from-
Mrs. Haley
on 5-25-06

HOLT

SCIENCE SPECTRUM

Physical Science

TEACHER EDITION

Ken Dobson, Ph.D.,
John Holman, Ph.D.,
Michael Roberts, Ph.D

Teacher Edition WALK-THROUGH

Student Edition CONTENTS IN BRIEF

HOLT, RINEHART AND WINSTON

A Harcourt Education Company

Orlando • **Austin** • New York • San Diego • Toronto • London

T1

Building a strong foundation for the future

Holt Science Spectrum: Physical Science introduces high school students to the broad spectrum of science study while developing reading and math skills. This program maintains a traditional emphasis on physical science while integrating physics, chemistry, Earth science, and space science.

STUDENTS OF ALL ABILITY LEVELS DEVELOP THE SKILLS THEY NEED FOR SUCCESS

- Math skills are developed through step-by-step examples, practice problems, and a ***Math Skills Workbook.***

- Reading support is found throughout the program with **Reading Skill Builders, Vocabulary Tips,** and reading strategies.

- Study skills are practiced and reviewed in both the chapter and the **Appendix.**

- Graphing skills are strengthened with in-text features throughout the chapter and the **Appendix.**

A FLEXIBLE LABORATORY PROGRAM BUILDS INQUIRY AND CRITICAL-THINKING SKILLS

- The laboratory program includes in-text labs for each chapter plus additional labs found in the ***Chapter Resource Files.***

- A variety of labs, from **QuickLABs** to **Inquiry Labs,** helps you meet your curriculum needs and work within the time constraints of your class schedule.

- Activities are labeled by difficulty level in the *Teacher Edition* to help you cater classroom instruction to the abilities of your students.

HOLT

Study Guide

HOLT

SCIENCE SPECTRUM

Physical Science

Integrating
✓ Chemistry
✓ Physics
✓ Earth Science
✓ Space Science
✓ Mathematics

Assessments

Includes
Concept Review
Worksheets

Pretests

CHAPTERS
1 Introduction to Science
2 Matter
3 States of Matter
4 Atoms and the Periodic Table
5 Chemical Reactions
6 Chemical Reactions
7 Solutions
8 Acids, Bases, and Salts
9 Nuclear Changes

HOLT

SCIENCE SPECTRUM

Physical Science

One-Stop Planner
with Test Generator
CD-ROM for Macintosh® and Windows®

Printable
Teaching Resources

Customizable
Lesson Plans

Powerful
Test Generator

HOLT

SCIENCE SPECTRUM

Physical Science

Integrating
✓ Chemistry
✓ Physics
✓ Earth Science
✓ Space Science
✓ Mathematics

Chapter Resource File 3

States of Matter

Skills Worksheets
Directed Reading 2
Reteaching 5
Concept Mapping 7
Concept Review 9

Labs and Activities
Datasheets for In-Text Health Labs
• Caffeine 35
• Drug Labels 39
Enrichment Activity

Teaching Transparencies

HOLT

SCIENCE SPECTRUM

Physical Science

includes
Bellringer Transparencies

Map Transparencies

Integrating
✓ Chemistry
✓ Physics
✓ Earth Science
✓ Space Science
✓ Mathematics

INTEGRATED TECHNOLOGY AND ONLINE RESOURCES EXPAND LEARNING BEYOND THE CLASSROOM

- Lighten the load with an interactive *Online Edition* or *CD-ROM Version* of the student text.

- **SciLinks,** a Web service developed and maintained by NSTA, contains current and prescreened links that engage students.

- The **Holt Physical Science Interactive Tutor CD-ROM** lets students advance at their own pace as they explore chemistry and physics concepts.

- All the resources you need are on the **One-Stop Planner CD-ROM with ExamView® Test Generator,** with worksheets, customizable lesson plans, and a powerful test generator.

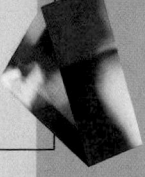

The Student Edition builds skills for success in science

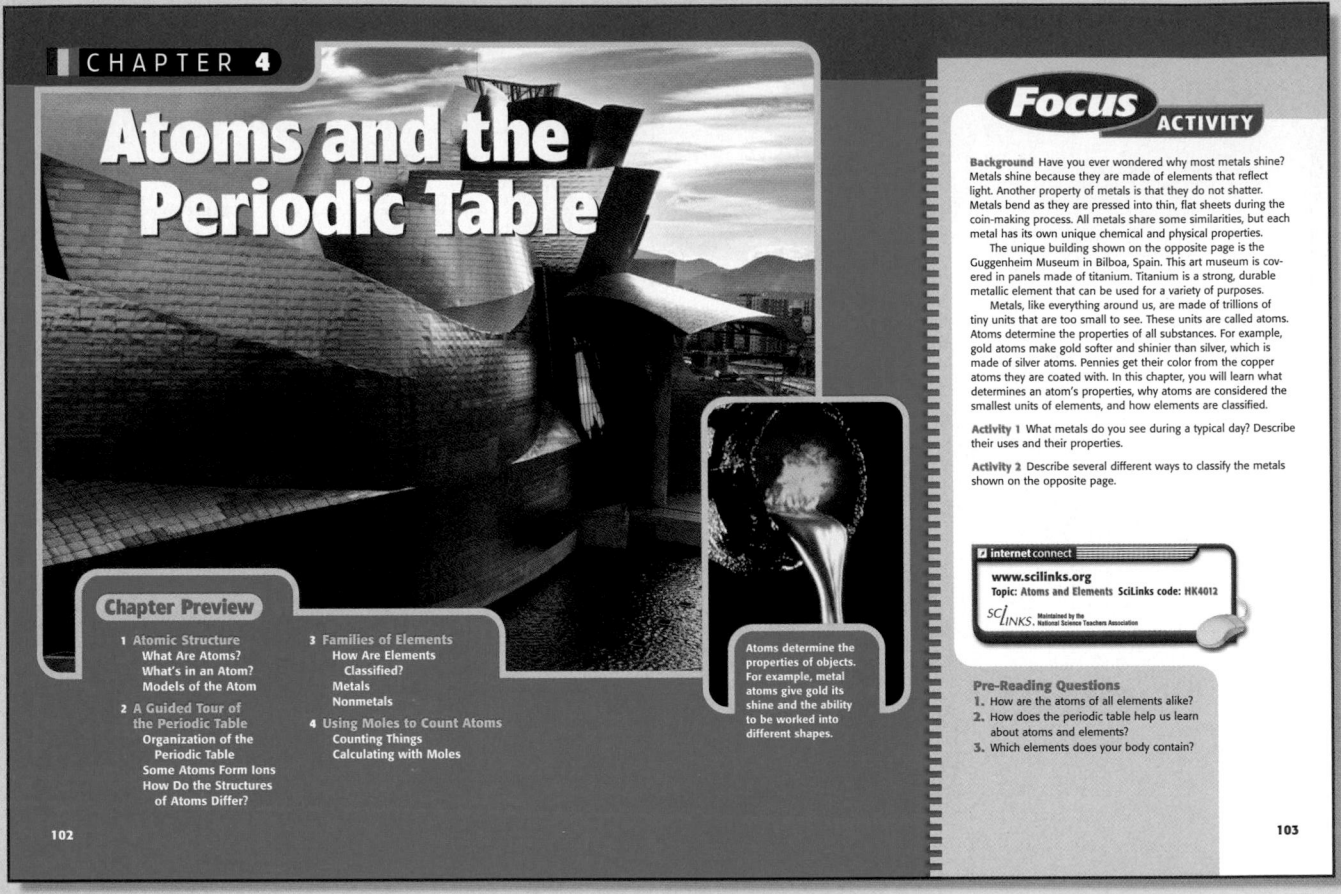

CHAPTER 4

Atoms and the Periodic Table

Chapter Preview

1 **Atomic Structure**
What Are Atoms?
What's in an Atom?
Models of the Atom

2 **A Guided Tour of the Periodic Table**
Organization of the Periodic Table
Some Atoms Form Ions
How Do the Structures of Atoms Differ?

3 **Families of Elements**
How Are Elements Classified?
Metals
Nonmetals

4 **Using Moles to Count Atoms**
Counting Things
Calculating with Moles

Atoms determine the properties of objects. For example, metal atoms give gold its shine and the ability to be worked into different shapes.

102

Focus ACTIVITY

Background Have you ever wondered why most metals shine? Metals shine because they are made of elements that reflect light. Another property of metals is that they do not shatter. Metals bend as they are pressed into thin, flat sheets during the coin-making process. All metals share some similarities, but each metal has its own unique chemical and physical properties.

The unique building shown on the opposite page is the Guggenheim Museum in Bilboa, Spain. This art museum is covered in panels made of titanium. Titanium is a strong, durable metallic element that can be used for a variety of purposes.

Metals, like everything around us, are made of trillions of tiny units that are too small to see. These units are called atoms. Atoms determine the properties of all substances. For example, gold atoms make gold softer and shinier than silver, which is made of silver atoms. Pennies get their color from the copper atoms they are coated with. In this chapter, you will learn what determines an atom's properties, why atoms are considered the smallest units of elements, and how elements are classified.

Activity 1 What metals do you see during a typical day? Describe their uses and their properties.

Activity 2 Describe several different ways to classify the metals shown on the opposite page.

internet connect
www.scilinks.org
Topic: Atoms and Elements SciLinks code: HK4012
SCILINKS Maintained by the National Science Teachers Association

Pre-Reading Questions
1. How are the atoms of all elements alike?
2. How does the periodic table help us learn about atoms and elements?
3. Which elements does your body contain?

103

Chapter Opener prepares students for the subject ahead with **Pre-Reading Questions, Focus Activity,** an engaging photo, and a **Chapter Preview** to help guide reading.

RELEVANT AND EXCITING FEATURES

Science and the Consumer creates ongoing debate by bringing up issues that analyze using risk-assessment models.

Viewpoint raises interesting issues and offers the opinions of students throughout the nation.

Did You Know? offers extra tidbits of information designed to get students more excited about the subject at hand.

Career Link features personal interviews that give students a realistic view of the education and training required for some fascinating careers.

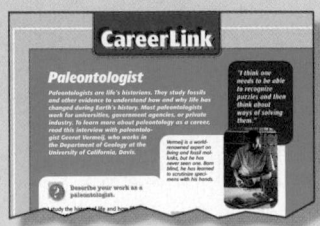

CareerLink

Paleontologist

LABS AND ACTIVITIES GRAB ATTENTION

QuickLAB and **Quick Activity** bring science to life with easy activities students can do themselves.

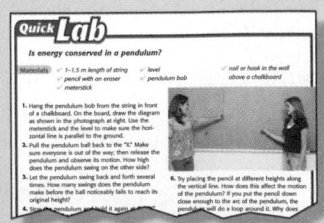

QuickLab
Is energy conserved in a pendulum?

Skills Practice Lab is more extensive, encouraging students to practice skills in order to more thoroughly understand the science being covered. Some also encourage students to design their own experiments.

SciLinks at point-of-use refers students to the NSTA Web site for up-to-date links, information, and interactive activities.

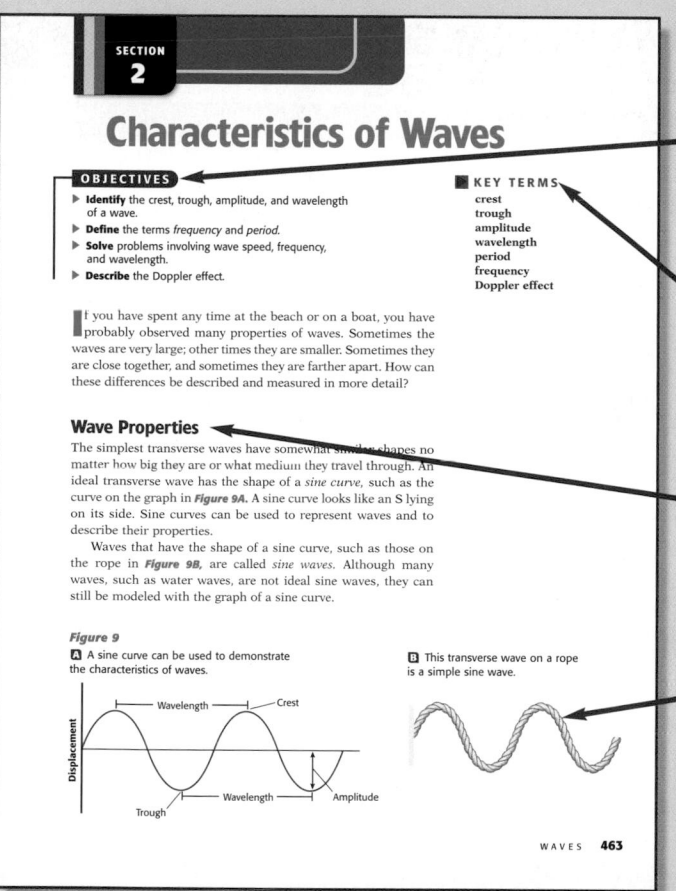

SECTION 2

Characteristics of Waves

OBJECTIVES

▶ **Identify** the crest, trough, amplitude, and wavelength of a wave.
▶ **Define** the terms *frequency* and *period*.
▶ **Solve** problems involving wave speed, frequency, and wavelength.
▶ **Describe** the Doppler effect.

KEY TERMS
crest
trough
amplitude
wavelength
period
frequency
Doppler effect

If you have spent any time at the beach or on a boat, you have probably observed many properties of waves. Sometimes the waves are very large; other times they are smaller. Sometimes they are close together, and sometimes they are farther apart. How can these differences be described and measured in more detail?

Wave Properties

The simplest transverse waves have somewhat similar shapes no matter how big they are or what medium they travel through. An ideal transverse wave has the shape of a *sine curve*, such as the curve on the graph in *Figure 9A.* A sine curve looks like an S lying on its side. Sine curves can be used to represent waves and to describe their properties.

Waves that have the shape of a sine curve, such as those on the rope in *Figure 9B,* are called *sine waves.* Although many waves, such as water waves, are not ideal sine waves, they can still be modeled with the graph of a sine curve.

Figure 9

A A sine curve can be used to demonstrate the characteristics of waves.

B This transverse wave on a rope is a simple sine wave.

WAVES **463**

Objectives set the groundwork for learning by alerting students to key concepts they need to look for.

Key Terms highlight vocabulary and present definitions at point-of-use, helping students focus on new vocabulary.

Accessible navigation engages students with outline-style headings, content grouped into small chunks, and text that doesn't break between pages.

Relevant graphics, tables, and photos enhance understanding with visual examples of concepts and topics.

CROSS-DISCIPLINARY FEATURES

Connection to... and **Integrating...** tie chapter topics to such diverse fields as language arts and social studies, as well as to other sciences.

Math Skills are integrated throughout the text, providing **Practice** for every skill and a **Practice Hint** feature to help students who struggle with math.

Math Skills, Study Skills, and **Graphing Skills** are mini-lessons presented as full-page features for more extended skills practice.

Connection to
SOCIAL STUDIES

ve used soap for thousands of years.
Egyptians took baths regularly with
om animal fats or vegetable oils
utions of alkali-metal compounds.

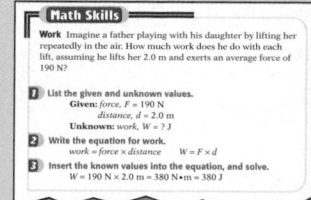

Math Skills

Work Imagine a father playing with his daughter by lifting her repeatedly in the air. How much work does he do with each lift, assuming he lifts her 2.0 m and exerts an average force of 190 N?

1 List the given and unknown values.
   Given: force, $F = 190$ N
   distance, $d = 2.0$ m
   Unknown: work, $W = ?$ J

2 Write the equation for work.
   work = force × distance $W = F \times d$

3 Insert the known values into the equation, and solve.
   $W = 190$ N × 2.0 m = 380 N·m = 380 J

REVIEW FOR TEST-READINESS

Section Review includes **Critical Thinking** and **Section Summaries** so students can apply what they've just learned.

Chapter Review covers every aspect of the chapter at a variety of skill levels with such features as **Understanding Concepts, Using Vocabulary, Building Math Skills,** and **Thinking Critically.**

Vocabulary Skills Tip improves reading by building language understanding.

Appendix features take learning further, with additional opportunities for practice and understanding—**Study Skills, Math Skills,** and **Graphing Skills** are also addressed here.

A Teacher Edition that makes planning easy

TEACHER EDITION

CHAPTER 14 — Waves
Chapter Planning Guide

Compression guide: To shorten your instruction because of time limitations, omit blocks 1, 8, and 9.

KEY
- TE Teacher Edition
- SE Student Edition
- OSP One-Stop Planner
- CRF Chapter Resource File
- TT Teaching Transparency
- TM Transparency Master
- * Also on One-Stop Planner
- ✦ Requires Advance Prep

PACING	CLASSROOM RESOURCES	LABS, ACTIVITIES, AND DEMONSTRATIONS	PROBLEM SOLVING AND PRACTICE	SECTION REVIEW AND ASSESSMENT	STANDARDS CORRELATION
BLOCK 1 · 45 min pp. 452–453 **Chapter Opener**		SE Activity 1, p. 453 SE Activity 2, p. 453		CRF Pretest* GENERAL	
BLOCKS 2 & 3 · 90 min pp. 454–462 **Section 1** Types of Waves	CRF Lesson Plan* TT Bellringer* TT Wave Model* TT Water Wave Model*	TE Demonstration Observing Wave Motion, p. 455 GENERAL TE Demonstration Harmonic Motion, p. 457 GENERAL SE Quick Lab How do particles move in a medium?, p. 460 CRF Datasheets for In-Text Labs How do particles move in a medium? GENERAL SE Quick Activity Polarization, p. 461 CRF Datasheets for In-Text Labs Polarization* GENERAL TE Demonstration Longitudinal Waves in Air, p. 461 GENERAL	CRF Cross-Disciplinary Worksheet Connection to Engineering—Wave Energy* ADVANCED SE Science and the Consumer Shock Absorbers: Why Are They Important?, p. 458 CRF Cross-Disciplinary Worksheet Science and the Consumer—Bicycle Design and Shock Absorption* ADVANCED	TE Quiz, p. 462 BASIC SE Section 1 Review, p. 462 CRF Concept Review* GENERAL CRF Quiz* BASIC	PS 4a, 5a, 5b, 5d UCP 1, 2, 4, 5 SAI 1 ST 1, 2 SPSP 5
BLOCKS 4 & 5 · 90 min pp. 463–471 **Section 2** Characteristics of Waves	CRF Lesson Plan* TT Bellringer* TT Transverse Wave* TM Longitudinal Wave* TT Frequency* TT Visible Light* TM The Electromagnetic Spectrum* TT Doppler Effect*	SE Quick Activity Wave Speed, p. 469 GENERAL CRF Datasheets for In-Text Labs Wave Speed* GENERAL TE Demonstration Doppler Effect, p. 470 GENERAL SE Skills Practice Lab Modeling Transverse Waves, pp. 484–485 ✦ GENERAL CRF Datasheets for SE Skills Practice Lab Modeling Transverse Waves* GENERAL CRF Observation Lab Creating and Measuring Standing Waves* ✦ BASIC	CRF Cross-Disciplinary Worksheet Connection to Language Arts—Writing a Plan for Wave Observation* ADVANCED CRF Cross-Disciplinary Worksheet Integrating Computers and Technology—Radio Waves* GENERAL CRF Cross-Disciplinary Worksheet Integrating Earth Science—Earthquake Waves* ADVANCED SE Math Skills Wave Speed, p. 468 GENERAL CRF Math Skills Wave Speed* GENERAL	TE Quiz, p. 471 BASIC SE Section 2 Review, p. 471 CRF Concept Review* GENERAL CRF Quiz* BASIC	PS 2e, 4a, 4e, 5a, 5d ES 3c UCP 1, 2, 3 ST 2 HNS 1 SPSP 5
BLOCKS 6 & 7 · 90 min pp. 472–478 **Section 3** Wave Interactions	CRF Lesson Plan* TT Bellringer* TT Interference* TM Beats* TT Concept Mapping*	TE Demonstration Observing Reflection and Refraction, p. 474 GENERAL CRF CBL™ Probeware Lab Tuning a Musical Instrument* ✦ ADVANCED	CRF Cross-Disciplinary Worksheet Integrating Math—Bending Light Waves to Magnify* ADVANCED SE Graphing Skills Interpreting Graphs, p. 479 SE CareerLink Ultrasonographer, pp. 486–487	TE Quiz, p. 478 BASIC SE Section 3 Review, p. 478 CRF Concept Review* GENERAL CRF Quiz* BASIC	PS 4a UCP 1, 2, 4 SAI 1 ST 1, 2 SPSP 5

BLOCKS 8 & 9 · 90 min

Chapter Review and Assessment Resources
- SE Chapter Review, pp. 480–483
- CRF Chapter Tests*
- OSP Test Generator
- CRF Standardized Test Practice with Guided Reading Development*
- CRF Test Item Listing for ExamView® Test Generator*
- OSP Scoring Rubrics and Classroom Management Checklists

Online Resources

go.hrw.com
Visit the HRW Web site for a variety of free resources related to the text. Just type in the keyword HK4 WAV.

Holt Online Learning
Holt Science Spectrum: Physical Science: Online Edition
Students can access interactive problem solving help and active visual concept development with the Holt Science Spectrum: Physical Science Online Edition available at www.hrw.com.

cnnstudentnews.com
Find the latest news, lesson plans, and activities related to important scientific events.

SciLINKS. www.scilinks.org
- Topic: Waves SciLinks code: HK4150
- Topic: Vibrations and Waves SciLinks code: HK4145
- Topic: Doppler Effect SciLinks code: HK4032
- Topic: Reflection, Refraction, Diffraction SciLinks code: HK4119
- Topic: Seismic Waves SciLinks code: HK4126
- Topic: Ultrasound SciLinks code: HK4143

Technology Resources

One-Stop Planner
All of your printable resources and the Test Generator are on this convenient CD-ROM.

Physical Science Interactive Tutor CD-Rom
- Disc Two, Module 12 Topic: Frequency and Wavelength
- Disc Two, Module 14 Topic: Refraction

452A Chapter 14 · Waves

452B Chapter 14 · Waves

TEACHING RESOURCES DESIGNED FOR CONVENIENCE

The **Chapter Planning Guide** breaks each chapter down into flexible 45-minute blocks, and offers a full listing of activities and classroom resources available for that lesson and how to use them. Look for guidance on:

- Pacing
- Classroom Resources
- Labs, Demonstrations, and Activities
- Enrichment and Skills Practice
- Section Review and Assessment
- National Science Standards Correlations
- Online and Technology Resources

A **Lesson Cycle** in the teacher's wrap builds structure around every lesson:

- **Focus** uses the objectives listed in the *Student Edition* to focus student attention on the upcoming content.

- **Motivate** uses demonstrations, discussions, and lively activities to get students excited about the material.

- **Teach** presents various techniques that help teach the section. Look for **Skill Builders, Teaching Tips,** and more in this part of the **Lesson Cycle.**

- Finally, **Close** with quiz questions to insure students understand the information covered.

ACTIVITIES AND DEMONSTRATIONS FOR EVERY LEARNING LEVEL

Activities are leveled by ability level in the teacher's wrap—**Basic, General,** and **Advanced**—helping you choose the activities you think your students are ready for.

Learning styles are addressed throughout—**Interpersonal, Intrapersonal, Auditory, Kinesthetic, Logical, Visual,** and **Verbal**—so you can adapt materials to different learning styles.

• **Bellringer** activities begin each section with an activity designed to get students thinking. Activities on transparency get them focused while you attend to administrative duties.

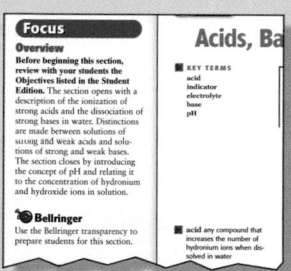

• **Quick Activity** experiments demonstrate concepts quickly and simply without the need for special equipment or resources.

• **Group Activity** encourages collaborative effort as students explore science concepts in pairs or larger groups.

Quick ACTIVITY

Materials (per group):
• sheet of wax paper
• dropper
• pin
• liquid detergent

Teacher's Notes: The detergent acts as a surfactant, or wetting agent, by reducing surface ten-

CREATING RELEVANCE AND UNDERSTANDING

On almost every page you will find exciting features to help ignite class discussion and keep students thinking.

• Misconception Alert
• Cross-Disciplinary Connection
• Connection To . . .
• Integrating . . .
• Demonstration
• Did You Know?
• Skill Builders for Reading, Writing, and Math

Teach, *continued*

Demonstration ——— GENERAL

A Natural pH Indicator

Did You Know?

Did you know that the concentration of hydronium ions and the concentration of hydroxide ions are related? In any solution made with water, the more hydronium ions there are (the more acidic the solution is), the fewer hydrox-

MISCONCEPTION ALERT

Science education research identified the following misconceptions about atoms.

• Students believe that atoms possess macro properties

Math Skills

Determining pH Determine the strong acid HCl dissolv

1 List the given and unknow
   Given: concentratio
   Unknown: pH

INCLUSION STRATEGIES MAKE MATERIAL ACCESSIBLE TO ALL

Written by professionals in the field of special needs education, **Inclusion Strategies** address many different learning exceptionalities in the classroom.

• Learning Disabled
• Developmentally Delayed
• Attention Deficit Disorder
• Behavior Control Issues
• Gifted and Talented
• English Language Learners

INCLUSION Strategies

• *Hearing Impaired* • *English Language Learners*

Have students visit a grocery store or provide ingredient labels from up to ten food or beverage products that contain an acid in their ingredients. The students should make

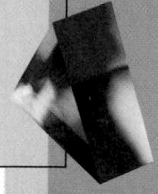

Assessment opportunities help you track students' progress

SECTION ASSESSMENT

- **Pre-Reading Questions,** preceding every chapter, get students thinking about what chapter content they already know, encouraging them to find connections between the topic and their own lives.

- **Reteaching** activities in the *Teacher Edition* help students understand section material by presenting a concept in a different way. Use these features to customize your lesson to your student population.

- **Quiz,** also in *Teacher Edition,* provides additional questions to assess student progress.

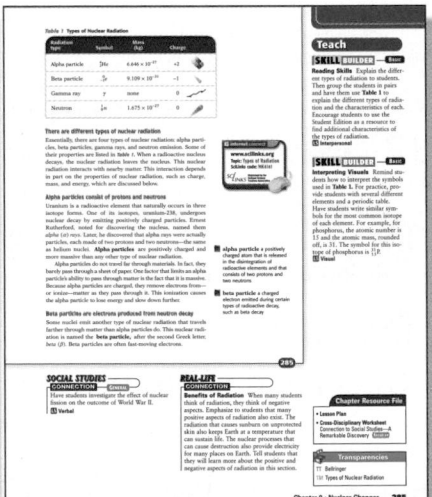

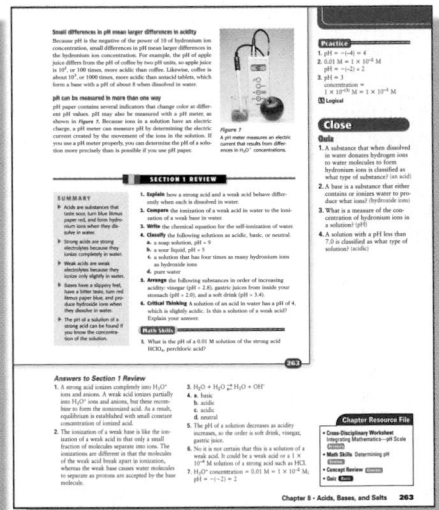

- Frequent reinforcement is provided by the **Section Review** while the information is still fresh, increasing retention and building on an overall understanding of the chapter.

SECTION 2 REVIEW

SUMMARY

▶ Elements are arranged in order of increasing atomic number so that elements with similar properties are in the same column, or group.

▶ Elements in the same group have the same number of

1. **Explain** how you can determine the number of protons, electrons, and neutrons an atom has from an atom's mass number and its atomic number.

2. **Calculate** how many neutrons a phosphorus-32 atom has.

3. **Name** the elements represented by the following symbols:

   a. Li   d. Br   g. Na

   b. Mg   e. He   h. Fe

   c. Cu   f. S   i. K

4. **Compare** the number of valence electrons an oxygen, O, atom

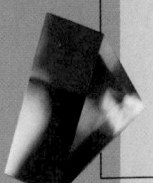

CHAPTER ASSESSMENT

More extensive **Chapter Reviews** prepare students for testing by approaching the material from a variety of conceptual levels. Features include:

- Understanding Concepts
- Using Vocabulary
- Building Math Skills
- Thinking Critically
- Developing Life/Work Skills
- Integrating Concepts

- **Assignment Guide** in the *Teacher Edition* lets you see which questions correlate with a specific section's content.

Assignment Guide

Section	Questions
1	1–3, 6, 8, 11, 16–19, 24, 28–31, 35–37, 41–44, 46
2	4, 9, 10, 12, 25–27, 32, 33, 38
3	5, 7, 13–15, 20–23, 34, 39, 40, 45

- **Concept Review** worksheets in the *Study Guide* help reinforce skills and concepts presented in the *Student Edition.*

- **Pretests, Chapter Tests,** and **Test Item Listing** are all together in the handy *Chapter Resource File* books, offering you a number of options for assessment.

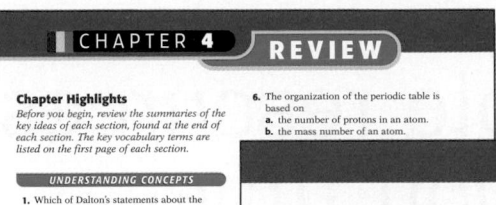

CHAPTER 4 REVIEW

Chapter Highlights
Before you begin, review the summaries of the key ideas of each section, found at the end of each section. The key vocabulary terms are listed on the first page of each section.

UNDERSTANDING CONCEPTS

1. Which of Dalton's statements about the atom was later proven false?
   a. Atoms cannot be subdivided.
   b. Atoms are tiny.
   c. Atoms of different elements are not identical.
   d. Atoms join to form molecules.

2. Which statement is not true of Bohr's model of the atom?
   a. The nucleus can be compared to the sun.
   b. Electrons orbit the nucleus.
   c. An electron's path is not known exactly.
   d. Electrons exist in energy levels.

3. According to the modern model of the atom,
   a. moving electrons form an electron cloud.
   b. electrons and protons circle neutrons.
   c. neutrons have a positive charge.
   d. the number of protons an atom has varies.

4. If an atom has a mass of 11 amu and contains five electrons, its atomic number must be
   a. 55.   c. 6.
   b. 16.   d. 5.

5. Which statement about ... the same group of the ...
   a. They have the same ...
   b. They have the same ...
   c. They have similar c...
   d. They have the same ... electrons.

6. The organization of the periodic table is based on
   a. the number of protons in an atom.
   b. the mass number of an atom.

USING VOCABULARY

12. How many *protons* and *neutrons* does a silicon, Si, atom have, and where are each of these subatomic particles located? How many *electrons* does a silicon atom have?

13. Identify the particles that make up an atom. How do these particles relate to the identity of an atom?

14. Draw two different types of *orbitals*, state their names, and describe how they are filled.

15. Describe the process of *ionization*, and give two different examples of elements that undergo this process.

16. Explain why different atoms of the same element always have the same *atomic number* but can have different *mass numbers*. What are these different atoms called?

17. Distinguish between the following:

BUILDING MATH SKILLS

24. **Graphing** Use a graphing calculator, a computer spreadsheet, or a graphing program to plot the atomic number on the x-axis and the average atomic mass in amu on the y-axis for the transition metals in Period 4 of the periodic table (from scandium to zinc). Do you notice a break in the trend near cobalt? Explain why elements with larger atomic numbers do not necessarily have larger atomic masses.

25. **Converting Mass to Amount** For an experiment you have been asked to do, you need 1.5 g of iron. How many moles of iron do you need?

26. **Converting Mass to Amount** James is holding a balloon that contains 0.54 g of helium gas. What amount of helium is this?

27. **Converting Amount to Mass** A pure gold bar is made of 19.55 mol of gold. What is ...

136 CHAPTER 4

---

31. **Evaluating Data** The figure below shows relative ionic radii for positive and negative ions of elements in Period 2 of the periodic table. Explain the trend in ion size as you move from left to right across the periodic table. Why do the negative ions have larger radii than the positive ions?

32. **Making Comparisons** Although carbon and lead are in the same group, some of their properties are very different. Propose a reason for this. (**Hint:** Look at the periodic table to locate each element and find out how each is classified.)

33. **Problem Solving** How does halving the amount of a sample of an element affect the sample's mass?

34. **Understanding Systems** When an atom loses an electron, what is the atom's charge? What do you think happens to the size of the atom?

35. **Applying Knowledge** What property do the noble gases share? How does this property relate to the electron configuration of the noble gases?

36. **Applying Knowledge** Write the chemical symbols for helium, carbon, gold, lead, sodium, potassium, and copper.

37. **Critical Thinking** Why is it difficult to measure the size of an atom?

38. **Critical Thinking** Particle accelerators are devices that speed up charged particles in order to smash the particles together. Sometimes the result of the collision is a new nucleus. How can scientists determine whether the nucleus formed is that of a new element or that of a new isotope of a known element?

39. **Problem Solving** What would happen to poisonous chlorine gas if the following alterations were made to the chlorine?
   a. A proton is added to each atom.
   b. An electron is added to each atom.
   c. A neutron is added to each atom.

DEVELOPING LIFE/WORK SKILLS

40. **Locating Information** Some "neon" signs contain substances other than neon to produce different colors. Design your own lighted sign, and find out which substances you could use to produce the colors you want your sign to be.

41. **Making Decisions** Suppose you have only 1.9 g of sulfur for an experiment and you must do three trials using 0.030 mol of S each time. Do you have enough sulfur?

42. **Communicating Effectively** The study of the nucleus produced a new field of medicine called nuclear medicine. Pretend you are writing an article for a hospital newsletter. Describe how radioactive substances called tracers are sometimes used to detect and treat diseases.

43. **Working Cooperatively** With a group of your classmates, make a list of 10 elements and their average atomic masses. Calculate the amount in moles for 6.0 g of each element. Rank your elements from the element with the greatest amount to the element with the least amount in a 6.0 g sample. Do you notice a trend in the amounts as atomic number increases? Explain why or why not.

44. **Applying Knowledge** You read a science fiction story about an alien race of silicon-based life-forms. Use information from the periodic table to hypothesize why the author chose silicon over other elements. (**Hint:** Life on Earth is carbon based.)

138 CHAPTER 4

... TABLE **137**

CUSTOM ASSESSMENT

Included on the convenient *One-Stop Planner CD-ROM* is the **ExamView® Test Generator,** allowing you to create customized assessment based on your teaching goals and the ability level of your class. See page T12 for more information.

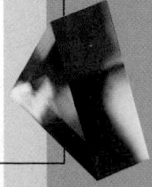

Flexible resources save you time

CHAPTER RESOURCE FILES

A *Chapter Resource File* accompanies each chapter of *Holt Science Spectrum: Physical Science.* Everything you need to plan and manage your lessons in a convenient timesaving format is included in each chapter book. Also included is a **Program Overview** booklet, your guide to the resources in each *Chapter Resource File.* Each chapter book includes:

Skills Worksheets
• Science Skills
• Math Skills
• Concept Review
• Cross-Disciplinary

Labs & Activities
• Datasheets for In-Text Activities
• Skills Practice Labs
• CBL™ Probeware Labs

Assessments
• Pretests
• Chapter Test
• Test Item Listing for ExamView® Test Generator

Teacher Resources
• Lesson Plans
• Lab Notes and Answers

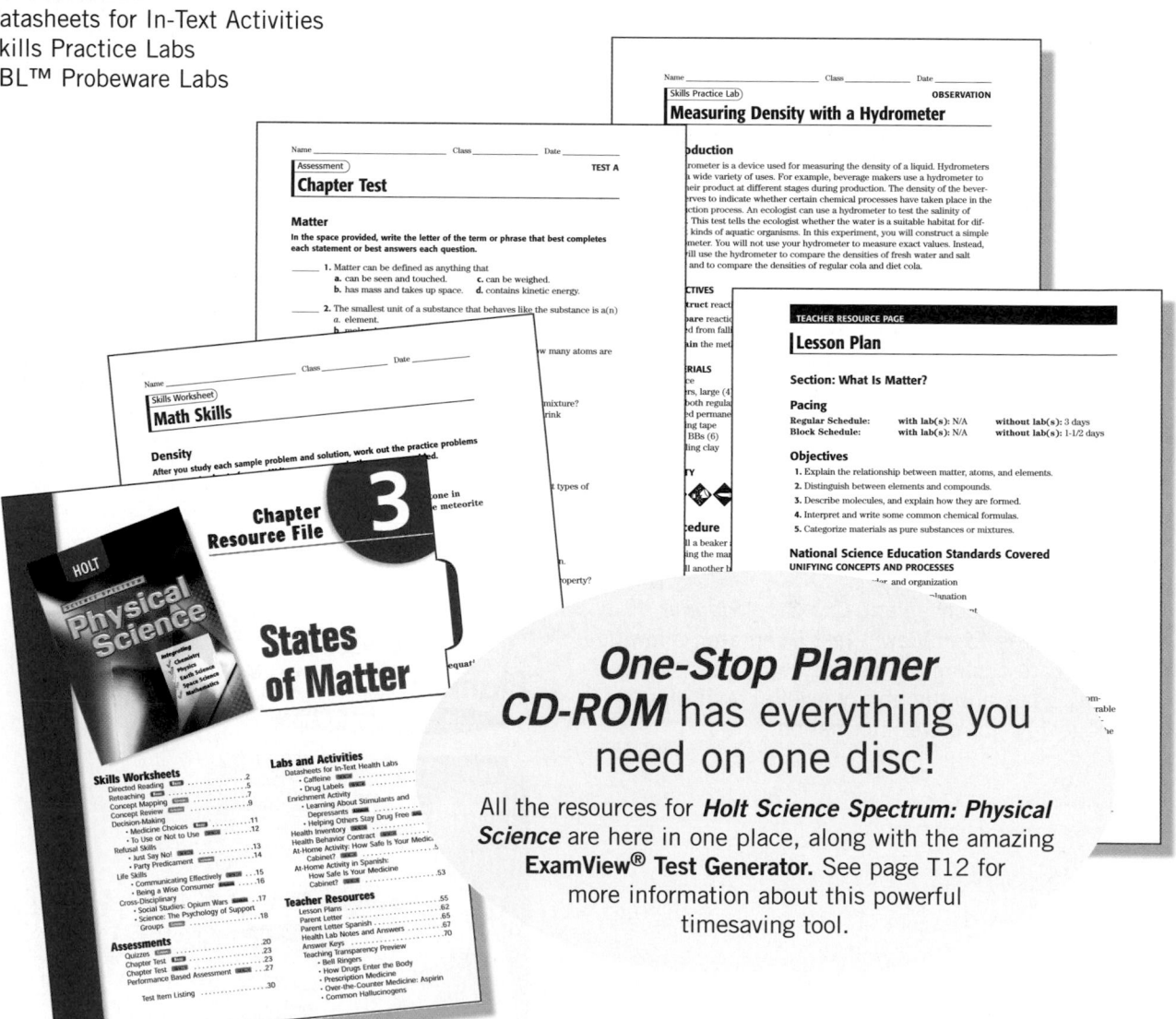

One-Stop Planner CD-ROM has everything you need on one disc!

All the resources for *Holt Science Spectrum: Physical Science* are here in one place, along with the amazing **ExamView® Test Generator**. See page T12 for more information about this powerful timesaving tool.

ADDITIONAL RESOURCES REINFORCE AND EXTEND LESSONS

- *Holt Science Skills Workshop: Reading in the Content Area* contains exercises that target the reading skills specific to the comprehension of science texts.

- *Holt Science Laboratory Manager's Professional Reference* is a well-organized reference that was created to help teachers understand risk management and the hazards that can occur in the classroom.

- *Math Skills Workbook* includes blackline master worksheets with problem-solving practice needed for chemistry and physics.

- *Study Guide* contains review worksheets to reinforce the skills and concepts presented in the *Student Edition.*

- Over 200 full-color **Teaching Transparencies** use graphics directly from the text to enhance classroom presentations.

- *Laboratory Manual* provides 20 traditional experiments that build students' lab skills. The *Teacher Guide* contains a master materials list, lab notes, and an answer key.

SPANISH RESOURCES BRING SCIENCE TO ENGLISH-LANGUAGE LEARNERS

- A **Spanish Glossary** is right at students' fingertips in the *Student Edition,* following the **Glossary.** It shows the English term, its Spanish equivalent, and a definition in Spanish.

- *Study Guide,* in Spanish, contains review worksheets that reinforce the skills and concepts presented in the *Student Edition.*

- *Assessments* in Spanish include **Section Quizzes** and **Chapter Tests.**

Technology that enhances teaching

One-Stop Planner CD-ROM®
with Test Generator

Planning and managing lessons has never been easier than with this convenient all-in-one CD-ROM that includes a variety of timesaving features, including:

Printable resources and worksheets

All the resources available for *Holt Science Spectrum: Physical Science* are in one place, including science skills development, concept practice, math practice, vocabulary development, Spanish materials, and transparency masters.

Customizable lesson plans

Tailor your lessons to your classroom's specific needs. Includes block-scheduling lesson plans in several word-processing formats.

Powerful ExamView® Test Generator

Contains test items organized by chapter, plus over a thousand editable questions, so you can put together your own tests and quizzes.

HOLT
SCIENCE SPECTRUM
Physical Science

One-Stop Planner®
with Test Generator
CD-ROM for Macintosh® and Windows®

Printable
Teaching Resources

Customizable
Lesson Plans

Powerful
Test Generator

HOLT PHYSICAL SCIENCE INTERACTIVE TUTOR CD-ROM

Help your students explore concepts at their own pace, build problem-solving skills, develop understanding, make observations, and much more with this interactive tutor.

GUIDED READING AUDIO CD PROGRAM

This direct read of *Holt Science Spectrum: Physical Science* on audio CD makes content more accessible, especially for auditory learners and reluctant readers.

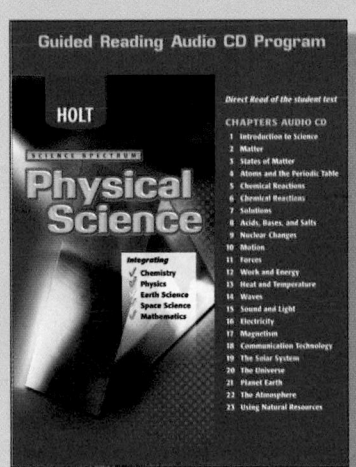

STUDENT EDITION, CD-ROM VERSION

Ideal for students who have limited access to the Internet, but who need to lighten the load of textbooks they carry home, the entire *Student Edition* is on one easy-to-navigate CD-ROM, page-for-page.

LESSON PRESENTATIONS ON CD-ROM

This CD-ROM is your guide to **Quick Concepts**— engaging, media-based presentations of core concepts. **Quick Concepts** can be projected for large group instruction or viewed by individuals or groups of students with a computer.

CNN PRESENTS SCIENCE IN THE NEWS VIDEOS

Chemistry Connections and **Physical Science Connections** videos include news segments that bring the relevance of science right into your classroom. Each news segment showcases useful applications of science concepts, often making cross-curricular connections to industry, careers, and a variety of other areas.

TECHNOLOGY RESOURCES

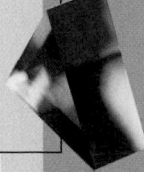

Online Resources are available anytime, anywhere!

THE ONLINE EDITION IS PORTABLE, EXPANDABLE, AND YET WEIGHS NOTHING AT ALL

The *Online Edition* of **Holt Science Spectrum: Physical Science** engages students in ways that were never before possible. And since it's all online, it's available anywhere you or your students connect to the Internet. You'll find:

- Interactive exercises and feedback
- Presentation materials
- Homework help
- And much more!

Contact your Holt sales representative or call 1-800-HRW-9799 for more information.

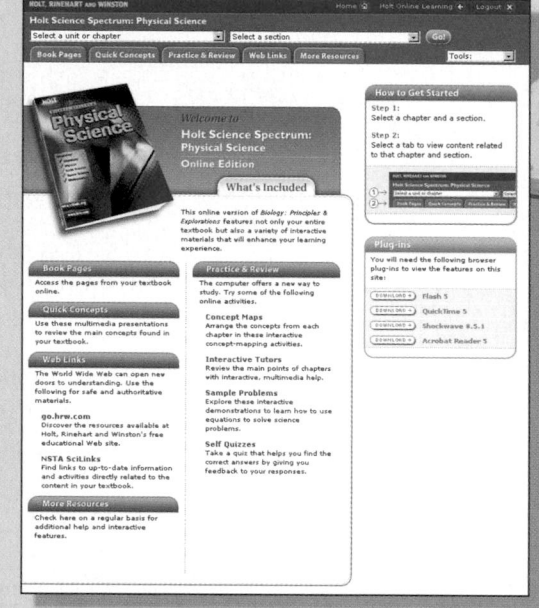

This Web service, developed and maintained by the National Science Teachers Association, contains a collection of prescreened links with current information and activities directly related to chapter topics.

- Saves you valuable time searching for relevant, up-to-date Web sites
- Sites are reviewed by science content experts and educators
- **Internet Connect** boxes within each chapter offer opportunities to enrich, enhance, and extend learning through **SciLinks**

student CNN News™

Go to **CNNStudentNews.com** for award-winning news and information for both teachers and students. You'll find a wealth of helpful features, including:

- News as it happens
- Classroom resources
- Student current events activities
- Lesson plans
- Projects and activities

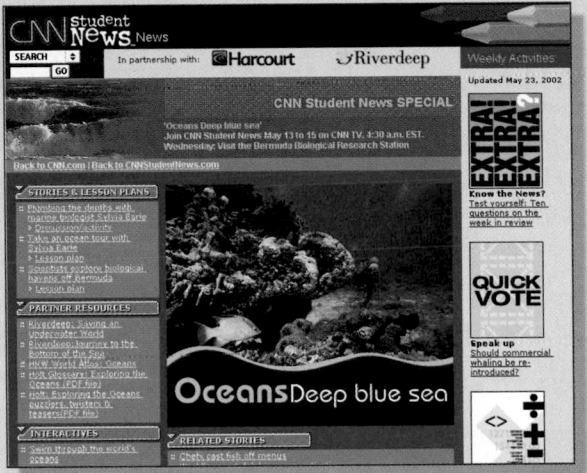

QUANTUM
INTELLIGENT
TUTORING
ENGINES

The first artificial-intelligence-based tutors give your students the help they need by coaching them through problems and giving them the confidence to work through challenging subjects.

- Easily accessible online, available whenever and wherever students need help
- Students can get help on the problem or concept of their choice
- Gives hints and step-by-step explanations
- Provides students with personalized feedback

Contact your Holt sales representative or call 1-800-HRW-9799 for more information.

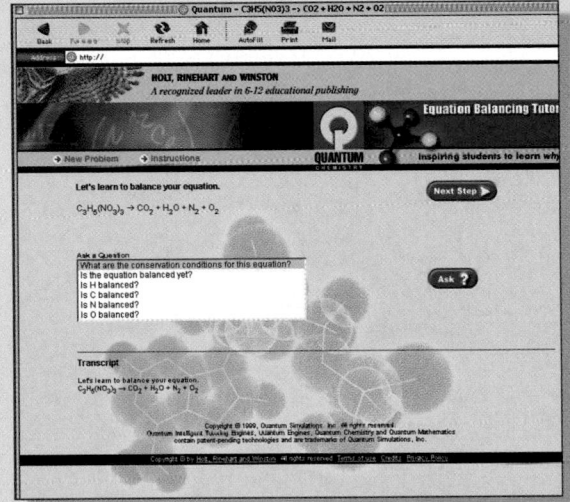

Holt, Rinehart and Winston's award-winning Web site, **go.hrw.com,** allows students to enrich their knowledge with a variety of worksheets, activities, projects, research articles and ideas, interactive quizzes, review activities, and teacher resources.

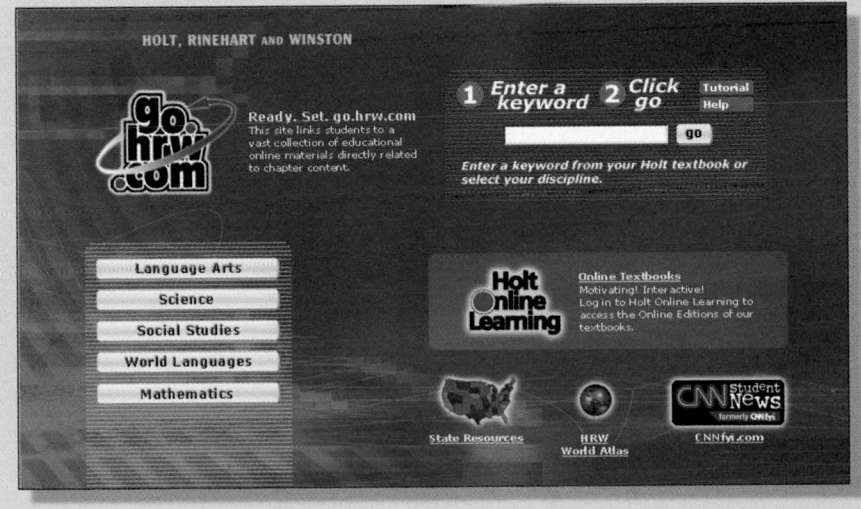

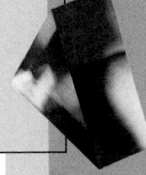

ONLINE RESOURCES

A variety of labs for every purpose

Holt Science Spectrum: Physical Science includes lab activities that meet the demands of your curriculum. This flexible laboratory program builds inquiry and critical-thinking skills.

All labs are **Bench-Tested** and **rated** to help you select the labs that suit your students' abilities.

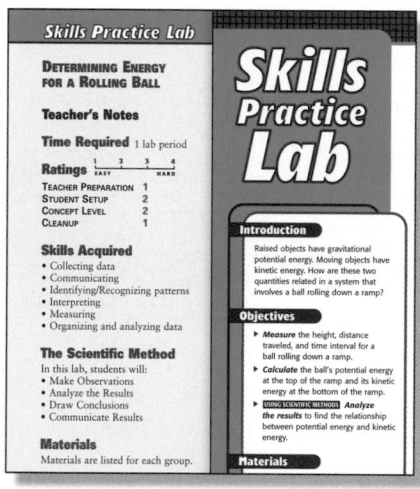

Skills Practice Lab

DETERMINING ENERGY FOR A ROLLING BALL

Teacher's Notes

Time Required 1 lab period

Ratings	$\frac{1}{EASY}$	2	3	$\frac{4}{HARD}$
TEACHER PREPARATION	1			
STUDENT SETUP	2			
CONCEPT LEVEL	2			
CLEANUP	1			

Skills Acquired
- Collecting data
- Communicating
- Identifying/Recognizing patterns
- Interpreting
- Measuring
- Organizing and analyzing data

The Scientific Method
In this lab, students will:
- Make Observations
- Analyze the Results
- Draw Conclusions
- Communicate Results

Materials
Materials are listed for each group.

Skills Practice Lab

Introduction
How can you distinguish metal elements by analyzing their physical properties?

Objectives
▶ USING SCIENTIFIC METHODS **Hypothesize** which physical properties can help you **distinguish** between different metals.

▶ **Identify** unknown metals by **comparing** the data you collect with reference information.

Materials
balance
beakers (several)
graduated cylinder
hot plate
ice
magnet
metal samples, unidentified (several)
metric ruler
stopwatch
water
wax

Comparing the Physical Properties of Elements

▶ **Procedure**

Identifying Metal Elements
1. In this lab, you will identify samples of unknown metals by comparing the data you collect with reference information listed in the table at right. Use at least two of the physical properties listed in the table to identify each metal.

Deciding Which Physical Properties You Will Analyze
2. Density is the mass per unit volume of a substance. If the metal is box-shaped, you can measure its length, width, and height, and then use these measurements to calculate the metal's volume. If the shape of the metal is irregular, you can add the metal to a known volume of water and determine what volume of water is displaced.

3. Relative hardness indicates how easy it is to scratch a metal. A metal with a higher value can scratch a metal with a lower value, but not vice versa.

4. Relative heat conductivity indicates how quickly a metal heats or cools. A metal with a value of 100 will heat or cool twice as quickly as a metal with a value of 50.

5. If a magnet placed near a metal attracts the metal, then the metal has been magnetized by the magnet.

Designing Your Experiment
6. With your lab partner(s), decide how you will use the materials provided to identify each metal you are given. There is more than one way to measure some of the physical properties that are listed, so you might not use all of the materials that are provided.

7. In your lab report, list each step you will perform in your experiment.

8. Have your teacher approve your plan before you carry out your experiment.

140 CHAPTER 4

BRIEF LABS

Focus Activity kicks off each chapter with engaging information and activities on real-world places, events, or phenomena to spark student interest in chapter content.

QuickLAB is an easy activity that can be completed in less than one class period.

Quick Activity experiments demonstrate concepts quickly and simply without the need for special equipment or resources.

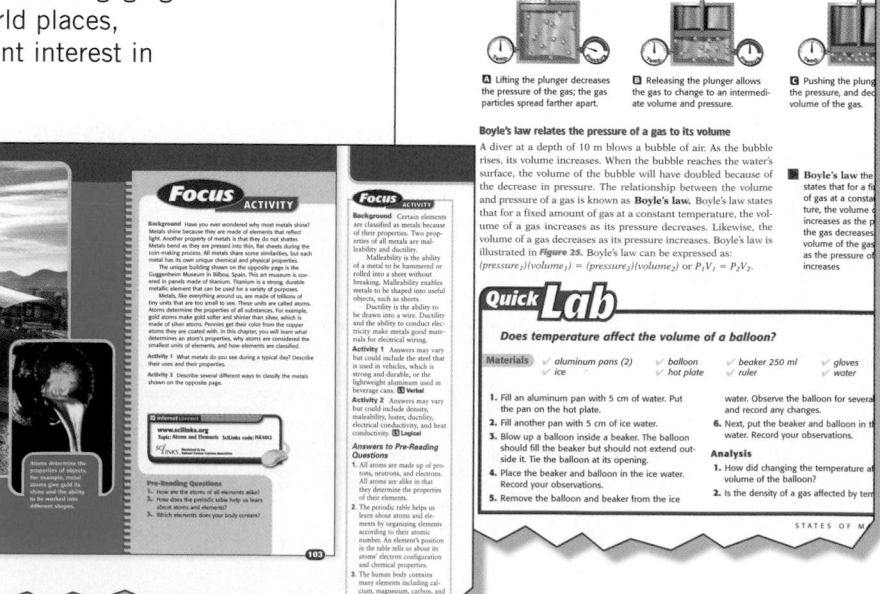

Design Your Own

Physical Properties of Some Metals

Metal	Density (g/mL)	Relative hardness	Relative heat conductivity	Magnetized by magnet?
Aluminum (Al)	2.7	28	100	no
Iron (Fe)	7.9	50	34	yes
Nickel (Ni)	8.9	67	38	yes
Tin (Sn)	7.3	19	28	no
Tungsten (W)	19.3	100	73	no
Zinc (Zn)	7.1	28	49	no

Performing Your Experiment

9. After your teacher approves your plan, carry out your experiment. Keep in mind that the more careful your measurements are, the easier it will be for you to identify the unknown metals.

10. Record all the data you collect and any observations you make in your lab report.

▶ Analysis

1. Make a table listing the physical properties you compared and the data you collected for each of the unknown metals.

2. Which metals were you given? Explain the reasoning you used to identify each metal.

3. Which physical properties were the easiest for you to measure and compare? Which were the hardest? Explain why.

4. What would happen if you tried to scratch aluminum foil with zinc?

5. Explain why it would be difficult to distinguish between iron and nickel unless you calculate each metal's density.

6. Suppose you find a metal fastener and determine that its density is 7 g/mL. What are two ways you could determine whether the unknown metal is tin or zinc?

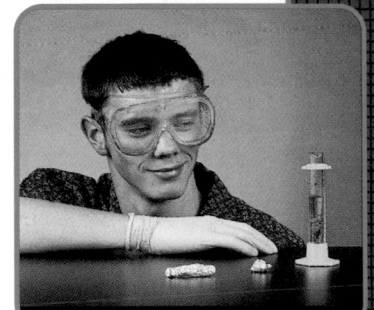

▶ Conclusions

7. Suppose someone gives you an alloy that is made of both zinc and nickel. In general, how do you think the physical properties of the alloy would compare with those of each individual metal?

ATOMS AND THE PERIODIC TABLE **141**

THE TEACHER EDITION OFFERS MORE ACTIVITIES

Additional activities found in the *Teacher Edition* also help illustrate science concepts. Examples include **Demonstration, Activity,** and **Group Activity.**

Motivate

Demonstration ── GENERAL

(Time: Approximately 30 minutes)

Materials:
• 25 g of solid
• 10 mL of eth

Group Activity ── BASIC

Interpreting the Periodic Table
Group students in pairs. Ask one student in each pair to choose an element. The other student should use the periodic table to identify the element's symbol, atomic num-

EXTENDED LABS

Skills Practice Labs are more extensive experiments and come in two flavors, traditional and **Design Your Own,** so your students can practice analysis and critical-thinking skills.

Chapter Resource Files include additional laboratory opportunities with:

• CBL™ Probeware Labs

• Skills Practice Labs

• Datasheets for In-Text Activities

Laboratory Manual provides 20 traditional experiments that build students' lab skills. The *Teacher Guide* contains a master materials list, lab notes, and an **Answer Key.**

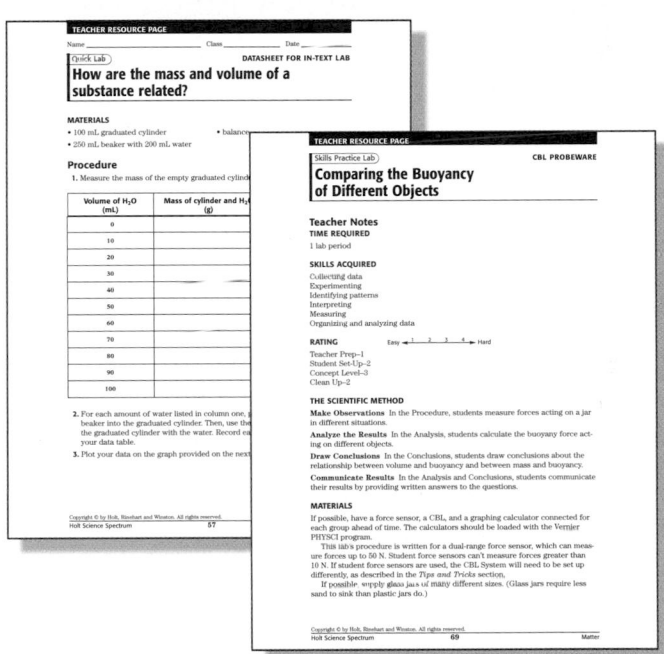

HOLT SCIENCE LABORATORY MANAGER'S PROFESSIONAL REFERENCE

Maintain safety in your classroom laboratory with information on risk management, checklists of typical hazards, and pertinent standards and regulations.

T17

Meeting individual needs

Students have a wide range of abilities and learning exceptionalities. These pages show you how *Holt Science Spectrum: Physical Science* provides resources and strategies to help you tailor your instruction to engage every student in your classroom.

Learning exceptionality	Resources and strategies	
Learning Disabilities and Slow Learners Students who have dyslexia or dysgraphia, students reading below grade level, students having difficulty understanding abstract or complex concepts, and slow learners	• Inclusion Strategies labeled *Learning Disabled* • Activities labeled *Basic* • Activities labeled *Visual* or *Kinesthetic*	• Hands-on activities or projects • Oral presentations instead of written tests or assignments
Developmental Delays Students who are functioning far below grade level because of mental retardation, autism, or brain injury; goals are to learn or retain basic concepts	• Inclusion Strategies labeled *Developmentally Delayed* • Activities labeled *Basic*	• Project-based activities • Observation Labs
Attention Deficit Disorders Students experiencing difficulty completing a task that has multiple steps, difficulty handling long assignments, or difficulty concentrating without sensory input from physical activity	• Inclusion Strategies labeled *Attention Deficit Disorder* • Activities labeled *Basic* • *Group Activities* • Activities labeled *Visual* or *Kinesthetic*	• Concepts broken into small chunks • Oral presentations instead of written tests or assignments
English as a Second Language Students learning English	• Inclusion Strategies labeled *English Language Learners* • Activities labeled *Basic*	• Activities labeled *Visual* • Observation Labs
Gifted and Talented Students who are performing above grade level and demonstrate aptitude in crosscurricular assignments	• Inclusion Strategies labeled *Gifted and Talented* • Activities labeled *Advanced* • *Connection* activities	• Activities that involve multiple tasks, a strong degree of independence, and student initiative

General Strategies The following strategies can help you modify instruction to help students who struggle with common classroom difficulties.

A student experiencing difficulty with ...	May benefit if you . . .	
Beginning assignments	• Assign work in small amounts • Have the student use cooperative or paired learning • Provide varied and interesting activities	• Allow choice in assignments or projects • Reinforce participation • Seat the student closer to you
Following directions	• Gain the student's attention before giving directions • Break up the task into small steps • Give written directions rather than oral directions • Use short, simple phrases • Stand near the student when you are giving directions	• Have the student repeat directions to you • Prepare the student for changes in activity • Give visual cues by posting general routines • Reinforce improvement in or approximation of following directions
Keeping track of assignments	• Have the student use folders for assignments • Have the student use assignment notebooks	• Have the student keep a checklist of assignments and highlight assignments when they are turned in
Reading the textbook	• Provide outlines of the textbook content • Reduce the length of required reading • Allow extra time for reading • Have the students read aloud in small groups	• Have the student use peer or mentor readers • Have the student use books on tape or CD • Discuss the content of the textbook in class after reading
Staying on task	• Reduce distracting elements in the classroom • Provide a task-completion checklist • Seat the student near you	• Provide alternative ways to complete assignments, such as oral projects taped with a buddy
Behavioral or social skills	• Model the appropriate behaviors • Establish class rules, and reiterate them often • Reinforce positive behavior • Assign a mentor as a positive role model to the student • Contract with the student for expected behaviors • Reinforce the desired behaviors or any steps toward improvement	• Separate the student from any peer who stimulates the inappropriate behavior • Provide a "cooling off" period before talking with the student • Address academic/instructional problems that may contribute to disruptive behaviors • Include parents in the problem-solving process through conferences, home visits, and frequent communication
Attendance	• Recognize and reinforce attendance by giving incentives or verbal praise • Emphasize the importance of attendance by letting the student know that he or she was missed when he or she was absent	• Encourage the student's desire to be in school by planning activities that are likely to be enjoyable, giving the student a preferred responsibility to be performed in class, and involving the student in extracurricular activities • Schedule problem-solving meeting with parents, faculty, or both
Test-taking skills	• Prepare the student for testing by teaching ways to study in pairs, such as using flashcards, practice tests, and study guides, and by promoting adequate sleep, nourishment, and exercise • During testing, allow the student to respond orally on tape or to respond using a computer; to use	notes; to take breaks; to take the test in another location; to work without time constraints; or to take the test in several short sessions • Decrease visual distraction by improving the visual design of the test through use of larger type, spacing, consistent layout, and shorter sentences

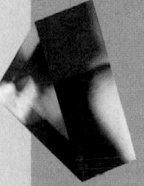

Build critical reading skills

HELP STUDENTS UNDERSTAND WHAT THEY READ

Like poles repel, and opposite poles attract

As you know, the closer two like electrical charges are brought together, the more they repel each other. The closer two opposite charges are brought together, the more they attract each other. A similar situation exists for **magnetic poles.**

Magnets have a pair of poles, a north pole and a south pole. The poles of magnets exert a force on one another. Two like poles, such as two south poles, repel each other. Two unlike poles, however, attract each other. Thus, the north pole of one magnet will attract the south pole of another magnet. Also, the north pole of one magnet repels the north pole of another magnet.

It is impossible to isolate a south magnetic pole from a north magnetic pole. If a magnet is cut, each piece will still have two poles. No matter how small the pieces of a magnet are, each piece still has both a north and a south pole.

Magnetic Fields

Try moving the south pole of one magnet toward the south pole of another that is free to move. As you do this, the magnet you are not touching will move away. A force is being exerted on the second magnet even though it never touches the magnet in your hand. The force is acting at a distance. This may seem unusual, but you are already familiar with other forces that act at a distance. Gravitational forces and the force between electric charges also act at a distance.

magnetic pole one of two points, such as the ends of a magnet, that have opposing magnetic qualities

VOCABULARY *Skills Tip*

The word pole *is used in physics for two related opposites that are separated by some distance along an axis. The word* polar, *used in chemistry, has the same origin.*

Vocabulary is called out at point-of-use throughout the text, along with **Vocabulary Skills Tips,** which provide tools and information on word origins for deriving the meanings of words.

Pre-Reading Questions, preceding every chapter, get students thinking about what chapter content they already know, encouraging them to find connections between the topic and their own lives.

Each chapter begins with **Objectives** and **Key Terms,** preparing students to pay attention to ideas that will later be tested.

SKILL BUILDER

This feature includes **Reading** and **Vocabulary** activities that help students understand the material that follows.

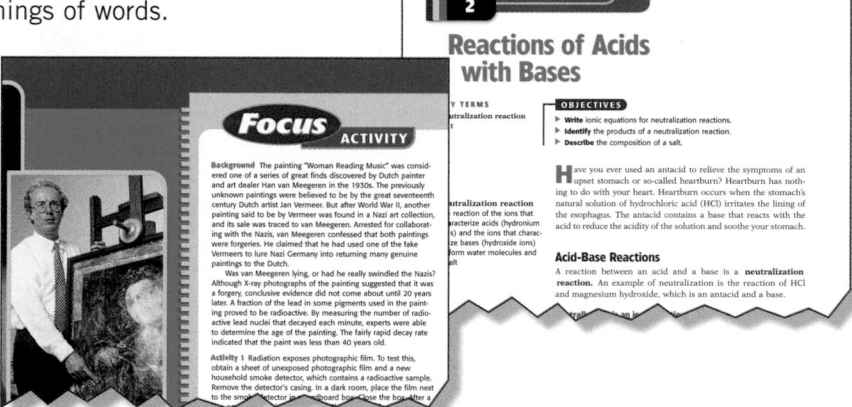

READING AND STUDY SKILLS APPENDIX

Located at the end of the *Student Edition,* this feature instructs students in a variety of strategies that can help them become better readers.

APPENDIX A

Reading and Study Skills

As a science student, you are expected to understand the information you read in this textbook and hear from your teacher. To be successful, you also need to be able to organize the information you receive and to take good notes. This appendix is designed to help you learn these skills, which will help you become a more successful student.

The first section, which contains reading skills and study skills, is followed by sections on graphing skills, a math refresher, and lab skills. At the end of every chapter in this book, there is a skills practice page. The skills practice pages are designed to help you develop your skills. If you have difficulty in one particular area, such as graphing skills, this appendix is an excellent resource for extra explanation and practice.

Recognizing Key Words

To begin improving your understanding of

The second reason that finding the key word or idea may be difficult is that its location within a paragraph or sentence often changes. Sometimes the key idea appears at the beginning of a sentence, and sometimes it is in the middle or at the end. To recognize key words in a sentence, ask yourself the following question about every word in a sentence.

If this word was taken out of the sentence, would I still understand what the sentence is trying to say?

Consider the following example.

Find the key words in the following sentence.

You are to report to the counselor's office at 4:00 P.M., and don't forget to take your books with you.

Key words: you, report, counselor's office, 4:00 P.M., take, books.

If you communicated the key words to someone, he or she would understand what to do.

T20

Additional Resources also help in reading Comprehension

GUIDED READING AUDIO CD PROGRAM

Auditory learners, visual learners, reluctant readers, and English-language learners will all find the critical reading support they need with a direct read of every chapter in *Holt Science Spectrum: Physical Science.*

HOLT SCIENCE SKILLS WORKSHOP: READING IN THE CONTENT AREA

Target the reading skills specific to the comprehension of science texts with these activities and exercises. Students learn to analyze text structures, recognize patterns, and organize information in ways that help them construct meaning.

Activities are flexible enough to be completed by students individually, or you can use the overhead transparencies provided in the *Teacher Edition* to teach analysis skills to a large group of students.

CONCEPT REVIEW

From straight recall to higher-order thinking, **Concept Review,** found in the *Chapter Resource Files,* helps reinforce what students have learned throughout a section.

Support for the development of critical math skills

MATH IS INTEGRATED THROUGHOUT THE TEXT

When math plays a critical part in understanding a principle discussed in the text, a **Math Skills** feature explains the concept and the math behind it. Then, the student is given opportunities to **Practice** the math, and a **Practice Hint** accompanies the exercise to provide additional assistance.

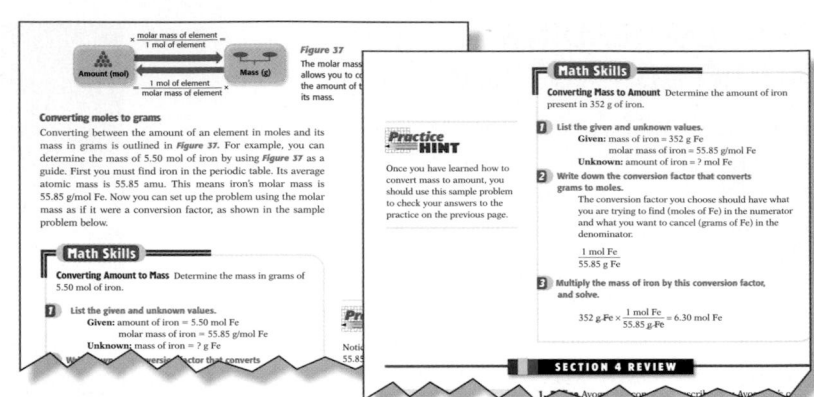

The **Chapter Review** offers more math practice. **Building Math Skills** revisits math covered throughout the chapter, while **Thinking Critically** encourages students to stretch their minds as they apply math to a variety of problems.

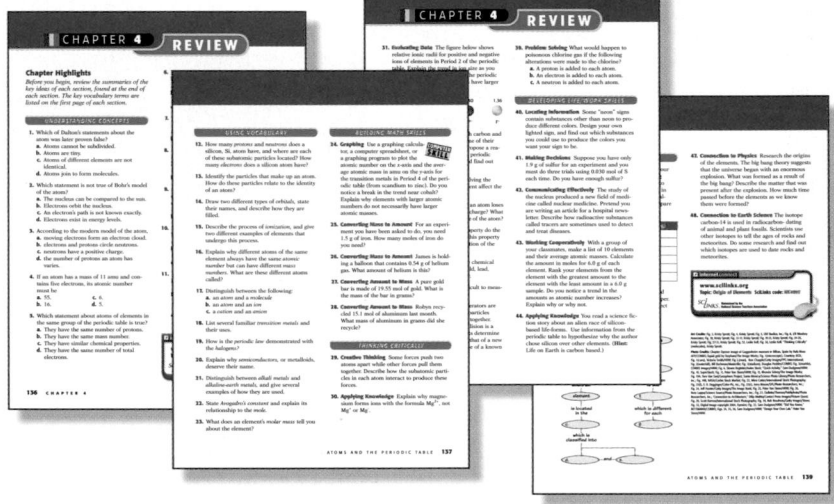

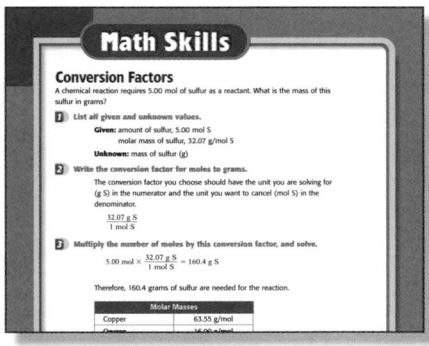

Math is also found in the **Appendix. Math Skills Refresher** covers basic topics such as graphing, exponents, fractions, percentages, and many more. **Problem Bank** contains more than 150 additional practice problems.

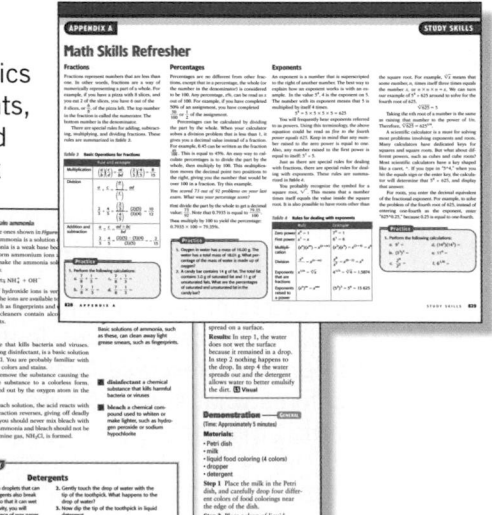

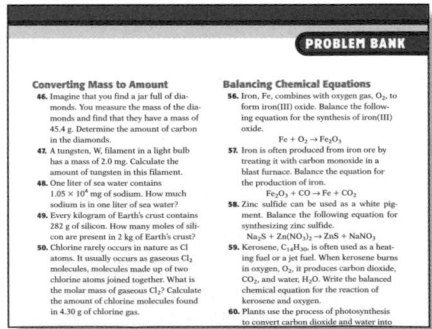

Skill Builder for math, in the *Teacher Edition,* provides additional tips and practice to improve math skills.

CHAPTER RESOURCE FILES

Math Skills Worksheets, located in the *Chapter Resource Files,*
focus on specific math skills that science students need.
Extensive problem-solving models and lots of practice problems
are included.

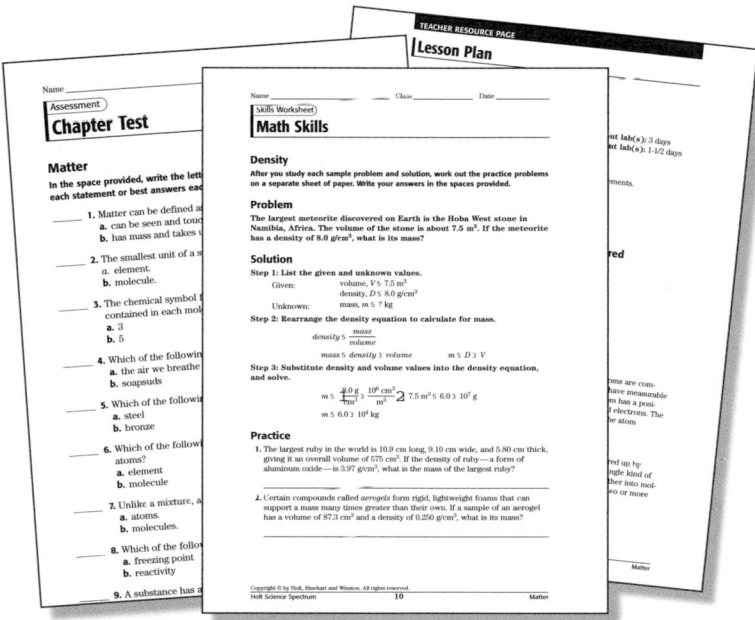

FULL-PAGE MATH FEATURES GET DEEPER INTO SKILLS

Math Skills, Graphing Skills, and **Study Skills** are all covered
in greater depth in these single page features.

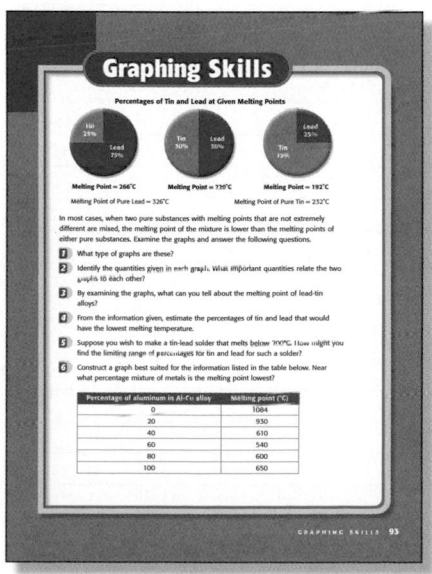

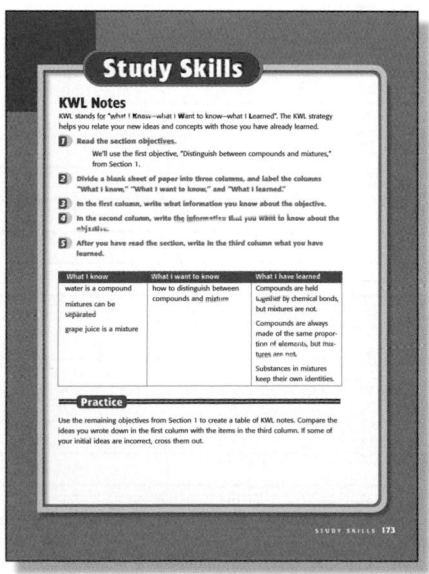

Pacing Guide

Today's chemistry classroom often requires a more flexible curriculum than ever before. ***Holt Science Spectrum: Physical Science*** can help you meet a variety of needs and challenges you and your students face in the classroom. The **Pacing Guide** below shows a number of ways to adapt the program to your teaching schedule.

This **Guide** can be further adapted, allowing you to mix and match or compress the material so you can spend more time on select topics, or to allow for special projects and activities.

- **General** provides the recommended course of study as indicated in the *Teacher Edition,* found in the individual chapter guides preceding each chapter. (This adds up to more than a year because of the advanced topics provided later in the book, which you may choose to omit, as indicated below.)

- **Compressed** indicates how you can still cover the essentials of physical science, even in the face of time constraints.
- **Basic** gives more time for the foundations of physical science, especially mathematical problem-solving, with less emphasis on some advanced topics from later in the course.
- **Advanced** moves quickly through the foundations of physical science for students who may be comfortable with the basics, to provide additional time for advanced topics.
- **Heavy Lab/Activity** indicates ways to streamline "lecture" time to provide hands-on experience for more than a third of the blocks in the school year. (Note: even this approach does not cover all of the labs and activities that are available with ***Holt Science Spectrum: Physical Science*** and its ***Chapter Resource Files.***)

	General	Compressed	Basic	Advanced	Heavy Lab/Activity
Chapter 1 *Introduction to Science*	9	6	11	6	9
Chapter Intro	1	–	1	–	1
1 Introduction to Science	2	1	2	1	1
2 The Way Science Works	2	2	3	2	2
3 Organizing Data	2	2	3	2	2
Resource File Labs	–	–	–	–	2
Chapter Review and Assessment	2	1	2	1	1
Chapter 2 *Matter*	9	6	10	7	9
Chapter Intro	1	–	1	–	1
1 What Is Matter	2	1	2	2	1
2 Properties of Matter	2	2	3	2	2
3 Changes of Matter	2	2	2	2	2
Resource File Labs	–	–	–	–	2
Chapter Review and Assessment	2	1	2	1	1
Chapter 3 *States of Matter*	9	4	8	7	7
Chapter Intro	1	–	1	–	1
1 Matter and Energy	2	2	2	2	2
2 Fluids	2	–	–	3	–
3 Behavior of Gases	2	1	3	2	1
Resource File Labs	–	–	–	–	2
Chapter Review and Assessment	2	1	2	–	1

	General	Compressed	Basic	Advanced	Heavy Lab/Activity
Chapter 4 Atoms and the Periodic Table	11	7	12	8	10
Chapter Intro	1	–	1	–	1
1 Atomic Sturcture	2	2	2	2	2
2 A Guided Tour of the Periodic Table	2	1	2	2	1
3 Families of Elements	2	1	1	1	1
4 Using Moles to Count Atoms	2	2	4	2	2
Resource File Labs	–	–	–	–	2
Chapter Review and Assessment	2	1	2	1	1
Chapter 5 The Structure of Matter	11	6	10	11	9
Chapter Intro	1	–	1	1	1
1 Compounds and Molecules	2	1	2	2	1
2 Ionic and Covalent Bonding	2	2	2	2	2
3 Compound Names and Formulas	2	2	3	2	2
4 Organic and Biochemical Compounds	2	–	–	3	–
Resource File Labs	–	–	–	–	2
Chapter Review and Assessment	2	1	2	1	1
Chapter 6 Chemical Reactions	11	5	11	10	8
Chapter Intro	1	–	1	–	1
1 Covalent Bonds	2	1	2	2	1
2 Drawing and Naming Molecules	2	2	3	2	2
3 Molecular Shapes	2	1	3	2	1
4 Rates of Change	2	–	–	3	
Resource File Labs	–	–	–	–	2
Chapter Review and Assessment	2	1	2	1	1
Chapter 7 Solutions	9	6	10	7	9
Chapter Intro	1	–	1	–	1
1 Solutions and Other Mixtures	2	2	2	2	2
2 How Substances Dissolve	2	2	2	2	2
3 Solubility and Concentration	2	1	3	2	1
Resource File Labs	–	–	–	–	2
Chapter Review and Assessment	2	1	2	1	1
Chapter 8 Acids, Bases, and Salts	9	6	11	7	9
Chapter Intro	1	–	1	–	1
1 Acids and Bases	2	2	4	2	2
2 Reactions of Acids with Bases	2	2	2	2	2
3 Acids, Bases, and Salts in the Home	2	1	2	2	1
Resource File Labs	–	–	–	–	2
Chapter Review and Assessment	2	1	2	1	1
Chapter 9 Nuclear Changes	9	6	9	7	9
Chapter Intro	1	–	1	–	1
1 What is Radioactivity?	2	2	2	2	2
2 Nuclear Fission and Fusion	2	2	4	2	2
3 Nuclear Radiation Today	2	1	2	2	1
Resource File Labs	–	–	–	–	2
Chapter Review and Assessment	2	1	–	1	1
Chapter 10 Motion	9	5	11	7	8
Chapter Intro	1	–	1	–	1
1 Measuring Motion	2	1	3	2	1
2 Acceleration	2	1	3	2	1
3 Motion and Force	2	2	2	2	2
Resource File Labs	–	–	–	–	2
Chapter Review and Assessment	2	1	2	1	1

	General	Compressed	Basic	Advanced	Heavy Lab/Activity
Chapter 11 *Forces*	9	5	10	7	8
Chapter Intro	1	–	1	–	1
1 Laws of Motion	2	2	4	2	2
2 Gravity	2	1	2	2	1
3 Newton's Third Law	2	1	1	2	1
Resource File Labs	–	–	–	–	2
Chapter Review and Assessment	2	1	2	1	1
Chapter 12 *Work and Energy*	11	7	13	9	10
Chapter Intro	1	–	1	–	1
1 Work, Power, and Machines	2	2	3	2	2
2 Simple Machines	2	1	2	2	1
3 What is Energy?	2	2	3	2	2
4 Conservation of Energy	2	1	2	2	1
Resource File Labs	–	–	–	–	2
Chapter Review and Assessment	2	1	2	1	1
Chapter 13 *Heat and Temperature*	9	6	10	6	8
Chapter Intro	1	–	–	–	–
1 Temperature	2	2	3	2	2
2 Energy Transfer	2	2	3	2	2
3 Using Heat	2	1	2	1	1
Resource File Labs	–	–	–	–	2
Chapter Review and Assessment	2	1	2	1	1
Chapter 14 *Waves*	9	6	10	6	9
Chapter Intro	1	–	1	–	1
1 Types of Waves	2	1	2	1	1
2 Characteristics of Waves	2	2	3	2	2
3 Wave Interactions	2	2	2	2	2
Resource File Labs	–	–	–	–	2
Chapter Review and Assessment	2	1	2	1	1
Chapter 15 *Sound and Light*	11	4	11	8	9
Chapter Intro	1	–	1	–	1
1 Sound	2	1	2	2	1
2 The Nature of Light	2	1	2	2	1
3 Reflection and Color	2	1	2	1	1
4 Refraction, Lenses and Prisms	2	–	2	2	2
Resource File Labs	–	–	–	–	2
Chapter Review and Assessment	2	1	2	1	1
Chapters 16 *Electricity*	9	5	10	9	8
Chapter Intro	1	–	1	–	1
1 Electric Charge and Force	2	1	2	2	1
2 Current	2	2	3	3	2
3 Circuits	2	1	3	3	1
Resource File Labs	–	–	–	–	2
Chapter Review and Assessment	2	1	1	1	1
Chapter 17 *Magnetism*	9	3	4	9	5
Chapter Intro	1	–	1	–	1
1 Magnets and Magnetic Fields	2	2	2	2	2
2 Magnetism from Electric Currents	2	–	–	3	–
3 Electric Currents from Magnetism	2	–	–	3	–
Resource File Labs	–	–	–	–	1
Chapter Review and Assessment	2	1	1	1	1

	General	Compressed	Basic	Advanced	Heavy Lab/Activity
Chapter 18 *Communication Technology*	9	4	9	9	6
Chapter Intro	1	–	1	–	1
1 Signals and Telecommunication	2	1	2	2	1
2 Telephone, Radio, and Television	2	1	2	3	1
3 Computers and the Internet	2	1	2	3	1
Resource File Labs	–	–	–	–	1
Chapter Review and Assessment	2	1	2	1	1
Chapter 19 *The Solar System*	9	3	0	7	5
Chapter Intro	1	–	–	–	1
1 Sun, Earth, and Moon	2	1	–	2	1
2 The Inner and Outer Planets	2	1	–	2	1
3 Formation of the Solar System	2	–	–	2	–
Resource File Labs	–	–	–	–	1
Chapter Review and Assessment	2	1	–	1	1
Chapter 20 *The Universe*	9	2	0	7	6
Chapter Intro	1	–	–	–	1
1 The Life and Death of Stars	2	1	–	2	1
2 The Milky Way and Other Galaxies	2	–	–	2	1
3 Origin of the Universe	2	–	–	2	1
Resource File Labs	–	–	–	–	1
Chapter Review and Assessment	2	1	–	1	1
Chapter 21 *Planet Earth*	11	5	0	9	7
Chapter Intro	1	–	–	–	1
1 Earth's Interior and Plate Technologies	2	1	–	2	1
2 Earthquakes and Volcanoes	2	1	–	2	1
3 Minerals and Rocks	2	1	–	2	1
4 Weathering and Erosion	2	1	–	2	1
Resource File Labs	–	–	–	–	1
Chapter Review and Assessment	2	1	–	1	1
Chapter 22 *The Atmosphere*	9	4	0	8	6
Chapter Intro	1	–	–	–	1
1 Characteristics of the Atmosphere	2	1	–	3	1
2 Water and Wind	2	1	–	2	1
3 Weather and Climate	2	1	–	2	1
Resource File Labs	–	–	–	–	1
Chapter Review and Assessment	2	1	–	1	1
Chapter 23 *Using Natural Resources*	9	4	0	9	6
Chapter Intro	1	–	–	–	1
1 Organisms and Their Environment	2	1	–	2	1
2 Energy and Resources	2	1	–	3	1
3 Pollution and Recycling	2	1	–	3	1
Resource File Labs	–	–	–	–	1
Chapter Review and Assessment	2	1	–	1	1
TOTAL	219	115	180	180	180

PACING GUIDE

Correlation to the National Science Education Standards

The following list shows the chapter correlation of *Holt Science Spectrum* with the National Science Education Standards (grades 9-12) for Physical Science content and Earth and Space Science content. For further detail, see the interleaf pages before each chapter.

UNIFYING CONCEPTS AND PROCESSES

Standard	Code
Systems, order, and organization	UCP 1
Evidence, models, and explanation	UCP 2
Change, consistency, and measurements	UCP 3
Evolution and equilibrium	UCP 4
Form and function	UCP 5

SCIENCE AS INQUIRY

Standard	Code
Abilities to do scientific inquiry	SAI 1
Understanding about scientific inquiry	SAI 2

SCIENCE AND TECHNOLOGY

Standard	Code
Abilities of technological design	ST 1
Understanding about science and technology	ST 2

HISTORY AND NATURE OF SCIENCE

Standard	Code
Science as a human endeavor	HNS 1
Nature of science	HNS 2
History of science	HNS 3

SCIENCE IN PERSONAL AND SOCIAL PERSPECTIVES

Standard	Code
Personal health	SPSP 1
Populations, resources, and environments	SPSP 2
Natural hazards	SPSP 3
Risks and benefits	SPSP 4
Science and technology in society	SPSP 5

PHYSICAL SCIENCE CONTENT STANDARDS

Standard	Code	Chapter Correlation
Structure of Atoms		
Matter is made of minute particles called atoms, and atoms are composed of even smaller components. These components have measurable properties, such as mass and electrical charge. Each atom has a positively charged nucleus surrounded by negatively charged electrons. The electric force between the nucleus and electrons holds the atom together.	PS 1a	Chapter 2 Chapter 4 Chapter 12 Chapter 16
The atom's nucleus is composed of protons and neutrons, which are much more massive than electrons. When an element has atoms that differ in the number of neutrons, these atoms are called different isotopes of the element.	PS 1b	Chapter 4 Chapter 9
The nuclear forces that hold the nucleus of an atom together, at nuclear distances, are usually stronger than the electric forces that would make it fly apart. Nuclear reactions convert a fraction of the mass of interacting particles into energy, and they can release much greater amounts of energy than atomic interactions. Fission is the splitting of a large nucleus into smaller pieces. Fusion is the joining of two nuclei at extremely high temperature and pressure, and is the process responsible for the energy of the sun and other stars.	PS 1c	Chapter 4 Chapter 9 Chapter 12
Radioactive isotopes are unstable and undergo spontaneous nuclear reactions, emitting particles and/or wavelike radiation. The decay of any one nucleus cannot be predicted, but a large group of identical nuclei decay at a predictable rate. This predictability can be used to estimate the age of materials that contain radioactive isotopes.	PS 1d	Chapter 9
Structure and Properties of Matter		
Atoms interact with one another by transferring or sharing electrons that are furthest from the nucleus. These outer electrons govern the chemical properties of the element.	PS 2a	Chapter 4 Chapter 5 Chapter 12
An element is composed of a single type of atom. When elements are listed in order according to the number of protons (called the atomic number), repeating patterns of physical and chemical properties identify families of elements with similar properties. This "Periodic Table" is a consequence of the repeating pattern of outermost electrons and their permitted energies.	PS 2b	Chapter 4 Chapter 12
Bonds between atoms are created when electrons are paired up by being transferred or shared. A substance composed of a single kind of atom is called an element. The atoms may be bonded together into molecules or crystalline solids. A compound is formed when two or more kinds of atoms bind together chemically.	PS 2c	Chapter 2 Chapter 3 Chapter 5 Chapter 12
The physical properties of compounds reflect the nature of the interactions among its molecules. These interactions are determined by the structure of the molecule, including the constituent atoms and the distances and angles between them.	PS 2d	Chapter 2 Chapter 3 Chapter 5 Chapter 7 Chapter 12

PHYSICAL SCIENCE CONTENT STANDARDS, *continued*

Standard	Code	Chapter Correlation
Structure and Properties of Matter, *continued*		
Solids, liquids, and gases differ in the distances and angles between molecules or atoms and therefore the energy that binds them together. In solids the structure is nearly rigid; in liquids molecules or atoms move around each other but do not move apart; and in gases molecules or atoms move almost independently of each other and are mostly far apart.	**PS 2e**	Chapter 2 Chapter 3 Chapter 5 Chapter 7 Chapter 12 Chapter 14 Chapter 15
Carbon atoms can bond to one another in chains, rings, and branching networks to form a variety of structures, including synthetic polymers, oils, and the large molecules essential to life.	**PS 2f**	Chapter 5
Chemical Reactions		
Chemical reactions occur all around us, for example in health care, cooking, cosmetics, and automobiles. Complex chemical reactions involving carbon-based molecules take place constantly in every cell in our bodies.	**PS 3a**	Chapter 2 Chapter 6
Chemical reactions may release or consume energy. Some reactions such as the burning of fossil fuels release large amounts of energy by losing heat and by emitting light. Light can initiate many chemical reactions such as photosynthesis and the evolution of urban smog.	**PS 3b**	Chapter 6 Chapter 12 Chapter 23
A large number of important reactions involve the transfer of either electrons (oxidation/reduction reactions) or hydrogen ions (acid/base reactions) between reacting ions, molecules, or atoms. In other reactions, chemical bonds are broken by heat or light to form very reactive radicals with electrons ready to form new bonds. Radical reactions control many processes such as the presence of ozone and greenhouse gases in the atmosphere, burning and processing of fossil fuels, the formation of polymers, and explosions.	**PS 3c**	Chapter 6 Chapter 8
Chemical reactions can take place in time periods ranging from the few femtoseconds (10^{-15} seconds) required for an atom to move a fraction of a chemical bond distance to geologic time scales of billions of years. Reaction rates depend on how often the reacting atoms and molecules encounter one another, on the temperature, and on the properties—including shape—of the reacting species.	**PS 3d**	Chapter 6
Catalysts, such as metal surfaces, accelerate chemical reactions. Chemical reactions in living systems are catalyzed by protein molecules called enzymes.	**PS 3e**	Chapter 6
Motion and Forces		
Objects change their motion only when a net force is applied. Laws of motion are used to calculate precisely the effects of forces on the motion of objects. The magnitude of the change in motion can be calculated using the relationship $F = ma$, which is independent of the nature of the force. Whenever one object exerts force on another, a force equal in magnitude and opposite in direction is exerted on the first object.	**PS 4a**	Chapter 10 Chapter 11 Chapter 12 Chapter 14

Standard	Code	Correlation
Motion and Forces, *continued*		
Gravitation is a universal force that each mass exerts on any other mass. The strength of the gravitational attractive force between two masses is proportional to the masses and inversely proportional to the square of the distance between them.	PS 4b	Chapter 1 Chapter 10 Chapter 11 Chapter 12 Chapter 16 Chapter 19
The electric force is a universal force that exists between any two charged objects. Opposite charges attract while like charges repel. The strength of the force is proportional to the charges, and, as with gravitation, inversely proportional to the square of the distance between them.	PS 4c	Chapter 16
Between any two charged particles, electric force is vastly greater than the gravitational force. Most observable forces such as those exerted by a coiled spring or friction may be traced to electric forces acting between atoms and molecules.	PS 4d	Chapter 16
Electricity and magnetism are two aspects of a single electromagnetic force. Moving electric charges produce magnetic forces, and moving magnets produce electric forces. These effects help students to understand electric motors and generators.	PS 4e	Chapter 14 Chapter 17
Conservation of Energy and the Increase in Disorder		
The total energy of the universe is constant. Energy can be transferred by collisions in chemical and nuclear reactions, by light waves and other radiations, and in many other ways. However, it can never be destroyed. As these transfers occur, the matter involved becomes steadily less ordered.	PS 5a	Chapter 12 Chapter 14 Chapter 15 Chapter 19
All energy can be considered to be either kinetic energy, which is the energy of motion; potential energy, which depends on relative position; or energy contained by a field, such as electromagnetic waves.	PS 5b	Chapter 3 Chapter 12 Chapter 14 Chapter 16 Chapter 17
Heat consists of random motion and the vibrations of atoms, molecules, and ions. The higher the temperature, the greater the atomic or molecular motion.	PS 5c	Chapter 1 Chapter 3 Chapter 13 Chapter 15
Everything tends to become less organized and less orderly over time. Thus, in all energy transfers, the overall effect is that the energy is spread out uniformly. Examples are the transfer of energy from hotter to cooler objects by conduction, radiation, or convection and the warming of our surroundings when we burn fuels.	PS 5d	Chapter 12 Chapter 13 Chapter 14 Chapter 15 Chapter 18
Interactions of Energy and Matter		
Waves, including sound and seismic waves, waves on water, and light waves, have energy and can transfer energy when they interact with matter.	PS 6a	Chapter 12 Chapter 14 Chapter 15 Chapter 17 Chapter 21

PHYSICAL SCIENCE CONTENT STANDARDS, *continued*

Standard	Code	Chapter Correlation
Interactions of Energy and Matter, *continued*		
Electromagnetic waves result when a charged object is accelerated or decelerated. Electromagnetic waves include radio waves (the longest wavelength), microwaves, infrared radiation (radiant heat), visible light, ultraviolet radiation, x-rays, and gamma rays. The energy of electromagnetic waves is carried in packets whose magnitude is inversely proportional to the wavelength.	**PS 6b**	Chapter 12 Chapter 15 Chapter 17 Chapter 18
Each kind of atom or molecule can gain or lose energy only in particular discrete amounts and thus can absorb and emit light only at wavelengths corresponding to these amounts. These wavelengths can be used to identify the substance.	**PS 6c**	Chapter 16 Chapter 20
In some materials, such as metals, electrons flow easily, whereas in insulating materials such as glass they can hardly flow at all. Semiconducting materials have intermediate behavior. At low temperatures some materials become superconductors and offer no resistance to the flow of electrons.	**PS 6d**	Chapter 17

EARTH AND SPACE SCIENCE CONTENT STANDARDS

Standard	Code	Chapter Correlation
Energy in the Earth System		
Earth systems have internal and external sources of energy, both of which create heat. The sun is the major external source of energy. Two primary sources of internal energy are the decay of radioactive isotopes and the gravitational energy from the Earth's original formation.	**ES 1a**	Chapter 12 Chapter 21 Chapter 22 Chapter 23
The outward transfer of Earth's internal heat drives convection circulation in the mantle that propels the plates comprising Earth's surface across the face of the globe.	**ES 1b**	Chapter 21
Heating of Earth's surface and atmosphere by the sun drives convection within the atmosphere and oceans, producing winds and ocean currents.	**ES 1c**	Chapter 13 Chapter 22
Global climate is determined by energy transfer from the sun at and near the Earth's surface. This energy transfer is influenced by dynamic processes such as cloud cover and the Earth's rotation, and static conditions such as the position of mountain ranges and oceans.	**ES 1d**	Chapter 22 Chapter 23
Geochemical Cycles		
The Earth is a system containing essentially a fixed amount of each stable chemical atom or element. Each element can exist in several different chemical reservoirs. Each element on Earth moves among reservoirs in the solid Earth, oceans, atmosphere, and organisms as part of geochemical cycles.	**ES 2a**	Chapter 22

Standard	Code	Chapter Correlation
Geochemical Cycles, *continued*		
Movement of matter between reservoirs is driven by the Earth's internal and external sources of energy. These movements are often accompanied by a change in the physical and chemical properties of the matter. Carbon, for example, occurs in carbonate rocks such as limestone, in the atmosphere as carbon dioxide gas, in water as dissolved carbon dioxide, and in all organisms as complex molecules that control the chemistry of life.	**ES 2b**	Chapter 22 Chapter 23
The Origin and Evolution of the Earth System		
The sun, the Earth, and the rest of the solar system formed from a nebular cloud of dust and gas 4.6 billion years ago. The early Earth was very different from the planet we live on today.	**ES 3a**	Chapter 19 Chapter 20
Geologic time can be estimated by observing rock sequences and using fossils to correlate the sequences at various locations. Current methods include using the known decay rates of radioactive isotopes present in rocks to measure the time since the rock was formed.	**ES 3b**	Chapter 21
Interactions among the solid Earth, the oceans, the atmosphere, and organisms have resulted in the ongoing evolution of the Earth system. We can observe some changes such as earthquakes and volcanic eruptions on a human time scale, but many processes such as mountain building and plate movements take place over hundreds of millions of years.	**ES 3c**	Chapter 14 Chapter 21 Chapter 22
Evidence for one-celled forms of life—the bacteria—extends back more than 3.5 billion years. The evolution of life caused dramatic changes in the composition of the Earth's atmosphere, which did not originally contain oxygen.	**ES 3d**	Chapter 22
The Origin and Evolution of The Universe		
The origin of the universe remains one of the greatest questions in science. The "big bang" theory places the origin between 10 and 20 billion years ago, when the universe began in a hot dense state; according to this theory, the universe has been expanding ever since.	**ES 4a**	Chapter 20
Early in the history of the universe, matter, primarily the light atoms hydrogen and helium, clumped together by gravitational attraction to form countless trillions of stars. Billions of galaxies, each of which is a gravitationally bound cluster of billions of stars, now form most of the visible mass in the universe.	**ES 4b**	Chapter 20
Stars produce energy from nuclear reactions, primarily the fusion of hydrogen to form helium. These and other processes in stars have led to the formation of all the other elements.	**ES 4c**	Chapter 20

NATIONAL SCIENCE EDUCATION STANDARDS

Safety in Your Laboratory

Risk Assessment

MAKING YOUR LABORATORY A SAFE PLACE TO WORK AND LEARN

Concern for safety must begin before any activity in the classroom and before students enter the lab. A careful review of the facilities should be a basic part of preparation for each school term. You should investigate the physical environment, identify any safety risks, and inspect your work areas for compliance with safety regulations.

The review of the lab should be thorough, and all safety issues must be addressed immediately. Keep a file of your review, and add to the list each year. This will allow you to continue to raise the standard of safety in your lab and classroom.

Many classroom experiments, demonstrations, and other activities are classics that have been used for years. This familiarity may lead to a comfort that can obscure inherent safety concerns. Review all experiments, demonstrations, and activities for safety concerns before presenting them to the class. Identify and eliminate potential safety hazards.

1. **Identify the Risks**

   Before introducing any activity, demonstration, or experiment to the class, analyze it and consider what could possibly go wrong. Carefully review the list of materials to make sure they are safe. Inspect the equipment in your lab or classroom to make sure it is in good working order. Read the procedures to make sure they are safe. Record any hazards or concerns you identify.

2. **Evaluate the Risks**

   Minimize the risks you identified in the last step without sacrificing learning. Remember that no activity you perform in the lab or classroom is worth risking injury. Thus, extremely hazardous activities, or those that violate your school's policies, must be eliminated. For activities that present smaller risks, analyze each risk carefully to determine its likelihood. If the pedagogical value of the activity does not outweigh the risks, the activity must be eliminated.

3. **Select Controls to Address Risks**

   Even low-risk activities require controls to eliminate or minimize the risks. Make sure that in devising controls you do not substitute an equally or more hazardous alternative. Some control methods include the following:

   - Explicit verbal and written warnings may be added or posted.

   - Equipment may be rebuilt or relocated, have parts replaced, or be replaced entirely by safer alternatives.

   - Risky procedures may be eliminated.

   - Activities may be changed from student activities to teacher demonstrations.

4. **Implement and Review Selected Controls**

   Controls do not help if they are forgotten or not enforced. The implementation and review of controls should be as systematic and thorough as the initial analysis of safety concerns in the lab and laboratory activities.

SOME SAFETY RISKS AND PREVENTATIVE CONTROLS

The following list describes several possible safety hazards and controls that can be implemented to resolve them. This list is not complete, but it can be used as a starting point to identify hazards in your laboratory.

Identified risk	Preventative control
Facilities and Equipment	
Lab tables are in disrepair, room is poorly lighted and ventilated, faucets and electrical outlets do not work or are difficult to use because of their location.	Work surfaces should be level and stable. There should be adequate lighting and ventilation. Water supplies, drains, and electrical outlets should be in good working order. Any equipment in a dangerous location should not be used; it should be relocated or rendered inoperable.
Wiring, plumbing, and air circulation systems do not work or do not meet current specifications.	Specifications should be kept on file. Conduct a periodic review of all equipment, and document compliance. Damaged fixtures must be labeled as such and must be repaired as soon as possible.
Eyewash fountains and safety showers are present but no one knows anything about their specifications.	Ensure that eyewash fountains and safety showers meet the requirements of the ANSI standard (Z358.1).
Eyewash fountains are checked and cleaned once at the beginning of each school year. No records are kept of routine checks and maintenance on the safety showers and eyewash fountains.	Flush eyewash fountains for 5 min. every month to remove any bacteria or other organisms from pipes. Test safety showers (measure flow in gallons per min) and eyewash fountains every 6 months and keep records of the test results.
Labs are conducted in multipurpose rooms, and equipment from other courses remains accessible.	Only the items necessary for a given activity should be available to students. All equipment should be locked away when not in use.
Students are permitted to enter or work in the lab without teacher supervision.	Lock all laboratory rooms whenever a teacher is not present. Supervising teachers must be trained in lab safety and emergency procedures.
Safety equipment and emergency procedures	
Fire and other emergency drills are infrequent, and no records or measurements are made of the results of the drills.	Always carry out critical reviews of fire or other emergency drills. Be sure that plans include alternate routes. Don't wait until an emergency to find the flaws in your plans.
Emergency evacuation plans do not include instructions for securing the lab in the event of an evacuation during a lab activity.	Plan actions in case of emergency: establish what devices should be turned off, which escape route to use, and where to meet outside the building.
Fire extinguishers are in out-of-the-way locations, not on the escape route.	Place fire extinguishers near escape routes so that they will be of use during an emergency.
Fire extinguishers are not maintained. Teachers are not trained to use them.	Document regular maintenance of fire extinguishers. Train supervisory personnel in the proper use of extinguishers. Instruct students not to use an extinguisher but to call for a teacher.

Identified risk	Preventative control
Safety equipment and emergency procedures, *continued*	
Teachers in labs and neighboring classrooms are not trained in CPR or first aid.	Teachers should receive training. The American Red Cross and other groups offer training. Certifications should be kept current with frequent refresher courses.
Teachers are not aware of their legal responsibilities in case of an injury or accident.	Review your faculty handbook for your responsibilities regarding safety in the classroom and laboratory. Contact the legal counsel for your school district to find out the extent of their support and any rules, regulations, or procedures you must follow.
Emergency procedures are not posted. Emergency numbers are kept only at the switchboard or main office. Instructions are given verbally only at the beginning of the year.	Emergency procedures should be posted at all exits and near all safety equipment. Emergency numbers should be posted at all phones, and a script should be provided for the caller to use. Emergency procedures must be reviewed periodically, and students should be reminded of them at the beginning of each activity.
Spills are handled on a case-by-case basis and are cleaned up with whatever materials happen to be on hand.	Have the appropriate equipment and materials available for cleaning up; replace them before expiration dates. Make sure students know to alert you to spilled chemicals, blood, and broken glass.
Work habits and environment	
Safety wear is only used for activities involving chemicals or hot plates.	Aprons and goggles should be worn in the lab at all times. Long hair, loose clothing, and loose jewelry should be secured.
There is no dress code established for the laboratory; students are allowed to wear sandals or open-toed shoes.	Open-toed shoes should never be worn in the laboratory. Do not allow any footwear in the lab that does not cover feet completely.
Students are required to wear safety gear but teachers and visitors are not.	Always wear safety gear in the lab. Keep extra equipment on hand for visitors.
Safety is emphasized at the beginning of the term but is not mentioned later in the year.	Safety must be the first priority in all lab work. Students should be warned of risks and instructed in emergency procedures for each activity.
There is no assessment of students' knowledge and attitudes regarding safety.	Conduct frequent safety quizzes. Only students with perfect scores should be allowed to work in the lab.
You work alone during your preparation period to organize the day's labs.	Never work alone in a science laboratory or a storage area.
Safety inspections are conducted irregularly and are not documented. Teachers and administrators are unaware of what documentation will be necessary in case of a lawsuit.	Safety reviews should be frequent and regular. All reviews should be documented, and improvements must be implemented immediately. Contact legal counsel for your district to make sure your procedures will protect you in case of a lawsuit.

Identified risk	Preventative control
Purchasing, storing, and using chemicals	
The storeroom is too crowded, so you decide to keep some equipment on the lab benches.	Do not store reagents or equipment on lab benches and keep shelves organized. Never place reactive chemicals (in bottles, beakers, flasks, wash bottles, etc.) near the edges of a lab bench.
You prepare solutions from concentrated stock to save money.	Reduce risks by ordering diluted instead of concentrated substances.
You purchase plenty of chemicals to be sure that you won't run out or to save money.	Purchase chemicals in class-size quantities. Do not purchase or have on hand more than one year's supply of each chemical.
You don't generally read labels on chemicals when preparing solutions for a lab, because you already know about a chemical.	Read each label to be sure it states the hazards and describes the precautions and first aid procedures (when appropriate) that apply to the contents in case someone else has to deal with that chemical in an emergency.
You never read the Material Safety Data Sheets (MSDSs) that come with your chemicals.	Always read the Material Safety Data Sheet (MSDS) for a chemical before using it and follow the precautions described. File and organize MSDSs for all chemicals where they can be found easily in case of an emergency.
The main stockroom contains chemicals that haven't been used for years.	Do not leave bottles of chemicals unused on the shelves of the lab for more than one week or unused in the main stockroom for more than one year. Dispose of or use up any leftover chemicals.
No extra precautions are taken when flammable liquids are dispensed from their containers.	When transferring flammable liquids from bulk containers, ground the container, and before transferring to a smaller metal container, ground both containers.
Students are told to put their broken glass and solid chemical wastes in the trash can.	Have separate containers for trash, for broken glass, and for different categories of hazardous chemical wastes.
You store chemicals alphabetically instead of by hazard class. Chemicals are stored without consideration of possible emergencies (fire, earthquake, flood, etc.), which could compound the hazard.	Use MSDSs to determine which chemicals are incompatible. Store chemicals by the hazard class indicated on the MSDS. Store chemicals that are incompatible with common fire-fighting media like water (such as alkali metals) or carbon dioxide (such as alkali and alkaline-earth metals) under conditions that eliminate the possibility of a reaction with water or carbon dioxide if it is necessary to fight a fire in the storage area.
Corrosives are kept above eye level, out of reach from anyone who is not authorized to be in the storeroom.	Always store corrosive chemicals on shelves below eye level. Remember, fumes from many corrosives can destroy metal cabinets and shelving.
Chemicals are kept on the stockroom floor on the days that they will be used so that they are easy to find.	Never store chemicals or other materials on floors or in the aisles of the laboratory or storeroom, even for a few minutes.

Safety Symbols and Safety Guidelines for Students

EYE PROTECTION

- Wear safety goggles, and know where the eyewash station is located and how to use it.
- Swinging objects can cause serious injury.
- Avoid directly looking at a light source, as this may cause permanent eye damage.

HAND SAFETY

- Wear latex or nitrile gloves to protect yourself from chemicals in the lab.
- Use a hot mitt to handle resistors, light sources, and other equipment that may be hot. Allow equipment to cool before handling it and storing it.

CLOTHING PROTECTION

- Wear a laboratory apron to protect your clothing.
- Tie back long hair, secure loose clothing, and remove loose jewelry to prevent their getting caught in moving parts or coming in contact with chemicals.

HEATING SAFETY

- When using a Bunsen burner or a hot plate, always wear safety goggles and a laboratory apron to protect your eyes and clothing. Tie back long hair, secure loose clothing, and remove loose jewelry.
- Never leave a hot plate unattended while it is turned on.
- If your clothing catches on fire, walk to the emergency lab shower, and use the shower to put out the fire.
- Wire coils may heat up rapidly during experiments. If heating occurs, open the switch immediately, and handle the equipment with a hot mitt.
- Allow all equipment to cool before storing it.

CHEMICAL SAFETY

- Do not eat or drink anything in the lab. Never taste chemicals.
- If a chemical gets on your skin or clothing or in your eyes, rinse it immediately with lukewarm water, and alert your teacher.
- If a chemical is spilled, tell your teacher, but do not clean it up yourself unless your teacher says it is OK to do so.

ELECTRICAL SAFETY

- Never close a circuit until it has been approved by your teacher. Never rewire or adjust any element of a closed circuit.
- Never work with electricity near water; be sure the floor and all work surfaces are dry.
- If the pointer of any kind of meter moves off the scale, open the circuit immediately by opening the switch.
- Light bulbs or wires that are conducting electricity can become very hot.
- Do not work with any batteries, electrical devices, or magnets other than those provided by your teacher.

GLASSWARE SAFETY

- If a thermometer breaks, notify your teacher immediately.
- Do not heat glassware that is broken, chipped, or cracked. Always use tongs or a hot mitt to handle heated glassware and other equipment because it does not always look hot when it is hot. Allow the equipment to cool before storing it.
- If a piece of glassware breaks, do not pick it up with your bare hands. Place broken glass in a specially designated disposal container.
- If a light bulb breaks, notify your teacher immediately. Do not remove broken bulbs from sockets.

WASTE DISPOSAL

- Use a dustpan, brush, and heavy gloves to carefully pick up broken glass, and dispose of it in a container specifically provided for this purpose.
- Dispose of any chemical waste only as instructed by your teacher.

HYGIENIC CARE

- Keep your hands away from your face and mouth.
- Always wash your hands thoroughly when you are done with an experiment.

Master Materials List

The following list indicates the quantities needed for 10 lab groups to perform the labs and quick activities in the textbook. If you have a different number of lab groups to plan for, scale the quantities accordingly.

A number alone indicates an end-of-chapter **Design Your Own Lab** or a **Skill-Builder Lab.** QA indicates a **Quick Activity** and QL indicates a **QuickLAB.** These codes are followed by numbers for the chapter and section in which they occur.

WARD'S is the supplier for *Holt Science Spectrum: Physical Science.* Catalog numbers for WARD'S are provided.

WARD'S Natural Science Establishment
5100 W. Henrietta Road
P.O. Box 92912
Rochester, New York, 14692-9012
1-800-962-2660

Chemicals and Consumable Materials	Qty.	Chapter, Quick Acivity, and/or QuickLAB	WARD'S Ordering Code
Aluminum foil pan, round, pkg. of 10	2	QA1.2, QL4.4	15 R 9890
Aluminum foil roll, 12″ X 25′	1	19	15 R 1009
Antacid tablets, pkg. of 75	1	QL8.3	37 R 1863
Apron, disposable polyethylene, box of 100	5	throughout	15 R 1050
Baking powder, 7 oz	1	QL8.1	37 R 2270
Balloons, pkg. of 12	3	QA16.1, QA20.3	15 R 0017
Battery, 6 V	20	QA16.3, QL16.2	15 R 3263
Battery, alkaline, 1.5 V, size D	40	16, 17, QL17.2	15 R 3247
Bleach, 1 pint	1	QL8.1	37 R 5554
Borax	1	QL5.4	38 R 4050
Box, corrugated, pkg. of 10	22	12, 19, QA21.1, QL16.2	18 R 1395
Cellophane, blue sheet, 20″ X 60″	1	QL15.2	15 R 9886
Cellophane, green sheet, 20″ X 60″	1	QL15.2	15 R 9885
Cellophane, red sheet, 20″ X 60″	1	QL15.2	15 R 9884
Chalk, white, box of 12	1	QL16.2	15 R 4637
Clay, red modeling, 1 lb	20	19, QL21.1	15 R 4640
Cloth, wool, 12″ X 24″	10	QA16.1	15 R 2537
Copper foil, 100 g	10	QL6.4	37 R 2513
Copper metal, piece	1	4, QA16.2	37 R 2202
Craft sticks, pkg. of 30	1	5	15 R 9893
Detergent, with phosphate, 50 g	1	QL8.1	15 R 1287
Distilled water, 1 gallon	3	4, QA5.4, QA6.4, QL2.3	88 R 7005
Dowel, wooden, 1″ X 8″, pkg. of 6	2	QA14.2	15 R 0082
Dowel, wooden, 1/4″ X 5″, pkg. of 12	2	18, QL16.2	15 R 0080

Chemicals and Consumable Materials, *continued*	Qty.	Chapter, Quick Acivity, and/or QuickLAB	WARD'S Ordering Code
Electrical tape	1	17	15 R 5020
Electrode, aluminum, 1 pkg. of 12	1	17	16 R 0081
Electrode, copper, pkg. of 12	1	QA16.2	16 R 0082
Electrode, iron, pkg. of 12	1	17	16 R 0083
Electrode, nickel, pkg. of 12	1	17	16 R 0085
Electrode, zinc, pkg. of 12	1	17, QA16.2	16 R 0086
Ethyl alcohol (ethanol), anhydrous	1	5, QA7.2	39 R 0273
Filter paper, pkg. of 100	1	QL2.3	15 R 2831
Gloves, latex disposable, medium, pkg. of 100	4	throughout	15 R 1071
Gloves, polyethylene disposable, medium	60	2, 11	15 R 1073
Glycerol (glycerin), 500 mL	1	22	39 R 1438
Hydrochloric acid, 1.0 M	2	6	37 R 8605
Hydrogen peroxide, 2% solution	1	QA6.4	37 R 8450
Index cards, 3 X 5, lined, pkg. of 1000	1	19, QA11.3	15 R 9807
Iodine, tincture of	1	QA7.2	37 R 2383
Iron filings, fine	1	QA6.4	37 R 2312
Latex solution, red, 1 L	1	5	37 R 2571
Light bulb, 150 W	20	19, QL13.2, QL23.1	36 R 4173
Light source for optical bench	10	15	16 R 0026
Limestone, gray, pkg. of 10	1	1	47 R 4602
Litmus paper, blue, pkg. of 12 vials	1	QL8.1, QL8.3	15 R 3105
Litmus paper, red, pkg. of 12 vials	1	QL8.1, QL8.3	15 R 3107
Magnesium, metal ribbon	1	QL6.4	37 R 2850

Chemicals and Consumable Materials, *continued*	Qty.	Chapter, Quick Acivity, and/or QuickLAB	WARD'S Ordering Code
Masking tape, 3/4″ X 60 yds, pkg. of 3	4	11, 12, 14, 16, 19, QA3.1, QA23.3	15 R 9828
Matches, wooden safety, box of 300	1	13, QA1.2, QL6.4	15 R 9427
Metal coffee can	10	QA3.1	17 R 2111
Miniature lamp, GE 458 type, pkg. of 10	1	QA13.3, QL13.2	16 R 0538
Nail, (8d, Common), pkg. of 35	1	14, QL16.2, QL17.2	15 R 9478
Olive oil, 500 mL	1	QL23.3	39 R 2514
Paper cups, 5 oz, pkg. of 100	1	4, 5	15 R 9830
Paper towels, 2-ply, 90-sheet roll	15	4, 5, 7, 14, QL2.3	15 R 9844
Paraffin candles, pkg. of 12	2	13, QA1.2	15 R 0055
Paraffin, refined, white wax, 500 g	1	4	39 R 2860
Petroleum jelly, 1 oz	5	22	15 R 9832
Plastic cup with lid, 9 oz, pkg. of 25	4	9, QA2.3, QA11.1, QL2.3	18 R 3676
Plastic foam cups, 8 oz, pkg. of 50	1	14, QL13.1	18 R 3677
Plastic spoons, pkg. of 100	1	QA13.2, QA15.1, QL2.3, QL8.3, QL16.2	15 R 9800
Plastic stirrers, pkg. of 500	1	QL8.3	15 R 9894
Potassium bromide, 100 g	1	QL6.3	37 R 2734
Potassium iodide, granular	1	QL6.3	37 R 4770
Rubber bands, assorted, 1/4 lb	1	11, QA5.4	15 R 9824
Sand, black, 32 oz	1	14	45 R 1987
Sandpaper, 180 grit, pkg. of 10	1	QL6.4	15 R 3010
Silver nitrate	1	QL6.3	37 R 5150
Soap, liquid antibacterial, 8 oz	5	QA8.3, QL8.1	18 R 1533
Sodium bicarbonate (baking soda), 1 lb	1	1, 2, QA2.3, QL8.1	37 R 5467
Sodium chloride	1	1, QL5.1, QL6.3	37 R 5487
Sodium silicate	1	5	36 R 5671
Steel wool, medium, pkg. of 16 pads	1	QL6.4	15 R 8798
String, 1/2 lb spool	5	1, 10, 14, QA12.2, QA15.1, QA17.1, QL13.1	15 R 9863
Sugar cubes	1	QL6.4	87 R 2203
Sugar, granular, 5 lb	1	QL5.1	39 R 3180
Swab applicator, pkg. of 100	1	QA7.2	14 R 5502
Syringe, 60 cc, disposable	22	10	14 R 1620

Chemicals and Consumable Materials, *continued*	Qty.	Chapter, Quick Acivity, and/or QuickLAB	WARD'S Ordering Code
Toothpicks, box of 800	1	23, QA8.3	15 R 4019
Twist ties, 1/4″ X 200′ roll	1	2, 7	14 R 0947
Vinegar, white, 1 pint	3	2, 5, QA2.3, QL6.4, QL8.1, QL8.3	39 R 0138
Wax paper, 12″ X 75′ roll	1	QA8.1, QL8.3	15 R 9857
Weighing papers, 3 1/8″ circle	1	2	15 R 2000
White glue	10	QL5.4	15 R 9806
Wire, annunciator (Bell), 1 lb spool	1	17, QA16.2	15 R 9425
Wire, bare copper magnet, 1 lb spool	1	13	15 R 9235
Wire, copper, PVC coated, 100′	1	16, 17, QA16.2, QA16.3, QL16.2, QL17.2	16 R 0549
Wire, nichrome, 18 gauge, 4 oz spool	1	13	15 R 9360
Wire, nichrome, 22 gauge, 4 oz spool	1	13	15 R 9370
Zinc, 100 pieces	1	3, QA16.2	37 R 6328
Zinc, sheet 30.48 cm X 30.48 cm	2	6, QL6.4	37 R 2347
Zipper resealable bags, pkg. of 10	4	2, 8, QA22.1, QL5.4	18 R 6921

Equipment and Reusable Materials	Qty.	Chapter, Quick Acivity, and/or QuickLAB	WARD'S Ordering Code
Alligator connector clip	10	QL16.2	15 R 9473
Aluminum metal sheet	1	4, 19	37 R 1883
Balance, triple-beam	10	1, 2, 3, 4, 9, 12, 23, QA3.1, QL2.2	15 R 6057
Bathroom scale, lb/kg	5	QL12.1	15 R 3800
Battery holder, double D cell	20	17	16 R 0603
Battery holder, single D cell	10	16	16 R 0602
Beaker, 1000 mL	10	7	17 R 4080
Beaker, 250 mL	20	1, 6, QA11.1, QL2.2, QL5.4, QL8.3	17 R 4040
Beaker, 400 mL	40	2, 3, 4, 18, 22, QL13.1	17 R 4050
Beaker, 50 mL	10	5, QL8.1	17 R 4010
Beaker, 500 mL	10	QA7.1	17 R 4060
Bottle, HDPE, 125 mL	20	23	18 R 0081

Equipment and Reusable Materials, *continued*	Qty.	Chapter, Quick Acivity, and/or QuickLAB	WARD'S Ordering Code
Box, hinged clear plastic	10	1	18 R 6560
Bunsen burner, standard	10	23, QL5.1, QL6.4	15 R 0612
Calculator, TI-1706	10	21, QA18.3	27 R 3055
Clamp, vinyl, 3-finger	20	23	15 R 0699
Clock, 7″ diameter wall	1	QL13.1, QL13.2	15 R 1492
Clothespin, large wooden, pkg. of 12	10	13	15 R 0018
Compass, ball-bearing, with pencil	10	21	15 R 4648
Compass, magnetic	10	QA17.1, QL17.2	12 R 0602
Cork sheet	1	QL16.2	14 R 7602
Dissection pan without wax	10	QA14.2, QA21.1, QL23.3	14 R 7010
Filter funnel	10	QL2.3	18 R 1462
Filter, polarizing	10	QA14.1	24 R 0891
Flasks, sidearm, 250 mL	20	6	17 R 3305
Galvanometer	10	QA16.2, QL17.2	16 R 0537
Glass plate, 8″ X 8″, pkg. of 12	2	23	15 R 3821
Glass tubing, flint, 6 mm, 1 lb	1	23	17 R 0941
Globe, world relief	10	QL23.1	80 R 5630
Gloves, Heat-Defier	10	4	15 R 1095
Goggles, general purpose safety	30	throughout	15 R 3046
Graduated cylinder, glass, 500 mL	10	18, 23	17 R 0580
Graduated cylinder, plastic, 10 mL	10	4, 6, QL6.3, QL6.4	18 R 1705
Graduated cylinder, plastic, 100 mL	10	2, 3, 4, QL2.2, QL13.1, QL13.2	18 R 1730
Graduated cylinder, plastic, 25 mL	20	1, 4, 6	18 R 1710
Graduated cylinder, plastic, 250 mL	10	QA7.1	18 R 1740
Hammer, rubber	10	18, QL15.1	16 R 0568
Hot plate, single burner	10	3, 4, QA21.1, QL13.1	15 R 7999
Jar, clear polystyrene, 16 oz	10	QA7.1	18 R 1635
Jar, clear polystyrene, 8 oz	10	QA9.1	18 R 1634
Jar, glass, wide-mouth, 1/2 gal	10	5	17 R 2070
Jar, glass, wide-mouth, 240 mL	10	19, QA22.1	17 R 2040
Jar lid, 53 mm	10	QA9.1	17 R 2133
Jar lid, 70 mm	10	19	17 R 2145
Jar lid, 89 mm	10	QA7.1	17 R 2153
Knife, plastic	20	QA10.2, QA16.2, QL21.1	25 R 8128

Equipment and Reusable Materials, *continued*	Qty.	Chapter, Quick Acivity, and/or QuickLAB	WARD'S Ordering Code
Lamp receptacle, plastic, pkg. of 12	2	QA16.3, QL16.2	16 R 0064
Lamp with reflector and clamp	10	19, QL13.2, QL23.1	36 R 4168
Lens and mirror support	10	15	16 R 0021
Lens, convex, 10 cm to 15 cm focal length	10	15	25 R 2005
Level, torpedo	10	QL12.4	12 R 0242
Magnet, steel bar	20	4, QA17.1, QL2.3, QL17.2	13 R 0115
Magnifying glass	10	20	24 R 1112
Map of U.S., raised relief	10	17	33 R 0460
Marbles, colored, pkg. of 25	1	QA7.1	15 R 3399
Marker, black lab	10	8, 19, QA20.3	15 R 3083
Marking pen for glass, black wax	10	QL6.3	15 R 1155
Mass, hooked, 2000 g	10	QL11.1	15 R 3724
Mass set, hooked, 10 g-200 g	10	11	15 R 3717
Meter stick, wood	10	1, 5, 12, 14, 18, 19, QL12.1, QL12.4	15 R 4065
Multimeter, economy analog	10	16	15 R 9426
Optical bench meter stick supports	10	15	16 R 0023
Optical bench screen, pkg. of 5	2	15	16 R 0027
Optical bench screen support	10	15	16 R 0028
Paper clips, #1, pkg. of 10 boxes	1	11, 17, QL6.4	15 R 9815
Pencils, no. 2, box of 12	2	9, 11, 12	15 R 9816
Pendulum clamp	10	QL12.4	15 R 3124
Ping-Pong balls, pkg. of 6	40	QA5.2	15 R 3636
Pipet bulbs, 1/4 oz, pkg. of 10	1	QL8.1, QL8.3	15 R 0511
Pipets, sterile plastic, 10 mL, pkg. of 100	2	8, QA8.3, QL8.1, QL8.3	17 R 4853
Plastic Bowl, 40 oz, pkg. of 5	6	QA13.2	18 R 7202
Pneumatic trough, polypropylene	10	23	15 R 7565
Resistors, carbon, 10 ohm, pkg. of 10	2	16	16 R 0001
Resonance tube, 45 X 4 cm	10	18	17 R 0179
Rods, metal or wooden	40	10	
Ruler, plastic	10	3, 4, 5, 7, 8, 11, 13, 14, 20, 21, QA4.1, QA15.1,	15 R 4655

Equipment and Reusable Materials, *continued*	Qty.	Chapter, Quick Acivity, and/or QuickLAB	WARD'S Ordering Code
Ruler, plastic, *continued*		QA16.2, QA21.1, QA22.2, QA23.3, QL21.1	
Scissors, nickel-plated	10	5, 6, 9, 10, 14, 19, QL21.1	14 R 0525
Solenoid	10	QL17.2	20 R 1395
Spring scale, pull type, 3 kg/30 N	20	10, QL11.1	15 R 3774
Spring, wave demonstration	10	QL14.1	16 R 0513
Stapler	10	QA12.2	15 R 1955
Stirring rods, glass, pkg. of 10	10	QL16.2	17 R 6005
Stopper, 1-hole black rubber, size 2	10	23	15 R 8482
Stopper, 2-hole black rubber, size 2	10	23	15 R 8512
Stoppers, black, size 6	20	6	15 R 8466
Stopwatch, digital	10	4, 6, 7, 12, 13, 14, QL5.1, QL12.1	15 R 0512
Support, rectangular, large, with base	10	8, 14, 23	15 R 0667
Support, rectangular, small, with base	10	QA17.1	15 R 0719
Support ring, zinc plated	10	QA17.1	15 R 0707
Switch, single pole, single throw	10	QA16.3	16 R 0547
Tape measure, metric/English	10	7	15 R 2541
Test tube rack, 16 mm	10	8, QL6.3	18 R 4213
Test tube with rim, Pyrex®, 13 X 100 mm	70	1, 8, QA6.4, QL5.1, QL6.3, QL6.4	17 R 0610
Test tube with rim, Pyrex®, 25 X 150 mm	20	23	17 R 0655
Thermometer –20°C to 110°C, red alcohol	20	5, 7, 19, 22, QA22.1, QL13.1, QL13.2	15 R 1416
Thermometer, wall model	1	1, 18	23 R 1400
Tongs	10	19, QL5.1, QL6.4, QL13.1	14 R 0960
Tubing, amber latex, 3/16″ X 1/16″	1	6, 23	15 R 1133
Tuning fork, 128 Hz	18	10	16 R 0556
Tuning fork, 384 Hz	10	18	16 R 0561
Tuning forks, set of 4	10	QL15.1	16 R 0565
Washers, pkg. of 12	40	QL13.1	15 R 0030
Wire stripper, crimping tool	10	17	15 R 0749
Wood block, 3″ X 5″ X 1-3/4″	10	QA9.2, QL16.2	16 R 0606

Materials Obtained Locally These common materials are not carried by WARD'S, but are easily obtained locally.	Qty.	Chapter, Quick Acivity, and/or QuickLAB
Basketball, volleyball or soccer ball	10	1
Board, 6 ft	10	12
Books	30	10, 12, QA5.2, QA12.2
Brass key	10	QL16.2
Candle holder	10	13
Coat hanger, metal	10	QA12.4
Comb or hairbrush	10	11
Dice	100	9
Dominoes	210	QA9.2
Flashlights, adjustable	30	QL18.2
Food coloring, box of 3	4	QA21.1
Golf ball	10	12
Graph paper	1	QL2.2
Ice		3, 4, 5, 6, 7, 18, QA13.2, QA22.1
Ice pick or awl	10	19
Lemon	10	QA16.2
Liver, raw		QA6.4
Mayonnaise		QL8.1
Metal hooks	20	QL16.2
Milk		QL8.1
Nail, aluminum	10	QL16.2
Paper ash		QL6.4
Paper, white	20	QA23.3, QL18.2, QL21.1
Plastic bags, small	20	7
Pennies	1300	QA9.1, QA11.1
Plastic rings from 6-pack holders	10	QA5.4
Popcorn kernels, bag	5	QA19.3
Powdered sugar, bag	1	QA2.3
Ribbon		QL14.1
Rolling pin	10	QL21.1
Shoebox	10	19
Sock	10	QA4.1
Soft drinks	40	7, QL8.1
Soup can, empty	30	QA22.2, QL13.2
Spoons, stainless steel	10	QA13.2
Tracing paper, pad	1	21
Water		3, 4, 14, 22, QA13.1, QL5.4, QL8.1, QL23.3
Water, mineral		QL8.1

About the Authors

Ken Dobson, Ph.D.

Dr. Dobson studied physics and education at Manchester and began teaching in 1956. At Wilson's grammar school in Camberwell he developed an early form of "self-study" for physics students. He has also served as head of science for Woodberry Down School in Hackney and Chief Awarder for Nuffield A-level Physics. He later was instrumental in developing the basic principles of "Suffolk Science," which became a national program for teaching science in over 600 schools in England. Dr. Dobson also served as a member of the Working Group that established the first National Curriculum in science. He has served as Honorary Editor of *Physics Education* from 1996 to 2000.

John Holman, Ph.D.

Dr. Holman studied natural sciences at the University of Cambridge and then taught chemistry at a number of independent and state schools. He has worked as a writer and science education specialist for Science and Technology in Society (SATIS), Salters Advanced Chemistry, Science Across the World, and other projects. He was head of Watford Grammar School for Boys, where he also taught chemistry. Currently, he is professor of chemical education and director of the Science Curriculum Centre at the University of York.

Michael Roberts, Ph.D.

Dr. Roberts was educated at Epsom College and Queens' College in England. He performed postgraduate research on the neurophysiology of annelids. He was a lecturer at the University of California at Santa Barbara and then head of biology at Marlborough College. During this time, he was also an external examiner in biology at the University of Botswana and Swaziland and the National University of Lesotho. Dr. Roberts was a research associate at Chelsea College (now King's College) and head of biology at Cheltenham College. He is now retired and focuses a great deal of time on writing textbooks.

Art Credits: Page v, vi (cl), vii (r), viii (molecules), Kristy Sprott; viii (tl), Morgan-Cain & Associates; ix (br), Kristy Sprott; xii (cl), Boston Graphics; xiii (tr), Uhl Studios, Inc.; xiv (tl), Tony Randazzo/American Artist's Rep., Inc.; xv (globe), Ortelius Design; xv (generator), Uhl Studios, Inc.

Photo Credits: Page v, Anglo-Australian Observatory/Photograph by David Malin; ix(br), Tom Myers/Photo Researchers, Inc; x, Bill Losh/Getty Images/Taxi. All other Table of Contents photos are credited within the chapter in which they appear.

Printed in the United States of America

ISBN 0-03-066471-3

2 3 4 5 6 048 07 06 05 04 03

Acknowledgements

Contributing Authors

Robert Davisson
Science Writer
Albuquerque, New Mexico

Mary Kay Hemenway, Ph.D.
Research Associate and Senior Lecturer
Department of Astronomy
The University of Texas at Austin
Austin, Texas

William G. Lamb, Ph.D.
Winningstad Chair in the Physical Sciences
Oregon Episcopal School
Portland, Oregon

Contributing Writers

David Bethel
Science Writer
Austin, Texas

Meredith Phillips
Science Writer
Brooklyn, New York

Rosemary Previte
Science Writer
Lexington, Massachusetts

Tracy Schagen
Science Writer
Austin, Texas

Inclusion Specialists

Joan A. Solorio
Special Education Director
Austin Independent School District
Austin, Texas

John A. Solorio
Multiple Technologies Lab Facilitator
Austin Independent School District
Austin, Texas

Feature Development

John M. Stokes
Science Writer
Socorro, New Mexico

Andrew Strickler
Science Writer
Oakland, California

Teacher Edition Development

Maria Hong
Science Writer
Austin, Texas

Bob Roth
Science Writer
Pittsburgh, Pennsylvania

Academic Reviewers

Mead Allison, Ph.D.
Associate Professor
Department of Geology and Earth Sciences
Tulane University
New Orleans, Louisiana

Eric Anslyn, Ph.D.
Professor of Chemistry
Department of Chemistry and Biochemistry
The University of Texas
Austin, Texas

Paul Asimow, Ph.D.
Assistant Professor of Geology and Geochemistry
Division of Geological and Planetary Sciences
California Institute of Technology
Pasadena, California

John A. Brockhaus, Ph.D.
Director of Mapping, Charting, and Geodesy Program
Department of Geography and Environmental Engineering
United States Military Academy
West Point, New York

Thomas Connolley, Ph.D.
Visiting Assistant Professor
Department of Mechanical Engineering
The University of Texas
San Antonio, Texas

Scott W. Cowley, Ph.D.
Associate Professor
Department of Chemistry and Geochemistry
Colorado School of Mines
Golden, Colorado

Nels F. Forsman, Ph.D.
Associate Professor of Geochemistry
Department of Physics and Astrophysics
University of North Dakota
Grand Forks, North Dakota

Gina Frey, Ph.D.
Professor of Chemistry
Department of Chemistry
Washington University
St. Louis, Missouri

Frank Guziec
Dishman Professor of Science
Department of Chemistry
Southwestern University
Georgetown, Texas

Vicki Hansen, Ph.D.
Professor of Geological Sciences
Department of Geology
Southern Methodist University
Dallas, Texas

Richard Hey, Ph.D.
Professor of Geophysics
School of Ocean and Earth
 Sciences Technology
University of Hawaii
Honolulu, Hawaii

Guy Indebetouw, Ph.D.
Professor of Physics
Department of Physics
Virginia Polytechnic Institute
 and State University
Blacksburg, Virginia

**Wendy L. Keeney-Kennicutt,
 Ph.D.**
*Associate Professor of
 Chemistry*
Chemistry Department
Texas A&M University
College Station, Texas

Samuel P. Kounaves
*Associate Professor of
 Chemistry*
Department of Chemistry
Tufts University
Medford, Massachusetts

David Lamp, Ph.D.
Associate Professor of Physics
Physics Department
Texas Tech University
Lubbock, Texas

Phillip LaRoe
Instructor
Department of Physics and
 Chemistry
Central Community College
Grand Isle, Nebraska

Joel Leventhal, Ph.D.
Emeritus Scientist
U.S. Geological Survey and
 Diversified Geochemistry
Lakewood, California

Joseph McClure, Ph.D.
Professor of Physics, Emeritus
Department of Physics
Georgetown University
Washington, D.C.

Gary Mueller, Ph.D.
*Associate Professor of Nuclear
 Engineering*
Department of Engineering
University of Missouri
Rolla, Missouri

Emily Neimeyer, Ph.D.
*Assistant Professor of
 Chemistry*
Department of Chemistry
Southwestern University
Georgetown, Texas

Hilary Olsen, Ph.D.
Research Scientist
Institute of Geophysics
The University of Texas
Austin, Texas

Brian Pagenkopf, Ph.D.
Professor of Chemistry
Department of Chemistry and
 Biochemistry
The University of Texas
Austin, Texas

Per F. Peterson, Ph.D.
Professor and Chair
Department of Nuclear
 Engineering
University of California
Berkeley, California

Barron Rector, Ph.D.
*Assistant Professor and
 Extension Range Specialist*
Texas Agricultural Extension
 Service
Texas A&M University
College Station, Texas

Dork Sahagian
*Research Professor,
 Stratigraphy and Basin
 Analysis, Geodynamics*
Global Analysis, Interpreta-
 tion, and Modeling Program
University of New Hampshire
Durham, New Hampshire

Charles Scaife, Ph.D.
Chemistry Professor
Department of Chemistry
Union College
Schenectady, New York

Fred Seaman, Ph.D.
Research Scientist and Chemist
Department of Pharmacologi-
 cal Chemistry
The University of Texas
Austin, Texas

Miles Silman, Ph.D.
Associate Professor of Biology
Department of Biology
Wake Forest University
Winston-Salem, North
 Carolina

Spencer Steinberg, Ph.D.
*Associate Professor, Environ-
 mental Organic Chemistry*
Chemistry Department
University of Nevada
Las Vegas, Nevada

Richard Storey, Ph.D.
*Dean of the Faculty and
 Professor of Biology*
Colorado College
Colorado Springs, Colorado

Jack B. Swift, Ph.D.
Associate Professor of Physics
Department of Physics
The University of Texas
Austin, Texas

***Acknowledgments continued
on page 902.***

Contents in Brief

Table of Contents

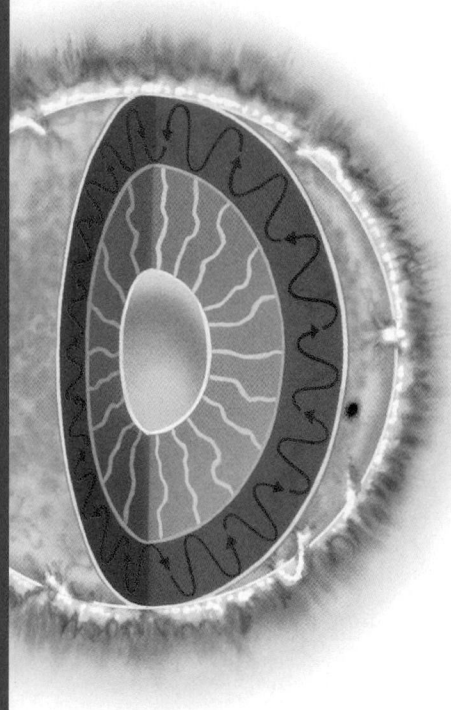

INTEGRATING
TECHNOLOGY
and *Society*

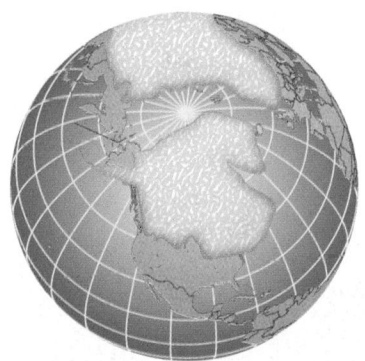

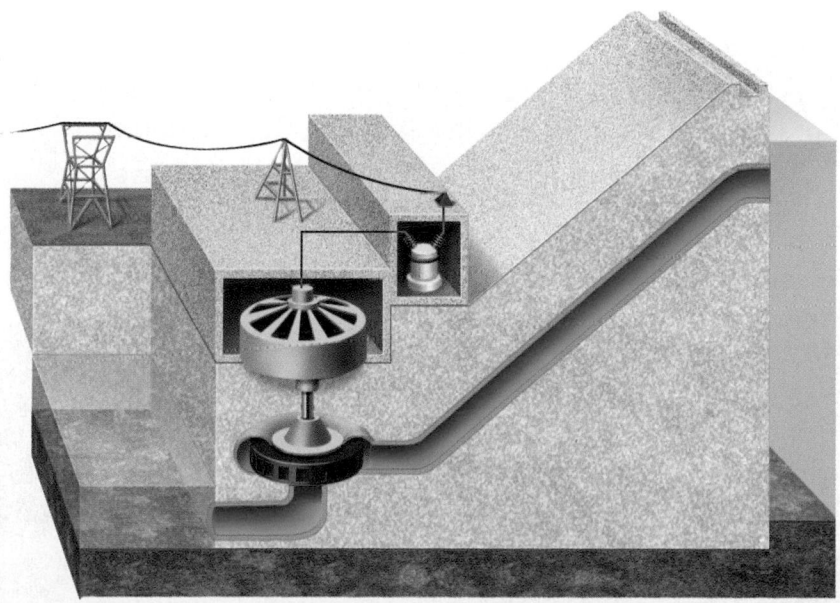

REFERENCE SECTION

LABORATORY EXPERIMENTS

Quick Lab

Quick Activity

How to Use YOUR TEXTBOOK

Your Roadmap for Success with Holt Science Spectrum

Read the Objectives
Objectives tell you what you'll need to know.
STUDY TIP Reread the objectives when studying for a test to be sure you know the material.

Study the Key Terms
Key Terms are listed for each section. Learn the definitions of these terms because you will most likely be tested on them. Use the on-page definitions to understand terms quickly as you read and study.
STUDY TIP If you don't understand a definition, reread the page where the term is introduced. The surrounding text should help make the definition easier to understand.

Take Notes and Get Organized
Keep a science notebook so that you are ready to take notes when your teacher reviews the material in class. Keep your assignments in this notebook so that you can review them when studying for the chapter test.

SECTION 3

Families of Elements

KEY TERMS
metal
nonmetal
semiconductor
alkali metal
alkaline-earth metal
transition metal
halogen
noble gas

internet connect
www.scilinks.org
Topic: Element Families
SciLinks code: HK4046
sciLINKS Maintained by the National Science Teachers Association

OBJECTIVES

▶ **Locate** alkali metals, alkaline-earth metals, and transition metals in the periodic table.
▶ **Locate** semiconductors, halogens, and noble gases in the periodic table.
▶ **Relate** an element's chemical properties to the electron arrangement of its atoms.

You may have wondered why groups in the periodic table are sometimes called families. Consider your own family. Though each member is unique, you all share certain similarities. All members of the family shown in **Figure 22A**, for example, have a similar appearance. Members of a family in the periodic table have many chemical and physical properties in common because they have the same number of valence electrons.

How Are Elements Classified?
Think of each element as a member of a family that is also related to other elements nearby. Elements are classified as metals or nonmetals, as shown in **Figure 22B**. This classification groups elements that have similar physical and chemical properties.

Figure 22
A Just like the members of this family,
B elements in the periodic table share certain similarities.

Note: Sometimes the boxed elements toward the right side of the periodic table are classified as a separate group and called semiconductors or metalloids.

120 CHAPTER 4

⬈ Be Resourceful, Use the Web

SCILINKS NSTA

Internet Connect boxes in your textbook take you to resources that you can use for science projects, reports, and research papers. Go to scilinks.org and type in the SciLinks code to get information on a topic.

go.hrw.com

Visit go.hrw.com
Find worksheets and other materials that go with your textbook at **go.hrw.com.** Click on the textbook icon and the table of contents to see all of the resources for each chapter.

Metals

Many elements are classified as metals. To further classify metals, similar metals are grouped together. There are four different kinds of metals. Two groups of metals are located on the left side of the periodic table. Other metals, like aluminum, tin, and lead, are located toward the right side of the periodic table. Most metals, though, are located in the middle of the periodic table.

The alkali metals are very reactive

Sodium is found in Group 1 of the periodic table, as shown in *Figure 23A*. Like other **alkali metals** it is soft and shiny and reacts violently with water. Sodium must be stored in oil, as in *Figure 23B*, to prevent it from reacting with moisture in the air.

An atom of an alkali metal is very reactive because it has one valence electron that can easily be removed to form a positive ion. You have already seen in Section 2 how lithium, another alkali metal, forms positive ions with a 1+ charge. Similarly, the valence electron of a sodium atom can be removed to form the positive sodium ion Na^+.

Because alkali metals such as sodium are so reactive, they are not found in nature as elements. Instead, they combine with other elements to form compounds. For example, the salt you use to season your food is actually the compound sodium chloride, NaCl.

conductor does

alkali metal one of the elements of Group 1 of the periodic table

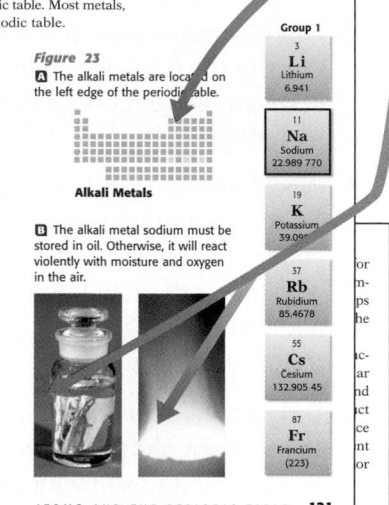

Figure 23
A The alkali metals are located on the left edge of the periodic table.

Alkali Metals

B The alkali metal sodium must be stored in oil. Otherwise, it will react violently with moisture and oxygen in the air.

Group 1

3
Li
Lithium
6.941

11
Na
Sodium
22.989 770

19
K
Potassium
39.09

37
Rb
Rubidium
85.4678

55
Cs
Cesium
132.905 45

87
Fr
Francium
(223)

ATOMS AND THE PERIODIC TABLE **121**

SECTION 3 REVIEW

SUMMARY

▶ Metals are shiny solids that conduct heat and electricity.

▶ Alkali metals, located in Group 1 of the periodic table, are very reactive.

▶ Alkaline-earth metals, located in Group 2, are less reactive than alkali metals.

▶ Transition metals, located in Groups 3–12, are not very reactive.

▶ Nonmetals usually do not conduct heat or electricity well.

▶ Nonmetals include the inert noble gases in Group 18, the reactive halogens in Group 17, and some elements in Groups 13–16.

▶ Semiconductors are nonmetals that are intermediate conductors of heat and electricity.

1. **Classify** the following elements as alkali, alkaline-earth, or transition metals based on their positions in the periodic table:
   a. iron, Fe   c. strontium, Sr
   b. potassium, K   d. platinum, Pt

2. **Predict** whether cesium forms Cs^+ or Cs^{2+} ions.

3. **Describe** why chemists might sometimes store reactive chemicals in argon, Ar. To which family does argon belong?

4. **Determine** whether the following substances are likely to be metals or nonmetals:
   a. a shiny substance used to make flexible bed springs
   b. a yellow powder from underground mines
   c. a gas that does not react
   d. a conducting material used within flexible wires

5. **Describe** why atoms of bromine, Br, are so reactive. To which family does bromine belong?

6. **Predict** the charge of a beryllium ion.

7. **Identify** which element is more reactive: lithium, Li, or barium, Ba.

8. **Creative Thinking** Imagine you are a scientist who has just discovered a new element. You have confirmed that the element is a metal but are unsure whether it is an alkali metal, an alkaline-earth metal, or a transition metal. Write a paragraph describing the additional tests you can do to further classify this metal.

WRITING SKILL

128 CHAPTER 4

Use the Illustrations and Photos

Art shows complex ideas and processes. Learn to analyze the art so that you better understand the material you read about in the text.

Tables and graphs display important information in an organized way to help you see relationships.

A picture is worth a thousand words. Look at the photographs to see relevant examples of the science concepts.

Answer the Section Reviews

Section Reviews test your knowledge over the main points of the section. Critical Thinking items challenge you to think about the material in greater depth and to find connections that you infer from the text.

STUDY TIP Use the summary to help you review the section. When you can't answer a question, reread the section. The answer is usually there.

Do Your Homework

Your teacher may assign worksheets to help you understand and remember the material in the chapter.

STUDY TIP Don't try to answer the questions without reading the text and reviewing your class notes. A little preparation up front will make your homework (and the test) a lot easier.

Holt Online Learning

Visit Holt Online Learning

If your teacher gives you a special password to log onto the **Holt Online Learning** site, you'll find your complete textbook on the Web. In addition, you'll find some great learning tools and practice quizzes. You'll be able to see how well you know the material from your textbook.

Visit CNN Student News

You'll find up-to-date events in science at **cnnstudentnews.com.**

Safety in the Laboratory

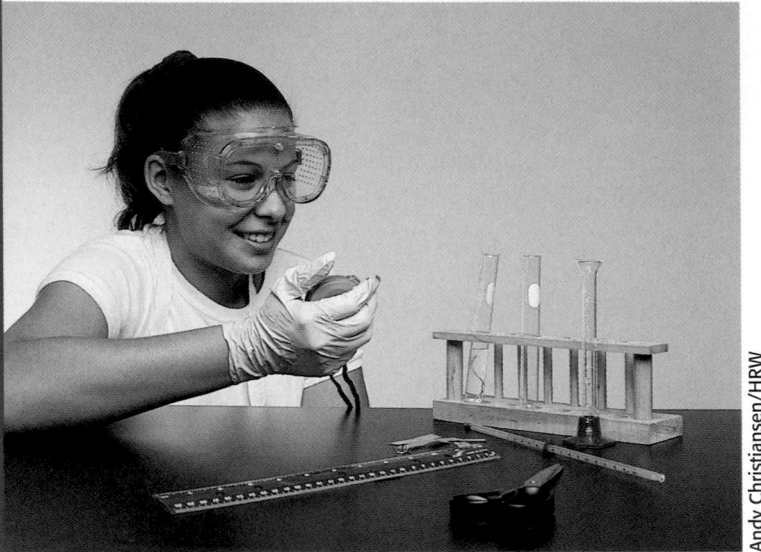

Andy Christiansen/HRW

Systematic, careful lab work is an essential part of any science program. The equipment and apparatus you will use involve various safety hazards, just as they do for working scientists. Your instructor will guide you in properly using the equipment and carrying out the experiments, but you must also take responsibility for your safety.

Anything can be dangerous if it is misused. Always follow the instructions and pay close attention to the safety notes.

These safety rules always apply in the lab

1. **Wear safety goggles, gloves, and a lab apron.**
   Wear these safety devices whenever you are in the lab. Keep the lab apron strings tied. If your safety goggles are uncomfortable or cloud up, ask for help.

2. **Do not wear contact lenses in the lab.**
   Contact lenses should not be worn during any investigations using chemicals (even if you are wearing goggles). In the event of an accident, chemicals can get behind contact lenses and cause serious eye damage. If your doctor requires that you wear contact lenses, you should wear eye-cup safety goggles in the lab.

3. **NEVER work alone in the lab.**
   Work in the lab only when supervised. Do not leave equipment unattended while it is in operation.

4. **Wear the right clothing for lab work.**
   Necklaces, neckties, dangling jewelry, long hair, and loose clothing can get caught in moving parts or catch on fire. Remove your wristwatch, and wear shoes that will protect your feet from chemical spills and falling objects.

5. **Only books and notebooks needed for the experiment should be in the lab.**
   Keep other items in your desk or locker.

6. **Read the entire experiment before entering the lab.**
   Memorize the safety precautions. Be familiar with the instructions for the experiment. Only authorized materials and equipment should be used. If you are not sure of something, ask your instructor about it.

7. **Always heed safety symbols and cautions listed in the experiments, listed on handouts, posted in the room, and given verbally by your instructor.**
   They are provided for a reason: YOUR SAFETY.

8. **Read chemical labels.**
   Follow the instructions and safety precautions stated on the labels.

9. **Be alert, and walk with care in the lab.**

10. **Know proper fire drill procedures and the location of fire exits and emergency equipment.**
    Make sure you know the procedures to follow in case of a fire or emergency.

11. **Know the location of and how to operate safety showers and eyewash stations.**

12. **If your clothing catches on fire, do not run; WALK to the safety shower, stand under it, and turn it on.**
    Alert your instructor while you do this.

13. **If you get a chemical in your eyes, walk immediately to the eyewash station, turn it on, and lower your head so that your eyes are in the running water.**
Hold your eyelids open with your thumbs and fingers, and roll your eyes. You have to flush your eyes continuously for at least 15 minutes. Alert your instructor while you do this.

14. **If you spill a chemical on your skin, wash it off with lukewarm water, and alert your instructor.**
If you spill a solid chemical on your clothing, brush it off carefully without scattering it. If you get liquid on your clothing, wash it off right away using the sink faucet. If the spill is on your pants or somewhere else that will not fit under the sink faucet, use the safety shower. Remove the affected clothing while under the shower.

15. **Report all accidents and spills, no matter how minor, to the instructor IMMEDIATELY.**
In addition, tell your instructor if you get a headache, feel sick to your stomach, or feel dizzy.

16. **The best way to prevent an accident is to stop it before it happens.**
If you have a close call, tell your instructor so that you can find a way to prevent it from happening again.

17. **DO NOT perform unauthorized experiments or use equipment and apparatus in a manner for which they were not intended.**
Use only materials and equipment listed in the instructions. Procedures should only be performed as described.

18. **Food, beverages, chewing gum, and tobacco products are NEVER permitted in the lab.**

19. **For all chemicals, take only what you need.**
However, if you happen to take too much and have some left over, DO NOT put it back into the container. Ask your instructor what to do with any leftover chemicals.

20. **NEVER taste chemicals. Do not touch chemicals or allow them to contact areas of bare skin.**

21. **Use a sparker to light a Bunsen burner.**
Do not use matches. Be sure that all gas valves are turned off when you leave the lab.

22. **Use extreme caution when working with hot plates or other heating devices.**
Keep your head, hands, hair, and clothing away from the flame or heating area, and turn the devices off when they are not in use. Remember that metal surfaces connected to the heated area will become hot by conduction. Remember that many metal, ceramic, and glass items do not always look hot when they are hot. Allow all items to cool before storing.

23. **Do not use electrical equipment with frayed or twisted wires.**

24. **Be sure your hands are dry before using electrical equipment.**
The area under and around electrical equipment should be dry; cords should not lie in puddles of spilled liquid.

25. **Do not let electrical cords dangle from work stations.**
Before plugging an electrical cord into a socket, be sure the equipment is turned OFF. When you are finished with the device, turn it off and unplug the device.

26. **Horseplay and fooling around in the lab are very dangerous.**

27. **Keep work areas and apparatus clean and neat.**
Always clean up any clutter made during the course of lab work, put away apparatus in an orderly manner, and report any damaged or missing items to your instructor.

28. **Always thoroughly wash your hands with soap and water at the conclusion of each lab.**

Safety in the Laboratory

Safety Symbols

The following symbols will appear in the laboratory experiments to emphasize important additional areas of caution.

EYE PROTECTION

▶ Wear safety goggles, and know where the eyewash station is located and how to use it. Contents under pressure may become projectiles and cause serious injury.

▶ Never look directly at the sun through any optical device or use direct sunlight to illuminate a microscope.

▶ If any substance gets into your eyes, notify your instructor immediately and flush your eyes with running water for at least 15 minutes.

CLOTHING PROTECTION

▶ Secure loose clothing, and remove dangling jewelry. Do not wear open-toed shoes or sandals in the lab.

▶ Wear an apron or lab coat to protect your clothing when you are working with chemicals.

▶ If a spill gets on your clothing, rinse it off immediately with water for at least 5 minutes while notifying your instructor.

CAUSTIC SUBSTANCES

▶ If a chemical gets on your skin, on your clothing, or in your eyes, rinse it immediately and alert your instructor.

▶ If a chemical is spilled on the floor or lab bench, alert your instructor, but do not clean it up yourself unless your instructor directs you to do so.

CHEMICAL SAFETY

▶ Always use caution when working with chemicals.

▶ Always wear appropriate protective equipment. Always wear eye goggles, gloves, and a lab apron or lab coat when you are working with any chemical or chemical solution.

▶ Never mix chemicals unless your instructor directs you to do so.

▶ Never taste, touch, or smell chemicals unless your instructor directs you to do so.

▶ Add an acid or base to water; never add water to an acid or base.

▶ Never return an unused chemical to its original container.

▶ Never transfer substances by sucking on a pipet or straw; use a suction bulb.

▶ Follow instructions for proper disposal.

ELECTRICAL SAFETY

▶ Never close a circuit until it has been approved by your instructor. Never rewire or adjust any element of a closed circuit.

▶ If the pointer of any kind of meter moves off the scale, open the circuit immediately by opening the switch.

▶ Do not place electrical cords in walking areas or let cords hang over a table edge in a way that could cause equipment to fall if the cord is accidentally pulled.

▶ Do not use equipment that has frayed electrical cords or loose plugs.

▶ Be sure that equipment is in the "off" position before you plug it in.

▶ Never use an electrical appliance around water or with wet hands or clothing.

▶ Be sure to turn off and unplug electrical equipment when you are finished using it.

HEATING SAFETY

▶ Avoid wearing hair spray or hair gel on lab days.

▶ Whenever possible, use an electric hot plate instead of an open flame as a heat source.

▶ When heating materials in a test tube, always angle the test tube away from yourself and others.

▶ Glass containers used for heating should be made of heat-resistant glass.

- Never leave a hot plate unattended while it is turned on.
- Wire coils may heat up rapidly during experiments. If heating occurs, open the switch immediately, and handle the equipment with heat-resistant gloves.
- Allow all equipment to cool before storing it.

SHARP OBJECTS
- Use knives and other sharp instruments with extreme care.
- Never cut objects while holding them in your hands. Place objects on a suitable work surface for cutting.
- Never use a double-edged razor in the lab.

HAND SAFETY
- To avoid burns, wear heat-resistant gloves whenever instructed to do so.
- Always wear protective gloves when working with an open flame, chemicals, and solutions.
- If you do not know whether an object is hot, do not touch it.
- Use tongs when heating test tubes. Never hold a test tube in your hand to heat the test tube.

EXPLOSION SAFETY
- Use flammable liquids only in small amounts.
- When working with flammable liquids, be sure that no one else in the lab is using or plans to use a lit Bunsen burner. Make sure that no other heat sources are present.

FIRE SAFETY
- Know the location of laboratory fire extinguishers and fire-safety blankets.
- Know your school's fire-evacuation routes.

GAS SAFETY
- Do not inhale any gas or vapor unless directed to do so by your instructor. Do not breathe pure gases.
- Handle materials prone to emit vapors or gases in a well-ventilated area. This work should be done in an approved chemical fume hood.

GLASSWARE SAFETY
- Check the condition of glassware before and after using it. Inform your instructor of any broken, chipped, or cracked glassware, because it should not be used.
- Do not pick up broken glass with your bare hands. Place broken glass in a specially designated disposal container.
- If a thermometer breaks, notify your instructor immediately.
- If a light bulb breaks, notify your instructor immediately. Do not remove broken bulbs from sockets.

WASTE DISPOSAL
- Clean and decontaminate all work surfaces and personal protective equipment as directed by your instructor.
- Dispose of all broken glass, contaminated sharp objects, and other contaminated materials in special containers as directed by your instructor.

HYGIENIC CARE
- Keep your hands away from your face and mouth.
- Always wash your hands thoroughly when you are done with an experiment.

Introduction to Science
Chapter Planning Guide

PACING	CLASSROOM RESOURCES	LABS, ACTIVITIES, AND DEMONSTRATIONS
BLOCK 1 · 45 min pp. 2–3 **Chapter Opener**		SE **Activity 1,** p. 3 SE **Activity 2,** p. 3
BLOCKS 2 & 3 · 90 min pp. 4–11 **Section 1** Introduction to Science	CRF **Lesson Plan*** TT **Bellringer*** TM **Branches of Science***	TE **Opening Discussion,** p. 4 TE **Group Activity** Models, p. 10 `BASIC`
BLOCKS 4 & 5 · 90 min pp. 12–19 **Section 2** The Way Science Works	CRF **Lesson Plan*** TT **Bellringer*** TM **The Scientific Method*** TM **SI Base Units*** TM **SI Prefixes***	TE **Demonstration** Are They the Same?, p. 13 SE **Quick Activity** Making Observations, p. 14 `GENERAL` CRF **Datasheets for In-Text Labs** Making Observations* TE **Group Activity** Standard Units of Measure, p. 16 `BASIC` CRF **CBL™ Probeware Lab** Designing a Pendulum Clock* ♦ `ADVANCED` CRF **Observation Lab** Comparing the Densities of Pennies* ♦ `BASIC`
BLOCKS 6 & 7 · 90 min pp. 20–26 **Section 3** Organizing Data	CRF **Lesson Plan*** TT **Bellringer*** TT **Line Graph*** TT **Bar Graph*** TT **Accuracy and Precision*** TT **Concept Mapping***	TE **Opening Activity,** p. 20 `GENERAL` SE **Skills Practice Lab** Making Measurements, pp. 32–35 ♦ `GENERAL` CRF **Datasheets for SE Skills Practice Lab** Making Measurements*

BLOCKS 8 & 9 · 90 min

Chapter Review and Assessment Resources

SE **Chapter Review,** pp. 28–31
CRF **Chapter Tests*** `GENERAL`
OSP **Test Generator**
CRF **Standardized Test Practice with Guided Reading Development***
CRF **Test Item Listing for ExamView® Test Generator***
OSP **Scoring Rubrics and Classroom Management Checklists**

Online Resources

Visit the HRW Web site for a variety of free resources related to the text. Just type in the keyword **HK4 INT**.

Holt Science Spectrum: Physical Science: Online Edition

Students can access interactive problem solving help and active visual concept development with the *Holt Science Spectrum: Physical Science* Online Edition available at **www.hrw.com**.

student CNN News

cnnstudentnews.com

Find the latest news, lesson plans, and activities related to important scientific events.

KEY

TE	Teacher Edition
SE	Student Edition
OSP	One-Stop Planner

CRF	Chapter Resource File
TT	Teaching Transparency
TM	Transparency Master

* Also on One-Stop Planner
◆ Requires Advance Prep

PROBLEM SOLVING AND PRACTICE	SECTION REVIEW AND ASSESSMENT	STANDARDS CORRELATION
	CRF **Pretest*** GENERAL	
CRF **Cross-Disciplinary Worksheet** Integrating Biology—Serendipity and Science* ADVANCED **CRF** **Science Skills** Creating a Concept Map* BASIC **CRF** **Cross-Disciplinary Worksheet** Integrating Chemistry—The Chemistry Connection* ADVANCED **CRF** **Science Skills** Compiling and Weighing Evidence* BASIC **CRF** **Cross-Disciplinary Worksheet** Integrating Mathematics—Using Quantitative Statements to Solve Problems* ADVANCED	**TE** **Quiz,** p. 11 BASIC **SE** **Section 1 Review,** p. 11 **CRF** **Concept Review*** GENERAL **CRF** **Quiz*** BASIC	PS 4b, 5c UCP 1, 2 HNS 1, 2, 3
SE **Math Skills** Conversions, p. 17 **CRF** **Math Skills** Conversions* GENERAL **CRF** **Science Skills** Forming a Hypothesis* BASIC **CRF** **Science Skills** Testing a Hypothesis* BASIC **CRF** **Cross-Disciplinary Worksheet** Connection to Language Arts—Medical Terminology* GENERAL **CRF** **Science Skills** Reading to Evaluate and to Identify Bias* BASIC **CRF** **Science Skills** Dimensional Analysis* BASIC **CRF** **Science Skills** SI Units and Conversions Between Them* BASIC **CRF** **Science Skills** Converting Between U.S. Conventional and SI Measurements* BASIC **TE** **Inclusion Strategies,** p. 18	**TE** **Quiz,** p. 19 BASIC **SE** **Section 2 Review,** p. 19 **CRF** **Concept Review*** GENERAL **CRF** **Quiz*** BASIC	SAI 2 UCP 3 ST 2 HNS 2, 3
CRF **Science Skills** Making a Line Graph* BASIC **CRF** **Science Skills** Making and Interpreting Bar Graphs and Pie Charts* BASIC **CRF** **Cross-Disciplinary Worksheet** Integrating Physics—Observing and Experimenting to Find Relationships* ADVANCED **TE** **Inclusion Strategies,** p. 22 **SE** **Math Skills** Writing Scientific Notation, p. 23 GENERAL **CRF** **Math Skills** Writing Scientific Notation* GENERAL **CRF** **Science Skills** Scientific Notation* BASIC **SE** **Math Skills** Using Scientific Notation, p. 24 GENERAL **CRF** **Math Skills** Using Scientific Notation* GENERAL **SE** **Math Skills** Significant Figures, p. 25 GENERAL **CRF** **Math Skills** Significant Figures* GENERAL **CRF** **Science Skills** Significant Figures* BASIC **SE** **Graphing Skills** Constructing a Pie Chart, p. 27	**TE** **Quiz,** p. 26 BASIC **SE** **Section 3 Review,** p. 26 **CRF** **Concept Review*** GENERAL **CRF** **Quiz*** BASIC	UCP 2 HNS 1

www.scilinks.org

Topic: Leonardo da Vinci
*Sci*Links code: HK4078

Topic: SI Units
*Sci*Links code: HK44128

Topic: New Discoveries in Science
*Sci*Links code: HK4093

Topic: Studying the Natural World
*Sci*Links code: HK4136

Topic: Presenting Scientific Data
*Sci*Links code: HK4109

Technology Resources

One-Stop Planner
All of your printable resources and the Test Generator are on this convenient CD-ROM.

 Science in the NEWS
each video segment is accompanied by a Critical Thinking Worksheet*

Segment 1
Amusement Park Physics

Overview

This chapter explores the nature of science, the scientific method of discovery, and the difference between scientific theories and laws. This chapter also describes how scientists use models and mathematics. Science skills, math skills, and units of measurement are taught next, followed by organization and presentation of data. The knowledge and skills gained in this chapter will serve as a foundation for science study as a whole.

Assessing Prior Knowledge

Students should be familiar with the basic skill of problem solving. Lead students in a discussion that allows them to explore the steps that a scientist might take to help them clearly identify a problem and look for a solution or to learn about a phenomenon.

MISCONCEPTION /// ALERT \\\

Science education research has identified the following misconceptions about science:

- Most students believe learning is passive and is, in fact, the canonical transfer of knowledge.
- Some students and teachers view science as a faithful copy of the world outside and not as a tentative human construction.

Introduction to Science

Chapter Preview

1 The Nature of Science
How Does Science Take Place?
Scientific Laws and Theories

2 The Way Science Works
Science Skills
Units of Measurement

3 Organizing Data
Presenting Scientific Data
Writing Numbers in Scientific Notation
Using Significant Figures

2

Standards Correlations

National Science Education Standards

The following descriptions summarize the National Science Standards that specifically relate to this chapter. For the full text of the standards, see the National Science Education Standards at the front of the book.

Section 1 The Nature of Science
Physical Science PS 4b, 5c
Unifying Concepts and Processes UCP 1, 2
History and Nature of Science HNS 1, 2, 3

Section 2 The Way Science Works
Science as Inquiry SAI 2
Unifying Concepts and Processes UCP 3
Science and Technology ST 2
History and Nature of Science HNS 2, 3

Section 3 Organizing Data
Unifying Concepts and Processes UCP 2
History and Nature of Science HNS 1

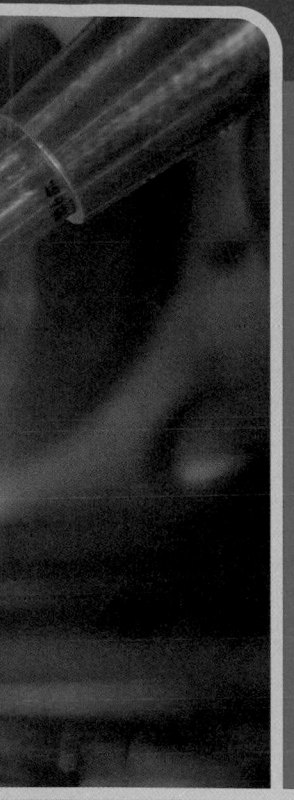

Focus ACTIVITY

Background Imagine that it is 1895 and you are a scientist working in your laboratory. Outside, people move about on foot, on bicycles, or in horse-drawn carriages. A few brave and rich people have purchased the new invention called an automobile. They ride along the street, but their auto sputters, pops, puffs smoke, and frightens both horses and people.

Your laboratory is filled with coils of wire, oddly shaped glass tubes, magnets of all sorts, and many heavy glass jars containing liquid and metal plates (batteries). A few dim electric bulbs are strung along the ceiling. Additional light comes from daylight through windows or from the old gas lamps along the wall.

It's an exciting time in science because new discoveries about matter and energy are being made almost every day. A few European scientists are even beginning to pay attention to those upstart scientists from America. However, some people believe that humans have learned nearly everything that is worth knowing about the physical world.

Activity 1 Interview someone old enough to have witnessed many technological changes. Ask the person what scientific discoveries have made the biggest differences in his or her life. Which changes do you think have been the most important?

Activity 2 Using a meterstick, measure the length and width of your desk surface. To what fraction of a unit can you reliably measure? Multiply your two measurements to calculate the surface area of your desk. Compare your results with those of other students. What might be the reasons for differences in calculations?

---

internet connect

www.scilinks.org
Topic: New Discoveries in Science
SciLinks code: HK4093

SCI LINKS. Maintained by the National Science Teachers Association

---

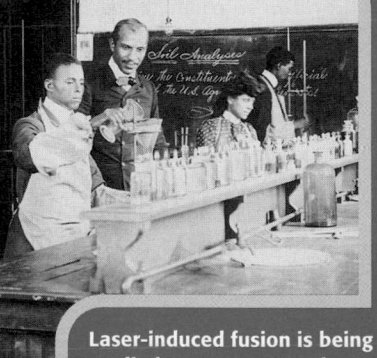

Laser-induced fusion is being studied as a way to produce energy for our growing needs. Lasers and fusion reactions were developed using the same approach to the scientific method used in 1896 by Dr. George Washington Carver at the Tuskegee Institute.

Pre-Reading Questions

1. How do scientific discoveries contribute to the development of new technology?
2. What are some problems that have been solved by science and technology in the last 10 years? What new problems have these technological changes caused?

3

Focus ACTIVITY

Background In the last 100 years, there have been incredible scientific advances in the fields of medicine, communication, and transportation. For example, advances in medicine include the discovery of antibiotics and many vaccines, understanding the roles that vitamins and minerals play in human health, and the development of more sophisticated diagnostic techniques. Ask students to think of examples of significant advances in the past century in the fields of communication and transportation.

Activity 1 Encourage students to ask neighbors or grandparents about technological changes and scientific discoveries they have witnessed. **LS Interpersonal**

Activity 2 Answers will vary. Students should use SI units (centimeters or millimeters) to measure their desks. Most students will be able to make measurements to one-tenth of a unit. Students may state that differences in calculations are due to differences in measurement tools. **LS Kinesthetic**

Answers to Pre-Reading Questions

1. Technology is the application of scientific principles to practical purposes. Thus, new discoveries in science pave the way for new technologies and for improvements in existing technologies.

See "Continuation of Answers" at the end of the chapter.

SECTION
1

Focus

Overview

Before beginning this section, review with your students the Objectives listed in the Student Edition. This section introduces students to the main branches of natural science. Students learn about confirming results by designing and repeating experiments. Scientific theories and laws are discussed, along with the role of models in both.

🔔 Bellringer

Use the Bellringer transparency to prepare students for this section.

Motivate

Opening Discussion —— GENERAL

It is important for students to understand that science is a process, not just a set of facts. Inform students that information about the world around us is always changing. Lead students in a discussion that explores the idea that science starts with a question. Ask students what they do when they have a question about the natural world, and discuss their responses. **LS Interpersonal**

Teach

SKILL BUILDER — GENERAL

Interpreting Visuals Have students examine **Figure 1.** Tell students that the illustration represents a technological advance. The television tube is a cathode-ray tube in which the direction of the rays is controlled by magnetic fields. **LS Visual**

The Nature of Science

■ **KEY TERMS**
science
technology
scientific law
scientific theory

OBJECTIVES

▶ **Describe** the main branches of natural science and relate them to each other.
▶ **Describe** the relationship between science and technology.
▶ **Distinguish** between scientific laws and scientific theories.
▶ **Explain** the roles of models and mathematics in scientific theories and laws.

Generally, scientists describe the universe using basic rules, which can be discovered by careful, methodical study. A scientist may perform experiments to find a new aspect of the natural world, to explain a known phenomenon, to check the results of other experiments, or to test the predictions of current theories.

How Does Science Take Place?

Imagine that it is 1895 and you are experimenting with cathode rays. These mysterious rays were discovered almost 40 years before, but in 1895 no one knows what they are. To produce the rays, you create a vacuum by pumping the air out of a sealed glass tube that has two metal rods at a distance from each other, as shown in *Figure 1A.* When the rods are connected to an electrical source, electric charges flow through the empty space between the rods, and the rays are produced.

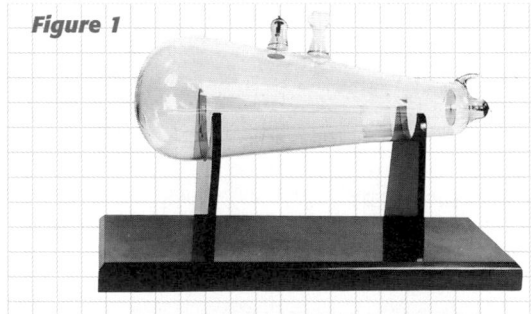

Figure 1

Ⓐ The cathode ray tube used in 1895 looked like this.

Ⓑ A television picture tube is a form of the same cathode ray tube.

4

LIFE SCIENCE
CONNECTION

Invite a biology teacher to be a guest lecturer in your classroom to explain the effect of X rays on the human body.

SOCIAL STUDIES
CONNECTION

The turn of the twentieth century was a time of changes in the United States. Ask a history teacher to lead a discussion of the impact of science and of technological changes on American culture between 1880 and 1920.

Scientists investigate

You have learned from the work of other scientists and have conducted experiments of your own. From these, you know that when certain minerals are placed inside the tube, the cathode rays make them fluoresce (glow). Pieces of cardboard coated with powder made from these minerals are used to detect the rays. With a very high voltage, even the glass tube itself glows.

Other scientists have found that cathode rays can pass through thin metal foils, but they travel in our atmosphere for only 2 or 3 cm. You wonder if the rays could pass through the glass tube. Others have tried this experiment and have found that cathode rays don't go through glass. But you think that the glow from the glass tube might have outshined any weak glow from the mineral-coated cardboard. So, you decide to cover the glass tube with heavy black paper.

Scientists plan experiments

Before experimenting, you write your plan in your laboratory notebook and sketch the equipment you are using. You make a table in which you can write down the electric power used, the distance from the tube to the fluorescent detector, the air temperature, and anything you observe. You state the idea you are going to test: At a high voltage, cathode rays will be strong enough to be detected outside the tube by causing the mineral-coated cardboard to glow.

Scientists observe

Everything is ready. You want to be sure that the black-paper cover doesn't have any gaps, so you darken the room and turn on the tube. The black cover blocks the light from the tube. Just before you switch off the tube, you glimpse a light nearby. When you turn on the tube again, the light reappears.

Then you realize that this light is coming from the mineral-coated cardboard you planned to use to detect cathode rays. The detector is already glowing, and it is on a table almost 1 m away from the tube. You know that 1 m is too far for cathode rays to travel in air. You suspect that the tube must be giving off some new rays that no one has seen before. What do you do now?

This is the question Wilhelm Roentgen had to ponder in Würzburg, Germany, on November 8, 1895, when all this happened to him. Should he call the experiment a failure because it didn't give the results he expected? Should he ask reporters to cover this news story? Maybe he should send letters about his discovery to famous scientists and invite them to come and see it.

VOCABULARY *Skills Tip*

Cathode rays *got their name because they come from the* cathode, *the rod connected to the negative terminal of the electricity source. The positive terminal is called the* anode.

INTEGRATING

BIOLOGY

In 1928, the Scottish scientist Alexander Fleming was investigating disease-causing bacteria when he saw that one of his cultures contained an area where no bacteria were growing. Instead, an unknown organism was growing in that area. Rather than discarding the culture as a failure, Fleming investigated the unfamiliar organism and found that it was a type of mold. This mold produced a substance that prevented the growth of many disease bacteria. What he found by questioning the results of a "failed" experiment became the first modern antibiotic, penicillin. Major discoveries are often made by accident when trying to find something else.

Reading Skills Have students silently read the paragraphs under the heading *Scientists observe*. Instruct students to mark on self-adhesive notes those portions of the text that they do not understand.

Group students together in pairs, and have one student in each pair discuss with the other student those parts of the passage that he or she found difficult. The listener should either clarify the difficult passages or help frame a common question for later explanation.
LS Interpersonal

INTEGRATING

BIOLOGY Fleming, who treated soldiers in France during World War I, was well aware of the lack of an effective treatment for many infections. After discovering penicillin, he learned that it could effectively treat many kinds of bacteria, including anthrax and the bacteria that causes meningitis. Although he published his results, penicillin's potential for treating infections was not well understood until about a decade later. World War II hastened the development of penicillin, which was quickly produced in large quantities by the United States, Britain, and later Russia. By 1944, there was enough penicillin to treat all of the Allied soldiers wounded on D-Day.

5

FINE ARTS
CONNECTION

Fine Arts Although photographs were fairly common by the late nineteenth century, photography was just beginning to emerge as an artistic medium. If your community has a museum that includes a photography collection, ask if the museum would give a survey lecture to the class, or ask an art teacher to show the class some images from this period.

Historical Perspective

Experimental Equipment Scientists at the turn of the twentieth century had to design and build most of their own experimental equipment. For example, a scientist studying cathode-rays may have had to be a glass blower to make the tube, an electrician to build the batteries and connect the rods in the tube to the batteries, and a mechanic to build and maintain the vacuum pump used to evacuate the cathode-ray tube. Although scientists today still build prototypes of new instruments, they can get assistance from specialists in technologies that were not available in 1895.

Chapter Resource File

- **Lesson Plan**
- **Cross-Disciplinary Worksheet**
  Integrating Biology—Serendipity and Science **ADVANCED**

Transparencies

TT Bellringer

Interpreting Visuals Have students examine **Figure 2.** Group students in pairs, and have them create an explanation of why the bones are dark and the surrounding areas are bright. (The bones block the X-rays and create a shadow.) **LS** Visual

SKILL BUILDER — BASIC

Reading Skills After students have read this section, have them work together as a group to brainstorm examples of how science occurs through the steps listed here (investigating, planning experiments, observing, and testing results). Ask a student volunteer to create a list on the board of all student contributions. Then have students create a diagram that summarizes the results of their brainstorming activity. **LS** Verbal

SKILL BUILDER — BASIC

Interpreting Diagrams Have students look at **Figure 3** and consider where they would place astronomy, analytical chemistry, archeology, geophysics, organic chemistry, paleontology, nuclear physics, and biochemistry. **LS** Logical

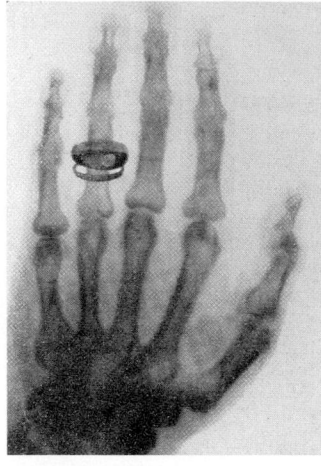

Figure 2
Roentgen included this X ray of his wife's hand in one of the first papers he wrote on X rays.

▶ **science** the knowledge obtained by observing natural events and conditions in order to discover facts and formulate laws or principles that can be verified or tested

Figure 3
This chart shows one way to look at science. Modern science has many branches and specialties.

Scientists always test results

Because Roentgen was a scientist, he first repeated his experiment to be sure of his observations. His results caused him to begin thinking of new questions and to design more experiments to find the answers.

He found that the rays passed through almost everything, although dense materials absorbed them somewhat. When he held his hand in the path of the rays, the bones were visible as shadows on the fluorescent detector, as shown in **Figure 2.** When Roentgen published his findings in December, he still did not know what the rays were. He called them *X rays* because *x* represents an unknown in a mathematical equation.

Within three months of Roentgen's discovery, a doctor in Massachusetts used X rays to help set properly the broken bones in a boy's arm. After a year, more than a thousand scientific articles about X rays had been published. In 1901, Roentgen received the first Nobel Prize in physics for his discovery.

Science has many branches

Roentgen's work with X rays illustrates how scientists work, but what is **science** about? Science is observing, studying, and experimenting to find the nature of things. You can think of science as having two main branches: social science, which deals with individual and group human behavior, and natural science. Natural science tries to understand how "nature," which really means "the whole universe," behaves. Natural science is usually divided into life science, physical science, and Earth science, as shown in **Figure 3.**

Life science is *biology*. Biology has many branches, such as *botany,* the science of plants; *zoology,* the science of animals; and *ecology,* the science of balance in nature. Medicine and agriculture are branches of biology too.

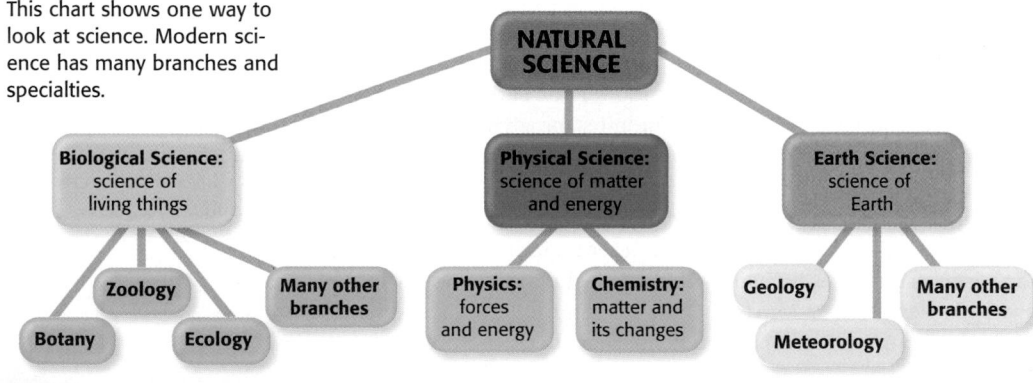

6

Alternative Assessment — ADVANCED

Planning an Experiment Ask students to imagine that they are Wilhelm Roentgen, and have them write their own plan for investigating cathode rays. Students should include a specific question to investigate, a diagram of their equipment setup, a data table for any information they plan to collect, and a detailed description of each step of the experiment.

Some students may have trouble coming up with their own plan. Tell these students that Roentgen's plan was to see if the cathode rays would pass through glass if the electric voltage applied to the rods was high enough.

Use the following criteria to assess students' plans:

- Did students pose thoughtful questions?
- Did they explain each step of the experiment they designed?
- Did they include a diagram and a data table?
- Is their design a reasonable approach for answering the proposed question?

LS Logical

Physical science has two main branches—*chemistry* and *physics*. Chemistry is the science of matter and its changes, and physics is the science of forces and energy. Both depend greatly on mathematics.

Some of the branches of Earth science are *geology*, the science of the physical nature and history of the Earth, and *meteorology*, the science of the atmosphere and weather.

This classification of science appears very tidy, like stacks of boxes in a shoe store, but there's a problem with it. As science has progressed, the branches of science have grown out of their little boxes. For example, chemists have begun to explain the workings of chemicals that make up living things, such as DNA, shown in *Figure 4*. This science is *biochemistry*, the study of the matter of living things. It is both a life science and a physical science. In the same way, the study of the forces that affect the Earth is *geophysics*, which is both an Earth science and a physical science.

Science and technology work together

Scientists who do experiments to learn more about the world are practicing *pure science*, also defined as the continuing search for scientific knowledge. Engineers look for ways to use this knowledge for practical applications. This application of science is called **technology**. For example, scientists who practice pure science want to know how certain kinds of materials, called superconductors, conduct electricity with almost no loss in energy. Engineers focus on how that technology can be best used to build high-speed computers.

Technology and science depend on one another, as illustrated by some of Leonardo da Vinci's drawings in *Figure 5*. For instance, scientists did not know that tiny organisms such as bacteria even existed until the technology to make precision magnifying lenses developed in the late 1600s.

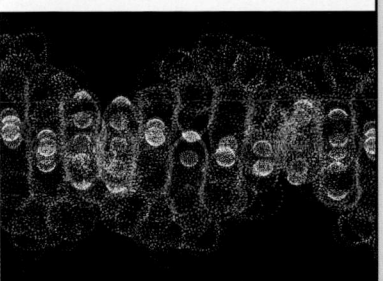

Figure 4
Our DNA (deoxyribonucleic acid) makes each of us unique.

▶ **technology** the application of science for practical purposes

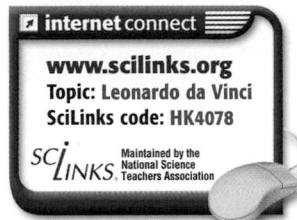

internet connect

www.scilinks.org
Topic: Leonardo da Vinci
SciLinks code: HK4078

SciLINKS Maintained by the National Science Teachers Association

Figure 5
Some of Leonardo da Vinci's ideas could not be built until twentieth-century technology developed. Some examples are: Ⓐ a design for a parachute and Ⓑ a design for a glider.

7

Teaching Tip
Modeling DNA Explain to students that the computer model shown in **Figure 4** uses different colors to show the locations and chemical compositions of the units that make up human DNA.

SKILL BUILDER
Vocabulary Be sure students understand the distinction between science and technology. The goal of science is to gain knowledge about the natural world. The goal of technology is to apply scientific understanding to solve problems.

Teaching Tip ——— GENERAL
Leonardo da Vinci's Diagrams Have students look at Leonardo da Vinci's parachute and glider in **Figure 5**. Then discuss how his original ideas might have worked and how their modern counterparts do work. Ask students if they think Leonardo da Vinci's parachute was designed to fold up and be deployed in the same way modern parachutes are or if it was to be used more like a modern parasail. His glider was powered by the arm and leg motion of a human. **LS Visual**

SKILL BUILDER — GENERAL
Interpreting Visuals Have students examine **Figure 5**. Allow students to brainstorm explanations for why the items shown could not be built until the twentieth century. **LS Logical**

Trends in Space Science

The International Space Station One recent example of the mutually beneficial relationship between science and technology is the international space station (ISS). The technology used to build the space station makes new scientific experiments possible. Scientists plan to use the ISS to conduct both pure and practical research in many branches of science and engineering, including gravity and microgravity research (physical science), life science, space science, earth science, space product development, and engineering research and technology. To learn more about the experiments planned for ISS, you can download a NASA brochure at the following web address: spacelink.nasa.gov/Instructional.Materials/NASA.Educational.Products/A.Key.to.Discovery/A.Key.to.Discovery.pdf. The brochure also contains student activities and provides web addresses for further information.

Chapter Resource File

- **Science Skills** Creating a Concept Map BASIC
- **Science Skills** Classifying Items BASIC
- **Cross-Disciplinary Worksheet** Integrating Chemistry—The Chemistry Connection ADVANCED

Transparencies

TM Branches of Science

Interpreting Visuals Have students examine **Figure 6B.** Explain that some of the effort of sawing is used to overcome friction—the saw does not slide smoothly through the wood. Friction results in the higher temperatures of both the saw and the wood. **LS Visual**

Teaching Tip

Theories and Laws The distinction between scientific theories and laws can be confusing to students. Stress that a scientific law is an observation about nature—a summary of a natural event. Many laws can be stated as mathematical formulas. A scientific law does not explain how or why something happens, but a scientific theory does. A scientific theory is a wide-ranging idea that explains many different laws. In some cases, the laws explained by a theory appeared to be unrelated before the theory was developed. Theories are always open to challenges and testing.

■ **scientific law** a summary of many experimental results and observations; a law tells how things work

■ **scientific theory** an explanation for some phenomenon that is based on observation, experimentation, and reasoning

Figure 6

The kinetic theory of heat explains many things that you can observe, such as why both the far end of the tube **A** and the saw blade **B** get hot.

Scientific Laws and Theories

People sometimes say things like, "My theory is that we'll see Jaime on the school bus," when they really mean, "I'm guessing that we'll find Jaime on the school bus." People use the word *theory* in everyday speech to refer to a guess about something. In science, a theory is much more than a guess.

Laws and theories are supported by experimental results

When you place a hot cooking pot in a cooler place, does the pot become hotter as it stands? No, it will always get cooler. This illustrates a **scientific law** that states that warm objects always become cooler when they are placed in cooler surroundings. A scientific law describes a process in nature that can be tested by repeated experiments. A law allows predictions to be made about how a system will behave under a wide range of conditions.

However, a law does not *explain* how a process takes place. In the example of the hot cooking pot, nothing in the law tells why hot objects become cooler in cooler surroundings. Such an explanation of how a natural process works must be provided by a **scientific theory.**

Scientific theories are always being questioned and examined. To be valid, a theory must continue to pass several tests.

▶ A theory must explain observations clearly and consistently. The theory that heat is the energy of particles in motion explains how the far end of a metal tube gets hot when you hold the tip over a flame, as shown in *Figure 6A.*

▶ Experiments that illustrate the theory must be repeatable. The far end of the tube always gets hot when the tip is held over a flame, whether it is done for the first time or the thirty-first time.

▶ You must be able to predict from the theory. You might predict that anything that makes particles move faster will make the object hotter. Sawing a piece of wood will make the metal particles in the saw move faster. If, as shown in *Figure 6B,* you saw rapidly, the saw will get hot to the touch.

Mathematics can describe physical events

How would you state the law of gravitation? You could say that something you hold will fall to Earth when you let go. This *qualitative* statement describes with words something you have seen many times. But many scientific laws and theories can be stated as mathematical equations, which are *quantitative* statements.

8

Alternative Assessment — **BASIC**

Distinguishing between Theories and Laws To help students understand the distinction between scientific theories and laws, have them analyze some theories and laws that they have studied in previous science courses. Instruct them to make a three-column chart with the column headings "Example," "Theory or Law?" and "Reasons." Tell them to list all of their examples in the first column and to classify each example as either a theory or a law in the second column. They should explain the reasons for their classification in the third column. Possible responses could include: Newton's laws of motion (laws); gas laws (laws); conservation of energy (law); universal law of gravitation (law); atomic theory (theory); theory of evolution (theory); kinetic theory (theory); plate tectonics (theory); relativity (theory). Be sure their reasons for each classification are accurate and logical. **LS Logical**

> **Rectangle Area Equation**
> $$A = l \times w$$

The rectangle area equation works for all rectangles, whether they are short, tall, wide, or thin.

> **Universal Gravitation Equation**
> $$F = G\frac{m_1 m_2}{d^2}$$

In the same way, the universal gravitation equation describes how big the force will be between two galaxies or between Earth and an apple dropped from your hand, as shown in *Figure 7*. Quantitative expressions of the laws of science make communicating about science easier. Scientists around the world speak and read many different languages, but mathematics, the language of science, is the same everywhere.

Theories and laws are always being tested

Sometimes theories have to be changed or replaced completely when new discoveries are made. Over 200 years ago, scientists used the *caloric theory* to explain how objects become hotter and cooler. Heat was thought to be an invisible fluid, called caloric, that could flow from a warm object to a cool one. People thought that fires were fountains of caloric, which flowed into surrounding objects, making them warmer. The caloric theory could explain everything that people knew about heat.

But the caloric theory couldn't explain why rubbing two rough surfaces together made them warmer. During the 1800s, after doing many experiments, some scientists suggested a new theory based on the idea that heat was a result of the motion of particles. The new theory was that heat is really a form of energy that is transferred when fast-moving particles hit others. Because this theory, the *kinetic theory*, explained the old observations as well as the new ones, it was kept and the caloric theory was discarded.

Models can represent physical events

When you see the word *model*, you may think of a small copy of an airplane or a person who shows off clothing. Scientists use models too. A scientific model is a representation of an object or event that can be studied to understand the real object or event. Sometimes, like a model airplane, models represent things that are too big, too small, or too complex to study easily.

What does this have to do with the force between two galaxies?

Figure 7
Gravitational attraction is described as a force that varies depending on the mass of the objects and the distance that separates them.

9

Alternative Assessment — ADVANCED

Revised Scientific Theories Have students research a scientific theory that is no longer accepted, such as the fluid theory of electricity or the phlogiston theory of combustion. Ask them to write a few paragraphs explaining why the theory was originally accepted, why it was later replaced, and what it was replaced by. They may wish to use the discussion of caloric theory on this page as a model. **LS Verbal**

Teaching Tip — ADVANCED

Making Predictions Scientific laws and theories can be used to make predictions about what will happen in different situations. If the results of an experiment match a prediction based on a scientific law or theory, the experiment supports that law or theory.

Challenge students to pose examples of hypothetical situations or experiments where a scientific law or theory may explain a given outcome. **LS Logical**

Teaching Tip

Laws as Approximations Tell students that many laws that are not correct under all circumstances are still used because they represent easy-to-understand models of things that we observe in the natural world. One example is the relationship between the current, resistance, and voltage in an electric circuit (resistance = voltage/ current), commonly called Ohm's Law. This equation does not apply to all materials and, when it does apply, it is only valid for a given range of voltages. But even though the "law" is not absolute, it is a very useful tool for many particular situations.

> ### Chapter Resource File
>
> - **Science Skills** Compiling and Weighing Evidence **BASIC**
> - **Science Skills** Understanding Symbols **BASIC**
> - **Cross-Disciplinary Worksheet** Integrating Mathematics—Using Quantitative Statements to Solve Problems **ADVANCED**

Group Activity ── BASIC

Models Point out to students that they probably use models every day without even realizing that they are doing so. When they plan a route to their next class via their locker, they are using a mental map that is a model of the school. A paper map is also a model of the real world. Any set of instructions on how to build something or put something together is a model of that object. If a student explains to a friend how to use a computer, that student is using a mental model of the operating rules of the computer. Have students work in small groups to brainstorm as many commonly used models as they can. Then have a volunteer from each group share their list with the class. **LS Interpersonal**

Figure 8

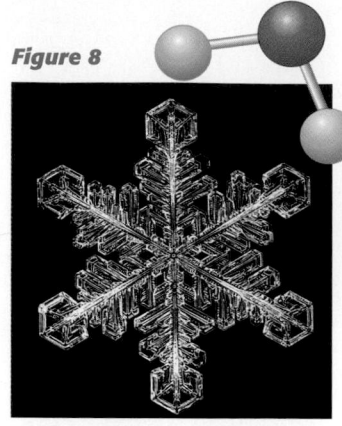

A Models can be used to describe a water molecule (top right) and to study how water molecules are arranged in a snowflake.

B Experiments show that this model depicts how a sound wave moves through matter.

A model of water is shown in *Figure 8A.* Chemists use models to study how water forms an ice crystal, such as a snowflake. Models can be drawings on paper. The spring shown in *Figure 8B* serves as a model of a sound wave moving through matter. Also, a model can be a mental "picture" or a set of rules that describes what something does. After you have studied atoms in Chapter 3, you will be able to picture atoms in your mind and use models to predict what will happen in chemical reactions.

Scientists and engineers also use computer models. These can be drawings such as the one shown in *Figure 9A;* more often, they are mathematical models of complex systems. Computer models can save time and money because long calculations are done by a machine to predict what will happen.

Figure 9

Crash tests give information that is used to make cars safer. Now, models **A** can replace some real-world crash tests **B**.

MISCONCEPTION ALERT

The Nature of Science Some students believe that science is simply an accumulation of facts. Use the discussion of theories and laws in this section to dispel this notion. Point out that science includes both the accumulation of observations and ever-changing interpretations of those observations. Remind them that a scientific theory can never be proved absolutely; there is always the possibility that it will be revised or even replaced by a new theory that explains additional observations or laws.

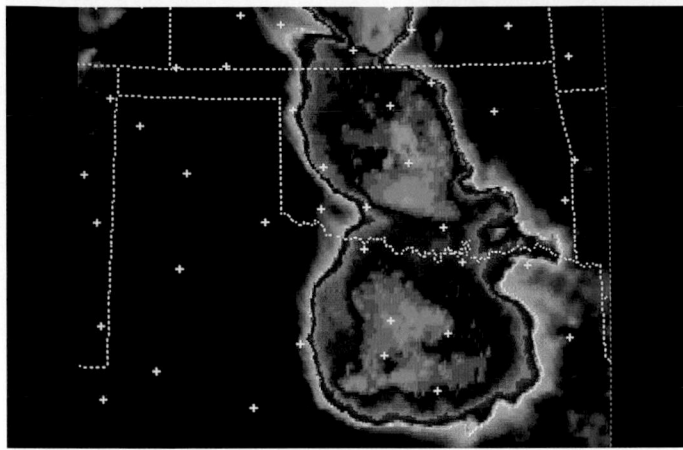

Figure 10
Models help forecast the weather and, in cases of dangerous storms, can help save lives.

Computer models have a variety of applications. For example, they can be used instead of expensive crash tests to study the effects of motion and forces in car crashes, as shown in *Figure 9.* Engineers use the predictions from the models to improve the design of cars. *Meteorologists* have computer models such as the one shown in *Figure 10,* which uses information about wind speed and direction, air temperature, moisture levels, and ground shape to help forecast the weather.

SECTION 1 REVIEW

SUMMARY

▶ A scientist makes objective observations.

▶ A scientist confirms results by repeating experiments and learns more by designing and conducting new experiments.

▶ Scientific laws and theories are supported by repeated experiments but may be changed when results are not consistent with predictions.

▶ Models are used to represent real situations and to make predictions.

1. Compare and Contrast the two main branches of physical science.

2. Explain how science and technology depend on each other.

3. Explain how a scientific theory differs from a guess or an opinion.

4. Define *scientific law* and give an example.

5. Compare and Contrast a scientific law and a scientific theory.

6. Compare quantitative and qualitative descriptions.

7. Describe how a scientific model is used, and give an example of a scientific model.

8. Creative Thinking How do you think Roentgen's training as a scientist affected the way he responded to his discovery?

9. Creative Thinking Pick a common happening, develop an explanation for it, and describe an experiment you could perform to test your explanation.

11

Answers to Section 1 Review

1. Chemistry is the study of matter and its changes. Physics is the study of forces and energy and their interaction with matter.

2. Technology is the application of science. Improving technology involves finding a use for a scientific discovery. However, some scientific discoveries cannot be made until a certain technology exists.

3. A guess or opinion is usually an unsupported statement. A scientific theory uses repeatedly tested results to explain observed events.

4. A scientific law describes how things work based on experimental observations. Examples

may vary but could include the laws of gravitation and the conservation of matter.

5. A law summarizes an observation; a theory explains observations.

6. Quantitative descriptions use numbers. Qualitative descriptions do not.

7. A model is used to study or make predictions about a situation the model represents. A computer simulation of the launch of a new kind of rocket is an example of a model.

8. Instead of being disappointed, he decided to learn more about his "failure."

9. Answers will vary.

Overview

Before beginning this section, review with your students the Objectives listed in the Student Edition. This section introduces students to many of the tools and skills they will use in science, including critical thinking skills, the steps of the scientific method, scientific tools, and the SI system of measurement.

🔊 Bellringer

Use the Bellringer transparency to prepare students for this section.

Motivate

Opening Discussion ——— GENERAL

Have students examine **Figure 11.** Have them brainstorm ideas about why the person is comparing peanut butter and why it might be important to do so. Be sure students consider differences other than price, such as ingredients (preservatives, sugar, salt), or size (how much is needed, will it go to waste).

For comparison, ask students to consider going to a bookstore. There are many books in the bargain rack, some for only $2 or $3. Have students brainstorm reasons why they would or would not want to buy the bargain books.

LS Logical

The Way Science Works

■ **KEY TERMS**
critical thinking
scientific method
variable
length
mass
volume
weight

■ **OBJECTIVES**

▶ **Understand** how to use critical thinking skills to solve problems.

▶ **Describe** the steps of the scientific method.

▶ **Know** some of the tools scientists use to investigate nature.

▶ **Explain** the objective of a consistent system of units, and identify the SI units for length, mass, and time.

▶ **Identify** what each common SI prefix represents, and convert measurements.

■ **critical thinking** the ability and willingness to assess claims critically and to make judgments on the basis of objective and supported reasons

If 16 ounces costs $3.59 and 8 ounces costs $2.19, then . . .

Figure 11
Making thoughtful decisions is important in scientific processes as well as in everyday life.

Throwing a spear accurately to kill animals for food or to ward off intruders was probably a survival skill people used for thousands of years. In our society, throwing a javelin is an athletic skill, and riding a bicycle or driving a car is considered almost a survival skill. The skills that we place importance on change over time, as society and technology change.

Science Skills

Although pouring liquid into a test tube without spilling is a skill that is useful in science, other skills are more important. Identifying problems, planning experiments, recording observations, and correctly reporting data are some of these more important skills. The most important skill is learning to think critically.

Critical thinking

If you were doing your homework and the lights went out, what would you do? Would you call the electric company immediately? A person who thinks like a scientist might first ask questions and make observations. Are lights on anywhere in the house? If so, what would you conclude? Suppose everything electrical in the house is off. Can you see lights in the neighbors' windows? If their lights are on, what does that mean? What if everyone's lights are off?

If you approach the problem this way, you are thinking logically. This kind of thinking is very much like **critical thinking.** You do this kind of thinking when you consider if the giant economy-sized jar of peanut butter is really less expensive than the regular size, as shown in **Figure 11,** or consider if a specific brand of soap makes you more attractive.

12

Teacher Resources

For the New Teacher Your students may think of the scientific method as a set of rigid steps for solving problems. When teaching this section, emphasize that there is no single scientific method. The scientific method is a way of thinking critically about a question and testing possible answers to that question by collecting data and making unbiased observations. Scientists approach problems from a variety of viewpoints. They conduct their research using available tools, data, time, and people. Research often leads to new problems and new hypotheses, which require further research and testing.

The Scientific Method

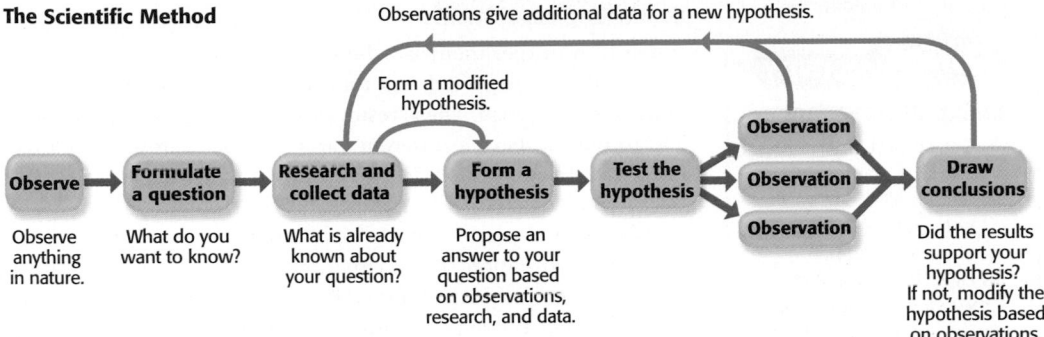

Figure 12
The scientific method is a general description of scientific thinking rather than an exact path for scientists to follow.

When the lights go out, if you get more facts before you call the power company, you're thinking critically. You're not making a reasonable conclusion if you immediately assume there is a citywide power failure. You can make observations and use logic.

Using the scientific method

In the **scientific method,** critical thinking is used to solve scientific problems. The scientific method is a general way to help organize your thinking about questions that you might think of as scientific. Using the scientific method helps you find and evaluate possible answers. The scientific method is often followed as a series of steps like those in **Figure 12.**

Most scientific questions begin with observations—simple things you notice. For example, you might notice that when you open a door, you hear a squeak. You ask the question: Why does this door make noise? You may gather data by checking other doors and find that the other doors don't make noise. So you form a *hypothesis,* a possible answer that you can test in some way. For instance, you may think that if the door makes a noise, the source of the noise is the doorknob.

Testing hypotheses

Scientists test a hypothesis by doing a *controlled experiment.* In a controlled experiment, all **variables** that can affect the outcome of the experiment are kept constant, or controlled, except for one. Only the results of changing one given variable are observed.

When you change more than one thing at a time, it's harder to make reasonable conclusions. If you remove the knob, sand the frame, and put oil on the hinges, you may stop the squeak, but you won't know what was causing the squeak. Even if you test one thing at a time, you may not find the answer on the first try. If you take the knob off the door and the door still makes noise, was your experiment a failure?

▶ **scientific method** a series of steps followed to solve problems including collecting data, formulating a hypothesis, testing the hypothesis, and stating conclusions

▶ **variable** a factor that changes in an experiment in order to test a hypothesis

13

Teacher Resources
Controlled Experiments A more detailed discussion of controlled experiments may be found online at **go.hrw.com.** Type in the keyword **HK4 CONTROLLED**.

Alternative Assessment ── BASIC
Applying the Scientific Method Point out to students that scientists do not necessarily use all of the steps of the scientific method for each problem. Have students work independently or in small groups to design a way to apply the scientific method to buying peanut butter, a CD, or a book. Ask volunteers to share their examples with the class, and discuss the differences between them. Point out that there can be more than one valid approach for any given problem. You could also emphasize that thinking like a scientist is a valuable skill in any profession. **LS Logical**

Quick ACTIVITY

Making Observations

Safety Caution: *Place the candles in a fireproof container, such as an aluminum pie plate, before lighting them.*

Materials (per group):
• candle
• fireproof container
• matches

Teacher's Notes: Encourage students to observe the changes as they occur.

Results: In step 2, students should note the color of the wax and the wick. Students should also record the length and/or thickness of the candle. Observations in step 4 may include relative assessments of the heat and light produced, height of flame, color of flame, changes in the wick, rate of consumption of the wax, and any color change as the solid wax melts. **LS Kinesthetic**

Teaching Tip

Learning Through Failure With experiments such as the discovery of X rays, it is easy to see that an unexpected result is not a failure. Point out to students that even experiments that do not work can be learning experiences. For example, when a chemical reaction fails to occur and the student or scientist can determine why the reaction did not occur, he or she will have learned more about the reaction being tested.

Quick ACTIVITY

Making Observations

1. Get an ordinary candle of any shape and color.
2. Record all the observations you can make about the candle.
3. Light the candle, and watch it burn for 1 minute.
4. Record as many observations about the burning candle as you can.
5. Share your results with your class, and find out how many different things were observed.

Figure 13

Computer models of Earth's crust help geologists understand how the continental plates (outlined in red) moved in the past and how they may move in the future.

Conducting experiments

In truth, no experiment is a failure. Experiments may not give the results you expected, but they are all observations of events in the natural world. These results are used to revise a hypothesis and to plan tests of a different variable. For example, once you know that the doorknob did not cause the squeak, you can revise your hypothesis to see if oiling the hinges stops the noise.

Scientists often do "what if" experiments to see what happens in a certain situation. These experiments are a form of data collection. Often, as with Roentgen's X rays, experimental results are surprising and lead scientists in new directions.

Scientists always have the question to be tested in mind. You can find out if ice is more dense than water just by thinking whether ice floats or sinks in water. The thinking that led to the law of gravitation began in 1666 when, according to legend, Isaac Newton saw an apple fall. He wondered why objects fall toward the center of Earth rather than in another direction.

Some questions, such as how Earth's continents have moved over millions of years, cannot be answered with experimental data. Instead of doing experiments, geologists make observations all over Earth. They also use models, such as the one shown in *Figure 13,* based on the laws of physics.

Using scientific tools

Of course, logical thinking isn't the only skill used in science. Scientists must make careful observations. Sometimes only the senses are needed for observations, as in the case of field botanists using their eyes to identify plants. At other times, special tools are provided through developments in technology. Scientists must know how to use these tools, what the limits of the tools are, and how to interpret data from them.

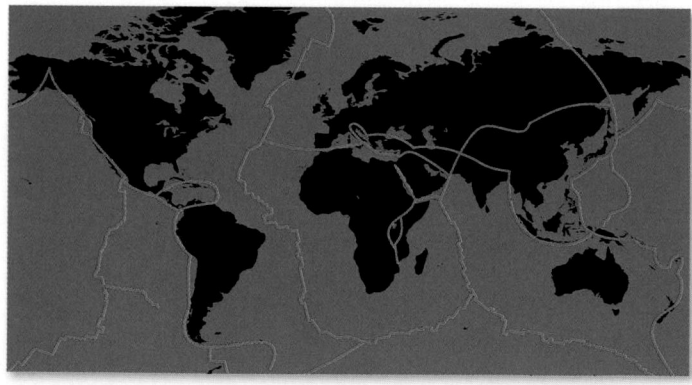

Historical Perspective

Plate Tectonics The theory of plate tectonics was developed in the 1960s and has replaced the older theory of continental drift. The theory of plate tectonics provides a way to reconstruct Earth's past geography. Volcanic activity, areas prone to earthquakes, and mountain ranges mark some of the boundaries between plates. Seismic studies, changes in Earth's magnetic field over time, and trails of volcanic activity on the ocean floor have contributed to our understanding of the boundaries of the plates (shown in **Figure 13**). Deep-sea drilling cores have been used to help determine the rate and direction of movement for the plates.

Did You Know ?

A Surprising Discovery Experiments do not always turn out as expected. In 1856, William Henry Perkin was experimenting to synthesize the antimalarial drug quinine from coal tar. He didn't succeed, but he accidentally made aniline purple (mauve), the first synthetic dye.

Figure 14

A The Gemini North observatory in Hawaii is a new tool for scientists. Its 8.1 m mirror is used to view distant galaxies.

B The Whirlpool galaxy (M51) and its companion NGC5195 are linked by a trail of gas and dust, which NGC5195 has pulled from M51 by gravitational attraction.

Connection to
LANGUAGE ARTS

Answers

A microscope is an instrument used to magnify small objects, such as bacteria. A retinoscope is an instrument used to see the retina of the eye. A kaleidoscope is a tube that contains bits of glass or plastic reflected by mirrors such that varied patterns are formed when the tube is held to the eye and rotated. A hygroscope is an instrument used to measure the changes in atmospheric humidity.

Astronomers, for example, use *telescopes* with lenses and mirrors such as the one shown in *Figure 14A* to magnify objects that appear small because they are far away, such as the distant galaxies shown in *Figure 14B*. Other kinds of telescopes do not form images from visible light. *Radio telescopes* detect the radio signals emitted by distant objects. Some of the oldest, most distant objects in the universe have been found with radio telescopes. Radio waves from those objects were emitted almost 15 billion years ago.

Several different types of *spectroscopes* break light into a rainbowlike *spectrum*. A chemist can learn a great deal about a substance from the light it absorbs or emits. Physicists use *particle accelerators* to make fragments of atoms move extremely fast and then let them smash into atoms or parts of other atoms. Data from these collisions give us information about the structure of atoms.

Units of Measurement

As you learned in Section 1, mathematics is the language of science, and mathematical models rely on accurate observations. But if your scientific measurements are in inches and gallons, some scientists may not understand because they do not use these units. For this reason scientists use the International System of Units, abbreviated SI, which stands for the French phrase *le Système Internationale d'Unités.*

Connection to
LANGUAGE ARTS

The word *scope* comes from the Greek word *skopein,* meaning "to see." Science and technology use many different scopes to see things that can't be seen with unaided eyes. For example, the telescope gets its name from the Greek prefix *tele-* meaning "distant" or "far." So a telescope is a tool for seeing far.

Making the Connection

Use a dictionary to find out what is seen by a microscope, a retinoscope, a kaleidoscope, and a hygroscope.

15

Trends in Space Science

The Gemini Project The Gemini North observatory (shown in **Figure 14**) is one of two new telescopes being built by an international partnership of seven nations. Its twin telescope, Gemini South, is currently under construction in northern Chile.

Both telescopes take advantage of new technologies that allow large, thin mirrors to collect and focus light with incredible precision. Together, they will be used to explore the entire northern and southern skies. Gemini North has already given astronomers some of the sharpest infrared images ever obtained by a telescope on Earth. Astronomers hope to use observations from both Gemini telescopes to learn more about the formation of stars and planets, the structure of galaxies (including the Milky Way), and the age and evolution of the universe.

Chapter Resource File

• **Quick Activity Datasheet** Making Observations GENERAL

• **Cross-Disciplinary Worksheet** Connection to Language Arts— Medical Terminology ADVANCED

• **Science Skills** Reading to Evaluate and to Identify Bias BASIC

• **Science Skills** Evaluating Data BASIC

Teaching Tip

Units of Measurement Be sure students understand the need to use appropriate units when measuring. Many (but not all) scientists use the SI units. Pose a theoretical question to students, such as "How far is it to the nearest bathroom?" Help them realize there is a big difference between 15 m (about 49 ft) and 15 km (about 9 mi), especially for the person who needs the information. Another possibility, "How far is it to the sun?" The answer could be 8.3, but 8.3 what? meters? kilometers? Actually, it is 8.3 light-minutes. A light-minute is the distance that light travels in 1 min—18 000 000 000 m (about 11 million miles).

Group Activity —— BASIC

Standard Units of Measure
Separate students into small groups. Ask each group to pick an object to use as a unit of measurement. Allow students to use anything, including their hands, feet, or a pencil. Have each group find the length of the classroom in their chosen unit of measure. Ask each group to share their result with the class. Then have a student volunteer write a list of all the units of measurements used on the chalkboard. Lead students into a discussion about why it is so important to use standard units of measure.
LS Interpersonal

Did You Know ?

SI started with the metric system in France in 1795. The meter was originally defined as 1/10 000 000 of the distance between the North Pole and the Equator.

Table 1 SI Base Units

Quantity	Unit	Abbreviation
Length	meter	m
Mass	kilogram	kg
Time	second	s
Temperature	kelvin	K
Electric current	ampere	A
Amount of substance	mole	mol
Luminous intensity	candela	cd

Table 2 Prefixes Used for Large Measurements

Prefix	Symbol	Meaning	Multiple of base unit
kilo-	k	thousand	1000
mega-	M	million	1 000 000
giga-	G	billion	1 000 000 000

Table 3 Prefixes Used for Small Measurements

Prefix	Symbol	Meaning	Multiple of base unit
deci-	d	tenth	0.1
centi-	c	hundredth	0.01
milli-	m	thousandth	0.001
micro-	μ	millionth	0.000 001
nano-	n	billionth	0.000 000 001

SI units are used for consistency

When all scientists use the same system of measurement, sharing data and results is easier. SI is based on the metric system and uses the seven SI base units that you see listed in **Table 1**.

Perhaps you noticed that the base units do not include area, volume, pressure, weight, force, speed, and other familiar quantities. Combinations of the base units, called *derived units*, are used for these measurements.

Suppose you want to order carpet for a floor that measures 8.0 m long and 6.0 m wide. You know that the area of a rectangle is its length times its width.

$$A = l \times w$$

The area of the floor can be calculated as shown below.

$$A = 8.0 \text{ m} \times 6.0 \text{ m} = 48 \text{ m}^2$$
(or 48 square meters)

The SI unit of area, m^2, is a derived unit.

SI prefixes are for very large and very small measurements

Look at a meterstick. How would you express the length of a bird's egg in meters? How about the distance you traveled on a trip? The bird's egg might be 1/100 m, or 0.01 m, wide. Your trip could have been 800 000 m in distance. To avoid writing a lot of decimal places and zeros, SI uses prefixes to express very small or very large numbers. These prefixes, shown in **Table 2** and **Table 3**, are all *multiples* of 10.

Using the prefixes, you can now say that the bird's egg is 1 cm (1 *centi*meter is 0.01 m) wide and your trip was 800 km (800 *kilo*meters are 800 000 m) long. Note that the base unit of mass is the *kilo*gram, which is already a multiple of the gram.

It is easy to convert SI units to smaller or larger units. Remember that to make a measurement, it takes more of a small unit or less of a large unit. A person's height could be 1.85 m, a fairly small number. In centimeters, the same height would be 185 cm, a larger number.

16

Historical Perspective

A Costly Mistake A $125-million space probe, the Mars Climate Observer, crashed on Mars in 1999 because key numbers were calculated by one group in English units while another group used metric units. Because critical maneuvers necessary to place the spacecraft in a proper Mars orbit relied on calculations involving numbers from both incompatible groups, the probe crashed.

Alternative Assessment —— ADVANCED

Evaluating Advertising Claims Have students bring in advertisements for products and services that contain claims that can be verified by measurement. Challenge students to identify the claim, describe the type of measurements that could verify or discredit the claim, and indicate which units they would use for the measurements. In addition, students should provide examples of sample data that would verify the claim and examples of sample data that would call the claim into question.
LS Logical

So, if you are converting to a smaller unit, multiply the measurement to get a bigger number. To write 1.85 m as *centi*meters, you multiply by 100, as shown below.

$$1.85 \ \cancel{m} \times \frac{100 \text{ cm}}{1 \ \cancel{m}} = 185 \text{ cm}$$

If you are converting to a larger unit, divide the measurement to get a smaller number. To change 185 cm to meters, divide by 100, as shown in the following.

$$185 \ \cancel{cm} \times \frac{1 \text{ m}}{100 \ \cancel{cm}} = 1.85 \text{ m}$$

internet connect

www.scilinks.org
Topic: SI Units
SciLinks code: HK4128

SciLINKS. Maintained by the National Science Teachers Association

Math Skills

Conversions A roll of copper wire contains 15 m of wire. What is the length of the wire in centimeters?

1 **List the given and unknown values.**
  Given: *length in meters, l* = 15 m
  Unknown: *length in centimeters* = ? cm

2 **Determine the relationship between units.**
  Looking at **Table 1-3**, you can find that 1 cm = 0.01 m. This also means that 1 m = 100 cm.
  You will multiply because you are converting from a larger unit (meters) to a smaller unit (centimeters).

3 **Write the equation for the conversion.**
  $$length \ in \ cm = m \times \frac{100 \text{ cm}}{1 \text{ m}}$$

4 **Insert the known values into the equation, and solve.**
  $$length \ in \ cm = 15 \ \cancel{m} \times \frac{100 \text{ cm}}{1 \ \cancel{m}}$$
  $$length \ in \ cm = 1500 \text{ cm}$$

Practice

Conversions
1. Write 550 *milli*meters as meters.
2. Write 3.5 seconds as *milli*seconds.
3. Convert 1.6 *kilo*grams to grams.
4. Convert 2500 *milli*grams to *kilo*grams.
5. Convert 4 *centi*meters to *micro*meters.
6. Change 2800 *milli*moles to moles.
7. Change 6.1 amperes to *milli*amperes.
8. Write 3 *micro*grams as *nano*grams.

Practice HINT

If you have done the conversions properly, all the units above and below the fraction will cancel except the units you need.

Did You Know?

A unit used for measuring the mass of precious metals and gems is the carat. The word *carat* comes from the word *carob*. Originally, the carat was the mass of one seed from the carob plant. It is now defined as 200 mg.

17

Did You Know?

Kelvins and Amperes The SI unit of temperature is called the kelvin and the unit of electric current is called the ampere. The kelvin was selected to honor William Thomson (later Lord Kelvin), a Scottish engineer, mathematician, and physicist. Lord Kelvin was a major contributor to the development of the laws of thermodynamics. The ampere was named to honor André-Marie Ampère, a French physicist, who by 1825 had laid the foundation of electromagnetic theory.

Math Skills

Additional Examples
Conversions Convert 15 m into:
a. mm
b. km

Answers
a. 15 000 mm
b. 0.015 km
LS Logical

Practice

Answers to Practice
1. $550 \text{ mm} \times \frac{1 \text{ m}}{1000 \text{ m}} = 0.55 \text{ m}$

2. $3.5 \text{ s} \times \frac{1000 \text{ ms}}{1 \text{ s}} = 3500 \text{ ms}$

3. $1.6 \text{ kg} \times \frac{1000 \text{ g}}{1 \text{ kg}} = 1600 \text{ g}$

4. $2500 \text{ mg} \times \frac{1 \text{ g}}{1000 \text{ mg}} \times \frac{1 \text{ kg}}{1000 \text{ g}}$
   $= 0.0025 \text{ kg}$

5. $4 \text{ cm} \times \frac{1 \text{ m}}{100 \text{ cm}}$
   $\frac{1000 \ 000 \ \mu m}{1 \text{ m}} = 40 \ 000 \ \mu m$

6. $2800 \text{ mmol} \times \frac{1 \text{ mol}}{1000 \text{ mmol}} =$
   2.8 mol

7. $6.1 \text{ A} \times \frac{1000 \text{ mA}}{1 \text{ A}} = 6100 \text{ mA}$

8. $3 \ \mu g \times \frac{1 \text{ g}}{1000 \ 000 \ \mu g} \times$
   $\frac{1000 \ 000 \ 000 \text{ ng}}{1 \text{ g}} = 3000 \text{ ng}$

LS Logical

Chapter Resource File

• **Science Skills** Dimensional Analysis **BASIC**
• **Science Skills** SI Units and Conversions Between Them **BASIC**
• **Science Skills** Converting Between U.S. Conventional and SI Measurements **BASIC**
• **Math Skills** Conversions **GENERAL**

Transparencies

TM SI Prefixes
TM SI Base Units

Weight and Mass Some students confuse weight and mass, believing that "felt weight" is a characteristic property of an object and that mass is something that "presses down." Emphasize that mass is how much matter an object has, while weight is how hard gravity is pulling on it.

An example can help illustrate this distinction. Tell students that when astronauts travel to the moon, they have the same mass. However, they weigh less because the moon is smaller (both in size and mass) than Earth. The change in weight is due to a change in gravitational attraction between the astronaut and Earth and the astronaut and the moon.

Teaching Tip ——— BASIC

Tools Have students consider some of the tools used for measuring time, length, mass, and volume and suggest other tools for those quantities. (An hourglass, a sundial, and an atomic clock can be used for measuring time. Length can also be measured with a surveyor's wheel or a car odometer. A double-pan balance and a single-beam balance that students may have seen in medical offices are both used to measure mass. Volume can be measured with pipettes and syringes.) **LS Logical**

Figure 15 Quantitative Measurements

	Time	Length
SI unit	second, s	meter, m
Other units	milliseconds, ms minutes, min hours, h	millimeter, mm centimeter, cm kilometer, km
Examples		91 m 2 cm 1 mm
Tools		

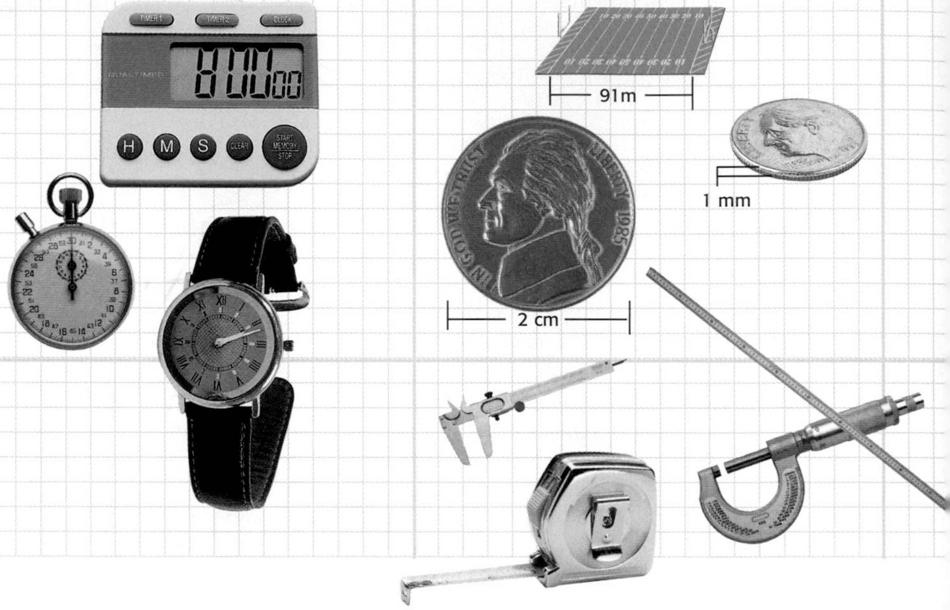

▶ **length** a measure of the straight-line distance between two points

▶ **mass** a measure of the amount of matter in an object

▶ **volume** a measure of the size of a body or region in three-dimensional space

▶ **weight** a measure of the gravitational force exerted on an object

Making measurements

Many observations rely on quantitative measurements. The most basic scientific measurements generally answer questions such as how much time did it take and how big is it?

Often, you will measure time, **length, mass,** and **volume.** The SI units for these quantities, examples of each quantity, and the tools you may use to measure them are shown in **Figure 15.**

Although you may hear someone say that he or she is "weighing" an object with a balance, **weight** is not the same as mass. Mass is the quantity of matter and weight is the force with which Earth's gravity pulls on that quantity of matter.

In your lab activities, you will use a graduated cylinder to measure the volume of liquids. The volume of a solid that has a specific geometric shape can be calculated from the measured lengths of its surfaces. Small volumes are usually expressed in cubic centimeters, cm³. One cubic centimeter is equal to 1 mL.

18

INCLUSION
Strategies

- *Learning Disabled*
- *English Language Learners*
- *Developmentally Delayed*

Ask students to cut pictures of common objects out of magazines. The objects should include solid objects, liquids, small objects, and large objects. Students may glue each picture to a piece of construction paper. Students can determine what properties of the object are to be measured, such as length,

mass, volume or weight. They may also choose a unit of measure that would be appropriate for measuring that property of the object. Students can label each picture with the property and unit of measure they chose. This activity can be done individually, in small groups, or with help from a teaching assistant.

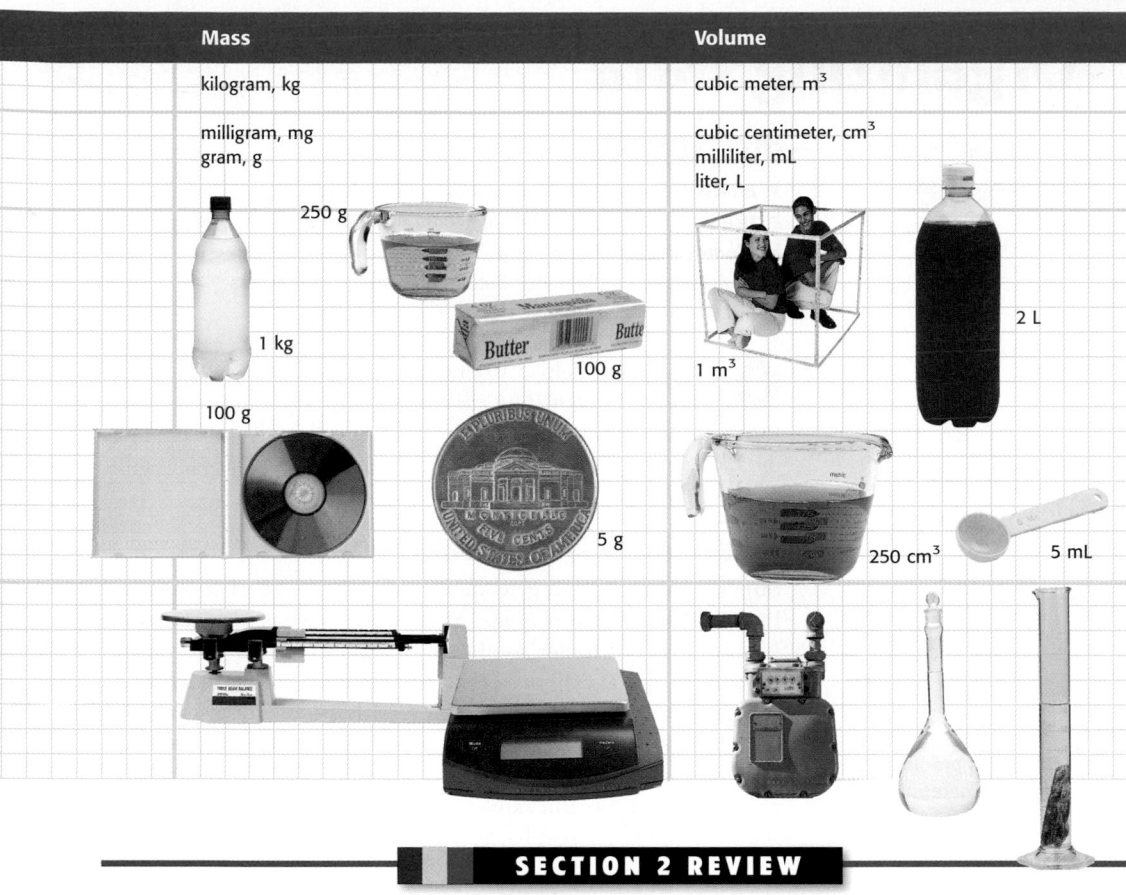

Mass	Volume
kilogram, kg	cubic meter, m³
milligram, mg	cubic centimeter, cm³
gram, g	milliliter, mL
	liter, L

250 g

1 kg

Butter   Butte
100 g

100 g

1 m³

2 L

5 g

250 cm³   5 mL

Close

Quiz — BASIC

1. The scientific method is a set of rigid steps that must always be followed in the same order. True or false? (false)

2. What are the SI base units of length, mass, and time? (meter, kilogram, second)

3. Convert 580 meters to kilometers. (580 m = 0.58 km)

LS Logical

SECTION 2 REVIEW

SUMMARY

▶ In the scientific method, a person asks a question, collects data about the question, forms a hypothesis, tests the hypothesis, draws conclusions, and if necessary, modifies the hypothesis based on results.

▶ In an ideal experiment, only one factor, the variable, is tested.

▶ SI has seven base units.

1. **List** three examples each of things that are commonly measured by mass, by volume, and by length.

2. **Explain** why the scientific method is said to involve critical thinking.

3. **Describe** a hypothesis and how it is used. Give an example of a hypothesis.

4. **Explain** why no experiment should be called a failure.

5. **Relate** the discussion of scientists' tools to how science and technology depend on each other.

6. **Explain** the difference between SI base units and derived units. Give an example of each.

7. **Critical Thinking** Why do you think it is wise to limit an experiment to test only one factor at a time?

19

Answers to Section 2 Review

1. Mass: solid food items, people, and mail; volume: liquid food items, gasoline; length: rope, distance, height.

2. The scientific method involves critical thinking in that it entails thinking about a problem and making objective judgments about results.

3. A hypothesis is a possible answer to a question that can be tested. An example would be, "I can pass the test if I study at least 5 hours."

4. No experiment should be called a failure because an experiment that produces unexpected results provides a chance to learn something new.

5. There are still scientific theories that have not been verified, because the tools needed to test these theories do not yet exist.

6. An SI base unit is a single unit while a derived unit is a combination of base units. Base units include: seconds, meters, kilograms, kelvins, amperes, moles, and candelas. Examples of derived units include meters squared (m²) and meters cubed (m³).

7. It is much easier to determine which factor your experiment depends on if you only check one factor at a time. If you change more than one thing and something unexpected happens, you will not know what caused the result.

Focus

Overview

Before beginning this section, review with your students the Objectives listed in the Student Edition. This section introduces students to techniques for organizing and interpreting data. They learn how to analyze line graphs, bar graphs, and pie charts, and how to use scientific notation and significant figures in problem solving. The difference between precision and accuracy is also covered.

🔔 Bellringer

Use the Bellringer transparency to prepare students for this section.

Motivate

Opening Activity — GENERAL

Ask students to think of something in their daily life to keep track of for one week. For example, they could track how many hours a day they spend on homework, how many hours they sleep every night, what percentage of their free time is spent watching television, or how much they weigh each day. Have them create a data table for the information. After they read this section, have students determine the appropriate type of graph and use it to display their data. They may wish to analyze their graphs for patterns and trends. **LS** Intrapersonal

Organizing Data

KEY TERMS
scientific notation
precision
significant figures
accuracy

OBJECTIVES

▶ **Interpret** line graphs, bar graphs, and pie charts.
▶ **Use** scientific notation and significant figures in problem solving.
▶ **Identify** the significant figures in calculations.
▶ **Understand** the difference between precision and accuracy.

One thing that helped Roentgen discover X rays was that he could read about the experiments other scientists had performed with the cathode ray tube. He was able to learn from their data. Organizing and presenting data are important science skills.

Presenting Scientific Data

Suppose you are trying to determine the speed of a chemical reaction that produces a gas. You can let the gas displace water in a graduated cylinder, as shown in **Figure 16.** You read the volume of gas in the cylinder every 10 seconds from the start of the reaction until there is no change in volume for four successive readings. *Table 4* shows the data you collect in the experiment.

Because you did the experiment, you saw how the volume changed over time. But how can someone who reads your report see it? To show the results, you can make a graph.

Figure 16

The volume of gas produced by a reaction can be determined by measuring the volume of water the gas displaces in a graduated cylinder.

20

Table 4 **Experimental Data**

Time (s)	Volume of gas (mL)	Time (s)	Volume of gas (mL)
0	0	90	116
10	3	100	140
20	6	110	147
30	12	120	152
40	25	130	154
50	43	140	156
60	58	150	156
70	72	160	156
80	100	170	156

Alternative Assessment — ADVANCED

Using a Spreadsheet Have students use a spreadsheet program to create graphs for the data shown in **Table 4.** Print the graphs, and have students label important points on the graph. Ask students if they have ever seen powers of ten, i.e., exponents, displayed in spreadsheet software or scientific calculators as E values. **LS** Logical

Line graphs are best for continuous changes

Many types of graphs can be drawn, but which one should you use? A *line graph* is best for displaying data that change. Our example experiment has two variables, time and volume. Time is the *independent variable* because you chose the time intervals to take the measurements. The volume of gas is the *dependent variable* because its value depends on what happens in the experiment.

Line graphs are usually made with the *x*-axis showing the independent variable and the *y*-axis showing the dependent variable. *Figure 17* is a graph of the data that is in *Table 4.*

A person who never saw your experiment can look at this graph and know what took place. The graph shows that gas was produced slowly for the first 20 s and that the rate increased until it became constant from about 40 s to 100 s. The reaction slowed down and stopped after about 140 s.

Bar graphs compare items

A *bar graph* is useful when you want to compare similar data for several individual items or events. If you measured the melting temperatures of some metals, your data could be presented in a way similar to that in *Table 5. Figure 18* shows the same values as a bar graph. A bar graph often makes clearer how large or small the differences in individual values are.

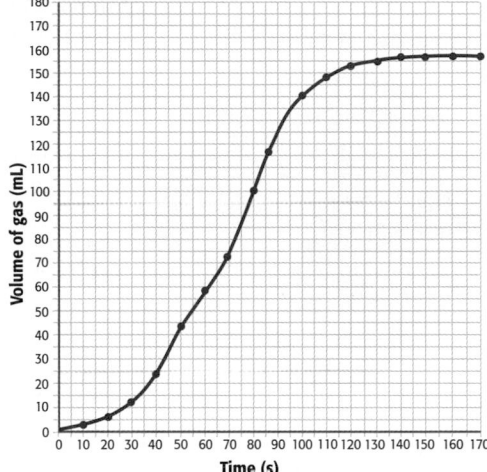

Volumes Measured over Time

Figure 17

Data that change over a range are best represented by a line graph. Notice that many in-between volumes can be estimated.

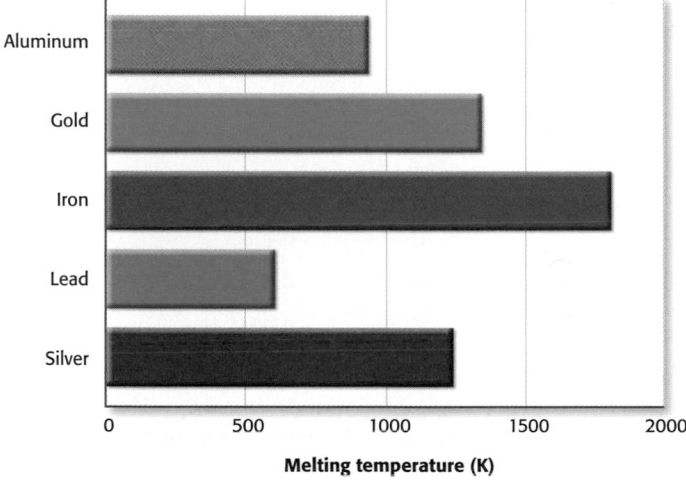

Graph of the Melting Points of Some Common Metals

Table 5 **Melting Points of Some Metals**

Element	Melting temp. (K)
Aluminum	933
Gold	1337
Iron	1808
Lead	601
Silver	1235

Figure 18

A bar graph is best for data that have specific values for different events or things.

Interpreting Visuals As students examine the pie chart in **Figure 19,** ask them the following questions: Which element makes up the greatest percent? (oxygen) About what fraction of the sample consists of this element? (1/2) What must the sum of percentages in a pie chart always add up to? (100) What additional data would you need to graph this information in the form of a bar graph instead of a pie chart? (To make a bar graph, you would need to know the total mass of the sample. With this value and the given percentages, you could calculate the mass of each element present in the sample and use that data to make a bar graph.) **LS** Visual

Teaching Tip

Scientific Notation Scientific notation is a shorthand way to represent where the decimal place is located in a measurement or value. Point out to students that a positive exponent, such as 10^4, means to move the decimal place to the right. So 5.4 m $\times$ 10^4 is the same as 54 000 m (move the decimal four places to the right). A negative exponent means move the decimal place to the left, so 2.54×10^{-3} cm is the same as 0.002 54 cm (move the decimal three places to the left).

Composition of Calcite

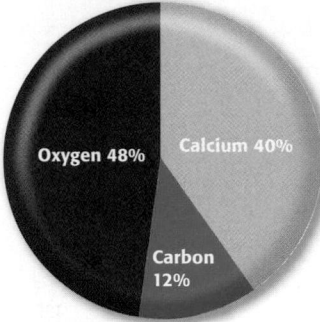

Figure 19

A pie chart is best for data that represent parts of a whole, such as the percentage of each element in the mineral calcite.

▶ **scientific notation** a method of expressing a quantity as a number multiplied by 10 to the appropriate power

internet connect

www.scilinks.org
Topic: Presenting Scientific Data
SciLinks code: HK4109

Maintained by the National Science Teachers Association

Pie charts show the parts of a whole

A *pie chart* is ideal for displaying data that are parts of a whole. Suppose you have analyzed a compound to find the percentage of each element it contains. Your analysis shows that the compound consists of 40 percent calcium, 12 percent carbon, and 48 percent oxygen. You can draw a pie chart that shows these percentages as a portion of the whole pie, the compound, as shown in *Figure 19*. To construct a pie chart, refer to the Graphing Skills Refresher in Appendix A and the skills page at the end of this chapter.

Writing Numbers in Scientific Notation

Scientists sometimes need to express measurements using numbers that are very large or very small. For example, the speed of light through space is about 300 000 000 m/s. Suppose you want to calculate the time required for light to travel from Neptune to Earth when Earth and Neptune are 4 500 000 000 000 m apart. To find out how long it takes, you would divide the distance between Earth and Neptune by the distance light travels in 1 s.

$$t = \frac{\text{distance from Earth to Neptune (m)}}{\text{distance light travels in 1 s (m/s)}}$$

$$t = \frac{4\ 500\ 000\ 000\ 000\ \text{m}}{300\ 000\ 000\ \text{m/s}}$$

This is a lot of zeros to keep track of when performing a calculation.

To reduce the number of zeros, you can express values as a simple number multiplied by a power of 10. This is called **scientific notation.** Some powers of 10 and their decimal equivalents are shown below.

$$10^4 = 10\ 000$$
$$10^3 = 1000$$
$$10^2 = 100$$
$$10^1 = 10$$
$$10^0 = 1$$
$$10^{-1} = 0.1$$
$$10^{-2} = 0.01$$
$$10^{-3} = 0.001$$

In scientific notation, 4 500 000 000 000 m can be written as 4.5×10^{12} m. The speed of light in space is 3.0×10^8 m/s. Refer to the Math Skills Refresher in Appendix A for more information on scientific notation.

Cultural Awareness

Ancient Numbers Numbers were expressed as powers during the Old Babylonian Empire almost 4000 years ago. Although the ancient Babylonians did not write numbers with exponents, they used squares and square roots to solve geometric problems. They expressed large numbers by positions with values of 60^0, 60^1, 60^2, and 60^3 just as we use 10^0, 10^1, 10^2, and 10^3 for ones, tens, hundreds, and thousands.

INCLUSION Strategies

• *Learning Disabled* • *English Language Learners*

To help students understand the concept of a pie chart, ask them to construct a pie chart that depicts the percentage of students of different ages in the class. Students' pie charts can be constructed of colored paper or colored on white paper. You may wish to have students try other pie charts as well, such as one showing the percentage of ingredients on a food product label.

Using scientific notation

When you use scientific notation in calculations, you follow the math rules for powers of 10. When you multiply two values in scientific notation, you add the powers of 10. When you divide, you subtract the powers of 10.

So the problem about Earth and Neptune can be solved more easily as shown below.

$$t = \frac{4.5 \times 10^{12} \text{ m}}{3.0 \times 10^8 \text{ m/s}}$$

$$t = \left(\frac{4.5}{3.0} \times \frac{10^{12}}{10^8}\right) \frac{\text{m}}{\text{m/s}}$$

$$t = (1.5 \times 10^{(12-8)})\text{s}$$

$$t = 1.5 \times 10^4 \text{ s}$$

Math Skills

Writing Scientific Notation The adult human heart pumps about 18 000 L of blood each day. Write this value in scientific notation.

1 List the given and unknown values.
  Given: *volume, V* = 18 000 L
  Unknown: *volume, V* = ? × 10$^?$ L

2 Write the form for scientific notation.
  $V = ? \times 10^? \text{ L}$

3 Insert the known values into the form, and solve.
  First find the largest power of 10 that will divide into the known value and leave one digit before the decimal point. You get 1.8 if you divide 10 000 into 18 000 L. So, 18 000 L can be written as (1.8 × 10 000) L.

  Then write 10 000 as a power of 10. Because 10 000 = 10^4, you can write 18 000 L as 1.8 × 10^4 L.
  $V = 1.8 \times 10^4 \text{ L}$

Practice

Writing Scientific Notation

1. Write the following measurements in scientific notation:
   a. 800 000 000 m
   b. 0.0015 kg
   c. 60 200 L
   d. 0.000 95 m
   e. 8 002 000 km
   f. 0.000 000 000 06 kg
2. Write the following measurements in long form:
   a. 4.5×10^3 g
   b. 6.05×10^{-3} m
   c. 3.115×10^6 km
   d. 1.99×10^{-8} cm

Practice HINT

▶ A shortcut for scientific notation involves moving the decimal point and counting the number of places it is moved. To change 18 000 to 1.8, the decimal point is moved four places to the left. The number of places the decimal is moved is the correct power of 10.

$$18\ 000 \text{ L} = 1.8 \times 10^4 \text{ L}$$

▶ When a quantity smaller than 1 is converted to scientific notation, the decimal moves to the right and the power of 10 is *negative*. For example, suppose an *E. coli* bacterium is measured to be 0.000 0021 m long. To express this measurement in scientific notation, move the decimal point to the right.

$$0.000\ 0021 \text{ m} = 2.1 \times 10^{-6} \text{ m}$$

Did You Know?

Big and Small The mass of an electron is 0.000 000 000 000 000 000 000 000 000 000 911 g.

The mass of the sun is about 2000 000 000 000 000 000 000 000 000 000 000 000 g.

These examples can be used to illustrate the value of scientific notation. In scientific notation, the electron's mass is expressed as 9.11×10^{-28} g, and the sun's mass is expressed as 2×10^{33} g.

Teach, continued

Math Skills

Additional Examples
Using Scientific Notation
Perform the following calculations:

a. $\dfrac{(5.2 \times 10^3 \text{ kg})(4.3 \times 10^3 \text{ m})}{(3.5 \times 10^2 \text{ s})(3.5 \times 10^2 \text{ s})}$

b. $\dfrac{(3.6 \times 10^3 \text{ kg})(6.5 \text{ m})}{(1.5 \times 10^2 \text{ m}^2)}$

Answers
a. 1.8×10^2 kg • m/s^2 (or N)
b. 1.6×10^2 kg/m

LS Logical

Practice

Answers to Practice

1. a. $(5.5 \times 10^4 \text{ cm}) \times (1.4 \times 10^4 \text{ cm}) = (5.5 \times 1.4)(10^{4+4})(\text{cm} \times \text{cm}) = 7.7 \times 10^8 \text{ cm}^2$

b. $(2.77 \times 10^{-5} \text{ m}) \times (3.29 \times 10^{-4} \text{ m}) = (2.77 \times 3.29)(10^{-5+-4})(\text{m} \times \text{m}) = 9.11 \times 10^{-9} \text{ m}^2$

c. $\left(4.34 \frac{\text{g}}{\text{mL}}\right) \times (8.22 \times 10^6 \text{ mL}) = (4.34 \times 8.22)(10^6)\left(\frac{\text{g}}{\text{mL}} \times \text{mL}\right) = 3.57 \times 10^7 \text{ g}$

d. $(3.8 \times 10^{-2} \text{ cm}) \times (4.4 \times 10^{-2} \text{ cm}) \times (7.5 \times 10^{-2} \text{ cm}) = (3.8 \times 4.4 \times 7.5)(10^{-2+-2+-2})(\text{cm} \times \text{cm} \times \text{cm}) = 1.3 \times 10^{-4} \text{ cm}^3$

2. a. $\dfrac{3.0 \times 10^4 \text{ L}}{62 \text{ s}} = 4.8 \times 10^2 \text{ L/s}$

b. $\dfrac{6.05 \times 10^7}{8.8 \times 10^6 \text{ cm}^3} = 6.9 \text{ g/cm}^3$

c. $\dfrac{5.2 \times 10^8 \text{ cm}^3}{9.5 \times 10^2 \text{ cm}} = 5.5 \times 10^5 \text{ cm}^2$

d. $\dfrac{3.8 \times 10^{-5} \text{ kg}}{4.6 \times 10^{-5} \text{ kg/cm}^3} = 8.3 \times 10^{-1} \text{ cm}^3$

LS Logical

Practice HINT

Because not all devices can display superscript numbers, scientific calculators and some math software for computers display numbers in scientific notation using E values. That is, 3.12×10^4 may be shown as 3.12 E4. Very small numbers are shown with negative values. For example, 2.637×10^{-5} may be shown as 2.637 E–5. The letter E signifies exponential notation. The E value is the exponent (power) of 10. The rules for using powers of 10 are the same whether the exponent is displayed as a superscript or as an E value.

▶ **precision** the exactness of a measurement

▶ **significant figure** a prescribed decimal place that determines the amount of rounding off to be done based on the precision of the measurement

Math Skills

Using Scientific Notation Your state plans to buy a rectangular tract of land measuring 5.36×10^3 m by 1.38×10^4 m to establish a nature preserve. What is the area of this tract in square meters?

1 List the given and unknown values.
  Given: *length*, $l = 1.38 \times 10^4$ m
  width, $w = 5.36 \times 10^3$ m
  Unknown: *area*, $A = ?$ m^2

2 Write the equation for area.
  $A = l \times w$

3 Insert the known values into the equation, and solve.
  $A = (1.38 \times 10^4 \text{ m})(5.36 \times 10^3 \text{ m})$
  Regroup the values and units as follows.
  $A = (1.38 \times 5.36)(10^4 \times 10^3)(\text{m} \times \text{m})$
  When multiplying, add the powers of 10.
  $A = (1.38 \times 5.36)(10^{4+3})(\text{m} \times \text{m})$
  $A = 7.3968 \times 10^7 \text{ m}^2$
  $A = 7.40 \times 10^7 \text{ m}^2$

Practice

Using Scientific Notation
1. Perform the following calculations.
  a. $(5.5 \times 10^4 \text{ cm}) \times (1.4 \times 10^4 \text{ cm})$
  b. $(2.77 \times 10^{-5} \text{ m}) \times (3.29 \times 10^{-4} \text{ m})$
  c. $(4.34 \text{ g/mL}) \times (8.22 \times 10^6 \text{ mL})$
  d. $(3.8 \times 10^{-2} \text{ cm}) \times (4.4 \times 10^{-2} \text{ cm}) \times (7.5 \times 10^{-2} \text{ cm})$
2. Perform the following calculations.
  a. $\dfrac{3.0 \times 10^4 \text{ L}}{62 \text{ s}}$
  c. $\dfrac{5.2 \times 10^8 \text{ cm}^3}{9.5 \times 10^2 \text{ cm}}$
  b. $\dfrac{6.05 \times 10^7 \text{ g}}{8.8 \times 10^6 \text{ cm}^3}$
  d. $\dfrac{3.8 \times 10^{-5} \text{ kg}}{4.6 \times 10^{-5} \text{ kg/cm}^3}$

Using Significant Figures

Suppose you measure a length of wire with two tape measures. One tape is marked every 0.001 m, and the other is marked every 0.1 m. The tape marked every 0.001 m gives you more **precision,** because with it you can report a length of 1.638 m. The other tape is only precise to 1.6 m.

To show the precision of a measured quantity, scientists use **significant figures.** The length of 1.638 m has four significant figures because the digits 1638 are known for sure. The measurement of 1.6 m has two significant figures.

24

Trends in Physics
Amazing Accuracy In April of 1999, scientists reported measuring the acceleration due to gravity with an accuracy of three parts in one billion. This is equivalent to measuring the width of Ireland with an accuracy within one millimeter! Scientists measured the acceleration by observing the effect of gravity on extremely cold cesium atoms falling in an "atom fountain." To prevent the atoms' thermal motion from interfering with the measurement, the atoms were cooled to ten billionths of a degree above absolute zero.

A Good accuracy (near post) and good precision (close together)

B Good accuracy (near post) and poor precision (spread apart)

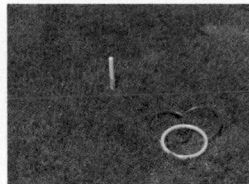
C Poor accuracy (far from post) and good precision (close together)

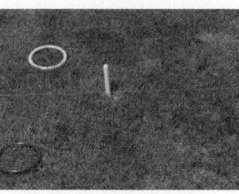
D Poor accuracy (far from post) and poor precision (spread apart)

Figure 20
A ring toss is a game of skill, but it is also a good way to visualize accuracy and precision in measurements.

If the tip of your tape measure has broken off, you can read 1.638 m precisely, but that number is not **accurate.** A measured quantity is only as accurate as the tool used to make the measurement. One way to think about the accuracy and precision of measurements is shown in *Figure 20*.

Math Skills

Significant Figures Calculate the volume of a room that is 3.125 m high, 4.25 m wide, and 5.75 m long. Write the answer with the correct number of significant figures.

1 **List the given and unknown values.**
   Given: *length, l* = 5.75 m
   width, w = 4.25 m
   height, h = 3.125 m
   Unknown: *Volume, V* = ? m³

2 **Write the equation for volume.**
   Volume, V = $l \times w \times h$

3 **Insert the known values into the equation, and solve.**
   $V = 5.75 \text{ m} \times 4.25 \text{ m} \times 3.125 \text{ m}$
   $V = 76.367\ 1875 \text{ m}^3$
   The answer should have three significant figures because the value with the smallest number of significant figures has three significant figures.
   $V = 76.4 \text{ m}^3$

Practice

Significant Figures
Perform the following calculations, and write the answer with the correct number of significant figures.
1. 12.65 m × 42.1 m
2. 3.02 cm × 6.3 cm × 8.225 cm
3. 3.7 g ÷ 1.083 cm³
4. 3.244 m ÷ 1.4 s

▮ **accuracy** a description of how close a measurement is to the true value of the quantity measured

Practice HINT

When rounding to get the correct number of significant figures, do you round up or down if the last digit is a 5? Your teacher may have other ways to round, but one very common way is to round to get an even number. For example, 3.25 is rounded to 3.2, and 3.35 is rounded to 3.4. Using this simple rule, half the time you will round up and half the time you will round down. See the Math Skills Refresher in Appendix A for more about significant figures and rounding.

1. What are the three different types of graphs? (line graph, bar graph, pie chart)

2. Write the following in scientific notation:

   a. 25 000 (2.5×10^4)

   b. 250 (2.5×10^2)

   c. 0.025 (2.5×10^{-2})

3. Round 50.76 to 2 significant figures. (51)

4. Which term describes how close a measurement approaches a true value: precision or accuracy? (accuracy)

5. Which term describes the degree of exactness of a measurement: precision or accuracy? (precision)

LS Logical

When you use measurements in calculations, the answer is only as precise as the least precise measurement used in the calculation—the measurement with the fewest significant figures. Suppose, for example, that the floor of a rectangular room is measured to the nearest 0.01 m (1 cm). The measured dimensions are reported to be 5.871 m by 8.14 m.

If you use a calculator to multiply 5.871 by 8.14, the display may show 47.789 94 as an answer. But you don't really know the area of the room to the nearest 0.000 01 m^2, as the calculator showed. To have the correct number of significant figures, you must round off your results. In this case the correct rounded result is $A = 47.8$ m^2, because the least precise value in the calculation had three significant figures.

When adding or subtracting, use this rule: the answer cannot be more precise than the values in the calculation. A calculator will add 6.3421 s and 12.1 s to give 18.4421 as a result. But the least precise value was known to 0.1 s, so round to 18.4 s.

SECTION 3 REVIEW

SUMMARY

▶ Representing scientific data with graphs helps you and others understand experimental results.

▶ Scientific notation is useful for writing very large and very small measurements because it uses powers of 10 instead of strings of zeros.

▶ Accuracy is the extent to which a value approaches the true value.

▶ Precision is the degree of exactness of a measurement.

▶ Expressing data with significant figures tells others how precisely a measurement was made.

1. **Describe** the kind of data that is best displayed as a line graph.

2. **Describe** the kind of data that is best displayed as a pie chart. Give an example of data from everyday experiences that could be placed on a pie chart.

3. **Explain** in your own words the difference between accuracy and precision.

4. **Critical Thinking** An old riddle asks, "Which weighs more, a pound of feathers or a pound of lead?" Answer the question, and explain why you think people sometimes answer incorrectly.

Math Skills

5. **Convert** the following measurements to scientific notation:

   a. 15 400 mm^3 **c.** 2050 mL

   b. 0.000 33 kg   **d.** 0.000 015 mol

6. **Calculate** the following:

   a. 3.16×10^3 m $\times$ 2.91×10^4 m

   b. 1.85×10^{-3} cm $\times$ 5.22×10^{-2} cm

   c. 9.04×10^5 g $\div$ 1.35×10^5 cm^3

7. **Calculate** the following, and round the answer to the correct number of significant figures.

   a. 54.2 cm^2 $\times$ 22 cm   **b.** 23 500 m $\div$ 89 s

26

Chapter Resource File

• Concept Review GENERAL

• Quiz BASIC

Answers to Section 3 Review

1. Line graphs are best for continuous changes.

2. Pie charts show the parts of a whole. An example is the percentages of types of CDs that make up a collection.

3. Accuracy is how close a measurement is to being correct. Precision indicates the reproducibility of the measurements.

4. The correct answer is that both weigh the same. A reason for an incorrect answer might be that mass is confused with volume, and a given volume of feathers would be much lighter than the same volume of lead.

Math Skills

5. **a.** 1.54×10^4 mm^3

   b. 3.3×10^{-4} kg

   c. 2.05×10^3 mL

   d. 1.5×10^{-5} mol

6. **a.** 9.20×10^7 m^2

   b. 9.66×10^{-5} cm^2

   c. 6.70 g/cm^3

7. **a.** 1.2×10^3 cm^3

   b. 2.6×10^2 m/s

Graphing Skills

Constructing a Pie Chart

Unlike line or bar graphs, pie charts require special calculations to accurately display data. The steps below show how to construct a pie chart from this data.

Wisconsin Hardwood Trees

Type of tree	Number found
Oak	600
Maple	750
Beech	300
Birch	1200
Hickory	150
Total	**3000**

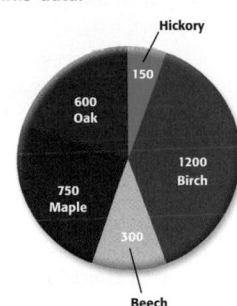

1 First, find the percentage of each type of tree. To do this, divide the number of each type of tree by the total number of trees and multiply by 100.

$$\frac{600\ oak}{3000\ trees} \times 100 = 20\%\ oak \qquad \frac{750\ maple}{3000\ trees} \times 100 = 25\%\ maple$$

Continuing these calculations for the rest of the trees, you find that 10% of the trees are beech, 40% are birch, and 5% are hickory. Check to make sure the sum is 100.

2 Now determine the size of the pie shapes that will make up the chart. Use the conversion factor 360°/100% to convert from percentage to degrees of a circle.

$$20\%\ oak \times \frac{360°}{100\%} = 72°\ oak \qquad 25\%\ maple \times \frac{360°}{100\%} = 90°\ maple$$

$$10\%\ beech \times \frac{360°}{100\%} = 36°\ beech \qquad 40\%\ birch \times \frac{360°}{100\%} = 144°\ birch$$

$$5\%\ hickory \times \frac{360°}{100\%} = 18°\ hickory$$

Check to make sure that the sum of all angles is 360°.

3 Use a compass to draw a circle and mark the circle's center. Then use a protractor to draw an angle of 144°. Mark this angle. From this mark, measure an angle of 90°. Continue marking angles from largest to smallest until all the angles have been marked. Finally, label each part of the chart, and choose an appropriate title for the graph.

Practice

1. A recipe for a loaf of bread calls for 474 g water, 9.6 g yeast, 28.3 g butter, 10 g salt, 10 g honey, and 907 g flour. Make a pie chart showing what percentage of the bread each of these ingredients is.

Graphing Skills

Teaching Tip

Before students begin the Study Skill, have them practice conversions between percentages or fractions and degrees. On the chalkboard, have them solve basic conversions, such as 50% of a circle equals 180°, 25% equals 90°, and so forth.

Point out that knowing some basic conversions can serve as a check for their work. For instance, if a quantity has a percentage equal to 28%, then you know that the angle of the circle for that quantity must be somewhat larger than 90° (25%).

Answers to Practice

1. The angles for the ingredients are as follows:

total mass = 1439 g
water = 119°
yeast = 2.4°
butter = 7.08°
salt = 2.5°
honey = 2.5°
flour = 227°

(The sum of the angles is slightly greater than 360° because of rounding.)

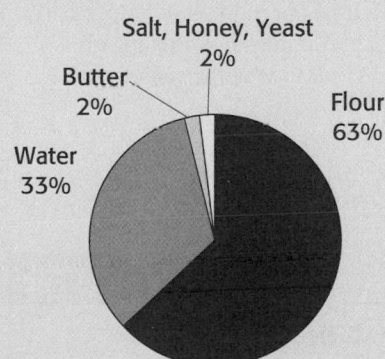

27

Chapter 1 • Introduction to Science 27

Understanding Concepts

1. d
2. b
3. a
4. d
5. c
6. b
7. d
8. a
9. c
10. c
11. b
12. c
13. d
14. b
15. a

Using Vocabulary

16. Physical science is no longer the study of only the nonliving world. As knowledge has increased, scientists have learned that the discoveries in one area are applicable to another. For example, chemistry, a physical science, applies to living beings. This field of study is called biochemistry, which is partly life science and partly physical science.

17. It has been observed repeatedly, and it does not attempt to explain why the sun sets in the west.

Chapter Resource File

- **Chapter Test** GENERAL
- **Chapter Test** ADVANCED
- **Test Item Listing for EvamView® Test Generator**

Chapter Highlights

Before you begin, review the summaries of key ideas of each section, found at the end of each section. The vocabulary terms are listed on the first page of each section.

UNDERSTANDING CONCEPTS

1. Which of the following is not included in physical science?
   - **a.** physics
   - **b.** chemistry
   - **c.** astronomy
   - **d.** zoology

2. What science deals most with energy and forces?
   - **a.** biology
   - **b.** physics
   - **c.** botany
   - **d.** agriculture

3. Using superconductors to build computers is an example of
   - **a.** technology.
   - **b.** applied biology.
   - **c.** pure science.
   - **d.** an experiment.

4. A balance is a scientific tool used to measure
   - **a.** temperature.
   - **b.** time.
   - **c.** volume.
   - **d.** mass.

5. Which of the following units is an SI base unit?
   - **a.** liter
   - **b.** cubic meter
   - **c.** kilogram
   - **d.** centimeter

6. The quantity 5.85×10^4 m is equivalent to
   - **a.** 5850 000 m.
   - **b.** 58 500 m.
   - **c.** 5840 m.
   - **d.** 0.000 585 m.

7. Which of the following measurements has two significant figures?
   - **a.** 0.003 55 g
   - **b.** 500 mL
   - **c.** 26.59 km
   - **d.** 2.3 cm

8. The composition of the mixture of gases that makes up our air is best represented on what kind of graph?
   - **a.** pie chart
   - **b.** bar graph
   - **c.** line graph
   - **d.** variable graph

9. Making sure an experiment gives the results you expect is an example of
   - **a.** the scientific method.
   - **b.** critical thinking.
   - **c.** unscientific thinking.
   - **d.** objective observation.

10. In a controlled experiment,
    - **a.** the outcome is controlled.
    - **b.** one variable is fixed while all others are changed.
    - **c.** one variable is changed while all others remain fixed.
    - **d.** results are obtained by computer models.

11. A line graph is best suited for
    - **a.** comparing electrical conductivities of different metals.
    - **b.** recording changes in a star's brightness over a 5 h period.
    - **c.** showing the proportion of different elements in an alloy.
    - **d.** comparing accelerations of automobiles.

12. The quantity 300 000 000 m/s is equivalent to
    - **a.** 3×10^6 m/s.
    - **b.** 3×10^7 m/s.
    - **c.** 3×10^8 m/s.
    - **d.** 3×10^9 m/s.

13. How many significant figures are in the quantity 6.022×10^{23} atoms/mol?
    - **a.** one
    - **b.** two
    - **c.** three
    - **d.** four

14. A 4.00 kg crowbar's mass is measured several times. Which set of measurements is precise but not accurate?
    - **a.** 3.5 kg, 2.5 kg, 4.5 kg
    - **b.** 3.55 kg, 3.58 kg, 3.56 kg
    - **c.** 3.99 kg, 4.02 kg, 4.00 kg
    - **d.** 4.0 kg, 4.2 kg, 3.9 kg

15. During a storm, a student measures rainwater depth every 15 min. The water's depth is a
    - **a.** dependent variable.
    - **b.** independent variable.
    - **c.** controlled variable.
    - **d.** significant figure.

18. The rotation of Earth could be considered a scientific theory because it is a tested, possible explanation of several natural events, including why the sun sets in the west. As Earth rotates on its axis every day, the sun's rays move westward across the surface of Earth. At night, stars appear to move across the horizon due to Earth's constant rotation.

19. Answers may vary. There are three significant figures, which means the volume was measured to the nearest milliliter. This is the precision of the measurement. The accuracy of the measurement is not known unless the measuring device has been compared to a standard, or has been calibrated.

20. The mass of an atom in units of kilograms would be a very small number. Scientific notation enables the measurer to write the measurement using powers of ten instead of a very long string of zeros.

21. Mass is a quantity of matter, whereas weight is the force with which Earth's gravity pulls on that quantity of matter. Weight would be expressed in units of force instead of mass units.

16. *Physical science* was once defined as the science of the nonliving world. Write a paragraph suggesting why that definition is no longer sufficient.

WRITING
SKILL

17. Explain why the observation that the sun sets in the west could be called a scientific law.

18. Explain why the rotation of Earth could be considered a *scientific theory*. Use it to account for the answer in item 16, as well as to explain the motion of stars in the night sky.

19. The volume of a bottle has been measured to be 465 mL. Use the terms *significant figures* and *precision* to explain what you know and do not know about the measured volume. How does the *accuracy* of the measurement affect the value?

20. Describe how *scientific notation* is useful in measuring the mass of an atom in units of kilograms.

21. Explain why *mass* and *weight* are not the same. How would the units in which they are measured differ?

BUILDING MATH SKILLS

22. Graphing The graph at right shows the changes in temperature during a chemical reaction. Study the graph and answer the following questions:

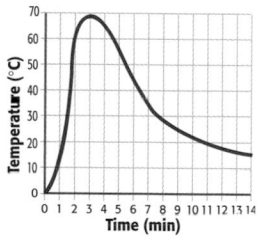

a. What was the highest temperature reached during the reaction?

b. How many minutes passed before the highest temperature was reached?

c. During what period of time was the temperature increasing?

d. Did heating or cooling occur faster?

23. Graphing Silver solder is a mixture of 40 percent silver, 40 percent tin, 14 percent copper, and 6 percent zinc. Draw a graph that shows the composition of silver solder.

24. Scientific Notation Write the following measurements in scientific notation:

a. 22 000 mg   **d.** 0.000 0037 kg
b. 0.005 km   **e.** 722 000 000 000 s
c. 65 900 000 m   **f.** 0.000 000 064 s

25. Scientific Notation Do the following calculations, and write the answers in scientific notation:

a. $37\ 000\ 000\ A \times 7\ 100\ 000\ s$
b. $0.000\ 312\ m^3 \div 486\ s$
c. $4.6 \times 10^4\ cm \times 7.5 \times 10^3\ cm$
d. $8.3 \times 10^6\ kg \div 2.5 \times 10^9\ cm^3$
e. $3.47 \times 10^4\ m \div 6.95 \times 10^{-3}\ s$

26. Significant Figures Round the following measurements to the number of significant figures shown in parentheses:

a. 7.376 m (2)   **c.** 0.087 904 85 g (1)
b. 48 794 km (3)   **d.** 362.003 06 s (5)

27. Significant Figures Do the following calculations, and write the answers with the correct number of significant figures:

a. $15.75\ m \times 8.45\ m$
b. $5650\ L \div 27\ min$
c. $0.0058\ km \times 0.228\ km$
d. $6271\ m \div 59.7\ s$
e. $3.5 \times 10^3\ cm^2 \times 2.11 \times 10^4\ cm$

28. SI Prefixes Express each of the following quantities using an appropriate SI prefix before the proper units.

a. 0.004 g
b. 75 000 m
c. 325 000 000 kg
d. 0.000 000 003 s
e. 4 570 000 s

29

Building Math Skills

22. a. 69°C
   b. 3 minutes
   c. the first 3 minutes
   d. heating

23. See pie chart below.

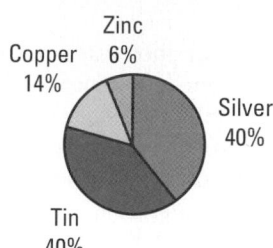

24. a. 2.2×10^4 mg
   b. 5×10^{-3} km
   c. 6.59×10^7 m
   d. 3.7×10^{-6} kg
   e. 7.22×10^{11} s
   f. 6.4×10^{-8} s

25. a. 2.6×10^{14} A•s
   b. 6.42×10^{-7} m³/s
   c. 3.4×10^8 cm²
   d. 3.3×10^{-3} kg/cm³
   e. 4.99×10^6 m/s

26. a. 7.4 m
   b. 48 800 km = 4.88×10^4 km
   c. 0.09 g
   d. 362.00 s

27. a. 133 m²
   b. 2.1×10^2 L/min
   c. 1.3×10^{-3} km²
   d. 105 m/s
   e. 7.4×10^7 cm³

28. a. 4 mg
   b. 75 km
   c. 325 Gg
   d. 3 ns
   e. 4.57 Ms

Section	Questions
Assignment Guide	
1	1–3, 16–18, 29, 31, 32, 37–39
2	4, 5, 9, 10, 21, 28, 30, 34
3	6–8, 11–15, 19, 20, 22–27, 33, 35, 36

Thinking Critically

29. On the back of the picture tube, there is a mineral coating that lights up in different colors when the cathode rays hit it.

30. under "research and collect data"

31. This description was a model in that it provided a mental picture of the universe, which could be used to study the apparent motion of the universe.

32. The law of gravitation states that objects fall to Earth; it even shows how to calculate the force. It does not explain why.

33. The type of fertilizer is the independent variable. Control factors are: the types of radishes, the amount of water, the amount of sunshine, etc. There are at least four things that could be used to determine the results: size, quantity, appearance, and taste.

34. Is 1200 cm^3 the volume of the glass used to make the container or the volume of the inside of the container or the volume that the container occupies? Is that 1200 with 4 significant figures or is that somewhere between 1100 and 1300 or between 1150 and 1250?

29. Applying Knowledge The picture tube in a television sends a beam of cathode rays to the screen. These are the same invisible rays that Roentgen was experimenting with when he discovered X rays. Use what you know about cathode rays to suggest what produces the light that forms the picture on the screen.

30. Applying Knowledge Today, scientists must do a search through scientific journals before performing an experiment or making methodical observations. Where would this step take place in the diagram of the scientific method?

31. Interpreting and Communicating Two thousand years ago, Earth was believed to be unmoving and at the center of the universe. The moon, sun, each of the known planets, and all of the stars were believed to be located on the surfaces of rotating crystal spheres. Explain how this description was a model of what ancient astronomers observed.

32. Creative Thinking At an air show, you are watching a group of skydivers when a friend says, "We learned in science class that things fall to Earth because of the law of gravitation." Tell what is wrong with your friend's statement, and explain your reasoning.

33. Applying Knowledge You have decided to test the effects of five different garden fertilizers by applying some of each to five separate rows of radishes. What is the independent variable? What factors should you control? How will you measure the results?

34. Interpreting and Communicating A person points to an empty, thick-walled glass bottle and says that the volume is 1200 cm^3. Explain why the person's statement is not as clear as it should be.

30

35. Interpreting Graphics A consumer magazine has tested several portable stereos and has rated them according to price and sound quality. The data are summarized in the bar graph shown below. Study the graph and answer the following questions:
  a. Which brand has the best sound?
  b. Which brand has the highest price?
  c. Which brand do you think has the best sound for the price?
  d. Do you think that sound quality corresponds to price?
  e. If you can spend as much as $150, which brand would you buy? Explain your answer.

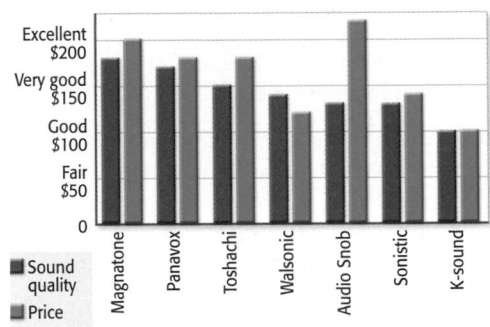

36. Making Decisions You have hired a painter to paint your room with a color that must be specially mixed. This color will be difficult to match if more has to be made. The painter tells you that the total length of your walls is 26 m and all walls are 2.5 m tall. You determine the area ($A = l \times w$) to be painted is 65 m^2. The painter says that 1 gal of paint will cover about 30 m^2 and that you should order 2 gal of paint. List at least three questions you should ask the painter before you buy the paint.

Developing Life/Work Skills

35. a. Magnatone
  b. Audio Snob
  c. Walsonic
  d. No
  e. Answers may vary. Walsonic costs about $125, and the sound quality is close to very good.

36. How accurate is the 30 m^2 per gallon measurement? How precise is the measurement? What is he going to do if he runs out of paint?

37. LASER stands for Light Amplification by Stimulated Emission of Radiation. Some examples of laser technology are: CD players, eye surgery, fiber-optic communications, and laser-guided targeting (military, space exploration, surveying).

37. Applying Technology Scientists discovered how to produce laser light in 1960. The substances in lasers emit an intense beam of light when electrical energy is applied. Find out what the word *laser* stands for, and list four examples of technologies that use lasers.

INTEGRATING CONCEPTS

38. Integrating Biology One of the most important discoveries involving X rays came in the early 1950s, when the work of Rosalind Franklin, a British scientist, provided evidence for the structure of a critical substance. Do library research to learn how Franklin used X rays and what her discovery was.

39. Concept Mapping Copy the unfinished concept map given below onto a sheet of paper. Complete the map by writing the correct word or phrase in the lettered box.

www.scilinks.org
Topic: Studying the Natural World
SciLinks code: HK4136

SCiLINKS. Maintained by the National Science Teachers Association

Art Credits: Fig. 3, Leslie Kell; Fig. 8, Kristy Sprott; Fig. 12, Leslie Kell; Fig. 15 (football stadium), Uhl Studios, Inc.; Fig. 18-17, Leslie Kell; Developing Life/Work Skills (graph), Leslie Kell.

Photo Credits: Chapter opener image of laser-induced fusion by Roger Ressmeyer/CORBIS; George Washington Carver in lab; Bettmann/CORBIS; Fig. 1A, Hulton Archive/Getty Images; Fig. 1B, Phil Degginger/Color-Pic, Inc.; Fig. 2, AIP Emilio Segrè Visual Archives; Fig. 4, Leonard Lessin/Peter Arnold, Inc.; Fig. 5A, Sheila Terry/Science Photo Library/Photo Researchers, Inc.; Fig. 5B, SuperStock; Fig. 5B(portrait), Bettmann/CORBIS; Fig. 6A, Peter Van Steen/HRW; Fig. 6B, Sam Dudgeon/HRW; Fig. 7, Peter Van Steen/HRW; Fig. 8, Kristian Hilsen/Getty Images/Stone; Fig. 9, BMW of North America, Inc.; Fig. 10, Chris Johns/Getty Images/Stone; Fig. 11, Peter Van Steen/HRW; "Quick Activity," Peter Van Steen/HRW; Fig. 13, Science Photo Library/Photo Researchers, Inc.; Fig. 14A, Roger Ressmeyer/CORBIS; Fig. 14B, Celestial Image Co./Science Photo Library/Photo Researchers, Inc.; Figure 15, all images HRW except, nickles, EyeWire, Inc.; calipers, tape measure, cd, Image Copyright ©2004 Photodisc, Inc.; Fig. 16, Peter Van Steen/HRW; Fig. 20, Sam Dudgeon/HRW; "Skills Practice Lab," (boy, box), Peter Van Steen/HRW; (girl), Sam Dudgeon/HRW.

Integrating Concepts

38. Rosalind Franklin used X rays to try to determine the double-helical structure of DNA. Had she not died at 37, she would have probably shared the Nobel Prize with Watson and Crick, who used her data. Nobel Prize recipients must be living.

39. a. life science (or biology)
b. physical science
c. Earth science
d. botany
e. zoology
f. ecology
g. physics
h. chemistry
i. geology
j. meteorology

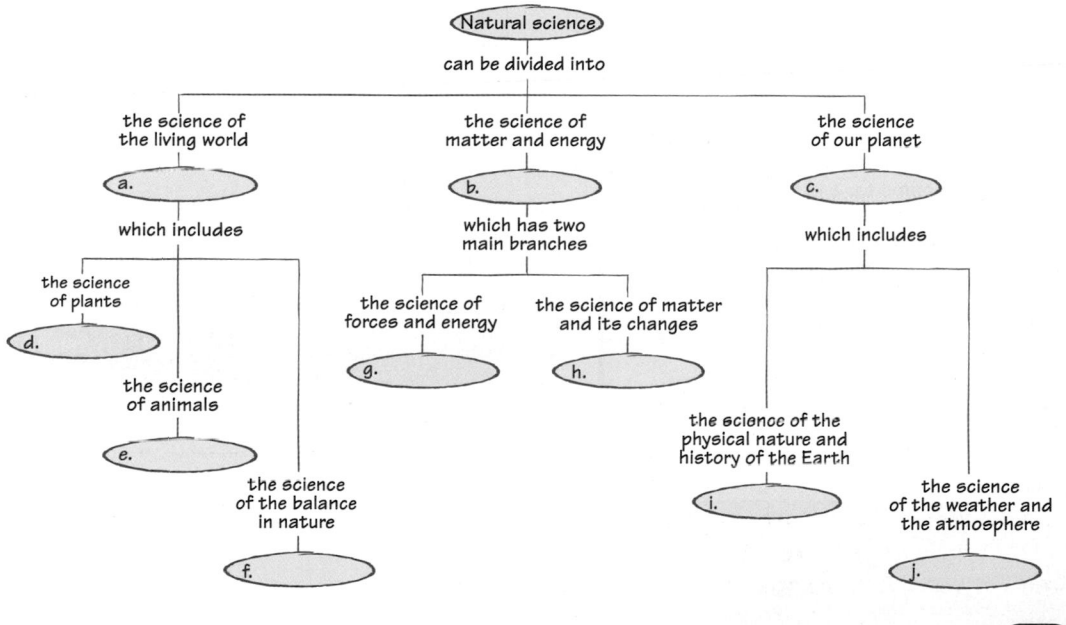

31

Transparencies

TT Concept Mapping

MAKING MEASUREMENTS

Teacher's Notes

Time Required 1 lab period

Ratings

1	2	3	4
EASY			HARD

TEACHER PREPARATION	2
STUDENT SETUP	2
CONCEPT LEVEL	1
CLEANUP	2

Skills Acquired

- Collecting data
- Communicating
- Experimenting
- Identifying/Recognizing Patterns
- Interpreting
- Measuring
- Organizing and analyzing data
- Predicting

The Scientific Method

In this lab, students will:
- Make Observations
- Analyze the Results
- Draw Conclusions
- Communicate Results

Materials

Some labs are equipped with test tubes having capacities of 35 mL or more. If the test tubes used are this size, a 50 mL graduated cylinder will be more appropriate than a 25 mL graduated cylinder.

You may want to use salt and baking soda purchased through a supply company or a grocery store instead of reagent-grade sodium chloride and sodium hydrogen carbonate.

Safety Cautions

Have students review safety guidelines before working in the lab.

Skills Practice Lab

32

Introduction

How can you use laboratory tools to measure familiar objects?

Objectives

- ▸ **Measure** mass, length, volume, and temperature.
- ▸ **Organize** data into tables and graphs.

Materials

balance, platform or triple-beam
basketball, volleyball, or soccer ball
graduated cylinder, 25 mL
meterstick or metric ruler marked with
　　centimeters and millimeters
small beaker
small block or box
small rock or irregularly-shaped object
sodium chloride (table salt)
sodium hydrogen carbonate
　　(baking soda)
string
test tubes
wall thermometer

Making Measurements

▸ Procedure

Preparing for Your Experiment

1. In this laboratory exercise, you will use a meterstick to measure length, a graduated cylinder to measure volume, a balance to measure mass, and a thermometer to measure temperature. You will determine volume by liquid displacement.

Measuring Temperature

2. At a convenient time during the lab, go to the wall thermometer and read the temperature. Be sure no one else is recording the temperature at the same time. On the chalkboard, record your reading and the time at which you read the temperature. At the end of your lab measurements, you will make a graph of the temperature readings made by the class.

Measuring Length

3. Measure the length, width, and height of a block or box in centimeters. Record the measurements in a table like *Table 6*, shown below. Using the equation below, calculate the volume of the block in cubic centimeters (cm^3), and write the volume in the table.

Volume = length (cm) × width (cm) × height (cm)
$$V = l \times w \times h$$
$$V = ?\ cm^3$$

4. Repeat the measurements twice more, recording the data in your table. Find the average of your measurements and the average of the volume you calculated.

Table 6 **Dimensions of a Rectangular Block**

	Length (cm)	Width (cm)	Height (cm)	Volume (cm^3)
Trial 1				
Trial 2				
Trial 3				
Average				

Sample Data Table 6 Dimensions of a Rectangular Block

	Length (cm)	Width (cm)	Height (cm)	Volume (cm^3)
Trial 1	8.15	4.25	2.20	76.2
Trial 2	8.20	4.25	2.25	78.4
Trial 3	8.15	4.20	2.20	75.3
Average	8.17	4.23	2.22	76.7

5. To measure the circumference of a ball, wrap a piece of string around the ball and mark the end point. Measure the length of the string using the meterstick or metric ruler. Record your measurements in a table like *Table 7,* shown below. Using a different piece of string each time, make two more measurements of the circumference of the ball, and record your data in the table.

6. Find the average of the three values and calculate the difference, if any, of each of your measurements from the average.

Table 7 Circumference of a Ball

	Circumference (cm)	Difference from average (cm)
Trial 1		
Trial 2		
Trial 3		
Average		

Measuring Mass

7. Place a small beaker on the balance, and measure the mass. Record the value in a table like *Table 8,* shown below. Measure to the nearest 0.01 g if you are using a triple-beam balance and to the nearest 0.1 g if you are using a platform balance.

8. Move the rider to a setting that will give a value 5 g more than the mass of the beaker. Add sodium chloride (table salt) to the beaker a little at a time until the balance just begins to swing. You now have about 5 g of salt in the beaker. Complete the measurement (to the nearest 0.01 or 0.1 g), and record the total mass of the beaker and the sodium chloride in your table. Subtract the mass of the beaker from the total mass to find the mass of the sodium chloride.

9. Repeat steps 7 and 8 two times, and record your data in your table. Find the averages of your measurements, as indicated in *Table 8.*

Table 8 Mass of Sodium Chloride

	Mass of beaker and sodium chloride (g)	Mass of beaker (g)	Mass of sodium chloride (g)
Trial 1			
Trial 2			
Trial 3			
Average			

Sample Data Table 7 Circumference of a Ball		
	Circumference (cm)	Difference from average (cm)
Trial 1	34.7	−0.2
Trial 2	35.1	+0.2
Trial 3	34.9	0
Average	34.9	—

Tips and Tricks

Use this lab as an opportunity to discuss significant figures and how they are estimated. Students are instructed to repeat the measuring process three times. This will help students become familiar with the measuring process and will also allow students to better understand the concepts of accuracy, precision, bias, and reproducibility.

Show students how to measure volume properly by reading the bottom of the meniscus. The concept of parallax can also be introduced with a discussion or demonstration of how it affects measurement.

Procedure

Measuring Temperature When students are ready to measure temperature, make sure they read the thermometer to the nearest 0.1°C or 0.5°C. Placing the thermometer near a window or a heating/cooling vent may produce greater variability.

Measuring Length

Students should be asked to vary the sequence of measurements of the block. The uncertainty or error in the measurement of each dimension of the block is multiplied when the volume of the block is calculated. The ball should be kept in a box or otherwise contained so that it is not a hazard. Remind students that while in the lab, everything used is a scientific tool, not a toy. Do not allow students to roll or throw the ball to each other.

Measuring Mass

The differences in behavior between the granular salt and the powdery baking soda may cause differences in students' ability to measure 5 g of each solid.

Measuring Volume

If the lab has large capacity test tubes, a 50 mL graduated cylinder will be appropriate instead of the 25 mL graduated cylinder. A demonstration or discussion of how to read a meniscus should be incorporated into this step.

Measuring Volume by Liquid Displacement

Caution students to dry the cylinder so that it does not slip from their hands. If a small rock or mineral sample is used for liquid displacement, encourage students to tilt the cylinder and gently slide the object down the cylinder's side instead of dropping the rock into the cylinder.

Disposal Information

The salt and soda can be rinsed down the sink with water.

10. Make a table like *Table 8,* substituting sodium hydrogen carbonate for sodium chloride. Repeat steps 7, 8, and 9 using sodium hydrogen carbonate (baking soda), and record your data.

Measuring Volume

11. Fill one of the test tubes with tap water. Pour the water into a 25 mL graduated cylinder.

12. The top of the column of water in the graduated cylinder will have a downward curve. This curve is called a *meniscus* and is shown in the figure at right. Take your reading at the bottom of the meniscus. Record the volume of the test tube in a table like *Table 9.* Measure the volume of the other test tubes, and record their volumes. Find the average volume of the three test tubes.

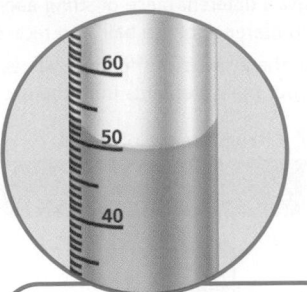

Table 9 Liquid Volume

	Volume (mL)
Test tube 1	
Test tube 2	
Test tube 3	
Average	

Measuring Volume by Liquid Displacement

13. Pour about 10 mL of tap water into the 25 mL graduated cylinder. Record the volume as precisely as you can in a table like *Table 10,* shown below.

Table 10 Volume of an Irregular Solid

	Total volume (mL)	Volume of water only (mL)	Volume of object (mL)
Trial 1			
Trial 2			
Trial 3			
Average			

Sample Data Table 8 Mass of Sodium Chloride

	Mass of beaker and sodium chloride (g)	Mass of beaker (g)	Mass of sodium chloride (g)
Trial 1	60.10	55.20	4.90
Trial 2	59.90	55.20	4.70
Trial 3	60.45	55.20	5.25
Average	60.15	55.20	4.95

14. Gently drop a small object, such as a stone, into the graduated cylinder; be careful not to splash any water out of the cylinder. You may find it easier to tilt the cylinder slightly and let the object slide down the side. Measure the volume of the water and the object. Record the volume in your table. Determine the volume of the object by subtracting the volume of the water from the total volume.

▶ Analysis

1. On a clean sheet of paper make a line graph of the temperatures that were measured with the wall thermometer over time. Did the temperature change during the class period? If it did, find the average temperature, and determine the largest rise and the largest drop.

▶ Conclusions

2. On a clean sheet of paper make a bar graph using the data from the three calculations of the mass of sodium chloride. Indicate the average value of the three determinations by drawing a line that represents the average value across the individual bars. Do the same for the sodium hydrogen carbonate masses. Using the information in your graphs, determine whether you measured the sodium chloride or the sodium hydrogen carbonate more precisely.

3. Suppose one of your test tubes has a capacity of 23 mL. You need to use about 5 mL of a liquid. Describe how you could estimate 5 mL.

4. Why is it better to align the meterstick with the edge of the object at the 1 cm mark rather than at the end of the stick?

5. Why do you think it is better to measure the circumference of the ball using string than to use a flexible metal measuring tape?

Answers to Analysis

1. Students' graphs should accurately represent their given data. It is unlikely that temperature changes over a few degrees will be experienced during the lab period. See graph on page 35A.

Answers to Conclusions

2. The precision of the measurements for salt and baking soda may vary because of the behavior of the solids. This is an opportunity to reinforce the difference between precision and accuracy. See graphs on page 35A.

3. Answers may vary. One option is to estimate the length of the test tube, divide it into five equal parts, and then fill it 1/5 full.

4. Metersticks may get damaged at the ends and not be accurate.

5. The circumference is more easily measured with the string because it makes contact all around the ball. Most flexible metal tapes do not conform to the surface well enough to make contact and stay in place easily.

See "Continuation of Answers" at the end of the chapter.

35

Sample Data Table 8 (additional) Mass of Sodium Hydrogen Carbonate			
	Mass of beaker and sodium hydrogen carbonate (g)	**Mass of beaker (g)**	**Mass of sodium hydrogen carbonate (g)**
Trial 1	60.35	55.20	5.15
Trial 2	60.20	55.20	5.00
Trial 3	59.90	55.20	4.70
Average	60.15	55.20	4.95

See "Continuation of Answers" at the end of the chapter.

Chapter Resource File

- **Datasheet** Making Measurements GENERAL
- **Observation Lab** Comparing the Densities of Pennies BASIC
- **CBL™ Probeware Lab** Designing a Pendulum Clock ADVANCED

Continuation of Answers

Continuation of Answers from p. 3
Answers to Pre-Reading Questions

2. Answers will vary. Students may state that discoveries in science and technology have provided treatments for disease, new methods of agriculture, and new means of communication and transportation in the last 10 years, and that many of these innovations have entailed their own medical, environmental, and ecological problems.

Continuation of Answers from p. 35
Lab Tables

Sample Data Table 9 Liquid Volume	
	Volume (mL)
Test Tube 1	16.5
Test Tube 2	15.8
Test Tube 3	16.0
Average	16.1

Sample Data Table 10 Volume of an Irregular Solid			
	Total volume (mL)	**Volume of water only (mL)**	**Volume of object (mL)**
Trial 1	17.5	10.5	7.0
Trial 2	14.5	9.0	5.5
Trial 3	21.5	10.0	11.5
Average	17.8	9.8	8.0

Continuation of Answers from p. 35
Answers to Analysis

1.

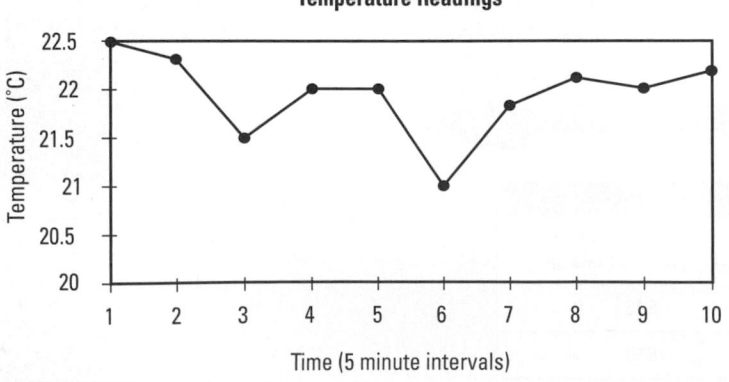

Continuation of Answers from p. 35
Answers to Conclusions

2.

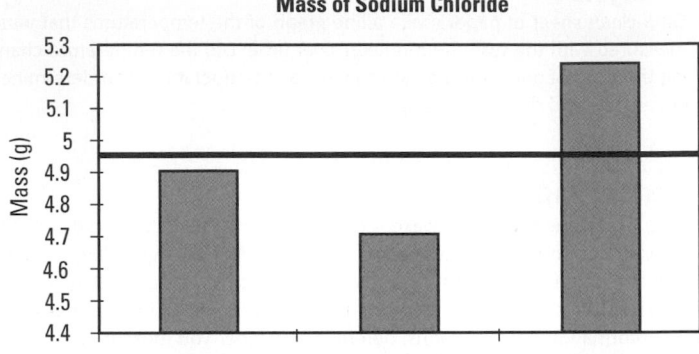

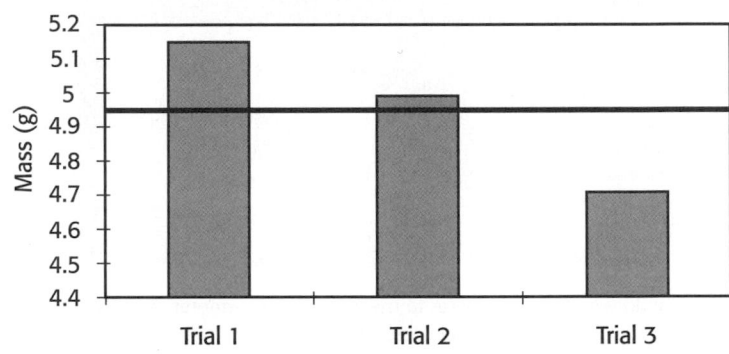

Matter
Chapter Planning Guide

PACING	CLASSROOM RESOURCES	LABS, ACTIVITIES, AND DEMONSTRATIONS
BLOCK 1 · 45 min pp. 36–37 **Chapter Opener**		SE **Activity 1,** p. 37 SE **Activity 2,** p. 37
BLOCKS 2 & 3 · 90 min pp. 38–44 **Section 1** What Is Matter?	CRF **Lesson Plan*** TT **Bellringer*** TT **Elements in the Human Body*** TM **Models of Water*** TM **Chemical Formula***	TE **Opening Demonstration,** p. 38
BLOCKS 4 & 5 · 90 min pp. 45–52 **Section 2** Properties of Matter	CRF **Lesson Plan*** TT **Bellringer*** TT **Three States of Water*** TM **Chemical and Physical Properties***	TE **Demonstration** Mass of Gaseous Matter, p. 45 TE **Demonstration** Temperature Effects on Density, p. 46 SE **Quick Lab** How are the mass and volume of a substance related?, p. 49 **GENERAL** TE **Demonstration** Properties of Iron, p. 50 CRF **Observation Lab** Measuring Density with a Hydrometer* ◆ **BASIC** CRF **CBL™ Probeware Lab** Comparing the Buoyancy of Different Objects* ◆ **ADVANCED**
BLOCKS 6 & 7 · 90 min pp. 53–58 **Section 3** Changes of Matter	CRF **Lesson Plan*** TT **Bellringer*** TT **Separating Mixtures*** TT **Chemical Changes*** TT **Concept Mapping***	TE **Opening Activity,** p. 53 SE **Quick Lab** How can physical properties separate a mixture?, p. 54 **GENERAL** TE **Demonstration** Chemical Change, p. 55 TE **Group Activity,** p. 55 TE **Demonstration** Separation, p. 55 TE **Demonstration** Physical or Chemical Change, p. 56 SE **Quick Activity** Compound Confusion, p. 58 **GENERAL** SE **Skills Practice Lab** Testing the Conservation of Mass, pp. 64–65 ◆ **GENERAL**

BLOCKS 8 & 9 · 90 min

Chapter Review and Assessment Resources

SE **Chapter Review,** pp. 60–63
CRF **Chapter Tests*** **GENERAL**
OSP **Test Generator**
CRF **Standardized Test Practice with Guided Reading Development***
CRF **Test Item Listing for ExamView® Test Generator***
OSP **Scoring Rubrics and Classroom Management Checklists**

Online Resources

Visit the HRW Web site for a variety of free resources related to the text. Just type in the keyword **HK4 MAT.**

Holt Online Learning

Holt Science Spectrum: Physical Science: Online Edition

Students can access interactive problem solving help and active visual concept development with the *Holt Science Spectrum: Physical Science* Online Edition available at **www.hrw.com.**

student CNN News

cnnstudentnews.com

Find the latest news, lesson plans, and activities related to important scientific events.

Compression guide:
To shorten your instruction because of time limitations, omit blocks 1, 8, and 9.

KEY

TE	Teacher Edition	CRF	Chapter Resource File	*	Also on One-Stop Planner
SE	Student Edition	TT	Teaching Transparency	◆	Requires Advance Prep
OSP	One-Stop Planner	TM	Transparency Master		

PROBLEM SOLVING AND PRACTICE	SECTION REVIEW AND ASSESSMENT	STANDARDS CORRELATION
	CRF Pretest* `GENERAL`	
CRF **Cross-Disciplinary Worksheet** RWA—Glassmaking* `ADVANCED` **CRF** **Science Skills** Making and Interpreting Bar Graphs and Pie Charts* `BASIC` **CRF** **Cross-Disciplinary Worksheet** Integrating Biology—What's Special about Indigo?* `ADVANCED` **CRF** **Science Skills** Ratios and Proportions* `BASIC` **SE** **Science and the Consumer** Dry Cleaning: How Are Stains Dissolved?, p. 43 `GENERAL` **CRF** **Cross-Disciplinary Worksheet** Science and the Consumer—Is Dry Cleaning Dangerous?* `ADVANCED` **CRF** **Cross-Disciplinary Worksheet** Integrating Earth Science—Uses of Pumice* `ADVANCED`	**TE** Quiz, p. 44 `BASIC` **SE** Section 1 Review, p. 44 **CRF** Concept Review* `GENERAL` **CRF** Quiz* `BASIC`	PS 1a, 2c UCP 1, 2, 3, 5 ST 2b
CRF **Science Skills** Rearranging Algebraic Equations* `BASIC` **CRF** **Math Skills** Density* `GENERAL` **CRF** **Cross-Disciplinary Worksheet** Real World Applications—Characteristic Properties* `ADVANCED` **CRF** **Cross-Disciplinary Worksheet** Real World Applications—Choosing Materials for Bicycle Frames* `ADVANCED` **CRF** **Cross-Disciplinary Worksheet** Integrating Environmental Science—Ozone Depletion* `ADVANCED`	**TE** Quiz, p. 52 `BASIC` **SE** Section 2 Review, p. 52 **CRF** Concept Review* `GENERAL` **CRF** Quiz* `BASIC`	PS 2d, 2e, 3a UCP 5 SAI 1 ST 2d
CRF **Cross-Disciplinary Worksheet** Connection to Language Arts—Hidden Meanings* `ADVANCED`	**TE** Quiz, p. 58 `BASIC` **SE** Section 3 Review, p. 58 **CRF** Concept Review* `GENERAL` **CRF** Quiz* `BASIC`	PS 3a UCP 2, 3, 5 ST 2d

SCI LINKS

www.scilinks.org

Topic: Chemical Changes
*Sci*Links code: HK4020

Topic: Density
*Sci*Links code: HK4031

Topic: Dry Cleaning
*Sci*Links code: HK4033

Topic: Glass
*Sci*Links code: HK4064

Topic: Origin of Elements
*Sci*Links code: HK4097

Topic: Physical/Chemical Changes
*Sci*Links code: HK4104

Technology Resources

 One-Stop Planner
All of your printable resources and the Test Generator are on this convenient CD-ROM.

 CNN Science in the NEWS
*each video segment is accompanied by a Critical Thinking Worksheet**

Segment 18
Chemical Separation Techniques for Plastics

Overview

Section 1 explains matter, atoms, and elements and distinguishes between elements and compounds. Molecules and chemical formulas are also discussed, as well as the differences between pure substances and mixtures. Section 2 covers physical and chemical properties and characteristic properties. The physical property of density, and calculations involving density, are introduced. Section 3 covers the physical and chemical changes of matter and explains how chemical changes can be detected.

Assessing Prior Knowledge

Be sure students understand the following concepts:

- scientific laws
- units of measurement
- using significant figures

MISCONCEPTION //ALERT \\\

Science education research has identified the following misconceptions about matter.

- Students confuse *pure substances* with things that have no additives or that are clean (or noncontaminated).
- Students confuse "Elements cannot be decomposed" with "Elements are always solid."
- Students confuse mass with weight, size, density, or volume, as well as density and weight.
- Students confuse changes of state with chemical changes.
- Students may fail to understand that N_2O_4 is one molecule, not an N_2 molecule bound to an O_4 molecule.
- Students often do not recognize that substances may fit into many categories (or new categories).

Matter

Chapter Preview

1 **What Is Matter?**
Composition of Matter
Pure Substances and Mixtures

2 **Properties of Matter**
Physical Properties
Chemical Properties
Comparing Physical and
Chemical Properties

3 **Changes of Matter**
Physical Changes
Chemical Changes

36

Standards Correlations

National Science Education Standards

The following descriptions summarize the National Science Standards that specifically relate to this chapter. For the full text of the standards, see the National Science Education Standards at the front of the book.

Section 1 What Is Matter?
Physical Science PS 1a, 2c
Unifying Concepts and Processes UCP 1, 2, 3, 5
Science and Technology ST 2b

Section 2 Properties of Matter
Physical Science PS 2d, 2e, 3a
Unifying Concepts and Processes UCP 5
Science as Inquiry SAI 1
Science and Technology ST 2d

Section 3 Changes of Matter
Physical Science PS 3a
Unifying Concepts and Processes UCP 2, 3, 5
Science and Technology ST 2d

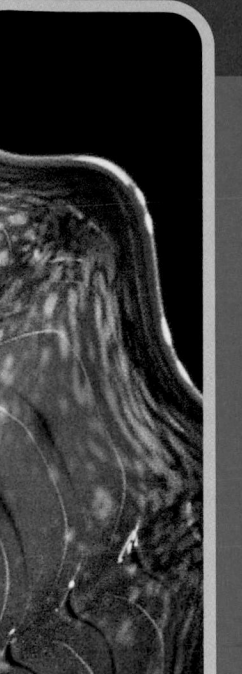

Dale Chihuly is the glass artist who created the sculptures in both of these photographs.

Focus ACTIVITY

Background People have been making glass for thousands of years. Glass making begins when sand is mixed with finely ground limestone and a powder called *soda ash.* When the mixture is heated to about 1500°C, the sand mixture becomes transparent and flows like honey.

A glass blower dips a hollow iron blowpipe into the red-hot mixture and picks up a gob of molten glass. By turning the sticky glob and blowing into the tube, the glass blower creates a hollow bulb that can be pulled, twisted, and blown into different shapes. When the finished shape is broken from the tube, a beautiful glass sculpture has been created. Through heating and cooling, glass changes from a solid to a liquid and back to a solid.

Activity 1 Your teacher will provide several samples of glass. Look at the different types of glass on display. Write down (a) the different characteristics of the glass (such as shape, color, texture, and density) and (b) possible uses for each type of glass.

Activity 2 Look at different types of plastic and plastic containers. List the differences you can observe between the examples of glass and plastic. Even though we have plastics and other materials to use in containers and other products, why do you think glass is still used?

---

internet connect

www.scilinks.org
Topic: Glass SciLinks code: HK4064

SCiLINKS® Maintained by the National Science Teachers Association

---

Pre-Reading Questions

1. Look around the room, and find several examples of matter. Can you find examples that are not matter?
2. Can matter ever change forms? Can one substance change into another? Explain how you think this change happens, and give examples.

37

Focus ACTIVITY

Background Glassmaking began about 1500 B.C.E. in Egypt and Mesopotamia. Glass consists of 75% silicate, a class of compounds containing silicon, oxygen, one or more metals, and possibly hydrogen. Sand is a familiar silicate. When glass is recycled, it is crushed into pieces, or cullet, which is melted and remolded again. Recycled glass thus undergoes many physical changes, including melting, solidifying, and crushing.

Activity 1 The different types of glass may include glass used in bottles and jars; high-heat ceramic glass for cooking; laminated glass (as used in car windshields); tempered glass (which breaks into chunks with few sharp edges); photosensitive glass that darkens when exposed to light; colored, stained, and decorative glass, etc. **LS Visual**

Activity 2 Responses may include that glass is rigid and heavy, whereas plastic is flexible and light. Glass is still used because it is made from a common material, is easily recyclable, and can be modified to have different properties (color, heat-resistivity, etc.). **LS Logical**

Answers to Pre-Reading Questions

1. Answers will vary. Examples of matter include anything that has mass and occupies space. Light and other forms of energy are examples of non-matter.

See "Continuation of Answers" at the end of the chapter.

Chapter Resource File

- **Cross-Disciplinary Worksheet** RWA—Glassmaking **ADVANCED**
- **Pretest** **GENERAL**
- **Teaching Transparency Preview**
- **Answer Keys**

Focus

Overview

Before beginning this section, review with your students the Objectives listed in the Student Edition. This section discusses the composition of matter and the relationship between matter, atoms, and elements. Students learn the differences between elements and compounds, and how molecules are formed. Chemical formulas and symbols are also introduced. The section concludes by comparing pure substances and mixtures, and by classifying mixtures as homogeneous or heterogeneous.

🔊 Bellringer

Use the Section 1 Bellringer transparency to prepare students for this section.

Motivate

Opening Demonstration —— GENERAL

(Time: Approximately 5 minutes)

Materials:

• charcoal
• granulated sugar
• vials with lids (2), one labeled "hydrogen" and one "oxygen"

Show students the four substances, then tell them that sugar is made from the other three substances. Ask students how sugar is like the substances that form it, and how it differs. (Few similarities exist. Differences include appearance, state, and solubility.) Ask students if they think sugar would form if you mixed charcoal, oxygen, and hydrogen. (No.) **LS Visual**

What Is Matter?

▶ **KEY TERMS**
chemistry
matter
element
atom
compound
molecule
chemical formula
pure substance
mixture

▶ **chemistry** the scientific study of the composition, structure, and properties of matter and the changes that matter undergoes

▶ **matter** anything that has mass and takes up space

OBJECTIVES

▶ **Explain** the relationship between matter, atoms, and elements.
▶ **Distinguish** between elements and compounds.
▶ **Describe** molecules, and explain how they are formed.
▶ **Interpret** and write some common chemical formulas.
▶ **Categorize** materials as pure substances or mixtures.

Making glass, as shown in *Figure 1,* is the process of changing the raw materials sand, limestone, and soda ash into a different substance. Such processes are what **chemistry** is all about: what things are made of, what their properties are, and how they interact and change. Chemistry is an important part of your daily life. Everything you use, from soaps to foods to carbonated drinks to books, you choose because of chemistry—what the object is made of, what its properties are, or how it changes.

Glass is used as a building material because its properties of being transparent, solid, and waterproof meet the needs we have for windows. The properties of sand, on the other hand, do not meet these needs. Chemistry helps you recognize how the differences in materials' properties relate to what the materials are composed of.

Composition of Matter

You are made of **matter.** This book is also matter. All the materials you can hold or touch are matter. Matter is anything that has mass and occupies space. The air you are breathing is matter even though you cannot see it. Light and sound are not matter. Unlike air, they have no mass or volume.

Figure 1

Glass blowers have been practicing their craft for more than 2000 years. Raw materials are changed into a new substance during the glass making process.

FINE ARTS
CONNECTION —— BASIC

Have students observe changes in state that occur as they work with various media (paint, ceramics, etc.). **LS Kinesthetic**

SOCIAL STUDIES
CONNECTION —— ADVANCED

Have students research how different elements have affected the materials for which they have been used throughout history. Have them share their findings in a class discussion or an oral report. **LS Verbal**

Atoms are matter

Wood is matter. Because it is rigid and lightweight, wood is a good choice for furniture and buildings. When wood gets hot enough, it chars—its surface turns black. The wood surface breaks down to form carbon, another kind of material that has different properties. The carbon in the charred remains will not decompose by further chemical reactions because carbon is an **element** and each element is made of only one kind of **atom.**

Diamonds, such as the one shown in *Figure 2,* are made of atoms of the element carbon. The shiny foil wrapped around a baked potato is made of atoms of the element aluminum. The elements that are most abundant on Earth and in the human body are shown in *Figure 3.* Each element is designated by a one- or two-letter symbol that is used worldwide. Symbols for elements are always a single capital letter or a capital letter followed by a lowercase letter. There are no exceptions! For example, the symbol for carbon is C, iron is Fe, copper is Cu, and aluminum is Al. Each of the more than 110 elements that we know of is unique and has different properties from the rest.

Elements combine chemically to form a compound

Many familiar substances, such as aluminum and iron, are elements. Nylon is a familiar substance, but it is not an element. Nylon is a **compound.** The basic unit that makes up nylon contains carbon, hydrogen, nitrogen, and oxygen atoms, but each strand contains many of these units linked together.

Figure 2
This diamond is made of carbon atoms.

▶ **element** a substance that cannot be seperated or broken down into simpler substances by chemical means

▶ **atom** the smallest unit of an element that maintains the properties of that element

▶ **compound** a substance made of atoms of two or more different elements that are chemically combined

Figure 3
Earth and the human body differ in the kind and proportion of elements they are composed of.

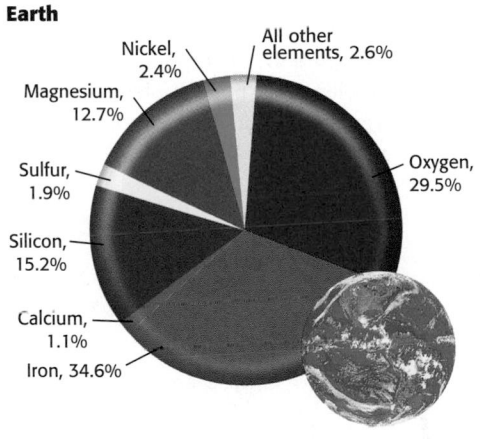

Earth
- Nickel, 2.4%
- All other elements, 2.6%
- Magnesium, 12.7%
- Sulfur, 1.9%
- Silicon, 15.2%
- Calcium, 1.1%
- Iron, 34.6%
- Oxygen, 29.5%

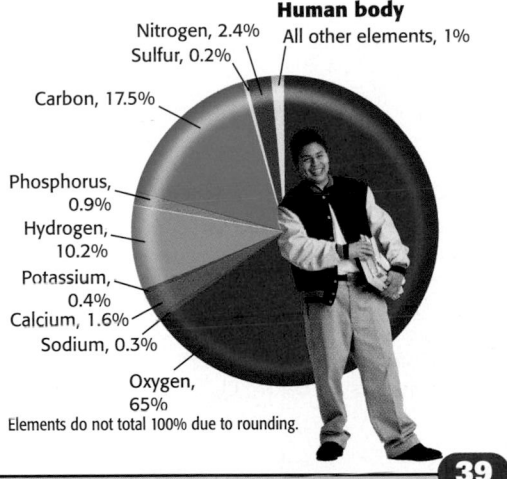

Human body
- Nitrogen, 2.4%
- Sulfur, 0.2%
- All other elements, 1%
- Carbon, 17.5%
- Phosphorus, 0.9%
- Hydrogen, 10.2%
- Potassium, 0.4%
- Calcium, 1.6%
- Sodium, 0.3%
- Oxygen, 65%

Elements do not total 100% due to rounding.

39

Teach

SKILL BUILDER — GENERAL

Reading Skills Have students read the definitions of *chemistry* and *matter.* Make two lists on the chalkboard, one labeled "matter" and one labeled "changes." Have students brainstorm a list of things they think are matter, as well as changes they have observed in each type of matter. **LS** Verbal

SKILL BUILDER — BASIC

Writing Skills Students may have preconceptions about what chemistry is. Have them list what they know about chemistry and what they want to know about it. After they have completed this section, have them look back at their lists and write what they have learned about chemistry.
LS Intrapersonal

SKILL BUILDER — GENERAL

Vocabulary Show students the derivation of the word *atom* from its Greek roots, and ask them what idea about matter is conveyed by that meaning.

a–: not *tomos*: cutting

Literal meaning: *indivisible.* The idea, which also originated in ancient Greece, is that matter can be subdivided only as small as an elemental particle: the atom.
LS Verbal

Historical Perspective

Elemental Discoveries In 1750, only 16 elements were known—antimony, arsenic, bismuth, carbon, cobalt, copper, gold, iron, lead, mercury, phosphorus, platinum, silver, sulfur, tin, and zinc. The next element identified was nickel, in 1751. None of the common gaseous elements, such as hydrogen, oxygen, and nitrogen, nor any of the halogens, had yet been identified, partly because scientists at the time were unclear about the nature of gases.

**MISCONCEPTION ///ALERT **

Elements Students may not realize that the human body, like Earth, is composed of the elements shown in **Figure 3**; they often think that animals are composed only of "skin, bones, and muscle." Clarify that the human body is composed of living cells, which in turn contain large molecules and compounds made up of elements.

Chapter Resource File

- **Lesson Plan**
- **Science Skills** Making and Interpreting Bar Graphs and Pie Charts **BASIC**

Transparencies

- TT Bellringer
- TT Elements in the Human Body

Definite Ratios If students do not understand what is meant by "combine in the same proportions," illustrate this concept with substances whose formulas are already familiar to them. For example, water exists in a ratio represented by the formula H_2O: two atoms of hydrogen to every one atom of oxygen. It is a definite ratio because all water molecules, regardless of their source, consist of these elements in the same ratio.

Teaching Tip

Interpreting Visuals Make sure students understand that a molecule may consist of two or more atoms of *different* elements or atoms of the *same* element. Tell them that examples of molecular elements other than those mentioned in the text and **Figure 5** include ozone, O_3, and several different molecules formed from carbon.

Figure 4

A water molecule can be represented by a formula, physical models, or computer images.

molecule the smallest unit of a substance that keeps all of the physical and chemical properties of that substance

Compounds have unique properties

Every compound is different from the elements it contains. For example, the elements hydrogen, oxygen, and nitrogen occur in nature as colorless gases. Yet when they combine with carbon to form nylon, the strands of nylon are a flexible solid.

When elements combine to make a specific compound, the elements always combine in the same proportions. For example, iron(III) oxide, which we see often as rust, always has two atoms of iron for every three atoms of oxygen.

A molecule acts as a unit

Atoms can join together to make millions of **molecules** like letters of the alphabet combine to form different words. A molecular substance you are familiar with is water. A water molecule is made of two hydrogen atoms and one oxygen atom, as shown in *Figure 4.*

When oxygen and hydrogen atoms form a molecule of water, the atoms combine and act as a unit. That is what a molecule is— the smallest unit of a substance that behaves like the substance. Most molecules are made of atoms of different elements, such as water. But a molecule may also be made of atoms of the same element, such as those shown in *Figure 5.* A compound is made of atoms of two or more different elements, but a molecule may be of the same elements or different elements.

Figure 5

The atoms of elements such as neon, Ne, are found singly in nature. Other elements, such as oxygen, hydrogen, chlorine, and phosphorus, form molecules that have more than one atom. Their unit molecules are O_2, H_2, Cl_2, and P_4.

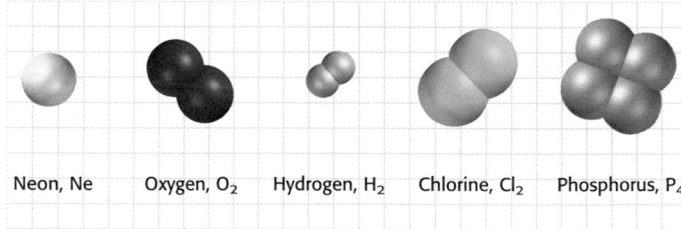

Neon, Ne   Oxygen, O_2 Hydrogen, H_2 Chlorine, Cl_2 Phosphorus, P_4

40

MISCONCEPTION ALERT

Polyatomic Forms of Elements Students may think that only single atoms are true "elements." As shown in **Figure 5**, explain that the atoms of some elements (e.g., neon) occur singly in nature, and the atoms of other elements combine to form polyatomic molecules (as in O_2 or H_2).

INCLUSION Strategies

• *Attention Deficit Disorders*

Using **Figure 3** and the Periodic Table, have the students construct a class poster showing the percentages of elements that make up Earth. The student should include the symbols for each element, the percentage of the element found on Earth, and definitions for matter, atom, element, and compound. The students may present the information to the class. They may tape record the information as though they were reporting the information on an education television channel.

Chemical formulas represent compounds and molecules

Indigo is the dye first used to turn jeans blue. The **chemical formula** for a molecule of indigo, $C_{16}H_{10}N_2O_2$, is shown in **Figure 6.** A chemical formula shows how many atoms of each element are in a unit of a substance. In a chemical formula, the number of atoms of each element is written after the element's symbol as a *subscript*. If only one atom of an element is present, no subscript number is used.

Numbers placed in front of the chemical formula show the number of molecules. So, three molecules of table sugar are written as $3C_{12}H_{22}O_{11}$. Each molecule of sugar contains 12 carbon atoms, 22 hydrogen atoms, and 11 oxygen atoms.

Figure 6

The chemical formula for a molecule of indigo shows that it is made of four elements and 30 atoms.

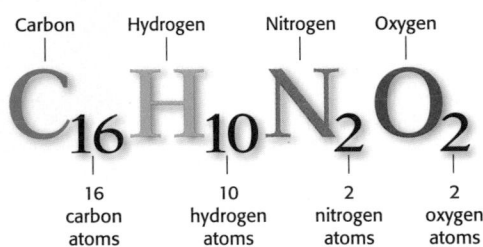

Carbon   Hydrogen   Nitrogen   Oxygen

$$C_{16}H_{10}N_2O_2$$

| 16 carbon atoms | 10 hydrogen atoms | 2 nitrogen atoms | 2 oxygen atoms |

Pure Substances and Mixtures

The word *pure* often means "not mixed with anything." For example, "pure grape juice" contains the juice of grapes and nothing else. In chemistry, the word *pure* has another meaning. A **pure substance** is matter that has a fixed composition and definite properties.

So, grape juice actually is not a pure substance. It is a **mixture** of many pure substances, such as water, sugars, and vitamins. The composition of grape juice is not fixed; it can have different amounts of water or sugar. Elements and compounds are pure substances, but mixtures are not. Many of the foods we eat are mixtures. The air we breathe is a mixture of gases.

Figure 7 shows a mixture and a pure substance. A mixture, such as grape juice, can be separated into its components. The components of water, a pure substance, are chemically combined and cannot be separated in the same way that the components of grape juice can be separated.

- ◼ **chemical formula** a combination of chemical symbols and numbers to represent a substance

- ◼ **pure substance** a sample of matter, either a single element or a single compound, that has definite chemical and physical properties

- ◼ **mixture** a combination of two or more substances that are not chemically combined

Figure 7

Grape juice is a mixture, and water is a pure substance. The components of grape juice, such as sugar and water, are not chemically combined. Water is a pure substance made up of the elements hydrogen and oxygen, which are chemically combined.

41

Vocabulary *Solution* and *homogeneous mixture* are synonymous. Homogeneous mixtures are mixed completely, all the way down to their most fundamental particles—atoms, molecules, or ions.

INTEGRATING

BIOLOGY The use of natural indigo dye decreased dramatically with its successful synthesis. The German chemist Adolf von Baeyer (1835–1917) analyzed indigo. While he was the first to accomplish the synthesis of indigo, others developed the methods used for its commercial production.

SKILL BUILDER — GENERAL

Vocabulary The words *homogeneous* and *heterogeneous* have the following Greek prefixes:

homo–: same

hetero–: different

So a homogeneous mixture is one that appears the same throughout, and a heterogeneous mixture is one in which differences can be seen.
LS Verbal

INTEGRATING

BIOLOGY
The pure substance indigo is a natural dye made from plants of the genus *Indigofera,* which is in the pea family. Before synthetic dyes were developed, indigo plants were widely grown in the East Indies, in India, and in the Americas. Most indigo species are shrubs 1 to 2 m tall. The leaves and branches are fermented to yield a paste, which is formed into blocks and then ground. The blue color develops when the material is exposed to air.

42

Figure 8

A Flour is suspended in water. **B** Powdered sugar is dissolved in water.

Mixtures are formed by mixing pure substances

While a compound is different from the elements that it is composed of, a mixture may have properties that are similar to the pure substances that form it. Although you cannot see the different pure substances in grape juice, the mixture has chemical and physical properties in common with its components. Grape juice is a liquid like the water that it contains, and it is sweet like the sugar that it also contains.

Mixtures are classified by how thoroughly the substances mix

Some mixtures are made by putting solids and liquids together. In *Figure 8,* two white, powdery solids—flour and powdered sugar—are each mixed with water. Although these solids look similar, the mixtures they form with water are different.

The flour and water form a cloudy white mixture. The flour does not dissolve in water. A mixture like this is called a *heterogeneous mixture.* The substances aren't mixed uniformly and are not evenly distributed.

The sugar-water mixture looks very different from the flour-water mixture. You cannot see the sugar, and the mixture is clear. Powdered sugar dissolves in water. If you leave the mixture for a long time, the sugar will not settle out. Sugar and water form a *homogeneous mixture* because the components are evenly distributed, and the mixture is the same throughout.

Gasoline is a liquid mixture—a homogeneous mixture of at least 100 liquids. Thus, gasoline is composed of *miscible* liquids.

If you shake a mixture of oil and water, the oil and water will not mix well together, and the water will settle out. Oil and water are *immiscible.* You can see two layers in the mixture.

MISCONCEPTION ALERT

Miscibility Many students believe that all liquids can "mix together." Clarify that only miscible liquids that are combined (e.g., ethylene glycol and water) can dissolve in each other. Such substances form a homogeneous liquid mixture called a solution. Other pairs of liquids, such as oil and water or toluene and water, are *immiscible*, meaning they are practically insoluble. Instead, they form two layers, with the denser liquid on the bottom.

Dry Cleaning: How Are Stains Dissolved?

Why do some clothes need to be dry cleaned, while others do not? Washing with water and detergents cleans most clothes. But if your clothes have a stubborn stain—such as ink or rust—if you have spilled something greasy on them, or if the label on the clothing recommends dry cleaning, then dry cleaning may be necessary. Dry cleaning is recommended for clothing made of fabrics that do not respond well to water. These fabrics, such as silk and wool, are usually cleaned without water because water causes them to shrink, to take on stubborn wrinkles, or to lose their shape.

Stain Removal

By knowing the composition of a stain, dry cleaners can decide how to treat the stain. Removing a stain that does not dissolve in water, such as oil or grease, requires two steps. First, the stain is treated with a substance that loosens the stain. Then, the stain is removed when the garment is washed in a mechanical dry-cleaning machine.

If a stain is water soluble, it will dissolve in water. A water-soluble stain is first treated with a stain remover that is specific to that stain. The stain is then flushed away with a steam gun. After the garment is dry, it is cleaned in a dry-cleaning machine to remove any stains that do not dissolve in water.

Once a fabric has been treated for tough stains, the garment is "washed" in a dry-cleaning machine.

Dry Cleaning Isn't Really Dry

In spite of its name, dry cleaning does involve liquids. But instead of water, another liquid is used to dissolve stains. It is always difficult to remove fats, greases, and oils from fabrics with water-based washing.

A good dry-cleaning substance must dissolve oil and grease, which can be trapped in the cloth fibers. The most commonly used dry-cleaning solvent is tetrachloroethylene, C_2Cl_4. Tetrachloroethylene dissolves oil, grease, and alcohols. Also, tetrachloroethylene is not flammable, and it evaporates easily, so it can be recycled by the process of distillation.

In distillation, the components of a liquid mixture are separated based on their rates of evaporation. Upon heating, the component that evaporates fastest is the first to vaporize and separate from the mixture. When the vapors are cooled, they condense to form a purified sample of that component.

Tetrachloroethylene is suspected of causing some kinds of cancer. To meet the standards of the U.S. Occupational Safety and Health Administration (OSHA) and other federal guidelines, dry-cleaning machines must be airtight so that no C_2Cl_4 escapes.

Your Choice

1. **Critical Thinking** Why is it difficult to remove greasy stains from fabrics with water-based cleaners?

2. **Critical Thinking** C_2Cl_4 evaporates faster than the fats and oils it dissolves. How can C_2Cl_4 be recycled by distillation?

☑ internet connect

www.scilinks.org
Topic: Dry Cleaning
SciLinks code: HK4033

SCi LINKS. Maintained by the National Science Teachers Association

43

Teaching Tip
Gas and Liquid Mixtures

Another example of a mixture of gases with liquids is lava. Explain that the molten rock in some types of volcanoes contains large quantities of gas. Pumice, a solid foam that occurs naturally on Earth, is a volcanic rock formed by the violent separation of these extremely hot gases from lava. As the exploding lava cools, it traps the gas bubbles. Some pumice is so soft that it is spongy, and some has such a low density that it floats on water.

Close

Quiz ——————— BASIC

Complete each statement with the correct term.

1. ___ has mass and occupies space. (matter)
2. Elements ___ be broken down into simpler substances. (cannot)
3. Pure substances include ___ and ___. (elements, compounds)
4. The formula H_2O represents a ___ of water. (molecule)
5. Mixtures are formed from ___ substances. (pure)

LS Logical

Chapter Resource File

- **Cross-Disciplinary Worksheet** Integrating Earth Science—Uses of Pumice ADVANCED
- **Concept Review** GENERAL
- **Quiz** BASIC

Figure 9
The meringue in this pie is a mixture of air and liquid egg white that has been beaten and then heated to form a solid foam.

Gases can mix with liquids

Air is a mixture of gases consisting mostly of nitrogen and oxygen. You inhale oxygen every time you breathe because the gases mixed in air form a homogeneous mixture. Carbonated drinks are also homogeneous mixtures. They contain sugar, flavorings, and carbon dioxide gas, CO_2, dissolved in water.

Even a liquid that is not carbonated can contain dissolved gases. For example, if you let a glass of cold water stand overnight, you may see bubbles on the sides of the glass the next morning. The bubbles form when some of the air that was dissolved in the cold water comes out of solution as the water warms up.

Carbonated drinks often have a foam on top. A foam is a kind of gas-liquid mixture. The gas is not dissolved in the liquid but has formed tiny bubbles in it. The bubbles join together to form bigger bubbles that escape from the foam, which causes the foam to collapse.

Other foams are stable and last for a long time. For example, if you whip egg whites with enough air, you get a foam. If you bake that foam in an oven, the liquid egg white dries and hardens, and you have a solid foam—meringue, shown in *Figure 9*.

SECTION 1 REVIEW

SUMMARY

- ► Matter has mass and occupies space.
- ► An element is a substance that cannot be broken down into simpler substances.
- ► An atom is the smallest unit of an element that has the properties of the element.
- ► Atoms can combine to form molecules or compounds.
- ► Chemical formulas represent the atoms in compounds and molecules.
- ► A mixture is a combination of two or more pure substances. Mixtures can be categorized as heterogeneous or homogeneous.

1. **State** the relationship between atoms and elements. Are both atoms and elements matter?
2. **List** the two types of pure substances.
3. **Describe** matter, and explain why light is not matter. Is light made of atoms and elements?
4. **Define** *molecule*, and give examples of a molecule formed by one element and a molecule formed by two elements.
5. **Classify** each of the following as an element or a compound.
   a. sulfur, S_8 c. carbon monoxide, CO
   b. methane, CH_4 d. cobalt, Co
6. **State** the chemical formula of water.
7. **Compare and Contrast** mixtures and pure substances. Give an example of each.
8. **Critical Thinking** David says, "'Pure honey' means it has nothing else added." Susan says, "The honey is not really pure. It is a mixture of many substances." Who is right? Explain your answer.

44

Answers to Section 1 Review

1. An atom is the smallest unit of an element that has that element's properties. Both atoms and elements are matter.
2. elements, compounds
3. Matter has mass and occupies space. Light has neither mass nor volume, nor is it made up of atoms.
4. A molecule is the smallest unit of a substance that keeps all of the physical and chemical properties of that substance. Oxygen (O_2) is an example of a molecule formed by one element. Water (H_2O) is an example of a molecule formed by two elements.

5. a. element
   b. compound
   c. compound
   d. element
6. Water contains two hydrogen atoms and one oxygen atom; H_2O.
7. A pure substance, such as table salt, has the same composition throughout and definite properties. A mixture, such as air, contains more than one pure substance.
8. David's statement reflects the common meaning of pure. Susan's use of the word "pure" is more correct scientifically because honey is a mixture of several different compounds.

Properties of Matter

OBJECTIVES

▶ **Distinguish** between the physical and chemical properties of matter, and give examples of each.

▶ **Perform** calculations involving density.

▶ **Explain** how materials are suited for different uses based on their physical and chemical properties, and give examples.

▶ **Describe** characteristic properties, and give examples.

■ **KEY TERMS**
melting point
boiling point
density
reactivity
flammability

The frame and engine of a car are made of steel. Steel is a mixture of iron, other metallic elements, and carbon. It is a strong solid that provides structure. The tires are made of a flexible solid that cushions your ride. You may not think of the cars you see in *Figure 10* as examples of chemistry. However, the properties and changes that make steel, gasoline, and other substances useful in cars are explained by chemistry.

Physical Properties

Some properties of matter, such as color and shape, are called *physical properties*. Physical properties are often very easy to observe. You rely on physical properties to identify things. You recognize your friends by their physical properties, such as height and hair color. When playing sports, you choose a ball that has the shape and mass suitable for your game. Mass, volume, and density are physical properties of matter. Matter can also be described in terms of the absence of a physical property. A physical property of air is that it is colorless.

Figure 10
The physical and chemical properties of substances determine how they are used in these cars.

45

Focus

Overview

Before beginning this section, review with your students the Objectives listed in the Student Edition. In this section, students learn about the physical and chemical properties of matter, and how materials are suited for different uses based on these properties. They also perform calculations involving the physical property of density. The section concludes with an explanation of characteristic properties.

🔊 Bellringer

Use the Bellringer transparency to prepare students for this section.

Motivate

Opening Discussion ———— GENERAL

Before students read about physical and chemical properties, have them choose items in the classroom and then identify and classify the properties of each item. Have students share and compare their classifications in a class discussion, which should include a justification for each classification. After students have completed this section, ask them to revisit their answers and to correct any that were inaccurate.

LS Interpersonal

MATH
CONNECTION

Review with students how to solve for a variable in a formula. Work through several examples of formulas, such as the area formula, $A = l \times w$, and show students how the same principles apply to the formula for density.

Chapter Resource File

• **Lesson Plan**

Transparencies

TT Bellringer

Mass of Gaseous Matter

(Time: Approximately 25 minutes)

Materials:

• balloon(s)

• balance

Disprove the misconception that air and gases have no mass.

Step 1 Have a pair of students measure the mass of a deflated balloon and record the mass on the board.

Step 2 Have one student blow up and tie the balloon.

Step 3 Have students measure the mass of the inflated balloon, and record the mass on the board.

Step 4 Discuss the discrepancy in mass. Ask students to explain where the extra mass of the inflated balloon came from. (The extra mass is the mass of the air inside the balloon.) **LS Logical**

SKILL BUILDER — GENERAL

Interpreting Visuals Point out the differences between the states of matter shown in **Figure 11** at the macroscopic and microscopic levels. In this text, there are many illustrations of micromodels that show macrophenomena. **LS Visual**

Figure 11

These models show water in three states. The molecules are close together in the solid and liquid states but far apart in the gas state. The molecules in the solid state are relatively fixed in position, but those in the liquid and gas states can flow around each other.

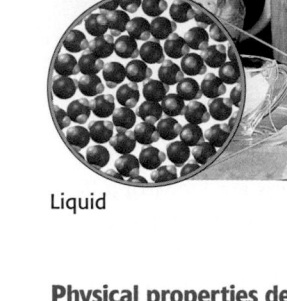

Solid

Gas

Liquid

▶ **melting point** the temperature and pressure at which a solid becomes a liquid

▶ **boiling point** the temperature at which a liquid becomes a gas

Physical properties describe matter

Many physical properties can be observed or measured to help identify a substance. You can use your senses to observe some of the basic physical properties of a substance: shape, color, odor, and texture. Other physical properties, such as **melting point, boiling point,** strength, hardness, and the ability to conduct electricity, magnetism, or heat, can be measured.

Because many physical properties remain constant for pure substances, you can use your observations or measurements of these properties to identify substances. At room temperature and atmospheric pressure, all samples of pure water are colorless and liquid; pure water is never a powdery green solid.

A characteristic of any pure substance is that its boiling point and its melting point are constant if the pressure remains the same. At sea level, water boils at 100°C and freezes at 0°C. It doesn't matter if you have a lot of water or a little water; these physical properties of the water are the same regardless of the mass or volume involved. This statement is true for all pure substances.

An easily observed physical property is *state*—the physical form in which a substance exists. Solids, liquids, and gases are three common states of matter. *Figure 11* shows the solid, liquid, and gas states of water at the molecular level.

MISCONCEPTION //// ALERT \\\\

Specific Temperatures of Phase Changes
Students often have the misconception that changes of state such as freezing, melting, or vaporizing are not connected to a specific temperature. Clarify the terms *melting point* and *boiling point*, explaining that they are physical properties that are constant for specific pure substances. Use the chart to the right for examples of the melting and boiling points of some common substances.

Substance	Melting Point (°C)	Boiling Point (°C)
Carbon	3550	4827
Gold	1063	2970
Iron	1535	2750
Mercury	−39	357
Nitrogen	−209	−196
Oxygen	−218	−183
NaCl	801	1413
Water	0	100

Density is a physical property

Density is a measurement of how much matter is contained in a certain volume of a substance. A substance that has a low **density** is "light" in comparison with something else of the same volume. The balloons in *Figure 12* float because they are less dense than the air around them. A substance that has a high density is "heavy" in comparison with another object of the same volume. A stone sinks to the bottom of a pond because the stone is more dense than the water around it.

You can compare the density of two objects of the same volume by holding one in each hand. The lighter one is less dense; the heavier one is more dense. If you held a brick in one hand and an equal-sized piece of sponge in the other hand, you would know instantly that the brick is more dense than the sponge. Remember that weight and density are different. Two pounds of feathers are heavier than one pound of steel. But the feathers are less dense than the steel, so two pounds of feathers have a greater volume than one pound of steel.

Density determines whether an object will float or sink. An object will float when placed in water if it is less dense than water. If an object is more dense than water, the object will sink.

Table 1 lists the densities of some common substances. The density of an object is calculated by dividing the object's mass by its volume.

Figure 12
Helium-filled balloons float upward because helium is less dense than air is. Similarly, hot-air balloons rise because hot air is less dense than cool air is.

Density Equation

$$D = m/V$$
$$density = mass/volume$$

> **density** the ratio of the mass of a substance to the volume of a substance

Table 1 Densities of Some Substances

Substance	Chemical formula	Density in g/cm³
Air, dry	mixture	0.00129
Brick, common	mixture	1.9
Gasoline	mixture	0.7
Helium	He	0.00018
Ice	H_2O	0.92
Iron	Fe	7.86
Lead	Pb	11.3
Nitrogen	N_2	0.00125
Steel	mixture	7.8
Water	H_2O	1.00

internet connect

www.scilinks.org
Topic: Density
SciLinks code: HK4031

SciLINKS. Maintained by the National Science Teachers Association

47

Historical Perspective

Blimps and Dirigibles Blimps and dirigibles are types of airships. An airship consists of an engine, a large, closed balloon that contains gas, and a gondola to carry passengers and crew. Airships float in air because the gases they contain are less dense than air. In the early 1900s, airships were commonly used for travel, including trans-Atlantic flights. Airships containing flammable hydrogen were used much less frequently after the 1937 explosion of the *Hindenburg* in New Jersey. Airships in use today contain helium, which is nonflammable.

Teaching Tip ——— GENERAL

Volume Review the concept of volume with students. Display several items and have students discuss in groups how the volume of each item can be determined. Some items should be regular in shape so that dimensions can be measured and the volume calculated. Other items should be irregular in shape so that students must estimate or use displacement to find the volume. **LS** Interpersonal

Math Skills

Additional Examples

Density The density of oak wood is generally 0.7 g/cm³. If a 35 cm³ piece of wood has a mass of 25 g, is the wood likely to be oak?

Answer: Wood with a density of 0.71 g/cm³ is probably oak.

The density of silver is 10.5 g/cm³. A bracelet made of silver has a volume of 1.12 cm³. What is the bracelet's mass?

Answer: 11.8 g
LS Logical

Practice

1. $D = 16.52 \text{ g}/2.26 \text{ cm}^3$
   $D = 7.31 \text{ g/cm}^3$
2. $D = 163 \text{ g}/50.0 \text{ cm}^3$
   $D = 3.26 \text{ g/cm}^3$
3. $m = 11.3 \text{ g/cm}^3 \times 6.7 \text{ cm}^3$
   $m = 76 \text{ g}$
LS Logical

Figure 13
A golf ball is denser than a table-tennis ball because the golf ball contains more matter in about the same volume.

Practice HINT

▶ When a problem requires you to calculate density, you can use the density equation.
$D = \frac{m}{V}$
▶ You can solve for mass by multiplying both sides of the density equation by volume.
$DV = \frac{mV}{V}$ $m = DV$
▶ You will need to use this form of the equation in Practice Problem 3.
▶ You can solve for volume by dividing both sides of the equation shown above by density.
$\frac{m}{D} = \frac{DV}{D}$ $V = \frac{m}{D}$

Density is often measured in units of g/cm³

A golf ball and a table tennis ball are shown in **Figure 13**. Which ball is more dense? The two balls have a similar volume, but the mass of a golf ball is 45.9 g and the mass of a table tennis ball is 2.5 g. The golf ball has more mass per unit of volume than a table tennis ball has, and therefore the golf ball is more dense.

The density of a liquid or a solid is usually reported in units of grams per cubic centimeter (g/cm³). For example, 10.0 cm³ of water has a mass of 10.0 g. Its density is 10.0 g for every 10.0 cm³, or 1.00 g/cm³. A cubic centimeter contains the same volume as a milliliter. You may see the density of water expressed as 1 g/mL.

Math Skills

Density If 10.0 cm³ of ice has a mass of 9.17 g, what is the density of ice?

1 List the given and the unknown values.

   Given: *mass, m* = 9.17 g
   volume, V = 10.0 cm³

   Unknown: *density, D* = ? g/cm³

2 Write the equation for density.
   $D = \frac{m}{V}$ or *density = mass/volume*

3 Insert the known values into the equation, and solve.
   $D = 9.17 \text{ g}/10.0 \text{ cm}^3$
   $D = 0.917 \text{ g/cm}^3$

Practice

Density

1. A piece of tin has a mass of 16.52 g and a volume of 2.26 cm³. What is the density of tin?

2. A man has a 50.0 cm³ bottle completely filled with 163 g of a slimy green liquid. What is the density of the liquid?

3. A piece of metal has a density of 11.3 g/cm³ and a volume of 6.7 cm³. What is the mass of this piece of metal?

INCLUSION Strategies ● GENERAL

• **Learning Disabled**
• **Attention Deficit Disorders**
• **English Language Learners**
• **Developmentally Delayed**

Using a reference book or the Internet, have the students create a Planet Notebook by listing the planets of the solar system on individual sheets of paper. Each page of this notebook should have a list of the elements or compounds found on each planet. The students will then identify the density of each planet and indicate whether that planet would or would not float in water. The students may add additional information about the planets and draw a picture representing the planet. A cover sheet would be designed for the front of the notebook and the pages could be bound with staples or string.

Physical properties help determine uses

Every day, you use physical properties to identify substances. Physical properties help you determine whether your socks are clean (odor), whether you can fit all your books into your backpack (volume), or whether your shirt matches your pants (color).

In industry, physical properties are used to select substances that may be useful. Copper is used in electrical power lines, telephone lines, and electric motors because it conducts electricity well. Antifreeze, which contains ethylene glycol (a poisonous liquid), remains a liquid at temperatures that would freeze or boil water in a car radiator. As shown in **Figure 14,** aluminum is used in foil because it is lightweight yet durable, water resistant, and flexible.

Can you think of other physical properties that help us determine how we can use a substance? Some substances have the ability to conduct heat, while others do not. Plastic-foam cups do not conduct heat well, so they are often used for holding hot drinks. What would happen if you poured hot tea into a metal cup?

Figure 14

Aluminum is light, strong, and durable, which makes it ideal for use in foil.

Quick Lab

How are the mass and volume of a substance related?

Analysis

1. Students should record the mass they measured in Step 2.
2. Predictions should be approximately 55 g and 100 g.
3. Predictions should be approximately 25 mL and 75 mL.
4. Divide the mass in the third column of the table by the corresponding volume from the first column. For any two points on the graph, divide the difference in the x values by the difference in the y values to determine density. Student answers will vary; accept any answer students can justify.

LS Logical

Quick Lab

How are the mass and volume of a substance related?

Materials
- 100 mL graduated cylinder
- 250 mL beaker with 200 mL water
- balance
- graph paper

1. Make a data table that has 3 columns and 12 rows. In the first row, label the columns "Volume of H_2O (mL)," "Mass of cylinder (g) and H_2O (g)," and "Mass of H_2O (g)." In the remaining spaces of the first column, write 0, 10, 20, 30, 40, 50, 60, 70, 80, 90, and 100.

2. Measure the mass of the empty graduated cylinder, and record it on a piece of paper.

3. For each amount of water listed in column one, pour the water from the beaker into the graduated cylinder. Then, use the balance to find the mass of the graduated cylinder with the water. Record each value in column two of your data table.

4. On graph paper, make a graph and label the horizontal x-axis "Mass of water (g)." Mark the x-axis in 10 equal increments for 10, 20, 30, 40, 50, 60, 70, 80, 90 and 100 g. Label the vertical

y-axis "Volume of water (mL)." Mark the y-axis in 10 equal increments for 10, 20, 30, 40, 50, 60, 70, 80, 90, and 100 mL.

5. Plot a graph of your data either on graph paper, on a graphing calculator, or by using a graphing/spreadsheet computer program.
COMPUTER SKILL

Analysis

1. What is the mass of the graduated cylinder?

2. Use your graph to estimate the mass of 55 mL of water and 100 mL of water.

3. Use your graph to predict the volume of 25 g of water and 75 g of water.

4. How could you use your data table or graph to calculate the density of water? Which method do you think gives better results? Why?

49

Alternative Assessment —ADVANCED

Soft Drink Densities Point out to students that the density of liquid mixtures depends on the mass of all matter in a given volume of liquid. Have students think about two cans of soda, one which is sweetened with sugar, and the other sweetened with an artificial sweetener. Ask students which can of soda has the greater density. Have them explain their reasoning. (Assuming the volumes of the two cans are identical, the soda with the greater mass has the greater density. Artificial sweeteners are stronger sweetening agents, so it takes a smaller amount of them to sweeten a can of sugar-free soda than it does sugar to sweeten a can of regular soda. Therefore, the regular soda has the greater mass, and therefore density, due to the large amount of sugar required to sweeten it.)

LS Logical

Chapter Resource File

- **Science Skills** Rearranging Algebraic Equations **BASIC**
- **Math Skills** Density **GENERAL**
- **Quick Lab Datasheet** How are the mass and volume of a substance related? **GENERAL**

Reading Skills Before students read this page, ask them: What is one unfavorable chemical property of iron?

a. its high melting point

b. its nonreactivity with oil and gasoline

c. its reactivity with oxygen

d. its nonflammability (c)

Ⓛ **Verbal**

Demonstration — GENERAL

Properties of Iron

(Time: Approximately 10 minutes)

Materials:

• magnet

• steel wool, rusted and nonrusted

Show students both pieces of steel wool. Then show them the effect a magnet has on both pieces of steel wool.

Analysis

1. Does steel wool have the same composition after it has rusted? (No.)

2. Is the ability to rust a chemical or a physical property? (chemical)

3. Does steel wool have the same composition after it is checked with the magnet? (Yes.)

4. Is attraction to a magnet a chemical or a physical property? (physical)

Ⓛ **Visual**

▶ **reactivity** the ability of a substance to combine chemically with another substance

▶ **flammability** the ability of a substance to react in the presence of oxygen and burn when exposed to a flame

Figure 15

Ⓐ This hole started as a small chip in the paint, which exposed the iron in the car to oxygen. The iron rusted and crumbled away.

Ⓑ Paint does not react with oxygen, so it provides a barrier between oxygen and the iron in the car's steel.

Ⓒ This bumper is rust free because it is coated with chromium, which is nonreactive with oxygen.

Chemical Properties

Some elements, such as sodium, react very easily with other elements and usually are found as compounds in nature. Other elements, like gold, are much less reactive and often are found uncombined in nature. Magnesium is so reactive that it is used to make emergency flares. Light bulbs are filled with argon gas because argon is not reactive, so the tungsten filament lasts longer. All of these are examples of *chemical properties.* Chemical properties are generally not as easy to observe as physical properties.

Chemical properties describe how a substance reacts

Although iron has many useful physical and chemical properties, one property that can cause problems for people is its reactivity with oxygen. When iron is exposed to oxygen, it rusts. You can see rust on the old car shown in *Figure 15.* The steel parts of a car rust when iron atoms in the steel react with oxygen in air to form iron(III) oxide. The painted and chromium parts of the car do not rust because they does not react with oxygen.

Chemical properties are related to the specific elements that make up substances. The elements in steel, paint, and chrome have different chemical properties. A chemical property describes how a substance changes into a new substance, either by combining with other elements or by breaking apart into new substances. Chemical properties include the **reactivity** of elements or compounds with oxygen, acid, water, or other substances.

Another chemical property is **flammability**—the ability to burn. For example, wood can be burned to create new substances (ash and smoke) with properties that are different from the original wood. A substance that does not burn, such as gold, has the chemical property of nonflammability. Remember that even when wood is not actually burning, it is still flammable because flammability is one of wood's chemical properties. A substance always has its chemical properties, even when you cannot observe them.

🌐 **Cultural Awareness**

Mercury is one of the oldest known elements and has been used by many cultures throughout history. It was found in an Egyptian tomb dating back to 1500 B.C.E. One chemical property of mercury is that it is toxic. In the 1960s, residents of Minamata City, in Japan, suffered mercury poisoning because they ate fish that were contaminated by high levels of mercury compounds in the water. As a result of this and other similar events, mercury contamination in the environment is closely monitored.

Comparing Physical and Chemical Properties

It is important to remember the differences between physical and chemical properties. You can observe physical properties without changing the identity of the substance. But you can observe chemical properties only in situations in which the identity of the substance changes.

Table 2 summarizes the physical and chemical properties of some common substances. As you can see, many substances have very similar physical properties but completely different chemical properties. For example, baking soda and powdered sugar are both white powders, but baking soda reacts with vinegar, whereas sugar does not.

Table 2 **Comparing Physical and Chemical Properties**

Substance	Physical property	Chemical property
Helium	less dense than air	nonflammable
Wood	grainy texture	flammable
Baking soda	white powder	reacts with vinegar to produce bubbles
Powdered sugar	white powder	does not react with vinegar
Rubbing alcohol	clear liquid	flammable
Red food coloring	red color	reacts with bleach and loses color
Iron	malleable	reacts with oxygen

REAL WORLD APPLICATIONS

Choosing Materials Materials are chosen because their properties are suitable for use. For example, white acrylic plastic can be used to make false teeth. Sometimes, porcelain is used. Metals are less commonly used, although gold teeth are still made sometimes. False teeth have a demanding job to do. They are constantly bathed in saliva, which is corrosive. They must withstand the forces from chewing hard objects, such as popcorn or hard candy. The material chosen has to be nontoxic, hard, waterproof, unreactive, toothlike in appearance, and affordable. Acrylic plastic satisfies these requirements.

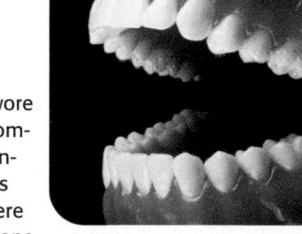

George Washington wore false teeth, which were common in the 1700s. But contrary to the legend that his teeth were wood, they were made of hippopotamus bone.

Applying Information

1. Compare the advantages and disadvantages of gold false teeth and Washington's bone teeth.
2. Identify some advantages of acrylic plastic teeth.

51

Science as a Human Endeavor

The *Hindenburg* The German zeppelin *Hindenburg*, which was filled with hydrogen, caught fire upon landing in 1937. The entire airship was engulfed in an orange fireball and burned in less than 32 seconds. For decades, most people believed the fire started when a spark ignited the flammable hydrogen. But hydrogen burns with a near-colorless flame, not an orange one. Scientists now think that the spark actually ignited the airship's highly flammable outer covering.

REAL WORLD APPLICATIONS

Choosing Materials In addition to being made from a single material, false teeth are also made from composite materials. Composite materials are those in which one material is embedded in another. The resulting material is stronger than either material by itself.

Applying Information

1. Advantages: Answers will vary, but could include that gold is easily shaped and is unreactive. Bone is similar in composition to actual teeth and would have similar strength.

   Disadvantages: Gold is a relatively soft metal and might become misshapen. Bone does not have the hard covering that teeth have, so bone might react with certain foods. Bone can break, and so can form splinters.
2. nontoxic, hard and durable, waterproof, noncorrosive, realistic appearance, reasonably priced

LS Verbal

Chapter Resource File

- **Cross-Disciplinary Worksheet** RWA—Characteristic Properties **ADVANCED**
- **Cross-Disciplinary Worksheet** RWA—Choosing Materials for Bicycle Frames **ADVANCED**
- **Cross-Disciplinary Worksheet** Integrating Environmental Science—Ozone Depletion **ADVANCED**

 Transparencies

TM Chemical and Physical Properties

Match each description with the correct term in the column below.

1. The temperature at which a solid becomes a liquid (a)

2. Ability of a substance to burn in the presence of oxygen (e)

3. Ability of a substance to combine chemically with another substance (d)

4. Mass per unit volume of a substance (c) **LS Logical**

   a. melting point
   b. boiling point
   c. density
   d. reactivity
   e. flammability

Figure 16
Helium is used in blimps because it is less dense than air and is nonflammable.

Characteristic properties help to identify and classify substances

You can describe matter by both physical and chemical properties. The properties that are most useful in identifying a substance, such as density, solubility (whether or not it dissolves), and reactivity with acids, are its *characteristic properties*. Characteristic properties include both types—physical and chemical properties. The characteristic properties of a substance are always the same whether the sample you are observing is large or small.

The blimp in *Figure 16* is filled with helium. The characteristic properties of helium, such as its density and nonflammability, make helium very useful for blimp flight.

SECTION 2 REVIEW

SUMMARY

▶ Physical properties can be observed or measured without changing the composition of matter.

▶ Physical properties help determine how substances are used.

▶ The density of a substance is equal to its mass divided by its volume.

▶ Chemical properties describe how a substance reacts; they can be observed when one substance reacts with another.

▶ Scientists use characteristic properties to identify and classify substances.

1. **Classify** the following as either chemical or physical properties.
   a. is shiny and silvery   **c.** has a density of 2.3 g/cm^3
   b. melts easily   **d.** tarnishes in moist air

2. **Identify** which of the following properties *are not* chemical properties.
   a. reacts with water
   b. boils at 100°C
   c. is red
   d. does not react with hydrogen

3. **Describe** several uses for plastic, and explain why plastic is a good choice for these purposes.

4. **Describe** characteristic properties, and explain why they are important. List some characteristic properties.

Math Skills

5. **Calculate** the density of a rock that has a mass of 454 g and a volume of 100 cm^3.

6. **Calculate** the density of a substance in a sealed 2500 cm^3 flask that is full to capacity with 0.36 g of a substance.

7. **Critical Thinking** Suppose you need to build a raft. Write a paragraph describing the physical and chemical properties of the raft that would be important to ensure your safety.

52

Answers to Section 2 Review

1. **a.** physical
   b. physical
   c. physical
   d. chemical

2. c

3. Answers will vary, but could include use of strong, rigid plastic in cases for delicate objects, such as compact discs, and use of flexible, transparent plastic as plastic food wrap. Plastic is a good choice for these purposes, because it can be hard, flexible, durable, opaque, or transparent.

4. They are physical and chemical properties that are most useful in identifying and classifying a substance; density, melting point, boiling point, solubility, and reactivity with acids.

See "Continuation of Answers" at the end of the chapter.

Changes of Matter

OBJECTIVES

▶ **Explain** physical change, and give examples of physical changes.

▶ **Explain** chemical change, and give examples of chemical changes.

▶ **Compare and contrast** physical and chemical changes.

▶ **Describe** how to detect whether a chemical change has occurred.

▶ **KEY TERMS**

physical change
chemical change

▶ **physical change** a change of matter from one form to another without a change in chemical properties

Some materials benefit us because they stay in the same state and do not change under normal conditions. Surgical steel pins are used to reinforce broken bones because surgical steel remains the same even after years in the human body. Concrete and glass are used as building materials because they change very little under most weather conditions. Other materials are valued for their ability to change states easily. Water is turned into steam to heat homes and factories. Liquid gasoline is changed into a gas so it can burn in car engines. The physical and chemical properties of a substance determine how the substances behave under different conditions.

Physical Changes

A **physical change** affects one or more physical properties of a substance without changing its identity. For example, if you break a piece of chalk in two, you change its physical properties of size and shape. But no matter how many times you break it, chalk would still be chalk and the chemical properties of the chalk would remain unchanged. Each piece of chalk would still produce bubbles if you placed it in vinegar.

Figure 17 shows a physical change taking place. The girl in the picture is getting her hair cut, but the chemical nature of her hair is not changing. The haircut will affect only the physical properties of her hair. Other examples of physical changes are dissolving sugar, melting ice, sanding a piece of wood, crushing an aluminum can, and mixing oil and vinegar.

Figure 17

Is this haircut a physical or a chemical change?

53

Alternative Assessment — ADVANCED

Physical Change Ask students to bring in different brands and types of plastic sandwich bags, along with packaging, ads, and other promotional materials. Have them design and perform an experiment to test how much mass each bag can hold before breaking. Pool the data and have students answer the following questions.

• What inferences can be drawn from the data? (Some claims of strength will be upheld, and others may not.)

• Was this a test of physical or chemical changes? (physical)

LS Logical

LANGUAGE ARTS

CONNECTION — GENERAL

Have students classify changes that occur to matter in materials they are reading in their language arts class. **LS** Logical

---

SECTION
3

Focus

Overview

Before beginning this section, review with your students the Objectives listed in the Student Edition. In this section, students learn about physical and chemical changes. They also learn how mixtures can be separated by physical changes, how compounds are broken down through chemical changes, and how chemical changes can be detected.

🔴 Bellringer

Use the Bellringer transparency to prepare students for this section.

Motivate

Opening Activity — GENERAL

You may perform this as a demonstration, or have students perform it in small groups. You will need a paper towel, a pie plate, vinegar, and 2 or 3 shiny pennies.

1. Place a folded paper towel in a small pie plate.

2. Pour vinegar into the pie plate until the paper towel is damp.

3. Place 2 or 3 shiny pennies on top of the paper towel.

4. Put the plate aside for 24 h.

Ask students if they think a physical or chemical change will occur (chemical), and to explain why. After 24 hours, have students observe the pennies and describe the change that occurred. (The shiny copper surface became coated with a dull, green substance.) Ask them to identify the type of change that occurred. (The change in the appearance of the coins indicates a chemical change.) **LS** Visual

---

Chapter Resource File

• Lesson Plan

Transparencies

TT Bellringer

Reading Skills Have students read the remainder of this section. Divide the class into pairs. Each student should point to an object and ask their partner to describe its physical properties and the physical changes that might occur. Next, students should ask each other to point out three materials that undergo chemical changes. Finally, students should evaluate whether their choices meet the criteria for physical and chemical changes. **LS** Interpersonal

Quick **Lab**

How can physical properties separate a mixture?

The mixture should contain 5 g samples of salt, sand, iron filings, and poppy seeds. Students should devise a procedure that uses solubility, density, and magnetism. Give them another sample of the mixture without telling them the exact masses of each component. Students can then run their procedures and determine the mass of each component.

Analysis

1. solubility, magnetism, low density
2. Salt is separated by solubility, poppy seeds by density, and iron filings by magnetism.
3. Sand, poppy seeds, and iron filings are insoluble; only the iron filings are magnetic.
4. salt, sand, iron filings, poppy seeds

LS Logical

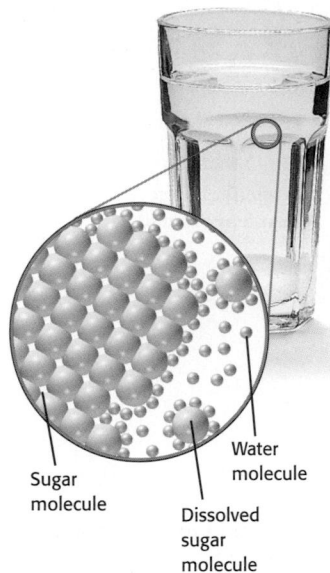

Figure 18

When sugar dissolves in water, water particles attract and pull apart sugar particles, so the sugar particles spread out in the water.

Sugar molecule

Water molecule

Dissolved sugar molecule

Physical changes do not change a substance's identity

Both quartz crystals and sand are made of SiO_2, but they look different. When quartz is crushed into sand, a physical change takes place. During physical changes, energy is absorbed or released. After a physical change, a substance may look different, but the arrangement of atoms that make up the substance are not changed.

Pounding a gold nugget into a ring results in physical changes. But physical changes do not change all the properties of a substance. For example, the color of the gold, its melting point, and its density do not change. Melting, freezing, and evaporating—all changes of state—are physical changes, too, because the identity of the substance does not change.

Dissolving is a physical change

When you stir sugar into water, the sugar dissolves and seems to disappear. But the sugar is still there; you can taste the sweetness when you drink the water. *Figure 18* shows sugar and water molecules dissolving. When sugar dissolves, it seems to disappear because the sugar particles become spread out between the particles of the water. The molecules of the sugar have not changed because dissolving is a physical change. Dissolving a solid in a liquid, a gas in a liquid, or a liquid in a liquid are all physical changes.

Quick **Lab**

How can physical properties separate a mixture?

Materials
- ✓ distilled water
- ✓ clear plastic cups
- ✓ filter funnel
- ✓ plastic spoon
- ✓ filter paper
- ✓ 5 g sample of mixture
- ✓ magnet
- ✓ paper towels

1. Design an experiment in which the given materials are used to separate the components of the sample mixture. (**Hint:** Consider physical properties such as solubility, density, and magnetism.)
2. Once you have separated the components of the sample mixture, describe them by their physical properties.

Analysis

1. What properties did you observe in each of the components of the mixture?
2. How did these properties help you to separate the components of the sample?
3. Did any of the components share similar properties?
4. Based on your observations, what do you think the mixture was composed of?

Alternative Assessment ── **GENERAL**

Changes in Matter Ask students to evaluate whether the following four processes are primarily physical or chemical changes.

- Baking soda releases CO_2 and H_2O when heated strongly. (chemical)
- Melting antimony and tin together produces pewter. (physical)
- Rubbing alcohol cools your skin as it evaporates. (physical)
- A piece of zinc is placed into a solution of hydrochloric acid, and bubbles begin to rise to the surface. (chemica)

LS Logical

MISCONCEPTION ///ALERT\\\

Physical and Chemical Changes

Students may have the idea that changes of state are chemical changes. For example, because solid ice and liquid water have some different properties, students may think they are different substances, and that melting is a chemical change. Similarly, they may confuse the combustion of a substance (e.g., alcohol) with the physical change of evaporation. Clarify that in changes of state, the identity of the substance does not change.

Figure 19

These pictures show ways that physical changes can be used to separate mixtures.

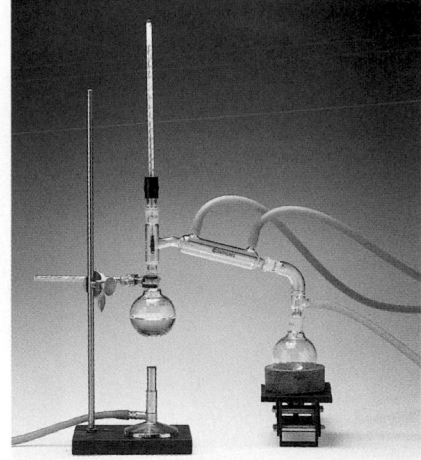

B The distillation device shown here can separate components of mixtures that have different boiling points. When heated, the component that boils and evaporates first, separates from the mixture and collects in the receiving flask.

A A centrifuge is a tool used to separate mixtures. It spins a sample of a mixture rapidly until the components of the mixture separate. You can see different layers in this sample of blood because it has been separated into its components.

Mixtures can be physically separated

Because mixtures are not chemically combined, each component of a mixture has the same chemical makeup it had before the mixture was formed. Each substance in a mixture keeps its identity. In some mixtures, such as a slice of pizza, you can easily see the individual components. In other mixtures, such as salt water, you cannot see all the components.

You can remove the mushrooms on a pizza, which results in a physical change. The identities of the substances in the pizza would not change. Unlike mixtures, compounds can be broken down only through chemical changes.

Not all mixtures are as easy to separate as a pizza. You cannot pick salt out of a saltwater mixture, but you can separate the salt from the water by heating the mixture. When the water evaporates, the salt remains behind. Several common techniques for separating mixtures are shown in *Figure 19*.

C Magnets can be used to separate mixtures that have components containing iron. In this mixture of nails, the magnet attracts and separates the nails containing iron from the nails that do not contain iron.

55

Alternative Assessment — GENERAL

Have students investigate why desalination is expensive, and determine under what conditions it should be used on a large scale. (Desalination is best used in locations where it is less expensive than piping or shipping fresh water from distant sources.) Have students summarize their findings in a brief written report. **LS** Verbal

Alternative Assessment — ADVANCED

Have students prepare posters that illustrate how the components of crude oil are separated by distillation. Encourage students to be creative. **LS** Visual

Demonstration — GENERAL
Chemical Change

(Time: Approximately 15 minutes)

Materials:

• beaker
• copper sulfate ($CuSO_4$) solution
• iron nail, large

Place the iron nail in a $CuSO_4$ solution for several minutes. Leave the nail in the solution overnight, and observe the nail and solution the next day.

Analysis

1. What change in the nail indicates that a chemical change occurred? (Copper from the solution appears on the nail as copper metal.)

2. How do you know that this change was not physical? (The copper was initially in a compound. It changed to a free element, which is a different substance.)
LS Visual

Group Activities — GENERAL

Have students work in small groups to identify a method to separate a mixture of sand, sawdust, and gravel into its components. Then, have them share their results with the class. **LS** Interpersonal

Demonstration

Separation Demonstrate the separation of salt (sodium chloride) and sand (silicon dioxide) using the method shown in **Figure 19A**. Ask students which compound dissolves in water (salt) and which does not (silicon dioxide). **LS** Visual

Physical or Chemical Change

(Time: Approximately 15 minutes)

Materials:

• hydrogen peroxide (H_2O_2) solution, 3%
• plastic bag, self-sealing
• plastic pill bottle, small
• steel wool

Safety Caution: *If students perform this, they should wear safety goggles, gloves, and an apron; they should handle the hydrogen peroxide carefully.*

Fill the pill bottle halfway with hydrogen peroxide. Then place a small piece of steel wool and the pill bottle into a plastic bag, being careful not to spill the hydrogen peroxide. Force the air out of the bag and seal it tightly. Tip the bottle over so that the hydrogen peroxide comes in contact with the steel wool. Invite students to feel the bag; ask them how they know a chemical change has occurred. (A gas was formed that inflated the bag, and the bag's contents became heated.) **LS Kinesthetic**

▶ **chemical change** a change that occurs when a substance changes composition by forming one or more new substances

internet connect

www.scilinks.org
Topic: Chemical Changes
SciLinks code: HK4020

SCiLINKS. Maintained by the National Science Teachers Association

---

Figure 20

Examples of Chemical Changes

A Soured milk smells bad because bacteria have formed new substances in the milk.

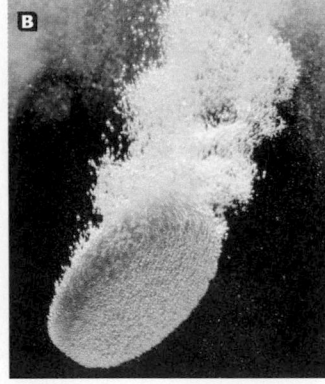

B Effervescent tablets bubble when the citric acid and baking soda in them react with water to produce CO_2.

C The Statue of Liberty is made of shiny, orange-brown copper. But the metal's interaction with carbon dioxide and water has formed a new substance, copper carbonate.

Chemical Changes

Some materials are useful because of their ability to change and combine to form new substances. For example, the compounds in gasoline burn in the presence of oxygen to form carbon dioxide and water, which releases energy. This is a **chemical change.** A chemical change occurs when one or more substances are changed into entirely new substances that have different properties.

Chemical changes happen everywhere

You see chemical changes happening more often than you may think. When a battery "dies," the chemicals inside the battery have changed, so the battery can no longer supply energy. The oxygen you inhale is used in a series of chemical reactions in your body. After it has undergone a chemical change by reacting with carbon, the oxygen is then exhaled as part of the compound carbon dioxide. Chemical changes occur when fruits and vegetables ripen and when the food you eat is digested. *Figure 20* shows some examples of other chemical changes that may be familiar to you.

56

Chemical changes form new substances that have different properties

A fun (and tasty) way to observe a chemical change is to bake a cake. When you bake a cake, you combine eggs, flour, sugar, butter, milk, baking powder, and other ingredients. Each ingredient has its own set of properties. For example, when baking powder combines with a liquid such as milk or water, it releases carbon dioxide, which causes the cake to rise. When you mix all of the ingredients and add heat by baking the cake batter, you get something completely different. The heat of the oven and the interaction of the ingredients cause chemical changes, which results in a cake with properties that are completely different from the properties of the original ingredients.

Chemical changes can be detected

When a chemical change takes place, there are often clues that suggest that a chemical change has happened. A change in odor or color is a good clue that a substance is changing chemically. When food burns, you can often smell the gases given off by the chemical changes. When paint fades, you can observe the effects of chemical changes in the paint. Chemical changes often cause color changes, fizzing or foaming, or the production of sound, heat, light, or odor.

Figure 21 shows table sugar being heated in a test tube. When sugar is heated to a high temperature, it breaks down into carbon and water. How do you know a chemical change is taking place in *Figure 21*? The sugar has changed color, bubbles are forming, and a caramel smell is filling the air.

Chemicals changes cannot be reversed by physical changes

Because new substances are formed in a chemical change, you cannot reverse chemical changes by using physical changes. You cannot "unbake" a cake by separating out each ingredient. Most of the chemical changes you observe in your daily life, such as a cake baking, milk turning sour, or iron rusting, are impossible to reverse. Imagine trying to unbake a cake! While some physical changes can be easily undone, chemical changes are often more difficult to undo.

However, some chemical changes can be reversed under the right conditions by other chemical changes. For example, the water formed in a space shuttle's rockets can be split back into the starting materials—hydrogen and oxygen—by using an electric current to initiate a reaction that separates the hydrogen and oxygen atoms in the water molecules.

Figure 21

Table sugar is a compound made of carbon, hydrogen, and oxygen. When table sugar is heated, it caramelizes. When heated to a high temperature, it breaks down completely into carbon and water.

internet connect

www.scilinks.org
Topic: Physical/Chemical Changes
SciLinks code: HK4104

SCI*LINKS*. Maintained by the National Science Teachers Association

ENVIRONMENTAL SCIENCE
CONNECTION — GENERAL

When fossil fuels are burned, a chemical change takes place involving sulfur (a substance in fossil fuels) and oxygen (from the air). This chemical change produces sulfur dioxide, a gas. When sulfur dioxide enters the atmosphere, it undergoes another chemical change by interacting with water and oxygen.

This chemical change produces sulfuric acid, a contributor to acid precipitation. Acid precipitation can kill trees and make ponds and lakes unable to support life. As an extension, have interested students research acid precipitation and present their findings in an oral or written report, or as a "TV News Bulletin." **LS Verbal**

Compound Confusion

Safety Caution: *Students should wear an apron, safety goggles, and gloves.*

Label the baking soda and sugar **A** and **B**. Dispose of the compounds in a sink with running water.

Answer

4. Baking soda reacts with vinegar by bubbling; powdered sugar does not react.

LS Visual

Close

Quiz ───────── BASIC

True or False Mark **T** if the statement is true, and **F** if it is false.

1. Water turning into steam and gasoline into a gas are both physical changes. (T)

2. Dissolving is a chemical change. (F)

3. In a chemical change, the identity of the substances stays the same. (F)

LS Logical

Chapter Resource File

- **Quick Activity Datasheet** Compound Confusion GENERAL
- **Concept Review** GENERAL
- **Quiz** BASIC

Quick ACTIVITY

Compound Confusion

1. Measure 4 g each of compounds A and B; place each in a clear plastic cup.
2. Observe the color and texture of each compound. Record your observations.
3. Add 5 mL of vinegar to each cup. Record your observations.
4. Baking soda reacts with vinegar, but powdered sugar does not. Which of these two substances is compound A, and which is B?

Compounds can be broken down through chemical changes

Some compounds can be broken down into elements through chemical changes. When the compound mercury(II) oxide is heated, it breaks down into the elements mercury and oxygen. If an electric current is passed through melted table salt, the elements sodium and chlorine are produced.

Other compounds undergo chemical changes to form simpler compounds. Carbonic acid is a compound that gives carbonated soda a tart taste and adds "fizz." In an unopened bottle of soda, you don't see bubbles because carbon dioxide is present in the form of carbonic acid. When you open a bottle of soda, the carbonic acid breaks down into carbon dioxide and water. The carbon dioxide escapes as bubbles. Through additional chemical changes, the carbon dioxide and water can be further broken down into the elements carbon, oxygen, and hydrogen.

SECTION 3 REVIEW

SUMMARY

▶ Physical changes are changes in the physical properties of a substance that do not change the identity of the substance.

▶ Changes of state are physical changes.

▶ Dissolving is a physical change.

▶ Physical changes are often easily reversed.

▶ Chemical changes form new substances that have new properties. Chemical changes can be reversed only through chemical reactions.

▶ Chemical changes often cause changes in color or produce sound, light, odor, or heat.

1. **Classify** the following as a chemical or a physical change.
   a. adding sugar to lemonade
   b. plants using CO_2 and H_2O to form O_2 and sugar
   c. boiling water
   d. frying an egg
   e. rust forming on metal
   f. fruit rotting
   g. removing salt from water by evaporation

2. **Explain** why changes of state are physical changes.

3. **Describe** how you would separate the components of a mixture, and state whether your methods would be physical or chemical changes.

4. **Define** physical change and chemical change, and give examples of each.

5. **Explain** why physical changes can easily be reversed but chemical changes cannot.

6. **Identify** two ways to break down a compound into simpler substances.

7. **List** three clues that indicate a chemical change.

8. **Critical Thinking** Describe the difference between physical and chemical changes in terms of what happens to the particles.

58

Answers to Section 3 Review

1. **a.** physical
   b. chemical
   c. physical
   d. chemical
   e. chemical
   f. chemical
   g. physical

2. The composition of the substances does not change in a change of state.

3. Mixtures can be separated by physical changes (evaporating, melting, distillation).

4. Physical changes alter the physical properties of a substance without changing its identity. Physical changes include pounding, breaking, and melting. Chemical changes form new substances that have new properties. Chemical changes include baking, rotting, and oxidation.

5. Physical changes do not change the identity of substances, and therefore are easy to undo. In chemical changes, new substances are formed, and cannot be reversed using physical means.

6. heating, electrolysis

See "Continuation of Answers" at the end of the chapter.

Study Skills

Two-Column Notes

Two-column notes help you learn or review details of specific concepts.

1 **Identify the main ideas using the section objectives.**

>The objectives from Section 1 will be used as a source for main ideas on matter.

2 **Divide a blank sheet of paper into two columns, write the main ideas and detail numbers in the left-hand column, and write the detail notes in the right-hand column.**

Main idea	Detail notes
Elements (2 characteristic properties)	contain one type of atom simplest form of substance
Compounds (2 characteristic properties)	made of two or more elements chemical properties differ from its elements
Pure substances (3 characteristic properties)	fixed composition definite properties examples: elements and compounds
Molecules (4 characteristic properties)	act as a unit smallest unit of a substance that has the same properties of the substance some molecules made of different elements are also compounds some molecules are made of atoms of the same element
Mixtures (3 characteristic properties)	combination of pure substances heterogeneous mixture: non-uniform homogeneous mixture: uniform

Practice

Use concepts from Section 3 to create a table of two-column notes. In the detail notes, include examples of physical and chemical changes, and explain how these changes can be distinguished.

Teaching Tip

Point out to students that the main ideas for two-column notes are obtained from reading all of the chapter objectives first. Students should then look at the headings for each subsection. These will provide a starting place for both finding the characteristic ideas and noting how many there are for each main idea.

Possible Answers

1. physical changes (no change to molecules; changes in physical form (size, shape, state); easily reversed; examples: crushing, evaporation, freezing, melting, crystallization, weathering, dissolving, mixture separation)

2. chemical changes: (change to molecules; new substances with new properties formed; not always easy to reverse; examples: burning fuel, reaction of effervescent tablets with water, rusting, baking a cake, souring milk, combining hydrogen and oxygen to form water, rotting eggs)

3. distinguishing between chemical and physical changes: (physical changes: no change to melting point, boiling point, or density; chemical changes: release of energy as heat or light; sound produced; appearance of odor; chemical changes form a gas or precipitate; chemical changes often cause change in color)

Understanding Concepts

1. c
2. d
3. d
4. a
5. c
6. a
7. c
8. d
9. a
10. c
11. c
12. a
13. b
14. a

Chapter Resource File

- **Chapter Tests** GENERAL
- **Test Item Listing for ExamView® Test Generator**
- **Cross-Disciplinary Worksheet** Connection to Language Arts— Hidden Meanings ADVANCED

Transparencies

TT Concept Mapping

Chapter Highlights

Before you begin, review the summaries of the key ideas of each section, found at the end of each section. The vocabulary terms are listed on the first page of each section.

UNDERSTANDING CONCEPTS

1. Matter is
   a. any visible solid that has mass.
   b. any liquid that takes up space and has mass.
   c. anything that takes up space and has mass.
   d. any liquid or solid that takes up space.

2. Which of the following is a compound?
   a. sodium, Na
   b. chlorine, Cl
   c. iodine, I
   d. water, H_2O

3. What is the chemical formula for iron(III) oxide?
   a. Fe^{2+}
   b. NaCl
   c. I_2
   d. Fe_2O_3

4. Which of the following is a mixture?
   a. air
   b. salt
   c. water
   d. sulfur

5. Compounds and elements are
   a. always solids.
   b. mixtures.
   c. pure substances.
   d. dense.

6. An element is a substance that
   a. cannot be broken down into simpler substances by chemical means.
   b. cannot react with another substance to create a third substance.
   c. is composed of two or more different atoms.
   d. is composed of two or more identical atoms.

7. What is the density of a piece of metal that has a volume of 8 cm³ and a mass of 64 g?
   a. 0.13 g/cm³
   b. 2.7 g/cm³
   c. 8.0 g/cm³
   d. 512 g/cm³

60

8. The chemical formula for indigo is $C_{16}H_{10}N_2O_2$, which indicates that each indigo molecule contains
   a. 4 elements.
   b. 30 atoms.
   c. 16 carbon atoms.
   d. All of the above

9. Which of the following is a physical change?
   a. melting ice cubes
   b. burning paper
   c. rusting iron
   d. burning gasoline

10. Which of the following is a pure substance?
    a. grape juice
    b. salt water
    c. table salt
    d. gasoline

11. If you add oil to water and shake the two liquids together, you will form a
    a. pure substance.
    b. miscible liquid.
    c. heterogeneous mixture.
    d. homogeneous mixture.

12. A carbonated drink is a
    a. mixture of gases and liquids.
    b. heterogeneous mixture.
    c. mixture of two immiscible liquids.
    d. compound.

13. A stone will sink in water because a stone
    a. is less dense than water.
    b. is denser than water.
    c. is denser than air.
    d. weighs a lot.

14. What percentage of the human body is composed of the element oxygen?
    a. 65%
    b. 29.5%
    c. 49.2%
    d. 17.5%

Using Vocabulary

15. Answers will vary. Student answers may include: melting point, boiling point, density, reactivity with acid, color, hardness, texture, flammability, malleability, thermal conductivity, and solubility.

16. a. An atom is the smallest unit of an element that has the properties of the element; a molecule is the smallest unit of a substance that exhibits all of the properties characteristic of that substance.
    b. A molecule is composed of 2 or more atoms of either the same or different elements; a compound is composed of atoms of two or more different elements.

    c. A compound is made of 2 or more different elements that are chemically combined; a mixture is a combination of 2 or more pure substances physically mixed together.

17. Wood has the chemical property of flammability, meaning it burns in the presence of oxygen. This is a chemical change, not a physical change, because new substances with properties that are different from wood are created; also heat (a sign of a chemical change) is present.

15. List four properties that can be used to classify elements.

16. Compare the following sets of terms:
  a. an *atom* and a *molecule*
  b. a *molecule* and a *compound*
  c. a *compound* and a *mixture*

17. When wood is burned, heat, ash, and smoke are produced. Describe this reaction, and explain what type of change is occurring. Use the terms *flammability, chemical property,* and *physical change* or *chemical change*.

18. When sugar is added to water, the sugar dissolves and the resulting liquid is clear. Have the combined sugar and water formed a *pure substance* or a *mixture*? Explain your answer.

19. When water and rubbing alcohol are mixed together, they completely dissolve. Are the two liquids *miscible* or *immiscible*? Explain the difference.

20. The figure below shows magnesium burning in the presence of oxygen. Give some evidence from the figure that a *chemical change* is occurring.

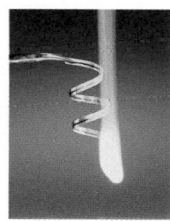

21. Make a table that has two columns. Label one column "Physical properties" and the other "Chemical properties." Put each of the following terms in the proper column: *color, density, reactivity, magnetism, melting point, corrosion, flammability, dissolving, conducting electricity,* and *boiling point*.

22. Graphing Make a graph that shows the relationship between the mass and volume of aluminum samples of different sizes. Use the *y*-axis to plot mass and the *x*-axis to plot volume. What does the shape of the graph tell you about the density of aluminum?

Block number	Mass (g)	Volume (cm³)
1	1.20	0.443
2	3.59	1.33
3	5.72	2.12
4	12.4	4.60
5	15.3	5.66
6	19.4	7.17
7	22.7	8.41
8	24.1	8.94
9	34.0	12.6
10	36.4	13.5

23. Calculating Density A piece of titanium metal has a mass of 67.5 g and a volume of 15 cm³. What is the density of titanium?

24. Calculating Density If a liquid has a volume of 620 cm³ and a mass of 480 g, what is its density?

25. Calculating Density A sample of a substance with a mass of 85 g occupies a volume of 110 cm³. What is the density of the substance? Will the substance float in water? Explain your answer.

26. Calculating Volume The density of a piece of brass is 8.4 g/cm³. If the mass of the brass is 510 g, find the volume of the brass.

61

Section	Questions
1	1–6, 8, 10–12, 14, 16, 18, 19, 40, 42, 43
2	7, 13, 15, 21–29, 32, 34–37, 39
3	9, 17, 20, 30, 31, 33, 38, 41

18. a mixture, because dissolving (a physical change) has occurred; It can be separated into its components, the sugar and water are mixed physically, it does not have a fixed composition or definite properties, it is formed from two pure substances (sugar and water).

19. miscible; miscible (two or more liquids that are able to dissolve), vs. immiscible (two or more liquids that do not dissolve)

20. Energy (as heat and light) is released; the properties of the magnesium change.

21. physical properties: color, density, magnetism, melting point, boiling point, dissolving, conducting electricity; chemical properties: reactivity, corrosion, flammability

Building Graphing Skills

22. The density of aluminum fluctuates somewhat, as volume and mass increase. These aluminum samples may not be pure aluminum, as density should remain constant in an element.

Building Math Skills

23. $D = 67.5 \text{ g}/15 \text{ cm}^3 = 4.5 \text{ g/cm}^3$

24. $D = 480 \text{ g}/620 \text{ cm}^3 = 0.77 \text{ g/cm}^3$

25. $D = 85 \text{ g}/110 \text{ cm}^3 = 0.77 \text{ g/cm}^3$
It will float because its density is less than that of water (1.00 g/cm³).

26. $V = 510 \text{ g}/(8.4 \text{ g/cm}^3) = 61 \text{ cm}^3$

27. $m = 1.00 \text{ g/cm}^3 \times (100.0 \text{ cm} \times 50.0 \text{ cm} \times 30.0 \text{ cm})$
$m = 1.50 \times 10^5 \text{ g}$

28. $V = 63.4 \text{ mL} - 40.0 \text{ mL}$
$V = 23.4 \text{ mL} = 23.4 \text{ cm}^3$
$m = 8.9 \text{ g/cm}^3 \times 23.4 \text{ cm}^3$
$m = 210 \text{ g}$

Thinking Critically

29. glycerin (because it is less dense); density of glycerin = 37.8 g/30.0 mL = 1.26 g/mL; density of corn syrup = 82.8 g/60.0 mL = 1.38 g/mL

30. Sample answer: Pass the mixture through a filter that allows the salt and pepper to pass through but traps the pebbles. Mix the salt and pepper with water to dissolve the salt. Filter the mixture to trap the pepper. Evaporate the water to recover the salt.

31. The powder is a compound. The change in color and the formation of a gas imply that a chemical change took place. Compounds can be broken down by chemical changes.

32. Sample answer: crushed shape, somewhat shiny, metallic

33. a physical change

34. The density before and after the change is the same because density is a characteristic property of matter.

35. No, chemical properties cannot be determined simply by looking at a substance. Chemical properties can only be observed when a chemical change occurs.

36. Answers could include high melting point, hardness, and strength to bore through different materials in Earth's crust and to endure the heat at Earth's core.

27. **Calculating Mass** What mass (in grams) of water will fill a tank that is 100 cm long, 50 cm wide, and 30 cm high?

28. **Calculating Volume and Mass** A graduated cylinder is filled with water to a level of 40.0 mL. When a piece of copper is lowered into the cylinder, the water level rises to 63.4 mL. Find the volume of the copper sample. If the density of the copper is 8.9 g/cm³, what is the copper's mass?

THINKING CRITICALLY

29. **Applying Knowledge** A jar contains 30 mL of glycerin (mass = 37.8 g) and 60 mL of corn syrup (mass = 82.8 g). Which liquid is on the top layer? Explain your answer.

30. **Applying Knowledge** Describe a procedure to separate a mixture of salt, finely ground pepper, and pebbles.

31. **Applying Knowledge** A light-green powder is heated in a test tube. A gas is given off while the solid becomes black. In which category of matter does the green powder belong? Explain your reasoning.

    Examine the photograph below, and answer the following questions.

32. **Interpreting Graphics** List three physical properties of this can.

33. **Interpreting Graphics** Was the change in the can's appearance caused by a chemical change or by a physical change?

34. **Applying Knowledge** How does the density of the metal in the can before the change compare with the density after the change?

35. **Applying Knowledge** Can you tell what the chemical properties of the can are by just looking at the picture? Explain your answer.

36. **Creative Thinking** Suppose you are planning a journey to the center of Earth in a self-propelled tunneling machine. List properties of the special materials that would be needed to build the machine, and explain why each property would be important.

DEVELOPING LIFE/WORK SKILLS

37. **Making Decisions** The frame of a tennis racket needs to be strong and stiff yet light. Tennis racket frames were once made of wood. But to be strong and stiff, the frame had to be thick and heavy. Now rackets can be made from different materials. Make a table of the advantages and disadvantages of each of the materials described in the graphs below.

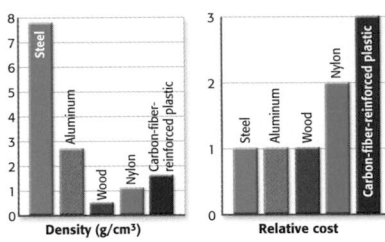

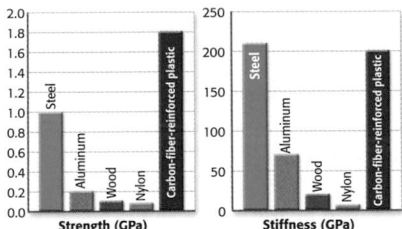

Developing Life/Work Skills

37. Tables should indicate that steel is strong, inexpensive, and stiff, but too heavy; aluminum and wood are lightweight and inexpensive but not very strong or stiff; nylon is lightweight and somewhat inexpensive but not strong or stiff; and carbon-fiber is strong, lightweight, and stiff, but expensive.

38. Student art will vary but may include one atom and a three-atom molecule forming two two-atom molecules.

39. Student answers will vary but may include testing to see if the material conducts electricity (metal), burns (wood), has a low melting point (plastic), or can be crushed and has a high melting point (glass).

38. Applying Technology Use a computer drawing program to illustrate a chemical change in which one atom and one molecule interact to form two molecules.

COMPUTER SKILL

39. Working Cooperatively Suppose you are given a piece of a material that is painted black so you cannot tell its normal appearance. Work in a small group to plan tests you would do on the material to decide whether it is metal, glass, plastic, or wood.

INTEGRATING CONCEPTS

40. Connection to Earth Science The air in Earth's atmosphere is a mixture. Research the atmosphere's contents. What are the main components of Earth's atmosphere? What is the most abundant substance in the mixture?

41. Connection to Biology Explain why the process of digestion involves mainly chemical changes. Research what the starting materials of digestion are and the final end products, and find out if physical changes are also involved. Why is digestion necessary?

42. Connection to Language Arts An element is sometimes named for one of its properties, an interesting fact about the element, or for the person who first discovered the element. Research the origin of the name of each of the following elements: promethium, oxygen, iridium, fermium, curium, tantalum, silver, polonium, ytterbium, and hafnium.

43. Concept Mapping Copy the unfinished concept map below onto a sheet of paper. Complete the map by writing the correct word or phrase in the lettered boxes.

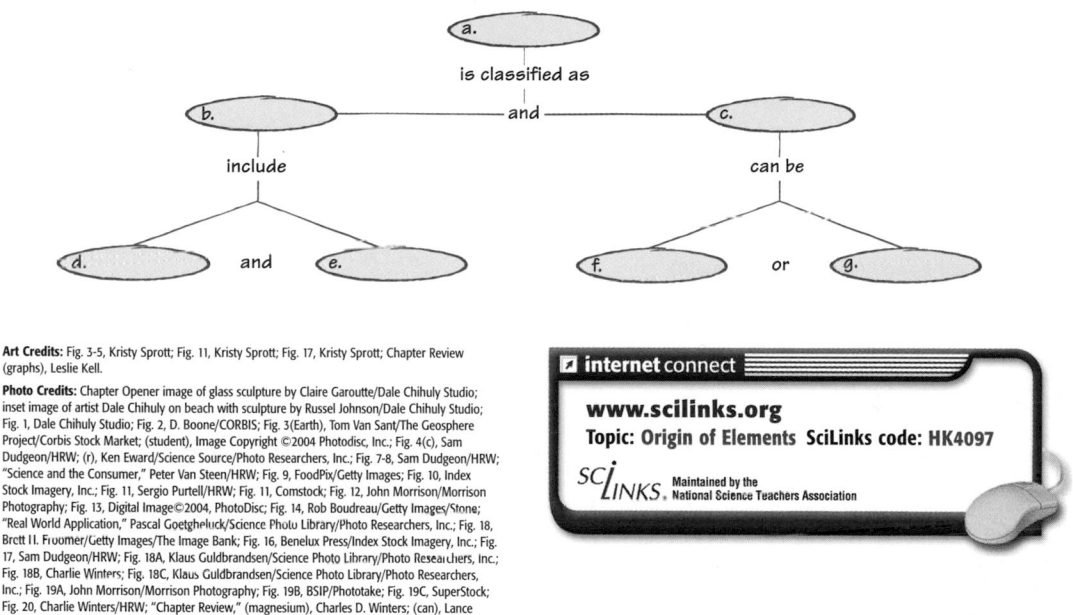

Art Credits: Fig. 3-5, Kristy Sprott; Fig. 11, Kristy Sprott; Fig. 17, Kristy Sprott; Chapter Review (graphs), Leslie Kell.

Photo Credits: Chapter Opener image of glass sculpture by Claire Garoutte/Dale Chihuly Studio; inset image of artist Dale Chihuly on beach with sculpture by Russel Johnson/Dale Chihuly Studio; Fig. 1, Dale Chihuly Studio; Fig. 2, D. Boone/CORBIS; Fig. 3(Earth), Tom Van Sant/The Geosphere Project/Corbis Stock Market; (student), Image Copyright ©2004 Photodisc, Inc.; Fig. 4(c), Sam Dudgeon/HRW; (r), Ken Eward/Science Source/Photo Researchers, Inc.; Fig. 7-8, Sam Dudgeon/HRW; "Science and the Consumer," Peter Van Steen/HRW; Fig. 9, FoodPix/Getty Images; Fig. 10, Index Stock Imagery, Inc.; Fig. 11, Sergio Purtell/HRW; Fig. 11, Comstock; Fig. 12, John Morrison/Morrison Photography; Fig. 13, Digital Image©2004, PhotoDisc; Fig. 14, Rob Boudreau/Getty Images/Stone; "Real World Application," Pascal Goetgheluck/Science Photo Library/Photo Researchers, Inc.; Fig. 18, Brett II. Froomer/Getty Images/The Image Bank; Fig. 16, Benelux Press/Index Stock Imagery, Inc.; Fig. 17, Sam Dudgeon/HRW; Fig. 18A, Klaus Guldbrandsen/Science Photo Library/Photo Researchers, Inc.; Fig. 18B, Charlie Winters; Fig. 18C, Klaus Guldbrandsen/Science Photo Library/Photo Researchers, Inc.; Fig. 19A, John Morrison/Morrison Photography; Fig. 19B, BSIP/Phototake; Fig. 19C, SuperStock; Fig. 20, Charlie Winters/HRW; "Chapter Review," (magnesium), Charles D. Winters; (can), Lance Schriner/HRW; "Skills Practice Lab," Peter Van Steen/HRW; "Viewpoints," HRW Photos.

internet connect

www.scilinks.org
Topic: Origin of Elements SciLinks code: HK4097

*SCI*LINKS. Maintained by the National Science Teachers Association

63

Integrating Concepts

40. nitrogen and oxygen; nitrogen

41. In digestion, nutrients (carbohydrates, fats, and proteins) are chemically broken down into the end products of simple sugars (monosaccharides), fatty acids and glycerol, and amino acids, respectively. Physical changes are also involved during mechanical digestion. (e.g., chewing, grinding, peristalsis, churning). Digestion is necessary to reduce nutrients to molecules small enough to be absorbed through intestinal walls to the blood and carried to body cells, where they are utilized for energy (metabolized).

42. promethium—named after Prometheus in Greek mythology; oxygen—named for Greek words that mean "acid former"; iridium—from the Greek word *iris*, meaning "rainbow"; fermium—named after Enrico Fermi; curium—named after Pierre and Marie Curie; tantalum—named after Tantalus, who in Greek mythology was the father of Niobe, because tantalum is closely related to niobium; silver—from the Anglo-Saxon word *siolfur*; polonium—named after Poland, the birthplace of Marie Curie; ytterbium—named after Ytterby, Sweden, where it was first discovered; hafnium—from the Latin word *Hafnia*, meaning "Copenhagen," because the element was discovered in Denmark.

43. a. matter
 b. pure substances
 c. mixtures
 d. elements
 e. compounds
 f. homogeneous
 g. heterogeneous

TESTING THE CONSERVATION OF MASS

Teacher's Notes

Time Required 1 lab period

Ratings

1	2	3	4
EASY			HARD

TEACHER PREPARATION	2
STUDENT SETUP	2
CONCEPT LEVEL	2
CLEANUP	2

Skills Acquired
- Collecting data
- Communicating
- Interpreting
- Measuring
- Organizing and analyzing data
- Predicting

The Scientific Method

In this lab, students will:
- Make Observations
- Analyze the Results
- Draw Conclusions
- Communicate Results

Materials

Materials listed are for each group.

Safety Cautions
- Safety goggles and a lab apron must be worn at all times.
- Read all safety cautions, and discuss them with your students.
- In case of an acid spill, first dilute the spill with water. Then mop while wearing disposable gloves. Designate separate cloths or mops for acid spills.

Introduction

How can you show that mass is conserved in a chemical reaction between two household substances—vinegar and baking soda?

Objectives

▶ USING SCIENTIFIC METHODS *Measure* the masses of reactants and products in a chemical reaction.

▶ *Design* an experiment to test the law of conservation of mass.

Materials

balance (with standard masses, if necessary)
baking soda (sodium bicarbonate)
beaker, 400 mL (optional)
clear plastic cups (capable of holding at least 150 mL each) (2)
graduated cylinder, 100 mL
plastic sandwich bag with zipper-type closure
twist tie
vinegar (acetic acid solution)
weighing papers (2)

64

Testing the Conservation of Mass

▶ Procedure

Observing the Reaction Between Vinegar and Baking Soda

1. On a blank sheet of paper, prepare a table like the one shown below.

	Initial mass (g)	Final mass (g)	Change in mass (g)
Trial 1			
Trial 2			

SAFETY CAUTION Put on a lab apron, safety goggles, and gloves. If you get a chemical on your skin or clothing, wash it off at the sink while calling to your teacher. If you get a chemical in your eyes, immediately flush it out at the eyewash station while calling to your teacher. When mixing chemicals, use a rimmed tray with a paper lining to catch and absorb spills.

2. Place a piece of weighing paper on the balance. Place about 4 to 5 g of baking soda on the paper. Carefully transfer the baking soda to a plastic cup.

3. Using the graduated cylinder, measure about 50 mL of vinegar. Pour the vinegar into the second plastic cup.

4. Place both cups on the balance, and determine the combined mass of the cups, baking soda, and vinegar to the nearest 0.01 g. Record the combined mass in the first row of your table under "Initial mass."

5. Take the cups off the balance. Carefully and slowly pour the vinegar into the cup that contains the baking soda. To avoid splattering, add only a small amount of vinegar at a time. Gently swirl the cup to make sure the reactants are well mixed.

6. When the reaction has finished, place both cups back on the balance. Determine the combined mass to the nearest 0.01 g. Record the combined mass in the first row of your table under "Final mass."

7. Subtract the final mass from the initial mass, and record the result in the first row of your table under "Change in mass."

Designing Your Experiment

8. Examine the plastic bag and the twist ties. With your lab partners, develop a procedure that will test the law of conservation of mass more accurately than Trial 1 did. Which products' masses were not measured? How can you be sure you measure the masses of all of the reaction products?

9. In your lab report, list each step you will perform in your experiment.

10. Before you carry out your experiment, your teacher must approve your plan.

Performing Your Experiment

11. After your teacher approves your plan, perform your experiment using approximately the same quantities of baking soda and vinegar you used in Trial 1.

12. Record the initial mass, final mass, and change in mass in your table.

▶ Analysis

1. Compare the changes in mass you calculated for the first and second trials. What value would you expect to obtain for a change in mass if both trials validated the law of conservation of mass?

2. Was the law of conservation of mass violated in the first trial? Explain your reasoning.

3. If the results of the second trial were different from those of the first trial, explain why.

▶ Conclusions

4. Suppose someone performs an experiment like the one you designed and finds that the final mass is much less than the initial mass. Would that prove that the law of conservation of mass is wrong? Explain your reasoning.

65

Answers to Procedure

8. The mass of the carbon dioxide produced is not measured. Any design students come up with must assure that the reaction system is truly closed. This will basically involve sealing baking soda into the corner of a large plastic bag using a twist tie, adding the vinegar to the bag, and sealing the top of the bag so that it is airtight. The bag should then be placed in the 400 mL beaker, so that the mass of the bag, beaker, and reactants can be measured. Removing the twist tie allows the reactants to mix. As the reaction occurs, the bag will inflate. After the reaction is complete, the mass of the bag, beaker, and products can be measured.

Note: Because the bag inflates, a mass loss of as much as 0.5–0.7 g will be found even if the bag is perfectly sealed, due to the effect of the buoyant force of air displaced. One way to avoid this error is to inflate the bag with air to the approximate volume it is expected to be filled by the CO^2 and seal it prior to measuring its mass the first time. In this way the volume change and the effects of the buoyant force it induces will be minimized. Then deflate the bag completely, reseal it, and allow the reactants to mix.

Answers to Analysis

1. Most students will observe that the mass change was greater in the first trial. The expected change in mass would be 0 g.

2. No. One of the products of the reaction is a gas, carbon dioxide. Since it escapes from the cup, its mass is not measured in the first test. Although students may not know that the reaction produces carbon dioxide, they can infer that some gas is produced by watching the bubbles that are formed.

3. The second test should have retained the gas within the plastic bag, so that the masses of all products could be determined.

Answers to Conclusions

4. No. The plastic bag may not have been sealed tight, allowing much of the gas produced by the reaction to escape, or one of the masses may have been measured incorrectly.

Chapter Resource File

- **Datasheet** Testing the Conservation of Mass **GENERAL**
- **Observation Lab** Measuring Density with a Hydrometer **BASIC**
- **CBL™ Probeware Lab** Comparing the Buoyancy of Different Objects **ADVANCED**

Continuation of Answers

Continuation of Answers from p. 37
Answers to Pre-Reading

2. Yes. Matter can undergo a physical change that alters its shape or form without changing its identity. In a chemical change, matter changes to become a new form of matter with different properties. Melting, evaporating, and freezing are physical changes. Combustion and corrosion are chemical changes.

Continuation of Answers from p. 52
Section 2 Review

Math Skills

5. $D = 454 \text{ g}/100.0 \text{ cm}^3 = 4.54 \text{ g/cm}^3$

6. $D = .36 \text{ g}/2500 \text{ cm}^3 = 1.4 \times 10^{-4} \text{ g/cm}^3$

7. Answers will vary. Physical properties could include a density less than that of water, enough strength to not bend or break when a load is placed on it, and insolubility in water. Chemical properties could include not reacting with water or not burning.

Continuation of Answers from p. 58
Section 3 Review

7. color change; bubbling; fizzing or foaming; production of heat, light, or sound

8. In a physical change, the material's particles do not change; in a chemical change, the particles recombine, and so have different properties.

PAPER OR PLASTIC AT THE GROCERY STORE?

The debate described in these pages is centered on a simple idea from Chapter 2—the selection of materials for specific uses based on their properties. Challenge students to list physical and chemical properties of paper and plastic grocery bags. The following are some possible responses:

Physical Properties

Paper: tears easily when wet; does not stretch

Plastic: does not break when wet; stretches

Chemical Properties

Paper: biodegradable

Plastic: breaks down when exposed to sunlight

Answers

1. Critiquing Viewpoints

Students' analyses of a single argument's weak points may vary. Be sure that students' responses clearly identify a weakness and respond in a way that provides support. Possible responses are listed below. (Note that students should analyze *only one of the following*.)

Jaclyn M.: Effective recycling requires that there is an effective use for the recycled materials. Not many businesses need reused bags.

But grocery stores might be better off without constantly re-ordering bags.

Eric S.: While paper is a renewable resource, the time it takes to grow a tree is very long.

---

viewpoints

Paper or Plastic at the Grocery Store?

As people focus more on the environment, there is a debate raging at the grocery store. It begins with a simple question asked at the checkout counter: "Paper or plastic?"

Some say that paper is a bad choice because making paper bags requires cutting down trees. Yet these bags are naturally biodegradable, and they recycle easily.

Others say that plastic is not a good choice because plastic bags are made from non-renewable petroleum products. But recent advances have made plastic bags that can break down when exposed to sunlight. Many stores collect used plastic bags and recycle them to make new ones.

How should people decide which bags to use? What do you think?

> FROM: Jaclyn M., Chicago, IL

I think people should choose paper bags because they can be recycled and reused. There should be a mandatory law that makes sure each community has a weekly recycling service for paper bags.

PAPER!

> FROM: Eric S., Rochester, MN

When it comes down to it, both types of bags can be recycled. However, as we know, not everybody recycles bags. Therefore, paper is a better choice because it is a renewable resource.

PLASTIC!

> FROM: Ashley A., Dyer, IN

Plastic is not necessarily better, but it is a lot more convenient. You can reuse plastic bags as garbage bags or bags to carry anything you need to take with you. Plastic is also easier to carry when you leave the store. Plastic bags don't get wet in the rain and break, causing you to drop your groceries on the ground.

> FROM: Christy M., Houston, TX

I believe we should use more plastic bags in grocery stores. By using paper, we are chopping down not only trees but also the homes of animals and plants.

66

But it is still better to use renewable resources.

Christy M.: Obtaining and refining petroleum for plastic bags can displace animals and plants from their habitats.

However, removal of trees can change an entire ecosystem, too.

Ashley A.: A plastic bag's handles can stretch and tear if it is too full.

But it's a simple matter to make sure you don't overload a plastic bag.

Andrew S.: People don't always make the right choices.

But if everyone were recycling their bags, fewer bags would get used in the first place.

Alice K.: Bringing canvas bags may not work if you don't always buy the same amount of groceries.

However, even if you use an extra bag every now and then, you're still saving more bags.

2. Critiquing Viewpoints

Students' analyses of strong points in opposing arguments may vary. Be sure that students' responses clearly identify the strong points and respond in a way that provides support for the opposite point of view. Possible responses are

> FROM: Andrew S., Bowling Green, KY

People should be able to use the bags they want. People that use paper bags should try to recycle them. People that use plastic bags should reuse them. We should be able to choose, as long as we recycle the bags in some way.

> FROM: Alicia K., Coral Springs, FL

Canvas bags would be a better choice than the paper or plastic bags used in stores. Canvas bags are made mostly of cotton, a very renewable resource, whereas paper bags are made from trees, and plastic bags are made from nonrenewable petroleum products.

BOTH or NEITHER!

> Your Turn

1. **Critiquing Viewpoints** Select one of the statements on this page that you *agree* with. Identify and explain at least one weak point in the statement. What would you say to respond to someone who brought up this weak point as a reason you were wrong?

2. **Critiquing Viewpoints** Select one of the statements on this page that you *disagree* with. Identify and explain at least one strong point in the statement. What would you say to respond to someone who brought up this point as a reason they were right?

3. **Creative Thinking** Make a list of at least 12 additional ways for people to reuse their plastic or paper bags.

4. **Life/Work Skills** Imagine that you are trying to decrease the number of bags being sent to the local landfill. Develop a presentation or a brochure that you could use to convince others to reuse or recycle their bags.

internet connect

TOPIC: Paper Vs. Plastic
GO TO: go.hrw.com
KEYWORD: HK4 Grocery Bag

Which kind of bag do you think is best to use? Why? Share your views on this issue and learn about other viewpoints at the HRW Web site.

67

Plastic Bag Reuses

1. Bring bags back to the store and reuse them.
2. Use bags to line household trash cans.
3. Use bags to carry lunches.
4. Use bags to carry gym clothes.
5. Use plastic bags to keep books and papers dry.
6. Use bags to organize clothes placed in storage.
7. Use bags to organize old papers, bills, and files.
8. Tie plastic bags shut to wrap food trash to prevent odors in the trash can.
9. Bring plastic bags to carry wet clothes and towels after swimming.
10. Tie plastic bags shut when camping to keep firewood dry.

4. Life/Work Skills

Evaluate students' presentations or brochures using a scale of 1–5 points for each of the following: creativity, factual accuracy, persuasiveness, attention to detail, and effort. (25 total possible points)

listed below. (Note that students should analyze *only one of the following*.)

Jaclyn M.: Mandatory recycling laws would increase recycling.

Community size may make such services hard to manage.

Eric S.: It's better to use renewable resources, because not everyone recycles bags.

Maybe people should change.

Christy M.: Logging can disrupt ecosystems.

However, paper bags are biodegradable.

Ashley A.: Plastic bags are sturdier than paper bags are when wet.

But should the environment be sacrificed for convenience's sake?

Andrew S.: Improve recycling of all bags instead of making laws.

But why continue to let people use a less effective kind of bag?

Alicia K.: Canvas bags can be reused for years without wearing out.

Canvas bags are paid for by the consumer instead of the seller.

3. **Creative Thinking**

Evaluate students' lists based on how creative and practical they are. Sample lists are provided below.

Paper Bag Reuses

1. Bring bags back to the store and reuse them.
2. Use bags to line household trash cans.
3. Use bags to carry lunches.
4. Use bags to carry gym clothes.
5. Keep unripe fruit in paper bags to ripen.
6. Use bags to organize clothes placed in storage.
7. Cut apart bags and use them as drop cloths for an art project.
8. Cut apart bags and place them on the garage floor to catch oil leaks from cars.
9. Use bags to organize old papers, bills, and files.
10. Cut apart bags and use them to make book covers for textbooks.

States of Matter
Chapter Planning Guide

PACING	CLASSROOM RESOURCES	LABS, ACTIVITIES, AND DEMONSTRATIONS
BLOCK 1 · 45 min pp. 68–69 **Chapter Opener**		SE **Activity 1,** p. 69 SE **Activity 2,** p. 69
BLOCKS 2 & 3 · 90 min pp. 70–78 **Section 1** Matter and Energy	CRF **Lesson Plan*** TT **Bellringer*** TT **States of Matter*** TT **Changes in State*** TM **Changes in State for Water***	TE **Opening Demonstration,** p. 70 GENERAL TE **Group Activity** Surface Tension, p. 72 GENERAL SE **Quick Activity** Hot or Cold?, p. 74 GENERAL CRF **Datasheets for In-Text Labs** Hot or Cold?* GENERAL
BLOCKS 4 & 5 · 90 min pp. 80–86 **Section 2** Fluids	CRF **Lesson Plan*** TT **Bellringer*** TM **Archimedes' Principle*** TT **Density***	TE **Opening Demonstration,** p. 80 TE **Group Activity,** p. 81 GENERAL TE **Group Activity** Making Models, p. 82 GENERAL
BLOCKS 6 & 7 · 90 min pp. 87–92 **Section 3** Behavior of Gases	CRF **Lesson Plan*** TT **Bellringer*** TM **Boyle's Law*** TM **Charles's Law*** TT **Concept Mapping***	TE **Opening Demonstration,** p. 87 SE **Quick Lab** Does temperature affect the volume of a balloon?, p. 89 GENERAL CRF **Datasheets for In-Text Labs** Does temperature affect the volume of a balloon?* GENERAL TE **Demonstration** Gas Pressure, p. 88 SE **Skills Practice Lab** Boiling and Freezing, pp. 98–99 ◆ GENERAL CRF **Datasheets for SE Skills Practice Lab** Boiling and Freezing* GENERAL CRF **CBL™ Probeware Lab** Investigating the Relationship Between Pressure and Volume* ◆ ADVANCED CRF **Observation Lab** Boyle's Law* ◆ BASIC

BLOCKS 8 & 9 · 90 min

Chapter Review and Assessment Resources

SE **Chapter Review,** pp. 94–97
CRF **Chapter Tests*** GENERAL
OSP **Test Generator**
CRF **Standardized Test Practice with Guided Reading Development***
CRF **Test Item Listing for ExamView® Test Generator***
OSP **Scoring Rubrics and Classroom Management Checklists**

Online Resources

Visit the HRW Web site for a variety of free resources related to the text. Just type in the keyword **HK4 SAM**.

Holt Science Spectrum: Physical Science: Online Edition

Students can access interactive problem solving help and active visual concept development with the *Holt Science Spectrum: Physical Science* Online Edition available at **www.hrw.com**.

student cnn News

cnnstudentnews.com

Find the latest news, lesson plans, and activities related to important scientific events.

KEY

TE	Teacher Edition	**CRF**	Chapter Resource File	*	Also on One-Stop Planner
SE	Student Edition	**TT**	Teaching Transparency	◆	Requires Advance Prep
OSP	One-Stop Planner	**TM**	Transparency Master		

PROBLEM SOLVING AND PRACTICE	SECTION REVIEW AND ASSESSMENT	STANDARDS CORRELATION
	CRF Pretest* [GENERAL]	
CRF **Cross-Disciplinary Worksheet** Integrating Physics—Plasma* [ADVANCED] **CRF** **Cross-Disciplinary Worksheet** Science and the Consumer—Dry Ice* [ADVANCED] **CRF** **Cross-Disciplinary Worksheet** Integrating Space Science—Our Changing Universe* [ADVANCED] **SE** **Science and the Consumer** Refrigeration, p. 79 **CRF** **Cross-Disciplinary Worksheet** Science and the Consumer—Refrigerants* [ADVANCED]	**TE** Quiz, p. 78 [BASIC] **SE** Section 1 Review, p. 78 **CRF** Concept Review* [GENERAL] **CRF** Quiz* [BASIC]	PS 2c, 2d, 2e, 5b, 5c UCP 1, 2, 3 ST 2d SAI 2 HNS 3
CRF **Cross-Disciplinary Worksheet** Real World Applications—Submarines* [ADVANCED] **CRF** **Cross-Disciplinary Worksheet** Integrating Biology—Density and Swim Bladders* [ADVANCED] **SE** **Math Skills** Pascal's principle, p. 84 **CRF** **Math Skills** Pascal's principle* [GENERAL]	**TE** Quiz, p. 86 [BASIC] **SE** Section 2 Review, p. 86 **CRF** Concept Review* [GENERAL] **CRF** Quiz* [BASIC]	PS 2e UCP 3, 4 ST 2d SAI 2 HNS 3
SE **Math Skills** Boyle's Law, p. 91 **CRF** **Math Skills** Boyle's Law* [GENERAL] **CRF** **Cross-Disciplinary Worksheet** Real World Applications—Gas Laws* [GENERAL] **SE** **Graphing Skills**, p. 93 **SE** **Science in Action** Plasma, pp. 100–101	**TE** Quiz, p. 92 [BASIC] **SE** Section 3 Review, p. 92 **CRF** Concept Review* [GENERAL] **CRF** Quiz* [BASIC]	PS 2e, 5c UCP 1, 3 SAI 1 HNS 3

www.scilinks.org

Topic: States of Matter
*Sci*Links code: HK4133

Topic: Law of Conservation of Energy
*Sci*Links code: HK4076

Topic: Refrigeration
*Sci*Links code: HK4121

Topic: Pascal's Principle
*Sci*Links code: HK4101

Topic: Gas Laws
*Sci*Links code: HK4062

Topic: Energy Transformations
*Sci*Links code: HK4049

Topic: Plasma
*Sci*Links code: HK4106

Technology Resources

One-Stop Planner
All of your printable resources and the Test Generator are on this convenient CD-ROM.

Overview

In this section, students learn about the kinetic theory of matter, the changes of state of matter, and the laws of conservation of mass and energy. This section then introduces the characteristics and behavior of fluids. Students learn about the relationship between a fluid's weight, buoyancy, and density. This section ends with the properties and behavior of gases, as well as Boyle's, Charles's, and Gay-Lussac's Laws.

Assessing Prior Knowledge

Be sure students understand the following concepts:

- scientific laws
- units of measurement
- using significant figures
- mass vs. weight
- density

MISCONCEPTION ///ALERT\\\

Science education research has identified the following misconceptions about matter:

- Students interpret "energy is neither created nor destroyed" to mean that energy is stored up and released in its original form.
- Students confuse heat and temperature and believe that temperature is a measurement of heat; they may not understand heating and cooling as processes of energy transfer.
- Students believe light objects float and heavy objects sink, and that mass is the factor determining whether an object sinks or floats.
- Students believe that matter is continuous, rather than particulate; they don't recognize intermolecular forces in solids, liquids, and gases.
- Students think that the addition of energy to a system always results in an increase in temperature, which is not so for phase changes.

States of Matter

Chapter Preview

1 Matter and Energy
Kinetic Theory
Energy's Role
Energy and Changes of State
Conservation of Mass and Energy

2 Fluids
Buoyant Force
Fluids and Pressure
Fluids in Motion

3 Behavior of Gases
Properties of Gases
Gas Laws

68

Standards Correlations

National Science Education Standards

The following descriptions summarize the National Science Standards that specifically relate to this chapter. For the full text of the standards, see the National Science Education Standards at the front of the book.

Section 1 Matter and Energy
Physical Science PS 2c, 2d, 2e, 5b, 5c
Unifying Concepts and Processes UCP 1, 2, 3
Science and Technology ST 2d
Science as Inquiry SAI 2
History and the Nature of Science HNS 3

Section 2 Fluids
Physical Science PS 2e
Unifying Concepts and Processes UCP 3, 4
Science and Technology ST 2d
Science as Inquiry SAI 2
History and the Nature of Science HNS 3

Section 3 Behavior of Gases
Physical Science PS 2e, 5c
Unifying Concepts and Processes UCP 1, 3
Science as Inquiry SAI 1
History and the Nature of Science HNS 3

Focus ACTIVITY

Background Do you notice something unique about the castle shown to the left? This castle is made of blocks of ice! Water exists in many forms, such as steam rising from a tea kettle, dew collecting on grass, crystals of frost forming on the windows in winter, and blocks of ice in an ice castle. But no matter what form you see, water always contains the same elements—hydrogen and oxygen.

In this chapter, you'll learn more about the many different properties of matter, such as the various forms water takes. Although you may not always be aware of them, changes of matter take place all around you. The atoms and molecules that make up all matter are in constant motion, whether the matter is a solid, a liquid, or a gas. This is true even in solid ice. Ice consists of water molecules held firmly in a rigid structure. Although they do not move about freely, ice molecules are able to vibrate back and forth.

Activity 1 Use the materials your teacher provides to make models of water molecules in ice.

Activity 2 Use the same materials to make models of liquid water molecules breaking away from ice crystals during melting.

🔲 internet connect

www.scilinks.org
Topic: States of Matter   SciLinks code: HK4133

SCiLINKS. Maintained by the National Science Teachers Association

Carbon dioxide, shown in this picture, changes directly from a solid to a gas under standard conditions.

Pre-Reading Questions

1. What three states of matter are you familiar with? Give examples of each.
2. Can you think of an example of matter changing from one state to another? What would cause a substance to change from one state to another?

69

Focus ACTIVITY

Background Have students discuss the importance of the properties of water in relation to life. For example, they might discuss the significance of water being liquid over such a wide range of temperatures, so that many areas of Earth are habitable. For further discussion, ask students to bring in newspaper and magazine articles and pictures that focus on the importance of water to life.
LS Interpersonal

Activity 1 Provide a wide assortment of materials (such as plastic foam balls, gumdrops, toothpicks, springs, tape, and dowels) and have groups of students make models that illustrate water molecules in ice. **LS Visual**

Activity 2 Have students use the same materials to make models that illustrate the process by which liquid water molecules break away from ice crystals during melting. The models should reflect the arrangements typical of sets of water molecules in both states. Have students compare their models from both activities. **LS Visual**

Answers to Pre-Reading Questions

1. solid, liquid, gas, and plasma; solids: glass, ice, plastic; liquids: water, milk, tea, mercury; gases: air, oxygen, helium; plasmas: lightning, solar wind, fire

2. Yes, matter can change from one state to another through processes such as melting, freezing, evaporation, condensation, and sublimation. Energy added to the matter causes bonds to break. Energy removed causes bonds to form.

Chapter Resource File

- **Pretest** GENERAL
- **Teaching Transparency Preview**
- **Answer Keys**

Focus

Overview

Before beginning this section, review with your students the Objectives listed in the Student Edition. This section reviews the kinetic theory of matter and compares the physical properties of solids, liquids, gases, and plasmas. The role of energy in changes of state is explained, as is temperature as a measure of average kinetic energy. The section concludes with an explanation of the laws of conservation of mass and energy.

Bellringer

Use the Bellringer transparency to prepare students for this section.

Motivate

Opening Demonstration —— GENERAL

(Time: Approximately 5 minutes)

Materials:

• balloon
• eye dropper
• vanilla extract

Before class, place five drops of vanilla extract into a balloon. Inflate the balloon and tie it shut. Have students smell the balloon.

Analysis

Ask students:

1. What do you smell at the surface of the balloon? (vanilla)

2. What does this smell tell about the particles of vanilla extract? (They move through the balloon's surface.) **LS Kinesthetic**

Matter and Energy

■ **KEY TERMS**

plasma
energy
thermal energy
evaporation
sublimation
condensation

OBJECTIVES

▶ **Summarize** the main points of the kinetic theory of matter.

▶ **Describe** how temperature relates to kinetic energy.

▶ **Describe** four common states of matter.

▶ **List** the different changes of state, and describe how particles behave in each state.

▶ **State** the laws of conservation of mass and conservation of energy, and explain how they apply to changes of state.

Figure 1

The ingredients in foods are chemicals. A skilled chef understands how the chemicals in foods interact and how changes of state affect cooking.

70

If you visit a restaurant kitchen, such as the one in *Figure 1,* you can smell the food cooking even if you are a long way from the stove. One way to explain this phenomenon is to make some assumptions. First, assume that the particles (atoms and molecules) within the food substances are always in motion and are constantly colliding. Second, assume that the particles move faster as the temperature rises. A theory based on these assumptions, called the kinetic theory of matter, can be used to explain things such as why you can smell food cooking from far away.

When foods are cooking, energy is transferred from the stove to the foods. As the temperature increases, some particles in the foods move very fast and actually spread through the air in the kitchen. In fact, the state, or physical form, of a substance is determined, partly by how its particles move.

Kinetic Theory

Here are the main points of the kinetic theory of matter:

▶ All matter is made of atoms and molecules that act like tiny particles.

▶ These tiny particles are always in motion. The higher the temperature of the substance, the faster the particles move.

▶ At the same temperature, more-massive (heavier) particles move slower than less massive (lighter) particles.

The kinetic theory helps you visualize the differences among the three common states of matter: solids, liquids, and gases.

Alternative Assessment —— GENERAL

Evaporation and Condensation Have students read this section and create a concept map that shows what happens during the evaporation and condensation cycle. The concept maps should also include where energy is absorbed or released in the cycle. **LS Visual**

Did You Know?

Absolute Zero Even the particles of a solid are constantly in motion. However, as matter is cooled to extremely cold temperatures, its particles move more slowly. Theoretically, matter can be cooled to a temperature at which all particle motion stops. This temperature is known as absolute zero, or 0 K (−273°C). A temperature of absolute zero has never been achieved in a laboratory.

Figure 2

Common States of Matter

A Here, the element sodium is shown as the solid metal.

B This is sodium melted as a liquid.

C Sodium exists as a gas in a sodium-vapor lamp.

The states of matter are physically different

The models for solids, liquids, and gases shown in *Figure 2* differ in the distances between the atoms or molecules and in how closely these particles are packed together. Particles in a solid, such as iron, are in fixed positions. In a liquid, such as cooking oil, the particles are closely packed, but they can slide past each other. Gas particles are in a constant state of motion and rarely stick together. Most matter found naturally on Earth is either a solid, a liquid, or a gas, although matter also exists in other states.

Solids have a definite shape and volume

Take an ice cube out of the freezer. The ice cube has the same volume and shape that it had in the ice tray. Unlike gases and liquids, a solid does not need a container in order to have a shape. The structure of a solid is rigid, and the particles have almost no freedom to change position. The particles in solids are held closely together by strong attractions, yet they vibrate.

Solids are often divided into two categories—crystalline and amorphous. *Crystalline solids* have an orderly arrangement of atoms or molecules. Examples of crystalline solids include iron, diamond, and ice. *Amorphous solids* are composed of atoms or molecules that are in no particular order. Each particle is in a particular place, but the particles are in no organized pattern. Examples of amorphous solids include rubber and wax. *Figure 3* and *Figure 4* illustrate the differences in these two types of solids.

Figure 3
Particles in a crystalline solid have an orderly arrangement.

Figure 4
The particles in an amorphous solid do not have an orderly arrangement.

Teach

Teaching Tip

Two-dimensional Models The drawings in **Figure 2** are two-dimensional models of the three principal states of matter. Be sure students note the spacing of the particles, especially liquid, which differs only slightly from solid.

MISCONCEPTION ALERT

Phase and State The terms *phase* and *state* are sometimes used interchangeably, but they do not have the same meaning. *State* refers to states of matter (solid, liquid, gas, plasma). *Phase* refers to a region of material with distinct boundaries and uniform properties for that region. For example, both corn oil and water are in the liquid state. A mixture of corn oil and water has two phases, the corn oil phase and the water phase.

SKILL BUILDER — GENERAL

Forms of Solids Provide students with a hand lens and samples of salt, flour, sugar, margarine or butter, and a rubber band. Give them time to observe and compare all of the samples. Encourage them to investigate and describe the visual differences between an amorphous solid and a crystalline solid.
LS Visual

MISCONCEPTION ALERT

Amorphous Solids A common misconception is that all solids are rigid, hard substances. Therefore, students may have difficulty understanding that amorphous solids are solids, and they may think they do not "hold their shape." Clarify that although the particles in an amorphous solid are not arranged in a definite pattern, each particle remains in position relative to the surrounding particles.

Chapter Resource File

• Lesson Plan

Transparencies

TT Bellringer

TT States of Matter

Group Activity GENERAL

Surface Tension

(Time: Approximately 10 minutes)

Materials:

• flour
• paper plate
• food coloring
• eye dropper
• a toothpick or common pin dipped in liquid dishwashing detergent
• a clean toothpick or pin

Safety Caution: *Students should wear safety goggles and aprons.*

Separate students into groups. Provide each group with the listed materials. Have students drop 1 to 2 drops of food coloring onto the flour. Ask them what they observe. (Surface tension causes beading of the water.)

Have students insert the toothpick with soap gently into one of the spheres and note any changes. (Spheres disappear and leave a wet spot on the powder.)

Have students insert the clean toothpick in a different sphere. Discuss any differences. (The clean toothpick should produce no changes in the sphere.) **LS Visual**

MISCONCEPTION ////ALERT

Gas Mass A common misconception is that gases do not have mass. Clarify that all gases have mass. For example, a cubic kilometer of air at sea level has a mass of about 1×10^9 kg.

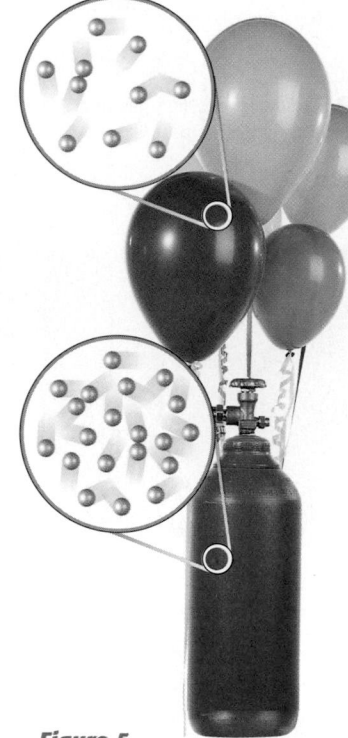

Figure 5

The particles of helium gas, He, in the cylinder are much closer together than the particles of the gas in the balloons.

▶ **plasma** a state of matter that starts as a gas and then becomes ionized

Figure 6

Auroras form when high-energy plasma collides with gas particles in the upper atmosphere.

72

Liquids change shape, not volume

Liquids have a definite volume, but they change shape. The particles of a liquid can slide past one another. And the particles in a liquid move more rapidly than those in a solid—fast enough to overcome the forces of attraction between them. This allows liquids to flow freely. As a result, the liquids are able to take the shape of the container they are put into. You see this every time you pour yourself a glass of juice. But even though liquids change shape, they do not easily change volume. The particles of a liquid are held close to one another and are in contact most of the time. Therefore, the volume of a liquid remains constant.

Another property of liquids is *surface tension*, the force acting on the particles at the surface of a liquid that causes a liquid, such as water, to form spherical drops.

Gases are free to spread in all directions

If you leave a jar of perfume open, particles of the liquid perfume will escape as gas, and you will smell it from across the room. Gas expands to fill the available space. And under standard conditions, particles of a gas move rapidly. For example, helium particles can travel 1200 m/s.

One cylinder of helium, as shown in *Figure 5,* can fill about 700 balloons. How is this possible? The volume of the cylinder is equal to the volume of only five inflated balloons. Gases change both their shape and volume. The particles of a gas move fast enough to break away from each other. In a gas, the amount of empty space between the particles changes. The helium atoms in the cylinder in *Figure 5,* for example, have been forced close together. But, as the helium fills the balloon, the atoms spread out, and the amount of empty space in the gas increases.

Plasma is the most common state of matter

Scientists estimate that 99% of the known matter in the universe, including the sun and other stars, is made of matter called **plasma.** Plasma is a state of matter that does not have a definite shape and in which the particles have broken apart.

Plasmas are similar to gases but have some properties that are different from the properties of gases. Plasmas conduct electric current, while gases do not. Electric and magnetic fields affect plasmas but do not affect gases. Natural plasmas are found in lightning, fire, and the aurora borealis, shown in *Figure 6.* The glow of a fluorescent light is caused by an artificial plasma, which is formed by passing electric currents through gases.

Did You Know?

Gels A gel is a liquid that has tiny particles of a solid suspended in it. Gels are best known for their elasticity. In a gel, the solid particles remain suspended, unaffected by gravity, giving gels their limited firmness. Examples of gels are gelatin, photographic film emulsions, and some kinds of toothpaste.

LIFE SCIENCE
CONNECTION

The tallest living organism is the giant sequoia tree, which grows primarily in California and can reach a height of 100 m. Surface tension helps water reach the top of these trees. As water evaporates from the leaves, a thin column of water molecules travels up through the tissues of the trunk and limbs. The surface tension of the water molecules, a result of cohesion, allows the molecules to be pulled to the top of the trees.

Figure 7
The particles in the steam have the most kinetic energy, but the ocean has the most total thermal energy because it contains the most particles.

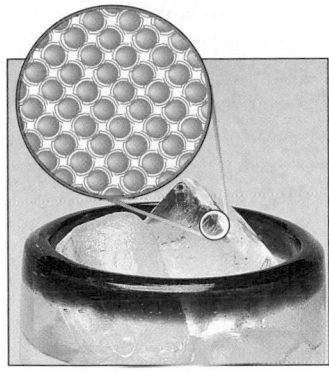

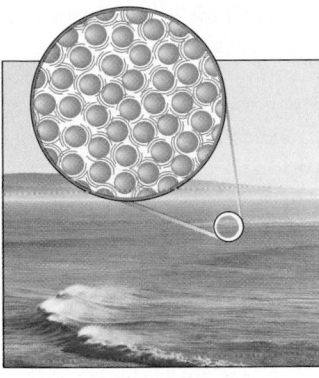

A The particles in an ice cube vibrate in fixed positions; therefore, they do not have a lot of kinetic energy.

B The particles in ocean water move around; therefore, they have more kinetic energy than the particles in an ice cube.

C The particles in steam move around rapidly; therefore, they have more kinetic energy than the particles in ocean water.

Energy's Role

What sources of energy would you use if the electricity were off? You might use candles for light and batteries to power a clock. Electricity, candles, and batteries are sources of energy. The food you eat is also a source of energy. Chemical reactions that release heat are another source of energy. You can think of **energy** as the ability to change or move matter. Later, you will learn how energy can be described as the ability to do work.

Thermal energy is the total kinetic energy of a substance

According to the kinetic theory, all matter is made of particles—atoms and molecules—that are constantly in motion. Because the particles are in motion, they have *kinetic energy*, or energy of motion. **Thermal energy** is the total kinetic energy of the particles that make up an object. The more kinetic energy the particles in the object have, the more thermal energy the object has. At higher temperatures, particles of matter move faster. The faster the particles move, the more kinetic energy they have, and the greater the object's thermal energy is. Thermal energy also depends on the number of particles in a substance. Look at **Figure 7.** Which substance do you think has the most thermal energy? The answer might surprise you.

▶ **energy** the capacity to do work

▶ **thermal energy** the kinetic energy of a substance's atoms

Teaching Tip
States of Matter Inform students that gas particles are approximately 10 times farther apart than the particles of a liquid or a solid. As a result, the volume of a substance as a gas is roughly 1000 ($10 \times 10 \times 10$) times the volume of the same amount of the substance as a liquid or a solid.

SKILL BUILDER — GENERAL
Reading Skills Make on the chalkboard three columns, labeled *heat*, *change*, and *move*. Based on what they have read, have students brainstorm to come up with a list of sources for each of these effects of energy. **LS** Logical

SKILL BUILDER — ADVANCED
Writing **Writing Skills** Explain to students that the thermal energy of Earth's oceans has a profound effect on climate and weather. An example of this is the phenomenon known as El Niño, and its counterpart, La Niña. Have students research these two phenomena and describe in a report or a poster how thermal energy is responsible for each. **LS** Verbal

Did You Know ?
Plasma Although solids, liquids, and gases are the most common states of matter on Earth, plasmas are the most common state of matter in the universe. Stars are made from plasma. The space between bodies of matter in the universe contains a thin plasma. In both cases, the particles in the plasmas have high kinetic energies. If scientists can find a way to contain a plasma without energy loss, the plasma could be a source for nuclear fusion, which produces great amounts of energy.

Chapter Resource File
• **Cross-Disciplinary Worksheet**
Integrating Physics—Plasma
ADVANCED

Quick ACTIVITY

Hot or Cold?

Materials (for class or small groups):
- bucket of warm water
- bucket of cold water
- bucket of hot water

Safety Caution: *Students should wear safety goggles and aprons for this activity.*

Teacher's Notes: Students should observe that the warm water felt warmer to the hand that had been in cold water, and cooler to the hand that had been in hot water. Students should conclude that their hands are unreliable temperature indicators. **LS Kinesthetic**

SKILL BUILDER — **BASIC**

Reading Skills Write the following on the board: *The temperature of boiling water is 100° on the Celsius scale and 212° on the Fahrenheit scale.* Then have students look at the following temperatures and decide whether they think that they are hot or cold: *60°F, 60°C, 37°F, 37°C, 0°C, 100°F, 70°F.* Discuss students' responses. **LS Logical**

SKILL BUILDER — **GENERAL**

Interpreting Visuals Have students look at **Figure 8.** Ask them: During which changes are the substances gaining energy? (melting, sublimation, and evaporation) Ask students how they are able to determine this. **LS Visual**

Quick ACTIVITY

Hot or Cold?

You will need three buckets: one with warm water, one with cold water, and one with hot water. **SAFETY:** Test a drop of the hot water to make sure it is not too hot.

Put both your hands into a bucket of warm water, and note how it feels. Now put one hand into a bucket of cold water and the other into a bucket of hot water. After a minute, take your hands out of the hot and cold water, and put them back in the warm water. Can you rely on your hands to determine temperature? Explain your observations.

Figure 8

This figure shows water undergoing five changes of state: freezing, melting, sublimation, evaporation, and condensation.

Temperature is a measure of average kinetic energy

Do you think of temperature as a measure of how hot or cold something is? Scientifically, temperature is a measure of the average kinetic energy of the particles in an object. The more kinetic energy the particles of an object have, the higher the temperature of the object is. Particles of matter are constantly moving, but they do not all move at the same speed. As a result, some particles have more kinetic energy than others have. So, when you measure an object's temperature, you measure the average kinetic energy of the particles in the object.

The temperature of a substance is not determined by how much of the substance you have. For example, a teapot contains more tea than a mug does, but the temperature, or average kinetic energy of the particles in the tea, is the same in both containers. However, the total kinetic energy of the particles in each container is different.

Energy and Changes of State

A change of state—the conversion of a substance from one physical form to another—is a physical change. The identity of a substance does not change during a change of state, but the energy of a substance does change. In *Figure 8,* the ice, liquid water, and steam are all the same substance—water, H_2O—but they all have different amounts of energy.

If energy is added to a substance, its particles move faster, and if energy is removed, its particles move slower. The temperature of a substance is a measure of its energy. Therefore, steam, for example, has a higher temperature than liquid water does, and the particles in steam have more energy than the particles in liquid water do. A transfer of energy known as *heat* causes the temperature of a substance to change, which can lead to a change of state.

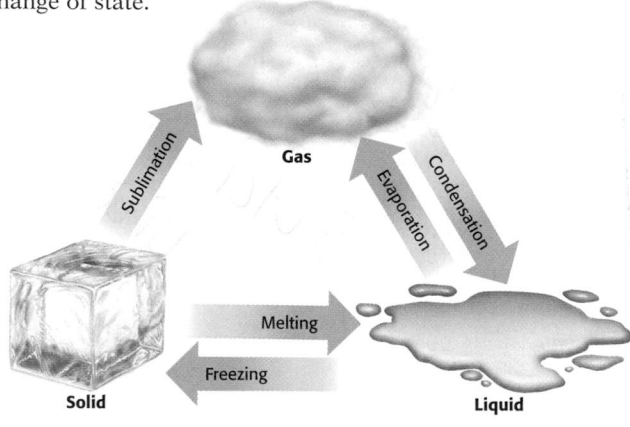

74

MISCONCEPTION ALERT

Heat and Temperature Students may confuse heat with temperature. Explain that heat is a transfer of energy from matter at a higher temperature to matter at a lower temperature. The form of the energy that is transferred is thermal energy.

INCLUSION Strategies

- *Learning Disabled* • *Attention Deficit Disorder*

Using **Figure 8,** ask students to create a class chart showing the four changes of state for water. Definitions for freezing, melting, evaporation, and condensation should be included in the chart. Additionally, the student can research and include the freezing, melting, and boiling points for water and alcohol to compare the differences for each liquid.

Some changes of state require energy

Changes, such as melting, that require energy are called *endothermic changes.* A solid changes to a liquid by melting. Heating a solid transfers energy to the atoms, which vibrate faster as they gain energy. Eventually, they break from their fixed positions, and the solid melts. The *melting point* is the temperature at which a substance changes from a solid to a liquid. The melting point of water is 0°C. Table salt has a melting point of 801°C.

Evaporation is the change of a substance from a liquid to a gas. Boiling is evaporation that occurs throughout a liquid at a specific temperature and pressure. The temperature at which a liquid boils is the liquid's *boiling point.* Like the melting point, the boiling point is a characteristic property of a substance. The boiling point of water at sea level is 100°C, and the boiling point of mercury is 357°C.

You can feel the effects of an energy change when you sweat. Energy from your body is transferred to sweat molecules as heat. When this transfer occurs, your body cools off. The molecules of sweat on your skin gain energy and move faster, as shown in ***Figure 9.*** Eventually, the fastest-moving molecules break away, and the sweat evaporates. Energy is needed to separate the particles of a liquid to form a gas.

Solids can also change to gases. ***Figure 10*** shows solid carbon dioxide undergoing **sublimation,** that is, the process by which a solid turns directly into a gas. Sometimes ice sublimes to form a gas. When left in the freezer for a while, ice cubes get smaller as the ice changes from a solid to a gas.

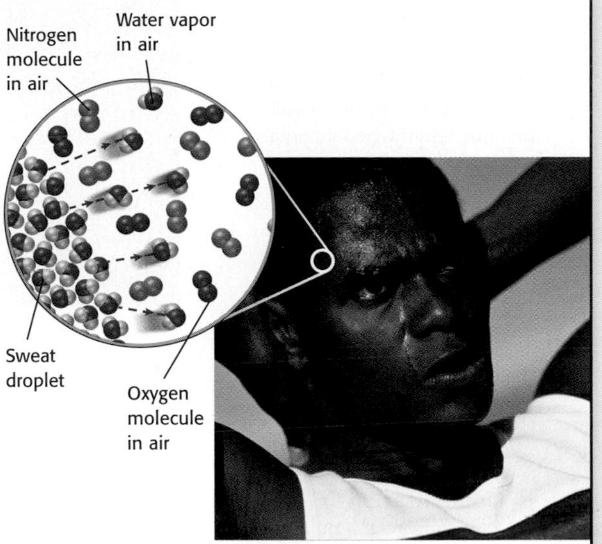

Nitrogen molecule in air
Water vapor in air
Sweat droplet
Oxygen molecule in air

Figure 9

Your body's heat provides the energy for sweat to evaporate.

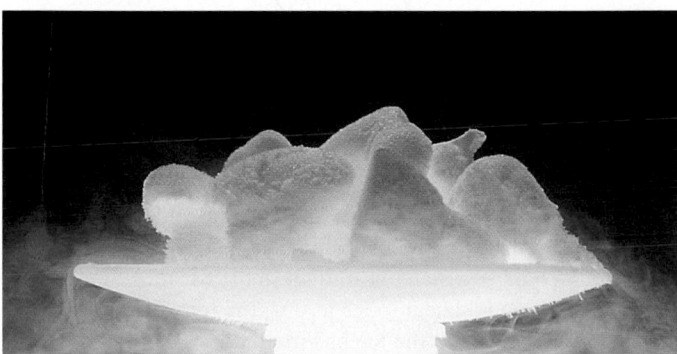

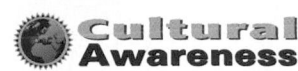

- **evaporation** the change of a substance from a liquid to a gas
- **sublimation** the process in which a solid changes directly into a gas (the term is sometimes also used for the reverse process)

Figure 10

Dry ice (solid carbon dioxide) sublimes to form gaseous carbon dioxide.

75

SKILL BUILDER — GENERAL

Reading Skills Discuss how different fabrics, colors, and styles of clothing can enhance evaporation and cooling, and point out how clothing varies among different cultures. People who live in warm climates wear light-weight and loose-fitting clothes that allow air to circulate near the body. This causes perspiration to evaporate, which cools the body. Examples of cultures that wear such clothing are those of the Middle East and Northern Africa. Separate students into groups to research other examples of how people dress for the climates where they live. Have them create posters showing the examples they find. **LS Interpersonal**

Teaching Tip

Sublimation Point out that many practical uses of sublimation exist. Dry ice (solid CO_2) is used when materials need to be kept cold but a liquid is undesirable. Wet clothes hung on a clothesline in below-freezing weather dry by sublimation. Certain foods and beverages, such as coffee, are "freeze-dried." But sublimation is not always helpful. If food is improperly wrapped or kept too long in a freezer, it dries out. This condition, called "freezer burn," is a result of sublimation.

Did You Know?

States of Water Water is the only substance that can be found as a solid, a liquid, and a gas at the normal surface temperatures on Earth.

Cultural Awareness

Synthesizing glass by adding energy to sand and melting it is not the only way to make glass. Glass can also be formed in nature by volcanic action or by a lightning bolt striking sand. American Indian cultures frequently used this naturally-formed glass to make arrowheads and spear points.

Chapter Resource File

- **Quick Activity Datasheet** Hot or Cold? GENERAL
- **Cross-Disciplinary Worksheet** Science and the Consumer—Dry Ice ADVANCED

Transparencies

TT Changes in State

Teaching Tip

Water Vapor Students might think that when they see a cloud they see water vapor. Emphasize that water vapor cannot be seen. If they see clouds or steam, they are actually seeing small droplets of water that have condensed from the water vapor. Even still, the amount of water vapor in the air is considerable. If all of the water vapor in Earth's atmosphere were to suddenly condense, the falling water would cover the United States to a depth of 8 m.

Ice and snow sometimes sublime directly to water vapor. If the air is dry after a snowstorm, a thin layer of ice on a street may disappear in a few hours, even though the temperature never rises above freezing.

MISCONCEPTION ALERT

Freezing Points Students may think that two different substances with the same mass and initial temperature release the same amount of energy in order to freeze, and that freezing only occurs at very cold temperatures. Point out that the amount of energy released by a substance during freezing and the substance's freezing point are physical properties unique to that substance. Depending on the substance, freezing can occur at high or low temperatures. For example, ammonia freezes at –77.7°C, and magnesium freezes at 650°C.

Figure 11
Gaseous water in the air will become liquid when it contacts a cool surface.

■ **condensation** the change of a substance from a gas to a liquid

If energy is added at 0°C, the ice will melt.

If energy is removed at 0°C, the liquid water will freeze.

Figure 12
Liquid water freezes at the same temperature that ice melts: 0°C.

Energy is released in some changes of state

When water vapor becomes a liquid, or when liquid water freezes to form ice, energy is released from the water to its surroundings. For example, the dew drops in *Figure 11* form as a result of **condensation,** which is the change of state from a liquid to a gas. During this energy transfer, the water molecules slow down. For a gas to become a liquid, large numbers of atoms clump together. Energy is released from the gas and the particles slow down.

Condensation sometimes takes place when a gas comes in contact with a cool surface. Have you ever noticed drops of water forming on the outside of a glass containing a cool drink? The *condensation point* of a substance is the temperature at which the gas becomes a liquid.

Energy is also released during freezing, which is the change of state from a liquid to a solid. The temperature at which a liquid changes into a solid is its *freezing point*. Freezing is the reverse of melting, so freezing and melting occur at the same temperature, as shown in *Figure 12*. For a liquid to freeze, the motion of its particles must slow down, and the attractions between the particles must overcome their motion. Like condensation, freezing is an *exothermic change* because energy is released from the substance as it changes state.

Temperature change verses change of state

When a substance loses or gains energy, either its temperature changes or its state changes. But the temperature of a substance does not change during a change of state, as shown in *Figure 13*. For example, if you add heat to ice at 0°C, the temperature will not rise until all the ice has melted.

The amount of gaseous water that air can hold decreases as the temperature of the air decreases.

As the air cools, some of the gaseous water condenses to form small drops of liquid water, which form clouds in the sky and fog near the ground.

Frederick McKinley Jones (1892–1961) was an African-American inventor with more than 60 patents to his name. One of his most important was a compact, automatic-refrigeration unit for trucks transporting produce. His invention, later adapted for trains, is still in use.

Figure 13

Changes of State for Water

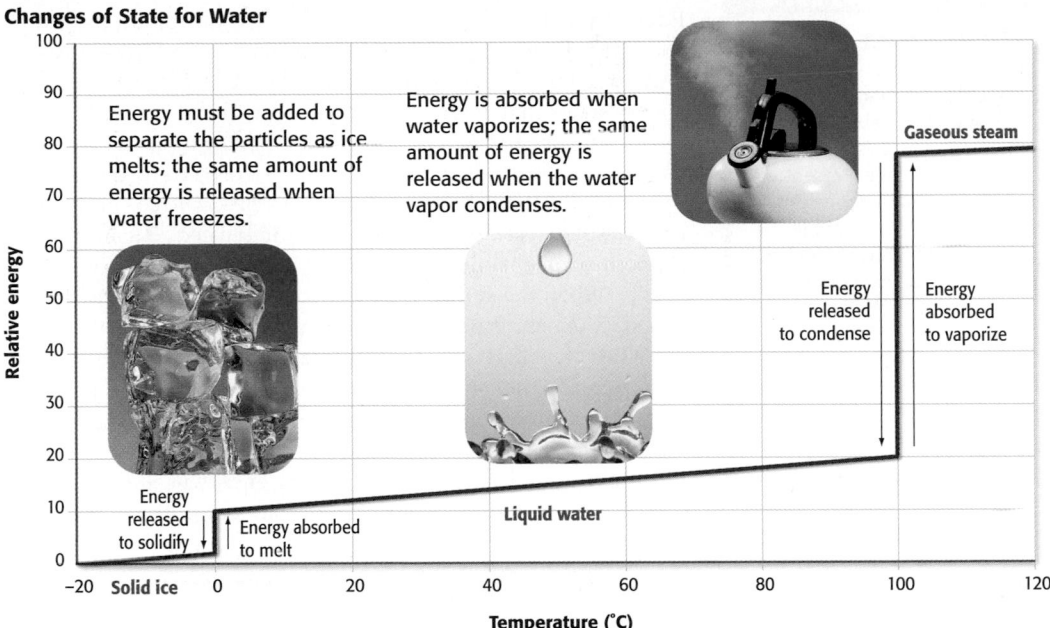

Energy must be added to separate the particles as ice melts; the same amount of energy is released when water freeezes.

Energy is absorbed when water vaporizes; the same amount of energy is released when the water vapor condenses.

Gaseous steam

Energy released to condense

Energy absorbed to vaporize

Liquid water

Energy released to solidify

Energy absorbed to melt

Solid ice

Relative energy

Temperature (°C)

Conservation of Mass and Energy

Look at the changes of state shown in *Figure 13*. Changing the energy of a substance can change the state of the substance, but it does not change the composition of a substance. Ice, water, and steam are all made of H_2O. When an ice cube melts, the mass of the liquid water is the same as the mass of the ice cube. When water boils, the number of molecules stays the same even as the liquid water loses volume. The mass of the steam is the same as the mass of the water that evaporated.

Mass cannot be created or destroyed

In chemical changes, as well as in physical changes, the total mass of matter stays the same before and after the change. Matter changes, but the total mass stays the same. The law of conservation of mass states that mass cannot be created or destroyed. For instance, when you burn a match, it seems to lose mass. The ash has less mass than the match. But there is also mass in the oxygen that reacts with the match, in the tiny smoke particles, and in the gases formed in the reaction. The total mass of the reactants (the match and oxygen) is the same as the total mass of the products (the ash, smoke, and gases).

internet connect

www.scilinks.org
Topic: Law of Conservation of Energy
SciLinks code: HK4076

SCiLINKS Maintained by the National Science Teachers Association

77

Integrating Space Science In astronomy, the term *nucleosynthesis* refers to the process by which all chemical elements are produced from hydrogen, the simplest element, by thermonuclear reactions within stars. During these reactions, a star forms elements from hydrogen and helium.

Close

Quiz   BASIC

Complete each statement with the correct term.

1. According to the ___ theory, all matter is made of particles that are in constant ___. (kinetic, motion)

2. The four states of matter are ___ ___ ___ ___. (solids, liquids, gases, plasmas)

3. The state of matter that changes shape but not volume is ___. (liquid)

4. The total kinetic energy of the particles of an object is ___. (thermal energy)

5. ___ is a measure of average kinetic energy. (temperature)

6. Condensation is a(n) ___ change because it involves a ___ of energy. (exothermic, release)

LS Logical

Chapter Resource File

- **Cross-Disciplinary Worksheet**
  Integrating Space Science—
  Our Changing Universe **ADVANCED**

- **Concept Review** **GENERAL**

- **Quiz** **BASIC**

INTEGRATING

SPACE SCIENCE
Studies of the chemical changes that stars and nebulae undergo are constantly adding to our knowledge. Present estimates are that hydrogen makes up more than 90% of the atoms in the universe and constitutes about 75% of the mass of the universe. Helium atoms make up most of the remainder of the mass. The total of all the other elements contributes very little to the total mass of the universe.

Energy cannot be created or destroyed

Energy may be converted to another form during a physical or chemical change, but the total amount of energy present before and after the change is the same. The law of conservation of energy states that energy cannot be created or destroyed.

Starting a car may seem to violate the law of conservation of energy. For the small amount of energy needed to turn the key in the ignition, a lot of energy results. But the car needs gasoline to run. Gasoline releases energy when it is burned. Because of the properties of chemicals that make up gasoline, it has stored energy. When the stored energy is considered, the energy before you start the car is equal to the energy that is produced.

When you drive a car, gasoline is burned to produce the energy needed to power the car. However, some of the energy from the gasoline is transferred to the surroundings as heat. That is why a car's engine gets hot. The total amount of energy released by the gasoline is equal to the energy used to move the car, plus the energy transferred to the surroundings as heat.

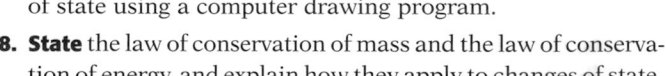

SECTION 1 REVIEW

SUMMARY

▶ The kinetic theory states that all matter is made of tiny, moving particles.

▶ Solids have a definite volume and shape. Liquids have a definite volume but a variable shape. Gases have a variable shape and volume.

▶ Thermal energy is the total kinetic energy of the particles of a substance.

▶ Temperature is a measure of average kinetic energy.

▶ A change of state is a physical change that requires or releases energy.

▶ Mass and energy are conserved in changes of state.

1. List three main points of the kinetic theory of matter.

2. Describe the relationship between temperature and kinetic energy.

3. State two examples for each of the four common states of matter.

4. Describe the following changes of state, and explain how particles behave in each state.
 a. freezing   **c.** sublimation
 b. boiling   **d.** melting

5. State whether energy is released or energy is required for the following changes of state to take place.
 a. melting   **c.** sublimation
 b. evaporation   **d.** condensation

6. Compare the shape and volume of solids, liquids, and gases.

7. Describe the role of energy when ice melts and when water vapor condenses to form liquid water. Portray each state of matter and the change of state using a computer drawing program. **COMPUTER SKILL**

8. State the law of conservation of mass and the law of conservation of energy, and explain how they apply to changes of state.

9. Critical Thinking Use the kinetic theory to explain how a dog could find you by your scent.

78

Answers to Section 1 Review

1. Atoms and molecules act like particles that are always in motion. Particles move faster at higher temperatures. At any temperature, heavier particles move slower.

2. Temperature is a direct measure of average kinetic energy.

3. solid—marble, brass; liquid—water, mercury; gas—oxygen, nitrogen; plasma—fire, lightning

4. a. freezing (liquid to solid)—particles slow down
 b. boiling (liquid to gas)—particles speed up
 c. sublimation (solid to gas)—particles speed up
 d. melting (solid to liquid)— particles speed up

5. a. endothermic
 b. endothermic
 c. endothermic
 d. exothermic

6. solids—definite volume and shape; liquids—definite volume but no definite shape; gases—neither definite volume nor definite shape

7. Drawings will vary. As energy is transferred to ice, the attraction between water molecules is broken and the ice melts. As water vapor gives up energy, attraction between molecules increases, and liquid water condenses.

See "Continuation of Answers" at the end of the chapter.

Refrigeration

Answers to Your Choice

1. Freon consists of chlorofluorocarbons (CFCs), which react with ozone (O_3). Ozone in the upper atmosphere absorbs harmful ultraviolet radiation from the sun. Depletion of this ozone by CFCs led to the banning of their manufacture in the United States.

2. As energy is transferred as heat to a refrigerant, the temperature of the refrigerant's surroundings drops. In order for these surroundings to become and remain cold, the refrigerant must evaporate at cold (low) temperatures.

LS Logical

Today the refrigerator is the most common kitchen appliance around. You can find one in more than 99% of American homes, but it wasn't always like this. Refrigerators didn't become widely available until 1916. Before that, people stored foods in "ice boxes" that held slabs of ice. This ice was often cut from frozen mountain lakes and carried long distances to be sold in cities. Ice was a luxury, but it played the same important role that refrigeration does today. Cold prevents the growth of bacteria that spoils food. If eaten, some bacteria can also cause sickness and even death.

responsible for several deaths in the 1920s. In 1928, a "miracle" refrigerant made from organic compounds called chlorofluorocarbons (CFCs) was introduced. Freon® not only was an efficient refrigerant but also was odorless and nonflammable.

Unfortunately, the "miracle" refrigerant was too good to be true. In the 1980s, scientists were alarmed to learn that the ozone layer, a protective layer of gases in Earth's atmosphere, was disappearing. Evidence linked CFCs to the ozone loss. Freon manufacture was banned in the United States, forcing companies to develop new, safer refrigerants.

Refrigerators keep food from spoiling

Ice Cold Science

Modern refrigeration systems use refrigerants to keep cool. A *refrigerant* is a substance that evaporates (and transfers energy) at a very low temperature. If a refrigerant can evaporate at a low temperature, it takes an input of less energy for the refrigerant to change from a liquid to a gas. On the back of a refrigerator, coiled tubes contain a refrigerant that alternately evaporates from a liquid into a gas and is then condensed back into a liquid. Through each cycle of evaporation, the refrigerant draws heat out of the air in the refrigerator, causing the air temperature in the refrigerator to go down.

Refrigerants

The first refrigerators used toxic gases such as ammonia as refrigerants, and leaking refrigerators were

Refrigerator Magnet

Scientists have recently discovered a way to use a magnet and the "magnetocaloric" element gadolinium to cool air. Magnetocaloric materials change temperature when in contact with a magnetic field. Because the device uses water instead of refrigerant, scientists are hopeful that the magnet refrigerator will one day be a safe, efficient form of refrigeration.

Your Choice

1. **Applying Knowledge** Why was the manufacture of Freon banned in the United States?

2. **Critical Thinking** Why is it important that a refrigerant evaporates at a low temperature?

internet connect

www.scilinks.org
Topic: Refrigeration   SciLinks code: HK4121

SCiLINKS® Maintained by the National Science Teachers Association

79

Focus

Overview

Before beginning this section, review with your students the Objectives listed in the Student Edition. In this section, students learn about buoyant force, and the relationship between an object's weight and its buoyant force. The role of density in an object's ability to float, and the concepts of fluid pressure and viscosity are explained. Students also learn three basic principles pertaining to fluids: Archimedes', Pascal's, and Bernoulli's principles.

⊙ Bellringer

Use the Bellringer transparency to prepare students for this section.

Motivate

Opening Demonstration —— GENERAL

(Time: Approximately 10 minutes)

Materials:
• graduated cylinder, 100 mL
• molasses, 20 mL
• cooking oil, 20 mL
• water, 20 mL

Add 20 mL each of molasses, cooking oil, and water to a 100 mL graduated cylinder. Either before students enter the classroom or while they observe, insert several different objects that will float on different layers. You might also try adding droplets of alcohol. Use the results of the demonstration to start a discussion about buoyant force. **LS Visual**

Fluids

▶ **KEY TERMS**

fluid
buoyant force
pressure
Archimedes' principle
pascal
Pascal's principle
viscosity

OBJECTIVES

▶ **Describe** the buoyant force and explain how it keeps objects afloat.
▶ **Define** Archimedes' principle.
▶ **Explain** the role of density in an object's ability to float.
▶ **State** and apply Pascal's principle.
▶ **State** and apply Bernoulli's principle.

What do liquids and gases have in common? Liquids and gases are states of matter that do not have a fixed shape. They have the ability to flow, and they are both referred to as **fluids.** Fluids are able to flow because their particles can move past each other easily. Fluids, especially air and water, play an important part in our lives. The properties of fluids allow huge ships to float, divers to explore the ocean depths, and jumbo jets to soar across the skies.

▶ **fluid** a nonsolid state of matter in which the atoms or molecules are free to move past each other, as in a gas or liquid

▶ **buoyant force** the upward force exerted on an object immersed in or floating on a fluid

Buoyant Force

Why doesn't a rubber duck sink to the bottom of a bath tub? Even if you push a rubber duck to the bottom, it will pop back to the surface when you release it. A force pushes the rubber duck to the top of the water. The force that pushes the duck up is the **buoyant force**—the upward force that fluids exert on matter. When you float on an air mattress in a swimming pool, the buoyant force keeps you and the air mattress afloat. A rubber duck and a large steel ship, such as the one shown in **Figure 14,** both float because they are less dense than the water that surrounds them and because the buoyant force pushes against them to keep them afloat.

Figure 14

Despite its large size and mass, this ship is able to float because its density is less than that of the water and because the buoyant force keeps it afloat.

80

MISCONCEPTION ///ALERT\\\

Buoyancy Students may think that all light objects will float, that all heavy objects sink, and that objects that float on water will float on any liquid. Clarify that whether or not an object floats depends on its density and the density of the fluid in which it is immersed. Only objects that are less dense than the fluid will float.

● INCLUSION Strategies

• **Attention Deficit Disorder** • **Learning Disabled**

Have students fill a small, clear plastic cup to the brim with water. Next, have students predict how many paperclips can be placed in the cup before the water overflows. Have students record their prediction, the actual number of paperclips, and the definition of surface tension of water. Additionally, you can ask students to prepare a glass of ice water to the brim and predict whether or not the water will overflow as the ice melts.

Buoyancy explains why objects float

The buoyant force, which keeps the ice in *Figure 15* floating, is a result of pressure. All fluids exert **pressure,** which is the amount of force exerted on a given area. The pressure of all fluids, including water, increases as the depth increases. The water exerts fluid pressure on all sides of each piece of ice. The pressure exerted horizontally on one side of the ice is equal to the pressure exerted horizontally on the opposite side. These equal pressures cancel one another. The only fluid pressures affecting the pieces of ice are above and below. Because pressure increases with depth, the pressure below the ice is greater than the pressure on top of the ice. Therefore, the water exerts a net upward force—the buoyant force—on the ice above it. Because the buoyant force is greater than the weight of the ice, the ice floats.

Determining buoyant force

Archimedes, a Greek mathematician in the third century BCE, discovered a method for determining buoyant force. **Archimedes' principle** states that the buoyant force on an object in a fluid is an upward force equal to the weight of the fluid that the object displaces. For example, imagine that you put a brick in a container of water, as shown in *Figure 16.* A spout on the side of the container at the water's surface allows water to flow out of the container. As the object sinks, the water rises and flows through the spout into another container. The total volume of water that collects in the smaller container is the displaced volume of water from the larger container. The weight of the displaced fluid is equal to the buoyant force acting on the brick. An object floats only when it displaces a volume of fluid that has a weight equal to the object's weight—that is, an object floats if the buoyant force on the object is equal to the object's weight.

Figure 16

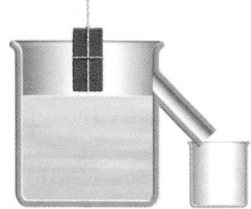

A An object is lowered into a container of water.

B The object displaces water, which flows into a smaller container.

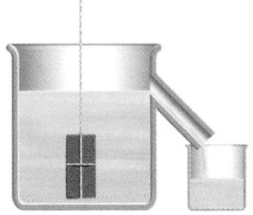

C When the object is completely submerged, the volume of the displaced water equals the volume of the object.

Figure 15

Ice floats in water because it is less dense than water and because of the upward buoyant force on the ice.

▶ **pressure** the amount of force exerted per unit area of a surface

▶ **Archimedes' principle** the principle that states that the buoyant force on an object in a fluid is an upward force equal to the weight of the volume of fluid that the object displaces

Historical Perspective
Archimedes and the Crown of Gold

Archimedes, a scientist and mathematician of ancient Greece, was born in Syracuse on the island of Sicily. According to legend, the king of Syracuse suspected that his crown was not pure gold, and he asked Archimedes to determine if it was pure gold without damaging it. When Archimedes figured out how to test the crown's authenticity using the buoyancy principle, he is reported to have exclaimed, "Eureka!" Archimedes submerged the crown and measured the increase in water level. He then repeated the procedure using an equal weight of pure gold, and he noted that the water level did not increase as much. He concluded that the goldsmith had substituted silver, which is cheaper and less dense, for the gold in the crown.

Teaching Tip

Buoyant Forces and Gases Use buoyant forces to explain why a helium balloon eventually will stop rising in the air. Helium is less dense than air, so a balloon filled with helium weighs less than the same volume of air. Air exerts a buoyant force on the balloon and pushes it up. The balloon will continue to rise as long as the weight of the displaced air is greater than the weight of the balloon. When the weight of the balloon equals the weight of the displaced air, the balloon will stop rising.

INTEGRATING

BIOLOGY Have interested students investigate the physical adaptations that enable sea organisms to utilize buoyant force. Have them select an organism and write a report or create a poster or presentation describing it. **LS Verbal**

Group Activity —— GENERAL

Making Models Have groups of students design models of a hot-air balloon. Beforehand, discuss how heating the air inside the balloon reduces the balloon's overall density. For their models, provide students with tissue paper, tape, glue, string, and other materials. To evaluate their models, have students hold them in place and fill them with hot air from a hair dryer, then release them to see if they fly. **LS Kinesthetic**

Figure 17
Helium in a balloon floats in air for the same reason a duck floats in water—the helium and the duck are less dense than the surrounding fluid.

INTEGRATING

BIOLOGY
Some fish can adjust their density so that they can stay at a certain water depth. Most fish have an organ called a *swim bladder,* which is filled with gases. The inflated swim bladder increases the fish's volume, decreases its overall density, and keeps it from sinking. The fish's nervous system controls the amount of gas in the bladder according to the fish's depth in the water. Some fish, such as sharks, do not have a swim bladder, so they must swim constantly to keep from sinking.

An object will float or sink based on its density

By knowing the density of a substance, you can determine if the substance will float or sink. For example, the density of a brick is 1.9 g/cm^3, and the density of water is 1.00 g/cm^3. The brick will sink because it is denser than the water.

One substance that is less dense than air is helium, a gas. Helium is about seven times less dense than air. A given volume of helium displaces a volume of air that is much heavier, so helium floats. That is why helium is used in airships and parade balloons, such as the one shown in *Figure 17.*

Steel is almost eight times denser than water. And yet huge steel ships cruise the oceans with ease, and they even carry very heavy loads. But hold on! Substances that are denser than water will sink in water. So, how does a steel ship float?

The shape of the ship allows it to float. Imagine a ship that was just a big block of steel, as shown in *Figure 18.* If you put that steel block into water, it would sink because it is denser than water. Ships are built with a hollow shape, as shown below. The amount of steel is the same, but the hollow shape decreases the boat's density. Water is denser than the hollow boat, so the boat floats.

Figure 18
A A block of steel is denser than water, so it sinks.

B Shaping the block into a hollow form increases the volume occupied by the same mass, which results in a reduced overall density. The ship floats because it is less dense than water.

Alternative Assessment —— GENERAL

Lifejacket Buoyancy Ask students to apply what they have learned to explain how a lifejacket helps prevent boaters who fall out of a boat from sinking. (Most lifejackets are made from porous material filled with air, which greatly increases the wearer's overall volume without greatly increasing his or her mass. Thus, the wearer's overall density decreases, enabling him or her to float.) **LS Logical**

GEOLOGY
CONNECTION

The rock that makes up the Earth's continents is about 15 percent less dense than the molten mantle rock below it. Because of this difference in densities, the continents are "floating" on the mantle.

Fluids and Pressure

You probably have heard the terms *air pressure, water pressure,* and *blood pressure.* Air, water, and blood are all fluids, and all fluids exert pressure. So, what is pressure? For instance, when you pump up a bicycle tire, you push air into the tire. Inside the tire, tiny air particles are constantly pushing against each other and against the walls of the tire, as shown in **Figure 19.** The more air you pump into the tire, the more the air particles push against the inside of the tire, and the greater the pressure against the tire is. Pressure can be calculated by dividing force by the area over which the force is exerted:

$$pressure = \frac{force}{area}$$

The SI unit for pressure is the **pascal.** One pascal (Pa) is the force of one newton exerted over an area of one square meter (1 N/m^2). You will learn more about newtons, but remember that a newton is a measurement of force. Weight is a force, and an object's weight can be given in newtons.

When you blow a soap bubble, you blow in only one direction. So, why does the bubble get rounder as you blow, instead of longer? The shape of the bubble is due partly to an important property of fluids: fluids exert pressure evenly in all directions. The air you blow into the bubble exerts pressure evenly in all directions, so the bubble expands in all directions and creates a round sphere.

Figure 19
The force of air particles inside the tire creates pressure, which keeps the tire inflated.

▶ **pascal** the SI unit of pressure; equal to the force of 1 N exerted over an area of 1 m^2 (abbreviation, Pa)

REAL WORLD APPLICATIONS

Density on the Move A submarine is a type of ship that can travel both on the surface of the water and underwater. Submarines have special ballast tanks that control their buoyancy. When the submarine dives, the tanks can be opened to allow sea water to flow in. This water adds mass and increases the submarine's density, so the submarine can descend into the ocean. The crew can control the amount of water taken in to control the submarine's depth. To bring the submarine through the water and to the surface,

compressed air is blown into the ballast tanks to force the water out. The first submarine, *The Turtle,* was used in 1776 against British warships during the American War of Independence. It was a one-person, hand-powered, wooden vessel. Most modern submarines are built of metals and use nuclear power, which enables them to remain submerged almost indefinitely.

Applying Information
1. Identify the advantages of using metals instead of wood in the construction of today's submarines.

83

Did You Know ?

Personal Weight The air in a large room in your house weighs about as much as an average adult male (about 737 N).

Teaching Tip

Pressure Students may assume that calculations involving pressure only involve the force of a fluid. Explain that because weight is a measure of gravitational force, anything that has weight exerts pressure. Thus, a crate or other object on a floor exerts pressure on the floor beneath it.

Math Skills

Additional Example

Pascal's Principle The pistons of a hydraulic lift have areas of 1.55 m^2 and 0.0065 m^2. How much force must be applied to the smaller piston if the larger piston is to lift a 45 000 N vehicle?

Answer

$F_2 = F_1A_2/A_1$

$F_2 = (45\ 000\ N)(0.0065\ m^2)/1.55\ m^2$

$F_2 = 190\ N$

LS Logical

Practice

1. $F_1/A_1 = F_2/A_2$

$F_2 = F_1A_2/A_1 = (150\ N)$
$(3.0\ cm^2)/0.5\ cm^2 = 900\ N$

LS Logical

internet connect

www.scilinks.org
Topic: Pascal's
Principle
SciLinks code: HK4101

SC*LINKS* Maintained by the National Science Teachers Association

▶ **Pascal's principle** the principle that states that a fluid in equilibrium contained in a vessel exerts a pressure of equal intensity in all directions

Practice HINT

Pressure, Force, and Area
The pressure equation

$$pressure = \frac{force}{area}$$

can be used to find pressure or can be rearranged to find force or area.

$$force = (pressure)(area)$$

$$area = \frac{force}{pressure}$$

Pascal's Principle

Have you ever squeezed one end of a tube of paint? Paint usually comes out the opposite end. When you squeeze the sides of the tube, the pressure you apply is transmitted throughout the paint. So, the increased pressure near the open end of the tube forces the paint out. This phenomenon is explained by Pascal's principle, which was named for the 17th-century scientist who discovered it. **Pascal's principle** states that a change in pressure at any point in an enclosed fluid will be transmitted equally to all parts of the fluid. Mathematically, Pascal's principle is stated as $p_1 = p_2$ or $pressure_1 = pressure_2$.

Math Skills

Pascal's principle
A hydraulic lift, shown in **Figure 20,** makes use of Pascal's principle, to lift a 19,000 N car. If the area of the small piston (A_1) equals 10.5 cm^2 and the area of the large piston (A_2) equals 400 cm^2, what force needs to be exerted on the small piston to lift the car?

1 **List the given and unknown values.**
Given: $F_2 = 19{,}000\ N$
$A_1 = 10.5\ cm^2$
$A_2 = 400\ cm^2$

Unknown: F_1

2 **Write the equation for Pascal's principle.**
According to Pascal's principle, $p_1 = p_2$

$$\frac{F_1}{A_1} = \frac{F_2}{A_2} \qquad F_1 = \frac{(F_2)(A_1)}{A_2}$$

3 **Insert the known values into the equation, and solve.**

$$F_1 = \frac{(19{,}000\ N)(10.5\ cm^2)}{400\ cm^2}$$

$$F_1 = 499\ N$$

Practice

Pascal's principle

1. In a car's liquid-filled hydraulic brake system, the master cylinder has an area of 0.5 cm^2, and the wheel cylinders each have an area of 3.0 cm^2. If a force of 150 N is applied to the master cylinder by the brake pedal, what force does each wheel cylinder exert on its brake pad?

84

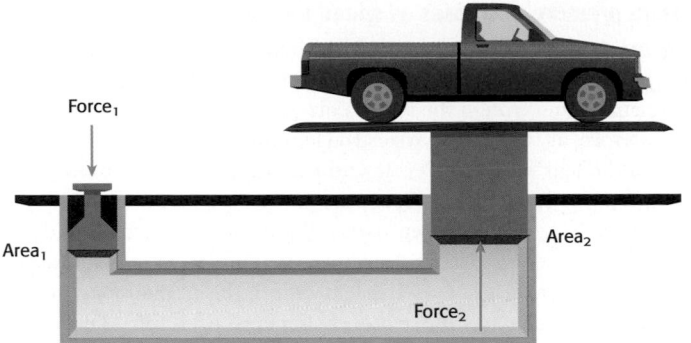

Force₁

Area₁

Area₂

Force₂

Figure 20

Because the pressure is the same on both sides of the enclosed fluid in a hydraulic lift, a small force on the smaller area (left) produces a much larger force on the larger area (right).

Hydraulic devices are based on Pascal's principle

Devices that use liquids to transmit pressure from one point to another are called *hydraulic devices*. Hydraulic devices use liquids because liquids cannot be compressed, or squeezed, into a much smaller space. This property allows liquids to transmit pressure more efficiently than gases, which can be compressed.

Hydraulic devices can multiply forces. For example, in *Figure 20*, a small downward force (F_1) is applied to a small area. This force exerts pressure on the liquid in the device, such as oil. According to Pascal's principle, this pressure is transmitted equally to a larger area, where it creates a force (F_2) larger than the initial force. Thus, the initial force can be multiplied many times.

Fluids in Motion

Examples of moving fluids include liquids flowing through pipes and air moving as wind. Have you ever used a garden hose? What happens when you place your thumb over the end of the hose? Your thumb blocks some of the area through which the water flows out of the hose, so the water exits at a faster speed. Fluids move faster through smaller areas than through larger areas, if the overall flow rate remains constant. Fluid speed is faster in a narrow pipe and slower in a wider pipe.

Viscosity is resistance to flow

Liquids vary in the rate at which they flow. For example, honey flows more slowly than lemonade. **Viscosity** is a liquid's resistance to flow. In general, the stronger the attraction between a liquid's particles the more viscous the liquid is. Honey flows more slowly than lemonade because it has a higher viscosity than lemonade. *Figure 21* shows a liquid that has a high viscosity.

Figure 21

The honey shown above has a higher viscosity than water.

▶ **viscosity** the resistance of a gas or liquid to flow

85

Did You Know ?

Pascal's Principle Each time you squeeze a tube of toothpaste, you experience Pascal's principle in action. The pressure you apply by squeezing the sides of the tube is transmitted throughout the toothpaste. The increased pressure near the open end of the tube forces the paste out and onto your toothbrush.

Historical Perspective

Pascal's Achievements Blaise Pascal (1623–1662), born in Clermont-Ferrand, France, published a geometry book when he was 16, and invented a mechanical calculator at the age of 19. Working with the mathematician Pierre Fermat, Pascal laid the foundation for the study of probability and statistics. In physics, he discovered that any pressure exerted on a fluid in a closed container would be transmitted throughout the fluid. Pascal eventually abandoned mathematics and physics for meditation and religious writing. He died at the age of 39.

Close

Quiz — BASIC

Fill in each blank with the correct term.

1. The tendency of a less dense substance to float in a denser liquid is due to ___. (buoyant force)

2. The amount of force exerted on a given area is called ___. (pressure)

3. The force of 1 N exerted over an area of 1 square meter equals 1 ___. (pascal)

4. ___ is a resistance to flow (viscosity)

5. ___ states that the buoyant force on an object immersed in a fluid equals the weight of the fluid that the object displaces. (Archimedes' Principle)

LS Logical

Chapter Resource File

- **Concept Review** GENERAL
- **Quiz** BASIC

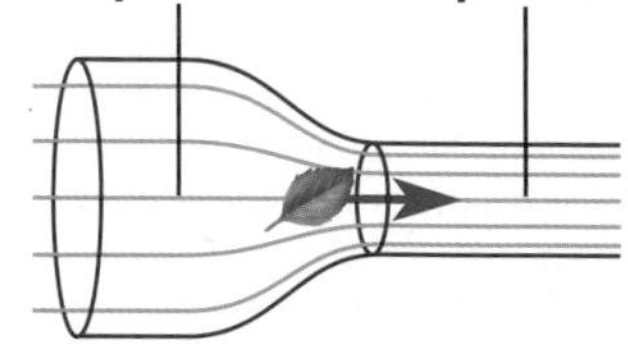

Fluid pressure decreases as speed increases

Figure 22 shows a water-logged leaf being carried along by water in a pipe. The water will move faster through the narrow part of the pipe than through the wider part, which is a property of fluids. Therefore, as the water carries the leaf into the narrow part of the pipe, the leaf moves faster. If you measure the pressure at point 1 and point 2, labeled in **Figure 22,** you would find that the water pressure in front of the leaf is less than the pressure behind the leaf. The pressure difference causes the leaf and the water around it to accelerate as the leaf enters the narrow part of the tube. This behavior illustrates a general principle, known as *Bernoulli's principle*, which states that as *the speed of a moving fluid increases, the pressure of the moving fluid decreases.* This property of moving fluids was first described in the 18th century by Daniel Bernoulli, a Swiss mathematician.

point 1   point 2

Figure 22

As a leaf passes through the drainage pipe, it speeds up. The water pressure on the right is less than the pressure on the left.

SECTION 2 REVIEW

SUMMARY

▶ Gases and liquids are fluids.

▶ Buoyancy is the tendency of a less dense substance to float in a denser liquid; buoyant force is the upward force exerted by fluids.

▶ Archimedes' principle states that the buoyant force on an object equals the weight of the fluid displaced by the object.

▶ Pressure is a force exerted on a given area; fluids exert pressure equally in all directions.

▶ Pascal's principle states that a change in pressure at any point in an enclosed fluid will be transmitted equally to all parts of the fluid.

▶ Bernoulli's principle states that fluid pressure decreases as the speed of a moving fluid increases.

1. Explain how differences in fluid pressure create buoyant force on an object.

2. State Archimedes' principle and give an example of how you could determine a buoyant force.

3. State Pascal's principle, and give an example of its use.

4. Compare the viscosity of milk and molasses.

5. Define the term *fluids*. What does Bernoulli's principle state about fluids?

6. Critical Thinking Two ships in a flowing river are sailing side-by-side with only a narrow space between them.
   a. What happens to the fluid speed between the two boats?
   b. What happens to the pressure between the boats?
   c. How could this change lead to a collision of the boats?

Math Skills

7. A water bed that has an area of 3.75 m^2 weighs 1025 N. Find the pressure that the water bed exerts on the floor.

8. An object weighs 20 N. It displaces a volume of water that weighs 15 N. (a) What is the buoyant force on the object? (b) Will the object float or sink? Explain.

9. Iron has a density of 7.9 g/cm^3. Mercury has a density of 13.6 g/cm^3. Will iron float or sink in mercury? Explain.

86

Answers to Section 2 Review

1. Water pressure is exerted on all sides of an object. The horizontal pressures cancel each other out. The pressure exerted at the bottom is greater than that exerted at the top because pressure increases with depth. This creates an overall upward, or buoyant, force.

2. The buoyant force on an object immersed in a fluid equals the weight of the fluid that the object displaces. Weigh the fluid displaced by an immersed object to find buoyant force.

3. Pressure applied to a fluid in a closed container is transmitted equally throughout the fluid. The principle is used in hydraulic lifts.

4. Molasses has a higher viscosity than milk.

5. Fluids are materials that flow and take the shape of their containers. Fluid pressure decreases as fluid speed increases.

6. a. It increases.
   b. It decreases.
   c. As fluid speed between the boats increases, fluid pressure decreases. The pressure on the outer sides of the boats exceeds the pressure between them, and so pushes the boats together.

See "Continuation of Answers" at the end of the chapter.

Behavior of Gases

OBJECTIVES

▶ **Explain** how gases differ from solids and liquids.

▶ **State and explain** the following gas laws: Boyle's law, Charles's law, and Gay-Lussac's law.

▶ **Describe** the relationship between gas pressure, temperature, and volume.

▶ **KEY TERMS**
Boyle's law
Charles's law
Gay-Lussac's law

Focus

Overview

Before beginning this section, review with your students the Objectives listed in the Student Edition. This section begins by reviewing the properties of gases and the ways in which gases differ from solids and liquids. Students learn the three basic gas laws: Boyle's Law, Charles's Law, and Gay-Lussac's Law, and perform calculations involving Boyle's Law.

🔔 Bellringer

Use the Bellringer transparency to prepare students for this section.

Because many gases are colorless and odorless, it is easy to forget that they exist. But, every day you are surrounded by gases. Earth's atmosphere is a gaseous mixture of elements and compounds. Some examples of gases in Earth's atmosphere are nitrogen, oxygen, argon, helium, and carbon dioxide, as well as methane, neon and krypton. In the study of chemistry, as in everyday life, gases are very important. In this section, you will learn how pressure, volume, and temperature affect the behavior of gases.

Properties of Gases

As you have already learned, the properties of gases are unique. Some important properties of gases are listed below.

▶ Gases have no definite shape or volume, and they expand to completely fill their container, as shown in *Figure 23.*

▶ Gas particles move rapidly in all directions.

▶ Gases are fluids.

▶ Gas molecules are in constant motion, and they frequently collide with one another and with the walls of their container.

▶ Gases have a very low density because their particles are so far apart. Because of this property, gases are used to inflate tires and balloons.

▶ Gases are compressible.

▶ Gases spread out easily and mix with one another. Unlike solids and liquids, gases are mostly empty space.

Figure 23
As you can see in this photo of chlorine gas, gases take the shape of their container.

Motivate

Opening Demonstration ── GENERAL

(Time: Approximately 5 minutes)

Materials:
• beaker, 1 L
• candle
• dry ice
• flask
• matches
• paper
• warm water

Demonstrate the fluid property of a gas using CO_2. Place a small piece of dry ice and a little warm water in a flask. As the container fills with CO_2, place the candle in the bottom of the large beaker and light the candle. Form a trough from a folded sheet of paper and pour the CO_2 down the trough into the beaker, extinguishing the candle. If the gas seems foggy, remind students that CO_2 is colorless. Water droplets suspended in CO_2 gas create the cloudy appearance. **LS Visual**

87

Historical Perspective

Demonstrating Air Pressure The effects of air pressure are noticed only if air pressure differs from one area to another. In the 1600s, Otto von Guericke, a German scientist who invented an air pump, demonstrated the effect of differences in air pressure to the King of Prussia. He placed two hemispherical shells together and pumped the air out of the resulting sphere. The atmospheric pressure on the outside of the hollow sphere was so much greater than the pressure inside the sphere that teams of horses were unable to pull the two halves apart. When Von Guericke opened a valve in the sphere to let air in and equalize the pressure, the hemispheres came apart easily.

Chapter Resource File

• Lesson Plan

Transparencies

TT Bellringer

Teach

Demonstration — GENERAL

Gas Pressure

(Time: Approximately 10 minutes)

Materials:

• baking soda, 4 teaspoons
• small plastic containers, 2 of different sizes with push-on lids
• vinegar, 4 tablespoons

Safety Caution: *Be sure to wear safety goggles and an apron when doing this demonstration.*

Place two teaspoonfuls of baking soda and two tablespoonfuls of vinegar into the smaller container. Quickly snap the lid in place. Shake the container once, and then leave it on the desk. The lid should pop off within seconds. Ask students why this happens. (Pressure from the gas caused the lid to pop off.)

Repeat the demonstration using the larger container, but use the same amount of reactants. It should take longer for the lid to pop off. Have students discuss why this is the case. (More gas molecules are needed to create the same pressure in a larger volume.) **LS Logical**

Did You Know?

The ability of gas particles to move and the relative sizes of gas particles have practical applications. Garrett Morgan was an African American born in Kentucky in 1875. Although he had only six years of formal education, he was an inventor of many useful products. For example, he invented a gas mask that allowed air, but not harmful gases, to pass through it. The mask received little recognition until Morgan used it to rescue 30 workers who were trapped underground in a tunnel that contained poisonous gases.

Gases exert pressure on their containers

A balloon filled with helium gas is under pressure. The gas in the balloon is pushing against the walls of the balloon. The kinetic theory helps to explain pressure. Helium atoms in the balloon are moving rapidly and they are constantly hitting each other and the walls of the balloon, as shown in *Figure 24*. Each gas particle's effect on the balloon wall is small, but the battering by millions of particles adds up to a steady force. The pressure inside the balloon is the measure of this force per unit area. If too many gas particles are in the balloon, the battering overcomes the force of the balloon holding the gas in, and the balloon pops.

If you let go of a balloon that you have held pinched at the neck, most of the gas inside rushes out and causes the balloon to shoot through the air. A gas under pressure will escape its container if possible. If there is a lot of pressure in the container, the gas can escape with a lot of force. For this reason, gases in pressurized containers, such as propane tanks for gas grills, can be dangerous and must be handled carefully.

Gas Laws

You can easily measure the volume of a solid or liquid, but how do you measure the volume of a gas? The volume of a gas is the same as the volume of its container but there are other factors, such as pressure, to consider.

The gas laws describe how the behavior of gases is affected by pressure and temperature. Because gases behave differently than solids and liquids, the gas laws will help you understand and predict the behavior of gases in specific situations.

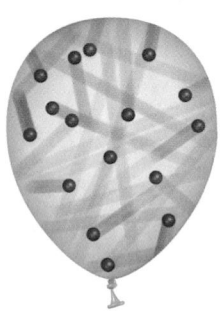 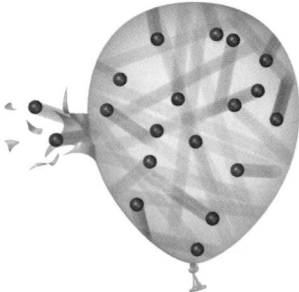

Figure 24

A Gas particles exert pressure by hitting the walls of the balloon.

B The balloon pops because the internal pressure is more than the balloon can hold.

88

Alternative Assessment — ADVANCED

Behavior of Gases Distribute a three-column chart to students with the following headings: *Observed Behavior of Gas, Property of Gas, Kinetic-Molecular Explanation.* Then, ask students to list several behaviors of gases, and have them complete columns two and three of the chart. **LS Logical**

Figure 25

Boyle's law Each illustration shows the same piston and the same amount of gas at the same temperature.

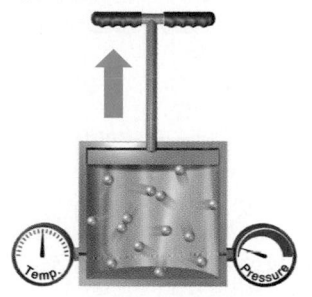  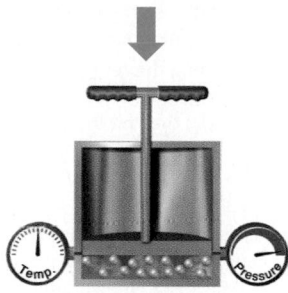

A Lifting the plunger decreases the pressure of the gas; the gas particles spread farther apart.

B Releasing the plunger allows the gas to change to an intermediate volume and pressure.

C Pushing the plunger increases the pressure, and decreases the volume of the gas.

Boyle's law relates the pressure of a gas to its volume

A diver at a depth of 10 m blows a bubble of air. As the bubble rises, its volume increases. When the bubble reaches the water's surface, the volume of the bubble will have doubled because of the decrease in pressure. The relationship between the volume and pressure of a gas is known as **Boyle's law.** Boyle's law states that for a fixed amount of gas at a constant temperature, the volume of a gas increases as its pressure decreases. Likewise, the volume of a gas decreases as its pressure increases. Boyle's law is illustrated in *Figure 25.* Boyle's law can be expressed as:
$(pressure_1)(volume_1) = (pressure_2)(volume_2)$ or $P_1V_1 = P_2V_2$.

▶ **Boyle's law** the law that states that for a fixed amount of gas at a constant temperature, the volume of the gas increases as the pressure of the gas decreases and the volume of the gas decreases as the pressure of the gas increases

Quick Lab

Does temperature affect the volume of a balloon?

Materials
- ✓ aluminum pans (2)
- ✓ ice
- ✓ balloon
- ✓ hot plate
- ✓ beaker 250 ml
- ✓ ruler
- ✓ gloves
- ✓ water

1. Fill an aluminum pan with 5 cm of water. Put the pan on the hot plate.
2. Fill another pan with 5 cm of ice water.
3. Blow up a balloon inside a beaker. The balloon should fill the beaker but should not extend outside it. Tie the balloon at its opening.
4. Place the beaker and balloon in the ice water. Record your observations.
5. Remove the balloon and beaker from the ice

water. Observe the balloon for several minutes, and record any changes.

6. Next, put the beaker and balloon in the hot water. Record your observations.

Analysis

1. How did changing the temperature affect the volume of the balloon?
2. Is the density of a gas affected by temperature?

SKILL BUILDER — BASIC

Interpreting Diagrams The pressure and volume of a gas are inversely proportional. Using **Figure 25,** discuss how changing the pressure applied to a gas will change its volume. Clarify that the temperature in this situation remains constant. Ask students to explain what would happen if they applied pressure to a solid or liquid. (There would be little to no change in volume, because solids and liquids are incompressible.) **LS Visual**

Quick Lab

Does temperature affect the volume of a balloon?

Safety Caution: *Remind students to wear goggles and aprons. They must also wear heat-resistant gloves when handling the hot beaker. Keep all power cords away from the beakers and pans of hot water. Be careful—hot plates may remain hot for a long time.*

Analysis

1. When the balloon cooled, it contracted. When heated, it expanded.
2. Yes. As temperature increases, the volume increases and the mass remains constant. Therefore, density decreases. Conversely, density increases when the temperature decreases.

LS Kinesthetic

Chapter Resource File

• **Quick Lab Datasheet** Does temperature affect the volume of a balloon? **GENERAL**

Transparencies

 TM Boyle's Law

Interpreting Diagrams Explain that the temperature and volume of a gas are proportional, and that they are directly proportional when the temperature is measured in kelvins. When the Kelvin temperature in **Figure 26** is doubled, the volume of the gas is also doubled. Have students predict what might happen to solids or liquids in a closed container when the temperature is changed. (There is little to no change.) **LS Visual**

Teaching Tip
Temperature Calculations

Remind students that temperature calculations with the gas laws must always be done with absolute temperatures (in K); the temperature can then be converted to other desired units, such as degrees Celsius (°C).

Reading Skills After students have finished reading this section, ask them why kelvins must be used when measuring temperature for gas law calculations. (Even though 1°C is the same size as 1 K, Celsius degrees may have negative values.) The Kelvin scale is the most widely used scale that uses only positive numbers to indicate temperature. Point out how the use of negative temperature values in gas law calculations leads to negative gas volume answers, an impossible condition. **LS Verbal**

Figure 26

Each illustration shows the same piston and the same amount of gas at the same pressure.

A Decreasing the temperature causes the gas particles to move more slowly; they hit the sides of the piston less often and with less force. As a result, the volume of the gas decreases.

B Raising the temperature of the gas causes the particles to move faster. As a result, the volume of the gas increases.

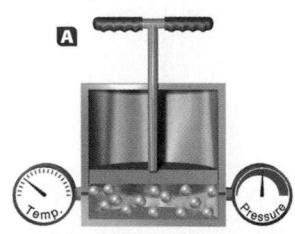
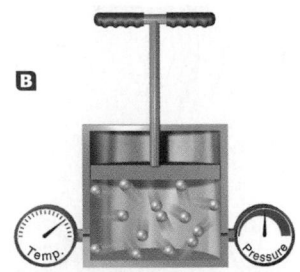

▶ **Charles's law** the law that states that for a fixed amount of gas at a constant pressure, the volume of the gas increases as the temperature of the gas increases and the volume of the gas decreases as the temperature of the gas decreases

Charles's law relates the temperature of a gas to its volume

An inflated balloon will also pop when it gets too hot which demonstrates another gas law—Charles's law. **Charles's law** states that for a fixed amount of gas at a constant pressure, the volume of the gas increases as its temperature increases. Likewise, the volume of the gas decreases as its temperature decreases. Charles's law is illustrated by the model in **Figure 26.** You can see Charles's law in action by putting an inflated balloon in the freezer and waiting about 10 minutes to see what happens!

As shown in **Figure 27,** if the gas in an inflated balloon is cooled (at constant pressure), the gas will decrease in volume and cause the balloon to deflate.

Figure 27
A Air-filled balloons are exposed to liquid nitrogen.
B The balloons shrink in volume.
C The balloons are removed from the liquid nitrogen and are warmed. The balloons expand to their original volume.

90

Ask students the following: "If one of your friends overinflated the tires on a bicycle, how would Charles's Law help you to explain why your friend should let out some of the air before going for a ride on a hot day?" (On a hot day, the gas in the tire will heat up and expand. If some of the air is not let out of the tire before going for a ride, the tire may burst due to the increased gas pressure.) **LS Logical**

Math Skills

Boyle's Law

The gas in a balloon has a volume of 7.5 L at 100 kPa. The balloon is released into the atmosphere, and the gas expands to a volume of 11 L. Assuming a constant temperature, what is the pressure on the balloon at the new volume?

1 List the given and unknown values.

   Given: V_1 = 7.5 L

   P_1 – 100 kPa

   V_2 = 11 L

   Unknown: P_2

2 Write the equation for Boyle's law, and rearrange the equation to solve for P_2.

   $P_1V_1 = P_2V_2$

   $P_2 = \dfrac{P_1V_1}{V_2}$

3 Insert the known values into the equation, and solve.

   $P_2 = \dfrac{(100\ \text{kPa})(7.5\ \text{L})}{11\ \text{L}}$

   $P_2 = 68$ kPa

Practice

Boyle's Law

1. A flask contains 155 cm³ of hydrogen collected at a pressure of 22.5 kPa. Under what pressure would the gas have a volume of 90.0 cm³ at the same temperature? (Recall that 1 cm³ 5 1 mL.)

2. If the pressure exerted on a 300.0 mL sample of hydrogen gas at constant temperature is increased from 0.500 atm to 0.750 atm, what will be the final volume of the sample?

3. A helium balloon has a volume of 5.0 L at a pressure of 101.3 kPa. The balloon is released and reaches an altitude of 6.5 km at a pressure of 50.7 kPa. If the gas temperature remains the same, what is the new volume of the balloon? Assume that the pressures are the same inside and outside of the balloon.

4. A sample of oxygen gas has a volume of 150 mL at a pressure of 0.947 atm. What will the volume of the gas be at a pressure of 1.000 atm if the temperature remains constant?

Practice HINT

Boyle's Law

The equation for Boyle's law can be rearranged to solve for volume in the following way. Use the equation $P_1V_1 = P_2V_2$ Divide both sides by P_2

$$\dfrac{P_1V_1}{P_2} = \dfrac{P_2V_2}{P_2}$$

$$\dfrac{P_1V_1}{P_2} = V_2$$

You will need to use this form of the equation in Practice Problems 2, 3, and 4.

internet connect

www.scilinks.org
Topic: Gas Laws
SciLinks code: HK4062

SCLINKS Maintained by the National Science Teachers Association

Math Skills

Additional Example

Boyle's Law A cylinder with a volume of 0.15 L contains a gas at a pressure of 150 kPa. If a piston compresses the gas at constant temperature to a volume of 0.025 L, what is the new pressure of the gas?

Answer

$P_2 = P_1V_1/V_2$

$P_2 = (150\ \text{kPa})(0.15\ \text{L})/0.025\ \text{L}$

$P_2 = 9.0 \times 10^2$ kPa

LS Logical

Practice

1. $P_2 = P_1V_1/V_2$

   $P_2 = (22.5\ \text{kPa})(155\ \text{cm}^3)/90.0\ \text{cm}^3$

   $P_2 = 38.8$ kPa

2. $V_2 = P_1V_1/P_2$

   $V_2 = (0.500\ \text{atm})(300.0\ \text{mL})/0.750\ \text{atm}$

   $V_2 = 2.00 \times 10^2$ mL

3. $V_2 = P_1V_1/P_2$

   $V_2 = (101.3\ \text{kPa})(5.00\ \text{L})/50.7\ \text{kPa}$

   $V_2 = 9.99$ L

4. $V_2 = P_1V_1/P_2$

   $V_2 = (0.947\ \text{atm})(150\ \text{mL})/1.000\ \text{atm}$

   $V_2 = 140$ mL

LS Logical

Teaching Tip

Ideal Gas Law Explain to students that each of the gas laws studied are special cases of a more general law called the ideal gas law. Write the mathematical form of this law on the chalkboard:

$$P_1V_1/T_1 = P_2V_2/T_2$$

Point out that by holding any one variable constant (temperature, pressure, or volume), the ideal gas law reduces to either Boyle's law (constant temperature), Charles's law (constant pressure), or Gay-Lussac's law (constant volume).

Close

Quiz ——————— BASIC

True or False Mark **T** if the statement is true, and **F** if it is false.

1. Gases are generally non-compressible. (F)

2. According to Boyle's law, the volume of a gas increases as the pressure decreases at constant temperature. (T)

3. Gay-Lussac's law describes the relationship between the volume and temperature of a gas. (F)

4. According to the kinetic theory, in contrast to solids and liquids, gases expand outward to fill space. (T)

LS Logical

Chapter Resource File

• Concept Review GENERAL

• Quiz BASIC

Gay-Lussac's law relates gas pressure to temperature

You have just learned about the relationship between the volume and temperature of a gas at constant pressure. What would you predict about the relationship between the pressure and temperature of a gas at constant volume? Remember that pressure is the result of collisions of gas molecules against the walls of their containers. As temperature increases, the kinetic energy of the gas particles increases. The energy and frequency of the collision of gas particles against their containers increases. For a fixed quantity of gas at constant volume, the pressure increases as the temperature increases.

Joseph Gay-Lussac is given credit for recognizing this property in 1802. **Gay-Lussac's law** states that the pressure of a gas increases as the temperature increases if the volume of the gas does not change. So if pressurized containers that hold gases, such as spray cans, are heated, they may explode. You should always be careful to keep containers of pressurized gas away from heat.

Gay-Lussac's law the law that states that the pressure of a gas at a constant volume is directly proportional to the absolute temperature

SECTION 3 REVIEW

SUMMARY

▶ Gases are fluids, their particles are in constant motion, they have low density, they are compressible, and they expand to fill their container.

▶ Gas pressure increases as the number of collisions of gas particles increases.

▶ Boyle's law states that the volume of a gas increases as the pressure decreases if the temperature does not change.

▶ Charles's law states that the volume of a gas increases as the temperature increases if the pressure does not change.

▶ Gay-Lussac's law states that the pressure of a gas increases as the temperature increases if the volume does not change.

1. List four properties of gases.

2. Explain why the volume of a gas can change.

3. Describe how gases are different from solids and liquids and give examples.

4. Identify what causes the pressure exerted by gas molecules on their container.

5. Restate Boyle's law, Charles's law, and Gay-Lussac's law.

6. Identify a real-life example for each of the three gas laws.

7. Critical Thinking When scientists record the volume of a gas, why do they also record the temperature and the pressure?

8. Critical Thinking Predict what would happen to the volume of a balloon left on a sunny windowsill. Which gas law predicts this result?

Math Skills

9. A partially inflated weather balloon has a volume of 1.56×10^3 L and a pressure of 98.9 kPa. What is the volume of the balloon when it is released to a height where the pressure is 44.1 kPa?

Answers to Section 3 Review

1. Gases flow, have low density, are compressible, and expand to fill their containers.

2. Attraction between gas particles is not strong, so the volume of a gas can easily change.

3. Unlike solids and liquids, gases have very low densities and can easily change volume. This makes them suitable for filling tires, balloons, and scuba tanks.

4. The pressure exerted on a container by a gas is caused by the gas particles colliding with the walls of the container.

5. Boyle's law states that the volume of a gas increases as pressure decreases at constant temperature. Charles's law states that the volume of a gas increases as temperature increases at constant pressure. Gay-Lussac's law states that gas pressure increases as temperature increases at constant volume.

6. Boyle's law: If you let air out of a tire, the air will expand outside of the tire, as pressure decreases. Charles's law: The air in a balloon will expand if the balloon is exposed to heat. Gay-Lussac's law: If you heat a spraycan it may explode due to the increase in pressure.

7. The volume of a gas can be changed by changing either the temperature or pressure.

See "Continuation of Answers" at the end of the chapter.

Graphing Skills

Percentages of Tin and Lead at Given Melting Points

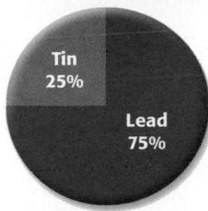
Tin 25%
Lead 75%

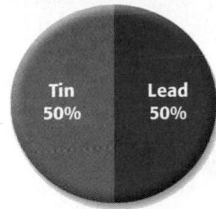
Tin 50%
Lead 50%

Lead 25%
Tin 75%

Melting Point = 266°C **Melting Point = 220°C** **Melting Point = 192°C**

Melting Point of Pure Lead = 326°C Melting Point of Pure Tin = 232°C

In most cases, when two pure substances with melting points that are not extremely different are mixed, the melting point of the mixture is lower than the melting points of either pure substances. Examine the graphs and answer the following questions.

1 What type of graphs are these?

2 Identify the quantities given in each graph. What important quantities relate the two graphs to each other?

3 By examining the graphs, what can you tell about the melting point of lead-tin alloys?

4 From the information given, estimate the percentages of tin and lead that would have the lowest melting temperature.

5 Suppose you wish to make a tin-lead solder that melts below 200°C. How might you find the limiting range of percentages for tin and lead for such a solder?

6 Construct a graph best suited for the information listed in the table below. Near what percentage mixture of metals is the melting point lowest?

Percentage of aluminum in Al-Cu alloy	Melting point (°C)
0	1084
20	930
40	610
60	540
80	600
100	650

Teaching Tip

Suggest to students that the data can be presented in other graphical forms. Ask them what data might be used to provide the same information in the form of a line graph (temperature and either the percentage of lead or tin).

Answers

1. pie charts

2. the percentages of lead and tin in a mixture; The graphs are related by different melting temperatures for different percentages of each metal in each mixture.

3. As the amount of tin in the mixture increases, the melting point decreases from that of pure lead, reaches a minimum value, and then increases until it equals the melting point of pure tin.

4. about 75% tin/25% lead

5. A series of such graphs over a narrow range of percentages (for instance, 10% increases in tin) would indicate a range of melting points from which the range under 200°C could be found.

6.

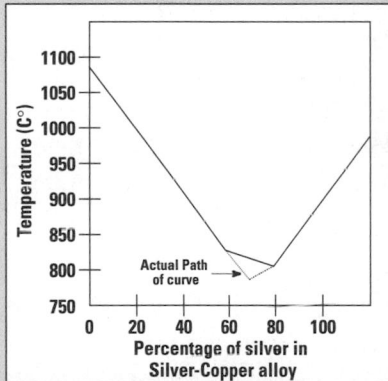

The lowest melting point occurs between 60% and 80% silver.

Understanding Concepts

1. d
2. c
3. b
4. b
5. b
6. b
7. a
8. c
9. d
10. c
11. d

Using Vocabulary

12. When the thermometer is placed in a warmer environment, the alcohol particles gain energy, and so move more quickly and farther apart. This increased movement causes the volume of the particles to increase, so the column of alcohol moves higher in the tube. In a cooler environment, the alcohol particles lose energy, and move more slowly. This reduced movement causes volume to decrease, so the alcohol moves lower in the tube.

13. Energy is absorbed in endothermic changes and released in exothermic changes.

14. exothermic: freezing, condensation; endothermic: melting, evaporation, sublimation (when solid becomes vapor)

Chapter Resource File

- **Chapter Test** GENERAL
- **Test Item Listing for ExamView® Test Generator**

Chapter Highlights

Before you begin, review the summaries of the key ideas of each section, found at the end of each section. The vocabulary terms are listed on the first page of each section.

UNDERSTANDING CONCEPTS

1. Which of the following assumptions is *not* part of the kinetic theory?
a. All matter is made up of tiny, invisible particles.
b. The particles are always moving.
c. Particles move faster at higher temperatures.
d. Particles are smaller at lower pressure.

2. Three common states of matter are
a. solid, water, and gas.
b. ice, water, and gas.
c. solid, liquid, and gas.
d. solid, liquid, and air.

3. Which of the following best describes the particles of a liquid?
a. The particles are far apart and moving fast.
b. The particles are close together but moving past each other.
c. The particles are far apart and moving slowly.
d. The particles are closely packed and vibrate in place.

4. Boiling points and freezing points are examples of
a. chemical properties.
b. physical properties.
c. energy.
d. matter.

5. During which change of state do atoms or molecules become more ordered?
a. boiling
b. condensation
c. melting
d. sublimation

94

6. Which of the following describes what happens as the temperature of a gas in a balloon increases?
a. The speed of the particles decreases.
b. The volume of the gas increases and the speed of the particles increases.
c. The volume decreases.
d. The pressure decreases.

7. Dew collects on blades of grass in the early morning. This is an example of
a. condensation.
b. evaporation.
c. sublimation.
d. melting.

8. Which of the following changes of state is exothermic?
a. evaporation
b. sublimation
c. freezing
d. melting

9. Fluid pressure is always directed
a. up.
b. down.
c. sideways.
d. in all directions.

10. Materials that can flow to fit their containers include
a. gases.
b. liquids.
c. both gases and liquids.
d. neither gases nor liquids.

11. If an object weighing 50 N displaces a volume of water with a weight of 10 N, what is the buoyant force on the object?
a. 60 N
b. 50 N
c. 40 N
d. 10 N

USING VOCABULARY

12. In an alcohol thermometer, the height of a constant amount of liquid alcohol in a thin glass tube increases or decreases as temperature changes. Using what you have learned about kinetic theory, explain the behavior of the alcohol using the following terms: *lose energy, gain energy, volume, movement, particles.*

13. Compare *endothermic* and *exothermic* changes.

14. Classify each change of state (*melting, freezing, evaporation, condensation,* and *sublimation*) as endothermic or exothermic.

15. Distinguish between *crystalline* and *amorphous solids,* and give examples of both.

16. For each pair of terms, explain the difference in meaning.
   a. *solid/liquid*
   b. *Boyle's law/Charles's law*
   c. *Gay-Lussac's law/Pascal's principle*

17. Describe four states of matter using the terms *solid, liquid, gas,* and *plasma.* Describe the behavior of particles in each state.

18. State the *law of conservation of energy* and the *law of conservation of mass* and explain what happens to energy and mass in a change of state.

19. Describe the *buoyant force* and explain how it relates to *Archimedes principle.*

20. Explain how fluid pressure is affected by the speed of the fluids by restating *Bernoulli's principle.*

21. Describe how *pressure* is exerted by fluids.

22. Why are liquids used in *hydraulic* brakes instead of gases?

BUILDING GRAPHING SKILLS

23. **Graphing** The graph below shows the effects of heating on ethylene glycol, the liquid commonly used as antifreeze. Until the temperature is 197°C, is the temperature increasing or decreasing? What physical change is taking place when the ethylene glycol is at 197°C? Describe what is happening to the ethylene glycol molecules at 197°C. How can you tell?

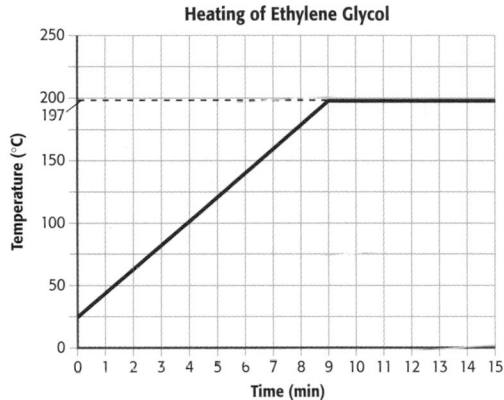

24. **Interpreting Data** Kate placed 100 mL of water in five different pans. She then placed the pans on a windowsill for a week, and measured how much water evaporated. Draw a graph of her data, shown below, with surface area on the *x*-axis. Is the graph linear or nonlinear? What does this tell you?

Pan number	1	2	3	4	5
Surface area (cm²)	44	82	20	30	65
Volume evaporated (mL)	42	79	19	29	62

95

15. Crystalline solids have an orderly arrangement of molecules, whereas amorphous solids are composed of particles that are in no particular order. Crystalline solids include iron and ice. Amorphous solids include wax and rubber.

16. a. Solid is the state of matter in which the substance has a definite shape and volume; liquid is the state in which the substance takes the shape of its container but has a definite volume.
   b. Boyle's law: when the pressure of a gas at constant temperature increases, its volume decreases; Charles's law: when the temperature of a gas at constant pressure increases, its volume increases.
   c. Gay-Lussac's law: when the temperature of a gas at constant volume increases, its pressure increases; Pascal's Principle: pressure applied to a fluid in a closed container is transmitted equally throughout the fluid.

17. In a solid, particles are held closely together and vibrate. In a liquid, particles are close together but slide past one another. In a gas, particles expand to fill available space. In plasma, particles expand to fill space but have also broken apart.

18. The law of conservation of energy states that energy cannot be created or destroyed. The law of conservation of mass states that mass cannot be created or destroyed. In a change of state, energy and mass may be transformed, but not destroyed.

19. Buoyant force is the upward force exerted on an object immersed in or floating on a fluid. Archimedes' principle states that the buoyant force on an object is equal to the weight of the volume of fluid that the object displaces.

20. Fluid pressure decreases as speed increases.

21. Fluids exert pressure evenly in all directions.

Assignment Guide

Section	Questions
1	1–5, 7, 8, 10, 12–15, 16a, 17, 18, 23–25, 32, 33, 38, 41–43
2	9, 11, 19–22, 26–28, 34, 36, 37
3	6, 16b, 16c, 29–31, 35, 39, 40

22. Liquids are used in hydraulic brakes because liquids cannot be compressed easily. Gases are easily compressible.

Building Graphing Skills

23. The temperature is increasing. Ethylene glycol undergoes a change of state at 197°C; it is changing from a liquid to a gas. At 197°C, ethylene glycol molecules are absorbing energy to vaporize, so there is no increase in temperature.

24.

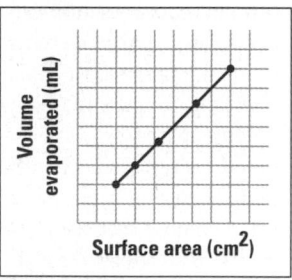

The graph is linear, which indicates that both variables (surface area and volume evaporated) increase together.

25. a. 80°C; 20°C
   b. liquid
   c. The temperature of the liquid will rise.

Building Math Skills

26. 3.0 m²; 250 Pa

27. 110 Pa

28. 5×10^3 N

29. 4.2 L

30. 12 L

31. 143.9 mL

25. Graphing Examine the graph below, and answer the following questions.
   a. What is the boiling point of the substance? What is the melting point?
   b. Which state is present at 30°C?
   c. How will the substance change if energy is added to the liquid at 20°C?

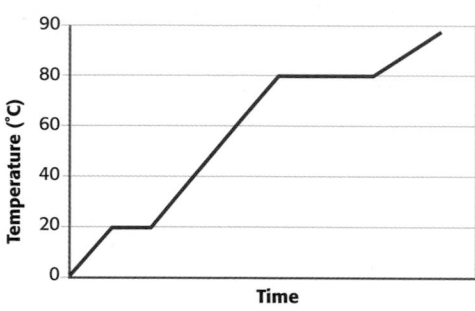

Time

BUILDING MATH SKILLS

26. Pressure Calculate the area of a 1500 N object that exerts a pressure of 500 Pa. Then calculate the pressure exerted by the same object over twice that area. Be sure to express your answer in the correct SI unit.

27. Pressure A box, half full of books, rests on the ground. The area where the box comes in contact with the floor is 1 m². The box weighs 110 N. How much pressure is the box exerting on the floor? Express your answer in pascals.

28. Pascal's principle One of the largest helicopters in the world weighs 1.0×10^6 N. If you were to place this helicopter on a large piston of a hydraulic lift, what force would need to be applied to the small piston, with an area of .7m², in order to lift the helicopter? The area of the large piston is 140 m².

29. Boyle's law A sample of neon gas occupies a volume of 2.8 L at 1.8 atm. What will its volume be at 1.2 atm?

30. Boyle's law 2.2 L of hydrogen at 6.5 atm pressure is used to fill a balloon at a final pressure of 1.15 atm. What is its final volume?

31. Boyle's law A sample of oxygen gas has a volume of 150 mL when its pressure is 0.947 atm. If the pressure is increased to 0.987 atm and the temperature remains constant, what will the new gas volume be?

THINKING CRITICALLY

32. Applying Knowledge After taking a shower, you notice that small droplets of water cover the mirror. Explain how this happens, describing where the water comes from and the changes it goes through.

33. Making Comparisons At sea level, water boils at 100°C, while methane boils at −161°C. Which of these substances has a stronger force of attraction between its particles? Explain your answer.

34. Understanding Systems An iceberg is floating, partially submerged, in the ocean. At what part of the iceberg is the water pressure the greatest?

35. Understanding Systems Use Boyle's Law to explain why "bubble wrap" pops when you squeeze it.

36. Applying Knowledge Compared with an empty ship, will a ship loaded with plastic-foam balls float higher or lower in the water? Explain.

37. Critical Thinking Inside all vacuum cleaners is a high-speed fan. Explain how this fan causes dirt to be picked up by the vacuum cleaner.

Thinking Critically

32. As you take a shower, some of the liquid water evaporates and becomes a gas. When the gaseous water touches the mirror, the water releases energy to the mirror and condenses into drops of liquid water.

33. Water has a stronger force of attraction between its particles. A higher temperature, and therefore more energy, is required to separate the water particles from one another than is needed to separate the methane particles from one another.

34. The pressure is greatest at the lowest depth.

35. As the bubble is squeezed to a smaller volume, the pressure of the gas inside the bubble increases until it is high enough to burst the plastic film.

36. The ship will float lower in the water because the plastic-foam balls will add to the ship's total mass, but will not increase its volume. Therefore, the overall density of the ship will increase, causing it to sink a little.

37. The fan causes the air inside the vacuum cleaner to move faster, which decreases pressure. The higher air pressure outside of the vacuum then pushes dirt into the vacuum cleaner.

38. Applying Knowledge In the photo below, water is being decomposed into hydrogen and oxygen. Is this a physical change? Explain your answer.

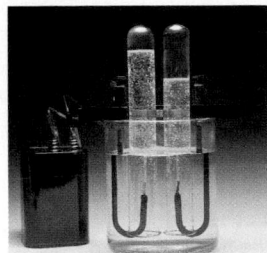

DEVELOPING LIFE/WORK SKILLS

39. Allocating Resources Use the Internet or library references to find out how using a pressure cooker can preserve the nutritional value of food. Prepare an illustrated report that explains the science behind pressure-cooking.

40. Locating Information Use the Internet or library references to find out why changes must be made in some recipes for cooking and baking at high elevations. Make a poster presentation that compares recipes for sea level and high altitude preparations.

INTEGRATING CONCEPTS

41. Connection to Biology Your body uses the food you eat to do work. However, some of the food energy is lost as heat. How does your body give off this heat?

42. Connection to Environmental Science Research the process of glass and metal recycling. Visit a metal salvage yard or recycling center. Find out what types of metal and glass are recycled and how the recycled material is bought and sold. Work with a group to set up a glass and metal collection site at your school.

43. Concept Mapping Copy the unfinished concept map below onto a sheet of paper. Complete the map by writing the correct word or phrase in the lettered boxes.

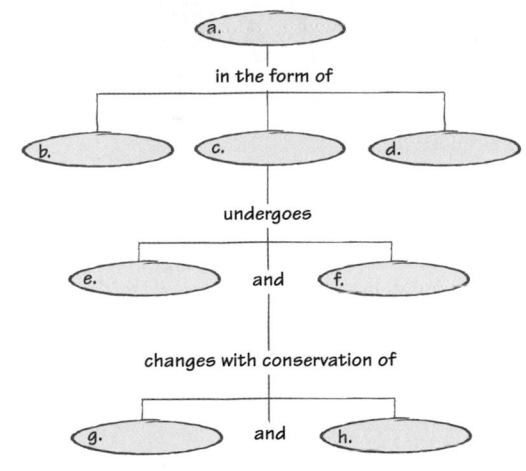

internet connect

www.scilinks.org
Topic: Energy Transformations
SciLinks code: HK4049

SCI/LINKS. Maintained by the National Science Teachers Association

Art Credits: Fig. 5, Stephen Durke/Washington Artists; Fig. 8, David Schleinkofer; Fig. 9, Kristy Sprott; Fig. 13, Leslie Kell; Fig. 16, Uhl Studios, Inc.; Fig. 18, Uhl Studios, Inc.; Fig. 19, Stephen Durke/Washington Artists; Fig. 24-26, Stephen Durke/Washington Artists.

Photo Credits: Chapter Opener photo of ice sculpture by Keren Su/CORBIS; dry ice photo by Matt Meadows/SPL/Photo Researchers, Inc.; Fig. 1, Getty Images/The Image Bank; Fig. 2(l,c), Sergio Purtell/Foca/HRW; (r), Rich Treptow/Photo Researchers, Inc.; Fig. 3, Digital Image copyright 2004, PhotoDisc; Fig. 4, Andy Sacks/Getty Images/Stone; Fig. 5, Scott Van Osdol/HRW; Fig. 6, Pekka Parvianen/SPL/Photo Researchers, Inc.; Fig. 7(l-r), John Langford/HRW; Paul A. Souders/CORBIS; Tony Freeman/PhotoEdit; Fig. 9, Lori Adamski Peek/Getty Images/Stone; Fig. 10, Charlie Winters/HRW; Fig. 11, Myrleen Ferguson/PhotoEdit; Fig. 12, Scott Van Osdol/HRW; Fig. 13(l-r), Steve Joester/Getty Images/Taxi; Michael Keller/Corbis Stock Market; E. R. Degginger/Color-Pic, Inc.; "Science and the Consumer," Sherman Clark/Getty Images; Fig. 14, Benelux Press/Getty Images/Taxi; Fig. 15, Roine Magnusson/Getty Images/Stone; Fig. 17, Joseph Sohm; ChromoSohm, Inc./CORBIS; "Real World Application," Chris Oxley/CORBIS; Fig. 21, Brian Hagiwara/FoodPix; Fig. 23, Charles D. Winters/Photo Researchers, Inc.; Fig. 27, Charlie Winters; "Chapter Review," Charles D. Winters/Photo Researchers, Inc.; "Skill Builder Lab," Victoria Smith/HRW; "Science in Action," (plasma sphere), Simon Terry/Photo Researchers, Inc.; (lightning), Rohan Sullivan/AP/Wide World Photos, (aurora borealis), Chris Madeley/Photo Researchers, Inc.

97

---

38. The decomposition of water into hydrogen and oxygen is not a physical change because the water does not keep its identity during the change; it is changed into two new substances: hydrogen and oxygen.

Developing Life/Work Skills

39. Students' answers may vary.

40. At higher elevations, air pressure decreases. The temperature at which food is cooked must be increased to compensate for concurrent decreases in temperature as pressure decreases.

Integrating Concepts

41. Answers could include energy is transferred by heat during breathing and through perspiration. Energy is also given up through infrared radiation from the skin.

42. Students' answers may vary.

43. a. matter
b. solid
c. liquid
d. gas
e. physical
f. chemical
g. mass
h. energy

Transparencies
TT Concept Mapping

BOILING AND FREEZING

Teacher's Notes

Time Required: 1 lab period

Ratings

	1	2	3	4
	EASY			HARD

TEACHER PREPARATION 2
STUDENT SETUP 2
CONCEPT LEVEL 2
CLEANUP 2

Skills Acquired

- Collecting data
- Communicating
- Identifying/Recognizing Patterns
- Interpreting
- Measuring
- Organizing and analyzing data
- Predicting

The Scientific Method

In this lab, students will:
- Make Observations
- Form a Hypothesis
- Analyze the Results
- Draw Conclusions
- Communicate Results

Materials

Materials are listed for each group.

Safety Cautions

Review glassware and heating safety before students begin the experiment. Remind students to wear heat-resistant gloves when glassware that may be hot. Caution students to handle thermometers carefully. They should gently place thermometers into glassware, and they should not stir with the thermometers.

Skills Practice Lab

Introduction

When you add or remove energy from a substance, does the substance's temperature always change? Investigate this question with a common substance—water.

Objectives

▶ **USING SCIENTIFIC METHODS** *Test your hypothesis* by measuring the temperature of water as it boils and freezes.

▶ *Graph* data, and interpret the slopes of the graphs.

Materials

beaker, 250 or 400 mL
coffee can, large
crushed ice
gloves, heat-resistant
graduated cylinder, 100 mL
graph paper
hot plate
rock salt
stopwatch
thermometer
water
wire-loop stirring device

98

Boiling and Freezing

▶ Procedure

Preparing for Your Experiment

1. Make a prediction: what happens to the temperature of boiling water as the boiling process continues?

2. Make a prediction: what happens to the temperature of freezing water as the freezing process continues?

3. Prepare two data tables like the one shown at right.

Boiling Water

4. Fill the beaker about one-third to one-half full with water.

5. Put on heat-resistant gloves. Turn on the hot plate, and put the beaker on the plate. Put the thermometer in the beaker.
   SAFETY CAUTION Be careful not to touch the hot plate. Also be careful not to break the thermometer.

6. In your first data table, record the temperature of the water every 30 seconds. Continue doing this until about one-fourth of the water boils away. Note the first temperature reading at which the water is steadily boiling.

7. Turn off the hot plate. Let the beaker cool for a few minutes, then use heat-resistant gloves to pick up the beaker. Pour the warm water out, and rinse the warm beaker with cool water.
   SAFETY CAUTION Even after cooling, the beaker may still be too hot to handle without gloves.

Freezing Water

8. Put approximately 20 mL of water in the graduated cylinder.

9. Put the graduated cylinder in the coffee can, and fill in around the graduated cylinder with crushed ice. Pour rock salt on the ice around the graduated cylinder. Slide the tip of the thermometer through the loop on the wire-loop stirring device and put the thermometer and the wire-loop stirring device in the graduated cylinder.

Answers to Procedure

1. Answers to the question may vary, with the most common responses being that temperature will remain constant while energy is absorbed by the water as it vaporizes, or that the temperature continues to rise as the water evaporates.

2. Answers to the question may vary, with the most common responses being that temperature will remain constant while energy is released by the water as it freezes, or that the temperature continues to drop as the water freezes.

Temperature of Water

Time (s)	30	60	90	120	150	180	etc.
Temperature (°C)							

10. As the ice melts and mixes with the rock salt, the level of ice will decrease. Add ice and rock salt to the can as needed.

11. In your second data table, record the temperature of the water in the graduated cylinder every 30 s. Stir the water occasionally by moving the wire-loop stirring device up and down along the thermometer.
SAFETY CAUTION Do not stir in a circular motion with the thermometer.

12. Once the water begins to freeze, stop stirring. Do not try to pull the thermometer out of the solid ice in the graduated cylinder.

13. Note the temperature when you first notice ice crystals forming in the water. Continue taking readings until the water in the graduated cylinder is frozen.

14. After you record your final reading, pour warm water into the can. Then wait until the ice in the graduated cylinder has melted. Pour the water out of the cylinder, and rinse the cylinder with water. Pour out the contents of the can, and rinse the can with water. Put away all equipment as directed by your instructor.

▶ Analysis

1. **Constructing Graphs** Make a graph of temperature (y-axis) versus time (x-axis) for the boiling-water data from the first table. Draw an arrow to the temperature reading at which the water started to boil.

2. **Constructing Graphs** Make a graph of temperature (y-axis) versus time (x-axis) for the freezing-water data from the second table. Draw an arrow to the temperature reading at which the water started to freeze.

3. What does the slope of the line on each graph represent?

4. In your first graph, how does the slope when the water is boiling compare to the slope before the water starts to boil?

5. In your second graph, how does the slope when the water is freezing compare to the slope before the water starts to freeze?

▶ Conclusions

6. Explain what happens to the energy that is added to the water while the water is boiling.

7. When water freezes, energy is removed from the water. What role do you think this energy played in the water before the energy was removed? Where does the energy go?

Analysis

1. The graphs will vary according to the data obtained, but should show an increase of temperature with time (positive slope) until 100°C is reached, at which point the graph becomes a nearly flat line with increasing time.

2. The graphs will vary according to the data obtained, but should show a decrease of temperature with time (negative slope) until 0°C is reached, at which point the graph becomes a nearly flat line with increasing time.

3. The slope represents change in water's temperature with time.

4. The slope becomes flat (zero) once the water starts to boil. Before this, the slope was positive (increasing).

5. The slope becomes flat (zero) once the water starts to freeze. Before this, the slope was negative (decreasing).

Conclusions

6. The energy added to the water during boiling is used to overcome the bonds between liquid water molecules, so that they can move freely as water vapor.

7. Before it was removed, the energy kept the liquid water molecules moving enough so that they did not form strong bonds, and thus crystallize. This energy, once removed from the water, is transferred to the surrounding air, causing its temperature to increase.

Chapter Resource File

- **Datasheet** Boiling and Freezing GENERAL
- **Observation Lab** Boyle's Law BASIC
- **CBL™ Probeware Lab** Investigating the Relationship between Pressure and Volume ADVANCED

Continuation of Answers

Continuation of Answers from p. 78
Section 1 Review

8. Mass cannot be created or destroyed. Energy cannot be created or destroyed. During a change of state, the total mass and total energy of the system remain constant.

9. Gas particles can move freely in all directions. The traces of chemicals a person leaves when he or she touches something absorb energy, vaporize, and spread outward, so that a dog can detect them.

Continuation of Answers from p. 86
Section 2 Review

Math Skills

7. $P = F/A = 1025 \text{ N}/3.75 \text{ m}^2 = 273 \text{ Pa}$

8. a. 15 N
   b. sink; weight exceeds buoyant force

9. It will float, being less dense than mercury.

Continuation of Answers from p. 92
Section 3 Review

8. Leaving the balloon on a sunny windowsill will cause the temperature of the gas in the balloon to increase. According to Charles's law, the volume will increase as the temperature increases at constant pressure.

Math Skills

9. $3.50 \times 10^3 \text{ L}$

Science in ACTION

Plasma

We are surrounded by matter in one of three states: solid, such as wood and plastic; liquid, such as milk and seawater; and gas, such as oxygen and helium. But almost everything in the universe (99.9% of all matter) including the sun and all other stars, is a *plasma*. Plasma is a strange state of matter that has some unexpected effects on your daily life.

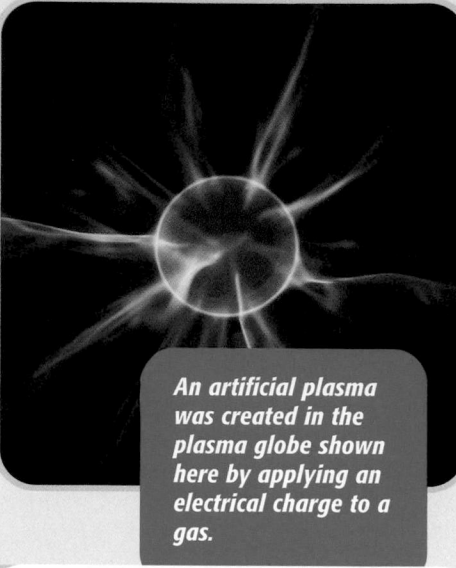

An artificial plasma was created in the plasma globe shown here by applying an electrical charge to a gas.

The lightning over Sydney, Australia, is an example of a natural plasma.

What Is Plasma?

Matter in the plasma state is a collection of free-moving electrons and ions (atoms that have lost electrons). Like gases, plasmas do not have a definite shape or volume. But unlike gases, plasmas conduct electricity and are affected by magnets. Plasmas require an energy source to exist. This energy may be a heat source, such as the heat of the sun; an electrical current; or a strong light, such as a laser. Some plasmas, including lightning and fire, do occur naturally on Earth. Artificial plasmas, including fluorescent and neon lights, are created by running an electrical current through a gas to change the gas into a plasma that emits light. When the current is removed the plasma becomes a gas again.

100

Space Weather

The sun is a giant ball of super heated plasma. On its surface, violent eruptions send waves of plasma streaming out into space at high speeds. As you read this book, Earth is bathed in waves of plasma known as solar wind. The most spectacular evidence of plasma in space is an aurora. When the highly charged plasma from the sun comes in contact with Earth's ionosphere (the uppermost region of Earth's atmosphere), the plasma produces an electrical discharge. This discharge, known as an aurora, lights up the sky.

After periods of disturbance on the surface of the sun, strong solar winds can disrupt radio and telephone communications, damage orbiting satellites, and cause electrical blackouts. Scientists are working to understand the forces behind solar wind and hope to better forecast damaging solar wind headed toward Earth.

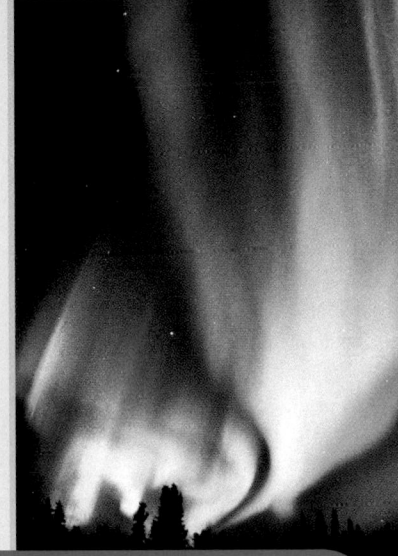

Aurora borealis, also known as the Northern Lights, results from solar wind interacting with the Earth's atmosphere.

Science and You

1. **Understanding Concepts** Describe one way that a plasma is similar to a gas. Describe one way plasma is different from gas.

2. **Understanding Concepts** Name one common technology that uses plasma.

3. **Critical Thinking** Do you think that auroras occur at the same time as eruptions on the surface of the sun? Explain.

4. **Critical Thinking** Why do you think scientists want to predict solar winds in Earth's atmosphere?

5. **Acquiring and Evaluating** Some scientists believe that plasma may be a key to source of energy in the future.

Research the Tokamak Fusion Reactor, and answer the following questions: What do scientists hope to achieve with this research? How are magnets involved in containing the plasma? What challenges do scientists face when studying plasma?

internet connect

www.scilinks.org
Topic: Plasma
SciLinks code: HK4106

SCiLINKS® Maintained by the National Science Teachers Association

101

CHAPTER 4

Atoms and the Periodic Table
Chapter Planning Guide

PACING	CLASSROOM RESOURCES	LABS, ACTIVITIES, AND DEMONSTRATIONS
BLOCK 1 · 45 min pp. 102–103 **Chapter Opener**		SE **Activity 1,** p. 103 SE **Activity 2,** p. 103
BLOCKS 2 & 3 · 90 min pp. 104–110 **Section 1** Atomic Structure	CRF **Lesson Plan*** TT Bellringer* TT Building a Model* TM Subatomic Particles*	SE **Quick Activity** Constructing a Model, p. 109 `GENERAL` CRF **Datasheets for In-Text Labs** Constructing a Model* `GENERAL` CRF **Observation Lab** Drawing Atomic Models* ◆ `BASIC`
BLOCKS 4 & 5 · 90 min pp. 111–119 **Section 2** A Guided Tour of the Periodic Table	CRF **Lesson Plan*** TT Bellringer* TM The Periodic Table* TT Nucleus* TT Isotopes*	TE **Opening Demonstration,** p. 111 `GENERAL` TE **Group Activity** Interpreting the Periodic Table, p. 113 `BASIC` TE **Group Activity** Protons and Neutrons, p. 116 `BASIC` SE **Quick Activity** Isotopes, p. 118 `GENERAL` CRF **Datasheets for In-Text Labs** Isotopes* `GENERAL`
BLOCKS 6 & 7 · 90 min pp. 120–128 **Section 3** Families of Elements	CRF **Lesson Plan*** TT Bellringer*	TE **Opening Demonstration,** p. 120 `GENERAL` SE **Quick Activity** Elements in Your Food, p. 122 `GENERAL` CRF **Datasheets for In-Text Labs** Elements in Your Food* `GENERAL` TE **Group Activity** Metal Coins, p. 123 `ADVANCED` SE **Quick Lab** Why do some metals cost more than others?, p. 124 `GENERAL` TE **Group Activity** Semiconductor Greeting, p. 127 `BASIC` SE **Skills Practice Lab** Comparing the Physical Properties of Elements, pp. 140–141 ◆ `GENERAL` CRF **CBL™ Probeware Lab** Predicting the Physical and Chemical Properties of Elements* ◆ `ADVANCED`
BLOCKS 8 & 9 · 90 min pp. 129–134 **Section 4** Using Moles to Count Atoms	CRF **Lesson Plan*** TT Bellringer* TM Mole-Mass Graph* TM Mole-Mass Conversion* TT Concept Mapping*	TE **Opening Activity,** p. 129 `GENERAL` TE **Group Activity** Conversion Factors, p. 131 `BASIC`

BLOCKS 10 & 11 · 90 min

Chapter Review and Assessment Resources

- SE **Chapter Review,** pp. 136–139
- CRF **Chapter Tests*** `GENERAL`
- OSP **Test Generator**
- CRF **Standardized Test Practice with Guided Reading Development***
- CRF **Test Item Listing for ExamView® Test Generator***
- OSP **Scoring Rubrics and Classroom Management Checklists**

Online Resources

Visit the HRW Web site for a variety of free resources related to the text. Just type in the keyword **HK4 ATO.**

Holt Online Learning

Holt Science Spectrum: Physical Science: Online Edition

Students can access interactive problem solving help and active visual concept development with the *Holt Science Spectrum: Physical Science* Online Edition available at **www.hrw.com.**

student CNN News

cnnstudentnews.com

Find the latest news, lesson plans, and activities related to important scientific events.

KEY

TE	Teacher Edition	**CRF**	Chapter Resource File	*****	Also on One-Stop Planner
SE	Student Edition	**TT**	Teaching Transparency	**♦**	Requires Advance Prep
OSP	One-Stop Planner	**TM**	Transparency Master		

PROBLEM SOLVING AND PRACTICE	SECTION REVIEW AND ASSESSMENT	STANDARDS CORRELATION
	CRF Pretest* GENERAL	
CRF **Cross-Disciplinary Worksheet** Integrating Technology—Seeing Atoms: the STM* GENERAL **CRF** **Cross-Disciplinary Worksheet** Integrating Physics—Atomic Fingerprints* GENERAL **CRF** **Cross-Disciplinary Worksheet** Real World Applications—How Do Scientists Find Cures for Diseases?* GENERAL	**TE** Quiz, p. 110 BASIC **SE** Section 1 Review, p. 110 **CRF** Concept Review* GENERAL **CRF** Quiz* BASIC	PS 1a, 1c, 2b UCP 2 HNS 1, 3
CRF **Cross-Disciplinary Worksheet** Integrating Biology—Elements in Your Body* GENERAL **CRF** **Cross-Disciplinary Worksheet** Connection to Language Arts—Chemical Symbols* GENERAL **CRF** **Cross-Disciplinary Worksheet** Connection to Fine Arts—Dating Masterpieces* GENERAL	**TE** Quiz, p. 119 BASIC **SE** Section 2 Review, p. 119 **CRF** Concept Review* GENERAL **CRF** Quiz* BASIC	PS 1c, 2a, 2b UCP 1, 5
CRF **Cross-Disciplinary Worksheet** Connection to Architecture—Buckyballs* GENERAL **CRF** **Cross-Disciplinary Worksheet** Integrating Earth Science—Magnesium: From Sea Water to Fireworks* GENERAL	**TE** Quiz, p. 128 BASIC **SE** Section 3 Review, p. 128 **CRF** Concept Review* GENERAL **CRF** Quiz* BASIC	PS 2a, 2b SAI 1 SPSP 2, 5
CRF **Science Skills** Scientific Notation* BASIC **CRF** **Math Skills** Conversion Factors* GENERAL **CRF** **Math Skills** Converting Amount to Mass* GENERAL **CRF** **Science Skills** Significant Figures* BASIC **CRF** **Science Skills** Dimensional Analysis* BASIC **CRF** **Math Skills** Converting Mass to Amount * GENERAL **SE** **Math Skills** Conversion Factors, p. 135	**TE** Quiz, p. 134 BASIC **SE** Section 4 Review, p. 134 **CRF** Concept Review* GENERAL **CRF** Quiz* BASIC	UCP 3 HNS 3

SCI LINKS®

www.scilinks.org

Topic: Atoms and Elements
*Sci*Links code: HK4012

Topic: Atomic Theory
*Sci*Links code: HK4011

Topic: Periodic Table
*Sci*Links code: HK4102

Topic: Element Families
*Sci*Links code: HK4046

Topic: Avogadro's Constant
*Sci*Links code: HK4013

Topic: Metals/Nonmetals
*Sci*Links code: HK4086

Topic: Origin of Elements
*Sci*Links code: HK4097

Technology Resources

 One-Stop Planner
All of your printable resources and the Test Generator are on this convenient CD-ROM.

Physical Science Interactive Tutor CD-Rom

Disc One, Module 1
Topic: Models of the Atom

Disc One, Module 3
Topic: Periodic Properties

 CNN. Science in the NEWS

each video segment is accompanied by a Critical Thinking Worksheet*

Segment 13
Atom Builders

Segment 15
Making Fullerenes

Overview

In this chapter, students learn what atoms are, what they are made up of, and how they are represented by models. This chapter also covers orbital levels and valence electrons. There is a guided tour of the periodic table, including organization of the table, periods, and groups. Students also learn about ionization, atomic number, mass number, isotopes, and average atomic mass. This chapter covers the classification of elements as metals (alkali, alkaline-earth, and transition) and nonmetals (halogens, noble gases, and semiconductors). Also, students learn how to use moles to count atoms.

Assessing Prior Knowledge

Be sure students understand the following concepts:

- units of measurement
- scientific notation
- significant figures
- atoms
- elements
- reactivity
- chemical and physical properties
- chemical changes

MISCONCEPTION ///ALERT\\\

Science education research has identified the following misconceptions about atoms.

- Students believe that atoms possess macro properties such as hardness, color, shape, or stickiness.
- Students believe that atoms contain no empty space and are static.
- Students do not understand how small molecules and atoms truly are.
- Students confuse molar mass with atomic mass.

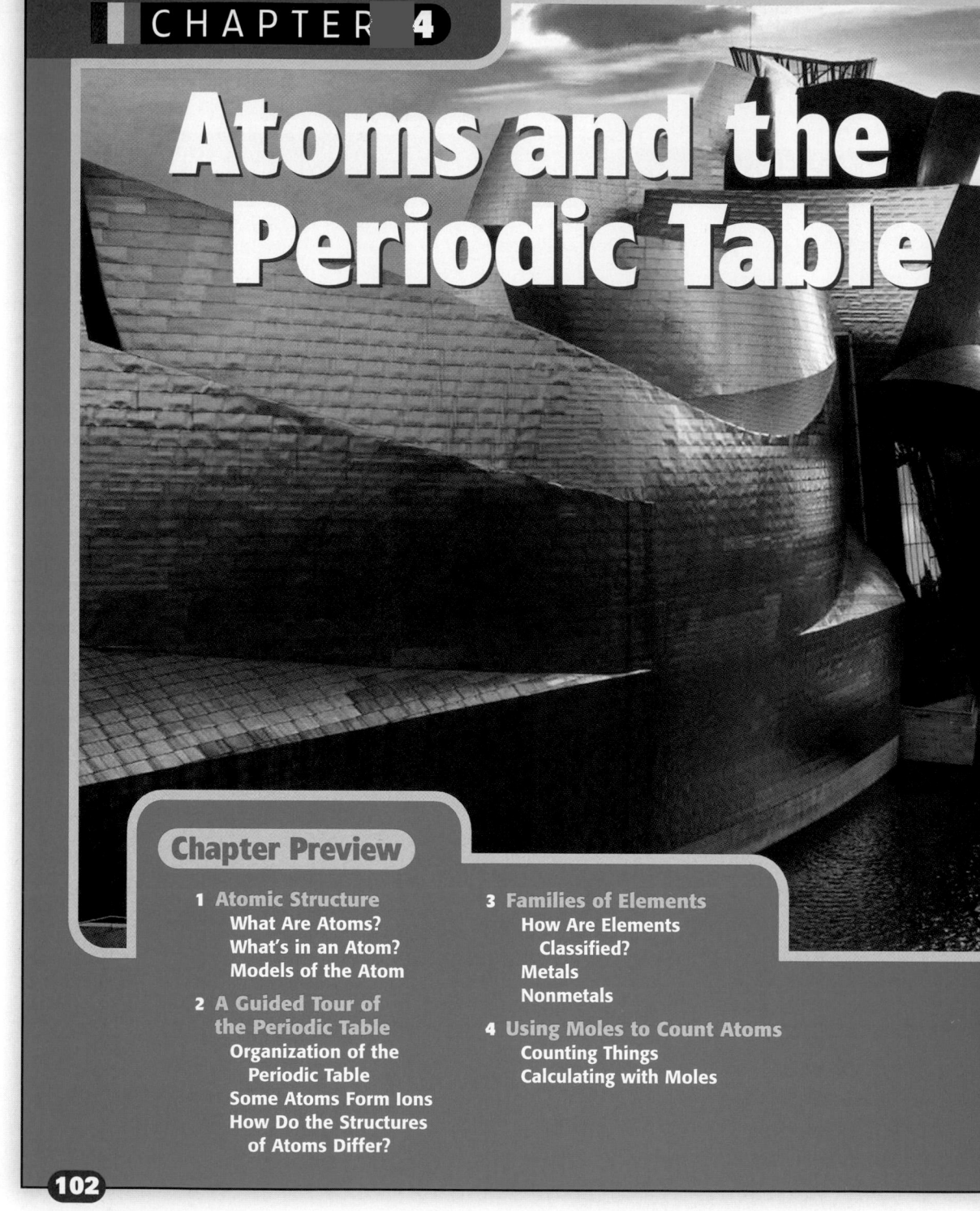

Atoms and the Periodic Table

Chapter Preview

1 Atomic Structure
What Are Atoms?
What's in an Atom?
Models of the Atom

2 A Guided Tour of the Periodic Table
Organization of the Periodic Table
Some Atoms Form Ions
How Do the Structures of Atoms Differ?

3 Families of Elements
How Are Elements Classified?
Metals
Nonmetals

4 Using Moles to Count Atoms
Counting Things
Calculating with Moles

102

Standards Correlations

National Science Education Standards

The following descriptions summarize the National Science Standards that specifically relate to this chapter. For the full text of the standards, see the National Science Education Standards at the front of the book.

Section 1 Atomic Structure
Physical Science PS 1a, 1b, 1c, 2b
Unifying Concepts and Processes UCP 2
History and Nature of Science HNS 1, 3

Section 2 A Guided Tour of the Periodic Table
Physical Science PS 1c, 2a, 2b
Unifying Concepts and Processes UCP 1, 5

Section 3 Families of Elements
Physical Science PS 2a, 2b
Science as Inquiry SAI 1
Science in Personal and Social Perspectives SPSP 2, 5

Section 4 Using Moles to Count Atoms
Unifying Concepts and Processes UCP 3
History and Nature of Science HNS 3

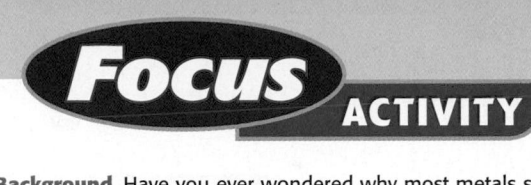

Focus ACTIVITY

Background Have you ever wondered why most metals shine? Metals shine because they are made of elements that reflect light. Another property of metals is that they do not shatter. Metals bend as they are pressed into thin, flat sheets during the coin-making process. All metals share some similarities, but each metal has its own unique chemical and physical properties.

The unique building shown on the opposite page is the Guggenheim Museum in Bilboa, Spain. This art museum is covered in panels made of titanium. Titanium is a strong, durable metallic element that can be used for a variety of purposes.

Metals, like everything around us, are made of trillions of tiny units that are too small to see. These units are called atoms. Atoms determine the properties of all substances. For example, gold atoms make gold softer and shinier than silver, which is made of silver atoms. Pennies get their color from the copper atoms they are coated with. In this chapter, you will learn what determines an atom's properties, why atoms are considered the smallest units of elements, and how elements are classified.

Activity 1 What metals do you see during a typical day? Describe their uses and their properties.

Activity 2 Describe several different ways to classify the metals shown on the opposite page.

Atoms determine the properties of objects. For example, metal atoms give gold its shine and the ability to be worked into different shapes.

internet connect

www.scilinks.org
Topic: Atoms and Elements SciLinks code: HK4012

sci LINKS Maintained by the National Science Teachers Association

Pre-Reading Questions

1. How are the atoms of all elements alike?
2. How does the periodic table help us learn about atoms and elements?
3. Which elements does your body contain?

103

Focus ACTIVITY

Background Certain elements are classified as metals because of their properties. Two properties of all metals are malleability and ductility.

Malleability is the ability of a metal to be hammered or rolled into a sheet without breaking. Malleability enables metals to be shaped into useful objects, such as sheets.

Ductility is the ability to be drawn into a wire. Ductility and the ability to conduct electricity make metals good materials for electrical wiring.

Activity 1 Answers may vary but could include the steel that is used in vehicles, which is strong and durable, or the lightweight aluminum used in beverage cans. **LS Verbal**

Activity 2 Answers may vary but could include density, maleability, luster, ductility, electrical conductivity, and heat conductivity. **LS Logical**

Answers to Pre-Reading Questions

1. All atoms are made up of protons, neutrons, and electrons. All atoms are alike in that they determinc the properties of their elements.

2. The periodic table helps us learn about atoms and elements by organizing elements according to their atomic number. An element's position in the table tells us about its atoms' electron configuration and chemical properties.

3. The human body contains many elements including calcium, magnesium, carbon, and many compounds of elements.

Chapter Resource File

• Pretest **GENERAL**
• Teaching Transparency Preview
• Answer Keys

Focus

Overview

Before beginning this section, review with your students the Objectives listed in the Student Edition. This section begins with a discussion of the atomic theories of Democritus and Dalton. Students then learn about the three basic subatomic particles (protons, electrons, neutrons) and explore modern models of the atom. The section concludes with a discussion of electron orbitals and valence electrons.

🎧 Bellringer

Use the Bellringer transparency to prepare students for this section.

Motivate

Opening Discussion ——— GENERAL

Write the following statements on the chalkboard:

- An atom cannot be broken down into smaller parts.
- An atom is the same throughout.
- An atom is made up of several different, smaller parts.

Ask students their opinions of the statements. Have them discuss the opinions and try to justify their own opinions. Save a list of the opinions for discussion after completing the section. **LS Verbal**

Atomic Structure

▶ KEY TERMS
  nucleus
  proton
  neutron
  electron
  orbital
  valence electron

OBJECTIVES

▶ **Explain** Dalton's atomic theory, and describe why it was more successful than Democritus's theory.
▶ **State** the charge, mass, and location of each part of an atom according to the modern model of the atom.
▶ **Compare** and contrast Bohr's model with the modern model of the atom.

Atoms are everywhere. They make up the air you are breathing, the chair you are sitting in, and the clothes you are wearing. This book, including this page you are reading, is also made of atoms.

Disc One, Module 2:
Models of the Atom
Use the Interactive Tutor to learn more about this topic.

What Are Atoms?

Atoms are tiny units that determine the properties of all matter. The aluminum cans shown in *Figure 1* are lightweight and easy to crush because of the properties of the atoms that make up the aluminum.

Our understanding of atoms required many centuries

In the fourth century BCE, the Greek philosopher Democritus suggested that the universe was made of invisible units called atoms. The word *atom* is derived from the Greek word meaning "unable to be divided." He believed movements of atoms caused the changes in matter that he observed.

Although Democritus's theory of atoms explained some observations, Democritus was unable to provide the evidence needed to convince people that atoms really existed. Throughout the centuries that followed, some people supported Democritus's theory. But other theories were also proposed. As the science of chemistry was developing in the 1700s, more emphasis was put on making careful and repeated measurements in scientific experiments. As a result, more-reliable data were collected and used to favor one theory over another.

Figure 1

The atoms in aluminum, seen here as an image from a scanning tunneling electron microscope, give these aluminum cans their properties.

Aluminum Atoms

104

SOCIAL STUDIES
CONNECTION ——— GENERAL

Have students do research and then make a timeline that shows the development of the modern atomic theory, identifying on the timeline the contributions of key physicists and chemists.

To put the dates in context, ask students to include information about other events in history that happened around these key dates. For example, Dalton proposed the atomic theory in 1808. This is 32 years after the U.S. Declaration of Independence (1776) and 5 years after the Louisiana Purchase

(1803). For recent events, students could also include important events in their family's history. Encourage them to interview parents and grandparents for this information.

Ask students to create their timelines on poster board. Students may wish to use different colors to distinguish between scientific, historical, and personal events. Encourage them to use illustrations as well. Display students' completed timelines in the classroom. **LS Interpersonal**

John Dalton developed an atomic theory

In 1808, an English schoolteacher named John Dalton proposed his own atomic theory. Dalton's theory was developed with a scientific basis, and some parts of his theory still hold true today. Like Democritus, Dalton proposed that atoms could not be divided. Today, we know that atoms are actually made up of even smaller particles! According to Dalton, all atoms of a given element were exactly alike. Dalton also stated that atoms of different elements could join to form compounds. Today, Dalton's theory is considered the foundation for modern atomic theory.

Atoms are the building blocks of molecules

An atom is the smallest part of an element that still has the element's properties. Imagine dividing a coin made of pure copper until the pieces were too small for you to see. If you were able to continue dividing these pieces, you would be left with the simplest units of the coin—copper atoms. All the copper atoms would be alike. Each copper atom would have the same chemical properties as the coin you started with.

You have learned that atoms can join. *Figure 2* shows atoms joined together to form molecules of water. The water we see is actually made of a very large number of water molecules. Whether it gushes downstream in a riverbed or is bottled for us to drink, water is always the same: each molecule is made of two hydrogen atoms and one oxygen atom.

internet connect

www.scilinks.org
Topic: Atomic Theory
SciLinks code: HK4011

SciLINKS. Maintained by the National Science Teachers Association

Figure 2

A The water that we see, no matter what its source, is made of many molecules.

B Each molecule of water is made of two hydrogen atoms and one oxygen atom.

Molecules of water

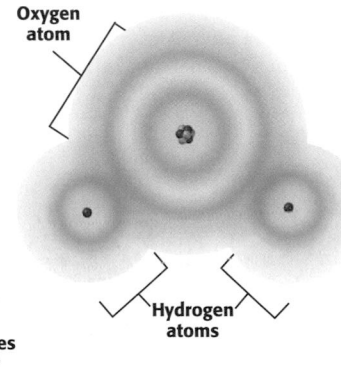

Oxygen atom

Hydrogen atoms

105

Teaching Tip

Neutral Atoms On the chalkboard, write a number line that contains both positive and negative numbers. Use the number line to show students that adding equal numbers of positive and negative charges results in no charge—zero on the number line.

SKILL BUILDER

Interpreting Visuals Students might ask why the protons in the nucleus stay together since positive charges repel each other.

Tell students that even though the protons in the nucleus do electrically repel each other, they are held together by a stronger force known as the *nuclear force,* also called the *strong nuclear force.* This force is unique to the nucleus and exists only over very short distances.

Gravity also attracts nuclear particles to each other. But gravity is not the main force that holds the nucleus together, as in **Figure 4,** because it is not strong enough to overcome the electrical repulsion.

Figure 3

If the nucleus of an atom were the size of a marble, the whole atom would be the size of a football stadium

▶ **nucleus** an atom's central region which is made up of protons and neutrons

▶ **proton** a subatomic particle that has a positive charge and that is found in the nucleus of an atom

▶ **neutron** a subatomic particle that has no charge and that is found in the nucleus of an atom

▶ **electron** a subatomic particle that has a negative charge

Figure 4

A helium atom is made of two protons, two neutrons, and two electrons ($2e^-$).

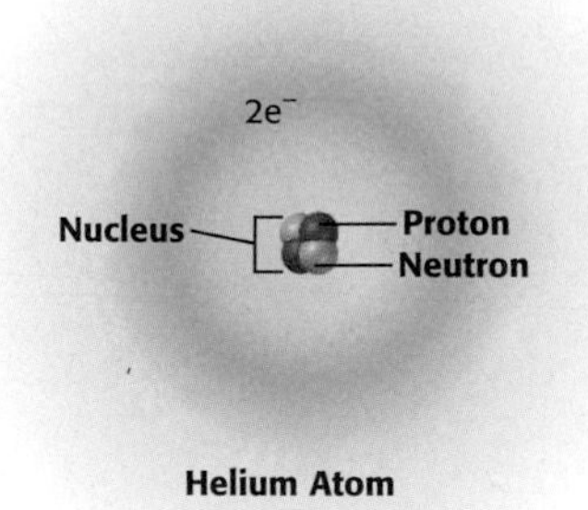

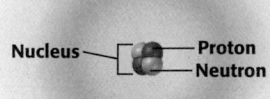

$2e^-$

Nucleus — Proton
— Neutron

Helium Atom

What's in an Atom?

Less than 100 years after Dalton published his atomic theory, scientists determined that atoms consisted of still smaller particles and could be broken down even further. While we now know that atoms are made up of many different subatomic particles, we need to study only three of these particles to understand the chemistry of most substances.

Atoms are made of protons, neutrons, and electrons

At the center of each atom is a small, dense **nucleus** with a positive electric charge. The nucleus is made of **protons** and **neutrons.** These two subatomic particles are almost identical in size and mass, but protons have a positive electric charge while neutrons have no electric charge at all. Moving around outside the nucleus is a cloud of very tiny negatively charged subatomic particles with very little mass. These particles are called **electrons.** To get an idea of how far from the nucleus an electron can be, see **Figure 3.** If the nucleus of an atom were the size of a marble, the whole atom would be the size of a football stadium! A helium atom, shown in **Figure 4,** has one more proton and one more electron than a hydrogen atom has. The number of protons and electrons an atom has is unique for each element.

Unreacted atoms have no overall charge

You might be surprised to learn that atoms are not charged even though they are made of charged protons and electrons. Atoms do not have a charge because they have an equal number of protons and electrons whose charges exactly cancel. A helium atom has two protons and two electrons. The atom is neutral because the positive charge of the two protons exactly cancels the negative charge of the two electrons.

Charge of two protons:	+2
Charge of two neutrons:	0
Charge of two electrons:	−2
Total charge of a helium atom:	0

Subatomic Particles

Particle	Charge	Mass (kg)	Location in the atom
Proton	+1	1.67×10^{-27}	In the nucleus
Neutron	0	1.67×10^{-27}	In the nucleus
Electron	−1	9.11×10^{-31}	Moving around outside the nucleus

Did You Know ?

Quarks and Leptons Most matter, including protons and neutrons, is made up of smaller particles, called quarks. There are six types of quarks: up, down, charm, strange, top (or truth), and bottom (or beauty). Protons and neutrons both consist of up and down quarks. The proton, for example, is made of two up quarks and one down quark. Electrons belong to a class of fundamental particles called leptons. Other leptons are the muon, the tau, and three types of neutrinos.

Trends in Physics

Particle Accelerators Today, scientists have identified more than 100 subatomic particles. Much of our knowledge of these particles comes from the study of collisions of protons and neutrons with atomic nuclei in particle accelerators. Most recently, in 1995, scientists at Fermi Lab experimentally proved the existence of the sixth quark—the elusive top quark, which had been theoretically predicted in 1977. This discovery may help physicists answer some of the unresolved questions in physics today.

Models of the Atom

Democritus in the fourth century BCE and later Dalton, in the nineteenth century, thought that the atom could not be split. That theory had to be modified when it was discovered that atoms are made of protons, neutrons, and electrons. Like most scientific models and theories, the model of the atom has been revised many times to explain such new discoveries.

Bohr's model compares electrons to planets

In 1913, the Danish scientist Niels Bohr suggested that electrons in an atom move in set paths around the nucleus much like the planets orbit the sun in our solar system. In Bohr's model, each electron has a certain energy that is determined by its path around the nucleus. This path defines the electron's energy level. Electrons can only be in certain energy levels. They must gain energy to move to a higher energy level or lose energy to move to a lower energy level.

One way to imagine Bohr's model is to compare an atom to the stairless building shown in **Figure 5.** Imagine that the nucleus is in a very deep basement and that the electronic energy levels begin on the first floor. Electrons can be on any floor of the building but not between floors. Electrons gain energy by riding up in the elevator and lose energy by riding down in the elevator. Bohr's description of energy levels is still considered accurate today.

Electrons act more like waves

By 1925, Bohr's model of the atom no longer explained electron behavior. So a new model was proposed that no longer assumed that electrons orbited the nucleus along definite paths like planets orbiting the sun. In this modern model of the atom, it is believed that electrons behave more like waves on a vibrating string than like particles.

Electrons cannot be between floors.

4th energy level

3rd energy level

2nd energy level

1st energy level

Basement

Nucleus

Figure 5

The energy levels of an atom are like the floors of the building shown here. The energy difference between energy levels decreases as the energy level increases.

107

Teaching Tip

Waves and Particles Scientists have discovered that particles act as waves, and waves act as particles. This concept is called the *wave-particle duality of nature.* The wave nature of most particles does not affect the particle much. However, the smaller the particle, the more the particle acts as both a wave and a particle.

MISCONCEPTION ///ALERT\\\

The Bohr Model The Bohr model leads some students to believe that the distance of an electron from the nucleus of an atom is directly proportional to the electron's energy. Be sure to emphasize that there is no direct proportion between the distance of a given level from the nucleus and the amount of energy needed to reach that energy level.

Historical Perspective

Wave-Particle Duality In 1923, Louis de Broglie, a French physicist, made a hypothesis that led to a statement of the wave-particle duality of nature and the present theory of how atoms are structured. De Broglie used research done by Albert Einstein and Max Planck to develop a mathematical equation that relates the mass and velocity of a particle to its wavelength.

Chapter Resource File

- **Cross-Disciplinary Worksheet** Integrating Physics—Atomic Fingerprints ADVANCED

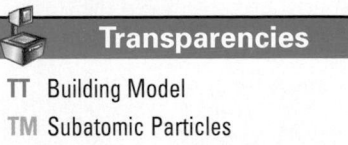

Transparencies

TT Building Model

TM Subatomic Particles

Reading Skills Have students revisit their active reading opinions from the Opening Discussion at the beginning of the section. Have them discuss whether their opinions have changed or remain the same. Have them cite passages in the text that account for the change in or reinforcement of their opinions. **LS Verbal**

Figure 6

The exact location of any of the blades of this fan is difficult to determine. The exact location of an electron is impossible to determine.

An electron's exact location cannot be determined

Imagine the moving blades of a fan, like the one shown in *Figure 6.* If you were asked where any one of the blades was located at a certain instant, you would not be able to give an exact answer. It is very difficult to know the exact location of any of the blades because they are moving so quickly. All you know for sure is that each blade could be anywhere within the blurred area you see as the blades turn.

It is impossible to determine both the exact location of an electron in an atom and the electron's speed and direction. The best scientists can do is calculate the chance of finding an electron in a certain place within an atom. One way to visually show the likelihood of finding an electron in a given location is by shading. The darker the shading, the better the chance of finding an electron at that location. The whole shaded region is called an electron cloud.

Electrons exist in energy levels

Within the atom, electrons with different amounts of energy exist in different energy levels. There are many possible energy levels that an electron can occupy. The number of filled energy levels an atom has depends on the number of electrons. The first energy level holds two electrons, and the second energy level holds eight. So, a lithium atom, with three electrons, has two electrons in its first energy level and one in the second. *Figure 7* shows how the first four energy levels are filled.

Figure 7

Electrons can be found in different energy levels. Each energy level holds a certain number of electrons.

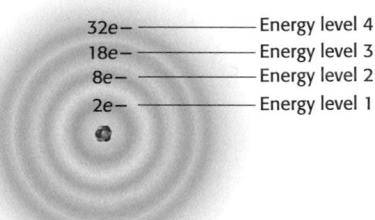

32e– ———— Energy level 4
18e– ———— Energy level 3
8e– ———— Energy level 2
2e– ———— Energy level 1

Alternative Assessment — ADVANCED

Modeling Atoms Ask students to create three-dimensional models of an atom of their choice. Encourage them to create more than one model of their atom, corresponding to different versions of the atomic theory throughout history. Instruct them to include information about which historical theories they are representing in each model, and which parts of those theories are no longer accepted today. Also ask them to include information about the limitations of their models.
LS Kinesthetic

Did You Know ?

A Tiny Mass The mass of an electron is more than 1,000 times less than the mass of a proton or neutron. For comparison, the ratio of masses between an electron and a proton or neutron is about equal to that between a domestic cat and an elephant.

Constructing a Model

A scientific model is a simplified representation based on limited knowledge that describes how an object looks or functions. In this activity, you will construct your own model.

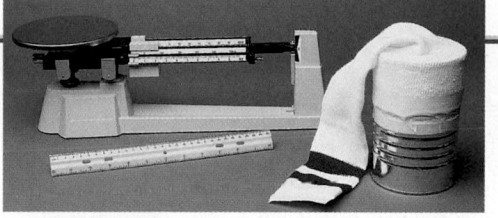

1. Obtain from your teacher a can that is covered by a sock and sealed with tape. An unknown object is inside the can.
2. Without unsealing the container, try to determine the characteristics of the object inside by examining it through the sock. What is the object's mass? What is its size, shape, and texture? Record all of your observations in a data table.
3. Remove the taped sock so that you can touch the object without looking at it. Record these observations as well.
4. Use the data you have collected to draw a model of the unknown object.
5. Finally remove the object to see what it is. Compare and contrast the model you made with the object it is meant to represent.

Electrons are found in orbitals within energy levels

The regions in an atom where electrons are likely to be found are called **orbitals.** Within each energy level, electrons occupy orbitals that have the lowest energy. The four different kinds of orbitals are the *s, p, d,* and *f* orbitals. The simplest kind of orbital is an *s* orbital. An *s* orbital can have only one possible orientation in space because it has a shape like a sphere, as shown in *Figure 8.* An *s* orbital has the lowest energy and can hold two electrons.

A *p* orbital, on the other hand, is dumbbell shaped and can be oriented three different ways in space, as shown in *Figure 9.* The axes on the graphs are drawn to help you picture how these orbitals look in three dimensions. Imagine the *y*-axis being flat on the page. Imagine the dotted lines on the *x*- and *z*-axes going into the page, and the darker lines coming out of the page. A *p* orbital has more energy than an *s* orbital has. Because each *p* orbital can hold two electrons, the three *p* orbitals can hold a total of six electrons.

The *d* and *f* orbitals are much more complex. There are five possible *d* orbitals and seven possible *f* orbitals. An *f* orbital has the greatest energy. Although all these orbitals are very different in shape, each can hold a maximum of two electrons.

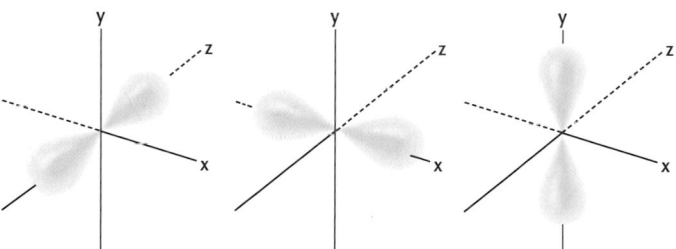

▶ **orbital** a region in an atom where there is a high probability of finding electrons

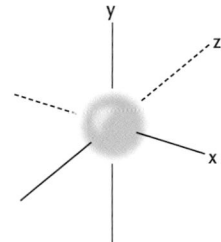

Figure 8

An *s* orbital is shaped like a sphere, so it has only one possible orientation in space. An *s* orbital can hold a maximum of two electrons.

Figure 9

Each of these *p* orbitals can hold a maximum of two electrons, so all three together can hold a total of six electrons.

109

Quiz

1. What part of an atom contains the protons and neutrons? (the nucleus)

2. What are the regions of an atom where electrons are found? (orbitals)

3. Who proposed the first widely accepted version of the atomic theory? (Dalton)

4. Who proposed a model in which electrons can only be in certain energy levels? (Bohr)

5. What are electrons in the outer-most energy level called? (valence electrons)

Figure 10
The neon atoms of this sign have eight valence electrons. The sign lights up because atoms first gain energy and then release this energy in the form of light.

▶ **valence electron** an electron that is found in the outermost shell of an atom and that determines the atom's chemical properties

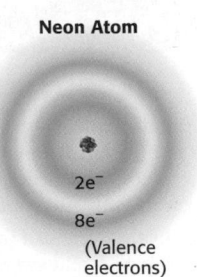

Neon Atom

2e⁻

8e⁻
(Valence electrons)

Every atom has between one and eight valence electrons

An electron in the outermost energy level of an atom is called a **valence electron.** Valence electrons determine an atom's chemical properties and its ability to form bonds. The single electron of a hydrogen atom is a valence electron because it is the only electron the atom has. The glowing red sign shown in *Figure 10* is made of neon atoms. In a neon atom, which has 10 electrons, 2 electrons fill the lowest energy level. Its valence electrons, then, are the 8 electrons that are farther away from the nucleus in the atom's second (and outermost) energy level.

SECTION 1 REVIEW

SUMMARY

▶ Elements are made of very small units called atoms.

▶ The nucleus of an atom is made of positively charged protons and uncharged neutrons.

▶ Negatively charged electrons surround the nucleus.

▶ Atoms have an equal number of protons and electrons.

▶ In Bohr's model of the atom, electrons orbit the nucleus in set paths.

▶ In the modern atomic model, electrons are found in orbitals within each energy level.

▶ Electrons in the outermost energy level are called valence electrons.

1. **Summarize** the main ideas of Dalton's atomic theory.

2. **Explain** why Dalton's theory was more successful than Democritus's theory.

3. **List** the charge, mass, and location of each of the three subatomic particles found within atoms.

4. **Explain** why oxygen atoms are neutral.

5. **Compare** an atom's structure to a ladder. What parts of the ladder correspond to the energy levels of the atom? Identify one way a real ladder is not a good model for the atom.

6. **Explain** how the path of an electron differs in Bohr's model and in the modern model of the atom.

7. **Critical Thinking** In the early 1900s, two associates of New Zealander Ernest Rutherford bombarded thin sheets of gold with positively charged subatomic particles. They found that most particles passed right through the sheets but some bounced back as if they had hit something solid. Based on their results, what do you think an atom is made of? What part of the atom caused the particles to bounce back? (**Hint:** Positive charges repel other positive charges.)

110

Answers to Section 1 Review

1. Elements are made of tiny, unique particles called atoms. Atoms cannot be divided. Atoms of the same element are identical. Atoms of different elements can join to form molecules.

2. Evidence existed to support Dalton's theory.

3. proton: $+1$, 1.67×10^{-27} kg, in the nucleus; neutron: 0, 1.67×10^{-27} kg, in the nucleus; electron: -1, 9.11×10^{-31} kg, outside the nucleus

4. Oxygen has eight negatively-charged electrons to balance out the charge of the eight positively-charged protons.

5. Each rung of the ladder represents an energy level. Electrons can only be on the rungs, not between rungs. A ladder is not a good model for an atom because electrons exist in three dimensional orbitals.

6. In Bohr's model, the electrons travel along fixed paths. In the modern model, electrons behave more like waves on a string than like particles.

7. empty space; the protons in the nucleus

A Guided Tour of the Periodic Table

OBJECTIVES

▶ **Relate** the organization of the periodic table to the arrangement of electrons within an atom.

▶ **Explain** why some atoms gain or lose electrons to form ions.

▶ **Determine** how many protons, neutrons, and electrons an atom has, given its symbol, atomic number, and mass number.

▶ **Describe** how the abundance of isotopes affects an element's average atomic mass.

▶ KEY TERMS
periodic law
period
group
ion
atomic number
mass number
isotope
atomic mass unit (amu)
average atomic mass

When you are in a store, are your favorite items placed randomly on the shelves? Usually similar items are grouped together, as shown in **Figure 11,** so that you can find what you need quickly. The periodic table organizes the elements in a similar way.

Organization of the Periodic Table

The periodic table groups similar elements together. This organization makes it easier to predict the properties of an element based on where it is in the periodic table. In the periodic table shown in **Figure 12,** elements are represented by their symbols and are arranged in a certain order. The order is based on the number of protons an atom of that element has in its nucleus.

A hydrogen atom has one proton, so hydrogen is the first element listed in the periodic table. A helium atom has two protons and is the second element listed. Elements are listed in this order in the periodic table because the **periodic law** states that when elements are arranged this way, similarities in their properties will occur in a regular pattern.

▶ **periodic law** the law that states that the repeating chemical and physical properties of elements change periodically with the atomic numbers of the elements

![Photograph of a store interior with signs reading SPORTSWEAR, PICTURES & FRAMES, HOUSEWARES]

Figure 11

In many stores, similar items are grouped so that they are easier to find.

111

Alternative Assessment — BASIC

Analyzing the Periodic Table Before students read this section, have them examine a periodic table and see what they can discover about its arrangement. Ask them to make a list of their observations. After reading the section, have students look over their lists again, correcting any errors and adding new information. (Students' initial lists could include the following: Elements are grouped into categories; elements in columns are often in the same category; elements are arranged horizontally in order of atomic number.) **LS Visual**

History of the Periodic Table — ADVANCED

Although the Russian chemist Dmitri Mendeleev is generally credited as being the "father" of the periodic table, his work was based on earlier versions of periodic tables by a number of scientists, including A. E. Béguyer de Chancourtois and John A. R. Newlands. A German scientist, Lothar Meyer, developed a periodic table very similar to Mendeleev's around the same time period. Have students conduct research and prepare a report detailing the contributions of these scientists and of Mendeleev. **LS Verbal**

Focus

Overview

Before beginning this section, review with your students the Objectives listed in the Student Edition. In this section, students study the organization of the periodic table. They learn about the process of ionization, and about how atomic number and atomic mass are used to represent properties of atoms. The section concludes with coverage of isotopes and the concept of average atomic mass.

🔊 Bellringer

Use the Bellringer transparency to prepare students for this section.

Motivate

Opening Demonstration — GENERAL

Show students a large, one-month calendar and ask them to make a list of things that they put on their calendars at home. (Students who don't keep calendars at home may list activities or events they have planned for the coming month.) Ask volunteers to read their lists and, as they do so, write the items on the calendar each time they occur. Ask students: What types of items did you list? (Answers will vary but could include appointments, weekly music lessons, and daily practices.) What items appear in a regular pattern, or are periodic? (Answers should include items that recur on a monthly, weekly, or daily basis.) **LS Intrapersonal**

Chapter Resource File

• **Lesson Plan**

• **Cross-Disciplinary Worksheet**
Integrating Biology—Elements in Your Body **ADVANCED**

Transparencies

TT Bellringer

Vocabulary Students should not worry if they do not understand the precise meaning of the periodic law. Assure them that its meaning will become clearer as they study the groups in the table. To help students gain a better understanding, ask them to name some periodicals and to define the word *periodical*. Use this idea to point out the arrangement of the periodic table. **LS Verbal**

SKILL BUILDER — **BASIC**

Interpreting Visuals Have students place bookmarks in their texts to mark the periodic table on these two pages. Encourage students to familiarize themselves with the information listed in the table and to make frequent reference to these pages as they read the rest of the chapter. **LS Visual**

Teaching Tip
Using the Periodic Table
Inform students of the information available in the periodic table, including symbol, name, atomic number, average atomic mass, group number, period number, and whether the element is a metal or a nonmetal. Students sometimes think that they have to memorize the information in the periodic table. Point out that the table exists as a scientific tool. The student's job is to learn to interpret the information on the table, not to memorize it.

Figure 12

The Periodic Table of the Elements

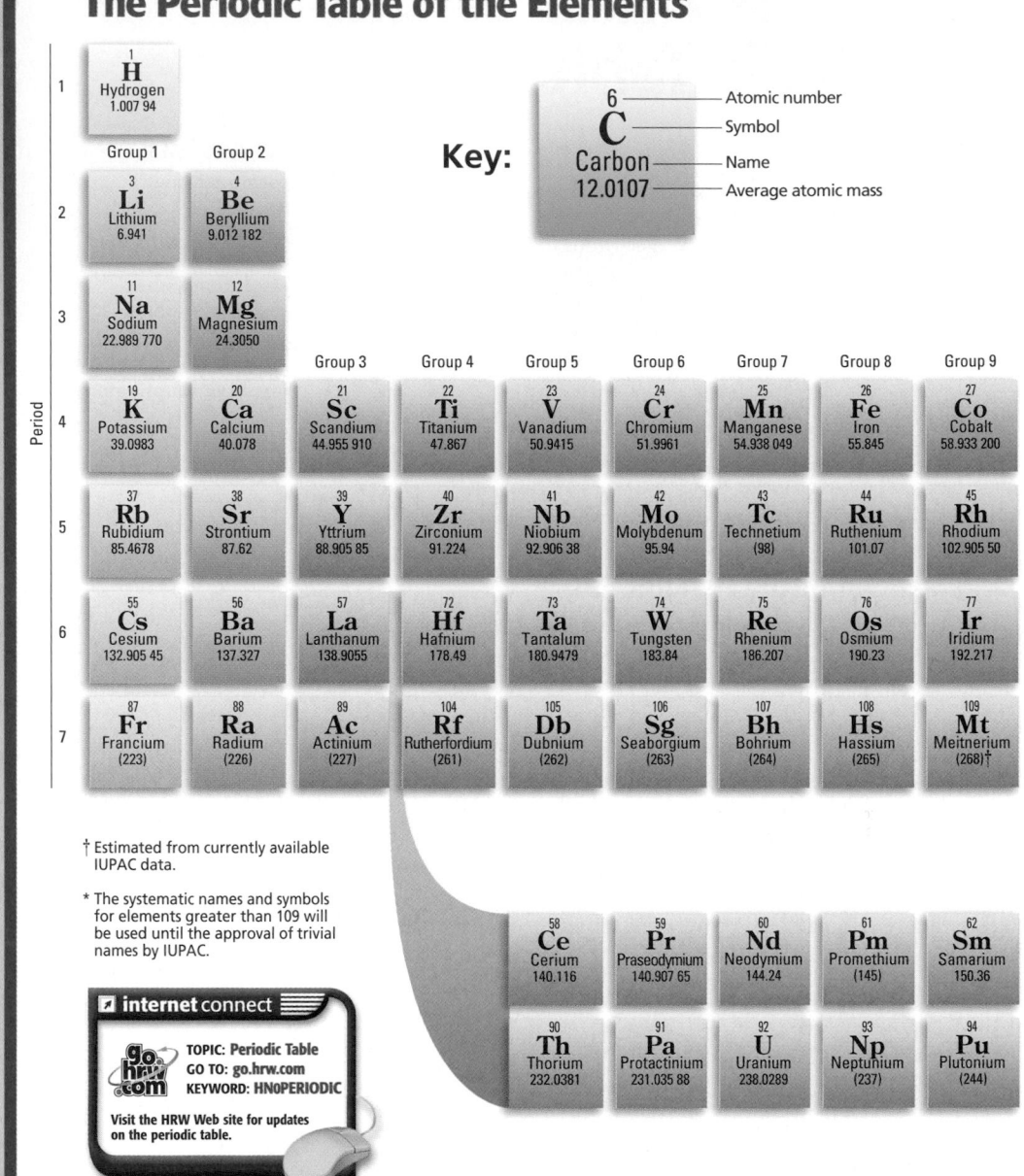

† Estimated from currently available IUPAC data.

* The systematic names and symbols for elements greater than 109 will be used until the approval of trivial names by IUPAC.

internet connect

go.hrw.com

TOPIC: Periodic Table
GO TO: go.hrw.com
KEYWORD: HN0PERIODIC

Visit the HRW Web site for updates on the periodic table.

Historical Perspective

Mendeleev's Periodic Table The modern periodic table is based on one presented by Dmitri Mendeleev, a Russian chemist, in 1869. Mendeleev realized that certain properties repeat periodically when the elements are arranged horizontally in order of atomic weight. He also placed chemically similar elements in vertical columns, leaving empty spaces as needed. From the empty spaces, Mendeleev deduced the existence of unknown elements and predicted some of their properties. For example, he left a space below silicon, to the left of arsenic. A few years later the element germanium was discovered; germanium fit into this spot based on its weight, and its properties matched Mendeleev's predictions.

When Mendeleev published his first table, scientists did not know about subatomic particles. Today, elements are arranged by atomic number (number of protons) instead of by atomic weight. This changed the order of a few elements, resolving some discrepancies between predicted and observed properties. Also, the modern periodic table has over 110 elements—many more than the 63 elements known to scientists in Mendeleev's time.

Metals
- Alkali metals
- Alkaline-earth metals
- Transition metals
- Other metals

Nonmetals
- Hydrogen
- Semiconductors (also known as *metalloids*)
- Other nonmetals
- Halogens
- Noble gases

Group 18

| 2 |
| He |
| Helium |
| 4.002 602 |

Group 13	Group 14	Group 15	Group 16	Group 17
5	6	7	8	9
B	C	N	O	F
Boron	Carbon	Nitrogen	Oxygen	Fluorine
10.811	12.0107	14.006 74	15.9994	18.998 4032

| 10 |
| Ne |
| Neon |
| 20.1797 |

13	14	15	16	17	18
Al	Si	P	S	Cl	Ar
Aluminum	Silicon	Phosphorus	Sulfur	Chlorine	Argon
26.981 538	28.0855	30.973 761	32.066	35.4527	39.948

Group 10	Group 11	Group 12
28	29	30
Ni	Cu	Zn
Nickel	Copper	Zinc
58.6934	63.546	65.39

31	32	33	34	35	36
Ga	Ge	As	Se	Br	Kr
Gallium	Germanium	Arsenic	Selenium	Bromine	Krypton
69.723	72.61	74.921 60	78.96	79.904	83.80

46	47	48	49	50	51	52	53	54
Pd	Ag	Cd	In	Sn	Sb	Te	I	Xe
Palladium	Silver	Cadmium	Indium	Tin	Antimony	Tellurium	Iodine	Xenon
106.42	107.8682	112.411	114.818	118.710	121.760	127.60	126.904 47	131.29

78	79	80	81	82	83	84	85	86
Pt	Au	Hg	Tl	Pb	Bi	Po	At	Rn
Platinum	Gold	Mercury	Thallium	Lead	Bismuth	Polonium	Astatine	Radon
195.078	196.966 55	200.59	204.3833	207.2	208.980 38	(209)	(210)	(222)

110	111	112		114
Uun*	Uuu*	Uub*		Uuq*
Unununilium	Unununium	Ununbium		Ununquadium
(269)†	(272)†	(277)†		(285)†

A team at Lawrence Berkeley National Laboratories reported the discovery of elements 116 and 118 in June 1999. The same team retracted the discovery in July 2001. The discovery of element 114 has been reported but not confirmed.

63	64	65	66	67	68	69	70	71
Eu	Gd	Tb	Dy	Ho	Er	Tm	Yb	Lu
Europium	Gadolinium	Terbium	Dysprosium	Holmium	Erbium	Thulium	Ytterbium	Lutetium
151.964	157.25	158.925 34	162.50	164.930 32	167.26	168.934 21	173.04	174.967

95	96	97	98	99	100	101	102	103
Am	Cm	Bk	Cf	Es	Fm	Md	No	Lr
Americium	Curium	Berkelium	Californium	Einsteinium	Fermium	Mendelevium	Nobelium	Lawrencium
(243)	(247)	(247)	(251)	(252)	(257)	(258)	(259)	(262)

The atomic masses listed in this table reflect the precision of current measurements. (Values listed in parentheses are those of the element's most stable or most common isotope.) In calculations throughout the text, however, atomic masses have been rounded to two places to the right of the decimal.

113

Teaching Tip ——— GENERAL
Common Chemical Symbols
Give students the names of several of the most common elements. Emphasize that while students should not try to memorize all of the chemical symbols on the periodic table, it is helpful to know the most common ones. Have students make a flashcard for each element on your list. Each card should contain the name of the element on one side and its symbol on the other side. After students study their cards, quiz them orally on the symbols. **LS Visual**

Group Activity ——— BASIC
Interpreting the Periodic Table
Group students in pairs. Ask one student in each pair to choose an element. The other student should use the periodic table to identify the element's symbol, atomic number, average atomic mass, group number, period number, and whether the element is a metal or a nonmetal, while the first student verifies their answers. Then have students switch roles and repeat the activity. **LS Interpersonal**

Cultural Awareness
Elements have connections to different cultures. Carbon, sulfur, tin, gold, and silver were named and used by many ancient civilizations. Platinum was introduced to Europe in 1750 from South America. The name for zirconium comes from the Arabic word *zargun*, which means "gold color." Vanadium was discovered in Mexico and was named after a Scandinavian goddess. Polonium was named after Poland to honor the birthplace of its discoverer, Marie Curie.

Did You Know?
Transuranium Elements Transuranium elements, which are those past uranium on the periodic table, have been difficult to study because they do not exist in nature. They must be created in a laboratory, and many exist for a very short period of time.

One particularly troublesome element is element 104. American scientists Albert Ghiorso and James Harris created this elusive element at the Lawrence Radiation Laboratory at the University of California at Berkeley in 1969.

Teaching Tip

Energy Levels Be sure students do not attempt to show how electrons are arranged in atoms past calcium in the periodic table. As electron energy levels increase, the difference in their energies decreases.

Starting with the third energy level, the levels overlap and the order in which orbitals fill becomes irregular. The pattern of this overlap is beyond the scope of this course, but more information can be found in a high school chemistry textbook.

Teaching Tip

Element Groups Be sure students understand that physical properties may or may not be similar among elements in a group. Although groups of elements may share certain physical properties, chemical properties are more likely to be similar.

SKILL BUILDER — BASIC

Interpreting Visuals Have students examine **Figure 13.** Ask: Why do elements in the same column have similar chemical properties? (because they have the same number of valence electrons, and the number of valence electrons determines how an element will chemically react with other elements) **LS Visual**

▶ **period** a horizontal row of elements in the periodic table

▶ **group (family)** a vertical column of elements in the periodic table

internet connect
www.scilinks.org
Topic: Periodic Table
SciLinks code: HK4102
SciLINKS Maintained by the National Science Teachers Association

Figure 13
The electronic arrangement of atoms becomes increasingly more complex as you move further right across a period and further down a group of the periodic table.

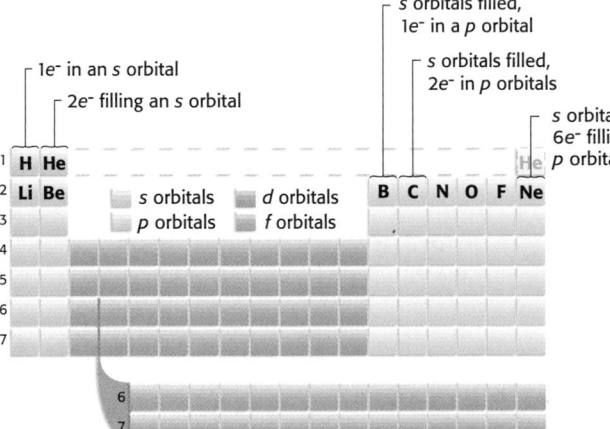

The periodic table helps determine electron arrangement

Horizontal rows in the periodic table are called **periods.** Just as the number of protons an atom has increases by one as you move from left to right across a period, so does its number of electrons. You can determine how an atom's electrons are arranged if you know where the corresponding element is located in the periodic table.

Hydrogen and helium are both located in Period 1 of the periodic table. *Figure 13* shows that a hydrogen atom has one electron in an *s* orbital, while a helium atom has one more electron, for a total of two. Lithium is located in Period 2. The electron arrangement for lithium is just like that for a helium atom, except that lithium has a third electron in an *s* orbital in the second energy level, as follows:

Energy level	Orbital	Number of electrons
1	s	2
2	s	1

As you continue to move to the right in Period 2, you can see that a carbon atom has electrons in s orbitals and *p* orbitals. The locations of the six electrons in a carbon atom are as follows:

Energy level	Orbital	Number of electrons
1	s	2
2	s	2
2	p	2

A nitrogen atom has three electrons in *p* orbitals, an oxygen atom has four, and a fluorine atom has five. *Figure 13* shows that a neon atom has six electrons in *p* orbitals. Each orbital can hold two electrons, so all three *p* orbitals are filled.

Elements in the same group have similar properties

Valence electrons determine the chemical properties of atoms. Atoms of elements in the same **group,** or column, have the same number of valence electrons, so these elements have similar properties. Remember that these elements are not exactly alike, though, because atoms of these elements have different numbers of protons in their nuclei and different numbers of electrons in their filled inner energy levels.

114

Alternative Assessment — GENERAL
Periodicity

Step 1 Divide the class into four groups. Assign each student group one of the following: Group 1, Group 2, Group 17, or Group 18 from the periodic table.

Step 2 Have students find out the properties and uses of the elements in their assigned group and determine what the elements have in common.

Step 3 Have each student group prepare a poster that shows what they learned so that they can present this information to the class.
LS Interpersonal

INCLUSION Strategies

• *Learning Disabled*
• *Attention Deficit Disorder*
• *English Language Learners*

Ask each student to select several elements from the Periodic Table. Label the front side of an index card with atomic number, symbol, name, and atomic mass. On the back of the card, give two examples of compounds that include this element. Students may present their own elements and exchange element cards. All of the completed cards may be placed on a poster or bulletin board to construct parts of the entire periodic table.

Some Atoms Form Ions

Atoms of Group 1 elements are reactive because their outermost energy levels contain only one electron. Atoms that do not have filled outer s and p orbitals may undergo a process called ionization. That is, they may gain or lose valence electrons so that they have a full outermost s and/or p orbital. If an atom gains or loses electrons, it no longer has the same number of electrons as it does protons. Because the charges do not cancel completely as they did before, the **ion** that forms has a net electric charge, as shown for the lithium ion in *Figure 14*. Sodium chloride, or table salt, shown in *Figure 15* is made of sodium and chloride ions.

A lithium atom loses one electron to form a 1+ charged ion

Lithium is located in Group 1 of the periodic table. It is so reactive that it even reacts with the water vapor in the air. An electron is easily removed from a lithium atom, as shown in *Figure 14*. The atomic structure of lithium explains its reactivity. A lithium atom has three electrons. Two of these electrons occupy the first energy level in the s orbital, but only one electron occupies the second energy level. This single valence electron makes lithium very reactive. Removing this electron forms a positive ion, or *cation*.

A lithium ion, written as Li$^+$, is much less reactive than a lithium atom because it has a full outer s orbital. Atoms of other Group 1 elements also have one valence electron. They are also reactive and behave similarly to lithium.

A fluorine atom gains one electron to form a 1– charged ion

Like lithium, fluorine is also very reactive. However, instead of losing an electron to become less reactive, an atom of the element fluorine gains one electron to form an ion with a 1– charge. Fluorine is located in Group 17 of the periodic table, and each atom has nine electrons. Two of these electrons occupy the first energy level, and seven valence electrons occupy the second energy level. A fluorine atom needs only one more electron to have a full outermost energy level. An atom of fluorine easily gains this electron to form a negative ion, or *anion*, as shown in *Figure 16*.

Ions of fluorine are called fluoride ions and are written as F$^-$. Because atoms of other Group 17 elements also have seven valence electrons, they are also reactive and behave similarly to fluorine.

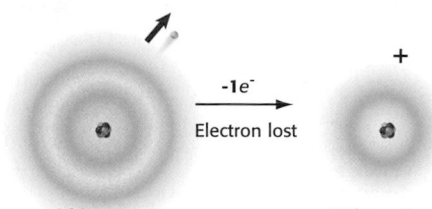

Figure 14

The valence electron of a reactive lithium atom may be removed to form a lithium ion, Li$^+$, with a 1+ charge.

▶ **ion** an atom or group of atoms that has lost or gained one or more electrons and has a negative or positive charge

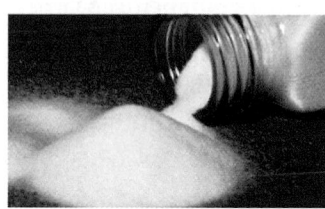

Figure 15

Table salt is made of sodium and chloride ions.

Figure 16

A fluorine atom easily gains one valence electron to form a fluoride ion, F$^-$, with a 1– charge.

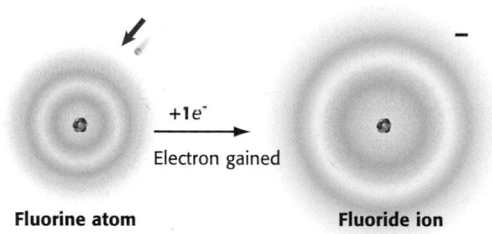

Teaching Tip

Charge of an Ion Students are sometimes confused by how the sign of an ion's charge is related to the loss and gain of electrons. Remind them that the *gain* of an electron means the gain of a negative charge, producing a negatively charged ion. Conversely, the *loss* of an electron means the loss of a negative charge, which leaves a positively-charged ion.

SKILL BUILDER

Vocabulary Inform students that because anions and cations are types of ions, they are pronounced accordingly. Anion is pronounced "an-ion," and cation is pronounced "cat-ion." Students who are familiar with batteries might relate these terms to the negative pole of a battery, the cathode, which attracts cations, and the positive pole of a battery, the anode, which attracts anions.

Did **You** Know ❓

Complete Levels The first electron energy level in an atom can contain only two electrons. If this level contains the valence electrons, it is full with two electrons. Energy levels two and higher are considered full when they have eight valence electrons, known as an octet. Atoms that ionize do so in a manner that will complete an octet. (The exceptions to this are H$^+$ and Li$^+$, which have 0 and 2 electrons in their outermost energy levels.)

Teaching Tip

Mass Number Students may find it hard to tell mass number from the average atomic masses given in the periodic table. Point out that the mass number is not an actual atomic mass, but rather a way of stating the total number of particles in the nucleus.

Group Activity ——— GENERAL

Protons and Neutrons Group students in pairs. List the symbols and mass numbers of 20 elements on the chalkboard. Have students create a table of number of protons (atomic number) and number of neutrons (mass number–atomic number) for all of the elements listed. Then, ask them to summarize the information in their table. (For light elements, the number of protons and the number of neutrons are approximately equal. For heavier elements, the number of neutrons increases faster than the number of protons in an atom.)
LS Logical

SKILL BUILDER ——— BASIC

Interpreting Visuals After students examine **Figure 17,** ask the following: Which colored spheres would you count to find the atomic number of this element? (pink) Which would you count to find the mass number? (blue and pink) Which color would increase or decrease for a different isotope of this same element? (blue) Are the number of pink and blue spheres necessarily the same? (no; the number of protons does not always equal the number of neutrons)
LS Visual

Disc One, Module 3:
Periodic Properties
Use the Interactive Tutor to learn more about these topics.

▶ **atomic number** the number of protons in the nucleus of an atom

▶ **mass number** the sum of the numbers of protons and neutrons in the nucleus of an atom

Figure 17
Atoms of the same element have the same number of protons and therefore have the same atomic number. But they may have different mass numbers, depending on how many neutrons each atom has.

How Do the Structures of Atoms Differ?

As you have seen with lithium and fluorine, atoms of different elements have their own unique structures. Because these atoms have different structures, they have different properties. An atom of hydrogen found in a molecule of swimming-pool water has properties very different from an atom of uranium in nuclear fuel.

Atomic number equals the number of protons

The **atomic number,** Z, tells you how many protons are in an atom. Remember that atoms are always neutral because they have an equal number of protons and electrons. Therefore, the atomic number also equals the number of electrons the atom has. Each element has a different atomic number. For example, the simplest atom, hydrogen, has just one proton and one electron, so for hydrogen, $Z = 1$. The largest naturally occurring atom, uranium, has 92 protons and 92 electrons, so $Z = 92$ for uranium. The atomic number for a given element never changes.

Mass number equals the total number of subatomic particles in the nucleus

The **mass number,** A, of an atom equals the number of protons plus the number of neutrons. A fluorine atom has 9 protons and 10 neutrons, so $A = 19$ for fluorine. Oxygen has 8 protons and 8 neutrons, so $A = 16$ for oxygen. This mass number includes only the number of protons and neutrons (and not electrons) because protons and neutrons provide most of the atom's mass. Although atoms of an element always have the same atomic number, they can have different mass numbers. *Figure 17* shows which subatomic particles in the nucleus of an atom contribute to the atomic number and which contribute to the mass number.

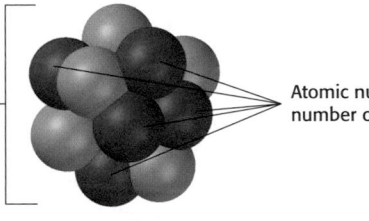

Mass number, A = number of protons + number of neutrons

Atomic number, Z = number of protons

Nucleus

Historical Perspective

Atomic Number The concept of atomic number was proposed by the British scientist Henry Moseley in 1915. Moseley found that the progression of the elements in the periodic table corresponded to an increase of one fundamental unit in the nucleus and that each element could be assigned an atomic number equal to the number of these units. In 1920, Rutherford announced the existence of the proton, Moseley's fundamental unit.

INCLUSION Strategies

• *Learning Disabled*
• *Attention Deficit Disorder*
• *English Language Learners*

Have students list three elements that are found in Group 2 of the Periodic Table. For each element, have the students describe some characteristics of the element. The student may also list some of the uses of the element. The information can be found in the textbook or reference books.

Figure 18

Protium has only a proton in its nucleus. Deuterium has both a proton and a neutron in its nucleus, while tritium has a proton and two neutrons.

Isotopes of Hydrogen

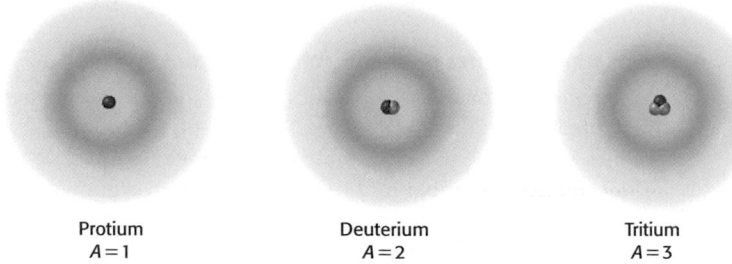

Protium	Deuterium	Tritium
$A = 1$	$A = 2$	$A = 3$

Isotopes of an element have different numbers of neutrons

Neutrons can be added to an atom without affecting the number of protons and electrons the atom is made of. Many elements have only one stable form, while other elements have different "versions" of their atoms. Each version has the same number of protons and electrons as all other versions but a different number of neutrons. These different versions, or **isotopes,** vary in mass but are all atoms of the same element because they each have the same number of protons.

The three isotopes of hydrogen, shown in **Figure 18,** have the same chemical properties because each is made of one proton and one electron. The most common hydrogen isotope, protium, has only a proton in its nucleus. A second isotope of hydrogen has a proton and a neutron. The mass number, A, of this second isotope is two, and the isotope is twice as massive. In fact, this isotope is sometimes called "heavy hydrogen." It is also known as deuterium, or hydrogen-2. A third isotope has a proton and two neutrons in its nucleus. This third isotope, tritium, has a mass number of three.

Some isotopes are more common than others

Hydrogen is present on both the sun and on Earth. In both places, protium (the hydrogen isotope without neutrons in its nucleus) is found most often. Only a very small fraction of the less common isotope of hydrogen, deuterium, is found on the sun and on Earth, as shown in **Figure 19.** Tritium is an unstable isotope that decays over time, so it is found least often.

▶ **isotope** an atom that has the same number of protons as other atoms of the same element do but that has a different number of neutrons

Figure 19

Ⓐ Hydrogen makes up less than 1% of Earth's crust. Only 1 out of every 6000 of these hydrogen atoms is a deuterium isotope.

Ⓑ Seventy-five percent of the mass of the sun is hydrogen, with protium isotopes outnumbering deuterium isotopes 50 000 to 1.

117

Trends in Physics

Deuterium as a Fuel for Fusion Deuterium is the fuel required for nuclear fusion, a potential energy source. The amount of energy released by fusing two deuterium atoms is about 10 times the amount of energy released by an equal mass of uranium during a nuclear fission reaction. Another advantage of fusion is that the waste products of certain fusion reactions are not radioactive. Scientists have not yet been able to create the conditions required for fusion in a laboratory, but many are working in this area with the hope of one day making fusion a practical energy source.

Did You Know ?

How Many Isotopes? Scientists have identified over 3500 isotopes! Many of these isotopes have very short half-lives and must be produced in laboratories by scientists who wish to study them.

Teaching Tip

Isotopes Make it clear to students that the identity of an atom is determined entirely by the number of protons in its nucleus. Atoms of the same element (same number of protons) may have varying numbers of neutrons. When an atom forms an ion, it may have more or fewer electrons than protons.

Some students may have the idea that all isotopes are radioactive and therefore dangerous. Point out that the water students drink every day contains two different isotopes of hydrogen (hydrogen-1 and hydrogen-2) and three isotopes of oxygen (oxygen-16, oxygen-17, and oxygen-18), none of which are radioactive or dangerous.

SKILL BUILDER — BASIC

Interpreting Visuals Ask students to draw a nucleus diagram like the one shown in **Figure 17** for the three isotopes of hydrogen shown in **Figure 18.** Tell them to include a key that distinguishes between protons and neutrons. (Student diagrams should illustrate the following: protium: 1 proton; deuterium: 1 proton, 1 neutron; tritium: 1 proton, 2 neutrons) **LS Visual**

Teaching Tip

Isomers Some isotopes exist in more than one form. The different forms, called isomers, have the same number of protons and neutrons, but they are arranged in a different ways. As a result, the isomers have different half-lives.

Chapter Resource File

- **Cross-Disciplinary Worksheet**
  Connection to Fine Arts—Carbon-Dating Masterpieces **ADVANCED**

Transparencies

TT Nucleus

Quick ACTIVITY

Teacher's Notes: To calculate the number of neutrons in the isotopes, students should subtract the atomic numbers (found in the periodic table) from the given mass numbers.

Results

1. $14 - 6 = 8$
2. $15 - 7 = 8$
3. $35 - 16 = 19$
4. $45 - 20 = 25$
5. $131 - 53 = 78$

LS Logical

SKILL BUILDER — BASIC

Interpreting Visuals Ask students to summarize orally the calculations shown in **Figure 20** and the example provided. Be sure students show understanding of atomic number and mass number and of how to use them to find the number of neutrons in an atom.

LS Verbal

MISCONCEPTION ///**ALERT**\\\

Atomic Masses Some students believe that atomic particles have no mass. Explain that although atomic masses are very small they are significant at a microscopic level. The masses of the proton and neutron are very large compared to the mass of the electron.

Figure 20
One isotope of chlorine has 18 neutrons, while the other isotope has 20 neutrons.

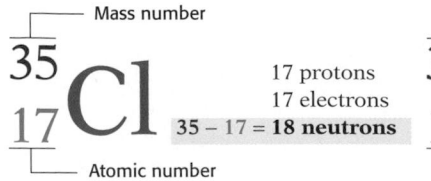

Mass number — 17 protons, 17 electrons, $35 - 17 = \mathbf{18}$ **neutrons** — Atomic number

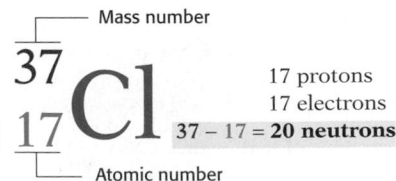

Mass number — 17 protons, 17 electrons, $37 - 17 = \mathbf{20}$ **neutrons** — Atomic number

▶ **atomic mass unit (amu)** a unit of mass that describes the mass of an atom or molecule; it is exactly one-twelfth of the mass of a carbon atom with mass number 12

▶ **average atomic mass** the weighted average of the masses of all naturally occurring isotopes of an element

Calculating the number of neutrons in an atom

Atomic numbers and mass numbers may be included along with the symbol of an element to represent different isotopes. The two isotopes of chlorine are represented this way in **Figure 20.** If you know the atomic number and mass number of an atom, you can calculate the number of neutrons it has.

Uranium has several isotopes. The isotope that is used in nuclear reactors is uranium-235, or $^{235}_{92}U$. Like all uranium atoms, it has an atomic number of 92, so it must have 92 protons and 92 electrons. It has a mass number of 235, which means its number of protons and neutrons together is 235. The number of neutrons must be 143.

Mass number (A):	235
Atomic number (Z):	− 92
Number of neutrons:	143

The mass of an atom

The mass of a single atom is very small. A single fluorine atom has a mass less than one trillionth of a billionth of a gram. Because it is very hard to work with such tiny masses, atomic masses are usually expressed in atomic mass units. An **atomic mass unit (amu)** is equal to one-twelfth of the mass of a carbon-12 atom. This isotope of carbon has six protons and six neutrons, so individual protons and neutrons must each have a mass of about 1.0 amu because electrons contribute very little mass.

Often, the atomic mass listed for an element in the periodic table is an average atomic mass for the element as it is found in nature. The **average atomic mass** for an element is a weighted average, so the more commonly found isotopes have a greater effect on the average than rare isotopes.

Figure 21 shows how the natural abundance of chlorine's two isotopes affects chlorine's average atomic mass. The average atomic mass of chlorine is 35.45 amu. This mass is much closer to 35 amu than to 37 amu. That's because the atoms of chlorine with masses of nearly 35 amu are found more often and therefore contribute more to chlorine's average atomic mass than chlorine atoms with masses of nearly 37 amu.

Quick ACTIVITY

Isotopes
Calculate the number of neutrons there are in the following isotopes. (Use the periodic table to find the atomic numbers.)

1. carbon-14
2. nitrogen-15
3. sulfur-35
4. calcium-45
5. iodine-131

118

Alternative Assessment — GENERAL

Weighted Average Tell students that a weighted average is based on both the number of items and the value of each.

Step 1 Provide students with the following example: a student received four As, 10 Bs, three Cs, and one F as grades. Using a 4-point grading scale, what is the student's average grade? $\{(4 \times 4) + (10 \times 3) + (3 \times 2) + (1 \times 0)\} \div (4 + 10 + 3 + 1) = 52/18 = 2.9$

Step 2 Have students work the following problem, then write problems of their own and share them. Juan had four quarters, six dimes, nine nickels, and 15 pennies. What is the average value of the coins? $\{(4 \times 25) + (6 \times 10) + (9 \times 5) + (15 \times 1)$ cents$\} \div (4 + 6 + 9 + 15) = 220$ cents$/34 = 6.5$ cents

LS Logical

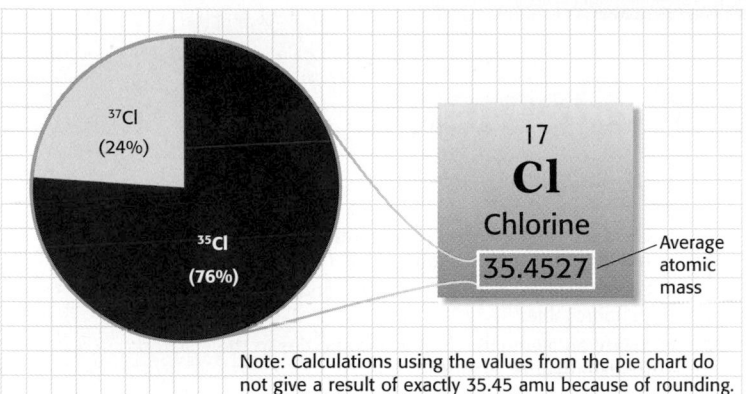

^{37}Cl (24%)

^{35}Cl (76%)

17
Cl
Chlorine
35.4527

Average atomic mass

Figure 21
The average atomic mass of chlorine is closer to 35 amu than it is to 37 amu because ^{35}Cl isotopes are found more often than ^{37}Cl isotopes.

Note: Calculations using the values from the pie chart do not give a result of exactly 35.45 amu because of rounding.

Close

Quiz

1. Elements in the same horizontal row, or period, have the same number of valence electrons and similar chemical properties. True or false? (false)

2. What is an ion? (a charged atom)

3. How are isotopes of an atom different from one another? (different numbers of neutrons)

4. What unit is commonly used to express atomic masses? (atomic mass unit, or amu)

SECTION 2 REVIEW

SUMMARY

▶ Elements are arranged in order of increasing atomic number so that elements with similar properties are in the same column, or group.

▶ Elements in the same group have the same number of valence electrons.

▶ Reactive atoms may gain or lose valence electrons to form ions.

▶ An atom's atomic number is its number of protons.

▶ An atom's mass number is its total number of protons and neutrons in the nucleus.

▶ Isotopes of an element have different numbers of neutrons, and therefore have different masses.

▶ An element's average atomic mass is a weighted average of the masses of its naturally occurring isotopes.

1. **Explain** how you can determine the number of protons, electrons, and neutrons an atom has from an atom's mass number and its atomic number.

2. **Calculate** how many neutrons a phosphorus-32 atom has.

3. **Name** the elements represented by the following symbols:
   a. Li   d. Br   g. Na
   b. Mg   e. He   h. Fe
   c. Cu   f. S   i. K

4. **Compare** the number of valence electrons an oxygen, O, atom has with the number of valence electrons a selenium, Se, atom has. Are oxygen and selenium in the same period or group?

5. **Explain** why some atoms gain or lose electrons to form ions.

6. **Predict** which isotope of nitrogen is more commonly found, nitrogen-14 or nitrogen-15. (**Hint:** What is the average atomic mass listed for nitrogen in the periodic table?)

7. **Describe** why the elements in the periodic table are arranged in order of increasing atomic number.

8. **Critical Thinking** Before 1937, all naturally occurring elements had been discovered, but no one had found any trace of element 43. Chemists were still able to predict the chemical properties of this element (now called technetium), which is widely used today for diagnosing medical problems. How were these predictions possible? Which elements would you expect to be similar to technetium?

119

Answers to Section 2 Review

1. Atomic number is the number of protons only. The number of electrons equals the number of protons in a neutral atom. Mass number is the total number of protons and neutrons. The number of neutrons is found by subtracting the atomic number from the mass number.

2. $A - Z = 32 - 15 = 17$ neutrons

3. a. lithium
   b. magnesium
   c. copper
   d. bromine
   e. helium
   f. sulfur
   g. sodium
   h. iron
   i. potassium

4. Both elements have six valence electrons. They are in the same group.

5. Atoms with unfilled outer s and p orbitals, such as atoms with one valence electron, often form ions. Atoms with one valence electron easily lose the valence electron to make a complete outermost energy level and form an ion.

6. The average atomic mass for nitrogen is 14.01 amu, so nitrogen-14 is more commonly found.

See "Continuation of Answers" at the end of the chapter.

Chapter Resource File

• **Quick Activity Datasheet** Isotopes
  GENERAL

• **Concept Review** GENERAL

• **Quiz** BASIC

Transparencies

π Isotopes

SECTION
3

Focus

Overview

Before beginning this section, review with your students the Objectives listed in the Student Edition. In this section, students learn how metals and nonmetals can be grouped into families with similar properties: alkali metals, alkaline-earth metals, transition metals, semiconductors, halogens, and noble gases. They learn about the general properties of each family, and also study specific examples.

🔊 Bellringer

Use the Bellringer transparency to prepare students for this section.

Motivate

Opening Demonstration —— GENERAL

For this demonstration, you will need a conductivity tester and samples of elements, such as a copper coin, an iron nail, charcoal, and sulfur. A conductivity tester can be made from insulated wire, a battery, and a flashlight bulb.

Use the conductivity tester to see which elements conduct an electric current. Ask students: Which elements are metals? How do you know? (Any elements that conduct a current under standard conditions are metals.) Which elements are nonmetals? How do you know? (Any elements that do not conduct a current under standard conditions are nonmetals.) **LS Logical**

Families of Elements

■ **KEY TERMS**
metal
nonmetal
semiconductor
alkali metal
alkaline-earth metal
transition metal
halogen
noble gas

◗ OBJECTIVES ◗

▶ **Locate** alkali metals, alkaline-earth metals, and transition metals in the periodic table.

▶ **Locate** semiconductors, halogens, and noble gases in the periodic table.

▶ **Relate** an element's chemical properties to the electron arrangement of its atoms.

📶 internet connect
www.scilinks.org
Topic: Element Families
SciLinks code: HK4046

SCLINKS. Maintained by the National Science Teachers Association

Figure 22

A Just like the members of this family,

B elements in the periodic table share certain similarities.

You may have wondered why groups in the periodic table are sometimes called families. Consider your own family. Though each member is unique, you all share certain similarities. All members of the family shown in *Figure 22A,* for example, have a similar appearance. Members of a family in the periodic table have many chemical and physical properties in common because they have the same number of valence electrons.

How Are Elements Classified?

Think of each element as a member of a family that is also related to other elements nearby. Elements are classified as metals or nonmetals, as shown in *Figure 22B*. This classification groups elements that have similar physical and chemical properties.

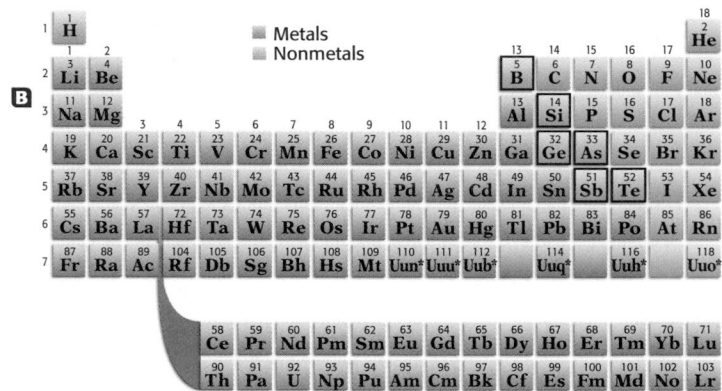

Note: Sometimes the boxed elements toward the right side of the periodic table are classified as a separate group and called semiconductors or metalloids.

Teacher Resources —— BASIC

For the New Teacher Remind students of what they have learned about the periodic table. They should know that it is an arrangement of elements in order of increasing atomic number (number of protons). They should also know that the table gives the average atomic mass of an element, which is the weighted average of the masses of the element's stable isotopes. Ask if students have learned anything that helps account for the table's arrangement. Lead them to note the relationship between electron configurations and the arrangement of elements in the table. **LS Verbal**

Alternative Assessment —— ADVANCED

What are Elements? Review with students the definition of an element. Have students brainstorm a list of substances that they think are elements. Have them use Sections 2 and 3 to confirm whether or not each of the substances on their list is an element. Have them use the properties of the elements to devise a classification system for elements. **LS Logical**

Elements are classified into three groups

As you can see in *Figure 22B,* most elements are **metals.** Most metals are shiny solids that can be stretched and shaped. They are also good conductors of heat and electricity. All **nonmetals,** except for hydrogen, are found on the right side of the periodic table. Nonmetals may be solids, liquids, or gases. Solid nonmetals are typically dull and brittle and are poor conductors of heat and electricity. But some elements that are classified as nonmetals can conduct under certain conditions. These elements are sometimes considered to be their own group and are called **semiconductors** or metalloids.

Metals

Many elements are classified as metals. To further classify metals, similar metals are grouped together. There are four different kinds of metals. Two groups of metals are located on the left side of the periodic table. Other metals, like aluminum, tin, and lead, are located toward the right side of the periodic table. Most metals, though, are located in the middle of the periodic table.

The alkali metals are very reactive

Sodium is found in Group 1 of the periodic table, as shown in *Figure 23A.* Like other **alkali metals** it is soft and shiny and reacts violently with water. Sodium must be stored in oil, as in *Figure 23B,* to prevent it from reacting with moisture in the air.

An atom of an alkali metal is very reactive because it has one valence electron that can easily be removed to form a positive ion. You have already seen in Section 2 how lithium, another alkali metal, forms positive ions with a 1+ charge. Similarly, the valence electron of a sodium atom can be removed to form the positive sodium ion Na$^+$.

Because alkali metals such as sodium are so reactive, they are not found in nature as elements. Instead, they combine with other elements to form compounds. For example, the salt you use to season your food is actually the compound sodium chloride, NaCl.

metal an element that is shiny and conducts heat and electricity well

nonmetal an element that conducts heat and electricity poorly

semiconductor an element or compound that conducts electric current better than an insulator but not as well as a conductor does

alkali metal one of the elements of Group 1 of the periodic table

Figure 23

A The alkali metals are located on the left edge of the periodic table.

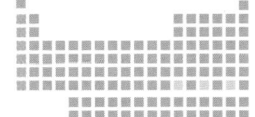

Alkali Metals

B The alkali metal sodium must be stored in oil. Otherwise, it will react violently with moisture and oxygen in the air.

Group 1
3 **Li** Lithium 6.941
11 **Na** Sodium 22.989 770
19 **K** Potassium 39.0983
37 **Rb** Rubidium 85.4678
55 **Cs** Cesium 132.905 45
87 **Fr** Francium (223)

121

Teach

SKILL BUILDER — BASIC

Reading Skills As students read each passage about a family of elements, ask for a volunteer to summarize the passage for the class. Then have the class, as listeners, ask for clarification of parts of the summary. All students may consult the text during the clarification process. **LS Interpersonal**

SKILL BUILDER

Vocabulary Make sure students distinguish between the scientific and everyday meanings of the word *metal.* The scientific term *metal* applies only to individual elements with certain properties. In common speech, the term *metal* is often applied to metallic objects that we encounter every day, which are almost always alloys, homogeneous mixtures of metals with other metals and/or nonmetals.

Teaching Tip
Uses of Alkali Metals
Compounds of alkali metals are used extensively, especially those of sodium and potassium. Sodium hydroxide is an important industrial compound that is used in the manufacture of paper, soap, and synthetic fabrics and in petroleum refining. Sodium chloride is common table salt, and potassium chloride is a table salt substitute. Potassium compounds are important components of chemical fertilizers.

LIFE SCIENCE
CONNECTION

Too much sodium can be harmful, but sodium ions and potassium ions are important for the proper functioning of nerves in the human body. They allow electrical impulses to pass from one nerve to another.

Did You Know?

Sodium in Power Lines Because sodium is a lightweight electrical conductor, it has been incorporated into high-voltage power lines. Sodium's low density reduces the weight of the cables without impairing their conductivity, which reduces sag and permits longer cable spans.

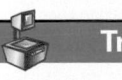

Teaching Tip

Reactivity Ionization energy is the amount of energy needed to remove an electron from an atom. The ionization energy for an alkali metal is low because one electron is easy to remove.

The first electron in an alkaline-earth atom is easy to remove also. However, the second electron requires more energy to remove. It must be removed from a positively charged particle, which does not lose an electron as easily as a neutral atom does. This is why alkaline-earth metals are less reactive than alkali metals.

Quick ACTIVITY

Elements in Your Food

Materials: none required

Teacher's Notes: If students need help performing this activity, give them the following hints. Ingredients that are not named as elements are probably compounds. Ingredients that are plant or animal products probably contain carbon, hydrogen, and oxygen. Commas separate ingredients. For example, if an ingredient is sodium citrate, it is a compound by that name, not the element sodium and then a compound.

Results: Student lists will vary.

LS Intrapersonal

Quick ACTIVITY

Elements in Your Food

1. For 1 day, make a list of the ingredients in all the foods and drinks you consume.
2. Identify which ingredients on your list are compounds.
3. For each compound on your list, try to figure out what elements it is made of.

▶ **alkaline-earth metal** one of the elements of Group 2 of the periodic table

Group 2

4 **Be** Beryllium 9.012 182
12 **Mg** Magnesium 24.3050
20 **Ca** Calcium 40.078
38 **Sr** Strontium 87.62
56 **Ba** Barium 137.327
88 **Ra** Radium (226)

Alkaline-earth metals form compounds that are found in limestone and in the human body

Calcium is in Group 2 of the periodic table, as shown in *Figure 24A,* and is an **alkaline-earth metal.** Atoms of alkaline-earth metals, such as calcium, have two valence electrons. Alkaline-earth metals are less reactive than alkali metals, but they may still react to form positive ions with a 2+ charge. When the valence electrons of a calcium atom are removed, a calcium ion, Ca^{2+}, forms. Alkaline-earth metals like calcium also combine with other elements to form compounds.

Calcium compounds make up the hard shells of many sea animals. When the animals die, their shells settle to form large deposits that eventually become limestone or marble, both of which are very strong materials used in construction. Coral is one example of a limestone structure. The "skeletons" of millions of tiny animals combine to form sturdy coral reefs that many fish rely on for protection, as shown in *Figure 24B.* Your bones and teeth also get their strength from calcium compounds.

Magnesium is another alkaline-earth metal that has properties similar to calcium. Magnesium is the lightest of all structural metals and is used to build some airplanes. Magnesium, as Mg^{2+}, activates many of the enzymes that speed up processes in the human body. Magnesium also combines with other elements to form many useful compounds. Two magnesium compounds are commonly used medicines—milk of magnesia and Epsom salts.

Figure 24

A The alkaline-earth metals make up the second column of elements from the left edge of the periodic table.

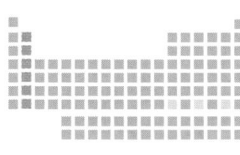

Alkaline-earth Metals

B Fish can escape their predators by hiding among the hard projections of limestone coral reefs that are made of calcium compounds.

Trends in Technology

Magnesium Alloys Magnesium and magnesium alloys are vital to the aerospace industry because of their low densities. But machining magnesium parts can be hazardous. Magnesium can catch fire and burn with an intensely hot, blinding white light. For this reason, magnesium is often machined in an inert atmosphere.

Did You Know?

Radioactive Waste Strontium-90 is a common radioactive waste product in nuclear reactors. Strontium-90 was released into the atmosphere as a result of the meltdown of the nuclear reactor at Chernobyl, Ukraine, in 1986. Smaller amounts of strontium-90 were released into the atmosphere during the first years of the cold war, until treaties limited weapons testing in the atmosphere.

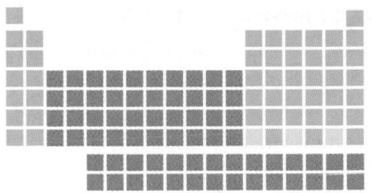

Transition Metals

Figure 25

A The transition metals are located in the middle of the periodic table.

B The transition metals platinum, gold, and silver are often shaped to make jewelry.

Gold, silver and platinum are transition metals

Gold is a valuable **transition metal.** *Figure 25A* shows that the transition metals are located in Groups 3–12 of the periodic table. Unlike most other transition metals, gold is not found combined with other elements as an ore but as the free metal.

Transition metals, like gold, are much less reactive than sodium or calcium, but they can lose electrons to form positive ions too. There are two possible cations that a gold atom can form. If an atom of gold loses only one electron, it forms Au^+. If the atom loses three electrons, it forms Au^{3+}. Some transition metals can form as many as four differently charged cations because of their complex arrangement of electrons.

All metals, including transition metals, conduct heat and electricity. Most metals can also be stretched and shaped into flat sheets, or pulled into wire. Because gold, silver, and platinum are the shiniest metals, they are often molded into different kinds of jewelry, as shown in *Figure 25B.*

There are many other useful transition metals. Copper is often used for electrical wiring or plumbing. Light bulb filaments are made of tungsten. Iron, cobalt, copper, and manganese play vital roles in your body chemistry. Mercury, shown in *Figure 26,* is the only metal that is a liquid at room temperature. It is often used in thermometers because it flows quickly and easily without sticking to glass.

▶ **transition metal** one of the elements of Groups 3–12 of the periodic table

VOCABULARY *Skills Tip*

The properties of transition metals *gradually transition, or shift, from being more similar to Group 2 elements to being more similar to Group 13 elements as you move from left to right across a period.*

Figure 26

Mercury is an unusual metal because it is a liquid at room temperature. Continued exposure to this volatile metal can harm you because if you breathe in the vapor, it accumulates in your body.

123

SKILL BUILDER

Interpreting Visuals The metals in **Figure 25B** are grouped together because of a common use. Other groups of transition elements are also grouped by use. Copper, silver, and gold, which are in the same family, are called coinage metals. Iron, cobalt, and nickel, which are in the same period, are called the iron triad and are the only elements that can be magnetized.

Group Activity ——— **ADVANCED**

Metal Coins Put students in small groups. Supply each group with samples of pennies, nickels, dimes, and quarters. Have students in each group list the metals that they think are used in each of the coins. Tell them to observe the edges of several coins. Discuss how a combination of value and properties determines what metals are used to make coins. Then have students in each group conduct research to find out more about the metals used in U.S. coins today and throughout history. You may wish to ask each group to research one particular type of coin, and then have them share their results with the class. **LS Interpersonal**

Teaching Tip

Variable Charges Because transition elements have different numbers of electrons they can gain or lose, many of them can form more than one kind of ion. Common transition element ions include Cu^+, Cu^{2+}, Fe^{2+}, Fe^{3+}, Hg^+, Hg^{2+}, Sn^{2+}, Sn^{4+}, Pb^{2+}, and Pb^{4+}.

REAL-LIFE CONNECTION

Gold Connectors Although silver is the best conductor, gold is also a very good conductor and has the advantage of not corroding or tarnishing under ordinary conditions. For this reason, gold is widely used on connectors in computers and other electronic devices.

Did You Know?

Colorful Gems The transition elements frequently form colorful compounds. Traces of transition elements provide the color in many gems, such as rubies and emeralds.

Chapter Resource File

• **Quick Activity Datasheet** Elements in Your Food **GENERAL**

1. Al, Fe, Cr, Zn, Cu, Sn, Ag, Au
2. Fe, Cr, Zn, Al, Cu, Sn, Ag, Au

Analysis

1. The lists match perfectly, except for the price of aluminum compared with its abundancy.

2. Aluminum is the most reactive metal on the list. While aluminum is also the most abundant, its extra cost is due to a necessary refining process.

3. Dividing $100 by the price per gram for each metal will give the number of grams of the metal. Because the prices listed in the table are given in dollars per kilogram, students should incorporate the conversion of kilograms to grams in their spreadsheets. **LS** Logical

---

SKILL BUILDER — **BASIC**

Vocabulary Ask students what is meant by the common use of the term decay. Student descriptions will probably include how plant and animal materials break down into other materials. From this description, ask them to hypothesize what it means for an element to decay. Student answers should indicate that these elements break down, forming other elements. **LS** Verbal

---

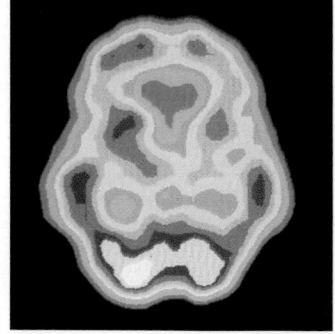

Figure 27
With the help of the radioactive isotope technetium-99, doctors are able to confirm that this patient has a healthy brain.

Technetium and promethium are synthetic elements

Technetium and promethium are both man-made elements. They are also both *radioactive*, which means the nuclei of their atoms are continually decaying to produce different elements. There are several different isotopes of technetium. The most stable isotope is technetium-99, which has 56 neutrons. Technetium-99 can be used to diagnose cancer as well as other medical problems in soft tissues of the body, as shown in **Figure 27**.

When looking at the periodic table, you might have wondered why part of the last two periods of the transition metals are placed toward the bottom. This keeps the periodic table narrow so that similar elements elsewhere in the table still line up. Promethium is one element located in this bottom-most section. Its most useful isotope is promethium-147, which has 86 neutrons. Promethium-147 is an ingredient in some "glow-in-the-dark" paints.

All elements with atomic numbers greater than 92 are also man-made and are similar to technetium and promethium. For example, americium, another element in the bottom-most section of the periodic table, is also radioactive. Tiny amounts of americium-241 are found in most household smoke detectors. Although even small amounts of radioactive material can affect you, americium-241 is safe when contained inside your smoke detector.

Quick **Lab**

Why do some metals cost more than others?

1. The table at right gives the abundance of some metals in Earth's crust. List the metals in order from most to least abundant.

2. List the metals in order of price, from the cheapest to the most expensive

3. Create a spreadsheet that can be used to calculate how many grams of each metal you could buy with $100.

Analysis

1. If the price of a metal depends on its abundance, you would expect the order to be the same on both lists. How well do the two lists match? Mention any exceptions.

2. The order of reactivity of these metals, from most reactive to least reactive, is aluminum, zinc, chromium, iron, tin, copper, silver, and gold. Use this information to explain any exceptions you noticed in item 3.

Metal	Abundance in Earth's crust (%)	Price ($/kg)
Aluminum (Al)	8.2	1.55
Chromium (Cr)	0.01	0.06
Copper (Cu)	0.0060	2.44
Gold (Au)	0.000 0004	11 666.53
Iron (Fe)	5.6	0.03
Silver (Ag)	0.000 007	154.97
Tin (Sn)	0.0002	6.22
Zinc (Zn)	0.007	1.29

124

Did You Know ?

Uses of Synthetic Elements Although a few synthetic elements have practical uses, many do not. The most common reason for this is that many synthetic elements have extremely short half-lives, sometimes measured in seconds. The half-life of a radioactive element is the amount of time it takes for half of the amount of element present to break down into other elements.

Historical Perspective

Americium-241 Americium-241, the key ingredient in most household smoke detectors, is a decay product of plutonium-241. It was discovered in 1945 by scientists working on the Manhattan Project—the U.S. top-secret effort to build an atomic weapon during World War II before Germany or Japan did so. The first sample was produced at the University of Chicago, when scientists bombarded plutonium with neutrons in a nuclear reactor.

Nonmetals

Except for hydrogen, nonmetals are found on the right side of the periodic table. They include some elements in Groups 13–16 and all the elements in Groups 17 and 18.

Carbon is found in three different forms and can also form many compounds

Carbon and other nonmetals are found on the right side of the periodic table, as shown in *Figure 28A*. Although carbon in its pure state is usually found as graphite (pencil "lead") or diamond, the existence of fullerenes, a third form, was confirmed in 1990. The most famous fullerene consists of a cluster of 60 carbon atoms, as shown in *Figure 28B*.

Carbon can also combine with other elements to form millions of carbon-containing compounds. Carbon compounds are found in both living and nonliving things. Glucose, $C_6H_{12}O_6$, is a sugar in your blood. A type of chlorophyll, $C_{55}H_{72}O_5N_4Mg$, is found in all green plants. Many gasolines contain isooctane, C_8H_{18}, while rubber tires are made of large molecules with many repeating C_5H_8 units.

Nonmetals and their compounds are plentiful on Earth

Oxygen, nitrogen, and sulfur are other common nonmetals. Each may form compounds or gain electrons to form the negative ions oxide, O^{2-}, sulfide, S^{2-}, and nitride, N^{3-}. The most plentiful gases in the air are the nonmetals nitrogen and oxygen. Although sulfur itself is an odorless yellow solid, many sulfur compounds, like those in rotten eggs and skunk spray, are known for their terrible smell.

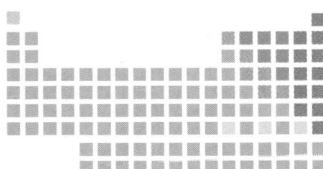

Figure 28

A Most nonmetals are located on the right side of the periodic table.

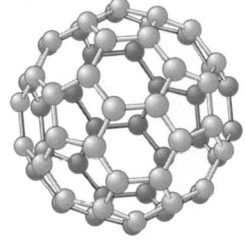

B The way carbon atoms are connected in the most recently discovered form of carbon resembles the familiar pattern of a soccer ball.

125

ORGANIC CHEMISTRY — CONNECTION

Organic chemistry is the study of carbon-containing compounds, called organic compounds. Carbon atoms have the ability to form molecule chains thousands of atoms long. There are over six million organic compounds, including all types of foods (carbohydrates, lipids, and proteins), plastics, synthetic and natural fibers, and petroleum products.

Teaching Tip

Halogen Elements Point out to students that halogen elements all have the same number of valence electrons, seven. This configuration is one electron short of the complete octet found in the noble gas atoms. As a result, halogen elements react by gaining one electron to form ions with a 1− charge.

INTEGRATING

EARTH SCIENCE One common way that magnesium and bromine are recovered from sea water is by the use of electrolysis. Bromide ions release an electron to the anode of the electrolysis setup, and free bromine is formed. At the cathode, magnesium ions accept electrons and become metallic magnesium.

SKILL BUILDER

Vocabulary The term *halogen* comes from the roots *hal* (salt) and *gen* (to form), so halogen means "salt former." The root *hal* is also used in halite, the term for rock salt. *Gen* is a common root used in many words, including generate and generation. Halogens were given this name because they form salts in compounds. Examples include common table salt (NaCl) and silver salt (AgBr), used in photography.

▶ **halogen** one of the elements of Group 17 of the periodic table

INTEGRATING

EARTH SCIENCE Eighty-one elements have been detected in sea water. Magnesium and bromine are two such elements. To recover an element from a sample of sea water, you must evaporate some of the water from the sample. Sodium chloride then crystallizes and the liquid that remains becomes more concentrated in bromide, magnesium, and other ions than the original sea water was, making their recovery easier.

Group 17

9
F
Fluorine
18.998 4032

17
Cl
Chlorine
35.4527

35
Br
Bromine
79.904

53
I
Iodine
126.904 47

85
At
Astatine
(210)

Chlorine is a halogen that protects you from harmful bacteria

Chlorine and other **halogens** are located in Group 17 of the periodic table, as shown in *Figure 29A*. You have probably noticed the strong smell of chlorine in swimming pools. Chlorine is widely used to kill bacteria in pools, like the one shown in *Figure 29B,* as well as in drinking-water supplies.

Like fluorine atoms, which you learned about in Section 2, chlorine atoms are very reactive. As a result, chlorine forms compounds. For example, the chlorine in most swimming pools is added in the form of the compound calcium hypochlorite, $Ca(OCl)_2$. Elemental chlorine is a poisonous yellowish green gas made of pairs of joined chlorine atoms. Chlorine gas has the chemical formula Cl_2. A chlorine atom may also gain an electron to form a negative chloride ion, Cl^-. The attractions between Na^+ ions and Cl^- ions form table salt, NaCl.

Fluorine, bromine, and iodine are other Group 17 elements. Fluorine is a poisonous yellowish gas, bromine is a dark red liquid, and iodine is a dark purple solid. Atoms of each of these elements can also form compounds by gaining an electron to become negative ions. A compound containing the negative ion fluoride, F^-, is used in some toothpastes and added to some water supplies to help prevent tooth decay. Adding a compound containing iodine as the negative ion iodide, I^-, to table salt makes "iodized" salt. You need this ion in your diet for your thyroid gland to function properly.

Figure 29

A The halogens are in the second column from the right of the periodic table.

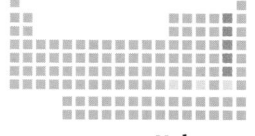

Halogens

B Chlorine keeps pool water bacteria-free for swimmers to enjoy.

REAL-LIFE CONNECTION

Chlorine Gas When certain cleaning agents are mixed together, toxic chlorine gas is released. The gas is quite hazardous to humans. This is one reason that different cleaning agents should never be mixed.

Did You Know ?

Fluorine The halogen fluorine is the most reactive element known. Fluorine gas (F_2) is so reactive that it is hard to find containers to store it in; most metals burst into flames when they come in contact with it, and it attacks glass and quartz. A compound of fluorine and carbon is used to make Teflon, a nonstick coating commonly used in cookware.

The noble gases are inert

Neon is one of the **noble gases** that make up Group 18 of the periodic table, as shown in *Figure 30A*. It is responsible for the bright reddish orange light of "neon" signs. *Figure 30B* shows how mixing neon with another substance, such as mercury, can change the color of a sign.

The noble gases are different from most elements that are gases because they exist as single atoms instead of as molecules. Like other members of Group 18, neon is inert, or unreactive, because its s and p orbitals are full of electrons. For this reason, neon and other noble gases do not gain or lose electrons to form ions. They also don't join with other atoms to form compounds under normal conditions.

Helium and argon are other common noble gases. Helium is less dense than air and is used to give lift to blimps and balloons. Argon is used to fill light bulbs because its lack of reactivity prevents filaments from burning.

Semiconductors are intermediate conductors of heat and electricity

Figure 31 shows that the elements sometimes referred to as semiconductors or metalloids are clustered toward the right side of the periodic table. Only six elements—boron, silicon, germanium, arsenic, antimony, and tellurium—are semiconductors. Although these elements are classified as nonmetals, each one also has some properties of metals. And as their name implies, semiconductors are able to conduct heat and electricity under certain conditions.

Boron is an extremely hard element. It is often added to steel to increase steel's hardness and strength at high temperatures. Compounds of boron are often used to make heat-resistant glass. Arsenic is a shiny solid that tarnishes when exposed to air. Antimony is a bluish white, brittle solid that also shines like a metal. Some compounds of antimony are used as fire retardants. Tellurium is a silvery white solid whose ability to conduct increases slightly with exposure to light.

Figure 30

A The noble gases are located on the right edge of the periodic table.

Noble Gases

B A neon sign is usually reddish orange, but adding a few drops of mercury makes the light a bright blue.

Group 18

| 2 |
| He |
| Helium |
| 4.002 602 |

| 10 |
| **Ne** |
| Neon |
| 20.1797 |

| 18 |
| Ar |
| Argon |
| 39.948 |

| 36 |
| Kr |
| Krypton |
| 83.80 |

| 54 |
| Xe |
| Xenon |
| 131.29 |

| 86 |
| Rn |
| Radon |
| (222) |

▶ **noble gas** an unreactive element of Group 18 of the periodic table

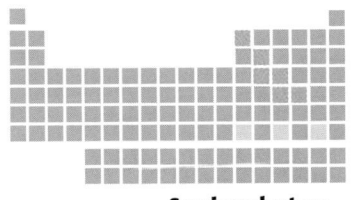

Semiconductors

Figure 31

Semiconductors are located toward the right side of the periodic table.

127

Group Activity ——— BASIC

Semiconductor Greeting This activity requires several musical greeting cards and hand lenses (at least one of each per student group). Play the greeting cards, then divide students into groups. Let each group of students take their card apart and examine the mechanism. Tell students that semiconductors play a part in the mechanism. Ask each group to answer the following questions: What parts of the mechanism can you recognize? (Answers will vary, but students will probably locate the speaker.) What do you think a semiconductor is? (Answers might include that it conducts electricity but not as well as a metal does.) **LS** Auditory

Extension ——— ADVANCED

Encourage interested students to learn more about why and how semiconductors are used in the greeting cards and to share their results with the class. **LS** Verbal

Teaching Tip

Doping The conductivity of semiconductors increases when certain impurities are added to them. This process is called *doping*.

The most common doping process is adding small amounts of arsenic or gallium to silicon. Because these elements contain different numbers of valence electrons, electrons flow more easily and conductivity increases.

SKILL BUILDER

Vocabulary Tell students the noble gases were named for their "nobility" in keeping to themselves and not readily combining with other elements.

Historical Perspective

The First Noble Gas Argon was the first noble gas to be identified. Its existence was proposed in 1785 because chemists could not account for all the major constituents of air. Argon makes up about 1 percent of air. It was not until 1894 that British chemist William Ramsay identified it. Because of its chemical inertness, it was given the name argon from the Greek *argos*, which means "lazy" or "inactive."

Trends in Chemistry

Noble Gas Compounds Group 18 gases were always considered to be completely inert. However, in 1962, the first Group 18 compound was formed from xenon. Currently, compounds of xenon, krypton, radon, and argon exist. In 2002, chemists at Ohio State University and the University of Virginia made the first compounds between three of the noble gases (argon, krypton, and xenon) and uranium.

Chapter Resource File

• **Cross-Disciplinary Worksheet**
  Integrating Earth Science—
  Magnesium: From Sea Water to
  Fireworks ADVANCED

Quiz
List the family each element belongs to.

1. neon (noble gases)
2. sodium (alkali metals)
3. chlorine (halogens)
4. silicon (semiconductors)
5. calcium (alkaline-earth metals)
6. gold (transition metals)

Silicon is the most familiar semiconductor

Silicon atoms, usually in the form of compounds, account for 28% of the mass of Earth's crust. Sand is made of the most common silicon compound, called silicon dioxide, SiO_2. Small chips made of silicon, like those shown in **Figure 32,** are used in the internal parts of computers.

Silicon is also an important component of other semiconductor devices such as transistors, LED display screens, and solar cells. Impurities such as boron, aluminum, phosphorus, and arsenic are added to the silicon to increase its ability to conduct electricity. These impurities are usually added only to the surface of the chip. This process can be used to make chips of different conductive abilities. This wide range of possible semiconductor devices has led to great advances in electronic technology.

Figure 32

Silicon chips are the basic building blocks of computers.

SECTION 3 REVIEW

SUMMARY

▶ Metals are shiny solids that conduct heat and electricity.

▶ Alkali metals, located in Group 1 of the periodic table, are very reactive.

▶ Alkaline-earth metals, located in Group 2, are less reactive than alkali metals.

▶ Transition metals, located in Groups 3–12, are not very reactive.

▶ Nonmetals usually do not conduct heat or electricity well.

▶ Nonmetals include the inert noble gases in Group 18, the reactive halogens in Group 17, and some elements in Groups 13–16.

▶ Semiconductors are nonmetals that are intermediate conductors of heat and electricity.

1. **Classify** the following elements as alkali, alkaline-earth, or transition metals based on their positions in the periodic table:
   a. iron, Fe
   b. potassium, K
   c. strontium, Sr
   d. platinum, Pt

2. **Predict** whether cesium forms Cs^+ or Cs^{2+} ions.

3. **Describe** why chemists might sometimes store reactive chemicals in argon, Ar. To which family does argon belong?

4. **Determine** whether the following elements are more likely to be a metal or nonmetal:
   a. a shiny substance used to make flexible bed springs
   b. a yellow powder from underground mines
   c. a gas that does not react
   d. a conducting material used within flexible wires

5. **Describe** why atoms of bromine, Br, are so reactive. To which family does bromine belong?

6. **Predict** the charge of a beryllium ion.

7. **Identify** which element is more reactive: lithium, Li, or beryllium, Be.

8. **Creative Thinking** Imagine you are a scientist who has just discovered a new element. You have confirmed that the element is a metal but are unsure whether it is an alkali metal, an alkaline-earth metal, or a transition metal. Write a paragraph describing the additional tests you can do to further classify this metal.

WRITING SKILL

Answers to Section 3 Review

1. **a.** transition metal
   b. alkali metal
   c. alkaline-earth metal
   d. transition metal

2. Cesium is an alkali metal, so its atom loses a valence electron to form a Cs^+ ion.

3. Reactive chemicals might react with oxygen or water vapor in the air. They will not react with argon because it is inert. Argon is a noble gas.

4. **a.** metal
   b. nonmetal
   c. nonmetal
   d. metal

5. A bromine atom is one electron short of a complete outermost energy level, and will react with an element that can supply that electron. Bromine is a halogen.

6. Beryllium is an alkaline-earth metal. Therefore, its atom has two valence electrons that can be removed to form a Be^{2+} ion.

7. Lithium is an alkali metal and is more reactive than an alkaline-earth metal such as beryllium.

8. Answers will vary, but one method is to check the reactivity of the metal against the reactivities of other alkali, alkaline-earth, and transition metals. The element belongs to the family that is most similar to the element chemically.

Using Moles to Count Atoms

OBJECTIVES

▶ **Explain** the relationship between a mole of a substance and Avogadro's constant.

▶ **Find** the molar mass of an element by using the periodic table.

▶ **Solve** problems converting the amount of an element in moles to its mass in grams, and vice versa.

▶ **KEY TERMS**

mole
Avogadro's constant
molar mass
conversion factor

Counting objects is one of the very first things children learn to do. Counting is easy when the objects being counted are not too small and there are not too many of them. But can you imagine counting the grains of sand along a stretch of beach or the stars in the night-time sky?

Counting Things

When people count out large numbers of small things, they often simplify the job by using counting units. For example, when you order popcorn at a movie theater, the salesperson does not count out the individual popcorn kernels to give you. Instead, you specify the size of container you want, and that determines how much popcorn you get. So the "counting unit" for popcorn is the size of the container: small, medium, or large.

There are many different counting units

The counting units for popcorn are only an approximation and are not exact. Everyone who orders a large popcorn will not get exactly the same number of popcorn kernels. Many other items, however, require more-exact counting units, as shown in **Figure 33.** For example, you cannot buy just one egg at the grocery store. Eggs are packaged by the dozen. Copy shops buy paper in reams, or 500-sheet bundles.

An object's mass may sometimes be used to "count" it. For example, if a candy shopkeeper knows that 10 gumballs have a mass of 21.4 g, then the shopkeeper can assume that there are 50 gumballs on the scale when the mass is 107 g (21.4 g × 5).

Figure 33

Eggs are counted by the dozen, and paper is counted by the ream.

129

Focus

Overview

Before beginning this section, review with your students the **Objectives listed in the Student Edition.** In this section, students learn what a mole is, how to find molar mass on the periodic table, and how to convert between moles and grams.

Bellringer

Use the Bellringer transparency to prepare students for this section.

Motivate

Opening Activity — GENERAL

Put students into groups. Each group will need a large container, identical items to fill the container (such as beans), and a balance. Ask students to find the following: the mass of the empty container (m_{cont}), the mass of a few (x) of the items (m_{xitems}), and the mass of the container with all of the items in it (m_{tot}). Tell each group to use their data to calculate the number of small items without counting them. (First find the mass of one item by dividing the mass of a given number of items by that number. Then find the mass of all of the items by subtracting the mass of the container from the total mass. Finally, the number of unknown items (n) equals the mass of all of the items divided by the mass of one item.

After students finish, let them count the items to check their results. Tell students that in this section they will see how this method can be used to "count atoms." LS **Kinesthetic**

LIFE SCIENCE
CONNECTION

Students may think that extremely large and small numbers do not apply to their daily lives. These concepts seem abstract because students cannot directly measure the quantities or count the numbers involved. Use the following example that applies to the human body to emphasize that extremely large and small numbers do affect students.

The average human body contains about 10^{14} cells. Each cell membrane has a thickness of 7×10^{-9} m. Of the cells in the human body, about 2.5×10^{12} are red blood cells. Each red blood cell has a diameter of about 7.5×10^{-6} m and a thickness of about 2×10^{-6} m. Each milliliter of blood contains about 5×10^{9} red blood cells. Because some cells die every moment, the human body replenishes itself with red blood cells at a rate of about 3×10^{6} each second.

Chapter Resource File

• **Lesson Plan**

Transparencies

TT Bellringer

Reading Skills Have students read the section and then organize the ideas presented in the section in the form of a concept map. Concept maps might show how to convert from moles to grams or grams to moles. Have students share their maps after you have checked them for accuracy.
LS Verbal

Did You Know?

Although Avogadro did not develop Avogadro's constant, he made many important contributions that dealt with particles of matter. He was the first scientist to make the distinction between atoms and molecules. This distinction led to the determination of an accurate table of atomic masses. He also hypothesized that equal volumes of gases at the same temperature and pressure contain equal numbers of particles.

Teaching Tip — BASIC

Counting Groups On the chalkboard, make a table that has one column labeled *Counting unit* and another column labeled *Number of units*. Have students brainstorm and record counting units that could be used for grouping students in the class. Possible units include basketball teams (5), baseball teams (9), dozens (12), and pairs (2). **LS Logical**

Did You Know?

Did you know that Avogadro never knew his own constant? Count Amedeo Avogadro (1776–1856) was a lawyer who became interested in mathematics and physics. Avogadro's constant was actually determined by Joseph Loschmidt, a German physicist in 1865, nine years after Avogadro's death.

▶ **mole** the SI base unit used to measure the amount of a substance whose number of particles is the same as the number of atoms of carbon in 12 g of carbon-12

▶ **Avogadro's constant** equals 6.022×10^{23}/mol; the number of particles in 1 mol

▶ **molar mass** the mass in grams of 1 mol of a substance

The mole is useful for counting small particles

Because chemists often deal with large numbers of small particles, they use a large counting unit—the **mole,** abbreviated *mol*. A mole is a collection of a very large number of particles.

About 602 213 670 000 000 000 000 000!

This number is usually written as 6.022×10^{23}/mol and is referred to as **Avogadro's constant.** The constant is named in honor of the Italian scientist Amedeo Avogadro. Avogadro's constant is defined as the number of particles, 6.022×10^{23}, in exactly 1 mol of a pure substance.

One mole of gumballs is 6.022×10^{23} gumballs. One mole of popcorn is 6.022×10^{23} kernels of popcorn. This amount of popcorn would cover the United States and form a pile about 500 km (310 mi) high! It is unlikely that you will ever come in contact with this much gum or popcorn, so it does not make sense to use moles to count either of these items. The mole is useful, however, for counting atoms.

You might wonder why 6.022×10^{23} represents the number of particles in 1 mol. The mole has been defined as the number of atoms in 12.00 grams of carbon-12. Experiments have shown that 6.022×10^{23} is the number of carbon-12 atoms in 12.00 g of carbon-12. One mole of carbon consists of 6.022×10^{23} carbon atoms, with an average atomic mass of 12.01 amu.

Moles and grams are related

The mass in grams of 1 mol of a substance is called its **molar mass.** For example, 1 mol of carbon-12 atoms has a molar mass of 12.00 g. But a mole of an element will usually include atoms of several isotopes. So the molar mass of an element in grams is the same as its average atomic mass in amu, which is listed in the periodic table. The average atomic mass for carbon is 12.01 amu. One mole of carbon, then, has a mass of 12.01 g. *Figure 34* demonstrates this idea for magnesium.

Figure 34

One mole of magnesium (6.022×10^{23} Mg atoms) has a mass of 24.30 g. Note that the balance is only accurate to one-tenth of a gram, so it reads 24.3 g.

12
Mg
Magnesium
24.3050

REAL-LIFE CONNECTION

From Stars to the Sea Astronomers estimate that there is about one mole of stars in the universe. One mole is also about the number of milliliters of water in the Pacific Ocean.

Did You Know?

A Mole of Textbooks One mole of high school textbooks would cover the entire surface area of the United States with a depth of about 200 miles!

Calculating with Moles

Because the amount of a substance and its mass are related, it is often useful to convert moles to grams, and vice versa. You can use **conversion factors** to relate units.

Using conversion factors

How did the shopkeeper mentioned on the previous page know the mass of 50 gumballs? He multiplied by a conversion factor to determine the number of gumballs on the scale from their combined mass. Multiplying by a conversion factor is like multiplying by 1 because both parts of the conversion factor are always equal.

The shopkeeper knows that exactly 10 gumballs have a combined mass of 21.4 g. This relationship can be written as two equivalent conversion factors, both of which are shown below.

$$\frac{10 \text{ gumballs}}{21.4 \text{ g}} \qquad \frac{21.4 \text{ g}}{10 \text{ gumballs}}$$

The shopkeeper can use one of these conversion factors to determine the mass of 50 gumballs because mass increases in a predictable way as more gumballs are added to the scale, as you can see from **Figure 35**.

▶ **conversion factor** a ratio that is derived from the equality of two different units and that can be used to convert from one unit to another

internet connect

www.scilinks.org
Topic: Avogadro's Constant
SciLinks code: HK4013

SciLINKS Maintained by the National Science Teachers Association

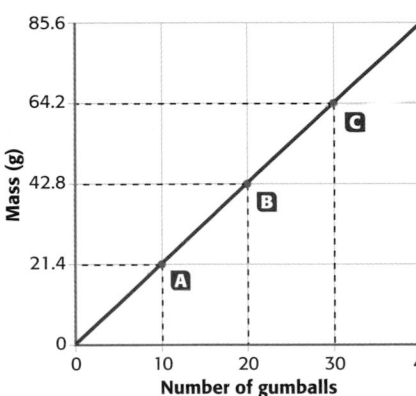

Figure 35
There is a direct relationship between the number of gumballs and their mass. Ten gumballs have a mass of 21.4 g, 20 gumballs have a mass of 42.8 g, and 30 gumballs have a mass of 64.2 g.

131

SOCIAL STUDIES
CONNECTION ─ **GENERAL**
From the business section of a major newspaper, obtain current costs of precious metals. Have students find the current cost of 1 mol of each metal. **LS Verbal**

MISCONCEPTION
ALERT

Molar Mass vs. Atomic Mass
Some students confuse molar mass and atomic mass, thinking that one atom of hydrogen has a mass of one gram, or one atom of magnesium has a mass of 24.3 grams. Be sure to emphasize the importance of units. Atomic mass is measured in amu, while molar mass in measured in grams. The mass of one *atom* of magnesium is 24.3 amu. With units of grams, 24.3 refers to the mass of one *mole* of magnesium, or 6.02×10^{23} Mg atoms.

Group Activity ─ **BASIC**

Conversion Factors To give students practice with conversion factors, divide the class into four groups. Assign one of the following to each group: length, mass and weight, volume, and area. Have students in each group make a set of cards of units that can be used to measure that property. For example, length cards can include inches, feet, meters, kilometers, and so on.

Have one student in each group shuffle the cards, and have another student draw two of them. Have students place those cards side-by-side. Ask students to figure out the conversion factors needed to change from the unit on the left to the unit on the right. Tell students that sometimes more than one conversion factor is needed. Have them repeat the activity for additional cards. **LS Interpersonal**

Chapter Resource File

• **Science Skills** Scientific Notation **BASIC**

Additional Examples

Conversion Factors

The mileage for Maria's car is 21 mi/gal of gasoline. If she needs to drive 97 mi, how much gas will her car use?

Answer: 4.6 gal

A school bus seat can hold three students. How many seats must the bus have if it is to haul 52 students on a field trip?

Answer: 18 seats

A bicycle travels at a speed of 30.0 km/h. How fast does the bicycle travel in m/s? (More than one conversion factor must be used.)

Answer: 8.33 m/s

LS Logical

Practice

1. 150 gumballs × (21.4 g/10 gumballs) 5 321 g

2. 50 eggs ÷ (1 dozen/12 eggs) = 4.2 dozen; 5 dozen eggs must be bought; 5 dozen × (12 eggs/ 1 dozen) − 50 eggs = 10 extra eggs

3. 1.7 ft × (1 yd/3 ft) = 0.57 yd

LS Logical

Math Skills

Conversion Factors What is the mass of exactly 50 gumballs?

1 List the given and unknown values.

Given: mass of 10 gumballs = 21.4 g

Unknown: mass of 50 gumballs = ? g

2 Write down the conversion factor that converts number of gumballs to mass.

The conversion factor you choose should have the unit you are solving for (g) in the numerator and the unit you want to cancel (number of gumballs) in the denominator.

$$\frac{21.4 \text{ g}}{10 \text{ gumballs}}$$

3 Multiply the number of gumballs by this conversion factor, and solve.

$$50 \text{ gumballs} \times \frac{21.4 \text{ g}}{10 \text{ gumballs}} = 107 \text{ g}$$

Practice

Conversion Factors

1. What is the mass of exactly 150 gumballs?

2. If you want 50 eggs, how many dozens must you buy? How many extra eggs do you have to take?

3. If a football player is tackled 1.7 ft short of the end zone, how many more yards does the team need to get a touchdown?

Relating amount to mass

Just as in the gumball example, there is also a relationship between the amount of an element in moles and its mass in grams. This relationship is graphed for iron nails in *Figure 36*. Because the amount of iron and the mass of iron are directly related, the graph is a straight line.

An element's molar mass can be used as if it were a conversion factor. Depending on which conversion factor you use, you can solve for either the amount of the element or its mass.

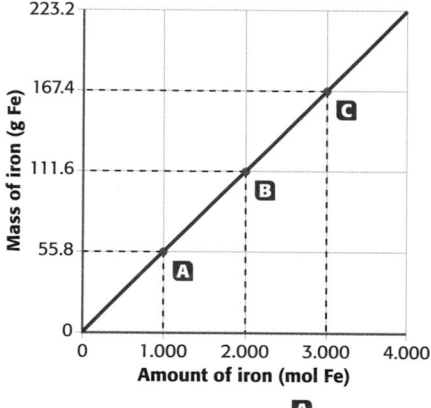

Figure 36

There is a direct relationship between the amount of an element and its mass.

132

Alternative Assessment ── BASIC

Graphing Have students study the graph in **Figure 36.** After they have completed this section, have them arbitrarily choose at least three different amounts, in moles, of an element other than iron. Then have them create a graph similar to that in **Figure 36** to confirm the direct relationship between mass and amount. **LS Logical**

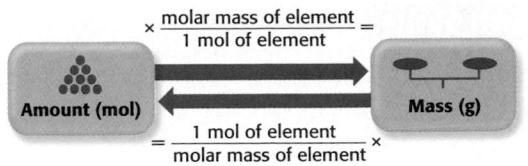

$$\times \frac{\text{molar mass of element}}{1 \text{ mol of element}} =$$

Amount (mol)

$$= \frac{1 \text{ mol of element}}{\text{molar mass of element}} \times$$

Mass (g)

Figure 37
The molar mass of an element allows you to convert between the amount of the element and its mass.

Converting moles to grams

Converting between the amount of an element in moles and its mass in grams is outlined in **Figure 37**. For example, you can determine the mass of 5.50 mol of iron by using **Figure 37** as a guide. First you must find iron in the periodic table. Its average atomic mass is 55.85 amu. This means iron's molar mass is 55.85 g/mol Fe. Now you can set up the problem using the molar mass as if it were a conversion factor, as shown in the sample problem below.

Math Skills

Converting Amount to Mass Determine the mass in grams of 5.50 mol of iron.

1 List the given and unknown values.
   Given: amount of iron = 5.50 mol Fe
   molar mass of iron = 55.85 g/mol Fe
   Unknown: mass of iron = ? g Fe

2 Write down the conversion factor that converts moles to grams.

   The conversion factor you choose should have what you are trying to find (grams of Fe) in the numerator and what you want to cancel (moles of Fe) in the denominator.

   $$\frac{55.85 \text{ g Fe}}{1 \text{ mol Fe}}$$

3 Multiply the amount of iron by this conversion factor, and solve.

   $$5.50 \text{ mol Fe} \times \frac{55.85 \text{ g Fe}}{1 \text{ mol Fe}} = 307 \text{ g Fe}$$

Practice HINT

Notice how iron's molar mass, 55.85 g/mol Fe, includes units (g/mol) and a chemical symbol (Fe). The units specify that this mass applies to 1 mol of substance. The symbol for iron, Fe, clearly indicates the substance. Remember to always include units in your answers and make clear the substance to which these units apply. Otherwise, your answer has no meaning.

Practice

Converting Amount to Mass
What is the mass in grams of each of the following?
1. 2.50 mol of sulfur, S
2. 1.80 mol of calcium, Ca
3. 0.50 mol of carbon, C
4. 3.20 mol of copper, Cu

Moles Emphasize that moles are a measure of the number of particles, not the mass. For example, 16.00 g of oxygen is the mass of 1 mol, or 6.022×10^{23} of oxygen atoms.

SKILL BUILDER

Interpreting Visuals Figure 37 illustrates how to convert between the amount of an element and its mass. Tell students to look at the periodic table to determine an element's average atomic mass, then arrange the conversion factor in the correct orientation.

Math Skills

Additional Example
Converting Amount to Mass
What is the mass in grams of 1.93 mol of cobalt, Co?
Answer: 114 g Co

Practice

1. 2.50 mol S × 32.07 g S/1 mol S = 80.2 g S
2. 1.80 mol Ca × 40.08 g Ca/ 1 mol Ca = 72.1 g Ca
3. 0.50 mol C × 12.01 g C/ 1 mol C = 6.0 g C
4. 3.20 mol Cu × 63.55 g Cu/ 1 mol Cu = 203 g Cu

LS Logical

Teacher Resources

For the New Teacher Many students benefit from the use of graphic organizers. For example, use this graphic organizer to visually illustrate how to convert back and forth between mass in grams and number of moles.

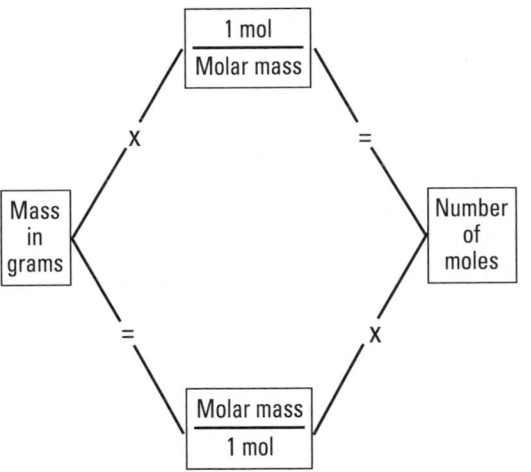

Chapter Resource File

- **Math Skills** Conversion Factors **GENERAL**
- **Math Skills** Converting Amount to Mass **GENERAL**
- **Science Skills** Significant Figures **BASIC**
- **Science Skills** Dimensional Analysis **BASIC**

Transparencies

TM Mole-Mass Graph
TM Mole-Mass Conversion

Quiz

1. What is the relationship between one mole and Avogadro's constant? (Avogadro's constant is the number of particles in one mole.)

2. What is the difference between molar mass and atomic mass? (Atomic mass is the mass of one atom; molar mass is the mass of one mole of atoms.)

3. Molar mass is measured in amu. True or false? (false)

4. The molar mass of iron is 55.85 g/mol Fe. What is the mass in grams of 2 mol of iron? (111.7 g Fe)

Practice HINT

Once you have learned how to convert mass to amount, you should use this sample problem to check your answers to the practice on the previous page.

Math Skills

Converting Mass to Amount Determine the amount of iron present in 352 g of iron.

1 **List the given and unknown values.**
　　Given: mass of iron = 352 g Fe
　　　　　　molar mass of iron = 55.85 g/mol Fe
　　Unknown: amount of iron = ? mol Fe

2 **Write down the conversion factor that converts grams to moles.**
　　The conversion factor you choose should have what you are trying to find (moles of Fe) in the numerator and what you want to cancel (grams of Fe) in the denominator.

$$\frac{1 \text{ mol Fe}}{55.85 \text{ g Fe}}$$

3 **Multiply the mass of iron by this conversion factor, and solve.**

$$352 \text{ g Fe} \times \frac{1 \text{ mol Fe}}{55.85 \text{ g Fe}} = 6.30 \text{ mol Fe}$$

SECTION 4 REVIEW

SUMMARY

▶ One mole of a substance has as many particles as there are atoms in exactly 12.00 g of carbon-12.

▶ Avogadro's constant, 6.022×10^{23}/mol, is equal to the number of particles in 1 mol.

▶ Molar mass is the mass in grams of 1 mol of a substance.

▶ An element's molar mass in grams is equal to its average atomic mass in amu.

▶ An element's molar mass can be used to convert from amount to mass, and vice versa.

1. **Define** Avogadro's constant. Describe how Avogadro's constant relates to a mole of a substance.

2. **Determine** the molar mass of the following elements:
   a. manganese, Mn
   b. cadmium, Cd
   c. arsenic, As
   d. strontium, Sr

3. **List** the two equivalent conversion factors for the molar mass of silver, Ag.

4. **Explain** why a graph showing the relationship between the amount of a particular element and the element's mass is a straight line.

5. **Critical Thinking** Which has more atoms: 3.0 g of iron, Fe, or 2.0 g of sulfur, S?

Math Skills

6. What is the mass in grams of 0.48 mol of platinum, Pt?

7. How many moles are present in 620 g of mercury, Hg?

8. How many moles are present in 11 g of silicon, Si?

9. How many moles are present in 205 g of helium, He?

Chapter Resource File

- **Math Skills** Converting Mass to Amount **GENERAL**
- **Concept Review** **GENERAL**
- **Quiz** **BASIC**

Answers to Section 4 Review

1. Avogadro's constant is 6.022×10^{23}/mol. It is the number of particles in one mole of anything.

2. **a.** 54.94 g/mol Mn
   b. 112.41 g/mol Cd
   c. 74.92 g/mol As
   d. 87.62 g/mol Sr

3. 1 mol Ag/107.87 g Ag; 107.87 g Ag/1 mol Ag

4. A direct relationship exists between the amount of an element in moles and the element's mass in grams.

5. 3.0 g Fe × 1 mol Fe/55.85 g Fe = 0.054 mol Fe; 2.0 g S × 1 mol S/32.07 g S = 0.062 mol S. Because the number of moles of sulfur is greater, the number of atoms of sulfur is greater.

See "Continuation of Answers" at the end of the chapter.

Math Skills

Conversion Factors

A chemical reaction requires 5.00 mol of sulfur as a reactant. What is the mass of this sulfur in grams?

1 **List all given and unknown values.**

    Given: amount of sulfur, 5.00 mol S

              molar mass of sulfur, 32.07 g/mol S

    Unknown: mass of sulfur (g)

2 **Write the conversion factor for moles to grams.**

The conversion factor you choose should have the unit you are solving for (g S) in the numerator and the unit you want to cancel (mol S) in the denominator.

$$\frac{32.07 \text{ g S}}{1 \text{ mol S}}$$

3 **Multiply the number of moles by this conversion factor, and solve.**

$$5.00 \text{ mol} \times \frac{32.07 \text{ g S}}{1 \text{ mol S}} = 160.4 \text{ g S}$$

Therefore, 160.4 grams of sulfur are needed for the reaction.

Practice

Molar Masses	
Copper	63.55 g/mol
Oxygen	16.00 g/mol
Sulfur	32.07 g/mol
Lead	207.2 g/mol

Using the example above, calculate the following:

1. If 4.00 mol of copper are needed as a reactant, what is the copper's mass in grams?

2. A combustion reaction requires 352 g of oxygen. What is the amount of oxygen atoms in moles?

3. If 622 g of lead are required for a reaction, what is the amount of the lead in moles?

Math Skills

Answers
1. 254 g Cu
2. 22.0 mol O
3. 3.00 mol Pb

135

Understanding Concepts

1. a
2. c
3. a
4. d
5. c
6. a
7. b
8. b
9. b
10. d
11. c

Using Vocabulary

12. In the nucleus, a silicon atom has 14 protons, and it usually has 14 neutrons. A silicon atom has 14 electrons, four of which are valence electrons.

13. The particles that make up an atom are protons, neutrons, and electrons. The number of protons determines the atom's atomic number. In a stable atom, this number also equals the number of electrons.

14. Drawings should resemble the orbitals in Figures 8 and 9 on page 109. Each orbital can hold up to two electrons, however an *s* orbital has only one orientation, whereas a *p* orbital has three; a *d* orbital has five; and an *f* orbital has seven.

Chapter Resource File

• **Chapter Tests** GENERAL

• **Test Item Listing for ExamView® Test Generator**

Chapter Highlights

Before you begin, review the summaries of the key ideas of each section, found at the end of each section. The key vocabulary terms are listed on the first page of each section.

UNDERSTANDING CONCEPTS

1. Which of Dalton's statements about the atom was later proven false?
   a. Atoms cannot be subdivided.
   b. Atoms are tiny.
   c. Atoms of different elements are not identical.
   d. Atoms join to form molecules.

2. Which statement is not true of Bohr's model of the atom?
   a. The nucleus can be compared to the sun.
   b. Electrons orbit the nucleus.
   c. An electron's path is not known exactly.
   d. Electrons exist in energy levels.

3. According to the modern model of the atom,
   a. moving electrons form an electron cloud.
   b. electrons and protons circle neutrons.
   c. neutrons have a positive charge.
   d. the number of protons an atom has varies.

4. If an atom has a mass of 11 amu and contains five electrons, its atomic number must be
   a. 55.   c. 6.
   b. 16.   d. 5.

5. Which statement about atoms of elements in the same group of the periodic table is true?
   a. They have the same number of protons.
   b. They have the same mass number.
   c. They have similar chemical properties.
   d. They have the same number of total electrons.

6. The organization of the periodic table is based on
   a. the number of protons in an atom.
   b. the mass number of an atom.
   c. the number of neutrons in an atom.
   d. the average atomic mass of an element.

7. The majority of elements in the periodic table are
   a. nonmetals.   c. synthetic.
   b. conductors.   d. noble gases.

8. Elements with some properties of metals and some properties of nonmetals are known as
   a. alkali metals.   c. halogens.
   b. semiconductors.   d. noble gases.

9. An atom of which of the following elements is unlikely to form a positively charged ion?
   a. potassium, K   c. barium, Ba
   b. selenium, Se   d. silver, Ag

10. Atoms of Group 18 elements are inert because
    a. they combine to form molecules.
    b. they have no valence electrons.
    c. they have filled inner energy levels.
    d. they have filled outermost energy levels.

11. Which of the following statements about krypton is not true?
    a. Its molar mass is 83.80 g/mol Kr.
    b. Its atomic number is 36.
    c. One mole of krypton atoms has a mass of 41.90 g.
    d. It is a noble gas.

⧉ **internet** connect

www.scilinks.org

Topic: Metals/Nonmetals SciLinks code: HK4086

SCI**LINKS**₀ Maintained by the National Science Teachers Association

---

15. In ionization, an atom whose outermost energy level is only partially filled may gain or lose a valence electron. For example, lithium may lose a valence electron to form a lithium ion with a 1+ charge, and fluorine may gain an electron to form a fluoride ion with a 1− charge.

16. Atomic number tells the number of protons in the atom. All atoms of the same element have the same number of protons, so they have the same atomic number. The mass number is the total number of protons and neutrons in an atom. The number of neutrons can vary, so the mass number can vary

among atoms of the same element. These different atoms are called isotopes.

17. a. An atom is the smallest unit of an element, and a molecule consists of two or more atoms.
    b. A neutral atom has no net charge. An ion is a charged atom or molecule.
    c. A cation has a positive charge, and an anion has a negative charge.

18. Answers will vary. Students may cite the following transition metals: gold, silver, platinum, copper, nickel, and zinc. Transition metals are often used in wiring, jewelry, and as conductors of heat and electricity.

USING VOCABULARY

12. How many *protons* and *neutrons* does a silicon, Si, atom have, and where are each of these subatomic particles located? How many *electrons* does a silicon atom have?

13. Identify the particles that make up an atom. How do these particles relate to the identity of an atom?

14. Draw two different types of *orbitals*, state their names, and describe how they are filled.

15. Describe the process of *ionization*, and give two different examples of elements that undergo this process.

16. Explain why different atoms of the same element always have the same *atomic number* but can have different *mass numbers*. What are these different atoms called?

17. Distinguish between the following:
    a. an *atom* and a *molecule*
    b. an *atom* and an *ion*
    c. a *cation* and an *anion*

18. List several familiar *transition metals* and their uses.

19. How is the *periodic law* demonstrated with the *halogens*?

20. Explain why *semiconductors*, or metalloids, deserve their name.

21. Distinguish between *alkali metals* and *alkaline-earth metals*, and give several examples of how they are used.

22. State *Avogadro's constant* and explain its relationship to the *mole*.

23. What does an element's *molar mass* tell you about the element?

BUILDING MATH SKILLS

24. **Graphing** Use a graphing calcula- **COMPUTER SKILL** tor, a computer spreadsheet, or a graphing program to plot the atomic number on the *x*-axis and the average atomic mass in amu on the *y*-axis for the transition metals in Period 4 of the periodic table (from scandium to zinc). Do you notice a break in the trend near cobalt? Explain why elements with larger atomic numbers do not necessarily have larger atomic masses.

25. **Converting Mass to Amount** For an experiment you have been asked to do, you need 1.5 g of iron. How many moles of iron do you need?

26. **Converting Mass to Amount** James is holding a balloon that contains 0.54 g of helium gas. What amount of helium is this?

27. **Converting Amount to Mass** A pure gold bar is made of 19.55 mol of gold. What is the mass of the bar in grams?

28. **Converting Amount to Mass** Robyn recycled 15.1 mol of aluminum last month. What mass of aluminum in grams did she recycle?

THINKING CRITICALLY

29. **Creative Thinking** Some forces push two atoms apart while other forces pull them together. Describe how the subatomic particles in each atom interact to produce these forces.

30. **Applying Knowledge** Explain why magnesium forms ions with the formula Mg^{2+}, not Mg^+ or Mg^-.

137

Assignment Guide

Section	Questions
1	1–3, 13, 14, 29, 40, 42, 47
2	4–6, 9, 11, 12, 15–17, 24, 30–32, 34, 36, 39, 44, 46, 48
3	7, 8, 10, 18–21, 35, 38, 45
4	22, 23, 25–28, 33, 37, 41, 43

19. All halogens have the same number of valence electrons (7) and similar chemical properties.

20. Semiconductors deserve their name because, under certain conditions, they can conduct heat or electricity the way that metals can.

21. Alkali metals are highly reactive, metallic elements. Alkaline-earth metals are less reactive. Alkali metals such as sodium are used in compounds such as sodium chloride. Alkaline-earth metals such as calcium are often found in limestone and other strong building materials and in the human body.

22. Avogadro's constant expresses the number of particles in one mole. It is equal to 6.022×10^{23}/mol.

23. the mass of one mole, or 6.022×10^{23} particles, of the element

Building Math Skills

24. Graphs should show increasing atomic mass with increasing atomic number, with one exception. The atomic number of Ni is higher than that of Co, but the atomic mass is less. Ni has more protons than Co does, but Co has more neutrons.

25. 1.5 g Fe × 1 mol Fe/55.85 g Fe = 0.027 mol Fe

26. 0.54 g He × 1 mol He/4.00 g He = 0.14 mol He

27. 19.55 mol Au × 196.97 g Au/ 1 mol Au = 3851 g Au

28. 15.1 mol Al × 26.98 g Al/1 mol Al = 407 g Al

Thinking Critically

29. Because like charges repel each other and unlike charges attract, protons repel each other and attract electrons. Electrons repel each other and attract protons. Since protons in the nucleus repel each other, there must be a stronger force that holds them together in the nucleus.

30. Magnesium has two valence electrons. To achieve a full outermost energy level, it will lose both electrons, forming Mg^{2+}.

31. All atoms in a period have the same number of electron energy levels. When an atom in that period becomes a positive ion, it loses that outer level and becomes smaller. An atom that becomes a negative ion keeps the outer level and becomes larger. Within their positive and negative groups, ions become smaller from left to right because the number of protons increases, increasing the attraction of the nucleus for the electrons.

32. Carbon is a nonmetal and lead is a metal.

33. The mass becomes halved also because only half the amount is present.

34. When an atom loses an electron, the atom gains a positive charge. The size of the atom decreases as its outermost energy levels lose electrons.

35. The noble gases are all very chemically stable due to the fact that their outermost energy levels are completely filled with electrons.

36. He, C, Au, Pb, Na, K, Cu

37. It is difficult to measure the size of an atom because atoms are very small. In addition, some atoms do not exist for very long due to their reactivity.

38. Scientists can compare the new entity to existing elements to determine whether or not the entity behaves like a known element or not. If the new entity possesses similar chemical and physical properties to an existing element, it is probably a new isotope of that element.

31. Evaluating Data The figure below shows relative ionic radii for positive and negative ions of elements in Period 2 of the periodic table. Explain the trend in ion size as you move from left to right across the periodic table. Why do the negative ions have larger radii than the positive ions?

0.60	0.31	1.71	1.40	1.36
●	·	⬤	⬤	⬤
Li^+	Be^{2+}	N^{3-}	O^{2-}	F^-

32. Making Comparisons Although carbon and lead are in the same group, some of their properties are very different. Propose a reason for this. (**Hint:** Look at the periodic table to locate each element and find out how each is classified.)

33. Problem Solving How does halving the amount of a sample of an element affect the sample's mass?

34. Understanding Systems When an atom loses an electron, what is the atom's charge? What do you think happens to the size of the atom?

35. Applying Knowledge What property do the noble gases share? How does this property relate to the electron configuration of the noble gases?

36. Applying Knowledge Write the chemical symbols for helium, carbon, gold, lead, sodium, potassium, and copper.

37. Critical Thinking Why is it difficult to measure the size of an atom?

38. Critical Thinking Particle accelerators are devices that speed up charged particles in order to smash the particles together. Sometimes the result of the collision is a new nucleus. How can scientists determine whether the nucleus formed is that of a new element or that of a new isotope of a known element?

138

39. Problem Solving What would happen to poisonous chlorine gas if the following alterations were made to the chlorine?
  a. A proton is added to each atom.
  b. An electron is added to each atom.
  c. A neutron is added to each atom.

DEVELOPING LIFE/WORK SKILLS

40. Locating Information Some "neon" signs contain substances other than neon to produce different colors. Design your own lighted sign, and find out which substances you could use to produce the colors you want your sign to be.

41. Making Decisions Suppose you have only 1.9 g of sulfur for an experiment and you must do three trials using 0.030 mol of S each time. Do you have enough sulfur?

42. Communicating Effectively The study of the nucleus produced a new field of medicine called nuclear medicine. Pretend you are writing an article for a hospital newsletter. Describe how radioactive substances called tracers are sometimes used to detect and treat diseases.

43. Working Cooperatively With a group of your classmates, make a list of 10 elements and their average atomic masses. Calculate the amount in moles for 6.0 g of each element. Rank your elements from the element with the greatest amount to the element with the least amount in a 6.0 g sample. Do you notice a trend in the amounts as atomic number increases? Explain why or why not.

44. Applying Knowledge You read a science fiction story about an alien race of silicon-based life-forms. Use information from the periodic table to hypothesize why the author chose silicon over other elements. (**Hint:** Life on Earth is carbon based.)

39. a. each atom would gain a positive charge and become an Ar atom. It would then gain an electron so as to become electrically neutral and chemically inert.
  b. each atom would form the negative chloride ion, Cl^-
  c. each atom would gain mass, forming a different isotope

Developing Life/Work Skills
40. Answers may vary, but could include that argon gives off a purple glow, and krypton gives off a pale violet glow.

41. 1.9 g S × 1 mol S/32.07 g S = 0.059 mol S. The experiment requires 3 × 0.030 mol, or 0.090 mol, so there is not enough sulfur.

42. Tracers are radioactive isotopes that can be detected by certain medical instruments but have low enough radioactivity that they do not hurt human cells in low concentrations. After entering the body, some tracers accumulate in certain body tissues. They can then be used to detect problems, such as tumors, or treat problems by accumulating and destroying undesirable tissues.

INTEGRATING CONCEPTS

45. Connection to Health You can keep your bones healthy by eating 1200–1500 mg of calcium a day. Use the table below to make a list of the foods you might eat in a day to satisfy your body's need for calcium. How does your typical diet compare with this?

Item, serving size	Calcium (mg)
Plain lowfat yogurt, 1 cup	415
Ricotta cheese, 1/2 cup	337
Skim milk, 1 cup	302
Cheddar cheese, 1 ounce	213
Cooked spinach, 1/2 cup	106
Vanilla ice cream, 1/2 cup	88

46. Concept Mapping Copy the unfinished concept map below onto a sheet of paper. Complete the map by writing the correct word or phrase in the lettered boxes.

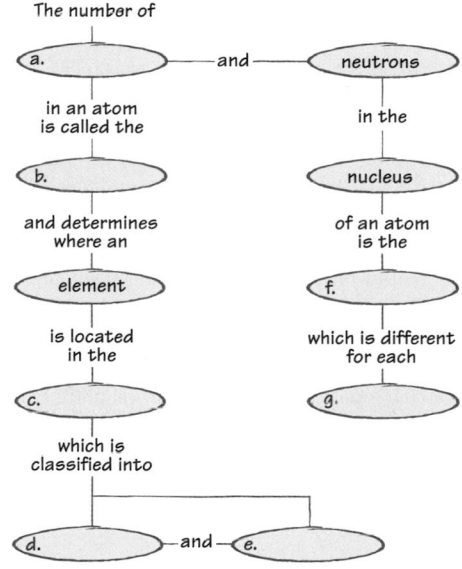

47. Connection to Physics Research the origins of the elements. The big bang theory suggests that the universe began with an enormous explosion. What was formed as a result of the big bang? Describe the matter that was present after the explosion. How much time passed before the elements as we know them were formed?

48. Connection to Earth Science The isotope carbon-14 is used in radiocarbon- dating of animal and plant fossils. Scientists use other isotopes to tell the ages of rocks and meteorites. Do some research and find out which isotopes are used to date rocks and meteorites.

www.scilinks.org
Topic: Origin of Elements SciLinks code: HK4097

SCI LINKS. Maintained by the National Science Teachers Association

43. Lists will vary. As atomic mass increases, the number of moles present decreases. In general, the number of moles decreases as the atomic number increases because atomic mass tends to increase as atomic number increases.

44. The author probably chose silicon, because silicon is in the same group as carbon, which is the basis of life forms on Earth. Thus, silicon could be construed as a possible life-forming element in other environments.

Integrating Concepts

45. Answers will vary. Any combination of the listed foods that has a total of 1200–1500 mg Ca is acceptable.

46. a. protons
b. atomic number
c. periodic table
d. metals
e. nonmetals
f. mass number
g. isotope

47. The big bang theory proposes that all matter was originally found at one point in the universe. An immense explosion occurred approximately 15 billion years ago, breaking this matter up into a cloud of energy and subatomic particles that expanded near the speed of light. These particles quickly fused to form mostly hydrogen and helium nuclei. Eventually, clouds of these materials formed, with further fusion forming elements through iron. Heavier elements are formed during supernovas.

48. Some isotopes used to date rocks and meteorites are Uranium-235/Lead-207, Potassium-40/Argon-40, Rubidium-87/Strontium-87, and Rhenium-187/Osmium-187.

139

Transparencies

TT Concept Mapping

COMPARING THE PHYSICAL PROPERTIES OF ELEMENTS

Teacher's Notes

Time Required
1 lab period

Ratings
1   2   3   4
EASY   HARD

TEACHER PREPARATION	2
STUDENT SETUP	2
CONCEPT LEVEL	1
CLEANUP	2

Skills Acquired
- Classifying
- Collecting data
- Communicating
- Designing Experiments
- Experimenting
- Identifying/Recognizing Patterns
- Inferring
- Measuring
- Organizing and analyzing data

The Scientific Method
In this lab, students will:
- Make Observations
- Form a Hypothesis
- Analyze the Results
- Draw Conclusions
- Communicate Results

Materials
Depending on which method students use to determine heat conductivity, the wax is used in solid or liquid form. Students could choose to test heat conductivity by adding hot melted wax to room temperature metals. In this case, melt the wax by placing it in a beaker on the hot plate and heating it. Caution students not to touch the hot beaker.

Skills Practice Lab

Introduction
How can you distinguish metal elements by analyzing their physical properties?

Objectives
▶ **USING SCIENTIFIC METHODS** *Hypothesize* which physical properties can help you *distinguish* between different metals.

▶ *Identify* unknown metals by *comparing* the data you collect with reference information.

Materials
balance
beakers (several)
graduated cylinder
hot plate
ice
magnet
metal samples, unidentified (several)
metric ruler
stopwatch
water
wax

140

Comparing the Physical Properties of Elements

▶ Procedure

Identifying Metal Elements
1. In this lab, you will identify samples of unknown metals by comparing the data you collect with reference information listed in the table at right. Use at least two of the physical properties listed in the table to identify each metal.

Deciding Which Physical Properties You Will Analyze
2. Density is the mass per unit volume of a substance. If the metal is box-shaped, you can measure its length, width, and height, and then use these measurements to calculate the metal's volume. If the shape of the metal is irregular, you can add the metal to a known volume of water and determine what volume of water is displaced.

3. Relative hardness indicates how easy it is to scratch a metal. A metal with a higher value can scratch a metal with a lower value, but not vice versa.

4. Relative heat conductivity indicates how quickly a metal heats or cools. A metal with a value of 100 will heat or cool twice as quickly as a metal with a value of 50.

5. If a magnet placed near a metal attracts the metal, then the metal has been magnetized by the magnet.

Designing Your Experiment
6. With your lab partner(s), decide how you will use the materials provided to identify each metal you are given. There is more than one way to measure some of the physical properties that are listed, so you might not use all of the materials that are provided.

7. In your lab report, list each step you will perform in your experiment.

8. Have your teacher approve your plan before you carry out your experiment.

Physical Properties of Some Metals

Metal	Density (g/mL)	Relative hardness	Relative heat conductivity	Magnetized by magnet?
Aluminum (Al)	2.7	28	100	no
Iron (Fe)	7.9	50	34	yes
Nickel (Ni)	8.9	67	38	yes
Tin (Sn)	7.3	19	28	no
Tungsten (W)	19.3	100	73	no
Zinc (Zn)	7.1	28	49	no

Performing Your Experiment

9. After your teacher approves your plan, carry out your experiment. Keep in mind that the more careful your measurements are, the easier it will be for you to identify the unknown metals.

10. Record all the data you collect and any observations you make in your lab report.

▶ **Analysis**

1. Make a table listing the physical properties you compared and the data you collected for each of the unknown metals.

2. Which metals were you given? Explain the reasoning you used to identify each metal.

3. Which physical properties were the easiest for you to measure and compare? Which were the hardest? Explain why.

4. What would happen if you tried to scratch aluminum foil with zinc?

5. Explain why it would be difficult to distinguish between iron and nickel unless you calculate each metal's density.

6. Suppose you find a metal fastener and determine that its density is 7 g/mL. What are two ways you could determine whether the unknown metal is tin or zinc?

▶ **Conclusions**

7. Suppose someone gives you an alloy that is made of both zinc and nickel. In general, how do you think the physical properties of the alloy would compare with those of each individual metal?

141

Students should wash their hands immediately and avoid touching their eyes if they accidentally touch a tin sample. This element poses a health risk if it enters the body.

Tips and Tricks

Discuss the work of Dmitri Mendeleev and the concept of periodicity with your students before starting this activity. In your discussion, emphasize the trends among groups of elements in the periodic table, but do not explain the chemical basis for these trends. Doing so might distract students from observing the various trends in the properties.

Minimize some of the initial frustration by directing students' attention to the trends in the properties among the alkali metals in the periodic table. Seeing the trends in the properties of lithium, sodium, and potassium should help students as they try to identify the unknown elements.

For each lab group, provide samples of several of the metals listed in the table **Physical Properties of Some Metals,** but do not reveal the identities of the metals. Emphasize to students that they may not use all the materials provided.

Students can measure heat conductivity in one of two ways. They could place the metal sample on a hot plate, add a drop of hardened wax, and then heat. (The wax on the metal with the greatest heat conductivity will melt first.) Or students could add hot melted wax to room temperature metals. (The wax on the metal with the greatest heat conductivity will harden first.)

Answers to Analysis

1. Student tables will vary depending on which unknown metals were analyzed.

2. Student answers will vary. Make sure answers are consistent with the properties of the unknown elements the student identifies.

3. Student answers will vary but should include a discussion of the measurement and comparison of each property.

4. Nothing. Neither aluminum nor zinc should be able to scratch the other one since they have such similar measurements of relative hardness.

5. Unless the densities of iron and nickel are calculated, it is difficult to distinguish between the two metals because their other physical properties are too similar.

6. Measuring the relative hardness and the relative heat conductivity of the metal fastener are two ways to determine whether the metal is tin or zinc.

Answers to Conclusions

7. The physical properties of an alloy of zinc and nickel would have intermediate values compared with the values of the properties of the metals alone.

Chapter Resource File

- **Datasheet** Comparing the Physical Properties of Elements GENERAL
- **Observation Lab** Drawing Atomic Models BASIC
- **CBL™ Probeware Lab** Predicting the Physical and Chemical Properties of Elements ADVANCED

CHAPTER 5

The Structure of Matter
Chapter Planning Guide

PACING	CLASSROOM RESOURCES	LABS, ACTIVITIES, AND DEMONSTRATIONS
BLOCK 1 · 45 min pp. 142–143 **Chapter Opener**		SE **Activity 1,** p. 143 SE **Activity 2,** p. 143
BLOCKS 2 & 3 · 90 min pp. 144–150 **Section 1** Compounds and Molecules	CRF **Lesson Plan*** TT Bellringer* TT Water Bonding*	TE **Opening Activity,** p. 144 GENERAL TE **Group Activity** Network Structures, p. 147 GENERAL SE **Quick Lab** Which melts more easily, sugar or salt?, p. 149 GENERAL CRF **Datasheets for In-Text Labs** Which melts more easily, sugar or salt?* GENERAL
BLOCKS 4 & 5 · 90 min pp. 151–158 **Section 2** Ionic and Covalent Bonding	CRF **Lesson Plan*** TT Bellringer* TT Multiple Bonds* TM Polyatomic Anions*	TE **Opening Demonstration,** p. 151 GENERAL SE **Quick Activity** Building a Close-Packed Structure, p. 154 GENERAL CRF **Datasheets for In-Text Labs** Building a Close-Packed Structure* GENERAL
BLOCKS 6 & 7 · 90 min pp. 159–164 **Section 3** Compound Names and Formulas	CRF **Lesson Plan*** TT Bellringer* TM Common Cations* TM Common Anions* TM Naming Prefixes*	TE **Opening Activity,** p. 159 GENERAL TE **Group Activity** Naming Covalent Compounds, p. 162 BASIC
BLOCKS 8 & 9 · 90 min pp. 165–172 **Section 4** Organic and Biochemical Compounds	CRF **Lesson Plan*** TT Bellringer* TT Six-Carbon Alkanes* TT Concept Mapping*	TE **Opening Demonstration,** p. 165 GENERAL SE **Quick Activity** Polymer Memory, p. 169 GENERAL CRF **Datasheets for In-Text Labs** Polymer Memory* GENERAL TE **Demonstration** Model a Polymer, p. 170 GENERAL SE **Quick Lab** What properties does a polymer have?, p. 171 GENERAL CRF **Datasheets for In-Text Labs** What properties does a polymer have?* SE **Skills Practice Lab** Comparing Polymers, pp. 178–179 ◆ GENERAL CRF **Datasheets for SE Skills Practice Lab** Comparing Polymers* GENERAL CRF **CBL™ Probeware Lab** Determining Which Household Solutions Conduct Electricity* ◆ ADVANCED CRF **Observation Lab** Extracting Iron from Cereal* ◆ BASIC

BLOCKS 10 & 11 · 90 min

Chapter Review and Assessment Resources

- SE **Chapter Review,** pp. 174–177
- CRF **Chapter Tests*** GENERAL
- OSP **Test Generator**
- CRF **Standardized Test Practice with Guided Reading Development***
- CRF **Test Item Listing for ExamView® Test Generator***
- OSP **Scoring Rubrics and Classroom Management Checklists**

Online Resources

go.hrw.com

Visit the HRW Web site for a variety of free resources related to the text. Just type in the keyword **HK4 STR**.

Holt Online Learning

Holt Science Spectrum: Physical Science: Online Edition

Students can access interactive problem solving help and active visual concept development with the *Holt Science Spectrum: Physical Science* Online Edition available at **www.hrw.com**.

student CNN News

cnnstudentnews.com

Find the latest news, lesson plans, and activities related to important scientific events.

KEY

TE Teacher Edition
SE Student Edition
OSP One-Stop Planner

CRF Chapter Resource File
TT Teaching Transparency
TM Transparency Master

* Also on One-Stop Planner
◆ Requires Advance Prep

PROBLEM SOLVING AND PRACTICE	SECTION REVIEW AND ASSESSMENT	STANDARDS CORRELATION
	CRF Pretest* GENERAL	
CRF Cross-Disciplinary Worksheet Connection to Fine Arts—What Happens in a Kiln?* GENERAL	**TE** Quiz, p. 150 BASIC **SE** Section 1 Review, p. 150 **CRF** Concept Review* GENERAL **CRF** Quiz* BASIC	PS 2c, 2d, 2e SAI 1 SPSP 5
CRF Cross-Disciplinary Worksheet Connection to Social Studies—Linus Pauling: A Life Well Spent* GENERAL **CRF** Cross-Disciplinary Worksheet Connection to Social Studies—Ion Propulsion in Deep Space 1* GENERAL	**TE** Quiz, p. 158 BASIC **SE** Section 2 Review, p. 158 **CRF** Concept Review* GENERAL **CRF** Quiz* BASIC	PS 2a, 2c UCP 2 HNS 3
SE Math Skills Writing Ionic Formulas, p. 161 **CRF** Math Skills Writing Ionic Formulas* GENERAL	**TE** Quiz, p. 164 BASIC **SE** Section 3 Review, p. 164 **CRF** Concept Review* GENERAL **CRF** Quiz* BASIC	SAI 1
CRF Cross-Disciplinary Worksheet Connection to Engineering—Fractions of Crude Oil* GENERAL **CRF** Cross-Disciplinary Worksheet Integrating Environmental Science—Plastics* GENERAL **CRF** Cross-Disciplinary Worksheet Integrating Math—Amino Acid Combinations* GENERAL **SE** Study Skills KWL Notes, p. 173 **SE** CareerLink Analytical Chemist, pp. 180–181	**TE** Quiz, p. 172 BASIC **SE** Section 4 Review, p. 172 **CRF** Concept Review* GENERAL **CRF** Quiz* BASIC	PS 2f SPSP 5

SCiLINKS®

www.scilinks.org

Topic: Properties of Substances
*Sci*Links code: HK4113

Topic: Structures of Substances
*Sci*Links code: HK4135

Topic: Chemical Bonding
*Sci*Links code: HK4021

Topic: Naming Compounds
*Sci*Links code: HK4092

Topic: Carbon Compounds
*Sci*Links code: HK4016

Topic: Vitamin C
*Sci*Links code: HK4147

Topic: Analytical Chemistry
*Sci*Links code: HK4006

Technology Resources

 One-Stop Planner
All of your printable resources and the Test Generator are on this convenient CD-ROM.

 Physical Science Interactive Tutor CD-Rom

 CNN Science in the NEWS
each video segment is accompanied by a Critical Thinking Worksheet*

Segment 12
Alloy Technology

Overview

This chapter distinguishes between compounds and mixtures, then covers compounds in more detail, including the use of models to visually represent compounds and the relationship between chemical structure and properties. This chapter explores why atoms join together in compounds and how they form ionic, covalent, and metallic bonds. Compound names and formulas for both ionic and covalent compounds are discussed. The concepts of the chapter are applied to a study of organic and biochemical compounds.

Assessing Prior Knowledge

Be sure students understand the following concepts:

- elements and compounds
- properties
- atomic structure
- the periodic table
- families of elements
- metals and nonmetals
- ions
- atomic mass
- molar mass

MISCONCEPTION ///ALERT\\\

Science education research has identified the following misconceptions about the structure of matter.

- Students fail to distinguish between mixtures and compounds, not recognizing the compound as a new substance rather than a mixture of the elements that compose it.
- Students confuse molecules and compounds.
- Students believe that, in covalent bonds, electrons are shared equally between atoms.

The Structure of Matter

Chapter Preview

1 Compounds and Molecules
   What Are Compounds?
   Models of Compounds
   How Does Structure Affect Properties?

2 Ionic and Covalent Bonding
   What Holds Bonded Atoms Together?
   Ionic Bonds
   Metallic Bonds
   Covalent Bonds
   Polyatomic Ions

3 Compound Names and Formulas
   Naming Ionic Compounds
   Writing Formulas for Ionic Compounds
   Naming Covalent Compounds
   Chemical Formulas for Covalent Compounds

4 Organic and Biochemical Compounds
   Organic Compounds
   Polymers
   Biochemical Compounds

INTEGRATING TECHNOLOGY and Society

142

Standards Correlations

National Science Education Standards

The following descriptions summarize the National Science Standards that specifically relate to this chapter. For the full text of the standards, see the National Science Education Standards at the front of the book.

Section 1 Compounds and Molecules
Physical Science PS 2c–2e
Science as Inquiry SAI 1
Science in Personal and Social Perspectives SPSP 5

Section 2 Ionic and Covalent Bonding
Physical Science PS 2a, 2c
Unifying Concepts and Processes UCP 2
History and Nature of Science HNS 3

Section 3 Compound Names and Formulas
Science as Inquiry SAI 1

Section 4 Integrating Technology and Society—Organic and Biochemical Compounds
Physical Science PS 2f
Science in Personal and Social Perspectives SPSP 5

Focus ACTIVITY

Background Suddenly, a glass object slips from your hand and crashes to the ground. You watch it break into many tiny pieces as you hear it hit the floor. Glass is a brittle substance. When enough force is applied, it breaks into many sharp, jagged pieces. Glass behaves the way it does because of its composition.

A glass container and a stained glass window have some similar properties because both are made mainly from silicon dioxide. But other compounds are responsible for the window's beautiful colors. Adding a compound of nickel and oxygen to the glass produces a purple tint. Adding a compound of cobalt and oxygen makes the glass deep blue, while adding a compound of copper and oxygen makes the glass dark red.

Activity 1 There are many different kinds of glass, each with its own use. List several kinds of glass that you encounter daily. Describe the ways that each kind of glass differs from other kinds of glass.

Activity 2 Research other compounds that are sometimes added to glass. Describe how each of these compounds changes the properties of glass. Write a report on your findings.

Glass is a brittle substance that is made from silicon dioxide, a compound with a very rigid structure. The addition of small amounts of other compounds changes the color of the glass, "staining" it.

> ☑ **internet** connect
>
> **www.scilinks.org**
> **Topic:** Properties of Substances
> **SciLinks code:** HK4113
>
> SCI*LINKS*. Maintained by the National Science Teachers Association

Pre-Reading Questions
1. How does atomic structure affect the properties of a substance?
2. Can bonds between atoms be broken?

143

Focus ACTIVITY

Background Glass is made primarily of silicon dioxide, the main ingredient in quartz and some types of sand. Have students compare the strength of glass and quartz. Emphasize that when materials are added to silicon dioxide, the structure of the compound might be affected. The resulting glass is usually more brittle than quartz or sand.

Activity 1 Answers may vary but might include glass used in drinking glasses and dishes, high-heat ceramics used in cooking, and safety (break-resistant) glass used in some windows and eyeglasses. **LS Intrapersonal**

Activity 2 Answers may vary. For example, glass that is used for optical fibers often contains arsenic, selenium, and tellurium. Be sure students understand that glass differs in properties because of the way the glass has been processed, not because of what has been added to it. **LS Verbal**

Answers to Pre-Reading Questions

1. The atoms in a substance and the way those atoms are arranged determine the properties of that substance, including the substance's hardness, ductility, and malleability.
2. Yes, bonds between atoms can be broken by heat, pressure, or dissolving them with a solution.

Chapter Resource File

- Pretest **GENERAL**
- Teaching Transparency Preview
- Answer Keys

Focus

Overview

Before beginning this section, review with your students the Objectives listed in the Student Edition. This section begins by comparing compounds and mixtures. Students learn that the atoms in a compound are bonded together chemically, and that compounds are represented by definite chemical formulas. They also learn how chemical structure can be described and modeled. The section concludes by discussing how chemical structure affects the properties of compounds.

📻 Bellringer

Use the Bellringer transparency to prepare students for this section.

Motivate

Opening Activity — GENERAL

Have students list as many uses of the term *compound* as they can. Lists might include compound sentences, chemical compounds, compound fracture, or compound interest rates. Ask them what the items on their lists have in common. Students should recognize that *compound* involves more than one thing. **LS Verbal**

Compounds and Molecules

▶ **KEY TERMS**
chemical bond
chemical structure
bond length
bond angle

◣ OBJECTIVES ◢

▶ **Distinguish** between compounds and mixtures.
▶ **Relate** the chemical formula of a compound to the relative numbers of atoms or ions present in the compound.
▶ **Use** models to visualize a compound's chemical structure.
▶ **Describe** how the chemical structure of a compound affects its properties.

If you step on a sharp rock with your bare foot, you feel pain. That's because rocks are hard substances; they don't bend. Many rocks are made of quartz. Table salt and sugar look similar; both are grainy, white solids. But they taste very different. In addition, salt is hard and brittle and breaks into uniform cube-like granules, while sugar does not. Quartz, salt, and sugar are all compounds. Their similarities and differences result from the way their atoms or ions are joined.

What Are Compounds?

Table salt is a compound made of two elements, sodium and chlorine. When elements combine to form a compound, the compound has properties very different from those of the elements that make it. **Figure 1** shows how the metal sodium combines with chlorine gas to form sodium chloride, NaCl, or table salt.

Figure 1

A The silvery metal sodium combines with **B** poisonous, yellowish green chlorine gas in a violent reaction **C** to form **D** white granules of table salt that you can eat.

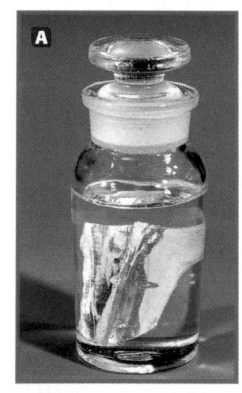  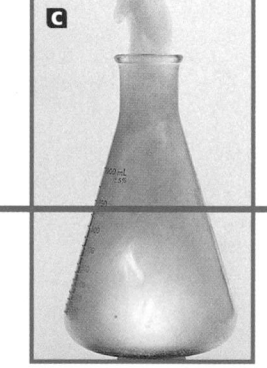 

144

MISCONCEPTION /// ALERT \\\

Mixtures and Compounds Many students have trouble distinguishing between mixtures and compounds. They do not recognize that a compound is a new substance with different properties. Discuss with students the two principal differences between mixtures and compounds covered in this section. The properties of a mixture reflect the properties of the substances it contains, but the properties of a compound often bear no resemblance to the properties of the elements that compose it. Also, compounds have a definite composition by mass of their combining elements, while the components of mixtures may be present in varying proportions. For example, water has a definite composition by mass. It is always 88.8% oxygen and 11.2% hydrogen. In contrast, substances in a mixture can have any mass ratio. The sand in a sand-and-gravel mixture may make up as much as 99% or as little as 1% of the overall mass.

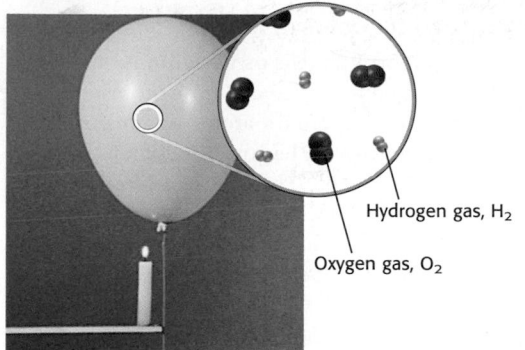

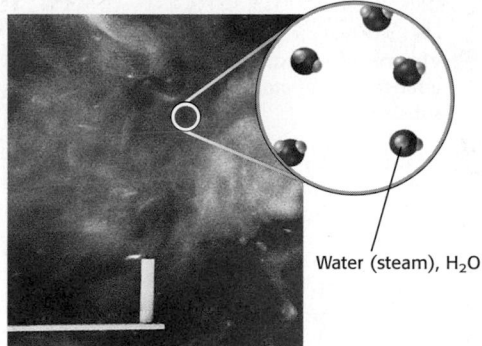

Figure 2

Ⓐ Placing a lit candle under a balloon containing hydrogen gas and oxygen gas causes the balloon to melt, releasing the mixed gases.

Hydrogen gas, H_2
Oxygen gas, O_2

Ⓑ The mixed gases are ignited by the candle flame, and water is produced.

Water (steam), H_2O

Chemical bonds distinguish compounds from mixtures

The attractive forces that hold different atoms or ions together in compounds are called **chemical bonds.** Recall how compounds and mixtures are different. Mixtures are made of different substances that are just placed together. Each substance in the mixture keeps its own properties.

For example, mixing blue paint and yellow paint makes green paint. Different shades of green can be made by mixing the paints in different proportions, but both original paints remain chemically unchanged.

Figure 2 shows that when a mixture of hydrogen gas and oxygen gas is heated, a violent reaction takes place and a compound forms. Chemical bonds are broken, and atoms are rearranged. New bonds form water, a compound with properties very different from those of the original gases.

A compound always has the same chemical formula

The chemical formula for water is H_2O, and that of table sugar is $C_{12}H_{22}O_{11}$. The salt you season your food with has the chemical formula NaCl. A chemical formula shows the types and numbers of atoms or ions making up the simplest unit of the compound.

There is another important way that compounds and mixtures are different. Compounds are always made of the same elements in the same proportion. A molecule of water, for example, is always made of two hydrogen atoms and one oxygen atom. This is true for all water. That means water frozen in a comet in outer space and water at 37°C (98.6°F) inside the cells of your body both have the same chemical formula—H_2O.

▶ **chemical bond** the attractive force that holds atoms or ions together

145

Be sure students realize the distinction between materials that bond for added strength when they dry and those materials that just dry. Have students think about each of their examples. Would the dried material return to its original consistency if the original solvent were added to it? If so, it probably just dried and did not change its bonding.

Making the Connection

1. Answers may vary but might include dental ceramics, adhesives, and some paints.
2. Paragraphs may vary. Each paragraph should refer to one substance and should include a discussion of how bonding changes in the substance.
LS Verbal

Teaching Tip

Molecular Models Discuss with students other examples of how something can be represented in more than one way. One example is found in music. The desired sounds can be represented as single notes on a staff or as letters that represent groups of notes, or chords.

In math, the process of multiplication can be represented by several different symbols—a multiplication cross, a dot, parentheses, or just two symbols written side by side. Emphasize to students that molecules can be represented in many different ways as well.

- ◼ **chemical structure** the arrangement of atoms in a substance

- ◼ **bond length** the average distance between the nuclei of two bonded atoms

- ◼ **bond angle** the angle formed by two bonds to the same atom

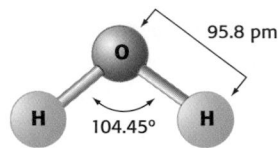

95.8 pm
104.45°

Figure 3
The ball-and-stick model in this figure is a giant representation of one molecule of water. A picometer (pm) is equal to 1×10^{-12} m.

Clay has a layered structure of silicon, oxygen, aluminum, and hydrogen atoms. Artists can mold wet clay into any shape because water molecules let the layers slide over one another. When clay dries, water evaporates and the layers can no longer slide. To keep the dry, crumbly clay from breaking apart, artists change the structure of the clay by heating it. The atoms in one layer bond to atoms in the layers above and below. When this happens, the clay hardens, and the artist's work is permanently set.

Making the Connection

1. Think of other substances that can be shaped when they are wet and that "set" when they are dried or heated.
2. Write a paragraph about one of these substances and why it has these properties.

146

Chemical structure shows the bonding within a compound

Although water's chemical formula tells us what atoms it is made of, it doesn't reveal anything about the way these atoms are connected. You can see how a compound's atoms or ions are connected by its **chemical structure.** The structure of a compound can be compared to that of a rope. The kinds of fibers used to make a rope and the way the fibers are intertwined determine how strong the rope is. Similarly, the atoms in a compound and the way the atoms are arranged determine many of the compound's properties.

Two terms are used to specify the positions of atoms relative to one another in a compound. A **bond length** gives the distance between the nuclei of two bonded atoms. And when a compound has three or more atoms, **bond angles** tell how these atoms are oriented in space. **Figure 3** shows the chemical structure of a water molecule. You can see that the way hydrogen and oxygen atoms bond to form water looks more like a boomerang than a straight line.

Models of Compounds

Figure 3 is a ball-and-stick model of a water molecule. Ball-and-stick models, as well as other kinds of models, help you "see" a compound's structure by showing you how the atoms or ions are arranged in the compound.

Some models give you an idea of bond lengths and angles

In the ball-and-stick model of water shown in **Figure 3,** the atoms are represented by balls. The bonds that hold the atoms together are represented by sticks. Although bonds between atoms aren't really as rigid as sticks, this model makes it easy to see the bonds and the angles they form in a compound.

Structural formulas can also show the structures of compounds. Notice how water's structural formula, which is shown below, is a lot like its ball-and-stick model. The difference is that only chemical symbols are used to represent the atoms.

$$H \diagup \overset{\displaystyle O}{} \diagdown H$$

ORGANIC CHEMISTRY
CONNECTION

Structural formulas are especially important in organic chemistry because many organic compounds have the same chemical formula but different structures. For example, butane and 2-methyl propane both have the chemical formula C_4H_{10}, but the atoms in each are arranged differently, so their structural formulas are different. Compounds with the same chemical formula but different structures are called isomers. Isomers can have different properties due to the different arrangement of atoms.

Isomers are common in organic chemistry because carbon is able to form many different kinds of chemical bonds. The more carbon atoms there are, the more isomers there can be. For example, C_8H_{18} has 18 isomers, $C_{20}H_{42}$ has 366,319 isomers, and $C_{40}H_{82}$ has approximately 6.25×10^{13} isomers!

Space-filling models show the space occupied by atoms

Figure 4 shows another way chemists picture a water molecule. It is called a space-filling model because it shows the space that is occupied by the oxygen and hydrogen atoms. The problem with this model is that it is harder to "see" bond lengths and angles.

How Does Structure Affect Properties?

Some compounds, such as the quartz found in many rocks, exist as a large network of bonded atoms. Other compounds, such as table salt, are also large networks, but of bonded positive and negative ions. Still other compounds, such as water and sugar, are made of many separate molecules. Different structures give these compounds different properties.

Compounds with network structures are strong solids

Quartz is sometimes found in the form of beautiful crystals, as shown in *Figure 5*. Quartz has the chemical formula SiO_2, and so does the less pure form of quartz, sand. *Figure 5* shows that every silicon atom in quartz is bonded to four oxygen atoms. The bonds that hold these atoms together are very strong. All of the $Si-O-Si$ and $O-Si-O$ bond angles are the same. That is, each one is 109.5°. This arrangement continues throughout the substance, holding the silicon and oxygen atoms together in a very strong, rigid structure.

This is why rocks containing quartz are hard and inflexible solids. Silicon and oxygen atoms in sand have a similar arrangement. It takes a lot of energy to break the strong bonds between silicon and oxygen atoms in quartz and sand. That's why the melting point and boiling point of quartz and sand is so high, as shown in *Table 1*.

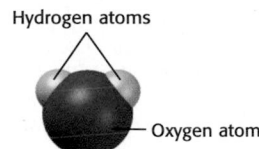

Figure 4
This space-filling model of water shows that the two hydrogen atoms take up much less space than the oxygen atom.

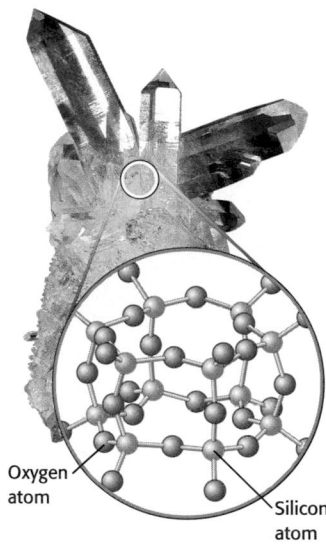

Figure 5
Quartz and sand are made of silicon and oxygen atoms bonded in a strong, rigid structure.

Table 1 Some Compounds with Network Structures

Compound	State (25°C)	Melting point (°C)	Boiling point (°C)
Silicon dioxide, SiO_2 (quartz)	solid	1700	2230
Magnesium fluoride, MgF_2	solid	1261	2239
Sodium chloride, NaCl (table salt)	solid	801	1413

internet connect
www.scilinks.org
Topic: Structures of Substances
SciLinks code: HK4135
SCiLINKS Maintained by the National Science Teachers Association

147

Did You Know ?

Network Solids Many network solids, such as diamond and many silicon compounds, are arranged in a tetrahedral pattern. For example, with the compound SiO_4, each oxygen atom is at one of the corners of a tetrahedron, with a silicon atom in the center of the tetrahedron. The tetrahedrons fit together to form a network solid.

Interpreting Visuals Sometimes a crystal of a network solid reveals the pattern made by the ions in the solid. Have students use a hand lens to examine some granulated salt and compare what they see with the network structure shown in **Figure 6.** Students should see that the salt granules are cubic, as is the arrangement of its ions.
LS Visual

Teaching Tip

States of Matter Students might think that a compound is, by nature, either a solid, a liquid, or a gas. Emphasize that most substances can be in any state, depending on temperature and sometimes, pressure. When a substance is classified as a solid, a liquid, or a gas, and no temperature is mentioned, the temperature is usually room temperature.

Figure 6

Each grain of table salt, or sodium chloride, is composed of a tightly packed network of Na^+ ions and Cl^- ions.

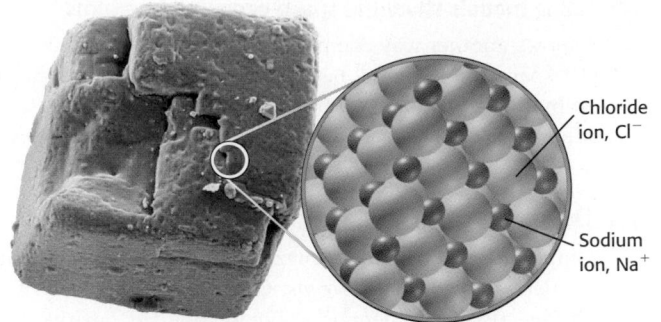

Chloride ion, Cl^-

Sodium ion, Na^+

Some compounds are made of networks of bonded ions

Like some quartz, table salt—sodium chloride—is found in the form of regularly shaped crystals. Crystals of sodium chloride are cube shaped. Like quartz and sand, sodium chloride is made of a repeating network connected by strong bonds. The network is made of tightly packed, positively charged sodium ions and negatively charged chloride ions, as shown in **Figure 6.** The strong attractions between the oppositely charged ions cause table salt and other similar compounds to have high melting points and boiling points, as shown in **Table 1.**

Some compounds are made of molecules

Salt and sugar are both white solids you can eat, but their structures are very different. Unlike salt, sugar is made of molecules. A molecule of sugar, shown in **Figure 7,** is made of carbon, hydrogen, and oxygen atoms joined by bonds. Molecules of sugar do attract each other to form crystals. But these attractions are much weaker than those that hold bonded carbon, hydrogen, and oxygen atoms together to make a sugar molecule.

We breathe nitrogen, N_2, oxygen, O_2, and carbon dioxide, CO_2, every day. All three substances are colorless, odorless gases made of molecules. Within each molecule, the atoms are so strongly attracted to one another that they are bonded. But the molecules of each gas have very little attraction for one another. Because the molecules of these gases are not very attracted to one another, they spread out as much as they can. That is why gases can take up a lot of space.

Figure 7

Sugar, $C_{12}H_{22}O_{11}$, is made of molecules.

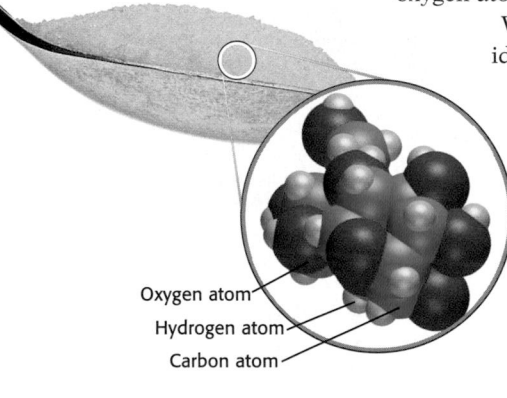

Oxygen atom

Hydrogen atom

Carbon atom

148

Alternative Assessment — ADVANCED

Growing Salt Crystals Have students use library or Internet resources to find out how to grow salt crystals. This is a relatively simple process that can be done with basic materials in the home, over a period of a few days. The following website links to several sites that have information about growing salt crystals: http://www.saltinstitute.org/15.html.

Tell students they can add food coloring to the water to make crystals of different colors.

They can also try the same process with sugar, baking soda, and cream of tartar. Have students bring their crystals to class so everyone can observe them. Tell students to hold the crystals up to the light to see how crystals reflect light. They can also use a magnifying glass to observe the crystal structure in more detail. **LS** Kinesthetic

Table 2 Comparing Compounds Made of Molecules

Compound	State (25°C)	Melting point (°C)	Boiling point (°C)
Sugar, $C_{12}H_{22}O_{11}$	Solid	185–186	———
Water, H_2O	Liquid	0	100
Dihydrogen sulfide, H_2S	Gas	–86	–61

The strength of attractions between molecules varies

Compare sugar, water, and dihydrogen sulfide in *Table 2*. Although all three compounds are made of molecules, their properties are very different. Sugar is a solid, water is a liquid, and dihydrogen sulfide is a gas. That means that sugar molecules have the strongest attractions for each other, followed by water molecules. Dihydrogen sulfide molecules have the weakest attractions for each other. The fact that sugar and water have such different properties probably doesn't surprise you. Their chemical structures are not at all alike. But what about water and dihydrogen sulfide, which do have similar chemical structures?

Which melts more easily, sugar or salt?

Materials
- ✓ table salt
- ✓ Bunsen burner
- ✓ stopwatch
- ✓ table sugar
- ✓ 2 test tubes
- ✓ tongs

SAFETY CAUTION Wear safety goggles and gloves. Tie back long hair, confine loose clothing, and use tongs to handle hot glassware. When heating a substance in a test tube, always point the open end of the test tube away from yourself and others.

1. Use your knowledge of structures to make a hypothesis about whether sugar or salt will melt more easily.

2. To test your hypothesis, place about 1 cm³ of sugar in a test tube.

3. Using tongs, position the test tube with sugar over the flame, as shown in the figure at right. Move the test tube back and forth slowly over the flame. Use a stopwatch to measure the time it takes for the sugar to melt.

4. Repeat steps 2 and 3 with salt. If your sample does not melt within 1 minute, remove it from the flame.

Analysis

1. Which compound is easier to melt? Was your hypothesis right?

2. How can you relate your results to the structure of each compound?

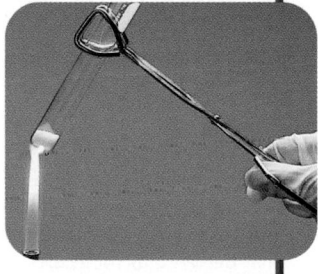

Quick Lab

Which melts more easily, sugar or salt?

Have test tube racks available so students can place hot test tubes in them until they cool. Remind students that all factors except what is in the test tube should be constant. Be sure students heat the sugar only until it melts. If it burns, the test tube will be difficult to clean.

Analysis

1. Sugar was easier to melt. The salt probably did not melt at all.

2. Sugar is molecular, and the particles have less attraction for each other than the salt particles do. Ions in salt form a network solid with strong attractions between the ions. **LS Logical**

Did You Know?

Hydrogen Bonds Why is water a liquid at room temperature when most molecules of similar mass are gases? Water molecules exhibit hydrogen bonding, which occurs between molecules formed from hydrogen and a strongly electronegative element, such as oxygen. Because oxygen pulls shared electrons away from the hydrogen atom, the proton of the hydrogen atom is exposed, and it attracts the negative parts of other polar molecules. This results in greater attraction between water molecules than exists between most other molecules.

Chapter 5 · The Structure of Matter **149**

SKILL BUILDER

Vocabulary Molecules are affected by both intermolecular and intramolecular forces. Intramolecular forces are those within the molecule, such as a covalent bond. Intermolecular forces are those between molecules such as the attraction between the positive part of one molecule and the negative part of another.

Close

Quiz

Choose the term or phrase that best completes each sentence.

1. The atoms in a compound are held together by chemical bonds/chemical structures. (chemical bonds)

2. A compound's chemical formula/bond angle shows which atoms or ions it is made up of. (chemical formula)

3. Chemical/structural formulas show how atoms are arranged in a compound. (structural)

4. Network solids usually have high/low melting and boiling points. (high)

Chapter Resource File

• Concept Review GENERAL

• Quiz BASIC

Transparencies

TT Water Bonding

Strong bonds *within* each water molecule

Weaker attractions *between* water molecules

Figure 8

Dotted lines indicate the *inter*molecular attractions that occur *between* water molecules, which is often referred to as "hydrogen bonding." Water is a liquid at room temperature because of these attractions.

Attractions between water molecules are called hydrogen bonds

The higher melting and boiling points of water suggest that water molecules attract each other more than dihydrogen sulfide molecules do. *Figure 8* shows how an oxygen atom of one water molecule is attracted to a hydrogen atom of a neighboring water molecule. This attraction is called a *hydrogen bond*. Water molecules attract each other, but these attractions are not as strong as the bonds holding oxygen and hydrogen atoms together within a molecule.

SECTION 1 REVIEW

SUMMARY

▸ Atoms or ions in compounds are joined by chemical bonds.

▸ A compound's chemical formula shows which atoms or ions it is made of.

▸ A model represents a compound's structure visually.

▸ Substances with network structures are usually strong solids with high melting and boiling points.

▸ Substances made of molecules have lower melting and boiling points.

▸ Whether a molecular substance is a solid, a liquid, or a gas at room temperature depends on the attractions between its molecules.

1. **Classify** the following substances as mixtures or compounds:
   a. air   **c.** SnF_2
   b. CO   **d.** pure water

2. **Explain** why silver iodide, AgI, a compound used in photography, has a much higher melting point than vanillin, $C_8H_8O_3$, a sweet-smelling compound used in flavorings.

3. **Draw** a ball-and-stick model of a boron trifluoride, BF_3, molecule. In this molecule, a boron atom is attached to three fluorine atoms. Each F−B−F bond angle is 120°, and all B−F bonds are the same length.

4. **Predict** which molecules have a greater attraction for each other, C_3H_8O molecules in liquid rubbing alcohol or CH_4 molecules in methane gas.

5. **Explain** why glass, which is made mainly of SiO_2, is often used to make cookware. (**Hint:** What properties does SiO_2 have because of its structure?)

6. **Critical Thinking** A picometer (pm) is equal to 1×10^{-12} m. O−H bond lengths in water are 95.8 pm, while S−H bond lengths in dihydrogen sulfide are 135 pm. Why are S−H bond lengths longer than O−H bond lengths? (**Hint:** Which is larger, a sulfur atom or an oxygen atom?)

Answers to Section 1 Review

1. **a.** mixture
   b. compound
   c. compound
   d. compound

2. Silver iodide has a network structure of positive and negative ions. Vanillin consists of molecules. The attraction between particles in silver iodide is much stronger than the attraction between particles of vanillin.

3. Student drawings should show a boron atom surrounded by three equally spaced fluorine atoms in the same plane. A line from each fluorine atom to the boron atom represents a bond. Lines should be of equal length.

4. Molecules in a liquid, such as C_3H_8O, have a greater attraction for each other because they are closer together and are moving more slowly than the molecules in a gas, such as CH_4.

5. SiO_2 has a network structure, resulting in a high melting point. So it does not melt when heated to high cooking temperatures.

6. Sulfur atoms are larger than oxygen atoms by one electron energy level. Their valence electrons are farther from the nucleus, so the nucleus-to-nucleus distance is greater.

Ionic and Covalent Bonding

OBJECTIVES

▶ **Explain** why atoms sometimes join to form bonds.

▶ **Explain** why some atoms transfer their valence electrons to form ionic bonds, while other atoms share valence electrons to form covalent bonds.

▶ **Differentiate** between ionic, covalent, and metallic bonds.

▶ **Compare** the properties of substances with different types of bonds.

▶ **KEY TERMS**
ionic bond
metallic bond
covalent bond
polyatomic ion

When two atoms join, a bond forms. You have already seen how bonded atoms form many kinds of substances. Atoms bond in different ways to form these many substances. The type of bonds that the atoms of a substance form affect the substance's properties.

What Holds Bonded Atoms Together?

Three different kinds of bonds describe the way atoms bond in most substances. In many of the models you have seen so far, the bonds that hold atoms together are represented by sticks. But what bonds atoms in a real molecule?

Bonded atoms usually have a stable electron configuration

Atoms bond when their valence electrons interact. You have learned that atoms with full outermost *s* and *p* orbitals are more stable than atoms with only partly filled outer *s* and *p* orbitals. Generally, atoms join to form bonds so that each atom has a stable electron configuration. When this happens, each atom has an electronic structure similar to that of a noble gas.

When two hydrogen atoms bond, as shown in *Figure 9,* the positive nucleus of one hydrogen atom attracts the negative electron of the other hydrogen atom, and vice versa. This attraction pulls the two atoms closer together. Soon the electron clouds of the hydrogen atoms cross each other. The shared electron cloud of the molecule that forms has two electrons (one from each atom). A hydrogen molecule, which consists of two hydrogen atoms bonded together, has an electronic structure similar to the noble gas helium. The molecule will not fall apart unless enough energy is added to break the bond.

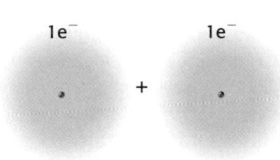
Hydrogen atom   Hydrogen atom

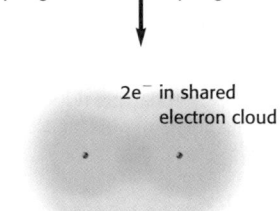
2e⁻ in shared electron cloud

Hydrogen molecule

Figure 9
When two hydrogen atoms are very close together, their electron clouds overlap, and a bond forms. The two electrons of the hydrogen molecule that forms are in the shared electron cloud.

151

Focus

Overview

Before beginning this section, review with your students the Objectives listed in the Student Edition. In this section, students learn why and how atoms bond together in ionic bonds, metallic bonds, covalent bonds, and polyatomic ions.

🎧 Bellringer

Use the Bellringer transparency to prepare students for this section.

Motivate

Opening Demonstration —— GENERAL

(Time: Approximately 10 minutes)

Materials
• foam ball (3), one larger
• pencil
• protractor
• copper wire, 2 50-cm lengths

Wrap the wire around the pencil to make two springs, each approximately 3 cm long. Leave about 4 cm of wire unwound at each end. Stick one end of a spring into each small ball. Stick the other ends into the large foam ball, forming a bond angle of 105°.

Explain to students that this is a model of a water molecule. Ask students: What element does the large ball represent? (oxygen) What element do the small balls represent? (hydrogen) Why do springs represent bonds better than sticks do? (Bonds bend and stretch.)
LS Visual

Alternative Assessment —— BASIC

Writing **Bonding** Explain the different types of bonding. Then have students form pairs and use the following figures to explain what is happening in terms of bonding. Have them use **Figure 11** to explain ionic bonding, **Figures 9, 14,** and **15** to explain nonpolar covalent bonding, **Figure 16** for polar covalent bonding, and **Figure 13** for metallic bonding. You may wish to have them recreate the illustrations on posterboard and add their explanations in written form. You could then display the posters around the classroom for students to refer to during the rest of this chapter.
LS Interpersonal

Chapter Resource File
• Lesson Plan

Transparencies
TT Bellringer

Connection to SOCIAL STUDIES

Tell students that, in general, electronegativity increases as you go left and up the periodic table. Excluding the noble gases, which have no electronegativity values, francium is the least electronegative element, and fluorine is the most electronegative.

Answers

1. Fluorine is more electronegative because it has a tendency to gain electrons to achieve a full outermost energy level. Calcium has a tendency to lose electrons, so it has little tendency to gain them.
2. Paragraphs will vary but should include that radiation disrupts the functions of cells and can destroy living tissue.

LS Logical

SKILL BUILDER — BASIC

Interpreting Visuals Have students examine **Figure 10**. Ask students why chemists use a solid bar to represent atomic bonds, even though the bonds are actually flexible like springs. (The model on the left is simpler to draw and use, and can serve the same purpose of representing the structure of a compound and the bonds that make it up.) Remind students that models are often used to represent things simply and quickly. Tell them that when working with such models, it is important to keep in mind their limitations. **LS** Logical

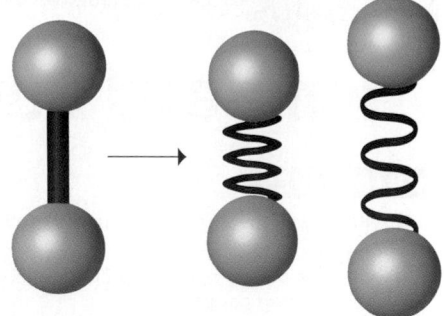

Figure 10
Chemists often use a solid bar to show a bond between two atoms, but real bonds are flexible, like stiff springs.

ionic bond a bond formed by the attraction between oppositely charged ions

Bonds can bend and stretch without breaking

Although some bonds are stronger and more rigid than others, all bonds behave more like flexible springs than like sticks, as *Figure 10* shows. The atoms move back and forth a little and their nuclei do not always stay the same distance apart. In fact, most reported bond lengths are averages of these distances. Although bonds are not rigid, they still hold atoms together tightly.

Connection to SOCIAL STUDIES

American scientist Linus Pauling studied how electrons are arranged within atoms. He also studied the ways that atoms share and exchange electrons. In 1954, he won the Nobel Prize in chemistry for his valuable research.

Later, Pauling fought to ban nuclear weapons testing. Pauling was able to convince more than 11 000 scientists from 49 countries to sign a petition to stop nuclear weapons testing. Pauling won the Nobel Peace Prize in 1962 for his efforts. A year later, a treaty outlawing nuclear weapons testing in the atmosphere, in outer space, and underwater went into effect.

Making the Connection

1. *Electronegativity* is an idea first thought of by Pauling. It tells how easily an atom accepts electrons. Which is more electronegative, a fluorine atom or a calcium atom? Why?
2. Nuclear weapons testing can harm living things because of the resulting radiation. Write a paragraph explaining how high levels of radiation can affect your body.

WRITING SKILL

Ionic Bonds

Ionic bonds are formed between oppositely charged ions. Atoms of metal elements, such as sodium and calcium, form the positively charged ions. Atoms of nonmetal elements, such as chlorine and oxygen, form the negatively charged ions.

Ionic bonds are formed by the transfer of electrons

Some atoms do not share electrons to fill their outermost energy levels completely. Instead, they transfer electrons. One of the atoms gains the electrons that the other atom loses. Both ions that form usually have stable electron configurations. The result is a positive ion and a negative ion, such as the Na^+ ion and the Cl^- ion in sodium chloride.

These oppositely charged ions attract each other and form an ionic bond. Each positive sodium ion attracts several negative chloride ions. These negative chloride ions attract more positive sodium ions, and so on. Soon a network of these bonded ions forms a crystal of table salt.

REAL-LIFE CONNECTION

Table Salt The table salt you buy is usually not pure NaCl. It often contains small amounts of finely divided insoluble substances such as silicates or complex salts of aluminum. Particles of these materials stick to the cubic NaCl crystals and keep them from clumping together in high humidity. In addition, much table salt is *iodized* and contains small amounts of KI, which provides dietary iodine for proper thyroid function.

Did You Know ?

Chemical Reactions Chemical reactions often depend on whether a compound is ionic or covalent. Certain types of reactions take place only if the reactants are one type or the other.

Ionic compounds are in the form of networks, not molecules

Because sodium chloride is a network of ions, it does not make sense to talk about "a molecule of NaCl." In fact, every sodium ion is next to six chloride ions, as shown in *Figure 6*. Instead, chemists talk about the smallest ratio of ions in ionic compounds. Sodium chloride's chemical formula, NaCl, tells us that there is one Na^+ ion for every Cl^- ion, or a 1:1 ratio of ions. This means the compound has a total charge of zero. One Na^+ ion and one Cl^- ion make up a *formula unit* of NaCl.

Not every ionic compound has the same ratio of ions as sodium chloride. An example is calcium fluoride, which is shown in *Figure 11*. The ratio of Ca^{2+} ions to F^- ions in calcium fluoride must be 1:2 to make a neutral compound. That is why the chemical formula for calcium fluoride is CaF_2.

When melted or dissolved in water, ionic compounds conduct electricity

Electric current is moving charges. Solid ionic compounds do not conduct electricity because the charged ions are locked into place, causing the melting points of ionic compounds to be very high—often well above 300°C. But if you dissolve an ionic compound in water or melt it, it can conduct electricity. That's because the ions are then free to move, as shown in *Figure 12*.

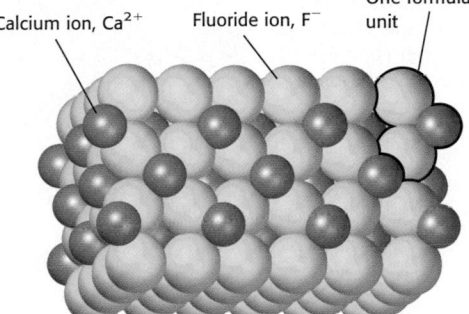

Figure 11

There are twice as many fluoride ions as calcium ions in a crystal of calcium fluoride, CaF_2. So one Ca^{2+} ion and two F^- ions make up one formula unit of the compound.

Figure 12

Like other ionic compounds, sodium chloride conducts electricity when it is dissolved in water.

Sodium ion, Na^+

Chloride ion, Cl^-

Water molecule, H_2O

153

HEALTH CONNECTION

The human body needs salt to regulate the balance of electrolytes in the body, both in and out of cells. Major salt deficiencies, such as a phenomenon known as "salt starvation" in India, can lead to serious health problems. In America, most people easily reach the minimum recommended daily amount—500 mg. (This is an average value; the actual amount needed by an individual depends on many factors, including lifestyle and genetic makeup.) In fact, Americans consume an average of 3,500 mg per day. In a healthy individual, the kidneys process excess salt not used by the body. Some scientists believe that excess salt intake is a risk factor for cardiovascular disease because there is a correlation in some individuals between salt intake and blood pressure. However, this is a controversial issue today, and the link is not fully understood.

Quick ACTIVITY

Building a Close-Packed Structure

Materials (per group):
• 3 books
• ping pong balls

Teacher's Notes: Be sure the spheres are small enough that several of them will fit into the triangle. Students will not see the pattern if too few balls are used. As an alternate activity, use equal numbers of balls of two different sizes, alternating them in the pattern. Students can also use a rectangular pattern instead of a triangle. Tell students that sometimes the arrangement of network ionic solids is called closest packing. Ask them to explain why this term is accurate. (The ions in upper layers fill in depressions in the lower levels. This arrangement is the closest the ions can be.) **LS** Kinesthetic

Teaching Tip

Metallic Bonding Students are used to thinking that electrons "belong" to one particular atom, ion, or molecule. Emphasize to them that in metallic bonding, metal ions are surrounded by what are known as delocalized electrons. Delocalized electrons are the valence electrons from the metal atoms. They move from one place to another and are not associated with a particular metal ion.

Figure 13
Copper is a flexible metal that melts at 1083°C and boils at 2567°C. Copper conducts electricity because electrons can move freely between atoms.

Copper

Metallic Bonds

Metals, like copper, shown in **Figure 13,** can conduct electricity when they are solid. Metals are also flexible, so they can bend and stretch without breaking. Copper, for example, can be hammered flat into sheets or stretched into very thin wire. What kind of bonds give copper these properties?

Electrons move freely between metal atoms

■ **metallic bond** a bond formed by the attraction between positively charged metal ions and the electrons around them

The atoms in metals like copper form **metallic bonds.** The attraction between one atom's nucleus and a neighboring atom's electrons packs the atoms closely together. This close packing causes the outermost energy levels of the atoms to overlap, as shown in **Figure 13.** Therefore, electrons are free to move from atom to atom. This model explains why metals conduct electricity so well. Metals are flexible because the atoms can slide past each other without their bonds breaking.

Quick ACTIVITY

Building a Close-Packed Structure

Copper and other metals have close-packed structures. This means their atoms are packed very tightly together. In this activity, you will build a close-packed structure using ping pong balls.

1. Place three books flat on a table so that their edges form a triangle.
2. Fill the triangular space between the books with the spherical "atoms." Adjust the books so that the atoms make a one-layer, close-packed pattern, as shown at right.
3. Build additional layers on top of the first layer. How many other atoms does each atom touch? Where have you seen other arrangements that are similar to this one?

154

Did You Know ?

Metals Many more metals exist than nonmetals. Most of their uses are based on properties that result from metallic bonding.

PHYSICS
CONNECTION

Electric current is the rate that electric charges move through a conductor. The charges can be negative, positive, or a combination. For example, in the human body, electric current in the brain and nerves consists of positive sodium and potassium ions in motion. In metals, current consists of the movement of negative electrons. This is why the ability of electrons to move in a metal affects the metal's ability to conduct current. Since electrons in a metal are free to move from atom to atom, metals are good conductors.

Covalent Bonds

Compounds that are made of molecules, like water and sugar, have **covalent bonds.** Compounds existing as networks of bonded atoms, such as silicon dioxide, are also held together by covalent bonds. Covalent bonds are often formed between non-metal atoms.

Covalent compounds can be solids, liquids, or gases. Except for silicon dioxide and other compounds with network structures, most covalent compounds have low melting points—usually below 300°C. In compounds that are made of molecules, the molecules are free to move when the compound is dissolved or melted. But most of these molecules remain intact and do not conduct electricity because they are not charged.

Atoms joined by covalent bonds share electrons

Some atoms, like the hydrogen atoms in *Figure 9,* bond to form molecules. *Figure 14A* shows how two chlorine atoms bond to form a chlorine molecule, Cl_2. Before bonding, each atom has seven electrons in its outermost energy level. The atoms don't transfer electrons to one another because each needs to gain an electron. If each atom shares one electron with the other atom, then both atoms together have a full outermost energy level. That is, both atoms together have eight valence electrons. The way electrons are shared depends on which atoms are sharing the electrons. Two chlorine atoms are exactly alike. When they bond, electrons are equally attracted to the positive nucleus of each atom. Bonds like this one, in which electrons are shared equally, are called *nonpolar covalent bonds.*

The structural formula in *Figure 14B* shows how the chlorine atoms are connected in the molecule that forms. A single line drawn between two atoms indicates that the atoms share two electrons and are joined by one covalent bond.

► **covalent bond** a bond formed when atoms share one or more pairs of electrons

internet connect

www.scilinks.org
Topic: Chemical Bonding
SciLinks code: HK4021

SCI LINKS. Maintained by the National Science Teachers Association

VOCABULARY *Skills Tip*

Co<u>v</u>alent bonds *form when atoms share pairs of* va<u>lence</u> *electrons.*

Figure 14

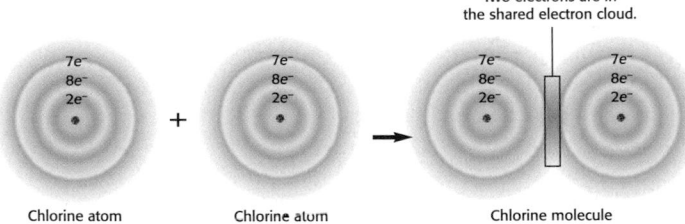

A Two chlorine atoms share electrons equally to form a *nonpolar covalent bond.*

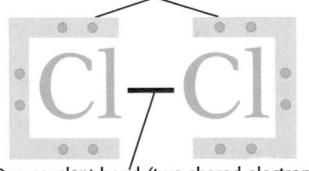

Each chlorine atom has six electrons that are not shared.

One covalent bond (two shared electrons)

B A single line drawn between two chlorine atoms shows that the atoms share two electrons. Dots represent electrons that are not involved in bonding.

155

Vocabulary The prefix *poly–* means "many." Ask students to think of other words with this prefix. (Responses could include polygon, polyester, polyglot, and polygamy.) Emphasize to students that polyatomic ions are ions that contain many atoms. **LS** Verbal

Teaching Tip — GENERAL

Polar Covalent Bond Have students think of a situation that is analogous to a polar covalent bond. Have them write a paragraph explaining their analogy and telling how the situation is similar to a polar covalent bond. For example, the analogy might be a tug-of-war between ten students who have been lifting weights and ten students who have not. Even though both groups have some pull on the rope, the pull will not be even. The weight lifters will have a greater pull for the shared rope, just as one atom in a polar bond has a greater attraction for the shared electrons. **LS** Verbal

---

Figure 15

The elements oxygen and nitrogen have covalent bonds. Electrons not involved in bonding are represented by dots.

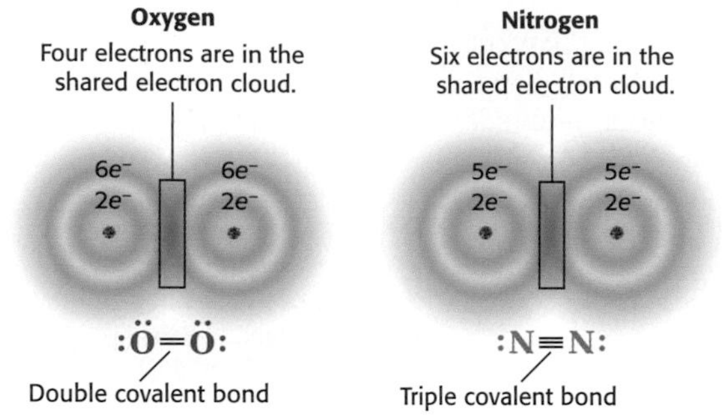

Oxygen
Four electrons are in the shared electron cloud.

Nitrogen
Six electrons are in the shared electron cloud.

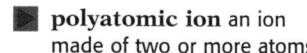

:Ö=Ö:
Double covalent bond

:N≡N:
Triple covalent bond

PHYSICAL SCIENCE INTERACTIVE TUTOR

Disc One, Module 4:
Chemical Bonding
Use the Interactive Tutor to learn more about this topic.

▶ **polyatomic ion** an ion made of two or more atoms

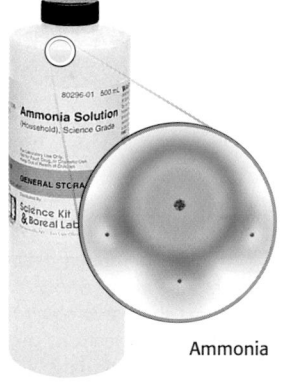

Ammonia

Atoms may share more than one pair of electrons

Figure 15 shows covalent bonding in oxygen gas, O_2 and nitrogen gas, N_2. Notice that the bond joining two oxygen atoms is represented by two lines. This means that two pairs of electrons (a total of four electrons) are shared to form a double covalent bond.

The bond joining two nitrogen atoms is represented by three lines. Two nitrogen atoms form a triple covalent bond by sharing three pairs of electrons (a total of six electrons).

The bond between two nitrogen atoms is stronger than the bond between two oxygen atoms. That's because more energy is needed to break a triple bond than to break a double bond. Triple and double bonds are also shorter than single bonds.

Atoms do not always share electrons equally

When two different atoms share electrons, the electrons are not shared equally. The shared electrons are attracted to the nucleus of one atom more than the other. An unequal sharing of electrons forms a *polar covalent bond*.

Usually, electrons are more attracted to atoms of elements that are located farther to the right and closer to the top of the periodic table. The shading in *Figure 16* shows that the shared electrons in the ammonia gas, NH_3, in the headspace of this container, are closer to the nitrogen atom than they are to the hydrogen atoms.

Polyatomic Ions

Until now, we have talked about compounds that have either ionic or covalent bonds. But some compounds have both ionic and covalent bonds. Such compounds are made of **polyatomic ions,** which are groups of covalently bonded atoms that have either lost or gained electrons. A polyatomic ion acts the same as the ions you have already encountered.

Figure 16

The darker shading around the nitrogen atom as compared to the hydrogen atoms shows that electrons are more attracted to nitrogen atoms than to hydrogen atoms. So the bonds in ammonia are *polar covalent bonds.*

156

Did You Know?

Chromatography A process called chromatography can be used to separate mixtures of covalent compounds. One type of chromatography, called paper chromatography, separates molecules based on their polarity. Water or some other liquid solvent carries the components of the mixture through a porous paper. The more polar the compound, the more it is attracted to the paper. Thus, the more polar the component, the less distance the component will travel in the paper.

MISCONCEPTION ///ALERT\\\

Polar Bonds Many students believe that electrons are always shared equally between atoms in covalent bonds. Use the discussion of polar covalent bonds on this page to emphasize that electrons are not always shared equally. In nonpolar bonding, the sharing is equal, while in polar bonding, the shared electrons are attracted to the nucleus of one atom more than to the nucleus of the other atom.

There are many common polyatomic ions

Many compounds you use either contain or are made from polyatomic ions. For example, your toothpaste may contain baking soda. Another name for baking soda is sodium hydrogen carbonate, $NaHCO_3$. Hydrogen carbonate, HCO_3^-, is a polyatomic ion. Sodium carbonate, Na_2CO_3, is often used to make soaps and other cleaners and contains the carbonate ion, CO_3^{2-}. Sodium hydroxide, $NaOH$, has hydroxide ions, OH^-, and is also used to make soaps. A few of these polyatomic ions are shown in **Figure 17.**

Oppositely charged polyatomic ions, like other ions, can bond to form compounds. Ammonium nitrate, NH_4NO_3, and ammonium sulfate, $(NH_4)_2SO_4$, both contain positively charged ammonium ions, NH_4^+. Nitrate, NO_3^-, and sulfate, SO_4^{2-}, are both negatively charged polyatomic ions.

Parentheses group the atoms of a polyatomic ion

You might be wondering why the chemical formula for ammonium sulfate is written as $(NH_4)_2SO_4$ instead of as $N_2H_8SO_4$. The parentheses around the ammonium ion are there to remind you that it acts like a single ion. Parentheses group the atoms of the ammonium ion together to show that the subscript 2 applies to the whole ion. There are two ammonium ions for every sulfate ion. Parentheses are not needed in compounds like ammonium nitrate, NH_4NO_3, because there is a 1:1 ratio of ions.

Always keep in mind that a polyatomic ion's charge applies not only to the last atom in the formula but to the entire ion. The carbonate ion, CO_3^{2-}, has a 2− charge. This means that CO_3, not just the oxygen atom, has the negative charge.

Some polyatomic anion names relate to their oxygen content

You may have noticed that many polyatomic anions are made of oxygen. Most of their names end with *-ite* or *-ate*. These endings do not tell you exactly how many oxygen atoms are in the ion, but they do follow a pattern. Think about sulfate (SO_4^{2-}) and sulfite (SO_3^{2-}), nitrate (NO_3^-) and nitrite (NO_2^-), and chlorate (ClO_3^-) and chlorite (ClO_2^-). The charge of each ion pair is the same. But notice how the ions have different numbers of oxygen atoms. Their names also have different endings.

An *-ate* ending is used to name the ion with one more oxygen atom. The name of the ion with one less oxygen ends in *-ite*. **Table 3,** on the next page, lists several common polyatomic anions. As you look at this table, you'll notice that not all of the anions listed have names that end in *-ite* or *-ate*. That's because some polyatomic anions, like hydroxide (OH^-) and cyanide (CN^-), are not named according to any general rules.

Hydroxide ion, OH^-

Carbonate ion, CO_3^{2-}

Ammonium ion, NH_4^+

Figure 17
The hydroxide ion (OH^-), carbonate ion (CO_3^{2-}), and ammonium ion (NH_4^+) are all polyatomic ions.

INTEGRATING

SPACE SCIENCE
Most of the ions and molecules in space are not the same as those that are found on Earth or in Earth's atmosphere. C_3H, C_6H_2, and HCO^+ have all been found in space. So far, no one has been able to figure out how these unusual molecules and ions form in space.

Close

Quiz

Match each term in the lower column with the correct description above.

1. bond formed between positively charged metal ions and electrons (c. metallic bond)

2. bond formed between oppositely charged ions (a. ionic bond)

3. bond formed between atoms that share electrons (b. covalent bond)

a. ionic bond

b. covalent bond

c. metallic bond

Chapter Resource File

- **Concept Review** GENERAL
- **Quiz** BASIC

Videos

CNN Presents Chemistry

- **Segment 12** Alloy Technology
  GENERAL

- **Critical Thinking** Worksheet 12

See the Science in the News video guide for more details.

Table 3 **Some Common Polyatomic Anions**

Ion name	Ion formula	Ion name	Ion formula
Acetate ion	$CH_3CO_2^-$	Hydroxide ion	OH^-
Carbonate ion	CO_3^{2-}	Hypochlorite ion	ClO^-
Chlorate ion	ClO_3^-	Nitrate ion	NO_3^-
Chlorite ion	ClO_2^-	Nitrite ion	NO_2^-
Cyanide ion	CN^-	Phosphate ion	PO_4^{3-}
Hydrogen carbonate ion	HCO_3^-	Phosphite ion	PO_3^{3-}
Hydrogen sulfate ion	HSO_4^-	Sulfate ion	SO_4^{2-}
Hydrogen sulfite ion	HSO_3^-	Sulfite ion	SO_3^{2-}

SECTION 2 REVIEW

SUMMARY

▸ Atoms bond when their valence electrons interact.

▸ Cations and anions attract each other to form ionic bonds.

▸ When ionic compounds are melted or dissolved in water, moving ions can conduct electricity.

▸ Atoms in metals are joined by metallic bonds.

▸ Metals conduct electricity because electrons can move from atom to atom.

▸ Covalent bonds form when atoms share electron pairs. Electrons may be shared equally or unequally.

▸ Polyatomic ions are covalently bonded atoms that have either lost or gained electrons. Their behavior resembles that of simple ions.

1. Determine if the following compounds are likely to have ionic or covalent bonds.
 a. magnesium oxide, MgO
 b. strontium chloride, $SrCl_2$
 c. ozone, O_3
 d. methanol, CH_3OH

2. Identify which two of the following substances will conduct electricity, and explain why.
 a. aluminum foil
 b. sugar, $C_{12}H_{22}O_{11}$, dissolved in water
 c. potassium hydroxide, KOH, dissolved in water

3. Draw the structural formula for acetylene. Atoms bond in the order HCCH. Carbon and hydrogen atoms share two electrons, and each carbon atom must have a total of four bonds. How many electrons do the carbon atoms share?

4. Predict whether a silver coin can conduct electricity. What kind of bonds does silver have?

5. Describe how it is possible for calcium hydroxide, $Ca(OH)_2$, to have both ionic and covalent bonds.

6. Explain why electrons are shared equally in oxygen, O_2, but not in carbon monoxide, CO.

7. Analyze whether dinitrogen tetroxide, N_2O_4, has covalent or ionic bonds. Describe how you reached this conclusion.

8. Critical Thinking *Bond energy* measures the energy per mole of a substance needed to break a bond. Which element has the greater bond energy, oxygen or nitrogen? (**Hint:** Which element has more bonds?)

Answers to Section 2 Review

1. a. ionic
 b. ionic
 c. covalent
 d. covalent

2. Aluminum foil will conduct, because it is a metal and its valence electrons are free to move. KOH dissolved in water will conduct, because its ions are free to move.

3. H−C≡C−H
 The carbon atoms share three pairs, or six, electrons.

4. Silver has metallic bonds so it has electrons that are free to move. The coin will conduct electricity.

5. The compound is comprised of Ca^{2+} and OH^- ions held together by ionic bonds. OH^- itself consists of covalently bonded oxygen and hydrogen atoms.

6. Because all atoms in a molecule of oxygen are the same, they attract and share electrons by the same amount. Carbon and oxygen attract electrons differently, so the sharing is unequal.

7. It has covalent bonds because both elements involved are nonmetals.

8. Nitrogen has greater bond energy. It takes more energy to break the three bonds in a nitrogen molecule than it takes to break the two bonds in an oxygen molecule.

Compound Names and Formulas

OBJECTIVES

▶ **Name** simple ionic and covalent compounds.

▶ **Predict** the charge of a transition metal cation in an ionic compound.

▶ **Write** chemical formulas for simple ionic compounds.

▶ **Distinguish** a covalent compound's empirical formula from its molecular formula.

■ **KEY TERMS**

empirical formula
molecular formula

Just like elements, compounds have names that distinguish them from other compounds. Although the compounds BaF_2 and BF_3 may appear to have similar chemical formulas, they have very different names. BaF_2 is *barium fluoride*, and BF_3 is *boron trifluoride*. When talking about these compounds, you have little chance for confusing their names. You can see that the names of these compounds reflect the elements from which the compounds are formed.

internet connect

www.scilnks.org
Topic: Naming Compounds
SciLinks code: HK4092

SCiLINKS Maintained by the National Science Teachers Association

Naming Ionic Compounds

Ionic compounds are formed by the strong attractions between cations and anions. Both ions are important to the compound's structure, so it makes sense that both ions are included in the name.

Names of cations include the elements of which they are composed

In many cases, the name of the cation is just like the name of the element from which it is made. You have already seen this for many cations. For example, when an atom of the element *sodium* loses an electron, a *sodium ion*, Na^+, forms. Similarly, when a *calcium* atom loses two electrons, a *calcium ion*, Ca^{2+}, forms. And when an *aluminum* atom loses three electrons, an *aluminum ion*, Al^{3+}, forms. These and other common cations are listed in *Table 4*. Notice how ions of Group 1 elements have a 1+ charge and ions of Group 2 elements have a 2+ charge.

Table 4 **Some Common Cations**

Ion name and symbol	Ion charge
Cesium ion, Cs^+	1+
Lithium ion, Li^+	
Potassium ion, K^+	
Rubidium ion, Rb^+	
Sodium ion, Na^+	
Barium ion, Ba^{2+}	2+
Beryllium ion, Be^{2+}	
Calcium ion, Ca^{2+}	
Magnesium ion, Mg^{2+}	
Strontium ion, Sr^{2+}	
Aluminum ion, Al^{3+}	3+

159

Focus

Overview

Before beginning this section, review with your students the Objectives listed in the Student Edition. In this section, students learn how to name ionic and covalent compounds, how to predict the charge of a cation in an ionic compound, how to write formulas for simple ionic compounds, and how to distinguish between the empirical and molecular formulas of a covalent compound.

● Bellringer

Use the Bellringer transparency to prepare students for this section.

Motivate

Opening Activity — GENERAL

For this activity, you will need a double-pan balance, pennies, and two index cards. Let stacks of pennies represent the amount of charge on ions. Place a card with a plus sign on it in front of the left-hand pan of the balance and a card with a minus sign in front of the right-hand pan. Choose several ionic compounds, and use the stacks to show that a correct formula balances charge. For example, if Fe^{3+} and S^{2-} ions are represented by stacks of three and two pennies, it will take two stacks of three pennies to balance three stacks of two pennies. The final formula is Fe_2S_3. Have students use the balance and pennies to find the formulas for all compounds that would form from: Au^{3+}, Pb^{2+}, O^{2-}, and Cl^- ions. (Au_2O_3, $AuCl_3$, $PbCl_2$, PbO) LS Logical

---

LANGUAGE-ARTS
CONNECTION — BASIC

Ask students to list abbreviations for common terms. Lists might include abbreviations for units of measurement and abbreviations for phrases, such as *etc.* and *e.g.* Relate the use of abbreviations in daily life to the use of chemical formulas to represent the composition of substances. LS **Verbal**

MATHEMATICS
CONNECTION — BASIC

To help students determine the charge of an ion in a compound, review with them how to add positive and negative integers. Provide a number line to students who have problems with this concept. LS **Logical**

Chapter Resource File

• Lesson Plan

Transparencies

TT Bellringer
TM Common Cations

SKILL BUILDER — BASIC

Reading Skills Have students read about how to name ionic compounds, write formulas for ionic compounds, name covalent compounds, and write formulas for covalent compounds. After reading each topic, ask for volunteers to summarize each process for the rest of the class. Have students ask for clarification of anything they do not understand once each summary is completed. **LS** Verbal

Teaching Tip

Order in Ionic Formulas Give examples, such as Einstein Albert, Marie Madame Curie, and Carver Washington George, to illustrate to students the importance of correct order and spelling of a name to accurately denote a person. Extend the discussion to include the naming of ionic compounds. Cations and anions can be thought of as the first and last names, respectively, of ionic compounds. Point out that writing the name in the correct order and with the correct spelling is important for accurately denoting an ionic compound.

Table 5 Some Common Anions

Element name and symbol	Ion name and symbol	Ion charge
Fluorine, F	Fluoride ion, F^-	1−
Chlorine, Cl	Chloride ion, Cl^-	
Bromine, Br	Bromide ion, Br^-	
Iodine, I	Iodide ion, I^-	
Oxygen, O	Oxide ion, O^{2-}	2−
Sulfur, S	Sulfide ion, S^{2-}	
Nitrogen, N	Nitride ion, N^{3-}	3−

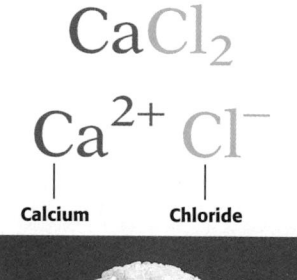

Figure 18

Ionic compounds are named for their positive and negative ions.

Names of anions are altered names of elements

An anion that is made of one element has a name similar to the element. The difference is the name's ending. **Table 5** lists some common anions and shows how they are named. Just like most cations, anions of elements in the same group of the periodic table have the same charge.

NaF is made of sodium ions, Na^+, and fluoride ions, F^-. Therefore, its name is *sodium fluoride*. **Figure 18** shows how calcium chloride gets its name.

Some cation names must show their charge

Think about the compounds FeO and Fe_2O_3. According to the rules you have learned so far, both of these compounds would be named *iron oxide*, even though they are not the same compound. Fe_2O_3, a component of rust, is a reddish brown solid that melts at 1565°C. FeO, on the other hand, is a black powder that melts at 1420°C. These different properties tell us that they are different compounds and should have different names.

Iron is a transition metal. Transition metals may form several cations—each with a different charge. A few of these cations are listed in **Table 6**. The charge of the iron cation in Fe_2O_3 is different from the charge of the iron cation in FeO. In cases like this, the cation name must be followed by a Roman numeral in parentheses. The Roman numeral shows the cation's charge. Fe_2O_3 is made of Fe^{3+} ions, so it is named *iron(III) oxide*. FeO is made of Fe^{2+} ions, so it is named *iron(II) oxide*.

Table 6 Some Transition Metal Cations

Ion name	Ion symbol	Ion name	Ion symbol
Copper(I) ion	Cu^+	Chromium(II) ion	Cr^{2+}
Copper(II) ion	Cu^{2+}	Chromium(III) ion	Cr^{3+}
Iron(II) ion	Fe^{2+}	Cadmium(II) ion	Cd^{2+}
Iron(III) ion	Fe^{3+}	Titanium(II) ion	Ti^{2+}
Nickel(II) ion	Ni^{2+}	Titanium(III) ion	Ti^{3+}
Nickel(III) ion	Ni^{3+}	Titanium(IV) ion	Ti^{4+}

160

Alternative Assessment — BASIC

Transition Metals On the chalkboard, write the following formulas for transition element compounds: CrO, $CrCl_3$, $Cr(SO_4)_3$, $CoBr_2$, Co_2O_3, $KMnO_4$, MnO_2, MnS, MnO_3, MnN. Have students use the formulas to make a table that gives the ions formed by chromium, cobalt, and manganese. Review with students how to use the periodic table to find charges of elements in certain families. The ions for chromium are Cr^{2+}, Cr^{3+}, and Cr^{6+}. The ions for cobalt are Co^{2+} and Co^{3+}. Those for manganese are Mn^{2+}, Mn^{3+}, Mn^{4+}, Mn^{6+}, and Mn^{7+}. **LS** Logical

Determining the charge of a transition metal cation

How can you tell that the iron ion in Fe_2O_3 has a charge of 3+? Like all compounds, ionic compounds have a total charge of zero. This means that the total positive charges must equal the total negative charges. An oxide ion, O^{2-}, has a charge of 2−. Three of them have a total charge of 6−. That means the total positive charge in the formula must be 6+. For two iron ions to have a total charge of 6+, each ion must have a charge of 3+.

Writing Formulas for Ionic Compounds

You have seen how to determine the charge of each ion in a compound if you are given the compound's formula. Following a similar process, you can determine the chemical formula for a compound if you are given its name.

Math Skills

Writing Ionic Formulas What is the chemical formula for aluminum fluoride?

1 **List the symbols for each ion.**
Symbol for an aluminum ion from *Table 4:* Al^{3+}
Symbol for a fluoride ion from *Table 5:* F^-

2 **Write the symbols for the ions with the cation first.**
$Al^{3+}F^-$

3 **Find the least common multiple of the ions' charges.**
The least common multiple of 3 and 1 is 3. To make a neutral compound, you need a total of three positive charges and three negative charges.
To get three positive charges: you need only one Al^{3+} ion because $1 \times 3+ = 3+$.
To get three negative charges: you need three F^- ions because $3 \times 1- = 3-$.

4 **Write the chemical formula, indicating with subscripts how many of each ion are needed to make a neutral compound.**
AlF_3

Practice HINT

Once you have determined a chemical formula, always check the formula to see if it makes a neutral compound. For this example, the aluminum ion has a charge of 3+. The fluoride ion has a charge of only 1−, but there are three of them for a total of 3−.

$(3-) + (3-) = 0$, so the charges balance, and the formula is neutral.

Practice

Writing Ionic Formulas

Write formulas for the following ionic compounds.

1. lithium oxide
2. beryllium chloride
3. titanium(III) nitride
4. cobalt(III) hydroxide

Teaching Tip — BASIC

Cation-anion Pairs Reinforce the idea of electroneutrality in the following drill. Make a vertical list on the board of the symbols of six cations, and make a parallel list of the symbols of six anions. The ions should have a variety of charges. Point to a cation-anion pair, and ask students how many of each will produce electroneutrality. You may want to ask students to write or state the correct formula, but the main purpose of the exercise is to help students understand how to construct an electroneutral combination. **LS Logical**

Math Skills

Additional Examples

Writing Ionic Formulas Write formulas for the following ionic compounds.

a. magnesium bromide
Answer: $MgBr_2$

b. rubidium oxide
Answer: Rb_2O

c. lithium nitride
Answer: Li_3N

d. potassium sulfate
Answer: K_2SO_4
LS Logical

Practice

1. Li_2O
2. $BeCl_2$
3. TiN
4. $Co(OH)_3$
LS Logical

Teacher Resources

For the New Teacher Students might notice what they consider to be a shortcut in writing ionic formulas. The numerical value of the charge on one ion is usually the subscript of the other ion in the formula for the compound. For example, if Sn^{2+} and PO_4^{3+} form a compound, the charge of 2 for Sn becomes the subscript of the PO_4 part of the formula, and vice versa. The final formula is $Sn_3(PO_4)_2$. If the charges have a common factor, divide by that factor to simplify the formula. For example, Mg^{2+} and O^{2-} form MgO, not Mg_2O_2. Although this procedure is a simple and quick way to write a formula, students should still follow the procedure listed in the Math Skills on this page. The Math Skills procedure emphasizes the concept of balanced charges and is not just a manipulation of numbers.

Chapter Resource File

• **Math Skills** Writing Ionic Formulas **GENERAL**

Transparencies

TM Common Anions

Chapter 5 · The Structure of Matter **161**

Prefixes Elicit from students examples of words containing prefixes that denote a number. Examples might include *bicycle, tricycle, octopus,* or *decade.* Other examples might include terms from mathematics, such as *triangle, pentagon, quadrilateral,* or *hexagon.* Tell students that such prefixes are also used to name molecular compounds. **LS** Verbal

Group Activity ——— BASIC

Naming Covalent Compounds
Separate students into teams of four. Provide two students on each team with flashcards with the prefixes listed in **Table 7** written on them. Supply one of the other students with cards containing names of nonmetals and the other student with *–ide* names based on nonmetals, such as *oxide.* Write the formula for a covalent compound on the chalkboard, and have students use the cards to name it. This assessment can be done as a learning exercise or as a group quiz. Before starting the activity, remind students of the following: *Mono–* is used only with the second part of the name, not the first. For example, CO is carbon monoxide, not monocarbon monoxide. Final *o's* and *a's* of prefixes can be dropped if it makes the name easier to pronounce. For example, tetroxide would be used, not tetraoxide. **LS** Interpersonal

Table 7 Prefixes Used to Name Covalent Compounds

Number of atoms	Prefix
1	*mono-*
2	*di-*
3	*tri-*
4	*tetra-*
5	*penta-*
6	*hexa-*
7	*hepta-*
8	*octa-*
9	*nona-*
10	*deca-*

$$N_2O_4$$
Dinitrogen tetroxide

Figure 19
One molecule of *di*nitrogen *tetr*oxide has *two* nitrogen atoms and *four* oxygen atoms.

■ **empirical formula** the composition of a compound in terms of the relative numbers and kinds of atoms in the simplest ratio

Naming Covalent Compounds

Covalent compounds, like SiO_2 (silicon dioxide) and CO_2 (carbon dioxide), are named using different rules than those used to name ionic compounds.

Numerical prefixes are used to name covalent compounds of two elements

For two-element covalent compounds, numerical prefixes tell how many atoms of each element are in the molecule. **Table 7** lists some of these prefixes. If there is only one atom of the first element, it does not get a prefix. Whichever element is farther to the right in the periodic table is named second and ends in *-ide.*

There are one boron atom and three fluorine atoms in *boron trifluoride*, BF_3. *Dinitrogen tetroxide,* N_2O_4, is made of two nitrogen atoms and four oxygen atoms, as shown in **Figure 19.** Notice how the *a* in *tetra* is dropped to make the name easier to say.

Chemical Formulas for Covalent Compounds

Emeralds, shown in **Figure 20,** are made of a mineral called beryl. The chemical formula for beryl is $Be_3Al_2Si_6O_{18}$. But how did people determine this formula? It took some experiments. Chemical formulas like this one were determined by first measuring the mass of each element in the compound.

A compound's simplest formula is its empirical formula

Once the mass of each element in a sample of the compound is known, scientists can calculate the compound's **empirical formula,** or simplest formula. An empirical formula tells us the smallest whole-number ratio of atoms that are in a compound. Formulas for most ionic compounds are empirical formulas.

Covalent compounds have empirical formulas, too. The empirical formula for water is H_2O. It tells you that the ratio of hydrogen atoms to oxygen atoms is 2:1. Scientists have to analyze unknown compounds to determine their empirical formulas.

Figure 20
Emerald gemstones are cut from the mineral beryl. Very tiny amounts of chromium(III) oxide impurity in the gemstones gives them their beautiful green color.

162

Alternative Assessment ——— BASIC

Common Names In addition to a scientific name, some compounds also have a common name. For example, few people would refer to dihydrogen monoxide, H_2O, as anything but water. Other compounds that have common names include vinegar (acetic acid) and rubbing alcohol (isopropanol).

Common names are especially frequent among compounds that have been known and used for years, before rules for naming were standard. For example, although many chemical salts exist, sodium chloride, NaCl, is what is commonly referred to as salt. Some compounds have common names resulting from old systems of naming or from the compound's effects, not the system currently used. For example, dinitrogen monoxide, N_2O, is commonly referred to as both nitrous oxide and laughing gas.

Have students research the common and standard names of some compounds they are familiar with. Students should include explanations for the origins of the common names when possible. You may wish to have students make a classroom poster of common and scientific names that compiles each student's research. **LS** Verbal

Determining empirical formulas

If a 142 g sample of an unknown compound contains only the elements phosphorus and oxygen and is found to contain 62 g of P and 80 g of O, its empirical formula is easy to calculate. This process is shown in **Figure 21**.

Different compounds can have the same empirical formula

It's possible for several compounds to have the same empirical formula because empirical formulas only represent a ratio of atoms. Formaldehyde, acetic acid, and glucose all have the empirical formula CH_2O, as shown in **Table 8**. These three compounds are not at all alike, though. Formaldehyde is sometimes used to keep dead organisms from decaying so that they can be studied. Acetic acid gives vinegar its sour taste and strong smell. And glucose is a sugar that plays a very important role in your body chemistry. Some other formula must be used to distinguish these three very different compounds.

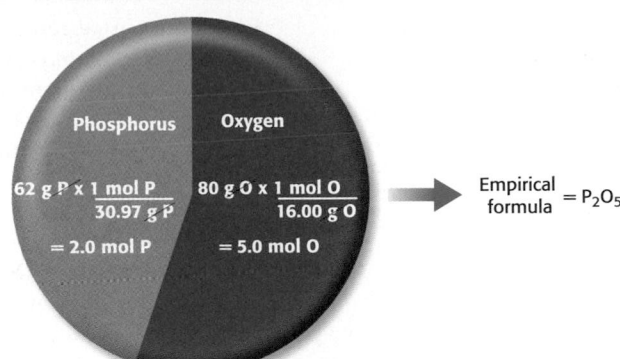

Exactly 142 g of Unknown Compound

Phosphorus

Oxygen

$$62 \text{ g P} \times \frac{1 \text{ mol P}}{30.97 \text{ g P}}$$

$$= 2.0 \text{ mol P}$$

$$80 \text{ g O} \times \frac{1 \text{ mol O}}{16.00 \text{ g O}}$$

$$= 5.0 \text{ mol O}$$

$$\text{Empirical formula} = P_2O_5$$

Figure 21

Once you determine the mass of each element in a compound, you can calculate the amount of each element in moles. The empirical formula for the compound is the ratio of these amounts.

Table 8 **Empirical and Molecular Formulas for Some Compounds**

Compound	Empirical formula	Molar mass	Molecular formula	Structure
Formaldehyde	CH_2O	30.03 g/mol	CH_2O	
Acetic acid	CH_2O	60.06 g/mol	$2 \times CH_2O = C_2H_4O_2$	
Glucose	CH_2O	180.18 g/mol	$6 \times CH_2O = C_6H_{12}O_6$	

Vocabulary Have students look up the meaning of the word *empirical* in the dictionary. Have them use the definition to explain where information is obtained about an empirical formula. Their explanations should reflect that empirical formulas are determined from experimental data. **LS** Verbal

Teaching Tip — ADVANCED

Finding Molecular Formulas

Show students how to find the molecular formula when given the empirical formula and the molar mass of the compound. Use the information in **Table 8** as an example. The molar mass of the empirical formula is found by adding the molar mass of each atom, expressed in g/mol. For CH_2O, the molar mass is $(1 \times 12.01) + (2 \times 1.01) + (1 \times 16.00)$, or 30.03. The molecular formula will be a multiple of the empirical formula, so divide the molar mass of the molecule by the molar mass of the empirical formula. For glucose, 180.18 g/mol ÷ 30.03 g/mol = 6. So, the molecule is made up of six units of the empirical formula. The molecular formula is $C_{1\times6}H_{2\times6}O_{1\times6}$, or $C_6H_{12}O_6$. Have students find the molecular formulas for the following molecules:

Molar mass (g/mol)	Empirical formula	Molecular formula
78.12	CH	C_6H_6
32.06	NH_2	N_2H_4
210.33	$C_3H_6N_2$	$C_9H_{18}N_6$

LS Logical

MISCONCEPTION ///ALERT\\\

Chemical Formulas Some students interpret chemical formulas incorrectly, believing, for example, that the chemical formula N_2O_4 represents an N_2 molecule bound to an O_4 molecule. Use examples in this section, including the structures illustrated in the last column of **Table 8**, to emphasize that this interpretation is incorrect.

Transparencies

TM Naming Prefixes

Close

Quiz

1. Which part of an ionic compound should be written first? (cation)

2. What does a Roman numeral in parentheses after a cation indicate? (the ion's charge)

3. What kind of formula tells the actual numbers of atoms in one molecule of a compound? (molecular formula)

4. What kind of formula gives the relative numbers of atoms of each element in a compound? (empirical formula)

▶ **molecular formula**
a chemical formula that shows the number and kinds of atoms in a molecule, but not the arrangement of atoms

Molecular formulas are determined from empirical formulas

Formaldehyde, acetic acid, and glucose are all covalent compounds made of molecules. They all have the same empirical formula, but each compound has its own **molecular formula.** A compound's molecular formula tells you how many atoms are in one molecule of the compound.

In some cases, a compound's molecular formula is the same as its empirical formula. The empirical and molecular formulas for water are both H_2O. You can see from *Table 8* on the previous page that this is also true for formaldehyde. In other cases, a compound's molecular formula is a small whole-number multiple of its empirical formula. The molecular formula for acetic acid is two times its empirical formula, and that of glucose is six times its empirical formula.

SECTION 3 REVIEW

SUMMARY

▶ To name an ionic compound, first name the cation and then the anion.

▶ If an element can form cations with different charges, the cation name must include the ion's charge. The charge is written as a Roman numeral in parentheses.

▶ Prefixes are used to name covalent compounds made of two different elements.

▶ An empirical formula tells the relative numbers of atoms of each element in a compound.

▶ A molecular formula tells the actual numbers of atoms in one molecule of a compound.

▶ Covalent compounds have both empirical and molecular formulas.

1. Name the following ionic compounds, specifying the charge of any transition metal cations.
   a. FeI_2 **c.** $CrCl_2$
   b. MnF_3 **d.** CuS

2. Name the following covalent compounds:
   a. As_2O_5 **c.** P_4S_3 **e.** SeO_2
   b. SiI_4 **d.** P_4O_{10} **f.** PCl_3

3. Explain why Roman numerals must be included in the names of MnO_2 and Mn_2O_7. Name both of these compounds.

4. Identify how many fluorine atoms are in one molecule of sulfur hexafluoride.

Math Skills

5. Critical Thinking An unknown compound contains 49.47% C, 5.20% H, 28.85% N, and a certain percentage of oxygen. What percentage of the compound must be oxygen? (**Hint:** The sum of the percentages should equal 100%.)

6. What is the charge of the cadmium cation in cadmium cyanide, $Cd(CN)_2$, a compound used in electroplating? Explain your reasoning.

7. Determine the chemical formulas for the following ionic compounds:
   a. magnesium sulfate   **c.** chromium(II) fluoride
   b. rubidium bromide   **d.** nickel(I) carbonate

164

Chapter Resource File

- Concept Review **GENERAL**
- Quiz **BASIC**

Answers to Section 3 Review

1. a. iron(II) iodide
   b. manganese(III) fluoride
   c. chromium(II) chloride
   d. copper(II) sulfide

2. a. diarsenic pentoxide
   b. silicon tetriodide
   c. tetraphosphorus trisulfide
   d. tetraphosphorus decoxide
   e. selenium dioxide
   f. phosphorus trichloride

3. Manganese is a transition metal, so its ions can vary in charge. Roman numerals indicate charge in a compound. MnO_2 is manganese(IV)

oxide; Mn_2O_7 is manganese(VII) oxide.

4. The prefix *hexa*– indicates six fluorine atoms.

Math Skills

5. $100.00\% - (49.47\% + 5.20\% + 28.85\%) = 16.48\%$ oxygen

6. The total charge must be zero. Each cyanide ion has a charge of 1–. The cadmium ion's charge must be 2+ to cancel the $2 \times (1-)$ charge from the cyanide ions.

7. a. $MgSO_4$ **c.** CrF_2
   b. $RbBr$ **d.** Ni_2CO_3

Organic and Biochemical Compounds

INTEGRATING TECHNOLOGY and Society

OBJECTIVES

▶ **Describe** how carbon atoms bond covalently to form organic compounds.

▶ **Identify** the names and structures of groups of simple organic compounds and polymers.

▶ **Identify** what makes up the polymers that are essential to life.

▶ **KEY TERMS**
organic compound
polymer
carbohydrate
protein
amino acid

The word *organic* has many different meanings. Most people associate the word *organic* with living organisms. Perhaps you have heard of or eaten organically grown fruits or vegetables. What this means is that they were grown using fertilizers and pesticides that come from plant and animal matter. In chemistry, the word *organic* is used to describe certain compounds.

Organic Compounds

An **organic compound** is a covalently bonded compound made of molecules. Organic compounds contain carbon and, almost always, hydrogen. Other atoms, such as oxygen, nitrogen, sulfur, and phosphorus, are also found in some organic compounds.

Many ingredients of familiar substances are organic compounds. The effective ingredient in aspirin is a form of the organic compound acetylsalicylic acid, $C_9H_8O_4$. Sugarless chewing gum also has organic compounds as ingredients. Two ingredients are the sweeteners sorbitol, $C_6H_{14}O_6$, and aspartame, $C_{14}H_{18}N_2O_5$, both of which are shown in *Figure 22*.

▶ **organic compound** a covalently bonded compound that contains carbon, excluding carbonates and oxides

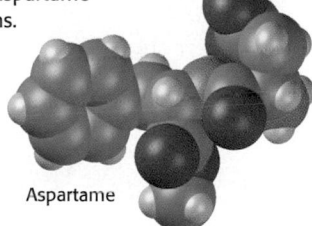

Figure 22
The organic compounds sorbitol and aspartame sweeten some sugarless chewing gums.

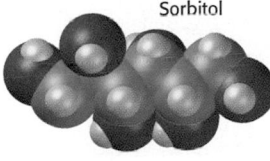
Sorbitol

Aspartame

165

REAL-LIFE CONNECTION — ADVANCED

Sugar Substitutes Have students find out more about the use of sugar substitutes, including sorbital and aspartame. You may wish to assign different topics to different students or groups of students. Possible topics include: Why do some people use sugar substitutes? What are the health risks associated with excess sugar consumption? Are there any health risks associated with sugar substitutes? What are the differences in chemical makeup and structure between

sugar and various sugar substitutes? Are some sugar substitutes better than others? Some students could also research individual sugar substitutes in more detail, including sorbitol, aspartame, saccharine, stevia, and the recently approved sucralose. Ask students to share their results with the class. You could also ask students to bring in examples of sugar substitutes they have researched to share with other students. **LS Verbal**

Focus

Overview
Before beginning this section, review with your students the Objectives listed in the Student Edition. In this section, students learn how carbon atoms bond covalently to form organic compounds, including alkanes, alkenes, and alcohols. Next, students study both man-made and natural polymers. The section concludes with a discussion of biochemical compounds, including carbohydrates, proteins, and DNA.

Bellringer
Use the Bellringer transparency to prepare students for this section.

Motivate

Opening Demonstration — GENERAL
This demonstration requires notebook paper, a test tube, a test-tube holder, a Bunsen burner, a sparker to light the burner, ice, a beaker, and tongs. Place a strip of notebook paper into the test tube. Place ice in the beaker. Using the test-tube holder, hold the test tube over the lit burner. At the same time, use tongs to hold the beaker just above the mouth of the test tube.

Ask students: What change did you notice in the paper? (The paper turned black.) What element could you see after the paper was heated? (carbon) What did you notice on the bottom of the beaker? (condensed water vapor) What other two elements must be present in the paper? (the hydrogen and oxygen that make up the water)
LS Logical

Chapter Resource File
• Lesson Plan

Transparencies

TT Bellringer

SKILL BUILDER — BASIC

Reading Skills Have students read the section and mark with gummed notes the passages they do not understand. Be sure students study all tables and figures to help clarify the relevant passages. After reading, pair up students and have the pairs discuss those passages that were difficult. For the passages that cannot be clarified by the pairs, have students ask questions for later class discussion or teacher explanation. **LS Interpersonal**

Teaching Tip

Inorganic Molecules Not all carbon-containing molecules are considered organic compounds. Carbon dioxide, carbon monoxide, and carbon disulfide are inorganic. Other carbon-containing compounds that are not organic include carbonic acid, carbides, carbonates, cyanides, and metal carbonyls.

SKILL BUILDER

Vocabulary Carbon is one of just a few elements capable of forming chains and rings by bonding to other carbon atoms. *Catenation* is the term that describes this tendency. Carbon exhibits *catenation* more than any other element does. Silicon also exhibits some catenation. The word catenation comes from the Latin word *catena*, which means "chain."

internet connect
www.scilinks.org
Topic: Carbon Compounds
SciLinks code: HK4016

SCiLINKS. Maintained by the National Science Teachers Association

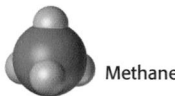

Methane

Figure 23

Methane is an alkane that has four C−H bonds.

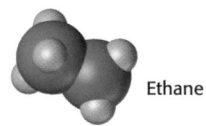

Ethane

Figure 24

Ethane, another alkane, has one C−C bond and six C−H bonds.

Figure 25

This camper is preparing his dinner on a gas grill fueled by propane. Propane is an alkane that has two C−C bonds and eight C−H bonds.

Propane

Carbon atoms form four covalent bonds in organic compounds

When a compound is made of only carbon and hydrogen atoms, it is called a *hydrocarbon*. Methane, CH_4, is the simplest hydrocarbon. Its structure is shown in *Figure 23*. Methane gas is formed when living matter, such as plants, decay, so it is often found in swamps and marshes. The natural gas used in Bunsen burners is also mostly methane. Carbon atoms have four valence electrons to use for bonding. In methane, each of these electrons forms a different C−H single bond.

A carbon atom may also share two of its electrons with two from another atom to form a double bond. Or a carbon atom may share three electrons to form a triple bond. However, a carbon atom can never form more than a total of four bonds.

Alkanes have single covalent bonds

Alkanes are hydrocarbons that have only single covalent bonds. *Figure 23* shows that methane, the simplest alkane, has only C−H bonds. But alkanes can also have C−C bonds. You can see from *Figure 24* that ethane, C_2H_6, has a C−C bond in addition to six C−H bonds. Notice how each carbon atom in both of these compounds bonds to four other atoms.

Many gas grills are fueled by another alkane, propane, C_3H_8. Propane is made of three bonded carbon atoms. Each carbon atom on the end of the molecule forms three bonds with three hydrogen atoms, as shown in *Figure 25*. Each of these end carbon atoms forms its fourth bond with the central carbon atom. The central carbon atom shares its two remaining electrons with two hydrogen atoms. You can see only one hydrogen atom bonded to the central carbon atom in *Figure 25* because the second hydrogen atom is on the other side.

166

Teacher Resources

For the New Teacher You may wish to use this graphic organizer to help students with alkane classification.

Classifying Alkanes

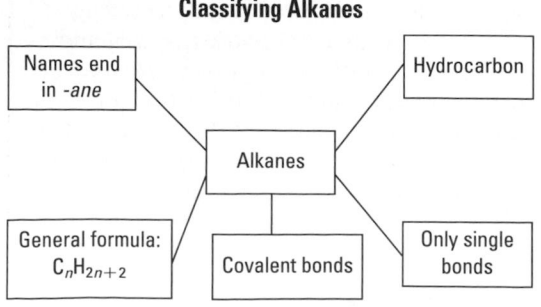

Names end in *-ane*

Hydrocarbon

Alkanes

General formula: C_nH_{2n+2}

Covalent bonds

Only single bonds

Arrangements of carbon atoms in alkanes

The carbon atoms in methane, ethane, and propane all line up in a row because that is their only possible arrangement. When there are more than three bonded carbon atoms, the carbon atoms do not always line up in a row. When they do line up, the alkane is called a *normal alkane,* or *n*-alkane for short. **Table 9** shows chemical formulas for the *n*-alkanes that have up to 10 carbon atoms. *Condensed structural formulas* are also included in the table to show how the atoms bond.

The carbon atoms in any alkane with more than three carbon atoms can have more than one possible arrangement. Carbon atom chains may be branched or unbranched, and they can even form rings. **Figure 26** shows some of the possible ways six carbon atoms can be arranged when they form hydrocarbons with only single covalent bonds.

Table 9 First 10 *n*-Alkanes

n-Alkane	Molecular formula	Condensed structural formula
Methane	CH_4	CH_4
Ethane	C_2H_6	CH_3CH_3
Propane	C_3H_8	$CH_3CH_2CH_3$
Butane	C_4H_{10}	$CH_3(CH_2)_2CH_3$
Pentane	C_5H_{12}	$CH_3(CH_2)_3CH_3$
Hexane	C_6H_{14}	$CH_3(CH_2)_4CH_3$
Heptane	C_7H_{16}	$CH_3(CH_2)_5CH_3$
Octane	C_8H_{18}	$CH_3(CH_2)_6CH_3$
Nonane	C_9H_{20}	$CH_3(CH_2)_7CH_3$
Decane	$C_{10}H_{22}$	$CH_3(CH_2)_8CH_3$

Alkane chemical formulas usually follow a pattern

Except for cyclic alkanes like cyclohexane, the chemical formulas for alkanes follow a special pattern. The number of hydrogen atoms is always two more than twice the number of carbon atoms. This pattern is shown by the chemical formula C_nH_{2n+2}.

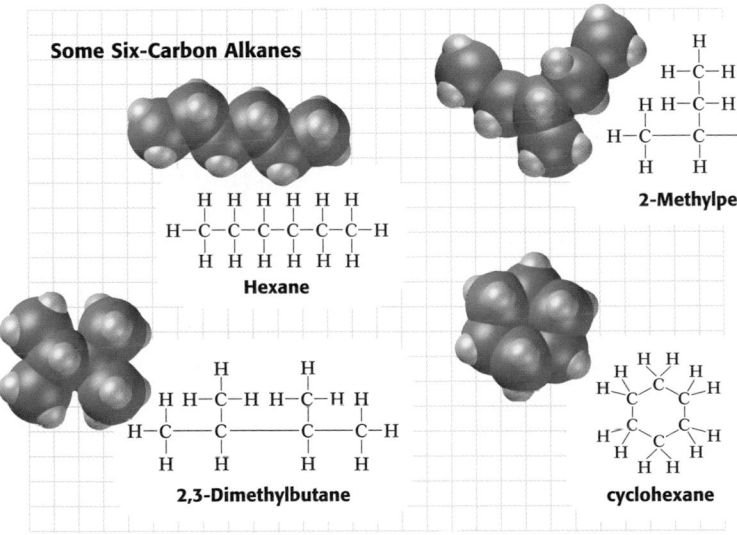

Some Six-Carbon Alkanes

Hexane

2-Methylpentane

2,3-Dimethylbutane

cyclohexane

Figure 26

Hexane, 2-methylpentane, 2,3-dimethylbutane, and cyclohexane are some of the forms six carbon atoms with single covalent bonds may take.

Teaching Tip

General Formulas Just as the general formula for an alkane is C_nH_{2n+2}, alkenes and alkynes have general formulas. The formula for an alkene is C_nH_{2n}, and that of an alkyne is C_nH_{2n-2}. An alkyne is an organic compound in which two carbon atoms share three pairs of electrons to form a triple bond.

Teaching Tip

Hydroxyl Group Be sure students understand that the hydroxyl group, —OH, found in alcohols, is not the same as the hydroxide ion, OH^-. The hydroxyl group is covalently bonded to an organic compound. The hydroxide ion is charged and is ionically bonded to a cation.

Figure 27

The peaches in this plastic container, which is made by joining propene molecules, release ethene gas as they ripen.

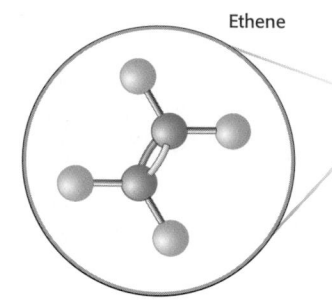

Ethene

Propene

Alkenes have double carbon-carbon bonds

Alkenes are also hydrocarbons. Alkenes are different from alkanes because they have at least one double covalent bond between carbon atoms. This is shown by C=C. Alkenes are named like alkanes but with the *-ane* ending replaced by *-ene*.

The simplest alkene is ethene (or ethylene), C_2H_4. Ethene is formed when fruit ripens. Propene (or propylene), C_3H_6, is used to make rubbing alcohol and some plastics. The structures of both compounds are shown in *Figure 27*.

Figure 28

Many products contain a mixture of the alcohols methanol and ethanol. This mixture is called "denatured alcohol."

Methanol

Ethanol

Alcohols have —OH groups

Alcohols are organic compounds that are made of oxygen as well as carbon and hydrogen. Alcohols have *hydroxyl*, or —OH, groups. The alcohol methanol, CH_3OH, is sometimes added to another alcohol ethanol, CH_3CH_2OH, to make denatured alcohol. Denatured alcohol is found in many familiar products, as shown in *Figure 28.* Isopropanol, which is found in rubbing alcohol, has the chemical formula C_3H_8O, or $(CH_3)_2CHOH$. You may have noticed how the names of these three alcohols all end in *-ol*. This is true for most alcohols.

Alcohol molecules behave similarly to water molecules

A methanol molecule is like a water molecule except that one of the hydrogen atoms is replaced by a methyl, or —CH_3, group. Just like water molecules, neighboring alcohol molecules are attracted to one another. That's why many alcohols are liquids at room temperature. Alcohols have much higher boiling points than alkanes of similar size.

168

Did You Know?

Substitutions Compounds known as substituted hydrocarbons form when an atom or group of atoms substitutes for a hydrogen atom in a hydrocarbon. For example, a halogen atom can replace a hydrogen atom. An alcohol is a substituted hydrocarbon, as is an organic acid. Other substituted hydrocarbons include ethers, aldehydes, ketones, and esters.

Polymers

What do the DNA inside the cells of your body, rubber, wood, and plastic milk jugs have in common? They are all made of large molecules called **polymers.**

Many polymers have repeating subunits

Some small organic molecules bond to form long chains called polymers. Polyethene, which is also known as polyethylene or polythene, is the polymer plastic milk jugs are made of. The name *polyethene* tells its structure. *Poly* means "many." *Ethene* is an alkene whose chemical formula is C_2H_4. Therefore, polyethene is "many ethenes," as shown in *Figure 29.* The original molecule, in this case C_2H_4, is called a *monomer.*

Some polymers are natural; others are man-made

Rubber, wood, cotton, wool, starch, protein, and DNA are all natural polymers. Man-made polymers are usually either plastics or fibers. Most plastics are flexible and easily molded, whereas fibers form long, thin strands.

Some polymers can be used as both plastics and fibers. For example, polypropene (polypropylene) is molded to make plastic containers, like the one shown in *Figure 27,* as well as some parts for cars and appliances. It is also used to make ropes, carpet, and artificial turf for athletic fields.

The elasticity of a polymer is determined by its structure

As with all substances, the properties of a polymer are determined by its structure. Polymer molecules are like long, thin chains. A small piece of plastic or a single fiber is made of billions of these chains. Polymer molecules can be likened to spaghetti. Like a bowl of spaghetti, the chains are tangled but can slide over each other. Milk jugs are made of polyethene, a plastic made of such noodlelike chains. You can crush or dent a milk jug because the plastic is flexible. Once the jug has been crushed, though, it does not return to its original shape. That's because polyethene is not elastic.

When the chains are connected to each other, or cross-linked, the polymer's properties change. Some become more elastic and can be likened to a volleyball net. Like a volleyball net, an elastic polymer can stretch. When the polymer is released, it returns to its original shape. Rubber bands are elastic polymers. As long as a rubber band is not stretched too far, it can shrink back to its original form.

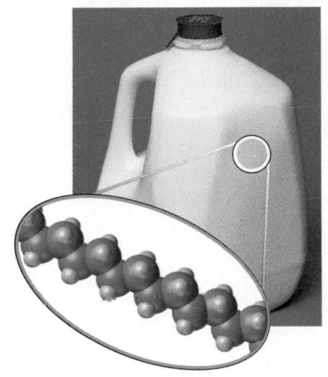

Figure 29   Polyethene

Polyethene is a polymer made of many repeating ethene units. As the polymer forms, ethene's double bonds are replaced by single bonds.

▶ **polymer** a large molecule that is formed by more than five monomers, or small units

Quick ACTIVITY

Polymer Memory

Polymers that return to their original shape after stretching can be thought of as having a "memory." In this activity, you will compare the memory of a rubber band with that of the plastic rings that hold a six-pack of cans together.

1. Which polymer stretches better without breaking?
2. Which one has better memory?
3. Warm the stretched six-pack holder over a hot plate, being careful not to melt it. Does it retain its memory?

169

Teaching Tip

Polymers Show students different types of polymers. Examples might include plastic wrap; rubber tubing; polyester thread; nylon pantyhose; foam cups; trash bags; a plastic that is firm but flexible, such as a margarine container; and a rigid, inflexible plastic, such as the material used to make a telephone. Tell students that all of the examples are made up of the same type of compound.

Quick ACTIVITY

Polymer Memory

Materials (per group):

- rubber band
- plastic rings from a six-pack of cans
- hot plate

Teacher's Notes: Caution students about not touching the hot surface of the hot plate and not melting the plastic. If the plastic accidentally does melt, tell students to notify you immediately. They should not touch the melted plastic and not breathe any of the fumes produced.

Results:

1. The rubber band stretches better without breaking.
2. The rubber band has better memory.
3. No. It changes shape and does not return to its original shape.

LS Kinesthetic

INCLUSION Strategies   GENERAL

- *Learning Disabled*
- *Attention Deficit Disorder*
- *English Language Learners*

Ask students to fold a piece of paper in half vertically. Label one side of the paper "Natural Polymers" and the other side "Synthetic Polymers." Under each title, have students write a definition for each term. Students should then write a list of different types of polymers or cut pictures from magazines showing the different types. Students can present their chart in class or use it as a study guide.

Did You Know?

Distinguishing Fibers Sometimes natural fibers, such as wool and cotton, are difficult to distinguish from synthetic fibers, such as nylon or rayon. One way to tell them apart is to hold them in a flame. The natural fiber will char, and the synthetic fiber will melt.

Chapter Resource File

- **Cross-Disciplinary Worksheet** Integrating Environmental Science— Plastics ADVANCED
- **Quick Activity Datasheet** Polymer Memory GENERAL

Teaching Tip

Carbohydrates Be sure students understand that not all carbohydrates are polymers, and not all carbohydrates provide energy to humans. Sugars are carbohydrates but not polymers. Cellulose is a carbohydrate, but humans are not able to digest it or use the energy in it. Animals, such as cows, contain in their digestive systems bacteria that break cellulose down into monomers the animal can use.

Demonstration —— GENERAL

Model a Polymer

(Time: About 10 minutes)

Materials:

• several spring clothespins

Attach one clothespin onto the leg of another one. Continue the process until a chain of clothespins is formed. Show students that the chain can bend and otherwise move, but it still remains a chain.

1. What does each clothespin represent? (a monomer)

2. What does the chain represent? (a polymer)

3. If the chain represents a protein, what does each clothespin represent? (an amino acid)

LS Visual

Figure 30
Athletes often eat lots of foods that are high in carbohydrates the day before a big event. This provides them with a ready supply of stored energy.

▶ **carbohydrate** any organic compound that is made of carbon, hydrogen, and oxygen and that provides nutrients to the cells of living things

▶ **protein** an organic compound that is made of one or more chains of amino acids and that is a principal component of all cells

▶ **amino acid** any one of 20 different organic molecules that contain a carboxyl and an amino group and that combine to form proteins

---

Biochemical Compounds

Biochemical compounds are naturally occurring organic compounds that are very important to living things. Carbohydrates give you energy. Proteins form important parts of your body, like muscles, tendons, fingernails, and hair. The DNA inside your cells gives your body information about what proteins you need. Each of these biochemical compounds is a polymer.

Many carbohydrates are made of glucose

The sugar glucose is a **carbohydrate.** Glucose provides energy to living things. Starch, also a carbohydrate, is made of many bonded glucose molecules. Plants store their energy as chains of starch.

Starch chains pack closely together in a potato or a pasta noodle. When you eat such foods, enzymes in your body break down the starch, making glucose available as a nutrient for your cells. Glucose that is not needed right away is stored as *glycogen*. When you become active, glycogen breaks apart and glucose molecules give you energy. Athletes often prepare themselves for their event by eating starchy foods. They do this so they will have more energy when they exert themselves later on, as shown in *Figure 30*.

Proteins are polymers of amino acids

Many polymers are made of only one kind of molecule. Starch, for example, is made of only glucose. **Proteins,** on the other hand, are made of many different molecules that are called **amino acids.** Amino acids are made of carbon, hydrogen, oxygen, and nitrogen. Some amino acids also contain sulfur. There are 20 amino acids found in naturally occurring proteins. The way these amino acids combine determines which protein is made.

Did You Know ?

Proteins Proteins in the food you eat are necessary to provide amino acids, especially the amino acids that human cells cannot synthesize.

LIFE SCIENCE
CONNECTION

Have a biology teacher visit the classroom and explain the importance of different compounds to a healthy human body. Have them emphasize the importance of trace minerals and how to obtain needed substances by eating a balanced diet.

Proteins are long chains made of amino acids. A small protein, insulin, is shown in **Figure 31.** Many proteins are made of thousands of bonded amino acid molecules. This means that millions of different proteins can be made with very different properties. When you eat foods that contain proteins, such as cheese, your digestive system breaks down the proteins into individual amino acids. Later, your cells bond the amino acids in a different order to form whatever protein your body needs.

DNA is a polymer with a complex structure

Your DNA determines your entire genetic makeup. It is made of organic molecules containing carbon, hydrogen, oxygen, nitrogen, and phosphorus.

Figuring out the complex structure of DNA was one of the greatest scientific challenges of the twentieth century. Instead of forming one chain, like many proteins and polymers, DNA is in the form of paired chains, or strands. It has the shape of a twisted ladder known as a *double helix.*

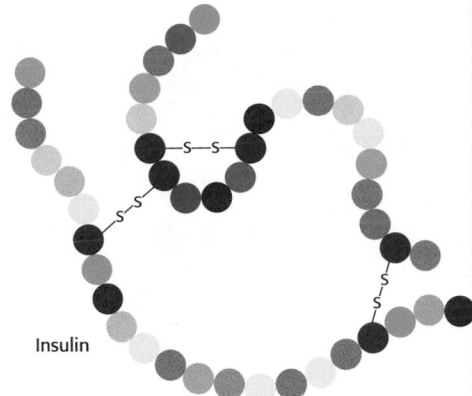

Insulin

Figure 31

Insulin controls the use and storage of glucose in your body. Each color in the chain represents a different amino acid.

Quick Lab

What properties does a polymer have?

Materials
- ✓ water
- ✓ white glue
- ✓ borax
- ✓ 250 mL beakers (2)
- ✓ plastic spoons
- ✓ plastic sandwich bags

SAFETY CAUTION Wear safety goggles, gloves, and a laboratory apron. Be sure to work in an open space and wear clothes that can be cleaned easily.

1. In one beaker, mix 4 g borax with 100 mL water, and stir well.

2. In the second beaker, mix equal parts of glue and water. This solution will determine the amount of new material made. The volume of diluted glue should be between 100 and 200 mL.

3. Pour the borax solution into the beaker containing the glue, and stir well using a plastic spoon.

4. When it becomes too thick to stir, remove the material from the cup and knead it with your fingers. You can store this new material in a plastic sandwich bag.

Analysis

1. What happens to the new material when it is stretched, or rolled into a ball and bounced?

2. Compare the properties of the glue with those of the new material.

3. The properties of the new material resulted from the bonds between the borax and the glue particles. If too little borax were used, in what way would the properties of the new material differ?

4. Does the new material have the properties of a polymer? Explain how you reached this conclusion.

171

Historical Perspective

Structure of DNA Scientists Francis Crick and James Watson determined the structure of DNA and were able to correctly predict how DNA replicates, or copies itself. This discovery has led the way for many genetic engineering applications. For their discovery, Crick and Watson won the 1963 Nobel Prize in chemistry.

Quiz

Fill in the blanks to complete each sentence.

1. Most organic compounds contain the elements ___ and ___. (carbon; hydrogen)

2. ___ are large organic molecules made up of many smaller bonded units. (polymers)

3. The 20 different organic molecules that combine to form proteins are called ___. (amino acids)

4. Genetic makeup is determined by ___, a polymer with a double helix structure. (DNA)

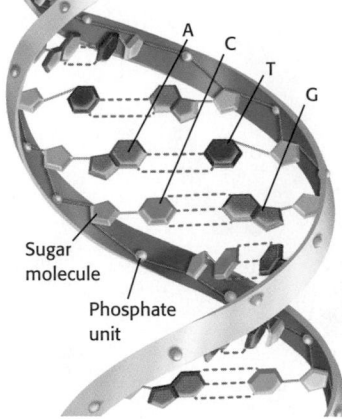

Figure 32

In DNA, cytosine, C, always pairs with guanine, G. Adenine, A, always pairs with thymine, T.

Your body has many copies of your DNA

Most cells in your body have a copy of your genetic material in the form of chromosomes made of DNA. For new cells to have the right amount of DNA, the DNA must be copied. Copying cannot happen unless the two DNA strands are first separated.

Proteins called helicases unwind DNA by separating the paired strands. Proteins called DNA polymerases then pair up new monomers with those already on the strand. At the end of this process, there are two strands of DNA.

DNA's structure resembles a twisted ladder

DNA's structure can be likened to a ladder. Alternating sugar molecules and phosphate units correspond to the ladder's sides, as shown in **Figure 32**. Attached to each sugar molecule is one of four possible DNA monomers—adenine, thymine, cytosine, or guanine. These DNA monomers pair up with DNA monomers attached to the opposite strand in a predictable way, as shown in **Figure 32**. Together, the DNA monomer pairs make up the rungs of the ladder.

SECTION 4 REVIEW

SUMMARY

▶ Alkanes have C—C and C—H bonds.

▶ Alkenes have C=C and C—H bonds.

▶ Alcohols have one or more —OH groups.

▶ Polymers form when small organic molecules bond to form long chains.

▶ Biochemical compounds are polymers important to living things.

▶ Sugars and starches are carbohydrates that provide energy.

▶ Amino acids bond to form polymers called proteins.

▶ DNA is a polymer shaped like a twisted ladder.

1. **Identify** the following compounds as alkanes, alkenes, or alcohols based on their names:
   a. 2-methylpentane   d. butanol
   b. 3-methyloctane   e. 3-heptene
   c. 1-nonene   f. cyclohexanol

2. **Explain** why the compound CBr_5 does not exist. Give an acceptable chemical formula for a compound made of only carbon and bromine.

3. **Determine** how many hydrogen atoms a compound has if it is a hydrocarbon and its carbon atom skeleton is C=C—C=C.

4. **Compare** the structures and properties of carbohydrates with those of proteins.

5. **Identify** which compound is an alkane: CH_2O, C_6H_{14}, or C_3H_4. Explain your reasoning.

6. **Critical Thinking** *Alkynes,* like alkanes and alkenes, are hydrocarbons. Alkynes have carbon-carbon triple covalent bonds, or C≡C bonds. Draw the structure of the alkyne that has the chemical formula C_3H_4. Can you guess the name of this compound?

172

Answers to Section 4 Review

1. a. alkane
   b. alkane
   c. alkene
   d. alcohol
   e. alkene
   f. alcohol

2. Carbon atoms form four bonds, not five. Bromine atoms form one bond, so a carbon-bromine compound would be CBr_4.

3. Each carbon atom must form four bonds. Each end carbon currently has two bonds, so each must further bond to two hydrogen atoms. The two interior carbon atoms each have three bonds, so each can bond to only one hydrogen atom. The total number of bonded hydrogen atoms is six.

4. Carbohydrates contain H, C, and O and may or may not be polymers. Proteins contain H, C, O, N, and S and are polymers of amino acids. In the human body, carbohydrates provide energy, and proteins provide the amino acids needed to create new proteins. Polymers of both must be broken down into their monomers before they can be used.

   See "Continuation of Answers" at the end of the chapter.

Study Skills

KWL Notes

KWL stands for "what I **K**now—what I **W**ant to know—what I **L**earned". The KWL strategy helps you relate your new ideas and concepts with those you have already learned.

1 **Read the section objectives.**

> We'll use the first objective, "Distinguish between compounds and mixtures," from Section 1.

2 **Divide a blank sheet of paper into three columns, and label the columns "What I know," "What I want to know," and "What I learned."**

3 **In the first column, write what information you know about the objective.**

4 **In the second column, write the information that you want to know about the objective.**

5 **After you have read the section, write in the third column what you have learned.**

What I know	What I want to know	What I have learned
water is a compound mixtures can be separated grape juice is a mixture	how to distinguish between compounds and mixture	Compounds are held together by chemical bonds, but mixtures are not. Compounds are always made of the same proportion of elements, but mixtures are not. Substances in mixtures keep their own identities.

Practice

Use the remaining objectives from Section 1 to create a table of KWL notes. Compare the ideas you wrote down in the first column with the items in the third column. If some of your initial ideas are incorrect, cross them out.

Practice

Answers may vary. Consult Appendix A for more information on the KWL strategy.

173

Understanding Concepts

1. d
2. a
3. a
4. b
5. c
6. a
7. c
8. a
9. d
10. b
11. b
12. d

Using Vocabulary

13. Both compounds have two identical atoms covalently bonded to a central atom of a different element. Carbon dioxide has the larger bond angle.
14. molecular formula
15. **a.** sulfur tetrafluoride
    b. dinitrogen monoxide
    c. phosphorus trichloride
    d. diphosphorus pentoxide
16. Metallic bonds are flexible so that metals can be bent and stretched, while ionic bonds are not flexible. Charges are free to move in a metal when it is solid, but the charges are held in place in an ionic solid. Metals are used in wiring because a solid that carries electrical current is needed.

Chapter Resource File

• **Chapter Test** GENERAL

• **Test Item Listing for ExamView®
  Test Generator**

Chapter Highlights

Before you begin, review the summaries of the key ideas of each section, found at the end of each section. The key vocabulary terms are listed on the first page of each section.

UNDERSTANDING CONCEPTS

1. Which of the following is not true of compounds made of molecules?
   a. They may exist as liquids.
   b. They may exist as solids.
   c. They may exist as gases.
   d. They always have very high melting points.

2. Compounds are different from mixtures because
   a. compounds are held together by chemical bonds.
   b. each substance in a compound maintains its own properties.
   c. each original substance in a compound remains chemically unchanged.
   d. mixtures are held together by chemical bonds.

3. What can be learned by looking at a model of compound?
   a. chemical structure
   b. the strength of attraction between molecules
   c. the electron configuration of the atoms involved
   d. the types of bonds formed between the atoms

4. Crystals of salt, called sodium chloride, are
   a. made of molecules.
   b. made of a network of ions.
   c. chemically similar to sugar crystals.
   d. weak solids.

174

5. Ionic solids
   a. are formed by networks of ions that have the same charge.
   b. melt at very low temperatures.
   c. have very regular structures.
   d. are sometimes found as gases at room temperature.

6. A chemical bond can be defined as
   a. a force that joins atoms together.
   b. a force blending nuclei together.
   c. a force caused by electric repulsion.
   d. All of the above

7. Which substance has ionic bonds?
   a. CO **c.** KCl
   b. CO_2 **d.** O_2

8. Covalent bonds
   a. join atoms in some solids, liquids, and gases.
   b. usually join one metal atom to another.
   c. are always broken when a substance is dissolved in water.
   d. join molecules in substances that have molecular structures.

9. A compound has an empirical formula CH_2. Its molecular formula could be
   a. CH_2.   **c.** C_4H_8.
   b. C_2H_4.   **d.** Any of the above

10. The chemical formula for calcium chloride is
    a. $CaCl$.   **c.** Ca_2Cl.
    b. $CaCl_2$.   **d.** Ca_2Cl_2.

11. The empirical formula of a molecule
    a. can be used to identify the molecule.
    b. is sometimes the same as the molecular formula for the molecule.
    c. is used to name the molecule.
    d. shows how atoms bond in the molecule.

12. All organic compounds
    a. come only from living organisms.
    b. contain only carbon and hydrogen.
    c. are biochemical compounds.
    d. have atoms connected by covalent bonds.

17. Proteins and most carbohydrates are made from many smaller molecules, or monomers, joined together. The monomers for proteins are amino acids, and glucose is the monomer for carbohydrates.
18. In nonpolar covalent bonds, electrons are shared equally between atoms. In polar covalent bonds, electrons are shared unequally and are attracted more strongly to one atom than another.
19. In ionic bonds, electrons are transferred from one atom to another, whereas in covalent bonds electrons are shared between atoms. Ionic bonds form by the attraction of

oppositely charged ions, whereas covalent bonds form between many different atoms.
20. Organic compounds contain carbon and usually hydrogen. Common organic compounds include acetylsalicyclic acid, sorbitol, methane, propane, and ethane.
21. Alkanes have single covalent bonds and include, propane, butane, and octane. Alkenes have double covalent bonds and include ethene and propene.
22. A hydroxyl group, or –OH group, is made of oxygen and hydrogen. Alcohols contain hydroxyl groups.

13. Compare the *chemical structure* of oxygen difluoride with that of carbon dioxide. Which compound has the larger *bond angle*?

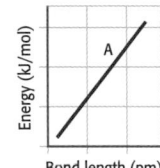

Carbon dioxide

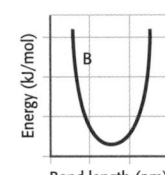

Oxygen difluoride

14. Determine whether the *chemical formula* $C_5H_5N_5$ is the *empirical formula* or *molecular formula* for adenine.

15. Name the following *covalent* compounds:
   a. SF_4
   c. PCl_3
   b. N_2O
   d. P_2O_5

16. Compare the *metallic bonds* of copper with the *ionic bonds* of copper sulfide. Why are metals rather than ionic solids used in electrical wiring?

17. Explain why *proteins* and *carbohydrates* are *polymers*. What is each polymer made of?

18. Discuss two ways that atoms share electrons using the terms *nonpolar covalent bonds* and *polar covalent bonds*.

19. Compare *ionic bonds* and *covalent bonds*, and list two differences between them.

20. What does an *organic compound* contain? List several organic compounds that can be found in your body or in your daily life.

21. Describe the type of bonds that *alkanes* and *alkenes* have. How are they different? Are there any alkanes or alkenes that you are familiar with?

22. What is a *hydroxyl* group? What organic compound contains a hydroxyl group?

23. What is a *hydrocarbon* made of? Name the most simple hydrocarbon.

24. **Graphing** Which of the graphs below shows how bond length and bond energy are related? Describe the flawed relationships shown by each of the other graphs.

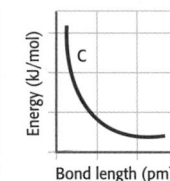

25. **Graphing** The melting points of elements in the same group of the periodic table follow a pattern. A similar pattern is also seen among the melting points of ionic compounds when the cations are made from elements that are in the same group. To see this, plot the melting point of each of the ionic compounds in the table below on the y-axis and the average atomic mass of the element that the cation is made from on the x-axis.
   a. What trend do you notice in the melting points as you move down Group 2?
   b. $BeCl_2$ has a melting point of 405°C. Is this likely to be an ionic compound like the others? Explain. (**Hint:** Locate beryllium in the periodic table.)
   c. Predict the melting point of the ionic compound $RaCl_2$. (**Hint:** Check the periodic table, and compare radium's location with the location of magnesium, calcium, strontium, and barium.)

Compound	Melting point (°C)
$MgCl_2$	714
$CaCl_2$	782
$SrCl_2$	875
$BaCl_2$	963

175

23. A hydrocarbon is made of hydrogen and carbon. The most simple hydrocarbon is methane.

Building Math Skills

24. Graph C accurately shows the relationship between bond length and bond energy: the shorter the bond length, the greater the bond energy. Graph A incorrectly illustrates that the longer the bond length, the greater the bond energy. Graph B incorrectly illustrates that both short and long bonds have high bond energy, while the relative middle bond length has the lowest bond energy.

25. a. The melting point of the compound increases as the atomic mass of the cation increases.
   b. $BeCl_2$ is an ionic compound because it is a compound of a metal and a nonmetal. It therefore has a relatively high melting point.
   c. Radium is in the same family as the other cations, and it has a larger atomic mass. Therefore, the melting point will be higher, somewhat over 1000°C.

26. a. $Sr(NO_3)_2$
   b. NaCN
   c. $Cr(OH)_3$
   d. AlN
   e. SnF_2
   f. K_2SO_4

Thinking Critically

27. Because it is a solid at room temperature, is only able to conduct electricity in the liquid state, and has a high melting point, this compound probably has ionic bonds.

28. The general formula for *n*–alkanes is C_nH_{2n+2}. With *n*=12, dodecane must have 2(12)+2, or 26 hydrogen atoms.

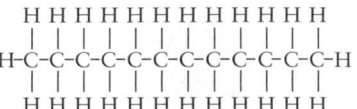

Assignment Guide	
Section	Questions
1	1–3, 6, 13, 24, 25, 30, 33
2	4, 5, 7, 8, 15, 16, 18, 19, 27, 31, 34–39, 41
3	9–11, 14, 26
4	12, 17, 20–23, 28, 29, 32, 40

29. Relatively, bond length increases from ethyne to ethene and from ethene to ethane. In the multiple bonds, more electrons are shared between the two carbon atoms, so the bonds are shorter.

30. Sodium gains a positive charge when it ionizes, because it loses a valence electron, not because it gains a proton.

31. In ionic bonds, electrons are transferred from one atom to another. In metallic bonds, electrons move freely from atom to atom. In covalent bonds, electrons are shared by atoms.

32. oxygen, nitrogen, sulfur, and phosphorous

33. Bonds between molecules must be overcome in order to melt ice.

34. a. one
b. two
c. three

35. Most metals are malleable and ductile, because their valence electrons move freely among different atoms, and thus atoms can slide past each other without their bonds breaking. In ionic compounds, electrons are transferred from atom to atom, creating a network of charged ions that are not free to move around.

26. Writing Ionic Formulas Determine the chemical formula for each of the following ionic compounds:
a. strontium nitrate, an ingredient in some fireworks, signal flares, and matches
b. sodium cyanide, a compound used in electroplating and treating metals
c. chromium(III) hydroxide, a compound used to tan and dye substances
d. aluminum nitride, a compound used in the computer-chip-making process
e. tin(II) fluoride, the source of fluoride for many toothpastes
f. potassium sulfate, a compound used in the glass-making process

THINKING CRITICALLY

27. Evaluating Data A substance is a solid at room temperature. It is unable to conduct electricity as a solid but can conduct electricity as a liquid. This compound melts at 755°C. Would you expect this compound to have ionic, metallic, or covalent bonds?

28. Creative Thinking Dodecane is a combustible organic compound used in jet fuel research. It is an *n*-alkane made of 12 carbon atoms. How many hydrogen atoms does dodecane have? Draw the structural formula for dodecane.

29. Applying Knowledge The length of a bond depends upon its type. Predict the relative lengths of the carbon-carbon bonds in the following molecules, and explain your reasoning.

```
  H  H                    H        H           
  |  |                     \      /            
H–C–C–H                     C=C               H–C≡C–H
  |  |                     /      \            
  H  H                    H        H           
 Ethane                  Ethene               Ethyne
```

30. Critical Thinking A classmate insists that sodium gains a positive charge when it becomes an ion because it gains a proton. Explain this student's error.

31. Applying Knowledge Compare the three types of bonds based on what happens to the valence electrons of the atoms.

32. Applying Knowledge In addition to carbon and hydrogen atoms, list four elements that can bond to carbon in organic compounds.

33. Critical Thinking Describe what attractive force(s) must be overcome to melt ice.

34. Applying Knowledge How many pairs of electrons are shared in the following types of bonds?
a. a single bond
b. a double bond
c. a triple bond

35. Understanding Systems Explain why most metals are malleable and ductile but ionic crystals are not.

DEVELOPING LIFE/WORK SKILLS

36. Working Cooperatively For one day, write down all of the ionic compounds listed on the labels of the foods you eat. Also write down the approximate mass you eat of each compound. As a class, make a master list in the form of a computer spreadsheet that includes all of the ionic compounds eaten by the whole class. Identify which compounds were eaten by the most people. Together, create a poster describing the dietary guidelines for the ionic compound that was eaten most often.

COMPUTER SKILL

Developing Life/Work Skills

36. Answers will vary. Sodium chloride (table salt) probably will be the most common compound. Posters for table salt should include the recommended dietary intake and emphasize that some salt is needed but too much should be avoided.

37. Answers will vary. Many salt substitutes contain KCl. Students who choose to discuss this substance should research the critical Na^+/K^+ balance in the body to learn why excess use of KCl can be dangerous. MSG is not strictly a salt substitute because it, too, contributes sodium ions to the diet.

38. code 1: PETE (polyethylene terephthalate), somewhat flexible and easily molded, soft-drink bottles; code 2: HDPE (high-density polyethylene), rigid but easily molded, milk containers; code 3: V (polyvinyl chloride), flexible but strong, garden hoses; code 4: LDPE (low-density polyethylene), rigid, cottage cheese containers; code 5: PP (polypropylene), somewhat rigid, shampoo bottles; code 6: PS (polystyrene), lightweight because it contains bubbles of air, foam cups and plates

39. Students' answers may vary.

37. Making Decisions People on low-sodium diets must limit their intake of table salt. Luckily, there are salt substitutes that do not contain sodium. Research different kinds of salt substitutes, and describe how each one affects your body. Determine which salt substitute you would use if you were on a low-sodium diet.

38. Locating Information Numerical recycling codes identify the composition of a plastic so that it can be sorted and recycled. For each of the recycling codes, 1–6, identify the plastic, its physical properties, and at least one product made of this plastic.

39. Interpreting and Communicating Covalently bonded solids, such as silicon, an element used in computer components, are harder than some pure metals. Research theories that explain the hardness of covalently bonded solids and their usefulness in the computer industry. Present your findings to the class.

INTEGRATING CONCEPTS

40. Connection to Health The figure below shows how atoms are bonded in a molecule of vitamin C. Which elements is vitamin C made of? What is its molecular formula? Write a paragraph explaining some of the health benefits of taking vitamin C supplements.

WRITING SKILL

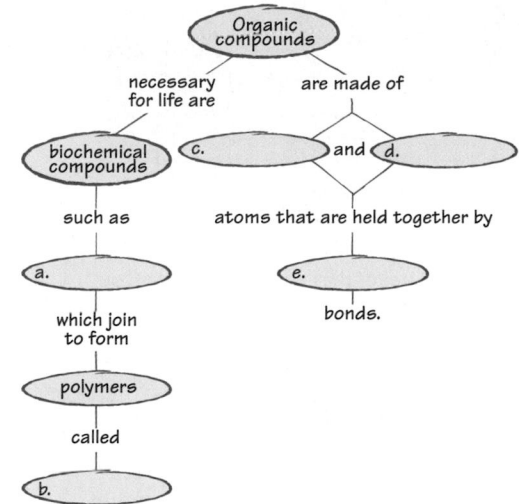

41. Concept Mapping Copy the unfinished concept map below onto a sheet of paper. Complete the map by writing the correct word or phrase in the lettered boxes.

internet connect

www.scilinks.org
Topic: Vitamin C SciLinks code: HK4147

SCiLINKS® Maintained by the National Science Teachers Association

Art Credits: Fig. 2-4, Kristy Sprott; Fig. 5 , J/B Woolsey Associates; Fig. 6-17, Kristy Sprott; Fig. 21, Leslie Kell; Table 8, Kristy Sprott; Fig. 22-29, Kristy Sprott; Fig. 31, Kristy Sprott; Fig. 32, Morgan-Cain & Associates.

Photo Credits: Chapter Opener photo of breaking bottle by John Bagley, stained glass by Robert Frerck/Getty Images/Stone; Fig. 1A, E. R. Degginger/Color-Pic, Inc.; Fig. 1B, Yoav Levy/Phototake; Fig. 1C, Charles D. Winters; Fig.1D, Sam Dudgeon/HRW; Fig. 2, Sergio Purtell/Foca/HRW; Fig. 5, E. R. Degginger/Color-Pic, Inc.; Fig. 6, Dr. Dennis Kunkel/Phototake; Fig. 7, Sergio Purtell/Foca/HRW; "Quick Lab," Sam Dudgeon/HRW; Fig. 8, Sergio Purtell/Foca/HRW; Fig. 12, Sam Dudgeon/HRW; Fig. 13, Erin Garvey/Index Stock Imagery, Inc.; "Quick Activity," Sam Dudgeon/HRW; Fig. 16, Randall Alhadef/HRW; Fig. 18, Peter Van Steen/HRW; Fig. 20, Getty Images/The Image Bank; Fig. 22, Peter Van Steen/HRW; Fig. 25, Marc Grimberg/Getty Images/The Image Bank; Figs. 27-29, Sam Dudgeon/HRW; Fig. 30, David J. Phillip/AP/Wide World Photos; "Skills Practice Lab," Sergio Purtell/Foca; "Career Feature," (portraits), Steve Fischbach/HRW; (barrels), W. & D. McIntyre/Photo Researchers, Inc.

Integrating Concepts

40. Vitamin C is made of carbon, hydrogen, and oxygen atoms. Its formula is $C_6H_8O_6$. Linus Pauling advocated large doses of vitamin C to treat the common cold and other conditions and diseases, such as cancer. Little scientific evidence supports his assertions.

41. a. amino acids
  b. proteins
  c. carbon
  d. hydrogen
  e. covalent

Transparencies

TT Concept Mapping

COMPARING POLYMERS

Teacher's Notes

Time Required 1 lab period

Ratings

```
        1     2     3     4
EASY                    HARD
```

TEACHER PREPARATION	2
STUDENT SETUP	3
CONCEPT LEVEL	1
CLEANUP	3

Skills Acquired

- Collecting data
- Communicating
- Interpreting
- Measuring
- Organizing and analyzing data

The Scientific Method

In this lab, students will:
- Make Observations
- Analyze the Results
- Draw Conclusions
- Communicate Results

Materials

For 5% acetic acid, use white vinegar. Do not dilute glacial acetic acid. Liquid latex may be purchased from Flinn-Scientific, Inc., in Batavia, IL. Sodium silicate solution, also known as water glass, can be purchased from many scientific supply companies.

Keep the bottles of latex and acetic acid solution in an operating hood because the vapors are unpleasant. Keep the bottle of ethanol in an operating fume hood because the vapors are flammable.

Skills Practice Lab

Introduction

Many polymers are able to "bounce back" after they are stretched, bent, or compressed. In this lab, you will compare the bounce heights of two balls made from different polymers.

Objectives

▶ **Synthesize** two different polymers, **shape** each into a ball, and **measure** how high each ball bounces.

▶ **USING SCIENTIFIC METHODS** *Conclude* which polymer would make a better toy ball.

Materials

acetic acid solution (vinegar), 5%
container, 2 L
ethanol solution, 50%
graduated cylinder, 10 mL
graduated cylinders, 25 mL (2)
liquid latex
meterstick
paper cups, medium-sized (2)
paper towels
sodium silicate solution
water, deionized
wooden craft sticks (2)

Comparing Polymers

▶ Procedure

1. Prepare a data table in your lab report similar to the one shown at right.

Making Latex Rubber

SAFETY CAUTION If you get a chemical on your skin or clothing, wash it  off with lukewarm water while calling to your teacher. If you get a chemical in your eyes, flush it out immediately at the eyewash station and alert your teacher.

2. Pour 1 L of deionized water into a 2 L container.

3. Use a 25 mL graduated cylinder to pour 10 mL of liquid latex into one of the paper cups.

4. Clean the graduated cylinder thoroughly with soap and water, then rinse it with deionized water and use it to add 10 mL of deionized water to the liquid latex.

5. Use the same graduated cylinder to add 10 mL of acetic acid solution to the liquid latex-water mixture.

6. Stir the mixture with a wooden craft stick. As you stir, a "lump" of the polymer will form around the stick.

7. Transfer the stick and the attached polymer to the 2 L container. While keeping the polymer underwater, gently pull it off the stick with your gloved hands.

8. Squeeze the polymer underwater to remove any unreacted chemicals, shape it into a ball, and remove the ball from the water.

9. Make the ball smooth by rolling it between your gloved hands. Set the ball on a paper towel to dry while you continue with the next part of the lab.

Bounce Heights of Polymers

Polymer	Bounce height (cm)					
	Trial 1	Trial 2	Trial 3	Trial 4	Trial 5	Average
Latex rubber						
Ethanol-silicate						

10. Wash your gloved hands with soap and water, then remove the gloves and dispose of them. Wash your hands again with soap and water.

Making an Ethanol-silicate Polymer

SAFETY CAUTION Put on a fresh pair of gloves. Ethanol is flammable, so make sure there are no flames or other heat sources anywhere in the laboratory.

11. Use a clean 25 mL graduated cylinder to pour 12 mL of sodium silicate solution into the clean paper cup.

12. Use a 10 mL graduated cylinder to add 3 mL of the ethanol solution to the sodium silicate solution.

13. Stir the mixture with the clean wooden craft stick until a solid polymer forms.

14. Remove the polymer with your gloved hands, and gently press it between your palms until you form a ball that does not crumble. This activity may take some time. Occasionally dripping some tap water on the polymer might be helpful.

15. When the ball no longer crumbles, dry it very gently with a paper towel.

16. Repeat step 10, and put on a fresh pair of gloves.

17. Examine both polymers closely. Record in your lab report how the two polymers are alike and how they are different.

18. Use a meterstick to measure the highest bounce height of each ball when each is dropped from a height of 1 m. Drop each ball five times, and record the highest bounce height each time in your data table.

▶ **Analysis**

1. Calculate the average bounce height for each ball by adding the five bounce heights and dividing by 5. Record the averages in your data table.

2. Based on only their bounce heights, which polymer would make a better toy ball?

▶ **Conclusions**

3. Suppose that making a latex rubber ball costs 22 cents and that making an ethanol-silicate ball costs 25 cents. Does this fact affect your conclusion about which polymer would make a better toy ball? Besides cost, what are other important factors that should be considered?

179

Safety Cautions

- Read all safety cautions, and discuss them with students. Excuse any student with latex allergies from this experiment.
- Wear safety goggles and a lab apron during the lab. Make sure students wear gloves throughout the lab and that they put on fresh gloves when the procedure tells them to.
- Be sure to have an MSDS for each of the following chemicals: acetic acid, liquid latex, ethanol, and sodium silicate.

- Promptly clean up all spills with paper towels.
- Keep the alcohol in a hood, use a container with a lid, and restrict the amount kept in the hood to the minimum needed by students.
- Do not allow students to take any ethanol-silicate polymer from the laboratory.

Tips and Tricks

Ethanol is flammable, so make sure no heat sources are present.

Sodium silicate can irritate if it comes into contact with the skin. Caution students to immediately wash off the affected area with soap and water and to remove contaminated clothing. If inhaled, sodium silicate can irritate the upper respiratory tract. If a student inhales sodium silicate, immediately get the student to fresh air.

Procedure
Techniques to Demonstrate
Students can use either deionized or distilled water when making the polymers. Show students how to roll the latex and the ethanol-silicate polymer into a ball. Students will have difficulty making a perfect sphere, but they should try to make it as regular as possible. The more irregular the shape is, the more difficulty students will have determining the bounce height because the ball will not bounce straight up.

Remind students to be patient with the ethanol-silicate polymer, which tends to crumble. If it crumbles too much, a few drops of water will rehydrate it so that it can be reshaped into a ball.

Disposal Information
Paper cups, paper towels, disposable gloves, latex, and ethanol-silicate polymer balls and fragments should be thrown in the trash can. Waste liquids from this lab can be poured down the drain.

See "Continuation of Answers" at the end of the chapter.

Chapter Resource File

- **Datasheet** Comparing Polymers **GENERAL**
- **Observation Lab** Extracting Iron from Cereal **BASIC**
- **Probeware Lab—CBL™** Determining Which Household Solutions Conduct Electricity **ADVANCED**

Continuation of Answers

Continuation of Answers from p. 172
Answers to Section 4 Review

5. An alkane has the general formula C_nH_{2n+2} and contains only H and C. C_6H_{14} is the alkane.

6.

```
              H
              |
  H−C≡C−C−H
              |
              H
```

Because the compound has three carbon atoms (*pro–*) and is an alkyne (*–yne*), the compound is propyne.

Continuation of Answers from p. 179
Skills Practice Lab
Answers to Analysis

1. Student answers will vary.

2. Student answers will vary depending on the measured bounce heights, but the ethanol-silicate polymer tends to bounce higher than the latex rubber.

Answers to Conclusions

3. Student answers will vary. The lower production cost of toy balls made of latex rubber compared with toy balls made of the ethanol-silicate polymer is a factor in determining which polymer would make a better toy ball. Other important factors that should be considered include texture of the polymer and how well the polymer resists crumbling.

CareerLink

Analytical Chemist

Have you ever looked at something and wondered what chemicals it contained? That's what analytical chemists do for a living. They use a range of tests to determine the chemical makeup of a sample. To find out more about analytical chemistry as a career, read the interview with analytical chemist Roberta Jordan, who works at the Idaho National Engineering and Environmental Laboratory, in Idaho Falls, Idaho.

In addition to working as an analytical chemist, Roberta Jordan mentors students regularly in the local schools.

"Chemistry is in everything we do. Just to take a breath and eat a meal involves chemistry."

 What is your work as an analytical chemist like?

We deal with radioactive waste generated by old nuclear power plants and old submarines, and we try to find a safe way to store the waste. I'm more like a consultant. A group of engineers that are working on a process will come to me. I tell them what things they need to analyze for and why they need to do that. On the flip side, I'll tell them what techniques they need to use.

 What do you like best about your work?

It forces me to stay current with any new techniques, new areas that are going on in analytical chemistry. And I like the team approach because it allows me to work on different projects.

 What do you find most interesting about your work?

Probably the most interesting thing is to observe how different industries and different labs conduct business. It gives you a broad feel for how chemistry is done.

 What qualities does a good chemist need?

I think you do need to be good at science and math and to like those subjects. You need to be fairly detail-oriented. You have to be precise. You need to be analytical in general, and you need to be meticulous.

180

What part of your education do you think was most valuable?

I think it was worthwhile spending a lot of energy on my lab work. With any science, the most important part is the laboratory experience, when you are applying those theories that you learn. I'm really a proponent of being involved in science-fair activities.

 What advice do you have for students who are interested in analytical chemistry?

It's worthwhile to go to the career center or library and do a little research. Take the time to find out what kinds of things you could do with your degree. You need to talk to people who have a degree in that field.

 Do you think chemistry has a bright future?

I think that there are a lot of things out there that need to be discovered. My advice is to go for it and don't think that everything we need to know has been discovered. Twenty to thirty years down the road, we will have to think of a new energy source, for example.

internet connect

www.scilinks.org
Topic: Analytical Chemistry
SciLinks code: HK4006

SCi LINKS. Maintained by the National Science Teachers Association

"One of the things necessary to be a good chemist is you have to be creative. You have to be able to think above and beyond the normal way of doing things to come up with new ideas, new experiments."
—Roberta Jordan

181

Historical Perspective

Originally, analytical chemists worked using tests in which they reacted the unknown substance with other substances. These methods often involved changing the original sample.

In the last 50 years, the field of instrumental analysis (in which instruments are used to analyze substances) has grown. Many of these instruments, such as infrared spectrophotometers and nuclear magnetic resonance spectrometers, are able to analyze a substance without changing it chemically.

Chemical Reactions
Chapter Planning Guide

PACING	CLASSROOM RESOURCES	LABS, ACTIVITIES, AND DEMONSTRATIONS
BLOCK 1 · 45 min pp. 182–183 **Chapter Opener**		SE **Activity 1**, p. 183 SE **Activity 2**, p. 183
BLOCKS 2 & 3 · 90 min pp. 184–189 **Section 1** The Nature of Chemical Reactions	CRF **Lesson Plan*** TT **Bellringer*** TT **Reaction Model*** TM **Exothermic and Endothermic***	TE **Opening Activity**, p. 184 GENERAL TE **Demonstration** Lemon Cell, p. 186 GENERAL
BLOCKS 4 & 5 · 90 min pp. 190–197 **Section 2** Reaction Types	CRF **Lesson Plan*** TT **Bellringer*** TT **Single-Displacement***	TE **Opening Demonstration**, p. 190 GENERAL
BLOCKS 6 & 7 · 90 min pp. 198–204 **Section 3** Balancing Chemical Equations	CRF **Lesson Plan*** TT **Bellringer*** TM **Balanced Equation***	TE **Opening Activity**, p. 198 GENERAL SE **Quick Lab** Can you determine the products of a reaction?, p. 204 GENERAL CRF **Datasheets for In-Text Labs** Can you determine the products of a reaction?* GENERAL CRF **Observation Lab** Combining Elements* ◆ BASIC
BLOCKS 8 & 9 · 90 min pp. 205–212 **Section 4** Rates of Change	CRF **Lesson Plan*** TT **Bellringer*** TT **Equilibrium*** TM **Changing Equilibrium*** TT **Concept Mapping***	TE **Opening Demonstration**, p. 205 TE **Demonstration** Increasing Concentration, p. 207 SE **Quick Lab** What affects the rates of chemical reactions?, p. 208 GENERAL CRF **Datasheets for In-Text Labs** What affects the rates of chemical reactions?* GENERAL SE **Quick Activity** Catalysts in Action, p. 209 GENERAL CRF **Datasheets for In-Text Labs** Catalysts in Action* GENERAL SE **Skills Practice Lab** Measuring the Rate of a Chemical Reaction, pp. 218–219 ◆ GENERAL CRF **Datasheets for SE Skills Practice Lab** Measuring the Rate of a Chemical Reaction* GENERAL CRF **CBL™ Probeware Lab** Investigating the Effect of Temperature on the Rate of a Reaction* ◆ ADVANCED

BLOCKS 10 & 11 · 90 min

Chapter Review and Assessment Resources

SE **Chapter Review**, pp. 214–217
CRF **Chapter Tests*** GENERAL
OSP **Test Generator**
CRF **Standardized Test Practice with Guided Reading Development***
CRF **Test Item Listing for ExamView® Test Generator***
OSP **Scoring Rubrics and Classroom Management Checklists**

Online Resources

Visit the HRW Web site for a variety of free resources related to the text. Just type in the keyword **HK4 CHE**.

Holt Online Learning

Holt Science Spectrum: Physical Science: Online Edition

Students can access interactive problem solving help and active visual concept development with the *Holt Science Spectrum: Physical Science* Online Edition available at **www.hrw.com**.

student CNN News

cnnstudentnews.com

Find the latest news, lesson plans, and activities related to important scientific events.

Compression guide:
To shorten your instruction because of time limitations, omit blocks 1, 10, and 11.

KEY

TE	Teacher Edition
SE	Student Edition
OSP	One-Stop Planner

CRF	Chapter Resource File
TT	Teaching Transparency
TM	Transparency Master

* Also on One-Stop Planner
◆ Requires Advance Prep

PROBLEM SOLVING AND PRACTICE	SECTION REVIEW AND ASSESSMENT	STANDARDS CORRELATION
	CRF Pretest* GENERAL	
CRF **Cross-Disciplinary Worksheet** Connection to Social Studies—Alchemists' Theory of the Elements* ADVANCED **CRF** **Cross-Disciplinary Worksheet** Real World Applications—Hot Meals on Hand* ADVANCED **CRF** **Cross-Disciplinary Worksheet** Integrating Biology—Organisms that Glow* ADVANCED	**TE** Quiz, p. 189 BASIC **SE** Section 1 Review, p. 189 **CRF** Concept Review* GENERAL **CRF** Quiz* BASIC	PS 3a, 3b LS 5b UCP 1, 2
SE **Science and the Consumer** Fire Extinguishers, p. 193 **CRF** **Cross-Disciplinary Worksheet** Science and the Consumer—The Right Fire Extinguisher for the Job* ADVANCED **TE** **Inclusion Strategies,** p. 194 BASIC **CRF** **Cross-Disciplinary Worksheet** Connection to Fine Arts—The Chemistry of Art* ADVANCED	**TE** Quiz, p. 197 BASIC **SE** Section 2 Review, p. 197 **CRF** Concept Review* GENERAL **CRF** Quiz* BASIC	PS 3c UCP 1, 2
TE **Inclusion Strategies,** p. 198 ADVANCED **CRF** **Science Skills** Balancing Chemical Equations* BASIC **CRF** **Cross-Disciplinary Worksheet** Connection to Social Studies—Fireworks* ADVANCED **SE** **Math Skills** Balancing Chemical Equations, pp. 201–202 GENERAL **CRF** **Science Skills** Ratios and Proportions* BASIC	**TE** Quiz, p. 204 BASIC **SE** Section 3 Review, p. 204 **CRF** Concept Review* GENERAL **CRF** Quiz* BASIC	SAI 1
CRF **Cross-Disciplinary Worksheet** Integrating Environmental Science—Fertilizers: Friend or Foe?* ADVANCED **SE** **Math Skills** Using Mole Ratios to Calculate Mass, p. 213 **SE** **Viewpoints** How Should Life-Saving Inventions Be Introduced?, pp. 220–221	**TE** Quiz, p. 212 BASIC **SE** Section 4 Review, p. 212 **CRF** Concept Review* GENERAL **CRF** Quiz* BASIC	PS 3d, 3e LS 4a UCP 1, 2, 5 SPSP 2

SCiLINKS®

www.scilinks.org

Topic: Fuels
*Sci*Links code: HK4059

Topic: Corrosion
*Sci*Links code: HK4029

Topic: Types of Reactions
*Sci*Links code: HK4142

Topic: Fire Extinguishers
*Sci*Links code: HK4052

Topic: Factors Affecting Reaction Rate
*Sci*Links code: HK4051

Topic: Catalysts
*Sci*Links code: HK4018

Topic: Biodegradable
*Sci*Links code: HK4015

Technology Resources

 One-Stop Planner
All of your printable resources and the Test Generator are on this convenient CD-ROM.

 Physical Science **Interactive Tutor CD-Rom**

Disc One, Module 5
Topic: Chemical Equations

Overview

In this chapter, students learn what chemical reactions are, how to recognize chemical reactions, and how energy is involved in chemical reactions. Next they study the five general types of reactions and the role of the electron in chemical bonding. Students also learn how to balance and interpret chemical equations, how to predict reaction amounts, and how to identify mole ratios. Finally, they study reaction rates and equilibrium systems.

Assessing Prior Knowledge

Be sure students understand the following concepts:

- conservation of mass
- formulas
- chemical changes
- the periodic table
- moles
- ions
- atomic structure
- atomic mass
- chemical structure
- bonding
- compound names
- formulas

MISCONCEPTION ///ALERT \\\

Science education research has identified the following misconceptions about chemical reactions.

- The failure to recognize products, especially gaseous ones, leads students to believe that reactants "disappear."
- Students think that reactions are a physical process that liberates "trapped" products from reactants.
- Students do not recognize that reactions involve the rearrangement of bonds among a set of atoms.
- Some students think most chemical changes are irreversible.

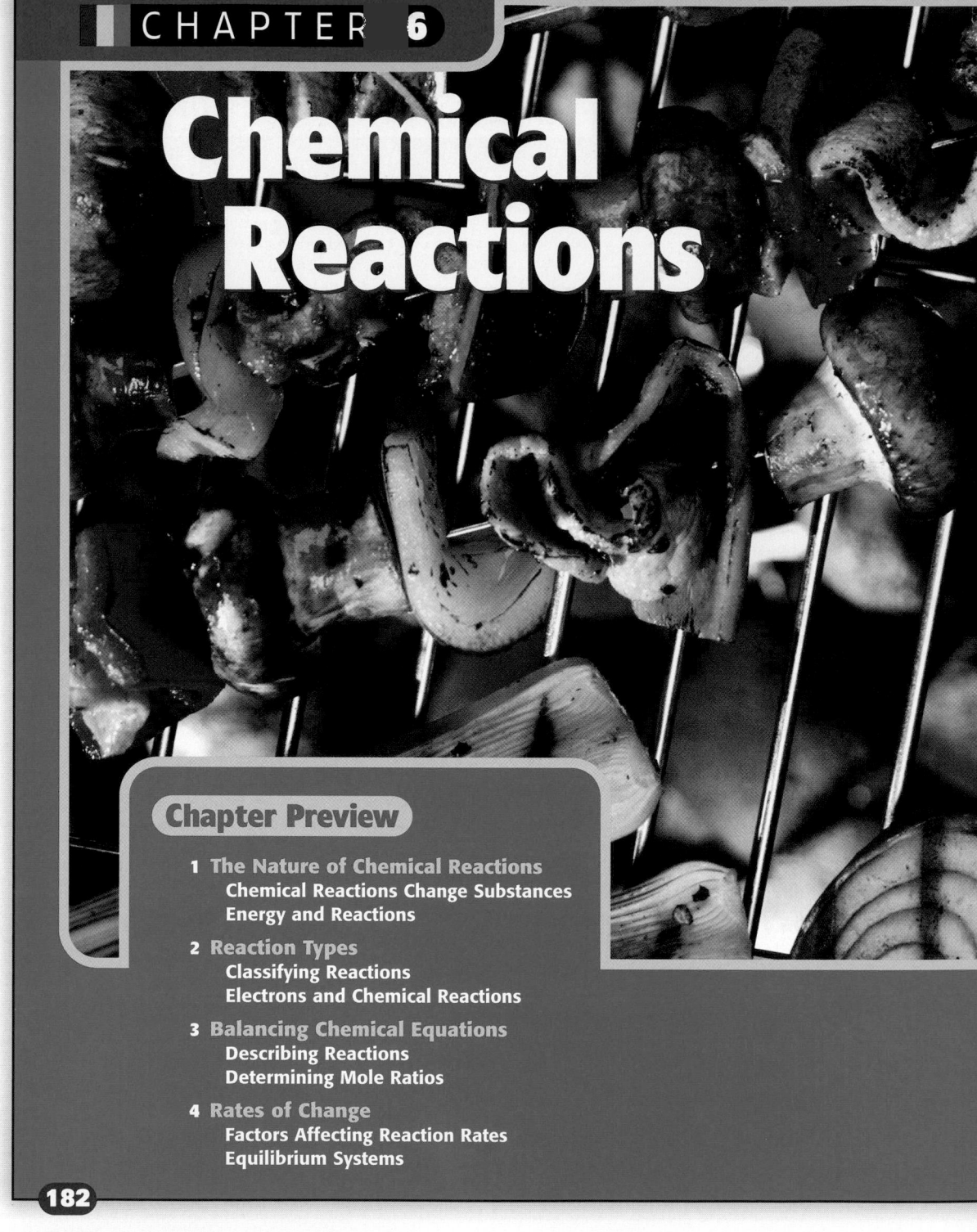

CHAPTER 6

Chemical Reactions

Chapter Preview

1 The Nature of Chemical Reactions
Chemical Reactions Change Substances
Energy and Reactions

2 Reaction Types
Classifying Reactions
Electrons and Chemical Reactions

3 Balancing Chemical Equations
Describing Reactions
Determining Mole Ratios

4 Rates of Change
Factors Affecting Reaction Rates
Equilibrium Systems

182

Standards Correlations

National Science Education Standards

The following descriptions summarize the National Science Standards that specifically relate to this chapter. For the full text of the standards, see the National Science Education Standards at the front of the book.

Section 1 The Nature of Chemical Reactions
Physical Science PS 3a, 3b
Life Science LS 5b
Unifying Concepts and Processes UCP 1, 2

Section 2 Reaction Types
Physical Science PS 3c
Unifying Concepts and Processes UCP 1, 2

Section 3 Balancing Chemical Equations
Science as Inquiry SAI 1

Section 4 Rates of Change
Physical Science PS 3d, 3e
Life Science LS 4a
Unifying Concepts and Processes UCP 1, 2, 5
Science in Personal and Social Perspectives SPSP 2

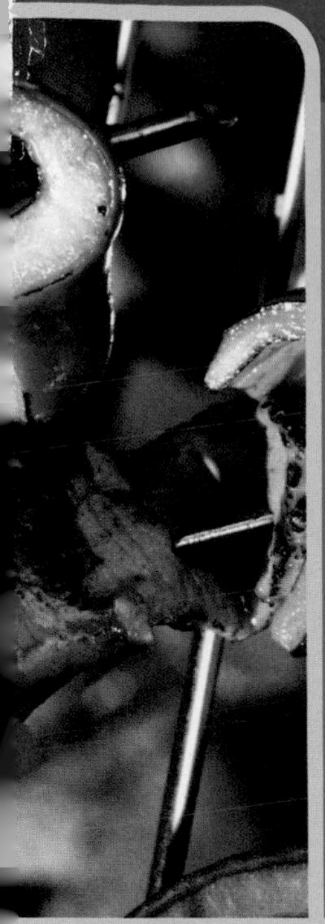

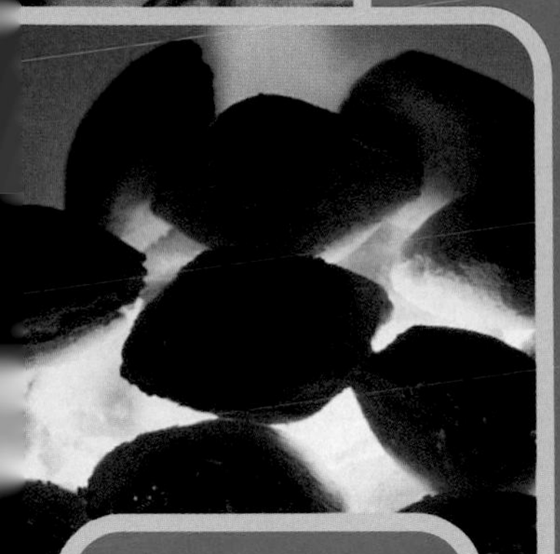

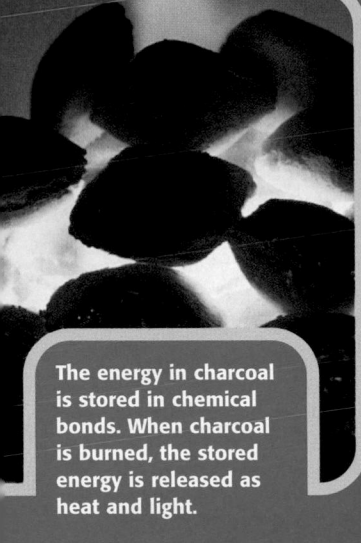

The energy in charcoal is stored in chemical bonds. When charcoal is burned, the stored energy is released as heat and light.

Focus ACTIVITY

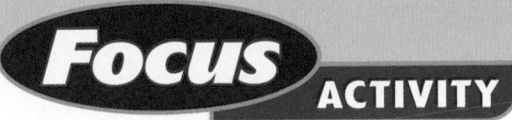

Background Many people look forward to summer as prime "grilling time." Although there are many ways to prepare food on a grill, the basic principle is the same: raw food is cooked to make it tastier, easier to digest, and safer to eat.

One method of grilling uses charcoal as a fuel. Charcoal is produced by heating wood or other plant matter to high temperatures in the absence of air. When charcoal is burned on a grill, the matter in the charcoal and oxygen in the air combine in a chemical reaction to produce light and heat energy that cooks the food.

Some food can be "cooked" using chemical reactions that are not caused by heat. For example, ceviche, which is usually served cold, contains fish that is "cooked" with lime juice. But in all cases, when food is cooked, chemical reactions make the energy in food easier to release when you eat it.

Activity 1 Obtain three freshly cut slices of apple. Cover one completely with water, and wrap another slice with clear plastic wrap. Allow the third slice to remain exposed to the air. Do the slices look the same after one hour? After six hours? Why or why not?

Activity 2 Sodium bicarbonate, also known as baking soda, is used to make pancakes, cookies, and other baked goods light and fluffy. Pour a small amount of vinegar into a cup and add a pinch of baking soda to the cup. What changes do you observe? How might the same reaction cause pancakes to rise?

internet connect

www.scilinks.org

Topic: Fuels SciLinks code: HK4059

SCILINKS. Maintained by the National Science Teachers Association

Pre-Reading Questions

1. When food is cooked, what are some signs to look for to indicate that the food is ready to eat?
2. Wood is sometimes used as a fuel to provide heat and light. What other things in addition to wood are required to start a fire?

Focus ACTIVITY

Background One chemical reaction that frequently occurs in cooking is the Maillard reaction. In this reaction, the molecules of amino acids and sugars combine to form new flavors. The Maillard reaction occurs at high temperatures, and can also occur at lower temperatures if the concentration of amino acids and sugars is high enough. This reaction creates the brown color associated with cooked foods.

Activity 1 The slice that is exposed to air will turn brown most rapidly, and the one that is covered in plastic will appear least changed. The differences in appearance are due to the different times the chemical reaction occurs.

Activity 2 When baking soda is added to vinegar, the release of carbon dioxide gas makes bubbles appear. This same reaction causes pancakes and other baked goods to rise.

Answers to Pre-Reading Questions

1. Indications could include changes in hardness, color, and temperature.
2. Oxygen and added energy in the form of an ignition source are also required to start a fire.

Chapter Resource File

- **Pretest** GENERAL
- **Teaching Transparency Preview**
- **Answer Keys**

Focus

Overview

Before beginning this section, review with your students the Objectives listed in the Student Edition. This section introduces students to chemical reactions. They learn how to recognize chemical reactions, what happens at the atomic level, and how energy is involved in chemical reactions.

🔔 Bellringer

Use the Bellringer transparency to prepare students for this section.

Motivate

Opening Activity — GENERAL

Write the following assertions on the chalkboard:

- In a chemical reaction, atoms change identity.
- A change of color indicates a chemical reaction.
- All chemical reactions give off energy.

Ask students to write down their opinion of each statement, then have students discuss their opinions in small groups. Tell students to keep a list of their opinions for discussion at the end of the section.

LS Interpersonal

The Nature of Chemical Reactions

▶ **KEY TERMS**
reactant
product
chemical energy
exothermic reaction
endothermic reaction

OBJECTIVES

▶ **Recognize** some signs that a chemical reaction may be taking place.

▶ **Explain** chemical changes in terms of the structure and motion of atoms and molecules.

▶ **Describe** the differences between endothermic and exothermic reactions.

▶ **Identify** situations involving chemical energy.

If someone talks about chemical reactions, you might think about scientists doing experiments in laboratories. But words like *grow, ripen, decay,* and *burn* describe chemical reactions you see every day. Even your own health is due to chemical reactions taking place inside your body. The food you eat reacts with the oxygen you inhale in processes such as respiration and cell growth. The carbon dioxide formed in these reactions is carried to your lungs, and you exhale it into the environment.

Chemical Reactions Change Substances

When sugar, water, and yeast are mixed into flour to make bread dough, a chemical reaction takes place. The yeast acts on the sugar to form new substances, including carbon dioxide and lactic acid. You know that a chemical reaction has happened because lactic acid and carbon dioxide are different from sugar.

Chemical reactions occur when substances undergo chemical changes to form new substances. Often you can tell that a chemical reaction is happening because you will be able to see changes, such as those in **Figure 1.**

Figure 1

Signs of a Chemical Reaction

A When the calcium carbonate in a piece of chalk reacts with an acid, bubbles of carbon dioxide gas are given off.

B When solutions of sodium sulfide and cadmium nitrate are mixed, a solid—yellow cadmium sulfide—settles out of the solution.

C When ammonium dichromate decomposes, energy is released as light and heat.

184

Teacher Resources — GENERAL

For the New Teacher As you work through this section, show students several examples of chemical reactions. This will generate student interest and help students apply the concepts discussed in the section. Include examples of familiar reactions, such as baking soda mixing with vinegar, an apple slice turning brown, or a match burning. For each reaction, have students compare and contrast the appearance of the reactants and the products and note any other signs that a reaction took place, such as the release of bubbles or energy. **LS** Visual

Production of gas and change of color are signs of chemical reactions

In bread making, the carbon dioxide gas that is produced expands the dough, causing the bread to rise. This release of gas is a sign that a chemical reaction may be happening.

As the dough bakes, old bonds break and new bonds form. Chemical reactions involving starch and protein make food turn brown when heated. A chemical change happens almost every time there is a change in color.

Chemical reactions rearrange atoms

When gasoline is burned in the engine of a car or boat, a lot of different reactions happen with the compounds that are in the mixture we call gasoline. In a typical reaction, isooctane, C_8H_{18}, and oxygen, O_2, are the **reactants.** They react and form two **products,** carbon dioxide, CO_2, and water, H_2O.

The products and reactants contain the same types of atoms: carbon, hydrogen, and oxygen. New product atoms are not created, and old reactant atoms are not destroyed. Atoms are rearranged as bonds are broken and formed. In all chemical reactions, mass is always conserved.

▶ **reactant** a substance or molecule that participates in a chemical reaction

▶ **product** a substance that forms in a chemical reaction

Energy and Reactions

Filling a car's tank with gasoline would be very dangerous if isooctane and oxygen could not be in the same place without reacting. Like most chemical reactions, the isooctane-oxygen reaction needs energy to get started. A small spark provides enough energy to start this reaction. That is why smoking or having any open flame near a gas pump is not allowed.

Energy must be added to break bonds

In each isooctane molecule, like the one shown in *Figure 2,* all the bonds to carbon atoms are covalent. In an oxygen molecule, a covalent bond holds the two oxygen atoms together. For the atoms in isooctane and oxygen to react, all of these bonds have to be broken. This takes energy.

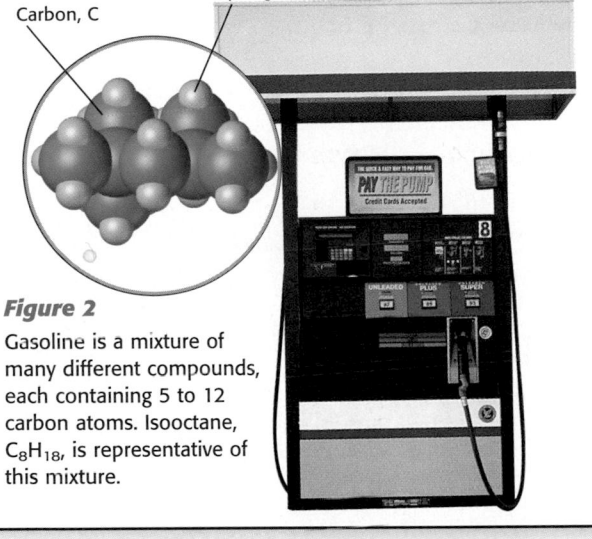

Carbon, C   Hydrogen, H

Figure 2

Gasoline is a mixture of many different compounds, each containing 5 to 12 carbon atoms. Isooctane, C_8H_{18}, is representative of this mixture.

Did You Know ?

Activation Energy Even if a reaction is exothermic, energy must be added to start the reaction. This energy is called activation energy. An example of activation energy is the flame needed to start a newspaper burning. Materials that will burn vary greatly in the amount of activation energy needed to start the reaction. Materials, such as gasoline, that have low activation energy are considered flammable. Other materials, such as paper, burn but require more energy to start the process and are classified as combustible.

Demonstration ——— GENERAL

Lemon Cell

(Time: About 10 minutes)

Materials

• lemon
• copper wire
• paper clip

Straighten the paper clip, and insert it and the copper wire into the lemon. Ask a volunteer to touch the ends of both wires with his or her tongue, and have the volunteer describe what happens. The volunteer should feel a tingling sensation on the tongue. Explain that this is because saliva provides an electrolytic solution that conducts current.

Analysis

1. What must be contained in the lemon for it to conduct a current? (ions)

2. Hydrogen ions in a lemon are part of the chemical reaction that produces the electricity. Hydrogen is present in almost all acids. Why do you think a lemon can be used to make a cell and an orange cannot? (A lemon is more acidic than an orange. As a result, the lemon contains more hydrogen ions to react.)

LS Logical

Figure 3

A Light passing through a camera lens causes silver bromide crystals on the film to form darker elemental silver on the negative.

B Light passing through the negative onto black and white photographic paper causes another reaction that forms the photograph.

A Negative

B Photo (positive image)

Many forms of energy can be used to break bonds. Sometimes the energy is transferred as heat, like the spark that starts the isooctane-oxygen reaction. Energy also can be transferred as electricity, sound, or light, as shown in **Figure 3.** When molecules collide and enough energy is transferred to separate the atoms, bonds can break.

Forming bonds releases energy

Once enough energy is added to start the isooctane-oxygen reaction, new bonds form to make the products, as shown in **Figure 4.** Each carbon dioxide molecule has two oxygen atoms connected to the carbon atom with a double bond. A water molecule is made when two hydrogen atoms each form a single bond with the oxygen atom.

When new bonds form, energy is released. When gasoline burns, energy in the form of heat and light is released as the products of the isooctane-oxygen reaction and other gasoline reactions form. Other chemical reactions can produce electrical energy.

Figure 4

The formation of carbon dioxide and water from isooctane and oxygen produces the energy used to power engines.

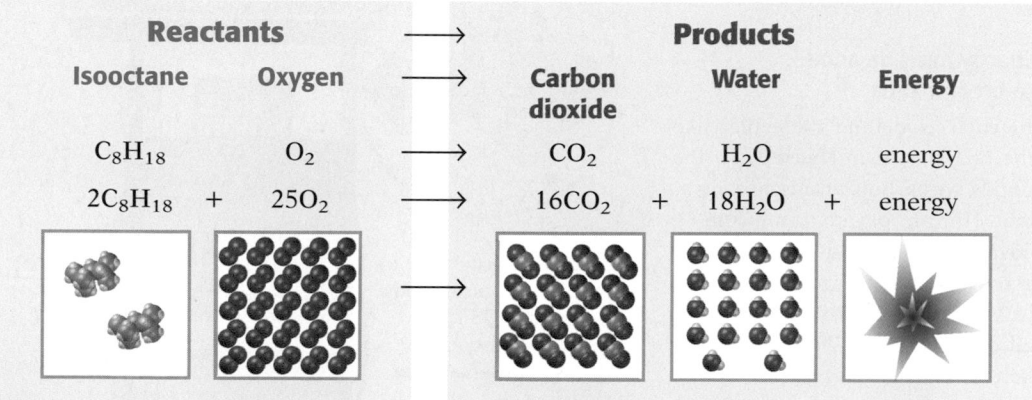

Reactants		→	Products		
Isooctane	Oxygen	→	Carbon dioxide	Water	Energy
C_8H_{18}	O_2	→	CO_2	H_2O	energy
$2C_8H_{18}$ +	$25O_2$	→	$16CO_2$ +	$18H_2O$ +	energy

REAL-LIFE —
CONNECTION

Voltaic Cells The two most common types of voltaic cells are the dry cell, such as a flashlight battery, and the wet cell. The reactants in a dry cell vary depending on what type of dry cell it is. The cell contains a shell of some relatively active metal, such as lithium or zinc, and compounds that will react with it. Wet cells are contained in the lead-storage battery that is used in an automobile. It produces electricity by the reaction involving lead, lead (IV) oxide, and sulfuric acid. Although a single voltaic cell, such as a dry cell, is commonly called a "battery," a battery is actually formed from several cells operating together as they do in an automobile battery.

Energy is conserved in chemical reactions

Energy may not appear to be conserved in the isooctane reaction. After all, a tiny spark can set off an explosion. The energy for that explosion comes from the bonds between atoms in the reactants. Often this stored energy is called **chemical energy.** The total energy of isooctane, oxygen, and their surroundings includes this chemical energy. The total energy before the reaction is equal to the total energy of the products and their surroundings.

Reactions that release energy are exothermic

In the isooctane-oxygen reaction, more energy is released as the products form than is absorbed to break the bonds in the reactants. Like all other combustion reactions, this is an **exothermic reaction.** After an exothermic reaction, the temperature of the surroundings rises because energy is released. The released energy comes from the chemical energy of the reactants.

Reactions that absorb energy are endothermic

If you put hydrated barium hydroxide and ammonium nitrate together in a flask, the reaction between them takes so much energy from the surroundings that water in the air will condense and then freeze on the surface of the flask. This is an **endothermic reaction** —more energy is needed to break the bonds in the reactants than is given off by forming bonds in the products.

internet connect
www.scilinks.org
Topic: Corrosion
SciLinks code: HK4029

SciLINKS Maintained by the National Science Teachers Association

■ **chemical energy** the energy released when a chemical compound reacts to produce new compounds

■ **exothermic reaction** a chemical reaction in which heat is released to the surroundings

■ **endothermic reaction** a chemical reaction that requires heat

REAL WORLD APPLICATIONS

Self-Heating Meals
Corrosion, the process by which a metal reacts with the oxygen in air or water, is not often desirable. However, corrosion is encouraged in self-heating meals so that the energy from the exothermic reaction can be used. Self-heating meals, as the name implies, have their own heat source.

Each meal contains a package of precooked food, a bag that holds a porous pad containing a magnesium-iron alloy, and some salt water. When the salt water is poured into the bag, the salt water soaks through the holes in the pad of metal alloy and begins to corrode the metals vigorously. Then the sealed food package is placed in the bag. The exothermic reaction raises the temperature of the food by 38°C in 14 minutes.

Applying Information
1. List some people for whom self-heating meals would be useful.
2. What other uses can you think of for this self-heating technology?

187

Reading Skills Review the list of opinions developed in the *Opening Activity* of this section. Ask students whether their opinions are the same or have changed. Have them cite passages in the text that account for the change in or reinforcement of their opinions.
LS Verbal

Teaching Tip

Respiration Although the exact chemical reactions are not the reverse, the reactants and products are opposite for the processes of photosynthesis and cellular respiration. Photosynthesis uses carbon dioxide, energy, and water to produce glucose and oxygen. Cellular respiration oxidizes glucose to produce carbon dioxide, energy, and water.

Graphing Have students work in pairs or small groups to use sentences to describe what is happening in the graphs shown in **Figure 5.** (The graph in Figure 5A illustrates the general form for exothermic reactions: energy as heat is released as the reactants form the products. The graph in Figure 5B illustrates the general form for endothermic reactions: energy as heat is absorbed as the reactants form the products.)
LS Interpersonal

Figure 5

Energy must be added to start both exothermic and endothermic reactions.

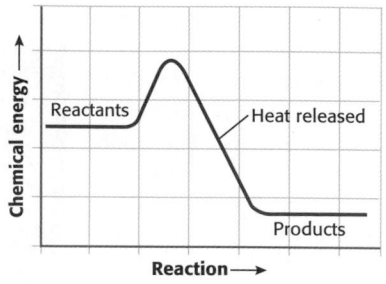

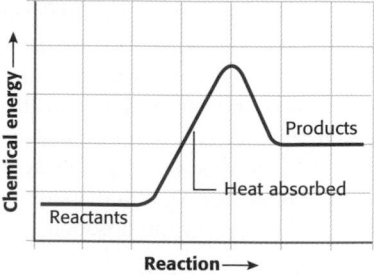

A In an exothermic reaction chemical energy is released, often as heat.

B In an endothermic reaction, energy from the surroundings is stored as chemical energy.

When an endothermic reaction occurs, you may be able to notice a drop in temperature. Some endothermic reactions cannot get enough energy as heat from the surroundings to happen; so energy must be added as heat to cause the reaction to take place. The changes in chemical energy for an exothermic reaction and for an endothermic reaction are shown in **Figure 5.**

Photosynthesis, like many reactions in living things, is endothermic. In photosynthesis, plants use energy from light to convert carbon dioxide and water to glucose and oxygen, as shown in **Figure 6.**

Figure 6

All of the food you eat comes directly or indirectly from the products of photosynthesis.

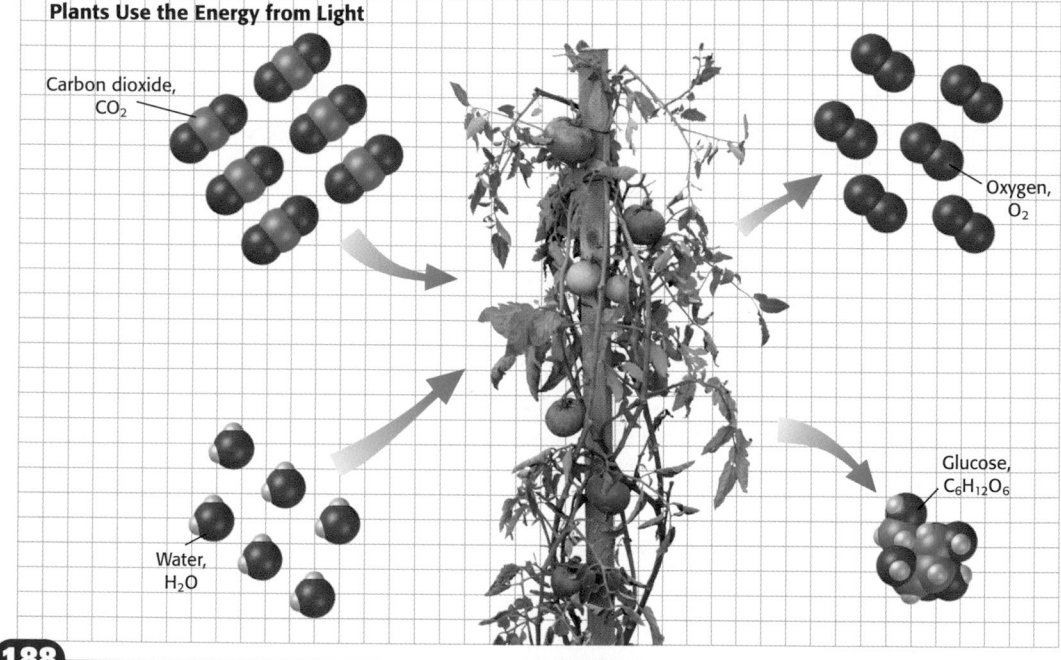

Plants Use the Energy from Light

Carbon dioxide, CO_2

Oxygen, O_2

Water, H_2O

Glucose, $C_6H_{12}O_6$

188

Alternative Assessment — **BASIC**

Concept Map Have students create a concept map that includes all of the key terms introduced in this section. Students should also include information about recognizing chemical reactions and about what happens to atoms and molecules during chemical reactions. Check student maps for completeness and accuracy. **LS** Logical

LIFE SCIENCE
CONNECTION — **ADVANCED**

Have students conduct research to learn about photosynthesis and respiration. Ask them to find out how each process works, how they are related, and why they are important to all living organisms. Have them present their results in a poster that includes the chemical equations and a detailed illustration of each process. **LS** Visual

Sometimes, reactions are described as exergonic or endergonic. These terms refer to the ease with which the reactions occur. In most cases in this book, exergonic reactions are exothermic and endergonic reactions are endothermic. Bioluminescence, shown in **Figure 7**, and respiration are exergonic reactions, and photosynthesis is an endergonic reaction.

Figure 7

A Some living things, such as this firefly, produce light through a chemical process called bioluminescence.

B The comb jelly (*Mnemiopsis leidyi*), shown above, is about 10 cm wide and is native to the Atlantic coast. Comb jellies are not true jellyfish.

INTEGRATING

BIOLOGY

People are charmed by fireflies because these common insects give off light. Scientists have found that fireflies are not alone in this. Some kinds of bacteria, worms, squids, and jellyfish also give off light. This process, called bioluminescence, involves an exothermic reaction made possible by the enzyme luciferase. Scientists can use bacteria that contain luciferase to track the spread of infection in the human body.

INTEGRATING

BIOLOGY You may wish to revisit this feature when students study redox reactions. Bioluminescence is produced from a redox reaction that involves luciferase. Bioluminescence is not limited to animals. For example, a type of mushroom exhibits bioluminescence.

Close

Quiz ─────────── BASIC

1. What is a substance that undergoes a chemical change called? (reactant)

2. What is the substance produced by a chemical change called? (product)

3. What happens to the atoms of a substance that undergoes a chemical reaction? (The atoms are rearranged as the original bonds are broken and new bonds are formed.)

4. What are two kinds of chemical reactions? (exothermic and endothermic)

LS Logical

SECTION 1 REVIEW

SUMMARY

▶ During a chemical reaction, atoms are rearranged.

▶ Signs of a chemical reaction include any of the following: a substance that has different properties than the reactants have; a color change; the formation of a gas or a solid precipitate; or the transfer of energy.

▶ Mass and energy are conserved in chemical reactions.

▶ Energy can be released or absorbed in a chemical reaction.

▶ Energy must be added to the reactants for bonds between atoms to be broken.

1. **Identify** which of the following is a chemical reaction:
   a. melting ice
   b. burning a candle
   c. rubbing a marker on paper
   d. rusting iron

2. **List** three signs that could make you think a chemical reaction might be taking place.

3. **List** four forms of energy that might be absorbed or released during a chemical reaction.

4. **Classify** the following reactions as exothermic or endothermic:
   a. paper burning with a bright flame
   b. plastics becoming brittle after being left in the sun
   c. a firecracker exploding

5. **Predict** which atoms will be found in the products of the following reactions:
   a. mercury(II) oxide, HgO, is heated and decomposes
   b. limestone, $CaCO_3$, reacts with hydrochloric acid, HCl
   c. table sugar, $C_{12}H_{22}O_{11}$, burns in air to form caramel

6. **Critical Thinking** Calcium oxide, CaO, is used in cement mixes. When water is added, heat is released as CaO forms calcium hydroxide, $Ca(OH)_2$. What signs are there that this is a chemical reaction? Which has more chemical energy, the reactants or the products? Explain your answer.

189

Answers to Section 1 Review

1. b, d

2. Students should list three of the following: gas given off, formation of a precipitate, energy change, change of color, change of identity of substances.

3. heat, light, electricity, sound

4. **a.** exothermic
   b. endothermic
   c. exothermic

5. **a.** Hg, O
   b. Ca, C, O, H, Cl
   c. C, H, O

6. Substances change identity, and energy is released. The reactants have more chemical energy because the energy of the reactants equals the energy of the products plus the energy released.

Chapter 6 • Chemical Reactions   189

Reaction Types

▶ **KEY TERMS**

synthesis reaction
decomposition reaction
electrolysis
combustion reaction
single-displacement
    reaction
double-displacement
    reaction
oxidation-reduction
    reaction
radical

▶ **synthesis reaction** a reaction in which two or more substances combine to form a new compound

OBJECTIVES

▶ **Distinguish** among five general types of chemical reactions.

▶ **Predict** the products of some reactions based on the reaction type.

▶ **Describe** reactions that transfer or share electrons between molecules, atoms, or ions.

In the last section, you saw how CO_2 is made from sugar by yeast, how isooctane from gasoline burns, and how photosynthesis happens. These are just a few examples of the many millions of possible reactions.

Classifying Reactions

Even though there are millions of unique substances and many millions of possible reactions, there are only a few general types of reactions. Just as you can follow patterns to name compounds, you also can use patterns to identify the general types of chemical reactions and to predict the products of the chemical reactions.

Synthesis reactions combine substances

Polyethene, a plastic often used to make trash bags and soda bottles, is produced by a **synthesis reaction** called polymerization. In polymerization reactions, many small molecules join together in chains to make larger structures called polymers. Polyethene, shown in *Figure 8,* is a polymer formed of repeating ethene molecules.

Hydrogen gas reacts with oxygen gas to form water. In a synthesis reaction, at least two reactants join to form a product. Synthesis reactions have the following general form.

$$A + B \longrightarrow AB$$

The following is a synthesis reaction in which the metal sodium reacts with chlorine gas to form sodium chloride, or table salt.

$$2Na + Cl_2 \longrightarrow 2NaCl$$

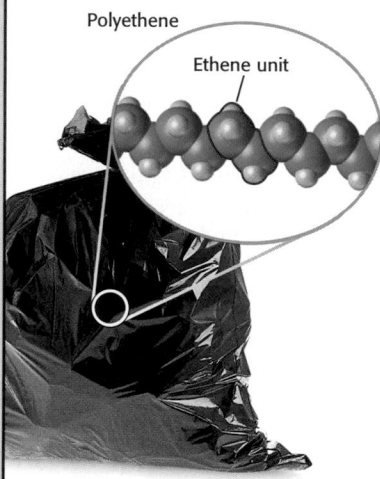
Polyethene
Ethene unit

Figure 8
A molecule of polyethene is made up of as many as 3500 units of ethene.

190

Historical Perspective

Synthetic Polymers In 1869, John Wesley Hyatt (1837–1920), in an attempt to find a substitute for the fragile ivory in billiard balls, discovered the first synthetic polymer. Called celluloid, it is a material that can be molded into different shapes. In 1884, Louis M. H. Berniguad pushed the same substance through tiny holes in a nozzle, making small threads that he called rayon.

REAL-LIFE
CONNECTION

A Popular Plastic Polyethylene, which has the simplest structure of all commercial polymers, is the most popular plastic in the world. It is used in a wide variety of products, including trash bags, grocery bags, soda bottles, shampoo bottles, and children's toys. It is even used to make some bullet-proof vests!

Synthesis reactions always join substances, so the product is a more complex compound than the reactants.

Photosynthesis is another kind of synthesis reaction—the synthesis reaction that goes on in plants. The photosynthesis reaction is shown in *Figure 9.*

Decomposition reactions break substances apart

Digestion is a series of reactions that break down complex foods into simple fuels your body can use. Similarly, in what is known as "cracking" crude oil, large molecules made of carbon and hydrogen are broken down to make gasoline and other fuels. Digestion and "cracking" oil are **decomposition reactions,** reactions in which substances are broken apart. The general form for decomposition reactions is as follows.

$$AB \longrightarrow A + B$$

The following shows the decomposition of water.

$$2H_2O \longrightarrow 2H_2 + O_2$$

The **electrolysis** of water is a simple decomposition reaction—water breaks down into hydrogen gas and oxygen gas when an electric current flows through the water.

Combustion reactions use oxygen as a reactant

Isooctane forms carbon dioxide and water during combustion. Oxygen is a reactant in every **combustion reaction,** so at least one product of such reactions always contains oxygen. Water is a common product of combustion reactions.

If the air supply is limited when a carbon-containing fuel burns, there may not be enough oxygen gas for all the carbon to form carbon dioxide. In that case, some carbon monoxide may form. Carbon monoxide, CO, is a poisonous gas that lowers the ability of the blood to carry oxygen. Carbon monoxide has no color or odor, so you can't tell when it is present. When there is not a good air supply during a combustion reaction, not all fuels are converted completely to carbon dioxide. In some combustion reactions, you can tell if the air supply is limited because the excess carbon is given off as small particles that make a dark, sooty smoke.

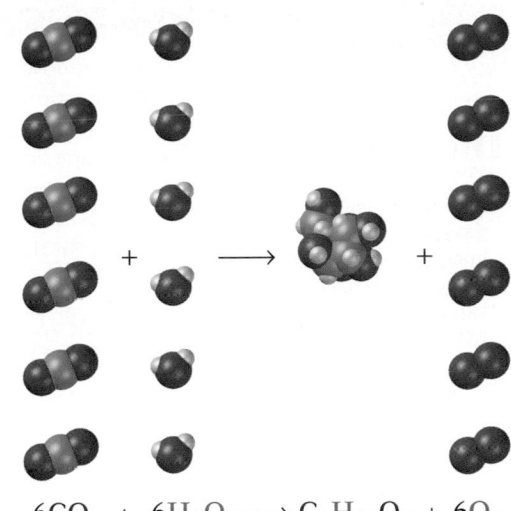

$$6CO_2 \ + \ 6H_2O \longrightarrow C_6H_{12}O_6 + 6O_2$$

Figure 9
Photosynthesis is the synthesis of glucose and oxygen gas from carbon dioxide and water.

internet connect

www.scilinks.org
Topic: Types of Reactions
SciLinks code: HK4142

SCiLINKS. Maintained by the National Science Teachers Association

▶ **decomposition reaction** a reaction in which a single compound breaks down to form two or more simpler substances

▶ **electrolysis** the process in which an electric current is used to produce a chemical reaction, such as the decomposition of water

▶ **combustion reaction** the oxidation reaction of an organic compound, in which heat is released

191

LIFE SCIENCE
CONNECTION

The hemoglobin in red blood cells is the protein that carries oxygen to all parts of the body. The reason that carbon monoxide is so dangerous to humans is that it bonds to hemoglobin more readily than oxygen does. When bonded to carbon monoxide, hemoglobin is incapable of carrying oxygen. Because carbon monoxide is colorless and odorless, it is important for homes to have carbon monoxide detectors.

MISCONCEPTION ///ALERT\\\

Combustion Some students think that burned steel wool will weigh less because iron or trapped air is "used up."

• Review the law of the conservation of mass and ask students to explain how this law applies to combustion.
• Students often see oxygen as required for combustion but do not consider it to be one of the reactants. Emphasize this point in your discussion of combustion reactions.

SKILL BUILDER — BASIC

Reading Skills Have students read the section and then organize what they learn in a concept map. Concept maps should start with the phrase *Reaction types,* and should include titles, descriptions, and examples for each type of reaction. **LS** Verbal

SKILL BUILDER — BASIC

Vocabulary Ask students to describe what the terms *synthesis, decomposition,* and *displacement* mean to them. All are terms that have common usage. Students should associate synthesis with making something, decomposition with things breaking down, and displacement with movement from one place to another. Have students compare their descriptions and discuss any similarities and any differences. **LS** Verbal

Teaching Tip

Polymerization Students might think that all polymers are made by simply combining monomers to form one large molecule. Many polymers, such as polyethylene, are formed that way in what is known as an addition polymerization reaction. When certain other types of polymers form, the reaction forms a small molecule—usually water—as well as the polymer. This type of reaction is condensation polymerization. An example of a condensation polymer is nylon.

Chapter Resource File

• **Lesson Plan**

Transparencies

TT Bellringer

INTEGRATING

EARTH SCIENCE Fossil fuels include not only coal, natural gas, and petroleum but also oil shale, peat, and bitumens. Oil shale is a sedimentary rock that releases hydrocarbons when heated above 500°C. Bitumens are dense, very viscous hydrocarbons similar to asphalt.

Natural gas is the cleanest burning fossil fuel. It is composed primarily of methane. Other fossil fuels do not burn as cleanly as natural gas. Oil and coal compounds pollute by releasing sulfur dioxide, nitrogen oxides, and carbon monoxide into the atmosphere.

Teaching Tip

Combustion Reactions You can show that the last two combustion reactions (on the student page) are incomplete by demonstrating that when the products of each reaction combine with more oxygen gas, carbon dioxide is produced. The product of the first incomplete reaction, carbon monoxide, forms carbon dioxide as follows:

$$2CO + O_2 \rightarrow 2CO_2$$

Similarly, for the second incomplete reaction, carbon (or soot) also forms carbon dioxide when more oxygen is added:

$$2C + 2O_2 \rightarrow 2CO_2$$

Did You Know ?

In the United States, natural gas supplies one-fifth of the energy used. The pipelines that carry this natural gas, if laid end-to-end, would stretch to the moon and back twice.

INTEGRATING

EARTH SCIENCE Compounds containing carbon and hydrogen are often called hydrocarbons. Most hydrocarbon fuels are fossil fuels, that is, compounds that were formed millions of years before dinosaurs existed. When prehistoric organisms died, they decomposed, and many were slowly buried under layers of mud, rock, and sand. During the millions of years that passed, the once-living material formed different fuels, such as oil, natural gas, or coal, depending on the kind of material present, the length of time the material was buried, and the conditions of temperature and pressure that existed when the material was decomposing.

In combustion the products depend on the amount of oxygen

To see how important a good air supply is, look at a series of combustion reactions for methane, CH_4. Because methane has only one carbon atom, it is the simplest carbon-containing fuel. Methane is the primary component in natural gas, the fuel often used in stoves, water heaters, and furnaces.

Methane reacts with oxygen gas to make carbon dioxide and water. In the balanced form of the chemical equation, four molecules of oxygen gas are needed for the combustion of two molecules of methane, as shown below.

$$2CH_4 + 4O_2 \longrightarrow 2CO_2 + 4H_2O$$

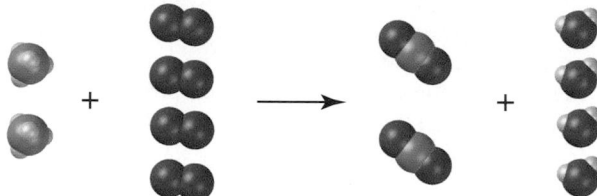

Now look at what happens when less oxygen gas is available. If there are only three molecules of oxygen gas for every two molecules of methane, water and carbon monoxide may form, as shown in the following reaction.

$$2CH_4 + 3O_2 \longrightarrow 2CO + 4H_2O$$

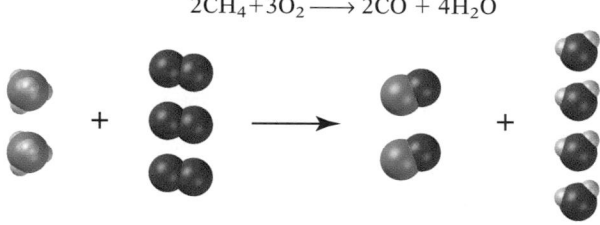

When the air supply is very limited and only two molecules of oxygen gas are available to react with two molecules of methane, water and tiny bits of carbon, or soot, are formed as follows.

$$2CH_4 + 2O_2 \longrightarrow 2C + 4H_2O$$

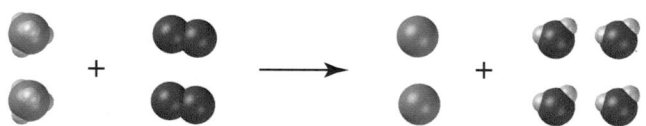

Did You Know ?

Nitrogen-Oxygen Compounds If air is used instead of pure oxygen in the combustion process, nitrogen-oxygen compounds—many of which are poisonous—are produced. Some nitrogen-oxygen compounds are also light sensitive and turn brown when exposed to sunlight, causing the brown smog that appears in some cities.

Historical Perspective —— ADVANCED

Phlogiston Theory Scientists of the early 18th century believed that all combustible substances contained a substance called *phlogiston*. Phlogiston was thought to be released when the substance burned. The phlogiston theory was disproved in the late 18th century by the chemist Antoine Lavoisier, who showed that combustion reactions require oxygen. Have interested students find out more about the theory of phlogiston and the experiments Lavoisier conducted to disprove the theory. Ask them to write a report summarizing the results of their research. **LS** Logical

Fire Extinguishers: Are They All The Same?

A fire is a combustion reaction in progress that is speeded up by high temperatures. Three things are needed for a combustion reaction to occur: a fuel, some oxygen, and an ignition source. If any of these three is absent, combustion cannot occur. So the goal of firefighting is to remove one or more of these parts. Fire extinguishers are effective in firefighting because they separate the fuel from the oxygen supply, which is most commonly air.

Fire extinguishers display codes indicating which types of fires they can put out.

Classes of Fires

A fire is classified by the type of fuel that combusts to produce it. Class A fires involve solid fuels, such as wood and paper. The fuel in a Class B fire is a flammable liquid, like grease, gasoline, or oil. Class C fires involve "live" electric circuits. And Class D fires are fueled by the combustion of flammable metals.

Types of Fire Extinguishers

Different types of fuels require different firefighting methods. Water extinguishers are used on Class A fires, which involve fuels such as most flammable building materials. The steam that is produced helps to displace the air around the fire, preventing the oxygen supply from reaching the fuel.

A Class B fire, in which the fuel is a liquid, is best put out by cold carbon dioxide gas, CO_2. Because carbon dioxide is more dense than air, it forms a layer underneath the air, cutting off the oxygen supply for the combustion reaction.

Class C fires, which involve a "live" electric circuit, can also be extinguished by CO_2. Liquid water cannot be used, or there will be a danger of electric shock. Some Class C fire extinguishers contain a dry chemical that smothers the fire. The dry chemical smothers the fire by reacting with the intermediates that drive the chain reaction that produces the fire. This stops the chain reaction and extinguishes the fire.

Finally, Class D fires, which involve burning metals, cannot be extinguished with CO_2 or water because these compounds may react with some hot metals. For these fires, nonreactive dry powders are used to cover the metal and keep it separate from oxygen. In many cases, the powders used in Class D extinguishers are specific to the type of metal that is burning.

Most fire extinguishers can be used with more than one type of fire. Check the fire extinguishers in your home and school to find out the kinds of fires they are designed to put out.

Your Choice

1. **Making Decisions** Aside from displacing the air supply, how does water or cold CO_2 gas reduce a fire's severity?

2. **Critical Thinking** How is the chain reaction in a Class C fire interrupted by the contents of a dry chemical extinguisher?

internet connect

www.scilinks.org
Topic: Fire Extinguishers SciLinks code: HK4052

SC*LINKS*® Maintained by the
National Science Teachers Association

193

REAL-LIFE CONNECTION

Operating a Fire Extinguisher The acronym PASS (Pull, Aim, Squeeze, Sweep) can be used to remember how to operate a fire extinguisher of any type. The four steps are:

1. Pull the pin at the top. The pin keeps the extinguisher from being pressed by accident.

2. Aim the nozzle toward the fire's base.

3. Squeeze the handle to discharge at a distance of about 8 feet from the fire.

4. Sweep the nozzle back and forth. Keep holding the handle down as you do so.

When the fire appears to be out, check carefully to see if the fire is fully out and watch to be sure it does not re-ignite.

Teaching Tip

Single Displacement Activity refers to how readily an element reacts. As shown on the student page, active elements replace less active elements in compounds. Some students may benefit from a second example of this concept. Tell students that fluorine is more active than chlorine. If fluorine gas, F_2, is bubbled through a sodium chloride, NaCl, solution, the fluorine will replace the chlorine in the compound. The products of the reaction are chlorine, Cl_2, and sodium fluoride, NaF. If chlorine gas is bubbled through a sodium fluoride solution, no reaction occurs because the more active nonmetal is already in the compound.

SKILL BUILDER — BASIC

Interpreting Visuals After students have studied **Figure 10,** have them answer the following: What are the reactants? (aluminum and copper(II) chloride) What are the products? (copper and aluminum chloride) Describe the reaction in words. (The chloride ions in the copper(II) chloride bond with the aluminum atoms, forming aluminum chloride. This leaves the copper atoms free.) **LS Visual**

In single-displacement reactions, elements trade places

Copper(II) chloride dissolves in water to make a bright blue solution. If you add a piece of aluminum foil to the solution, the color fades, and clumps of reddish brown material form. The reddish brown clumps are copper metal. Aluminum replaces copper in the copper(II) chloride, forming aluminum chloride. Aluminum chloride does not make a colored solution, so the blue color fades as the amount of blue copper(II) chloride decreases, as shown in **Figure 10.**

At first, the copper atoms are in the form of copper(II) ions, as part of copper(II) chloride, and the aluminum atoms are in the form of aluminum metal. After the reaction, the aluminum atoms become ions, and the copper atoms become neutral in the copper metal. Because the atoms of one element appear to move into a compound, and atoms of the other element appear to move out, this is called a **single-displacement reaction.** Single-displacement reactions have the following general form.

▶ **single-displacement reaction** a reaction in which one element or radical takes the place of another element or radical in a compound

$$AX + B \longrightarrow BX + A$$

The single-displacement reaction between copper(II) chloride and aluminum is shown as follows.

$$3CuCl_2 + 2Al \longrightarrow 2AlCl_3 + 3Cu$$

Generally, in a single-displacement reaction, a more reactive element will take the place of a less reactive one.

Figure 10
Aluminum undergoes a single-displacement reaction with copper(II) chloride to form copper and aluminum chloride.

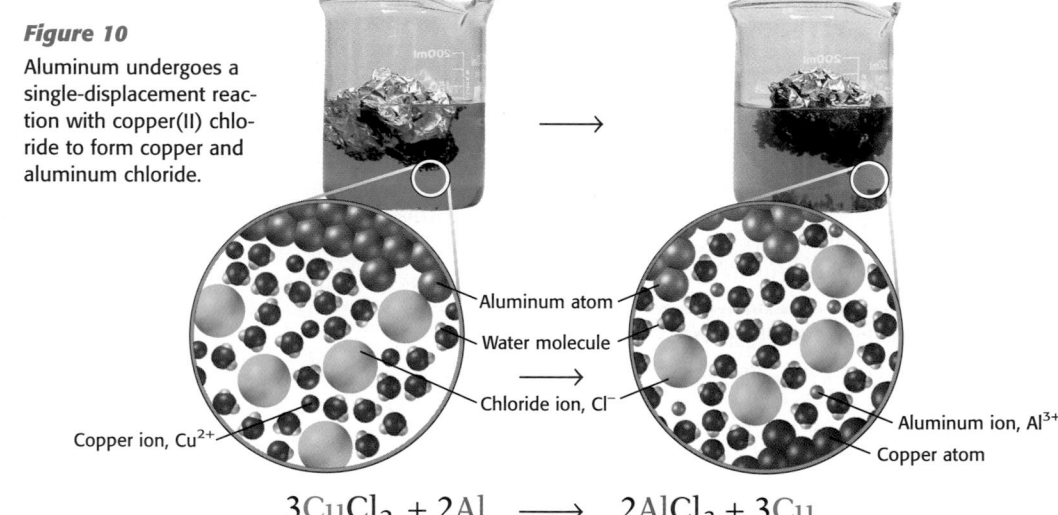

Copper ion, Cu^{2+}
Aluminum atom
Water molecule
Chloride ion, Cl^-
Aluminum ion, Al^{3+}
Copper atom

$$3CuCl_2 + 2Al \longrightarrow 2AlCl_3 + 3Cu$$

194

INCLUSION Strategies

• *English Language Learners*
• *Attention Deficit Disorder*
• *Learning Disabled*

Write the types of chemical reactions on five note cards (one type per card). Write definitions for each of the chemical reactions on five additional note cards. Have students work individually or in pairs to match each type of reaction with its definition. To create a more difficult matching activity, make additional cards with examples of each type of reaction. **LS Logical**

Did You Know ?

Fighting Fire with Sand If a class D fire is not on a vertical surface, an effective way to extinguish the fire is to cover it with sand.

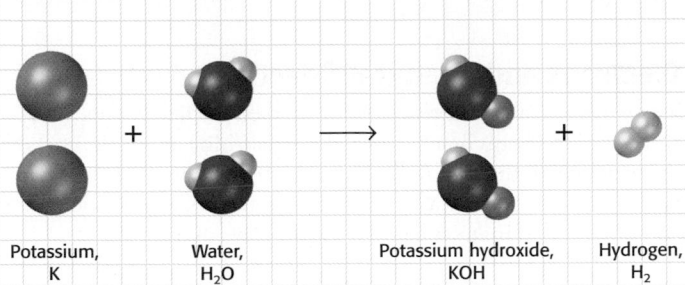

Potassium, K   +   Water, H_2O ⟶   Potassium hydroxide, KOH   +   Hydrogen, H_2

Alkali metals react with water to form ions

Potassium metal is so reactive that it undergoes a single-displacement reaction with water. A potassium ion appears to take the place of one of the hydrogen atoms in the water molecule. Potassium ions, K^+, and hydroxide ions, OH^-, are formed. The hydrogen atoms displaced from the water join to form hydrogen gas, H_2.

The potassium and water reaction, shown in **Figure 11,** is so exothermic that the H_2 may explode and burn instantly. All alkali metals and some other metals undergo single-displacement reactions with water to form hydrogen gas, metal ions, and hydroxide ions.

All of these reactions happen rapidly and give off heat but some alkali metals are more reactive than others. Lithium reacts steadily with water to form lithium ions, hydroxide ions, and hydrogen gas. Sodium and water react vigorously to make sodium ions, hydroxide ions, and hydrogen gas. Rubidium and cesium are so reactive that the hydrogen gas will explode as soon as they are put into water.

In double-displacement reactions, ions appear to be exchanged between compounds

The yellow lines painted on roads are colored with lead chromate, $PbCrO_4$. This compound can be formed by mixing solutions of lead nitrate, $Pb(NO_3)_2$, and potassium chromate, K_2CrO_4. In solution, these compounds form the ions Pb^{2+}, NO_3^-, K^+, and CrO_4^{2-}. When the solutions are mixed, the yellow lead chromate compound that forms doesn't dissolve in water, so it settles to the bottom. A **double-displacement reaction,** such as this one, occurs when two compounds appear to exchange ions. The general form of a double-displacement reaction is as follows.

$$AX + BY \longrightarrow AY + BX$$

The double-displacement reaction that forms lead chromate is as follows.

$$Pb(NO_3)_2 + K_2CrO_4 \longrightarrow PbCrO_4 + 2KNO_3$$

Figure 11
Potassium reacts with water in a single-displacement reaction.

▶ **double-displacement reaction** a reaction in which a gas, a solid precipitate, or a molecular compound forms from the apparent exchange of atoms or ions between two compounds

Teaching Tip
Activity Series In most single-displacement reactions, a more active element replaces a less active one. An *activity series*, which lists substances in order of relative activity, can be used to predict whether a replacement reaction will occur. An activity series lists the most active elements first (or at the top) and the least active ones last (or at the bottom). An activity series of common metals is as follows:

Li, K, Ca, Na, Mg, Al, Mn, Zn, Fe, Ni, Sn, Pb, Cu, Ag, Pt, Au

Each element is able to displace those listed after it from a solution. This series can usually be used to predict whether a single-displacement reaction will occur at room temperature in an aqueous solution, but there are some exceptions.

Teaching Tip
Double-Displacement Reactions One common example of a double-displacement reaction is an acid-base reaction. Acid-base reactions fall under the category of neutralization reactions because the two components neutralize each other's properties, producing a salt and water.

195

Historical Perspective

Pain Relievers From the time of Hippocrates, an ancient Greek healer, people knew that chewing willow bark would relieve pain. Native Americans and the Chinese also used it. But with the pain relief of willow bark came a problem: stomach discomfort. In the 1800s, the active ingredient in the bark was isolated. It is salicylic acid, which causes stomach discomfort because of its acidity. Felix Hoffman, a German chemist, used a displacement reaction to make the acid less acidic without destroying its pain-relieving properties. He replaced a hydrogen atom on the acid molecule with an acetyl group, $-OOCCH_3$. The result—acetylsalicylic acid—relieves pain with fewer stomach problems.

    Transparencies

TT Single-Displacement

Different metals corrode differently. Although the green coating, or *patina*, on the Statue of Liberty was caused by corrosion, the coating helped protect the copper from further corrosion. However, the rust formed on the iron framework did not protect it. The rust crumbled away, and the exposed iron corroded further. When repairing the framework of the statue, materials resistant to corrosion were used. Two materials were chosen—an extremely strong iron-aluminum alloy called *ferallium* and a more flexible stainless steel alloy made from iron, chromium, and nickel. Sealants were also used to protect the structure from the environment and to separate the copper and iron-based parts of the statue.

Answers

1. The presence of water speeds up corrosion.

2. They have the most surface area in contact with corroding agents.

Teaching Tip

Mnemonic Devices An easy mnemonic device for associating oxidation and reduction with electron transfer is **OIL RIG: O**xidation **I**s the **L**oss of electrons, **R**eduction **I**s the **G**ain of electrons. Some students may prefer **LEO** says **GER: L**oss of **E**lectrons is **O**xidation, **G**ain of **E**lectrons is **R**eduction.

▶ **oxidation-reduction reaction** any chemical change in which one species is oxidized (loses electrons) and another species is reduced (gains electrons); also called *redox reaction*

▶ **radical** an organic group that has one or more electrons available for bonding

Metal sculptures often corrode because of redox reactions. The Statue of Liberty, which is covered with 200 000 pounds of copper, was as bright as a new penny when it was erected. However, after more than 100 years, the statue had turned green. The copper reacted with the damp air of New York harbor. More importantly, oxidation reactions between the damp, salty air and the internal iron supports made the structure dangerously weak. The statue was closed for several years in the 1980s while the supports were cleaned and repaired.

Making the Connection

1. Metal artwork in fountains often rusts very quickly. Suggest a reason for this.

2. Why do you think the most detailed parts of a sculpture are the first to appear worn away?

Electrons and Chemical Reactions

The general classes of reactions described earlier in this section were used by early chemists, who knew nothing about the parts of the atom. With the discovery of the electron and its role in chemical bonding, another way to classify reactions was developed. We can understand many reactions as transfers of electrons.

Electrons are transferred in redox reactions

The following **oxidation-reduction reaction** is an example of electron transfer. When the metal iron reacts with oxygen to form rust, Fe_2O_3, each iron atom loses three electrons to form Fe^{3+} ions, and each oxygen atom gains two electrons to form the O^{2-} ions.

Substances that accept electrons are said to be *reduced*; substances that give up electrons are said to be *oxidized*. One way to remember this is that the gain of electrons will reduce the positive charge on an ion or will make an uncharged atom a negative ion. Reduction and oxidation are linked. In all redox reactions, one or more reactants is reduced and one or more is oxidized.

Some redox reactions do not involve ions. In these reactions, oxidation is a gain of oxygen or a loss of hydrogen, and reduction is the loss of oxygen or the gain of hydrogen. Respiration and combustion are redox reactions because oxygen gas reacts with carbon compounds to form carbon dioxide. Carbon atoms in CO_2 are oxidized, and oxygen atoms in O_2 are reduced.

Radicals have electrons available for bonding

Many synthetic fibers, as well as plastic bags and wraps, are made by polymerization reactions, as you have already learned. Polymerization reactions can occur when **radicals** are formed.

When a covalent bond is broken such that at least one unpaired electron is left on each fragment of the molecule, these fragments are called radicals. Because an uncharged hydrogen atom has one electron available for bonding, it is a radical. Radicals react quickly to form covalent bonds with other substances, making new compounds. Often, when you see chemical radicals mentioned in the newspaper or hear about them on the radio or television, they are called free radicals.

Alternative Assessment ── BASIC

Oxidation and Reduction Have students use a number line to explain why gaining electrons (adding a negative charge) during reduction results in a decrease in positive charge. Have them also explain why losing electrons (subtracting negative charge) during oxidation results in an increase in positive charge. Remind students that free elements have zero charge, and provide them with several examples of changes of charge. For each, have them identify the change as oxidation or reduction. Examples of oxidation include Fe to Fe^{2+}, Cr^{2+} to Cr^{3+}, and Cl^- to Cl. Examples of reduction include F to F^-, Pb^{4+} to Pb^{2+}, and Mg^{2+} to Mg. After mastering the concepts of oxidation and reduction, have groups of students create posters that use a number line to explain oxidation and reduction along with examples of each. **LS** Logical

Radicals are part of many everyday reactions besides the making of polymers, such as those shown in *Figure 12.* Radicals can also be formed when coal and oil are processed or burned. The explosive combustion of rocket fuel is another reaction involving the formation of radicals.

Figure 12
Radical reactions are used to make polystyrene. Polystyrene foam is often used to insulate or to protect things that can break.

SUMMARY

▸ Synthesis reactions make larger molecules.

▸ Decomposition breaks compounds apart.

▸ In combustion, substances react with oxygen.

▸ Elements appear to trade places in single-displacement reactions.

▸ In double-displacement reactions, ions appear to move between compounds, resulting in a solid that settles out of solution, a gas that bubbles out of solution, and/or a molecular substance.

▸ In redox reactions, electrons transfer from one substance to another.

1. **Classify** each of the following reactions by type:
   a. $S_8 + 8O_2 \longrightarrow 8SO_2 + heat$
   b. $6CO_2 + 6H_2O \longrightarrow C_6H_{12}O_6 + 6O_2$
   c. $2NaHCO_3 \longrightarrow Na_2CO_3 + H_2O + CO_2$
   d. $Zn + 2HCl \longrightarrow ZnCl_2 + H_2$

2. **Identify** which element is oxidized and which element is reduced in the following reaction.
$$Zn + CuSO_4 \longrightarrow ZnSO_4 + Cu$$

3. **Define** *radical.*

4. **Compare and Contrast** single-displacement and double-displacement reactions based on the number of reactants. Use the terms *compound, atom* or *element,* and *ion.*

5. **Explain** why charcoal grills or charcoal fires should never be used for heating inside a house. (**Hint:** Doors and windows are closed when it is cold, so there is little fresh air.)

6. **Contrast** synthesis and decomposition reactions.

7. **List** three possible results of a double-displacement reaction.

8. **Creative Thinking** Would you expect larger or smaller molecules to be components of a more viscous liquid? Which is likely to be more viscous, crude oil or oil after cracking?

197

Teaching Tip

Free Radicals Because they are so reactive, free radicals can be harmful. Ultraviolet radiation from the sun breaks chlorofluorocarbons (CFCs) from certain refrigerants down, forming a chlorine free radical. Each of these free radicals can destroy thousands of ozone molecules.

Close

Quiz ────────── **BASIC**

Determine which type of reaction each general equation describes.

1. $XA + B \rightarrow BA + X$ (single-displacement reaction)

2. $AB \rightarrow A + B$ (decomposition reaction)

3. $A + B \rightarrow AB$ (synthesis reaction)

4. $AX + BY \rightarrow AY + BX$ (double-displacement reaction)

LS Logical

Answers to Section 2 Review

1. **a.** synthesis
   b. synthesis
   c. decomposition
   d. single-displacement

2. Zinc is oxidized, and copper is reduced.

3. a fragment of a molecule that has at least one electron available for bonding

4. In single-displacement reactions, atoms of one element replace the atoms of another element in a compound. In a double-displacement reaction, positive ions in two compounds trade places, forming two different compounds.

5. Because oxygen might be limited, carbon monoxide might form instead of carbon dioxide. Carbon monoxide is harmful to animals, including humans.

6. In synthesis reactions, two or more substances combine to form a more complex substance. Decomposition is the opposite process.

7. formation of a gas; a precipitate; or a covalent molecule, such as water

8. You would expect larger molecules in a more viscous liquid, so crude oil would be more viscous. Cracking breaks large molecules down into smaller ones.

Chapter Resource File

• **Cross-Disciplinary Worksheet**
  Connection to Fine Arts—The Chemistry of Art **ADVANCED**

• **Concept Review** **GENERAL**

• **Quiz** **BASIC**

SECTION
3

Focus

Overview

Before beginning this section, review with your students the Objectives listed in the Student Edition. In this section, students learn how to balance and interpret chemical equations. The section also covers the law of definite proportions and mole ratios.

🔔 Bellringer

Use the Bellringer transparency to prepare students for this section.

Motivate

Opening Activity — GENERAL

For this activity, you will need 20 index cards and tape or thumbtacks. Divide the index cards into 4 groups of 5 cards. Write the term center on the cards of the first group, guard on the cards of the second group, forward on the cards of the third group, and basketball team on the cards of the fourth group. Fasten the cards on the chalkboard or a bulletin board. Have volunteers create as many basketball teams as they can from the cards. Ask students the following: Can the leftover cards be used? Explain. (No; there are not enough of all the positions to form another team.) Write an equation that shows the formation of a basketball team. (2 forwards + 1 center + 2 guards = 1 basketball team)
LS Interpersonal

▶ **KEY TERMS**
chemical equation
mole ratio

▶ **chemical equation** a representation of a chemical reaction that uses symbols to show the relationship between the reactants and the products

Figure 13

A A methane flame is used to polish the edges of these glass plates.

Balancing Chemical Equations

OBJECTIVES

▶ **Demonstrate** how to balance chemical equations.
▶ **Interpret** chemical equations to determine the relative number of moles of reactants needed and moles of products formed.
▶ **Explain** how the law of definite proportions allows for predictions about reaction amounts.
▶ **Identify** mole ratios in a balanced chemical equation.
▶ **Calculate** the relative masses of reactants and products from a chemical equation.

You may have seen a combustion reaction in the lab or at home if you have a gas stove. When natural gas burns, methane, the main component, reacts with oxygen gas to form carbon dioxide and water. Energy is also released as heat and light, as shown in *Figure 13A*.

Describing Reactions

You can describe this reaction in many ways. You could take a photograph or make a videotape. One way to record the products and reactants of this reaction is to write a word equation.

$$\text{methane} + \text{oxygen} \longrightarrow \text{carbon dioxide} + \text{water}$$

Chemical equations summarize reactions

In Section 1, you learned that all chemical reactions are rearrangements of atoms. This is shown clearly in *Figure 13B*. A better way to write the methane combustion reaction is as a **chemical equation,** using the formulas for each substance.

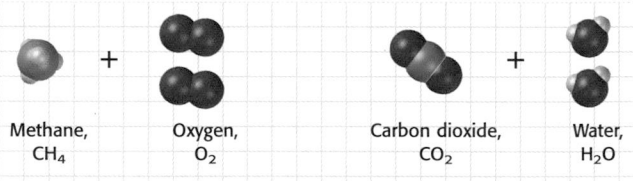

Methane, CH_4 Oxygen, O_2 Carbon dioxide, CO_2 Water, H_2O

B Methane burns with oxygen gas to make carbon dioxide and water.

198

PHYSICAL EDUCATION
CONNECTION — **BASIC**

Work with a physical education teacher to emphasize what balance is. Have students walk on a narrow board placed on the floor. Students should notice that they have the greatest balance when their arms are held out the same amount on both sides of their bodies. Ask the physical education teacher to have students do other activities that involve balance. **LS Kinesthetic**

🔵 INCLUSION Strategies

• *Gifted and Talented*

Arrange for a student to interview a chemical engineer. The interview should include questions regarding the educational requirements to become a chemical engineer, the different types of work involved in chemical engineering, the different types of companies or agencies that hire chemical engineers, and a list of local schools that offer a chemical engineer degree program. The student may present findings to the class and submit a written report. **LS Interpersonal**

Unbalanced Chemical Equation

$CH_4 + O_2$	$\longrightarrow$	$CO_2 + H_2O$
reactants	"give" or "yield"	products

In a chemical equation, such as the one above, the reactants, which are on the left-hand side of the arrow, form the products, which are on the right-hand side. When chemical equations are written, $\longrightarrow$ means "gives" or "yields." People all over the world write chemical equations the same way, as shown in **Figure 14.**

Balanced chemical equations account for the conservation of mass

The chemical equation shown above can be made more useful. As written, it does not tell you anything about the amount of the products that will be formed from burning a given amount of methane. When the number of atoms of each element on the right-hand side of the equation matches the number of atoms of each element on the left, then the chemical equation is said to be *balanced*. A balanced chemical equation is the standard way of writing equations for chemical reactions because it follows the law of conservation of mass.

How to balance chemical equations

In the previous equation, the number of atoms on each side of the arrow did not match for all of the elements in the equation. Carbon is balanced because one carbon atom is on each side of the equation. However, four hydrogen atoms are on the left, and only two are on the right. Also, two oxygen atoms are on the left, and three are on the right. This can't be correct because atoms can't be created or destroyed in a chemical reaction.

Remember that you cannot balance an equation by changing the chemical formulas. You have to leave the subscripts in the formulas alone. Changing the formulas would mean that different substances were in the reaction. An equation can be balanced only by putting numbers, called coefficients, in front of the chemical formulas.

Because there is a total of four hydrogen atoms in the reactants, a total of four hydrogen atoms must be in the products. Instead of a single water molecule, this reaction makes two water molecules to account for all four hydrogen atoms. To show that two water molecules are formed, a coefficient of 2 is placed in front of the formula for water.

$$CH_4 + O_2 \longrightarrow CO_2 + 2H_2O$$

Figure 14
This student is giving a talk on reactions that use copper. You can read the chemical equations even if you can't read Japanese.

By the fourteenth century, Europe had surpassed China's development of fireworks, or pyrotechnics. During the Renaissance, an Italian school emphasized fireworks for entertainment, and a German school promoted scientific advancement in pyrotechnics. As fireworks became more popular, improper manufacture and use resulted in injuries. As a result, the use of fireworks is limited in many countries.

Answers

1. Fort McHenry, located in Baltimore Harbor, protected the city in the past from invasion by sea. In 1814, the British were defeated as their leader was killed during an attempt to take the fort by land.

After the battle, the sight of a waving American Flag (a sign of victory) inspired Frances Scott Key to write a poem, which he titled *Defense of Fort McHenry.* The poem was later set to music, and the result was the song we now call the "Star Spangled Banner."

Although the song was immediately popular, it did not officially became our national anthem until 1931, 116 years after Key wrote the lyrics.

Connection to
SOCIAL STUDIES

No one can be sure when fireworks were first used. When the Mongols attacked China in 1232, the defenders used "arrows of flying fire," which some historians think were rockets fired by gunpowder. The Arabs probably used rockets when they invaded the Spanish peninsula in 1249. For hundreds of years, the main use of rockets was to add terror and confusion to battles. In the late 1700s, rockets were used with some success against the British in India. Because of this, Sir William Congreve began to design rockets for England. Congreve's rockets were designed to explode in the air or be fired along the ground.

Making the Connection

British forces used Congreve's rockets during the War of 1812. Research the battle of Fort McHenry. Find out what happened, who won the battle, and what lyrics the rockets inspired.

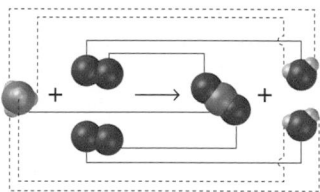

Disc One, Module 5:
Chemical Equations
Use the Interactive Tutor to learn more about this topic.

Figure 15

Magnesium in these fireworks gives off energy as heat and light when it burns to form magnesium oxide.

200

Next look at the oxygen. There is a total of four oxygen atoms in the products. Two are in the CO_2, and each water molecule contains one oxygen atom. To get four oxygen atoms on the left side of the equation, two oxygen molecules must react. That would account for all four oxygen atoms.

> **Balanced Chemical Equation**
>
> $$CH_4 + 2O_2 \longrightarrow CO_2 + 2H_2O$$

Now the numbers of atoms for each element are the same on each side, and the equation is balanced, as shown below.

Information from a balanced equation

You can learn a lot from a balanced equation. In our example, you can tell that each molecule of methane requires two oxygen molecules to react. Each methane molecule that burns forms one molecule of carbon dioxide and two molecules of water. Balanced chemical equations are the standard way chemists write about reactions to describe both the substances in the reaction and the amounts involved.

If you know the formulas of the reactants and products in a reaction, like the one shown in *Figure 15,* you can always write a balanced equation, as shown on the following pages.

Alternative Assessment ── BASIC

Balanced Chemical Equations Have students create a concept map that shows the steps necessary to write a balanced chemical equation. The map should include writing the formula for each reactant and product, using plus signs and an arrow where needed, and using coefficients to balance the number of each type of atom on each side of the equation. Maps might also include an initial step of writing a word equation or a final step of double-checking the equation for balance.
LS Logical

Did You Know?

Firework Colors Before the 1800s, fireworks displays were limited to orange and yellow, produced with steels and charcoals. In the late 1700s, the discovery of chlorates made two new firework colors possible: red and green. It was not until the 20th century that stable formulas were developed for white, blue, and purple fireworks. Although our modern firework displays are very colorful, they hardly ever contain green or turquoise; a simple, effective production method for the color green eludes the fireworks industry to this day.

Balancing Chemical Equations Write the equation that describes the burning of magnesium in air to form magnesium oxide.

1 **Identify the reactants and products.**

Magnesium and oxygen gas are the reactants that form the product, magnesium oxide.

2 **Write a word equation for the reaction.**

$$\text{magnesium} + \text{oxygen} \longrightarrow \text{magnesium oxide}$$

3 **Write the equation using formulas for the elements and compounds in the word equation.**

Remember that some gaseous elements, like oxygen, are molecules, not atoms. Oxygen in air is O_2, not O.

$$Mg + O_2 \longrightarrow MgO$$

4 **Balance the equation one element at a time.**

The same number of each kind of atom must appear on both sides. So far, there is one atom of magnesium on each side of the equation.

Atom	Reactants	Products	Balanced?
Mg	1	1	✔
O	2	1	✘

But there are two oxygen atoms on the left and only one on the right. To balance the number of oxygen atoms, you need to double the amount of magnesium oxide:

$$Mg + O_2 \longrightarrow 2MgO$$

Atom	Reactants	Products	Balanced?
Mg	1	2	✘
O	2	2	✔

This equation gives you two magnesium atoms on the right and only one on the left. So you need to double the amount of magnesium on the left, as follows.

$$2Mg + O_2 \longrightarrow 2MgO$$

Atom	Reactants	Products	Balanced?
Mg	2	2	✔
O	2	2	✔

Now the equation is balanced. It has an equal number of each type of atom on both sides.

Practice **HINT**

▶ Sometimes changing the coefficients to balance one element may cause another element in the equation to become unbalanced. So always check your work.

201

Teaching Tip

Real Equations If you or your students write equations to balance, be sure the reactions actually occur. Although an equation for a reaction that does not occur can be balanced, it can be confusing to students. If necessary, refer to a high school or college chemistry text for practice equations to balance.

Math Skills

Additional Examples

Balancing Chemical Equations

Write balanced equations for each of the following:

1. Potassium metal reacts violently with water to produce potassium hydroxide and hydrogen gas. ($2K + 2H_2O \rightarrow 2KOH + H_2$)

2. Calcium oxide, an ingredient in cement, combines with water to produce calcium hydroxide. ($CaO + H_2O \rightarrow Ca(OH)_2$)

3. Ethanol, C_2H_5OH, produces water and carbon dioxide when it burns. ($C_2H_5OH + 3O_2 \rightarrow 2CO_2 + 3H_2O$)

LS Logical

Alternative Assessment ———————— **BASIC**

Writing **Coefficients** Have students write a paragraph explaining why the sum of the coefficients on one side of a balanced equation doesn't necessarily equal the sum of the coefficients on the other side of the equation. Paragraphs should include the idea that the number of each type of atom must be the same on both sides of the equation. That number is determined by the formulas, including subscripts, not just coefficients. **LS** Logical

Chapter Resource File

• **Cross-Disciplinary Worksheet** Connection to Social Studies— Fireworks **ADVANCED**

• **Math Skills** Balancing Chemical Equations **GENERAL**

1. $3CuSO_4 + 2Al \rightarrow Al_2(SO_4)_3 + 3Cu$

2. $Na_2S + 2AgNO_3 \rightarrow 2NaNO_3 + Ag_2S$

3. $2H_2O_2 \rightarrow 2H_2O + O_2$

LS Logical

Teaching Tip

Balancing Equations Remind students to watch for elements contained in more than one reactant or more than one product. For example, H is contained in both reactants in the reaction $NH_3 + HCl \rightarrow NH_4Cl$. Both the hydrogen locations must be considered in balancing the equation.

Tell students that polyatomic ions in reactants that remain the same in the products can be balanced as units. For example, in $2Fe + 3H_2SO_4 \rightarrow Fe_2(SO_4)_3 + 3H_2$, the sulfate ion is on both sides. It can be balanced as sulfate ions instead of as oxygen and sulfur atoms. If the ion is changed at all, it cannot be considered a unit.

Practice

Balancing Chemical Equations

1. Copper(II) sulfate, $CuSO_4$, and aluminum react to form aluminum sulfate, $Al_2(SO_4)_3$, and copper. Write the balanced equation for this single-displacement reaction.

2. In a double-displacement reaction, sodium sulfide, Na_2S, reacts with silver nitrate, $AgNO_3$, to form sodium nitrate, $NaNO_3$, and silver sulfide, Ag_2S. Balance this equation.

3. Hydrogen peroxide, H_2O_2, is sometimes used as a bleach or as a disinfectant. Hydrogen peroxide decomposes to give water and molecular oxygen. Write a balanced equation for the decomposition reaction.

Determining Mole Ratios

Look at the reaction of magnesium with oxygen to form magnesium oxide.

$$\text{magnesium} + \text{oxygen} \longrightarrow \text{magnesium oxide}$$

$$2Mg + O_2 \longrightarrow 2MgO$$

The single molecule of oxygen in the equation might be shown as $1O_2$. However, a coefficient of 1 is never written.

Balanced equations show the conservation of mass

Other ways of looking at the amounts in the reaction are shown in *Figure 16*. Notice that there are equal numbers of magnesium and oxygen atoms in the product and in the reactants. The total mass of the reactants is always the same as the total mass of the products.

Figure 16   **Information from the Balanced Equation: $2Mg + O_2 \longrightarrow 2MgO$**

Equation:	2Mg	+	O_2	$\longrightarrow$	2MgO
Amount (mol)	2		1	$\longrightarrow$	2
Molecules	$(6.02 \times 10^{23}) \times 2$		$(6.02 \times 10^{23}) \times 1$	$\longrightarrow$	$(6.02 \times 10^{23}) \times 2$
Mass (g)	24.3 g/mol $\times$ 2 mol		32.0 g/mol $\times$ 1 mol	$\longrightarrow$	40.3 g/mol $\times$ 2 mol
Total mass (g)	48.6		32.0	$\longrightarrow$	80.6
Model	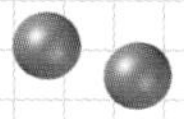		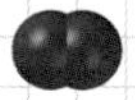	$\longrightarrow$	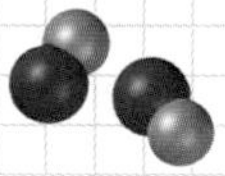

202

MATHEMATICS
CONNECTION — **BASIC**

Review with students how to write and use a ratio, which is a comparison of two numbers. Then have them create a proportion by setting two ratios equal to each other. Show them how to use a proportion to find the value of a variable in the proportion if the other three numbers are known, as in $x/4 = 9/12$; $x = 3$. **LS** Logical

The law of definite proportions

What if you want 4 mol of magnesium to react completely? If you have twice as much magnesium as the balanced equation calls for, you will need twice as much oxygen. Twice as much magnesium oxide will be formed. No matter what amounts of magnesium and oxygen are combined or how the magnesium oxide is made, the balanced equation does not change. This follows the law of definite proportions, which states:

A compound always contains the same elements in the same
proportions, regardless of how the compound is made or
how much of the compound is formed.

Mole ratios can be derived from balanced equations

Whether the magnesium-oxygen reaction starts with 2 mol or 4 mol of magnesium, the proportions remain the same. One way to understand this is to look at the **mole ratios** from the balanced equation. For 2 mol of magnesium and 1 mol of oxygen, the ratio is 2:1. If 4 mol of magnesium is present, 2 mol of oxygen is needed to react. The ratio is 4:2, which reduces to 2:1.

The mole ratio for any reaction comes from the balanced chemical equation. For example, in the following equation for the electrolysis of water, the mole ratio for $H_2O:H_2:O_2$, using the coefficients, is 2:2:1.

$$2H_2O \longrightarrow 2H_2 + O_2$$

As you can see in **Figure 17**, the hydrogen gas produced occupies twice the volume of the oxygen gas. That is because there are twice as many molecules of hydrogen gas produced in electrolysis as there are molecules of oxygen gas.

Mole ratios allow you to calculate the mass of the reactants

If you know the mole ratios of the substances involved in a reaction, you can determine the relative masses of the substances required to react completely.

The most convenient way to determine the relative masses is by multiplying the molecular mass of each substance by the mole ratio from the balanced equation. For example, for the reaction shown in **Figure 16,** the atomic mass of magnesium, 24.3 g/mol, is multiplied by 2 to get a total mass of 48.6 g. The mass of molecular oxygen, 32.0 g/mol, is multiplied by 1. This means that in order for magnesium to react completely with oxygen, there must be 32 g of oxygen available for every 48.6 g of magnesium.

▶ **mole ratio** the relative number of moles of the substances required to produce a given amount of product in a chemical reaction

Figure 17
Electrical energy causes the decomposition of water into oxygen (in the test tube on the left) and hydrogen (on the right).

203

Teacher Resources

For the New Teacher You may need to explain to students the reason why coefficients in a balanced equation can represent either single units or moles. To illustrate this concept, you can write a simple balanced chemical equation on the chalkboard. Multiply all coefficients by a small integer, such as 3. Show that even though the coefficients are not as simple as they could be, the equation is still balanced.

Using multiple examples can help students who are having difficulty understanding a new concept. Repeat this demonstration several times, using different integers. Emphasize that a chemical equation is like a mathematical equation in that you can multiply through the entire equation by the same number without changing the relationship among the formulas. For coefficients to represent moles, the number used to multiply the coefficients in a balanced equation is Avogadro's constant.

Answers

1. the formation of a precipitate
2. reactants: NaCl, $AgNO_3$
   products: AgCl, $NaNO_3$;
   reactants: KBr, $AgNO_3$
   products: AgBr, KNO_3;

See "Continuation of Answers" at the end of the chapter.

Close

Quiz ——— BASIC

1. Which can you adjust to balance a chemical equation, coefficients or subscripts? (coefficients)

2. Any amount of a compound always contains the same elements in the same proportions. True or false? (true)

3. What can you use to describe the relative amounts of two or more substances in a chemical reaction? (a mole ratio)

4. Determine whether each equation is balanced or unbalanced.

   a. $3P_4 + 10KClO_3 \rightarrow 10KCl + 6P_2O_5$ (balanced)

   b. $H_2SO_4 + Al \rightarrow Al_2(SO_4)_3 + 3H_2$ (unbalanced)

   LS Logical

Chapter Resource File

- **Quick Lab Datasheet** Can You Determine the Products of a Reaction? GENERAL
- **Concept Review** GENERAL
- **Quiz** BASIC

Quick **Lab**

Can you determine the products of a reaction?

Materials
- ✓ 7 test tubes
- ✓ test-tube rack
- ✓ labels or wax pencil
- ✓ 10 mL graduated cylinder
- ✓ bottles of the following solutions: sodium chloride, NaCl; potassium bromide, KBr; potassium iodide, KI; and silver nitrate, $AgNO_3$

SAFETY CAUTION Wear safety goggles and an apron. Silver nitrate will stain your skin and clothes.

1. Label three test tubes, one each for NaCl, KBr, and KI.
2. Using the graduated cylinder, measure 5 mL of each solution into the properly labeled test tube. Rinse the graduated cylinder between each use.
3. Add 1 mL of $AgNO_3$ solution to each of the test tubes. Record your observations.

Analysis

1. What did you observe as a sign that a double-displacement reaction was occurring?
2. Identify the reactants and products for each reaction.
3. Write the balanced equation for each reaction.
4. Which ion(s) produced a solid with silver nitrate?
5. Does this test let you identify all the ions? Why or why not?

SECTION 3 REVIEW

SUMMARY

▸ A chemical equation shows the reactants that combine and the products that result from the reaction.

▸ Balanced chemical equations show the proportions of reactants and products needed for the mass to be conserved.

▸ A compound always contains the same elements in the same proportions, regardless of how the compound is made or how much of the compound is formed.

▸ A mole ratio relates the amounts of any two or more substances involved in a chemical reaction.

1. **Identify** which of the following is a complete and balanced chemical equation:
   a. $H_2O \longrightarrow H_2 + O_2$ **c.** $Fe + S \longrightarrow FeS$
   b. $NaCl + H_2O$ **d.** $CaCO_3$

2. **Balance** the following equations:
   a. $KOH + HCl \longrightarrow KCl + H_2O$
   b. $Pb(NO_3)_2 + KI \longrightarrow KNO_3 + PbI_2$
   c. $NaHCO_3 \longrightarrow H_2O + CO_2 + Na_2CO_3$
   d. $NaCl + H_2SO_4 \longrightarrow Na_2SO_4 + HCl$

3. **Explain** why the numbers in front of chemical formulas, not the subscripts, must be changed to balance an equation.

4. **Describe** the information needed to calculate the mass of a reactant or product for the following balanced equation:

   $$FeS + 2HCl \longrightarrow H_2S + FeCl_2$$

5. **Critical Thinking** Ammonia is manufactured by the Haber process in the reaction shown below:

   $$N_2 + 3H_2 \rightleftharpoons 2NH_3 + heat$$

   This involves the reaction of nitrogen with hydrogen. What mass of nitrogen is needed to make 34 g of ammonia?

Answers to Section 3 Review

1. The complete and balanced equation is (c).
2. **a.** $KOH + HCl \rightarrow KCl + H_2O$
   b. $Pb(NO_3)_2 + 2KI \rightarrow 2KNO_3 + PbI_2$
   c. $2NaHCO_3 \rightarrow H_2O + CO_2 + Na_2CO_3$
   d. $2NaCl + H_2SO_4 \rightarrow Na_2SO_4 + 2HCl$
3. Changing subscripts changes the identities of the substances in the reaction. Changing coefficients does not change the identity of a substance, just the number of units of it.
4. You need to know the molar mass of the substance and the mass and molar mass of one other substance in the reaction.

5. molar mass of $N_2 = 2 \cdot 14$ g/mol $= 28$ g/mol; molar mass of NH3 $= 14$ g/mol $+ (31$ g/mol$) = 17$ g/mol

   $$\frac{mass_1}{coefficient_1 \times molar\ mass_1} =$$

   $$\frac{mass_2}{coefficient_2 \times molar\ mass_2}$$

   $$\frac{x}{1\ mol \times 28\ g/mol} = \frac{34\ g}{2\ mol \times 17\ g/mol}$$

   $$x = 28\ g$$

Rates of Change

OBJECTIVES

▶ **Describe** the factors affecting reaction rates.

▶ **Explain** the effect a catalyst has on a chemical reaction.

▶ **Explain** chemical equilibrium in terms of equal forward and reverse reaction rates.

▶ **Apply** Le Châtelier's principle to predict the effect of changes in concentration, temperature, and pressure in an equilibrium process.

■ **KEY TERMS**

catalyst
enzyme
substrate
chemical equilibrium

Chemical reactions can occur at different speeds or rates. Some reactions, such as the explosion of nitroglycerin, shown in **Figure 18,** are very fast. Other reactions, such as the burning of carbon in charcoal, are much slower. But what if you wanted to slow down the nitroglycerin reaction to make it safer? What if you wanted to speed up the reaction by which yeast make carbon dioxide, so bread would rise in less time? If you think carefully, you may already know some things about how to change reaction rates.

internet connect

www.scilinks.org
Topic: Factors Affecting
Reaction Rate
SciLinks code: HK4051

SciLINKS Maintained by the National Science Teachers Association

Factors Affecting Reaction Rates

Think about the following observations:

▶ A potato slice takes 5 minutes to fry in oil at 200°C but takes 10 minutes to cook in boiling water at 100°C. Therefore, potatoes cook faster at higher temperatures.

▶ Potato slices take 10 minutes to cook in boiling water, but whole potatoes take about 30 minutes to cook. Therefore, potatoes cook faster if you cut them up into smaller pieces.

These observations relate to the speed of chemical reactions. For any reaction to occur, the particles of the reactants must collide with one another. In each situation where the potatoes cooked faster, the contact between particles was greater, so the cooking reaction went faster.

Oxygen Hydrogen Carbon Nitrogen

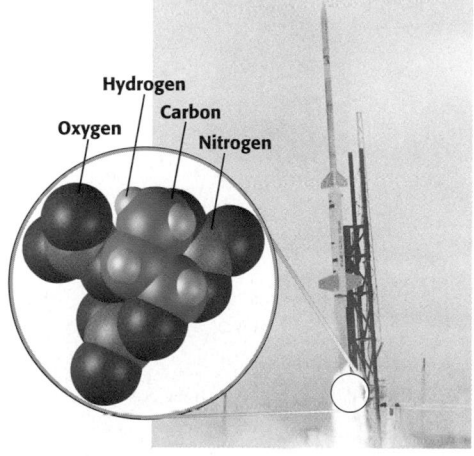

Figure 18

Nitroglycerin can be used as a rocket fuel as well as a medicine for people with heart ailments.

205

Teacher Resources

For the New Teacher Before beginning this section, point out to students that chemical reactions occur when reactant particles collide with each other with sufficient energy to cause a reaction. Ask students to suggest ways that you could increase the frequency of collisions, the energy of the collisions, or both to increase the reaction rate. Lead students to realize that increasing concentration, temperature, or surface area will increase the reaction rate. Ask students to explain how each of these changes would affect the frequency or energy of the collisions.

Focus

Overview

Before beginning this section, review with your students the Objectives listed in the Student Edition. In this section, students learn how temperature, surface area, concentration, pressure, molecule shape and size, and the use of catalysts can affect reaction rates. The section concludes with a discussion of equilibrium systems and Le Châtelier's principle.

🔊 Bellringer

Use the Bellringer transparency to prepare students for this section.

Motivate

Opening Demonstration

For this demonstration, you will need 2 100 mL graduated cylinders, rock salt, table salt, weighing paper, a balance, and 1 can of carbonated soda.

Label one graduated cylinder "rock salt" and the other cylinder "table salt." Pour 75 mL of carbonated soda into each graduated cylinder. Measure the mass of one large crystal of rock salt. Obtain an equivalent mass of table salt. Simultaneously dump each salt sample into the appropriately labeled graduated cylinder. Tell students that since the masses of the salt samples are identical, the only variable between the two samples is the amount of surface area. The increased surface area of the table salt provides many more reactive sites for the dissolved CO_2 to form, resulting in a greater amount of foam being formed.

Chapter Resource File

• Lesson Plan

Transparencies

TT Bellringer

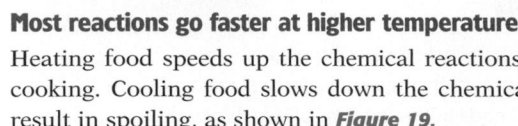

Demonstration ——— GENERAL

The Effect of Temperature on Dissolving

(Time: About 30 minutes)

Materials

• 400 mL beakers (2)
• beaker tongs
• 2 tea bags
• tap water
• ice
• hot plate

Add 300 mL of water to each 400 mL beaker. Place the first beaker on a hot plate and heat the water until it is almost boiling. Add ice to the second beaker and allow the water to cool to almost 0°C. Remove any remaining ice. Add a tea bag to each beaker. The hot water quickly becomes brown as the tea dissolves into it. The cold water remains colorless because the rate of dissolving is much slower at low temperatures.

Analysis

1. Is it important to keep the volumes of water constant in this demonstration? (yes, because then temperature will be the only variable)

2. How does temperature relate to the rate of molecular motion? (As temperature increases, so does the rate of molecular motion.)

LS Visual

Figure 19

A Mold will grow on bread stored at room temperature. **B** Bread stored in the freezer for the same length of time will be free of mold when you take it out.

Figure 20

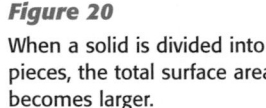

When a solid is divided into pieces, the total surface area becomes larger.

206

Most reactions go faster at higher temperatures

Heating food speeds up the chemical reactions that happen in cooking. Cooling food slows down the chemical reactions that result in spoiling, as shown in **Figure 19.**

The kinetic theory states that particles move faster at higher temperatures. The faster moving particles collide more often, and there are more chances for the particles to react. Therefore, the reaction will be faster.

A large surface area speeds up reactions

When a whole potato is placed in boiling water, only the outside is in direct contact with the boiling water. The energy transferred from the water takes longer to reach the center of the potato than it would if the potato were sliced. As **Figure 20** shows, cutting potatoes into pieces allows parts that were inside the potato to be exposed. In other words, the *surface area* of the potato is increased. The surface area of a solid is the amount of the surface that is exposed. Generally solids that have a large surface area react more rapidly because more particles can come in contact with the other reactants.

Concentrated solutions react faster

Think about a washing machine full of clothes with grass stains on them. If you put a drop of bleach in the water, little will happen to the dirty clothes. If you pour a bottle of bleach into the washing machine, the stained clothes will be clean. The more concentrated solution has more bleach particles. This means a higher chance for particle collisions with the stains.

Reactions are faster at higher pressure

The concentration of a gas can be thought of as the number of particles in a given volume. A gas at high pressure is more concentrated than the same amount of a gas at a low pressure because the gas at high pressure has been squeezed into a smaller volume. Gases react faster at higher pressures; the particles have less space, so they have more collisions.

Teacher Resources ——— BASIC

For the New Teacher Concept maps can help students clarify the relationships between ideas. Ask students to make a concept map for the main ideas in the section "Factors Affecting Reaction Rates." You can use the following example to get them started. Students should add information that explains how each factor affects reaction rates.

LS Logical

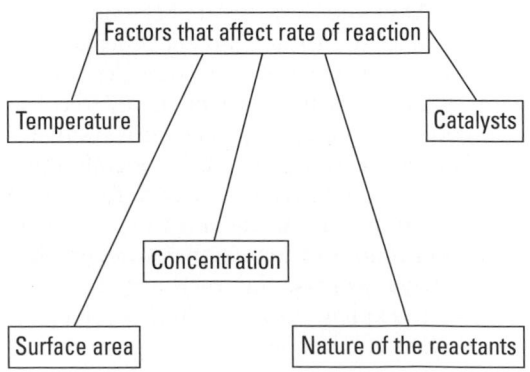

Massive, bulky molecules react slower

The size and shape of the reactant molecules affect the rate of reaction. You know from the kinetic theory of matter that massive molecules move more slowly than less massive molecules at the same temperature. This means that for equal numbers of massive and "light" molecules of about the same size, the molecules with more mass collide less often with other molecules.

Some molecules, such as large biological compounds, must fit together in a particular way to react. They can collide with other reactants many times, but if the collision occurs on the wrong end of the molecule, they will not react. Generally these compounds react very slowly because many unsuccessful collisions may occur before a successful collision begins the reaction.

Catalysts change the rates of chemical reactions

Why add a substance to a reaction if the substance may not react? This is done all the time in industry when **catalysts** are added to make reactions go faster. Catalysts are not reactants or products. They speed up or slow reactions. Catalysts that slow reactions are called *inhibitors*. Catalysts are used to help make ammonia, to process crude oil, and to accelerate making plastics. Catalysts can be expensive and still be profitable because they can be cleaned or renewed and reused. Sometimes the name of the catalyst is written over the reaction arrow of a chemical equation when a catalyst is present.

Catalysts work in different ways. Most solid catalysts, such as those in car exhaust systems, speed up reactions by providing a surface where the reactants can collect and react. Then the reactants can form new bonds to make the products. Most solid catalysts are more effective if they have a large surface area.

Enzymes are biological catalysts

Enzymes are proteins that are catalysts for chemical reactions in living things. Enzymes are very specific. Each enzyme controls one reaction or set of similar reactions. Some common enzymes and the reactions they control are listed in *Table 1*. Most enzymes are fragile. If they are kept too cold or too warm, they tend to decompose. Most enzymes stop working above 45°C.

☑ internet connect ≡

www.scilinks.org
Topic: Catalysts
SciLinks code: HK4018

SCiLINKS Maintained by the National Science Teachers Association

▶ **catalyst** a substance that changes the rate of a chemical reaction without being consumed or changed significantly

▶ **enzyme** a type of protein that speeds up metabolic reactions in plants and animals without being permanently changed or destroyed

Table 1   Common Enzymes and Their Uses

Enzyme	Substrate	What the enzyme does
Amylase	starch	breaks down long starch molecules into sugars
Cellulase	cellulose	breaks down long cellulose molecules into sugars
DNA polymerase	nucleic acid	builds up DNA chains in cell nuclei
Lipase	fat	breaks down fat into smaller molecules
Protease	protein	breaks down proteins into amino acids

Demonstration

Increasing Concentration

(Time: About 10 minutes)

Materials

- plastic CD case
- 50–100 metal BBs

You can easily model the effect of increasing concentration on reaction rate. Remove the plastic insert that holds the CD in the case to get a clear shallow box. You may wish to cover the outer $\frac{3}{4}$ in. of the box so that students will not be distracted by the BBs hitting the sides. Place 25 to 50 metal BBs in the case, and place the case on an overhead projector. Swirl and jerk the case so that the frequency of collisions between the BBs can be seen. Next double the number of BBs in the box. This doubles the concentration of BBs. Swirl the box again; students should be able to see the increase in the frequency of collisions.

SKILL BUILDER

Vocabulary Although inhibitors appear to retard reactions, they actually bind reactants so that they are not available for other reactions. The term *inhibitor* comes from the Latin *in*, which means "in," and *habere*, which means "to have or to hold." Thus, an inhibitor is a substance that holds a reactant, keeping it from reacting with another substance.

207

LIFE SCIENCE
CONNECTION

Few chemical reactions in the human body would occur at a rate fast enough to sustain life without the presence of enzymes. Not all enzymes break down large molecules. Some enzymes catalyze the formation of larger molecules, including large biopolymers such as starch and proteins. These enzymes carry out biosyntheses that are usually endothermic. For this reason they use a large amount of energy stored in the form of ATP, adenosine triphosphate, to carry out their functions. This is one of the reasons why living things require a constant supply of energy.

Teaching Tip

Types of Catalysts Two types of catalysts exist. Heterogeneous catalysts provide a surface where reactants concentrate, increasing reaction rate. A heterogeneous catalyst is used in catalytic converters. Homogeneous catalysts, such as enzymes, mix with the reactants and form an intermediate compound that reacts more readily. After the final reaction, the catalyst is released unchanged.

Quick Lab

Answers to Analysis

1. *Observations:* In step 1, Mg reacted most quickly, forming bubbles of H_2. Zn reacted but not as quickly, and Cu did not react. In step 2, the paper clip showed no evidence of reaction. The steel wool had rust, indicating a reaction with oxygen. In step 3, Mg reacted most readily in A, less readily in B, and least in C. In step 4, the sugar cube did not burn.

   Interpretations: The reactivity of the metals in step 1 differed. In step 2, the increased surface area of iron in the steel wool increased the rate of reaction. In step 3, rate of reaction decreased as the concentration of vinegar decreased. The ash was a catalyst in step 4.

2. nature of the reactants; increased surface area; concentration; presence of a catalyst
   LS Logical

▶ **substrate** a part, substance, or element that lies beneath and supports another part, substance, or element; the reactant in reactions catalyzed by enzymes

Figure 21
The enzyme hexokinase catalyzes the addition of phosphate to glucose. This model shows the enzyme, in blue, before **A** and after **B** it fits with a glucose molecule, shown in red.

Catalase, an enzyme produced by humans and most other living organisms, breaks down hydrogen peroxide. Hydrogen peroxide is the **substrate** for catalase.

$$2H_2O_2 \xrightarrow{\text{catalase}} 2H_2O + O_2$$

For an enzyme to catalyze a reaction, the substrate and the enzyme must fit exactly—like a key in a lock. This fit is shown in **Figure 21.** Enzymes are very efficient. In 1 minute, one molecule of catalase can catalyze the decomposition of 6 million molecules of hydrogen peroxide.

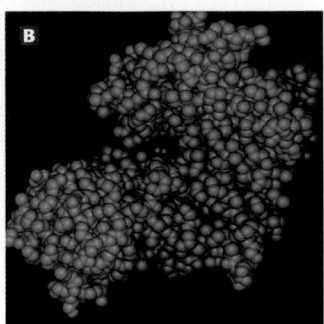

Quick Lab

What affects the rates of chemical reactions?

Materials
- ✔ Bunsen burner
- ✔ paper clip
- ✔ 6 test tubes
- ✔ paper ash
- ✔ sandpaper
- ✔ tongs
- ✔ matches
- ✔ 2 sugar cubes
- ✔ steel wool ball, 2 cm diameter
- ✔ graduated cylinder
- ✔ vinegar
- ✔ magnesium ribbon, copper foil strip, zinc strip; each 3 cm long, uniform width

SAFETY CAUTION Wear safety goggles and an apron.

1. Label three test tubes 1, 2, and 3. Place 10 mL of vinegar in each test tube. Sandpaper the metals until they are shiny. Then add the magnesium to test tube 1, the zinc to test tube 2, and the copper to test tube 3. Record your observations.

2. Using tongs, hold a paper clip in the hottest part of the burner flame for 30 s. Repeat with a ball of steel wool. Record your observations.

3. Label three more test tubes A, B, and C. To test tube A, add 10 mL of vinegar; to test tube B, add 5 mL of vinegar and 5 mL of water; and to test tube C, add 2.5 mL of vinegar and 7.5 mL of water. Add a piece of magnesium ribbon to each test tube. Record your observations.

4. Using tongs, hold a sugar cube and try to ignite it with a match. Rub paper ash on another cube and try again. Record your observations.

Analysis

1. Describe and interpret your results.

2. For each step, list the factor(s) that influenced the rate of reaction.

208

REAL-LIFE CONNECTION

Catalytic Converters Catalytic converters have one section where nitrogen oxides are reduced to N_2 and O_2 using catalysts, such as platinum. The second section uses the same catalysts to completely oxidize unoxidized or only partially oxidized hydrocarbons and carbon monoxide in exhaust gases.

Did You Know?

Hydrogen Peroxide Hydrogen peroxide foams up when placed on a cut or scrape because blood and cells in the cut both contain the enzyme catalase. The bubbles are oxygen gas created by the reaction stimulated by the enzyme. Since catalase is not found on the surface of skin, hydrogen peroxide does not react when placed directly on skin, and no foam is observed.

Equilibrium Systems

When nitroglycerin explodes, not much nitroglycerin is left. When an iron nail rusts, given enough time, all the iron is converted to iron(III) oxide and only the rust remains. Even though an explosion occurs rapidly and rusting occurs slowly, both reactions go to completion. Most of the reactants are converted to products, and the amount that is not converted is not noticeable and usually is not important.

Some changes are reversible

You may get the idea that all chemical reactions go to completion if you watch a piece of wood burn or see an explosion. However, reactions don't always go to completion; some are reversible.

For example, carbonated drinks, such as the soda shown in **Figure 22,** contain carbon dioxide. These drinks are manufactured by dissolving carbon dioxide in water under pressure. To keep the carbon dioxide dissolved, you need to maintain the pressure by keeping the top on the bottle. Opening the soda allows the pressure to decrease. When this happens, some of the carbon dioxide comes out of solution, and you see a stream of carbon dioxide bubbles. This carbon dioxide change is reversible.

$$\text{CO}_2 \text{ (gas above liquid)} \underset{\substack{\text{decrease} \\ \text{pressure}}}{\overset{\substack{\text{increase} \\ \text{pressure}}}{\rightleftharpoons}} \text{CO}_2 \text{ (gas dissolved in liquid)}$$

The physical change can go in either direction. The $\rightleftharpoons$ sign indicates a reversible change. Compare it with the arrow you normally see in chemical reactions, $\longrightarrow$, which indicates a change that goes in one direction—toward completion.

Figure 22

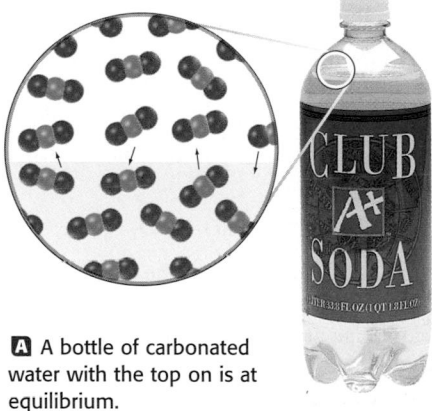

A A bottle of carbonated water with the top on is at equilibrium.

B When the top is removed, the carbonated water is no longer at equilibrium.

Historical Perspective

"Cracking" molecules for gasoline In the early part of the twentieth century, a rapid increase in use of vehicles that contained internal combustion engines resulted in an equally rapid increase in demand for gasoline. Distillation of petroleum resulted in production of gasoline, but the amount produced was not nearly adequate. Other, heavier petroleum products were available from the distillation, and few uses existed for them.

In 1912, gasoline producers started using high temperatures to perform large-scale "cracking" of these larger molecules. The resulting smaller molecules were suitable for gasoline. In 1936, a cracking procedure that used catalysts instead of high temperatures was developed. This procedure was instrumental in providing Allied forces with needed gasoline during World War II.

Vocabulary The word *dynamic* comes from the Greek word *dynamis*, which means "power." Ask students to brainstorm related terms. Lists will probably include such terms as *dynamite* and *dynamo*. All relate to power and/or motion. *Dynamic*, as it relates to equilibrium, indicates that the situation is not static. Constant, opposite reactions occur at equal rates.
LS Verbal

Teaching Tip — GENERAL

Writing **Dissolving Gases and Pressure** Have students bring in bottles of carbonated soft drinks. Hold a discussion to establish that there is carbon dioxide in the space above the liquid in the bottle, and that there is carbon dioxide dissolved within the liquid. Explain that the carbon dioxide is in equilibrium. Ask students to identify whether an opened carbonated soft drink is at higher or lower pressure than a sealed one. (The opened bottle is at lower pressure.)

Have students write a paragraph predicting how the amount of dissolved gas will be influenced by decreasing the pressure. When students have made predictions, they can demonstrate the influence of decreasing pressure on the dissolved gas by opening their bottles. In a second paragraph, ask students to discuss how well their observations match their predictions.
LS Kinesthetic

▶ **chemical equilibrium** a state of balance in which the rate of a forward reaction equals the rate of the reverse reaction and the concentrations of products and reactants remain unchanged

VOCABULARY *Skills Tip*

Equilibrium comes from the Latin aequilibris *meaning equally balanced. In Latin,* aequil *means equal, and* libra *means a balance scale. You may have seen the constellation called Libra. The stars in the constellation roughly represent a balance.*

Figure 23
Cement for ancient buildings, like this one in Limeni, Greece, probably contained lime made from seashells.

210

Equilibrium results when rates balance

When a carbonated drink is in a closed bottle, you can't see any changes. The system is in **chemical equilibrium** —a balanced state. This balanced state is dynamic. No changes are apparent, but changes are occurring. If you could see individual molecules in the bottle, you would see continual change. Molecules of CO_2 are coming out of solution constantly. However, CO_2 molecules from the air above the liquid are dissolving at the same time and the same rate.

The result is that the amount of dissolved and undissolved CO_2 doesn't change, even though individual CO_2 molecules are moving in and out of the solution. This is similar to the number of players on the field for a football team. Although different players can be on the field at any time, eleven players are always on the field for each team.

Systems in equilibrium respond to minimize change

When the top is removed from a carbonated drink, the drink is no longer at equilibrium, and CO_2 leaves as bubbles. For equilibrium to be reached, none of the reactants or products can escape.

The conversion of limestone, $CaCO_3$, to lime, CaO, is a chemical reaction that can lead to equilibrium. Limestone and seashells, which are also made of $CaCO_3$, were used to make lime more than 2000 years ago. By heating limestone in an open pot, lime was produced to make cement. The ancient buildings in Greece and Rome, such as the one shown in *Figure 23,* were probably built with cement made by this reaction.

$$CaCO_3 + heat \longrightarrow CaO + CO_2$$

Because the CO_2 gas can escape from an open pot, the reaction proceeds until all of the limestone is converted to lime.

However, if some dry limestone is sealed in a closed container and heated, the result is different. As soon as some CO_2 builds up in the container, the reverse reaction starts. Once the concentrations of the $CaCO_3$, CaO, and CO_2 stabilize, equilibrium is established.

$$CaCO_3 \rightleftharpoons CaO + CO_2$$

If there aren't any changes in the pressure or the temperature, the forward and reverse reactions continue to take place at the same rate. The concentration of CO_2 and the amounts of $CaCO_3$ and CaO in the container do not change.

Alternative Assessment — ADVANCED

Writing **Chemical Equilibrium** Have students write a paragraph that explains why most reactions in the human body go to completion and don't reach equilibrium. If students have difficulty, allow them to discuss their ideas in small groups before writing their paragraphs. Paragraphs should include the idea that energy produced by exothermic reactions is used or released by the body, shifting the reaction toward completion. Also, for reactions such as cellular respiration, the circulatory system constantly supplies reactants of oxygen and glucose to cells and removes carbon dioxide, shifting the reaction to completion. **LS Verbal**

Table 2 The Effects of Change on Equilibrium

Condition	Effect
Temperature	Increasing temperature favors the reaction that absorbs energy.
Pressure	Increasing pressure favors the reaction that produces fewer molecules of gas.
Concentration	Increasing the concentration of one substance favors the reaction that produces less of that substance.

Le Châtelier's principle predicts changes in equilibrium

Le Châtelier's principle is a general rule that describes the behavior of equilibrium systems.

> **If a change is made to a system in chemical equilibrium,
> the equilibrium shifts to oppose the change
> until a new equilibrium is reached.**

The effects of different changes on an equilibrium system are shown in **Table 2.**

Ammonia is a chemical building block used to make fertilizers, dyes, plastics, cosmetics, cleaning products, and fire retardants, such as those you see being applied in **Figure 24.** The Haber process, which is used to make ammonia industrially, is exothermic; it releases energy.

$$\text{nitrogen} + \text{hydrogen} \rightleftarrows \text{ammonia} + \text{heat}$$

$$N_2 \text{ (gas)} + 3H_2 \text{ (gas)} \rightleftarrows 2NH_3 \text{ (gas)} + \text{heat}$$

At an ammonia-manufacturing plant production chemists must choose the conditions that favor the highest yield of NH_3. In other words, the equilibrium should favor the production of NH_3.

211

INTEGRATING

ENVIRONMENTAL SCIENCE

All living things need nitrogen, which cycles through the environment. Nitrogen gas, N_2, is changed to ammonia by bacteria in soils. Different bacteria in the soil change the ammonia to nitrites and nitrates. Nitrogen in the form of nitrates is needed by plants to grow. Animals eat the plants and deposit nitrogen compounds back in the soil. When plants or animals die, nitrogen compounds are also returned to the soil. Additional bacteria change the nitrogen compounds back to nitrogen gas, and the cycle can start again.

Figure 24
Ammonium sulfate and ammonium phosphate are being dropped from the airplane as fire retardants. The red dye used for identification fades away after a few days.

Teaching Tip

Equilibrium Tell students that chemical equilibrium means that reaction rates are equal. It does not mean that amounts of reactants and products are equal. The percentage of reactants that produce products at equilibrium depends on the reaction itself and the conditions, such as temperature, that affect the equilibrium.

INTEGRATING

ENVIRONMENTAL SCIENCE
Soil often lacks sufficient amounts of the organisms that change nitrogen gas to ammonia, so fertilizers containing usable nitrogen compounds are added. To reduce the amount of chemical fertilizers needed, agriculturists usually rotate crops, alternating a nitrogen-fixing legume crop, such as clover, beans, peas, or peanuts, with other crops.

Nodules in the roots of legumes contain nitrogen-fixing bacteria that convert atmospheric nitrogen to ammonia. Recently, researchers at the University of Sydney, in Australia, have succeeded in adding nitrogen-fixing bacteria to the roots of wheat plants. Similar research is ongoing with other crops, such as corn and rice.

Historical Perspective

The Haber Process In 1909, Fritz Haber used what is now known as the Haber process to manufacture ammonia from nitrogen and hydrogen. Carl Bosch, who built an operating ammonia plant in Germany in 1913, further refined the process. Throughout World War I, the Bosch factory supplied Germany with ammonia that was needed for explosives. As a result, controversy existed regarding Haber receiving the Nobel Prize in Chemistry in 1918 and Bosch receiving the same award in 1931.

Chapter Resource File

• **Cross-Disciplinary Worksheet**
Integrating Environmental Science—
Fertilizers: Friend or Foe? **ADVANCED**

Transparencies

TM Changing Equilibrium

Quiz ——— BASIC

Determine whether each change will increase or decrease the reaction rate (in general).

1. a decrease in temperature (decrease in reaction rate)

2. an increase in surface area (increase in reaction rate)

4. an increase in concentration (increase in reaction rate)

3. the addition of an inhibitor (decrease in reaction rate)

LS Logical

Figure 25

Ammonia, which is manufactured in plants such as this, is used to make ammonium perchlorate—one of the space shuttle's fuels.

Le Châtelier's principle can be used to control reactions

If you raise the temperature, Le Châtelier's principle indicates that the equilibrium will shift to the left, the direction that absorbs energy and makes less ammonia. If you raise the pressure, the equilibrium will move to reduce the pressure according to Le Châtelier's principle. One way to reduce the pressure is to have fewer gas molecules. This means the equilibrium moves to the right—more ammonia—because there are fewer gas molecules on the right side. So to get the most ammonia from this reaction, you need to use a high pressure and a low temperature. The Haber process is a good example of balancing equilibrium conditions to make the most product. A manufacturing plant that uses the Haber process to produce ammonia is shown in **Figure 25.**

SECTION 4 REVIEW

SUMMARY

▶ Increasing the temperature, surface area, concentration, or pressure of reactants may speed up chemical reactions.

▶ Catalysts alter the rate of chemical reactions. Most catalysts speed up chemical reactions. Others, called inhibitors, slow reactions down.

▶ In a chemical reaction, chemical equilibrium is achieved when reactants change to products and products change to reactants at the same time and the same rate.

▶ At chemical equilibrium, no changes are apparent even though individual particles are reacting.

▶ Le Châtelier's principle states that for any change made to a system in equilibrium, the equilibrium will shift to minimize the effects of the change.

1. List five factors that may affect the rate of a chemical reaction.

2. Describe what can happen to the reaction rate of a system that is heated and then cooled.

3. Compare and Contrast a catalyst and an inhibitor.

4. Analyze the error in reasoning in the following situation: A person claims that because the overall amounts of reactants and products don't change, a reaction must have stopped.

5. Decide which way an increase in pressure will shift the following equilibrium system involving ethane, C_2H_6, oxygen, O_2, water, H_2O, and carbon dioxide, CO_2.

$$2C_2H_6 \text{ (gas)} + 7O_2 \text{ (gas)} \rightleftarrows 6H_2O \text{ (liquid)} + 4CO_2 \text{ (gas)}$$

6. Identify and Explain an example of Le Châtelier's principle.

7. Identify the effect of the following changes on the system in which the reversible reaction shown below is taking place:

$$4HCl \text{ (gas)} + O_2 \text{ (gas)} \rightleftarrows 2Cl_2 \text{ (gas)} + 2H_2O \text{ (gas)} + \text{heat}$$

a. the pressure of the system is increased

b. the pressure of the system is decreased

c. the concentration of O_2 is decreased

d. the temperature of the system is increased

8. Critical Thinking Consider the decomposition of solid calcium carbonate to solid calcium oxide and carbon dioxide gas.

$$\text{heat} + CaCO_3 \rightleftarrows CaO + CO_2 \text{ (gas)}$$

What conditions of temperature and pressure would you choose to get the most decomposition of $CaCO_3$? Explain.

212

Answers to Section 4 Review

1. surface area, concentration, temperature, presence of a catalyst, pressure, nature of the reactants

2. Most reactions speed up when heated and slow down when cooled.

3. A catalyst alters the rate of a reaction, usually by speeding it up. An inhibitor is a type of catalyst that ties up a reactant, the effect being to slow the rate of reaction.

4. If the reaction is in equilibrium, it is proceeding in both directions at equal rates.

5. Pressure changes affect gases only. Because there are more moles of gas on the left side of

the equation, increasing the pressure forces the reaction to the right.

6. If a change is made to a system in chemical equilibrium, the equilibrium shifts to oppose the change until a new equilibrium is reached. Carbonic acid, H_2CO_3, decomposes into CO_2 and H_2O. ($H_2CO_3 = CO_2 + H_2O$) If the CO_2 is removed as it is produced, the equilibrium will shift to the right to produce more products.

7. a. the reaction will be forced to the right

b. the reaction will be forced to the left

c. the reaction will be forced to the left

d. the reaction will be forced to the left

See "Continuation of Answers" at the end of the chapter.

Math Skills

Using Mole Ratios to Calculate Mass

Determine the mass of hydrogen gas, H_2, and oxygen gas, O_2, produced by 4 mol of water, H_2O, in the following chemical reaction:

$$2H_2O \longrightarrow 2H_2 + O_2$$

1 Write down the mole ratio for the balanced equation and multiply the ratio to obtain the number of moles of H_2O.

There are 4 mol of H_2O, so multiply each number in the ratio by 2.

Equation	$2H_2O$	$\longrightarrow$	$2H_2$	+	O_2
Mole ratio	2	:	2	:	1
Amount (mol)	4		4		2

2 Determine the mass per mol of each substance.

Look up the atomic mass of each element first. Since there are 2 hydrogen atoms and 1 oxygen atom in each molecule of H_2O, the mass per mol of H_2O is 2×1 g/mol $+$ 16 g/mol $=$ 18 g/mol. Similarly, the mass of H_2 is 2 g/mol, and the mass of O_2 is 32 g/mol.

3 Multiply the number of moles by the mass per mol of each substance.

The total mass of the reactants should match the total mass of the products.

Equation	$2H_2O$		$\longrightarrow$	$2H_2$	+	O_2
Mole ratio	2		:	2	:	1
Amount (mol)	4			4		2
Mass per mol	18 g/mol			2 g/mol		32 g/mol
Mass	18 g/mol $\times$ 4 mol	=		2 g/mol $\times$ 4 mol	+	32 g/mol $\times$ 2 mol
Total mass	72 g	=		8 g	+	64 g

4 mol of H_2O (72 g) will produce 8 g of H_2 and 64 g of O_2.

Practice

1. Determine the mass of H_2SO_4 produced when 1 mol of H_2O reacts with 1 mol of SO_3 in the following reaction:

$$H_2O + SO_3 \longrightarrow H_2SO_4$$

2. Determine the mass of $ZnSO_4$ produced in the following reaction if 2 mol of Zn reacts with 2 mol of $CuSO_4$.

$$Zn + CuSO_4 \longrightarrow ZnSO_4 + Cu$$

Teaching Tip

Balancing Equations Remind students that calculating the mole fraction depends on having a balanced chemical equation. All chemical equations need to be checked to be sure that the number of all reactant atoms equals the number of all product atoms.

If the equation is not balanced, point out that there are a few methods that can make the trial-and-error process of equation-balancing easier. As a rule, the atoms in complex molecules should be balanced first, followed by simpler molecules, and ending with the simplest molecules.

For example, consider the following reaction:

$$P_4 + KClO_3 \rightarrow KCl + P_2O_5$$

The first step is to balance the oxygen atoms in the $KClO_3$ and P_2O_5 molecules with coefficients of 5 and 3, respectively. Further balancing indicates that there must be 5 KCl molecules. However, the balancing coefficient for P_4 is 1.5, which is not a whole number.

$$1.5P_4 + 5KClO_3 \rightarrow$$
$$5KCl + 3P_2O_5$$

By multiplying the equation by 2, the correct coefficients (3, 10, 10, and 6) are obtained.

Answers

1. 98 g

2. 322 g

213

Understanding Concepts

1. c
2. d
3. c
4. a
5. c
6. b
7. c
8. b
9. c
10. d
11. c
12. b

Using Vocabulary

13. It means that a stress that favors formation of products has been placed on the system. The system will shift to produce more products to relieve that stress.

14. Wood undergoes combustion, which is an exothermic reaction. The total energy released in the reaction plus the chemical energy in the bonds of the products is the same chemical energy contained in the chemical bonds of the reactants. Energy was neither created nor destroyed.

15. Accept any accurate sentence. Sample sentence: 1 mol of methane reacts with 2 mol of oxygen gas to yield 1 mol of carbon dioxide gas and 2 mol of water.

Chapter Resource File

- **Chapter Test** GENERAL
- **Test Item Listing for ExamView® Test Generator**

Chapter Highlights

Before you begin, review the summaries of the key ideas of each section, found at the end of each section. The key vocabulary terms are listed on the first page of each section.

UNDERSTANDING CONCEPTS

1. When a chemical reaction occurs, atoms are never
   a. ionized.   c. destroyed.
   b. rearranged.   d. vaporized.

2. In an exothermic reaction,
   a. energy is conserved.
   b. the formation of bonds in the product releases more energy than is required to break the bonds in the reactants.
   c. energy is released as bonds form.
   d. All of the above

3. Which of the following is an endothermic reaction?
   a. fireworks exploding in the sky
   b. water boiling
   c. photosynthesis
   d. respiration

4. A + B $\longrightarrow$ AB is an example of a
   a. synthesis reaction.
   b. decomposition reaction.
   c. single-displacement reaction.
   d. double-displacement reaction.
   e. redox reaction.

5. Which of the following reactions is not an example of a redox reaction?
   a. combustion   c. dissolving in salt water
   b. rusting   d. cellular respiration

6. Radicals
   a. form ionic bonds with other ions.
   b. result from broken covalent bonds.
   c. usually break apart to form smaller components.
   d. bind molecules together.

214

7. In any chemical equation, the arrow means
   a. "equals."
   b. "is greater than."
   c. "yields."

8. Hydrogen peroxide, H_2O_2, decomposes to produce water and oxygen gas. The balanced equation for this reaction is
   a. $H_2O_2 \longrightarrow H_2O + O_2$.
   b. $2H_2O_2 \longrightarrow 2H_2O + O_2$.
   c. $2H_2O_2 \longrightarrow H_2O + 2O_2$.
   d. $2H_2O_2 \longrightarrow 2H_2O + 2O_2$.

9. Which of the following chemical equations is balanced?
   a. $Fe + O_2 \longrightarrow Fe_2O_3$
   b. $Ca + SbCl_3 \longrightarrow Si + Sb + 3CaCl_2$
   c. $3CuCl_2 + 2Al \longrightarrow 2AlCl_3 + 3Cu$
   d. $CS_2 + 2O_2 \longrightarrow CO_2 + SO_2$

10. Most reactions speed up when
    a. the temperature is lowered.
    b. equilibrium is achieved.
    c. the concentration of the products is increased.
    d. the reactants are in small pieces.

11. Enzymes
    a. can be used to speed up almost any chemical reaction.
    b. rely on increased surface area to catalyze reactions.
    c. catalyze specific biological reactions.
    d. always work faster at higher temperatures.

12. A system in chemical equilibrium
    a. has particles that don't move.
    b. responds to minimize change.
    c. is undergoing visible change.
    d. is stable only when all of the reactants have been used.

16. In an exothermic reaction, such as combustion, energy is released to the surroundings. In an endothermic reaction, such as photosynthesis, energy is absorbed or transferred to the reactants from the surroundings.

17. In a synthesis reaction, two or more substances combine to form a new compound, whereas in a decomposition reaction, a single compound breaks down to form two or more substances.

18. A combustion reaction differs from other reactions in that oxygen is always involved as a reactant and product.

19. According to kinetic theory, all matter consists of particles in constant motion.

Increasing the surface area of a reactant increases the rate of chemical reaction by increasing the chance of collisions between particles of the reactants. Increasing temperature increases the speed at which particles move, and thus also increases the likelihood of collisions between particles.

20. a. forward
    b. forward
    c. no effect
    d. forward
    e. reverse

13. Explain what it means when a system in equilibrium shifts to favor the products.

14. When wood is burned, energy is released in the forms of heat and light. Describe the reaction, and explain why this change does not violate the law of conservation of energy. Use the terms *combustion*, *exothermic*, and *chemical energy*.

15. Translate the following chemical equation into a sentence.

$$CH_4 + 2O_2 \longrightarrow CO_2 + 2H_2O$$

16. Explain the difference between an *exothermic reaction* and an *endothermic reaction*, and give an example of each.

17. How is a *synthesis reaction* different from a *decomposition reaction*?

18. How does a *combustion* reaction differ from other chemical reactions?

19. Use the *kinetic theory* to explain how an increase in the surface area of a reactant and higher temperatures can increase the rate of a chemical reaction.

20. For each of the following changes to the equilibrium system below, predict which reaction will be favored—forward (to the right), reverse (to the left), or neither.

$$H_2 \text{ (gas)} + Cl_2 \text{ (gas)} \rightleftharpoons 2HCl \text{ (gas)} + heat$$

   a. addition of Cl_2
   b. removal of HCl
   c. increased pressure
   d. decreased temperature
   e. removal of H_2

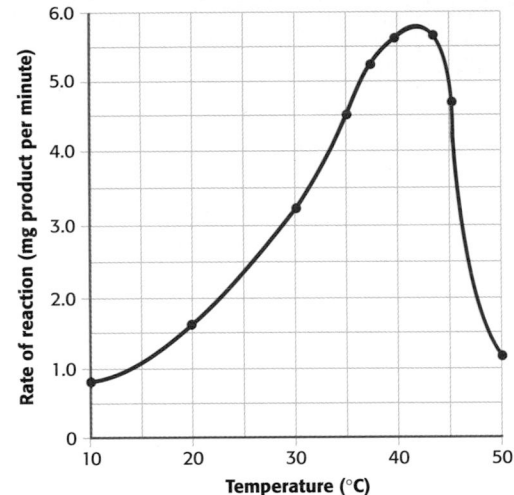

21. **Graphing** A technician carried out an experiment to study the effect of increasing temperature on a certain reaction. Her results are shown in the graph above.
   a. Between which temperatures does the rate of the reaction rise?
   b. Between which temperatures does the rate of the reaction slow down?
   c. At what temperature is the rate of the reaction fastest?

22. **Chemical Equations** In 1774, Joseph Priestly discovered oxygen when he heated solid mercury(II) oxide, HgO, and produced the element mercury and oxygen gas. Write and balance this equation.

23. **Chemical Equations** Write the balanced chemical equation for methane, CH_4, reacting with oxygen gas to produce water and carbon dioxide.

215

Building Graphing Skills
21. a. from 10°C to about 42°C
    b. from about 42°C to 50°C
    c. about 42°C

Building Math Skills
22. $2HgO \rightarrow 2Hg + O_2$
23. $CH_4 + 2O_2 \rightarrow 2H_2O + CO_2$
24. a. The mole ratio of $Al_2(SO_4)_3$ and H_2SO_4 is 1:3. If 6 mol of H_2SO_4 are used, 2 mol of $Al_2(SO_4)_3$ are produced.
    b. The mole ratio of Al_2O_3 and H_2O is 1:3. Therefore, 3 mol of Al_2O_3 are needed for 9 mol of H_2O.
    c. 588 mol of $Al_2(SO_4)_3$ and 1764 mol of H_2O
25. 1:12 for $C_{12}H_{22}O_{11}$: O_2 and $C_{12}H_{22}O_{11}$: CO_2; 1:11 for $C_{12}H_{22}O_{11}$:H_2O; 1:1 for O_2:CO_2; 12:11 for O_2:H_2O and CO_2:H_2O

26. a. $64 \text{ g S} \times \dfrac{1 \text{ mol S}}{32.07 \text{ g S}} \times$

$\dfrac{1 \text{ mol } SO_2}{1 \text{ mol S}} \times \dfrac{64.07 \text{ g } SO_2}{1 \text{ mol } SO_2} =$

$130 \text{ g } SO_2$

   b. $256 \text{ g } SO_2 \times \dfrac{1 \text{ mol } SO_2}{64.07 \text{ g } SO_2} \times$

$\dfrac{1 \text{ mol S}}{1 \text{ mol } SO_2} \times \dfrac{32.07 \text{ g S}}{1 \text{ mol S}} =$

128 g S

27. $Zn + 2HCl \rightarrow ZnCl_2 + H_2$
28. 450 g

Thinking Critically
29. Answers will vary but might include adding water to shredded newspaper to make a slurry, then adding enzymes that break down cellulose. Problems might include interference from inks or other materials in the paper, using the wrong amount of enzyme, or being unable to separate the sugars from other materials. Precautions might include controlling the temperature so the enzyme stays active.

Assignment Guide

Section	Questions
1	1–3, 14, 16
2	4–6, 17, 18, 31–34, 37
3	7–9, 15, 22–28, 30, 36
4	10–13, 19–21, 29, 35

30. Answers will vary. One general form for the answers follows.

   a. $2A + 3B_2 \rightarrow 2AB_3$
   b. $3A_2 + B_2 \rightarrow 2A_3B$
   c. $AB_4 + 2C_2 \rightarrow$
       $C_2A + 2CB_2$

31. In single-displacement reactions, the more reactive element will take the place of the less reactive one. Thus, when a reactive metal reacts with water, the metal displaces hydrogen atoms in the water and hydrogen is released as gas.

32. a. synthesis
   b. single-displacement
   c. decomposition
   d. combustion
   e. decomposition
   f. synthesis
   g. double-displacement

Developing Life/Work Skills

33. In a cigarette, tobacco combusts with a limited supply of oxygen. In addition to CO, cigarette smoke contains other toxic gases, nicotine, and tars, all of which can damage the lungs and circulatory system.

34. Answers will vary. Biodegradation reactions rely on bacteria or other organisms to degrade materials to more basic forms, such as the biodegradation of hydrocarbons to CO_2 and H_2O.

Integrating Concepts

35. Answers will vary. Students' five paragraphs should explain how the following enzymes act as catalysts: amylase, cellulase, DNA polymerase, lipase, and protease.

---

24. Chemical Equations Aluminum sulfate, $Al_2(SO_4)_3$, is used to fireproof fabrics and to make antiperspirants. It can be formed from a reaction between aluminum oxide, Al_2O_3, and H_2SO_4.

$$Al_2O_3 + 3H_2SO_4 \longrightarrow Al_2(SO_4)_3 + 3H_2O$$

   a. How many moles of $Al_2(SO_4)_3$ would be produced if 6 mol of H_2SO_4 reacted with an unlimited amount of Al_2O_3?
   b. How many moles of Al_2O_3 are required to make 9 mol of H_2O?
   c. If 588 mol of Al_2O_3 reacts with unlimited H_2SO_4, how many moles of each of the products will be produced?

25. Chemical Equations Sucrose, $C_{12}H_{22}O_{11}$, is a sugar used to sweeten many foods. Inside the body, it is broken down to produce H_2O and CO_2.

$$C_{12}H_{22}O_{11} + 12O_2 \longrightarrow 12CO_2 + 11H_2O$$

List all of the mole ratios that can be determined from this equation.

26. Chemical Equations Sulfur burns in air to form sulfur dioxide.

$$S + O_2 \longrightarrow SO_2$$

   a. What mass of SO2 is formed from 64 g of sulfur?
   b. What mass of sulfur is necessary to form 256 g of SO2?

27. Chemical Equations Zinc metal will react with hydrochloric acid, HCl, to produce hydrogen gas and zinc chloride, $ZnCl_2$. Write and balance the chemical equation for this reaction.

28. Chemical Formulas What is the mass of 25 moles of water, H_2O?

THINKING CRITICALLY

29. Designing Systems Paper consists mainly of cellulose, a complex compound made up of simple sugars. Suggest a method for turning old newspapers into sugars using an enzyme. What problems would there be? What precautions would need to be taken?

30. Applying Knowledge Molecular models of some chemical reactions are pictured below. Correct the drawings by adding coefficients or drawing molecules with a computer drawing program to reflect balanced equations.

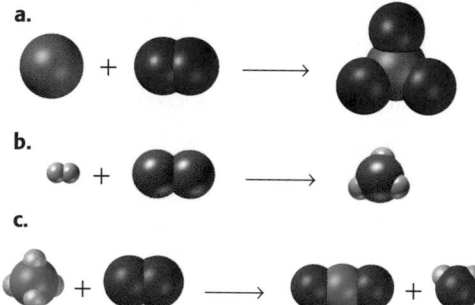

31. Creative Thinking Explain why hydrogen gas is given off when a reactive metal undergoes a single-displacement reaction with water.

32. Applying Knowledge Classify each of the following reactions as synthesis, decomposition, single-displacement, double-displacement, or combustion:

   a. $N_2 + 3H_2 \longrightarrow 2NH_3$
   b. $2Li + 2H_2O \longrightarrow 2LiOH + H_2$
   c. $2NaNO_3 \longrightarrow 2NaNO_2 + O_2$
   d. $2C_6H_{14} + 19O_2 \longrightarrow 12CO_2 + 14H_2O$
   e. $NH_4Cl \longrightarrow NH_3 + HCl$
   f. $BaO + H_2O \longrightarrow Ba(OH)_2$
   g. $AgNO_3 + NaCl \longrightarrow AgCl + NaNO_3$

---

36. A balanced equation shows that the total mass of reactants equals the total mass of products, illustrating the law of conservation of mass.

37. a. exothermic
   b. endothermic
   c. bond
   d. single-displacement reaction
   e. element
   f. activity
   g. double-displacement reaction
   h. positive ions in two ionic compounds
   i. precipitate
   j. gas
   k. molecule

33. Making Decisions Cigarette smoke contains carbon monoxide. Why do you think carbon monoxide is in the smoke? Why is smoking bad for your health?

34. Interpreting and Communicating Choose several items labeled "biodegradable," and research the decomposition reactions involved. Write balanced chemical equations for the decomposition reactions. Be sure to note any conditions that must occur for the substance to biodegrade. Present your information to the class to inform the students about what products are best for the environment.

35. Integrating Biology Research the enzymes listed in the table called "Common Enzymes and Their Uses" in Section 4. Write a paragraph on each one, describing in what way it acts as a catalyst.

36. Integrating Physics Explain how a balanced chemical equation illustrates that mass is never lost or gained in a chemical reaction.

□ internet connect

www.scilinks.org
Topic: Biodegradable   SciLinks code: HK4015

SCiLINKS. Maintained by the National Science Teachers Association

37. Concept Mapping Copy the unfinished concept map given below onto a sheet of paper. Complete the map by writing the correct word or phrase in the lettered box.

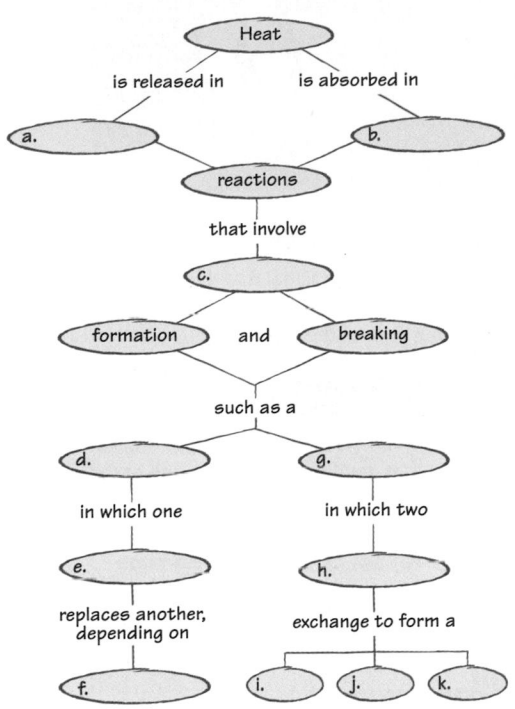

Art Credits: Fig. 2, Kristy Sprott; Fig. 4, Kristy Sprott; "Real World Applications", Uhl Studios, Inc.; Fig. 5-6, Kristy Sprott; Fig. 8-9, Kristy Sprott; p. 156, Kristy Sprott; Fig. 10-11, Kristy Sprott; Fig. 13, Kristy Sprott; p. 164, Kristy Sprott; Fig. 16, Kristy Sprott; Fig. 18, Kristy Sprott; Fig. 22, Kristy Sprott; "Thinking Critically", Kristy Sprott.

Photo Credits: Chapter opener image of grilled vegetables by Rita Moss/Getty Images/The Image Bank; burning charcoals by Charles O' Rear/CORBIS; Fig. 1A, 1B Charlie Winters/HRW; Fig. 1C, Charlie Winters; Fig. 2, Peter Van Steen/HRW; Fig. 3, E. R. Degginger/Color-Pic, Inc.; Fig. 6, Peter Van Steen/HRW; Fig. 7A, E. R. Degginger/Color-Pic, Inc.; Fig. 7B, Runk/Schoenberger/Grant Heilman Photography; Fig. 8, Sergio Purtell/Foca/HRW; "Science and the Consumer," Peter Van Steen/HRW; Fig. 10, 11, Sergio Purtell/Foca/HRW; "Connection to Fine Arts," Andy Levin/Photo Researchers, Inc.; Fig. 12, Charlie Winters; Fig. 13, Hank Morgan/Photo Researchers, Inc.; Fig. 14, Sam Dudgeon/HRW; Fig. 15, Visuals Unlimited/John Sohldon; Fig. 17, Charlie Winters; Fig. 18, NASA; Fig. 19, Sergio Purtell/Foca/HRW; Fig. 20, Peter Van Steen/HRW; Fig. 21, Dr. Thomas A. Steitz/Yale University, Fig. 22, Sam Dudgeon/HRW; Fig. 23, SuperStock; Fig. 24, Visuals Unlimited/Steve McCutcheon; Fig. 25, Visuals Unlimited/Tom J. Ulrich; "Skills Practice Lab," Sam Dudgeon/HRW.

217

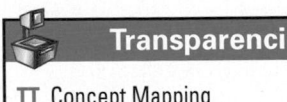

Transparencies

TT Concept Mapping

MEASURING THE RATE OF A CHEMICAL REACTION

Teacher's Notes

Time Required 1 lab period

Ratings

1	2	3	4
EASY			HARD

TEACHER PREPARATION	2
STUDENT SETUP	3
CONCEPT LEVEL	3
CLEANUP	3

Skills Acquired

- Collecting data
- Communicating
- Designing experiments
- Experimenting
- Identifying/Recognizing patterns
- Inferring
- Interpreting
- Measuring
- Organizing and analyzing data

The Scientific Method

In this lab, students will:
- Make observations
- Ask a question
- Form a hypothesis
- Analyze the results
- Draw conclusions
- Communicate results

Safety Cautions

Polyethylene gloves must be worn when handling hydrochloric acid. Hydrogen gas is extremely flammable. Be sure that no open flame or spark source is present in the lab during this experiment.

Skills Practice Lab

Introduction

How can you show that the rate of a chemical reaction depends on the temperature of the reactants?

Objectives

▶ **Measure** the volume of gas evolved to determine the average rate of the reaction between zinc and hydrochloric acid.

▶ **USING SCIENTIFIC METHODS** *Determine* how the rate of this reaction depends on the temperature of the reactants.

Materials

beaker to hold a 10 mL graduated cylinder
graduated cylinder, 10 mL
graduated cylinder, 25 mL
heavy scissors
hydrochloric acid, 1.0 M
ice
metric ruler
rubber tubing
sidearm flasks with rubber stoppers (2)
stopwatch
strips of thick zinc foil, 10 mm wide
thermometer
water bath to hold a sidearm flask

Measuring the Rate of a Chemical Reaction

▶ ## Procedure

Observing the Reaction Between Zinc and Hydrochloric Acid

1. On a blank sheet of paper, prepare a table like the one shown at right.
   SAFETY CAUTION Hydrochloric acid can cause severe burns. Wear a lab apron, gloves, and safety goggles. If you get acid on your skin or clothing, wash it off at the sink while calling to your teacher. If you get acid in your eyes, immediately flush it out at the eyewash station while calling to your teacher. Continue rinsing for at least 15 minutes or until help arrives.

2. Fill a 10 mL graduated cylinder with water. Turn the cylinder upside down in a beaker of water, taking care to keep the cylinder full. Place one end of the rubber tubing under the spout of the graduated cylinder. Attach the other end of the tubing to the arm of the flask. Place the flask in a water bath at room temperature. Record the initial gas volume of the cylinder and the temperature of the water bath in your data table.

3. Cut a piece of zinc about 50–75 mm long. Measure the length, and record this in your data table. Place the zinc in the sidearm flask.

4. Measure 25 mL of hydrochloric acid in a graduated cylinder.

5. Carefully pour the acid from the graduated cylinder into the flask. Start the stopwatch as you begin to pour. Stopper the flask as soon as the acid is transferred.

6. Record any signs of a chemical reaction you observe.

7. After 15 minutes, determine the amount of gas given off by the reaction. Record the volume of gas in your data table.

	Length of zinc strip (mm)	Initial gas volume (mL)	Final gas volume (mL)	Temperature (°C)	Reaction time (s)
Reaction 1					
Reaction 2					

Designing Your Experiment

8. With your lab partners decide how you will answer the question posed at the beginning of the lab. By completing steps 1–7, you have half the data you need to answer the question. How can you collect the rest of the data?

9. In your lab report, list each step you will perform in your experiment. Because temperature is the variable you want to test, the other variables in your experiment should be the same as they were in steps 1–7.

10. Before you carry out your experiment, your teacher must approve your plan.

Performing Your Experiment

11. After your teacher approves your plan, carry out your experiment. Record your results in your data table.

12. How do the two reactions differ?

▶ Analysis

1. Express the rate of each reaction as mL of gas evolved in 1 minute.

2. Which reaction was more rapid?

3. Divide the faster rate by the slower rate, and express the reaction rates as a ratio.

4. According to your results, how does decreasing the temperature affect the rate of a chemical reaction?

▶ Conclusions

5. How could you test the effect of temperature on this reaction without using an ice bath?

6. How can you express the rate of each of the two reactions you conducted as a function of the surface area of the zinc?

7. How would you design an experiment to test the effect of surface area on this reaction?

219

See "Continuation of Answers" at the end of the chapter.

Tips and Tricks

You may want to review how to collect gas by volumetric displacement. Ask students to write and balance the equation for the reaction.

$$Zn + 2HCl \rightarrow ZnCl_2 + H_2$$

The reaction will not go to completion in the 15 minutes allotted for the reaction.

Sidearm flasks (i.e., filter flasks) are specified rather than test tubes to avoid the risk of students inserting glass tubing through a test-tube stopper. Because of the large head space in a sidearm flask, several minutes may pass before enough hydrogen flows to the graduated cylinder. If filter tubes are available, this experiment can be conducted with 10 mL of 1 M HCl and a piece of zinc 5 mm wide and 25 cm in length. After 15 minutes, about 0.5 mL of H_2 will be evolved at 0°C and about 2 mL of H_2 at room temperature.

In the second half, students need to devise a way to conduct the same reaction at a different temperature. Although student answers may vary, having ice available will point them toward conducting the experiment in an ice bath. Because of the risk of igniting the hydrogen, the second half of the experiment should be performed at an elevated temperature only if no spark source or open flame is present.

Disposal Information

Neutralize the acid with 0.1 M NaOH, and filter the solution. The filtrate may be poured down the drain. After the $Zn(OH)_2$ precipitate has dried, it may be wrapped in newspaper and discarded in the trash.

Sample Data Table

	Length of zinc strip (mm)	Initial gas volume (mL)	Final gas volume (mL)	Temperature of water bath (°C)	Reaction time(s)
Reaction 1	50	2.6	10.3	21	900
Reaction 2	50	3.2	4.8	0	900

Chapter Resource File

- **Datasheet** Measuring the Rate of a Chemical Reaction GENERAL
- **Observation Lab** Combining Elements BASIC
- **CBL™ Probeware Lab** Investigating the Effect of Temperature on the Rate of a Reaction ADVANCED

Continuation of Answers

Continuation of Answers from p. 204
Answers

    reactants: KI, AgNO$_3$
    products: AgI, KNO$_3$

3. NaCl + AgNO$_3$ → AgCl + NaNO$_3$

KBr + AgNO$_3$ → AgBr + KNO$_3$

KI + AgNO$_3$ → AgI + KNO$_3$

4. Cl$^-$, Br$^-$, and I$^-$ all form solids with AgNO$_3$.

5. Solid AgCl is white, AgBr is pale yellow, and AgI is also yellow. Although AgI is darker than AgBr, it might be difficult to distinguish the two by color.

LS Kinesthetic

Continuation of Answers from p. 212
Answers to Section 4 Review

8. CO$_2$ is the only substance affected by pressure. Decreasing pressure would increase decomposition of CaCO$_3$. Because the reaction is endothermic, increasing temperature would also increase the rate of decomposition.

Continuation of Answers from p. 219

Answers to Procedure

11. See sample data table.

12. The two reactions differ in that the reaction at an elevated temperature gave off more hydrogen gas than the reaction at the lower temperature.

Answers to Analysis

1. Answers will vary. Sample data show 0.51 mL H$_2$/min at room temperature and 0.11 mL H$_2$/min at 0C.

2. The reaction at an elevated temperature is more rapid than the reaction at the lower temperature.

3. The ratio of the sample data is 4.6:1.

4. Decreasing the temperature slows the rate of a chemical reaction.

Answers to Conclusions

5. The effect of temperature on this reaction could be determined by comparing the results at room temperature with the results at a higher temperature.

6. The rate of the reactions can be given in mL of hydrogen gas/mm^2 of zinc.

7. To test the effect of surface area, the reaction can be conducted with equal amounts of zinc but one sample folded tightly in half so that mass is not a variable.

HOW SHOULD LIFE-SAVING INVENTIONS BE INTRODUCED?

Answers to Your Turn

1. Critiquing Viewpoints

Students' analyses of a single argument's weak points may vary. Be sure that student responses clearly identify a weakness and respond in some way that provides support. Possible responses are listed below. (Note that students should analyze *only one of the following.*)

Stacey F.: Some companies might conceal their ability to pay to avoid having to do it. But that would be better than placing an undue burden on airplane manufacturers.

Emily B.: When people make purchases, they don't always consider safety as one of the factors involved. But airlines could remind people of safety in their advertisements.

Virginia M.: The FAA does not pass laws. But such a law, if passed by Congress, could help companies make this decision, if they stand to lose even more money without safety.

April R.: For some, the price of a plane ticket will determine whether they travel. But people should consider safety.

Carlene de C.: Airline passengers already have the right to sue airlines if they've been negligent. But stronger regulations about safer materials could help make it easier for such suits.

viewpoints

How Should Life-Saving Inventions Be Introduced?

Researchers are developing better fireproof materials to use inside passenger airplanes. But the new materials are much more expensive than the ones currently used.

Should the Federal Aviation Administration (FAA) require that the new materials be used on all new and old planes, or should it be up to the plane manufacturers and airlines to decide whether to use the new materials?

A similar debate occurs whenever life-saving inventions are introduced, from automobile airbags to better child-safety seats. If the inventions should be used, who should bear the cost? Should it be the federal government, an insurance company, a manufacturer, or the customers?

If the device shouldn't be required at all times, how do you decide when it should be used? When are the risks so small that it doesn't make sense to spend money on another safety device?

What do you think?

> FROM: Stacey F., Rochester, MN.
>
> It should be up to the plane manufactuers because not all companies would be able to afford the cost. The FAA should look into the budgets of all plane companies and companies that can afford it should be required to use the new material.

> FROM: Emily B., Coral Springs, FL
>
> I think it should be up to the plane manufacturers and airlines. The new materials shouldn't be required on planes that are already built or on planes that are being built, because of expenses. However, it would be to an airline's advantage to have the best safety material possible for their customers' sake.

> FROM: Virginia M., Houston, TX
>
> The airlines are responsible for the lives of their passengers, so they should decide. But the FAA should pass a law stating that if the airlines refuse new safety measures, the airlines will accept total responsibility for any accidents that occur.

Leave the Decisions to the Companies Involved

220

Shannon B.: The airlines may have other ways to keep people safe besides fireproof materials, such as better fire extinguishers and smoke detectors. Even so, these new materials will do no harm.

2. Critiquing Viewpoints

Students' analyses of strong points in opposing arguments may vary. Be sure that student responses clearly identify the strong points and respond in some way that provides support for the opposite point of view. Possible responses are listed below. (Note that students should analyze *only one of the following.*)

Stacey F. and Emily B.: Providing this safety equipment on all planes could be a substantial financial burden. A company should not place its profits above the lives of its customers.

Virginia M.: Airline companies should be responsible for safety. But companies may not realize the risks they are running without regulations to serve as guidelines.

April R.: Lives should be valued more than profits. But you can't be sure that any safety improvement will solve all problems.

Carlene de C.: Airline companies should minimize their liabilities. But each company should be able to decide which strategies it wants to use to keep liabilities low.

> FROM: April R., Coral Springs, FL

If it can save just one life, it's worth spending money and time on. Eventually the technology will be required on all planes anyway. If an airline chose not to use these materials and there were an accident, there would be liability cases because lives might have been saved. Most people will have no problem spending more for a plane ticket if their safety is ensured.

Require Safety Immediately

> FROM: Carlene de C., Chicago, IL

The FAA should require that all planes—those currently in use and those being built—have fireproof materials. Otherwise, passengers could sue the airline company if they were hurt in a fire and it could have been prevented.

> FROM: Shannon B., Bowling Green, KY.

They should put the new fireproof materials on all planes, even the ones that have already been built. The public's health is at risk if a plane malfunctions, and the airlines should want to keep everybody safe. Otherwise they will lose customers.

Your Turn

1. **Critiquing Viewpoints** Select one of the statements on this page that you agree with. Explain at least one weak point in the statement. How would you respond to someone who used this point as a reason you were wrong?

2. **Critiquing Viewpoints** Select one of the statements on this page that you disagree with. Explain one strong point in the statement. How would you respond to someone who used this point as a reason they were right?

3. **Life/Work Skills** Imagine that you are preparing to testify in a congressional hearing about this matter. Choose the four most important points you'd make, and draft a statement that explains all of them persuasively.

4. **Working Cooperatively** With your teacher's help, stage a role-playing exercise, with students serving as the panel of congressional representatives preparing to vote on this issue and as witnesses for the airlines, the airplane manufacturers, insurance companies, safety organizations, and a passenger's rights group.

internet connect

TOPIC: Lifesaving Technology
GO TO: go.hrw.com
KEYWORD: HK4 Lifesavers

What do you think should be done? Why? Share your views on this issue and learn about other viewpoints at the HRW Web site.

221

4. **Working Cooperatively**

This classroom activity is a flexible one and can take as little or as much time as you choose. Students should be given time prior to the role-playing exercise to consider what questions and arguments they will want to state, given their different roles.

Evaluate students' performances using a scale of 1–5 points for each of the following: clarify of presentation, use of logic, attention to role, persuasiveness, and effort (25 total possible points).

Shannon B.: Public health should be protected in the event of a malfunction. Malfunctions that don't cause fires can also cause death, and the addition of these fireproof materials will not solve those problems.

3. **Life/Work Skills**

Evaluate student responses based on whether the four points chosen are pertinent and are persuasively explained in student statements. The following sample statements include some arguments students may use.

Leave the Decisions to the Companies Involved

1. The costs and logistics of adding these materials to existing planes will be burdensome to businesses.

2. The forces of the marketplace should be used rather than regulatory matters. Customers that want safety will reward companies providing it.

3. In the unfortunate event of an accident, customers can still resort to lawsuits if the company has truly been negligent.

4. While these materials resist fires, it is possible that an accident will still result in deaths even with the materials present.

Require Safety Immediately

1. The lives of citizens should be more important than the profits of a private company.

2. Government action can help guide companies to do the right thing.

3. Fireproof materials can help airlines minimize their liability and the damage to their aircraft as well as save lives.

4. Airlines can use the presence of these materials as a demonstration of their commitment to customer safety.

	PACING	CLASSROOM RESOURCES	LABS, ACTIVITIES, AND DEMONSTRATIONS
BLOCK 1 · 45 min pp. 222–223 **Chapter Opener**			SE **Activity 1**, p. 223 SE **Activity 2**, p. 223
BLOCKS 2 & 3 · 90 min pp. 224–231 **Section 1** Solutions and Other Mixtures		CRF **Lesson Plan*** TT **Bellringer*** TT **Suspension*** TT **Homogeneous Mixtures***	TE **Opening Activity**, p. 224 GENERAL TE **Group Activity** Suspensions, p. 225 GENERAL TE **Demonstration** Solutions or Not?, p. 226 GENERAL SE **Quick Activity** Making Butter, p. 227 GENERAL CRF **Datasheets for In-Text Labs** Making Butter* GENERAL CRF **Observation Lab** Separating Substances in a Mixture* ♦ BASIC
BLOCKS 4 & 5 · 90 min pp. 232–238 **Section 2** How Substances Dissolve		CRF **Lesson Plan*** TT **Bellringer*** TT **Hydrogen Bonding*** TT **Surface Area***	SE **Quick Activity** What dissolves a nonpolar substance?, p. 235 GENERAL CRF **Datasheets for In-Text Labs** What dissolves a nonpolar substance?* GENERAL TE **Demonstration** Effect of Stirring on Dissolving, p. 236 GENERAL TE **Demonstration** Temperature and Solubility, p. 237 GENERAL
BLOCKS 6 & 7 · 90 min pp. 239–244 **Section 3** Solubility and Concentration		CRF **Lesson Plan*** TT **Bellringer*** TM **Solubilities of Some Common Compounds*** TT **Concept Mapping***	SE **Skills Practice Lab** Investigating How Temperature Affects Gas Solubility, pp. 250–251 ♦ GENERAL CRF **Datasheets for SE Skills Practice Lab** Investigating How Temperature Affects Gas Solubility* GENERAL CRF **CBL™ Probeware Lab** Determining the Concentration of an Ionic Solution* ♦ BASIC

BLOCKS 8 & 9 · 90 min

Chapter Review and Assessment Resources

- SE **Chapter Review**, pp. 732–735
- CRF **Chapter Tests*** GENERAL
- OSP **Test Generator**
- CRF **Standardized Test Practice with Guided Reading Development***
- CRF **Test Item Listing for ExamView® Test Generator***
- OSP **Scoring Rubrics and Classroom Management Checklists**

Online Resources

Visit the HRW Web site for a variety of free resources related to the text. Just type in the keyword **HK4 SLT**.

Holt Science Spectrum: Physical Science: Online Edition

Students can access interactive problem solving help and active visual concept development with the *Holt Science Spectrum: Physical Science* Online Edition available at **www.hrw.com**.

student CNN News

cnnstudentnews.com

Find the latest news, lesson plans, and activities related to important scientific events.

PROBLEM SOLVING AND PRACTICE	SECTION REVIEW AND ASSESSMENT	STANDARDS CORRELATION
	CRF Pretest* GENERAL	
CRF Cross-Disciplinary Worksheet Integrating Physics—The Centrifuge* GENERAL **CRF Science Skills** SI Units and Conversions Between Them* GENERAL **TE Inclusion Strategies,** p. 225 **CRF Cross-Disciplinary Worksheet** Integrating Biology—Phospholipids* GENERAL **TE Inclusion Strategies,** p. 227 **CRF Cross-Disciplinary Worksheet** Connection to Engineering—Inks* GENERAL **CRF Cross-Disciplinary Worksheet** Real World Applications—Chromatography* GENERAL	**TE Quiz,** p. 231 BASIC **SE Section 1 Review,** p. 231 **CRF Concept Review*** GENERAL **CRF Quiz*** BASIC	SAI 2
CRF Cross-Disciplinary Worksheet Integrating Biology—Diffusion* GENERAL	**TE Quiz,** p. 238 BASIC **SE Section 2 Review,** p. 238 **CRF Concept Review*** GENERAL **CRF Quiz*** BASIC	SAI 1
CRF Science Skills Percentages* GENERAL **SE Math Skills** Molarity, p. 243 **CRF Math Skills** Molarity* GENERAL **SE Graphing Skills,** p. 245 **SE Science in Action** In Search of a Blood Substitute, pp. 252–253	**TE Quiz,** p. 244 BASIC **SE Section 3 Review,** p. 244 **CRF Concept Review*** GENERAL **CRF Quiz*** BASIC	

SCILINKS.

www.scilinks.org

Topic: Vaccines
*Sci*Links code: HK4144

Topic: Colloids
*Sci*Links code: HK4023

Topic: Chromatography
*Sci*Links code: HK4022

Topic: Diffusion
*Sci*Links code: HK4160

Topic: Solubility
*Sci*Links code: HK4129

Topic: Hydrogen Bonding
*Sci*Links code: HK4070

Topic: Artificial Blood
*Sci*Links code: HK4161

Topic: Paleontology
*Sci*Links code: HK4157

Technology Resources

 One-Stop Planner
All of your printable resources and the Test Generator are on this convenient CD-ROM.

 **Physical Science
Interactive
Tutor CD-Rom**

Disc One, Module 8
Solutions

 CNN. Science in the NEWS
each video segment is accompanied by a Critical Thinking Worksheet *

Segment 14
Harvesting Salt

Overview

In this section, students learn about the characteristics of heterogeneous mixtures and homogeneous mixtures. This section then introduces the dissolving process and factors that affect it. Finally, students learn about solubility and quantitative means of expressing concentration.

Assessing Prior Knowledge

Be sure students understand the following concepts:

- moles
- equilibrium
- Le Châtelier's principle
- ions
- bonding
- chemical formulas

MISCONCEPTION //ALERT\\\

Science education research has identified the following misconceptions about solutions.

- Students fail to recognize that dissolving involves two substances.
- Students may believe that the solution has less mass than the masses of the uncombined solute and solvent because the solute is no longer "pressing down" on the beaker.
- Students fail to recognize that the disappearance of the solute-solvent boundary implies a single phase and therefore think that a solution can still be filtered or settled.
- Many students, when asked to draw a sugar solution remember to draw sugar molecules, but rarely include water molecules.

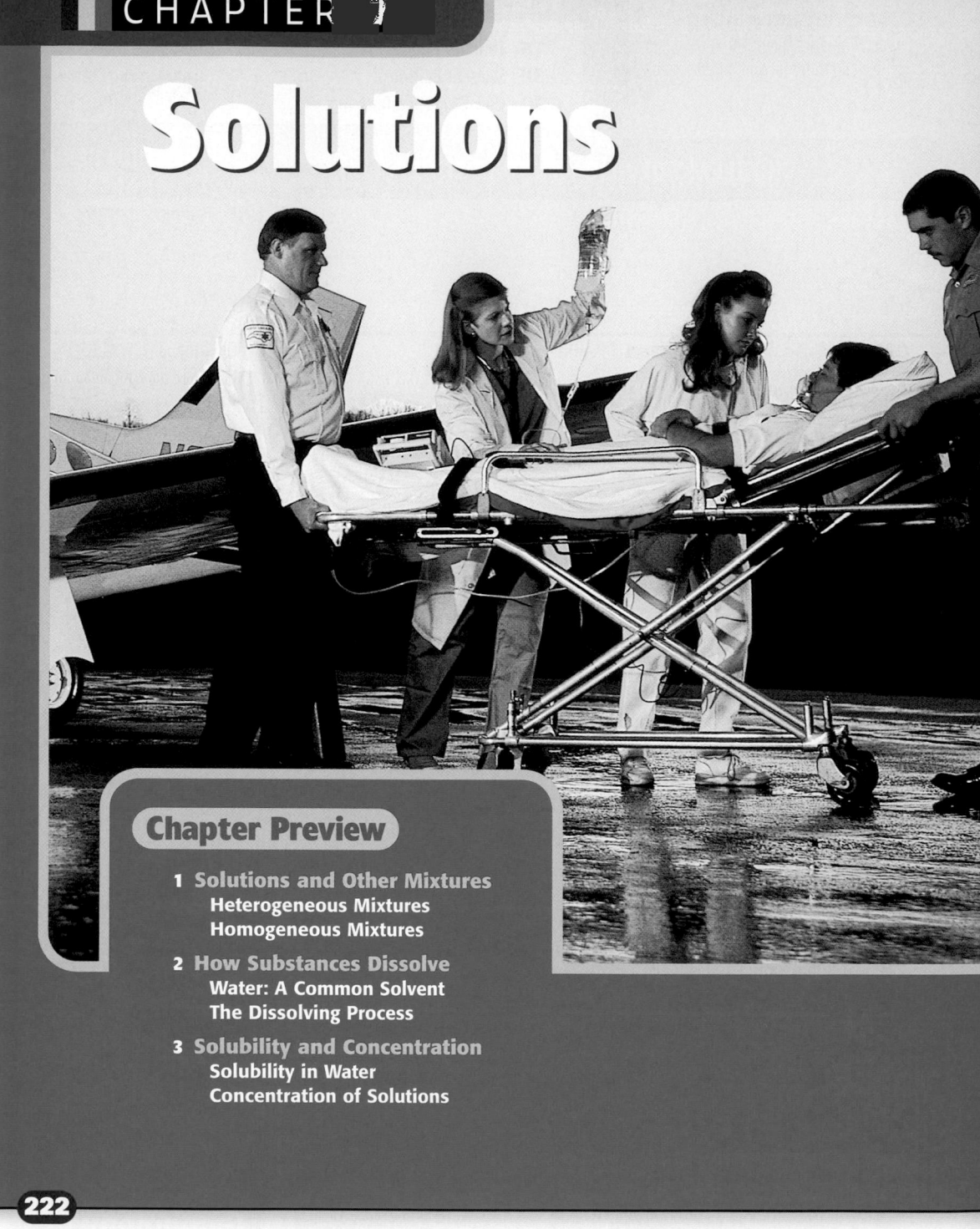

Solutions

Chapter Preview

1 Solutions and Other Mixtures
Heterogeneous Mixtures
Homogeneous Mixtures

2 How Substances Dissolve
Water: A Common Solvent
The Dissolving Process

3 Solubility and Concentration
Solubility in Water
Concentration of Solutions

222

Standards Correlations

National Science Education Standards

The following descriptions summarize the National Science Standards that specifically relate to this chapter. For the full text of the standards, see the National Science Education Standards at the front of the book.

Section 1 Solutions and Other Mixtures
Science as Inquiry SAI 2

Section 2 How Substances Dissolve
Science as Inquiry SAI 1

Section 3 Solubility and Concentration

Focus ACTIVITY

Background Paramedics rush to the scene of an accident. Someone has been injured, and the person's blood pressure has become dangerously low. Paramedics pump a *saline solution*, a mixture of water and sodium chloride that is similar to blood, into the person's veins. This mixture maintains the blood pressure that is needed to keep the person alive on the way to the hospital.

Shots called *vaccines* are mixtures that help protect you from many diseases. Vaccines have a tiny amount of the disease-causing organism you are trying to protect yourself from. The shot you get is harmless because the organism contained in it is dead, or inactivated. But the shot keeps you from getting the disease because your body can now recognize this harmful bacterium or virus again and fight it.

Activity 1 Look up the word *saline* in the dictionary. Which group of elements in the periodic table form ionic compounds that can be described by the word *saline*? Explain how the word *saline* applies to sodium chloride.

Activity 2 Fill a clear plastic cup with water. After the water settles, add table salt one teaspoon at a time to the water. Stir after you add each spoonful until all of the salt dissolves. How much salt are you able to dissolve before it stops dissolving and settles to the bottom of the cup? Perform this activity again, but this time use sugar instead of salt. Does the same amount of sugar dissolve? If not, what might explain the difference?

 **internet** connect

www.scilinks.org
Topic: Vaccines SciLinks code: HK4144

SCi LINKS. Maintained by the National Science Teachers Association

Pre-Reading Questions
1. The label on some drinks reads "Shake well before serving." Why do you need to shake these drinks? Why don't you need to shake all drinks?
2. Frozen orange juice must be mixed with water before you can drink it. Why is frozen orange juice referred to as *orange juice concentrate*?

Many solutions can be life-saving. Some solutions replace vital fluids in your body if you are injured, while others protect you from deadly diseases.

223

Focus ACTIVITY

Background Blood is a mixture called a suspension. The suspended particles, mainly red blood cells, white blood cells, and platelets, are actually suspended in a liquid called plasma, which is a solution, another type of mixture. Plasma is 90 percent water and 10 percent dissolved solutes, including sugar, vitamins, and proteins.

Activity 1 Saline refers to salts. The group of elements related to this meaning would be the alkali metals. Sodium chloride contains ions of sodium, an alkali metal.

Activity 2 More salt will dissolve in the water than sugar. Salt or sodium chloride dissolves readily, because it is ionic compound. Not as much sugar will dissolve in the same amount of water, because sugar crystals are made of larger molecular compounds (sucrose).

Answers to Pre-Reading Questions

1. Students should infer that in drinks that need to be shaken, the components of the drink may have separated and that in drinks that do not have to be shaken, the components stay mixed.
2. Students may infer that the term concentrate means that water has been removed from the orange juice before it was frozen.

Chapter Resource File

- **Pretest** GENERAL
- **Teaching Transparency Preview**
- **Answer Keys**

Chapter 7 · Solutions   223

Solutions and Other Mixtures

■ KEY TERMS

suspension
colloid
emulsion
solution
solute
solvent
alloy

OBJECTIVES

▶ **Distinguish** between heterogeneous mixtures and homogeneous mixtures.
▶ **Compare** the properties of suspensions, colloids, and solutions.
▶ **Give** examples of solutions that contain solids or gases.

Any sample of matter is either a pure substance or a mixture of pure substances. You can easily tell that fruit salad is a mixture because it is a blend of different kinds of fruit. But some mixtures look like they are pure substances. For example, a mixture of salt and water looks the same as pure water. Air is a mixture of several gases, but you cannot see different gases in the air.

Heterogeneous Mixtures

The amount of each substance in different samples of a *heterogeneous mixture* varies, just as the amount of each kind of fruit varies in each spoonful of fruit salad, as shown in *Figure 1*. If you compared two shovels full of dirt from a garden, they would not be exactly the same. Each shovelful would have a different mixture of rock, sand, clay, and decayed matter.

Another naturally occurring heterogeneous mixture is granite, a type of igneous rock shown in *Figure 2*. Granite is a mixture of crystals of the minerals quartz, mica, and feldspar. Because a mixture has no fixed composition, samples of granite from different locations can vary greatly in appearance because the samples have different proportions of minerals.

Figure 1

Fruit salad is a heterogeneous mixture. Each spoonful has a different composition of fruit because the fruits are not distributed evenly throughout the salad.

Figure 2

These paperweights are made from granite, which is a mixture of quartz, black mica, and feldspar.

224

Figure 3 **Orange Juice: A Heterogeneous Mixture**

A The pulp in the orange juice is spread throughout the mixture right after the orange juice is shaken.

B Over time, the pulp does not stay mixed with the water molecules. The pulp settles to the bottom, and two layers form.

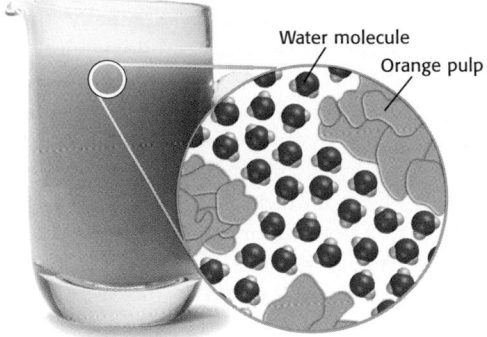

Water molecule
Orange pulp

Particles in a suspension are large and eventually settle out

Have you ever forgotten to shake the orange juice carton before pouring yourself a glass of juice? The juice probably tasted thin and watery. Natural orange juice is a **suspension** of orange pulp in a clear liquid that is mostly water, as shown in *Figure 3A*. A property of a suspension is that the particles settle out when the mixture is allowed to stand. So if the orange juice carton is not shaken, the top layer of the juice in the carton is mostly water because all the pulp has settled to the bottom.

Orange juice is clearly a heterogeneous mixture because after it settles, the liquid near the top of the container is not the same as the liquid near the bottom. Shaking the container mixes the pulp and water, but the pulp pieces are big enough that they will eventually settle out again, as shown in *Figure 3B*.

Particles in a suspension may be filtered out

Particles in suspensions are usually the size of or larger than the tip of an extremely sharp pencil, which has a diameter of about 1000 nm, or about the size of a bacterial cell. Particles of this size are large enough that they can be filtered out of the mixture. For example, a filter made of porous paper can be used to catch the suspended pulp in orange juice. That is, the pulp stays in the filter, while water molecules pass through the filter easily. You can classify a mixture as a suspension if the particles settle out or can be filtered out.

 suspension a mixture in which particles of a material are more or less evenly dispersed throughout a liquid or gas

225

Demonstration —— GENERAL

Solution or Not?

(Time: Approximately 10 minutes)

Materials

• hot distilled water
• gelatin
• table sugar
• flashlight
• 2 clear, colorless tumblers or beakers
• 2 spoons

Add equal amounts of hot water to the two tumblers. To one, add some dry gelatin and stir. To the other, add the same quantity of table sugar and stir to dissolve. Darken the room. Shine the flashlight through both tumblers from the side. Have students observe. Identify the first mixture as a colloid and the second as a solution. Let the tumblers sit undisturbed until near the end of class. Then, shine the light through them; have students make observations.

Analysis

1. What did you observe? (The light beam could not be seen in the solution but could be seen in colloid.)

2. Particles can reflect light. In which mixture do you think the particles are larger? Explain. (Light was reflected from the gelatin and not from the sugar in the other tumbler. The gelatin particles are larger.)

3. Did anything settle out in either tumbler? (no)

4. What can you state about the size of the particles in a colloid? (The particles are large enough to scatter light but also small enough to stay suspended.) **LS** Visual

Figure 4

Some salad dressings are made with oil and vinegar, which form a suspension when shaken. Oil and vinegar mixtures separate after standing for a few minutes.

 colloid a mixture consisting of tiny particles that are intermediate in size between those in solutions and those in suspensions and that are suspended in a liquid, solid, or gas

Figure 5

The liquid in the jar on the right is a colloid. Colloids exhibit the Tyndall effect, in which light is scattered by the invisible particles.

internet connect

www.scilinks.org
Topic: Colloids
SciLinks code: HK4023

SciLINKS. Maintained by the National Science Teachers Association

226

Some mixtures of two liquids will separate

Oil, vinegar, and flavorings can be shaken together to make salad dressing. But when the dressing stands for a few minutes, two layers form, as shown in *Figure 4.* The two liquids separate because they are *immiscible,* which means they do not mix. Eventually, the oil, which is less dense, rises and floats on the vinegar, which is denser.

One way to separate two immiscible liquids is to carefully pour the less dense liquid off the top. Some cooks use this technique to separate melted fat from meat juices. The cook removes the fat by pouring or spooning it off the meat juices, which are denser than the fat. The process of pouring a lighter liquid off of a heavier liquid is called *decanting.*

Particles in a colloid are too small to settle out

Latex paint is an example of another kind of heterogeneous mixture called a **colloid.** Latex paint is a thick combination of solid particles of pigment dispersed in water and other substances that make the pigment stick to a surface. The difference between colloids and suspensions is that the particles in colloids are smaller than those in suspensions—ranging from only 1 to 100 nm in diameter. Because the particles in colloids are so small, they pass through ordinary filters and stay dispersed throughout the mixture. However, the particles are large enough to scatter light that passes through the colloid, even though the colloid may look like clear water, as shown in *Figure 5.* This scattering of light is called the *Tyndall effect.*

Did You Know?

Colloid Classifications The two parts of a colloid are the dispersed substance and the continuous substance. For example, in fog, the dispersed substance is liquid water and the continuous substance is air. The states of the dispersed and continuous substances determine the classification of the colloid. For example, fog classifies as an aerosol because it consists of a liquid dispersed in a gas. Smog is also an aerosol and consists of a solid dispersed in a gas. Foam is a gas dispersed in a liquid (whipped cream) and solid foam is a gas dispersed in a solid (packaging peanuts). A gel is a liquid dispersed in a solid (soft contact lenses).

Making Butter

Cream is a lipid-in-water emulsion. Churning or shaking cream causes lipid droplets to stick to one another, which forms butter. You can make your own butter in the following way.

1. Pour 250 mL (about 1/2 pint) of heavy cream into an empty 500 mL container.
2. Add a clean marble, and seal the container tightly so that it does not leak.
3. Take turns shaking the container. When the cream becomes very thick, you will no longer hear the marble moving.
4. Open the container to look at the substance that formed. Record your observations.
5. If the butter is made of joined lipid droplets, what must make up most of the liquid that is left behind?

Other familiar materials are also colloids

The particles in most colloids are composed of many atoms, ions, or molecules, but individual protein molecules are also large enough to form colloids. Examples of protein colloids include gelatin desserts, egg whites, and blood plasma. These colloids consist of protein molecules dispersed in a liquid.

Other examples of colloids include whipped cream, which is made by dispersing a gas in a liquid, and marshmallows, which are made by dispersing a gas in a solid. Fog consists of small droplets of water dispersed in air, and smoke contains small solid particles dispersed in air.

Some immiscible liquids can form colloids

Mayonnaise is a colloid consisting of very small droplets of oil suspended in vinegar. Vinegar-and-oil salad dressings separate into two layers, but the vinegar and oil in mayonnaise do not separate. Mayonnaise has another ingredient that keeps the oil and vinegar together—egg yolk. Egg yolk coats the oil droplets, which keeps them from joining to form a separate layer. Mayonnaise is an **emulsion,** a colloid in which liquids that normally do not mix are spread throughout each other. Another example of an emulsion is found in your body. Bile salts cause fats to form an emulsion in the small intestine. Then digestive enzymes are able to break down the smaller fat particles.

Like other colloids, an emulsion has particles so small that it may appear to be uniform. But a closer look shows that it is not. For example, cream does not form separate layers, so it looks like a pure substance. Cream is really a mixture of oily fats, proteins, and carbohydrates dispersed in water. The lipid droplets are coated with a protein. The protein is an *emulsifier* that keeps the lipid droplets dispersed in the water so that they can spread throughout the entire mixture, as shown in *Figure 6.*

▶ **emulsion** any mixture of two or more immiscible liquids in which one liquid is dispersed in the other

Figure 6

Cream may look like a pure substance, but when cream is magnified, you can see that it is an emulsion of droplets of fats, or lipids, dispersed in water.

227

Making Butter
Materials (per group):
- jar with tight fitting lid
- 250 ml (1/2 pt) heavy cream
- clean marble

Teachers' Notes: If the butter were to be prepared for actual use, the liquid would be drained off and any liquid remaining in the butter would be squeezed out. A small amount of salt might be added to improve the taste.

Results: Students will find that the activity produces butter and a liquid that is mostly water and contains a few milk solids. Try to get students to realize that the lipids that formed the emulsion (cream) clumped together to form the butter. **LS Visual**

SKILL BUILDER

Vocabulary Tell students that one of the first characteristcs of colloids noticed by scientists was their consistency. The word *colloid* comes from the Greek word *kolla,* which means "glue." Colloids were thought to be glue-like materials.

Teaching Tip ——— BASIC

Stabilizing Emulsions An ingredient that stabilizes an emulsion, such as the egg yolk in mayonnaise, is called an emulsifier. One common emulsifier found in many foods is the group of compounds known as lecithins. Lecithins are found in egg yolks and soybeans. Have students check for lecithin on the ingredient labels of food items such as salad dressings, sprays for cooking pans, soft vegetable spreads, and pancake mix. **LS Verbal**

Chapter Resource File

- **Quick Activity Datasheet** Making Butter GENERAL
- **Cross-Disciplinary Worksheet** Integrating Biology– Phospholipids GENERAL

• Gifted and Talented

Ask students to research bile and how it works in the body. Included in the research should be: where bile is produced and stored, the function of bile, and why this substance is considered to be an emulsifier. The research should include what health problems occur from bile related actions. Students may also interview a medical professional involved with the function of bile and related organs. **LS Verbal**

PHYSICS
CONNECTION —— ADVANCED

Ask students what color a material that can reflect all colors of light appears when illuminated with white light. (white) Have students compare the color of cream with that of the magnified view in **Figure 6.** Ask students why the cream appears white even though the magnified view shows colorless droplets in a clear liquid. (The suspended particles completely reflect all the light that strikes them.) **LS Visual**

The ink used in ball-point pens can be traced back to the Middle Ages, when it was called iron-gall ink. The term *gall* refers to a tumor that is caused by bacteria, fungi, or the eggs of insects laid in the bark of trees. The galls used in iron-gall ink are from oak trees that have been infested by the gall-fly and are rich in tannic acid.

Answers

1. The two major types of inks are water-based inks and [organic] solvent-based inks. The former is used on porous material, in which it is absorbed into the material and dries. The latter is used on materials that have nonporous surfaces. Here the ink depends on the rapid evaporation of the solvent allowing the ink to adhere to the surface.
LS Verbal

SKILL BUILDER — **GENERAL**

Interpreting Visuals Have students refer to **Figure 7.** Ask students what the smallest particles are that make up water (water molecules); what the smallest particles are that make up sodium chloride (sodium ions and chloride ions); and what the smallest particles are that make up a solution of sodium chloride and water. (water molecules, sodium ions, and chloride ions). Also ask if any new substances are formed when sodium chloride dissolves in water (no)
LS Visual

Ink is a complicated mixture of substances. Some inks, such as those used in printing books and magazines, contain *pigments* that give the ink most of its color. Pigments are added to a liquid as finely ground, solid particles to form a suspension. The ink is then applied to the paper on a printing press and allowed to dry.

The ink used in some ballpoint pens is different from ink used in printing. The ink in pens contains a dissolved iron salt, such as ferrous sulfate, and an organic substance called *tannic acid.* Tannic acid and the iron salt mix to form a dark blue solution that gives the ink its blue-black color.

Making the Connection

1. In a library or on the Internet, research two different types of inks used in printing. How do the properties of the inks differ? What properties make them good choices for printing different materials?

Figure 7

A Plain water is homogeneous because it is a single substance.
B Salt water is also homogeneous because it is a uniform mixture of water molecules, sodium ions, and chloride ions.

Homogeneous Mixtures

Homogeneous mixtures not only look uniform but *are* uniform. Salt water is an example of a homogeneous mixture. If you add pure salt to a glass of water and stir, the mixture soon looks like pure water. The mixture looks uniform even when you examine it under a microscope. That's because the individual components of the mixture are too small to be seen. The mixture is made of sodium ions and chloride ions surrounded by water molecules, as shown in *Figure 7.* Because the number of ions will be the same throughout the salt water, the mixture is homogeneous.

When salt and water are mixed, no chemical reaction occurs. For this reason, the two substances can be easily separated by evaporating the water. As the water evaporates, the sodium and chloride ions of the salt begin to rejoin to form salt crystals like those that originally dissolved. When all the water evaporates, only salt is left.

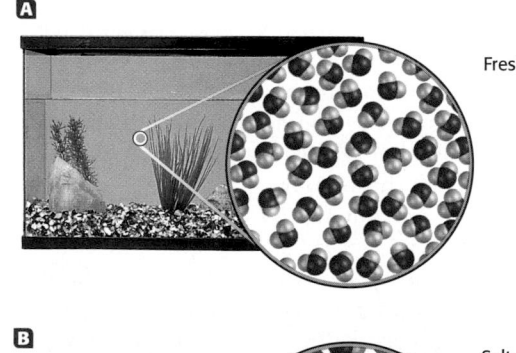

Fresh water

Saltwater solution

Sodium ion, Na$^+$

Chloride ion, Cl$^-$

Water

Teaching Tip ——— **ADVANCED**
Conservation of Mass but not Volume

Obtain four identical 100-mL graduated cylinders. Fill two each with 50 mL of ethanol and the other two each with 50 mL of distilled water. Place a cylinder each of ethanol and water on one pan of a double pan balance and the other two cylinders on the opposite pan. Adjust the balance screw so that the two exactly balance. On one of the pans, add the ethanol to the water and replace the cylinder on the pan. Have students observe that the pans remain balanced. Ask students what they can infer about the mass of separate water and ethanol and the mass of the ethanol-water solution. (They are the same.) Ask students to predict the volume of the solution. (Most students will say 100 mL.) Have students observe that the volume is less than 100 mL (The final volume depends on the concentration of ethanol.) Have interested students research why the volume of the ethanol-water solution is less than the volumes of ethanol and water. (The ethanol molecules distort the liquid structure of water making the water molecules move closer together and therefore reduce the volume.) **LS** Logical

Homogeneous mixtures are solutions

When you add aquarium salt to water and stir, the solid seems to disappear. What really happens is that salt has *dissolved* in water to form a **solution.** In this particular solution, aquarium salt is the **solute,** the substance that dissolves. Water is the **solvent,** the substance in which the solute dissolves. When a solute has dissolved in a solvent, the dissolved particles are so small that you can't see them, even with a powerful microscope. They are invisible because they are the smallest particles of the substance—atoms, ions, or molecules. Mixtures are homogeneous when the smallest particles of one substance are uniformly spread among similar particles of another substance. This description of a homogeneous mixture is also a description of a solution, so all homogeneous mixtures are also solutions.

Miscible liquids mix to form solutions

Two or more liquids that form a single layer when mixed are said to be miscible. Some solutions of two liquids are useful. For example, water mixed with isopropanol makes a solution called *rubbing alcohol,* and some skin lotions contain a solution of glycerol and water.

Chemists often have to separate miscible liquids when purifying substances in the laboratory. Because miscible liquids do not separate into layers, they are not as easy to separate as immiscible liquids are. One way to separate miscible liquids is by a process called *distillation,* which involves an apparatus such as the one shown in *Figure 8.* Distillation works only when the two miscible liquids have different boiling points. For example, a mixture of methanol and water can be separated by distillation because methanol boils at 64.5°C and water boils at 100.0°C. When this mixture is heated in a distillation apparatus, the methanol boils away first, leaving most of the water behind.

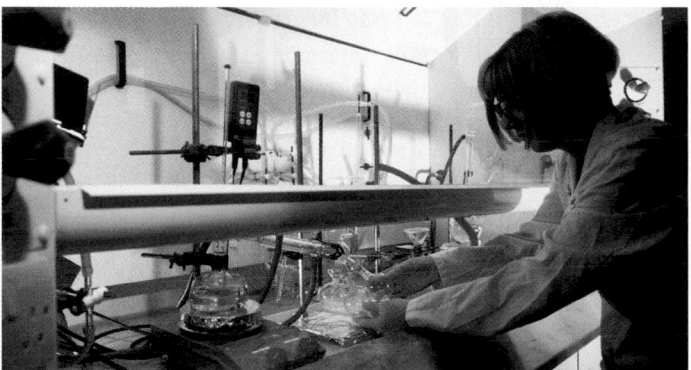

solution a homogeneous mixture of two or more substances uniformly dispersed throughout a single phase

solute in a solution, the substance that dissolves in the solvent

solvent in a solution, the substance in which the solute dissolves

PHYSICAL SCIENCE INTERACTIVE TUTOR

Disc One, Module 8:
Solutions
Use the Interactive Tutor to learn more about this topic.

Figure 8

A solution of miscible liquids that have different boiling points may be separated by distillation in an apparatus such as this one.

Teaching Tip

Solutes and Solvents Point out that water is not always the solvent when it is part of a solution. The solvent is the part of the solution that is in the greatest concentration. The solute exists in smaller concentration. For example, if 25 mL of ethanol is mixed with 10 mL of water, ethanol is the solvent, and water is the solute. Nitrogen, which makes up approximately 79 percent of air, is the solvent in air, and the other gases present are solutes.

MISCONCEPTION ALERT   BASIC

Separating Solutions
Remind students that filtration utilizes a difference in the sizes of the particles of the substances in a mixture to separate the mixture.

Ask students why filtration cannot be used to separate a solution. (The particles of each substance in a solution are of similar size; therefore; a difference in particle size cannot be used to separate them.) **LS Logical**

229

Did You Know ?

Solar Water Stills Solar stills are devices that distill fresh water from saltwater or brackish water. They are often found as survival equipment in maritime boats. A commercial solar still consists of a black solar collector with a reservoir for salt water beneath and a clear dome above. Saltwater is fed onto the collector where it evaporates and then condenses on the interior of the dome. The purified water runs down the interior of dome into a gutter and then a collecting bag.

Chapter Resource File

• **Cross-Disciplinary Worksheet:** Connection to Engineering— Inks GENERAL

Transparencies

TT Homogeneous Mixtures

Chromatography Inks can be pure substances or mixtures. Most ballpoint pen inks are complex mixtures containing pigments or dyes that can be separated by paper chromatography.

Answers

1. The blue dye has a structure more like that of the solvent because it is near the top of the paper.

2. All parts of the ink would travel at the same rate, and the color on the paper would be the color of the ink.

LS Visual

Teaching Tip — **ADVANCED**

Chromatograph Several different types of chromatography are used under different circumstances, but all chromatography is based on differences in polarity. Have interested students investigate the difference between column chromatography, gas chromatography, ion chromatography, and thin layer chromatography and then present their findings to the class. **LS Verbal**

internet connect

www.scilinks.org
Topic: Chromatography
SciLinks code: HK4022

SCILINKS Maintained by the National Science Teachers Association

230

REAL WORLD APPLICATIONS

Chromatography Chromatography is often used to separate mixtures that can't be separated by simpler methods. The figure at right shows how paper chromatography can be used to separate colored dyes in three different samples of black ink.

First, ink marks are made on absorbent paper. Then the paper is put in a jar holding a small volume of solvent. The solvent travels upward through the paper, carrying the ink with it. The finished *chromatogram* reveals which dyes make up each of the inks.

Each dye has a different chemical structure. Dyes that have structures more like that of the paper than that of the solvent stick to the paper and travel slower. Dyes that have structures more like that of the solvent move upward with the solvent and therefore travel farther.

Applying Information

1. Does the blue dye in each sample have a structure more like that of the paper or the solvent? Explain your answer.
2. How would the result differ if the inks were made from a single dye instead of a mixture of several dyes?

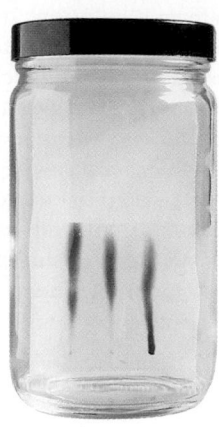

Liquid solutions sometimes contain no water

Many examples of solutions of liquids in other liquids do not contain water. For example, some kinds of fingernail-polish removers and paint strippers are mixtures of liquid substances that contain no water. Fuels such as gasoline, diesel fuel, and kerosene are homogeneous mixtures of several different liquid substances. These fuels are made from a liquid solution called petroleum, which is also called *crude oil*. Petroleum is a solution of many different carbon compounds. Fuels and other useful materials are made from petroleum by distillation. For example, gasoline is a solution containing several liquid substances distilled from petroleum.

Other states of matter can also form solutions

Like the water in a saltwater aquarium, many common solutions are solids dissolved in liquids. However, solutes and solvents can be in any state. For instance, vinegar is a solution of acetic acid dissolved in water, both of which are liquids. The air you breathe is a solution of nitrogen, oxygen, argon, and other gases. Gases can also dissolve in liquids. For example, a soft drink contains carbon dioxide gas dissolved in liquid water. Mothballs, which are made of a solid substance called *naphthalene,* slowly give off vapor that forms a solution with air. The liquid element mercury dissolves in solid silver to form a solution called an *amalgam,* which can be used to fill cavities in teeth.

Alternative Assessment — **ADVANCED**

The Advantages of Alloys Students can work individually or in small groups when performing this activity. Some students are most familiar with liquid solutions. Use this activity to make sure all students become familiar with solid solutions. Have each student choose an alloy, research its makeup and its uses, and report his or her findings to the class. The report can consist of an oral or written report, a poster, or some other method of communicating the information. Emphasize that the report should include the advantages the alloy has over the use of the individual components. Possible alloys include brass, bronze, dental amalgam, and many different types of iron (steel) and aluminum alloys. Checking recent sources of information on steel and aluminum should provide many choices of iron and aluminum alloys. **LS Verbal**

Solids can dissolve in other solids

The musical instrument shown in *Figure 9* is made of brass, a solution of zinc metal dissolved in copper metal. Brass is an example of an **alloy**, a homogeneous mixture that is usually composed of two or more metals. Of course, the metals must be melted to liquids and then mixed, but when the mixture cools, the result is a solid solution of one metal in another metal.

Alloys are important because they have properties that the individual metals do not have. For example, pure copper is too soft and bends too easily to be used to make a sturdy musical instrument. When zinc is dissolved in copper, the resulting brass is harder and tougher than copper, but the brass is still easy to form into complicated shapes. Bronze, an alloy of tin in copper, is harder than tin or copper alone. Bronze resists corrosion, so bronze has been used since ancient times to make sculptures and other objects meant to last a long time. Not all alloys contain only metals. Some types of steel are alloys containing the nonmetal element carbon.

■ **alloy** a solid or liquid mixture of two or more metals

Figure 9
This cornet is made of the alloy brass, which is a solid solution.

Close

Quiz

1. A difference in what physical property is often used to separate two immiscible liquids? (density)

2. Fog and smoke are examples of what type of heterogeneous mixture? (colloid)

3. What is another term for a homogeneous mixture? (a solution)

4. What is an alloy? (a solid solution of two or more metals)

SECTION 1 REVIEW

SUMMARY

▶ A heterogeneous mixture is a nonuniform blend of two or more substances.

▶ The particles in a suspension settle out of the mixture or may be filtered out.

▶ The dispersed particles in a colloid are too small to settle out or to be filtered out.

▶ A homogeneous mixture, or solution, is a uniform blend of substances.

▶ In a solution, the solute is dissolved in the solvent.

▶ Solutions may be formed from solids, liquids, or gases.

1. Classify the following mixtures as heterogeneous or homogeneous:
 a. orange juice without pulp   **c.** cinnamon sugar
 b. sweat   **d.** concrete

2. Explain the difference between a suspension and a colloid.

3. List three examples of solutions that are not liquids.

4. Identify the solvent and solute in a solution made by dissolving a small quantity of baking soda in water.

5. Arrange the following mixtures in order of increasing particle size: muddy water, sugar water, and egg white.

6. Explain why distillation would not be an effective way to separate a mixture of the miscible liquids formic acid, which boils at 100.7°C, and water, which boils at 100.0°C.

7. Critical Thinking A small child watches you as you stir a spoonful of sugar into a glass of clear lime-flavored drink. The child says she believes that the sugar went away because it seemed to disappear. How would you explain to the child what happened to the sugar, and how could you show her that you can get the sugar back?

231

Answers to Section 1 Review

1. a. homogeneous
 b. homogeneous
 c. heterogeneous
 d. heterogeneous

2. The particles in a suspension are larger than those in a colloid. As a result, the particles in a suspension settle out and can be filtered out, but those in a colloid do not settle out and cannot be filtered out.

3. Answers may include air, mixtures of gases used in diving, dental amalgam, sterling silver, brass, bronze, other metal alloys.

4. Water is the solvent and baking soda or sodium hydrogen carbonate is the solute.

5. sugar water, egg white, muddy water

6. Distillation can separate liquids that have different boiling points. The boiling points of formic acid and water are too close together for separation by distillation.

7. You can show her that the drink is sweet which is evidence that the sugar is still present. You could tell her that the sugar mixed with the water, coming apart into pieces so small that they cannot be seen. You could allow the water to evaporate, showing that the crystalline sugar is regained, although contaminated with the drink flavoring.

Chapter Resource File

• **Cross-Disciplinary Worksheet** Real World Applications—Chromatography GENERAL

• **Concept Review** GENERAL

• **Quiz** BASIC

SECTION
2

Focus

Overview

Before beginning this section, review with your students the Objectives listed in the Student Edition. This section begins with a discussion of the structure of the water molecule and its role in the dissolving process. The ability of a solvent to dissolve a solute is then related to the relative strength of molecular forces. The section continues by identifying three factors that influence solubility. The section concludes with a discussion of the effects of a solute on the freezing point and the boiling point of a solution.

🎙️ Bellringer

Use the Bellringer transparency to prepare students for this section.

Motivate

Opening Discussion ——— GENERAL

Have students recall [from section on "Matter"] how the kinetic theory accounts for sugar dissolving in water. (When sugar crystals are placed in water, water molecules attract sugar molecules in the crystals and pull them apart. The random collisions of water molecules with the individual sugar molecules disperse the sugar molecules homogeneously throughout the water.) Have students recall the different types of forces (gravitational, electrical, magnetic, and nuclear) and ask them which one might cause the force of attraction between water and sugar molecules. (electrical force) **LS Logical**

How Substances Dissolve

▶ **KEY TERMS**
polar compound
hydrogen bonding
nonpolar compound

▶ **OBJECTIVES**

▶ **Explain** how the polarity of water enables it to dissolve many different substances.

▶ **Relate** the ability of a solvent to dissolve a solute to the relative strengths of forces between molecules.

▶ **Describe** three ways to increase the rate at which a solute dissolves in a solvent.

▶ **Explain** how a solute affects the freezing point and boiling point of a solution.

▶ **polar compound** a molecule that has an uneven distribution of electrons

Suppose you and a friend are drinking iced tea. You add one spoonful of loose sugar to your glass of tea, stir, and all of the sugar dissolves quickly. Your friend adds a sugar cube to her tea and finds that she must stir longer than you did to dissolve all of the sugar cube. Why does the sugar cube take longer to dissolve? Why does sugar dissolve in water at all?

Water: A Common Solvent

Two-thirds of Earth's surface is water. The liquids you drink are mostly water, and three-fourths of your body weight is water. Many different substances can dissolve in water. For this reason, water is sometimes called the *universal solvent*.

Water can dissolve ionic compounds because of its structure

To understand what makes water such a good solvent, consider the structure of water. A water molecule is made up of two hydrogen atoms bonded to one oxygen atom by covalent bonds. But electrons are not evenly distributed throughout a water molecule because oxygen atoms strongly attract electrons. The oxygen atom pulls electrons away from the hydrogen atoms, giving them a partial positive charge. The electrons are then bunched around the oxygen atom, giving it a partial negative charge. This uneven distribution of electrons, combined with a water molecule's bent shape, causes it to be a **polar compound,** as shown in *Figure 10.* A polar molecule has distinct positively and negatively charged sides, which are indicated by δ+ and δ− in *Figure 10.* Water molecules attract both the positive and negative ions of an ionic compound.

Figure 10

Water is a polar molecule because the oxygen atom strongly attracts electrons, which leaves the hydrogen atoms with a positive charge.

232

Caramel Ask students if they like caramel. Point out caramel is made by changing sugar from a solid state to a liquid state. Ask students: What is this change of state called? (melting) How do you melt sugar? (by heating it) Remind students that another way of changing solid sugar is by adding water and making a sugar solution. Ask: Is liquid sugar or a sugar solution a pure substance? (pure substance) Which is a mixture? (the sugar solution) Which change required large amounts of heat? (melting) Point out that when sugar melts, sugar molecules move farther from neighboring sugar molecules. In dissolving, neighboring sugar molecules are replaced by water molecules. **LS Logical**

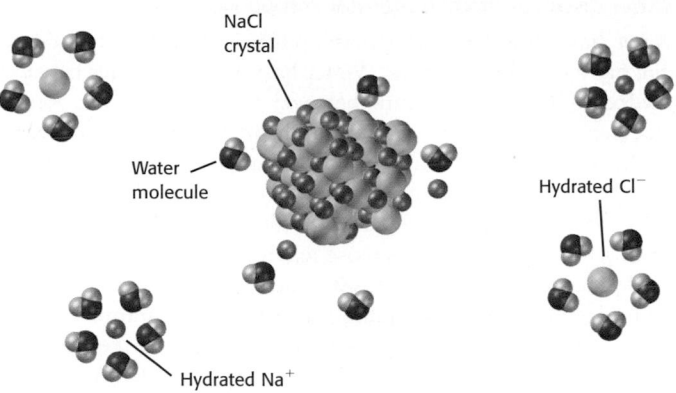

NaCl crystal

Water molecule

Hydrated Cl⁻

Hydrated Na⁺

Figure 11

Figure 11
Polar water molecules attract the positive sodium ions and negative chloride ions, pulling the ions away from the crystal of NaCl. Notice that the negative oxygen atoms of the water molecules are attracted to the positive sodium ion and that the positive hydrogen atoms are attracted to the negative chloride ions.

Polar water molecules pull ionic crystals apart

Figure 11 shows how a sodium chloride, NaCl, crystal dissolves in water. First, the partially negative oxygen atoms of water molecules attract the positively charged sodium ions at the surface of the sodium chloride crystal. The partially positive hydrogen atoms of water molecules also attract the negatively charged chloride ions. As more water molecules attract the ions, the force of attraction between the ions and the water molecules increases. Finally, this force becomes stronger than the force of attraction between the sodium and chloride atoms in the crystal. Then, the ions are pulled away from the crystal and surrounded by water molecules. Eventually, all of the ions in the crystal are pulled into solution, and the substance is completely dissolved.

Dissolving depends on forces between particles

Water dissolves baking soda and many other ionic compounds in exactly the same way that water dissolves sodium chloride. The attraction of water molecules pulls the crystals apart into individual ions. Many other ionic compounds, though, do not dissolve in water. Silver chloride, unlike sodium chloride, is an ionic compound that does not dissolve in water.

Why does one ionic compound dissolve in water, but another ionic compound does not? The answer is related to forces of attraction. To dissolve an ionic substance, water molecules must exert a force on the ions that is more attractive than the force between the ions in the crystal. This principle applies not only to water molecules and ionic compounds but also to any solvent and any substance. To dissolve a substance, solvent molecules must exert more force on the particles of the substance than the particles exert on one another.

INTEGRATING

BIOLOGY
When a solute dissolves in water, the random movement of particles called *diffusion* ensures that the solute spreads out evenly through the solution. Cells rely on diffusion to transport molecules. When the concentration of a solute is greater inside a cell than it is outside, the solute moves out of the cell through the cell membrane.

Not all substances can diffuse across a cell membrane. Sodium ions and potassium ions, for example, are transported into and out of cells through structures called *sodium-potassium pumps*. These ions give many cells an electrical charge that makes it possible for the cells to send electrical signals to different parts of the body.

SKILL BUILDER — BASIC

Interpreting Visuals Perform the following demonstration to emphasize the polarity of water as shown in **Figure 10.** Use a plastic comb to rapidly comb someone's hair. Hold the comb near a small stream of water running from a faucet or poured slowly from a beaker. Show students that the stream bends toward the comb. The comb is charged, and the partial charges of the water molecules are attracted to it. **LS Visual**

INTEGRATING

BIOLOGY The diffusion of a solvent across a selectively permeable membrane (a membrane that allows the passage of some molecules but not others) because of a difference in concentration is known as osmosis. In biology the osmosis of water through the cell membrane is important in maintaining the shape and structure of a cell. Crisping celery stalks by placing them in water is one example of osmosis. The concentration of water outside the cells is higher than the concentration within the cells. As a result, water diffuses into the cells, plumping them. If the celery stalk is left in a dry environment, the reverse occurs; water diffuses from the cells by osmosis causing the cells to become flaccid.

233

Did You Know?

Electrolytes and Nonelectrolytes Pure water does not conduct an electric current. However, some water solutions do. The ability of a solution to conduct an electric current can be used to classify the solutes in the solution. Solutes whose water solutions conduct an electric current are classified as electrolytes. These substances exit as ions in solution. Solutes such as Na⁺ and Cl⁻ are electrolytes. Solutes whose solutions do not conduct electric current are classified as nonelectrolytes. These substances, such as sugar, exist as molecules in solution. A simple apparatus can be use to classify solutes. It consists of light bulb, battery, and two leads connected in series. The end of each lead is immersed in the solution and then bulb is observed. Unless the solution is very dilute, the bulb will glow indicating the solute contains electrolytes. If the bulb doesn't glow, the solution contains nonelectrolytes.

Chapter Resource File

- **Lesson Plan**
- **Science Skills** Surface Area **GENERAL**
- **Cross-Disciplinary Worksheet** Integrating Biology— Diffusion **GENERAL**

Transparencies

TT Bellringer

Reading Skills Write the title of this passage, *Water dissolves many molecular compounds*, on the chalkboard. Then have students write brief statements describing how water can dissolve molecular substances, such as sugar. Have students refer to these statements after they have read the passage. **LS** Verbal

Interpreting Visuals Have students refer to **Figure 13** and count the number of −OH groups in each sucrose molecule.(8) Point out that sugar easily dissolves in water because the water molecules hydrogen-bond with the sucrose molecules and pull them into the solution. **LS** Visual

Teaching Tip

Hydrogen Bond A hydrogen bond is considered a *partial* covalent bond because the bond is formed by the attraction between a covalently bonded hydrogen atom in one molecule and the *partial* charge of another polar molecule.

Figure 12

As water slowly dissolves the sugar cube, streams of denser sugar solution move downward. Table sugar, or sucrose, is a molecular compound.

■ **hydrogen bonding** the intermolecular force occurring when a hydrogen atom that is bonded to a highly electronegative atom of one molecule is attracted to two unshared electrons of another molecule

Figure 13

Water molecules form hydrogen bonds with the −OH groups of a sucrose molecule. These forces pull the sucrose molecule away from the sugar crystal and into solution.

234

Water dissolves many molecular compounds

Water has a low molecular mass, but it is a fairly dense liquid that has a high boiling point. Water has these properties because **hydrogen bonding** occurs between water molecules. Recall that electrons are pulled away from the hydrogen atoms of water by the oxygen atom. As a result, the hydrogen atom of one water molecule attracts electrons from the oxygen atom of another water molecule to form a partial covalent bond. This hydrogen bond pulls water molecules close together.

Besides ionic compounds, water also dissolves many molecular compounds, such as ethanol, ascorbic acid (vitamin C), and table sugar, as shown in *Figure 12*. These molecular compounds are polar because they have hydrogen atoms bonded to oxygen, which attract electrons strongly. Ethanol, for example, has a polar −OH group in its structure, CH_3CH_2OH. As a result, the negative oxygen atom of a water molecule attracts the positive hydrogen atom of an ethanol molecule. The positive hydrogen atom of a water molecule also attracts the negative oxygen atom of an ethanol molecule. This attraction is one force that pulls an ethanol molecule into water solution.

Hydrogen bonding plays a large part in the dissolving of other molecular compounds such as sucrose, $C_{12}H_{22}O_{11}$ as shown in *Figure 13*. Water molecules form hydrogen bonds with the −OH groups in the sucrose molecule and easily pull the sucrose molecules away from the sugar crystal and into solution.

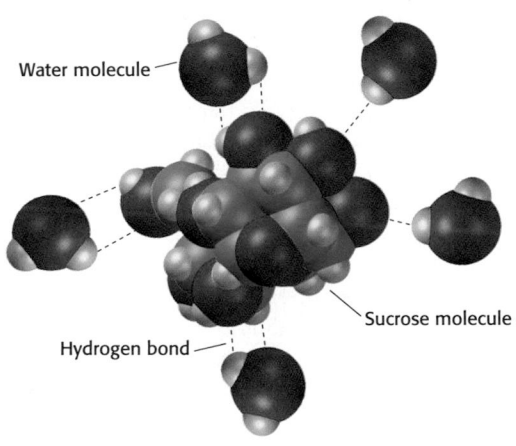

Water molecule

Sucrose molecule

Hydrogen bond

Did You Know?

Intra- and Intermolecular Forces
Covalent, metallic, and ionic bonds are caused by intramolecular forces, that is, forces within a molecule, between ions, or between a metallic ion and surrounding electrons. Hydrogen bonds are caused by intermolecular forces (forces between molecules). An ion-polar bond is another example of an intermolecular bond. This bond occurs because of the attraction between an ion and the opposite partial charge of a polar molecule. A third example of a bond caused by intermolecular forces is a polar-polar bond and it is caused by the attraction between opposite partial charges of two polar molecules. The table on the facing page compares the relative strengths of these bonds. Copy the table on the chalkboard and discuss with students the types of bonds formed by intramolecular and intermolecular forces. Have students compare the relative strengths of the bonds. Point out to students that the ion-polar bond, which forms between water molecules and ions of dissolved substances, are strong enough to overcome bonds in some ionic substances.

What dissolves a nonpolar substance?

Some elements, such as oxygen, nitrogen, and chlorine exist as diatomic molecules. These molecules are completely nonpolar because neither atom in the molecule attracts electrons more strongly. Iodine, I_2, is a nonpolar molecule. In this activity, you will determine whether water or ethanol dissolves iodine better.

1. Dip a cotton swab in tincture of iodine, and make two small spots on the palm of your hand. Let the spots dry. The spots that remain are iodine.

2. Dip a second cotton swab in water, and wash one of the iodine spots with it. What happens to the iodine spot?

3. Dip a third cotton swab in ethanol, and wash the other iodine spot with it. What happens to the iodine spot?

4. Did water or ethanol dissolve the iodine spot better? Is water more polar or less polar than ethanol? Write a paragraph explaining your reasoning. **WRITING SKILL**

Like dissolves like

A rule of thumb in chemistry is that like dissolves like. This rule means that a solvent will dissolve substances that are like the solvent in molecular structure. For example, water is a polar molecule because it has positive and negative ends. So, water dissolves ions and polar molecules, which also have charges.

Nonpolar compounds usually will not dissolve in water. Nonpolar molecules do not have charges on opposite sides, like water molecules have. For example, olive oil is not soluble in water because it is a mixture of nonpolar compounds. Water does not attract nonpolar molecules enough to overcome the attractive forces between the nonpolar molecules. So, nonpolar solvents must be used to dissolve nonpolar materials, as shown in *Figure 14.* Nonpolar solvents that are distilled from petroleum are often used to dissolve nonpolar materials.

▶ **nonpolar compound** a compound whose electrons are equally distributed among its atoms

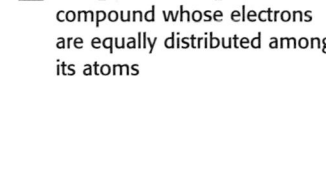
Figure 14
Nonpolar substances, such as oil-based paint, must be dissolved by a like solvent. This means that a nonpolar solvent must be used.

235

Ask students if hydrogen bonds are strong enough to overcome polar-polar bonds. (yes) Ask students to use the relative strength information to explain why water can dissolve some polar substances. (By water forming hydrogen bonds with molecules of the polar substance, the hydrogen bonds can overcome the polar-polar bonds between the polar molecules of the substance and pull the molecules into solution.) **LS Logical**

What dissolves a nonpolar substance?
Materials (per group):
• three cotton swabs
• tincture of iodine
• water
• ethanol

Teacher's Notes: Be sure no students in the class have skin reactions to iodine before performing this activity. Warn students that tincture of iodine stains clothing and that iodine is poisonous. Have students wear safety goggles and laboratory aprons. Properly dispose of cotton swabs containing iodine according to your local environmental guidelines.

Results: Students will find that ethanol removes iodine better than water does. Because iodine is nonpolar, it will dissolve better in a solvent that has little or no polarity. Because iodine is more soluble in ethanol than water, ethanol must be less polar than water is. **LS Visual**

Teaching Tip — BASIC

Nonpolar molecules Molecules of substances, such as methane (CH_4), are nonpolar because of the arrangement of the atoms that make up the molecules. Each methane molecule consists of a central carbon atom surrounded by four covalently bonded hydrogen atoms arranged symmetrically around the carbon atom. Point out that carbon tetrachloride, CCl_4, is also a nonpolar substance. Have students predict the shape of the CCl_4 molecule. (A central carbon atom surrounded by four covalently bonded hydrogen atoms arranged symmetrically around it.) **LS Logical**

Chapter Resource File

• **Quick Activity Datasheet** What dissolves a nonpolar substance? **GENERAL**

Demonstration ── GENERAL

Effect of Stirring on Dissolving

(Time: Approximately 15 minutes)

Materials
• 500 mL beakers (2)
• stirring plate
• magnetic stirring bar
• tap water
• powdered drink mix

Before doing the demonstration, attach white paper to the back of the beakers to increase the effectiveness of the demonstration by making the color changes more dramatic.

Step 1 Fill each beaker with room temperature tap water.

Step 2 Add the magnetic stirring bar to one of the beakers. Turn on the stirring plate and adjust the rate of spinning so that a slight vortex is observed at the surface of the water.

Step 3 Have students predict with a show of hands in which beaker—stirred or unstirred—a powdered drink mix will dissolve more quickly.

Step 4 Carefully drop a small amount of the mix into each beaker. Have students observe the contents of each beaker.

Analysis

1. In which beaker did the drink mix dissolve more quickly? (the stirred beaker)

2. How might stirring increase solubility? (Stirring moves the dissolved sugar away from the crystal so that more water molecules can interact with the solid.) **LS** Visual

The Dissolving Process

According to the kinetic theory of matter, water molecules in a glass of tea are always moving. When sugar is poured into the tea, water molecules collide and transfer energy to the sugar molecules at the surface of the sugar crystal. This energy, as well as the attractive forces between water and sugar molecules, causes molecules at the surface of the crystal to dissolve.

Every time a layer of sugar molecules leaves the crystal, another layer of sugar molecules is uncovered. Sugar molecules break away from the crystal layer by layer in this way until the crystal completely dissolves.

Solutes with a larger surface area dissolve faster

When a solid is crushed into small pieces, it dissolves faster than the same substance in large pieces. Breaking the solid into smaller pieces exposes more surface area, uncovering more of the solute. More molecules exposed to a solvent leads to more solute-solvent collisions. Therefore, the solid dissolves faster.

Figure 15 shows how breaking a solid into many smaller pieces increases the total exposed surface area. The large crystal is a cube that is 1 cm on each edge, so each face of the crystal has an area of 1 cm^2. A cube has six surfaces, so the total surface area is 6 cm^2. Suppose the large cube is cut into 1000 cubes that are 0.1 cm on each edge. The face of each small cube has an area of 0.1 cm $\times$ 0.1 cm = 0.01 cm^2, so each cube has a surface area of 0.06 cm^2. The total surface area of the small cubes is 1000 $\times$ 0.06 cm^2 = 60 cm^2, which is ten times the surface area of the large cube.

Figure 15

 This salt crystal has a small surface area compared to its total volume.

 When the salt crystal is crushed into small pieces, more of the salt can be exposed to a solute, which allows the salt to dissolve faster.

0.1 cm

1 cm

236

Did You Know?

Brining Some food products are prepared by soaking them in brine, a solution of sodium chloride. The technique dehydrates the food by causing water to leave the product by osmosis and enter the brine. Cheeses of the Swiss variety are soaked in brine solution sometimes up to two weeks. The brining not only dehydrates the curd but also inhibits the activity of the starter bacteria. Brining also cures meats and fish. Lox, a variety of smoked salmon, is prepared by first brining the cleaned salmon, then air drying it, and finally smoking it.

Historical Perspective

Deep Drilling for Brine By the first century B.C., the Chinese had developed methods of drilling to depths of about 1.5 km to obtain brine. The brine was brought to the surface by natural pressure of the wells or later by pumps. At the surface the salt was recovered by evaporation or by heating the brine in large cast-iron pans that were heated by burning natural gas, which was also obtained from the wells.

Figure 16

How Stirring Affects the Dissolving Process

A If the solution is not stirred, crystals of table sugar, or sucrose, become surrounded by dissolved sugar molecules, which prevent water molecules from reaching the undissolved sugar.

B Stirring moves the dissolved sugar molecules away from the undissolved sugar, and more water molecules can reach the undissolved sugar.

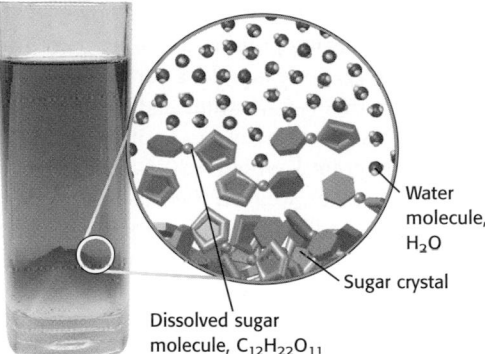

Water molecule, H_2O

Sugar crystal

Dissolved sugar molecule, $C_{12}H_{22}O_{11}$

Stirring or shaking a solution helps the solute dissolve faster

If you pour sugar in a glass of water and let it sit without stirring, the sugar will take a long time to dissolve completely. The sugar takes a long time to dissolve because it is at the bottom of the glass surrounded by dissolved sugar molecules, as shown in **Figure 16A.** Dissolved sugar molecules will slowly *diffuse,* or spread out, throughout the entire solution. But until that happens, the dissolved sugar molecules cluster near the surface of the crystal. These dissolved molecules keep water molecules from reaching the sugar that has not yet dissolved.

Stirring or shaking the solution moves the dissolved sugar away from the sugar crystals. So, more water molecules can interact with the solid, as shown in **Figure 16B,** and the sugar crystals dissolve faster.

Solutes dissolve faster when the solvent is hot

Solutes dissolve faster in a hot solvent than in a cold solvent. The kinetic theory states that when matter is heated, its particles move faster. As a result of heating, particles of solvent collide with undissolved solute more often. These collisions also transfer more energy than collisions that occur when the solvent is cold. The greater frequency and energy of the collisions help "knock" undissolved solute particles away from each other and spread them throughout the solution.

⊿ internet connect

www.scilinks.org
Topic: Diffusion
SciLinks code: HK4160

SCI Maintained by the
$LINKS$ National Science
Teachers Association

237

Chapter 7 • Solutions   **237**

Teaching Tip

Boiling points/freezing points of solutions Adding a nonvolatile solute (one that does not readily evaporate) to a solvent "stretches" the temperature range in which the solvent can still remain a liquid.

Close

Quiz

1. Why is a water molecule a polar molecule? (The oxygen atom attracts electrons strongly leaving the hydrogen atoms with a partial positive charge.)

2. A hydrogen bond is a weak bond that results from the attraction between a hydrogen atom and another atom that is rich in what? (electrons)

3. How does grinding a solid affect the rate at which it dissolves? (Grinding the solid increases the surface area of the solid to the solvent, allowing the solid to dissolve faster.)

4. In most cases, what happens to the solubility of solids as the temperature of the solvent increases. (The solubility increases with increased temperature.)

Chapter Resource File

- **Concept Review** GENERAL
- **Quiz** BASIC

Figure 17
The coolant mixture of ethylene glycol and water keeps the radiator fluid from freezing in winter and boiling in summer.

Solutes affect the physical properties of a solution

The boiling point of pure water is 100°C. But, if you dissolve 12 g of sodium chloride in 100 mL of water, you will find that the boiling point of the solution is increased to about 102°C. Also, the freezing point of the solution will be lowered to about −7°C. Many solutes increase the boiling point of a solution above that of the pure solvent. Likewise, the same solutes lower the freezing point of the solution below that of the pure solvent.

The effect of a solute on freezing point and boiling point can be useful, as shown in **Figure 17**. For example, a car's cooling system often contains a mixture that is 50% water and 50% ethylene glycol, a type of alcohol. This solution acts as antifreeze in cold weather because its freezing point is about −30°C. The solution also helps prevent boiling in hot weather because its boiling point is about 109°C. Adding sodium chloride to ice can lower the freezing point of the mixture to about −15°C, which is cold enough to freeze ice cream.

SECTION 2 REVIEW

SUMMARY

▶ Water dissolves ionic compounds and polar molecular compounds because water molecules are polar.

▶ To dissolve a solute, a solvent must attract solute particles more strongly than solute particles attract each other.

▶ Nonpolar compounds do not dissolve in water but may be dissolved by nonpolar solvents.

▶ Solids with a greater surface area dissolve faster.

▶ Stirring or heating a solvent dissolves solutes faster.

▶ Solutes can lower the freezing point of a solution and raise its boiling point.

1. **Explain** how water can dissolve some ionic compounds, such as ammonium chloride, NH_4Cl, as well as some molecular compounds, such as methanol.

2. **Describe** the relationship of attractive forces between molecules and the ability of a solvent to dissolve a substance.

3. **Explain** why large crystals of coarse sea salt take longer to dissolve in water than crystals of fine table salt.

4. **Use** the like dissolves like rule to predict whether glycerol, which is a polar molecular compound, is soluble in water.

5. **Describe** three methods you could use to make a spoonful of salt dissolve faster in water.

6. **Critical Thinking** You have made some strawberry-flavored drink from water, sugar, and drink mix. You decide to freeze the mixture into ice cubes, so you pour the liquid into an ice-cube tray and place it in the freezer along with another tray of plain water. Two hours later, you find the water has frozen, but the fruit drink has not. How can you explain this result?

238

Answers to Section 2 Review

1. Water is a polar molecule, which means that one end has a slight positive charge and the other end has a slight negative charge. These charged ends are attracted to opposite charges on other polar molecules, such as methanol and to charged ions in ionic substances, such as ammonium chloride. As a result water molecules can pull polar molecules and ions into solution.

2. Solvents can dissolve a substance only if the attraction of the solvent molecules for the particles of the substance is greater than the attraction between the particles of the substance.

3. The large crystals of salt have less surface area exposed to water than the same mass of smaller crystals. Therefore, more sodium and chloride ions are attracted by water molecules at the surface of the smaller crystals.

4. Water will dissolve glycerol because the polar water molecules attract the polar groups of glycerol and pull it into solution.

5. Raising the temperature of the water, shaking, or stirring the mixture would increase the rate of dissolving of the salt.

See "Continuation of Answers" at the end of the chapter.

Solubility and Concentration

OBJECTIVES

▶ **Explain** the meaning of solubility and compare the solubilities of various substances.

▶ **Describe** dilute, concentrated, saturated, unsaturated, and supersaturated solutions.

▶ **Relate** changes in temperature and pressure to changes in solubility of solid and gaseous solutes.

▶ **Express** the concentration of a solution as molarity, and calculate the molarity of a solution given the amount of solute and the volume of the solution.

▶ **KEY TERMS**
solubility
concentration
unsaturated solution
saturated solution
supersaturated solution
molarity

How much would you have to shake, stir, or heat the olive oil and water mixture shown in *Figure 18* to dissolve the oil in the water? The answer is that the oil would never dissolve in the water no matter what you did. Some substances are *insoluble* in water, meaning they do not dissolve. Other substances such as sugar and baking soda are said to be *soluble* in water because they dissolve easily in water. However, there is often a limit to how much of a substance will dissolve.

Solubility in Water

Have you ever tried to dissolve several spoonfuls of salt in a glass of water? If you have tried, you may have observed that some of the salt did not dissolve, no matter how much you stirred. Unlike olive oil, salt is soluble in water, but the amount of salt that will dissolve is limited. The maximum amount of salt that can be dissolved in 100 g of water at room temperature is 36 g, or about two tablespoonfuls. The maximum mass of a solute that will dissolve in 100 g of water at 20°C and standard atmospheric pressure is known as the **solubility** of the substance in water.

Some substances such as acetic acid, methanol, ethanol, glycerol, and ethylene glycol are completely soluble in water. This means that any amount of the substance will mix with water to form a solution. Some ionic compounds, such as silver chloride, AgCl, are almost completely insoluble in water. Only 0.000 19 g of AgCl will dissolve in 100 g of water.

▶ **solubility** the maximum amount of a solute that will dissolve in a given quantity of solvent at a given temperature and pressure

Figure 18

Olive oil and water form two layers when they are mixed because olive oil is *insoluble* in water.

239

Science as a Human Endeavor

Student Involvement in Space Station Crystal Growth Experiments More than 100 high school students prepared and loaded samples of protein crystal solutions that the Space Shuttle, *Atlantis*, carried to the International Space Station on September 8, 2000. The students packed hundreds of tiny tubes each containing a frozen solution of a protein, virus, or macromolecule (very large molecules) and a precipitant in various concentrations into a Thermos-like system that keeps the samples frozen until they reach the space station. In the space station each sample was thawed and combined with the precipitant. The reaction of the precipitant with the solution caused crystals to form in the weightless environment of the space station. Research into protein crystal growth may one day lead to improved drug design and treatment of diseases.

Conservation of Mass Have students refer to Table 1 and ask them how they would make a saturated solution of calcium chloride using 100 g of water. (Dissolve 75 g CaCl$_2$ in the water.) Ask them what the mass of this solution would be. (75 g + 100 g = 175 g) If students hesitate answering, remind them that the mass of a solution is equal to the combined masses of the solute and solvent. **LS** Logical

SKILL BUILDER — **BASIC**

Interpreting Tables Have students refer to **Table 1** and ask them the following questions. How many characteristics of each substance are given in the table? (3—the substance name, chemical formula, and solubility) How many substances are listed? (10) What is the solubility of calcium chloride at 20°C and what does this value mean? (75 g CaCl$_2$/100 g H$_2$O; 75 g of calcium chloride will dissolve in 100 g of water at a temperature of 20°C.) What substance is the most soluble in water at 20°C? (silver nitrate) least soluble at 20°C? (silver chloride) **LS** Logical

Table 1 Solubilities of Some Ionic Compounds in Water

Substance	Formula	Solubility in g/100 g H$_2$O at 20°C
Silver nitrate	AgNO$_3$	216
Silver chloride	AgCl	0.000 19
Sodium fluoride	NaF	4.06
Sodium chloride	NaCl	35.9
Sodium iodide	NaI	178
Calcium chloride	CaCl$_2$	75
Calcium sulfate	CaSO$_4$	0.32
Calcium fluoride	CaF$_2$	0.0015
Sodium sulfide	Na$_2$S	26.3
Iron(II) sulfide	FeS	0.0006

▶ **concentration** the amount of a particular substance in a given quantity of a mixture, solution, or ore

▶ **unsaturated solution** a solution that contains less solute than a saturated solution does and that is able to dissolve additional solute

☑ **internet** connect

www.scilinks.org
Topic: Solubility
SciLinks code: HK4129

SC*LINKS* Maintained by the National Science Teachers Association

240

Different substances have different solubilities
Even closely related compounds can have very different solubilities. Compare the solubilities of some of the ionic compounds in *Table 1.* All compounds listed that contain sodium also contain one other element, but sodium iodide is much more soluble than either sodium chloride or sodium fluoride.

The solubility of any substance in water depends on the strength of the forces acting between water molecules and solute particles compared to the forces acting between the solute particles. Sodium iodide is more soluble than many other compounds that also contain sodium because the forces between water molecules and particles of sodium iodide are much greater than the forces between sodium iodide particles.

How much of a substance can dissolve in a solvent?
Because substances vary greatly in solubility, you need a way to specify how much solute is dissolved in a given solution. One way is to refer to a solution as *weak* if only a small amount of solute is dissolved or *strong* if a large amount of solute is dissolved. However, *weak* and *strong* mean different things to different people. For example, most people would describe the sulfuric acid solution found in automobile batteries as strong. The solution can injure the skin and react with the fiber in textiles to create holes in clothing. A chemist, though, might describe battery acid as a weak solution of sulfuric acid because the chemist knows that much stronger solutions of sulfuric acid can be prepared.

The terms *weak* and *strong* do not specify the **concentration** of a solution. Concentration is the quantity of solute dissolved in a given volume of solution. Scientists refer to a solution that contains a large amount of solute as *concentrated*. A solution that contains only a small amount of solute is *dilute*.

Unsaturated solutions can dissolve more solute
The terms *concentrated* and *dilute* give no information about the actual quantity of solute dissolved in a solution. An **unsaturated solution** contains less than the maximum amount of solute that will dissolve in the solvent under the same conditions. The solution of sodium acetate in *Figure 19A* is unsaturated. A solution is unsaturated as long as it is able to dissolve more solute.

REAL WORLD APPLICATIONS

Le Châtelier's Principle and Plumping Raisins (Prior to discussing this topic, obtain a few raisins and plump half of them by placing them in some warm water for about 15 minutes.) Have students examine the plumped and unplumped raisins and compare them. Ask them what the difference is. (The plumped raisins seem to have more water.) Explain to students that when a raisin is placed in water, stress is placed on the system because the concentration of the materials within the raisin is less than the concentration of materials without. Normally, the solutes would move from a region of higher concentration within the raisin to one of lower concentration outside the raisin. However, the solutes cannot pass easily through the membranes of the cells within the raisin, but the solvent can. Thus, water moves at a greater rate into the raisins than solutes leave. As a result the raisin swells. Ask students when a new equilibrium will be reached. (When the rates of the water entering and the solutes leaving the raisins are equal.) **LS** Visual

Figure 19
How Concentration Affects the Dissolving Process

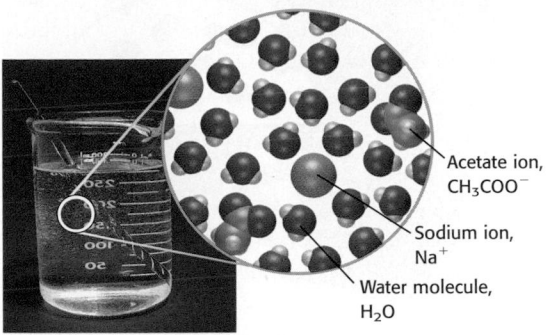

Acetate ion,
CH₃COO⁻

Sodium ion,
Na⁺

Water molecule,
H₂O

A Additional sodium acetate can dissolve when added to this unsaturated solution.

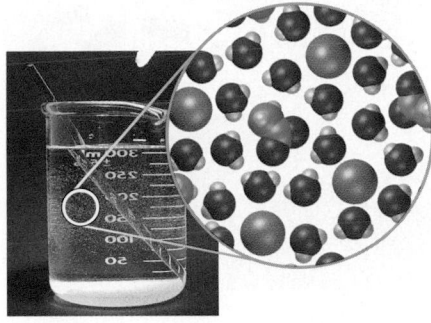

B No more sodium acetate will dissolve in this saturated solution. Any additional sodium acetate that is dissolved causes an equal amount to settle out of the solution.

At some point, most solutions become saturated with solutes

If you keep adding sodium acetate to the solution in *Figure 19A,* the added sodium acetate dissolves until the solution becomes saturated, as shown in *Figure 19B.* A **saturated solution** can dissolve no more solute. The dissolved solute is in equilibrium with undissolved solute. To be in equilibrium means that the dissolved solute re-forms crystals at the same rate that the undissolved solute dissolves. So if you add more solute, it just settles to the bottom of the container. No matter how much you stir, no more sodium acetate will dissolve in a saturated solution.

Heating a saturated solution usually dissolves more solute

The solubility of most solutes increases as the temperature of the solution increases. If you heat a saturated solution of sodium acetate, more sodium acetate can dissolve until the solution becomes saturated at the higher temperature.

But something interesting happens when the temperature of the solution decreases again. The extra sodium acetate does not re-form crystals, but remains in solution. At the cooler temperature, this unstable **supersaturated solution** holds more solute than it normally can. Adding a small crystal of sodium acetate to the solution provides the surface that the excess solute needs to crystallize, as shown in *Figure 20.* The solute crystallizes out of the solution until the solution is saturated at the cooler temperature.

▶ **saturated solution** a solution that cannot dissolve any more solute under the given conditions

▶ **supersaturated solution** a solution that holds more dissolved solute than is required to reach equilibrium at a given temperature

Figure 20
Adding a single crystal of sodium acetate to a supersaturated solution causes the excess sodium acetate to quickly crystallize out of the solution.

241

Interpreting Visuals Have students refer to **Figure 21.** Ask students if there is any evidence in Figure 21A that a gas is dissolved in the soda. (no) Then ask students what evidence in Figure 21B allows them to infer that a soda contains a dissolved gas. (the formation of bubbles within the soda) **LS Visual**

Teaching Tip

Saturation Conditions Be sure students understand that "the given conditions" mentioned in the definition of a saturated solution refer to temperature and, if a gas is involved, pressure. Because most substances have a unique solubility for each temperature, expressing solubility values are meaningless unless temperature is stated. Because pressure affects the solubility of gases, the existing pressure of a gas must be stated for the solubility value to have meaning.

CO₂ under high pressure above solvent

Dissolved CO₂ molecules

A Carbon dioxide gas is dissolved in this unopened bottle of soda.

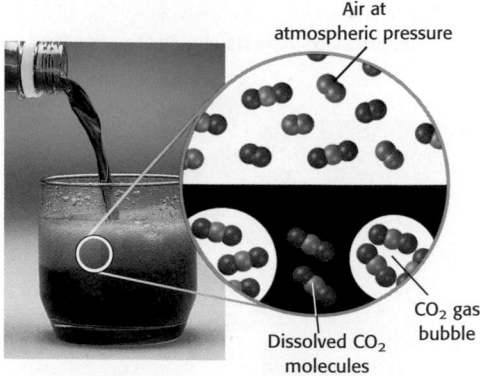

Air at atmospheric pressure

Dissolved CO₂ molecules

CO₂ gas bubble

B When the bottle is opened, the pressure inside the bottle decreases. Carbon dioxide gas then forms bubbles as it comes out of solution.

Figure 21
The solubility of a gas in water depends on the pressure of the gas.

Temperature and pressure affect the solubility of gases

Gases can also dissolve. For instance, some household cleaners contain ammonia gas dissolved in water. Soda is a solution of carbon dioxide gas, sweetener, and flavorings in water. Unlike solid solutes, gaseous solutes are less soluble in warmer water. For example, soda goes flat quickly at room temperature but stays fizzy for a longer period of time when it is cold.

The solubility of gases also depends on pressure, as shown in *Figure 21.* Carbon dioxide is dissolved in the soda under high pressure, and the bottle is sealed. When the bottle is opened, the gas pressure decreases to atmospheric pressure and the soda fizzes as the carbon dioxide comes out of solution.

Divers must understand the solubility of gases. Increased pressure underwater causes nitrogen gas to dissolve in the blood. If the diver returns to the surface too quickly, nitrogen comes out of solution and forms bubbles in blood vessels. This condition, called the *bends,* is extremely painful and dangerous.

Concentration of Solutions

Describing a solution as concentrated, dilute, saturated, or unsaturated still does not reveal the quantity of dissolved solute. For example, only 0.173 g of calcium hydroxide will dissolve in 100 g of water. This solution is saturated but is still dilute because so little solute is present. Scientists express the quantity of solute in a solution in several ways, but one of the most useful forms of expressing this measurement is molarity.

Trends in Sports Medicine

Treating Injuries with Oxygen A new tool to help heal sports injuries is hyperbaric oxygen (HBO). HBO is 100% oxygen that is delivered to the patient at pressure greater than atmospheric pressure. Treatment usually takes place in a chamber that is filled with HBO. One of the principles of HBO therapy is that the solubility of oxygen gas in blood increases with increased gas pressure. A patient breathing pure oxygen at three times normal air pressure (equivalent to the pressure experienced by a diver at a depth of about 30 m) will have a 15-fold increase in the amount of dissolved oxygen delivered to tissues. In the treatment of an injury, the increased oxygen would accelerate wound healing by promoting the growth of new capillaries in the area of the wound. Professional sports team such as the Detroit Red Wings, Vancouver Canucks, and Dallas Cowboys are using HBO chambers to rehabilitate injuries, such as sprains and injuries to soft tissues.

Molarity is a precise way of measuring concentration

You have already seen that the solubility of a substance can be expressed as grams of solute per 100 grams of solvent. Concentration can also be expressed as **molarity.**

$$\text{Molarity} = \frac{\text{moles of solute}}{\text{liters of solution}}, \text{ or } M = \frac{\text{mol}}{L}$$

Note that molarity is expressed as moles per liter of solution, not per liter of solvent.

A 1.0 M, which is read as "one molar," solution of NaCl contains 1.0 mol of dissolved NaCl in every 1.0 L of solution. To find the molar concentration of a solute, divide the number of moles of solute by the volume of the solution in liters.

▶ **molarity** an expression of the concentration of a solution in moles of dissolved solute per liter of solution

---

Math Skills

Molarity Calculate the molarity of sodium carbonate, Na_2CO_3, in a solution of 38.6 g of solute in 0.500 L of solution.

1 List the given and unknown values.

> **Given:** mass of sodium carbonate = 38.6 g
> volume of solution = 0.500 L
> **Unknown:** molarity, amount of Na_2CO_3 in 1 L of solution

2 Write the equation for moles Na_2CO_3 and molarity.

$$moles\ Na_2CO_3 = \frac{mass\ Na_2CO_3}{molar\ mass\ Na_2CO_3}$$

$$molarity = \frac{moles\ Na_2CO_3}{volume\ of\ solution}$$

3 Find the number of moles of Na_2CO_3 and calculate molarity.

$$molar\ mass\ Na_2CO_3 = 106\ g$$

$$moles\ Na_2CO_3 = \frac{38.6\ g}{106\ g} = 0.364\ mol\ Na_2CO_3$$

$$molarity\ of\ solution = \frac{0.364\ mol\ Na_2CO_3}{0.500\ L\ solution} = 0.728\ M$$

Practice HINT

▶ When calculating molarity, remember that molarity is moles per liter of solution, not moles per liter of solvent.

▶ If volume is given in milliliters, you must multiply by 1 L/1000 mL to change milliliters to liters. You will need to do this in Practice problems 1c and 1d.

Practice

1. a. $molarity = \dfrac{2\ mol\ CaCl_2}{1\ L\ solution} = 2\ M$

b. $molarity = \dfrac{0.75\ mol\ CuSO_4}{1.5\ L\ solution} = 0.50\ M$

c. $725\ mL\ solution \times \dfrac{1\ L}{1000\ mL} = 0.725\ L\ solution$

$molarity = \dfrac{2.25\ mol\ H_2SO_4}{0.725\ L\ solution} = 3.10\ M$

d. molar mass $Pb(NO_3)_2 = 207.2\ g + 2 \times (14\ g + (3 \times 16\ g)) = 331.2\ g$

$moles\ Pb(NO_3)_2 = \dfrac{525\ g}{331.2\ g} = 1.59\ mol\ Pb(NO_3)_2$

$molarity = \dfrac{1.59\ mol\ Pb(NO_3)_2}{1.25\ L\ solution} = 1.27\ M$

Practice

Molarity

1. Determine the molarity of each of the following solutions:

a. 2 mol of calcium chloride, $CaCl_2$, dissolved in 1 L of solution

b. 0.75 mol of copper(II) sulfate, $CuSO_4$, dissolved in 1.5 L of solution

c. 2.25 mol of sulfuric acid, H_2SO_4, dissolved in 725 mL of solution

d. 525 g of lead(II) nitrate, $Pb(NO_3)_2$, dissolved in 1250 mL of solution

243

MATH

CONNECTION — ADVANCED

Write the following values on the chalkboard.

molarity of Na^+ in seawater: 0.481 M
molarity of Cl^- in seawater: 0.560 M

Tell students that dissolved sodium chloride accounts for the sodium ions in seawater. Have students explain if dissolved sodium chloride in seawater account for all the chloride ions. (No. If sodium chloride were the only source of chloride ions, the molarity of Na^+ and Cl^- would be the same because the chemical formula of sodium chloride, NaCl, shows that the ratio of the ions is 1 to 1.) Point out that the students are correct because seawater contains dissolved magnesium chloride. **LS Logical**

Chapter Resource File

• **Math Skills** Molarity GENERAL

Additional Examples — GENERAL

Mass Percent Show students how to calculate mass percent.

$$\text{mass percent} = \frac{\text{mass of solute}}{\text{mass of solute} + \text{mass of solvent}} \times 100\%$$

Have students calculate the mass percent of a solution of 5 g NaCl dissolved in 95 g of water.

$$\text{mass percent} = \frac{5\text{g (NaCl)}}{5\text{g (NaCl)} + 95\text{ g (H}_2\text{O)}}$$

$$\text{mass percent} = \frac{5\text{ g}}{100\text{ g}} \times 100\% =$$

$$0.05 \times 100\% = 5\% \quad \text{LS Logical}$$

Close

Quiz

1. What is solubility? (the maximum amount of a solute that will dissolve in a given quantity of solvent at a given temperature and pressure)

2. In what unit is solubility measured? (g of solute per 100 g solvent)

3. Does the solubility of a gas in a liquid increase or decrease with increased pressure? (increase)

Chapter Resource File

- **Concept Review** GENERAL
- **Quiz** BASIC

Other measures of solution concentration can be used

Concentration is sometimes expressed as mass percent, which is grams of solute per 100 g of solution. To make a 5.0% solution of sodium chloride, you would dissolve 5 g of sodium chloride in 95 g of water. The concentration of ordinary rubbing alcohol solution is usually 70%, as shown in **Figure 22.**

The concentrations of solutions that contain extremely small amounts of solutes are sometimes given in parts per million, ppm, or parts per billion, ppb. Sea water contains slightly more than 1 ppm of fluoride ions, which means that every 1 000 000 g of seawater (about 1000 L) contains 1 g of dissolved fluoride ions. One million grams of water is about the mass of water in a child's plastic wading pool. Sea water also contains about 4 ppb of arsenic, or 4 g arsenic in every 1 000 000 000 g of sea water. The U.S. Environmental Protection Agency has set a maximum safe level of lead in drinking water at 15 ppb.

Figure 22

There are 70 grams of alcohol in every 100 g of this solution.

SECTION 3 REVIEW

SUMMARY

▶ The solubility of a substance is the maximum amount that can dissolve in a solvent at a certain temperature and pressure.

▶ The concentration of a solution is the quantity of solute dissolved per unit volume of solution.

▶ An unsaturated solution can dissolve more solute.

▶ A saturated solution cannot dissolve more solute.

▶ A solute's solubility is exceeded in a supersaturated solution.

▶ The solubility of a gas in water depends on both pressure and temperature.

▶ Molarity is a useful way of expressing the concentration of a solution. Molarity is moles of solute per liter of solution.

1. Explain how a solution can be both saturated and dilute at the same time. Use an example from *Table 1*.

2. Describe how a saturated solution can become supersaturated.

3. Propose a way to determine whether a saltwater solution is unsaturated, saturated, or supersaturated.

4. Compare the solubility of olive oil and acetic acid in water. What is the solubility of olive oil? of acetic acid?

5. Determine whether your sweat would evaporate more quickly if the humidity were 92% or 37%. (**Hint:** When the humidity is 100%, the air is saturated with dissolved water vapor.)

6. Express the molarity of a solution that contains 0.5 mol of calcium acetate per 1.0 L of solution.

7. Critical Thinking When you fill a glass with cold water from a faucet and then let the glass sit undisturbed for two hours, you will see small bubbles sticking to the glass. What are the bubbles? Why did they form?

Math Skills

8. Math Skills Calculate the molarity of a solution that contains 35.0 g of barium chloride, $BaCl_2$, dissolved in 450.0 mL of solution.

Answers to Section 3 Review

1. Some substances are only slightly soluble in water. For example, only 0.32 g of calcium sulfate dissolves in 100 g of water. Such a solution is saturated because it contains the maximum amount of solute, but still dilute because it contains only a little solute per unit volume.

2. The solution can be heated and additional solute dissolved. When the solution is cooled to the original temperature, it will be supersaturated.

3. Student responses should include the idea of adding a very small amount of additional

solute. If it dissolves, the solution is unsaturated. If it remains undissolved, the solution is saturated. If new crystals form in the solution, the solution is supersaturated.

4. Acetic acid is completely soluble in water, while olive oil is completely insoluble in water. The solubility of olive oil is zero, and the solubility of acetic acid is effectively infinite.

5. Evaporation would take place faster at 37% humidity because the solution of water vapor in air is less saturated than it is at 92% humidity.

See "Continuation of Answers" at the end of the chapter.

Graphing Skills

Maximum Percentage of Ammonium Nitrate Dissolved in Water

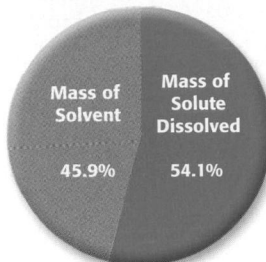

Mass of Solvent 45.9% Mass of Solute Dissolved 54.1%

Solution Temperature = 0°C

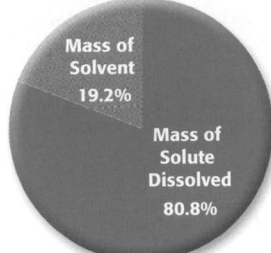

Mass of Solvent 19.2% Mass of Solute Dissolved 80.8%

Solution Temperature = 60°C

Examine the graphs, and answer the following questions. (See Appendix A for help interpreting a graph.)

1 What type of graphs are these?

2 Identify the quantities that are given in each graph. What important quantities relate the two graphs to each other?

3 By examining the graphs, what can you tell about the solubility of ammonium nitrate?

4 Suppose 500.0 g of water is used to dissolve the ammonium nitrate. What is the maximum mass of ammonium nitrate that can be dissolved at 0°C? What is the maximum mass of ammonium nitrate that can be dissolved at 60°C?

5 From the two graphs, what might the percentage of solute dissolved be if the temperature of the solution is 30°C?

6 The information in the table below shows the solubility of calcium acetate dihydrate in water at various temperatures. Construct a graph best suited for the information listed. How does solubility vary with temperature?

Temperature (°C)	Maximum mass of solute dissolved in 100 g of water (g)
0	37.4
20	34.7
60	32.7
100	29.7

245

Answers

1. pie charts
2. The percentage of solute and solvent by mass are shown in each graph. The graphs are related by different temperatures for the solution, and the changes in the solubility of the solute.
3. Ammonium nitrate dissolves easily in water, and its solubility greatly increases with solution temperature.
4. 590 g; 2100 g
5. 67.4%, which can be obtained by assuming that the percentage of solute dissolved at 30°C will be halfway between the percentages shown in the charts. Because the solubility increases with temperature, the percentage is closer to 70%.

6. The line graph of mass versus temperature should indicate that solubility of calcium acetate dihydrate in water decreases as temperature increases. See graph below.

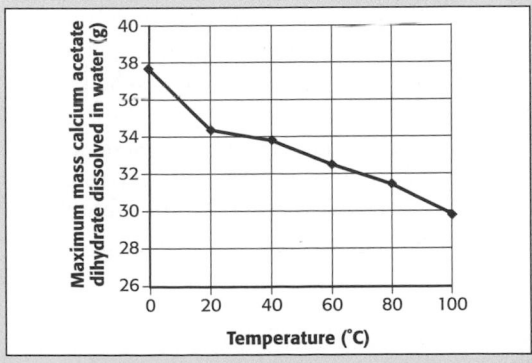

Understanding Concepts

1. c
2. b
3. c
4. a
5. b
6. b
7. a
8. a
9. d
10. c
11. b
12. a
13. b
14. c
15. d
16. a

Using Vocabulary

17. **a.** saturated

    b. unsaturated

18. An alloy is a solution because it contains particles of one or more substances, usually metals, dispersed uniformly throughout another metal. The composition of an alloy is homogeneous. Alloys are used because their properties are usually more suitable for practical applications than those of pure metals.

19. The oil in the furniture polish probably consists of hydrocarbons, which are not polar molecules. Therefore you would need a material, such as a light petroleum solvent, whose molecules are nonpolar like those of the oil. The dried spill of the lemon drink consists of substances that were dissolved in water, such as polar molecular substances and ionic substances. A solvent such

Chapter Highlights

Before you begin, review the summaries of key ideas of each section, found at the end of each section. The vocabulary terms are listed on the first page of each section.

UNDERSTANDING CONCEPTS

1. Which of the following is a homogeneous mixture?
   a. tossed salad   **c.** salt water
   b. soil   **d.** vegetable soup

2. If the label on a bottle of medicine reads "shake well before using," the medicine is probably a
   a. solution.   **c.** colloid.
   b. suspension.   **d.** gel.

3. Which of the following affects the solubility of a solute in a solvent?
   a. the surface area of the solute
   b. stirring the solution
   c. the temperature of the solvent
   d. All of the above

4. Suppose you add a teaspoon of table salt to a cool saltwater solution and stir until all of the salt dissolves. The solution you started with was
   a. unsaturated.   **c.** saturated.
   b. supersaturated.   **d.** concentrated.

5. Which of the following materials is an example of a heterogeneous mixture?
   a. air   **c.** water
   b. granite   **d.** aluminum

6. Which of the following materials is an example of a solid dissolved in another solid?
   a. smoke   **c.** mayonnaise
   b. bronze   **d.** ice

7. The dispersed particles of a suspension are _____ than the particles of a colloid.
   a. larger   **c.** lighter
   b. smaller   **d.** less dense

8. Water attracts the ions of ionic compounds because water molecules are
   a. polar.   **c.** magnetic.
   b. ions.   **d.** nonpolar.

9. To dissolve a substance, a solvent must attract particles of the substance more strongly than the _____ attract each other.
   a. solvent particles   **c.** ions
   b. water molecules   **d.** solute particles

10. To cause a solute to dissolve faster, you could
    a. add more solute.   **c.** stir the solution.
    b. cool the solution.   **d.** saturate the solution.

11. To increase the solubility of a solid substance in a solvent, you could
    a. add more solute.   **c.** stir the solution.
    b. heat the solution.   **d.** lower the pressure.

12. A solution is _____ if it contains as much dissolved solute as it will hold under a given set of conditions.
    a. saturated   **c.** supersaturated
    b. dilute   **d.** concentrated

13. Gases are more soluble in liquids when the pressure is _____ and the temperature is _____.
    a. high, high   **c.** low, high
    b. high, low   **d.** low, low

14. A solution that contains 2.0 mol of dissolved magnesium sulfate in 1.0 L of solution has a concentration of
    a. 0.5 M.   **c.** 2.0 M.
    b. 1.0 M.   **d.** 4.0 M.

15. The maximum amount of a substance that dissolves in 100 g of water is the _____ of the substance.
    a. unsaturation   **c.** dilution
    b. molarity   **d.** solubility

16. The boiling point of a solution of sugar in water is _____ the boiling point of water.
    a. higher than   **c.** the same as
    b. lower than   **d.** not related to

as water, which consists of polar molecules, should dissolve the substances in the spill.

20. Molecules of ethylene glycol have –OH groups like those of water are polar because the oxygen atom attracts electrons strongly from the hydrogen atoms. This attraction results in a partial negative charge on the oxygen atom and a partial positive charge on the hydrogen atom. Polar water molecules attract these –OH groups and pull ethylene glycol molecules into solution. Also, hydrogen bonds can form between water and ethylene glycol. Dichloroethane molecules do not have polar groups and so are not strongly attracted by water molecules.

21. A saturated solution of ethanol in water is not possible. The term *miscible* is taken to mean that the two substances can mix in any and all proportions.

22. The solubility of a gas in a liquid decreases with increasing temperature; on the other hand, the solubility of a solid in a liquid increases with increasing temperature.

23. muddy water: suspension; salt water: solution; mayonnaise: emulsion; vinegar: solution; fog: colloid; dry air: solution; cream: emulsion

USING VOCABULARY

17. A small amount of *solute* is added to two different solutions. Based on the figures below, which solution was *unsaturated*? Which solution was *saturated*? Explain your answer.

a. 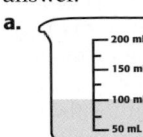 **b.**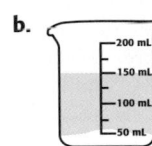

18. Explain why an *alloy* is a type of solution. Why are alloys used more often than pure metals?

19. You need to clean up a spill of oil-based liquid furniture polish and another spot where some lemon-lime soda has spilled and dried. Explain how you would apply the like dissolves like rule to clean up these two spills. Use the words *polar* and *nonpolar* in your explanation.

20. The chemical formula of ethylene glycol is $HOCH_2CH_2OH$, and the chemical formula of dichloromethane is $ClCH_2CH_2Cl$. Ethylene glycol is *miscible* in water, but dichloromethane is almost completely *immiscible* in water. Explain this difference using two properties of water.

21. Ethanol is completely *miscible* in water. Is it possible to prepare a *saturated solution* of ethanol in water? Explain your answer.

22. How does temperature affect the *solubility* of a gas solute in a liquid solvent? How does the solubility of a gas solute differ from the solubility of a solid solute in a liquid solvent?

23. Classify the following as either a *suspension*, a *colloid*, an *emulsion*, or a *solution*: muddy water, salt water, mayonnaise, vinegar, fog, dry air, and cream.

BUILDING MATH SKILLS

24. Molarity How many moles of lithium chloride are dissolved in 3.00 L of a 0.200 M solution of lithium chloride?

25. Molarity What is the molarity of 250 mL of a solution that contains 12.5 g of dissolved zinc bromide, $ZnBr_2$?

26. Solubility The solubility of lead(II) chloride, $PbCl_2$, in water at 20°C is 1.00 g $PbCl_2$/100 g of water. If you stirred 7.50 g $PbCl_2$ in 400 g of water at 20°C, what mass of lead(II) chloride would remain undissolved?

27. Solubility The solubility of sodium fluoride, NaF, is 4.06 g NaF/100 g H_2O at 20°C. What mass of NaF would you have to dissolve in 1000 g H_2O to make a saturated solution?

BUILDING GRAPHING SKILLS

28. Graphing Make a solubility graph for $AgNO_3$ from the data in the table below. Plot temperature on the *x*-axis, and plot solubility on the *y*-axis. Answer the following questions.
   a. How does the solubility of $AgNO_3$ vary with the temperature of water?
   b. Estimate the solubility of $AgNO_3$ at 35°C, at 55°C, and at 75°C.
   c. At what temperature would the solubility of $AgNO_3$ be 512 g per 100 g of H_2O?
   d. If 100 g $AgNO_3$ were added to 100 g H_2O at 10°C, would the solution be saturated or unsaturated?

Temperature (°C)	Solubility of $AgNO_3$ in g $AgNO_3$/100 g H_2O
0	122
20	216
40	311
60	440
80	585

247

Assignment Guide

Section	Questions
1	1, 2, 5–7, 18, 23, 33, 35, 40
2	8–11, 16, 19, 20, 31, 32, 34, 36, 37, 39
3	3, 4, 12–15, 17, 21, 22, 24–30, 38

Building Math Skills

24. $\text{molarity} = \dfrac{\text{moles of solute}}{\text{liters of solution}}$

moles of solute = molarity × liters of solution

moles LiCl = 0.200 M LiCl × 3.00 L solution = 0.600 mol LiCl

25. moles $ZnBr_2$ =

$\dfrac{\text{mass } ZnBr_2}{\text{molar mass } ZnBr_2} =$

$\dfrac{12.5 \text{ g } ZnBr_2}{225 \text{ g } ZnBr_2/\text{mol } ZnBr_2} =$

0.0556 mol $ZnBr_2$

$\dfrac{0.0556 \text{ mol } ZnBr_2}{0.250 \text{ L}} =$

0.222 M $ZnBr_2$

26. solubility of $PbCl_2 =$

$\dfrac{1.00 \text{ g } PbCl_2}{100 \text{ g } H_2O}$

$\dfrac{1.00 \text{ g } PbCl_2}{100 \text{ g } H_2O} = \dfrac{x \text{ g } PbCl_2}{400 \text{ mL } H_2O}$

$x = 4.00$ g $PbCl_2$

undissolved $PbCl_2$ =
7.50 g $PbCl_2$ − 4.00 g $PbCl_2$ =
3.50 g $PbCl_2$

27. $\dfrac{4.06 \text{ g NaF}}{100 \text{ g } H_2O} = \dfrac{x}{1000 \text{ g } H_2O}$

$x = 4.06 \text{ g NaF} \dfrac{1000 \text{ g } H_2O}{100 \text{ g } H_2O}$

$x = 40.6$ g NaF

Building Graphing Skills

28. a. The solubility of $AgNO_3$ increases with increasing temperature. Students may also note that solubility increases slightly faster with increasing temperature.
   b. 35°C: 285g; 55°C: 410 g; 75°C: 560 g
   c. about 93°C
   d. The solution would be unsaturated because the solubility of silver nitrate is about 160 g/100g H_2O at 10°C.

Chapter Resource File

• **Chapter Test** GENERAL
• **Test Item Listing for ExamView® Test Generator**

29. a. at about 30°C

b. at about 56°C

c. The solubility of $CoCl_2$ increases with increasing temperature, but increases more slowly above 60°C. The solubility of $CdSeO_4$ decreases steadily with increasing temperature. The solubility trend of $CdSeO_4$ in that the solubility decreases with increasing temperature. The plateau reached in the solubility curve of $CoCl_2$ is a bit unusual, too.

d. Student responses will vary but should include the idea of making a saturated solution at a cooler temperature, such as 30°C, and then heating it gently to 50°C.

e. According to the graph, the solubility of $CoCl_2$ at 20°C is about 55 g/100 g H_2O. So, about 25 g of $CoCl_2$ would remain undissolved.

30. The solubility of a gas in a liquid decreases with increasing temperature, so the graph is most likely to represent the solubility curve of a gas.

Thinking Critically

31. Methane is a nonpolar molecule, so attractive forces between methane molecules are very weak because there is no dipole interaction or opportunity for hydrogen bonding. Thus, methane remains a gas until very low temperatures have slowed down the molecules enough for them to condense to a liquid. Water is a polar molecule, so attractive forces between water molecules are much stronger than those between methane molecules.

29. Interpreting Graphs The graph below shows the solubilities of two different substances, cadmium selenate, $CdSeO_4$, and cobalt(II) chloride, $CoCl_2$, over a range of temperatures. Use the graph to answer the following questions.

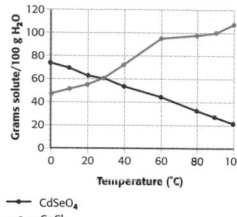

a. At what temperature do both substances have the same solubility?

b. At what temperature is the solubility of $CdSeO_4$ equal to 40 g $CdSeO_4$/100 g H_2O?

c. Describe how the solubility of each substance changes with increasing temperature. Which substance has an unusual solubility trend? How is the trend unusual?

d. Propose a way to make a supersaturated solution from a saturated solution of $CdSeO_4$ at 50°C.

e. Suppose that you add 80 g $CoCl_2$ to 100 g of water at 20°C and stir until no more $CoCl_2$ dissolves. What mass of $CoCl_2$ remains undissolved at 20°C?

30. Interpreting Graphs The line in the graph below represents the change in solubility of a substance with the temperature of the solvent. Is the substance most likely to be a solid, liquid, or gas? Explain your reasoning.

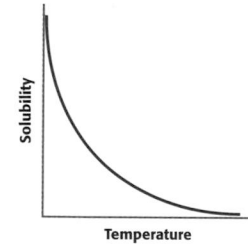

31. Applying Knowledge Substances that have similar molecular masses tend to have similar boiling points if no other factors affect the molecules of the substances. The molecular mass of methane, CH_4, is 16 amu, and the molecular mass of water is 18 amu. However, the boiling point of methane is −162°C, and the boiling point of water is 100°C. How can you account for the difference of 262°C in the boiling points of methane and water?

Methane   Water

32. Creative Thinking Sea water is a solution that contains dissolved sodium chloride, dissolved magnesium chloride, and many other dissolved ions. Suppose you have a sample of seawater that contains mud and sand. What is the simplest way to get clear sea water from this mixture? Could you use distillation to separate the sea water from the mud? Explain your answer.

33. Applying Knowledge You have been investigating the nature of suspensions, colloids, and solutions and have made the observations on four unknown mixtures in the table below. From your data, decide whether each mixture is a solution, a suspension, or a colloid.

Sample	Clarity	Settles out	Scatters light
1	clear	no	yes
2	clear	no	no
3	cloudy	yes	yes
4	cloudy	no	yes

Also, water molecules can hydrogen-bond to each other when an electron-deficient hydrogen comes near an electron-rich oxygen atom. As a result of these factors, the boiling point of water is very high for its mass.

32. Student responses may vary, but the most common means of obtaining clear seawater is by filtration. It is not possible to obtain clear seawater by distillation, because only the water would boil away leaving dissolved salts behind.

33. Sample 1: colloid; Sample 2: solution; Sample 3: suspension; Sample 4: colloid

34. Student presentations should show a crystal of an ionic compound with its positive and negative ions arranged in a cubic pattern, surrounded by water molecules. The partially positive parts of the water molecules attract the negative ions of the crystal, and the negative parts of the water molecules attract the positive ions of the crystal. As more water molecules attract each ion, the ions are pulled away from the surface of the crystal and surrounded by water molecules. The ions are then in solution. The diagram should show some of these surrounded ions with water molecules in the correct orientation for the charge on the ion.

34. Designing Systems Use what you have learned about the polarity of water molecules and how ionic substances dissolve to make a three-step diagram or computer presentation showing how water molecules attract, surround, and dissolve an ionic substance.

COMPUTER SKILL

35. Problem Solving Sometimes, powerful searchlights are pointed toward the sky to draw the public's attention to grand openings of stores and restaurants. Even when the air appears clear, you can see the searchlight beam shining into the sky. Explain why these beams are visible in air.

WRITING SKILL

BUILDING LIFE/WORK SKILLS

36. Working Cooperatively The salt you use to flavor foods is usually mined from the ground. But when salt is mined, it is often mixed with dirt and minerals. Work with a partner to design a method to yield pure salt from this mixture. Write a plan that describes your method.

37. Making Presentations Research the technique of *reverse osmosis*. Develop a presentation explaining how reverse osmosis produces drinking water from sea water. Include information about the areas of the world where this technology is commonly used and about the energy sources that are used to power the process. Demonstrate a reverse osmosis device from a store that deals in outdoor equipment.

38. Interpreting Data Use a reference book to look up the solubilities of the fluoride compounds and chloride compounds made from alkali metals and alkaline earth metals. What conclusions can you draw about the solubilities of the compounds of these metals?

INTEGRATING CONCEPTS

39. Connection to Biology Hydrogen bonding is extremely important to the function of substances within organisms. Use biology books to find three examples in which hydrogen bonds play a crucial role in the life processes of a cell.

40. Concept Mapping Copy the unfinished concept map below onto a sheet of paper. Complete the map by writing the correct word or phrase in the lettered boxes.

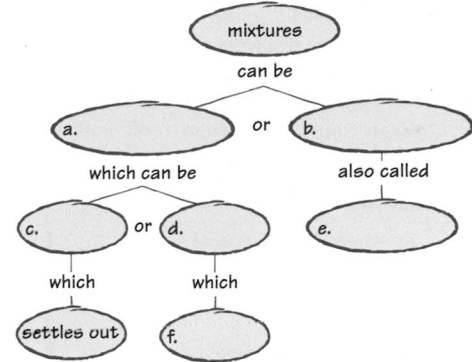

www.scilinks.org
Topic: Hydrogen Bonding SciLinks code: HK4070

SCiLINKS Maintained by the National Science Teachers Association

Art Credits: Fig. 3, Kristy Sprott; Fig. 8, Kristy Sprott; Fig. 11-12, Kristy Sprott; Fig. 14-16, Kristy Sprott; Fig. 19, Kristy Sprott; Fig. 21, Kristy Sprott.

Photo Credits: Chapter Opener image of rescue medics by Will & Deni McIntyre/Photo Researchers, Inc.; doctor preparing a shot by Jose Luis Palacz, Inc./CORBIS; Fig. 1, Peter Van Steen/HRW; Fig. 2, Victoria Smith/HRW; Fig. 3-4, Sam Dudgeon/HRW; Fig. 5, Richard Haynes/HRW; Fig. 6, Peter Van Steen/HRW; Fig. 6 (inset), Bruce Iverson; Fig. 7, Sergio Purtell/Foca/HRW; Colin Cuthbert/SPL/Photo Researchers, Inc.; "Real World Application," Sergio Purtell/Foca/HRW; Fig. 9, Image Copyright ©2004 PhotoDisc, Inc./HRW; Fig. 10, Mark Douet/Getty Images/Stone; Fig. 12, Richard Megna/Fundamental Photographs, New York; Fig. 14, Sam Dudgeon/HRW; Fig. 15, Eugene Hecht; Fig. 16, Peter Van Steen/HRW; Fig. 17, Sam Dudgeon/HRW; Fig. 18, Charlie D. Winters/Photo Researchers, Inc.; Fig. 19, Peter Van Steen/HRW; Fig. 20, Charlie Winters/Photo Researchers, Inc.; Fig. 21, Charlie Winters/HRW; Fig. 22, Sam Dudgeon/HRW; "Skills Practice Lab," Peter Van Steen/HRW; "Science in Action," (blood cells), Microworks/Phototake; (blood bank), Jerry Mason/SPL/Photo Researchers, Inc.; (mouse), Black Star.

249

---

38. Students should discover that, among the alkaline earth metal salts, fluorides tend to be significantly less soluble than the corresponding chloride. Among alkali metal salts, fluorides increase in solubility with respect to chlorides, becoming more soluble than chlorides at potassium and below.

Integrating Concepts

39. Examples may include the hydrogen bonding between the base pairs of DNA and RNA. This bonding protects the genetic code and makes exact duplication possible since each base hydrogen bonds only with its complementary base.

Hydrogen bonding between amino acids in a peptide chain is one kind of bonding that keeps proteins in the correct three-dimensional conformation to perform their functions.

Hydrogen bonding contributes to the interaction between enzymes and their substrate molecules when the enzymes put together or break apart substances

40. a. heterogeneous
   b. homogeneous
   c. a suspension
   d. a colloid
   e. a solution
   f. does not settle out

---

35. The beam reflects from materials suspended or dispersed in the air. These include suspended particles that have not settled out, but often, the air contains colloidal condensed water droplets that you cannot see. These droplets exhibit the Tyndall effect, scattering light that strikes them.

Building Life/Work Skills

36. Most plans will include making a concentrated salt solution (brine), filtering it to remove dirt particles, and then evaporating it to reclaim purified salt crystals.

37. Students should demonstrate how ordinary osmosis works, showing how water diffuses through a membrane from regions of high water concentration (dilute solutions or plain water) into regions of low water concentration (more concentrated solutions) until the water concentrations on both sides of the membrane are equal. Osmosis would normally let freshwater diffuse into saltwater, but, by applying pressure to the saltwater side of the membrane, freshwater can be made to diffuse through the membrane from the saltwater. Energy must be used to maintain the pressure as the water diffuses and to replenish the saltwater. Hand-powered reverse osmosis devices are sometimes included in life-raft equipment. Typically this method of water purification is feasible only where there is no freshwater available and where sun is abundant to produce electric power for the pumps from solar cells. This would include sunny areas near saltwater such as coastal areas in the Middle East, Africa, Mexico, Central America, and South America.

Transparencies

TT Concept Mapping

INVESTIGATING HOW TEMPERATURE AFFECTS GAS SOLUBILITY

Teacher's Notes

Time Required 1 lab period

Ratings

	1	2	3	4
	EASY			HARD

TEACHER PREPARATION	1
STUDENT SETUP	2
CONCEPT LEVEL	2
CLEANUP	1

Skills Acquired

- Collecting data
- Communicating
- Inferring
- Measuring
- Organizing and analyzing data

The Scientific Method

In this lab, students will:
- Make Observations
- Analyze the Results
- Draw Conclusions
- Communicate Results

Materials

Use thin plastic bags like the kind found in the produce section of a supermarket.

Safety Cautions

Have students review safety guidelines before working in the lab.

Skills Practice Lab

Introduction

In general, a solid solute dissolves faster in a liquid solvent if the liquid is warm. In this lab, you will determine whether this is true of carbon dioxide, a gaseous solute, dissolved in a soft drink.

Objectives

▶ **Compare** the volume of carbon dioxide released from a warm soft drink with that released from a cold soft drink.

▶ **USING SCIENTIFIC METHODS** **Draw conclusions** to relate carbon dioxide's solubility in each soft drink to the temperature of each soft drink.

Materials

beaker, 1 L
carbonated soft drinks in plastic bottles, 2
crushed ice
flexible metric tape measure
paper towels
plastic bags, 2 small
stopwatch
thermometer
twist ties, 4

Investigating How Temperature Affects Gas Solubility

▶ Procedure

Preparing for Your Experiment

1. Prepare a data table in your lab report similar to the one shown at right.

Testing the Solubility of Carbon Dioxide in a Warm Soft Drink

SAFETY CAUTION Wear safety goggles, gloves, and a laboratory apron.

2. Obtain a bottle of carbonated soft drink that has been stored at room temperature, and carry it to your lab table. Try not to disturb the liquid.

3. Use a thermometer to measure the temperature in the laboratory. Record this temperature in your data table.

4. Remove the bottle's cap, and quickly place the open end of a deflated plastic bag over the bottle's opening. Seal the bag tightly around the bottle's neck with a twist tie. Begin timing with a stopwatch.

5. When the bag is almost fully inflated, stop the stopwatch. Very carefully remove the plastic bag from the bottle, making sure to keep the bag sealed so the carbon dioxide inside does not escape. Seal the bag tightly with another twist tie.

6. Gently mold the bag into the shape of a sphere. Measure the bag's circumference in centimeters by wrapping the tape measure around the largest part of the bag. Record the circumference in your data table.

Soft Drink Data

	Temperature (°C)	Circumference of bag (cm)	Radius of of bag (cm)	Volume of of bag (cm³)
Soft drink at room temp				
Chilled soft drink				

Testing the Solubility of Carbon Dioxide in a Cold Soft Drink

7. Obtain a second bottle of carbonated soft drink that has been chilled. Place the bottle in a 1 L beaker, and pack crushed ice around the bottle. Use paper towels to dry any water on the outside of the beaker, and then carefully move the beaker to your lab table.

8. Repeat step 4. Let the second plastic bag inflate for the same length of time that the first bag was allowed to inflate. Very carefully remove the bag from the bottle as you did before. Seal the plastic bag tightly with a twist tie.

9. Wait for the bag to warm to room temperature. While you are waiting, use the thermometer to measure the temperature of the cold soft drink. Record the temperature in your data table.

10. When the bag has warmed to room temperature, repeat step 6.

▶ **Analysis**

1. Calculate the radius in centimeters of each inflated plastic bag by using the following equation. Record the results in your data table.

$$\text{radius (in cm)} = \frac{\text{circumference (in cm)}}{2\pi}$$

2. Calculate the volume in cubic centimeters of each inflated bag by using the following equation. Record the results in your data table.

$$\text{volume (in cm}^3) = \frac{4}{3}\pi \times [\text{radius (in cm)}]^3$$

3. Compare the volume of carbon dioxide released from the two soft drinks. Use your data to explain how the solubility of carbon dioxide in a soft drink is affected by temperature.

▶ **Conclusions**

4. Suppose someone says that your conclusion is not valid because a soft drink contains many other solutes besides carbon dioxide. How could you verify that your conclusion is correct?

Procedure

Techniques to Demonstrate

Show students how to seal the plastic bag with a twist tie without allowing the carbon dioxide gas inside the bag to escape in Step 5 of the Procedure.

Tips and Tricks

An alternative method of collecting the gas in Step 4 of the Procedure is to stopper the soda bottles with a one-hole stopper and collect the gas in a graduated cylinder under a bath of warm water. This method can be used to accelerate the results of the room temperature bottle.

Disposal Information

Dispose of soft drinks by pouring them down the sink. Discard the plastic bottles, bags, twist ties, and paper towels in the trash.

Answers to Analysis

1. Answers will vary.

2. Answers will vary.

3. The volume of the bag attached to the room-temperature drink should be greater than the volume of the bag attached to the cold drink. This would mean that more carbon dioxide came out of the room-temperature drink, indicating that the solubility of carbon dioxide in the soft drink decreases as temperature increases.

Answers to Conclusions

4. Students could repeat the experiment using a solution such as seltzer water.

Chapter Resource File

- **Datasheet** Investigating How Temperature Affects Gas Solubility GENERAL
- **Observation Lab** Separating Substances in a Mixture BASIC
- **CBL™ Probeware Lab** Determining the Concentration of an Ionic Solution ADVANCED

Continuation of Answers

Continuation of Answers from p. 238
Section 2 Review

6. The presence of a solute lowers the freezing point of a solution below that of the pure solvent. So, the freezing point of the drink mix is lower than that of water, and the drink mix has not become cold enough to freeze, although the water has.

Continuation of Answers from p. 244
Section 3 Review

6. 0.5 M

7. Student hypotheses should include the idea that gases, such as nitrogen and oxygen, from the air are dissolved in the cold water. Gas solubility decreases with increasing temperature, so the dissolved gas has come out of solution, forming bubbles as the temperature of the tap water increased.

Math Skills

8. 0.373 M

In Search of a Blood Substitute

Teaching Tip

Blood Have students recall the main functions of blood is to transport oxygen and nutrients to the cells and carry carbon dioxide and waste products from the cells and transport disease fighting white cells and platelets throughout the body. Explain that that blood is a suspension of solid substances (red blood cells, white blood cells, and platelets) in a liquid, called plasma and that plasma is a water solution of dissolved nutrients.

Point out that only about one percent of the oxygen necessary for respiration is dissolved in the plasma. The remainder is hydrogen-bonded to hemoglobin molecules in the red blood cells.

Unlike hemoglobin-based oxygen transport, some artificial blood substitutes use an emulsion of oxygen (up to 15%) in inert substances called perfluorocarbons to transport oxygen throughout the body.

Science in ACTION

In Search of a Blood Substitute

When patients lose a lot of blood, doctors give them a blood transfusion, or an infusion of replacement blood. This replacement blood comes from either the patient or a blood donor. Blood donation has declined in recent years, leaving many areas with severe blood shortages. For several years, sick or injured dogs have received transfusions of a new blood substitute. Some day, blood substitutes could save human lives by alleviating blood shortages and by helping people who have rare blood types and other conditions that make traditional blood transfusions difficult.

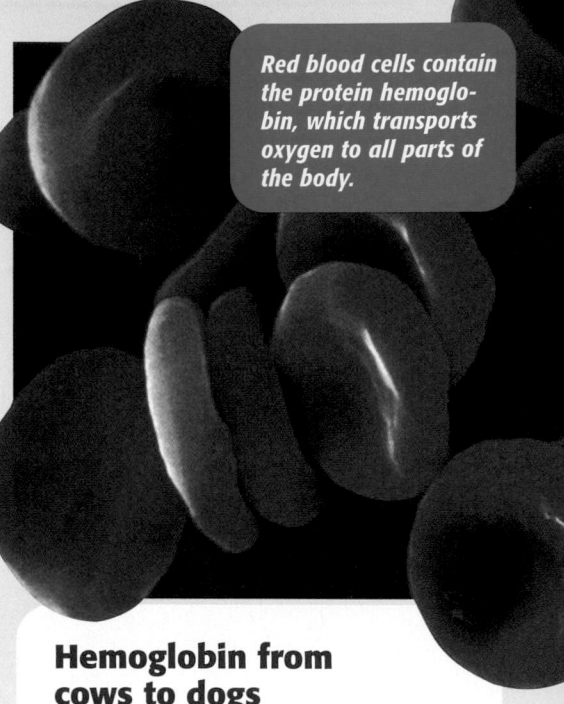

Red blood cells contain the protein hemoglobin, which transports oxygen to all parts of the body.

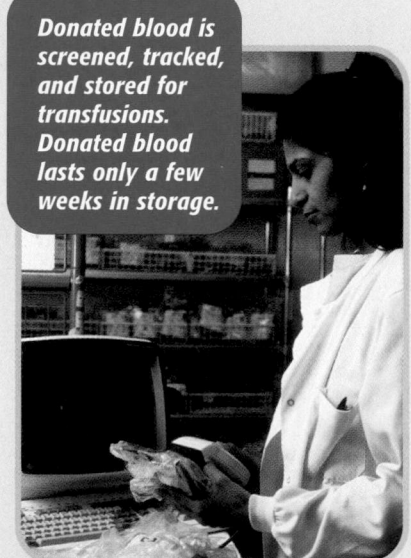

Donated blood is screened, tracked, and stored for transfusions. Donated blood lasts only a few weeks in storage.

Hemoglobin from cows to dogs

When an animal or human is injured, the number of red blood cells in the blood drops. This drop may cause the body's tissues to be starved for oxygen. The artificial blood now available for injured dogs reduces the risk of anemia by replacing the natural hemoglobin, the protein that allows blood to carry oxygen through the body. The artificial blood given to dogs is made from hemoglobin recycled from slaughtered cows. Because this hemoglobin lacks the complex, natural covering normally present in red blood cells, the hemoglobin is often absorbed by the dog's body in less than a day. These products are saving dogs' lives by giving dogs time to replace the blood they have lost and to recover from their injuries.

252

Artificial blood for humans

Researchers are now racing to develop a blood substitute for humans. One promising new family of substances known as *perfluorocarbons* is made from fluorine and carbon. Perfluorocarbons are oily substances that are coated with a bonding chemical that allows them to mix with a water-based saline solution. Perfluorocarbons have a major advantage over real blood because they don't have to match the patient's blood type. Perfluorocarbons also carry twice as much oxygen as human hemoglobin does. And unlike donated human blood, which must be thrown out after six weeks, artificial blood can be stored for up to two years. However, artificial blood does have one major limitation—it lacks white blood cells and platelets, a critical part of the body's defense and healing systems. For this reason, artificial blood will probably never be a permanent replacement for real blood. If approved for humans, blood substitutes will most likely be used for short-term needs (such as during surgery) until matching human blood can be found.

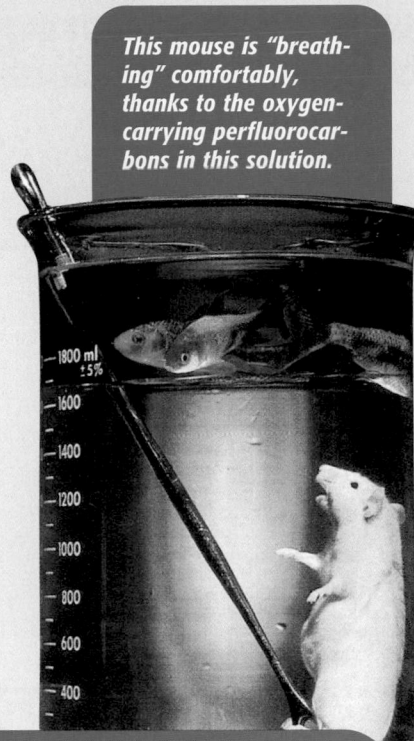

This mouse is "breathing" comfortably, thanks to the oxygen-carrying perfluorocarbons in this solution.

> ## Science and You

1. **Applying Knowledge** Describe one medical condition that can be cured or helped with artificial blood.

2. **Understanding Concepts** If perfluorocarbons are more efficient at carrying oxygen through the body than natural blood is, why don't doctors expect to use perfluorocarbons as a permanent blood replacement?

3. **Critical Thinking** Do you think that the ability of artificial blood to be stored for long periods of time will help reduce blood shortages? Why or why not?

4. **Creative Thinking** Why do you think doctors don't use cow hemoglobin for human use?

5. **Acquiring and Evaluating** Research the medical condition called *sickle cell anemia,* and write a short paper that answers the following questions: What is sickle cell anemia? How are sickle cells different from healthy cells? How is the condition normally treated? How might artificial blood be used to treat people who have sickle cell anemia?

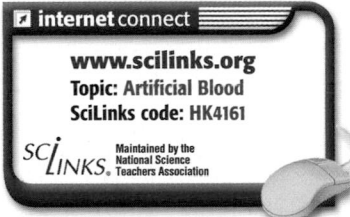

internet connect

www.scilinks.org
Topic: Artificial Blood
SciLinks code: HK4161

SCiLINKS. Maintained by the National Science Teachers Association

253

Science and You

1. Artificial blood can be used during surgery to avoid anemia.

2. Artificial blood lacks white blood cells and platelets.

3. Present types of artificial blood will alleviate blood shortages for situations that require short-term use of blood.

4. Possible answer: cow's hemoglobin may be attacked by human immune system.

5. Sickle cell anemia is a genetically inherited disease in which the amino acid sequence of the hemoglobin molecule is altered causing the cell to have an altered shape and a shorter life span than normal red blood cells. The curved, sickle-shape of the red blood cells differs from the disk-shape of normal blood cells, and causes them to lodge in capillaries which impairs circulation causing damage to bones and kidneys. The condition is normally treated with transfusions and drugs. The artificial blood could transport oxygen more efficiently than the misshapen cells.

Acids, Bases, and Salts
Chapter Planning Guide

PACING	CLASSROOM RESOURCES	LABS, ACTIVITIES, AND DEMONSTRATIONS
BLOCK 1 · 45 min pp. 254–255 **Chapter Opener**		**SE Activity 1,** p. 255 **SE Activity 2,** p. 255
BLOCKS 2 & 3 · 90 min pp. 256–263 **Section 1** Acids and Bases	**CRF Lesson Plan*** **TT Bellringer*** **TT pH Scale*** **TM Some Common Acids*** **TM Some Common Bases***	**SE Quick Lab** Which household substances are acidic, which are basic, and which are neither?, p. 260 `GENERAL` **CRF Datasheets for In-Text Labs** Which household substances are acidic, which are basic, and which are neither?* `GENERAL` **TE Demonstration** A Natural pH Indicator, p. 262 `GENERAL` **CRF CBL™ Probeware Lab** Determining the Concentration of an Acid Solution* ◆ `ADVANCED` **CRF Observation Lab** Properties of Acids and Bases* ◆ `BASIC`
BLOCKS 4 & 5 · 90 min pp. 264–268 **Section 2** Reactions of Acids with Bases	**CRF Lesson Plan*** **TT Bellringer*** **TT Neutralization Reaction***	**TE Opening Activity,** p. 264 `GENERAL` **TE Demonstration** Spectator Ions, p. 264 `GENERAL` **SE Skills Practice Lab** Measuring Quantities in an Acid-Base Reaction, pp. 280–281 ◆ `GENERAL` **CRF Datasheets for SE Skills Practice Lab** Measuring Quantities in an Acid-Base Reaction* `GENERAL`
BLOCKS 6 & 7 · 90 min pp. 269–274 **Section 3** Acids, Bases, and Salts in the Home	**CRF Lesson Plan*** **TT Bellringer*** **TT Concept Mapping***	**TE Opening Demonstration,** p. 269 `GENERAL` **SE Quick Activity** Detergents, p. 271 `GENERAL` **CRF Datasheets for In-Text Labs** Detergents* `GENERAL` **SE Quick Lab** What does an antacid do?, p. 272 `GENERAL` **CRF Datasheets for In-Text Labs** What does an antacid do?* `GENERAL` **TE Demonstration** Do Detergents Emulsify Liquids?, p. 271

BLOCKS 8 & 9 · 90 min

Chapter Review and Assessment Resources

- **SE Chapter Review,** pp. 276–279
- **CRF Chapter Tests*** `GENERAL`
- **OSP Test Generator**
- **CRF Standardized Test Practice with Guided Reading Development***
- **CRF Test Item Listing for ExamView® Test Generator***
- **OSP Scoring Rubrics and Classroom Management Checklists**

Online Resources

Visit the HRW Web site for a variety of free resources related to the text. Just type in the keyword **HK4 ABS**.

Holt Science Spectrum: Physical Science: Online Edition

Students can access interactive problem solving help and active visual concept development with the *Holt Science Spectrum: Physical Science* Online Edition available at **www.hrw.com**.

student CNN News

cnnstudentnews.com

Find the latest news, lesson plans, and activities related to important scientific events.

KEY

TE	Teacher Edition
SE	Student Edition
OSP	One-Stop Planner

CRF	Chapter Resource File
TT	Teaching Transparency
TM	Transparency Master

*	Also on One-Stop Planner
◆	Requires Advance Prep

PROBLEM SOLVING AND PRACTICE	SECTION REVIEW AND ASSESSMENT	STANDARDS CORRELATION
	CRF Pretest* GENERAL	
SE **Math Skills** Determining pH, p. 262 **CRF** **Science Skills** Scientific Notation* GENERAL **CRF** **Science Skills** Operations with Exponents* GENERAL **TE** **Inclusion Strategies,** p. 256 **CRF** **Cross-Disciplinary Worksheet** Integrating Mathematics—pH Scale* GENERAL **CRF** **Math Skills** Determining pH* GENERAL	**TE** **Quiz,** p. 263 BASIC **SE** **Section 1 Review,** p. 263 **CRF** **Concept Review*** GENERAL **CRF** **Quiz*** BASIC	SAI 1
CRF **Science Skills** Balancing Chemical Equations* GENERAL **TE** **Inclusion Strategies,** p. 266	**TE** **Quiz,** p. 268 BASIC **SE** **Section 2 Review,** p. 268 **CRF** **Concept Review*** GENERAL **CRF** **Quiz*** BASIC	PS 3c
CRF **Science Skills** Classifying Items* GENERAL **CRF** **Cross-Disciplinary Worksheet** Connection to Social Studies—Detergents: Helpful or Harmful?* GENERAL **CRF** **Cross-Disciplinary Worksheet** Integrating Biology—Balance in the Body* GENERAL **SE** **Graphing Skills,** p. 275	**TE** **Quiz,** p. 274 BASIC **SE** **Section 3 Review,** p. 274 **CRF** **Concept Review*** GENERAL **CRF** **Quiz*** BASIC	PS 3a SAI 2

www.scilinks.org

Topic: Acids and Bases
***Sci*Links code:** HK4163

Topic: Hydronium Ion
***Sci*Links code:** HK4071

Topic: pH
***Sci*Links code:** HK4103

Topic: Acids and Bases at Home
***Sci*Links code:** HK4003

Topic: Baking Soda/Baking Powder
***Sci*Links code:** HK4164

Technology Resources

One-Stop Planner
All of your printable resources and the Test Generator are on this convenient CD-ROM.

 Science in the NEWS
*each video segment is accompanied by a Critical Thinking Worksheet**

Segment 26
Acids in the Environment

Overview

In this chapter, students first learn about the characteristics of acids and bases and the relationship between acid or base concentration and pH. Neutralization reactions and the formation of salts are introduced. Students also learn about cleaning products and household uses of acids, bases, and salts.

Assessing Prior Knowledge

Be sure students understand the following concepts:

- moles
- ions
- writing formulas
- bonding
- chemical reactions
- chemical equations

MISCONCEPTION /// ALERT \\\

Science education research has identified the following misconceptions about acids.

- Students believe that the primary activity of acids is "eating something away," and use this concept to explain and categorize all acid reactions.
- Students believe that "strong acids" are always more caustic than "weak acids" regardless of concentration.

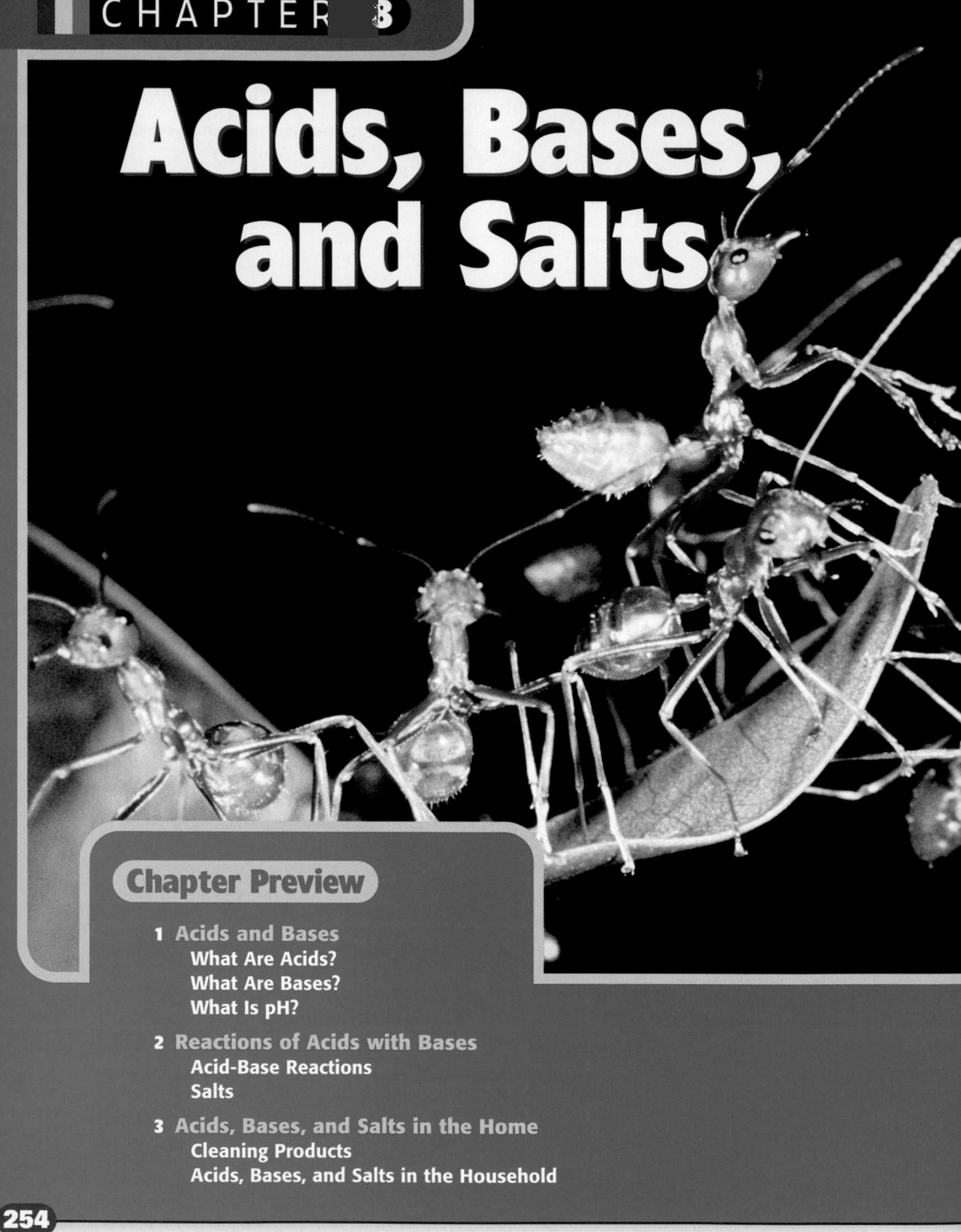

CHAPTER 8

Acids, Bases, and Salts

Chapter Preview

1 Acids and Bases
What Are Acids?
What Are Bases?
What Is pH?

2 Reactions of Acids with Bases
Acid-Base Reactions
Salts

3 Acids, Bases, and Salts in the Home
Cleaning Products
Acids, Bases, and Salts in the Household

254

Standards Correlations

National Science Education Standards

The following descriptions summarize the National Science Standards that specifically relate to this chapter. For the full text of the standards, see the National Science Education Standards at the front of the book.

Section 1 Acids and Bases
Science as Inquiry SAI 1

Section 2 Reactions of Acids With Bases
Physical Science PS 3a

Section 3 Acids, Bases, and Salts in the Home
Physical Science PS 3c
Science as Inquiry SAI 2

Background Some kinds of ants can defend themselves with a quick squirt of highly irritating formic acid solution. These ants are often called *stinging ants,* but in fact the ants bite and then squirt the acid into the wound. Formic acid was identified in 1670 by a chemist who heated ants in a flask and collected the vapors given off. The name *formic acid* is from the Latin *formica,* meaning ant. Many other acids are also found in living things.

Acids also react with a type of chemical called *bases.* In many ways, acids and bases are chemical opposites. For example, the base calcium hydroxide can be used to treat lakes that are too acidic. The reaction that neutralizes the lake is similar to the reaction that happens when you take an antacid for an upset stomach.

Activity 1 Cut a lemon in half. Squeeze the lemon over a clean dish to get about a teaspoon of juice. Dip a clean finger into the juice, and taste it. Describe the taste. Do you think that lemon juice is acidic or basic? Give reasons for your decision.

Activity 2 After you have tasted the lemon juice in Activity 1, add a teaspoon of water to it, and stir with your finger. With a clean, dry spoon, add 1/2 teaspoon of baking soda to the diluted lemon juice. What happens to the juice and baking soda? Baking soda is a basic substance. What evidence do you see that a chemical reaction takes place?

> **internet** connect
>
> **www.scilinks.org**
> Topic: **Acids and Bases** SciLinks code: **HK4163**
>
> **SCI**LINKS. Maintained by the National Science Teachers Association

Some species of ants produce formic acid and inject it into their victims when they bite. The helicopter is adding a base to an acidic lake to neutralize it.

Pre-Reading Questions

1. The orange is known as a citrus fruit because it contains *citric acid.* What other foods may contain citric acid?
2. Bee venom is also acidic. How might a solution of baking soda in water reduce the pain of a bee sting?

255

Focus ACTIVITY

Background Ants that produce formic acid are used as fumigants by birds, which place the ants in their feathers to rid themselves of mites. Beekeepers use a gel of formic acid to fumigate hives. Formic acid is lethal to parasites, such as tracheal and varroa mites which can devastate a hive, but does not harm the bees.

Activity 1 Answers may vary. Most students will describe the taste as sour. Some students may attribute the astringent sensation on the tongue as being caused by an acid "burning" the tongue. Caution students that taste should never be used as a test for acids in the laboratory.

Activity 2 Students observe that bubbles form. Students should recognize that the formation of bubbles indicates that a gas is being evolved, which, in turn, indicates a chemical reaction.

Answers to Pre-Reading Questions

1. Students should infer that other citrus fruits, such as limes, lemons, and grapefruits, contain citric acid.
2. Students may infer that the easing of the pain is caused by a neutralization reaction between the acidic bee venom and the basic baking soda solution.

Chapter Resource File

- Pretest GENERAL
- Teaching Transparency Preview
- Answer Keys

Acids, Bases, and pH

▶ **KEY TERMS**

acid
indicator
electrolyte
base
pH

OBJECTIVES

▶ **Describe** the ionization of strong acids in water and the dissociation of strong bases in water.

▶ **Distinguish** between solutions of weak acids or bases and solutions of strong acids or bases.

▶ **Relate** pH to the concentration of hydronium ions and hydroxide ions in a solution.

Does the thought of eating a lemon make your mouth pucker and your saliva flow? You know to expect that sour, piercing taste that can sometimes make you shudder. Eating a lime or a dill pickle may cause you to have a similar response.

What Are Acids?

Each of the foods shown in *Figure 1* tastes sour because it contains an **acid.** Several fruits, including lemons and limes, contain citric acid. Dill pickles are soaked in vinegar, which contains acetic acid. Other acidic foods include apples, which contain malic acid, and grapes, which contain tartaric acid.

When acids dissolve in water, they *ionize*, which means that they form ions. Hydrogen ions, H^+, attach to water molecules to form hydronium ions, H_3O^+. These hydronium ions are responsible for the sour taste you experience. **Indicators** respond to the concentration of hydronium ions in water by changing color. Blue litmus paper contains an indicator that can help you determine if a substance is an acid. Acids turn blue litmus paper red, as shown in *Figure 1.*

▶ **acid** any compound that increases the number of hydronium ions when dissolved in water

▶ **indicator** a compound that can reversibly change color depending on the pH of the solution or other chemical change

Figure 1

Lemons, limes, and dill pickles taste sour because they contain acids. Acids, such as the citric acid in lemon juice, turn blue litmus paper red.

256

Nitric acid Water Hydronium ion Nitrate ion

Strong acids ionize completely

All acids ionize when dissolved in water. The ionization process shown in *Figure 2A* occurs when nitric acid is added to water. The single arrow pointing to the right shows that nitric acid ionizes completely in water. When the acid ionizes, it forms hydronium ions and nitrate ions. These charged ions are able to move around in the solution and conduct electricity, as you see in *Figure 2B.* A substance that conducts electricity when dissolved in water is an **electrolyte.**

Solutions of some acids, such as nitric acid, conduct electricity well. Nitric acid, HNO_3, is a *strong acid* because it ionizes completely in water. Other strong acids behave similarly to nitric acid when dissolved in water. A solution of sulfuric acid in water, for example, conducts electric current in car batteries. Strong acids are strong electrolytes because solutions of these acids have as many hydronium ions as the acid can possibly form.

Weak acids do not ionize completely

Solutions of *weak acids,* such as acetic acid, CH_3COOH, do not conduct electricity as well as nitric acid. When acetic acid is added to water, the equilibrium shown in *Figure 3A* is reached.

When acetic acid is dissolved in water, some molecules of acetic acid combine with water molecules to form ions. Many of the ions then recombine to form molecules of acetic acid. Because there are fewer charged ions in a solution of acetic acid, it does not conduct electricity very well, as shown in *Figure 3B.* Acetic acid and other weak acids are weak electrolytes.

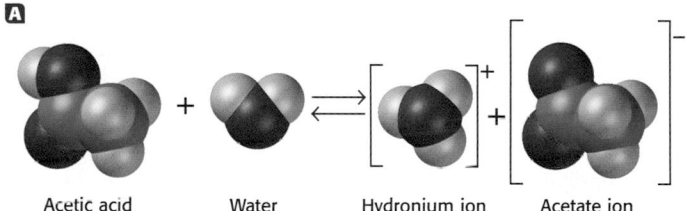

Acetic acid Water Hydronium ion Acetate ion

Figure 2

B Nitric acid, HNO_3, is a strong electrolyte and a strong acid because it ionizes completely in water to form hydronium, ions H_3O^+, and nitrate ions, NO_3^-.

▮ **electrolyte** a substance that dissolves in water to give a solution that conducts an electric current

Figure 3

B Acetic acid, CH_3COOH, is a weak acid and a weak electrolyte because it ionizes only partially in water to form hydronium ions, H_3O^+, and acetate ions, CH_3COO .

257

Teach

SKILL BUILDER — BASIC

Reading Skills Have students read the section silently and mark on self-adhesive notes those passages that they do not understand. Be sure students study figures and tables to help clarify the relevant passages. Pair students and have them discuss the passages each found difficult. If passages remain unclear, have the pair keep track of their questions for later class discussion or teacher explanation.
LS Interpersonal

SKILL BUILDER — GENERAL

Interpreting Visuals Ask students how an acid behaves in solution. (An acid donates hydrogen ions to form hydronium ions when dissolved in water.) Have students look at **Figures 2A** and **3A.** Ask them how the compounds HNO_3 and CH_3COOH can be classified as acids. (Both substances transfer hydrogen ions, H^+, to water to form hydronium ions.) Have students refer to **Figures 2B** and **3B** and ask them what evidence indicates that nitric acid is a strong electrolyte while acetic acid is a poor electrolyte. (In **Figure 2B,** the brightly lit bulb indicates that a large current is moving through the circuit. Thus the solution is a good conductor of electrical current and contains strong electrolytes. In **Figure 3B** the dimly lit bulb indicates little current is moving through the circuit. Thus the solution contains weak electrolytes.) **LS Visual**

MISCONCEPTION ALERT

Hydrogen Atoms and Acid Strength

Students might think that the more hydrogen atoms an acid has, the stronger it is. Acids that have more than one acidic hydrogen atom per molecule ionize by losing the atoms one at a time. Each atom is more difficult to lose than the one before because it is being lost from a negative ion. Phosphoric acid, H_3PO_4, loses one hydrogen atom relatively easily, leaving an $H_2PO_4^-$ ion. Fewer hydrogen atoms leave the $H_2PO_4^-$ ion because it is negatively charged. Few HPO_4^{2-} ions lose their final hydrogen ion. Phosphoric acid contains three acidic hydrogen atoms, but it does not have a high degree of ionization and is therefore a weak acid.

Chapter Resource File

• **Lesson Plan**
• **Science Skills** Scientific Notation GENERAL
• **Science Skills** Operations with Exponents GENERAL

Transparencies

TT Bellringer

Interpreting Tables Have students examine **Table 1.** Point out that acetic, formic, and citric acids are classified as organic acids while hydrochloric, sulfuric, and nitric acids are classified as inorganic (non-organic) acids. Ask students if the organic acids listed in the table form strong or weak acids. (weak acids) Ask student if the inorganic or organic acids listed in the table ionize more completely. (inorganic acids) Point out that organic acids tend to form strong acids in solution while organic acids tend to form weak acids in solution.
LS Visual

Teaching Tip

Indicators Indicators change color over a range of H_3O^- concentrations. Some indicators, such as crystal violet, change color in solutions that are quite acidic. Others, such as alizarin yellow, change color in a solution with low H_3O^- concentration.

Teaching Tip

Slippery Bases Explain to students that bases feel slippery because they react with fats and oils in the skin. Weak bases, such as soap, do not harm skin, but strong bases will damage human tissue.

Table 1 Some Common Acids

Acid	Formula	Strength	Uses for the acid
Hydrochloric acid	HCl	strong	cleaning masonry; treating metal before plating or painting; adjusting the pH of swimming pools
Sulfuric acid	H_2SO_4	strong	manufacturing fertilizer and chemicals; most used industrial chemical; the electrolyte in car batteries
Nitric acid	HNO_3	strong	manufacturing fertilizers and explosives
Acetic acid	CH_3COOH	weak	manufacturing chemicals, plastics, and pharmaceuticals; the acid in vinegar
Formic acid	HCOOH	weak	dyeing textiles; the acid in stinging ants
Citric acid	$H_3C_6H_5O_7$	weak	manufacturing flavorings and soft drinks; the acid in citrus fruits (oranges, lemons, limes)

Figure 4

These household items are bases because they produce OH^- ions in water solution.

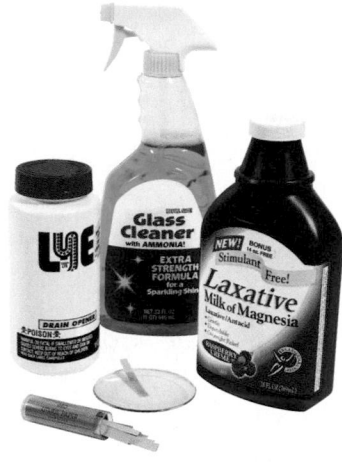

■ **base** any compound that increases the number of hydroxide ions when dissolved in water

Any acid can be dangerous in a concentrated form

Some examples of common strong and weak acids and their uses are listed in *Table 1.* Acids are used in many manufacturing processes and are necessary to many organisms even though strong acids can damage living tissue. For example, your stomach normally contains a dilute solution of hydrochloric acid that helps you digest food, but concentrated hydrochloric acid can burn your skin.

Even weak acids are not always safe to handle. Most vinegar is a 5% solution of acetic acid in water, but concentrated acetic acid can damage the skin, and the vapors are harmful to the eyes, mouth, and lungs. To be safe, always wear safety goggles, gloves, and a laboratory apron when working with acids.

What Are Bases?

Like acids, all **bases** share common properties. Bases have a bitter, soapy taste, and solutions of bases feel slippery. *Figure 4* shows some common household substances that contain bases. Like solutions of acids, solutions of bases contain ions and can conduct electricity. Some bases contain hydroxide ions, OH^-, but others do not. Bases that do not contain hydroxide ions will react with water molecules to form hydroxide ions. Bases cause indicators to change color, such as turning red litmus paper blue.

Historical Perspective

Inorganic Acids Inorganic acids, such as HCl and H_2SO_4, were first reported in medieval times. The importance of the discovery and use of these acids is considered by some scientific historians to be second only to the manufacture of iron items 3000 years earlier.

Cultural Awareness

Pickling Acids have been used in many cultures for food preservation in a process known as pickling. Pickling inhibits growth of bacteria and other microorganisms. Ancient Greeks ate pickled birds. Pickled fish have been an important food supply in Scandinavia since the Maglemosian culture there used the pickling technique as early as 10 000 B.C.E.

Table 2 Some Common Bases

Base	Formula	Strength	Uses for the base
Potassium hydroxide (potash)	KOH	strong	manufacturing soap; absorbing carbon dioxide from flue gases; dyeing products
Sodium hydroxide (lye)	$NaOH$	strong	manufacturing soap; refining petroleum; cleaning drains; manufacturing synthetic fibers
Calcium hydroxide	$Ca(OH)_2$	strong	treating acidic soil; treating lakes polluted by acid precipitation; making mortar, plaster, and cement
Ammonia	NH_3	weak	fertilizing soil; manufacturing other fertilizers; manufacturing nitric acid; making cleaning solutions
Methylamine	CH_3NH_2	weak	manufacturing dyes and medicines; tanning leather
Aniline	$C_6H_5NH_2$	weak	manufacturing dyes and varnishes; used as a solvent

Many common bases contain hydroxide ions

Strong bases are ionic compounds that contain a metal ion and a hydroxide ion. These strong bases are also known as *metal hydroxides*. When a metal hydroxide is dissolved in water, the metal ions and the hydroxide ions *dissociate,* or separate.

For example, sodium hydroxide, NaOH, is a metal hydroxide that is found in some drain cleaners. Solutions of sodium hydroxide conduct electricity well, so sodium hydroxide is a strong electrolyte. The dissociation of sodium hydroxide in water is shown below.

$$NaOH \rightarrow Na^+ + OH^-$$

Some metal hydroxides, such as calcium hydroxide and magnesium hydroxide, are not very soluble in water, but the ions in the part of the metal hydroxide that does dissolve separate completely. Calcium hydroxide is used to treat soil that is too acidic. Other useful bases are listed in *Table 2.*

Like acids, bases can be very dangerous in concentrated form, and in the case of bases such as sodium hydroxide and potassium hydroxide, bases can be dangerous even in fairly dilute form. Because bases attack living tissue very rapidly, bases are in some ways more dangerous than acids. To protect yourself when working with bases in the laboratory, always wear safety goggles, gloves, and a laboratory apron. If possible, work with very dilute bases instead of concentrated ones.

259

Interpreting Visuals Write the chemical equation for dissociation of potassium hydroxide in solution.

$$KOH = K^+ + OH^-$$

Ask students what process forms the hydroxide ion in solution. (dissociation) Now have students examine **Figure 5.** Ask: What role does a water molecule play in the formation of a hydroxide ion when ammonia dissolves in water? (A water molecule donates a hydrogen ion to the ammonia molecule.) What process forms ammonium ions and hydroxide ions in solution? (ionization). Does water behave like an acid or a base in this ionization process? (acid) Why? (It donates a hydrogen ion.) **LS** Visual

Quick Lab

Analysis

1. The acidic materials are soft drinks, milk, white vinegar, and mayonnaise because the liquids turned blue litmus red. The basic materials are baking soda, bleach, mineral water, dishwashing liquid, and laundry detergent because they turned red litmus blue.

2. Depending on the sample tested, baking powder and tap water are probably neutral substances because they did not cause either litmus paper to change color.
LS Visual

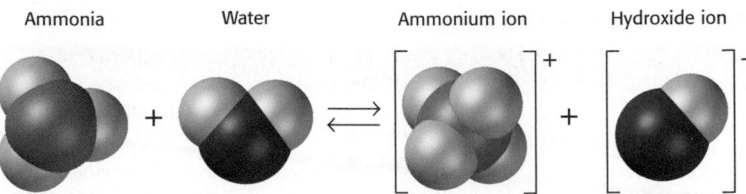

Ammonia   Water   Ammonium ion   Hydroxide ion

Figure 5
Ammonia produces hydroxide ions in water through ionization. An ammonia molecule accepts H^+ ions from water to form ammonium ions, NH_4^+, and hydroxide ions, OH^-.

Other bases ionize in water to form hydroxide ions

Ammonia, like other bases, forms hydroxide ions when it dissolves in water. But ammonia does not contain hydroxide ions. Instead, it forms hydroxide ions with water through an ionization process, shown in **Figure 5.** In this process, water acts as an acid and donates a hydrogen ion to ammonia to form an ammonium ion, NH_4^+, and leaves a hydroxide ion, OH^-, behind.

A solution of ammonia in water is a poor conductor of electricity. This shows that only some of the ammonia molecules actually become ammonium ions when the ammonia dissolves. So, an ammonia solution consists mostly of water and dissolved ammonia, along with a few ammonium ions and hydroxide ions. Ammonia is a much weaker base than potassium hydroxide, which is a metal hydroxide that dissociates completely.

Quick Lab

Which household substances are acidic, which are basic, and which are neither?

Materials
- baking powder
- baking soda
- several 50 mL beakers
- pipet bulbs
- milk
- mineral water
- bleach
- blue litmus paper
- white vinegar
- dishwashing liquid
- soft drinks
- mayonnaise
- red litmus paper
- disposable pipets or eyedroppers
- laundry detergent
- tap water

SAFETY CAUTION Wear safety goggles, gloves, and a laboratory apron. Never pipet anything by mouth.

1. Prepare a sample of each substance you will test. If the substance is a liquid, pour about 5 mL of it into a small beaker. If the substance is a solid, place a small amount of it in a beaker, and add about 5 mL of water. Label each beaker clearly with the name of the substance that is in the beaker.

2. Use a pipet to transfer a drop of liquid from one of the samples to red litmus paper. Then transfer another drop of liquid from the same sample to blue litmus paper. Record your observations.

3. Repeat step 2 for each sample. Be sure to use a clean pipet to transfer each sample.

Analysis

1. Which substances are acidic? Which are basic? How did you determine this?

2. Which substances are not acids or bases? How did you determine this?

260

Alternative Assessment — ADVANCED
Properties of Acids and Bases

Step 1 Have groups of students brainstorm lists of properties, examples, and uses of acids and bases.

Step 2 After checking the lists for accuracy, have students use the lists to write books for younger students that explain in simple language what acids and bases are and how they are used.

Step 3 Have students create outlines and rough drafts for the books. Check these items

for clarity of explanation and scientific accuracy before students finalize their books. Students can use pictures that they draw themselves or acquire from magazines to illustrate their written copy.

Step 4 If possible, have students use the completed books to teach another class about acids and bases.
LS Verbal

What Is pH?

You can tell if a solution is acidic or basic by using an indicator, such as litmus paper. But to determine exactly how acidic or basic a solution is, you must measure the concentration of hydronium (H_3O^+) ions. The **pH** of a solution indicates its concentration of H_3O^+ ions. The pH of a solution is often critical. For example, enzymes in your body work only in a narrow pH range.

pH values correspond to the concentration of hydronium ions

pH is a measure of the H_3O^+ concentration in a solution, but pH also indicates hydroxide ion (OH^-) concentration. So a pH value can tell you how acidic or basic a solution is. A pH value can even tell you if a solution is neutral, or neither an acid nor a base.

Typically, the pH of solutions ranges from 0 to 14, as shown in *Figure 6*. In neutral solutions, or in substances such as pure water, the concentration of hydronium ions equals the concentration of hydroxide ions, and the pH is 7. Solutions that have a pH of less than 7 are acidic. In acidic solutions, such as apple juice, the concentration of hydronium ions is greater than the concentration of hydroxide ions. Solutions that have a pH of greater than 7 are basic. In basic solutions, the concentration of hydroxide ions is greater than the concentration of hydronium ions.

Figure 6

The pH of a solution is easily measured by moistening a piece of pH paper with the solution and then comparing the color of the pH paper with the color scale on the dispenser of the pH paper.

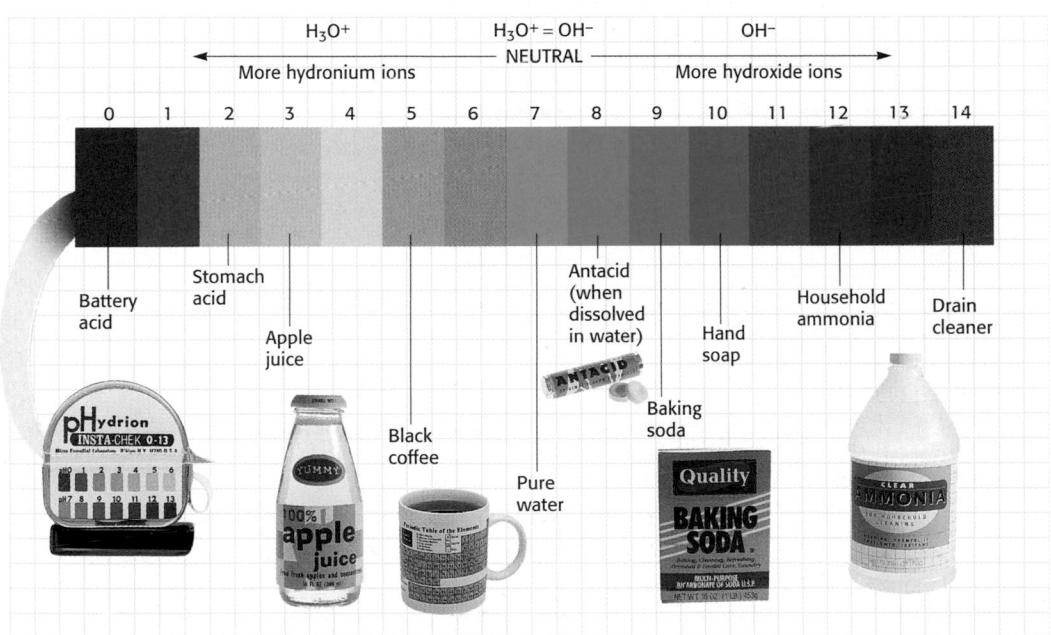

261

Demonstration —— GENERAL

A Natural pH Indicator

Materials:

- red cabbage juice
- 24-well spot plate
- pipets
- various household products (ammonia, vinegar, detergent, etc.)
- overhead projector

Before the demonstration prepare the red cabbage juice by placing a half-head of red cabbage into a blender. Fill the blender with water and chop for a minute. Strain the liquid into a jar that has a tight-fitting lid.

Step 1 Place the spot plate on the overhead projector. Pipet a small amount of each of the household products into the wells on the plate.

Step 2 Pipet a few drops of red cabbage juice into each of the wells.

Step 3 Have students observe the color changes of the solutions.

Step 4 Copy the table below on the chalkboard and have students use it to determine the pH of each of the solutions.

pH	Cabbage Juice Color
2–3	Red
3–4	Pink
4–6	Purple
6–8	Blue-Green
8–12	Green

LS Visual

internet connect

www.scilinks.org
Topic: pH
SciLinks code: HK4103

SCiLINKS Maintained by the National Science Teachers Association

Did You Know ?

Did you know that the concentration of hydronium ions and the concentration of hydroxide ions are related? In any solution made with water, the more hydronium ions there are (the more acidic the solution is), the fewer hydroxide ions there are (the less basic the solution is).

Practice HINT

► If a solution contains a base, you should expect the pH to be greater than 7. If the solution contains an acid, the pH will be less than 7.

► To find the concentration of a solution of strong acid from its pH, multiply the pH value by –1. Then use the result as a power of 10. The result is the concentration of the acid in moles per liter (mol/L).

The concentration of a strong acid allows you to calculate pH

When you describe the concentration of a substance in a solution, you probably write the concentration as a *molarity* (M), or the number of moles of the substance per liter of solution. For example, the hydronium ion (H_3O^+) concentration of pure water at 25°C is 0.000 000 1 mol/L, or 10^{-7} M.

When the H_3O^+ concentration of a solution can be written as a power of 10, the pH is the negative of the power of 10 used to describe the concentration of hydronium ions. For example, the pH of pure water is 7, so the concentration of hydronium ions in water is 10^{-7} M. The pH of apple juice is about 3, so the concentration of H_3O^+ in apple juice is 10^{-3} M.

If you know the concentration of a solution of a strong acid, you can calculate the pH of the solution. When a strong monoprotic acid ionizes in a solution, one hydronium ion is formed for each particle of acid that dissolves. So the concentration of hydronium ions in a solution of strong acid is the same as the concentration of the acid itself, and this information allows you to find the pH of the solution.

Math Skills

Determining pH Determine the pH of a 0.0001 M solution of the strong acid HCl dissolved in water.

1 List the given and unknown values.
   Given: *concentration of HCl in solution* = 0.0001 M
   Unknown: pH

2 Determine the molar concentration of hydroxide ions.
   concentration of HCl in solution = 0.0001 M
   HCl is completely ionized into H_3O^+ and Cl^- ions.
   concentration of H_3O^+ ions in solution = 0.0001 M = 1×10^{-4} M

3 Convert the H_3O^+ concentration to pH.
   concentration of H_3O^+ ions = 1×10^{-4} M
   pH = –(–4) = 4

Practice

Determining pH

1. Calculate the pH of a 1×10^{-4} M solution of HBr, a strong acid.

2. Determine the pH of a 0.01 M solution of HNO_3, a strong acid.

3. Nitric acid, HNO_3, is a strong acid. The pH of a solution of HNO_3 is 3. What is the concentration of the solution?

INTEGRATING

MATHEMATICS

Adding Exponents Illustrate the rule of adding exponents when multiplying quantities in scientific notation by having students calculate the following: 100×10; $1/100 \times 1/10$ (1000; 1/1000)

Have them write the quantities in scientific notation as well as the answers.
($(1 \times 10^2) \times (1 \times 10^1) = 1 \times 10^3$
$(1 \times 10^{-2}) \times (1 \times 10^{-1}) = 1 \times 10^{-3}$)

LS Logical

Historical Perspective

Acids and Bases Most acids and bases in this chapter fit the definitions of acids and bases that Svante Arrhenius, a Swedish chemist, formulated in 1887 which defined an acid as a substance that produces hydrogen ions in water, and a base as a substance that produces hydroxide ions in water. In 1923, the Brønsted-Lowry theory, proposed by a Danish and an English chemist, defined an acid as a proton donor and a base as a proton acceptor. Gilbert Lewis expanded the definition further by defining an acid as an electron-pair acceptor and a base as an electron-pair donor which included substances that do not include hydrogen.

Small differences in pH mean larger differences in acidity

Because pH is the negative of the power of 10 of hydronium ion concentration, small differences in pH mean larger differences in the hydronium ion concentration. For example, the pH of apple juice differs from the pH of coffee by two pH units, so apple juice is 10^2, or 100 times, more acidic than coffee. Likewise, coffee is about 10^3, or 1000 times, more acidic than antacid tablets, which form a base with a pH of about 8 when dissolved in water.

pH can be measured in more than one way

pH paper contains several indicators that change color at different pH values. pH may also be measured with a pH meter, as shown in *Figure 7*. Because ions in a solution have an electric charge, a pH meter can measure pH by determining the electric current created by the movement of the ions in the solution. If you use a pH meter properly, you can determine the pH of a solution more precisely than is possible if you use pH paper.

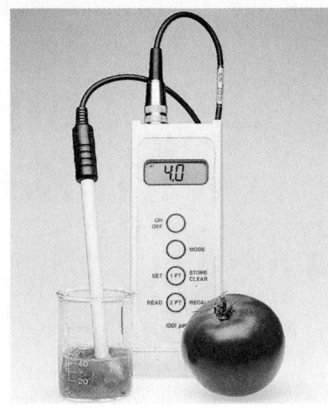

Figure 7

A pH meter measures an electric current that results from differences in H_3O^+ concentrations.

Practice

1. pH = $-(-4) = 4$
2. 0.01 M = 1×10^{-2} M
   pH = $-(-2) = 2$
3. pH = 3
   concentration =
   $1 \times 10^{-(3)}$ M = 1×10^{-3} M

LS Logical

Close

Quiz

1. A substance that when dissolved in water donates hydrogen ions to water molecules to form hydronium ions is classified as what type of substance? (an acid)

2. A base is a substance that either contains or ionizes water to produce what ions? (hydroxide ions)

3. What is a measure of the concentration of hydronium ions in a solution? (pH)

4. A solution with a pH less than 7.0 is classified as what type of solution? (acidic)

SECTION 1 REVIEW

SUMMARY

▶ Acids are substances that taste sour, turn blue litmus paper red, and form hydronium ions when they dissolve in water.

▶ Strong acids are strong electrolytes because they ionize completely in water.

▶ Weak acids are weak electrolytes because they ionize only slightly in water.

▶ Bases have a slippery feel, have a bitter taste, turn red litmus paper blue, and produce hydroxide ions when they dissolve in water.

▶ The pH of a solution of a strong acid can be found if you know the concentration of the solution.

1. **Explain** how a strong acid and a weak acid behave differently when each is dissolved in water.

2. **Compare** the ionization of a weak acid in water to the ionization of a weak base in water.

3. **Write** the chemical equation for the self-ionization of water.

4. **Classify** the following solutions as acidic, basic, or neutral.
   a. a soap solution, pH = 9
   b. a sour liquid, pH = 5
   c. a solution that has four times as many hydronium ions as hydroxide ions
   d. pure water

5. **Arrange** the following substances in order of increasing acidity: vinegar (pH = 2.8), gastric juices from inside your stomach (pH = 2.0), and a soft drink (pH = 3.4).

6. **Critical Thinking** A solution of an acid in water has a pH of 4, which is slightly acidic. Is this a solution of a weak acid? Explain your answer.

Math Skills

7. What is the pH of a 0.01 M solution of the strong acid $HClO_4$, perchloric acid?

Answers to Section 1 Review

1. A strong acid ionizes completely into H_3O^+ ions and anions. A weak acid ionizes partially into H_3O^+ ions and anions, but these recombine to form the nonionized acid. As a result, equilibrium is established with small constant concentration of ionized acid.

2. The ionization of a weak base is like the ionization of a weak acid in that only a small fraction of molecules separate into ions. The ionizations are different in that the molecules of the weak acid break apart in ionization, whereas the weak base causes water molecules to separate as protons are accepted by the base molecule.

3. $H_2O + H_2O \rightleftarrows H_3O + OH^-$
4. a. basic
   b. acidic
   c. acidic
   d. neutral

5. The pH of a solution decreases as acidity increases, so the order is soft drink, vinegar, gastric juice.

6. No it is not certain that this is a solution of a weak acid. It could be a weak acid or a 1×10^{-4} M solution of a strong acid such as HCl.

7. H_3O^+ concentration = 0.01 M = 1×10^{-2} M; pH = $-(-2) = 2$

Chapter Resource File

• **Cross-Disciplinary Worksheet** Integrating Mathematics—pH Scale **ADVANCED**

• **Math Skills** Determining pH **GENERAL**

• **Concept Review** **GENERAL**

• **Quiz** **BASIC**

SECTION
2

Focus

Overview

Before beginning this section, review with your students the **Objectives listed in the Student Edition.** The section opens with the chemical equations for acid-base reactions and defines a neutralization reaction and its products. It then describes titrations and the equivalence point in a neutralization reaction. The section closes with a discussion of the composition of a salt.

Bellringer

Use the Bellringer transparency to prepare students for this section.

Motivate

Opening Activity — GENERAL

Have pairs of students brainstorm words and phrases they associate with the term *salt*. Have each pair compile a list of their responses. After the students have completed this section, have them select a response from the list and use it to evaluate what they have learned.

LS Interpersonal

Reactions of Acids with Bases

▶ **KEY TERMS**
neutralization reaction
salt

OBJECTIVES

▶ **Write** ionic equations for neutralization reactions.
▶ **Identify** the products of a neutralization reaction.
▶ **Describe** the composition of a salt.

▶ **neutralization reaction**
the reaction of the ions that characterize acids (hydronium ions) and the ions that characterize bases (hydroxide ions) to form water molecules and a salt

Have you ever used an antacid to relieve the symptoms of an upset stomach or so-called heartburn? Heartburn has nothing to do with your heart. Heartburn occurs when the stomach's natural solution of hydrochloric acid (HCl) irritates the lining of the esophagus. The antacid contains a base that reacts with the acid to reduce the acidity of the solution and soothe your stomach.

Acid-Base Reactions

A reaction between an acid and a base is a **neutralization reaction.** An example of neutralization is the reaction of HCl and magnesium hydroxide, which is an antacid and a base.

Neutralization is an ionic reaction

A solution of a strong acid, such as hydrochloric acid, ionizes completely, as shown below.

$$HCl + H_2O \rightarrow H_3O^+ + Cl^-$$

In a similar way, a solution of a strong base, such as sodium hydroxide, dissociates completely, as shown below.

$$NaOH \rightarrow Na^+ + OH^-$$

If the two solutions of equal concentrations and equal volumes are combined, the following neutralization reaction takes place:

$$H_3O^+ + Cl^- + Na^+ + OH^- \rightarrow Na^+ + Cl^- + 2H_2O$$

The Na^+ and Cl^- ions are called *spectator ions* because they are like spectators watching on the sidelines. These ions do not change during the reaction between H_3O^+ and OH^-. As you can see in *Figure 8,* energy is also released in the reaction of Na and Cl.

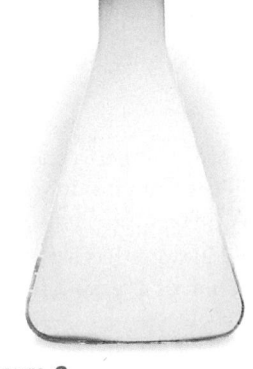

Figure 8
When HCl reacts with NaOH, sodium chloride is produced and energy is released.

Demonstration —————— GENERAL

Spectator Ions

(Time: Approximately 10 minutes)

Materials:

• 2 pipets
• beaker
• 6 M HCl
• 6 M NaOH

Pipet 1 mL each of 6 M HCl and 6 M NaOH into the beaker. Heat the beaker on a hot plate until the water evaporates. Have students observe the crystals.

Analysis:

1. What ions are in the hydrochloric acid, HCl, solution? (H_3O^+, Cl^-) in the sodium hydroxide, NaOH, solution? (Na^+, OH^-)

2. What ions make up the crystals remaining in the beaker after the reaction? (Na^+, Cl^-)

3. Did the sodium and chloride ions participate in the neutralization reaction? (no)

4. What are these sodium and chloride ions called? (spectator ions)

LS Visual

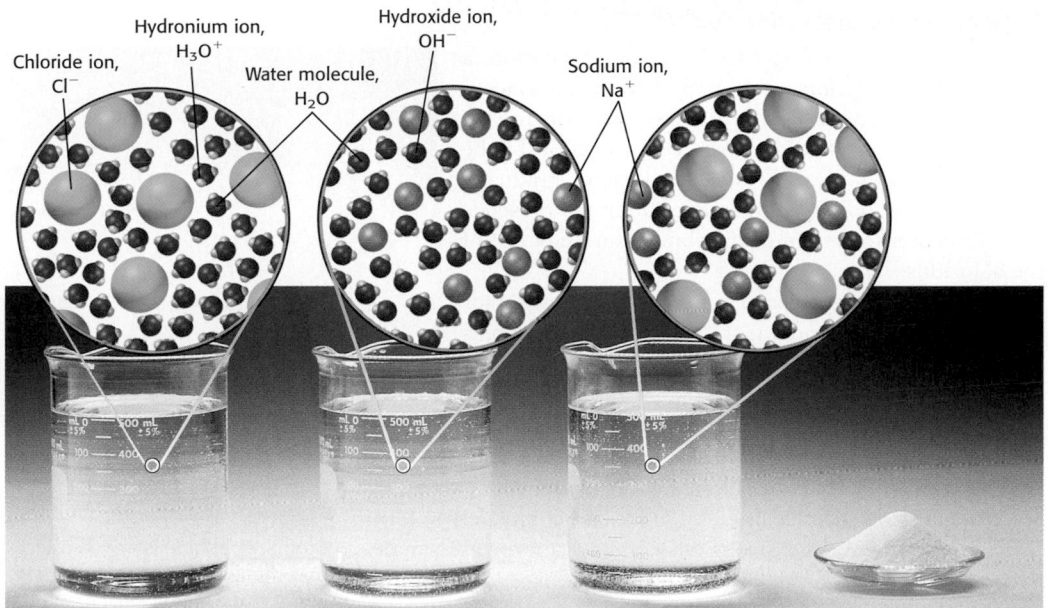

Chloride ion, Cl⁻

Hydronium ion, H₃O⁺

Water molecule, H₂O

Hydroxide ion, OH⁻

Sodium ion, Na⁺

Teach

Teaching Tip

Neutral Versus Neutralization Reaction Be sure students distinguish the term *neutral* from the term *neutralization reaction*. *Neutral* refers to having a pH of 7. *Neutralization reaction* refers to an acid-base reaction—a chemical reaction that occurs when an acid is mixed with a base. The resulting solution of a neutralization reaction can be acidic, basic, or neutral.

Strong acids and bases react to form water and a salt

If you include only the substances that react during neutralization, the equation can be written as follows:

$$H_3O^+ + OH^- \rightarrow 2H_2O$$

When an acid reacts with a base, hydronium ions react with hydroxide ions to form water. The other ions—positive ions from the base and negative ions from the acid—form an ionic compound called a **salt,** such as sodium chloride. Salts are ionic compounds that are often soluble in water, as you can see in **Figure 9.**

Not all neutralization reactions produce neutral solutions

Reactions between acids and bases do not always produce neutral solutions. The final pH of the solution depends on the amounts of acid and base that are combined. The pH also depends on whether the acid and base are strong or weak.

If a strong acid, such as nitric acid, reacts with an equal amount of a weak base, such as sodium hydrogen carbonate from an antacid tablet, the resulting solution will still be acidic. A similar situation occurs when a strong base reacts with a weak acid. When a strong acid reacts with an equal amount of a weak base, the resulting solution will be acidic.

Figure 9

When a solution of HCl reacts with a solution of NaOH, the reaction produces water and leaves sodium and chloride ions in solution. When the water is evaporated, the sodium and chloride ions crystallize to form pure sodium chloride.

▶ **salt** an ionic compound that forms when a metal atom or a positive radical replaces the hydrogen of an acid

265

BIOLOGY
CONNECTION — **BASIC**

If possible, set up an aquarium of freshwater fish in the classroom. Discuss with students what factors might cause the pH of the water to change. Factors might include increased acidity from the carbon dioxide in the air bubbled through the tank or increased basicity from ammonia released in fish wastes. Each day, assign a student to check the pH of

the water and to decide whether it needs to be adjusted or not. If it does, have the student propose what needs to be added to make the adjustment. Approve student plans before any action is taken. After approval, have the student adjust and recheck the pH. **Visual**

Chapter Resource File

• **Lesson Plan**
• **Science Skills** Balancing Chemical Equations **GENERAL**

Transparencies

TT Bellringer
TT Neutralization Reaction

Teaching Tip

Hydrolysis The pH of the solution after a neutralization reaction depends on the *hydrolysis* of the salt produced. Hydrolysis is the reaction of a salt with water. Water ionizes to a small extent, forming a few OH^- ions and an equal number of H_3O^+ ions. Suppose the salt produced is sodium chloride, NaCl. Because NaOH is a strong base and HCl is a strong acid, the ions from the salt do not combine with the ions from the water, and the solution is neutral. However, if the salt formed is aluminum nitrate, $Al(NO_3)_3$, the Al^{3+} ion from the salt will combine with the OH^- from the water because $Al(OH)_3$ is a weak base. The H_3O^- ions stay in solution because HNO_3 is a strong acid. An $Al(NO_3)_3$ solution is therefore acidic. Similarly, a salt formed from a strong base and a weak acid, such as sodium carbonate, Na_2CO_3, will be basic. The result of a neutralization reaction between a weak acid and a weak base depends on the relative strengths of the two reactants.

Titrations are neutralization reactions

When an acid solution is added to a basic solution, a neutralization reaction occurs. If you know the concentration of the acid solution or the basic solution, a *titration* can help you determine the concentration of the other solution. A titration is the process of gradually adding one solution to another solution in the presence of an indicator to determine the concentration of one of the solutions.

In a titration, an indicator is used that changes color when the original amount of the base in solution is equal to the amount of the acid added to the solution. For example, when a strong acid is titrated with a strong base, an indicator called *bromthymol blue* is used because bromthymol blue changes color when the solution reaches a pH of about 7, as shown in *Figure 10.*

When a strong acid is dissolved, it ionizes completely to form hydronium ions. When a strong base dissolves, it forms as many hydroxide ions as possible. And as you have learned, hydronium ions and hydroxide ions combine in a neutralization reaction. If the number of hydronium ions is equal to the number of hydroxide ions in a solution, the product of the reaction will be neutral. The *equivalence point* in a titration of a strong acid with a strong base is reached when the original amount of the acid equals the original amount of the base and occurs at pH 7. *Figure 11* shows the change in pH during the titration of nitric acid with sodium hydroxide.

The equivalence point is not always neutral

Titrations can also be carried out with a strong acid and a weak base, or with a weak acid and a strong base. In these cases, however, the equivalence point will not be at pH 7. For example, when 1 mol of acetic acid, a weak acid, is dissolved in water, only some of the molecules of the acid ionize to form H_3O^+ ions. When 1 mol of sodium hydroxide is added to water, it dissociates to form 1 mol of OH^- ions because sodium hydroxide is a strong base. When the neutralization reaction takes place between acetic acid and sodium hydroxide, there are OH^- ions left over.

When there are OH^- ions left over, a neutralization reaction does not produce a neutral solution. Neutralization occurs when water is formed from H_3O^+ ions and OH^- ions, but if there are any hydroxide ions left over, the solution will still be basic. A similar situation occurs when a strong acid is titrated with a weak base, but in this case the product is acidic and has a pH of less than 7.

Figure 10

Bromthymol blue is an indicator that changes color between a pH of 6.0 and 7.6. It is ideal for a titration involving a strong acid and a strong base.

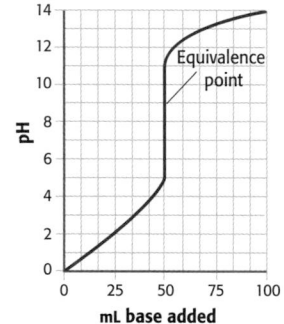

Figure 11

When a strong acid, such as nitric acid, is titrated with a strong base, the pH of the solution changes rapidly when the equivalence point is reached.

Salts

When you hear the word *salt*, you probably think of white crystals that you sprinkle on food. But to a chemist, a salt can be almost any combination of cations and anions, except for hydroxides and oxides, which are bases.

Sodium chloride has many different uses

Common table salt contains sodium chloride, NaCl, which is an ionic compound that can be formed from the reaction of hydrochloric acid with sodium hydroxide. NaCl is the source of most of the sodium in your diet. It is widely used to season and preserve food. Most NaCl in the United States comes from underground deposits that were left when ancient seas dried up.

NaCl is also used in ceramic glazes, soap manufacturing, home water softeners, highway de-icing, and fire extinguishers. Many other salts also contain sodium, as you can see in *Table 3* below.

Salts are all around us

Salts can be formed by acid-base neutralization, but more often, they are formed from other salts. Another familiar example of a salt is baking soda, sodium hydrogen carbonate. Photographic film contains the salts silver bromide and silver iodide, which are sensitive to light. Ordinary soaps and detergents are also examples of salts. *Figure 12* shows a salt that is used in medical diagnosis.

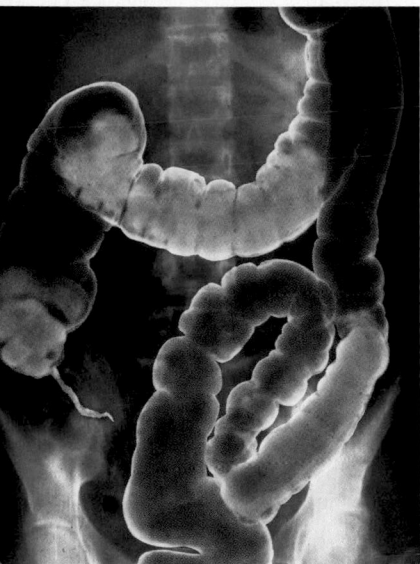

Figure 12

The salt barium sulfate, BaSO₄, is a highly insoluble salt that blocks X rays. After barium sulfate is placed into the large intestine, the form of the intestine shows up lighter on an X-ray photo.

SKILL BUILDER — BASIC

Modeling a Neutralization Reaction Label five large sheets of construction paper, H⁺, Cl⁻, H₂O, Na⁺, and OH⁻, respectively. Give each sheet to a student. Walk them through the neutralization reaction of hydrochloric acid and sodium hydroxide by first showing HCl ionizing in water to form Cl⁻, H⁺, with H⁺ migrating to the H₂O molecule to form the hydronium ion. Similarly show that NaOH dissociates to form Na⁺ and OH⁻ ions. Show that the hydronium ion transfers the hydrogen ion to the hydroxide ion. As a result two molecules of water are formed.
LS Visual

Table 3 **Some Common Salts**

Salt	Formula	Uses
Aluminum sulfate	Al₂(SO₄)₃	purifying water; used in antiperspirants
Ammonium sulfate	(NH₄)₂SO₄	flameproofing fabric; used as fertilizer
Calcium chloride	CaCl₂	de-icing streets and highways; used in some kinds of concrete
Potassium chloride	KCl	treating potassium deficiency; used as table-salt substitute
Sodium carbonate	Na₂CO₃	manufacturing glass; added to wash to soften water
Sodium hydrogen carbonate	NaHCO₃	treating upset stomach; ingredient in baking powder; used in fire extinguishers
Sodium stearate	NaOOCC₁₇H₃₄	typical example of a soap
Sodium lauryl sulfonate	NaSO₃C₁₂H₂₅	typical example of a detergent

267

INCLUSION Strategies

• *Gifted and Talented*

Ask students to research the use of barium sulfate (BaSO₄) for medical purposes. The students should create a report that includes uses, procedures, and benefits of barium sulfate. Additionally, you can arrange for the student to interview a medical professional involved with the use of barium sulfate and/or persons that use barium sulfate in medical procedures. The students may report their findings by tape recording the information as though they were broadcasting from a science channel on television.

Quiz

1. What is a neutralization reaction? (A reaction in which hydronium ions from an acid combine with hydroxide ions from a base to form water molecules.)

2. What will be the pH of a solution resulting from the neutralization reaction of a strong acid and a weak base? (less than 7)

3. What is a salt? (an ionic compound composed of cations and anions other than oxide and hydroxide anions.)

4. Minerals that are necessary for good nutrition are usually obtained in what form from their mineral salts? (ions)

Figure 13
These butterflies can obtain the salt they need from the dried sweat on an old sneaker.

Salts are useful substances

You have probably seen a lot of chalk since you entered school, but did you know that chalk is a salt? Chalk is one form of the salt calcium carbonate, $CaCO_3$, which also makes up limestone and marble. It is likely that the walls in your house are made of slabs of gypsum, which is one form of the salt calcium sulfate, $CaSO_4$.

You often hear that a healthful diet should include minerals such as potassium, sodium, calcium, magnesium, iron, phosphorus, and iodine. However, ingesting these nutrients in the form of free elements is not very common. Instead, you get ions of these elements in their ionic form from salts. You need calcium ions, Ca^{2+}, for strong bones and teeth and for proper function of nerves and muscles. The correct proportion of potassium ions, K^+, and sodium ions, Na^+, is crucial for transmission of nerve impulses, even in insects, such as those shown in **Figure 13.** Phosphorus, in the form of phosphate ions, PO_4^{3-}, is needed for many processes in living cells, from transporting energy to the reproduction of the genetic code.

SECTION 2 REVIEW

SUMMARY

▶ Acids and bases react with each other in a process called *neutralization*.

▶ Neutralization is a reaction between an acid and a base to form water and a salt.

▶ Neutralization reactions between weak acids and strong bases result in basic solutions.

▶ Neutralization reactions between strong acids and weak bases result in acidic solutions.

▶ Salts are ionic substances composed of cations and anions other than oxide or hydroxide.

1. **Write** the chemical equation for the neutralization of nitric acid, HNO_3, with magnesium hydroxide, $Mg(OH)_2$, first with spectator ions and then without spectator ions.

2. **Determine** which acid and which base you would combine to form the salt aluminum sulfate, $Al_2(SO_4)_3$.

3. **Identify** the spectator ions in the neutralization of lithium hydroxide, LiOH, with hydrobromic acid, HBr.

4. **Predict** whether the reaction of each of the following acids and bases will yield an acidic, a basic, or a neutral solution. Explain your answer for each.
   a. sulfuric acid, H_2SO_4, and ammonia, NH_3
   b. formic acid, HCOOH, and potassium hydroxide, KOH
   c. nitric acid, HNO_3, and calcium hydroxide, $Ca(OH)_2$

5. **Critical Thinking** A classmate observes a neutralization reaction between an acid and a base. After the reaction is complete, your classmate is surprised to find that the pH of the resulting solution is 4, not 7, the pH of a neutral solution. What can you tell your classmate to help them understand what happened?

Answers to Section 2 Review

1. $Mg^{2+} + 2OH^- + 2H_3O^+ + 2NO_3^- \rightarrow Mg^{2+} + 2NO_3^- + 4H_2O$
   $2OH^- + 2H_3O^+ \times 4 H_2O$

2. The Al^{3+} ion would have to come from the base, aluminum hydroxide, $Al(OH)_3$. The sulfate ion, SO_4^{2-} would have to come from the acid sulfuric acid, H_2SO_4.

3. $Li^+ + OH^- + H_3O^+ + Br^- \times Li^+ + Br^- + 2H_2O$
   The spectator ions are the lithium ion Li^+ and the bromide ion Br^-.

4. **a.** The reaction will yield an acidic solution because sulfuric acid is a strong acid, and ammonia is a weak base.

   b. The reaction will yield a basic solution because formic acid is a weak acid, and potassium hydroxide is a strong base.

   c. The reaction will yield a neutral solution because nitric acid is a strong acid and calcium hydroxide is a strong base.

5. If the neutralization was done correctly, the reaction must have occurred between a strong acid and a weak base, such as ammonia. The resulting ammonium ions combine with water in an equilibrium that yields extra hydronium ions, resulting in an acidic solution.
   $$NH_4^+ + H_2O \rightleftarrows NH_3 + H_3O^+$$

Acids, Bases, and Salts in the Home

INTEGRATING TECHNOLOGY and Society

▶ **Describe** the chemical structures of soaps and detergents and explain how they work.

▶ **Describe** the chemical composition of bleach and its uses.

▶ **Describe** how an antacid reduces stomach acid.

▶ **Identify** acidic and basic household products and their uses.

▶ **KEY TERMS**
soap
detergent
disinfectant
bleach
antacid

As you have seen, you won't find acids, bases, and salts only in a laboratory. Many items in your own home, such as soaps, detergents, shampoos, antacids, vitamins, sodas, and juices in your kitchen are examples of household products that contain acids, bases, and salts.

▶ **soap** a substance that is used as a cleaner and that dissolves in water

Cleaning Products

If you work on an oily bicycle chain or if you've been eating potato chips, water alone will not remove the greasy film from your hands. Water will not work because it doesn't mix with grease or oil. Something else must be added to water to improve its ability to clean.

Soaps allow oil and water to mix

Soap improves water's ability to clean because it can dissolve in both oil and in water. This property allows oil and water to form an emulsion that can be washed away by rinsing. For example, when you are washing your face with soap, as the girl in *Figure 14* is, the oil on your face is emulsified by the soapy water. The water you rinse with carries away both the soap and unwanted oil to leave your face clean.

Soaps are salts of sodium or potassium and fatty acids, which have long hydrocarbon chains. Soaps are made through a reaction of animal fats or vegetable oils with a solution of sodium hydroxide or potassium hydroxide. The products of the reaction are soap and an alcohol called *glycerol*.

Figure 14

When you wash with soap, you create an emulsion of oil droplets spread throughout water.

269

BIOLOGY
CONNECTION

Most garden plants thrive in soil that is neutral or slightly acidic. (Azaleas, rhododendrons, and some conifers tolerate acidic soils.) Nutrients necessary for plant growth are more soluble or available in acidic soil.

Several factors tend to increase soil acidity. Rainwater removes basic ions, such as calcium, potassium, and sodium. Carbon dioxide from root respiration dissolves in water, producing carbonic acid. Decaying plant matter and the oxidation of ammonium and sulfur fertilizers add nitric and sulfuric acid

to the soil. To increase soil pH, many gardeners apply hydrated lime, powdered calcium hydroxide to their garden soil.

(Calcium hydroxide is made by reacting calcium oxide (lime) with water.

$$CaO + H_2O \rightarrow Ca(OH)_2)$$

Because calcium hydroxide is slightly soluble in water, it dissociates at a slow rate making it a long-term source of hydroxide ions.

$$Ca(OH)_2 \rightarrow Ca^{2+} + 2OH^-$$

Focus

Overview

Before beginning this section, review with your students the Objectives listed in the Student Edition. A description of detergents opens the section. Chemical composition of bleach as well as acidic and basic household products are then discussed.

🔔 Bellringer

Use the Bellringer transparency to prepare students for this section.

Motivate

Demonstration —— GENERAL

(Time: Approximately 30 minutes)

Materials:

• 25 g of solid vegetable shortening
• 10 mL of ethanol
• 1.2 g of NaOH
• 30 mL of distilled water
• 25 mL saturated NaCl solution
• cheesecloth

Safety Caution: *There should be no open flames while using ethanol.*

Step 1 Mix the shortening, ethanol, NaOH, and 5 mL of water in a 250 mL beaker.

Step 2 Warm it on a hot plate while stirring for 15 minutes. Cool the mixture in ice water.

Step 3 Add the rest of the water and the NaCl solution.

Step 4 Filter using cheesecloth and press the soap into a mold to shape. Dry for several days.

Analysis

Which of the materials formed the soap? (NaOH, shortening) **LS** Visual

Soap was also used in Sumer, which is now Iran and Iraq, as early as 2500 B.C. Around 600 B.C., the Phoenicians made soap from wood ashes and goat fat and found it to be valuable for barter or trade. Ancient Mediterranean and British cultures also made soap.

Answers

1. The term is based on the Latin word for soap, *sapo*, which came from the name of the mountain that was the source of soapy water.

2. Ashes provided the base, and the hog fat provided the hydrocarbon chain with a water-soluble end.

LS Verbal

Teaching Tip

Solid and Liquid Soaps Tell students that in a saponification reaction, sodium hydroxide produces solid soaps and potassium hydroxide produces liquid soaps.

Teaching Tip

Cleaner Sooner Clothes wash cleaner if they are cleaned soon after they are soiled. Oils from the skin react with oxygen soon after they are deposited on clothes, making the molecules more polar and thus more easily removed by soaps and detergents. However, certain oils change over time, forming large molecules that tend to bond with molecules in fabric. These changes cause yellowing in fabrics, especially cotton, and the yellowing is not easily removed.

Connection to
SOCIAL STUDIES

People have used soap for thousands of years. Ancient Egyptians took baths regularly with soap made from animal fats or vegetable oils and basic solutions of alkali-metal compounds. According to Roman legend, people discovered that the water in the Tiber River near Mount Sapo was good for washing. Mount Sapo was used for elaborate animal-sacrifice rituals, and the combination of animal fat and the basic ash that washed down the mountain made the river soapy.

Making the Connection

1. The process of making soap is sometimes referred to as *saponification*. How does this word relate to the Roman soap legend?

2. Homemade soap can be made from hog fat and ashes. Which material provides the base needed to make the soap?

▶ **detergent** a water-soluble cleaner that can emulsify dirt and oil

Figure 15
The charged ends of soap or detergent dissolve in water, and the hydrocarbon chains dissolve in oil, keeping the oil droplet in suspension.

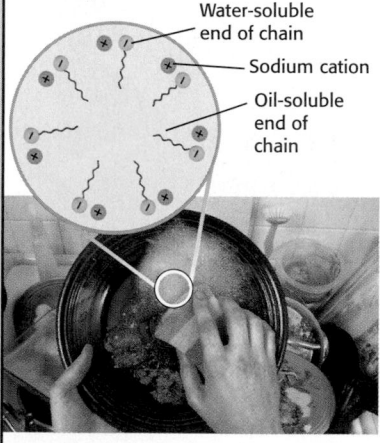

Water-soluble end of chain

Sodium cation

Oil-soluble end of chain

How soap removes grease

Soap is an ionic compound. Its negative ion is a long hydrocarbon chain of the soap anion with a carboxylate group ($-COO^-$) at one end. For every negatively charged end, there is a positive sodium or potassium ion.

Soap is able to remove grease and oil because the cations and the negatively charged ends of the chains ($-COO^-$) dissolve in water, while the hydrocarbon chains dissolve in oil. Soap acts as an emulsifier by surrounding droplets of oil, as shown in *Figure 15*. This action causes the droplets of oil to stay suspended in water. When you are washing your hands with soap, you probably rub them together. Rubbing your hands together actually helps clean them. When you do this, you lift most of the emulsion of grease and water into the lather where it can be rinsed into the sink.

Detergents have replaced soap in many uses

As useful as soap is for cleaning, it does not work well in hard water, that is, water containing the dissolved ions Mg^{2+}, Ca^{2+}, and Fe^{3+}. These cations combine with the fatty acid anions of soap to form an insoluble salt called *soap scum*. This soap scum settles out on clothing, dishes, your skin, and your hair. The scum also makes a ring around the bathtub or washbasin. To prevent this problem, **detergents** are used instead of soap to wash clothes and dishes. Most shampoos, liquid hand soaps, and body washes are actually detergents, not soap.

Detergents are sodium, potassium, and sometimes ammonium salts. Like anions in soaps, the anion in detergents are composed of a long hydrocarbon chain that has a negatively charged end. But the charged end of a detergent is a sulfonate group ($-SO_3^-$), not a carboxylate group. These sulfonate ions do not form scum with the ions in hard water. Detergents are also different from soaps because their hydrocarbon chains are made from petroleum products instead of from animal fats or plant oils.

Because soaps and detergents act in the same way, *Figure 15* represents detergents as well as soaps. The long hydrocarbon chains are soluble in oil or grease. The sulfonate ends are highly soluble in water. Water molecules attract the charged sulfonate group and keep the oil droplet suspended among the water molecules.

Did You Know?

Toilet Bowl Cleaners One result of hard water is the build up of insoluble calcium carbonate, $CaCO_3$, on the interior of toilet bowls. These deposits appear as discoloration and scale and can be removed by toilet bowl cleaners. Liquid toilet bowl cleaners usually contain hydrochloric acid. The reaction between the hydrochloric acid and calcium carbonate deposits produces soluble calcium and chlorine ions and carbon dioxide gas.

$$2HCl + CaCO_3 \rightarrow Ca^{2+} + 2Cl^- + H_2O + CO_2$$

Bleaches and toilet bowl cleaners should never be used in tandem with bleach to whiten the surface of the toilet bowls. The reaction between the hydrochloric acid in the cleaner and the bleach yields toxic chlorine gas.

$$2HCl + NaClO \rightarrow Na^+ + Cl^- + H_2O + Cl_2$$

Many household cleaners contain ammonia

Ammonia solutions, such as the ones shown in *Figure 16*, are also effective cleaners. Household ammonia is a solution of ammonia gas in water. Recall that ammonia is a weak base because it ionizes only slightly in water to form ammonium ions and hydroxide ions. The hydroxide ions make the ammonia solution basic, as shown in the reaction below.

$$NH_3 + H_2O \leftrightarrows NH_4^+ + OH^-$$

Although the concentration of hydroxide ions is very low in an ammonia solution, enough of the ions are available to help emulsify thin layers of oily dirt, such as fingerprints and oily smears. In addition, many ammonia cleaners contain alcohols, detergents, and other cleaning agents.

Bleach can eliminate stains

A **disinfectant** is a substance that kills bacteria and viruses. Household **bleach,** a very strong disinfectant, is a basic solution of sodium hypochlorite, NaOCl. You are probably familiar with the ability of bleach to remove colors and stains.

Bleach does not actually remove the substance causing the stain. Instead, it changes the substance to a colorless form. This bleaching action is carried out by the oxygen atom in the hypochlorite ion, ClO^-.

If an acid is added to a bleach solution, the acid reacts with the hydroxide ions, and the reaction reverses, giving off deadly chlorine gas. For this reason, you should never mix bleach with an acid, such as vinegar. Also, ammonia and bleach should not be mixed because noxious chloramine gas, NH_2Cl, is formed.

Figure 16
Basic solutions of ammonia, such as these, can clean away light grease smears, such as fingerprints.

▶ **disinfectant** a chemical substance that kills harmful bacteria or viruses

▶ **bleach** a chemical compound used to whiten or make lighter, such as hydrogen peroxide or sodium hypochlorite

Quick ACTIVITY

Materials (per group):
• sheet of wax paper
• dropper
• pin
• liquid detergent

Teacher's Notes: The detergent acts as a surfactant, or wetting agent, by reducing surface tension so that the droplet will spread on a surface.

Results: In step 1, the water does not wet the surface because it remained in a drop. In step 2 nothing happens to the drop. In step 4 the water spreads out and the detergent allows water to better emulsify the dirt. **LS** Visual

Demonstration ── GENERAL

(Time: Approximately 5 minutes)

Materials:
• Petri dish
• milk
• liquid food coloring (4 colors)
• dropper
• detergent

Step 1 Place the milk in the Petri dish, and carefully drop four different colors of food colorings near the edge of the dish.

Step 2 Place a drop of liquid detergent in the middle of the dish as students observe.

Analysis

1. What did you observe after the detergent was added (The colors mixed.)

2. Recall that milk contains lipids. Do detergents emulsify lipids? (yes)

LS Visual

Quick ACTIVITY

Detergents

Detergents help break up oil into droplets that can be washed away by water. Detergents also break up the surface tension of water so that it can wet materials more easily. In this activity, you will demonstrate this effect using a piece of wax paper, a drop of water, a toothpick, and liquid detergent.

1. Lay some wax paper on a flat surface, and put a drop of water on it. Does the water wet the surface of the wax paper? How can you tell?

2. Gently touch the drop of water with the tip of the toothpick. What happens to the drop of water?

3. Now dip the tip of the toothpick in liquid detergent.

4. Gently touch the drop of water with the tip of the toothpick after it has been dipped in detergent. What happens to the drop of water? How could this action help water clean away dirt?

271

Did You Know ?

Color-Safe Bleaches Not all bleaches are disinfectants. Color-safe bleaches contain nonchlorine-oxidizing agents and do not disinfect items.

Historical Perspective

Sal Ammoniac Medieval physicians prescribed sal ammoniac (from the Latin *sal ammoniacus* meaning "salt of Ammon") to patients to relieve congestion. Sal ammoniac is the ammonia salt, ammonium chloride (NH_4Cl). Noted as early as the 8th century, sal ammoniac was reported to have been produced at the oracle of Jupiter Ammon in modern day Libya from the soot of burnt camel dung and exported throughout the Middle East. Sal ammoniac gave rise to the term *ammonia*.

Chapter Resource File

• **Cross-Disciplinary Worksheet**
Connection to Social Studies— Detergents: Helpful or Harmful? GENERAL

• **Quick Activity Datasheet**
Detergents GENERAL

internet connect

www.scilinks.org
Topic: Acids and Bases
at Home
SciLinks code: HK4003

SCiLINKS. Maintained by the National Science Teachers Association

Acids, Bases, and Salts in the Household

You probably have taken many of the acidic and basic materials in your home for granted. For example, many of the clothes in your closet get their color from acidic dyes. These dyes are sodium salts of organic compounds that contain the sulfonic acid group ($-SO_3H$) or the carboxylic acid group ($-COOH$). If you have ever had an upset stomach because of excess stomach acid, you may have taken an antacid tablet to feel better. The antacid made you feel better because it neutralized the excess stomach acid. Many other useful products in your home are also acids or bases.

Many healthcare products are acids or bases

In the morning before school, you may drink a glass of orange juice that contains vitamin C. Ascorbic acid is the chemical name for vitamin C, which your body needs to grow and repair bone and cartilage. Both sodium hydrogen carbonate and magnesium hydroxide (milk of magnesia) can be used as **antacids.** Antacids are basic substances that you swallow to neutralize stomach acid when you have an upset stomach. **Figure 17** shows how adding an antacid tablet to an acidic solution changes the pH of the solution. A similar reaction (without the color change) takes place in your stomach when you take an antacid.

antacid a weak base that neutralizes stomach acid

Teaching Tip

Buffers Tell students that buffers are systems that can dissolve moderate amounts of acids or bases without significantly changing pH. Buffered aspirin minimizes the effects of the acid in aspirin on the stomach. Blood must maintain a narrow pH range, and the hydrogen carbonate ion, HCO_3^-, acts as a buffer in blood. If blood becomes too acidic, the ion acts as a base. If blood becomes too basic, the ion acts as an acid.

Quick Lab

Answers

1. The antacid neutralizes the stomach acid by reacting with it.
2. Answers may vary. The brand that needed the most acid to neutralize it is probably the best.

LS Visual

Quick Lab

What does an antacid do?

Materials
- plastic stirrer
- wax paper
- several varieties of antacid tablets
- red litmus paper
- 150 to 200 mL beakers (2)
- pipet bulbs
- vinegar
- spoon
- blue litmus paper
- disposable pipets

1. Pour 100 mL of water in a beaker. Add vinegar one drop at a time while stirring. Test the solution with litmus paper after each drop is added. Record the number of drops it takes for the solution to turn blue litmus paper bright red.

2. Use the back of a spoon to crush an antacid tablet to a fine powder on a piece of wax paper. Pour 100 mL of water in the second beaker, add the powdered tablet, and stir until a suspension forms.

3. Use litmus paper to find out whether the mixture is acidic, basic, or neutral. Record your results.

4. Now add vinegar to the antacid mixture. Record the number of drops it takes to react with the antacid and turn the blue litmus paper bright red. Compare this solution with the solution that has only vinegar and water. Compare the brand of antacid you tested with the brands of other groups.

Analysis

1. How does an antacid work to relieve the pain caused by excess stomach acid?

2. Of the brands that were tested, which brand worked the best? Explain your reasoning.

272

Did You Know?

Antacids Two types of antacids currently exist. One type is insoluble metallic hydroxides, such as magnesium hydroxide, $Mg(OH)_2$. These undergo a neutralization reaction with stomach acid, forming chloride salts and water. Another type is a carbonate-based antacid, such as calcium carbonate, $CaCO_3$, or sodium hydrogen carbonate, $NaHCO_3$. These react with stomach acid to produce a chloride salt and carbonic acid, which breaks down into carbon dioxide and water.

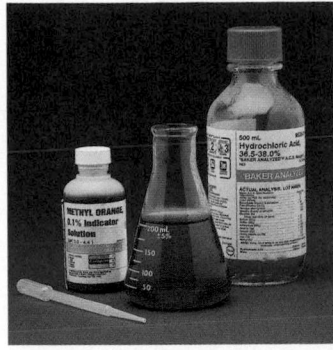
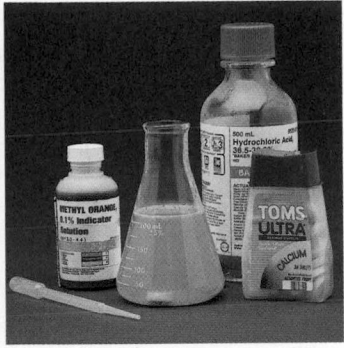

Figure 17

Stomach acid has about the same concentration of HCl as the solution in the flask in the left photo. When an antacid tablet reacts with the acid, the pH increases to a less acidic level, as shown in the photo on the right.

Shampoos are adjusted for an ideal pH

Shampoos can be made from soap. But if they are, they can leave sticky soap scum on your hair if you happen to live in an area that has hard water. Most shampoos today are made from detergents and are able to remove dirt as well as most of the oil from your hair without leaving soap scum, even when they are used with hard water. Shampoo is not meant to remove all of the oil from your hair. Some oil is needed to give your hair shine and to keep it from becoming dry and brittle.

The appearance of your hair is greatly affected by the pH of the shampoo you use. Hair—which consists of strands of a protein called *keratin*—looks best when it is kept at either a slightly acidic pH or very close to neutral. If a shampoo is too basic, it can cause strands of hair to swell, which gives them a dull, lifeless appearance. Shampoos are usually pH balanced, which means that they are made to be in a specific pH range. The pH of most shampoos is between 5 and 8. Shampoos that have higher pH values are more effective in cleaning oil from your hair. Shampoos that have lower pH values protect dry hair.

Acids keep fruit fresh longer

Some cut fruits slowly turn brown when they are exposed to air, such as the right side of the cut apple shown in *Figure 18.* This happens because certain molecules in the apple are oxidized to form darker substances. Both sides of the apple in *Figure 18* were cut at the same time, so why does the left side of the apple look like it was just cut? The left side was moistened with lemon juice shortly after it was cut. The citric acid in lemon juice helps *antioxidants* in the apple that react with oxygen before the oxygen can react with other substances in the apple. Vitamin C is another example of a natural antioxidant.

Did You Know?

The fibrous protein keratin builds up in the outermost cells of your epidermis, the outer layer of your skin. Keratin in these cells makes the skin tough and almost completely waterproof. Keratin forms callouses in places on the skin where it is rubbed.

The horns, hoofs, claws, feathers, and scales of animals grow from the same type of tissue that makes up your epidermis and also consist mainly of keratin.

Figure 18

The left side of this cut apple was coated with lemon juice. Citric acid in the lemon juice kept the surface of the apple looking fresh.

SKILL BUILDER — BASIC

Reading Skills Ask for volunteers to summarize parts of the section for the class. After each summary is completed, allow questions. Allow the student answering the questions to consult the text. Clarify any confusing concepts that the student presenter is not able to address. **LS Verbal**

SKILL BUILDER — GENERAL

Interpreting Visuals Have students inspect **Figure 17** closely. Tell students that methyl orange, like most indicators, is a weak organic acid that changes color when it loses a proton. It appears orange in acidic solutions that have a pH of 3.1–4.4. It appears red in solutions with pH values less than 3.1, and yellow in solutions witih pH values above 4.4. Ask students the pH of the solution on the left (3.0 or less) and the pH of the solution on the right. (between 3.1 and 4.4) Ask students why methyl orange would not be useful to differentiate water from a solution of baking soda. (Both would appear yellow because each has a pH value greater than 4.5)

Have students use **Figure 6** to determine the color the pH test paper if it were used to test the two. (left: golden yellow; right: yellow) **LS Logical**

273

Trends in Biochemistry

Free Radicals and Antioxidants

Compounds called oxygen-centered free radicals, designated RO•, can cause damage to cells at the molecular level. These substances each contain a single unpaired electron, which makes them very reactive. A free radical can easily strip away an electron from another molecule making it a free radical. This cascading effect can damage cell structures, such as the cell membrane and DNA.

Antioxidants are substances that donate electrons to the free radicals before they can react with other substances in the cell. For example, the antioxidants vitamin E and C work in tandem to protect the cell membrane from damage by free radicals. Vitamin E is lipid soluble and resides within the cell membrane while vitamin C is water-soluble and exists within the cell. Vitamin E donates an electron to the free radical and becomes a free radical. Vitamin C within the cell donates an electron to the vitamin E radical converting the radical to vitamin E and itself becoming a radical. Within the cell, the vitamin C radical is converted to vitamin C by the action of enzymes.

Chapter Resource File

• **Quick Lab Datasheet** What does an antacid do? GENERAL

• **Cross-Disciplinary Worksheet** Integrating Biology—Balance in the Body GENERAL

Videos

CNN. Presents Chemistry

• **Segment 26** Acids in the Environment

• **Critical Thinking** Worksheet 26

See the Science in the News video guide for more details.

Quiz

1. Fatty acids have long chains of what? (hydrocarbons)

2. What is a soap? (a substance that dissolves in water or oil and is a sodium or potassium salt of a fatty acid, usually derived from plant or animal fat)

3. A substance that dissolves in both fats and oils and is a sodium salt of a hydrocarbon sulfonic acid is called a what? (a detergent)

4. Is lye a strong acid or a strong base? (strong base)

Figure 19

Adding vinegar to milk causes the milk to curdle, because casein, the main protein in milk, becomes denatured by the acid.

Acids, bases, and salts in the kitchen

Acids have other uses in the kitchen. Acidic marinades made of vinegar or wine can be used to tenderize meats because they can *denature* proteins in the meat. That is, the acids cause the protein molecules to unravel and lose their characteristic shapes. As a result, the meat becomes more tender.

Figure 19 shows that milk curdles if you add vinegar to it. This reaction may seem undesirable, but a similar reaction occurs in the formation of yogurt. Bacteria convert lactose, a sugar in milk, into lactic acid. The lactic acid denatures the protein casein in milk and changes the milk into a thick gel known as yogurt.

There are many bases and salts in the kitchen. You can unclog a drain by using the strong base sodium hydroxide, also called *lye*. Baking soda, or sodium hydrogen carbonate, is a salt that forms carbon dioxide gas at high temperatures, which makes cookies rise when they are baked. Baking powder consists of baking soda and an acidic substance that react to release CO_2, which makes light, fluffy batter for cakes.

SECTION 3 REVIEW

SUMMARY

▶ Soaps and detergents can dissolve in oil and water. They are usually sodium or potassium salts of carboxylic or sulfonic acids, which have long hydrocarbon chains.

▶ Detergents do not form an insoluble scum in hard water as soap does.

▶ Bleach is an alkaline solution of sodium hypochlorite, NaOCl. Bleach is a disinfectant and oxidizes stains to a colorless form.

▶ Antacids are basic substances that react with hydrochloric acid in the stomach.

▶ Acids, bases, and salts have many practical uses in the kitchen, both in cleaning and cooking.

1. **Describe** how soap can dissolve in both oil and water. How does soap work with water to remove oily dirt?

2. **Explain** why soap scum might form in hard water that contains Mg^{2+} ions when soap is used instead of detergent to wash dishes.

3. **Explain** why the agitation of a washing machine helps a detergent clean your clothes. (**Hint:** Compare this motion to rubbing your hands together when you wash them.)

4. **Explain** why it is not necessary for bleach to actually remove the substance that causes a stain.

5. **Explain** how milk of magnesia, an antacid, can reduce acidity in stomach acid.

6. **List** three acidic household substances and three basic household substances. How are the substances most often used?

7. **Critical Thinking** Crayon companies recommend treating wax stains on clothes by spraying the stains with an oily lubricant, applying dishwashing liquid, and then washing the clothes. Explain in a paragraph why this treatment would remove the stain.

Chapter Resource File

• Concept Review **GENERAL**

• Quiz **BASIC**

Answers to Section 3 Review

1. Soap is a salt of a fatty acid. The anion consists of a hydrocarbon chain with a carboxyl group , COO⁻, attached. The hydrocarbon chain is similar to the molecular structure of oil so that end of the molecule dissolves in oil, while the ionic carboxyl end dissolves in water. When an oil droplet has enough soap hydrocarbon chains dissolved in it, the attraction between water and the carboxyl groups pulls the droplet into suspension where it can be washed away by rinsing.

2. Insoluble magnesium stearate is formed, which precipitates forming soap scum.

3. The agitation moves suspended oily dirt particles away from the clothing enabling new detergent ions to reach the remaining dirt. The dirt becomes suspended in water and suds and can be rinsed away.

4. Usually, bleach can decolorize the stain by oxidizing the colored substance to a colorless form.

5. As a basic solution, milk of magnesia can neutralize the stomach acid, and reduce its acidity.

See "Continuation of Answers" at the end of the chapter.

Graphing Skills

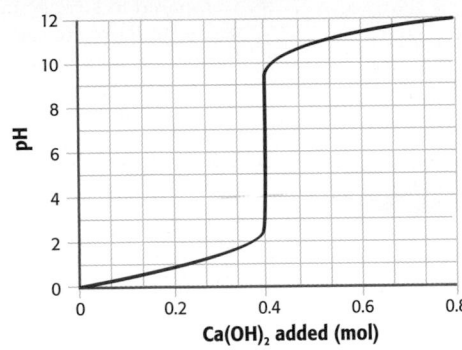

Examine the above graph, and answer the following questions. (See Appendix A for help interpreting a graph.)

1 Does the solution's acidity increase or decrease as calcium hydroxide is added? Explain your answer.

2 Identify the independent and dependent variables. What is the relationship between the two variables?

3 At what point on the graph are there equal moles of acid and base? Explain your answer.

4 Use your answer to the previous question to calculate the number of moles of acid present before the calcium hydroxide is added to the solution.

5 What is the pH of the calcium hydroxide solution added to the acid? How did you reach this conclusion?

6 A person who has a stomach disorder is advised to avoid acidic foods. Construct the type of graph best suited for the data given in the table below. Which substance is most acidic? Which substance has a pH closest to the pH of pure water?

Substance	Average pH
Bananas	4.6
Dill pickles	3.4
Eggs	7.8
Salmon	6.2
Soda crackers	7.5

Answers

1. It decreases, as indicated by the increase in the pH value. Larger pH indicates a more basic, less acidic solution.

2. Moles of calcium hydroxide is the independent variable; pH is the dependent variable. As more moles of calcium hydroxide are added, the pH of the solution becomes less acidic and more basic.

3. When the pH value of the solution equals 7 (the curve is vertical at this point). A pH value of 7 indicates that there are equal amounts of acid and base, or H_3OH^+ and OH^-, in the solution.

4. 0.4 moles of acid

5. 12. As more calcium hydroxide is added, the curve levels off at a pH of 12, indicating that this is the maximum pH value of the calcium hydroxide solution.

6. dill pickles; soda crackers

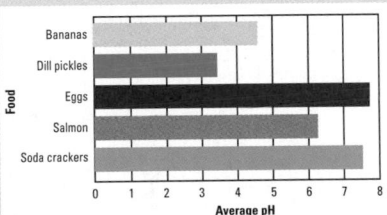

Understanding Concepts

1. b
2. c
3. d
4. a
5. c
6. c
7. b
8. d
9. a
10. d
11. b
12. a
13. a
14. b
15. a

Using Vocabulary

16. Strong acids ionize completely in water to hydronium ions and the anion of the acid. Weak acids ionize only slightly in water, establishing an equilibrium between ionized and non-ionized forms.

    Student choices of acids will vary. Examples are shown below.

    Strong acid:

    $HNO_3 + H_2O \rightarrow H_3O^+ + NO_3^-$

    Weak acid:

    $HCOOH + H_2O \rightleftarrows H_3O^+ + HCOO^-$

17. calcium bromide, $CaBr_2$

Chapter Resource File

- **Chapter Tests** GENERAL
- **Test Item Listing for ExamView™ Test Generator**

Chapter Highlights

Before you begin, review the summaries of key ideas of each section, found at the end of each section. The vocabulary terms are listed on the first page of each section.

UNDERSTANDING CONCEPTS

1. Which ions does an acid form in solution?
   a. oxygen
   b. hydronium
   c. hydroxide
   d. sulfur

2. Which ions does a base form in solution?
   a. oxygen
   b. hydronium
   c. hydroxide
   d. sulfur

3. A substance with a pH of 9 has
   a. the same number of H_3O^+ ions and OH^- ions.
   b. more H_3O^+ ions than OH^- ions.
   c. no H_3O^+ ions, but many OH^- ions.
   d. more OH^- ions than H_3O^+ ions.

4. When a solution of nitric acid is added to a solution of calcium hydroxide, the salt formed has the formula
   a. $Ca(NO_3)_2$.
   b. $Ca(OH)_2$.
   c. H_2O.
   d. CaH.

5. An antacid relieves an overly acidic stomach because antacids are
   a. acidic.
   b. neutral.
   c. basic.
   d. dilute.

6. Any substance that conducts electricity when it dissolves in water is called a(n)
   a. salt.
   b. antacid.
   c. electrolyte.
   d. weak base.

7. Detergents have replaced soap in many uses because detergents
   a. are made from animal fat.
   b. do not form insoluble substances.
   c. are milder than soap.
   d. contain ammonia.

8. Compared to strong acids, weak acids
   a. ionize more completely in water.
   b. are less soluble in water.
   c. do not react with bases.
   d. ionize only slightly in water.

9. Which of the following ions could be present in a salt?
   a. Br^-
   b. OH^-
   c. H_3O^+
   d. H^+

10. Which of the following ionic equations best represents a neutralization reaction?
    a. $Na + H_2O \rightarrow Na^+ + OH^- + H_2$
    b. $HNO_3 + H_2O \rightarrow H_3O^+ + NO_3^-$
    c. $2OH^- + NH_4Cl \rightarrow Cl^- + H_2O + NH_3$
    d. $OH^- + H_3O^+ \rightarrow 2H_2O$

11. An increase in the hydronium ion concentration of a solution _____ the pH.
    a. raises
    b. lowers
    c. does not affect
    d. doubles

12. A complete neutralization of a weak acid by a strong base yields a solution that is
    a. basic.
    b. neutral.
    c. acidic.
    d. saturated.

13. Bleach removes stains by
    a. changing the color of the stain.
    b. covering the stain.
    c. removing the stain-causing substances.
    d. disinfecting the stain.

14. Which of the following is *not* a property of soap?
    a. It is a salt.
    b. It is made from petroleum.
    c. It dissolves in both oil and water.
    d. It is an ionic substance.

15. Which of the following is *not* an acidic material found in the kitchen?
    a. baking soda
    b. lemon juice
    c. vinegar
    d. vitamin C

276

18. A solution that is acidic will turn blue litmus paper red. A solution that is basic will turn red litmus paper blue. A neutral solution will have little effect on either type of litmus paper.

19. Strong bases are metal hydroxides. The most abundant ions in the solution of a metal hydroxide are the metal cation, M^+, and the hydroxide ion, OH^-.

20. The pH is the negative of the exponent, or power, of 10 in the expression of the molar concentration of hydronium ions. Because pH is the negative of the exponent, the pH decreases as the exponent becomes larger (less negative).

21. Soaps and detergents are salts of acids that have a hydrocarbon chain. The hydrocarbon chain is similar to the molecular structure of oil so that end of the molecule dissolves in oil, while the ionic end of the molecule dissolves in water. When an oil droplet has enough hydrocarbon chains dissolved in it, the attraction between water and the ionic groups pulls the droplet into suspension where it can be washed away by rinsing.

22. The active substance in bleach is sodium hypochlorite, which is made by passing chlorine gas into sodium hydroxide solution.

16. Explain how the *ionization* of a strong acid differs from the ionization of a weak acid in a solution. Give an example of a strong acid and a weak acid. Show which ions form when each is dissolved in water.

17. Give both the name and the formula of the salt produced in the following *neutralization reaction*:

$$2H_3O^+ + 2Br^- + Ca^{2+} + 2OH^- \rightarrow Ca2^+ + 2Br^- + 4H_2O$$

18. Explain how you can use the *indicator* litmus, in the form of litmus paper, to determine whether a solution is *acidic, basic,* or *neutral.*

19. List the two kinds of ions that are in greatest concentration in a solution of a strong *base.*

20. How is the pH of a solution related to its *hydronium ion* concentration? What happens to pH as this concentration changes?

21. Explain how the molecular structure of *soaps* and *detergents* can cause water to wash away oil and grease.

22. What is the active substance in *bleach*? How is bleach made?

23. Why are most shampoos made from *detergents* rather than *soaps*?

24. Microbiologists often wipe down work areas with a *bleach* solution before working with bacterial cultures. What is the purpose of using bleach in this way?

25. Explain why a solution of a *strong acid* is a good conductor of electricity.

26. What is a *neutralization reaction*? How might the product of a neutralization reaction have a pH of less than 7?

27. How would you find the concentration of a strong acid in a *titration*? Use the terms *indicator* and *equivalence point* in your answer.

28. **Determining pH** What is the pH of a 0.001 M solution of rubidium hydroxide, RbOH, a strong base?

29. **Determining pH** What is the pH of a solution that contains 0.10 mol of HCl in a volume of 100.0 L?

30. **Using pH** What is the molar concentration of hydroxide ions in a solution with a pH of 6?

31. **Determining pH** The concentration of hydronium ions in a certain acid solution is 100 times the concentration of hydronium ions in a second acid solution. If the second solution has a pH of 5, what is the pH of the first solution?

32. **Interpreting Graphs** The point at which equal amounts of an acid and a base have reacted in a neutralization reaction is called the equivalence point. Study the graph of pH versus volume of base added below, and note the pH at the equivalence point. Classify both the acid and the base in this neutralization reaction as either weak or strong. Explain your answer.

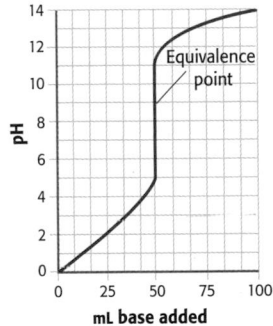

23. The fatty acid anions in soap form insoluble salts with Ca^{2+} and Mg^{2+}, which precipitate as soap scum. If soaps were used in shampoo, this soap scum would precipitate on the hair. The sulfonate anions in detergents do not form precipitates with these metal ions.

24. Bleach acts as a disinfectant, which means that it kills bacteria and other microorganisms. Wiping the work surface kills stray bacteria that could contaminate the cultures with which the microbiologist is working.

25. Strong acids ionize completely, producing abundant ions in water solution. These ions are able to move through the solution conducting an electric current.

26. A neutralization reaction is one in which hydronium and hydroxide ions react to form water molecules and a salt. When a strong acid reacts with a weak base, the product of the reaction will be acidic and have a pH of less than 7.

27. To determine the concentration of a strong acid in a titration, you could gradually add a strong base to the solution until the equivalence point is reached and the equal amounts of acid and base have reacted. An indicator immersed in the solution will change color when the equivalence point is reached.

Building Math Skills

28. As a strong base, RbOH should be completely dissociated in water solution. So, the concentration of OH⁻ ions is 0.001 M, or 1×10^{-3} M. pOH = $-(-3)$ = 3. pH = 14 − 3 = 11

29. Because HCl is a strong acid, the concentration of hydronium ions in the solution is the same as the molar concentration of HCl. 0.10 mol HCl/100.0 L = 0.0010 mol/L = 1×10^{-3} M H_3O^+ pH = $-(-3)$ = 3

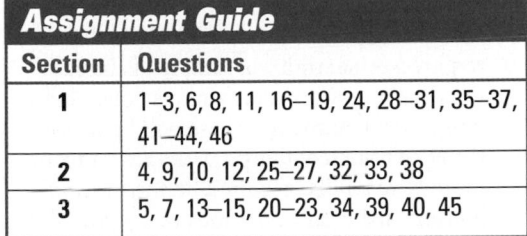

30. pH + pOH = 14; pOH = 14 − 6 = 8; concentration of OH− ions is $1 \times 10^{-(pOH)}$ M = 1×10^{-8} M

31. The hydronium ion concentration in the second solution of pH 5 is 1×10^{-5} M. So, the H_3O^+ concentration of the first solution is $100 \times 1 \times 10^{-5} = 10^2 \times 10^{-5} = 10^{(-5 + 2)} = 10^{-3}$ M; so, the pH is −(−3) = 3.

Building Graphing Skills

32. In this neutralization, the base is strong and the acid is weak. When equal amounts of the acid and base have reacted, the pH is 9, a basic pH. This pH indicates that the solution contains the anions of a weak acid, which react with water to produce extra OH− ions.

33. a. The pH increases slowly because nearly all of the added OH− ions are reacting with H_3O^+ ions from the acid.
    b. When 50 mL of base has added, the equivalence point is reached. No additional OH− ions are reacting with H_3O^+ ions. Instead, the concentration of OH− ions increases rapidly and the pH increases rapidly.
    c. The concentration of OH− continues to increase. However each additional pH unit must come from a tenfold increase in OH− ions, so the graph moves steadily upward, but at only a fraction of the rate that it increased at around 50 mL.

33. **Interpreting Graphs** The graph below shows how pH changes as a 0.1 M solution of NaOH is added to 50 mL of a 0.1 M solution of HCl. Use the graph to answer the following questions.
    a. Describe how pH changes as the first 30 mL of NaOH solution is added. What takes place in the solution during this addition?
    b. What is happening in the solution just as 50 mL of NaOH has been added? Why is the pH changing so rapidly at this point? (**Hint:** When 50 mL of the NaOH solution has been added, equal amounts of acid and base have combined.)
    c. What is happening in the solution as more than 50 mL of NaOH solution is added?

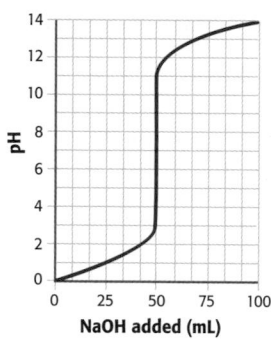

THINKING CRITICALLY

34. **Applying Knowledge** Baking soda, sodium hydrogen carbonate ($NaHCO_3$), is useful in the kitchen for baking and to absorb odors in the refrigerator. Baking soda can also be tossed onto a grease fire to extinguish it. How can baking soda extinguish fires?

35. **Creative Thinking** If you wish to change the pH of a solution very slightly, should you add a strong acid or a weak acid? Explain your answer.

36. **Problem Solving** Insect bites hurt because the insect injects a toxin into the victim. When certain kinds of ants bite, they inject a small amount of highly irritating formic acid. Suggest a treatment that might stop an ant bite from itching or hurting.

37. **Designing Systems** Suppose you measure the pH of a clear solution in a beaker and find that it has a pH of 3. You are asked to determine whether the solution is a very dilute solution of a strong acid or a stronger solution of a weak acid. Propose a method to answer the question.

38. **Creative Thinking** You need several grams of the substance ammonium bromide, NH_4Br, for an experiment, but you do not have any. You do, however, have a solution of hydrobromic acid, HBr, and a solution of ammonia. Suggest a way to use an acid-base reaction to make a small quantity of NH_4Br.

39. **Applying Knowledge** Pure water is a poor conductor of electricity. But it is still dangerous to have any sort of plugged-in appliances near the bathtub or shower. Why does this danger exist? Explain your reasoning by discussing the composition of tap water.

BUILDING LIFE/WORK SKILLS

40. **Communicating Effectively** When there is an oil spill in the ocean, emergency-response teams use the properties of oil and water along with solubility principles to clean spills and prevent them from spreading. Describe the research behind these techniques, and evaluate the impact this research has had on the environment.

41. **Locating Information** Research the invention of the pH meter by Dr. Arnold O. Beckman. Why was the pH meter invented? How does it work? Prepare a poster to present your results.

Thinking Critically

34. Baking soda decomposes and releases CO_2 when it is heated. The CO_2 from the baking soda thrown on a fire deprives the fire of oxygen and puts it out.

35. A weak acid would be better because it changes the pH more gradually. Because of a weak acid's poor ionization, fewer hydronium ions would be added.

36. You would neutralize the acid by applying a basic material such as ammonia solution or a paste of baking soda. A strong base would also neutralize the acid, but would also damage the skin.

37. Student responses may vary. One possible method is to measure the amount of base required to neutralize a sample of the solution. If the solution contains a strong acid, only a small amount of base will be needed. If the solution contains a weak acid, much more base will be required because the acid ionizes further as H_3O^+ ions react with OH− ions in the base.

38. You would neutralize an amount of ammonia in solution with an equal amount of hydrobromic acid solution. After the water is evaporated, crystals of the salt ammonium bromide will be left.

42. Applying Knowledge Design an experiment to measure the pH of four types of shampoo: baby shampoo, shampoo for extra body, shampoo for oily hair, and shampoo that contains conditioner. Also compare two brands of pH-balanced shampoo. Write a paragraph summarizing your results.

INTEGRATING CONCEPTS

43. Connection to Biology The pH of human blood is about pH 7.4 and must be kept within a few tenths of a pH unit of the normal pH. Reactions within the human body ensure that a proper pH is maintained. The equilibrium shown below between carbonic acid, H_2CO_3, and the hydrogen carbonate ion, HCO_3^-, is important to maintain the blood's pH.

$$H_2CO_3 + H_2O \rightleftharpoons HCO_3^- + H_3O^+$$

Find out what happens to keep the pH from decreasing as extra hydronium ions enter the blood. Also find out how the pH is kept from increasing as extra OH^- ions enter the blood.

44. Connection to History In the 18th century, the French chemist Antoine Lavoisier experimented with substances containing oxygen, such as CO_2 and SO_2, that formed acidic solutions when dissolved in water. His observations led him to infer that for a solution to be acidic, it must contain oxygen. Provide evidence to disprove Lavoisier's conclusion.

45. Locating Information A reaction between baking soda and a baking batter that is made of acidic ingredients produces CO_2 gas. The reaction makes the batter fluffier. Some recipes call for baking powder instead of baking soda. Find out what regular baking powder and double-acting baking powder are made of. How do they each differ from baking soda?

46. Concept Mapping Copy the unfinished concept maps below onto a sheet of paper. Complete the maps by writing the correct word or phrase in the lettered boxes.

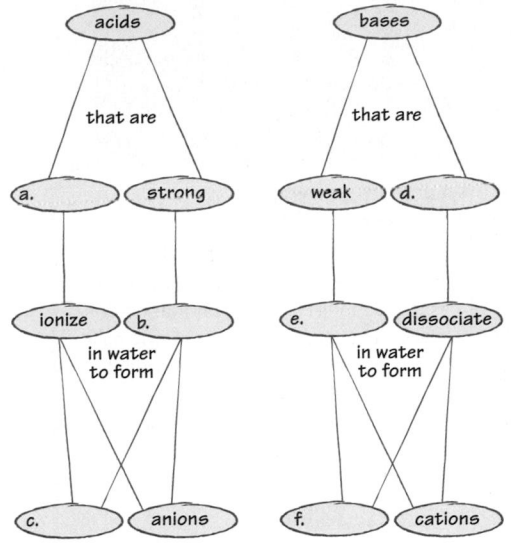

internet connect

www.scilinks.org
Topic: Baking Soda/Baking Powder
SciLinks code: HK4164

SCI*LINKS*® Maintained by the National Science Teachers Association

Art Credits: Fig. 2A, Kristy Sprott; Fig. 3B, Kristy Sprott; Fig. 5-7, Kristy Sprott; Fig. 11, Kristy Sprott.

Photo Credits: Chapter Opener image of Weaver Ants by Philip Chapman/Taxi/Getty Images; helicopter dropping lime by Peter Arnold, Inc.; Fig. 1, Sam Dudgeon/HRW; Fig. 2B, 3B, Sergio Purtell/Foca/HRW; Fig. 4, Sam Dudgeon/HRW; Fig. 6, Charles D. Winters; Fig. 7, Charles D. Winters/Photo Researchers, Inc.; Fig. 8, Yoav Levy/Phototake; Fig. 9, Charlie Winters; Fig. 10, CNRI/Photo Researchers, Inc.; Fig. 11, Charlie Winters; Fig. 13, Frans Lanting/Minden Pictures; Fig. 14, Bob Thomason/Getty Images/Stone; Fig. 15, Sam Dudgeon/HRW; Figs. 16-19, Peter Van Steen/HRW; "Skill Builder Lab," Charlie Winters/HRW.

39. Pure water is a poor conductor because it contains very few ions. Tap water is often hard water containing dissolved salts of calcium and magnesium, making it much more conductive than pure water. Soaps and detergents are also ionic salts that dissociate into ions when dissolved in water. So, the danger exists because you are almost never in contact with pure water.

Developing Life/Work Skills

40. Collecting the oil is the best method. Booms are used to surround and contain the oil before it is collected using skimmers or absorbents. Chemicals that emulsify and disperse the oil are also used along with microorganisms that feed on oil.

41. Students will find that Dr. Beckman built the first pH meter in 1935 at the request of a friend who worked at a citrus processing plant. The original purpose of the instrument was to measure the acidity of lemon juice so that a uniform product could be shipped. The device was first called an acidimeter. Posters describing the pH meter should explain the operation of the glass electrode, the crucial part of the pH meter.

42. Students' experiments should designate means of measuring pH of shampoos (most likely with litmus paper.)

279

43. Maintaining blood pH is a complicated process involving enzymes, red blood cells , and other factors. However the main processes can be summarized in two chemical equations involving the hydrogen carbonate ion, HCO_3^-.

When H_3O^+ ions begin to build up, they are absorbed as the equilibrium reaction

$$HCO_3^- + H_3O^+ \rightleftharpoons H_2CO_3 + H_2O$$

shifts to the right, producing more carbonic acid, H_2CO_3, molecules.

When OH^- ions begin to build up, they are absorbed as the equilibrium reaction

$$H_2CO_3 + OH- \rightleftharpoons HCO_3^- + H_2O$$

shifts to the right, producing more hydrogen carbonate ions, HCO_3^-.

44. Some acids can be formed by a reaction that occurs when an oxide of a nonmetal dissolves in water. Many acids, though, do not contain oxygen atoms. Examples are HF, HCl, H_2S, HBr, HCN, and others.

See "Continuation of Answers" at the end of the chapter

Transparencies

TT Concept Mapping

MEASURING QUANTITIES IN ACID-BASE REACTIONS

Teacher's Notes

Time Required 1 lab period

Ratings

EASY 1 2 3 4 HARD

TEACHER PREPARATION	2
STUDENT SETUP	3
CONCEPT LEVEL	2
CLEANUP	2

Skills Acquired
- Collecting Data
- Communicating
- Experimenting
- Interpreting
- Measuring
- Organizing and Analyzing Data
- Predicting

The Scientific Method
In this lab, students will:
- Make observations
- Form a hypothesis
- Analyze the results
- Draw conclusions
- Communicate results

Safety Cautions
Have students review safety guidelines before working in the lab.

Skills Practice Lab

Introduction

Acids and bases neutralize each other to form a salt and water. Phenolphthalein is a good indicator to use in the neutralization of a strong acid by a strong base. It is a good indicator because phenolphthalein changes color at a pH very near the neutral point of a reaction of a strong base and a strong acid.

Objectives

▶ **Determine** the volume of a base solution needed to neutralize a given volume of acid solution.

▶ USING SCIENTIFIC METHODS **Analyze the results** to compare the volume of base solution needed to neutralize a given volume of HCl solution with the volume needed to neutralize the same volume of H_2SO_4 solution.

Materials

0.1 M H_2SO_4 solution
0.1 M HCl solution
0.1 M NaOH solution
marker
phenolphthalein indicator solution
plastic pipets, disposable
test-tube rack
test tubes

280

Measuring Quantities in an Acid-Base Reaction

Preparing for Your Experiment

1. On a sheet of paper, prepare a data table similar to the one shown at right.

▶ Procedure

Neutralizing HCl with NaOH

SAFETY CAUTION Wear an apron or lab coat to protect your clothing when working with chemicals. If a spill gets on your clothing, rinse it off immediately with water for at least 5 minutes, while. Wear safety goggles and gloves when handling chemicals. If any substance gets in your eyes, immediately flush your eyes with running water for at least 15 minutes and notify your instructor. Always use caution when working with chemicals. Add an acid or a base to water; never do the opposite.

2. Use the marker to write "HCl" on the bulb of one pipet. This pipet should be used only for hydrochloric acid solution. Mark a second pipet NaOH. This pipet should be used only for sodium hydroxide solution.

3. Add 40 drops of 0.1 M HCl solution to a clean test tube at a steady rate. Do not let the tip of the pipet touch the sides of the test tube. Hold the long tube of the pipet with the other hand, if necessary.

4. Add two drops of phenolphthalein indicator to the test tube. Gently shake the test tube from side to side to mix the liquid in the tube. Be careful not to spill or splash the liquid.

5. Note the concentrations of the HCl and NaOH solutions. Predict how many drops of NaOH solution will be required to neutralize the 40 drops of HCl. Record your prediction in the data table.

6. Add 25 drops of 0.1 M NaOH solution to the test tube. You will probably see a pink color develop temporarily. This is the color of phenolphthalein in a basic solution. Remember this color. Gently swirl the test tube to mix the liquid. The pink color should disappear.

Neutralization Reaction Data

	Number of drops	Drops NaOH needed (predicted)	Drops NaOH needed (measured)
HCl			
H_2SO_4			

7. Add more NaOH solution to the test tube two drops at a time, and mix the liquids after each addition. As the pink color starts to disappear more slowly when you mix the liquids, start adding the NaOH solution one drop at a time, and mix the solution with each addition. When the mixture remains slightly pink after the addition of a drop and does not change within 10 seconds, you have reached the end of the neutralization reaction. Record in the data table the total number of drops of NaOH solution you added.

Neutralizing H_2SO_4 with NaOH

8. Use the marker to label a third pipet "H_2SO_4." Use this pipet only for sulfuric acid solution.

9. Repeat steps 3–6, but start with 40 drops of 0.1 M H_2SO_4 solution instead of 40 drops of HCl solution. Make and record your prediction as in step 4.

▶ Analysis

1. In the neutralization of HCl with NaOH, how close was your predicted number of drops to the actual number of drops of NaOH solution needed? If there is a large difference, explain the reasoning that led to your prediction.

2. Write a complete nonionic chemical equation for the reaction of HCl and NaOH. Then, write the ionic equation for the reaction without spectator ions.

3. In the neutralization of H_2SO_4 with NaOH, how close was your predicted number of drops to the actual number of drops of NaOH solution needed? If there is a large difference, explain the reasoning that led to your prediction.

4. Write a complete nonionic chemical equation for the reaction of H_2SO_4 and NaOH. Then, write the ionic equation for the reaction without spectator ions.

▶ Conclusions

5. Suppose someone tries to explain your results by saying that H_2SO_4 is twice as strong an acid as HCl. How could you explain that this person's reasoning is incorrect?

Techniques to Demonstrate

Show students how to "stir" small amounts of liquids in a test tube by holding the tube at the lip with one hand and tapping the other end gently against the palm of the other as in Step 4 of the Procedure.

Tips and Tricks

If students agitate the contents of the tube for too long a time in Step 6, the mixture will change from pink back to colorless as the solution absorbs atmospheric CO_2 and becomes slightly more acidic.

To save time in step 9 students may start by adding 60 drops of NaOH solution.

Discrepancies between the number of drops of NaOH and HCl needed for neutralization in question 1 of the Analysis can be often explained by the fact that students tend to become impatient and add more base than is necessary to neutralize the acid.

Disposal Information

Neutralize both acids with the sodium hydroxide solution. Dilute with excess water and discard in the drain. Wrap pipettes in paper toweling and dispose in the trash.

Analysis

1. Since both solutions are of the same concentration and HCl and NaOH react in a 1 : 1 ratio, the number of drops of HCl and NaOH should be equal.

2. HCl + NaOH → NaCl + H_2O
   H_3O^+ + OH^- → $2H_2O$

3. Student should infer from their experimental data and the formula of H_2SO_4 that two hydrogen ions are available from H_2SO_4. Therefore, twice as many drops of NaOH is needed to neutralize the H_2SO_4.

4. H_2SO_4 + 2NaOH → Na_2SO_4 + $2H_2O$
   $2H_3O^+$ + $2OH^-$ → $4H_2O$

Conclusions

5. Point out that the person is confusing strength of acids with concentration of hydronium ions. The strength of an acid refers to its degree of ionization. Strong acids are completely ionized in solution. What the person is referring to as "stronger" is that, per mole of acid, sulfuric acid produces a greater concentration of H_3O^+ ions in solution than HCl does.

Chapter Resource File

- **Datasheet** Measuring Quantities in an Acid-Base Reaction **GENERAL**
- **Observation Lab** Properties of Acids and Bases **BASIC**
- **CBL™ Probeware Lab** Determining the Concentration of an Acid Solution **ADVANCED**

Continuation of Answers

Continuation of Answers from p. 274
Section 3 Review

6. Answers will vary. Acidic materials could include vinegar (acetic acid), lemon juice (citric acid), Vitamin C (ascorbic acid), aspirin (acetylsalicylic acid), buttermilk or sour cream (lactic acid), acidic salts in baking powder (sodium aluminum sulfate and calcium dihydrogen phosphate), muriatic acid (hydrochloric acid in a concentration used to clean masonry), and others.

    Basic materials could include ammonia solution, drain cleaner made with lye (sodium hydroxide), milk of magnesia (magnesium hydroxide), other antacids (basic salts including calcium carbonate and sodium hydrogen carbonate, aluminum hydroxide), potash (potassium carbonate), lime (calcium hydroxide), washing soda (sodium carbonate), and others.

7. Oil and wax are both nonpolar materials, so wax will dissolve in the oily lubricant. The nonpolar ends of the detergent molecules dissolve in the droplets of oil containing the wax, while the ionic ends dissolve in water. The agitation of the washing machine disperses the droplets throughout the water, away from the clothes and the oil rinses away.

Continuation of Answers from p. 279
Chapter Review
Integrating Concepts

45. Single-acting baking powders contain sodium hydrogen carbonate and a dry acidic salt such as cream of tartar, potassium hydrogen tartrate, $KHC_4H_4O_6$. These two ingredients react at room temperature to release CO_2 gas when they become wet. The released CO_2 puffs the uncooked batter. Some baking powders use other dry acidic substances. Double acting baking powders contain ingredients similar to single-acting powders, but they also include a second dry acidic salt, usually sodium aluminum sulfate, $NaAl(SO_4)_2$. This salt reacts with the baking soda only at higher temperatures. This reaction is the second "action" in double-acting powder.

46. a. weak
   b. ionize
   c. hydronium ions
   d. strong
   e. dissociate
   f. hydroxide ions

Nuclear Changes
Chapter Planning Guide

PACING	CLASSROOM RESOURCES	LABS, ACTIVITIES, AND DEMONSTRATIONS
BLOCK 1 · 45 min pp. 282–283 **Chapter Opener**		**SE Activity 1,** p. 283
BLOCKS 2 & 3 · 90 min pp. 284–292 **Section 1** What Is Radioactivity?	**CRF Lesson Plan*** **TT Bellringer*** **TM Types of Nuclear Radiation*** **TM Half-Life***	**TE Opening Demonstration,** p. 284 **SE Quick Activity** Modeling Decay and Half-Life, p. 290 GENERAL **CRF Datasheets for In-Text Labs** Modeling Decay and Half-Life* GENERAL
BLOCKS 4 & 5 · 90 min pp. 293–298 **Section 2** Nuclear Fission and Fusion	**CRF Lesson Plan*** **TT Bellringer*** **TT Chain Reaction***	**TE Opening Activity,** p. 293 GENERAL **TE Demonstration** Modeling Nuclear Forces, p. 294 GENERAL **TE Demonstration** A Chain Reaction, p. 296 GENERAL **SE Quick Activity** Modeling Chain Reactions, p. 297 GENERAL **CRF Datasheets for In-Text Labs** Modeling Chain Reactions* GENERAL
BLOCKS 6 & 7 · 90 min pp. 299–306 **Section 3** Nuclear Radiation Today	**CRF Lesson Plan*** **TT Bellringer*** **TT MRI Images of Healthy Brain*** **TT MRI Image of Brain with Alzheimer's*** **TT Concept Mapping***	**TE Opening Activity,** p. 299 GENERAL **SE Skills Practice Lab** Simulating Nuclear Decay Reactions, pp. 312–313 ◆ GENERAL **CRF Datasheets for SE Skills Practice Lab** Simulating Nuclear Decay Reactions* GENERAL **CRF Observation Lab** Modeling Radioactive Decay With Pennies* ◆ BASIC **CRF CBL™ Probeware Lab** Determining the Effective Half-Life of Iodine-131 in the Human Body* ◆ ADVANCED

BLOCKS 8 & 9 · 90 min

Chapter Review and Assessment Resources

SE **Chapter Review,** pp. 308–311
CRF **Chapter Tests*** GENERAL
OSP **Test Generator**
CRF **Standardized Test Practice with Guided Reading Development***
CRF **Test Item Listing for ExamView® Test Generator***
OSP **Scoring Rubrics and Classroom Management Checklists**

Online Resources

go.hrw.com

Visit the HRW Web site for a variety of free resources related to the text. Just type in the keyword **HK4 NUC**.

Holt Online Learning

Holt Science Spectrum: Physical Science: Online Edition

Students can access interactive problem solving help and active visual concept development with the *Holt Science Spectrum: Physical Science* Online Edition available at **www.hrw.com**.

student CNN News

cnnstudentnews.com

Find the latest news, lesson plans, and activities related to important scientific events.

KEY

TE Teacher Edition	**CRF** Chapter Resource File
SE Student Edition	**TT** Teaching Transparency
OSP One-Stop Planner	**TM** Transparency Master

* Also on One-Stop Planner
◆ Requires Advance Prep

PROBLEM SOLVING AND PRACTICE	SECTION REVIEW AND ASSESSMENT	STANDARDS CORRELATION
	CRF Pretest* GENERAL	
CRF **Cross-Disciplinary Worksheet** Connection to Social Studies—A Remarkable Discovery* ADVANCED **CRF** **Cross-Disciplinary Worksheet** Connection to Language Arts—Marie Curie and the Naming of a Unit* ADVANCED **SE** **Math Skills** Nuclear Decay, p. 288 GENERAL **CRF** **Math Skills** Nuclear Decay* GENERAL **CRF** **Cross-Disciplinary Worksheet** Integrating Chemistry—Radiochemistry* ADVANCED **CRF** **Cross-Disciplinary Worksheet** Integrating Earth Science—Radioactivity Within the Earth* ADVANCED **SE** **Math Skills** Half-Life, p. 291 GENERAL **CRF** **Math Skills** Half-Life* GENERAL	**TE** Quiz, p. 292 BASIC **SE** Section 1 Review, p. 292 **CRF** Concept Review* GENERAL **CRF** Quiz* BASIC	PS 1d UCP 2
	TE Quiz, p. 298 BASIC **SE** Section 2 Review, p. 298 **CRF** Concept Review* GENERAL **CRF** Quiz* BASIC	PS 1c UCP 2
CRF **Cross-Disciplinary Worksheet** Integrating Environmental Science* ADVANCED **CRF** **Cross-Disciplinary Worksheet** Real World Applications—Radiation and Medicine* ADVANCED **CRF** **Cross-Disciplinary Worksheet** Integrating Space Science—Nuclear-Powered Space Probes* ADVANCED **CRF** **Cross-Disciplinary Worksheet** Integrating Space Science—The Life Cycle of a Star* ADVANCED **SE** **Math Skills** Calculating Times of Decay, p. 307 **SE** **CareerLink** Science Reporter, pp. 314–315	**TE** Quiz, p. 306 BASIC **SE** Section 3 Review, p. 306 **CRF** Concept Review* GENERAL **CRF** Quiz* BASIC	SAI 2 ST 2 SPSP 1, 4, 5, 6

www.scilinks.org

Topic: Radioactive Isotopes
*Sci*Links code: HK4114

Topic: Types of Radiation
*Sci*Links code: HK4141

Topic: Fission
*Sci*Links code: HK4053

Topic: Fusion
*Sci*Links code: HK4060

Topic: Nuclear Power
*Sci*Links code: HK4096

Topic: Radioactive Tracers
*Sci*Links code: HK4115

Topic: Science Writer
*Sci*Links code: HK4125

Technology Resources

 One-Stop Planner
All of your printable resources and the Test Generator are on this convenient CD-ROM.

 CNN Science in the NEWS
each video segment is accompanied by a Critical Thinking Worksheet*

Segment 32
Radioisotopes in Medicine

Segment 31
Nuclear Wastes

Overview

This chapter introduces students to radioactivity. The students will learn of four different kinds of radiation. Students also learn how to balance nuclear decay equations and calculate half-life. Nuclear fission and nuclear fusion will also be explored. Students study the interaction between matter and energy and explore how chain reactions occur. Some applications of radioactivity and the associated risks will be introduced, including the use of radioactivity in medicine and the benefits and drawbacks of nuclear power. Students also learn about natural background radiation.

Assessing Prior Knowledge

Be sure students understand the following concepts:

- scientific notation
- elements
- matter and energy
- atomic structure
- isotopes
- mass number
- atomic number

CHAPTER 9

Nuclear Changes

Chapter Preview

1 What Is Radioactivity?
Nuclear Radiation
Nuclear Decay
Radioactive Decay Rates

2 Nuclear Fission and Fusion
Nuclear Forces
Nuclear Fission
Nuclear Fusion

3 Nuclear Radiation Today
Where Is Radiation?
Beneficial Uses of Nuclear Radiation
Possible Risks of Nuclear Radiation
Nuclear Power

INTEGRATING TECHNOLOGY and Society

282

Standards Correlations

National Science Education Standards

Section 1 What is Radioactivity?
Physical Science PS 1d
Unifying Concepts and Processes UCP 2

Section 2 Nuclear Fission and Fusion
Physical Science PS 1c
Unifying Concepts and Processes UCP 2

Section 3 Nuclear Radiation Today
Science as Inquiry SAI 2
Science and Technology ST 2
Science in Personal and Social Perspectives SPSP 1, 4, 5, 6

Focus ACTIVITY

Background The painting "Woman Reading Music" was considered one of a series of great finds discovered by Dutch painter and art dealer Han van Meegeren in the 1930s. The previously unknown paintings were believed to be by the great seventeenth century Dutch artist Jan Vermeer. But after World War II, another painting said to be by Vermeer was found in a Nazi art collection, and its sale was traced to van Meegeren. Arrested for collaborating with the Nazis, van Meegeren confessed that both paintings were forgeries. He claimed that he had used one of the fake Vermeers to lure Nazi Germany into returning many genuine paintings to the Dutch.

Was van Meegeren lying, or had he really swindled the Nazis? Although X-ray photographs of the painting suggested that it was a forgery, conclusive evidence did not come about until 20 years later. A fraction of the lead in some pigments used in the painting proved to be radioactive. By measuring the number of radioactive lead nuclei that decayed each minute, experts were able to determine the age of the painting. The fairly rapid decay rate indicated that the paint was less than 40 years old.

Activity 1 Radiation exposes photographic film. To test this, obtain a sheet of unexposed photographic film and a new household smoke detector, which contains a radioactive sample. Remove the detector's casing. In a dark room, place the film next to the smoke detector in a cardboard box. Close the box. After a day, open the box in a dark room. Place the film in a thick envelope. Have the film processed. How does the image differ from the rest of the film? How can you tell that the image is related to the radioactive source?

⊿ internet connect

www.scilinks.org
Topic: **Radioactive Isotopes SciLinks code: HK4114**

SCi*LINKS* Maintained by the National Science Teachers Association

Pre-Reading Questions

1. What are some applications of nuclear radiation?
2. How does nuclear power compare to other sources of power?

283

Focus ACTIVITY

Background In 1968, an American chemist used uranium-lead dating to prove that van Meegeren was a forger. The parent material for the lead isotope found in paint is uranium-238. The series of reactions that form lead includes steps that form radium-226 and polonium-210. The ratio of these two elements in the paint determines the age of the paint. The higher the ratio, the older the paint.

Activity 1 The film will show a scattering of spots on the otherwise unaffected film. The spots will be more numerous near the radioactive source.

LS Visual

Answers to Pre-Reading Questions

1. Answers may vary. Students may state that nuclear radiation is used in smoke detectors, radioactive tracers in medicine and agricultural research, radiotherapy to treat cancerous tumors, and nuclear reactors to generate energy.

2. Answers may vary. Students may state that nuclear power differs from other sources of power in that it does not produce gaseous pollutants, but that it poses other risks to the environment in the form of radioactive products and nuclear waste, which require special safeguards.

Chapter Resource File

- Pretest GENERAL
- Teaching Transparency Preview
- Answer Keys

What Is Radioactivity?

▶ **KEY TERMS**

radioactivity
nuclear radiation
alpha particle
beta particle
gamma ray
half-life

▶ OBJECTIVES ◀

▶ **Identify** four types of nuclear radiation and their properties.
▶ **Balance** equations for nuclear decay.
▶ **Calculate** the half-life of a radioactive isotope.

Our lives are affected by radioactivity in many ways. Technology using radioactivity has helped to detect disease and dysfunction, kill cancer cells, generate electricity, and design smoke detectors. On the other hand, there are also risks associated with too much nuclear radiation, so it is important to know where it may exist and how to counteract it. What exactly is radioactivity?

Nuclear Radiation

▶ **radioactivity** the process by which an unstable nucleus emits one or more particles or energy in the form of electromagnetic radiation

▶ **nuclear radiation** the particles that are released from the nucleus during radioactive decay

Many elements change through **radioactivity.** Radioactive materials have unstable nuclei, which go through changes by emitting particles or releasing energy to become stable, as shown in *Figure 1.* This nuclear process is called *nuclear decay*. After the changes in the nucleus, the element can transform into a different isotope of the same element or into an entirely different element. Recall that isotopes of an element are atoms that have the same number of protons but different numbers of neutrons in their nuclei. Different elements are distinguished by having different numbers of protons in their nuclei.

The released energy and matter are called **nuclear radiation.** Just as radioactivity changes the materials that undergo nuclear decay, nuclear radiation has effects on other materials. These effects depend on the type of radiation and on the properties of the materials that nuclear radiation encounters. (Note that the term *radiation* can refer to light or to energy transfer. *To avoid confusion, the term* nuclear radiation *will be used to describe radiation associated with nuclear changes.*)

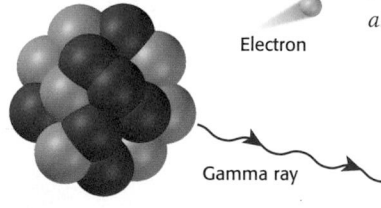

Electron

Gamma ray

Figure 1

During radioactivity an unstable nucleus emits one or more particles or high-energy electromagnetic radiation.

Table 1 Types of Nuclear Radiation

Radiation type	Symbol	Mass (kg)	Charge	
Alpha particle	$^{4}_{2}\text{He}$	6.646×10^{-27}	+2	
Beta particle	$^{0}_{-1}e$	9.109×10^{-31}	−1	
Gamma ray	γ	none	0	
Neutron	$^{1}_{0}n$	1.675×10^{-27}	0	

There are different types of nuclear radiation

Essentially, there are four types of nuclear radiation: alpha particles, beta particles, gamma rays, and neutron emission. Some of their properties are listed in **Table 1.** When a radioactive nucleus decays, the nuclear radiation leaves the nucleus. This nuclear radiation interacts with nearby matter. This interaction depends in part on the properties of nuclear radiation, such as charge, mass, and energy, which are discussed below.

Alpha particles consist of protons and neutrons

Uranium is a radioactive element that naturally occurs in three isotope forms. One of its isotopes, uranium-238, undergoes nuclear decay by emitting positively charged particles. Ernest Rutherford, noted for discovering the nucleus, named them *alpha* (α) *rays.* Later, he discovered that alpha rays were actually particles, each made of two protons and two neutrons—the same as helium nuclei. **Alpha particles** are positively charged and more massive than any other type of nuclear radiation.

Alpha particles do not travel far through materials. In fact, they barely pass through a sheet of paper. One factor that limits an alpha particle's ability to pass through matter is the fact that it is massive. Because alpha particles are charged, they remove electrons from— or ionize—matter as they pass through it. This ionization causes the alpha particle to lose energy and slow down further.

Beta particles are electrons produced from neutron decay

Some nuclei emit another type of nuclear radiation that travels farther through matter than alpha particles do. This nuclear radiation is named the **beta particle,** after the second Greek letter, *beta* (β). Beta particles are often fast-moving electrons.

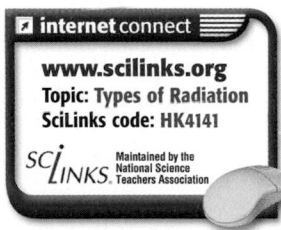

internet connect

www.scilinks.org
Topic: Types of Radiation
SciLinks code: HK4141

SC**LINKS** Maintained by the National Science Teachers Association

▶ **alpha particle** a positively charged atom that is released in the disintegration of radioactive elements and that consists of two protons and two neutrons

▶ **beta particle** a charged electron emitted during certain types of radioactive decay, such as beta decay

285

Reading Skills Write the word *radiation* on the chalkboard. Under the word, list the terms *positive* and *negative* as column heads. Have students brainstorm positive and negative perceptions that they think are true about radiation. Keep a copy of the list, and use it as a discussion tool when students complete this section. **LS** Verbal

Teaching Tip

Other Radiation Tell students that the main types of radiation are discussed in the student text. However, other types exist. For example, chromium-49 decays forming vanadium-49 and what is known as a *positron*.

A positron ($\updownarrow^+$) is the same as an electron except that it is positively charged. Positrons exist for an extremely short period of time because they are the antiparticles of electrons. When an electron and a positron collide, all the mass in both particles converts to energy in the form of gamma rays.

Figure 2
In 1898, Marie Curie discovered the element radium, which was later found to emit gamma rays.

▶ **gamma ray** the high-energy photon emitted by a nucleus during fission and radioactive decay

Negative particles coming from a positively charged nucleus puzzled scientists for years. However, in the 1930s, another discovery helped to clear up the mystery: neutrons, which are not charged, decay to form a proton and an electron. The electron, having very little mass, is then ejected at a high speed from the nucleus as a beta particle.

Beta particles easily go through a piece of paper, but most are stopped by 3 mm of aluminum or 10 mm of wood. This greater penetration occurs because beta particles aren't as massive as alpha particles and therefore move faster. But like alpha particles, beta particles can easily ionize other atoms. As they ionize atoms, beta particles lose energy. This property prevents them from penetrating matter very deeply.

Gamma rays are very high energy

In 1898, Marie Curie, shown in *Figure 2,* and her husband, Pierre, isolated the radioactive element radium. In 1900, studies of radium by Paul Villard revealed that the element emitted a previously undetected form of nuclear radiation. This radiation was much more penetrating than even beta particles. Following the pattern established by Rutherford, this new kind of nuclear radiation was named the **gamma ray,** after the third Greek alphabet letter, *gamma* (γ).

Unlike alpha or beta particles, gamma rays are not made of matter and do not have an electrical charge. Instead, gamma rays consist of a form of electromagnetic energy called photons, like visible light or X rays. Gamma rays, however, have more energy than light or X rays.

Although gamma rays have no electrical charge, they can easily ionize matter. High-energy gamma rays can cause damage in matter. They can penetrate up to 60 cm of aluminum or 7 cm of lead. They are not easily stopped by clothing or most building materials and therefore pose a greater danger to health than either alpha or beta particles.

Neutron radioactivity may occur in an unstable nucleus

Like alpha and beta radiation, *neutron emission* consists of matter that is emitted from an unstable nucleus. In fact, scientists first discovered the neutron by detecting its emission from a nucleus.

Because neutrons have no charge, they do not ionize matter as alpha and beta particles do. Because neutrons do not use their energy ionizing matter, they are able to travel farther through matter than either alpha or beta particles. A block of lead about 15 cm thick is required to stop most fast neutrons emitted during radioactive decay.

286

Did You Know?

Nobel Prize Winners In 1903, Marie Curie, her husband, Pierre, and Henri Becquerel were jointly awarded a Nobel Prize in physics for their studies in the field of radioactivity.

Nuclear Decay

Anytime an unstable nucleus emits alpha or beta particles, the number of protons or neutrons changes. An example would be radium-226 (an isotope of radium with the mass number 226), which changes to radon-222 by emitting an alpha particle.

A nucleus gives up two protons and two neutrons during alpha decay

Nuclear decay equations are similar to those for chemical reactions. The nucleus before decay is like a reactant and is placed on the left side of the equation. Products are placed on the right side. The process of the alpha decay of radium-226 is written as follows.

$$\,^{226}_{88}\text{Ra} \longrightarrow \,^{222}_{86}\text{Rn} + \,^{4}_{2}\text{He} \qquad \begin{array}{l} 226 = 222 + 4 \\ 88 = 86 + 2 \end{array}$$

The mass number of the atom before decay is 226 and equals the sum of the mass numbers of the products, 222 and 4. The atomic numbers follow the same principle. The 88 protons in radium before the nuclear decay equals the 86 protons in the radon-222 nucleus and 2 protons in the alpha particle.

A nucleus gains a proton and loses a neutron during beta decay

With beta decay, the form of the equation is the same except the symbol for a beta particle is used. This symbol, with the appropriate mass and atomic numbers, is $\,^{0}_{-1}e$.

Of course, an electron is not an atom and should not have an atomic number, which is the number of positive charges in a nucleus. But for the sake of convenience, since an electron has a single negative charge, an electron is given an atomic number of –1 when you write a nuclear decay equation. Similarly, the beta particle's mass is so much less than that of a proton or neutron that it can be regarded as having a mass number of 0.

A beta decay process occurs when carbon-14 decays to nitrogen-14 by emitting a beta particle.

$$\,^{14}_{6}\text{C} \longrightarrow \,^{14}_{7}\text{N} + \,^{0}_{-1}e \qquad \begin{array}{l} 14 = 14 + 0 \\ 6 = 7 + (-1) \end{array}$$

In all cases of beta decay, the mass number before and after the decay does not change. Note that the atomic number of the product nucleus increases by 1. This occurs because a neutron decays into a proton, causing the positive charge of the nucleus to increase by 1.

Did You Know?

Ernest Rutherford showed that alpha particles are helium nuclei by trapping alpha particles from radon-222 decay in a glass tube. He then applied a high electric voltage across the gas, causing it to glow. The glow was identical to the glow produced by helium atoms, indicating that the two substances were the same.

Teaching Tip

Decay Make sure students are familiar with the word *decay* indicating the breakdown of a substance, such as when food or a tooth decays. Be sure students know that *radioactive decay* is the release of radiation by isotopes that are radioactive.

SKILL BUILDER — GENERAL

Interpreting Visuals To help students understand how nuclear decay processes are represented by equations, have them compare a nuclear equation with a chemical equation. Write the equation for the alpha decay of radium-226 on the chalkboard, and write the following chemical equation directly beneath it:

$$2\text{H}_2 + \text{O}_2 \rightarrow 2\text{H}_2\text{O}$$

Ask students: How are the two equations similar? (In both examples, the "reactants" are on the left, and the "products" are on the right. Both use an arrow to separate the terms. Both use numbers to show the quantities of items involved in the reactions.)

How are they different? [In the chemical equation, *atoms* are conserved. Both sides of the equation have 4 hydrogen atoms and 2 oxygen atoms. In the nuclear equation, *protons* and *neutrons* are conserved, but the atoms are not the same. The radium atom (88 protons) emits an alpha particle (2 protons), and thereby becomes a radon atom (86 protons).] **LS Logical**

287

Historical Perspective

Becquerel's Discovery In 1896, the French scientist Henri Becquerel discovered that uranium gives off penetrating radiation. He tried to prove that uranium's radiation resulted from its absorbing sunlight and releasing this energy in the form of X rays. But after using uranium that had not been exposed to sunlight to expose a photographic plate, he discovered that uranium spontaneously emits radiation.

Pierre and Marie Curie used Becquerel's findings to conclude that a nuclear change takes place within the uranium atoms, resulting in what they first called *radioactivity*.

Chapter Resource File

• **Cross-Disciplinary Worksheet**
Connection to Language Arts—Marie Curie and the Naming of a Unit **ADVANCED**

Practice

Nuclear Decay Complete the following radioactive-decay equations by identifying the nuclide X. Indicate whether alpha or beta decay takes place.

$^{14}_{6}C \rightarrow {}^{A}_{Z}X + {}^{0}_{-1}e$

Answer: $^{14}_{7}N$, beta

$^{238}_{92}U \rightarrow {}^{234}_{90}Th + {}^{A}_{Z}X$

Answer: $^{4}_{2}He$, alpha

$^{40}_{19}K \rightarrow {}^{40}_{20}Ca + {}^{A}_{Z}X$

Answer: $^{0}_{-1}e$, beta

$^{219}_{86}Rn \rightarrow {}^{A}_{Z}X + {}^{4}_{2}He$

Answer: $^{215}_{84}Po$, alpha

LS Logical

Figure 3

A nucleus that undergoes beta decay has nearly the same atomic mass afterward, except that it has one more proton and one less neutron.

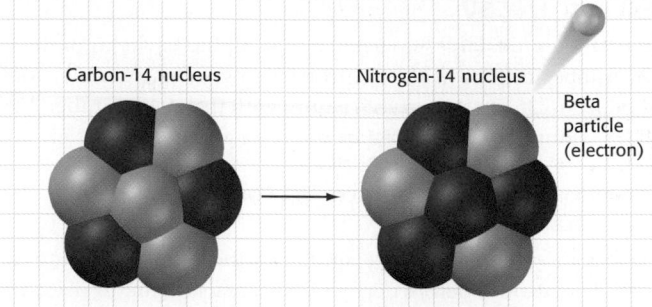

Carbon-14 nucleus   Nitrogen-14 nucleus

Beta particle (electron)

Figure 3 shows how the positive charge of the nucleus increases by 1 when a neutron decays into a proton. When the nucleus undergoes nuclear decay by gamma rays, there is no change in the atomic number of the element. This is because the number of protons does not change. The atomic number is the number of protons in the nucleus of the atom. The only change is in the energy content of the nucleus.

Math Skills

Nuclear Decay Actinium-217 decays by releasing an alpha particle. Write the equation for this decay process, and determine what element is formed.

1 Write down the equation with the original element on the left side and the products on the right side.

Use the letter X to denote the unknown product. Note that the mass and atomic numbers of the unknown isotope are represented by the letters A and Z.

$$^{217}_{89}Ac \longrightarrow {}^{A}_{Z}X + {}^{4}_{2}He$$

2 Write math equations for the atomic and mass numbers.

$217 = A + 4 \qquad 89 = Z + 2$

3 Rearrange the equations.

$A = 217 - 4 \qquad Z = 89 - 2$

4 Solve for the unknown values, and rewrite the equation with all nuclei represented.

$A = 213 \qquad Z = 87$

The unknown decay product has an atomic number of 87, which is francium, according to the periodic table. The element is therefore $^{213}_{87}Fr$.

$$^{217}_{89}Ac \longrightarrow {}^{213}_{87}Fr + {}^{4}_{2}He$$

288

Trends in Particle Physics

Finding Neutrino Mass Some scientists are hoping to use beta decay to shed light on one of the most mysterious particles of modern physics—the neutrino. The neutrino is a fundamental particle in our universe, but we don't know much about it. Learning more about the neutrino might help scientists answer some of the most fundamental questions of modern physics today.

The neutrino, whose existence was first predicted by Wolfgang Pauli in 1931, has no electric charge and almost no mass. (Current estimates suggest that a neutrino's mass is about 10,000 times less than an electron's mass.) As a result, neutrinos are very difficult to detect and measure. Many current attempts to detect neutrinos involve elaborate set-ups, such as huge detectors in underground caves.

In 1986, scientists first observed a process that is now known as double beta decay: a nucleus emits two electrons simultaneously. Today, scientists are studying to learn more about the neutrino. They plan to use measurements of these double beta decay reactions to make the first measurement of the absolute mass of an electron neutrino.

Practice

Nuclear Decay

Complete the following radioactive-decay equations by identifying
the isotope *X*. Indicate whether alpha or beta decay takes place.

1. $^{12}_{5}B \longrightarrow ^{12}_{6}C + ^{A}_{Z}X$

3. $^{63}_{28}Ni \longrightarrow ^{A}_{Z}X + ^{0}_{-1}e$

2. $^{225}_{89}Ac \longrightarrow ^{221}_{87}Fr + ^{A}_{Z}X$

4. $^{212}_{83}Bi \longrightarrow ^{A}_{Z}X + ^{4}_{2}He$

Radioactive Decay Rates

If you were asked to pick up a rock and determine its age, you
would probably not be able to do so. After all, old rocks do not
look much different from new rocks. How, then, would you go
about finding the rock's age? Likewise, how would a scientist find
out the age of cloth found at the site of an ancient village?

One way to do it involves radioactive decay. Although it is
impossible to predict the moment when any particular nucleus will
decay, it is possible to predict the time it takes for half the nuclei in
a given radioactive sample to decay. The time in which half a
radioactive substance decays is called the substance's **half-life.**

After the first half-life of a radioactive sample has passed, half
the sample remains unchanged, as indicated in **Figure 4** for
carbon-14. After the next half-life, half the remaining half decays,
leaving only a quarter of the sample undecayed. Of that quarter,
half will decay in the next half-life. Only one-eighth will remain
undecayed then.

▶ **half-life** the time required
for half of a sample of a
radioactive substance to
disintegrate by radioactive
decay or by natural processes

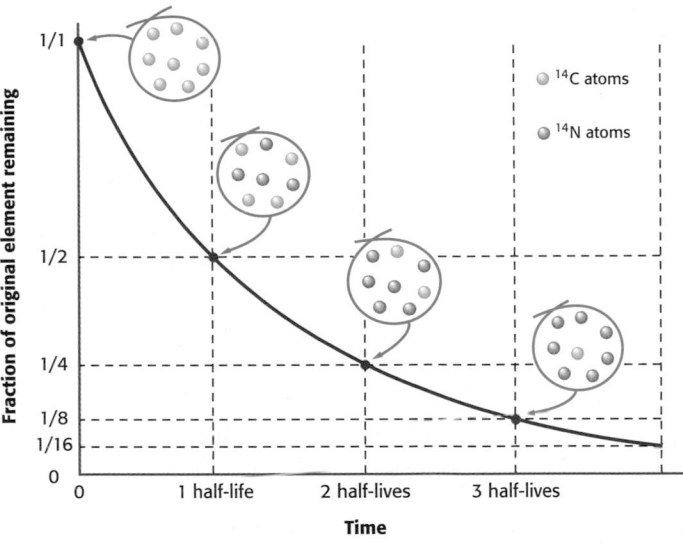

Figure 4

With each successive half-life, half
the remaining sample decays to
form another element.

Practice

1. $12 = 12 + A$; $A = 0$
$5 = 6 + Z$; $Z = -1$
$X = e$
Beta decay occurs, and $^{0}_{-1}e$ is
produced.

2. $225 = 221 + A$; $A = 4$
$89 = 87 + Z$; $Z = 2$
$X = He$
Alpha decay occurs, and $^{4}_{2}He$ is
produced.

3. $63 = A + 0$; $A = 63$
$28 = Z + (-1)$; $Z = 29$
$X = Cu$
Beta decay occurs, and $^{63}_{29}Cu$ is
produced.

4. $212 = A + 4$; $A = 208$
$83 = Z + 2$; $Z = 81$
$X = Tl$
Alpha decay occurs, and $^{208}_{81}Tl$ is
produced.

Teaching Tip

Radioactive Dating Students
might think that all items can be
dated using radioisotopes. Make
sure students understand that after
several half-lives, the amount of
radioactive material present
becomes so small that it is not
measurable.

Teacher Resources ——— GENERAL

For the New Teacher Students who are having difficulty understanding a new concept
will often benefit from a physical demonstration of the idea. For example, to help students
understand half-life, put all the students on
one side of the room. Instruct half the students to move to the opposite side of the
room, and then instruct half of the remaining
students to also move to the opposite side of
the room. Have one student record the
number of students remaining after each half-
life. Complete this process three or four times,
each time representing one half-life. The number of students represents the number of
atoms of a particular isotope. Notice that the
students do not disappear. When the activity
is complete, have students graph the number
of students remaining after each half-life.
LS Kinesthetic

Modeling Decay and Half-life

Materials (per group):
• jar with a lid
• 128 pennies
• pencil
• paper
• flat work surface

Teacher's Notes: This activity differs from an actual half-life in that eventually all the pennies are removed. Even in a small sample of a radioactive isotope, the number of atoms is so large that theoretically, some parent atoms would always remain. The number would decrease to the point that the parent atoms would not be detectable, but they would be present.

Results: Students should find that the ratio of heads-up pennies to the total number of pennies in the trial is approximately 1:2, or 1/2. Point out that if the actual number of atoms in a sample were considered, the ratio would be even closer to 1:2 because of the large sample size. **LS** Visual

Table 2 **Half-lives of Selected Isotopes**

Isotope	Half-life	Nuclear radiation emitted
Thorium-219	1.05×10^{-6} s	$^{4}_{2}He$
Hafnium-156	2.5×10^{-2} s	$^{4}_{2}He$
Radon-222	3.82 days	$^{4}_{2}He$, γ
Iodine-131	8.1 days	$^{0}_{-1}e$, γ
Radium-226	1599 years	$^{4}_{2}He$, γ
Carbon-14	5715 years	$^{0}_{-1}e$
Plutonium-239	2.412×10^{4} years	$^{4}_{2}He$, γ
Uranium-235	7.04×10^{8} years	$^{4}_{2}He$, γ
Potassium-40	1.28×10^{9} years	$^{0}_{-1}e$, γ
Uranium-238	4.47×10^{9} years	$^{4}_{2}He$, γ

Half-life is a measure of how quickly a substance decays

Different radioactive isotopes have different half-lives, as indicated in *Table 2.* Half-lives can last from nanoseconds to billions of years, depending on the stability of the nucleus.

Using half-lives, scientists can predict how old an object is. Using the half-lives of long-lasting isotopes, such as potassium-40, geologists calculate the age of rocks. Potassium-40 decays to argon-40, so the ratio of potassium-40 to argon-40 is smaller for older rocks than it is for younger rocks.

Modeling Decay and Half-life

For this exercise, you will need a jar with a lid, 128 pennies, pencil and paper, and a flat work surface.

1. Place the pennies in the jar, and place the lid on the jar. Shake the jar, and then pour the pennies onto the work surface.

2. Separate pennies that are heads up from those that are tails up. Count and record the number of heads-up pennies, and set these pennies aside. Place the tails-up pennies back in the jar.

3. Repeat the process until all pennies have been set aside.

4. For each trial, divide the number of heads-up pennies set aside by the total number of pennies used in the trial. Are these ratios nearly equal to each other? What fraction are they closest to?

SKILL BUILDER ───────── ADVANCED

Graphing Have students graph their results of the Quick Activity. Graphs should show the number of remaining parent pennies on the *y*-axis and the trial number (number of half-lives) on the *x*-axis for each trial. A trial number of zero represents the starting number of pennies.

Make sure student graphs show a curve with a *y*-intercept at 128 that curves downward asymptotic to the *x*-axis. **LS** Logical

Carbon-14 is used to date materials

Archaeologists use the half-life of radioactive carbon-14 to date more recent materials, such as the remains of an animal or fibers from ancient clothing. All of these materials came from organisms that were once alive. When plants absorb carbon dioxide during photosynthesis, a tiny fraction of the CO_2 molecules contains carbon-14 rather than the more common carbon-12. While the plant is alive, the ratio of the carbon isotopes remains constant. This is also true for animals that eat plants.

When a plant or animal dies, it no longer takes in carbon-14. The amount of carbon-14 decreases through beta decay, while the amount of carbon-12 remains constant. Thus, the ratio of carbon-14 to carbon-12 decreases with time. By measuring this ratio and comparing it with the ratio in a living plant or animal, scientists can estimate the age of the once-living organism.

Math Skills

Half-life Radium-226 has a half-life of 1599 years. How long would it take seven-eighths of a radium-226 sample to decay?

1 **List the given and unknown values.**

Given: half-life = 1599 years

fraction of sample decayed $= \frac{7}{8}$

Unknown: fraction of sample remaining = ?

total time of decay = ?

2 **Calculate the fraction of radioactive sample remaining.**

To find the fraction of sample remaining, subtract the fraction that has decayed from 1.

fraction of sample remaining = 1 − fraction decayed

fraction of sample remaining $= 1 - \frac{7}{8} = \frac{1}{8}$

3 **Calculate the number of half-lives.**

Amount of sample remaining after one half-life $= \frac{1}{2}$

Amount of sample remaining after two half-lives

$= \frac{1}{2} \times \frac{1}{2} = \frac{1}{4}$

Amount of sample remaining after three half-lives

$= \frac{1}{2} \times \frac{1}{2} \times \frac{1}{2} = \frac{1}{8}$

Three half-lives are needed for one-eighth of the sample to remain undecayed.

4 **Calculate the total time required for the radioactive decay.**

Each half-life lasts 1599 years.

total time of decay $= 3 \text{ half-lives} \times \frac{1599 \text{ y}}{\text{half-life}} = 4797 \text{ years}$

EARTH SCIENCE
Earth's interior is extremely hot. One reason is because uranium and the radioactive elements produced by its decay are present in amounts of about 3 parts per million beneath the surface of Earth, and their nuclear decay produces energy that escapes into their surroundings.

The long half-lives of uranium-238 and -235 allow their radioactive decay to heat Earth for billions of years. The very large distance this heat energy must travel to reach Earth's surface keeps the interior of Earth much hotter than its surface.

291

INTEGRATING

EARTH SCIENCE Emphasize to students the great amount of heat released on Earth by radioactive elements. Tell them that 1.0 g of radium during its lifetime releases by natural radiation the same amount of energy released when half a ton of coal is burned.

Math Skills

Additional Examples

Half-life The half-life of tritium, $^{3}_{1}\text{H}$, is 12.3 years. How long will it take for 7/8 of a sample to decay?

Answer: $1 - 7/8 = 1/8$ remains

It will take three half-lives.

3 half-lives = 12.3 years/half-life = 36.9 years

The half-life of cobalt-60 is 5.3 years. How much of a 20.0 g sample will remain after 21.2 years?

Answer: 21.2 years/5.3 years per half-life = 4 half-lives.

$20.0 \text{ g} \times 1/2 \times 1/2 \times 1/2 \times 1/2 = 1.25 \text{ g}$

LS Logical

Alternative Assessment ——— GENERAL
Radiation

Step 1 Have students review the list of ideas about radiation that they created in the *Skill Builder: Reading Skills* at the beginning of the section.

Step 2 Discuss each item, evaluating whether each item is accurate and in the correct category of positive and negative.

Step 3 Have students adjust items based on what they learned about radiation in this section. They should use the text to justify any changes. **LS** Verbal

Alternative Assessment ——— ADVANCED

Carbon Dating Have interested students investigate how carbon-14 dating was used to date the Shroud of Turin. Ask a volunteer to give a class presentation with the results of his or her research. **LS** Verbal

---

Chapter Resource File

- **Quick Activity Datasheet** Modeling Decay and Half-life GENERAL
- **Cross-Disciplinary Worksheet** Integrating Earth Science— Radioactivity Within the Earth ADVANCED
- **Math Skills** Half-life GENERAL

Transparencies

TM Half-life

Practice

1. $1 - 3/4 = 1/4$, or $1/2 \times 1/2$;
   2 half-lives remain
   2×8.1 days = 16 days
2. $1 - 15/16 = 1/16$, or $1/2 \times 1/2 \times 1/2 \times 1/2$; 4 half-lives remain; 4×3.82 days = 15.3 days
3. 13.4 billion years/4.47 billion years/half-life = 3 half-lives;
   $1/2 \times 1/2 \times 1/2 = 1/8 = 0.125$;
   $0.125 \times 100\% = 12.5\%$
4. $1/8 = 1/2 \times 1/2 \times 12$; 3 half-lives;
   87.3 years/3 half-life = 29.1 years/half-life
5. $1/16 = 1/2 \times 1/2 \times 1/2 \times 1/2$;
   4 half-lives; 80 min/4 half-lives = 20 min/half-life

Close

Quiz ——————— BASIC

1. What happens in nuclear decay? (An unstable nucleus emits particles or energy. The element transforms into either a different isotope of the element or a new element.)
2. The half-life of iodine-131, a radioactive isotope of iodine, is 8.1 days. What fraction of a sample of iodine-131 would remain *unchanged* after 16.2 days? (1/4) **LS Logical**

Practice

Half-life

1. The half-life of iodine-131 is 8.1 days. How long will it take for three-fourths of a sample of iodine-131 to decay?
2. Radon-222 is a radioactive gas with a half-life of 3.82 days. How long would it take for fifteen-sixteenths of a sample of radon-222 to decay?
3. Uranium-238 decays very slowly, with a half-life of 4.47 billion years. What percentage of a sample of uranium-238 would remain after 13.4 billion years?
4. A sample of strontium-90 is found to have decayed to one-eighth of its original amount after 87.3 years. What is the half-life of strontium-90?
5. A sample of francium-212 will decay to one-sixteenth its original amount after 80 minutes. What is the half-life of francium-212?

SECTION 1 REVIEW

SUMMARY

▶ Nuclear radiation includes alpha particles, beta particles, gamma rays, and neutron emissions.

▶ Alpha particles are helium-4 nuclei.

▶ Beta particles are electrons emitted by neutrons decaying in the nucleus.

▶ Gamma radiation is an electromagnetic wave like visible light but with much greater energy.

▶ In nuclear decay, the sums of the mass numbers and the atomic numbers of the decay products equal the mass number and atomic number of the decaying nucleus.

▶ The time required for half a sample of radioactive material to decay is called its half-life.

1. **Identify** which of the four common types of nuclear radiation correspond to the following descriptions.
   a. an electron
   c. can be stopped by a piece of paper
   b. uncharged particle
   d. high-energy light
2. **Describe** what happens when beta decay occurs.
3. **Explain** why charged particles do not penetrate matter deeply.

Math Skills

4. **Determine** the product denoted by X in the following alpha decay.
$$^{212}_{86}\text{Rn} \longrightarrow {}^{A}_{Z}X + {}^{4}_{2}\text{He}$$

5. **Determine** the isotope produced in the beta decay of iodine-131, an isotope used to check thyroid-gland function.
$$^{131}_{53}\text{I} \longrightarrow {}^{A}_{Z}X + {}^{0}_{-1}e$$

6. **Calculate** the time required for three-fourths of a sample of cesium-138 to decay given that its half-life is 32.2 minutes.

7. **Calculate** the half-life of cesium-135 if seven-eighths of a sample decays in 6×10^6 years.

8. **Critical Thinking** An archaeologist discovers charred wood whose carbon-14 to carbon-12 ratio is one-sixteenth the ratio measured in a newly fallen tree. How old does the wood seem to be, given this evidence?

292

Answers to Section 1 Review

1. a. beta   c. alpha
   b. neutron   d. gamma
2. A neutron decays, forming a proton and an electron. The electron is released as a beta particle.
3. They ionize the materials they pass through. Each ionization transfers energy from the alpha or beta particle to the ionized particle. Less energy means less penetration.

Math Skills

4. $212 = A + 4$, $A = 208$; $86 = Z + 2$, $Z = 84$; $X = \text{Po}$; The product is $^{208}_{84}\text{Po}$.
5. $131 = A + 0$, $A = 131$; $53 = Z + (-1)$, $Z = 54$; $X = \text{Xe}$; The product is $^{131}_{54}\text{Xe}$.
6. $1 - 3/4 = 1/4$, or $1/2 \times 1/2$, remains. Two half-lives take 2×32.2 minutes, or 64.4 minutes.
7. $1 - 7/8 = 1/8$, or $1/2 \times 1/2 \times 1/2$, remains 6×10^6 years/3 half-lives = 2×10^6 years/half-life
8. $1/16 = 1/2 \times 1/2 \times 1/2 \times 1/2$ Four half-lives have passed. The charred wood is 4 half-lives $\times$ 5730 years/half-life, or 22 920 years old.

Nuclear Fission and Fusion

OBJECTIVES

▶ **Describe** how the strong nuclear force affects the composition of a nucleus.

▶ **Distinguish** between fission and fusion, and provide examples of each.

▶ **Recognize** the equivalence of mass and energy, and why small losses in mass release large amounts of energy.

▶ **Explain** what a chain reaction is, how one is initiated, and how it can be controlled.

▶ **KEY TERMS**

fission
nuclear chain reaction
critical mass
fusion

In 1939, German scientists Otto Hahn and Fritz Strassman conducted experiments in the hope of forming heavy nuclei. Using the apparatus shown in *Figure 5*, they bombarded uranium samples with neutrons, expecting a few nuclei to capture one or more neutrons. The new elements they made had chemical properties they could not explain.

It wasn't until their colleague Lise Meitner and her nephew Otto Frisch read the results of Hahn and Strassman's work that an explanation was offered. Meitner and Frisch believed that instead of making heavier elements, the uranium nuclei had split into smaller elements.

Nuclear Forces

Protons and neutrons are tightly packed in the tiny nucleus of an atom. As we saw in the previous section, certain nuclei are unstable and undergo decay by emitting nuclear radiation. Also, an element can have both stable and unstable isotopes. For instance, carbon-12 is a stable isotope, while carbon-14 is unstable and radioactive. The stability of a nucleus depends on the nuclear forces that hold the nucleus together. These forces act between the protons and the neutrons.

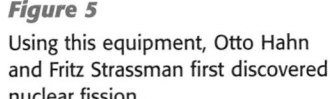

Figure 5

Using this equipment, Otto Hahn and Fritz Strassman first discovered nuclear fission.

293

SOCIAL STUDIES
CONNECTION

One of the most controversial events of recent history is President Truman's decision to drop atomic bombs on two Japanese cities during World War II. A single atomic bomb was dropped on Hiroshima on August 6, 1945; a second was dropped on Nagasaki three days later. Japan surrendered to the Allied forces less then one month afterwards.

Many scientists whose discoveries played a role in the development of the bomb were personally affected by the events of World War II. Albert Einstein faced Nazi persecution in Germany before fleeing to the United States. Einstein, who had been a fervent pacifist before the rise of Nazism in Germany, was instrumental in urging President Roosevelt to fund the research that produced the bomb. Lise Meitner, who offered the first explanation of nuclear fission (with her nephew Otto Frisch), was also forced to leave Nazi Germany. She fled to Switzerland in 1938.

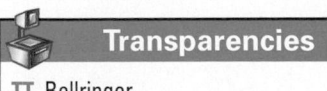

Demonstration ── GENERAL

Modeling Nuclear Forces

(Time: Approximately 5 minutes)

Materials:

- 2 bar magnets
- rubber band

Using two bar magnets, you can model the nuclear force between two protons. Show students how like poles of magnets repel each other. Place the magnets so that the like poles are together and the magnets touch. Wrap the rubber band tightly around them so that the magnets are held together.

Analysis

1. What part of a nucleus do the magnets represent? Explain. (protons; they repel each other)

2. What does the rubber band represent? (a force strong enough to hold objects together even though they repel each other)

LS Visual

Teaching Tip

Nuclear Fission Students might think that fission in a cell and in a nucleus are identical processes because both involve splitting. The product of cell fission is two cells that are identical to each other and to the parent cell. Nuclear fission produces particles that differ from the parent atom and from each other.

Nuclei are held together by a special force

Like charges repel, so how can so many positively charged protons fit into an atomic nucleus without flying apart?

The answer lies in the existence of the *strong nuclear force*. This force causes protons and neutrons in the nucleus to attract each other. The attraction is much stronger than the electric repulsion between protons. However, this attraction due to the strong nuclear force occurs over a very short distance, less than 3×10^{-15} m, or about the width of three protons.

Neutrons contribute to nuclear stability

Due to the strong nuclear force, neutrons and protons in a nucleus attract other protons and neutrons. Because neutrons have no charge, they do not repel each other or the protons. On the other hand, the protons in a nucleus both repel and attract each other, as shown in *Figure 6*. In stable nuclei, the attractive forces are stronger than the repulsive forces.

Too many neutrons or protons can cause a nucleus to become unstable and decay

While more neutrons can help hold a nucleus together, there is a limit to how many neutrons a nucleus can have. Nuclei with too many or too few neutrons are unstable and undergo decay.

Nuclei with more than 83 protons are always unstable, no matter how many neutrons they have. These nuclei will always decay, releasing large amounts of energy and nuclear radiation. Some of this released energy is transferred to the various particles ejected from the nucleus, the least massive of which move very fast as a result. The rest of the energy is emitted in the form of gamma rays. The radioactive decay that takes place results in a more stable nucleus.

Figure 6

The nucleus is held together by the attractions among protons and neutrons. These forces are greater than the electric repulsion among the protons alone.

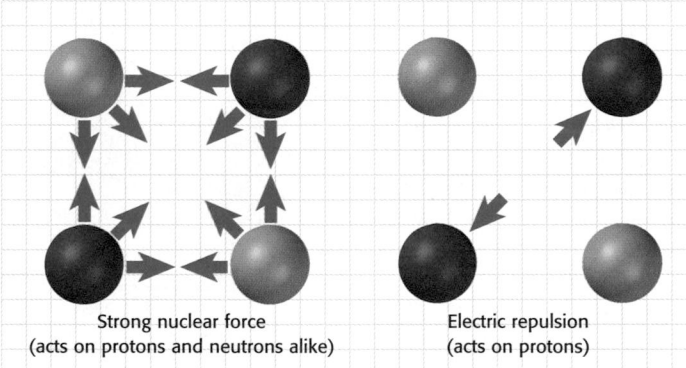

Strong nuclear force
(acts on protons and neutrons alike)

Electric repulsion
(acts on protons)

294

Scientists Around the World

Hideki Yukawa Hideki Yukawa, a Japanese scientist, first theorized in 1935 that a strong force exists that holds the nucleus together. He described the force as being stronger over short distances than the electrical repulsion that repels protons. He thought that this strong force was the result of the transfer of some unknown particle.

In 1947, the particle Yukawa predicted was found. This particle is the pion, which exists for only about a billionth of a second. After his theory was confirmed by the discovery of this particle, Yukawa received the Nobel Prize in 1949. He was the first Japanese person ever to receive this prestigious award.

Nuclear Fission

The process of splitting heavier nuclei into lighter nuclei, which Hahn and Strassman observed, is called **fission.** In their experiment, uranium-235 was bombarded by neutrons. The products of this fission reaction included two lighter nuclei barium-137 and krypton-84, together with neutrons and energy.

$$^{235}_{92}U + ^{1}_{0}n \longrightarrow ^{137}_{56}Ba + ^{84}_{36}Kr + 15^{1}_{0}n + energy$$

Notice that the products include 15 neutrons. Uranium-235 can also undergo fission by producing different pairs of lighter nuclei with a different number of neutrons. For example, a different fission of uranium-235 produces strontium-90, xenon-143, and three neutrons. So, in either fission process when the nucleus splits, both neutrons and energy are released.

Energy is released during nuclear fission

During fission, as shown in *Figure 7,* the nucleus breaks into smaller nuclei. The reaction also releases large amounts of energy. Each dividing nucleus releases about 3.2×10^{-11} J of energy. By comparison, the chemical reaction of one molecule of the explosive trinitrotoluene (TNT) releases only 4.8×10^{-18} J.

In their experiment, Hahn and Strassman determined the masses of all the nuclei and particles before and after the reaction. They found that the overall mass had decreased after the reaction. The missing mass had changed into energy.

The equivalence of mass and energy observed in nature is explained by the special theory of relativity, which Albert Einstein presented in 1905. This equivalence means that matter can be converted into energy and energy into matter. This equivalence is expressed by the following equation.

Mass-Energy Equation

$$Energy = mass \times (speed\ of\ light)^2$$
$$E = mc^2$$

Because c, which is constant, has such a large value, 3.0×10^8 m/s, the energy associated with even a small mass is immense. The mass-equivalent energy of 1 kg of matter is 9×10^{16} J. This is more than the chemical energy of 22 million tons of TNT.

Obviously, it would be devastating if objects around us changed into their equivalent energies. Under ordinary conditions of pressure and temperature, matter is very stable. Objects, such as chairs and tables, never spontaneously change into energy.

■ **fission** the process by which a nucleus splits into two or more fragments and releases neutrons and energy

Figure 7

When the uranium-235 nucleus is bombarded by a neutron the nucleus breaks apart. It forms smaller nuclei, such as barium-137 and krypton-84, and releases energy through fast neutrons.

Did You Know?

Enrico Fermi and his associates achieved the first controlled nuclear reaction in December 1942. The reactor was built on a racquetball court under the unused football stadium at the University of Chicago. The reactor consisted of blocks of uranium for fuel and graphite to slow the neutrons so that they could be captured by the uranium nuclei and cause fission.

Did You Know?

A Natural Nuclear Reactor Mother Nature created the first nuclear reactor some two billion years ago. Oklo uranium deposits in Gabon, Africa, divided in a fission process and sustained a chain reaction that lasted hundreds of thousands of years.

SKILL BUILDER — BASIC

Interpreting Visuals Have students examine **Figure 8** and make a concept map that summarizes what happens when a chain reaction occurs.

Maps should start with a neutron hitting a uranium-235 atom, which then splits into barium-140 and krypton-93 atoms. Three other neutrons are released and each neutron hits another uranium-235 atom. **LS Visual**

Demonstration — GENERAL

A Chain Reaction

(Time: Approximately 10 minutes)

Materials:

• 8 books of matches
• ring stand
• tape

Safety Caution: *Work in a well-ventilated area. Have water available to extinguish the matches.*

Open the books of matches. Starting at the top of the ring stand, use tape to attach the matches so that the heads of the matches form a vertical row on the stand. Light the bottom match.

Analysis

1. Have students explain their observations. (Each match supplied enough energy to the next match to light it.)

2. How is this demonstration a model of a nuclear chain reaction? (Just as each match supplies energy to the next match, each neutron emitted in one fission reaction causes another fission reaction.) **LS Visual**

internet connect
www.scilinks.org
Topic: Fission
SciLinks code: HK4053

SciLINKS Maintained by the National Science Teachers Association

▶ **nuclear chain reaction** a continuous series of nuclear fission reactions

Figure 8
A nuclear chain reaction may be triggered by a single neutron.

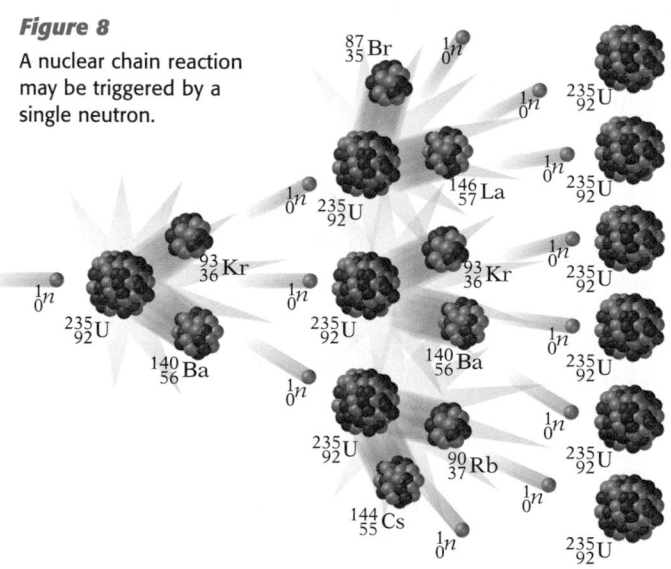

When the total mass of any nucleus is measured, it is less than the individual masses of the neutrons and protons that make up the nucleus. This missing mass is referred to as the *mass defect*. But what happens to the missing mass? Einstein's equation provides an explanation—it changes into energy. However, the mass defect of a nucleus is very small.

Another way to think about mass defect is to imagine constructing a nucleus by bringing individual protons and neutrons together. During this process a small amount of mass changes into energy, as described by $E = mc^2$.

Neutrons released by fission can start a chain reaction

Have you ever played marbles with lots of marbles in the ring? When one marble is shot into the ring, the resulting collisions cause some of the marbles to scatter. Some nuclear reactions are like this, where one reaction triggers another.

A nucleus that splits when it is struck by a neutron forms smaller product nuclei. These smaller nuclei need fewer neutrons to be held together. Therefore, excess neutrons are emitted. One of these neutrons can collide with another large nucleus, triggering another nuclear reaction. This reaction releases more neutrons, and so it is possible to start a chain reaction.

When Hahn and Strassman continued experimenting, they discovered that each dividing uranium nucleus, on average, produced between two and three additional neutrons. Therefore, two or three new fission reactions could be started from the neutrons ejected from one reaction.

If each of these three new reactions produce three additional neutrons, a total of nine neutrons become available to trigger nine additional fission reactions. From these nine reactions, a total of 27 neutrons are produced, setting off 27 new reactions, and so on. You can probably see from *Figure 8* how the reaction of uranium-235 nuclei would very quickly result in an uncontrolled **nuclear chain reaction.** Therefore, the ability to create a chain reaction partly depends on the number of neutrons released.

REAL-LIFE
CONNECTION

Power from Fission The United States receives approximately 20 percent of its electrical power from power plants that use nuclear fission as an energy source. The country that currently receives the greatest percentage of its electrical power from nuclear fission is France. France receives more than 75% of its electricity from nuclear power.

Did You Know?

Reactors Around the World Today, there are more than 1100 nuclear reactors in use around the world, including over 430 large reactors that generate electricity, over 400 reactors powering ships (primarily submarines), and about 280 small reactors that are used for research and for producing isotopes for medicine and industry.

Quick ACTIVITY

Modeling Chain Reactions

1. To model a fission chain reaction, you will need a small wooden building block and a set of dominoes.
2. Place the building block on a table or counter. Stand one domino upright in front of the block and parallel to one of its sides, as shown at right.
3. Stand two more dominoes vertically, parallel, and symmetrical to the first domino. Continue this process until you have used all the dominoes and a triangular shape is created, as shown at right.
4. Gently push the first domino away from the block so that it falls and hits the second group. Note how more dominoes fall with each step.

Chain reactions can be controlled

Energy produced in a controlled chain reaction can be used to generate electricity. Particles released by the splitting of the atom strike other uranium atoms, splitting them. The particles that are given off split still other atoms. A chain reaction is begun, which gives off heat energy that is used to boil water. The boiling water heats another set of pipes filled with water to make steam. The steam then rotates a turbine to generate electricity. So, energy released by the chain reaction changes the atomic energy into heat energy.

The chain-reaction principle is also used in the nuclear bomb. Two or more masses of uranium-235 are contained in the bomb. These masses are surrounded by a powerful chemical explosive. When the explosive is detonated, all of the uranium is pushed together to create a **critical mass.** The critical mass refers to the minimum amount of a substance that can undergo a fission reaction and can also sustain a chain reaction. If the amount of fissionable substance is less than the critical mass, a chain reaction will not continue. Fortunately, the concentration of uranium-235 in nature is too low to start a chain reaction naturally. Almost all of the escaping neutrons are absorbed by the more common and more stable isotope uranium-238.

In nuclear power plants, control rods are used to regulate splitting, slowing the chain reaction. In nuclear bombs, reactions are not controlled, and almost pure pieces of the element uranium-235 or plutonium of a precise mass and shape must be brought together and held together with great force. These conditions are not present in a nuclear reactor.

▶ **critical mass** the minimum mass of a fissionable isotope that provides the number of neutrons needed to sustain a chain reaction

297

Alternative Assessment —— ADVANCED

Fission Debate Have students prepare for a debate about which is a better way to generate electricity: fossil fuels or nuclear fission. Ask them to use library materials and the Internet to learn about the pros and cons of each method. Then divide the class in half, and set aside 20 minutes for a class debate.

Afterwards, ask students to choose a side and write a paper convincing an opponent of the benefits of his or her choice. Remind students that a successful position paper discusses both the pros and cons of that side, explaining how the cons can be addressed. **LS Interpersonal**

Teaching Tip

Containing Fusion Explain to students that one of the most difficult problems with nuclear fusion is containing the reaction. For the reaction to readily occur, it must happen at temperatures of approximately 10^8 °C, where the fuel is in the plasma state. No known material could contain such a reaction without melting. Because plasma consists of charged particles, some success has been achieved by containing the reaction in a magnetic field. However, the moving, charged particles produce magnetic fields that interfere with the magnetic field.

Close

Quiz BASIC

1. What force holds protons and neutrons in the nucleus? (the strong nuclear force)

2. What is the difference between fission and fusion? (In fission, one nucleus splits into two or more smaller fragments. In fusion, two or more light nuclei combine.)

3. Which type of nuclear reaction is used to generate electricity in nuclear power plants? (fission)

4. Which type of nuclear reaction produces energy in our sun? (fusion)

Chapter Resource File

- Concept Review GENERAL
- Section Quiz BASIC

---

fusion the process in which light nuclei combine at extremely high temperatures, forming heavier nuclei and releasing energy

internet connect

www.scilinks.org
Topic: Fusion
SciLinks code: HK4060

SCiLINKS. Maintained by the National Science Teachers Association

---

Nuclear Fusion

Just as energy is obtained when heavy nuclei break apart, energy can also be obtained when very light nuclei are combined to form heavier nuclei. This type of nuclear process is called **fusion.**

In stars, including the sun, energy is primarily produced when hydrogen nuclei combine, or fuse together, and release tremendous amounts of energy. However, a large amount of energy is needed to start a fusion reaction. This is because all nuclei are positively charged, and they repel each other with an electrical force. Energy is required to bring the hydrogen nuclei close together until the electrical forces are overcome by the attractive nuclear forces between two protons. In stars, the extreme temperatures provide the energy needed to bring hydrogen nuclei together.

Four hydrogen atoms fuse together in the sun to produce a helium atom and enormous energy in the form of gamma rays. This occurs in a multistep process that involves two isotopes of hydrogen: ordinary hydrogen ($^1_1$H), and deuterium ($^2_1$H).

$$^1_1\text{H} + ^1_1\text{H} \longrightarrow ^2_1\text{H} + \text{two particles}$$
$$^2_1\text{H} + ^1_1\text{H} \longrightarrow ^3_2\text{He} + ^0_0\gamma$$
$$^3_2\text{He} + ^3_2\text{He} \longrightarrow ^4_2\text{He} + ^1_1\text{H} + ^1_1\text{H}$$

SECTION 2 REVIEW

SUMMARY

▶ Neutrons and protons in the nucleus are held together by the strong nuclear force.

▶ Nuclear fission takes place when a large nucleus divides into smaller nuclei.

▶ Nuclear fusion occurs when light nuclei combine.

▶ Mass is converted into energy during fusion reactions of light elements and fission reactions of heavy elements.

1. **Explain** why most isotopes of elements with a high atomic number are radioactive.

2. **Indicate** whether the following are fission or fusion reactions.
   a. $^1_1\text{H} + ^2_1\text{H} \longrightarrow ^3_2\text{He} + \gamma$
   b. $^1_0n + ^{235}_{92}\text{U} \longrightarrow ^{146}_{57}\text{La} + ^{87}_{35}\text{Br} + 3^1_0n$
   c. $^{21}_{10}\text{Ne} + ^4_2\text{He} \longrightarrow ^{24}_{12}\text{Mg} + ^1_0n$
   d. $^{208}_{82}\text{Pb} + ^{58}_{26}\text{Fe} \longrightarrow ^{265}_{108}\text{Hs} + ^1_0n$

3. **Predict** whether the total mass of the 26 protons and 30 neutrons that make up the iron nucleus will be more, less, or equal to 55.847 amu, the mass of an iron atom, $^{56}_{26}\text{Fe}$. If it is not equal, explain why.

4. **Critical Thinking** Suppose a nucleus captures two neutrons and decays to produce one neutron; is this process likely to produce a chain reaction? Explain your reasoning.

Answers to Section 2 Review

1. The strong force acts over such a small distance that large nuclei are difficult to hold together. These unstable nuclei undergo nuclear reactions that make them more stable.

2. **a.** fusion
   b. fission
   c. fusion
   d. fusion

3. The total mass of this nucleus—56 amu—will be greater than 55.847 amu. The mass is not equal to 55.847 amu because that value is an average of the masses of several different isotopes.

4. A continued (critical) chain reaction will not occur. Each step of fission reactions requires more neutrons than the previous fission releases.

Nuclear Radiation Today

INTEGRATING TECHNOLOGY and Society

OBJECTIVES

▶ **Describe** sources of nuclear radiation, including where it exists as background radiation.

▶ **List** and explain three beneficial uses and three possible risks of nuclear radiation.

▶ **Compare** and contrast the advantages and disadvantages of nuclear energy as a power source.

■ **KEY TERMS**
background radiation
rem
radioactive tracer

It may surprise you to learn that you are exposed to some form of nuclear radiation every day. Some forms of nuclear radiation are beneficial. Others present some risks. This section will discuss both the benefits and the possible risks of nuclear radiation.

Where Is Radiation?

Nuclear radiation is all around you. This form of nuclear radiation is called **background radiation.** Most of it comes from natural sources, such as the sun, heat, soil, rocks, and plants, as shown in **Figure 9.** The living tissues of most organisms are adapted to survive these low levels of natural nuclear radiation.

■ **background radiation**
the nuclear radiation that arises naturally from cosmic rays and from radioactive isotopes in the soil and air

Figure 9
Sources of background radiation are all around us.

Did You Know ?

Levels of Background Radiation
Although there are small variations in background radiation levels at different locations, the average levels in the United States and around the world are generally in the same range. However, due to high concentrations of radioactive minerals in the soil, a few areas in Brazil, China, and India have much higher levels. For example, the background radiation in certain black-sand beaches of Brazil is almost 400 times greater than the normal level in the United States.

Graphing Have students use the Internet or library sources to find out the average level of radiation exposure in their area. Ask them to also find other values for comparison, such as the highest and lowest levels in the United States and in the world. Then ask them to create a bar graph that illustrates these comparisons. They could also include some average values for other U.S. cities (given in **Table 3**). Remind them that all values must have the same units for an accurate comparison. **LS Logical**

Table 3

Radiation Exposure Per Location

Location	Radiation Exposure (millirems/year)
Tampa, FL	63.7
Richmond, VA	64.1
Las Vegas, NV	69.5
Los Angeles, CA	73.6
Portland, OR	86.7
Rochester, NY	88.1
Wheeling, WV	111.9
Denver, CO	164.6

Source: United States Department of Energy, Nevada Operations Office

■ **rem** the quantity of ionizing radiation that does as much damage to human tissue as 1 roentgen of high-voltage X rays does

Radiation is measured in units of rems

Levels of radiation absorbed by the human body are measured in **rems** or millirems (1 rem = 1000 millirems).

In the United States, many people work in occupations involving nuclear radiation. Nuclear engineering, health physics, radiology, radiochemistry, X-ray technology, magnetic resonance imaging (MRI), and other nuclear medical technology all involve nuclear radiation. A safe limit for these workers has been set at 5000 millirems annually, in addition to natural background exposures.

Exposure varies from one location to another

People in the United States receive varying amounts of natural radiation. Those in higher altitudes receive more exposure to nuclear radiation from space than those in lower altitudes do. People in areas with many rocks have higher nuclear radiation exposure than people in areas without many rocks do. Because of large differences both in altitude and background radiation sources, exposure varies greatly from one location to another, as illustrated in **Table 3**.

Some activities add to the amount of nuclear radiation exposure

Another factor that affects levels of exposure is participation in certain activities. **Table 4** shows actual exposure to nuclear radiation for just a few activities. There are more activities that add to the amount of nuclear radiation exposure than those in this table, but these listed are at least a few of the activities that will add nuclear radiation to the air, affecting all those in the area around these activities.

Table 4 **Radiation Exposure Per Activity**

Activity	Radiation (millirems/year)
Smoking 1 1/2 packs of cigarettes per day	8,000
Flying for 720 hours (airline crew)	267
Inhaling radon from the environment	360
Giving or receiving medical X rays	100

Source: United States Department of Energy, Nevada Operations Office

300

Did You Know?

What is a rem? *Rem* stands for *r*oentgen *e*quivalent in *m*an. Roentgen is a German physicist who discovered X-rays in 1895. One rem is the amount of radiation that causes the same effect on a human being as a given amount (1 rad) of X rays.

REAL-LIFE CONNECTION

Radioactive Foods Natural radioactivity is all around us—in the air we breathe, in the rocks and soil that make up our planet, and even in the foods we eat. All foods have small amounts of radioactivity. The most common radioactive elements found in foods are potassium-40, radium-226, and uranium-238. For example, some foods that contain potassium-40 are bananas, brazil nuts, carrots, white potatoes, red meat, and lima beans.

Beneficial Uses of Nuclear Radiation

Radioactive substances have a wide range of applications. In these applications, nuclear radiation is used in a controlled way to take advantage of its effects on other materials.

Smoke detectors help to save lives

Small radioactive sources are present in smoke alarms, as shown in *Figure 10.* They release alpha particles, which are charged and produce an electric current. Smoke particles in the air reduce the flow of the current. The drop in current sets off the alarm before levels of smoke increase.

Nuclear radiation is used to detect diseases

Several types of nuclear radiation procedures have been very helpful to medical science. The digital computer, ultrasound scanning, CT scanning, PET, and magnetic resonance imaging (MRI) have combined to create a large variety of diagnostic imaging techniques. Using these procedures, doctors can view images of parts of the organs and can detect dysfunction or disease.

An X ray once was the primary imaging technique used in medicine. An image was created by focusing X rays for 11 minutes through a part of the body and onto a single piece of film. Today, X-ray imaging is done in milliseconds.

The MRI, an imaging process, as in *Figure 11,* uses radio frequency pulses to provide images of even small bodily structures.

Radioactive tracers are widely used in medicine. Tracers are short-lived isotopes that tend to concentrate in affected cells and are used to locate tumors.

Figure 10

In a smoke alarm, a small alpha-emitting isotope detects smoke particles in the air.

▶ **radioactive tracer** a radioactive material that is added to a substance so that its distribution can be detected later

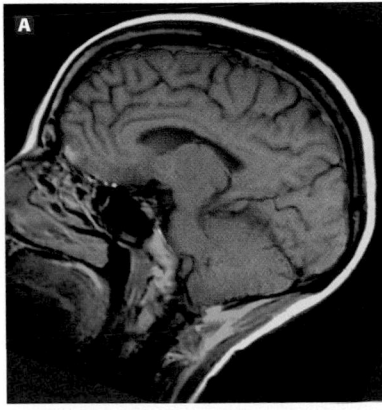 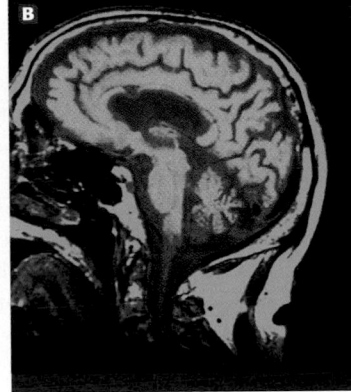

Figure 11

A This is an image of a healthy brain obtained with magnetic resonance imaging (MRI).

B Magnetic resonance imaging reveals that this brain has Alzheimer's disease.

301

Did You Know ?

Nuclear Medicine Today, one out of every three people admitted to a U.S. hospital undergoes some form of nuclear medical procedure for either diagnosis or therapy.

Nuclear medicine is the use of radioactive materials for the diagnosis and treatment of patients as well as for the study of disease. Nuclear medicine utilizes radioactive tracers to study physiology rather than anatomy.

The reason this is important is that biochemical and physiological changes occur in disease before anatomical changes can be identified. The techniques used in nuclear medicine are non-invasive and usually require no more than an intravenous injection. The radiation dosage is similar to and often far less than the dosage given in an equivalent radiological procedure.

Teaching Tip

Irradiation Tell students that another beneficial use of nuclear radiation is the irradiation of food. Some foods are irradiated by gamma rays from cobalt-60 or cesium-137 to retard spoilage. Irradiation is FDA approved and most commonly used on fruits, vegetables, and spices. Irradiation kills bacteria, so irradiated food requires fewer chemicals to keep it fresh. Students might think that such foods become radioactive. Point out that irradiated food is not radioactive and is safe to eat.

Teaching Tip

Example of Agricultural Application of Radioisotopes

A solution of phosphate, containing radioactive phosphorus-32, is injected into the root system of a plant. Phosphorus-32 behaves identically to that of phosphorus-31, which is the more common and non-radioactive form of the element. Therefore, the two isotopes of phosphorus are used by the plant in the same way. Movement of the radioactive phosphorus-32 throughout the plant can be followed by an instrument like a Geiger counter. The information that is obtained helps scientists to understand the detailed mechanism of how plants utilize phosphorus to grow and to reproduce.

Figure 12
Research farms use radioactive tracers to reveal water movement and other biochemical processes.

Nuclear radiation therapy is used to treat cancer

Radiotherapy is treatment that uses controlled doses of nuclear radiation for treating diseases such as cancer. For example, certain brain tumors can be targeted with small beams of gamma rays.

Radiotherapy is also used for treating thyroid cancer, using an iodine isotope. Treatment of leukemia also uses radiotherapy. The defective bone marrow is first killed with a massive dose of nuclear radiation and then replaced with healthy bone marrow from a donor.

Agriculture uses radioactive tracers and radioisotopes

On research farms, as in *Figure 12,* radioactive tracers in flowing water can show how fast water moves through the soil or through stems and leaves of crops. They help us to understand biochemical processes in plants. Radioisotopes are chemically identical with other isotopes of the same element. Because of that similarity, they are substituted in chemical reactions. Radioactive forms of the element can then be easily located with sensors.

Possible Risks of Nuclear Radiation

While nuclear radiation has many benefits, there are also risks. It is important to know what they are so that you can make informed decisions and exercise caution.

Nuclear radiation can ionize atoms

Nuclear radiation interacts with living tissue. This radiation includes charged particles (alpha and beta) as well as gamma rays and X rays. Alpha and beta particles, as well as gamma and X rays, can change the number of electrons in atoms in living materials. This is known as *ionization*. Molecules containing ionized atoms may form substances that are harmful to life.

The ability to penetrate matter differs among different types of nuclear radiation. A layer of clothing or an inch of air can stop alpha particles, which are heavy and slow moving. Beta particles are lighter and faster than alpha particles. Beta particles can penetrate a fraction of an inch in solids and liquids and can travel several feet in air. The ability of gamma rays to penetrate a material depends upon their energy. Several feet of material may protect you from high-energy gamma rays.

302

Alternative Assessment —— ADVANCED

Radioactive Tracer Ask students to write a short paragraph explaining why the half-life of a medical radioactive tracer is so important. The half-life of a medical radioactive tracer must be long enough to reach its destination and be detected but short enough to minimize the length of time healthy cells are exposed to radiation. Paragraphs might include the effects on the human body if either of these requirements is not met. **LS Logical**

The risk depends upon amounts of radiation

The effects of low levels of nuclear radiation on living cells are so small that they may not be detected. However, studies have shown a relationship between exposure to high levels of nuclear radiation and cancer. Cancers associated with high-dose exposure include leukemia as well as breast, lung, and stomach cancers.

Radiation sickness results from high levels of nuclear radiation

Radiation sickness is an illness resulting from excessive exposure to nuclear radiation. This sickness may occur from a single massive exposure, such as a nuclear explosion, or repeated exposures to very high nuclear radiation levels.

Individuals working with nuclear radiation must protect themselves with shields and special clothing. A person working in radioactive areas should wear a *dosimeter*, a device for measuring the amount of nuclear radiation exposure.

internet connect

www.scilinks.org
Topic: Nuclear Power
SciLinks code: HK4096

SCiLINKS Maintained by the National Science Teachers Association

REAL WORLD APPLICATIONS

Medical Radiation Exposure Graves' disease causes the thyroid gland to produce excess hormones. This excess induces increase in metabolism, weight loss (despite a healthy appetite), and irregular heartbeat.

Graves' disease and similar illnesses can be treated in several ways. Parts of the thyroid gland can be surgically removed, or patients can be treated with radioactive iodine-131. The thyroid cells need iodine to make hormones. When they take in the radioactive iodine-131, the overactive cells are destroyed, and hormone levels drop.

Examine the table below, which shows radiation exposures for different situations and the resulting increased risks in leukemia rates.

Applying Information
1. Given that the typical exposure for radioisotope therapy is about 10 rems, mostly delivered at once, do you think leukemia rates are likely to go up for this group? If so, estimate the expected risk.
2. The link of low-level nuclear radiations to cancers such as leukemia is still in question. Describe what other information would help you evaluate the risks.

WRITING SKILL

Person tested	Radiation exposure	Measured increased leukemia risk
Hiroshima atomic bomb survivor	27 rem at once	6%
U.S. WW II radiology technician	50 rem over 2 years	0%
Austrian citizen after the nuclear accident at Chernobyl	0.025 rem	0%

303

Teaching Tip

Testing for Radon Tell students that radon tests are either short term or long term, depending on whether the testing period is more or less than 90 days. If a short-term test shows high levels of radon, it should be followed by another test. The advantage of long-term testing is that it allows for seasonal variations in radon levels.

Did You Know ?

Reducing Radon Contractors who are experienced in reducing radon levels should make most building corrections for radon. Different venting methods are used based on whether the building has a basement, a crawl space, or a slab and how porous the soil is at that location.

Historical Perspective

On December 2, 1942, physicist Enrico Fermi and his colleagues at the University of Chicago demonstrated that a self-sustaining nuclear fission reaction could be produced from uranium and uranium oxide that was shielded by numerous blocks of graphite. Although the reactor produced only a few watts of power, this was sufficient to show that nuclear fission could be used to produce energy.

Did You Know ?

Radon-222 problems in homes or offices can be eliminated by sealing cracks in foundations or by installing vents that draw air out of the building.

High concentrations of radon gas can be hazardous

Colorless and inert, *radon gas* is produced by the radioactive decay of uranium-238 present in soil and rock. Radon gas emits alpha and beta particles and gamma rays. Tests have shown a correlation between lung cancer and high levels of exposure to radon gas, especially for smokers. Some areas have higher radon levels than others do. Tests for radon gas in buildings are widely available.

Nuclear Power

Today, nuclear reactors, as shown in *Figure 13*, are used in dozens of countries to generate electricity. Energy produced from fission is used to light the homes of millions of families. There are numerous advantages to this source of energy. There are also disadvantages.

Nuclear fission has both advantages and disadvantages

One advantage of nuclear fission is that it does not produce gaseous pollutants, and there is much more energy in the known uranium reserves than in the known reserves of coal and oil.

In nuclear fission reactors, energy is produced by triggering a controlled fission reaction in uranium-235. However, the products of fission reactions are often radioactive isotopes. Therefore, serious safety concerns must be addressed. Radioactive products of fission must be handled carefully so they do not escape in the environment and release nuclear radiation.

Another safety issue involves the safe operation of the nuclear reactors in which the controlled fission reaction is carried out. A nuclear reactor must be equipped with many safety features. The reactor requires considerable shielding and must meet very strict safety requirements. Thus, nuclear power plants are expensive to build.

Figure 13

Nuclear reactors like this are used over much of the world to generate electricity.

304

Chernobyl On April 26, 1986, the operators of a nuclear reactor at Chernobyl, in Ukraine, improperly managed the water coolant levels. Overheating and an explosion resulted in radioactive materials being blown high into the air. Thirty-one people died, and over two hundred others were seriously injured.

Some people cite this accident as an argument against nuclear power, but proper safety precautions could have prevented the disaster. Had the explosion occurred in a containment vessel, such as that required for any nuclear power plant in the United States, no nuclear materials would have escaped into the environment.

Nuclear waste must be safely stored

Besides the expenses that occur during the life of a nuclear power plant, there is the expense of storing radioactive materials, such as the fuel rods used in the reactors. After their use they must be placed in safe facilities that are well shielded, as shown in ***Figure 14.*** These precautions are necessary to keep nuclear radiation from leaking out and harming living things. The facilities must also keep nuclear radiation from contacting ground water.

Ideal places for such facilities are sparsely populated areas with little water on the surface or underground. These areas must be free from earthquakes.

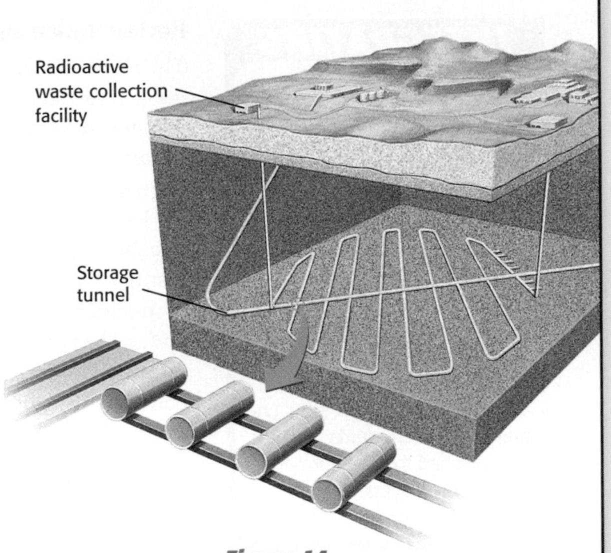

Figure 14
Storage facilities for nuclear waste must be designed to contain radioactive materials safely for thousands of years.

Nuclear fusion reactors are being tested

Another option that holds some promise as an energy source is nuclear fusion. Fusion means joining (fusing) smaller nuclei to make a larger nucleus. The sun uses the nuclear fusion of hydrogen atoms; this fusion results in larger helium atoms. This process of fusion gives off heat, light, and other radiation, otherwise known as solar energy. Solar energy can be captured by solar panels or other means to provide energy for homes and other types of buildings.

Recall from the last section that the process of fusion takes place when light nuclei, such as hydrogen, are forced together to produce heavier nuclei, such as helium, producing large amounts of energy. Some scientists estimate that 1 pound of hydrogen in a fusion reactor could release as much energy as 16 million pounds of burning coal. Nuclear fusion releases very little waste or pollution.

Because fusion requires that the electrical repulsion between protons be overcome, these reactions are difficult to produce in the laboratory. However, successful experiments have been conducted in the United States when researchers took a major step toward exploiting a safe, clean source of power that uses fuels extracted from ordinary water. Other experiments for power generated in a nuclear fusion reactor have also been carried out near Oxford, England.

INTEGRATING

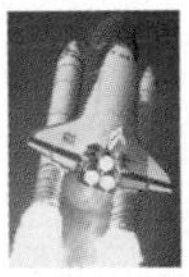

SPACE SCIENCE Unmanned space probes have greatly increased our knowledge of the solar system. Nuclear-powered probes can venture far from the sun without losing power, as solar-powered probes do. *Cassini,* which has been sent to explore Saturn, has been powered by the heat generated by the radioactive decay of plutonium.

Teaching Tip

To strengthen their understanding of the issues involved in the containment and disposal of nuclear waste, students may want to check out the following web sites.

- Nuclear Regulatory Commission http://www.nrc.gov
- Environmental Protection Agency http://www.epa.gov/radiation/
- Department of Energy http://www.ymp.gov
- Nuclear Information and Resource Service http://www.nirs.org

INTEGRATING

SPACE SCIENCE Nuclear-powered probes eventually lose power. The older the plutonium in the probe, the less energy the plutonium releases. As time passes, it will no longer be able to release enough energy to operate all the instruments, and some of the instruments will be turned off. Eventually, the plutonium will not release enough energy for the probe's vital functions.

Trends in Physics

Learning from Nature Nuclear power has several benefits. The raw materials are more plentiful than fossil fuels, and the process does not pollute the air. But nuclear power poses another environmental problem. The fission process generates radioactive waste that must be safely stored for thousands of years. The question is this: What type of storage is safe?

In the United States, some leaders have proposed storing waste deep under Yucca Mountain, in Nevada. The consequences of this storage method are not fully understood. However, the discovery of the natural, two-billion year old nuclear reactor in Gabon, Africa, may change that. Scientists are studying the site in Africa to learn how far buried nuclear waste travels. They hope to apply their results toward the safe disposal of radioactive waste in the United States and other countries.

Videos

CNN Presents Chemistry

- **Segment 31** Nuclear Waste
- **Critical Thinking** Worksheet 31

See the Science in the News video guide for more details.

INTEGRATING

SPACE SCIENCE In regular fusion reactions in stars, elements lighter than and including iron are formed. Elements heavier than iron are not formed because they consume instead of release energy when they form. When fuel in the star is spent, the star collapses, ending in a tremendous explosion called a supernova. Supernovas release enough energy to produce elements heavier than iron.

Close

Quiz ———————— BASIC

1. It is unsafe to be exposed to background radiation without wearing a dosimeter.
True or false? (false)

2. What is one beneficial application of nuclear radiation? (Accept any response from text.)

3. Why must products of nuclear fission be stored safely? (Products of fission are radioactive isotopes that could release nuclear radiation into the environment.)

Chapter Resource File

- **Concept Review** Section 3 GENERAL
- **Cross-Disciplinary Worksheet** Integrating Space Science—The Life Cycle of a Star ADVANCED
- **Section Quiz** BASIC

INTEGRATING

SPACE SCIENCE All heavy elements, from cobalt to uranium, are made when massive stars explode. The pressure produced in the explosion causes nearby nuclei to fuse together, in some cases more than once.

The explosion carries the newly created elements into space. These elements later become parts of new stars and planets. The elements of Earth are believed to have formed in the outer layers of an exploding star.

Nuclear fusion also has advantages and disadvantages

The most attractive feature of fusion is that the fuel for it is abundant. Hydrogen is the most common element in the universe and is plentiful in many compounds on Earth, such as water. Earth's oceans could provide enough hydrogen to meet current world energy demands for millions of years.

Unfortunately, practical fusion-based power is far from being a reality. Fusion reactions have some drawbacks. They can produce fast neutrons, a highly energetic and potentially dangerous form of nuclear radiation. Shielding material in the reactor would have to be replaced periodically, increasing the expense of operating a fusion power plant. Lithium can be used to slow down these neutrons, but it is chemically reactive and rare, making its use impractical.

Nuclear fusion is still in its infancy. Successful experiments are just beginning. Who can say what the future may hold? Perhaps scientists yet to come will find the answers to the nagging questions that plague the government today concerning the perfect fuel for United States citizens.

SECTION 3 REVIEW

SUMMARY

- Background radiation comes from natural sources and is everywhere. Living tissue adapts to background radiation in most cases.

- Beneficial uses of nuclear radiation include smoke detectors, X rays, CT, MRI, radioactive tracers, PET, radiotherapy, radioactive tracers, and radioisotopes.

- Risks of high levels of nuclear radiation include cancers and radiation sickness. High levels of radon gas can be harmful. Tests for radon gas are widely available.

- Nuclear fission is an alternative to fossil fuels as a source of energy.

1. **List** three sources of background radiation.

2. **Identify** three activities that add to background radiation under normal circumstances.

3. **Describe** how smoke detectors use alpha particles and what sets off the alarm.

4. **Name** three nuclear radiation diagnostic imaging techniques that help detect diseases.

5. **Explain** how radioactive tracers help locate tumors.

6. **Describe** how gamma rays are used in cancer therapy.

7. **Compare** and contrast the benefits and risks of radiation therapy in general.

8. **Explain** why it is important to use low levels of nuclear radiation for detection and treatment of disease.

9. **Summarize** why the testing of buildings for radon gas levels may be important, especially for smokers.

10. **Critical Thinking** Suppose uranium-238 could undergo fission as easily as uranium-235. Predict how that would change the advantages and drawbacks of fission reactors.

Answers to Section 3 Review

1. Answers may include sun, heat, soil, water, plants, and air

2. Answers may include: smoking, flying, inhaling radon from environment, giving or receiving medical X rays, etc.

3. Accept any answers similar to text.

4. Accept X rays, CT, MRI, PET, or other similar techniques.

5. They can be used to trace water underground, through both soil and crops. In the human body, tracers locate tumors and trace the path of drugs.

6. Several small beams target the cancerous tissue but not the tissue around it.

7. Radiation therapy enables doctors to treat cancer by targeting and exposing tumors to radiation. However, high doses of radiotherapy may harm healthy tissue.

8. Risks of high levels of nuclear radiation include cancers and radiation sickness.

9. Accept answers similar to text.

10. Advantages might include that fuel would be more abundant, thus lowering expense. The relative radioactivity and usefulness of waste products would need to be evaluated.

Math Skills

Calculating Times of Decay

A sample of francium-223 has a half-life of 22 minutes.

a. What fraction of francium-223 remains if 93.75 percent of it has undergone radioactive decay?

b. How many half-lives does it take for the sample to decay?

c. How long does it take for the sample to decay?

1 **List all given and unknown values.**

> **Given:** fraction of sample decayed, 93.75 percent
>
> half-life, 22 min
>
> **Unknown:** fraction of sample remaining
>
> number of half-lives (n)
>
> time of decay

2 **Write down the equation relating the fraction of the sample remaining to the percentage of sample decayed, and the equation relating the time of decay to the number of half-lives.**

$$\text{fraction of sample remaining} = 1 - \text{fraction of sample decayed}$$
$$= 1 - \frac{\text{percentage of sample decayed}}{100}$$
$$= \left(\frac{1}{2}\right)^n$$

$$\text{time of decay} = n \times \text{half-life}$$

3 **Calculate the unknown quantities.**

> **a.** $\text{fraction of sample remaining} = 1 - \dfrac{93.75}{100} = 1 - 0.9375 = 0.0625$

To express this as a fraction, divide the answer into 1 to find the denominator of the fraction. $1/0.0625 = 16$, so the fraction of sample remaining is 1/16.

> **b.** $\left(\dfrac{1}{2}\right)^n = \dfrac{1}{16} = \left(\dfrac{1}{2}\right) \times \left(\dfrac{1}{2}\right) \times \left(\dfrac{1}{2}\right) \times \left(\dfrac{1}{2}\right) = \left(\dfrac{1}{2}\right)^4$
>
> $\text{number of half-lives} = n = 4$

> **c.** $\text{time of decay} = 4 \times 22 \text{ min} = 88 \text{ min}$

Practice

Following the example above, calculate the following:

1. What fraction of iodine-132 remains if 87.5% has undergone radioactive decay?

2. How many half-lives does it take for the sample to decay?

3. Iodine-132 has a half-life of 2.3 hours. How long does it take for the sample to decay?

Teaching Tip

Percentages and Fractions In calculating half-lives, students may find it helpful if they can associate certain fractions and percentage values as multiples of 1/2. As a calculation aid, write the following table on the chalkboard, showing the relationship between these numbers.

n	$(1/2)^n$	$(1/2)^n \times 100\%$
1	1/2	50%
2	1/4	25%
3	1/8	12.5%
4	1/16	6.25%
5	1/32	3.125%

Explain that whenever the fraction or percentage of remaining radioactive material equals one of the values in the middle or right columns, respectively, the number of half-lives that has passed is on the same line in the left column.

Also explain that, to determine the number of half-lives that have passed when the fraction or percentage is given of radioactive material that has already decayed, simply subtract that value from 1 (or 100 percent) to obtain the value in the table above, and solve as before.

Answers

1. 1/8

2. 3

3. 9.2 h

Understanding Concepts

1. d
2. b
3. c
4. a
5. d
6. b
7. d
8. c
9. d
10. c
11. a
12. b
13. a
14. d
15. b

Using Vocabulary

16. Radioactivity can increase atomic number ($\updownarrow$), decrease it ($\mapsto$), or leave it unaffected (). It can either decrease mass number ($\mapsto$) or not affect it ($\updownarrow$ and). When mass number is decreased, the atom is more stable because the strong nuclear force is greater in a smaller nucleus.

Chapter Resource File

- Test Item Listing for ExamView® Test Generator
- Chapter Tests **GENERAL**

Chapter Highlights

Before you begin, review the summaries of the key ideas of each section, found at the end of each section. The key vocabulary terms are listed on the first page of each section.

UNDERSTANDING CONCEPTS

1. When a heavy nucleus decays, it may emit
   a. alpha particles.
   c. gamma rays.
   b. beta particles.
   d. All of the above

2. A neutron decays to form a proton and a(n)
   a. alpha particle.
   c. gamma ray.
   b. beta particle.
   d. emitted neutron.

3. Alpha particles
   a. are negatively charged rays emitted from uranium-238.
   b. are equivalent to lithium nuclei.
   c. are too massive to pass through paper.
   d. gain energy as they ionize matter.

4. Beta particles
   a. are actually electrons emitted from a decayed neutron.
   b. are neutral in charge because they came from a neutron.
   c. are negatively charged but cannot ionize other atoms.
   d. gain energy as they ionize and then deeply penetrate matter.

5. Gamma rays
   a. have a positive charge and can therefore ionize matter.
   b. have a negative charge and can therefore ionize matter.
   c. have no electrical charge and cannot therefore ionize matter.
   d. have no electrical charge but can ionize matter.

6. Neutrons
   a. cannot travel as far through matter as alpha and beta particles can.
   b. can travel farther through matter than either alpha or beta particles.
   c. can ionize matter, although they have no charge.
   d. cannot ionize matter, although they do have a charge.

7. After three half-lives, _____ of a radioactive sample remains.
   a. all
   c. one-third
   b. one-half
   d. one-eighth

8. Carbon dating can be used to measure the age of each of the following except
   a. a 7000-year-old human body.
   b. a 1200-year-old wooden statue.
   c. a 2600-year-old iron sword.
   d. a 3500-year-old piece of fabric.

9. Of the following elements, only the isotopes of _____ are all radioactive.
   a. nitrogen
   c. sulfur
   b. gold
   d. uranium

10. The strong nuclear force
    a. attracts protons to electrons.
    b. holds molecules together.
    c. holds the atomic nucleus together.
    d. attracts electrons to neutrons.

11. The process in which a heavy nucleus splits into two lighter nuclei is called
    a. fission.
    c. alpha decay.
    b. fusion.
    d. a chain reaction.

12. The amount of energy produced during nuclear fission is related to
    a. the temperature in the atmosphere during nuclear fission.
    b. the masses of the missing nuclei and particles released.
    c. the volume of the nuclear reactor.
    d. the square of the speed of sound.

17. An alpha particle is a helium nucleus with a charge of +2 and is only slightly penetrating. A beta particle is an electron emitted when a neutron in a nucleus changes into a proton and an electron; its charge is −1 and it is moderately penetrating. A gamma ray is high-energy, penetrating, electromagnetic radiation. Neutron emission occurs as an uncharged particle is emitted when an unstable nucleus undergoes fission.

18. Alpha particles travel less distance than neutrons, because they are massive and charged.

19. Beta particles come from decayed neutrons.

20. Gamma rays have no mass, because they are a form of electromagnetic energy and are not made of matter.

21. No; the half-life of the tracer must be long enough to reach its destination and be detected.

22. Enough of the radioactive substance must be present for fission to release enough neutrons to cause other nuclei to undergo fission in a continuing process.

13. Which condition is not necessary for a chain reaction to occur?
  a. The radioactive sample must have a short half-life.
  b. The neutrons from one split nucleus must cause other nuclei to divide.
  c. The radioactive sample must be at critical mass.
  d. Not too many neutrons must be allowed to leave the radioactive sample.

14. Exposure to nuclear radiation varies from location to location because
  a. the altitude varies from location to location.
  b. the activities in certain areas vary from those in other locations.
  c. the amount of rock varies with location.
  d. All of the above

15. Which of the following is *not* a use for radioactive isotopes?
  a. as tracers for diagnosing disease
  b. as an additive to paints to increase their durability
  c. as a way of treating forms of cancer
  d. as a way to study biochemical processes in plants

USING VOCABULARY

16. How can *nuclear radioactivity* affect the atomic number and mass number of a nucleus that changes after undergoing decay?

17. Describe the main differences between the four main types of nuclear *radiation: alpha particles, beta particles, gamma rays,* and *neutron emission.*

18. What are two factors that cause alpha particles to lose energy and travel less distance than neutrons travel?

19. Where do beta particles come from?

20. Why do gamma rays have no mass at all?

21. Would a substance with a one-second *half-life* be effective as a *radioactive tracer?*

22. For the nuclear *fission* process, how is *critical mass* important in a *chain reaction?*

23. How does nuclear *fusion* account for the energy produced in stars?

24. What is *background radiation,* and what are its sources?

25. The amount of nuclear radiation exposure that is received into a human body is measured in *rems.* How does the amount of exposure in rems per year in Denver, Colorado, compare with the amount that has been set as a safe limit for workers in occupations with relatively high radiation exposure?

26. How can a *radioactive tracer* be used to locate tumors?

BUILDING MATH SKILLS

27. Nuclear Decay Bismuth-212 undergoes a combination of alpha and beta decays to form lead-208. Depending on which decay process occurs first, different isotopes are temporarily formed during the process. Identify these isotopes by completing the equations given below:
  a. $^{212}_{83}\text{Bi} \longrightarrow {}^{\square}_{\square}\text{X} + {}^{4}_{2}\text{He}$

  $^{\square}_{\square}\text{X} \longrightarrow {}^{208}_{82}\text{Pb} + {}^{0}_{-1}e$

  b. $^{212}_{83}\text{Bi} \longrightarrow {}^{\square}_{\square}\text{Y} + {}^{0}_{-1}e$

  $^{\square}_{\square}\text{Y} \longrightarrow {}^{208}_{82}\text{Pb} + {}^{4}_{2}\text{He}$

28. Nuclear Decay The longest-lived radioactive isotope yet discovered is the beta-emitter tellurium-130. It has been determined that it would take 2.5×10^{21} years for 99.9% of this isotope to decay. Write the equation for this reaction, and identify the isotope into which tellurium-130 decays.

309

23. Nuclear fusion occurs when two small nuclei combine or fuse to form a larger more stable nucleus. This process occurs at extremely high temperatures with the release of energy. The energy in stars is produced when hydrogen nuclei fuse together and release tremendous amounts of energy.

24. Background radiation is radiation that arises naturally from cosmic rays and from radioactive isotopes in the soil and in the air.

25. The average amount of radiation exposure in the United States is 360 millirems per year. This average is much lower than the safe limit set for workers in high-radiation jobs, which is 5 000 millirems per year.

26. Some radioactive tracers tend to concentrate in tumors and thus can be used to locate these tumors with radioactive detectors.

Building Math Skills

27. a. $-212 - 4 = 208$
  $^{212}_{83}\text{Bi} \rightarrow {}^{208}_{81}\text{Tl} + {}^{4}_{2}\text{He}$
  $83 - 2 = 81$
  $^{208}_{81}\text{Tl} \rightarrow {}^{208}_{82}\text{Pb} + {}^{0}_{-1}e$
  b. $-212 - 0 = 212$
  $^{212}_{83}\text{Bi} \rightarrow {}^{212}_{84}\text{Po} + {}^{0}_{-1}e$
  $83 - (-1) = 84$

28. $^{130}_{52}\text{Te} \rightarrow {}^{130}_{53}\text{I} + {}^{0}_{-1}e$

29. $^{149}_{62}\text{Sm} \rightarrow {}^{145}_{60}\text{Nd} + {}^{4}_{2}\text{He}$

30. $\frac{1}{8} = \frac{1}{2} \times \frac{1}{2} \times \frac{1}{2}$; three half-lives
  $3 \times 5730 \text{ years} = 1.72 \times 10^{4}$ years

31. $15.2 \text{ days} \times \frac{1 \text{ half-life}}{3.82 \text{ days}} =$
  about 4 half-lives
  $\frac{1}{2} \times \frac{1}{2} \times \frac{1}{2} \times \frac{1}{2} = \frac{1}{16}$
  $\frac{1}{16} \times 4.38 \text{ μg} = 0.274 \text{ μg}$

Building Graphing Skills

32. This graph should be a straight line extending from the zero point to somewhere in the upper right corner of the graph. Point out to the students that this graph represents a direct relationship.

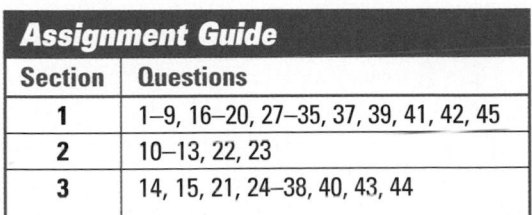

Assignment Guide	
Section	**Questions**
1	1–9, 16–20, 27–35, 37, 39, 41, 42, 45
2	10–13, 22, 23
3	14, 15, 21, 24–38, 40, 43, 44

33. Graphs may vary. One possibility is to graph number of half-lives on the *x*-axis and percentage of sample remaining on the *y*-axis.

   a. 4 days/8.1 days/half-life = 0.494 half-lives; 29%
   b. 64.5%
   c. 82%
   d. 8.8%

Thinking Critically

34. During fission, the mass of the nucleus is slightly more than the mass of the fission fragments. A small amount of mass is transformed to energy. The total amount of mass and energy is the same in the nucleus and in the products of the fission reaction.

35. They are alike in mass and charge. Beta particles possess more energy and are not part of a particular atom.

36. Where radioactivity strikes the plate, it will expose it. The amount and pattern of exposure indicates the amount of radiation exposure.

37. Answers might include that it is calculated mathematically using an equation based on current and original amounts of radioactivity and lapsed time.

38. The time lapsed is approximately 10 half-lives.

39. The object must have come from a living thing. The amounts of C-12 and C-14 in the sample are measured. The fraction of C-14 remaining tells how many half-lives have passed since the carbon was incorporated into the object.

29. **Nuclear Decay** It takes about 10^{16} years for just half the samarium-149 in nature to decay by alpha-particle emission. Write the decay equation, and find the isotope that is produced by the reaction.

30. **Half-life** The ratio of carbon-14 to carbon-12 in a prehistoric wooden artifact is measured to be one-eighth of the ratio measured in a fresh sample of wood from the same region. The half-life of carbon-14 is 5730 years. Determine its age.

31. **Half-life** Health officials are concerned about radon levels in homes. The half-life of radon-222 is 3.82 days. If a sample of gas taken from a basement contains 4.38 mg of radon-222, how much will remain in the sample after 15.2 days?

BUILDING GRAPHING SKILLS

32. **Graphing** Draw a graph representing stable nuclei. Entitle it "Number of protons versus number of neutrons for stable nuclei." Let the *x*-axis (horizontal axis) represent number of protons. Let the *y*-axis (vertical axis) represent number of neutrons. Remember that in a stable nucleus, there are equal numbers of protons and neutrons.

33. **Graphing** Using a graphing calculator or computer graphing program, create a graph for the decay of iodine-131, which has a half-life of 8.1 days. Use the graph to answer the following questions:

   COMPUTER SKILL

   a. Approximately what percentage of the iodine-131 has decayed after 4 days?
   b. Approximately what percentage of the iodine-131 has decayed after 12.1 days?
   c. What fraction of iodine-131 has decayed after 2.5 half-lives have elapsed?
   d. What percentage of the original iodine-131 remains after 3.5 half-lives?

310

THINKING CRITICALLY

34. **Applying Knowledge** Explain how the equivalence of mass and energy accounts for the small difference between the mass of a uranium-235 nucleus and the masses of the nuclei of its fission fragments.

35. **Applying Knowledge** Describe the similarities and differences between atomic electrons and beta particles.

36. **Critical Thinking** Why do people working around radioactive waste in a radioactive storage facility wear badges containing strips of photographic film?

37. **Creative Thinking** Many radioactive isotopes have half-lives of several billion years. Other radioactive isotopes have half-lives of billionths of a second. Suggest a way in which the half-lives of such isotopes are measured.

38. **Problem Solving** A radioactive tracer can be used to measure water movement through soil. In order to avoid contamination of ground water, 99.9% of the tracer must decay between the time that it is introduced into the soil and the time that it reaches the ground water supply. Estimate this time and calculate the half-life of an ideal tracer that could be used in this particular application.

39. **Critical Thinking** Explain the concept of why carbon-14 is used to determine the age of an object.

40. **Critical Thinking** Why would carbon-14 not be a good choice to use in household smoke detectors?

41. **Critical Thinking** Would an emitter of alpha particles be useful in measuring the thickness of a brick? Explain your answer.

42. Allocating Resources An archeologist has collected seven samples from a site: two scraps of fabric, two strips of leather, and three bone fragments. The age of each item must be determined, but the budget for carbon-14 dating is only $4500. Carbon-14 mass spectrometry is an accurate way to find a sample's age, but it costs $820 per sample. Carbon-14 dating by liquid scintillation costs only $400 a sample, but is less reliable. How would you apply either or both of these techniques to the samples to obtain the most reliable information and still stay within your budget?

43. Making Decisions Suppose you are an energy consultant who has been asked to evaluate a proposal to build a power plant in a remote area of the desert. Investigate the requirements for and possible hazards of nuclear-fission power plants, coal-burning power plants, and solar-energy farms. Study research about their environmental impacts. Using this information and what you have learned from this chapter, write a paragraph supporting your decision about which of these power plants would be best for its surroundings.

WRITING SKILL

44. Working Cooperatively Read the following, and research with a group of classmates possible solutions that make use of radioactivity. Report your findings.

A person believed to be suffering from cancer has been admitted to a hospital. What are some possible methods of diagnosing the patient's conditions? Assuming that cancer is found, how might the disease be treated? Suppose you suspect that another patient is suffering from radiation poisoning. How would you be able to tell?

45. Concept Mapping Copy the unfinished concept map below onto a sheet of paper. Complete the map by writing the correct word or phrase in the lettered boxes.

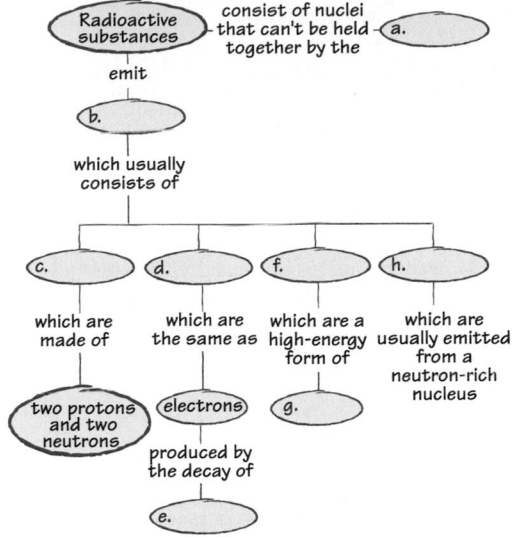

internet connect

www.scilinks.org
Topic: **Radioactive Tracers** SciLinks code: **HK4115**

SCiLINKS® Maintained by the National Science Teachers Association

Art Credits: Fig. 1, Kristy Sprott; Table 1, Kristy Sprott; Figs. 3, 4, 6, 7, 8, Kristy Sprott; Fig. 14, Uhl Studios, Inc.

Photo Credits: Chapter Opener image of the painting "Woman Reading Music", courtesy Rijksmuseum, Amsterdam; man next to X-ray of forgery, Volker Steger/Science Photo Library/Photo Researchers, Inc.; Fig. 2, Bettmann/CORBIS; "Quick Activity" (pennies), Peter Van Steen/HRW; Fig. 5, American Institute of Physics/AIP Emilio Segrè Visual Archives; "Quick Activity" (dominos) Peter Van Steen/HRW, Fig. 9, SuperStock; Fig. 10, Peter Van Steen/HRW; Fig. 11, Howard J. Radzyner/Phototake; Fig. 12, Larry Lefever/Grant Heilman Photography; Fig. 13, Cuthbert/SPL/Photo Researchers, Inc.; "Career Feature" (portraits), Peter Cutts; (article), Reprinted with permission from Science News, ©1998 by Science Service.

311

40. Carbon-14 decays by emitting beta particles, which are more penetrating than the alpha particles emitted by the isotopes that are usually used.

41. No. Alpha particles are relatively slow, have a large mass, and can be stopped by a sheet of paper. Therefore, they will not penetrate a brick.

Developing Life/Work Skills

42. Answers might include assuming that all samples of each type are of approximately the same age. Date one fabric sample, one leather sample, and two bone samples by mass spectrometry for a cost of 4 × $820, or $3280. Confirm these ages by dating the other samples by liquid scintillation for a cost of 3 × $400, or $1200. The total cost is $3280 + $1200, or $4480, which is within the budget.

43. Answers might include the following: Nuclear fission power plants require large amounts of water, which are not present in the desert. Coal would have to be shipped in, which would not be practical. However, solar-energy technology may need more development to be practical and cost-efficient.

44. A tracer that concentrates in fast-growing cells, such as tumors, could be used to detect any cancer. Depending on the type of cancer, it could be treated by focused gamma radiation. Symptoms of radiation poisoning include lower white cell count in the blood, hair loss, sterility, unhealthy bones, and cancer.

Integrating Concepts

45. a. strong nuclear force
b. radiation
c. alpha particles
d. beta particles
e. neutrons
f. gamma rays
g. electromagnetic radiation
h. neutrons

Transparencies

TT Concept Mapping

SIMULATING NUCLEAR DECAY REACTIONS

Teacher's Notes

Time Required 1 lab period

Ratings

1	2	3	4
EASY			HARD

TEACHER PREPARATION 2
STUDENT SETUP 1
CONCEPT LEVEL 3
CLEANUP 1

Skills Acquired
- Classifying
- Collecting data
- Communicating
- Identifying/Recognizing Patterns
- Interpreting
- Measuring
- Organizing and analyzing data

The Scientific Method
In this lab, students will:
- Make Observations
- Analyze the Results
- Draw Conclusions
- Communicate Results

Safety Cautions
Have students review safety guidelines before working in the lab.

Skills Practice Lab

Introduction

In this lab you will simulate the decay of lead-210 into its isotope lead-206. This decay of lead-210 into its isotope lead-206 occurs in a multistep process. Lead-210, $^{210}_{82}Pb$, first decays into bismuth-210, $^{210}_{83}Bi$, which decays into polonium-210, $^{210}_{84}Po$, which finally decays into the isotope lead-206, $^{206}_{82}Pb$.

Objectives

▶ **USING SCIENTIFIC METHODS** *Simulate* the decay of radioactive isotopes by throwing a set of dice, and observe the results.

▶ *Graph* the results to identify patterns in the amounts of isotopes present.

Materials

10 dice
large paper cup with plastic lid
roll of masking tape
scissors

Simulating Nuclear Decay Reactions

▶ Procedure

1. On a sheet of paper, prepare a table as shown below. Leave room to add extra rows at the bottom, if necessary.

Throw #	# of dice representing each Isotope			
	$^{210}_{82}Pb$	$^{210}_{83}Bi$	$^{210}_{84}Po$	$^{206}_{82}Pb$
0 (start)	10	0	0	0
1				
2				
3				
4				

2. Place all 10 dice in the cup. Each die represents an atom of $^{210}_{82}Pb$, a radioactive isotope.

3. Put the lid on the cup, and shake it a few times. Then remove the lid, and spill the dice. In this simulation, each throw represents a *half-life*.

4. All the dice that land with 1, 2, or 3 up represent atoms of $^{210}_{82}Pb$ that have decayed into $^{210}_{83}Bi$. The remaining dice still represent $^{210}_{82}Pb$ atoms. Separate the two sets of dice. Count the dice, and record the results in your data table.

5. To keep track of the dice representing the decayed atoms, you will make a small mark on them. On a die, the faces with *1*, *2*, and *3* share a corner. With a pencil, draw a small circle around this shared corner, and this die represents the $^{210}_{83}Bi$ atoms.

6. Put all the dice back in the cup, shake them and roll them again. In a decay process, there are two possibilities: some atoms decay and some do not. See the diagram below to track your results.

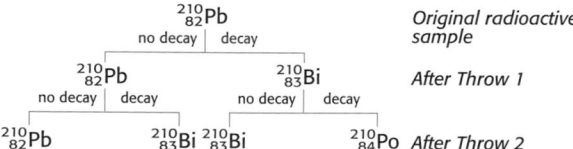

Original radioactive sample — $^{210}_{82}Pb$
After Throw 1 — $^{210}_{82}Pb$ ... $^{210}_{83}Bi$
After Throw 2 — $^{210}_{82}Pb$ ... $^{210}_{83}Bi$ $^{210}_{83}Bi$ ... $^{210}_{84}Po$

Tips and Tricks
Some students may have difficulty understanding the concept of probability. Begin the lab with the following activity to illustrate the concept of probability.

Have each student write his or her name on a strip of paper. Collect all the strips and place them in a container. Ask students how likely it is that his or her name will be picked if someone were to draw a name from the container.

Isotope type	Decays into	Signs of decay	Identifying the atoms in column 2
$^{210}_{82}Pb$	$^{210}_{83}Bi$	Unmarked dice land on *1*, *2*, or *3*	Mark $^{210}_{83}Bi$ by drawing a circle around the corner where faces *1*, *2*, and *3* meet.
$^{210}_{83}Bi$	$^{210}_{84}Po$	Dice with one loop land on *1*, *2*, or *3*	Draw a circle around the corner where faces *4*, *5*, and *6* meet.
$^{210}_{84}Po$	$^{206}_{82}Pb$	Dice with two loops land on *1*, *2*, or *3*	Put a small piece of masking tape over the two circles.
$^{206}_{82}Pb$	Decay ends		

7. After the second throw, we have three types of atoms. Sort the dice into three sets.

   a. The first set consists of dice with a circle drawn on them that landed with *1*, *2*, or *3* facing up. These represent $^{210}_{83}Bi$ atoms that have decayed into $^{210}_{84}Po$.

   b. The second set consists of two types of dice: the dice with one circle that did not land on *1*, *2*, or *3* (undecayed $^{210}_{83}Bi$) and the unmarked dice that landed with *1*, *2*, or *3* facing up (representing the decay of original $^{210}_{82}Pb$ into $^{210}_{83}Bi$).

   c. The third set includes unmarked dice that did not land with *1*, *2*, or *3* facing up. These represent the original undecayed $^{210}_{82}Pb$ atoms.

8. After each throw, do the following: separate the different types of atoms in groups, count the atoms in each group, record your data in your table, and mark the dice to identify each isotope. Use the table above as a guide.

9. For your third throw, put all the dice back into the cup. After the third throw, some of the $^{210}_{84}Po$ will decay into the stable isotope $^{206}_{82}Pb$. Use the table above and step 8 to figure out what else happens after the third throw.

10. Continue throwing the dice until all the dice have decayed into $^{206}_{82}Pb$, which is a stable isotope. Hence, these dice will remain unchanged in all future throws.

▶ Analysis

1. Write nuclear decay equations for the nuclear reactions modeled in this lab.

2. In your lab report, prepare a graph like the one shown at right. Using a different color or symbol for each atom, plot the data for all four atoms on the same graph.

3. What do your results suggest about how the amounts of $^{210}_{82}Pb$ and $^{206}_{82}Pb$ on Earth are changing over time?

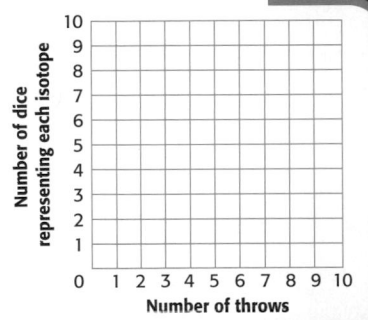

▶ Conclusions

4. $^{210}_{82}Pb$ is continually produced through a series of nuclear decays that begin with $^{238}_{92}U$. Does this information cause you to modify your answer to item 3? Explain why.

313

Answers to Analysis

1. $^{210}_{82}Pb \rightarrow {}^{210}_{83}Bi + {}^{0}_{-1}e$
   $^{210}_{83}Bi \rightarrow {}^{210}_{84}Po + {}^{0}_{-1}e$
   $^{210}_{84}Po \rightarrow {}^{206}_{82}Pb + {}^{4}_{2}He$

2. Students' graphs will vary. The graphs will depend on the results of throwing 10 dice (governed by probability) as described in the lab. A graph of the sample data in Data Table 1 is shown below.

3. Lead-210 decays into lead-206. The amount of lead-210 gradually decreases, while lead-206 increases. Eventually, all the lead-210 will be used up.

Answers to Conclusions

4. If lead-210 is continually produced through the decay of uranium-238, then the amount of lead-210 will not necessarily decrease. This depends on how much new lead-210 is produced in a given time and how much lead-210 decays into lead-206.

Data Table 1

Throw	$^{210}_{82}Pb$	$^{210}_{83}Bi$	$^{210}_{84}Po$	$^{206}_{82}Pb$
0	10	0	0	0
1	6	4	0	0
2	3	5	2	0
3	2	3	4	1
4	0	4	3	3
5	0	2	4	4
6	0	0	3	7
7	0	0	2	8
8	0	0	0	10

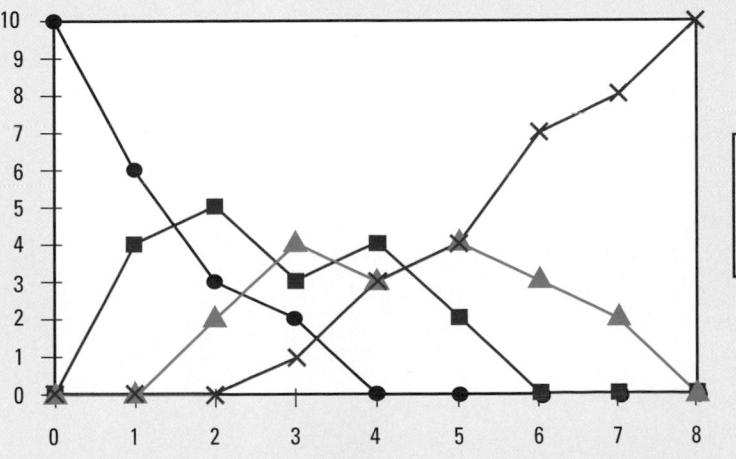

- ● Lead-210
- ■ Bismuth-210
- ▲ Polonium-210
- ✕ Lead-206

Chapter Resource File

- **Datasheet** Simulating Nuclear Decay Reactions **GENERAL**
- **Observation Lab** Modeling Radioactive Decay With Pennies **BASIC**
- **Skills Practice Lab—CBL™** Determining the Effective Half-life of Iodine-131 in the Human Body **ADVANCED**

SCIENCE REPORTER

Corinna Wu reports on matters involving chemistry and materials science, including the topics covered in the chemistry chapters of this text.

CareerLink

Science Reporter

Science reporters are usually among the first people to hear about scientific discoveries. News organizations hire science reporters to explain these discoveries to the general public in a clear, understandable, and entertaining way. To learn more about science reporting as a career, read the interview with science reporter Corinna Wu, who writes for Science News magazine, in Washington, D.C.

Corinna Wu describes scientific research and discovery in the articles she writes.

"I think writing is something you can learn— it's a craft. Lots of people talk about talents, but I think it's something you can do if you work at it."

 What does a science reporter do?

I write and report news and feature articles for a weekly science news magazine. That entails finding news stories— generally about research. I have to call the researchers and ask them questions about how they did their work and the significance of the work. Then I write a short article explaining the research to ordinary people.

 What is your favorite part of your work?

I like learning about a new subject every week. I get to ask all the stupid questions I was afraid to ask in school.

 How did you become interested in science reporting as a career?

After college, I had a summer internship at NASA, at the Johnson Space Center in Houston, Texas, doing materials research there. I had lots of time to read space news magazines. It was at that time that I realized, "Hey, people write this stuff."

 What kinds of skills are important for a science reporter?

One thing that is really important is to really love writing. If you don't like to write already, it's pretty hard to make yourself do it every day. It helps to have a creative bent, too. It also helps to enjoy explaining things. Science writing by nature is explanatory, more so than other kinds of journalism.

 You have a science background. How does that help you do your job?

I majored in chemistry as an undergraduate and got a master's degree in materials science. I find that I draw on that academic background a lot, in terms of understanding the research.

314

Do you think a science reporter needs a science background?

Ideally, you should be studying science while writing on the side. But if you have to do one or the other, I'd do science first. It's harder to pick up the science later. Science builds on itself. It takes years to really get a grasp of it.

Why do you think science reporting is important?

Science and technology are becoming part of our everyday lives. It's important for people to keep up on research in these areas. There is an element of education in everything you write.

What advice do you have for students who are interested in science reporting?

Read as much as you can—newspapers, magazines, books. Nothing beats getting real experience writing. If you have a newspaper or magazine at school, get involved in that. You draw on academic experiences—you don't know when they will become useful.

🔲 **internet** connect

www.scilinks.org
Topic: Science Writer
SciLinks code: HK4125

SCI LINKS® Maintained by the National Science Teachers Association

"Science is a strong tool, a strong way of looking at the world. I feel that trying to introduce people to that way of looking at the world is very important."
—**Corinna Wu**

315

SCIENCE REPORTER

Historical Perspective Science Service, the organization which publishes *Science News* magazine, has been in existence since 1921. It was originally founded to provide reliable science news for newspapers in the United States. The first editors of the service were concerned that most other science reporting was overly sensationalized.

The weekly *Science News* magazine, originally entitled *Science News Letter,* was first published on March 13, 1922. The publication was an outgrowth of Science Service's wire reports for newspapers. It was intended to explain scientific news to nonscientists. While the wire service was discontinued when many newspapers began to assign reporters to cover science in the 1960s, the magazine has been published ever since.

In 1941, Science Service began the Science Talent Search, an annual competition that showcases original research work by high school students.

Motion
Chapter Planning Guide

PACING	CLASSROOM RESOURCES	LABS, ACTIVITIES, AND DEMONSTRATIONS
BLOCK 1 · 45 min pp. 316–317 **Chapter Opener**		**SE** Activity 1, p. 317 **SE** Activity 2, p. 317
BLOCKS 2 & 3 · 90 min pp. 318–324 **Section 1** Measuring Motion	**CRF** Lesson Plan* **TT** Bellringer* **TT** Distance vs. Displacement* **TM** Distance-Time Graph* **TM** Resultant Velocity*	**SE** Math Skills, p. 323
BLOCKS 4 & 5· 90 min pp. 325–330 **Section 2** Acceleration	**CRF** Lesson Plan* **TT** Bellringer* **TT** Changing Speed to Accelerate* **TM** Acceleration Graphs*	**TE** Opening Demonstration, p. 325 **GENERAL** **TE** Group Activity Acceleration, p. 326 **ADVANCED** **CRF** Observation Lab Testing Reaction Time* ◆ **BASIC**
BLOCKS 6 & 7 · 90 min pp. 331–336 **Section 3** Motion and Force	**CRF** Lesson Plan* **TT** Bellringer* **TT** Frictional Forces and Acceleration* **TT** Concept Mapping*	**TE** Demonstration Static vs. Kinetic Friction, p. 333 **GENERAL** **TE** Group Activity Researching Tires, p. 334 **ADVANCED** **TE** Group Activity Ice Racing, p. 335 **ADVANCED** **SE** Skills Practice Lab Static, Sliding, and Rolling Friction, pp. 342–343 ◆ **GENERAL** **CRF** Datasheets for SE Skills Practice Lab Static, Sliding, and Rolling Friction **GENERAL** **CRF** CBL™ Probeware Lab Determining the Coefficients of Static and Kinetic Friction* ◆ **ADVANCED**

BLOCKS 8 & 9 · 90 min

Chapter Review and Assessment Resources

- **SE** Chapter Review, pp. 331–336
- **CRF** Chapter Tests* **GENERAL**
- **OSP** Test Generator
- **CRF** Standardized Test Practice with Guided Reading Development*
- **CRF** Test Item Listing for ExamView® Test Generator*
- **OSP** Scoring Rubrics and Classroom Management Checklists

Online Resources

Visit the HRW Web site for a variety of free resources related to the text. Just type in the keyword **HK4 MOT**.

Holt Science Spectrum: Physical Science: Online Edition

Students can access interactive problem solving help and active visual concept development with the *Holt Science Spectrum: Physical Science* Online Edition available at **www.hrw.com**.

student CNN News

cnnstudentnews.com

Find the latest news, lesson plans, and activities related to important scientific events.

KEY

TE Teacher Edition	**CRF** Chapter Resource File	***** Also on One-Stop Planner
SE Student Edition	**TT** Teaching Transparency	**◆** Requires Advance Prep
OSP One-Stop Planner	**TM** Transparency Master	

PROBLEM SOLVING AND PRACTICE	SECTION REVIEW AND ASSESSMENT	STANDARDS CORRELATION
	CRF Pretest* GENERAL	
CRF **Cross-Disciplinary Worksheet** Real World Applications—Hiking in Yellowstone* GENERAL **CRF** **Cross-Disciplinary Worksheet** Connection to Fine Arts—The Motions of Dance* GENERAL **CRF** **Science Skills** Slope of a Line* GENERAL **CRF** **Cross-Disciplinary Worksheet** Integrating Health—Energy Costs of Walking and Running* GENERAL **CRF** **Cross-Disciplinary Worksheet** Integrating Biology—Speedy Dinosaurs* GENERAL **SE** **Math Skills** Velocity, p. 323 **CRF** **Science Skills** Integrating Mathematics—Instantaneous Rates of Change* GENERAL **CRF** **Math Skills** Velocity* GENERAL **CRF** **Cross-Disciplinary Worksheet** Connection to Social Studies—Expanding City* GENERAL	**TE** Quiz, p. 324 BASIC **SE** Section 1 Review, p. 324 **CRF** Concept Review* GENERAL **CRF** Quiz* BASIC	SAI 1
SE **Math Skills** Acceleration, p. 328 **CRF** **Cross-Disciplinary Worksheet** Integrating Math—Jesse Owens in the 100-Meter Dash* GENERAL **CRF** **Math Skills** Acceleration* BASIC	**TE** Quiz, p. 330 BASIC **SE** Section 2 Review, p. 330 **CRF** Concept Review* GENERAL **CRF** Quiz* BASIC	PS 4b UCP 2
CRF **Cross-Disciplinary Worksheet** Real World Application—Designing Race Cars* GENERAL **CRF** **Cross-Disciplinary Worksheet** Connection to Language Arts—Friction in Fiction* GENERAL **SE** **Graphing Skills,** p. 337	**TE** Quiz, p. 336 BASIC **SE** Section 3 Review, p. 336 **CRF** Concept Review* GENERAL **CRF** Quiz* BASIC	PS 4a UCP 2

SCiLINKS

www.scilinks.org

Topic: Motion
*Sci*Links code: HK4091

Topic: Measuring Motion
*Sci*Links code: HK4084

Topic: Acceleration
*Sci*Links code: HK4001

Topic: Force and Friction
*Sci*Links code: HK4122

Topic: Graphing Speed, Velocity, Acceleration
*Sci*Links code: HK4066

Technology Resources

 One-Stop Planner
All of your printable resources and the Test Generator are on this convenient CD-ROM.

 Physical Science **Interactive Tutor CD-Rom**

Disc Two, Module 9
Speed and Acceleration

 CNN Science in the NEWS

each video segment is accompanied by a Critical Thinking Worksheet*

Segment 2
Land Speed Record

Overview

This chapter introduces the concept of motion, and distinguishes between speed and velocity. Students also learn how to calculate speed. The chapter also covers acceleration, including how to calculate acceleration and how to determine acceleration from a velocity-time graph. The concept of force is also introduced, including balanced and unbalanced forces and the force of friction.

Assessing Prior Knowledge

Be sure students understand the following concept:

- units of measurement

CHAPTER 10

Motion

Chapter Preview

1 Measuring Motion
Observing Motion
Speed and Velocity

2 Acceleration
Acceleration and Motion
Calculating Acceleration

3 Motion and Force
Balanced and Unbalanced Forces
The Force of Friction
Friction and Motion

316

Standards Correlations

National Science Education Standards

The following descriptions summarize the National Science Standards that specifically relate to this chapter. For the full text of the standards, see the National Science Education Standards at the front of the book.

Section 1 Measuring Motion
Science as Inquiry SAI 1

Section 2 Acceleration
Physical Science PS 4b
Unifying Concepts and Processes UCP 2

Section 3 Motion and Force
Physical Science PS 4a
Unifying Concepts and Processes UCP 2

Focus ACTIVITY

Background A skier such as the one shown here is obviously in motion, very fast motion. In fact, skiers have set world records by reaching speeds of over 240 km/h. So, it is easy to understand that a skier coming down a mountain is very much in motion. But what is motion? What is the correct way to describe motion?

To describe motion in the language of science, you need to answer questions such as the following. Does an object make patterns with its motion? How fast does the object move? In what direction does the object move? Does its speed change? Does the object change its direction? Does it repeat its motion? Do you need to compare the motion to another object in motion or to an object standing still? Is motion a relative term?

Activity 1 Choose a windup or battery-operated toy. You will also need a meterstick, a stopwatch, paper, and a pencil. Measure and record time, distance, direction, and any pattern of the toy's motion. Record your findings.

Activity 2 Use the same windup or battery-operated toy you used for Activity 1. After putting the toy into motion, set it on a toy truck or train. Then, put that toy truck or train in motion. Measure and record the distance traveled by the toy on the truck or train and the corresponding time interval. Repeat this activity several times and record all of your findings.

How is an athlete's speed calculated? When instruments report the speed, they do so by measuring both distance and time in small increments and then dividing the distance by the time.

internet connect

www.scilinks.org
Topic: Motion SciLinks code: HK4091

SCI LINKS. Maintained by the National Science Teachers Association

Pre-Reading Questions
1. What are some of the ways that objects move?
2. How can you tell when something is moving?

317

Focus ACTIVITY

Background Downhill skiers can reach incredibly fast speeds. Olympic skiers average over 60 miles per hour (100 km/h). The 1999 speed skiing world record, set by Harry Egger, was 153.760 miles per hour (248.000 km/h).

Activity 1 Students should be able to answer most of the questions just by observing the toy's motion. They should notice any patterns and repetitions of motion, the initial direction and changes in direction, and changes in speed. They will need to use tools, such as the meter stick and stopwatch, to find out how fast the toy moves.

Activity 2 This activity introduces the concept of a frame of reference. The first object's speed could be determined relative to the second object (the toy truck or train) or relative to the ground.

Answers to Pre-Reading Questions

These should be open-ended questions for students at this point. Any reasonable answer should be accepted.

1. Answers could include the following: Objects can move forward or backward in a straight or curved line; they can stay in the same place while spinning around; they can have moving parts; and they can speed up or slow down.

2. You can measure how far the object travels in a certain amount of time.

Chapter Resource File

- Pretest GENERAL
- Teaching Transparency Preview
- Answer Keys

Measuring Motion

◼ **KEY TERMS**

motion
displacement
speed
velocity

◼ **OBJECTIVES**

▶ **Explain** the relationship between motion and a frame of reference.

▶ **Relate** speed to distance and time.

▶ **Distinguish** between speed and velocity.

▶ **Solve** problems related to time, distance, displacement, speed, and velocity.

◼ **motion** an object's change in position relative to a reference point

We are surrounded by moving things. From a car moving in a straight line to a satellite traveling in a circle around Earth, objects move in many ways. In everyday life, **motion** is so common that it seems very simple. But understanding and describing motion scientifically requires some advanced concepts. To begin, how do we know when an object is moving?

Observing Motion

You may think that the motion of an object is easy to detect—just observe the object. But you actually must observe the object in relation to another object that stays in place, called a *stationary* object. The stationary object is a *reference point*, sometimes called a *reference frame*. Earth is a common reference point. In *Figure 1*, a mountain is used as a reference point.

When an object changes position in comparison to a reference point, the object is in motion. You can describe the direction of an object in motion with a reference direction. Typical reference directions are north, south, east, west, up, or down.

Figure 1

During the time required to take these two photographs, the hot-air balloon changed position compared with a stationary reference point—the mountain. Therefore, the balloon was in motion.

EARTH SCIENCE
CONNECTION

Earth—our most common frame of reference for describing motion—rotates around its axis. Because of this rotation, something traveling in a straight line from a pole to the equator—such as a jetliner, wind currents, or ocean currents—appears to curve. This is called the *Coriolis Effect*.

For someone in the northern hemisphere looking toward the equator, Earth appears to rotate counterclockwise. As a result of this rotation, ocean and wind currents traveling straight appear to curve to the right. Someone in the southern hemisphere observes the opposite direction of rotation (clockwise), and wind and ocean currents appear to curve to the left.

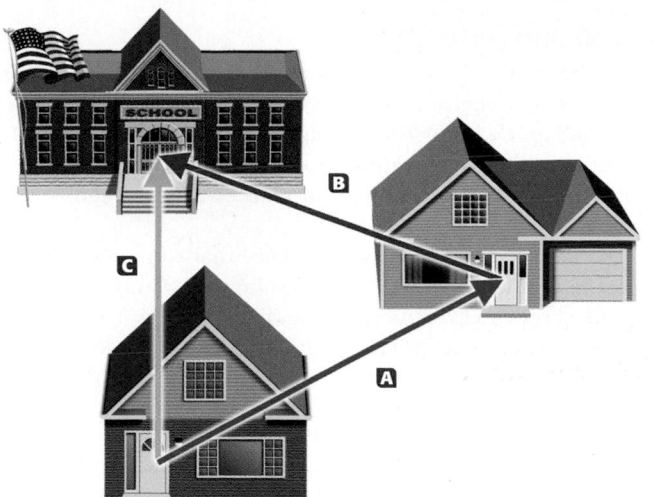

Figure 2
A student walks from his house to his friend's house (A), and then from his friend's house to the school (B). Line (A) plus line (B) equals the total distance he traveled. Line (C) is the displacement he traveled.

Distance measures the path taken

In addition to direction, you also need to know how far an object moves if you want to accurately describe its motion. To measure distance, you measure the actual path you took. If you started at your home and wandered around your neighborhood for a while by changing directions a few times, a string that followed your path would be as long as the distance you traveled.

Displacement is the change of an object's position

If you stretched a string in a straight line from your home directly to your final destination, the length of that string would be your **displacement.** This concept is illustrated in **Figure 2** above. In that illustration, the total of line (A) plus line (B) represents the actual distance traveled. Line (C) represents displacement, which is the change in position.

There are two differences between distance and displacement: straightness and direction. Distance can be a straight line, but it doesn't have to be. Displacement must be a straight line. So, displacement is shorter than the actual distance traveled unless the actual distance traveled is a straight line from the initial position to the final position.

Also, displacement must be in a particular direction. The distance between your home and school may be twelve blocks, but that information doesn't indicate whether you are going toward or away from school. Displacement must always indicate the direction, such as twelve blocks *toward school*.

displacement the change in position of an object

internet connect
www.scilinks.org
Topic: **Measuring Motion**
SciLinks code: **HK4084**
SC/*LINKS* Maintained by the National Science Teachers Association

Describing Motion During a student's first exposure to ideas such as speed, velocity, and acceleration, research shows that very few students grasp the subtleties of describing motion. The definitions of average speed, average velocity, and average acceleration are deceivingly simple. But the idea of speed at a particular moment (instantaneous speed) involves sophisticated mathematical ideas.

To help students grasp the concept of speed, first introduce constant motion. The definition of constant motion—an object covering equal distances in equal amounts of time—naturally motivates the concept of speed.

Describing non-constant motion—an object that does not cover equal distances in equal intervals of time—provides the motivation to develop the concept of average speed.

SKILL BUILDER

Interpreting Diagrams Explain to students that the scale in **Figure 3** is adjusted so that objects with vastly different speeds can be shown on a page spread. If the figure were drawn in a linear scale using the beginning speed, the position for the jet's speed (257 m/s) would be 5.65 m, or about 18 ft, to the right of the starting point. Therefore, a logarithmic scale is used to arrange the different objects of varying speeds shown in **Figure 3.**

Speed and Velocity

As has been stated, an object is moving if its position changes against some background that stays the same. In *Figure 3,* a horse is seen galloping against the background of stationary trees. The change in position as compared to a reference frame or reference point is measured in terms of an object's displacement from a fixed point.

You know from everyday experience that some objects move faster than others. **Speed** describes how fast an object moves. *Figure 3* shows speeds for some familiar things. A speeding race car moves faster than a galloping horse. But how do we determine speed?

Speed measurements involve distance and time

To find speed, you must measure two quantities: the distance traveled by an object and the time it took to travel that distance. Notice that all the speeds shown in *Figure 3* are expressed as a distance unit divided by a time unit. The SI unit for speed is meters per second (m/s). Speed is sometimes expressed in other units, such as kilometers per hour (km/h) or miles per hour (mi/h). The captions for *Figure 3* express speed in all three of these units of measurement.

When an object covers equal distances in equal amounts of time, it is moving at a *constant speed*. For example, if a race car has a constant speed of 96 m/s, the race car travels a distance of 96 meters every second, as shown in *Table 1.* So, the term *constant speed* means that the speed does not change. As you probably know, most objects do not move with constant speed.

> ▶ **speed** the distance traveled divided by the time interval during which the motion occurred

Table 1
Distance-Time Values for a Race Car

Time (s)	Distance (m)
0	0
1	96
2	192
3	288
4	384

Figure 3
We encounter a wide range of speeds in our everyday life.

1.4 m/s 5.0 km/h 3.1 mi/h	7.3 m/s 26 km/h 16 mi/h	19 m/s 68 km/h 42 mi/h
Walking person	**Wheelchair racer**	**Galloping horse**

HEALTH CONNECTION — GENERAL

In many types of aerobic exercise, the number of calories burned depends on the speed at which the exercise is performed. For example, a 150-pound person burns 7.35 calories per minutes bicycling at 15 mi/h, or 20.85 calories per minute bicycling at 25 mi/h. This is because when you move faster, you cover more distance in a given time interval. Have students use Internet or library resources to find out how many calories someone of their weight burns for different types of aerobic exercise at various speeds. Alternately, give students the following data for their calculations: walking at 3.5 mph burns 0.035 cal/lb/min; walking at 4.5 mph burns 0.048 cal/lb/min; jogging at 5 mph burns 0.061 cal/lb/min; running at 10 mph burns 0.114 cal/lb/min. **LS Verbal**

Speed can be studied with graphs and equations

You can investigate the relationship between distance and time in many ways. You can plot a graph with distance on the vertical axis and time on the horizontal axis, you can use mathematical equations and calculations, or you can combine these two approaches. Whatever method you use, your measurements are always either distances or displacements and time intervals during which the distances or displacements occur.

Speed can be determined from a distance-time graph

In a distance-time graph the distance covered by an object is noted at regular intervals of time, as shown on the line graph in **Figure 4.** Line graphs are usually made with the *x*-axis (horizontal axis) representing the independent variable and the *y*-axis (vertical axis) representing the dependent variable.

On our graph, time is the independent variable because time will pass whether distance is traveled or not. Distance is the dependent variable because the distance traveled depends upon the amount of time the object is moving. So, time is plotted on the *x*-axis and distance is plotted on the *y*-axis.

For a race car moving at a constant speed, the distance-time graph is a straight line. The speed of the race car can be found by calculating the slope of the line. The slope of any distance-time graph gives the speed of the object.

Suppose all objects in **Figure 3** are moving at a constant speed. The distance-time graph of each object is drawn in **Figure 4.** Notice that the distance-time graph for a faster moving object is steeper than the graph for a slower moving object. An object at rest, such as a parked car, has a speed of 0 m/s. Its position does not change as time goes by. So, the distance-time graph of a resting object is a flat line with a slope of zero.

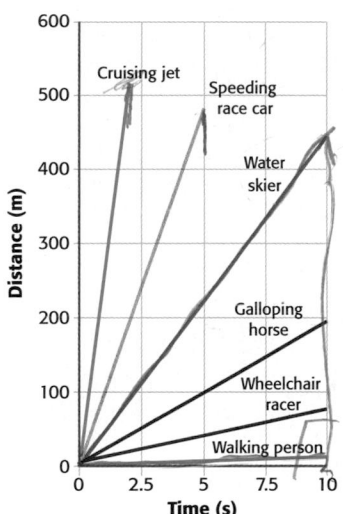

Figure 4

When an object's motion is graphed by plotting distance on the *y*-axis and time on the *x*-axis, the slope of the graph is speed.

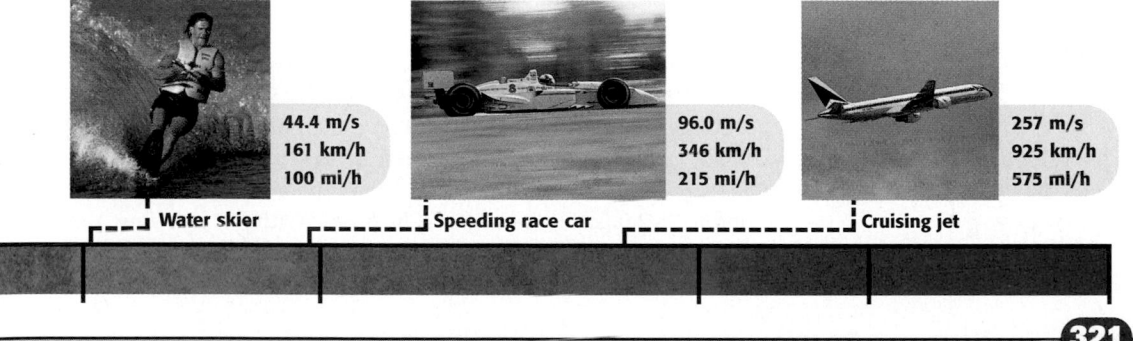

Water skier	Speeding race car	Cruising jet
44.4 m/s 161 km/h 100 mi/h	96.0 m/s 346 km/h 215 mi/h	257 m/s 925 km/h 575 mi/h

321

PHYSICS CONNECTION

Einstein discovered that there is a universal speed limit: no physical object in the universe can travel at or faster than the speed of light, 3.0×10^8 m/s. If an object with mass begins traveling close to that speed, its mass starts increasing and would become infinite when it reached the speed of light! Particles of light, called photons, are able to travel at this speed because they have no mass.

FINE ARTS CONNECTION

The technology for creating motion films was developed when a way was invented to photograph a live action many times per second. The photo frames are played very rapidly. In regular films, 24 frames per second are projected to produce the perception of continuous motion.

Teaching Tip

How are Speed and Velocity Different? Help students distinguish between speed and velocity by describing a car moving at constant speed as it rounds a curve. The speed remains constant, but its velocity has changed because the direction in which the car is moving has changed.

You can also use the example of a car race to help students understand the difference. Point out that cars have *speedometers*, not "velocitometers." A car in a race could spend hours going around a track very fast (high speed) but getting nowhere (zero velocity). In order to have a nonzero velocity, the car must finish at some point other than the starting point.

Figure 5

A wheelchair racer's speed can be determined by timing the racer on a set course.

▶ **velocity** the speed of an object in a particular direction

Average speed is calculated as distance divided by time

Most objects do not move at a constant speed. The speed of an object can change from one instant to another. One way to describe the motion of an object moving at changing speeds is to use *average speed*. Average speed is simply the distance traveled by an object divided by the time the object takes to travel that distance. Average speed can also be expressed as a simple mathematical formula.

> **Equation for Average Speed**
>
> $$speed = \frac{distance}{time} \qquad v = \frac{d}{t}$$

Suppose a wheelchair racer, such as the one shown in **Figure 5,** finishes a 132 m race in 18 s. By inserting the time and distance measurements into the formula, you can calculate the racer's average speed.

$$v = \frac{d}{t} = \frac{132 \text{ m}}{18 \text{ s}} = 7.3 \text{ m/s}$$

The racer's average speed over the entire distance is 7.3 m/s. But the racer probably did not travel at this speed for the whole race. For instance, the racer's pace may have been faster near the start of the race and slower near the end as the racer became tired.

Instantaneous speed is the speed at a given point in time

You could find the racer's speed at any given point in time by measuring the distance traveled in a shorter time interval. The smaller the time interval, the more accurate the measurement of speed would be. Speed measured in an infinitely small time interval is called *instantaneous speed*. Although it is impossible to measure an infinitely small time interval, some devices measure speed over very small time intervals. For practical purposes, a car's speedometer gives the instantaneous speed of the car.

Velocity describes both speed and direction

Sometimes, describing the speed of an object is not enough. You may also need to know the direction in which the object is moving. In 1997, a 200 kg (450 lb) lion escaped from a zoo in Florida. The lion was located by searchers in a helicopter. The helicopter crew was able to guide searchers on the ground by reporting the lion's **velocity,** which is its speed *and* direction of motion. The escaped lion's velocity may have been reported as 4.5 m/s *to the north* or 2.0 km/h *toward the highway*. Without knowing the direction of the lion's motion, it would have been impossible to predict the lion's position.

322

Did You Know?

Average Speed Calculating speed as distance divided by time always yields average speed. Because the time by which you divide is some finite amount (even if very small), the speed is always the average speed for that time interval.

Alternative Assessment ── GENERAL

Tracking Speed Ask students to keep track of their speed the next time they are in a car. They should use a watch or stopwatch to measure time intervals, and note the speedometer reading every 5 seconds for a period of about a minute. (They may wish to create a data table before beginning this experiment.) Ask them to use their data to plot line graphs of speed vs. time. Have them note where each graph represents the car's speed increasing (positive slope), decreasing (negative slope), and remaining constant (zero slope). **LS** Logical

The direction of motion can be described in various ways, such as east, west, south, or north of a fixed point. Or, it can be an angle from a fixed line. Also, direction can be described as positive or negative along the line of motion. So, if a body is moving in one direction, it has positive velocity. If it is moving in the opposite direction, it has negative velocity. In this book, velocity is considered to be positive in the direction of motion.

Math Skills

Velocity Metal stakes are sometimes placed in glaciers to help measure a glacier's movement. For several days in 1936, Alaska's Black Rapids glacier surged as swiftly as 89 meters per day down the valley. Find the glacier's velocity in m/s. Remember to include direction.

1 List the given and the unknown values.

Given: time, t = 1 day

displacement, d = 89 m down the valley

Unknown: velocity, v = ? (m/s and direction)

2 Perform any necessary conversions.

To find the velocity in meters per second, the value for time must be in seconds.

$$t = 1 \text{ day} = 24 \text{ h} \times \frac{60 \text{ min}}{1 \text{ h}} \times \frac{60 \text{ s}}{1 \text{ min}}$$

$$t = 86\,400 \text{ s} = 8.64 \times 10^4 \text{ s}$$

3 Write the equation for speed.

$$\text{speed} = \frac{displacement}{time} = \frac{d}{t}$$

4 Insert the known values into the equation, and solve.

$$v = \frac{d}{t} = \frac{89 \text{ m}}{8.64 \times 10^4 \text{ s}} \quad \text{(For velocity, include direction.)}$$

$$v = 1.0 \times 10^{-3} \text{ m/s down the valley}$$

Practice

Velocity

1. Find the velocity in m/s of a swimmer who swims 110 m toward the shore in 72 s.

2. Find the velocity in m/s of a baseball thrown 38 m from third base to first base in 1.7 s.

3. Calculate the displacement in meters a cyclist would travel in 5.00 h at an average velocity of 12.0 km/h to the southwest.

Practice HINT

▶ When a problem requires you to calculate velocity, you can use the speed equation. Remember to specify direction.

▶ The speed equation can also be rearranged to isolate distance or displacement on the left side of the equation in the following way.

$$v = \frac{d}{t}$$

Multiply both sides by t

$$v \times t = \frac{d}{t} \times t$$

$$vt = d$$

$$d = vt$$

You will need to use this form of the equation in Practice Problem 3. Remember to specify direction when you are asked for a displacement.

323

Historical Perspective

Describing Motions with Mathematics

Galileo Galilei (1564–1642) demonstrated many aspects of motion that we take for granted today. For example, his study of the vertical and horizontal components of projectile motion contributed to our modern understanding of velocity and vector addition.

Almost two thousand years before Galileo, philosophers such as Aristotle and Zeno described motion philosophically. Galileo believed that the natural world—including motions—could be explored and described with precise experiments and mathematics, and he began the attempt to do so in his work *Two New Sciences*. Galileo conducted a number of experiments with inclined planes, projectiles, and falling bodies. He used mathematics in his demonstrations and explanations, and his results—which contradicted some of Aristotle's long-accepted views—are the foundation of modern dynamics.

Math Skills

Additional Examples

Velocity Suppose the lion in the previous discussion moves due east at different speeds so that it travels 25 km in 4.0 hours. What is the lion's average speed? What is its average velocity?

Answer: average speed: 6.2 km/h = 1.7 m/s; average velocity: 6.2 km/h east = 1.7 m/s east

What would the lion's average velocity be if it traveled 15 km due north in 2 hours and 15 minutes?

Answer: 6.7 km/h north = 1.9 m/s north

LS Logical

Practice

1. $v = d/t = 110 \text{ m}/72 \text{ s} = 1.5 \text{ m/s toward shore}$

2. $v = d/t = 38 \text{ m}/1.7 \text{ s} = 22 \text{ m/s toward first base}$

3. $d = vt = (12.0 \text{ km/h})(5.00 \text{ h}) = 60.0 \text{ km}$

$(60.0 \text{ km})(1000 \text{ m/km}) = 6.00 \times 10^4 \text{ m}$

LS Logical

Chapter Resource File

- **Science Skills** Integrating Mathematics—Instantaneous Rates of Change **BASIC**
- **Math Skills** Velocity **GENERAL**
- **Cross-Disciplinary Worksheet** Connection to Social Studies—An Expanding City **GENERAL**

Videos

 Presents Physical Science

- **Segment 2** Land Speed Record
- **Critical Thinking** Worksheet 2

Determine whether each statement is true or false. If false, replace the underlined term with the correct word or phrase.

1. Distance divided by time is a measure of average velocity. (false; speed)

2. <u>Displacement</u> must be straight and must include a direction. (true)

3. Displacement divided by time is a measure of <u>speed</u>. (false; velocity)

4. <u>Instantaneous speed</u> is a speed measured in an infinitely small time interval. (true)

5. The SI unit of speed is <u>miles per hour</u>. (false; meters per second)

Figure 6
Determining Resultant Velocity

Person's resultant velocity
| 15 m/s east + 1 m/s east = 16 m/s east |

A When you have two velocities that are in the same direction, add them together to find the resultant velocity, which is in the direction of the two velocities.

Person's resultant velocity
| 15 m/s east + (−1 m/s west) = 14 m/s east |

B When you have two velocities that are in opposite directions, add the positive velocity to the negative velocity to find the resultant velocity, which is in the direction of the larger velocity.

Combine velocities to determine resultant velocities

If you are riding in a bus traveling east at 15 m/s, you and all the other passengers are traveling at a velocity of 15 m/s east. But suppose you stand up and walk down the bus's aisle while it is moving. Are you still moving at the same velocity as the bus? No! *Figure 6* shows how you can combine velocities to determine the *resultant velocity*.

SECTION 1 REVIEW

SUMMARY

▶ When an object changes position in comparison to a stationary reference point, the object is in motion.

▶ The average speed of an object is defined as the distance the object travels divided by the time of travel.

▶ The distance-time graph of an object moving at constant speed is a straight line. The slope of the line is the object's speed.

▶ The velocity of an object consists of both its speed and its direction of motion.

1. Describe the measurements necessary to find the average speed of a high school track athlete.

2. Determine the unit of a caterpillar's speed if you measure the distance in centimeters (cm) and the time it takes to travel that distance in minutes (min).

3. Identify the following measurements as speed or velocity.
 a. 88 km/h   **c.** 18 m/s down
 b. 19 m/s to the west   **d.** 10 m/s

4. Critical Thinking Imagine that you could ride a baseball that is hit high enough and far enough for a home run. Using the baseball as a reference frame, what does the Earth appear to do?

Math Skills

5. How much time does it take for a student running at an average speed of 5.00 m/s to cover a distance of 2.00 km?

Chapter Resource File

- Concept Review **GENERAL**
- Quiz **BASIC**

Transparencies

TT Resultant Velocity

Answers to Section 1 Review

1. Measure the distance he/she travels, and the time required to travel that distance.

2. cm/min

3. a. speed
 b. velocity
 c. velocity
 d. speed

4. Earth appears to recede below you for a distance equal to the height of the baseball above Earth. Earth also appears to travel backwards for the distance the baseball travels and at the speed at which the baseball travels.

Math Skills

5. 2 km = 2000 m

$$t = d/v = \frac{2000 \text{ m}}{5 \text{ m/s}} = 400. \text{ s or } 6.67 \text{ min}$$

LS Logical

Acceleration

OBJECTIVES

▶ **Describe** the concept of acceleration as a change in velocity.

▶ **Explain** why circular motion is continuous acceleration even when the speed does not change.

▶ **Calculate** acceleration as the rate at which velocity changes.

▶ **Graph** acceleration on a velocity-time graph.

▶ KEY TERMS

acceleration

When you increase speed, your velocity changes. Your velocity also changes if you decrease speed or if your motion changes direction. For example, your velocity changes when you turn a corner. Any time you change velocity, you are accelerating. Any change in velocity is called **acceleration.**

■ **acceleration** the rate at which velocity changes over time; an object accelerates if its speed, direction, or both change

Acceleration and Motion

Imagine that you are a race car driver. You press on the accelerator. The car goes forward, moving faster and faster each second. Like velocity, acceleration has direction. When the car is speeding up, it is accelerating positively. Positive acceleration is in the same direction as the motion and increases velocity.

Acceleration can be a change in speed

Suppose you are facing south on your bike and you start moving and speed up as you go. Every second, your southward velocity increases, as shown in **Figure 7.** After 1 s, your velocity is 1 m/s south. After 2 s, your velocity is 2 m/s south. Your velocity after 5 s is 5 m/s south. Your acceleration can be expressed as an increase of one meter per second per second (1 m/s/s) or 1 m/s² south.

Figure 7

You are accelerating whenever your speed changes. This cyclist's speed increases by 1 m/s every second.

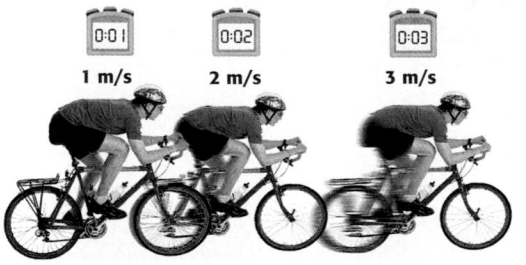
1 m/s    2 m/s    3 m/s    4 m/s   5 m/s

325

HISTORY
CONNECTION — ADVANCED

One of the scientists who first described motion quantitatively was Galileo Galilei. Galileo was born in Italy in 1564 and lived until 1642. Ask students to use the library or the Internet to find out what famous English writer was born the same year as Galileo. (William Shakespeare) Ask them to speculate and conduct research to find out about what similar historical and political forces in Europe at that time might have encouraged both the sciences and the humanities.
LS Verbal

Alternative Assessment — GENERAL
Observing Acceleration Ask students to look around in their daily lives for examples of acceleration. Have them make a list of at least 10 examples that they observe. Also have students find a few examples of objects in motion that have zero acceleration. Then ask students to classify each example of acceleration as an increase in speed, a decrease in speed, a change in direction, or a combination. You may also wish to have students list the examples in the approximate order of least to greatest acceleration. **LS** Logical

Focus

Overview

Before beginning this section, review with your students the Objectives listed in the Student Edition. In this section, students learn what acceleration is, how to calculate acceleration, and how to determine acceleration from a velocity-time graph.

Bellringer

Use the Bellringer transparency to prepare students for this section.

Motivate

Opening Demonstration — GENERAL

For this demonstration you will need a marble, an inclined plane, tape, a stopwatch, a meter stick, a protractor, and a metal or glass cup. Have a student volunteer release the marble from rest at the top of the plane. Ask students to observe the motion of the ball. Have them identify the locations of lowest speed (top) and highest speed (bottom). Repeat the demonstration at a different angle, and have students compare the two trials.

Add quantitative values by timing the runs. Place a tape marker near the top of the plane to serve as a start line. Use a metal or glass cup at the end of the plane as a sound cue for stopping the stopwatch. Measure the length of the plane. Have students record the time and angle measurements for several trials. Discuss the results.
LS Logical

Chapter Resource File

• Lesson Plan

Transparencies

TT Bellringer
TT Changing Speed to Accelerate

Reading Skills Have students read aloud or silently the first two pages of this section. Start a class discussion about scenarios that involve acceleration. Be sure to emphasize that any change in velocity (including change of direction) is acceleration. **LS** Verbal

Teaching Tip

Changes in Direction Students may wonder why a change in direction is considered to be an acceleration. At this point, emphasize that since velocity involves both magnitude and direction, a change in either is a change in velocity, or an acceleration.

You can return to this concept when students study Newton's laws. A force can produce a change in speed, a change in direction, or both. According to Newton's second law, force is proportional to acceleration. Any object experiencing a net force—such as a satellite in orbit around Earth acted upon by the force of gravity—must have an acceleration.

Group Activity — ADVANCED

Acceleration Group students in pairs and have each group apply the idea of acceleration to the design of roller coasters. Students should research a particular roller coaster and create a drawing or diagram of its features. Have them describe the motion of the roller coaster in terms of its velocity and acceleration. **LS** Interpersonal

Figure 8

These skaters accelerate when changing direction, even if their speed doesn't change.

internet connect

www.scilinks.org

Topic: Acceleration
SciLinks code: HK4001

SCLINKS Maintained by the National Science Teachers Association

Figure 9

The blades of these windmills are constantly changing direction as they travel in a circle. So, centripetal acceleration is occurring.

Acceleration can also be a change in direction

Besides being a change in speed, acceleration can also be a change in direction. The skaters in *Figure 8* are accelerating because they are changing direction. Why is changing direction considered to be an acceleration? Acceleration is defined as the rate at which velocity changes over time. Velocity includes both speed and direction, so an object accelerates if its speed, direction, or both change. This idea leads to the seemingly strange but correct conclusion that you can constantly accelerate while never speeding up or slowing down.

If you travel at a constant speed in a circle, even though your speed is never changing, your direction is always changing. So, you are always accelerating. The moon is constantly accelerating in its orbit around Earth. A motorcyclist who rides around the inside of a large barrel is constantly accelerating. When you ride a Ferris wheel at an amusement park, you are accelerating. All these examples have one thing in common—change in direction as the cause of acceleration.

Uniform circular motion is constant acceleration

Are you surprised to find out that as you stand on Earth you are accelerating? After all, you are not changing speed, and you are not changing direction—or are you? In fact, you are traveling in a circle as Earth revolves. An object traveling in a circular motion is always changing its direction. As a result, its velocity is always changing, even if its speed does not change. Thus, acceleration is occurring. The acceleration that occurs in uniform circular motion is known as centripetal acceleration. Another example of *centripetal acceleration* is shown in *Figure 9.*

Historical Perspective

Motion of the Planets The ancient Greek philosopher Aristotle believed that since the heavens are perfect, the motion of the planets must be eternal and unchanging. For this reason, he believed that the planets were in uniform circular motion around Earth, the center of the universe. However, this theory was unable to explain certain observations, such as the varying brightness of some of the planets.

The principle of uniform circular motion was so well accepted that the astronomer Ptolemy devised an elaborate theory to account for the variations in brightness and other observed discrepancies while maintaining uniform circular motion. Ptolemy posited that the planets actually move in smaller circles called *epicycles*, which themselves move in spherical orbits around Earth.

In our modern theory—based on the work of Copernicus, Kepler, and Newton—the planets do not have uniform circular motion. All of the planets, including Earth, travel around the Sun in elliptical paths with varying speeds.

Calculating Acceleration

To find the acceleration of an object moving in a straight line, you need to measure the object's velocity at different times. The average acceleration over a given time interval can be calculated by dividing the change in the object's velocity by the time in which the change occurs. The change in an object's velocity is symbolized by Δv.

Acceleration Equation (for straight-line motion)

$$acceleration = \frac{final\ velocity - initial\ velocity}{time} \qquad a = \frac{\Delta v}{t}$$

If the acceleration is small, the velocity is increasing very gradually. If the acceleration has a greater value, the velocity is increasing more rapidly. For example, a human can accelerate at about 2 m/s². On the other hand, a sports car that goes from 0 km/h to 96 km/h (60 mi/h) in 3.7 s has an acceleration of 7.2 m/s².

Because we use only positive velocity in this book, a positive acceleration always means the object's velocity is increasing—the object is speeding up. Negative acceleration means the object's velocity is decreasing—the object is slowing down.

Acceleration is the rate at which velocity changes

People often use the word *accelerate* to mean "speed up," but in science it describes any change in velocity. Imagine that you are skating down the sidewalk. You see a large rock in your path. You slow down and swerve to avoid the rock. A friend says, "That was great acceleration. I'm amazed that you could slow down and turn so quickly!" You accelerated because your velocity changed. The velocity decreased in speed, and you changed directions. So, your velocity changed in two different ways.

The student in *Figure 10* is accelerating to a stop. Suppose this student was originally going at 20 m/s and stopped in 0.50 s. The change in velocity is 0 m/s − 20 m/s = −20 m/s, which is negative because the student is slowing down. The student's acceleration is

$$\frac{0\ m/s - 20\ m/s}{0.50\ s} = -40\ m/s^2$$

INTEGRATING

MATHEMATICS
In the seventeenth century, both Sir Isaac Newton and Gottfried Leibniz studied acceleration and other rates of change. Independently, each created calculus, a branch of math that allows for describing rates of change of a quantity like velocity.

Figure 10
The rate of velocity change is acceleration, whether it is direction or speed that changes.

Teaching Tip ——— BASIC

Illustrating Acceleration You can use a slow-motion "strobe" example on the chalkboard to illustrate acceleration. Draw a horizontal line labeled "distance" across the bottom of the board with equally-marked intervals. First, in a horizontal row across the top of the board, illustrate zero acceleration. To do this, draw a car in several positions, with equal spaces (distances) between each position. Beneath this example, in a second row, illustrate an accelerating car. For this example, each successive distance interval should be longer than the previous interval. Tell students to imagine that each "snapshot" image was taken after equal time intervals. Ask students if either or both cars are accelerating, and have them explain their answer. (The car in the first row is not accelerating because it covers equal distances in equal times; this car has a uniform velocity. The car in the second row is accelerating because it covers successively more distance in equal time intervals.) **LS Visual**

SKILL BUILDER

Vocabulary Some students may associate the term *deceleration* with a negative acceleration. In this book, acceleration can be either positive or negative. Speeding up is a positive acceleration, while slowing down is a negative acceleration. The term deceleration is not used.

327

Did You Know?

How Fast? The record for the fastest bird is held by the peregrine falcon, with a diving speed of 217 mi/h. The fastest jet-powered aircraft to date, the Lockheed SR-71, has a record speed of 2,193.16 mi/h—more than three times the speed of sound!

REAL-LIFE ———
CONNECTION ——— GENERAL

Automobile Accelerations Have students conduct research to find out the time period (in seconds) in which different kinds of cars can accelerate from 0 to 60 mi/h. For example, the Ford Escape, an SUV with a 3.0-liter V-6 engine, can accelerate from 0 to 60 mi/h in 8.5 s. Students should include a variety of models in their research, such as full-size cars, compact cars, trucks, SUVs, gas-electric hybrids, and sports cars. Ask them to create a bar graph to compare their data.

Chapter Resource File

• **Cross-Disciplinary Worksheet**
Integrating Math—Jesse Owens in the 100-Meter Dash GENERAL

Math Skills

Additional Examples

Acceleration A car accelerates from 0 m/s to 45 m/s northward in 15 s. What is the acceleration of the car?
Answer: 3.0 m/s² northward

After reaching 45 m/s, the car slows down to 0 m/s in 10.0 s. What is the acceleration of the car?
Answer: 4.5 m/s² southward (or −4.5 m/s² northward)
LS Logical

Practice

1. $\dfrac{(4.0 \text{ m/s} - 0 \text{ m/s})}{2.5 \text{ s}} =$
   1.6 m/s² along her path

2. $\dfrac{(0.80 \text{ m/s} - 0.50 \text{ m/s})}{4.0 \text{s}} =$
   0.075 m/s² toward the shore

3. $\dfrac{(9.6 \text{ m/s} - 12 \text{ m/s})}{0.8 \text{ s}} =$
   −3 m/s² north = 3 m/s² south

4. $t = \Delta v/a =$
   $\dfrac{(26.8 \text{ m/s} - 24.6 \text{ m/s})}{2.6 \text{ m/s}^2} = 0.85$ s

5. $v_f = v_i + at = 4.5$ m/s $+$
   $(2.3 \text{ m/s}^2)(5.0 \text{ s}) = 16$ m/s
   LS Logical

PHYSICAL SCIENCE

Disc Two, Module 9:
Speed and Acceleration
Use the Interactive Tutor to learn more about these topics.

Practice HINT

▶ When a problem asks you to calculate acceleration, you can use the acceleration equation.
$$a = \frac{\Delta v}{t}$$
To solve for other variables, rearrange it as follows.

▶ To isolate *t*, first multiply both sides by *t*.
$$a \times t = \frac{\Delta v}{t} \times t$$
$$\Delta v = at$$
Next divide both sides by *a*.
$$\frac{\Delta v}{a} = \frac{at}{a}$$
$$t = \frac{\Delta v}{a}$$
You will need to use this form of the equation in Practice Problem 4.

▶ In Practice Problem 5, isolate final velocity.
$$v_f = v_i + at$$

When you press on the gas pedal in a car, you speed up. Your acceleration is in the direction of the motion and therefore is positive. When you press on the brake pedal, your acceleration is opposite the direction of motion. You slow down, and your acceleration is negative. When you turn the steering wheel, your velocity changes because you are changing direction.

Math Skills

Acceleration A flowerpot falls off a second-story windowsill. The flowerpot starts from rest and hits the sidewalk 1.5 s later with a velocity of 14.7 m/s. Find the average acceleration of the flowerpot.

1 List the given and unknown values.
   Given: *time*, $t = 1.5$ s
   initial velocity, $v_i = 0$ m/s
   final velocity, $v_f = 14.7$ m/s down
   Unknown: *acceleration*, $a = ?$ m/s² (and direction)

2 Write the equation for acceleration.
$$acceleration = \frac{final\ velocity - initial\ velocity}{time} = \frac{v_f - v_i}{t}$$

3 Insert the known values into the equation, and solve.
$$a = \frac{v_f - v_i}{t} = \frac{14.7 \text{ m/s} - 0 \text{ m/s}}{1.5 \text{ s}}$$
$$a = \frac{14.7 \text{ m/s}}{1.5 \text{ s}} = 9.8 \text{ m/s}^2 \text{ down}$$

Practice

Acceleration

1. Natalie accelerates her skateboard along a straight path from 0 m/s to 4.0 m/s in 2.5 s. Find her average acceleration.

2. A turtle swimming in a straight line toward shore has a speed of 0.50 m/s. After 4.0 s, its speed is 0.80 m/s. What is the turtle's average acceleration?

3. Find the average acceleration of a northbound subway train that slows down from 12 m/s to 9.6 m/s in 0.8 s.

4. Marisa's car accelerates at an average rate of 2.6 m/s². Calculate how long it takes her car to speed up from 24.6 m/s to 26.8 m/s.

5. A cyclist travels at a constant velocity of 4.5 m/s westward, and then speeds up with a steady acceleration of 2.3 m/s². Calculate the cyclist's speed after accelerating for 5.0 s.

328

Trends in Physics

Particle Accelerators Some of today's physicists use particle accelerators to learn about the fundamental particles that make up our universe. The most powerful accelerator in the world is the Tevatron at Fermilab in Batavia, Illinois. The Tevatron is a 4-mile-long underground ring in which protons and other particles are accelerated to incredible speeds, and then put into collisions against one another. Scientists study these collisions to learn about the fundamental particles of matter and about conditions in the very early universe.

Some of this theoretical research has practical applications as well. For example, Magnetic Resonance Imaging (MRI) is a technique used in medicine to image the inside of the human body. The powerful superconducting magnets used in MRI technology were first developed in the 1970s for Fermilab's Tevatron.

Acceleration can be determined from a velocity-time graph

You have learned that an object's speed can be determined from a distance-time graph of its motion. You can also make a velocity-time graph by plotting velocity on the vertical axis and time on the horizontal axis.

A straight line on a velocity-time graph means that the velocity changes by the same amount over each time interval. This is called *constant acceleration*. The slope of a line on a velocity-time graph gives you the value of the acceleration. A line with a positive slope represents an object that is speeding up. A line with a negative slope represents an object that is slowing down. A straight horizontal line represents an object that has an unchanging velocity and therefore has no acceleration.

The bicyclists in *Figure 11A* are riding in a straight line at a constant speed of 13.00 m/s, as shown by the data in *Table 2*. *Figure 11B* is a distance-time graph for the cyclists. Because the velocity is constant, the graph is a straight line. The slope of the line equals the cyclists' velocity. *Figure 11C* is a velocity-time graph for the same cyclists. The slope of this line represents the cyclists' acceleration. In this case, the slope is zero (a horizontal line) because the acceleration is zero.

Did You Know?

The faster a car goes, the longer it takes a given braking force to bring the car to a stop. *Braking distance* describes how far a car travels between the moment the brakes are applied and the moment the car stops. As a car's speed increases, so does its braking distance. For example, when a car's speed is doubled, its braking distance is four times as long.

SKILL BUILDER — GENERAL

Graphing Have students create a distance-time graph from the data in **Table 2**. To do this, have them first create a new data table that contains the information in **Table 2** in the first two columns. Have students add two more columns—the distance traveled in the time interval and the total distance traveled.

To calculate the distance traveled in a time interval, students should use $d = vt$. To calculate the total distance traveled, students should add the distance traveled in the time interval (column 3) to the previous total distance.

For an extension exercise, give students a data table with an increasing velocity (3 m/s, 6 m/s, 9 m/s, 12 m/s corresponding to the times 1 s, 2 s, 3 s, 4 s). Have them create both a velocity–time graph and a distance–time graph. Discuss the differences between the two graphs. **LS Logical**

Figure 11

A When you ride your bike straight ahead at constant speed, you are not accelerating, because neither your velocity nor your direction changes.

Table 2
Data for a Bicycle with Constant Speed

Time (s)	Speed (m/s)
0	13.00
1	13.00
2	13.00
3	13.00
4	13.00

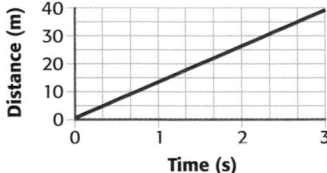

B If you plot the distance traveled against the time it takes, the resulting graph is a straight line with a slope of 13.00 m/s.

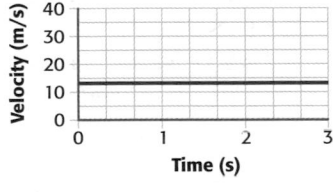

C Plotting the velocity against time results in a horizontal line because the velocity does not change. The acceleration is 0 m/s^2.

329

Trends in Technology

Space Acceleration Measurement System

NASA has developed an acceleration measurement and recording system called the Space Acceleration Measurement System, or SAMS. SAMS can measure, condition, and record low-gravity accelerations for microgravity experiments in space. It can be used with up to three separate experiments simultaneously. By June of 1998, SAMS had been used on over 20 Shuttle missions, as well as on the MIR space station.

Did You Know?

A Sprinter's Acceleration In a 200-m sprint, a top runner accelerates very rapidly at first, and continues accelerating for the first 50–70 m (50–60 m for females, and 60–70 m for males). The sprinter continues at his or her highest speed—about 10 m/s for a female and 11 m/s for a male—for the next 20–30 m. In the final 100 m, the sprinter has a negative acceleration while his or her speed decreases.

Chapter Resource File

• **Math Skills** Acceleration GENERAL

Transparencies

 TM Acceleration Graphs

Close

Quiz

1. What is an acceleration? (a change in velocity, which means change in speed or direction)

2. How is acceleration calculated? ($a = \Delta v/t$)

3. What are the SI units of acceleration? (m/s^2)

4. Does an object moving at a constant speed in uniform circular motion have acceleration? (yes)

5. Given a velocity-time graph, how can you find acceleration? (acceleration = the slope of the line)

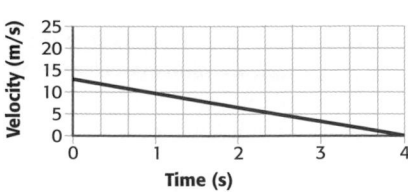

Figure 12

A When you slow down, your velocity changes. Your acceleration is negative because you are decreasing your velocity.

B If you plot the distance you travel against the time it takes you, the distance you travel each second becomes shorter and shorter until you finally stop.

C Plotting the velocity against time results in a line that has a negative slope, which means the acceleration is negative.

Table 3
Data for a Slowing Bicycle

Time (s)	Speed (m/s)
0	13.00
1	9.75
2	6.50
3	3.25
4	0

The rider in **Figure 12A** is slowing down from 13.00 m/s to 3.25 m/s over a period of 3.00 s, as shown by the data in **Table 3.** You can find out the rate at which velocity changes by calculating the acceleration.

$$a = \frac{3.25 \text{ m/s} - 13.00 \text{ m/s}}{3.00 \text{ s}} = -3.25 \text{ m/s}^2$$

The rider's velocity decreases by 3.25 m/s each second. The acceleration has a negative sign because the rider is slowing down. **Figure 12B** is a distance-time graph of the rider's motion, and **Figure 12C** is a velocity-time graph.

SECTION 2 REVIEW

SUMMARY

▶ Acceleration is a change in an object's velocity. Accelerating means speeding up, slowing down, or changing direction.

▶ For straight-line motion, average acceleration is defined as the change in an object's velocity per unit of time.

▶ Circular motion is acceleration because of the constant change of direction.

▶ A velocity-time graph can be used to determine acceleration.

1. Identify the straight-line accelerations below as either speeding up or slowing down.
   a. 5.7 m/s^2 **c.** -2.43 m/s^2
   b. -9.8 m/s^2 **d.** 9.8 m/s^2

2. Critical Thinking Joshua skates in a straight line at a constant speed for one minute, then begins going in circles at the same rate of speed, and then finally begins to increase speed. When is he accelerating? Explain your answer.

Math Skills

3. What is the final speed of a skater who accelerates at a rate of 2.0 m/s^2 from rest for 3.5 s?

4. Graph the velocity of a car accelerating at a uniform rate from 7.0 m/s to 12.0 m/s in 2.0 s. Calculate the acceleration.

330

Answers to Section 2 Review

1. a. speeding up
   b. slowing down
   c. slowing down
   d. speeding up

2. He is accelerating both when he is going in circles at the same rate of speed and when he is increasing speed. One involves a change in direction, while the other involves a change in speed. Both speed and direction are a part of velocity, and any change of velocity is acceleration.

Math Skills

3. $v_f = v_i + at = 0 + (2.0 \text{ m/s}^2)(3.5 \text{ s}) = 7.0$ m/s

4. 2.5 m/s^2 is the acceleration. The graph should be a straight line from 7.0 m/s at the zero line of time to 12 m/s at the 2.0 s line of time (with time on the x-axis and velocity on the y-axis).

Motion and Force

OBJECTIVES

▶ **Explain** the effects of unbalanced forces on the motion of objects.

▶ **Compare and contrast** static and kinetic friction.

▶ **Describe** how friction may be either harmful or helpful.

▶ **Identify** ways in which friction can be reduced or increased.

▶ **KEY TERMS**
force
friction
static friction
kinetic friction

Focus

Overview

Before beginning this section, review with your students the Objectives listed in the Student Edition. This section begins by defining force and distinguishing between balanced and unbalanced forces. Next, students study friction, including static friction, kinetic friction, fluid friction, ways to decrease or increase friction, and the relationship between friction and acceleration.

You often hear the word **force** in everyday conversation: "That storm had a lot of force!" "Our basketball team is a force to be reckoned with." But what exactly is a force? In science, force is defined as anything that changes the state of rest or motion of an object. This section will explore how forces change motions.

▶ **force** an action exerted on a body in order to change the body's state of rest or motion; force has magnitude and direction

🎧 Bellringer

Use the Bellringer transparency to prepare students for this section.

Balanced and Unbalanced Forces

When you throw or catch a ball, you exert a force to change the ball's velocity. What causes an object to change its velocity, or accelerate? Usually, many forces are acting on an object at any given time. The *net force* is the combination of all of the forces acting on the object. Whenever there is a net force acting on an object, the object accelerates in the direction of the net force. An object will not accelerate if the net force acting on it is zero.

Motivate

Identifying Preconceptions —— GENERAL

Write on the chalkboard the words *movement* and *force*. Ask students to come up with words or phrases that relate to the concepts of movement and force. This exercise should help identify your students' preconceptions about motion and forces. Encourage students to write their ideas down.

Balanced forces do not change motion

When the forces applied to an object produce a net force of zero, the forces are balanced. *Balanced forces* do not cause an object at rest to start moving. Furthermore, balanced forces do not cause a change in the motion of a moving object.

Many objects have only balanced forces acting on them. For example, a light hanging from the ceiling does not move up or down, because an elastic force due to tension pulls the light up and balances the force of gravity pulling the light down. A hat resting on your head is also an example of balanced forces. In **Figure 13,** the opposing forces on the piano are balanced. Therefore, the piano remains at rest.

Figure 13

The forces applied by these two students balance each other, so the piano does not move.

Based on their ideas, ask them to predict: if a force (acting on an object) were removed, would the object *necessarily* come to a stop? Ask them the same question when the section is finished. Based on everyday experiences that are dominated by frictional forces, many of us develop inaccurate preconceptions about motion and forces. **LS** Verbal

331

MISCONCEPTION ///ALERT\\\

Forces and Motion Many students will conclude that the speed of a body is directly related to the force currently applied, and that constant motion requires a constant force. They may also believe that if a body is at rest, then there is no force acting on it.

List each of these misconceptions on the chalkboard. Then, when you have finished this section, use examples from the text to address how each misconception is incorrect.

CHEMISTRY
CONNECTION —— ADVANCED

Molecules are held together by balanced forces; otherwise, the atoms that make them up would accelerate apart. Have students find out what kind of forces hold molecules together. **LS** Verbal

Chapter Resource File

• Lesson Plan

Transparencies

TT Bellringer

Teaching Tip

Balanced Forces Be sure to emphasize that balanced forces do not *change* the motion of an object. That does not imply that the object is not moving. For example, a car moving at a constant speed in a straight line has balanced forces acting on it. The force exerted by the engine is balanced by the frictional forces within the car's components, by the road, and by the air. Because the car is not accelerating, we can tell that the forces are balanced. If the car speeds up, slows down, or turns, then the forces are no longer balanced.

Teaching Tip ——— BASIC

Forces Because the direction of a force is important, the description of a force includes both a size and direction, just like velocity and acceleration. To emphasize the importance of direction, pose the following question to students: What is the net (or total) force acting on a piano if two students are each putting a force of 100 N on it?

After a brief discussion, draw two scenarios on the board. The first scenario should show two forces (drawn as arrows) acting on an object in the same direction. The net force in this case would be 200 N in the direction of the forces. The second scenario should show the forces acting in opposite directions. The net force in this case would be zero. **LS** Logical

Figure 14
When two opposite forces acting on the same object are unequal, the forces are unbalanced. A change in motion occurs in the direction of the greater force.

Unbalanced forces do not cancel completely

In *Figure 14,* another student pushes on one side of the piano. In this case, there are two students pushing against the piano on one side and only one student pushing against the piano on the other side. If the students all have the same mass and are all pushing with the same force, there is an *unbalanced force:* two students pushing against one student. Because the net force on the piano is greater than zero, the piano will begin to accelerate in the direction of the greater force.

What happens if forces act in different directions that are not opposite each other? In this situation, the combination of forces acts like a single force on the object, which causes acceleration in a direction that combines the directions of the applied forces. If you push eastward on a box, and your friend pushes northward, the box will accelerate in a northeasterly direction.

The Force of Friction

Imagine a car that is rolling along a flat, evenly paved street. Experience tells you that the car will keep slowing down until it eventually stops. This steady change in the car's speed gives you a clue that a force must be acting on the car. The unbalanced force that acts against the car's direction of motion is **friction.**

Friction occurs because the surface of any object is rough. Even surfaces that look or feel very smooth are actually covered with microscopic hills and valleys. When two surfaces are in contact, the hills and valleys of one surface stick to the hills and valleys of the other surface.

VOCABULARY *Skills Tip*

The word force *comes from the Latin word* fortis, *which means "strength." The word* fortress *comes from the same root.*

■ **friction** a force that opposes motion between two surfaces that are in contact

Historical Perspective
Aristotle and Galileo: Theories of Force
Aristotle believed that a moving object must have a force acting on it or it would stop moving. His belief was suggested by everyday experiences. If you slide an object and then remove your hand, it will stop sliding. What Aristotle did not realize was that when you remove your hand, you remove a force that is balancing another opposing force—friction. So the object stops moving because friction is still acting on it.

The force of friction was first demonstrated by Galileo. Galileo used two inclined planes facing each other to support his conviction that in the absence of friction, no force is needed to keep an object in motion. He found that with smooth surfaces, a ball rolled down one plane rolled almost to the same height on the other plane, whereas it did not roll as high when the inclined planes had two rough surfaces. (You may want to demonstrate this using a ball on a bent piece of carpet and then on a bent piece of poster board.)

Friction opposes the applied force

Because of friction, a constant force must be applied to a car just to keep it moving. The force pushing the car forward must be greater than the force of friction opposing the car's motion, as shown in *Figure 15A*. Once the car reaches its desired speed, the car will maintain this speed if the forces acting on the car are balanced, as shown in *Figure 15B*.

Friction also affects objects that aren't moving. For example, when a truck is parked on a hill with its brakes set, as shown in *Figure 15C*, friction provides the force of gravity along the hill and prevents the truck from rolling away.

Static friction is greater than kinetic friction

The friction between surfaces that are stationary is called **static friction.** The friction between moving surfaces is called **kinetic friction.** Because of forces between molecules of the two surfaces, the force required to make a stationary object start moving is usually greater than the force necessary to keep it moving. In other words, static friction is usually greater than kinetic friction.

Not all kinetic friction is the same

There are different kinds of kinetic friction. The type of friction depends on the motion and the nature of the objects. For example, when objects slide past each other, the friction that occurs is called *sliding friction.* If a round object rolls over a flat surface, the friction is called *rolling friction.* Rolling friction is usually less than sliding friction.

▶ **static friction** the force that resists the initiation of sliding motion between two surfaces that are in contact and at rest

▶ **kinetic friction** the force that opposes the movement of two surfaces that are in contact and are sliding over each other

Figure 15

Frictional Forces and Acceleration

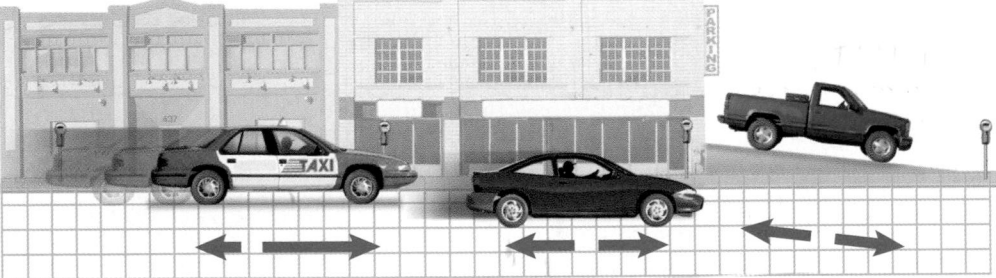

| Unbalanced forces: acceleration | Balanced forces: constant speed | Balanced forces: no motion |

A When a car is accelerating, the forces are unbalanced. The force moving the car forward is greater than the opposing force of friction.

B When a car is cruising at constant speed, the force moving the car forward is balanced by the force of friction.

C This truck does not roll, because the force of friction between the brakes and the wheels balances the gravity.

333

Demonstration —— GENERAL

Static vs. Kinetic Friction

(Time: about 15 minutes)

Materials:

- rectangular block
- hook
- spring scale

Step 1 Use the spring scale to measure the force required to start the rectangular block moving.

Step 2 Use the spring scale to measure the frictional force for constant velocity.

Step 3 Perform several trials. Have students record all data and find the average for each.

Analysis

1. What might account for any difference between the two average values? (Since the contact force and surfaces remain the same, the only difference in the two average values is due to motion.)

2. Which is greater, kinetic or static friction? (static friction)

LS Kinesthetic

Teaching Tip

Static Friction The force of static friction is a reaction force. Static frictional forces do not exist if there is no opposing force. For example, when trying to slide a crate to the left, you have an opposing frictional force to the right. However, if you stop pushing the crate, the frictional force stops.

Also, consider a sliding crate. If you stop pushing, the crate will slow down because of friction. As the relative motion between the crate and the floor decreases, so does the frictional force. So, when the crate stops, the frictional force is zero.

Chapter Resource File

- **Cross-Disciplinary Worksheet**
  Real World Application—Designing Race Cars

 Transparencies

TT Frictional Forces and Acceleration

Vocabulary Students may not understand why air resistance is classified as fluid friction. Explain that in scientific terminology, both liquids and gases are classified as fluids. Thus, air is a fluid, and so air resistance is an example of fluid friction.

MISCONCEPTION ///ALERT\\\

Friction A common misconception is that friction is a force of "directionless resistance" that occurs between solids only. Emphasize that, like all forces, friction acts in a particular direction. Use the idea of fluid friction, including air resistance, to counter the misconception that friction can occur only between solids.

Group Activity ——— ADVANCED

Researching Tires Vehicle tires are designed to increase grip. Have students work in small groups to find information about as many different kinds of tires, tire compounds, and tread designs as they can. Have them make a poster or other project showing some of the types of tires and treads they have learned about. **LS Interpersonal**

Figure 16

With the need for better fuel efficiency and increased speed, car designs have been changed to reduce air resistance. Modern cars are much more aerodynamic than cars of earlier eras.

---

internet connect

www.scilinks.org
Topic: Force and Friction
SciLinks code: HK4054

SCiLINKS. Maintained by the National Science Teachers Association

---

Air resistance also opposes motion

Any object moving through a fluid such as air encounters friction between the air and the surface of the moving object. That friction is called *fluid friction.* Air slides past a car as it moves, which causes fluid friction. Fluid friction can be minimized by very smooth surfaces.

In addition to fluid friction, another factor involved in air resistance is the displacement of air. For example, as a car moves, it must push air out of the way. The car must displace a certain volume of air for each car length that it moves. Air resistance to the car's motion increases as the car travels faster, because more air must be moved each second. This effect is very different from kinetic friction. The amount of air moved depends on the shape of the car. Designing the shape of the car so that less air must be displaced, as shown in **Figure 16,** is called *streamlining.*

Friction and Motion

Without friction, the tires of a car would not be able to push against the ground and move the car forward, the brakes would not be able to stop the car, and you would not even be able to grip the door handle to get inside! Without friction, a car is useless. Friction between your pencil and your paper is necessary for the pencil to leave a mark. Without friction, balls and other sports equipment would slip from your fingers when you tried to pick them up, and you would slip and fall when you tried to walk.

However, friction can cause some problems, too. In a car, friction between moving engine parts increases their temperature and causes the parts to wear down. Motor oil must be regularly added to the engine to keep it from overheating due to friction, and engine parts need to be changed as they wear out.

BIOLOGY
CONNECTION ——— GENERAL

When a fish swims in water, it experiences "water resistance" similar to air resistance. Ask students to find out how fish have adapted to their water environment so that water resistance during swimming is minimized. **LS Verbal**

SPACE SCIENCE
CONNECTION

Objects in space do not experience air resistance. This is why Earth continually orbits the sun without slowing down. (Earth's velocity is actually decreasing because of collisions with small masses such as meteoroids, but this is a minor effect.)

Harmful friction can be reduced

Because friction can be both harmful and helpful, it is sometimes desirable to reduce or increase friction. One way to reduce harmful friction is to use *lubricants*. Lubricants are substances that are applied to surfaces to reduce the friction between them. Some examples of common lubricants are motor oil, wax, and grease. *Figure 17* shows why lubricants are important to maintaining car parts.

Lubricants are usually liquids, but they can be solids or gases, too. An example of a lubricant gas is the air that comes out of the tiny holes of an air-hockey table.

Friction can also be reduced by replacing sliding friction with rolling friction. Ball bearings are placed between the wheels and axles of in-line skates and bicycles to reduce friction and thereby make the wheels turn more easily.

Another way to reduce friction is to make the surfaces smoother. For example, sliding across rough wood on a park bench can be uncomfortable if there is a large amount of friction between your legs and the bench. Rubbing the bench with sandpaper makes it smoother and therefore more comfortable for sitting, because the friction between the bench and your legs is reduced.

Competitive swimmers and bikers reduce the amount of fluid friction by wearing clothes that fit closely. Even their headgear is designed to decrease fluid friction in both the air and the water.

Helpful friction can be increased

One way to increase helpful friction is to make surfaces rougher. For example, sand scattered on icy roads keeps cars from skidding. Baseball players sometimes wear textured batting gloves to increase the friction between their hands and the bat so that the bat does not slide or fly out of their hands.

Another way to increase friction is to increase the force pushing the surfaces together. For example, you can ensure that your magazine will not blow away at the park by putting a heavy rock on it. The added mass of the rock increases the friction between the magazine and the ground or park bench. If you are sanding a piece of wood, you can sand the wood faster by pressing harder on the sandpaper. *Figure 18* gives another example of a way to increase helpful friction.

Figure 17

Motor oil is used as a lubricant in car engines. Without oil, engine parts would wear down quickly, like the connecting rod shown in the bottom of this photograph.

Figure 18

No one enjoys cleaning pans with baked-on food! To make the chore pass quickly, press down on the pan with the scrubber to increase friction.

335

Trends in Materials Science
Studying the Atomic Origin of Friction

Scientists at the Lawrence Berkeley National Laboratory are using a new kind of Atomic Force Microscope (AFM) to study the atomic origin of frictional forces. As the tip of this special AFM moves across the surface of a material, it moves up and down over individual atoms to map surface topography. At the same time, the AFM also measures friction by recording the sideways motion of the flexible tip. The researchers are able to study wear at an atomic level by measuring how much force is needed to dislodge a single atom.

Scientists plan to use the AFM to learn more about what causes friction and what factors increase or decrease friction. They also hope to determine how variations in atomic topography affect friction. The research might one day lead to the development of better lubricants, which in turn will improve the energy efficiencies of many kinds of mechanical devices.

Quiz

Select the term that best completes each statement.

1. A(n) balanced/unbalanced force must be present to cause a change in motion. (unbalanced)

2. Friction is the force between the surfaces of two objects that accelerates/opposes the motion of either object. (opposes)

3. Of static and kinetic friction, static/kinetic friction is always greater. (static)

4. Friction is always/sometimes harmful. (sometimes)

5. Putting salt on a snowy road is a way to increase/decrease friction. (increase)

Figure 19
Without friction, a car cannot be controlled.

Cars could not move without friction

What causes a car to move? A car's wheels turn, and they push against the road. The road pushes back on the car and causes the car to accelerate. Without friction between the tires and the road, the tires would not be able to push against the road, and the car would not experience a net force. Friction, therefore, causes the acceleration (whether speeding up, slowing down, or changing direction).

Water, snow, and ice provide less friction between the road and the car than usual. Normally, as a car moves slowly over water on the road, the water is pushed out from under the tires. However, if the car moves too quickly, the water becomes trapped and cannot be pushed out from under the tires. The water trapped between the tires and the road may lift the car off the road, as shown in *Figure 19*. This is called *hydroplaning*. When hydroplaning occurs, there is very little friction between the tires and the water, and the car becomes difficult to control. This dangerous situation is an example of the need for friction.

SECTION 3 REVIEW

SUMMARY

▶ Objects subjected to balanced forces either do not move or move at constant velocity.

▶ An unbalanced force must be present to cause any change in an object's state of motion or rest.

▶ Friction is a force that opposes motion between the surfaces of objects moving, or attempting to move, past each other.

▶ Static friction opposes motion between two stationary surfaces. Kinetic friction opposes motion between two surfaces that are moving past one another.

▶ Friction can be helpful or harmful. There are many ways to decrease or increase friction.

1. **Describe** a situation in which unbalanced forces are acting on an object. What is the net force on the object, and how does the net force change the motion of the object?

2. **Identify** the type of friction in each situation described below.
   a. Two students are pushing a box that is at rest.
   b. The box pushed by the students is now sliding.
   c. The students put rollers under the box and push it forward.

3. **Explain** why friction is necessary to drive a car on a road. How could you increase friction on an icy road?

4. **Describe** three different ways to decrease the force of friction between two surfaces that are moving past each other.

5. **Critical Thinking** When you wrap a sandwich in plastic food wrap to protect it, you must first unroll the plastic wrap from the container, and then wrap the plastic around the sandwich. In both steps you encounter friction. In each step, is friction helpful or harmful? Explain your answer.

6. **Critical Thinking** The force pulling a truck downhill is 2000 N. What is the size of the static friction acting on the truck if the truck doesn't move?

336

Answers to Section 3 Review

1. Answers may vary. A baseball falling to Earth is an example of a situation involving unbalanced forces. The net force is the force of gravity pulling the baseball toward the ground combined with kinetic friction.

2. a. static friction
   b. kinetic friction
   c. rolling friction

3. Friction between the tires and the road enables the tires to push against the road and vice versa so that the car experiences a net force and drives forward. To increase friction and to melt ice on an icy road, you can salt or sand the road.

4. Lubricate the surfaces. Smooth the surfaces. Place bearings between the surfaces.

5. When unrolling the plastic wrap, friction between your hand and the wrap is helpful, because it enables you pull the wrap; but the friction between two parts of the wrap is harmful, because it slows the flow of the wrap. When wrapping the sandwich, friction between edges of the plastic is helpful, because it keeps the plastic sticking to itself.

6. If there is no acceleration, the net force is zero. The force of static friction is equal to, but in the opposite direction of the force pulling the truck downhill: 2000 N uphill.

Graphing Skills

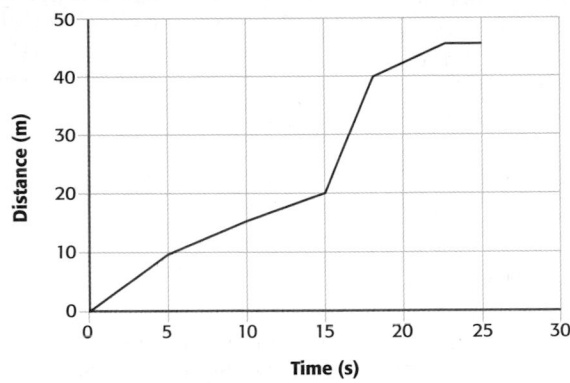

Time (s)

Examine the above graph, and answer the following questions. (See Appendix A for help in interpreting a graph.)

1 Does the graph indicate an increase or decrease of the quantities? Explain your answer.

2 Identify the independent and dependent variables. What is the relationship between the two variables?

3 What information about the runner's speed can be determined from the graph? Is the speed constant during the run?

4 What is the runner's maximum speed? During what 5-second time interval does the runner reach this speed? What is the runner's minimum speed?

5 What is the total distance traveled by the runner? What trend suggests that this is the total distance run even when the graph is continued beyond the 25.0 s mark?

6 How is this graph similar to any graph showing distance traveled in a single direction over a given time interval?

7 Construct a graph best suited for the information in the table below. Assuming all measurements are made in 7.0 s, which car has the greatest acceleration? If the time interval for car B is 8.0 s instead of 7.0 s, which car has the greatest acceleration?

Car type	Maximum speed (m/s)
A	23.3
B	28.0
C	26.2

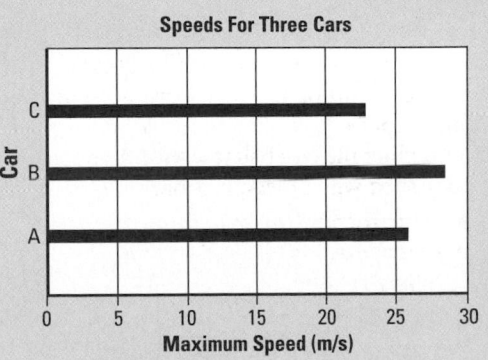

Speeds For Three Cars

Car

Maximum Speed (m/s)

Teaching Tip

Interpreting Graphs Indicate how the shape of a distance versus time graph can provide qualitative information about velocity and acceleration. A horizontal line indicates zero velocity and acceleration, a straight line with a positive slope has positive velocity and zero acceleration, and a straight line with a negative slope has negative velocity and zero acceleration. A curved line indicates non-zero acceleration.

Answers

1. increasing

   As time increases, so does the distance traveled.

2. Time is the independent variable; distance traveled is the dependent variable. During a time interval, the runner travels farther, according to the runner's speed (slope of line).

3. The runner's speed can be calculated from the graph. No, the speed is not constant, but varies over the entire run.

4. 8.0 m/s; from $t = 15.0$ s to $t = 17.5$ s; 0 m/s

5. 45.0 m

   The runner has stopped moving after 22.5 s, as indicated by the horizontal line. It may therefore be inferred that the run is over at 22.5 s.

6. All such graphs indicate an increase in distance with time, regardless of changes in speed.

7. The best suited graph is a bar graph (see sample below). The acceleration over 7.0 s for each car is:

   Car A. 3.3 m/s^2
   Car B. 4.0 m/s^2
   Car C. 3.7 m/s^2

   Car B has the greatest acceleration. If the time for car B is increased to 8.0 s, car C has the greatest acceleration.

Understanding Concepts

1. b

2. d

3. a

4. a

5. c

6. b

7. a

8. d

9. c

10. c

11. a

Using Vocabulary

12. both; the speed is 30 m/s, and the velocity is 30 m/s westward.

13. At the beginning of a race, a cyclist uses friction between her tires and the road to create a net force, which changes the bicycle's velocity and enables acceleration forward.

14. acceleration

15. The reference frame defines the starting, ending, and comparison points.

16. Distance is the length of the path that you travel even if you change direction. Displacement is the straight-line distance between the starting point and the ending point.

Chapter Resource File

- **Chapter Tests** GENERAL
- **Test Item Listing for ExamView® Test Generator**

Chapter Highlights

Before you begin, review the summaries of the key ideas of each section, found at the end of each section. The vocabulary terms are listed on the first page of each section.

UNDERSTANDING CONCEPTS

1. If you jog for 1 h and travel 10 km, 10 km/h describes your
- **a.** momentum.
- **c.** displacement.
- **b.** average speed.
- **d.** acceleration.

2. _____ is a speed in a certain direction.
- **a.** Acceleration
- **c.** Momentum
- **b.** Friction
- **d.** Velocity

3. An object's speed is a measure of
- **a.** how fast the object is moving.
- **b.** the object's direction.
- **c.** the object's displacement per unit of time.
- **d.** All of the above

4. A car travels a distance of 210 mi in exactly 4 h. The driver calculates that he traveled 52.5 mi/h. Which of the following terms most nearly describes his calculation?
- **a.** average speed
- **b.** instantaneous speed
- **c.** instantaneous acceleration
- **d.** displacement

5. Which of the quantities below represents a velocity?
- **a.** 25 m/s
- **c.** 15 mi/h eastward
- **b.** 10 km/min
- **d.** 3 mi/h

6. Which of the following is *not* accelerating?
- **a.** a ball being juggled
- **b.** a woman walking at 2.5 m/s along a straight road
- **c.** a satellite circling Earth
- **d.** a braking cyclist

7. At the end of a game, a basketball player on the winning team throws the basketball straight up as high as he can throw it. At the top of its path, the basketball's velocity is
- **a.** 0 m/s.
- **b.** 10 m/s up.
- **c.** 10 m/s down.
- **d.** Not enough information is given to determine its velocity.

8. Which one of the following is *not* caused by a net force?
- **a.** starting up a bicycle that was previously not moving
- **b.** changing a bicycle's speed while it is moving in a straight line
- **c.** changing a bicycle's direction while it is moving at constant speed
- **d.** keeping a bicycle going in a straight line at constant speed

9. A book is sitting still on your desk. Which of the following best describes this situation?
- **a.** There are no forces acting on the book.
- **b.** The book is moving compared to the reference frame.
- **c.** There are balanced forces acting on the book.
- **d.** There are unbalanced forces acting on the book.

10. When you graph displacement vs. time, velocity is represented by
- **a.** the *x*-intercept of the graph.
- **b.** the *y*-intercept of the graph.
- **c.** the slope of the graph.
- **d.** the curve of the graph.

11. If a track athlete runs an 800 m race at a constant speed of 2 m/s, how long will it take her to run the race?
- **a.** 6.7 min
- **b.** 16 min
- **c.** 26.7 min
- **d.** 400 min

338

USING VOCABULARY

12. State whether 30 m/s westward represents a *speed*, a *velocity*, or both.

13. Describe the motion of a cyclist at the start of a race. In your answer, use the terms *velocity*, *acceleration*, and *friction*.

14. What does the slope of a velocity-time graph tell you about an object?

15. Why is identifying the *reference frame* important in describing motion?

16. What is the difference between *distance* and *displacement*?

17. Why is traveling in a circle at a constant speed called *acceleration*?

18. What is *uniform circular motion*?

19. How are *friction* and *air resistance* alike? How are they different?

20. What is the difference between a *force* and an *unbalanced force*?

21. How do *static friction* and *kinetic friction* differ from each other?

BUILDING MATH SKILLS

22. Interpreting Data Bob straps on his in-line skates and pushes down a hill. His velocity changes from 0 m/s at the start to 4.5 m/s exactly 15 s later. What is Bob's average acceleration?

23. Interpreting Data A baseball is hit straight up at an initial velocity of 30 m/s. If the ball has a negative acceleration of about 10 m/s², how long does the ball take to reach the top of its path?

24. Velocity An airplane traveling from San Francisco northeast to Chicago travels 1260 km in 3.5 h. What is the airplane's average velocity?

25. Velocity Heather and Matthew take 45 s to walk eastward along a straight road to a store 72 m away. What is their average velocity?

26. Velocity Simpson drives his car with an average velocity of 85 km/h eastward. How long will it take him to drive 560 km on a perfectly straight highway?

27. Acceleration A driver is traveling eastward on a dirt road when she spots a pothole ahead. She slows her car from 14.0 m/s to 5.5 m/s in 6.0 s. What is the car's acceleration?

28. Acceleration How long will it take a cyclist with a forward acceleration of -0.50 m/s² to bring a bicycle with an initial forward velocity of 13.5 m/s to a complete stop?

BUILDING GRAPHING SKILLS

29. Graphing The following graphs describe the motion of four different balls—*a*, *b*, *c*, and *d*. Use the graphs to determine whether each ball is accelerating, sitting still, or moving at a constant velocity.

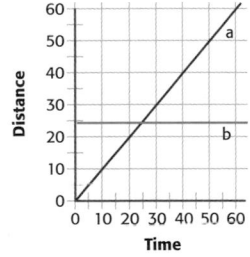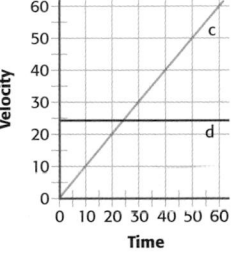

17. Acceleration is the change in velocity per unit of time. Velocity includes both speed and direction, so a change in direction is a change in velocity. Traveling in a circle requires constantly changing direction, therefore constantly changing velocity, therefore accelerating.

18. Uniform circular motion is motion at constant speed in a circle.

19. They are alike in that they oppose motion. In addition, part of air resistance is friction between the air and the surface of the moving object. They are different because air resistance also involves pushing air out of the way. The more air that must be pushed out of the way each second, the greater the air resistance. That's why air resistance increases as speed increases but friction does not.

20. A force is an action exerted on a body in order to change the body's state of rest or motion. If a force is canceled by another force in the opposite direction, the forces are balanced. However, if there is no force acting in the opposite direction, then the force is unbalanced, and acceleration results.

21. Static means "not moving" (stationary) and kinetic means "moving." Static friction is greater than kinetic friction, because when two surfaces are not moving past each other, the irregularities in one surface can stick to the irregularities of the other surface more easily than when the surfaces are in motion.

339

Assignment Guide

Section	Questions
1	1–5, 7, 10–12, 15, 16, 24–26, 30, 31, 36, 38, 40
2	6, 14, 17, 18, 22, 23, 27–29, 32–35
3	8, 9, 13, 19–21, 37, 39

Building Math Skills

22. $a = \dfrac{\Delta v}{t} = \dfrac{4.5 \text{ m/s}}{15 \text{ s}} = 0.3 \text{ m/s}^2$

23. Velocity at the top of the path is 0 m/s.

$t = \dfrac{\Delta v}{a} = \dfrac{0 \text{ m/s} - 30 \text{ m/s}}{-10 \text{ m/s}^2} = 3.0 \text{ s}$

24. $v = \dfrac{d}{t} = \dfrac{1260 \text{ km}}{3.5 \text{ h}} = 360 \text{ km/h}$ northeast

25. $v = \dfrac{d}{t} = \dfrac{72 \text{ m}}{45 \text{ s}} = 1.6 \text{ m/s}$ eastward

26. $t = \dfrac{d}{v} = \dfrac{560 \text{ km}}{85 \text{ km/h}} = 6.6 \text{ h}$

27. $a = \dfrac{5.5 \text{ m/s} - 14.0 \text{ m/s}}{6.0 \text{ s}} = \dfrac{-8.5 \text{ m/s}}{6.0 \text{ s}} = -1.4 \text{ m/s}^2$ eastward $= 1.4 \text{ m/s}^2$ westward

28. $t = \dfrac{\Delta v}{a} = \dfrac{v_f - v_i}{a} = \dfrac{0 \text{ m/s} - 13.5 \text{ m/s}}{-0.50 \text{ m/s}^2} = 27 \text{ s}$

Building Graphing Skills

29. Objects *a* and *d* are moving with constant velocity; object *b* is at rest; object *c* is moving with constant acceleration.

30.

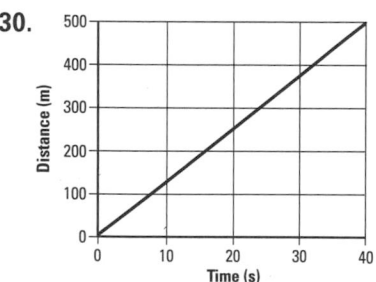

a. Find the slope of the line, about 13 m/s.

b. about 19 s

30. Graphing A cyclist was observed riding from her home to a location 500 m away. Her distance was measured at varying times, as indicated below. Graph the data points shown in the chart below, and draw the straight line that best fits the data. Then, use your graph to answer the questions below.

Time	Distance
0 s	0 m
8 s	100 m
15 s	200 m
23 s	300 m
30 s	400 m
38 s	500 m

a. How can you use the graph to find the cyclist's average speed? What is the value of her average speed?

b. About how long did it take the cyclist to travel 250 m?

31. Interpreting Graphics Two cars are traveling eastward on a highway, as shown in the left figure below. After 5.0 s, the cars are side by side at the next telephone pole, as shown on the right. The distance between each of the poles is 70.0 m. Determine the following quantities:

a. the distance the red car has traveled during the 5.0 s interval

b. the distance the blue car has traveled during the 5.0 s interval

c. the average velocity of the red car during this 5.0 s time interval

d. the average velocity of the blue car during this 5.0 s time interval

32. Graphing A rock is dropped from a bridge, and the distance it travels and the speed at which it is falling are measured every second until it hits the water. The data are shown in the chart below. Make two graphs of the data, a distance-time graph and a velocity-time graph. Use your graphs to answer the questions below.

Time	Distance traveled	Downward speed
0 s	0 m	0 m/s
1 s	5 m	10 m/s
2 s	20 m	20 m/s
3 s	45 m	30 m/s

a. Why is the distance-time graph curved?

b. Why is the velocity-time graph a straight line?

c. Use the velocity-time graph to figure out the rock's acceleration.

THINKING CRITICALLY

33. Drawing Conclusions What can you conclude about the forces acting on an object traveling in uniform circular motion?

34. Applying Knowledge When you drive, you will sometimes have to decide in a brief moment whether to stop for a yellow light. Discuss the variables you must consider in making your decision. Use the concepts of force, acceleration, and velocity in your discussion.

35. Applying Knowledge An instructor suggested that a driver stop accelerating while turning a corner. Using the definition of *acceleration* found in this chapter, explain why the driver will not be able to comply.

---

31. a. 140 m (red car)

   b. 70 m (blue car)

   c. $v = \dfrac{d}{t} = \dfrac{140 \text{ m}}{5.0 \text{ s}} = 28 \text{ m/s east}$

   d. $v = \dfrac{70 \text{ m}}{5.0 \text{ s}} = 14 \text{ m/s east}$

32. a. The distance traveled in each succeeding time interval is bigger than the distance traveled in the preceding time interval because the rock is speeding up. That is why the distance vs. time graph is curved.

b. The velocity vs. time graph is a straight line because the rock speeds up by the same amount in each time interval.

c. 10 m/s² downward

Thinking Critically

33. An object traveling in uniform circular motion must be accelerating, so there must be a net force. The acceleration is centripetal—toward the center of the circle—so the net force causing the centripetal acceleration must also be toward the center of the circle.

36. Working Cooperatively For one day, write down a brief description of the different kinds of motions you see. Work with a group to generate a common list of the different kinds of motions observed. What reference points did you use to detect the motion? How did these reference points help you determine motion? Compare your list with those from other groups in your class.

37. Applying Information Visit a local hardware store or interview a carpenter to investigate various textures of sandpaper. Write a report describing the kinds of surfaces for which various sandpapers are appropriate, and how they are used. What is a "grit number?" Explain how to choose the *best* grit number for a particular wood surface. Incorporate information about friction from this chapter.

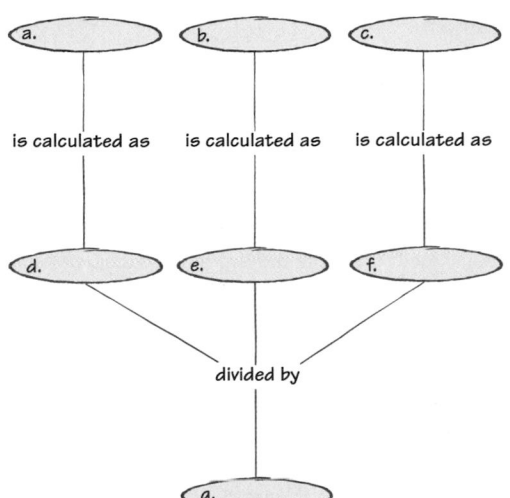

38. Connection to Physical Education A track athlete ran the 50 m dash. She ran 4 m/s during the first 25 m and 5 m/s during the last 25 m. What was her average speed? How long did it take her to run the 50 m dash?

39. Creative Thinking What are some of the ways that competitive swimmers can decrease the amount of friction or drag between themselves and the water they are swimming through? How does each method work to decrease friction?

40. Concept Mapping Copy the unfinished concept map below onto a sheet of paper. Complete the map by writing the correct word or phrase in the lettered boxes.

internet connect

www.scilinks.org
Topic: Graphing Speed, Velocity, Acceleration
SciLinks code: HK4066

SCI*LINKS* Maintained by the
National Science Teachers Association

Art Credits: Fig. 2, Uhl Studios, Inc.; Fig. 6, Marty Roper/Planet Rep; Fig. 7, Mike Carroll/Steve Edsey & Sons; Fig. 15, Uhl Studios, Inc.; Chapter Review ("Thinking Critically"), Uhl Studios, Inc.
Photo Credits: Chapter Opener image of ski jumper by Mike Powell/Getty Images; scoreboard by Rudi Blaha/AP/Wide World Photos; Fig. 1, SuperStock; Fig. 3(l-r), David Madison/Getty Images/Stone; Alan Levenson/Getty Images/Stone; Index Stock Photography, Inc.; Anne Powell/Index Stock Imagery, Inc.; Michael H. Dunn/Corbis Stock Market; Brian Stablyk/Getty Images/Stone; Fig. 5, Steve Coleman/AP/Wide World Photos; Fig. 7, Sergio Purtell/Foca; Fig. 8, Clive Brunskill/Getty Images; Fig. 9, Wernher Krutein/CORBIS Images/HRW; Fig. 10, Roberto SCHMIDT/AFP PHOTO/CORBIS; Fig. 11, Dennis Curran/Index Stock Imagery, Inc.; Fig. 12, Phil Cole/Getty Images; Fig. 13, Sam Dudgeon/HRW; Fig. 16(l-r), SuperStock; SuperStock; ©2004 Ron Kimball Studios; Fig. 17, Sam Dudgeon/HRW; Fig. 18, Michelle Bridwell/HRW; Fig. 19, Digital Image ©2004, PhotoDisc; "Design Your Own Lab," Sam Dudgeon/HRW

34. Answers will vary. Some of the variables include the length of time the light is yellow, the force that can be applied by the brakes compared with the force that can be applied with the engine, the distance to the intersection when the light turns yellow, and the width of the intersection. To stop successfully, the brakes must provide enough force for negative acceleration to give zero velocity before the intersection is entered. To get through the intersection safely, the engine must apply enough positive acceleration to travel the distance to the intersection plus the width of the intersection before the light turns red.

35. The driver will not be able to comply, because when he/she turns the corner, he/she will be changing direction and thus will be accelerating.

Developing Life/Work Skills

36. Student responses will vary. In their descriptions of various motions, students should answer at least some of the questions posed in the student text.

37. Answers will vary.

Integrating Concepts

38. avg $v = t_1 + t_2 = t_{\text{total}}$

$t_1 = \dfrac{d_1}{v_1} = \dfrac{25 \text{ m}}{4 \text{ m/s}} = 6.25$ s
(rounded to 6 s)

$t_2 = \dfrac{d_2}{v_2} = \dfrac{25 \text{ m}}{5 \text{m/s}} = 5$ s

avg $v = 6$ s $+ 5$ s $= 11$ s

39. Answers will vary but they should all involve streamlining (e.g., shaving heads, wearing bathing caps, wearing a helmet similar to racing bike helmets) and/or lubrication

40. a. average speed
b. average velocity
c. acceleration
d. distance
e. displacement
f. change in velocity
g. time

 Transparencies

TT Concept Mapping

Teacher's Notes

Time Required 1 lab period

Ratings

1	2	3	4
EASY			HARD

TEACHER PREPARATION	1
STUDENT SETUP	1
CONCEPT LEVEL	2
CLEANUP	1

Skills Acquired

- Collecting data
- Experimenting
- Inferring
- Measuring
- Organizing and analyzing data
- Predicting

The Scientific Method

In this lab, students will:
- Ask Questions
- Test the Hypothesis
- Analyze the Results
- Draw Conclusions
- Communicate Results

Materials

If the spring scale is not very sensitive, students may record a force of zero for rolling friction. Encourage students to discuss whether this is realistic and what would be causing them to get such a result. You may want to provide more sensitive spring scales for steps 10 and 11 to avoid this problem. Textbooks should be covered with paper to provide an even sliding surface and to protect the book covers. Also make sure tabletops are clean and smooth.

Skills Practice Lab

Introduction

In this experiment, you will investigate three types of friction—static, sliding, and rolling—to determine which is the largest force and which is the smallest force.

Objectives

- ▶ **USING SCIENTIFIC METHODS** *Form a hypothesis* to predict which type of friction force—static, sliding, or rolling—will be greatest and which will be smallest.

- ▶ *Measure* the static, sliding, and, rolling friction when pulling a textbook across a table.

- ▶ *Calculate* average values from multiple trials.

- ▶ *Compare* results to initial predictions.

Materials

scissors
spring scale
string
textbook (covered)
wooden or metal rods (4)

Tips and Tricks

For best results, students should keep spring scales parallel to the table as they pull. They should also pull as gently and gradually as possible; quick pulls will give incorrect readings.

Your spring scales may or may not show force in units of newtons. The data table calls for forces in units of newtons. Students do not learn about units of force in this chapter. You may tell students to leave off units until they study force units. The units used are not important, as the purpose of the lab is to compare the relative values for different types of friction.

Static, Sliding, and Rolling Friction

▶ Procedure

Preparing for Your Experiment

1. Which type of friction do you think is the largest force: static, sliding, or rolling? Which is the smallest?

2. Form a hypothesis by writing a short paragraph that answers the question above. Explain your reasoning.

3. Prepare a data table like the one shown at right. **SAFETY CAUTION** Secure loose clothing and remove dangling jewelry. Don't wear open-toed shoes or sandals in the lab. Use knives and other sharp instruments with extreme care. Never cut objects while holding them in your hands. Place objects on a suitable work surface for cutting.

Collecting Data and Testing the Hypothesis

4. Cut a piece of string, and tie it in a loop that fits inside a textbook. Hook the string to the spring scale as shown.

5. To measure the static friction between the book and the table, pull the spring scale very slowly. Gradually increase the force with which you pull on the spring scale until the book starts to slide across the table. Pull very gently. If you pull too hard, the book will start lurching, and you will not get accurate results.

6. Practice pulling the book as in step 5 several times. On a smooth trial, note the largest force that appears on the scale *before* the book starts to move. Record this result in your data table as *static friction*.

7. Repeat step 6 two times, and record the results in your data table.

8. After the textbook begins to move, you can determine the sliding friction. Start pulling the book as in step 5. Once the book starts to slide, continue applying just enough force to keep the book sliding at a slow, constant speed. Practice this several times. On a smooth trial, note the force that appears on the scale as the book is sliding at a slow, constant speed. Record this force in your data table as *sliding friction*.

	Static friction (N)	Sliding friction (N)	Rolling friction (N)
Trial 1			
Trial 2			
Trial 3			
Average			

9. Repeat step 8 two times, and record the results in your data table.

10. Place two or three rods under the textbook to act as rollers. Make sure the rods are evenly spaced. Place another rod in front of the book so that the book will roll onto it. Pull the spring scale slowly so that the book rolls across the rods at a slow, constant speed. Practice this several times, repositioning the rods each time. On a smooth trial, note the force that appears on the scale as the book is moving at a slow, constant speed. Record this force in your data table as *rolling friction*.

11. Repeat step 10 two times, and record the results in your data table.

▶ Analysis

1. **Organizing Data** For each type of friction, add the results of the three trials and divide by three to get an average. Record these averages in your data table.

2. **Analyzing Data** Which of the three types of friction was the largest force on average?

3. **Analyzing Data** Which of the three types of friction was the smallest force on average?

▶ Conclusions

4. **Evaluating Results** Did your answers to Analysis questions 2 and 3 agree with the hypotheses you made before collecting data? If not, explain how your results differed from what you predicted.

5. **Applying Conclusions** Imagine that you are an engineer at a construction site. You are planning to drag a heavy load of building materials on a palette by using a cable attached to a truck. When will the force exerted by the cable be greatest, before the palette starts moving or while it is moving? How could you reduce the amount of force needed to move the palette?

6. **Evaluating Methods** In each trial, the force that you measured was actually the force that you were exerting on the spring scale, which was in turn exerted on the book. Why could you assume that this was equal to the force of friction in each case?

343

Safety Cautions
Remind students not to wear jewelry or open-toed shoes in the lab. Jewelry can get caught in the spring scale, and falling books can pose a hazard to unprotected feet. Also remind students to be cautious when using scissors.

Answers to Analysis
1. Answers will vary based on data. The correct answers should be the averages of the three trials for each of the three types of friction.
2. Answers may vary. Students should find that static friction is the largest force.
3. Answers may vary. Students should find that rolling friction is the smallest force.

Answers to Conclusions
4. Answers will vary based on the initial predictions and results. Students should explain why their results might have differed from their predictions.
5. The force exerted by the cable will be greatest just before the palette starts moving. The required force could be reduced by putting wheels or rolling bars under the palette.
6. Because the book is moving at a constant speed (or at rest), the forces on the book are balanced. Therefore, the force that the spring scale exerts on the book is equal to the force of friction.

Chapter Resource File
- **Datasheets** Skills Practice Lab— Static, Sliding, and Rolling Friction GENERAL
- **Observation Lab** Testing Reaction Time BASIC
- **CBL™ Probeware Lab** Determining the Coefficients of Static and Kinetic Friction ADVANCED

Forces
Chapter Planning Guide

PACING	CLASSROOM RESOURCES	LABS, ACTIVITIES, AND DEMONSTRATIONS
BLOCK 1 • 45 min pp. 344–345 **Chapter Opener**		SE **Activity 1,** p. 345 SE **Activity 2,** p. 345
BLOCKS 2 and 3 • 90 min pp. 346–351 **Section 1** Laws of Motion	CRF **Lesson Plan*** TT **Bellringer*** TT **Newton's Second Law***	TE **Opening Demonstration,** p. 346 GENERAL SE **Quick Activity** Newton's First Law, p. 347 GENERAL CRF **Datasheets for In-Text Labs** Newton's First Law* GENERAL
BLOCKS 4 and 5• 90 min pp. 352–359 **Section 2** Gravity	CRF **Lesson Plan*** TT **Bellringer*** TM **Law of Universal Gravitation*** TT **Terminal Velocity*** TT **Two Motions Cause Orbiting*** TM **Projectile Motion***	TE **Group Activity** Artificial Satellites, p. 357 ADVANCED SE **Skills Practice Lab** Measuring Forces, pp. 372–373 ◆ GENERAL
BLOCKS 6 and 7 • 90 min pp. 360–366 **Section 3** Newton's Third Law	CRF **Lesson Plan*** TT **Bellringer*** TM **Rocket Propulsion*** TT **Concept Mapping***	TE **Opening Demonstration,** p. 360 GENERAL SE **Quick Lab** How are action and reaction forces related?, p. 366 GENERAL CRF **Datasheets for In-Text Labs** How are action and reaction forces related?* GENERAL CRF **Datasheets for SE Skills Practice Lab** Measuring Forces* GENERAL CRF **CBL™ Probeware Lab** Determining Your Acceleration on a Bicycle* ◆ ADVANCED CRF **Observation Lab** Observing the Conservation of Momentum* ◆ BASIC

BLOCKS 8 & 9 • 90 min

Chapter Review and Assessment Resources

- SE **Chapter Review,** pp. 368–371
- CRF **Chapter Tests*** GENERAL
- OSP **Test Generator**
- CRF **Standardized Test Practice with Guided Reading Development***
- CRF **Test Item Listing for ExamView® Test Generator***
- OSP **Scoring Rubrics and Classroom Management Checklists**

Online Resources

go hrw .com

Visit the HRW Web site for a variety of free resources related to the text. Just type in the keyword **HK4 FOR.**

TOPIC: **Bicycle Helmets**

Keyword: **MK4Helmet**

Holt Online Learning

Holt Science Spectrum: Physical Science: Online Edition

Students can access interactive problem solving help and active visual concept development with the *Holt Science Spectrum: Physical Science* Online Edition available at **www.hrw.com.**

student CNN News

cnnstudentnews.com

Find the latest news, lesson plans, and activities related to important scientific events.

Compression guide:
To shorten your instruction because of time limitations, omit blocks 1, 8, and 9.

KEY

TE	Teacher Edition	**CRF**	Chapter Resource File
SE	Student Edition	**TT**	Teaching Transparency
OSP	One-Stop Planner	**TM**	Transparency Master

* Also on One-Stop Planner
◆ Requires Advance Prep

PROBLEM SOLVING AND PRACTICE	SECTION REVIEW AND ASSESSMENT	STANDARDS CORRELATION
	CRF Pretest* GENERAL	
SE **Science and the Consumer** Should a Car's Air Bags Be Disconnected?, p. 348 **CRF** **Cross-Disciplinary Worksheet** Real World Applications—Car Seat Safety* GENERAL **CRF** **Cross-Disciplinary Worksheet** Integrating Mathematics—Using Force Diagrams* GENERAL **SE** **Math Skills** Newton's Second Law, pp. 350–351 **CRF** **Math Skills** Newton's Second Law* GENERAL	**TE** Quiz, p. 351 BASIC **SE** Section 1 Review, p. 351 **CRF** Concept Review* GENERAL **CRF** Quiz* BASIC	PS 4a
CRF **Science Skills** Equations Involving a Constant* GENERAL **CRF** **Cross-Disciplinary Worksheet** Integrating Biology—Blood Pressure in Space* ADVANCED **CRF** **Cross-Disciplinary Worksheet** Connection to Social Studies—The Great Plague and Isaac Newton* GENERAL **CRF** **Cross-Disciplinary Worksheet** Integrating Space Science—Gravity and the Planets* ADVANCED **CRF** **Cross-Disciplinary Worksheet** Integrating Biology—How Fish Maintain Neutral Buoyancy* ADVANCED	**TE** Quiz, p. 359 BASIC **SE** Section 2 Review, p. 359 **CRF** Concept Review* GENERAL **CRF** Quiz* BASIC	PS 4b
CRF **Cross-Disciplinary Worksheet** Integrating Technology—Hydraulic Lift Force* ADVANCED **SE** **Math Skills** Momentum, p. 363 **CRF** **Math Skills** Momentum* GENERAL **CRF** **Cross-Disciplinary Worksheet** Connection to Fine Arts—Momentum of Line in Art* ADVANCED **CRF** **Cross-Disciplinary Worksheet** Applications—Driving Safety* ADVANCED **SE** **Math Skills** Algebraic Rearrangements, p. 367 **CRF** **Science Skills Worksheet** Ordering Multiple Operations* GENERAL **SE** **Viewpoints** Should Bicycle Helmets Be Required by Law?, pp. 374–375	**TE** Quiz, p. 366 BASIC **SE** Section 3 Review, p. 366 **CRF** Concept Review* GENERAL **CRF** Quiz* BASIC	PS 4a

SCiLINKS®
www.scilinks.org

Topic: Forces
*Sci*Links code: HK4055

Topic: Intertia
*Sci*Links code: HK4072

Topic: Newton's Laws of Motion
*Sci*Links code: HK4094

Topic: Momentum
*Sci*Links code: HK4088

Topic: Rocket Technology
*Sci*Links code: HK4123

Technology Resources

One-Stop Planner
All of your printable resources and the Test Generator are on this convenient CD-ROM.

Physical Science
Interactive
Tutor CD-Rom

 CNN. Science in the NEWS
each video segment is accompanied by a Critical Thinking Worksheet *

Segment 4
Crash-Test
Dummies

Segment 7
Zero-Gravity
Plane

Segment 6
Egg Drop Contest

Overview

This chapter covers Newton's first and second laws of motion, including problem-solving with the second law. It also discusses the law of universal gravitation, free fall, and projectile motion. Finally, the chapter explores Newton's third law of motion and also covers momentum.

Assessing Prior Knowledge

Be sure students understand the following concepts:

- mass
- acceleration
- balanced and unbalanced forces

MISCONCEPTION ///ALERT \\\

Science education research has identified the following misconceptions about forces and free fall.

- Students believe that when there is no motion, there is no force.
- Students characterize equilibrium as the result of the strongest forces winning and believe that all forces cease to act at equilibrium.
- Students believe that heavier objects fall faster, and they are unable to separate the effects of air resistance.

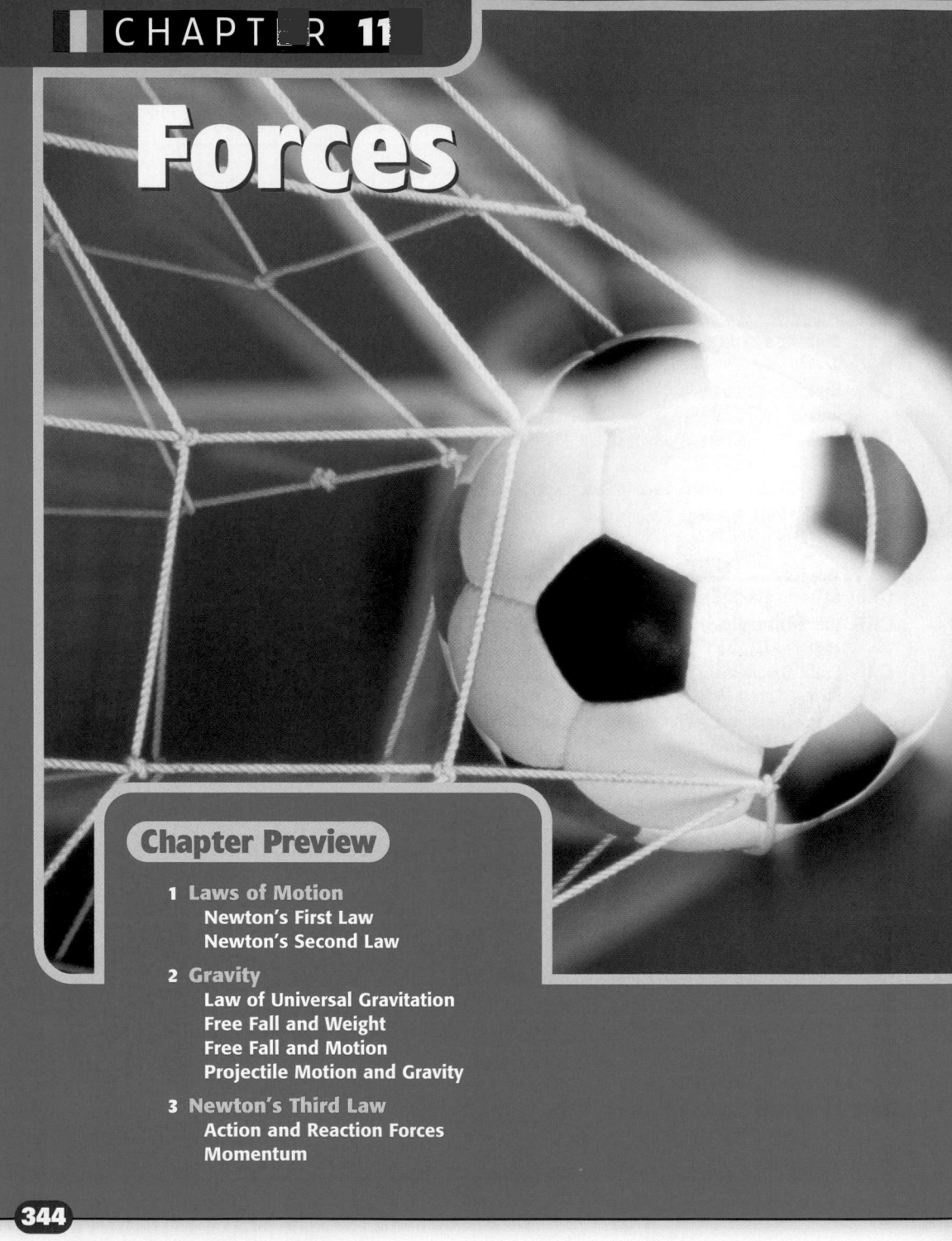

Forces

Chapter Preview

1 Laws of Motion
Newton's First Law
Newton's Second Law

2 Gravity
Law of Universal Gravitation
Free Fall and Weight
Free Fall and Motion
Projectile Motion and Gravity

3 Newton's Third Law
Action and Reaction Forces
Momentum

344

Standards Correlations

National Science Education Standards

The following descriptions summarize the National Science Standards that specifically relate to this chapter. For the full text of the standards, see the National Science Education Standards at the front of the book.

Section 1 Laws of Motion
Physical Science PS4a

Section 2 Gravity
Physical Science PS4b

Section 3 Newton's Third Law
Physical Science PS4a

Focus ACTIVITY

Background When you kick a soccer ball, you are applying force to the ball. At the same time, the ball is also applying force to your foot. Force is involved in soccer in many ways. Soccer players have to know what force to apply to the ball. They must be able to anticipate the effects of force so that they will know where the ball is going and how to react. They sometimes experience force in the form of collisions with other players.

Another illustration of force in soccer is a change in the ball's direction. Soccer players need to know how to use force to kick the ball in a new direction or to continue to kick it in the same direction it had been going.

Activity 1 Hold this textbook at arm's length in front of your shoulders. Move the book from left to right and back again. Repeat these actions with a piece of paper. What differences do you notice between the effort needed to change the direction of the paper and the effort needed to change the direction of the textbook? Why would there be a difference?

Activity 2 You can investigate Earth's pull on objects by using a stopwatch, a board, and two balls of different masses. Set one end of the board on a chair and the other end on the floor. Time each ball as it rolls down the board. Then, several more times, roll both balls down the board at different angles by using books under one end of the board and changing the number of books to change the angle. Does the heavier ball move faster, move more slowly, or take the same amount of time as the lighter one? What factors do you think may have affected the motion of the two balls?

🖳 internet connect

www.scilinks.org
Topic: Forces SciLinks code: HK4055

SC*LINKS*. Maintained by the
National Science Teachers Association

Pre-Reading Questions

1. How is force used in other sports, such as basketball, baseball, and hockey? Give examples.
2. Give as many examples as you can think of about how force is involved in driving a car.

345

Soccer gives us many examples of the use of force.

Focus ACTIVITY

Background Soccer is the most popular sport in the world, and the World Cup— a national soccer competition that occurs every four years— may be the world's most popular athletic event. A regulation soccer ball has a mass of 397g–454g when inflated properly. When students study Newton's second law, give them this range of values, and have them calculate what forces are required for various accelerations of balls with masses in this range. **LS Logical**

Activity 1 This activity allows students to feel the effects of inertia. Students should note that the paper requires much less effort to move than the textbook. They should realize that this is because the textbook has a greater mass.

Activity 2 Do not allow students to use hollow balls for this activity. (This is because solid balls act more nearly like ideal point masses. Hollow thin-walled balls roll differently, and do not behave ideally.) The weight of the balls is not important. The closer the board is to vertical, the faster the ball accelerates.

Answers to Pre-Reading Questions

1. Accept all answers in which students describe examples of forces in various sports.
2. Accept any examples of forces involved in driving a car.

Chapter Resource File

• Pretest **GENERAL**
• Teaching Transparency Preview
• Answer Keys

SECTION 1

Focus

Overview

Before beginning this section, review with your students the Objectives listed in the Student Edition. In this section, students learn Newton's first law of motion, the law of inertia, and they learn that mass is a measure of inertia. Next, students study and apply Newton's second law of motion ($F = ma$).

Bellringer

Use the Bellringer transparency to prepare students for this section.

Motivate

Opening Demonstration — GENERAL

You will need a toy dump truck, a small doll, and a textbook. Set the doll in the back of the truck. Place the truck and the book on a flat surface, with the truck facing the book. Give the truck a push toward the book. When the truck hits the book and stops, the doll will fly out. Ask students to explain this in terms of inertia. (Since the doll and truck are in motion, they both have a tendency to stay in motion until experiencing an unbalanced force. The book exerts an unbalanced force on the truck; the truck stops. This force does not affect the doll, so the doll continues moving and flies out of the truck.) LS **Visual**

Transparencies

TT Bellringer

SECTION 1

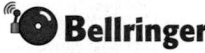

Laws of Motion

KEY TERMS
inertia

OBJECTIVES

- **Identify** the law that says that objects change their motion only when a net force is applied.
- **Relate** the first law of motion to important applications, such as seat belt safety issues.
- **Calculate** force, mass, and acceleration by using Newton's second law.

inertia the tendency of an object to resist being moved or, if the object is moving, to resist a change in speed or direction until an outside force acts on the object

Every motion you observe or experience is related to a force. Sir Isaac Newton described the relationship between motion and force in three laws that we now call Newton's laws of motion. Newton's laws apply to a wide range of motion—a caterpillar crawling on a leaf, a person riding a bicycle, or a rocket blasting off into space.

Newton's First Law

If you slide your book across a rough surface, such as carpet, the book will soon come to rest. On a smooth surface, such as ice, the book will slide much farther before stopping. Because there is less frictional force between the ice and the book, the force must act over a longer time before the book comes to a stop. Without friction, the book would keep sliding forever. This is an example of Newton's first law, which is stated as follows.

An object at rest remains at rest and an object in motion maintains its velocity unless it experiences an unbalanced force.

In a moving car, you experience the effect described by Newton's first law. As the car stops, your body continues forward, as the crash-test dummies in ***Figure 1*** do. Seat belts and other safety features are designed to counteract this effect.

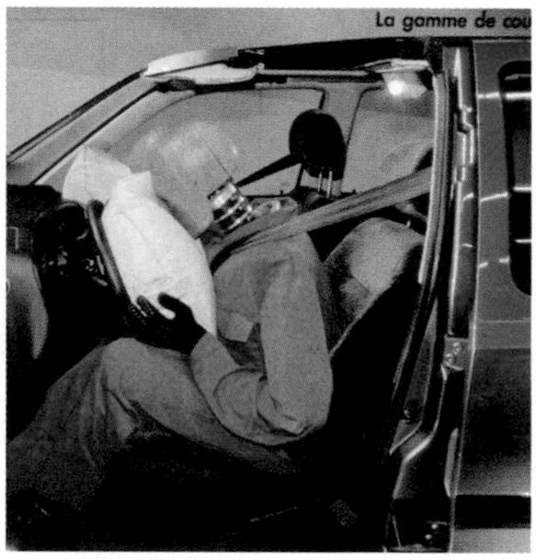

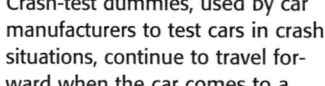

Figure 1
Crash-test dummies, used by car manufacturers to test cars in crash situations, continue to travel forward when the car comes to a sudden stop, in accordance with Newton's first law.

Objects tend to maintain their state of motion

Inertia is the tendency of an object at rest to remain at rest or, if moving, to continue moving at a constant velocity. All objects resist changes in motion, so all objects have inertia. An object with a small mass, such as a baseball, can be accelerated with a small force. But a much larger force is required to accelerate a car, which has a relatively large mass.

REAL-LIFE CONNECTION

Crash-Test Dummies Crash-test dummies come in various sizes and shapes. Each dummy is outfitted with sensors that record how the dummy moves and how hard it presses against different parts of the car during a crash when the dummy is strapped in by a seatbelt. Automobile manufacturers use this information to develop and improve seat belts and other safety devices, such as air bags and padded dashboards.

Inertia is related to an object's mass

Newton's first law of motion is often summed up in one sentence: Matter resists any change in motion. An object at rest will remain at rest until something makes it move. Likewise, a moving object stays in motion at the same velocity unless a force acts on it to change its speed or direction. Since this property of matter is called inertia, Newton's first law is sometimes called the *law of inertia.*

Mass is a measure of inertia. An object with a small mass has less inertia than an object with a large mass. Therefore, it is easier to change the motion of an object with a small mass. For example, a softball has less mass and less inertia than a bowling ball does. Because the softball has a small amount of inertia, it is easy to pitch, and the softball's direction will change easily when it is hit with a bat. Imagine how difficult playing softball with a bowling ball would be! The bowling ball would be hard to pitch, and changing its direction with a bat would be very difficult.

Seat belts and car seats provide protection

Because of inertia, you slide toward the side of a car when the driver makes a sharp turn. Inertia is also why it is impossible for a plane, car, or bicycle to stop instantaneously. There is always a time lag between the moment the brakes are applied and the moment the vehicle comes to rest.

When the car you are riding in comes to a stop, your seat belt and the friction between you and the seat stop your forward motion. They provide the unbalanced rearward force needed to bring you to a stop as the car stops.

Babies are placed in special backward-facing car seats, as shown in *Figure 2.* With this type of car seat, the force that is needed to bring the baby to a stop is safely spread out over the baby's entire body.

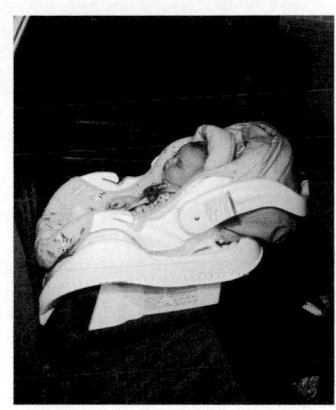

Figure 2

During an abrupt stop, this baby would continue to move forward. The backward-facing car seat distributes the force that holds the baby in the car.

Quick ACTIVITY

Newton's First Law

1. Set an index card over a glass. Put a coin on top of the card.
2. With your thumb and forefinger, quickly flick the card sideways off the glass. Observe what happens to the coin. Does the coin move with the index card?
3. Try again, but this time slowly pull the card sideways and observe what happens to the coin.
4. Use Newton's first law of motion to explain your results.

347

Alternative Assessment — GENERAL

Performing Magic Ask students to perform the dishes-under-the-tablecloth magic trick at home. Instruct them to start with a single, non-breakable dish and to use caution. If a student is able to master the trick, have him or her perform it for the class with non-breakable dishes. Be sure students are seated at a safe distance. Ask a volunteer to explain how inertia is involved in the demonstration. **LS Kinesthetic**

Teach

SKILL BUILDER — GENERAL

Interpreting Visuals Have students examine **Figure 2.** Explain to students that very young babies have little muscle control and are unable to even hold their head up. Have students discuss why infant car seats for small babies face backward while car seats for older children face forward.
LS Visual

Quick ACTIVITY

Newton's First Law

Materials:
- index card
- glass
- coin

Teacher's Notes: In step 2, students should observe the coin fall into the glass. The coin may move slightly sideways with the card. If students are having difficulty, emphasize that they must flick the card very quickly.

Make sure that students understand that in step 3, the coin should not fall into the glass; it should remain resting on the card. The force of friction caused the coin to move sideways with the card.
LS Visual

Chapter Resource File

- **Lesson Plan**
- **Quick Activity Datasheet** Newton's First Law GENERAL

Videos

CNN. Presents Physical Science

- **Segment 4** Crash-Test Dummies
- **Critical Thinking** Worksheet 4

See the Science in the News video guide for more details.

Teach, continued

Science and the Consumer ——— GENERAL

Should a Car's Air Bags Be Disconnected? The key to an air bag's success during a crash is the speed at which it inflates. Inside the bag is a gas generator that contains the compounds sodium azide, potassium nitrate, and silicon dioxide. At the moment of a crash, an electronic sensor in the vehicle detects the sudden decrease in speed. The sensor sends a small electric current to the gas generator, providing the energy needed to start the chemical reaction.

The force that triggers the inflation of an air bag is approximately the same as that of hitting a solid barrier head-on at 20 km/h. The chemicals sodium azide and potassium nitrate react to form nontoxic, nonflammable nitrogen gas, which is what actually inflates the bag. (Several toxic chemicals are also formed, but they quickly react with other substances to make them less hazardous.)

The rate at which the reaction occurs is very fast. In 0.04 s—less than the blink of an eye—the gas formed in the reaction inflates the bag. By filling the space between the person and the car's dashboard, the air bag protects the person from injury.

Answers to Your Choice

1. No, because your body moves toward the seat back and not toward the dashboard, where the air bag deploys.

2. Students' reports may vary.

LS Verbal

Science and the Consumer

Should a Car's Air Bags Be Disconnected?

Air bags are standard equipment in every new automobile sold in the United States. These safety devices are credited with saving almost 1700 lives between 1986 and 1996. However, air bags have also been blamed for the deaths of 36 children and 20 adults during the same period. In response to public concern about the safety of air bags, the National Highway Traffic Safety Administration has proposed that drivers be allowed to disconnect the air bags on their vehicles.

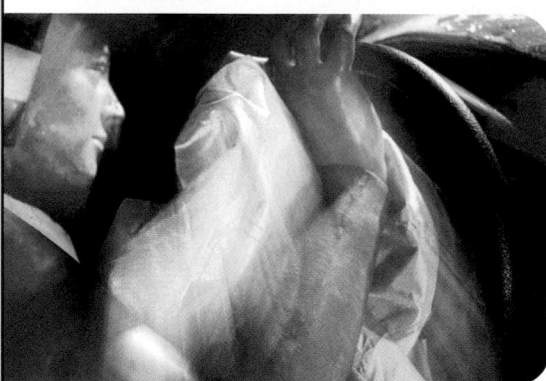

In a collision, air bags explode from a compartment to cushion the passenger's upper body and head.

How Do Air Bags Work?

When a car equipped with air bags comes to an abrupt stop, sensors in the car detect the sudden change in speed (negative acceleration) and trigger a chemical reaction inside the air bags. This reaction very quickly produces nitrogen gas, which causes the bags to inflate and explode out of their storage compartments in a fraction of a second. The inflated air bags cushion the head and upper body of the driver and the passenger in the front seat, who keep moving forward at the time of impact because of their inertia. Also, the inflated air bags increase the amount of time over which the stopping forces act. So, as the riders move forward, the air bags absorb the impact.

What Are the Risks?

Because an air bag inflates suddenly and with great force, it can cause serious head and neck injuries in some circumstances. Seat belts reduce this risk by holding passengers against the seat backs. This allows the air bag to inflate before the passenger's head comes into contact with it. In fact, most of the people killed by air bags either were not using seat belts or had not adjusted the seat belts properly.

Two groups of people are at risk for injury by air bags even with seat belts on: drivers shorter than about 157 cm (5 ft 2 in) and infants riding next to the driver in a rear-facing safety seat.

Alternatives to Disconnecting Air Bags

Always wearing a seat belt and placing child safety seats in the back seat of the car are two easy ways to reduce the risk of injury from air bags. Shorter drivers can buy pedal extenders that allow them to sit farther back and still safely reach the pedals. Some vehicles without a back seat have a switch that can deactivate the passenger-side air bag. Automobile manufacturers are also working on air bags that inflate less forcefully.

Your Choice

1. **Critical Thinking** Are air bags useful if your car is struck from behind by another vehicle?

2. **Locating Information** Research "smart" air-bag systems, and prepare a report.

internet connect

www.scilinks.org
Topic: Inertia SciLinks code: HK4072

SCI**LINKS** Maintained by the National Science Teachers Association

348

Did You Know?

Reduced Risk In purely frontal crashes, drivers protected by front air bags have a 31% reduction in fatality risk. The reduction in fatality risk for all types of crashes (purely frontal, partially frontal, and other) is 11%.

Newton's Second Law

Newton's first law describes what happens when the net force acting on an object is zero: the object either remains at rest or continues moving at a constant velocity. What happens when the net force is not zero? Newton's second law describes the effect of an unbalanced force on the motion of an object.

Force equals mass times acceleration

Newton's second law, which describes the relationship between mass, force, and acceleration, can be stated as follows.

> **The unbalanced force acting on an object equals the object's mass times its acceleration.**

Mathematically, Newton's second law can be written as follows.

Newton's Second Law
$$force = mass \times acceleration$$
$$F = ma$$

Consider the difference between pushing an empty shopping cart and pushing the same cart filled with groceries, as shown in *Figure 3*. If you push with the same amount of force in each situation, the empty cart will have a greater acceleration because it has a smaller mass than the full cart does. The same amount of force produces different accelerations because the masses are different. If the masses are the same, a greater force produces a greater acceleration, as shown in *Figure 4* on the next page.

www.scilinks.org
Topic: Newton's Laws of Motion
SciLinks code: HK4094

Figure 3

Because the full cart has a larger mass than the empty cart does, the same force gives the empty cart a greater acceleration.

349

MISCONCEPTION ALERT

Forces and Motions Many students have the misconception that when there is no unbalanced force, there is no motion. In other words, forces cause motions, and constant motion requires a constant force. You may have addressed this misconception when studying balanced and unbalanced forces. When studying Newton's laws, you can address this misconception again.

As seen with Newton's first law, an object experiencing no unbalanced force maintains its state of rest or motion. An unbalanced force is not needed to *keep* an object in motion, but to *change* its state of motion or rest. Newton's second law emphasizes this further. Unbalanced forces cause changes of *motion*; they cause *accelerations*.

Additional Examples — GENERAL

Newton's Second Law A 1200 kg car has a force of 1500 N from the engine pushing it forward. The car also has a combined frictional force of 1100 N pushing it backward. What is the acceleration of the car? (**Hint:** You will need to calculate the net force first.)

Answer: 0.33 m/s² forward

If the driver eases up on the gas pedal, the frictional force remains the same, and the engine is exerting a force of only 950 N, what is the new acceleration?

Answer: −0.13 m/s² forward = 0.13 m/s² backward (The car is slowing down.) **LS Logical**

Teaching Tip

Mass versus Weight The introduction of the pound as a unit of force may bring up the issue of weight. Remind students that mass and weight are not the same. In the SI system, mass is measured in kilograms, and weight—a force—is measured in newtons. The weight of an apple is about 1 newton. Tell students they will learn more about weight in this section, when they study gravity.

Figure 4

A A small force on an object causes a small acceleration.

B A larger force causes a larger acceleration.

Force is measured in newtons

Newton's second law can be used to derive the SI unit of force, the newton (N). One newton is the force that can give a mass of 1 kg an acceleration of 1 m/s², expressed as follows.

$$1 \text{ N} = 1 \text{ kg} \times 1 \text{ m/s}^2$$

The pound (lb) is sometimes used as a unit of force. One newton is equivalent to 0.225 lb. Conversely, 1 lb is equal to 4.448 N.

Did You Know?

Sir Isaac Newton (1642–1727) was a central figure in the Scientific Revolution during the seventeenth century. He was born in 1642, the same year that Galileo died.

Math Skills

Newton's Second Law Zookeepers lift a stretcher that holds a sedated lion. The total mass of the lion and stretcher is 175 kg, and the lion's upward acceleration is 0.657 m/s². What is the unbalanced force necessary to produce this acceleration of the lion and the stretcher?

1 List the given and unknown values.
  Given: *mass, m* = 175 kg
  acceleration, a = 0.657 m/s²
  Unknown: *force, F* = ? N

2 Write the equation for Newton's second law.
  force = mass × acceleration
  $F = ma$

3 Insert the known values into the equation, and solve.
  $F = 175 \text{ kg} \times 0.657 \text{ m/s}^2$
  $F = 115 \text{ kg} \times \text{m/s}^2 = 115 \text{ N}$

350

Alternative Assessment

Isaac Newton — ADVANCED

It is generally accepted that Isaac Newton was one of the greatest scientists in history. In addition to his work with forces, he made significant advancements in mathematics, astronomy, and other branches of physics. Have students research some aspect of his life or work. Possible topics include: his studies of light and color, his invention of Calculus, his personal history, or his personality and character. Ask each student to prepare a brief presentation for the class. **LS Verbal**

Concept Map — BASIC

Have students create a concept map for Newton's first and second laws. Instruct them to include all of the following terms in their maps: Newton's first law, inertia, mass, unbalanced force, balanced force, Newton's second law, acceleration, newtons. **LS Verbal**

Newton's Second Law

1. What is the net force necessary for a 1.6×10^3 kg automobile to accelerate forward at 2.0 m/s^2?

2. A baseball accelerates downward at 9.8 m/s^2. If the gravitational force is the only force acting on the baseball and is 1.4 N, what is the baseball's mass?

3. A sailboat and its crew have a combined mass of 655 kg. Ignoring frictional forces, if the sailboat experiences a net force of 895 N pushing it forward, what is the sailboat's acceleration?

Newton's second law can also be stated as follows:

The acceleration of an object is proportional to the net force on the object and inversely proportional to the object's mass.

Therefore, the second law can be written as follows.

$$acceleration = \frac{force}{mass}$$

$$a = \frac{F}{m}$$

Practice HINT

▶ When a problem requires you to calculate the unbalanced force on an object, you can use Newton's second law ($F = ma$).

▶ The equation for Newton's second law can be rearranged to isolate mass on the left side as follows.

$$F = ma$$

Divide both sides by a.

$$\frac{F}{a} = \frac{ma}{a}$$

$$m = \frac{F}{a}$$

You will need this form in Practice Problem 2.

▶ In Practice Problem 3, you will need to rearrange the equation to isolate acceleration on the left.

Practice

1. $F = ma = (1.6 \times 10^3 \text{ kg})$
   $(2.0 \text{ m/s}^2) = 3.2 \times 10^3$ N

2. $m = \dfrac{F}{a} = \dfrac{1.4 \text{ N}}{9.8 \text{ m/s}^2} =$
   0.14 kg

3. $a = \dfrac{F}{m} = \dfrac{895 \text{ N}}{655 \text{ kg}} = 1.37$ m/s^2
   in the direction of the force

LS Logical

Close

Quiz

1. What is Newton's first law of motion? (When there is no unbalanced force, an object at rest remains at rest, and an object in motion maintains its velocity.)

2. How is inertia related to mass? (Mass is a measure of inertia. A smaller mass has less inertia than a larger mass does.)

3. Express Newton's second law mathematically. ($F = ma$, or $a = F/m$)

4. What acceleration would be produced if a 25 N force acts on a 5.0 kg mass? ($a = F/m = 25 \text{ N}/5.0 \text{ kg} = 5.0$ m/s^2)

SECTION 1 REVIEW

SUMMARY

▶ An object at rest remains at rest and an object in motion maintains a constant velocity unless it experiences an unbalanced force (Newton's first law).

▶ Inertia is the property of matter that resists change in motion.

▶ Properly used seat belts protect passengers.

▶ The unbalanced force acting on an object equals the object's mass times its acceleration, or $F = ma$ (Newton's second law).

1. **State** Newton's first law of motion in your own words, and give an example that demonstrates that law.

2. **Explain** how the law of inertia relates to seat belt safety.

3. **Critical Thinking** Using Newton's laws, predict what will happen in the following situations:
   a. A car traveling on an icy road comes to a sharp bend.
   b. A car traveling on an icy road has to stop quickly.

Math Skills

4. What is the acceleration of a boy on a skateboard if the unbalanced forward force on the boy is 15 N? The total mass of the boy and the skateboard is 58 kg.

5. What force is necessary to accelerate a 1250 kg car at a rate of 40 m/s^2?

6. What is the mass of an object if a force of 34 N produces an acceleration of 4 m/s^2?

Answers to Section 1 Review

1. Answers may vary. Students should say in their own words that an object at rest remains at rest and an object in motion maintains its velocity unless it experiences an unbalanced force. They should each give an example of this law.

2. Answers may vary. Students should say something like this: When you have on a seat belt and the vehicle stops suddenly, the seat belt applies a force that stops you and keeps you from flying forward, as the force of inertia keeps you in motion after the car has stopped.

3. **a.** The car may be unable to turn. Newton's first law states that the object (car) will continue to travel in a straight line unless an unbalanced force acts on the object. Since the road is icy, the friction between the tires and the ice may not be large enough to turn the car.

   b. The car will slide for the same reasons in (a). Also, Newton's second law states that acceleration is proportional to force. Since the friction force is much smaller on an icy road, the negative acceleration (deceleration) is much smaller.

See "Continuation of Answers" at the end of the chapter.

Chapter Resource File

• **Math Skills** Newton's Second Law **GENERAL**

• **Concept Review** **GENERAL**

• **Quiz** **BASIC**

Focus

Overview

Before beginning this section, review with your students the Objectives listed in the Student Edition. In this section, students study gravitational force and the law of universal gravitation, free-fall acceleration, mass versus weight, terminal velocity, and projectile motion.

🔊 Bellringer

Use the Bellringer transparency to prepare students for this section.

Motivate

Opening Discussion ——— GENERAL

Ask students if they would weigh the same at sea level, on Mount Everest, and on the moon. Some students may already know that their weight would be less on the moon. Write the universal gravitation equation on the chalkboard and challenge students to use the equation to explain why weight would vary. (Be sure to explain what each symbol in the equation represents.) Students should also be able to deduce from the equation that they would weigh slightly less on Mount Everest than at sea level. **LS** Logical

Transparencies

TT Bellringer

Gravity

▶ **KEY TERMS**
free fall
terminal velocity
projectile motion

OBJECTIVES

▶ **Explain** that gravitational force becomes stronger as the masses increase and rapidly becomes weaker as the distance between the masses increases, $F = G\frac{m_1 m_2}{d^2}$.

▶ **Evaluate** the concept that free-fall acceleration near Earth's surface is independent of the mass of the falling object.

▶ **Demonstrate** mathematically how free-fall acceleration relates to weight.

▶ **Describe** orbital motion as a combination of two motions.

Have you ever seen a videotape of the first astronauts on the moon? When they tried to walk on the lunar surface, they bounced all over the place! Why did the astronauts—who were wearing heavy spacesuits—bounce so easily on the moon, as shown in *Figure 5*?

Law of Universal Gravitation

For thousands of years, two of the most puzzling scientific questions were "Why do objects fall toward Earth?" and "What keeps the planets in motion in the sky?" A British scientist, Sir Isaac Newton (1642–1727), realized that they were two parts of the same question. Newton generalized his observations on gravity in a law now known as the *law of universal gravitation.* The law states that all objects in the universe attract each other through gravitational force.

Universal Gravitation Equation

$$F = G\frac{m_1 m_2}{d^2}$$

This equation says that the gravitational force increases as one or both masses increase. It also says that the gravitational force decreases as the distance between the masses increases. In fact, because distance is squared in the equation, even a small increase in distance can cause a large decrease in force. The symbol G in the equation represents a constant.

Figure 5
Because gravity is less on the moon than on Earth, the Apollo astronauts bounced as they walked on the moon's surface.

352

SOCIAL STUDIES
CONNECTION

For your discussion of gravity, rent and play the video of Neil Armstrong's first step on the moon during the Apollo 11 mission.

Did You Know ❓

Universal Gravitation Equation The symbol G in the universal gravitation equation is a constant, sometimes called the *constant of universal gravitation.* Scientists have used experiments to determine the value of G: 6.673×10^{-11} N•m²/kg².

All matter is affected by gravity

Whether two objects are very large or very small, there is a gravitational force between them. When something is very large, such as Earth, the force is easy to detect. However, we do not notice that something as small as a toothpick exerts gravitational force. Yet no matter how small or how large the objects are, every object exerts a gravitational force, as illustrated by both parts of **Figure 6.** The force of gravity between two masses is easier to understand if you consider it in two parts: (1) the size of the masses and (2) the distance between them. So, these two ideas will be considered separately.

Gravitational force increases as mass increases

Gravity is given as the reason why an apple falls down from a tree. When an apple breaks its stem, it falls down because the gravitational force between Earth and the apple is much greater than the gravitational force between the apple and the tree.

Imagine an elephant and a cat. Because an elephant has a larger mass than a cat does, the gravitational force between an elephant and Earth is greater than the gravitational force between a cat and Earth. That is why a cat is much easier to pick up than an elephant! There is also gravitational force between the cat and the elephant, but it is very small because the cat's mass and the elephant's mass are so much smaller than Earth's mass. The gravitational force between most objects around you is relatively very small.

Figure 6

The arrows indicate the gravitational force between objects. The length of the arrows indicates the strength of the force.

A Gravitational force is small between objects that have small masses.

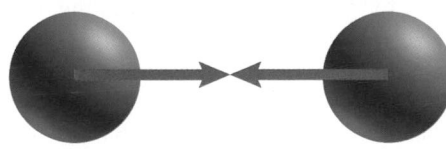

B Gravitational force is larger when one or both objects have larger masses.

LIFE SCIENCE
CONNECTION

In the reduced gravity of space, astronauts lose bone and muscle mass, even after a very short time. Sleep patterns may be affected, and so may cardiovascular strength and the immune response. These same effects happen more gradually as people age on Earth. Scientists are interested in studying the effects of microgravity so they can find ways to counteract them in space and here on Earth.

MISCONCEPTION ALERT

Gravity Some students think of gravity as "holding," or as "that which keeps things from floating away." Students also believe that gravity requires air; these students think that gravity is not present in space. Another common misconception is that gravity stops acting on an object when it has finished falling. Ask specific questions to target these misconceptions, such as "Can gravity occur in space?" and "Does gravity continue acting on a fallen object after it hits Earth's surface?"

Teach

INTEGRATING

BIOLOGY The human body makes use of gravity. One example is blood pressure. Gravity pulls blood to the lower part of the body; without gravity, blood would remain closer to the heart. Muscles then work as pumps to return blood back to the heart.

Teaching Tip

Weight Influences Shape The dramatic necessity of having a shape correct for your size can be shown by comparing an ant's legs to the legs of an elephant. An ant has very thin legs relative to the size of its body, while an elephant has thick legs. Students may have seen science fiction movies with giant ants, but in reality these giant ants could never exist. If an ant were somehow enlarged to the size of an elephant, it wouldn't even be able to walk. This is because strength is proportional to the cross-sectional area—how big the muscle is. In order to support its new, larger body, the ant would have to grow thick, strong legs like an elephant!

Chapter Resource File

- **Science Skills Worksheet** Equations Involving a Constant **BASIC**
- **Cross-Disciplinary Worksheet** Integrating Biology—Blood Pressure in Space **ADVANCED**
- **Cross-Disciplinary Worksheet** Connection to Social Studies—The Great Plague and Isaac Newton **GENERAL**

Transparencies

TM Law of Universal Gravitation

Videos

CNN Presents Physical Science

- **Segment 6** Egg Drop Contest
- **Critical Thinking** Worksheet 6

See the Science in the News video guide for more details.

Free-Fall Acceleration On the Apollo 15 mission, astronaut David Scott released a hammer and a feather on the moon at precisely at the same moment. Ask students to hypothesize about what happened in the experiment based on their knowledge of free-fall acceleration. (Since there is no air resistance on the moon, the hammer and feather fell with the same acceleration, and both hit the moon at the same instant.) Explain that since the moon's mass and radius are smaller than Earth's, free-fall acceleration on the moon is about 1/6 what it is on Earth. As a result, the acceleration of the hammer and feather was less than it would be in a vacuum on Earth, making the motions easier to observe.

If you have Internet access, you can show students a video of this experiment at the following NASA website: http://nssdc.gsfc.nasa.gov/planetary/lunar/apollo.html
LS Logical

SKILL BUILDER ━ **GENERAL**

Interpreting Visuals Have students look at **Figure 8**. Ask: What would happen if this experiment were performed outside, on Earth's surface? (The feather would fall slower than the apple because of air resistance.) Assuming that the vacuum is on Earth and that the apple and feather are dropped from a high enough distance that they do not hit the ground for 5 seconds, what would their speed be after 1 s, 2 s, and 3 s? (9.8 m/s, 19.6 m/s, 29.4 m/s) **LS** Logical

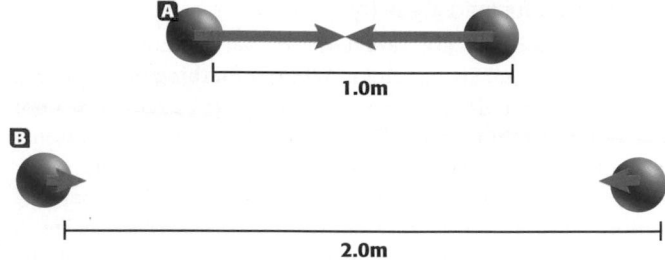

Figure 7

A Gravitational force rapidly becomes stronger as the distance between two objects decreases.

B Gravitational force rapidly becomes weaker as the distance between two objects increases.

1.0m

2.0m

▶ **free fall** the motion of a body when only the force of gravity is acting on the body

Gravitational force decreases as distance increases

Gravitational force also depends on the distance between two objects, as shown in **Figure 7**. The force of gravity changes as the distance between the balls changes. If the distance between the two balls is doubled, the gravitational force between them decreases to one-fourth its original value. If the original distance is tripled, the gravitational force decreases to one-ninth its original value. Gravitational force is weaker than other types of forces, even though it holds the planets, stars, and galaxies together.

Free Fall and Weight

When gravity is the only force acting on an object, the object is said to be in **free fall.** The free-fall acceleration of an object is directed toward the center of Earth. Because free-fall acceleration results from gravity, it is often abbreviated as the letter g. Near Earth's surface, g is approximately 9.8 m/s^2.

Free-fall acceleration near Earth's surface is constant

In the absence of air resistance, all objects near Earth's surface accelerate at the same rate, regardless of their mass. As shown in **Figure 8,** the feather and the apple, dropped from the same height, would hit the ground at the same moment. In this book, we disregard air resistance for all calculations. We assume that all objects on Earth accelerate at exactly 9.8 m/s^2.

Why do all objects have the same free-fall acceleration? Newton's second law shows that acceleration depends on both force and mass. A heavier object experiences a greater gravitational force than a lighter object does. But a heavier object is also harder to accelerate because it has more mass. The extra mass of the heavy object exactly compensates for the additional gravitational force.

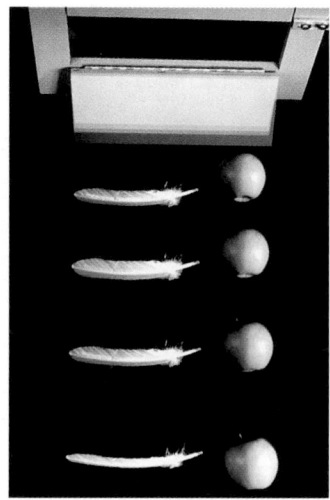

Figure 8

In a vacuum, a feather and an apple fall with the same acceleration because both are in free fall.

354

MATHEMATICS
CONNECTION ━ **ADVANCED**

Challenge students to use the universal gravitation equation and Newton's second law to demonstrate mathematically that free-fall acceleration is independent of the mass of the object, disregarding air resistance.

$F = Gm_1m_2/d^2 = Gm_{obj}m_{Earth}/(radius_{Earth})^2$

$F = ma$, so $a = F/m$

$a_{obj} = F/m_{obj} = Gm_{obj}m_{Earth}/(radius_{Earth})^2/m_{obj}$

$a_{obj} = G{\cancel{m_{obj}}}m_{Earth}/(radius_{Earth})^2/{\cancel{m_{obj}}}$

$a_{obj} = Gm_{Earth}/(radius_{Earth})^2$

Students can also substitute the values for G, m_{Earth}, and $radius_{Earth}$ to see that $a_{obj} = 9.8$ m/s^2. ($G = 6.673 \times 10^{-11}$ N•m^2/kg^2; $m_{Earth} = 5.98 \times 10^{24}$ kg; $r_{Earth} = 6.37 \times 10^6$ m)

$a_{obj} = Gm_{Earth}/(radius_{Earth})^2$

$a_{obj} = (6.673 \times 10^{-11}$ N•m^2/kg$^2)(5.98 \times 10^{24}$ kg$)/(6.37 \times 10^6$ m$)^2$

$a_{obj} = 9.8$ m/s^2

LS Logical

Weight is equal to mass times free-fall acceleration

The force on an object due to gravity is called its *weight*. On Earth, your weight is simply the amount of gravitational force exerted on you by Earth. If you know the free-fall acceleration, *g*, acting on a body, you can use $F = ma$ (Newton's second law) to calculate the body's weight. Weight equals mass times free-fall acceleration. Mathematically, this is expressed as follows.

$$weight = mass \times free\text{-}fall\ acceleration$$
$$w = mg$$

Note that because weight is a force, the SI unit of weight is the newton. For example, a small apple weighs about 1 N. A 1.0 kg book has a weight of 1.0 kg × 9.8 m/s² = 9.8 N.

You may have seen pictures of astronauts floating in the air, as shown in **Figure 9**. Does this mean that they don't experience gravity? In orbit, astronauts, the space shuttle, and all objects on board experience free fall because of Earth's gravity. In fact, the astronauts and their surroundings all accelerate at the same rate. Therefore, the floor of the shuttle does not push up against the astronauts and the astronauts appear to be floating. This situation is referred to as *apparent weightlessness*.

Weight is different from mass

Mass and weight are easy to confuse. Although mass and weight are directly proportional to one another, they are not the same. Mass is a measure of the amount of matter in an object. Weight is the gravitational force an object experiences because of its mass.

The weight of an object depends on gravity, so a change in an object's location will change the object's weight. For example, on Earth, a 66 kg astronaut weighs 66 kg × 9.8 m/s² = 650 N (about 150 lb), but on the moon's surface, where *g* is only 1.6 m/s², the astronaut would weigh 66 kg × 1.6 m/s², which equals 110 N (about 24 lb). The astronaut's mass remains the same everywhere, but the weight changes as the gravitational force acting on the astronaut changes in each place.

Weight influences shape

Gravitational force influences the shapes of living things. On land, large animals must have strong skeletons to support their mass against the force of gravity. The trunks of trees serve the same function. For organisms that live in water, however, the downward force of gravity is balanced by the upward force of the water. For many of these creatures, strong skeletons are unnecessary. Because a jellyfish has no skeleton, it can drift gracefully through the water but collapses if it washes up on the beach.

Figure 9

In the low-gravity environment of the orbiting space shuttle, astronauts experience apparent weightlessness.

INTEGRATING

SPACE SCIENCE
Planets in our solar system have different masses and different diameters. Therefore, each has its own unique value for *g*. Find the weight of a 58 kg person on the following planets:

Earth, where *g* = 9.8 m/s²

Venus, where *g* = 8.8 m/s²

Mars, where *g* = 3.7 m/s²

Neptune, where *g* = 11.8 m/s²

355

Interpreting Visuals Have students examine **Figure 10.** Ask: Since the two forces are balanced, is the sky diver at rest? Use Newton's laws to explain your answer. (No. According to Newton's first law, the sky diver will maintain his state of rest or motion if there are no unbalanced forces acting on him. Since he was in motion before the forces became balanced, he remains in motion at the same velocity that was reached at the moment the forces became balanced.) **LS** Logical

Teaching Tip

Terminal Velocity Students have learned that, in the *absence* of air resistance, all objects accelerate at the same rate. You can use the concept of terminal velocity to explain why, in the *presence* of air resistance, heavy objects fall faster.

A sky diver continues to accelerate until the force of gravity (the sky diver's weight) is balanced by the force of air resistance, which increases as speed increases. The moment the two forces are equal will occur later for a diver with a greater weight. For example, a 100 kg diver will accelerate until the force of air resistance is equal to the diver's weight of 980 N ($w = mg = 100 \text{ kg} \times 9.8 \text{ m/s}^2 = 980$ N), while a 50 kg diver would accelerate only until air resistance equals 490 N ($50 \text{ kg} \times 9.8 \text{ m/s}^2 = 490$ N). Since the heavier diver accelerates longer and then falls with a greater terminal velocity, that diver will reach Earth first.

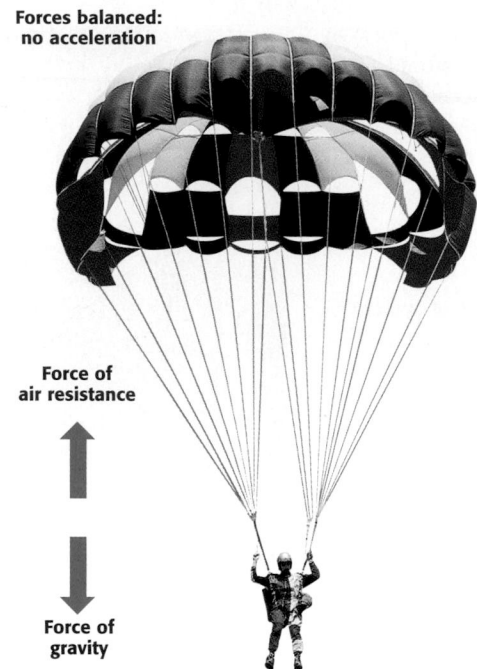

Forces balanced: no acceleration

Force of air resistance

Force of gravity

Figure 10

When a skydiver reaches terminal velocity, the force of gravity is balanced by air resistance.

▶ **terminal velocity** the constant velocity of a falling object when the force of air resistance is equal in magnitude and opposite in direction to the force of gravity

Velocity is constant when air resistance balances weight

Both air resistance and gravity act on objects moving through Earth's atmosphere. A falling object stops accelerating when the force of air resistance becomes equal to the gravitational force on the object (the weight of the object), as shown in *Figure 10.* This happens because the air resistance acts in the opposite direction to the weight. When these two forces are equal, the object stops accelerating and reaches its maximum velocity, which is called the **terminal velocity.**

When skydivers start a jump, their parachutes are closed, and they are accelerated toward Earth by the force of gravity. As their velocity increases, the force they experience because of air resistance increases. When air resistance and the force of gravity are equal, skydivers reach a terminal velocity of about 320 km/h (200 mi/h). But when they open the parachute, air resistance increases greatly. For a while, this increased air resistance slows them down. Eventually, they reach a new terminal velocity of several kilometers per hour, which allows them to land safely.

Free Fall and Motion

Skydivers are often described as being in free fall before they open their parachutes. However, that is an incorrect description, because air resistance is always acting on the skydiver. An object is in free fall only if gravity is pulling it down and no other forces are acting on it. Because air resistance is a force, free fall can occur only where there is no air—in a vacuum (a place in which there is no matter) or in space. Thus, a skydiver falling to Earth is not in free fall.

Because there is no air resistance in space, objects in space are in free fall. Consider a group of astronauts riding in a spacecraft. When they are in space, gravity is the only force acting on the spacecraft and the astronauts. As a result, the spacecraft and the astronauts are in free fall. They all fall at the same rate of acceleration, no matter how great or small their individual masses are.

MISCONCEPTION /// ALERT \\\

Falling Objects Many students believe that heavier objects fall faster, and they have trouble separating the effects of air resistance. Use discussion questions and the examples discussed throughout this section to address this misconception.

Did You Know ?

Spacediving In the 1950s, the Air Force began conducting experiments to learn about the hazards faced by flight crews bailing from high-altitude aircraft. Their experiments included three high-altitude "spacedives" by Captain Joseph Kittinger. His final jump, in August of 1960, was from the incredible height of 31,334 meters (102,800 feet, or almost 20 miles)! Even today Kittinger holds the record for the highest jump and the fastest speed attained by a human being in the atmosphere.

Orbiting objects are in free fall

Why do astronauts appear to float inside a space shuttle? Is it because they are "weightless" in space? You may have heard that objects are weightless in space, but this is not true. It is impossible to be weightless anywhere in the universe.

As you learned earlier in this section, weight—a measure of gravitational force—depends on the masses of objects and the distances between them. If you traveled in space far away from all the stars and planets, the gravitational force acting on you would be almost undetectable because the distance between you and other objects would be great. But you would still have mass, and so would all the other objects in the universe. Therefore, gravity would still attract you to other objects—even if just slightly—so you would still have weight.

Astronauts "float" in orbiting spaceships, not because they are weightless but because they are in free fall. The moon stays in orbit around Earth, as in **Figure 11,** and the planets stay in orbit around the sun, all because of free fall. To better understand why these objects continue to orbit and do not fall to Earth, you need to learn more about what orbiting means.

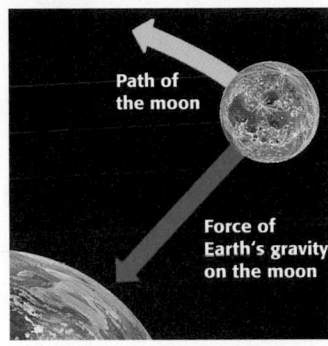

Figure 11

The moon stays in orbit around Earth because Earth's gravitational force provides a pull on the moon.

Two motions combine to cause orbiting

An object is said to be orbiting when it is traveling in a circular or nearly circular path around another object. When a spaceship orbits Earth, it is moving forward but it is also in free fall toward Earth. **Figure 12** shows how these two motions combine to cause orbiting.

Figure 12

How an Orbit Is Formed

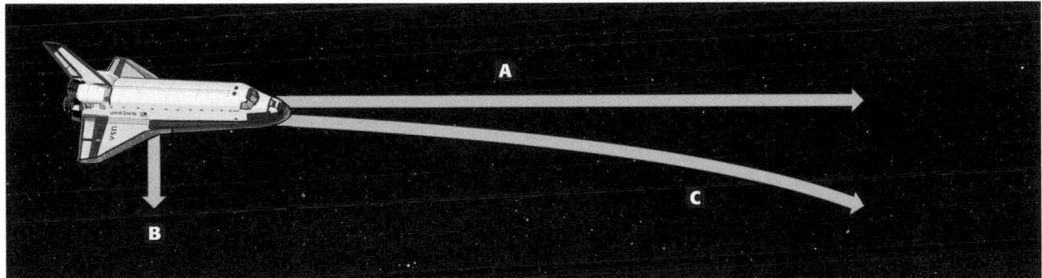

A The shuttle moves forward at a constant speed. This would be its path if there were no gravitational pull from Earth.

B The shuttle is in free fall because gravity pulls it toward Earth. This would be its path if it were not traveling forward.

C When the forward motion combines with free fall, the shuttle follows the curve of Earth's surface. This is known as *orbiting*.

357

Historical Perspective

Newton's Idea of Gravity One of Isaac Newton's amazing scientific achievements was his recognition that the force that pulls an apple toward Earth is the same force that holds the moon in its orbit around Earth. The legend about Newton first realizing this while watching an apple fall to the ground stems from an account written down by John Conduitt, Newton's assistant at the royal mint. (John Conduitt was also Newton's niece's husband.) Conduitt wrote:

In the year 1666 . . . while he was musing in a garden it came into his thought that the power of gravity (which brought an apple from a tree to the ground) was not limited to a certain distance from earth, but that this power must extend much further than was usually thought. Why not as high as the Moon thought he to himself & that if so, that must influence her motion & perhaps retain her in her orbit, whereupon he fell a-calculating what would be the effect of that superposition . . .

Teaching Tip —— BASIC

Newton's Thought Experiment

To help students understand how an orbiting space station is an example of projectile motion, you can use Newton's thought experiment. Newton imagined putting a cannon on top of a tall mountain. When the cannon is fired, the cannonball has projectile motion, and it eventually falls to Earth. The greater the ball's speed, the more distance it covers before hitting the ground.

Draw a few examples of this scenario on a chalkboard or overhead projector, with different speeds. Ask students what would eventually happen when the speed reached a high enough value. (Remind them that this is a thought experiment, and the setup is purely hypothetical.) By extrapolating from the previous examples, students should realize that, if the cannonball's speed were great enough, it would continually "fall" around Earth, never touching the ground; in other words, it would be in orbit around Earth. This is to demonstrate that an object orbiting Earth is simply an example of projectile motion. **LS Logical**

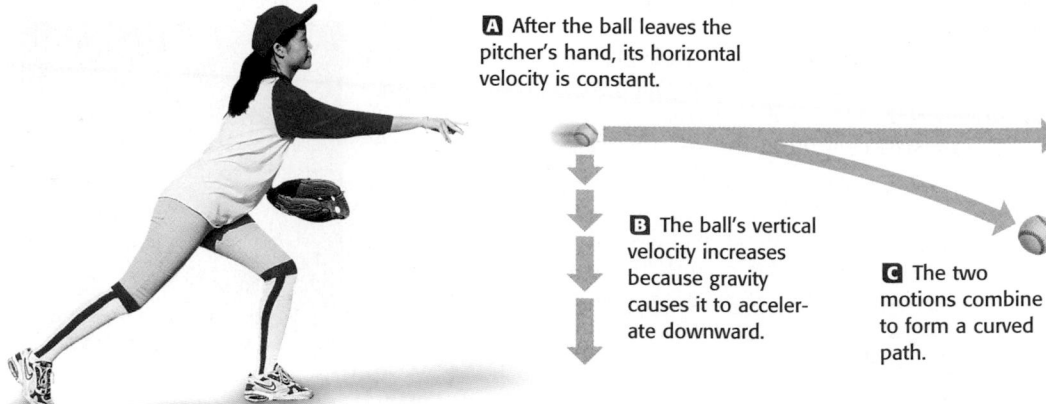

A After the ball leaves the pitcher's hand, its horizontal velocity is constant.

B The ball's vertical velocity increases because gravity causes it to accelerate downward.

C The two motions combine to form a curved path.

Figure 13
Two motions combine to form projectile motion.

▶ **projectile motion** the curved path that an object follows when thrown, launched, or otherwise projected near the surface of Earth; the motion of objects that are moving in two dimensions under the influence of gravity

Projectile Motion and Gravity

The orbit of the space shuttle around Earth is an example of **projectile motion.** Projectile motion is the curved path an object follows when thrown, launched, or otherwise projected near the surface of Earth. The motions of leaping frogs, thrown balls, and arrows shot from a bow are all examples of projectile motion. Projectile motion has two components—horizontal and vertical. The two components are independent; that is, they have no effect on each other. In other words, the downward acceleration due to gravity does not change a projectile's horizontal motion, and the horizontal motion does not affect the downward motion. When the two motions are combined, they form a curved path, as shown in **Figure 13.**

Projectile motion has some horizontal motion

When you throw a ball, your hand and arm exert a force on the ball that makes the ball move forward. This force gives the ball its horizontal motion. Horizontal motion is motion that is perpendicular (90°) to Earth's gravitational field.

After you have thrown the ball, there are no horizontal forces acting against the ball (if you ignore air resistance). Therefore, there are no forces to change the ball's horizontal motion. So, the horizontal velocity of the ball is constant after the ball leaves your hand, as shown in **Figure 13.**

Ignoring air resistance allows you to simplify projectile motion so that you can understand the horizontal and then the vertical components of projectile motion. Then, you can put them together to understand projectile motion as a whole.

358

SPACE SCIENCE — CONNECTION

When a rocket is sent into space, its velocity must be great enough that it can escape Earth's gravitational field. With a lower velocity, the rocket would either fall back toward Earth in projectile motion or fall into orbit around Earth. The velocity required to escape Earth's gravity—called *escape velocity*—depends on Earth's mass. Since the force of gravity is proportional to mass, a greater mass corresponds to a greater escape velocity. For example, the sun has a greater escape velocity than Earth.

Black holes are objects so massive that the speed required to escape their gravitational pull is greater than the speed of light. Einstein found that no mass can travel at or greater than the speed of light. As a result, no mass can escape the gravitational pull of a black hole. Even light—which always travels at the same speed, the fastest speed in the universe—cannot travel fast enough to escape a black hole!

Projectile motion also has some vertical motion

In addition to horizontal motion, vertical motion is involved in the movement of a ball that has been thrown. If it were not, the ball would continue moving in a straight line, never falling. Again imagine that you are throwing the ball as in *Figure 13*. When you let go of the ball, gravity pulls it downward, which gives the ball vertical motion. Vertical motion is motion in the direction in which the force of Earth's gravity acts.

In the absence of air resistance, gravity on Earth pulls downward with an acceleration of 9.8 m/s^2 on objects that are in projectile motion, just as it does on all falling objects. *Figure 14* shows that the downward accelerations of a thrown object and of a falling object are identical.

Because objects in projectile motion accelerate downward, you should always aim above a target if you want to hit it with a thrown or propelled object. This is why archers point their arrows above the bull's eye on a target. If they aimed an arrow directly at a bull's eye, the arrow would strike below the center of the target rather than the middle.

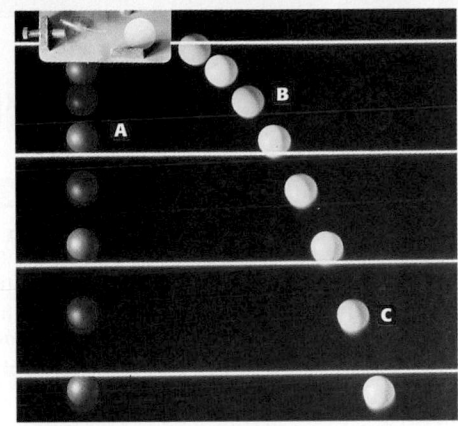

Figure 14

A The red ball was dropped without a horizontal push.

B The yellow ball was given a horizontal push off the ledge at the same time it was dropped. It follows a projectile-motion path.

C The two balls have the same acceleration downward because of gravity. The horizontal motion of the yellow ball does not affect its vertical motion.

Teaching Tip

Microgravity When an object such as a satellite or a space station is in free fall around Earth, a condition of microgravity is said to exist. Although astronauts in *microgravity* are not weightless, they appear to be so because they are falling around Earth at the same rate as their surroundings.

Close

Quiz

Choose the word that completes each sentence correctly.

1. The force of gravity is inversely proportional to mass/distance squared. (distance squared)

2. Disregarding air resistance, free-fall acceleration near Earth's surface does/does not depend on mass. (does not)

3. The mass/weight of an astronaut on the moon is the same as it is on Earth. (mass)

4. Terminal velocity is reached when the forces on a skydiver become balanced/unbalanced. (balanced)

SECTION 2 REVIEW

SUMMARY

▶ Gravitational force between two masses strengthens as the masses become more massive and rapidly weakens as the distance between them increases.

▶ Gravitational acceleration results from gravitational force, is constant, and does not depend on mass.

▶ Mathematically, *weight = mass × free-fall acceleration*, or *w = mg*.

▶ Projectile motion is a combination of a downward free-fall motion and a forward horizontal motion.

1. **State** the law of universal gravitation, and use examples to explain the effect of changing mass and changing distance on gravitational force.

2. **Explain** why your weight would be less on the moon than on Earth even though your mass would not change. Use the law of universal gravitation in your explanation.

3. **Describe** the difference between mass and weight.

4. **Name** the two components that make up orbital motion, and explain how they do so.

5. **Critical Thinking** Using Newton's second law, explain why the gravitational acceleration of any object near Earth is the same no matter what the mass of the object is.

Math Skills

6. The force between a planet and a spacecraft is 1 million newtons. What will the force be if the spacecraft moves to half its original distance from the planet?

359

Answers to Section 2 Review

1. Sample answer: The law of universal gravitation says that the force of gravitational acceleration is proportional to the attracting masses and inversely proportional to the square of the distance between the masses. If you make the mass of one or both of the attracting masses larger, then the force will be larger. (Weight on Jupiter is larger than weight on Earth because the mass of Jupiter is so much larger than the mass of the earth.) The attraction of Earth for a satellite gets smaller as the satellite gets farther away from Earth.

2. The moon's mass is much smaller than Earth's mass. Since the force of attraction depends on the mass of both attracting objects, the force of attraction between you and the moon would therefore be smaller than the force between you and Earth.

3. Mass is the resistance to change in motion of one object, whereas weight is a pull between two objects. Weight is proportional to the mass of those two objects and inversely proportional to the square of the distance between them. Mass is the amount of stuff, and doesn't change; weight is gravitational pull, and depends on location.

See "Continuation of Answers" at the end of the chapter.

Chapter Resource File

• **Concept Review** GENERAL

• **Quiz** BASIC

Transparencies

TM Projectile Motion

Videos

 Presents Physical Science

• **Segment 7** Zero-Gravity Plane

• **Critical Thinking** Worksheet 7

See the Science in the News video guide for more details.

SECTION
3

Focus

Overview

Before beginning this section, review with your students the objectives listed in the Student Edition. In the first part of this section, students learn about Newton's third law, also called the law of action and reaction. Next, they study momentum, including problem solving with momentum and the conservation of momentum. The section concludes with a discussion of rocket propulsion and additional examples of action-reaction pairs.

Bellringer

Use the Bellringer transparency to prepare students for this section.

Motivate

Opening Demonstration —— GENERAL

Use two identical horseshoe magnets and an overhead projector to demostrate Newton's third law. Place the two magnets close together on the overhead projector, with the poles facing one another. Release the magnets at the same moment, and have students observe their motion and final positions. The magnets should move away from one another and come to rest in a symmetrical pattern. **LS Visual**

Transparencies

TT Bellringer

Newton's Third Law

▶ **KEY TERMS**
momentum

OBJECTIVES

▶ **Explain** that when one object exerts a force on a second object, the second object exerts a force equal in size and opposite in direction on the first object.
▶ **Show** that all forces come in pairs commonly called *action* and *reaction pairs.*
▶ **Recognize** that all moving objects have momentum.

When you kick a soccer ball, as shown in *Figure 15,* you notice the effect of the force exerted by your foot on the ball. The ball experiences a change in motion. Is this the only force present? Do you feel a force on your foot? In fact, the soccer ball exerts an equal and opposite force on your foot. The force exerted on the ball by your foot is called the *action force*, and the force exerted on your foot by the ball is called the *reaction force.*

Action and Reaction Forces

Notice that the action and reaction forces are applied to different objects. These forces are equal and opposite. The action force acts on the ball, and the reaction force acts on the foot. This is an example of Newton's third law of motion, also called the *law of action and reaction.*

For every action force, there is an equal and opposite reaction force.

Figure 15

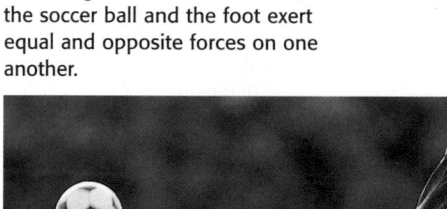

According to Newton's third law, the soccer ball and the foot exert equal and opposite forces on one another.

Forces always occur in pairs

Newton's third law can be stated as follows: All forces act in pairs. Whenever a force is exerted, another force occurs that is equal in size and opposite in direction. Action and reaction force pairs occur even when there is no motion. For example, when you sit on a chair, your weight pushes down on the chair. This is the action force. The chair pushing back up with a force equal to your weight is the reaction force.

BIOLOGY
CONNECTION

Newton's three laws of motions were published in the *Principia* in 1686. There were many significant developments in biology around the same time.

In 1665, Robert Hooke published *Micrographia*, which described his detailed observations of insects, sponges, and similar objects with one of the best compound microscopes of the day. Observing the cell walls in cork tissue, Hooke marveled at the regular pattern of repeating structures, which he also observed in other woods and in

plants. Since the box-like structures reminded him of the cells of a monastery, he coined the term "cells" to describe them.

Soon afterward, Antony van Leeuwenhoek— inspired by Hooke's work—learned how to make simple microscopes which produced greater clarity and brightness than other current microscopes. He observed a vast array of items over the next 50 years, and made a number of significant discoveries, including little "animalcules" (now known as bacteria and protozoa), red blood cells, and sperm cells.

Figure 16

A The action force is the swimmer pushing the water backward.

B The reaction force is the water pushing the swimmer forward.

Force pairs do not act on the same object

Newton's third law indicates that forces always occur in pairs. In other words, every force is part of an action and reaction force pair. Although the forces are equal and opposite, they do not cancel one another because they are acting on different objects. In the example shown in **Figure 16,** the swimmer's hands and feet exert the action force on the water. The water exerts the reaction force on the swimmer's hands and feet. In this and all other examples, the action and reaction forces do not act on the same object. Also note that action and reaction forces always occur at the same time.

Equal forces don't always have equal effects

Another example of an action-reaction force pair is shown in **Figure 17.** If you drop a ball, the force of gravity pulls the ball toward Earth. This force is the action force exerted by Earth on the ball. But the force of gravity also pulls Earth toward the ball. That force is the reaction force exerted by the ball on Earth.

It's easy to see the effect of the action force—the ball falls to Earth. Why don't you notice the effect of the reaction force—Earth being pulled upward? Remember Newton's second law: an object's acceleration is found by dividing the force applied to the object by the object's mass. The force applied to Earth is equal to the force applied to the ball. However, Earth's mass is much larger than the ball's mass, so Earth's acceleration is much smaller than the ball's acceleration.

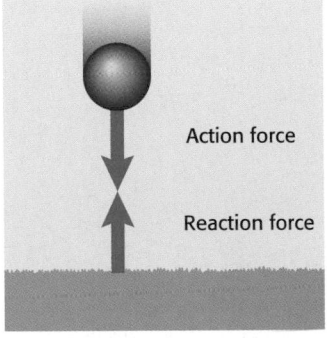

Action force

Reaction force

Figure 17

The two forces of gravity between Earth and a falling object are an example of a force pair.

361

Teaching Tip — BASIC

Momentum Introduce the subject of momentum by asking students for examples of non-scientific uses of the term, such as, "The baseball team started gaining momentum near the end of the season." After students read this page, discuss how the scientific use of the term *momentum* relates to its use in common speech. Also discuss how the two are different. LS **Verbal**

SKILL BUILDER — BASIC

Interpreting Visuals Have students examine **Figure 18.** Ask students: How could you increase the *momentum* of the bowling ball? (increase its velocity) Would a less massive ball at the same velocity have more or less momentum? (less) Do the pins have any momentum before the ball hits them? How do you know? (No, the pins do not have any momentum because they do not have any velocity.) LS **Logical**

internet connect
www.scilinks.org
Topic: Momentum
SciLinks code: HK4088

SCI LINKS. Maintained by the National Science Teachers Association

■ **momentum** a quantity defined as the product of the mass and velocity of an object

Momentum

If a compact car and a large truck are traveling with the same velocity and the same braking force is applied to each, the truck takes more time to stop than the car does. Likewise, a fast-moving car takes more time to stop than a slow-moving car with the same mass does. The truck and the fast-moving car have more **momentum** than the compact car and the slow-moving car do. Momentum is a property of all moving objects, which is equal to the product of the mass and the velocity of the object.

Moving objects have momentum

For movement along a straight line, momentum is calculated by multiplying an object's mass by its velocity. The SI unit for momentum is kilograms times meters per second (kg•m/s).

> **Momentum Equation**
>
> $$momentum = mass \times velocity$$
> $$p = mv$$

The momentum equation shows that for a given velocity, the more mass an object has, the greater its momentum is. A massive semi truck on the highway, for example, has much more momentum than a sports car traveling at the same velocity has. The momentum equation also shows that the faster an object is moving, the greater its momentum is. For instance, a fast-moving train has much more momentum than a slow-moving train with the same mass has. If an object is not moving, its momentum is zero.

Like velocity, momentum has direction. An object's momentum is in the same direction as its velocity. The momentum of the bowling ball shown in *Figure 18* is directed toward the pins.

Figure 18

Because of the large mass and high speed of this bowling ball, it has a lot of momentum and is able to knock over the pins easily.

 362

Alternative Assessment — ADVANCED
Momentum and Newton's Second Law
If students have read Newton's laws in the *Principia*, they have probably noticed that Newton actually uses momentum rather than acceleration in his second law. If students did not read this, explain that in his second law of motion, Newton states that force is proportional to the "change of motion." Also explain that he defines motion as velocity multiplied by the "quantity of matter," or velocity times mass (momentum). Thus, Newton's own version of the second law states that force is proportional to

change in momentum. Challenge students to mathematically demonstrate how this is equivalent to our modern formulation of $f = ma$.

force = change in momentum $(F = \Delta p)$
momentum = mass × velocity $(p = mv)$
force = mass × change in velocity $(F = m\Delta v)$
change in velocity = acceleration $(\Delta v = a)$
force = mass × acceleration $(F = ma)$

Symbolically:
$F = \Delta p = \Delta mv = m\Delta v = ma$
LS **Logical**

Momentum Calculate the momentum of a 6.00 kg bowling ball moving at 10.0 m/s down the alley toward the pins.

1 **List the given and unknown values.**
Given: *mass, m* = 6.00 kg
velocity, v = 10.0 m/s down the alley
Unknown: *momentum, p* = ? kg • m/s (and direction)

2 **Write the equation for momentum.**
momentum = mass × velocity, p = mv

3 **Insert the known values into the equation, and solve.**
$p = mv = 6.00 \text{ kg} \times 10.0 \text{ m/s}$
$p = 60.0$ kg • m/s down the alley

Practice

Momentum

1. Calculate the momentum of the following objects.
   a. a 75 kg speed skater moving forward at 16 m/s
   b. a 135 kg ostrich running north at 16.2 m/s
   c. a 5.0 kg baby on a train moving eastward at 72 m/s
   d. a seated 48.5 kg passenger on a train that is stopped
2. Calculate the velocity of a 0.8 kg kitten with a momentum of 5 kg • m/s forward.

Force is related to change in momentum

To catch a baseball, you must apply a force on the ball to make it stop moving. When you force an object to change its motion, you force it to change its momentum. In fact, you are actually changing the momentum of the ball over a period of time.

As the time period of the momentum's change becomes longer, the force needed to cause this change in momentum becomes smaller. So, if you pull your glove back while you are catching the ball, as in *Figure 19,* you are extending the time for changing the ball's momentum. Extending the time causes the ball to put less force on your hand. As a result, the sting to your hand is reduced.

As another example, when pole-vaulters, high jumpers, and gymnasts land after jumping, they move in the direction of the motion. This motion extends the time of the momentum change. As a result, the impact force decreases.

Practice HINT

▶ When a problem requires that you calculate velocity when you know momentum and mass, you can use the momentum equation.
▶ You may rearrange the equation to isolate velocity on the left side, as follows: $v = \frac{p}{m}$. You will need this form of the momentum equation for Practice Problem 2.

Figure 19

Moving the glove back during the catch increases the time of the momentum's change and decreases the impact force.

363

Additional Examples

Momentum An athlete with a mass of 73.0 kg runs with a constant forward velocity of 1.50 m/s. What is the athlete's momentum?

Answer: 110 kg • m/s forward (rounded to correct significant figures)

If a car with a mass of 925 kg has the same momentum as the athlete, what is the car's speed?

Answer: 0.119 m/s

LS Logical

Practice

1. $p = mv$
   a. (75 kg)(16 m/s) = 1200 kg•m/s forward
   b. (135 kg)(16.2 m/s) = 2187 kg•m/s north (using significant figures: 2190 kg•m/s north)
   c. (5.0 kg)(72 m/s) = 360 kg•m/s eastward
   d. (48.5 kg)(0 m/s) = 0 kg • m/s

2. $v = \frac{p}{m} = \frac{5 \text{ kg•m/s}}{0.8 \text{ kg}} =$
   6.3 m/s forward (using significant figures)

LS Logical

REAL-LIFE CONNECTION

Air Bags Air bags in automobiles take advantage of the fact that force decreases when the time period of a change in momentum increases. If a driver is in a collision, the driver keeps moving forward (in accordance with Newton's first law of motion), even after the vehicle stops. If the driver hits a window or dashboard, his change in momentum occurs very quickly, so the force on the driver is great. Air bags increase the time interval over which momentum changes and, as a result, the force on the driver is greatly reduced.

Chapter Resource File

- **Math Skills** Momentum GENERAL
- **Cross-Disciplinary Worksheet** Connection to Fine Arts—Momentum of Line in Art ADVANCED
- **Cross-Disciplinary Worksheet** Applications—Driving Safety ADVANCED

Newton's Cradle You can use a toy called "Newton's cradle" to demonstrate the conservation of momentum. Newton's cradle—a common office toy—consists of five or seven steel balls suspended from a horizontal bar, hanging close enough together that the balls are all touching when they are at rest. If a ball from one side is picked up and then released, a ball from the other side flies out, and the motion continues back and forth. If two balls are used, two balls move on the other side, and so on. Show students what happens when you pick up one, two, and three balls, and discuss how the toy demonstrates the conservation of momentum. You can also discuss the demonstration in terms of action/reaction forces.

SKILL BUILDER — GENERAL

Interpreting Visuals Ask students how they could calculate the velocity of the billiard ball in **Figure 20C,** given the mass of each ball and the initial velocity of the cue ball. (Since momentum is conserved, the total momentum before the collision (mass of the cue ball × initial velocity of the cue ball) equals the total momentum after the collision (mass of the billiard ball × final velocity of the billiard ball). This equation can be rearranged to find the billiard ball's velocity after the collision:

$$m_c v_{c(initial)} = m_b v_{b(final)}$$
$$v_{b(final)} = m_c v_{c(initial)} / m_b)$$

LS Logical

Figure 20

A The cue ball is moving forward with momentum. The billiard ball's momentum is zero.

B When the cue ball hits the billiard ball, the action force makes the billiard ball move forward. The reaction force stops the cue ball.

C Because the cue ball's momentum was transferred to the billiard ball, the cue ball's momentum after the collision is zero.

Momentum is conserved in collisions

Imagine that two cars of different masses traveling with different velocities collide head on. Can you predict what will happen after the collision? The momentum of the cars after the collision can be predicted. This prediction can be made because, in the absence of outside influences, momentum is conserved. Some momentum may be transferred from one vehicle to the other, but the total momentum remains the same. This principle is known as the law of conservation of momentum.

The total amount of momentum in an isolated system is conserved.

The total momentum of the two cars before a collision is the same as the total momentum after the collision. This law applies whether the cars bounce off each other or stick together. In some cases, cars bounce off each other to move in opposite directions. If the cars stick together after a collision, the cars will continue in the direction of the car that originally had the greater momentum.

Momentum is transferred

When a moving object hits a second object, some or all of the momentum of the first object is transferred to the second object. If only some of the momentum is transferred, the rest of the momentum stays with the first object.

Imagine you hit a billiard ball with a cue ball so that the billiard ball starts moving and the cue ball stops, as shown in *Figure 20.* The cue ball had a certain amount of momentum before the collision. During the collision, all of the cue ball's momentum was transferred to the billiard ball. After the collision, the billiard ball moved away with the same amount of momentum the cue ball had originally. This example illustrates the law of conservation of momentum. Any time two or more objects interact, they may exchange momentum, but the total momentum stays the same.

Newton's third law can explain conservation of momentum. In the example of the billiard ball, the cue ball hit the billiard ball with a certain force. This force was the action force. The reaction force was the equal but opposite force exerted by the billiard ball on the cue ball. The action force made the billiard ball start moving, and the reaction force made the cue ball stop moving.

364

Historical Perspective

Descartes and Huygens The concept of momentum was first proposed by the French philosopher Rene Descartes (1596–1650). Descartes defined momentum as "amount of motion," or "mass times speed." Descartes thought momentum was a useful quantity because it is conserved in some cases. However, according to this definition, momentum would not always be conserved. For example, if two bowling balls both come to a stop after a head-on collision, momentum would not be conserved. Descartes was unable to reconcile this discrepancy.

This was left to the Dutch scientist Christian Huygens (1629–1695), who realized that momentum would be conserved in all examples if direction is taken into account. With the example of the two bowling balls, if the momentum of one is positive and that of the other negative (since the directions are opposite), the total momentum is the same before and after the collision. In other words, Huygens realized that in order to be conserved, momentum must be the product of mass and *velocity*, rather than of mass and *speed*.

Conservation of momentum explains rocket propulsion

Newton's third law and the conservation of momentum are used in rocketry. Rockets have many different sizes and designs, but the basic principle remains the same. The push of the hot gases through the nozzle is matched by an equal push in the opposite direction on the combustion (burning) chamber, which accelerates the rocket forward, as shown in *Figure 21*.

Many people wrongly believe that rockets work because the hot gases flowing out the nozzle push against the atmosphere. If this were true, rockets couldn't travel in outer space where there is no atmosphere. What really happens is that momentum is conserved. Together, the rocket and fuel form a system. When the fuel is pushed out the back, they remain a system. The change in the fuel's momentum must be matched by a change in the rocket's momentum in the opposite direction for the overall momentum to stay the same. This example shows the conservation of momentum. Also, the upward force on the rocket and the downward force on the fuel are an action-reaction pair.

Action and reaction force pairs are everywhere

Figure 22 gives more examples of action-reaction pairs. In each example, notice which object exerts the action force and which object exerts the reaction force. Even though we are concentrating on just the action and reaction force pairs, there are other forces at work, each of which is also part of an action-reaction pair. For example, when the bat and ball exert action and reaction forces, the bat does not fly toward the catcher, because the batter is exerting yet another force on the bat.

Figure 22

Examples of Third Law Forces

A The rabbit's legs exert a force on Earth. Earth exerts an equal force on the rabbit's legs, which causes the rabbit to accelerate upward.

B The bat exerts a force on the ball, which sends the ball into the outfield. The ball exerts an equal force on the bat.

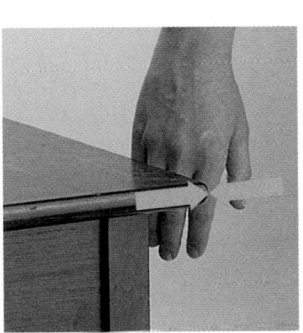

C When your hand hits the table with a force, the table's force on your hand is equal in size and opposite in direction.

Gases push rocket forward

— Hydrogen

— Oxygen

— Combustion chamber

Rocket pushes gases backward

Figure 21

All forces occur in action-reaction pairs. In this case, the upward push on the rocket equals the downward push on the exhaust gases.

365

Quick Lab

How are action and reaction forces related?

Analysis

1. If students have difficulty with item 1, encourage them to list the forces acting on each object separately. For example, the forces acting on the middle spring scale are: its weight, the downward pull of the 2 kg mass, and the upward pull of the top spring scale.

See "Continuation of Answers" at the end of the chapter.

Close

Quiz

Choose the term that best completes each sentence.

1. Action-reaction forces are equal/unequal in size and opposite in direction. (equal)

2. Momentum is equal to mass times speed/velocity. (velocity)

3. Momentum is always/sometimes conserved. (always)

Quick Lab

How are action and reaction forces related?

Materials ✓ 2 spring scales ✓ 2 kg mass

1. Hang the 2 kg mass from one of the spring scales.

2. Observe the reading on the spring scale.

3. While keeping the mass connected to the first spring scale, link the two scales together. The first spring scale and the mass should hang from the second spring scale, as shown in the figure at right.

4. Observe the readings on each spring scale.

Analysis

1. What are the action and reaction forces involved in the spring scale–mass system you have constructed?

2. How did the readings on the two spring scales in step 4 compare? Explain how this is an example of Newton's third law.

SECTION 3 REVIEW

SUMMARY

▶ When one object exerts an action force on a second object, the second object exerts a reaction force on the first object. Forces always occur in action-reaction force pairs.

▶ Action-reaction force pairs are equal in size and opposite in direction.

▶ Every moving object has momentum, which is the product of the object's mass and velocity.

▶ Momentum can be transferred in collisions, but the total momentum before and after a collision is the same. This is the law of conservation of momentum.

▶ Rocket propulsion can be explained in terms of conservation of momentum.

1. **State** Newton's third law of motion, and give an example that shows how this law works.

2. **List** three examples of action-reaction force pairs that are not mentioned in the chapter.

3. **Define** momentum, and explain what the law of conservation of momentum means.

4. **Explain** why, when a soccer ball is kicked, the action and reaction forces don't cancel each other.
   a. The force of the player's foot on the ball is greater than the force of the ball on the player's foot.
   b. They act on two different objects.
   c. The reaction force happens after the action force.

5. **Critical Thinking** The forces exerted by Earth and a skier become an action-reaction force pair when the skier pushes the ski poles against Earth. Explain why the skier accelerates while Earth does not seem to move at all. (**Hint:** Think about the math equation for Newton's second law of motion for each of the forces.)

Math Skills

6. **Calculate** the momentum of a 1.0 kg ball traveling eastward at 12 m/s.

366

Answers to Section 3 Review

1. Newton's third law states that any time one object applies a force to a second object, the second object applies a force on the first object that is equal in size and opposite in direction. For example, when you sit in a chair, you push down on the chair (action force), and the chair pushes up on you (reaction force).

2. Answers may vary. They can include any action-reaction force pairs, such as fingers and keyboard (on computer or piano, etc.), cheese and cheese cutter, person and floor, etc.

3. Momentum is mass × velocity. The law of conservation of momentum says that in any system of group of objects, the momentum will not

change if a net outside force does not act on the system or group of objects.

4. b

5. The forces exerted by Earth and a skier are an action-reaction force pair because Earth exerts an equal and opposite force on the skier as the skier had exerted on Earth. Since $F = ma$ and F is equal for both, the smaller mass of the skier (when compared to the mass of Earth) will mean the skier's acceleration away from the skier will be very much greater than Earth's unnoticeable acceleration away from the skier.

See "Continuation of Answers" at the end of the chapter.

Math Skills

Algebraic Rearrangements

A car's engine exerts a force of 1.5×10^4 N in the forward direction, while friction exerts an opposing force of 9.0×10^3 N. If the car's mass is 1.5×10^3 kg, what is the magnitude and direction of the car's net acceleration?

1 **List all given and unknown values.**

> **Given:** *forward force, $F_1 = 1.5 \times 10^4$ N*
> *opposing force, $F_2 = 9.0 \times 10^3$ N*
> *mass, $m = 1.5 \times 10^3$ kg*

> **Unknown:** acceleration (m/s^2)

2 **Write down the equation for force, and rearrange it for calculating the acceleration.**

$$F = ma$$
$$a = \frac{F}{m}$$

3 **Solve for acceleration.**

> The net acceleration must be obtained from the net force, which is the overall, unbalanced force acting on the car.

$$F_{net} = F_1 - F_2 \text{, so}$$
$$ma_{net} = F_1 - F_2$$

> The net acceleration is therefore

$$a_{net} = \frac{(F_1 - F_2)}{m} = \frac{(1.5 \times 10^4 \text{ N} - 9.0 \times 10^3 \text{ N})}{1.5 \times 10^3 \text{ kg}} = \frac{6.0 \times 10^3 \text{ N}}{1.5 \times 10^3 \text{ kg}} = 4.0 \text{ m/s}^2$$

> The net acceleration is 4.0 m/s^2, and like the net force, it is in the forward direction.

Practice

Following the example above, calculate the following:

1. A car has a mass of 1.50×10^3 kg. If the net force acting on the car is 6.75×10^3 N to the east, what is the car's acceleration?

2. A bicyclist decelerates with a force of 3.5×10^2 N. If the bicyclist and bicycle have a total mass of 1.0×10^2 kg, what is the acceleration?

3. Roberto and Laura are studying across from each other at a wide table. Laura slides a 2.2 kg book toward Roberto with a force of 11.0 N straight ahead. If the force of friction opposing the movement is 8.4 N, what is the magnitude and direction of the book's acceleration?

Practice ——— GENERAL
Algebraic Rearrangements

Teaching Tip
Net Forces and Acceleration
Remind students that the net acceleration describes the change in the speed of an object due to the unbalanced forces acting on it. While there is an acceleration associated with each force, these are in the same direction of the force. Thus, any force in the opposite direction provides an acceleration that is also in the opposite direction, and which must cancel with the original acceleration. The net acceleration is the result of this cancellation, and accounts for the motion that is actually observed.

Be sure that students understand that, in problems where more than one force or acceleration value are given, they must determine whether these are in the same or opposite direction to each other, and thus add or subtract them accordingly. Once the net force or acceleration has been found, the net acceleration or force, respectively, can be calculated.

Answers
1. 4.5 m/s^2 to the east
2. 3.5 m/s^2 in a backward direction (deceleration)
3. $a_{net} = (F_1 - F_2)/m =$
   11.0 N − 8.4 N / 2.2 kg =
   1.2 m/s^2 (using significant figures)

LS Logical

Chapter Resource File

• **Science Skills Worksheet** Ordering Multiple Operations

Understanding Concepts

1. d
2. c
3. a
4. b
5. c
6. d
7. b
8. d
9. b
10. d
11. a
12. d
13. c
14. c
15. b

Using Vocabulary

16. Inertia is the behavior of matter that defines mass in the laws of motion. It is important because inertia resists the effects of net force. The more mass or inertia an object has, the smaller the acceleration for a given net force.

17. The *newton* is the unit of measure for force in the SI system. 1 newton = 1 kg•m/s^2

Chapter Highlights

Before you begin, review the summaries of the key ideas of each section, found at the end of each section. The vocabulary terms are listed on the first page of each section.

UNDERSTANDING CONCEPTS

1. Newton's first law of motion states that an object
   a. at rest remains at rest unless it experiences an unbalanced force.
   b. in motion maintains its velocity unless it experiences an unbalanced force.
   c. will tend to maintain its motion unless it experiences an unbalanced force.
   d. All of the above

2. The first law of motion applies to
   a. only objects that are moving.
   b. only objects that are not moving.
   c. all objects, whether moving or not.
   d. no object, whether moving or not.

3. A measure of inertia is an object's
   a. mass.
   b. weight.
   c. velocity.
   d. acceleration.

4. Automobile seat belts are necessary for safety because of a passenger's
   a. weight.
   b. inertia.
   c. speed.
   d. gravity.

5. The newton is a measure of
   a. mass.
   b. length.
   c. force.
   d. acceleration.

6. Any change in an object's velocity is caused by
   a. the object's mass.
   b. the object's direction.
   c. a balanced force.
   d. an unbalanced force.

7. _____ is a force that opposes the motion between two objects in contact with each other.
   a. Motion   c. Acceleration
   b. Friction   d. Velocity

8. An object's acceleration is
   a. directly proportional to the net force.
   b. inversely proportional to the object's mass.
   c. in the same direction as the net force.
   d. All of the above

9. Suppose you are pushing a car with a certain net force. If you then push with twice the net force, the car's acceleration
   a. becomes four times as much.
   b. becomes two times as much.
   c. stays the same.
   d. becomes half as much.

10. Gravitational force between two masses _____ as the masses increase and rapidly _____ as the distance between the masses increases.
    a. strengthens, strengthens
    b. weakens, weakens
    c. weakens, strengthens
    d. strengthens, weakens

11. According to Newton's third law, when a 450 N teacher stands on the floor, that teacher applies a force of 450 N on the floor, and
    a. the floor applies a force of 450 N on the teacher.
    b. the floor applies a force of 450 N on the ground.
    c. the floor applies a force of 450 N in all directions.
    d. the floor applies an undetermined force on the teacher.

368

18. Free fall literally means falling freely. Falling straight down is free fall. An object in orbit falls freely toward Earth but is also propelled forward. The combination of those two motions is what keeps the object in orbit. Because we experience our weight by pressing against Earth or an object supported by Earth, when we fall freely, we don't feel our weight. We feel "weightless." However, weightless actually means that there is no force of gravity acting. (Weight is the force of gravity acting on an object.) So there's no such thing as actual weightlessness. Weight of objects that are very far from any other objects is very small, but it is not zero.

19. The wrestler will weigh less on the moon than he does on Earth, because the *force* exerted on him will be different at these locations. Since he has the same *mass* in both places, his *weight* will depend on the acceleration due to *gravity* at each location. The moon causes a much smaller acceleration due to *gravity*, so the wrestler will weigh less there.

12. An example of action-reaction forces is
 a. air escaping from a toy balloon.
 b. a rocket traveling through the air.
 c. a ball bouncing off a wall.
 d. All of the above

13. A baseball player catches a line-drive hit. If the reaction force is the force of the player's glove stopping the ball, the action force is
 a. the force of the player's hand on the glove.
 b. the force applied by the player's arm.
 c. the force of the ball pushing on the glove.
 d. the force of the player's shoes on the dirt.

14. Which of the following objects has the smallest amount of momentum?
 a. a loaded tractor-trailer driven at highway speeds
 b. a track athlete running a race
 c. a motionless mountain
 d. a child walking slowly on the playground

15. The force that causes a space shuttle to accelerate is exerted on the shuttle by
 a. the exhaust gases pushing against the atmosphere.
 b. the exhaust gases pushing against the shuttle.
 c. the shuttle's engines.
 d. the shuttle's wings.

USING VOCABULARY

16. What is *inertia*, and why is it important in the laws of motion?

17. What is the unit of measure of force, and how is it related to other measurement units?

18. What is the difference between *free fall* and *weightlessness*?

19. A wrestler weighs in for the first match on the moon. Will the athlete weigh more or less on the moon than he does on Earth? Explain your answer by using the terms *weight, mass, force,* and *gravity*.

20. Describe a skydiver's jump from the airplane to the ground. In your answer, use the terms *air resistance, gravity,* and *terminal velocity.*

21. Give an example of *projectile motion,* and explain how the example demonstrates that a vertical force does not influence horizontal motion.

BUILDING MATH SKILLS

22. Force The net force acting on a 5 kg discus is 50 N. What is the acceleration of the discus?

23. Force A 5.5 kg watermelon is pushed across a table. If the acceleration of the watermelon is 4.2 m/s^2 to the right, what is the net force exerted on the watermelon?

24. Force A block pushed with a force of 13.5 N accelerates at 6.5 m/s^2 to the left. What is the mass of the block?

25. Force The net force on a 925 kg car is 37 N as it pulls away from a stop sign. Find the car's acceleration.

26. Force What is the unbalanced force on a boy and his skateboard if the total mass of the boy and skateboard is 58 kg and their acceleration is 0.26 m/s^2 forward?

27. Force A student tests the second law of motion by accelerating a block of ice at a rate of 3.5 m/s^2. If the ice has a mass of 12.5 kg, what force must the student apply to the ice?

28. Weight A bag of sugar has a mass of 2.26 kg. What is its weight in newtons on the moon, where the acceleration due to gravity is one-sixth of that on Earth? (**Hint:** On Earth, g = 9.8 m/s^2.)

29. Weight What would the 2.26 kg bag of sugar from the previous problem weigh on Jupiter, where the acceleration due to gravity is 2.64 times that on Earth?

369

Assignment Guide	
Section	**Questions**
1	1–9, 16–17, 22–27, 33, 37, 39, 45
2	10, 18–21, 28–29, 34–36, 38, 41–42, 44
3	11–15, 30–32, 40, 43

20. As a skydiver jumps from a plane, *gravity* pulls her downward. *Air resistance* pushes upward against the downward motion. The skydiver accelerates downward until the force of *air resistance* equals the downward force of *gravity*. Then the skydiver stops accelerating and falls downward at a constant speed. This is called *terminal velocity*.

21. A ball thrown or an orbit would be examples. (Accept all reasonable answers.) *Projectile motion* combines a constant horizontal velocity with downward vertical acceleration due the force of gravity. The curved path of projectiles fits this assumption, providing proof that vertical force does not affect horizontal motion. The force of gravity downward cannot change horizontal velocity.

Building Math Skills

22. $a = \dfrac{F}{m} = \dfrac{50 \text{ N}}{5 \text{ kg}} = 10 \text{ m/s}^2$

23. $F = ma = (5.5 \text{ kg})(4.2 \text{ m/s}^2) = 23 \text{ N}$

24. $m = \dfrac{F}{a} = \dfrac{13.5 \text{ N}}{6.5 \text{ m/s}^2} = 2.1 \text{ kg}$

25. $a = \dfrac{F}{m} = \dfrac{37 \text{ N}}{925 \text{ kg}} = 0.04 \text{ m/s}^2$ away from the stop sign = 4.0×10^{-2} m/s^2

26. $F = ma = 58 \text{ kg} \times 0.26 \text{ m/s}^2 = 15 \text{ N}$

27. $F = ma = 12.5 \text{ kg} \times 3.5 \text{ m/s}^2 = 43.75 \text{ N}$

28. $W = \dfrac{mg}{6} = \dfrac{2.26 \text{ kg} \times 9.8 \text{ m/s}^2}{6} = \dfrac{22.1 \text{ N}}{6} = 3.69 \text{ N}$

29. $W = (mg)(2.64) = (2.26 \text{ kg} \times 9.8 \text{ m/s}^2)(2.64) = 58.5 \text{ N}$

30. $p = mv = (85 \text{ kg})(2.65 \text{ m/s}) = 225 \text{ kg}\bullet\text{m/s}$ north (or 230 kg•m/s north using significant figures)

31. $p = mv = (9.1 \text{ kg})(89 \text{ km/hr}) = \left(\dfrac{1000 \text{ m}}{1 \text{ km}}\right)\left(\dfrac{1 \text{ h}}{60 \text{ min}}\right)\left(\dfrac{1 \text{ min}}{60 \text{ s}}\right) = 225 \text{ kg}\bullet\text{m/s}$ east (or 220 kg•m/s east using significant figures)

32. $p = mv$

 a. (65 kg)(3.0 m/s forward) = 195 kg•m/s forward
 b. (20.0 kg)(22 m/s west) = 440 kg•m/s west
 c. (16 kg)(0 m/s) = 0 kg•m/s
 d. (2.5 kg)(4.8 m/s to the right) = 12 kg•m/s to the right

Building Graphing Skills

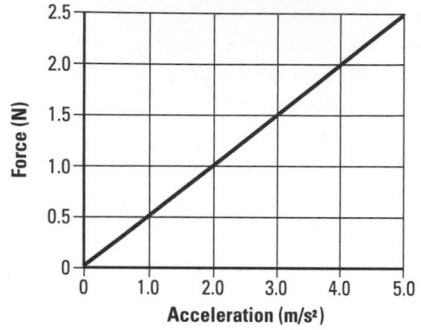

33. The slope of the line represents mass. From $F = ma$, and using numbers from the table, we learn that mass = 0.50 kg

Thinking Critically

34. The gravitational force between the two objects would become 1/16 as much.

35. The gravitational force between the two objects would become 1.5 times greater.

36. The statement is false. Even in outer space, objects exert gravitational force on each other.

37. The acceleration will be 1/6 what it was.

38. C is correct. When the mass halves, the mass in the numerator on both sides of the equation halves. So, the gravitational force would half.

---

30. Momentum Calculate the momentum of an 85 kg man jogging north along the highway at a rate of 2.65 m/s.

31. Momentum Calculate the momentum of a 9.1 kg toddler who is riding in a car moving east at 89 km/h.

32. Momentum Calculate the momentum of the following objects:
 a. a 65 kg skateboarder moving forward at the rate of 3.0 m/s
 b. a 20.0 kg toddler in a car traveling west at the rate of 22 m/s
 c. a 16 kg penguin at rest
 d. a 2.5 kg puppy running to the right at the rate of 4.8 m/s

BUILDING GRAPHING SKILLS

33. Graphing An experiment is done with a lab cart. Varying forces are applied to the cart and measured while the cart is accelerating. These forces are all in the direction of the movement of the cart, such as pushes to make the cart accelerate. Each push is made after the cart has stopped from the previous push. The following data were obtained in the experiment.

Trial	Acceleration	Applied force
1	0.70 m/s²	0.35 N
2	1.70 m/s²	0.85 N
3	2.70 m/s²	1.35 N
4	3.70 m/s²	1.85 N
5	4.70 m/s²	2.35 N

Graph the data in the table. Place acceleration on the x-axis and applied force on the y-axis. Because $F = ma$ and therefore $m = \frac{F}{a}$, what does the line on the graph represent? Use your graph to determine the mass of the lab cart.

THINKING CRITICALLY

34. Critical Thinking What happens to the gravitational force between two objects if their masses do not change but the distance between them becomes four times as much?

35. Critical Thinking What will happen to the gravitational force between two objects if one of the objects gains 50% in mass while the mass of the other object does not change? Assume that the distance between the two objects does not change.

36. Critical Thinking There is no gravity in outer space. Write a paragraph explaining whether this statement is true or false.

37. Problem Solving How will acceleration change if the mass being accelerated is multiplied by three but the net force is cut in half?

38. Problem Solving If Earth's mass decreased to half its current mass but its radius (your distance to its center) did not change, your weight would be
 a. twice as big as it is now.
 b. the same as it is now.
 c. half as big as it is now.
 d. one-fourth as big as it is now.

39. Applying Knowledge If you doubled the net force acting on a moving object, how would the object's acceleration be affected?

40. Applying Knowledge For each pair, determine whether the objects have the same momentum. If the objects have different momentums, determine which object has more momentum.
 a. a car and train that have the same velocity
 b. a moving ball and a still bat
 c. two identical balls moving at the same speed in the same direction
 d. two identical balls moving at the same speed in opposite directions

---

39. The object's acceleration would be doubled.

40. a. train
 b. ball
 c. equal
 d. equal but opposite

41. Locating Information Use the library and/or the Internet to find out about the physiological effects of free fall. Also find out how astronauts counter the effects of living in a free-fall environment, sometimes called *microgravity*. Use your information to propose a minimum acceleration for a space shuttle to Mars that would minimize physiological problems for the crew, both during flight and upon return to Earth's atmosphere.

42. Interpreting Data At home, use a garden hose to investigate the laws of projectile motion. Design experiments to investigate how the angle of the hose affects the range of the water stream. (Assume that the initial speed of water is constant.) How can you make the water reach the maximum range? How can you make the water reach the highest point? What is the shape of the water stream at each angle? Present your results to the rest of the class, and discuss the conclusions with regard to projectile motion and its components.

43. Working Cooperatively Read the following arguments about rocket propulsion. With a small group, determine which of the following statements is correct. Use a diagram to explain your answer.
- **a.** Rockets cannot travel in space because there is nothing for the gas exiting the rocket to push against.
- **b.** Rockets can travel because gas exerts an unbalanced force on the front of the rocket. This net force causes the acceleration.
- **c.** Argument b can not be true. The action and reaction forces will be equal and opposite. Therefore, the forces will balance, and no movement will be possible.

44. Connection to Social Studies Research Galileo's work on falling bodies. What did he want to demonstrate? What theories did he try to refute?

45. Concept Mapping Copy the concept map below onto a sheet of paper. Write the correct phrase in each lettered box.

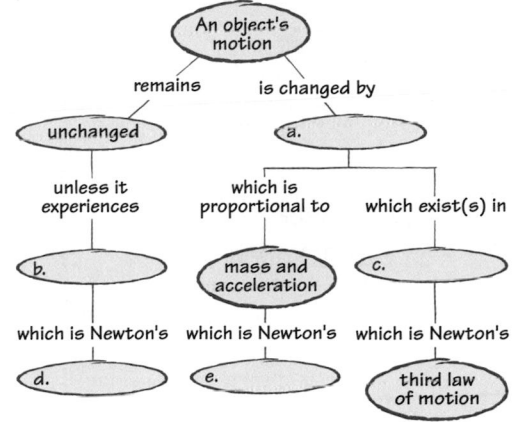

internet connect

www.scilinks.org
Topic: Rocket Technology   SciLinks code: HK4123

SC*I*NKS. Maintained by the National Science Teachers Association

Art Credits: Fig. 6-7, Stephen Durke/Washington Artists; Fig. 11-12, Craig Attebery/Jeff Lavaty Artist Agent; Fig. 18, Gary Ferster; Fig. 22, Uhl Studios, Inc.; Lab (scales), Uhl Studios, Inc.

Photo Credits: Chapter Opener image of soccer ball by Japack Company/CORBIS; players colliding by Andy Lyons/Allsport/Getty Images; Fig. 1, SuperStock; Fig. 2, Sam Dudgeon/HRW; "Quick Activity," Sam Dudgeon/HRW; "Science and the Consumer," Nicholas Pinturas/Stone; Fig. 3, Peter Van Steen/HRW ; Fig. 4, Sam Dudgeon/HRW; Fig. 5, NASA; Fig. 6, James Sugar/Black Star; Fig. 7, NASA; Fig. 8, Toby Rankin/Masterfile; Fig. 11, Michelle Bridwell/Frontera Fotos; Fig. 12, Richard Megna/Fundamental Photographs; Fig. 16-17, David Madison; Fig. 19, Rene Sheret/Stone; Fig. 20, Yellow Dog Productions/Getty Images/Stone; Fig. 21, Michelle Bridwell/Frontera Fotos; Fig. 23A, Gerard Lacz/Animals Animals/Earth Sciences; Fig. 23B, Image Copyright ©(2004 PhotoDisc, Inc./HRW and Sam Dudgeon/HRW; Fig. 23C, Lance Schriner/HRW; "Design Your Own Lab," and "Viewpoints," HRW Photos.

371

Developing Life/Work Skills

41. Answers may vary. Accept all reasonable answers.

42. Answers may vary. Accept all reasonable answers.

43. a. cannot be correct because rockets *do* travel in outer space
b. is true; the gas in the tank is pushing in all directions. Some pushes forward on the rocket, and some exits through the back of the rocket. This is an imbalance that pushes the rocket forward.
c. cannot be true because action-reaction pairs never cancel

Integrating Concepts

44. He wanted to demonstrate that the gravitational acceleration is the same for all objects. He was trying to disprove the theory that heavier bodies fall faster. He conducted experiments with rolling and dropping balls. He used logic, observations, and experiments.

45. a. force
b. unbalanced force
c. action-reaction pairs
d. first law of motion
e. second law of motion

MEASURING FORCES

Teacher's Notes

In 1660, Robert Hooke (1635–1703) stated that for small deformations, the stretching of a solid body is directly proportional to the applied force. Under these conditions, the object will return to the original configuration after the load is removed. Hooke's law is given as $F=kx$, where F is the applied force; k is a constant, which depends on the material, its dimensions, and shape; and x is the change in length.

Time Required 1 lab period

Ratings

| 1 | 2 | 3 | 4 |
| EASY | | | HARD |

TEACHER PREPARATION	2
STUDENT SETUP	2
CONCEPT LEVEL	2
CLEANUP	2

Skills Acquired

- Collecting data
- Communicating
- Designing experiments
- Experimenting
- Identifying/Recognizing patterns
- Inferring
- Interpreting
- Measuring
- Organizing and analyzing data

The Scientific Method

In this lab, students will:
- Make observations
- Form a hypothesis
- Analyze the results
- Draw conclusions
- Communicate results

Safety Cautions

Have students review safety guidelines before working in the lab.

Skills Practice Lab

Introduction

How can you use a rubber band to measure the force necessary to break a human hair?

Objectives

▶ USING SCIENTIFIC METHODS *Design* an experiment to test a hypothesis.

▶ *Build* and calibrate an instrument that measures force.

▶ *Use* your instrument to measure how much force it takes to stretch a human hair until it breaks.

Materials

comb or hairbrush
metal paper clips, large and small
metric ruler
pen or pencil
rubber bands of various sizes
standard hooked masses ranging
    from 10 to 200 g

Measuring Forces

▶ Procedure

Testing the Strength of a Human Hair

1. Obtain a rubber band and a paper clip.

2. Carefully straighten the paper clip so that it forms a double hook. Cut the rubber band and tie one end to the ring stand and the other end to one of the paper clip hooks. Let the paper clip dangle.

3. In your lab report, prepare a table as shown below.

4. Measure the length of the rubber band. Record this length in *Table 1*.

5. Hang a hooked mass from the lower paper clip hook. Supporting the mass with your hand, allow the rubber band to stretch downward slowly. Then remove your hand carefully so the rubber band does not move.

6. Measure the stretched rubber band's length. Record the mass that is attached and the rubber band's length in *Table 1*. Calculate the change in length by subtracting your initial reading of the rubber band's length from the new length.

7. Repeat steps 5 and 6 three more times using different masses each time.

8. Convert each mass (in grams) to kilograms using the following equation.

$$mass \text{ (in kg)} = mass \text{ (in g)} \div 1000$$

Record your answers in *Table 1*.

9. Calculate the force (weight) of each mass in newtons using the following equation.

$$force \text{ (in N)} = mass \text{ (in kg)} \times 9.8 \text{ m/s}^2$$

Record your answers in *Table 1*.

Table 1 Calibration

Rubber-band length (cm)	Change in length (cm)	Mass on hook (g)	Mass on hook (kg)	Force (N)
	0	0	0	0

Table 1: Sample Data Table Calibration			
Rubber-band length (cm)	Mass on hook (g)	Mass on hook (kg)	Force (N)
0	0	0	0
2.1	200	0.200	2.0
7.1	500	0.500	4.9
11.2	1000	1.000	9.8

Design Your Own

Designing Your Experiment

10. With your lab partner(s), devise a plan to measure the force required to break a human hair using the instrument you just calibrated. How will you attach the hair to your instrument? How will you apply force to the hair?

11. In your lab report, list each step you will perform in your experiment.

12. Have your teacher approve your plan before you carry out your experiment.

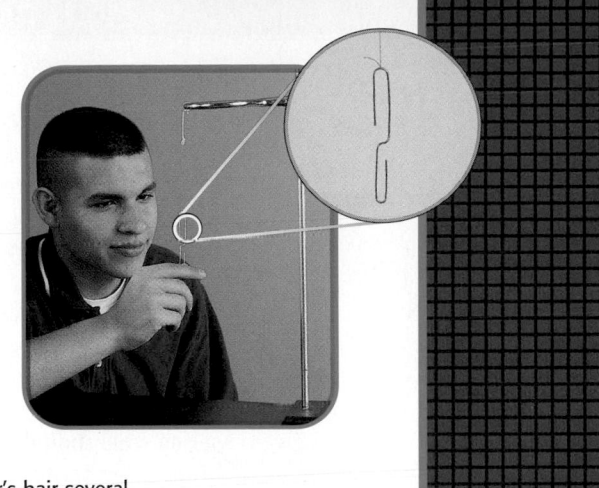

Performing Your Experiment

13. After your teacher approves your plan, gently run a comb or brush through a group member's hair several times until you find a loose hair at least 10 cm long that you can test.

14. In your lab report, prepare a data table similar to the one shown at right to record your experimental data.

15. Perform your experiment on three different hairs from the same person. Record the maximum rubber-band length before the hair snaps for each trial in Table 2.

Table 2 Experimentation

Trial	Rubber-band length (cm)	Force (N)
Hair 1		
Hair 2		
Hair 3		

▶ Analysis

1. Plot your calibration data in your lab report in the form of a graph like the one shown at right. On your graph draw the line or smooth curve that fits the points best.

2. Use the graph and the length of the rubber band for each trial of your experiment to determine the force that was necessary to break each of the three hairs. Record your answers in *Table 2*.

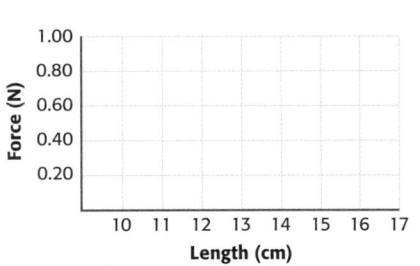

▶ Conclusions

3. Suppose someone tells you that your results are flawed because you measured length and not force. How can you show that your results are valid?

373

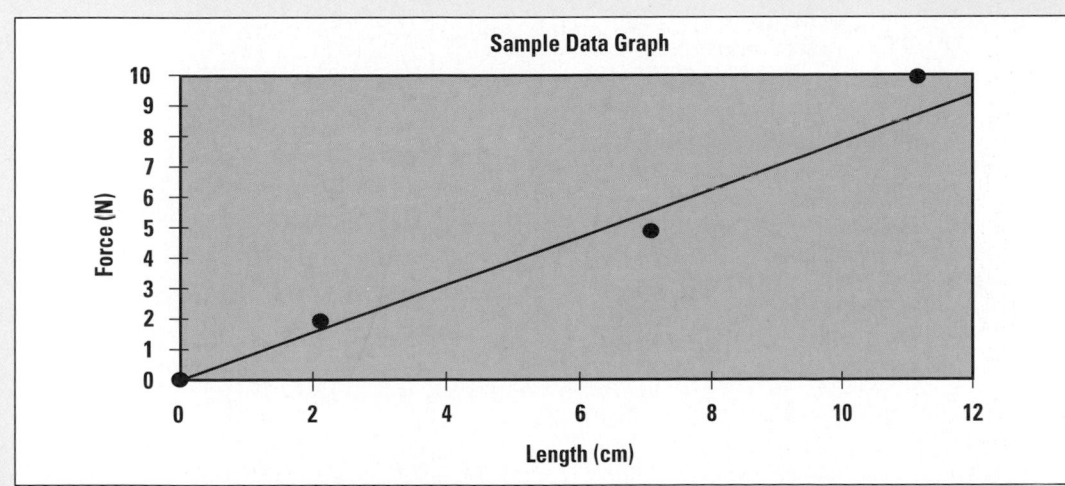

Sample Data Graph

Chapter 11 • Forces **373**

Continuation of Answers

Continuation of Answers from p. 351
Section 1 Review

Math Skills

4. $a = \dfrac{F}{m} = \dfrac{15 \text{ N}}{58 \text{ kg}} = 0.26 \text{ m/s}^2$ forward

5. $F = ma = 1250 \text{ kg} \times 40 \text{ m/s}^2 = 50,000 \text{ N} = 5.0 \times 10^4 \text{ N}$

6. $m = \dfrac{F}{a} = \dfrac{34 \text{ N}}{4 \text{ m/s}^2} = 8.5 \text{kg}$

LS Logical

Continuation of Answers from p. 359
Section 2 Review

4. Orbital motion has two components—horizontal and vertical. Horizontal motion propels the object forward, and vertical free fall pulls the object downward toward the center of gravity of the larger mass.

5. Answers may vary, but they should state something similar to the following: Newton's second law shows that acceleration depends on both force and mass. A heavier object experiences a greater gravitational force than a lighter object (as you can see from the law of universal gravitation). But a heavier object is also harder to accelerate because it has more mass. The extra mass of the heavy object exactly compensates for the additional gravitational force. Since $F = ma$ (or $a = F/m$), if F is increased at the same rate as m, then a remains the same.

Math Skills

6. The force of gravity is inversely proportional to the square of distance. Since the distance is made twice as close, the force of gravity will be four times as great (the square of 2), or 4 million N.

Continuation of Answers from p. 366
Quick Lab

The reaction forces that correspond with each force are: the upward force on the Earth from the mass of the spring scale, the upward force on the 2 kg mass, and the downward pull on the top spring scale.

The pairs are much easier to identify if students list all forces acting on each object separately. Scan the lists for pairs like: A pulls down on B, and B lifts up on A.

2. Students should be able to determine that the reading on the top spring scale should be the weight of the 2 kg mass plus the weight of the other spring scale. The reading on the bottom spring scale should be the weight of the 2 kg mass. Since the bottom spring scale is pulling down on the top spring scale with a force equal to its weight plus the weight of the 2 kg mass, the top scale must exert an upward force on the bottom scale that is equal to the weight of the bottom scale plus the 2 kg mass.

LS Logical

Section 3 Review

Math Skills

6. $p = mv = (1 \text{ kg})(12 \text{ m/s}) = 12 \text{ kg} \cdot \text{m/s}$ eastward

SHOULD BICYCLE HELMETS BE REQUIRED BY LAW?

1. Critiquing Viewpoints

Students' analyses of a single argument's weak points may vary. Be sure that student responses clearly identify a weakness and respond in some way that provides support. Possible responses for each of the viewpoints in the feature are listed below. (Note that students should analyze *only one of the following.*)

Chad A.: Some objections to helmet use are more serious than appearance, such as decreases in peripheral vision and ability to hear.

But bicycle helmets can still protect you from serious head injuries.

Laurel R.: Ticketing children is not likely to change matters.

But children are at greater risk because they are more likely to be riding bicycles and have less experience.

Jocelyn B.: The unfortunate accident described happened to one person. That does not necessarily mean everyone's at risk.

But this single incident does show that accidents that could be less tragic with bicycle helmets do occur, even when people don't expect them.

viewpoints

viewpoints

Should Bicycle Helmets Be Required by Law?

In some communities, bicyclists are required by law to wear a helmet and can be ticketed if they do not. Few people dispute the fact that bicycle helmets can save lives when used properly.

But others say that it is a matter of private rights and that the government should not interfere. Should it be up to bicyclists to decide whether or not to wear a helmet and to suffer any consequences?

But are the consequences limited to the rider? Who will pay when the rider gets hurt? Should the rider bear the cost of an injury that could have been prevented?

Is this an issue of public health or private rights? What do you think?

> FROM: Chad A., Rochester, MN

More and more people are getting head injuries every year because they do not wear a helmet. Nowadays helmets look so cool—I wouldn't be ashamed to wear one.

Require Bicycle Helmets

> FROM: Laurel R., Coral Springs, FL

I believe that this is a public issue only for people under the age of 12. Children 12 and under still need guidance and direction about safety, and they are usually the ones riding their bicycles out in the road or in traffic. Often they don't pay attention to cars or other motor vehicles around them.

> FROM: Jocelyn B., Chicago, IL

They should treat helmets the same way they treat seat belts. I was in a tragic bike accident when I was 7. I was jerked off my bike, and I slid on the glass-laden concrete. To make a long story short, I think there should be a helmet law because people just don't know the danger.

374

Megan J. and Heather R.: The argument that individual rights exceed the public health benefits do not take into account the fact that preventable injuries are a waste of money for insurance companies, hospitals, the government, and taxpayers.

But laws limiting individual freedom can have unintended consequences.

Melissa F.: If people can't afford to bicycle safely and with helmets, perhaps they shouldn't bicycle at all.

But is it right to deny someone access to such a simple vehicle because of their financial situation?

2. Critiquing Viewpoints

Students' analyses of strong points in opposing arguments may vary. Be sure that student responses clearly identify the strong points and respond in some way that provides support for the opposite point of view. Possible responses for each of the viewpoints in the feature are listed below. (Note that students should analyze *only one of the following.*)

Chad A. and Jocelyn B.: More and more injuries that are preventable are occurring.

But the government should not get involved in every risk someone takes.

> FROM: Megan J., Bowling Green, KY

Although wearing a bicycle helmet can be considered a matter of public health, the rider is the one at risk. It is a personal choice, no matter what the public says.

> FROM: Melissa F., Houston, TX

Bicycle helmets shouldn't be required by law. Helmets are usually a little over $20, and if you have five kids, the helmets alone cost $100. You'd still have to buy the bikes.

Don't Require Bicycle Helmets

> FROM: Heather R., Rochester, MN

It has to do with private rights. The police have more serious issues to deal with, like violent crimes. Bicycle riders should choose whether or not they want to risk their life by riding without a helmet.

Your Turn

1. **Critiquing Viewpoints** Select one of the statements on this page that you agree with. Identify and explain at least one weak point in the statement. What would you say to respond to someone who brought up this weak point as a reason you were wrong?

2. **Critiquing Viewpoints** Select one of the statements on this page that you disagree with. Identify and explain at least one strong point in the statement. What would you say to respond to someone who brought up this point as a reason they were right?

3. **Creative Thinking** Suppose you live in a community that does not have a bicycle helmet law. Design a campaign to persuade people to wear helmets, even though it isn't required by law. Your campaign could include brochures, posters, and newspaper ads.

4. **Acquiring and Evaluating Data** When a rider falls off a bicycle, the rider continues moving at the speed of the bicycle until the rider strikes the pavement and slows down rapidly. For bicycle speeds ranging from 5.0 m/s to 25.0 m/s, calculate what acceleration would be required to stop the rider in just 0.50 s. How large is the force that must be applied to a 50.0 kg rider to cause this acceleration? Organize your data and results in a series of charts or graphs.

☑ internet connect

TOPIC: Bicycle Helmets
GO TO: go.hrw.com
KEYWORD: HK4 Helmet

Should helmets be required by law? Why or why not? Share your views on this issue and learn about other viewpoints at the HRW Web site.

375

Laurel R.: Children need to be protected the most from such accidents.

But because they are children, they can't force their parents to get them helmets anyway.

Megan J.: Individual rights are important.

But when an individual decides to risk his life, others do have the right to try to stop him.

Melissa F.: Bicycle helmets are not cheap.

But far more money can be saved by preventing a costly accident.

Heather R.: This is not likely to be a high-priority for police enforcement.

Many other laws are rarely enforced, but have been instituted so society can go on record in favor of or opposing something.

3. **Creative Thinking**
Student answers will vary. Evaluate students' plans for campaigns using a scale of 1–5 points for each of the following: thought behind the planning of the advertising strategy, clarity in the advertising message, persuasiveness, creativity, and effort. (25 total possible points)

4. **Acquiring and Evaluating Data**
Student answers should indicate that an acceleration of 10 m/s^2 (nearly the same as gravity) and a force of 500 N would be required to stop the slowest bicyclist, and an acceleration of 50 m/s^2 and 2500 N would be required to stop the fastest one.

Students should probably choose a bar graph to display their results if they evaluated specific speeds. Some may choose to show all values in between the maximum and minimum by plotting a line graph.

Evaluate students' charts or graphs using a scale of 1–5 points for each of the following: accuracy of calculation, accuracy of graphing, neatness of graph, explanation of significance of data, and effort. (25 total possible points)

Work and Energy
Chapter Planning Guide

PACING	CLASSROOM RESOURCES	LABS, ACTIVITIES, AND DEMONSTRATIONS
BLOCK 1 • 45 min pp. 376–377 **Chapter Opener**		SE **Activity 1,** p. 377 SE **Activity 2,** p. 377
BLOCKS 2 & 3 • 90 min pp. 378–384 **Section 1** Work, Power, and Machines	CRF **Lesson Plan*** TT **Bellringer***	SE **Quick Lab** What is your power output when you climb the stairs?, p. 380 `GENERAL` CRF **Datasheets for In-Text Labs** What is your power output when you climb the stairs?* `GENERAL` TE **Demonstration** Work, p. 379
BLOCKS 4 & 5 • 90 min pp. 385–390 **Section 2** Simple Machines	CRF **Lesson Plan*** TT **Bellringer*** TT **Levers*** TT **Pulleys***	TE **Opening Activity,** p. 385 `GENERAL` TE **Demonstration** Pulley Power, p. 387 `GENERAL` SE **Quick Activity** A Simple Inclined Plane, p. 388 `GENERAL` CRF **CBL™ Probeware Lab** Determining Which Ramp is More Efficient* ◆ `ADVANCED`
BLOCKS 6 & 7 • 90 min pp. 391–399 **Section 3** What Is Energy?	CRF **Lesson Plan*** TT **Bellringer*** TT **Kinetic Energy Graph*** TM **Kinetic Energy Graph***	CRF **Observation Lab** Exploring Work and Energy* ◆ `BASIC`
BLOCKS 8 & 9 • 90 min pp. 400–408 **Section 4** Conservation of Energy	CRF **Lesson Plan*** TT **Bellringer*** TT **Energy Graphs*** TT **Concept Mapping***	TE **Opening Activity,** p. 400 `GENERAL` SE **Quick Activity** Energy Transfer, p. 403 `GENERAL` SE **Quick Lab** Is energy conserved in a pendulum?, p. 405 `GENERAL` SE **Skills Practice Lab** Determining Energy for a Rolling Ball, pp. 414–415 ◆ `GENERAL`

BLOCKS 10 & 11 • 90 min

Chapter Review and Assessment Resources

- SE **Chapter Review,** pp. 410–413
- CRF **Chapter Tests*** `GENERAL`
- OSP **Test Generator**
- CRF **Standardized Test Practice with Guided Reading Development***
- CRF **Test Item Listing for ExamView® Test Generator***
- OSP **Scoring Rubrics and Classroom Management Checklists**

Online Resources

Visit the HRW Web site for a variety of free resources related to the text. Just type in the keyword **HK4 WOR.**

Holt Science Spectrum: Physical Science: Online Edition

Students can access interactive problem solving help and active visual concept development with the *Holt Science Spectrum: Physical Science* Online Edition available at **www.hrw.com.**

student CNN News

cnnstudentnews.com

Find the latest news, lesson plans, and activities related to important scientific events.

Compression guide:
To shorten your instruction because of time limitations, omit blocks 1, 10, and 11.

KEY

TE	Teacher Edition	**CRF**	Chapter Resource File	* Also on One-Stop Planner
SE	Student Edition	**TT**	Teaching Transparency	◆ Requires Advance Prep
OSP	One-Stop Planner	**TM**	Transparency Master	

PROBLEM SOLVING AND PRACTICE	SECTION REVIEW AND ASSESSMENT	STANDARDS CORRELATION
	CRF Pretest* GENERAL	
SE Math Skills Work, p. 379 **CRF Cross-Disciplinary Worksheet** Integrating Biology—Muscles and Work* ADVANCED **CRF Math Skills** Work* GENERAL **SE Math Skills** Power, p. 381 **CRF Math Skills** Power* GENERAL **SE Math Skills** Mechanical Advantage, pp. 383–384 **CRF Math Skills** Mechanical Advantage* GENERAL **CRF Science Skills** Equations with Three Parts* BASIC	**TE Quiz**, p. 384 BASIC **SE Section 1 Review**, p. 384 **CRF Concept Review*** GENERAL **CRF Quiz*** BASIC	PS 4a UCP 1–3, 5 SAI 1
CRF Cross-Disciplinary Worksheet Social Studies Connection—The Pyramids* ADVANCED	**TE Quiz**, p. 390 BASIC **SE Section 2 Review**, p. 390 **CRF Concept Review*** GENERAL **CRF Quiz*** BASIC	PS 4a UCP 1–3, 5 ST 2
CRF Cross-Disciplinary Worksheet Connection to Language Arts—The Concept of Energy* ADVANCED **SE Math Skills** Gravitational Potential Energy, p. 393 **SE Math Skills** Kinetic Energy, p. 395 **CRF Science Skills** Squares and Square Roots* BASIC **CRF Cross-Disciplinary Worksheet** Real World Applications—Calories and Nutrition* ADVANCED **CRF Cross-Disciplinary Worksheet** Integrating Chemistry—Chemical Reactions* ADVANCED	**TE Quiz**, p. 399 BASIC **SE Section 3 Review**, p. 399 **CRF Concept Review*** GENERAL **CRF Quiz*** BASIC	PS 1a, 1c, 2a–2e, 3b, 4b, 5a–5d, 6a, 6b LS 1b, 1f, 4a, 5a–5c, 5f ES 1a UCP 1–3 SPSP 2
CRF Cross-Disciplinary Worksheet Integrating Technology—Batteries and Emerging Technology* ADVANCED **CRF Cross-Disciplinary Worksheet** Integrating Environmental Science—The Conservation of Energy* ADVANCED **SE Math Skills** Efficiency, p. 407 **CRF Science Skills** Percentages* BASIC	**TE Quiz**, p. 408 BASIC **SE Section 4 Review**, p. 408 **CRF Concept Review*** GENERAL **CRF Quiz***	BAS PS 5a, 5d LS 5f UCP 1–3

SCILINKS

www.scilinks.org

Topic: Machines
*Sci*Links code: HK4081

Topic: Mechanical Advantage
*Sci*Links code: HK4085

Topic: Potential Energy
*Sci*Links code: HK4108

Topic: Kinetic Energy
*Sci*Links code: HK4075

Topic: Energy Transformations
*Sci*Links code: HK4049

Topic: Energy and Sports
*Sci*Links code: HK4047

Topic: Engineer
*Sci*Links code: HK1999

Technology Resources

 One-Stop Planner
All of your printable resources and the Test Generator are on this convenient CD-ROM.

 Physical Science Interactive **Tutor CD-Rom**
Disc Two, Module 10 Work

 CNN Science in the NEWS
*each video segment is accompanied by a Critical Thinking Worksheet**

Segment 3
Trebuchet Design

Chapter 12 · Work and Energy   **376B**

Overview

This chapter covers work, power, and the mechanical advantage of machines. This chapter then explores the six simple machines and relates work to energy, and distinguishes between different forms of energy. This chapter finally covers energy transformations, the conservation of energy, and the efficiency of machines.

Assessing Prior Knowledge

Be sure students understand the following concepts:

- speed
- velocity
- forces
- gravity
- mass versus weight

MISCONCEPTION
///ALERT \\\

Science education research has identified the following misconceptions about work and energy.

- Students often use energy, force, and work interchangeably.
- Some students think energy is only associated with inanimate objects, or only with humans, or that it is a fluid, ingredient, or fuel.
- Students believe that energy transformations involve only one form of energy at a time.
- The idea of energy conservation is counter-intuitive to many students.

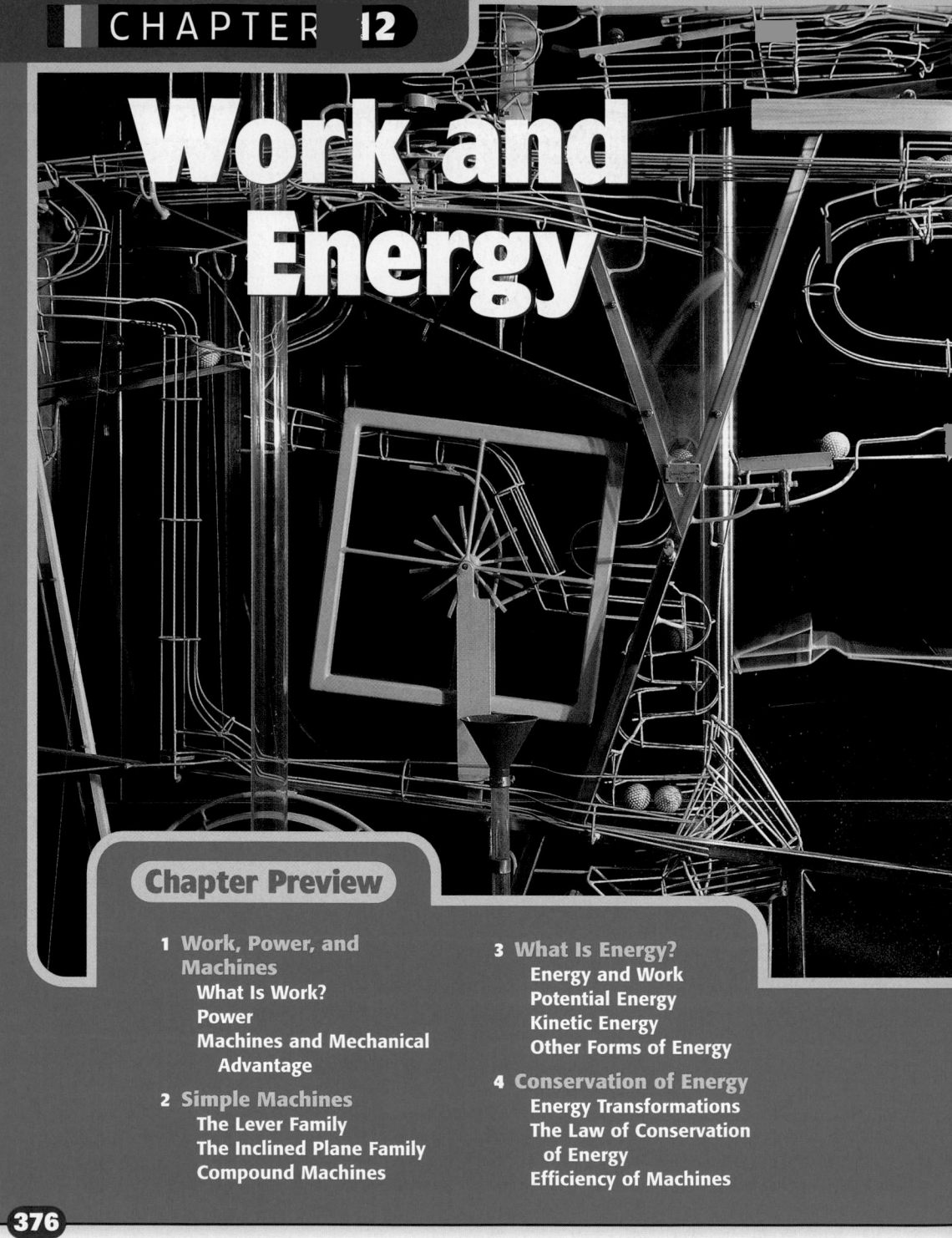

CHAPTER 12

Work and Energy

Chapter Preview

376

Standards Correlations

National Science Education Standards

The following descriptions summarize the National Science Standards that specifically relate to this chapter. For the full text of the standards, see the National Science Education Standards at the front of the book.

Section 1 Work, Power, and Machines
Physical Science PS 4a
Unifying Concepts and Processes UCP 1–3, 5
Science as Inquiry SAI 1

Section 2 Simple Machines
Physical Science PS 4a

Unifying Concepts and Processes UCP 1–3, 5
Science and Technology ST 2

Section 3 What Is Energy?
Physical Science PS 1a, 1c, 2a–2e, 3b, 4b, 5a–5d, 6a, 6b
Life Science LS 1b, 1f, 4a, 5a–5c, 5f
Earth and Space Science ES 1a
Unifying Concepts and Processes UCP 1–3
Science in Personal and Social Perspectives SPSP 2

Section 4 Conservation of Energy
Physical Science PS 5a, 5d
Life Science LS 5f
Unifying Concepts and Processes UCP 1–3

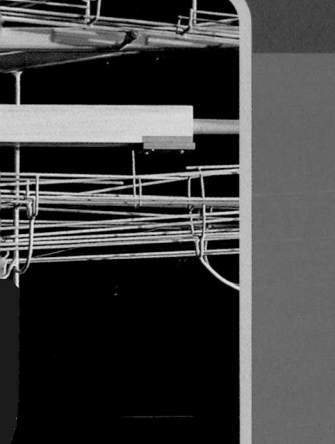

Kinetic sculptures are sculptures that have moving parts. The changes in the motion of different parts of a kinetic sculpture can be explained in terms of forces or in terms of energy transformations.

Focus ACTIVITY

Background The collection of tubes, tracks, balls, and blocks of wood shown at left is an audio-kinetic sculpture. A conveyor belt lifts the balls to a point high on the track, and the balls wind their way down as they are pulled by the force of gravity and pushed by various other forces. They twist through spirals, drop straight down tubes, and sometimes go up and around loops as if on a roller coaster. Along the way, the balls trip levers and bounce off elastic membranes. The sculpture uses the energy of the falling balls to produce sounds in wood blocks and metal tubes.

This kinetic sculpture can be considered a machine or a collection of many small machines. Other kinetic sculptures may incorporate simple machines such as levers, wheels, and screws. The American artist Alexander Calder, shown at left, is well known for his hanging mobiles that move in response to air currents.

This chapter introduces the basic principles of energy that explain the motions and interactions of machines.

Activity 1 Look around your kitchen or garage. What kinds of tools or utensils do you see? How do these tools help with different kinds of projects? For each tool, consider where force is applied to the tool and how the tool may apply force to another object. Is the force transferred to another part of the tool? Is the force that the tool can exert on an object larger or smaller than the force exerted on the tool?

Activity 2 Any piece of artwork that moves is a kinetic sculpture. Design one of your own. Some ideas for materials include hangers, rubber bands, string, wood and metal scraps, and old toys.

▸ internet connect

www.scilinks.org
Topic: Machines SciLinks code: HK4081

SCI LINKS. Maintained by the National Science Teachers Association

Pre-Reading Questions

1. How would you define work and energy? Do these words have the same meaning in everyday speech and in science?
2. What different types of energy do you know about?

377

Focus ACTIVITY

Background The artist Alexander Calder was born in Philadelphia, Pennsylvania, in 1898. He is best known for inventing the mobile, a type of kinetic sculpture in which balanced weights could move in elaborate ways when only lightly touched. Calder died at the age of 78, after creating 16 000 works of art.

Activity 1 Answers may vary. Student lists should include tools around the home and should indicate some thought to the function of each tool, where forces act on the tools, and how the tools apply forces to other objects. **LS Kinesthetic**

Activity 2 Encourage students to begin modeling their sculpture using the ideas of simple machines from this chapter. Students may then construct a kinetic sculpture by combining simple machines into a compound machine. **LS Kinesthetic**

Answers to Pre-Reading Questions

1. Answers will vary. Discuss student definitions of work and energy and emphasize that in science work refers specifically to the quantity of energy produced by a force when it is applied to a body and causes that body to move in the direction of the force.

2. Answers will vary. Students may say that they know about chemical, nuclear, electrical, sound, and solar energy.

Chapter Resource File

• Pretest **GENERAL**
• Teaching Transparency Preview
• Answer Keys

SECTION
1

Focus

Overview

Before beginning this section, review with your students the objectives listed in the Student Edition. This section introduces work and the specific conditions in which work is done. Power, the rate at which work is done, is also discussed. The section concludes with a discussion of the mechanical advantage of various simple machines.

🎧 Bellringer

Use the Bellringer transparency to prepare students for this section.

Motivate

Opening Discussion ——— GENERAL

Have students brainstorm all the words, phrases, and ideas that they associate with the term *work*. Write each term on the board, and lead students in a discussion about the listed terms. Allow students to ask you and other students for clarification of the listed ideas. As part of the discussion, ask the following: Which of the examples involve work in the scientific sense? Which examples use a different definition of *work*? **LS Verbal**

Work, Power, and Machines

▶ **KEY TERM**
work
power
mechanical advantage

OBJECTIVES

▶ **Define** *work* and *power*.
▶ **Calculate** the work done on an object and the rate at which work is done.
▶ **Use** the concept of mechanical advantage to explain how machines make doing work easier.
▶ **Calculate** the mechanical advantage of various machines.

I f you needed to change a flat tire, you would probably use a car jack to lift the car. Machines—from complex ones such as a car to relatively simple ones such as a car jack, a hammer, or a ramp—help people get things done every day.

What Is Work?

▶ **work** the transfer of energy to a body by the application of a force that causes the body to move in the direction of the force

Imagine trying to lift the front of a car without using a jack. You could exert a lot of force without moving the car at all. Exerting all that force might seem like hard work. In science, however, the word **work** has a very specific meaning.

Work is done only when force causes a change in the position or the motion of an object in the direction of the applied force. Work is calculated by multiplying the force by the distance over which the force is applied. We will always assume that the force used to calculate work is acting along the line of motion of the object.

Figure 1
As this weightlifter holds the barbell over her head, is she doing any work on the barbell?

Work Equation

$$work = force \times distance$$
$$W = F \times d$$

In the case of trying to lift the car, you might apply a large force, but if the distance that the car moves is equal to zero, the work done on the car is also equal to zero.

However, once the car moves even a small amount, you have done some work on it. You could calculate how much by multiplying the force you have applied by the distance the car moves.

The weightlifter in **Figure 1** is applying a force to the barbell as she holds it overhead, but the barbell is not moving. Is she doing any work on the barbell?

378

Alternative Assessment ——— BASIC

Work Students may confuse the scientific meaning of *work* with the common, everyday meaning. In order to clarify the proper scientific meaning, point out that when used in the scientific sense, *work* is always done *by* a force, *on* an object, changing the motion of the object. Have students come up with sentences using *work* in both the common sense and the scientific sense. **LS Verbal**

Historical Perspective

James Prescott Joule The unit of work and energy was named in honor of the English physicist James Prescott Joule (1818–1889). Joule was one of the first scientists to recognize that mechanical work and non-mechanical energy could be interchangeable, which was the foundation of the law of conservation of energy.

Work is measured in joules

Because work is calculated as force times distance, it is measured in units of newtons times meters, N•m. These units are also called *joules* (J). In terms of SI base units, a joule is equivalent to $1 \text{ kg} \cdot \text{m}^2/\text{s}^2$.

$$1 \text{ N} \cdot \text{m} = 1 \text{ J} = 1 \text{ kg} \cdot \text{m}^2/\text{s}^2$$

Because these units are all equal, you can choose whichever unit is easiest for solving a particular problem. Substituting equivalent units will often help you cancel out other units in a problem.

You do about 1 J of work when you slowly lift an apple, which weighs about 1 N, from your arm's length down at your side to the top of your head, a distance of about 1 m.

PHYSICAL SCIENCE
Disc Two, Module 10: **Work**
Use the Interactive Tutor to learn more about this topic.

Math Skills

Work Imagine a father playing with his daughter by lifting her repeatedly in the air. How much work does he do with each lift, assuming he lifts her 2.0 m and exerts an average force of 190 N?

1 List the given and unknown values.
  Given: *force, F* = 190 N
    distance, d = 2.0 m
  Unknown: *work, W* = ? J

2 Write the equation for work.
  work = force × distance $W = F \times d$

3 Insert the known values into the equation, and solve.
  $W = 190 \text{ N} \times 2.0 \text{ m} = 380 \text{ N} \cdot \text{m} = 380 \text{ J}$

Practice

Work

1. A crane uses an average force of 5200 N to lift a girder 25 m. How much work does the crane do on the girder?
2. An apple weighing 1 N falls through a distance of 1 m. How much work is done on the apple by the force of gravity?
3. The brakes on a bicycle apply 125 N of frictional force to the wheels as the bicycle travels 14.0 m. How much work have the brakes done on the bicycle?
4. While rowing in a race, John uses his arms to exert a force of 165 N per stroke while pulling the oar 0.800 m. How much work does he do in 30 strokes?
5. A mechanic uses a hydraulic lift to raise a 1200 kg car 0.5 m off the ground. How much work does the lift do on the car?

Practice **HINT**

▶ In order to use the work equation, you must use units of newtons for force and units of meters for distance. Practice Problem 5 gives a mass in kilograms instead of a weight in newtons. To convert from mass to force (weight), use the definition of weight from Section 3:

$$w = mg$$

where *m* is the mass in kilograms and $g = 9.8 \text{ m/s}^2$. Then plug the value for weight into the work equation as the force.

Teach

Demonstration
Work
(Time: About 15 minutes)
Materials:
• spring scale
• string
• textbook

Step 1 Hang the book from the scale with string. Have students note the scale reading. Is there a force acting on the book? (Yes, gravity pulls down and the spring exerts an upward force.) Is work being done on the book? (No, because the book's displacement is zero.)

Step 2 Now lift the book (using the scale) about 1 m at a constant velocity. Does the scale reading change? (Yes, at the beginning and the end of the motion) Is work being done on the book? (Yes, because a force moves the book through a distance.)

Step 3 Now hold the scale at shoulder height and carry the book at a constant speed across the room. Does the reading on the scale change? (No) Is work being done on the book? (No, because the motion is perpendicular to the force.)
LS Visual

Practice
1. $W = (5200 \text{ N})(25 \text{ m}) = 1.3 \times 10^5 \text{ J}$
2. $W = (1 \text{ N})(1 \text{ m}) = 1 \text{ J}$
3. $W = (125 \text{ N})(14.0 \text{ m}) = 1750 \text{ J}$
4. $W = (30)(165 \text{ N})(0.800 \text{ m}) = 3960 \text{ J}$
5. $W = (1200 \text{ kg})(9.8 \text{ m/s}^2)(0.5 \text{ m}) = 6000 \text{ J}$
LS Logical

Chapter Resource File
• **Lesson Plan**
• **Cross-Disciplinary Worksheet** Integrating Biology—Muscles and Work **ADVANCED**
• **Math Skills** Work **GENERAL**

Transparencies
TT Bellringer

Work versus Power Use the following analogy to help students understand the difference between work and power. Consider a summer lawn-mowing job. If you have to mow 30 lawns in one month, you could mow six lawns each day and finish in five days, or mow one lawn each day and take the entire month. The total number of mowed lawns (30) is the same, but the rate is different (either 6 per day or 1 per day). Ask students what is analogous to the total number of lawns mowed (work) and to the rate of mowing (power).
LS Verbal

Quick Lab

What is your power output when you climb the stairs?

Safety Caution: *Instruct students not to run on the stairs. Also caution students not to overexert themselves. Emphasize to students that this is not a contest.*

Analysis

1. Your power output would be greater if you walked up the stairs faster.

2. Answers will vary depending on data. Answer should be slightly larger than the power output calculated in item 5 of the procedure.

3. Students are lifting themselves up the stairs against the force of gravity. The gravitational force is equivalent to their weight.
LS Kinesthetic

▶ **power** a quantity that measures the rate at which work is done or energy is transformed

Power

Running up a flight of stairs doesn't require more work than walking up slowly does, but it is more exhausting. The amount of time it takes to do work is an important factor when considering work and machines. The quantity that measures work in relation to time is **power.** Power is the rate at which work is done, that is, how much work is done in a given amount of time.

Power Equation

$$power = \frac{work}{time} \qquad P = \frac{W}{t}$$

Running takes less time than walking does. How does reducing the time in this equation affect the power if the amount of work stays the same?

Power is measured in watts

Power is measured in SI units called *watts* (W). A watt is the amount of power required to do 1 J of work in 1 s, about as much power as you need to lift an apple over your head in 1 s. Do not confuse the abbreviation for watts, W, with the symbol for work, *W*. You can tell which one is meant by the context in which it appears and by whether it is in italics.

Quick Lab

What is your power output when you climb the stairs?

Materials ✓ flight of stairs   ✓ stopwatch   ✓ meterstick

1. Determine your weight in newtons. If your school has a scale that reads in kilograms, multiply your mass in kilograms by 9.8 m/s² to determine your weight in newtons. If your school has a scale that weighs in pounds, multiply your weight by a factor of 4.45 N/lb.

2. Divide into pairs. Have your partner use the stopwatch to time how long it takes you to walk quickly up the stairs. Record the time. Then switch roles and repeat.

3. Measure the height of one step in meters. Multiply the number of steps by the height of one step to get the total height of the stairway.

4. Multiply your weight in newtons by the height of the stairs in meters to get the work you did in

joules. Recall the work equation: *work = force × distance,* or *W = F × d.*

5. To get your power in watts, divide the work done in joules by the time in seconds that it took you to climb the stairs.

Analysis

1. How would your power output change if you walked up the stairs faster?

2. What would your power output be if you climbed the same stairs in the same amount of time while carrying a stack of books weighing 20 N?

3. Why did you use your weight as the force in the work equation?

380

Scientists Around the World ———— GENERAL

James Watt The unit of power was named for James Watt (1736–1819), a Scottish inventor who played an important role in the development of the steam engine. Watt was born in Greenock, Scotland in 1736. Although he did not invent the steam engine, as is commonly stated, his contributions—which dramatically improved its efficiency—were fundamental to its development. Inventions on his first patent of 1769 included a separate condensing chamber for the steam engine (which prevented huge amounts of steam loss), oil lubrication,

and cylinder insulation. He later developed a steam indicator to measure the amount of steam pressure in an engine.

Watt also made contributions to other fields throughout his lifetime. In addition to his physical science work, he conducted surveys of canal routes as a civil engineer. He also invented an adaptor for telescopes to aid in distance measuring. Watt died in England in 1819.

Power It takes 100 kJ of work to lift an elevator 18 m. If this is done in 20 s, what is the average power of the elevator during the process?

1 List the given and unknown values.
  Given: *work*, $W = 100$ kJ $= 1 \times 10^5$ J
  time, $t = 20$ s
  The distance of 18 m will not be needed to calculate power.
  Unknown: *power*, $P = ?$ W

2 Write the equation for power.
  $$power = \frac{work}{time} \qquad P = \frac{W}{t}$$

3 Insert the known values into the equation, and solve.
  $$P = \frac{1 \times 10^5 \text{ J}}{20 \text{ s}} = 5 \times 10^3 \text{ J/s}$$
  $$P = 5 \times 10^3 \text{ W}$$
  $$P = 5 \text{ kW}$$

Practice

Power

1. While rowing across the lake during a race, John does 3960 J of work on the oars in 60.0 s. What is his power output in watts?

2. Every second, a certain coal-fired power plant produces enough electricity to do 9×10^8 J (900 MJ) of work. What is the power output of this power plant in units of watts (or in units of megawatts)?

3. Using a jack, a mechanic does 5350 J of work to lift a car 0.500 m in 50.0 s. What is the mechanic's power output?

4. Suppose you are moving a 300 N box of books. Calculate your power output in the following situations:
  a. You exert a force of 60.0 N to push the box across the floor 12.0 m in 20.0 s.
  b. You lift the box 1 m onto a truck in 3 s.

5. Anna walks up the stairs on her way to class. She weighs 565 N and the stairs go up 3.25 m vertically.
  a. Calculate her power output if she climbs the stairs in 12.6 s.
  b. What is her power output if she climbs the stairs in 10.5 s?

Did You Know ?

Another common unit of power is horsepower (hp). This originally referred to the average power output of a draft horse. One horsepower equals 746 W. With that much power, a horse could raise a load of 746 apples, weighing 1 N each, by 1 m every second.

Practice HINT

▸ In order to calculate power in Practice Problems 4 and 5, you must first use the work equation to calculate the work done in each case.

Additional Examples

Power A student lifts a 12 N textbook 1.5 m in 1.5 s and carries the book 5 m across the room in 7 s.

a. How much work does the student do on the textbook? (18 J)

b. What is the power output of the student? (12 W)

Compare the work and power used in the following cases:

a. A 43 N force is exerted through a distance of 2.0 m over a time of 3.0 s. (W = 86 J, P = 29 W)

b. A 43 N force is exerted through a distance of 3.0 m over a time of 2.0 s. (W = 130 J, P = 65 W)

LS Logical

Practice

1. $P = \dfrac{W}{t} = \dfrac{3960 \text{ J}}{60.0 \text{ s}} = 66.0$ W

2. $P = \dfrac{900 \text{ MJ}}{1 \text{ s}} = 900$ MW

3. $P = \dfrac{5350 \text{ J}}{50.0 \text{ s}} = 107$ W

4. a. $W = (60.0 \text{ N})(12.0 \text{ m}) = 720$ J
  $P = \dfrac{W}{t} = \dfrac{720 \text{ J}}{20.0 \text{ s}} = 36$ W

  b. $W = (300 \text{ N})(1 \text{ m}) = 300$ J
  $P = \dfrac{W}{t} = \dfrac{300 \text{ J}}{3 \text{ s}} = 100$ W

5. a. $P = \dfrac{W}{t} = \dfrac{(565 \text{ N})(3.25 \text{ m})}{12.6 \text{ s}} = 146$ W

  b. $P = \dfrac{(565 \text{ N})(3.25 \text{ m})}{10.5 \text{ s}} = 175$ W

LS Logical

REAL-LIFE CONNECTION — BASIC

Electric Power Most utility companies bill customers in units of kilowatt-hours, which is actually a unit of energy, not power. One kilowatt-hour, which equals 3 600 000 joules, is the amount of energy used at the rate of one kilowatt (1000 watts, or 1000 joules/second) over a time period of one hour (1000 J/s × 3600 s/h × 1 h = 3 600 000 J). Have students obtain a recent power bill and determine the average number of kilowatt-hours used each day. Also ask them to convert this value to joules. **LS** Intrapersonal

Chapter Resource File

• **Math Skills** Power **GENERAL**

• **Quick Lab Datasheet** What is your power output when you climb the stairs? **GENERAL**

Teaching Tip

Machines Be sure to emphasize that machines do not increase the quantity of work that one can do. Given a specific amount of work to be done, a machine takes advantage of the fact that force and distance are inversely proportional. Either one can be increased by decreasing the other.

As an example, you can use the equation $W = F \times d$ to show that a longer distance implies a smaller force for the same amount of work. For instance, to lift a 225 N box into the back of a truck that is 1.00 m off the ground requires 225 J of work. If you lift the box straight up into the truck, the force needed is 225 N, but if you use a 3.00 m ramp, the force needed is only 75 N (ignoring friction).

SKILL BUILDER — GENERAL

Interpreting Visuals Use **Figure 3** with the concrete example above to help students learn the concept of mechanical advantage. Be sure to point out that whether or not you use the ramp, the box ends up in the same place—a sure sign that the work done on the box is the same in either case. **LS Logical**

Figure 2

A jack makes it easier to lift a car by multiplying the input force and spreading the work out over a large distance.

Figure 3

A When lifting a box straight up, a mover applies a large force over a short distance.

B Using a ramp to lift the box, the mover applies a smaller force over a longer distance.

Machines and Mechanical Advantage

Which is easier, lifting a car yourself or using a jack as shown in *Figure 2*? Which requires more work? Using a jack is obviously easier. But you may be surprised to learn that using a jack requires the same amount of work. The jack makes the work easier by allowing you to apply less force at any given moment.

Machines multiply and redirect forces

Machines help us do work by redistributing the work that we put into them. Machines can change the direction of an input force. Machines can also increase or decrease force by changing the distance over which the force is applied. This process is often called multiplying the force.

Different forces can do the same amount of work

Compare the amount of work required to lift a box straight onto the bed of a truck, as shown in *Figure 3A,* with the amount of work required to push the same box up a ramp, as shown in *Figure 3B.* When the mover lifts straight up, he must apply 225 N of force for a short distance. Using the ramp, he can apply a smaller force over a longer distance. But the work done is about the same in both cases.

Both a car jack and a loading ramp make doing work easier by increasing the distance over which force is applied. As a result, the force required at any point is reduced. Therefore, a machine allows the same amount of work to be done by either decreasing the distance while increasing the force or by decreasing the force while increasing the distance.

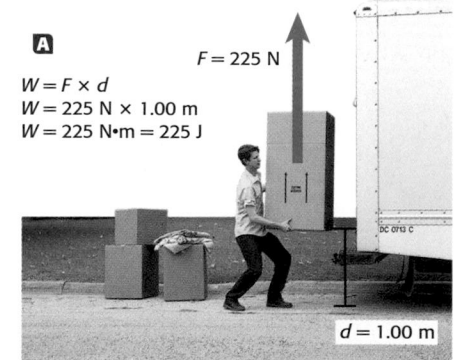

A

$F = 225$ N

$W = F \times d$
$W = 225$ N $\times$ 1.00 m
$W = 225$ N•m $= 225$ J

$d = 1.00$ m

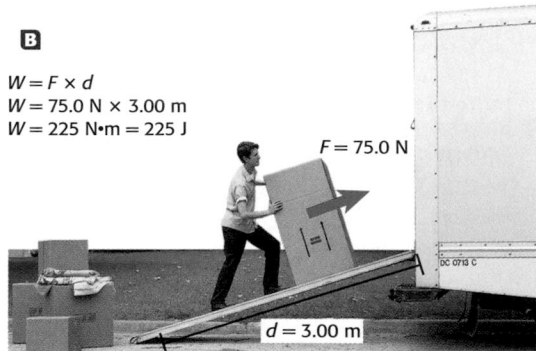

B

$W = F \times d$
$W = 75.0$ N $\times$ 3.00 m
$W = 225$ N•m $= 225$ J

$F = 75.0$ N

$d = 3.00$ m

382

Alternative Assessment — BASIC

Conservation of Mechanical Energy

The statement that different forces can do the same amount of work is really an expression of the law of conservation of energy. The law is implicit throughout this chapter, but it is not stated explicitly until Section 4. Have students recall how the conservation of energy in Chapter 3 indicated that energy took different forms or was transferred in different ways. Ask students to write a paragraph explaining how energy transferred to objects by work is conserved in machines. Be sure they give special attention to how changes in force and displacement, the factors affecting work, vary so that energy is conserved. **LS Verbal**

Mechanical advantage tells how much a machine multiplies force or increases distance

A ramp makes doing work easier by increasing the distance over which force is applied. But how long should the ramp be? An extremely long ramp would allow the mover to use very little force, but he would have to push the box a long distance. A very short ramp, on the other hand, would be too steep and would not help him very much.

To solve problems like this, scientists and engineers use a number that describes how much the force or distance is multiplied by a machine. This number is called the **mechanical advantage,** and it is defined as the ratio between the output force and the input force. It is also equal to the ratio between the input distance and the output distance.

internet connect

www.scilinks.org
Topic: Mechanical Advantage
SciLinks code: HK4085

SCi LINKS Maintained by the National Science Teachers Association

Mechanical Advantage Equation

$$mechanical\ advantage = \frac{output\ force}{input\ force} = \frac{input\ distance}{output\ distance}$$

A machine with a mechanical advantage greater than 1 multiplies the input force. Such a machine can help you move or lift heavy objects, such as a car or a box of books. A machine with a mechanical advantage of less than 1 does not multiply force, but increases distance and speed. When you swing a baseball bat, your arms and the bat together form a machine that increases speed without multiplying force.

▶ **mechanical advantage**
a quantity that measures how much a machine multiplies force or distance

Math Skills

Mechanical Advantage Calculate the mechanical advantage of a ramp that is 5.0 m long and 1.5 m high.

1 List the given and unknown values.
   Given: *input distance* = 5.0 m
   output distance = 1.5 m
   Unknown: *mechanical advantage* = ?

2 Write the equation for mechanical advantage.
   Because the information we are given involves only distance, we only need part of the full equation:
   $$mechanical\ advantage = \frac{input\ distance}{output\ distance}$$

3 Insert the known values into the equation, and solve.
   $$mechanical\ advantage = \frac{5.0\ \cancel{m}}{1.5\ \cancel{m}} = 3.3$$

INTEGRATING

BIOLOGY
You may not do any work on a car if you try to lift it without a jack, but your body will still get tired from the effort because you are doing work on the muscles inside your body.

When you try to lift something, your muscles contract over and over in response to a series of electrical impulses from your brain. With each contraction, a tiny bit of work is done on the muscles. In just a few seconds, this can add up to thousands of contractions and a significant amount of work.

383

Did You Know?

Conservation of Energy No machine can increase both the force and the distance at the same time. An increase in one always corresponds to a decrease in the other. This is another way to express the idea that you cannot ever get more work out of a machine than you put in. This is because energy is always conserved; it cannot be created or destroyed. Machines are not useful because

they increase the amount of work done, but because they make work easier.

If there were no friction, the work output would be exactly equal to the work input. However, because friction acts against work input, the actual work output is always less than the work input.

1. *MA = input distance/output distance* = 6.0 m/1.5 m = 4.0
2. *MA = output force/input force* = 9900 N/150 N = 66
3. *MA* = 140 N/140 N = 1.0
4. *output force* = (*MA*)(*input force*) = (5.2)(15 N) = 78 N
5. *output distance* = *input distance/MA* = 0.80 m/1.5 = 0.53 m

LS Logical

Close

Quiz ———————— BASIC

Match each term at the top with the correct description below. Then write the equation and the SI unit for each term.

left column:

1. work (c; force × distance; joules)
2. power (b; work/time; watts)
3. mechanical advantage (a; input distance/output distance; no units)

right column:

a. the amount that a machine multiplies force or distance
b. the rate at which work is done
c. what is done when a force makes an object move

LS Logical

Chapter Resource File

- Concept Review GENERAL
- Quiz BASIC

Practice HINT

▶ The mechanical advantage equation can be rearranged to isolate any of the variables on the left.
▶ For practice problem 4, you will need to rearrange the equation to isolate output force on the left.
▶ For practice problem 5, you will need to rearrange to isolate output distance. When rearranging, use only the part of the full equation that you need.

Practice

Mechanical Advantage

1. Calculate the mechanical advantage of a ramp that is 6.0 m long and 1.5 m high.
2. Determine the mechanical advantage of an automobile jack that lifts a 9900 N car with an input force of 150 N.
3. A sailor uses a rope and pulley to raise a sail weighing 140 N. The sailor pulls down with a force of 140 N on the rope. What is the mechanical advantage of the pulley?
4. Alex pulls on the handle of a claw hammer with a force of 15 N. If the hammer has a mechanical advantage of 5.2, how much force is exerted on a nail in the claw?
5. While rowing in a race, John pulls the handle of an oar 0.80 m on each stroke. If the oar has a mechanical advantage of 1.5, how far does the blade of the oar move through the water on each stroke?

SECTION 1 REVIEW

SUMMARY

▶ Work is done when a force causes an object to move. This meaning is different from the everyday meaning of *work*.
▶ Work is equal to force times distance. The most commonly used SI unit for work is joules.
▶ Power is the rate at which work is done. The SI unit for power is watts.
▶ Machines help people by redistributing the work put into them. They can change either the size or the direction of the input force.
▶ The mechanical advantage of a machine describes how much the machine multiplies force or increases distance.

1. Define work and power. How are work and power related to each other?
2. Determine if work is being done in these situations:
   a. lifting a spoonful of soup to your mouth
   b. holding a stack of books motionless over your head
   c. letting a pencil fall to the ground
3. **Describe** how a ramp can make lifting a box easier without changing the amount of work being done.
4. **Critical Thinking** A short ramp and a long ramp both reach a height of 1 m. Which has a greater mechanical advantage?

Math Skills

5. How much work in joules is done by a person who uses a force of 25 N to move a desk 3.0 m?
6. A bus driver applies a force of 55.0 N to the steering wheel, which in turn applies 132 N of force on the steering column. What is the mechanical advantage of the steering wheel?
7. A student who weighs 400 N climbs a 3 m ladder in 4 s.
   a. How much work does the student do?
   b. What is the student's power output?
8. An outboard engine on a boat can do 1.0×10^6 J of work in 50.0 s. Calculate its power in watts. Convert your answer to horsepower (1 hp = 746 W).

Answers to Section 1 Review

1. Work is the quantity of energy transferred by a force applied to an object that moves in the direction of the force. Power is a quantity that measures the rate at which work is done. Power equals work divided by time.
2. **a.** yes
   b. no
   c. yes (by gravity)
3. A ramp allows the use of a smaller input force exerted over a longer distance, so that work is unchanged.
4. The long ramp has a greater mechanical advantage than the short ramp.

Math Skills

5. *W* = 75 J
6. *MA* = 2.40
7. **a.** *W* = 1200 J
   b. *P* = 300 W
8. *P* = 2.0×10^4 W (27 hp)

Simple Machines

OBJECTIVES

▶ **Name** and describe the six types of simple machines.

▶ **Discuss** the mechanical advantage of different types of simple machines.

▶ **Recognize** simple machines within compound machines.

▶ **KEY TERMS**
simple machines
compound machines

The most basic machines of all are called **simple machines.** Other machines are either modifications of simple machines or combinations of several simple machines. *Figure 4* shows examples of the six types of simple machines. Simple machines are divided into two families, the lever family and the inclined plane family.

▶ **simple machine** one of the six basic types of machines, which are the basis for all other forms of machines

The Lever Family

To understand how levers do work, imagine using a claw hammer to pull out a nail. As you pull on the handle of the hammer, the head turns around the point where it meets the wood. The force you apply to the handle is transferred to the claw on the other end of the hammer. The claw then does work on the nail.

Figure 4

The Six Simple Machines

The lever family

Simple lever	Pulley	Wheel and axle

The inclined plane family

Simple inclined plane	Wedge	Screw

385

Focus

Overview

Before beginning this section, review with your students the Objectives listed in the Student Edition. This section discusses each of the six simple machines in detail, then explains what a compound machine is.

🔔 Bellringer

Use the Bellringer transparency to prepare students for this section.

Motivate

Opening Activity — GENERAL

Have students read this section before class. Write the following statement on the board: "All machines are simple machines." Separate students into pairs or small groups and have the groups debate the statement. Each group should decide whether it agrees or disagrees with the statement and should provide some evidence to support its opinion. Have a class discussion to compare opinions and rationales.

Students who disagree with the statement will usually point out that compound machines are more complicated because they are combinations of simple machines working intricately together. On the other hand, students who agree with the statement may point out that a compound machine is really nothing more than a collection of simple machines. **LS Interpersonal**

FINE ARTS

CONNECTION — BASIC

Have students examine the chapter opening photographs and identify as many simple machines in the kinetic sculptures as they can. Students may refer to **Figure 4** for examples of the six simple machines. **LS Visual**

HISTORY

CONNECTION — ADVANCED

Have students research the history of one of the simple machines shown in **Figure 4.** You may wish to divide students into six groups and assign one machine to each group. Research topics could include first known uses of the machine and interesting inventions throughout history that have relied on the machine. **LS Verbal**

Chapter Resource File

• Lesson Plan

Transparencies

TT Bellringer

Interpreting Visuals Figure 5 shows the three classes of levers. Ask students: For each of the three examples, where is force input and where is force output? Do the levers multiply force or increase distance on the output side? (Second-class levers always multiply force, third-class levers always increase distance, and first-class levers may either multiply force or increase distance.) **LS** Visual

Figure 5

The Three Classes of Levers

A All **first-class levers** have a fulcrum located between the points of application of the input and output forces.

B In a **second-class lever**, the fulcrum is at one end of the arm and the input force is applied to the other end. The wheel of a wheelbarrow is a fulcrum.

C **Third-class levers** multiply distance rather than force. As a result, they have a mechanical advantage of less than 1. The human body contains many third-class levers.

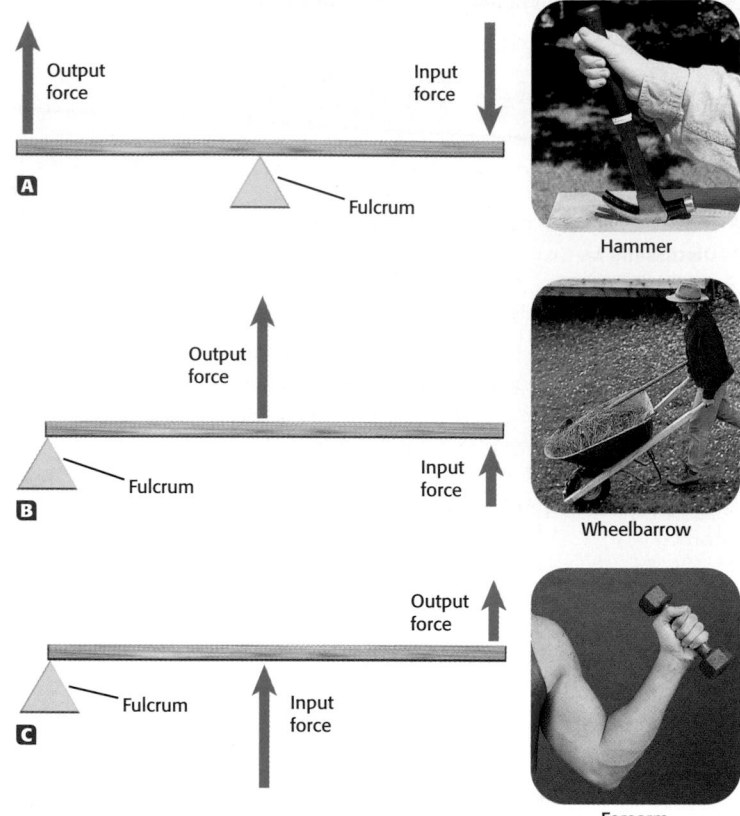

Hammer

Wheelbarrow

Forearm

Levers are divided into three classes

All levers have a rigid *arm* that turns around a point called the *fulcrum*. Force is transferred from one part of the arm to another. In that way, the original input force can be multiplied or redirected into an output force. Levers are divided into three classes depending on the location of the fulcrum and of the input and output forces.

Figure 5A shows a claw hammer as an example of a first-class lever. First-class levers are the most common type. A pair of pliers is made of two first-class levers joined together.

Figure 5B shows a wheelbarrow as an example of a second-class lever. Other examples of second-class levers include nutcrackers and hinged doors.

Figure 5C shows the human forearm as an example of a third-class lever. The biceps muscle, which is attached to the bone near the elbow, contracts a short distance to move the hand a large distance.

386

Double Levers Many common levers, such as scissors (first-class), nutcrackers (second-class), and tweezers (third-class), are actually compound machines combining two levers together. Ask students to look for everyday examples of both single and compound (double) levers. Have students add their examples to a list on the chalkboard, noting the class of the lever and whether the lever is single or double. Have other students check the classifications for accuracy. (Responses could include the following: hammer claw, 1st class, single; see-saw, 1st class, single; scissors, 1st class, double; pliers, 1st class, double; wheelbarrow, 2nd class, single; bottle opener, 2nd class, single; nutcracker, 2nd class, double; fishing rod, 3rd class, single; tweezers, 3rd class, double; tongs, 3rd class, double.) **LS** Intrapersonal

Pulleys are modified levers

You may have used pulleys to lift things, as when raising a flag to the top of a flagpole or hoisting a sail on a boat. A pulley is another type of simple machine in the lever family.

Figure 6A shows how a pulley is like a lever. The point in the middle of a pulley is like the fulcrum of a lever. The rest of the pulley behaves like the rigid arm of a first-class lever. Because the distance from the fulcrum is the same on both sides of a pulley, a single, fixed pulley has a mechanical advantage of 1.

Using moving pulleys or more than one pulley at a time can increase the mechanical advantage, as shown in *Figure 6B* and *Figure 6C*. Multiple pulleys are sometimes put together in a single unit called a *block and tackle*.

Figure 6
The Mechanical Advantage of Pulleys

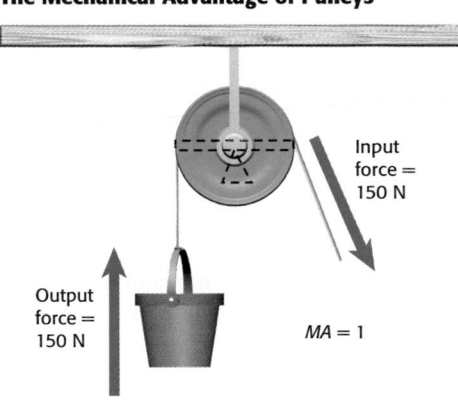

A Lifting a 150 N weight with a single, fixed pulley, the weight must be fully supported by the rope on each side of the pulley. This type of pulley has a mechanical advantage of 1.

B Using a moving pulley, the 150 N force is shared by two sections of rope pulling upward. The input force on the right side of the pulley has to support only half of the weight. This pulley system has a mechanical advantage of 2.

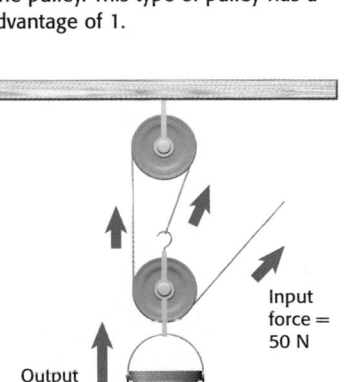

C In this arrangement of multiple pulleys, all of the sections of rope pull up against the downward force of the weight. This gives an even higher mechanical advantage.

387

Interpreting Visuals Have students examine **Figure 7**. Lead a discussion about the similarities between a wheel and axle and a lever. Have students note that the small input force acts through a large distance—the radius of the large wheel, while the large output force acts through a small distance—the radius of the axle. The wheel's center is the fulcrum around which the wheel and axle pivot. The mechanical advantage of a wheel and axle is equal to the ratio of the wheel radius to the axle radius. **LS** Visual

Quick ACTIVITY

A Simple Inclined Plane

Materials (per group):
• board
• stack of books
• string
• heavy object with low friction

Teacher's Notes: The object used in this activity can be any heavy rolling object (such as a wheeled dynamics cart) or simply a heavy object with low friction. A roll of wire with string tied through the hollow shaft so that the roll moves freely when dragged up the ramp will work well.

Answer

4. Lifting the object straight up requires more force than dragging it up the ramp. The work done is the same in either case, if friction is ignored.
LS Kinesthetic

Figure 7

How is a wheel and axle like a lever? How is it different from a pulley?

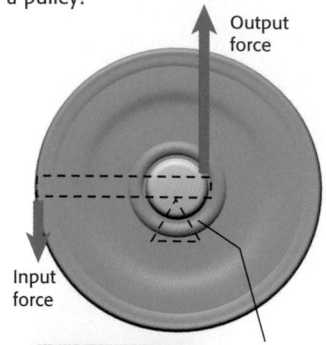

Output force

Input force

Fulcrum

Quick ACTIVITY

A Simple Inclined Plane

1. Make an inclined plane out of a board and a stack of books.
2. Tie a string to an object that is heavy but has low friction, such as a metal toy car or a roll of wire. Use the string to pull the object up the plane.
3. Still using the string, try to lift the object straight up through the same distance.
4. Which action required more force? In which case did you do more work?

388

A wheel and axle is a lever or pulley connected to a shaft

The steering wheel of a car is another kind of simple machine: a wheel and axle. A wheel and axle is made of a lever or a pulley (the wheel) connected to a shaft (the axle), as shown in **Figure 7**. When the wheel is turned, the axle also turns. When a small input force is applied to the steering wheel, the force is multiplied to become a large output force applied to the steering column, which turns the front wheels of the car. Screwdrivers and cranks are other common wheel-and-axle machines.

The Inclined Plane Family

Earlier we showed how pushing an object up a ramp requires less force than lifting the same object straight up. A loading ramp is another type of simple machine, an inclined plane.

Inclined planes multiply and redirect force

When you push an object up a ramp, you apply a force to the object in a direction parallel to the ramp. The ramp then redirects this force to lift the object upward. This is why the output force of the ramp is shown in **Figure 8A** as an arrow pointing straight up. The output force is the force needed to lift the object straight up.

An inclined plane turns a small input force into a large output force by spreading the work out over a large distance. Pushing something up a long ramp that climbs gradually is easier than pushing something up a short, steep ramp.

Figure 8 The Inclined Plane Family

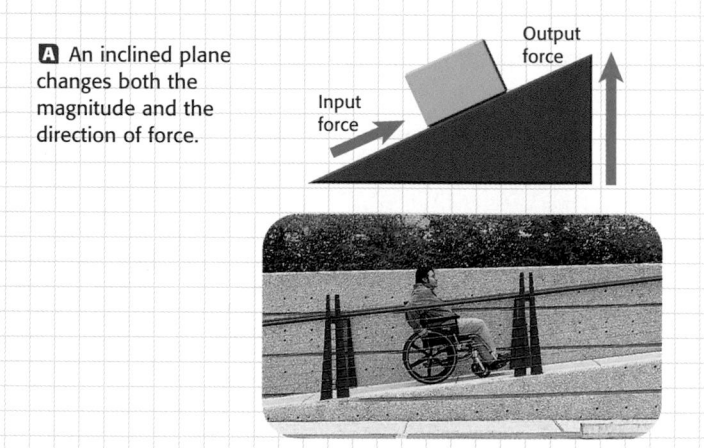

A An inclined plane changes both the magnitude and the direction of force.

Output force

Input force

Historical Perspective

Archimedes' Machines Archimedes was a Greek mathematician (287 B.C.–212 B.C.) of Syracuse, Sicily. Although his first love was pure mathematics, he also figured out how to use simple machines in a variety of practical applications. In response to a challenge by the King, Archimedes developed a system of levers and pulleys to launch a heavy ship. The King was astonished to see the fully-loaded ship glide into the water with a simple pull of a rope.

Later, when the Romans were attacking Syracuse, Archimedes designed machines to defend the city. He made cranes that would pick up Roman ships and smash them against rocks. He also created catapults to hurl huge stones at ships and soldiers.

The Romans were forced to withdraw, but they eventually gained control by starving the citizens. The Roman leader Marcellus ordered his men to "Spare that mathematician" but, tragically, Archimedes was killed. Encourage interested students to find out more about the life, death, and scientific contributions of Archimedes.

A wedge is a modified inclined plane

When an ax blade or a splitting wedge hits a piece of wood, it pushes through the wood and breaks it apart, as shown in *Figure 8B.* An ax blade is an example of a wedge, another kind of simple machine in the inclined plane family. A wedge functions like two inclined planes back to back. Using a wedge is like pushing a ramp instead of pushing an object up the ramp. A wedge turns a single downward force into two forces directed out to the sides. Some types of wedges, such as nails, are used as fasteners.

A screw is an inclined plane wrapped around a cylinder

A type of simple machine that you probably use often is a screw. The threads on a screw look like a spiral inclined plane. In fact, a screw is an inclined plane wrapped around a cylinder, as shown in *Figure 8C.* Like pushing an object up a ramp, tightening a screw with gently sloping threads requires a small force acting over a large distance. Tightening a screw with steeper threads requires more force. Jar lids are screws that people use every day. Spiral staircases are also common screws.

The ancient Egyptians built dozens of large stone pyramids as tombs for the bodies of kings and queens. The largest of these is the pyramid of Khufu at Giza, also called the Great Pyramid. It is made of more than 2 million blocks of stone. These blocks have an average weight of 2.5 tons, and the largest blocks weigh 15 tons. How did the Egyptians get these huge stones onto the pyramid?

Making the Connection

1. The Great Pyramid is about 140 m tall. How much work would be required to raise an average-sized pyramid block to this height? (2.5 tons = 2.2×10^4 N)

2. If the Egyptians used ramps with a mechanical advantage of 3, then an average block could be moved with a force of 7.3×10^3 N. If one person can pull with a force of 525 N, how many people would it take to pull an average block up such a ramp?

There is some uncertainty about how the blocks were lifted, although simple machines must have been involved. One hypothesis posits that the Egyptians primarily used ramps coated with mud to reduce friction. The exercise here is based on that method. Another hypothesis holds that the Egyptians used levers, mounted on either side of each block, to lift the blocks to successively higher levels.

Answers

1. $W = Fd = (2.2 \times 10^4 \text{ N})(140 \text{ m})$
   $W = 3.1 \times 10^6 \text{ J} = 3.1 \text{ MJ}$
2. $7.3 \times 10^3 \text{ N}/(525 \text{ N/person}) = 14$ people

LS Logical

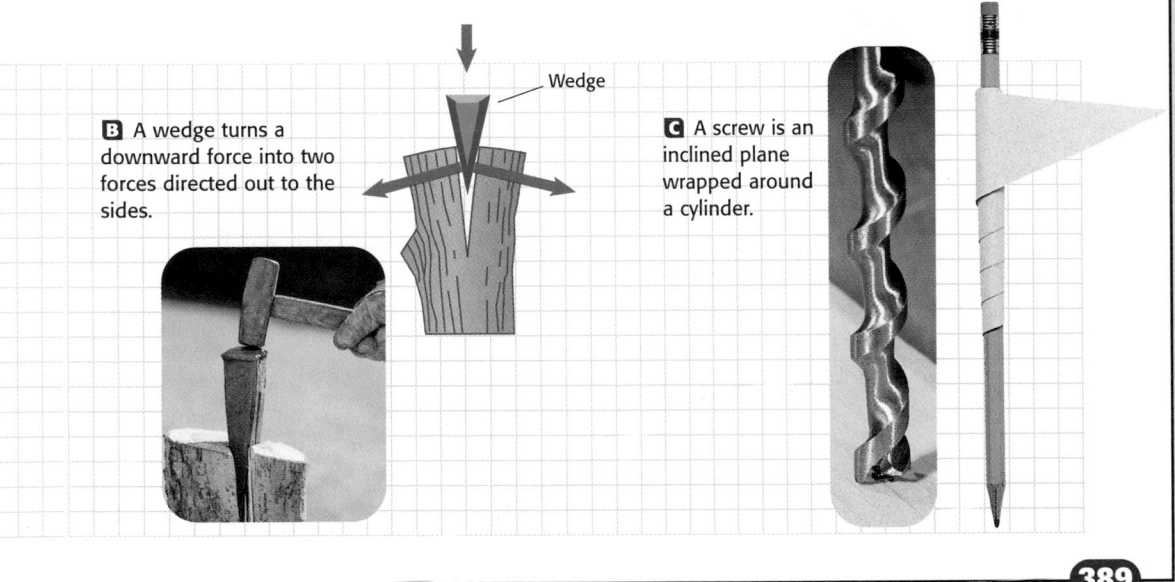

B A wedge turns a downward force into two forces directed out to the sides.

C A screw is an inclined plane wrapped around a cylinder.

389

Alternative Assessment — BASIC

Simple Machine Acrostic Ask students to create an acrostic they can use as a memory aid for the six simple machines. Use the following example (or create one of your own) to give students the idea:

Louis **P**lank **W**as an **I**nteresting, **W**itty **S**cholar.

(**L**ever, **P**ulley, **W**heel and axle, **I**nclined plane, **W**edge, **S**crew)

Tell students to create their own example because they will probably remember their own words better than an acrostic written by someone else. Also remind them that they can list the machines in any order. **LS** Verbal

Chapter Resource File

- **Quick Activity Datasheet** A Simple Inclined Plane GENERAL
- **Cross-Disciplinary Worksheet** Social Studies Connection—The Pyramids ADVANCED

Chapter 12 • Work and Energy **389**

List the two families of simple machines and the type of simple machines that belong to each family. Then describe what a compound machine is.

1. Lever family: levers, pulleys, and wheel-and-axle machines

2. Inclined plane family: inclined planes, wedges, and screws

3. A compound machine is any machine made up of two or more simple machines.

(Answers 1 and 2 could also be in the reverse order.) **LS** Logical

Chapter Resource File

• **Concept Review** GENERAL

• **Quiz** BASIC

Videos

CNN Presents Physical Science

• **Segment 3** Trebuchet Design

• **Critical Thinking** Worksheet 3

See the Science in the News video guide for more details.

Compound Machines

▶ **compound machine**
a machine made of more than one simple machine

Many devices that you use every day are made of more than one simple machine. A machine that combines two or more simple machines is called a **compound machine.** A pair of scissors, for example, uses two first class levers joined at a common fulcrum; each lever arm has a wedge that cuts into the paper. Most car jacks use a lever in combination with a large screw.

Of course, many machines are much more complex than these. How many simple machines can you identify in the bicycle shown in **Figure 9**? How many can you identify in a car?

Figure 9
A bicycle is made of many simple machines.

SECTION 2 REVIEW

SUMMARY

▶ The most basic machines are called simple machines. There are six types of simple machines in two families.

▶ Levers have a rigid arm and a fulcrum. There are three classes of levers.

▶ Pulleys and wheel-and-axle machines are also in the lever family.

▶ The inclined plane family includes inclined planes, wedges, and screws.

▶ Compound machines are made of two or more simple machines.

1. List the six types of simple machines.

2. Identify the kind of simple machine represented by each of these examples:
   a. a drill bit   **b.** a skateboard ramp   **c.** a boat oar

3. Describe how a lever can increase the force without changing the amount of work being done.

4. Explain why pulleys are in the lever family.

5. Compare the mechanical advantage of a long, thin wedge with that of a short, wide wedge. Which is greater?

6. Critical Thinking Can an inclined plane have a mechanical advantage of less than 1?

7. Critical Thinking Using the principle of a lever, explain why it is easier to open a door by pushing near the knob than by pushing near the hinges. What class of lever is a door?

8. Creative Thinking Choose a compound machine that you use every day, and identify the simple machines that it contains.

390

Answers to Section 2 Review

1. lever, pulley, wheel and axle, inclined plane, wedge, screw

2. a. screw
   b. inclined plane
   c. lever

3. A lever can increase the force without increasing the work done, because the output force will be exerted through a smaller distance.

4. The middle of the pulley is the fulcrum; the wheel of the pulley is like a lever-arm extended into a circle. Pulleys are different from ordinary levers because you can change the location of the fulcrum of a lever.

5. A long, thin wedge has a greater mechanical advantage than a short, wide wedge because the ratio of its length to its height is greater.

6. If the *MA* were 1, the plane would be as long as it is tall. Because you cannot travel a distance shorter than the actual height you need to lift an object, you cannot build an inclined plane with a *MA* less than 1.

See "Continuation of Answers" at the end of the chapter.

What Is Energy?

OBJECTIVES

▶ **Explain** the relationship between energy and work.
▶ **Define** *potential energy* and *kinetic energy.*
▶ **Calculate** kinetic energy and gravitational potential energy.
▶ **Distinguish** between mechanical and nonmechanical energy.

▶ **KEY TERMS**
potential energy
kinetic energy
mechanical energy

The world around you is full of energy. When you see a flash of lightning and hear a thunderclap, you are observing light and sound energy. When you ride a bicycle, you have energy just because you are moving. Even things that are sitting still have energy waiting to be released. We use other forms of energy, like nuclear energy and electrical energy, to power things in our world, from submarines to flashlights. Without energy, living organisms could not survive. Our bodies use a great deal of energy every day just to stay alive.

Energy and Work

When you stretch a slingshot, as shown in *Figure 10,* you are doing work, and you transfer energy to the elastic band. When the elastic band snaps back, it may in turn transfer that energy again by doing work on a stone in the slingshot. Whenever work is done, energy is transformed or transferred to another system. In fact, one way to define energy is as the ability to do work.

Energy is measured in joules

While work is done only when an object experiences a change in its position or motion, energy can be present in an object or a system when nothing is happening at all. But energy can be observed only when it is transferred from one object or system to another, as when a slingshot transfers the energy from its elastic band to a stone in the sling.

The amount of energy transferred from the slingshot can be measured by how much work is done on the stone. Because energy is a measure of the ability to do work, energy and work are expressed in the same units—joules.

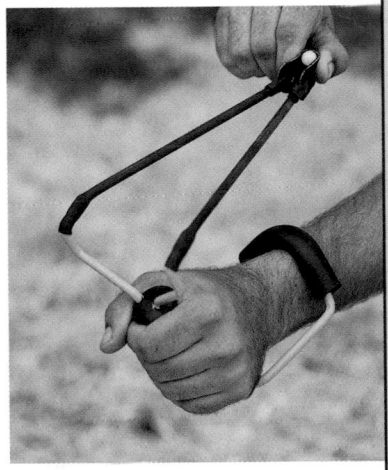

Figure 10
A stretched slingshot has the ability to do work.

391

Figure 11

This apple has gravitational potential energy. The energy results from the gravitational attraction between the apple and Earth.

▶ **potential energy** the energy that an object has because of the position, shape, or condition of the object

☑ internet connect ▤

www.scilinks.org

Topic: Potential Energy
SciLinks code: HK4108

SCILINKS Maintained by the National Science Teachers Association

Potential Energy

Stretching a rubber band requires work. If you then release the stretched rubber band, it will fly away from your hand. The energy used to stretch the rubber band is stored as potential energy so that it can do work at a later time. But where is the energy between the time you do work on the rubber band and the time you release it?

Potential energy is stored energy

A stretched rubber band stores energy in a form called **potential energy.** Potential energy is sometimes called energy of position because it results from the relative positions of objects in a system. The rubber band has potential energy because the two ends of the band are far away from each other. The energy stored in any type of stretched or compressed elastic material, such as a spring or a bungee cord, is called *elastic potential energy.*

The apple in **Figure 11** will fall if the stem breaks off the branch. The energy that could potentially do work on the apple results from its position above the ground. This type of stored energy is called *gravitational potential energy.* Any system of two or more objects separated by a distance contains gravitational potential energy resulting from the gravitational attraction between the objects.

Gravitational potential energy depends on both mass and height

An apple at the top of the tree has more gravitational potential energy with respect to the Earth than a similar apple on a lower branch. But if two apples of different masses are at the same height, the heavier apple has more gravitational potential energy than the lighter one.

Because it results from the force of gravity, gravitational potential energy depends both on the mass of the objects in a system and on the distance between them.

Gravitational Potential Energy Equation

$$grav.\ PE = mass \times free\text{-}fall\ acceleration \times height$$
$$PE = mgh$$

In this equation, notice that *mg* is the weight of the object in newtons, which is the same as the force on the object due to gravity. So this equation is really just a calculation of force times distance, like the work equation.

Did You Know ❓

Potential Energy Potential energy is a term that encompasses many types of energy. Other forms of potential energy include elastic, chemical, electrical, and magnetic. These may seem like very different concepts to students, but all forms of potential energy deal with position. Just as gravitational potential energy depends on distances between masses, elastic potential energy depends on the elongation of elastic materials. Chemical potential energy depends on bonds (position) between atoms within molecules. Electrical potential energy depends on distances between charged particles, and magnetic potential depends on the orientation of magnetic particles in a magnetic field.

Height can be relative

The height used in the equation for gravitational potential energy is usually measured from the ground. However, in some cases, a relative height might be more important. For example, if an apple were in a position to fall into a bird's nest on a lower branch, the apple's height above the nest could be used to calculate the apple's potential energy relative to the nest.

Math Skills

Gravitational Potential Energy A 65 kg rock climber ascends a cliff. What is the climber's gravitational potential energy at a point 35 m above the base of the cliff?

1 List the given and unknown values.

   Given: *mass, m* = 65 kg
   height, h = 35 m
   free-fall acceleration, g = 9.8 m/s^2
   Unknown: *gravitational potential energy, PE* = ? J

2 Write the equation for gravitational potential energy.

   $PE = mgh$

3 Insert the known values into the equation, and solve.

   $PE = (65 \text{ kg})(9.8 \text{ m/s}^2)(35 \text{ m})$
   $PE = 2.2 \times 10^4 \text{ kg} \cdot \text{m}^2/\text{s}^2 = 2.2 \times 10^4 \text{ J}$

Practice

Gravitational Potential Energy

1. Calculate the gravitational potential energy in the following systems:
   a. a car with a mass of 1200 kg at the top of a 42 m high hill
   b. a 65 kg climber on top of Mount Everest (8800 m high)
   c. a 0.52 kg bird flying at an altitude of 550 m
2. Lake Mead, the reservoir above Hoover Dam, has a surface area of approximately 640 km^2. The top 1 m of water in the lake weighs about 6.3×10^{12} N. The dam holds that top layer of water 220 m above the river below. Calculate the gravitational potential energy of the top 1 m of water in Lake Mead.
3. A science student holds a 55 g egg out a window. Just before the student releases the egg, the egg has 8.0 J of gravitational potential energy with respect to the ground. How far is the student's arm from the ground in meters? (**Hint:** Convert the mass to kilograms before solving.)
4. A diver has 3400 J of gravitational potential energy after stepping up onto a diving platform that is 6.0 m above the water. What is the diver's mass in kilograms?

Practice HINT

▶ The gravitational potential energy equation can be rearranged to isolate height on the left.

   $mgh = PE$

   Divide both sides by mg, and cancel.

   $$\frac{\cancel{m}gh}{\cancel{m}g} = \frac{PE}{mg}$$

   $$h = \frac{PE}{mg}$$

▶ You will need this version of the equation for practice problem 3.

▶ For practice problem 4, you will need to rearrange the equation to isolate mass on the left. When solving these problems, use g = 9.8 m/s2.

Math Skills

Additional Examples

Gravitational Potential Energy
A spider has 0.080 J of gravitational potential energy as it reaches the halfway point climbing up a 2.8 m wall. What is the potential energy of the spider at the top of the wall? (**HINT:** You do not need to find the mass first.) (0.16 J)

A 0.50 g leaf falls from a branch 4.0 m off the ground to a bird's nest 2.5 m off the ground. How much gravitational potential energy did the leaf lose? (7.4×10^{-3} J)
LS Logical

Practice

1. **a.** $PE = mgh = (1200 \text{ kg})$
      $(9.8 \text{ m/s}^2)(42 \text{ m}) = 4.9 \times 10^5 \text{ J}$
   b. $PE = (65 \text{ kg})(9.8 \text{ m/s}^2)$
      $(8800 \text{ m}) = 5.6 \times 10^6 \text{ J}$
   c. $PE = (0.52 \text{ kg})(9.8 \text{ m/s}^2)$
      $(550 \text{ m}) = 2.8 \times 10^3 \text{ J}$
2. $PE = (6.3 \times 10^{12} \text{ N})(220 \text{ m}) = 1.4 \times 10^{15} \text{ J}$
3. $h = \dfrac{PE}{mg} =$

   $\dfrac{8.0 \text{ J}}{(0.055 \text{ kg})(9.8 \text{ m/s}^2)} = 15 \text{ m}$

4. $m = \dfrac{PE}{gh} = \dfrac{3400 \text{ J}}{(9.8 \text{ m/s}^2)(6.0 \text{ m})} = 58 \text{ kg}$

LS Logical

Did You Know?

Potential Energy in Food Whenever we eat food, we are supplying our bodies with energy. The body constantly uses that energy to perform actions and to stay alive, for example to breathe and to pump blood. A person who exercises regularly needs more calories of energy each day than someone who is inactive. In fact, each individual has a daily caloric requirement based on their size and activity level. The daily requirements listed in nutrition labels on food are for a person who needs 2000 calories per day, but many individuals actually need more or less than this average value.

If someone eats more calories than their daily requirement, the body stores the extra energy in the form of fat. The energy—which is now in the form of potential energy—can be used at a later time when it is needed. Losing body fat requires using up this potential energy by using more calories than are taken in every day. This is best accomplished by a combination of calorie reduction and exercise.

Chapter Resource File

• **Math Skills** Gravitational Potential Energy **GENERAL**

Kinetic Energy Kinetic energy, like all other kinds of energy, is an ability to do work. Kinetic energy is probably the most obvious form of energy. If students have had problems grasping energy, now is a good time to drive it home. When you think of an "energetic" person, you may think of someone who moves around a lot. Something moving has, by nature of its motion, an ability to do work.

SKILL BUILDER — **BASIC**

Interpreting Graphs Have students examine the graph in **Figure 12B.** The graph shows kinetic energy versus speed for a falling apple that weighs 1 N. Point out that as the speed increases, the kinetic energy increases rapidly. That is because kinetic energy depends on the square of the speed. The curve on this graph is half of a parabola. **LS** Visual

Teaching Tip — **BASIC**

Kinetic Energy and Momentum Some students may be confused about the distinction between kinetic energy and momentum. For those students, write the equations for momentum and for kinetic energy side-by-side. Have students consider how they are similar and how they are different. (Both quantities depend on mass and velocity, but kinetic energy has a more sensitive dependence on velocity because of the squared quantity.) **LS** Logical

---

▶ **kinetic energy** the energy of a moving object due to the object's motion

VOCABULARY *Skills Tip*

Kinetic *comes from the Greek word* kinetikos, *which means "motion."*

Figure 12

A A falling apple can do work on the ground underneath—or on someone's head.

B A small increase in the speed of an apple results in a large increase in kinetic energy.

394

Kinetic Energy

Once an apple starts to fall from the branch of a tree, as in *Figure 12A,* it has the ability to do work. Because the apple is moving, it can do work when it hits the ground or lands on the head of someone under the tree. The energy that an object has because it is in motion is called **kinetic energy.**

Kinetic energy depends on mass and speed

A falling apple can do more work than a cherry falling at the same speed. That is because the kinetic energy of an object depends on the object's mass.

An apple that is moving at 10 m/s can do more work than an apple moving at 1 m/s can. As an apple falls, it accelerates. The kinetic energy of the apple increases as it speeds up. In fact, the kinetic energy of a moving object depends on the square of the object's speed.

Kinetic Energy Equation

$$kinetic\ energy = \frac{1}{2} \times mass \times speed\ squared$$
$$KE = \frac{1}{2}mv^2$$

Figure 12B shows a graph of kinetic energy versus speed for a falling apple that weighs 1.0 N. Notice that kinetic energy is expressed in joules. Because kinetic energy is calculated using both mass and speed squared, the base units are $kg \bullet m^2/s^2$, which are equivalent to joules.

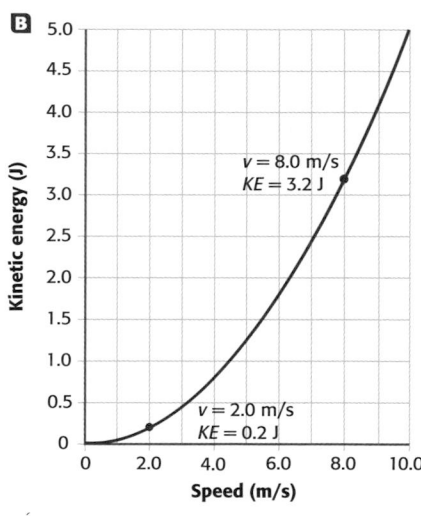

$v = 8.0$ m/s
$KE = 3.2$ J

$v = 2.0$ m/s
$KE = 0.2$ J

Alternative Assessment — **BASIC**

Potential and Kinetic Energy Have students write a brief paragraph describing the changes in potential and kinetic energy in a roller coaster as it sits motionless at the top of a rise, then begins rolling down the rise until it reaches the bottom. (At the top of the rise, the energy is all PE_1 based on the coaster's height. Because the coaster is motionless, it has no KE. As it starts rolling down the hill, its height decreases so it loses PE. At the same time, its speed increases, so it gains KE. At the bottom of the hill, the relative height is zero, so there is no longer any PE. The speed has been gradually increasing, so KE is at a maximum at this point.)

After students have read Section 4, ask them to reread their paragraphs in light of their new understanding of energy conversions and conservation. (At that point they should recognize that, disregarding friction, the sum of PE and KE at any given time is a constant, so that the PE at the top equals the KE at the bottom.) **LS** Verbal

---

Kinetic energy depends on speed more than mass

The line on the graph of kinetic energy versus speed curves sharply upward as speed increases. At one point, the speed is 2.0 m/s and the kinetic energy is 0.20 J. At another point, the speed has increased four times to 8.0 m/s. But the kinetic energy has increased 16 times, to 3.2 J. In the kinetic energy equation, speed is squared, so a small increase in speed produces a large increase in kinetic energy.

You may have heard that car crashes are much more dangerous at speeds above the speed limit. The kinetic energy equation provides a scientific reason for that fact. Because a car has much more kinetic energy at higher speeds, it can do much more work—which means much more damage—in a collision.

internet connect

www.scilinks.org
Topic: Kinetic Energy
SciLinks code: HK4075

SC*LINKS*. Maintained by the National Science Teachers Association

---

> ## Math Skills
>
> **Kinetic Energy** What is the kinetic energy of a 44 kg cheetah running at 31 m/s?
>
> **1** List the given and unknown values.
>   Given: *mass*, $m = 45$ kg
>   *speed*, $v = 31$ m/s
>   Unknown: *kinetic energy*, $KE = ?$ J
>
> **2** Write the equation for kinetic energy.
> $kinetic\ energy = \frac{1}{2} \times mass \times speed\ squared$
> $KE = \frac{1}{2}mv^2$
>
> **3** Insert the known values into the equation, and solve.
> $KE = \frac{1}{2}(44\ \text{kg})(31\ \text{m/s})^2$
> $KE = 2.1 \times 10^4\ \text{kg} \cdot \text{m}^2/\text{s}^2 = 2.1 \times 10^4\ \text{J}$

Practice

Kinetic Energy

1. Calculate the kinetic energy in joules of a 1500 kg car moving at the following speeds:
  a. 29 m/s
  b. 18 m/s
  c. 42 km/h (**Hint:** Convert the speed to meters per second before substituting into the equation.)

2. A 35 kg child has 190 J of kinetic energy after sledding down a hill. What is the child's speed in meters per second at the bottom of the hill?

3. A bowling ball traveling 2.0 m/s has 16 J of kinetic energy. What is the mass of the bowling ball in kilograms?

Practice **HINT**

▶ The kinetic energy equation can be rearranged to isolate speed on the left.

$$\frac{1}{2}mv^2 = KE$$

Multiply both sides by $\frac{2}{m}$.

$$\left(\frac{2}{m}\right) \times \frac{1}{2}mv^2 = \left(\frac{2}{m}\right) \times KE$$

$$v^2 = \frac{2KE}{m}$$

Take the square root of each side.

$$\sqrt{v^2} = \sqrt{\frac{2KE}{m}}$$

$$v = \sqrt{\frac{2KE}{m}}$$

You will need this version of the equation for Practice Problem 2.

▶ For Practice Problem 3, you will need to use the equation rearranged with mass isolated on the left:

$$m = \frac{2KE}{v^2}$$

395

Teaching Tip

Nonmechanical Energy The distinction between mechanical energy and nonmechanical energy is vague by nature. Nonmechanical energy is often called "internal energy," also a vague concept. In most cases, nonmechanical energy can be reduced to some kind of mechanical energy (for instance, the kinetic energy of atoms in a gas as the basis for thermal energy). For each type of energy introduced in the next few pages, discuss how it can also be considered as either kinetic or potential energy.

Teaching Tip

Approximations in Science
Explain to students that the practice of considering only major effects and ignoring small effects is a very common practice in science. As students read this page they may ask how scientists can consider some energy to be nonmechanical at some times and not at others. Explain that scientists are quite often investigating one certain aspect of things.

For example, a physicist trying to explain the flight of a horseshoe could include the motion of the molecules (with the help of a computer), but the physicist would probably ignore the motion of the molecules because that motion is so small compared to the horseshoe flying through the air.

■ **mechanical energy**
the amount of work an object can do because of the object's kinetic and potential energies

Figure 13

The atoms in a hot object, such as this horseshoe, have kinetic energy. The kinetic energy is related to the object's temperature.

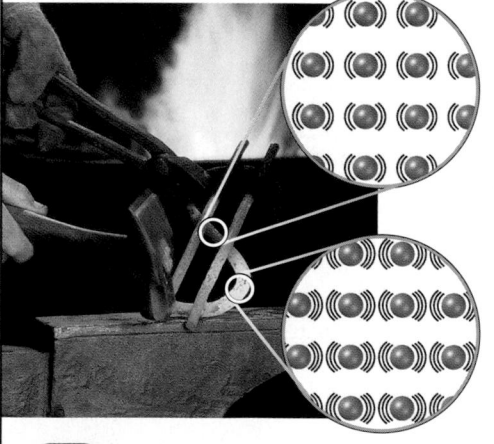

Other Forms of Energy

Apples have potential energy when they are hanging on a branch above the ground, and they have kinetic and potential energy when they are falling. The sum of the potential energy and the kinetic energy in a system is called **mechanical energy.** Mechanical energy can also be thought of as the amount of work an object can do because of the object's kinetic and potential energies.

Apples can also give you energy when you eat them. What kind of energy is that? In almost every system, there are hidden forms of energy that are related to the motion and arrangement of atoms that make up the objects in the system.

Energy that lies at the level of atoms and that does not affect motion on a large scale is sometimes called *nonmechanical energy.* However, a close look at the different forms of energy in a system usually reveals that they are in most cases just special forms of kinetic or potential energy.

Atoms and molecules have kinetic energy

You have learned that atoms and molecules are constantly in motion. Therefore, these tiny particles have kinetic energy. Like a bowling ball hitting pins, kinetic energy is transferred between particles through collisions. The average kinetic energy of particles in an object increases as the object gets hotter and decreases as it cools down. In another chapter, you will learn more about how the kinetic energy of particles relates to heat and temperature.

Figure 13 shows the motion of atoms in two parts of a horseshoe at different temperatures. In both parts, the iron atoms inside the horseshoe are vibrating. The atoms in the hotter part of the horseshoe are vibrating more rapidly than the atoms in the cooler part, so they have greater kinetic energy.

If a scientist wanted to analyze the motion of a horseshoe in a game of "horseshoes," the motion of particles inside the shoes would not be important. For the sake of that study, the energy due to the motion of the atoms would be considered nonmechanical energy.

However, if the same scientist wanted to study the change in the properties of iron when heated in a blacksmith's shop, the motion of the atoms would become significant to the study, and the kinetic energy of the particles within the horseshoe would then be viewed as mechanical energy.

396

Teacher Resources

For the New Teacher The fundamental concepts about energy that are explored in this section will be used and built upon in the coverage of many other physics topics throughout the book. Topics that are especially related include the following: the kinetic theory of matter; heat and temperature; chemical reactions; energy derived from natural resources; nuclear energy; electricity; energy processes in stars.

When working with those topics, be sure students remember the information covered in this section. For topics you have already studied, you could have students review the topics again with their new understanding of energy. For upcoming topics, you may wish to briefly review this fundamental material before covering the related topic.

Chemical reactions involve potential energy

In a chemical reaction, bonds between atoms break apart. When the atoms bond together again in a new pattern, a different substance is formed. Both the formation of bonds and the breaking of bonds involve changes in energy. The amount of *chemical energy* associated with a substance depends in part on the relative positions of the atoms it contains.

Because chemical energy depends on position, it is a kind of potential energy. Reactions that release energy involve a decrease in the potential energy within substances. For example, when a match burns, as shown in **Figure 14,** the release of stored energy from the match head produces light and a small explosion of hot gas.

Living things get energy from the sun

Where do you get the energy you need to live? It comes in the form of chemical energy stored in the food you eat. But where did that energy come from? When you eat a meal, you are eating either plants or animals, or both. Animals also eat plants or other animals, or both. At the bottom of the food chain are plants and algae that derive their energy directly from sunlight.

Plants use *photosynthesis* to turn the energy in sunlight into chemical energy. This energy is stored in sugars and other organic molecules that make up cells in living tissue. When your body needs energy, some of these organic molecules are broken down through *respiration*. Respiration releases the energy your body needs to live and do work.

Figure 14

When a match burns, the chemical energy stored inside the head of the match is released, producing light and a small explosion of hot gas.

REAL WORLD APPLICATIONS

The Energy in Food You can introduce the idea that a food Calorie is actually what scientists would call a kilocalorie, or 1000 calories. So, to a scientist, one calorie (with a small "c") equals 4.186 J. This is important if you are comparing the work you do exercising to the energy in the food you eat.

Applying Information

1. 230 Cal

2. 230 Cal × 4186 J/Cal = 9.6×10^5 J

3. 1×10^7 J/9.6×10^5 J = 10

LS Logical

REAL WORLD APPLICATIONS

The Energy in Food
We get energy from the food we eat. This energy is often measured by another unit, the Calorie. One Calorie is equivalent to 4186 J.

Applying Information
1. Look at the nutrition label on this "energy bar." How many Calories of energy does the bar contain?

2. Calculate how many joules of energy the bar contains by multiplying the number of Calories by the conversion factor of 4186 J/Cal.

3. An average person needs to take in about 10 million joules of energy every day. How many energy bars would you have to eat to get this much energy?

397

Teach, *continued*

Teaching Tip

Nuclear Fission Tell students to imagine a large water balloon filled almost to the breaking point. You have to carry a balloon like that very carefully because the slightest bump will cause it to break. Now have students imagine two smaller water balloons that are not as full. There is still a large amount of flexibility in the "wrapper" (or balloon). Large nuclei are like an overfilled water balloon. The wrapper (strong nuclear forces that hold the nucleus together) is stretched to the breaking point. When a nucleus undergoes fission, it splits into smaller nuclei, where the nuclear forces are not "stretched" as much. The energy that is released is the difference in the energy required to hold the nuclei together. It takes less energy to contain two small nuclei than it does to contain one very large one.

MISCONCEPTION /// ALERT \\\

What is Energy? Students commonly believe that energy is a fluid, an ingredient, or a fuel. Emphasize that energy is not a material substance. The definition of energy as the ability to do work may seem vague to some students; remind them that the amount of energy associated with an object or system can be precisely quantified, as for example with the *KE* and *PE* equations they have learned in this section.

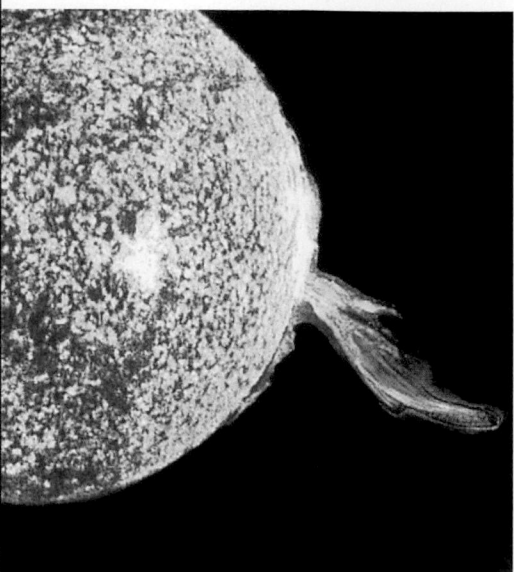

Figure 15

The nuclei of atoms contain enormous amounts of energy. The sun is fueled by nuclear fusion reactions in its core.

The sun gets energy from nuclear reactions

The sun, shown in *Figure 15*, not only gives energy to living things but also keeps our whole planet warm and bright. And the energy that reaches Earth from the sun is only a small portion of the sun's total energy output. How does the sun produce so much energy?

The sun's energy comes from nuclear fusion, a type of reaction in which light atomic nuclei combine to form a heavier nucleus. Nuclear power plants use a different process, called nuclear fission, to release nuclear energy. In fission, a single heavy nucleus is split into two or more lighter nuclei. In both fusion and fission, small quantities of mass are converted into large quantities of energy.

You have learned that mass is converted to energy during nuclear reactions. This nuclear energy is a kind of potential energy stored by the forces holding subatomic particles together in the nuclei of atoms.

Electricity is a form of energy

The lights and appliances in your home are powered by another form of energy, electricity. Electricity results from the flow of charged particles through wires or other conducting materials. Moving electrons can increase the temperature of a wire and cause it to glow, as in a light bulb. Moving electrons also create magnetic fields, which can do work to power a motor or other devices. The lightning shown in *Figure 16* is caused by electrons traveling through the air between the ground and a thundercloud.

Figure 16

Electrical energy is derived from the flow of charged particles, as in a bolt of lightning or in a wire. We can harness electricity to power appliances in our homes.

398

Did You Know ?

Tons of TNT Nuclear energy, especially as contained in nuclear weapons, is sometimes expressed in units of tons of TNT. One ton of TNT is equal to the amount of energy released from the explosion of 1 ton of TNT explosive. 1 ton of TNT = 4.2×10^9 J.

Light can carry energy across empty space

An asphalt surface on a bright summer day is hotter where light is shining directly on it than it is in the shade. Light energy travels from the sun to Earth across empty space in the form of *electromagnetic waves*.

A beam of white light can be separated into a color spectrum, as shown in **Figure 17**. Light toward the blue end of the spectrum carries more energy than light toward the red end.

Figure 17
Light is made of electromagnetic waves that carry energy across empty space.

SECTION 3 REVIEW

SUMMARY

▶ Energy is the ability to do work.

▶ Like work, energy is measured in joules.

▶ Potential energy is stored energy.

▶ Elastic potential energy is stored in any stretched or compressed elastic material.

▶ The gravitational potential energy of an object is determined by its mass, its height, and *g*, the free-fall acceleration due to gravity. *PE = mgh*.

▶ An object's kinetic energy, or energy of motion, is determined by its mass and speed. $KE = \frac{1}{2}mv^2$.

▶ Potential energy and kinetic energy are forms of mechanical energy.

▶ In addition to mechanical energy, most systems contain nonmechanical energy.

▶ Nonmechanical energy does not usually affect systems on a large scale.

1. **List** three different forms of energy.

2. **Explain** how energy is different from work.

3. **Explain** the difference between potential energy and kinetic energy.

4. **Determine** what form or forms of energy apply to each of the following situations, and specify whether each form is mechanical or nonmechanical:
   a. a Frisbee flying though the air
   b. a hot cup of soup
   c. a wound clock spring
   d. sunlight
   e. a boulder sitting at the top of a cliff

5. **Critical Thinking** Water storage tanks are usually built on towers or placed on hilltops. Why?

6. **Creative Thinking** Name one situation in which gravitational potential energy might be useful, and name one situation where it might be dangerous.

=== **Math Skills** ===

7. Calculate the gravitational potential energy of a 93.0 kg sky diver who is 550 m above the ground.

8. What is the kinetic energy in joules of a 0.02 kg bullet traveling 300 m/s?

9. Calculate the kinetic or potential energy in joules for each of the following situations:
   a. a 2.5 kg book held 2.0 m above the ground
   b. a 15 g snowball moving through the air at 3.5 m/s
   c. a 35 kg child sitting at the top of a slide that is 3.5 m above the ground
   d. an 8500 kg airplane flying at 220 km/h

399

Answers to Section 3 Review

1. Answers may include: kinetic energy, potential energy (gravitational or elastic), mechanical energy, nonmechanical energy, chemical energy, electrical energy, nuclear energy, light energy.

2. Energy is the ability to do work. Or when work is done, energy is transferred from one object to another.

3. Potential energy is energy due to position. Kinetic energy is the energy of motion.

4. a. gravitational *PE* and *KE* (both mechanical)
   b. kinetic energy of the molecules (nonmechanical), chemical energy of the molecules (nonmechanical)

   c. elastic potential energy, kinetic energy as the spring unwinds (both mechanical)
   d. light energy (nonmechanical)
   e. gravitational *PE* (mechanical)

5. Storing the water up high gives the water gravitational potential energy, so the water will naturally flow out of the tank if needed.

6. The water tank in item 5 is a case where gravitational *PE* is useful. Gravitational *PE* is dangerous to people hanging from the side of a cliff or building.

See "Continuation of Answers" at the end of the chapter.

Conservation of Energy

Overview

Before beginning this section, review with your students the Objectives listed in the Student Edition. In this section, students learn about energy transformations and the law of energy conservation. The section concludes with a discussion of the efficiency of machines.

🔊 Bellringer

Use the Bellringer transparency to prepare students for this section.

Opening Activity — GENERAL

Have students devise a hypothesis about the subject of the section based on the objectives and headings. Have students create a concept map or graphic organizer to show the structure of the section. Ask students to add details to their concept map or graphic organizer as they read the section. **LS Verbal**

SKILL BUILDER — BASIC

Interpreting Visuals Have students discuss how the roller-coaster car in **Figure 18** behaves going up and down the hills. Ask them to describe the types and magnitudes of energy the car has at the highest and lowest points. **LS Visual**

▶ **KEY TERM**
efficiency

OBJECTIVES

▶ **Identify** and describe transformations of energy.
▶ **Explain** the law of conservation of energy.
▶ **Discuss** where energy goes when it seems to disappear.
▶ **Analyze** the efficiency of machines.

Imagine you are sitting in the front car of a roller coaster, such as the one shown in *Figure 18.* The car is pulled slowly up the first hill by a conveyor belt. When you reach the crest of the hill, you are barely moving. Then you go over the edge and start to race downward, speeding faster and faster until you reach the bottom of the hill. The wheels are roaring along the track. You continue to travel up and down through a series of smaller humps, twists, and turns. Finally, you climb another hill almost as big as the first, drop down again, and then coast to the end of the ride.

Figure 18

The tallest roller coaster in the world is the Fujiyama, in Fujikyu Highland Park, Japan. It spans 70 m from its highest to lowest points.

Energy Transformations

In the course of a roller coaster ride, energy changes form many times. You may not have noticed the conveyor belt at the beginning, but in terms of energy it is the most important part of the ride. All of the energy required for the entire ride comes from work done by the conveyor belt as it lifts the cars and the passengers up the first hill.

The energy from that initial work is stored as gravitational potential energy at the top of the first hill. After that, the energy goes through a series of transformations, or changes, turning into kinetic energy and turning back into potential energy. A small quantity of this energy is transferred as heat to the wheels and as vibrations that produce a roaring sound in the air. But whatever form the energy takes during the ride, it is all there from the very beginning.

SOCIAL STUDIES
CONNECTION

Invite a history teacher to be a guest lecturer during your discussion of energy transformation. Have the history teacher explain the impact of new sources of power during the Industrial Revolution.

LIFE SCIENCE
CONNECTION

Plants use energy from the sun to carry out photosynthesis. Invite a biology teacher to be a guest lecturer during your discussion of energy transformations. Have the biology teacher explain the energy transformations that take place during the process of photosynthesis.

Figure 19

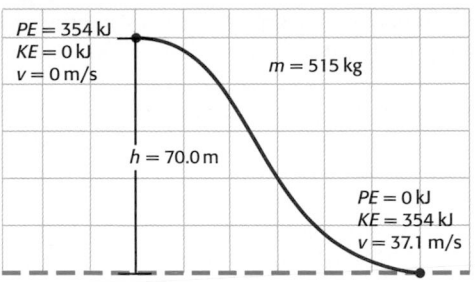

A As a car goes down a hill on a roller coaster, potential energy changes to kinetic energy.

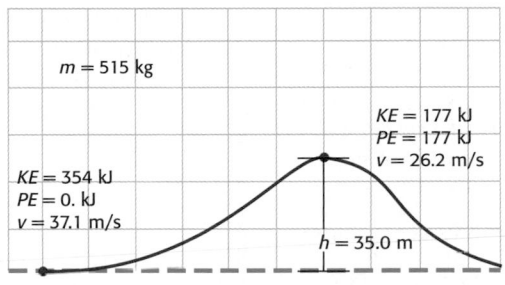

B At the top of this small hill, half the kinetic energy has become potential energy. The rest of the kinetic energy carries the car over the crest of the hill at high speed.

Potential energy can become kinetic energy

Almost all of the energy of a car on a roller coaster is potential energy at the top of a tall hill. The potential energy gradually changes to kinetic energy as the car accelerates downward. At the bottom of the lowest hill, the car has a maximum of kinetic energy and a minimum of potential energy.

Figure 19A shows the potential energy and kinetic energy of a car at the top and the bottom of the biggest hill on the Fujiyama roller coaster. Notice that the system has the same amount of energy, 354 kJ, whether the car is at the top or the bottom of the hill. That is because all of the gravitational potential energy at the top changes to kinetic energy as the car goes down the hill. When the car reaches the lowest point, the system has no potential energy because the car cannot go any lower.

Kinetic energy can become potential energy

When the car is at the lowest point on the roller coaster, it has no more potential energy, but it has a lot of kinetic energy. This kinetic energy can do the work to carry the car up another hill. As the car climbs the hill, the car slows down, decreasing its kinetic energy. Where does that energy go? Most of it turns back into potential energy as the height of the car increases.

At the top of a smaller hill, the car will still have some kinetic energy, along with some potential energy, as shown in *Figure 19B.* The kinetic energy will carry the car forward over the crest of the hill. Of course, the car could not climb a hill taller than the first one without an extra boost. The car does not have enough energy.

401

Teaching Tip

Friction Point out to students that in a world with no friction, the exchange of energy—from potential to kinetic and back again—could go on forever. In the real world, some energy is lost to friction as the car rolls along the track, so the roller coaster could not continue forever without some energy input (such as that provided by the motor that pulls the cars up the first hill).

SKILL BUILDER — BASIC

Interpreting Diagrams Figure 19 shows diagrams of two different hills on a roller coaster. Walk students through the energy quantities, which are generated by the equations learned in Section 3. What assumption is made in each diagram about the energy of the roller-coaster car? (The total mechanical energy of the car at a later time is equal to the total mechanical energy of the car at any earlier time. This is based on the law of conservation of energy.)
LS Logical

Historical Perspective

The First Roller Coasters The first roller coasters were giant ice slides made in Russia. The first of these ice slides, which was in St. Petersburg, consisted of a 70-foot wooden frame packed with watered-down snow that turned into ice. The sleds were made of two-foot ice blocks with seats carved into them.

Because ice has a low coefficient of friction, the ice "cars" easily rolled down the ice slide. The seats were lined with straw or fur for insulation. Sand was placed near the end of the slide. Friction between the sand and the sleds slowed them down near the bottom of the slide.

Teaching Tip — **BASIC**

Flight of a Ball On the board, draw the flight of a ball thrown from one person to another. This should be drawn as an inverted parabola. Ask students to identify where the ball has the maximum kinetic energy (at the bottom on either side of the curve) and where the ball has the maximum potential energy (at the top). Point out to students that the ball you have just drawn on the board does not have 0 J *KE* at the top, because it is traveling sideways as well as up and down. Therefore, the ball still has some *KE* at the top. If it did not, it would fall straight down, because there would be no energy at the top of the path to carry the ball sideways. **LS Visual**

SKILL BUILDER — **BASIC**

Interpreting Visuals Ask students if a greater initial kinetic energy from the tennis player would make the ball go higher or lower (higher). Then have them calculate how high the tennis ball would go if the tennis player gives it 0.8 J of initial kinetic energy (about 1.6 m). **LS Logical**

www.scilinks.org
Topic: Energy
Transformations
SciLinks code: HK4049

Energy transformations explain the flight of a ball

The relationship between potential energy and kinetic energy can explain motion in many different situations. Let's look at some other examples.

A tennis player tosses a 0.05 kg tennis ball into the air to set up for a serve, as shown in *Figure 20.* He gives the ball 0.5 J of kinetic energy, and it travels straight up. As the ball rises higher, the kinetic energy is converted to potential energy. The ball will keep rising until all the kinetic energy is gone. At its highest point, the ball has 0.5 J of potential energy. As the ball falls down again, the potential energy changes back to kinetic energy.

Imagine that a tennis trainer wants to know how high the ball will go when it is given 0.5 J of initial kinetic energy by a tennis player. The trainer could make a series of calculations using force and acceleration, but in this case using the concept of energy transformations is easier. The trainer knows that the ball's initial kinetic energy is 0.5 J and that its mass is 0.05 kg. To find out how high the ball will go, the trainer has to find the point where the potential energy equals its initial kinetic energy, 0.5 J. Using the equation for gravitational potential energy, the height turns out to be 1 m above the point that the tennis player releases the ball.

Figure 20

The kinetic energy of the ball at the bottom of its path equals the potential energy at the top of the path.

$$PE = mgh$$

$$h = \frac{PE}{mg}$$

$$h = \frac{0.5 \text{ J}}{(0.05 \text{ kg})(9.8 \text{ m/s}^2)}$$

$$h = 1 \text{ m}$$

$$KE = 0.5 \text{ J}$$

Did You Know?

Tennis, anyone? When tossing a ball for a serve in tennis, the height of the ball toss is important. This is because the tennis player wants to hit the ball near the top of the arc, when it is not moving or is barely moving. If the ball is thrown too high, it will be harder to time the serve and hit the ball at the right point. If the ball is too low, the tennis player will not be able to extend his or her arm and therefore will not hit the ball with as much force.

Energy transformations explain a bouncing ball

Before a serve, a tennis player usually bounces the ball a few times while building concentration. The motion of a bouncing ball can also be explained using energy principles. As the tennis player throws the ball down, he adds kinetic energy to the potential energy the ball has at the height of her hand. The kinetic energy of the ball then increases steadily as the ball falls because the potential energy is changing to kinetic energy.

When the ball hits the ground, there is a sudden energy transformation as the kinetic energy of the ball changes to elastic potential energy stored in the compressed tennis ball. The elastic potential energy then quickly changes back to kinetic energy as the ball bounces upward.

If all of the kinetic energy in the ball changed to elastic potential energy, and that elastic potential energy all changed back to kinetic energy during the bounce, the ball would bounce up to the tennis player's hand. Its speed on return would be exactly the same as the speed at which it was thrown down. If the ball were dropped instead of thrown down, it would bounce up to the same height from which it was dropped.

Mechanical energy can change to other forms of energy

If changes from kinetic energy to potential energy and back again were always complete, then balls would always bounce back to the same height they were dropped from and cars on roller coasters would keep gliding forever. But that is not the way things really happen.

When a ball bounces on the ground, not all of the kinetic energy changes to elastic potential energy. Some of the kinetic energy compresses the air around the ball, making a sound, and some of the kinetic energy makes the ball, the air, and the ground slightly hotter. Because these other forms of energy are not directly due to the motion or position of the ball, they can be considered nonmechanical energy. With each bounce, the ball loses some mechanical energy, as shown in *Figure 21*.

Likewise, a car on a roller coaster cannot keep moving up and down the track forever. The total mechanical energy of a car on a roller coaster constantly decreases due to friction and air resistance. This energy does not just disappear though. Some of it increases the temperature of the track, the car's wheels, and the air. Some of the energy compresses the air, making a roaring sound. Often, when energy seems to disappear, it has really just changed to a nonmechanical form.

Figure 21

With each bounce of a tennis ball, some of the mechanical energy changes to nonmechanical energy.

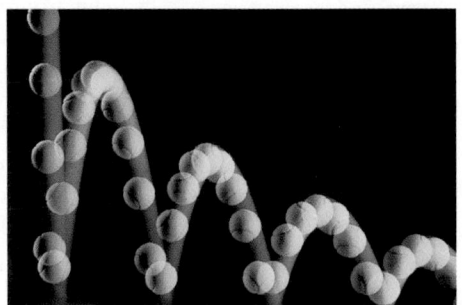

403

Alternative Assessment — ADVANCED

Writing **Accounting for Energy** Many examples in the chapter so far have assumed "ideal" circumstances, disregarding friction and air resistance. However, once the idea that mechanical energy can change to nonmechanical energy is introduced, deviations from the ideal in "real world" situations can be explained.

Have students, after they have finished reading this page, return to earlier examples in the chapter to consider if energy would be transferred from the system in realistic circumstances (for instance, pushing a box up a ramp, or a roller coaster). Have them write a few paragraphs about a given example, making sure that they account for as much of the "lost" energy as possible.

Demonstration—GENERAL

Bouncing Balls

(Time: About 10 minutes)

Materials

• several different types of balls (super ball, tennis ball, racquet ball, squash ball, ball made of clay, steel ball bearing, etc.)

Rotating through all the types of balls, hold two at a time approximately 1 m above the floor or a desk. Drop the balls simultaneously.

Lead a discussion about why each ball bounces to a different height. Discuss the energy conversions mentioned in the text. Ask students to hypothesize about the "disappearance" of the energy. Explain to students that each ball has a different ability to store elastic potential energy—the super ball is very elastic, while the ball of clay is definitely not. Students should be able to conclude that some energy is lost in the collision with the floor or table. Lead students to the realization that some energy is released as sound, and some is stored internally as the temperature of the ball rises. **LS Visual**

INTEGRATING

COMPUTERS AND TECHNOLOGY

In order for a flashlight to work, there must be a supply of energy.

A flashlight battery contains different chemicals that can react with each other to release energy. When the flashlight is turned on, chemical potential energy changes to electrical energy, and electrons begin to flow through a wire attached to the battery. Inside the bulb, the wire filament begins to glow, and the energy is transformed into light energy.

After the flashlight has been used for a certain amount of time, the battery will run out of energy. It will have to be replaced or recharged.

The Law of Conservation of Energy

In our study of machines, we saw that the work done on a machine is equal to the work that it can do. Similarly, in our study of the roller coaster, we found that the energy present at the beginning of the ride is present throughout the ride and at the end of the ride, although the energy continually changes form. The energy in each system does not appear out of nowhere and never just disappears.

This simple observation is based on one of the most important principles in all of science—the law of conservation of energy. Here is the law in its simplest form.

Energy cannot be created or destroyed.

In a mechanical system such as a roller coaster or a swinging pendulum, the energy in the system at any time can be calculated by adding the kinetic and potential energy to get the total mechanical energy. The law of conservation of energy requires that at any given time, the total energy should be the same.

Energy doesn't appear out of nowhere

Energy cannot be created from nothing. Imagine a girl jumping on a trampoline. After the first bounce, she rises to a height of 0.5 m. After the second bounce, she rises to a height of 1 m. Because she has greater gravitational potential energy after the second bounce, we must conclude that she added energy to her bounce by pushing with her legs. Whenever the total energy in a system increases, it must be due to energy that enters the system from an external source.

Energy doesn't disappear

Because mechanical energy can change to nonmechanical energy due to friction, air resistance, and other factors, tracing the flow of energy in a system can be difficult. Some of the energy may leak out of the system into the surrounding environment, as when the roller coaster produces sound as it compresses the air. But none of the energy disappears; it just changes form.

Scientists study energy systems

Energy has many different forms and can be found almost everywhere. Accounting for all of the energy in a given situation can be complicated. To make studying a situation easier, scientists often limit their view to a small area or a small number of objects. These boundaries define a system.

404

MISCONCEPTION ALERT

Conservation of Energy The idea of energy conservation is counter-intuitive to many students. Students do not understand that this law can be used to explain phenomena. Also, students often interpret the phrase "energy is neither created or destroyed" to mean that energy is stored up in a system and then released in its original form.

Apply the law to various examples to show students how it can be used to interpret observed phenomena and to make accurate predictions. Also explain that the law of energy conservation does not make any stipulations about the forms of energy involved; the energy can change form multiple times, but the sum of all forms of energy (in a closed system) will not change. Introducing the concept of energy dissipation while teaching the conservation law may also help alleviate confusion.

Systems may be open or closed

A system might include a gas burner and a pot of water. A scientist could study the flow of energy from the burner into the pot and ignore the small amount of energy going into the pot from the lights in the room, from a hand touching the pot, and so on.

When the flow of energy into and out of a system is small enough that it can be ignored, the system is called a *closed system.* Most systems are *open systems,* which exchange energy with the space that surrounds them. Earth is an open system, as shown in *Figure 22.* Is your body an open or closed system?

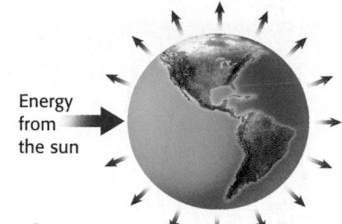

Figure 22
Earth is an open system because it receives energy from the sun and radiates some of its own energy out into space.

Quick Lab

Is energy conserved in a pendulum?

Materials
- ✓ 1–1.5 m length of string
- ✓ pencil with an eraser
- ✓ meterstick
- ✓ level
- ✓ pendulum bob
- ✓ nail or hook in the wall above a chalkboard

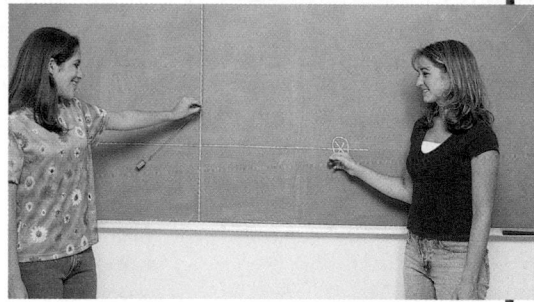

1. Hang the pendulum bob from the string in front of a chalkboard. On the board, draw the diagram as shown in the photograph at right. Use the meterstick and the level to make sure the horizontal line is parallel to the ground.

2. Pull the pendulum ball back to the "X." Make sure everyone is out of the way; then release the pendulum and observe its motion. How high does the pendulum swing on the other side?

3. Let the pendulum swing back and forth several times. How many swings does the pendulum make before the ball noticeably fails to reach its original height?

4. Stop the pendulum and hold it again at the "X" marked on the board. Have another student place the eraser end of a pencil on the intersection of the horizontal and vertical lines. Make sure everyone is out of the way again, especially the student holding the pencil.

5. Release the pendulum again. This time its motion will be altered halfway through the swing as the string hits the pencil. How high does the pendulum swing now? Why?

6. Try placing the pencil at different heights along the vertical line. How does this affect the motion of the pendulum? If you put the pencil down close enough to the arc of the pendulum, the pendulum will do a loop around it. Why does that happen?

Analysis

1. Use the law of conservation of energy to explain your observations in steps 2–6.

2. If you let the pendulum swing long enough, it will start to slow down, and it won't rise to the line any more. That suggests that the system has lost energy. Has it? Where did the energy go?

405

Useful Work Compare the last sentence on this page, "The useful work output of a machine never equals—and certainly cannot exceed—the work input," with the following phrase from the sub-section in Section 1 titled *Machines and Mechanical Advantage*: "Therefore, a machine allows the same amount of work to be done . . ." How can these both be true? (The law of conservation of energy requires that the energy that goes into the machine does not disappear, although it may change form. If the energy changes form, it may no longer be able do the work that the machine is designed to do, so the useful work output decreases.)
LS Verbal

Figure 23
Like all machines, the pulleys on a sailboat are less than 100 percent efficient.

efficiency a quantity, usually expressed as a percentage, that measures the ratio of useful work output to work input

Efficiency of Machines

If you use a pulley to raise a sail on a sailboat like the one in *Figure 23,* you have to do work against the forces of friction in the pulley. You also have to lift the added weight of the rope and the hook attached to the sail. As a result, only some of the energy that you transfer to the pulley is available to raise the sail.

Not all of the work done by a machine is useful work

Because of friction and other factors, only some of the work done by a machine is applied to the task at hand; the machine also does some incidental work that does not serve any intended purpose. In other words, there is a difference between the total work done by a machine and the *useful* work done by the machine, that is, work that the machine is designed or intended to do.

Although all of the work done on a machine has some effect on the output work that the machine does, the output work might not be in the form that you expect. In lifting a sail, for example, some of the work available to lift the sail, which would be useful work, is transferred away as heat that warms the pulley because of friction. This warming is not a desired effect. The amount of useful work might decrease slightly more if the pulley squeaks, because some energy is "lost" as it dissipates into forces that vibrate the pulley and the air to produce the squeaking sound.

Efficiency is the ratio of useful work out to work in

The **efficiency** of a machine is a measure of how much useful work it can do. Efficiency is defined as the ratio of useful work output to total work input.

> **Efficiency Equation**
>
> $$efficiency = \frac{useful\ work\ output}{work\ input}$$

Efficiency is usually expressed as a percentage. To change an answer found using the efficiency equation into a percentage, just multiply the answer by 100 and add the percent sign, "%."

A machine with 100 percent efficiency would produce exactly as much useful work as the work done on the machine. Because every machine has some friction, no machine has 100 percent efficiency. The useful work output of a machine never equals—and certainly cannot exceed—the work input.

406

REAL-LIFE CONNECTION

EnergyGuide Labels In the United States, consumers can easily determine the relative efficiency of many common appliances, including refrigerators, freezers, clothes washers, dishwashers, and room air conditioners. All they need to do is consult the mandatory yellow *EnergyGuide* label placed on appliances by the manufacturers. These labels indicate the lowest and highest amounts of average yearly energy use for appliances of this type. The energy use for the appliance in question is indicated on

this scale. This shows consumers the relative efficiency of the appliance.

For example, a clothes washer label might show that washers of this type use from 312 kW•h/y to 1306 kW•h/y, with the washer in question operating at 860 kW•h/y. Thus, this washing machine's efficiency is about average. The *EnergyGuide* labels also show the approximate yearly operating cost for the appliance. (The labels became mandatory in the 1970s; appliances made before then do not have the labels.)

Perpetual motion machines are impossible

Figure 24 shows a machine designed to keep on going forever without any input of energy. These theoretical machines are called *perpetual motion machines*. Many clever inventors have devoted a lot of time and effort to designing such machines. If such a perpetual motion machine could exist, it would require a complete absence of friction and air resistance.

Math Skills

Efficiency A sailor uses a rope and an old, squeaky pulley to raise a sail that weighs 140 N. He finds that he must do 180 J of work on the rope in order to raise the sail by 1 m (doing 140 J of work on the sail). What is the efficiency of the pulley? Express your answer as a percentage.

1 List the given and unknown values.

Given: *work input* = 180 J
useful work output = 140 J

Unknown: *efficiency* = ? %

2 Write the equation for efficiency.

$$efficiency = \frac{useful\ work\ output}{work\ input}$$

3 Insert the known values into the equation, and solve.

$$efficiency = \frac{140\ J}{180\ J} = 0.78$$

To express this as a percentage, multiply by 100 and add the percent sign, "%."

$$efficiency = 0.78 \times 100 = 78\%$$

Practice

Efficiency

1. Alice and Jim calculate that they must do 1800 J of work to push a piano up a ramp. However, because they must also overcome friction, they actually must do 2400 J of work. What is the efficiency of the ramp?

2. It takes 1200 J of work to lift the car high enough to change a tire. How much work must be done by the person operating the jack if the jack is 25 percent efficient?

3. A windmill has an efficiency of 37.5 percent. If a gust of wind does 125 J of work on the blades of the windmill, how much output work can the windmill do as a result of the gust?

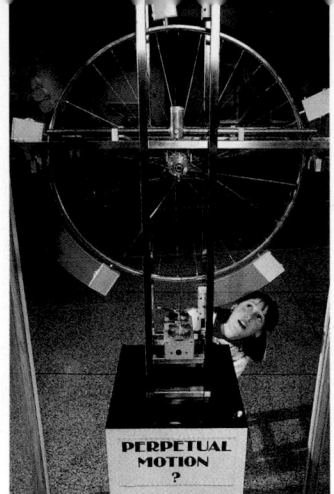

Figure 24
Theoretically, a perpetual motion machine could keep going forever without any energy loss or energy input.

Practice HINT

▶ The efficiency equation can be rearranged to isolate any of the variables on the left

▶ For practice problem 2, you will need to rearrange the equation to isolate work input on the left side.

▶ For practice problem 3, you will need to rearrange to isolate useful work output.

▶ When using these rearranged forms to solve the problems, you will have to plug in values for efficiency. When doing so, do not use a percentage, but rather convert the percentage to a decimal by dropping the percent sign and dividing by 100.

407

1. What is one example of an energy transformation? (Accept all accurate responses. Examples would include gravitational potential energy transferred to kinetic energy, or kinetic energy transformed to thermal energy.)

2. What is the law of the conservation of energy? (Energy cannot be created or destroyed.)

3. What is the efficiency of a machine, and how can it be calculated? (Efficiency is a measure of how much useful work a machine can do for a given amount of input work; *efficiency = useful work output/work input*.)

LS Logical

Machines need energy input

Because energy always leaks out of a system, every machine needs at least a small amount of energy input to keep going. Unfortunately, that means that perpetual motion machines are impossible. But new technologies, from magnetic trains to high speed microprocessors, reduce the amount of energy leaking from systems so that energy can be used as efficiently as possible.

SECTION 4 REVIEW

SUMMARY

▶ Energy readily changes from one form to another.

▶ In a mechanical system, potential energy can become kinetic energy, and kinetic energy can become potential energy.

▶ Mechanical energy can change to nonmechanical energy as a result of friction, air resistance, or other means.

▶ Energy cannot be created or destroyed, although it may change form. This is called the law of conservation of energy.

▶ A machine cannot do more work than the work required to operate the machine. Because of friction, the work output of a machine is always somewhat less than the work input.

▶ The efficiency of a machine is the ratio of the useful work performed by the machine to the work required to operate the machine.

1. List three situations in which potential energy becomes kinetic energy and three situations in which kinetic energy becomes potential energy.

2. State the law of conservation of energy in your own words. Give an example of a situation in which the law of conservation of energy is demonstrated.

3. Describe the rise and fall of a basketball using the concepts of kinetic energy and potential energy.

4. Explain why machines are not 100 percent efficient.

5. Applying Knowledge Use the concepts of kinetic energy and potential energy to describe the motion of a child on a swing. Why does the child need a push from time to time?

6. Creative Thinking Using what you have learned about energy transformations, explain why the driver of a car has to continuously apply pressure to the gas pedal in order to keep the car cruising at a steady speed, even on a flat road. Does this situation violate the law of conservation of energy? Explain.

Math Skills

7. Efficiency When you do 100 J of work on the handle of a bicycle pump, it does 40 J of work pushing the air into the tire. What is the efficiency of the pump?

8. Efficiency and Power A river does 6500 J of work on a water wheel every second. The wheel's efficiency is 12 percent.
   a. How much work in joules can the axle of the wheel do in a second?
   b. What is the power output of the wheel?

9. Efficiency and Work John is using a pulley to lift the sail on his sailboat. The sail weighs 150 N and he must lift it 4.0 m.
   a. How much work must be done on the sail?
   b. If the pulley is 50 percent efficient, how much work must John do on the rope in order to lift the sail?

Chapter Resource File

• Concept Review GENERAL

• Quiz BASIC

Answers to Section 4 Review

1. *PE* to *KE*: a falling ball, anything rolling downhill, a pendulum on the downswing; *KE* to *PE*: a rising ball, anything rolling uphill, a pendulum on the upswing

2. Energy can neither be created nor destroyed. In a swinging pendulum, energy is constantly transformed from potential to kinetic energy and back again. In all of these transformations, the total mechanical energy remains the same.

3. The player throws the ball, giving it *KE*. The ball begins to rise and slow down as *KE* is transformed into *PE*. At its peak, the ball has maximum *PE*, then begins to fall, transforming *PE* into *KE*.

4. Friction prevents machines from being 100 percent efficient.

5. A child on a swing undergoes energy transformations from maxium *PE* at the top (both sides) to maximum *KE* at the bottom and back to maximum *PE* at the top of the opposite side. The child needs a push every now and then to make up for the energy lost to friction between the rope and the support, as well as some energy lost to air resistance.

See "Continuation of Answers" at the end of the chapter.

Graphing Skills

Total Mechanical Energy of a Bouncing Ball

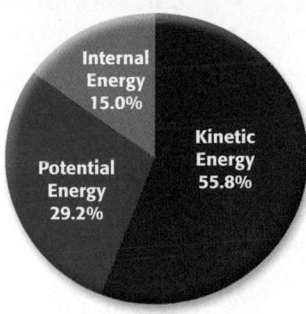

t = 0.75 s

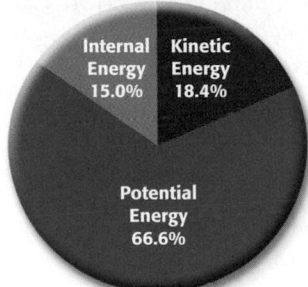

t = 1.5 s

Examine the graphs above and answer the following questions.

1 What type of graphs are these?

2 Identify the information provided by each graph.

3 Does the total mechanical energy change between 0.75 s and 1.5 s? What does change in this time interval?

4 Assume that the internal energy of the ball increases only when it bounces off the floor. What can you tell about the number of times the ball has bounced between 0.75 s and 1.5 s?

5 In which graph is the ball moving fastest? In which is the ball higher above the ground? Explain your answers.

6 Suppose you are asked to design a ball that bounces to nearly the same height as that from which it is dropped. In terms of energy, what property would this ball require?

7 Construct the type of graph best suited for the data given in the table below. Is mechanical energy conserved in this process? Explain your answer.

Time (s)	Potential energy (J)	Kinetic energy (J)	Internal energy (J)
0	30.0	0	0
0.50	15.0	12.0	3.0
1.00	5.0	20.0	5.0
1.50	0	24.0	6.0

Teaching Tip

Internal Energy Explain that internal energy is that portion of energy absorbed by the atoms of an object, and accounts for the "lost" mechanical energy.

Answers

1. pie charts

2. the potential, kinetic, and internal energy associated with the ball after a particular time interval; The three types of energy are related by conservation of energy.

3. No. Total mechanical energy is conserved. *KE* decreases, *PE* increases, internal energy remains constant

4. The ball has not bounced in that time, as internal energy has not changed.

5. The first graph has higher kinetic energy, so the ball is moving faster at 0.75 s. Potential energy is greater in the second graph, indicating that the ball is at a greater height with respect to the ground.

6. For the ball to bounce to nearly the same height, it must not change much mechanical energy to internal energy.

7. A bar graph would be best suited for the data. Mechanical energy is not conserved, because the sum of the *PE* and *KE* values steadily decreases. Total energy (*PE*, *KE*, and internal energy) is conserved in the process, because it equals 30.0 J at all times.

Understanding Concepts

1. c
2. a
3. d
4. c
5. d
6. c
7. d
8. c
9. c
10. a
11. d
12. a
13. a

Using Vocabulary

14. Answers will vary. Work used in the scientific sense should imply a force acting on an object and changing the object's motion, while work in other contexts may have other meanings.

15. Answers should contain some statement that the word kinetic relates to motion.

16. wheel and axle, wedge, lever

17. a. *PE*
b. *KE*
c. *PE*
d. *KE*

Chapter Highlights

Before you begin, review the summaries of the key ideas of each section, found at the end of each section. The key vocabulary terms are listed on the first page of each section.

UNDERSTANDING CONCEPTS

1. _____ is defined as force acting over a distance.
 a. Power
 b. Energy
 c. Work
 d. Potential energy

2. The quantity that measures how much a machine multiplies force is called
 a. mechanical advantage.
 b. leverage.
 c. efficiency.
 d. power.

3. Scissors are an example of
 a. a lever.
 b. a wedge.
 c. a wheel and axle.
 d. a compound machine.

4. The unit that represents 1 J of work done each second is the
 a. power.
 b. newton.
 c. watt.
 d. mechanical advantage.

5. Units of joules could be used when measuring
 a. the work done in lifting a bowling ball.
 b. the potential energy of a bowling ball held in the air.
 c. the kinetic energy of a rolling bowling ball.
 d. All of the above

6. Which of the following situations does *not* involve potential energy being changed into kinetic energy?
 a. an apple falling from a tree
 b. shooting a dart from a spring-loaded gun
 c. pulling back on the string of a bow
 d. a creek flowing downstream

7. _____ is determined by both mass and velocity.
 a. Work
 b. Power
 c. Potential energy
 d. Kinetic energy

8. Energy that does not involve the large-scale motion or position of objects in a system is called
 a. potential energy.
 b. mechanical energy.
 c. nonmechanical energy.
 d. conserved energy.

9. The law of conservation of energy states that
 a. the energy of a system is always decreasing.
 b. no machine is 100 percent efficient.
 c. energy is neither created nor destroyed.
 d. Earth has limited energy resources.

10. Power is measured in
 a. watts.
 b. joules.
 c. newtons.
 d. kilograms.

11. Which of the following can a machine not do?
 a. change the direction of a force
 b. multiply or increase a force
 c. redistribute work
 d. increase the total amount of work done

12. A machine with a mechanical advantage of less than one
 a. increases speed and distance.
 b. multiplies force.
 c. increases output force.
 d. reduces distance and speed.

13. A perpetual motion machine is impossible because
 a. machines require energy input.
 b. machines do not require energy input.
 c. machines have become too efficient.
 d. friction is negligible.

18. Because work is force times distance, the elephant does much more work than the mouse; the distance is the same, but the elephant weighs much more. Power is work divided by time, and because the mouse beat the elephant by only a small amount of time, the fact that the elephant did much more work means that the power of the elephant is much greater than the power of the mouse.

19. Energy is the ability to do work. Doing work is transferring or transforming energy. Work is exerting a force through a distance to change the motion, and thus the energy, of an object. An object that has energy has the ability to exert a force through a distance. The rate of changing energy, or work, per unit time is power.

20. Answers will vary. Students may say that electrical energy supplies the power for computers, light bulbs, air conditioners, refrigerators, and many other appliances and machines, and that light energy provides plants with the energy that is converted by photosynthesis into the chemical energy that sustains living things.

14. Write one sentence using *work* in the scientific sense, and write another sentence using it in a different, nonscientific sense. Explain the difference in the meaning of *work* in the two sentences. **WRITING SKILL**

15. The first page of this chapter shows an example of *kinetic sculpture*. You have now also learned the definition of *kinetic energy*. Given your knowledge of these two terms, what do you think the word *kinetic* means?

16. A can opener is a *compound machine*. Name three *simple machines* that it contains.

17. For each of the following, state whether the system contains primarily *kinetic energy* or *potential energy*:
a. a stone in a stretched slingshot
b. a speeding race car
c. water above a hydroelectric dam
d. the water molecules in a pot of boiling water

18. An elephant and a mouse race up the stairs. The mouse beats the elephant by a full second, but the elephant claims, "I am more powerful than you are, and this race has proved it." Use the definitions of *work* and *power* to support the elephant's claim.

19. How is *energy* related to *work*, *force*, and *power*?

20. List several examples of how *electrical energy* and *light energy* are useful to you.

21. You and two friends apply a force of 425 N to push a piano up a 2.0 m long ramp.
a. Work How much work in joules has been done when you reach the top of the ramp?
b. Power If you make it to the top in 5.0 s, what is your power output in watts?
c. Mechanical Advantage If lifting the piano straight up would require 1700 N of force, what is the mechanical advantage of the ramp?

22. A crane uses a block and tackle to lift a 2200 N flagstone to a height of 25 m.
a. Work How much work is done on the flagstone?
b. Efficiency In the process, the crane's hydraulic motor does 110 kJ of work on the cable in the block and tackle. What is the efficiency of the block and tackle?
c. Potential Energy What is the potential energy of the flagstone when it is 25 m above the ground?

23. A 2.0 kg rock sits on the edge of a cliff 12 m above the beach.
a. Potential Energy Calculate the potential energy in the system.
b. Energy Transformations The rock falls off the cliff. How much kinetic energy will it have just before it hits the beach? (Ignore air resistance.)
c. Kinetic Energy Calculate the speed of the rock just before it hits the beach. (For help, see Practice Hint on page 301.)
d. Conservation of Energy What happens to the energy after the rock hits the beach?

411

Assignment Guide

Section	Questions
1	1, 2, 4, 10–12, 14, 18, 21, 29, 33, 37
2	3, 16, 28, 31, 34–36, 41
3	5–8, 15, 17, 19, 20, 26, 38–40
4	9, 13, 22–25, 27, 30, 32

Building Math Skills

21. a. 850 J
b. 170 W
c. 4.0
22. a. 55 000 J (or 55 kJ)
b. 0.50 (or 50%)
c. 55 kJ
23. a. 240 J
b. 240 J
c. 15 m/s
d. It is transferred into the kinetic energy of the sand, sound energy (*KE* of molecules in air), and increased temperature (*KE* of the molecules in the rock and sand).

Thinking Critically

24. a. A, E, B, D, C
b. C, D, B, E, A
c. The lists are identical, except in reverse order.

25. Because energy cannot be created, the machine can only put out an amount of work equal to or less than the energy within the machine, which is equal to or less than the work input.

26. nine times

27. The work done by the hammer is converted into kinetic energy of the nail and then into useful work on the wood, splitting it open so the nail can enter. Much of the energy goes into heating the hammer, nail, and wood. Some of the energy goes into the air as sound. This does not violate the law of conservation of energy.

28. The work done on the lever will be greater than the work done on the rock by the lever, because some energy is dissipated or "lost" as nonmechanical energy every time energy is transferred from one object to another.

29. The advantage of using a machine lies in its ability to redistribute work by changing the direction of an input force or changing the distance over which the force is applied.

30. No, the design will not be successful, because the car will not have enough kinetic energy to climb a hill that is taller than the first one, without receiving an additional input of energy.

31. Since work equals force multiplied times distance, machines can be used to decrease or increase force by changing the distance over which the force is applied. A second-class lever, for example, multiplies input force by the decreasing the distance over which the work occurs, whereas a third-class lever, decreases input force by increasing distance.

32. Levers have a greater mechanical efficiency than other simple machines do, because there is less opportunity for energy to be transformed into unuseful non-mechanical energy in levers than in the other types of simple machines. In most cases, using pulleys, wheel and axles, inclined planes, wedges, and screws involves more friction than using levers.

Developing Life/Work Skills

33. You should use a longer screwdriver. The output length remains the same (the distance from the fulcrum to the output force), but the input length increases with a longer screwdriver, creating a larger mechanical advantage and therefore a larger output force.

24. Interpreting Graphics The diagram below shows five different points on a roller coaster.

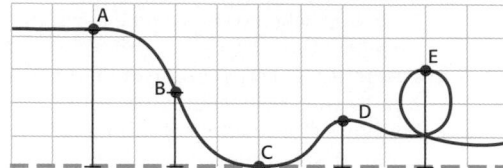

a. List the points in order from the point where the car would have the greatest potential energy to the point where it would have the least potential energy.

b. Now list the points in order from the point where the car would have the greatest kinetic energy to the point where it would have the least kinetic energy.

c. How are your two lists related to each other?

25. Critical Thinking Use the law of conservation of energy to explain why the work output of a machine can never exceed the work input.

26. Applying Knowledge If a bumper car triples its speed, how much more work can it do on a bumper car at rest? (**Hint:** Use the equation for kinetic energy.)

27. Understanding Systems When a hammer hits a nail, there is a transfer of energy as the hammer does work on the nail. However, the kinetic energy and potential energy of the nail do not change very much. What happens to the work done by the hammer? Does this violate the law of conservation of energy?

28. Critical Thinking You are attempting to move a large rock using a long lever. Will the work you do on the lever be greater than, the same as, or less than the work done by the lever on the rock? Explain your answer.

29. Applying Knowledge If a machine cannot multiply the amount of work, then what is the advantage of using a machine?

30. Applying Knowledge You are designing a roller coaster in which a car will be pulled to the top of a hill and then will be released to roll freely down the hill toward the the top of the next hill. The next hill is twice as high. Will your design be successful?

31. Applying Knowledge In two or three sentences, explain the force-distance trade-off that occurs when a machine is used to make work easier. Use the lever as an example of one type of trade-off.

32. Applying Knowledge Why do you think that levers have a greater mechanical efficiency than other simple machines do?

DEVELOPING LIFE/WORK SKILLS

33. Applying Knowledge You are trying to pry the lid off a paint can with a screwdriver, but the lid will not budge. Should you try using a shorter screwdriver or a longer screwdriver? Explain.

34. Designing Systems Imagine you are trying to move a piano into a second-floor apartment. It will not fit through the stairwell, but it will fit through a large window 3.0 m off the ground. The piano weighs 1740 N and you can exert only 290 N of force. Design a system of machines you could use to lift the piano to the window.

35. Teaching Others Prepare a poster or a series of models of common machines that explains their uses and how they work. Include a diagram next to each sample labeling parts of each machine. Add your own examples of simple machines to the following list: nail clipper, wheelbarrow, can opener, nutcracker, electric drill, screwdriver, tweezers, and a key in a lock.

34. Answers may vary. One option is to use a ramp that is 18 m long, but that is not practical. Another option is a block and tackle with a mechanical advantage of 6 (three moving and three fixed pulleys).

35. Answers will vary. Student posters should reflect information from Section 2 of this chapter.

36. Mechanical energy is not conserved because of the longer distance travelled. However, the zigzag design acts as a series of inclined planes, so it provides the advantage of spreading the work over a larger distance,

and thus decreasing the input force required at any given moment. The amount of power provided is lower because energy is spread out over more time.

37. Work is done when the bag is lifted, because force applied to the bag has moved the bag a vertical distance in the direction of the force. Work, in the scientific sense, is not done when the bag is carried, because the motion of the bag is perpendicular to the direction of the force.

36. Designing Systems Many mountain roads are built so that they zigzag up a mountain rather than go straight up toward the peak. Discuss the advantage of such a design from the viewpoint of energy conservation and power. Think of a winding road as a series of inclined planes.

37. Applying Knowledge Explain why you do work on a bag of groceries when you pick it up, but not when you are carrying it.

INTEGRATING CONCEPTS

38. Connection to Sports A baseball pitcher applies a force to the ball as his arm moves a distance of 1.0 m. Using a radar gun, the coach finds that the ball has a speed of 18 m/s after it is released. A baseball has a mass of 0.15 kg. Calculate the average force that the pitcher applied to the ball. (**Hint:** You will need to use both the kinetic energy equation and the work equation.)

39. Concept Mapping Copy the unfinished concept map below onto a sheet of paper. Complete the map by writing the correct word or phrase in the lettered boxes.

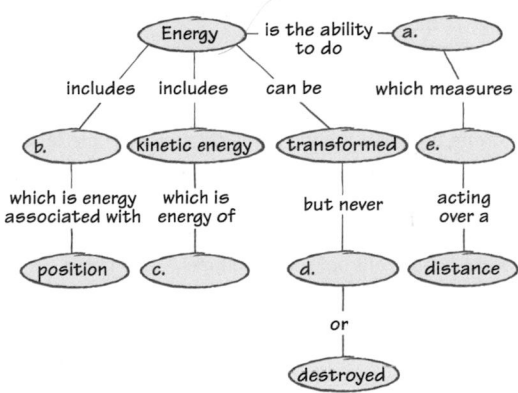

40. Connection to Earth Science Many fuels come from fossilized plant and animal matter. How is the energy stored in these fuels? How do you think that energy got into the fuels in the first place?

41. Connection to Biology When lifting an object using the biceps muscle, the forearm acts as a lever with the fulcrum at the elbow. The input work is provided by the biceps muscle pulling up on the bone. Assume that the muscle is attached 1.0 cm from the elbow and that the total length of the forearm from elbow to palm is 32 cm. How much force must the biceps exert to lift an object weighing 12 N? What class of lever is the forearm in this example?

📶 **internet** connect

www.scilinks.org
Topic: Energy and Sports **SciLinks code:** HK4047

SCiLINKS Maintained by the
National Science Teachers Association

Art Credits: Figs. 6,7, 8A, 8B, 13, Stephen Durke/Washington Artists; Fig. 22, Uhl Studios, Inc.

Photo Credits: Chapter Opener photo of George Rhodes' auto-kinetic sculpture by Wayne Source; portrait of Alexander Calder by Tony Vaccaro/AKG Photo, London; Fig. 1, Amwell/Getty Images/Stone; Inquiry Lab, Sam Dudgeon/HRW; Figs. 2-3, Peter Van Steen/HRW; Fig. 4 (hammer), Michelle Bridwell/HRW; (pulley), Visuals Unlimited/A. J. Copley; (wheel), Peter Van Steen/HRW; (ramp), Sam Dudgeon/HRW; (wedge), Superstock; (drill bit), Dr. E. R. Degginger/Color-Pic, Inc.; Fig. 5 (hammer) Michelle Bridwell/HRW; (wheelbarrow), John P. Kelly/Getty Images/The Image Bank; (arm), Sam Dudgeon/HRW; Fig. 7, Peter Van Steen/HRW, Fig. 8 (ramp), Sam Dudgeon/HRW; (wedge), Superstock; (drill bit), Dr. E. R. Degginger/Color-Pic, Inc.; "Connection to Social Studies," M. O'Neill/Getty Images/The Image Bank; Figs. 9-10, Peter Van Steen/HRW; Fig. 11, Picture Perfect; Fig. 12, Kindra Clineff/The Picture Cube/Index Stock; Fig. 13, Randy Ury/Corbis Stock Market; Fig. 14, Al Francekevich/Corbis Stock Market; "Real World Applications," Peter Van Steen/HRW; Fig. 15, NASA/Phototake; Fig. 16, Steve Bloom/Picture Perfect; Fig. 17, E. R. Degginger/Color-Pic, Inc.; Fig. 18, AFP/CORBIS; Fig. 20, Getty Images/FPG International; Fig. 21, Henry Groskinsky/Peter Arnold, Inc.; "Inquiry Lab," Sam Dudgeon/HRW; Fig. 23, Visuals Unlimited/Albert J. Copley; Fig. 37, Hank Morgan/Rainbow Inc.; "Chapter Review," (can opener), Peter Van Steen/HRW; "Skill Builder Lab," Peter Van Steen/HRW, "Career Link," Ken Kinzie/Latent Image Photography/HRW.

413

Integrating Concepts

38. 24 N;

$$W = KE, F \times d = \frac{1}{2}mv^2$$

$$F = \frac{mv^2}{2d} = \frac{(0.15 \text{ kg})(18 \text{ m/s})^2}{2(1.0 \text{ m})}$$

$$= 24 \text{ N}$$

39. a. work
   b. potential energy
   c. motion
   d. created
   e. force

40. chemical energy; Light (solar) energy was converted into chemical energy through photosynthesis

41. 380 N; The forearm is acting as a third-class lever.

work input = work output;
$$F_i \times d_i = F_o \times d_o$$

$$F_i = F_o \frac{d_o}{d_i} = 12 \text{ N} \left(\frac{32 \text{ cm}}{1.0 \text{ cm}} \right)$$

$$= 380 \text{ N}$$

Transparencies

TT Concept Mapping

DETERMINING ENERGY FOR A ROLLING BALL

Teacher's Notes

Time Required 1 lab period

Ratings

1	2	3	4
EASY			HARD

TEACHER PREPARATION	1
STUDENT SETUP	2
CONCEPT LEVEL	2
CLEANUP	1

Skills Acquired

- Collecting data
- Communicating
- Identifying/Recognizing patterns
- Interpreting
- Measuring
- Organizing and analyzing data

The Scientific Method

In this lab, students will:
- Make Observations
- Analyze the Results
- Draw Conclusions
- Communicate Results

Materials

Materials are listed for each group.

Safety Cautions

The balls used for the experiment could cause trip and fall hazards. Be sure that students use a catch box at the end of the ramp.

Skills Practice Lab

Introduction

Raised objects have gravitational potential energy. Moving objects have kinetic energy. How are these two quantities related in a system that involves a ball rolling down a ramp?

Objectives

▶ **Measure** the height, distance traveled, and time interval for a ball rolling down a ramp.

▶ **Calculate** the ball's potential energy at the top of the ramp and its kinetic energy at the bottom of the ramp.

▶ **USING SCIENTIFIC METHODS** *Analyze the results* to find the relationship between potential energy and kinetic energy.

Materials

balance
board, at least 90 cm (3 ft) long
box
golf ball, racquet ball, or handball
masking tape
meterstick
stack of books, at least 60 cm (2 ft) high
stopwatch

Determining Energy for a Rolling Ball

▶ Procedure

Preparing for Your Experiment

1. On a blank sheet of paper, prepare a table like the one shown below.

Table I **Potential Energy and Kinetic Energy**

	Height 1	Height 2	Height 3
Mass of ball (kg)			
Length of ramp (m)			
Height of ramp (m)			
Time ball traveled, first trial (s)			
Time ball traveled, second trial (s)			
Time ball traveled, third trial (s)			
Average time ball traveled (s)			
Final speed of ball (m/s)			
Final kinetic energy of ball (J)			
Initial potential energy of ball (J)			

2. Measure the mass of the ball, and record it in your table.

3. Place a strip of masking tape across the board close to one end, and measure the distance from the tape to the opposite end of the board. Record this distance in the row labeled "Length of ramp."

4. Make a catch box by cutting out one side of a box.

5. Make a stack of books approximately 30 cm high. Build a ramp by setting the taped end of the board on top of the books, as shown in the photograph on the next page. Place the other end in the catch box. Measure the vertical height of the ramp at the tape, and record this value in your table as "Height of ramp."

Making Time Measurements

6. Place the ball on the ramp at the tape. Release the ball, and measure how long it takes the ball to travel to the bottom of the ramp. Record the time in your table.

7. Repeat step 6 two more times and record the results in your table. After three trials, calculate the average travel time and record it in your table.

8. Repeat steps 5–7 with a stack of books approximately 45 cm high, and repeat the steps again with a stack approximately 60 cm high.

► Analysis

1. Calculate the average speed of the ball using the following equation:

$$average\ speed = \frac{length\ of\ ramp}{average\ time\ ball\ traveled}$$

2. Multiply average speed by 2 to obtain the final speed of the ball, and record the final speed.

3. Calculate and record the final kinetic energy of the ball by using the following equation:

$$KE = \frac{1}{2} \times mass\ of\ ball \times (final\ speed)^2$$

$$KE = \frac{1}{2}mv^2$$

4. Calculate and record the initial potential energy of the ball by using the following equation:

$$grav.\ PE = mass\ of\ ball \times (9.8\ m/s^2) \times height\ of\ ramp$$
$$PE = mgh$$

► Conclusions

5. For each of the three heights, compare the ball's potential energy at the top of the ramp with its kinetic energy at the bottom of the ramp.

6. How did the ball's potential and kinetic energy change as the height of the ramp was increased?

7. Suppose you perform this experiment and find that your kinetic energy values are always just a little less than your potential energy values. Does that mean you did the experiment wrong? Why or why not?

415

Sample Data Table Potential Energy and Kinetic Energy			
	Height 1	Height 2	Height 3
Mass of ball (kg)	0.045	0.045	0.045
Length of ramp (m)	1.513	1.513	1.513
Height of ramp (m)	0.28	0.445	0.583
Time ball traveled, first trial (s)	1.59	1.31	1.09
Time ball traveled, second trial (s)	1.62	1.28	1.06
Time ball traveled, third trial (s)	1.56	1.25	1.04
Average time ball traveled (s)	1.59	1.28	1.06
Final speed of ball (m/s)	1.90	2.36	2.86
Final kinetic energy of ball (J)	0.081	0.125	0.184
Initial potential energy of ball (J)	0.123	0.196	0.257

Continuation of Answers

Continuation of Answers from p. 390
Section 2 Review

7. A door is normally a second-class lever. Pushing near the knob is easier because the input distance is longer. If you push near the hinges, the input arm is shorter than the output arm, and the door becomes a third-class lever, with an *MA* of less than 1.

8. Answers will vary. A pencil sharpener, for example, is a compound machine that consists of a couple of screws (the blades to sharpen the pencil), wedges (the edges of those blades), and a wheel and axle (the crank).

Continuation of Answers from p. 399
Section 3 Review

Math Skills

7. $PE = mgh = (93.0 \text{ kg})(9.8 \text{ m/s}^2)(550 \text{ m}) = 5.0 \times 10^5 \text{ J}$

8. $KE = \frac{1}{2} mv^2 = (\frac{1}{2})(0.02 \text{ kg})(300 \text{ m/s})^2 = 900 \text{ J}$

9. **a.** $PE = mgh = (2.5 \text{ kg})(9.8 \text{ m/s}^2)(2.0 \text{ m}) = 49 \text{ J}$
   b. $KE = \frac{1}{2} mv^2 = (\frac{1}{2})(0.015 \text{ kg})(3.5 \text{ m/s})^2 = 0.092 \text{ J}$
   c. $PE = mgh = (35 \text{ kg})(9.8 \text{ m/s}^2)(3.5 \text{ m}) = 1200 \text{ J}$
   d. $KE = \frac{1}{2} mv^2 = (\frac{1}{2})(8500 \text{ kg})[(220 \text{ km/h}) (1000 \text{ m/km})(1 \text{ h/3600 s})]^2 = 1.6 \times 10^7 \text{ J}$

Continuation of Answers from p. 408
Section 4 Review

6. The driver must keep transferring potential energy from the gas to the kinetic energy of the car to make up for the losses due to friction within the car's mechanism and between the tires and the road. This does not violate the law of conservation of energy because mechanical energy is transformed into nonmechanical forms.

Math Skills

7. *efficiency = useful work output/work input* = (40 J)/(100 J) = 0.4 or 40%

8. **a.** *useful work output = (efficiency)(work input)* = (0.12)(6500 J) = 780 J
   b. $P = W/t = 780 \text{ J/1 s} = 780 \text{ W}$

9. **a.** $W = Fd = (150 \text{ N})(4.0 \text{ m}) = 6.0 \times 10^2 \text{ J}$
   b. *work input = useful work output/efficiency* = $6.0 \times 10^2 \text{ J}/0.50 = 1200 \text{ J}$

CIVIL ENGINEER

Teaching Tip

Civil engineers rely on many of the principles of physics to plan their projects and be certain they are appropriate. In Grace Pierce's work with Traffic Systems, Inc., she uses the concepts of speed, acceleration, and force that were all discussed in this textbook

CareerLink

Civil Engineer

In a sense, civil engineering has been around since people started to build structures. Civil engineers plan and design public projects, such as roads, bridges, and dams, and private projects, such as office buildings. To learn more about civil engineering as a career, read the profile of civil engineer Grace Pierce, who works at Traffic Systems, Inc., in Orlando, Florida.

"I get to help in projects that provide a better quality of life for people. It's a good feeling."

As a civil engineer, Grace Pierce designs roads and intersections.

 What do you do as a civil engineer?

I'm a transportation engineer with a bachelor's degree in civil engineering. I do a lot of transportation studies, transportation planning, and engineering—anything to do with moving cars. Right now, my clients are about a 50-50 mix of private and public.

 What part of your job do you like best?

Transportation planning. On the planning side, you get to be involved in developments that are going to impact the community . . . being able to tap into my creative sense to help my clients get what they want.

 What do you find most rewarding about your job?

Civil engineering in civil projects. They are very rewarding because I get to see my input on a very fast time scale.

 What kinds of skills do you think a good civil engineer needs?

You need a good solid academic background. You need communication skills and writing ability. Communication is key. You should get involved in activities or clubs like Toastmasters, which can help you with your presentation skills. You should get involved with your community.

What part of your education do you think was most important?

Two years before graduation, I was given the opportunity to meet with the owner of a company who gave me a good preview of what he did. It's really important to get out there and get the professional experience as well as the academic experience before you graduate.

416

Did You Know?

According to the Southern California Association of Governments, by the year 2020, traffic in Los Angeles will move twice as slow as it does now.

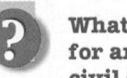 **You didn't enter college immediately after high school. Did you have to do anything differently from a younger student?**

I went to school as an older student. I didn't go back to college until age 27. I knew that because I was competing with younger folks, I really had to hustle.

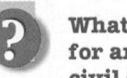 **What advice do you have for anyone interested in civil engineering?**

Have a vision. Have a goal, whatever that might be, and envision yourself in that arena. Work as hard as you can to realize that vision. Find out what you want to do, and find someone who can mentor you. Use every resource available to you in high school and college, including professors and people in the community. And in the process, have fun. It doesn't have to be dreary.

☑ internet connect

www.scilinks.org
Topic: Engineer
SciLinks code: HK4156

SCiLINKS® Maintained by the National Science Teachers Association

INTEGRATING

COMPUTERS AND TECHNOLOGY One fast-growing area of traffic research is building new models that can predict traffic flow patterns. Many of these models are computer simulations of traffic situations, either generated with a few key equations or with simulations of independent vehicles whose movement is governed by a few basic rules. Another promising field of inquiry compares the creation and dilution of traffic jams by individual drivers' behaviors to the behavior of the particles in a liquid as the liquid undergoes a phase change to become a solid.

" *I think my industry is going toward the 'smart' movement of vehicles and people. The future is intelligent transportation systems using automated systems.* "
—Grace Pierce

Heat and Temperature
Chapter Planning Guide

PACING	CLASSROOM RESOURCES	LABS, ACTIVITIES, AND DEMONSTRATIONS
BLOCK 1 · 45 min pp. 418–419 **Chapter Opener**		SE **Activity 1,** p. 419 SE **Activity 2,** p. 419
BLOCKS 2 & 3 · 90 min pp. 420–426 **Section 1** Temperature	CRF **Lesson Plan*** TT **Bellringer*** TT **Temperature Scales***	TE **Opening Demonstration,** p. 420 `GENERAL` SE **Quick Activity** Sensing Hot and Cold, p. 421 `GENERAL` CRF **Datasheets for In-Text Labs** Sensing Hot and Cold* `GENERAL` TE **Demonstration** Building a Thermometer, p. 423 SE **Quick Lab** How do temperature and energy relate?, p. 425 `GENERAL` CRF **Datasheets for In-Text Labs** How do temperature and energy relate?* `GENERAL`
BLOCKS 4 & 5 · 90 min pp. 427–434 **Section 2** Energy Transfer	CRF **Lesson Plan*** TT **Bellringer*** TM **Specific Heats***	TE **Opening Demonstration,** p. 427 `GENERAL` SE **Quick Activity** Convection, p. 429 `GENERAL` CRF **Datasheets for In-Text Labs** Convection* `GENERAL` TE **Demonstration** Energy Transfer in Boiling Water, p. 429 `BASIC` SE **Quick Lab** What color absorbs more radiation?, p. 430 `GENERAL` CRF **Datasheets for In-Text Labs** What color absorbs more radiation?* `GENERAL` SE **Quick Activity** Conductors and Insulators, p. 431 `GENERAL` CRF **Datasheets for In-Text Labs** Conductors and Insulators* `GENERAL` CRF **CBL™ Probeware Lab** Determining the Better Insulator for Your Feet* ◆ `ADVANCED` SE **Skills Practice Lab** Investigating Conduction of Heat, pp. 450–451 ◆ `GENERAL`
BLOCKS 6 & 7 · 90 min pp. 435–444 **Section 3** Using Heat	CRF **Lesson Plan*** TT **Bellringer*** TT **Air Conditioner*** TM **Insulation *R*-Values*** TM **Refrigerant Flow*** TT **Internal Combustion Engine*** TT **Concept Mapping***	TE **Demonstration** Heat Pumps, p. 441 TE **Group Activity** Solar Houses, p. 438 `ADVANCED` CRF **Datasheets for SE Skills Practice Lab** Investigating Conduction of Heat* `GENERAL` CRF **Observation Lab** Energy Transfer and Specific Heat* ◆ `BASIC`

BLOCKS 8 & 9 · 90 min

Chapter Review and Assessment Resources

SE **Chapter Review,** pp. 446–449
CRF **Chapter Tests*** `GENERAL`
OSP **Test Generator**
CRF **Standardized Test Practice with Guided Reading Development***
CRF **Test Item Listing for ExamView® Test Generator***
OSP **Scoring Rubrics and Classroom Management Checklists**

Online Resources

Visit the HRW Web site for a variety of free resources related to the text. Just type in the keyword **HK4 HAT.**

Holt Science Spectrum: Physical Science: Online Edition

Students can access interactive problem solving help and active visual concept development with the *Holt Science Spectrum: Physical Science* Online Edition available at **www.hrw.com.**

student CNN News

cnnstudentnews.com

Find the latest news, lesson plans, and activities related to important scientific events.

KEY

TE Teacher Edition	**CRF** Chapter Resource File
SE Student Edition	**TT** Teaching Transparency
OSP One-Stop Planner	**TM** Transparency Master

* Also on One-Stop Planner
◆ Requires Advance Prep

PROBLEM SOLVING AND PRACTICE	SECTION REVIEW AND ASSESSMENT	STANDARDS CORRELATION
	CRF Pretest* GENERAL	
CRF Cross-Disciplinary Worksheet Integrating Space Science—Starlight, Star Heat* GENERAL **SE** Math Skills Temperature Scale Conversion, p. 424 **CRF** Math Skills Temperature Conversions* GENERAL **CRF** Cross-Disciplinary Worksheet Integrating Health—Skin Temperature* GENERAL	**TE** Quiz, p. 426 BASIC **SE** Section 1 Review, p. 426 **CRF** Concept Review* GENERAL **CRF** Quiz* BASIC	SAI 1
SE Math Skills Specific Heat, pp. 433–434 **CRF** Cross-Disciplinary Worksheet Integrating Earth Science—Land and Sea Breezes* GENERAL **CRF** Math Skills Specific Heat* GENERAL	**TE** Quiz, p. 434 BASIC **SE** Section 2 Review, p. 434 **CRF** Concept Review* GENERAL **CRF** Quiz* BASIC	PS 5d PS 5c
CRF Cross-Disciplinary Worksheet Connection to Social Studies—Early Central Heating* GENERAL **CRF** Cross-Disciplinary Worksheet Connection to Social Studies—The Little Ice Age* GENERAL **CRF** Cross-Disciplinary Worksheet Integrating Biology—Hibernation and Torpor* GENERAL **CRF** Cross-Disciplinary Worksheet Integrating Environmental Science—Thermal Pollution* GENERAL **CRF** Cross-Disciplinary Worksheet Real World Applications—Appliance Energy Use and Cost* GENERAL **SE** Math Skills Order of Operations, p. 445	**TE** Quiz, p. 444 BASIC **SE** Section 3 Review, p. 444 **CRF** Concept Review* GENERAL **CRF** Quiz* BASIC	PS 5d ES 1c

SCiLINKS®

www.scilinks.org

Topic: Electromagnetic Spectrum
*Sci*Links code: HK4043

Topic: Temperature Scales
*Sci*Links code: HK4138

Topic: Energy Transfer
*Sci*Links code: HK4048

Topic: Insulators
*Sci*Links code: HK4073

Topic: Heating and Cooling Systems
*Sci*Links code: HK4067

Topic: Conduction, Convection, and Radiation
*Sci*Links code: HK4027

Technology Resources

 One-Stop Planner
All of your printable resources and the Test Generator are on this convenient CD-ROM.

 Physical Science Interactive Tutor CD-Rom
Disc One, Module 7
Heat

 CNN. **Science in the NEWS**
each video segment is accompanied by a Critical Thinking Worksheet*

Segment 12
Urban Heat Islands

Overview

This chapter covers temperature and heat, including temperature conversions. It also explores methods of energy transfer (conduction, convection, and radiation) and conductors and insulators. Students also learn how to use specific heat in calculations. Finally, this chapter applies the concepts learned to the study of heating and cooling systems.

Assessing Prior Knowledge

Be sure students understand the following concepts:

- units of measurement
- energy
- conservation of energy

MISCONCEPTION ALERT

Science education research has identified the following misconceptions about temperature and heat.

- Students do not distinguish between temperature and heat.
- Students believe that temperature depends on volume, or that it is related to size and mass.
- Students think that there are different types of heat.
- Students do not understand heating and cooling as processes of energy transfer; many think that *hot* and *cold* are transferred.

CHAPTER 13

Heat and Temperature

Chapter Preview

1 Temperature
Temperature and Energy
Relating Temperature to Energy Transfer

2 Energy Transfer
Methods of Energy Transfer
Conductors and Insulators
Specific Heat

3 Using Heat
Heating Systems
Cooling Systems
Heat Engines

INTEGRATING TECHNOLOGY and Society

418

Standards Correlations

National Science Education Standards

The following descriptions summarize the National Science Standards that specifically relate to this chapter. For the full text of the standards, see the National Science Education Standards at the front of the book.

Section 1 Temperature
Science as Inquiry SAI 1

Section 2 Energy Transfer
Physical Science PS 5c
Physical Science PS 5d
Earth Science ES 1c

Section 3 Using Heat
Physical Science PS 5d

Unlike visible light, infrared radiation from the fire passes through the smoke, making an otherwise invisible fire easy to see and locate. In infrared images, the high-temperature fire is brighter than its cooler surroundings.

Background The fire started at night. By the time firefighters arrived the next morning, the forest was filled with thick smoke. The firefighters knew the fire was raging, but they had to see through the smoke to find the fire's location.

Fortunately, firefighters have instruments that detect infrared radiation, which is a form of light that is invisible to the eye. It is given off by hot objects, such as burning wood. Infrared radiation passes through the smoke and is picked up by infrared detectors. The images formed by these instruments are converted into pictures. From these pictures, the fire's location can be determined, and the firefighters can fight the fire.

Activity 1 Use a prism to separate a beam of sunlight into its component colors, and project these onto a sheet of paper. Use a thermometer to record the temperature of the air in the room, and then place the thermometer bulb in each colored band for 3 minutes. Record the final temperature of each colored band. Place the thermometer on the dark side of the red band, where infrared radiation is found, for 3 minutes. How do the final temperature readings differ? Do your results suggest why infrared radiation is associated with hot objects?

Activity 2 Obtain several cups that are the same size but are made of different materials (glass, metal, ceramic, plastic foam). Fill one cup with hot tap water. Measure the time it takes for the outside of the cup to feel hot (about 35°C). Repeat this for each cup. List the cups by their materials, with the one that warms fastest listed first. Note any differences such as cup thickness, cup volume, or changes in the temperature of your hand.

internet connect

www.scilinks.org
Topic: Electromagnetic Spectrum
SciLinks code: HK4043

SC*i*LINKS. Maintained by the National Science Teachers Association

Pre-Reading Questions
1. Write a paragraph summarizing what you know about heat as energy.
2. List three ways that temperature has affected you recently.

419

Background Infrared detectors are also useful to firefighters trying to find people trapped in smoke-filled buildings. Since the temperature of a living human body is higher than the temperature of most objects in a room, a firefighter can use the detector to find the people.

Activity 1 This activity will work best with a sheet of white paper.

Students should find that the temperature of the bands increases as they move from violet to red and that the infrared region (the dark side of the red band) is the highest temperature.

Activity 2 Answers will vary depending on which materials are used and on the thickness of each. Students should find that metal heats up faster, followed by glass, ceramic, and plastic foam.

Answers to Pre-Reading Questions
1. Answers may vary.
2. Answers may vary.

Chapter Resource File

• Pretest GENERAL
• Teaching Transparency Preview
• Answer Keys

Chapter 13 • Heat and Temperature   **419**

Focus

Overview

Before beginning this section, review with your students the Objectives listed in the Student Edition. In this section, students learn about temperature and heat energy; conversions between Fahrenheit, Celsius, and Kelvin scales; and energy transfer.

🔔 Bellringer

Use the Bellringer transparency to prepare students for this section.

Motivate

Opening Demonstration — GENERAL

Safety Caution: *Do not handle dry ice with bare hands. Use gloves and/or tongs.*

This demonstration requires dry ice, a beaker, water, a hot plate, and a thermometer. Before class begins, place the beaker on the hot plate. Be sure that the hot plate is turned off and that the plug or power light is not visible. Fill the beaker halfway with water and add dry ice. Place the thermometer in the water. When students enter the class, the water should be bubbling as if boiling.

Analysis

1. Ask students to observe the beaker and write their observations.

2. Have students write down their guess of the water's temperature.

3. Have a student read the thermometer.

4. Discuss the importance of using instruments to make observations. **LS** Visual

Temperature

■ **KEY TERMS**
temperature
thermometer
absolute zero
heat

▶ **OBJECTIVES**

▶ **Define** *temperature* in terms of the average kinetic energy of atoms or molecules.
▶ **Convert** temperature readings between the Fahrenheit, Celsius, and Kelvin scales.
▶ **Recognize** heat as a form of energy transfer.

■ **temperature** a measure of how hot (or cold) something is; specifically, a measure of the average kinetic energy of the particles in an object

People use **temperature** readings, such as those shown in *Figure 1,* to make a wide variety of decisions every day. You check the temperature of the outdoor air to decide what to wear. The temperature of a roasting turkey is monitored to see if it is properly cooked. A nurse monitors the condition of a patient by checking the patient's body temperature. But what exactly is it that you, the cook, and the nurse are measuring? What does the temperature indicate?

Temperature and Energy

When you touch the hood of an automobile, you sense how hot or cold it is. In everyday life, we associate this sensation of hot or cold with the temperature of an object. However, this sensation serves only as a rough indicator of temperature. The Quick Activity on the next page illustrates this point.

Figure 1
Many decisions are made based on temperature.

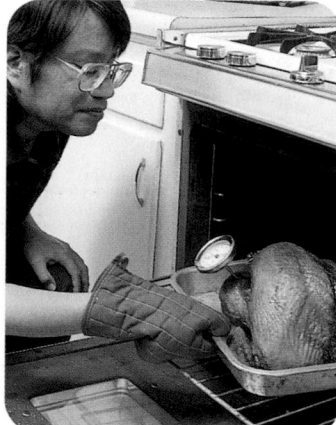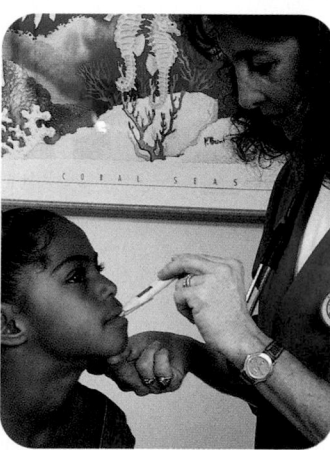

420

🔵 INCLUSION Strategies

• *Developmentally Delayed*

Using **Figure 1** as a guide, have students cut pictures from magazines and newspapers showing ways temperature is used to make decisions in daily life. The pictures can be glued to construction paper and labeled with a description of the way temperature is used in that picture. Pictures will include types of weather and the clothing used in each extreme, cooking and the variety of products and times required for safe cooking, body temperature and the need for medicine or medical attention, outdoor temperature and its effect on plants or the use of heat or air conditioning.

Sensing Hot and Cold

For this exercise you will need three bowls.

1. Put an equal amount of water in all three bowls. In the first bowl, put some cold tap water. Put some hot tap water in the second bowl. Then, mix equal amounts of hot and cold tap water in the third bowl.

2. Place one hand in the hot water and the other hand in the cold water. Leave them there for 15 s.

3. Place both hands in the third bowl, which contains the mixture of hot and cold water. How does the water temperature feel to each hand? Explain.

As you know, all particles in a substance are constantly moving. Like all moving objects, each particle has kinetic energy. If we average the kinetic energy of all the particles in an object, it turns out that this average kinetic energy is proportional to the temperature of the object.

In other words, as the average kinetic energy of an object increases, its temperature will increase. Compared to a cool car hood, the particles in a hot hood move faster because they have more kinetic energy. But how do we measure the temperature of an object? It is impossible to find the kinetic energy of every particle in an object and calculate its average. Actually, nature provides a very simple way to measure temperature directly.

Common thermometers rely on expansion

Icicles forming on trees, flowers wilting in the sun, and the red glow of a stove-top burner are all indicators of certain temperature ranges. You feel these temperatures as hot or cold. How you sense hot and cold depends not only on an object's temperature but also on other factors, such as the temperature of your skin.

To measure temperature, we rely on a simple physical property of substances: most objects expand when their temperatures increase. Ordinary **thermometers** are based on this principle and use liquid substances such as mercury or colored alcohol that expand as their temperature increases and contract as their temperature falls, because of energy exchange.

For example, the thermometer shown in *Figure 2* can measure the temperature of air on a sunny day. As the temperature rises, the particles in the liquid inside the thermometer gain kinetic energy and move faster. With this increased motion, the particles in the liquid move farther apart causing it to expand and rise up the narrow tube.

▶ **thermometer** an instrument that measures and indicates temperature

Figure 2

A liquid thermometer uses the expansion of a liquid, alcohol or mercury, to indicate changes in temperature.

Historical Perspective

The First Thermometer It is generally believed that Galileo built the first thermometer around 1592. Galileo's thermometer consisted of a long-necked glass vessel that was slightly heated, inverted, and placed in a trough of water to partially submerge the open-mouthed end of the glass vessel in the water. As the temperature of the air inside the glass vessel changed, the air expanded or contracted, changing the water level in the glass vessel. The change in the water level indicated change in temperature, which was measured against a temperature scale.

REAL-LIFE
CONNECTION

Bridges and Roads Because most substances expand when their temperature increases, bridges and roads are built to account for this expansion when the weather gets hotter. Expansion joints are built between sections of the road to prevent it from breaking when the road expands.

Teach

Quick ACTIVITY

Sensing Hot and Cold

Safety Caution *Monitor "hot" tap water carefully to ensure students do not burn themselves.*

Materials:

- 3 bowls
- cold tap water
- hot tap water

Teacher's Notes: If tap water is not cold, have students add one or two ice cubes to the "cold" bowl. Be sure students grasp the idea that terms like *hot* and *cold* are qualitative measures of temperature. Professionals in all disciplines prefer measurements that are quantitative, so a scientist uses a thermometer to measure temperature.

Results: The main conclusion from the Quick Activity is that the hot and cold sensations are relative and can be misleading.

LS Kinesthetic

Chapter Resource File

- **Lesson Plan**
- **Quick Activity Datasheet** Sensing Hot and Cold GENERAL

Transparencies

TT Bellringer

Teaching Tip

Temperature Scale Draw two horizontal number lines on the chalkboard, one above the other. At the left end of the lines, draw a vertical line and write "Water Freezes." At the right end, draw another vertical line and write "Water Boils." Label the top line °F and the bottom line °C. Label the left end of the top line "32" and the bottom line "0." Label the right ends "212" and "100."

Show students that the temperatures mean the same thing but they have different units, like 2.54 cm and 1 inch. Also point out that there are 180 spaces in between water freezing and boiling on the top line, but there are only 100 spaces on the bottom line. That means that a °C must be 1.8 times as big as a °F.

SKILL BUILDER — ADVANCED

Graphing Tell students that there is a linear relationship between the Celsius and Fahrenheit scales. Then have them use the freezing and boiling points of water, (0°C, 32°F) and (100°C, 212°F), to draw a line graph of °F versus °C. Ask students to determine the slope of the line graph (1.8/1) and the y–intercept (32°). Students can use their graphs to convert between the two scales. Also ask students to find the equation of the line. (Using the equation $y = mx + b$, where m = slope and b = y intercept, $y = 1.8x + 32$.)
LS Logical

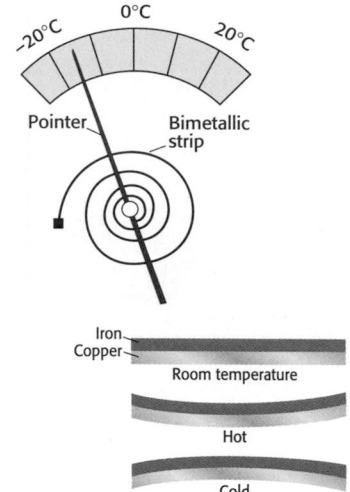

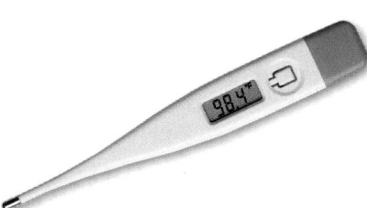

Figure 3
A refrigerator thermometer uses the bending of a strip made from two metals to indicate the correct temperature.

Figure 4
A digital thermometer uses changes in electricity to measure temperature.

internet connect

www.scilinks.org
Topic: Temperature Scales
SciLinks code: HK4138

SCiLINKS Maintained by the National Science Teachers Association

422

Thermometers can use different methods

Liquid thermometers can measure only temperatures within a certain range. This is because below a certain temperature, the liquid used in the thermometer freezes. Also, above a certain temperature the liquid boils. Therefore, different types of thermometers are designed to measure extreme temperatures.

A refrigerator thermometer is based on the expansion of metal, as shown in **Figure 3.** The thermometer contains a coil made from two different metal strips pressed together. Both strips expand and contract at different rates as the temperature changes. As the temperature falls, the coil unwinds moving the pointer to the correct temperature. As the temperature rises, the coil winds up moving the pointer in the opposite direction.

A digital thermometer, shown in **Figure 4,** is designed to measure temperature by noting the change in current. Changes in temperature also cause electric current to change.

Fahrenheit and Celsius are common scales used for measuring temperatures

The units on the Fahrenheit scale are called degrees Fahrenheit, or °F. On the Fahrenheit scale, water freezes at 32°F and boils at 212°F.

Most countries other than the United States use the Celsius (or centigrade) scale. This scale is widely used in science. The Celsius scale gives a value of 0°C to the freezing point of water and a value of 100°C to the boiling point of water at standard atmospheric pressure. The difference between these two points is divided into 100 equal parts, called degrees Celsius, or °C.

A degree Celsius is 1.8 times as large as a degree Fahrenheit. Also, the temperature at which water freezes differs for the two scales by 32 degrees. To convert from one scale to the other, use one of the following formulas.

Conversion Equations

Fahrenheit temperature = $\left(1.8 \times \text{Celsius temperature}\right) + 32.0$

$$T_F = 1.8t + 32.0$$

Celsius temperature = $\dfrac{(\text{Fahrenheit temperature} - 32.0)}{1.8}$

$$t = \dfrac{(T_F - 32.0)}{1.8}$$

LIFE SCIENCE
CONNECTION

The human body maintains a fairly constant body temperature of about 98.6°F. This is in contrast to reptiles, like snakes and lizards, whose temperature adjusts according to the temperature of their surroundings.

MATHEMATICS
CONNECTION — GENERAL

Show students the origin of the factor 1.8 in the Fahrenheit-to-Celsius conversions.

100°C divisions =

180°F divisions; therefore,

1°C division = 180/100°F divisions

1°C division = 1.8°F divisions

The Kelvin scale is based on absolute zero

You have probably heard of negative temperatures, such as those reported on extremely cold winter days in the northern United States and Canada. Remember that temperature is a measure of the average kinetic energy of the particles in an object. Even far below 0°C these particles are moving and therefore have some kinetic energy. But how low can the temperature fall? Physically, the lowest possible temperature is −273.16°C. This temperature is referred to as **absolute zero** At absolute zero the energy of an object is zero. That is, the energy of the object cannot be any lower.

Absolute zero is the basis for another temperature scale called the Kelvin scale. On this scale, 0 kelvin, or 0 K, is absolute zero. Since the lowest possible temperature is assigned a zero value, there are no negative temperature values on the Kelvin scale. The Kelvin scale is used in many fields of science, especially those involving low temperatures. The three temperature scales are compared in *Figure 5*.

In magnitude, a unit of kelvin is equal to a degree on the Celsius scale. Therefore, the temperature of any object in kelvins can be found by simply adding 273 to the object's temperature in degrees Celsius. The equation for this conversion is given below.

Celsius-Kelvin Conversion Equation

$$\text{Kelvin temperature} = \text{Celsius temperature} + 273$$
$$T = t + 273$$

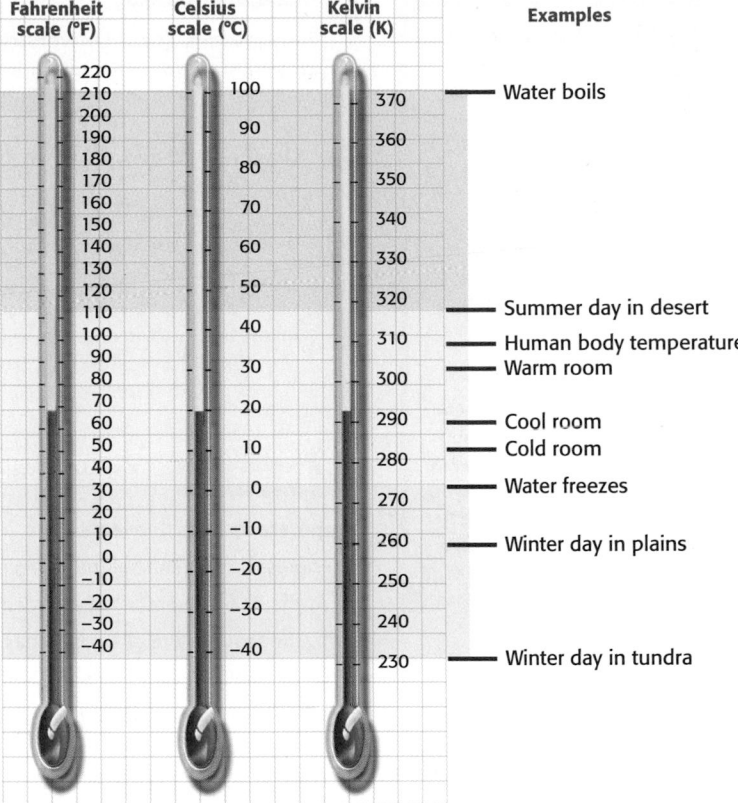

Temperature Values on Different Scales

Fahrenheit scale (°F)	Celsius scale (°C)	Kelvin scale (K)	Examples
	100	370	Water boils
220			
210			
200	90	360	
190			
180	80	350	
170			
160	70	340	
150			
140	60	330	
130			
120	50	320	
110			Summer day in desert
100	40	310	Human body temperature
90			Warm room
80	30	300	
70			
60	20	290	Cool room
50	10	280	Cold room
40			
30	0	270	Water freezes
20			
10	−10	260	Winter day in plains
0	−20	250	
−10			
−20	−30	240	
−30			
−40	−40	230	Winter day in tundra

Figure 5

Temperature on the Celsius scale can be converted to both Fahrenheit and Kelvin scales. Note that all Kelvin temperatures are positive.

▶ **absolute zero** the temperature at which molecular energy is at a minimum (0 K on the Kelvin scale or −273.16°C on the Celsius scale)

423

Historical Perspective

Temperature Scales The Kelvin scale is named after the British physicist Lord Kelvin (1824–1907). The Fahrenheit scale is named after the German physicist Gabriel Fahrenheit (1686–1736). The Celsius scale is named for the Swedish astronomer Anders Celsius (1701–1744). The Celsius scale was originally called the centigrade scale because there are 100 degrees between the freezing and boiling points of water. In Latin, *centi* means "100" and *gradus* means "degree."

Practice

1. a. $1.8(-252.87) + 32.0 = -423.2\ °F$
      $-252.87 + 273 = 20.13\ K$
   b. $1.8(-40.0) + 32.0 = -40.0°F$
      $-40.0 + 273 = 233\ K$
   c. $1.8(1064) + 32.0 = 1947°F$
      $1064 + 273 = 1337\ K$

2. $1.8(21) + 32 = 70°F$
   $21 + 273 = 294\ K$

   $388 - 273 = 115°C$
   $1.8(115) + 32.0 = 239°F$

   $1.8(-200.0) + 32.0 = -328°F$
   $-200.0 + 273 = 73\ K$

   $\dfrac{(110.0 - 32.0)}{1.8} = 43°C$

   $43 + 273 = 316\ K$

3. d
4. c

INTEGRATING

SPACE SCIENCE
From cold deep space to hot stars, astronomers measure a wide range of temperatures of objects in the universe. All objects produce different types of electromagnetic waves depending on their temperature. By identifying the distribution of wavelengths an object radiates, astronomers can estimate the object's temperature.

Light (an electromagnetic wave) received from the sun indicates that the temperature of its surface is 6000 K. If you think that is hot, try the center of the sun, where the temperature increases to 15 000 000 K!

PHYSICAL SCIENCE INTERACTIVE TUTOR

Disc One, Module 7: **Heat**
Use the Interactive Tutor to learn more about these topics.

Math Skills

Temperature Scale Conversion The highest atmospheric temperature ever recorded on Earth was 57.8°C. Express this temperature both in degrees Fahrenheit and in kelvins.

1 List the given and unknown values.
   Given: $t = 57.8°C$
   Unknown: $T_F = ?°F$, $T = ?K$

2 Write down the equations for temperature conversions.
   $T_F = 1.8t + 32.0$
   $T = t + 273$

3 Insert the known values into the equations, and solve.
   $T_F = (1.8 \times 57.8) + 32.0 = 104 + 32.0 = 136°F$
   $T = 57.8 + 273 = 331\ K$

Practice

Temperature Scale Conversion

1. Express these temperatures in degrees Fahrenheit and in kelvins.
   a. the boiling point of liquid hydrogen (–252.87°C)
   b. the temperature of a winter day at the North Pole (–40.0°C)
   c. the melting point of gold (1064°C)

2. Make the necessary conversions to complete the table below.

Example	Temp. (°C)	Temp. (°F)	Temp. (K)
Air in a typical living room	21	?	?
Metal in a running car engine	?	?	388
Liquid nitrogen	–200	?	?
Air on a summer day in the desert	?	110	?

3. Use **Figure 5** to determine which of the following is a likely temperature for ice cubes in a freezer.
   a. – 20°C   c. 253 K
   b. – 4°F   d. all of the above

4. Use **Figure 5** to determine which of the following is the nearest value for normal human body temperature.
   a. 50°C   c. 310 K
   b. 75°F   d. all of the above

424

SPACE SCIENCE
CONNECTION — BASIC

Solar System Have the students create a table of the planets in our solar system, comparing the temperature of each planet with the classroom's temperature. Record the table on the board or on an overhead transparency.

Develop some general principles that can guide the students to classify which planets are "hotter" or "colder." General principles that should be suggested to students are the distance of a planet from the sun and the type of atmosphere of the planet.

As an extension, have students graph the relationship between the temperature of the planets and their distance from the sun. For a dramatic effect, you could prepare a large wall-sized graph on butcher paper or a bulletin board, using a cutout thermometer for the y-axis of the graph, and have students plot the points using cutouts of the planets.
LS Logical

Relating Temperature to Energy Transfer

When you touch a piece of ice, it feels very cold. When you step into a hot bath, the water feels very hot. Clasping your hands together usually produces neither sensation. These three cases can be explained by comparing the temperatures of the two objects that are making contact with each other.

The feeling associated with temperature difference results from energy transfer

Imagine that you are holding a piece of ice. The temperature of ice is lower than the temperature of your hand; therefore, the molecules in the ice move slowly compared with the molecules in your hand. As the molecules on the surface of your hand collide with those on the surface of the ice, energy is transferred to the ice. As a result, the molecules in the ice speed up and their kinetic energy increases. This causes the ice to melt.

How do temperature and energy relate?

Materials
- ✓ glass beaker
- ✓ tongs
- ✓ 2 pieces of string, 20 cm each
- ✓ thermometer
- ✓ clock
- ✓ electric hot plate
- ✓ graduated cylinder
- ✓ 40 identical small metal washers
- ✓ 2 plastic-foam cups

1. Tie 10 washers on one piece of string and 30 washers on another piece of string.
2. Fill the beaker two-thirds full with water, lower the washers in, and set the beaker on the hot plate.
3. Heat the water to boiling.
4. While the water heats, put exactly 50 mL of cool water in each plastic-foam cup.
5. Use a thermometer to measure and record the initial temperature of water in each cup.
6. When the water in the beaker has boiled for about 3 minutes, use tongs to remove the group of 30 washers. Gently shake any water off the washers back into the beaker, and quickly place the washers into one of the plastic-foam cups.
7. Observe the change in temperature of the cup's water. Record the highest temperature reached.

8. Repeat steps 6 and 7 by placing the 10 washers in the other plastic-foam cup.

Analysis

1. Which cup had the higher final temperature?
2. Both cups had the same starting temperature. Both sets of washers started at 100°C. Why did one cup reach a higher final temperature?

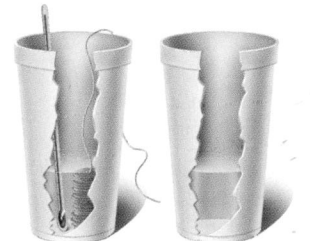

425

Quick Lab

How do temperature and energy relate?

Analysis

The cup with 30 washers reaches a higher final temperature because a greater mass (of 30 washers as opposed to 10 washers) can transfer more energy. If students have trouble answering item 2, guide them with further questions such as the following:

- Why did the water temperature rise?
- Which set of washers had more energy available to transfer?

LS Logical

LIFE SCIENCE CONNECTION

Animals Some animals, like lizards and frogs, do not have an internal mechanism to regulate their body temperature. Lizards and frogs bask in the sun to increase their body temperature by absorbing energy as heat. When it is very cold outside, these animals become inactive because they do not have a lot of extra energy to spend.

REAL-LIFE CONNECTION

Burns The painful sensation of a burn is caused by a rapid transfer of energy. This is usually caused by touching a very hot object, but it can also be caused by touching a very cold object. When you touch a hot object, energy flows from the object into your skin cells very rapidly, causing damage. When you touch a very cold object, energy transfers from your skin into the object, also resulting in tissue damage.

Chapter Resource File

- **Cross-Disciplinary Worksheet** Integrating Space Science—Starlight, Star Heat **GENERAL**
- **Math Skills** Temperature Scale Conversions **GENERAL**
- **Quick Lab Datasheet** How do temperature and energy relate? **GENERAL**

INTEGRATING

HEALTH Active people use more energy when they exercise than less active people, so they need to take in more than 2000 Calories per day.

Close

Quiz

1. Temperature is a measure of what? (how hot (or cold) something is; the average kinetic energy of the particles in an object)

2. What are three temperature scales? (Fahrenheit, Celsius, Kelvin)

3. What is the temperature at which an object's energy is minimal? (absolute zero)

4. Convert 55°F to °C and K. (13°C; 286 K)

5. What happens when two objects at different temperatures come into contact with each other are? (Energy is transferred from the object at a higher temperature to the object at a lower temperature.)

Chapter Resource File

• **Cross-Disciplinary Worksheet** Integrating Health—Skin Temperature GENERAL

• **Concept Review** GENERAL

• **Quiz** BASIC

■ **heat** the energy transferred between objects that are at different temperatures

INTEGRATING

HEALTH Food supplies the human body with energy. An active 120 lb teenager on a typical diet takes in and expends about 2400 Calories (4.187×10^7J) per day, or 48.5 J/s. Much of this energy is eventually transferred away as heat, which is why a full classroom feels hotter toward the end of class.

SUMMARY

▶ Temperature is a measure of the average kinetic energy of an object's particles.

▶ On the Celsius temperature scale, water freezes at 0° and boils at 100°.

▶ A kelvin is the same size as a degree Celsius. The lowest temperature possible—absolute zero—is 0 K.

▶ At absolute zero, particle kinetic energy is minimal.

▶ Heat is the energy transferred between objects with different temperatures.

Temperature changes indicate an energy transfer

The energy transferred between the particles of two objects due to a temperature difference between the two objects is called **heat.** This transfer of energy always takes place from a substance at a higher temperature to a substance at a lower temperature. For example, if you hold a glass of ice water in your hands, energy will be transferred as heat from your hand to the glass. However, if you hold a very hot cup of water, energy will be transferred as heat from the cup to your hand.

Because temperature is an indicator of the average kinetic energy of internal particles, you can use temperature to predict which way energy will be transferred. Internal kinetic energy will be transferred as heat from the warmer object to the cooler object. So, when this energy is transferred from the hot water in the cup to your skin, the temperature of the water falls while the temperature of your skin rises.

When both your skin and the cup in your hand approach the same temperature, less energy is transferred from the cup to your skin. To continue the transfer of energy, enough energy must be added to the water as heat to keep the water's temperature higher than the skin's temperature. The greater the difference in the temperatures of the two objects is, the greater the amount of energy that will be transferred as heat is.

SECTION 1 REVIEW

1. **Define** *absolute zero* in terms of kinetic energy of particles.

2. **Predict** which molecules will move faster on average: water molecules in hot soup or water molecules in iced lemonade.

3. **Predict** whether a greater amount of energy will be transferred as heat between 1 kg of water at 10°C and a freezer at –15°C or between 1 kg of water at 60°C and an oven at 65°C.

4. **Critical Thinking** Determine which of the following has a higher temperature and which contains a larger amount of total kinetic energy: a cup of boiling water or Lake Michigan.

Math Skills

5. Convert the temperature of the air in an air-conditioned room, 20.0°C, to equivalent values on the Fahrenheit and Kelvin temperature scales.

6. Convert the coldest outdoor temperature ever recorded, –128.6°F, to equivalent Celsius and Kelvin temperatures.

Answers to Section 1 Review

1. Absolute zero is the temperature at which particles have minimal kinetic energy.

2. The higher temperature of the hot soup means that the water molecules in the soup will move faster on average than the water molecules in iced lemonade.

3. More energy would be transferred between water at 10°C and a freezer at –15°C because the temperature difference is greater.

4. A cup of boiling water has a higher temperature than Lake Michigan, but Lake Michigan has more total kinetic energy since it has more particles.

Math Skills

5. 1.8 (20.0) + 32.0 = 68.0 °F

   20 + 273 = 293K

6. $\dfrac{(-128.6 - 32.0)}{1.8} = -89.2$ °C

   –89.2 + 273 = 184K

Energy Transfer

OBJECTIVES

▶ **Investigate** and demonstrate how energy is transferred by conduction, convection, and radiation.
▶ **Identify** and distinguish between conductors and insulators.
▶ **Solve** problems involving specific heat.

▶ **KEY TERMS**
thermal conduction
convection
convection current
radiation
specific heat

While water is being heated for your morning shower, your breakfast food is cooking. In the freezer, water in ice trays becomes solid after the freezer cools the water to 0°C. Outside, the morning dew evaporates soon after light from the rising sun strikes it. These are all examples of energy transfers from one object to another.

Methods of Energy Transfer

The transfer of heat energy from a hot object can occur in three ways. Roasting marshmallows around a campfire, as shown in *Figure 6*, provides an opportunity to experience each of these three ways.

internet connect

www.scilinks.org
Topic: Energy Transfer
SciLinks code: HK4048

SciLINKS Maintained by the National Science Teachers Association

Figure 6 **Ways of Transferring Energy**

A Conduction transfers energy as heat along the wire and into the hand.

B Embers swirl upward in the convection currents that are created as warmed air above the fire rises.

C Electromagnetic waves emitted by the hot campfire transfer energy by radiation.

427

Focus

Overview

Before beginning this section, review with your students the Objectives listed in the Student Edition. In this section, students learn about conduction, convection, and radiation; conductors and insulators; and problem solving with specific heat.

🔔 Bellringer

Use the Bellringer transparency to prepare students for this section.

Motivate

Opening Demonstration ── GENERAL

You will need 1 tablespoon of baking soda, a 2 L soda bottle, 2 500 mL graduated cylinders, 200 mL of distilled water, 200 mL of vinegar, a cork, a coffee filter, and a twist tie (optional).

Safety Caution: *Wear safety goggles. Do not lean over the bottle or have it pointing toward anyone once the baking soda is placed inside.*

This demonstration is an example of energy transfer. Put a tablespoon of baking soda into the center of a coffee filter. Twist the ends of the filter tightly shut, using a twist tie if necessary. Pour 200 mL of water into an empty 2 L soda bottle. Pour 200 mL of vinegar into the same soda bottle. Place the bottle upright on newspaper. Drop the coffee filter into the soda bottle; quickly place a cork in the mouth of the bottle. Step back from the bottle. Instruct students to observe what happens both inside the bottle and to the cork.
LS Visual

Alternative Assessment ── BASIC

Interpreting the Opening Demonstration
After performing the opening demonstration, have students answer these questions:

1. Do the water, vinegar, and baking soda have energy? (Answers will vary, but students should infer that the water, vinegar, and baking soda have energy because they produced a reaction that popped the cork.)

2. How does the cork reaction show that the cork has energy? (Answers will vary, but students should mention that the movement of the cork indicates that it has energy.)
LS Verbal

Chapter Resource File

• Lesson Plan

Transparencies

TT Bellringer

Interpreting Visuals Have students examine **Figure 8.** Have them brainstorm explanations for why you can put your hand close to the side of a candle flame and not get burned while putting your hand above the flame will burn your hand. Much of the candle's energy is transferred by convection as hot air rises. Therefore, the air next to the flame is not as hot as the air above the flame. **LS Logical**

Teaching Tip ——— BASIC

Convection Essentially, a convection current is generated when a layer of fluid becomes hot and rises as it becomes less dense. Suppose a glass tube containing water is heated by putting a flame next to the top portion of the water layer. Ask the students if energy transfer will occur by convection, causing the bottom layer of water to get hot. **LS Logical**

Figure 7

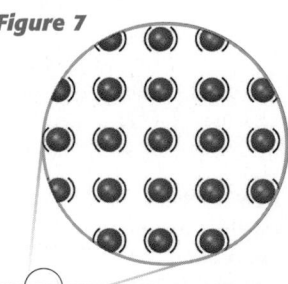

A Before conduction takes place, the average kinetic energy of the particles in the metal wire is the same throughout.

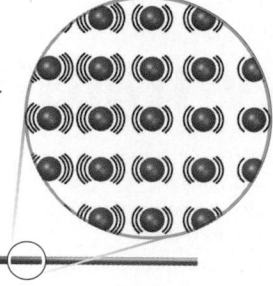

B During conduction, the rapidly moving particles in the wire transfer some of their energy to slowly moving particles nearby.

▶ **thermal conduction** the transfer of energy as heat through a material

▶ **convection** the movement of matter due to differences in density that are caused by temperature variations

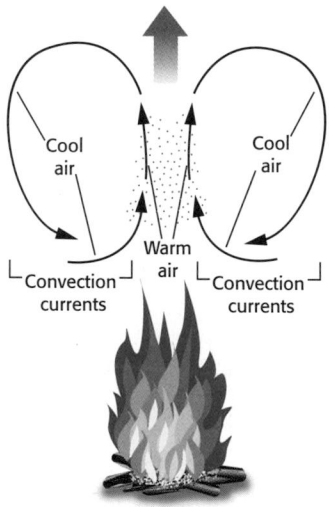

Cool air   Cool air

Warm air

Convection currents   Convection currents

Figure 8

During convection, energy is carried away by a heated fluid that expands and rises above cooler, denser fluids.

428

Conduction involves objects in direct contact

Imagine you place a marshmallow on one end of a wire made from a metal coat hanger. Then you hold the other end of the wire while letting the marshmallow cook in the campfire flame. Soon, the end of the wire you are holding will get warmer. This is an example of energy transfer by **thermal conduction.**

Conduction is one of the methods of energy transfer. Conduction takes place when two objects that are in contact are at unequal temperatures. It also takes place between particles within an object. In the case of the wire in the campfire, the rapidly moving air molecules close to the flame collide with the atoms at the end of the wire. The energy transferred to the atoms in the wire causes them to vibrate rapidly. As shown in *Figure 7,* these rapidly vibrating atoms collide with slowly vibrating atoms, transferring energy as heat all along the wire. The energy is then transferred to you as the wire's atoms collide with the molecules in your skin, creating a hot sensation in your hand.

Convection results from the movement of warm fluids

While roasting your marshmallow, you may notice that tiny glowing embers from the fire rise and begin to swirl, as shown in *Figure 6.* They are following the movement of air away from the fire. The air close to the fire becomes hot and expands so that there is more space between the air particles. As a result, the air becomes less dense and moves upward, carrying its extra energy with it, as shown in *Figure 8.* The rising warm air is replaced by cooler, denser air. The cooler air then becomes hot by the fire until it also expands and rises. Eventually, the rising hot air cools, contracts, becomes denser, and sinks. This is an example of energy transfer by **convection.**

Convection involves the movement of the heated substance itself. This is possible only if the substance is a fluid—either a liquid or a gas—because particles within solids are not as free to move.

REAL-LIFE CONNECTION

Popping Popcorn Cooking popcorn is a familiar example of energy transfer, as well as a dramatic example of what happens when water rapidly undergoes a phase change to become steam. The hard kernels absorb energy until, at a high temperature, superheated water inside the kernel suddenly turns to steam and rushes outward, and the kernels burst open to form popcorn.

Heated fluids have convection currents

The cycle of a heated fluid that rises and then cools and falls is called a **convection current.** When a pan of water is heated, the molecules of water at the bottom of the pan gradually rise and heat the molecules toward the top. The proper heating and cooling of a room requires the use of convection currents. Warm air expands and rises from vents near the floor. It cools and contracts near the ceiling and then sinks back to the floor. Eventually, the temperature of all the air in the room is increased by convection currents.

Radiation does not require physical contact between objects

As you stand close to a campfire, you can feel its warmth. This warmth can be felt even when you are not in the path of a convection current. The energy that is transferred as heat from the fire in this case is in the form of *electromagnetic waves,* which include infrared radiation, visible light, and ultraviolet rays. The energy that is transferred as electromagnetic waves is called **radiation.** You will learn more about electromagnetic radiation later.

When you stand near a fire, your skin absorbs the energy radiated by the fire. As the molecules in your skin absorb this energy, the average kinetic energy of these molecules—and thus the temperature of your skin—increases. A hot object radiates more energy than a cool object, as shown in **Figure 9.**

Radiation differs from conduction and convection in that it does not involve the movement of matter. Radiation is therefore the only method of energy transfer that can take place in a vacuum, such as outer space. Much of the energy we receive from the sun is transferred by radiation.

▶ **convection current** the vertical movement of air currents due to temperature variations

▶ **radiation** the energy that is transferred as electromagnetic waves, such as visible light and infrared waves

Figure 9
Changes in Radiated Energy

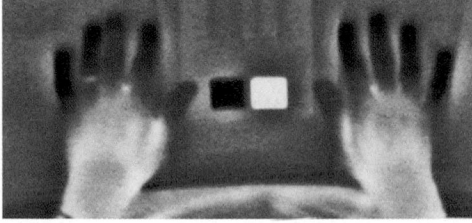

A Before surgery, as seen in the infrared photo, the fingers are cooler than the rest of the hand. This results from poor blood flow in this patient's fingers.

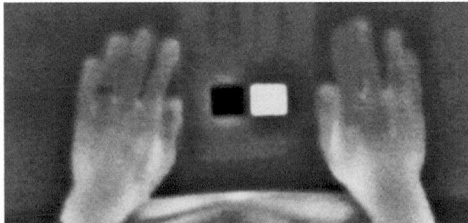

B After surgery, the blood flow has been restored, so the temperature of the fingers increases. The amount of energy they radiate also increases.

429

EARTH SCIENCE — CONNECTION

Earth's Temperature The electromagnetic waves from the sun provide energy, causing the temperature of our planet to increase. It is often assumed that the temperature of the atmosphere rises because of the direct energy transfer between the electromagnetic waves and the air molecules. In fact, very little energy is absorbed by the air molecules as the electromagnetic waves pass through the atmosphere. First, the electromagnetic waves transfer energy to the ground, raising its temperature. The hot ground then transfers energy to the layer of air next to the ground by conduction. This is followed by the generation of convection currents, which rise up and transfer energy to subsequent layers of the atmosphere. The hot ground also produces infrared radiation, which travels upward in the air. The infrared radiation is mostly absorbed by water vapor and CO_2 in the air.

Quick **Lab**

What color absorbs more radiation?

Have students bring cans from home, but be sure each group gets two cans that are matched in size. If you decide to use sunlight, try to take students outside, since the glass in the classroom windows will reflect or absorb the infrared and ultraviolet light. If you go outside, you will need stopwatches. If you use lamps, try to use 100 W bulbs.

Analysis

1. Both sets of data should show an increase in the temperature of the water.

2. The black can should absorb more energy.

3. **a.** The starting temperature of the water in each can should be the same.

   b. The volume of the water was controlled (50 mL in each can).

   c. The distance should be the same for each can, although some groups might not have been careful about this.

   d. The size of each can should be the same.

4. Black absorbs more energy, which is why solar panels are often black.

5. In winter, a black car would absorb wanted energy, and in summer, a white or silver car would reflect unwanted energy.

LS Kinesthetic

Quick **Lab**

What color absorbs more radiation?

Materials
- ✓ empty soup can, painted black inside and out, label removed
- ✓ empty soup can, label removed
- ✓ 2 thermometers
- ✓ clock
- ✓ graduated cylinder
- ✓ bright lamp or sunlight

1. Prepare a data table with three columns and at least seven rows. Label the first column "Time," the second column "Temperature of painted can (°C)," and the third column "Temperature of unpainted can (°C)."

2. Pour 50 mL of cool water into each can.

3. Place a thermometer in each can, and record the temperature of the water in each can at the start. Leave the thermometers in the cans. Aim the lamp at the cans, or place them in sunlight.

4. Record the temperature of the water in each can every 3 minutes for at least 15 minutes.

Analysis

1. Prepare a graph. Label the *x*-axis "Time" and the *y*-axis "Temperature." Plot your data for each can of water.

2. Which color absorbed more radiation?

3. Which variables in the lab were controlled (unchanged throughout the experiment)? For each of the following variables, explain your answer.

   a. starting temperature of water in cans

   b. volume of water in cans

   c. distance of cans from light

   d. size of cans

4. Use your results to explain why panels used for solar heating are often painted black.

5. Based on your results, what color would you want your car to be in the winter? in the summer? Justify your answer.

Conductors and Insulators

When you are cooking, the pan must conduct energy to heat the food, but the handle must be insulated from the heat so that you can hold it. If you are using conduction to increase the temperature of a substance, you must use materials through which energy can be quickly transferred as heat. Cooking pans are usually made of metal because energy is passed quickly between the particles in most metals. Any material through which energy can be easily transferred as heat is called a *conductor*.

Many people try to avoid wasting energy. It is most often wasted by energy transfer through the roof or the walls of your home. You can reduce this energy transfer by using poor conductors, called *insulators* or *insulation*. Insulation in the attic or walls of homes helps to prevent unwanted energy transfer.

REAL-LIFE CONNECTION

Superconductors Copper and aluminum, like most metals, are good conductors of electric current, which is the flow of electrons in a substance. At very low temperatures, certain substances lose all their resistance to electric current and become excellent conductors of electric current. These substances are called superconductors. The temperature at which a substance becomes a superconductor is close to absolute zero. For example, aluminum becomes a superconductor at 1.2 K.

◕ INCLUSION *Strategies*

- *Learning Disabled*
- *Attention Deficit Disorder*
- *English as a Second Language*

Ask students to fold a piece of paper in half vertically. Label one side of the paper "Conductors" and the other side "Insulators." Under each term, write a definition. Using information from the chapter, students can list materials that are good conductors of energy and poor conductors (insulators). Discussion in small groups about the practical uses of these materials would help students understand the concept.

Energy transfers through particle collisions

Gases are extremely poor conductors because their particles are far apart, and transfer of energy is less likely to occur. The particles in liquids are more closely packed. However, while liquids conduct better than gases, they are not effective conductors.

Some solids, such as rubber and wood, conduct energy about as well as liquids. So, rubber and wood are good insulators. Some solids are better conductors than other solids. Metals, such as copper and silver, conduct energy as heat very well. Metals, in general, are better conductors than nonmetals.

Examples of conductors and insulators are shown in **Figure 10.** The skillet is made of iron, a good conductor, so energy is transferred effectively as heat to the food. Wood is an insulator, so the energy from the hot skillet won't reach your hand through the wooden spoon or the wooden handle.

Figure 10
The skillet conducts energy from the stove element to the food. The wooden spoon and handle insulate the hands from the energy of the skillet.

internet connect

www.scilinks.org
Topic: Insulators
SciLinks code: HK4073

SciLINKS Maintained by the National Science Teachers Association

Quick ACTIVITY

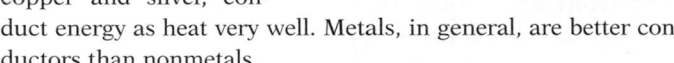

Conductors and Insulators

For this activity you will need several flatware utensils. Each one should be made of a different material, such as stainless steel, aluminum, and plastic. You will also need a bowl and ice cubes.

1. Place the ice cubes in the bowl. Position the utensils in the bowl so that an equal length of each utensil lies under the ice.
2. Check the utensils' temperature by briefly touching each utensil at the same distance from the ice every 20 s. Which utensil becomes colder first? What variables might affect your results?

431

Interpreting Visuals Have students examine **Table 1.** Point out that the numbers in the chart represent how much energy is required to raise the temperature of 1 kg of the substance by 1 K (or 1°C). Have students make a list ranking the substances from the ones that absorb energy the easiest (gold and lead) to the one that absorbs the least energy (water). **LS Logical**

Graphing Have students prepare a graph of the following data (for a 1.0 kg sample) with energy on the *y*-axis and the temperature change on the *x*-axis. Students should note the constant nature of the relationship. You may have advanced students calculate the slope (140 J/kg•K) and use **Table 1** to determine the substance (mercury).

Energy (J)	Temperature change (K)
700	5
1400	10
2100	15
2800	20
3500	25

LS Logical

Figure 11
The spoon's temperature increases rapidly because of the spoon's low specific heat.

▶ **specific heat** the quantity of heat required to raise a unit mass of homogenous material 1 K or 1°C in a specified way given constant pressure and volume

Specific Heat

You have probably noticed that a metal spoon, like the one shown in **Figure 11,** becomes hot when it is placed in a cup of hot liquid. You have also probably noticed that a spoon made of a different material, such as plastic, does not become hot as quickly. The difference between the final temperatures of the two spoons depends on whether they are good conductors or good insulators. But what makes a substance a good or poor conductor depends in part on how much energy a substance requires to change its temperature by a certain degree.

Specific heat describes how much energy is required to raise an object's temperature

Not all substances behave the same when they absorb heat energy. For example, a metal spoon left in a metal pot becomes hot seconds after the pot is placed on a hot stovetop burner. This is because a few joules of energy are enough to raise the spoon's temperature substantially. However, if an amount of water with the same mass as the spoon is placed in the same pot, that same amount of energy will produce a much smaller temperature change in the water.

For all substances, **specific heat** is a characteristic physical property, which is denoted by c. In this book, we will think of specific heat of any substance as the amount of energy required to raise 1 kg of that substance by 1 K.

Some values for specific heat are given in *Table 1.* They are in units of J/kg•K, meaning that each is the amount of energy in J needed to raise the temperature of 1 kg of the substance by exactly 1 K.

Table 1 **Specific Heats at 25°C**

Substance	c (J/kg•K)	Substance	c (J/kg•K)
Water (liquid)	4186	Copper	385
Steam	1870	Gold	129
Ammonia (gas)	2060	Iron	449
Ethanol (liquid)	2440	Mercury	140
Aluminum	897	Lead	129
Carbon (graphite)	709	Silver	234

432

Did You Know?

Liquids: Are They Conductors? Liquids, in general, are poor thermal conductors. An exception is molten or liquid metals, which are very good conductors.

MISCONCEPTION ALERT

Rate of Energy Transfer When one substance feels colder than another, is it because the first substance has a lower temperature? Actually, the difference is not in the temperature of the two, but in the rate at which they transfer energy. For example, a tile floor might feel much colder to bare feet than a rug in the same room does. This is not because the tile floor and the rug have different temperatures, but because the tile, which is a better conductor than the rug, transfers energy to the bare feet much faster than the rug does.

On a hot summer day, the temperature of the water in a swimming pool remains much lower than the air temperature and the temperature of the concrete around the pool. This is due to water's relatively high specific heat as well as the large mass of water in the pool. Similarly, at night, the concrete and the air cool off quickly, while the water changes temperature only slightly.

Specific heat can be used in calculations

Because specific heat is a ratio, it can be used to predict the effects of larger temperature changes for masses other than 1 kg. For example, if it takes 4186 J to raise the temperature of 1 kg of water by 1 K, twice as much energy, 8372 J, will be required to raise the temperature of 2 kg of water by 1 K. But about 25 120 J will be required to raise the temperature of the 2 kg of water by 3 K. This relationship is summed up in the equation below.

> **Specific Heat Equation**
> energy = (specific heat) × (mass) × (temperature change)
> $$energy = cm\Delta t$$

Specific heat can change slightly with changing pressure and volume. However, problems and questions in this chapter will assume that specific heat does not change.

Math Skills

Specific Heat How much energy must be transferred as heat to the 420 kg of water in a bathtub in order to raise the water's temperature from 25°C to 37°C?

1 List the given and unknown values.

> **Given:** $\Delta t = 37°C - 25°C = \Delta 12°C = \Delta 12$ K
> $\Delta T = 12$ K
> $m = 420$ kg
> $c = 4186$ J/kg•K

> **Unknown:** energy = ? J

2 Write down the specific heat equation from this page.

> $energy = cm\Delta t$

3 Substitute the specific heat, mass, and temperature change values, and solve.

> $energy = \left(\dfrac{4186 \text{ J}}{\text{kg}\bullet\text{K}}\right) \times (420 \text{ kg}) \times (12 \text{ K})$
> $energy = 21\ 000\ 000$ J $= 2.1 \times 10^4$ kJ

433

INTEGRATING

EARTH SCIENCE Sea breezes result from both convection currents in the coastal air and differences in the specific heats of water and sand or soil. During the day, the temperature of the land increases more than the temperature of the ocean water, which has a larger specific heat. As a result, the temperature of the air over land increases more than the temperature of air over the ocean. This causes the warm air over the land to rise and the cool ocean air to move inland to replace the rising warm air. At night, the temperature of the dry land drops below that of the ocean, and the direction of the breezes is reversed.

Practice

1. *energy* = (449 J/kg•K)(0.755 kg) (403 K − 283 K) = 40 700 J

2. *energy* = (4186 J/kg•K)(0.225 kg) (35°C − 5°C) = 28 000 J

3. *energy* = (449 J/kg•K)(144 kg) (35°C − 25°C)(1 kJ/1000 J) = 650 kJ

4. ΔT = (2.4 × 10⁴ J)/[(897 J/kg•K) (0.225 kg)]

   ΔT = 120 K = 120°C

   T_f − 25°C = 120°C

   T_f = 145°C

5. *m* = energy/($c \times \Delta T$) = (4.7 × 10⁵ J)/[(4186 J/kg•K)(355 K − 298 K)] = 2.0 kg

6. *c* = 1124 J/[93 g (1 kg/1000 g) 25K] = 480 J/kg•K

LS Logical

Close

Quiz

Determine what type of energy transfer is described by each phrase below.

1. transfer of energy by electromagnetic waves (radiation)

2. transfer of energy between colliding particles (conduction)

3. transfer of energy by the movement of fluids (convection)

Practice HINT

▶ To rearrange the equation to isolate temperature change, divide both sides of the equation by *mc*.

$$\frac{energy}{mc} = \left(\frac{\cancel{mc}}{\cancel{mc}}\right)\Delta t$$

$$\Delta t = \frac{energy}{mc}$$

▶ Use this version of the equation for Practice Problem 4.

▶ For Practice Problems 5 and 6, you will need to isolate *m* and *c*.

SUMMARY

▶ Conduction is the transfer of energy as heat between particles as they collide within a substance or between objects in contact.

▶ Convection currents are the movement of gases and liquids as they become heated, expand, and rise, then cool, contract, and fall.

▶ Radiation is energy transfer by electromagnetic waves.

▶ Conductors are materials through which energy is easily transferred as heat.

▶ Insulators are materials that conduct energy poorly.

▶ Specific heat is the energy required to heat 1 kg of a substance by 1 K.

Practice

Specific Heat

1. How much energy is needed to increase the temperature of 755 g of iron from 283 K to 403 K?

2. How much energy must a refrigerator absorb from 225 g of water so that the temperature of the water will drop from 35°C to 5°C?

3. A 144 kg park bench made of iron sits in the sun, and its temperature increases from 25°C to 35°C. How many kilojoules of energy does the bench absorb?

4. An aluminum baking sheet with a mass of 225 g absorbs 2.4 x 10⁴ J from an oven. If its temperature was initially 25°C, what will its new temperature be?

5. What mass of water is required to absorb 4.7 x 10⁵ J of energy from a car engine while the temperature increases from 298 K to 355 K?

6. A vanadium bolt gives up 1124 J of energy as its temperature drops 25 K. If the bolt's mass is 93 g, what is its specific heat?

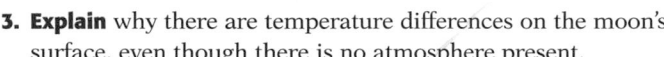

SECTION 2 REVIEW

1. **Describe** how energy is transferred by conduction, convection, and radiation.

2. **Predict** whether the hottest part of a room will be near the ceiling, in the center, or near the floor, given that there is a hot-air vent near the floor. Explain your reasoning.

   WRITING SKILL

3. **Explain** why there are temperature differences on the moon's surface, even though there is no atmosphere present.

4. **Critical Thinking** Explain why cookies baked near the turned-up edges of a cookie sheet receive more energy than those baked near the center.

Math Skills

5. When a shiny chunk of metal with a mass of 1.32 kg absorbs 3250 J of energy, the temperature of the metal increases from 273 K to 292 K. Is this metal likely to be silver, lead, or aluminum?

6. A 0.400 kg sample of glass requires 3190 J for its temperature to increase from 273 K to 308 K. What is the specific heat for this type of glass?

Answers to Section 2 Review

1. Energy transfer by conduction involves direct contact. Higher-energy molecules transfer energy to lower-energy molecules. Energy transfer by convection involves the movement of the higher-energy molecules from one place to another. Energy transfer by radiation involves the emission and absorption of electromagnetic waves.

2. The hottest part of the room should be near the ceiling because hot air rises.

3. The temperature differences on the moon's surface are due to two factors. One is the varying composition of the moon's surface.

The other factor is the location of the spot, whether in the sun or in the shadow.

4. The cookies near the turned-up edge receive conduction energy from the cookie sheet, just as the cookies in the middle do. However, those near the edge also receive energy from air convection currents and radiation from the side.

Math Skills

5. $c = \dfrac{3250 \text{ J}}{(1.32\text{kg})(292\text{K} - 273\text{K})}$ = 130J/kg•K;

   lead

6. 228 J/kg•K

Using Heat

INTEGRATING TECHNOLOGY and Society

OBJECTIVES

▶ **Describe** the concepts of different heating and cooling systems.

▶ **Compare** different heating and cooling systems in terms of their transfer of usable energy.

▶ **Explain** how a heat engine uses heat energy to do work.

▶ **KEY TERMS**
refrigerant
heat engine

Heating a house in the winter, cooling an office building in the summer, or preserving food throughout the year is possible because of machines that transfer energy as heat from one place to another. An example of one of these machines, an air conditioner, is shown in **Figure 12.** An air conditioner does work to remove energy as heat from the warm air inside a room and then transfers the energy to the warmer air outside the room. An air conditioner can do this because of two principles about energy that you have already studied.

The first principle is that the total energy used in any process—whether that energy is transferred as a result of work, heat, or both—is conserved. This principle of conservation of energy is called the first law of thermodynamics.

The second principle is that the energy transferred as heat always moves from an object at a higher temperature to an object at a lower temperature.

Figure 12

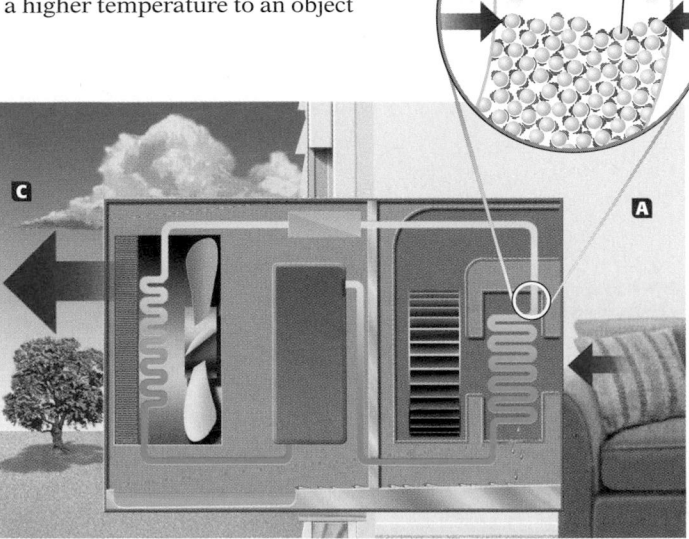

Gaseous refrigerant   Liquid refrigerant

A A substance that easily evaporates and condenses is used in air conditioners to transfer energy from a room to the air outside.

B When the liquid evaporates, it absorbs energy from the surrounding air, thereby cooling it.

C Outside, the air conditioner causes the gas to condense, releasing energy.

435

Did You Know?

Radiators The radiator in a hot-water heating system will effectively transfer energy by both conduction and radiation if it has a dark surface. This is because dark surfaces become hotter by absorbing radiation more effectively than surfaces with light colors. Dark objects are also very effective emitters of radiation. In fact, all good absorbers of radiation are also good emitters of radiation. Therefore, most radiators consist of a dark surface, particularly, a dull black surface.

Focus

Overview

Before beginning this section, review with your students the Objectives listed in the Student Edition. In this section, students explore different heating and cooling systems in terms of how they work, advantages and drawbacks, and the associated decrease of usable energy.

Bellringer

Use the Bellringer transparency to prepare students for this section.

Motivate

Opening Discussion —— GENERAL

Ask students for examples of heating and cooling systems they have in their homes. They can include any device that heats or cools the entire home, a single room, or a smaller enclosed area (such as a refrigerator). List all examples on the board. Tell students that they will learn about how some of these systems work in this section.
LS Verbal

Teach

SKILL BUILDER —— GENERAL

Interpreting Visuals Discuss the energy transfers in **Figure 12** with students. Have them work in pairs to describe how a heat pump would work to heat a room. Discuss proposed solutions with the class. **LS** Interpersonal

Chapter Resource File

• Lesson Plan

Transparencies

TT Bellringer

TT Air Conditioner

Reading Skills Have students list what they know or think they know about heating and cooling systems, then have them list what they want to know. After the students read this section, have them write down what they have learned about heating and cooling systems.
LS Verbal

Connection to
SOCIAL STUDIES

In honor of James Watt's contribution, the metric unit of power was named the watt (W). One watt is one joule per second.

Making the Connection

1. Answers will vary but may include: steam-driven turbines in power plants, WWII boats and submarines, nuclear reactors, trains, and vegetable steamers.

2. radiators
LS Verbal

Teaching Tip — BASIC

Keeping Warm Discuss with students their own experiences in keeping themselves warm in cold weather. Discussion topics could include the following: Why are a hat and gloves especially important in cold weather? Why is layered clothing useful? What happens to the body when you exercise in cold weather? Encourage interested students to research these and related topics in more detail.
LS Intrapersonal

Connection to
SOCIAL STUDIES

In 1769, a Scottish engineer named James Watt patented a new design that made steam engines more efficient. During the next 50 years, the improved steam engines were used to power trains and ships. Previously, transportation had depended on the work done by horses or the wind.

Watt's new steam engines were used in machines and factories of the industrial revolution. In 1784, Watt used steam coils to heat his office. This was the first practical use of steam for heating.

Making the Connection

1. Old steam-powered riverboats are popular tourist attractions in many cities. Make a list of at least three other instances in which the energy in steam is used for practical purposes.

2. What devices in older buildings function like the steam coils Watt used for heating his office?

☑ internet connect

www.scilinks.org
Topic: Heating and
Cooling Systems
SciLinks code: HK4067

SCLINKS Maintained by the
National Science
Teachers Association

436

Did You Know?

Detecting Energy Electromagnetic radiation emitted by the human body cannot be seen by humans because the radiation's wavelengths are primarily in the infrared portion of the electromagnetic spectrum. However, this radiation can be seen by some snakes. By detecting infrared radiation, these snakes seek out their prey in what appears to humans to be complete darkness. A similar concept is used in some types of night-vision goggles.

Heating Systems

People generally feel and work their best when the temperature of the air around them is in the range of 21°C–25°C (70°F–77°F). To raise the indoor temperature on colder days, energy must be transferred into a room's air by a *heating system.* Most heating systems use a source of energy to raise the temperature of a substance such as air or water.

Work can increase average kinetic energy

When you rub your hands together, they become warmer. The energy you transfer to your hands by work is transferred to the molecules of your hands, and their temperature increases. Processes that involve energy transfer by work are called mechanical processes.

Another example of a mechanical heating process is a device used in the past by certain American Indian tribes to start fires. The device consists of a bow with a loop in the bowstring that holds a pointed stick. The sharp end of the stick is placed in a small indentation in a stone. A small pile of wood shavings is then put around the place where the stick and stone make contact. A person then does work to move the bow back and forth. This energy is transferred to the stick, which turns rapidly. The friction between the stick and stone causes the temperature to rise until the shavings are set on fire.

Some of the energy from food is transferred as heat to blood moving throughout the human body

You may not think of yourself as a heating system. But unless you are sick, your body maintains a temperature of about 37°C (98.6°F), whether you are in a place that is cool or hot. Maintaining this temperature in cool air requires your body to function like a heating system.

If you are surrounded by cold air, energy will be transferred as heat from your skin to the air, and the temperature of your skin will drop. To compensate, stored nutrients are broken down by your body to provide energy, and this energy is transferred as heat to your blood. The warm blood circulates through your body, transferring energy as heat to your skin and increasing your skin's temperature. In this way your body can maintain a constant temperature.

LANGUAGE-ARTS
CONNECTION — ADVANCED

Since ancient times, fire has stirred the imagination of people all over the world. Many myths and legends about fire have originated in various civilizations. Have your students interview their parents, grandparents, or other friends or relatives to learn about any traditional stories of fire. After students have conducted these interviews, ask them to share the stories with the class. Then have students write their own legend describing the origin of fire and how it came to be used on Earth.
LS Verbal

Heated water or air transfers energy as heat in central heating systems

Most modern homes and large buildings have a central heating system. As is the case with your body, when the building is surrounded by cold air, energy is transferred as heat from the building to the outside air. The temperature of the building begins to drop.

A central heating system has a furnace that burns coal, fuel oil, or natural gas. The energy released in the furnace is transferred as heat to water, steam, or air, as shown in ***Figure 13.*** The steam, hot water, or hot air is then moved to each room through pipes or ducts. Because the temperature of the pipe is higher than that of the air, energy is transferred as heat to the air in the room.

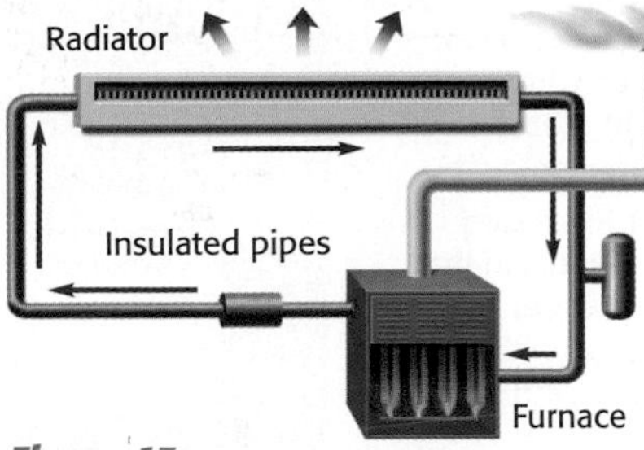

Figure 13
Hot-water, steam, and hot-air systems heat buildings by circulating heated fluids to each room.

Solar heating systems also use warmed air or water

Cold-blooded animals, such as lizards and turtles, increase their body temperature by using external sources, such as the sun. You may have seen these animals sitting motionless on rocks on sunny days, as shown in ***Figure 14.*** During such behavior, called basking, energy is absorbed by the reptile's skin through conduction from the warmer air and rocks and by radiation from sunlight. This absorbed energy is then transferred as heat to the reptile's blood. As the blood circulates, it transfers this energy to all parts of the reptile's body.

Solar heating systems, such as the one illustrated in ***Figure 15,*** use an approach similar to that of a basking reptile. A solar collector uses panels to gather energy radiated from the sun. This energy is used to heat water. The hot water is then moved throughout the house by the same methods other hot-water systems use.

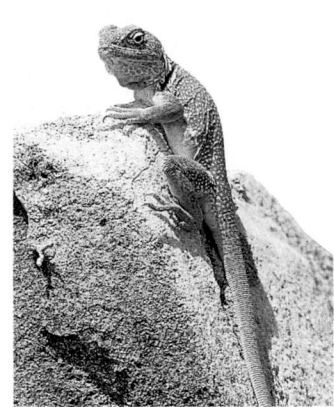

Figure 14
Reptiles bask in the sun to raise their body temperature.

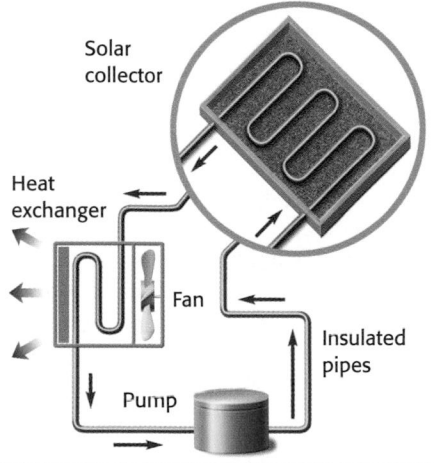

Figure 15
An active solar heating system moves solar-heated water through pipes and a heat exchanger.

437

SKILL BUILDER — BASIC

Graphing In the United States, 55% of central heating systems are powered by natural gas; 26% by electricity; 11% by fuel oil; and the rest by other sources. Have students draw a pie graph that displays this data. **LS Logical**

Teaching Tip

Energy Efficiency Discuss with students the importance of efficient heating and cooling systems in both economic and environmental terms. Tell students that in an average household, 44% of the money spent on utilities goes toward heating and cooling the home. Point out that in addition to an energy-efficient system, energy use can also be reduced by having good insulation in the walls and ducts. For example, leaks in duct systems can cost hundreds of dollars a year. Temperature regulation can also make a significant difference: setting back a thermostat 10% to 15% for 8 hours each night in winter can reduce the year's energy bills by 10%.

SKILL BUILDER — BASIC

Interpreting Visuals Have students examine **Figure 15.** Discuss the various energy transfers that are occurring. Have students work in pairs to describe what would happen on a cold, cloudy day when the temperature inside the house is higher than the temperature outside the house. Have them brainstorm ways to prevent energy loss. **LS Interpersonal**

Trends in Technology

Heating From the Ground Up As the earliest cave dwellers knew, a good way to stay warm in the winter is to go underground. Now scientists and engineers are using the same concept to heat aboveground homes for a fraction of the cost of conventional air-conditioning and heating systems.

Although the average specific heat capacity of Earth has a smaller value than the specific heat capacity of air, Earth has a greater density. That means there are more kilograms of earth than there are of air near a house and that a 1°C change in temperature involves transferring more energy to or from the ground than to or from the air. Thus in wintertime, the ground will probably have a higher temperature than the air above it, while in the summer the ground will likely have a lower temperature than the air. An earth-coupled heat pump enables homeowners to tap Earth's below-ground temperature to heat their homes in the winter or cool them in the summer.

Chapter Resource File

- **Cross-Disciplinary Worksheet** Connection to Social Studies—Early Central Heating GENERAL

Teach, *continued*

Group Activity —— **ADVANCED**

Solar Houses Divide students into groups. Have each group learn more about the design principles involved in solar heating, such as building orientation, placement of windows, shading and overhangs, thermal mass materials, and the use of wing walls. You may wish to assign a different topic to each group. Instruct groups to prepare a brief presentation to share their information with the class.

As an extension, you could have students use the information compiled by all of the groups to make a basic design for a solar house.

LS Interpersonal

Teaching Tip

Loss of Useful Energy Point out to students that transferring energy is like pouring water from one cup into another, and then another, etc. Each time you pour into a new cup, some leftover water remains in the "empty" cup. Even if you hold the cup upside down and wait, some of the water still remains in the "empty" cup. Each time you pour water into a cup, the amount of water will decrease.

Figure 16

A In a passive solar heating system, energy from sunlight is absorbed in a rooftop panel.

B Pipes carry the hot fluid that exchanges heat energy with the air in each room.

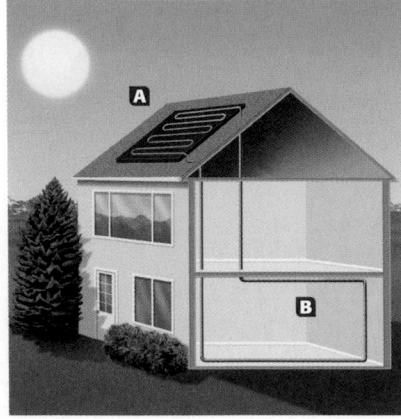

internet connect

www.scilinks.org
Topic: Conduction, Convection, and Radiation
SciLinks code: HK4027

SCI**LINKS** Maintained by the National Science Teachers Association

The warm water can also be pumped through a device called a heat exchanger, which transfers energy from the water to a mass of air by conduction and radiation. The warmed air is then blown through ducts as with other warm-air heating systems.

Both of these types of solar heating systems are called active solar heating systems. They require extra energy from another source, such as electricity, in order to move the heated water or air around.

Passive solar heating systems, as shown in **Figure 16,** require no extra energy to move the hot fluids through the pipe. In this type of system, energy transfer is accomplished by radiation and convection currents created in heated water or air. In warm, sunny climates, passive solar heating systems are easy to construct and maintain and are clean and inexpensive to operate.

Usable energy decreases in all energy transfers

When energy can be easily transformed and transferred to accomplish a task, such as heating a room, we say that the energy is in a usable form. After this transfer, the same amount of energy is present, according to the law of conservation of energy. Yet less of it is in a form that can be used.

The energy used to increase the temperature of the water in a hot-water tank should ideally stay in the hot water. However, it is impossible to keep some energy from being transferred as heat to parts of the hot-water tank and its surroundings. The amount of usable energy decreases even in the most efficient heating systems.

Due to conduction and radiation, some energy is lost to the tank's surroundings, such as the air and nearby walls. Cold water in the pipes that feed into the water heater also draws energy from some of the hot water in the tank. When energy from electricity is used to heat water in the hot-water heater, some of the energy is used to increase the temperature of the electrical wire, the metal cover of the water heater, and the air around the water heater. All of these portions of the total energy can no longer be used to heat the water. Therefore, that energy is no longer in a usable form. In general, the amount of usable energy always decreases whenever energy is transferred or transformed.

438

HISTORY
CONNECTION —— **ADVANCED**

Have students research the history of a modern heating or cooling system, such as air conditioning or refrigerators. Instruct them to draw a timeline of the major events in the development of their chosen system. Then have students find out what other historical events were happening around the same times, and ask them to include these events in their timeline for context. **LS** Verbal

LANGUAGE-ARTS
CONNECTION —— **BASIC**

Ask students to imagine what life would be like without the heating or cooling system they researched for their *History* timeline (left). Have them write a short story set in a time before their chosen system was invented, or before it was in common use. For example, a student who researched the history of air conditioning could write about a family visiting a movie theater with air conditioning for the first time. **LS** Verbal

Insulation minimizes undesirable energy transfers

During winter, some of the energy from the warm air inside a building is lost to the cold outside air. Similarly, during the summer, energy from warm air outside seeps into an air-conditioned building, raising the temperature of the cool inside air. Good insulation can reduce, but not entirely eliminate, the unwanted transfer of energy to and from the building's surroundings. As shown in *Figure 17,* insulation material is placed in the walls and attics of homes and other buildings to reduce the unwanted transfer of energy as heat.

A standard rating system has been developed to measure the effectiveness of insulation materials. This rating, called the *R-value,* is determined by the type of material used and the material's thickness. *R*-values for several common building and insulating materials of a given thickness are listed in *Table 2.* The greater the *R*-value, the greater the material's ability to decrease unwanted energy transfers.

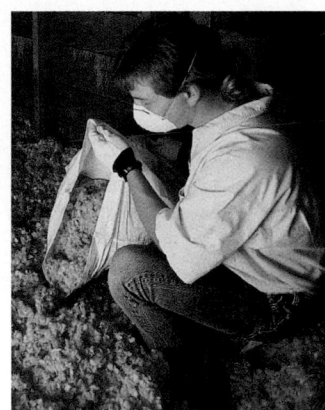

Figure 17

Insulating materials, such as fiberglass and cellulose, are used in most buildings to reduce the loss of heat energy.

Cooling Systems

If you quickly let the air out of a compressed-air tank like the one used by scuba divers, the air from the tank and the tank's nozzle feel slightly cooler than they did before the air was released. This is because the molecules in the air lose some of their kinetic energy as the air's pressure and volume change and the temperature of the air decreases. This process is a simple example of a *cooling system.* In all cooling systems, energy is transferred as heat from one substance to another, leaving the first substance with less energy and thus a lower temperature.

Table 2 **R-Values for Some Common Building Materials**

Substance	*R*-value
Drywall, 1.3 cm (0.50 in.)	0.45
Wood shingles, (overlapping)	0.87
Flat glass, 0.318 cm (0.125 in.)	0.89
Hardwood siding, 2.54 cm (1.00 in.)	0.91
Vertical air space, 8.9 cm (3.5 in.)	1.01
Insulating glass, 0.64 cm (0.25 in.)	1.54
Cellulose fiber, 2.54 cm (1.00 in.)	3.70
Brick, 10.2 cm (4.00 in.)	4.00
Fiberglass batting, 8.9 cm (3.5 in.)	10.90

439

BIOLOGY Elephants flap their ears to increase the air circulation and thus increase the energy transfer from the blood vessels in their ears to the air.

SKILL BUILDER — **BASIC**

Reading Skills Have students read the section on refrigerators. Ask students to brainstorm reasons why leaving the refrigerator door open will not cool down the house. Be sure to guide the discussion to the correct explanation.

The refrigerator removes energy from the air inside the refrigerator and then releases that energy to the air outside the refrigerator (in the house). If the refrigerator did nothing else, the house would still not cool because the energy removed from the air in the refrigerator is released into the air in the house. The refrigerator transfers extra energy from the motor to the air in the house. The result of leaving the refrigerator door open would be to heat up the house! **LS** **Verbal**

▶ **refrigerant** a material used to cool an area or an object to a temperature that is lower than the temperature of the environment

INTEGRATING

BIOLOGY
In hot regions, the ears of many mammals serve as cooling systems. Larger ears provide more area for energy to be transferred from blood to the surrounding air, helping the animals to maintain their body temperature. Rabbits and foxes that live in the desert have much longer ears than rabbits and foxes that live in temperate or arctic climates.

Cooling systems often use evaporation to transfer energy from their surroundings

In the case of a refrigerator, the temperature of the air and food inside is lowered. But because the first law of thermodynamics requires energy to be conserved, the energy inside the refrigerator must be transferred to the air outside the refrigerator. If you place your hand near the rear or base of a refrigerator, you will feel warm air being discharged. Much of the energy in this air was removed from inside the refrigerator.

Hidden in the back wall of a refrigerator is a set of coiled pipes through which a substance called a **refrigerant** flows, as shown in *Figure 18*. During each operating cycle of the refrigerator, the refrigerant evaporates into a gas and then condenses back into a liquid.

Recall from the beginning of this section that evaporation produces a cooling effect. Changes of state always involve the transfer of relatively large amounts of energy. In liquids that are good refrigerants, evaporation occurs at a much lower temperature than that of the air inside the refrigerator. When the liquid refrigerant is in a set of pipes near the inside of the refrigerator, heat energy is transferred from the air to the refrigerant. This exchange causes the air and food to cool.

Figure 18

A Liquid refrigerant flowing through the pipes inside a refrigerator cools the compartment by evaporation.

B Energy is removed by the outside coils as the warmed refrigerant vapor cools and condenses back into a liquid.

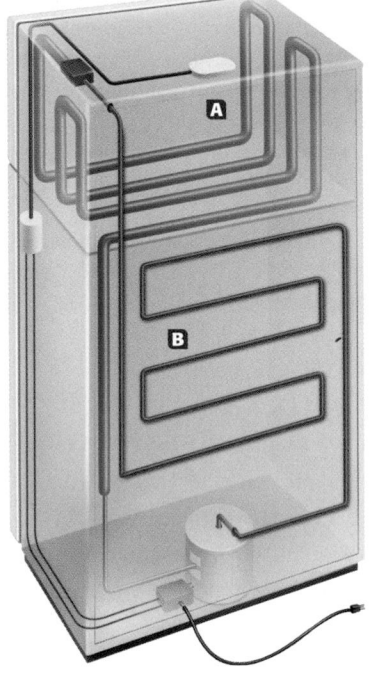

ENVIRONMENTAL SCIENCE
CONNECTION

When refrigerators were first developed, many toxic gases were used as refrigerants, including ammonia, methyl chloride, and sulfur dioxide. In the 1920s, methyl chloride leaks caused a number of fatal accidents. This initiated a search for safer refrigerants that resulted in the discovery of Freon, a chlorofluorocarbon (CFC). Soon Freon was used as a refrigerant in all household refrigerators.

Today, we know that CFCs such as Freon pose another danger—they deplete the ozone layer, which shields us from some of the sun's ultraviolet radiation. As a result, the Environmental Protection Agency carefully monitors and regulates the use of CFCs. Their goal is to eventually phase out the use of all CFCs, replacing them with safer alternatives.

Condensation transfers energy to the surroundings

The refrigerant has become a gas by absorbing energy. This gas moves to the section of coils outside the refrigerator, where electrical energy is used to power a compressor. Pressure is used to condense the refrigerant back into a liquid. Because condensation involves transferring heat energy from the vapor, the temperature of the air outside the refrigerator increases. This explains why the outside coils stay warm.

Air-conditioning systems in homes and buildings use the same process that refrigerators use. As air near the evaporation coils is cooled, a fan blows this air through ducts into the rooms and hallways. Convection currents in the room then allow the cool air to circulate as displaced warmer air flows into return ducts.

Heat pumps can transfer energy to or from rooms

Heat pumps use the evaporation and condensation of a refrigerant to provide heating in the winter and cooling in the summer. A heat pump is a refrigeration unit in which the cooling cycle can be reversed.

As shown in *Figure 19A,* the liquid refrigerant travels through the outdoor coils during the winter and absorbs enough energy from the outside air to evaporate. Work is done on the gas by a compressor, increasing the refrigerant's energy. Then the refrigerant moves through the coils inside the house, as shown in *Figure 19B.* The hot gas transfers heat energy to the air inside the house. This process warms the air while cooling the refrigerant gas enough for it to condense back into a liquid.

In the summer, the refrigerant is pumped in the opposite direction, so that the heat pump functions like a refrigerator or an air conditioner. The liquid refrigerant absorbs energy from the air inside the house as it evaporates. The hot refrigerant gas is then moved to the coils, which are outside the house. The refrigerant then condenses, transferring energy as heat to the outside air.

Figure 19

A Liquid refrigerant evaporates in the outdoor coils as energy is transferred from the air.

B The hot refrigerant gas moves through the coils into the indoor portion of the pump, where the refrigerant condenses back into a liquid and transfers energy as heat into the room.

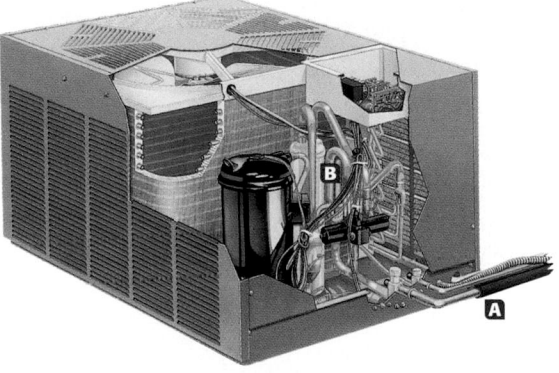

Teaching Tip

It may help the students as they learn about internal combustion engines to have a little more explanation about the parts of the engine discussed.

Crank shaft The crank shaft turns the piston's up-and-down motion into circular motion.

Connecting rod The connecting rod connects the piston to the crankshaft. It can rotate at both ends so that its angle can change as the piston moves and the crankshaft rotates.

Spark plug The spark plug supplies the spark that ignites the air-fuel mixture so that combustion can occur.

Valves The intake and exhaust valves open at the proper time to let in air and fuel and to let out exhaust. Both valves are closed during compression and combustion so that the combustion chamber is sealed.

Piston A piston is a piece of metal shaped like the cylinder but smaller in diameter than the cylinder. It moves up and down inside the cylinder.

Combustion chamber The combustion chamber is the area where compression and combustion take place. As the piston moves up and down, the size of the combustion chamber changes.

■ **heat engine** a machine that transforms heat into mechanical energy, or work

Heat Engines

Heat engines convert potential chemical energy and internal kinetic energy to mechanical energy by using the process of combustion. The two main types of heat engines—internal combustion engines and external combustion engines—are named for where combustion takes place (inside the engine or outside the engine). Examples of internal engines are the engines in cars and trucks. An example of an external engine is a steam engine.

Internal combustion engines burn fuel inside the engine

In an internal combustion engine, fuel burns in cylinders within the engine. There are pistons inside the cylinders, as shown in *Figure 20*. Up and down movements, or strokes, of the pistons cause the crankshaft to turn. The motion of the crankshaft is transferred to the wheels of the car or truck, for example.

An automobile engine is a four-stroke engine, because four strokes take place for each cycle of the piston. The four strokes are called *intake, compression, power,* and *exhaust* strokes.

Figure 21 illustrates the four-stroke cycle of the pistons in an engine with a carburetor. A *carburetor* is another part of the engine, in which gasoline liquid becomes vaporized.

Some engines have fuel injectors instead of carburetors. In fuel-injected engines, only air enters the cylinder during the intake stroke. During the compression stroke, fuel vapor is injected directly into the compressed air in the cylinder. The other steps are the same as in an engine with a carburetor.

Figure 20

The pistons move within the cylinders of the four-stroke engine to turn the crankshaft, which transfers motion to the wheels of the car or truck.

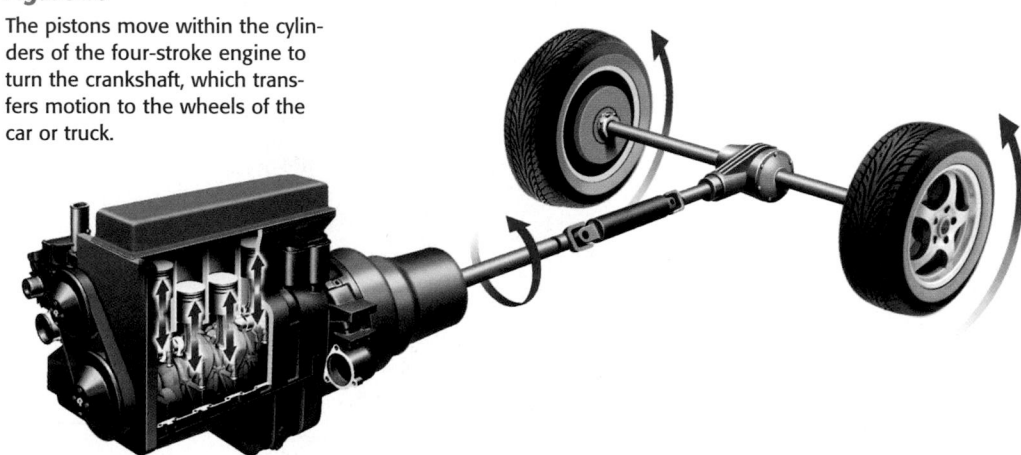

442

Historical Perspective

Benjamin Thompson, Count Rumford

Benjamin Thompson, Count Rumford (1753–1814) was a physicist who made an early connection between work and heat. At the time, most scientists believed that heat was a fluid called *caloric*. While supervising cannon boring at a military workshop in Munich, Rumford noticed that the boring produced a great deal of heat in a brass gun and its metal shavings. He conducted a number of experiments with the cannon borer. He found that the friction between two metallic surfaces produced a constant stream of heat in all

directions without any sign of lessening, even over a long period of time. His observations could not be explained by the caloric theory and, after eliminating all other variables, Rumford came to the conclusion that heat is simply a form of motion.

It took some time for Rumford's idea of heat to gain acceptance. Rumford also made many practical innovations during his lifetime, including central heating, thermal underwear, the smokeless chimney, the kitchen oven, and the pressure cooker.

Not all internal combustion engines work alike

Diesel engines are also internal combustion engines, but they work differently. A diesel engine has no spark plugs. Instead, the fuel-air mixture is compressed so much that it becomes hot enough to ignite without a spark from a spark plug.

In an internal combustion engine, only part of the potential chemical energy is converted to mechanical energy. As engine parts move, friction and other forces cause much of the energy to be lost to the atmosphere as heat. In fact, an internal combustion engine becomes so hot that a cooling system is used to cool the engine.

Internal combustion engines vary in number of pistons

Most motorcycle engines have two cylinders. Automobile engines usually have four, six, or eight cylinders. Because of the four-stroke cycle, a four-piston engine can run efficiently with each piston at a different stroke of the cycle. However, engines with six or eight cylinders have more power than four-piston engines.

Figure 21

A In the *intake* stroke, a mixture of fuel vapor and air is brought into the cylinder from the carburetor as the piston moves downward.

B In the *compression* stroke, the piston moves up and compresses the fuel-air mixture.

C At the beginning of the *power* stroke, a spark from the spark plug ignites the compressed mixture and causes the mixture to expand quickly and move the piston down to turn the crankshaft.

D The *exhaust* stroke takes place when the piston moves up again and forces the waste products to move out of the exhaust valve.

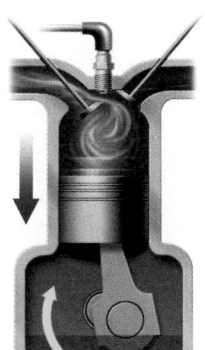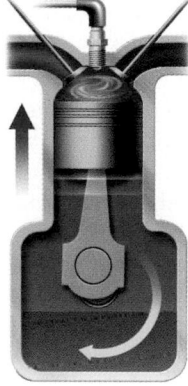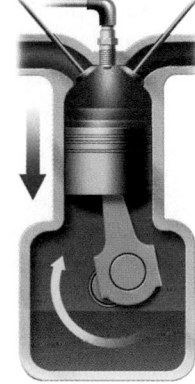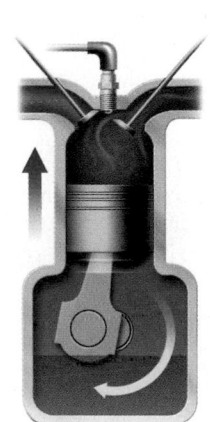

Did You Know ?

Automobiles, jet aircrafts, and even rockets all are powered by internal combustion engines operated under high pressures.

Chapter 13 • Heat and Temperature **443**

 **APPLICATIONS**

Buying Appliances

Applying Information

1. (4939 kW•h/year)(1 year/ 365 days)(1 day/24 h) = 0.5638 kW•h/h

2. ($415/year)(1 year/365 days) = $1.14/day

LS Logical

Close

Quiz

1. When energy is transferred between two objects, what happens to the amount of usable energy? (It decreases.)

2. What two energy sources are often used for heating systems? (fuel-burning furnaces, sunlight)

3. What do refrigerators and air conditioners use for cooling? (the evaporation of a refrigerant)

Chapter Resource File

• **Cross-Disciplinary Worksheet** Real World Applications—Appliance Energy Use and Cost **GENERAL**

• **Concept Review** **GENERAL**

• **Quiz** **BASIC**

REAL WORLD APPLICATIONS

Buying Appliances Most major appliances, including those that involve the transfer of energy as heat, are required by law to have an *Energyguide* label attached to them.

The label indicates the average amount of energy used by the appliance in a year. It also gives the average cost of using the appliance based on a national average of cost per energy unit.

The *Energyguide* label provides consumers a way to compare various brands and models of appliances.

Applying Information

1. Use the *Energyguide* label shown to find how much energy the appliance uses each hour.

2. What is the daily operating cost of the appliance?

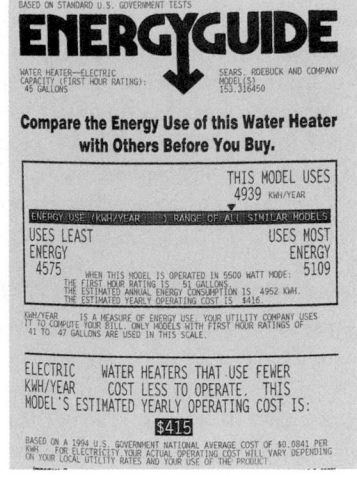

ENERGYGUIDE

Compare the Energy Use of this Water Heater with Others Before You Buy.

SECTION 3 REVIEW

SUMMARY

▶ Heating and cooling systems regulate temperature by transferring energy.

▶ Usable energy decreases during any process in which energy is transferred.

▶ The total amount of energy, both usable and unusable, is constant in any process.

▶ In heating systems, energy is transferred to a fluid, which then transfers its energy to the air in rooms.

▶ Refrigerators and air conditioners use the evaporation of a refrigerant for cooling.

▶ Heat engines use heat to do work.

1. **Explain** how evaporation is a cooling process.

2. **List** one type of home heating system, and describe how it transfers energy to warm the air inside the rooms.

3. **Describe** how energy changes from a usable form to a less usable form in a building's heating system.

4. **Compare** the advantages and disadvantages of using a solar heating system in your geographical area.

5. **Search** the Internet to find information on how *R*-values of insulation affect the environment.

6. **Critical Thinking** Water has a high specific heat, meaning it takes a good deal of energy to raise its temperature. For this reason, the cost of heating water may be a large part of a monthly household energy bill. Describe two ways the people in your household could change their routines, without sacrificing results, in order to save money and energy by using less hot water.

7. **Critical Thinking** Draw and describe each of the strokes of an automobile engine. Explain how the spark-plug ignition of compressed gas results in work done by the engine.

444

Answers to Section 3 Review

1. When a liquid changes state to become a gas, energy must be absorbed. This absorption cools the system.

2. A natural gas heater uses heat from burning natural gas to transfer energy as heat to the air, which is then circulated throughout the house.

3. The warm exhaust from the natural gas heater is the result of taking usable energy (in the natural gas), using some of it, then releasing the less usable form (exhaust).

4. Answers will depend on your area, but some possibilities are: advantages—free energy from sun, less pollution; disadvantages—not much sun available, expensive to set up, difficult to install

5. Answers may vary. Accept all reasonable responses.

6. Accept all reasonable responses.

7. Drawings should resemble **Figure 21.** The spark plug ignition uses electrical energy to convert the chemical energy of gas into mechanical energy, which is also called work.

Math Skills

Order of Operations

A plate with a temperature of 95.0°C is placed in a vat of water with a temperature of 26.0°C. The equilibrium temperature of the plate and water is 28.2°C. The mass of the plate is 1.5 kg, and the mass of the water is 3.0 kg. What is the plate's specific heat? To calculate this, first calculate the energy transferred as heat to the water. Then use energy conservation, and rearrange the equation to calculate the plate's specific heat.

1 **List all the given and unknown values.**

Use this step to perform the first operation, which is calculating the temperature change.

Given: temperature change of plate (Δt_{plate}) = 95.0°C − 28.2°C
$= \Delta66.8°C = \Delta66.8$ K
temperature change of water (Δt_{water}) = 28.2°C − 26.0°C
$= \Delta2.2°C = \Delta2.2$ K
mass of plate (m_{plate}) = 1.5 kg
mass of water (m_{water}) = 3.0 kg
specific heat of water (c_{water}) = 4186 J/kg·K

Unknown: specific heat of plate (c_{plate}) (J/kg·K)

2 **Write down the specific heat equation, and then rearrange it to calculate the specific heat of the plate.**

$$energy = cm\Delta t = c_{water}m_{water}\Delta t_{water}$$

$$c_{plate} = \frac{energy}{m_{plate}\Delta t_{plate}}$$

3 **Solve for energy, and then calculate the specific heat of the plate.**

$$energy = \left(\frac{4186 \, J}{kg \cdot K}\right) \times (3.0 \, kg) \times (2.2 \, K) = 2.8 \times 10^4 \, J$$

$$c_{plate} = \frac{2.8 \times 10^4 \, J}{(1.5 \, kg) \times (\Delta66.8 \, K)} = \frac{2.8 \times 10^4 \, J}{1.0 \times 10^2 \, kg \cdot K} = 280 \, J/kg \cdot K$$

Practice

Follow the example above to calculate the following:

1. Suppose in the example problem that the water's initial temperature was 29.0°C and that the equilibrium temperature of the plate and water was 35.0°C. Assuming that the plate's properties are the same as those in the example, what would the mass of the water be?

Teaching Tip

Energy Conservation Because the energy transferred as heat is not specifically called for in solving the problem, students can use the principle of energy conservation to equate $c_{plate}m_{plate}\Delta t_{plate}$ with $c_{water}m_{water}\Delta t_{water}$. This not only simplifies calculations, but also makes use of the important property that the energy leaving the plate must be transferred to other matter, in this case the water.

Answer

1. 1.0 kg

SKILL BUILDER

Temperature Conversions
Remind students that, when they calculate the temperature difference, that the magnitude of a degree Celsius is identical to the magnitude of a kelvin. Thus, the difference between two Celsius temperatures can be expressed in kelvins, and vice versa. Have students convince themselves of this by taking the difference between two Celsius temperatures (for instance, 45°C and 14°C). Then have them express these in kelvins using the Celsius-Kelvin Conversion Formula (45 + 273 and 14 + 273). By taking the difference, the conversion factors 273 cancel out, so that $\Delta T = 31°C = \Delta t = 31$ K.

Understanding Concepts

1. c
2. b
3. b
4. b
5. a
6. d
7. c
8. d
9. d
10. b
11. c
12. d
13. a
14. b

Using Vocabulary

15. An object at 0 K cannot transfer energy to its surroundings because 0 K is absolute zero, which is the temperature at which the molecules of an object have no transferrable energy. An object at 0°C (273 K) has particles with a much higher average kinetic energy and can transfer that energy to its surroundings.

16. The difference would be how the thermometer is numbered because $T = t + 273$.

Chapter Resource File

- **Test Item Listing for ExamView®**
  Test Generator
- **Chapter Tests** GENERAL

Chapter Highlights

Before you begin, review the summaries of the key ideas of each section, found at the end of each section. The vocabulary terms are listed on the first page of each section.

UNDERSTANDING CONCEPTS

1. Temperature is proportional to the average kinetic energy of particles in an object. Thus an increase in temperature results in a(n)
   a. increase in mass.
   b. decrease in average kinetic energy.
   c. increase in average kinetic energy.
   d. decrease in mass.

2. As measured on the Celsius scale, the temperature at which ice melts is
   a. −27°C.   c. 32°C.
   b. 0°C.   d. 100°C.

3. As measured on the Fahrenheit scale, the temperature at which water boils is
   a. 32°F.   c. 100°F.
   b. 212°F.   d. 451°F.

4. The temperature at which the particles of a substance have no more kinetic energy to transfer is
   a. −273 K.   c. 0°C.
   b. 0 K.   d. 273 K.

5. The type of energy transfer that takes place between objects in direct contact is
   a. conduction.   c. contraction.
   b. convection.   d. radiation.

6. Which type of energy transfer can occur in empty space?
   a. convection   c. conduction
   b. contraction   d. radiation

7. An *R*-value is a rating for materials used as
   a. conduction.   c. insulation.
   b. convection.   d. condensation.

---

8. Campfires transfer energy as heat to their surroundings by methods of
   a. convection and conduction.
   b. convection and radiation.
   c. conduction and radiation.
   d. convection, conduction, and radiation.

9. Which of the following would be an example of a very good conductor of heat energy?
   a. liquid   c. air
   b. wood   d. metal

10. Which of the following would be an example of a very good insulator?
    a. metal   c. wood
    b. air   d. liquid

11. The amount of energy required to raise the temperature of 1 kg of a substance by 1 K is determined by its
    a. *R*-value.
    b. usable energy.
    c. specific heat.
    d. convection current.

12. The amount of usable energy decreases when
    a. systems are used only for heating.
    b. systems are used only for cooling.
    c. systems are used for heating or cooling.
    d. the heating or cooling system's design allows loss of heat energy.

13. A refrigerant in a cooling system cools the surrounding air
    a. as it evaporates.
    b. as it condenses.
    c. both as it evaporates and as it condenses.
    d. when it neither evaporates nor condenses.

14. Solar heating systems are classified as
    a. positive and negative.
    b. active and passive.
    c. AC and DC.
    d. active and indirect.

---

17. Water can transfer energy by conduction when an object is placed in the water. For example, a raw egg placed in 100°C water will absorb energy by conduction and become a boiled egg. Water can transfer energy by convection as it flows around an object. For example, a warm can of soda in a river transfers energy to the cold water, making the water warm. Then the warm water moves away, and more cold water surrounds the can of soda.

18. During the day, the hot desert sand raises the temperature of the surrounding air, which rises up along the mountain wall. As the air rises up, it loses energy through conduction. After reaching the top, this air, having

cooled, becomes dense and sinks down along the opposite wall of the mountain, creating a downdraft.

19. Metal is a conductor with a low specific heat, so pouring a hot beverage in a metal cup would make the cup hot also. This cup would be unpleasant to hold. A china cup is a better insulator with a higher specific heat, so it will not become hot as rapidly.

20. The dark clothing absorbs energy transferred from the sun by radiation.

21. The ammonia would evaporate very easily in a room, absorbing energy from the room. Even a cold room is warmer than −33.4°C, so the ammonia would still evaporate, taking

15. Use the concepts of average particle kinetic energy, *temperature,* and *absolute zero* to predict whether an object at 0°C or an object at 0 K will transfer more energy as heat to its surroundings.

16. How would a *thermometer* that measures temperatures using the Kelvin scale differ from one that measures temperatures using the Celsius scale?

17. Explain how water can transfer energy by *conduction* and by *convection.*

18. Explain how *convection currents* form updrafts near tall mountain ranges along deserts, as shown in the figure below.

19. Use the differences between a *conductor* and an *insulator* and the concept of *specific heat* to explain whether you would rather drink a hot beverage from a metal cup or from a china cup.

20. If you wear dark clothing on a sunny day, the clothing will become hot after a while. Use the concept of *radiation* to explain this.

21. Explain why ammonia, which has a boiling point of –33.4°C, is sometimes used as a *refrigerant* in a cooling system. Why would ammonia be less effective in a heating system?

22. Describe how a *heat engine* works, including the four strokes of the heat-engine cycle.

23. Temperature Scale Conversion A piece of dry ice, solid CO_2, has a temperature of –100°C. What is its temperature in kelvins and in degrees Fahrenheit?

24. Temperature Scale Conversion The temperature in deep space is thought to be around 3 K. What is 3 K in degrees Celsius? in degrees Fahrenheit?

25. Specific Heat How much energy is needed to raise the temperature of a silver necklace chain with a mass of 22.5 g from room temperature, 25°C, to body temperature, 37°C? (**Hint:** Refer to **Table 1** on p. 432)

26. Specific Heat How much energy would be absorbed by 550 g of copper when it is heated from 24°C to 45°C? (**Hint:** Refer to **Table 1** on p. 432.)

27. Interpreting Graphics Graph the Celsius-Fahrenheit conversion equation, plotting Celsius temperature along the *x*-axis and Fahrenheit temperature on the *y*-axis. Use an *x*-axis range from –100°C to 100°C, then use the graph to find the following values:
  a. the Fahrenheit temperature equal to 77°C
  b. the Fahrenheit temperature equal to –40°C
  c. the Celsius temperature equal to 23°F
  d. the Celsius temperature equal to –17°F

energy from the room rather than adding energy to the room.

22. Answers should contain information similar to the text in the spread about heat engines.

Building Math Skills

23. –100 + 273 = 173 K
1.8 (–100.) + 32.0 = –148°F

24. 3 – 273 = –270°C
1.8 (–270) + 32.0 = –454°F

25. (234 J/kg•K)(0.0225 kg)
(12 K) = 63 J

26. (385 J/kg•K)(0.55 kg)
(21 K) = 4400 J

Building Graphing Skills

27. a. 170°F
   b. –40°F
   c. –5°C
   d. –27°C

Thinking Critically

28. Answers should include the fact that the liquids in the thermometer expand as they take on more energy and, therefore, increase in temperature.

29. Answers should include the facts that two metals are bound together to form a coil, that each metal has a different rate of expansion and contraction, and that the resultant bending moves a needle that indicates the temperature.

30. Answers should describe the increasing particle movement and resulting temperature increases as energy is added.

31. No net transfer of energy; the amount leaving an object will equal the amount entering the object.

32. Their temperatures will be closer together than they were at the start.

33. Answers may vary in complexity but should mention taking in both heat and chemical energy.

34. Answers may include the fact that the materials vary in specific heat or may mention conductors and insulators, both of which are part of the same concept.

35. The metal is a better conductor than the carpet, so energy flows more easily away from you.

36. to let hot air escape

Assignment Guide	
Section	**Questions**
1	1–4, 15–16, 23–24, 27–31
2	5–6, 8–11, 17–20, 25–26, 32, 34–39, 48
3	7, 12–14, 21–22, 33, 40–47

447

37. The metal spoon is a good conductor, so energy flows into the spoon. The spoon transfers energy to the surrounding air.

38. The crust is an insulator, whereas the filling is a conductor, so energy flows easily out of the filling.

39. The thicker glass is a better insulator than thin glass and the layer of air (or other gas) is an even better insulator.

40. The air conditioner must exhaust some energy. If the air conditioner were contained entirely within the room, the exhaust energy would also be in the room, heating the room up.

41. Water would be better because it has a higher specific heat. For a certain amount of water or ethanol, the water would absorb more energy from the engine per degree of temperature change.

42. In good refrigerants, evaporation occurs at a lower temperature than that of the air inside the refrigerator. Heat energy is transferred from the air to the refrigerant. (Answers should agree with the text.)

Developing Life/Work Skills

43. Statement (b) is correct. Energy can never be lost; it is always conserved.

44. Answers may vary.

45. Answers may vary.

46. Answers may vary.

47. Many websites and encyclopedias have articles about Benjamin Thompson.

28. Applying Knowledge Explain how the common thermometer works by expansion. What expands, and how does that expansion indicate the temperature?

29. Critical Thinking Describe and illustrate how you think that a thermometer might indicate temperature when the thermometer uses the bending of a strip made from two metals that expand at different rates.

30. Creative Thinking Imagine the particles within an object that is receiving energy as heat. Describe what is happening to individual particles and how what's happening would be related to temperature.

31. Applying Knowledge If two objects that have the same temperature come into contact with each other, what can you say about the amount of energy that will be transferred between them as heat?

32. Applying Knowledge If two objects that have different temperatures come into contact with each other, what can you say about their temperatures after several minutes of contact?

33. Critical Thinking Search the Internet to find two types of heat engines, and answer the following questions about each of them. Does the engine take in heat energy and convert it to mechanical energy? Or, does the engine take in another form of energy and convert it to mechanical energy? If it was another form, what type of energy is taken in, and why do you think the engines are called heat engines?

34. Critical Thinking When you get into your car on a very hot day and the car windows have been up, why do the buckles of the seat belt feel very hot, while the seat belt itself does not feel hot?

35. Creative Thinking Why does a metal doorknob feel cooler to your hand than a carpet feels to your bare feet?

36. Creative Thinking Why do the metal shades of desk lamps have small holes at the top?

37. Creative Thinking Why does the temperature of hot chocolate decrease faster if you place a metal spoon in the liquid?

38. Creative Thinking If you bite into a piece of hot apple pie, the pie filling might burn your mouth while the crust, at the same temperature, will not. Explain why.

39. Applying Technology Glass can conduct some energy. Double-pane windows consist of two plates of glass separated by a small layer of insulating air. Explain why a double-pane window prevents more energy from escaping your house than a single-pane window.

40. Understanding Systems Explain why window unit air conditioners always have the back part of the air conditioner hanging outside. Why is it that the entire air-conditioner cannot be in the room?

41. Making Decisions If the only factor considered were specific heat, which would make a better coolant for automobile engines: water or ethanol? Explain your answer.

42. Critical Thinking Explain why a refrigerant must have a very low boiling point. Why is it important that the refrigerant evaporates?

448

48. a. kinetic energy
b. Celsius
c. Kelvin
d. Fahrenheit
e. conduction
f. convection
g. radiation

43. Working Cooperatively Read the following statements, and discuss with a group of classmates which statement is correct. Explain your answer.
   a. Energy is lost when water is boiled.
   b. The energy used to boil water is still present, but it is no longer in a usable form unless you use work or heat to make it usable.

44. Allocating Resources In one southern state the projected yearly costs for heating a home were $463 using a heat pump, $508 using a natural-gas furnace, and $1220 using electric radiators. Contact your local utility company to determine the projected costs for the three different systems in your area. Make a table comparing the costs of the three systems.

45. Interpreting and Communicating Suppose that an internal combustion engine has a 25% efficiency, meaning that 25% of the energy put into the engine is converted to usable energy. Search the Internet for alternative energy sources that would have a greater efficiency than you found from an internal combustion engine. Report which alternative energy source you would recommend. Explain why you would recommend that energy source.

46. Interpreting and Communicating In a store, look at actual ENERGYGUIDE labels attached to three different models of one brand of any appliance you choose. From the information provided on the labels, compare those three models. Report to the class which of the three models you found to be the most energy efficient, according to the information on the ENERGYGUIDE labels.

47. Connection to Social Studies Research the work of Benjamin Thompson. What was the prevailing theory of heat during Thompson's time? What observations led to Thompson's theory?

48. Concept Mapping Copy the unfinished concept map below onto a sheet of paper. Complete the map by writing the correct word or phrase in the lettered boxes.

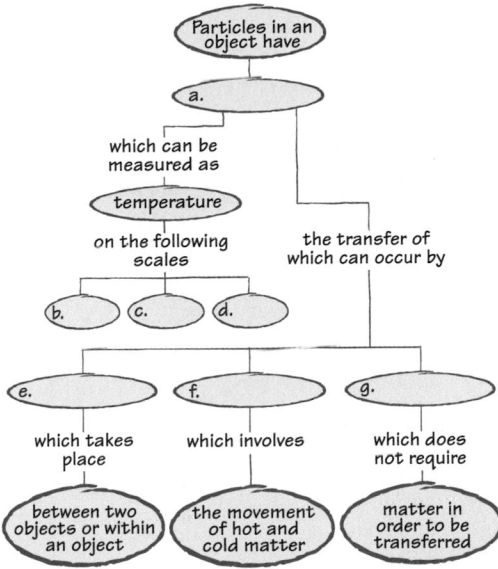

Art Credits: "Inquiry Lab," Uhl Studios, Inc.; Fig. 7, Stephen Durke/Washington Artists; Fig. 8, Stephen Durke/Washington Artists; Fig. 12B, Kristy Sprott; Fig. 12A, C, Uhl Studios, Inc.; Fig. 13, Uhl Studios, Inc.; Fig. 15–16, Uhl Studios, Inc.; Fig. 18, Uhl Studios, Inc.; Fig. 19, Uhl Studios, Inc.; Chapter Review (mountain range), Uhl Studios, Inc.; Fig. 20–21, Uhl Studios, Inc.

Photo Credits: Chapter Opener image of forest fire by John M. Roberts/Corbis Stock Market; infrared aerial photo courtesy NOAA; Fig. 1 (l), Peter Van Steen/HRW; (c), Sam Dudgeon/HRW; (r), Blair Seitz/Photo Researchers, Inc.; "Quick Activity," Sam Dudgeon/HRW; Fig. 2–3, Sam Dudgeon/HRW; Fig. 6, Peter Van Steen/HRW; Fig. 9, CECOM Night Vision/U. S. Army; "Quick Lab," Sam Dudgeon/HRW; Fig. 10, Sam Dudgeon/HRW; "Quick Activity," Sam Dudgeon/HRW; Fig. 11, Charles D. Winters; Fig. 14, Rod Planck/Photo Researchers, Inc.; Fig. 17, John Maher/Corbis Stock Market; Fig. 19, Courtesy Rheem; "Real World Application," Peter Van Steen/HRW; "Chapter Review," Sam Dudgeon/HRW.

449

Transparencies

TT Concept Mapping

INVESTIGATING CONDUCTION OF HEAT

Teacher's Notes

Time Required 1 lab period

Ratings

1	2	3	4
EASY			HARD

TEACHER PREPARATION	1
STUDENT SETUP	2
CONCEPT LEVEL	3
CLEANUP	1

Skills Acquired

- Collecting data
- Communicating
- Designing experiments
- Experimenting
- Identifying/recognizing patterns
- Inferring
- Interpreting
- Measuring
- Organizing and analyzing data
- Predicting

The Scientific Method

In this lab, students will:
- Make observations
- Ask a question
- Analyze the results
- Draw conclusions
- Communicate results

Materials

The wires should be bare, solid rather than braided, and made of the same metal. Copper wires of various diameters are available in electronics stores, electrical supply stores, and hardware stores. Select candle holders that are designed to catch wax drips.

Skills Practice Lab

Introduction

How can you determine whether the thickness of a metal wire affects its ability to conduct energy as heat?

Objectives

▶ **USING SCIENTIFIC METHODS** *Develop* a plan to measure how quickly energy is transferred as heat through a metal wire.

▶ *Compare* the speed of heat conduction in metal wires of different thicknesses.

Materials

candle
candle holder
clothespin
lighter or matches
metal wires of different thicknesses, each about 30 cm long (3)
metric ruler
stopwatch

450

Investigating Conduction of Heat

▶ ## Procedure

Demonstrating Conduction in Wires

1. Obtain three wires of different thicknesses. Clip a clothespin on one end of one of the wires. Lay the wire and attached clothespin on the lab table.

2. Light the candle and place it in the holder.

   SAFETY CAUTION Tie back long hair and confine loose clothing. Never reach across an open flame. Always use the clothespin to hold the wire as you heat it and move it to avoid burning yourself. Remember that the wires will be hot for some time after they are removed from the flame.

3. Hold the lighted candle in its holder above the middle of the wire, and tilt the candle slightly so that some of the melted wax drips onto the middle of the wire.

4. Wait a couple of minutes for the wire and dripped wax to cool completely. The dripped wax will harden and form a small ball. Using the clothespin to hold the wire, place the other end of the wire in the candle's flame. When the ball of wax melts, remove the wire from the flame, and place it on the lab table. Think about what caused the wax on the wire to melt.

Designing Your Experiment

5. With your lab partner(s), decide how you will use the materials available in the lab to compare the speed of conduction in three wires of different thicknesses. Form a hypothesis about whether a thick wire will conduct energy more quickly or more slowly than a thin wire.

6. In your lab report, list each step you will perform in your experiment.

7. Have your teacher approve your plan before you carry out your experiment.

Sample Data Table Heat Conductivity for Copper Wire				
Wire	**Time to melt wax (s)**			
diameter (mm)	Trial 1	Trial 2	Trial 3	Average time
Wire 1 1.0	39.1	37.8	38.4	38.4
Wire 2 1.5	35.0	36.1	34.8	35.3
Wire 3 2.0	34.0	33.3	30.9	32.7

Performing Your Experiment

8. After your teacher approves your plan, you can carry out your experiment.

9. Prepare a data table in your lab report that is similar to the one shown below.

10. Record in your table how many seconds it takes for the ball of wax on each wire to melt. Perform three trials for each wire, allowing the wires to cool to room temperature between trials.

Conductivity Data

Wire diameter (mm)	Time to melt wax (s)			
	Trial 1	Trial 2	Trial 3	Average time
Wire 1				
Wire 2				
Wire 3				

▶ Analysis

1. Find the diameter of each wire you tested. If the diameter is listed in inches, convert it to millimeters by multiplying by 25.4. If the diameter is listed in mils, convert it to millimeters by multiplying by 0.0254. In your data table, record the diameter of each wire in millimeters.

2. Calculate the average time required to melt the ball of wax for each wire. Record your answers in your data table.

3. Plot your data in your lab report in the form of a graph like the one shown. On your graph, draw the line or smooth curve that fits the points best.

4. **Reaching Conclusions** Based on your graph, does a thick wire or a thin wire conduct energy more quickly?

5. When roasting a large cut of meat, some cooks insert a metal skewer into the meat to make the inside cook more quickly. If you were roasting meat, would you insert a thick skewer or a thin skewer? Why?

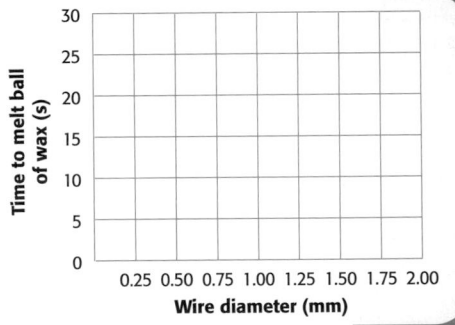

▶ Conclusions

6. Suppose someone tells you that your conclusion is valid only for the particular metal you tested. How could you show that your conclusion is valid for other metals as well?

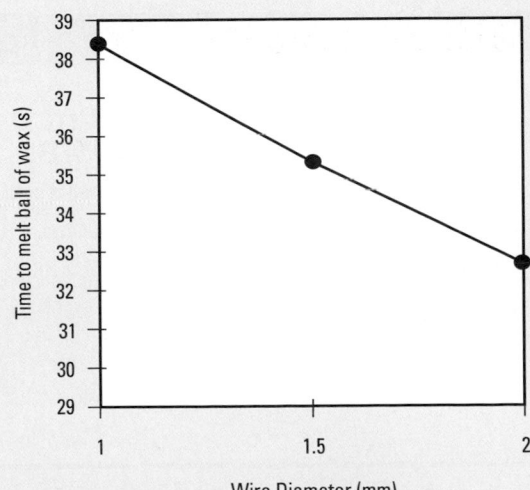

Procedure

Students should control all variables except for wire diameter. A distance from the end of the wire that is consistent among the three wires and across the three trials should be specified. The end of each wire should be held in the same part of the flame, so the temperature at the end will be the same for all wires. The wax balls on the wires should be close to the same size and be the same distance from the flame. If the thinnest wires are too flexible, students may hold both ends of the wire (with two clamps or clothespins) and heat the wire somewhere between the clamps.

Safety Cautions

Clothespins can be a fire hazard with students. As an alternative, wires can be strung between two test-tube clamps on ring stands or lab supports. Be sure students always use a clamp or clothespin to hold the wire as it heats. Remind students that the wires will be hot for some time after they are removed from the flame.

Answers to Analysis

1. Answers will vary. See sample data.

2. Answers will vary. See sample data.

3. See sample graph. Results will vary, but times should be greater for smaller-diameter wires. The line that best fits the data points should decrease as diameter increases.

4. Students should find that thicker wires conduct heat energy more quickly than thin wires.

5. A thicker skewer will conduct heat energy more quickly than a thin skewer.

Answers to Conclusions

6. Students could test wires made from different kinds of metals.

Chapter Resource File

- **Datasheet** Investigating Conduction of Heat **GENERAL**
- **Observation Lab** Energy Transfer and Specific Heat **BASIC**
- **CBL™ Probeware Lab** Determining the Better Insulator for Your Feet **ADVANCED**

Waves
Chapter Planning Guide

PACING	CLASSROOM RESOURCES	LABS, ACTIVITIES, AND DEMONSTRATIONS
BLOCK 1 · 45 min pp. 452–453 **Chapter Opener**		SE **Activity 1**, p. 453 SE **Activity 2**, p. 453
BLOCKS 2 & 3 · 90 min pp. 454–462 **Section 1** Types of Waves	CRF **Lesson Plan*** TT **Bellringer*** TT **Wave Model*** TT **Water Wave Model***	TE **Demonstration** Observing Wave Motion, p. 455 **GENERAL** TE **Demonstration** Harmonic Motion, p. 457 **GENERAL** SE **Quick Lab** How do particles move in a medium?, p. 460 **GENERAL** CRF **Datasheets for In-Text Labs** How do particles move in a medium?* **GENERAL** SE **Quick Activity** Polarization, p. 461 **GENERAL** CRF **Datasheets for In-Text Labs** Polarization* **GENERAL** TE **Demonstration** Longitudinal Waves in Air, p. 461 **GENERAL**
BLOCKS 4 & 5 · 90 min pp. 463–471 **Section 2** Characteristics of Waves	CRF **Lesson Plan*** TT **Bellringer*** TM **Transverse Wave*** TM **Longitudinal Wave*** TT **Frequency*** TT **Visible Light*** TM **The Electromagnetic Spectrum*** TT **Doppler Effect***	SE **Quick Activity** Wave Speed, p. 469 **GENERAL** CRF **Datasheets for In-Text Labs** Wave Speed* **GENERAL** TE **Demonstration** Doppler Effect, p. 470 **GENERAL** SE **Skills Practice Lab** Modeling Transverse Waves, pp. 484–485 ◆ **GENERAL** CRF **Datasheets for SE Skills Practice Lab** Modeling Transverse Waves* **GENERAL** CRF **Observation Lab** Creating and Measuring Standing Waves* ◆ **BASIC**
BLOCKS 6 & 7 · 90 min pp. 472–478 **Section 3** Wave Interactions	CRF **Lesson Plan*** TT **Bellringer*** TT **Interference*** TM **Beats*** TT **Concept Mapping***	TE **Demonstration** Observing Reflection and Refraction, p. 474 **GENERAL** CRF **CBL™ Probeware Lab** Tuning a Musical Instrument* ◆ **ADVANCED**

BLOCKS 8 & 9 · 90 min

Chapter Review and Assessment Resources

SE **Chapter Review**, pp. 480–483
CRF **Chapter Tests*** **GENERAL**
OSP **Test Generator**
CRF **Standardized Test Practice with Guided Reading Development***
CRF **Test Item Listing for ExamView® Test Generator***
OSP **Scoring Rubrics and Classroom Management Checklists**

Online Resources

Visit the HRW Web site for a variety of free resources related to the text. Just type in the keyword **HK4 WAV**.

Holt Science Spectrum: Physical Science: Online Edition

Students can access interactive problem solving help and active visual concept development with the *Holt Science Spectrum: Physical Science* Online Edition available at **www.hrw.com**.

student **CNN** **News**

cnnstudentnews.com

Find the latest news, lesson plans, and activities related to important scientific events.

KEY

TE	Teacher Edition
SE	Student Edition
OSP	One-Stop Planner

CRF Chapter Resource File
TT Teaching Transparency
TM Transparency Master

* Also on One-Stop Planner
◆ Requires Advance Prep

PROBLEM SOLVING AND PRACTICE	SECTION REVIEW AND ASSESSMENT	STANDARDS CORRELATION
	CRF Pretest* GENERAL	
CRF **Cross-Disciplinary Worksheet** Connection to Engineering—Wave Energy* ADVANCED SE **Science and the Consumer** Shock Absorbers: Why Are They Important?, p. 458 CRF **Cross-Disciplinary Worksheet** Science and the Consumer—Bicycle Design and Shock Absorption* ADVANCED	TE **Quiz**, p. 462 BASIC SE **Section 1 Review**, p. 462 CRF **Concept Review*** GENERAL CRF **Quiz*** BASIC	PS 4a, 5a, 5b, 5d UCP 1, 2, 4, 5 SAI 1 ST 1, 2 SPSP 5
CRF **Cross-Disciplinary Worksheet** Connection to Language Arts—Writing a Plan for Wave Observation* ADVANCED CRF **Cross-Disciplinary Worksheet** Integrating Computers and Technology—Radio Waves* ADVANCED CRF **Cross-Disciplinary Worksheet** Integrating Earth Science—Earthquake Waves* ADVANCED SE **Math Skills** Wave Speed, p. 468 GENERAL CRF **Math Skills** Wave Speed* GENERAL	TE **Quiz**, p. 471 BASIC SE **Section 2 Review**, p. 471 CRF **Concept Review*** GENERAL CRF **Quiz*** BASIC	PS 2e, 4a, 4e, 5a, 5d ES 3c UCP 1, 2, 3 ST 2 HNS 1 SPSP 5
CRF **Cross-Disciplinary Worksheet** Integrating Math—Bending Light Waves to Magnify* ADVANCED SE **Graphing Skills** Interpreting Graphs, p. 479 SE **CareerLink** Ultrasonographer, pp. 486–487	TE **Quiz**, p. 478 BASIC SE **Section 3 Review**, p. 478 CRF **Concept Review*** GENERAL CRF **Quiz*** BASIC	PS 4a UCP 1, 2, 4 SAI 1 ST 1, 2 SPSP 5

SCiLINKS®

www.scilinks.org

Topic: Waves
*Sci*Links code: HK4150

Topic: Vibrations and Waves
*Sci*Links code: HK4145

Topic: Doppler Effect
*Sci*Links code: HK4032

Topic: Reflection, Refraction, Diffraction
*Sci*Links code: HK4119

Topic: Seismic Waves
*Sci*Links code: HK4126

Topic: Ultrasound
*Sci*Links code: HK4143

Technology Resources

One-Stop Planner
All of your printable resources and the Test Generator are on this convenient CD-ROM.

Physical Science Interactive Tutor CD-Rom

Disc Two, Module 12
Topic: Frequency and Wavelength

Disc Two, Module 14
Topic: Refraction

Overview

This chapter introduces students to waves. This chapter first defines waves and distinguishes between different wave types. The chapter then explores wave characteristics such as amplitude, wavelength, frequency, and period. Students also calculate wave speed and learn about the Doppler effect. The chapter finally discusses wave behaviors and interactions.

Assessing Prior Knowledge

Be sure students understand the following concepts:

- Interactions of matter and energy
- motion
- force
- work
- forms of energy
- energy transformations

MISCONCEPTION ALERT

Science education research has identified the following misconceptions about waves.

- The motion of the medium (i.e. water) is frequently confused with the motion of the wave itself.
- Students often confuse the independent aspects of waves, such as amplitude, frequency, and velocity. Some students believe that a rapid oscillation ensures a large amplitude and fast velocity.
- Students intuit that wave collisions result in the permanent cancellation of both waves as if they were mechanical objects.

Waves

Chapter Preview

1 Types of Waves
What Is a Wave?
Vibrations and Waves
Transverse and Longitudinal Waves

2 Characteristics of Waves
Wave Properties
Wave Speed
The Doppler Effect

3 Wave Interactions
Reflection, Diffraction, and Refraction
Interference
Standing Waves

452

Standards Correlations

National Science Education Standards

Section 1 Types of Waves
Physical Science PS 4a, 5a, 5b, 5d, 6a
Unifying Concepts and Processes UCP 1, 2, 4, 5
Science as Inquiry SAI 1
Science and Technology ST 1, 2
Science in Personal and Social Perspectives SPSP 5

Section 2 Characteristics of Waves
Physical Science PS 2e, 4a, 4e, 5a, 5d
Earth and Space Science ES 3c

Unifying Concepts and Processes UCP 1, 2, 3
Science and Technology ST 2
History and Nature of Science HNS 1
Science in Personal and Social Perspectives SPSP 5

Section 3 Wave Interactions
Physical Science PS 4a
Unifying Concepts and Processes UCP 1, 2, 4
Science as Inquiry SAI 1
Science and Technology ST 1, 2
Science in Personal and Social Perspectives SPSP 5

Focus ACTIVITY

Background The energy in an ocean wave can lift a surfboard up into the air and carry the surfer into shore. Ocean waves get most of their energy from the wind. A wave may start as a small ripple in a calm sea, then build up as the wind pushes it along. Waves that start on the coast of northern Canada may be very large by the time they reach a beach in the Hawaiian Islands.

The winds that create ocean waves are caused by convection currents in the atmosphere, which are driven by energy from the sun. Energy travels across empty space from the sun to Earth—in the form of light waves.

Waves are all around us. As you read this book, you are depending on light waves. Light bounces off the pages and into your eyes. When you talk with a friend you are depending on sound waves traveling through the air. Sometimes the waves can be gentle, such as those that rock a canoe in a pond. Other times waves can be very destructive, such as those created by earthquakes.

Activity 1 Fill a long, rectangular pan with water. Experiment with making waves in different ways. Try making waves by sticking the end of a pencil into the water, by moving a wide stick or board back and forth, and by striking the side of the pan. Place wooden blocks or other obstacles into the pan, and watch how the waves change when they encounter the obstacles.

Activity 2 With some of your classmates, practice the "wave" as is often performed by fans at football games. How does it resemble an ocean wave? How does it differ from other forms of motion?

---

internet connect

www.scilinks.org
Topic: Waves SciLinks code: HK4150

SciLINKS Maintained by the National Science Teachers Association

---

Pre-Reading Questions

1. What things do you do every day that depend on waves? What kinds of waves do you think are involved?
2. What properties do you think these different kinds of waves have in common?

453

A surfer takes advantage of the energy in ocean waves. Energy travels from the sun to Earth in the form of waves.

Focus ACTIVITY

Background Ask students to describe what they think sound waves and light waves would look like if they could be observed directly. Many students will have trouble getting away from the up-and-down undulations they associate with water waves.

Activity 1 See the Demonstration *Observing Wave Motion* for additional information about observing waves in a pan. **LS Visual**

Activity 2 Students may note that doing the "wave" resembles an ocean wave in that both motions are continuous and rolling. They may also observe that both types of waves involve the transfer of motion. With an ocean wave, motion is transferred between water particles; with students in this activity, motion is transferred between students. **LS Kinesthetic**

Answers to Pre-Reading Questions

1. Answers may vary. Students may state that daily activities such as reading, listening to music, talking, cooking, and watching TV depend on light, sound, heat, and radio waves.
2. Students may state that different kinds of waves seem to move up and down or back and forth.

Chapter Resource File

- Pretest **GENERAL**
- Teaching Transparency Preview
- Answer Keys

Chapter 14 • Waves 453

Types of Waves

◀ **KEY TERMS**

wave
medium
mechanical wave
electromagnetic wave
transverse wave
longitudinal wave

OBJECTIVES

▶ **Recognize** that waves transfer energy.

▶ **Distinguish** between mechanical waves and electromagnetic waves.

▶ **Explain** the relationship between particle vibration and wave motion.

▶ **Distinguish** between transverse waves and longitudinal waves.

When a stone is thrown into a pond, it creates ripples on the surface of the water, as shown in *Figure 1*. If there is a leaf floating on the water, the leaf will bob up and down and back and forth as each ripple, or wave, disturbs it. But after the waves pass, the leaf will almost return to its original position on the water.

What Is a Wave?

◀ **wave** a periodic disturbance in a solid, liquid, or gas as energy is transmitted through a medium

Like the leaf, individual drops of water do not travel outward with a wave. They move only slightly from their resting place as each ripple passes by. If drops of water do not move very far as a wave passes, and neither does a leaf on the surface of the water, then what moves along with the wave? Energy does. A wave is not just the movement of matter from one place to another. A **wave** is a disturbance that carries energy through matter or space.

Figure 1
A stone thrown into a pond creates waves.

Properties of Waves:

Type of Wave	Mechanical		Electromagnetic
Form	Longitudinal	Transverse	Modeled as transverse
Medium	Solids, liquids, gases	Solids and liquids	None required
Description	Compressions and rarefactions in matter	Back-and-forth (or up-and-down) movement of matter	Oscillating electric and magnetic fields
Examples	Sound waves, some earthquake waves	Water waves, rope waves, some earthquake waves	Visible light waves radio waves, X rays

Most waves travel through a medium

The waves in a pond are disturbances traveling through water. Sound also travels as a wave. The sound from a stereo is a pattern of changes in the air between the stereo speakers and your ears. Earthquakes create waves, called *seismic waves*, that travel through Earth.

In each of these examples, the waves involve the movement of some kind of matter. The matter through which a wave travels is called the **medium.** In the example of the pond, the water is the medium. For sound from a stereo, air is the medium. And in earthquakes, Earth itself is the medium.

Waves that require a medium are called **mechanical waves.** Almost all waves are mechanical waves, with one important exception: light waves.

Light does not require a medium

Light can travel from the sun to Earth across the empty space between them. This is possible because light waves do not need a medium through which to travel. Instead, light waves consist of changing electric and magnetic fields in space. For that reason, light waves are also called **electromagnetic waves.**

Visible light waves are just one example of a wide range of electromagnetic waves. Radio waves, such as those that carry signals to your radio or television, are also electromagnetic waves. Other kinds of electromagnetic waves will be introduced in Section 2. In this book, the terms *light* and *light wave* may refer to any electromagnetic wave, not just visible light.

Waves transfer energy

Energy is the ability to exert a force over a certain distance. It is also known as the ability to do *work.* We know that waves carry energy because they can do work. For example, water waves can do work on a leaf, on a boat, or on a beach. Sound waves can do work on your eardrum. Light waves can do work on your eye or on photographic film.

A wave caused by dropping a stone in a pond might carry enough energy to move a leaf up and down several centimeters. The bigger the wave is, the more energy it carries. A cruise ship moving through water in the ocean could create waves big enough to move a fishing boat up and down a few meters.

▶ **medium** a physical environment in which phenomena occur

▶ **mechanical wave** a wave that requires a medium through which to travel

▶ **electromagnetic wave** a wave that consists of oscillating electric and magnetic fields, which radiate outward at the speed of light

455

Teaching Tip

Tsunamis—Deep and Shallow

It may surprise students to know that if they were on a ship in the deep part of the ocean when a tidal wave passed, it would cause only a smooth rise and fall of a few inches. In other words, no one would notice. However, a lot of energy is required to lift a great depth of water a few inches. When that energy becomes concentrated in shallow water, it will produce a fast and fearsome tsunami.

MISCONCEPTION ALERT

Wave Motion Stress at this point that wave motion transfers only energy, not matter. Even though mechanical waves are transferred by the motion of matter, no matter itself is transferred.

Teaching Tip

Spherical Wave Fronts

Mechanical waves spread out through the medium evenly and spherically from the source if the source is open to the medium in all directions. If the source, such as a speaker, is pointed in a particular direction, the sound waves will spread spherically but will be stronger in the direction the speaker points. Note that the spherical wave fronts are represented as circles in **Figure 3.**

Figure 2
This portrait of a tsunami was created by the Japanese artist Hokusai in 1830.

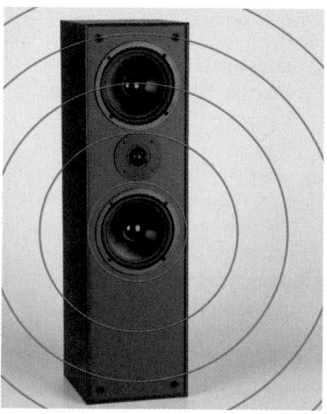

Figure 3
Sound waves from a stereo speaker spread out in spherical wave fronts.

Figure 2 shows a woodblock print of a *tsunami,* a huge ocean wave caused by earthquakes. A tsunami may be as high as 30 m when it reaches shore, taller than a 10-story building. Such waves carry enough energy to cause a lot of damage to coastal towns and shorelines. Normal-sized ocean waves do work on the shore, too, breaking up rocks into tiny pieces to form sandy beaches.

Energy may spread out as a wave travels

If you stand next to the speakers at a rock concert, the sound waves may damage your ears. Likewise, if you look at a bright light bulb from too close, the light may damage your eyes. But if you are 100 m away, the sound of the rock band or the light from the bulb is harmless. Why?

Think about waves created when a stone falls into a pond. The waves spread out in circles that get bigger as the waves move farther from the center. Each of these circles, called a *wave front,* has the same amount of total energy. But as the circles get larger, the energy spreads out over a larger area.

When sound waves travel in air, the waves spread out in spheres, as shown in **Figure 3.** These spheres are similar to the circular ripples on a pond. As they travel outward, the spherical wave fronts get bigger, so the energy in the waves spreads out over a larger area. This is why large amplifiers and speakers are needed to fill a concert hall with sound, even though the same music can sound just as loud if it is played on a portable radio and listened to with a small pair of headphones.

456

Cultural Awareness

People who live on Pacific islands or in Pacific Coast countries, including Chile and Australia, have great respect for the power of tsunamis. These tidal waves occur more frequently around the Pacific than in other oceans because of earthquakes and underwater volcanic activity associated with the movement of the continental plates in that region.

Did You Know ?

Chilean Tsunami One of the most destructive tsunamis in recent history occurred in Chile in 1960. The tsunami was triggered by a huge earthquake off the coast of South Central Chile. Together, the tsunami and earthquake caused over half a billion dollars worth of damage.

Vibrations and Waves

When a singer sings a note, vocal cords in the singer's throat move back and forth. That motion makes the air in the throat vibrate, creating sound waves that eventually reach your ears. The vibration of the air in your ears causes your eardrums to vibrate. The motion of the eardrum triggers a series of electrical pulses to your brain, and your brain interprets them as sounds.

Waves are related to vibrations. Most waves are caused by a vibrating object. Electromagnetic waves may be caused by vibrating charged particles. In a mechanical wave, the particles in the medium also vibrate as the wave passes through the medium.

internet connect

www.scilinks.org
Topic: Vibrations and Waves
SciLinks code: HK4145

SciLINKS Maintained by the National Science Teachers Association

Vibrations involve transformations of energy

Figure 4 shows a mass hanging on a spring. If the mass is pulled down slightly and released, it will begin to move up and down around its original resting position. This vibration involves transformations of energy, much like those in a swinging pendulum.

When the mass is pulled away from its resting place, the mass-spring system gains elastic potential energy. The spring exerts a force that pulls the mass back to its original position.

As the spring moves back toward the original position, the potential energy in the system changes to kinetic energy. The mass moves beyond its original resting position to the other side.

At the top of its motion, the mass has lost all its kinetic energy. But the system now has both elastic potential energy and gravitational potential energy. The mass moves downward again, past the resting position, and back to the beginning of the cycle.

Figure 4

When a mass hanging on a spring is disturbed from rest, it starts to vibrate up and down around its original position.

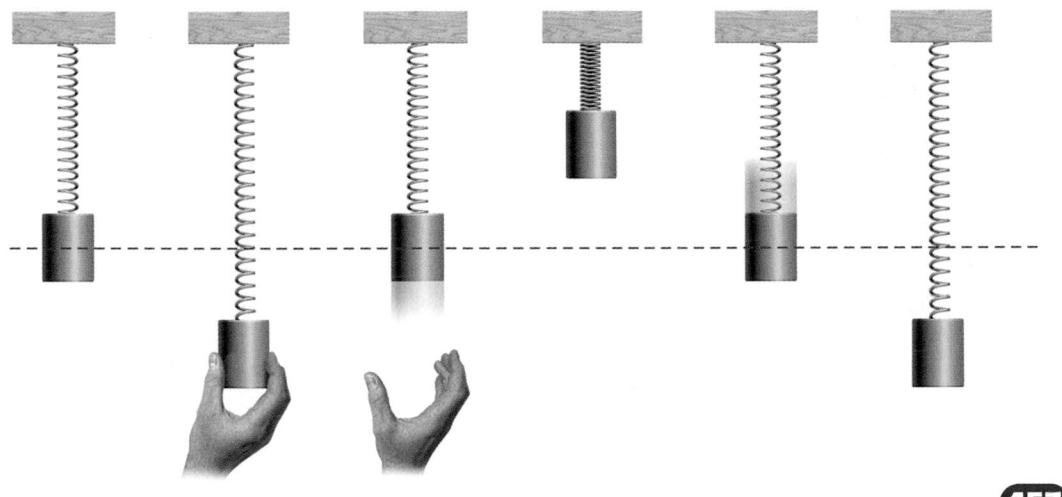

Demonstration —— GENERAL

Harmonic Motion

(Time: About 5 minutes)

Materials:

• spring
• support for spring
• chalkboard
• chalk
• mass

Step 1 Set up a demonstration of the system illustrated in **Figure 4.** The mass must be heavy enough to stretch the spring significantly without permanently deforming it. The spring should not completely collapse at the top of the vibration.

Step 2 Demonstrate the up-and-down vibration (oscillation) of the mass when it is pulled down or pushed up from its resting position.

Step 3 Draw a set of *x*- and *y*-axes on the chalkboard. Extend the *x*-axis out along the board to the right, and label it "Elapsed time." Label the *y*-axis "Displacement." Zero on the *x*-axis represents the resting position of the mass.

Step 4 Set the mass in motion and walk along the board to the right at a steady pace while reproducing the up-and-down motion of the mass with chalk. The resulting curve should be a reasonable approximation of a sine curve that describes the harmonic motion of the weight. Point out that any vibration can be graphed in this way. **LS** Visual

Alternative Assessment —— ADVANCED

Simple Harmonic Motion Have students apply the principles of simple harmonic motion to a playground swing. After studying **Figure 4,** ask them to draw a swing at four different points of one cycle (up in one direction, back through the starting point, up in the other direction, back through the starting point). Tell them to disregard friction and assume that the swing is in motion the entire time. Have them label the illustrations *A, B, C,* and *D.* For each, have students list whether potential and kinetic energy are at maximum or minimum. Have students make a single graph of displacement (from the starting point) versus time. Finally, ask students what type of potential energy is involved in this example. (At *A* and *C*, potential energy is maximum and kinetic energy is minimum. At *B* and *D*, kinetic energy is maximum and potential energy is minimum. The graph is a sine curve. The high and low points correspond to *A* and *C*, and the curve crosses the horizontal axis at *B* and *D*. The potential energy is gravitational potential energy.) **LS** Visual

Science and the Consumer — GENERAL

Shock Absorbers: Why Are They Important? This feature puts the distinction between simple harmonic motion and damped harmonic motion into a real-world context. Ask students if they have ever ridden in a car with poor suspension. What did it feel like? Did the car continue bouncing up and down after going over a bump? Did the shock from hitting a bump transfer more directly to the passengers in the car?

Students may also be familiar with shock absorbers on a bicycle. Ask students if they have or know someone who has a bike with shock absorbers. Have them find out if the shock absorbers contain springs or a fluid such as oil or pressurized air. **LS Kinesthetic**

Answers to Your Choice

1. Students' answers should indicate that they would look for a vehicle with leaf springs, because these types of springs can bear heavier loads than coil springs can.

2. Shock absorbers stop an automobile from continually bouncing by dampening the vibrations of the springs.

Science and the Consumer

Shock Absorbers: Why Are They Important?

Bumps in the road are certainly a nuisance, but without strategic use of damping devices, they could also be very dangerous. To control a car going 100 km/h (60 mi/h), a driver needs all the wheels of the vehicle on the ground. Bumps in the road lift the wheels off the ground and may rob the driver of control of the car.

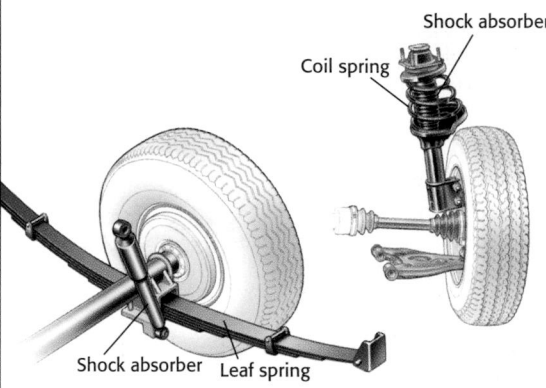

Coil spring

Shock absorber

Shock absorber   Leaf spring

Springs Absorb Energy

To solve this problem, cars are fitted with springs at each wheel. When the wheel of a car goes over a bump, the spring absorbs kinetic energy so that the energy is not transferred to the rest of the car. The energy becomes elastic potential energy in the spring, which then allows the spring to push the wheel back down onto the road.

Springs Alone Prolong Vibrations

Once a spring is set in motion, it tends to continue vibrating up and down in simple harmonic motion. This can create an uncomfortable ride, and it may also affect the driver's control of the car. One way to cut down on unwanted vibrations is to use stiff springs that compress only a few centimeters with thousands of newtons of force. However, the stiffer the spring is, the rougher the ride is and the more likely the wheels are to come off the road.

Shock Absorbers Dampen Vibrations

Modern automobiles are fitted with devices known as shock absorbers that absorb energy without prolonging vibrations. Shock absorbers are fluid-filled tubes that turn the simple harmonic motion of the springs into a damped harmonic motion. In a damped harmonic motion, each cycle of stretch and compression of the spring is much smaller than the previous cycle. Modern auto suspensions are set up so that all a spring's energy is absorbed by the shock absorbers in just one up-and-down cycle.

Shock Absorbers and Springs Come in Different Arrangements

Different types of springs and shock absorbers are combined to give a wide variety of responses. For example, many passenger cars have coil springs with shock absorbers parallel to the springs, or even inside the springs, as shown at near left. Some larger vehicles have heavy-duty leaf springs made of stacks of steel strips. Leaf springs are stiffer than coil springs, but they can bear heavier loads. In this type of suspension system, the shock absorber is perpendicular to the spring, as shown at far left.

The stiffness of the spring can affect steering response time, traction, and the general feel of the car. Because of the variety of combinations, your driving experiences can range from the luxurious "floating-on-air" ride of a limousine to the bone-rattling feel of a true sports car.

Your Choice

1. **Making Decisions** If you were going to haul heavy loads, would you look for a vehicle with coil springs or leaf springs? Why?

2. **Critical Thinking** How do shock absorbers stop an automobile from continually bouncing?

458

INCLUSION Strategies

- *Learning Disabled*
- *Attention Deficit Disorder*
- *English Language Learners*

Students can create a basic outline of the chapter by first listing the chapter headings. Under each heading, students will find key terms in bold lettering. Key terms should be listed under the heading where they are found.

1. Types of Waves
   A. What is a Wave?
      1. wave

B. Most waves travel through a medium
   1. medium
   2. mechanical waves
C. Light does not require a medium
   1. electromagnetic waves

This basic outline can be used as a study guide for individuals or small groups of students.

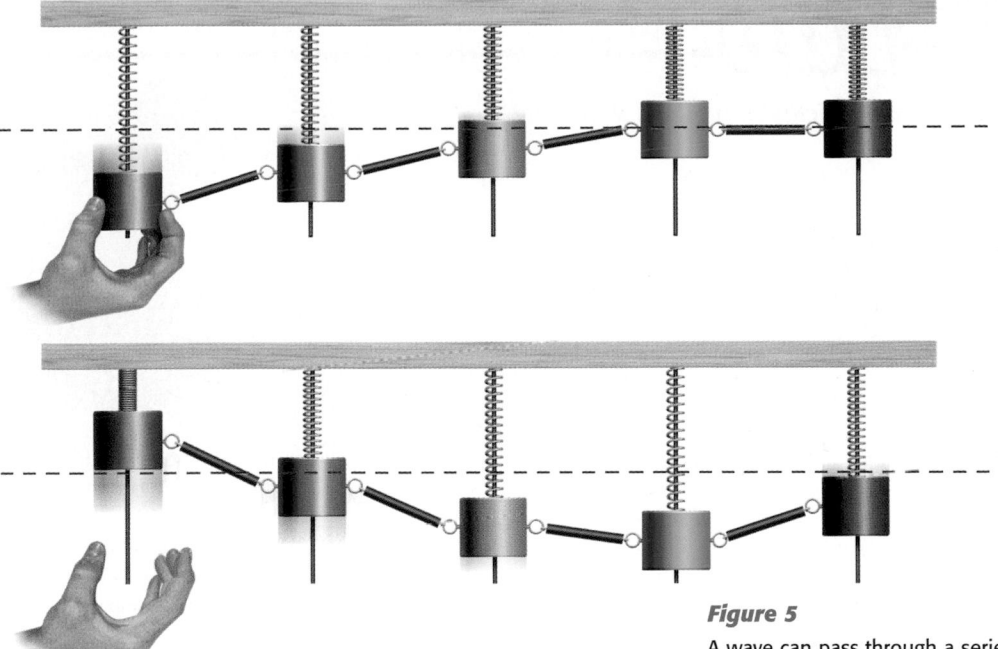

Figure 5

A wave can pass through a series of masses on springs. The masses act like the particles in a medium.

Interpreting Visuals **Figure 5** shows a simple model of wave motion in a medium. Have students look at the top part of the figure and predict what will happen. (You may wish to use a transparency, covering up the bottom portion of the diagram.) Then have students look at the bottom part and see if their predictions were correct. Ask students: What will happen when the wave reaches the end of the series? (The wave will be reflected.) **LS Visual**

Whenever the spring is expanded or compressed, it is exerting a force that pushes the mass back almost to the original resting position. As a result, the mass will continue to bounce up and down. This type of vibration is called *simple harmonic motion*.

A wave can pass through a series of vibrating objects

Imagine a series of masses and springs tied together in a row, as shown in *Figure 5*. If you pull down on a mass at the end of the row, that mass will begin to vibrate up and down. As the mass on the end moves, it pulls on the mass next to it, causing that mass to vibrate. The energy in the vibration of the first mass, which is a combination of kinetic energy and elastic potential energy, is transferred to the mass-spring system next to it. In this way, the disturbance that started with the first mass travels down the row. This disturbance is a wave that carries energy from one end of the row to the other.

If the first mass were not connected to the other masses, it would keep vibrating up and down on its own. However, because it transfers its energy to the second mass, it slows down and then returns to its resting position. A vibration that fades out as energy is transferred from one object to another is called *damped harmonic motion*.

459

Trends in Acoustical Engineering

Recording studios typically have soundproofed booths and rooms. An effective soundproofing or acoustic deadening material must be able to absorb the energy of sound waves of all frequencies. This energy is usually absorbed by mats of fibers, which move slightly and rub against each other, changing the energy of the sound wave to heat. The amount of absorbed energy is small in relation to the amount of fiber, so the mats do not get hot.

Alternative Assessment ⎯ ADVANCED

Soundproofing Have interested students learn more out how rooms and buildings are soundproofed and what sorts of materials are used. They may wish to find and interview contractors who specialize in this kind of work. Have students prepare a presentation to share their results with the class. **LS Verbal**

Chapter Resource File

• **Cross-Disciplinary Worksheet** Science and the Consumer— Bicycle Design and Shock Absorption ADVANCED

 Transparencies

TT Wave Model

Quick Lab

How do particles move in a medium? Be sure students do not stretch the spring enough to permanently deform it. A metal spring toy is preferable to a plastic one.

Students should first send single wave pulses along the spring and observe the ribbon as this wave passes. In step 3, students can also make longitudinal waves by rapidly pushing and pulling on the end of the spring.

Answers to Analysis

1. Students should observe the ribbon move from the resting position to one side, to the other side, and then back to the resting position.
2. The energy was able to set matter in motion along the length of the spring. The energy came from the motion (vibration) of the hand holding the spring. **LS Visual**

SKILL BUILDER — **BASIC**

Interpreting Visuals Have students examine **Figure 6**. Ask students if the ribbon's position will change after the wave passes. Ask students to relate the ribbon to other examples they may have seen, such as a leaf floating on the surface of a pond or corks floating in the ripple tank in the *Harmonic Motion* demonstration. **LS Visual**

Quick Lab

How do particles move in a medium?

Materials ✓ long, flexible spring ✓ colored ribbon

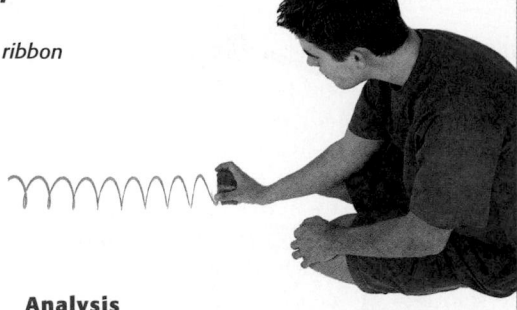

1. Have two people each grab an end of the spring and stretch it out along a smooth floor. Have another person tie a small piece of colored ribbon to a coil near the middle of the spring.
2. Swing one end of the spring from side to side. This will start a wave traveling along the spring. Observe the motion of the ribbon as the wave passes by.
3. Take a section of the spring and bunch it together as shown in the figure at right. Release the spring. This will create a different kind of wave traveling along the spring. Observe the motion of the ribbon as this wave passes by.

Analysis

1. How would you describe the motion of the ribbon in step 2? How would you describe its motion in step 3?
2. How can you tell that energy is passing along the spring? Where does that energy come from?

The motion of particles in a medium is like the motion of masses on springs

If you tie one end of a rope to a doorknob, pull it straight, and then rapidly move your hand up and down once, you will generate a single wave along the rope, as shown in *Figure 6*. A small ribbon tied to the middle of the rope can help you visualize the motion of a single particle of matter in the rope.

As the wave approaches, the ribbon moves up in the air, away from its resting position. As the wave passes farther along the rope, the ribbon drops below its resting position. Finally, after the wave passes by, the ribbon returns to its original starting point. Like the ribbon, each part of the rope moves up and down as the wave passes by.

The motion of each part of the rope is like the vibrating motion of a mass hanging on a spring. As one part of the rope moves, it pulls on the part next to it, transferring energy. In this way, a wave passes along the length of the rope.

Figure 6

As this wave passes along a rope, the ribbon moves up and down while the wave moves to the right.

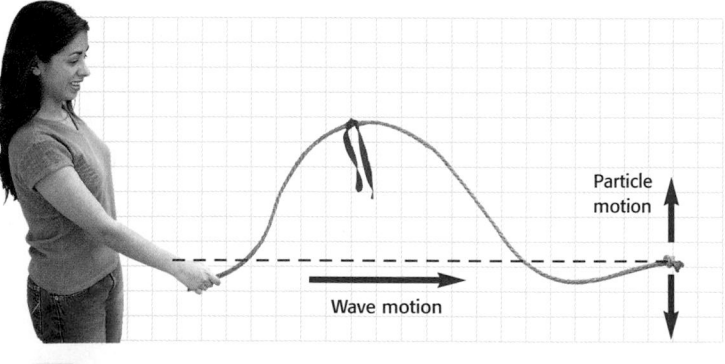

Particle motion

Wave motion

460

Did You Know?

A Matter of Attraction The transfer of energy by transverse mechanical waves depends on the attraction or bonds of the molecules of matter to each other, so that the motion of one particle is easily transferred to the next. This explains why transverse waves in nature occur commonly in the condensed states of matter—solid and liquid—but seldom in gases.

Transverse and Longitudinal Waves

Particles in a medium can vibrate either up and down or back and forth. Waves are often classified by the direction that the particles in the medium move as a wave passes by.

Transverse waves have perpendicular motion

When a crowd does "the wave" at a sporting event, people in the crowd stand up and raise their hands into the air as the wave reaches their part of the stadium. The wave travels around the stadium in a circle, but the people move straight up and down. This is similar to the wave in the rope. Each particle in the rope moves straight up and down as the wave passes by from left to right.

In these cases, the motion of the particles in the medium (in the stadium, the people in the crowd) is perpendicular to the motion of the wave as a whole. Waves in which the motion of the particles is perpendicular to the motion of the wave as a whole are called **transverse waves.**

Light waves are another example of transverse waves. The fluctuating electric and magnetic fields that make up a light wave are perpendicular to one another and are also perpendicular to the direction the light travels.

Longitudinal waves have parallel motion

Suppose you stretch out a long, flexible spring on a table or a smooth floor, grab one end, and move your hand back and forth, directly toward and directly away from the other end of the spring. You would see a wave travel along the spring as it bunches up in some spots and stretches in others, as shown in *Figure 7.*

As a wave passes along the spring, a ribbon tied to one of the coils of the spring will move back and forth, parallel to the direction that the wave travels. Waves that cause the particles in a medium to vibrate parallel to the direction of wave motion are called **longitudinal waves.**

Sound waves are an example of longitudinal waves that we encounter every day. Sound waves traveling in air compress and expand the air in bands. As sound waves pass by molecules in the air move backward and forward parallel to the direction that the sound travels.

▶ **transverse wave** a wave in which the particles of the medium move perpendicular to the direction the wave is traveling

▶ **longitudinal wave** a wave in which the particles of the medium vibrate parallel to the direction of wave motion

Figure 7

As a longitudinal wave passes along this spring, the ribbon tied to the coils moves back and forth, parallel to the direction the wave is traveling.

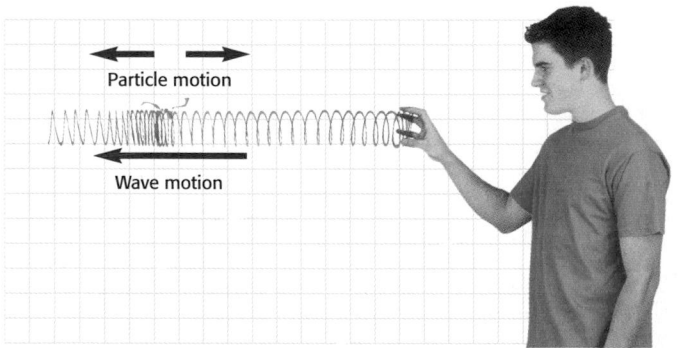

Particle motion

Wave motion

461

Did You Know?

Compression Waves The vibration of a mass on a spring is also a model of how longitudinal waves are created. As the weight moves down, the bottom surface compresses the air in front of it. As it moves up, the bottom surface "stretches" the air behind it. This is why longitudinal waves are often called *compression waves.* We don't hear this wave as sound because its frequency is much too low for human hearing.

1. What does a wave carry through matter or space? (energy)

2. Most waves are caused by vibrating objects. True or false? (true)

3. Give one example of each of the following:

   a. mechanical wave (sample answer: water wave)

   b. electromagnetic wave (sample answer: visible light)

   c. transverse wave (sample answer: wave traveling through a rope)

   d. longitudinal wave (sample answer: wave traveling through a spring)

LS Logical

Chapter Resource File

• **Concept Review** GENERAL

• **Quiz** BASIC

Transparencies

TT Water Wave Model

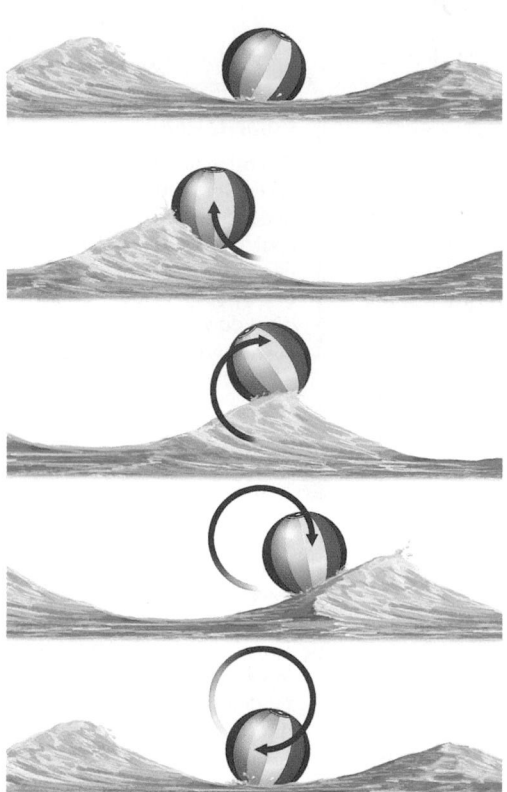

In a surface wave, particles move in circles

Waves on the ocean or in a swimming pool are not simply transverse waves or longitudinal waves. Water waves are an example of *surface waves*. Surface waves occur at the boundary between two different mediums, such as between water and air. The particles in a surface wave move both perpendicularly and parallel to the direction that the wave travels.

Follow the motion of the beach ball in *Figure 8* as a wave passes by traveling from left to right. At first, the ball is in a trough. As the crest approaches, the ball moves to the left and upward. When the ball is very near the crest, it starts to move to the right. Once the crest has passed, the ball starts to fall back downward, then to the left. The up and down motions combine with the side to side motions to produce a circular motion overall.

The beach ball helps to make the motion of the wave more visible. Particles near the surface of the water also move in a similar circular pattern.

Figure 8
Ocean waves are surface waves at the boundary between air and water.

SECTION 1 REVIEW

SUMMARY

▶ A wave is a disturbance that carries energy through a medium or through space.

▶ Mechanical waves require a medium through which to travel. Light waves, also called electromagnetic waves, do not require a medium.

▶ Particles in a medium may vibrate perpendicularly to or parallel to the direction a wave is traveling.

1. **Identify** the medium for the following waves:
   a. ripples on a pond
   b. the sound waves from a stereo speaker
   c. seismic waves

2. **Name** the one kind of wave that does not require a medium.

3. **Describe** the motion of a mass vibrating on a spring. How does this relate to wave motion?

4. **Explain** the difference between transverse waves and longitudinal waves. Give an example of each type.

5. **Describe** the motion of a water molecule on the surface of the ocean as a wave passes by.

6. **Critical Thinking** Describe a situation that demonstrates that water waves carry energy.

462

Answers to Section 1 Review

1. **a.** water
   b. air
   c. Earth's crust and interior

2. electromagnetic (light) waves

3. The mass vibrates up and down from a low point to a high point. Particles in a medium vibrate like masses on springs when a wave passes.

4. In a transverse wave, particles in the medium move back and forth at right angles to the direction the wave is moving. A wave along a rope is a transverse wave. In a longitudinal wave, particles in the medium move back and forth in the same direction the wave is moving. A sound wave is a longitudinal wave.

5. The wave causes the molecule to move up and down as well as forward and backward at the same time. This combination causes the molecule to move in a circular path.

6. Answers will vary. Students should cite any instance in which waves move objects from their positions. Examples include tsunamis, beach erosion during storms, and waves destroying waterfront buildings in a hurricane.

Characteristics of Waves

OBJECTIVES

▶ **Identify** the crest, trough, amplitude, and wavelength of a wave.

▶ **Define** the terms *frequency* and *period*.

▶ **Solve** problems involving wave speed, frequency, and wavelength.

▶ **Describe** the Doppler effect.

▶ KEY TERMS

crest
trough
amplitude
wavelength
period
frequency
Doppler effect

If you have spent any time at the beach or on a boat, you have probably observed many properties of waves. Sometimes the waves are very large; other times they are smaller. Sometimes they are close together, and sometimes they are farther apart. How can these differences be described and measured in more detail?

Wave Properties

The simplest transverse waves have somewhat similar shapes no matter how big they are or what medium they travel through. An ideal transverse wave has the shape of a *sine curve*, such as the curve on the graph in *Figure 9A*. A sine curve looks like an S lying on its side. Sine curves can be used to represent waves and to describe their properties.

Waves that have the shape of a sine curve, such as those on the rope in *Figure 9B*, are called *sine waves*. Although many waves, such as water waves, are not ideal sine waves, they can still be modeled with the graph of a sine curve.

Figure 9

A A sine curve can be used to demonstrate the characteristics of waves.

B This transverse wave on a rope is a simple sine wave.

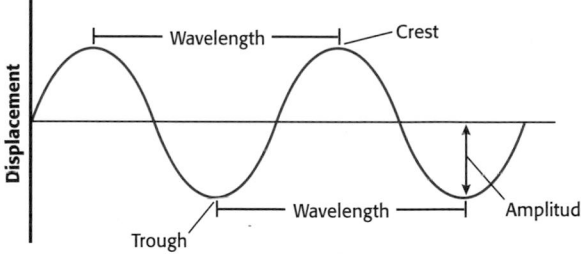

463

Focus

Overview

Before beginning this section, review with your students the Objectives listed in the Student Edition. This section introduces many of the terms used to describe wave characteristics: crest, trough, amplitude, wavelength, frequency, and period. Students learn about the relationship between wavelength, frequency, and wave speed. The section concludes with a discussion of the Doppler effect.

🔊 Bellringer

Use the Bellringer transparency to prepare students for this section.

Motivate

Opening Discussion/Question —— GENERAL

Ask students to define the word *sinuous* and to give examples of things that are described as sinuous. (Sample answers: climbing vines, snakes.) Tell students that both sinuous and sine are derived from the Latin *sinus*, which means "curve." It can also mean "hollow," as in the nasal sinus cavities. Explain to students that in this section, they will see how sine curves can be used to model many different types of waves. **LS Verbal**

Chapter Resource File

• Lesson Plan

Transparencies

TT Bellringer

TM Transverse Wave

Interpreting Visuals Have students examine **Figure 10.** Ask students to explain the meaning of the sine-curve representation of a longitudinal wave. Ask students which parts of each wave represent energy maximums and minimums. **LS Visual**

Teaching Tip — **BASIC**

Representing Sound Waves
Students often see sound waves and electrical signals represented as transverse waves. In fact, neither is a transverse wave, but sine curves can represent sound waves in a useful way. Ask students to suppose they needed to represent several different sound waves on paper. Would it be better to depict longitudinal waves, as in **Figure 10A,** or sine waves, as in **Figure 10B?** **LS Visual**

SKILL BUILDER — **BASIC**

Reading Skills As students read Section 2, have them add to their table, *Properties of Waves*, from Section 1. (See the sample below.) Information may vary slightly for each student. **LS Visual**

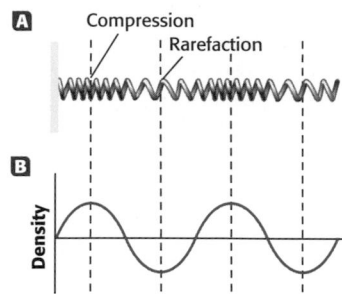

Disc Two, Module 12:
Frequency and Wavelength
Use the Interactive Tutor to learn more about these topics.

Figure 10

A A longitudinal wave has compressions and rarefactions.

B The high and low points of this sine curve correspond to compressions and rarefactions in the spring.

▶ **crest** the highest point of a wave

▶ **trough** the lowest point of a wave

▶ **amplitude** the maximum distance that the particles of a wave's medium vibrate from their rest position

▶ **wavelength** the distance from any point on a wave to an identical point on the next wave

464

Amplitude measures the amount of particle vibration

The highest points of a transverse wave are called **crests.** The lowest parts of a transverse wave are called **troughs.** The greatest distance that particles are displaced from their normal resting positions because of a wave is called the **amplitude.** The amplitude is also half the vertical distance between a crest and a trough. Larger waves have bigger amplitudes and carry more energy.

But what about longitudinal waves? These waves do not have crests and troughs because they cause particles to move back and forth instead of up and down. If you make a longitudinal wave in a spring, you will see a moving pattern of areas where the coils are bunched up alternating with areas where the coils are stretched out. The crowded areas are called *compressions.* The stretched-out areas are called *rarefactions. Figure 10A* illustrates these properties of a longitudinal wave.

Figure 10B shows a graph of a longitudinal wave. Density of the medium is plotted on the vertical axis; the horizontal axis represents the distance along the spring. The result is a sine curve. The amplitude of a longitudinal wave is the maximum deviation from the normal density or pressure of the medium, which is shown by the high and low points on the graph.

Wavelength measures the distance between two equivalent parts of a wave

The crests of ocean waves at a beach may be separated by several meters, while ripples in a pond may be only a few centimeters apart. Crests of a light wave may be separated by only billionths of a meter.

The distance from one crest of a wave to the next crest, or from one trough to the next trough, is called the **wavelength.** In a longitudinal wave, the wavelength is the distance between two compressions or between two rarefactions. The wavelength is the distance between any two successive identical parts of a wave.

Not all waves have a single wavelength that is easy to measure. Most sound waves have a very complicated shape, so the wavelength may be difficult to determine. If the source of a wave vibrates in an irregular way, the wavelength may change over time.

When used in equations, wavelength is represented by the Greek letter lambda, λ. Because wavelength is a distance measurement, it is expressed in the SI unit meters.

Properties of Waves:

Type of Wave	Mechanical		Electro-magnetic
Form	Longitudinal	Transverse	Modeled as transverse
Wavelength	Distance between successive compressions or rarefactions	Distance between successive crests or troughs	Distance between successive crests or troughs
Amplitude	Difference in pressure between maximum compression and the resting state	Difference in height between a crest and the resting state	Modeled as the difference between maximum field strength and zero

Teacher Resources

For the New Teacher You can use simple demonstrations to illustrate wave characteristics. Use a spring toy to represent longitudinal waves. Have students identify the compressions and rarefactions, and point out that they represent energy maximums and minimums.

You can use the ripple tank (from the first demonstration) to demonstrate period and wavelength. First, use the dowel to generate regular waves. Then vary the frequency of vibration of the dowel to produce waves of varying wavelength and period.

The period measures how long it takes for waves to pass by

If you swim out into the ocean until your feet can no longer touch the bottom, your body will be free to move up and down as waves come into shore. As your body rises and falls, you can count off the number of seconds between two successive wave crests.

The time required for one full wavelength of a wave to pass a certain point is called the **period** of the wave. The period is also the time required for one complete vibration of a particle in a medium—or of a swimmer in the ocean. In equations, the period is represented by the symbol T. Because the period is a time measurement, it is expressed in the SI unit seconds.

Frequency measures the rate of vibrations

While swimming in the ocean or floating in an inner tube, as shown in *Figure 11,* you could also count the number of crests that pass by in a certain time, say in 1 minute. The **frequency** of a wave is the number of full wavelengths that pass a point in a given time interval. The frequency of a wave also measures how rapidly vibrations occur in the medium, at the source of the wave, or both.

The symbol for frequency is f. The SI unit for measuring frequency is hertz (Hz), named after Heinrich Hertz, who in 1888 became the first person to experimentally demonstrate the existence of electromagnetic waves. Hertz units measure the number of vibrations per second. One vibration per second is 1 Hz, two vibrations per second is 2 Hz, and so on. You can hear sounds with frequencies as low as 20 Hz and as high as 20 000 Hz. When you hear a sound at 20 000 Hz, there are 20 000 compressions hitting your ear every second.

The frequency and period of a wave are related. If more vibrations are made in a second, each one takes a shorter amount of time. In other words, the frequency is the inverse of the period.

■ **period** the time that it takes a complete cycle or wave oscillation to occur

■ **frequency** the number of cycles or vibrations per unit of time

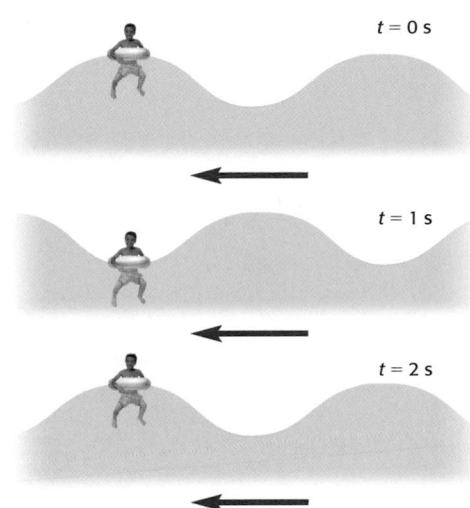

t = 0 s

t = 1 s

t = 2 s

Figure 11

A person floating in an inner tube can determine the period and frequency of the waves by counting off the number of seconds between wave crests.

Frequency-Period Equation

$$frequency = \frac{1}{period} \qquad f = \frac{1}{T}$$

In the inner tube example, a wave crest passes the inner tube every 2 s, so the period is 2 s. The frequency can be found by using the frequency-period equation above. Because 0.5 (1/2) is the inverse of 2, the frequency is 0.5 Hz, or half a wave per second.

465

Did You Know ?

Elephant Sounds Elephants communicate with low-frequency sounds that humans cannot hear. Their low-frequency calls travel much further than higher-frequency sounds. Under the right conditions, an elephant call can carry over thirty square kilometers or more! Males use these calls to find fertile females. Elephant families separated by several kilometers use the calls to coordinate their movements. Other animals that communicate with low-frequency sounds include blue and finback whales, hippos, and rhinos.

INCLUSION Strategies

- *Learning Disabled*
- *Attention Deficit Disorder*
- *English Language Learners*

Using **Figure 9,** students can create a model showing the characteristics of a wave. Have students first form the shape of the wave using pipe cleaners or string, and then glue their wave to a piece of cardboard. Have students label the following: wavelength, crest, trough, and amplitude. This model can be used as a study guide by covering the labels and asking students to recall which wave characteristic is hidden.

Teaching Tip

Frequency One wave every 5 s is the equivalent of 0.2 waves per second. Point out that there is no such thing as 0.2 waves, but wave frequencies are always expressed as waves per second, or Hz.

SKILL BUILDER — BASIC

Interpreting Visuals Figure 11 shows a boy floating in an inner tube as waves pass by. Each successive image is one second later than the previous image. Have students determine the frequency and period of these waves. ($T = 2$ s; $f = 0.5$ Hz) **LS** Logical

Teaching Tip — GENERAL

Period and Frequency To help students understand the inverse relationship between wave period and wave frequency, try this analogy. Suppose you are at a railroad crossing waiting for a freight train to pass. All of the boxcars are the same length, and you find that 120 cars pass in 5.0 minutes. Calculate the period and frequency at which the cars pass. ($T = 2.5$ s; $f = 0.4$ Cars per second) **LS** Logical

Chapter Resource File

- **Cross-Disciplinary Worksheet** Connection to Language Arts— Writing a Plan for Wave Observation **ADVANCED**

Transparencies

TM Longitudinal Wave

TT Frequency

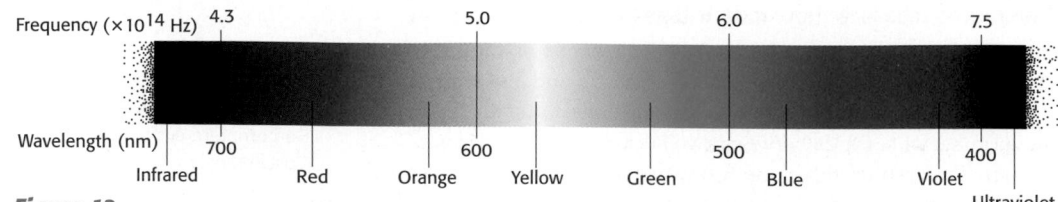

Frequency (×10^{14} Hz) 4.3   5.0   6.0   7.5

Wavelength (nm) 700   600   500   400
Infrared   Red   Orange   Yellow   Green   Blue   Violet   Ultraviolet

Figure 12

The part of the electromagnetic spectrum that we can see is called visible light.

Teach, *continued*

SKILL BUILDER — BASIC

Interpreting Visuals Have students examine **Figure 12**. Ask students to formulate a general statement comparing the wavelengths and frequencies at the red and violet extremes of the visible spectrum. (Sample answer: The radiation at the violet end of the spectrum has the shortest wavelengths and the highest frequencies. Radiation at the red end has the longest wavelengths and the lowest frequencies.) **LS Visual**

Teaching Tip

Visible Light Help students understand that visible light has no special properties when compared to other electromagnetic waves. In addition, visible light makes up only a tiny fraction of the entire electromagnetic spectrum.

SKILL BUILDER

Interpreting Visuals Table 1 shows the ranges of the different parts of the electromagnetic spectrum. Point out that the range for visible light is very small compared to some of the other ranges. Also point out that as frequency increases, wavelength decreases.

Light comes in a wide range of frequencies and wavelengths

Our eyes can detect light with frequencies ranging from about 4.3×10^{14} Hz to 7.5×10^{14} Hz. Light in this range is called *visible light*. The differences in frequency in visible light account for the different colors we see, as shown in **Figure 12**.

Electromagnetic waves also exist at other frequencies that we cannot see directly. The full range of light at different frequencies and wavelengths is called the *electromagnetic spectrum*. **Table 1** lists several different parts of the electromagnetic spectrum, along with some real-world applications of the different kinds of waves.

Table 1   The Electromagnetic Spectrum

Type of wave	Range of frequency and wavelength	Applications
Radio wave	$f < 1 \times 10^9$ Hz $\lambda > 30$ cm	AM and FM radio; television broadcasting; radar; aircraft navigation
Microwave	1×10^9 Hz $< f < 3 \times 10^{11}$ Hz 30 cm $> \lambda > 1$ mm	Atomic and molecular research; microwave ovens
Infrared (IR) wave	3×10^{11} Hz $< f < 4.3 \times 10^{14}$ Hz 1 mm $> \lambda > 700$ nm	Infrared photography; remote-control devices; heat radiation
Visible light	4.3×10^{14} Hz $< f < 7.5 \times 10^{14}$ Hz 700 nm (red) $> \lambda > 400$ nm (violet)	Visible-light photography; optical microscopes; optical telescopes
Ultraviolet (UV) light	7.5×10^{14} Hz $< f < 5 \times 10^{15}$ Hz 400 nm $> \lambda > 60$ nm	Sterilizing medical instruments; identifying fluorescent minerals
X ray	5×10^{15} Hz $< f < 3 \times 10^{21}$ Hz 60 nm $> \lambda > 1 \times 10^{-4}$ nm	Medical examination of bones, teeth, and organs; cancer treatments
Gamma ray	3×10^{18} Hz $< f < 3 \times 10^{22}$ Hz 0.1 nm $> \lambda > 1 \times 10^{-5}$ nm	Food irradiation; studies of structural flaws in thick materials

466

Historical Perspective

The Discovery of Electromagnetic Waves

In 1819, Hans Christian Oersted observed that a current-carrying wire placed near a magnetic compass moved the compass needle. This demonstrated that there is a physical relationship between electricity and magnetism.

The exact nature of the relationship remained unclear until 1871, when James Clerk Maxwell published a comprehensive theory of electromagnetism. Maxwell described electromagnetic waves mathematically, and discovered that they travel at the speed of light. This suggested that visible light itself is an electromagnetic wave—an idea that was further confirmed when Maxwell's equations correctly predicted the properties of light.

In 1886, Heinrich Hertz verified Maxwell's theoretical idea empirically. Hertz used a current vibrating in a wire to produce radio signals and transmit them across his lab. In the twentieth century, scientists began developing practical applications for electromagnetic waves, including radios, televisions, and microwave ovens.

Wave Speed

Imagine watching as water waves move past a post at a pier such as the one in **Figure 13.** If you count the number of crests passing the post for 10 s, you can determine the frequency of the waves by dividing the number of crests you count by 10 s. If you measure the distance between crests, you can find the wavelength of the wave. But how fast are the water waves moving?

Wave speed equals frequency times wavelength

The speed of a moving object is found by dividing the distance the object travels by the time it takes to travel that distance. Speed can be calculated using the speed equation:

$$speed = \frac{distance}{time}$$

$$v = \frac{d}{t}$$

If SI units are used for measuring distance and time, speed is expressed as meters per second (m/s). The *wave speed* is simply how fast a wave moves. Finding the speed of a wave is just like finding the speed of a moving object: you need to measure how far the wave travels in a certain amount of time.

For a wave, it is most convenient to use the wavelength as the distance traveled. The amount of time it takes the wave to travel a distance of one wavelength is the period. Substituting these into the speed equation gives an equation that can be used to calculate the speed of a wave.

$$speed = \frac{wavelength}{period}$$

$$v = \frac{\lambda}{T}$$

Because the period is the inverse of the frequency, dividing by the period is equivalent to multiplying by the frequency. Therefore, the speed of a wave can also be calculated by multiplying the wavelength by the frequency.

> **Wave Speed Equation**
> $$wave\ speed = frequency \times wavelength$$
> $$v = f \times \lambda$$

Suppose that waves passing by a post at a pier have a wavelength of 10 m, and two waves pass by in 5 s. The period is therefore 2.5 s, and the frequency is the inverse of 2.5 s, or 0.4 Hz. The waves in this case travel with a wave speed of 4 m/s.

Figure 13

By observing the frequency and wavelength of waves passing a pier, you can calculate the speed of the waves.

INTEGRATING

EARTH SCIENCE
Earthquakes create waves, called *seismic waves,* that travel through Earth. There are two main types of seismic waves, *P waves* (primary waves) and *S waves* (secondary waves).

P waves travel faster than S waves, so the P waves arrive at a given location first. P waves are longitudinal waves that tend to shake the ground from side to side.

S waves move more slowly than P waves but also carry more energy. S waves are transverse waves that shake the ground up and down, often damaging buildings and roads.

Teaching Tip
Demonstrating Wave Speed

Use a ripple tank or shallow baking dish to show differences in wave speed. Adjust the water depth for the best effect.

The speed of water waves changes in shallow water. Place a block with a large horizontal surface in the tank so that it is submerged less than 1 cm. Students will observe that the ripples change speed as they pass over the block.

INTEGRATING

EARTH SCIENCE P waves and S waves travel through Earth from the focus of an earthquake. In addition, earthquakes produce L waves and R waves. L waves are up-and-down transverse waves that travel along Earth's surface. R waves are side-to-side transverse waves that also travel along Earth's surface.

OCEANOGRAPHY
CONNECTION — **GENERAL**

Ask students if they have observed waves breaking at the beach. When a wave approaches the shore, its height increases as the water depth decreases. The crests become peaked, and the wave loses stability. Eventually, the speed of the crest exceeds the speed of the wave, and the wave "breaks."

The steepness of the ocean floor determines how the waves break. When the floor is very steep, the waves roll onto the beach before actually breaking. These waves—called *surging breakers*—are the most destructive. A moderately steep slope produces *plunging breakers*, a favorite of surfers. Plunging breakers have a characteristic curl at the top of the crest. Gentle slopes produce *spilling breakers*, which break far from the shore. Foam slides down the surface as these waves slowly approach the shore.

Chapter Resource File

- **Cross-Disciplinary Worksheet** Integrating Computers and Technology—Radio Waves ADVANCED

- **Cross-Disciplinary Worksheet** Integrating Earth Science— Earthquake Waves ADVANCED

  Transparencies

TT Visible Light

TM The Electromagnetic Spectrum

Math Skills

Additional Examples

Wave Speed The speed of sound in dry air at 40°C is 355 m/s. The speed at 0°C is 331 m/s. Will a sound of a given frequency have a longer or shorter wavelength at 40°C than at 0°C? (longer; the waves will be farther apart because the speed is greater.)

Calculate the frequency of microwaves that have a wavelength of 0.0085 m and a speed of 3.00×10^8 m/s. (3.5×10^{10} Hz)

The speed of sound in sea water is 1530 m/s at 25°C. A sonar device emits a pulse of sound of 922 Hz. What is the wavelength of sound at this frequency in sea water? If an echo from the sonar device returns 4.76 s after the pulse is sent, how deep is the object that reflected the sonar? (1.66 m; 3640 m in depth) **LS Logical**

Practice

1. $v = f \times \lambda$
   $v = 0.100 \, \text{Hz} \times 15.0 \, \text{m} = 1.50 \, \text{m/s}$

2. $v = f \times \lambda$
   $v = (9.45 \times 10^7 \, \text{Hz})(3.17 \, \text{m}) =$
   $3.00 \times 10^8 \, \text{m/s}$

3. $f = \dfrac{v}{\lambda}$

   $f = \dfrac{3.00 \times 10^8 \, \text{m/s}}{5.20 \times 10^{-7} \, \text{m}} =$

   $5.77 \times 10^{14} \, \text{Hz}$

4. $\lambda = \dfrac{v}{f}$

   $\lambda = \dfrac{340 \, \text{m/s}}{200 \text{Hz}} = 1.7 \, \text{m}$

Practice HINT

▶ When a problem requires you to calculate wave speed, you can use the wave speed equation on the previous page.

▶ The wave speed equation can also be rearranged to isolate frequency on the left in the following way:

$$v = f \times \lambda$$

Divide both sides by λ.

$$\frac{v}{\lambda} = \frac{f \times \lambda}{\lambda}$$

$$f = \frac{v}{\lambda}$$

You will need to use this form of the equation in Practice Problem 3.

▶ In Practice Problem 4, you will need to rearrange the equation to isolate wavelength on the left.

Math Skills

Wave Speed The string of a piano that produces the note middle C vibrates with a frequency of 264 Hz. If the sound waves produced by this string have a wavelength in air of 1.30 m, what is the speed of sound in air?

1 List the given and unknown values.
   Given: *frequency, f* = 264 Hz
   wavelength, λ = 1.30 m
   Unknown: *wave speed, v* = ? m/s

2 Write the equation for wave speed.
   $v = f \times \lambda$

3 Insert the known values into the equation, and solve.
   $v = 264 \, \text{Hz} \times 1.30 \, \text{m} = 264 \, \text{s}^{-1} \times 1.30 \, \text{m}$
   $v = 343 \, \text{m/s}$

Practice

Wave Speed

1. The average wavelength in a series of ocean waves is 15.0 m. A wave crest arrives at the shore on average every 10.0 s, so the frequency is 0.100 Hz. What is the average speed of the waves?

2. An FM radio station broadcasts electromagnetic waves at a frequency of 94.5 MHz (9.45×10^7 Hz). These radio waves have a wavelength of 3.17 m. What is the speed of the waves?

3. Green light has a wavelength of 5.20×10^{-7} m. The speed of light is 3.00×10^8 m/s. Calculate the frequency of green light waves with this wavelength.

4. The speed of sound in air is about 340 m/s. What is the wavelength of a sound wave with a frequency of 220 Hz (on a piano, the A below middle C)?

The speed of a wave depends on the medium

Sound waves can travel through air. If they couldn't, you would not be able to have a conversation with a friend or hear music from a radio across the room. Because sound travels very fast in air (about 340 m/s), you don't notice a time delay in most normal situations.

If you swim with your head underwater, you may hear certain sounds very clearly. Sound waves travel better—and three to four times faster—in water than in air. Dolphins, such as those in *Figure 14,* use sound waves to communicate with one another over long distances underwater. Sound waves travel even faster in solids than in air or in water. Sound waves have speeds 15 to 20 times faster in rock or metal than in air.

468

MISCONCEPTION ALERT

Students confuse independent properties of waves, such as amplitude, frequency, and wave speed. Some students believe that a rapid vibration will always produce a large amplitude and a fast wave speed. Emphasize that wave speed depends on the medium. A vibration in air will produce a different wave speed than the same vibration in water. Temperature also affects wave speed. Sound travels faster through hot air than through cool air.

If someone strikes a long steel rail with a hammer at one end and you listen for the sound at the other end, you might hear two bangs. The first sound comes through the steel rail itself and reaches you shortly before the second sound, which travels through the air.

The speed of a wave depends on the medium. In a given medium, though, the speed of waves is constant; it does not depend on the frequency of the wave. No matter how fast you shake your hand up and down to create waves on a rope, the waves will travel the same speed. Shaking your hand faster just increases the frequency and decreases the wavelength.

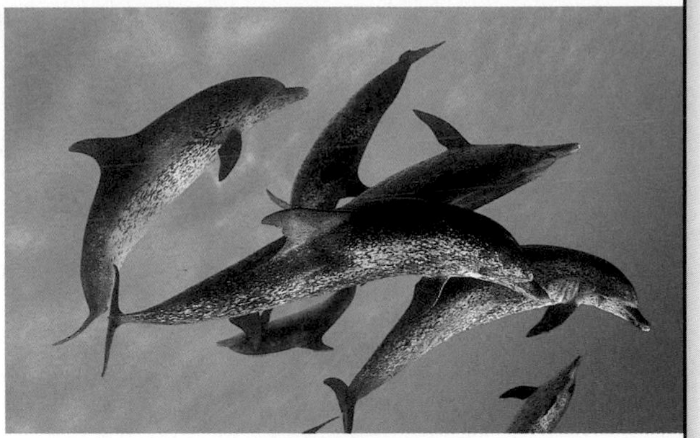

Figure 14
Dolphins use sound waves to communicate with one another. Sound travels three to four times faster in water than in air.

Kinetic theory explains differences in wave speed

The arrangement of particles in a medium determines how well waves travel through it. The different states of matter are due to different degrees of organization at the particle level.

In gases, the molecules are far apart and move around randomly. A molecule must travel through a lot of empty space before it bumps into another molecule. Waves don't travel as fast in gases.

In liquids, such as water, the molecules are much closer together. But they are also free to slide past one another. As a result, vibrations are transferred more quickly from one molecule to the next than they are in a gas. This situation can be compared to vibrating masses on springs that are so close together that the masses rub against each other.

In a solid, molecules are not only closer together but also tightly bound to each other. The effect is like having vibrating masses that are glued together. When one mass starts to vibrate, all the others start to vibrate almost immediately. As a result, waves travel very quickly through most solids.

Light has a finite speed

When you flip a light switch, light seems to fill the room instantly. However, light does take time to travel from place to place. All electromagnetic waves in empty space travel at the same speed, the speed of light, which is 3.00×10^8 m/s (186 000 mi/s). The speed of light in empty space is a constant that is often represented by the symbol c. Light travels slower when it has to pass through a medium such as air or water.

Quick ACTIVITY

Wave Speed

1. Place a rectangular pan on a level surface, and fill the pan with water to a depth of about 2 cm.
2. Cut a wooden dowel (3 cm in diameter or thicker) to a length slightly less than the width of the pan, and place the dowel in one end of the pan.
3. Move or roll the dowel slowly back and forth, and observe the length of the wave generated.
4. Now move the dowel back and forth faster (increased frequency). How does that affect the wavelength?
5. Do the waves always travel the same speed in the pan?

Did You Know?

Speed of Light Although light requires no medium for transmission, it does interact with matter that it passes through by slowing down somewhat, depending on the material. For example, light slows down from 3.00×10^8 m/s to around 2.25×10^8 m/s when it enters water. The speed of light in diamond is only about 40% the speed of light in air.

Teaching Tip

Elasticity The speed of sound in a material depends on the elasticity of the material. Elasticity is the tendency of a material to restore itself or "bounce back" after being deformed by an external force. A sound wave is a compression wave, so it is reasonable that a material that quickly bounces back from a compression can transmit sound at a high velocity.

Very elastic materials include steel, glass, and aluminum. Lead, on the other hand, is very inelastic by comparison because it deforms semi-permanently rather than bouncing back. The speed of sound in lead is about one-third of the speed of sound in steel. Materials thought of as elastic, such as rubber, actually have relatively poor elasticity. Materials such as dough and modeling clay are very inelastic.

Quick ACTIVITY

Wave Speed

Materials (per group):

- rectangular pan
- water
- wooden dowel (3 cm diameter or thicker)

Teacher's Notes: You may also use the setup from the first Demonstration in Section 1.

Results: Students should observe that the velocity of the waves stays the same because it is determined by the medium. Waves of higher frequency will have shorter wavelengths, and waves of lower frequency will have longer wavelengths. LS **Kinesthetic**

Chapter Resource File

- **Math Skills** Wave Speed GENERAL
- **Quick Activity Datasheet** GENERAL

Demonstration ——— GENERAL

Doppler Effect

(Time: About 5 minutes)

Materials:

• strong mesh bag
• battery-powered alarm clock
• 1 m of heavy cord

Step 1 Adjust the clock to start the alarm. Place the clock in a mesh bag tied securely to a sturdy cord about 1 m long.

Step 2 Swing the bag containing the clock in a circle over your head. To hear the changes in pitch, students should stand just below the plane of the circle. Do not allow students to stand in the plane in case the bag or cord should break. If necessary, add mass to the bag to slow the circular motion. **LS Auditory**

Teaching Tip

The Doppler Effect Reinforce the idea that the pitch of sound you hear depends entirely on the frequency at which compressions strike the eardrum, even if the source of vibration is at a higher or lower frequency. Emphasize that the frequency of vibration of the sound source remains constant. However, the waves are pushed closer together ahead of the moving vehicle and stretched farther apart behind it. Also stress the point that the speed of the waves does not change; it is determined by the nature of the medium.

The Doppler Effect

Imagine that you are standing on a corner as an ambulance rushes by. As the ambulance passes, the sound of the siren changes from a high pitch to a lower pitch. Why? Do the sound waves produced by the siren change as the ambulance goes by? How does the motion of the ambulance affect the sound?

Pitch is determined by the frequency of sound waves

The *pitch* of a sound, how high or low it is, is determined by the frequency at which sound waves strike the eardrum in your ear. A higher-pitched sound is caused by sound waves of higher frequency. As you know from the wave speed equation, frequency and wavelength are also related to the speed of a wave.

Suppose you could see the sound waves from the ambulance siren when the ambulance is at rest. You would see the sound waves traveling out from the siren in circular wave fronts, as shown in *Figure 15A.* The distance between two successive wave fronts shows the wavelength of the sound waves. When the sound waves reach your ears, they have a frequency equal to the number of wave fronts that strike your eardrum each second. That frequency determines the pitch of the sound that you hear.

Figure 15

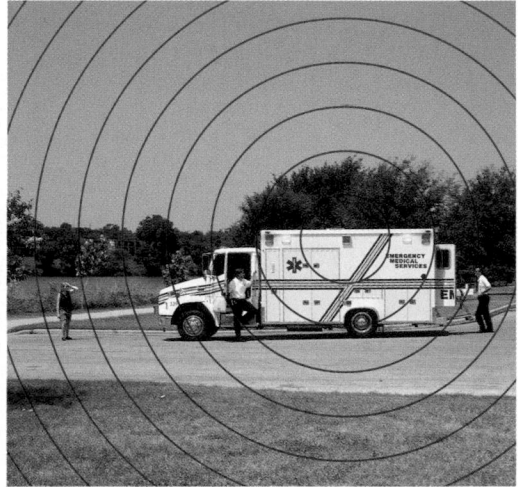

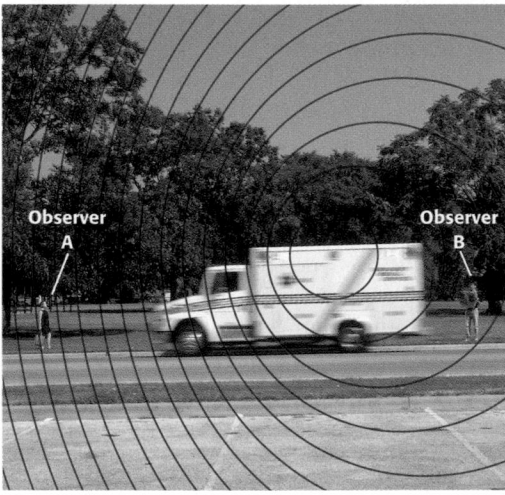

A When an ambulance is not moving, the sound waves produced by the siren spread out in circles. The frequency of the waves is the same at any location.

B When an ambulance is moving, the sound waves produced by the siren are closer together in front and farther apart behind. Observer A hears a higher-pitched sound than Observer B hears.

470

Did You Know ?

Ambulance Siren A siren sound does not become noticeably higher as an ambulance approaches. It will be high when it is first heard. The only observable change in pitch occurs as the sound source passes.

Alternative Assessment ——— BASIC

Wave Review Have students draw examples of two transverse waves, making the second wave have a higher frequency and lower amplitude than the first. Ask students to label wavelength and amplitude on the two waves. Ask students to explain the nature of a sound wave and how it travels through matter (as compressions and rarefactions). Also ask them what measure constitutes wavelength in a sound wave (distance between compressions or rarefactions). **LS Visual**

Frequency changes when the source of waves is moving

If the ambulance is moving toward you, the sound waves from the siren are compressed in the direction of motion, as shown in **Figure 15B.** Between the time that one sound wave and the next sound wave are emitted by the siren, the ambulance moves a short distance. This shortens the distance between wave fronts, while the wave speed remains the same. As a result, the sound waves reach your ear at a higher frequency.

Because the waves now have a higher frequency, you hear a higher-pitched sound than you would if the ambulance were at rest. Similarly, if the ambulance were moving away from you, the frequency at which the waves reached your ear would be less than if the ambulance were at rest, and you would hear the sound of the siren at a lower pitch. This change in the observed frequency of a wave resulting from the motion of the source or observer is called the **Doppler effect.** The Doppler effect occurs for light and other types of waves as well.

internet connect

www.scilinks.org
Topic: Doppler Effect
SciLinks code: HK4032

SciLINKS Maintained by the National Science Teachers Association

▶ **Doppler effect** an observed change in the frequency of a wave when the source or observer is moving

Teaching Tip
Observing the Doppler Effect

Ask students to relate experiences in which they heard the Doppler effect, such as along highways, at airports, or near railroad tracks. Ask students to explain the characteristic sound that allows you to identify an auto race on TV, even if you can't see the picture. Consider having students use an audio recorder or a video camera to record examples of the Doppler effect. **LS Auditory**

SECTION 2 REVIEW

SUMMARY

▶ The highest points of a transverse wave are called crests; the lowest parts are called troughs.

▶ The amplitude of a transverse wave is half the vertical distance between a crest and a trough.

▶ The wavelength is the distance between two identical parts of a wave.

▶ The period of a wave is the time it takes a wavelength to pass a certain point.

▶ The frequency of a wave is the number of vibrations that occur in a given amount of time. (1 Hz = 1 vibration/s)

▶ The speed of a wave equals the frequency times the wavelength. ($v = f \times \lambda$)

1. **Draw** a sine curve, and label a crest, a trough, and the amplitude.

2. **State** the SI units used for wavelength, period, frequency, and wave speed.

3. **Describe** how the frequency and period of a wave are related.

4. **Explain** why sound waves travel faster in liquids or solids than in air.

5. **Critical Thinking** What happens to the wavelength of a wave when the frequency of the wave is doubled but the wave speed stays the same?

6. **Critical Thinking** Imagine you are waiting for a train to pass at a railroad crossing. Will the train whistle have a higher pitch as the train approaches you or after it has passed you by?

Practice

7. A wave along a guitar string has a frequency of 440 Hz and a wavelength of 1.5 m. What is the speed of the wave?

8. The speed of sound in air is about 340 m/s. What is the wavelength of sound waves produced by a guitar string vibrating at 440 Hz?

9. The speed of light is 3×10^8 m/s. What is the frequency of microwaves with a wavelength of 1 cm?

471

Answers to Section 2 Review

1. Students' drawings should resemble **Figure 9A.**

2. wavelength, m; period, s; frequency, Hz; wave speed, m/s.

3. Frequency and period are the inverse of each other.

4. Sound travels faster in liquids and solids because the particles of the medium are closer together than particles in air.

5. The wavelength is halved.

6. The pitch will sound higher as the train approaches and will become lower as the whistle passes.

Practice

7. $v = f \times \lambda$
$v = (440 \text{ Hz})(1.5 \text{ m}) = 660 \text{ m/s}$

8. $\lambda = \dfrac{v}{f}$

$\lambda = \dfrac{340 \text{ m/s}}{440 \text{ Hz}} = 0.77 \text{ m}$

9. $f = \dfrac{v}{\lambda}$

$f = \dfrac{3 \times 10^8 \text{ m/s}}{0.01 \text{ m}}$

$f = 3 \times 10^{10} \text{ Hz}$

SECTION
3

Focus

Overview

Before beginning this section, review with your students the Objectives listed in the Student Edition. This section covers wave behaviors and interactions. Students learn about reflection, diffraction, refraction, and constructive and destructive interference. They also learn how interference can create standing waves.

🎧 Bellringer

Use the Bellringer transparency to prepare students for this section.

Motivate

Identifying Preconceptions ── BASIC

Ask students what will happen when two compression waves traveling toward each other on a toy spring collide. Some students may predict that each wave will block the other from passing, effectively canceling both waves. Have two volunteers use a toy spring to demonstrate this example. Be sure the students stand far enough apart, and instruct each to generate a single wave with one quick motion. Point out that the waves did not cancel one another; instead, each passed through the other and continued traveling in its original direction. **LS** Kinesthetic

Wave Interactions

■ **KEY TERMS**
reflection
diffraction
refraction
interference
constructive interference
destructive interference
standing wave

▶ **OBJECTIVES**

▶ **Describe** how waves behave when they meet an obstacle or pass into another medium.

▶ **Explain** what happens when two waves interfere.

▶ **Distinguish** between constructive interference and destructive interference.

▶ **Explain** how standing waves are formed.

PHYSICAL SCIENCE INTERACTIVE TUTOR

Disc Two, Module 13: Reflection
Use the Interactive Tutor to learn more about this topic.

■ **reflection** the bouncing back of a ray of light, sound, or heat when the ray hits a surface that it does not go through

W hen waves are simply moving through a medium or through space, they may move in straight lines like waves on the ocean, spread out in circles like ripples on a pond, or spread out in spheres like sound waves in air. But what happens when a wave meets an object or another wave in the medium? And what happens when a wave passes into another medium?

Reflection, Diffraction, and Refraction

You probably already know what happens when light waves strike a shiny surface: they reflect off the surface. Other waves reflect, too. *Figure 16* shows two ways that a wave on a rope may be reflected. **Reflection** is simply the bouncing back of a wave when it meets a surface or boundary.

Figure 16

A If the end of a rope is free to slide up and down a post, a wave on the rope will reflect from the end.

B If the end of the rope is fixed, the reflected wave is turned upside down.

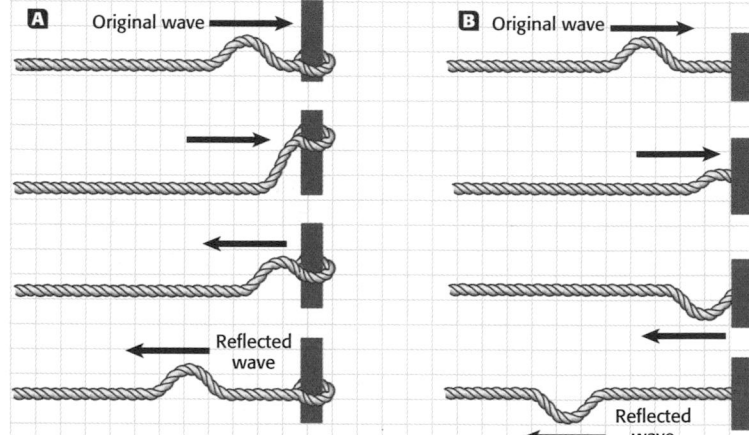

Did You Know ?

Electromagnetic Waves Mechanical waves will always spread out from a source into the available medium. However, spreading out is not an inherent property of electromagnetic waves. If light waves are given out from a point source in space, they will travel in all directions. But light waves can also be focused into a beam that does not spread out. For example, a laser beam consists of light waves traveling parallel to each other and not spreading out.

Waves reflect at a free boundary

Figure 16A shows the reflection of a single wave traveling on a rope. The end of the rope is free to move up and down on a post. When the wave reaches the post, the loop on the end moves up and then back down. This is just what would happen if someone were shaking that end of the rope to create a new wave. The reflected wave in this case is exactly like the original wave except that the reflected wave is traveling in the opposite direction to the direction of the original wave.

At a fixed boundary, waves reflect and turn upside down

Figure 16B shows a slightly different situation. In this case, the end of the rope is not free to move because it is attached to a wall. When the wave reaches the wall, the rope exerts an upward force on the wall. The wall is too heavy to move, but it exerts an equal and opposite downward force on the rope, following Newton's third law. The force exerted by the wall causes another wave to start traveling down the rope. This reflected wave travels in the opposite direction and is turned upside down.

Diffraction is the bending of waves around an edge

If you stand outside the doorway of a classroom, you may be able to hear the sound of voices inside the room. But if the sound waves cannot travel in a straight line to your ear, how are you able to hear the voices?

When waves pass the edge of an object or pass through an opening, such as an open window or a door, they spread out as if a new wave were created there. In effect, the waves seem to bend around an object or opening. This bending of waves as they pass an edge is called **diffraction.**

Figure 17A shows waves passing around a block in a tank of water. Before they reach the block, the waves travel in a straight line. After they pass the block, the waves near the edge bend and spread out into the space behind the block. Diffraction is the reason that shadows never have perfectly sharp edges.

The tank in *Figure 17B* contains two blocks placed end to end with a small gap in between. In this case, waves bend around two edges and spread out as they pass through the opening. Sound waves passing through a door behave the same way. Because sound waves spread out into the space beyond the door, a person near the door on the outside can hear sounds from inside the room.

internet connect

www.scilinks.org
Topic: Reflection, Refraction, Diffraction
SciLinks code: HK4119

SCiLINKS. Maintained by the National Science Teachers Association

▶ **diffraction** a change in the direction of a wave when the wave finds an obstacle or an edge, such as an opening

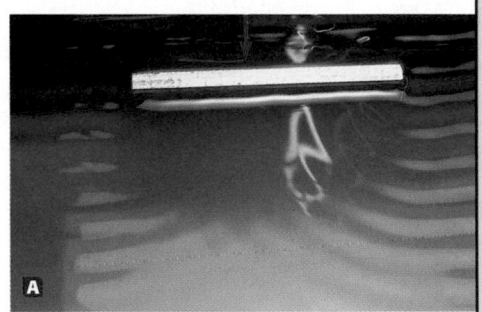

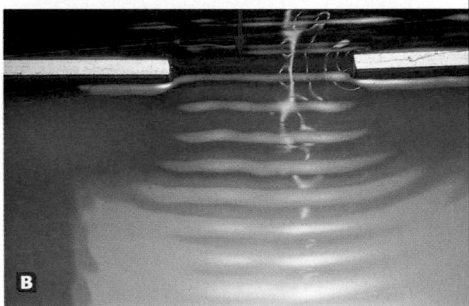

Figure 17

🅰 Waves bend when they pass the edge of an obstacle.

🅱 When they pass through an opening, waves bend around both edges.

473

Teach

SKILL BUILDER — BASIC

Reading Skills Have students read the material on reflection, diffraction, and refraction, one topic at a time. After each topic, ask for volunteers to summarize each phenomenon, giving particular attention to what happens to the waves and the conditions that are necessary to make this happen. **LS Verbal**

Teaching Tip

Bending Waves Caution students that the word *bend*, when applied to light waves, seldom means that they bend in a curving path like a highway. In diffraction and refraction, light waves change direction abruptly.

Alternative Assessment — BASIC

Wave Phenomena Ask students to draw diagrams showing reflection, refraction, and diffraction of waves. Students do not need to draw waves but simply sketch the geometry of each phenomenon. Tell them to label each of the diagrams clearly as to which phenomenon it represents. Students should also label the objects and boundaries involved and use arrows to show the direction of wave travel both before and after the encounter between the waves and the objects. **LS Visual**

Chapter Resource File

• Lesson Plan

Transparencies

TT Bellringer

Observing Reflection and Refraction

(Time: About 5 minutes)

Materials:

• aquarium tank (5 or 10 gal) or other large transparent container
• milk
• mirror
• laser pointer or focusing flashlight

Step 1 Fill the aquarium with water. Leave enough space above the water so that you can reach to the bottom of the tank without it spilling over. Add three drops of milk to the water, and stir. Shine the laser into the water to see if the beam is visible. Add more milk as needed.

Step 2 Demonstrate refraction by directing the laser beam straight down into the water and then moving it in an arc to change the angle at which the beam enters the water. Students will be able to see the beam bend toward the normal (vertical) as it enters the water.

Step 3 Demonstrate reflection by placing the small mirror on the bottom of the tank. **LS** Visual

Figure 18

Because light waves bend when they pass from one medium to another, this spoon looks like it is in two pieces.

▶ **refraction** the bending of a wavefront as the wavefront passes between two substances in which the speed of the wave differs

▶ **interference** the combination of two or more waves of the same frequency that results in a single wave

Figure 19

Water waves passing through each other produce interference patterns.

474

Waves can also bend by refraction

Figure 18 shows a spoon in a glass of water. Why does the spoon look like it is broken into two pieces? This strange sight results from light waves bending, but not because of diffraction. This time, the waves are bending because of **refraction.** Refraction is the bending of waves when they pass from one medium into another. All waves are refracted when they pass from one medium to another at an angle.

Light waves from the top of the spoon handle pass straight through the air and the glass from the spoon to your eyes. But the light waves from the rest of the spoon start out in the water, then pass into the glass, then into the air. Each time the waves enter a new medium, they bend slightly because of a change in speed. By the time those waves reach your eyes, they are coming from a different angle than the waves from the top of the spoon handle. But your eyes just see that one set of light waves are coming from one direction, and another set of waves are coming from a different direction. As a result, the spoon appears to be broken.

Interference

What would happen if you and another person tried to walk through the exact same place at the same time? You would run into each other. Material objects, such as a human body, cannot share space with other material objects. More than one wave, however, can exist in the same place at the same time.

Waves in the same place combine to produce a single wave

When several waves are in the same location, the waves combine to produce a single, new wave that is different from the original waves. This is called **interference.** *Figure 19* shows interference occurring as water waves pass through each other. Once the waves have passed through each other and moved on, they return to their original shape.

You can show the interference of two waves by drawing one wave on top of another on a graph, as in *Figure 20.* The resulting wave can be found by adding the height of the waves at each point. Crests are considered positive, and troughs are considered negative. This method of adding waves is sometimes known as the *principle of superposition.*

Did You Know ?

Change in Direction Waves change direction when they pass at an angle from one medium into another (unless both mediums have exactly the same refractive qualities). However, if the waves meet the boundary between mediums at a right angle, the waves will cross the boundary without changing direction.

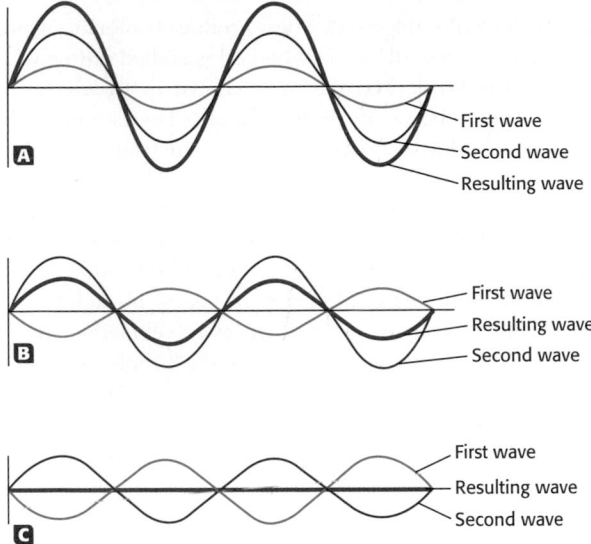

A

First wave
Second wave
Resulting wave

B

First wave
Resulting wave
Second wave

C

First wave
Resulting wave
Second wave

Figure 20

Constructive and Destructive Interference

A When two waves line up so their crests overlap, they add together to make a larger wave.

B When the crest of a large wave overlaps with the trough of a smaller wave, subtraction occurs.

C Two waves of the same size may completely cancel each other out.

PHYSICAL SCIENCE **INTERACTIVE TUTOR**

Disc Two, Module 14: **Refraction**
Use the Interactive Tutor to learn more about this topic.

Constructive interference increases amplitude

When the crest of one wave overlaps the crest of another wave, the waves reinforce each other, as shown in *Figure 20A.* Think about what happens at the particle level. Suppose the crest of one wave would move a particle up 4 cm from its original position, and another wave crest would move the particle up 3 cm.

When both waves hit at the same time, the particle moves up 4 cm due to one wave and 3 cm due to the other for a total of 7 cm. The result is a wave whose amplitude is the sum of the amplitudes of the two individual waves. This is called **constructive interference.**

Destructive interference decreases amplitude

When the crest of one wave meets the trough of another wave, the resulting wave has a smaller amplitude than the larger of the two waves, as shown in *Figure 20B.* This is called **destructive interference.**

To understand how this works, imagine again a single particle. Suppose the crest of one wave would move the particle up 4 cm, and the trough of another wave would move it down 3 cm. If the waves hit the particle at the same time, the particle would move in response to both waves, and the new wave would have an amplitude of just 1 cm. When destructive interference occurs between two waves that have the same amplitude, the waves may completely cancel each other out, as shown in *Figure 20C.*

▶ **constructive interference** any interference in which waves combine so that the resulting wave is bigger than the original waves

▶ **destructive interference** any interference in which waves combine so that the resulting wave is smaller than the largest of the original waves

475

Teach, continued

Teaching Tip

Color by Interference Other examples of color in nature may be attributed to light interference rather than colored pigments. Ask students to give examples of where they have seen colors similar to those in soap bubbles. The colors are typically jewel-like colors that we associate with the term *iridescent*. Students may think of thin oil slicks they see after a rain or the colors of liquid-crystal thermometers that you press to the forehead. The colors on the wings of butterflies and moths and the colors of many beetles' shells are produced by interference. Interference also produces colors in bird feathers, as in the tail of a peacock.

SKILL BUILDER — BASIC

Interpreting Diagrams **Figure 22** shows two waves interfering constructively and destructively to produce beats. Have students add the amplitudes of the top two waves to see how the resulting "beat wave" is produced. **LS Visual**

Figure 21

The colorful swirls on a bubble result from the constructive interference of some colors and the destructive interference of other colors.

Figure 22

A When two waves of slightly different frequencies interfere with each other, they produce beats.

B A piano tuner can listen for beats to tell if a string is out of tune.

Interference of light waves creates colorful displays

The interference of light waves often produces colorful displays. You can see a rainbow of colors when oil is spilled onto a watery surface. Soap bubbles, like the ones shown in *Figure 21,* have reds, blues, and yellows on their surfaces. The colors in these examples are not due to pigments or dyes. Instead, they are due to the interference of light.

When you look at a soap bubble, some light waves bounce off the outside of the bubble and travel directly to your eye. Other light waves travel into the thin shell of the bubble, bounce off the inner side of the bubble's shell, then travel back through the shell, into the air and to your eye. Those waves travel farther than the waves reflected directly off the outside of the bubble. At times the two sets of waves are out of step with each other. The two sets of waves interfere constructively at some frequencies (colors) and destructively at other frequencies (colors). The result is a swirling rainbow effect.

Interference of sound waves produces beats

The sound waves from two tuning forks of slightly different frequencies will interfere with each other as shown in *Figure 22A.* Because the frequencies of the tuning forks are different, the compressions arrive at your ear at different rates.

When the compressions from the two tuning forks arrive at your ear at the same time, constructive interference occurs, and the sound is louder. A short time later, the compression from one and the rarefaction from the other arrive together. When this happens, destructive interference occurs, and a softer sound is heard. After a short time, the compressions again arrive at the same time, and again a loud sound is heard. Overall, you hear a series of loud and soft sounds called *beats*.

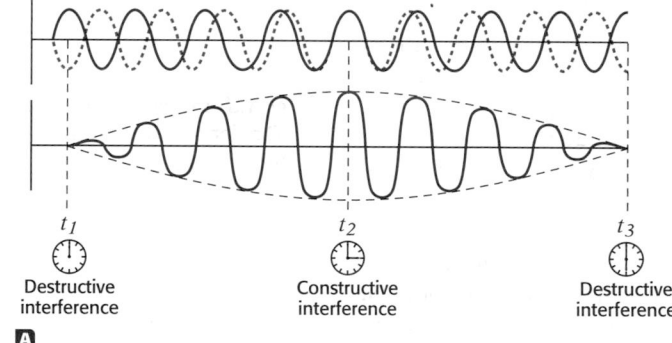

t_1
Destructive interference
A

t_2
Constructive interference

t_3
Destructive interference

B

476

Alternative Assessment — GENERAL

Standing Waves on a Rope Students can experience standing waves by using a rope. They can also observe the basics of harmonics and resonance.

Step 1 Have two students stand about 5 m apart holding a rope or piece of rubber tubing between them. The rope should be several meters long. Adjust the rope so that there is only a little slack between the students.

Step 2 Ask one student to make the entire rope vibrate in a single standing wave by moving the end up and down continuously. The rope should vibrate in a single standing wave only at a certain frequency.

Step 3 After students get the rhythm of this vibration, tell them to make the rope vibrate in two segments, with a node in the middle. They will discover that this will happen only at twice the frequency of the previous wave.

Figure 22B shows a piano tuner tuning a string. Piano tuners listen for beats between a tuning fork of known frequency and a string on a piano. By adjusting the tension in the string, the tuner can change the pitch (frequency) of the string's vibration. When no beats are heard, the string is vibrating with the same frequency as the tuning fork. In that case, the string is said to be in tune.

Standing Waves

Waves can also interfere in another way. Suppose you send a wave through a rope tied to a wall at the other end. The wave is reflected from the wall and travels back along the rope. If you continue to send waves down the rope, the waves that you make will interfere with those waves that reflect off the wall and travel back toward you.

Interference can cause standing waves

Standing waves can form when a wave is reflected at the boundary of a medium. In a standing wave, interference of the original wave with the reflected wave causes the medium to vibrate in a stationary pattern that resembles a loop or a series of loops. Although it appears as if the wave is standing still, in reality waves are traveling in both directions.

Standing waves have nodes and antinodes

Each loop of a standing wave is separated from the next loop by points that have no vibration, called *nodes*. Nodes lie at the points where the crests of the original waves meet the troughs of the reflected waves, causing complete destructive interference.

One of the nodes on a fixed rope lies at the point of reflection, where the rope cannot vibrate. Another node is near your hand. If you shake the rope up and down at the right frequency, you can create standing waves with several nodes along the length of the string.

Midway between the nodes lie points of maximum vibration, called *antinodes*. Antinodes form where the crests of the original waves line up with the crests of the reflected waves so that complete constructive interference occurs.

standing wave a pattern of vibration that simulates a wave that is standing still

Caution students to hold the rope tightly as considerable energy can build up in the standing waves.

Step 4 Ask students to make the rope vibrate at a frequency somewhere between the two observed frequencies. They will find this impossible to do. Ask students to find the next higher natural frequency of the rope. This will occur at three times the original frequency, and the rope will exhibit a standing wave of three parts with two nodes.

Step 5 Have students suggest ways to make the rope vibrate at different frequencies. Almost any change in length or tension will result in standing waves of different frequencies. Point out that the rope is a model for a string on a violin, guitar, piano, or other stringed instrument. **LS Kinesthetic**

1. What happens to a wave when it reaches a boundary? (It bounces back, or is reflected.)

2. What happens to a wave when it passes from one medium to another at an angle? (It bends, or is refracted.)

3. Interference always results in a smaller wave. True or false? (false)

4. In a standing wave, what happens at a point where a crest of the original wave meets a trough of the reflected wave? (Complete destructive interference produces a node at this point.)

LS Logical

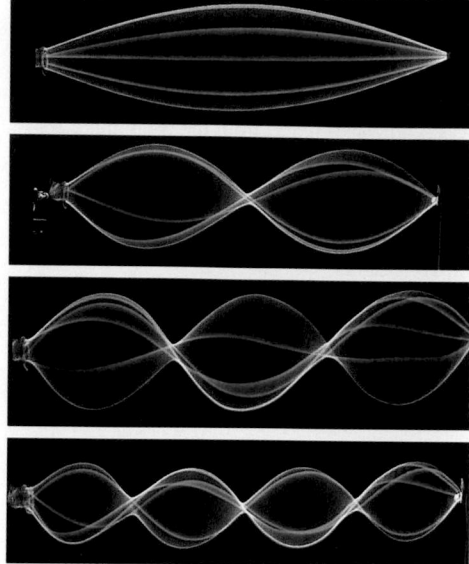

Figure 23
These photos of standing waves were captured using a strobe light that flashes different colors at different times.

Standing waves can have only certain wavelengths

Figure 23 shows several different possible standing waves on a string fixed at both ends. Only a few waves with specific wavelengths can form standing waves in any given string.

The simplest standing waves occur when the wavelength of the waves is twice the length of the string. In that case, it just looks like the entire string is shaking up and down. The only nodes are on the two ends of the string.

If the string vibrates with a higher frequency, the wavelength becomes shorter. At a certain frequency, the wavelength is exactly equal to the length of the string. In the middle of the string, complete destructive interference occurs, producing a node.

In general, standing waves can exist whenever a multiple of half-wavelengths will fit exactly in the length of the string. It is even possible for standing waves of more than one wavelength to exist on a string at the same time.

SECTION 3 REVIEW

SUMMARY

▸ Waves bouncing off a surface is called reflection.

▸ Diffraction is the bending of waves as they pass an edge or corner.

▸ Refraction is the bending of waves as they pass from one medium to another.

▸ Interference results when two waves exist in the same place and combine to make a single wave.

▸ Interference may cause standing waves.

1. **Describe** what may happen when ripples on a pond encounter a large rock in the water.

2. **Explain** why you can hear two people talking even after they walk around a corner.

3. **Name** the conditions required for two waves to interfere constructively.

4. **Explain** why colors appear on the surface of a soap bubble.

5. **Draw** a standing wave, and label the nodes and antinodes.

6. **Critical Thinking** What conditions are required for two waves on a rope to interfere completely destructively?

7. **Critical Thinking** Imagine that you and a friend are trying to tune the lowest strings on two different guitars to the same pitch. Explain how you could use beats to determine if the strings are tuned to the same frequency.

8. **Critical Thinking** Determine the longest possible wavelength of a standing wave on a string that is 2 m long.

Answers to Section 3 Review

1. The waves will reflect off of the rock.

2. Sound waves diffract when they pass the edge of a barrier, spreading into the medium beyond the barrier.

3. The waves must be in the same place and the crest of one wave must overlap the crest of the other.

4. When light strikes a bubble, some light reflects off the outer surface of the bubble, while other light passes through the thickness of the bubble membrane and reflects off the inner surface. The two sets of reflected waves interfere. Some wavelengths are reinforced while others are canceled, producing colors.

5. See the answer to question 28 in the Chapter Review.

6. The waves must match in amplitude and frequency, and the crests of one wave must exactly overlap the troughs of the other.

7. Decide which guitar will be used as the standard pitch. Pluck the strings of both guitars and listen for beats as the tones fade. Adjust the tension of the string on the second guitar until no beats are heard.

8. The longest possible wavelength is twice the length of the string, 4 m.

Graphing Skills

Interpreting Graphs

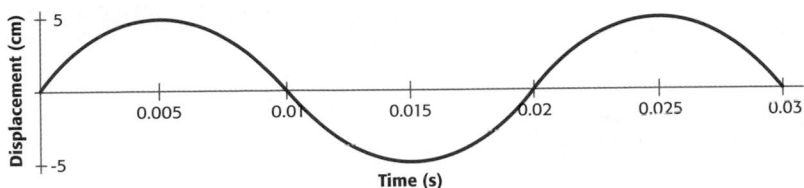

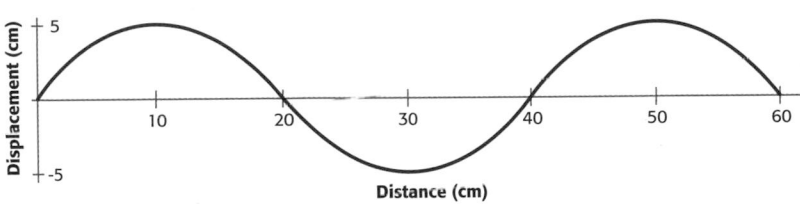

The graphs above show the behavior of a single transverse wave. Study the graphs, and answer the following questions.

1 What types of graphs are these?

2 What variable is described in the *x*-axis of the first graph? What variable is described in the *x*-axis of the second graph?

3 What information about the wave is indicated by the first graph? What information is indicated by the second graph?

4 Determine from the graphs the period, amplitude, and wavelength of the wave.

5 Using the frequency-period equation, calculate the frequency of the wave. Use the wave speed equation to calculate the speed of the wave.

6 Why are both graphs needed to provide complete information about the wave?

7 Plot the data given in the table to the right. From the graph, calculate the wavelength and amplitude of the wave.

x-axis (cm)	*y*-axis (cm)
0	0.93
0.25	2.43
0.50	3.00
0.75	2.43
1.00	0.93
1.25	−0.93
1.50	−2.43
1.75	−3.00
2.00	−2.43
2.25	−0.93
2.50	0.93
2.75	2.43
3.00	3.00

Teaching Tip

To help students understand the graphs, draw three or four transverse waves on the board, each showing a wave at different moments. Indicate that each graph is like a snapshot of the wave taken at different times. Point out that if the graphs could be stacked, the time axis would be at a right angle to the chalkboard.

Answers to Practice

1. line graphs
2. time; distance
3. amplitude and period; amplitude and wavelength
4. $T = 0.02$ s, amplitude = 5 cm, $\lambda = 40$ cm
5. $f = 5 \times 10^1$ Hz; $v = 2.00 \times 10^3$ cm/s = 20.0 m/s
6. One graph describes the change in the wave with time, while the other describes its motion through the medium. The first graph depicts how a single point on the wave (x = constant) changes its transverse position with time; the second graph depicts how the motion of the wave in the x-direction appears at a single moment (t = constant).
7. The amplitude of this sinusoidal graph is 3.00 cm, and its wavelength is 2.5 cm.

Understanding Concepts

1. c

2. a

3. a

4. c

5. c

6. a

7. c

8. c

9. d

10. a

11. c

12. c

13. b

14. a

15. b

Using Vocabulary

16. The amplitude of a wave is half the vertical distance between the crest and the trough.

17. Waves bend due to refraction when they pass from one medium into another. Waves bend due to diffraction when they pass an edge or an opening.

18. If the rope is shaken more quickly, the wave speed will remain constant; the frequency will increase; and the wavelength will decrease.

19. When a mass is pulled away from its resting place, the mass-spring system will gain elastic potential energy. As the spring moves back to its original position, the elastic potential energy changes to kinetic energy, and the mass moves beyond its original resting place. At the top of its motion, the mass loses its kinetic energy but gains elastic potential energy and gravitational potential energy. The mass moves downward again, and the cycle repeats.

Chapter Highlights

Before you begin, review the summaries of the key ideas of each section, found at the end of each section. The key vocabulary terms are listed on the first page of each section.

UNDERSTANDING CONCEPTS

1. A wave is a disturbance that transmits
   a. matter.   **c.** energy.
   b. particles.   **d.** a medium.

2. Electromagnetic waves
   a. are transverse waves.
   b. require a medium.
   c. are mechanical waves.
   d. are longitudinal waves.

3. The speed of a wave depends on the
   a. medium.   **c.** amplitude.
   b. frequency.   **d.** wavelength.

4. Waves that need a medium in which to travel are called
   a. longitudinal waves.
   b. transverse waves.
   c. mechanical waves.
   d. All of the above.

5. Most waves are caused by
   a. velocity.   **c.** a vibration.
   b. amplitude.   **d.** earthquakes.

6. For which type of waves do particles in the medium vibrate perpendicular to the direction in which the waves are traveling?
   a. transverse waves
   b. longitudinal waves
   c. P waves
   d. none of the above

7. A sound wave is an example of
   a. an electromagnetic wave.
   b. a transverse wave.
   c. a longitudinal wave.
   d. a surface wave.

8. In an ocean wave, the molecules of water
   a. move perpendicular to the direction of wave travel.
   b. move parallel to the direction of wave travel.
   c. move in circles.
   d. don't move at all.

9. Half the vertical distance between the crest and trough of a wave is called the
   a. frequency.   **c.** wavelength.
   b. crest.   **d.** amplitude.

10. The number of waves passing a given point per unit of time is called the
   a. frequency.   **c.** wavelength.
   b. wave speed.   **d.** amplitude.

11. The Doppler effect of a passing siren results from an apparent change in
   a. loudness.   **c.** frequency.
   b. wave speed.   **d.** interference.

12. The combining of waves as they meet is known as
   a. a crest.   **c.** interference.
   b. noise.   **d.** the Doppler effect.

13. Waves bend when they pass through an opening. This is called
   a. interference.   **c.** refraction.
   b. diffraction.   **d.** the Doppler effect.

14. Refraction occurs whenever
   a. a wave passes from one medium to another at an angle.
   b. two waves interfere with one another.
   c. a wave is reflected at a free boundary.
   d. standing waves occur.

15. The Greek letter λ is often used to represent a wave's
   a. period.   **c.** frequency.
   b. wavelength.   **d.** amplitude.

20. Kinetic theory states that different states of matter are due to different degrees of organization at the particle level. Since molecules are far apart in gases, one molecule must travel through a lot of empty space before it bumps into another molecule, and thus waves travel relatively slowly in gases. In liquids, molecules are closer together and slide past one another, so that vibrations are transferred more quickly than they are in gases, and wave speed increases. Waves speed is fastest in solids, where molecules are close together and bound to each other.

21. At a free boundary, a wave is reflected exactly as it was in the original wave. At a fixed boundary, the reflected wave is turned upside down and travels in the opposite direction from the original wave.

22. Beats occur when sound waves from different sources vibrate at different frequencies. When compressions in the two waves arrive at a listener's ear at the same time, constructive interference occurs, creating a loud beat. When the compression from one wave and the rarefaction from the other arrive together, destructive interference occurs and a soft beat is heard. When no beats are heard, the waves are vibrating at the same frequency.

16. How would you describe the *amplitude* of a wave using the words *crest* and *trough*?

17. Explain the difference between waves bending due to *refraction* and *diffraction*.

18. Imagine you are shaking the end of a rope to create a series of waves. What will you observe if you begin shaking the rope more quickly? Use the terms *wave speed, frequency,* and *wavelength* in your answer.

19. Describe the changes in *elastic potential energy* and *kinetic energy* that occur when a mass vibrates on a spring.

20. Use the *kinetic theory* to explain the difference in *wave speed* in solids, liquids, and gases.

21. Why is the reflection of a wave at a *free boundary* different from reflection at a *fixed boundary*?

22. How do *beats* help determine whether two sound waves are of the same *frequency*? Use the terms *constructive interference* and *destructive interference* in your answer.

23. How is an *electromagnetic wave* different from a *mechanical wave*?

24. You have a long metal rod and a hammer. How would you hit the metal rod to create a *longitudinal wave*? How would you hit it to create a *transverse wave*?

25. Identify each of the following as a distance measurement, a time measurement, or neither.
   a. amplitude   **d.** frequency
   b. wavelength   **e.** wave speed
   c. period

26. Explain the difference between *constructive interference* and *destructive interference*.

27. Imagine a train approaching a crossing where you are standing safely behind the gate. Explain the changes in sound of the horn that you may hear as the train passes. Use the following terms in your answer: *frequency, wavelength, wave speed,* and *Doppler effect*.

28. Draw a picture of a *standing wave,* and label a *node* and an *antinode*.

BUILDING MATH SKILLS

29. Wave Speed Suppose you tie one end of a rope to a doorknob and shake the other end with a frequency of 2 Hz. The waves you create have a wavelength of 3 m. What is the speed of the waves along the rope?

30. Wave Speed Ocean waves are hitting a beach at a rate of 2.0 Hz. The distance between wave crests is 1.5 m. Calculate the speed of the waves.

31. Wavelength All electromagnetic waves have the same speed in empty space, 3.00×10^8 m/s. Using that speed, find the wavelengths of the electromagnetic waves at the following frequencies:
   a. radio waves at 530 kHz
   b. visible light at 6.0×10^{14} Hz
   c. X rays at 3.0×10^{18} Hz

32. Frequency Microwaves range in wavelength from 1 mm to 30 cm. Calculate their range in frequency. Use 3.00×10^8 m/s as the speed of electromagnetic waves.

33. Wavelength The frequency of radio waves range from about 3.00×10^5 Hz to 3.00×10^7 Hz. What is the range of wavelengths of these waves? Use 3.00×10^8 m/s as the speed of electromagnetic waves.

34. Frequency The note A above middle C on a piano emits a sound wave with wavelength 0.77 m. What is the frequency of the wave? Use 340 m/s as the speed of sound in air.

481

Assignment Guide

Section	Questions
1	1, 2, 4–8, 19, 23, 24, 44, 49, 51, 52
2	3, 9–11, 15, 16, 18, 20, 25, 27, 29–40, 43, 45, 50
3	12–14, 17, 21, 22, 26, 28, 41, 42, 46–48

23. Mechanical waves involve the movement of matter. Electromagnetic waves do not require a material medium.

24. To cause a longitudinal wave in a rod, you would strike the end of the rod, along the axis of the rod. To cause a transverse wave, you would strike the rod at right angles to its axis.

25. a. distance in a transverse mechanical wave; neither in other waves
   b. distance
   c. time
   d. neither (inverse time)
   e. neither (distance and time)

26. In constructive interference, the crests of one wave align with the crests of another wave. Their energies combine to give a wave with energy greater than either wave alone. In destructive interference, the crests of one wave align with the troughs of the other. Their energies combine to give a wave with less energy than either of the waves alone.

27. As the train approaches, you hear a sound with a frequency slightly higher than the actual pitch of the train's horn. The waves emitted by the train's horn travel at a fixed speed through air. The motion of the train squeezes the waves closer together in the forward direction, shortening their wavelength. The waves strike your ear with a frequency greater than that emitted by the horn. As the train moves past you, the wavelength of the waves increases, and as it moves away, the waves are stretched farther apart and strike your ear at a lower frequency. This is called the Doppler effect.

Chapter Resource File

- Concept Review
- Chapter Test
- Test Item Listing for EvamView® Test Generator

28. Antinode

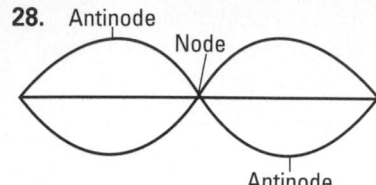

Node

Antinode

Building Math Skills

29. $v = f \times \lambda$
$v = (2\ \text{Hz})(3\ \text{m}) = 6\ \text{m/s}$

30. $v = f \times \lambda$
$v = (2.0\ \text{Hz})(1.5\ \text{m})$
$v = 3.0\ \text{m/s}$

31. a. $\lambda = \dfrac{v}{f} = \dfrac{3.0 \times 10^8\ \text{m/s}}{530\ 000\ \text{Hz}}$
$\lambda = 570\ \text{m}$

b. $\lambda = \dfrac{v}{f} = \dfrac{3.0 \times 10^8\ \text{m/s}}{6.0 \times 10^{14}\ \text{Hz}}$
$\lambda = 5.0 \times 10^{-7}\ \text{m}$

c. $\lambda = \dfrac{v}{f} = \dfrac{3.0 \times 10^8\ \text{m/s}}{3.0 \times 10^{18}\ \text{Hz}}$
$\lambda = 1.0 \times 10^{-10}\ \text{m}$

32. $f = \dfrac{v}{\lambda} = \dfrac{3 \times 10^8\ \text{m/s}}{1\ \text{mm}} \times \dfrac{1000\ \text{mm}}{1\ \text{m}} = 3 \times 10^{11}\ \text{Hz}$

$f = \dfrac{v}{\lambda} = \dfrac{3 \times 10^8\ \text{m/s}}{30\ \text{cm}} \times \dfrac{100\ \text{cm}}{1\ \text{m}} = 1 \times 10^{9}\ \text{Hz}$

f range $= 1 \times 10^9\ \text{Hz}$ to $3 \times 10^{11}\ \text{Hz}$

33. $\lambda = \dfrac{v}{f} = \dfrac{3.00 \times 10^8\ \text{m/s}}{3.00 \times 10^5\ \text{Hz}} = 1.00 \times 10^3\ \text{m}$

$\lambda = \dfrac{v}{f} = \dfrac{3.00 \times 10^8\ \text{m/s}}{3.00 \times 10^7\ \text{Hz}} = 1.00 \times 10^1\ \text{m}$

λ range $= 1.00 \times 10^1\ \text{m}$ to $1.00 \times 10^3\ \text{m}$

35. Graphing Draw a sine curve, and label a crest, a trough, and the amplitude.

36. Interpreting Graphics The wave shown in the figure below has a frequency of 25.0 Hz. Find the following values for this wave:
a. amplitude   **c.** speed
b. wavelength   **d.** period

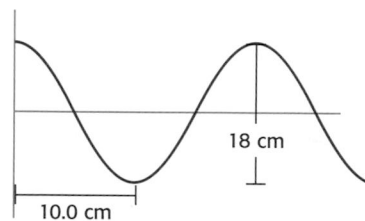

18 cm

10.0 cm

37. Understanding Systems A friend standing 2 m away strikes two tuning forks at the same time, one at a frequency of 256 Hz and the other at 240 Hz. Which sound will reach your ear first? Explain.

38. Applying Knowledge When you are watching a baseball game, you may hear the crack of the bat a short time after you see the batter hit the ball. Why does this happen? (**Hint:** Consider the relationship between the speed of sound and the speed of light.)

39. Understanding Systems You are standing on a street corner, and you hear a fire truck approaching. Does the pitch of the siren stay constant, increase, or decrease as it approaches you? Explain.

40. Applying Knowledge If you yell or clap your hands while standing at the edge of a large rock canyon, you may hear an echo a few seconds later. Explain why this happens.

482

41. Interpreting Graphics Draw the wave that results from interference between the two waves shown below.

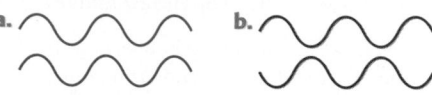

a.   **b.**

42. Understanding Systems An orchestra is playing in a huge outdoor amphitheater, and thousands of listeners are sitting on a hillside far from the stage. To help those listeners hear the concert, the amphitheater has speakers halfway up the hill. How could you improve this system? A computer delays the signal to the speakers by a fraction of a second. Why is this computer used? Explain what might happen if the signal were not delayed at all.

43. Applying Knowledge Dolphins use sound waves to detect other organisms close by. How can a dolphin use the Doppler effect to determine whether an organism is moving towards it? (**Hint:** the sound waves dolphins use can reflect off objects in the water and create an echo.)

44. Applying Knowledge Describe how you interact with waves during a typical school day. Document the types of waves you encounter. Document also how often you interact with each type of wave. Decide whether one type of wave is more important in your life than the other types of waves.

45. Applying Technology With your teacher's help, use a microphone and an oscilloscope or a CBL interface to obtain an image of a sound. Determine the frequency and wavelength of the sound.

COMPUTER SKILL

---

34. $f = \dfrac{v}{\lambda} = \dfrac{340\ \text{m/s}}{0.77\ \text{m}} = 440\ \text{Hz}$

Building Graphing Skills

35.

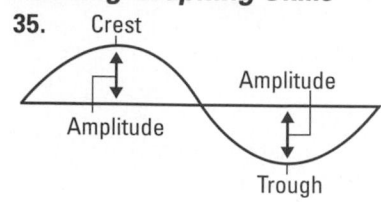

Crest

Amplitude

Amplitude

Trough

36. a. $9\ \text{cm} = 0.09\ \text{m}$
b. $20.0\ \text{cm} = 0.200\ \text{m}$
c. $v = f \times \lambda$
$v = (25.0\ \text{Hz})(0.200\ \text{m})$
$v = 5.00\ \text{m/s}$

d. $T = \dfrac{1}{f} = \dfrac{1}{25.0\ \text{Hz}}$
$T = 0.0400\ \text{s}$

46. Making Decisions A new car is advertised as having antinoise technology. The manufacturer claims that inside the car any sounds are negated. Evaluate the possibility of such a claim. What would have to be created to cause destructive interference with any sound in the car? Do you believe that the manufacturer is correct in its statement?

47. Working Cooperatively Work with other classmates to research architectural acoustics that would affect a restaurant. Investigate acoustics problems in places where many people gather. How do odd-shaped ceilings, decorative panels, and glass windows affect echoes? Prepare a model of your school cafeteria showing what changes you would make to reduce the level of noise.

48. Applying Knowledge A piano tuner listens to a tuning fork vibrating at 440 Hz to tune the string of a piano. He hears beats between the tuning fork and the piano string. Is the string in tune? Explain your answer.

INTEGRATING CONCEPTS

49. Connection to Space Science The Doppler effect occurs for light waves as well as for sound waves. Research some of the ways in which the Doppler effect has helped astronomers understand the motion of distant galaxies and other objects in deep space.

50. Connection to Earth Science What is the medium for seismic waves?

51. Connection to Architecture Explore and describe research on earthquake-proof buildings and materials. Evaluate the impact of this research on society in terms of building codes and architectural styles in earthquake-prone areas such as Los Angeles, San Francisco, and Tokyo.

52. Concept Mapping Copy the unfinished concept map below onto a sheet of paper. Complete the map by writing the correct word or phrase in the lettered boxes.

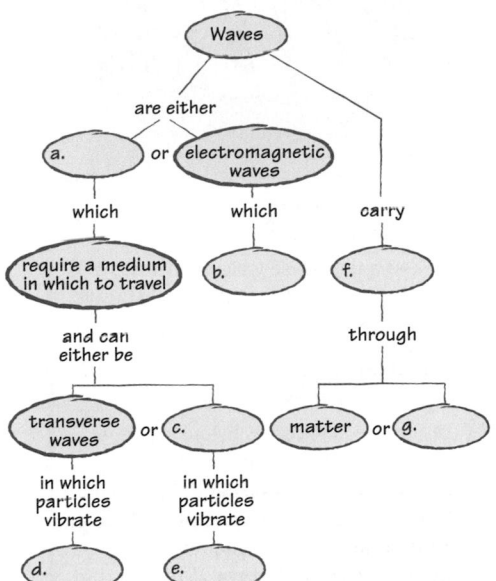

internet connect

www.scilinks.org
Topic: Seismic Waves   SciLinks code: HK4126

SCiLINKS. Maintained by the National Science Teachers Association

Art Credits: Fig. 4-5, Uhl Studios, Inc.; Fig. 8, Uhl Studios, Inc.; Fig. 9, Mark Schroeder; Fig. 11, Uhl Studios, Inc.; Fig. 16, Mark Schroeder; Fig. 20, Uhl Studios, Inc.

Photo Credits: Chapter Opener image of surfer, International Stock Photography; sunset over beach, Bill Brooks/Masterfile; Fig. 1, SuperStock; Fig. 2, Art Resource, NY; Fig. 3, "Quick Lab," Fig. 6-7, Peter Van Steen/HRW; Fig. 13, Joe Devenney/Getty Images/The Image Bank; Fig. 14, Jeff Hunter/Getty Images/The Image Bank; Fig. 15, Peter Van Steen/HRW; Fig. 17, E. R. Degginger/Color-Pic, Inc.; Fig. 18, Peter Van Steen/HRW; Fig. 19, E. R. Degginger/Color-Pic, Inc.; Fig. 21, Michael Freeman/Bruce Coleman, Inc.; Fig. 22, Neil Nissing/Getty Images/FPG International; Fig. 23, Richard Megna/Fundamental Photographs; "Skills Practice Lab," Sam Dudgeon/HRW; "Career Link," Peter Van Steen/HRW.

483

Thinking Critically

37. Both sounds will reach your ear at the same time. The speed of sound in a given medium is constant.

38. You see the bat strike the ball before you hear the crack of the bat because the speed of light is much faster than the speed of sound. The light from the bat reaches your eyes almost instantly, but the sound takes a discernable, although small, time to reach your ears.

39. As a fire truck approaches, the pitch of the siren is higher than it would be if the fire truck were at rest. However, the pitch of the truck does not change noticeably until the fire truck is passing right in front of you, when the pitch rapidly drops. In fact, as the truck approaches, the pitch from the siren is always dropping to some degree.

40. You hear an echo because the sound waves you produced by clapping or shouting have reflected off the walls of the canyon and returned to your ears.

41. a. Drawings should have three crests (high points), two troughs (low points) and be twice as large (tall) as the waves shown.
  b. Drawings should show a straight line (the waves completely cancel each other.)

42. To avoid cancellation, the amplifier delays the sound so that the sound waves will be in phase with those directly from the orchestra.

43. By emitting sound waves that reflect off objects in the water around it, a dolphin can use the Doppler effect to determine whether an organism is moving closer to it. When objects are approaching the dolphin (or the dolphin is approaching the object), the echo from the reflection of sound waves off the object will occur at higher frequency and

---

thus at a higher pitch than they would when the object is at rest or moving away from the dolphin. Increases in pitch would alert the dolphin to approaching organisms.

Developing Life/Work Skills

44. Students will probably think of sound waves, visible light waves, and maybe water waves. Encourage them to look at the entire electromagnetic spectrum, including infrared light, microwaves, and radio waves.

45. The oscilloscope image will be a transverse wave. Ask students to explain how that image can represent a longitudinal wave.

46. Ask students to consider how they would listen to music if the antinoise device canceled all interior noise. They should consider the effect of such cancellation on safety. Students may be interested in exploring the current status of these devices and which cars have them. The technology is called Active Noise Reduction or ANR.

47. Have students also learn about materials used for acoustic control including the use of carpet and upholstered furniture.

See "Continuation of Answers" at the end of the chapter.

 Transparencies

TT Concept Mapping

MODELING TRANSVERSE WAVES

Teacher's Notes

Time Required 1 lab period

Ratings
```
      1    2    3    4
      |----|----|----|
     EASY           HARD
```

TEACHER PREPARATION	2
STUDENT SETUP	2
CONCEPT LEVEL	3
CLEANUP	2

Skills Acquired

- Collecting data
- Constructing models
- Experimenting
- Identifying/Recognizing patterns
- Interpreting
- Measuring
- Organizing and Analyzing data
- Predicting

The Scientific Method

In this lab, students will:
- Form a Hypothesis
- Analyze the Results
- Draw Conclusions

Tips and Tricks

Review Section 2 *Characteristics of Waves* with students before beginning this lab.

Skills Practice Lab

Introduction

When a transverse wave model is created with a sand pendulum, what wave characteristics can you measure?

Objectives

▶ **Create** sine curves by pulling paper under a sand pendulum.

▶ **Measure** the amplitude, wavelength, and period of transverse waves using sine curves as models.

▶ **USING SCIENTIFIC METHODS** *Form a hypothesis* about how changes to the experiment may change the amplitude and wavelength.

▶ **Calculate** frequency and wave speed using your measurements.

Materials

colored sand
masking tape
meterstick
nail
paper or plastic-foam cup
ring stand or other support
rolls of white paper, about 30 cm wide
stopwatch
string and scissors

Modeling Transverse Waves

▶ **Procedure**

Making Sine Curves with a Sand Pendulum

1. Review the discussion in Section 2 on the use of sine curves to represent transverse waves.

2. On a blank sheet of paper, prepare a table like the one shown at right.

3. Use a nail to puncture a small hole in the bottom of a paper cup. Also punch two holes on opposite sides of the cup near the rim. Tie strings of equal length through the upper holes. Make a pendulum by tying the strings from the cup to a ring stand or other support. Clamp the stand down at the end of a table, as shown in the photograph at right. Cover the bottom hole with a piece of tape, then fill the cup with sand. **SAFETY CAUTION** Wear gloves while handling the nails and punching holes.

4. Unroll some of the paper, and mark off a length of 1 m using two dotted lines. Then roll the paper back up, and position the paper under the pendulum, as shown in the photograph at right.

5. Remove the tape over the hole. Start the pendulum swinging as your lab partner pulls the paper perpendicular to the cup's swing. Another lab partner should loosely hold the paper roll. Try to pull the paper in a straight line with a constant speed. The sand should trace a sine curve on the paper, as in the photograph at right.

6. As your partner pulls the paper under the pendulum, start the stopwatch when the sand trace reaches the first dotted line marking the length of 1 m. When the sand trace reaches the second dotted line, stop the watch. Record the time in your table.

7. When you are finished making a curve, stop the pendulum and cover the hole in the bottom of the cup. Be careful not to jostle the paper; if you do, your trace may be erased. You may want to tape the paper down.

Procedure

If students find it difficult punching holes in a plastic or paper cup without ruining the entire cup, they can melt holes in a cup with a nail that has been slightly heated in the flame of a laboratory burner. As always, use tongs and gloves when working with the Bunsen burner. You may want to prepare the cups before students perform the lab.

To make sure the hole in the bottom of the cup is the right size: cover the bottom hole with a piece of tape, then fill the cup with sand. Uncover the hole and swing the cup in a small arc over a sheet of test paper. If the sand does not leak fast enough to make a continuous track on the sheet, increase the diameter of the hole until it does. Once the flow seems right, cover the hole again and pour the sand back into the cup. Students may use a whisk broom and dustpan when returning the sand to the cup or make a crimp at one end of the paper and pour the sand into the cup.

Have students perform a trial run before doing the actual experiment. Students should identify the following on the wave trace: crests, troughs, amplitude, wavelength.

Length along paper = 1 m	Time (s)	Average wavelength (m)	Twice average amplitude (m)
Curve 1			
Curve 2			
Curve 3			

8. For the part of the curve between the dotted lines, measure the distance from the first crest to the last crest, then divide that distance by the total number of crests. Record your answer in the table under "Average wavelength."

9. For the same part of the curve, measure the vertical distance between the first crest and the first trough, between the second crest and the second trough, and so on. Add the distances together, then divide by the number of distances you measured. Record your answer in the table under "Twice average amplitude."

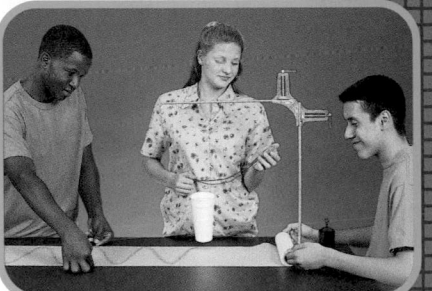

Designing Your Experiment

10. With your lab partners, form a hypothesis about how to make two additional sine curve traces, one with a different average wavelength than the first trace and one with a different average amplitude.

11. In your lab report, write down your plan for changing these two factors. Before you carry out your experiment, your teacher must approve your plan.

Performing Your Experiment

12. After your teacher approves your plan, carry out your experiment. For each curve, measure and record the time, the average wavelength, and the average amplitude.

13. After each trace, return the sand to the cup and roll the paper back up.

▶ Analysis

1. For each of your three curves, calculate the average speed at which the paper was pulled by dividing the length of 1 m by the time measurement. This is equivalent to the speed of the wave that the curve models or represents.

2. For each curve, use the wave speed equation to calculate average frequency.

$$\text{average frequency} = \frac{\text{average wave speed}}{\text{average wavelength}} \qquad f = \frac{v}{\lambda}$$

▶ Conclusions

3. What factor did you change to alter the average wavelength of the curve? Did your plan work? If so, did the wavelength increase or decrease?

4. What factor did you change to alter the average amplitude? Did your plan work?

485

Students will not pull the paper at exactly the same rate for the entire time; as a result all the values are calculated as averages. (The pendulum will slow down slightly over time.)

Designing Your Experiment

To make a curve with a different average amplitude, students simply have to change the initial displacement of the pendulum (so it swings through a smaller or larger arc).

To make a curve with a different average wavelength, students will have to pull the paper at a different average speed.

Performing Your Experiment

If students want to change the length of the pendulum, make sure the strings are still tied securely to the ringstand.

Disposal Information The sand can be collected and reused.

Answers to Analysis

1. Students should calculate the average speed of the paper by dividing the length measurement by the time measurement.

2. Students should use the wave speed equation to calculate the average frequency. Another way to calculate frequency is to count the number of times the cup swings back and forth during a 10 s interval. Divide the number of swings by 10 to calculate the average period of the pendulum. Then use the frequency-period equation to calculate the frequency of the wave.

   The frequency will be close to the same in every case, unless the length of the pendulum was changed. (A pendulum of given length has a more or less fixed period. Increasing the length increases the period.)

Answers to Conclusions

3. To make a curve with a different average wavelength, students should have pulled the paper at a different average speed, and/or have changed the frequency of the pendulum (by altering the length of the pendulum).

4. To make a curve with a different average amplitude, students should have changed the initial displacement of the pendulum (so it swings through a smaller or larger arc).

   Additional question to pose to students: How could the same apparatus be used to model longitudinal waves? (The sand pendulum could be used to model a longitudinal wave if the paper were moved parallel to the motion of the pendulum. Compressions would appear as areas with more sand and rarefactions as areas with less sand.)

Chapter Resource File

• **Skills Practice Lab Datasheet** Modeling Transverse Waves GENERAL

• **Observation Lab** Creating and Measuring Standing Waves BASIC

• **CBL™ Probeware Lab** Tuning a Musical Instrument ADVANCED

Continuation of Answers

Continuation of Answers from p. 483

48. If beats are heard, the string is not in tune. The string and the tuning fork vibrate at slightly different frequencies. As a result, the two sounds interfere constructively and destructively in succession. These successive reinforcements and cancellations are heard as beats. When the frequencies are exactly equal, no beats are heard.

Integrating Concepts

49. Answers will vary, but students may note that the Doppler effect has helped astronomers understand a phenomenon called red shift, which is a shift toward the red end of the spectrum in the observed spectral lines of stars and galaxies. The red shift shows that all galaxies are spreading apart from one another thus that the universe is expanding.

50. Earth, or the ground

51. Building codes are often updated when engineers study the damage to buildings after an earthquake. Computer models allow scientists and engineers to evaluate how new building designs may be affected by earthquakes. Students may be surprised to learn that many modern buildings in earthquake-prone regions are built to be able to sway, slide, or move horizontally. Some even have giant rollers, springs, or rubber shock absorbers that let the ground move beneath the building.

52. a. mechanical waves
   b. can travel through empty space or require no medium
   c. longitudinal waves
   d. at right angles to the path of the wave
   e. parallel to the path of the wave
   f. energy
   g. space

TEACHING TIP

Ultrasonographers use the properties of sound waves in order to generate images of structures within the body. When the ultrasound waves pass from one material to a different material, such as from muscle tissues to the lungs, some of the waves are reflected back toward the source. In addition, the speed of sound is slightly different in these different materials, so the time it takes for a reflection to arrive at the detector depends on the material the reflection passes through. A computer is able to build an image based on the patterns of reflection detected.

SKILL BUILDER — GENERAL

Vocabulary The word sonograph comes from two different root words. The Latin word *sonos* means "sound," and the Greek word *graphein* means "to write."
LS Verbal

CareerLink

Ultrasonographer

Most people have seen a sonogram showing an unborn baby inside its mother's womb. Ultrasound technologists make these images with an ultrasound machine, which sends harmless, high-frequency sound waves into the body. Ultrasonographers work in hospitals, clinics, and doctors' offices. Besides checking on the health of unborn babies, ultrasonographers use their tools to help diagnose cancer, heart disease, and other health problems. To find out more about this career, read this interview with Estela Zavala, a registered diagnostic medical sonographer who works at Austin Radiological Association in Austin, Texas.

"Ultrasound is a helpful diagnostic tool that allows you to see inside the body without the use of X rays."

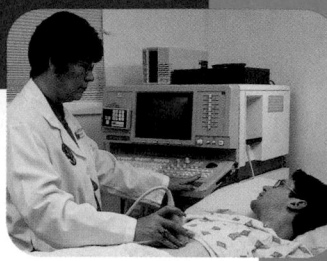

Estela Zavala uses an ultrasound machine to check this man's kidneys.

 What does an ultrasonographer do?

We scan various organs of the body with an ultrasound machine to see if there are any abnormalities. For example, we can check a gallbladder to see if there are any stones in it. We already know what healthy organs look like, so anything unusual shows up in these pictures. This can help a doctor make a diagnosis.

 How does the ultrasound machine make pictures?

The machine creates high-frequency sound waves. When the sound waves reflect off of the organs in the body, the waves strike a piezoelectric crystal in the detector. Piezoelectric crystals can convert the pressure energy from the ultrasound wave into an electrical signal. The ultrasound system processes the signal to create an image.

 What part of your job do you find most satisfying?

I like helping people find out what is wrong with them in a noninvasive way. Before ultrasound, invasive surgery was often the only option.

 What is challenging about your job?

The technology is constantly changing. We are using ultrasound in more ways than ever before. For example, we can now create images of veins and arteries.

 What skills does an ultrasonographer need?

First you need excellent hand-eye coordination, so you can move the equipment over the parts of the body you need to image. You also have to know a lot about all of the organs in the body. It is very important to be able to work quickly, because sometimes patients are uncomfortable.

486

Trends in Medicine

Ultrasound Cauterization Scientists at the University of Washington in Seattle have developed a technique that focuses ultrasound waves in one spot to cause clots in some blood vessels in a certain organ. This technique can be used before surgery on organs such as the spleen, liver, and brain. Usually, surgery on these organs is very difficult because their tissue is so fragile, and doctors have trouble stopping the bleeding after an operation.

A related technology could allow doctors to use ultrasound to detect internal bleeding, which otherwise could lead to death in minutes.

How much training and education did you receive before becoming an ultrasonographer?

After graduating from high school, I went to an X-ray school to be licensed as an X-ray technologist. A medical background like this is necessary for entering ultrasound training. First I went to an intensive 1-month training program, which involved 2 weeks of hands-on work and 2 weeks of classwork. After that, I worked for a licensed radiologist for about a year. Finally, I attended an accredited year-long ultrasound program at a local community college before becoming fully licensed.

 What part of your education do you think was the most valuable?

The best part was the on-the-job training that was involved. You need to do a lot of ultrasounds before you can become proficient.

 Would you recommend ultrasound technology as a career to students?

Yes, I would recommend it. Just remember that you must have some medical experience first, such as being a nurse or an X-ray technician, if you want to continue into other areas of radiology, like ultrasound.

internet connect

www.scilinks.org
Topic: Ultrasound
SciLinks code: HK4143

SCI*LINKS* Maintained by the National Science Teachers Association

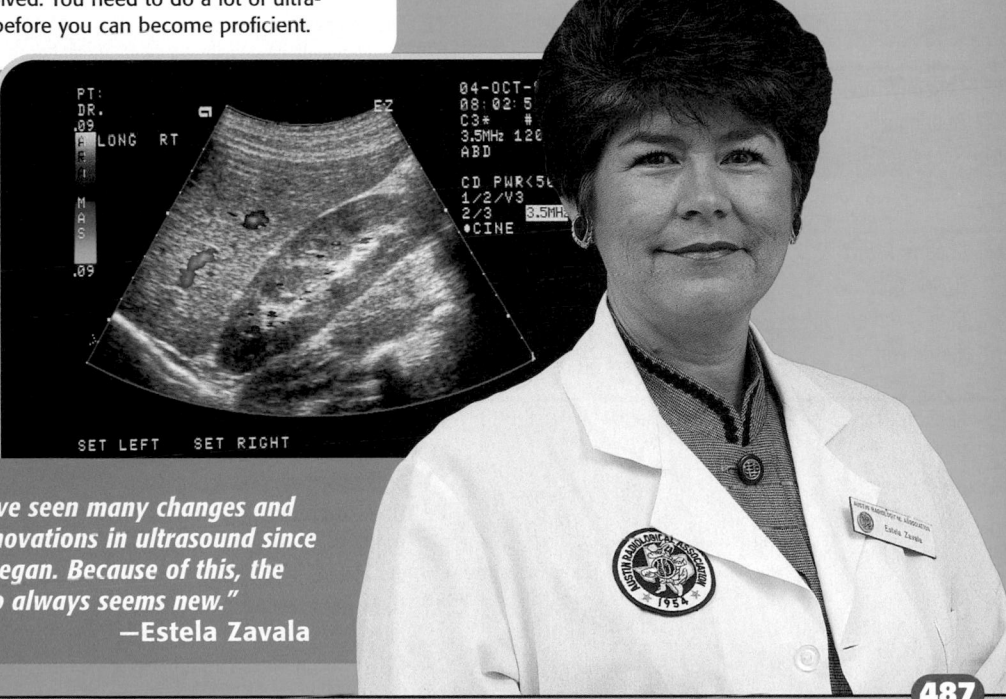

"I've seen many changes and innovations in ultrasound since I began. Because of this, the job always seems new."
—Estela Zavala

487

Sound and Light
Chapter Planning Guide

PACING	CLASSROOM RESOURCES	LABS, ACTIVITIES, AND DEMONSTRATIONS
BLOCK 1 • 45 min pp. 488–489 **Chapter Opener**		**SE Activity 1,** p. 489 **SE Activity 2,** p. 489
BLOCKS 2 & 3 • 90 min pp. 490–498 **Section 1** Sound	**CRF Lesson Plan*** **TT Bellringer*** **TT The Ear***	**TE Opening Demonstration,** p. 490 `GENERAL` **SE Quick Activity** Sound in Different Mediums, p. 491 `GENERAL` **CRF Datasheets for In-Text Labs** Sound in Different Mediums* `GENERAL` **SE Quick Activity** Frequency and Pitch, p. 492 `GENERAL` **CRF Datasheets for In-Text Labs** Frequency and Pitch* `GENERAL` **TE Demonstration** Showing Waves with a Stroboscope, p. 494 `GENERAL` **SE Quick Lab** How can you amplify the sound of a tuning fork?, p. 495 `GENERAL` **CRF Datasheets for In-Text Labs** How can you amplify the sound of a tuning fork?* `GENERAL` **TE Demonstration** Resonance, p. 496 `GENERAL`
BLOCKS 4 & 5 • 90 min pp. 499–505 **Section 2** The Nature of Light	**CRF Lesson Plan*** **TT Bellringer*** **TM Spectrum Diagram***	**TE Opening Activity,** p. 499 `GENERAL`
BLOCKS 6 & 7 • 90 min pp. 506–511 **Section 3** Reflection and Color	**CRF Lesson Plan*** **TT Bellringer*** **TM Law of Reflection*** **TM Flat Mirror***	**TE Demonstration** Observing Reflection, pp. 507–508 `GENERAL` **TE Demonstration** Additive Color TV, p. 510 `GENERAL` **CRF Observation Lab** Mirror Images* ◆ `BASIC`
BLOCKS 8 & 9 • 90 min pp. 512–518 **Section 4** Refraction, Lenses, and Prisms	**CRF Lesson Plan*** **TT Bellringer*** **TT Refraction*** **TT The Eye*** **TT Concept Mapping***	**TE Opening Demonstration,** p. 512 **TE Demonstration** Total Internal Reflection, p. 514 `GENERAL` **TE Demonstration** Dispersion, p. 517 `GENERAL` **SE Skills Practice Lab** Forming Images with Lenses, pp. 524–525 ◆ `GENERAL` **CRF Datasheets for SE Skills Practice Lab** Forming Images with Lenses* `GENERAL` **CRF CBL™ Probeware Lab** Choosing Pair of Sunglasses* ◆ `ADVANCED`

BLOCKS 10 & 11 • 90 min

Chapter Review and Assessment Resources

- **SE Chapter Review,** pp. 520–523
- **CRF Chapter Tests*** `GENERAL`
- **OSP Test Generator**
- **CRF Standardized Test Practice with Guided Reading Development***
- **CRF Test Item Listing for ExamView® Test Generator***
- **OSP Scoring Rubrics and Classroom Management Checklists**

Online Resources

Visit the HRW Web site for a variety of free resources related to the text. Just type in the keyword **HK4 SAL**.

Holt Online Learning

Holt Science Spectrum: Physical Science: Online Edition

Students can access interactive problem solving help and active visual concept development with the *Holt Science Spectrum: Physical Science* Online Edition available at **www.hrw.com**.

student CNN news

cnnstudentnews.com

Find the latest news, lesson plans, and activities related to important scientific events.

Compression guide:
To shorten your instruction because of time limitations, omit blocks 1, 10, and 11.

KEY

TE	Teacher Edition	CRF	Chapter Resource File	*	Also on One-Stop Planner
SE	Student Edition	TT	Teaching Transparency	◆	Requires Advance Prep
OSP	One-Stop Planner	TM	Transparency Master		

PROBLEM SOLVING AND PRACTICE	SECTION REVIEW AND ASSESSMENT	STANDARDS CORRELATION
	CRF Pretest* GENERAL	
TE Teaching Tip Using Sonar, p. 497	**TE** Quiz, p. 498 BASIC **SE** Section 1 Review, p. 498 **CRF** Concept Review* GENERAL **CRF** Quiz* BASIC	PS 3e, 5a, 5c, 5d LS 6a UCP 2, 3, 5 ST 2 SPSP 2, 4, 5
CRF Cross-Disciplinary Worksheet Real World Applications—How Does Sunscreen Work?* ADVANCED	**TE** Quiz, p. 505 BASIC **SE** Section 2 Review, p. 505 **CRF** Concept Review* GENERAL **CRF** Quiz* BASIC	PS 5d UCP 1, 2 ST 2 HNS 1–3 SPSP 2–5
CRF Cross-Disciplinary Worksheet Integrating Space Science—Telescopes* ADVANCED	**TE** Quiz, p. 511 BASIC **SE** Section 3 Review, p. 511 **CRF** Concept Review* GENERAL **CRF** Quiz* BASIC	UCP 2, 5
SE Math Skills Fractions, p. 519 **SE** Science in Action Holography, pp. 526–527	**TE** Quiz, p. 518 BASIC **SE** Section 4 Review, p. 518 **CRF** Concept Review* GENERAL **CRF** Quiz* BASIC	LS 6a UCP 2 ST 2

SCiLINKS®

www.scilinks.org

Topic: Transfer of Sound and Light Energy
*Sci*Links code: HK4140

Topic: Properties of Sound
*Sci*Links code: HK4112

Topic: The Ear
*Sci*Links code: HK4034

Topic: Properties of Light
*Sci*Links code: HK4110

Topic: Reflection
*Sci*Links code: HK4118

Topic: Refraction
*Sci*Links code: HK4120

Topic: Microwaves
*Sci*Links code: HK4087

Topic: Holography
*Sci*Links code: HK4068

Technology Resources

 One-Stop Planner
All of your printable resources and the Test Generator are on this convenient CD-ROM.

 CNN® Science in the NEWS
each video segment is accompanied by a Critical Thinking Worksheet *

Segment 14
Virtual Practice Room

Segment 16
Color Deficiency Lenses

Overview

This chapter begins by exploring sound, including the properties of sound, musical instruments, the human ear, and ultrasound and sonar. Next, it explores the wave and particle properties of light and the electromagnetic spectrum. It also covers the reflection of light, including mirrors, and explains how we see colors. Finally, the chapter covers the refraction of light, including the use of lenses in microscopes, telescopes, and the human eye. The chapter concludes with a discussion of dispersion and prisms.

Assessing Prior Knowledge

Be sure students understand the following concepts:

• electrons
• structure of atoms
• energy transformation
• types of waves
• relationship of vibrations and waves
• reflection
• diffraction
• interference
• standing waves
• waves and energy

MISCONCEPTION //// ALERT \\\\

Science education research has identified the following misconceptions about sound and light.

• Students intuit that sound cannot travel through solids and liquids, and can travel through a vacuum and space.
• Students believe that sound can be produced without using any materials, and that hitting an object harder will change the pitch.
• Students do not think of light as traveling from one place to another.
• Most students believe that white light is colorless and pure, and that a color filter adds color to a white beam.

Sound and Light

Chapter Preview

1 Sound
Properties of Sound
Musical Instruments
Hearing and the Ear
Ultrasound and Sonar

2 The Nature of Light
Waves and Particles
The Electromagnetic Spectrum

3 Reflection and Color
Reflection of Light
Mirrors
Seeing Colors

4 Refraction, Lenses, and Prisms
Refraction of Light
Lenses
Dispersion and Prisms

488

Standards Correlations

National Science Education Standards

The following descriptions summarize the National Science Standards that specifically relate to this chapter. For the full text of the standards, see the National Science Education Standards at the front of the book.

Section 1 Sound
Physical Science PS 2e, 5a, 5c, 5d
Life Science LS 6a
Unifying Concepts and Processes UCP 2, 3, 5
Science and Technology ST 2
Science in Personal and Social Perspectives SPSP 2, 4, 5

Section 2 The Nature of Light
Physical Science PS 5d
Unifying Concepts and Processes UCP 1, 2
Science and Technology ST 2
History and Nature of Science HSN 1, 2, 3
Science in Personal and Social Perspectives SPSP 2, 3, 4, 5

Section 3 Reflection and Color
Unifying Concepts and Processes UCP 2, 5

Section 4 Refraction, Lenses, and Prisms
Life Science LS 6a
Unifying Concepts and Processes UCP 2
Science and Technology ST 2
History and Nature of Science HNS 1

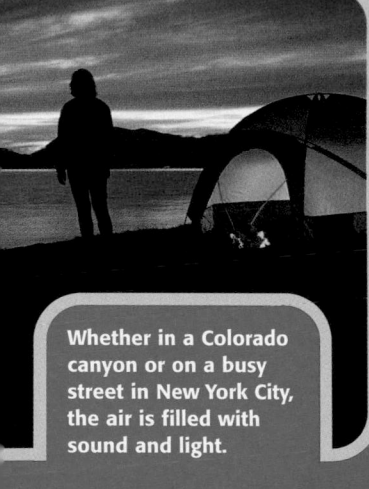

Whether in a Colorado canyon or on a busy street in New York City, the air is filled with sound and light.

Focus ACTIVITY

Background Imagine that you are walking through a canyon at sunset. As the light of the sun fades, you see the first stars of the night. Your footsteps make an echo that bounces around the canyon walls. As you approach your destination, you hear the sounds of people talking around a campfire.

Now imagine that you are walking down a street in a big city. You see the flash of neon signs and the colors of street lights. You hear the sound of cars, and you hear music from a car radio. Through the open door of a restaurant, you hear dishes clanking and people laughing.

Sound and light carry information about the world around us. This chapter will focus on the behavior of sound waves and light waves. You will learn how sound is produced, how mirrors and lenses work, how we see and hear, and how sound and light are used in different applications, ranging from music to medicine.

Activity 1 Stand outside in front of a large wall, and clap your hands. Do you hear an echo? How much time passes between the time you clap your hands and the time you hear the echo? Use this estimated time to estimate the distance to the wall. You will need two other pieces of information: the speed of sound in air, about 340 m/s, and the speed equation, $v = d/t$.

Activity 2 Find a crosswalk with a crossing signal. Watch as the signal changes from "Walk" to "Don't Walk" and back again. Does the crossing signal ever produce a sound? If so, why? If not, why would it be a good idea for the signal to produce a sound?

---

internet connect

www.scilinks.org
Topic: Transfer of Sound and Light Energy
SciLinks code: HK4140

SCI**LINKS** Maintained by the National Science Teachers Association

---

Pre-Reading Questions

1. How do you hear sounds? Are there sounds you can't hear?
2. Does a reflection of an object look exactly like the object? Why or why not?
3. How high is the speed of light? How would you measure it?

489

Focus ACTIVITY

Background Ask students to imagine hiking in an area similar to the one in the small photo, and have them describe information they might receive from their sense of sound. What are the sources of the sound? How are those sounds produced? What are we responding to when we hear sounds? Repeat this exercise for the information we receive by sight. In this chapter, students will learn how sound and light waves are produced, how their properties are determined, and how we detect and interpret them.

Activity 1 Encourage students to find a location where echoes return in a second or more. This would require a distance of 170 m from the wall. This may be possible only outdoors with a large building or a geological feature. **LS Auditory**

Activity 2 Some crosswalks use buzzers or other audible signals. These devices help the visually impaired cross the street in a straight line. **LS Logical**

Answers to Pre-Reading Questions

1. Sound waves are transmitted as vibrations through the ear. Nerve fibers then send impulses to the brain to be interpreted as sound. Human beings cannot hear sounds with frequencies below 20 Hz or above 20 000 Hz.
2. No. A reflection of an object in a flat mirror appears to be flipped from left to right.
3. In a vacuum, light travels at 3.0×10^8 m/s. You would need sophisticated instruments to measure the speed of light, because light travels so fast.

Chapter Resource File

- Pretest **GENERAL**
- Teaching Transparency Preview
- Answer Keys

Focus

Overview

Before beginning this section, review with your students the Objectives listed in the Student Edition. In this section, students relate properties of sound to wave characteristics. The section also covers musical instruments, the human ear, and applications of ultrasound.

Bellringer

Use the Bellringer transparency to prepare students for this section.

Motivate

Opening Demonstration —— GENERAL

Try to borrow a practice piano that you can take to your classroom on rollers. If no piano is available, use an acoustic guitar or other musical instruments students volunteer to bring in. First, demonstrate differences in loudness using notes of the same pitch. Use a sound level meter to reinforce the judgment of students. Then demonstrate difference in pitch. Students may be able to see that the lowest string of a guitar vibrates with noticeably lower frequency than the highest string. Having students touch the strings lightly as they vibrate will also help convey this idea. **LS** Auditory

Sound

▶ **KEY TERMS**

sound wave
pitch
infrasound
ultrasound
resonance
sonar

▶ **OBJECTIVES**

▶ **Recognize** what factors affect the speed of sound.

▶ **Relate** loudness and pitch to properties of sound waves.

▶ **Explain** how harmonics and resonance affect the sound from musical instruments.

▶ **Describe** the function of the ear.

▶ **Explain** how sonar and ultrasound imaging work.

▶ **sound wave** a longitudinal wave that is caused by vibrations and that travels through a material medium

When you listen to your favorite musical group, you hear a variety of sounds. You may hear the steady beat of a drum, the twang of guitar strings, the wail of a saxophone, chords from a keyboard, or human voices.

Although these sounds all come from different sources, they are all longitudinal waves produced by vibrating objects. How does a musical instrument or a stereo speaker make sound waves in the air? What happens when those waves reach your ears? Why does a guitar sound different from a violin?

Properties of Sound

When a drummer hits a drum, the head of the drum vibrates up and down, as shown in *Figure 1A*. Each time the drumhead moves upward, it compresses the air above it. As the head moves back down again, it leaves a small region of air that has a lower pressure. As this happens over and over, the drumhead creates a series of compressions and rarefactions in the air, as shown in *Figure 1B*.

The **sound waves** from a drum are longitudinal waves, in which the particles of air vibrate in the same direction the wave travels. Sound waves are caused by vibrations, and carry energy through a medium. Sound waves in air spread out in all directions away from the source. When sound waves from the drum reach your ears, the waves cause your eardrums to vibrate. Other sounds are produced in different ways from the sound waves produced by a drum, but in all cases a vibrating object sets the medium around it in motion.

Figure 1

A The head of a drum vibrates up and down when it is struck by the drummer's hand.
B The vibrations of the drumhead create sound waves in the air.

490

FINE ARTS
CONNECTION

Ask the school's instrumental music director to conduct a lesson on pitch and quality of sound by demonstrating or having players demonstrate instruments that have similar and contrasting sounds.

MISCONCEPTION ALERT —— GENERAL

Sources of Sound Some students believe that sound can be produced without any materials. Ask students to name examples of sounds, and write them in a column on the board. For each example, have students determine the source of the sound. Write these next to each example, in a second column. Use these examples to illustrate that all sounds can be traced back to a vibrating source. **LS** Logical

Table 1 Speed of Sound in Various Mediums

Medium	Speed of sound (m/s)	Medium	Speed of sound (m/s)
Gases		**Liquids at 25°C**	
Air (0°C)	331	Water	1490
Air (25°C)	346	Sea water	1530
Air (100°C)	386	**Solids**	
Helium (0°C)	972	Copper	3813
Hydrogen (0°C)	1290	Iron	5000
Oxygen (0°C)	317	Rubber	54

The speed of sound depends on the medium

If you stand a few feet away from a drummer, it may seem that you hear the sound from the drum at the same time that the drummer's hand strikes the drum head. Sound waves travel very fast, but not infinitely fast. The speed of sound in air at room temperature is about 346 m/s (760 mi/h).

Table 1 shows the speed of sound in various materials and at various temperatures. The speed of sound in a particular medium depends on how well the particles can transmit the compressions and rarefactions of sound waves. In a gas, such as air, the speed of sound depends on how often the molecules of the gas collide with one another. At higher temperatures, the molecules move around faster and collide more frequently. An increase in temperature of 10°C increases the speed of sound in a gas by about 6 m/s.

Sound waves travel faster through liquids and solids than through gases. In a liquid or solid the particles are much closer together than in a gas, so the vibrations are transferred more rapidly from one particle to the next. However, some solids, such as rubber, dampen vibrations so that sound does not travel well. Materials like rubber can be used for soundproofing.

Loudness is determined by intensity

How do the sound waves change when you increase the volume on your stereo or television? The loudness of a sound depends partly on the energy contained in the sound waves. The *intensity* of a sound wave describes the rate at which a sound wave transmits energy through a given area of the medium. Intensity depends on the amplitude of the sound wave as well as your distance from the source of the waves. *Loudness* depends on the intensity of the sound wave. The greater the intensity of a sound, the louder the sound will seem.

internet connect

www.scilinks.org
Topic: Properties of Sound
SciLinks code: HK4112

SciLINKS Maintained by the National Science Teachers Association

Quick ACTIVITY

Sound in Different Mediums

1. Tie a spoon or other utensil to the middle of a 1–2 m length of string.
2. Wrap the loose ends of the string around your index fingers and place your fingers against your ears.
3. Swing the spoon so that it strikes a tabletop, and compare the volume and quality of the sound received with those received when you listen to the sound directly through the air.
4. Does sound travel better through the string or through the air?

491

PHYSICS CONNECTION

The speed of sound in a gas depends on the velocity of the molecules. Temperature is a measure of the average kinetic energy of molecules, so at 0°C, molecules of air have the same average kinetic energy as molecules of hydrogen or helium. However, the molecules in air have a mass of about 29 amu (H = 2 amu, He = 4 amu) and kinetic energy depends on the mass and velocity ($KE = \frac{1}{2}mv^2$). Therefore, in order to have the same kinetic energy, molecules of hydrogen and helium must be moving much faster than molecules of air.

MISCONCEPTION ALERT

Sound Mediums Some students think that sound cannot travel through solids and liquids. Use the Quick Activity on this page to emphasize that sound can travel through all three types of matter, and that it actually travels through solids and liquids faster than it does through gases.

Teach

Teaching Tip

Sound Waves Make clear that sound waves are not transverse waves even though they are usually represented in a transverse wave form. Amplitude in a sound wave is determined by the degree of compression and rarefaction compared to the normal pressure of the medium.

Quick ACTIVITY

Sound in Different Mediums
Materials (per group):
- spoon or other utensil
- string (1–2 m long)

Teacher's Notes: The utensil should not be one with parts that rattle. It is better to use hard kite string instead of yarn. Students should wrap the string near their fingertips and try to arrange the string so that the last loop before the downward part passes over the ends of their fingers ensuring that the string is pushed against their ears. Heavier objects, such as wrenches, make the loudest sounds. A welded oven rack is truly impressive. Try striking with a soft object such as a rubber stopper.

Results: Students should find that the object produces a ringing, bell-like sound that is louder when transmitted by the string than when transmitted through air. **LS Auditory**

Chapter Resource File

- **Lesson Plan**
- **Quick Activity Datasheet** Sound in Different Mediums **GENERAL**

Transparencies

TT Bellringer

Chapter 15 · Sound and Light **491**

Teaching Tip

Pitch Versus Sound Students unfamiliar with the language of music may confuse the idea of high and low pitch with loud and soft sounds. All students should understand these words before proceeding. Also, students should understand that the words *fast* and *slow*—when applied to vibration—refer to the frequency, or "oftenness," of the back and forth movement.

SKILL BUILDER — BASIC

Interpreting Visuals Figure 2 shows the decibel levels of several common sounds. Have students examine the figure. What is the decibel level of normal conversation? (50 dB) What is the level of a vacuum cleaner? (70 dB) How many times louder would a vacuum cleaner sound than a normal conversation? (Four times as loud; an increase of 10 dB makes a sound seem twice as loud.) **LS Visual**

Quick ACTIVITY

Frequency and Pitch

Materials (per student):

• flexible metal or plastic ruler

Teacher's Notes: Have students relate their observations to the idea of pitch.

Results: A possible conclusion is that, given a longer and a shorter object of the same form and material, the shorter object will vibrate at a higher frequency (pitch). **LS Auditory**

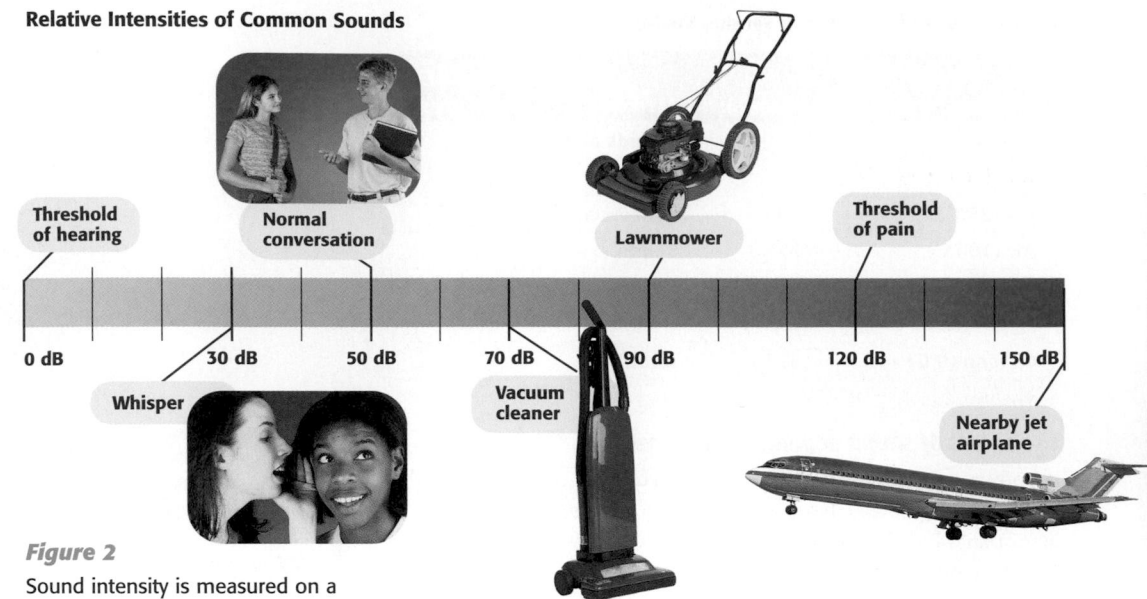

Relative Intensities of Common Sounds

Threshold of hearing — 0 dB
Whisper — 30 dB
Normal conversation — 50 dB
Vacuum cleaner — 70 dB
Lawnmower — 90 dB
Threshold of pain — 120 dB
Nearby jet airplane — 150 dB

Figure 2
Sound intensity is measured on a logarithmic scale of decibels.

▶ **pitch** a measure of how high or low a sound is perceived to be depending on the frequency of the sound wave

Quick ACTIVITY

Frequency and Pitch

1. Hold one end of a flexible metal or plastic ruler on a desk with about half of the ruler hanging off the edge. Bend the free end of the ruler and then release it. Can you hear a sound?
2. Try changing the position of the ruler so that less hangs over the edge. How does that change the sound produced?

However, a sound with twice the intensity of another sound does not seem twice as loud. Humans perceive loudness on a logarithmic scale. This means that a sound seems twice as loud when its intensity is 10 times the intensity of another sound.

The *relative intensity* of sounds is found by comparing the intensity of a sound with the intensity of the quietest sound a person can hear, the threshold of hearing. Relative intensity is measured in units called *decibels*, dB. A difference in intensity of 10 dB means a sound seems about twice as loud. *Figure 2* shows some common sounds and their decibel levels.

The quietest sound a human can hear is 0 dB. A sound of 120 dB is at the threshold of pain. Sounds louder than this can hurt your ears and give you headaches. Extensive exposure to sounds above 120 dB can cause permanent deafness.

Pitch is determined by frequency

The **pitch** of a sound is related to the frequency of sound waves. A high-pitched note is made by something vibrating very rapidly, like a violin string or the air in a flute. A low-pitched sound is made by something vibrating more slowly, like a cello string or the air in a tuba.

In other words, a high-pitched sound corresponds to a high frequency, and a low-pitched sound corresponds to a low frequency. Trained musicians are capable of detecting subtle differences in frequency, even as slight as a change of 2 Hz.

REAL-LIFE CONNECTION — BASIC

Sound Levels If you have or can borrow a sound level meter, let students measure the intensity of various sounds around school. Have them take note of the decibel (dB) scale on the meter. These meters are inexpensive and available at local electronic stores. **LS Auditory**

MISCONCEPTION ///ALERT \\\

Pitch Some students believe that hitting an object harder will change the pitch of the sound. This actually affects the amplitude, and thus the intensity or loudness of the sound. The pitch depends on the frequency of the vibration, which is not affected by how hard an object is struck. Use examples to emphasize this point.

Ranges of Hearing for Various Mammals

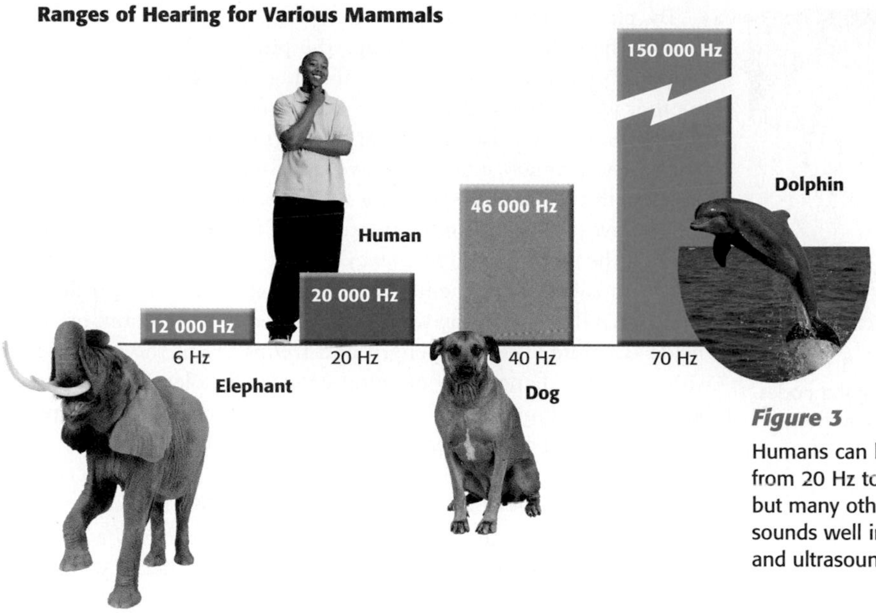

150 000 Hz	**Dolphin**
46 000 Hz	
20 000 Hz	**Human**
12 000 Hz	
6 Hz	20 Hz
Elephant	40 Hz
	Dog
	70 Hz

Figure 3

Humans can hear sounds ranging from 20 Hz to about 20 000 Hz, but many other animals can hear sounds well into the infrasound and ultrasound ranges.

Humans hear sound waves in a limited frequency range

The human ear can hear sounds from sources that vibrate as slowly as 20 vibrations per second (20 Hz) and as rapidly as 20 000 Hz. Any sound with a frequency below the range of human hearing is known as **infrasound;** any sound with a frequency above human hearing range is known as **ultrasound.** Many animals can hear frequencies of sound outside the range of human hearing, as shown in *Figure 3.*

▶ **infrasound** slow vibrations of frequencies lower than 20 Hz

▶ **ultrasound** any sound wave with frequencies higher than 20 000 Hz

Musical Instruments

Musical instruments, from deep bassoons to twangy banjos, come in a wide variety of shapes and sizes and produce a wide variety of sounds. But musical instruments can be grouped into a small number of categories based on how they make sound. Most instruments produce sound through the vibration of strings, air columns, or membranes.

Musical instruments rely on standing waves

When you pluck the string of a guitar, particles in the string start to vibrate. Waves travel out to the ends of the string, and then reflect back toward the middle. These vibrations cause a standing wave on the string, as shown in *Figure 4.* The two ends of the strings are nodes, and the middle of the string is an antinode.

Figure 4

Vibrations on a guitar string produce standing waves on the string. These standing waves in turn produce sound waves in the air.

493

SKILL BUILDER — BASIC

Interpreting Visuals Have students examine **Figure 3.** Ask them which of these mammals can hear the highest sounds (dolphins) and which can hear the lowest (elephants). Do the hearing ranges of mammals overlap? (yes) What frequency of sounds can be heard by all of the mammals shown? (70 Hz to 12 000 Hz) **LS** Visual

Teaching Tip

Standing Waves Stress to students that only certain frequencies of standing waves are possible because these are the "natural" frequencies of that particular rope at a certain state of tension and length.

SKILL BUILDER — BASIC

Reading Skills Write the words *sound* and *music* on the chalkboard. Ask students to list words and phrases that apply to both terms and that are exclusive to one term or the other. Students should generally conclude that musical sounds have organized or regular properties, such as pitch and quality, that other sounds do not have. **LS** Verbal

INCLUSION Strategies

• *Learning Disabled*
• *Attention Deficit Disorder*
• *English Language Learners*

Using **Table 1,** have students rearrange the speed of sound in various materials from fastest to slowest. Ask students what materials could be used for soundproofing the classroom. Student can also be asked to explain, in writing, orally, or dictated, why sound waves travel faster in one material than another.

Did You Know?

Elephant Sounds Several years ago, scientist Katherine Payne of Cornell University discovered that elephants can make and hear sounds having pitches well below the range of human hearing, which ends at about 20 Hz. She found that elephants in the wild communicate over long distances with these sounds, which were in the range of 6 to 18 Hz. Few objects or materials can absorb the energy of sound waves of these frequencies, so the sounds can be heard by other elephants long distances away.

Chapter Resource File

• **Science Skills** Equations Involving a Constant BASIC
• **Quick Activity Datasheet** Frequency and Pitch GENERAL

Videos

 Presents Physical Science

• **Segment 14** Virtual Practice Room
• **Critical Thinking** Worksheet 14

See the Science in the News video guide for more details.

Fundamental Frequency To explain how standing waves relate to fundamental frequency, point out that any object, when disturbed, will vibrate at certain natural frequencies that are characteristic of that object. Certain objects vibrate better than others and produce sounds of recognizable pitch. Ask students to name objects that fit into this category. (Responses may include bells, stretched strings, xylophone bars (metal), marimba bars (wood), crystal glassware.) **LS Logical**

Demonstration

Showing Waves with a Stroboscope If your school has an adjustable strobe lamp, show students the standing wave on a vibrating string of a guitar. Adjust the strobe until students can see that the string vibrates not only in one segment but in two or more at the same time. A bright flashlight shining through the blades of a fan is a makeshift substitute, but the fan's speed must be continuously adjustable.

Figure 5

Colored dust lies along the nodes of the standing waves on the head of this drum.

By placing your finger on the string somewhere along the neck of the guitar, you can change the pitch of the sound. This happens because a shorter length of string vibrates more rapidly. In other words, the standing wave has a higher frequency.

Standing waves can exist only at certain wavelengths on a string. The primary standing wave on a vibrating string has a wavelength that is twice the length of the string. The frequency of this wave, which is also the frequency of the string's vibrations, is called the *fundamental frequency*.

All musical instruments use standing waves to produce sound. In a flute, standing waves are formed in the column of air inside the flute. The wavelength and frequency of the standing waves can be changed by opening or closing holes in the flute body, which changes the length of the air column. Standing waves also form on the head of a drum, as shown in **Figure 5**.

Harmonics give every instrument a unique sound

If you play notes of the same pitch on a tuning fork and a clarinet, the two notes will sound different from each other. If you listen carefully, you may be able to hear that the clarinet is actually producing sounds at several different pitches, while the tuning fork produces a pure tone of only one pitch.

A tuning fork vibrates only at its fundamental frequency. The air column in a clarinet, however, vibrates at its fundamental frequency and at certain whole-number multiples of that frequency, called *harmonics*. **Figure 6** shows the harmonics present in a tuning fork and a clarinet when each sounds the note A-natural.

Figure 6

The note A-natural on a clarinet sounds different from the same note on a tuning fork due to the relative intensity of harmonics.

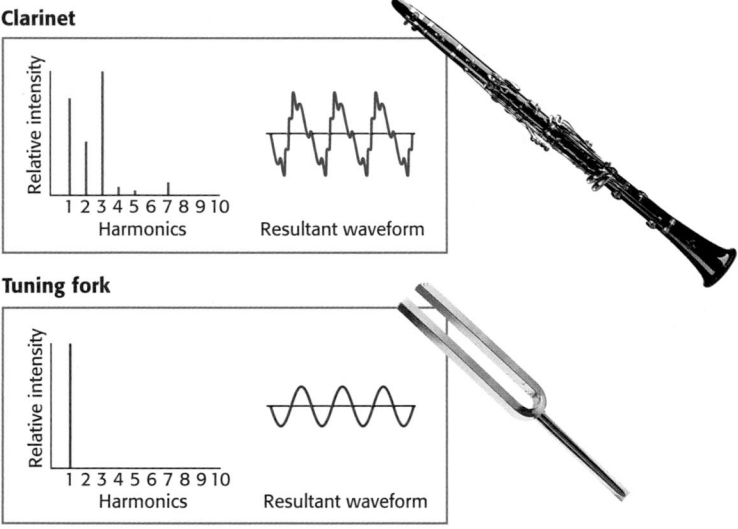

A musical interval (two notes played together) can sound either consonant (harmonious or pleasing) or dissonant (inharmonious or harsh). In many consonant intervals, there is a whole-number ratio between the two frequencies, for example the octave (2:1), the perfect fifth (3:2), and the major third (5:4). Many dissonant intervals have irrational ratios. Encourage students to learn more about the relationship between mathematics and music and to share their results with the class. **LS Logical**

Alternative Assessment ——— **GENERAL**

Instruments of Other Cultures Ask students to research instruments from other cultures. Students may wish to focus on the instruments of one culture, or on one general type of instrument. Ask them to design a sales brochure or catalog for the instruments they learn about, including illustrations, information about what the instruments sound like and how the sounds are produced, and comparisons to familiar instruments. **LS Interpersonal**

In the clarinet, several harmonics combine to make a complex wave. Note, however, that this wave still has a primary frequency that is the same as the frequency of the wave produced by the tuning fork. This is the fundamental frequency, which makes the note sound a certain pitch. Every musical instrument has a characteristic sound quality resulting from the mixture of harmonics.

Instruments use resonance to amplify sound

When you pluck a guitar string, you can feel that the bridge and the body of the guitar also vibrate. These vibrations, which are a response to the vibrating string, are called *forced vibrations*. The body of the guitar is more likely to vibrate at certain specific frequencies called *natural frequencies*.

The sound produced by the guitar will be loudest when the forced vibrations cause the body of the guitar to vibrate at a natural frequency. This effect is called **resonance.** When resonance occurs, the sound is amplified because both the string and the guitar itself are vibrating at the same frequency.

resonance a phenomenon that occurs when two objects naturally vibrate at the same frequency

Quick Lab

How can you amplify the sound of a tuning fork?

Materials
✓ tuning forks of various frequencies
✓ rubber block for activating forks
✓ various objects made of metal and wood

1. Activate a tuning fork by striking the tongs of the fork against a rubber block.
2. Touch the base of the tuning fork to different wood or metal objects, as shown in the figure at right. Listen for any changes in the sound of the tuning fork.
3. Activate the fork again, but now try touching the end of the tuning fork to the ends of other tuning forks (make sure that the tines of the forks are free to vibrate, not touching anything). Can you make another tuning fork start vibrating in this way?
4. If you find two tuning forks that resonate with each other, try activating one and holding it near the tongs of the other one. Can you make the second fork vibrate without touching it?

Analysis

1. What are some characteristics of the objects that helped to amplify the sound of the tuning fork in step 2?
2. What is the relationship between the frequencies of tuning forks that resonate with each other in steps 3 and 4?

495

Alternative Assessment — ADVANCED

Comparing Frequency and Intensity Have students use common objects or musical instruments to demonstrate the following:

- two sounds of different intensities but the same frequency; two sounds of different frequencies but about the same intensity
- two sounds of different pitches but about the same amplitude; two sounds of different amplitudes but the same pitch
- two sounds of different wavelengths but the same amplitude

LS Auditory

Quick Lab

How can you amplify the sound of a tuning fork?

You should provide each group with at least one pair of forks of the same frequency. If you have a hollow resonance box, you may include it along with blocks of wood, metal bars, and so on. Caution students not to strike the tuning forks against anything hard. Dents and dings will eventually make a tuning fork useless. Tuning forks may be activated with a rubber stopper if rubber blocks are not available. In step 3, students should touch the bases, not the tines, of the two forks. In step 4, have students try bringing the tines of the active fork very near, but not touching, the tines of another fork of the same frequency. Students can also get the second fork to vibrate by touching the bases of both forks simultaneously to a resonant surface.

Analysis

1. Answers may vary. Generally, objects will be made of hard, thin materials with a relatively large surface area, such as desktops, boxes, tabletops, and sheets of poster board.
2. The forks will have the same pitch. If one fork has a pitch double (one octave above) that of another, the forks may also resonate.

LS Kinesthetic

Chapter Resource File

- **Quick Lab Datasheet** How can you amplify the sound of a tuning fork? GENERAL

Resonance

(Time: About 10 minutes)

Materials:

• piano, guitar, or other string
  instruments

Step 1 When one object vibrates at the natural frequency of another object, vibrations can be induced in the second object. This is an example of sympathetic vibration caused by resonance. A piano works best for this demonstration, but other acoustic, string instruments may work also.

Step 2 Press down the loud pedal of the piano to lift the felt dampers from all the strings. Sing or have a student sing a single note very loudly for 2 s and then stop. When the singing stops, you will hear the same note coming from the piano. This is caused by the induced vibration of the string of the same pitch (along with its harmonics). Try several other pitches. With a guitar, match the pitch of the sung note to the pitch of one of the strings.

Step 3 Students will also hear evidence that the human voice has several strong harmonics, or overtones, because several other strings will also sound along with the fundamental. In addition, all strings will vibrate a little due to forced vibrations.

LS **Auditory**

SKILL BUILDER

Interpreting Visuals **Figure 7** shows the different parts of the ear. As students read the description of vibrations passing through the ear, have them follow the path of the vibrations in the figure.

internet connect

www.scilinks.org
Topic: The Ear
SciLinks code: HK4034

SCiLINKS. Maintained by the National Science Teachers Association

Figure 7

Sound waves are transmitted as vibrations through the ear. Vibrations in the cochlea stimulate nerves that send impulses to the brain.

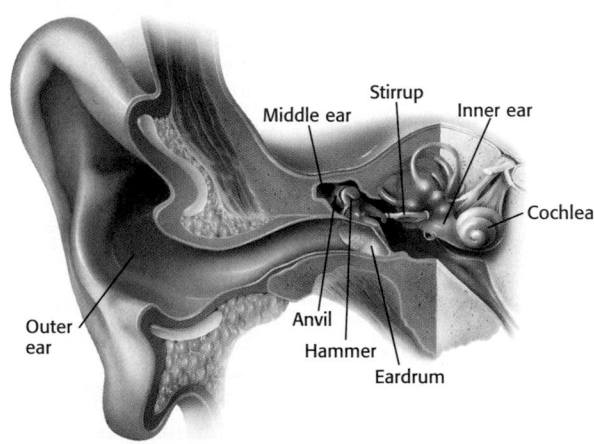

Middle ear   Stirrup   Inner ear

Cochlea

Outer ear

Anvil

Hammer

Eardrum

The natural frequency of an object depends on its shape, size, and mass, as well as the material it is made of. Complex objects such as a guitar have many natural frequencies, so they resonate well at many different pitches. However, some musical instruments, such as an electric guitar, do not resonate well and must be amplified electronically.

Hearing and the Ear

The head of a drum or the strings and body of a guitar vibrate to create sound waves in air. But how do you hear these waves and interpret them as different sounds?

The human ear is a very sensitive organ that senses vibrations in the air, amplifies them, and then transmits signals to the brain. In some ways, the process of hearing is the reverse of the process by which a drum head makes a sound. In the ear, sound waves cause membranes to vibrate.

Vibrations pass through three regions in the ear

Your ear is divided into three regions—outer, middle, and inner—as shown in **Figure 7.** Sound waves are funneled through the fleshy part of your outer ear and down the ear canal. The ear canal ends at the eardrum, a thin, flat piece of tissue.

When sound waves strike the eardrum, they cause the eardrum to vibrate. These vibrations pass from the eardrum through the three small bones of the middle ear—known as the hammer, the anvil, and the stirrup. When the vibrations reach the stirrup, the stirrup strikes a membrane at the opening of the inner ear, sending waves through the spiral-shaped cochlea.

Resonance occurs in the inner ear

The cochlea contains a long, flexible membrane called the *basilar membrane*. Different parts of the basilar membrane vibrate at different natural frequencies. As waves pass through the cochlea, they resonate with specific parts of the basilar membrane.

A wave of a particular frequency causes only a small portion of the basilar membrane to vibrate. Hair cells near that part of the membrane then stimulate nerve fibers that send an impulse to the brain. The brain interprets this impulse as a sound with a specific frequency.

496

Trends in Medicine

Cochlear Implants Since the late 1980s, scientists around the world have used cochlear implants in over 70,000 people. While hearing aids merely amplify sound, cochlear implants actually bypass damaged parts of the inner ear, thereby restoring some level of hearing to people with profound deafness. Although the implants cannot fully restore hearing, they provide enough improvement to drastically improve the recipient's ability to communicate fluently in verbal conversation.

The recipient of a cochlear implant wears a microphone in a hearing case behind the ear, and a sound processor about the size of a personal radio on a belt or in a pocket. The microphone picks up sounds and send them to the processor, which selects and arranges the sounds. A transmitter converts these sounds to electrical signals, which are sent to electrodes implanted in the cochlea and then to the brain. Researchers are currently working to improve hearing for cochlear-implant recipients and to find additional candidates for the use of this new technology.

Ultrasound and Sonar

If you shout over the edge of a rock canyon, you may hear the sound reflected back to you in an echo. Like all waves, sound waves can be reflected. The reflection of sound waves can be used to determine distances and to create maps and images.

Sonar is used for underwater location

How can a person on a ship measure the distance to the ocean floor, which may be thousands of meters from the surface of the water? One way is to use **sonar.**

A sonar system determines distance by measuring the time it takes for sound waves to be reflected back from a surface. A sonar device on a ship sends a pulse of sound downward, and measures the time, t, that it takes for the sound to be reflected back from the ocean floor. Using the average speed of the sound waves in water, v, one can calculate the distance, d, by using a form of the speed equation that solves for distance.

$$d = vt$$

If a school of fish or a submarine passes under the ship, the sound pulse will be reflected back much sooner.

Ultrasound waves—sound waves with frequencies above 20 000 Hz—work particularly well in sonar systems because they can be focused into narrow beams and can be directed more easily than other sound waves. Bats, like the one in *Figure 8,* use reflected ultrasound waves to navigate in flight and to locate insects for food.

Ultrasound imaging is used in medicine

The echoes of very high frequency ultrasound waves, between 1 million and 15 million Hz, are used to produce computerized images called *sonograms*. Using sonograms, doctors can safely view organs inside the body without having to perform surgery. Sonograms can be used to diagnose problems, to guide surgical procedures, or even to view unborn fetuses, as shown in *Figure 9.*

Figure 8
Bats use ultrasound echoes to navigate in flight.

▶ **sonar s**ound **n**avigation **a**nd **r**anging, a system that uses acoustic signals and echo returns to determine the location of objects or to communicate

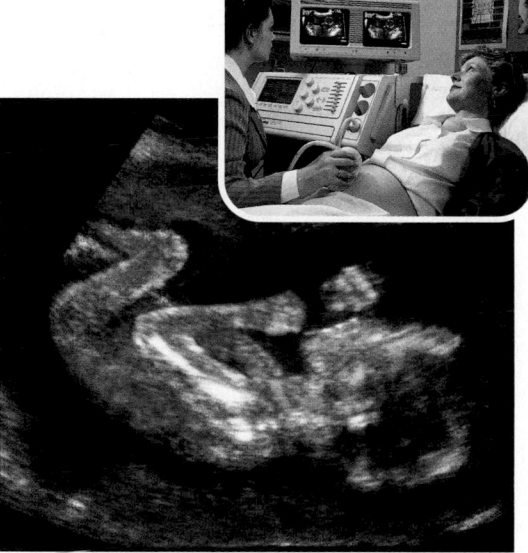

Figure 9
An image of an unborn fetus can be generated from reflected ultrasound waves.

497

LIFE SCIENCE
CONNECTION

Bats use ultrasound to piece together detailed information about the location, size, and movement of insects and other prey. By comparing the sound received by its right and left ears, a bat can determine location. Folds of the ear help the bat determine the insect's precise vertical position. The intensity of the echo provides information about the insect's size. Because of the Doppler effect, there is a change in pitch that tells the bat if the insect is moving toward it or away from it.

Some bats emit the sound waves from their mouth; others use their nose. The reflected sound is received by the bat's ears and the information is processed by the brain. All of this is done unconsciously, just like we process aural information gathered by our ears. The result is an "echolocation image" which bats use in conjunction with their sense of vision to track down and capture prey.

Close

Quiz ————————— BASIC

1. Sound travels through solids, liquids, and gases at the same speed. True or false? (false)

2. What are the units of relative sound intensity? (decibel)

3. What is the pitch of a note determined by? (the frequency of the sound waves)

4. What are the three regions of the ear? (outer, middle, and inner)

5. What type of sound waves are used in sonar, bat echolocation, and sonograms? (ultrasound)

LS Logical

Some ultrasound waves are reflected at boundaries

At high frequencies, ultrasound waves can travel through most materials. But some sound waves are reflected when they pass from one type of material into another. How much sound is reflected depends on the density of the materials at each boundary. The reflected sound waves from different boundary surfaces are compiled into a sonogram by a computer and displayed on a screen.

The advantage of using sound to see inside the human body is that it doesn't harm living cells as X rays may do. However, to see details, the wavelengths of the ultrasound must be slightly smaller than the smallest parts of the object being viewed. The higher the frequency of waves in a given medium are, the shorter the wavelength will be. Sound waves with a frequency of 15 million Hz have a wavelength of less than 1 mm when they pass through soft tissue.

SECTION 1 REVIEW

SUMMARY

▶ The speed of sound waves depends on temperature, density, and other properties of the medium.

▶ Pitch is determined by the frequency of sound waves.

▶ Infrasound and ultrasound lie beyond the range of human hearing.

▶ The loudness of a sound depends on intensity. Relative intensity is measured in decibels (dB).

▶ Musical instruments use standing waves and resonance to produce sound.

▶ The ear converts vibrations in the air into nerve impulses to the brain.

▶ Reflection of sound or ultrasound waves can be used to determine distances or to create sonograms.

1. Identify two factors that affect the speed of sound.

2. Explain why sound travels faster in water than in air.

3. Distinguish between infrasound and ultrasound waves.

4. Determine which of the following must change when pitch gets higher.
   a. amplitude   **d.** intensity
   b. frequency   **e.** speed of the sound waves
   c. wavelength

5. Determine which of the following must change when a sound gets louder.
   a. amplitude   **d.** intensity
   b. frequency   **e.** speed of the sound waves
   c. wavelength

6. Explain why the note middle C played on a piano sounds different from the same note played on a violin.

7. Explain why an acoustic guitar generally sounds louder than an electric guitar without an electronic amplifier.

8. Describe the process through which sound waves in the air are translated into nerve impulses to the brain.

9. Critical Thinking Why are sonograms made with ultrasound waves instead of audible sound waves?

10. Creative Thinking Why do most pianos contain a large *sounding board* underneath the strings? (**Hint:** The piano would be harder to hear without it.)

Answers to Section 1 Review

1. Students may list state of matter, nature of the medium, or temperature.

2. In air, molecules are very far apart and collide less often than in water, where molecules are packed closely together.

3. Audible sound can be heard by the human ear. Its frequency range is about 20 Hz to 20 000 Hz. Sounds with frequencies lower than 20 Hz make up infrasound. Ultrasound has frequencies higher than 20 000 Hz.

4. b and c: Frequency increases and wavelength decreases.

5. a and d: Amplitude and intensity both increase.

6. Although the two notes have the same fundamental pitch, they sound different because they emphasize different harmonics.

7. The acoustic guitar is constructed so that the hollow body vibrates in resonance with the string. Electric guitars usually have a solid body that vibrates very little.

See "Coninuation of Answers" at the end of the chapter.

The Nature of Light

OBJECTIVES

▶ **Recognize** that light has both wave and particle characteristics.
▶ **Relate** the energy of light to the frequency of electromagnetic waves.
▶ **Describe** different parts of the electromagnetic spectrum.
▶ **Explain** how electromagnetic waves are used in communication, medicine, and other areas.

▶ **KEY TERMS**
photon
intensity
radar

Most of us see and feel light almost every moment of our lives, from the first rays of dawn to the warm glow of a campfire. Even people who cannot see can feel the warmth of the sun on their skin, which is an effect of infrared light. We are very familiar with light, but how much do we understand about what light really is?

Waves and Particles

It is difficult to describe all of the properties of light with a single scientific model. The two most commonly used models describe light either as a wave or as a stream of particles.

Light produces interference patterns like water waves

In 1801, the English scientist Thomas Young devised an experiment to test the nature of light. He passed a beam of light through two narrow openings and then onto a screen on the other side. He found that the light produced a striped pattern on the screen, like the pattern in *Figure 10A*. This pattern is similar to the pattern caused by water waves interfering in a ripple tank, as shown in *Figure 10B*.

internet connect
www.scilinks.org
Topic: Properties of Light
SciLinks code: HK4110
*SCi*LINKS® Maintained by the National Science Teachers Association

Figure 10

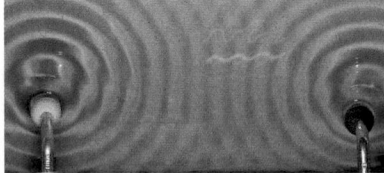

A Light passed through two small openings produces light and dark bands on a screen.

B Two water waves in a ripple tank also produce an interference pattern with light and dark bands.

499

EARTH SCIENCE
CONNECTION

Invite an Earth Science teacher to be a guest lecturer in your class, and ask the teacher to display maps of the sea floor. If a relief globe is available, have the teacher show the sea floors and continental shelves. They could also show satellite images of Earth with vegetation growth or other features detected with infrared light or other wavelengths. Discuss how these maps were obtained using both sonar (sound waves) and satellite radar (electromagnetic waves).

Focus

Overview
Before beginning this section, review with your students the Objectives listed in the Student Edition. In this section, students learn about the wave and particle characteristics of light. They also study the electromagnetic spectrum in detail, including the relationship of energy to frequency and applications of electromagnetic waves in communication, medicine, and other areas.

💿 Bellringer
Use the Bellringer transparency to prepare students for this section.

Motivate

Opening Activity — GENERAL
Write the words *sound* and *light* on the chalkboard. Have students brainstorm words and phrases commonly used to describe sound and light. Allow a student volunteer to write the student contributions on the board under the appropriate term. (Students should provide words such as *waves, strong, bright, dim, colored, loud, soft, musical, high-pitched,* and *low-pitched.*) Next have students pair a word used for sound with a word used for light when the words have the same meaning in terms of wave motion. (For example, *loud* would pair with *bright* because both are related to the energy of the wave. *Color* would pair with *pitch* because both depend on frequency.)
LS Verbal

Chapter Resource File

• **Lesson Plan**

Transparencies

TT Bellringer

Reading Skills The speed of light and the wave-particle duality of light are fundamental ideas of science. To help students grasp these concepts, proceed in a stepwise fashion by having students read the subsections under "Waves and Particles" one at a time. At the end of each subsection, stop and ask for volunteers to summarize that part. After the student finishes, ask students whether they agree on the summary given. If not, ask for other volunteers to offer their insights. **LS Interpersonal**

Interpreting Diagrams To help students understand the photon concept pictured in **Figure 11,** have them visualize a brick wall being struck by streams of thousands of ping-pong balls. No single ball has enough kinetic energy to chip the bricks. Then imagine the same wall being struck at the same speed by only a few steel bolts. Even though just a few strike the wall, each has enough energy to break chips out of the bricks. The many ping-pong balls represent bright red light and the few bolts represent dim blue light. The brightness of light corresponds to the number of photons striking a surface per unit time, and the color of light corresponds to the energy per photon.

Figure 11

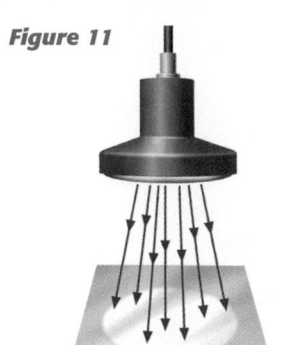

A Bright red light cannot knock electrons off this metal plate.

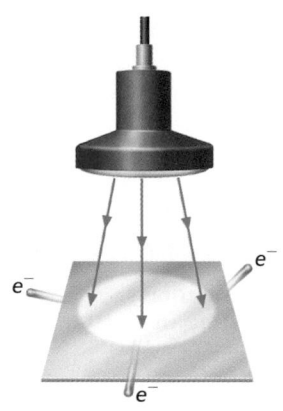

B Dim blue light can knock electrons off the plate. The wave model of light cannot explain this effect, but the particle model can.

photon a unit or quantum of light

Light can be modeled as a wave

Because the light in Young's experiment produced interference patterns, Young concluded that light must consist of waves. The model of light as a wave is still used today to explain many of the basic properties of light and its behavior.

This model describes light as transverse waves that do not require a medium in which to travel. Light waves are also called electromagnetic waves because they consist of changing electric and magnetic fields. The transverse waves produced by these fields can be described by their amplitude, wavelength, and frequency.

The wave model of light explains much of the observed behavior of light. For example, light waves may reflect when they meet a mirror, refract when they pass through a lens, or diffract when they pass through a narrow opening. Light waves also interfere with one another, and they can even form standing waves.

The wave model of light cannot explain some observations

In the early part of the 20th century, physicists began to realize that some observations could not be explained with the wave model of light. For example, when light strikes a piece of metal, electrons may fly off the metal's surface. Experiments show that in some cases, dim blue light may knock some electrons off a metal plate, while very bright red light cannot knock off any electrons, as shown in *Figure 11.*

According to the wave model, very bright red light should have more energy than dim blue light because the waves in bright light should have greater amplitude. But this does not explain how the blue light can knock electrons off the plate while the red light cannot.

Light can be modeled as a stream of particles

One way to explain the effects of light striking a metal plate is to assume that the energy of the light is contained in small packets. A packet of blue light carries more energy than a packet of red light, enough to knock an electron off the plate. Bright red light contains many packets, but no single one has enough energy to knock an electron off the plate.

In the particle model of light, these packets are called **photons,** and a beam of light is considered to be a stream of photons. Photons are considered particles, but they are not like ordinary particles of matter. Photons do not have mass; they are more like little bundles of energy. But unlike the energy in a wave, the energy in a photon is located in a particular place.

500

Historical Perspective

Photons of Light Photons were first proposed by Albert Einstein in 1905, the same year he published his theory of special relativity. Einstein used photons to explain the photoelectric effect, which is the ejection of electrons from a metal plate by light. Einstein's photon theory was based on the earlier (1900) quantum theory of Max Planck, who said that atoms absorb and release energy in certain discrete amounts or quanta. This work by Planck and Einstein laid the foundation for a major revolution in physics that led to our modern theory of quantum mechanics.

Did You Know?

Photoelectric Effect The photoelectric effect was once widely used in devices called "electric eyes," which detected when a light beam was broken and caused an action such as setting off an alarm or opening a door.

The model of light used depends on the situation

Light can be modeled as either waves or particles; so which explanation is correct? The success of any scientific theory depends on how well it can explain different observations. Some effects, such as the interference of light, are more easily explained with the wave model. Other cases, like light knocking electrons off a metal plate, are explained better by the particle model. The particle model also easily explains how light can travel across empty space without a medium.

Most scientists currently accept both the wave model and the particle model of light, and use one or the other depending on the situation that they are studying. Some believe that light has a "dual nature," so that it actually has different characteristics depending on the situation. In many cases, using either the wave model or the particle model of light gives good results.

The energy of light is proportional to frequency

Whether modeled as a particle or as a wave, light is also a form of energy. Each photon of light can be thought of as carrying a small amount of energy. The amount of this energy is proportional to the frequency of the corresponding electromagnetic wave, as shown in **Figure 12.**

A photon of red light, for example, carries an amount of energy that corresponds to the frequency of waves in red light, 4.5×10^{14} Hz. A photon with twice as much energy corresponds to a wave with twice the frequency, which lies in the ultraviolet range of the electromagnetic spectrum. Likewise, a photon with half as much energy, which would be a photon of infrared light, corresponds to a wave with half the frequency.

The speed of light depends on the medium

In a vacuum, all light travels at the same speed, called c. The speed of light is very large, 3×10^8 m/s (about 186 000 mi/s). Light is the fastest signal in the universe. Nothing can travel faster than the speed of light.

Light also travels through transparent mediums, such as air, water, and glass. When light passes through a medium, it travels slower than it does in a vacuum. *Table 2* shows the speed of light in several different mediums.

Figure 12

The energy of photons of light is related to the frequency of electromagnetic waves.

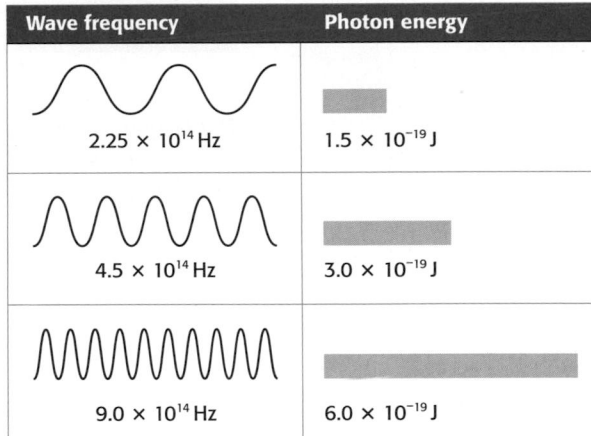

Wave frequency	Photon energy
2.25×10^{14} Hz	1.5×10^{-19} J
4.5×10^{14} Hz	3.0×10^{-19} J
9.0×10^{14} Hz	6.0×10^{-19} J

Table 2 **Speed of Light in Various Mediums**

Medium	Speed of light ($\times 10^8$ m/s)
Vacuum	2.997925
Air	2.997047
Ice	2.29
Water	2.25
Quartz (SiO_2)	2.05
Glass	1.97
Diamond	1.24

SKILL BUILDER

Interpreting Visuals Use **Figure 12** to remind students that light color relates to the energy per photon. In the wave model, color relates to frequency, as pointed out in the next subsection. Therefore, frequency and energy per photon correspond to each other.

Teaching Tip

Speed of Light Unlike sound, which travels better in liquids and solids than in air, light travels more slowly as the density of the medium increases. When light passes through a medium, it encounters the many atoms that make up the medium. If light hits an atom, the light is absorbed and emitted again, which takes a (very) small amount of time. As light passes through the empty space between atoms in the medium, the light moves at its full speed of 3.0×10^8 m/s. The speed of light in the medium is really an average speed that takes into account the stopping and starting of the light as it encounters atoms.

SKILL BUILDER

Interpreting Visuals The speed of light in air is slower than the speed of light in a vacuum, but the difference is very small, as students can see in **Table 2**. The speed of light in a diamond is less than half the speed of light in a vacuum but still on the same order of magnitude ($\times 10^8$ m/s).

Historical Perspective

Particle versus Wave Isaac Newton believed that light is made up of tiny particles, which he called corpuscles. He published this theory in his book *Optics* in 1704. One of Newton's contemporaries, Christian Huygens, argued that light is a wave. Both Newton and Huygens offered convincing evidence, and the subject was a matter of great debate. Later experiments (such as Young's double-split experiment) seemed to verify Huygens' theory, which became widely accepted by scientists until Einstein proposed his photon theory in 1905. Einstein did not dispute the wave nature of light, but argued that in some cases, a particle model is more useful.

Alternative Assessment — ADVANCED

The Newton-Huygens Debate Have students research the arguments used by Newton and Huygens for their theories of light. Ask them to assess the validity of the arguments, and evaluate which arguments still hold today and which have been overturned by our modern understanding of light.
LS Logical

Teaching Tip
Electromagnetic Spectrum

Remind students that radiation in other parts of the electromagnetic spectrum is qualitatively no different from visible light. However, our eyes are sensitive to the frequencies in the visible spectrum. The unique properties mentioned in the text depend less on the nature of the radiation and more on the way the radiation interacts with matter. Emphasize the continuity of the visible spectrum by pointing out that it consists not only of red, orange, yellow, green, blue, indigo, and violet, but all intermediate colors as well.

SKILL BUILDER — BASIC

Interpreting Visuals Figure 14 shows the different parts of the electromagnetic spectrum. Ask students: Which end of the spectrum has higher frequency? (Frequency increases toward the top end of the spectrum.) Which end of the spectrum has longer wavelengths? (Wavelength increases toward the bottom end of the spectrum.) Which end of the spectrum has higher energy? (The upper end has higher energy; the energy of electromagnetic waves is proportional to frequency.)
LS Visual

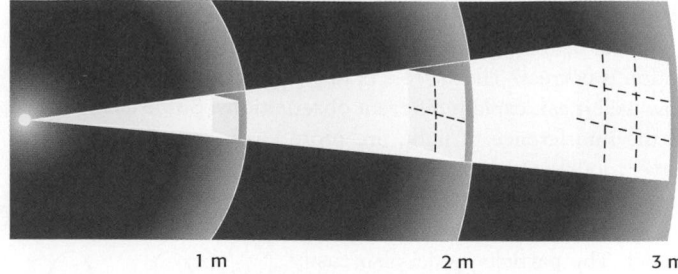

1 m   2 m   3 m

Figure 13

Less light falls on each unit square as the distance from the source increases.

▶ **intensity** the rate at which energy flows through a given area of space

Figure 14

The electromagnetic spectrum includes all possible kinds of light.

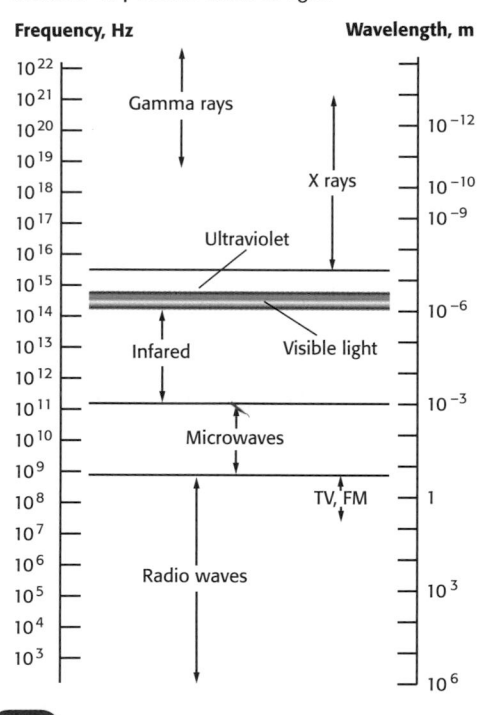

The brightness of light depends on intensity

You have probably noticed that it is easier to read near a lamp with a 100 W bulb than near a lamp with a 60 W bulb. That is because a 100 W bulb is brighter than a 60 W bulb. The quantity that measures the amount of light illuminating a surface is called **intensity**, and it depends on the amount of light—the number of photons, or power—that passes through a certain area of space.

Like the intensity of sound, the intensity of light from a light source decreases as the light spreads out in spherical wave fronts. Imagine a series of spheres centered around a source of light, as shown in *Figure 13.* As light spreads out from the source, the number of photons or power passing through a given area on a sphere, say 1 cm^2, decreases. An observer farther from the light source will therefore see the light as dimmer than will an observer closer to the light source.

The Electromagnetic Spectrum

Light fills the air and space around us. Our eyes can detect light waves ranging from 400 nm (violet light) to 700 nm (red light). But the visible spectrum is only one small part of the entire electromagnetic spectrum, shown in *Figure 14.* We live in a sea of electromagnetic waves, ranging from the sun's ultraviolet light to radio waves transmitted by television and radio stations.

The electromagnetic spectrum consists of light at all possible energies, frequencies, and wavelengths. Although all electromagnetic waves are similar in certain ways, each part of the electromagnetic spectrum also has unique properties. Many modern technologies, from radar guns to cancer treatments, take advantage of the different properties of electromagnetic waves.

502

Did You Know?

Inverse Square Illumination is an extension of the inverse square law that applies to most fields. Another way to state it is to say that as distance from a light source doubles, illumination decreases to one-fourth its original value.

Alternative Assessment — ADVANCED

Electromagnetic Wave Applications Have students research electromagnetic wave applications. Ask each student to focus on one application that is not discussed in the student text. Have them prepare a poster that illustrates the application, including how it works, what type of electromagnetic waves are used, and why that type works best. Display student posters around the classroom. **LS Verbal**

Sunlight contains ultraviolet light

The invisible light that lies just beyond violet light falls into the *ultraviolet* (UV) portion of the spectrum. Ultraviolet light has higher energy and shorter wavelengths than visible light does. Nine percent of the energy emitted by the sun is UV light. Because of its high energy, some UV light can pass through thin layers of clouds, causing you to sunburn even on overcast days.

X rays and gamma rays are used in medicine

Beyond the ultraviolet part of the spectrum lie waves known as *X rays,* which have even higher energy and shorter wavelengths than ultraviolet waves. X rays have wavelengths less than 10^{-8} m. The highest energy electromagnetic waves are gamma rays, which have wavelengths as short as 10^{-14} m.

An X-ray image at the doctor's office is made by passing X rays through the body. Most of them pass right through, but a few are absorbed by bones and other tissues. The X rays that pass through the body to a photographic plate produce an image such as the one in *Figure 15.*

X rays are useful tools for doctors, but they can also be dangerous. Both X rays and gamma rays have very high energies, so they may kill living cells or turn them into cancer cells. However, gamma rays can also be used to treat cancer by killing the diseased cells.

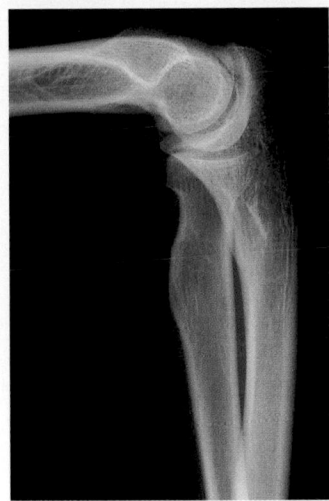

Figure 15
X-ray images are negatives. Dark areas show where the rays passed through, while bright areas show denser structures in the body.

REAL WORLD APPLICATIONS

Sun Protection Short-term exposure to UV light can cause sunburn; prolonged or repeated exposure may lead to skin cancer. To protect your skin, you should shield it from UV light whenever you are outdoors by covering your body with clothing, wearing a hat, and using a sunscreen.

Sunscreen products contain a chemical that blocks some or all UV light, preventing it from penetrating your skin. The Skin Protection Factor (SPF) of sunscreens varies as shown in the table at right.

Applying Knowledge

1. A friend is taking an antibiotic, and his doctor tells him to avoid UV light while on the medication. What SPF factor should he use, and why?

2. You and another friend decide to go hiking on a cloudy day. Your friend claims that she does not need any sunscreen because the sun is not out. What is wrong with her reasoning?

SPF factor	Effect on skin
None	Offers no protection from damage by UV
4 to 6	Offers some protection if you tan easily
7 to 8	Offers extra protection but still permits tanning
9 to 10	Offers excellent protection if you burn easily but would still like to get a bit of a tan
15	Offers total protection from burning
22	Totally blocks UV

503

Trends in Medicine

Digital X Rays Dentists have started using digital X rays, which use electronic sensors in place of photographic film. There are many advantages to this new technology. The images are higher resolution, and they can be stored on a computer. Also, the chemical waste used to develop film is eliminated. Another advantage is that the exposure is about one-tenth the typical dosage today. Researchers hope to apply this technique to other areas of medicine, such as mammography and heart imaging.

Teaching Tip

Incandescence Infrared radiation is sometimes called radiant heat because it increases the velocity of molecules when it is absorbed by matter. The warm matter also radiates energy as infrared waves. As a sample of matter becomes hotter, the average frequency of radiated energy becomes higher. At some temperature, the matter begins to radiate visible light (become incandescent) and appears to be dull red. As matter is heated further, it becomes orange and then yellow. When it is hot enough that significant amounts of blue light are emitted, the object appears white and we say it is "white-hot." In fact, all matter above 0 K emits radiation.

Teaching Tip

Division Points Be sure students understand that the division point between microwaves and radio waves is arbitrarily set, as are other division points on the electromagnetic spectrum. There is no significant difference between the waves on either side of a division point. The exception is visible light, which is limited by the eye's sensitivity.

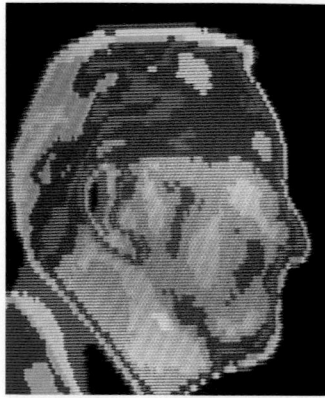

Figure 16
An infrared camera reveals the temperatures of different parts of an object.

▶ **radar** radio detection and ranging, a system that uses reflected radio waves to determine the velocity and location of objects

Did You Know?

Because microwaves reflect off the inside walls of a microwave oven, they may form standing waves. Food lying at the antinodes, where the vibrations are at a maximum, gets cooked more than food lying at the nodes, where there are no vibrations. For that reason, most microwave ovens rotate food items to ensure even heating.

504

Infrared light can be felt as warmth

Electromagnetic waves with wavelengths slightly longer than red light fall into the *infrared* (IR) portion of the spectrum. Infrared light from the sun, or from a heat lamp, warms you. Infrared light is used to keep food warm. You might have noticed reddish lamps above food in a cafeteria. The energy provided by the infrared light is just enough to keep the food hot without continuing to cook it.

Devices and photographic film that are sensitive to infrared light can reveal images of objects like the one in **Figure 16**. An infrared sensor can be used to measure the heat that objects radiate and then create images that show temperature variations. By detecting infrared radiation, areas of different temperature can be mapped. Remote sensors on weather satellites that record infrared light can track the movement of clouds and record temperature changes in the atmosphere.

Microwaves are used in cooking and communication

Electromagnetic waves with wavelengths in the range of centimeters, longer than infrared waves, are known as *microwaves*. The most familiar application of microwaves today is in cooking.

Microwave ovens in the United States use microwaves with a frequency of 2450 MHz (12.2 cm wavelength). Microwaves are reflected by metals and are easily transmitted through air, glass, paper, and plastic. However, water, fat, and sugar all absorb microwaves. Microwaves can travel about 3–5 cm into most foods.

As microwaves penetrate deeper into food, they are absorbed along with their energy. The rapidly changing electric field of the microwaves causes water and other molecules to vibrate. The food warms up as the energy of these vibrations is delivered to other parts of the food.

Microwaves are also used to carry telecommunication signals. Most mobile phones use microwaves to transmit information, and space probes transmit signals back to Earth with microwaves.

Radio waves are used in communications and radar

Electromagnetic waves longer than microwaves are classified as *radio waves*. Radio waves have wavelengths that range from tenths of a meter to millions of meters. This portion of the electromagnetic spectrum includes TV signals, AM and FM radio signals, and other radio waves.

Air-traffic control towers at airports use **radar** to determine the locations of aircraft. Antennas at the control tower emit radio waves, or sometimes microwaves, out in all directions.

REAL-LIFE
CONNECTION

Microwave Ovens Although microwaves are reflected by metals, powdered metals absorb a significant amount of microwaves and become very hot. This technology is used in packaging microwavable foods. A paper or plastic sheet containing metal powder becomes hot and browns or fries the food.

Did You Know?

Microwaves from the Big Bang In the 1960s, scientists detected microwave radiation of 7.3 cm wavelength (or 4.1 GHz) coming from all directions. It is theorized that this is the residual radiation from the big bang, the birth of the universe.

When the signal reaches an airplane, a transmitter on the plane sends another radio signal back to the control tower. This signal gives the plane's location and elevation above the ground.

At shorter range, the original signal sent by the antenna may reflect off the plane and back to a receiver at the control tower. A computer then calculates the distance to the plane using the time delay between the original signal and the reflected signal. The locations of various aircraft around the airport are displayed on a screen like the one shown in *Figure 17*.

Radar is also used by police to monitor the speed of vehicles. A radar gun fires a radar signal of known frequency at a moving vehicle and then measures the frequency of the reflected waves. Because the vehicle is moving, the reflected waves will have a different frequency, according to the Doppler effect. A computer chip converts the difference in frequency into a speed and shows the result on a digital display.

Figure 17

The radar system in an air traffic control tower uses reflected radio waves to monitor the location and speed of airplanes.

Close

Quiz — BASIC

1. What are two different ways light can be modeled? (as a wave and as a particle)

2. What is the energy of light proportional to? (frequency)

3. What is the range of light waves from radio waves to gamma rays called? (the electromagnetic spectrum)

4. Match each type of electromagnetic wave with the correct application.
   a. cooking (microwaves) X rays
   b. radar (radio waves) microwaves
   c. medicine (X rays) radio waves

LS Logical

SECTION 2 REVIEW

SUMMARY

▶ Light can be modeled as electromagnetic waves or as a stream of particles called photons.

▶ The energy of a photon is proportional to the frequency of the corresponding light wave.

▶ The speed of light in a vacuum, c, is 3.0×10^8 m/s. Light travels more slowly in a medium.

▶ The electromagnetic spectrum includes light at all possible values of energy, frequency, and wavelength.

1. **State** one piece of evidence supporting the wave model of light and one piece of evidence supporting the particle model of light.

2. **Name** the regions of the electromagnetic spectrum from the shortest wavelengths to the longest wavelengths.

3. **Determine** which photons have greater energy, those associated with microwaves or those associated with visible light.

4. **Determine** which band of the electromagnetic spectrum has the following:
   a. the lowest frequency   c. the greatest energy
   b. the shortest wavelength   d. the least energy

5. **Critical Thinking** You and a friend are looking at the stars, and you notice two stars close together, one bright and one fairly dim. Your friend comments that the bright star must emit much more light than the dimmer star. Is he necessarily right? Explain your answer.

505

Answers to Section 2 Review

1. Interference, reflection, and refraction support the wave model. The fact that blue light can knock electrons off a metal plate while red light cannot (the photoelectric effect) supports the particle model.

2. gamma rays, X rays, ultraviolet light, visible light (from violet to red), infrared light, microwaves, radio waves

3. Photons of visible light have higher energy than photons associated with microwaves because their frequency is greater.

4. a. radio waves
   b. gamma rays
   c. gamma rays
   d. radio waves

5. He may be wrong because there is no way to judge. The star that appears brighter may actually be intrinsically dimmer but much closer than the less-bright star.

Chapter Resource File

• Concept Review GENERAL
• Quiz BASIC

SECTION
3

Overview

Before beginning this section, review with your students the Objectives listed in the Student Edition. This section begins with the law of reflection. Next, students learn how flat and concave mirrors form virtual and real images. The section concludes with a discussion of color, including why objects appear different colors and how colors can be added or subtracted.

🔊 Bellringer

Use the Bellringer transparency to prepare students for this section.

Motivate

Opening Discussion ——— GENERAL

Write the following questions on the chalkboard and conduct a short discussion on possible responses.

• How do you know things reflect light?

• What kinds of things reflect light best?

• Why can you see your image in some reflectors and not in others?

• Is there any way to predict what direction a light ray will travel after it hits a reflecting surface?

After students have read the first four pages of this section, discuss the questions again. **LS** Verbal

Reflection and Color

■ **KEY TERMS**
light ray
virtual image
real image

■ **light ray** a line in space that matches the direction of the flow of radiant energy

Figure 18

This solar collector in the French Pyrenees uses mirrors to reflect and focus light into a huge furnace, which can reach temperatures of 3000°C.

506

▶ **OBJECTIVES**

▶ **Describe** how light reflects off smooth and rough surfaces.
▶ **Explain** the law of reflection.
▶ **Show** how mirrors form real and virtual images.
▶ **Explain** why objects appear to be different colors.
▶ **Describe** how colors may be added or subtracted.

You may be used to thinking about light bulbs, candles, and the sun as objects that send light to your eyes. But everything else that you see, including this textbook, also sends light to your eyes. Otherwise, you would not be able to see them.

Of course, there is a difference between the light from the sun and the light from a book. The sun emits its own light. The light that comes from a book is created by the sun or a lamp, then bounces off the pages of the book to your eyes.

Reflection of Light

Every object reflects some light and absorbs some light. Mirrors, such as those on the solar collector in *Figure 18,* reflect almost all incoming light. Because mirrors reflect light, it is possible for you to see an image of yourself in a mirror.

Light can be modeled as a ray

It is useful to use another model for light, the **light ray,** to describe reflection, refraction, and many other effects of light at the scale of everyday experience. A light ray is an imaginary line running in the direction that the light travels. It is the same as the direction of wave travel in the wave model of light or the path of photons in the particle model of light.

Light rays do not represent a full picture of the complex nature of light but are a good approximation of light in many cases. The study of light in cases in which light behaves like a ray is called *geometrical optics.* Using light rays, one can trace the path of light in geometrical drawings called *ray diagrams.*

🌐 Cultural Awareness

In ancient times, people believed that vision was a result of mysterious rays sent out from the eyes. The idea still survives in language, for example when we "cast a glance" at something, or in "She threw him a disgusted look." Even recent studies show that many people still believe this idea, probably because they have never been asked to analyze it.

Questioning may reveal some students who interpret vision in this way. In the eleventh century, the Arabian scholar Alhazen proved that light was not created by the eyes. He pointed out that light must come from the sun because people who look directly at the sun become blinded.

Figure 19

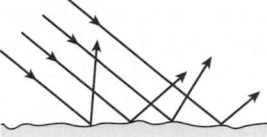

A Light rays reflected from a rough surface are reflected in many directions.

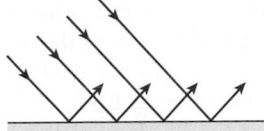

B Light rays reflected from a smooth surface are reflected in the same direction.

Rough surfaces reflect light rays in many directions

Many of the surfaces that we see every day, such as paper, wood, cloth, and skin, reflect light but do not appear shiny. When a beam of light is reflected, the path of each light ray in the beam changes from its initial direction to another direction. If a surface is rough, light striking the surface will be reflected at all angles, as shown in **Figure 19A**. Such reflection of light into random directions is called *diffuse reflection*.

Smooth surfaces reflect light rays in one direction

When light hits a smooth surface, such as a polished mirror, it does not reflect diffusely. Instead, all the light hitting a mirror from one direction is reflected together into a single new direction, as shown in **Figure 19B**.

The new direction of the light rays is related to the old direction in a definite way. The angle of the light rays reflecting off the surface, called the *angle of reflection*, is the same as the angle of the light rays striking the surface, called the *angle of incidence*. This is called the *law of reflection*.

The angle of incidence equals the angle of reflection.

Both of these angles are measured from a line perpendicular to the surface at the point where the light hits the mirror. This line is called the *normal*. **Figure 20** is a ray diagram that illustrates the law of reflection.

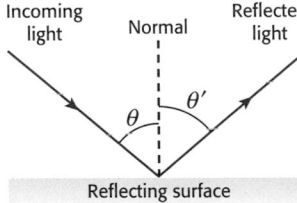

Incoming light   Normal   Reflected light

θ θ'

Reflecting surface

Figure 20

When light hits a smooth surface, the angle of incidence (θ) equals the angle of reflection (θ').

INTEGRATING

SPACE SCIENCE
There are two primary types of telescopes, refracting and reflecting. Refracting telescopes use glass lenses to focus light into an image at the eyepiece; reflecting telescopes use curved mirrors to focus light.

The lens of a refracting telescope cannot be very large, because the weight of the glass would cause the lens to bend out of shape. Curved mirrors are thinner and lighter than lenses, so they are stable at larger diameters.

The largest refracting telescope, at the Yerkes Observatory in Wisconsin, has a lens that is 1 m in diameter. The Mauna Kea Observatory in Hawaii houses four of the largest reflecting telescopes. Two of them have single mirrors that are over 8 m in diameter, and two use multiple mirrors for a total diameter of 10 m.

507

Did You Know ?

Law of Reflection Light rays striking any surface always obey the law of reflection on a microscopic level. But on a macroscopic level, most objects show diffuse reflection—they reflect light in all directions. For the law of reflection to work on a macroscopic level, the irregularities on the surface light is striking must be smaller than the wavelength of the incoming light.

Teach

SKILL BUILDER

Interpreting Diagrams Explain that the small arrows on the rays in **Figure 19** indicate direction of travel. Point out that in both cases, the incoming light rays are parallel. After reflecting from the rough surface, the rays are no longer parallel, but the rays reflected from the smooth surface remain parallel.

Demonstration — GENERAL

Observing Reflection

(Time: About 15 minutes)

Materials:

• aquarium tank (5 or 10 gal) or other large transparent container
• milk
• small mirror
• aluminum foil
• laser pointer or focusable flashlight
• transparent plastic protractor
• concave mirror
• convex mirror

Step 1 Fill the aquarium about two-thirds full with water. Add three drops of milk to the water, and stir. Shine the laser into the water to see if the beam is visible. Add more milk as needed.

Step 2 Simulate **Figure 19A** with the shiny side of crumpled aluminum foil and **Figure 19B** with a flat mirror lying at the bottom of the tank.

Continued on the next page

Demonstration, *continued*

Step 3 Demonstrate the angles of incidence and reflection with the flat mirror on the bottom of the tank. Move the light source to achieve various angles to show the validity of the law of reflection. Tape a transparent protractor to the outside of the tank to measure angles.

Step 4 Place the convex ("wide-angle") mirror at the bottom of the tank. Shine the laser pointer vertically downward onto the mirror, and move it from side to side over the mirror. Rays will reflect increasingly outward as you move the beam in any direction away from the center of the mirror. Reverse the rays by shining the beam onto the mirror from many different angles. Prompt students to conclude why this type of mirror is called a wide-angle mirror.

Step 5 Place a concave mirror at the bottom of the tank. Again shine the beam straight down at the mirror, and move it back and forth while keeping it vertical. Students will observe that all of the rays reflect inward through a common point. Use two beams if you can to show that they cross at a common point no matter where the beams strike the mirror.

Step 6 Place both mirrors in the tank and adjust a flashlight to produce a straight beam. Shine the beam into the mirrors. Students will see the light spread from the convex mirror and focus to a point from the concave mirror.
LS Visual

Mirrors

When you look into a mirror, you see an image of yourself that appears to be behind the mirror, as in *Figure 21A.* It is like seeing a twin or copy of yourself standing on the other side of the glass, although flipped from left to right. You also see a whole room, a whole world of space beyond the mirror, even if the mirror is placed against a wall. How is this possible?

Flat mirrors form virtual images by reflection

The ray diagram in *Figure 21B* shows the path of light rays striking a flat mirror. When a light ray is reflected by a flat mirror, the angle it is reflected is equal to the angle of incidence, as described by the law of reflection. Some of the light rays reflect off the mirror into your eyes.

However, your eyes do not know where the light rays have been. They simply sense light coming from certain directions, and your brain interprets the light as if it traveled in straight lines from an object to your eyes. As a result, you perceive an image of yourself behind the mirror.

Of course, there is not really a copy of yourself behind the mirror. If someone else looked behind the mirror, they would not see you, an image, or any source of light. The image that you see results from the apparent path of the light rays, not an actual path. An image of this type is called a **virtual image.** The virtual image appears to be as far behind the mirror as you are in front of the mirror.

▶ **virtual image** an image that forms at a location from which light rays appear to come but do not actually come

Figure 21

🅰 A virtual image appears behind a flat mirror.

🅱 A ray diagram shows where the light actually travels as well as where you perceive that it has come from.

508

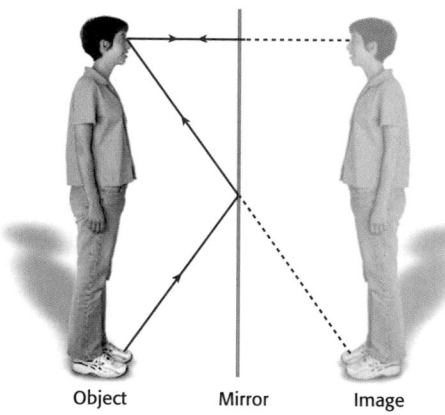

Object   Mirror   Image

MISCONCEPTION
ALERT

Light Many students associate light with its source, and do not understand that light travels from one place to another. This confusion makes it hard to for students to understand reflection. To emphasize the fact that light travels from place to place, tell students that it takes light from the sun about 8 minutes to reach Earth.

Curved mirrors can distort images

If you have ever been to the "fun house" at a carnival, you may have seen a curved mirror like the one in *Figure 22.* Your image in a curved mirror does not look exactly like you. Parts of the image may be spread out, making you look wide or tall. Other parts may be compressed, making you look thin or short. How does such a mirror work?

Curved mirrors still create images by reflecting light according to the law of reflection. But because the surface is not flat, the line perpendicular to the mirror (the normal) points in different directions for different parts of the mirror.

Where the mirror bulges out, two light rays that start out parallel are reflected into different directions, making an image that is stretched out. Mirrors that bulge out are called *convex mirrors.*

Similarly, parts of the mirror that are indented reflect two parallel rays in toward one another, making an image that is compressed. Indented mirrors are called *concave mirrors.*

Concave mirrors can create real images

Concave mirrors are used to focus reflected light. A concave mirror can form one of two kinds of images. It may form a virtual image behind the mirror or a **real image** in front of the mirror. A real image results when light rays from a single point of an object are focused onto a single point or small area.

If a piece of paper is placed at the point where the light rays come together, the real image appears on the paper. If you tried this with a virtual image, say by placing a piece of paper behind a mirror, you would not see the image on the paper. That is the primary difference between a real and a virtual image. With a real image, light rays really exist at the point where the image appears; a virtual image appears to exist in a certain place, but there are no light rays there.

Telescopes use curved surfaces to focus light

Many reflecting telescopes use curved mirrors to reflect and focus light from distant stars and planets. Radio telescopes, such as the one in *Figure 23,* gather radio waves from extremely distant objects, such as galaxies and quasars.

Different materials will reflect certain wavelengths of light and will allow other wavelengths to pass through them. Because radio waves reflect off almost any solid surface, these telescopes do not need to use mirrors. Instead, parallel radio waves bounce off a curved dish, which focuses the waves onto another, smaller curved surface poised above the dish. The waves are then directed into a receiver at the center of the dish.

Figure 22

A curved mirror produces a distorted image.

> **real image** an image of an object formed by light rays that actually come together at a specific location

Figure 23

A radio telescope dish reflects and focuses radio waves into the receiver at the center of the dish.

509

Color Perception When the brain receives signals from certain combinations of photoreceptor cells in the retina, it interprets them as the color green. These receptors are three kinds of cone cells, one each for red, green, and blue.

Be sure students understand that there is nothing inherently green about electromagnetic waves in one part of the visible spectrum. "Green" is just the way our brains interpret certain signals. Some students may be interested in conducting a more in-depth project on vision and how cells in the retina respond to light.

Teaching Tip ——— **BASIC**

Lighting and Color Ask students if they know of any instances in which light color is used to make something look different from how it would look in white light. Examples include the lights over produce displays to make the produce look lush green and deep red, the lighting at makeup counters and some clothing sales areas, and stage lighting to achieve a wide variety of effects. **LS** Verbal

Demonstration ——— **BASIC**

Additive Color TV If some students have never used a magnifying glass to look at the glowing phosphors of a color TV or monitor, this would be an ideal time because it will help reinforce the idea of additive colors. **LS** Visual

Seeing Colors

The different wavelengths of visible light correspond to many of the colors that you perceive. When you see light with a wavelength of about 550 nm, your brain interprets it as *green*. If the light comes from the direction of a leaf, then you may think, "That leaf is green."

A leaf does not emit light on its own; in the darkness of night, you may not be able to see the leaf at all. So where does the green light come from?

Objects have color because they reflect certain wavelengths

If you pass the light from the sun through a prism, the prism separates the light into a rainbow of colors. White light from the sun actually contains light from all the visible wavelengths of the electromagnetic spectrum.

When white light strikes a leaf, as shown in *Figure 24A,* the leaf reflects light with a wavelength of about 550 nm, corresponding to the color green. The leaf absorbs light at other wavelengths. When the light reflected from the leaf enters your eyes, your brain interprets it as *green*. Transparent objects such as color filters work in a similar way. A green filter transmits green light and absorbs other colors.

Likewise, the petals of a rose reflect red light and absorb other colors, so the petals appear to be red. If you view a rose and its leaves under red light, as shown in *Figure 24B,* the petals will still appear red but the leaves will appear black. Why?

Figure 24

A Rose in White and Red Light

A Under white light, the petals of a rose reflect red light, while the leaves reflect green light.

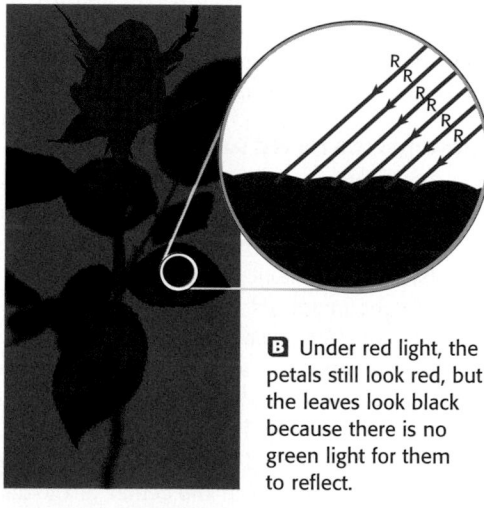

B Under red light, the petals still look red, but the leaves look black because there is no green light for them to reflect.

510

Colors of Light Most students believe that white light is colorless and pure, and that a color filter adds color to a white beam. These students have trouble understanding that white light is actually a mixture of colored light. You can show students a prism separating white light into a rainbow of colors to dispel this misconception.

Did You Know?

Reflection and Color Another way to think of subtractive colors is as minus colors. Each subtractive color is the full spectrum minus a color. Magenta is minus green, cyan is minus red, and yellow is minus blue. When all three are mixed, all colors are taken away, and the result is black.

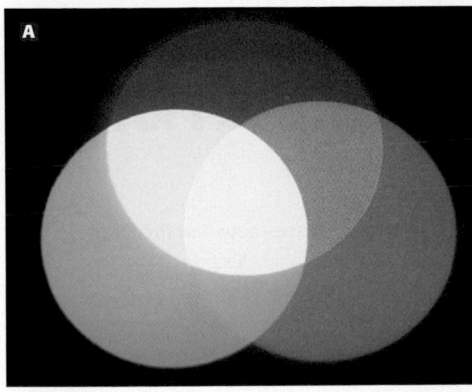

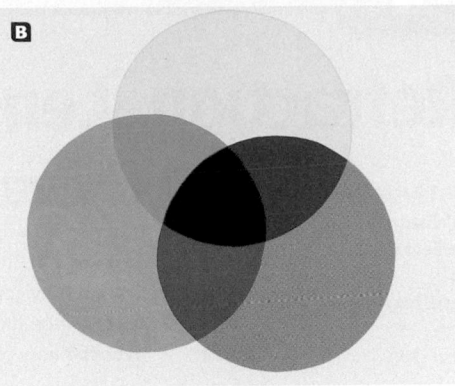

Quiz ──────────── BASIC

1. What is the law of reflection (angle of incidence equals the angle of reflection)

2. What type of image can be formed by a flat mirror? (virtual image)

3. What type can be formed by a concave mirror? (virtual image or real image)

4. White light from the sun contains wavelengths of all colors of visible light. True or false? (true)

LS Logical

Colors may add or subtract to produce other colors

Televisions and computer monitors display many different colors by combining light of the *additive primary colors*, red, green, and blue. Adding light of two of these colors together can produce the secondary colors yellow, cyan, and magenta, as shown in *Figure 25A*. Mixing all three additive primary colors makes white.

In the reverse process, pigments, paints, or filters of the *subtractive primary colors*, yellow, cyan, and magenta, can be combined to create red, green, and blue as shown in *Figure 25B*. If filters or pigments of all three colors are combined in equal proportions, all the light is absorbed, leaving black. Black is not really a color at all; it is the absence of color.

Figure 25

A Red, green, and blue lights can combine to produce yellow, magenta, cyan, or white.

B Yellow, magenta, and cyan filters can be combined to produce red, green, blue, or black.

SECTION 3 REVIEW

SUMMARY

▶ Light is reflected when it strikes a boundary between two different mediums.

▶ When light reflects off a surface, the angle of reflection equals the angle of incidence.

▶ Mirrors form images according to the law of reflection.

▶ The color of an object depends on the wavelengths of light that the object reflects.

1. **List** three examples of the diffuse reflection of light.

2. **Describe** the law of reflection in your own words.

3. **Draw** a diagram to illustrate the law of reflection.

4. **Discuss** how a plane mirror forms a virtual image.

5. **Discuss** the difference, in terms of reflection, between objects that appear blue and objects that appear yellow.

6. **Explain** why a plant may look green in sunlight but black under red light.

7. **Critical Thinking** A friend says that only mirrors and other shiny surfaces reflect light. Explain what is wrong with this reasoning.

8. **Creative Thinking** A convex mirror can be used to see around a corner at the intersection of hallways. Draw a simple ray diagram illustrating how this works.

511

Answers to Section 3 Review

1. Examples may include bicycle reflectors, clothing, and paper. Almost anything that is visible, except a direct source of light, reflects light. Only polished surfaces, such as mirrors, reflect light non-diffusely.

2. Answers should restate the law that states that the angle of incidence of a light ray equals the angle of reflection.

3. Diagrams should resemble **Figure 20**.

4. Light that reaches your eyes from the mirror comes from in front of the mirror. But because the rays come from the mirror, you see an image that appears to be behind the mirror.

5. Students should express the idea that the blue object reflects blue light while absorbing colors in the rest of the visible spectrum. A yellow object reflects yellow light while absorbing other colors.

6. The appearance of an object partly depends on the color of light striking it. A green leaf reflects green light only if green light is in the light shining on it. Light from the sun contains all colors, so the leaf looks green. Red light contains no green, so the leaf reflects no light and appears black.

See "Continuation of Answers" at the end of the chapter.

Chapter Resource File

• **Concept Review** GENERAL

• **Quiz** BASIC

 Videos

CNN Presents Physical Science

• **Segment 16** Color Deficiency Lenses

• **Critical Thinking** Worksheet 16

See the Science in the News video guide for more details.

Refraction, Lenses, and Prisms

▶ **KEY TERMS**

total internal
  reflection
lens
magnification
prism
dispersion

OBJECTIVES

▶ **Describe** how light is refracted as it passes between mediums.
▶ **Explain** how fiber optics use total internal reflection.
▶ **Explain** how converging and diverging lenses work.
▶ **Describe** the function of the eye.
▶ **Describe** how prisms disperse light and how rainbows form.

Light travels in a straight line through empty space. But in our everyday experience, we encounter light passing through various mediums, such as the air, windows, a glass of water, or a pair of eyeglasses. Under these circumstances, the direction of a light wave may be changed by refraction.

Figure 26

Light refracts when it passes from one medium into another.

Refraction of Light

Light waves bend, or refract, when they pass from one medium to another. If light travels from one transparent medium to another at any angle other than straight on, the light changes direction when it meets the boundary, as shown in **Figure 26.** Light bends when it changes mediums because the speed of light is different in each medium.

Imagine pushing a lawn mower at an angle from a sidewalk onto grass, as in **Figure 27.** The wheel that enters the grass first will slow down due to friction. If you keep pushing on the lawn mower, the wheel on the grass will act like a moving pivot, and the lawn mower will turn to a different angle.

Figure 27

This lawn mower changes direction as it passes from the sidewalk onto the grass.

MATHEMATICS
CONNECTION

The amount that a light ray will bend when passing into a medium can be described by the medium's index of refraction. The larger the index of refraction, the more light will bend. The index of refraction (n) can be calculated with the following ratio, where $c =$ the speed of light in a vacuum, and $v =$ the speed of light in the given medium:

$$n = \frac{c}{v}$$

Since c and v are both measures of velocity, n is a dimensionless number. Also, since c is the maximum speed of light, n is always greater than one.

To determine whether light rays will bend toward or away from the normal when passing from one medium to another, you can compare the index of refraction for the two mediums. When light moves from a medium with a lower index to one with a higher index, light rays bend toward the normal. Conversely, when it moves from a higher to a lower index, light rays bend away from the normal.

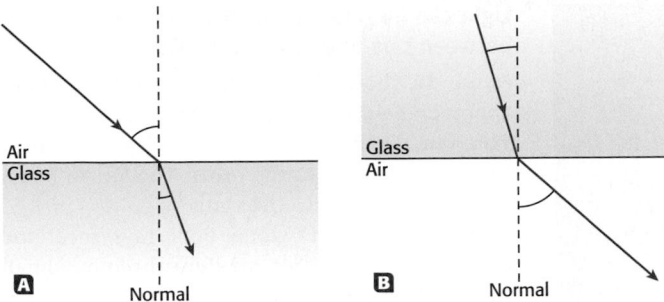

A Normal

B Normal

Air
Glass

Glass
Air

Figure 28

Figure 28
A When the light ray moves from air into glass, its path is bent toward the normal.

B When the light ray passes from glass into air, its path is bent away from the normal.

When light moves from a material in which its speed is higher to a material in which its speed is lower, such as from air to glass, the ray is bent toward the normal, as shown in *Figure 28A.* This is like the lawnmower moving from the sidewalk onto the grass. If light moves from a material in which its speed is lower to one in which its speed is higher, the ray is bent away from the normal, as shown in *Figure 28B.*

Refraction makes objects appear to be in different positions

When a cat looks at a fish underwater, the cat perceives the fish as closer than it actually is, as shown in the ray diagram in *Figure 29A.* On the other hand, when the fish looks at the cat above the surface, the fish perceives the cat as farther than it really is, as shown in *Figure 29B.*

The misplaced images that the cat and the fish see are virtual images like the images that form behind a mirror. The light rays that pass from the fish to the cat bend away from the normal when they pass from water to air. But the cat's brain doesn't know that. It interprets the light as if it traveled in a straight line, and sees a virtual image. Similarly, the light from the cat to the fish bends toward the normal, again causing the fish to see a virtual image.

Refraction in the atmosphere creates mirages

Have you ever been on a straight road on a hot, dry summer day and seen what looks like water on the road? If so, then you may have seen a *mirage.* A mirage is a virtual image caused by refraction of light in the atmosphere.

Light travels at slightly different speeds in air of different temperatures. Therefore, when light from the sky passes into the layer of hot air just above the asphalt on a road, it refracts, bending upward away from the road. Because you see an image of the sky coming from the direction of the road, your mind may assume that there is water on the road causing a reflection.

Figure 29

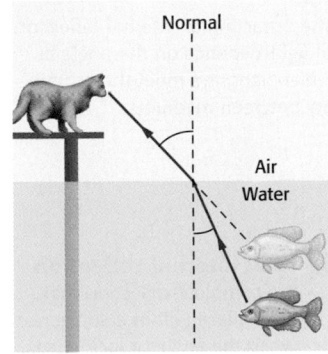

Normal

Air
Water

A To the cat on the pier, the fish appears to be closer than it really is.

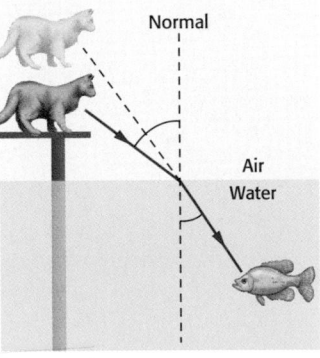

Normal

Air
Water

B To the fish, the cat seems to be farther from the surface than it actually is.

513

Teach

Teach

SKILL BUILDER — **BASIC**

Reading Skills Have students read the first three pages of this section and write notes about concepts they do not understand. Students should work in pairs to discuss the difficult passages that each person found difficult.
LS Interpersonal

Teaching Tip

Does Light Bend? Remind students that when we say light bends we mean that it changes direction abruptly.

Teaching Tip

Light Refraction Steering bulldozers and tanks is analogous to light refraction. Ask students if they have ever steered a track-driven vehicle. A tracked vehicle changes direction by the driver either applying a brake to slow down the track on one side or by speeding up the track on the other side (or both actions for sharp turns). Another analogy is making a turn while rowing a boat.

SKILL BUILDER — **BASIC**

Interpreting Diagrams **Figure 29,** like **Figure 20** earlier in the chapter, shows both light rays and dotted lines. The dotted lines show the perceived, or apparent, path of the light rays. Ask students whether the images in the figure are virtual images or real images. (They are virtual images.) **LS Logical**

Alternative Assessment — **ADVANCED**

Index of Refraction After discussing the Mathematics connection on the facing page, have students use a reference source to find the speed of light in various mediums. (Alternately, you may wish to provide them with this data.) Ask students to calculate the index of refraction for each medium. Then have them determine whether light will bend toward or away from the normal for different combinations of mediums. You may wish to have students illustrate this with diagrams similar to **Figure 28.** Also remind them that the higher the index, the more the light bends. **LS Logical**

Did You Know?

Mirages There are two types of mirages, inferior and superior. With an inferior mirage, an inverted object is seen *below* the original object. An apparent pool of water on desert sand is an inferior mirage. In a superior mirage, the inverted reflection appears *above* the original image. In special cases, a third, upright image may appear above the inverted image. If conditions are right, the curvature of Earth can hide the original and inverted images, creating an eerie image of a ship at sea or land on the horizon.

Chapter Resource File

• Lesson Plan

Transparencies

TT Bellringer
TT Refraction

Demonstration

Total Internal Reflection

(Time: About 10 minutes)

Materials:

• aquarium (5 or 10 gal)
• milk
• laser pointer or focusable flashlight
• regular flashlight

Step 1 Fill the aquarium two-thirds full with water. Add three drops of milk and stir. Check to see if the beam is visible and add more milk if needed.

Step 2 Demonstrate total internal reflection by shining the laser pointer upward toward the water surface. Slowly increase the angle at which the beam strikes the interface. Just as the critical angle is reached, the beam will shine across the interface, which seems to light up. Beyond that angle, the beam reflects back into the water.

Step 3 Repeat using a wider beam from a flashlight.

Teaching Tip ─── BASIC

Underwater Reflection Ask the class if anyone has ever looked up at the air/water interface while swimming underwater. Those who were observant will recall that they could see things above them, but as they looked farther off, the surface looked like a mirror. This happens when the angle of view is greater than the critical angle. Beyond this point, one can no longer see above the water. **LS Kinesthetic**

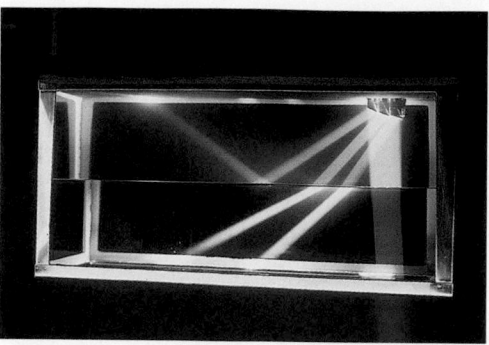

Figure 30

The refraction or internal reflection of light depends on the angle at which light rays meet the boundary between mediums.

▶ **total internal reflection**
the complete reflection that takes place within a substance when the angle of incidence of light striking the surface boundary is less than the critical angle

Light can be reflected at the boundary between two transparent mediums

Figure 30 shows four different beams of light approaching a boundary between air and water. Three of the beams are refracted as they pass from one medium to the other. The fourth beam is reflected back into the water.

If the angle at which light rays meet the boundary between two mediums becomes small enough, the rays will be reflected as if the boundary were a mirror. This angle is called the *critical angle*, and this type of reflection is called **total internal reflection.**

Fiber optics use total internal reflection

Fiber-optic cables are made by fusing bundles of transparent fibers together, as shown in *Figure 31A*. Light inside a fiber in a fiber-optic cable bounces off of the walls of the fiber due to total internal reflection, as shown in *Figure 31B*.

If the fibers are arranged in the same pattern at both ends of the cable, the light that enters one end can produce a clear image at the other end. Fiber-optic cables of that sort are used to produce images of internal organs during surgical procedures.

Because fiber-optic cables can carry many different frequencies at once, they transmit computer data or signals for telephone calls more efficiently than standard metal wires. Many long-distance phone calls are now transmitted over optical fibers.

Figure 31

A A fiber optic cable consists of several glass or plastic fibers bundled together.

B Light is guided along an optic fiber by multiple internal reflections.

Did You Know?

Fiber-Optic Cables There are two primary types of fiber-optic cables. In those used for imaging purposes, the fibers at each end have the same arrangement so that the images will not be jumbled. In those used for transmitting signals, the arrangement of the fibers is less important.

REAL-LIFE ─── CONNECTION

Advantages of Fiber Optics Although fiber-optic cables are much smaller than conventional copper cables, they are able to transmit much more information. Another advantage is that electromagnetic interference (as from weather conditions or other wires) does not affect data transmission through fiber-optic cables. Many phone companies now use fiber optics to connect between switching stations, but in most cases the last connection (from the local station to individual homes) is still made by conventional wires.

Figure 32

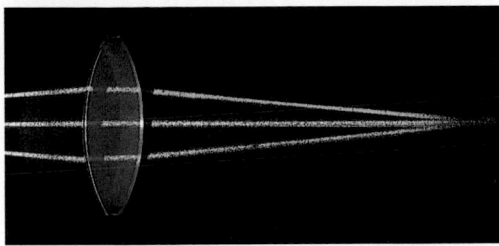

A When rays of light pass through a converging lens (thicker at the middle), they are bent inward.

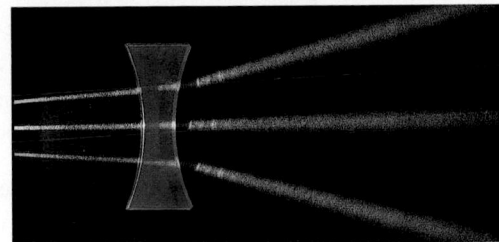

B When they pass through a diverging lens (thicker at the ends), they are bent outward.

Lenses

You are probably already very familiar with one common application of the refraction of light: lenses. From cameras to microscopes, eyeglasses to the human eye, lenses change the way we see the world.

Lenses rely on refraction

Light traveling at an angle through a flat piece of glass is refracted twice—once when it enters the glass and again when it reenters the air. The light ray that exits the glass is still parallel to the original light ray, but it has shifted to one side.

On the other hand, when light passes through a curved piece of glass, a **lens**, there is a change in the direction of the light rays. This is because each light ray strikes the surface of a curved object at a slightly different angle.

A *converging lens*, as shown in *Figure 32A*, bends light inward. A lens that bends light outward is a *diverging lens*, as shown in *Figure 32B*.

A converging lens can create either a virtual image or a real image, depending on the distance from the lens to the object. A diverging lens, however, can only create a virtual image.

Lenses can magnify images

A magnifying glass is a familiar example of a converging lens. A magnifying glass reveals details that you would not normally be able to see, such as the pistils of the flower in *Figure 33*. The large image of the flower that you see through the lens is a virtual image. **Magnification** is any change in the size of an image compared with the size of the object. Magnification usually produces an image larger than the object, but not always.

▶ **lens** a transparent object that refracts light waves such that they converge or diverge to create an image

▶ **magnification** a change in the size of an image compared with the size of an object

Figure 33
A magnifying glass makes a large virtual image of a small object.

(515)

Interpreting Visuals Figure 34 shows that light from a point on the specimen is focused to a point somewhere inside the body tube. If a screen were placed at this point, a small image of the specimen would appear on it. However, a real image can be viewed without a screen. A magnifying lens—the eyepiece—at the top of the tube, lets you view a magnified image of the already-magnified real image inside the tube. A telescope works in the same way as a microscope, except that the objective lens is large to capture a lot of light, and it brings light to a focus much farther from the lens.

Teaching Tip ── GENERAL

The Eye Lens Some students were probably taught in the past that the lens of the eye is solely responsible for forming an image on the retina. Help them overcome this error by calling their attention to the fact that the curved cornea is a lens too. The lens is important, because muscles change its curvature to adjust the focus for nearby and distant objects. Some students may be interested in exploring the use of surgery to reshape the cornea to correct vision defects that would otherwise be corrected by contact lenses or eyeglasses.
LS Verbal

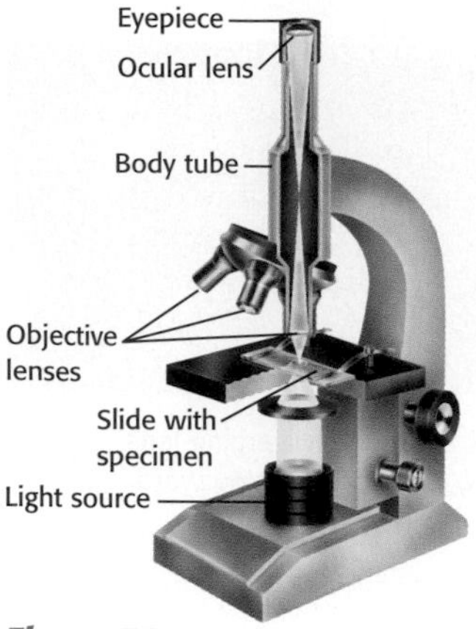

Eyepiece
Ocular lens
Body tube
Objective lenses
Slide with specimen
Light source

Figure 34

A compound microscope uses several lens to produce a highly magnified image.

Figure 35

The cornea and lens refract light onto the retina at the back of the eye.

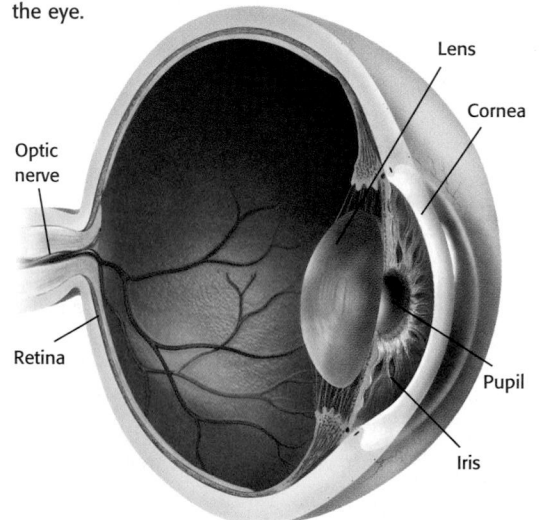

Lens
Cornea
Optic nerve
Retina
Pupil
Iris

If you hold a magnifying glass over a piece of paper in bright sunlight, you can see a real image of the sun on the paper. By adjusting the height of the lens above the paper, you can focus the light rays together into a small area, called the *focal point*. At the focal point, the image of the sun may contain enough energy to set the paper on fire.

Microscopes and refracting telescopes use multiple lenses

A compound microscope uses multiple lenses to provide greater magnification than a magnifying glass. *Figure 34* illustrates a basic compound microscope. The objective lens first forms a large real image of the object. The eyepiece then acts like a magnifying glass and creates an even larger virtual image that you see when you peer through the microscope.

Section 3 explained how reflecting telescopes use curved mirrors to create images of distant objects such as planets and galaxies. Refracting telescopes work more like a microscope, focusing light through several lenses. Light first passes through a large lens at the top of the telescope, then through another lens at the eyepiece. The eyepiece focuses an image onto your eye.

The eye depends on refraction and lenses

The refraction of light by lenses is not just used in microscopes and telescopes. Without refraction, you could not see at all.

The operation of the human eye, shown in *Figure 35,* is in many ways similar to that of a simple camera. Light enters a camera through a large lens, which focuses the light into an image on the film at the back of the camera.

Light first enters the eye through a transparent tissue called the cornea. The cornea is responsible for 70 percent of the refraction of light in the eye. After the cornea, light passes through the pupil, a hole in the colorful iris.

From there, light travels through the lens, which is composed of glassy fibers situated behind the iris. The curvature of the lens determines how much further the lens refracts light. Muscles can adjust the curvature of the lens until an image is focused on the back layer of the eye, the retina.

The retina is composed of tiny structures, called rods and cones, that are sensitive to light. When light strikes the rods and cones, signals are sent to the brain where they are interpreted as images.

516

LIFE SCIENCE ── CONNECTION

If a biology teacher has done dissections of a sheep's or a cow's eye, ask that teacher to display some dissected eyes and conduct a lesson on their structure and function.

Cones are concentrated in the center of the retina, while rods are mostly located on the outer edges. The cones are responsible for color vision, but they only respond to bright light. That is why you cannot see color in very dim light. The rods are more sensitive to dim light, but cannot resolve details very well. That is why you can glimpse faint movements or see very dim stars out of the corners of your eyes.

internet connect

www.scilinks.org
Topic: Refraction
SciLinks code: HK4120

SCi LINKS Maintained by the National Science Teachers Association

Dispersion and Prisms

A **prism,** like the one in *Figure 36,* can separate white light into its component colors. Water droplets in the air can also do this, producing a rainbow. But why does the light separate into different colors?

Different colors of light are refracted differently

Although light waves of all wavelengths travel at the same speed $(3.0 \times 10^8 \text{ m/s})$ in a vacuum, when a light wave travels through a medium the speed of the light wave *does* depend on its wavelength. In the visible spectrum, violet light travels the slowest and red light travels the fastest.

Because violet light travels slower than red light, violet light refracts more than red light when it passes from one medium to another. When white light passes from air to the glass in the prism, violet bends the most, red the least, and the rest of the visible spectrum appears in between. This effect, in which light separates into different colors because of differences in wave speed, is called **dispersion.**

▶ **prism** in optics, a system that consists of two or more plane surfaces of a transparent solid at an angle with each other

▶ **dispersion** in optics, the process of separating a wave (such as white light) of different frequencies into its individual component waves (the different colors)

Figure 36

A prism separates white light into its component colors. Notice that violet light is bent more than red light.

517

Demonstration

Dispersion

(Time: About 10 minutes)

Materials:

• aquarium
• prism
• slide projector
• focusable halogen flashlight

Step 1 Darken the room for this demonstration. Focus the flashlight to produce a beam that does not diverge. Shine the beam through a prism, as shown in **Figure 36.** Arrange the prism and beam so that a light spectrum appears on a light-colored wall or on white paper.

Step 2 Fill the aquarium with clear water. Shine the projector beam through the side of the tank so that it refracts, as shown in the diagram below. Arrange the aquarium so that a bright spectrum falls on a light-colored wall. You will need to adjust the lens of the projector to produce the narrowest possible beam. If you do not get a good spectrum, place a sheet of aluminum foil with a 1/8 inch wide vertical slit over the projector lens. This should produce a narrow beam. You may have to adjust the placement of the projector for the best results.

Step 3 Place colored plastic sheets in front of the projector lens to see which colors they absorb and transmit. You can also place colored objects in the light of the projected spectrum to see how the colors of the objects appear under various colors of light.

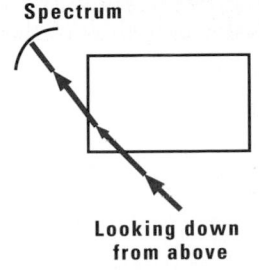

Spectrum

Looking down from above

PHYSICS
CONNECTION

Violet light slows down more than red light because, as you move toward the violet end of the spectrum, the light's frequency (energy) becomes closer to the natural frequency of electron transitions between energy levels of the atom. Thus, photons of violet light are more likely to be absorbed temporarily by atoms than are photons of lower frequency (less energy). As a result, the higher the frequency of light, the more likely it will be delayed.

Alternative Assessment ── BASIC

Light Rays Give students a sketch showing a light ray passing from air into water at an angle. Have students predict and draw the path of the light ray in the water. Do the same thing with sketches of a convex lens, a concave lens, and a prism. With the prism, have students show the dispersion of white light as well as the path of the ray. Allow leeway in judging student responses. The criterion in each case should be whether the ray bends in the correct direction. **LS Visual**

Transparencies

TT The Eye

1. Why does light bend when it moves from one medium to another? (because the speed of light is different in different mediums)

2. What are the two types of lenses? (converging and diverging)

3. What happens when white light passes through a prism? (the light separates into its component colors)

LS Logical

Figure 37

Sunlight is dispersed and internally reflected by water droplets to form a rainbow.

Rainbows are caused by dispersion and internal reflection

Rainbows, like the one in **Figure 37,** may form any time water droplets are in the air. When sunlight strikes a droplet of water, the light is dispersed into different colors as it passes from the air into the water. Some of the light then reflects off the back surface of the droplet by total internal reflection. The light disperses further when it passes out of the water back into the air.

When light finally leaves the droplet, violet light emerges at an angle of 40 degrees, red light at 42 degrees, with the other colors in between. We see light from many droplets as arcs of color, forming a rainbow. Red light comes from droplets higher in the air and violet light comes from lower droplets.

SECTION 4 REVIEW

SUMMARY

▶ Light may refract when it passes from one medium to another.

▶ Light rays may also be reflected at a boundary between mediums.

▶ Lenses form real or virtual images by refraction.

▶ Converging lenses cause light rays to converge to a point. Diverging lenses cause light rays to spread apart, or diverge.

▶ A prism disperses white light into a color spectrum.

1. **Explain** why a lawn mower turns when pushed at an angle from a sidewalk onto the grass.

2. **Draw** a ray diagram showing the path of light when it travels from air into glass.

3. **Explain** how light can bend around corners inside an optical fiber.

4. **Explain** how a simple magnifying glass works.

5. **Describe** the path of light from the time it enters the eye to the time it reaches the retina.

6. **Critical Thinking** Which color of visible light travels the slowest through a glass prism?

7. **Critical Thinking** A spoon partially immersed in a glass of water may appear to be in two pieces. Is the image of the spoon in the water a real image or a virtual image?

8. **Creative Thinking** If light traveled at the same speed in raindrops as it does in air, could rainbows exist? Explain.

518

Answers to Section 4 Review

1. The wheels turn slower in the grass. When one wheel is in the grass, the wheel on the sidewalk turns faster, pushing the lawnmower into a turn.

2. Diagrams should resemble **Figure 28A.**

3. When an optical fiber bends, light rays will encounter a wall of the fiber. The rays bounce off the walls of the fiber by total internal reflection.

4. A magnifying lens refracts light rays from an object near the lens so that the light from points on the object appear to be coming from points that are farther apart. This produces an enlarged virtual image.

5. A light ray is first refracted as it passes into the cornea of the eye. From there it passes through the pupil and then on to the lens. The lens refracts the light further so that light is focused on the retina, forming a real image.

6. violet light

7. It is a virtual image because the displaced part of the spoon only seems to be where it is seen.

8. No, rainbows are seen as a result of the dispersion of colors, resulting from refraction. Refraction results from the change in the speed of light as it passes from one medium to another. If light traveled the same speed in water as in air, no refraction would occur.

Math Skills

Fractions

The intensity of light at a distance r from a source with power output P is described by the following equation:

$$intensity = \frac{power}{4\pi(distance)^2} = \frac{P}{4\pi r^2}$$

What is the intensity of a 100.0 W light bulb at a distance of 5.00 m?

1 List all given and unknown values.

    Given: power (P), 100.0 W

          distance (r), 5.00 m

    Unknown: intensity (W/m^2)

2 Write down the equation for intensity.

$$intensity = \frac{P}{4\pi r^2}$$

3 Solve for intensity.

$$intensity = \frac{100.0 \text{ W}}{4 \times \pi \times (5.00 \text{ m})^2}$$

$$intensity = \frac{100.0 \text{ W}}{4 \times 3.14 \times 25 \text{ m}^2}$$

$$intensity = \frac{100.0 \text{ W}}{314 \text{ m}^2} = 0.318 \text{ W/m}^2$$

Because a watt (W) is the amount of power required to do 1 joule (J) of work in 1 s, 0.318 J of energy pass through each square meter of area at a distance of 5.00 m from the light bulb each second.

Practice

Follow the example above to answer the following questions.

1. What is the intensity of the 100.0 W light bulb in the example above at a distance of 1.00 m from the light bulb? How many times greater is this intensity than the intensity at 5.00 m?

2. The total power output of the sun is 3.84×10^{26} W. What is the intensity of sunlight at a distance of 1.49×10^{11} m (the average distance of Earth from the sun)?

3. When the sun is directly above the clouds that form the "surface" of Saturn, the intensity of sunlight is 15.13 W/m^2. What is the distance from the sun to Saturn? How many times farther is this than the distance between the sun and Earth?

Teaching Tip

Energy Conservation To help students visualize the variation of intensity with the square of distance, have them think in terms of energy conservation. Have them consider a light that is at the center of several clear spheres. Each sphere has a slightly greater radius (1 m, 2 m, 3 m, etc.). If the amount of energy the light emits each second (its power) is constant, the amount of energy that must go through each sphere each second must be the same. Because the surface area of the sphere through which the light passes increases as it moves further from the source, a smaller fraction of energy passes through each unit of area on each sphere. The surface area of each sphere is proportional to the square of its radius (the distance from the light), and so for an observer at a point on each sphere, the light's intensity decreases with the square of the distance to the light.

Answers to Practice

1. 7.96 W/m^2, 25.0 times greater
2. 1.38×10^3 W/m^2
3. 1.42×10^{12} m , 9.54 times further

519

Understanding Concepts

1. a
2. d
3. b
4. a
5. d
6. c
7. d
8. a
9. d
10. a
11. d
12. c
13. d
14. c

Using Vocabulary

15. Both amplitude and intensity are directly related to the energy of sound waves. The more energy the waves contain, the louder the sound is.

16. Pitch is the way the brain interprets the frequency of a sound wave. Higher frequency corresponds to higher pitch, and lower frequency corresponds to lower pitch.

17. When a guitar string is plucked, it vibrates in standing waves, producing a fundamental pitch plus harmonics. The body of the guitar is resonant at all the frequencies produced by the strings and vibrates at these same frequencies, moving more air and thus amplifying the sound.

Chapter Resource File

- **Concept Review** GENERAL
- **Chapter Test** GENERAL
- **Test Item Listing for EvamView® Test Generator**

Chapter Highlights

Before you begin, review the summaries of the key ideas of each section, found at the end of each section. The key vocabulary terms are listed on the first page of each section.

UNDERSTANDING CONCEPTS

1. All sound waves are
   a. longitudinal waves.
   b. transverse waves.
   c. electromagnetic waves.
   d. standing waves.

2. The speed of sound depends on
   a. the temperature of the medium.
   b. the density of the medium.
   c. how well the particles of the medium transfer energy.
   d. All of the above

3. A sonar device can use the echoes of ultrasound underwater to find the
   a. speed of sound.
   b. depth of the water.
   c. temperature of the water.
   d. height of waves on the surface.

4. Relative intensity is measured in units abbreviated as
   a. dB.   c. J.
   b. Hz.   d. V.

5. During a thunderstorm, you see lightning before you hear thunder because
   a. the thunder occurs after the lightning.
   b. the thunder is farther away than the lightning.
   c. sound travels faster than light.
   d. light travels faster than sound.

6. A flat mirror forms an image that is
   a. smaller than the object.   c. virtual.
   b. larger than the object.   d. real.

7. Which situation does not involve ultrasound?
   a. bats navigating
   b. a machine generating sonograms of a fetus
   c. ships determining the depth of the ocean floor
   d. a person tuning a piano

8. The wave interaction most important for echolocation is
   a. reflection.   c. diffraction.
   b. interference.   d. resonance.

9. The speed of light
   a. depends on the medium.
   b. is fastest in a vacuum.
   c. is the fastest speed in the universe.
   d. All of the above

10. Which of the following forms of light has the most energy?
    a. X rays   c. infrared light
    b. microwaves   d. ultraviolet light

11. Light can be modeled as
    a. electromagnetic waves.
    b. a stream of particles called photons.
    c. rays that travel in straight lines.
    d. All of the above

12. The energy of light is proportional to
    a. amplitude.   c. frequency.
    b. wavelength.   d. the speed of light.

13. Which of the following wavelengths of visible light bends the most when passing through a prism?
    a. red   c. green
    b. yellow   d. blue

14. When you look at yourself in a plane mirror, you see a
    a. real image behind the mirror.
    b. real image on the surface of the mirror.
    c. virtual image that appears to be behind the mirror.
    d. virtual image that appears to be in front of the mirror.

520

18. Both are producing sounds of the same fundamental frequency, but the clarinet's sound includes several strong harmonics, whole-number multiples of the fundamental frequency. The tuning fork produces almost no harmonics.

19. Infrasonic waves occur below the range of human hearing. Ultrasonic waves occur above the human hearing range.

20. Sound waves are funneled through the outer ear down the ear canal to eardrum, which vibrates. These vibrations pass through the anvil, hammer, and stirrup of the middle ear to a membrane at the opening of the inner ear. The waves then pass through the cochlea, which contains the basilar membrane. Hairs near the membrane stimulate nerve fibers that send impulses to the brain.

21. Sonar is a system that uses reflected sound waves to determine the distance to objects, whereas radar is a system that uses reflected radio waves to determine distance.

USING VOCABULARY

15. How is the loudness of a sound related to *amplitude* and *intensity*?

16. How is the *pitch* of a sound related to its *frequency*?

17. Explain how a guitar produces sound. Use the following terms in your answer: *standing waves, resonance.*

18. Why does a clarinet sound different from a tuning fork, even when played at the same pitch? Use these terms in your answer: *fundamental frequency, harmonics.*

19. Describe *infrasonic* and *ultrasonic* waves, and explain why humans cannot hear them.

20. Describe the anatomy of the ear, and explain how the ear senses vibrations. Use the terms *outer ear, middle ear,* and *inner ear.*

21. Define *sonar* and *radar*. State their differences and their similarities, and describe a situation in which each technique would be useful.

22. Explain the difference between a *virtual image* and a *real image*. Give an example of each type of image.

23. Explain why a leaf may appear green in white light but black in red light. Use the following terms in your answer: *wavelength, reflection.*

24. What is the difference between a *converging lens* and a *diverging lens*?

25. Draw a figure that illustrates what happens when light rays hit a rough surface and a smooth surface. Explain how your illustration demonstrates *diffuse reflection* and the *law of reflection.*

BUILDING GRAPHING SKILLS

26. Graphing As a ship travels across a lake, a sonar device on the ship sends out pulses of ultrasound and detects the reflected pulses. The table below gives the ship's distance from the shore and the time for each pulse to return to the ship. Construct a graph of the depth of the lake as a function of distance from the shore.

Distance from shore (m)	Time to receive pulse ($\times 10^{-2}$ s)
100	1.7
120	2.0
140	2.6
150	3.1
170	3.2
200	4.1
220	3.7
250	4.4
270	5.0
300	4.6

BUILDING MATH SKILLS

You may use the following for items 27–30:

▶ the wave speed equation, $v = f \times \lambda$

▶ a rearranged form of the speed equation, $d = vt$

▶ the average speed of sound in water or soft tissue, 1500 m/s

▶ the speed of light, 3.0×10^8 m/s

27. Sound Waves Calculate the wavelength of ultrasound used in medical imaging if the frequency is 15 MHz.

28. Sonar Calculate the distance to the bottom of a lake when a ship using sonar receives the reflection of a pulse in 0.055 s.

22. When you view a virtual image, you are seeing light rays that only appear to be coming from the location of the image. Examples include a reflection in a flat mirror and the image seen through a magnifying glass. When you view a real image, you are seeing light rays actually coming from the location of the image. A real image can be projected on a screen. Examples are projected movies and the images formed in the tubes of microscopes and telescopes.

23. The leaf appears green because it reflects green light and absorbs other wavelengths. Red light contains no green wavelengths, so the leaf can reflect no light and appears black.

24. A converging lens bends light inward, while a diverging lens bends light outward.

25. Figures should resemble the drawings in **Figure 19**. Diffuse reflection occurs when light strikes a rough surface sending light rays into random directions. When light hits a smooth surface, the law of reflection applies, and the angle of incidence equals the angle of reflection.

26. Students' graphs should show a plot of the following data points:

Distance from shore (m) (x-axis)	Depth (m) (y-axis)
100	13
120	15
140	20
150	23
170	24
200	31
220	28
250	33
270	38
300	35

Assignment Guide

Section	Questions
1	1–4, 15–20, 26–28, 31–36, 41
2	5, 9–12, 21, 29, 30, 38, 40, 42–44, 46, 48
3	6, 14, 22, 23, 25, 37, 45, 47
4	13, 24, 39

Building Math Skills

See pp. 525A–525B for worked-out solutions.

27. 1.0×10^{-4} m (0.10 mm)

28. 41 m

29. 250 m

30. 5.5×10^{14} Hz

Thinking Critically

31. dolphin

32. Different instruments in the orchestra will transmit sound at different frequencies, since each instrument has its unique fundamental frequency and associated harmonic range. However, the sounds of the different instruments would reach your ears at the same time, proving that the speed of sound is the same at all frequencies.

33. Dogs can hear sounds with higher frequencies those that human beings are able to hear.

34. Cupping a hand around the outer ear helps a person hear better by increasing the surface area of the outer ear. Similarly, having large ears enables some animals to hear better.

35. A narrow beam remains focused, so that it still has most of its energy when it returns to the ship. A narrow beam is also less likely to hit interfering objects on the way down to the ocean's floor. A wider beam is more useful for detecting closer objects such as fish.

29. Electromagnetic Waves Calculate the wavelength of radio waves from an AM radio station broadcasting at 1200 kHz.

30. Electromagnetic Waves Waves composing green light have a wavelength of about 550 nm. What is their frequency?

THINKING CRITICALLY

31. Interpreting Graphics Review *Figure 3,* which illustrates ranges of frequencies of sounds that different animals can hear. Which animal on the chart can hear sounds with the highest pitch?

32. Understanding Systems By listening to an orchestra, how can you determine that the speed of sound is the same for all frequencies?

33. Applying Knowledge As you are walking through a park, you see someone blowing a whistle. You can't hear the whistle, but you notice that several dogs are responding to it. Why can't you hear the dog whistle?

34. Applying Knowledge When people strain to hear something, they often cup a hand around the outer ear. How does this help them hear better? Some animals, such as rabbits and foxes, have very large outer ears. How does this affect their hearing?

35. Thinking Critically Sonar devices on ships use a narrow ultrasonic beam for determining depth. A wider beam is used to locate fish. Why?

36. Aquiring and Evaluating Data A guitar has six strings, each tuned to a different pitch. Research what pitches the strings are normally tuned to and what frequencies correspond to each pitch. Then calculate the wavelength of the sound waves that each string produces. Assume the speed of sound in air is 340 m/s.

522

37. Creative Thinking Imagine laying this page flat on a table, then standing a mirror upright at the top of the page. Using the law of reflection, draw the image of each of the following letters of the alphabet in the mirror.

A B C F W T

38. Understanding Systems The glass in greenhouses is transparent to certain wavelengths and opaque to others. Research this type of glass, and write a paragraph explaining why it is an ideal material for greenhouses. *WRITING SKILL*

39. Applying Knowledge Why is white light not dispersed into a spectrum when it passes through a flat pane of glass like a window?

40. Interpreting Graphics Examine the chart of the electromagnetic spectrum in *Figure 14.* Which type of electromagnetic wave has the shortest wavelength? Which has the highest frequency? How does wavelength change as frequency increases?

DEVELOPING LIFE/WORK SKILLS

41. Teaching Others Your aunt is scheduled for an ultrasound examination of her gall bladder and she is worried that it will be painful. All she knows is that the examination has something to do with sound. How would you explain the procedure to her?

42. Applying Technology Meteorologists use Doppler radar to measure the speed of approaching storms and the velocity of the swirling air in tornadoes. Research this application of radar, and write a short paragraph describing how it works. *WRITING SKILL*

36. Notes: E–A–D–G–B–E
Frequencies: 165 Hz, 220 Hz, 294 Hz, 392 Hz, 494 Hz, and 659 Hz
Wavelengths: 2.06 m, 1.55 m, 1.16 m, 0.867 m, 0.688 m, 0.516 m

37. Student drawings should resemble the following:

∀ B C Ⅎ Ѡ T

38. Students should find that it allows most wavelengths of visible light to pass through. When the light is absorbed by objects inside

the greenhouse, it warms those objects. The objects reradiate the energy they have absorbed as long-wave infrared light, which cannot pass through the glass. The energy stays in the greenhouse, keeping it warm.

39. It is dispersed slightly as it refracts when entering the glass. However, when it emerges, it is refracted in exactly the opposite way and is recombined into white light.

40. Gamma rays have the shortest wavelength and the highest frequency. Wavelength decreases as frequency increases.

43. Connection to Earth Science Landsat satellites are remote sensing satellites that can detect electromagnetic waves at a variety of wavelengths to reveal hidden features on Earth. Research the use of Landsat satellites to view Earth's surface. What kind of electromagnetic waves are detected? What features do Landsat images reveal that cannot be seen with visible light?

44. Connection to Health Research the effect of UV light on skin. Are all wavelengths of UV light harmful to your skin? What problems can too much exposure to UV light cause? Is UV light also harmful to your eyes? If so, how can you protect your eyes?

45. Connection to Biology While most people can see all the colors of the spectrum, people with *colorblindness* are unable to see at least one of the primary colors. What part of the eye do you think is malfunctioning in colorblind people?

46. Connection to Fine Arts Describe how you can make red, green, blue, and black paint with a paint set containing only yellow, magenta, and cyan paint.

47. Connection to Space Science Telescopes can produce images in several different regions of the electromagnetic spectrum. Research photos of areas of the galaxy taken with infrared light, microwaves, and radio waves. What features are revealed by infrared light that are hidden in visible light? What kinds of objects are often studied with radio telescopes?

48. Concept Mapping Copy the unfinished concept map below onto a sheet of paper. Complete the map by writing the correct word or phrase in the lettered boxes.

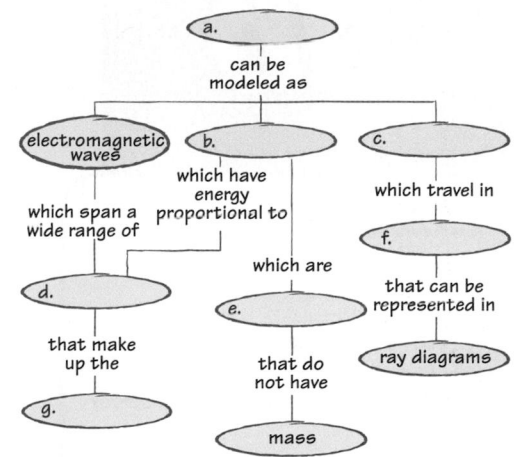

www.scilinks.org
Topic: Microwaves SciLinks code: HK4087

SCiLINKS Maintained by the National Science Teachers Association

Art Credits: Fig. 1B, Uhl Studios, Inc.; Fig. 2 (graph), Leslie Kell; Fig. 7, Keith Kasnot; Fig. 11, Uhl Studios, Inc.; Fig. 12, Leslie Kell; Fig. 13, Boston Graphics; Fig. 14, Boston Graphics; Fig. 27, Uhl Studios, Inc.; Fig. 34, Morgan-Cain & Associates; Fig. 35, Keith Kasnot; "Skills Practice Lab", Uhl Studios, Inc.

Photo Credits: Chapter Opener photo of Times Square by Paul Hardy/CORBIS; camper by Ron Watts/CORBIS; Fig. 1A, A. Ramey/PhotoEdit; Fig. 2(students speaking, lawnmower, vacuum, dog), Peter Van Steen/HRW; (jet), Peter Gridley/Getty Images/FPG International; (elephant, human), Image Copyright (c)2004 PhotoDisc, Inc.; (dolphin), Doug Perrine/Masterfile; Fig. 4, Tom Pantages Photography; Fig. 5, Thomas D. Rossing; Fig. 6(clarinet), SuperStock; (tuning fork), Sam Dudgeon/HRW; "Quick Lab," Peter Van Steen/HRW; Fig. 8, Benny Odeur/Wildlife Pictures/Peter Arnold, Inc.; Fig. 9(t), Telegraph Colour Library/Getty Images/FPG International; (b), Saturn Stills/Science Photo Library/Photo Researchers; Fig. 10B, E. R. Degginger/Bruce Coleman, Inc.; Fig. 15, Ron Chapple/Getty Images/FPG International; Fig. 16, E. R. Degginger/Animals Animals/Earth Scenes; Fig. 17, Telegraph Colour Library/Getty Images/FPG International; Fig. 18, Claude Gazuit/Photo Researchers, Inc.; Fig. 21, 22, Peter Van Steen/HRW; Fig. 23, Telegraph Colour Library/Getty Images/FPG International; Fig. 24, Peter Van Steen/HRW; Fig. 25A, Leonard Lessin/Peter Arnold, Inc.; Fig. 25B, Sam Dudgeon/HRW; Fig. 26, Richard Megna/Fundamental Photographs; Fig. 30, Ken Kay/Fundamental Photographs; Fig. 30, Guntram Gerst/Peter Arnold, Inc.; Fig. 32, Richard Menga/Fundamental Photographs; Fig. 33, Peter Van Steen/HRW; Fig. 36, Science Photo Library/Photo Researchers, Inc.; Fig. 37, Graham French/Masterfile; "Science In Action" (lasers), Zefa Visual Media/Index Stock Imagery, Inc.; (bog man), Sam Dudgeon/HRW; (butterflies), Yves Gentet.

523

45. Colors are perceived primarily by the cones of the eye. On the retina, there is one set of cones for red light, one set for green light, and one set for blue light. Different types of colorblindness are caused by malfunctions of one or more of these sets of cones.

46. Paints are subtractive pigments. Red paint can be made by combining yellow and magenta. Green paint can be made by combining yellow and cyan. Blue paint can be made by combining magenta and cyan. Black paint can be made by combining yellow, magenta, and cyan. If students have trouble, refer them to **Figure 25** in this chapter.

See "Continuation of Answers" at the end of the chapter.

Developing Life/Work Skills

41. Answers will vary. They should reflect the idea that the procedure is non-invasive, that the sound is above the range of human hearing, and that the sound will reflect from the gallstones to produce an image showing their size and location.

42. Answers will vary. Students will find that meteorologists can measure the speed of approach in the same way as police radar determines the speed of cars. In addition, Doppler radar can show the velocities on either side of a storm cell. If one side is moving forward and the other side is moving away, the storm must be rotating.

Integrating Concepts

43. Landsat uses visible light and infrared wavelengths. The Landsat program monitors agricultural activities, forests, population change, urbanization, water resources, desert growth, disasters, geology of Earth's surface, and wildlife habitat.

44. The shortest wavelengths of UV radiation are absorbed or reflected before reaching Earth's surface. Medium wavelengths that are not absorbed by the ozone layer reach Earth's surface. Their high energy gives them penetrating power, and they can damage DNA and the eyes, but they also help the body produce vitamin D. UV rays in this range cause the most severe sunburns. In order to protect your skin and eyes, you should wear sunscreen and sunglasses with a protective UV coating when you spend long periods of time outside. The longest wavelengths contribute to smog formation and cause fading of paints and dyes. Otherwise, they are not that harmful to humans.

FORMING IMAGES WITH LENSES

Teacher's Notes

Time Required 1 lab period

Ratings

EASY 1 — 2 — 3 — 4 HARD

TEACHER PREPARATION	2
STUDENT SETUP	2
CONCEPT LEVEL	3
CLEANUP	1

Skills Acquired
- Collecting data
- Communicating
- Interpreting
- Measuring
- Organizing and analyzing data

The Scientific Method
In this lab, students will:
- Make Observations
- Analyze the Results
- Draw Conclusions
- Communicate Results

Materials
A 15 W bulb works well for the experiments. Each setup can accommodate 2–3 students.

Safety Cautions
Students should be cautioned that the light bulb will become hot to the touch during the course of the lab. Only devices that are UL-listed should be used. The condition of the wiring and adequate grounding should be checked prior to use.

Skills Practice Lab

Introduction
How can you find the focal length of a lens and verify the value?

Objectives
▶ **Observe** images formed by a convex lens.

▶ **Measure** the distance of objects and images from the lens.

▶ USING SCIENTIFIC METHODS **Analyze** your results to determine the focal length of the lens.

Materials
cardboard screen, 10 cm × 20 cm
convex lens, 10 cm to 15 cm
    focal length
lens holder
light box with light bulb
meterstick
screen holder
supports for meterstick

524

Forming Images with Lenses

▶ Procedure

Preparing for Your Experiment

1. The shape of a lens determines the size, position, and types of images that it may form. When parallel rays of light from a distant object pass through a converging lens, they come together to form an image at a point called the focal point. The distance from this point to the lens is called the focal length. In this experiment, you will find the focal length of a lens, and then verify this value by forming images, measuring distances, and using the lens formula,

$$\frac{1}{d_o} + \frac{1}{d_i} = \frac{1}{f}$$

where d_o = object distance,
d_i = image distance, and
f = focal length.

2. On a clean sheet of paper, make a table like the one shown at right.

3. Set up the equipment as illustrated in the figure below. Make sure the lens and screen are securely fastened to the meterstick.

Determining Focal Length

4. Stand about 1 m from a window, and point the meterstick at a tree, parked car, or similar object. Slide the screen holder along the meterstick until a clear image of the distant object forms on the screen. Measure the distance between the lens and the screen in centimeters. This distance is very close to the focal length of the lens you are using. Record this value at the top of your data table.

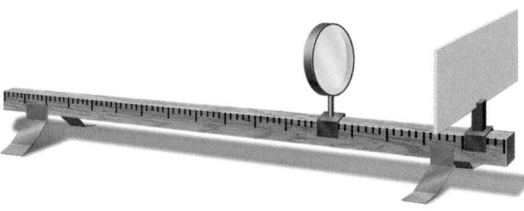

Tips and Tricks
Pre-Lab Be sure that students understand that light is refracted and results in an image; the image is called a *refracted image*.

In step 4, a partially darkened room produces a closer approximation of *f*. The actual point of focus is a judgment call. As a result, the students may have some difficulty measuring the actual size of the image in step 7.

Stress that the value of *f* obtained in step 4 is an approximation. Therefore, if students expect an image at exactly 2*f* in step 8, they will see some discrepancy. Another step can be added to show the effects of the object when it is inside *f*. Emphasize the relationship between *f* measured in step 1 and $\frac{1}{f}$ calculated in the analysis of the results.

Focal length of lens, f: _____ cm	Object distance, d_o (cm)	Image distance, d_i (cm)	$\frac{1}{d_o}$	$\frac{1}{d_i}$	$\frac{1}{d_o}+\frac{1}{d_i}$	$\frac{1}{f}$	Size of object (mm)	Size of image (mm)
Trial 1								
Trial 2								
Trial 3								

Forming Images

5. Set up the equipment as illustrated in the figure at right. Again, make sure that all components are securely fastened.

6. Place the lens more than twice the focal length from the light box. For example, if the lens has a focal length of 10 cm, place the lens 25 or 30 cm from the light.

7. Move the screen along the meterstick until you get a good image. Record the distance from the light to the lens, d_o, and the distance from the lens to the screen, d_i, in centimeters as Trial 1 in your data table. Also record the height of the object and of the image in millimeters. The object in this case may be either the filament of the light bulb or a cut-out shape in the light box.

8. For Trial 2, place the lens exactly twice the focal length from the object. Slide the screen along the stick until a good image is formed, as in step 7. Record the distances from the screen and the sizes of the object and image as you did in step 7.

9. For Trial 3, place the lens at a distance from the object that is greater than the focal length but less than twice the focal length. Adjust the screen, and record the measurements as you did in step 7.

▶ Analysis

1. Perform the calculations needed to complete your data table.
2. How does $\frac{1}{d_o}+\frac{1}{d_i}$ compare with $\frac{1}{f}$ in each of the three trials?

▶ Conclusions

3. If the object distance is greater than the image distance, how will the size of the image compare with the size of the object?

Procedure

You may find that a partially darkened room creates a clearer image on the card and therefore, a better approximation of *f*.

In step 5, a 15 W light bulb works best with the lettering on the bulb acting as the object. The bulb can be arranged such that the lettering is upright. This creates an inverted image on the card.

Answers to Analysis

1. Results will vary. Sample data are given in the sample data table.
2. The value of $\frac{1}{d_o}+\frac{1}{d_i}$ is generally very close to the value of $\frac{1}{f}$.

Answers to Conclusions

3. If the object distance is greater than the image distance, the image will be smaller than the object.

Sample Data

Focal length of lens: 10.2 cm	Object distance d_o (cm)	Image distance d_i (cm)	$\frac{1}{d_o}$	$\frac{1}{d_i}$	$\frac{1}{d_o}+\frac{1}{d_i}$	$\frac{1}{f}$	Size of object (mm)	Size of image (mm)
Trial 1	22.0	18.2	0.045	0.055	0.100	0.100	19	16
Trial 2	20.4	21.6	0.049	0.046	0.095	0.095	19	18
Trial 3	16.5	28.0	0.061	0.036	0.097	0.097	19	32

Chapter Resource File

- **Datasheet** Forming Images with Lenses GENERAL
- **Observation Lab** Mirror Images BASIC
- **CBL™ Probeware Lab** Choosing a Pair of Sunglasses ADVANCED

Continuation of Answers

Continuation of Answers from p. 498
Section 1 Review

8. Compressions and rarefactions in sound waves strike the eardrum, causing it to move move back and forth. This vibration is transmitted through three small bones of the middle ear, then to the basilar membrane in the cochlea, a fluid-filled, snail-shaped organ. As the vibrations pass through the cochlea, they stimulate hair cells, which, in turn, stimulate nerves leading to the brain.

9. Ultrasound vibrations travel easily through tissue, but audible waves do not.

10. As the strings vibrate, the sound board vibrates at the same frequency (through resonance), causing stronger compressions and rarefactions in the air than the vibrating string alone could cause. This produces a louder sound.

Continuation of Answers from p. 511
Section 3 Review

7. You see (opaque) objects by light reflected from them. Therefore, anything you can see, except for direct sources of light, reflects light.

8. Diagrams should show how reflected light is directed outward.

Continuation of Answers from p. 522
Chapter Review
Building Math Skills

27. $\lambda = \dfrac{v}{f} = \dfrac{1500 \text{ m/s}}{1.5 \times 10^7 \text{ Hz}} = 1.0 \times 10^{-4} \text{ m}$

28. $d = vt = (1500 \text{ m/s}) \left(\dfrac{0.055 \text{ s}}{2}\right) = 41 \text{ m}$

29. $\lambda = \dfrac{v}{f} = \dfrac{3.0 \times 10^8 \text{ m/s}}{1.2 \times 10^6 \text{ Hz}} = 250 \text{ m}$

30. $f = \dfrac{v}{\lambda} = \dfrac{3.0 \times 10^8 \text{ m/s}}{5.5 \times 10^{-7} \text{ m}} = 5.5 \times 10^{14} \text{ Hz}$

Continuation of Answers from p. 523

Chapter Review

47. Infrared light can penetrate interstellar matter, so telescopes that can detect infrared light are used to study the interiors of interstellar clouds and to view the center of the galaxy (which is surrounded by interstellar matter). Radio telescopes are often used to study extremely distant objects, such as galaxies and quasars. Light from these objects may be red shifted entirely into radio frequencies.

48. a. light
 b. photons
 c. light rays
 d. frequencies
 e. particles
 f. straight lines
 g. electromagnetic spectrum

Science in ACTION

Holography

Have you ever been watching TV or looking at a magazine and seen something that made you think, "Wait a minute—is that real?" Sometimes, special effects and images made with computers are so lifelike that you can't believe your eyes. But even the most realistic TV images or photographs can't fool you into thinking that they are real, solid objects. Images on a movie screen have height and width but no depth. Therefore, these images always appear flat. Solid objects, however, are three-dimensional, which means they have height, width, and depth. Holograms are three-dimensional images so real that many people try to reach out and pick them up!

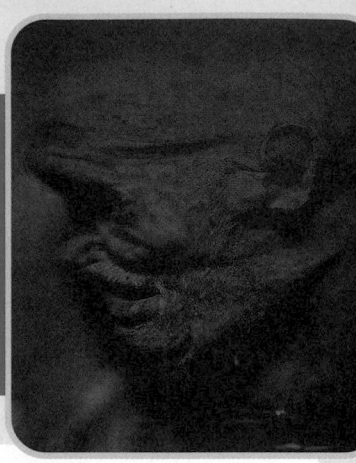

"Lindow Man" lived about 2300 years ago. This hologram allows researchers anywhere in the world to study his form.

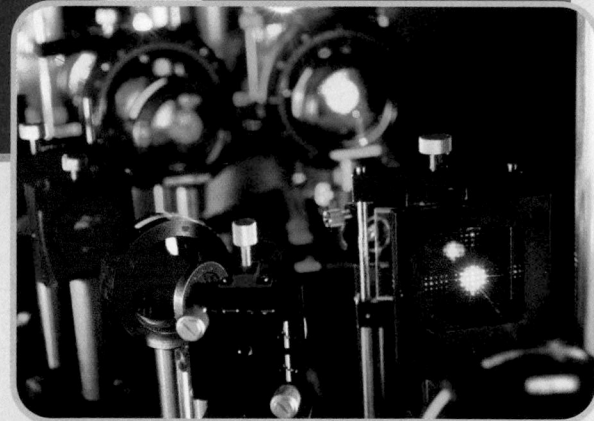

Laser light helps capture the information required to form a three-dimensional image on holographic film.

The Science of Holograms

To create a hologram, a person places an object behind or near a sheet of transparent holographic film. Laser light is shown through the film and onto the object. The light reflects off the object and back onto the film, where it meets the original light beam. The colliding beams create an image in the light-sensitive film. Like regular photographic film, the holographic film records the light's wavelength (which determines color) and amplitude (which determines intensity, or brightness). But holographic film also captures the light's direction. After the film is developed and light is shown on the holographic film, tiny reflectors in the film bounce back light at exactly the same directions at which it originally came from the object. The result is a three-dimensional image that appears to float in space. Amazingly, holographic images are so complete that the viewer can view the image from different angles and see different sides of the image, just like a solid object can be viewed.

Putting Holograms to Work

Although simple holograms printed on reflective plastic are a common security device on credit cards, holograms that truly appear three-dimensional are not yet widely used. But with recent advances in holographic film, high-definition holograms may soon appear in some unexpected places. Holographic "clones" of art or cultural artifacts may one day be displayed in museums to avoid risk to the originals. Because holographic film contains no pigments or dyes, they do not fade over time and can therefore be used as a permanent three-dimensional record of rare and fragile objects, such as a specimen of a nearly extinct insect. The technology also exists to create hologram-like images of moving objects—including people—that appear to move through thin air.

This full-color hologram is brighter and more realistic thanks to recent advances in holographic film and techniques.

> ### Science and You

1. **Applying Knowledge** Why don't images on a TV screen or in photographs appear as real as solid objects?

2. **Applying Knowledge** What two characteristics of light does regular photographic film capture?

3. **Applying Knowledge** What third characteristic of light captured by holographic film gives the holographic image depth?

4. **Critical Thinking** Why do you think some people may object to using holographic images in museums in place of originals?

5. **Critical Thinking** Can you think of a practical application for holographic images not mentioned here? Write a short paragraph explaining how a particular problem or challenge could be solved by using a hologram.

internet connect

www.scilinks.org
Topic: Holography
SciLinks code: HK4068

SC*i*LINKS. Maintained by the National Science Teachers Association

Answers to Science and You

1. Images on a TV screen and in photographs are two-dimensional, while solid objects are three-dimensional.

2. It captures the light's wavelengths, which are responsible for color, and the light's wave heights (amplitude), which are responsible for the light's intensity.

3. Holographic film records the light's direction, which adds depth to the image.

4. Answers will vary. Sample answer: If museums display holographic images, people may stop valuing original works of art, and as a result they may stop collecting and preserving art and artifacts.

5. Answers will vary. Students' paragraphs should explore a practical application of holograms not covered in the text and/or explore how holograms might be used to solve a particular problem.

Electricity
Chapter Planning Guide

PACING	CLASSROOM RESOURCES	LABS, ACTIVITIES, AND DEMONSTRATIONS
BLOCK 1 · 45 min pp. 528–529 **Chapter Opener**		SE **Activity 1,** p. 529 SE **Activity 2,** p. 529
BLOCKS 2 & 3 · 90 min pp. 530–536 **Section 1** Electric Charge and Force	CRF **Lesson Plan*** TT **Bellringer*** TT **Charging by Contact*** TT **Induced Charges*** TM **Point Charges*** TM **Electric Fields***	TE **Opening Demonstration,** pp. 530–531 GENERAL TE **Demonstration** Charging by Friction and Induction, p. 532 SE **Quick Activity** Charging Objects, p. 534 GENERAL CRF **Datasheets for In-Text Labs** Charging Objects* GENERAL
BLOCKS 4 & 5 · 90 min pp. 537–545 **Section 2** Current	CRF **Lesson Plan*** TT **Bellringer*** TM **Electrical Potential Energy***	TE **Opening Demonstration,** p. 537 TE **Group Activity** Comparing Batteries, p. 540 GENERAL SE **Quick Activity** Using a Lemon as a Cell, p. 541 GENERAL CRF **Datasheets for In-Text Labs** Using a Lemon as a Cell* GENERAL SE **Quick Lab** How can materials be classified by resistance?, p. 544 GENERAL CRF **Datasheets for In-Text Labs** How can materials be classified by resistance?* GENERAL CRF **CBL™ Probeware Lab** Investigating How the Length of a Conductor Affects Resistance* ◆ ADVANCED
BLOCKS 6 & 7 · 90 min pp. 546–552 **Section 3** Circuits	CRF **Lesson Plan*** TT **Bellringer*** TT **Series and Parallel*** TM **Circuit and Diagram*** TT **Concept Mapping***	TE **Opening Activity,** p. 546 GENERAL SE **Quick Activity** Series and Parallel Circuits, p. 549 GENERAL CRF **Datasheets for In-Text Labs** Series and Parallel Circuits* GENERAL TE **Group Activity** Energy Use in Home Appliances, p. 550 BASIC SE **Skills Practice Lab** Constructing Electric Circuits, pp. 558–559 ◆ GENERAL CRF **Datasheets for SE Skills Practice Lab** Constructing Electric Circuits* GENERAL CRF **Observation Lab** Converting Wind Energy into Electricity* ◆ BASIC

BLOCKS 8 & 9 · 90 min

Chapter Review and Assessment Resources

SE **Chapter Review,** pp. 554–557
CRF **Chapter Tests*** GENERAL
OSP **Test Generator**
CRF **Standardized Test Practice with Guided Reading Development***
CRF **Test Item Listing for ExamView® Test Generator***
OSP **Scoring Rubrics and Classroom Management Checklists**

Online Resources

Visit the HRW Web site for a variety of free resources related to the text. Just type in the keyword **HK4 ELE**.

Holt Science Spectrum: Physical Science: Online Edition

Students can access interactive problem solving help and active visual concept development with the *Holt Science Spectrum: Physical Science* Online Edition available at **www.hrw.com.**

student CNN News

cnnstudentnews.com

Find the latest news, lesson plans, and activities related to important scientific events.

PROBLEM SOLVING AND PRACTICE	SECTION REVIEW AND ASSESSMENT	STANDARDS CORRELATION
	CRF Pretest* GENERAL	
CRF Cross-Disciplinary Worksheet Connection to Social Studies—Incandescent Light Bulbs* ADVANCED **CRF** Cross-Disciplinary Worksheet Integrating Biology—Electric Eels* ADVANCED	**TE** Quiz, p. 536 BASIC **SE** Section 1 Review, p. 536 **CRF** Concept Review* GENERAL **CRF** Quiz* BASIC	PS 1a, 4b, 4d, 4c, 5b, 6c UCP 1, 2, 3 SAI 1, 2 ST 1, 2 SPSP 5
SE Science and the Consumer Which Is the Best Type of Battery?, p. 540 **TE** Inclusion Strategies, p. 540 **CRF** Cross-Disciplinary Worksheet Science and the Consumer—Battery Issues* ADVANCED **CRF** Cross-Disciplinary Worksheet Integrating Chemistry—Rechargeable Ni-Cd Batteries* ADVANCED **CRF** Cross-Disciplinary Worksheet Real World Applications—Electric Shock: Caution!* ADVANCED **SE** Math Skills Resistance, p. 543 **CRF** Math Skills Resistance* GENERAL	**TE** Quiz, p. 545 BASIC **SE** Section 2 Review, p. 545 **CRF** Concept Review* GENERAL **CRF** Quiz* BASIC	PS 5b, 6c UCP 1, 2, 3 SAI 1, 2 ST 1, 2 SPSP 4, 5
SE Math Skills Electric Power, pp. 550–551 **CRF** Math Skills Electric Power* GENERAL **CRF** Cross-Disciplinary Worksheet Integrating Health—Recording Electricity in the Brain* ADVANCED **SE** Graphing Skills, p. 553 **SE** CareerLink Physicist, pp. 560–561	**TE** Quiz, p. 552 BASIC **SE** Section 3 Review, p. 552 **CRF** Concept Review* GENERAL **CRF** Quiz* BASIC	PS 5b UCP 1, 2, 3 SAI 1, 2 ST 1, 2 SPSP 5

SCiLINKS®

www.scilinks.org

Topic: Applications of the Electric Spark
*Sci*Links code: HK4008

Topic: Static Electricity
*Sci*Links code: HK4134

Topic: Batteries
*Sci*Links code: HK4014

Topic: Semiconductors and Insulators
*Sci*Links code: HK4162

Topic: Electric Circuits
*Sci*Links code: HK4042

Topic: Electrostatic Precipitators
*Sci*Links code: HK4045

Topic: Physicist
*Sci*Links code: HK4105

Technology Resources

 One-Stop Planner
All of your printable resources and the Test Generator are on this convenient CD-ROM.

 Physical Science Interactive Tutor CD-Rom

Disc Two, Module 15
Topic: Force Between Charges

Disc Two, Module 8
Topic: Batteries and Cells

 CNN. Science in the NEWS
*each video segment is accompanied by a Critical Thinking Worksheet**

Segment 4
Student Semiconductors

Segment 20
Eagle Electrocution

Overview

In this chapter, students learn about electric charge, the electric force, and the electric field. Next they study electric current and related concepts, including potential difference, resistance, and the classification of materials as conductors, insulators, superconductors, and semiconductors. Finally, they learn about electric circuits, including schematic diagrams, series and parallel circuits, electrical power, and fuses and circuit breakers.

Assessing Prior Knowledge

Be sure students understand the following concepts:

- scientific notation
- atomic structure
- force
- acceleration
- potential energy

MISCONCEPTION /// ALERT \\\

Science education research has identified the following misconceptions about electricity.

- Some students think that electricity travels from a battery to a bulb, where it is used up in an "electrical sink." Others believe that electricity leaves both terminals of the battery and races to the bulb, where it smashes together.
- Students believe that current flowing through a battery grows weaker with distance, and that a bulb farther away from a battery will be dimmer than one that is closer.
- Students believe that batteries store current.

CHAPTER 16

Electricity

Chapter Preview

1 Electric Charge and Force
 Electric Charge
 Electric Force

2 Current
 Voltage and Current
 Electrical Resistance

3 Circuits
 What Are Circuits?
 Series and Parallel Circuits
 Electric Power and Electrical Energy
 Fuses and Circuit Breakers

Electricity arcs across the fusion chamber at Sandia National Laboratory in the large photo above. Video games and all other electrical appliances use the movement of electrons to operate.

528

Standards Correlations

National Science Education Standards

The following descriptions summarize the National Science Standards that specifically relate to this chapter. For the full text of the standards, see the National Science Education Standards at the front of the book.

Section 1 Electric Charge and Force
Physical Science PS 1a, 4b, 4d, 4c, 5b, 6c
Unifying Concepts and Processes UCP 1, 2, 3
Science as Inquiry SAI 1, 2
Science and Technology ST 1, 2
Science in Personal and Social Perspectives SPSP 5

Section 2 Current
Physical Science PS 5b, 6c
Unifying Concepts and Processes UCP 1, 2, 3
Science as Inquiry SAI 1, 2
Science and Technology ST 1, 2
Science in Personal and Social Perspectives SPSP 4, 5

Section 3 Circuits
Physical Science PS 5b
Unifying Concepts and Processes UCP 1, 2, 3
Science as Inquiry SAI 1, 2
Science and Technology ST 1, 2
Science in Personal and Social Perspectives SPSP 5

Background A race car rounds a curve and speeds to the finish line in first place. Afterward, the screen darkens and the driver's score is displayed. Video games are complex pieces of electrical equipment with a detailed video display and computer chips that use electric power supplied by a power plant miles away. And in turn, that energy comes from burning fossil fuels, falling water, the wind, or nuclear fission.

At the Sandia National Laboratory, in New Mexico, powerful electrical arcs are generated in a split second when scientists fire a fusion device. Each electrical arc is similar to a bolt of lightning. A huge number of electrons move across the chamber with each arc. Although they cannot be seen, electrons move inside all electrical devices, including video games. Electricity is involved in many interactions between everyday objects, and is a vital part of the natural world and of every living organism.

Activity 1 Use the bulb and battery from a flashlight, and some wire or aluminum foil to make the light bulb light up. Try connecting the light bulb to the battery in several different ways. What works? What doesn't?

Activity 2 Find your electric meter at home. Observe how the horizontal gear moves and the numbers on the dials change. If you have an electric clothes dryer or air conditioner, observe the dials on the meter when one of these appliances is operating. Compare this with the rate of movement of the dials when all the electrical appliances and lights are turned off. Based on your results, what do you think the electric meter measures?

🔲 **internet** connect

www.scilinks.org
Topic: Applications of the Electric Spark
SciLinks code: HK4008

SCLINKS® Maintained by the National Science Teachers Association

---

Pre-Reading Questions

1. Why are power outages more common during thunderstorms?
2. Make a list of all the electrical devices in your home. What do they all have in common? How do they differ?

529

Focus ACTIVITY

Background Sandia National Laboratory uses X rays to implode a tiny capsule containing deuterium and tritium in order to produce nuclear fusion. In the implosion, the capsule is compressed to a density of about 200 g/cm³ and reaches a temperature of 5.5×10^7 K.

Three electron beams strike the screen in a carefully controlled manner to produce the picture in a video game. Video games contain many other electrical components, such as switches, computer chips, and speakers.

Activity 1 To light the bulb, students should connect the bottom of the bulb to one terminal of the battery and the side of the bulb's base to the other terminal. They can then hold the base of the bulb to one of the battery's terminals and use the wire or aluminum foil to connect the side of the bulb's base to the other terminal.
LS Kinesthetic

Activity 2 Refrigerators and electric heaters use the most electricity of home appliances. Ask students to bring in their electric bills for the month. Have students compare their family's electrical usage over a given billing period with that of other students' families and with the class average.
LS Logical

Answers to Pre-Reading Questions

1. During thunderstorms, lightning can strike houses, power lines, and other objects, delivering up to 1 million joules of

See "Continuation of Answers" at the end of the chapter.

Chapter Resource File

- Pretest **GENERAL**
- Teaching Transparency Preview
- Answer Keys

Focus

Overview

Before beginning this section, review with your students the Objectives listed in the Student Edition. In this section students learn how pairs of charges repel and attract. Next, students learn how objects become charged and how charges move within objects. The section ends with a discussion of electric force and electric fields.

🔊 Bellringer

Use the Bellringer transparency to prepare students for this section.

Motivate

Opening Demonstration —— GENERAL

For this demonstration, you will need 2 pith balls, string, PVC pipe 6 inches long, wool, aluminum pipe, and plastic cling wrap.

Step 1 Construct an electroscope by attaching each pith ball to a piece of string with a plastic hook or other fastening device. Then hang the pith balls together from a hook so they touch.

Step 2 Ask students to rub the PVC pipe with the wool. Have students touch the charged PVC pipe to the pith balls. The charged pith balls will repel each other.

Step 3 Remove both of the pith balls, with the strings still attached, and discharge them by touching them to a faucet. Hang them back on the hook.

Continued on the next page

Electric Charge and Force

KEY TERMS
electric charge
electrical conductor
electrical insulator
electric force
electric field

OBJECTIVES

▶ **Indicate** which pairs of charges will repel and which will attract.

▶ **Explain** what factors affect the strength of the electric force.

▶ **Describe** the characteristics of the electric field due to a charge.

PHYSICAL SCIENCE INTERACTIVE TUTOR

Disc Two, Module 15:
Force Between Charges
Use the Interactive Tutor to learn more about this topic.

■ **electric charge** an electrical property of matter that creates electric and magnetic forces and interactions

Figure 1

A If you rub a balloon across your hair on a dry day, the balloon and your hair become charged and are attracted to each other.

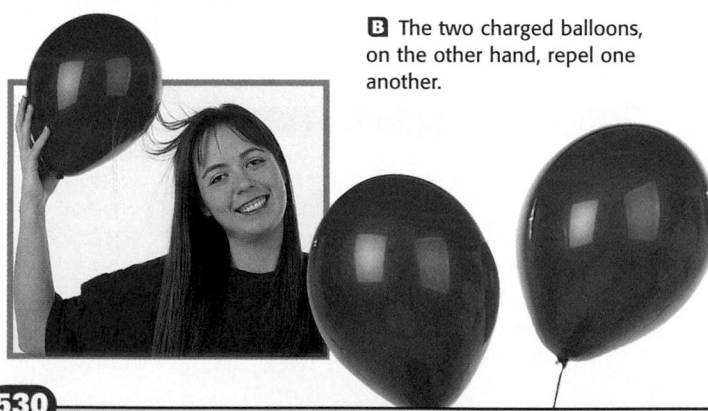

530

B The two charged balloons, on the other hand, repel one another.

When you speak into a telephone, the microphone in the handset changes your sound waves into electric signals. Light shines in your room when you flip a switch. And if you step on a pin with bare feet, your nerves send messages back and forth between your brain and your muscles so that you react quickly. These messages are carried by electric pulses moving through your nerve cells.

Electric Charge

You have probably been shocked from touching a doorknob after walking across a rug on a dry day. This happens because your body picks up **electric charge** as your shoes move across the carpet. Although you may not notice these charges when they are spread throughout your body, you notice them as they pass from your finger to the metal doorknob. You experience this movement of charges as a shock.

Like charges repel, and opposite charges attract

One way to observe charge is to rub a balloon back and forth across your hair. You may find that the balloon is attracted to your hair, as shown in *Figure 1A.* If you rub two balloons across your hair and then gently bring them near each other, as shown in *Figure 1B,* the balloons will push away from, or repel, each other.

Teacher Resources

For the New Teacher You can use the triboelectric series shown at right to determine the charge on two objects that are rubbed together for the *Opening Demonstration.* When two materials are rubbed together, the material that is closer to the positive end of the series charges positively, and the material that is closer to the negative end charges negatively.

Triboelectric Series

Positive	Negative
glass	hard rubber
nylon	nickel and copper
wool	brass and silver
silk	synthetic rubber
aluminum	Orlon™
paper	Saran™
cotton	polyethylene
steel	Teflon™
	silicone rubber

After this experiment, the balloons and your hair have some kind of charge on them. Your hair is attracted to both balloons, yet the two balloons are repelled by each other. This means there must be two types of charges—the type on the balloons and the type on your hair.

The two balloons must have the same kind of charge because each became charged in the same way. Because the two charged balloons repel each other, we see that like charges repel. However, a rubbed balloon and your hair, which did not become charged in the same way, are attracted to one another. This is because unlike charges attract.

The two types of charges are called *positive* and *negative*. When you rub a balloon on your hair, the charge on your hair is positive and the charge on the balloon is negative. When there is an equal amount of positive and negative charges on an object, it has no net charge.

An object's electric charge depends on the imbalance of its protons and electrons

All matter, including you, is made up of atoms. Atoms in turn are made up of even smaller building blocks—electrons, protons, and neutrons. Electrons are negatively charged, protons are positively charged, and neutrons are neutral (no charge).

Objects are made up of an enormous number of neutrons, protons, and electrons. Whenever there is an imbalance in the number of protons and electrons in an atom, molecule, or other object, it has a net electric charge. The difference in the numbers of protons and electrons determines an object's electric charge. Negatively charged objects have more electrons than protons. Positively charged objects have fewer electrons than protons.

The SI unit of electric charge is the *coulomb*, C. The electron and proton have exactly the same amount of charge, 1.6×10^{-19} C. Because they are oppositely charged, a proton has a charge of $+1.6 \times 10^{-19}$ C, and an electron has a charge of -1.6×10^{-19} C. An object with a total charge of -1.0 C has 6.25×10^{18} excess electrons. Because the amount of electric charge on an object depends on the numbers of protons and electrons, the net electric charge of a charged object is always a multiple of 1.6×10^{-19} C.

internet connect

www.scilinks.org
Topic: Static Electricity
SciLinks code: HK4134

SCiLINKS Maintained by the National Science Teachers Association

531

Demonstration

Charging by Friction and Induction

(Time: About 10 minutes)

Materials:

• PVC pipe, 6 inches long
• wool
• aluminum pipe
• plastic cling wrap
• electroscope

Step 1 Using the wool, rub the PVC pipe. The pipe will become negatively charged by friction.

Step 2 Touch the end of the PVC pipe to the top of the electroscope. The negative charges on the PVC pipe will move downward towards the leaves. The leaves will now repel each other and separate.

Step 3 Repeat the preceeding steps with the aluminum pipe. Rub the aluminum with the plastic. Touch the aluminum pipe to the top of the electroscope. This time the electroscope leaves become positively charged, so they initially fall together again, then repel each other.

Step 4 Try this demonstration with other materials from the triboelectric series (listed under Teacher Resources on page 530).

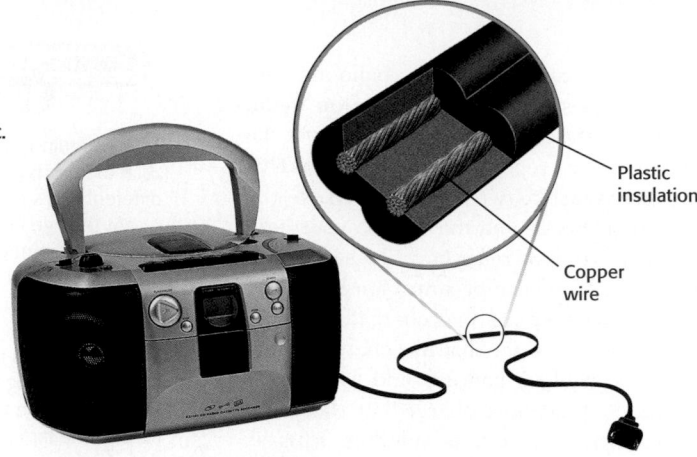

Figure 2
Appliance cords are made of metal wire surrounded by plastic. Electric charges move easily through the wire, but the plastic insulation prevents them from leaking into the surroundings.

Plastic insulation

Copper wire

■ **electrical conductor** a material in which charges can move freely and that can carry an electric current

■ **electrical insulator** a material that does not transfer current easily

INTEGRATING

BIOLOGY
Atoms or molecules with a net electric charge are known as *ions*. All living cells contain ions. Most cells also need to be bathed in solutions of ions to stay alive. As a result, most living things are fairly good conductors.

Dry skin can be a good insulator. But if your skin gets wet it becomes a conductor, and charge can move through your body more easily. So there is a greatly increased risk of electrocution when your skin is wet.

Conductors allow charges to flow; insulators do not

Have you ever noticed that the electric cords attached to appliances, such as the stereo shown in *Figure 2,* are plastic? These cords are not plastic all the way through, however. The center of an electric cord is made of thin copper wires twisted together. Cords are layered like this because of the electric properties of each material.

Materials like the metal in cords are called **conductors.** Conductors allow electric charges to move relatively freely. The plastic in the cord, however, does not allow electric charges to move freely. Materials that do not transfer charge easily are called **insulators.** Cardboard, glass, silk, and plastic are insulators.

Charges in the electric cord attached to an appliance can move through the conducting center but cannot escape through the surrounding insulator. This design makes the appliances more efficient and helps protect people from dangerous electric shock.

Objects can be charged by the transfer of electrons

Protons and neutrons are relatively fixed in the nucleus of the atom, but the outermost electrons can be easily transferred from one atom to another. When different materials are rubbed together, electrons can be transferred from one material to the other. The direction in which the electrons are transferred depends on the materials.

For example, when you slide across a fabric car seat, some electrons are transferred between your clothes and the car seat. Depending on the types of materials involved, the electrons can be transferred from your clothes to the seat or from the seat to your clothes. One material gains electrons and becomes negatively charged, and the other loses electrons and becomes positively charged. This is an example of *charging by friction.*

532

MATERIALS SCIENCE
CONNECTION

Optical engineers make optical components to be used for laser experiments. The mirrors being used with the lasers must be extremely good reflectors. Scientists have found that materials that are good conductors of electricity are also good reflectors of light. One of the best conductors of electricity is gold, so gold mirrors tend to be very good reflectors.

Gold mirrors are made by cooling a flat piece of glass, then holding it over molten gold that is starting to boil. The evaporating gold condenses onto the flat glass, just like steam condensing on a bathroom mirror, coating it with a smooth layer of gold.

Figure 3

A When a negative rod touches a neutral doorknob, electrons move from the rod to the doorknob.

B The transfer of electrons to the metal doorknob gives the doorknob a net negative charge.

Objects can also be charged without friction. One way to charge a neutral object without friction is by touching it with a charged object. As shown in *Figure 3A,* when the negatively charged rubber rod touches a neutral object, like the doorknob, some electrons move from the rod to the doorknob. The doorknob then has a net negative charge, as shown in *Figure 3B.* The rubber rod still has a negative charge, but the charge is smaller. If a positively charged rod touches a neutral doorknob, electrons move into the rod from the neutral doorknob, giving the doorknob a positive charge. Objects charged in this manner are said to be charged by *contact.*

Charges move within uncharged objects

The charges in a neutral conductor can be redistributed without contacting a charged object. If you just bring a negatively charged rubber rod close to the doorknob, the movable electrons in the doorknob will be repelled. Because the doorknob is a conductor, the electrons will move away from the rod. As a result, the portion of the doorknob closest to the negatively charged rod will have an excess of positive charge. The portion farthest from the rod will have a negative charge. But the doorknob will be neutral. Although the total charge on the doorknob will be zero, the opposite sides will have an *induced* charge, as shown in *Figure 4.*

Figure 4

A negatively charged rod brought near a metal doorknob induces a positive charge on the side of the doorknob closest to the rod and a negative charge on the side farthest from the rod.

Historical Perspective

Discovery of the Electron Sir Joseph John Thomson is credited for being the scientist that discovered the electron. He was the first to recognize that negatively charged beams, known as cathode rays, were actually made up of electrons. He received the Nobel Prize for his discovery in 1906.

REAL-LIFE
CONNECTION

Helicopters When helicopters are flying, the propellers rub across the air many times per second. If the humidity is low, the friction can charge the propeller blades, which in turn charge the entire helicopter. For this reason, the electronics of the helicopter must be insulated from the outer skin of the helicopter so that they cannot be burned out by this built-up charge. When the helicopter lands, the excess charge goes into the ground and the helicopter becomes neutral. When the helicopter is flying in humid air, the helicopter continually discharges into the atmosphere.

Interpreting Visuals Ask students to explain why the tissue paper is hanging from the comb in **Figure 5.** (Molecules in the tissue paper are polarized, producing an induced positive charge on the surface nearest the comb and an induced negative charge on the surface farthest from the comb. As a result, the pieces of tissue paper are electrically attracted to each other, and they hang from the comb.)
LS Logical

Quick ACTIVITY

Materials (per group):
• two balloons
• wool

Teacher's Notes: In step 4, be sure that the faucet is turned on low.

Results: When the balloons are rubbed on wool, they become negatively charged and the wool becomes positively charged. The two balloons repel each other. A negatively charged balloon will polarize molecules in the wall, creating a positively charged area to which the balloon is attracted. The side of the water stream nearest the balloon will be positively charged. This is due to the polar nature of water molecules—which rotate so the positive hydrogen end faces the balloon—and the presence of ions in the water.
LS Kinesthetic

Figure 5
The negatively charged comb induces a positive charge on the surface of the tissue paper closest to the comb, so the comb and the paper are attracted to each other.

▎**electric force** the force of attraction or repulsion between objects due to charge

Quick ACTIVITY

Charging Objects
1. Rub two air-filled balloons vigorously on a piece of wool.
2. Hold your balloons near each other.
3. Now try to attach one balloon to the wall.
4. Turn on a faucet, and hold a balloon near the stream of tap water.
5. Explain what happens to the charges in the balloons, wool, water, and wall.

How can the negatively charged comb in **Figure 5** pick up pieces of neutral tissue paper? The electrons in tissue paper cannot move about freely because the paper is an insulator. But when a charged object is brought near an insulator, the positions of the electrons within the individual molecules of the insulator change slightly. One side of a molecule will be slightly more positive or negative than the other side. This *polarization* of the atoms or molecules of an insulator produces an induced charge on the surface of the insulator. The surface of the tissue paper nearest the comb has an induced positive charge. The surface farthest from the comb has an induced negative charge.

Electric Force

The attraction of tissue paper to a negatively charged comb and the repulsion of the two balloons are examples of **electric force.** It is also the reason clothes sometimes cling to each other when you take them out of the dryer. Such pushes and pulls between charges are all around you. For example, a table feels solid, even though its atoms contain mostly empty space. The electric force between the electrons in the table's atoms and your hand is strong enough to prevent your hand from going through the table. In fact, the electric force at the atomic and molecular level is responsible for most of the common forces we can observe, such as the force of a spring and the force of friction.

The electric force is also responsible for effects that we can't see; it is part of what holds an atom together. The bonding of atoms to form molecules is also due to the electric force. The electric force plays a part in the interactions among molecules, such as the proteins and other building blocks of our bodies. Without the electric force, life itself would be impossible.

534

Alternative Assessment — BASIC

Graphing Force and Distance At right is a list of the distance between two positively or negatively charged objects and the electric force between them. Have students graph the data to see the shape of the curve that describes the interaction between charges. Using their graphs, students should be able to predict the force when the distance is 6 mm. (about 6.75 N) **LS** Visual

Distance (mm) (x-axis)	Force (N) (y-axis)
1	243
3	27
5	10
7	5
9	3

Electric force depends on charge and distance

The electric force between two charged objects varies depending on the amount of charge on each object and the distance between them. The electric force between two objects is proportional to the product of the charges on the objects. If the charge on one object is doubled, the electric force between the objects will also be doubled.

The electric force is also inversely proportional to the square of the distance between two objects. For example, if the distance between two small charges is doubled, the electric force between them decreases to one-fourth its original value. If the distance between two small charges is quadrupled, the electric force between them decreases to one-sixteenth its original value.

Electric force acts through a field

As described earlier, electric force does not require that objects touch. How do charges interact over a distance? One way to model this property of charges is with the concept of an **electric field.** A charged particle produces an electric field in the space around it. Another charged particle in that field will experience an electric force. This force is due to the electric field associated with the first charged particle.

One way to show an electric field is by drawing *electric field lines*. Electric field lines point in the direction of the electric force on a positive charge. Because two positive charges repel one another, the electric field lines around a positive charge point outward, as shown in *Figure 6A*. In contrast, the electric field lines around a negative charge point inward, as shown in *Figure 6B*. Regardless of the charge, electric field lines never cross one another.

▶ **electric field** a region in space around a charged object that causes a stationary charged object to experience an electric force

Figure 6

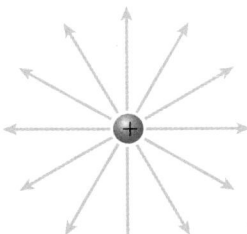 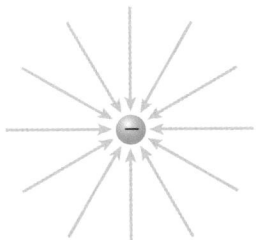

A The electric field lines show that a positive charge placed in the electric field due to a positive charge would be pushed away.

B A positive charge placed in the electric field due to a negative charge would be pulled in.

Did You Know ?

Electric force and gravitational force both depend on a physical property of objects—charge and mass, respectively—and the distance between the objects. They have the same mathematical form. But gravitational force is attractive, while electric force is both attractive and repulsive. Also, the electric force between two charged particles separated by a given distance is much greater than the gravitational force between the particles.

Teaching Tip

Electric Field Lines Some students may think that electric field lines are a physical phenomenon. Students should be told explicitly that electric field lines do not actually exist. The lines are a visual representation of the field that would be experienced by a test charge.

|SKILL| BUILDER — BASIC

Interpreting Diagrams Figure 6 shows the electric fields for a positive source charge and a negative source charge. Be sure students understand the similarities and differences between the two cases shown in this figure. Ask students what can be concluded about the charges in **Figure 6A** and **B** by comparing the electric field lines for each case. (Because the number of field lines in **A** is equal to the number of field lines approaching the charge in **B**, the charges must be equal in magnitude. The field lines also show that the charge in **A** is positive because they are pointing away from the charge. Because the field lines point toward the charge in **B**, they show that the charge is negative.) **LS Visual**

MATH
CONNECTION

Have a math teacher visit your class and discuss the inverse square law, or $1/r^2$ dependence. Discuss the asymptotic behavior of this law and its implications as to how far you must be away from an electrical source for the effects of that source to be zero. Have the math teacher discuss how changing the distance between charged objects changes the electric force between them.

BIOLOGY
CONNECTION

Sharks have specialized organs (ampullae of Lorenzini) that can sense the minute electrical fields generated by all marine animals and that assist in the detection of prey. Scientists have developed a device called the Shark POD (Protective Oceanic Device) for scuba divers to use against sharks. The Shark POD produces an electric field so strong that sharks cannot come near the divers. This would be analogous to humans being unable to look into extremely bright lights. It is believed that the Shark POD causes mild pain in the sharks' ampullae, repelling the sharks from distances of 1–7 m.

Chapter Resource File

• **Quick Activity Datasheet** Charging Objects GENERAL

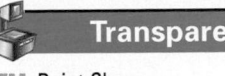
Transparencies

TM Point Charges

Teaching Tip
Drawing Electric Field Lines
Students may have difficulty remembering that the electric field points in the direction of the electric force on a positive charge. Draw a positive test charge at the location of the electric field. The electric field will point in the direction that this positive charge would move. Negatively charged particles will move in the opposite direction of the electric field.

Close

Quiz ———— BASIC
Complete the sentences with the appropriate terms.
1. _____ charges repel, while _____ charges attract. (Like; unlike)
2. The electric force between two charged objects is proportional to the product of the _____ and inversely proportional to the _____ squared. (charges; distance between them)
3. All charged objects are surrounded by an electric _____. (field)

LS Logical

Chapter Resource File
- Concept Review GENERAL
- Quiz BASIC

Transparencies
TM Electric Fields

Figure 7

A The electric field lines for two positive charges show the repulsion between the charges.

B Half the field lines starting on the positive charge end on the negative charge because the positive charge is twice as great as the negative charge.

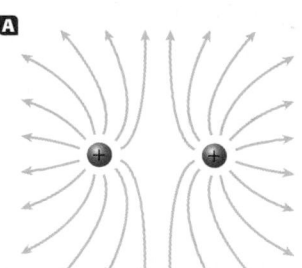
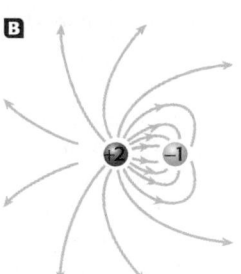

You can see from **Figure 7** that the electric field between two charges can be represented using these rules. The field lines in **Figure 7A** point away from the positive charges, showing that the positive charges repel each other. Field lines show not only the direction of an electric field but also the relative strength due to a given charge. As shown in **Figure 7B,** there are twice as many field lines pointing outward from the +2 charge as there are ending on the −1 charge. More lines are drawn for greater charges to indicate greater force.

SECTION 1 REVIEW

SUMMARY
- ▶ There are two types of electric charge, positive and negative.
- ▶ Like charges repel; unlike charges attract.
- ▶ The electric force between two charged objects is proportional to the product of the charges and inversely proportional to the square of the distance between the objects.
- ▶ Electric force acts through electric fields.
- ▶ Electric fields surround charged objects. Any charged object that enters an electric field experiences an electric force.

1. **Identify** the electric charge of each of the following atomic particles: a proton, a neutron, and an electron.

2. **Describe** the interaction between two like charges. Is the interaction the same between two unlike charges?

3. **Diagram** what will happen if a positively charged rod is brought near the following objects:
   a. a metal washer   **b.** a plastic disk

4. **Categorize** the following as conductors or insulators:
   a. copper wire
   b. your body when your skin is wet
   c. a plastic comb

5. **Explain** how the electric force between two positive charges changes if
   a. the distance between the charges is tripled.
   b. the amount of one charge is doubled.

6. **Critical Thinking** What missing electric charge would produce the electric field shown at right?

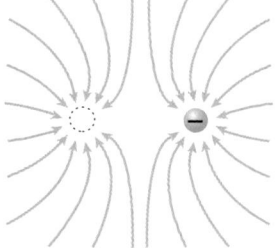

Answers to Section 1 Review

1. proton—positive; neutron—neutral, no charge; electron—negative

2. Like charges repel each other. No, two unlike charges attract each other.

3. **a.** The drawing should show negative charges on the side of the metal washer nearest the rod and positive charges on the side of the washer farthest from the rod.
   b. The drawing should show "polarized molecules" (refer to **Figure 5**) with the side of

each "molecule" nearest the rod negative and the side farthest from the rod positive.

4. **a.** conductor
   b. conductor
   c. insulator

5. **a.** The force will become one-ninth as large as the original force.
   b. The force will double.

6. a negative charge of equal amount

Current

OBJECTIVES

▶ **Describe** how batteries are sources of voltage.
▶ **Explain** how a potential difference produces a current in a conductor.
▶ **Define** *resistance*.
▶ **Calculate** the resistance, current, or voltage, given the other two quantities.
▶ **Distinguish** between conductors, superconductors, semiconductors, and insulators.

KEY TERMS
electrical potential
    energy
potential difference
cell
current
resistance

W hen you wake up in the morning, you reach up and turn on the light switch. The light bulb is powered by moving charges. How do charges move through a light bulb? And what causes the charges to move?

Disc Two, Module 8:
Batteries and Cells
Use the Interactive Tutor to learn more about these topics.

Focus

Overview
Before beginning this section, review with your students the Objectives listed in the Student Edition. This section covers current and many related concepts, including electrical potential energy, potential difference, batteries, conventional current, and resistance. Students also learn how materials can be classified as conductors, superconductors, insulators, and semiconductors.

🔊 Bellringer
Use the Bellringer transparency to prepare students for this section.

Voltage and Current

Gravitational potential energy depends on the relative position of the ball, as shown in *Figure 8A.* A ball rolling downhill moves from a position of higher gravitational potential energy to one of lower gravitational potential energy. An electric charge also has potential energy— **electrical potential energy** —that depends on its position in an electric field.

Just as a ball will roll downhill, a negative charge will move away from another negative charge. This is because of the first negative charge's electric field. The electrical potential energy of the moving charge decreases, as shown in *Figure 8B,* because the electric field does work on the charge.

▶ **electrical potential energy** the ability to move an electric charge from one point to another

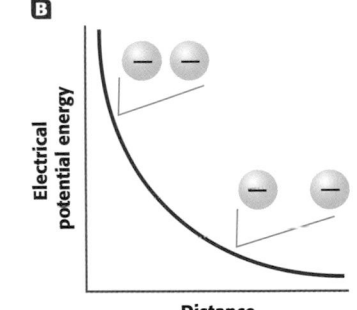

A More gravitational potential energy

Less gravitational potential energy

B Electrical potential energy

Distance

Figure 8
A The gravitational potential energy of a ball decreases as it rolls downhill.

B The electrical potential energy between two negative charges decreases as the distance between them increases.

Motivate

Opening Demonstration
Safety Caution: *Use a plastic fluorescent tube cover to protect students from possible shattering glass. Be careful to discharge the generator before touching it.*

Borrow a Van de Graaff generator from the physics teacher. You will also need a small, tube-shaped fluorescent light bulb. Turn on the generator and allow it to build up a charge. The fluorescent tube can be taped to the end of a meterstick for better visibility. Hold the tube near the generator so that it points out radially, but do not touch the ball of the generator. Turn the lights off in the room. There will be a potential difference across the tube, and it will light up slightly. Tell students they will learn about potential difference in this section.

537

Alternative Assessment — ADVANCED

✏️ **Electrical Potential Energy** Ask students to write a paragraph describing how the electrical potential energy of a charged object depends on the object's position relative to another charged object. Also ask them to explain what energy conversion occurs if either charge is free to move. (Sample answer: If two charged objects repel each other,

the electrical potential energy will be greatest when the objects are near each other, as shown in **Figure 8.** If two charged objects attract each other, the electrical potential energy will be greatest when they are far apart. In either case, the electrical potential energy can be converted to kinetic energy if either charge is free to move.)
LS **Verbal**

Chapter Resource File

• **Lesson Plan**

Transparencies

TT Bellringer

Reading Strategy — BASIC

Reading Organizer Have students read the first three pages of this section and then organize the ideas presented in an outline. The outline should be organized in terms of the details presented for each heading. **LS** Verbal

SKILL BUILDER — BASIC

Interpreting Visuals Have students look at **Figure 9,** and ask them when the electrical potential energy would be the lowest for an electron in the electric field due to a positively charged object. (It would be lowest when the electron and positively charged object are very close together.) **LS** Visual

MISCONCEPTION //ALERT\\\

Batteries Some students believe that batteries store current. Stress that a battery does not supply the charges that move through a circuit. Rather, the battery does work on charges to increase their electrical potential energy. Tell students this is analogous to lifting an object: the lifter does work on the object and thereby increases the object's gravitational potential energy.

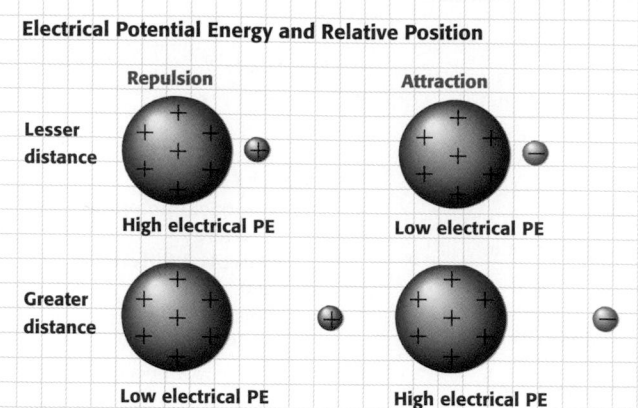

Figure 9

The electrical potential energy of a charge depends on its position in an electric field.

▶ **potential difference** between any two points, the work that must be done against electric forces to move a unit charge from one point to the other

▶ **cell** a device that is a source of electric current because of a potential difference, or voltage, between the terminals

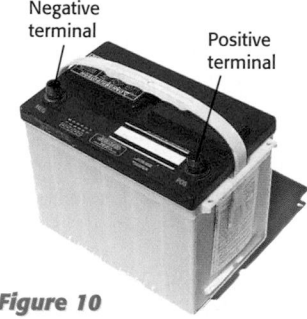

Negative terminal

Positive terminal

Figure 10

For a typical car battery, there is a voltage of 12 V across the negative (black) terminal and the positive (red) terminal.

538

You can do work on a ball to move it uphill. This will increase the ball's gravitational potential energy. In the same way, a force can push a charge in the opposite direction of the electric force. This increases the electrical potential energy associated with the charge's relative position. **Figure 9** shows how the electrical potential energy depends on the distance between two charged objects for both an attractive and a repulsive electric force.

Potential difference is measured in volts

Usually, it is more practical to consider the **potential difference** than electrical potential energy. Potential difference is the change in the electrical potential energy of a charged particle divided by its charge. This change occurs as a charge moves from one place to another in an electric field.

The SI unit for potential difference is the *volt*, V, which is equivalent to one joule per coulomb (1 J/C). For this reason, potential difference is often called *voltage*.

There is a voltage across the terminals of a battery

The voltage across the two *terminals* of a battery can range from about 1.5 V for a small battery to about 12 V for a car battery, as shown in **Figure 10.** Most common batteries are an electric **cell**—or a combination of connected electric cells—that convert chemical energy into electrical energy. One terminal is positive, and the other is negative. A summary of various types of electric cells is given in **Table 1.**

Electrochemical cells contain an *electrolyte,* a solution that conducts electricity, and two *electrodes,* each a different conducting material. These cells can be dry cells or wet cells. Dry cells, such as those used in flashlights, contain a paste-like electrolyte. Wet cells, such as those used in almost all car batteries, contain a liquid electrolyte. An average cell has a potential difference of 1.5 V between the positive and negative terminals.

A voltage sets charges in motion

When a flashlight is switched on, the terminals of the battery are connected through the light bulb. Electrons move through the light bulb from the negative terminal to the positive terminal.

REAL-LIFE CONNECTION

Car Batteries The battery in a car contains two different metals as well as a strong acid. The acid allows electrons from one metal to be attracted to the other metal. This flow of electrons is used by the car in the form of electricity.

As the battery uses its energy, the voltage stays the same but the internal resistance rises until the battery cannot supply any more current. Attached to the motor is a generator that recharges the battery as the car is driven. The motor converts the chemical potential energy of gasoline into kinetic energy. The generator then converts this kinetic energy into electrical energy.

Table 1 **Types of Electric Cells**

Electrical cell	Basic principle	Uses
Electrochemical	Electrons are transferred between different metals immersed in an electrolyte.	Common batteries, automobile batteries
Photoelectric and photovoltaic	Electrons are released from a metal when struck by light of sufficient energy.	Artificial satellites, calculators, streetlights
Thermoelectric	Two different metals are joined together, and the junctions are held at different temperatures, causing electrons to flow.	Thermostats for furnaces and ovens
Piezoelectric	Opposite surfaces of certain crystals become electrically charged when under pressure.	Crystal microphones and headsets, computer keypads, record cartridge

When charges are accelerated by an electric field to move to a position of lower potential energy, an electric **current** is produced. Current is the rate at which these charges move through a conductor. The SI unit of current is the *ampere*, A. One ampere, or *amp*, equals 1 C of charge moving past a point in 1 second.

A battery is a *direct current* source because the charges always move from one terminal to the other in the same direction. Current can be made up of positive, negative, or a combination of both positive and negative charges. In metals, moving electrons make up the current. In gases and many chemical solutions, current is the result of both positive and negative charges in motion.

In our bodies, current is mostly positive charge movement. Nerve signals are in the form of a changing voltage across the nerve cell membrane. *Figure 11A* shows that a resting cell has more negative charges on the inside than on the outside. *Figure 11B* shows how a nerve impulse moves along the cell membrane. As one end of the cell is stimulated, channels nearby in the cell membrane open, allowing Na^+ ions to enter. Later, potassium channels open, and K^+ ions exit the cell, restoring the original voltage across the cell membrane.

Conventional current is defined as movement of positive charge

A negative charge moving in one direction has the same effect as a positive charge moving in the opposite direction. *Conventional current* is defined as the current made of positive charge that would have the same effect as the actual motion of charge in the material. *In this book, the direction of current will always be given as the direction of positive charge movement that is equivalent to the actual motion of charges in the material.* So the direction of current in a wire is opposite the direction that electrons move in that wire.

▶ **current** the rate that electric charges move through a conductor

Figure 11

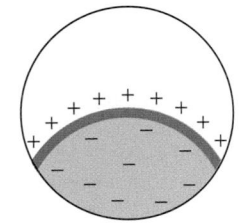

A A resting nerve cell is more negatively charged than its surroundings.

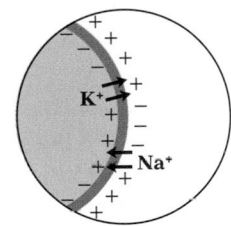

B As a nerve impulse moves along the cell membrane, the voltage across it changes.

539

REAL-LIFE CONNECTION

Lightning Protection If a house is hit by lightning, the electrical charges flow through any metal material throughout the house until the charges reach Earth. Lightning can explode brick walls and concrete and start fires. A lightning protection system provides a designated path for the current. The system neither attracts nor repels a lightning strike, but simply intercepts the current and guides it harmlessly to the ground.

Did You Know?

Lightning Around the world, lightning strikes about 100 times every second, or 8 million times every day. A typical flash of lightning has about 1 billion volts of potential difference and between 10,000 and 200,000 amps of current.

Teaching Tip

Lightning Lightning is the flow of charge from a cloud to Earth, from Earth to a cloud, or between clouds. In the first situation, negative charges build up in the cloud. As the charge increases, an electric field forms between Earth and the cloud. When the potential difference between the cloud and the ground is great enough, the air acts as a conductor. At that point, the lightning strikes the closest point on the ground, delivering up to 1 million joules of electrical energy.

Teaching Tip

Current Currents that consist of both positive and negative charge carriers in motion include those that exist in batteries, in the human body, in the ocean, and in the ground. Electric currents in the brain and nerves of the human body consist of moving sodium ions and potassium ions. Remind students that current refers to the movement of matter, not energy.

 Transparencies

TM Electrical Potential Energy

Group Activity —— GENERAL

Comparing Batteries You will need AA or AAA batteries (different brands of the same voltage), battery holders, holiday lights, wire connectors with alligator clips, and wire strippers.

Cut the lights apart and strip the insulation off the ends. Use the wire connectors to make a complete circuit using the battery, battery holder, and lights. Give each student group a different brand of battery. Have the students start all the circuits at the same time and check their bulbs periodically. By the end of class, some of the light bulbs will be dimmer, demonstrating which batteries last the longest.
LS **Interpersonal**

Answers

1. rechargeable alkaline; they hold a charge like a regular alkaline and are cheapest in the long run.

2. NiCads lose about 1% of their energy every day. A smoke detector needs a battery that will last for a long time.

3. Answers may vary. Lead acid storage batteries are used in gasoline-powered cars. They use lead-antimony electrodes, and sulfuric acid as the electrolyte. Electric vehicles can use several types of rechargeable batteries, such as Ni-Cd, nickel-metal hydride, lead acid, or sodium-sulfur.

Science and the Consumer

Which Is the Best Type of Battery?

"Heavy-duty," "long-lasting alkaline," and "environmentally friendly rechargeable" are some of the labels that manufacturers put on batteries. But how do you know which type to use?

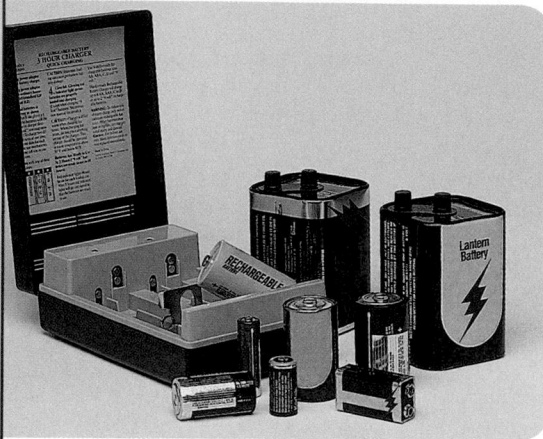

The answer depends on how you will use the battery. Some batteries are used continuously, but others are turned off and on frequently, such as those used in a stereo. Still other batteries must be able to hold a charge without being used, such as those used in smoke detectors and flashlights.

Heavy-Duty Batteries Are Inexpensive

In terms of price, a heavy-duty battery typically costs the least but lasts only about 30 percent as long as an alkaline battery. This makes heavy-duty batteries impractical for most uses and an unnecessary source of landfill clutter.

Regular Alkaline Batteries Are Expensive but Long Lasting

Regular alkaline batteries are more expensive but have longer lives, lasting up to 6 hours with continuous use and up to 18 hours with intermittent use. They hold a full charge for years, making them good for use in flashlights and similar devices. They are less

of an environmental problem than they previously were because manufacturers have stopped using mercury in them. However, because they are single-use batteries, they also end up in landfills very quickly.

Rechargeable Batteries Don't Clutter Landfills

Rechargeable batteries are the most expensive to purchase initially. If recycled, however, they are the most economical in the long run and are the most environmentally sound choice. The most common rechargeable cells are either NiCads—containing nickel, Ni, and cadmium, Cd, metals—or alkaline. Either type of rechargeable battery can be recharged hundreds of times. Although rechargeable batteries last only about half as long on one charge as regular alkaline batteries, the energy to recharge them costs pennies. Rechargeable NiCads lose about 1% of their stored energy each day they are not used and should therefore never be used in smoke detectors or flashlights.

Your Choice

1. **Making Decisions** Which type of battery would you use in a portable stereo? Explain your reasoning.

2. **Critical Thinking** Why is it important not to use NiCads in smoke detectors?

3. **Locating Information** Use library resources or the Internet to learn more about batteries used in gasoline-powered and electric cars. Prepare a summary of the types of rechargeable car batteries available.

internet connect

www.scilinks.org
Topic: Batteries SciLinks code: HK4014

SCiLINKS. Maintained by the National Science Teachers Association

Historical Perspective

Volta's Battery The first battery was developed and constructed by Alessandro Volta in about 1800. He was a professor of natural philosophy at the University of Pavia, in Italy. His device was made of a series of silver and zinc disks in pairs, each separated with a sheet of pasteboard saturated with salt water. This type of battery is called a voltaic pile. A current is generated when the first silver disk is connected with a wire to the last zinc disk.

INCLUSION Strategies

• *Learning Disabled*
• *Attention Deficit Disorder*
• *English Language Learners*

Using the information in the "Which is the Best Type of Battery" activity in the textbook, ask students to create television commercials for Heavy-duty Batteries, Regular Alkaline Batteries, and Rechargeable Batteries. Each commercial should advertise the positive qualities and uses of each type of battery. Students can present their commercials or tape-record them.

Using a Lemon as a Cell

Because lemons are very acidic, their juice can act as an electrolyte. If various metals are inserted into a lemon to act as electrodes, the lemon can be used as an electrochemical cell.

SAFETY CAUTION Handle the wires only where they are insulated.

1. Using a knife, make two parallel cuts 6 cm apart along the middle of a juicy lemon. Insert a copper strip into one of the cuts and a zinc strip the same size into the other.

2. Cut two equal lengths of insulated copper wire. Use wire strippers to remove the insulation from both ends of each wire. Connect one end of each wire to one of the terminals of a galvanometer.

3. Touch the free end of one wire to the copper strip in the lemon. Touch the free end of the other wire to the zinc strip, as shown in the figure at right. Record the galvanometer reading for the zinc-copper cell.

4. Replace the strips of copper and zinc with equally sized strips of different metals. Record the galvanometer readings for each pair of electrodes. Which pair of electrodes resulted in the largest current?

5. Construct a table of your results.

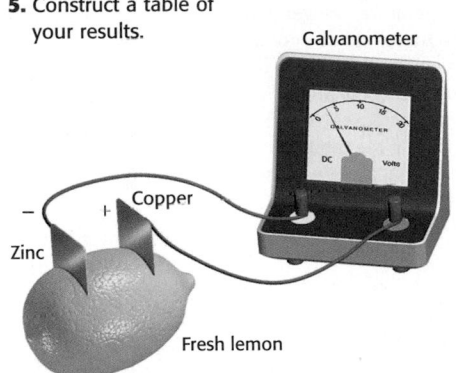

Galvanometer
Copper
Zinc
Fresh lemon

Electrical Resistance

Most electrical appliances you plug into an outlet are designed for the same voltage: 120 V. But light bulbs come in many varieties, from dim 40 W bulbs to bright 100 W bulbs. These bulbs shine differently because they have different amounts of current in them. The difference in current between these bulbs is due to their **resistance.** Resistance is caused by internal friction, which slows the movement of charges through a conducting material. Because it is difficult to measure the internal friction directly, resistance is defined by a relationship between the voltage across a conductor and the current through it.

The resistance of the *filament* of a light bulb, as shown in *Figure 12,* determines how bright the bulb is. The filament of a dim 40 W light bulb has a higher resistance than the filament of a bright 100 W light bulb.

resistance the opposition posed by a material or a device to the flow of current

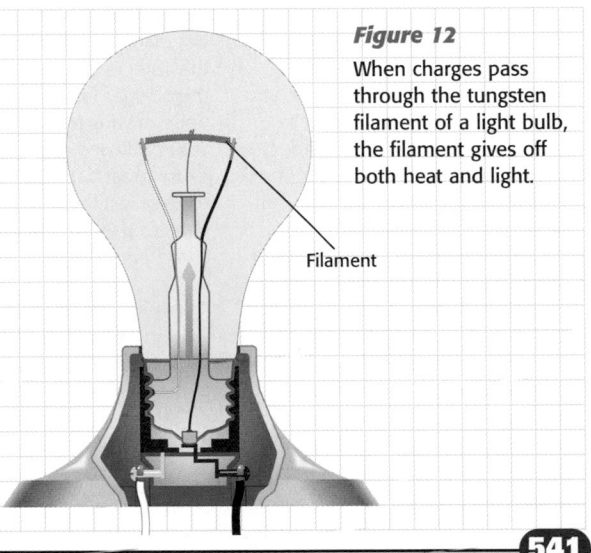

Figure 12

When charges pass through the tungsten filament of a light bulb, the filament gives off both heat and light.

Filament

541

Historical Perspective

Electrical Lighting The first commercially viable incandescent light bulb was made by Thomas Edison on October 21, 1879. Between October 21 and December 31, Edison and his crew worked day and night to construct the huge number of components necessary for a complete electric lighting system—including wires, insulators, bulb sockets, and connectors. This electrical system was then used for the first public demonstration of electrical lighting on New Year's Eve in 1879.

Materials (per group):
- knife
- juicy lemon
- copper strip
- zinc strip
- insulated copper wire
- wire strippers
- galvanometer
- strips of other metals

Teacher's Notes: The electrodes need to be of a uniform size. It is important that the electrodes are inserted as deeply as possible into the lemon so that there is maximum contact between the metal and the lemon juice. If zinc is not available, use magnesium. Be sure to review any safety concerns regarding the magnesium. Any fruit with an acidic juice, such as an orange or grapefruit, will work. Potatoes will also work.

Results: Answers will vary depending on the metals used as electrodes. Larger electrodes will produce greater current. Consult an electrochemical series to determine which pair of electrodes should produce the greatest current.
LS Kinesthetic

Chapter Resource File

- **Cross-Disciplinary Worksheet** Science and the Consumer—Battery Issues ADVANCED
- **Quick Activity Datasheet** Using a Lemon as a Cell GENERAL
- **Cross-Disciplinary Worksheet** Integrating Chemistry—Rechargeable Ni-Cd Batteries ADVANCED

Did You Know?

Resistance Length and Width Use the following two analogies to help students understand why resistance decreases with increasing area and increases with increasing length. The fire department uses very thick hoses so that the resistance to the flow of water is less. They also use hoses that can be connected. In this way the length of the hose can be kept as short as possible, thus minimizing resistance.

Speaker wire is used to connect speakers to a stereo. If the stereo is to be played at high volume, then you want a low resistance in the speaker wires. The thicker the wire, the louder and the better quality the sound will be. Longer speaker wires will cause the speakers to play a little softer.

REAL WORLD APPLICATIONS

Answers

1. $I = \dfrac{V}{R} = \dfrac{24V}{10^5 \Omega} = 2.4 \times 10^{-4}$ A
This much current could not even be felt.

2. $I = \dfrac{V}{R} = \dfrac{24V}{10^3 \Omega} = 2.4 \times 10^{-2}$ A
This amount of current is enough to cause you to lose muscle control and probably enough to cause a bad burn.

Did You Know?

Resistance depends on the material used as well as the material's length, cross-sectional area, and temperature. Longer pieces of a material have greater resistance. Increasing the cross-sectional area of a material decreases its resistance. Lowering the temperature of a material also decreases its resistance.

REAL WORLD APPLICATIONS

The Danger of Electric Shock

If you are in contact with the ground or with water, you can receive an electric shock by touching an uninsulated conducting, or "live," wire. An electric shock from such a wire can result in serious burns or even death.

The degree of damage to your body by an electric shock depends on several factors. Large currents are more dangerous than smaller currents. A current of 0.1 A is often fatal. But the amount of time you are exposed to the current also matters. If the current is larger than about 0.01 A, the muscles in the

hand touching the wire contract, and you may be unable to let go of the wire. In this case, the charges will continue moving through your body and can cause great damage, especially if the charges pass through a vital organ, such as the heart.

Applying Information

1. You can use the definition of *resistance* to calculate the amount of current that would be in a body, given the voltage and resistance. Using the table above as a reference, determine the effect of touching the terminals of a 24 V battery. Assume that your body is dry and has a resistance of 100 000 Ω.

2. If your skin is moist, your body's resistance is only about 1000 Ω. How would touching the terminals of a 24 V battery affect your body if your skin is moist?

Current (A)	Effect
0.001	Slight tingle
0.005	Pain
0.010	Muscle spasms
0.015	Loss of muscle control
0.070	Probably fatal (if contact is more than 1 second)

Resistance can be calculated from current and voltage

You have probably noticed that electrical devices such as televisions or stereos become warm after they have been on for a while. As moving electrons collide with the atoms of the material, some of their kinetic energy is transferred to the atoms. This energy transfer causes the atoms to vibrate, and the material warms up. In most materials, some of the kinetic energy of electrons is lost as heat.

A conductor's resistance indicates how much the motion of charges within it is resisted because of collisions. Resistance is found by dividing the voltage across the conductor by the current.

Resistance Equation

$$resistance = \frac{voltage}{current} \qquad R = \frac{V}{I}$$

The SI unit of resistance is the *ohm*, Ω, which is equal to volts per ampere. If a voltage across a conductor of 1 V produces a current of 1 A, then the resistance of the conductor is 1 Ω.

A *resistor* is a special type of conductor used to control current. Every resistor is designed to have a specific resistance. For example, for any applied voltage, the current in a 10 Ω resistor is half the current in a 5 Ω resistor.

Did You Know?

Resistor Color Codes Resistors have three colored bands that signify their resistance. The first two bands are read as digits, and the third band is the multiplier. For example, the color red represents the digit 2 and the multiplier 100. So, 3 red bands would represent 22 multiplied by 100, or 2200 Ω. Resistors can also have a fourth band which indicates whether the resistance value is accurate within 5%, 10%, or 20%. If the example resistor were accurate within 5%, its resistance is actually between 2090 and 2310 Ω.

Teacher Resources

For the New Teacher Be sure students understand that the resistance in a circuit is a physical constant. Resistance cannot be changed by increasing or decreasing voltage or current. To change the resistance, you must change the components or resistors in the circuit.

Resistance The headlights of a typical car are powered by a 12 V battery. What is the resistance of the headlights if they draw 3.0 A of current when turned on?

1 List the given and unknown values.
 Given: *current, I = 3.0 A*
 voltage, V = 12 V
 Unknown: *resistance, R = ? Ω*

2 Write the equation for resistance.
$$resistance = \frac{voltage}{current} \qquad R = \frac{V}{I}$$

3 Insert the known values into the equation, and solve.
$$R = \frac{V}{I} = \frac{12\ V}{3.0\ A}$$
$$R = 4.0\ \Omega$$

Practice

Resistance

1. Find the resistance of a portable lantern that uses a 24 V power supply and draws a current of 0.80 A.
2. The current in a resistor is 0.50 A when connected across a voltage of 120 V. What is its resistance?
3. The current in a handheld video game is 0.50 A. If the resistance of the game's circuitry is 12 Ω, what is the voltage produced by the battery?
4. A 1.5 V battery is connected to a small light bulb with a resistance of 3.5 Ω. What is the current in the bulb?

Conductors have low resistances

Whether or not charges will move in a material depends partly on how tightly electrons are held in the atoms of the material. A good conductor is any material in which electrons can flow easily under the influence of an electric field. Metals, like the copper found in wires, are some of the best conductors because electrons can move freely throughout them. Certain metals, conducting alloys, or carbon are used in resistors.

When you flip the switch on a flashlight, the light seems to come on immediately. But the electrons don't travel that rapidly. The electric field is directed through the conductor at almost the speed of light when a voltage source is connected to the conductor. Electrons everywhere throughout the conductor simultaneously experience a force due to the electric field and move in the opposite direction of the field lines. This is why the light comes on so quickly in a flashlight.

Practice **HINT**

▶ When a problem requires you to calculate the resistance of an object, you can use the resistance equation as shown on the previous page.
▶ The resistance equation can also be rearranged to isolate voltage on the left in the following way:
$$R = \frac{V}{I}$$
Multiply both sides by I.
$$IR = \frac{VI}{I}$$
$$V = IR$$
You will need this version of the equation for Practice Problem 3.
▶ For Practice Problem 4, you will need to rearrange the equation to isolate current on the left.

543

Additional Examples

Resistance A battery-operated CD player uses 12 V from the wall socket and draws a current of 2.5 A. Calculate the resistance of the CD player.
$$R = \frac{V}{I} = \frac{12\ V}{2.5\ A} = 4.8\ \Omega$$

A light bulb has a resistance of 12 Ω. It is attached to a battery that has a voltage of 24 V. Calculate the current in the light bulb.
$$I = \frac{V}{R} = \frac{24\ V}{12\ \Omega} = 2.0\ A$$
LS Logical

Practice

1. $R = \dfrac{V}{I} = \dfrac{24\ V}{0.80\ A} = 3.0 \times 10^1\ \Omega$

2. $R = \dfrac{V}{I} = \dfrac{120\ V}{0.50\ A} = 240\ \Omega$

3. $V = IR = (0.50\ A)(12\ \Omega) = 6.0\ V$

4. $I = \dfrac{V}{R} = \dfrac{1.5\ V}{3.5\ \Omega} = 0.43\ A$
LS Logical

Did You Know ?

Electron Movement in a Conductor Free electrons in a conductor move about randomly until there is a potential difference across the conductor. Then the electrons move through the conductor in the opposite direction of the electric field. The electrons do not move in straight lines, but collide repeatedly with the vibrating metal atoms of the conductor. Despite these collisions, electrons move slowly along the conductor toward the positive terminal of the battery.

MISCONCEPTION ///ALERT\\\

Electricity in a Circuit Some students believe that electricity "travels" from a battery to a bulb in a circuit. Emphasize that the charges are already present throughout the wire between the battery and bulb, as explained in the student text on this page. The charges begin to move slowly when acted on by an electric field. It is the field that travels rapidly through the wire (at the speed of light) when a switch is activated to turn the circuit on.

Chapter Resource File

• **Cross-Disciplinary Worksheet** Real World Application—Electric Shock: Caution! **ADVANCED**
• **Math Skills** Resistance **GENERAL**

Teaching Tip

Variable Resistors The resistance of electrical circuits is not always fixed. This can be demonstrated by changing the volume control on a portable stereo. The volume control is an example of a variable resistor. Ask students to speculate how a variable resistor works based on their knowledge of resistance. Have students think of other devices that also have variable resistors. (Responses could include indoor light-dimmer switches and electric oven controls.)

Quick Lab

Safety Caution: *Students should be careful not to touch both terminals of the battery.*

Analysis

1. The conductivity tester lights up.

2. The iron nail, copper wire, aluminum nail, and brass key were good conductors.

3. The glass stirring rod, wooden dowel, chalk, cardboard, plastic utensil, and cork were poor conductors.

4. Good conductors have low resistance and allow charges to pass through, while good insulators have high resistance.

LS Kinesthetic

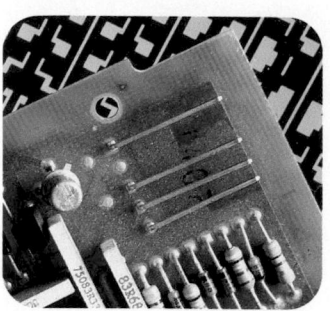

Figure 13

Most electrical devices contain conductors, insulators, and semiconductors.

Some materials become superconductors below a certain temperature

Certain metals and compounds have zero resistance when their temperature falls below a certain temperature called the *critical temperature*. These types of materials are called *superconductors*. The critical temperature varies among materials, from less than $-272°C$ ($-458°F$) to as high as $-123°C$ ($-189°F$).

Metals such as niobium, tin, and mercury and some metallic compounds containing barium, copper, and oxygen become superconductors below their respective critical temperatures. Superconductors have been used in electrical devices such as filters, powerful magnets, and Maglev high-speed express trains.

Semiconductors are intermediate to conductors and insulators

Semiconductors belong to a third class of materials with electrical properties between those of insulators and conductors. In their pure state, semiconductors are insulators. The controlled addition of specific atoms of other materials as impurities dramatically increases a semiconductor's ability to conduct electric charge. Silicon and germanium are two common semiconductors. Complex electrical devices, like the computer board shown in *Figure 13,* are made of conductors, insulators, and semiconductors.

Quick Lab

How can materials be classified by resistance?

Materials
- ✓ 6 V battery
- ✓ flashlight bulb in base holder
- ✓ 2 wire leads with alligator clips
- ✓ 2 metal hooks
- ✓ block of wood

- ✓ glass stirring rod
- ✓ iron nail
- ✓ wooden dowel
- ✓ copper wire
- ✓ piece of chalk

- ✓ strip of cardboard
- ✓ plastic utensil
- ✓ aluminum nail
- ✓ brass key
- ✓ strip of cork

1. Construct a conductivity tester, as shown in the diagram.
2. Test the conductivity of various materials by laying the objects one at a time across the hooks of the conductivity tester.

Analysis

1. What happens to the conductivity tester if a material is a good conductor?
2. Which materials were good conductors?
3. Which materials were poor conductors?
4. Explain the results in terms of resistance.

Wooden base
Alligator clip
Metal screw hooks
Alligator clip
Wire leads
6 V battery
Flashlight bulb in base holder

Alternative Assessment — ADVANCED

Writing **Maglev Trains** Have students research the Japanese or German Maglev trains and write a report summarizing their results. These Maglev trains levitate on a magnetic field provided by superconducting electromagnets. Have students discuss whether Maglev trains would be practical in the United States. **LS Verbal**

Historical Perspective

Superconductivity Superconductivity was first observed in 1911 when H. K. Onnes cooled mercury to below 3 K, or –270°C, and found that the resistance went almost to zero (3×10^{-6} Ω, which is one ten-millionth its resistance at 0°C). Onnes was able to cool the mercury using liquid helium because he had developed the method to liquefy helium three years earlier.

Insulators have high resistance

Insulators have high resistance to charge movement. So insulating materials are used to prevent electric current from leaking. For example, plastic coating around the copper wire of an electric cord keeps the current from escaping into the floor or your body.

Sometimes it is important to provide a pathway for current to leave a charged object. So a conducting wire is run between the charged object and the ground, thereby *grounding* the object. Grounding is an important part of electrical safety.

Many electrical sockets are wired with three connections: two current-carrying wires and the ground wire. If there is any charge buildup, or if the live wire contacts an appliance, the ground wire conducts the charge to Earth. The excess charge can spread over the planet safely.

internet connect

www.scilinks.org
Topic: Semiconductors and Insulators
SciLinks code: HK4162

SCiLINKS Maintained by the National Science Teachers Association

Close

Quiz ──────────────── **BASIC**

1. What unit measures potential difference? (volts)

2. What is the rate that electric charges move through a conductor called? (current)

3. What is the ratio of voltage to current called? (resistance)

4. What is the resistance equation? ($R = V/I$)

5. What are four categories of materials, based on their resistance? (conductor, superconductor, insulator, semiconductor)

LS Logical

SECTION 2 REVIEW

SUMMARY

▶ A charged object has electrical potential energy due to its position in an electric field.

▶ Potential difference, or voltage, is the difference in electrical potential energy per unit charge.

▶ A voltage causes charges to move, producing a current.

▶ Current is the rate of charge movement.

▶ Electrical resistance can be calculated by dividing voltage by current.

▶ Conductors are materials in which electrons flow easily.

▶ Superconductors have no resistance below their critical temperature.

▶ Insulators are materials with high resistance.

1. Describe the motion of charges through a flashlight, from one terminal of a battery to the other.

2. Identify which of the following could produce current:
  a. a wire connected across a battery's terminals
  b. two electrodes in a solution of positive and negative ions
  c. a salt crystal, whose ions cannot move
  d. a sugar-water mixture

3. Predict which way charges are likely to move between two positions of different electrical potential energy, one high and one low.
  a. from low to high
  b. from high to low
  c. back and forth between high and low

4. Define resistance, and state the quantities needed to calculate an object's resistance.

5. Classify the following materials as conductors or insulators: wood, paper clip, glass, air, paper, plastic, steel nail, rubber.

6. Critical Thinking Recent discoveries have led some scientists to hope that a material will be found that is superconducting at room temperature. Why would such a material be useful?

Math Skills

7. If the current in a certain resistor is 6.2 A and the voltage across the resistor is 110 V, what is its resistance?

8. If the voltage across a flashlight bulb is 3 V and the bulb's resistance is 6 Ω, what is the current through the bulb?

545

Answers to Section 2 Review

1. Electrons move away from the negative terminal of the battery and travel through the filament in the light bulb. They heat the filament, causing it to give off light. The electrons then travel toward the positive terminal of the battery.

2. a, b

3. b

4. Resistance is the opposition posed by a material or a device to the flow of charge. To calculate resistance, you need to know the voltage across the object and the current.

5. wood—insulator
paper clip—conductor

glass—insulator
air—insulator
paper—insulator
plastic—insulator
steel nail—conductor
rubber—insulator

6. Such a material would enable people to use devices that rely on extremely efficient conductors of electricity (such as Maglev trains) without having to cool the conducting material to extremely cold temperatures.

See "Continuation of Answers" at the end of the chapter.

Chapter Resource File

• **Quick Lab Datasheet** How can materials be classified by resistance? GENERAL

• **Concept Review** GENERAL

• **Quiz** BASIC

Videos

CNN. Presents Chemistry

• **Segment 4** Student Semiconductors

• **Critical Thinking** Worksheet 4

See the Science in the News video guide for more details.

Focus

Overview

Before beginning this section, review with your students the Objectives listed in the Student Edition. In this section, students learn what electric circuits are, how they are represented by schematic diagrams, and how they can be classified as series or parallel. They also learn how to use voltage and current to calculate electric power. The section concludes with a discussion of fuses and circuit breakers.

🔔 Bellringer

Use the Bellringer transparency to prepare students for this section.

Motivate

Opening Activity — GENERAL

Ask students to read the first three pages of this section before class. Draw different basic circuit diagrams having 4, 5, or 6 light bulbs on the board. Have groups of students wire these diagrams using a 6 V battery, wire connectors with alligator clips on the ends, and lights cut off of a set of holiday lights.

Next, have students work in pairs and take turns drawing simple schematic diagrams and wiring the circuits. **LS Visual**

Circuits

▶ **KEY TERMS**

electric circuit
schematic diagram
series
parallel
electrical energy
fuse
circuit breaker

PHYSICAL SCIENCE INTERACTIVE TUTOR

Disc Two, Module 16:
Frequency and Wavelength
Use the Interactive Tutor to learn more about these topics.

▶ **electric circuit** a set of electrical components connected such that they provide one or more complete paths for the movement of charges

Figure 14

When this battery is connected to a light bulb, the voltage across the battery generates a current that lights the bulb.

OBJECTIVES

▶ **Use** schematic diagrams to represent circuits.
▶ **Distinguish** between series and parallel circuits.
▶ **Calculate** electric power using voltage and current.
▶ **Explain** how fuses and circuit breakers are used to prevent circuit overload.

Think about how you would get the bulb shown in *Figure 14* to light up. Would the bulb light if the bulb were not fully screwed into the socket? How about if one of the clips were removed from the battery?

What Are Circuits?

When a wire connects the terminals of the battery to the light bulb, as shown in *Figure 14*, charges that built up on one terminal of the battery have a path to follow to reach the opposite charges on the other terminal. Because there are charges moving uniformly, a current exists. This current causes the filament inside the light bulb to give off heat and light.

An electric circuit is a path through which charges can be conducted

Together, the bulb, battery, and wires form an **electric circuit.** In the circuit shown in *Figure 14,* the path from one battery terminal to the other is complete. Because of the voltage of the battery, electrons move through the wires and bulb from the negative terminal to the positive terminal. Then the battery adds energy to the charges as they move within the battery from the positive terminal back to the negative one.

In other words, there is a closed-loop path for electrons to follow. The conducting path produced when the light bulb is connected across the battery's terminals is called a *closed circuit.* Without a complete path, there is no charge flow and therefore no current. This is called an *open circuit.*

The inside of the battery is part of the closed path of current through the circuit. The voltage source, whether a battery or an outlet, is always part of the conducting path of a closed circuit.

MATERIALS SCIENCE
CONNECTION

Light bulbs are slightly damaged each time they are turned on and then turned off. When the filament becomes hot, it expands slightly. When the light bulb is turned off, it cools and contracts slightly. The process of expansion and contraction stresses the surface of the filament, causing it to crack slightly over time. Repeated expansion and contraction causes the filament to crack even more. Eventually, the filament cracks too much and breaks, causing the light bulb to burn out. If the light bulb were kept on all the time, it would last much longer.

Switches interrupt the flow of charges in a circuit

If a device called a *switch* is added to the circuit, as shown in **Figure 15,** you can use the switch to open and close the circuit. You have used a switch many times. The switches on your wall at home are used to turn lights on and off. Although they look different from the switch in **Figure 15,** their function is the same. When you flip a switch at home, you either close or open the circuit to turn a light on or off.

The switch shown in **Figure 15** is called a knife switch. The metal bar is a conductor. When the bar is touching both sides of the switch, as shown in **Figure 15,** the circuit is closed. Electrons can move through the bar to reach the other side of the switch and light the bulb. If the metal bar on the switch is lifted, the circuit is open. Then there is no current, and the bulb does not glow.

Schematic diagrams are used to represent circuits

Suppose you wanted to describe to someone the contents and connections in the photo of the light bulb and battery in **Figure 15.** How might you draw each element? Could you use the same representations of the elements to draw a bigger circuit?

A diagram that depicts the construction of an electrical circuit or apparatus is called a **schematic diagram.** **Figure 16** shows how the battery and light bulb can be drawn as a schematic diagram. The symbols that are used in this figure can be used to describe any other circuit with a battery and one or more bulbs. All electrical devices, from toasters to computers, can be described using schematic diagrams. Because schematic diagrams use standard symbols, they can be read by people all over the world.

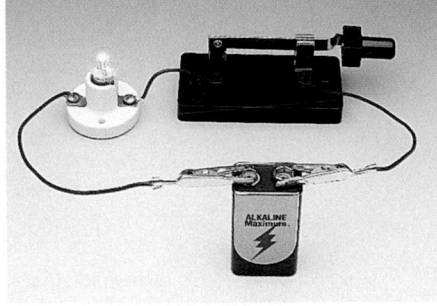

Figure 15

When added to the circuit, a switch can be used to open and close the circuit.

■ **schematic diagram** a graphical representation of a circuit that uses lines to represent wires and different symbols to represent components

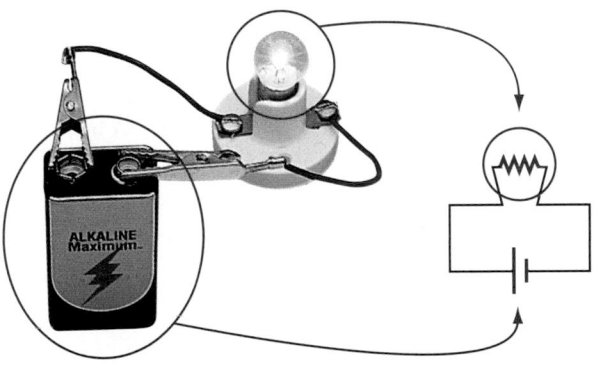

Figure 16

The connections between the light bulb and battery can be represented by symbols. This type of illustration is called a schematic diagram.

internet connect

www.scilinks.org
Topic: Electric Circuits
SciLinks code: HK4042

SCiLINKS. Maintained by the National Science Teachers Association

547

Historical Perspective

Computer Chips The first computer chips, called integrated circuits, were invented by Jack Kilby and Robert Noyce in the late 1950s. Since 1962, the number of electrical components on a chip has nearly doubled every year. Both the size of the components and the distance between them has become much smaller. Discuss with your class whether this trend is likely to continue.

SKILL BUILDER

Interpreting Diagrams Tell students that if you were to draw out the entire schematic diagram for a computer chip using components the same size as in **Table 2,** the circuit schematic would probably cover the entire wall of your classroom.

Teaching Tip

Fluid Model of Electric Current
Many teachers use a fluid model of electric current. In this model, charges moving due to potential difference are analogous to water moving to a level of lower gravitational potential energy. Wires are analogous to horizontal pipes, and resistors are analogous to water wheels, which transform the energy to another form. Batteries and generators act like pumps in that they lift water upward, increasing its potential energy.

As shown in **Table 2,** each element used in a piece of electrical equipment is represented by a symbol that reflects the element's construction or function. For example, the schematic-diagram symbol that represents an open switch resembles the open-knife switch shown in the corresponding photograph. Any circuit can be drawn using a combination of these and other, more complex schematic diagram symbols.

Table 2 Schematic Diagram Symbols

Component	Symbol used in this book	Explanation
Wire or conductor		Wires that connect elements are conductors.
Resistor		Resistors are shown as wires with multiple bends, indicating resistance to a straight path.
Bulb or lamp		The winding of the filament indirectly indicates that the light bulb is a resistor, something that impedes the movement of electrons or the flow of charge.
Battery or other direct current source		The difference in line height indicates a voltage between positive and negative terminals of the battery. The taller line represents the positive terminal of the battery.
Switch Open Closed	Open Closed	The small circles indicate the two places where the switch makes contact with the wires. Most switches work by breaking only one of the contacts, not both.

548

TECHNOLOGY CONNECTION

Mercury is the only metal that is a liquid at room temperature. Some switches make use of mercury in what is called a rocker switch. When the switch is rocked in one direction, the liquid mercury flows downhill and will cover two electrodes. Because mercury is a conductor, it makes an electrical connection with these two electrodes and closes the circuit. When the switch is rocked in the opposite direction, the liquid flows in the opposite direction and moves off the electrodes. This breaks the connection between the electrodes and opens the circuit, stopping charges from flowing.

Series and Parallel Circuits

Section 2 showed that the current in a circuit depends on voltage and the resistance of the device in the circuit. What happens when there are two or more devices connected to a battery?

Series circuits have a single path for current

When appliances or other devices are connected in a **series** circuit, as shown in *Figure 17A,* they form a single pathway for charges to flow. Charges cannot build up or disappear at a point in a circuit. For this reason, the amount of charge that enters one device in a given time interval equals the amount of charge that exits that device in the same amount of time. Because there is only one path for a charge to follow when devices are connected in series, the current in each device is the same. Even though the current in each device is the same, the resistances may be different. Therefore, the voltage across each device in a series circuit can be different.

If one element along the path in a series circuit is removed, the circuit will not work. For example, if either of the light bulbs in *Figure 17A* were removed, the other one would not glow. The series circuit would be open. Several kinds of breaks may interrupt a series circuit. The opening of a switch, the burning out of a light bulb, a cut wire, or any other interruption can cause the whole circuit to fail.

Parallel circuits have multiple paths for current

When devices are connected in **parallel,** rather than in series, the voltage across each device is the same. The current in each device does not have to be the same. Instead, the sum of the currents in all of the devices equals the total current. A simple parallel circuit is shown in *Figure 17B.* The two lights are connected to the same points. The electrons leaving one end of the battery can pass through either bulb before returning to the other terminal. If one bulb has less resistance, more charge moves through that bulb because the bulb offers less opposition to the movement of charges.

Even if one of the bulbs in the circuit shown in *Figure 17B* were removed, charges would still move through the other loop. Thus, a break in any one path in a parallel circuit does not interrupt the flow of electric charge in the other paths.

Figure 17

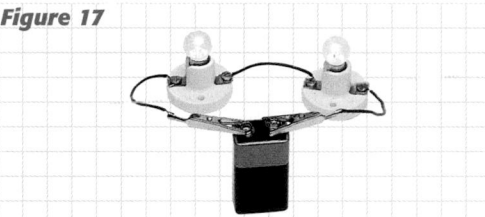

A When bulbs are connected in series, charges must pass through both light bulbs to complete the circuit.

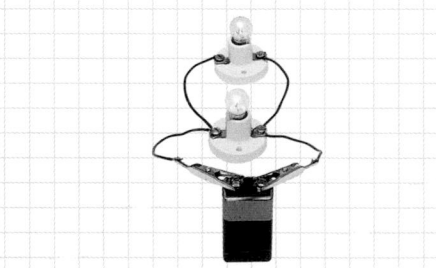

B When devices are connected in parallel, charges have more than one path to follow. The circuit can be complete even if one light bulb burns out.

549

REAL-LIFE CONNECTION

Holiday Lights When holiday lights became popular in the 1970s, they were usually sold in strands of 48 bulbs. Each strand had 48 2.5-volt lights wired into a series circuit. Since the lights were wired in series, a single burned-out bulb opened the circuit and caused the entire strand to go dark.

Today, most holiday lights have a combination of series and parallel circuits. For strands with 100 or 150 lights, groups of 50 lights connected in series are wired together in parallel. But fortunately, a single burned-out bulb is not the hassle it once was. This is because today's bulbs contain an internal shunt wire below the light bulb filament. If a bulb burns out, the shunt wire activates, allowing current to continue flowing through the circuit. However, a loose or removed bulb will cause a 50-bulb portion of the strand to go dark, because this opens the series circuit the bulb belongs to.

Energy Use in Home Appliances

Safety Caution: *Unplug appliances and use caution when handling electrical equipment.*

Determine the power ratings for a few large appliances, such as a refrigerator, an air conditioner, or an electric heater. Bring a household electric-company bill (optional) and a few small appliances to class, such as a toaster, a clock/radio, and a hand-held vacuum cleaner. Look for a label on the back or bottom of each appliance. Record the power rating, which is given in units of watts (W). Use the billing statement to find the cost of energy per kilowatt-hour.

Put students into groups. Ask the students in each group to calculate the cost of running each appliance for 1 hour. Then have them estimate how many hours a day each appliance is used, and calculate the monthly cost of using each appliance based on their estimate. After students have finished, compare the estimates from each group. **LS** Interpersonal

▶ **electrical energy** the energy that is associated with charged particles because of their positions

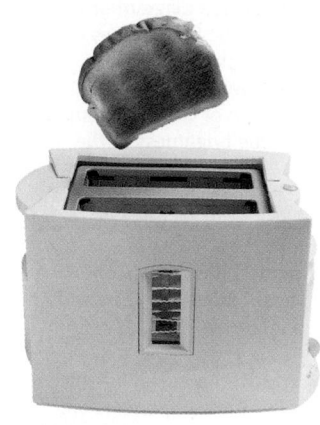

Figure 18
Household appliances use electrical energy to do useful work. Some of that energy is lost as heat.

Vocabulary *Skills Tip*

The SI unit of power, the watt, *was named after the Scottish inventor James Watt in honor of his important work on steam engines.*

Electric Power and Electrical Energy

Many of the devices you use on a daily basis, such as the toaster shown in *Figure 18,* require **electrical energy** to run. The energy for these devices may come from a battery or from a power plant miles away.

Electric power is the rate at which electrical energy is used in a circuit

When a charge moves in a circuit, it loses energy. This energy is transformed into useful work, such as the turning of a motor, and is lost as heat in a circuit. The rate at which electrical work is done is called *electric power*. Electric power is the product of total current (I) in and voltage (V) across a circuit.

> **Electric Power Equation**
> $$power = current \times voltage$$
> $$P = IV$$

The SI unit for power is the watt (W). A watt is equivalent to $1\ A \times 1\ V$. Light bulbs are rated in terms of watts. For example, a typical desk lamp uses a 60 W bulb. A typical hair dryer is rated at about 1800 W.

If you combine the electric power equation above with the equation $V = IR$, the power lost, or *dissipated*, by a resistor can be calculated.

$$P = I^2R = \frac{V^2}{R}$$

> ### Math Skills
>
> **Electric Power** When a hair dryer is plugged into a 120 V outlet, it has a 9.1 A current in it. What is the hair dryer's power rating?
>
> **1** List the given and unknown values.
>   **Given:** *voltage,* V = 120 V
>   *current,* I = 9.1 A
>   **Unknown:** *electric power,* P = ? W
>
> **2** Write the equation for electric power.
> $power = current \times voltage$
> $P = IV$
>
> **3** Insert the known values into the equation, and solve.
> $P = (9.1\ A)(120\ V)$
> $P = 1.1 \times 10^3\ W$

550

Trends in Space Science
Powering the International Space Station
The international space station will be powered by large solar panels. These panels will convert solar energy from the sun into electrical energy to run the station. The solar panels will be able to generate 89 000 W of electrical power for the station. Some of this power will go directly to the station, while some will be directed to the battery. When the space station is in the shadow of Earth, the solar panels will not work. The space station will then rely upon the batteries to continue operation.

Practice

Electric Power

1. An electric space heater requires 29 A of 120 V current to adequately warm a room. What is the power rating of the heater?
2. A graphing calculator uses a 6.0 V battery and draws 2.6×10^{-3} A of current. What is the power rating of the calculator?
3. A color television has a power rating of 320 W. How much current is in the television when it is connected across 120 V?
4. The operating voltage for a light bulb is 120 V. The power rating of the bulb is 75 W. Find the current in the bulb.
5. The current in the heating element of an electric iron is 5.0 A. If the iron dissipates 590 W of power, what is the voltage across it?

Electric companies measure energy consumed in kilowatt-hours

Power companies charge for energy used in the home, not power. The unit of energy that electric companies use to track consumption of energy is the kilowatt-hour (kW•h). One kilowatt-hour is the energy delivered in 1 hour at the rate of 1 kW. In SI units, 1 kW•h = 3.6×10^6 J.

Depending on where you live, the cost of energy ranges from 5 to 20 cents per kilowatt-hour. All homes and businesses have an electric meter, like the one shown in *Figure 19*. Electric meters are used by an electric company to determine how much electrical energy is consumed over a certain time interval.

Fuses and Circuit Breakers

When too many appliances, lights, CD players, televisions, and other devices are connected across a 120 V outlet, the overall resistance of the circuit is lowered. That means the electrical wires carry more than a safe level of current. When this happens, the circuit is said to be *overloaded*. The high currents in overloaded circuits can cause fires.

Worn insulation on wires can also be a fire hazard. If a wire's insulation wears down, two wires may touch, creating an alternative pathway for current. This is called a *short circuit*. The decreased resistance greatly increases the current in the circuit. Short circuits can be very dangerous. Grounding appliances reduces the risk of electric shock from a short circuit.

Practice *HINT*

▶ When a problem requires you to calculate power, you can use the power equation as shown on the previous page.
▶ The electric power equation can also be rearranged to isolate current on the left in the following way:

$$P = IV$$

Divide both sides by V.

$$\frac{P}{V} = \frac{IV}{V}$$

$$I = \frac{P}{V}$$

You will need this version of the equation for Practice Problems 3 and 4.
▶ For Practice Problem 5, you will need to rearrange the equation to isolate voltage on the left.

Figure 19

An electric meter, such as the one shown here, records the amount of energy consumed.

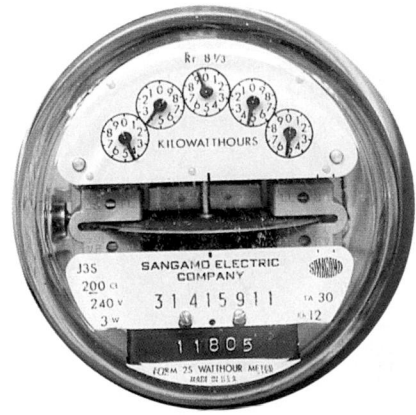

Practice

1. $P = IV = (29 \text{ A})(120 \text{ V}) = 3.5 \times 10^3$ W
2. $P = IV = (2.6 \times 10^{-3} \text{ A})(6.0 \text{ V}) = 1.6 \times 10^{-2}$ W
3. $I = \dfrac{P}{V} = \dfrac{320 \text{ W}}{120 \text{ V}} = 2.7$ A
4. $I = \dfrac{P}{V} = \dfrac{75 \text{ W}}{120 \text{ V}} = 0.62$ A
5. $V = \dfrac{P}{I} = \dfrac{590 \text{ W}}{5.0 \text{ A}} = 120$ V

LS Logical

Math Skills

Additional Examples

Electric Power A certain electrical motor needs a 9.0 V battery to operate and has a power output of 1.6 W. Calculate the current in the motor.

$$I = \frac{P}{V} = \frac{1.6 \text{ W}}{9.0 \text{ V}} = 0.18 \text{ A}$$

A single solar panel for home use puts out 550 W of electrical power. If the electrical current produced by this panel is 4.2 A, calculate the voltage generated by the panel.

$$V = \frac{P}{I} = \frac{550 \text{ W}}{4.2 \text{ A}} = 130 \text{ V}$$

LS Logical

Teaching Tip

Light Bulbs The rate of energy a lamp uses depends on the type of bulb in it. You may want to discuss the advantages and disadvantages of incandescent, fluorescent, and lower-wattage bulbs with students.

551

REAL-LIFE
CONNECTION

Ground Fault Interrupters Ground Fault Circuit Interrupters and Ground Fault Interrupters are mounted in electrical outlets and in certain appliances to prevent electrocution. GFCIs and GFIs are usually installed in the kitchen, bathroom, and outdoor outlets. They function by comparing the current in both wires of a socket. If there is a difference in current, the device opens the circuit within a few milliseconds. If you were to touch a bare wire, the device would detect the change in current and open the circuit; you would get only a small shock. Some circuit breakers are equipped with a GFI.

Chapter Resource File

- **Math Skills** Electric Power [GENERAL]
- **Cross-Disciplinary Worksheet**
  Integrating Health—Recording
  Electricity in the Brain [ADVANCED]

Videos

CNN. Presents Physical Science

- **Segment 20** Eagle Electrocution
- **Critical Thinking** Worksheet 20

See the Science in the News video guide for more details.

Quiz **BASIC**

1. Which type of circuit has a single conducting path? (series)

2. Which type of circuit has multiple conducting paths? (parallel)

3. What is the relationship between the power, current, and voltage in a circuit? ($P = IV$)

4. What are two devices that prevent overloaded circuits? (fuses and circuit breakers)

LS Logical

fuse an electrical device that contains a metal strip that melts when current in the circuit becomes too great

circuit breaker a switch that opens a circuit automatically when the current exceeds a certain value

Fuses melt to prevent circuit overloads

To prevent overloading in circuits, **fuses** are connected in series along the supply path. A fuse is a ribbon of wire with a low melting point. If the current in the line becomes too large, the fuse melts and the circuit is opened.

Fuses "blow out" when the current in the circuit reaches a certain level. For example, a 20 A fuse will melt if the current in the circuit exceeds 20 A. A blown fuse is a sign that a short circuit or a circuit overload may exist somewhere in your home. It is best to find out what made a fuse blow out before replacing it.

Circuit breakers open circuits with high current

Many homes are equipped with **circuit breakers** instead of fuses. A circuit breaker uses a magnet or *bimetallic strip,* a strip with two different metals welded together, that responds to current overload by opening the circuit. The circuit breaker acts as a switch. As with blown fuses, it is wise to determine why the circuit breaker opened the circuit. Unlike fuses, circuit breakers can be reset by turning the switch back on.

SECTION 3 REVIEW

SUMMARY

▶ An electric circuit is a path charges can move along.

▶ In a series circuit, devices are connected along a single pathway. A break anywhere along the path will stop the current.

▶ In a parallel circuit, two or more paths are connected to the voltage source. A break along one path will not stop the movement of charges in the other paths.

▶ Electric power supplied to a circuit or dissipated in a circuit is calculated as the product of the current and voltage.

▶ Circuit breakers and fuses protect circuits from current overloads.

1. Identify the types of elements in the schematic diagram at right and the number of each type.

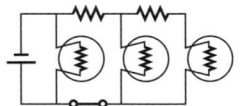

2. Describe the advantage of using a parallel arrangement of decorative lights rather than a series arrangement.

3. Draw a schematic diagram with four lights in parallel.

4. Draw a schematic diagram of a circuit with two light bulbs in which you could turn off either light and still have a complete circuit. (**Hint:** You will need to use two switches.)

5. Contrast how a fuse and a circuit breaker work to prevent overloading in circuits.

6. Critical Thinking Predict whether a fuse will work successfully if it is connected in parallel with the device it is supposed to protect.

Math Skills

7. When a VCR is connected across a 120 V outlet, the VCR has a 9.5 A current in it. What is the power rating of the VCR?

8. A 40 W light bulb and a 75 W light bulb are in parallel across a 120 V outlet. Which bulb has the greater current?

552

Answers to Section 3 Review

1. battery = 1, switch = 1, resistor = 2, light bulb = 3

2. In a parallel arrangement, if one light burns out, the rest will still keep working. In a series arrangement, if one burns out, all of the lights will stop working.

3.

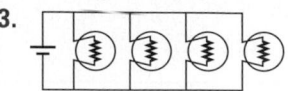

4.

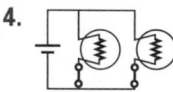

5. Fuses melt when the current exceeds their current rating, while circuit breakers break the connection when the current exceeds their current rating. Circuit breakers can also be reset and reused.

6. If a fuse is attached in parallel, it will not protect the intended device because even if the fuse blows out, there will still be current in the device.

See "Continuation of Answers" at the end of the chapter.

Graphing Skills

Resistance of Metal Wires

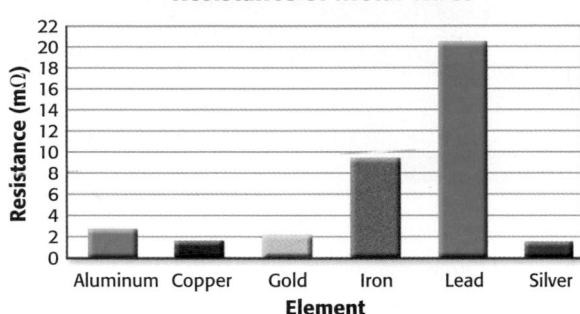

(All wire samples are 0.1 cm² in area and 1.0 m in length.
Resistances are measured at 20.0°C.)

Examine the graph above and answer the following questions (See Appendix A for help interpreting a graph.)

1 What type of graph is this?

2 What variables are shown in this graph?

3 Identify the dependent variable. What is the relationship between the two variables?

4 The information provided by the graph is restricted to certain conditions. What are those conditions? How would the data vary if these conditions changed?

5 Which element listed would make the poorest conductor? Which listed element conducts electricity best?

6 Suppose aluminum, copper, iron, and silver cost $0.50, $0.95, $0.18, and $135 per kilogram, respectively. Based on this information and the graph above, which element would you choose as a conductor of electricity over a long distance? Explain your answer.

7 Use the data in the table below to construct the type of graph best suited for the data. How does carbon's electrical behavior differ from that of most conducting metals?

Temperature (°C)	Resistance of Carbon Wire (1.0 m long; 1 cm² in area) (Ω)
0	0.354
200	0.318
400	0.284
600	0.248
800	0.214

553

Teaching Tip

Metals and Resistance Students may think that metals always conduct electricity better than nonmetals. Explain that, while this is often the case, there are many exceptions. Selenium and silicon are comparable to platinum and iron as conductors, while metals like mercury and bismuth are relatively poor conductors.

Answers

1. a bar graph

2. element and resistance

3. Resistance is the dependent variable because resistance is characteristic to each element, and so varies with the element.

4. All elements are in the form of 1.0 m-long wires with cross-sectional areas of 0.1 cm², and are at a temperature of 20.0°C. Resistance increases with temperature and length of the conducting material, and decreases with increasing area.

5. lead; silver

6. aluminum; Silver has the least resistance, but it is far too expensive. Iron is 2.8 times as cheap as aluminum, but has 3.6 times more resistance. Aluminum's resistance is 1.6 times that of copper, but it is 1.9 times cheaper, making it the best value for large-scale use.

7. Students' should plot a line graph with resistance as the dependent variable (on the vertical axis). Carbon's resistance decreases with increasing temperature, which is opposite the behavior of conducting metals.

Understanding Concepts

1. c
2. b
3. c
4. d
5. c
6. a
7. b
8. c
9. a
10. d
11. d
12. c and d

Using Vocabulary

13. As a positive charge moves toward a negatively charged object, the charge's kinetic energy increases and its electrical potential energy decreases. The electric field due to the negatively charged object does work on the positive charge.

14. Resistance is caused by internal friction, which slows the movement of charges through a material. Resistance is measured by dividing the voltage across a conductor by the current.

15. Due to the potential difference across the wire, there is a current in the wire. The wire is made of a conductor covered with an insulator, which prevents the current from leaking to the surroundings.

Chapter Resource File

- **Chapter Test** GENERAL
- **Test Item Listing for ExamView® Test Generator**

Chapter Highlights

Before you begin, review the summaries of the key ideas of each section, found at the end of each section. The key vocabulary terms are listed on the first page of each section.

UNDERSTANDING CONCEPTS

1. Which of the following particles is electrically neutral?
   a. a proton
   c. a hydrogen atom
   b. an electron
   d. a hydrogen ion

2. Which of the following is not an example of charging by friction?
   a. sliding over a plastic-covered car seat
   b. scraping food from a metal bowl with a metal spoon
   c. walking across a woolen carpet
   d. brushing dry hair with a plastic comb

3. The electric force between two objects depends on all of the following except
   a. the distance between the objects.
   b. the electric charge of the first object.
   c. how the two objects became electrically charged.
   d. the electric charge of the second object.

4. A positive charge placed in the electric field of a second positive charge will
   a. experience a repulsive force.
   b. accelerate away from the second positive charge.
   c. have greater electrical potential energy when near the second charge than when farther away.
   d. All of the above

5. If two charges attract each other,
   a. both charges must be positive.
   b. both charges must be negative.
   c. the charges must be different.
   d. the charges must be the same.

6. In the figure below,
   a. the positive charge is greater than the negative charge.
   b. the negative charge is greater than the positive charge.
   c. both charges are positive.
   d. both charges are negative.

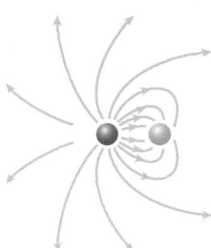

7. The _____ is the change in the electrical potential energy of a charged particle per unit charge.
   a. circuit
   c. induction
   b. voltage
   d. power

8. The type of electrical cell in a common battery is
   a. piezoelectric.
   c. electrochemical.
   b. thermoelectric.
   d. photoelectric.

9. In order to produce a current in a cell, the terminals must
   a. have a potential difference.
   b. be exposed to light.
   c. be in a liquid.
   d. be at two different temperatures.

10. An electric current does not exist in a(n)
    a. closed circuit.
    c. parallel circuit.
    b. series circuit.
    d. open circuit.

11. Which of the following can help prevent a circuit from overloading?
    a. a fuse
    c. a circuit breaker
    b. a switch
    d. both (a) and (c)

16. The conventional current is moving in a clockwise direction.

17. You can ground an electrical appliance by running a conducting wire between the object and the ground.

18. Check students' drawings for accuracy. In a series circuit, every charge must travel through every component in the circuit. In a parallel circuit, there is more than one conducting path for charges to move through. As a charge moves through the circuit, it may travel through any one of the complete paths.

19. The bulbs are probably connected in a series circuit because if one element in a series cir-

cuit is not working, the circuit will be open and will not work.

20. Electric power is the rate at which electrical energy is used in a circuit. Electric meters read the amount of electric energy consumed in kilowatt-hours rather than electric power.

21. A fuse is an electrical device that melts when current in a circuit becomes too great, whereas a circuit breaker is a switch that opens a circuit when the current becomes too great. In designing a circuit for a reading lamp, a circuit breaker would work better than a fuse because a circuit breaker can be reset after it has opened an overloaded circuit.

12. Which of the following schematic diagrams represent circuits that cannot have current in them as drawn?

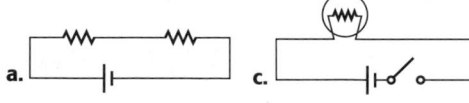

a. c.

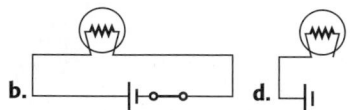

b. d.

USING VOCABULARY

13. Explain the energy changes involved when a positive charge moves because of a nearby, negatively charged object. Use the terms *electrical potential energy, work,* and *kinetic energy* in your answer.

14. What causes *resistance* in an electric circuit? How is resistance measured?

15. How do charges move through an insulated wire connected across a battery? Use the terms *potential difference, current, conductor,* and *insulator* in your answer.

16. If electrons in a circuit are moving in a counterclockwise direction, in what direction is the *conventional current* moving?

17. How would you *ground* an electrical appliance?

18. Contrast the movement of charges in a *series circuit* and in a *parallel circuit*. Use a diagram to aid in your explanation.

19. If a string of lights goes out when one of the bulbs is removed, are the lights probably connected in a *series circuit* or a *parallel circuit*? Explain your answer.

20. What is *electric power*? Do electric meters measure electric power? If not, what do they measure?

21. Explain the difference between a *fuse* and a *circuit breaker*. If you were designing a circuit for a reading lamp, would you include a fuse, a circuit breaker, or neither? Explain your answer.

BUILDING MATH SKILLS

22. Electric Force The electric force is proportional to the product of the charges and inversely proportional to the square of the distance between them. If q_1 and q_2 are the charges on two objects, and d is the distance between them, which of the following represents the electric force, F, between them?

a. $F \propto \dfrac{q_1 q_2}{d}$ **c.** $F \propto \dfrac{d^2}{q_1 q_2}$

b. $F \propto \dfrac{q_1 q_2}{d^2}$ **d.** $F \propto \dfrac{(q_1 q_2)^2}{d}$

23. Resistance A potential difference of 12 V produces a current of 0.30 A in a piece of copper wire. What is the resistance of the copper wire?

24. Resistance What is the voltage across a 75 Ω resistor with 1.6 A of current?

25. Resistance A nickel wire with a resistance of 25 Ω is connected across the terminals of a 3.0 V flashlight battery. How much current is in the wire?

26. Power A portable cassette player uses 3.0 V (two 1.5 V batteries in series) and has 0.33 A of current. What is its power rating?

27. Power Find the current in a 2.4 W flashlight bulb powered by a 1.5 V battery.

28. Power A high-voltage transmission line carries 1.0×10^3 A of current. The power transmitted is 7.0×10^8 W. Find the voltage of the transmission line.

555

Assignment Guide

Section	Questions
1	1–6, 22, 29, 32, 35, 41–43
2	7–9, 13–17, 23–25, 28, 30, 31, 34, 38–40
3	10–12, 18–21, 26, 27, 33, 36, 37

Building Math Skills

22. b

23. $R = \dfrac{V}{I} = \dfrac{12 \text{ V}}{0.30 \text{ A}} = 4.0 \times 10^1 \, \Omega$

24. $V = IR = (1.6 \text{ A})(75 \, \Omega) = 120 \text{ V}$

25. $I = \dfrac{V}{R} = \dfrac{3.0 \text{ V}}{25 \, \Omega} = 0.12 \text{ A}$

26. $P = IV = (0.33 \text{ A})(3.0 \text{ V}) = 0.99 \text{ W}$

27. $I = \dfrac{P}{V} = \dfrac{2.4 \text{ W}}{1.5 \text{ V}} = 1.6 \text{ A}$

28. $V = \dfrac{P}{I} = \dfrac{7.0 \times 10^8 \text{ W}}{1.0 \times 10^3 \text{ A}} = 7.0 \times 10^5 \text{ V}$

Building Graphing Skills

29. The second charge is positive.

Thinking Critically

30. Protons are trapped in the nucleus and cannot escape. As a result, electrons are the only subatomic particles that can be transferred. Metals transfer electrons most easily. In gases and certain chemical solutions, current can be the result of positive charge movement.

31. Your skin's resistance decreases when it gets wet because the ions on your skin dissolve in the water, making your skin a good conductor.

32. Decreases—shocks from static electricity would be worse in dry air. Dryer air is more insulating, allowing more charge to build up on an object.

33. Masses are always positive, while charges can be either positive or negative.

34. four (ignoring the position of the battery and as long as the light bulbs are identical): all in series, all in parallel, two in parallel in series with the third, two in series in parallel with the third

35. $R = \dfrac{V^2}{P} = \dfrac{V^2}{200 \text{ W}}$

$R = \dfrac{V^2}{P} = \dfrac{V^2}{75 \text{ W}}$

The 75 W bulb has a greater resistance.

Alternatively, students may argue that a dimmer bulb (75 W) has a greater resistance than a brighter bulb (200 W) does.

Developing Life/Work Skills

36. The positively charged ball induces a negative charge on the surface of the can closest to it. The ball is then attracted to the can and swings toward t. Upon contact, some of the electrons from the can are transferred to the ball, leaving the can positively charged. The ball is still positively charged, but the amount of charge is less. The ball is unlikely to touch the can after this.

37. Your local fire department will have electrical safety brochures. You can also find information in the public library as well as on the Internet.

38.
$$\frac{\text{average electric energy}}{\text{day}} =$$

$$\frac{471 \text{ kWh}}{33 \text{ days}} = \frac{14 \text{ kWh}}{1 \text{ day}}$$

$$\frac{\text{cost of fuel}}{\text{day}} = \frac{\$6.91}{33 \text{ days}} = \frac{\$0.21}{\text{day}}$$

Integrating Concepts

39. a. current
b. resistance
c. insulators
d. friction
e. contact

---

29. Interpreting Graphs The graph below shows how electrical potential energy changes as the distance between two charges changes. Is the second charge positive or negative?

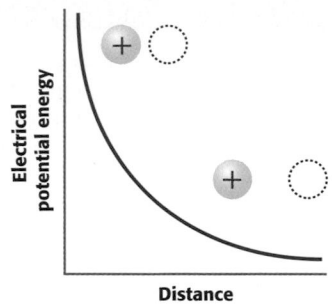

THINKING CRITICALLY

30. Understanding Systems Why is charge usually transferred by electrons? Which materials transfer electrons most easily? In what situations can positive charge move?

31. Applying Knowledge Why does the electrical resistance of your body decrease if your skin gets wet?

32. Problem Solving Humid air is a better electrical conductor because it has a higher water content than dry air. Do you expect shocks from static electricity to be worse as the humidity increases or as it decreases? Explain your answer.

33. Understanding Systems The gravitational force is always attractive, while the electric force is both attractive and repulsive. What accounts for this difference?

34. Designing Systems How many ways can you connect three light bulbs in a circuit with a battery? Draw a schematic diagram of each circuit.

---

35. Applying Knowledge At a given voltage, which light bulb has the greater resistance, a 200 W light bulb or a 75 W light bulb?

DEVELOPING LIFE/WORK SKILLS

36. Interpreting and Communicating A metal can is placed on a wooden table. If a positively charged ball suspended by a thread is brought **COMPUTER SKILL** close to the can, the ball will swing toward the can, make contact, then move away. Explain why this happens, and predict what will happen to the ball next. Use presentation software or a drawing program to make diagrams showing the charges on the ball and on the can at each phase.

37. Working Cooperatively With a small group of classmates, make a chart about electrical safety in the home and outdoors. Use what you have learned in this chapter and information from your local fire department. Include how to prevent electric shock.

38. Allocating Resources Use the electric bill shown below to calculate the average amount of electrical energy used per day and the average cost of fuel to produce the electricity per day.

```
New England Electric    1-888-555-5555
                                    471 KWH
IN 33 DAYS YOU USED       METER # 00790510
READ DATE                            60591
01/21/00                             60120
12/19/99                               471
DIFFERENCE

RATE CALCULATION:
RESIDENTIAL SERVICE RATE, MULTI-FUEL
CUSTOMER CHARGE:                   $ 6.00
ENERGY: 471 KWH AT  $.03550/KWH     16.72
FUEL:   471 KWH AT  $.01467/KWH      6.91
SUBTOTAL ELECTRIC CHARGES         $ 29.63
SALES TAX                             .30
TOTAL COST FOR ELECTRIC SERVICE   $ 29.93
FOR THIS 33 DAY PERIOD, YOUR
AVERAGE DAILY COST FOR ELECTRIC
SERVICE WAS     $.91

DETACH                            DETACH
HERE                              HERE
PLEASE NOTIFY US 10 DAYS BEFORE MOVING
```

---

40. Alessandro Volta (1745–1827) was an Italian physicist who served as chair of physics at the University of Pavia. He invented the battery and was the first to discover and isolate methane gas.

André-Marie Ampère (1775–1836) was a French physicist and mathematics professor at the Ecole Polytechnique, in Paris. He founded and named electrodynamics. He developed techniques for measuring electricity that were later refined to produce the galvanometer. He is responsible for recognizing several of the fundamental principles of electromagnetism. Ampère's law describes the magnetic field produced by electric currents.

George Simon Ohm (1789–1854) was a German physicist and professor of mathematics at the Polytechnic School of Nürnberg. He discovered that the current in a conductor is proportional to the potential difference across the conductor and inversely proportional to its resistance.

41. Refer to **Table 1** for the basic principle for each type of electrical cell.

INTEGRATING CONCEPTS

39. Concept Mapping Copy the unfinished concept map below onto a sheet of paper. Complete the map by writing the correct word or phrase in the lettered boxes.

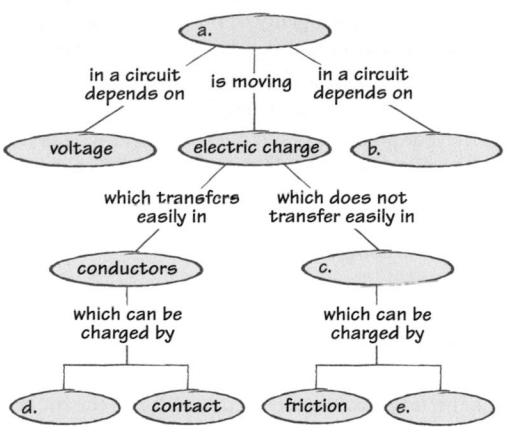

40. Connection to Social Studies The units of measurement you learned about in this chapter were named after three famous scientists—Alessandro Volta, André-Marie Ampère, and Georg Simon Ohm. Create a presentation about one of these scientists. Research the life, work, discoveries, and contributions of the scientist. The presentation can be in the form of a report, poster, short video, or computer presentation.

41. Connection to Engineering Research one of the four types of electrical cells. Write a report describing how it works.

42. Connection to Environmental Science Research how an *electrostatic precipitator* removes smoke and dust particles from the polluting emissions of fuel-burning industries. Find out what industries in your community use a precipitator. What are the advantages and costs of using this device? What alternatives are available? Summarize your findings in a brochure, poster, or chart.

43. Connection to Chemistry Atoms are held together partly because of the electric force between electrons and protons. Chemical bonding is also explained by the attraction between positive and negative particles. Prepare a poster that explains the types of bonding within substances using information from this book and the library. Give examples of common substances that contain these bonds. Describe the relative strengths of the bonds and the types of atoms these bonds form between.

44. Connection to Engineering The common copying machine was designed in the 1960s, after the American inventor Chester Carlson developed a practical device for attracting carbon black to paper by causing a charge imbalance on the paper. Research how this process works and determine why the last copy made when several hundred copies are made can be noticeably less sharp than the first copy. Create a report, poster, or brochure that summarizes your findings.

internet connect

www.scilinks.org
Topic: Electrostatic Precipitators
SciLinks code: HK4045
Maintained by the
National Science Teachers Association

Art Credits: Fig. 2 , Uhl Studios, Inc.; Fig. 3, Kristy Sprott; Fig. 4, Kristy Sprott; Fig. 5-7, Kristy Sprott; Section 1 Review, Kristy Sprott; Fig. 8, Kristy Sprott; Quick Activity, Stephen Durke/Washington Artists; Fig. 14, Boston Graphics; Quick Lab, Stephen Durke/Washington Artists; Chapter Review, "Developing Life/Work Skills"; Chapter Review, "Understanding Concepts", Kristy Sprott; Chapter Review, "Building Graphing Skills", Kristy Sprott.

Photo Credits: Chapter Opener photo of fusion chamber at Sandia National Laboratory by Walter Dickenman, Sandia Lab/David R. Frazier Photolibrary; inset photo of video arcade by DiMaggio/Kalish/Corbis Stock Market; Fig. 1, Michelle Bridwell/HRW; Fig. 2, Peter Van Steen/HRW; Fig. 5, Fundamental Photographs, New York; "Quick Activity," Charles D. Winters/Photo Researchers, Inc.; Fig. 10, Sam Dudgeon/HRW; "Science and the Consumer," Peter Van Steen/HRW; Fig. 13, SuperStock; Figs. 14-16, Sam Dudgeon/HRW; "Schematic Diagram Symbols" Table, HRW Photos; Fig. 17, Sam Dudgeon/HRW; Fig. 18, Peter Van Steen/HRW; Fig. 19, Michelle Bridwell/HRW; "Skills Practice Lab," Peter Van Steen/HRW; "Career Link," Sam Dudgeon/HRW.

557

44. Electrostatic photocopying involves using static electricity to create a positively charged image that corresponds to the dark areas of the original. Negatively charged toner sticks to the positively charged areas, and then heat melts the toner to create a permanent copy. Repeated charging can make the pattern spread, causing blurry copies.

42. Students' answers will vary. Electrostatic precipitators can remove either solid particles or liquid drops that are suspended in a gas. They can handle large volumes of gas, even at high temperatures, and can remove particles in the micrometer range. However, they must be designed and tested for each application, making them more expensive than other methods. Alternatives include scrubbers (gas passes through streams of water), air filters (gas passes through coated fiber filters or dry filters), packed-bed particle separation devices (gas passes through layers of materials such as sand, glass, or steel wool), gravity settling chambers (velocity of gas is decreased so particles settle out), cloth collectors (gas passes through layers of fabric), and inertial separators (direction of gas motion is changed suddenly and particles settle out due to their increased inertia).

43. The three types of bonds are covalent bonds, ionic bonds, and metallic bonds. Many substances have covalent bonds, such as water, sand, and many organic compounds. Covalent bonds form between nonmetal atoms and are usually very strong. Table salt, NaCl, is an example of a compound with ionic bonds. Ionic bonds form between metal and nonmetal atoms and are also very strong. Other examples of compounds with ionic bonds are CaF_2 and KI. All metals have metallic bonds. Examples include copper, aluminum, silver, and gold. Metallic bonds form between metal atoms. They are fairly strong but flexible so that metals can be bent and stretched.

Transparencies

TT Concept Mapping

CONSTRUCTING ELECTRIC CIRCUITS

Teacher's Notes

Time Required 1 lab period

Ratings
EASY 1 2 3 4 HARD

TEACHER PREPARATION	1
STUDENT SETUP	2
CONCEPT LEVEL	3
CLEANUP	2

Skills Acquired

- Experimenting
- Recognizing patterns
- Inferring
- Interpreting
- Measuring
- Organizing and analyzing data
- Predicting

The Scientific Method

In this lab, students will:
- Make Observations
- Analyze the Results
- Draw Conclusions
- Communicate Results

Materials

Most new batteries will have 1.55–1.6 V. Battery holders that have tabs or wires attached to the positive and negative terminals will make it easier for students to connect the batteries in a circuit. To connect components, students may use miniature test leads with spring clips at each end, or wires approximately 10 cm long with small alligator clips soldered to each end. The test leads attached to the multimeter should also have alligator clips at their ends.

Skills Practice Lab

Introduction

How can you show how the current that flows through an electric circuit depends on voltage and resistance?

Objectives

▶ **Construct** parallel and series circuits.

▶ USING SCIENTIFIC METHODS **Predict** voltage and current by using the resistance law.

▶ **Measure** voltage, current, and resistance.

Materials

battery holder
connecting wires (3)
dry-cell battery
masking tape
multimeter
resistors (2)

558

Safety Cautions

Warn students to be careful not to short the batteries. The resistors may become warm during the lab. Check students' circuits before the final connection is made.

Constructing Electric Circuits

▶ Procedure

1. In this laboratory exercise, you will use an instrument called a multimeter to measure voltage, current, and resistance. Your teacher will demonstrate how to use the multimeter to make each type of measurement.

2. As you read the steps listed below, refer to the diagrams for help making the measurements. Write down your predictions and measurements in your lab notebook. **SAFETY CAUTION** Handle the wires only where they are insulated.

Circuits with a Single Resistor

3. Measure the resistance in ohms of one of the resistors. Write the resistance on a small piece of masking tape, and tape it to the resistor. Repeat for the other resistor.

4. Use the resistance equation to predict the current in amps that will be in a circuit consisting of one of the resistors and one battery. (**Hint:** You must rearrange the equation to solve for current.)

5. Test your prediction by building the circuit. Do the same for the other resistor.

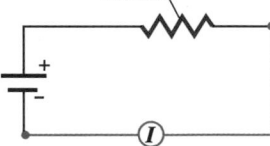

Circuits with Two Resistors in Series

6. Measure the total resistance across both resistors when they are connected in series.

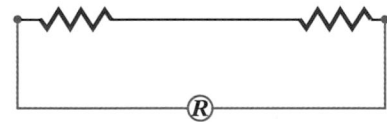

7. Using the total resistance you measured, predict the current that will be in a circuit consisting of one battery and both resistors in series. Test your prediction.

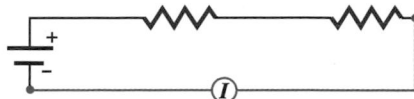

8. Using the current you measured, predict the voltage across each resistor in the circuit you just built. Test your prediction.

Circuits with Two Resistors in Parallel

9. Measure the total resistance across both resistors when they are connected in parallel.

10. Using the total resistance you measured, predict the total current that will be in an entire circuit consisting of one battery and both resistors in parallel. Test your prediction.

11. Predict the current that will be in each resistor individually in the circuit you just built. Test your prediction.

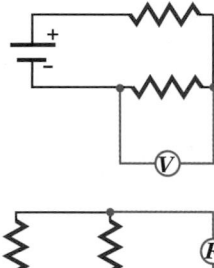

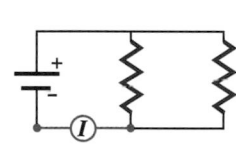

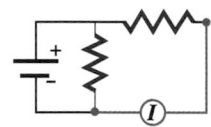

▶ Analysis

1. If you have a circuit consisting of one battery and one resistor, what happens to the current if you double the resistance?

2. What happens to the current if you add a second, identical battery in series with the first battery?

3. What happens to the current if you add a second resistor in parallel with the first resistor?

4. **Reaching Conclusions** Suppose you have a circuit consisting of one battery plus a 10 Ω resistor and a 5 Ω resistor in series. Which resistor will have the greater voltage across it?

5. **Reaching Conclusions** Suppose you have a circuit consisting of one battery plus a 10 Ω resistor and a 5 Ω resistor in parallel. Which resistor will have more current in it?

▶ Conclusions

6. Suppose someone tells you that you can make the battery in a circuit last longer by adding more resistors in parallel. Is that correct? Explain your reasoning.

559

Answers to Procedure

Circuits with a Single Resistor

3. The resistances should be about 100 Ω and 200 Ω.

4. $I = \dfrac{V}{R} = \dfrac{1.5\,\text{V}}{100\,\Omega} = 0.015$ A in the 100 Ω circuit

   $I = \dfrac{V}{R} = \dfrac{1.5\,\text{V}}{200\,\Omega} = 0.0075$ A in the 200 Ω circuit

5. About 0.015 A should be in the 100 Ω circuit and 0.0075 A should be in the 200 Ω circuit.

Circuits with Two Resistors in Series

6. The total resistance should be about 300 Ω.

7. The total current should be about 0.005 A.

8. The voltage should be about 0.5 V across the 100 Ω resistor and 1.0 V across the 200 Ω resistor.

Circuits with Two Resistors in Parallel

9. The total resistance should be about 67 Ω.

10. The total current should be about 0.022 A.

11. About 0.015 A should be in the 100 Ω resistor and 0.0075 A should be in the 200 Ω resistor.

See "Continuation of Answers" at the end of the chapter.

Continuation of Answers

Continuation of Answers from p. 529
Answers to Pre-Reading Questions

electrical energy every time it strikes. The electrical charges in light-ning flow through any metal material until the charges reach the Earth, and can explode brick walls and concrete and start fires. The excess charges along power lines can also overload circuits and cause power outages.

2. Answers will vary. Students may list devices such as refrigerators, hair dryers, televisions, stereos, blenders, dishwashers, etc. and state that some electrical devices use far more electrical energy than others.

Continuation of Answers from p. 545
Section 2 Review

Math Skills

7. $R = \dfrac{V}{I} = \dfrac{110\,V}{6.2\,A} = 18\,\Omega$

8. $I = \dfrac{V}{R} = \dfrac{3\,V}{6\,\Omega} = 0.5\,A$

Continuation of Answers from p. 552
Section 3 Review

Math Skills

7. $P = IV = (9.5\,A)(120\,V) = 1.1 \times 10^3\,W$

8. $I = \dfrac{P}{V} = \dfrac{40\,W}{120\,V} = 0.3\,A$

$I = \dfrac{P}{V} = \dfrac{75\,W}{120\,V} = 0.62\,A$

The 75 W bulb has more current in it.

Continuation of Answers from p. 559
Skills Practice Lab

Answers to Analysis

1. The current will be half as much as the original current.

2. The total current will be double the original current.

3. The total current will increase.

4. The 10 Ω resistor will have the greater voltage across it.

5. The 5 Ω resistor will have more current in it.

Answers to Conclusions

6. No, adding more resistors will increase the amount of current in the circuit, which will drain the battery more quickly.

Career Link

Teaching Tip

Background Dr. Martinez's research work involves exploring the electrical and magnetic properties of individual molecules and how these molecules interact with the electromagnetic fields associated with light.

CareerLink

Physicist

Physicists are scientists who are trying to understand the fundamental rules of the universe. Physicists pursue these questions at universities, private corporations, and government agencies. To learn more about physics as a career, read the interview with physicist Robert Martinez, who works at the University of Texas in Austin, Texas.

Robert Martinez uses a microscope that he has developed to identify single molecules.

"I think of our current project a little bit like the nineteenth century explorers did. They didn't know what they would find on the other side of the ridge or the other side of the ocean, but they had to go look."

 What kinds of problems are you studying?

We're working on a technique that will allow us to study single molecules. We could look at, say, molecules on the surface of a cell. What we're doing is building a kind of microscope for optical spectroscopy, which is a way to find out the colors of molecules. Studying the colors of molecules can tell us what those molecules are made of.

 How does this allow you to identify molecules?

Atoms act as little beams, and the bonds act as little springs. By exciting them with light, we can get them to vibrate and give off different colors of light. It's a little bit like listening to a musical instrument and telling from the overtones that a piano is different from a trumpet or a clarinet.

 What facets of your work do you find most interesting?

The thing that I like about what we're doing is that it's very practical, very hands-on. Also, the opportunity exists to explore whole new areas of physics and chemistry that no one has explored before. What we are doing has the promise of giving us new tools—new "eyes"—to look at important problems.

 What qualities do you think a physicist needs?

You've got to be innately curious about how the world works, and you've got to think it's understandable and you are capable of understanding it. You've got to be courageous. You've got to be good at math.

560

Historical Perspective

Spectroscopy While Dr. Martinez's work with molecular spectroscopy is cutting-edge research, other types of spectroscopy have been used for many years.

Isaac Newton first studied the spectrum of sunlight in 1672. In 1802, W. H. Wollaston, and in 1814, Joseph Fraunhofer independently demonstrated that by passing sunlight through a slit first, a spectrum could be made that contained a series of dark lines.

In 1859, G. R. Kirchoff explained that the dark lines Fraunhofer had found in the solar spectrum years before were due to elements in the sun's atmosphere absorbing wavelengths of light.

In the 1860s, working with R. Bunsen, Kirchoff demonstrated that spectral analysis of light from sparks or flames could be used to identify specific elements. In 1861, Bunsen and Kirchoff co-discovered the elements cesium and rubidium as the sources for previously undescribed lines in some spectra of alkali metals.

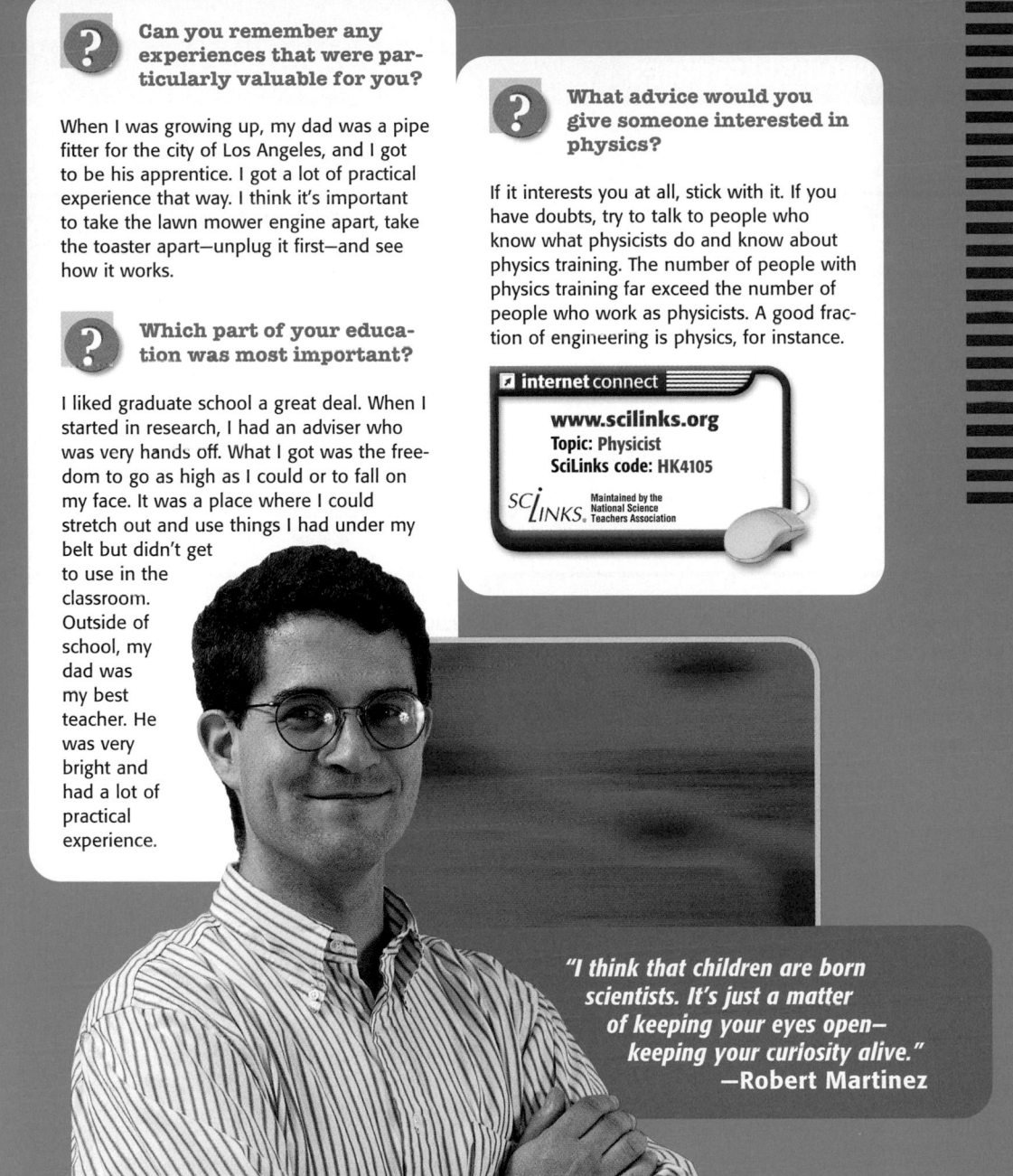

Can you remember any experiences that were particularly valuable for you?

When I was growing up, my dad was a pipe fitter for the city of Los Angeles, and I got to be his apprentice. I got a lot of practical experience that way. I think it's important to take the lawn mower engine apart, take the toaster apart—unplug it first—and see how it works.

Which part of your education was most important?

I liked graduate school a great deal. When I started in research, I had an adviser who was very hands off. What I got was the freedom to go as high as I could or to fall on my face. It was a place where I could stretch out and use things I had under my belt but didn't get to use in the classroom. Outside of school, my dad was my best teacher. He was very bright and had a lot of practical experience.

What advice would you give someone interested in physics?

If it interests you at all, stick with it. If you have doubts, try to talk to people who know what physicists do and know about physics training. The number of people with physics training far exceed the number of people who work as physicists. A good fraction of engineering is physics, for instance.

internet connect

www.scilinks.org
Topic: Physicist
SciLinks code: HK4105

SCI**LINKS** Maintained by the National Science Teachers Association

"I think that children are born scientists. It's just a matter of keeping your eyes open— keeping your curiosity alive."
—Robert Martinez

561

PACING	CLASSROOM RESOURCES	LABS, ACTIVITIES, AND DEMONSTRATIONS
BLOCK 1 · 45 min pp. 562–563 **Chapter Opener**		SE **Activity 1**, p. 563 SE **Activity 2**, p. 563
BLOCKS 2 & 3 · 90 min pp. 564–569 **Section 1** Magnets and Magnetic Fields	CRF **Lesson Plan*** TT **Bellringer*** TM **Magnetic Field***	TE **Opening Demonstration**, p. 564 `GENERAL` SE **Quick Activity** Test Your Knowledge of Magnetic Poles, p. 566 `GENERAL` CRF **Datasheets for In-Text Labs** Test Your Knowledge of Magnetic Poles* `GENERAL` TE **Demonstration** Shapes of Magnetic Fields, p. 567 `GENERAL` SE **Quick Activity** Magnetic Field of a File Cabinet, p. 568 `GENERAL` CRF **Datasheets for In-Text Labs** Magnetic Field of a File Cabinet* CRF **Observation Lab** Constructing and Using a Compass* ◆ `BASIC`
BLOCKS 4 & 5 · 90 min pp. 570–575 **Section 2** Magnetism from Electric Currents	CRF **Lesson Plan*** TT **Bellringer*** TM **Right-Hand Rule*** TM **Solenoid*** TM **Galvanometer*** TT **Electric Motor***	TE **Opening Demonstration**, p. 570 SE **Quick Lab** How can you make an electromagnet?, p. 572 `GENERAL` CRF **Datasheets for In-Text Labs** How can you make an electromagnet?* `GENERAL` TE **Demonstration** Galvanometers, p. 573 TE **Demonstration** Magnetic Force on a Current Loop, p. 574 SE **Skills Practice Lab** Making a Better Electromagnet, pp. 588–589 ◆ `GENERAL` CRF **Datasheets for SE Skills Practice Lab** Making a Better Electromagnet* `GENERAL` CRF **CBL™ Probeware Lab** Testing Magnets from an Electric Motor* ◆ `ADVANCED`
BLOCKS 6 & 7 · 90 min pp. 576–582 **Section 3** Electric Currents from Magnetism	CRF **Lesson Plan*** TT **Bellringer*** TT **AC Generator*** TT **Induced Current*** TT **Transformers*** TT **Concept Mapping***	SE **Quick Lab** Can you demonstrate electromagnetic induction?, p. 577 `GENERAL` CRF **Datasheets for In-Text Labs** Can you demonstrate electromagnetic induction?* `GENERAL` TE **Demonstration** Mechanical Energy to Electrical Energy, p. 578 TE **Demonstration** Alternating Current, p. 579 TE **Demonstration** Measuring Energy, p. 580 TE **Demonstration** DC Transformer, p. 581

BLOCKS 8 & 9 · 90 min

Chapter Review and Assessment Resources

- SE **Chapter Review**, pp. 584–587
- CRF **Chapter Tests*** `GENERAL`
- OSP **Test Generator**
- CRF **Standardized Test Practice with Guided Reading Development***
- CRF **Test Item Listing for ExamView® Test Generator***
- OSP **Scoring Rubrics and Classroom Management Checklists**

Online Resources

Visit the HRW Web site for a variety of free resources related to the text. Just type in the keyword **HK4 MAG**.

Holt Online Learning

Holt Science Spectrum: Physical Science: Online Edition

Students can access interactive problem solving help and active visual concept development with the *Holt Science Spectrum: Physical Science* Online Edition available at **www.hrw.com**.

student CNN News

cnnstudentnews.com

Find the latest news, lesson plans, and activities related to important scientific events.

KEY

TE	Teacher Edition	**CRF**	Chapter Resource File	***** Also on One-Stop Planner
SE	Student Edition	**TT**	Teaching Transparency	**◆** Requires Advance Prep
OSP	One-Stop Planner	**TM**	Transparency Master	

Compression guide:
To shorten your instruction because of time limitations, omit blocks 1, 8, and 9.

PROBLEM SOLVING AND PRACTICE	SECTION REVIEW AND ASSESSMENT	STANDARDS CORRELATION
	CRF Pretest* [GENERAL]	
TE Inclusion Strategies, p. 567 **CRF** Cross-Disciplinary Worksheet Connection to Social Studies—The Natural Force and Laws of Compasses* [GENERAL]	**TE** Quiz, p. 569 [BASIC] **SE** Section 1 Review, p. 569 **CRF** Concept Review* [GENERAL] **CRF** Quiz* [BASIC]	PS 5b UCP 1–3, 5 SAI 1, 2 ST 1, 2 HNS 1–3 SPSP 5
CRF Cross-Disciplinary Worksheet Integrating Chemistry—Molecular Magnetism* [GENERAL]	**TE** Quiz, p. 575 [BASIC] **SE** Section 2 Review, p. 575 **CRF** Concept Review* [GENERAL] **CRF** Quiz* [BASIC]	PS 4e, 5b UCP 1–3, 5 SAI 1, 2 ST 1, 2 HNS 1–3 SPSP 5
TE Inclusion Strategies, p. 581 **CRF** Cross-Disciplinary Worksheet Integrating Technology—Magnetic Resonance Imaging* [GENERAL] **SE** Study Skills Interpreting Scientific Illustrations, p. 583	**TE** Quiz, p. 582 [BASIC] **SE** Section 3 Review, p. 582 **CRF** Concept Review* [GENERAL] **CRF** Quiz* [BASIC]	PS 4e, 5b, 6b UCP 1–3, 5 SAI 1, 2 ST 1, 2 HNS 1–3 SPSP 5

www.scilinks.org

Topic: Maglev Trains
*Sci*Links code: HK4082

Topic: Properties of Magnets
*Sci*Links code: HK4111

Topic: Earth's Magnetic Field
*Sci*Links code: HK4036

Topic: Electromagnetism
*Sci*Links code: HK4044

Topic: Generators
*Sci*Links code: HK4063

Topic: Magnetic Fields of Power Lines
*Sci*Links code: HK4083

Technology Resources

 One-Stop Planner
All of your printable resources and the Test Generator are on this convenient CD-ROM.

 Physical Science Interactive Tutor CD-Rom
Disc Two, Module 17
Magnetic Field of a Wire

 CNN Science in the NEWS
each video segment is accompanied by a Critical Thinking Worksheet *

Segment 21
Magnetic Attractions

Overview

This chapter begins with a discussion of magnets, magnetic force, magnetic fields, compasses, and Earth's magnetic field. It then explores how electric currents can produce magnetic fields, and discusses galvanometers, electric motors, and stereo speakers. Electromagnetic induction and Faraday's law are explained so that students can learn how generators and transformers work.

Assessing Prior Knowledge

Be sure students understand the following concepts:

- charged particles
- electromagnetic waves
- electrical currents
- electric fields
- circuits

MISCONCEPTION ///ALERT\\\

Science education research has identified the following misconceptions about magnetism.

- Students believe that the size of a magnet determines its strength.
- Students believe that anything metal, or anything silver, will be attracted to a magnet.
- Students guess that magnetism will pass through paper but not through wood, books, and similar materials.
- Students believe that only magnets produce magnetic fields, and that magnetic fields only exist outside the magnet in two-dimensional lines.

Magnetism

Chapter Preview

1 Magnets and Magnetic Fields
  Magnets
  Magnetic Fields

2 Magnetism from Electric Currents
  Magnetism from Electric Currents
  Electromagnetic Devices

3 Electric Currents from Magnetism
  Electromagnetic Induction
    and Faraday's Law
  Transformers

562

Standards Correlations

National Science Education Standards

The following descriptions summarize the National Science Standards that specifically relate to this chapter. For the full text of the standards, see the National Science Education Standards at the front of the book.

Section 1 Magnets and Magnetic Fields
Physical Science PS 5b
Unifying Concepts and Processes UCP 1–3, 5
Science as Inquiry SAI 1, 2
Science and Technology ST 1, 2
History and Nature of Science HNS 1–3
Science in Personal and Social Perspectives SPSP 5

Section 2 Magnetism from Electric Currents
Physical Science PS 4e, 5b
Unifying Concepts and Processes UCP 1–3, 5
Science as Inquiry SAI 1, 2
Science and Technology ST 1, 2
History and Nature of Science HNS 1–3
Science in Personal and Social Perspectives SPSP 5

Section 3 Electric Currents from Magnetism
Physical Science PS 4e, 5b, 6b
Unifying Concepts and Processes UCP 1–3, 5
Science as Inquiry SAI 1, 2
Science and Technology ST 1, 2
History and Nature of Science HNS 1–3
Science in Personal and Social Perspectives SPSP 5

Focus ACTIVITY

Background Just as a magnet exerts a force on the iron filings in the small photo at left, a modern type of train called a *Maglev* train is levitated and accelerated by magnets. A Maglev train uses magnetic forces to lift the train off the track, reducing the friction and allowing the train to move faster. These trains, in fact, have reached speeds of more than 500 km/h (310 mi/h).

In addition to enabling the train to reach high speeds, the lack of contact with the track provides a smoother, quieter ride. With improvements in the technologies that produce the magnetic forces used in levitation, these trains may become more common in high-speed transportation.

Activity 1 You can see levitation in action with two ring-shaped magnets and a pencil. Drop one of the ring magnets over the tip of the pencil so that it rests on your hand. Now drop the other magnet over the tip of the pencil. If the magnets are oriented correctly, the second ring will levitate above the other. If the magnets attract, remove the second ring, flip it over, and again drop it over the tip of the pencil.

The magnetic force exerted on the levitating magnet is equal to the magnet's weight. Use a scale to find the magnet's mass; then use the weight equation $w = mg$ to calculate the magnetic force necessary to levitate this magnet.

Activity 2 Place two bar magnets flat on a table with the N poles about 2 cm apart. Cover the magnets with a sheet of plain paper. Sprinkle iron filings on the paper. Tap the paper gently until the filings line up. Make a sketch showing the orientation of the filings. Where does the magnetic force seem to be the strongest?

internet connect

www.scilinks.org
Topic: **Maglev Trains** SciLinks code: **HK4082**

SCiLINKS. Maintained by the
National Science Teachers Association

Pre-Reading Questions

1. Magnets can exert a force on objects without touching the objects. What other forces behave the same way? Do all these forces attract the same kinds of objects?

2. Do all parts of a bar magnet attract a paperclip equally? Why or why not?

563

The iron filings in the photo above are moved into a pattern by the magnetic force of the magnet. Maglev trains, like the one shown above, levitate above their tracks using magnetic force.

Focus ACTIVITY

Background Modern Maglev trains are levitated above guideways by on-board superconducting electromagnets and coils on the guideway. The electromagnets move at a high speed very close to the coils, inducing an electric current within the coils, which then act as electromagnets temporarily. The interaction between the coils and the on-board magnets levitates the train, keeps it centered on the guideway, and propels it forward.

Activity 1 The top magnet will levitate if the magnets' north poles are next to each other or if the south poles are next to each other. If the north and south poles are next to each other, the magnets will be attracted, and the top magnet will not levitate.

Activity 2 When you try this experiment, it is advisable to put the magnet into a plastic sandwich bag. Have students try many different arrangements of magnets to see the resulting magnetic fields.

Answers to Pre-Reading Questions

1. Gravity and the force between electric charges also act a distance, however these forces all attract different kinds of objects.

2. No, the ends or poles of a bar magnet will attract a paperclip more strongly than other parts of the magnet will, because the poles are where the magnetic force is strongest.

Chapter Resource File

- Pretest **GENERAL**
- Teaching Transparency Preview
- Answer Keys

Chapter 17 • Magnetism   **563**

Focus

Overview

Before beginning this section, review with your students the Objectives listed in the Student Edition. In the first part of this section, students learn what magnets are. Next they study magnetic fields and learn how compasses work. The section concludes with a discussion of Earth's magnetic field.

Bellringer

Use the Bellringer transparency to prepare students for this section.

Motivate

Opening Demonstration —— GENERAL

For this demonstration, you will need a bar magnet, an iron nail, and paper clips. Magnetize the nail with the bar magnet by rubbing the magnet on the end of the nail. Rub the nail in only one direction, and do not change direction. If you rub back and forth in two directions, the nail will not become magnetized. Have a student rub the magnet on the nail 10 times and then determine how many paper clips the nail will lift.

Next, rub the nail 10 more times and see how many paper clips the nail will lift. If you continue to rub the nail, the magnetic force will increase. At some point, additional rubbing with the bar magnet will not increase the strength of the magnetized nail.

LS Kinesthetic

Magnets and Magnetic Fields

■ **KEY TERMS**
magnetic pole
magnetic field

internet connect
www.scilinks.org
Topic: Properties
of Magnets
SciLinks code: HK4111
SCILINKS. Maintained by the
National Science
Teachers Association

OBJECTIVES

▸ **Recognize** that like magnetic poles repel and unlike poles attract.
▸ **Describe** the magnetic field around a permanent magnet.
▸ **Explain** how compasses work.
▸ **Describe** the orientation of Earth's magnetic field.

You may think of magnets as devices used to attach papers or photos to a refrigerator door. But magnets are involved in many different devices, such as alarm systems like the one shown in *Figure 1.* This type of alarm system uses the simple magnetic attraction between a piece of iron and a magnet to alert home-owners that a window or door has been opened.

When the window is closed, as shown in *Figure 1A,* the iron switch is attracted to the magnet. This attraction keeps the electrical contacts in the switch closed, which completes the circuit. Thus, a current is in the system when it is turned on. When the window slides open, as shown in *Figure 1B,* the magnet is no longer close enough to the iron to attract it strongly. The spring pulls the switch open, which breaks the circuit, and sounds the alarm.

Figure 1

A When the window is closed, the magnet holds the switch closed so that current is in the circuit.

B If the window is opened, the switch will open, and the alarm will sound.

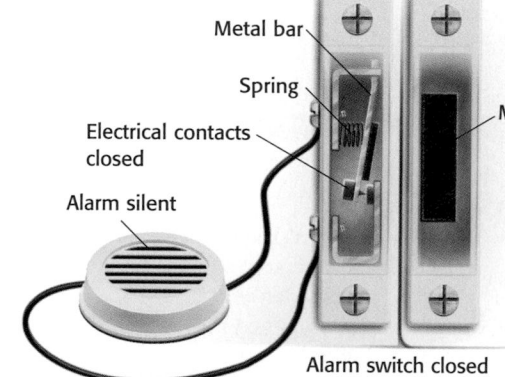

Metal bar
Spring
Electrical contacts closed
Magnet
Alarm silent
Alarm switch closed

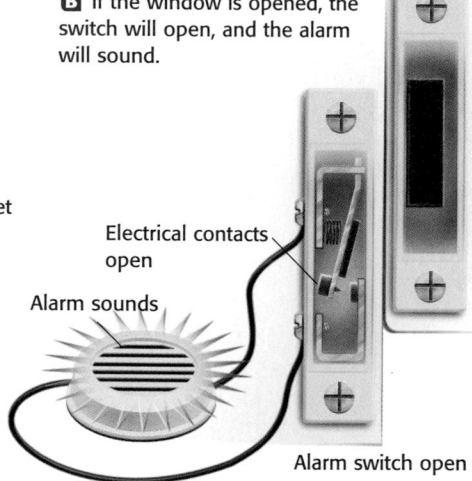

Electrical contacts open
Alarm sounds
Alarm switch open

564

MATERIAL SCIENCE
CONNECTION

Bar magnets do not start out magnetic. They must be magnetized during the manufacturing process. Magnets are made by melting steel and pouring the molten steel into a mold. As the molten steel is cooling, it is exposed to a strong magnetic field. This magnetic field makes all the magnetic fields of the iron atoms inside the molten steel reorient themselves to point in the same direction.

As a result, the magnetic field from each iron atom combines with the magnetic fields

of all the other iron atoms to generate an overall magnetic field. Once the molten iron has hardened, the external magnetic field is turned off.

One way to demagnetize a bar magnet is to hit it hard with a hammer many times. Striking with a hammer can change the orientation of the individual atoms' magnetic fields. This has the effect of reducing the strength of the bar magnet.

Magnets

Magnets got their name from the region of Magnesia, which is now part of modern-day Greece. The first naturally occurring magnetic rocks, called *lodestones*, were found in this region almost 3000 years ago. A lodestone, shown in *Figure 2*, is composed of an iron-based material called *magnetite*.

Some materials can be made into permanent magnets

Some substances, such as lodestones, are magnetic all the time. These types of magnets are called *permanent magnets*. You can change any piece of iron, such as a nail, into a permanent magnet by stroking it several times with a permanent magnet. A slower method is to place the piece of iron near a strong magnet. Eventually the iron will become magnetic and will remain magnetic even when the original magnet is removed.

Although a magnetized piece of iron is called a "permanent" magnet, its magnetism can be weakened or even removed. Possible ways to do this are to heat or hammer the piece of iron. Even when this is done, some materials retain their magnetism longer than others.

Scientists classify materials as either magnetically *hard* or magnetically *soft*. Iron is a soft magnetic material. Although a piece of iron is easily magnetized, it also tends to lose its magnetic properties easily. In contrast, hard magnetic materials, such as cobalt and nickel, are more difficult to magnetize. Once magnetized, however, they don't lose their magnetism easily.

Magnets exert magnetic forces on each other

As shown in *Figure 3*, a magnet lowered into a bucket of nails will often pick up several nails. As soon as a nail touches the magnet, the nail acts as a magnet and attracts other nails. More than one nail is lifted because each nail in the chain becomes temporarily magnetized and exerts a *magnetic force* on the nail below it. This ability disappears when the chain of nails is no longer touching the magnet, although the nails may become slightly magnetized after they have been in contact with the permanent magnet. In contrast, the aluminum bucket is not attracted to the magnet at all.

There is a limit to how long the chain of nails can be. The length of the chain depends on the ability of the nails to become magnetized and the strength of the magnet. The farther from the magnet each nail is, the smaller its magnetic force. Eventually, the magnetic force between the two lowest nails is not strong enough to overcome the force of gravity, and the bottom nail falls.

Figure 2

A naturally occurring magnetic rock, called a lodestone, will attract a variety of iron objects.

Figure 3

When a magnet is lowered into a bucket of nails, it can pick up a chain of nails. Each nail is temporarily magnetized by the nail above it.

565

Did You Know?

Magnetite Magnetite, Fe_3O_4, is a lustrous black magnetic mineral that occurs in crystals with a cubic structure. It is one of the important ores of iron and is a common constituent of igneous and metamorphic rocks. It is found in Norway, Sweden, the Urals, and various parts of the United States. One variety of magnetite—lodestone—has been noted for its natural magnetism since antiquity.

MISCONCEPTION ///ALERT\\\

Magnets Some students believe that anything metal, or anything silver, will be attracted to a magnet. Show students some examples to dispel this misconception. Use a bar magnet and a variety of metal objects, including some that are attracted to the magnet and others that are not attracted.

Materials (per group):
- bar magnets (2)
- tape
- string
- support stand

Teacher Notes: The string should be tied so that the magnet will be balanced and hang parallel to the ground. A small piece of tape will keep the string from moving while the magnet is suspended. Be sure you perform this experiment away from metal objects like metal cabinets and faucets which may attract the magnet.

Results: Like poles repel, and opposite poles attract. If left alone, the hanging magnet will align itself with Earth's magnetic field, with its N pole pointing north. Using this fact, you can determine the unknown poles of both magnets. **LS Kinesthetic**

Teaching Tip
Magnetic Fields and Field Lines
Students often think the poles of a magnet are named for the North and South Poles of Earth. Point out that the N pole of a magnet is actually the North-seeking pole. Also, many students believe that magnetic field lines show the direction in which the magnet will push another magnet. But the magnetic field lines indicate how a second magnet will align itself in the magnetic field of the first magnet.

▶ **magnetic pole** one of two points, such as the ends of a magnet, that have opposing magnetic qualities

VOCABULARY *Skills Tip*

The word pole *is used in physics for two related opposites that are separated by some distance along an axis. The word* polar, *used in chemistry, has the same origin.*

Like poles repel, and opposite poles attract
As you know, the closer two like electrical charges are brought together, the more they repel each other. The closer two opposite charges are brought together, the more they attract each other. A similar situation exists for **magnetic poles.**

Magnets have a pair of poles, a north pole and a south pole. The poles of magnets exert a force on one another. Two like poles, such as two south poles, repel each other. Two unlike poles, however, attract each other. Thus, the north pole of one magnet will attract the south pole of another magnet. Also, the north pole of one magnet repels the north pole of another magnet.

It is impossible to isolate a south magnetic pole from a north magnetic pole. If a magnet is cut, each piece will still have two poles. No matter how small the pieces of a magnet are, each piece still has both a north and a south pole.

Magnetic Fields
Try moving the south pole of one magnet toward the south pole of another that is free to move. As you do this, the magnet you are not touching will move away. A force is being exerted on the second magnet even though it never touches the magnet in your hand. The force is acting at a distance. This may seem unusual, but you are already familiar with other forces that act at a distance. Gravitational forces and the force between electric charges also act at a distance.

Quick ACTIVITY

Test Your Knowledge of Magnetic Poles
1. Tape the ends of a bar magnet so that its pole markings are covered.
2. Tie a piece of string to the center of the magnet and suspend it from a support stand, as shown in the figure at right.
3. Use another bar magnet to determine which pole of the hanging magnet is the north pole and which is the south pole. What happens when you bring one pole of your magnet near each end of the hanging magnet?
4. Now try to identify the poles of the hanging magnet using the other pole of your magnet.
5. After you have decided the identity of each pole, remove the tape to check. Can you determine which are north poles and which are south poles if you cover the poles on both magnets?

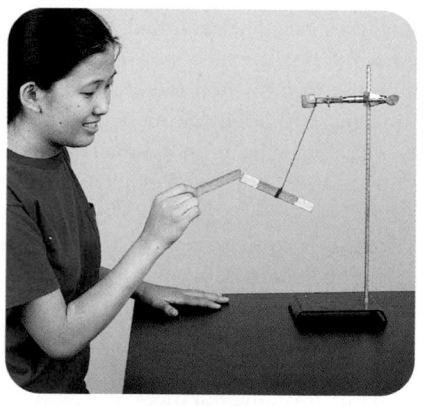

Did You Know ?
Ceramic Magnets Some permanent magnets are made out of ceramic materials, such as barium or strontium ferrite. These magnets can be extremely strong, are resistant to demagnetization, and are inexpensive, but they are also extremely brittle. Ceramic magnets must be machined when they are unmagnetized. Then they can be magnetized in the desired direction. These magnets are used for a wide range of applications from motors and cellular phones to toys.

Alternative Assessment — ADVANCED
Comparing Forces Some students might think that the forces exerted by gravity, by electric charges, and by magnets are identical. Have students write a few paragraphs that compare and contrast these three forces. Be sure students include both similarities and differences in their paragraphs. For example, one similarity is that all three forces act at a distance. One difference is that electric and magnetic forces can be attractive or repulsive, while the force of gravity is always an attractive force. **LS Verbal**

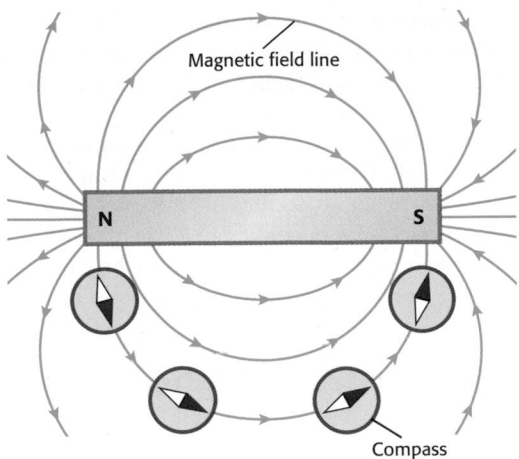

Magnetic field line

Compass

Figure 4

Figure 4
The magnetic field of a bar magnet can be traced with a compass. Note that the north pole of each compass points in the direction of the field lines from the magnet's north pole to its south pole.

Magnets are sources of magnetic fields

Magnetic force is a field force. When magnets repel or attract each other, it is due to the interaction of their **magnetic fields.**

All magnets produce a magnetic field. Some magnetic fields are stronger than others. The strength of the magnetic field depends on the material from which the magnet is made and the degree to which it has been magnetized.

Electric field lines are used to represent an electric field. Similarly, magnetic field lines are used to represent the magnetic field of a bar magnet, as shown in *Figure 4.* These field lines all form closed loops. *Figure 4* shows only the field near the magnet. The field also exists within the magnet and farther away from the magnet. The magnetic field, however, gets weaker with distance from the magnet. As with electric field lines, magnetic field lines that are close together indicate a strong magnetic field. Field lines that are farther apart indicate a weaker field. Knowing this, you can tell from *Figure 4* that a magnet's field is strongest near its poles.

Compasses can track magnetic fields

One way to analyze a magnetic field's direction is to use a compass, as shown in *Figure 4.* A compass is a magnet suspended on top of a pivot so that the magnet can rotate freely. You can make a simple compass by hanging a bar magnet from a support with a string tied to the magnet's midpoint.

magnetic field a region where a magnetic force can be detected

SKILL BUILDER

Interpreting Visuals The orientation of Earth's magnetic field, shown in **Figure 5,** can be confusing for many students. Emphasize that the N and S poles of a magnet are named for the geographical poles they point to.

Quick ACTIVITY

Materials:
• compass
• metal file cabinet

Teacher's Notes: If you are performing this lab in class, test the file cabinet before the lab. Possible substitutes include iron flagpoles and iron fence posts, such as those around tennis courts. Remind students that Earth's magnetic field has both a vertical and horizontal component. The file cabinet is magnetized by the vertical component. Have them try to find an object that has been magnetized by the horizontal component of Earth's magnetic field.

Results: The file cabinet can become magnetized because of prolonged exposure to Earth's magnetic field. However, the amount of magnetization that occurs is dependent upon the kind of metal from which the cabinet is made. If the cabinet is aluminum, for example, it will not be magnetic at all. If the cabinet is steel, it is probably slightly magnetic and will cause the compass needle to be deflected. **LS Kinesthetic**

Quick ACTIVITY

Magnetic Field of a File Cabinet

1. Stand in front of a metal file cabinet, and hold a compass face up and parallel to the ground.
2. Move the compass from the top of the file cabinet to the bottom, and check to see if the direction of the compass needle changes. If the compass needle changes direction, the file cabinet is magnetized.
3. Can you explain what might have caused the file cabinet to become magnetized? Remember that Earth's magnetic field not only points horizontal to Earth but also points up and down.

A compass aligns with Earth's magnetic field just as iron filings align with the field of a bar magnet. The compass points in a direction that lies along, or is tangent to, the magnetic field line at that point.

The first compasses were made using lodestones. A lodestone was placed on a small plank of wood and floated in calm water. Sailors then watched as the wood turned and pointed toward the north star. In this way, sailors could gauge their direction even during the day, when stars were not visible. Later, sailors found that a steel or iron needle rubbed with lodestone acted in the same manner.

Earth's magnetic field is like that of a bar magnet

A compass can be used to determine direction because Earth acts like a giant bar magnet. As shown in *Figure 5,* Earth's magnetic field has both direction and strength. If you were to move northward along Earth's surface with a compass whose needle could point up and down, the needle of the compass would slowly tilt forward. At a point in northeastern Canada, the needle would point straight down. This point is one of Earth's magnetic poles. There is an opposite magnetic pole in Antarctica.

The source of Earth's magnetism is a topic of scientific debate. Although Earth's core is made mostly of iron, the iron in the core is too hot to retain any magnetic properties. Instead, many researchers believe that the circulation of ions or electrons in the liquid layer of Earth's core may be the source of the magnetism. Others believe it is due to a combination of several factors.

Earth's magnetic field has changed direction throughout geologic time. Evidence of more than 20 reversals in the last 5 million years is preserved in the magnetization of sea-floor rocks.

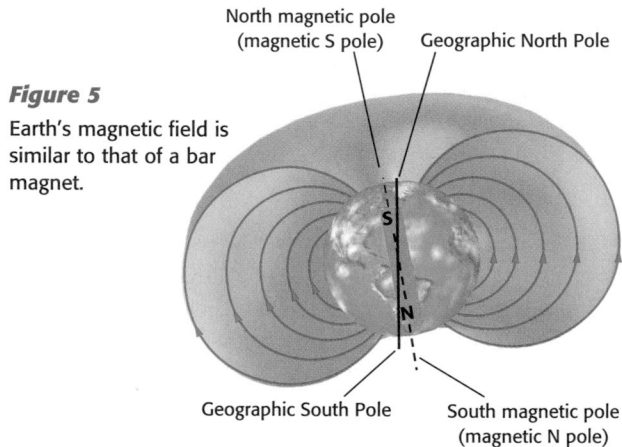

Figure 5
Earth's magnetic field is similar to that of a bar magnet.

North magnetic pole (magnetic S pole)   Geographic North Pole

Geographic South Pole   South magnetic pole (magnetic N pole)

SPACE SCIENCE CONNECTION

Space Science The sun also has a magnetic field and ejects charged particles into space. This stream of particles is called the *solar wind*. Earth's magnetic field deflects most of these particles so that the charged particles can only enter Earth's atmosphere near the magnetic poles. These particles collide with atoms in the atmosphere, resulting in the emission of visible light. This light causes the sky at the North and South Poles to glow with different colors. These colors are called the Northern and Southern Lights, or the aurora borealis and aurora australis.

INCLUSION Strategies

• Developmentally Delayed   • Learning Disabled
• English Language Learners   • Attention Deficit Disorder

To help students understand the difference between materials that are attracted to magnets and those that are not, give students a magnet and ask them to discover materials in the classroom that are attracted to magnets. Have students create two lists of objects, one list of objects that are attracted to magnets and the other list of objects that are not. Discuss the materials of which the objects in each category are made. **LS Kinesthetic**

Earth's magnetic poles are not the same as its geographic poles

One of the interesting things about Earth's magnetic poles, as shown in **Figure 5,** is that they are not in the same place as the geographic poles. Another important feature of Earth's magnetic poles is the orientation of their magnetic field. Earth's magnetic field points from the geographic South Pole to the geographic North Pole. This orientation is similar to a bar magnet, like the one shown in **Figure 5.** The magnetic pole in Antarctica is actually a magnetic N pole, and the magnetic pole in northern Canada is actually a magnetic S pole.

For historical reasons, the poles of magnets are named for the geographic pole they point toward. Thus, the end of the magnet labeled *N* is a "north-seeking" pole, and the end of the magnet labeled *S* is a "south-seeking" pole.

Close

Quiz — BASIC

Choose the word that best completes each sentence.

1. All magnets have one/two pole(s). (two)
2. Opposite magnetic poles attract/repel. (attract)
3. The magnetic field of a magnet is strongest/weakest near the magnet's poles. (strongest)
4. Earth's magnetic poles are/ are not at the same location as Earth's geographic poles. (are not)

LS Logical

SECTION 1 REVIEW

SUMMARY

▶ All magnets have two poles that cannot be isolated.

▶ Like poles repel each other, and unlike poles attract each other.

▶ The magnetic force is the force due to interacting magnetic fields.

▶ The magnetic field of a magnet is strongest near its poles and gets weaker with distance.

▶ The direction of a magnetic field can be traced using a compass.

▶ Earth's magnetic field has both north and south poles.

▶ Earth's magnetic poles are not at the same location as the geographic poles. The magnetic N pole is in Antarctica, and the magnetic S pole is in northern Canada.

1. **Determine** whether the magnets will attract or repel each other in each of the following cases.

   a. S N | N S
   b. S N | S N
   c. N S / S N

2. **State** how many poles each piece of a magnet will have when you break it in half.

3. **Identify** which of the compass-needle orientations in the figure below correctly describe the direction of the bar magnet's magnetic field.

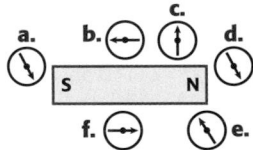

4. **Describe** the direction a compass needle would point if you were in Australia.

5. **Critical Thinking** The north pole of a magnet is attracted to the geographic North Pole, yet like poles repel. Explain why.

569

EARTH SCIENCE CONNECTION

Evidence for reversals of Earth's magnetic field is provided by basalt (an iron-containing rock) on the ocean floor. Volcanic activity along mid-ocean ridges produces new ocean floor. As the lava cools and solidifies, it retains a record of Earth's magnetic field direction at that time. Radioactive dating of these rocks provides evidence for periodic reversals of Earth's magnetic field.

Answers to Section 1 Review

1. a. repel
   b. attract
   c. attract
2. Each piece of the magnet will have two poles. It is impossible for a magnet to have only one pole.
3. a, b
4. The compass needle would point toward the geographic North Pole.
5. The north pole of a magnet is attracted to Earth's magnetic south pole, which is near the geographic North Pole.

SECTION
2

Focus

Overview

Before beginning this section, review with your students the Objectives listed in the Student Edition. In this section, students learn how electric currents produce magnetic fields and what causes magnetism. Then they apply these concepts to galvanometers, electric motors, and stereo speakers.

🔔 Bellringer

Use the Bellringer transparency to prepare students for this section.

Motivate

Opening Demonstration

For this demonstration you will need a 6 V battery, a spool of insulated wire, a compass, and 2 ring stands. First, make a large coil of wire, perhaps 1 ft in diameter. Use many loops around the coil so that the magnetic field is strong. An easy way to make the coil is to wrap the wire around a soccer ball or basketball. Use a ring stand to hold the loop off the table. The loop must be oriented perpendicular to the ground in order to use the compass. Attach the wire to the battery so that a current is going through the loop, creating a magnetic field at its center. Place the compass at the center of the loop and turn the circuit on and off. The compass needle should move when the circuit is activated. The compass does not need to be at the exact center; however, that is where the magnetic field is the strongest. Use the compass to trace the magnetic field around the loop.

Magnetism from Electric Currents

▶ **KEY TERMS**
solenoid
electromagnet
galvanometer
electric motor

OBJECTIVES

▶ **Describe** how magnetism is produced by electric currents.
▶ **Interpret** the magnetic field of a solenoid and of an electromagnet.
▶ **Explain** the magnetic properties of a material in terms of magnetic domains.
▶ **Explain** how galvanometers and electric motors work.

During the eighteenth century, people noticed that a bolt of lightning could momentarily change the direction of a compass needle. They also noticed that iron pans sometimes became magnetized during lightning storms. These observations suggested a relationship between electricity and magnetism, but it wasn't until 1820 that the relationship was understood.

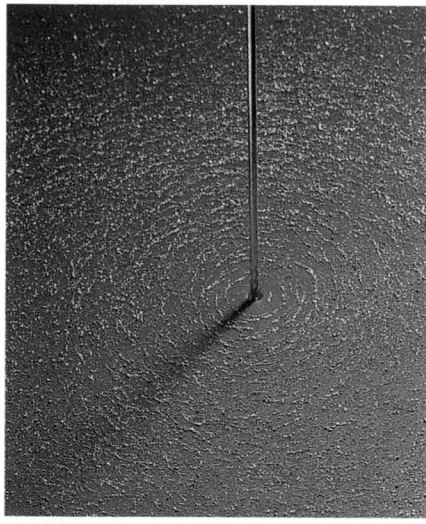

Figure 6

The iron filings show that the magnetic field of a current-carrying wire forms concentric circles around the wire.

570

Magnetism from Electric Currents

In 1820, a Danish science teacher named Hans Christian Oersted first experimented with the effects of an electric current on the needle of a compass. He found that magnetism is produced by moving electric charges.

Electric currents produce magnetic fields

The experiment shown in *Figure 6* uses iron filings to demonstrate that a current-carrying wire creates a magnetic field. Because of this field, the iron filings make a distinct pattern around the wire.

If pieces of iron are free to move, they will align with a magnetic field. The pattern of the filings in *Figure 6* suggests that the magnetic field around a current-carrying wire forms concentric circles around the wire. If you were to bring a compass close to a current-carrying wire, as Oersted did, you would find that the needle points in a direction tangent to the circles of iron filings. When the current stops flowing, the magnetic field disappears.

REAL-LIFE CONNECTION

Wire Taps Telephone wire taps are possible because the current that is carrying the conversation produces a magnetic field. Because magnetic fields travel through most materials, coils are placed around the wire to measure the magnetic field. The pattern of changes in the magnetic field is converted to current (a current is induced by the changing magnetic field), which is converted to sound by a speaker, reproducing the telephone conversation. Fiber-optic cables cannot be tapped in this way because they do not produce an external magnetic field.

MISCONCEPTION //// ALERT \\\\

Producing Magnetic Fields Some students believe that only magnets produce magnetic fields. Emphasize that magnetic fields can be produced by magnets or by moving electric charges, and tell students that there is no qualitative difference between the magnetic fields produced by either one.

Use the right-hand rule to find the direction of the magnetic field produced by a current

Is the direction of the wire's magnetic field clockwise or counterclockwise? Repeated measurements have shown an easy way to predict the direction of the field; this method is called the *right-hand rule*. The right-hand rule is explained below.

> **If you imagine holding the wire in your right hand with your thumb pointing in the direction of the positive current, the direction your fingers would curl is in the direction of the magnetic field.**

Figure 7 illustrates the right-hand rule. Pretend the wire is grasped with the right hand with the thumb pointing upward, in the direction of the current. When the hand holds the wire, the fingers encircle the wire with the fingertips pointing in the direction of the magnetic field, counterclockwise in this case. If the current were toward the bottom of the page, the thumb would point downward, and the magnetic field would point clockwise. *Remember—never grasp or touch an uninsulated wire. You could be electrocuted.*

The magnetic field of a coil of wire resembles that of a bar magnet

As Oersted demonstrated, the magnetic field of a current-carrying wire exerts a force on a compass needle. This force causes the needle to turn in the direction of the wire's magnetic field. However, this force is very weak. One way to increase the force is to increase the current in the wire, but large currents can be fire hazards. A safer way to create a strong magnetic field that will provide a greater force is to wrap the wire into a coil, as shown in **Figure 8.** This device is called a **solenoid.**

In a solenoid, the magnetic field of each loop of wire adds to the strength of the magnetic field of the loop next to it. The result is a strong magnetic field similar to the magnetic field produced by a bar magnet. A solenoid even has a north and south pole, just like a magnet.

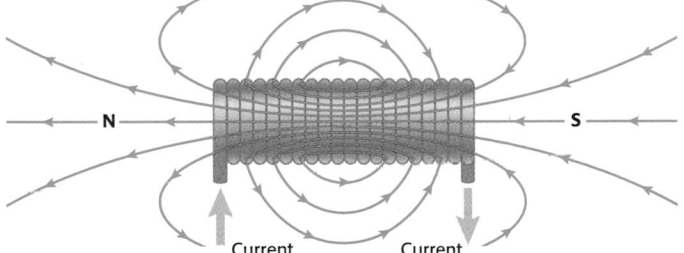

Current   Current

PHYSICAL SCIENCE INTERACTIVE TUTOR

Disc Two, Module 17:
Magnetic Field of a Wire
Use the Interactive Tutor to learn more about this topic.

Figure 7

Use the right-hand rule to find the direction of the magnetic field around a current-carrying wire.

Current

Magnetic Field

■ **solenoid** a coil of wire with an electric current in it

Figure 8

The magnetic field of a solenoid resembles the magnetic field of a bar magnet.

Trends in Medicine

Magnetic Resonance Imaging Prior to 1973, the most common way that doctors looked inside the human body was with X rays. In 1973, Mansfield and Grannell invented magnetic resonance imaging (MRI). When an MRI is done, the patient is placed inside a large coil of wire. A current is sent through the coil, exposing the patient to a strong magnetic field. This magnetic field, used in conjunction with low-frequency radio waves, can cause hydrogen atoms to emit low-frequency electromagnetic waves. Molecules of water contain hydrogen, and water is prevalent in soft tissue. The waves emitted by the hydrogen atoms are used to do the imaging.

Teach

SKILL BUILDER — BASIC

Reading Skills Have students read this section and then organize the ideas presented in the form of a concept map or other reading organizer. The organization can be in terms of the order presented in the chapter or in terms of detailed characteristics listed for each idea.
LS Verbal

SKILL BUILDER — BASIC

Interpreting Visuals Have students practice using the right-hand rule by confirming that the direction of the magnetic field shown in **Figure 8** is correctly drawn for the solenoid. Ask them which end of the solenoid would be the N pole if the current were reversed. (the right side) **LS** Visual

Chapter Resource File

• **Lesson Plan**

Transparencies

TT Bellringer
TM Right-Hand Rule
TM Solenoid

Quick Lab

The electromagnet will work better if the wire is wound very tightly together.

Safety Caution: The wires will get hot. Students should use heat-resistant gloves to handle the electromagnet. Be sure students disconnect the battery after making their observations.

Answers to Analysis

1. an electromagnet—the coil of wire is the solenoid, and the nail is the magnetic core
2. The direction of the compass needle should reverse when the current is in the opposite direction.
3. Hit the nail on its side sharply with a hammer.

LS Kinesthetic

Teaching Tip ——— BASIC

Magnetic Domains Have students use the discussion of magnetic domains on these two pages to explain what happened in the Opening Demonstration at the beginning of the chapter. (When the magnet was rubbed on the iron nail, the domains in the iron became aligned, and the iron became magnetic. The overall magnetic field increased as more domains became aligned.) Ask students what would happen to the nail if it were dropped after being rubbed by the magnet. (Some domains would lose their alignment, and the nail's magnetic field would weaken.) **LS Verbal**

▶ **electromagnet** a coil that has a soft iron core and that acts as a magnet when an electric current is in the coil

internet connect
www.scilinks.org
Topic: **Electromagnetism**
SciLinks code: **HK4044**

SciLINKS Maintained by the National Science Teachers Association

---

The strength of the magnetic field of a solenoid depends on the number of loops of wire and the amount of current in the wire. In particular, more loops or more current can create a stronger magnetic field.

The strength of a solenoid's magnetic field can be increased by inserting a rod made of iron (or some other potentially magnetic metal) through the center of the coils. The resulting device is called an **electromagnet.** The magnetic field of the solenoid causes the rod to become a magnet as well. The magnetic field of the rod then adds to the coil's field, creating a stronger magnet than the solenoid alone.

Magnetism can be caused by moving charges

The movement of charges causes all magnetism. The magnetic field of a bar magnet is an example.

But what charges are moving in a bar magnet? Negatively charged electrons moving around the nuclei of all atoms make magnetic fields. Atomic nuclei also have magnetic fields because protons move within the nucleus. Each electron has a property called *electron spin*, which also produces a tiny magnetic field.

In most cases the various sources of magnetic fields in an element cancel out and leave the atom essentially nonmagnetic. However, in some materials such as iron, nickel, and cobalt, not all of the fields cancel. Thus, each atom in those metals has its own magnetic field.

Quick Lab

How can you make an electromagnet?

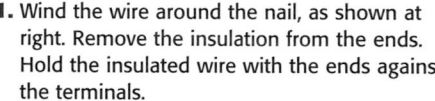 **Materials**
✓ D-cell   ✓ 1 m length of insulated wire
✓ compass   ✓ large iron or steel nail

1. Wind the wire around the nail, as shown at right. Remove the insulation from the ends. Hold the insulated wire with the ends against the terminals.
2. Move the compass toward the nail to determine whether the nail is magnetized. If it is magnetized, the compass needle will spin to align with the nail's magnetic field.
3. Switch connections to the cell so the current is reversed. Again bring the compass toward the same part of the nail.

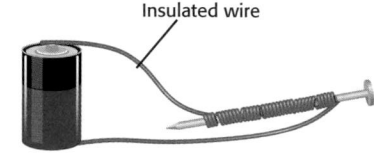
Insulated wire

Analysis

1. What type of device have you produced? Explain your answer.
2. What happens to the direction of the compass needle after you reverse the direction of the current? Why does this happen?
3. After detaching the coil from the cell, what can you do to make the nail nonmagnetic?

572

REAL-LIFE CONNECTION

Electromagnets Electromagnets are used in many different industries. Perhaps their best-known use is for lifting junk cars at a junkyard. These electromagnets are suspended at the end of a crane. Large currents are used to power the electromagnet so that its magnetic field is strong enough to lift a car off the ground.

These heavy-duty electromagnets are also used to lift machine parts that are being cast in foundries. After parts have been cast, they must cool in their molds. Once they solidify but are not completely cool, they must be removed from the mold for additional processing. Electromagnets are used to lift the cooling parts and continue the manufacturing process because the parts are still too hot to touch.

Figure 9

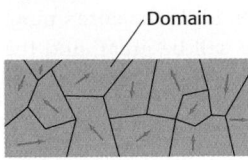

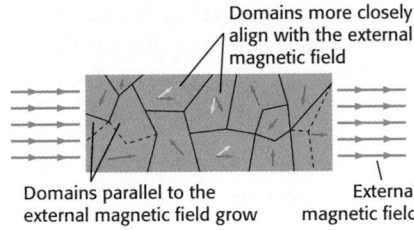

Domains more closely align with the external magnetic field

Domain

Domains parallel to the external magnetic field grow

External magnetic field

A When a potentially magnetic substance is unmagnetized, its domains are randomly oriented.

B When in an external magnetic field, the direction of the domains becomes more uniform, and the material becomes magnetized.

Just as a compass needle rotates to align with a magnetic field, magnetic atoms rotate to align with the magnetic fields of nearby atoms. The result is small regions within the material called *domains*. The magnetic fields of atoms in a domain point in the same direction.

As shown in *Figure 9A,* the magnetic fields of the domains inside an unmagnetized piece of iron are not aligned. When a strong magnet is brought nearby, the domains line up more closely with the magnetic field, as shown in *Figure 9B.* The result of this reorientation is an overall magnetization of the iron.

▶ **galvanometer** an instrument that detects, measures, and determines the direction of a small electric current

Electromagnetic Devices

Many modern devices make use of the magnetic field produced by coils of current-carrying wire. Devices as different as hair dryers and stereo speakers function because of the magnetic field produced by these current-carrying conductors.

Galvanometers detect current

Galvanometers are devices used to measure current in *ammeters* and voltage in *voltmeters*. The basic construction of a galvanometer is shown in *Figure 10.* In all cases, a galvanometer detects current, or the movement of charges in a circuit.

A galvanometer consists of a coil of insulated wire wrapped around an iron core that can rotate between the poles of a permanent magnet. When the galvanometer is attached to a circuit, a current will exist in the coil of wire. The coil and iron core will act as an electromagnet and produce a magnetic field. This magnetic field will interact with the magnetic field of the surrounding permanent magnet. The resulting forces will turn the core.

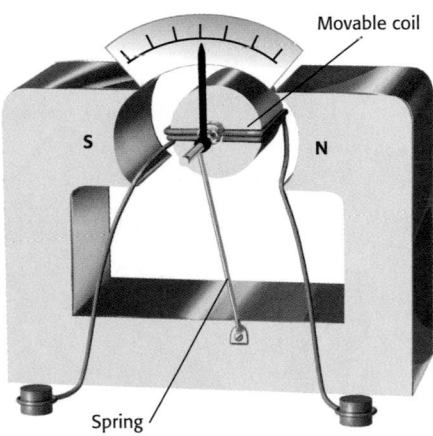

Movable coil

S   N

Spring

Figure 10

When there is current in the coil of a galvanometer, magnetic repulsion between the coil and the magnet causes the coil to twist.

573

Did You Know?

Particle Accelerators CERN, the Center for European Nuclear Research, studies subatomic particles. This is done by accelerating charged particles to speeds near the speed of light and then smashing them into other particles. Large wire coils are used to generate strong magnetic fields that help accelerate and steer the particles.

REAL-LIFE
CONNECTION

Electromagnetic Keys Many hotels use electromagnetic card keys instead of traditional keys. Within the door lock mechanism are small solenoids with many loops. These solenoids read the magnetic signature on a key to determine whether the lock should unlock. The same kind of readers are also used in ATMs and credit card reading machines.

Interpreting Diagrams Have students trace the path of electric charge through the motor in **Figure 11.** In which direction does the magnetic field of the loop point? (toward the upper right of the page) **LS** Visual

Demonstration —— **GENERAL**

Magnetic Force on a Current Loop

(Time: About 20 minutes)

Materials:

• strong horseshoe magnet
• wire
• ring stands and supports (2)
• variable DC power supply

Step 1 Lay the horseshoe magnet on its side so that one pole is above the other. Connect the wire to the power supply, and support the wire so that it passes through the center of the poles of the horseshoe magnet.

Step 2 Turn on the power supply, and gradually increase the current in the wire until the wire is forced to one side or the other. Ask students what will happen if the direction of the current is reversed. (The wire will move in the opposite direction.)

Step 3 Turn off the power supply, detach the wires, and attach them to the opposite terminals. Test the students' hypotheses. Have students sketch the magnetic fields of the magnet and the wire for both cases.
LS Visual

electric motor a device that converts electrical energy into mechanical energy

As stated earlier in this section, the greater the current in the electromagnet, the stronger its magnetic field. If the core's magnetic field is strong, the force on the core will be great, and the core will rotate through a large angle. A needle extends upward from the core to a scale. As the core rotates, the needle moves across the scale. The greater the movement across the scale, the larger the current.

Electric motors convert electrical energy to mechanical energy

Electric motors are another type of device that uses magnetic force to cause motion. *Figure 11* is an illustration of a simple direct current, or DC, motor.

As shown by the arrow in *Figure 11,* the coil of wire in a motor turns when a current is in the wire. But unlike the coil in a galvanometer, the coil in an electric motor keeps spinning. If the coil is attached to a shaft, it can do work. The end of the shaft is connected to some other device, such as a propeller or wheel. This design is often used in mechanical toys.

A device called a *commutator* is used to make the current change direction every time the flat coil makes a half revolution. This commutator is two half rings of metal. Devices called *brushes* connect the wires to the commutator. Because of the slits in the commutator, charges must move through the coil of wire to reach the opposite half of the ring. As the coil and commutator spin, the current in the coil changes direction every time the brushes come in contact with a different side of the ring.

So the magnetic field of the coil changes direction as the coil spins. In this way, the coil is repelled by both the north and south poles of the magnet surrounding it. Because the current keeps reversing, the loop rotates in one direction. If the current did not keep changing direction, the loop would just bounce back and forth in the magnetic field until the force of friction caused it to come to rest.

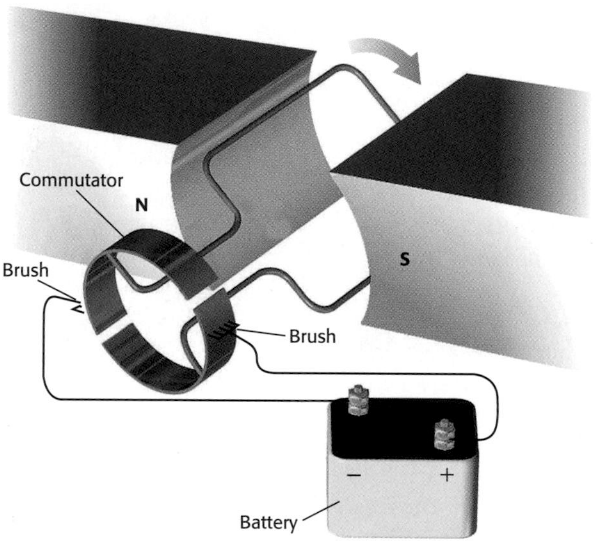

Commutator

N

Brush

S

Brush

Battery

574

Figure 11

In an electric motor, the current in the coil produces a magnetic field that interacts with the magnetic field of the surrounding magnet, causing the coil to turn.

Alternative Assessment —— **ADVANCED**

Electric Motors Obtain some small electric motors. It doesn't matter if these motors work or not. Remove the outer casing with a pair of pliers. Once this casing is removed, the inner components of the motor will be visible to students. Let them investigate the inner components. Students will see the wire coils, which have hundreds of turns each, as well as the permanent magnets in the casing. They will also be able to see the brushes and commutator. Ask students to draw the inner components and to write a paragraph describing what happens during one rotation of the coil.
LS Kinesthetic

Stereo speakers use magnetic force to produce sound

Motion caused by magnetic force can even be used to produce sound waves. This is how most stereo speakers work. The speaker shown in *Figure 12* consists of a permanent magnet and a coil of wire attached to a flexible paper cone. When a current is in the coil, a magnetic field is produced. This field interacts with the field of the permanent magnet, causing the coil and cone to move in one direction. When the current reverses direction, the magnetic force on the coil also reverses direction. As a result, the cone accelerates in the opposite direction.

This alternating force on the speaker cone makes it vibrate. Varying the magnitude of the current changes how much the cone vibrates. These vibrations produce sound waves. In this way, an electric signal is converted to a sound wave.

Figure 12

In a speaker, when the direction of the current in the coil of wire changes, the paper cone attached to the coil moves, producing sound waves.

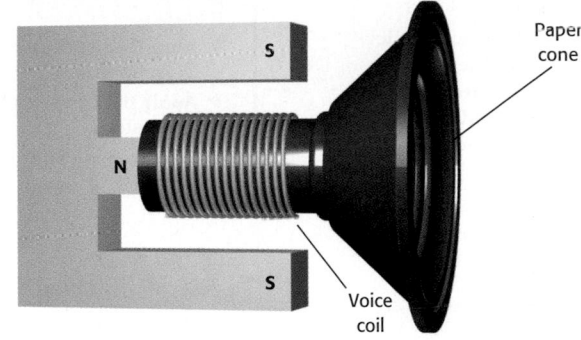

Paper cone

Voice coil

Interpreting Visuals In **Figure 12**, the voice coil makes the speaker move forward and backward in a line. When a speaker is turned up too loud, it tries to move in another direction, side to side. The speaker generates distortion in the sound, and the paper cone begins to tear, causing the speaker to buzz.

Close

Quiz — GENERAL

1. What is produced around a current-carrying wire? (a magnetic field)

2. What is a solenoid? (a coiled wire)

3. What is an electromagnet? (a current-carrying solenoid with an iron core)

4. What type of electromagnetic device uses the magnetic field to measure current? (galvanometer)

5. What type of energy conversion occurs in an electric motor? (An electric motor converts electrical energy to mechanical energy.)

LS Logical

SECTION 2 REVIEW

SUMMARY

▶ A magnetic field is produced around a current-carrying wire.

▶ A current-carrying solenoid has a magnetic field similar to that of a bar magnet.

▶ An electromagnet consists of a current-carrying solenoid with an iron core.

▶ A domain is a group of atoms whose magnetic fields are aligned.

▶ Galvanometers measure the current in a circuit using the magnetic field produced by a current in a coil.

▶ Electric motors convert electrical energy to mechanical energy.

1. **Describe** the shape of the magnetic field produced by a straight current-carrying wire.

2. **Determine** the direction in which a compass needle will point when held above a wire with positive charges moving west. (**Hint:** Use the right-hand rule.)

3. **Identify** which of the following would have the strongest magnetic field. Assume the current in each is the same.
   a. a straight wire
   b. an electromagnet with 30 coils
   c. a solenoid with 20 coils
   d. a solenoid with 30 coils

4. **Explain** why a very strong magnet attracts both poles of a weak magnet. Use the concept of magnetic domains in your explanation.

5. **Predict** whether a solenoid suspended by a string could be used as a compass.

6. **Critical Thinking** A friend claims to have built a motor by attaching a shaft to the core of a galvanometer and removing the spring. Can this motor rotate through a full rotation? Explain your answer.

575

Answers to Section 2 Review

1. The magnetic field generated by a straight current-carrying wire is in the shape of concentric rings, with the wire at the center.

2. Using the right-hand rule, point your thumb west. Your fingers curl in the direction of the magnetic field. Above the wire, your fingers would point north.

3. b

4. A very strong magnet is able to change the orientation of the domains in a weak magnet. As a result, the strong magnet is able to attract both poles of a weak magnet.

5. A solenoid suspended by a string could be used as a compass. The magnetic field of the solenoid would align itself with the magnetic field of Earth just as a bar magnet would. The N pole of the solenoid would point to Earth's magnetic south pole in northern Canada.

6. This motor will not be able to make a complete rotation because of the lack of a commutator. This motor will only oscillate back and forth.

SECTION
3

Focus

Overview

Before beginning this section, review with your students the Objectives listed in the Student Edition. In this section, students learn what conditions are required for electromagnetic induction. Next, they apply this concept to generators. They also learn how transformers change voltage across power lines.

🔊 Bellringer

Use the Bellringer transparency to prepare students for this section.

Motivate

Opening Discussion ——— GENERAL

Remind students that in the previous section, they learned that an electric current produces a magnetic field. Ask students to speculate about what might happen when a loop of wire moves through a magnetic field. List all student ideas on the board, and discuss whether the ideas are accurate. Explain that an electric current is induced in the wire. The current exists as long as the magnetic field is changing relative to the wire. Some students may have difficulty accepting this idea at first. Viewing this experimentally, as is done in the Quick Lab on the next page, may help convince them. **LS Verbal**

Electric Currents from Magnetism

■ **KEY TERMS**
electromagnetic
induction
generator
alternating current
transformer

● OBJECTIVES

▶ **Describe** the conditions required for electromagnetic induction.

▶ **Apply** the concept of electromagnetic induction to generators.

▶ **Explain** how transformers increase or decrease voltage across power lines.

Can you have current in a wire without a battery or some other source of voltage? In 1831, Michael Faraday discovered that a current can be produced by pushing a magnet through a coil of wire. In other words, moving a magnet in and out of a coil of wire causes charges in the wire to move. This process is called **electromagnetic induction.**

■ **electromagnetic induction** the process of creating a current in a circuit by changing a magnetic field

Electromagnetic Induction and Faraday's Law

Electromagnetic induction is so fundamental that it has become one of the laws of physics—*Faraday's law.* Faraday's law states the following:

> **An electric current can be produced in a circuit by a changing magnetic field.**

Consider the loop of wire moving between the two magnetic poles in ***Figure 13.*** As the loop moves in and out of the magnetic field of the magnet, a current is *induced* in the circuit. As long as the wire continues to move in or out of the field in a direction that is not parallel to the field, an induced current will exist in the circuit.

Rotating the circuit or changing the strength of the magnetic field will also induce a current in the circuit. In each case, there is a changing magnetic field passing through the loop. You can predict whether a current will be induced using the concept of magnetic field lines. A current will be induced if the number of field lines that pass through the loop changes.

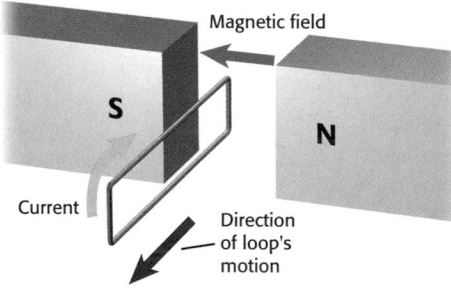

Figure 13

When the loop moves in or out of the magnetic field, a current is induced in the wire.

576

REAL-LIFE CONNECTION

Traffic Signals The operation of detectors at traffic signals is based on Faraday's law. A wire is placed in a groove cut into the road so that the wire makes a loop where the first car will sit. The ends of the wire go back to the computer controller. Earth's magnetic field goes through this loop. When a car moves over the loop, the strength of the magnetic field going through the loop decreases. This change in the magnetic field induces a current in the loop and tells the computer that a car is waiting.

PHYSICS CONNECTION

Physicists first studied protons, neutrons, and electrons by firing them through magnetic fields. They found that when a charged particle goes through a constant magnetic field, the particle moves in a circular path. The radius of the path is directly related to the mass of the particle and inversely related to the particle's charge. Electrons and protons fired through a constant magnetic field curve in different directions, indicating that they are oppositely charged. Neutrons travel straight, indicating no charge.

It would seem that electromagnetic induction creates energy from nothing, but this is not true. Electromagnetic induction does not violate the law of conservation of energy. Pushing a loop through a magnetic field requires work. The greater the magnetic field, the stronger the force required to push the loop through the field. The energy required for this work comes from an outside source, such as your muscles pushing the loop through the magnetic field. So while electrical energy is produced by electromagnetic induction, energy is required to move the loop.

Moving electric charges experience a magnetic force when in a magnetic field

When studying electromagnetic induction, it is helpful to imagine the individual charges in a wire. A charged particle moving in a magnetic field will experience a force due to the magnetic field. Experiments have shown that this magnetic force is zero when the charge moves along or opposite the direction of the magnetic field lines. The force is at its maximum value when the charge moves perpendicular to the field. As the angle between the charge's direction and the direction of the magnetic field decreases, the force on the charge decreases.

INTEGRATING

BIOLOGY
Many types of bacteria contain magnetic particles of iron oxide and iron sulfide. These particles are encased in a membrane within the cell, forming a magnetosome. The magnetosomes in a bacterium spread out in a line and align with Earth's magnetic field. In this way, as the cell uses its flagella to swim, it travels along a north-south axis. Recently, magnetite crystals have been found in human brain cells, but the role these particles play remains uncertain.

Teach

SKILL BUILDER — BASIC

Interpreting Visuals Have students look at **Figure 13.** Ask them to state the condition required for current to exist in the loop in the figure. (There must be relative motion between the loop and the magnetic field.) Some students may have trouble distinguishing between the external magnetic field and the magnetic field that is created by the induced current. Point out that only when the loop is moved will there be a magnetic field created by the induced current in the loop. Use the right hand rule to show that it will point in the opposite direction to the magnetic field between the pole pieces.
LS Logical

Quick Lab

Can you demonstrate electromagnetic induction?

Materials
✓ galvanometer
✓ solenoid
✓ 2 insulated wire leads
✓ 2 bar magnets

1. Set up the apparatus as shown in the photo at right. With this arrangement, current induced in the solenoid will pass through the galvanometer.

2. Holding one of the bar magnets, insert its north pole into the solenoid while observing the galvanometer needle. What happens?

3. Pull the magnet out of the solenoid, and record the movement of the galvanometer needle.

4. Turn the magnet around, and move the south pole in and out of the solenoid. What happens?

5. Vary the speed of the magnet. What happens if you do not move the magnet at all?

6. Try again using two magnets alongside each other with north poles and south poles together. How does the amount of current induced depend on the strength of the magnetic field?

Analysis

1. What evidence did you find that current is induced by a changing magnetic field?

2. Compare the current induced by a south pole with that induced by a north pole.

3. What two observations did you make that show that more current is induced if the magnetic field changes rapidly?

577

Quick Lab

Answers to Analysis

1. As the magnet is being brought toward the coil, the strength of the magnetic field in the coil increases, current is induced in the coil, and the galvanometer needle deflects.

2. in opposite directions

3. The induced current increases with speed and with the number of magnets.

Trends in Space Science

Tethered Satellites A tethered satellite is a satellite connected to a spacecraft through a long cable called a tether. As the tether moves through Earth's magnetic field, the charges in the tether experience a force due to the magnetic field and try to move to one end of the tether. This motion produces a voltage across the tether. The tether can then be used as an energy source. A tethered satellite should be able to generate more than 1 A of current.

NASA conducted the first test of a tethered satellite in 1992. The Space Shuttle Atlantis released a spherical satellite with a 21-km

long tether into space. Unfortunately, the tethered satellite jammed on the reel, and could not be deployed further. NASA conducted a second test in 1996 with the Space Shuttle Columbia. This time, the tether broke after deployment, and the satellite drifted away. However, before the break, the satellite had been generating 3,500 volts and up to 0.5 amps of current. So the mission did meet its objective, by demonstrating that tethered satellites can use Earth's magnetic field to generate electric current.

Chapter Resource File

• **Lesson Plan**

• **Quick Lab Datasheet** Can you demonstrate electromagnetic induction? **GENERAL**

Transparencies

TT Bellringer

Interpreting Visuals Have students use **Figure 14** to explain why a current is not induced in a loop that is moving parallel to a magnetic field. (There is no magnetic force on a charge moving in the direction of the magnetic field. So no current is induced in Figure 14B.)
LS Visual

Demonstration

Mechanical Energy to Electrical Energy

(Time: About five minutes)

Materials

• hand crank generator
• light bulb and socket

Use the generator to light up a light bulb. This demonstrates that different types of energy can be converted into electrical energy. The energy flow for this device is the following: chemical energy in your muscles is converted into kinetic energy to make your hand crank the motor. This kinetic energy is converted into electrical energy when a magnet or coil rotates inside the generator. The magnetic field through the wire loop generates an electrical current, lighting the bulb.

Figure 14

A When the wire in a circuit moves perpendicular to a magnetic field, the current induced in the wire is at a maximum.
B When the wire moves parallel to a magnetic field, there is zero current induced in the wire.

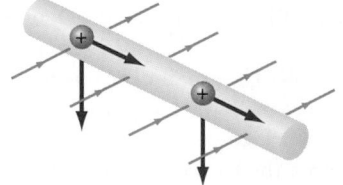

A Maximum current when the wire moves perpendicular to the magnetic field

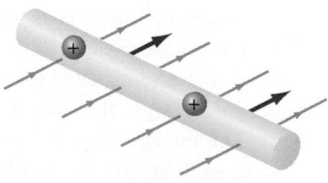

B Zero current when the wire moves parallel to the magnetic field

Now apply this concept to current. Imagine the wire in a circuit as a tube full of charges, as shown in **Figure 14.** When the wire is moving perpendicular to a magnetic field, the force on the charges is at a maximum. In this case, there will be a current in the wire and circuit, as shown in **Figure 14A.** When a wire is moving parallel to the field, as in **Figure 14B,** no current is induced in the wire. Because the charges are moving parallel to the field, they experience no magnetic force.

generator a machine that converts mechanical energy into electrical energy

alternating current an electric current that changes direction at regular intervals (abbreviation, AC)

Generators convert mechanical energy to electrical energy

Generators are similar to motors except that they convert mechanical energy to electrical energy. If you expend energy to do work on a simple generator, like the one in **Figure 15,** the loop of wire inside turns within a magnetic field and current is produced. For each half rotation of the loop, the current produced by the generator reverses direction.

This type of generator is therefore called an **alternating current,** or *AC,* generator. The generators that produce the electrical energy that you use at home are alternating current generators. The current supplied by the outlets in your home and in most of the world is alternating current.

As can be seen by the glowing light bulb in **Figure 15,** the coil turning in the magnetic field of the magnet creates a current. The magnitude and direction of the current that results from the coil's rotation vary depending on the orientation of the loop in the field.

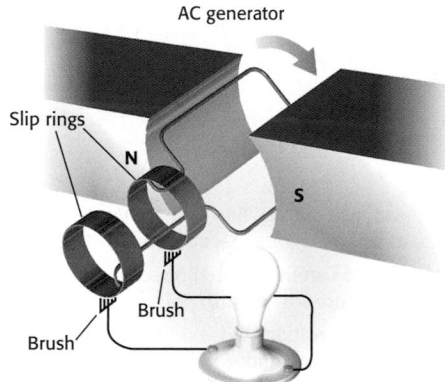

AC generator

Slip rings

N

S

Brush

Brush

Figure 15

In an alternating current generator, the mechanical energy of the loop's rotation is converted to electrical energy when a current is induced in the wire. The current lights the light bulb.

Did You Know?

Types of Current There are two types of electric current: direct current (DC) and alternating current (AC). In DC, the net current flow is constant, while in AC, the current oscillates between a negative and a positive value. Batteries always produce DC, because the potential difference between the terminals of a battery is fixed.

AC is supplied to homes and businesses, because it can be transmitted more easily and efficiently than DC. This is because power transmission works best at high voltages, but high voltages are not safe in the home. Transformers can be used to change the voltage of the current before and after transmission, but it is easier to change the voltage of AC than it is to change the voltage of DC. Another advantage of AC is that some devices, such as washing machines, would have to be much more complex (and thus more expensive) to use DC.

Table 1 Induced Current in a Generator

Position of loop	Amount of current	Graph of current versus angle of rotation
Magnetic field	Zero current	Current vs Rotation angle: 0° 90° 180° 270° 360° (flat at zero)
Magnetic field	Maximum current	Current vs Rotation angle: 0° 90° 180° 270° 360° (rising curve to peak)
Magnetic field	Zero current	Current vs Rotation angle: 0° 90° 180° 270° 360° (arch returning to zero)
Magnetic field	Maximum current (opposite direction)	Current vs Rotation angle: 0° 90° 180° 270° 360° (sine curve going negative)
Magnetic field	Zero current	Current vs Rotation angle: 0° 90° 180° 270° 360° (full sine wave)

Table 1 shows how the magnitude of the current produced by an AC generator varies with time. When the loop is perpendicular to the field, the current is zero. Recall that a charge moving parallel to a magnetic field experiences no magnetic force. This is the case here. The charges in the wire experience no magnetic force, so no current is induced in the wire.

As the loop continues to turn, the current increases until it reaches a maximum. When the loop is parallel to the field, charges on either side of the wire move perpendicular to the magnetic field. Thus, the charges experience the maximum magnetic force, and the current is large. Current decreases as the loop rotates, reaching zero when it is again perpendicular to the magnetic field. As the loop continues to rotate, the direction of the current reverses.

internet connect

www.scilinks.org
Topic: Generators
SciLinks code: HK4063

SCLINKS Maintained by the National Science Teachers Association

579

Historical Perspective
The Battle of the Currents Thomas Edison opened America's first power plant in New York City in 1882. Edison used underground cables to transmit DC, which he considered to be safer and more manageable than AC. He worked tirelessly to spread demand for electricity and to convince the public that it was safe. His campaign was successful, and by 1887 there were 121 of his power stations—owned by the Edison General Electric Company—around the United States.

However, around this time, Edison began facing competition from another inventor,

George Westinghouse, who championed AC over DC. Edison claimed that AC was extremely dangerous, and a fierce debate began.

Westinghouse incorporated the Westinghouse Electric Company in 1886. While Edison's DC could only travel about a mile, Westinghouse could send AC hundreds of miles with little power loss. Another great advantage was a very efficient AC motor and generator invented by Nicola Tesla. Westinghouse was eventually victorious, and today all power stations generate AC.

Demonstration
Measuring Energy
(Time: About 30 minutes)
Materials
- small electrical motor
- galvanometer
- string
- various weights
- ring stand
- test-tube clamp

Attach the motor to the ring stand with the test-tube clamp. Raise it to the highest position. Put something heavy on top of the ring stand base to keep it from falling over. The shaft of the motor should hang over the edge of the table. Tape or glue the end of the string to the shaft of the motor and wind it around the shaft. Tie a small mass to the string. The mass should be heavy enough so that, when released, the weight of the mass will cause the motor to turn. Attach the galvanometer. Release the weight and record the reading of the galvanometer when the mass reaches the floor or some fixed position. Repeat the experiment with different weights. Discuss the energy transformations involved.

Did You Know?

Although the light from a fluorescent light bulb appears to be constant, the current in the bulb actually varies, changing direction 60 times each second. The light appears to be steady because the changes are too rapid for our eyes to perceive.

Figure 16

An electromagnetic wave consists of electric and magnetic field waves at right angles to each other.

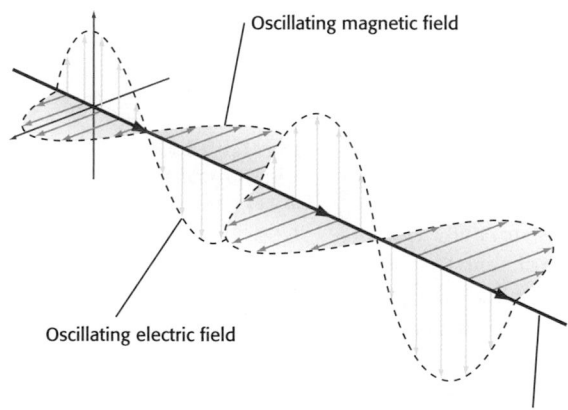

Oscillating magnetic field

Oscillating electric field

Direction of the electromagnetic wave

Generators produce the electrical energy you use in your home

Large power plants use generators to convert mechanical energy to electrical energy. The mechanical energy used in a commercial power plant comes from a variety of sources. One of the most common sources is running water. Dams are built to harness the kinetic energy of falling water. Water is forced through small channels at the top of the dam. As the water falls to the base of the dam, it turns the blades of large turbine fans. The fans are attached to a core wrapped with many loops of wire that rotate within a strong magnetic field. The end result is electrical energy.

Coal power plants use the heat from burning coal to make steam that eventually turns the blades of the turbines. Other sources of energy are nuclear fission, wind, hot water from geysers (geothermal), and solar power.

Some mechanical energy is always lost as heat, and the available electrical energy is reduced due to resistance in the wires of the generator. Many power plants are not very efficient. More efficient and safer methods of producing energy are continually being sought.

Electricity and magnetism are two aspects of a single electromagnetic force

So far you've read that a moving charge produces a magnetic field and that a changing magnetic field causes an electric charge to move. The energy that results from these two forces is called electromagnetic energy.

Light is a form of electromagnetic energy. Visible light travels as electromagnetic waves, or *EM* waves, as do other forms of radiation, such as radio signals and X rays. These waves are also called *EMF* (electromagnetic frequency) waves. As shown in **Figure 16,** EM waves are made up of oscillating electric and magnetic fields that are perpendicular to each other. This is true of any type of EM wave, regardless of the frequency.

Both the electric and magnetic fields in an EM wave are perpendicular to the direction the wave travels. So EM waves are transverse waves. As the wave moves along, the changing electric field generates the magnetic field. The changing magnetic field generates the electric field. Each field regenerates the other, allowing EM waves to travel through empty space.

REAL-LIFE
CONNECTION
Electric Guitars When an electric guitar is played, current goes through the strings. Beneath each string is a series of solenoids called pickups. As the string moves across the pickups, the magnetic field generated by the strings goes through the solenoids. The moving string also induces a current. This current is sent to the amplifier to be amplified and then to the speakers to be converted into sound. Musicians can change the sound of their guitars by changing how the solenoids are wound.

Did You Know?
Fish Ladders Many dams and hydroelectric power plants stop or inhibit the migration of salmon and other fish on the waterway. This prevents the fish from migrating upstream to spawn. As a result of legislation, power companies in many states are required to build a "fish ladder" to enable the fish to swim past the dam in order to reproduce.

Figure 17

A transformer uses the alternating current in the primary circuit to induce an alternating current in the secondary circuit.

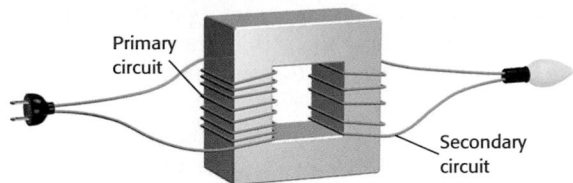

Primary circuit

Secondary circuit

▶ **transformer** a device that increases or decreases the voltage of alternating current

Transformers

You may have seen metal cylinders on power line poles in your neighborhood. These cylinders hold EM devices called **transformers**. *Figure 17* is a simple representation of a transformer. Two wires are coiled around opposite sides of a closed iron loop. In this transformer, one wire is attached to a source of alternating current, such as a power outlet in your home. The other wire is attached to an appliance, such as a lamp.

When there is current in the primary wire, this current creates a changing magnetic field that magnetizes the iron core. The changing magnetic field of the iron core then induces a current in the secondary coil. The direction of the current in the secondary coil changes every time the direction of the current in the primary coil changes.

Transformers can increase or decrease voltage

The voltage induced in the secondary coil of a transformer depends on the number of loops, or *turns,* in the coil. As shown in *Figure 18A,* both the primary and secondary wires are coiled only once around the iron core. If the incoming current has a voltage of 5 V, then the voltage measured in the other circuit will be close to 5 V. When the number of turns in the two coils is equal, the voltage induced in the secondary coil is about the same as the voltage in the primary coil.

In *Figure 18B,* two secondary coils with just one turn each are placed on the iron core. In this case, a voltage of slightly less than 5 V is induced in each coil. If these turns are joined together to form one coil with two turns, as shown in *Figure 18C,* the voltmeter will measure an induced voltage of slightly less than twice as much as the voltage produced by one coil.

Figure 18

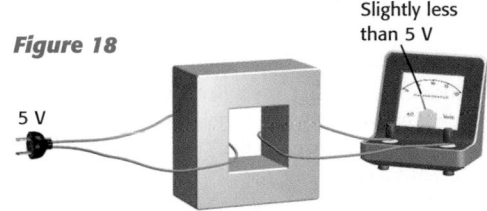

5 V

Slightly less than 5 V

A When the primary and secondary circuits in a transformer each have one turn, the voltage across each is about equal.

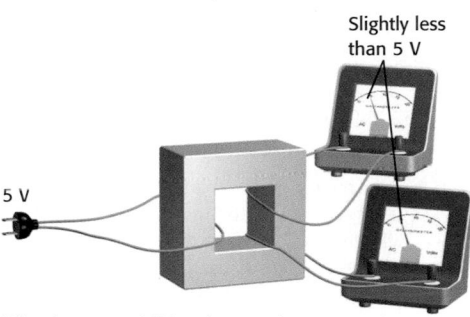

5 V

Slightly less than 5 V

B When an additional secondary circuit is added, the voltage across each is again about equal.

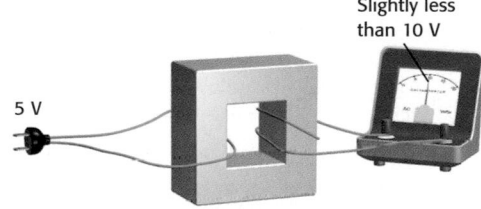

5 V

Slightly less than 10 V

C When the two secondary circuits are combined, the secondary circuit has about twice the voltage of the primary circuit. Actual transformers may have thousands of turns.

581

Interpreting Visuals Voltages are transformed down near homes and businesses with step-down transformers on utility poles, like the one shown in **Figure 19**, and at substations. Have students locate their neighborhood transformers or substations. In areas with underground wiring, the transformers may be installed on the ground.
LS Kinesthetic

Close

Quiz ———————— BASIC

Choose the term that best completes each sentence.

1. Electricity and magnetism are/are not part of a single force. (are)

2. A transformer/generator can increase or decrease voltage. (transformer)

3. An electric current can be produced in a circuit by a changing electric/magnetic field. (magnetic)

4. A transformer/generator uses electromagnetic induction to convert mechanical energy to electrical energy. (generator)
LS Logical

Figure 19
Step-down transformers like this one are used to reduce the voltage across power lines so that the electrical energy supplied to homes and businesses is safer to use.

Thus, the voltage across the secondary coil is about twice as large as the voltage across the primary coil. This device is called a *step-up transformer* because the voltage across the secondary coil is greater than the voltage across the primary coil.

If the secondary coil has fewer loops than the primary coil, then the voltage is lowered by the transformer. This type of transformer is called a *step-down transformer*.

Step-up and step-down transformers are used in the transmission of electrical energy from power plants to homes and businesses. A step-up transformer is used at or near the power plant to increase the voltage of the current to about 120 000 V. At this high voltage, less energy is lost due to the resistance of the transmission wires. A step-down transformer, like the one shown in *Figure 19,* is then used near your home to reduce the voltage of the current to about 120 V. This lower voltage is much safer. Many appliances in the United States operate at 120 V.

SECTION 3 REVIEW

SUMMARY

▶ A current is produced in a circuit by a changing magnetic field.

▶ In a generator, mechanical energy is converted to electrical energy by a conducting loop turning in a magnetic field.

▶ Electromagnetic waves consist of magnetic and electric fields oscillating at right angles to each other.

▶ In a transformer, the magnetic field produced by a primary coil induces a current in a secondary coil.

▶ The voltage across the secondary coil of a transformer is proportional to the number of loops, or turns, it has relative to the number of turns in the primary coil.

1. **Identify** which of the following will *not* increase the current induced in a wire loop moving through a magnetic field.
   a. increasing the strength of the magnetic field
   b. increasing the speed of the wire
   c. rotating the loop until it is perpendicular to the field

2. **Explain** how hydroelectrical power plants use moving water to produce electricity.

3. **Determine** whether the following statement describes a step-up transformer or a step-down transformer: The primary coil has 7000 turns, and the secondary coil has 500 turns.

4. **Predict** the movement of the needle of a galvanometer attached to a coil of wire for each of the following actions. Assume that the north pole of a bar magnet has been inserted into the coil, causing the needle to deflect to the right.
   a. pulling the magnet out of the coil
   b. letting the magnet rest in the coil
   c. thrusting the south pole of the magnet into the coil

5. **Critical Thinking** A spacecraft orbiting Earth has a coil of wire in it. An astronaut measures a small current in the coil, even though there is no battery connected to it and there are no magnets on the spacecraft. What is causing the current?

Answers to Section 3 Review

1. c

2. The moving water rotates a turbine. The turbine is attached to a core wrapped with many loops of wire that rotates within a magnetic field. Alternating current is induced in the wire loop as it turns.

3. step-down transformer

4. a. The needle will be deflected to the left.
   b. The needle will fall back to the zero reading because the magnetic field is no longer changing.

   c. The needle will be deflected to the left. The south end of the magnet going into the loop is the same as the north end of the magnet going out of the loop.

5. As the spacecraft orbits, the coil encounters a change in both the strength and orientation of Earth's magnetic field, inducing a current in the coil.

Study Skills

Interpreting Scientific Illustrations

Illustrations, figures, and photographs can be useful for understanding a scientific concept that is difficult to visualize. In the case of magnetism, keeping track of the directions of field lines and currents can be made easier with proper understanding of illustrations and their relation to the *right-hand rule*.

1 **Determine what the figure is trying to show.**

We'll use *Figure 20A* below, which shows how an electric current is induced in a wire loop moved out of a magnetic field. The caption reads, "When the loop moves in or out of the magnetic field, a current is induced in the wire." Examine the directions of the arrows in the figure, as these indicate the relationship between the loop's motion, the magnetic field, and the induced current.

2 **Examine the illustration's labels and art to learn general information.**

Note that the red arrow indicates the direction in which the loop is moved. Take your right hand, palm outstretched and thumb extended out, and point the thumb in the direction of the loop's motion. Now move your hand to align your fingers with the direction of the magnetic field, which is indicated by the blue arrow. (Do not curl your fingers.) The direction of the current will be out of the palm of your hand, along whichever part of the wire loop is still in the magnetic field. As this is the far, short end of the loop, the current points downward in that part of the wire, and so the current moves around the loop in the direction of the yellow arrows.

Figure 20

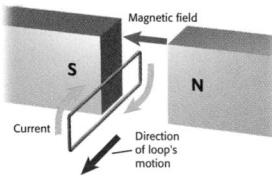

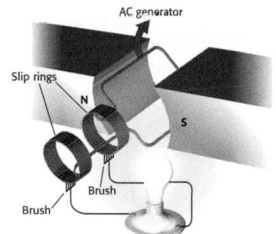

A When the loop moves in or out of the magnetic field, a current is induced in the wire.

B In an alternating current generator, the mechanical energy of the loop's rotation is converted into electrical energy.

Practice

1. Apply the right-hand rule to the generator shown in *Figure 20B.* In what direction is the current flowing?

2. Does the current in *Figure 20B* always flow in this direction? If not, why not?

583

Understanding Concepts

1. d
2. c
3. b
4. c
5. c
6. b
7. d
8. d
9. d
10. b
11. b
12. a
13. a

Using Vocabulary

14. Earth's magnetic S pole is located in northern Canada. The magnetic field of the Earth points from the magnetic N pole in Antarctica (South Magnetic Pole) to the magnetic S pole in northern Canada (North Magnetic Pole). The N pole of a compass needle is attracted to the magnetic S pole and ends up pointing to the north.

15. A magnetic compass aligns with Earth's magnetic field and points in the direction that lies along the magnetic field toward Earth's North magnetic pole (magnetic S pole). Using a magnetic compass is useful for determining location relative to the North magnetic pole.

Chapter Resource File

- **Concept Review** GENERAL
- **Chapter Test** GENERAL
- **Test Item Listing for Exam View®️ Test Generator**

Chapter Highlights

Before you begin, review the summaries of the key ideas of each section, found at the end of each section. The key vocabulary terms are listed on the first page of each section.

UNDERSTANDING CONCEPTS

1. If the poles of two magnets repel each other,
   a. both poles must be south poles.
   b. both poles must be north poles.
   c. one pole is a south pole and the other is a north pole.
   d. the poles are the same type.

2. Cutting a bar magnet in half will result in
   a. one magnet with a north pole only, and one magnet with a south pole only.
   b. two unmagnetized bars.
   c. two smaller magnets, each with a north pole and a south pole.
   d. two magnets with a north pole only.

3. The part of a magnet where the magnetic field and forces are strongest is called a magnetic
   a. field.   c. attraction.
   b. pole.   d. repulsion.

4. A _____ magnetic material is easy to magnetize but loses its magnetism easily.
   a. hard
   b. magnetically unstable
   c. soft
   d. No such material exists.

5. An object's ability to generate a magnetic field depends on its
   a. size.   c. composition.
   b. location.   d. direction.

6. A straight current-carrying wire produces
   a. an electric field.
   b. a magnetic field.
   c. beams of white light.
   d. All of the above.

7. A compass held directly below a current-carrying wire with a positive current moving north will point
   a. east.   c. south.
   b. north.   d. west.

8. An electric motor uses an electromagnet to change
   a. mechanical energy to electrical energy.
   b. magnetic fields in the motor.
   c. magnetic poles in the motor.
   d. electrical energy to mechanical energy.

9. An electric generator is a device that can convert
   a. nuclear energy to electrical energy.
   b. wind energy to electrical energy.
   c. energy from burning coal to electrical energy.
   d. All of the above.

10. The process of producing an electrical current by moving a magnet in and out of a coil of wire is called
    a. magnetic deduction.
    b. electromagnetic induction.
    c. magnetic reduction.
    d. magnetic production.

11. _____ law states that an electric current can be produced in a circuit by a changing magnetic field.
    a. Wien's   c. Boyle's
    b. Faraday's   d. Kepler's

12. In a generator, the current produced is _____ when the loop is parallel to the surrounding magnetic field.
    a. at a maximum   c. zero
    b. very small   d. average

13. In a transformer, the voltage of a current will be increased if the secondary circuit
    a. has more turns than the primary circuit.
    b. has fewer turns than the primary circuit.
    c. has the same number of turns as the primary circuit.
    d. is parallel to the primary circuit.

584

However, the magnetic fields of other objects may interfere with the functioning of a magnetic compass.

16. One could imagine holding the wire in one's right hand with the thumb pointing north. The fingers would then curl eastward on top of the wire, westward under the wire, downward to the east of the wire and upward on the other side. One's fingers would then curl in the direction of the resulting magnetic field.

17. An electromagnet is made by inserting an iron core into a solenoid.

18. The magnetic domains align with the external magnetic field.

19. AC stands for alternating current.

20. A galvanometer consists of a coil of insulated wire wrapped around an iron core that spins between the poles of a permanent magnet. When attached to a circuit, the coil and core act as an electromagnet, producing a magnetic field, which causes the core to rotate. The greater the rotation, the stronger the magnetic field and the greater the electric current. It is similar to an electric motor in that both use magnetic force to cause motion. However, unlike the coil in a galvanometer, the coil in a motor keeps spinning.

USING VOCABULARY

14. Use the terms *magnetic pole* and *magnetic field* to explain why the N pole of a compass points toward northern Canada.

15. Write a paragraph explaining some of the advantages and disadvantages of using a magnetic compass to determine direction. Use the terms *magnetic pole* and *magnetic field* in your answer. **WRITING SKILL**

16. A wire is carrying a positive electric current north. Describe how you would use the *right-hand rule* to predict the direction of the resulting magnetic field.

17. What is made by inserting an iron core into a *solenoid*?

18. What happens to the magnetic *domains* in a material when it is placed in a strong magnetic field?

19. What does the abbreviation *AC* stand for?

20. How does a *galvanometer* measure electric current? How is it similar to an *electric motor*? How does it differ?

21. What is the purpose of a *commutator* in an *electric motor*?

22. Use the terms *generator* and *electromagnetic induction* to explain how *electrical energy* can be produced using the *kinetic energy* of falling water.

23. What does the abbreviation *EM* stand for?

24. Describe how a *step-down transformer* reduces the voltage across a power line. Use the terms *primary circuit, secondary circuit,* and *electromagnetic induction* in your answer.

BUILDING GRAPHING SKILLS

25. Graphing The figure below is a graph of current versus rotation angle for the output of an alternating-current generator.
   a. At what point(s) does the generator produce no current?
   b. Is less or more current being produced at point *B* than at points *C* and *E*?
   c. Is less or more current being produced at point *D* than at points *C* and *E*?
   d. What does the negative value for the current at *D* signify?

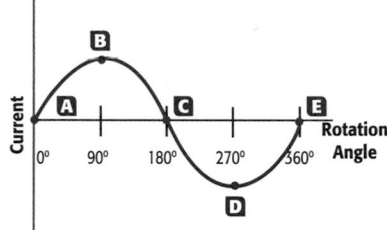

26. Interpreting Graphics If the coil of the generator referred to in item 25 were like the one shown in *Table 1,* what would the coil orientation be relative to the magnetic field in order to produce the maximum current at *B*?

THINKING CRITICALLY

27. Problem Solving How could you use a compass with a magnetized needle to determine if a steel nail were magnetized?

28. Applying Knowledge If you place a stethoscope on an unmagnetized iron nail and then slowly move a strong magnet toward the nail, you can hear a faint crackling sound. Use the concept of domains to explain this sound.

585

21. The commutator makes the current change direction every time the coil makes a half revolution, and thus it keeps the coil rotating in one direction.

22. Falling water can be used to generate electricity by converting the kinetic energy of the water into electrical energy. This is done by having the water turn a coil of wire that is inside a magnetic field, inducing a current in the wire. This is called electromagnetic induction, which is the basis of all power generators.

23. EM stands for electromagnetic.

24. In a transformer, electromagnetic induction occurs when current in a primary coil magnetizes an iron core and induces electric current in secondary coil. A step-down transformer decreases the voltage across a power line, by reducing the number of loops from the primary to the secondary coil.

Building Graphing Skills

25. a. *A, C, E*
   b. more
   c. more
   d. Current at *D* is moving in the opposite direction to current at *B*.

26. parallel

Thinking Critically

27. If a steel nail is magnetized, then the compass needle will point to the S end of the nail. As you move the compass from one end of the nail to the other, the compass should rotate to align with the nail's magnetic field. If it does not rotate, the nail is not magnetized. Refer to **Figure 4** for the orientation of the compass needle as it is moved around a magnetized object.

28. The faint crackling sound is the sound of the domains in the iron nail aligning with the magnetic field of the strong magnet.

29. Hold your arm still with the bracelet parallel to the magnetic field.

30. Electromagnets are used so that the doors can be closed automatically in the event of a fire.

Assignment Guide

Section	Questions
1	1–5, 14, 15, 27, 30, 33, 35, 42
2	6–8, 16–21, 28, 29, 32, 34, 36, 37, 40
3	9–12, 22–26, 31, 38, 43–46

31. Usually, a transformer would not work if DC current were used; however, it would work if the current were turned off and on repeatedly.

32. b

33. Hold the bars together in a **T** configuration. If there is no attraction, the bar at the top of the **T** is the magnet. If there is attraction, the bar making up the stem is the magnet.

34. The left-hand rule would state that you imagine holding the wire in your left hand with your thumb pointing in the opposite direction of the current. Your fingers would then curl in the direction of the magnetic field.

Developing Life/Work Skills

35. Students' plans should be logical, based on principles from the text or previous experiences. A compass could be used to determine if a material is magnetic. The needle would point toward the magnetic south pole of the metal.

36. Check students' drawings. They may include a combination of the following: a source of a magnetic field, wire loops, solenoids, electromagnets, and a voltage source.

37. a. A hair dryer uses an electric motor to blow air. The air is warmed by resistance coils.

  b. A doorbell uses an electromagnet to pull a striker that hits a bell.

  c. A tape consists of a thin coating of magnetic particles on a plastic tape. The record head magnetizes the tape in the pattern of the changing magnetic field

29. Problem Solving You walk briskly into a strong magnetic field while wearing a copper bracelet. How should you hold your wrist relative to the magnetic field lines to avoid inducing a current in the bracelet?

30. Understanding Systems Fire doors are doors that can slow the spread of fire from room to room when they are closed. In some buildings, fire doors are held open by electromagnets. Explain why electromagnets are used instead of permanent magnets.

31. Understanding Systems Transformers are usually used to raise or lower the voltage across an alternating-current circuit. Could a transformer be used in a direct-current circuit? How about if the direct current were pulsating (turning on and off)?

32. Understanding Systems Which of the following might be the purpose of the device shown below?

  a. to measure the amount of voltage across the wire
  b. to determine the direction of the current in the wire
  c. to find the resistance of the wire

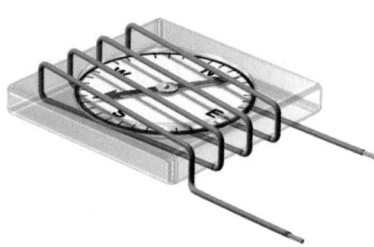

33. Problem Solving You are an astronaut stranded on a planet with no equipment or minerals. The planet does not have a magnetic field. You have two iron bars in your possession; one bar is magnetized, and one is not magnetized. How can you determine which bar is magnetized?

that corresponds to the intensity of sound being recorded. The playback head reads the pattern and produces an output voltage, which is used to control the speakers.

38. Adapters change the voltage and the current to different levels in order to power electrical devices. Most adapters contain step-down transformers because they reduce household AC to a lower voltage.

34. Creative Thinking The right-hand rule allows you to find the direction of the magnetic field around a current-carrying wire. What would the left-hand rule be? (**Hint:** Hold your hands so that the fingers of both hands curl in the same direction. What direction are your thumbs pointing?)

DEVELOPING LIFE/WORK SKILLS

35. Working Cooperatively During a field trip, you find a round chunk of metal that attracts iron objects. In groups of three, design a procedure to determine whether the object is magnetic and, if so, to locate its poles. What materials would you need? How would you draw your conclusions? List all the possible results and the conclusions you could draw from each result.

36. Applying Technology Use your imagination and your knowledge of electromagnetism to invent a useful electromagnetic device. Use a computer-drawing program to make sketches of your invention, and write a description of how it works. **COMPUTER SKILL**

37. Interpreting and Communicating Research one of the following electromagnetic devices: a hair dryer, a doorbell, and a tape recorder. Write a half-page description of how electromagnetism is used in the device, using diagrams where appropriate. **WRITING SKILL**

38. Applying Knowledge What do adapters do to voltage and current? Examine the input/output information on several adapters to find out. Do they contain step-up or step-down transformers?

39. Researching Information A transformer is needed to plug American appliances into wall sockets in other countries. Research how these transformers work. Why are they necessary?

39. Because the household voltage in some countries is greater than 120 volts, a transformer is required to reduce the voltage to a level that is safe for American appliances to use. Electrical outlets that carry AC at higher voltages are of a different shape to prevent damage to electrical devices that are only designed to use 120 volts of current.

40. Concept Mapping Copy the unfinished concept map below onto a sheet of paper. Complete the map by writing the correct word or phrase in the lettered boxes.

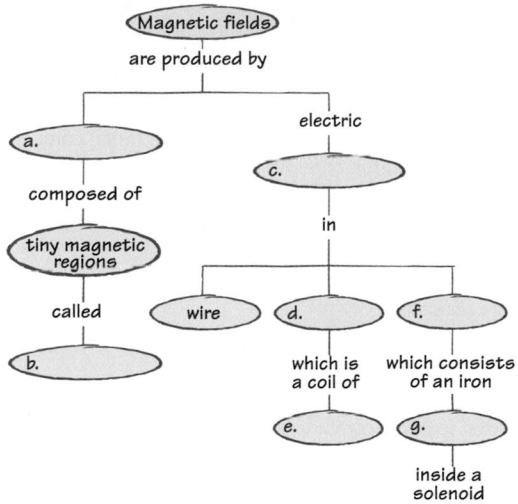

41. Concept Mapping Create your own concept map by using the content in one of the sections of this chapter. Write three propositions from your completed concept map.

42. Connection to Social Studies Why was the discovery of lodestones in Greece important to navigators hundreds of years later?

43. Connection to Health Some studies indicate that magnetic fields produced by power lines may contribute to leukemia among children who grow up near high-voltage power lines. Research the history of scientific studies of the connection between leukemia and power lines. What experiments show that growing up near power lines increases risk of leukemia? What evidence is there that there is no relation between leukemia and the magnetic fields produced by power lines?

44. Connection to Physics Find out how electromagnetism is used in containing nuclear fusion reactions. Write a report on your findings. **WRITING SKILL**

45. Connection to Social Studies Research the debate between proponents of alternating current and proponents of direct current in the 1880s and 1890s. How were Thomas Edison and George Westinghouse involved in the debate? What advantages and disadvantages did each side claim? What kind of current was finally generated in the Niagara Falls hydroelectric plant? If you had been in a position to fund the projects at that time, which projects would you have funded? Prepare your arguments so that you can reenact a meeting of businesspeople in Buffalo in 1887.

46. Connection to Fine Arts Electric guitars use electromagnetic induction to produce sound. Research electric guitars to find out how they work, and draw a diagram that illustrates the process electric guitars use to create sound. What other musical instruments could be modified to work the same way?

internet connect

www.scilinks.org
Topic: Magnetic Fields of Power Lines
SciLinks code: HK4083

SCI**LINKS** Maintained by the National Science Teachers Association

Art Credits: Fig. 1, Uhl Studios, Inc.; Fig. 4, Boston Graphics; Fig. 5, Stephen Durke/Washington Artists; Fig. 7, Mark Persyn; Fig. 8, Boston Graphics; "Quick Lab" (electromagnet), Uhl Studios, Inc.; Fig. 9 10, Stephen Durke/Washington Artists; Fig. 11, Uhl Studios, Inc.; Fig. 12-13, Stephen Durke/Washington Artists; Fig. 14, Kristy Sprott; Fig. 15, Stephen Durke/Washington Artists; Table 1, Kristy Sprott; Fig. 16, Kristy Sprott; Fig. 17-18, Stephen Durke/Washington Artists; "Understanding Concepts" (compass), Stephen Durke/Washington Artists; "Skills Practice Lab" (electromagnet), Uhl Studios, Inc.; "Study Skills", Stephen Durke/Washington Artists.

Photo Credits: Chapter Opener image of maglev train by Alex Bartel/Science Photo Library/Photo Researchers, Inc.; iron fillings reacting to magnet by Yoav Levy/Phototake; Fig. 2, Breck P. Kent/Animals Animals/Earth Scenes; Fig. 3, "Quick Activities," Peter Van Steen/HRW; Fig. 6, Richard Megna/Fundamental Photographs; "Quick Lab", Mike Fager/HRW; Fig. 19, Peter Van Steen/HRW; "Skills Practice Lab", Sam Dudgeon/HRW.

587

---

44. The plasmas used in nuclear fusion reactions have extremely high temperatures, usually more than 100 000 000°C, so they cannot be contained in a vessel. Because plasmas are made up of charged particles (atomic nuclei and electrons), they can be contained by magnetic fields.

45. Thomas Edison and General Electric wanted DC, while George Westinghouse and Nikola Tesla wanted AC. Edison claimed that DC was safer. Westinghouse claimed that AC would have less loss in transmission and that AC was more appropriate for use in large motors, such as in washing machines and refrigerators. The Niagara Falls power plant produced alternating current. Students should be able to defend their arguments based on their research.

46. The Real Life Connection in the ATE of Section 3 provides information on how electric guitars work. Other musical instruments that could be modified in this way include violins and pianos.

MAKING A BETTER ELECTROMAGNET

Teacher's Notes

Time Required 1 lab period

Ratings

1 — 2 — 3 — 4
EASY ——————— HARD

TEACHER PREPARATION	2
STUDENT SETUP	2
CONCEPT LEVEL	3
CLEANUP	2

Skills Acquired
- Collecting data
- Designing experiments
- Inferring
- Interpreting
- Measuring
- Organizing and analyzing data
- Predicting

The Scientific Method

In this lab, students will:
- Make observations
- Ask a question
- Form a hypothesis
- Analyze the results
- Draw conclusions
- Communicate results

Safety Cautions

Safety goggles and heat-resistant gloves must be worn at all times. Instruct students on the safe use of the wire strippers. Warn students to be careful not to short the batteries. Do not let metal rods roll on the floor; they create a slip hazard.

Students should wash their hands after the lab and avoid touching their eyes if they touched the tin and lead samples. These elements pose a health risk.

Skills Practice Lab

Introduction

How can you build the strongest electromagnet from a selection of batteries, wires, and metal rods?

Objectives

▶ **Build** several electromagnets.

▶ **Determine** how many paper clips each electromagnet can lift.

▶ USING SCIENTIFIC METHODS **Analyze your results** to identify the features of a strong electromagnet.

Materials

battery holders (2)
D-cell batteries (2)
electrical tape
extra insulated wire
metal rods (1 each of iron, tin, aluminum, and nickel)
small paper clips (1 box)
thick insulated wire, 1 m
thin insulated wire, 1 m
wire stripper

Making a Better Electromagnet

▶ Procedure

Building an Electromagnet

1. Review the Inquiry Lab in Section 2 on the basic steps in making an electromagnet.

2. On a blank sheet of paper, prepare a table like the one shown at right.

3. Wind the thin wire around the thickest metal core. Carefully pull the core out of the center of the thin wire coil. Repeat the above steps with the thick wire. You now have two wire coils that can be used to make electromagnets.

   SAFETY CAUTION Handle the wires only where they are insulated.

Designing Your Experiment

4. With your lab partners, decide how you will determine what features combine to make a strong electromagnet. Think about the following before you predict the features that the strongest electromagnet would have.

   a. Which metal rod would make the best core?

   b. Which of the two wires would make a stronger electromagnet?

   c. How many coils should the electromagnet have?

   d. Should the batteries be connected in series or in parallel?

Answers to Analysis

1. Answers will depend on the diameter of the copper conductor in each wire, the thickness of each wire's insulation, and the number of coils made with each wire. A larger-diameter conductor can carry more current, but a thinner wire (conductor plus insulator) allows more coils to be made. Both effects increase the strength of an electromagnet.

2. The iron and nickel cores should make the strongest electromagnets. Both metals contain domains that are aligned in the same direction.

3. The electromagnet should pick up more paper clips with the batteries connected in series.

With a series connection, the voltages of the two batteries add together. A larger voltage produces a greater current in the wire, making the magnetic field stronger.

4. The strongest electromagnet should have resulted from using the iron or nickel core, the series arrangement of batteries, and the thicker wire (as long as the difference in number of coils for the thin and thick wires is small).

Electromagnet number	Wire (thick or thin)	# of coils	Core (iron, tin, alum., or nickel)	Batteries (series or parallel)	# of paper clips lifted
1					
2					
3					
4					
5					
6					

5. In your lab report, list each step you will perform in your experiment.

6. Before you carry out your experiment, your teacher must approve your plan.

Performing Your Experiment

7. After your teacher approves your plan, carry out your experiment. You should test all four metal rods, both thicknesses of wire, and both series and parallel battery connections. Count the number of coils of wire in each electromagnet you build.

8. Record your results in your data table.

▶ Analysis

1. Did the thick wire or the thin wire make a stronger electromagnet? How can you explain this result?

2. Which metal cores made the strongest electromagnets? Why?

3. Could your electromagnet pick up more paper clips when the batteries were connected in series or in parallel? Explain why.

4. What combination of wire, metal core, and battery connection made the strongest electromagnet?

▶ Conclusions

5. Suppose someone tells you that your conclusion is invalid because each time you tested a magnet on the paper clips, the paper clips themselves became more and more magnetized. How could you show that your conclusion is valid?

589

Tips and Tricks

Review the construction of an electromagnet with the students before beginning the activity. Discuss the parts of an electromagnet, the solenoid and metal core, and their functions in an electromagnet. Remind the students that the strength of the magnetic field produced by a solenoid increases with increasing current in the solenoid and with the number of coils. Review the relationship between current, voltage, and resistance. Remind students that a thin wire will have a greater resistance than a thick wire. Encourage them to discuss how the diameter of a wire will affect how many times it can be wound around a core. Also review series and parallel circuits. Ask students to draw schematic diagrams for two batteries and a resistor in series and in parallel. Have them trace the path of a charge through each circuit and discuss the energy changes involved.

Procedure

Students should propose a systematic testing procedure, changing only one variable (wire thickness, type of metal core, or battery connection) from one test to the next.

Disposal Information

The metal cores can be kept for future use. The wire can be unwound and stored. The batteries should be tested. Discharged batteries should be taken to hazardous waste disposal.

Answers to Conclusions

5. Answers may vary. To account for the magnetism of the paper clips, students could determine how many paper clips stick to each electromagnet when the batteries are disconnected. They could also use a different set of paper clips for each electromagnet design that they test.

Communication Technology
Chapter Planning Guide

PACING	CLASSROOM RESOURCES	LABS, ACTIVITIES, AND DEMONSTRATIONS
BLOCK 1 · 45 min pp. 590–591 **Chapter Opener**		SE **Activity 1**, p. 591 SE **Activity 2**, p. 591
BLOCKS 2 & 3 · 90 min pp. 592–600 **Section 1** Signals and Telecommunication	CRF **Lesson Plan*** TT **Bellringer*** TT **Transducers*** TM **Binary Code***	TE **Opening Activity**, p. 592 GENERAL TE **Demonstration** Building a Communication System, p. 594 BASIC TE **Demonstration** Model of Radio Wave Transmission, p. 599 SE **Skills Practice Lab** Determining the Speed of Sound, pp. 624–625 ◆ GENERAL CRF **Datasheets for SE Skills Practice Lab** Determining the Speed of Sound* GENERAL
BLOCKS 4 & 5 · 90 min pp. 601–609 **Section 2** Telephone, Radio, and Television	CRF **Lesson Plan*** TT **Bellringer*** TT **Telephone*** TT **Television***	TE **Demonstration** Fiber Optics Cable, p. 602 TE **Demonstration** Internal Reflection, p. 604 TE **Demonstration** Deflecting Electron Beams, p. 606 SE **Quick Lab** How do red, blue, and green TV phosphors produce other colors?, p. 609 GENERAL CRF **Datasheets for In-Text Labs** How do red, blue, and green TV phosphors produce other colors?* GENERAL CRF **Observation Lab** Constructing a Radio Receiver* ◆ BASIC
BLOCKS 6 & 7 · 90 min pp. 610–618 **Section 3** Computers and the Internet	CRF **Lesson Plan*** TT **Bellringer*** TM **Logic Gates*** TT **Concept Mapping**	TE **Opening Demonstration**, p. 610 SE **Quick Activity** How Fast Are Digital Computers?, p. 612 GENERAL CRF **Datasheets for In-Text Labs** How Fast Are Digital Computers?* GENERAL CRF **CBL™ Probeware Lab** Transmitting and Receiving a Message Using a Binary Code* ◆ ADVANCED

BLOCKS 8 & 9 · 90 min

Chapter Review and Assessment Resources

- SE **Chapter Review**, pp. 620–623
- CRF **Chapter Tests*** GENERAL
- OSP **Test Generator**
- CRF **Standardized Test Practice with Guided Reading Development***
- CRF **Test Item Listing for ExamView® Test Generator***
- OSP **Scoring Rubrics and Classroom Management Checklists**

Online Resources

Visit the HRW Web site for a variety of free resources related to the text. Just type in the keyword **HK4 COM**.

Holt Online Learning

Holt Science Spectrum: Physical Science: Online Edition

Students can access interactive problem solving help and active visual concept development with the *Holt Science Spectrum: Physical Science* Online Edition available at **www.hrw.com**.

student CNN News

cnnstudentnews.com

Find the latest news, lesson plans, and activities related to important scientific events.

KEY

TE	Teacher Edition	**CRF**	Chapter Resource File	***** Also on One-Stop Planner
SE	Student Edition	**TT**	Teaching Transparency	**◆** Requires Advance Prep
OSP	One-Stop Planner	**TM**	Transparency Master	

PROBLEM SOLVING AND PRACTICE	SECTION REVIEW AND ASSESSMENT	STANDARDS CORRELATION
	CRF Pretest* GENERAL	
CRF Cross-Disciplinary Worksheet Social Studies Connection—Morse Code and Computers* GENERAL **CRF** Cross-Disciplinary Worksheet Integrating Math—Converting Binary Numbers* GENERAL	**TE** Quiz, p. 600 BASIC **SE** Section 1 Review, p. 600 **CRF** Concept Review* GENERAL **CRF** Quiz* BASIC	PS 5d, 6b ST 2
CRF Cross-Disciplinary Worksheet Integrating Biology—The Brain's Signals* GENERAL **SE** Science and the Consumer TV by the Numbers: High-Definition Digital TV, p. 608 **CRF** Cross-Disciplinary Worksheet Science and the Consumer—HDTV: Why Make the Switch?* GENERAL	**TE** Quiz, p. 609 BASIC **SE** Section 2 Review, p. 609 **CRF** Concept Review* GENERAL **CRF** Quiz* BASIC	PS 5d, 6b SPSP 5 ST 2
CRF Cross-Disciplinary Worksheet Integrating Physics—Building a Computer* GENERAL **CRF** Cross-Disciplinary Worksheet Architecture Connection—Computers and Design Fields* GENERAL **CRF** Cross-Disciplinary Worksheet Real World Applications—World Wide Web Robots* GENERAL **CRF** Cross-Disciplinary Worksheet Fine Arts Connection—Arts and the Internet* GENERAL **CRF** Science Skills Basic Exercises in Logic* GENERAL **SE** Study Skills Pattern Puzzle, p. 619 **SE** Science in Action Is there life outside our solar system, pp. 626–627	**TE** Quiz, p. 618 BASIC **SE** Section 3 Review, p. 618 **CRF** Concept Review* GENERAL **CRF** Quiz* BASIC	PS 5d, 6b SPSP 5 ST 2

Technology Resources

SCI LINKS

www.scilinks.org

Topic: Space Messages
*Sci*Links code: HK4132

Topic: Morse Code
*Sci*Links code: HK4090

Topic: Analog/Digital Signals
*Sci*Links code: HK4005

Topic: RCommunication Satellites
*Sci*Links code: HK4026

Topic: Television Technology
*Sci*Links code: HK4137

Topic: Internet
*Sci*Links code: HK4074

Topic: Communication Technology
*Sci*Links code: HK4025

Topic: Search for Extraterrestrial Life
*Sci*Links code: HK4062

One-Stop Planner
All of your printable resources and the Test Generator are on this convenient CD-ROM.

Overview

This chapter begins with a discussion of how signals and codes are used in comunication. Next students learn about telecommunication, including the use of optical fibers and communication satellites. The chapter also explores how telephones, televisions, and radios transmit and receive signals. Basic functions and workings of a computer are also covered. The chapter concludes with a discussion of computer networks and the Internet.

Assessing Prior Knowledge

Be sure students understand the following concepts:
- electromagnetic waves
- electric current
- circuits

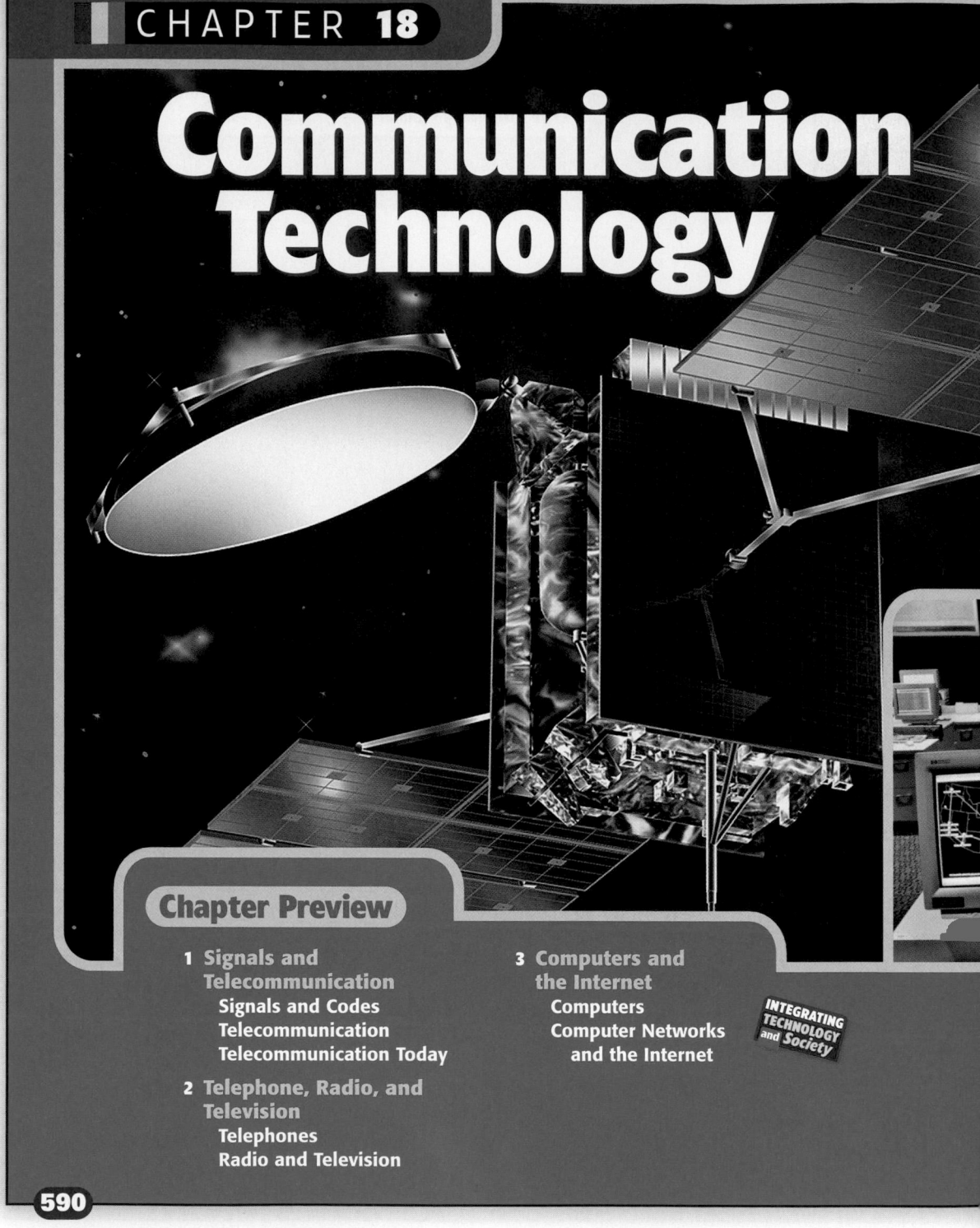

CHAPTER 18

Communication Technology

Chapter Preview

1 Signals and Telecommunication
   Signals and Codes
   Telecommunication
   Telecommunication Today

2 Telephone, Radio, and Television
   Telephones
   Radio and Television

3 Computers and the Internet
   Computers
   Computer Networks and the Internet

INTEGRATING TECHNOLOGY and Society

590

Standards Correlations

National Science Education Standards

The following descriptions summarize the National Science Education Standards that specifically relate to this chapter. For the full text of the standards, see the National Science Education Standards at the front of the book.

Section 1 Signals and Telecommunications
Physical Science PS 5d, 6b
Science and Technology ST 2

Section 2 Telephone, Radio, and Television
Physical Science PS 5d, 6b
Science in Personal and Social Perspectives SPSP 5
Science and Technology ST 2

Section 3 Integrating Technology and Society—Computers and the Internet
Physical Science PS 5d, 6b
Science in Personal and Social Perspectives SPSP 5
Science and Technology ST 2

Focus ACTIVITY

Background When you write a letter, you assume that the recipient can read and understand what you have written. But how would you send a message to intelligent life on another planet?

NASA had to consider this question in the early 1970s when it began sending spacecraft to the outer regions of the solar system. Like bottles drifting on an ocean, these probes would eventually drift out of the solar system into deep space. With messages attached to these spacecraft, any extraterrestrial beings that might discover a craft could learn where the craft came from if they could understand the message.

When the *Voyager 1* and *2* spacecraft were launched, large gold-plated copper disks were sent with them. Each disk was a large phonograph record consisting of sounds of nature, music from various nations, and greetings in all modern languages.

Activity 1 Suppose you are chosen to develop a visual message to be sent with a probe into deep space. Make a list of information you think would be most important to convey to intelligent extraterrestrial beings. What assumptions have you made about the receivers of this information?

Activity 2 On a piece of plain paper, draw the design for the space message you developed in Activity 1. Share your design with your classmates. See if they understand what you tried to communicate. Are there parts of your design that your classmates have trouble understanding? What might you do to remedy this?

☑ internet connect

www.scilinks.org
Topic: Space Messages SciLinks code: HK4132

SC*i*LINKS. Maintained by the
National Science Teachers Association

Pre-Reading Questions

1. Give examples of situations in which observing written communication may save someone's life.
2. List three specific ways in which communications have changed in your lifetime.

591

Improvements in communications satellites (left) make it possible for more telephone, radio, and television signals to travel from one place to another. These advancements, in both satellites and other communication equipment, are largely the result of improvements in the speed and storage capacity of modern computers.

Focus ACTIVITY

Background Students can research the *Voyager 2* satellite launched by NASA. On board this satellite is a gold record, which has visual and audio messages for any alien culture that may eventually find this space craft.

Activity 1 You may suggest images of humans, of different plants and animals, and of our solar system, for example. Students may also put sample languages on their messages and perhaps images of sample mathematical theorems. Extraterrestrial beings could receive this information, assuming they have sight and hearing or some other way to understand the messages.

Activity 2 Have students research what image is etched into the gold record that is attached to the *Voyager 2* satellite and compare it to their own designs. See if the students can understand the message of *Voyager 2*.

Answers to Pre-Reading Questions

1. Answers may include the traffic stop signs, warning signs on the road, warnings of poisons, warnings about small children and front-seat air bags, reminders to fasten seat belts, instructions about taking medications, and more.

2. Students may mention greater dependence upon e-mail, the Internet, cell phones, cell phone Internet service, cell phone text messaging, or other types of communications.

Chapter Resource File

- Pretest **GENERAL**
- Teaching Transparency Preview
- Answer Keys

SECTION
1

Focus

Overview

Before beginning this section, review with your students the Objectives listed in the Student Edition. This section begins with a general introduction to communication, including the use of signals and codes and the definition of telecommunication. Next students study analog and digital signals, including binary digital signals. The section concludes with a discussion of telecommunication today, including the use of optical fibers and communication satellites.

Bellringer

Use the Bellringer transparency to prepare students for this section.

Motivate

Opening Activity — GENERAL

Have students brainstorm words that they believe are related to communication and communication technology. Write these relevant words on the chalkboard. Have students suggest how some of these words affect their daily lives. They encounter many of these words on a daily basis, whether they know it or not. The students should write down the brainstormed terms so that they can define them at the end of the chapter, using what they have learned about communication and communication technology. **LS Verbal**

Signals and Telecommunication

■ **KEY TERMS**
signal
code
telecommunication
analog signal
digital signal
optical fiber

▶ **OBJECTIVES**

▶ **Distinguish** between signals and codes.
▶ **Define** and give an example of telecommunication.
▶ **Compare** analog signals with digital signals.
▶ **Describe** two advantages of optical fibers over metal wires for transmitting signals.
▶ **Describe** how microwave relays transmit signals using Earth-based stations and communications satellites.

You communicate with people every day. Each time you talk to a friend, wave goodbye or hello, or give someone a "thumbs up," you are sending and receiving information. Even actions such as shaking someone's hand and frowning are forms of communication.

Signals and Codes

All of the different forms of communication just mentioned use **signals.** A signal is any sign or event that conveys information. People often use nonverbal signals along with words to communicate. Some signals, such as those shown in *Figure 1,* are so common that almost everyone in the United States recognizes their meaning. Signals can be sent in the form of gestures, flags, lights, shapes, colors, or even electric current.

■ **signal** anything that serves to direct, guide, or warn

Figure 1

A A handshake indicates friendship or good will.
B A green light means "go."
C A football referee's raised hands tell the crowd and the score-keeper that the kick was good.

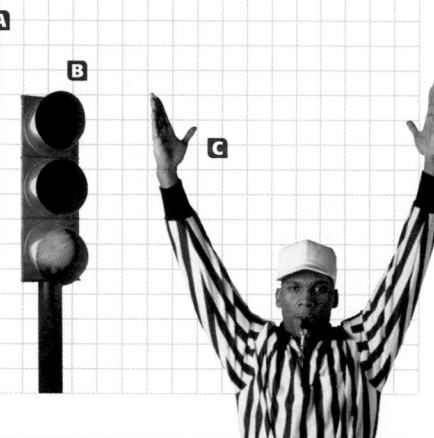

592

Did You Know ?

Infant Communication Typically, infants are not able to speak until they reach the age of 12–14 months. This is partly because they have not learned communication codes, or language. However, they are able to communicate with their parents. Infants use nonverbal communication skills, and they interpret the nonverbal reactions that their parents present back to them.

Codes are used to send signals

In a baseball game, the catcher often sends signals to the pitcher. These signals can tell the pitcher what type of pitch to throw. For the catcher's signals to be understood by the pitcher, the two players must work out the meaning of the signals, or **code,** before the game starts.

You hear and use codes every day, perhaps without even being aware of it. The language you speak is a code. Not everybody in the world understands it. An idea or message can be represented in different languages using very different symbols. The phrase "thank you" in English, for instance, is expressed as *gracias* in Spanish, شكران in Arabic, and 謝謝 in Chinese.

Some codes are used by particular groups. For example, chemists around the world use Au, Pb, and O as symbols for the elements gold, lead, and oxygen, respectively. Also, all mathematicians recognize =, −, and + as symbols that mean "equals", "minus", and "plus."

In addition to signals and codes, communication requires a sender and a receiver. A sender transmits, or sends, a message to a receiver.

Signals are sent in many different forms

Signals such as waving or speaking can be received only if the person at the other end can see or hear the signal. As a result, these signals cannot be sent very far. To send a message over long distances, the signal needs to be converted into a form that can travel long distances easily. Both electricity and electromagnetic waves offer excellent ways to send such signals.

The first step in using electricity to send sound is to convert the sound into an electric current. This electrical signal is produced by using a microphone. The microphone matches the changes in sound waves with comparable changes in electric current. You can imagine the microphone making a copy of the sound in the form of electricity. Next, this electrical signal travels along a wire over longer distances. At the other end, the electrical signal is amplified and converted back into sound by using a speaker.

> **code** a set of rules used to interpret data that convey information

Connection to SOCIAL STUDIES

In 1837, an American named Samuel Morse received a patent on a device called the electric telegraph. The telegraph uses a code made of a series of pulses of electric current to send messages. A machine at the other end marks a paper tape—a dot in response to a short pulse and a dash in response to a long pulse. Morse code, as shown below, represents letters and numbers as a series of dashes and dots.

A ·—	N —·	1 ·————
B —···	O ———	2 ··———
C —·—·	P ·——·	3 ···——
D —··	Q ——·—	4 ····—
E ·	R ·—·	5 ·····
F ··—·	S ···	6 —····
G ——·	T —	7 ——···
H ····	U ··—	8 ———··
I ··	V ···—	9 ————·
J ·———	W ·——	0 —————
K —·—	X —··—	
L ·—··	Y —·——	
M ——	Z ——··	

Making the Connection

1. Write a simple sentence, such as "I am here."
2. Translate it into Morse code, and using sounds, tapping, or a flashlight, send it to a partner.
3. Have your partner write down the code and try to translate the message using Morse code.

593

Demonstration ——— BASIC

Building a Communication System

(Time: About 30 minutes)

Materials

- string
- paper clips
- plastic cups with small holes in the bottom

Step 1 Putting holes in the bottom of the cups can be difficult with scissors. The easiest way is to use a large paper clip, a pair of pliers, and a Bunsen burner. Heat the paper clip and melt a small hole in the bottom of each cup.

Step 2 Divide the class into groups of 2–3 students per group. Give each group a long piece of string (10 ft) and a cup and a paper clip for each student.

Step 3 Have students make a communication system with these materials. To attach the string to the cup, students should thread the string through the hole in the bottom of the cup. The string will stay connected to the cup better if the string is tied to the paper clip inside the cup; the paper clip will keep the knot in the string from pulling through the hole. Students can experiment with having only two cups on the string, or they can attach many cups by connecting multiple strings together at the center so that the strings form a star pattern.

Step 4 You can also have students experiment with different kinds of string, such as fishing line, twine, or guitar strings. Some strings will transmit vibrations better than others.

LS Kinesthetic

---

internet connect

www.scilinks.org
Topic: Morse Code
SciLinks code: HK4090

SCI**LINKS**. Maintained by the National Science Teachers Association

■ **telecommunication** the sending of visible or audible information by electromagnetic means

Figure 2

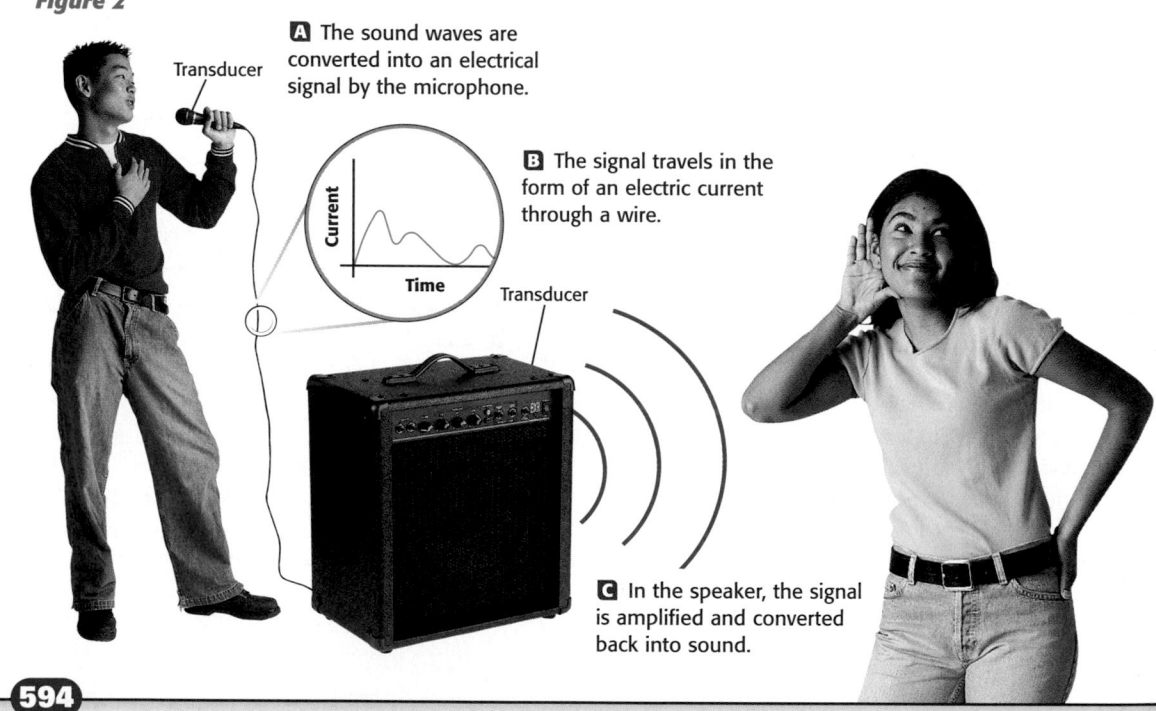

A The sound waves are converted into an electrical signal by the microphone.

Transducer

B The signal travels in the form of an electric current through a wire.

Current

Time

Transducer

C In the speaker, the signal is amplified and converted back into sound.

594

A transducer converts signals

A speaker is a type of *transducer*, which is a device that converts a signal from one form to another. A speaker converts an incoming electrical signal into sound. After the conversion, the original sound is heard once again. Two types of transducers, a speaker and a microphone, are shown in **Figure 2.** The microphone is a transducer that converts a sound signal into an electrical signal.

Telecommunication

Not long after the discovery of electric current, people tried to find ways of using electricity to send messages over long distances. In 1844, the first telegraph line provided a faster way to send messages between Baltimore and Washington, D.C. More telegraph lines were then installed. By 1861, messages could be sent rapidly between the West Coast and the East Coast.

About 30 years after the first electric telegraph service was provided, the telephone was developed. In another 25 years, the wireless telegraph was invented. With wireless technology, a telegraph message could be sent by radio waves without the use of wires and cables. Sending and receiving signals by using electromagnetic means is referred to as **telecommunication.**

Historical Perspective

Communication by Fire The Romans had a sophisticated communication system for their time. They could send messages over long distances by the use of fire.

They would build towers approximately 20 mi apart and place large piles of dry wood and kindling near the towers. Generals in the field would decide what message the fire indicated.

When it was time to pass along the message, the general would order the pile of wood to be lit. The lookout in the next tower, 20 mi away, would see the fire, and then set his wooden pile on fire for the next tower to see.

Because the wooden piles were large, the fires could be seen over great distances, carrying a message faster than a horse and rider could. The drawback of this system was that only one message could be sent at a time.

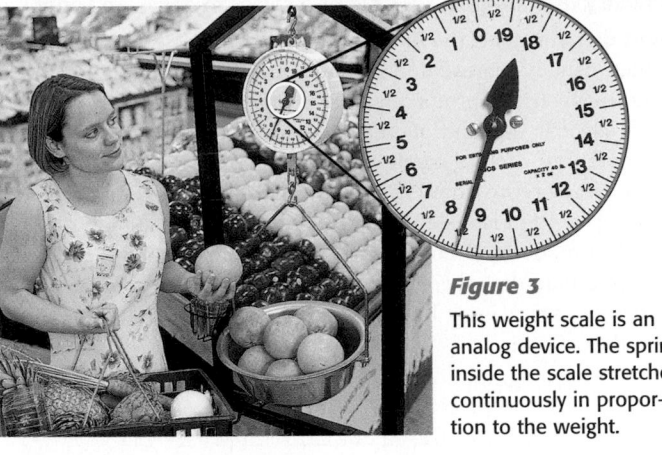

Figure 3

This weight scale is an analog device. The spring inside the scale stretches continuously in proportion to the weight.

An analog signal varies continuously within a range

What do a thermometer, a speedometer, and a spring scale have in common? They are analog devices, which means that their readings change continuously as the quantity they are measuring—temperature, wheel rotation, or weight—changes. A reading given by each of these measuring devices is an **analog signal.**

An example of an analog device is shown in **Figure 3.** As the weight on the scale increases, the needle moves in one direction. As the weight decreases, the needle moves in the opposite direction. The position of the needle on this scale can have any possible value between 0 and 20.0 lb (0 N and 89 N).

The audio signal from the microphone in **Figure 2** is an analog signal in the form of a changing electric current. Analog signals consisting of radio waves can be used to transmit picture, sound, and telephone messages.

Digital signals consist of separate bits of information

Unlike an analog signal, which can change continuously, a **digital signal** consists of only discrete, or fixed, values. The binary number system consists of two discrete values, 0 and 1. The combination for a lock, shown in **Figure 4,** is in a digital form. It is composed of discrete values, or digits, each of which can have one of six values—1, 2, 3, 4, 5, or 6.

A simple type of digital signal uses a flashing light. Sailors sometimes use a flashing *signal lamp* to send Morse code for ship-to-ship and ship-to-shore communication. Morse code, which was developed by Samuel Morse for transmitting information by telegraph, uses three "digits": a short interval between clicks, a long interval between clicks, and no click at all.

■ **analog signal** a signal whose properties, such as amplitude and frequency, can change continuously in a given range

■ **digital signal** a signal that can be represented as a sequence of discrete values

Figure 4

The code to open this lock is in a digital form, consisting of a series of whole numbers.

595

---

internet connect

www.scilinks.org
Topic: Analog/Digital Signals
SciLinks code: HK4005

SCI LINKS. Maintained by the National Science Teachers Association

---

Did You Know ?

Scanners A scanner used with a computer can digitize an image. It looks at every part of the picture, subdivides the picture into small squares, and assigns a number to each square. This number, or digit, represents the color and brightness of the image in the square.

REAL-LIFE CONNECTION

Bar Codes Bar codes are a digital system used to identify merchandise. The black strips of a bar code do not reflect light, while the white section between the bars does reflect light. When laser light hits the bar code, the light detectors see a pattern of light, no light, light, no light. The thickness and the number of the bars change the pattern, identifying the merchandise to the computer.

Vocabulary The word *binary* means characterized by two parts. Students should recognize the prefix *bi–*, as in bicycle (two wheels) or biannual (two times per year). Challenge students to think of other words with this prefix. Then ask students what the two parts of binary signals are. (zeros and ones, or on and off) **LS** Verbal

Interpreting Visuals **Figure 6** shows a CD and the pattern of pits on the CD. The pits etched on the CD surface are microscopically small, just one-thousandth of a millimeter wide. Each CD is marked by roughly 3 billion pits in a spiral track. The area between two pits is called a flat. The digital copy of the sound or any other type of data is etched on the CD in a pattern of pits and flats. **Figure 6** shows the playback mechanism of a CD player. The data are read by a laser beam. When the light hits the pit, light is dispersed and no light is reflected back to the detector. But when the light hits a flat, the light is reflected back to the detector. As the laser reads the pattern of pits and flats, the digital copy is converted back into analog signal, which is used to recreate the sound by the speakers.

A binary digital signal consists of a series of zeros and ones

Most digital signals use *binary digital code*, which consists of two values, usually represented as 0 and 1. Each binary digit is called a bit. In electrical form, 0 and 1 are represented by the two states of an electric current: *off* (no current present) and *on* (current present). Information such as numbers, words, music, and pictures can be represented in binary code. *Figure 5* shows a binary digital code that is used to represent the English alphabet.

Most modern telecommunication systems transmit and store data in binary digital code. A compact disc (CD) player, shown in *Figure 6*, uses a laser beam to read the music that is digitally stored on the disc.

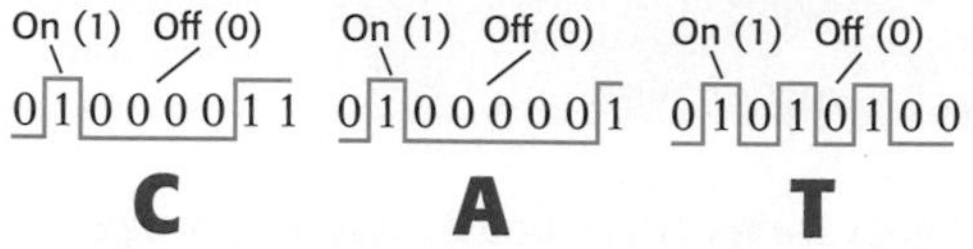

Figure 5
The English alphabet can be represented by combinations of the binary digits 1 and 0.

On (1) Off (0)   On (1) Off (0)   On (1) Off (0)
0 1 0 0 0 0 1 1   0 1 0 0 0 0 0 1   0 1 0 1 0 1 0 0
C   **A**   **T**

Alphabetic Characters and Their Binary Codes			
A 01000001	**H** 01001000	**O** 01001111	**V** 01010110
B 01000010	**I** 01001001	**P** 01010000	**W** 01010111
C 01000011	**J** 01001010	**Q** 01010001	**X** 01011000
D 01000100	**K** 01001011	**R** 01010010	**Y** 01011001
E 01000101	**L** 01001100	**S** 01010011	**Z** 01011010
F 01000110	**M** 01001101	**T** 01010100	
G 01000111	**N** 01001110	**U** 01010101	

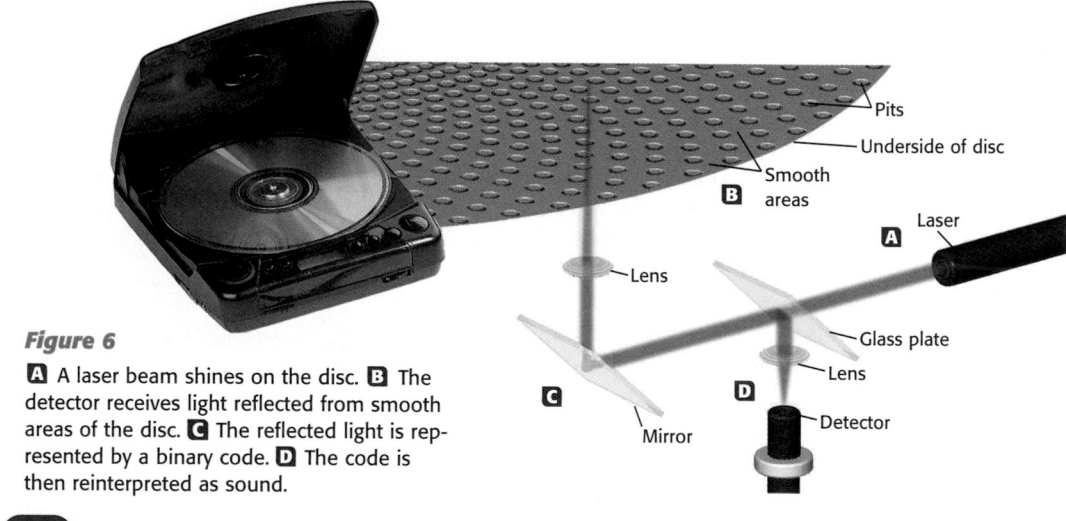

Figure 6
A A laser beam shines on the disc. **B** The detector receives light reflected from smooth areas of the disc. **C** The reflected light is represented by a binary code. **D** The code is then reinterpreted as sound.

Alternative Assessment — **ADVANCED**

Digital Technology In recent years, there has been an explosion of new digital applications, such as DVD players, portable MP3 players, and digital cameras. Have students choose one of these recent technologies to research. Ask them to make a poster that demonstrates how the device works, including an illustration and descriptive text. Their posters should also explain how digital signals are used in the device. **LS** Verbal

Did You Know?

MP3 Format Today, many people enjoy downloading songs from the Internet. Some use portable music players to store and listen to the songs. These songs are usually in MP3 format, a compression system that reduces the number of bytes in a song. A 3-minute song, which would take up 32 MB on a CD, takes up just 2–3 MB in MP3 format. This makes it possible to download songs much faster, and to store more songs in less space on a memory card or hard drive. The MP3 technology was patented in the U.S. by a German company in 1996.

Sound can be stored digitally

Sound is a wave of compressions (high air pressure) and rarefactions (low air pressure). Therefore, a sound can be described by noting the air pressure changes. The air pressure is measured in numbers and represented in binary digits.

How is the air pressure measured in numbers? This process is indirect. First, a microphone is used to convert the sound into an analog signal as a changing electric current. Then, an electronic device measures this changing current in numbers or digits at regular intervals. In fact, for CD sound recordings, the current is measured 44 100 times every second! The air pressure measurement is converted into binary digits in terms of 16 bits. For instance, 0000000010000010 is the digital representation of air pressure at a particular moment. This conversion process is basically the same for creating digital signals from analog signals. This conversion also occurs in a digital telephone.

Digital signals can be sent quickly and accurately

Digital signals have many advantages over analog signals. Some digital "switches," consisting of electronic components, can be turned on and off up to a billion times per second. This allows a digital signal to send a lot of data in a small amount of time to receivers such as the audio player in *Figure 7.*

Noise and static have less effect on digital transmissions. Most digital signals include codes that constantly check the pattern of the received signal and correct any errors that may occur in the signal. By contrast, analog signals must be received, amplified, and retransmitted several times by components along the transmission route. Each time, the signal can get a little more distorted.

Telecommunication Today

Many telecommunication devices, such as telephones, transmit signals along metal wires. But, other ways are more efficient. Metal wires are being replaced with glass fibers that carry signals using pulses of light. Radio waves also carry signals. A call may at times involve sending a signal by way of a communication satellite.

Optical fibers are more efficient than metal wires

A thin glass or plastic fiber, called an **optical fiber,** can be used to carry a beam of light. The light is reflected by the inside walls of the fiber, so it does not escape. Instead of carrying signals that are coded into electric currents, these fibers carry signals that are represented by pulses of light emitted by a laser.

Figure 7

The world's first wrist-wear audio player can be used with a computer to download MP3 music files.

▶ **optical fiber** a transparent thread of plastic or glass that transmits light

Teaching Tips

Digital Coding Morse code was among the first digital coding systems. This system of dots and dashes is similar to the 1's and 0's used in computers. Additional uses of digital signals are in video tape recording, computer information storage, and CD-ROMs. Bar codes in retail stores are also an example of digital systems.

Optical-Fiber Cables Today, optical-fiber cables are used to transmit some (but not all) telephone and Internet signals. They are also used to transmit digital cable-television signals. One advantage of using optical-fiber cable in the telecommunication industry is that a light signal traveling through the cable loses very little energy. With standard electrical cable, an electrical signal loses energy as the signal travels. If the signal is being sent over a large distance, then signal repeaters are needed. Because light can travel farther without losing energy in an optical fiber, fewer repeaters are needed to boost the original signal.

REAL-LIFE CONNECTION

New Uses for Radio Signals Gasoline companies are beginning to distribute to their customers key chains with a small radio transmitter attached. This radio transmitter is as thick as a pencil and approximately 1 in. long. When placed near the gasoline pump, the computer in the pump receives the code being transmitted by the device and searches its files for the proper account. In this way, a credit card is not needed to purchase gasoline.

The same kind of technology is being used in states that require drivers to pay tolls on the highways. Drivers that frequently use toll highways can purchase a device that broadcasts a radio signal to the toll booths. When the car drives through a toll booth, the computer in the booth receives the signal, identifies the account number, and charges the account the proper amount. In this way, the driver does not need to stop and does not need to have money in the car at all times.

SKILL BUILDER

Interpreting Visuals Be sure students understand that each hair-thin strand shown in **Figure 8B** is a single optical fiber. Tell students that each fiber has three parts: an inner core, where light travels; cladding, which reflects light in the core; and a buffer coating, which protects the fiber from damage and moisture. The fibers are bundled together in optical-fiber cables. One cable holds hundreds or even thousands of optical fibers. The cable has an outer covering, called a jacket, which gives the fibers an added layer of protection.

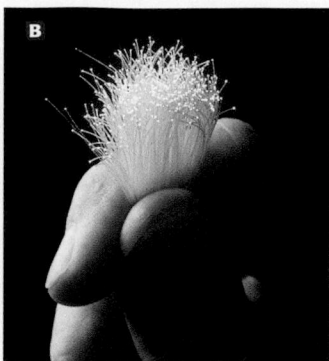

Figure 8

A single standard metal-wire cable **A** is much thicker than an optical-fiber cable **B**, yet it carries much less information than an optical fiber does.

Many telephone lines now in use in the United States consist of optical fibers. The optical-fiber system is lighter and smaller than the wire-cable system, as shown in *Figure 8,* making it much easier to put in place. A standard metal-wire cable, which is about 7.6 cm in diameter, can carry up to 1000 coded conversations at one time. A single optical fiber can carry 11 000 conversations at once using the present coding system.

As the use of the Internet and telephones dramatically increases, telephone companies are busy expanding fiber-optic networks. The materials used to make the optical fibers are so pure that a half-mile-thick slab made from them would transmit as much light as a clean windowpane.

Figure 9

A microwave relay tower picks up a signal, amplifies it, and relays it to the next tower.

Relay systems make it possible to send messages across the world

If you've traveled around the United States, you may have noticed tall steel towers with triangular or cone-shaped boxes and perhaps some dish-shaped antennas. These are microwave relay towers. They use microwave frequencies to transmit and relay signals over land.

As shown in *Figure 9,* a tower picks up a signal transmitted by another tower, amplifies the signal, and retransmits it toward the next tower. The next tower repeats the process, passing the signal along until it reaches its destination. Microwave transmission is often used to connect distant places with telephone signals.

598

Historical Perspective

The First Transatlantic Cable Originally, telephone cables were strung from city to city, and from state to state. Eventually, people wanted to be able to call people who were in different countries. As a result, a telephone cable that would span the Atlantic Ocean was needed. The transatlantic cable, completed in July 1866, marked the beginning of rapid communication across the seas. The establishment of this cable is credited to Massachusetts merchant and financier Cyrus W. Field. He first proposed running the 2000 mi long copper cable from Newfoundland to Ireland in 1854. The first three attempts ended in broken cables, but eventually Field succeeded. Messages that would have taken weeks took hours to reach Europe. Cyrus Field was showered with honors and recognition for his accomplishment.

Microwave towers should be tall

Microwaves are a form of electromagnetic waves. For the microwave signals to be sent from one tower to the next as in *Figure 10,* each tower must be almost visible from the top of the other. A tower built high in the Rocky Mountains would be able to relay signals for 80 to 160 km. However, a tower built in the plains can relay only a little farther than the horizon, or about 40 km.

Microwave transmission allows you to make telephone calls across land without wires or fiber-optic networks. But how could you call a friend who lives across the ocean in Australia?

In the past, your call would have been carried by one of the cables that run along the ocean floor between continents. Because there are so many telephones, online computers, and fax machines today, the demand is too much for these cables. Communication satellites that orbit Earth help send these messages.

Communications satellites receive and transmit electromagnetic waves

These satellites use solar power to generate electricity. This allows them to operate receivers, transmitters, and antennas. These satellites receive and send microwaves just like the towers described earlier. Because they are so high above the ground, these satellites can relay signals between telephone exchanges thousands of kilometers apart.

A satellite receives a microwave signal, called an *uplink,* from a ground station on Earth. The satellite then processes and transmits a *downlink* signal to another ground station. To keep the signals separate, the uplink signal consists of electromagnetic waves with a frequency of around 6 GHz (gigahertz, or 10^9 cycles per second), while the downlink signal typically has a lower frequency of about 4 GHz.

The transmitting antenna of a communications satellite must be aimed so that it covers the largest land area without the signal becoming too weak. This area is called a *satellite footprint* and increases as the distance between the satellite and Earth's surface increases. With several such satellites, a signal from one location can be transmitted and received anywhere in the world.

Many communications satellites have geostationary orbits

If you live in an area where people receive television signals from satellites by using dish-shaped antennas, you may have noticed that the dish always points in one direction. If a satellite orbits Earth, its position would change. Why does the dish not have to be moved in order to stay pointed at the orbiting satellite?

Figure 10
A microwave relay tower picks up a signal, amplifies it, and relays it to the next tower.

internet connect

www.scilinks.org
Topic: Communications Satellites
SciLinks code: HK4026

SciLINKS. Maintained by the National Science Teachers Association

599

SPACE SCIENCE
CONNECTION

Satellites A satellite cannot stay in orbit around Earth forever. The orbit of the satellite gradually degrades. This is because the Earth is not a perfectly smooth ball. The planet is inhomogeneous, which means that the gravitational force generated by Earth can fluctuate slightly. As a result, satellites must have the ability to correct their orbits over time. They accomplish this by two methods: the satellites may have thrusters that can correct the orbit, or the satellite may require a visit from the space shuttle in order to correct its orbit. If it is cheaper, however, to construct and launch a new satellite, then the old satellite will be left alone. The old satellite's orbit will eventually degrade and the satellite will finally spiral into Earth's atmosphere, burning up on re-entry.

Demonstration
Model of Radio Wave Transmission
(Time: About 15 minutes)
Materials
• large beach ball
• string
• plastic straws
• tape

Step 1 Cut a plastic straw in half and tape it to the beach ball so that it stands straight up.

Step 2 Tie or tape the end of a long piece of string to the end of the straw that is pointed away from the center of the beach ball. The beach ball represents Earth, and the straw represents a transmission tower. The string represents the radio wave that is being transmitted by the tower.

Step 3 Pull the string so that it is straight. Notice that the straight string can reach only so far from the tower before the curvature of the Earth (beach ball) blocks the string. A radio tower can basically transmit only as far as the horizon. Similarly, if someone were on the opposite side of the Earth, he or she could not receive this radio signal.

Step 4 Add another transmission tower by taping another straw to the beach ball so that it stands straight out from the ball.

Step 5 Tie another string to that tower. As long as the towers can "see" each other, they can transmit a signal to one another. The signal is transmitted from the first tower to the second tower. The second tower then transmits the original signal, giving the original signal a greater transmission range. With multiple towers, it should be possible to transmit a signal completely around Earth.

Quiz

Choose the term that best completes each sentence.

1. An analog/digital signal varies continuously. (analog)

2. A signal/code is any sign or event that represents information. (signal)

3. A CD player uses analog/digital signals. (digital)

4. Metal wires/optical fibers use light to transmit signals. (optical fibers)

5. Communication satellites receive and transmit electromagnetic/ sound waves. (electromagnetic)

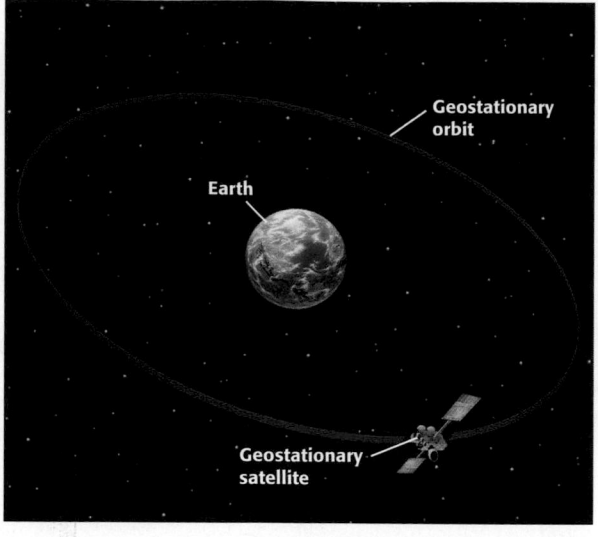

The answer is that these satellites orbit Earth every 24 hours, the same amount of time it takes for Earth to rotate once. Therefore the position of the satellite relative to the ground doesn't change. The orbit of this type of satellite is called a *geostationary orbit*, or a *geosynchronous orbit*. To be in a geostationary orbit, a satellite must be 35 880 km directly over Earth's equator and have a speed of 11 050 km/h, as shown in **Figure 11.**

Figure 11

A geostationary satellite appears to stay in a fixed position above the same spot on Earth. Once a dish is aimed at one of these satellites, it does not have to be moved again.

SECTION 1 REVIEW

SUMMARY

▶ A signal conveys a message that can be sent using gestures, shapes, colors, electricity, or light.

▶ An analog signal varies continuously.

▶ A digital signal represents information in the form of discrete digits.

▶ A two-number code, called a binary digital code, represents the signal conditions of "on" or "off" by either a 1 or a 0.

▶ Telecommunication sends a signal long distances by means of electricity or light.

▶ Satellites are used to relay microwave signals around the world.

1. List five examples of telecommunication.

2. Explain why talking to your friend on the telephone is an example of telecommunication but talking to her face to face is not an example of telecommunication.

3. Describe how a sound is translated into an analog signal.

4. Indicate which of the following are analog signals and which are digital:
   a. music recorded on a compact disc
   b. speed displayed by the needle on a speedometer dial
   c. time displayed on a clock with three or four numerals
   d. time displayed on a clock with hands and a circular dial

5. Discuss two advantages of optical fibers over metal wires as media for carrying signals.

6. Explain how communications satellites transmit messages around the world.

7. Explain what a geostationary orbit is and why many communications satellites are put in geostationary orbits.

8. Critical Thinking Explain why a taller microwave relay tower on Earth's surface has a longer transmission range than a shorter relay tower has.

Chapter Resource File

• **Concept Review** GENERAL
• **Quiz** BASIC

Answers to Section 1 Review

1. telephone communication, television transmission, radio communication, satellite communication, fiber-optic communication

2. Telecommunication is communication with someone by electronic means. While talking face to face does not fit this definition, talking on the telephone does.

3. Sound can be converted into an analog signal through the use of a microphone. Sound waves cause a diaphragm to vibrate, which also vibrates the voice coil. The vibrating voice coil generates an electrical current that vibrates in the same way that the sound wave does, converting it to an electrical analog signal.

4. a. digital
   b. analog
   c. digital
   d. analog

5. Compared to metal wires, optical-fiber cables (a) carry more telephone calls, (b) transmit signals over a greater distance, and (c) result in less energy loss.

6. Communication satellites can transmit signals around the world by passing the signal by satellite. The signal reaches the satellite through an uplink, then passes from satellite to satellite until it reaches its destination, then is transmitted to the ground through a downlink.

See "Continuation of Answers" at the end of the chapter.

Telephone, Radio, and Television

SECTION
2

OBJECTIVES

▶ **Describe** how a telephone converts sound waves to electric current during a phone call.

▶ **Distinguish** between physical transmission and atmospheric transmission for telephone, radio, and television signals.

▶ **Explain** how radio and television signals are broadcast using electromagnetic waves.

▶ **Explain** how radio and television signals are received and changed into sound and pictures.

▶ **KEY TERMS**
atmospheric transmission
carrier
modulate
cathode-ray tube
pixel

Focus

Overview

Before beginning this section, review with your students the Objectives listed in the Student Edition. In this section, students learn how telephone, television, and radio signals are transmitted and received. They also learn how telephones convert sound waves to electric current, and how radio and television signals are converted to and from electromagnetic waves before and after transmission.

Bellringer

Use the Bellringer transparency to prepare students for this section.

What sort of information do communication satellites relay around the world? Some information is vital business information, and some is secret government and military communication. Much of the information, however, consists of radio and television programming and telephone conversations.

Telephones

When you talk on the telephone, the sound waves of your voice are converted to an electrical signal by a transducer, a microphone in the mouthpiece of the telephone. As you hear the voice from the earpiece, a speaker, another transducer, is changing an electrical signal back into sound waves.

The electret microphone vibrates with sound waves, creating an analog signal

Most newer telephones use an *electret microphone*. In this type of microphone, an electrically charged membrane is mounted over an *electret*, which is a material that has a constant electric charge. The membrane vibrates up and down with the sound waves of your voice, as shown in *Figure 12.*

This motion causes a changing electric field so that an analog electrical signal that corresponds to your voice is produced. This signal is then transmitted as variations in an electric current between your telephone and the telephone of the person to whom you are talking.

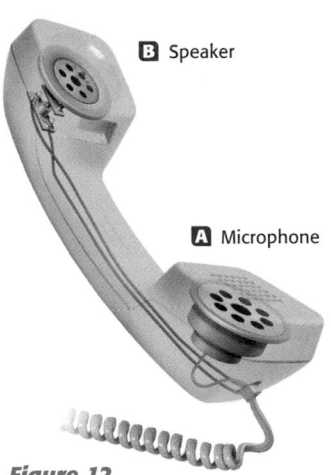

B Speaker

A Microphone

Figure 12

The sound waves from your voice are transformed by the microphone **A** into an analog electrical signal. A speaker **B** converts the analog electrical signal back to sound waves.

601

Motivate

Opening Discussion ——— GENERAL

Ask students to think about what their lives would be like without telephones, televisions, and radios. Ask them to describe ways these technologies improve our lives. For example, radios and televisions provide entertainment and news. They also warn us of dangerous weather conditions. Ask students if there are any disadvantages of these technologies. Some students might suggest that overuse of any of these technologies can use up time that could be spent for other activities such as reading, conversing with family and friends, exercising or playing sports, and so on.

LS Verbal

Did You Know ?

Electret Membranes The membrane of an electret microphone is only 1×10^{-6} m (0.000 001 m) thick! In other words, one million electret membranes stacked on top of each other would only be as tall as a meterstick.

PHYSICS
CONNECTION

Electrets Electret membranes can also be used as particle detectors, primarily for detecting radioactive alpha particles and radioactive radon gas. When these small particles hit the membrane, they alter the total charge. Detectors near the membrane can measure this change in charge and identify the captured particles. In addition, these membranes are excellent for filtering very small particles out of the air.

Chapter Resource File

• Lesson Plan

 Transparencies

TT Bellringer

INTEGRATING

BIOLOGY The nervous system extends throughout the body. Sometimes a person can become injured and lose feeling in a part of his or her body. This loss of feeling may be the result of damaged or torn nerve cells. Damaged nerve cells break the pathway for the nerve impulses. This is similar to breaking a wire in an electrical circuit, which interrupts the electrical current.

Demonstration

Fiber Optics Cable

(Time: About 15 minutes)

Materials

• laser demonstration
• optical-fiber cable
• blank white paper

Step 1 Borrow a helium neon laser and a thick demonstration optical-fiber cable (in the shape of a spiral) from the physics department. Show the optical fiber to students, and explain that this one is thick so that it can be easily seen by students in a classroom. The actual size of a piece of optical cable is approximately the same thickness as a human hair.

Step 2 Shine the laser beam into one end of the cable and let students gather around so that they can see the beam bouncing inside the cable. Project the light that is coming out of the end of the cable onto a piece of white paper, so that they can see that the light is being transmitted.

INTEGRATING

BIOLOGY Biologists have discovered that information is transmitted through the human body by the nervous system. The nervous system contains billions of nerve cells that form bundles of cordlike fibers. Nerve signals, known as *impulses,* can travel along nerve fibers at speeds ranging from about 1 m/s to 90 m/s. Nerve signals are relayed through the body by a combination of electrical and chemical processes.

▶ **atmospheric transmission** the passage of an electromagnetic wave signal through the atmosphere between a transmitter and a receiver

602

The movement of the speaker cone converts the analog signal back into sound waves

When you get a telephone call, the electrical signal enters your telephone. The incoming electrical signal travels through a coil of wire that is fastened to a thin membrane called a *speaker cone.*

The wire coil is placed in a constant magnetic field and can move back and forth. The varying electric current of the incoming signal creates a varying magnetic field that interacts with the constant magnetic field. This causes the coil to move back and forth, which in turn causes the speaker cone to move in the same way. The movement of the speaker cone creates sound waves in the air that match the sound of your caller's voice. Speakers in radios, televisions, and stereo systems work the same way.

Telephone messages are sent through a medium in physical transmission

Telephone messages can be voice calls, faxes, or computer data. But how do the messages arrive at the right place? When you make a call, the signal is sent along wires to a local station. Telephone wires arrive at and leave the station in bundles called cables that are strung along poles or run underground. The station's switching equipment detects the number called.

If you are calling someone who lives nearby, such as a neighbor, the switching equipment sends the signal down wires that connect your phone through the station to your neighbor's phone. When the signal reaches your neighbor's phone, the phone rings. When your neighbor picks up the phone, the circuit is completed.

Sometimes telephone conversations travel a short distance by wire and then are carried by light through fiber-optic cables. In this case, the varying current is fed to a laser *diode*, a device used to convert alternating current, causing the laser light to brighten and dim. In this way, the electrical signal is converted into a light or optical signal. This light passes through an optical fiber to its destination, where a sensor changes light back to an electrical signal. Transmission of signals by wires or optical fibers is called *physical transmission.*

Messages traveling longer distances are sent by atmospheric transmission

Long-distance calls may be transmitted over wire or fiber-optic cables, or they may be sent through the atmosphere using microwave radiation. The transfer of information by means of electromagnetic waves through the atmosphere or space is called **atmospheric transmission.** The use of microwaves for telephone signals is one example of atmospheric transmission.

Did You Know ?

Modems Modems were developed to allow computers to communicate with each other. Originally, the first modems converted the digital signals in the computer into analog sound waves, to be transmitted through the phone lines. Later it was found that digital signals worked better than analog signals. As a result, modems today convert the digital electrical signals in the computer into digital sound waves.

Historical Perspective

Manual Switching Before the development of automatic switches, operators had to connect a call to the next circuit manually so that the call would be routed to its final destination. This would increase the time it took to have a call connected and increase the number of wrong phone connections.

Computers help route calls

When you make a call, computers are used to find the most direct form of routing. Either physical or atmospheric transmission or a combination of the two may be used for long-distance calls, as shown in *Figure 13*. If the telephone system is very busy, computers may route your call indirectly through a combination of cables and microwave links. Your call to someone 100 mi away could actually travel for thousands of miles.

Cellular phones transmit messages in the form of electromagnetic waves

A cellular phone is just a small radio transmitter/receiver, or *transceiver*. Cellular phones communicate with one of an array of antennas mounted on towers or tall buildings. The area covered by each antenna is called a *cell*.

As the user moves from one cell to another, the phone switches to communicate with the next antenna. As long as the telephone is not too far from a cellular antenna, the user can make and receive calls.

A cordless phone is also a radiowave transceiver. The phone communicates with its base station, which is also a transceiver. The base station is connected to a standard phone line.

Figure 13

Your telephone call **A** arrives at a local switching station **B**. Depending on its destination, the call is routed through a wire cable **C**, fiber-optic cable **D**, microwave towers **E**, or communication satellites **F**. The telephone signal then arrives at another switching station **G** where it travels to your friend's house, and the phone rings.

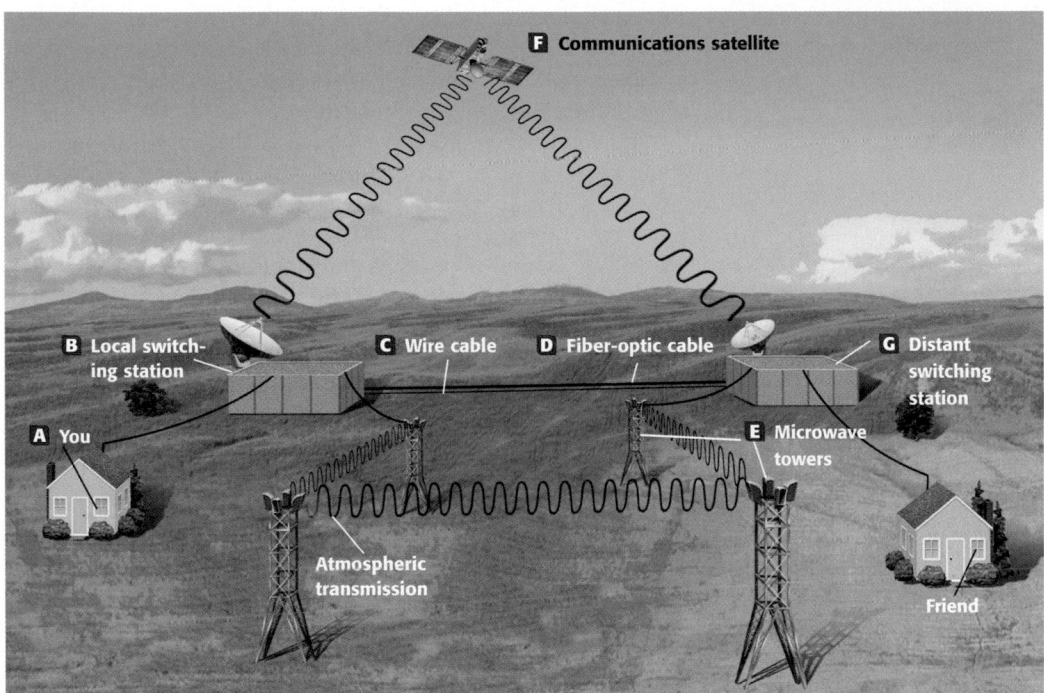

603

Demonstration

Internal Reflection

(Time: About 15 minutes)

Materials

• laser
• salt water
• round bottom flask
• ring stand
• test-tube clamp

Step 1 Fill the flask with water and put a little salt into the water.

To test the amount of salt, turn out the lights and shine the laser beam into the salt water. You should be able to easily see the laser beam in the salt water. The laser beam should emerge from the other side of the flask without much distortion.

When salt concentration is correct, attach the flask to a tall ring stand with a test-tube clamp.

Step 2 Turn off the lights and shine the laser beam into the flask. Now aim the beam from underneath the flask so that the beam hits the boundary between the water and air at an angle.

The most common occurrence is that light bends (refracts) when it passes from one medium to another. But if you increase the angle of incidence (from an imaginary line perpendicular to the water's surface) until the beam is very close to the water's surface, the light will reflect into the water instead of refracting into the air. This phenomenon is called *total internal reflection.*

This principle is used to send data using light pulses through optical fibers. Due to total internal reflection, the light beam stays inside the cable.

Figure 14

Morse code was a method of transmitting communication signals before the time of radio waves.

▶ **carrier** a wave that can be modulated to send a signal

▶ **modulate** to change a wave's amplitude or frequency in order to send a signal

Figure 15

Amplitude Modulation

An audio signal carrying sound information modulates a carrier wave.

Radio and Television

The first long-distance transmission of a signal using radio waves was made across the English Channel in 1899. At the time, the signals were sent in Morse code, as in *Figure 14.* For the next 20 years, all radio transmissions were sent this way. It was not until 1918 that voice messages could be sent over the air using radio waves. In 1920, the first commercial radio station, KDKA, in Pittsburgh, Pennsylvania, went on the air, broadcasting sound signals by means of radio waves.

Sound waves are converted to electromagnetic waves for radio broadcast

A radio signal begins as a sound, or audio, signal that is first converted into a varying electric current from a microphone, tape deck, or CD player. This varying current is the analog of the sound waves from a voice or music source, as shown in *Figure 15A.*

A microphone is capable of producing only a weak signal, which has to be amplified, or increased in power, using an electronic device called an amplifier.

Now the signal is ready to be broadcast using a transmitter at the radio station. The visible part of the transmitter is an antenna, and the transmitter also contains different electric circuits including an oscillator. The oscillator produces a **carrier,** which is a signal of constant frequency and amplitude, as shown in *Figure 15B.* The numbers you see on your radio dial correspond to the carrier wave's frequency.

You can imagine the carrier wave as the wave on which the audio signal to be broadcasted will ride. The audio signal contains the sound information in the frequency range of the human voice, from about 100 to 3000 Hz. Also, the change in the loudness of the sound appears in the signal in terms of changing amplitude. The sound signal and the carrier signal meet in a specialized circuit in the transmitter. Here they combine and the audio signal changes, or **modulates,** the carrier wave. The result is a signal of constant frequency with an amplitude that is shaped by the audio signal, as shown in *Figure 15C.*

604

Historical Perspective

Crystal Radios At the turn of the twentieth century, an American engineer, Greenleaf Whittier Pickard, found that a number of naturally occurring crystalline materials, particularly silicon, could be used to detect radio signals. The detection occurs at the contact point between one of these crystals and the tip of a piece of wire.

Radios employing this type of detector became known as crystal radios. In the typical early radiowave crystal detector, the crystal rock was fixed into a brass cup and the radio operator found the loudest signal by touching the wire, called a cat's whisker, to various points on the surface of the crystal.

Modulation can be either AM or FM

Most broadcast carrier waves are modulated either by *amplitude modulation* (AM) or by *frequency modulation* (FM). In amplitude modulation, the audio signal increases and decreases the amplitude of the carrier wave in a pattern that matches the audio signal. In frequency modulation, the audio signal affects the frequency of the carrier wave, changing it in a pattern that matches the audio signal.

The modulated signal generated in the transmitter causes electric charges to move up and down along the length of the antenna. The resulting motion of the charges produces radio waves corresponding to the modulated signal.

The path that radio waves follow depends on the transmission frequency. Higher frequency transmissions can follow only a simple straight line. This is called *line-of-sight transmission.* To receive a signal from an FM radio station, which can broadcast at frequencies between 88 and 108 MHz, your radio must be located no farther than just over the horizon from the broadcasting antenna (usually about 40 to 80 km).

You can receive AM stations that are much farther than 80 km away. AM frequencies between 540 and 1700 kHz can travel as *ground waves,* which can follow the curvature of the Earth for some distance, unlike line-of-sight transmissions.

Radio stations use sky waves to broadcast long distances

Another way AM radio stations can broadcast farther is by using *sky waves.* Sky waves spread out from the antenna into the sky and are reflected in the upper atmosphere, which contains charged particles. Sky waves are reflected back to Earth by these particles.

Some radio broadcasting uses sky waves to reach distant locations around the world. Certain powerful AM signals that use sky waves can be received thousands of miles away. These stations are often limited to using sky waves at night, when stations that have interfering signals may be off the air.

Radio receivers convert electromagnetic waves back into sound

The antenna of your radio receiver works as a transducer. When radio waves strike it, they produce very weak electric currents that match the original radio signal. But radio waves from many stations with different frequencies are striking the antenna. Fortunately, each station broadcasts with a different carrier frequency. Like the girl in **Figure 16,** you have to adjust the antenna circuit with a *tuner* so that the radio responds to only the frequency of the station you want to hear.

605

Figure 16

Even when the antenna is hidden, it is responding to radio waves.

SKILL BUILDER — GENERAL

Reading Skills Have students draw flowcharts that illustrate how radio and television signals are broadcast and received. Their charts should show how the signals are transmitted and should also include all conversions that occur on both ends (before they are sent and after they are received). Encourage students to consult the appropriate sections in the text as they create their charts. **LS Visual**

Teaching Tip

Deep Space Network NASA is able to receive transmissions from its satellites in space by the use of extremely large receiving antennas. These antennas make up a system called the Deep Space Network. This network consists of three deep-space communication facilities that are located approximately 120° apart around the world in the Mojave Desert; in an area near Madrid, Spain; and in an area near Canberra, Australia. The placement of these antennas permits the constant observation of satellites as Earth rotates on its axis. During the *Voyager 1* and *Voyager 2* missions, the Deep Space Network kept a vital link with the two spacecraft. The network received thousands of hours of radio transmissions from the satellites as well as transmitted instructions to the spacecraft.

Did You Know ?

Radio Waves Radio and television transmissions travel through the atmosphere to our antennas. These transmissions also travel into space. Radio waves travel at the speed of light, which is 1.86×10^5 miles per second.

REAL-LIFE CONNECTION

Short Wave Radio Amateur short wave radio operators, or ham radio operators, make use of Earth's atmosphere to communicate over large distances. Ham radio is limited by Earth's curvature just as microwave communications are. One way to get past the horizon is to bounce the short wave radio waves off the atmosphere. This works the same as when light bounces inside a piece of optical-fiber cable. As a result, the radio wave can travel farther than the horizon.

Demonstration

Deflecting Electron Beams

(Time: About 15 minutes)

Materials

• oscilloscope or CRT
• magnet
• PVC pipe
• wool

You can probably borrow an oscilloscope or CRT from the physics department. These devices work just like televisions. A beam of electrons is fired at the screen from inside the device. Magnetic and electric fields deflect the beam to form the different images on the screen.

Step 1 Turn on the oscilloscope or CRT, and position the screen so that students can easily see it.

Step 2 Place a magnet near the screen. This should have the effect of distorting the image seen on the screen.

Step 3 It may also be possible to distort the image with a strong electric field. Charge a piece of PVC pipe by rubbing it with a piece of wool. Depending on the strength of the electron beam, this may also deflect the beam and distort the image.

Figure 17
After the detector removes the audio signal from the carrier wave, the signal is amplified and sent to a speaker. (Note that the amplifiers and detector shown as boxes correspond to different circuits that are part of the radio.)

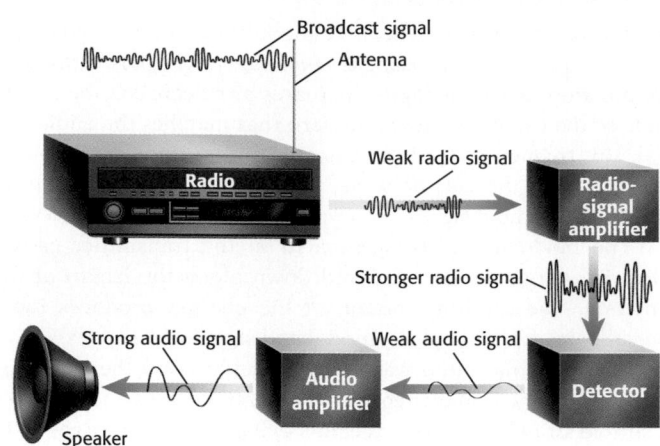

☐ internet connect

www.scilinks.org
Topic: Television
Technology
SciLinks code: HK4137
SCILINKS. Maintained by the National Science Teachers Association

■ **cathode-ray tube** a tube that uses an electron beam to create a display on a phosphorescent screen

Next the modulated signal from the antenna is sent to a detector, as shown in *Figure 17.* The carrier wave has a very high frequency compared with the original electrical signal, so the two can be separated easily. The electrical signal then goes to an amplifier, which increases the signal's power. Finally, the amplified signal is sent to a speaker, where the sound that was originally broadcast is recreated.

Television sets convert electromagnetic waves back into images and sound

Television signals are also received by an antenna. By selecting a channel, you tune the television to the carrier frequency of the station of your choice. The carrier wave is passed to a detector that separates the audio and video electrical signals from the carrier. The audio electrical signals are sent to an audio amplifier and speaker, just as in a radio. The video electrical signal, which contains the color and brightness information, is used to create an image on the face of a picture tube.

The picture tube of a black-and-white television is a large **cathode-ray tube,** or CRT. A CRT makes a beam (ray) of electrons from a negatively charged cathode. The beam is directed toward the face of the tube that is covered with *phosphors,* which glow when an electron beam strikes them. Electromagnets arranged around the neck of the tube deflect the beam, causing it to move across the phosphor-coated face. The moving beam lights up the phosphors in a pattern that recreates the shot taken by the television camera. Each pass of the beam is called a *scan line.* In the United States, each complete image is made up of 525 scan lines.

606

REAL-LIFE CONNECTION

Television Sets A television works by the firing of a beam of electrons at the screen. The screen glows where the beam of electrons hits. The electrons are guided by strong electric and magnetic fields so that the beam of electrons spreads across the screen, recreating the image that is contained in the video signal.

The television signal that is broadcast to your TV tells the set how to change the electric and magnetic fields. A magnet placed near a TV screen will cause the electron beam to be deflected and will change the color on the screen. Never leave magnets near a television set because they can damage it with permanent distortion.

Color picture tubes produce electron beams

Color picture tubes in some televisions, like the one shown in *Figure 18,* produce three electron beams, one for each of the primary colors of light: red, blue, and green. The phosphors on the face of the tube are arranged in groups of three dots, one of each color. Each group of three dots is a **pixel,** the smallest piece of an electronically produced picture.

To make sure the beam for red strikes only red phosphors, two different approaches can be used. In one, a screen with holes, called a *shadow mask,* lies just behind the face of the tube. The beam for each color passes through a hole in the shadow mask at an angle so that the beam strikes only the phosphor dot that glows the correct color. Another approach in some televisions use a single electron beam deflected toward the phosphor of the correct color by a charged wire grid.

▆ **pixel** the smallest element of a display image

VOCABULARY *Skills Tip*

The term pixel *is derived from the phrase* **pic**ture **el**ement.

SKILL BUILDER — **BASIC**

Interpreting Diagrams Figure 18 illustrates the path of a video signal from an electromagnetic wave to an image on a television screen. Ask students to describe the path in their own words. You may wish to point to the appropriate parts of the image on an overhead projector during their descriptions. **LS** Verbal

Teaching Tip
Computer Scanners A computer scanner operates in a manner similar to a television set. It scans the entire image and breaks it up into individual pixels. It then identifies the color and brightness of each pixel and digitally records the location of the pixel and the color and brightness of the pixel. When it reforms the image on the screen, the series of small colored dots that are generated on the screen form a continuous image to our eyes.

Figure 18

A The video signals modulating the television carrier waves are detected and are then used to control the electron beams in the cathode-ray tube. The sound signal is amplified and sent to a speaker, while video signals vary the intensity of the three electron beams.

B Electromagnets sweep the beams across the face of the screen. The intensity of each beam determines how bright the phosphor dots light up.

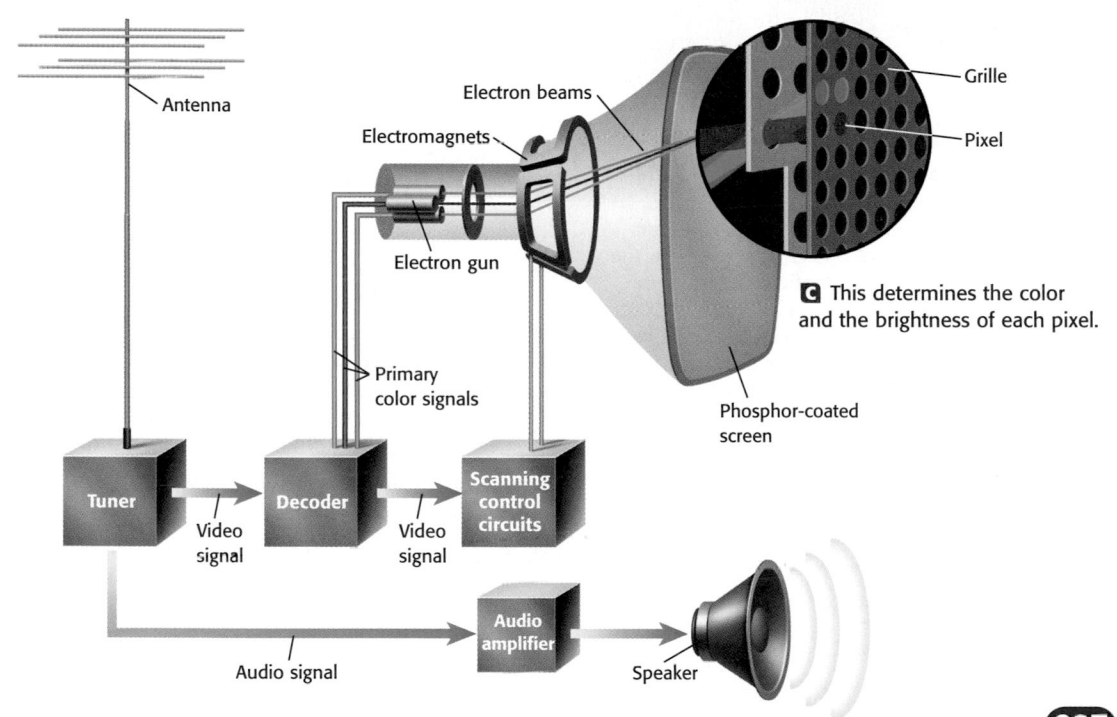

Grille

Pixel

C This determines the color and the brightness of each pixel.

Antenna

Electron beams

Electromagnets

Electron gun

Primary color signals

Phosphor-coated screen

Tuner — Video signal — Decoder — Video signal — Scanning control circuits

Audio signal — Audio amplifier — Speaker

607

Did You Know?

Satellite Dishes The large curved dish of a satellite is called the reflector dish, and the part that sticks out of the center of the dish is the antenna. Electromagnetic (EM) waves are emitted from a satellite that is in orbit. When these waves hit the dish, they are reflected to the antenna. The EM waves generate an electrical signal. The electrical signal then goes into your house via a cable and is amplified by an amplifier. This amplified signal is then sent to your television.

Transparencies

TT Television

Science and the Consumer ── GENERAL

TV by the Numbers: High-Definition Digital TV High Definition Television (HDTV) represents a radical change in the television industry. This new format will require broadcasters to transmit their television signals in a different way. This is because the number of scan lines and the number of pixels are increased to create sharper images. In addition to increased video picture quality, the HDTV systems will also incorporate six audio channels for surround sound systems. This further changes the requirements for broadcasting companies to implement this technology in their businesses.

Consumers are not yet willing to purchase HDTVs because HDTVs are currently extremely expensive, and there are not enough shows in the HDTV format to make them a worthwhile purchase.

Answers to Your Choice

1. Accept all reasonable answers.

2. Have students search the Internet or call an electronics store in the area to determine the current cost of an HDTV. The cost of VCRs dropped by 76% in the 5 years after their introduction. To project the cost of an HDTV in 5 years, calculate 76% of the current cost of an HDTV, and subtract that value from the current cost of an HDTV.

LS Verbal

TV by the Numbers: High-Definition Digital TV

When you turn on a television in the year 2006, it may look a lot different than the one you watch now. That's because the Federal Communications Commission (FCC) has decided that all television stations in the United States must broadcast only digital, high-definition television, called HDTV for short, by 2006.

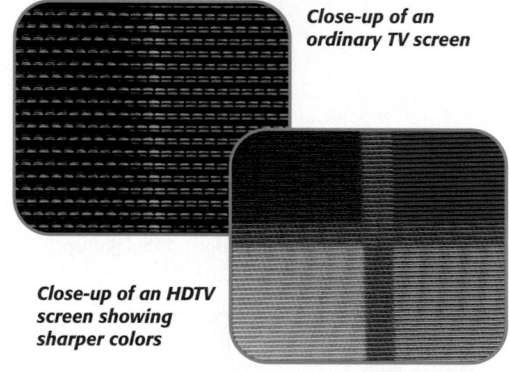

Close-up of an ordinary TV screen

Close-up of an HDTV screen showing sharper colors

Comparing HDTV with Ordinary TV

The HDTV picture looks very detailed and sharp compared with an ordinary television picture. You can even see the faces of fans at a sports event. The picture has a width-to-height ratio of 16:9, similar to many movies that you see in a theater. By 2006, you may have a TV with a large flat screen that hangs on the wall like a painting or mirror. HDTV sound is clear, digital sound, like that recorded on a CD.

However, you won't have to throw out your old television set in 2006. A converter box will let your old TV show pictures that are broadcast in HDTV. However, the picture won't look any better than if it is a regular broadcast.

History of HDTV

The development of HDTV began in the early 1980s when engineers realized that newer microprocessors

would be able to both send and decode data fast enough to transmit a detailed television picture digitally. In 1988, 23 different HDTV systems were proposed to the FCC. In 1993, several companies joined the Massachusetts Institute of Technology in what was called the Grand Alliance. Its purpose was to create HDTV standards for broadcasters. In 1996, the FCC approved an entirely digital system, and by late 1998, the first commercial HDTV receivers were on sale at prices between $10 000 and $20 000.

HDTV Technology

HDTV achieves its sharp picture by using almost 1200 scan lines, compared with 525 on analog TV. The digital signal can also be continuously checked for accuracy, so the picture remains clear.

The movie industry is very interested in HDTV. It will be able to release new HDTV tapes and discs of movies already released on other video formats. Another possibility is that some movies can be released to pay-per-view HDTV at the same time they are released in theaters. The HDTV picture and sound quality should be so good that some people may prefer to stay at home to watch a new movie.

Your Choice

1. **Making Decisions** What effect do you think HDTV will have on movie theaters, especially if studios make movies available on HDTV for the same price as a theater ticket? Explain whether you think people will still want to go to theaters.

2. **Critical Thinking** When home VCRs were introduced in the mid-1970s, they cost about $2500. By 1980, the price was about $600. By 1985, the price was about $450. By the mid-1990s, you could buy a good-quality VCR for about $250. Check the current price of an HDTV, and use the VCR example to project what one will cost in 5 years.

608

Quick Lab

How do red, blue, and green TV phosphors produce other colors?

Materials ✓ three adjustable flashlights with bright halogen bulbs ✓ white paper
✓ several pieces of red, blue, and green cellophane

1. Adjust the focus of each flashlight so that it produces a circle of light about 15 cm in diameter on a white sheet of paper. Turn off the flashlights.

2. Place a piece of red cellophane over the lens of one flashlight, green cellophane over the lens of another, and blue over the lens of the third.

3. Turn on the flashlights, and shine the three beams on white paper so the circles of light overlap slightly.

4. Adjust the distance between the flashlights and the paper until the area where all three circles overlap appears white. Add more cellophane if necessary.

Analysis

1. Describe the three colors formed where two of the beams overlap.

2. What combinations of light produced the colors yellow and cyan?

SECTION 2 REVIEW

SUMMARY

▶ Telephones change sound to electrical signals and electrical signals to sound.

▶ Signals can be sent by physical transmission or by atmospheric transmission.

▶ Signals modulate carrier waves by amplitude modulation (AM) and frequency modulation (FM).

▶ Television converts electromagnetic waves.

1. **Describe** how telephones convert sound to electrical signals and electrical signals to sound.

2. **List** three ways that telephone signals can travel.

3. **Identify** which have a higher frequency—AM or FM signals.

4. **Describe** the function of phosphors in a cathode-ray tube.

5. **Describe** how a radio receiver converts a broadcast signal into sound waves.

6. **Describe** how television sets convert electromagnetic waves into images and sound.

7. **Critical Thinking** Why do you think there is increasing interest in using fiber-optic cables to provide homes with cable television service?

609

Quick Lab

How do red, blue, and green TV phosphors produce other colors? Color filters can be purchased inexpensively at theatrical supply stores. You want to choose primary red, primary green, and primary blue. These will give the best color mixing.

In addition, you can also purchase colored flood lights from a hardware store. You can then project these three lights onto the overhead projector screen and the screen will appear white. If you separate the lights by approximately 1 m and then shine the light onto the projector screen, you will get interesting shadowing effects. Have a student stand in front of these lights, and you will see six different-colored shadows. Each shadow is the mixture of two different colors. **LS** Visual

Close

Quiz

Determine whether each statement is true or false. If false, replace the underlined term with the correct term.

1. <u>Atmospheric</u> transmission involves wires, cables, or optical fibers. (false; physical)

2. Cell phones transmit messages as <u>sound</u> waves. (false; electromagnetic)

3. Television sets convert <u>electromagnetic</u> waves into images and sound. (true)

Answers to Section 2 Review

1. Sound waves cause a thin membrane in the microphone in the mouthpiece of a telephone to vibrate. This membrane is attached to an electronic component that generates an electrical current. The electrical signal is then transmitted through the telephone lines to its destination. The process works in the reverse in the speaker to change the electrical signal back into sound.

2. Telephone signals can travel through conventional cable wires, through optical-fiber cable, and through orbiting satellites.

3. FM radio waves have higher frequencies than AM radio waves.

4. The purpose of phosphors in a cathode-ray tube is to glow when struck by the electron beam.

5. The broadcast signal is an electromagnetic wave that causes the electrons in a radio antenna to vibrate at the same frequency as the radio wave. The vibrating electrons carry the signal into a radio to be amplified and filtered. The final signal then goes to the speakers, where it recreates the broadcast sound.

See "Continuation of Answers" at the end of the chapter.

Chapter Resource File

- **Cross-Disciplinary Worksheet** Science and the Consumer—HDTV: Why Make the Switch? GENERAL
- **Quick Lab Datasheet** How do red, blue, and green TV phosphors produce other colors? GENERAL
- **Concept Review** GENERAL
- **Quiz** BASIC

Focus

Overview

Before beginning this section, review with your students the Objectives listed in the Student Edition. In this section, students learn about the four basic functions of a computer, the binary nature of computer data, the use of logic gates in computers, and the distinction between hardware and software. The section concludes with a discussion of what the Internet is and what tools are needed to use the Internet.

🔊 Bellringer

Use the Bellringer transparency to prepare students for this section.

Motivate

Opening Demonstration

Purchase some vacuum tubes from a TV repair shop or an electronics store. Hand them out to students to examine. Tell students that the first computers consisted of thousands of such types of vacuum tubes, which filled entire rooms. The development of the transistor eliminated the need for these tubes.

Computers and the Internet

▶ **KEY TERMS**
computer
random-access
  memory
read-only memory
hardware
software
operating system
Internet

▶ **OBJECTIVES**

▶ **Describe** a computer, and list its four basic functions.
▶ **Describe** the binary nature of computer data and the use of logic gates.
▶ **Distinguish** between hardware and software, and give examples of each.
▶ **Explain** how the Internet works.
▶ **Define** how technological tools are applied to address personal and societal needs.

▶ **computer** an electronic device that can accept data and instructions, follow the instructions, and output the results

Figure 19
ENIAC, the world's first practical digital computer, like the one shown here, used 18 000 vacuum tubes. The modern microprocessor has thousands of times ENIAC's computing power.

610

D id you heat a bagel in the microwave for breakfast? Have you ever inserted a card in a slot to pay your fare on a bus or subway? Maybe you rode to school in a car. Did you stop for a traffic light? Was the temperature in your classroom comfortable? Did a clerk scan a bar code on an item you bought at the store?

All of these situations involve computers or the use of computers to function. The computer that controls traffic lights may be large and complex, while the one in the microwave oven is likely to be small and simple.

Computers

A **computer** is a machine that can receive data, perform high-speed calculations or logical operations, and output the results. Although computers operate automatically, they do only what they are programmed to do. Computers respond to commands that humans give them, even though they sometimes may appear to "think" on their own.

Computers have been changing greatly since the 1940s

The first electronic computer was the Electronic Numerical Integrator and Computer (ENIAC), shown in *Figure 19.* It was developed during World War II. ENIAC was as big as a house and weighed 30 tons. Its 18 000 vacuum tubes consumed 180 000 W of electric power. During the late 1940s, computers began to be used in business and industry. As they became smaller, faster, and cheaper, their use in offices and homes dramatically increased.

Alternative Assessment —— ADVANCED

Researching Computer Technology Have students research the historical development of the computer. There is much information available both in books at the public library and on the Internet. Have students make a timeline showing when certain developments took place. Make the timeline long enough to stretch across the entire length of the classroom. Once the timeline is made, you can assign different aspects of the timeline to different groups of students for detailed research. Then students can give oral reports in chronological order on the historical development of the computer. **LS** Verbal

Today computers are so common that we hardly notice them. Try to imagine what computer developers in the 1940s would think if they could see a modern personal computer, or PC, which fits on a desk and computes thousands of times faster than the earlier cumbersome computers like ENIAC.

Computers carry out four functions

Digital computers perform four basic functions: input, storage, processing, and output. The input function can be carried out using any number of devices, as shown in *Figure 20*. When you use a personal computer, you can use a keyboard or a mouse to input data and instructions for the computer. You may use a mouse to draw or select text in a document.

Other input devices include a scanner, which can enter drawings or photographs. A modem connected to a telephone line can be both an input and an output device.

Microphones, musical instruments, and cameras can be used as input devices. Once the data are processed, the result, or output, may be displayed on a monitor. You can also send output to a printer. Sound output goes to speakers or to a recording device. Both input and output data can be stored in storage devices.

Figure 20

A computer can receive data from many devices, store information on a hard drive, process data as needed, and store results or send them to an output device.

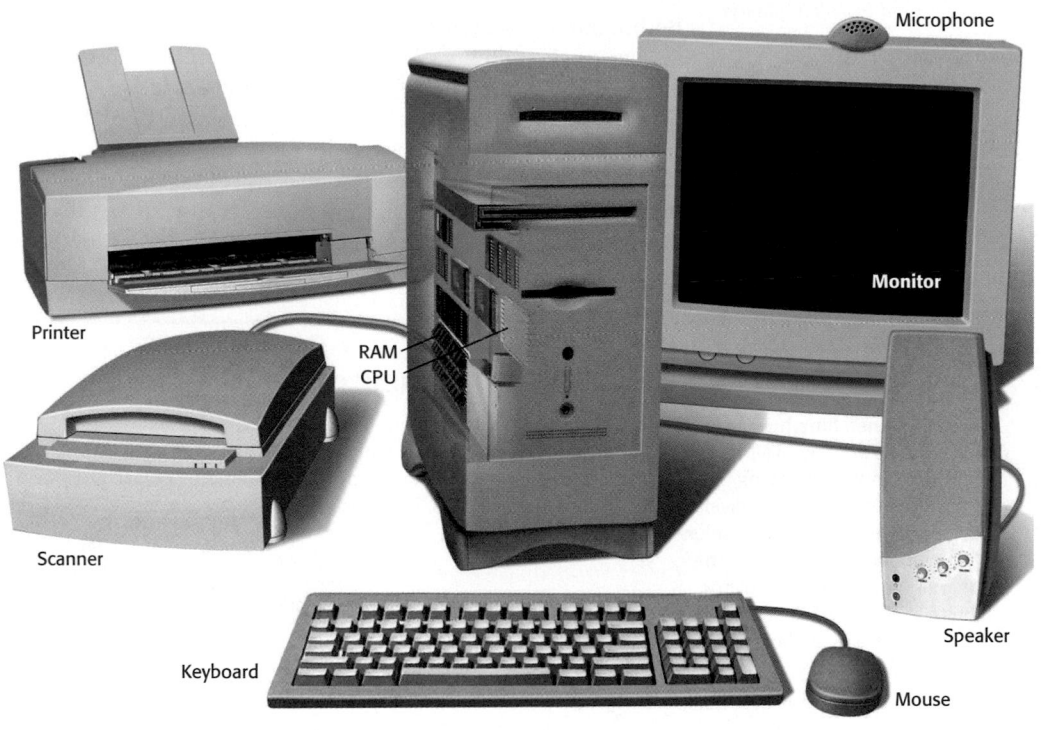

Printer
Scanner
RAM
CPU
Keyboard
Monitor
Microphone
Speaker
Mouse

Historical Perspective

Moon Landing In 1969, Neil Armstrong and Buzz Aldrin were the first men to land on the moon. The computers that were used on the landing craft, the *Eagle*, were advanced for the time but were less powerful than the most basic game computer today. During the descent, the amount of data that was received by the computer overloaded many of the circuits. The computer could not handle the total number of tasks, and it gave error messages to the two astronauts. The computer had to shut down some of its less important systems. Computer system errors occurred multiple times during the landing.

Teaching Tip
Sound Cards and Digital Images

Computer engineers have developed computer interface cards that allow musicians to play music into the computer. These cards allow a musical keyboard to communicate directly with the computer. A musician can play music on the keyboard, and the computer will not only record the music and play it back, but will also write out the notes with musical symbols in standard five-staff notation. This music can then be printed out for other musicians to use. In addition, the computer can be used to mix different sounds together and to add special effects. This makes it possible to have a basic recording studio in the home.

Computers have integrated not only musical sounds but also color images. The recent advances in image capturing and image enhancing have made it affordable for people to design and print graphics on their computers. Scanners scan the image, and photo enhancement programs make it possible to touch up the image. This software makes it possible to alter images in many different ways. Digital cameras make it possible to capture images and copy them directly into the computer for printing or enhancement. Video capture cards make it possible to capture video images and store them on a computer or CD-ROM.

611

Chapter Resource File

• Lesson Plan

Transparencies

TT Bellringer

PHYSICS Electricity is used in the current generation of analog and digital computers to activate circuits and resistors. The next generation is expected to be optical computers. The optical computer will use light instead of electricity as an energy source.

Quick ACTIVITY

How Fast Are Digital Computers?

Materials (per pair):
• hand-held calculator
• paper and pencil
• watch or stopwatch

Teacher's Notes: Some calculators show the entire calculation on their screens, rather than just the answer. As a result, you may find a difference in the calculation speed depending upon the type of calculator being used. Compare these calculator speed variances to an average computer, which can do millions of calculations per second. **LS Logical**

INTEGRATING

PHYSICS All PCs are digital computers, but analog computers also exist. An example of an analog computer in a car is the gasoline gauge, whose needle moves in response to a voltage sent from a sensor in the gas tank.

VOCABULARY *Skills Tip*

Most disks used for computer work today are not actually floppy. The name floppy disk *originally referred to a larger disk that was encased in softer, plastic sleeves.*

Computer input is in the form of binary code

All input devices provide data to the computer in the form of binary code. For example, a keyboard contains a small processor that detects which key is pressed and sends the computer a binary code that represents the character you typed. Devices such as temperature sensors, pressure sensors, and light sensors provide information in the form of varying voltage. This information is analog; that is, it changes continuously over the range of the quantity being measured. Such information must be passed through an analog-to-digital converter (A to D converter) before the data can be used by a computer.

Computers process binary data, including numbers, letters, and other symbols, in groups of eight *bits*. Each bit can have only one of two values, usually represented as 1 and 0. A group of eight bits is called a *byte*.

As shown in *Figure 5,* when you type the capital letter *W*, the computer receives the data byte 01010111. The lowercase letter *e* is received as 01100101. So, if you type the word *We*, the computer recognizes the word as 0101011101100101, a combination of the *W* and *e* bytes.

Computers must have a means of storing data

Both input and output data can be stored on long-term storage devices, such as the *hard-disk drive,* sometimes called the hard drive. Hard-drive storage capacity has increased very rapidly. From 1999 to 2001, available storage increased from approximately 20 to 80 billion bytes (gigabytes or Gigs). Hard drives are so called to distinguish them from disk drives that use removable "floppy" disks and drives that use compact discs (CDs). Floppy disks can be removed from one computer and used in another.

Quick ACTIVITY

How Fast Are Digital Computers?

1. With a partner, time how long it takes for each of you to solve problems involving adding, subtracting, multiplying, and dividing large numbers. Do each problem first by hand and then with the help of a hand-held calculator, which is a form of digital computer. Solve at least five problems using each method.

2. Find the average amount of time spent doing the problems by hand and with a calculator. Compare the two averages, and discuss your results.

Historical Perspective

Batch-Card Processing Initially, inputting data on a computer was a difficult task. A popular method in the 1970s was batch-card processing. This method used computer cards with holes punched in them for the computer to read. Programmers would type a command into a batch console. The console would punch holes into the cards, and each card represented only one command. Large programs might have 10 000 cards. When the card entered the computer, it would pass between a light source and a light detector. The computer "read" the holes by measuring the light that passed through the holes to the detector.

Figure 21
The head of the hard drive moves over the surface of the disk, reading and recording data in narrow tracks.

Disk coated with magnetizable substance

Read-write head

Both hard drives and floppy drives use disks coated with a magnetizable substance. Disks of this type are generally referred to as *magnetic media*. A small read-write head, similar to the record-play head in a cassette tape recorder, transfers data to and from the disk, as shown in *Figure 21.* Each data bit consists of a very small area that is magnetized in one direction for 0 and in the opposite direction for 1. These magnetized areas are arranged in tracks around the disk. When data are being read, the disk spins and the head detects the magnetic direction of each area that passes. When data are being recorded, or "written" on the disk, a current passes through a small coil of wire in the head. The direction of the current at any time creates a magnetic field in one direction or the other. This magnetic field allows the head to record information on the disks in bits of 0s and 1s.

On a disk, the time required to access (read or write) data depends on where the information is stored on the disk and the position of the read-write head.

Random-access memory is used for short-term storage of data and instructions

For working memory, the computer needs to be able to access data quickly. This type of memory is contained on microchips, tiny integrated circuits, as shown in *Figure 22,* and is called **random-access memory,** or RAM.

Each RAM microchip is covered with millions of tiny transistors, electronic devices that transfer current across resistors. Like a light switch, each transistor can be placed in one of two electrical states: *on* or *off*. Each transistor represents either a 0 or a 1 and can thus store one data bit. This memory is called random-access because any of the data stored in RAM can be accessed at the same time. Unlike accessing data stored on the disk, accessing information in RAM doesn't depend on location.

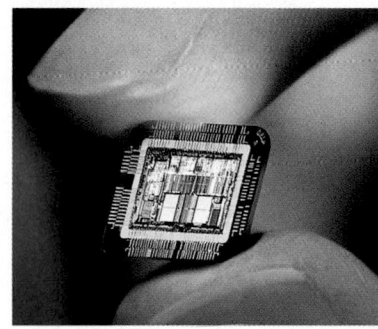

Figure 22
This chip is covered with tiny transistors that function as two-position switches. This feature allows the computer to operate as a binary machine.

▶ **random-access memory**
a storage device that allows a computer user to write and read data; it is the amount of data that the memory chips can hold at one time (abbreviation, RAM)

613

REAL-LIFE
CONNECTION

Virtual Memory The amount of programs that can be run at one time on a computer depends on how much RAM is installed. Some computer users install extra RAM cards. Another option is using virtual memory. If you turn on your computer's virtual memory, you will be able to run more programs than you could otherwise. However, this is useful only up to a certain point; using too much virtual memory can significantly slow down a computer.

Did You Know?
Memory in Electronic Devices
Computers are not the only electronic devices that use memory. Cell phones, personal digital assistants (PDAs), game consoles, car radios, VCRs, and TVs all use some form of memory.

Teaching Tip
Rewritable CD-ROMs The first kind of recordable CD-ROM to enter the market was called a WORM drive. WORM stands for Write Once, Read Many. This means that you could write to the CD-ROM once, but you could not erase and rewrite the information like you could with a floppy diskette. Currently, rewritable CD-ROM drives, which allow you to write to the CD-ROM many times, have entered the market. You can erase the information many times and reuse the CD. These CDs can hold up to 600 MB of information, which is a significant amount.

Teaching Tip ── BASIC
Operating Systems Ask students for examples of operating systems they are familiar with. Many older computers used a DOS operating system. Today, most PCs use some form of Windows, such as Windows 98 or Windows 2000. Macintosh computers use a different operating system, such as OS 9.2 or OS X. **LS Verbal**

▶ **read-only memory** a memory device that contains data that can be read but cannot be changed (abbreviation, ROM)

▶ **hardware** the parts or pieces of equipment that make up a computer

▶ **software** a set of instructions or commands that tells a computer what to do; a computer program

▶ **operating system** the software that controls a computer's activities

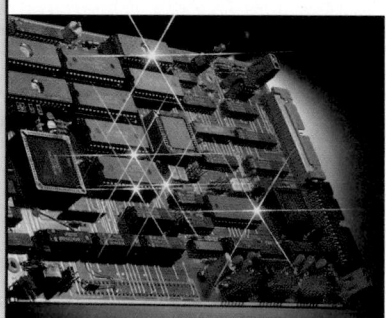

Figure 23
The motherboard is like the nervous system of a computer. It contains the CPU, memory chips, and logic circuits.

614

Read-only memory is for long-term storage of operating instructions

Another type of memory is called **read-only memory,** or ROM. The information in ROM is permanently stored when the chip is manufactured. As a result, it can be read but not changed. When you first turn on a computer, instructions that are stored in ROM set up the computer so that it is ready to receive input data from the keyboard or the hard drive.

Optical storage devices can be more permanent than magnetic disks

Information can also be stored on *compact discs* (CDs) and *digital versatile discs* (DVDs). These discs are called optical media because the information on them is read by a laser light. When they are used to store computer data, they are referred to as CD-ROMs and DVD-ROMs because the data they hold are permanently recorded on them.

Operating systems control hardware

All of the physical components of a computer are called **hardware.** The hardware of the computer can compute and store data only if we provide it with the necessary instructions. These instructions are called computer programs, or **software.**

When a computer is turned on, one of the first programs executed by the computer is the **operating system,** or OS. The OS coordinates the computer hardware—memory, keyboard, disks, printer, mouse, and monitor. It also handles the transfer of computer files to and from disks and organizes the files. The operating system provides the environment in which other computer programs run. These other programs are called applications. Applications include word processors, drawing programs, spreadsheet programs, and programs to organize and manipulate large amounts of information, such as a store's inventory or polling data. Applications also include computer games and programs that allow you to browse the Internet, as this section will explain.

The processing function is the primary operation of a computer

The processing function is where computing actually takes place. Computing or data processing is carried out by the *central processing unit,* or CPU. The CPU of a personal computer usually consists of one microchip, which is not much larger than a postage stamp. The CPU is one of the many chips located on the motherboard, as shown in *Figure 23*.

Chips have many components

This chip, or microprocessor, consists of millions of tiny electronic parts, including resistors, transistors, and capacitors (devices for storing electric charges), most of which act as switches. These components form huge numbers of circuits on the surface of the chip.

Logic circuits in the CPU make decisions

The heart of the CPU is an *arithmetic/logic unit,* or ALU, which performs calculations and logic decisions. The CPU also contains temporary data storage units, called *registers,* which hold results from previous calculations and other data waiting to be processed. A control section coordinates all of the processor activities. Finally, there are conductors that connect the various parts of the CPU to one another and to the rest of the computer.

When you start a program, the program first loads into random-access memory. Next the CPU performs a "fetch" operation, which brings in the first program instruction. Then it carries out that instruction and fetches the next instruction. The CPU proceeds in this fashion, fetching new instructions and obtaining data from the keyboard, mouse, disk, or other input device. Then it processes the data and creates output that is sent to the monitor or printer.

The CPU's logic gates can be built up to evaluate data and make decisions

As with memory chips, transistors in the CPU act as switches. The switches can operate as devices called *logic gates.* Just as a real gate can be open or shut, a logic gate can open or close a circuit depending on the condition of two inputs. One kind of logic gate, called an AND gate, closes the circuit and allows current to pass only when both inputs are in the "on" position.

You could use a similar device to alert you when it is both cold and raining so that you would know how to dress. You could connect moisture and temperature sensors to an AND gate and arrange to have it close a circuit and ring a bell. The bell would ring only when the temperature fell below 40°F and it was raining. If it were cold but dry, the bell would not ring. Similarly, the bell would not ring if it were warm but raining.

615

Trends in Technology

Magnetic RAM Most of today's memory is dynamic memory, or DRAM, which requires a large power supply at all times. Within a few years, the DRAM in computers may be replaced by a new technology called Magnetic RAM, or MRAM. MRAM uses magnetism rather than electrical power to store data. In addition to computers, MRAM might also be used to replace Flash memory in MP3 players, cell phones, and other devices

MRAM has a number of advantages. It will use much less power, store more data,

and access the data faster than any current memory technologies. Also, computers will no longer have to "boot up" when first turned on, since MRAM does not rely on electrical power. A computer will be ready as soon as it is turned on, just like a television set. Another benefit of reducing electrical power is that battery life for laptop computers and other portable devices that use MRAM will be greatly extended.

Teaching Tip

Computer Networks Computer networks can be connected in many different ways. When a computer series is connected together in an office or a library, it is called a Local Area Network, or LAN. This terminology is used when the computers are in the same building. Multiple LANs in different buildings can be connected together, perhaps at a college or a university. This type of network is called a Wide Area Network, or WAN. The Internet is a WAN because it connects computers in different locations over great distances.

SKILL BUILDER —ADVANCED

Interpreting Diagrams A logic system controls circuits based on the kinds of logic gates used and the conditions of the inputs received. Ask students to draw a logic system similar to the one shown in **Figure 24** that would ring a bell when it is warm and dry outside, but only during daylight hours. (Student drawings should match Figure 24 with the following differences: the OR gate at D is replaced with an AND gate, and each of the switches in sensors A, B, and C is reversed. For example, in B "Dry weather" becomes ON and "Rain or snow" becomes OFF, and likewise for the other two sensors.)
LS Logical

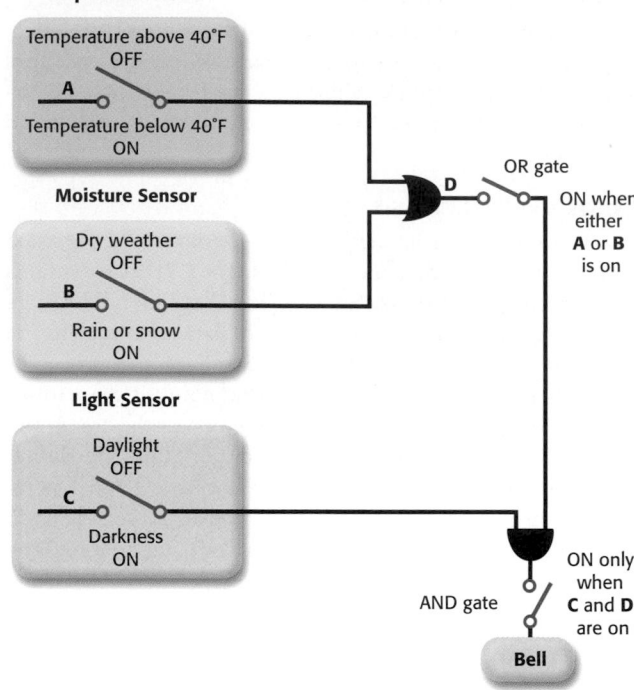

Figure 24

This logic system evaluates three variables—temperature, moisture, and light—in order to make a decision.

Figure 25

LANs have given schools, businesses, and government great communication possibilities.

If you use a type of logic gate called an OR gate, as shown in *Figure 24,* the bell will ring when it is cold or raining. If you want the bell to ring when it is cold or when it is raining, but only if it is dark outside, you can use an OR gate followed by an AND gate.

Computer Networks and the Internet

As the use of desktop personal computers became common in the 1980s, people looked for ways to link all of the computers within a single business, university, or government agency. The development of local area networks, or LANs, was the solution. In a LAN, as in *Figure 25,* all PCs are connected by cables to a central computer called a *server*. A server consists of a computer with lots of memory and several hard-disk drives for storing huge amounts of information.

This system allows workers to share data files that are stored on the server. One can also send a document to another person on the network. Soon after LANs were established, people were exchanging memos and documents over the network. This type of communication is called electronic mail, or E-mail.

TECHNOLOGY CONNECTION

Internet Ask the technology teacher to come into class and discuss digital communications and the Internet. Discuss how the use of the Internet has changed over the past few years, including how some companies do "e-business" entirely over the Internet.

The Internet is a worldwide network of computers

As the number of powerful computers increased, especially in government and universities, the U.S. Department of Defense wanted to connect them in a nationwide network. However, the department's computer experts worried about setting up a network that depended on only a few computers. If anything went wrong, the entire network would stop working.

Instead, a network in which every computer could communicate with every other computer was created. If part of the network were destroyed, the remainder would still be able to transmit information. This was the beginning of the **Internet.**

Because many companies had set up internal networks that used the same communication methods as the Defense Department's network, it was easy for them to connect to the network by telephone lines. Many other governments and corporations around the world joined to form a worldwide network that we now call the Internet, which is really a network of other networks.

If you have used the Internet, you are probably most familiar with the part known as the *World Wide Web,* or WWW, or just the Web. The Web was created in Europe in 1989 as a way for scientists to use the Internet to share data and other information.

The Web was mostly a resource for scientific information. It has since exploded into a vast number of sites created by individuals, government agencies, companies, and other groups. The Web is meeting many needs of individuals and of society.

internet connect

www.scilinks.org
Topic: Internet
SciLinks code: HK4074

SciLINKS Maintained by the National Science Teachers Association

► **Internet** a large computer network that connects many local and smaller networks all over the world

REAL WORLD APPLICATIONS

Using a Search Engine
Search engines provide a way to find specific information in the vast amount of information that is available on the Internet. Finding information successfully depends on several things, one of which is picking appropriate keywords for your search.

Applying Information
1. Pick a science topic that interests you. Write down a few keywords that you think will occur in information about the topic.

2. Use an Internet search engine to find three Web sites that have information on the topic. Experiment with keywords until you find the kinds of sites you want.

3. Try the same keywords on other search engines. Do they all find the same sites?

4. How did the results differ? Can you detect whether an engine specializes in certain types of information?

SKILL BUILDER — BASIC

Vocabulary The prefix *inter–* in Internet means "between or among." This is appropriate because the Internet is a shared network between many different computers.

Some students may have heard the term intranet. This word uses the prefix *intra–,* which means within or inside. An intranet is a network within an organization or company. All organization members or employees have access, but people outside the organization do not. Other examples include intrastate and intramural. Challenge students to think of or look up other words that contain these two prefixes. **LS Verbal**

REAL WORLD APPLICATIONS

Using a Search Engine The most common way to connect to the Internet is to use a modem. Modems come in various speeds, measured in BAUD rates. The higher the BAUD rate, the faster the rate of communication. If a cable or broadband (such as DSL) modem is used instead of a telephone modem, then the communication speed is much faster. For example, cable modems run at 10 MB, or 10 000 000 bits, per second, which is much faster than the 56 000 bits per second of standard modems. **LS Verbal**

Alternative Assessment — ADVANCED

Student Web Sites Encourage students with access to a computer and the Internet to design and create a simple Web site. Students can learn some basic commands in HTML or use a Web-design program. Their sites could explore a topic they are interested in or give information about themselves, a favorite hobby, a vacation, and so on. If your school has a scanner, allow students to scan in photographs or illustrations for their sites.

Many Internet service providers offer their users space to post Web sites. Students without Internet accounts can also find hosts that offer free Web space in exchange for displaying advertisement banners along the top or bottom of the page. Ask students to bring in the Web address of their completed sites to share with other students. If enough students participate, you may wish to hold a contest in which students vote for the best site. **LS Intrapersonal**

Chapter Resource File

• **Cross-Disciplinary Worksheet** RWA—World Wide Web Robots GENERAL

• **Cross-Disciplinary Worksheet** Fine Arts Connection—Arts and the Internet GENERAL

• **Science Skills** Basic Exercises in Logic GENERAL

Transparencies

TT Logic Gates

Close

Quiz

Match each term on the right with the correct description.

[left column:]

1. instructions, data, and programming used by a computer (c. software)
2. a large computer network (d. Internet)
3. software that controls a computer's activities (b. operating system)
4. equipment that makes up a computer system (a. hardware)

[right column:]

a. hardware
b. operating system
c. software
d. Internet

Chapter Resource File

- **Concept Review** GENERAL
- **Quiz** BASIC

Figure 26
The Internet opens doors of opportunity in mere seconds.

You need three things to use the Internet

To use the Internet, as in ***Figure 26,*** you need a computer with a modem to connect the computer to a telephone line. The word *modem* is short for modulator/demodulator, a device that codes the output data of your computer and uses it to modulate a carrier wave that is transmitted over telephone lines. The modem also extracts data from an incoming carrier wave and sends that data to your computer.

Next you need a software program called an Internet, or Web, browser. This program interprets signals received from the Internet and shows the results on your monitor. It also changes your input into signals that can be sent out.

Finally, you need a telephone connection to an Internet service provider, or ISP. An ISP is usually a company that connects the modem signal of your computer to the Internet for a monthly fee.

Internet communication uses transmission pathways like those used to relay television, radio, and telephone signals. With the Internet system, you can communicate with a Web site anywhere on Earth in just seconds.

SECTION 3 REVIEW

SUMMARY

- Computers perform four functions: input, storage, processing, and output.
- Input is in the form of binary code, grouped in eight bits called a *byte.*
- The physical components of computers are called hardware.
- Programs and instructions are called software.
- Computing activity takes place in a central processing unit, which also carries out logic functions.
- The Internet is a worldwide network of computers that can store and transmit vast amounts of data.

1. **List** three computer input devices and three output devices.
2. **Describe** the four main functions of a digital computer.
3. **Explain** how data are stored on and read from a magnetic hard-disk drive.
4. **Indicate** how ROM and RAM differ.
5. **Distinguish** between the ways that optical media and magnetic media function.
6. **Identify** which of the following components are part of a computer's hardware and which are part of its software.
   - **a.** a CPU microchip
   - **b.** a program to calculate when a car needs an oil change
   - **c.** the instructions for the computer clock to be displayed
   - **d.** RAM memory
7. **Explain** the purpose of an operating system.
8. **Compare** AND gates with OR gates.
9. **Restate** the three things, in addition to a computer, that you need to use the Internet.
10. **Creative Thinking** Describe how technological tools might be used in the future both for personal needs and for the needs of society.

618

Answers to Section 3 Review

1. input devices: mouse, keyboard, modem; output devices: monitor screen, printer, speaker
2. Computers perform input, storage, process, and output.
3. Data is stored on magnetic media through the use of a current-carrying coil. Current is sent through the coil, generating a magnetic field. This magnetic field causes the magnetic domains within the diskette to change, storing the information in those magnetic domains.
4. ROM is "read-only memory," while RAM is "random-access memory." Programs are stored on CD-ROMs, where the computer can read the program but not rewrite it. RAM stores the program when the computer is running it. Without RAM, the computer program would not run.
5. Optical media use lasers to read the information while magnetic media use magnetic fields and solenoids to detect the information stored on diskettes.

See "Continuation of Answers" at the end of the chapter.

Study Skills

Pattern Puzzle

Pattern puzzles help you remember information in the correct order and can help you understand scientific processes.

1 **Write down the steps of a process in your own words.**

We'll use the chapter's description of how a computer's four functions apply to a typed sentence. In your own words, write these steps on a sheet of notebook paper, one step per line.

- Words are typed on a keyboard (input).
- The input words are sent as digital data bytes to a central processing unit (CPU).
- The CPU stores data bytes in a file on the hard-disk drive.
- The words are taken from the stored file and are displayed as output on a monitor.

2 **Cut the sheet of paper into thin strips with one step per strip. Shuffle the strips so that they are out of sequence.**

- The CPU stores data bytes in a file on the hard disk drive.
- The words are taken from the stored file and displayed as output on a monitor.
- The input words are sent as digital data bytes to a central processing unit (CPU).
- Words are typed on a keyboard (input).

3 **Place the strips in the correct sequence. Confirm the order of the process by checking your text or class notes.**

- Words are typed on a keyboard (input).
- The input words are sent as digital data bytes to a central processing unit (CPU).
- The CPU stores data bytes in a file on a hard-disk drive.
- The words are taken from the stored file and are displayed as output on a monitor.

Practice

Use concepts from the chapter to properly arrange the following pattern puzzle:

- The call undergoes atmospheric transmission by microwaves.
- The call is received at a switching station and sent to its final destination.
- A telephone call is transmitted by wire cable to the local switching station.
- The microwaves are amplified and relayed between transmission towers.

619

Teaching Tip

Pattern Puzzle Solving As an additional exercise, have students add the following steps in the correct order to the four steps in the example.

- Activate "Print" function by using keyboard commands with mouse.
- Word is printed on paper by printer (output).
- Open document with stored words.
- Move the mouse toward icon on the monitor screen for stored document and click.

The fourth and third of these steps should be placed between the third and fourth steps in the example. The first and second of the above steps should be placed at the end. Write all four lines out of order on the chalkboard, and have students decide what the proper order should be.

Answers

1. A telephone call is transmitted by wire cable to the local switching station.
2. The call undergoes atmospheric transmission by microwaves.
3. The microwaves are amplified and relayed between transmission towers.
4. The call is received at a switching station and sent to its final destination.

Understanding Concepts

1. d
2. b
3. b
4. a
5. a
6. c
7. c
8. c
9. c
10. a
11. b
12. b
13. c
14. d
15. c

Using Vocabulary

16. Telecommunication occurs over large distances using electromagnetic means, while other forms of communication occur without electromagnetic means.

17. Analog signals can vary continuously within a range, whereas digital signals consist of separate bits of information. Digital signals cannot vary continuously; rather, they are discrete values.

18. Physical transmission involves sending signals through a medium like a telephone wire or a cable. Atmospheric transmission involves transmitting signals through the air over larger distances. Microwave communica-

Chapter Resource File

- **Chapter Tests** GENERAL
- **Test Item Listing for Exam View®
  Test Generator**

Chapter Highlights

Before you begin, review the summaries of key ideas of each section, found at the end of each section. The vocabulary terms are listed on the first page of each section.

UNDERSTANDING CONCEPTS

1. A _____ is necessary in order to interpret a signal.
   a. CPU
   b. modulation
   c. operating system
   d. code

2. A microphone uses a transducer to
   a. convert digital signals into analog signals.
   b. convert sound waves into an electric current.
   c. convert a digital signal into an analog signal.
   d. amplify a sound wave.

3. The microphone in the mouthpiece of a telephone produces a(n) _____ signal.
   a. microwave
   b. analog
   c. light
   d. digital

4. A communications signal from a ground station to a satellite is an example of _____ transmission.
   a. atmospheric
   b. cellular
   c. physical
   d. ground-wave

5. The up-and-down movement of electrons in the wire of a transmitting antenna produces _____ waves.
   a. electromagnetic
   b. visible light
   c. sound
   d. television

6. Telephone signals sent down wires rely on
   a. electromagnetic waves.
   b. digital signals.
   c. electric current.
   d. atmospheric transmission.

7. If the _____ is adjusted, a radio can receive a certain station.
   a. amplifier voltage
   b. speaker circuit
   c. tuner circuit
   d. carrier frequency

8. FM radio waves do *not* rely on
   a. line-of-sight transmission.
   b. carrier waves.
   c. amplitude modulation.
   d. a transducer.

9. Cathode-ray tubes are used in
   a. telephones.
   b. telegraphs.
   c. televisions.
   d. radios.

10. Materials that glow when struck by an electron beam are called
    a. phosphors.
    b. pixels.
    c. cathode rays.
    d. transistors.

11. Computers rely on RAM for
    a. long-term storage.
    b. working memory.
    c. processing instructions.
    d. "fetch" operations.

12. Memory in a computer that is permanent and cannot be added to is called
    a. RAM.
    b. ROM.
    c. CPU.
    d. CRT.

13. A computer program that coordinates all of the computer hardware is a(n)
    a. read-only memory.
    b. application.
    c. operating system.
    d. browser.

14. A modem connects a computer to a
    a. printer.
    b. transmitting antenna.
    c. hard-disk drive.
    d. telephone line.

15. The most common data-storage device on a modern personal computer is the
    a. ROM.
    b. keyboard.
    c. hard-disk drive.
    d. CPU.

tions are an example of atmospheric transmission.

19. A carrier wave can be modulated by changing either the frequency or the amplitude of the carrier wave. Frequency modulation is used in FM radio signals and amplitude modulation is used in AM radio signals.

20. b; Light travels through an optical fiber by total internal reflection. A binary signal is sent through a piece of optical fiber by pulsing the light—that is, turning it on and off. "On" represents a 1 and "off" represents a 0.

21. Electromagnetic waves received by the television set's antenna are passed to a detector that separates the audio and video electrical

signals from the carrier waves. The audio signals are sent to an audio amplifier and speaker, as in a radio. The video signals are used to control electron beams in a cathode-ray tube. The electron beams are directed toward a screen covered with phosphors, which glow in a pattern that recreates the image taken by the television camera. Each pass of the electron beam is called a scan line. In color televisions, a shadow mask may be used to control the color of each phosphor dot.

22. The arithmetic/logic unit or ALU performs calculations and makes logic decisions. Registers hold results from previous calculations and other data. Conductors connect

16. How does *telecommunication* differ from ordinary communication?

17. Describe the differences between *analog signals* and *digital signals*.

18. How does *physical transmission* differ from *atmospheric transmission*?

19. Describe two ways that a broadcasting station can *modulate* a carrier wave.

20. Which of the following diagrams correctly represents the path of a light beam through an *optical fiber*? Explain your choice. How is a binary digital signal sent through an optical fiber?

a.

b.

21. Describe how a television converts a television signal into sounds and images that you can hear and see. Use the terms *cathode-ray tube, phosphor, scan line,* and *shadow mask* in your answer.

22. List three parts of a computer's CPU, and explain the functions of those three parts.

23. List two examples of computer *hardware* that are input devices. List two examples of computer hardware that are output devices.

24. RAM stands for *random-access memory.* Why is this kind of computer memory called *random access?*

25. **Graphing** In 1965, an engineer named Moore stated that the number of transistors on integrated-circuit chips would double every 18 to 24 months. This idea became known as Moore's law. The data in the table below show the actual numbers of transistors on the CPU chips that have been introduced since 1972. Make a graph with "Years" on the *x*-axis and "Number of transistors" on the *y*-axis. Describe the shape of the graph. Does your graph support Moore's law? Does the projected value for the year 2010 seem realistic?

Year	Microprocessor	Number of transistors
1972	4004	2300
1973	8008	3500
1974	8080	6000
1978	8086	29 000
1982	80286	134 000
1986	80386	275 000
1989	80486	1,200,000
1993	Pentium	3,100,000
1996	Pentium Pro	5,500,000
1997	Pentium II	7,500,000
1999	Pentium III	9,500,000
2010	?	800,000,000 (estimated)

26. **Applying Knowledge** Your basketball team and coach have a meeting in which you decide that certain hand gestures and finger positions will convey certain messages such as pass, stall, or play zone defense, etc. Use this example to explain the difference between a signal and a code.

621

parts of the CPU to one another and to the rest of the computer.

23. Answers will vary. Two input devices might be a mouse and a keyboard, and two output devices might be a printer and a monitor.

24. It is called "random" because the time to access—read or write—information does not depend on the location of the information on a memory chip.

Building Graphing Skills

25. The number of transistors on an integrated circuit chip for the year 2010 seems rather unrealistic. What will probably happen is that a change in technology will occur so that the number of transistors is not as great.

Thinking Critically

26. The signals are the hand gestures being sent from the coach to the players. The code is the meaning of these signals—pass, stall, etc.

27. A microphone converts sound waves to electrical signals, whereas a speaker converts electrical signals into sound waves. Both are transducers.

28. **a.** B
    b. A
    c. C

29. As the height of a transmitter tower is increased, the FM signal can be broadcast over a greater distance. This is because the paths of the wave transmission are above the heights of many hills and other obstructions. The same principle is true for higher TV antennas.

30. If a laser beam were amplitude modulated, you would see the beam get brighter and dimmer as it transmits the signal. If the laser beam were frequency modulated, you would see the beam change color as the frequency changes. Frequency modulation is probably more efficient because of energy losses that occur over the distance traveled. If the beam is to be sent over a large distance, then losses in beam intensity would alter an amplitude-modulated signal but would not affect a frequency-modulated signal.

Assignment Guide

Section	Questions
1	1, 16, 17, 20, 26
2	2–10, 18, 19, 21, 27–30, 33, 34, 38
3	11–15, 22–25, 31, 32, 35–37, 39–41

31. You would need a light sensor outside that can tell the system that it is dark or light outside. In addition, you would need an AND gate that reads the information from the proximity sensor and the light sensor. These two would form the inputs for the AND gate. If both are true, it is dark <u>and</u> someone is nearby, then the system activates.

32. Student answers will vary but might include that the computer does not have a modem or an Internet browser software program.

Developing Life/Work Skills

33. $\lambda = \dfrac{v}{f} = \dfrac{3.0 \times 10^8 \text{ m/s}}{6.5 \times 10^5 \text{ Hz}} = $

460 m

length of antenna $= \dfrac{1}{4} \lambda = $

$\dfrac{1}{4} \times 460$ m $= 115$ m

(If rounding off properly, the answer above is 120 m or 1.2×10^2.)

34. $\lambda = \dfrac{v}{f} = \dfrac{3.0 \times 10^8 \text{ m/s}}{1.055 \times 10^8 \text{ Hz}} = 2.8$ m

length of antenna $= \dfrac{1}{2} \lambda = $

$\dfrac{1}{2} \times 2.8$ m $= 1.4$ m

FM antenna towers are taller than AM towers because the FM station can broadcast farther with a taller tower.

35. Have students make use of the references section of a public library and the Internet to find information on this subject.

27. Understanding Systems How do a speaker and a microphone differ? How are they similar?

28. Interpreting Graphics Identify the diagram that represents each of the following:
a. a carrier wave
b. an audio signal
c. an amplitude-modulated carrier

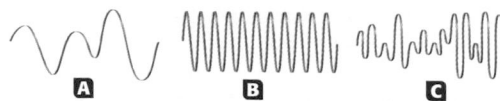

A   **B**   **C**

29. Applying Knowledge Use words or draw a diagram to explain why an FM radio signal can be received farther away as the height of the transmitting tower is increased. Also explain why you can receive more-distant television stations by using a higher television antenna.

30. Applying Knowledge Visible light from a laser can be used as a signal carrier. Describe what you would see if laser light is amplitude modulated and frequency modulated, assuming the modulation is slow enough for you to see the result. Which type of modulation do you think is more practical for visible light? Explain.

31. Problem Solving Suppose you want a light to come on automatically when someone comes to your door but only if it is dark outside. You have a proximity sensor, which is a device that closes an electric circuit when a person comes close to the door. What other sensor and what kind of logic gate do you need?

32. Applying Knowledge Suppose you are attempting to connect your computer to the World Wide Web, but it is not working. List two possible reasons why your computer is not able to connect, and explain how you could check for each one.

622

33. Applying Technology AM radio antennas are usually as tall as one-fourth the wavelength of the carrier. How tall is the antenna of a radio station transmitting at 650 kHz? Use the equation relating wavelength, velocity, and frequency from Chapter 11.

34. Applying Technology FM stations usually broadcast with antennas that are 1/2 the wavelength of the carrier. What is the length of an antenna for a station broadcasting at 105.5 MHz? Compare your answer with the answer in item 23, and explain why FM stations broadcast from towers much taller than the actual antenna length.

35. Interpreting and Communicating ENIAC was among the world's first digital computers. At your library or on the Internet, research the construction and early uses of ENIAC. Write a paragraph that summarizes your findings.

36. Applying Technology What computer-input device would work best in each of the following situations? Justify your choices.
a. You want to use a picture from a magazine in a report for history class.
b. You want to play a computer game in which you fly a plane.
c. You want to compose an E-mail message and send it to a friend.
d. You want to copy parts of several different documents on the Internet and put them all into one document.

37. Researching and Communicating Microchips consist of many components. At your library or on the Internet, research the construction and functions of microchips. Communicate those findings in a written report or in a sketch with labels and captions. When you report your findings, explain why you chose the method of communication that you used.

36. a. scanner
  b. joy stick
  c. keyboard
  d. mouse
37. Have students make use of the references section of a public library and the Internet to find this information.
38. Have students make use of the references section of a public library and the Internet to find information on these people.

Integrating Concepts

39. a. telephone
  b. signal
  c. analog
  d. physical transmission
  e. atmospheric transmission
  f. satellites
40. Student answers will vary. Students can research specific cases involving free speech and privacy issues on the Internet or at the library.
41. Student answers will vary. Students can research the development of computer animation and effects on the Internet or at the library.

38. Working Cooperatively Working with a group of classmates, research the achievements of the following people in the fields of communication and computer technology. Construct a classroom display that includes a picture of each person along with a summary of his or her contributions to the advancement of communication and computer technology.

a. Edwin Armstrong
b. Grace Murray Hopper
c. An Wang
d. Lewis Latimer
e. Vladimir Zwyorkin
f. John W. Mauchly

INTEGRATING CONCEPTS

39. Concept Mapping Copy the unfinished concept map below onto a sheet of paper. Complete the map by writing the correct word or phrase in each of the lettered boxes.

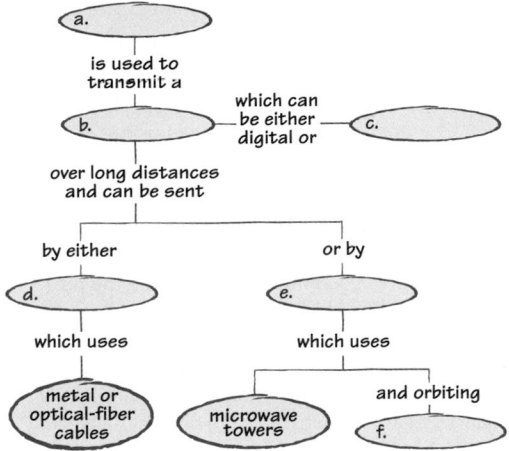

40. Connection to Social Studies Internet access has led to problems involving free speech and privacy. Should there be controls to prevent the spread of potentially dangerous or offensive information? Should others have free access to information about you? Research specific examples of these problems and find out what laws have been passed to address them. What arguments are made for and against free speech? An important fact to consider is that the Internet is not limited to one country.

41. Connection to Fine Arts Many animated motion pictures and television shows are now produced with the help of computers. Research the development of computer animation and special effects over the last several years. When were the first computerized special effects used in a motion picture? Based on your findings, what advances in computer effects and animation do you predict will occur in the next 10 years?

www.scilinks.org
Topic: Communication Technology
SciLinks code: HK4025

SCILINKS Maintained by the
National Science Teachers Association

Art Credits: Fig. 6, 9, 11-13, 15, 17, 18, 20, Uhl Studios, Inc.; Fig. 24, Leslie Kell; "Understanding Concepts" (light beam paths), Uhl Studios, Inc.; "Thinking Critically" (diagrams), Uhl Studios, Inc.; "Skills Practice Lab," Uhl Studios, Inc.

Photo Credits: GE Americom communication satellite artwork courtesy Lockheed Martin; telecommunications network room by Sam Dudgeon/HRW, courtesy Broadwing, Inc.; Fig. 1A, Michael Newman/PhotoEdit; Figs. 1B-C, Image Copyright ©2004 PhotoDisc, Inc.; Fig. 2-3 Peter Van Steen/HRW; Fig. 4, Andy Christiansen/HRW; Fig. 6, Peter Van Steen/HRW; Fig. 7, Reuters NewMedia Inc./CORBIS; Fig. 8A, Peter Van Steen/HRW; Fig. 8B, Don Mason/Corbis Stock Market; Fig. 10, Lester Lefkowitz/CORBIS; Fig. 14, Bettmann/CORBIS; Fig. 16, Mary Kate Denny/PhotoEdit; "Science and the Consumer," Alexander Tsiaras/Science Source/Photo Researchers, Inc.; "Quick Lab," Peter Van Steen/HRW; Fig. 19(t), Tom Pantages Photography; Fig. 19(b), Los Alamos National Laboratory/Science Photo Library/Photo Researchers, Inc.; "Quick Activity," Peter Van Steen/HRW; Fig. 21, Robert Mathena/Fundamental Photographs; Fig. 22, Telegraph Colour Library/Getty Images/FPG International; Fig. 23, Image Copyright ©2004 Photodisc, Inc.; "Connection to Architecture," Dennis Hallinan/Getty Images/FPG International; Fig. 25, William Taufic/CORBIS; "Real World Applications," HRW; Fig. 26, Michael Newman/PhotoEdit; "Science in Action"(alien parking), Brad Stockton/CORBIS; (telescopes), Dr. Seth Shostak/SPL/Photo Researchers, Inc.

623

Transparencies

TT Concept Mapping

DETERMINING THE SPEED OF SOUND

Teacher's Notes

Time Required 1 lab period

Ratings

```
        1     2     3     4
EASY ├─────┼─────┼─────┤ HARD
```

TEACHER PREPARATION	2
STUDENT SETUP	2
CONCEPT LEVEL	2
CLEANUP	1

Skills Acquired

- Collecting data
- Communicating
- Experimenting
- Identifying/recognizing patterns
- Inferring
- Measuring
- Organizing and analyzing data

The Scientific Method

In this lab, students will:
- Make observations
- Analyze the results
- Draw conclusions
- Communicate results

Safety Cautions

A plastic tube may be used instead of the glass tube to reduce the risk of the glass tube shattering if struck by the tuning fork.

Skills Practice Lab

Introduction

Can you determine the speed of sound in air?

Objectives

▶ USING SCIENTIFIC METHODS *Observe* the reinforcement of sound in a column of air.

▶ *Determine* the speed of sound in air by calculating the wavelength of sound at a known frequency.

Materials

glass cylinder, tall
glass tube, about 4 cm in diameter and 50 cm in length
meterstick
rubber stopper, large
thermometer
tuning forks of known frequency (2)
wood dowel or wire handle for stopper

Determining the Speed of Sound

▶ Procedure

1. The speed of sound is equal to the product of frequency and wavelength. The frequency is known in this experiment, and the wavelength will be determined.

2. If you hold a vibrating tuning fork above a column of air, the note or sound produced by the fork is strongly reinforced when the air column in the glass tube is just the right length. This reinforcement is called *resonance,* and the length is called the *resonant length*. The resonant length of a closed tube is about one-fourth the wavelength of the note produced by the fork.

3. On a paper, copy *Table 1* at right.

Determining the Speed of Sound

SAFETY CAUTION Make sure the tuning fork does not touch the glass tube or cylinder, as the glass may shatter from the vibrations.

4. Set up the equipment as shown in the figure at right.

5. Record the frequency of the tuning fork as the number of vibrations per second (vps) in your *Table 1*.

6. Make the tuning fork vibrate by striking it with a large rubber stopper mounted on a dowel or heavy wire.

7. Hold the tube in a cylinder nearly full of water, as illustrated in the figure.

8. Hold the vibrating fork over the open end of the tube. Adjust the air column by moving the tube up and down until you find the point where the resonance causes the loudest sound. Then hold the tube in place while your partner measures the distance from the top of the glass tube to the surface of the water (which is the part of the tube sticking out of water). Record this length to the nearest millimeter as Trial 1 in *Table 1*.

9. Repeat steps 6–8 two more times using the same tuning fork, and record your data in *Table 1*.

10. Using a different tuning fork, repeat steps 4 through 8.

Table 1 Data Needed to Determine the Speed of Sound in Air

	Tuning fork 1	Tuning fork 2
Vibration rate of fork (vps)		
Length of tube above water (mm), Trial 1		
Length of tube above water (mm), Trial 2		
Length of tube above water (mm), Trial 3		

▶ **Analysis**

1. On a clean sheet of paper, make a table like the one shown below.

2. Measure the inside diameter of your tube and record this measurement in your **Table 2**. The reflection of sound at the open end of a tube occurs at a point about 0.4 of its diameter above the end of the tube. Calculate this value and record it in **Table 2**. This distance is added to the length to get the resonant length. Record the resonant length in **Table 2**.

3. Complete the calculations shown in **Table 2**.

4. Measure the air temperature, and calculate the speed of sound using the information shown below. Record your answer in **Table 2**.

   Speed of sound = 5 332 m/s at 0°C + 0.6 m/s for every degree above 0°C

Table 2 Calculating the Speed of Sound

	Tuning fork 1	Tuning fork 2
Average measured length of air column (mm)		
Inside diameter (mm)		
Inside diameter $\times$ 0.4 (mm)		
Resonant length		
Wavelength of sound (mm), 4 $\times$ resonant length		
Wavelength of sound (m), wavelength of sound (mm) $\times \dfrac{1}{1000}$		
Speed of sound (m/s), wavelength $\times$ vibration rate of fork		
Speed of sound (m/s), calculated from step 4		

▶ **Conclusions**

5. Should the speed of sound determined with the two tuning forks be the same?

6. How does the value for the speed of sound you calculated compare with the speed of sound you determined by measuring the air column?

7. How could you determine the frequency of a tuning fork that had an unknown value?

Tuning fork

Open-ended glass tube

625

Tips and Tricks

To obtain the greatest sound volume, the tuning fork tines should be aligned for the maximum amplitude to enter the tube. Have students start with the tube totally withdrawn from the water.

In this experiment, the frequency of the tuning forks is described as vibrations per second (vps). The SI unit for frequency is the hertz (Hz). Students may be familiar with megahertz (MHz) in reference to the clock speed of personal computers.

Procedure

Each setup can accommodate 2–3 students. Students may try to match harmonics of the tuning fork to what they hear in the resonance tube rather than the fundamental frequency.

Frequencies of 250 to 700 vps work best to stay within convenient measurements. Not all tuning forks are equal in amplitude (volume). Each tuning fork should be tested prior to the lab.

If a strobe light is available, students can observe the vibrations of the tines on the tuning fork and estimate their amplitudes. As an extension, ask students why temperature affects the speed of sound. (Temperature affects the movement of the molecules of the medium.)

Answers to Analysis

1.–4. Sample answers are given in Sample Data Table 1.

Answers to Conclusions

5. The speed of sound should be the same regardless of tuning fork frequency.

See "Continuation of Answers" at the end of the chapter.

Calculating the Speed of Sound

	Tuning fork 1	Tuning fork 2
Average measured length of air column (mm)	179.6	213.0
Inside diameter (mm)	35	35
Inside diameter 3 0.4 (mm)	14	14
Resonant length (mm)	193.6	227.0
Wavelength of sound (mm), 4 3 resonant length	774.4	908.0
Speed of sound (m/s), wavelength 3 vibration rate of fork	340.7	348.7
Speed of sound (m/s), calculated from step 4	344.6	344.6

Chapter Resource File

• **Datasheet** Determining the Speed of Sound GENERAL

• **Observation Lab** Constructing a Radio Receiver BASIC

• **CBL™ Probeware Lab** Transmitting and Receiving a Message Using a Binary Code ADVANCED

Continuation of Answers

Continuation of Answers from p. 600
Section 1 Review

7. A geostationary satellite is one that stays fixed in its position over Earth and rotates as Earth rotates, thereby maintaining its position comparative to Earth. This is important in satellite communication because the satellite does not appear to move in the sky. As a result, a satellite dish remains pointed at the orbiting satellite over time.

8. Taller microwave relay towers have a longer transmission range because the taller the tower is the less it is influenced by hills and tall buildings. Its transmission path is above many of them. As a result, the taller tower can transmit farther than the shorter transmission tower.

Continuation of Answers from p. 609
Section 2 Review

6. Electromagnetic waves received by the television set's antenna are passed to a detector that separates the audio and video electrical signals from the carrier waves. The audio signals are sent to an audio amplifier and speaker, as in a radio. The video signals are used to control electron beams in a cathode-ray tube. The electron beams are directed toward a screen covered with phosphors, which glow in a pattern that recreates the image taken by the television camera.

7. Optical-fiber cable offers cleaner signals in transmission, fewer signal losses, and fewer power losses.

Continuation of Answers from p. 618
Section 3 Review

6. a. hardware
   b. software
   c. software
   d. hardware

7. The operating system controls the computer hardware, memory, keyboard, disks, printer, mouse, and monitor.

8. An AND gate gives a TRUE response (1) if both inputs are on. An OR gate gives a TRUE response (1) if only one of the inputs is on.

9. a modem, an Internet browser, and an Internet service provider (ISP)

10. Answers will vary. Students may include voice-activated personal computers, especially useful for people with disabilities. They may list more up-to-date personal computers in the classroom, cell phones eventually replacing land-line phones, cell phone Internet service and cell phone text messaging becoming common place, and other types of technology used for communications.

Continuation of Answers from p. 625
Skills Practice Lab

6. The speed of sound calculated from step 4 is similar to the lab calculations.

7. The frequency of the tuning fork can be determined by first determining the wavelength of the sound produced by the tuning fork and then applying the appropriate equation, using the value for the speed of sound at room temperature.

SHOULD WE SEARCH FOR EXTRATERRESTRIAL LIFE?

The search for extraterrestrial intelligence represents a challenge for communication technology. All of the different aspects of communications described in Chapter 15 are required here. Radio frequencies are scanned by computers that search for unexpected regularities that could be signs of a signal.

1. Critiquing Viewpoints

Students' analyses of a single argument's weak points may vary. Be sure that student responses clearly identify a weakness and respond in some way that provides support. Possible responses for each of the viewpoints in the feature are listed below. (Note that students should analyze *only one of the following.*)

Lauren L.: It is also possible that if we find extraterrestrial life, we may find out that it is not friendly.

Either way, it is better to find out and be prepared to meet other forms of life than to act as if they aren't there.

Ghautam P. and Thuy N.: Although the search is not too expensive, it isn't free either. Those resources could be doing something better.

Advances in technology will make it cheaper and cheaper to continue pursuing evidence of extraterrestrial life.

Samuel S.: Perhaps the reason there's no evidence yet is because we haven't looked hard enough.

But how will we ever know when we've looked hard enough. Better to abandon the search now.

Science in ACTION

Is there life outside our solar system?

Although the great distances make travel to other solar systems in today's spacecraft impossible, can we 'hear' evidence of intelligent life in the universe? A unique organization of scientists known as the Search for Extra-Terrestrial Intelligence (SETI) is working to answer that question.

Will this Roswell, NM, "Alien Parking" sign become a common sight?

History of SETI

In 1959, two young physicists named Philip Morrison and Giuseppe Cocconi noticed that because radio waves travel at the speed of light and require little power, they were the perfect way to communicate across the vastness of space. Morrison and Cocconi reasoned that if intelligent civilizations did exist somewhere in the universe, it would be possible to use radio telescopes to eavesdrop on their radio transmissions, or even detect a signal deliberately sent into space. They even proposed an exact frequency to start the search: 1420 MHz, the emission frequency of the hydrogen atom, the most common element in the universe.

626

The Search Is On

The very next year, the first attempts were made to listen in on a handful of nearby stars. Although modest, these early attempts paved the way for much broader investigations. Today, SETI uses some of the world's biggest radio telescopes, including the mammoth Arecibo dish in Puerto Rico, to monitor millions of radio channels over an ever-expanding area of the universe. Likewise, the SETI organization has grown to hundreds of astronomers, physicists, and communications specialists from all over the world. SETI now also includes scientists from the relatively new discipline of astrobiology, the study of the conditions and environments necessary for life on other planets.

This 305 m–diameter aluminum dish at the Arecibo Observatory in Puerto Rico receives signals from outer space.

Zack J.: While requiring extraterrestrial life to have advanced technology makes it less likely to find, that doesn't mean it will never be found.

But why spend resources on something this unlikely?

Samantha J.: The amount being spent on this search is small enough that it wouldn't have much of an impact on problems here on Earth.

Every little bit of money that can be used to address real problems will help make them less difficult.

2. Critiquing Viewpoints

Students' analyses of strong points in opposing arguments may vary. Be sure that student responses clearly identify the strong points and respond in some way that provides support for the opposite point of view. Possible responses for each of the viewpoints in the feature are listed below. (Note that students should analyze *only one of the following.*)

Lauren L.: Thousands of planets in the universe could have intelligent life.

Just because intelligent life is possible does not mean that it is definitely out there.

The Future of SETI

It's now known that many millions of stars in the universe are orbited by planets, and there is growing evidence that water, an essential ingredient of life, exists on other planets. Based on these and other discoveries, SETI recently received funding to build the first large-scale radio telescope dedicated to this research. The Allen Telescope Array will consist of approximately 350 small satellites linked together to make a collecting area equal to a 100-meter telescope. SETI scientists are optimistic that this new telescope will eventually help them search more than 1 million stars. SETI will also look for new kinds of alien evidence and will use conventional optical telescopes in California to search the night sky for flashes of light (such as from high-powered lasers) signaling us from other solar systems. Although SETI has not yet found a confirmed extra-terrestrial signal, improved technologies and greater interest from the public continue to support SETI in their search for proof that we are not alone.

The Allen Telescope Array, designed by astronomers and engineers with SETI and the University of California–Berkeley, is a significant step in the exploration of the cosmos in the search for extraterrestrial intelligence.

Science and You

1. **Understanding Concepts** Most SETI researchers monitor radio frequencies in the microwave range of the electromagnetic spectrum because it is almost free of natural interference. Why is this a good idea?

2. **Critical Thinking** Explain one possible flaw in concentrating on radio transmissions for evidence of alien life.

3. **Acquiring and Evaluating** A unique project called SETI@Home uses the power of thousands of personal computers in homes, schools, and offices around the world to process the enormous amount of data gathered by SETI. Using the Internet, research the SETI@Home project and write a short paper explaining how the project works and why it's a good idea to utilize personal computers to process scientific data. If you like, ask your teacher for permission to use a computer at school for SETI@Home.

internet connect

www.scilinks.org
Topic: **Search for Extraterrestrial Life**
SciLinks code: **HK4061**

SCiLINKS. Maintained by the National Science Teachers Association

627

Ghautam P.: It would be a great discovery if life was found elsewhere in the universe.

But life may not be found. Then, all of the work and expense would be in vain.

Thuy N.: We just won't know if extraterrestrial life has tried to contact us if we don't continue the search.

But if we continue the search and find nothing, we won't know if that means there really is no life out there or that we just didn't find it.

Samuel S. and Zack J.: There has been no evidence yet, and the odds are so small that we will find anything.

Perhaps that's because we haven't been looking long enough or hard enough.

Samantha J.: There are many other things the money could be spent on.

There are always many more things to spend money on than there is money. The amount being spent on the search is fairly small.

3. **Life/Work Skills**

Student answers will vary. Evaluate students' presentations or brochures proposing further projects in searching for extraterrestrial life using a scale of 1–5 points for each of the following: organization, clarity in explanation, persuasiveness, creativity, and effort. (25 total possible points)

4. **Creative Thinking**

Student answers about how to communicate with living beings from other planets will vary. Answers should involve using items that we are likely to share with other living things. For example, a code could be worked out using samples of different elements because the same elements should be found throughout the universe. Another way could be to use waves of varying frequencies because presumably the same electromagnetic spectrum occurs throughout the universe.

Evaluate students' suggestions using a scale of 1–5 points for each of the following: practicality, clarity in explanation, probable ease of use, creativity, and effort. (25 total possible points)

The Solar System
Chapter Planning Guide

PACING	CLASSROOM RESOURCES	LABS, ACTIVITIES, AND DEMONSTRATIONS
BLOCK 1 • 45 min pp. 628–629 **Chapter Opener**		SE **Activity 1**, p. 629 SE **Activity 2**, p. 629
BLOCKS 2 & 3 • 90 min pp. 630–636 **Section 1** Sun, Earth, and Moon	CRF **Lesson Plan*** TT **Bellringer*** TM **Planetary Orbits*** TM **The Solar System*** TT **Phases of the Moon*** TT **Eclipses***	TE **Demonstration** Orbital Speed, p. 632 GENERAL TE **Demonstration** Phases of the Moon, p. 634 GENERAL SE **Skills Practice Lab** Estimating the Size and Power Output of the Sun, pp. 660–661 ◆ GENERAL CRF **Datasheets for SE Skills Practice Lab** Estimating the Size and Power Output of the Sun* CRF **CBL™ Probeware Lab** Determining the Speed of an Orbiting Moon* ◆ ADVANCED
BLOCKS 4 & 5 • 90 min pp. 637–645 **Section 2** The Inner and Outer Planets	CRF **Lesson Plan*** TT **Bellringer***	TE **Opening Activity**, p. 637 GENERAL TE **Demonstration** The Reflection of Light from Venus, p. 638 CRF **Observation Lab** Explaining the Motion of Mars* ◆ BASIC
BLOCKS 6 & 7 • 90 min pp. 646–654 **Section 3** Formation of the Solar System	CRF **Lesson Plan*** TT **Bellringer*** TT **The Nebular Model*** TM **Formation of the Moon*** TT **Concept Mapping***	SE **Quick Activity** Estimating 4.6 Billion, p. 649 GENERAL CRF **Datasheets for In-Text Labs** Estimating 4.6 Billion*

BLOCKS 8 & 9 • 90 min

Chapter Review and Assessment Resources

- SE **Chapter Review**, pp. 656–659
- CRF **Chapter Tests*** GENERAL
- OSP **Test Generator**
- CRF **Standardized Test Practice with Guided Reading Development***
- CRF **Test Item Listing for ExamView® Test Generator***
- OSP **Scoring Rubrics and Classroom Management Checklists**

Online Resources

go hrw .com

Visit the HRW Web site for a variety of free resources related to the text. Just type in the keyword **HK4 SOL**.

Holt Online Learning

Holt Science Spectrum: Physical Science: Online Edition

Students can access interactive problem solving help and active visual concept development with the *Holt Science Spectrum: Physical Science* Online Edition available at **www.hrw.com**.

student CNN News

cnnstudentnews.com

Find the latest news, lesson plans, and activities related to important scientific events.

KEY

TE	Teacher Edition
SE	Student Edition
OSP	One-Stop Planner

CRF	Chapter Resource File
TT	Teaching Transparency
TM	Transparency Master

*	Also on One-Stop Planner
◆	Requires Advance Prep

PROBLEM SOLVING AND PRACTICE	SECTION REVIEW AND ASSESSMENT	STANDARDS CORRELATION
	CRF Pretest* `GENERAL`	
TE Inclusion Strategies, p. 630 **CRF** Science Skills Classifying Items* `BASIC` **CRF** Cross-Disciplinary Worksheet Integrating Mathematics—Using Comparisons to Understand Space Statistics* `ADVANCED`	**TE** Quiz, p. 636 `BASIC` **SE** Section 1 Review, p. 636 **CRF** Concept Review* `GENERAL` **CRF** Quiz* `BASIC`	ES 3a UCP 1, 2, 3, 4, 5 SAI 2 ST 2 HNS 1, 2, 3 SPSP 5
CRF Cross-Disciplinary Worksheet Connection to Social Studies—Egyptian Calendars* `ADVANCED` **TE** Inclusion Strategies, p. 640 **CRF** Cross-Disciplinary Worksheet Connection to Language Arts—Science Fiction and Fact* `GENERAL` **CRF** Cross-Disciplinary Worksheet Real World Applications—Deep Space 1* `ADVANCED`	**TE** Quiz, p. 645 `BASIC` **SE** Section 2 Review, p. 645 **CRF** Concept Review* `GENERAL` **CRF** Quiz* `BASIC`	ES 3a UCP 1, 2, 3, 4, 5 SAI 2 ST 2 HNS 1, 2, 3 SPSP 5
CRF Cross-Disciplinary Worksheet Real World Applications—Weather Satellites* `GENERAL` **SE** Study Skills Concept Mapping, p. 655 **SE** Science in Action Mining the Moon, pp. 662–663	**TE** Quiz, p. 654 `BASIC` **SE** Section 3 Review, p. 654 **CRF** Concept Review* `GENERAL` **CRF** Quiz* `BASIC`	PS 4a, 4b, 5a UCP 1, 2, 3, 4, 5 SAI 2 ST 2 HNS 1, 2, 3 SPSP 5

SCiLINKS®

www.scilinks.org

Topic: Comets, Asteroids, and Meteoroids
*Sci*Links code: HK4024

Topic: Moons of Other Planets
*Sci*Links code: HK4089

Topic: Eclipses
*Sci*Links code: HK4039

Topic: The Nine Planets
*Sci*Links code: HK4095

Topic: Origins of the Solar System
*Sci*Links code: HK4099

Topic: Astronomy
*Sci*Links code: HK4009

Topic: Lunar Mining
*Sci*Links code: HK4080

Technology Resources

One-Stop Planner
All of your printable resources and the Test Generator are on this convenient CD-ROM.

 CNN. Science in the NEWS

each video segment is accompanied by a Critical Thinking Worksheet *

Segment 29
Looking for Life

Segment 30
Europa Pics

Segment 19
Asteroid Extinction

Overview

This chapter explores the solar system and includes the study of planet Earth and Earth's moon. The chapter describes each planet in the solar system, including information on important space missions. This chapter also presents models for the formation of the solar system and of Earth's moon.

Assessing Prior Knowledge

Be sure students understand the following concepts:

- properties of matter
- atoms
- the periodic table
- the force of gravity
- temperature
- light, reflection, and refraction

MISCONCEPTION /// ALERT \\\

Science education research has identified the following misconceptions about the solar system:

- Most students do not understand that Earth is older than 10^9 years old. This may be due to a general confusion about large numbers, and students' general inability to distinguish between millions and billions.
- Some students believe there is an absolute down direction (as in a flat Earth). Students also believe day and night are caused by daily orbiting of the sun, or by clouds or the moon blocking the sun.
- Students are confused about the relative sizes and motions of the sun, Earth, and moon as well as the causes of both solar and lunar eclipses.
- Most students think the seasons are caused by the changing distances between Earth and the sun (so it is summer when the sun is closest and winter when the sun is farthest).
- Very few students understand that the sun is a star, and many regard it as a planet.

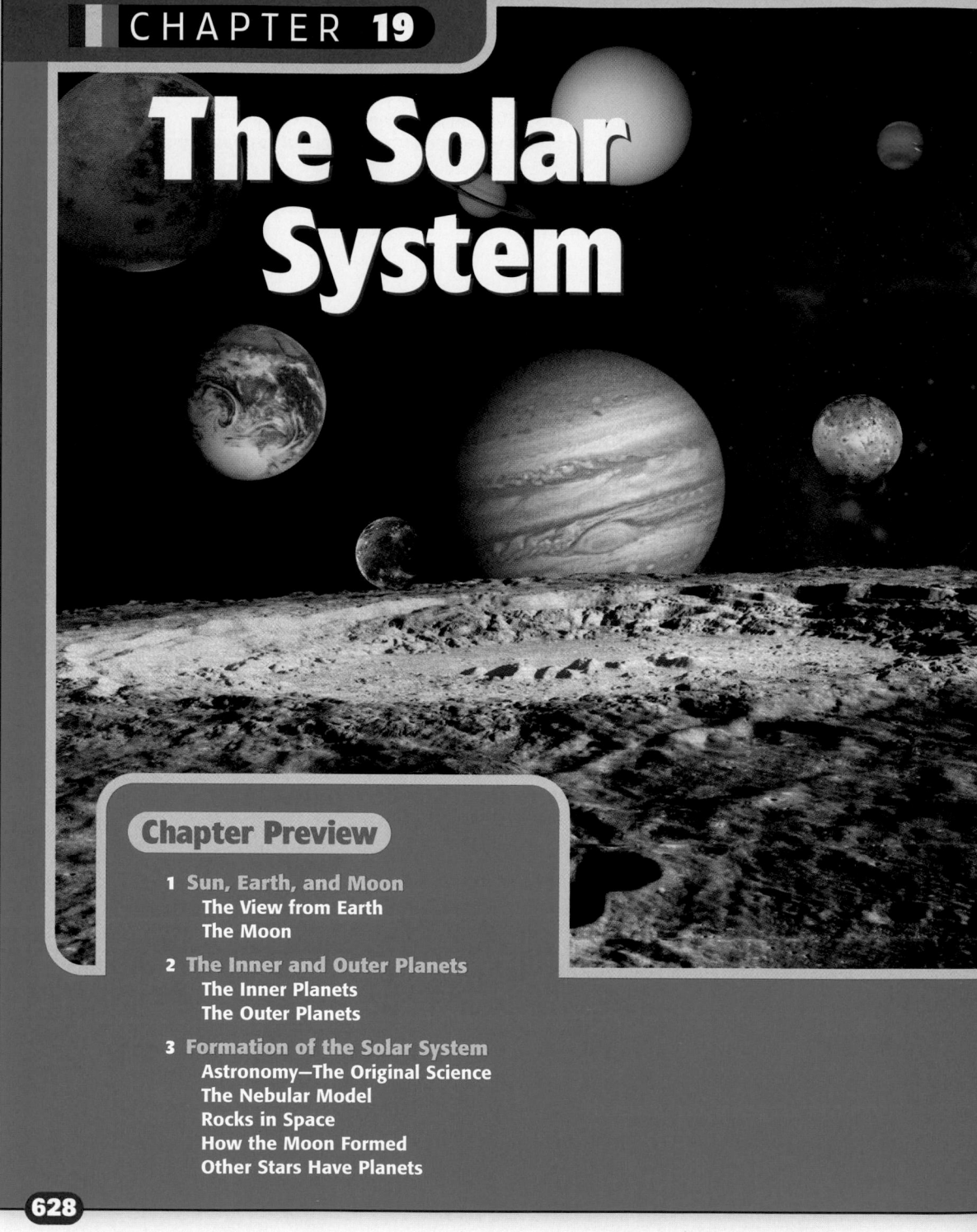

The Solar System

Chapter Preview

1 Sun, Earth, and Moon
The View from Earth
The Moon

2 The Inner and Outer Planets
The Inner Planets
The Outer Planets

3 Formation of the Solar System
Astronomy—The Original Science
The Nebular Model
Rocks in Space
How the Moon Formed
Other Stars Have Planets

628

Standards Correlations

National Science Education Standards

The following descriptions summarize the National Science Education Standards that specifically relate to this chapter. For the full text of the standards, see the National Science Education Standards at the front of the book.

Section 1 Sun, Earth, and Moon
Earth and Space Science ES 3a
Unifying Concepts and Processes UCP 1, 2, 3, 4, 5
Science as Inquiry SAI 2

Science and Technology ST 2
History and Nature of Science HNS 1, 2, 3
Science in Personal and Social Perspectives SPSP 5

Section 2 The Inner and Outer Planets
Earth and Space Science ES 3a
Unifying Concepts and Processes UCP 1, 2, 3, 4, 5
Science as Inquiry SAI 2
Science and Technology ST 2
History and Nature of Science HNS 1, 2, 3
Science in Personal and Social Perspectives SPSP 5

Focus ACTIVITY

Background How do you describe to people where you live? If they live nearby, just naming your neighborhood may be enough. What if the people lived very far away? You may have to tell them the name of the town, state, country, or continent. But what if they lived outside of our solar system? You may have to explain to them that you live on Earth, the third planet from the sun and one of thousands of celestial bodies that orbit around an average star in the Milky Way galaxy.

Our solar system contains a vast diversity of objects—from small, rocky asteroids to huge, gas giant planets. So far, we have discovered life as we know it only in one place—Earth. Could there be other life in our solar system or on planets around other stars? This is a big question for the 21st century. Learning about life means understanding what makes our planet special and exploring the characteristics of the other bodies in the solar system.

Activity 1 Make a list of 20 items you would take on a camping trip. Add a spacesuit and oxygen, and examine your list to see if it has what you need to survive on Mars for a month. Explain what items you would eliminate from your list and what items would you add?

Activity 2 Go outside on a clear evening at dusk, and look up at the sky. Spend 20 min counting stars as they become visible. Describe the first stars you see. How many did you count in 20 min? Do you see any lights in the sky that might not be stars? If so, what do you think they might be, and how would you find out what they are?

internet connect

www.scilinks.org
Topic: Comets, Asteroids, and Meteoroids
SciLinks code: HK4024

SCI LINKS Maintained by the National Science Teachers Association

This artist's composite shows the nine planets of our solar system and four of Jupiter's moons. Other objects, such as the comet above, whirl through our dynamic solar system.

Pre-Reading Questions

1. How is the moon different from a planet such as Earth?
2. When you look up at the sky, how can you tell the difference between stars and planets?

629

Section 3 Formation of the Solar System
Physical Science PS 4a, 4b, 5a
Unifying Concepts and Processes UCP 1, 2, 3, 4, 5
Science as Inquiry SAI 2
Science and Technology ST 2
History and Nature of Science HNS 1, 2, 3
Science in Personal and Social Perspectives SPSP 5

Focus ACTIVITY

Background The detection of planets around other stars is one of the most important areas of research today. Like the Copernican revolution, in which we came to realize that Earth is just one of several planets orbiting the sun, the current research is proving that our solar system is just one among many. By studying other solar systems, we may come to understand more about our own. Understanding comets, asteroids, and meteoroids is also important. The impact of comet Shoemaker-Levy with Jupiter and research on impacts in Earth's history have increased awareness of the potential risks to Earth.

Activity 1 You may also ask what might be needed for a longer stay. Possibilities include renewable sources of food, water, and oxygen. **LS Logical**

Activity 2 An evening field trip can be a rewarding experience for students who may never have seen the Milky Way or shooting stars. Even in the city, students should be able to see the brightest stars and Venus and Jupiter. **LS Visual**

Answers to Pre-Reading Questions

1. The moon orbits Earth, while planets orbit the sun. Also, the moon is smaller than most planets and lacks an atmosphere.

2. Observed over time, planets will change positions relative to the fixed background of stars. When magnified, planets may appear to be disks.

Chapter Resource File

• Pretest
• Teaching Transparency Preview
• Answer Keys

Focus

Overview

Before beginning this section, review with your students the Objectives listed in the Student Edition. This section introduces students to the sun, moon, and planets as viewed from Earth. Students will get an introduction to the nine planets in the solar system and their satellites. The section ends with a focus on Earth's moon, including explanations of phases, eclipses, and tides.

Bellringer

Use the Bellringer transparency to prepare students for this section.

Motivate

Identifying Preconceptions — BASIC

Is The Sun A Star? Ask students to name the star closest to Earth. (the sun; the closest star other than the sun is Proxima Centauri, about 4.3 light-years away)

Many people do not realize that the sun is a star. In fact, the sun is a very ordinary star. It appears bright only because it is much closer to Earth than other stars are. One simple piece of evidence that the sun is close to Earth is the fact that it has an observable diameter. Most stars are so far away that they appear as mere points of light, even through powerful telescopes. Some stars appear larger than mere points in the sky due to the dispersion of starlight in the atmosphere. **LS** Logical

Sun, Earth, and Moon

KEY TERMS
planet
solar system
satellite
phase
eclipse

OBJECTIVES

▶ **Recognize** Earth as one of many planets that orbit the sun.
▶ **Explain** how gravity works within the solar system.
▶ **Describe** eclipses and phases of the moon.
▶ **List** two characteristics of the moon, and show how the moon affects Earth's tides.

planet any of the nine primary bodies that orbit the sun; a similar body that orbits another star

You know the sun, moon, and stars appear to rise and set each day because Earth spins on its axis. The stars that are visible at night revolve throughout the year as Earth orbits the sun. These two motions affect our view of the sky.

The View from Earth

Like ancient viewers of the sky, you can go outside and watch stars cross the sky on any clear night. Over time, you may notice one of the brighter objects changing its position and crossing the paths of stars. The Greeks called these objects *planets,* which means "wanderers." We now think of a **planet** as any large object that orbits the sun or another star. Five planets (in addition to Earth) are visible to the unaided eye: Mercury, Venus, Mars, Jupiter, and Saturn.

The sun is our closest star

There are billions of stars, but one is special to us. It took thousands of years for scientists to realize that the sun is a star. Because the sun is so close to us, it is very bright. As you see in *Figure 1,* our atmosphere scatters the sun's light and makes the daytime sky so bright that we can't see the other stars. The sun is an average star, not particularly hot or cool, and of average size. Its diameter is 1.4 million kilometers, about 110 times the diameter of Earth. Its mass is over 300 000 times the mass of Earth.

The solar system is the sun, planets, and other objects that orbit the sun. The system includes objects of all sizes—large planets, small satellites, asteroids, comets, gas, and dust. Astronomers are discovering that many other stars in our galaxy have planets, but we don't know any system as well as we know our own.

Figure 1
People on Earth are very familiar with one star—the sun.

630

APPLICATIONS ....

Protection Against Ultraviolet Rays
While sunlight is beneficial to life on Earth, ultraviolet rays from the sun are harmful to people. Have your students research the role that ultraviolet rays from the sun play in the lives of humans. What are some of the ways that people protect themselves from the sun's UV rays? (sunscreen, clothing) How does Earth's atmosphere protect us from UV rays? (The ozone layer blocks the most energetic and harmful UV light.) **LS** Verbal

INCLUSION Strategies

• *Learning Disabled*
• *Developmentally Delayed*
• *English Language Learners*

Using a calendar that shows the moon's phases, count how many days there are between one full moon and the next full moon. Students can then try to predict when the full moon will occur in the same month of the next calendar year. Students can check their predictions by counting off the days on a calendar for the next year. Have students explain how they predicted the date of the next full moon to the class.

Everything revolves around the sun

As the largest member of our solar system, the sun is not just the object that all the planets orbit. It is also the source of heat and light for the entire system. As Earth spins on its axis every 24 h, we see the sun rise and set. Many patterns of human life such as rising in the morning, eating meals at certain times throughout the day, and sleeping at night follow the sun's cycle. Most animals have patterns of activity and sleep. But unlike us, animals don't have school, television, or electric lights to interrupt their patterns.

As each year progresses, you can watch the growing seasons of plants change. Some plants, such as the morning glories in ***Figure 2,*** are very sensitive to light and move to face the sun as it rises and travels across the daytime sky.

Heat from the sun is a main cause of weather patterns on Earth. Another type of weather, space weather, is caused by energetic particles that leave the sun during solar flares and storms. When these particles reach Earth, they can zap communication satellites and cause blackouts.

Figure 2

The sun is important to all life on Earth. These morning glories turn their faces to the morning sun.

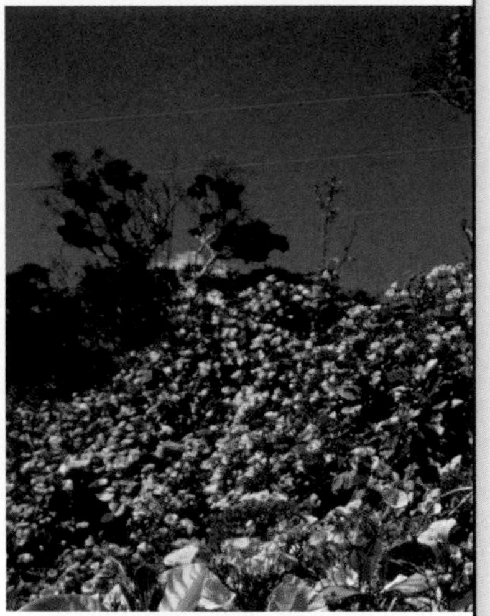

Planets and distant stars are visible in the night sky

Ancient peoples looked at the night sky and saw patterns that reminded them of their myths. When we pick out the shapes of constellations, we use some of the same patterns the Greeks saw and named. These ancient scientists also watched the five bright planets wander in regular paths among the stars. ***Figure 3*** shows Saturn wandering through the constellation Leo, which is named for its lionlike shape. By watching the sky for many years, the ancient Greeks calculated that the stars were more distant than the planets were. Over a thousand years later, after the invention of the telescope, people found other objects in the night sky, including many faint stars and three more planets: Uranus, Neptune, and Pluto.

August 2007
September 2007
October 2007

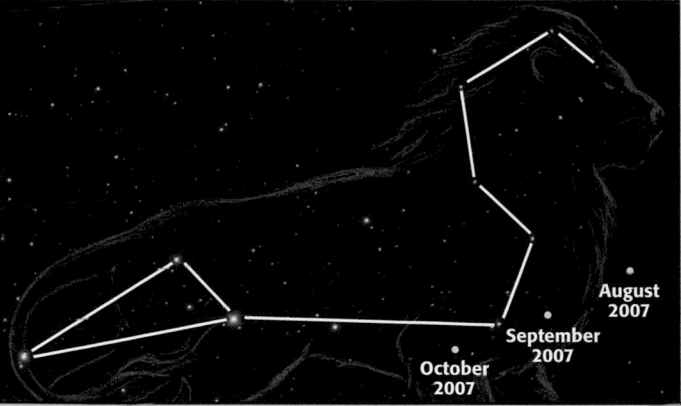

Figure 3

The planet Saturn moves against the background of stars in the constellation Leo.

631

Trends in Astronomy

Discovering Planets The first three planets discovered outside our solar system were found in 1994 by Alexander Wolszczan. These first planets were all orbiting a pulsar—not an ordinary star like the sun—in the constellation Virgo. Since then, close to 80 new planets have been discovered orbiting around other stars, and this number is rapidly growing. However, most of these planets have been large, gaseous planets, generally referred to as "Jupiter-like" planets. They also have orbits very close to their parent stars, sometimes even closer than Mercury in our solar system. Large planets in close orbits cause the greatest "wobble" in the motion of their parent stars. This wobble is what tells astronomers that the planet is there. Currently, no planet outside of our solar system has been viewed directly.

Astronomers are in hot pursuit of evidence of "Earth-like" planets—small, rocky planets at about the same distance as Earth's distance from the sun. NASA's Space Interferometry Mission, scheduled to launch in 2009, will put an array of telescopes in space capable of detecting the wobble caused by an Earth-sized planet.

Teach

SKILL BUILDER — BASIC

Reading Skills Have students list what they know about the solar system, and then have them list what they would like to learn. After students have read the chapter, tell them to look at their lists and fill in what they have learned. Students should write down any new questions that they have. Encourage students to do further reading on the solar system in other sources. **LS Verbal**

MISCONCEPTION ALERT

Earth's Rotation Some students believe day and night are caused by daily orbiting of the sun, or by clouds, the moon or other objects blocking the sun.

Demonstrate how the daily rotation of Earth on its axis can cause day and night. Turn out the lights, and shine a flashlight on a white sphere, such as a plastic-foam ball or a golf ball. Mark a point on the sphere with a marker. Then rotate the sphere, showing how the marked point passes from light to dark and back again as the sphere rotates. Ask students to identify at which points in the rotation the marked point is at "sunset" and "sunrise."

Ask students: What unit of time is determined by Earth's rotation about its axis? (a day) What unit of time is determined by Earth's rotation around the sun? (a year) **LS Logical**

Chapter Resource File

- **Lesson Plan**
- **Science Skills** Classifying Items BASIC

Transparencies

TT Bellringer

Teaching Tip

Circular and Elliptical Orbits
The orbits of the planets may look circular, but they are in fact elliptical. Differentiate the two by drawing both shapes on the board. Explain that an ellipse can appear very nearly circular if its major axis is nearly equal to its minor axis. In fact, a circle can be defined as an ellipse with both axes equal.

MISCONCEPTION ///ALERT\\\

Gravity in Space Many students think that gravity is not present in space. On the contrary, gravity holds the moon in orbit around Earth, Earth and other planets in orbit around the sun, and the solar system in orbit around the galactic center.

Astronauts in space are affected by Earth's gravity. They are in free-fall, with Earth's gravity pulling them toward Earth. But their forward motion along their orbit keeps them from getting closer to Earth. The sense of "weightlessness" comes from the spaceship and everything inside the spaceship falling at the same rate. This is like being in free-fall on a roller coaster; you have a sense of floating, but you are under the influence of gravity at all times.

▶ **solar system** the sun and all of the planets and other bodies that travel around it

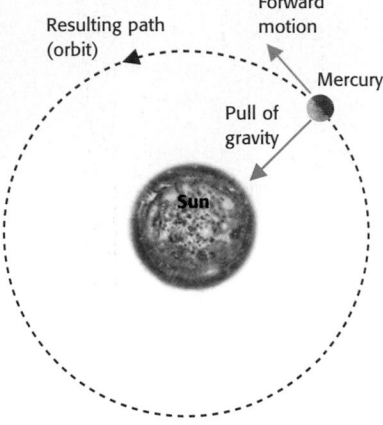

Figure 4
The pull of gravity causes Mercury to fall toward the sun, changing what would be a straight line into a curved orbit.

Earth is a part of a solar system

A school system has parts such as students, teachers, and administrators, which all interact according to a set of rules. The solar system also has its parts and its own set of rules. The **solar system** is the sun and all of the objects that orbit it. The sun is the most important part of our system and makes up 99% of the total mass of the solar system. The nine planets and their moons make up most of the remaining 1%. The solar system has many other smaller objects—meteoroids, asteroids, comets, gas, and dust. Although smaller objects don't have much mass, they help us understand how the solar system is organized.

Gravity holds the solar system together

The force of gravity between two objects depends upon their masses and the distance between them. The greater the mass, the larger the gravitational force an object exerts on another if they are equally distant. The closer two objects are to each other, the stronger the gravitational pull is between them. The sun exerts the largest force in the solar system because its mass is so large. *Figure 4* shows the pull of the sun, which keeps Mercury in its orbit. Imagine swinging a ball on a string. If you let go of the string, the ball flies off in a straight line. Without the sun's pull, Mercury and all the other planets would similarly shoot off into space. Gravity is also the force that keeps moons orbiting around planets. You experience gravity as the force that keeps you on Earth. Every object in the solar system pulls on every other object. Even though Jupiter is more massive than Earth, you don't notice its pull on you because it is too far away.

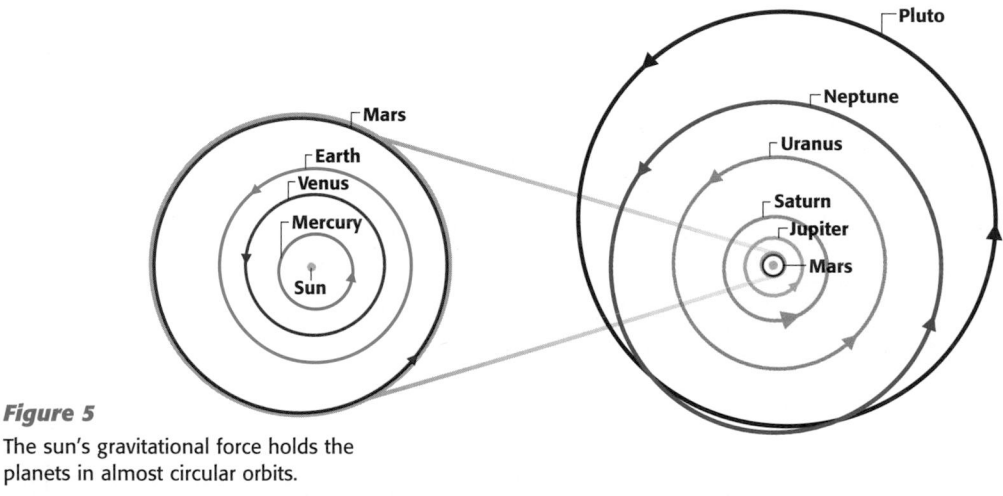

Figure 5
The sun's gravitational force holds the planets in almost circular orbits.

Demonstration ——— GENERAL

Orbital Speed

(Time: Approximately 5 minutes)

Materials

• gravity well
• marble (or coin)

Step 1 Have students stand around the gravity well. Start the marble rolling near the outer edge of the well. The marble should slowly move around the well in a nearly circular orbit.

Step 2 Let the marble continue until it reaches the middle. Prompt students to pay attention to the speed of the marble as it approaches the center of the well. You may even want to dip the marble in water-soluble paint to mark its path.

Step 3 Ask students how the speed of the ball changed as it got closer to the center? (the speed increases) How does this correspond to the speed of the planets? (The closer a planet is to the sun, the faster it moves in its orbit.)

Nine planets orbit the sun

Planets can be seen because their surfaces or atmospheres reflect sunlight. A planet's distance from the sun determines how long the planet takes to orbit the sun. *Figure 5* shows the orbits of the planets in order of distance from the sun. Mercury, the closest, takes 88 days to orbit the sun, which is the shortest time of all the planets. Earth takes one year, or 365.25 days. Pluto, the most distant planet, takes 248 years or over 90 000 days. For part of its orbit, Pluto is closer to the sun than Neptune is, but its average distance from the sun is the farthest.

A **satellite** is an object in orbit around a body that has a larger mass. The moon is Earth's satellite because Earth has the larger mass. *Figure 6* shows the relative diameters of the planets. The four closest to the sun are small and rocky and have few or no satellites. The next four are large and gaseous and have many satellites. Pluto has one satellite.

Satellites orbit planets

All of the planets in our solar system have moons except Mercury and Venus. Currently, we know of 102 natural satellites, or moons, orbiting the planets in our solar system. In 1970, we only knew of 33. Space missions have discovered many small satellites, and more could be found in the future. The smallest satellites are less than 3 km in diameter, while the largest, including Jupiter's Ganymede and Saturn's Titan, are larger than the planet Mercury. All satellites are held in their orbits by the gravitational forces of their planets. Like planets, satellites reflect sunlight. A few satellites have atmospheres, but most do not.

Figure 6

The planets in the solar system are shown in relative scale. The sun's diameter is almost 10 times larger than Jupiter's. Distances between the planets are not shown to scale.

▶ **satellite** a natural or artificial body that revolves around a planet

■ **internet** connect ≡
www.scilinks.org
Topic: Moons of Other Planets
SciLinks code: HK4089

SciLINKS Maintained by the National Science Teachers Association

633

Would a year on Mars be longer or shorter than a year on Earth? (longer, because Mars is farther from the sun than Earth is)

Step 4 You may also ask a more difficult question: Why don't the orbits of the planets move slowly toward the sun, like the marble did? (Actually, they do, but there is much less friction in space than there is between the ball and the surface of the well. Because there is less friction, the planets do not lose energy and move towards the center as noticeably as the marble did in the experiment.)

Tip: If you do not have access to a gravity well, you can make one by using a sheet of stretchy material. Have four students hold the corners of the sheet and stretch the sheet taut. Place a heavy weight in the center of the sheet. This will cause the sheet to curve downward in a funnel shape. Experiment with the mass of the object and the tautness of the sheet to get the best effect. **LS Visual**

Teaching Tip

Dark Side of the Moon Some
students may have the misconception that one side of the moon is
always dark. It is true that the same
side of the moon always faces
Earth, so we cannot ever observe
the far side of the moon from
Earth. However, when the moon is
in new phase as viewed from Earth,
the far side is bathed in sunlight,
the moon equivalent of daytime.
The only time the far side of the
moon is completely dark is when
the moon is in full phase as seen
from Earth.

SKILL BUILDER — GENERAL

Interpreting Diagrams Have students examine **Figure 8.** Follow the
path of the moon counterclockwise, starting with the new moon.
At each phase, refer students to the
corresponding boxed pictures of
the moon outside the circle. These
images show how the moon would
appear to an observer on Earth.

To reinforce understanding—
and for fun—you may ask students
at each phase to describe or draw
on the board what Earth would
look like to an astronaut on the
moon. (At new moon, Earth would
appear "full." At full moon, Earth
would appear "new." When the
moon is gibbous, Earth would appear
crescent, and vice-versa.) **LS Visual**

Figure 7

The moon has dark maria and
light highlands and craters.

▶ **phase** the change in the
illuminated area of one
celestial body as seen from
another celestial body;
phases of the moon are
caused by the positions of
Earth, the sun, and the moon

Figure 8

As the moon changes
position relative to
Earth and the sun, it
goes through different
phases. (The figure is
not to scale.)

The Moon

The moon does not orbit the sun directly; it orbits Earth at a distance of 385 000 km. The moon's surface is covered with craters,
mostly caused by asteroid collisions early in the history of the
solar system. The *maria,* or large, dark patches on the moon,
shown in **Figure 7,** are seas of lava that flowed out of the moon's
interior, filled the impact craters, and cooled to solid rock.

The moon has phases because it orbits Earth

The moon appears to have different shapes throughout the
month that are called **phases.** The relative positions of Earth,
the moon, and the sun determine the phases of the moon, as
shown in **Figure 8.** At any given time, the sun illuminates half
the moon's surface, just as at any given time it is day on one half
of Earth and night on the other half. As the moon revolves
around Earth, the illuminated portion of the side of the moon
facing Earth changes. When the moon is full, the half that is facing you is lit. When the moon is new, the side that is facing you
is dark, so you can't see it. Quarter phases occur when you can
see half of the sunlit side. The time from one full moon to the
next is 29.5 days, or about one calendar month.

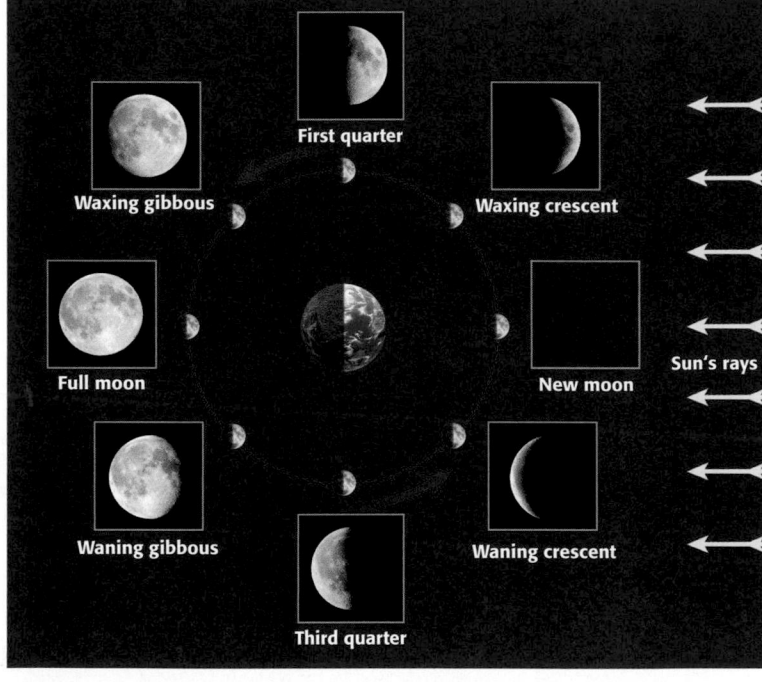

First quarter

Waxing gibbous

Waxing crescent

Full moon

New moon

Sun's rays

Waning gibbous

Waning crescent

Third quarter

Demonstration ——————— GENERAL

Phases of the Moon

(Time: Approximately 15 minutes)

Materials

- lamp
- a bright flashlight
- round ball (plastic-foam)
- 2 ft of string
- tape
- a meterstick

Step 1 Attach one end of the string to the
ball, and the other end to the meterstick. Turn
out the lights in the room. Have one student
hold the ball on the meterstick in the middle
of the room and have another student point
the flashlight so that it illuminates one side
of the ball.

Step 2 Students should take turns slowly
walking around the ball and observing the
shape of the illuminated portion of the ball
that they can see. Have students determine
where they can see a crescent moon, a
quarter moon, and a gibbous moon. Ask
students where they would have to stand to
simulate a solar eclipse or a lunar eclipse.
LS Visual

Phases of the moon are not caused by Earth's shadow

The relative sizes of and distance between Earth and the moon are not shown to scale in *Figure 8.* Earth and the moon may seem to make shadows that fall on each other all the time. But Earth, which has a diameter that is four times the diameter of the moon, is a distance of 30 Earth-diameters from the moon. Therefore, the moon's shadow is so small that it hits only a small part of Earth even when Earth, the moon, and the sun line up exactly.

Eclipses occur when Earth, the moon, and the sun line up

While exploring Jamaica in 1504, Christopher Columbus impressed the native people by consulting a table of astronomical observations and predicting that the sky would darken. The event he predicted was an **eclipse.** Eclipses can be predicted and can happen when Earth, the sun, and the moon are in a straight line. An eclipse occurs when one object moves into the shadow cast by another object.

During a new moon, the moon may cast a shadow onto Earth, as shown in *Figure 9A.* Observers within that small shadow on Earth see the sky turn dark as the moon blocks out the sun. This event is called a solar eclipse. On the other hand, when the moon is full, it may pass into the shadow of Earth, as shown in *Figure 9B.* All the observers on the nightside of Earth can see the full moon darken as the moon passes through Earth's shadow. This event is called a lunar eclipse. Because the moon's orbit is slightly tilted compared with Earth's orbit around the sun, the moon is usually slightly above or below the line between Earth and the sun. So, eclipses are relatively rare.

Figure 9

Eclipses occur when Earth, the sun, and the moon are in a line. **A** shows a solar eclipse, and **B** shows a lunar eclipse.

▶ **eclipse** an event in which the shadow of one celestial body falls on another

internet connect

www.scilinks.org
Topic: Eclipses
SciLinks code: HK4039

SCiLINKS. Maintained by the National Science Teachers Association

635

Did You Know ?

Earthshine On a dark night during the crescent phase of the moon, you can sometimes see the dark part of the moon's disk with the naked eye. This is due to *earthshine,* light reflected off Earth and shining on the part of the moon that is in darkness. To an astronaut on the moon at that time, Earth would appear bright and nearly full in the night sky.

Historical Perspective ── GENERAL

Eclipses Have students research some of the more recent solar and lunar eclipses. Where were they visible? When are the next eclipses predicted to occur, and what kind will they be?
LS Verbal

Teaching Tip

Viewing the Sun Remind students that they should never look directly at the sun. This is also true during a solar eclipse. The solar viewer constructed during the end-of-chapter lab can be used for observing the sun. Solar filters are available for viewing the sun through a telescope. Before using a filter, make sure it has no holes, not even small ones.

MISCONCEPTION
///**ALERT**\\\

Eclipses Students are often confused about the sizes and motions of the sun, Earth, and moon as well as the causes of eclipses. Have students examine **Figure 9** carefully. Point out that these figures are not to scale; the distances between the moon, Earth, and the sun are much greater than can be shown in these diagrams.

Ask students what phase the moon would appear to be in during a solar eclipse? (new) What phase would the moon appear to be in during a lunar eclipse? (full) Why can a solar eclipse only be viewed from certain locations on Earth? (The shadow of the moon covers only a small area on Earth's surface. As the eclipse progresses, the shadow traces out a narrow band across Earth; the eclipse can only be viewed from within this band. Outside this band, the sun appears as usual.)

You may also ask why solar and lunar eclipses don't occur every month? (The orbit of the moon around Earth is tilted slightly with respect to the orbit of Earth. Most of the time, during a new or full moon, the shadow of the moon on Earth or of Earth on the moon falls above or below the other body.)
LS Visual

📦 **Transparencies**

TT Phases of the Moon
TT Eclipses

Chapter 19 · The Solar System **635**

Quiz ———— BASIC

1. What is the star closest to Earth? (the sun)

2. What object do all the planets orbit? What object does the moon orbit? (the sun; Earth)

3. What phase must the moon be in for a solar eclipse to be possible? (new)

4. What celestial object is most responsible for tides on Earth? (the moon)

LS Logical

Figure 10

The gravitational pull of the moon is the main cause of tides on Earth.

The moon affects Earth's tides

Coastal areas on Earth, such as the one shown in *Figure 10*, have two high tides and two low tides each day. Even though tides are affected by Earth's landscape, tides are mainly a result of the gravitational influence of the moon. The moon's gravitational pull is strongest on the side of Earth nearest the moon. On the side near the moon, the water and land is pulled toward the moon, which creates a bulge. The movement of water is more noticeable than the movement of land because water is more changeable. The pull of the moon is weaker on the side of Earth that is farthest from the moon.

Earth rotates and so one area on Earth will have two maximum, or high, tides and two minimum, or low, tides in one day. Because the moon is also orbiting Earth, the times of these tides change throughout the month.

The sun has a minor effect on tides. When the sun is on the same side of Earth as the moon, the gravitational forces are at their strongest, and tides are at their highest for the month.

SECTION 1 REVIEW

SUMMARY

▶ The sun and the nine planets make up our solar system.

▶ Planets are visible because they reflect sunlight.

▶ Gravity holds the solar system together and keeps planets in orbit around the sun.

▶ The moon's surface has meteor-impact craters and maria from lava flows.

▶ Eclipses and phases of the moon are caused by the relative positions of Earth, the sun, and the moon.

▶ Tides are caused by differences in the pull of the moon's gravity on different areas of Earth.

1. **Identify** what makes planets and satellites shine.

2. **Explain** how gravity keeps planets in orbit around the sun.

3. **Predict** which satellite experiences the larger gravitational force if two satellites have the same mass, but one is twice as far away from the planet as the other.

4. **Explain** what happens during a lunar eclipse. What phase is the moon in during a lunar eclipse?

5. **Describe** two features of the moon, and explain how they formed.

6. **Explain** what causes tides.

7. **Describe** the positions of the sun, Earth, and the moon during a full moon and during a quarter phase.

8. **Critical Thinking** At what phase of the moon will tides be the highest? Explain.

9. **Critical Thinking** Examine *Figure 8* closely. If the moon were a crescent as seen from Earth, what would Earth look like to an astronaut on the moon?

10. **Critical Thinking** The Greeks thought that there were five planets visible with the unaided eye: Mercury, Venus, Mars, Jupiter, and Saturn. What other planet is visible with the unaided eye?

Answers to Section 1 Review

1. Planets and satellites appear to shine because they reflect sunlight.

2. Gravity pulls planets toward the sun because the sun has the greatest mass of any object in the solar system and therefore exerts the strongest gravitational force. The sun's gravity pulls planets into curved orbits that would otherwise travel off into space.

3. The satellite at the greater distance will experience a gravitational force 1/4 as large as the closer one. Because they are of equal mass, the gravitational pull is inversely proportional to the distance squared, and 1/2 squared is 1/4.

4. During a lunar eclipse, the Earth's shadow falls on the moon. The moon is at full phase.

5. Craters on the moon were formed by the impact of asteroids or other rocks from space. Maria were formed by the moon's volcanic activity.

6. Tides are caused by the gravitational influence of the moon and sun on Earth. The moon pulls more strongly on the side of the Earth nearest to it, and less strongly on the side farthest from the moon.

See "Continuation of Answers" at the end of the chapter.

The Inner and Outer Planets

OBJECTIVES

▶ **Identify** the planets of the solar system and their features.

▶ **Distinguish** between the inner and outer planets and their relative distances from the sun.

▶ **State** two characteristics that allow Earth to sustain life.

▶ **Describe** two characteristics of a gas giant.

▶ **KEY TERMS**

terrestrial planet
hydrosphere
asteroid
gas giant

The solar system has inner planets close to the sun and more-distant outer planets, too. The inner ones are called **terrestrial planets** because they are rocky like Earth. They receive more of the sun's energy and have higher temperatures than the outer planets do.

The Inner Planets

Figure 11 shows the orbits of the terrestrial planets: Mercury, Venus, Earth, and Mars. They are small and have solid, rocky surfaces. Using telescopes, satellites, and surface probes, scientists can study the geologic features of these planets.

Mercury has extreme temperatures

Until we sent space missions, such as *Mariner 10*, to investigate Mercury, we did not know much about it. The photograph in *Figure 12* shows that Mercury, much like Earth's moon, is pocked with craters. Because Mercury has such a small orbit around the sun, it is never very far from the sun. The best times to observe Mercury are just before sunrise or just after sunset, but even then it is difficult to see, even with a telescope.

▶ **terrestrial planet** one of the highly dense planets nearest to the sun; Mercury, Venus, Earth, and Mars

Figure 11

The four terrestrial inner planets—Mercury, Venus, Earth, and Mars—are closest to the sun.

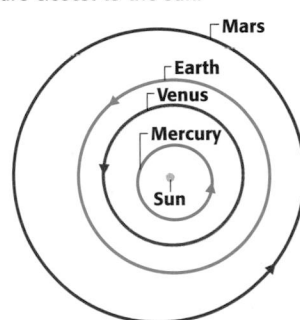

Figure 12

Mercury is pocked with craters.

637

Focus

Overview

Before beginning this section, review with your students the Objectives listed in the Student Edition. This section introduces students to key properties and features of the planets in our solar system, starting with Mercury and working outward.

🔊 Bellringer

Use the Bellringer transparency to prepare students for this section.

Motivate

Opening Activity — GENERAL

Writing Have students write either a description of how they imagine life on another planet would be, or a short story that takes place on another planet. Encourage students to consider the properties of the planet and how those properties would affect the life forms living there. For example, what would life be like on a gas giant? Because there is no solid surface, the life forms would have to float in the gaseous atmosphere. Other interesting settings that you may suggest include oceans underneath the ice-covered surface of a moon, the mouth of an active volcano, or the low-gravity environment on the surface of an asteroid. **LS Verbal**

INTEGRATING

BIOLOGY Life on Other Planets? Have students research the conditions on at least three other planets in our solar system. Tell students that research sources may include their textbook, encyclopedias, and Internet Web sites. Encourage them to find data including surface temperature and pressure, composition of the atmosphere, and the amount of water on each planet. For each planet, have students list at least two reasons why the planet probably could not support life as we know it. **LS Verbal**

SOCIAL STUDIES
CONNECTION — **GENERAL**

The English names for the planets come from the names of Roman gods. Have students research the origins of the names of several planets, moons, and asteroids in our solar system. Are the names appropriate for each object? Why or why not? **LS Verbal**

Chapter Resource File

• Lesson Plan

Transparencies

TT Bellringer

SKILL BUILDER — GENERAL

Reading Skills Have students read the section on the inner planets and then organize what they learned in a chart. One heading should be *Similarities* and the other should be *Differences*. Have students fill in their chart by describing each planet's similarities and differences to the other inner planets. **LS Logical**

Connection to
SOCIAL STUDIES

Answers

1. Answers may vary but may include that the sun is the most visible or that it has the greatest impact to life on Earth.

2. It suggests the Egyptians carefully observed the sky.

SKILL BUILDER — BASIC

Interpreting Visuals Have students examine **Figure 13.** If Venus is obscured, how could this map have been created? (by using radar; radio waves can pass through the clouds) What do the colors on the map represent? (Yellow and brown are higher elevations; blue and green are lower elevations.) Remind students that the colors are computer generated.

Prompt students to look for features on the surface of Venus. (The large brown areas are like continents. The large blue and green patches are comparable to Earth's oceans, but because they contain no liquid water, they are more like low plains.) **LS Visual**

Connection to
SOCIAL STUDIES

All civilizations in the world have used the motions of celestial objects to keep time. Most calendars are based on Earth's orbit around the sun and have a 365-day year. Some cultures, such as the Chinese, based their calendars on the moon. The Maya of ancient Mexico had several sophisticated calendars. One was based on the sun, just as ours is today, but their special interest was Venus. They carefully observed and measured the 584 days it takes for Venus to return to the same place in the sky and calculated that five of these cycles took eight of the 365-day years. The huge Aztec calendar stone shown below shows the Aztec's belief in the cyclical nature of change in the cosmos.

Making the Connection

1. Why are more calendars based on the sun than on other celestial objects?

2. The Egyptians built pyramids that were aligned with celestial objects. What does this tell you about their interests?

Figure 13

The Magellan spacecraft used radar to measure the surface below Venus' clouds. Brown and yellow show high ground, and green and blue show low ground.

Distances in the solar system are often measured in terms of the distance from Earth to the sun, which is one astronomical unit (AU), or 150 million km. Mercury is 0.4 AU from the sun. Mercury is so close to the sun that its surface temperature is over 670 K, which is hot enough to melt tin. The temperature on Mercury's night side drops to 103 K, which is far below the freezing point of water. Mercury spins slowly on its axis, with three spins on its axis for every two orbits Mercury makes around the sun. Its day is 58 Earth days; its year is 0.24 Earth years. Mercury is not a likely place to find life because it has almost no atmosphere and no water.

Thick clouds on Venus cause a runaway greenhouse effect

Venus is 0.7 AU from the sun. It is only seen near sunset or sunrise and is called either the morning star or the evening star. From Earth, Venus shows phases. Photos taken by *Mariner 10* show thick layers of clouds made mostly of carbon dioxide. These cloud layers make Venus very reflective.

Radar maps that measure the surface of Venus through the clouds, like the map shown in *Figure 13,* indicate that the surface of Venus has mountains and plains. Venus spins in the opposite direction from the other planets and the sun. One day on Venus is 243 Earth days long, and one year on Venus is 0.6 Earth years long.

Venus does not provide an environment that can support life. Venus is hot, and its atmosphere contains large amounts of sulfuric acid. In addition, the atmospheric pressure at the surface of Venus is more than 90 times the pressure on Earth. Venus' thick atmosphere prevents the release of energy by radiation, creating a "runaway" greenhouse effect that keeps the surface temperature of Venus over 700 K. A greenhouse effect occurs when radiation from the sun is trapped and heat builds up, as in a greenhouse on Earth. The "runaway" greenhouse effect that is taking place on Venus means that the more heat is built up, the more efficient the atmosphere becomes at trapping radiation. This effect causes unrelenting high temperatures.

Demonstration

The Reflection of Light from Venus

(Time: Approximately 15 minutes)

Materials

- flashlight
- 2 golf balls
- flat black paint

Venus is bright because its atmosphere reflects a large amount of the sunlight that it receives. Use this demonstration to illustrate this fact. This demonstration works only in a room that is very dark.

Step 1 Paint one golf ball black and leave the other white (do this well before class so the paint can dry).

Step 2 Put the golf balls on a table, and turn out the classroom lights. Ask students if they can see the golf balls. If the balls are visible then the room is not dark enough.

Step 3 Shine the flashlight on the balls. The white golf ball should be visible because it reflects light. The black ball should not be visible because it does not reflect light.

Earth has ideal conditions for living creatures

Earth, our home, is the third planet from the sun. We measure other planets in the solar system in relation to Earth. Earth rotates on its axis in one Earth day. It revolves around the sun at a distance of 1 AU in one Earth year. It has a mass of one Earth mass.

Earth is the only planet we know that sustains life. It is also the only planet that has large amounts of liquid water on its surface. Water floats when it freezes, so life can continue under the ice. Other substances, such as carbon dioxide, freeze from the bottom up. All the water on Earth's surface, both liquid and frozen, is called the **hydrosphere.** The continents and the hydrosphere hold an amazing diversity of life, as shown in *Figure 14.* Because water takes a long time to heat or cool, the hydrosphere moderates the temperature of Earth. One example of this effect is that areas near coasts seldom have temperatures as cold as inland areas on a continent.

The atmosphere protects Earth from radiation and sustains life

Earth, shown in *Figure 15,* has an atmosphere composed of 78% nitrogen, 21% oxygen, and 1% carbon dioxide and other gases. Like the hydrosphere, the atmosphere helps moderate temperatures between day and night. The greenhouse effect traps heat in the atmosphere, so Earth's surface doesn't freeze at night. Compare Earth with Mercury, a planet that doesn't have an atmosphere to protect it. Mercury is extremely hot during the day, and it gets very cold at night.

Earth's atmosphere protects us from some harmful ultraviolet radiation and high-energy particles from the sun. The radiation and particles are blocked in our upper atmosphere before they can damage life on Earth. The atmosphere also protects us from space debris, which can be leftover portions of artificial satellites or small rocks from space. As they speed through the atmosphere toward Earth's surface, these objects heat up and vaporize or shatter. Only very large objects can survive the trip through Earth's atmosphere.

Earth's original atmosphere changed over time as gases were released from volcanoes and by plants during photosynthesis. Earth is the only planet we know of that has enough oxygen in its atmosphere to sustain complex life as we know it.

Figure 14
Oceans on Earth host a vast diversity of life.

■ **hydrosphere** the portion of Earth that is water

Figure 15
A photo of Earth taken from space shows clouds, oceans, and land.

639

Did You Know ?

The Moon and Life Earth's moon may play an important role in Earth's ability to support life. The moon's gravity helps to keep Earth's axis from changing its orientation, which makes the environment on Earth more stable. Also, the moon causes tides which may have been important in the origin of life. Several theories on how life originated on Earth state that tides provided energy and the transport of molecules needed for chemical reactions in the first life forms.

Teaching Tip ———— GENERAL

Mars Missions NASA has sent several missions to Mars, including four Mariner missions, two Viking missions, Mars Observer, Pathfinder, Mars Global Surveyor, and Mars Odyssey 2001. Have students choose one of these missions to research. What did the mission set out to do? What did the mission discover? Students may also choose to research a future mission to Mars or to another planet. **LS** Verbal

|SKILL| BUILDER — GENERAL

Interpreting Visuals Have students examine **Figure 17.** Besides the polar caps, what other large features are visible on the surface of Mars? (possible answers include: the volcano Olympus Mons, volcanic craters, impact craters, canyons or cracks in the surface) **LS** Visual

Figure 16

The small *Sojourner* robot rover from the *Pathfinder* mission to Mars moved on the surface to examine rocks; this rock was named *Yogi.*

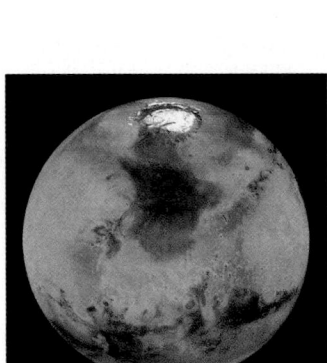

Figure 17

The *Viking I* and *II* probes took many images of Mars to make this composite. Note the polar ice caps.

Many missions have been sent to Mars

Although humans have yet to visit Mars, many probes have landed on its surface. *Viking 1* and *2* each sent a lander to the surface in 1976. In 1997, the *Pathfinder* mission reached Mars and deployed a rover named *Sojourner*, shown in **Figure 16,** which explored the surface using robotic navigation systems.

Figure 17 shows a white region at the poles of Mars. This is one of the polar icecaps of carbon dioxide that contain small amounts of frozen water. People who dream of colonizing Mars hope to harvest water from the ice caps. Features on other parts of the planet suggest that water used to flow across the surface as a liquid. Mars has a very thin atmosphere, composed mostly of carbon dioxide. Mars is 1.5 AU from the sun and has two small satellites, Phobos and Deimos. Mars's mass is 10% of Earth's. It orbits the sun in 1.9 Earth years and its day is 24.6 Earth hours. Mars is very cold; its surface temperature ranges from 144 K to 300 K.

Mars has the largest volcanoes in the solar system

Orbiting space missions detected some of Mars's unique features. The Martian volcano Olympus Mons is the largest mountain in the solar system. It is almost three times the height of Mount Everest. The Martian volcanoes grew from lava flows. Because Mars has low gravity, the weight of the lava was lower than on other planets, and volcanoes could grow very large.

Like the moon, Mars has many impact craters. Its thin atmosphere doesn't burn up objects from space, and so they often impact the surface. The surface of Mars is red from iron oxide. It has frequent dust storms stronger than those in the Sahara desert. These dust storms form large red dunes.

INCLUSION Strategies — GENERAL

- **Developmentally Delayed**
- **Learning Disabled**
- **English Language Learners**

Ask students to write a story entitled "How I Spent My Solar System Vacation" describing a planet they visited on an imaginary vacation away from Earth. They may use the text, reference books, and/or Internet searches for details about a specific planet. Students should include the following in their descriptions:

- Order of the planet from the sun
- How far the planet is from the sun
- List moons (if any)
- How long a day lasts on the planet
- How long a year is in the planet
- General information about the atmosphere on the planet

Students may write down or tape-record their stories to be shared with the class.

The asteroid belt divides the inner and outer planets

Between Mars and Jupiter lie hundreds of smaller, rocky objects that range in diameter from 3 km to 700 km. These objects are called **asteroids,** or minor planets. There are probably thousands of other asteroids too small to see from Earth. *Figure 18* shows the asteroid Ida as photographed in 1993. Most asteroids remain between Mars and Jupiter, but some wander away from this region. Some pass close to the sun and sometimes cross Earth's orbit. The odds of a large asteroid hitting Earth are fortunately very small, but many research programs are keeping track of them. Some pieces of asteroids have hit Earth as meteorites. As a portion of the rock burns up in the atmosphere it makes a bright streak in the sky, which we call a meteor.

The Outer Planets

Figure 19 shows the orbits of the planets most distant from the sun: are Jupiter, Saturn, Uranus, Neptune, and Pluto. Except for Pluto, the outer planets are much larger than the inner planets and have thick, gaseous atmospheres, many satellites, and rings. These large planets are called the **gas giants.**

Because the gas giants have no solid surface, a spaceship cannot land on them. However, the *Pioneer* missions, launched in 1972 and 1973, the *Voyager 1* and *2* missions, launched in 1977, and the *Galileo* spacecraft, launched in 1989, flew to the large outer planets. *Galileo* even dropped a probe into the atmosphere of Jupiter in 1995. A mission called *New Horizons* is planned to investigate Pluto before the year 2020.

Figure 18

Asteroid Ida was photographed by the Galileo spacecraft. It is 56 km long.

▶ **asteroid** a small, rocky object that orbits the sun, usually in a band between the orbits of Jupiter and Mars

▶ **gas giant** a planet that has a deep, massive atmosphere, such as Jupiter, Saturn, Uranus, or Neptune

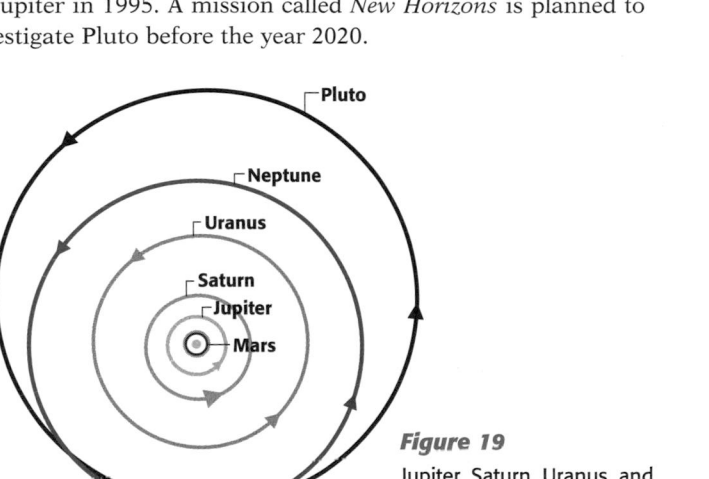

Figure 19

Jupiter, Saturn, Uranus, and Neptune—the gas giants—and little Pluto are the most distant planets known in the solar system.

641

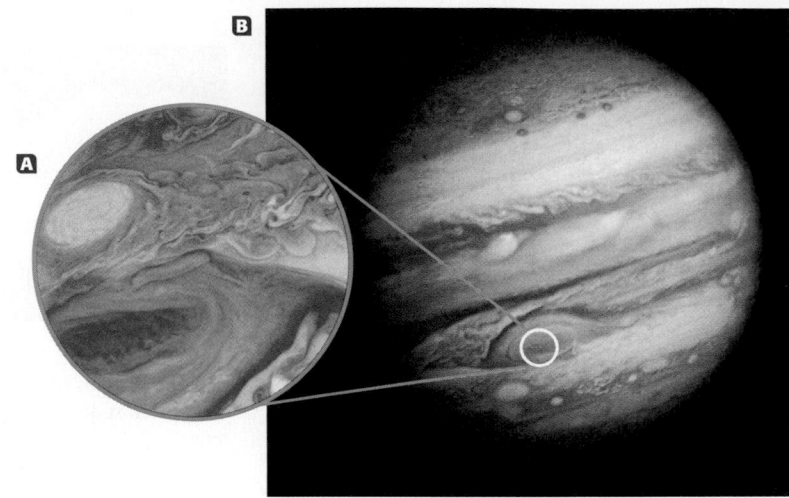

Teach, continued

Teaching Tip

Observing the Planets Jupiter is one of the brightest objects in the night sky. Encourage students to look up the current position of Jupiter and the other planets, and then to try to observe them. Venus, Mars, Jupiter, and even Saturn can be seen even in urban areas if the sky is relatively clear and the planets are above the horizon.

Jupiter is especially spectacular when viewed through binoculars. With binoculars you, like Galileo, can observe as many as four moons that appear to lie on a line passing through Jupiter. These moons are called the Galilean moons. You can watch these satellites change positions over the course of a few nights. Earth's moon also takes on new character when viewed through binoculars. You can see individual craters, especially near the inner edge of a crescent moon.

Figure 20

A The Great Red Spot is a huge, hurricane-like storm on Jupiter.

B Jupiter is the largest planet in the solar system.

internet connect

www.scilinks.org
Topic: The Nine Planets
SciLinks code: HK4095

SC*LINKS*. Maintained by the National Science Teachers Association

Jupiter is the largest planet in the solar system

Jupiter, shown in *Figure 20B,* is the first planet beyond the asteroid belt. Jupiter is big enough to hold 1300 Earths. If it were 80 times more massive than it is, it could have become a star. At a distance of 5 AU, Jupiter takes about 12 Earth years (y) to orbit the sun and rotates once around its axis in less than 10 hours. Images of Jupiter's atmosphere show swirling clouds of hydrogen, helium, methane, and ammonia. Complex features in Jupiter's atmosphere appear to be jet streams and huge storms. One of these storms, the Great Red Spot, shown in *Figure 20A,* is a huge hurricane that measures over twice the diameter of Earth. The Great Red Spot has existed for hundreds of years.

In 1610, Galileo discovered the four largest of Jupiter's 39 satellites, which he named Io, Europa, Ganymede, and Callisto. Using binoculars or a small telescope, you can see them near Jupiter. Io has a thin atmosphere and active volcanoes. Europa may have liquid water under its icy surface.

All the gas giant planets have rings and satellites

Although the vast rings of Saturn were recognized in 1659, it took modern technology to discover the thin, faint rings around the other gas giants. Uranus' rings were not discovered until 1977. Space missions have discovered most of the satellites known in the solar system. Jupiter has 39, Saturn has 30, Uranus has 21, and Neptune has 8. Most are cratered, and many have interesting surface features. Some satellites are thought to have thin atmospheres.

642

Did You Know?

Atmospheres and Mass The thickness of a planet's atmosphere is closely related to the planet's mass. Larger planets, such as the gas giants, have stronger gravitational influences, so they can hold light gases such as helium and hydrogen. Medium-sized planets, such as Earth and Venus, can only hold heavier gases, such as nitrogen, oxygen, and carbon dioxide. Small planets such as Mercury and Pluto cannot hold even the lightest gases and so do not sustain an atmosphere.

Saturn has the most extensive ring system

Saturn is 95 times the mass of the Earth and takes over 29 y to orbit the sun. It rotates in 10.7 h. Like Jupiter, it is a gas giant and rotates fastest at the equator and slower near the poles. In addition to its many satellites, Saturn has a spectacular system of rings, as shown in *Figure 21*.

These rings are narrow bands of tiny particles of dust, rock, and ice. There is a range of sizes of the particles which measure from millimeters to meters. Most are probably the size of a large snowball on Earth. Competing gravitational forces from Saturn and its many satellites hold the particles in place around the planet. The rings are rather thin in comparison to their diameter. Many are only 10 or 20 m thick and stretch around the entire planet. Scientists aren't sure exactly how the rings formed. One hypothesis is that they came from a smashed satellite. Others believe the rings formed from leftover material when Saturn and its satellites formed long ago.

Saturn may still be forming

Saturn radiates three times more energy than it receives from the sun. Scientists believe helium in Saturn's outer layers is condensing and falling inward. As the helium nears the central core, it heats up. Think about pumping air into a bicycle tire. As you pump, the air inside the tire compresses, which causes the tire to heat up. Eventually, for both the tire and Saturn, the extra energy is radiated away. When Saturn uses up its atmospheric helium, this process will stop and Saturn will reach a state of equilibrium. Until then, Saturn is considered to still be forming.

Did You Know?

The density of Saturn is so low that it would float in water—if you could find a big enough bathtub.

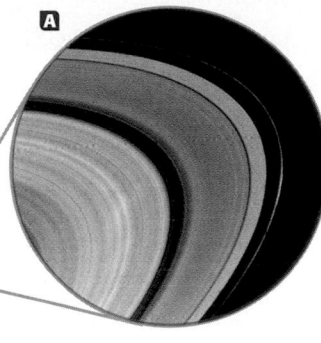

Figure 21

A This photo shows a close-up view of the structure of Saturn's rings, made of thousands of small rocks and ice.

B You can see the shadow of the rings on Saturn's surface if you look closely.

643

SKILL BUILDER — BASIC

Math Skills A day on Saturn is just 10.7 Earth hours long, but a year on Saturn is 29 Earth years long. Have students calculate how many days there are in a year on Saturn. (23 700 days) **LS** Logical

Teaching Tip

Rings of Saturn Some students may think that the rings of Saturn are rigid, solid rings. Remind students that the rings are made of separate particles of dust, rock, and ice. The material for the rings may have come from a moon that was shattered, or it may be material from the early solar system that never clumped together to form a moon. These particles are held in orbits at specific distances from the planet by the complex tidal forces from Saturn's many moons.

INTEGRATING

GEOLOGY **Tidal Forces** Tides on Earth are caused by the gravitational pull of the moon. The moon's gravity is strongest on the side of the Earth closer to the moon, and weakest on the side opposite the moon. As a result, the surface of the seas is stretched into an ellipsoid shape, like an egg. The long axis of the ellipsoid creates one bulge pointing toward the moon and another bulge pointing away from the moon. As the moon orbits Earth, these bulges rotate around the planet, in effect raising the level of the oceans wherever the bulges are.

Because Earth has only one moon, the tides on Earth are relatively simple. Saturn's 30 moons create much more complex tidal forces, which are responsible for holding Saturn's rings in place.

Three primary rings, called A, B, and C, can be seen in Saturn's rings using telescopes on Earth. The division between ring A and ring B, called the Cassini division, is the easiest to see. Tidal effects caused by Saturn's seventh moon, Mimas, are believed to be responsible for the Cassini division.

 **Videos**

CNN **Presents Earth Science**

• **Segment 30** Europa Pics [GENERAL]

• **Critical Thinking** Worksheet 30

See the Science in the News video guide for more details.

Applying Information

1. Images taken by probes in space are clearer because the light used to make the pictures does not have to pass through Earth's atmosphere. Also, probes can get closer to their subjects.

2. When the ion engine is turned on, the probe goes in a direction opposite to the exhaust. The probe will go slower than the ions because it has more mass. However, there is almost no friction or air resistance in space, so the probe can keep accelerating as long as the engine is on.

SKILL BUILDER

Interpreting Visuals Figure 23 shows Pluto with its moon, Charon. Because it is so small and so distant, Pluto was not discovered until 1930. Like Neptune, the planet Pluto was predicted before it was actually observed.

Charon was discovered in 1978. Even with many powerful telescopes, Pluto and Charon cannot be seen separately. This image taken by the Hubble Space Telescope is one of the clearest ever obtained of Pluto and Charon together.

Figure 22

A Methane in the atmosphere of Uranus gives it a blue color.

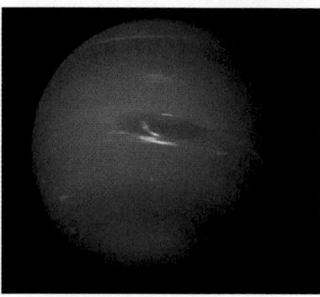

B The Great Dark Spot is a huge storm in Neptune's blue atmosphere.

Uranus and Neptune are blue giants

Beyond Saturn lie the planets Uranus and Neptune, which are shown in *Figure 22A* and *Figure 22B.* These two gas giants are similar to each other in size and color. Although they are smaller than Saturn and Jupiter, they are still large enough to hold thick, gaseous atmospheres composed of hydrogen, helium, and methane. The methane gives both planets a bluish color.

William Herschel discovered Uranus by accident in 1781. He wanted to name it after King George III, but another astronomer suggested that it be given a name from mythology, as the other planets were. Uranus is 14 Earth masses, and it takes approximately 84 y to orbit the sun at its distance of 19 AU.

After Uranus was discovered, astronomers used what they knew about gravity to guide their search for other planets. Because every mass attracts every other mass, changes in the expected orbit of Uranus could be used to predict the existence and position of other planets. Predicted independently by John Adams and Urbain Leverrier, Neptune was discovered in 1846 by Johann Galle. It is 17 Earth masses, and takes approximately 164 y to orbit the sun at a distance of 30 AU.

Uranus and Neptune are far away from the sun. The gas in their upper atmospheres is very cold, about 58 K. Uranus rotates in 17 h, but its pole is tilted over on its side at a 98° angle. Because of this tilt, Uranus has the most extreme seasons in the solar system. The few clouds in the atmosphere of Uranus show wind speeds of 200 to 500 km/h. Neptune rotates in 16 h. Neptune also has storm systems similar to Jupiter's.

REAL WORLD APPLICATIONS

Deep Space Exploration
Deep space missions help astronomers get much closer to and take more detailed images of objects in our solar system.

Ion power was tested on a probe called *Deep Space 1.* Ions are charged particles much like those that make your clothes stick when they come out of the dryer. Ions made from the element xenon race out of the probe at a speed of 100 000 km/h. Each one that is pushed out propels the probe forward, just as a balloon moves forward when some of its air is released. This new propulsion lets probes travel into very deep space.

Applying Information
1. Why do probes get better images than those taken from telescopes on Earth?
2. What direction will the probe go when the ion engine is on? What determines its speed?

644

Historical Perspective

Discovery of the Outer Planets Observers have known of the existence of Jupiter and Saturn for many centuries. The other outer planets were discovered much later.

Uranus was discovered in 1781 by an English music teacher named William Herschel. He originally named the planet *Georgium Sidus* (George's Star), in honor of King George III. The name Uranus was later officially chosen to continue the tradition of naming the planets after Roman gods.

Astronomers in the nineteenth century observed discrepancies in the orbit of Uranus that could not be explained by its attraction to the sun and to other known planets. They predicted that another large planet existed beyond Uranus. In 1846, the position of the planet, Neptune, was predicted by the English astronomer John Adams and the French astronomer Joseph Leverrier. John Galle is also credited with the discovery of Neptune because he located the planet in the sky.

Pluto is an oddball planet

After the discovery of Neptune, American astronomer Percival Lowell used fluctuations in Neptune's orbit to predict yet another planet. In 1930, Clyde Tombaugh found a planet very close to where Lowell had predicted one might be. This new planet Pluto, shown in *Figure 23* with its satellite Charon, is not like the other outer planets. It has only a thin, gaseous atmosphere and a solid, icy surface. Pluto orbits the sun in a long ellipse and its orbit is at a different angle than the rest of the solar system. For these reasons, some scientists believe Pluto was captured by the gravity of the sun some time after the formation of the solar system.

Pluto isn't always the farthest planet in the solar system. Its orbit sometimes cuts inside the orbit of Neptune. Pluto's average distance from the sun is almost 40 AU. It takes 248 y to complete one orbit. Its mass is only 0.002 Earth's mass, closer to the mass of a satellite than that of a planet. Some scientists even refuse to classify it as a planet. They think Pluto might be an ejected satellite of Neptune or simply a leftover piece of debris from when the solar system formed.

Figure 23

Pluto's satellite, Charon, has a diameter almost half as large as Pluto. Pluto was discovered in 1930, and Charon was discovered in 1978.

Close

Quiz ——————————— BASIC

1. What substance makes Earth the most likely place in the solar system for life to have developed? (liquid water)

2. What effect does the thick atmosphere of Venus have on the temperature of the planet? (The atmosphere causes a greenhouse effect, which makes the planet very hot.)

3. What lies in the region between the inner and outer planets? (the asteroid belt)

4. What primary feature do all of the outer planets—except Pluto—have in common? (They are all gas giants with very thick, gaseous atmospheres.)

LS Logical

SECTION 2 REVIEW

SUMMARY

▶ The solar system has the planets Mercury, Venus, Earth, Mars, Jupiter, Saturn, Uranus, Neptune, and Pluto.

▶ The inner planets are relatively close to the sun and have few satellites.

▶ Earth sustains life. It has water, an atmosphere, and oxygen.

▶ The inner and outer planets are separated by the asteroid belt.

▶ The gas giants are farther from the sun, are relatively cold, and have rings and many satellites.

1. **List** the planets in order of distance from the sun.

2. **Describe** one feature of each inner planet.

3. **Explain** why the surface of Venus is hotter than the surface of Mercury.

4. **Describe** one feature of each outer planet.

5. **Compare** the terrestrial planets with the gas giants.

6. **Explain** why most satellites and planets have many craters.

7. **Describe** how the hydrosphere and atmosphere help make Earth a good place for life as we know it.

8. **Explain** why distance from the sun is an important factor for a planet to support life.

9. **Critical Thinking** Why are space missions needed to learn about other planets?

10. **Critical Thinking** What characteristics would a planet around another star need to sustain life as we know it?

645

Answers to Section 2 Review

1. Mercury, Venus, Earth, Mars, Jupiter, Saturn, Uranus, Neptune, Pluto.

2. Answers may vary. Students may list the following features: Mercury has craters; Venus is the hottest planet, has a dense atmosphere, and rotates backwards; Earth: has hydrosphere, oxygen, life, and one moon; Mars is small, rocky, and has craters, thin atmosphere, icecaps on poles, and two satellites.

3. Venus is hotter than Mercury, although it is more distant from the sun, because its atmosphere traps heat very well.

4. Answers may vary. Students may list the following features: Jupiter is the largest planet, has 28 satellites, the Great Red spot, and fast rotation; Saturn has the largest ring system, and 30 satellites; Uranus has 21 satellites, is blue and cold; Neptune has 8 satellites, is blue and cold; Pluto is not a gas giant like other outer planets, has 1 satellite, and is the most distant.

See "Continuation of Answers" at the end of the chapter.

Chapter Resource File

• **Concept Review**

• **Cross Disciplinary Worksheet** Real World Applications— Deep Space 1 **ADVANCED**

SECTION
3

Focus

Overview

Before beginning this section, review with your students the Objectives listed in the Student Edition. This section introduces students to ancient models of the universe and solar system, and to the modern nebular model for the formation of the solar system. This section also discusses smaller objects in the solar system, including comets, asteroids, and meteors. The section closes by discussing a model for the formation of Earth's moon, and a description of the current state of discovery of planets around stars other than the sun.

Bellringer

Use the Bellringer transparency to prepare students for this section.

Motivate

Opening Discussion ——— GENERAL

Students may have some understanding of the origin of the universe according to the big bang theory. However, few students will have prior knowledge of theories of the formation of the solar system. Ask students to brainstorm about possible ways that the solar system could have formed. Direct student models by reminding students that their models should offer explanations for certain key features of the solar system, such as the fact that all the planets orbit in the same direction around the sun and all the planets orbit in nearly the same plane. **LS Logical**

Formation of the Solar System

■ **KEY TERMS**
nebula
nebular model
accretion
comet

OBJECTIVES

▶ **Contrast** ancient models of the solar system with the current model.
▶ **Estimate** the age of our solar system.
▶ **Summarize** two points of the nebular model, and describe how it can explain astronomical observations.
▶ **Explain** how scientists think the moon was formed.

The oldest record of human interest in astronomy was found in Nabta, Egypt. Scientists and historians believe a group of stones were specially arranged 6000 to 7000 years ago to line up with the sun on the longest day of the year, called the summer solstice. How ancient people in Egypt used this ancient observatory and how they understood the sky remains a mystery.

Astronomy—The Original Science

Historians believe that many ancient peoples watched the changing sky. *Figure 24* shows Stonehenge, a structure thousands of years old, that was probably used for keeping time. Its stones are also aligned with the summer and winter solstices.

Some ancient peoples used stories or myths to explain star movements. Eventually, mathematical tools began to be used to make models of observed astronomical objects. But ancient curiosity was the beginning of astronomy. Ancient questions about the universe gave rise to science and led to the scientific method which is used throughout the sciences today.

internet connect

www.scilinks.org
Topic: Origins of the Solar System
SciLinks code: HK4099

SCiLINKS. Maintained by the National Science Teachers Association

Figure 24
Stonehenge, located in England, is one of the world's oldest observatories.

Cultural Awareness

Ancient Astronomers Virtually every culture that has developed writing has also left records of astronomical observations. The Babylonians kept very detailed records of celestial events. Ptolemy used data from the Babylonians to track the motion of the heavens far back in history as he developed his models of the motions of the heavens. The ancient Chinese also kept very good astronomical records. The only report of the supernova in 1054 A.D. (the supernova that produced what is now the Crab Nebula) comes from China. The Maya also tracked the motions of the planets, especially Venus, with astonishing accuracy.

The first model put Earth in the center

Like many people who came before them, the ancient Greeks observed the sky to keep track of time. But they took a new approach in trying to understand Earth's place in the universe. They used logic and mathematics, especially geometry. The Greek philosopher Aristotle explained the phases of the moon and eclipses by using a model of the solar system with Earth in the center. His *geocentric* or "Earth-centered" model is shown in *Figure 25*.

This model was expanded by Ptolemy in 140 CE. Ptolemy thought that the sun, moon, and planets orbited Earth in perfect circles. His theory described what we see in day-to-day life, including motions of the sun and planets. Because it predicted many astronomical events well, Ptolemy's model was used for over a thousand years.

Copernicus moved the sun to the center

In 1543, Nicolaus Copernicus proposed a *heliocentric,* or "sun-centered," model. He realized that many adjustments used to make Ptolemy's model work would not be needed in a model in which the sun was in the center. In this new model, Earth and the other planets orbit the sun in perfect circles. Although Copernicus's model was not perfect, it explained the motion of the planets more simply than Ptolemy's model did. In 1605, Johannes Kepler improved the model by proposing that the orbits around the sun are ellipses, or ovals, rather than circles.

Newton explained it all

The heliocentric model was useful for time keeping and for navigation. However, no one had explained why planets orbited the sun in elliptical orbits. In 1687, Isaac Newton explained that the force that keeps the planets in orbit around the sun, and satellites in orbit around planets is gravity. His theory states that the gravitational force that keeps planets in orbit around the sun is the same force we experience when things fall to Earth. His theory also states that every object in the universe exerts a gravitational force on every other object. Newton was the first to propose that everything in the universe follows the same rules and acts in a predictable way. All classical physics, including much of astronomy, is built on this assumption.

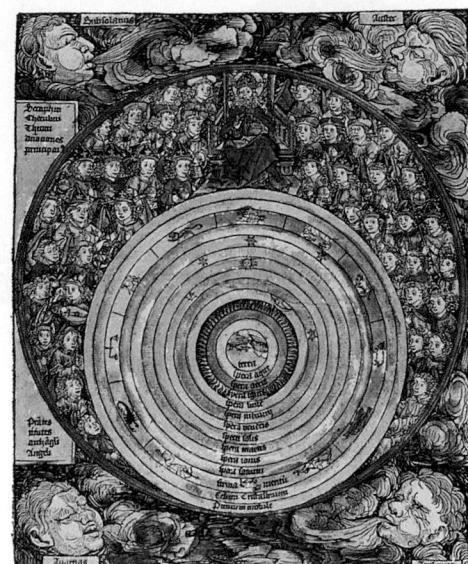

Figure 25
Before Copernicus, people believed Earth was in the center of the universe.

Did You Know ?

The astronomer Johannes Kepler proposed that planets orbit the sun in elliptical paths. However, some scientists continued to believe that Earth was the center of the solar system until Isaac Newton showed that elliptical orbits could be predicted using his laws of motion.

647

Historical Perspective

Galileo's Telescopes Italian astronomer Galileo Galilei was the first astronomer to work with a telescope. In the summer of 1609, Galileo made his first three-power telescope (which magnified objects to three times their size). He went on to create a twenty-power telescope that he used to observe the moon and the four largest moons of Jupiter.

At the time of Galileo's observations, Copernicus's sun-centered model of the universe was still not widely accepted; most people still used Ptolemy's Earth-centered model, which theorized that every object in the heavens orbits Earth. When Galileo saw four moons orbiting Jupiter, he realized that objects can orbit around other heavenly bodies as well. Galileo also observed that Venus has phases like the moon. Galileo's observations were of key importance in establishing that Earth and the other planets orbit the sun.

Figure 26

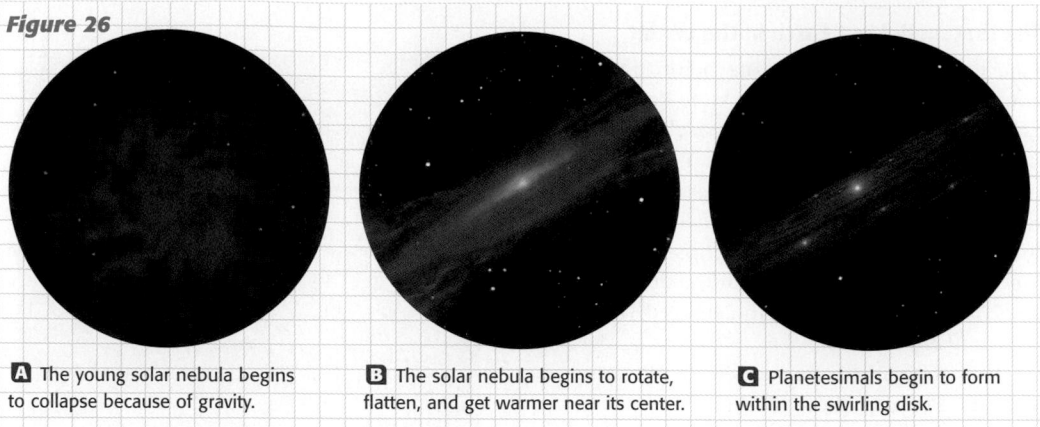

A The young solar nebula begins to collapse because of gravity.

B The solar nebula begins to rotate, flatten, and get warmer near its center.

C Planetesimals begin to form within the swirling disk.

Teach, continued

SKILL BUILDER —ADVANCED

Writing Skills After students have read the section on the nebular model, have them write a description of the formation of the solar system from the point of view of a grain of carbon dust. Encourage them to follow the dust grain through the collapse of the disk, accretion into a planet, and perhaps even into the origin of a carbon-based life form. **LS** Verbal

Teach, continued

SKILL BUILDER —ADVANCED

Writing Skills After students have read the section on the nebular model, have them write a description of the formation of the solar system from the point of view of a grain of carbon dust. Encourage them to follow the dust grain through the collapse of the disk, accretion into a planet, and perhaps even into the origin of a carbon-based life form. **LS** Verbal

Teaching Tip

Before the Solar System All elements heavier than hydrogen and helium are made by the nuclear processes in the interiors of stars. Elements heavier than iron are made only in supernovae. Every atom on Earth, including all of the atoms in every rock and mineral, in the air, in every plant, and in every person's body, was in the cloud of gas and dust that collapsed to form the solar system. Because there are many elements on Earth heavier than iron (such as gold, silver, and lead) the solar system must have formed from some material that came from a star that blew up in a supernova sometime in the distant past. Your body is made of stardust!

▶ **nebula** a large cloud of dust and gas in interstellar space; a region in space where stars are born or where stars explode at the end of their lives

▶ **nebular model** a model for the formation of the solar system in which the sun and planets condense from a cloud (or nebula) of gas and dust

The Nebular Model

According to dating of rocks, scientists believe the solar system is approximately 4.6 billion years old. Scientists trying to build a good model to explain how the solar system formed started with the following questions: Why are the planets so far away from each other? Why are they almost in the same plane? Why are their orbits nearly circular? Why do they orbit in the same direction? Why are the terrestrial planets different from the gas giants? A good model would also have to explain the presence and behavior of objects such as satellites, comets, and asteroids.

The solar system may have begun as a nebula

A **nebula** is a large cloud of dust and gas in space. The most widely accepted model of the formation of the solar system is the **nebular model.** According to the nebular model, the sun, like every star, formed from a cloud of gas and dust that collapsed because of gravity, as shown in *Figure 26A.*

The nebula then formed a rotating disk

As this cloud collapsed, it formed into a flat, rotating disk, as shown in *Figure 26B.* In the center, where the material became denser and hotter, a star began to form. As the cloud collapsed, it spun faster and faster, just as ice skaters spin faster when they pull their arms in. Because spinning bodies tend to change shape as they collapse, the nebula flattened. Scientists find this model helpful because objects that form out of a disk will lie in the same plane, have almost circular orbits, and orbit in the same direction as the material in the center.

648

INTEGRATING

PHYSICS Conservation of Angular Momentum As gas and dust in a rotating cloud collapses in toward the center, the gas and dust rotates faster and faster. This is due to the fact that angular momentum—which is proportional to the product of radius and angular speed—is conserved. As the radius of orbit decreases, the angular speed increases, so the quantity is conserved.

Because the clouds that collapse to form a solar system are extremely large, the clouds may start out with an almost imperceptible rotation. As the cloud collapses, this slow rotation speeds up until the cloud is rotating at a speed similar to that of the planets orbiting in our solar system today.

D Because of their greater gravitational attraction, the largest planetesimals begin to collect the dust and gas of the solar nebula.

E Smaller planetesimals collide with the larger ones, and the planets begin to grow.

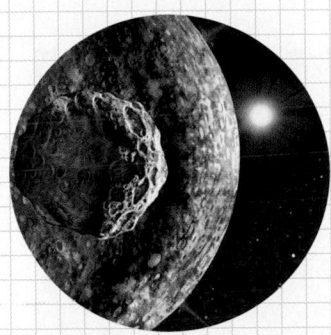

F The remaining dust and gas are gradually removed from the solar nebula, which leaves planets around the sun—a new solar system.

Planets formed by the accretion of matter in the disk

As in *Figure 26C, planetesimals*, or particles that become planets, began to form in the disk. They formed mostly through the process of **accretion,** which is when small particles collide and stick together. As shown in *Figure 26D,* these planetesimals grew bigger as they collected more material. *Figure 26E* shows the planets beginning to grow. Each planet swept up material in its region of the disk, which explains why the planetary orbits are separate from each other.

The nebular theory explains why the terrestrial planets are different from the gas giants. The gas and dust close to the sun did not join together easily. Radiation from the newly formed sun exerted pressure on the rest of the gas and dust in the disk. The planets near the sun lost their lighter materials, leaving behind rocky planets. Colder gas and dust in the outer part of the disk joined easily and became the gas giants. These planets were large enough and cold enough to hold light nebular gases, such as hydrogen and helium, in their atmospheres.

The nebular model also explains smaller rocks in space

Satellites may have formed around gas giants in the same way planets formed around the sun. Another possibility is that planetesimals were captured by the gravitational pull of the gas giants. Some smaller satellites may have broken off from larger ones. Most satellites orbit planets in the same direction that the planets orbit the sun. *Figure 26F* shows the solar system as it looks today. Asteroids and other small rocks are most likely leftover planetesimals from solar-system formation.

▶ **accretion** the accumulation of matter

Quick ACTIVITY

Estimating 4.6 Billion

1. Find a small box, and measure its length, width, and height
2. Multiply the height by the width by the length to find the volume of the box.
3. Fill half the box with popcorn kernels.
4. Count the number of kernels in the box, and multiply that number by two.
5. Divide 4 600 000 000 by the total number from step 4.
6. How big would your box need to be to hold 4.6 billion popcorn kernels?
7. If you have time, estimate the weight of 4.6 billion kernels.

649

Did You Know?

Planetary Oddballs One strength of the nebular model is that it explains why all the planets are orbiting the sun in the same direction. Most planets also spin in this same direction on their axes, and most satellites orbit planets in this same direction as well. However, there are exceptions. Venus spins in the completely opposite direction on its axis, and Uranus spins almost perpendicularly to the plane of its orbit around the sun. Ask students how these unusual cases might be explained. (Planets may change their direction of spin if they collide with a large object. A satellite that orbits out of the plane of the solar system may have been a stray rock that was captured by the planet.) **LS** Logical

Reading Skills Instruct students to make a table to organize the information presented in this section about rocks in space. Headings for the table should include *Type of Rock, Composition,* and *Location.* Have them fill out the table as they read through the section. They should have information about at least 4 different types of rocks. After they are finished reading, have them look through the chapter to find the information missing from their tables.
LS Logical

Teaching Tip ——— GENERAL

Comet Missions NASA and the European Space Agency have sent several missions to intercept comets and study them at close distance. Several more such missions are planned for the next few decades. Have students research one or more of these missions to find out how they worked (or will work) and what type of information the missions found (or are intended to find). **LS** Verbal

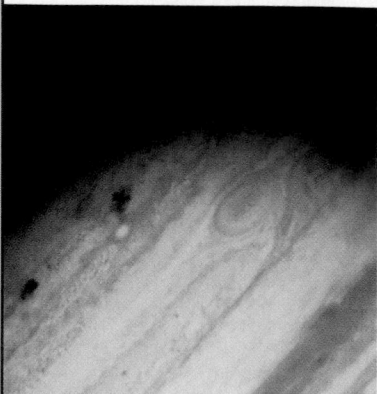

Figure 27

The many pieces of comet Shoemaker-Levy 9 hit Jupiter.

■ **comet** a small body of ice, rock, and cosmic dust loosely packed together that follows an elliptical orbit and that gives off gas and dust in the form of a tail as it passes close to the sun

Figure 28

In this photo, you can see the two tails of Comet Hale-Bopp. The blue streak is the ion tail, and the white streak is the dust tail.

Rocks in Space

There are many types of small bodies in our solar system, including satellites, asteroids, meteoroids, and comets. Satellites orbit planets, and most asteroids can be found between Mars and Jupiter. Meteoroids are small pieces of rock that enter Earth's atmosphere. Most meteoroids burn up in the atmosphere, and we see them as meteors streaking through the night sky. If a meteoroid survives the atmosphere and hits the ground, it is called a meteorite. **Comets** are probably composed of leftover material from when the solar system formed.

Comets may give us clues to the origin of the solar system

By studying comets, scientists have gained important information about the material that made the solar system. Comets are composed of dust and of ice made from methane, ammonia, carbon dioxide, and water. In 1994, pieces of the comet Shoemaker-Levy 9 plowed into the planet Jupiter, as shown in *Figure 27.* The comet had been pulled into many pieces by Jupiter's gravity. The impacts showed that the comet also contained silicon, magnesium, and iron. We will learn more about comets in 2006 when the *Stardust* mission returns to Earth with comet samples.

Comets have long tails and icy centers

Because of their composition, comets are sometimes called "dirty snowballs." When a comet passes near the sun, solar radiation heats the ice so the comet gives off gases in the form of a long tail. Some comets, such as the one in *Figure 28,* have two tails—an ion tail made of charged particles that is blown by the solar wind and a dust tail that follows the comet's orbit.

A comet's orbit is usually very long. Although some of the comet is lost with each passage by the sun, the icy center, or nucleus, continues its journey around the sun. When its tail eventually disappears, a comet becomes more difficult to see. It will brighten only when it passes by the sun again.

650

Historical Perspective

Halley's Comet The first written account of Halley's Comet came from China in 240 BCE. The comet also made a high-profile appearance during the Battle of Hastings in 1066.

The American writer Mark Twain was born while Halley's Comet was visible from Earth in 1835. In 1909, Twain predicted that he would die when it returned the following year, and his prediction turned out to be correct.

Halley's Comet was most recently visible from Earth in 1986. It will visit the inner solar system again in 2061.

Where do comets come from?

During the formation of our solar system, some planetesimals did not join together. These leftovers strayed far from the sun. The gravitational force of the gas giants pulled on these small pieces, and over hundreds of millions of years they moved into distant orbits. These far-flung pieces make up the Oort cloud of comets, shown in *Figure 29,* which may be up to 100 000 AU wide and extend in all directions. Planetesimals that remained in the nebular disk formed the Kuiper belt beyond the orbit of Neptune. Some scientists believe that Pluto may simply be the largest object in the Kuiper belt.

Halley's comet is one of the most famous comets. It travels in a highly elliptical orbit that brings it near the sun every 76 years. It appears in Earth's sky once every 76 years. Compared with the rest of the solar system it orbits backward, which suggests that its orbit was probably greatly altered by a planet's gravity.

Asteroids can be made of many different elements

We can study asteroids by studying meteorites. Meteor showers can occur when Earth passes through a comet tail, but larger rocks that make it through our atmosphere come from asteroids. As shown in *Figure 30,* there are three major types of asteroids. Stony meteorites include carbon-rich specimens that contain organic materials and water. Metallic meteorites are made of iron and nickel. Stony-iron meteorites are a combination of the two types. Most meteorites that have been collected are stony and have compositions like those of the inner planets and the moon.

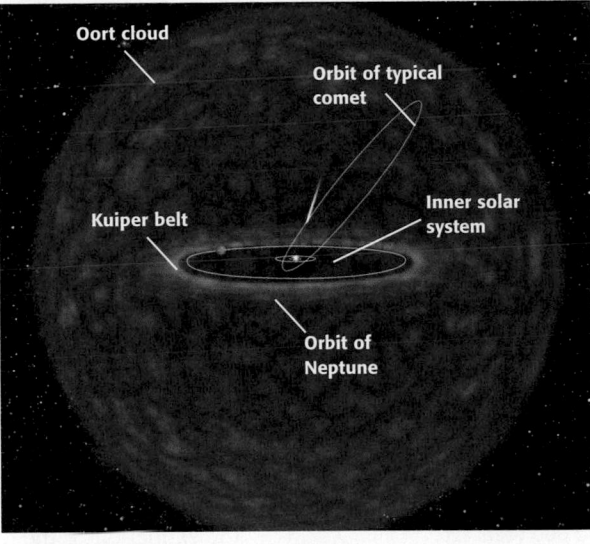

Figure 29

Comets come from the Kuiper belt, a disk-shaped region beyond the orbit of Neptune, and the Oort cloud, a spherical region in the outer solar system beyond the orbit of Pluto.

Stony
Rocky material

Metallic
Iron and nickel

Stony-iron
Rocky, iron, and nickel

Figure 30
There are three major types of meteorites.

SKILL BUILDER

Vocabulary "Oort cloud" and "Kuiper belt" can be difficult to pronounce. "Oort" is pronounced with the "oo-" as in "food" followed by "-rt" as in "dirt." "Kuiper" is pronounced like "kie" (rhymes with "pie") + "purr."

SKILL BUILDER — **Basic**

Interpreting Visuals Figure 29 shows a view of the solar system on a very large scale. Have students examine the figure. What shape is the Oort cloud? (spherical) What shape is the Kuiper belt? (disk-shaped) Which of these two is farther from the sun? (the Oort cloud) **LS Visual**

651

Answers to Artificial Satellites

1. Each satellite is in its own orbit. Two satellites on the same orbital path at the same distance from Earth would always stay at the same distance. The only chance for impact would be between satellites on orbits that cross. Even then, the enormous sizes of the orbits compared to the sizes of the satellites make an impact very unlikely.

2. If a satellite breaks up, the pieces continue to orbit Earth (or whatever it was orbiting before it broke up). Any piece that falls back to Earth melts or burns up as it passes through Earth's atmosphere.

3. To reach all of Earth, you need more than one satellite. Just as you can't see the moon from all locations on Earth at the same time, you could not have a clear view to a single satellite (in order to access it for communication) at all times.

Figure 31

Barringer Crater in Arizona is due to an impact that happened over 50 000 years ago.

Meteorites sometimes strike Earth

Most meteoroids were once part of asteroids, although a few came from Mars or from our moon. Objects less than 10 m across usually burn up in the atmosphere. *Figure 31* shows Barringer Crater, which was made by the impact of a meteoroid that probably weighed 200 000 metric tons (200 000 000 kg). The crater is over 1 km wide and 175 m deep. As you recall, when a meteoroid strikes Earth, it is called a meteorite. Twenty-five metric tons (25 000 kg) of iron meteorite fragments have been found. The rest vaporized on impact, or were scattered, broken by erosion, or buried. Earth has nearly 100 craters larger than 0.1 km. Many craters look like circular lake basins. Erosion and volcanic activity can erase traces of the craters, but new technologies, including satellite photos, can help locate them. From this data, scientists estimate that a large-scale collision happens only once every few hundred thousand years.

Many scientists believe that an asteroid or comet that was 10 to 15 km wide struck Earth about 65 million years ago caused the extinction of the dinosaurs. The energy released was probably as much as the energy of 10 million hydrogen bombs. Large amounts of dust may have swirled into Earth's atmosphere and darkened the sky for years. Plants died, and eventually the dinosaurs, which depended on plants for food, died. Today, government programs track asteroids that come near Earth.

Artificial satellites Sputnik 1 in 1957 was the first of thousands of artificial satellites launched into Earth orbit. Satellites today are used for weather monitoring, communications, espionage, and monitoring changes in oceans and land-use on Earth. Astronomers use satellites to look out into space. Most satellites are only a few hundred kilometers above Earth's surface. To get a satellite to orbit Earth requires a vehicle such as the space shuttle. Rockets, which are also used to launch satellites, follow Newton's third law of motion: for every action force, there is an equal and opposite reaction force. Rockets throw gas out the bottom end of the rocket that, in turn, gives motion in the opposite direction and pushes the rocket upward.

Applying Information

1. There are over 2000 satellites orbiting Earth. Why don't they run into each other?

2. If a satellite breaks apart, what happens to the pieces?

3. To set up a communications network that would reach all of Earth, would you need more than one satellite? Explain.

LANGUAGE ARTS

CONNECTION — GENERAL

Have students write a short story that describes a meteor shower. The students should write the story as if they are a scientist or an astronaut collecting data about the size, color, shape, and composition of the falling meteors. **LS Verbal**

How the Moon Formed

Before the moon's composition was known, there were several theories for how it formed. Some thought a separate body was captured by Earth's gravity. If this theory were true, the moon's composition would be very different from Earth's. Others thought the moon formed at the same time as Earth, which would mean its composition would be identical to Earth's. When the moon's composition was learned to be similar to Earth's, but not identical, a third theory emerged about how the moon was formed.

Earth collided with a large body

When Earth was still forming, it was *molten*, or heated to an almost liquid state. The heavy material was sinking to the center to form the core, and the lighter material floated to form the mantle and crust. A Mars-sized body struck Earth at an angle and was deflected, as shown in **Figure 32A.** At impact, a large part of Earth's mantle was blasted into space.

The ejected material clumped together

The debris began to clump together to form the moon, as shown in **Figure 32B.** The debris consisted of the iron core of the body mantle material from Earth and from the impacting body. The iron core became the core of the moon. This theory explains why moon rocks brought back to Earth from the *Apollo* mission share some characteristics of Earth's mantle.

The moon began to orbit Earth

Figure 32C shows the material that formed the moon revolving around Earth because of Earth's gravitational pull. After the moon cooled, impacts created basins on the surface. Lava flooded the basins to make the maria. Smaller impacts made craters on the lunar surface. Today, lava flows on the moon have essentially ceased.

Some scientists doubt this theory because it involves a "unique" event. A unique event has only happened a few times in history. But, the moon itself is unique. It is the only large satellite around a terrestrial planet, and except for Charon, it is the largest moon with respect to its planet.

Figure 32

A Impact A large body collided with Earth and blasted part of the mantle into space.

B Ejection The resulting debris began to revolve around Earth.

C Formation The material began to join together to form the moon.

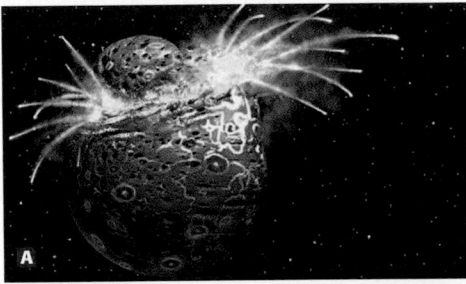

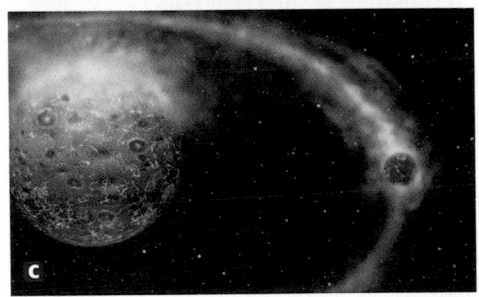

653

INTEGRATING

GEOLOGY Craters Students may not realize that the craters on the moon are due to impacts. Explain that all large objects in the solar system, including all the planets and their moons, are struck by meteorites and even asteroids from time to time. The rate of such collisions was much higher during the first billion years or so of the solar system's history. During that time, there was much more debris floating freely around the solar system.

Planets or satellites that have very little geological activity, such as Mercury or Earth's moon, retain craters as evidence of impacts for a long time, sometimes for billions of years. Almost all of the impact craters on Earth have disappeared due to erosion and the continual reshaping of Earth's surface by plate tectonics.

Quiz ━━━━━━━━━━ **BASIC**

1. What object did Copernicus think was at the center of the universe? (the sun)

2. Out of what did the solar system form, according to the nebular model? (a cloud of gas and dust)

3. Name three types of rocks that can be found in space. (Answers may include: asteroids, meteoroids, comets, satellites, or planets)

4. Describe the primary event leading to the formation of Earth's moon. (Molten Earth collided with a Mars-sized body.)

LS Logical

Other Stars Have Planets

Astronomers have discovered over 90 planets by measuring the small gravitational effects they have on their parent stars. As a planet orbits its star, it causes the star to wobble back and forth. Imagine an adult and a child on a seesaw. The adult sits very close to the center and doesn't move much. The child sits away from the center and moves up and down much farther than the adult. The massive star is like the adult, and the small planet like the child. We have no images of these newly discovered planets. Because planets shine by reflected starlight, they are too faint to see. Astronomers use special techniques and technology such as the Keck telescopes, shown in *Figure 33,* to observe the movements of the parent star over time.

Figure 33

The Keck telescopes and others like them help astronomers find planets outside our solar system.

Almost all of the newly discovered planets have masses close to the mass of Jupiter or Saturn. Many of them have noncircular orbits that bring them close to 1 AU from their star. Only a few are in systems that have more than one planet. Modern detection methods favor finding large planets. Although many are around stars like our sun, these systems are not like our solar system.

SECTION 3 REVIEW

SUMMARY

▶ The sun is in the center of the solar system. Most planets are in the same plane and orbit in the same direction that the sun rotates.

▶ The solar system is approximately 4.6 billion years old.

▶ In the nebular model, a disk formed from a cloud of gas and dust. Planets formed by accretion of matter in the disk.

▶ The moon may have formed after a Mars-sized object struck Earth.

1. **Explain** how our current model of the solar system differs from Ptolemy's model.

2. **State** the approximate age of the solar system.

3. **Describe** two steps of solar system formation.

4. **Describe** how comets change when they near the sun.

5. **List** the three types of meteorites.

6. **Explain** why the study of comets, asteroids, and meteoroids is important to understand the formation of the solar system.

7. **Contrast** the theory of the formation of our moon with how satellites may have formed around the gas giants.

8. **Critical Thinking** Do you think the government should spend money on programs to search for asteroids that may strike Earth? Explain.

9. **Critical Thinking** Do you think planets like Earth exist around other stars? Do you think they may contain life? Explain.

654

Answers to Section 3 Review

1. Our current model of the solar system has the sun in the center, as opposed to Ptolemy's Earth-centered model.

2. 4.6 billion years

3. Correct answers should include two of the following: asteroids, comets, and satellites.

4. A comet has a nucleus, a dust tail, and an ion tail. The nucleus brightens as the comet nears the sun. The dust tail follows the comet's orbital path around the sun. The ion tail points directly away from the sun.

5. stony, metallic, stony-iron

6. They are made of the leftover building blocks of the solar system from 4.6 billion years ago.

7. Our moon may have formed when a Mars-sized object hit the young molten Earth and tore off some of the mantle. The object and the mantle material formed the moon. The satellites of the gas giants probably were leftover planetesimals.

See "Continuation of Answers" at the end of the chapter.

Study Skills

Concept Mapping

Concept mapping is an effective tool for helping you organize the important material in a chapter. It also is a means for checking your understanding of key terms and concepts.

1 **Select a main concept for the map.**

> We will use *planets* as the main concept of this map.

2 **List all the important concepts.**

> We'll use the terms: *planets, satellites, phases,* and *eclipses.*

3 **Build the map by placing the concepts according to their importance under the main concept and adding linking words to give meaning to the arrangement of concepts.**

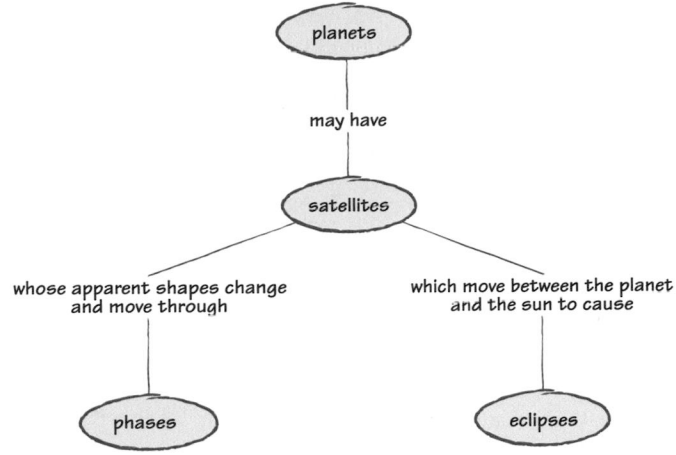

> From this completed concept map we can write the following propositions:
>
> Planets may have satellites whose apparent shapes change and move through phases.
>
> Planets may have satellites, which move between the planet and the sun to cause eclipses.

Practice

1. Draw your own concept map by using the main topic *outer planets.*
2. Write three propositions from your completed concept map.

655

Understanding Concepts

1. d

2. d

3. c

4. a

5. b

6. b

7. c

8. b

9. a

10. b

11. a

12. b

13. c

14. a

15. d

16. b

17. b

18. d

Chapter Highlights

Before you begin, review the summaries of the key ideas of each section, found at the end of each section. The key vocabulary terms are listed on the first page of each section.

UNDERSTANDING CONCEPTS

1. Which outer planet is not a gas giant?
   a. Jupiter   **c.** Saturn
   b. Neptune   **d.** Pluto

2. Earth is gravitationally attracted to
   a. Mercury.   **c.** the sun.
   b. the moon.   **d.** All of the above.

3. The nebular theory explains
   a. why Venus rotates backwards.
   b. why Halley's comet has a very elongated orbit.
   c. why the planets lie in almost the same plane.
   d. why Io has volcanoes.

4. Meteor showers occur when
   a. Earth passes through the orbit of a comet.
   b. a large asteroid comes near Earth.
   c. Earth enters the asteroid belt.
   d. Earth enters the Oort cloud.

5. The current theory states that our moon formed when
   a. a Jupiter-sized object collided with Earth.
   b. a Mars-sized object collided with Earth.
   c. planetesimals came together in a disk around the Earth.
   d. Earth's gravity captured a stray object in space, and it began to orbit Earth.

6. The planet Jupiter has a day that is
   a. longer than one Earth day.
   b. shorter than one Earth day.
   c. equal to one Earth day.
   d. equal to one Jupiter year.

7. The theory that the sun and planets formed out of the same cloud of gas and dust is called the _____ theory.
   a. big bang   **c.** nebular
   b. planetary   **d.** nuclear

8. A total solar eclipse can occur only when the moon is
   a. full.   **c.** rising.
   b. new.   **d.** setting.

9. The greenhouse effect causes
   a. Venus to get hot.   **c.** Plants to grow on Mars
   b. our Moon to be hot.   **d.** Saturn's rings.

10. Which planet was not known to ancient observers?
   a. Venus   **c.** Mars
   b. Neptune   **d.** Jupiter

11. Which planet does not have rings?
   a. Venus   **c.** Uranus
   b. Neptune   **d.** Jupiter

12. Which planet has the hottest surface?
   a. Mercury   **c.** Earth
   b. Venus   **d.** Mars

13. The center of our solar system is
   a. Earth.   **c.** the sun.
   b. the moon.   **d.** the Oort cloud.

14. Most calendars are based on
   a. Earth's orbit of sun.
   b. Moon's orbit of Earth.
   c. Jupiter's orbit of sun.
   d. Venus' position in sky.

15. The solar system is about _____ years old.
   a. 0.5 million   **c.** 0.5 billion
   b. 4.6 million   **d.** 4.6 billion

16. One reason that Earth is a good home for life as we know it is that Earth has
   a. volcanoes.   **c.** cities.
   b. an atmosphere.   **d.** animals.

656

Chapter Resource File

• **Chapter Test** GENERAL

• **Test Item Listing for ExamView® Test Generator**

17. Why are the volcanoes on Mars larger than the volcanoes on other planets?
   a. Mars has small satellites compared with the planet's mass.
   b. Mars has low gravity.
   c. Mars has very cold surface temperatures.
   d. Mars has a carbon dioxide atmosphere.

18. Planets near other stars are found by
   a. making images of the planet and the star.
   b. detecting the temperature of the planet.
   c. detecting the atmosphere of the planet.
   d. detecting the gravitational effect of the planet.

USING VOCABULARY

19. Arrange the following from largest to smallest in mass: *terrestrial planet, satellite, meteorite, sun,* and *solar system.*

20. Name two distinguishing characteristics of a *satellite.*

21. Write a paragraph explaining the origin of the solar system according to the nebular theory. Use the following terms: *planet, accretion, nebula,* and *nebular model.* **WRITING SKILL**

22. Describe the arrangement of the sun, the moon, and Earth during a *solar eclipse* and a *lunar eclipse.*

23. Name the three major types of *meteorites.*

24. List the names of the *terrestrial planets* in order of distance from the sun.

25. List the names of the *gas giant* planets in order of mass.

26. Explain why Earth is the only planet with a *hydrosphere.*

27. Explain why Earth has ideal conditions for life by using the terms *AU, atmosphere,* and *hydrosphere.*

28. Describe the difference between an *asteroid* and a *satellite,* and explain where most *asteroids* in our solar system are found.

29. Sketch two different *phases* of the moon, and describe the arrangement of the sun, Earth, and the moon that causes them.

30. Describe two different models of the solar system using the terms *heliocentric* and *geocentric.* Which model do we use today?

BUILDING MATH SKILLS

31. **Applying Technology** The free-fall acceleration, g, near the surface of a planet is given by the equation below.

$$g = \left(6.67 \times 10^{-11} \; \frac{m^3}{kg \cdot s^2}\right) \times \frac{(planet \; mass)}{(planet \; radius)^2}$$

Use this equation to create a spreadsheet that will calculate the free-fall acceleration near the surface of each of the planets in the table below.

Name	Mass (kg)	Radius (m)
Mercury	3.3×10^{23}	2.4 million
Venus	4.9×10^{24}	6.1 million
Earth	6.0×10^{24}	6.1 million
Mars	6.4×10^{23}	3.4 million

32. **Applying Technology** Add a column to the spreadsheet you created in item 31 to record the weight on each of the inner planets of a person who has a mass of 70 kg. Calculate the weight using the equation *weight = mass × acceleration.* Which of the planets has gravity most like Earth's?

Assignment Guide

Section	Questions
1	2, 8, 13, 19, 20, 22, 29, 38–41, 44, 50
2	1, 6, 9–12, 14, 16, 17, 24–28, 32–37, 42, 43, 45, 46, 48
3	3–5, 7, 15, 18, 21, 23, 30, 47, 49, 53

Using Vocabulary

19. solar system, sun, terrestrial planet, satellite, meteorite

20. Satellites are objects that orbit around bodies that have larger masses than they do. Satellites reflect sunlight, and some have atmospheres although most do not.

21. Answers may vary, but should include the concepts that the nebular model has the sun and planets forming from a nebula, a cloud of gas and dust. The planetesimals that form in the disk as it collapses become larger through accretion.

22. In a solar eclipse, the sun, moon, and Earth are lined up with the moon in the middle and in new phase. During a lunar eclipse, the sun, moon, and Earth are lined up with Earth in the middle and the moon in full phase.

23. stony, metallic, and stony-iron

24. Mercury, Venus, Earth, Mars

25. Jupiter, Saturn, Neptune, Uranus

26. Earth is the only planet that receives the right amount of sunlight to keep water in liquid form on the surface of the planet.

27. Earth has ideal conditions for life, because its distance of 1 AU from the sun enables it to sustain a hydrosphere and an atmosphere. The hydrosphere and atmosphere help moderate temperatures between day and night promoting the growth of many life forms.

28. An asteroid orbits the sun directly while a satellite orbits a planet. Most asteroids are found in the asteroid belt between Mars and Jupiter.

29. Sketches should show the arrangement of sun, Earth, and moon as in Figure 8 in Section 1.

30. In the heliocentric model, the sun is at the center of the solar system, whereas in the geocentric model, Earth is at the solar system's center. We use the heliocentric model today.

Building Math Skills

Answers to 31 and 32 are shown below.

Planet	g in m/s^2	Wt in N
Mercury	3.8	270
Venus	8.8	620
Earth	9.8	680
Mars	3.7	260

Venus' gravity is most like Earth's.

Building Graphing Skills

33. Students may make line graphs or bar graphs of the data. Bar graphs will best show the differences in mass and density among the planets.

34. Students may want to use a line graph to show the relationship between distance and period.

35. From the graph, about 5 years (student answers may vary from 3 to 8).

36. From the graph, about 18 AU (student answers may vary from 14 to 21).

Thinking Critically

37. Venus is very bright because it reflects a high percentage of the sunlight that strikes its cloudy atmosphere. At the times we can see Venus, it is the closest planet to Earth. The planet's proximity also makes it bright.

38. midnight

39. No, the far side of the moon is fully illuminated when the moon is in new phase as seen from Earth.

40. The moon moves 12.2 degrees per day.

BUILDING GRAPHING SKILLS

33. Graphing The table below gives the density and mass for the planets. Using the data in the table make two different kinds of graphs. What kind of graph best displays the information? Explain.

Planet	Mass (Earth mass)	Density (gm/cm^3)
Mercury	0.055	5.43
Venus	0.82	5.24
Earth	1.00	5.52
Mars	0.10	3.94
Jupiter	318	1.33
Saturn	95.2	0.70
Neptune	17.2	1.76
Uranus	14.5	1.30
Pluto	0.0025	1.1

34. Graphing The table below shows the average distance from the sun for each planet and its orbital period. Plot the data in a graph and explain why you used that kind of graph.

Planet	Distance (AU)	Period (y)
Mercury	0.4	0.2
Venus	0.7	0.6
Earth	1.0	1
Mars	1.5	2
Jupiter	5.2	12
Saturn	9.6	30
Neptune	19.2	83
Uranus	30	164
Pluto	39	246

35. Graphing The asteroid Ceres has an average distance of 2.8 AU from the sun. Using your graph from item 34, estimate how long Ceres will take to orbit the sun.

658

36. Graphing Halley's comet has a period of 76 years. Using your graph from item 34, determine its average distance from the sun.

THINKING CRITICALLY

37. Critical Thinking How can Venus be the third brightest object in the sky when it doesn't produce any visible light of its own?

38. Understanding Systems You are the first astronaut to go to Mars. You look into the sky and see Deimos directly overhead. It is in full phase. What time of day is it on Mars?

39. Applying Knowledge The moon rotates at the same rate as it revolves, so one side never faces Earth. Is the far side of the moon always dark? Explain your answer.

40. Applying Knowledge It takes the moon 29.5 days to make its orbit around Earth. How many degrees does it move in one day?

41. Applying Knowledge The moon is constantly moving in the sky, but the times of tides are different every day. How much time is between high tide and low tide?

42. Critical Thinking The moon and Earth both average 1 AU from the sun. The moon is covered with craters. Earth has very few craters. Explain this difference.

43. Applying Knowledge If Mars were the same distance from the sun as Earth is, how would it be different than it is today?

44. Critical Thinking Pluto rotates on its axis in 6.387 days. Charon orbits Pluto in 6.387 days. Explain how Charon would look from the surface of Pluto.

41. There are 24 hours and 50 minutes between the first high tide of one day to the first high tide of the next day. So, a high tide to a low tide is 6 hours 12 minutes.

42. The Earth's atmosphere protects it from meteorites so that most burn up during their passage towards the surface. Also, the Earth may have had more craters in the past, but geological processes have covered them up.

43. Student answers may vary. Sample answers include: Mars' year would shorten to 365 days, it would be warmer, and the greenhouse effect might slowly increase in its atmosphere over time.

44. Charon does not rise and set. It is visible from only one side of Pluto.

Developing Life/Work Skills
45. Answers may vary.
46. Answers may vary.
47. Answers may vary.

DEVELOPING LIFE/WORK SKILLS

45. Interpreting and Communicating In your library or on the Internet, research the size of Jupiter and its satellites and the distances between different objects in the Jupiter system. Create a poster, booklet, computer presentation, or other presentation that can be used to teach fifth-graders about making and comparing scale models of the Jupiter system the Earth-moon system.

46. Working Cooperatively Working in teams, research the question "Should astronauts go to Mars?" Consider the financial costs and the technical challenges for the development of the mission as well as the psychological challenges for the astronauts.

47. Applying Technology Use a computer-art program to illustrate how a comet looks at different parts of its elliptical orbit.

INTEGRATING CONCEPTS

48. Connection to Music In 1914 to 1916, the composer Gustav Holst composed *The Planets*, a symphonic suite that portrays each of the planets according to its role in mythology. Listen to a recording of *The Planets*. Write a paragraph describing which parts of the music seem to match scientific facts about the planets and which do not. Why do you think Pluto is missing from *The Planets*?

49. Connection to Art Many artists have used astronomical themes or objects in their works. Two examples are Vermeer's 1668 painting *The Astronomer* and Alexander Calder's 1940s mobile entitled *Constellation*. Examine one of these works, and describe how the work relates to astronomy, or create a work of art that relates your feelings about the solar system.

50. Concept Mapping Copy the unfinished concept map below onto a sheet of paper. Complete the map by writing the correct word or phrase in the lettered boxes.

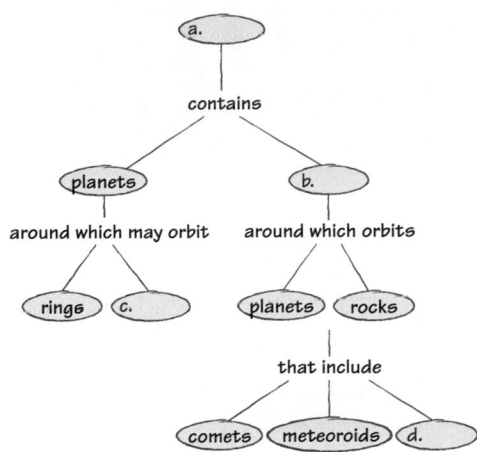

internet connect

www.scilinks.org
Topic: **Astronomy** SciLinks code: **HK4009**

SCI*LINKS* Maintained by the
National Science Teachers Association

659

Integrating Concepts

48. Answers may vary. Pluto is missing from *The Planets*, because it had not yet been discovered when the piece was written.

49. Answers may vary.

50. a. the solar system
   b. the sun
   c. satellites
   d. asteroids

ESTIMATING THE SIZE AND POWER OUTPUT OF THE SUN

Teacher's Notes

Time Required 1 lab period

Ratings

1	2	3	4
EASY			HARD

TEACHER PREPARATION	3
STUDENT SETUP	2
CONCEPT LEVEL	2
CLEANUP	2

Skills Acquired

- Collecting data
- Constructing models
- Experimenting
- Inferring
- Measuring
- Organizing and analyzing data

The Scientific Method

In this lab, students will:
- Make observations
- Analyze the results
- Draw conclusions
- Communicate results

Materials

Poke holes in the jar lids using an ice pick or an awl, and cut the metal strips to size. Smooth any rough edges.

Shape the metal strips into the wings by bending the strips around a pencil. This will reduce the risk of students breaking a thermometer when constructing the collectors.

To save time and minimize risks, you may choose to construct the solar collectors yourself beforehand. A CBL temperature probe can be used to remove the risk of breaking a glass thermometer.

Skills Practice Lab

Introduction

The sun is 1.496×10^{11} m from Earth. How can you use this distance and a few simple measurements to find the size and power output of the sun?

Objectives

▶ **Construct** devices for observing and measuring properties of the sun.

▶ **Measure** an image of the sun, and temperatures in sunlight and in light from a light bulb.

▶ **USING SCIENTIFIC METHODS** **Analyze the results** by calculating the size of the sun and the power output of the sun.

Materials

aluminum foil and pin
black paint or magic marker
Celsius thermometer
glass jar with a hole in the lid
index card
lamp with a 100 W bulb
masking tape
meterstick
modeling clay
shoe box
scissors
2×8 cm sheet of very thin metal

660

Estimating the Size and Power Output of the Sun

▶ Procedure

1. Construct a solar viewer in the following way.
   a. Cut a round hole in one end of the shoe box.
   b. Tape a piece of aluminum foil over the hole. Use the pin to make a tiny hole in the center of the foil.
   c. Tape the index card inside the shoebox on the end opposite the hole.

2. Construct a solar collector in the following way.
   a. Gently fit the sheet metal around the bulb of the thermometer. Bend the edges out so that they form "wings," as shown at right. Paint the wings black. **SAFETY CAUTION Thermometers are fragile.** Do not squeeze the bulb of the thermometer or let the thermometer strike any solid objects.
   b. Slide the top of the thermometer through the hole in the lid of the jar. Use modeling clay and masking tape to hold the thermometer in place.
   c. Place the lid on the jar. Adjust the thermometer so the metal wings are centered.

3. Place the lamp on one end of a table that is not in direct sunlight. Remove any shade or reflector from the lamp.

Measurements with the Solar Viewer

4. Stand in direct sunlight, with your back to the sun, and position the solar viewer so that an image of the sun appears on the index card. **SAFETY CAUTION Never look directly at the sun.** Permanent eye damage or even blindness may result.

5. Carefully measure and record the diameter of the image of the sun. Also measure and record the distance from the image to the pinhole.

Tips and Tricks

The first part of the experiment may be performed with an index card and a piece of paper. Poke a pinhole in the card, then have a student hold the card in the sun. Have a second student position the paper so that an image of the sun focuses on it. Have a third student measure the diameter of the image and the distance from the card to the paper.

Instead of constructing the solar collector, the second part of this lab may be performed with any device that measures a quantity dependent on light energy. The units of measurement are not important, because this measurement is only used to calibrate the distance from the sun to the distance from the light bulb.

Measurements with the Solar Collector

6. Place the solar collector in sunlight. Tilt the jar so that the sun shines directly on the metal wings. Watch the temperature reading rise until it reaches a maximum value. Record that value. Place the collector in the shade to cool.

7. Now place the solar collector about 30 cm from the lamp on the table. Tilt the jar so that the light shines directly on the metal wings. Watch the temperature reading rise until it reaches a stable value.

8. Move the collector toward the lamp in 2 cm increments. At each position, let the collector sit until the temperature reading stabilizes. When you find a point where the reading on the thermometer matches the reading you observed in step 6, measure and record the distance from the solar collector to the light bulb.

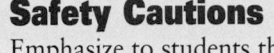

▶ Analysis

1. The ratio of the sun's actual diameter to its distance from Earth is the same as the ratio of the diameter of the sun's image to the distance from the pinhole to the image.

$$\frac{diameter\ of\ the\ sun,\ S}{Earth - sun\ distance,\ D} = \frac{diameter\ of\ image,\ i}{pinhole - image\ distance,\ d}$$

Solving for the sun's diameter, S, gives the following equation.

$$S = \frac{D}{d} \times i$$

Substitute your measured values, and $D = 1.5 \times 10^{11}$ m, into this equation to calculate the value of S. Remember to convert all distance measurements to units of meters.

2. The ratio of the power output of the sun to the sun's distance from Earth squared is the same as the ratio of the power output of the light bulb to the solar collector's distance from the bulb squared.

$$\frac{power\ of\ sun,\ P}{(earth - sun\ distance,\ D)^2} = \frac{power\ of\ light\ bulb,\ b}{(bulb - collector\ distance,\ d)^2}$$

Solving for the sun's power output, P, gives the following equation.

$$P = \frac{D^2}{d^2} \times b$$

Substitute your measured distance for d, the known wattage of the bulb for b, and $D = 1.5 \times 10^{11}$ m, into this equation to calculate the value of P in watts. Remember to convert all distance measurements to units of meters.

▶ Conclusions

3. How does your S compare with the accepted diameter of the sun, 1.392×10^9 m?

4. How does your P compare with the accepted power output of the sun, 3.83×10^{26} W?

661

If the lab is conducted indoors, turn off any artificial lights in the room while collecting data on the sun.

If you are using a device other than the solar collector, ask the students to look at the device that they used to determine relative intensity. In what units does the device give data? What kind of energy is the light energy converted into by the device? What other kinds of devices could be used to perform this experiment?

To extend the discussion of the sun's characteristics, give the students the diameter of Earth as 1.2×10^4 km. Ask how many Earths, placed side-by-side, would fit on a line as long as the sun's diameter. (116) Also, the volume of a sphere with radius r is equal to $(4/3)\pi r^3$. How many Earths would fit inside a space with the same volume as the sun? (1.5×10^6)

Safety Cautions
Emphasize to students the danger of looking directly at the sun. Gloves should be worn when working with sheet metal. There is a high risk of breaking thermometers in this activity. Caution students never to push the thermometers against anything. This is particularly important when sliding thermometers through the lids of jars. Caution should also be used when shaping the metal strips around the thermometers, or when carrying the collectors.

Disposal Information
Both the solar viewers and the solar collectors can be saved and reused when other classes perform the experiment. The solar viewers can also be used for viewing the sun at other times, such as during a solar eclipse.

Answers to Analysis
1. Answers will vary. Values should be fairly close to 1.4×10^9 m.
2. Answers will vary. Values should be fairly close to 3.8×10^{26} W.

Answers to Conclusions
1. Answers will vary based on the values obtained in the analysis.
2. Answers will vary based on the values obtained in the analysis.

Chapter Resource File

- **Datasheets** Estimating the Size and Power Output of the Sun **GENERAL**
- **Observation Lab** Explaining the Motion of Mars **BASIC**
- **CBL™ Probeware Lab** Determining the Speed of an Orbiting Moon **ADVANCED**

Continuation of Answers from p. 636

Section 1 Review

8. Pro: It could help protect Earth if we had enough warning and better technology to change its path. Keeping track of them would help predict a collision. Con: Since we don't know how to destroy or deflect an Earth-crossing asteroid, we should not bother spending money trying to track them.

7. At full moon, the moon is in a straight line with the sun and Earth. The entire face of the moon is illuminated. At quarter phase, the moon is at a 90-degree angle from the line that joins the sun and Earth.

8. Tides will be highest when the sun and moon are pulling in the same direction at new moon.

9. If the moon were seen as crescent on Earth, from the moon Earth would appear to be gibbous.

10. Planet Earth is also visible to the unaided eye.

Continuation of Answers from p. 645

Section 2 Review

5. Terrestrial planets are small, dense, and rocky, with few or no satellites. Gas giants are large, gaseous, with many satellites and rings. The gas giants are colder and rotate more quickly than the terrestrial planets.

6. Most terrestrial satellites and planets have many craters because they don't have atmospheres to protect them from incoming rocks from space.

7. The hydrosphere provides water that moderates the temperature on Earth and provides a place for life to grow and develop. The atmosphere also moderates the temperature, provides oxygen for us to breath, and protects us from dangerous radiation and rocks from space.

8. Planets very close to the sun will be too hot to support life as we know it, and planets very far from the sun will be too cold.

9. Although telescopes in space are smaller than those on Earth, they don't suffer from the disturbing effects of Earth's atmosphere blurring the view. Also, space probes get close to the planet or object and thus have a better view.

10. For a planet to sustain life as we know it, it needs an atmosphere, a hydrosphere, and moderate temperatures.

Continuation of Answers from p. 654

Section 3 Review

9. Earth-sized planets probably exist around other stars, but the current discovery techniques are not precise enough to find them. Whether or not these planets are "Earth-like" (with water and oxygen) is unknown and difficult to predict. Accept all reasonable answers.

Science in ACTION

Mining the Moon

On December 14, 1972, the crew of Apollo 17 climbed from the surface of the moon back on board the spacecraft for the return to Earth. Since then, NASA has turned its attention to more-distant goals and no one has set foot on the moon. But today, a rare isotope of helium known as helium-3 is fueling new interest in returning to the moon—and not just for scientific discovery.

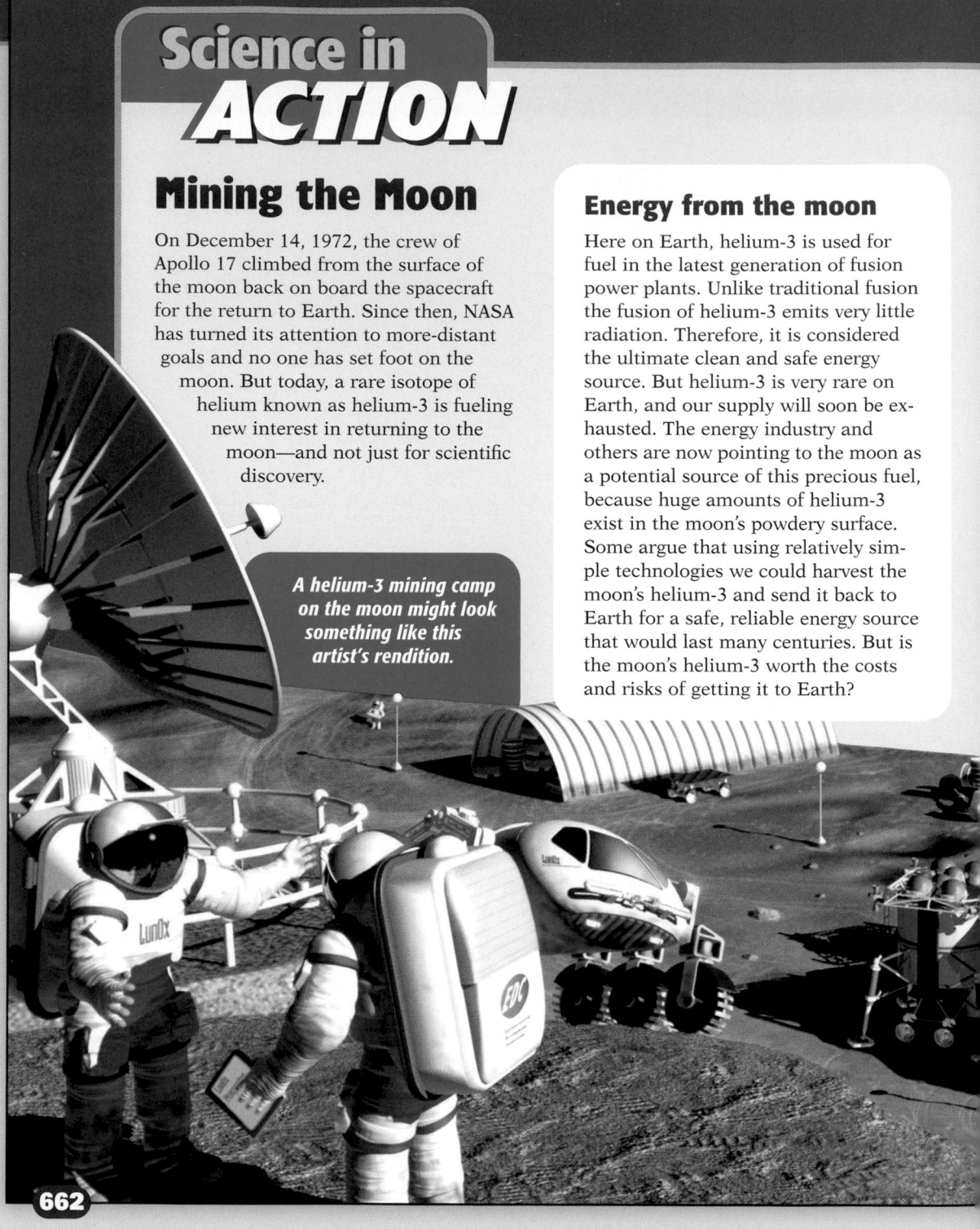

A helium-3 mining camp on the moon might look something like this artist's rendition.

Energy from the moon

Here on Earth, helium-3 is used for fuel in the latest generation of fusion power plants. Unlike traditional fusion the fusion of helium-3 emits very little radiation. Therefore, it is considered the ultimate clean and safe energy source. But helium-3 is very rare on Earth, and our supply will soon be exhausted. The energy industry and others are now pointing to the moon as a potential source of this precious fuel, because huge amounts of helium-3 exist in the moon's powdery surface. Some argue that using relatively simple technologies we could harvest the moon's helium-3 and send it back to Earth for a safe, reliable energy source that would last many centuries. But is the moon's helium-3 worth the costs and risks of getting it to Earth?

662

This false-color mosaic shows differences in the moon's composition.

The challenges of lunar mining

Extreme temperatures, solar radiation, the lack of atmosphere, and the constant barrage of small meteorites make the moon a dangerous place to live and work. Advocates for moon mining argue that much of the mining work could be carried out by robots, which would reduce the risk to human life. Also, the knowledge gained by building a permanent mining outpost on the moon would be valuable to any future manned space missions. So, how do we fund an expensive lunar mining program? Plans to minimize costs range from the practical (such as gathering solar energy on the moon to power mining equipment) to the highly imaginative (such as using huge slingshots to launch packages of helium-3 to orbiting spacecraft for transport). Some space entrepreneurs envision the moon as a destination for tourists who would pay high fees to bounce across the Sea of Tranquility and climb lunar mountains. The money gathered from these enterprises could be used to fund mining and other space-related industry. Although moon mining may be many years off, some people are already looking ahead to mining other space resources, including precious metals and rare-earth elements from asteroids that pass near Earth.

Science and You

1. **Applying Knowledge** Why is helium-3 such a valuable resource?

2. **Applying Knowledge** Name two major challenges to mining the moon.

3. **Understanding Concepts** Why do you think NASA has not sent astronauts to the moon since 1972? Give two reasons.

4. **Critical Thinking** In 1998, the unmanned spacecraft Lunar Prospector gathered evidence that large amounts of frozen water exist in deep craters on the moon. Do you think the presence of frozen water on the moon could help lunar mining efforts?

5. **Critical Thinking** The United Nations has declared that no country can lay claim to the moon or the resources that exists there. However, the rule does not apply to private companies. Do you think private companies should be allowed to "own" mining rights on the moon? Explain.

internet connect

www.scilinks.org
Topic: Lunar Mining
SciLinks code: HK4080

SCiLINKS Maintained by the National Science Teachers Association

663

The Universe
Chapter Planning Guide

PACING	CLASSROOM RESOURCES	LABS, ACTIVITIES, AND DEMONSTRATIONS
BLOCK 1 · 45 min pp. 664–665 **Chapter Opener**		**SE** Activity 1, p. 665 **SE** Activity 2, p. 665
BLOCKS 2 & 3 · 90 min pp. 666–673 **Section 1** The Life and Death of Stars	**CRF** Lesson Plan* **TT** Bellringer* **TM** Distances to Stars in Orion* **TT** Structure of the Sun* **TT** Starlight Intensity Graph* **TT** H-R Diagram*	**SE** Quick Activity Using a Star Chart, p. 667 **GENERAL** **CRF** Datasheets for In-Text Labs Using a Star Chart* **TE** Demonstration Understanding Stellar Brightness, p. 668 **CRF** Observation Lab The Sun's Yearly Trip Through the Zodiac ♦ **BASIC**
BLOCKS 4 & 5 · 90 min pp. 674–679 **Section 2** The Milky Way and Other Galaxies	**CRF** Lesson Plan* **TT** Bellringer* **TT** Investigating Different Types of Galaxies*	**SE** Skills Practice Lab Investigating Different Types of Galaxies, pp. 694–695 ♦ **GENERAL** **CRF** Datasheets for SE Skills Practice Lab Investigating Different Types of Galaxies*
BLOCKS 6 & 7 · 90 min pp. 680–688 **Section 3** Origin of the Universe	**CRF** Lesson Plan* **TT** Bellringer* **TM** Timeline of Universe* **TM** Fates of the Universe* **TT** Concept Mapping*	**SE** Quick Activity Modeling the Expanding Universe, p. 684 **GENERAL** **CRF** Datasheets for In-Text Labs Modeling the Expanding Universe*

BLOCKS 8 & 9 · 90 min

Chapter Review and Assessment Resources

- **SE** Chapter Review, pp. 690–693
- **CRF** Chapter Tests* **GENERAL**
- **OSP** Test Generator
- **CRF** Standardized Test Practice with Guided Reading Development*
- **CRF** Test Item Listing for ExamView® Test Generator*
- **OSP** Scoring Rubrics and Classroom Management Checklists

Online Resources

Visit the HRW Web site for a variety of free resources related to the text. Just type in the keyword **HK4 GAL**.

Holt Online Learning

Holt Science Spectrum: Physical Science: Online Edition

Students can access interactive problem solving help and active visual concept development with the *Holt Science Spectrum: Physical Science* Online Edition available at **www.hrw.com**.

student CNN News

cnnstudentnews.com

Find the latest news, lesson plans, and activities related to important scientific events.

KEY

TE	Teacher Edition	CRF	Chapter Resource File	*	Also on One-Stop Planner
SE	Student Edition	TT	Teaching Transparency	◆	Requires Advance Prep
OSP	One-Stop Planner	TM	Transparency Master		

PROBLEM SOLVING AND PRACTICE	SECTION REVIEW AND ASSESSMENT	STANDARDS CORRELATION
	CRF Pretest* GENERAL	
TE Inclusion Strategies, p. 670 **CRF** Cross-Disciplinary Worksheet Connection to Social Studies—Ancient Chinese Astronomy*	**TE** Quiz, p. 673 BASIC **SE** Section 1 Review, p. 673 **CRF** Concept Review* GENERAL **CRF** Quiz* BASIC	PS 6e UCP 1, 2, 3 SAI 1, 2 ST 1, 2 HNS 1, 2, 3 SPSP 5
CRF Cross-Disciplinary Worksheet Real World Application—Telescope Interference* ADVANCED	**TE** Quiz, p. 679 BASIC **SE** Section 2 Review, p. 679 **CRF** Concept Review* GENERAL **CRF** Quiz* BASIC	UCP 1, 2 ST 1, 2 HNS 1, 2, 3 SPSP 5
CRF Cross-Disciplinary Worksheet Integrating Health—Exercise in Space* ADVANCED **CRF** Cross-Disciplinary Worksheet Integrating Chemistry—Red Shift, Blue Shift* ADVANCED **CRF** Science Skills Ratios and Proportions* BASIC **SE** Graphing Skills Bar Graphs, p. 689 **SE** CareerLink Joanne Cohn, Cosmologist, pp. 696–697	**TE** Quiz, p. 688 BASIC **SE** Section 3 Review, p. 688 **CRF** Concept Review* GENERAL **CRF** Quiz* BASIC	ES 3a, 4a, 4b, 4c UCP 1, 2, 4 ST 1, 2 HNS 1, 2, 3 SPSP 5

SCiLINKS®

www.scilinks.org

Topic: Radioastronomy
*Sci*Links code: HK4116

Topic: How Stars Evolve
*Sci*Links code: HK4069

Topic: Origin of the Universe
*Sci*Links code: HK4098

Topic: Dark Matter
*Sci*Links code: HK4030

Topic: Formation of the Elements
*Sci*Links code: HK4056

Topic: Astrophysicist
*Sci*Links code: HK4010

Technology Resources

 One-Stop Planner
All of your printable resources and the Test Generator are on this convenient CD-ROM.

 CNN Science in the NEWS
each video segment is accompanied by a Critical Thinking Worksheet

Segment 24
Wisp of Creation

Overview

The chapter first discusses the nature of stars, how scientists study stars, and the life of stars from birth to death. The chapter then discusses galaxies, including the Milky Way Galaxy, different types of galaxies, and how galaxies evolve. Finally, the chapter discusses the universe as a whole, including theories of the origin of the universe and the possible futures of the universe.

Assessing Prior Knowledge

Be sure students understand the following concepts:

- properties of matter
- atoms
- radioactivity
- nuclear fusion
- the force of gravity
- heat and energy transfer
- the electromagnetic spectrum

MISCONCEPTION ///ALERT\\\

Science education research has identified the following misconceptions:

- Many students associate light with its source (that light comes from a lightbulb) and do not understand light as something that travels from one place to another.
- Most students do not understand that Earth is older than 10^9 years old. This may be due to a general confusion about large numbers.
- The action of gravity is often characterized by students as "holding." It is often confused with atmospheric pressure, and described as *that which keeps things from floating away*. Gravity is also thought to require air, and students believe it stops acting on an object when it has finished falling.
- Very few students describe the sun as a star, and some even regard it as a planet.

The Universe

Chapter Preview

1 The Life and Death of Stars
What Are Stars?
Studying Stars
The Fate of Stars

2 The Milky Way and Other Galaxies
Galaxies
Types of Galaxies
How Galaxies Evolve

3 Origin of the Universe
What Is the Universe?
What Happened at the Beginning?
Predicting the Future of the Universe

664

Standards Correlations

National Science Education Standards

The following descriptions summarize the National Science Standards that specifically relate to this chapter. For the full text of the standards, see the National Science Education Standards at the front of the book.

Section 1 The Life and Death of Stars
Physical Science PS 6c
Unifying Concepts and Processes UCP 1, 2, 3
Science as Inquiry SAI 1, 2
Science and Technology ST 1, 2
History and Nature of Science HNS 1, 2, 3
Science in Personal and Social Perspectives SPSP 5

Section 2 The Milky Way and Other Galaxies
Unifying Concepts and Processes UCP 1, 2
Science and Technology ST 1, 2
History and Nature of Science HNS 1, 2, 3
Science in Personal and Social Perspectives SPSP 5

Section 3 Origin of the Universe
Earth and Space Science ES 3a, 4a, 4b, 4c
Unifying Concepts and Processes UCP 1, 2, 4
Science and Technology ST 1, 2
History and Nature of Science HNS 1
Science in Personal and Social Perspectives SPSP 5

Focus ACTIVITY

Background Optical telescopes take pictures of objects by collecting visible light. Visible light is only a small part of the electromagnetic spectrum. Radio waves have longer wavelengths than the light waves that allow us to see with our eyes. In 1932, scientists discovered that astronomical objects emit radio waves. Radio telescopes are sophisticated systems that collect the radio waves emitted by astronomical objects. One of the largest radio telescopes is the Very Large Array in New Mexico. It is composed of 27 dish antennas that work together as a single instrument. Unlike your radio at home, a radio telescope does not convert radio waves to sound. The radio waves are processed by a computer to form a picture such as the image of the Crab Nebula shown below. When scientists combine the data coming from the 27 radio telescopes, the data are as precise as if they had come from an antenna 36 km in diameter!

Activity 1 Examine the photograph of the Very Large Array. Look at the sky behind the antennas—do you see star trails due to the rotation of Earth? Knowing that Earth rotates once every 24 h, estimate how many degrees a star travels in the night sky in 1 h. (**Hint:** One rotation is 360°.)

Activity 2 Turn to Appendix B, and locate the star maps. Pick a constellation and brainstorm how it got its name. You may need to look up the name in a dictionary to find out what the name means in its original language. Write a few paragraphs that explain your hypothesis.

internet connect

www.scilinks.org
Topic: **Radioastronomy** SciLinks code: **HK4116**

SCiLINKS. Maintained by the
National Science Teachers Association

Pre-Reading Questions

1. Name as many celestial objects as you can think of. Which objects are outside our solar system?
2. Where would you find information about the locations of the objects that you named in question 1?

The Very Large Array in New Mexico can make detailed images of astronomical objects such as the Crab Nebula, which scientists think is the remains of a supernova explosion.

665

Focus ACTIVITY

Background Astronomers observe the sky in all parts of the electromagnetic spectrum. Optical telescopes, which use lenses to produce images that you can observe directly with your eyes, have been around since the time of Galileo. Other types include radio telescopes, infrared telescopes, and X-ray telescopes. In order to view images from these telescopes, the signals they receive must be processed by a computer.

Activity 1 The center of the motion of star trails is very near Polaris, the North Star. Earth rotates through 15 degrees of arc, or 1/24 of a circle, every hour. **LS** Logical

Activity 2 Many constellations are named for characters in Greek or Roman mythology. Others are named for animals. Encourage students to come up with their own stories about the constellations, or even to make up new constellations in the night sky. **LS** Verbal

Answers to Pre-Reading Questions

1. Answers may vary, but could include the sun, the moon, comets, asteroids, any of the planets, stars, galaxies, the Milky Way, black holes, pulsars, and quasars.
2. In addition to their textbook, students can find information about celestial objects in a variety of places. The World Wide Web has a huge amount of astronomical information, in many cases well organized and from reliable sources. Other possible sources include books from the library or science magazines.

Chapter Resource File

- **Pretest** GENERAL
- **Teaching Transparency Preview**
- **Answer Keys**

Focus

Overview

Before beginning this section, review with your students the Objectives listed in the Student Edition. This section describes stars and the methods that scientists use to study stars. The section traces the life of the sun, and also describes the life cycle of other stars.

Bellringer

Use the Bellringer transparency to prepare students for this section.

Motivate

Identifying Preconceptions

Many students may have heard the term "light-year," but may be confused about its meaning. Ask students what a light-year (ly) measures: time, distance, or speed? (distance; A ly is the distance the light travels in one year.)

Remind students that while light moves extremely fast, it does not move infinitely fast. Light from the sun takes about 8 min to reach Earth. The distance from the sun to Earth is equal to 8 light-minutes.

Light from the star nearest the sun takes over 4 y to reach Earth. In this chapter, students will learn about galaxies so far away that light from them takes millions or billions of years to reach Earth.

The Life and Death of Stars

KEY TERMS

star
light-year
red giant
white dwarf
supernova
black hole

OBJECTIVES

▶ **Describe** the basic structure and properties of stars.

▶ **Explain** how the surface temperature of a star is measured.

▶ **Recognize** that all normal stars are powered by fusion reactions that form elements.

▶ **Identify** the stages in the evolution of stars.

On a clear night, you can see about 6000 stars. People have observed stars for thousands of years, but only in the last 100 years have we begun to understand the life of stars.

star a large celestial body that is composed of gas and that emits light; the sun is a typical star

light-year the distance that light travels in one year; about 9.5 trillion kilometers

What Are Stars?

Stars are huge spheres of very hot gas that emit light and other radiation. The nearest star to Earth is the sun. Ancient Greek scientists thought that the stars were attached to a large, invisible sphere. The Greeks also grouped stars into shapes and patterns called constellations. Today, we still use constellations to group stars, such as those in the constellation Orion, shown in *Figure 1.* Since ancient times, we have learned that stars are located at different distances from Earth. We use the unit **light-year** (ly) to describe a star's distance from Earth. One light-year is the distance that light travels in one year, or 9.5×10^{15} m.

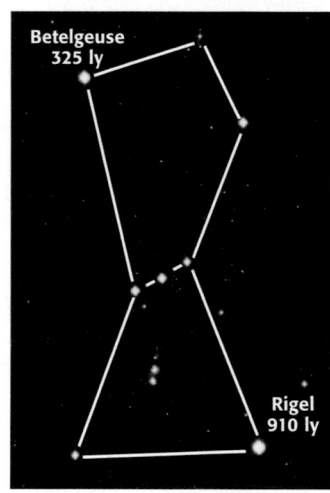

Betelgeuse
325 ly

Rigel
910 ly

Figure 1

A The stars in the constellation Orion looked like the shape of a hunter to the ancient Greeks.

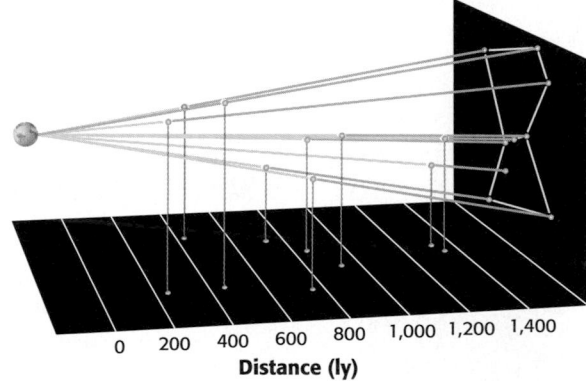

Distance (ly)

B The stars in Orion, which appear close together when viewed from Earth, are located at different distances from us and from each other.

666

Transparencies

TM Distance to Stars in Orion

SOCIAL STUDIES
CONNECTION

The constellations are often modeled after characters from Greek and Roman mythology. Have students pick a constellation from the star chart in Appendix B (or another constellation that they know) and research the origin of its name.

REAL WORLD APPLICATIONS

Nuclear Power Plants Have students research nuclear power plants and answer the following questions: How do these facilities produce energy? Compare nuclear power plants with the sun. How similar or different are the two methods of producing energy? (Nuclear power plants rely on nuclear fission reactions, while the sun derives energy from nuclear fusion reactions. In both cases, a small amount of mass is converted into large amounts of energy.)
LS Verbal

Stars are driven by nuclear fusion reactions

A star is a huge sphere of very hot hydrogen and helium gas that emits light. A star is held together by the enormous gravitational forces that result from its own mass. Inside the core, or middle, of a star, these forces create a harsh environment. The pressure is more than a billion times the atmospheric pressure on Earth. The temperature is hotter than 15 million kelvins, and the density is more than 13 times the density of lead.

Nuclear fusion takes place in the core. Fusion combines the nuclei of hydrogen atoms into helium. Positively charged particles, such as the nuclei of hydrogen atoms, normally repel each other, but inside a star, where the temperature and pressure are very high, these particles collide at high speeds. When they collide, they fuse together to form new nuclei called *deuterons*, which have one proton and one neutron. Next, two deuterons collide to form the nucleus of a helium atom. When two particles fuse, energy is released. The energy from these fusion reactions creates outward pressure that balances the inward pull of gravity.

Energy moves slowly through the layers of a star

Figure 2 shows the layers of the sun. Other stars have similar structures, although the temperatures and depths of the layers may differ. Energy moves through the layers of a star by a combination of radiation and convection. During convection, rising hot gas moves upward, away from the star's center, and cooler, denser gas sinks toward the center. During radiation, energy is transferred to individual atoms. The atoms absorb the energy and then transfer it to other atoms in random directions. Atoms near the star's surface radiate energy into space.

The energy from a nuclear fusion reaction may take millions of years to work its way through a star. When the energy finally reaches the surface, it is released into space as radiation and light.

Once light leaves the surface of a star, it radiates across space at the speed of light in a vacuum, 3×10^8 m/s. At this speed, it takes light from the sun about eight minutes to reach Earth.

Quick ACTIVITY

Using a Star Chart
Locate the following stars on the star chart in Appendix B: Betelgeuse, Rigel, Sirius, Capella, and Aldebaran. Name the constellation to which each star belongs. Which of these stars appears closest in the sky to Polaris, the North Star?

Figure 2

Energy released by fusion reactions in the core slowly works its way through the layers of the sun by the forces of radiation and convection.

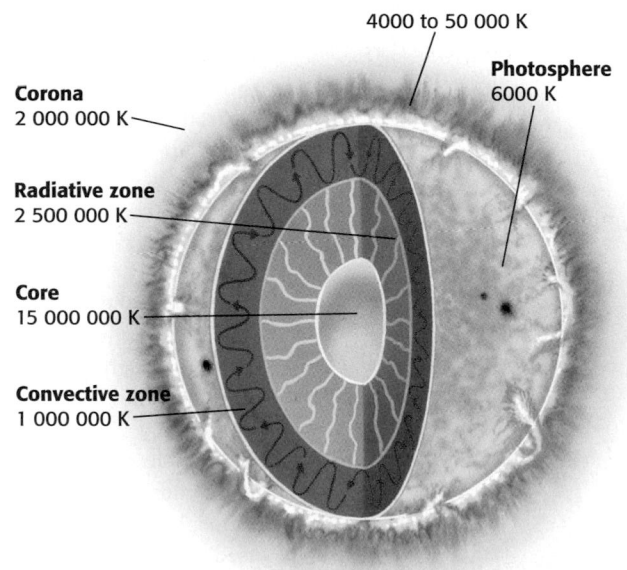

4000 to 50 000 K

Photosphere
6000 K

Corona
2 000 000 K

Radiative zone
2 500 000 K

Core
15 000 000 K

Convective zone
1 000 000 K

667

Teach

SKILL BUILDER — BASIC

Interpreting Visuals Figure 1 shows the constellation Orion. Have the students examine the figure. Do they recognize Orion? Have they seen it in the sky before? What key features make it easy to locate in the sky?

Orion represents a human figure, a hunter. The red star Betelgeuse is Orion's shoulder, and the blue star Rigel represents Orion's foot. What might the three stars across the middle represent? (Traditionally, they represent a belt. The cloudy smudge below the belt, called the Orion nebula, represents a sword hanging from the belt.)
LS Visual

SKILL BUILDER

Interpreting Visuals Have students examine **Figure 2,** and ask them to use the picture to help explain the movement of energy through the layers of the sun. Where does nuclear fusion take place? (the core) Where is energy transferred by radiation? (the radiative zone) Where does convection take place? (the convective zone) What is the temperature of the photosphere? (6000 K) **LS** Visual

Did You Know ?

The part of the sun that we see from Earth is actually a layer of the sun's atmosphere called the photosphere. This layer is composed of gas clouds hundreds of miles thick and grouped in cells called granules. Each of these cells is roughly the size of Earth.

INCLUSION Strategies

- *Learning Disabled*
- *Developmentally Delayed*
- *Attention Deficit Disorder*
- *English Language Learners*

Using **Figure 2,** ask students to make a colorful poster, bulletin board, or computer picture showing the layers of the sun. Students should label the layers' names and temperatures with large print. Students can present their finished products to small or large groups.

Chapter Resource File

- **Lesson Plan**

Transparencies

TT Bellringer

TT Structure of the Sun

668

Quick ACTIVITY

Using a Star Chart The star charts in Appendix B shows the sky as it appears in northern latitudes at 8 P.M. on March 1 or midnight on January 1. To use the chart when observing the sky, hold the page overhead and face north (for the upper half of the chart) or south (for the lower half). Betelgeuse and Rigel are in the constellation Orion; Sirius is in Canis Major; Capella is in Auriga; and Aldebaran is in Taurus. These constellations are in the southern sky, except Auriga. Of the five stars, Capella is the closest to Polaris, the North Star. Polaris is in the constellation Ursa Minor, and it marks the end of the "handle" of the Little Dipper. **LS Visual**

Teaching Tip ——— GENERAL

Photographing Star Trails To make star trail photographs you need a camera mounted on a tripod with a shutter that can be held open for several minutes. Reduce the aperture of the camera to its minimum size, then take an exposure including part of the sky, holding the shutter open for about 5 minutes. Experiment by taking several pictures, using different settings and exposure times. Be sure to write down the settings so you can see which ones produce the best results. **LS Kinesthetic**

Figure 3
Sirius is the brightest star in the night sky, and it is shown on the mouth of the larger dog on this star chart that dates back to 1725.

Studying Stars

Although the ancient Greeks noticed that stars had color and divided stars by their apparent brightness, astronomers did not really begin to learn about the nature of stars until after the invention of the optical telescope.

Why do some stars appear brighter than others?

The brightness of a star depends on the star's temperature, size, and distance from Earth. The brightest star in the night sky is Sirius in the constellation Canis Major, which is shown in *Figure 3*. Sirius appears so bright because it is relatively close to Earth, only about 9 ly away. The surface temperature of Sirius is about 10 000 K. The sun's surface is only 6000 K, but the sun is so close to Earth that it dominates the sky during the day.

We learn about stars by studying light

When we look with our eyes or use binoculars, as in *Figure 4,* we detect only light in the visible part of the spectrum. But stars also produce other wavelengths of electromagnetic radiation, from high-energy X rays to low-energy radio waves. Scientists use optical telescopes to study visible light and radio telescopes to study radio waves emitted from astronomical objects. Earth's atmosphere blocks other wavelengths, so telescopes in space are used to study a wider range of the electromagnetic spectrum.

Figure 4
You can observe some stars and constellations more easily with binoculars than with the unaided eye.

Demonstration

Understanding Stellar Brightness

(Time: Approximately 10 minutes)

Materials

• 1 large, bright flashlight
• 3 small, dim flashlights

Step 1 Have the students line up at one end of the classroom, and turn out the lights. Give one student the large flashlight and give another student one of the small flashlights.

Step 2 Send those two students to the other end of the classroom. Ask which light appears to be brightest.

Step 3 Now have the student with the dimmer light walk across the room until the lights appear equally bright. Ask students how this situation relates to the stars Sirius and Rigel. Which flashlight corresponds to which star? (The large flashlight corresponds to Rigel, the small to Sirius.)

Step 4 Repeat the procedure, but this time have one student hold all three dim flashlights bundled together. (In this case, the large flashlight still corresponds to Rigel, and the bundle of dim flashlights corresponds to Betelgeuse.)

A star's color is related to its temperature

When light from a glowing hot object passes through a prism, it generates a spectrum of many colors. This spectrum changes with temperature in a definite way: hotter objects glow with light that is more intense and that has shorter wavelengths (closer to the blue end of the spectrum), while the light from cooler objects has greater intensity and longer wavelengths (closer to red).

Although the light from a glowing object contains many colors, the color that we see when we look directly at a hot object is determined mainly by the wavelength at which the object emits the most light. ***Figure 5*** is a graph that shows the intensity, or brightness, of light at different wavelengths for three stars. The sun appears yellow because the peak wavelength of the sun is near the color yellow. Yellow also corresponds to a temperature near 6000 K. Hot stars emit more energy at every wavelength than cooler stars do.

Figure 5
This graph shows the intensity of light at different wavelengths for the sun and two other stars.

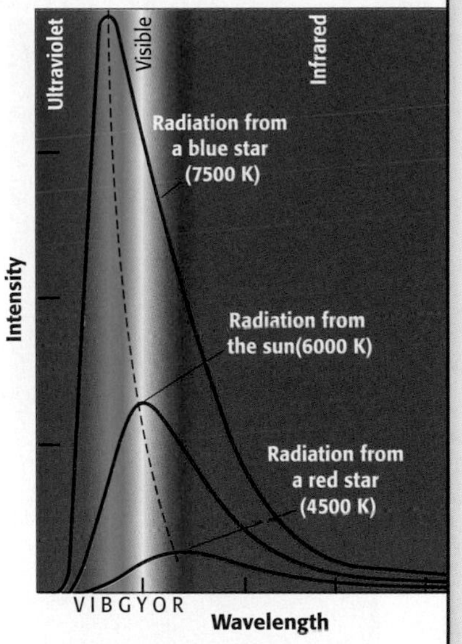

Spectral lines reveal the composition of stars

How do we know what stars are made of? The spectra of most stars have dark lines. These dark lines are caused by gases in the outer layers of the stars that absorb the light at these wavelengths. The temperature of these outer layers determines which gases produce spectral lines. For example, cool hydrogen has no spectral lines. Because each element produces a unique pattern of spectral lines, astronomers can match the dark lines in starlight to the known lines of elements found on Earth. ***Figure 6*** shows how the spectral lines of both hydrogen and helium can be found in a star's spectrum.

Astronomers have analyzed more than 20 000 lines in the sun's spectrum to find the composition of its atmosphere. Like the composition of most stars of its age, the sun's mass is 71% hydrogen, 27% helium, and 2% other elements.

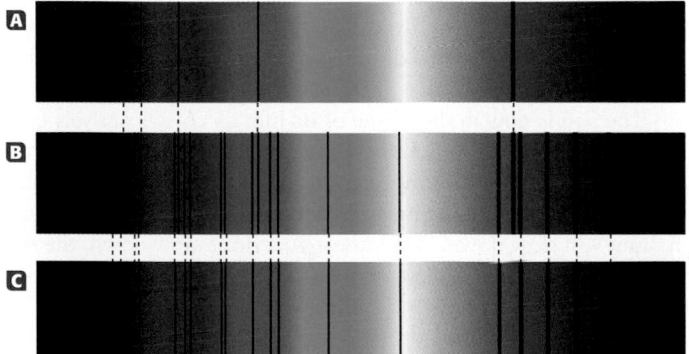

Figure 6
When light is passed through hydrogen gas **A**, or helium gas **C**, then through a slit and prism, dark lines appear in the spectrum. If both hydrogen and helium are present, both sets of lines appear **B**.

SKILL BUILDER — BASIC

Interpreting Visuals Have students examine **Figure 5.** Explain that the three curves represent light from three different stars. Each star emits light across a wide range of the electromagnetic spectrum.

The color of the star corresponds to the peak of the curve. For the curve representing the sun, what color of the spectrum is the highest point of the graph? (yellow)

The color of a star also corresponds to the surface temperature of the star. Which of the stars has the highest temperature? (the blue star) Which has the lowest temperature? (the red star)

According to Wien's law, a surface temperature of 6000 K actually corresponds to the color green, but the sun appears yellow because our eyes blend the red, blue and green light emitted.

The total area under each curve corresponds to the total energy output of the star. Which star produces the most energy? (the blue star) **LS Visual**

Historical Perspective

Instruments of the Scientific Revolution

The telescope was one of many instruments that came out of the Scientific Revolution of the seventeenth century. Have students research some of the other instruments produced during this era. You may assign students to write a paragraph describing one instrument from this era, including its invention and any contribution to science or society in which the instrument played a role.

Chapter Resource File

- **Quick Activity Datasheet**
  Using a Star Chart GENERAL

Transparencies

TT Starlight Intensity Graph

Reading Skills Have students read the section on the fate of stars. Then have them organize the ideas presented in the section in a timeline showing the sequence of events in the life of a star. The timeline should divide into two branches, showing the different l ife cycles for sun-like stars and for stars much more massive than the sun. **LS** Verbal

|SKILL BUILDER

Interpreting Visuals **Figure 7** shows a region of star birth in the Eagle Nebula. Although some stars are visible in the picture, the stars that are being born are hidden deep within the pillars of gas. The pillars form as lighter gases are blown away by radiation from nearby stars. The stars form as the denser material is drawn together due to gravity.

Figure 7

New stars are constantly being formed in clouds of gas and dust such as these columns in the Eagle Nebula.

internet connect

www.scilinks.org

Topic: How Stars Evolve

SciLinks code: HK4069

SCiLINKS. Maintained by the National Science Teachers Association

670

The Fate of Stars

Figure 7 shows stars being formed in a cloud of gas and dust called a *nebula.* Stars are born, go through different stages of development, and eventually die. Stars appear different from one another in part because they are at different stages in their life cycles. Nearly 90% of all stars in our galaxy, including the sun, are in midlife, still converting hydrogen into helium in their interiors.

Some stars, such as Rigel, are younger than the sun, while others, such as Betelgeuse, are farther along in their life cycles. Some objects in the universe are remnants of very old stars that died long ago. But how do stars form? And how do they keep on shining for billions of years?

The sun formed from a cloud of gas and dust

About 5 billion years ago, in an arm of the Milky Way galaxy, a thin, invisible cloud of gas and dust collapsed inward, pulled by the force of the cloud's own gravity. As the cloud fell together, it began to spin. The smaller the cloud became, the faster it spun. About 30 million years after the cloud started to collapse, the center of the cloud reached a temperature of 15 million kelvins.

Electrons were then stripped from hydrogen atoms to leave hydrogen nuclei, which are positively charged protons. Recall that positively charged particles repel each other. But at very high temperatures, protons may get as close to each other as 10^{-15} m. At such a small distance, the strong nuclear force overpowers the electrical repulsion. Through this process of nuclear fusion, the protons combine to form helium. Scientists think that once this process of nuclear fusion started in the core of the cloud, the star we call the sun turned on.

The sun now has a balance of inward and outward forces

The fusion reactions in the core of the sun produce an outward force that balances the inward force due to gravity. With these two forces evenly balanced, the sun has maintained an equilibrium for 5 billion years.

The sun is now in the prime of its life; its core is actively converting hydrogen into helium. Over time, the percentage of the core that is helium becomes larger. Eventually, the core will run out of hydrogen, and the fusion reactions that turn hydrogen into helium will slow down. When these reactions slow down, the sun will begin to die. Scientists estimate that the sun can continue nuclear fusion for another 5 billion years.

 INCLUSION *Strategies* GENERAL

- *Learning Disabled*
- *Attention Deficit Disorder*
- *English Language Learners*

Ask students to write a classified ad looking for an astronomer to direct the "new observatory" at your school. The advertisement must include the different kinds of objects the astronomer will observe in the universe, types of tools to be used in the observatory, and types of experience and education required for the position. Students may use a computer to "typeset" the classified ad. They may refer to the classified section of a newspaper to model the layout and language used in their ad.

The sun will become a red giant before it dies

As fusion slows, the pressure in the core of the sun will drop and the core will contract, which will cause the core temperature to rise. The sun's outer layers will expand, and the sun will become a **red giant** like the one shown in *Figure 8*. The star is red because its surface is cooler, but the core is hot enough to convert helium into carbon and oxygen.

After about 100 million years, the core of the red giant sun will run out of helium and will contract further, which will cause the outer layers to expand again. At this point, the temperature at the core is not high enough to fuse these heavier elements. The outer layers will continue to expand out from the core and will eventually leave the star. The remnant will become a **white dwarf,** a small and very dense star about the size of Earth. White dwarfs no longer fuse elements, so they slowly cool. Stars with a mass of 1.4 solar masses or smaller will have a similar life cycle. Most stars in our galaxy will end as white dwarfs.

Supergiant stars explode in supernovas

Massive stars evolve faster than smaller stars do. They also develop hotter cores that create heavier elements through fusion. Forming an iron core signals the beginning of a supergiant star's violent death because fusing iron atoms to make heavier elements requires adding rather than releasing energy. When a core becomes mostly iron, fusion stops. When fusion stops, there is no longer any outward pressure to balance the gravitational force. The core collapses because of its own gravity and then rebounds with a shock wave that violently blows the star's outer layers away from the core. The resulting huge, bright explosion is called a Type II **supernova,** shown in *Figure 9.* Elements heavier than iron (such as gold and lead) form during a supernova. A Type I supernova occurs when a white dwarf in a binary system (a system composed of two stars) collects enough mass from its companion to exceed 1.4 solar masses.

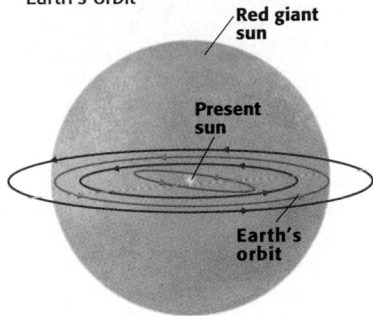

Figure 8

When the sun becomes a red giant, it will expand out past Earth's orbit

Red giant sun

Present sun

Earth's orbit

▶ **red giant** a large, reddish star late in its life cycle

▶ **white dwarf** a small, hot, dim star that is the leftover center of an old star

▶ **supernova** a gigantic explosion in which a massive star collapses and throws its outer layers into space; plural *supernovae*

Figure 9

Supernova 1987A, a Type II supernova, was the first supernova visible to the unaided eye in 400 years. The first image shows what the original star looked like before the explosion.

Teaching Tip

Black Holes Some students may think that black holes are completely empty. Quite the contrary, black holes are the densest objects in the universe. They only appear black because light cannot escape from them. They are only holes in the sense that things fall into them and cannot get back out.

|SKILL BUILDER — BASIC

Interpreting Visuals Have students examine **Figure 11.** Discuss with students what the axes of the diagram represent. The vertical axis is brightness, expressed as either intensity or magnitude (if the axis is in units of magnitude, beware that higher numbers mean dimmer stars). The horizontal axis may be temperature, color, or spectral type, all of which are closely related.

Point out to students the main sequence, the part of the H-R diagram, which shows the state of most stars during the prime of their lives. Then point out the giant region, above and to the right of the main sequence. This is where giant stars are found. The average surface temperature of a giant star is cooler than that of a main sequence star, but giant stars are brighter overall because they have a larger surface area. **LS Visual**

Connection to SOCIAL STUDIES

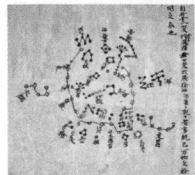

On July 4, 1054, a bright supernova appeared in the constellation Taurus. It was visible for three weeks. Imperial Chinese astronomers named it a "guest" star because it was new to the sky. These astronomers told the emperor that the star's brightness meant that the emperor was a person of great worth. This supernova may have also been observed by Native Americans in New Mexico and Arizona where rock paintings have been found. Later, the remnants of this supernova gained the name "Crab Nebula."

▶ **black hole** an object so massive and dense that not even light can escape its gravity

Figure 10
The Crab Nebula is the remains of a supernova seen by Chinese observers in the year 1054.

672

After a Type II supernova, either a neutron star or a black hole forms

Figure 10 shows a nebular remnant of a supernova. If the core that remains after a supernova has a mass of 1.4 to 3 solar masses, the remnant can become a *neutron star*. Neutron stars are only a few kilometers in diameter, but they are very massive. A neutron star is as dense as matter in the nucleus of an atom, about 10^{17} kg/m^3. A thimbleful of a neutron star would weigh more than 100 million tons on Earth. Neutron stars can be detected as *pulsars*, or sources of pulsating radio waves.

If the leftover core has a mass that is greater than three solar masses, it will collapse to form an even stranger object—a **black hole,** which consists of matter so massive and compressed that nothing, not even light, can escape its gravity. Because no light can escape, a black hole cannot be seen directly. Black holes have a powerful gravitational pull, so they can be detected indirectly by observing the radiation of light and X rays from objects that revolve rapidly around them.

The H-R diagram shows how stars evolve

In 1911, Ejnar Hertzsprung compared the temperature and brightness of stars and carefully plotted his data on a graph. In 1913, Henry Norris Russell made similar plots. Together, the two graphs form the Hertzsprung-Russell diagram, or H-R diagram, which is shown in *Figure 11.* The vertical axis indicates brightness. Absolute magnitude indicates how bright stars would be if they were all the same distance from Earth. The horizontal axis indicates surface temperature of the stars, with hotter temperatures on the left.

When stars are born, they appear as *protostars* on a diagonal line called the *main sequence.* Most stars are main sequence stars. None of them are old enough to have evolved off the main sequence. The position of a star on the main sequence depends on the initial mass of the star. As stars age and pass through different stages in their life cycles, their positions on the H-R diagram change. Because most stars spend most of their lives in midlife, more stars appear on the main sequence than on other parts of the H-R diagram. Red giant stars are both cool and bright, so they appear in the upper right. White dwarf stars are both faint and hot, so they appear in the lower left.

Science as a Human Endeavor

The Discovery of Pulsars In 1967, Jocelyn Bell detected several radio signals that were unusual in that they flashed on and off very rapidly, with periods of just a few seconds. Because the pulses were so regular, she and her colleagues thought at first that the signals might actually have been transmitted by extraterrestrial (ET) civilizations. They called the signals "LGM's," with "LGM" standing for "Little Green Men." After they ruled out the possibility that the signals were from ET, they started calling them "pulsars." Eventually, pulsars were shown to be rapidly spinning neutron stars.

Alternative Assessment — BASIC

Have students draw pictures or make paintings of what a star would look like at several stages of its life cycle. Drawings should reflect the true properties of the stars at each stage, such as size and color. **LS Visual**

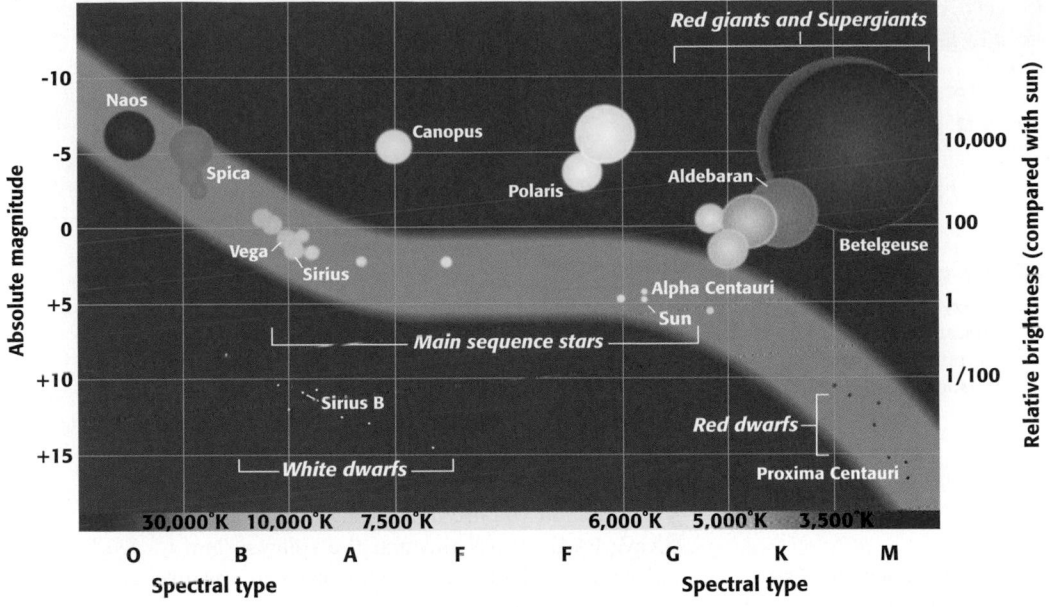

Red giants and Supergiants

Naos

Spica

Canopus

Polaris

Aldebaran

Vega

Sirius

Betelgeuse

Alpha Centauri

Sun

Main sequence stars

Sirius B

Red dwarfs

White dwarfs

Proxima Centauri

Absolute magnitude: -10, -5, 0, +5, +10, +15

Relative brightness (compared with sun): 10,000, 100, 1, 1/100

30,000°K 10,000°K 7,500°K   6,000°K 5,000°K 3,500°K

O   B   A   F   F   G   K   M

Spectral type   **Spectral type**

Our sun went from a protostar to a main sequence star in tens of millions of years. It will stay on the main sequence for about 10 billion years. As it becomes a red giant, it will become brighter, cooler, and redder; it will move up and to the right on the H-R diagram for about 100 million years. The sun will become a white dwarf, in the lower left, about 50 million years later.

Figure 11

The H-R diagram is a tool that astronomers use to help them understand how stars change over time.

SECTION 1 REVIEW

SUMMARY

▶ Stars are spheres of gas that produce energy by fusion.

▶ The composition of stars is measured using spectra.

▶ In most stars, outward pressure balances the inward pull of the star's gravity.

▶ Stars smaller than 1.4 solar masses become red giants and then white dwarfs.

▶ Massive stars become supergiants and explode in supernovae to become neutron stars or black holes.

1. **Determine** the distance between Polaris and Earth in meters. Polaris is 431 ly from Earth. The speed of light is 3.0×10^8 m/s.

2. **Arrange** the following from smallest to largest: sun, supernova, red giant, and white dwarf.

3. **Describe** the stages in the life of a star of 1 solar mass and in the life of a star of 20 solar masses.

4. **Critical Thinking** Which of the following elements is not likely to be formed in the sun at some time during its life?
   a. helium   **c.** oxygen
   b. carbon   **d.** iron

5. **Critical Thinking** You and a friend are looking at the stars, and your friend says, "Stars must be shrinking because gravity is constantly pulling their particles together." Explain what is wrong with this reasoning.

673

Answers to Section 1 Review

1. 4.1×10^{18} m

2. white dwarf, sun, red giant, supernova

3. A one solar mass star stays on the main sequence about 10 billion years, and then becomes a red giant. After a hundred million years, it becomes a white dwarf. A star of 20 solar masses will stay a short time on the main sequence, and then become a supergiant star.

When it starts to fuse iron in its core, there will be an explosion and the remnant will be a black hole.

4. d

5. Although gravity is always pulling the star together, the forces caused by fusion in the core are pushing it apart. A main sequence star balances these inward and outward forces.

Close

Quiz

1. Is a light-year a unit of time, distance, or speed? (distance)

2. What type of reactions provides fuel for stars? (nuclear fusion)

3. What physical property of a star most directly determines the star's color? (temperature)

4. Name two objects that may remain after a supernova. (neutron star, black hole; pulsar is also a correct answer)

Chapter Resource File

• Concept Review GENERAL

• Quiz BASIC

Transparencies

TT H–R Diagram

Focus

Overview

Before beginning this section, review with your students the Objectives listed in the Student Edition. This section introduces students to our galaxy, the Milky Way, and to galaxies in general. The section also describes the three main types of galaxies, and explains how galaxies change over time.

🔊 Bellringer

Use the Bellringer transparency to prepare students for this section.

Motivate

Opening Discussion ——— GENERAL

Ask students if they have ever seen the Milky Way. Students may look ahead to **Figure 15.** Prompt students to discuss what they think the Milky Way is. Why is it a narrow band? Are the other stars in the sky part of the Milky Way?

Confusion may arise during this discussion about the distinction between the Milky Way and the Milky Way Galaxy. Point out that both can be called the Milky Way, and that they are related. Tell students that the Milky Way appears as a bright band in the sky because you are looking along the plane of the galaxy, which is full of gas and dust, especially in the spiral arms. The other stars in the sky are also part of the Milky Way Galaxy, but do not lie in the direction of the plane. **LS Interpersonal**

The Milky Way and Other Galaxies

▶ **KEY TERMS**
galaxy
cluster
interstellar matter
quasar

OBJECTIVES

▶ **Define** galaxy, and identify Earth's home galaxy.
▶ **Describe** two characteristics of a spiral galaxy.
▶ **Distinguish** between the three types of galaxies.
▶ **Describe** two aspects of a quasar, and identify the tools scientists use to study quasars.

Imagine that you are in a special space ship that allows you to leave Earth, travel through the solar system to nearby stars, and explore all of space. What do you imagine you will see beyond our solar system?

galaxy a collection of stars, dust, and gas bound together by gravity

Galaxies

While the nearest stars are a few light-years away, the nearest galaxy to our own is millions of light-years from Earth. A **galaxy** is a collection of millions or billions of stars. The deeper scientists look into space, the more galaxies they find. There may be more than 100 billion galaxies. If you counted 1000 galaxies per night, it would take 275 000 years to count all of them.

Figure 12
The Andromeda Galaxy is 2.2 million ly from Earth. From a dark location, this galaxy is visible to the unaided eye as a faint blur.

Galaxies contain millions or billions of stars

Galaxies, such as the Andromeda Galaxy shown in **Figure 12,** contain millions to billions of stars bound together by gravity. Because stars age at different rates, a galaxy may contain many types of stars. Young stars are often found near the nebular gas and dust where they were born. Older stars may be throughout the galaxy or in regions that contain no gas and dust. Although galaxies contain many stars, scientists do not expect to be able to observe stellar systems within other galaxies. The distances to other galaxies are so large that searches for other planets focus on nearby stars, usually within our own galaxy.

674

Did You Know ?

Collision Course Although the Andromeda galaxy is an incredible 2.2 million ly from Earth, it is actually the nearest full-sized galaxy to the Milky Way. The mass of Andromeda is roughly twice the mass of the Milky Way. The two galaxies are drawn to each other by gravity. Some evidence suggests that the Andromeda galaxy and the Milky Way have passed near each other several times. This galactic dance may someday end in a collision in which the two galaxies merge to form one huge galaxy.

Gravity holds galaxies together in clusters

Without gravity, everything in space might be a veil of gas spread out through space. With gravity, clouds of gas come together and collapse to form stars. After the first stars in a galaxy age, throwing off gas and dust or becoming supernovae, new stars form. The gas, dust, and stars collapse into galaxies because of gravity.

Galaxies are not spread out evenly through space. They are grouped together in **clusters** like the one shown in *Figure 13.* The members of a cluster of galaxies are bound together by gravity. The Milky Way galaxy and the Andromeda galaxy are two of the largest members of the Local Group, a cluster of more than 30 galaxies. New members of the Local Group are being discovered as new telescopes, such as the Hubble Space Telescope shown in *Figure 14,* become available to astronomers.

Clusters of galaxies can form even larger groups called *superclusters.* A typical supercluster contains thousands of galaxies containing trillions of stars in individual clusters. Superclusters can be as large as 100 million ly across. They are the largest structures in the universe.

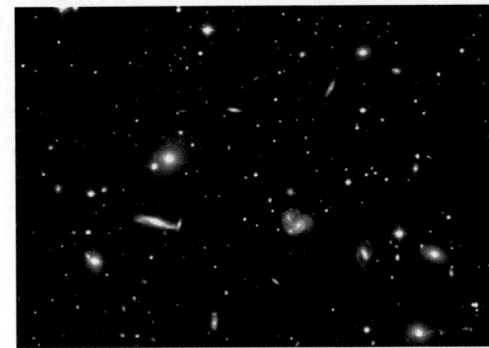

Figure 13
The Hercules Cluster of galaxies is 650 million ly from Earth.

cluster a group of stars or galaxies bound by gravity

Figure 14

A Edwin Hubble used the telescopes at Mount Wilson Observatory in California to explore galaxies beyond the Milky Way galaxy.

B The *Hubble Space Telescope,* shown here being launched from the space shuttle, now probes the depths of the universe from its orbit above Earth.

675

Teach

SKILL BUILDER — **BASIC**

Interpreting Visuals Have students look at **Figure 12** and answer the following question: If you are looking through a telescope at the Andromeda galaxy—which is 2.2 million ly from Earth—how old is the light you are seeing? (2.2 million years) **LS** Logical

SKILL BUILDER — **GENERAL**

Interpreting Visuals **Figure 13** shows the Hercules Cluster, a collection of galaxies held together by gravity. The figure also shows some stars in the foreground. These stars are part of our own galaxy.

Have students examine the figure and identify which objects are galaxies and which objects are stars. Also ask students if they can detect any differences between the individual galaxies. Do there seem to be recurring shapes? Can they make out different colors? Recognizing these differences will help prepare students for learning about the different kinds of galaxies. **LS** Visual

Did You Know ?

The Magellanic Clouds The Milky Way galaxy's nearest neighbors are two small, irregular galaxies, the Large Magellanic Cloud and the Small Magellanic Cloud. These two galaxies are joined by a long, diffuse stream of stars that have been pulled out of the galaxies as they have passed near each other and near the Milky Way. Eventually, they may be swallowed up entirely by our galaxy.

Chapter Resource File

• Lesson Plan

 Transparencies

TT Bellringer

Vocabulary In ancient times, people thought that the Milky Way looked like a radiant band of light flowing like a river of milk. Thus, our galaxy came to be known as the Milky Way. **LS** Verbal

SKILL BUILDER — BASIC

Interpreting Visuals Have students examine **Figure 16** and imagine standing within the Milky Way galaxy but outside the central bulge. In what direction would they need to look to see the Milky Way as a band of light? (The students could look left, right, or straight ahead along the plane of the galaxy to see the band of light. If they looked up or down out of the plane, they would see stars but no continuous band of light.) **LS** Visual

Teaching Tip

Interstellar Matter In very dark, clear skies, interstellar matter can be seen clearly, but indirectly, by looking at the Milky Way. The Milky Way contains dark, splotchy patches, interspersed throughout its glowing band. These dark patches are caused by interstellar clouds blocking the light from the many stars that make up the band of the Milky Way.

Figure 15

When we see the band of light called the *Milky Way*, we are looking along the plane of our galaxy, the Milky Way galaxy.

▶ **interstellar matter** the gas and dust located between the stars in a galaxy

Figure 16

An idea of what the Milky Way galaxy might look like from the outside can be pieced together from astronomical data.

Types of Galaxies

Edwin Hubble divided all galaxies into three major types: spiral, elliptical, and irregular. All three types have many stars, but they have different structures. Spiral galaxies have spiral arms made of gas, dust, and stars. Elliptical galaxies have little gas or dust. Irregular galaxies do not have a particular shape.

We live in the Milky Way galaxy

If you live away from bright outdoor lights, you may be able to see the Milky Way, a faint, narrow band of light and dark patches across the sky. This band, shown in *Figure 15,* consists of stars, gas, and dust in our galaxy, the Milky Way galaxy.

Most of the objects you can see in the night sky are part of the Milky Way galaxy. Because our solar system is inside the Milky Way galaxy, we cannot see all of it at once. But scientists can use astronomical data to piece together a picture of the Milky Way galaxy, such as the one shown in *Figure 16.* Our solar system is located within a spiral arm, about 26 000 ly from the center, or about half of the distance to the edge.

The Milky Way is a spiral galaxy

Our galaxy is a huge spiraling disk of stars, gas, and dust. Like most spiral galaxies, the Milky Way galaxy has a huge bulge in the center. The nucleus of the galaxy is very dense and has many old stars. The gas and dust have been used up to form stars. Many astronomers think that a large black hole is at the very center of our galaxy. Spiral galaxies, such as Messier 74 (M74), which is shown in *Figure 17A,* have gas and dust between the stars. This gas and dust is called **interstellar matter.** Clouds of interstellar matter provide materials that allow new stars to form. Because hot young stars are blue, the spiral arms often appear bluish. Because old stars are often red, the bulge in the middle appears reddish. The arms have both old and new stars as well as gas and dust.

676

Cultural Awareness

Constellations in the Dust Most ancient cultures saw patterns or constellations among the stars. The Incas, however, were unique in that they saw and named constellations in the dark patches of the Milky Way, now known to be interstellar dust. Careful observations of the details of the Milky Way were possible because of the extremely dark skies the Incas enjoyed from their location in the Andes Mountains of South America.

Figure 17

A Seen from above, the Milky Way galaxy might look like this spiral galaxy, named *Messier 74.*

B Unlike the Milky Way galaxy, elliptical galaxies such as *Messier 87* do not have spiral arms.

C The Magellanic Cloud is a large irregular galaxy that is easily seen in the Southern Hemisphere.

Elliptical galaxies have no spiral arms

Elliptical galaxies have no spiral arms and are spherical or egg shaped. They contain mostly older stars and have little interstellar matter. Older stars are red, so elliptical galaxies, such as M87 in *Figure 17B* often have a reddish color. Elliptical galaxies are found in a wide range of sizes. Giant elliptical galaxies contain trillions of stars and can be up to 200 000 ly in diameter. Dwarf elliptical galaxies contain a few million stars and are much smaller.

A spiral galaxy can be recognized even when it is tilted at an angle, but because an elliptical galaxy has no regular features, scientists have trouble knowing whether an elliptical galaxy is head-on or sideways in relation to Earth.

All other galaxies are irregular galaxies

Edwin Hubble named the third category irregular galaxies because they lack regular shapes and do not have a well-defined structure. Some irregular galaxies contain little interstellar matter, while others have large amounts and contain mostly young blue stars. *Figure 17C* shows the large irregular galaxy that is nearest to the Milky Way, the Large Magellanic Cloud. This galaxy is a part of the Local Group of galaxies.

There are many more dwarf irregular galaxies than large ones. Dwarf galaxies are often found near larger galaxies. Some irregular galaxies may be oddly shaped because the gravitational influence of nearby galaxies distorts their spiral arms.

Did You Know?

Many new technologies have come out of the space program. A company has recently starting selling a jacket made out of the same material that NASA uses to insulate spacecraft. The material, called *aerogel,* can withstand temperatures from –45°C to 1650°C (–50°F to 3000°F), and keeps the person inside the jacket very warm even in the coldest weather.

677

Applying Information

1. Cell phones affect radio telescopes more than conventional phones because cell phones transmit signals using radio waves.

2. Any source of electromagnetic radiation in the radio region might bother a radio telescope. Some examples are automobile ignition (spark plugs), electrical storms, aircraft or ground-based radar stations, commercial radio or TV towers, and man-made satellites. To keep it safe, a radio telescope should be built far from civilization or in a valley where the natural protection of the mountains may shield it from radio pollution.

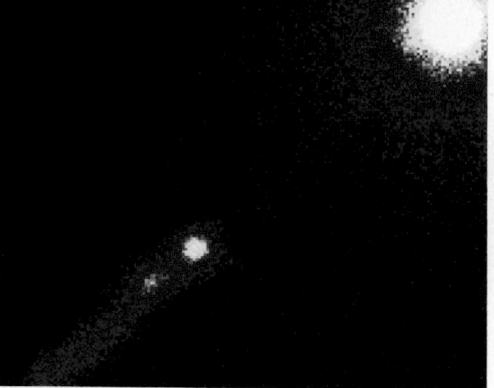

Figure 18
This quasar, which is named 3C273, has a powerful jet shooting from it and produces more energy than the sun.

quasar quasi-stellar radio sources; very luminous objects that produce energy at a high rate and that are thought to be the most distant objects in the universe

How Galaxies Evolve

When scientists observe distant galaxies, they are looking back in time. When astronomers observe a galaxy that is 1 billion ly away, they are observing light that left the galaxy one billion years ago. Scientists do not know what such a galaxy is doing now, but by studying closer galaxies that might be similar to ancient ones, they can slowly piece together the puzzle of how galaxies change over time.

Quasars may be infant galaxies

When astronomers first detected radio waves in space, they had difficulty finding stars that accompanied some of the radio sources. In 1960, a faint object was finally matched with a strong radio signal. This object was the first **quasar,** or quasi-stellar object, named for its starlike appearance, as shown in *Figure 18.* On further study, scientists discovered that quasars are the most distant and most radiant objects in space. One explanation for the strong radiation is that each quasar has a huge central black hole (about a billion solar masses) and a large disk of gas and dust around it. Friction in the disk releases energy into space at many wavelengths, especially radio waves. Optical telescopes show quasars embedded in faint galaxies. Quasars may be the central parts of distant galaxies, seen as they were when very young.

Radio Telescopes and Cell Phones

Electromagnetic radiation comes in many wavelengths. Radio telescopes detect radio waves from space. Radio waves are especially important in studying distant objects such as quasars and the interstellar matter within galaxies. But just as optical telescopes are affected by light pollution, radio telescopes are affected by unwanted radio signals. Cell phones can easily drown out the distant signals of many galaxies. For this reason, cell phones are banned near many radio telescopes, such as the Arecibo in Puerto Rico.

Applying Information
1. Why do cell phones affect telescopes more than phones that use wires do?
2. What other sorts of pollution may affect radio telescopes? Where might you build a radio telescope to avoid radio pollution?

678

Did You Know?

Galactic Collisions Because they are attracted to each other, galaxies sometimes orbit each other and occasionally collide. These collisions can last millions of years and span over a trillion cubic ly of space. The distance between stars is so great that a galactic collision produces very few star collisions. But the motion of the stars in each galaxy is affected by the gravitational pull of the stars in the other galaxy, so the galaxies are transformed by the encounter.

Galaxies change over time

All stars change over time. Massive stars explode in supernovae, and lower-mass stars become red giants and, eventually, white dwarfs. Because gas, dust, and stars make up galaxies, entire galaxies also change over time. As galaxies consume their gas and dust, they become unable to make new stars. Many galaxies in the Hubble Deep Field, shown in *Figure 19,* are blue, indicating that we are viewing them when they were young, before they used their stores of gas and dust.

Galaxies also change as a result of collisions. Stars within a galaxy are far apart and can easily pass each other if two galaxies collide. But as galaxies approach each other, the mutual gravitational attraction changes their shapes. While the stars rarely hit each other, the collision of gas and dust sets off rapid bursts of new star formation. Scientists are trying to discover why elliptical galaxies do not contain young stars and how they used up their gas and dust. One possibility is that gas and dust were stripped away in collisions that also stripped away many of the younger stars.

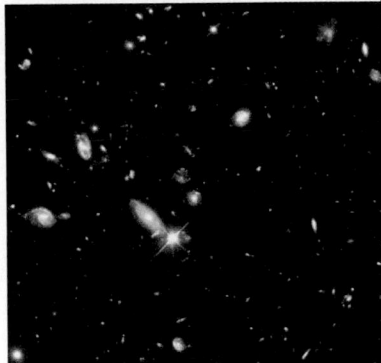

Figure 19
The Hubble Deep Field project discovered many faint and very distant galaxies.

Close

Quiz

1. What do you call a collection of clusters bound together by gravity? (a supercluster)

2. What are the three main types of galaxies? (spiral, elliptical, and irregular)

3. What type of galaxy is the Milky Way galaxy? (a spiral galaxy)

4. What color do elliptical galaxies tend to be, and why? (red; they contain relatively few young stars)

SECTION 2 REVIEW

SUMMARY

▶ A galaxy is a collection of millions or billions of stars bound together by gravity.

▶ Our solar system lies in a spiral arm in the Milky Way galaxy.

▶ Spiral galaxies have a bulge near the center and spiral arms made of gas, dust, and stars.

▶ Elliptical galaxies have little gas and dust and a spherical or oval shape.

▶ Irregular galaxies have no regular shape.

▶ Quasars are stellar in appearance but emit large amounts of radiation, especially radiowaves.

1. **List** the types of galaxies and describe one important feature of each type.

2. **Explain** why the Milky Way appears as a narrow band of light in the night sky.

3. **Draw** a sketch of the Milky Way galaxy. Label the nucleus and the central bulge, and indicate the position of our solar system.

4. **Compare** the colors of spiral galaxies with the colors of elliptical galaxies.

5. **Describe** how quasars got their name.

6. **Arrange** these structures from largest to smallest: solar system, sun, spiral galaxy, dwarf elliptical galaxy, and cluster of galaxies.

7. **Explain** how scientists know that elliptical galaxies do not contain many young stars?

8. **Critical Thinking** Why do stars rarely collide during galactic collisions?

679

Answers to Section 2 Review

1. Spiral galaxies have spiral arms; elliptical galaxies are spherical or elongated in shape; irregular galaxies have no regular features.

2. The Milky Way looks like a narrow band because we are looking along the plane of the galaxy from inside a spiral arm.

3. Sketch should resemble Figure 16 or 17a.

4. Spiral galaxies often have blue stars in the spiral arms and red/yellow stars in the core. Elliptical galaxies contain mostly red and yellow stars.

5. Quasars don't look quite like stars and have a strange radio emission, unlike stars. Because

they were "quasi" stellar, they were called "quasars."

6. cluster of galaxies, spiral galaxy, dwarf elliptical galaxy, solar system, sun

7. Scientists know that elliptical galaxies don't contain many young stars because elliptical galaxies are mostly reddish in color.

8. Stars rarely collide in galactic collisions because the distance between stars in a galaxy is extremely large compared to the size of the stars.

Chapter Resource File

• **Cross-Disciplinary Worksheet** Real World Applications—Telescope Interference **ADVANCED**

• **Concept Review** **GENERAL**

• **Quiz** **BASIC**

Focus

Overview

Before beginning this section, review with your students the Objectives listed in the Student Edition. This section introduces students to the concept of the universe as a whole. The section also describes the evidence for one model of the origin of the universe, the big bang theory, and discusses the possible future of the universe.

Bellringer

Use the Bellringer transparency to prepare students for this section.

Motivate

Opening Discussion — GENERAL

Engage students in a class discussion about all the different components that make up the universe. Encourage them to name things that are part of the universe, and challenge them to try to think of things that are not a part of the universe. This activity should reinforce in students' minds the idea that the universe contains everything that exists. **LS Verbal**

Origin of the Universe

KEY TERMS

universe
red shift
blue shift
big bang theory

OBJECTIVES

▶ **Describe** the basic structure of the universe.

▶ **Describe** red shift, and explain what it tells scientists about our universe.

▶ **State** the main features of the big bang theory, and explain the evidence supporting the expansion of the universe.

▶ **Explain** how scientists are using tools and models to hypothesize what may happen to the universe in the future.

universe the sum of all space, matter, and energy that exist, that have existed in the past, and that will exist in the future

Just imagine the following: colliding galaxies that rip stars from each other, a dead star so dense that one thimbleful of its matter would weigh more than 100 million tons on Earth, a volcano on Mars that is nearly three times taller than Mount Everest and that has a base larger than Louisiana. All of these things are part of the universe.

What Is the Universe?

By the term **universe,** scientists mean everything physical that exists in space and time. The universe consists of all space, matter, and energy that exists—now, in the past, and in the future. There is only one observable universe. **Figure 20** shows objects in the universe and their relative sizes.

Figure 20

The sizes of astronomical objects are so great that measuring units such as the light-year are needed to describe these objects.

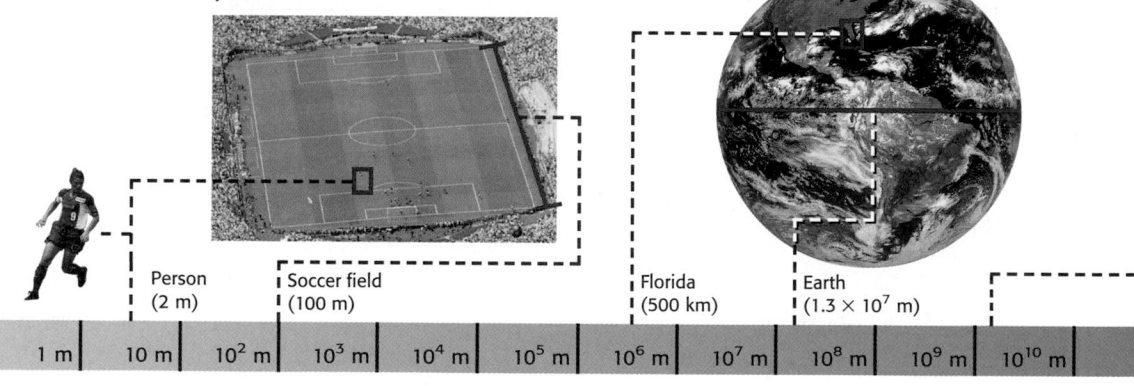

| Person (2 m) | Soccer field (100 m) | | Florida (500 km) | Earth (1.3×10^7 m) |

| 1 m | 10 m | 10^2 m | 10^3 m | 10^4 m | 10^5 m | 10^6 m | 10^7 m | 10^8 m | 10^9 m | 10^{10} m |

Did You Know?

The ancient Greeks proposed a question about the size of the universe. What would happen if you went to the edge of the universe and then shot an arrow beyond the boundary? Because the universe by definition contains everything, it would have to stretch further to encompass the space into which the arrow was flying. The Greeks used this puzzle to argue that the universe is infinite.

A modern model of the universe portrays it as the surface of a four-dimensional sphere. Imagine a three-dimensional sphere and then imagine drawing a line on the surface of a sphere. You could draw the line forever and never reach the end of the surface. The surface is not infinite in size, but it has no boundary. In this model, the sphere is expanding.

You are part of the universe, as is Earth and everything on it. With the unaided eye, we can see about 6000 stars, 5 of the planets, our moon, and several nebulae, star clusters, and galaxies. As shown in *Figure 20,* huge distances are involved in studying the universe. Perhaps you can imagine the size of a soccer field, your city, your country, or even Earth, but the comparisons become difficult as we imagine objects on scales beyond Earth, such as the solar system. The solar system is only a small part of our galaxy, which is but one of many galaxies in one of many clusters of galaxies.

We see the universe now as it was in the past

Astronomers need large units of measure to express distances. As you recall, a light-year is the distance that light travels in one year, or 9.5×10^{15} m. This distance is so long that driving it in a car moving at highway speed would take more than 10 million years. Remember that while a year is a unit of time, a light-year is a unit of distance.

It takes time for light to travel in space. The farther away an object is, the older the light that we get from that object is. When we say the sun is 8 light-minutes away, we are not only expressing its distance, but also the fact that we see it as it was eight minutes ago. We never see it as it really is, right now. The same is true for stars, planets (Pluto is more than 5 light-hours away), galaxies, or clusters of galaxies. When we see very distant objects, we see them as they were when they were younger. Astronomers can compare how galaxies age by looking at many galaxies at different distances, and therefore at different ages.

☑ **internet** connect

www.scilinks.org
Topic: Origin of the
Universe
SciLinks code: HK4098

SC*I*LINKS. Maintained by the National Science Teachers Association

Did **You** Know ?

Because light travels at a constant speed of 3×10^8 m/s, we can measure how long it takes starlight to reach Earth. Light from Sirius travels 8 years and 7 months. If you look at Altair (in the constellation Aquila), you see light that left the star 16 years and 9 months ago. You could even find a star whose light left the year you were born.

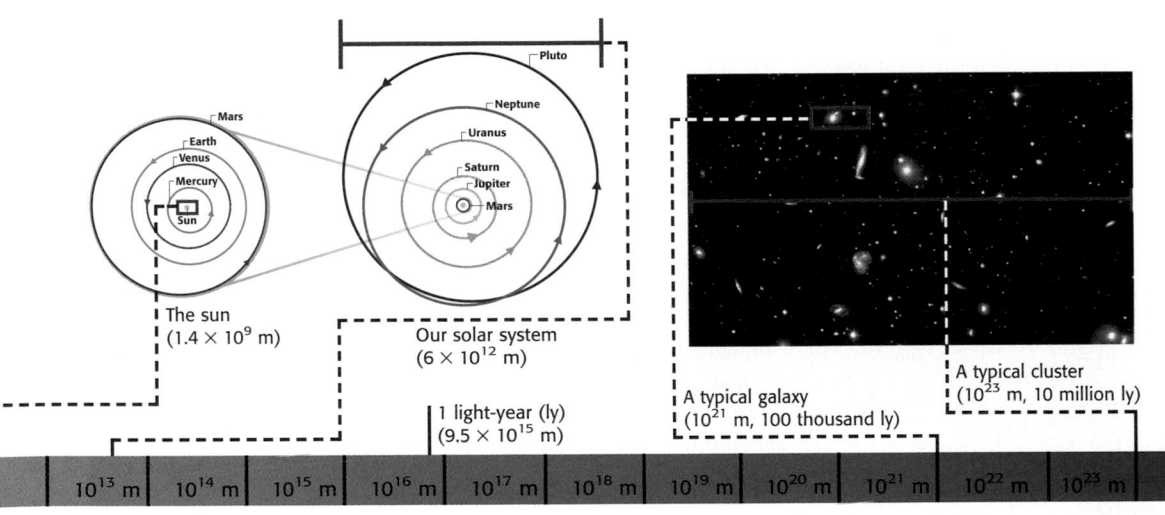

The sun
(1.4×10^9 m)

Our solar system
(6×10^{12} m)

1 light-year (ly)
(9.5×10^{15} m)

A typical galaxy
(10^{21} m, 100 thousand ly)

A typical cluster
(10^{23} m, 10 million ly)

| 10^{13} m | 10^{14} m | 10^{15} m | 10^{16} m | 10^{17} m | 10^{18} m | 10^{19} m | 10^{20} m | 10^{21} m | 10^{22} m | 10^{23} m |

681

Did **You** Know ?

Age of the Universe Scientists estimate that some quasars, the most distant objects that have been detected, are at least 12 billion ly away. If this is true, then the universe must be at least 12 billion years old in order for the light from those quasars to have reached Earth.

Reading Organizer As students read the section entitled *What Happened at the Beginning?* and have them make a list of evidence that could support the big bang theory or the idea of an expanding universe. **LS** Verbal

Teaching Tip ——— **GENERAL**

Hubble and The Hubble Space Telescope Have students go to their local library and research the life and work of Edwin Powell Hubble and the Hubble Space Telescope. If some students need guidance to start their research, have them answer the following questions: What significant contributions did Edwin Powell Hubble make in the field of astronomy? When was the Hubble Space Telescope first launched into space? What problems did it encounter? What role does the Hubble Telescope play in astronomical research? **LS** Verbal

Figure 21

Astronaut Bruce McCandless maneuvers through space in a suit specially designed to allow him to propel himself. His space-suit also protects him from the conditions of space.

▶ **red shift** an apparent shift toward longer wavelengths of light caused when a luminous object moves away from the observer

▶ **blue shift** an apparent shift toward shorter wavelengths of light caused when a luminious object moves toward the observer

Most of the universe is empty space

Despite the variety of objects in the universe, such as interstellar matter, stars, and galaxies, there is almost nothing between objects. *Figure 21* shows an astronaut in a suit designed to help a person survive in space. It is extremely hot facing the sun in space, and it is very cold facing away from it. Space is a vacuum with no air and no air pressure. The suit provides the insulation, breathable air, and air pressure that the human body needs to survive. In this case, the astronaut is bathed in particles streaming from the sun. Farther out, there is so little between stars that the space can truly be called "empty."

What Happened at the Beginning?

How the universe came to be is an age-old question. Ancient cultures had myths to explain the origin of the universe. Today, scientists study stars and galaxies for clues by using new tools and techniques. Scientists interested in the early history of our universe use large telescopes to study the most distant objects, whose light was emitted billions of years ago.

The universe is expanding

In 1929, Edwin Hubble announced that the universe is expanding. Hubble based his conclusion on observations of the spectral lines in light from other galaxies. He found that these spectral lines were almost always shifted toward the red end of the spectrum. This effect, called **red shift,** can be explained by the Doppler effect. The Doppler effect states that when an object is moving away from us, waves emitted from the object stretch out. The faster a light source moves away, the more that light stretches to longer wavelengths and shifts toward the red end of the spectrum, as shown in *Figure 22.*

When an object is approaching us, the shift is toward shorter wavelengths at the spectrum's blue end and is called **blue shift.** *Table 1* shows the distance, velocity, and frequency shift of several galaxies. Hubble found that most galaxies have red shifts and that galaxies that are farther away have greater red shifts. Hubble explained this by proposing that almost every galaxy is moving away from Earth. Therefore, galaxies are also moving away from each other, and the universe is expanding.

682

Trends in Technology

Space Telescopes The Hubble Space Telescope, which is primarily an optical telescope, was the first major telescope to be put into orbit around Earth. Space telescopes have the advantage of being above Earth's atmosphere and away from the radiation pollution caused by cities. Other space telescopes now or soon to be in orbit include infrared telescopes (SIRTF) and X-ray telescopes (CHANDRA). The Next Generation telescope, which will take over the Hubble Space Telescope's role as the primary space telescope, will be sensitive to infrared radiation.

Table 1 Velocity, Frequency Shift, and Distance from Earth of Several Galaxies

Name of galaxy	Type	Velocity (km/s)	Red shift or blue shift	Distance (ly)
Andromeda galaxy (M31)	spiral	–10	blue	2.4×10^6
Barnard's galaxy (NGC 6822)	irregular	15	red	2.2×10^6
NGC 55 (in Sculptor)	spiral	115	red	1.0×10^7
Sunflower galaxy (M 63)	spiral	550	red	3.6×10^7
Virgo A (M87)	elliptical	822	red	7.2×10^7
Fornax A (NGC 1316)	spiral	1713	red	9.8×10^7

Expansion implies that the universe was once smaller

Although galaxies that are close to each other are gravitationally attracted to each other, galaxies are moving away from each other in general. Imagine time running backward, like a movie being rewound. If every galaxy normally moves away from every other galaxy, then as time goes backward, the galaxies move closer together. Long ago, the entire universe might have been contained in an extremely small space, effectively a point.

If time moves forward again from that point, all of the matter in the universe appears to expand rapidly outward like a gigantic explosion. Scientists call this hypothetical explosion the *big bang*. If the expansion has been constant since the big bang, we can estimate the age of the universe. Velocity is equal to distance divided by time. The velocities of galaxies can be measured by using red shift and the Doppler effect, but the distances to these faint objects are difficult to measure. Using estimates for distance and velocity, scientists have estimated that the age of the universe is between 10 and 20 billion years.

INTEGRATING

CHEMISTRY
One of the major discoveries of the 1800s was that spectral lines found in chemistry labs were also found in celestial objects. This discovery showed that objects act predictably everywhere in the universe and allowed scientists to identify elements found in space as identical to elements found on Earth. Astronomers began to study how atoms react in conditions that differed from those on Earth.

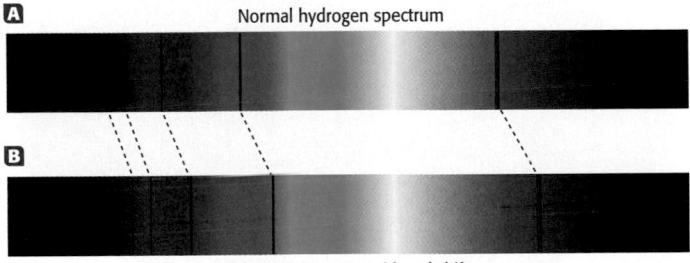

Normal hydrogen spectrum

Hydrogen spectrum with red shift

Figure 22

A The spectral lines of hydrogen gas can be seen and measured in a laboratory.

B When this pattern appears in starlight, we know that the star contains hydrogen. In this case, the lines show a red shift, suggesting that the star is moving away from us.

683

Trends in Physics

The Doppler effect describes how the frequency of sound waves increases when the source of the sound is moving toward you and decreases when the source is moving away from you. The same is true of the frequency of light waves. Would a decrease in the frequency of light from a galaxy make the galaxy appear redder or bluer? (redder)

Did You Know ?

Even though the universe is expanding not every galaxy exhibits a red shift. Some nearby galaxies, including the Andromeda galaxy, show a blue shift because galaxies in our local cluster, called the Local Group, are gravitationally attracted to each other. While the universe is expanding on a very large scale, galaxies within a single cluster may move toward each other and even collide.

■ **big bang theory** the theory that all matter and energy in the universe was compressed into an extremely small volume that 10 to 20 billion years ago exploded and began expanding in all directions

Quick ACTIVITY

Modeling the Expanding Universe

1. Inflate a round balloon to about half full, and then pinch it closed to keep the air inside.
2. Use a marker to draw several dots close together on the balloon. Mark one of the dots with an *M* to indicate the Milky Way galaxy.
3. Now continue inflating the balloon. How do the dots move relative to each other?
4. How is this inflating balloon a good model for the expanding universe?
5. In what ways might this balloon not be a good model for the expanding universe?

Figure 23

The colors in this computerized map of cosmic background radiation across the entire sky represent slight differences in temperature above and below 2.7 K.

684

Did the universe start with a big bang?

Although scientists have proposed several different theories to explain the expansion of the universe, the most complete and widely accepted is the big bang theory. The **big bang theory** states that the universe began with a gigantic explosion 10 billion to 20 billion years ago. *In this book, we will assume that the universe is 15 billion years old.*

According to this theory, nothing existed before the big bang. There was no time and no space. But out of this nothingness came the vast system of space, time, matter, and energy that now makes up the universe. The explosion released all of the matter and energy that still exist in the universe today.

Cosmic background radiation supports the big bang theory

In 1965, Arno Penzias and Robert Wilson were making adjustments to a new radio antenna that they had built. They could not explain a steady but very dim signal from all over the sky in the form of radiation at microwave wavelengths. They realized that the signal they were receiving was the *cosmic background radiation* predicted by the big bang theory.

Imagine the changes in color that occur as the burner on an electric stove cools off. First, the hot burner glows yellow or white. As the burner cools, it becomes dimmer and glows red. It may still be rather hot when it finally looks black. The color you see corresponds to the wavelength at which the burner radiates the most light. In outer space, the burner would cool until it reached the temperature of space.

Many scientists believe that the microwaves detected by Penzias and Wilson are dim remains of the radiation produced during the big bang. Using maps of cosmic background radiation, such as the one shown in *Figure 23,* scientists have found that the universe has an overall temperature of about 2.7 K.

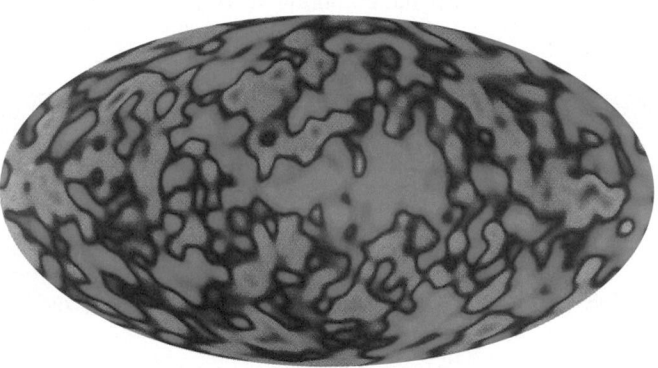

Trends in Astronomy

Cosmic Background Radiation Cosmic background radiation can tells us more than just the temperature of the universe. One important question in astronomy today is why the universe evolved so that the matter is clumped in galaxies and clusters, rather than spread evenly throughout the universe. Detailed maps of cosmic background radiation, such as the one shown in **Figure 23,** show slight fluctuations in temperature above and below the average 2.7 K. These slight fluctuations might reflect slight deviations from the nearly smooth distribution of matter in the early universe. These slight deviations might have led to the formation of larger structures, such as galaxies, clusters, and superclusters.

Figure 24

This timeline shows major events in the evolution of the universe. The first 1 million (10^6) years are based on the big bang theory.

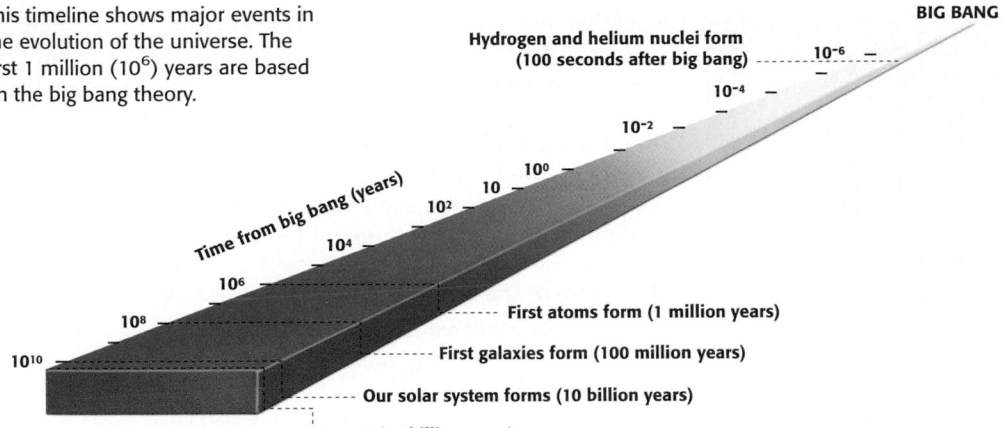

BIG BANG

Hydrogen and helium nuclei form (100 seconds after big bang) ---- 10^{-6}

10^{-4}

10^{-2}

10^0

10

10^2

Time from big bang (years)

10^4

10^6

10^8

10^{10}

First atoms form (1 million years)

First galaxies form (100 million years)

Our solar system forms (10 billion years)

Present (15 billion years)

Radiation dominated the early universe

According to the big bang theory, expansion cooled the universe enough for matter such as protons, neutrons, and electrons to form from the radiation within a few seconds after the big bang. Hydrogen and helium nuclei and other particles were present, but the temperature was still too high for entire atoms to form and remain stable. The universe was dominated by radiation, which immediately overcame the attraction between electrons and nuclei. **Figure 24** shows key points in the evolution of the universe as predicted by the big bang theory. Note that the timeline uses a logarithmic scale, so the last 5 billion years can be found in a small area near the end.

Processes in stars lead to bigger atoms

In a million years, the universe had expanded and cooled enough for hydrogen and helium atoms to form. Hydrogen comprised 75% of the matter, and helium comprised 25%. Hydrogen fuels stars and acts as a building block for other elements. Once hydrogen atoms formed, stars and galaxies began to form, too. Our solar system is thought to be 4.6 billion years old, forming 10 billion years after the big bang.

All elements other than hydrogen and helium form in stars. Nuclear fusion in stars produces helium and elements up to the atomic number of iron. Heavier elements form during supernovae. **Figure 25** shows helium and lead that were produced in a star. The lead is in the form of galena, or lead sulfide.

Figure 25

Helium is found in stars, but heavier elements, such as lead, are the result of supernovae.

685

Did You Know?

Although the nuclei of atoms formed just 100 s after the big bang, whole atoms, with electrons orbiting the nuclei, did not form for another million years. The energy in the early universe was so concentrated that electrons could only stay close to any given nucleus for only a tiny fraction of a second.

Interpreting Graphics **Figure 27** is a graph showing how the size of the universe might change with time. Have students examine the figure, and ask them the following questions. What does the vertical axis represent? (the size of the universe) What does the horizontal axis represent? (time) How do you interpret the straight, diagonal line on the graph? (The line represents the universe expanding at a constant rate.) If the universe were not expanding at all, how would this appear on the graph? (as a horizontal line) If the expansion of the universe was actually accelerating, how would this appear on the graph? (the line would curve upward) What does it mean if the value of the vertical coordinate is zero? (it means the universe has a size of zero; in other words, the whole universe is compressed into a point) **LS** Visual

Figure 26
Astronomers observe the universe by using modern telescopes, such as the telescopes at the Cerro Tololo Inter-American Observatory in Chile.

Figure 27
There are three possible fates for the universe.

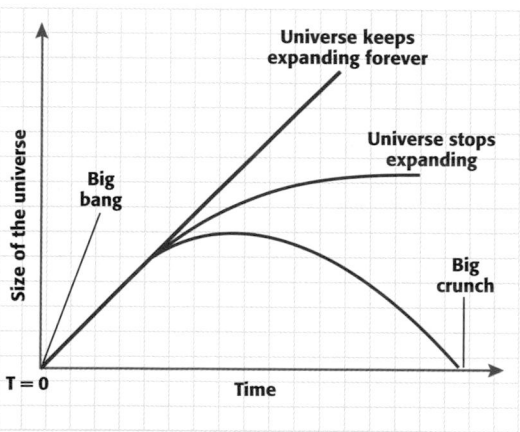

Predicting the Future of the Universe

Scientists use their ever-increasing knowledge to hypothesize what might happen in the future. They depend on a mixture of theory and precise observations of very faint objects. These observations depend on technology, such as the telescopes shown in *Figure 26.* New space telescopes that collect infrared radiation and X rays are being built and launched. Data in these regions of the electromagnetic spectrum may provide important clues about the beginning and future of the universe.

The future of the universe is uncertain

The universe is still expanding, but it may not do so forever. The combined gravity of all of the mass in the universe is also pulling the universe inward, in the direction opposite to the expansion. The competition between these two forces leaves three possible outcomes for the universe:

1. The universe will keep expanding forever.
2. The expansion of the universe will gradually slow down, and the universe will approach a limit in size.
3. The universe will stop expanding and start to fall back in on itself.

The fate of the universe depends on mass

Figure 27 shows three possible fates of the universe. Which one occurs depends on the amount of matter in the universe. If there is not enough mass, the gravitational force will be too weak to stop the expansion, so the universe will keep expanding forever. If there is just the right amount of mass, the expansion will continually slow down but will never stop completely. If there is more mass than this right amount, gravity will eventually overcome the expansion and the universe will start to contract. Eventually, a contracting universe could collapse back to a single point in a "big crunch." As things drew closer to each other, galaxies and stars would collide. The universe would become extremely hot and very small. At this point, the universe may end, or another big bang may start the cycle all over again.

It is hard to predict what will happen in this very distant future. Much of the mass in the universe is very difficult to detect, so we do not yet know the total mass of the universe.

686

Trends in Astronomy

Accelerating Universe? Several recent studies of the evolution of the universe have focused on observations of supernovas in extremely distant galaxies. Some of these studies have found evidence that the expansion of the universe may actually be accelerating, rather than staying constant or slowing down. This surprising result seems to contradict one of the fundamental laws of physics, the conservation of energy. If the expansion of the universe is accelerating, then where is the extra energy coming from? The question has not yet been resolved, and may serve to significantly change the way we understand our universe and the laws that make it work.

New technology helps scientists test theories

Predictions of the future of the universe rest on theories of the past. Scientists test theories by making observations to see whether the theories make accurate predictions. If observations do not agree with theory, new theories are needed. To make these important observations, very powerful telescopes and other sensitive equipment are needed.

One example of new, more sensitive technology is the Chandra X-Ray Observatory, shown in *Figure 28A*. The presence of X rays indicates matter at temperatures of more than one million degrees. The Crab Nebula emits radiation at many wavelengths, including X rays, as shown in *Figure 28B*. Compare this image with the visible light picture in *Figure 10* and the radio image on the first page of this chapter. Observations in each wavelength region tell us something about the Crab Nebula and about supernovae and their release of elements in general.

There is debate about dark matter

Astronomers estimate the mass of the universe by measuring stars, galaxies, and matter in the interstellar medium. But observations of gravitational interactions between galaxies, such as the interaction shown in *Figure 29* indicate that there is more matter than what is visible. Some scientists call this undetectable matter *dark matter*. Dark matter may be planets, black holes, or brown dwarfs. Brown dwarfs are starlike objects that lack enough mass to begin fusion. Dark matter could also be exotic atomic particles that no one knows how to observe. As much as 90% of the universe may be composed of dark matter. What it is, where it is, and how to detect it remain a mystery.

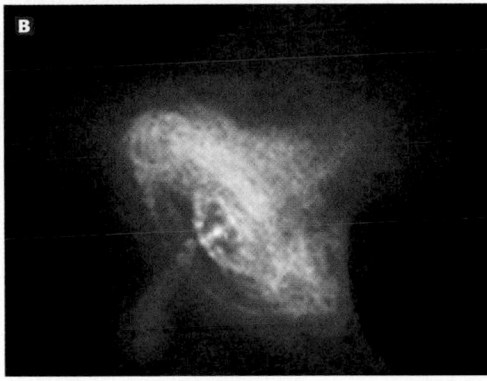

Figure 28

A The Chandra X-Ray Observatory collects information from matter at very high temperatures.

B The Chandra X-Ray Observatory created this image of the Crab Nebula.

internet connect

www.scilinks.org
Topic: Dark Matter
SciLinks code: HK4030

SCiLINKS. Maintained by the National Science Teachers Association

Figure 29

Spiral galaxies NCG 2207 and IC 2163 are colliding.

687

Trends in Physics

Neutrinos Dark matter may not all be in the form of massive objects such as planets or black holes. In fact, one significant source of dark matter may turn out to be neutrinos, some of the tiniest particles in the universe.

Neutrinos are elusive. Many neutrinos pass right through the entire Earth without interacting with any of Earth's atoms. For this reason, they are very hard to study. So hard, in fact, that scientists are not even sure whether or not neutrinos have mass.

Determining the mass of neutrinos could provide an important clue to the possible fate of the universe. If they have no mass, then they cannot contribute to the tally of dark matter in the universe. But there are an extremely high number of neutrinos in the universe. If they do have mass, no matter how small, then they could make a large contribution to the total mass of the universe—large enough to mean the difference between an ever-expanding universe and a big crunch.

Close

Quiz

1. If the spectral lines of hydrogen in a star are seen to be shifted toward the blue end of the spectrum, is the star moving toward us or away from us? (toward us)

2. If you observed that the light from nearly every galaxy was shifted toward the blue end of the spectrum, what could you conclude about the universe? (You could conclude that the universe is shrinking.)

3. What would happen to the universe if it was shrinking, and continued to shrink for a very long time? (Eventually, the universe would contract into a single point. This type of event is called a big crunch.)

4. Name two examples of dark matter. (Possible answers include: planets, black holes, brown dwarfs, neutrinos.)

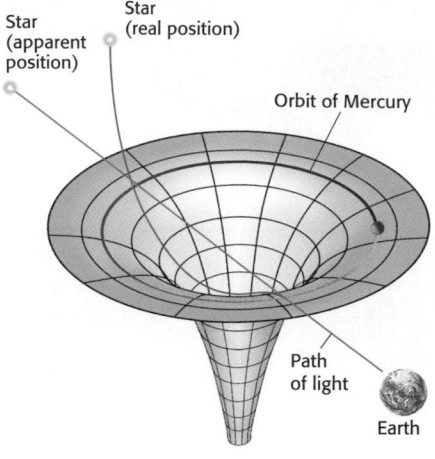

Star (apparent position)
Star (real position)
Orbit of Mercury
Path of light
Earth

Figure 30

According to Einstein's theory of relativity, the mass of the sun curves the space near the sun.

Scientists use mathematics to build better models

People easily accept Newton's theory of gravity because it corresponds to our experiences of the world. Newton wrote his theory in a mathematical form that can be applied to many circumstances on Earth and in space. In 1916, Albert Einstein expanded on Newton's theories by developing the general theory of relativity, which he also expressed in a mathematical form. Einstein's theory is hard to understand, in part because its effects are only noticeable at a very large scale.

According to Einstein's theory, mass curves space, much in the same way that your body curves a mattress when you sit on it. In 1919, observations of a total solar eclipse showed that Einstein was correct. Stars in the direction of the sun, which could be seen only during the eclipse, were in slightly different positions than expected. The mass of the sun had curved space, causing light to come from a slightly different location, as shown in **Figure 30**. Larger masses, such as galaxies, will distort space even more. In this way, a mathematical model was tested and supported by observation.

SECTION 3 REVIEW

SUMMARY

▶ The universe is all of the space, matter, and energy that exist, have existed, and will exist.

▶ The big bang theory states that the universe began 10 billion to 20 billion years ago as an explosion.

▶ The discovery of cosmic background radiation supports the big bang theory.

▶ Red shift shows that most galaxies are moving away from each other and that the universe is expanding.

▶ Astronomers use mathematical models and observations to discern the past and future of the universe.

1. Define the word *universe*, and list three things that are found in the universe.

2. Define the terms *red shift* and *blue shift*.

3. Describe the evidence that the universe is expanding.

4. Explain why the microwave background radiation is now less than 3 K even though the universe was originally very hot.

5. Compare the features that you see in the three images of the Crab Nebula in this chapter. Make a list of similarities and a list of differences.

6. Critical Thinking Why didn't the first stars to form have solar systems with Earth-like planets and satellites?

7. Critical Thinking If an object is moving away from us at a high speed and is observed in the radio region of the spectrum, what does red shift mean? Explain.

8. Critical Thinking Why is it unlikely that dark matter is composed mostly of stars?

688

Answers to Section 3 Review

1. The universe is everything physical that exists in space and time. (Examples of things in the universe will vary.)

2. Red shift is a shift in wavelength to longer wavelengths of light. Blue shift is a shift in wavelength to shorter wavelengths of light.

3. Observations of the red shifts of galaxies show that most galaxies are moving away from us and from each other.

4. As the universe expanded, it cooled. The temperature is a measure of the original energy, now spread over a much larger volume.

5. Answers may vary but could point out that there are colors shown in the radio and optical images but none present in the X-ray image shown. Also, different details are shown by all three images.

6. The first stars were made from an interstellar medium of only hydrogen and helium. There were no heavier elements (like carbon, iron, or gold) to make solid planets like Earth.

See "Continuation of Answers" at the end of the chapter.

Graphing Skills

Bar Graphs

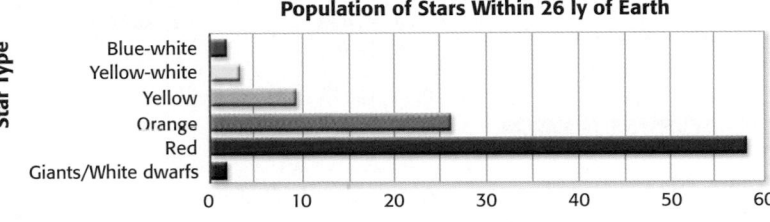

Population of Stars Within 26 ly of Earth

Star Type:
- Blue-white
- Yellow-white
- Yellow
- Orange
- Red
- Giants/White dwarfs

Number of Stars: 0, 10, 20, 30, 40, 50, 60

Stars have different colors depending on their masses and ages. The graph above indicates the colors and types of the 100 stars nearest Earth (within 26 ly). Examine the graph and answer the following questions. (See Appendix A for help interpreting a graph.)

1 What variables are shown in this graph?

2 What is the most common type of star nearest Earth? What stars are least common?

3 Most stars are main sequence stars. The more massive a main sequence star is, the greater its surface temperature and brightness. Using the graph and H-R diagram, what can you conclude about the brightness, temperature, and mass of the stars nearest Earth? Which form most easily: stars with large or small masses?

4 If you were to construct a similar graph from a list of the 100 brightest stars as seen from Earth, the number of red and orange main sequence stars would decrease and the number of giant stars would increase. What can you conclude from this information?

5 Construct a graph best suited for the information listed below. How many elliptical galaxies are among the nearest (within 26 million ly) to Earth?

Galaxy Type	Percentage among 190 nearest galaxies
Elliptical	10.0
Spiral	38.4
Irregular	51.6

6 Beyond 26 million ly, the number of irregular galaxies does not increase as much as the numbers of the other types of galaxies. What can you conclude from this information?

689

Graphing Skills

Answers

1. star types and the number of stars

2. Orange stars are most common. Blue-white, giant, and white dwarf stars are the least common.

3. Red main sequence stars are less bright, cooler, less massive, and less common than other stars. The more massive a star is (orange through blue-white), the brighter, hotter, more massive, and less common it is. Stars with small masses form more easily.

4. The percentage of stars should remain the same if the distribution is uniform. However, because red and orange stars are less bright than hotter main sequence stars or giant stars, they are not among the brightest stars, and therefore their numbers decrease. Brighter stars can be seen at greater distances, so their numbers increase.

5. A pie chart is best suited for the data given. Nineteen elliptical galaxies are within 26 million ly of Earth.

6. At great distances, irregular galaxies are not as easily seen as spiral and elliptical galaxies. This suggests that irregular galaxies are on average smaller and less bright than other types of galaxies.

Understanding Concepts

1. a
2. d
3. d
4. c
5. d
6. a
7. a
8. b
9. c
10. b
11. d
12. d
13. d
14. d
15. d
16. c
17. b
18. a

Using Vocabulary

19. cluster of galaxies, spiral galaxy, dwarf elliptical galaxy, red giant, pulsar

20. nuclear fusion combines the nuclei of smaller atoms into larger nuclei; fusion takes place in the core of a star

21. Spectral lines are dark lines that appear at specific wavelengths in the electromagnetic spectra of stars. Spectral lines can reveal the composition of stars because different elements have unique patterns of spectral lines.

Chapter Resource File

• **Chapter Tests** GENERAL
• **Test Item Listing for ExamView®
Test Generator**

Chapter Highlights

Before you begin, review the summaries of the key ideas of each section, found at the end of each section. The key vocabulary terms are listed on the first page of each section.

UNDERSTANDING CONCEPTS

1. What are the three basic types of galaxies?
   a. spiral, elliptical, and irregular
   b. closed, elliptical, and open
   c. spiral, quasar, and pulsar
   d. open, binary, and globular

2. A pattern of stars seen from Earth is a
   a. galaxy.
   b. nebula.
   c. Milky Way.
   d. constellation.

3. By studying starlight, astronomers may learn
   a. the elements that are in the star.
   b. the surface temperature of the star.
   c. the speed at which the star is moving toward or away from Earth.
   d. All of the above

4. The core of a star that remains after a supernova may be any of the following *except*
   a. a black hole.
   b. a neutron star.
   c. a red giant.
   d. a pulsar.

5. A light-year is a unit of
   a. time.
   b. mass.
   c. temperature.
   d. distance.

6. The spectral lines of galaxies that are moving away from us shift toward the _____ end of the spectrum.
   a. red
   b. yellow
   c. green
   d. blue

7. If two stars have the same diameter and are at the same distance from Earth, the brighter star is
   a. hotter.
   b. colder.
   c. faster.
   d. slower.

8. If two stars have the same temperature and are the same distance from Earth, the brighter star is
   a. faster.
   b. larger.
   c. slower.
   d. smaller.

9. A star like the sun will end its life as a
   a. pulsar.
   b. black hole.
   c. white dwarf.
   d. supernova.

10. What kind of galaxy is the Milky Way galaxy?
    a. elliptical
    b. spiral
    c. cluster
    d. irregular

11. Giant elliptical galaxies have _____ of stars.
    a. dozens
    b. thousands
    c. hundreds
    d. millions

12. A Type II supernova explodes when it begins to fuse _____ in its core.
    a. hydrogen
    b. carbon
    c. helium
    d. iron

13. Most astronomers agree that quasars are
    a. very old.
    b. very distant.
    c. very bright.
    d. All of the above

14. Which of the following are in the universe?
    a. Mars
    b. stars
    c. Milky Way
    d. All of the above

15. Which of the following is a possible age of the universe, according to the big bang theory?
    a. 4.6 million years
    b. 15 million years
    c. 4.6 billion years
    d. 15 billion years

16. If the big bang theory is correct, what percentage of the universe is helium?
    a. 10% to 25%
    b. exactly 25%
    c. more than 25%
    d. 0%

17. Dark matter is detected because it
    a. is bright.
    b. has gravity.
    c. is dark.
    d. is hot.

18. According to Einstein's theory of relativity, space is curved by a great _____ nearby.
    a. mass
    b. comet
    c. vacuum
    d. satellite

690

22. A supernova is the explosion of a supergiant star. Type II supernovae occur when fusion stops in the core of a supergiant star. Type I supernovae occur when a white dwarf star collects enough mass from a companion star to exceed 1.4 solar masses.

23. Answers may vary. A sample answer: The universe began with an explosion called the big bang. As the universe expanded outward, it rapidly cooled to the point where hydrogen and helium nuclei could form. After about a million years, atoms formed. These are the building blocks for stars that formed from interstellar matter. Galaxies began to form at

an age of about 100 million years. We can measure how the universe is expanding by using the red shift of other galaxies with respect to us.

24. The Milky Way has a bulge in the center and spiral arms in the disk. The spiral arms contain interstellar matter from which new stars form. The spiral arms have lots of young, blue stars. The bulge contains older, redder stars.

19. Arrange the following from largest to smallest: *dwarf elliptical galaxy, spiral galaxy, pulsar, red giant,* and *cluster of galaxies.*

20. Describe *nuclear fusion,* and identify in which part of a star it takes place.

21. Describe *spectral lines,* and explain how they help scientists study the composition of stars.

22. Define *supernova,* and describe the difference between a Type I supernova and a Type II supernova.

23. Write a paragraph that explains the origin of the universe as presented in this chapter. Use the following terms: *big bang theory, red shift, galaxy, interstellar matter,* and *star.*

24. Describe the arrangement of the components of the Milky Way galaxy. Use the terms *interstellar matter, stars, bulge,* and *spiral arms.*

25. Describe the various stages in the life of a star like the sun. Use the terms *white dwarf, red giant,* and *star.*

26. Explain the current theory of what a *quasar* is.

27. Write a paragraph that describes how the mass of a star determines the death of the star. Include the terms *star, white dwarf, supernova, neutron star,* and *black hole.*

WRITING SKILL

28. Describe the difference between *blue shift* and *red shift.*

29. Explain what makes up a *cluster,* and approximate how large a cluster may be.

30. Describe how *cosmic background radiation* was discovered.

31. Using Graphics The figure below shows the intensity of radiation at different wavelengths for three stars. Draw the curve for a star whose surface temperature is 20 000 K.

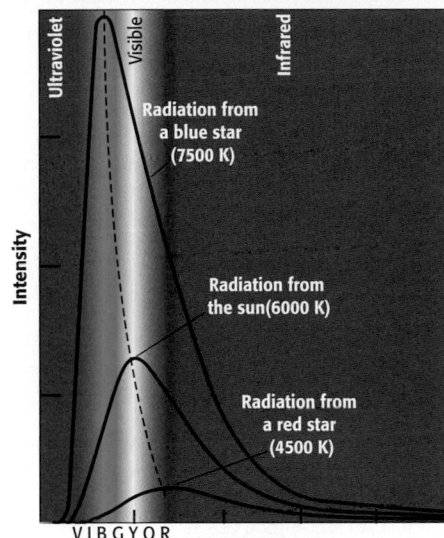

Radiation from a blue star (7500 K)

Radiation from the sun (6000 K)

Radiation from a red star (4500 K)

V I B G Y O R

Wavelength

32. Graphing Graph the following data. Put distance on the horizontal axis.

Object	Distance (thousand ly)	Velocity (km/s)
Andromeda galaxy	224	−10
Centaurus A	2116	251
M66	3680	593
M49	6746	822
Fornax A	9200	1713

691

Assignment Guide

Section	Questions
1	2–5, 7–9, 12, 20–22, 25, 27, 31, 35–37, 40–43, 48, 50, 51
2	1, 10, 11, 13, 19, 24, 26, 29, 44, 45
3	6, 14–18, 23, 28, 30, 32–34, 38, 39, 46, 47, 49

25. The sun formed from a cloud of gas and dust. It will be a main sequence star for 10 billion years, balancing the inward force of gravity with the outward pressure from fusion reactions in its core. When it runs out of hydrogen in the core, the sun will become a red giant, with a larger diameter and a cooler exterior. Eventually, the sun will lose its outer atmosphere and its core will collapse to make a white dwarf star. The white dwarf will slowly cool off and become dark.

26. Quasars may be infant galaxies. We see them as they were when they first formed. They may have large central black holes embedded in faint galaxies.

27. The initial mass of a star determines how long it will stay on the main sequence and what its eventual fate will be. Stars whose cores are less than 1.4 solar masses after they puff off their outer layers in the red giant stage become white dwarfs. More massive stars that can fuse elements up to iron will undergo supernova explosions. If there is a remnant left, it may become a neutron star, or if even more massive, a black hole.

28. Blue shift is a change in wavelength toward shorter wavelengths due to the relative motion of an object towards the observer. Red shift is a change in wavelength toward longer wavelengths of light due to the relative motion of an object away from the observer.

29. A cluster is a collection of galaxies bound by gravity. Clusters may be tens of millions of light-years in diameter.

30. Cosmic background radiation was discovered by accident, when Penzias and Wilson found interference in a new radio antenna. This radiation had already been predicted by the big bang theory, so its discovery helped to support the theory.

Building Math Skills

31. The curve should be higher than the three indicated and its peak should be more to the blue (to the left) than the peaks of the three curves shown in the figure. Also, the area under the curve should be larger than those of the other three.

Building Graphing Skills

32. Graphs should be roughly linear, with the line running from lower left to upper right.

33. Answers may vary. 12,300 plus or minus about 1,000 light years is a reasonable estimate.

Thinking Critically

34. The Andromeda galaxy is moving towards Earth, and has a blue shift.

35. Most of the stars in the Milky Way are red dwarfs. They are still on the main sequence. Other stars that were once on the main sequence are now red giants. There are very few massive blue stars in our galaxy at any one time since they are blue for only a very short time before they evolve.

36. The most massive stars have the shortest lifetimes because they use up their fuel for fusion faster than less massive stars.

37. A star would have a different brightness if it were a different diameter or a different temperature.

38. Fusion reactions in the core release radiation. Pressure from this radiation balances the inward force of gravity.

39. They will have more blue stars. Younger galaxies contain more young stars, and young stars tend to be bluer than old stars.

33. Estimate the distance between Earth and a galaxy whose velocity is 2000 km/s.

34. The Andromeda galaxy has a negative value for velocity. What does this mean physically?

THINKING CRITICALLY

35. Critical Thinking Why are most of the stars in the Milky Way galaxy red?

36. Understanding Systems Given that very massive stars fuse hydrogen into helium at a faster rate than less massive stars, explain why the most massive stars have the shortest lifetimes.

37. Critical Thinking Name two ways that two stars that are the same distance from Earth can have different brightnesses.

38. Understanding Systems What keeps a star from collapsing under its own weight?

39. Critical Thinking When looking at very distant galaxies, astronomers see the galaxies as they were when the galaxies were very young. Will these galaxies have more blue stars or fewer blue stars than nearby galaxies do? Explain your answer.

40. Applying Knowledge If Edwin Hubble had observed that the spectral lines from every galaxy were blue shifted, what might he have concluded about the universe? What could we conclude about the fate of the universe?

41. Critical Thinking Where in a galaxy would a black hole most likely be?

42. Applying Knowledge Could a black hole consume an entire galaxy? Explain your answer in a paragraph. Use concepts you learned in the chapter.

43. Interpreting Graphics The spectra shown below were taken for hydrogen, helium, and lithium in a laboratory on Earth. The spectra labeled as "Star 1" and "Star 2" were taken from starlight. What elements are found in Star 1 and Star 2.

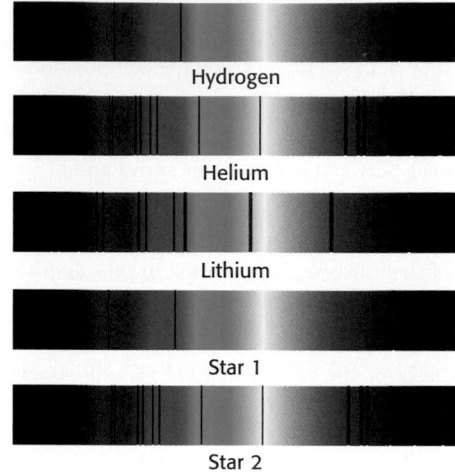

Hydrogen

Helium

Lithium

Star 1

Star 2

DEVELOPING LIFE/WORK SKILLS

44. Interpreting and Communicating Research how astronomers find distances to stars and galaxies. Make a poster or presentation that describes at least three methods.

45. Working Cooperatively Working in teams, research how ancient cultures around the world explained the Milky Way. Present your findings to the class.

46. Applying Technology Use a computer art program to illustrate different types of galaxies.

47. Communicating Effectively Write an article for your school newspaper in which you explain why developing theories about the future of the universe is important.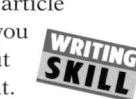

40. He might have concluded that the universe is shrinking.

41. in the center

42. It is unlikely since the black hole can only consume nearby objects. Eventually, the center would run out of things for the black hole to consume.

43. Star 1 contains only hydrogen. Star 2 contains hydrogen and helium.

Developing Life/Work Skills

44. Answers may vary.

45. Answers may vary.

46. Answers may vary. Illustrations should resemble models of galaxies shown in the chapter.

47. Answers may vary. Students may state that we should develop theories about the future of the universe, in order to predict and prepare for events that may have a direct impact on life on Earth.

Integrating Concepts

48. A big crunch, which would be unimaginably hot, would be like the world ending in fire. A universe that expanded forever would eventually cool to near absolute zero. This would be like the world ending in ice.

49. Answers may vary.

48. Connection to Literature Robert Frost wrote a poem entitled "Fire and Ice" that begins "Some say the world will end in fire, Some say in ice." Explain how these lines relate to possible fates of the universe.

49. Connection to Literature In *Following the Equator*, Mark Twain wrote, "Constellations have always been troublesome things to name. If you give one of them a fanciful name, it will always refuse to live up to it; it will always persist in not resembling the thing it has been named for." Choose a constellation that you have tried to observe or have seen on a star map. Draw the stars and connect them in a new way. Give the constellation a new name, and explain how you arrived at the name.

50. Connection to Chemistry Based on the descriptions within this chapter, how did hydrogen and helium first form? What are the possible sources of the elements found on the periodic table from lithium to carbon? What are the possible sources of elements from carbon to iron? How could atoms heavier than iron form?

51. Connection to Science Fiction Many authors, such as Poul Anderson, Isaac Asimov, David Brin, Larry Niven, and Fred Pohl, have incorporated black holes, neutron stars, or supernovae into their stories. Read one of their stories and compare the author's use of scientific concepts with information that you learned in this chapter.

internet connect

www.scilinks.org
Topic: **Formation of the Elements**
SciLinks code: HK4056

SCiLINKS Maintained by the National Science Teachers Association

52. Concept Mapping Copy the unfinished concept map below onto a sheet of paper. Complete the map by writing the correct word or phrase in the lettered boxes.

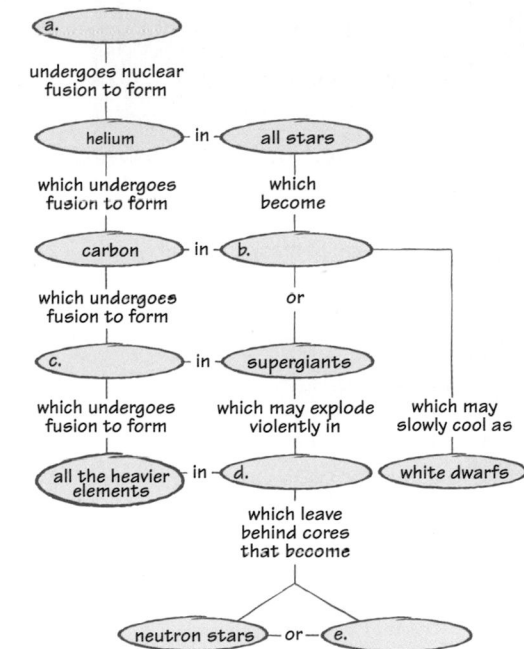

Art Credits: Fig. 1B, Stephen Durke/Washington Artists; Fig. 2, Tony Randazzo/American Artist's Rep., Inc.; Fig. 8, Uhl Studios, Inc.; Fig. 11, Stephen Durke/Washington Artists; Fig. 16, Uhl Studios, Inc.; Fig. 20 (sun and solar system), Sidney Jablonski (meter bar), Uhl Studios, Inc.; Fig. 24, Uhl Studios, Inc.; Fig. 30, Craig Attebery/Jeff Lavaty Artist Agent.

Photo Credits: Chapter Opener photo of Very Large Array radio telescopes by Roger Ressmeyer/COR-BIS; radio image of Crab Nebula by R.A. Perely/NRAO; Fig. 1A, Roger Ressmeyer/CORBIS; Fig. 3, Stapelton Collection/CORBIS; Fig. 4, Victoria Smith/HRW; Fig. 7, Jeff Hester and Paul Scowen (Arizona St. University)/NASA; Fig. 9, Anglo-Australian Observatory; "Connection to Social Studies," Tha British Library Picture Library; Fig. 10, Malin/Pasachoff/Caltech/Anglo-Australian Observatory; Fig. 12, John Gleason/Celestial Images; Fig. 13, Dr. Victor Anderson (University of Alabama, KPNO), courtesy W. Keel; Fig. 14A, The Observatories of the Carnegie Institution of Washington; Fig. 14B, NASA; Fig. 15, Jerry Schad/Photo Researchers, Inc.; Fig. 17A, Gemini Observatory, GMOS team; Fig. 17B, Anglo-Australian Observatory; Fig. 17C, Dennis Di Cicco/Peter Arnold, Inc.; Fig. 18, NASA/CXC/SAO/H. Marshall et. al.; "Real World Applications," Arecibo Observatory, Courtesy Dr. Jose Alonso; Fig. 19, R. Williams and the HDF Team (ST Scl)/NASA; Fig. 20(soccer player), Doug Pensinger/Allsport/Getty Images; (field), Bob Long/AP/Wide World Photos; (earth), ESA/PLI/Corbis Stock Market; (galaxies), Dr. Victor Anderson (University of Alabama, KPNO), courtesy W. Keel; Fig. 21, Digital image (c) 1996 CORBIS; Original image courtesy of NASA/CORBIS; NASA/SPL/Photo Researchers, Inc.; Fig. 25, Sam Dudgeon/HRW/Courtesy Dr. Ann Molineux, Texas Memorial Museum, University of Texas at Austin; Fig. 26, Roger Ressmeyer/CORBIS; Fig. 28A, TRW; Fig. 28B, NASA/CXC/SAO; Fig. 29, NASA; "Skills Practice Lab," Dr. Victor Anderson (University of Alabama, KPNO), courtesy W. Keel; "Career Link," (Dr. Cohn portraits), Peg Skorpinski/HRW; (galaxy "chalkboard"), Digital Image copyright (c)2004 PhotoDisc; (nebula), D. Walter(South Carolina State University) and P. Scowen (Arizona State University)/NASA.

693

50. The nuclei of hydrogen and helium were formed about 100 seconds after the big bang. Atoms of hydrogen and helium formed about one million years later. Lithium and carbon are formed in red giant stars. Elements from carbon to iron form in supergiant stars. Elements heavier than iron form in supernovae.

51. Answers will vary.

52. a. hydrogen
   b. red giants
   c. iron
   d. supernovae
   e. black holes

 Transparencies

TT Concept Mapping

INVESTIGATING DIFFERENT TYPES OF GALAXIES

Teacher's Notes

Time Required 1 lab period

Ratings 1 2 3 4
EASY HARD

TEACHER PREPARATION	1
STUDENT SETUP	1
CONCEPT LEVEL	2
CLEANUP	1

Skills Acquired

- Classifying
- Collecting data
- Identifying/recognizing patterns
- Measuring
- Organizing and analyzing data

The Scientific Method

In this lab, students will:
- Make observations
- Analyze the results
- Draw conclusions
- Communicate results

Materials

A large black-and-white negative of the photo is provided in the datasheets.

Safety Cautions

Have students review safety guidelines before working in the lab.

Skills Practice Lab

Introduction

Galaxies are of many sizes and types. How can you tell the differences in type among galaxies by simple measurements of their images?

Objectives

▶ **Recognize** the orientation of galaxies.

▶ **Classify** galaxies according to type.

▶ **Measure** the diameters of galaxies.

▶ USING SCIENTIFIC METHODS **Analyze the results** by calculating the ratio of spiral galaxies to elliptical galaxies within a cluster.

Materials

magnifying glass
metric ruler, clear plastic
photograph of a galaxy cluster

Investigating Different Types of Galaxies

▶ Procedure

Preparing for Your Experiment

1. Examine the photographs of galaxies in this chapter. Make sketches of what each galaxy might look like if you rotated it from a "top-down" view to a "side" view.

2. Examine the large photograph of the Hercules Cluster of galaxies in *Figure 31* on the next page. Your teacher may also provide you with a larger version of the photograph. The photograph contains both stars that are between us and the cluster and galaxies that are within the cluster. Write your criteria for distinguishing between a nearby star and a galaxy.

Classifying Galaxies

3. Set up a classification system that divides different galaxies into categories. Ignore the individual identity of stars. You should have at least three different types of galaxies. Discuss in your group what types you will use, and what characteristics define each type.

4. Find at least one example of each type in the photograph and write down the coordinates of each example. Compare your examples with others in your group.

5. Classify each galaxy you see in the photograph, and note the coordinates of each galaxy. If necessary, use a magnifying glass to view the picture more clearly. If you can identify something as a galaxy but are unclear of its type, classify it as "uncertain" galaxy.

Measuring Galaxies

6. Locate the largest and smallest galaxy for each of your classification types.

7. Measure the sizes of these galaxies in millimeters. This process may be easier to do if you use a magnifying glass and a clear ruler.

694

Answers to Analysis

1. Answers may vary. More than 50 galaxies can be resolved in the photo, although students may not recognize all of them as galaxies.

2. Answers may vary depending on classification types and the galaxies that are actually observed. The numbers of spiral galaxies and elliptical galaxies are roughly equal. There are relatively few irregular galaxies. Cautious students, however, will have more "uncertain" galaxies, as it is often hard to distinguish between types.

3. Answers may vary as described above.

Answers to Conclusions

4. Answers may vary. Elliptical galaxies are probably the most common, but spiral galaxies or uncertain galaxies are also acceptable answers.

5. Answers may vary. There are several very large elliptical galaxies, but also many dwarf ellipticals. For that reason, students may find that spirals are larger on average.

6. Yes. It is very hard to distinguish the smallest galaxies from faint stars. There may also be galaxies in the cluster that are too faint or too small to be seen at all in the photograph.

7. (Dwarf) elliptical galaxies are easiest to confuse with stars in the foreground.

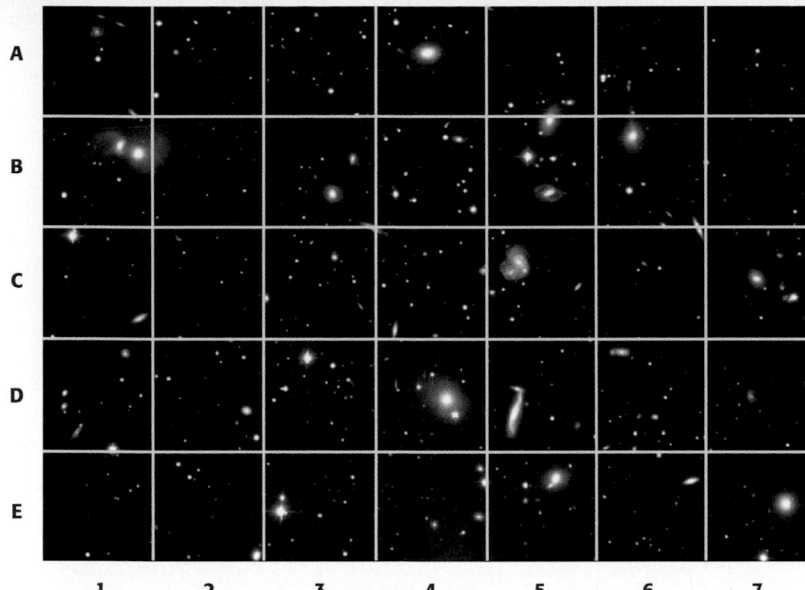

Figure 31
The Hercules Cluster

► Analysis

1. Count how many galaxies you have of each type. Add up the totals for all types to get the total number of galaxies observed.

2. Calculate the ratio of each type to the other types by dividing the total of one type by the total number of galaxies observed. For example if you have 33 of type A and 100 total galaxies, then the ratio would be 33 divided by 100, or about one-third.

3. Make a table showing the types of galaxies, their total numbers, and their calculated ratios.

► Conclusions

4. What type of galaxy is most common in this cluster?

5. Which of your classification types typically has larger galaxies?

6. Do you think there may be smaller galaxies that you missed seeing? Explain.

7. Which type of galaxy is easiest to confuse with foreground stars?

8. All the galaxies in a cluster are about the same distance from Earth. Therefore, differences in cluster size are due to differences in the sizes of the galaxies. If the largest spiral you measured was the same size as the Milky Way galaxy, estimate the total diameter of this cluster of galaxies.

695

8. Answers may vary. Assuming the largest spiral measured was 15 mm and the total diameter is measured as 200 mm, and using 100 000 ly as the diameter of the Milky Way galaxy, the diameter of the cluster is found as follows: $D = (200 \text{ mm}/15 \text{ mm})$ *100 000 ly = 1×10^6 ly.

Tips and Tricks

Students may choose to classify galaxies into the 3 types described in the chapter, or they may invent their own classification schemes. Encourage students not to get to caught up in being "right" about what type of galaxies they are seeing. The classification types of galaxies are to some extent arbitrary, and even astronomers debate how galaxies should be classified. Furthermore, it can be hard to distinguish a galaxy's type from a single image. Encourage students to consider what other types of images they might need in order to be more sure of their classifications (for example, higher resolution images, images at different wavelengths).

Bright foreground stars can be distinguished from galaxies because the stars have cross-shaped patterns of lines coming out from them. Faint foreground stars may be hard to distinguish from small galaxies.

Continuation of Answers

Continuation of Answers from p. 688

Section 3 Review

7. In the radio region, going toward "red" would mean going towards a shorter wavelength. However, the term "red shift" is used to denote always a shift to longer wavelength, even when it doesn't make physical sense. So, when someone says "red shift" in the radio region, they mean a shift to longer radio wavelengths.

8. If dark matter were composed of stars, then it would not be "dark." Dark matter is matter that, for the most part, cannot be observed directly because it does not emit light.

SKILL BUILDER — GENERAL

Vocabulary The word *cosmology* comes from two different root words. The Greek word *cosmos* means universe, and *logos* means word. In the sciences, words ending with –logy refer to the study of a particular area. Therefore cosmology literally means the study of the universe. **LS Verbal**

CareerLink

Joanne Cohn, Cosmologist

Cosmologists study the origin, evolution, and future of the universe. They use observations and scientific theory to try to answer some of science's most fundamental questions: How old is the universe? What happened during the Big Bang? What are Black Holes? Is the universe expanding? And if so, how fast? Read on to hear from Joanne Cohn, a cosmologist at the University of California in Berkeley, California.

Joanne Cohn is a cosmologist at the University of California in Berkeley, California.

"I'm still amazed that the laws of physics that we use on Earth can help us measure and describe the entire known universe!"

 What does a cosmologist do?

A cosmologist studies the universe as a physical system. To study a system, you want to know what it is made of, and that's one of the important questions in the field today—what is the universe filled with? We want to know what ingredients make up our universe so we can understand various scenarios, like how matter assembles into large collapsed objects such as galaxies, how light travels through what is out there to us, and how light is given off by stars. It is very much like a detective story, where you try to figure out the players (types of matter in the universe) and the story (how they interacted in the past to get where they are today).

 Why do you think your work is important?

My work is important to people who want to know what the universe is made of and how it is evolving. The universe contains us and all physical phenomena we know about, so it is of interest to people who want to know what is out there and what is happening and what has happened.

 What question about the universe are you most interested in answering?

I'm interested in knowing how gravity from a galaxy changes the way light rays that travel near it behave. Light that travels near a galaxy is deflected and 'lensed' by the galaxy, and it's possible to use this lensing to learn about the matter in the galaxy.

What kinds of tools and models do you use in your work?

I use theoretical descriptions of how gravity and other physical forces work, and then use computers to make calculations and simulations. I also use simple models of galaxy shapes as a starting point for comparisons with real observations from galaxies.

696

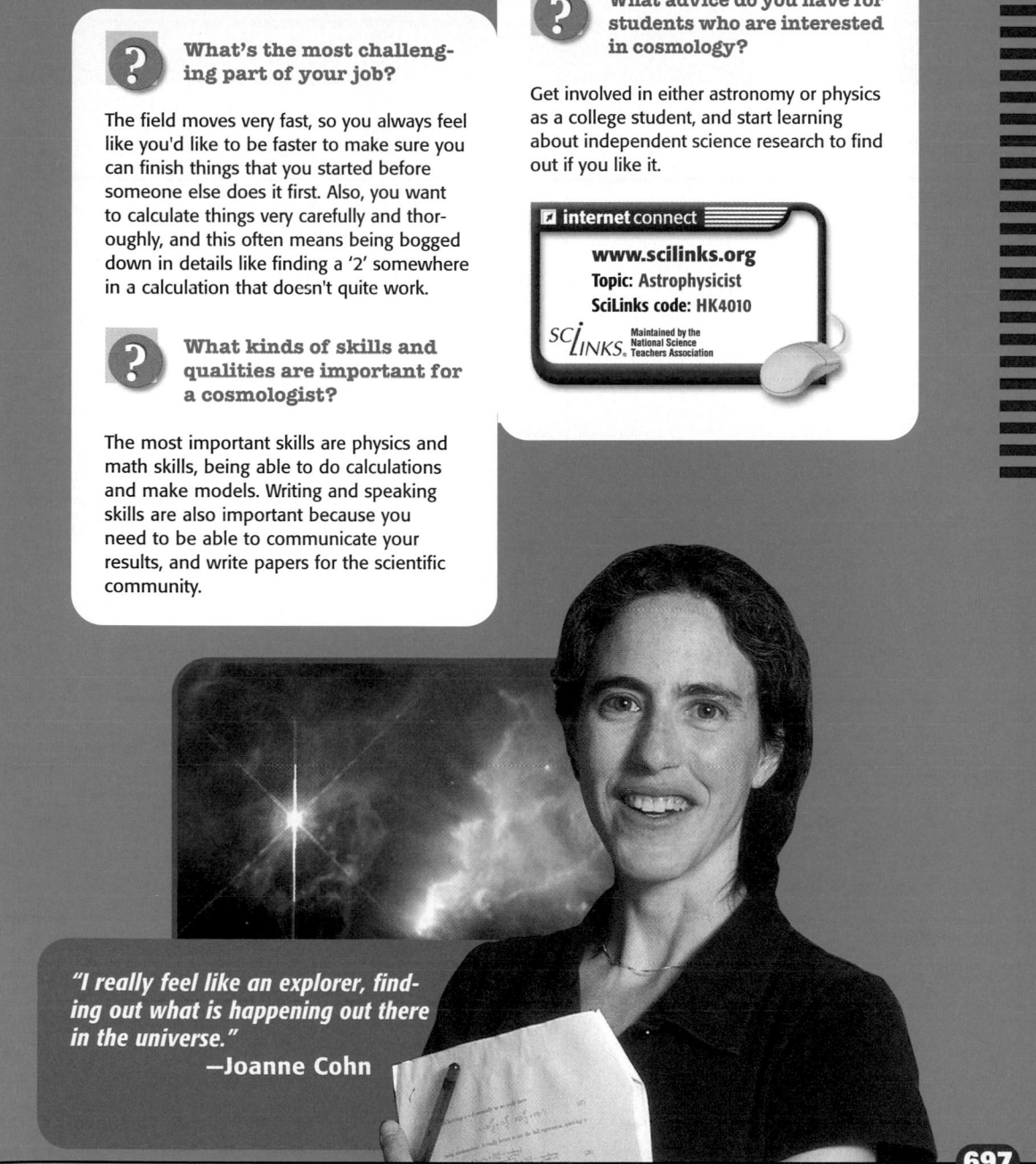

What's the most challenging part of your job?

The field moves very fast, so you always feel like you'd like to be faster to make sure you can finish things that you started before someone else does it first. Also, you want to calculate things very carefully and thoroughly, and this often means being bogged down in details like finding a '2' somewhere in a calculation that doesn't quite work.

What kinds of skills and qualities are important for a cosmologist?

The most important skills are physics and math skills, being able to do calculations and make models. Writing and speaking skills are also important because you need to be able to communicate your results, and write papers for the scientific community.

What advice do you have for students who are interested in cosmology?

Get involved in either astronomy or physics as a college student, and start learning about independent science research to find out if you like it.

☑ internet connect

www.scilinks.org
Topic: Astrophysicist
SciLinks code: HK4010

SCI**LINKS**® Maintained by the National Science Teachers Association

"I really feel like an explorer, finding out what is happening out there in the universe."
—Joanne Cohn

697

Planet Earth
Chapter Planning Guide

PACING	CLASSROOM RESOURCES	LABS, ACTIVITIES, AND DEMONSTRATIONS
BLOCK 1 · 45 min pp. 698–699 **Chapter Opener**		SE **Activity 1**, p. 699 SE **Activity 2**, p. 699
BLOCKS 2 & 3 · 90 min pp. 700–708 **Section 1** Earth's Interior and Plate Tectonics	CRF **Lesson Plan*** TT **Bellringer*** TT **Tectonic Plates*** TT **Divergent and Convergent Boundaries***	TE **Opening Activity**, p. 700 GENERAL SE **Quick Activity** Convection and Plate Tectonics, p. 704 GENERAL CRF **Datasheets for In-Text Labs** Convection and Plate Tectonics* GENERAL SE **Quick Lab** Can you model tectonic plate boundaries with clay?, p. 708 GENERAL CRF **Datasheets for In-Text Labs** Can you model tectonic plate boundaries with clay?* GENERAL CRF **CBL™ Probeware Lab** Relating Convection to the Movement of Tectonic Plates* ♦ ADVANCED
BLOCKS 4 & 5 · 90 min pp. 709–717 **Section 2** Earthquakes and Volcanoes	CRF **Lesson Plan*** TT **Bellringer** TM **Epicenter and Focus*** TM **Locating an Epicenter*** TT **Volcanoes***	TE **Opening Demonstration**, p. 709 TE **Demonstration** Earthquake!, p. 710 TE **Demonstration** Eruption Simulation, p. 714
BLOCKS 6 & 7 · 90 min pp. 718–724 **Section 3** Minerals and Rocks	CRF **Lesson Plan*** TT **Bellringer*** TT **Rock Cycle***	TE **Opening Activity**, p. 718 GENERAL TE **Demonstration** Temperature Affects Crystal Growth, p. 720 CRF **Observation Lab** Classifying Rocks* ♦ BASIC
BLOCKS 8 & 9 · 90 min pp. 725–730 **Section 4** Weathering and Erosion	CRF **Lesson Plan*** TT **Bellringer*** TT **Concept Mapping***	TE **Opening Demonstration**, p. 725 TE **Demonstration** Erosion, p. 728 TE **Group Activity** Sandstone Arches, p. 729 GENERAL SE **Skills Practice Lab** Analyzing Seismic Waves, pp. 736–737 ♦ GENERAL CRF **Datasheets for SE Skills Practice Lab** Analyzing Seismic Waves* GENERAL

BLOCKS 10 & 11 · 90 min

Chapter Review and Assessment Resources

SE **Chapter Review**, pp. 732–735
CRF **Chapter Tests*** GENERAL
OSP **Test Generator**
CRF **Standardized Test Practice with Guided Reading Development***
CRF **Test Item Listing for ExamView® Test Generator***
OSP **Scoring Rubrics and Classroom Management Checklists**

Online Resources

Visit the HRW Web site for a variety of free resources related to the text. Just type in the keyword **HK4 PLA**.

Holt Science Spectrum: Physical Science: Online Edition

Students can access interactive problem solving help and active visual concept development with the *Holt Science Spectrum: Physical Science* Online Edition available at **www.hrw.com**.

student CNN News

cnnstudentnews.com

Find the latest news, lesson plans, and activities related to important scientific events.

KEY

TE	Teacher Edition
SE	Student Edition
OSP	One-Stop Planner

CRF	Chapter Resource File
TT	Teaching Transparency
TM	Transparency Master

*	Also on One-Stop Planner
◆	Requires Advance Prep

PROBLEM SOLVING AND PRACTICE	SECTION REVIEW AND ASSESSMENT	STANDARDS CORRELATION
	CRF Pretest* `GENERAL`	
CRF Science Skills Compiling and Weighing Evidence* `GENERAL` **TE Inclusion Strategies,** p. 701 **CRF Cross-Disciplinary Worksheet** Integrating Physics—High Up in the Himalayas* `GENERAL` **CRF Cross-Disciplinary Worksheet** Integrating Biology—Mountaineering: How Our Bodies Acclimatize* `GENERAL`	**TE Quiz,** p. 708 `BASIC` **SE Section 1 Review,** p. 708 **CRF Concept Review*** `GENERAL` **CRF Quiz*** `BASIC`	ES 1a, 1b LS 2a, 2b UCP 1, 2, 4, 5 SAI 1, 2 HNS 2
CRF Cross-Disciplinary Worksheet Integrating Environmental Science—The Eruption of Mount Pinatubo* `GENERAL` **CRF Cross-Disciplinary Worksheet** Connection to Social Studies—Plinian Eruptions* `ADVANCED`	**TE Quiz,** p. 715 `BASIC` **SE Section 2 Review,** p. 717 **CRF Concept Review*** `GENERAL` **CRF Quiz*** `BASIC`	PS 6a ES 3c UCP 1, 2, 3, 4 SAI 2 ST 1, 2 HNS 1, 2 SPSP 3, 5
CRF Cross-Disciplinary Worksheet Connection to Social Studies—Human Tools: From Stone to Iron* `ADVANCED` **TE Inclusion Strategies,** p. 721 **CRF Cross-Disciplinary Worksheet** Integrating Chemistry—From Granite to Paper* `ADVANCED` **CRF Cross-Disciplinary Worksheet** Connection to Language Arts—Names of Rocks* `ADVANCED`	**TE Quiz,** p. 724 `BASIC` **SE Section 3 Review,** p. 724 **CRF Concept Review*** `GENERAL` **CRF Quiz*** `BASIC`	PS 1d ES 2b, 3b UCP 1, 4 SAI 2 HNS 2 SPSP 5
CRF Cross-Disciplinary Worksheet Integrating Biology—Living Sources of Weathering* `GENERAL` **SE Study Skills** Concept Mapping, p. 731 **SE CareerLink** Paleontologist, pp. 738–739	**TE Quiz,** p. 730 `BASIC` **SE Section 4 Review,** p. 730 **CRF Concept Review*** `GENERAL` **CRF Quiz*** `BASIC`	UCP 1, 4 HNS 1, 2

SCiLINKS
www.scilinks.org

Topic: Volcanoes
*Sci*Links code: HK4148

Topic: Earth's Geologic Layers
*Sci*Links code: HK4035

Topic: Earthquakes
*Sci*Links code: HK4038

Topic: Earthquake Measurement
*Sci*Links code: HK4037

Topic: Rock Types
*Sci*Links code: HK4122

Topic: Weathering
*Sci*Links code: HK4152

Topic: Erosion
*Sci*Links code: HK4050

Topic: Sonar
*Sci*Links code: HK4131

Topic: Paleontology
*Sci*Links code: HK4157

Technology Resources

One-Stop Planner
All of your printable resources and the Test Generator are on this convenient CD-ROM.

CNN Science in the NEWS
each video segment is accompanied by a Critical Thinking Worksheet *

Segment 4
Seattle Quake

Segment 5
Earthquake Seekers

Segment 6
LA Quake House

Segment 7
Earth's Core

Segment 16
Beach Erosion Tools

Overview

In this chapter, students learn about Earth's interior and study the theory of plate tectonics. Next they learn about the causes and classification of earthquakes and volcanoes. This chapter then covers types and properties of rocks and discusses the rock cycle. Finally, this chapter discusses physical weathering, chemical weathering, and erosion.

Assessing Prior Knowledge

Be sure students understand the following concepts:

- chemical reactions
- acids
- pH
- radiation
- radioactive dating
- waves
- magnetic fields

MISCONCEPTION /// ALERT \\\

Science education research has identified the following misconceptions about Earth.

- Students believe that Earth has always been the same and looked the same.
- Students think that the changes in Earth that have occurred were sudden and comprehensive.
- Students do not realize that the Earth is older than 10^9 years old.

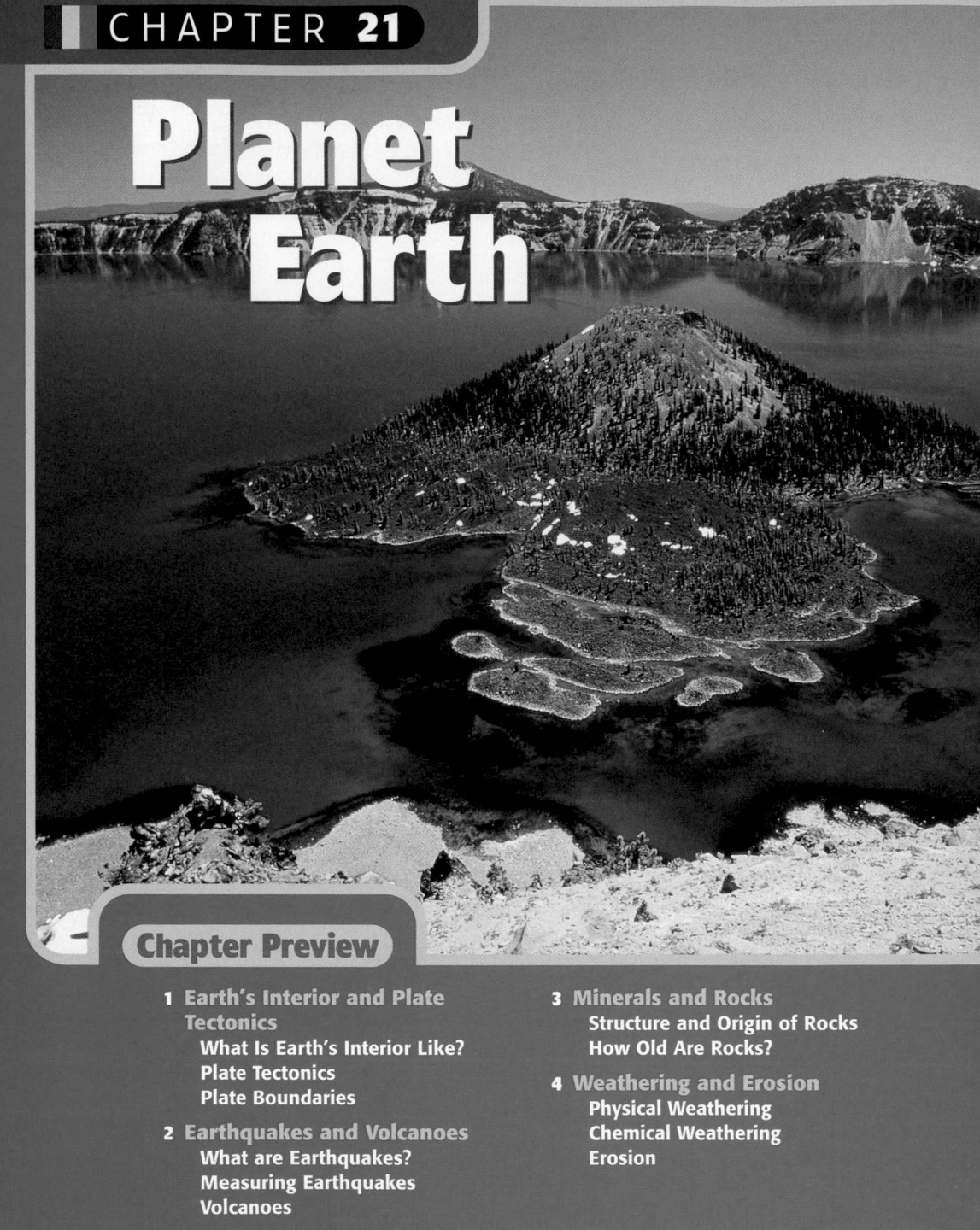

Planet Earth

Chapter Preview

698

Standards Correlations

National Science Education Standards

The following descriptions summarize the National Science Educational Standards that specifically relate to this chapter. For the full text of the standards, see the National Science Education Standards at the front of this book.

Section 1 Earth's Interior and Plate Tectonics
Earth Science ES 1a, 1b
Life Science LS 2a, 2b
Unifying Concepts and Processes UCP 1, 2, 4, 5
Science as Inquiry SAI 1, 2
History and Nature of Science HNS 2

Section 2 Earthquakes and Volcanoes
Physical Science PS 6a
Earth Science ES 3c
Unifying Concepts and Processes UCP 1, 2, 3, 4
Science as Inquiry SAI 2
Science and Technology ST 1, 2
History and Nature of Science HNS 1, 2
Science in Personal and Social Perspectives SPSP 3, 5

Section 3 Minerals and Rocks
Physical Science PS 1d
Earth Science ES 2b, 3b

Focus ACTIVITY

Background Crater Lake is the deepest lake in the United States, measuring 589 m (1932 ft) at its deepest point. The lake is inside a volcano called Mount Mazama. As Mount Mazama erupted around 6800 years ago, the molten rock and volcanic ash that helped to support the cone of the volcano were ejected. The top then collapsed, creating a big hole. As the hole filled with rainwater and melted snow, Crater Lake was formed. A secondary eruption produced a small volcanic cone, which rose above the water's surface and became Wizard Island, the small island seen in the photo at left.

Activity 1 Imagine you are an early explorer who has just discovered Crater Lake. Examine the photos at left, and describe what you see, explaining how the lake may have formed. When you are finished, write down possible weaknesses for your explanation. Share your results with your class.

Activity 2 Collect a handful of rocks of different sizes. Examine them using a magnifying glass, and make notes about each rock's shape and surface texture. Place the rocks in a plastic container with a tight-fitting lid, add enough water to cover the rocks, and close the container. Shake the container 100 times, and drain the water into a glass jar. Examine the rocks and the water carefully, and report any changes in either. If you have time, repeat shaking the container another 100 times, and write down your observations. What forces does this activity mimic?

📲 **internet** connect

www.scilinks.org
Topic: Volcanoes SciLinks code: HK4148

SC*LINKS*. Maintained by the National Science Teachers Association

Crater Lake, in Oregon, sits within the top of a collapsed volcano.

Pre-Reading Questions
1. Think about the area where you live, and try to describe what it looked like one year ago. What part of the landscape has changed in a year? Brainstorm on what changes may take place over 100 years.
2. Have the continents always looked exactly as they do today? If not, what happened?

699

Unifying Concepts and Processes UCP 1, 4
Science as Inquiry SAI 2
History and Nature of Science HNS 2
Science in Personal and Social Perspectives SPSP 5

Section 4 Weathering and Erosion
Unifying Concepts and Processes UCP 1, 4
History and Nature of Science HNS 1, 2

Focus ACTIVITY

Background Because of the constant movement of Earth, landforms such as mountains, canyons, and, in this instance, a volcano, have been transformed in different ways over time. Make sure that students understand that Earth has been and will continue to be in motion and that those motions can cause dramatic changes in Earth's surface.

Activity 1 Responses may vary, but explanations about how the lake formed may include the following: (1) the lake is a crater formed by the impact of a meteor or (2) Mount Mazuma was pushed up by colliding continental plates, and the top of the mountain became a lake by collecting rainwater. **LS** Logical

Activity 2 This activity models physical weathering. Student observations may include the following: the amount of residue increases, the particle size decreases, the rocks show rounding and breaking, and the water becomes dirtier as small rock particles accumulate and are suspended in it. **LS** Visual

Answers to Pre-Reading Questions
1. Answers may vary. Encourage students to anticipate changes to the environment and to consider what effect those changes have.
2. The continents have changed over time. Have students talk about the changes that have occurred as well as the mechanisms for change.

Chapter Resource File

- Pretest **GENERAL**
- Teaching Transparency Preview
- Answer Keys

Focus

Overview

Before beginning this section, review with your students the Objectives listed in the Student Edition. This section begins with a discussion of Earth's geologic layers. The theory of plate tectonics including supporting evidence is covered in detail. The section concludes with a discussion of the types of plate boundaries and the structures that form at each type.

Bellringer

Use the Bellringer transparency to prepare students for this section.

Motivate

Opening Activity — GENERAL

Write the words *crust, mantle,* and *core* on the board. Have students work in pairs to hypothesize what the words refer to. Students should then draw a cross section of Earth on the board, and label the diagram based on their hypothesis. After students read the section, have them make corrections to their diagrams if necessary.
LS Interpersonal

Chapter Resource File

- **Lesson Plan**
- **Science Skills** Compiling and Weighing Evidence GENERAL

Transparencies

TT Bellringer

Earth's Interior and Plate Tectonics

▶ **KEY TERMS**
crust
mantle
core
lithosphere
plate tectonics
magma
subduction
fault

OBJECTIVES

▶ **Identify** Earth's different geologic layers.

▶ **Explain** how the presence of magnetic bands on the ocean floor supports the theory of plate tectonics.

▶ **Describe** the movement of Earth's lithosphere using the theory of plate tectonics.

▶ **Identify** the three types of plate boundaries and the principal structures that form at each of these boundaries.

▶ **crust** the thin and solid outermost layer of Earth above the mantle

▶ **mantle** the layer of rock between Earth's crust and core

You know from experience that Earth's surface is solid. You walk on it every day. You may have even dug into it and found that it is often more solid once you dig and reach rock. However, Earth is not solid all the way to the center.

What Is Earth's Interior Like?

Figure 1 shows Earth's major compositional layers. We live on the topmost layer of Earth—the **crust.** Because the crust is relatively cool, it is made up of hard, solid rock. The crust beneath the ocean is called oceanic crust and has an average thickness of 5–8 km (3.1–4.9 mi). Continental crust is less dense and thicker, with an average thickness of about 20–40 km (12–25 mi). The continental crust is deepest beneath high mountains, where it commonly reaches depths of 70 km or more.

Beneath the crust lies the **mantle,** a layer of rock that is denser than the crust. Almost 2900 km (1800 mi) thick, the mantle makes up about 80% of Earth's volume. Because humans have never drilled all the way to the mantle, we do not know for sure what it is like. However, geologic events, such as earthquakes and volcanoes, provide evidence of the mantle's consistency.

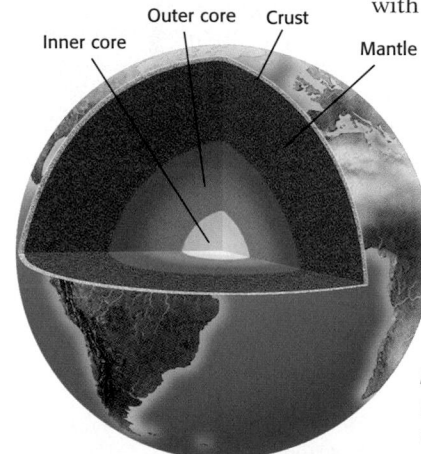
Inner core   Outer core   Crust
                            Mantle

Figure 1

Earth is composed of an inner core, an outer core, a mantle, and a crust. Though it is difficult to see, the oceanic crust is thinner than the continental crust.

700

LANGUAGE ARTS
CONNECTION — ADVANCED

Have students read *Journey to the Center of the Earth* by Jules Verne. After completing the study of this chapter, ask students to comment on connections between concepts in the book and some of the concepts they learned in the chapter. **LS Visual**

MATHEMATICS
CONNECTION — GENERAL

Tell students to assume that the average thickness of the oceanic crust is about 5.5 km and that of the continental crust is about 35 km. Have students calculate the ratio of the thickness of the mantle to that of the oceanic crust, and the ratio of the thickness of the mantle to that of the continental crust. (The mantle is about 530 times the thickness of the oceanic crust, 2900 km/5.5 km = 530; the mantle is about 83 times the thickness of the continental crust, 2900 km/35 km = 83) **LS Logical**

For the most part, the mantle is solid. The outermost part is also rigid, like the crust. Deeper than a few hundred kilometers, however, it is extremely hot, and said to be "plastic"—soft and easily deformed, like a piece of gum.

The center of Earth, the **core,** is believed to be composed mainly of iron and nickel. It has two layers. The *inner core,* which is solid metal, is surrounded by the liquid metal *outer core.*

Earth's interior gets warmer with depth

If you have ever been in a cave, you may have noticed that the temperature in the cave was cool. That's because the air and rocks beneath Earth's surface are shielded from the warming effects of the sun. However, if you were to travel far beneath the surface, such as into a deep mine, you would find that the temperature becomes uncomfortably hot. South African gold mines, for instance, reach depths of up to 3 km (2 mi), and their temperatures approach 50°C (120°F). The high temperatures in these mines are caused not by the sun but by energy that comes from Earth's interior.

Geologists believe the mantle is much hotter than the crust, as shown in *Figure 2.* These high temperatures cause the rocks in the mantle to behave plastically. This is the reason for the inner mantle's deformable, gumlike consistency.

The core is hotter still. On Earth's surface, the metals contained in the core would boil at the temperatures shown in *Figure 2.* Iron boils at 2750°C (4982°F), and nickel boils at 2732°C (4950°F). But in the outer core, these metals remain liquid because the pressure due to the weight of the mantle and crust is so great that the substances in the outer core are prevented from changing to their gaseous form. Similarly, pressure in the inner core is so great that the atoms are forced together as a solid despite the intense heat.

Radioactive elements contribute to Earth's high internal temperature

Earth's interior contains radioactive isotopes. These radioactive isotopes (mainly those of uranium, thorium, and potassium) are quite rare. Their nuclei break up, releasing energy as they become more stable. Because Earth is so large, it contains enough atoms of these elements to produce a huge quantity of energy. This energy is one of the major factors contributing to Earth's high internal temperature.

www.scilinks.org
Topic: Earth's Geologic layers
SciLinks code: HK4035

▶ **core** the center part of Earth below the mantle

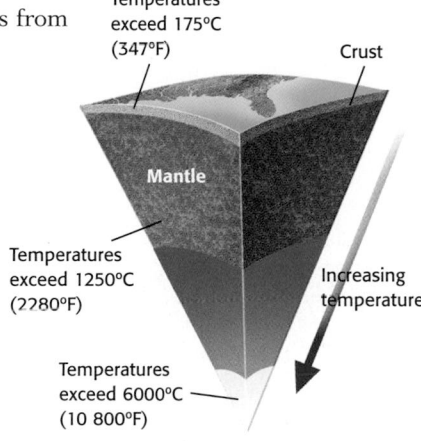

Temperatures exceed 175°C (347°F)

Crust

Mantle

Temperatures exceed 1250°C (2280°F)

Increasing temperature

Temperatures exceed 6000°C (10 800°F)

Figure 2

Temperatures in Earth's interior increase with depth. Temperatures near the center of the core can be as hot as the surface of the sun.

701

REAL-LIFE
CONNECTION ⬤ **ADVANCED**

Driving Through Earth Have students calculate the distance from the surface of the Earth to the core in kilometers. Then, have them convert that figure into miles. Ask students how long driving that distance at a rate of 89 km/h (55 mi/h) would take. (Average distance from the surface to the mantle is 30 km (18 mi). The distance from the mantle to the core is 2900 km (1800 mi). Therefore the average distance from the surface to the core is 2930 km (1818 mi). At a rate of 89 km/h, 2930 km would take approximately 33 h to traverse.) **LS** Logical

Reading Skills Before students read the section on plate tectonics, ask them if the positions of the continents have ever been different from their current positions. Write the words *Pangaea* and *continental drift* on the board, and ask students to guess the meanings of these words. Have students discuss opinions, and make a list of hypotheses. After students have read the section, discuss the list again. **LS Verbal**

Vocabulary Tell students that the term *Pangaea* comes from two Greek roots: *pan*, which means "all," and *gaea*, which means "goddess of the Earth" or simply "Earth." Thus, *Pangaea* means "all Earth." Have students find other words that contain the root *pan*, such as *panacea, panorama, pandemic,* and *panoply*. **LS Verbal**

Teaching Tip — GENERAL

Continental Drift Have students find South Africa and Argentina on a map, and ask students to identify similarities or differences between the two countries. Let the students know that the two countries have similar mineral deposits, particularly diamond deposits. Have students hypothesize how the two countries, which are separated by the Atlantic Ocean, have similar mineral deposits. **LS Visual**

Figure 3

This map shows Pangaea as Alfred Wegener envisioned it.

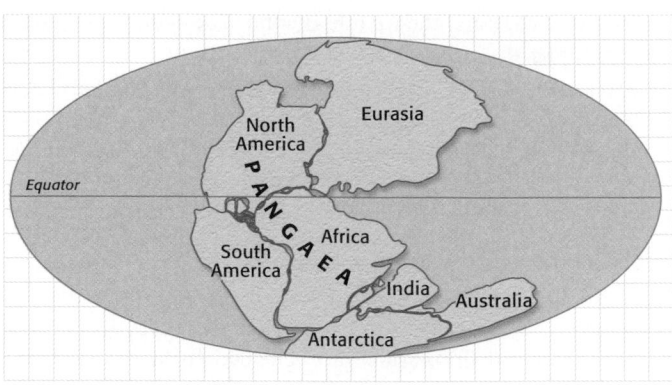

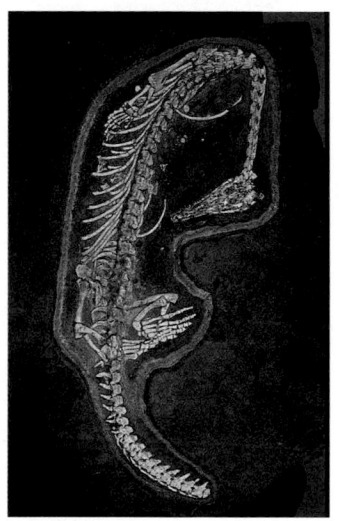

Figure 4

These Mesosaurus bones were discovered in Sao Paulo, Brazil.

702

Plate Tectonics

Around 1915, a German scientist named Alfred Wegener noticed that the eastern coast of South America and the western coast of Africa appeared to fit together like pieces of a puzzle. By studying maps, Wegener found that several other continents' coastlines also seemed to fit together. He pieced all the continents together to form a supercontinent that he called *Pangaea* (pan GEE uh). *Figure 3* shows what Pangaea might have looked like approximately 200 million years ago.

Using fossil evidence, Wegener showed that 200 million years ago the same kinds of animals lived on continents that are now oceans apart. He argued that the animals could not have evolved on separate continents. *Figure 4* shows the fossil of a Mesosaurus found in Brazil. Identical fossils were found in western Africa, giving scientists strong evidence for a past connection between the continents.

Evidence for Wegener's ideas came later

The evidence for *continental drift* or the theory that Earth's surface is made up of large moving plates, was compelling. However, scientists did not have an explanation of how continents could move. Wegener's theory was ignored until the mid-1960s, when structures discovered on the ocean floor gave evidence of a mechanism for the slow movement of continents, or continental drift.

In the 1960s, evidence was discovered in the middle of the oceans that helped explain the mechanisms of continental drift. New technology provided images of "bands" of rock on the ocean floor with alternating magnetic polarities, like the bands illustrated in *Figure 6*. These bands differ from one another in the alignment of the magnetic minerals in the rocks they contain.

Scientists Around the World — GENERAL

Alfred Lothar Wegener Alfred Wegener was born in Germany in 1880. He obtained a degree in astronomy in 1905 and later developed an interest in meteorology. Wegener, a soldier in World War I, developed the theory of continental drift while recuperating from a war injury. The theory was not accepted by geologists, who considered Wegener a meddling outsider. Immune to his critics, Wegener spent much of his life searching for more evidence to support his theory.

In 1930, a friend asked Wegener to coordinate a meteorological expedition to Greenland to help establish a weather station. Despite extreme weather conditions, Wegener went ahead with the trip because he knew that scientists at the station desperately needed the 4,000 pounds of supplies he was carrying. Wegener survived the five-week trip but, tragically, froze to death on the return journey. His theory gained acceptance 30 years after his death as new technologies and research led to additional evidence in support of the theory.

Alignment of oceanic rocks supports the theory of moving plates

As molten rock pours out onto the ocean floor, as shown in *Figure 5,* iron minerals such as magnetite align themselves parallel to Earth's magnetic field, just as compass needles do. After the rocks cool to about 550°C (1020°F), the alignment of these magnetic regions in the iron minerals becomes fixed like the stripes shown in *Figure 6.* The result is a permanent record of Earth's magnetic field as it was just before the rock cooled.

So why are there differently oriented magnetic bands of rock? Earth's magnetic field has reversed direction many times during its history, with the north magnetic pole becoming the south magnetic pole and the south magnetic pole becoming the north magnetic pole. This occurs on average once every 200 000 years. This process is recorded in the rocks as bands. These magnetic bands are symmetrical on either side of the Mid-Atlantic Ridge. The rocks are youngest near the center of the ridge. The farther away from the ridge you go, the older the rocks appear. This suggests that the crust was moving away from the plate boundary.

Earth has plates that move over the mantle

The **lithosphere** is approximately 100 km (60 mi) thick and is made up of the crust and the upper portion of the mantle. The lithosphere is made up of about seven large pieces (and several smaller pieces) called *tectonic plates.* The word *tectonic* refers to the structure of the crust of a planet. The continents are embedded into these plates, which fit together like pieces of a puzzle and move in relation to one another. The theory describing the movement of plates is called **plate tectonics.**

Tectonic plates move at speeds ranging from 1 to 16 cm (0.4 to 6.3 in) per year. Although this speed may seem slow, tectonic plates have moved a considerable distance because they have been moving for hundreds of millions or billions of years.

▪ **lithosphere** the solid, outer layer of Earth, that consists of the crust and the rigid upper mantle

▪ **plate tectonics** the theory that explains how the outer parts of Earth change through time, and that explains the relationships between continental drift, sea-floor spreading, seismic activity, and volcanic activity

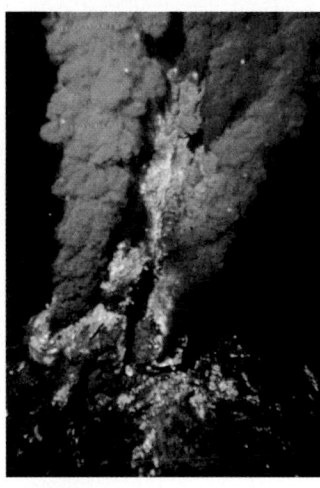

Figure 5

Hydrothermal vents are driven by heat from the eruption of fresh lava on the sea floor.

Figure 6

The stripes illustrate Earth's alternating magnetic field. Light stripes represent when Earth's polarity was the same way it is today, while the darker stripes show reversed polarity.

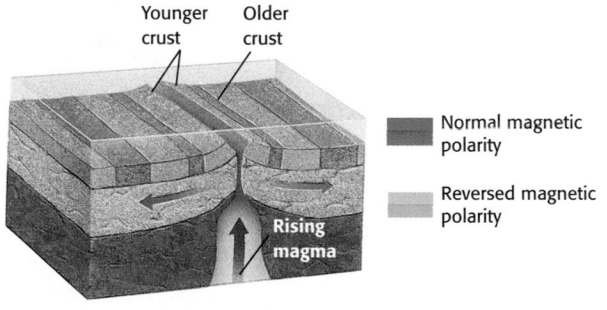

Younger crust — Older crust

Normal magnetic polarity

Reversed magnetic polarity

Rising magma

703

LIFE SCIENCE
CONNECTION — ADVANCED

Before Pangaea broke up, dinosaurs roamed the entire continent. As Pangaea began to break up, dinosaur populations became fragmented and isolated on the new continents. As a result, dinosaurs began to evolve divergently. By the time dinosaurs became extinct, about 65 million years ago, there was great biological diversity among them. Have students research the biodiversity among dinosaurs and plants before and after Pangaea broke up. **LS Logical**

Did You Know ?

Related Dinosaurs In 1991, scientists found fossil evidence of a new dinosaur species, *Cryolophosaurus ellioti*, 600 km from the South Pole. The fossil evidence suggests that these dinosaurs lived about 200 million years ago. They were probably similar in appearance to the *Allosaurus*, discovered at the southeastern tip of mainland Australia in Dinosaur Cove. The similarity can be explained by the presence of a common ancestor dating back to when modern-day Antarctica and southern Australia were neighboring areas of Pangaea.

Teaching Tip

A Powerful Theory Remind students that a scientific theory often brings together many apparently unrelated facts or observations. The theory of plate tectonics is a great example of this because it explains why the continent shapes "fit together" and why earthquakes and volcanoes occur in specific areas of the world. It also explains how great mountain ranges formed and why similar fossils and mineral deposits are found on continents separated by oceans.

SKILL BUILDER — BASIC

Reading Skills Have students create an outline for the sections entitled "Plate Tectonics" and "Plate Boundaries." Each heading in the text should correspond to an outline heading. Students will need to use additional subheadings. Remind students that an outline should summarize information and should show the relationships between ideas. Outlines do not need to cover every single piece of information but should include all of the main points. **LS Visual**

Teaching Tip — ADVANCED

From New York to Africa Have students calculate the number of years that it took New York and the west coast of Africa to reach their current locations, 6.76×10^8 cm apart, if the sea floor is spreading an average of 4 cm per year. (about 169 million years, 6.76×10^8 cm / (4 cm/y) = 1.69×10^8 y. This is fairly close to the estimate based on geologic and fossil evidence of the beginning of the breakup of Pangaea 180 million years ago.) **LS Logical**

Chapter Resource File

• CBL™ Probeware Lab Relating Convection to the Movement of Tectonic Plates ADVANCED

Materials (per group):

• water
• shallow rectangular pan
• stove or hot plate
• dark food coloring,
• five cardboard pieces (such as shirt-packaging cardboard)

Teacher Notes: This model represents a *hypothesis* about why the tectonic plates move, not an accepted theory. The water in the pan represents the asthenosphere, the cardboard pieces represent tectonic plates, and the heat source represents the heat coming from Earth's core. One limitation of the model is that heat from Earth's core is more evenly distributed than is represented. Some cardboard pieces may begin moving before others due to proximity to the heat source.

Results: In step 2, food coloring moves in a circular pattern showing the convection currents which result from heat circulating in the pan. In step 4, cardboard pieces nearest the heat source move in a circular pattern. Eventually, all of the pieces move in a circular pattern, modeling the movement of tectonic plates. **LS** Visual

Quick ACTIVITY

Convection and plate tectonics

1. Fill a shallow pan with water until it is 3 cm from the top.
2. Heat the water over low heat for 30 s. Add a few drops of food coloring to the pan, and watch what happens.
3. Turn off the heat, and place 5 cardboard pieces as close together as possible in the center of the pan.
4. Turn on the heat, and sketch the movement of the cardboard.

5. What do the water and the cardboard pieces represent? What did you observe in step 2?
6. How was the movement in step 4 like continental drift? How could you make a more accurate model of plate tectonics?

VOCABULARY *Skills Tip*

The word tectonic *originates from the Greek word* tektonikos, *meaning "construction." In everyday usage, the word* tectonics *also relates to architecture.*

Figure 7

Earth's lithosphere is made up of several large tectonic plates. Plate boundaries are marked in red, and arrows indicate plate movement.

It is unknown exactly why tectonic plates move

Figure 7 shows the edges of Earth's tectonic plates. The arrows indicate the direction of each plate's movement. Note that plate boundaries do not always coincide with continental boundaries. Some plates move toward each other, some move away from each other, and still others move alongside each other. One hypothesis suggests that plate movement results from convection currents in the *asthenosphere*, the hot, plastic portion of the mantle. The plates of the lithosphere "float" on top of the asthenosphere.

Some scientists believe that the plates are pieces of the lithosphere that are being moved around by convection currents. The soft rock in the asthenosphere circulates by convection, similar to the way mushy oatmeal circulates as it boils, and this slow movement of rock might push the plates of the lithosphere along. Other scientists believe that the forces generated by convection currents are not sufficient to move the plates, and that instead plates are driven by the force of gravity acting on their own weight.

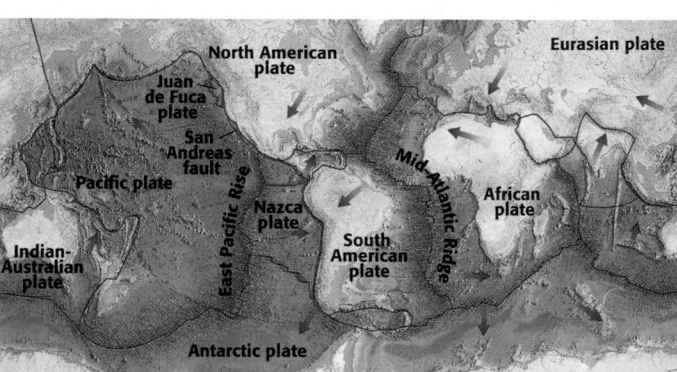

Transparencies

TT Tectonic Plates

Trends in Geophysics

Measuring Plate Movement Geophysicists use the Global Positioning System (GPS) to measure the rate of tectonic plate movement. Radio signals are continuously beamed from satellites to GPS ground stations, which record the distance between the satellites and the ground station. By calculating the time that the GPS ground stations take to move a given distance, scientists can measure the rate of motion of each tectonic plate.

Historical Perspective

Catastrophism Versus Uniformitarianism
Until the 1700s, most scientists believed in *catastrophism,* a theory that states that all changes to Earth are based on sudden, catastrophic events. In 1785, James Hutton proposed an alternative theory, called *uniformitarianism.* According to this theory, only gradual forces change Earth and the processes observed today are the same processes that occurred in the past to make our planet the way it is today. The two theories were debated for many years, but by the mid-19th century, uniformitarianism became dominant.

Plate Boundaries

The theory of plate tectonics helps scientists study and sometimes predict volcanic eruptions and has provided information on earthquakes. Volcanoes and earthquakes, such as the one that caused the damage shown in **Figure 8,** often occur where tectonic plates come together. At these plate boundaries, many other dramatic geological features, such as mountains and rift valleys, can occur.

Mid-oceanic ridges result from divergent boundaries

A *divergent boundary* occurs where two plates move apart, creating a gap between them. When this happens, hot rock rises from the asthenosphere and cools, forming new lithospheric rock. The two diverging plates then pull the newly formed lithosphere away from the gap. The drop in pressure also causes the rising asthenosphere to partially melt, forming **magma,** which separates to form new oceanic crust.

Mid-oceanic ridges are mountain ranges at divergent boundaries in oceanic crust. Unlike most mountains on land, which are formed by the bending and folding of continental crust, mid-oceanic ridges are mountain ranges created by magma rising to Earth's surface and cooling. **Figure 9B** shows how a mid-oceanic ridge forms. As the plates move apart, magma rises from between the diverging plates and fills the gap. The new oceanic crust forms a large valley, called a *rift valley,* surrounded by high mountains. The most studied mid-oceanic ridge is called the Mid-Atlantic Ridge which is shown in **Figure 9A.** This ridge runs roughly down the center of the Atlantic Ocean from the Arctic Ocean to an area off the southern tip of South America.

Figure 8

An earthquake, which occurred in 1999, damaged this running track in Taiwan.

► **magma** liquid rock produced under Earth's surface

Figure 9

A When divergent boundaries occur in the oceanic crust they form a mid-oceanic ridge.
B Tectonic plates move apart at divergent boundaries, forming rift valleys and mountain systems.

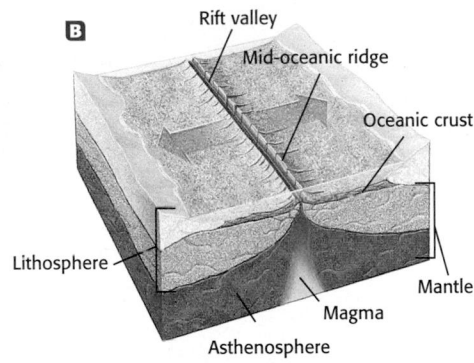

Rift valley
Mid-oceanic ridge
Oceanic crust
Lithosphere
Mantle
Magma
Asthenosphere

705

Teaching Tip

Magma Be sure that students understand the difference between magma and lava. Magma is hot, molten rock within the ground, and it does not become lava until it reaches the surface of Earth.

SKILL BUILDER — BASIC

Vocabulary Before reading the section on plate boundaries, write the words *divergent* and *convergent* on the board, and ask students if they know what these words mean. Having students think about the prefixes *di-* and *con-* may help them. **LS** Visual

SKILL BUILDER — BASIC

Interpreting Visuals Have students use **Figure 9** to explain the forces that pull rocks outward from mid-oceanic ridges. Guide their explanations by asking the following questions: "Why does molten rock from the mantle come to the surface at the ridges? Why does the ocean floor spread apart at the ridges? Why is rock at the ridge "new" rock?"

Be sure students understand that it was not until the 1950s, when researchers used sonar to study the ocean floor, that the mid-oceanic ridge was discovered. Stress that mid-oceanic ridges are not always in the middle of an ocean. **Figure 9A** is from a large section of the Mid-Atlantic Ridge in Iceland. **LS** Visual

Transparencies

TT Divergent and Convergent Boundaries

Interpreting Visuals Have students use **Figure 10** to explain how mountains and volcanoes form at convergent boundaries. (Mountains can form when crust buckles as two plates collide. Volcanoes form when subducting rock melts and rises back to the surface.) **LS** Visual

INTEGRATING

PHYSICS This transfer of energy by heat is an example of a general law of physics: when two objects are at different temperatures, energy is transferred from the warmer object to the cooler object until the temperatures of the objects are the same. This energy transfer occurs when atoms of the hotter material transfer some of their motion, though collisions, to atoms of the cooler material. A simpler example is a cup of hot coffee in a room. The coffee gradually warms the air around it until the coffee and the air have reached equilibrium and are at the same temperature.

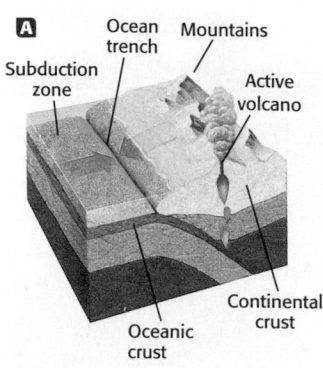

Figure 10

A Ocean trenches, volcanoes, and mountains, such as those shown in **B** form near the boundary where oceanic and continental plates collide.

▶ **subduction** the process by which one lithospheric plate moves beneath another as a result of tectonic forces

Oceanic plates dive beneath continental plates at convergent boundaries

Knowing that lithosphere is being created, you may wonder why Earth isn't expanding. The reason is that while new lithosphere is formed at divergent boundaries, older lithosphere is destroyed at *convergent boundaries.* The Andes Mountains, which are shown in *Figure 10B,* formed along a convergent boundary between an oceanic plate and the South American continental plate. The oceanic plate, which is denser, dives beneath the continental plate and drags the oceanic crust along with it. This process is called **subduction.** As shown in *Figure 10A,* ocean trenches, mountains, and volcanoes are formed at *subduction zones.*

Ocean trenches form along the boundary between two oceanic plates or between an oceanic plate and a continental plate. These trenches can be very deep. The deepest is the Mariana Trench in the Pacific Ocean. Located off the coast of Asia, the deepest point in the trench is more than 11 km (6.8 mi) beneath the ocean surface. The Peru-Chile Trench is associated with the formation of the Andes Mountains and is more than 7 km (4.3 mi) deep.

Subduction of ocean crust generates volcanoes

Chains of often-explosive volcanic mountains form on the overriding plate at subduction zones—where oceanic crust meets continental crust. As the water-bearing rocks and sediments of the oceanic plate are heated by surrounding mantle, they release water into the overlying mantle. Water is very effective at lowering the melting point of rock at high presure and so magma is formed and rises into the crust. The magma cools, and the accumulation of low-density magnetic rock over time forms a chain of high mountains and plateaus.

706

Alternative Assessment ── **ADVANCED**

Mountains in Space Astronomers give extraterrestrial mountains the name *mons,* while extraterrestrial mountain ranges are called either *montes,* or *high-lands.* Encourage students to find out more about the formation of mountains on Mercury, Mars, Earth's moon, or one of the moons of Jupiter or Saturn. Have students compare the mountains that they study with mountains on Earth in terms of size and formation. **LS** Logical

Volcanic mountains also form at convergent boundaries. Magma rises to the surface and cools, forming new rock. These volcanoes are formed far inland from their associated oceanic trenches. Aconcagua (ah kawng KAH gwah), the tallest mountain in the Western Hemisphere, is a volcanic mountain in the Andes. At a height of 6959 m (22 831 ft), the peak of Aconcagua is more than 13.8 km (8.6 mi) above the bottom of the Peru-Chile Trench.

Colliding tectonic plates create mountains

The Himalayas, shown in *Figure 11,* are the tallest mountains. They formed during the collision between the continental tectonic plate containing India and the Eurasian continental plate. They continue to grow in both width and height as the two plates continue to collide. Mount Everest, the highest mountain in the world, is part of this range. Mount Everest's peak is 8850 m (29 034 ft) above sea level.

Transform fault boundaries can crack Earth

Plate movement can cause breaks in the lithosphere. Once a break occurs, rock in the lithosphere continues to move, scraping past nearby rock. The crack where rock moves is called a **fault.** Faults can occur in any area where forces are great enough to break rock. When rock moves horizontally at faults along plate boundaries, the boundary is called a *transform fault boundary* as shown in *Figure 12A.*

Plate movement at transform fault boundaries is one cause of earthquakes. You may have heard of earthquakes along the San Andreas fault which is shown in *Figure 12B,* and which runs from Mexico through California and out to sea north of San Francisco. Transform fault boundaries occur in many places across Earth, including the ocean floor.

Figure 11
The Himalayas are still growing today as the tectonic plate containing Asia and the plate containing India continue to collide.

▶ **fault** a crack in Earth created when rocks on either side of a break move

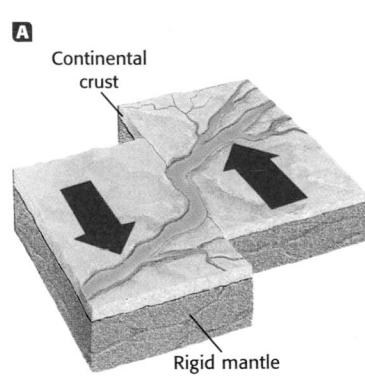

Ⓐ
Continental crust

Rigid mantle

Ⓑ

Figure 12
Ⓐ The change in the course of the river and the fault results from plate movement.

Ⓑ The San Andreas fault system is over 800 mi long.

707

Quick **Lab**

Answers to Analysis

1. convergent boundary
2. transform fault boundary
3. Answers may vary but for step 3 may include words such as *rough, wrinkled,* and *bumpy.* Mountains, volcanoes, and trenches form at convergent plate boundaries. Answers for step 4 may include words such as *smoother, flatter,* and *less bumpy.*

Close

Quiz

1. What are the three main geologic layers of Earth in order from the topmost layer to the center? (crust, mantle, and core)
2. According to the theory of plate tectonics, what is Earth's surface made of? (large moving plates)
3. What are the three types of plate boundaries? (convergent, divergent, and transform fault)
4. True or false: Volcanoes and earthquakes frequently occur where tectonic plates come together. (true)

Chapter Resource File

- **Datasheet for QuickLab** BASIC
- **Concept Review** GENERAL
- **Quiz** BASIC

Quick **Lab**

Can you model tectonic plate boundaries with clay?

Materials
- ✓ ruler
- ✓ plastic knife
- ✓ paper
- ✓ lab apron
- ✓ scissors
- ✓ 2–3 lb modeling clay
- ✓ rolling pin or rod

Procedure

1. Use a ruler to draw two 10 cm × 20 cm rectangles on your paper, and cut them out.
2. Use a rolling pin to flatten two pieces of clay until they are each about 1 cm thick. Place a paper rectangle on each piece of clay. Using the plastic knife, trim each piece of clay along the edges to match the shape of the paper.
3. Flip the two clay rectangles so that the paper is at the bottom, and place them side by side on a flat surface. Slowly push the models toward each other until the edges of the clay begin to buckle and rise off the table.
4. Turn the models around so that the unbuckled edges are touching. Place one hand on each. Slide one clay toward you and the other away from you. Apply only slight pressure toward the seam where the two pieces of clay touch.

Analysis

1. What type of plate boundary are you demonstrating with the model in step 3?
2. What type of plate boundary are you demonstrating in step 4?
3. Compare the appearances of the facing edges of the models in the two processes. How do you think similar processes might affect Earth's surface?

SECTION 1 REVIEW

SUMMARY

▶ The layers of Earth are the crust, mantle, and core.

▶ Earth's outer layer is broken into several tectonic plates, which ride on top of the mantle beneath.

▶ The alignment of iron in oceanic rocks supports the theory of plate tectonics.

▶ Plates spread apart at divergent boundaries, collide at convergent boundaries, and slide past each other at transform fault boundaries.

1. **Explain** why the inner core remains a solid even though it is very hot.
2. **Describe** how the gap is filled when two tectonic plates move away from each other.
3. **Determine** whether each of the following is likely to occur at convergent or divergent boundaries:
   a. rift valley
   b. continental mountains
   c. mid-oceanic ridge
   d. ocean trench
4. **Explain** how magnetic bands provide evidence that tectonic plates are moving apart at mid-oceanic ridges.
5. **Predict** what type of plate boundary exists along the coastline near Japan's volcanic mountain ranges.
6. **Critical Thinking** The oldest continental rocks are 4 billion years old, whereas the oldest sea-floor rocks are 200 million years old. Explain the difference in these ages.

Answers to Section 1 Review

1. The inner core remains a solid even though it is very hot because pressure due to the weight of the mantle and crust is so great that the atoms are forced together as a solid.
2. When two tectonic plates move away from each other, the gap is filled by hot rock that rises from the asthenosphere, partially melts, and cools. This cooled magma forms new oceanic crust.
3. a. divergent plate boundary
   b. convergent plate boundary
   c. divergent plate boundary
   d. convergent plate boundary
4. During the Earth's history, Earth's magnetic field has reversed directions many times, meaning that the north magnetic pole has become the south magnetic pole and the south magnetic pole has become the north magnetic pole. This process of switching poles is recorded in rocks as the iron in the rocks cooled and became magnetically fixed. These magnetic bands are symmetrical on either side of the Mid-Atlantic Ridge, with the rocks nearest the center of the ridge being the youngest, and the rocks farthest away from the ridge being the oldest.

   See "Continuation of Answers" at the end of the chapter.

Earthquakes and Volcanoes

OBJECTIVES

▶ **Identify** the causes of earthquakes.
▶ **Distinguish** between primary, secondary, and surface waves in earthquakes.
▶ **Describe** how earthquakes are measured and rated.
▶ **Explain** how and where volcanoes occur.
▶ **Describe** the different types of common volcanoes.

▶ **KEY TERMS**

focus
epicenter
surface waves
seismology
Richter scale
vent

Imagine rubbing two rough-sided rocks back and forth against each other. The movement won't be smooth. Instead, the rocks will create a vibration that is transferred to your hands. The same thing happens when rocks slide past one another at a fault. The resulting vibrations are called earthquakes.

What Are Earthquakes?

Compare the occurrence of earthquakes, shown as red dots in *Figure 13,* with the plate boundaries, marked by black lines. Each red dot marks the occurrence of an earthquake sometime between 1985 and 1995. You can see that earthquakes occur mostly at the boundaries of tectonic plates, where the plates shift with respect to one another.

internet connect

www.scilinks.org
Topic: Earthquakes
SciLinks code: HK4038

SC/LINKS Maintained by the National Science Teachers Association

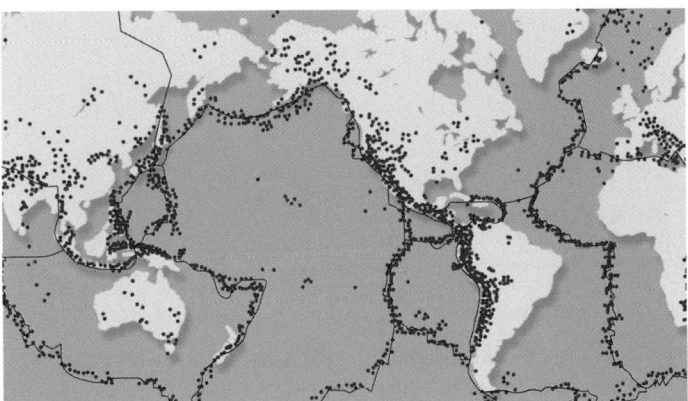

Figure 13
Each red dot in this illustration marks the occurrence of a moderate to large earthquake sometime between 1985 and 1995.

709

Did You Know ?

Millions of Earthquakes Earthquakes are not rare. In fact, more than 3 million earthquakes happen each year, which is about 1 every 10 seconds! Most earthquakes are too weak for humans to feel. The Ring of Fire, a volcanic zone that lies along the plate boundaries surrounding the Pacific Ocean, is the world's largest and most active earthquake zone.

Focus

Overview
Before beginning this section, review with your students the Objectives listed in the Student Edition. In this section, students learn what causes earthquakes, how earthquakes are measured and rated. Students also learn about how and where volcanoes occur and how volcanoes are classified.

Bellringer
Use the Bellringer transparency to prepare students for this section.

Motivate

Opening Demonstration
You will need two smooth wooden blocks, coarse sandpaper, and glue. Glue or staple the sandpaper onto one side of each block. Firmly hold the sandpaper-covered sides against each other while pushing them in opposite directions until there is a sudden movement. Explain that this shows how rock slides along a fault. As the rock slides, it releases energy that travels as seismic waves.

Teach

SKILL BUILDER — BASIC

Reading Skills As students read the section entitled "What Are Earthquakes?" have them write down difficult concepts and words on self-adhesive notes. They should then work together through discussion and note taking to clarify the difficult portions of the section.
LS Interpersonal

Chapter Resource File

• **Lesson Plan**

Transparencies

TT Bellringer

Interpreting Visuals Have students examine **Figure 14.** Ask them where this earthquake initiated (at the focus), what point on land will probably have the most damage (the epicenter), and what limitation this illustration has (Sample answer: It shows the seismic waves spreading out from the focus in two dimensions, while in reality they spread out in three dimensions). **LS Visual**

Demonstration

Earthquake!

(Time: About 10 minutes)

Materials:

• two large pieces of clay of different colors

Shape the clay into flat sheets of varying thicknesses, and stack the layers. Make two stacks of clay to represent two plates. Use the stacks of clay to demonstrate the motions of each type of boundary, and ask students to observe the effect of the motion on the clay. Tell the students that earthquakes at certain types of boundaries are shallow and that earthquakes at other boundaries are usually deep. Ask students to hypothesize which earthquakes occur at which boundaries and to predict whether the earthquakes would be weak, moderate, or strong. (transform fault boundary: moderate and shallow; divergent boundary: weak and shallow; convergent boundary: strong and deep)

▶ **focus** the area along a fault at which the first motion of an earthquake occurs

▶ **epicenter** the point on Earth's surface directly above an earthquake's focus

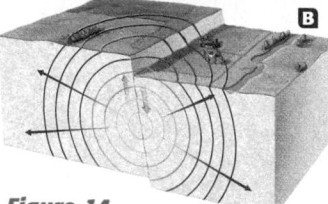

Figure 14

A Earthquakes cause rock to break apart.

B The epicenter of an earthquake is the point on the surface directly above the focus.

Earthquakes occur at plate boundaries

As plates move, their edges experience immense pressure. Eventually, the stress becomes so great that it breaks rock along a fault line. Energy is released as *seismic waves*. As the seismic waves travel through Earth, they create the shaking that we experience during an earthquake.

The exact point inside Earth where an earthquake originates is called the **focus.** Earthquake waves travel in all directions from the focus, which is often located far below Earth's surface. The point on the surface immediately above the focus is called the **epicenter,** as shown in *Figure 14A.* Because the epicenter is the point on Earth's surface that is closest to the focus, the damage there is usually greatest although damage can occur many miles from the epicenter, as shown in *Figure 14B.*

Energy from earthquakes is transferred by waves

The energy released by an earthquake is measured as shock waves. Earthquakes generate three types of waves. *Longitudinal* waves originate from an earthquake's focus. Longitudinal waves move faster through rock than other waves do and are the first waves to reach recording stations. For this reason, longitudinal waves are also called *primary,* or *P waves.*

A longitudinal wave travels by compressing Earth's crust in front of it and stretching the crust in back of it. You can simulate longitudinal waves by compressing a portion of a spring and then releasing it, as shown in *Figure 15A.* Energy will travel through the coil as a longitudinal wave.

The second type of wave is a *transverse wave.* Transverse waves move more slowly than longitudinal waves. Thus, these slower waves are called *secondary* or *S waves.* The motion of a transverse wave is similar to that of the wave created when a rope is shaken up and down, as shown in *Figure 15B.*

Figure 15

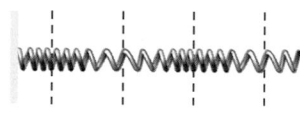

Longitudinal wave

A P waves can be modeled by compressing and releasing a spring.

Transverse wave

B S waves can be modeled by shaking a rope.

710

Alternative Assessment — ADVANCED

Tsunamis Underwater earthquakes, if strong enough, can create a giant wave called a *tsunami.* Have students research the origin of the word *tsunami* and the damaging effects of tsunamis. Ask students to prepare a public education brochure designed for people who live in potentially affected areas. They may wish to include some specific examples of damaging tsunamis in history and information about the areas of the world that are affected by tsunamis. **LS Visual**

Waves move through Earth and along its surface

Both P waves and S waves spread out from the focus in all directions, like light from a light bulb. In contrast, the third type of wave moves only across Earth's surface. These waves, called **surface waves,** are the result of Earth's entire mass shaking like a bell that has been rung. Earth's surface bends and reshapes as it shakes. The resulting rolling motion of Earth's surface is a combination of up-and-down motion and back-and-forth motion. In this type of wave, points on Earth's surface have a circular motion, like the movement of ocean waves far from shore.

Surface waves, such as the ones shown in *Figure 16,* cause more destruction than either P waves or S waves. P waves and S waves shake buildings back and forth or up and down at relatively high frequencies. But the rolling action of surface waves, with their longer wavelengths, can cause buildings to collapse.

Measuring Earthquakes

Because energy from earthquakes is transferred by waves, scientists can measure the waves to learn about earthquakes, and about the interior of the Earth through which the waves travel. Scientists hope that learning more will give them tools to predict earthquakes and save lives.

Seismologists detect and measure earthquakes

Seismology is the study of earthquakes. Seismologists use sensitive machines called *seismographs* to record data about earthquakes, including P waves, S waves, and surface waves. Seismographs use inertia to measure ground motion during an earthquake. Examine the seismograph in *Figure 17.* A stationary pendulum hangs from a support fastened to Earth as a drum of paper turns beneath the pendulum with a pen at its tip. When Earth does not shake, the seismograph records an almost straight line. If Earth shakes, the base of the seismograph moves, but the pendulum is protected from Earth's movement by the string. The pendulum draws zigzag lines on the paper that indicate an earthquake has occurred. Records of seismic activity are called seismograms. *Figure 17* shows a typical seismogram.

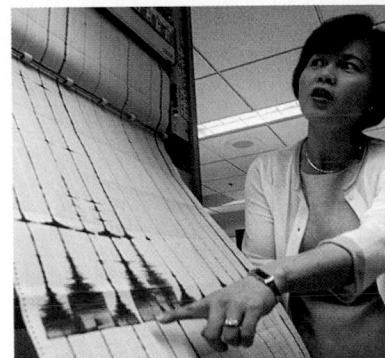

Figure 16

A seismologist points out a surface wave that was measured during a large earthquake.

▶ **surface wave** a seismic wave that can move only through solids

▶ **seismology** the study of earthquakes including their origin, propagation, energy, and prediction

internet connect

www.scilinks.org
Topic: Earthquake Measurement
SciLinks code: HK4037

SCI LINKS. Maintained by the National Science Teachers Association

Figure 17

When the ground shakes, the pendulum remains still while a rotating drum of paper records Earth's movement.

Cultural Awareness

References to predicting earthquakes by observing animal behavior date back 3000 years in Chinese literature. This behavior includes dogs howling, chickens leaving their roosts, fish thrashing about in ponds, and snakes awakening from hibernation to leave their holes. The Chinese government still monitors animal behavior as part of an early warning system for earthquakes. On February 4, 1975, the Chinese government issued an earthquake warning in Liaoning province based on observations of animal behavior, and more than 3 million people were evacuated from their homes. That night, an earthquake with a magnitude of 7.3 occurred and many buildings were destroyed. But, because of the evacuation, only around 300 people died.

Chapter 21 • Planet Earth   711

SKILL BUILDER — BASIC

Interpreting Visuals Ask students why the intersection of the circles in **Figure 18** is the epicenter. (Each circle shows all possible locations of the epicenter based on data from that station. The intersection of the circles is the point that matches the data from all three stations.) **LS** Visual

Teaching Tip

Discovering Earth's Layers The speed of seismic waves increases sharply at the Moho, which is the boundary between the crust and mantle. The evidence for a liquid core is based on the shadow zones, the areas of Earth's surface where no direct seismic waves can be detected. In 1936, Inge Lehmann demonstrated that Earth's core had two parts: an outer, liquid part and an inner, solid part. Her discovery was based on observations of the reflection and refraction of seismic waves generated by deep-focus earthquakes.

SKILL BUILDER — BASIC

Interpreting Visuals Have students study **Figure 19.** Ask them, "Why do scientists believe that the S waves (blue arrows) are absorbed by Earth's core?" (because they are not detected further than 105° from the epicenter) **LS** Logical

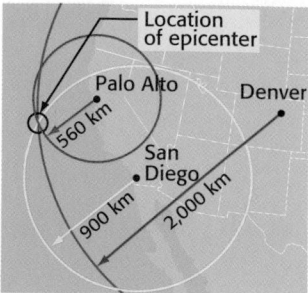

Figure 18

The point where the three circles intersect is the epicenter of the earthquake.

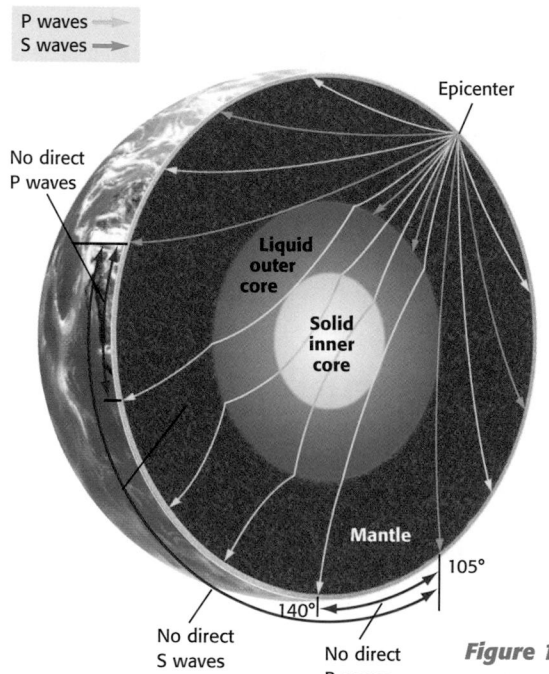

Figure 19

No direct S waves can be detected more than 105° from the epicenter. No direct P waves can be detected between 105° and 140° from the epicenter.

Three seismograph stations are necessary to locate the epicenter of an earthquake

There are more than 1000 seismograph stations across the world. At each station, three seismographs are used to measure different motions: north to south, east to west, and up and down. P waves, the first to be recorded by seismographs, make a series of small, zigzag lines. S waves arrive later, appearing as larger, more ragged lines. Surface waves arrive last and make the largest lines. The difference in time between the arrival of P waves and the arrival of S waves enables seismologists to calculate the distance between the seismograph station and the earthquake's epicenter. When the distances from three seismograph stations are calculated, three circles can be drawn on a map as shown in **Figure 18.** Each circle has the geographic locations of a seismograph station as its center, with the radius being the distance to the epicenter. Therefore, the epicenter can be found by finding where all three circles intersect.

Geologists use seismographs to investigate Earth's interior

Seismologists have found that S waves do not reach seismographs on the side of Earth's core opposite the focus, as shown in **Figure 19.** This is evidence that part of the core is liquid because S waves, which are transverse waves, cannot travel through a liquid.

By comparing seismograms recorded during earthquakes, seismologists have noticed that the velocity of seismic waves varies depending on where the waves are measured. Waves change speed and direction whenever the density of the material they are traveling through changes. The differences in velocity suggest that Earth's interior consists of several layers of different densities. By comparing data, scientists have constructed the model of Earth's interior described in Section 1.

712

LIFE SCIENCE CONNECTION

Prior to an earthquake, changes in Earth's crust occur such as changes in magnetic field, and the sinking, tilting, and bulging of the surface. These changes can be monitored by scientific instruments. Many studies have shown that electromagnetic fields can affect the behavior of living organisms. For example, migrating birds and fish navigate by using magnetic fields, and fish and sharks use electroreceptors to detect objects around them. Have students research the relationship between animal behavior and earthquakes.

LANGUAGE ARTS CONNECTION

Have students write a newspaper article detailing events before, during, and after an earthquake. Tell students to be creative, but specify that their articles should include some scientific information that they have been learning.

Table 1 Earthquake Magnitude, Effects, and Frequency

Magnitude (Richter scale)	Characteristic effects of shallow earthquakes	Estimated number of earthquakes recorded each year
2.0 to 3.4	Not felt but recorded	More than 150,000
3.5 to 4.2	Felt by a few people in the affected area	30,000
4.3 to 4.8	Felt by most people in the affected area	4800
4.9 to 5.4	Felt by everyone in the affected area	1400
5.5 to 6.1	Moderate to slight damage	500
6.2 to 6.9	Widespread damage to most structures	100
7.0 to 7.3	Serious damage	15
7.4 to 7.9	Great damage	4
8.0 to 8.9	Very great damage	Occur infrequently
9.0	Would be felt in most parts of the Earth	Possible but never recorded
10.0	Would be felt all over the Earth	Possible but never recorded

The Richter scale is a measure of the magnitude of earthquakes

The **Richter scale** is a measure of the energy released by an earthquake. The 1964 Alaskan earthquake, with a magnitude of 8.4, is the largest earthquake of recent times. Each step on the Richter scale represents a 30-fold increase in the energy released. So an earthquake of magnitude 8 releases 30^4, or 810,000, times as much energy as one of magnitude 4! *Table 1* summarizes the effects and number of earthquakes with varying magnitudes. Notice that low magnitude earthquakes occur frequently.

The Richter scale cannot predict how severe an earthquake will be. The amount of damage depends on several factors, such as the distance between populated areas and the epicenter, and the type of construction used in area buildings. The Armenian earthquake of 1988 and the San Francisco earthquake of 1989 both had a magnitude of 7 on the Richter scale. In Armenia, there was devastating property damage, and more than 25,000 people died. In contrast, only 70 people died in San Francisco.

Why was there such a big difference between the effects of these two earthquakes? The focus of the Armenian earthquake was 5 km down, but in San Francisco it was 19 km down. The deeper the focus, the less the effects will be felt at the surface.

Also, the rock in San Francisco is harder than that of Armenia. Softer rock breaks apart and changes position more easily than rigid rock. The difference in fatalities was mainly due to building construction. Other factors that determine how destructive an earthquake will be are the time of day that the event occurs, soil composition, and how saturated the ground is.

▶ **Richter scale** a scale that expresses the magnitude of an earthquake

Did You Know ?

The effect of an earthquake on Earth's surface is called the earthquake's intensity. The modified Mercalli scale is the most commonly used intensity scale. An earthquake is assigned a lower number if people felt the quake but it didn't cause much damage. Earthquakes that cause structural damage are assigned a higher number. The scale has been used to develop intensity maps for planners, building officials, and insurance companies.

Did You Know ?

Earthquake Damage The best structures for resisting earthquake damage are wood-framed buildings because they are not very rigid and can flex without collapsing. Structures built on waterlogged or unconsolidated sediment such as sand are more likely to suffer intense damage than structures built on bedrock.

Demonstration

Eruption Simulation

(Time: About 15 minutes)

Materials:

- 10 mL of baking soda
- 50 mL of vinegar
- 200 mL beaker or measuring cup
- a few sheets of bathroom tissue
- large plate or pan
- modeling clay
- funnel
- red food coloring
- liquid dish soap
- stirring stick

Step 1 Place 10 mL of baking soda in the center of a piece of bathroom tissue. Fold the corners of the tissue over the baking soda, and press the edges until the ends stay in place. Place the tissue packet in the middle of a large plate or pan.

Step 2 Put some modeling clay around the top edge of a funnel. Turn the funnel upside down over the tissue packet in the bottom of the pan. The clay should form a watertight seal between the base of the funnel and the plate or pan. Press down on the clay to make a tight seal.

Step 3 Add 50 mL of vinegar, two drops of red food coloring, and several drops of liquid dish soap to a 200 mL beaker or measuring cup, and stir.

Step 4 Carefully pour the liquid into the spout of the upturned funnel. Have students predict how much time will elapse before the volcano erupts. Have one student record the time of the final eruption.

▶ **vent** an opening at the surface of Earth through which volcanic material passes

Figure 20

Volcanoes build up into hills or mountains as lava and ash explode from openings in Earth called vents.

Scientists are trying to predict earthquakes

In the past, people would try to predict earthquakes by watching animals for strange behavior. Today, scientists are trying to measure changes in Earth's crust that can signal an earthquake. Scientists might someday be able to warn people of an impending earthquake and save lives by learning to observe rock for signs of stress and strain. The random nature of earthquake rupture makes prediction extremely difficult, but finding a reliable system could save tens of thousands of lives in the future.

Volcanoes

A volcano is any opening in Earth's crust through which magma has reached Earth's surface. These openings are called **vents.** Volcanoes often form hills or mountains as materials pour or explode from the vent, as shown in *Figure 20.* Volcanoes release molten rock, ash, and a variety of gases that result from melting in the mantle or in the crust.

Volcanoes generally have one central vent, but they can also have several smaller vents. Magma from inside a volcano can reach Earth's surface through any of these vents. When magma reaches the surface, its physical behavior changes, and it is called *lava.*

Shield volcanoes have mild eruptions

Magma rich in iron and magnesium is very fluid and forms lava that tends to flow great distances. The eruptions are usually mild and can occur several times. The buildup of this kind of lava produces a gently sloping mountain, called a *shield volcano.* Shield volcanoes are some of the largest volcanoes. Mauna Loa, in Hawaii, is a shield volcano, as shown in *Figure 21A.* Mauna Loa's summit is more than 4000 m (13 000 ft) above sea level and more than 9020 m (29 500 ft) above the sea floor.

Composite volcanoes have trapped gas

Composite volcanoes are made up of alternating layers of ash, cinders, and lava. Their magma is rich in silica and therefore is much more viscous than the magma of a shield volcano. Gases are trapped in the magma, causing eruptions that alternate between flows and explosive activity that produces cinders and ash. Composite volcanoes are typically thousands of meters high, with much steeper slopes than shield volcanoes. Japan's Mount Fuji, shown in *Figure 21B,* is a composite volcano. Mount St. Helens, Mount Rainier, Mount Hood, and Mount Shasta, all in the western United States, are also composite volcanoes.

PHYSICS

CONNECTION

When the water dissolved in magma changes from liquid to steam, the volume of the water increases dramatically. This change causes pressure that creates a great deal of explosive force. Have students think of other examples in which water can be an explosive force, such as in a car's radiator or in popcorn.

Alternative Assessment — ADVANCED

Volcanic Eruptions Have students research several aspects of a volcanic eruption, including the name and date of the eruption, the type of volcano, the geographic location, and other disruptions in the surrounding ecosystem. Have students present their findings in a report. **LS Logical**

Cinder cones are the most abundant volcano

Cinder cones are the smallest and most abundant volcanoes. When large amounts of gas are trapped in magma, violent eruptions occur—vast quantities of hot ash and lava are thrown from the vent. These particles then fall to the ground around the vent, forming the cone. Cinder cones tend to be active for only a short time and then become dormant. As shown in **Figure 21C,** Parícutin (pah REE koo teen), in Mexico, is a cinder cone. Parícutin erupted in 1943. After 2 years, the volcano's cone had grown to a height of 450 m (1480 ft). The eruptions finally ended in 1952. Volcanoes form not only on land but also under the oceans. In shallow water, volcanoes can erupt violently, forming clouds of ash and steam. An underwater volcano is called a *seamount* and looks like a composite volcano.

Types of Volcanoes

A Shield volcano

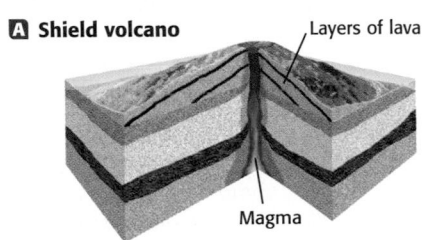
Layers of lava
Magma

Mauna Loa, Hawaii

B Composite volcano

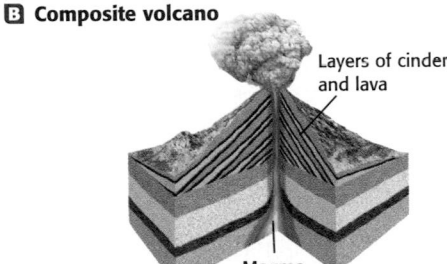
Layers of cinders and lava
Magma

Mount Fuji, Japan

C Cinder cone

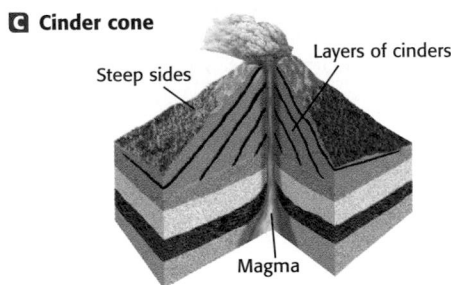
Steep sides
Layers of cinders
Magma

Parícutin, Mexico

Figure 21

The type of volcano that forms depends largely on the makeup of the magma. Differences in the fluidity of the magma determine the type of eruption that occurs.

715

Did You Know?

Nonexplosive Eruptions Although explosive volcanoes get the most attention, nonexplosive extrusions play a much more significant role in shaping our world. Much of the ocean floor is covered by lava from nonexplosive eruptions at the mid-oceanic ridges, and nonexplosive volcanoes formed many of the Pacific islands. In fact, there are 10 times as many active volcanoes at the bottom of the ocean as there are on land. The ocean floor, roughly three-fourths of Earth's surface, was produced primarily by volcanic activity.

Alternative Assessment

Birth and Death of a Volcano The Paricutín volcano did not exist before its 1943 eruption in the flat cornfield of a farmer named Dominic Pulido. By 1952, the cinder cone was 424 m high! Pulido and his wife were the first people to witness the birth of a volcano in the 20th century. Scientists had the unique opportunity of observing the volcano throughout its entire nine-year lifetime. Have students research this spectacular event, including the initial eruption, the effects on the area, and the observations of the Pulidos and of scientists.

Teaching Tip

Diverging Tectonic Plates Point out to students that most volcanic activity takes place on the ocean floor, where vast amounts of lava rise through rifts or volcanoes at diverging plate boundaries. Ask students to identify a spot where plates are diverging. (the Mid-Atlantic Ridge) Then, ask students what landmass formed as a result of volcanic activity along this diverging plate boundary. (the island of Iceland)

Remind students that Iceland is merely a visible part of the Mid-Atlantic Ridge. Also explain that volcanic eruptions along diverging tectonic plates are generally much less violent than volcanoes typical of convergent boundaries are.

Connection to
SOCIAL STUDIES

Answers

1. Answers may vary but should include the fact that gases and magma may have been building up for an extended amount of time before the eruption. Mount St. Helens is a composite volcano and has very viscous magma in which gases are trapped.

2. Winds in the atmosphere were responsible for carrying ash from Washington to Montana.

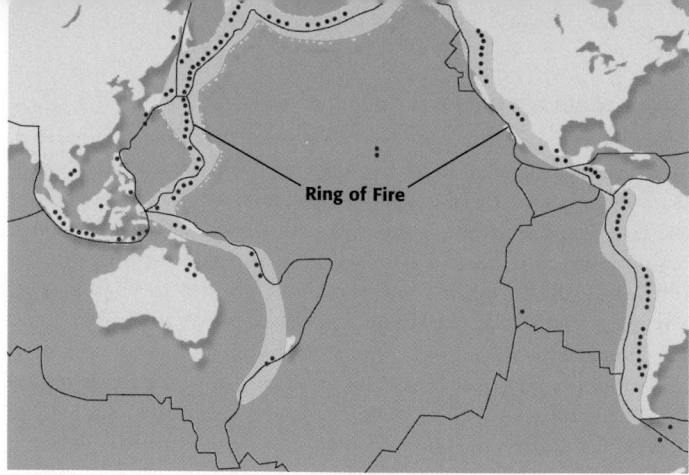

Figure 22
Seventy-five percent of the active volcanoes on Earth occur along the edges of the Pacific Ocean. Together these volcanoes form the Ring of Fire.

Connection to
SOCIAL STUDIES

Mount St. Helens, in the Cascade Range in Washington, erupted explosively on May 18, 1980. Sixty people and thousands of animals were killed, and 10 million trees were blown down by the air blast created by the explosion. During the eruption, the north side of the mountain was blown away. Gas and ash were ejected upward, forming a column more than 19.2 km (11.9 mi) high. The ash was reported to have fallen as far east as central Montana.

Since the May 18 explosion, Mount St. Helens has had several minor eruptions. As a result, a small volcanic cone is now visible in the original volcano's crater.

Making the Connection

1. What might have caused the eruption of Mount St. Helens to be so explosive?

2. The force of the blast didn't push the ashes all the way to Montana. What other natural force might have transported the ashes that far?

716

Most volcanoes occur at convergent plate boundaries

Like earthquakes, volcanoes are linked to plate movement. Volcanoes are common all around the edges of the Pacific Ocean, where oceanic tectonic plates collide with continental plates. In fact, 75% of the active volcanoes on Earth are located in these areas. As seen in **Figure 22,** the volcanoes around the Pacific Ocean lie in a zone known as the Ring of Fire.

As a plate sinks at a convergent boundary, it causes melting in the mantle and magma rises to the surface. The volcanoes that result form the edges of the Ring of Fire. These volcanoes tend to erupt cooler, less-fluid lava and clouds of ash and gases. The high-viscosity lava makes it difficult for the gases to escape. Gas pressure builds up, causing explosive eruptions.

Underwater volcanoes occur at divergent plate boundaries

As plates move apart at divergent boundaries, magma rises to fill in the gap. This magma creates the volcanic mountains that form the ridges around a central rift valley.

The volcanic island of Iceland, in the North Atlantic Ocean, is on the Mid-Atlantic Ridge. The island is continuously expanding from its center; the eastern and western sides of the island are growing outward in opposite directions. As a result, a great deal of geologic activity, such as volcanoes and hot springs, occurs on the island.

Alternative Assessment — BASIC

Volcano Illustration Have students draw a cross section of a volcano in which they show what is inside the volcano and what is in the ground underneath. The picture should include information that shows details of how the volcano was created and the forces that led to eruptions. **LS Visual**

Tectonic Plate Boundaries Post on the bulletin board a map of the world that shows the location of tectonic plates. Pair students together, and have partners explain how tectonic plate boundaries and volcanoes are related. Each partner should evaluate the other's understanding by assessing his or her descriptions of converging and diverging tectonic plates, subduction, hot spots, and magma formation. **LS Interpersonal**

Volcanoes occur at hot spots

Some volcanoes occur in the middle of plates. They occur because mushroom-shaped trails of hot rock, called *mantle plumes*, rise from deep inside the mantle, melt as they rise, and erupt from volcanoes at *hot spots* at the surface.

When mantle plumes form below oceanic plates, lava and ash build up on the ocean floor. If the resulting volcanoes grow large enough, they break through the water's surface and become islands. As the oceanic plate continues moving, however, the mantle plume does not move along with it. The plume continues to rise under the moving oceanic plate, and a new volcano is formed at a different point. A "trail" in the form of a chain of extinct volcanic islands is left behind.

The Hawaiian Islands lie in a line that roughly corresponds to the motion of the Pacific plate. The island of Hawaii is the most recently formed in the chain, and contains the active volcanoes situated over the mantle plume. Volcanic activity produces fertile soil which helps tropical plants, like those shown in **Figure 23,** grow.

Figure 23
Tropical plants often grow on the fertile ground that results from volcanoes.

Teaching Tip

Hot Spot Be sure students understand that a hot spot remains relatively stationary and that volcanic activity occurs as a tectonic plate moves over a hot spot. You may want to compare a hot spot to a candle flame and a tectonic plate to a piece of paper passing over the flame.

Close

Quiz

1. True or false: Earthquakes usually occur in the center of tectonic plates. (false)
2. What kind of machine do scientists use to record data about earthquakes? (seismographs)
3. What is an opening through which molten rock flows onto Earth's surface? (volcano or vent)
4. What are the three types of volcanoes? (shield volcano, composite volcano, and cinder cone)

SECTION 2 REVIEW

SUMMARY

▶ Earthquakes occur as a result of sudden movement within Earth's lithosphere.

▶ P waves are longitudinal waves, and they travel the fastest.

▶ S waves are transverse waves, and they travel more slowly.

▶ Surface waves travel the slowest. They result from Earth's vibrating like a bell.

▶ Volcanoes are formed when magma rises and penetrates the surface of Earth.

▶ The three types of volcanoes are shield volcanoes, cinder cones, and composite volcanoes.

1. **Identify** which type of seismic wave is described in each of the following:
   a. cannot travel through the core
   b. cause the most damage to buildings
   c. are the first waves to reach seismograph stations

2. **Select** which of the following describes a shield volcano:
   a. formed from violent eruptions
   c. formed from hot ash
   b. has gently sloping sides
   d. has steep sides

3. **Identify** whether volcanoes are likely to form at the following locations:
   a. hot spot
   b. transform fault boundary
   c. divergent plate boundary
   d. convergent boundary between continental and oceanic plates

4. **Differentiate** between the focus and the epicenter of an earthquake.

5. **Explain** how a mid-oceanic ridge is formed.

6. **Explain** why Iceland is a good place to use hydrothermal power, which is power produced from heated water.

7. **Critical Thinking** Are quiet eruptions or explosive eruptions more likely to increase the height of a volcano? Why?

717

Answers to Section 2 Review

1. **a.** S waves
   b. surface waves
   c. P waves

2. b

3. **a.** yes
   b. no
   c. yes
   d. yes

4. The focus of an earthquake is the area along a fault where an earthquake originates, while the epicenter of an earthquake is the point on the surface of the Earth directly above the focus.

5. A mid-oceanic ridge forms as plates move apart at divergent boundaries. Mantle rises to fill the gap between the plates, partially melts, and this magma creates the volcanic mountains that form ridges.

6. Iceland is a volcanic island on a mid-oceanic ridge. It is an excellent place to use hydrothermal power because of the heat that results from the large amount of geologic activity that occurs on the island.

See "Continuation of Answers" at the end of the chapter.

Chapter Resource File

• **Cross-Disciplinary Worksheet**
  Connection to Social Studies—
  Plinian Eruptions ADVANCED

• **Concept Review** GENERAL

• **Quiz** BASIC

SECTION
3

Focus

Overview

Before beginning this section, review with your students the Objectives listed in the Student Edition. This section discusses the types and properties of rocks, the rock cycle, and methods used to determine the relative and absolute age of rocks.

 Bellringer

Use the Bellringer transparency to prepare students for this section.

Motivate

Opening Activity — GENERAL

You will need magnifying glasses and several examples of each type of rock. Rock samples could include: igneous rock, such as pumice, granite, obsidian, basalt, and rhyolite; sedimentary rock, such as shale, limestone, conglomerate, and sandstone; and metamorphic rock, such as slate, marble, quartzite, gneiss, and schist. Have students read this section before class. Organize the rocks by type, and ask the class to describe how each type forms. Invite students to examine the rocks with the magnifying glasses and to list some of their characteristics. Then, mix the rocks up. Have students work in small groups to try to reorganize the rocks. **LS Kinesthetic**

Transparencies

TT Bellringer

Minerals and Rocks

▶ **KEY TERMS**

mineral
igneous rock
weathering
sedimentary rock
metamorphic rock

OBJECTIVES

▶ **Identify** the three types of rock.
▶ **Explain** the properties of each type of rock based on physical and chemical conditions under which the rock formed.
▶ **Describe** the rock cycle and how rocks change form.
▶ **Explain** how the relative and absolute ages of rocks are determined.

▶ **mineral** a natural, usually inorganic solid that has a characteristic chemical composition, an orderly internal structure, and a characteristic set of physical properties

Figure 24

Devils Tower, in northeastern Wyoming, is the solidified core of a volcano.

718

Devils Tower, in Wyoming, shown in *Figure 24,* rises 264 m (867 ft) above its base. According to an American Indian legend, the tower's jagged columns were formed by a giant bear scraping its claws across the rock. The tower is actually the solidified core of a volcano. Over millions of years, the surrounding softer rock was worn away by the Belle Fourche River finally exposing the core. Volcanic pipes, which are similar to volcanic cores, can be a source of diamonds. They contain solidified magma that extends from the mantle to Earth's surface.

Structure and Origins of Rocks

All rocks are composed of **minerals.** Minerals are naturally occurring, nonliving substances that have a composition that can be expressed by a chemical formula. Minerals also have a definite internal structure. Quartz, for example, is a mineral made of silicon dioxide, SiO_2. It is composed of crystals, as are most minerals. Coal, on the other hand, is not a mineral because it is formed from decomposed plant matter. *Granite* is not a mineral either; it is a rock composed of different minerals.

There are about 3500 known minerals in Earth's crust. However, no more than 20 of these are commonly found in rocks. Together, these 20 or so minerals make up more than 95% of all the rocks in Earth's crust. Some of the most common of these *rock-forming minerals* are feldspar, pyroxene, mica, olivine, dolomite, quartz, amphibole, and calcite.

Each combination of rock-forming minerals results in a rock with a unique set of properties. Rocks may be porous, granular, or smooth; they may be soft or hard and have different densities or colors. The appearance and characteristics of a rock reflect its mineral composition and the way it formed.

Alternative Assessment — GENERAL

Devils Tower Have students describe all of the characteristics of Devils Tower (shown in **Figure 24**). Have students imagine that they are geologists exploring the tower. Ask them to write a short story, script, or poem describing the formation of Devils Tower based on their reading. Encourage students to speculate about the original height of the volcano and the history of rivers in the area. **LS Verbal**

Minerals Ask students to brainstorm a list of minerals. Have them use Section 3 and a geology field guide to confirm or deny the items on their lists. Ask them to list reasons why each item listed is or is not a mineral. **LS Logical**

Figure 25

A Notice the coarse-grained texture of this sample of granite, an intrusive igneous rock.

B Obsidian, an extrusive igneous rock, cools much more quickly than granite.

Molten rock cools to form igneous rock

When molten rock cools and solidifies it forms **igneous rock.** Nearly all igneous rocks are made of crystals of various minerals, such as those shown in the granite in *Figure 25A*. As the rock cools, the minerals in the rock crystallize and grow. In general, the more quickly the rock cools, the less the crystals grow. For instance, obsidian, a smooth stone used by early American Indians to make tools, is similar to granite in composition, but it cools much more quickly. As a result, obsidian has either very small crystals or no crystals at all and is mostly glass. *Figure 25B* shows a piece of obsidian.

Obsidian is categorized as an *extrusive* igneous rock because it cools on Earth's surface. *Basalt*, a fine-grained, dark-colored rock, is the most common extrusive igneous rock. Granite, on the other hand, is called an *intrusive* igneous rock because it forms from magma that cools while trapped beneath Earth's surface. Because the magma is insulated by the surrounding rocks, it takes a very long time to cool—sometimes millions of years. Because of this long cooling period, the crystals in intrusive igneous rocks are larger than those in extrusive igneous rocks. The crystals of granite, for example, are easy to see with the naked eye. They are much lighter in color than those of basalt. Both rocks contain feldspar, but granite also has quartz, while basalt has pyroxene.

▶ **igneous rock** rock that forms when magma cools and solidifies

Connection to
SOCIAL STUDIES

Throughout history, humans have used rocks and minerals to fashion tools. During the Stone Age, the Bronze Age, and the Iron Age people used stone, bronze, and iron, respectively, to make tools and weapons. The industrial revolution began when humans learned to burn coal to run machinery. After humans learned to extract oil from Earth's crust, gasoline-powered vehicles were invented, and we entered the automobile age.

Making the Connection

1. Research minerals that have been mined for their iron content. Where are the mines that were first used to harvest these minerals?

2. Scientists have divided the Stone Age into three phases—Paleolithic, Mesolithic, and Neolithic—on the basis of toolmaking techniques. Research these phases, and distinguish between the techniques used in each.

719

Cultural Awareness

Obsidian is a smooth stone that was used by many Native Americans to make tools. Have students research other types of stones that were used by Native Americans. What were these tools used for, and what do these rocks imply about the environments in which the Native Americans lived?

Teach

SKILL BUILDER — **BASIC**

Reading Skills Students should read the section entitled "Structure and Origins of Rocks" and then create a table with the following headings: "Formation process," "Texture," and "Examples." Have students fill in the table by using information on the three types of rocks that they have just read about. **LS** Logical

Teaching Tip

Minerals and Rocks Be sure students understand the difference between minerals and rocks. A mineral has a definite chemical composition and a regular crystal structure. A mineral can be a single element, such as copper, or a compound, such as quartz. Rocks are made up of one or more minerals in varying proportions.

SKILL BUILDER — **BASIC**

Interpreting Visuals Ask students to identify some of the differences between the intrusive igneous rock and the extrusive igneous rock shown in **Figure 25.** (The intrusive igneous rock is lighter in color, and its crystals are larger and easier to see with the naked eye.) **LS** Visual

Connection to
SOCIAL STUDIES

Answers

1. Answers may vary according to the sources used.
2. Answers may vary.

Chapter Resource File

• **Lesson Plan**

• **Cross-Disciplinary Worksheet** Connection to Social Studies—Human Tools: From Stone to Iron **ADVANCED**

Demonstration

Temperature Affects Crystal Growth

(Time: About 45 minutes)

Materials

- heat-resistant gloves
- 400 mL beaker
- hot plate
- Celsius thermometer
- magnesium sulfate (Epsom salts)
- hand lens
- pointed laboratory scoop
- medium test tube
- distilled water
- clock
- aluminum foil
- test-tube tongs

Safety Caution: *In step 3, be sure to direct the opening of the test tube away from you and the students. Always use the test-tube tongs to handle the hot test tube.*

Step 1 Fill the beaker halfway with tap water, and place it on the hot plate. The temperature of the water should be between 40°C and 50°C.

Step 2 Have students use a hand lens to examine crystals of magnesium sulfate. Then, fill the test tube about halfway with the magnesium sulfate, and add an equal amount of distilled water. Use one finger to tap the test tube gently to mix the contents. Place the test tube in the beaker of hot water, and heat it for 3 minutes.

Step 3 Shape the aluminum foil into two small, boatlike containers, and label the boats. You can also grow the crystals on a watch glass.

Step 4 If all of the magnesium sulfate does not dissolve after 3 min,

Continued on the next page

Figure 26
Sedimentary rock can have many distinct layers.

▶ **weathering** the natural process by which atmospheric and environmental agents, such as wind, rain, and temperature changes, disintegrate and decompose rocks

▶ **sedimentary rock** a rock formed from compressed or cemented layers of sediment

Remains of older rocks and organisms form sedimentary rocks

Even very hard rock with large crystals will break down over thousands of years. The process by which rocks are broken down is called **weathering.** Pieces of rock fall down hillsides due to gravity or get washed down by wind and rain. Rivers then carry the pieces down into deltas, lakes, or the sea. Chemical processes also knock pieces of rock away. The action of physical and chemical weathering eventually breaks the pieces into pebbles, sand, and even smaller pieces.

As pieces of rock accumulate, they can form another type of rock— **sedimentary rock.** Think of sedimentary rocks like those shown in *Figure 26* as recycled rocks. The sediment they are made of contains fragments of older rocks and, in some cases, fossils.

Loose sediment forms rock in two ways

There are two ways sediment can become rock; and both require precipitation. In one, layers of sediment get compressed from weight above, forming rock. In the second way, minerals dissolved in water seep between bits of sediment and "glue" them together. In *Figure 27A,* the bits of rock in the conglomerate are fused together with material containing mostly quartz.

Sedimentary rocks are named according to the size of the fragments they contain. As mentioned, a rock made of pebbles is called a conglomerate. A rock made of sand is called sandstone. A rock made of fine mud is usually called mudstone, but if it is flaky and breaks easily into layers, it is called shale. Limestone, another kind of sedimentary rock, is often made of the fossils of organisms that lived in the water, as shown in *Figure 27B.* Sometimes the fossilized skeletons are so small or are broken up into such small fragments that they can't be seen with the naked eye. Places where limestone is found were once beneath water.

Figure 27

A Conglomerate rock is composed of rounded, pebble-sized fragments of weathered rock.

B Limestone is made mostly of fossils of sea creatures.

LIFE SCIENCE
CONNECTION ▶ **BASIC**

Limestone is about 95% calcite, or calcium carbonate. Many mollusks remove calcium and carbonate from sea water and combine them in special tissues that then harden to form a calcium carbonate shell. When the mollusk dies, its shell either dissolves in the water or becomes part of the sediment on the ocean floor. If the shell is deposited, it may become a fossil. Have students research the Mazon Creek Deposits in Kansas to learn more about these fossil beds. **LS Logical**

FINE ARTS
CONNECTION ▶ **BASIC**

One type of printing used to reproduce fine art is called *lithography*. Lithography uses a flat piece of fine-grained porous limestone. Interestingly, many important fossil beds were discovered while people quarried for lithographic limestone. The same qualities that make some limestone good for lithography also allow the preservation of extremely detailed fossils. Ask students to find out more about lithography and lithographic limestone beds around the world. **LS Visual**

Rocks that undergo pressure and heating without melting form metamorphic rock

Heat and pressure within Earth cause changes in the texture and mineral content of rocks. These changes produce **metamorphic rocks.** The word *metamorphic* comes from the Greek word *metamorphosis*, which means "to change form."

Limestone, a sedimentary rock, will turn into marble, a metamorphic rock, under the effects of heat and pressure. Marble is a stone used in buildings, such as the Taj Mahal, in India. *Figure 28* is a photo of the exterior of the Taj Mahal. Notice the swirling, colored bands that make marble so attractive. These bands are the result of impurities that existed in the limestone before it was transformed into marble.

Rocks may be changed, or *metamorphosed,* in two ways: by heat alone or, more commonly, by a combination of heat and pressure. In both cases, the solid rock undergoes a chemical change over millions of years, without melting. As a result, new minerals form in the rocks. The texture of the rocks is changed too, and any fossils in sedimentary rocks are transformed and destroyed.

The most common types of metamorphic rock are formed by heat and pressure deep in the crust. *Slate* forms in this way. It metamorphoses from mudstone or shale, as shown in *Figure 29.* Slate is a hard rock that can be split very easily along planes in the rock, creating large, flat surfaces.

Figure 28

The Taj Mahal, in India, is made of marble, a metamorphic rock often used in buildings.

▶ **metamorphic rock** a rock that forms from other rocks as a result of intense heat, pressure, or chemical processes

Figure 29

A Mudstone is composed of silt- or clay-sized particles. Its characteristics can be seen in some examples of slate.

B Slate is a metamorphic rock that is transformed under heat and pressure from sedimentary shale rocks.

721

INCLUSION *Strategies*

- Developmentally Delayed
- Learning Disabled
- English Language Learners

Provide or ask students to collect at least 10 rocks. Have students categorize the sedimentary rocks as conglomerate, sandstone, mudstone, limestone, or shale according to the size of the fragments they contain. Have students then create a display using manila file folders. Each folder tab should be labeled with a rock category. Their displays should include descriptions of each category and the rock specimens they collected.

Did You Know?

The Rocks of Earth's Surface Ninety-five percent of the outer 10 km of Earth is igneous and metamorphic rock. Although sedimentary rock makes up less than 5% of Earth's crust, it is spread thinly over 75% of Earth's continental surfaces.

Demonstration, *continued*

tap the test tube again and heat it for 3 more minutes.

Step 5 Place the first boat on the hot plate, and turn the hot plate off. Place the second boat on the table, away from the hot plate.

Step 6 Using the test-tube tongs, remove the test tube from the beaker of water and evenly distribute the contents among the boats. Do not move or disturb the boats. Ask students how the temperature of the solution will affect the size of the crystals and the rate at which the crystals form.

Step 7 Record the time it takes for the first crystals to appear. When the crystals form, have the students examine them using a hand lens.

Teaching Tip

Foliated Versus Nonfoliated
Texture can be used to classify metamorphic rock. A metamorphic rock can be either foliated or nonfoliated. Foliated metamorphic rock consists of minerals that are aligned or sorted into layers, while minerals in nonfoliated metamorphic rock are randomly arranged. Many metamorphic rocks break easily along planes because the minerals are aligned along those planes.

Teaching Tip

Continental Rock Continental crust is much older than the surrounding oceanic crust. The core of a continent, called a *craton,* is generally composed of ancient, crystalline igneous and metamorphic rock. Cratons are relatively thick and make up the most stable part of continents. They range from 200 million to 3.9 billion years old.

Chapter Resource File

- **Cross-Disciplinary Worksheet** Integrating Chemistry— From Granite to Paper ADVANCED

- **Cross-Disciplinary Worksheet** Connection to Language Arts— Names of Rocks ADVANCED

Interpreting Visuals Have students study **Figure 30,** which illustrates the rock cycle. Ask students to orally summarize the rock cycle, using only the diagram. **LS Verbal**

Teaching Tip

Rock Cycle Be sure students understand that rocks rarely undergo the complete process shown in the rock-cycle diagram. Sedimentary rocks can become igneous rocks, and metamorphic rocks can become sedimentary rocks. Some students may not realize the length of time it takes for changes to occur in the rock cycle. The processes shown in the diagram can take millions of years.

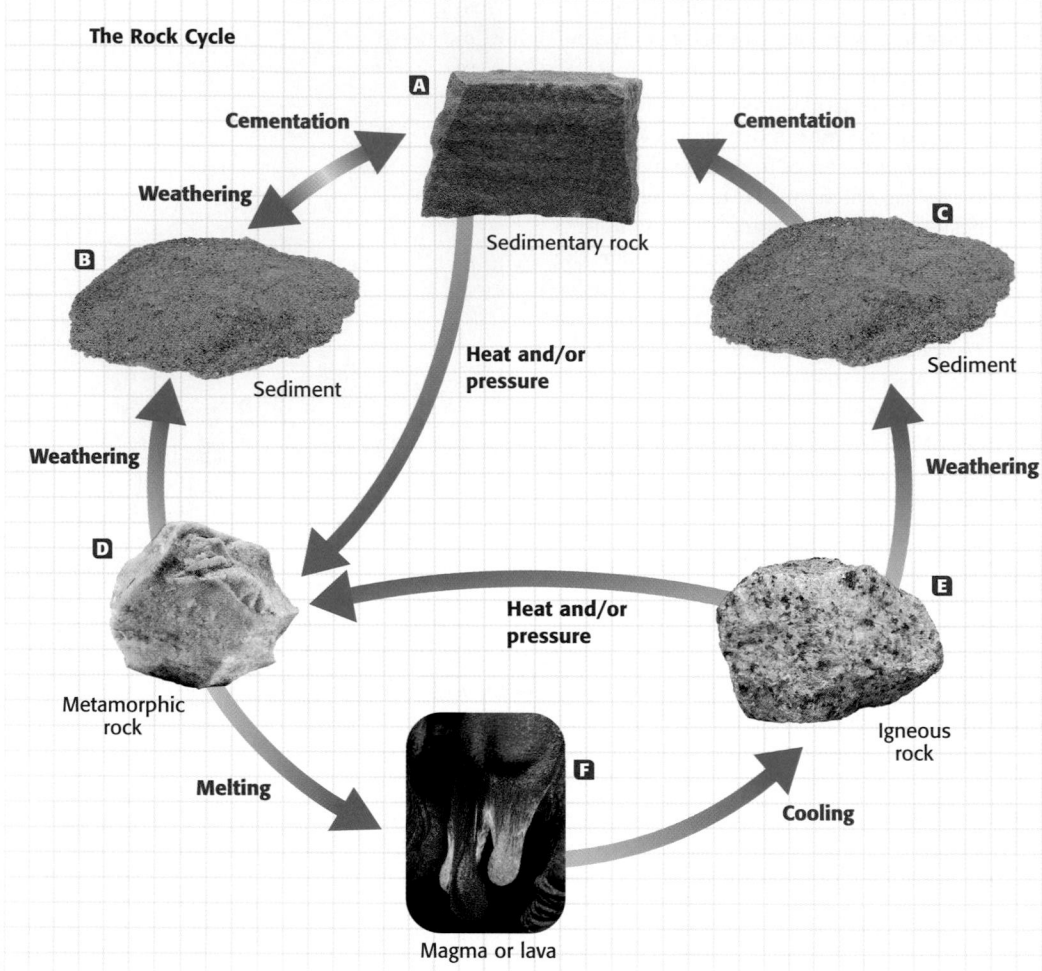

The Rock Cycle

A Sedimentary rock
Cementation
Weathering
B Sediment
C Sediment
Cementation
Heat and/or pressure
Weathering
Weathering
D Metamorphic rock
E Igneous rock
Heat and/or pressure
Melting
F Magma or lava
Cooling

Figure 30
The rock cycle illustrates the changes that sedimentary, igneous, and metamorphic rocks undergo.

Old rocks in the rock cycle form new rocks

So far, you have seen some examples of one type of rock becoming another. For instance, limestone exposed to heat and pressure becomes marble. Exposed rocks are weathered, forming sediments. These sediments may be cemented together to make sedimentary rock. The various types of rock are all a part of one rock system. The sequence of events in which rocks can be weathered, melted, altered, and formed is described by the *rock cycle*.

Figure 30 illustrates the stages of the rock cycle. Regardless of which path is taken, rock formation occurs very slowly, often over tens of thousands to millions of years.

722

Scientists Around the World

Avicenna The Islamic scholar Avicenna (980–1037) contributed immensely to our knowledge of medicine, astronomy, mathematics, and geology. In the *Book of Minerals*, he describes how rivers and seas lay down sediment that eventually becomes rock. Avicenna's theories contributed to the foundations of Western geology. Many of his controversial ideas did not gain acceptance in Europe until the 1600s.

Did You Know ?

Ancient Rocks The oldest known fossils are from rocks that were deposited 3.5 billion years ago! This is about 1 billion years after Earth's formation, which occurred approximately 4.6 billion years ago.

As magma or lava (F) cools underground, it forms igneous rock (E), such as granite. If the granite is heated and put under pressure, it may become metamorphic rock (D); if it is exposed at the surface of Earth, it may be weathered and become sand (B, C). The sand may be transported, deposited, and cemented to become the sedimentary rock (A) sandstone. As more time passes, several other layers of sediment are deposited above the sandstone. With enough heat and pressure, the sandstone becomes a metamorphic rock (D). This metamorphic rock (D) may then be forced deep within Earth, where it melts, forming magma (F).

internet connect ≡

www.scilinks.org
Topic: Rock types
SciLinks code: HK4122

SCiLINKS. Maintained by the National Science Teachers Association

How Old Are Rocks?

Rocks form and change over millions of years. It is difficult to know the exact time when a rock formed. To determine the age of rocks on a geological time scale, several techniques have been developed.

The relative age of rocks can be determined using the principle of superposition

Think about your hamper of dirty clothes at home. If you don't disturb the stack of clothes in the hamper, you can tell the relative time the clothes were placed in the hamper. In other words, you may not know how long ago you placed a particular red shirt in the hamper, but you can tell that the shirts above the red shirt were placed there more recently. In a similar manner, the *relative age* of rocks can be determined using the *principle of superposition*. The principle of superposition states the following:

Assuming no disturbance in the position of the rock layers, the oldest will be on the bottom, and the youngest will be on top.

The principle of superposition is useful in studying the sequence of life on Earth. For instance, the cliffside in *Figure 31* shows several sedimentary layers stacked on top of one another. The layers on the bottom are older than the layers above them.

Although the various layers of sedimentary rock are most visible in cliffsides and canyon walls, you would also find layering if you dug down anywhere there is sedimentary rock. By applying the principle of superposition, scientists know that fossils in the upper layers are the remains of animals that lived more recently than the animals that were fossilized in lower layers.

Figure 31

According to the principle of superposition, the layers of sedimentary rock on top are the most recent layers if the rocks have not been disturbed.

723

Historical Perspective

Dating Rocks The relative positions of sedimentary rocks were studied as early as the mid-1600s by the Danish scientist Nicholas Steno. Steno formulated the Law of Original Horizontality, which states that rock layers originally form horizontally. Slanted or vertical layers either formed in an unusual way or became tilted during a geologic event, such as mountain building.

Relative rock dating was a common practice in the 18th and 19th centuries, when many geologists began studying rock layers.

Geologists divided Earth's history into different eras, but they were unable to assign numeric values to these time periods. They believed that Earth was very old but had no precise way to estimate Earth's age.

Absolute dating was not practiced until the 20th century, when scientists began to understand the process of radioactive decay. Absolute dating allowed geologists to determine the absolute time periods of the geologic eras and to estimate the age of Earth.

SKILL BUILDER — GENERAL

Interpreting Visuals Have students analyze **Figure 31.** Make sure that students can see the different layers of sedimentary rock. Remind students that rock layers do not have to be parallel to the ground. Rock layers may be curved and folded because of compressive stress or angled relative to the ground. Tell students that the rocks that make up the walls of the Grand Canyon range from about 200 million years old to almost 2 billion years old. Does the principle of superposition tell us that the rocks formed continuously over this time? (No, it states only that the deepest rocks are the oldest and that the rocks higher up in the canyon are younger.) **LS Visual**

Teaching Tip ——— ADVANCED

Radioactive Dating Review the concepts of radioactive decay and half-lives with the students. Then, give them the following problem: "Potassium-40 decays to become argon-40 within a half-life of 1.3 billion years. A sample of volcanic rock has been analyzed for potassium-40 to determine the rock's absolute age. The rock contains 8.0×10^{-5} g of potassium-40 and 2.4×10^{-4} g of argon-40. How old is the rock?" (2.6 billion years old; the ratio of daughter isotope to parent isotope is 3:1, so the rock contains a mixture of 75% daughter isotope and 25% parent isotope, thus, the rock's age is two half-lives, or 2.6 billion years)

Ask students to explain which isotopes (those with long half-lives or short half-lives) should be used to date older rocks and which should be used to date younger rocks. (Isotopes with long half-lives give accurate ages for older rocks. Isotopes with short half-lives give accurate ages for younger rocks.) **LS Logical**

Quiz

Determine which type of rock is described by each statement. Some types may be used more than once.

1. This rock forms from compressed or cemented deposits. (sedimentary)

2. This rock forms from heat, pressure, or chemical processes. (metamorphic)

3. This rock forms from cooled and hardened magma or lava. (igneous)

4. This rock may contain fossils. (sedimentary)

5. This rock often forms deep in Earth's crust. (metamorphic)

Did You Know?

Radioactive dating is not always accurate. For instance, as heat and pressure are applied to a rock and water flows through it, soluble radioactive materials can escape from the minerals in the rock. Because there is often no method for measuring how much radioactive material is lost, it is difficult to accurately date some older rocks that have been heated and put under pressure or that are partly weathered.

Radioactive dating can determine a more exact, or absolute, age of rocks

The chapter on nuclear changes showed that the nuclei of some isotopes decay, emitting energy at a fairly constant rate. These isotopes are said to be radioactive. The radioactive elements that make up minerals in rocks decay over billions of years. Physicists have determined the rate at which these elements decay, and geologists can use this data to determine the age of rocks. They measure both the amount of the original radioactive material left undecayed in the rock and the amount of the product of the radioactive material's decay. The amount of time that passed since the rock formed can be calculated from this ratio.

Many different isotopes can be analyzed when rocks are dated. Some of the most reliable are isotopes of potassium, argon, rubidium, strontium, uranium, and lead.

While the principle of superposition gives only the relative age of rocks, radioactive dating gives the *absolute age* of a rock.

SECTION 3 REVIEW

SUMMARY

▸ Igneous rocks are formed from cooling molten rock.

▸ Sedimentary rocks form by the deposition of pieces of other rocks and the remains of living organisms.

▸ Metamorphic rocks form after exposure to heat and/or pressure for an extended time.

▸ Rocks can change type, as described by the rock cycle.

▸ The relative age of rock can be determined using the principle of superposition. Unless the layers are disturbed, the layers on the bottom are the oldest.

▸ Radioactive dating is used to determine the absolute age of rocks.

1. **Modify** the following false statement to make it a true statement: Fossils are found in igneous rock.

2. **Explain** how the principle of superposition is used by geologists to compare the ages of rocks.

3. **Determine** the type of rock that will form in each of the following scenarios:
   a. Lava pours onto the ocean floor and cools.
   b. Minerals cement small pieces of sand together.
   c. Mudstone is subjected to great heat and pressure over a long period of time.

4. **Explain** why a construction worker who uses a jackhammer on a rock does not produce a metamorphic rock.

5. **Identify** what type of rock might have a lot of holes in it due to the formation of gas bubbles. Explain your answer.

6. **Critical Thinking** A paleontologist who is researching extinctions notices that certain fossils are never found above a layer of sediment containing the radioactive isotope rubidium-87 or below another layer containing the same isotope. To determine when these animals became extinct, should the paleontologist use relative dating, absolute dating, or a combination of the two? Explain your answer.

724

Answers to Section 3 Review

1. "Fossils are found in sedimentary rock." It is also true to say, "Fossils are not found in igneous rock."

2. Because of the principle of superposition, geologists know that the oldest rocks will always be in the bottom layers and that the youngest rocks will be in the top layers.

3. a. igneous
   b. sedimentary
   c. metamorphic

4. Metamorphic rocks form when heat and pressure cause changes in a rock's texture and mineral content. A jackhammer breaks up rock, which may change the rock's texture, but it does not change the mineral content of the rock.

5. An extrusive igneous rock may have a lot of holes in it because of the formation of gas bubbles. Igneous rocks form from cooled molten rock from a volcanic eruption. Gases trapped in the magma form bubbles in the molten rock because the pressure decreases when the volcano erupts.

See "Continuation of Answers" at the end of the chapter.

Weathering and Erosion

OBJECTIVES

▶ **Distinguish** between chemical and physical weathering.
▶ **Explain** how chemical weathering can form underground caves in limestone.
▶ **Describe** the importance of water to chemical weathering.
▶ **Identify** three different physical elements that can cause erosion.

▶ KEY TERMS

acid precipitation
erosion
deposition

Compared to the destructive power of an earthquake or a volcano, the force exerted by a river may seem small. But, over time, forces such as water and wind can make vast changes in the landscape. Parunaweep Canyon, shown in *Figure 32,* is one of the most magnificent examples of how water can shape Earth's surface.

Physical Weathering

There are two types of weathering processes: physical and chemical. Physical, or mechanical, weathering breaks rocks into smaller pieces but does not alter their chemical compositions. Erosion by water or wind are examples of physical weathering. Chemical weathering breaks down rock by changing its chemical composition.

Ice can break rocks

Ice can play a part in the physical or mechanical weathering of rock. A common kind of mechanical weathering is called *frost wedging.* This occurs when water seeps into cracks or joints in rock and then freezes. When the water freezes, its volume increases by about 10%, pushing the rock apart. Every time the ice thaws and refreezes, it wedges farther into the rock, and the crack in the rock widens and deepens. This process eventually breaks off pieces of the rock or splits the rock apart.

Plants can also break rocks

The roots of plants can also act as wedges as the roots grow into cracks in the rocks. As the plant grows, the roots exert a constant pressure on the rock. The crack continues to deepen and widen, eventually causing a piece of the rock to break off.

Figure 32
Parunaweep Canyon, in Zion National Park, Utah, is a striking example of the effect of water on Earth's surface.

725

Focus

Overview

Before beginning this section, review with your students the Objectives listed in the Student Edition. In this section, students will study examples of weathering, including frost wedging, underground cave formation, and acid precipitation. They will also explore three physical causes of erosion: water, glacier, and wind.

🔘 Bellringer

Use the Bellringer transparency to prepare students for this section.

Motivate

Opening Demonstration

To demonstrate the role of ice in weathering, place some small pebbles in a plastic container. (Note: Do not use glass.) Fill the container with water, and freeze overnight. Have students generate hypotheses about what will happen to the water and pebbles as they freeze. Let students observe the container the following day to see if their hypotheses were correct.

Teach

SKILL BUILDER — BASIC

Interpreting Visuals Have students examine **Figure 32,** and ask them to describe what they see. Encourage students to hypothesize what forces caused the canyon to look the way that it does. **LS Visual**

Did You Know ?

Kolob Arch Parunaweep Canyon is not the only spectacular sight in Zion National Park. Park visitors can also view Kolob Arch, the largest arch in the world. It has a span of 310 feet!

Alternative Assessment — ADVANCED

Examples of Weathering Have students conduct research to find examples of physical and chemical weathering in your area. Examples could include major landforms, such as canyons and caves, or the effects of acid precipitation on local statues and structures. Ask students to choose one example to illustrate in a poster. The poster should use text and illustrations to explain what kind of weathering occurred and what changes were produced. **LS Logical**

Chapter Resource File

• **Cross-Disciplinary Worksheet**
Integrating Biology—Living Sources of Weathering **GENERAL**

• **Lesson Plan**

Transparencies

TT Bellringer

SKILL BUILDER — BASIC

Reading Skills As students read the sections about weathering, ask them to create a table that has the following headings: "Description of process," "Agents of weathering", and "Examples." Have students fill in the table by using information on the two types of weathering (physical and chemical). LS Logical

SKILL BUILDER — GENERAL

Interpreting Visuals Have students look at **Figure 34** and identify the features along the ceiling of the cave. Ask students how the stalactites could have formed if limestone was dissolved away by acidic rainwater to form the cave. These stalactites are made of a type of limestone called *dripstone*. Water and dissolved limestone that drip downward from cracks in a cave's ceiling leave behind calcite deposits that form the stalactite. At the same time, water drops that fall to the cave's floor build the stalagmites. LS Visual

REAL-LIFE
CONNECTION — BASIC

Oxidation and Rusting Tell your students that rusting is an example of oxidation. Challenge your students to think of common household items that may be subject to oxidation. What do these items have in common? LS Visual

Figure 33
Red sedimentary layers in Badlands National Park contain iron that has reacted with oxygen to form hematite.

Figure 34
Carbonic acid dissolved the calcite in the sedimentary rock limestone to produce this underground cavern.

Chemical Weathering

Figure 33 shows the sedimentary layers in Badlands National Park, in South Dakota. They appear red because they contain hematite. Hematite, Fe_2O_3, is one of the most common minerals and is formed as iron reacts with oxygen in an oxidation reaction. When certain elements, especially metals, react with oxygen, they become oxides and their properties change. When these elements are in minerals, oxidation can cause the mineral to decompose or form new minerals. This is an example of *chemical weathering*. The results of chemical weathering are not as easy to see as those of physical weathering, but chemical weathering can have a great effect on the landscape over millions of years.

Carbon dioxide can cause chemical weathering

Another common type of chemical weathering occurs when carbon dioxide from the air dissolves in rainwater. The result is water that contains carbonic acid, H_2CO_3. Although carbonic acid is a weak acid, it reacts with some minerals. As the slightly acidic water seeps into the ground, it can weather rock underground.

For example, calcite, the major mineral in limestone, reacts with carbonic acid to form calcium bicarbonate. Because the calcium bicarbonate is dissolved in water, the decomposed rock is carried away in the water, leaving underground pockets. The cave shown in **Figure 34** resulted from the weathering action of carbonic acid on calcite in underground layers of limestone.

726

LIFE SCIENCE
CONNECTION — GENERAL

Show students the effect of acid precipitation on plant growth. Plot on the board a graph of the data in the table on the opposite page. Plant A was not exposed to acid precipitation, while Plant B was. Relate these results and the graph to the section on acid precipitation. Ask students to identify the trend of the data. (The average leaf length of the plant that was not exposed to acid precipitation is increasing, while the average leaf length of the plant that was exposed to acid precipitation is decreasing.) LS Logical

Time (days)	Average leaf length for Plant A (cm)	Average leaf length for Plant B (cm)
1	12	12
3	12	11.5
5	12.5	11
7	13	10.5
9	14	10
11	14	10
13	14.5	9.5

Water plays a key role in chemical weathering

Minerals react chemically with water. This reaction changes the physical properties of minerals, and often changes entire landscapes. Other times, minerals dissolve completely into water and are carried to a new location. Often minerals are transported to lower layers of rock. This process is called *leaching*. Some mineral ore deposits, like those mined for aluminum, are deposited by leaching.

Water can also carry dissolved oxygen that reacts with minerals that contain metals such as iron. This type of chemical weathering is called *oxidation*. When oxygen combines with the iron found in rock, it forms iron oxide, or rust. The red color of soil in some areas of the southeastern United States is mainly caused by the oxidation of minerals containing iron.

Acid precipitation can slowly dissolve minerals

Rain and other forms of precipitation have a slightly acidic pH, around 5.7, because they contain carbonic acid. When fossil fuels, especially coal, are burned, sulfur dioxide and nitrogen oxides are released and may react with water in clouds to form nitric acid, or nitrous acid, and sulfuric acid. These clouds form precipitation that falls to Earth as **acid precipitation.** The pH value of rainwater in some northeastern United States cities between 1940 and 1990 averaged between 4 and 5. In some individual cases, the pH dropped below 4, to levels nearly as acidic as vinegar.

Acid precipitation causes damage to both living organisms and inorganic matter. Acid rain can erode metal and rock, such as the statue in Brooklyn, New York, shown in *Figure 35.* Marble and limestone dissolve relatively rapidly even in weak acid.

In 1990, the Acid Rain Control Program was added to the Clean Air Act of 1970. According to the program, power plants and factories were given 10 years to decrease the release of sulfur dioxide to about half the amount they emitted in 1980. The acidity of rain has been greatly reduced since power plants have installed *scrubbers* that remove the sulfur oxide gases.

internet connect

www.scilinks.org
Topic: Weathering
SciLinks code: HK4152

SciLINKS Maintained by the National Science Teachers Association

■ **acid precipitation** precipitation, such as rain, sleet, or snow, that contains a high concentration of acids, often because of pollution in the atmosphere

Figure 35

Acid precipitation weathers stone structures, such as this marble statue in Brooklyn, New York.

Teaching Tip

Chemistry Review You may wish to review chemical reactions and acids with your students. Explain that calcite (calcium carbonate, $CaCO_3$) slowly dissolves in rain water, which contains some dissolved CO_2, to form calcium bicarbonate, $Ca(HCO_3)_2$. This process involves three steps.

Step 1: CO_2 dissolves in rain water.
$$CO_2 + H_2O = H_2CO_3$$

Step 2: Some of the carbonic acid ionizes in the rain water.
$$H_2CO_3 + H_2O = H_3O^+ + HCO_3^-$$

Step 3: Calcite dissolves in the carbonic acid solution to form dissolved calcium and bicarbonate.
$$CaCO_3 + H_3O^+ = Ca^{2+} + HCO_3^- + H_2O$$

This reaction is reversible. Some of the dissolved calcium and bicarbonate may precipitate very slowly over many years to form solid calcium carbonate (calcite), which makes the stalactites and stalagmites found in limestone caves. Therefore, in limestone country, water contains dissolved calcium and bicarbonate, which makes the water "hard."

727

Demonstration

Erosion

(Time: About 15 minutes)

Materials:

- erosion tray (or a rain gutter or baking pan)
- soil or sand
- pebbles
- water

Put the pebbles in a layer on the bottom of the erosion tray, and add a layer of soil or sand on top. Tilt one end of the tray upward by placing some books underneath one end. Slowly drip water over the top of the tray. Have students observe the action and flow of the water as it cuts a path and carries soil to the bottom of the tray. Ask students to relate what they have seen in the demonstration to the words *erosion* and *deposition*. Ask students to hypothesize what will happen if the tray is tilted further upward or is placed on a level surface.

Teaching Tip

River Erosion A river or stream can carry particles in three ways. A river can bounce large materials, such as pebbles and boulders, along the riverbed. Small rocks and soil can be carried in suspension. This suspended material makes a river look muddy. When the current slows to a point that the particles can no longer be carried in suspension, the rocks and soil are deposited. Some ions, such as sodium and calcium, are carried in solution.

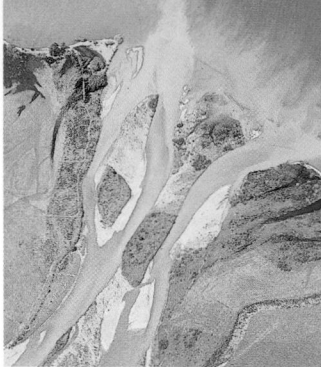

Figure 36

Deltas, such as this one in New Zealand, are formed by deposition.

▶ **erosion** a process in which the materials of the Earth's surface are loosened, dissolved, or worn away and transported from one place to another by a natural agent, such as wind, water, ice, or gravity

▶ **deposition** the process in which material is laid down

Erosion

Erosion is the removal and transportation of weathered and non-weathered materials by running water, wind, waves, ice, underground water, and gravity.

Water erosion shapes Earth's surface

Water is the most effective physical weathering agent. Have you ever seen a murky river? Muddy rivers carry sediment in their water. As sediment moves along with the water, it scrapes the riverbanks and the river bottom. As the water continues to scour the surface, it carries the new sediment away. This process of loosening and moving sediments is known as **erosion.**

There is a direct relationship between the velocity of the water and the size and amount of sediment it can carry. Quickly moving rivers can carry away a lot of sediment, and create extraordinary canyons.

As a river becomes wider or deepens, it flows more slowly and cannot carry as much sediment. As a result, sediment is deposited on the floor of these calmer portions of the river or stream. The process of depositing sediment is called **deposition.** Rivers eventually flow into large bodies of water, such as seas and oceans, where the sediment is deposited along the continental shores. As rivers slow at the continental boundary, large deposits of sediment are laid down. These areas, called deltas, often have rich, fertile soils, making them excellent agricultural areas. *Figure 36* shows the Greenstone River delta, in New Zealand.

Oceans also shape Earth

The oceans also have a dramatic effect on Earth's landscape. On seashores, the waves crash onto land, creating tall cliffs and jagged coastlines. The Cliffs of Moher, in western Ireland, shown in *Figure 37,* reach heights of 204 m (669 ft) above the water. The cliffs were formed partially by the force of waves in the Atlantic Ocean eroding the rocky shale and sandstone coast.

Figure 37

The action of waves slowly tearing away at the rocky coast formed the Cliffs of Moher.

728

Did You Know ?

Erosion and Deposition When water flows down a river channel, it does not flow straight down the channel. Water flowing in a river or stream channel moves in a helical, or corkscrew, motion. This motion causes erosion on the bank where the water is rising and causes deposition on the bank where the water is falling. The helical flow of water helps explain the formation of bends or meanders in the channel.

Glaciers erode mountains

Large masses of ice, such as the glacier shown in *Figure 38A,* can exert tremendous forces on rocks. The constantly moving ice mass carves the surface it rests on, often creating U-shaped valleys, such as the one shown in *Figure 38B.* The weight of the ice and the forward movement of the glacier cause the mass to act like a huge scouring pad. Immense boulders that are carried by the ice scrape across other rocks, grinding them to a fine powder. Glacial meltwater streams carry the fine sediment away from the glacier and deposit it along the banks and floors of streams or at the bottom of glacier-formed lakes.

Wind can also shape the landscape

Just as water or glaciers can carry rocks along, scraping other rocks as they pass, wind can also weather the Earth's surfaces. Have you ever been in a dust storm and felt your skin "burn" from the swirling dust? This happens because fast-moving wind can carry sediment, just as water can. Wind that carries sediment creates a sandblaster effect, smoothing Earth's surface and eroding the landscape.

The sandstone arches of Arches National Park, in Utah, are formed partly by wind erosion. Look at *Figure 39.* Can you guess how these arches might have formed? Geologists have struggled to find a good explanation for the formation of arches.

The land in and around Arches National Park is part of the Colorado Plateau, an area that was under a saltwater sea more than 300 million years ago. As this sea evaporated, it deposited a thick layer of salt that has since been covered by many layers of sedimentary rock. The salt layer deforms more easily than rock layers. As the salt layers warped and deformed over the years, they created surface depressions and bulges. Arches formed where the overlying sedimentary rocks were pushed upward by the salt.

Figure 38

A Tustamena Glacier, in Alaska, has slowly pushed its way through these mountains.

B Glaciers are capable of carving out large U-shaped valleys, such as this valley in Alaska.

internet connect

www.scilinks.org
Topic: Erosion
SciLinks code: HK4050

*sci*LINKS. Maintained by the National Science Teachers Association

Figure 39

This sandstone arch in Arches National Park, in Utah, was created as high-speed winds weathered the terrain.

729

Trends in Technology

Discovering Ruins The Taka Makan Desert in China's arid northwest is so inhospitable that its name in the local language means "Go in, and you don't come out." The desert is covered with treacherous dunes of fine, dry sand. Buried under those dunes are the remains of cities that prospered along the ancient Silk Road. The Silk Road was a trade route that connected China to civilizations in the West. NASA's Spaceborne Imaging Radar (SIR-C), which flew on space shuttles twice in 1994, is being used to examine the desert. The radar-imaging technology has already helped archeologists locate some buried cities and promises to help them find other ruins.

Quiz

For questions 1–4, determine whether the statement is an example of physical weathering or chemical weathering.

1. Acid precipitation damages a marble statue. (chemical)

2. Plant roots split a rock into two pieces. (physical)

3. Oxygen in the air reacts with iron to form hematite. (chemical)

4. Freezing water expands in a crack of a rock, which breaks the rock. (physical)

5. What are three agents of erosion? (water, ice, and wind)

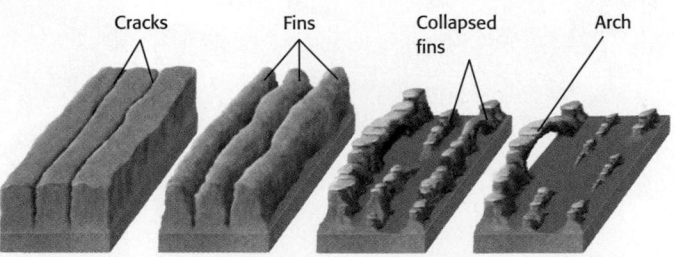

Figure 40
Fins are formed when sandstone is pushed upward, and cracks are slowly eroded. Wind, water, and ice erode the fins until they collapse or form arches.

Cracks   Fins   Collapsed fins   Arch

Figure 40 shows how one theory explains the formation of arches. As land is pushed upward in places, small surface cracks form. These cracks are eroded by water, ice, and wind until narrow free-standing rock formations, called *fins*, are formed. When these fins are exposed along their sides, the wind wears away at the cement that holds the sediment together, causing large pieces of the rock to fall away. Some fins collapse completely; others that are more sturdy and balanced form arches.

SECTION 4 REVIEW

SUMMARY

▶ Physical weathering breaks down rock by water erosion, ice wedging, wind abrasion, glacial abrasion, and other forces.

▶ In chemical weathering, rock is altered as minerals in rock react chemically.

▶ Carbonic acid acts as a chemical weathering agent and is responsible for the formation of underground limestone caves.

▶ Water plays an important role in shaping Earth's landscape.

▶ Acid precipitation can weather rock and harm living organisms.

1. **List** two agents of physical weathering that might occur in the mountains in northern Montana.

2. **Explain** how the wind may be involved in the formation of sandstone arches.

3. **Distinguish** between physical weathering, chemical weathering, and erosion in the following examples:
   a. Rock changes color as it is oxidized.
   b. Rock shatters as it freezes.
   c. Wind erodes the sides of the Egyptian pyramids in Giza.
   d. An underground cavern is formed as water drips in from Earth's surface.

4. **Explain** why the following statement is incorrect: Acid precipitation is any precipitation that has a pH less than 7.

5. **Predict** which will experience more weathering, a rock in the Sonora Desert, in southern Arizona, or a rock on a beach in North Carolina.

6. **Critical Thinking** On many coastlines, erosion is wearing the beach away and threatening to destroy homes. How would you prevent this destruction?

730

Chapter Resource File

• **Concept Review** GENERAL

• **Quiz** BASIC

Answers to Section 4 Review

1. Two agents of physical weathering that may occur in the mountains in northern Montana are ice and plants.

2. After the narrow, free-standing rocks (fins) form, exposure to the wind leaves them vulnerable to erosion. As the wind wears away at the cement that holds the sediments together, large pieces of rock fall away, while others are more sturdy and form balanced arches.

3. **a.** chemical weathering
   b. physical weathering
   c. erosion
   d. chemical weathering

4. The statement is incorrect because in most areas, natural rain has a slightly acidic pH of about 5.7.

5. A rock on a beach in North Carolina will experience more weathering because it is exposed to more physical and chemical weathering agents—water, wind, and sea water.

6. Answers may vary but may include ideas such as jetties, groins, sea walls, beach nourishment (bringing sand from other places and pumping it onto the beach), and moving homes away from the water.

Study Skills

Concept Mapping

Concept mapping is an important study guide and a good way to check your understanding of key terms and concepts.

1 **Select a main concept for the map.**

We will use tectonic plates as the main concept of this map.

2 **List all the important concepts.**

We'll use the terms: lithosphere, divergent boundaries, convergent boundaries, and transform fault boundaries.

3 **Build the map by placing the concepts according to their importance under the main concept, tectonic plates, and add linking words to give meaning to the arrangement of concepts.**

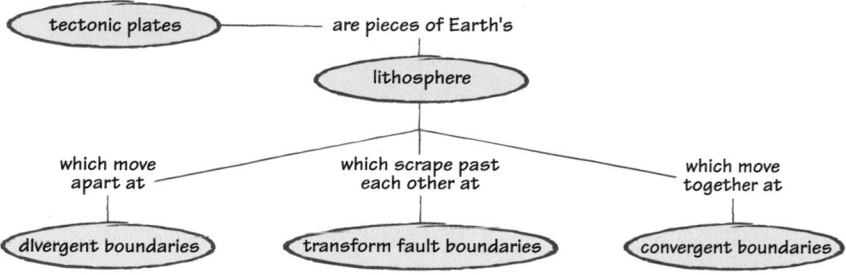

From this completed concept map we can write the following propositions:

Tectonic plates are pieces of Earth's lithosphere which move apart at divergent boundaries.

Tectonic plates scrape past each other at transform fault boundaries.

Tectonic plates move together at convergent boundaries.

Practice

1. Add on to the concept map using the words *earthquake, volcano,* and *fault,* as well as the appropriate linking words.

2. Use as the main concept erosion and create your own concept map.

3. Write two propositions from your completed concept map.

731

Teaching Tip

Alternative Arrangements Show that those concepts that relate in a cause-and-effect way can also be organized in a process flow chart. For instance, the concept map in Figure 41 would form three linear process flow charts. The first one of these would connect the phrase *tectonic plates are pieces of Earth's lithosphere,* to the phrase *plates at divergent boundaries move apart,* which would then connect to the phrase *forming mid-oceanic mountain ridges and volcanoes.*

Have students construct process flow charts for the other two cases given in the example, adding a third step for results. (*Tectonic plates are pieces of Earth's lithosphere* connects to *plates at transform fault boundaries scrape past each other,* which connects to *causing earthquakes to occur. Tectonic plates are pieces of Earth's lithosphere* connects to *plates at convergent boundaries move together,* which connects to *causing mountain ranges, volcanoes, and ocean trenches to form.*)

Answers

1. The phrases *which leads to the formation of a, along which occur,* and *which causes the formation of* connect *divergent boundaries* to *volcano, transform fault boundaries* to *earthquakes,* and *convergent boundaries* to *mountain ranges,* respectively.

2. Answer may vary. See the sample concept map below.

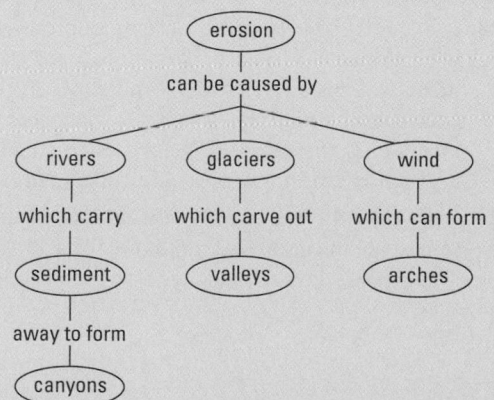

3. Answers may vary. For the sample map in item 2, the propositions would be two of the following: "Erosion can be caused by rivers, which carry sediment away to form canyons."

"Erosion can be caused by glaciers, which carve out valleys."

"Erosion can be caused by wind, which can form arches."

Understanding Concepts

1. a
2. d
3. b
4. a
5. d
6. c
7. b
8. a
9. c
10. d
11. c
12. a
13. a
14. a
15. d

Using Vocabulary

16. The crust is Earth's topmost layer, the mantle lies beneath the crust, and the core is the center of Earth. There is a liquid outer core and a solid inner core.

17. The relative age of a rock tells whether a rock is older or younger than another rock, based on the apparent order in which the rocks were deposited in sedimentary layers. The absolute age of the rock is the actual age of the rock in years. Absolute age can be determined using radioactive dating methods.

18. seismology

Chapter Highlights

Before you begin, review the summaries of the key ideas of each section, found at the end of each section. The key vocabulary terms are listed on the first page of each section.

UNDERSTANDING CONCEPTS

1. The thin layer of crust and upper mantle that makes up the tectonic plates is called the
   a. lithosphere.
   b. oceanic crust.
   c. asthenosphere.
   d. tectonic plate boundary.

2. Two tectonic plates moving away from each other form a(n)
   a. transform fault boundary.
   b. convergent boundary.
   c. ocean trench.
   d. divergent boundary.

3. Vibrations in Earth caused by sudden movements of rock are called
   a. epicenters.   c. faults.
   b. earthquakes.   d. volcanoes.

4. Using the difference in the time it takes for P waves and S waves to arrive at three different seismograph stations, seismologists can find an earthquake's
   a. epicenter.   c. fault zone.
   b. surface waves.   d. intensity.

5. The Richter scale expresses an earthquake's
   a. damage.   c. duration.
   b. location.   d. magnitude.

6. High pressure and high temperature cause igneous rocks to become
   a. sedimentary rocks.
   b. limestone.
   c. metamorphic rocks.
   d. clay.

7. The sequence of events in which rocks change from one type to another and back again is described by
   a. a rock family.
   b. the rock cycle.
   c. metamorphism.
   d. deposition.

8. _____ rock is formed from magma.
   a. Igneous   c. Sedimentary
   b. Metamorphic   d. Schist

9. A common kind of mechanical weathering is called
   a. oxidation.   c. frost wedging.
   b. carbonation.   d. leaching.

10. Underground caves in limestone can be formed by a reaction including
    a. sulfuric acid.   c. ice crystals.
    b. hematite.   d. carbonic acid.

11. Which of the following is not part of Earth's interior?
    a. core   c. lava
    b. crust   d. mantle

12. Which of the following is not a type of plate boundary?
    a. oceanic   c. convergent
    b. divergent   d. transform fault

13. The most common type of volcano is the
    a. cinder cone.
    b. composite volcano.
    c. shield volcano.
    d. giant volcano.

14. What causes the area around the Pacific Ocean to have abundant volcanoes?
    a. convergent plates   c. magnetic fields
    b. divergent plates   d. lava fields

15. Which one of these forces shapes Earth's surface?
    a. glaciers   c. winds
    b. oceans   d. all of the above

732

19. The point inside Earth where an earthquake originates is called the *focus*. The epicenter is the point on the surface of Earth immediately above the focus. An earthquake releases energy through three types of waves. The first type is a P wave, which is the first to reach seismology recording stations. The second type of wave originating from an earthquake's focus is an S wave, which travels at a slower rate than a P wave does. Finally, surface waves move across the surface of Earth, and cause the most destruction.

20. sedimentary

21. sedimentary

22. weathering

23. Erosion is the process of loosening and moving sediments. A river carries these loose sediments, and as a river widens and slows at the continental boundary, it deposits these sediments onto the shores. This process is called *deposition*, and these areas are called *deltas*.

24. Answers will vary but may include: feldspar, pyroxene, mica, olivine, dolomite, quartz, amphibole, clay, and calcite.

16. Using the terms *crust, mantle,* and *core,* describe Earth's internal structure.

17. Explain the difference between the relative and the absolute age of a rock.

18. What is the name of the field of study concerning earthquakes?

19. Use the terms *focus, epicenter, P waves, S waves,* and *surface waves* to describe what happens during an earthquake.

20. What type of rock is formed when small rock fragments are cemented together?

21. In what type of rock—igneous, sedimentary, or metamorphic—are you most likely to find a fossil?

22. What is the name of the process of breaking down rock?

23. Explain how deltas form at the continental boundary using the terms *erosion* and *deposition.*

24. Name two minerals that form rock.

25. Explain what role water plays in chemical weathering, and in physical weathering.

26. Define acid precipitation and describe how it forms.

27. Restate the theory of plate tectonics using the terms: *plate tectonics, lithosphere,* and *continental drift.*

28. Explain what happens to a continental plate during subduction.

29. Explain why earthquakes happen near faults.

30. Describe the Richter Scale, including information on its purpose and what kind of phenomena it describes.

31. What kind of rock are marble and slate?

32. Describe how wind can erode rock and shape Earth's surface, and list two other elements that can cause erosion.

33. Describe the difference between a divergent and convergent plate boundary. Name the features found near each type of boundary.

34. **Graphing** Examine the graph in the figure below, and answer the questions that follow.

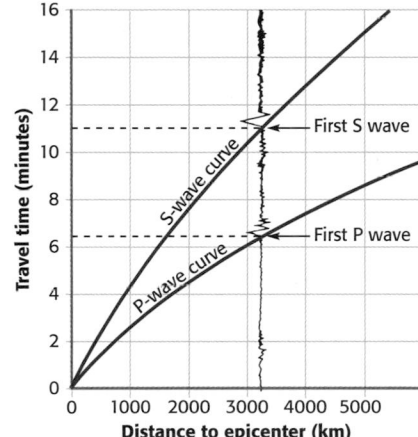

a. Approximately how far, in kilometers, was the epicenter of the earthquake from the seismograph that recorded the seismogram shown here?
b. When did the first P wave reach the seismograph?
c. When did the first S wave reach the seismograph?
d. What is the difference in travel time between the two waves?

25. Water contributes to chemical weathering as it carries solutes, such as oxygen or carbonic acid, that can react with minerals. Minerals can also react chemically with water itself. Water contributes to physical weathering as it slowly erodes or breaks up rock, as on a beach. When water freezes and expands, it may also break up rock.

26. Acid precipitation is precipitation that has an unusually high concentration of sulfuric or nitric acids resulting from chemical pollution in the air. The pollutants react with water in clouds to form precipitation that includes nitric or sulfuric acid(s).

27. The theory of plate tectonics states that Earth's surface is made up of large moving tectonic plates, which together comprise the lithosphere or thin outer shell of Earth. This theory is supported by evidence of continental drift, or the movement of continents and tectonic plates.

28. During subduction an oceanic plate dives beneath a continental plate, dragging the hydrated oceanic crust along with it. Rising magma from the resulting melting mantle intrudes and erupts on the continental crust, forming mountains. The magma addition changes the composition, thermal structure, and density of the continental crust.

29. Earthquakes happen near faults, because the moving plates along faults experience immense pressure, which can eventually break rock along the fault line. The energy released when rocks shift along a pre-existing fault creates earthquakes.

733

Assignment Guide

Section	Questions
1	1, 2, 11, 12, 16, 17, 27, 28, 33, 37, 41a, 41d, 44, 49
2	3–5, 13, 14, 18, 19, 29, 30, 34, 38, 41d, 43, 45–47
3	6–8, 20, 21, 24, 31, 35, 36, 39, 40, 48
4	9, 10, 15, 22, 23, 25, 26, 32, 41b, 41c

Chapter Resource File

• **Chapter Test** GENERAL
• **Text Item Listing for Exam View®** Test Generator

30. The purpose of the Richter scale is to express the relative magnitude of an earthquake. The Richter scale describes the measure of energy released by an earthquake. It does not predict how severe an earthquake will be in terms of its damage to Earth's surface.

31. metamorphic

32. Wind erodes rock and shapes Earth's surface by carrying sediment, creating a sandblaster effect. Water and glaciers can also cause erosion.

33. A divergent boundary occurs where two plates move apart. A convergent boundary occurs where two plates collide. Mountains, mid-oceanic ridges, and rift valleys form at divergent boundaries. Ocean trenches, mountains, and volcanoes form at convergent boundaries.

Building Math Skills

34. **a.** about 3200 km
    b. about 6.5 min after the earthquake started
    c. about 11 min after the earthquake started
    d. about 4.5 min

THINKING CRITICALLY

35. **Critical Thinking** Why are mathematics and theory the only practical ways to determine the temperature of Earth's core?

36. **Applying Knowledge** Is a tall building more likely to be damaged by an earthquake if it is on a mountain of granite or in a valley of sediment? Explain.

37. **Applying Knowledge** You are on a field trip to the top of a mountain. Your teacher tells you that the rocks are metamorphic. What can you tell about the geologic history of the area?

38. **Problem Solving** Imagine you are on a dinosaur dig and your team finds two sets of dinosaur bones entwined as if they died while engaged in battle. You know one of the dinosaurs lived 180 million to 120 million years ago and the other lived 130 million to 115 million years ago. What can you say about the age of the rock the dinosaur fossils were found in?

39. **Interpreting Diagrams** The figure below shows a cross-sectional diagram of a portion of a plate boundary. It shows plates, the ocean, mountains, and rivers. The mouths of the rivers reach the shore, where they deposit sediments. Using the diagram, answer the following questions:
    a. What type of plate boundary is shown? How do you know?
    b. What type of physical weathering processes are probably acting on the mountains?
    c. What type of land-shaping processes are occurring on and near the beach?
    d. What forces might have formed this coastal mountain range?

734

DEVELOPING LIFE/WORK SKILLS

40. **Interpreting Graphics** Use a map of the United States to plan a car trip to study the geology of three national parks. How many miles will you have to drive roundtrip? How many days will you stay in each park? How long will the trip take?

41. **Making Decisions** Pretend you are the superintendent of the Washington State Police Department. Seismologists in the area of Mount St. Helens predict that the volcano will erupt in one week. Write a report describing the evacuation procedures. In your report, describe the area you will evacuate and a plan for how people will be contacted.

42. **Applying Technology** Sonar is the best method for identifying features under water. Sound waves are emitted from one device, and the time it takes for the wave to bounce off an object and return is calculated by another. This information is used to determine how far away a feature is. Draw a graph that could be made by a ship's sonar device as it tracks the ocean bottom from the shoreline to a deep ocean trench and farther out to sea.

43. **Applying Technology** Use a computer drawing program to create cutaway diagrams of a fault before, during, and after an earthquake. COMPUTER SKILL

44. **Acquiring and Evaluating Data** Mount Mazama exploded with a great deal of force. Go to the library and research Mount Mazama. How big was the eruption? How might the eruption have affected Earth's global climate?

Thinking Critically

35. There is no way to measure the temperature of Earth's core directly because the core is too deep in Earth's interior for drilling to reach.

36. Because sediment is looser and not as compact or sturdy as granite is, a building in a valley of sediment is more likely to be damaged by an earthquake.

37. Over time, these rocks have been transformed by heat and pressure within Earth, and currently they have different textures and mineral content than they did in the past. The layer of rock may have been pushed upward by tectonic movement and then exposed by weathering and erosion.

38. The rock is between 120 million and 130 million years old.

39. **a.** convergent plate boundary; an oceanic plate is subducting under a continental plate to form an ocean trench and mountains
    b. frost wedging and physical weathering due to plant growth
    c. erosion and deposition
    d. Answers may vary. The continental crust may have buckled to the force of the collision between the plates. Magma produced during subduction of the oceanic plate may have risen to the surface to form a volcanic mountain range.

45. Connection to Biology Charles Darwin, known for his theory of evolution by natural selection, was the first to explain how atolls (AT ahls) form. What are atolls? What was his explanation?

46. Connection to Social Studies The Grand Canyon is one of the most geologically informative and beautiful sites in the United States. Read about the people who have lived along the Colorado River and in the Grand Canyon. How do we know Paleo-Indians lived there? Who were the Anasazi? Who was John Wesley Powell? Write a one-page essay about the history of these people and this region.

47. Concept Mapping Copy the unfinished concept map on this page onto a sheet of paper. Complete the map by writing the correct word or phrase in the lettered boxes.

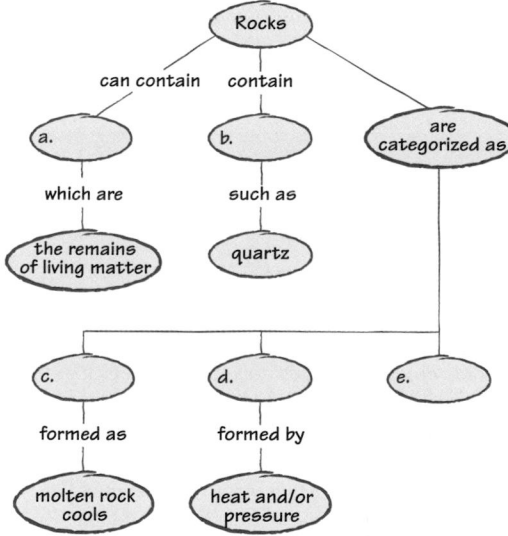

www.scilinks.org
Topic: Sonar   SciLinks code: HK4131

SCLINKS. Maintained by the National Science Teachers Association

Art Credits Fig. 1–3, Ortelius Design; Fig. 6, Uhl Studios, Inc.; Fig. 7, (arrows) Function Thru Form; Fig. 9–10, Uhl Studios, Inc.; Fig. 12, Uhl Studios, Inc.; Fig. 13, Function Thru Form; Fig. 14–15, Uhl Studios, Inc.; Fig. 17, Uhl Studios, Inc.; Fig. 18, Doug Walston; Fig. 19, Uhl Studios, Inc.; Fig. 21, Uhl Studios, Inc.; Fig. 22, Function Thru Form; Fig. 40, Uhl Studios, Inc.; "Thinking Critically" (diagram), Uhl Studios, Inc.; "Building Math Skills," (seismograph), Doug Walston; (pie graph), HRW; "Skill Builder Lab," (map), Doug Walston.

Photo Credits Chapter Opener photo of Wizard Island at Crater Lake, A & L Sinibaldi/Getty Images/Stone; wide shot of Crater Lake, Georg Gerster/Photo Researchers, Inc.; Fig. 4, Charles Palek/Animals Animals/Earth Scenes; Fig. 5, OAR/National Undersea Research Program (NURP)/NOAA; "Quick Activity," Sam Dudgeon/HRW; Fig. 7, Marie Tharp; Fig. 8, Wang Yuan-mao/AP/Wide World Photos; Fig. 9, Bernhard Edmaier/SPL/Photo Researchers, Inc.; Fig. 10, Galen Rowell/CORBIS; Fig. 11, Robin Prange/Corbis Stock Market; Fig. 12, David Parker/Science Photo Library/Photo Researchers, Inc.; "Quick Lab," Sam Dudgeon/HRW; Fig. 14, Mark Downey/Lucid Images; Fig. 16, Simon Kwong/Reuters/TimePix; Fig. 20, Greg Vaughn/Getty Images/Stone; Fig. 21(top), Michael T. Sedam/CORBIS; Fig. 21(center), Orion Press/Getty Images/Stone; Fig. 21(bottom), SuperStock; Fig. 23, Gary Braasch/CORBIS; Fig. 24, John M. Roberts/Corbis Stock Market; Fig. 25A, Breck P. Kent/Animals Animals/Earth Scenes, Fig. 25B, G.R. Roberts Photo Library; "Connection to Social Studies," National Geographic Image Collection/Kenneth Garrett; Fig. 26, Tom Bean/CORBIS; Fig. 27A, Grant Heilman Photography; Fig. 27B, Breck P. Kent/Animals Animals/Earth Scenes; Fig. 28, Harvey Lloyd/Corbis Stock Market; Fig. 29A-B, Fig. 30A, Sam Dudgeon/HRW; Fig. 30B-C, Image Copyright ©2004 PhotoDisc, Inc.; Fig. 30D, Grant Heilman Photography; Fig. 30E, Breck P. Kent/Animals Animals/Earth Scenes, Fig. 30F, SuperStock; Fig. 31, Laurence Parent; Fig. 32, James Kay; Fig. 33, Larry Ulrich/Getty Images/Stone; Fig. 34, Mark Gibson Photography; Fig. 35, Ray Pfortner/Peter Arnold, Inc.; Fig. 36, G.R. Roberts Photo Library; Fig. 37, SuperStock; Fig. 38A, Grant Heilman Photography; Fig. 38B, Visuals Unlimited/Bill Kamin, Fig. 39, Mark Tomalty/Masterfile; "Career Link" (all), Max Aguilera-Hellweg/GEO.

735

Developing Life/Work Skills

40. Answers may vary.

41. Answers may vary.

42. Check students' graphs for accuracy.

43. Check students' drawings for accuracy. The fault should be in noticeably different positions before and after the earthquake. The fault should be moving during the earthquake.

44. Mount Mazama was formerly a 12,000-foot-tall volcanic mountain. The eruption leveled the mountain, leaving a 7-mile-wide caldera, which is now known as Crater Lake. The ash and dust blown into the atmosphere might have blocked sunlight over a large portion of Earth's surface, causing a period of cooler temperatures.

Integrating Concepts

45. Atolls are rings of coral reefs that form on the top of underwater volcanic islands. Darwin correctly theorized that atolls form as volcanic islands slowly subside over time.

46. Students answers may vary. Paleo-Indians lived in the region about 9000 to 11 000 years ago. They left pictographs and rock paintings. The Anasazi (circa AD 100–AD 1300) were a Native American agricultural people of the southwestern U.S. who lived in adobe villages and great cliff dwellings. They are the predecessors of the Pueblo Indians. John Wesley Powell (1834–1902) was an American geologist and anthropologist who studied the Rocky Mountains, the canyons of the Green and Colorado rivers (including the Grand Canyon), and Native American languages. He was the first United States explorer to map the Colorado river. The Mormons were some of the first United States settlers in the region and continue to live there today. Today, the Havasupai live in villages and farm on the floor of the Grand Canyon.

47. a. fossils
   b. minerals
   c. igneous
   d. metamorphic
   e. sedimentary

Transparencies

TT Concept Mapping

ANALYZING SEISMIC WAVES

Teacher's Notes

Time Required 45 minutes

Ratings
1 — EASY — 2 — 3 — 4 — HARD

TEACHER PREPARATION	1
STUDENT SETUP	1
CONCEPT LEVEL	3
CLEANUP	1

Skills Acquired

- Classifying
- Communicating
- Constructing models
- Identifying and recognizing patterns
- Measuring
- Predicting

The Scientific Method

In this lab, students will:
- Analyze the Results
- Draw Conclusions
- Communicate Results

Materials

Materials listed are for each group. To save time, you may want to provide photocopies of the map for each lab group to work on.

Safety Cautions

Warn students about the sharp point on a drawing compass. If any students have never used a drawing compass, demonstrate how to draw a circle with the compass. Emphasize that the point should be held still and that the paper should be held to keep it from sliding.

Skills Practice Lab

Introduction

During an earthquake, seismic waves travel through Earth in all directions from the earthquake's focus. How can you find the location of the epicenter by studying seismic waves?

Objectives

▶ **Calculate** the distance from an earthquake's epicenter to surrounding seismographs.

▶ **Find** the location of the earthquake's epicenter.

▶ USING SCIENTIFIC METHODS **Draw conclusions** by explaining the relationship between seismic waves and the location of an earthquake's epicenter.

Materials

calculator
drawing compass
ruler
tracing paper

Analyzing Seismic Waves

▶ Procedure

Preparing for Your Experiment

1. In this lab, you will examine seismograms showing two kinds of seismic waves: primary waves (P waves) and secondary waves (S waves).

2. P waves have an average speed of 6.1 km/s. S waves have an average speed of 4.1 km/s.
   a. How long does it take P waves to travel 100 km?
   b. How long does it take S waves to travel 100 km?
   (**Hint:** You will need to use the equation for velocity and rearrange it to solve for time.)

3. Because S waves travel more slowly than P waves, S waves will reach a seismograph after P waves arrive. This difference in arrival times is known as the lag time.

4. Use the time intervals found in step 2 to calculate the lag time you would expect from a seismograph located exactly 100 km from the epicenter of an earthquake.

Measuring the Lag Time from Seismographic Records

5. On a blank sheet of paper, prepare a table like the one shown below.

City	Lag time(s)	Distance to epicenter
Austin, TX		
Portland, OR		
Bismarck, ND		

6. The illustration at the top of the next page shows the records produced by seismographs in three cities following an earthquake.

7. Using the time scale at the bottom of the illustration, measure the lag time for each city. Be sure to measure from the start of the P wave to the start of the S wave. Enter your measurements in your table.

736

Answers to Procedure

2. **a.** For the P waves

$$speed = \frac{distance}{time}$$

$$time = \frac{distance}{speed}$$

$$t = \frac{100 \text{ km}}{6.1 \text{ km/s}} = 16 \text{ s}$$

b. For the S waves

$$t = \frac{100 \text{ km}}{4.1 \text{ km/s}} = 24 \text{ s}$$

4. $lag\ time = time_{Swave} - time_{Pwave}$
   Substituting the time each wave takes to travel 100 km from step 2 yields
   $lag\ time = 24 \text{ s} - 16 \text{ s} = 8 \text{ s}$

7. See Data Table 1.

8. *distance = (measured lag time/lag time for 100 km) × 100 km*
   Austin:
   $d = (150 \text{ s}/8 \text{ s}) \times 100 \text{ km} = 1875 \text{ km}$
   Bismarck:
   $d = (170 \text{ s}/8 \text{ s}) \times 100 \text{ km} = 2125 \text{ km}$
   Portland:
   $d = (120 \text{ s}/8 \text{ s}) \times 100 \text{ km} = 1500 \text{ km}$

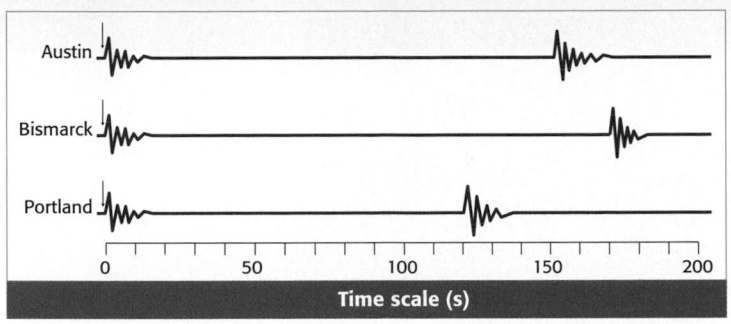

Time scale (s)

8. Using the lag time you found in step 4 and the formula below, calculate the distance from each city to the epicenter of the earthquake. Enter your results in your table.

distance = (measured lag time ÷ lag time for 100 km) × 100 km

► Analysis

1. Trace the map at the bottom of this page on a blank sheet of paper. Using the scale below your map, adjust the drawing compass so that it will draw a circle whose radius equals the distance from the epicenter of the earthquake to Austin. Then put the point of the compass on Austin, and draw a circle on your map. How is the location of the epicenter related to the circle?

2. Repeat the process in item 1 using the distance from Portland to the epicenter. This time put the point of the compass on Portland, and draw the circle. Where do the two circles intersect? The epicenter is one of these two sites.

3. **Reaching Conclusions** Repeat the process once more for Bismarck, and find that city's distance from the epicenter. The epicenter is located at the site where all three circles intersect. What city is closest to that site?

► Conclusions

4. Why is it necessary to use seismographs in three different locations to find the epicenter of an earthquake?

5. Would it be possible to use this method for locating an earthquake's epicenter if earthquakes produced only one kind of seismic wave? Explain your answer.

6. Someone tells you that the best way to determine the epicenter is to find a seismograph where the P and S waves occur at the same time. What is wrong with this reasoning?

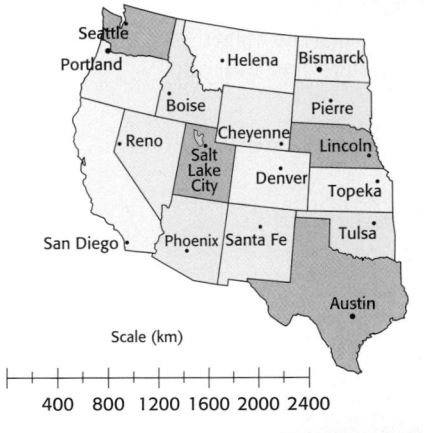

Scale (km)

400 800 1200 1600 2000 2400

Answers to Analysis

1. The circle centered on Austin should have a radius of 1875 km. The epicenter of the earthquake will be on the circumference of the circle.

2. The circle centered on Portland should have a radius of 1500 km.

3. The circle centered on Bismarck should have a radius of 2125 km. The circles intersect at San Diego, California.

Answers to Conclusions

4. The three circles can intersect at only one point.

5. This technique would not work if an earthquake produced only one kind of wave, because one must know the lag time between P waves and S waves to determine the location of the epicenter.

6. If the P waves and S waves arrived at the same time, the lag time between them would be 0 s. A lag time is needed to locate the earthquake's epicenter. Also, there would be no such seismograph because it would have to be right at the focus, which is underground.

Data Table 1

Measuring the Lag Time from Seismographic Records

City	Lag time (s)	Distance from city to epicenter (km)
Austin, TX	150	1875
Portland, OR	120	1500
Bismarck, ND	170	2125

Continuation of Answers

Continuation of Answers from p. 708
Section 1 Review

5. A convergent plate boundary exists along the coastline near Japan's volcanic mountain ranges because convergent plate boundaries form volcanic mountains.

6. There is a difference in the age of continental rocks and sea floor rocks because crust on the sea floor is created and destroyed at a faster rate than crust on land.

Continuation of Answers from p. 717
Section 2 Review

7. Accept both answers. Students should support their choice with reasoning based on the chapter material. Answers may include that explosive eruptions tend to produce volcanoes with steeper slopes, and so they will build more height than quiet eruptions. However, explosive eruptions are more likely to destroy the build-up of materials around the vent of a volcano, preventing an increase in height. Over time, quiet eruptions can build tall volcanoes with gently sloped sides.

Continuation of Answers from p. 724
Section 3 Review

6. The paleontologist should use a combination of the two methods. Relative dating reveals that the animals existed during the time between when the lower layer was formed and when the higher layer was formed. Absolute dating can give absolute ages to those two bands of rock.

PALEONTOLOGIST

The recent history of our planet has been greatly affected by the presence of life, as will be described in the section about Earth's early atmosphere in "The Atmosphere" chapter. At the same time, paleontologists such as Dr. Vermeij need to know a lot about Earth's history and the nature of geologic changes on Earth, as described in the chapter "Planet Earth," to be able to understand the significance of their finds.

|SKILL BUILDER

Vocabulary The word *paleontology* comes from three root words. The Greek word *palaios* means "ancient," and the Greek words *ontos* and *logos* mean "being" and "word," respectively. Thus, paleontology is the study of ancient beings.

CareerLink

Paleontologist

Paleontologists are life's historians. They study fossils and other evidence to understand how and why life has changed during Earth's history. Most paleontologists work for universities, government agencies, or private industry. To learn more about paleontology as a career, read this interview with paleontologist Geerat Vermeij, who works in the Department of Geology at the University of California, Davis.

"I think one needs to be able to recognize puzzles and then think about ways of solving them."

Vermeij is a world-renowned expert on living and fossil mollusks, but he has never seen one. Born blind, he has learned to scrutinize specimens with his hands.

? **Describe your work as a paleontologist.**

I study the history of life and how life has changed from its beginning to today. I'm interested in long-term trends and long-term patterns. My work involves everything from field studies of how living organisms live and work to a lot of work in museum collections. I work especially on shell-bearing mollusks, but I have thought about and written about all of life.

? **What questions are you particularly interested in?**

How enemies have influenced the evolution of plants and animals. I study arms races (evolutionary competitions) over geological time. And I study how the physical history of Earth has affected evolution.

? **What is your favorite part of your work?**

That's hard to say. I enjoy doing the research and writing. I'd say it was a combination of working with specimens, reading for background, and writing (scientific) papers and popular books. I have written four books.

? **What qualities make a good paleontologist?**

First and foremost, hard work. The second thing is you need to know a lot. You have to have a lot of information at hand to put what you observe into context. And you need to be a good observer.

? **What skills does a paleontologist need?**

To me, the curiosity to learn a lot is essential. You have to have the ability to understand and do science and to communicate it.

738

What attracted you to a career in paleontology?

It's a love of natural history in general and shells in particular that led inexorably to my career. As long as I can remember, I have been interested in natural history. I knew pretty much what I wanted to be from the age of 10.

 What education and experiences have been most useful to you?

I think it was very good for me to start early. I started school when I was just shy of my fourth birthday. I started reading the scientific literature in high school.

 What advice do you have for students who are interested in paleontology?

People should work on their interests and not let them slide. They should pursue their interests outside of school. If they live near a museum, for example, getting involved in the museum's activities, getting to know the people there, and so forth is a good idea.

 Why do you think paleontology is important, and did that influence your choice of career?

It gives us a window on life and the past, which like history in general, can provide lessons on what we are doing to the Earth. It gives us perspective on crises and opportunities. The main reason people should pursue interests is for their own sake. I just love the things I work on. It can be utilitarian, but that's not the rationale for my work.

🔲 **internet** connect

www.scilinks.org
Topic: Paleontology
SciLinks code: HK4157

SCiLINKS Maintained by the National Science Teachers Association

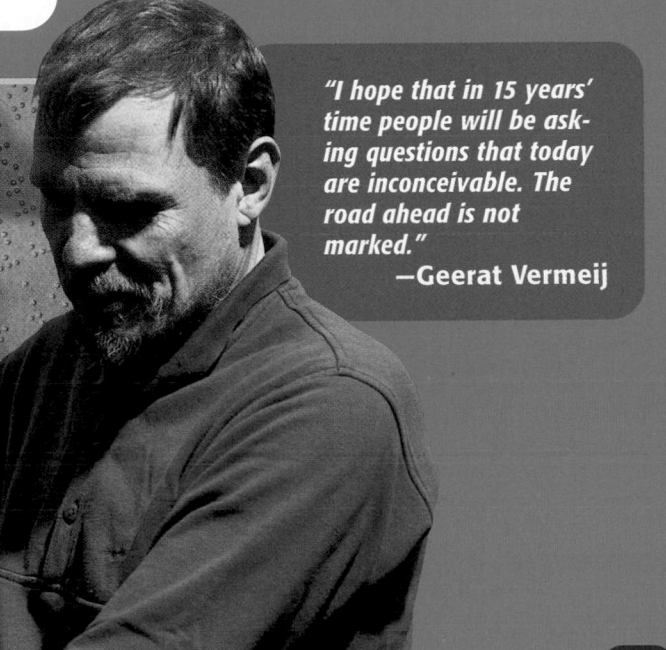

"I hope that in 15 years' time people will be asking questions that today are inconceivable. The road ahead is not marked."
—Geerat Vermeij

PALEONTOLOGIST

In September 1996, W. H. Freeman and Company published Vermeij's memoirs, entitled *Privileged Hands: A Scientific Life*.

In this book, Vermeij describes the current state of evolutionary theory and discusses his contributions through his study of marine mollusks. He also explores what it is like to have science as a career, how discoveries are made, and what it is like to visit places from Baja California to New Zealand in search of scientific facts about living things.

Vermeij has also written a book entitled *A Natural History of Shells*.

739

PACING	CLASSROOM RESOURCES	LABS, ACTIVITIES, AND DEMONSTRATIONS
BLOCK 1 • 45 min pp. 740–741 **Chapter Opener**		SE **Activity 1**, p. 741 SE **Activity 2**, p. 741
BLOCKS 2 & 3 • 90 min pp. 742–749 **Section 1** Characteristics of the Atmosphere	CRF **Lesson Plan*** TT **Bellringer*** TT **Layers of the Atmosphere*** TT **Greenhouse Effect***	TE **Opening Demonstration**, p. 742 SE **Quick Activity** The Greenhouse Effect, p. 748 `GENERAL` CRF **Datasheets for In-Text Labs** The Greenhouse Effect* `GENERAL`
BLOCKS 4 & 5 • 90 min pp. 750–756 **Section 2** Water and Wind	CRF **Lesson Plan*** TT **Bellringer*** TM **Humidity*** TT **Cloud Types*** TM **Barometer***	TE **Opening Demonstration**, p. 750 TE **Demonstration** Transpiration, p. 751 TE **Demonstration** How to Make a Cloud, p. 752 SE **Quick Activity** Measuring Rainfall, p. 753 `GENERAL` CRF **Datasheets for In-Text Labs** Measuring Rainfall* `GENERAL` TE **Demonstration** Air Pressure and Barometers, p. 753 TE **Demonstration** Windsocks, p. 754 CRF **Observation Lab** Building a Cup Anemometer* ◆ `BASIC` CRF **CBL™ Probeware Lab** Predicting Coastal Winds* `ADVANCED`
BLOCKS 6 & 7 • 90 min pp. 757–762 **Section 3** Weather and Climate	CRF **Lesson Plan*** TT **Bellringer*** TM **Hurricane*** TT **Concept Mapping***	TE **Opening Activity**, p. 757 TE **Demonstration** A Cold Front Meets a Warm Front, p. 758 TE **Demonstration** Creating a Tornado, p. 759 TE **Group Activity**, p. 760 TE **Group Activity**, p. 761 SE **Skills Practice Lab** Measuring Temperature Effects, pp. 768–769 ◆ `GENERAL` CRF **Datasheets for SE Skills Practice Lab** Measuring Temperature Effects* `GENERAL`

BLOCKS 8 & 9 • 90 min

Chapter Review and Assessment Resources

- SE **Chapter Review**, pp. 764–767
- CRF **Chapter Tests*** `GENERAL`
- OSP **Test Generator**
- CRF **Standardized Test Practice with Guided Reading Development***
- CRF **Test Item Listing for ExamView® Test Generator***
- OSP **Scoring Rubrics and Classroom Management Checklists**

Online Resources

Visit the HRW Web site for a variety of free resources related to the text. Just type in the keyword **HK4 ATM**.

Holt Science Spectrum: Physical Science: Online Edition

Students can access interactive problem solving help and active visual concept development with the *Holt Science Spectrum: Physical Science* Online Edition available at **www.hrw.com**.

student CNN News

cnnstudentnews.com

Find the latest news, lesson plans, and activities related to important scientific events.

KEY

TE	Teacher Edition	**CRF** Chapter Resource File	* Also on One-Stop Planner
SE	Student Edition	**TT** Teaching Transparency	◆ Requires Advance Prep
OSP	One-Stop Planner	**TM** Transparency Master	

PROBLEM SOLVING AND PRACTICE	SECTION REVIEW AND ASSESSMENT	STANDARDS CORRELATION
	CRF Pretest* GENERAL	
CRF **Cross-Disciplinary Worksheet** Integrating Physics—The Tropopause* ADVANCED **CRF** **Cross-Disciplinary Worksheet** Connection to Language Arts—The Layers of the Atmosphere* ADVANCED **TE** Inclusion Strategies, p. 745	**TE** Quiz, p. 749 BASIC **SE** Section 1 Review, p. 749 **CRF** Concept Review* GENERAL **CRF** Quiz* BASIC	ES 1a, 3d UCP 1, 2, 3 SAI 1, 2 ST 1 HNS 1, 3 SPSP 1, 2, 4, 5
CRF **Cross-Disciplinary Worksheet** Integrating Chemistry—Cloud Seeding* ADVANCED **CRF** **Cross-Disciplinary Worksheet** Integrating Health—Why Your Ears Pop* ADVANCED **TE** Inclusion Strategies, p. 752	**TE** Quiz, p. 756 BASIC **SE** Section 2 Review, p. 756 **CRF** Concept Review* GENERAL **CRF** Quiz* BASIC	ES 1c, 2a, 2b UCP 1, 3 SAI 1, 2 ST 1, 2 SPSP 1, 2, 3
CRF **Cross-Disciplinary Worksheet** Integrating Technology—Doppler Weather Radar* ADVANCED **CRF** **Cross-Disciplinary Worksheet** Real World Applications—Understanding Thunderstorms* ADVANCED **CRF** **Cross-Disciplinary Worksheet** Integrating Physics—Adobe* ADVANCED **SE** Viewpoints Should Laws Require Zero-Emission Cars?, pp. 770–771	**TE** Quiz, p. 762 BASIC **SE** Section 3 Review, p. 762 **CRF** Concept Review* GENERAL **CRF** Quiz* BASIC	ES 1d, 3c UCP 1, 3 SAI 1 HNS 1 SPSP 1, 5

www.scilinks.org

Topic: Coriolis Effect
*Sci*Links code: HK4028

Topic: Global Warming
*Sci*Links code: HK4065

Topic: Layers of the Atmosphere
*Sci*Links code: HK4077

Topic: Ozone Depletion
*Sci*Links code: HK4100

Topic: Visible Light
*Sci*Links code: HK4146

Topic: Water Cycle
*Sci*Links code: HK4149

Topic: Weather Maps
*Sci*Links code: HK4151

Topic: Zero-emission vehicles
*Sci*Links code: HK4158

Technology Resources

 One-Stop Planner
All of your printable resources and the Test Generator are on this convenient CD-ROM.

Overview

This chapter first discusses the characteristics of each layer of Earth's atmosphere and changes in the atmosphere, including ozone depletion, the greenhouse effect, and global warming. The chapter then discusses the water cycle, air pressure, the causes of winds, and global wind patterns. This chapter finally explores weather and climate, including discussions of weather maps, fronts, severe weather conditions, and local and global climate.

Assessing Prior Knowledge

Be sure students understand the following concepts:

- properties of matter
- compounds and molecules
- reflection and color
- the solar system

MISCONCEPTION ///ALERT\\\

Science education research has identified the following misconceptions about the atmosphere and weather.

- Students believe air and oxygen are the same thing.
- Students believe pressure is present in wind but not in still air.
- Students rarely account for winds in terms of pressure differences between regions of the atmosphere; instead they account for winds in terms of visible moving objects (such as cars, clouds, or tides), the movement of Earth, or the coldness of the poles.
- Students link wind speed with temperature, believing that a high speed corresponds to cold winds, while warm winds are gentler or slower.

The Atmosphere

Chapter Preview

740

Standards Correlations

National Science Education Standards

The following descriptions summarize the National Science Standards that specifically relate to this chapter. For the full text of the standards, see the National Science Education Standards at the front of the book.

Section 1 Characteristics of the Atmosphere
Earth and Space Science ES 1a, 3d
Unifying Concepts and Processes UCP 1, 2, 3
Science as Inquiry SAI 1, 2
Science and Technology ST 1
History and Nature of Science HNS 1, 2, 3
Science in Personal and Social Perspectives SPSP 1, 2, 4, 5

Section 2 Water and Wind
Earth and Space Science ES 1c, 2a, 2b
Unifying Concepts and Processes UCP 1, 3
Science as Inquiry SAI 1, 2
Science and Technology ST 1, 2
Science in Personal and Social Perspectives SPSP 1, 2, 3

Section 3 Weather and Climate
Earth and Space Science ES 1d, 3c
Unifying Concepts and Processes UCP 1, 3
Science as Inquiry SAI 1
History and Nature of Science HNS 1
Science in Personal and Social Perspectives SPSP 1, 3, 5

Background Like many other weather phenomena, rainbows are caused by water in Earth's atmosphere. Rainbows are visible when the air is filled with water droplets. Sunlight striking the droplets passes through their front surface and is partially reflected back toward the viewer from the back of the droplet.

But why do we see the rainbow of colors? A rainbow occurs when sunlight is bent as it passes from air to water and back to air again.

Activity 1 Look at the two rainbows in the smaller photo at left. One of these rainbows, called a secondary rainbow, results when light is reflected a second time in the raindrops. The second reflection causes the order of the colors to be reversed. Compare the order of the colors in the two rainbows with that of the rainbow in the photo of the irrigation trucks. Can you tell which rainbow is the secondary rainbow? How?

Activity 2 Rainbows are most commonly seen as arches because the ends of the rainbow disappear at the horizon. But if an observer is at an elevated vantage point, such as on an airplane or at the rim of a canyon, a complete circular rainbow can be seen.

You can create a circular rainbow in your yard. On a warm, clear day when the sun is overhead, turn on a water hose and spray water into the air above you. If the mist is fine enough, you should be able to create a rainbow that encircles your body.

internet connect

www.scilinks.org
Topic: Visible Light SciLinks code: HK4146

SCi LINKS. Maintained by the National Science Teachers Association

Rainbows can be seen when the air is filled with water droplets.

Pre-Reading Questions
1. What would Earth be like without an atmosphere?
2. How do scuba divers breathe under water?

741

Focus ACTIVITY

Background Ask your students if they know why and how rainbows occur. Have them hypothesize reasons why rainbows form.

Rainbows are produced by sunlight reflecting and refracting off of droplets of water in the air. The droplets act as mirrors and prisms that cause the sunlight to separate into a spectrum of visible colors. When sunlight is reflected twice, a secondary rainbow that is much dimmer than the first may be produced.

Activity 1 The first rainbow shown in the photo is actually the secondary rainbow. When light from the sun is reflected twice in the same set of raindrops, two rainbows are produced. The primary rainbow is more intense in color than the dimmer secondary rainbow.
LS Visual

Activity 2 If your school has access to a water hose, you may want to use this activity as a demonstration. Make sure the sun is positioned behind you before spraying the water in front and over your head.
LS Kinesthetic

Answers to Pre-Reading Questions
1. There would be no oxygen to support life, and temperatures on Earth would be much more extreme.
2. Their diving equipment provides the oxygen needed to breathe.

Chapter Resource File
• Pretest **GENERAL**
• Teaching Transparency Preview
• Answer Keys

SECTION
1

Focus

Overview

Before beginning this section, review with your students the Objectives listed in the Student Edition. This section discusses characteristics of the primary layers of the atmosphere, then covers the oxygen-carbon dioxide cycle and recent atmospheric changes such as ozone depletion and global warming.

🔔 Bellringer

Use the Bellringer transparency to prepare students for this section.

Motivate

Opening Demonstration — GENERAL

This demonstration requires 2 glass beakers, hot tap water, cold tap water, food coloring, and a spoon. Fill one of the beakers halfway with cold water. Fill the other beaker with hot water, and stir two drops of food coloring in it. Slowly and carefully use the spoon to layer the colored warm water on top of the cold water. Ask students to relate this demonstration to what occurs in Earth's atmosphere when cool air gets trapped beneath a layer of warm air in the troposphere. (A temperature inversion results.) Alternatively, you can set this up as a lab, give students the materials, and challenge them to demonstrate a temperature inversion without any other guidance. **LS Logical**

Characteristics of the Atmosphere

▶ **KEY TERMS**

troposphere
temperature inversion
stratosphere
ozone
mesosphere
thermosphere
greenhouse effect

OBJECTIVES

▶ **Identify** the primary layers of the atmosphere.
▶ **Describe** how the atmosphere has evolved over time.
▶ **Describe** how the oxygen–carbon dioxide cycle works, and explain its importance to living organisms.
▶ **Discuss** the recent changes in Earth's atmosphere.

If you were to see Earth's atmosphere from space, it would look like a thin blue halo of light around Earth. This fragile envelope provides the air we breathe, regulates global temperature, and filters out dangerous solar radiation.

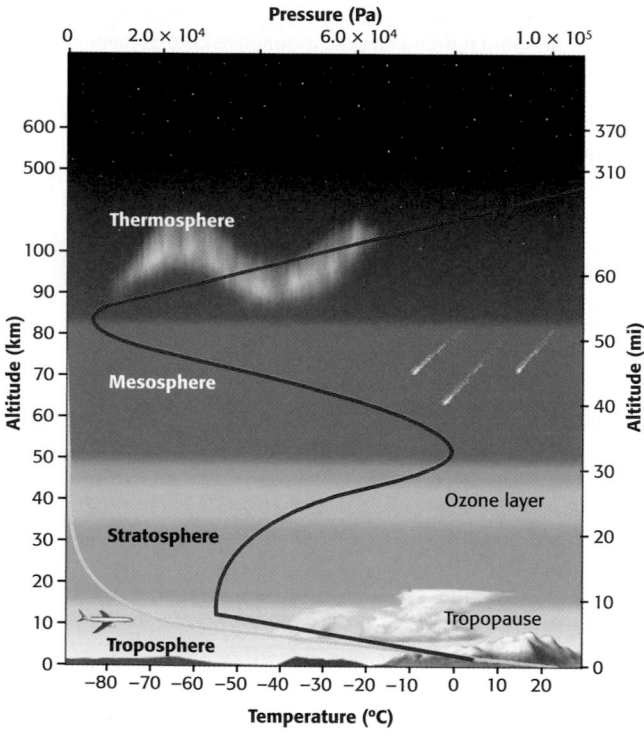

Layers of the Atmosphere

Without the atmosphere, you would have no oxygen to breathe. The atmosphere, however, does not contain oxygen alone. Earth's atmosphere consists of a variety of gases. The two main gases in the atmosphere are nitrogen (about 78%) and oxygen (about 21%). The remaining elements exist only in very small amounts and are called *trace gases*.

The atmosphere has several layers. These layers differ in temperature, density, and amount of certain gases present. **Figure 1** shows the position and relative thicknesses of Earth's atmospheric layers.

Figure 1

The layers of the atmosphere are marked by differences in temperature and pressure. Red indicates temperature. Yellow indicates pressure.

742

LIFE SCIENCE
CONNECTION — **ADVANCED**

Ask a biology teacher or a health professional to visit your classroom and explain some of the respiratory conditions such as asthma, bronchitis, and emphysema that can result from breathing air with pollutants in it. Have students research the harmful effects air pollutants have on other organisms. **LS Verbal**

CHEMISTRY
CONNECTION — **GENERAL**

Have students look at **Figure 2.** Tell them that the polluted air trapped in the Los Angeles Basin is called smog. Ask a chemistry teacher to visit your classroom and explain the principles behind the formation of smog. **LS Visual**

Almost all weather occurs in the troposphere

You live in the layer closest to Earth's surface. This layer is called the **troposphere.** Clouds, wind, rain, and snow occur mostly in the troposphere.

The troposphere is the densest of the atmospheric layers. Consider the weight of all the other gas layers pressing down on the gases in the troposphere. The weight causes the gas molecules to squeeze together into a smaller volume. The result is a greater density than exists in higher layers of the atmosphere.

The troposphere gets cooler with increasing altitude

If you were to climb a mountain, you would notice that the air is much colder at the top of the mountain than it is at the base. The air closest to the mountain's base is warmed by the ground and oceans, which absorb solar energy during the day and heat the atmosphere by radiation and conduction. Air at higher altitudes is less dense and is not as close to those sources of heat. As you travel higher into the troposphere, the temperature decreases by about 6°C for every kilometer of altitude.

At the top of the troposphere, the temperature stops decreasing. The boundary where this occurs is called the *tropopause.* The low temperature (–55°C) at the tropopause keeps the water vapor and clouds in the troposphere. This is the reason that clouds, rain and snow are restricted to the troposphere. The tropopause acts like a lid on the water and water vapor in the troposphere. The altitude of the tropopause is different at different places on Earth. At the poles, it occurs at about 8 km (5 mi). Near the equator, it rises to nearly 18 km (11 mi).

Cold air can become trapped beneath warm air

Although the troposphere is generally warmer close to Earth's surface, cool air sometimes gets trapped beneath the warm air. This is called a **temperature inversion.**

When a temperature inversion occurs, cooler air cannot rise and the air can become thick with pollution. This is true especially in areas surrounded by mountains, which also prevent the polluted air from escaping. The Los Angeles Basin in California is often filled with a brown haze as shown in *Figure 2.* Cool air from the Pacific Ocean blows into the basin, becomes trapped by the overriding warm air, and fills with pollutants.

▶ **troposphere** the lowest layer of the atmosphere, in which temperature drops at a constant rate as altitude increases

▶ **temperature inversion** the atmospheric condition in which warm air traps cooler air near Earth's surface

VOCABULARY *Skills Tip*

The names of all of Earth's atmospheric layers end in the root word sphere, *implying their spherical shape.*

internet connect

www.scilinks.org
Topic: Layers of the Atmosphere
SciLinks code: HK4077

SC*i*LINKS. Maintained by the National Science Teachers Association

Figure 2

A temperature inversion traps polluted air in the Los Angeles Basin, in California.

743

Did You Know?

Clouds Since temperature decreases with increasing altitude, clouds at the bottom of the troposphere are made of drops of water, while clouds at the top of the troposphere are made of ice crystals.

REAL-LIFE
CONNECTION

Pressure Altimeter A pressure altimeter uses atmospheric pressure as an indicator of altitude. Airplane pilots use this device which can look like a wristwatch to track the changing conditions of Earth's atmosphere. But because standard air pressure at a specific elevation may vary from the actual air pressure, pilots must adjust their altimeter before and during flights.

Teaching Tip ——— GENERAL

The Troposphere and Stratosphere Earth's diameter is 12 756 km (7911 mi), and the troposphere and stratosphere extend about 50 km (about 30 mi) from the surface. To give your students a visual image of how thin these layers are, have them go to a paved area of the school's parking area. Using colored chalk, draw a circle with a diameter of 4 m (about 13 ft) to represent Earth. This is easy if you tie the chalk to a string 2 m (about 6.5 ft) long and have one person hold the end of the string on the ground while another person moves around drawing the circle.

Calculate how thick the combined troposphere and stratosphere need to be on your drawing to be in proportion to your "Earth" (for a radius of 2 m, the layers would extend about 15.7 cm outward). Then use white chalk to draw another circle around the first one to represent these two layers. Ask your students if the layers are thinner than they thought. This demonstration could also be done on a smaller scale on a classroom blackboard. **LS** Visual

SKILL BUILDER ——— BASIC

Reading Skills Have students create a table like the one shown below and to the right. Once the table is set up, have students fill the table with appropriate information from **Figure 1** and other areas of their book. **LS** Logical

---

▶ **stratosphere** the upper layer of the atmosphere, which lies immediately above the troposphere and extends from 10 km to about 50 km above Earth's surface

▶ **ozone** a gas molecule that is made up of three oxygen atoms

Figure 3
A heated cabin in this aircraft allows its pilot to do high-altitude atmospheric research, such as collecting air and particle samples after a volcanic eruption.

As long as a temperature inversion remains, trapped pollutants cannot escape. A person breathing these toxins can become ill. While these conditions exist, it is not healthy for people to exercise outside because they inhale a greater amount of pollutants as they breathe heavily.

The stratosphere gets *warmer* with increasing altitude

In 1892, unmanned balloons were built that could record temperatures in the **stratosphere,** the layer above the tropopause. Later, humans further explored this cold, low-pressure layer by using balloons and airplanes with enclosed, heated cabins such as the WB-57F shown in *Figure 3.* These explorers found that the temperature in the lower stratosphere remains fairly constant, staying around –55°C (–67°F) from near the tropopause to an altitude of about 25 km (about 16 mi). At 25 km, the temperature begins to increase with altitude until it reaches about 0°C (32°F).

The stratosphere extends to about 50 km (31 mi). In addition to getting warmer instead of cooler, the stratosphere differs from the troposphere in composition, weather, and density. Unlike the troposphere, the stratosphere has little water vapor—the gaseous form of water because water vapor cannot get through the cold tropopause. Because of this lack of water vapor, the stratosphere contains few clouds and no storms.

The increase in temperature in the upper stratosphere occurs in the atmospheric layer known as the *ozone layer.* The ozone layer is warmer because it contains a form of oxygen called **ozone** that absorbs solar radiation. Whereas the oxygen we breathe is a molecule that consists of two oxygen atoms, as shown in *Figure 4A,* ozone molecules have three oxygen atoms, as shown in *Figure 4B.* Ozone is important because it absorbs much of the sun's ultraviolet radiation. The ozone layer shields life on Earth's surface from ultraviolet-radiation damage. Ozone will be discussed later in this section.

Figure 4

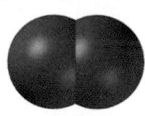 

A Oxygen molecules have two atoms of oxygen.

B Ozone molecules have three atoms of oxygen.

Layers of the Atmosphere

	Troposphere	Stratosphere	Mesosphere	Thermosphere
Height	8–18 km	about 50 km	50–80 km	80–480 km
Temperature	255°C–10°C	255°C–0°C	280°C–0°C	average 5 980°C
Pressure	2.0×10^4–1.0×10^5 Pa	$< 2.0 \times 10^4$ Pa	$< 2.0 \times 10^4$ Pa	$< 2.0 \times 10^4$ Pa

The mesosphere and thermosphere exhibit extremes of temperature

Temperature begins to fall again in the **mesosphere.** As in the troposphere, temperatures in the mesosphere, 50–80 km (31–50 mi) above Earth's surface, decrease with increasing altitude. Near the top of this layer, temperatures fall to below –80°C (–112°F), the coldest temperature in Earth's atmosphere.

Beyond the mesosphere, temperatures begin to rise again. This layer, at an altitude of about 80–480 km (50–298 mi), is called the **thermosphere.** The main gases are still nitrogen and oxygen, but the molecules are very far apart. This may lead you to think that the thermosphere is very cold, but it is actually very hot. Temperatures in this layer average around 980°C (1796°F) because the small amount of molecular oxygen in the thermosphere heats up as it absorbs intense solar radiation.

The outermost portion of the thermosphere, at about 480 km, is known as the *exosphere.* In the exosphere, some gases escape from the gravitational pull of Earth and exit into space. In addition, gases in space are captured by Earth's gravity and added to Earth's atmosphere.

The ionosphere is used in radio communication

When solar energy is absorbed in the lower thermosphere and upper mesosphere, electrically charged ions are formed. The area where this occurs is sometimes called the *ionosphere.*

Electrons in the ionosphere reflect radio waves, as shown in **Figure 5,** allowing them to be received over long distances. Without the ionosphere, most radio signals would travel directly into space, and only locations very close to a transmitter could receive the signals.

Because these ions require solar radiation in order to form, their number in the lower layers of the ionosphere decreases at night. This means the radio waves can travel higher into the atmosphere before being reflected. As a result, the radio waves return to Earth's surface farther from their source than they do in the daytime, as shown at right.

▶ **mesosphere** the coldest layer of the atmosphere, between the stratosphere and the mesopause

▶ **thermosphere** the uppermost layer of the atmosphere, in which temperature increases as altitude increases

Figure 5

Radio waves can be received from far away because they are reflected by the ionosphere. At night, when ion density decreases in the lower atmosphere, transmissions can be received farther away.

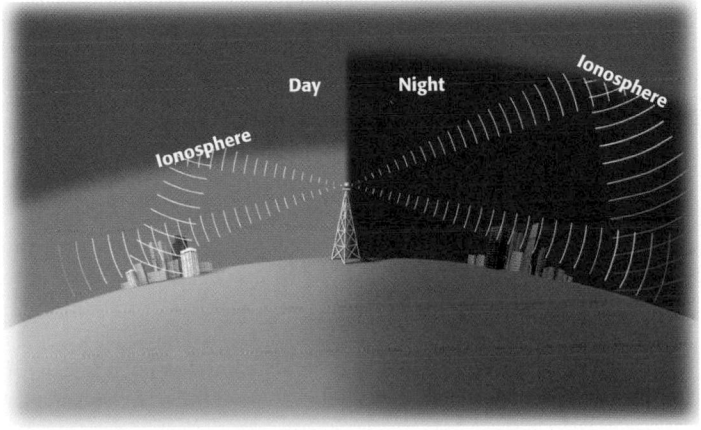

745

Vocabulary Have students find out the meaning of the prefixes used to describe the four primary layers of the atmosphere. Then ask them to explain why each prefix is appropriate (when possible). You could also ask them to find other words with the same prefixes. (*Tropo:* indicates turning or change, especially of temperature or condition; the troposphere is where conditions frequently change and produce different kinds of weather. *Strato:* from stratum, which means horizontal layer. *Meso:* indicates middle or intermediate; the mesosphere is one of the middle layers of the atmosphere. *Thermo:* from therm, which indicates heat; the thermosphere is the hottest layer of the atmosphere. Other words with these prefixes include tropic, tropism; stratify, strata; mesoderm, Mesolithic; thermal and thermodynamic.) **LS Verbal**

INCLUSION Strategies

• **Gifted and Talented**

Using **Figure 1,** have students estimate the temperatures and pressures an imaginary aircraft must withstand as it reaches altitudes of 15, 30, 45, 60, 75, and 90 km. Ask students to create a table showing the changes in temperature and pressure with differing altitude. Students may also write a description of the aircraft's journey through 90 km that includes specific information regarding temperature and pressure.

Alternative Assessment — BASIC

Temperature and Pressure Review with students the topics of temperature and pressure. Remind them that each layer of the atmosphere is classified according to changes in temperature and pressure. Ask students to write a few paragraphs explaining how temperature and pressure affect the different layers of the atmosphere. **LS Verbal**

Chapter Resource File

• **Cross-Disciplinary Worksheet**
Connection to Language Arts—The Layers of the Atmosphere **GENERAL**

Reading Skills Write the word *outgassing* on the board. Have students list all the words, phrases, and ideas that they associate with the term. Begin a discussion in which students inquire about clarification of the listed ideas. Form a definition of outgassing and its role in the creation of an atmosphere of gases. **LS** Verbal

Teaching Tip

Formation of the Ozone Layer

Explain to students that when organisms began using sunlight for photosynthesis, oxygen gas (O_2) started accumulating in the atmosphere. Energy from the sun's UV rays split some of the O_2 molecules into individual oxygen atoms. These free oxygen atoms began combining with other O_2 molecules to form O_3 molecules, or ozone. Ozone molecules are able to absorb some of the sun's UV rays. Gradually a thin layer of ozone formed around Earth. By about 600 million years ago, the ozone layer shielded enough UV light to allow organisms to develop and live on land. At that time the atmosphere contained just 10% of the oxygen found in our current atmosphere.

MISCONCEPTION ALERT

Oxygen and Air Some students believe oxygen and air are the same thing. Explain that the air we breathe contains many different substances, including nitrogen, oxygen, and a number of other elements.

Figure 6
Auroras, such as this one seen above mountains in Alaska, occur in the ionosphere.

Figure 7

A Early in Earth's existence, the atmosphere contained mostly carbon dioxide, nitrogen, and a few other trace gases.

B As Earth changed, so did the gases in the atmosphere.

The ionosphere is where auroras take place

The ionosphere is also where colorful light displays called *auroras* can be seen encircling Earth's magnetic poles. Auroras form when energetic ions from the sun hit atoms and molecules in the ionosphere, causing photons to be emitted. The *aurora borealis*, shown in **Figure 6**, appears in the sky above the Northern magnetic pole. A similar phenomenon, the *aurora australis*, is observed in the south, above Antarctica.

Changes in Earth's Atmosphere

When Earth began to solidify about 4.4 billion years ago, volcanic eruptions released a variety of gases. This process, called *outgassing*, created an atmosphere of gases, some of which would be poisonous to us today. As shown in **Figure 7A,** these gases included hydrogen, H_2, water vapor, H_2O, ammonia, NH_3, methane, CH_4, carbon monoxide, CO, carbon dioxide, CO_2, and nitrogen, N_2, but not O_2, oxygen.

Photosynthetic plants contribute oxygen to the atmosphere

Amazingly, life-forms evolved that were comfortable in this early atmosphere. Bacteria and other single-celled organisms lived in the oceans. Around 2.5 billion years ago, some cells evolved a method of capturing energy from the sun and converting it to sugar that could be used as a food source. This process, called *photosynthesis*, also produced oxygen as a byproduct. These organisms needed only sunlight, water, and carbon dioxide for their survival, so they thrived and multiplied in this environment. Gradually, the oxygen content of the atmosphere increased to about 20%, as shown in **Figure 7B.** About 350 million years ago, the concentration of oxygen reached a level similar to what it is today.

746

LANGUAGE ARTS CONNECTION

Aurora was the Roman goddess of dawn, and *borealis* is the Latin word for "northern." Aurora borealis refers to the Northern Lights, which are visible only in the Northern Hemisphere. Similarly, *australis* is the Latin word for "southern," and the aurora australis is visible only in the Southern Hemisphere.

CHEMISTRY CONNECTION — GENERAL

Photosynthetic plants use carbon dioxide, water, and light energy to produce carbohydrates and oxygen. Write the following photosynthesis reaction on the board, and have students balance the equation.

$CO_2 + H_2O \xrightarrow{\text{light energy}} C_6H_{12}O_6 + O_2$

$(6CO_2 + 6H_2O \xrightarrow{\text{light energy}} C_6H_{12}O_6 + 6O_2)$

LS Logical

Animals produce carbon dioxide necessary for photosynthesis

As *aerobic*, or oxygen-breathing, organisms evolved, they joined plants in a balance that led to our present atmosphere. The steps of the oxygen–carbon dioxide cycle describe this balance. These steps are summarized in *Figure 8,* which shows a simple depiction of the series of chemical reactions that take place. Plants need carbon dioxide, CO_2, for photosynthesis and food production. Oxygen, O_2, is then released as a waste product of photosynthesis. Animals breathe oxygen during a process called *respiration* and release carbon dioxide as waste. The carbon dioxide they exhale is then used by plants and other photosynthetic organisms, and the process is repeated.

Man-made chemicals can deplete the ozone layer

Recall that the stratosphere contains a layer of ozone molecules. Ozone is formed when the sun's ultraviolet rays strike molecules of O_2. The energy splits the molecules, and the single atoms of oxygen bond with O_2 molecules to make O_3, ozone. These O_3 molecules in turn absorb much of the sun's damaging ultraviolet radiation. Without the ozone layer, ultraviolet rays would cause serious damage to the cells of living things. Thus, scientists were concerned when they found lower than expected concentrations of ozone in the stratosphere in 1985.

Ozone destruction was caused mainly by chemicals known as chlorofluorocarbons, or CFCs. CFCs were widely used in the last 65 years of the twentieth century as refrigerants and in spray cans. Persuaded by evidence of a connection between CFCs and ozone destruction, most industrialized countries stopped production of CFCs on January 1, 1996. These bans have drastically decreased the amount of CFCs entering the stratosphere.

VOCABULARY *Skills Tip*

The word ozone *comes from the Greek word* ozein, *which means "to smell." Ozone gas has a strong odor. You may have smelled ozone after a thunderstorm when atmospheric oxygen was converted to ozone by the electrical energy of lightning.*

internet connect

www.scilinks.org
Topic: Ozone Depletion
SciLinks code: HK4100

*sci*LINKS. Maintained by the National Science Teachers Association

Figure 8

In the oxygen–carbon dioxide cycle, plants produce oxygen, which is used by animals for respiration. Animals produce carbon dioxide, which is used by plants for photosynthesis.

747

Teaching Tip

Ozone Layer Many students may have heard about the ozone layer, probably in connection to ozone depletion. Stress to students that the ozone layer, found in the stratosphere, protects us from harmful UV radiation. Ozone can also form in the lower troposphere as a result of human activities. In this layer, ozone is a pollutant and can be harmful.

SKILL BUILDER — BASIC

Interpreting Visuals Have students examine **Figure 8,** and discuss the steps of the oxygen–carbon dioxide cycle. Have students write a brief paragraph explaining the roles that plants and animals play in the cycle.
LS Verbal

Teaching Tip

CFC Molecules Inform students that CFC molecules remain active in the stratosphere for decades. CFCs released thirty years ago are still destroying ozone today.

Did You Know?

Earth's Magnetic Field In addition to the ozone layer, there is a magnetic field around Earth that also protects us from harmful radiation. High-energy particles are caught in the Van Allen Belt that surrounds our planet and deflects the particles from our planet's surface. There is geological evidence that this magnetic field is disrupted occasionally, which results in dangerous particles and rays hitting unprotected organisms.

Teaching Tip

Greenhouse Gases Tell students that greenhouse gases are necessary for life to survive on Earth. An atmosphere without any carbon dioxide or water vapor in it would result in a cold, inhospitable Earth.

SKILL BUILDER —ADVANCED

Interpreting Visuals Ask students to brainstorm possible solutions to global warming. Have them identify which step shown in **Figure 9** is interrupted by their solution. **LS** Logical

Quick ACTIVITY

The Greenhouse Effect
Materials (per group):
• 2 identical glass jars
• 1000 mL water
• 10 ice cubes
• 2 thermometers
• resealable plastic bag
• light source (2 lamps or the sun)
• watch or stopwatch

Results: The jar in the plastic bag becomes warmer first because the bag traps the heat. Students' graphs should show both temperatures rising, but the temperature of the covered jar rises at a faster rate.

Teacher's Notes: After students have completed this activity, ask them to explain which jar models the greenhouse effect. (The covered jar models the greenhouse effect; the plastic traps heat in the jar just as gases in Earth's atmosphere trap heat on Earth.) **LS** Logical

---

Figure 9
The greenhouse effect is a process in which atmospheric gases trap some of the energy from the sun in the troposphere.

A Solar radiation warms Earth's surface and is radiated back into the atmosphere as heat radiation.

B Greenhouse gases, such as CO_2 and H_2O, receive this heat radiation and radiate some of it back toward Earth's surface.

C CO_2 is added to the air in the burning of fossil fuels and in forest fires, possibly causing global warming.

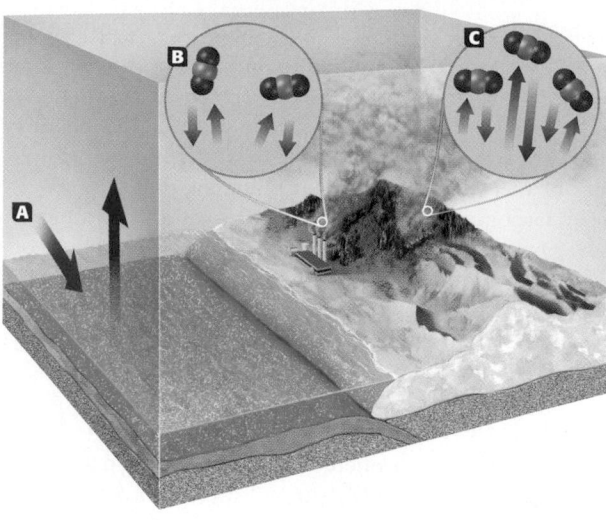

▶ **greenhouse effect** the warming of the surface and lower atmosphere of Earth that occurs when carbon dioxide, water vapor, and other gases in the air absorb and reradiate infrared radiation

The greenhouse effect keeps Earth warm

Have you ever been in a greenhouse or opened a car on a sunny day? It is surprisingly warm inside. Although some greenhouses are heated, much of the warmth results from the sun's energy entering and becoming trapped inside the glass or plastic walls of the greenhouse.

Unlike a greenhouse, Earth's atmosphere has no walls, but certain atmospheric gases act like glass walls by keeping Earth much warmer than it would be without an atmosphere. As shown in *Figure 9,* energy released from the sun as radiation is absorbed by Earth's surface. Then some of this energy is transferred back toward space as radiation. Carbon dioxide, water vapor, and other gases absorb some of this energy, making the atmosphere warmer. The warm atmosphere releases some of this energy in the form of radiation, some of which is directed back toward Earth's surface. This effect is called the **greenhouse effect.**

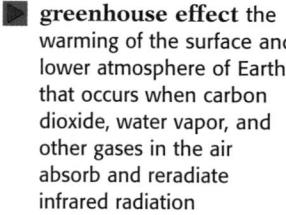
Quick ACTIVITY

The Greenhouse Effect

1. Pour 500 mL of water into two identical glass jars. (If the jars are small, use 200 mL of water and fewer ice cubes.)
2. Add five ice cubes and a thermometer to each jar, and wrap one jar in a resealable plastic bag.
3. Put both jars in the sun or under lamps.
4. Record the temperature of the water in each jar every five minutes, and record your observations.
5. Make a line graph from your results.
6. Which jar warmed up faster? Explain why.

748

REAL-LIFE
CONNECTION

CFCs People who work with chlorofluorocarbons, such as air conditioner repair personnel, must be trained and licensed in the proper handling of CFCs. Strict guidelines must be followed. Invite someone who repairs air conditioners to be a guest lecturer in your class. Have the person explain how the handling of CFCs has changed in recent years.

Increased levels of carbon dioxide may lead to global warming

Without the greenhouse effect, Earth would have a colder average temperature than it does. But too much of the greenhouse effect can cause problems. If too much heat is trapped, the global temperature will rise. This *global warming* could cause the icecaps to melt, ocean levels to rise, and droughts to occur in some areas.

Carbon dioxide occurs naturally and is necessary for plant photosynthesis. In the last 100 years, the burning of coal, oil, and gas for power plants, machinery, and cars has added excess carbon dioxide to the air. Recently, scientists have hypothesized that this increase in the amount of carbon dioxide is the reason the troposphere's average temperature has risen 0.5°C in the past 100 years. Whether carbon dioxide is responsible for global warming and what to do about it continues to be debated around the world.

internet connect
www.scilinks.org
Topic: Global Warming
SciLinks code: HK4065

sciLINKS. Maintained by the National Science Teachers Association

Close

SECTION 1 REVIEW

SUMMARY

▶ The layers of Earth's atmosphere are the troposphere, stratosphere, mesosphere, and thermosphere.

▶ The oxygen–carbon dioxide cycle produces the oxygen we breathe. Plants release oxygen. Animals breathe this oxygen and release carbon dioxide, which is used by plants.

▶ The ozone layer protects life on Earth by absorbing much of the ultraviolet radiation entering Earth's atmosphere.

▶ CFCs are linked to the deterioration of the ozone layer. For this reason, their use has been banned in most countries.

▶ The addition of CO_2 to the atmosphere by the burning of fossil fuels may cause global warming. This issue continues to be debated.

1. **Identify** the two atmospheric layers that contain air as warm as 25°C.

2. **Identify** which characteristic is true of the ionosphere.
   a. It gets warmer with altitude.
   b. It is used in radio communication.
   c. It is where auroras take place.
   d. It exhibits extremes in temperature.

3. **Identify** which of the following gases is most abundant in Earth's atmosphere today.
   a. argon   c. oxygen
   b. nitrogen   d. carbon dioxide

4. **Compare** Earth's early atmosphere with its present one.

5. **Describe** the role that plants play in the oxygen–carbon dioxide cycle.

6. **Explain** why the following statement is incorrect:
   Global warming could cause oceans to rise, so the greenhouse effect must be eliminated completely.

7. **Predict** how much colder it is at the top of Mount Everest, which is almost 9 km above sea level, than it is at the Indian coastline. (**Hint:** The temperature in the troposphere decreases by 6°C/km.)

8. **Critical Thinking** In 1982, Larry Walters rose to an altitude of approximately 4900 m on a lawn chair attached to 45 helium-filled weather balloons. Give two reasons why Walters's efforts were dangerous.

749

Answers to Section 1 Review

1. troposphere and thermosphere

2. c

3. b

4. Earth's early atmosphere was composed of many gases that would be poisonous to us today. As forms of life evolved and began to photosynthesize, oxygen was produced. Once animals adapted to breathing oxygen, they began to give off carbon dioxide, balancing the production of oxygen.

5. Plants use carbon dioxide to produce oxygen through photosynthesis. Through respiration, animals give off the carbon dioxide that plants need, and use the oxygen that plants produce.

6. If we were to eliminate the greenhouse effect completely, the world's climate would become too cold for humans to survive.

7. About 54°C colder

8. The oxygen content of the air is considerably lower and the temperature is extremely cold at that high altitude.

Focus

Overview

Before beginning this section, review with your students the Objectives listed in the Student Edition. This section covers the water cycle, temperature and humidity, cloud types, air pressure, the cause of winds, the Coriolis effect, and global wind patterns.

🔊 Bellringer

Use the Bellringer transparency to prepare students for this section.

Motivate

Opening Demonstration — GENERAL

You will need a glass container, ice water, and food coloring. Fill the glass container with ice water. Add a few drops of food coloring to the water to distinguish it from liquid water that condenses on the glass surface. Allow the glass to sit for a few minutes. Have students describe what they observe, and explain how this demonstration proves that water is in the air.

• Does the water seep through the glass?
• Does the water come from the air?
• Why don't the water beads form on a warm container?

LS Visual

Water and Wind

▶ **KEY TERMS**

water cycle
transpiration
precipitation
humidity
dew point
barometric pressure
Coriolis effect

internet connect

www.scilinks.org
Topic: Water Cycle
SciLinks code: HK4149

SCI LINKS. Maintained by the National Science Teachers Association

■ **water cycle** the continuous movement of water from the ocean to the atmosphere to the land and back to the ocean

Figure 10
Evaporation, transpiration, condensation, and precipitation make up the continuous process called the water cycle.

OBJECTIVES

▶ **Describe** the three phases of the water cycle.
▶ **Explain** how temperature and humidity are related.
▶ **Identify** various cloud types by their appearance and the altitudes at which they typically occur.
▶ **Use** the concept of pressure gradients to explain how winds are created, and explain how Earth's rotation affects their direction.

You come in contact with water throughout every day, not just when you drink it or when you shower or wash your hands. You experience water in the air because water exists as an invisible gas in the air. It also exists in air as a liquid, suspended in the atmosphere as clouds or fog, or falling as rain or snow. All of this water in the atmosphere affects the weather on Earth.

The Water Cycle

Water is continuously being moved through the troposphere in a process described by the **water cycle**. *Figure 10* shows the main processes that take place in the water cycle.

The major part of the water cycle occurs between the oceans and the continents. Solar energy strikes ocean water, causing water molecules to escape from the liquid and rise as gaseous water vapor. This process is known as evaporation.

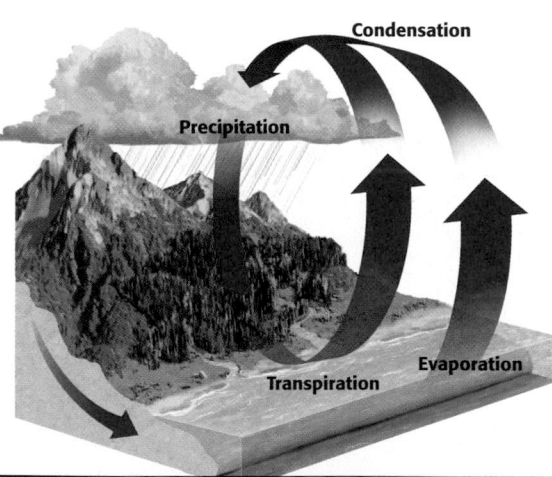

LIFE SCIENCE
CONNECTION

When the air is humid, hair becomes frizzy. Hair is made of a protein called keratin and each hair fiber has a scaly outer cuticle, which you can feel by running your fingers up and down a single hair. These scales allow moisture to enter the inner part of the hair fiber.

When the air is humid, hair absorbs moisture and becomes longer, which makes it frizzy. Hair dries out and becomes shorter when the air is dry. Humidity can cause hair length to change by as much as 2.5 percent. A device called a hair hygrometer can very accurately measure changes in humidity.

MISCONCEPTION
ALERT

Evaporation Some students think that weight or mass is lost when a liquid changes to a gas or vapor because material substance seems to disappear. Be sure to emphasize that this is not the case; the mass changes form, but does not disappear.

Evaporated water vapor condenses to form precipitation

Evaporation occurs when sunlight strikes water in lakes, rivers, and the soil. Additional water travels to the air through **transpiration.** During transpiration, plants lose moisture through small pores in their leaves. While transpiration may seem insignificant, the addition of water vapor to the atmosphere can be large. For example, a 1 km² cornfield typically transpires 3 40 0000 L (900 000 gal) of water per day.

In the atmosphere, water vapor rises until the air is cool enough to condense the vapor into tiny droplets of water. We observe these droplets as clouds. As clouds cool and condense more vapor, they often release their moisture in the form of **precipitation.** Most precipitation falls as rain, snow, hail, or sleet. Precipitation can occur over land or water. When precipitation falls on land, it stays on the surface until it evaporates, flows into a larger body of water, or is absorbed into the ground to become *ground water.*

Air contains varying quantities of water vapor

You have probably noticed that air doesn't always feel the same. Sometimes it is thick and moist, while at other times it is crisp and dry. The air around you contains varying amounts of water vapor, and you experience the effects of changing **humidity,** or the quantity of moisture in the atmosphere. Some animals depend on this humidity as shown in *Figure 11.*

Relative humidity is the actual quantity of water vapor present in the air compared with the maximum quantity of water vapor that can be present at that particular temperature. In weather forecasts, the relative humidity is usually given as a percentage. A relative humidity of 85% means that the air contains 85 percent of the water that it can contain at that temperature. Air that has a relative humidity of 100% is said to be *saturated.*

Warmer temperatures evaporate more water

As illustrated in *Figure 12,* the temperature of the air determines the air's maximum water vapor content. At warm temperatures, molecules move very quickly and are farther apart. Thus, the water is more likely to exist as a gas. As the temperature decreases, the water molecules slow down. At these slower speeds, the attractive forces between water molecules have a greater effect, and the molecules may condense into a liquid.

Figure 11

This frog depends on the humidity in its environment.

▶ **transpiration** the process by which plants release water vapor into the air through stomata

▶ **precipitation** any form of water that falls to Earth's surface from the clouds

▶ **humidity** the amount of water vapor in the air

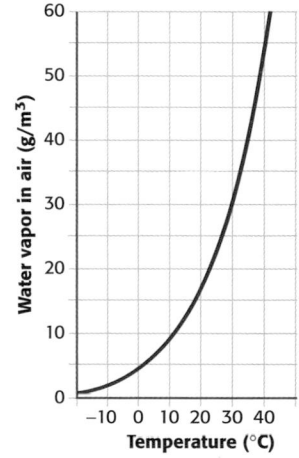

Figure 12

Warm air can hold more water vapor than cold air can.

751

Teach

SKILL BUILDER ─ **BASIC**

Reading Skills Have students read a passage out loud, then ask for volunteers to summarize it for the class. After the summary is completed, have members of the class ask for clarification of parts of the summary they did not understand. **LS Interpersonal**

Teaching Tip

Water Vapor Versus Water Droplets Emphasize to students that water vapor is invisible because it is a gas. Water droplets are condensed water vapor, and are visible because they are liquid.

Demonstration ─ GENERAL

Transpiration

(Time: About 20 minutes [over 3–4 days])
Materials:

• plastic bag
• rubberband
• healthy plant with large leaves

Ask students how they might capture some of the moisture that passes through a plant's pores during transpiration. Allow students to discuss ideas among themselves. To demonstrate, take a plastic bag and tie it around a large leaf of a healthy plant. Watch the plant over several days. Moisture should collect on the bag after a few days. **LS Visual**

Demonstration

How to Make a Cloud

(Time: About 20 minutes)

Materials:

• 2 L soda bottle
• matches
• water

Step 1 Pour a small amount of water into the soda bottle.

Step 2 Light one of the matches and hold it in the bottle. Try to capture as much smoke in the bottle as possible.

Step 3 Replace the lid and shake (to moisten the air), and then squeeze the bottle to release it quickly.

If the cloud is difficult to see, turn off the classroom lights and have a student shine a flashlight at the bottle, so the beam of light is passing through the bottle.

Squeezing the bottle increases the temperature of the air, enabling it to hold more moisture. Releasing the bottle reduces the temperature and promotes condensation of the water vapor. Clouds need condensation nuclei, or small bits of dust or dirt, to condense around. In this case, the smoke from the match served as the condensation nuclei.

▶ **dew point** the temperature at which air or a gas begins to condense to a liquid

Figure 13

Clouds are classified by their form and the altitude at which they occur.

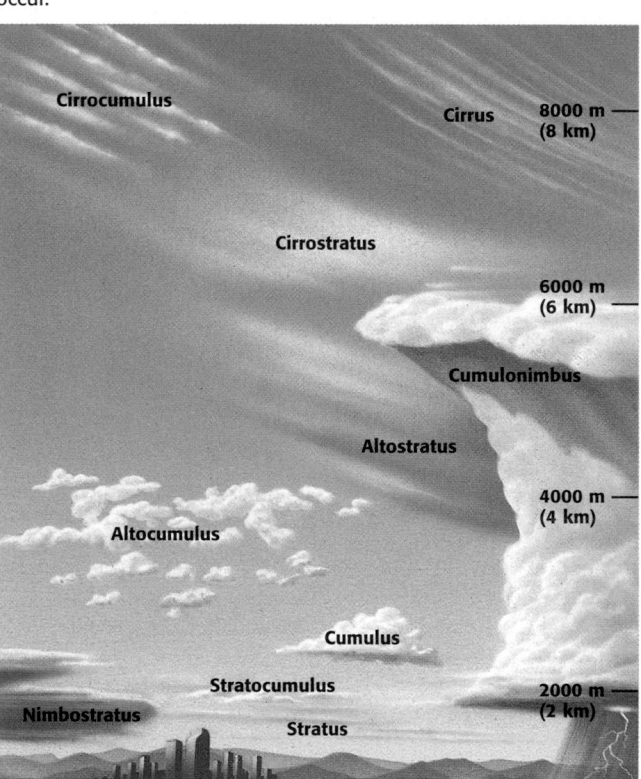

Cirrocumulus

Cirrus   8000 m (8 km)

Cirrostratus

6000 m (6 km)

Cumulonimbus

Altostratus

4000 m (4 km)

Altocumulus

Cumulus

Stratocumulus   2000 m (2 km)

Nimbostratus

Stratus

752

Water vapor becomes liquid at dew point

The exact temperature at which water vapor molecules slow enough to form liquid water is called the **dew point.** Dew point depends on humidity. When the humidity is high, there are more molecules of water in the air and it is easier for them to form liquid water. So the higher the humidity is, the higher the dew point is. In fact, we can measure humidity by finding the dew point. You may have seen drops of water, or condensation, on a glass of ice water. The cold surface of the glass provides a place where the dew point is reached, and water shifts from gas to liquid.

Clouds form as warm, moist air rises

You may know from walking through fog—a low-lying cloud—that clouds are wet. Clouds are formed when warm air rises and water vapor condenses into tiny droplets of liquid as it cools. This process usually occurs in the troposphere. Clouds are made up of tiny droplets of liquid water and, at higher altitudes, small ice crystals. Depending on where clouds form, they can have many different shapes and characteristics.

Cloud names describe their shape and altitude

Figure 13 shows the different kinds of clouds that can occur. Cloud names describe both their appearance and the altitude at which they occur. Different types are named using combinations of the three root words *cirrus, stratus,* and *cumulus.*

Cirrus clouds are thin and wispy, and they occur at high altitudes—between 6 km and 11 km (3.7–6.8 mi) above Earth. Stratus clouds are sheetlike and layered. These clouds typically form at lower altitudes—less than 6 km. Cumulus clouds are white and fluffy with somewhat flat bottoms. The flat base is the point at which rising air begins to condense. Cumulus clouds form at various altitudes—anywhere from about 500 m to about 12 km (7.5 mi) above Earth.

REAL-LIFE CONNECTION

Condensation Why is it possible to "see one's breath" on a cold day but not on a warm day? Water vapor in your breath condenses when it hits the cold air outside. This does not happen in warm air because condensation only occurs in low temperatures.

Evaporation Why might you shiver when you get out of a swimming pool on a windy day? As you leave the water, air blowing over you evaporates the water on your skin. Because evaporation is a cooling process, it chills your skin.

INCLUSION Strategies

• *Learning Disabled*   • *English Language Learners*

Have students label five index cards with the following: cirrus, stratus, cumulus, cumulonimbus, and nimbostratus. Using their textbooks, have students record the altitude, associated weather conditions, and a description of each type of cloud on the corresponding index card.

Lower level students can draw a scene illustrating each type of clouds and their altitudes. Make sure that students also label their illustrations.

Cloud names reflect combined characteristics

Cloud names reflect the combined characteristics of each cloud type. *Cirrostratus* clouds are high, layered clouds that form a thin white veil over the sky. *Altostratus* and *altocumulus* clouds are simply stratus and cumulus clouds that occur at middle altitudes. When a cloud name includes the root *nimbo* or *nimbus*, the cloud is the type that produces precipitation. Cumulonimbus clouds are towering rain clouds that often produce thunderstorms. Nimbostratus clouds are large, gray clouds that often produce steady precipitation.

You may have seen a halo around the sun or moon. This halo results from the refraction of light as it passes through ice crystals in cirrostratus clouds. Sometimes the presence of a halo is the only way to tell that a very thin, transparent layer of cirrostratus clouds is present.

Air Pressure

The term **barometric pressure** is often used in weather reports in the newspaper and on television. Changes in barometric pressure often accompany changes in the weather. Falling pressure may indicate a large air mass is leaving the area, while rising pressure can mean an air mass is moving in. The barometric pressure, also called *atmospheric pressure* or *air pressure*, is the pressure that results from the weight of a column of air extending from the top of the thermosphere to the point of measurement.

Instruments used to measure air pressure are called barometers. *Figure 14A* is a photo of a *mercury barometer*, and *Figure 14B* shows an *aneroid barometer*. Aneroid barometers do not contain liquid and are more portable than mercury barometers, but mercury barometers are more accurate.

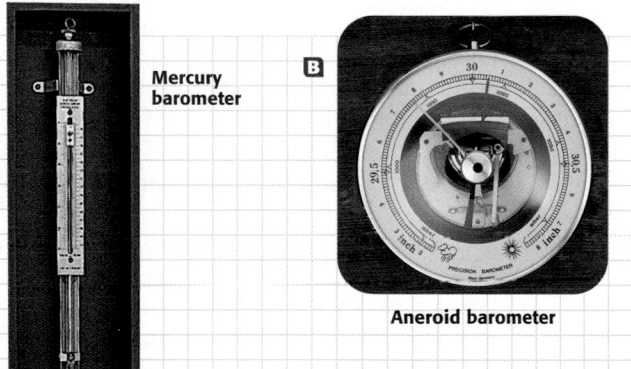

A Mercury barometer
B Aneroid barometer

▶ **barometric pressure** the pressure due to the weight of the atmosphere; also called air pressure or atmospheric pressure

Figure 14

A Mercury barometers measure the weight of the air, which is also called *barometric pressure*.

B Aneroid barometers do not contain liquid and are less accurate than mercury barometers.

Demonstration

Windsocks

(Time: About 40 minutes)

Materials:

• bed sheets
• pillowcases or fabric remnants
• scissors
• stapler
• dowels (1 m long)
• wire
• wire cutters

Safety Caution: *Be careful when handling wire pieces because the ends may be sharp.*

Step 1 Cut the wire into pieces long enough to bend into a 20 cm diameter circle and still wrap the ends around one end of the dowel.

Step 2 Draw a square with 63 cm sides on the fabric.

Step 3 Draw a line from each lower corner to the midpoint of the top, forming a large triangle.

Step 4 Cut out this triangle with scissors. Fold the triangle so the sides meet and overlap about 1 cm.

Step 5 Staple the edges together, and cut off the tip of the triangle.

Step 6 Bend the wire into a 20 cm diameter circle, and wrap the extra wire around the end of the dowel.

Step 7 Roll the bottom edge of the windsock around the wire and staple it.

Step 8 Once you have a finished windsock, take your students outside and use a directional compass to determine the direction of the wind.

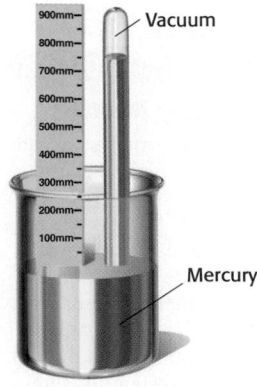

Figure 15

The height of the mercury in the tube of a mercury barometer indicates the barometric pressure in millimeters.

INTEGRATING

HEALTH

Just as air exerts pressure on the objects around it, blood in your body exerts pressure against the walls of your arteries. To measure a person's blood pressure, a cuff is wrapped around the upper arm and a stethoscope is placed over the arteries of the forearm. Air is pumped into the cuff until the pressure exerted by the cuff stops the flow of blood.

The doctor or nurse then listens to the person's pulse as air is let out slowly until the pressure in the cuff is less than the blood pressure when the heart contracts. This pressure is called the *systolic* pressure.

More air is released until the pulsing of the heart is no longer audible. The pressure at this point, called the *diastolic* pressure, is the blood pressure when the heart relaxes.

754

Barometers measure air pressure

Figure 15 shows how a simple mercury barometer works. The mercury barometer contains a long tube that is open at one end and closed at the other. The tube is filled with mercury and then inverted into a small container of mercury. Some but not all of the mercury spills out of the tube and into the container. The atmosphere exerts a pressure on the mercury in the container, holding some of the mercury in the tube to a height above the mercury in the container. Any change in the height of the column of mercury means that the atmosphere's pressure has changed.

At sea level, the barometric pressure of air at 0°C is around 760 mm of mercury. This amount of pressure is defined as 1 atmosphere (1 atm) of pressure. The SI unit for pressure is the pascal (Pa), which is equivalent to one newton per square meter.

Aneroid barometers are more commonly used. The word *aneroid* means "without liquid." This type of barometer contains a sealed metal chamber from which part of the air has been removed. When the air pressure changes, the chamber expands or contracts, moving a needle on a dial.

Wind

Have you ever seen a movie in which an airplane window gets broken? When the window breaks, all the loose objects in the plane are pushed out the window. Although it is unlikely that an airplane window would actually break, the portrayal of objects flying out the window is correct.

Differences in pressure create winds

Commercial airplanes fly very high in the troposphere, between 10 km and 13 km above Earth. At this altitude, the air is not very dense and the atmospheric pressure is very low. However, because of pressurization by pumps, the pressure inside the airplane is relatively high—similar to the air pressure at Earth's surface.

If an airplane window were to break, the dense air in the plane's cabin would spread out into the less-dense air outside the cabin. The airflow produced in this situation would push loose objects out the window.

Just as a difference in air pressure would create airflow from inside the airplane to the outside, differences in pressure in the atmosphere can cause wind. When air pressure varies from one place to another, a pressure gradient exists. The air in a *pressure gradient* moves from areas of high pressure to areas of low pressure. This movement of air from a high-pressure area to a low-pressure area is called wind.

BIOLOGY

CONNECTION

Although airplane cabins are pressurized, passengers still feel the pressure decrease as the plane climbs, and the pressure increase as the plane descends. *Middle ear barotrauma* is an earache caused by the difference in pressure between the air and a person's middle ear. The trauma occurs when the Eustachian tube, a passageway between the middle ear and the throat, fails to open wide enough to equalize the pressure. Some scuba divers also experience this kind of earache. Chewing gum, yawning, or swallowing often alleviates the condition.

Earth's rotation affects the direction of winds

The direction in which wind moves is influenced by Earth's rotation. The effect of Earth's rotation on the direction of wind is described by the **Coriolis effect.** To understand how the Coriolis effect works, you must first understand that points at different latitudes on Earth move at different speeds as Earth rotates. Consider two houses at different latitudes. A house located on the equator travels faster than a house located near one of the poles. Can you see why this must be true? Earth goes through one full rotation in 24 hours. During this time, the house at the equator must travel the distance of Earth's circumference. Closer to the poles, a circle of latitude is smaller. Therefore, the house closer to the pole moves through a shorter distance in the same amount of time. Thus, this house moves more slowly.

Imagine a cannonball in a cannon at the equator. The cannonball is moving along with Earth as it rotates. The cannonball's speed at this time is a little greater than 1610 km/h (1000 mi/h) to the east—the speed at which all points on the equator move because of Earth's rotation. When the cannonball is fired to the north, it continues to move east at about 1610 km/hr. As the cannonball moves farther north, however, the portion of Earth beneath the cannonball is moving more slowly. The result is a flight path like the one shown in *Figure 16* where the cannonball's path appears to curve eastward.

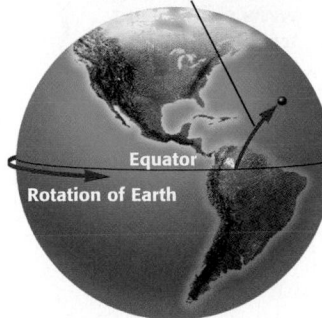

Figure 16

Because of Earth's rotation, a cannonball fired directly north from the equator will curve to the east relative to Earth's surface.

► **Coriolis effect** the curving of the path of a moving object from an otherwise straight path due to Earth's rotation

Predictable air circulation forms wind patterns

The movement of winds is analogous to the cannonball's movement. When moving north in the Northern Hemisphere, winds curve to the right. Conversely, winds moving south in the Southern Hemisphere curve to the left. Next consider air moving south from the North Pole. The wind would lag behind the rotation of Earth and travel west, or to the left on the figure at right because it has a slower speed than the spinning Earth. Similarly, wind moving north from the South Pole would travel west, or to the left, because of its slower speed.

In summary, *winds in the Northern Hemisphere curve clockwise, and winds in the Southern Hemisphere curve counterclockwise.* The resulting circulation patterns are so regular that meteorologists have named them as shown in *Figure 17*.

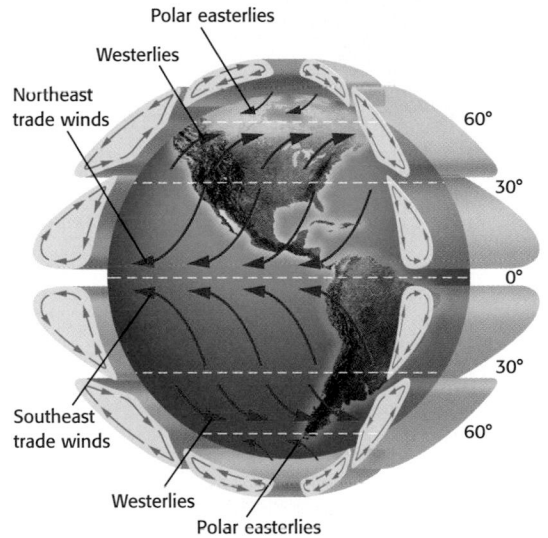

Figure 17

Both the Northern Hemisphere and the Southern Hemisphere have three wind belts.

SKILL BUILDER — **BASIC**

Reading Skills Have students read the passage on the Coriolis effect, then ask for a volunteer to summarize the passage. Allow students to ask for clarification and to consult their text. **LS Verbal**

MISCONCEPTION ALERT

The Coriolis Effect It is a common misconception that the Coriolis effect causes the water in a sink or toilet to rotate one way in the northern hemisphere, and the opposite way in the southern hemisphere. The Coriolis effect can influence the direction of long-lasting vortices, such as those associated with hurricanes and large mid-latitude storms, but it does not affect the direction that water in a sink or toilet drains. While the rotation of the Earth happens once every 24 h, water in a toilet might rotate once every few seconds— a rate at least 10,000 times greater than Earth's rotation. As a result, the Coriolis effect is much too small to have any influence.

The direction of rotation in a sink is actually caused by the way the sink was filled, or by vortices created while using the water to wash. Have students pay attention to the direction of rotation for a variety of sinks in the school and at their homes, and they should find examples of both directions of rotation. **LS Visual**

LANGUAGE ARTS
CONNECTION — **ADVANCED**

Writing Have students research names given to winds in various parts of the world and write a one-page report including answers to the following questions: Why do these winds have the names they do? How do these winds affect parts of the world? Winds to research include but are not limited to: monsoons, jet streams, williwaws, Pineapple Express, Chinooks, mistrals, and Santa Ana. **LS Verbal**

Did You Know?
Wind and Ocean Currents For the same reason that winds are deflected clockwise in the Northern Hemisphere and counterclockwise in the Southern Hemisphere, ocean currents are deflected in the same directions.

Chapter Resource File

• **Cross-Disciplinary Worksheet** Integrating Health—Why Your Ears Pop **GENERAL**

Transparencies

TM Barometer

Quiz ——————— BASIC

1. What is the water cycle? (continuous movement of water between Earth and the atmosphere)

2. What two characteristics are used to classify clouds? (form or shape and altitude)

3. What does barometric pressure result from? (the weight of the atmosphere)

4. What causes winds? (a difference in air pressure)

5. What is the Coriolis effect? (the effect of Earth's rotation on the direction of wind)

☑ internet connect ▤

www.scilinks.org
Topic: Coriolis effect
SciLinks code: HK4028

SC_**LINKS**_ Maintained by the National Science Teachers Association

Global wind patterns form circulation cells

Note that the wind patterns shown in *Figure 17* move in vertical loops. Because temperatures at the equator tend to be warmer than at other latitudes, the air there rises and creates a low-pressure belt. As this warm air rises, it moves toward the poles.

In the Northern Hemisphere, much of the northward-moving air sinks at about 30° latitude, forming a high pressure area near Earth's surface. Flowing from a high-pressure area to a low-pressure area, air flows both north and south. At about 60° latitude, air flowing along the surface from the polar high and the high pressure band at 30° converge. As these air masses converge, air rises, forming a low-pressure belt.

A similar circulation pattern, in which rising warm air is coupled with sinking cold air, occurs in the Southern Hemisphere. Thus, air in each of the hemispheres completes three loops, called *cells*.

SECTION 2 REVIEW

SUMMARY

▶ In the water cycle, water from oceans and lakes evaporates and rises in the atmosphere. After it cools and condenses, the water falls back to Earth as precipitation.

▶ Warm air can contain more water vapor than cold air.

▶ Clouds are classified according to their appearance and the altitude at which they occur. *Figure 13* summarizes the various types of clouds.

▶ Wind is caused by the air in a pressure gradient moving from a high-pressure area to a low-pressure area.

▶ Earth's rotation affects the direction of winds. This phenomenon is described by the Coriolis effect.

1. **Identify** which one of the following processes is not a step in the water cycle:
   a. evaporation
   c. photosynthesis
   b. condensation
   d. precipitation

2. **Determine** whether wind moving north from the equator will curve eastward or westward due to the Coriolis effect.

3. **Distinguish** between humidity and relative humidity.

4. **Select** which one of the following describes a cumulus cloud:
   a. sheetlike
   b. fluffy, white, and flat-bottomed
   c. wispy and feathery
   d. high altitude

5. **Identify** which one of the following statements describes a mercury barometer:
   a. less accurate than an aneroid barometer
   b. contains a chamber that has a lot of water inside
   c. the height of the mercury in the tube indicates the barometric pressure
   d. more commonly used than aneroid barometers

6. **Predict** which of the following would be a more humid area: the Sahara Desert on a cold night or the Florida coast on a warm day.

7. **Critical Thinking** Which has a lower pressure, the air in your lungs as you inhale or the air outside your body?

756

Chapter Resource File

• **Concept Review** GENERAL

• **Quiz** BASIC

Answers to Section 2 Review

1. c

2. eastward

3. Humidity is the quantity of water vapor (moisture) in the air, while relative humidity is the quantity of water vapor in the air compared with the maximum quantity of water vapor that can be present at a given temperature.

4. b

5. c

6. The Florida coast on a warm day would be more humid because along the coastline there is water that can evaporate into the air, unlike in the Sahara Desert. Also, the cold Sahara air cannot hold as much moisture as the warm Florida air.

7. During inhalation the air inside the lungs is less densely packed, and thus is under less pressure.

Weather and Climate

OBJECTIVES

▶ **Describe** the formation of cold fronts and warm fronts.

▶ **Describe** various severe weather situations, including thunderstorms, tornadoes, and hurricanes.

▶ **Distinguish** between climate and weather.

▶ **Identify** factors that affect Earth's climate.

▶ **KEY TERMS**
air mass
front
climate
topography

How is a weather forecast made? *Meteorologists,* people who study weather, gather data about weather conditions in different areas. By using weather maps, meteorologists can try to predict weather by tracking the movement of air pockets called **air masses.** Interactions between air masses have predictable effects on the weather in a given location.

Fronts and Severe Weather

You have probably heard about *cold fronts* and *warm fronts*. A **front** is the place where a cold air mass and a warm air mass meet. Examine the weather map in *Figure 18A.* Cold fronts are shown as a blue line with blue triangles. Warm fronts are shown as a red line with red semicircles. Clouds, rain, and sometimes snow can occur at fronts. When fronts move through an area, the result is usually precipitation, as shown in *Figure 18B,* and a change in wind direction and temperature.

▶ **air mass** a large body of air where temperature and moisture content are similar throughout

▶ **front** the boundary between air masses of different densities and usually different temperatures

Figure 18

A Weather maps can give you a large-scale view of the weather.

B People also experience weather on a very local level.

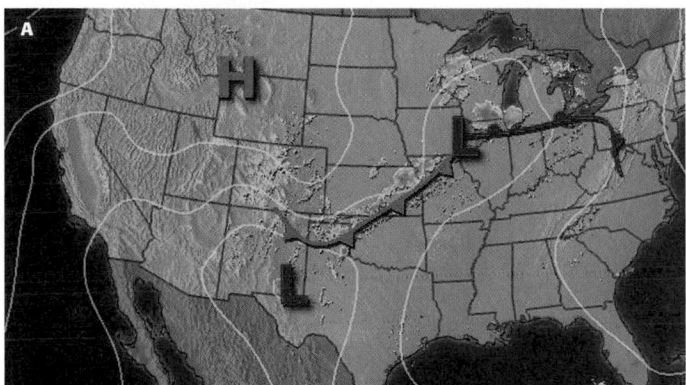

757

Focus

Overview
Before beginning this section, review with your students the Objectives listed in the Student Edition. In this section, students study weather maps, cold and warm air fronts, and severe weather conditions including tornadoes, and hurricanes. They also learn the difference between weather and climate and study factors that affect climate.

Bellringer
Use the Bellringer transparency to prepare students for this section.

Motivate

Opening Activity — GENERAL
Obtain the weather section of the local newspaper. Challenge students to decipher the weather map with the help of the key. Ask them to describe all aspects of the weather for various geographical points across the country. **LS** Visual

Alternative Assessment — ADVANCED

Weather Prediction Throughout history, people have predicted approaching weather by interpreting natural signs. Animals and plants are usually more sensitive to changes in the atmosphere—such as changes in air pressure, humidity, and temperature—than are humans.

Have students research natural signs for predicting weather and write a short report which includes an answer to the following question: If you did not have access to the weather forecast on the news, radio, or television, how would you forecast the weather? **LS** Verbal

Chapter Resource File

• **Lesson Plan**

• **Cross-Disciplinary Worksheet**
Integrating Technology—Doppler Weather Radar GENERAL

Transparencies

TT Bellringer

Interpreting Visuals Have students use **Figure 19** to answer the following questions: What kinds of clouds are associated with warm fronts? (cirrus, cirrostratus, altostratus, and nimbostratus) What kind of weather is associated with warm fronts? (clouds, rain, and sometimes snow) What kinds of clouds are associated with cold fronts? (cumulonimbus) **LS** Visual

Demonstration

A Cold Front Meets a Warm Front

(Time: About 10 minutes)

Materials:

- cold water
- warm water dyed with food coloring
- a transparent container

Fill the container with the warm, dyed water. Next, slowly pour a small amount of the cold water down the side the container. Have your students comment on the results and apply them to cold and warm fronts. (The cold water represents a cold front. When these types of fronts meet, the cold air slides under the warm front and pushes it up because the cold air is denser.)

INTEGRATING

PHYSICS
Large bodies of water regulate local climates because water has a high specific heat. Even though Minneapolis, Minnesota, and Portland, Oregon, are at about the same latitude, they have very different climates. The difference is caused by the Pacific Ocean. During winter months, the ocean does not get as cold as the surrounding air. As a result, Pacific winds warm the Oregon coastline.

In summertime, the Pacific Ocean does not get as warm as the surrounding air, and Pacific winds cool Portland.

There are three types of fronts

In a warm front, a mass of warm air moves toward and over a slower mass of cold air, as shown in **Figure 19A.** As the warm air is pushed up over the cool air, it cools and forms clouds.

Cirrus and cirrostratus clouds can be seen high in the sky as a warm front approaches. As time passes, lower-lying clouds move overhead. Often, nimbostratus clouds release steady rain or snow for one to two days.

With a cold front, the forward edge of a mass of cold air moves under a slower mass of warm air and pushes it up, as shown in **Figure 19B.** Note that the front edge of the cold front is steeper than that of the warm front shown in **Figure 19A.** Because of this steep edge, warm air rises quickly, forming cumulonimbus clouds. High winds, thunderstorms, and sometimes tornadoes accompany this type of front.

A *stationary front* occurs when two air masses meet but neither is displaced. Instead, the air masses move side by side along the front. The weather conditions near a stationary front are similar to those near a warm front.

Figure 19

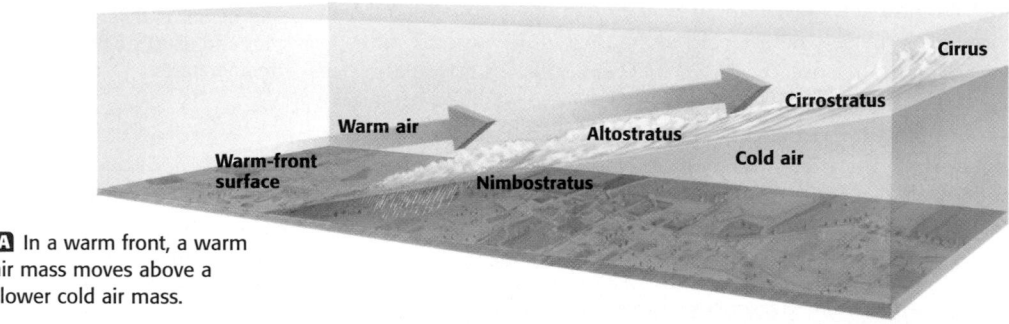

A In a warm front, a warm air mass moves above a slower cold air mass.

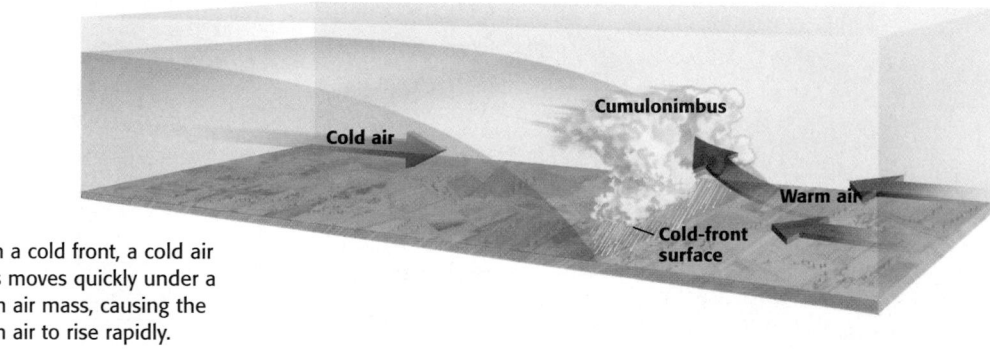

B In a cold front, a cold air mass moves quickly under a warm air mass, causing the warm air to rise rapidly.

758

Alternative Assessment — ADVANCED

Fronts After performing the Demonstration above, ask students if an airline pilot should avoid flying into an area where two different fronts are meeting. Have students write a paragraph explaining their answer. **LS** Logical

Did You Know?

Coriolis Effect The two air masses that move side by side along a stationary front move in opposite directions. The winds blow in opposite directions as a result of the Coriolis effect.

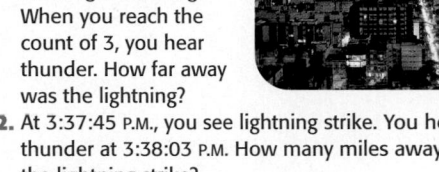

Calculating the Distance to a Thunderstorm

How can you tell if lightning is close by? The distance can be determined by counting the seconds between the lightning flash and the sound of thunder. This time lag occurs because light travels faster than sound. Count the seconds between the flash of lightning and the thunder, and use the following calculation:

time (s)/3 = distance (km)

time (s)/5 = distance (mi)

Applying Information

1. You see lightning and begin counting. When you reach the count of 3, you hear thunder. How far away was the lightning?

2. At 3:37:45 P.M., you see lightning strike. You hear thunder at 3:38:03 P.M. How many miles away did the lightning strike?

Lightning is a discharge of atmospheric electrical energy

Lightning is a big spark. Water droplets and ice crystals in thunderclouds build up electrical charges. Often, sparks jump between clouds or between clouds and Earth to equalize the charge. Thunderstorms can be exciting, but they can also be very dangerous if lightning strikes the ground near you. Lightning superheats the air so fast that the air expands faster than the speed of sound. The shockwave created is thunder.

Tornadoes are funnels of high-speed wind

Tornadoes like the one shown in *Figure 20* are high-speed, rotating winds. Tornadic winds are the most violent winds on Earth with speeds thought to be as great as 500 km/h (about 310 mi/h). Tornadoes occur most commonly in the United States, especially in the Midwest and along the Gulf of Mexico. They occur during spring and early summer, and typically form along a front between cool, dry air and warm, humid air. As warm air rises, more air rushes in to replace it. This air sometimes begins to rotate as it rises and can spawn a tornado.

Typically a tornado begins as a column of water droplets, called a *funnel cloud,* that reaches down from dark storm clouds with heavy rain and lightning. As the funnel reaches the ground, it begins sucking objects upward. The rotating winds on the outer edge of a tornado can tear apart homes and trees. Tornadoes are too fast and unpredictable to attempt to outrun, even if you are in a car. If you see a tornado, move to a storm cellar or basement, or lie flat under a table at the center of a room with few windows. If you are outside, lie in a ditch or low-lying area, and cover your head with your hands.

Figure 20

A tornado such as this one in Pampa, Texas, is seen as a rapidly spinning funnel cloud.

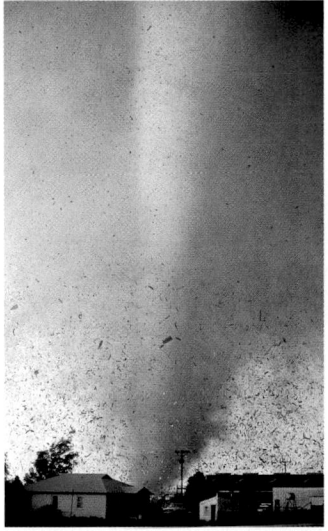

759

Demonstration

Creating a Tornado Use this demonstration to model a tornado vortex. You will need a clean, empty jar with its lid, water, food coloring, a teaspoon of liquid dishwashing detergent, and a teaspoon of vinegar. Fill the jar about three-quarters full of water. Add a few drops of food coloring, the detergent, and vinegar to the water. Cap the jar tightly and shake it vigorously. Once the solution is mixed, you can create a vortex by giving the jar a quick twist with a flick of the wrist.

FINE ARTS

CONNECTION — GENERAL

Ask students to write a creative story about an object that becomes trapped in a tornado. Be sure that their stories have some of the scientific elements in them that are described in the text. **LS Verbal**

Did You Know ?

Raining Fish Did you know that fish have been known to fall from the sky? Some scientists think the phenomenon of raining fish is caused by waterspouts. A waterspout is a tornado that occurs over water. When the funnel comes into contact with the surface of the water, it causes the water and its contents to spray several meters upward.

Chapter Resource File

- **Cross-Disciplinary Worksheet** Real World Applications—Understanding Thunderstorms GENERAL

Videos

 Presents Earth Science

- **Segment 24** Twister Mysteries

- **Critical Thinking** Worksheet 24

See the Science in the News video guide for more details.

Group Activity —— GENERAL

Major Hurricanes Put students into pairs or small groups and have each group choose a major hurricane, cyclone, or typhoon to research in detail. Students should do preliminary research before choosing a hurricane to focus on. Their research should focus on the duration of the hurricane, areas affected, diameter of the hurricane, wind speeds, and destruction caused by the hurricane. Ask students to create a poster, or some other visual aid, with the information they gathered. Have students view the posters and compare hurricanes. **LS** Visual

Teaching Tip

Hurricane Warnings In 1900, a hurricane at Galveston killed 6000 people. But in 1992, only about 60 deaths were attributed to the destructive Hurricane Andrew. Why this huge difference?

Today, a sophisticated detection system makes it possible to predict hurricanes far enough in advance to evacuate dangerous areas when needed. At the National Hurricane Center in Miami, Florida, weather forecasters constantly monitor the oceans, receiving new satellite images every 30 minutes. If these meteorologists see a major storm developing, they alert an air force unit called the *Hurricane Hunters,* who fly into storms to measure wind speed, moisture levels, and pressure. Meteorologists use this data to predict when and where a storm is likely to hit land.

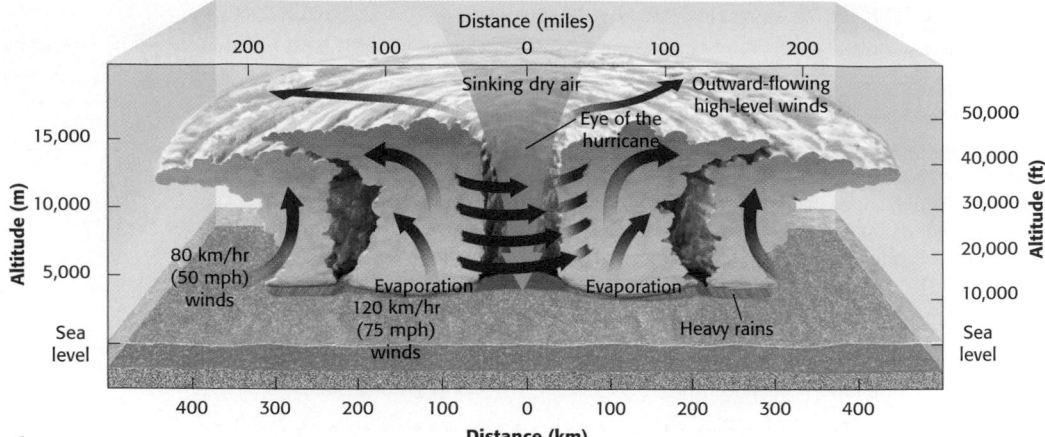

Figure 21
Hurricanes are nearly circular in shape and rotate around a center called the eye as shown in this cross section.

■ **climate** the average weather conditions in an area over a long period of time

Did **You Know?**

In the Northern Hemisphere, the winds and clouds in a hurricane rotate counterclockwise around the center, or eye, of the hurricane. In the Southern Hemisphere, hurricanes rotate clockwise. The direction of rotation is caused by Earth's rotation as described by the Coriolis effect.

Hurricanes are large storm systems

Hurricanes are similar to thunderstorms but are much larger. These storms are called *hurricanes* in North America and the Caribbean, *cyclones* in the Indian Ocean, and *typhoons* in the western Pacific. In the Northern Hemisphere, hurricanes occur in late summer and early fall, when the oceans are warmest. As the warm water evaporates and the water vapor rises, intense low-pressure areas called *tropical depressions* form. These tropical depressions can build strength and become hurricanes. As shown in **Figure 21,** hurricanes are large circulating masses of clouds, wind, and rain with diameters of about 600 km (373 mi).

Hurricanes are powered by the energy released as water vapor condenses to form clouds. As the vapor condenses, it releases heat into the air, the air heats and expands, and the pressure inside the clouds decreases. Warm, moist air continues to rise and condense, releasing more energy. This rising air creates fierce winds, shown by the red arrows in **Figure 21.** The storm gradually weakens as it moves over land or cool water.

Although hurricanes move fairly slowly, they are extremely powerful. Winds in a hurricane reach speeds from 118 km/h (73 mi/h) to greater than 250 km/h (155 mi/h). The eye of the hurricane is very calm. This can be very dangerous because people often believe the storm has passed and leave the protective cover of their homes only to be caught in the storm again.

Climate

Weather changes day to day, but **climate** does not change. Climate is the *average* weather of a region, often measured over many years.

760

HISTORY
CONNECTION

Hurricanes played a significant role in early colonial history. In 1609, a fleet of ships with settlers from England bound for Virginia was blown off course by a hurricane. Some of the ships landed in Bermuda, and the settlers started the first European colony there.

ASTRONOMY
CONNECTION

Wind speeds on Jupiter reach up to 540 km/h. Storms last for decades, and one—the Great Red Spot—has been observed swirling around since 1664. The Great Red Spot has a diameter of more than one and a half times that of Earth.

Temperatures tend to be higher close to the equator

Figure 22 shows sunlight striking Earth. Rays striking farther from the equator spread out over a greater area and are less concentrated than rays that strike the equator. At the poles, rays pass parallel to Earth's surface and do not warm the atmosphere as much. Hence, the poles are very cold. Earth is not always oriented so that the equator is perpendicular to incoming solar radiation. However, the equator is close to perpendicular throughout the year. Because of this, areas close to the equator have warmer climates.

Earth's tilt and rotation account for our seasons

In summer months, the days are longer and warmer. In winter months, the days are shorter and colder. Why do we experience these differing conditions? Earth's orbit is an ellipse, and you might expect that summer is when Earth is closest to the sun. This is not true. In fact, Earth is farthest from the sun on July 4 and closest to the sun about January 3. Earth's seasons are actually caused by the tilt of Earth on its axis. As shown in *Figure 23,* Earth's axis is tilted 23.5° from the perpendicular to the plane of the planet's orbit about the sun. Because of this tilt, the sun seems to rise to different heights in different seasons.

When the North Pole is tilted toward the sun, as in position *A* in *Figure 23,* the sun rises higher in the Northern Hemisphere than when the North Pole is pointed away from the sun. The days become longer and the temperature increases. This is summer in the Northern Hemisphere and winter in the Southern Hemisphere. Our longest day of the year is the *summer solstice,* which occurs around June 21.

When the South Pole is tilted toward the sun, as in position *C* in *Figure 23,* the sun rises higher in the Southern Hemisphere. It is summer in the Southern Hemisphere and winter in the Northern Hemisphere. Our shortest day of the year occurs on December 21, which is called the *winter solstice.*

At positions *B* and *D,* Earth's axis is tilted neither away from nor toward the sun. Day and night are of equal length all over Earth. The day on which this happens is called an equinox. Position *D* corresponds to the *vernal* (spring) *equinox,* which occurs around March 21. Position *B* occurs about September 22 and is called the *autumnal* (fall) *equinox.*

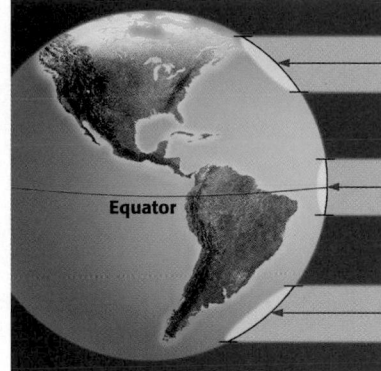

Figure 22

Solar energy is more concentrated at the equator than at the poles because Earth's surface is curved.

Figure 23

This illustration (not to scale) shows that Earth's axis is tilted 23.5° from the perpendicular to the orbital plane. The direction of tilt of Earth's axis remains the same throughout Earth's orbit.

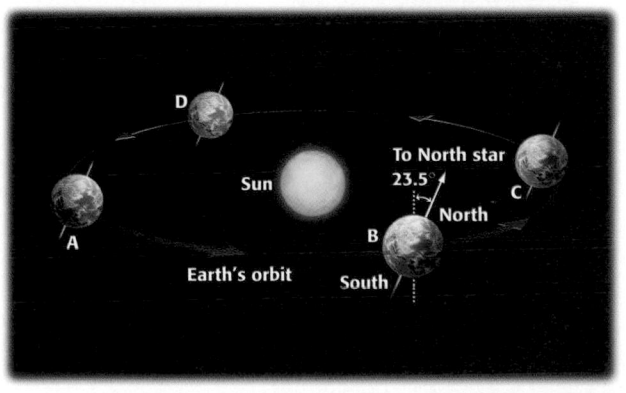

Travel Posters Group students in pairs and have each choose a location near the poles or the equator. Instruct them to find out about the weather and then to design a travel poster which describes the weather conditions and includes suggested activities appropriate to those conditions. **LS Verbal**

MISCONCEPTION ALERT

Seasons Some students believe that the changing distance between Earth and the sun causes the seasons, so that summer occurs when the Earth is closest to the sun and winter occurs when Earth is farthest. To emphasize that this is not the case, remind students that the Northern and Southern hemispheres undergo summer and winter at opposite times.

SKILL BUILDER

Vocabulary The word *solstice* comes from the Latin word *solstitium,* with the prefix *sol–,* which means sun, and the root *–stitium,* which refers to a standing. Thus, solstice means "standing of the sun." In the term *equinox,* the prefix *equi–* means equal, while the stem *–nox* comes from *–noct,* which means night. Thus, equinox means "equal nights."

Cultural Awareness GENERAL

Weather and climate have inspired rhymes, greetings, and other folklore. In the hot, wet climate of Venezuela, indigenous people sometimes greet each other by saying, "How have the mosquitoes used you?" Russia's cold climate inspired the saying, "There's no bad weather, only bad clothing." Invite students to interview relatives or research weather folklore in another country. Have them share their findings with the class. **LS Interpersonal**

Chapter Resource File

• **Cross-Disciplinary Worksheet** Integrating Physics—Adobe GENERAL

Transparencies

TM Hurricane

Videos

CNN Presents Earth Science

• **Segment 25** Hurricane Double Whammy GENERAL

• **Critical Thinking** Worksheet 25

See the Science in the News video guide for more details.

Topography Remind students that topography is an important factor in the formation of a temperature inversion in areas like the Los Angeles Basin. In this way, topography can affect air quality as well as climate.

SKILL BUILDER — BASIC

Interpreting Visuals Have students examine **Figure 24** and explain all aspects of the diagram. Make sure they incorporate the words *evaporation* and *condensation* in their explanations. **LS** Visual

Close

Quiz ———— BASIC

Determine whether each statement is true or false. If false, replace the underlined term with the correct term.

1. An air mass is a large body of air with <u>varying</u> temperature and moisture content. (false; uniform)

2. <u>Cold fronts</u> are often associated with high winds and thunderstorms. (true)

3. <u>Climate</u> is a region's general weather conditions over a period of many years. (true)

4. Earth's tilt and rotation account for <u>tornadoes and hurricanes.</u> (false; our seasons)

Chapter Resource File

• Concept Review GENERAL

• Quiz BASIC

---

topography the configuration of a land surface, including its relief

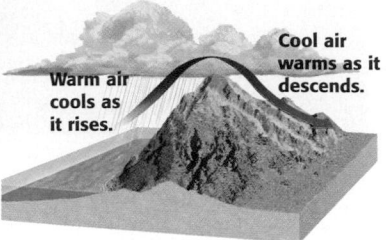

Figure 24

Air passing over a mountain loses moisture as it rises and cools. This dry air warms as it descends.

Warm air cools as it rises.

Cool air warms as it descends.

SUMMARY

▶ A warm front forms as warm air moves over slower cold air. A cold front forms as cold air moves under slower warm air.

▶ Lightning is a discharge of electrical energy.

▶ Tornadoes are high-speed rotating winds that form around rising warm air.

▶ Hurricanes are large storm systems that have highspeed winds and very low pressures.

▶ Climate is the average weather of a region over time.

▶ Some factors that affect climate are latitude, seasons, and topography.

Earth's surface features affect climate

The rise and fall of a land surface is called **topography**. Variations in Earth's topography affect the climate of a region. Mountains can have a profound effect on the climate of an area. As air rises over a mountain, it cools, and clouds form. When mountains are near oceans, as shown in *Figure 24,* the air is so humid that the clouds cannot hold all of the water vapor, and precipitation results. On the other side of the mountain, cool, dry air warms as it descends. Deserts often form and are said to lie in a rain shadow. Broad flat surfaces, such as the Great Plains, do not stop wind flow. Winds can come from several directions and merge on the plains. This mixing of wind produces thunderstorms and even tornadoes.

Global climate changes over time

From the early atmosphere to the many ice ages, Earth's climate has varied greatly over time. Many factors, such as the movement of the continents and slight changes in Earth's tilt, have produced these changes. The greenhouse effect may increase and temperatures could rise. Volcanic eruptions may produce gases that reflect solar energy, causing cooling. Because of all these factors, Earth's climate is likely to continue to change over the millennia to come.

SECTION 3 REVIEW

1. **Determine** whether each of the following statements describes a warm front or a cold front:
   a. A warm air mass moves above a slower cold air mass.
   b. It is characterized by high winds and thunderstorms.
   c. A cold air mass moves quickly under a slower-moving warm air mass.
   d. It is characterized by steady rain.

2. **Distinguish** between thunder and lightning.

3. **Identify** which of the following would not have an effect on the climate of a nearby region:
   a. a mountain range   c. a thunderstorm
   b. the equator   d. the Atlantic Ocean

4. **Determine** whether a tornado is more likely to form along a cold front or a warm front.

5. **Critical Thinking** Grapes grow well in areas where the climate is generally mild. Would you recommend planting grapes on the California coast or on the plains of North Dakota? Explain your answer.

762

Answers to Section 3 Review

1. **a.** warm front
   b. cold front
   c. cold front
   d. warm front

2. Lightning is a large spark that results from the buildup of electrical charges in the atmosphere, and thunder is the noise that results from the rapid heating of the air through which the spark travels.

3. c

4. A tornado is more likely to form along a cold front, in which warm, moist air rises more rapidly than it does in a warm front.

5. You should recommend planting grapes on the California coast because large bodies of water can regulate local climates. Plains do not stop wind flow, which can result in severe weather.

Math Skills

Percentages

Relative humidity is the amount of water vapor in the air divided by the maximum amount of water vapor that can remain in air at the same temperature. Relative humidity is usually expressed as a percentage. If air at 35°C can contain 40.0 g/m³ of water vapor, what is the amount of water vapor actually present if the relative humidity is 65%?

1 **List all given and unknown values.**

    Given: relative humidity, 65%

    maximum amount of water in air, 40.0 g/m³

    Unknown: actual amount of water in air (g/m³)

2 **Write down the equation for relative humidity, and rearrange it so you can calculate the actual amount of water in the air.**

$$relative\ humidity = \frac{actual\ water\ in\ air}{maximum\ water\ in\ air} \times 100$$

$$actual\ water\ in\ air = \frac{relative\ humidity \times maximum\ water\ in\ air}{100}$$

3 **Solve for the actual amount of water in the air.**

$$actual\ water\ in\ air = \frac{65 \times 40.0\ g/m^3}{100} = 26\ g/m^3$$

Therefore, 26 g of water vapor is actually present in each cubic meter of air at a temperature of 35°C.

Practice

Following the example above, calculate the following:

1. If the maximum amount of water vapor in air at 15°C is 12.5 g/m³, what is the actual amount of water vapor in the air if the relative humidity is 85.0%?

2. If, at a relative humidity of 60.0%, the actual amount of water vapor in air at 30°C is 18.0 g/m³, what is the maximum amount of water vapor in the air?

3. If the maximum amount of water vapor in air at 20°C is 17.0 g/m³, and the actual amount of water vapor in the air is 14.0 g/m³, what is the relative humidity?

Teaching Tip

Humidity Point out that, for a given temperature, the relative humidity is much greater for regions where there is more water available for evaporation than in regions where there is relatively little water. Because more water is available in regions where there are rivers and lakes, there is enough water to evaporate until the air is nearly saturated. In arid climates, like the desert, there is not enough water available to saturate the air.

Answers to Practice

1. 10.6 g/m³
2. 30.0 g/m³
3. 82.4%

SKILL BUILDER

Applying Knowledge Students can use the equation for relative humidity to estimate how much water vapor is actually in the air. By checking local weather reports, both the temperature and relative humidity can be obtained. **Figure 12** can then be used to estimate the maximum amount of water vapor that can remain in air for the particular temperature. From these quantities, the actual amount of water vapor can be determined.

Understanding Concepts

1. b
2. c
3. d
4. d
5. c
6. b
7. d
8. c
9. b
10. c
11. c
12. a
13. b
14. d
15. d

Using Vocabulary

16. Carbon dioxide is a greenhouse gas, and it helps to trap solar energy within the Earth's atmosphere. High levels of carbon dioxide may trap too much heat, leading to increased global temperatures.

17. Ozone is formed when the sun's ultraviolet rays strike molecules of O_2. The energy splits the molecules, and then the single atoms of oxygen bond with other O_2 molecules to make ozone.

Chapter Resource File

- Chapter Test GENERAL
- Test Item Listing for ExamView® Test Generator

Transparencies

TT Concept Mapping

Chapter Highlights

Before you begin, review the summaries of the key ideas of each section, found at the end of each section. The key vocabulary terms are listed on the first page of each section.

UNDERSTANDING CONCEPTS

1. Around Los Angeles, frequent temperature inversions are the result of cool, polluted air being trapped by
   a. acid rain.
   b. a layer of warmer air.
   c. a thunderstorm.
   d. the ocean.

2. The _____ is the process in which the atmosphere traps warming solar energy near Earth's surface.
   a. summer solstice
   b. Coriolis effect
   c. greenhouse effect
   d. water cycle

3. Almost all the water vapor in the atmosphere is in the
   a. exosphere.
   b. ionosphere.
   c. stratopause.
   d. troposphere.

4. The addition of _____ to the atmosphere by the burning of fossil fuels for cars, machinery, and power plants may lead to global warming.
   a. gasoline
   b. CFCs
   c. oxygen
   d. carbon dioxide

5. CFCs, chemicals that are used as refrigerants and propellants in spray cans, are partly to blame for the reduction of _____ in the stratosphere.
   a. carbon dioxide
   b. oxygen
   c. ozone
   d. clouds

6. Clouds form when water vapor in the air condenses as
   a. the air is heated.
   b. the air is cooled.
   c. snow falls.
   d. snow forms.

7. When air temperature drops, the air's ability to contain water vapor is
   a. slightly higher.
   b. much higher.
   c. about the same.
   d. lower.

8. Winds in the Northern Hemisphere curve to the right because of
   a. CO_2.
   b. climate.
   c. the Coriolis effect.
   d. CFCs.

9. Cumulonimbus and nimbostratus clouds both
   a. appear white and fluffy.
   b. produce precipitation.
   c. occur at high altitudes.
   d. look thin and wispy.

10. When a moving warm air mass encounters a mountain range, it
    a. stops moving.
    b. slows and sinks.
    c. rises and cools.
    d. reverses direction.

11. If you hear on the radio that a tornado is approaching, you should
    a. head to high ground.
    b. attempt to drive away from the tornado.
    c. sit in the center of a basement.
    d. hold onto a solid object, such as a tree.

12. Which of these layers of the atmosphere is closest to the ground?
    a. troposphere
    b. thermosphere
    c. mesosphere
    d. stratosphere

13. Electrons in the _____ reflect radio waves.
    a. mesosphere
    b. ionosphere
    c. radiosphere
    d. mesosphere

14. Which one of the following is not a part of the water cycle?
    a. precipitation
    b. condensation
    c. transpiration
    d. humidity

15. _____ are created when air masses of different temperatures meet.
    a. Wind storms
    b. Weather systems
    c. Air masses
    d. Fronts

764

18. Evaporation occurs as sunlight strikes water in lakes, rivers, oceans, and soil. The water that evaporates goes into the atmosphere. Transpiration, in which plants lose moisture through their pores, also adds water to the atmosphere. As the temperature cools, condensation occurs as the moisture in the air condenses and eventually precipitates back to the Earth in the form of precipitation or rain, snow, or sleet.

19. Humidity is the amount of moisture in the air. As the temperature drops overnight, the moisture in the air will condense on a surface—in this case, the lawn. The

temperature at which the moisture in the air condenses and forms water droplets is known as the dew point.

20. A warm air mass moves above a slower cold air mass that is moving in the same direction or is stationary.

21. Weather changes daily, whereas climate is the average weather of a region measured over many years.

22. The temperature of air determines the air's maximum water vapor content or maximum humidity. The higher the temperature, the higher the humidity can be.

16. Explain how carbon dioxide in the atmosphere relates to the *greenhouse effect.*

17. How is *ozone* formed?

18. Describe the *water cycle* using the terms *precipitation, condensation, transpiration,* and *evaporation.*

19. Using the terms *humidity* and *dew point,* explain why you might find small droplets of water on your lawn in the morning.

20. Describe the formation of a warm *front.*

21. Explain the difference between *weather* and *climate.*

22. Describe the relationship between air temperature and *humidity.*

23. Define *barometric pressure,* and describe how it affects weather.

24. Distinguish between a *tornado* and a *funnel cloud.*

25. Explain how a storm becomes a hurricane using the terms *hurricane, tropical depression,* and *winds.*

26. Write a paragraph which discusses how seasons change using the terms *tilt, rotation, solstice,* and *equinox.* **WRITING SKILL** Explain the differences between seasons in the Northern and Southern hemispheres.

27. Describe two factors that can cause global changes in *climate.*

28. Draw a picture of a molecule of *ozone.* How is it different from a molecule of atmospheric oxygen?

29. Explain what causes *lightning.* Why is lightning always followed by thunder, and why is there a delay in between?

30. Explain *respiration* and its role in photosynthesis.

31. **Climatography** Visual aids called climatographs are used to display information about the climate of a specific region. Using the climatograph for Moscow, Idaho, in the figure below, answer the following questions:
    a. What was the average temperature in Moscow during August?
    b. What was the total precipitation in the Moscow area for the month of January?
    c. What was the approximate total precipitation for the year?

Climatograph for Moscow, Idaho

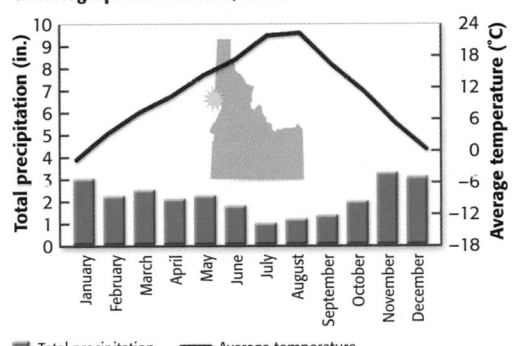

☐ Total precipitation ━━ Average temperature

32. Research the average rainfall in your city, area, or state for each month of the year. Using this data, create two different kinds of graphs. Decide which graph is better for clearly representing the information. Include both of your graphs with a short paragraph explaining which one you would use in a presentation.

765

23. Barometric pressure is the pressure due to the weight of the atmosphere. Changes in barometric pressure reflect changes in the weather, as falling and rising barometric pressure indicate when an air mass is moving in or out of an area.

24. A tornado is a high-speed, rotating wind, which begins as a funnel cloud, or a column of water droplets that reach down from a storm cloud.

25. A tropical depression occurs when warm water evaporates and rises creating an intense low-pressure area, which can build strength and become a large circulating mass of wind, rain, and clouds, known as a hurricane.

26. Earth's seasons derive from the fact that Earth's axis is tilted from the perpendicular plane of its orbit around the sun. Because of this tilt, the sun seems to rise to different heights in the sky at different points in Earth's rotation around the sun or at different times of the year. When the North Pole is tilted toward the sun, the sun rises higher, temperature increases, and the days are longer in the Northern hemisphere, creating summer there and winter in the Southern hemisphere. When the South Pole is tilted toward the sun, it is summer in the Southern hemisphere and winter here. At solstice, one of the poles is at its closest point to the sun. At equinox, Earth's axis is tilted neither toward nor away from the sun, and day and night are of equal length all over Earth.

27. Volcanic eruptions and the movement of continents may cause global changes in climate.

28. Ozone molecules have three atoms of oxygen, whereas atmospheric oxygen has only two atoms of oxygen. **Figure 4** shows a three-dimensional representation of Ozone.

29. Lightning is a big spark, which occurs when electrical charges in clouds build up to the point where the charge is significantly different from that of surrounding clouds or Earth. Thunder accompanies lightning, because lightning superheats the air, making it expand faster than the speed of sound, creating a shockwave we perceive as thunder. The delay between lightning and thunder occurs, because light travels more quickly than sound.

30. Animals breathe oxygen and release carbon dioxide during respiration. During photosynthesis, plants use carbon dioxide and release oxygen. In this way, the two processes are complimentary.

Building Math Skills

31. a. 22°C
b. 3 in.
c. 26 in.

Building Graphing Skills

32. Answers may vary. Students should create both a bar and a line graph to represent their data and give clear explanations for choosing one type of graph to best represent their information.

Thinking Critically

33. The peaks are covered with snow because they are tall, and temperatures are cold at high altitudes in the troposphere.

34. The CFC ban helps the stratosphere because that is where the protective layer of ozone is located.

33. Critical Thinking Explain in a paragraph why some of the mountain peaks located near the equator are covered with snow.

34. Applying Knowledge All use of CFCs has been banned in the United States for environmental reasons. Which one of the four layers of the atmosphere does this ban help protect? Explain your answer.

35. Creative Thinking Describe how your life would be changed if global temperatures were to increase by several degrees.

36. Creative Thinking In what ways would a knowledge of the global wind belts have helped a sixteenth-century explorer sailing between Spain and the northern part of South America?

37. Applying Knowledge One body of air has a relative humidity of 97 percent. Another has a relative humidity of 44 percent. If they are at the same temperature, which body of air is closer to its dew point? Explain your answer.

38. Making Comparisons Where would the air contain the most moisture, over Panama or over Antarctica? Explain your answer.

39. Critical Thinking Is it safe to be on the street in an automobile during a tornado? Explain your answer.

40. Acquiring and Evaluating Data Find out the local high and low temperatures for each day of a 2 week period. Using a computer spreadsheet program, graph the data and find the average high and low temperatures. **COMPUTER SKILL**

41. Interpreting Graphics Use the weather map shown below to name three states that have a cold front moving through them. Predict what kind of weather they can expect.

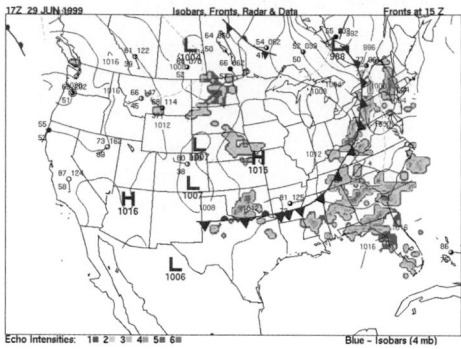

DataStreme Project, American Meteorological Society

42. Creative Thinking Predict how the strength of a Northern Hemisphere hurricane will change as it moves northward in the Atlantic Ocean, and explain why.

43. Critical Thinking What evidence do scientists have that Earth's climate has changed over time? How can this evidence help scientists study changes our climate is making now?

44. Acquiring and Evaluating Data Research the debate concerning global warming. What evidence supports global warming, and what evidence calls it into question? Write one paragraph describing each side of the debate. Then, write a third paragraph examining the limitations of the scientific data and describing how scientists are trying to predict what will happen in the future. **WRITING SKILL**

35. Answers may vary but might include: having to move away from coastal areas, different and more intense climate and weather patterns, increased severity of storms, and higher energy costs.

36. Knowledge of the wind belts would help an explorer chart the quickest route between Spain and South America.

37. The body of air with a relative humidity of 97% would be closer to its dew point because the water vapor in the more humid air will condense at a warmer temperature than the less humid air sample would.

38. Air over Panama would contain the most moisture because it is warm, and warmer air holds more water vapor.

39. No, because a tornado contains winds strong enough to destroy the car. Also, cars can be very dangerous during flood conditions. Flood waters can easily sweep a car off the road and trap those inside.

40. Answers may vary.

41. Texas, Arkansas, and Tennessee have cold fronts moving through them. They might expect high winds, thunderstorms, and perhaps even tornadoes.

45. Allocating Resources You are planning an expedition to Mount Everest, the tallest mountain in the world. Identify which four of the following items you will need the most, and explain why.

    a. inflatable raft
    b. insulated clothes
    c. life vest
    d. television
    e. oxygen equipment
    f. fire-starting equipment
    g. raincoat

46. Working Cooperatively Obtain a week's worth of local or national weather maps from the paper or the Internet. In a small group, prepare a weather forecast. Interpret the daily weather, and follow any trends. Explain pressure areas and the resulting fronts, any precipitation, and average temperatures. Have a volunteer from your group present the forecast to the class.

47. Acquiring and Evaluating Data
WRITING SKILL
Contact or research a news source, and find out about the satellites it uses to predict the weather. Write a short paragraph describing how satellites help predict the weather and climate.

Art Credits: Fig. 4, Kristy Sprott; Fig. 5, Uhl Studios, Inc.; Fig. 7, Uhl Studios, Inc.; Fig. 9, Uhl Studios, Inc.; Fig. 10, Uhl Studios, Inc.; Fig. 13, Uhl Studios, Inc.; Fig. 15, Stephen Durke/Washington Artists; Fig. 16, Uhl Studios, Inc.; Fig. 17, Uhl Studios, Inc.; Fig. 19, Craig Attebery/Jeff Lavaty Artist Agent; Fig. 21, Uhl Studios, Inc.; Fig. 22, Uhl Studios, Inc.; Fig. 23, Uhl Studios, Inc.; Fig. 24, Uhl Studios, Inc.; Building Math Skills (Climatograph), Leslie Kell; Analyzing Your Results (graph), Leslie Kell.

Photo Credits: Chapter Opener image of irrigation truck, Cameron Davidson/Getty Images/Stone; double rainbow, Craig Tuttle/Corbis Stock Market; Fig. 2, Ned Haines/Photo Researchers, Inc.; Fig. 3, NASA; Fig. 6, George Lepp/Getty Images/Stone; Fig. 8, Larry Lefever/Grant Heilman Photography; Fig. 11, Bernard Photo Productions/Animals Animals/Earth Scenes; Fig. 13A, Tom Pantages Photography; Fig. 13B, Runk/Schoenberger/Grant Heilman Photography; Fig. 18A, Photo Courtesy of The Weather Channel; Fig. 18B, Ali Jarekji/REUTERS/TimePix; "Real World Application" (lightning), Telegraph Colour Library/Getty Images/FPG International; Fig. 20, Alan R. Moller/Getty Images/Stone; Chapter Review (weather map), DataStreme Project, American Meteorological Society; "Viewpoints," HRW.

48. Concept Mapping Copy the unfinished concept map below onto a sheet of paper. Complete the map by writing the correct word or phrase in the lettered boxes.

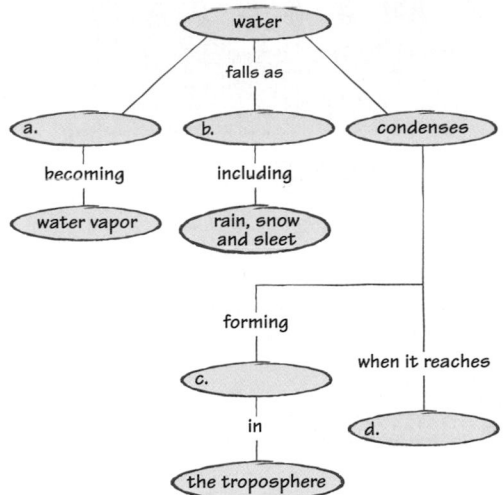

49. Concept Mapping Create your own concept map using one or two sections of this chapter. Copy your concept map on a piece of paper, leaving some cells blank. Swap concept maps with members of your small group, and discuss any difficulties that may arise.

50. Connection to Chemistry Research the effects of methane on the atmosphere. How is methane produced? What are some sources of methane in the atmosphere? What is its effect on global warming? Can we eliminate methane from the atmosphere?

internet connect

www.scilinks.org
Topic: Weather Maps SciLinks code: HK4151

SC*i*LINKS. Maintained by the National Science Teachers Association

767

42. The hurricane will become weaker as it moves northward over cooler waters because less water vapor is present at lower temperatures.

43. Changes in Earth's atmosphere and crust (such as the increase in average temperature of the troposphere and the movement of glaciers) provide evidence that Earth's climate has changed over time. By monitoring changes in atmosphere and crust, scientists can study climatic changes that are occurring now.

44. Answers may vary.

Developing Life/Work Skills

45. b. because the temperatures at high altitudes are very low;
e. because the levels of oxygen in the air decrease with altitude;
f. because you might need to build a fire to keep warm in an emergency while approaching the mountain; and
g. because a raincoat can provide you with waterproof protection from rain, snow and sleet.

46. Answers may vary depending upon the source used.

47. Answers may vary.

Integrating Concepts

48. a. evaporates
    b. precipitation
    c. clouds
    d. the dew point

49. Answers may vary.

50. Methane is the main component of natural gas. It is produced by the decomposition of plants. Some sources are ruminant animals, swamps, and wetlands. Methane is a greenhouse gas that may promote global warming. No, we cannot eliminate it from the atmosphere.

MEASURING TEMPERATURE EFFECTS

Teacher's Notes

Have the students review the concept of density. Boyle's law, $PV = k$, shows the inverse relationship between the pressure and the volume of a gas. Charles's law, $V = kT$, shows the direct relationship between the temperature and the volume of a gas. The combined gas law, $PV / T = k$, expresses Boyle's law and Charles's law in one equation. Although students may not know these relationships by name, their observation of the world around them probably has given them some familiarity with these concepts.

Time Required 1 lab period

Ratings
$$\underset{\text{EASY}}{1} \quad 2 \quad 3 \quad \underset{\text{HARD}}{4}$$

TEACHER PREPARATION	2
STUDENT SETUP	2
CONCEPT LEVEL	3
CLEANUP	1

Skills Acquired
- Collecting data
- Designing experiments
- Experimenting
- Inferring
- Interpreting
- Measuring
- Organizing and analyzing data

The Scientific Method

In this lab, students will:
- Form a Hypothesis
- Analyze the Results
- Draw Conclusions
- Communicate Results

Skills Practice Lab

Introduction

Air rises or sinks in Earth's atmosphere due to differences in the density of air that are caused by differences in temperature. How can you determine the effect of temperature on the density of air?

Objectives

▶ Measure the volume of a constant mass of air at different temperatures.

▶ **USING SCIENTIFIC METHODS** *Draw conclusions* by inferring changes in density from changes in volume.

Materials

400 mL beaker
60 mL disposable syringe
glycerin
hot and cold tap water
ice
petroleum jelly
thermometer

Measuring Temperature Effects

▶ **Procedure**

Preparing for Your Experiment

1. On a blank sheet of paper, prepare a table like the one shown below.

Temp. (°C)	Pull volume (mL)	Push volume (mL)	Average volume (mL)

2. Measure the air temperature in the room, and record the temperature in your data table.

3. Remove the cap from the tip of the syringe, and move the plunger. If the plunger does not move smoothly and easily, lubricate the inside wall of the syringe with a few drops of glycerin.

4. Adjust the position of the plunger until the syringe is about two-thirds full of air. Add a dab of petroleum jelly to the tip of the syringe, and replace the cap.

Measuring the Volume of Air

5. Gently pull on the plunger, and then release it. When the plunger stops, read the volume of air inside the syringe. Record the volume in your data table in the column labeled "Pull volume."

6. With your finger on the cap, gently push on the plunger and then release it. When the plunger stops, read the volume of air inside the syringe. Record the volume in your data table in the column labeled "Push volume."

   SAFETY CAUTION Do not point the syringe at anyone while you push on the plunger. Wear safety goggles.

Safety Cautions

Read all safety cautions and discuss them with your students. Safety goggles must be worn at all times. Caution the students not to depress the syringe plunger with excessive force. The syringe can break.

Tips and Tricks

Review the density equation, $D = \frac{m}{V}$ and the rearrangement to $V = \frac{m}{D}$

Ask students what changes they would expect in the density of the air if a known mass occupied a larger volume (the density would be less) or a smaller volume. (the density would be greater) Ask students what would happen if a balloon is placed in a freezer (volume would decrease, density would increase) or in a warm oven. (volume would increase, density would decrease)

Designing Your Experiment

7. With your lab partners, decide how you will use the materials available in the lab to determine the effect of temperature on air density. Test at least two temperatures below room temperature and two temperatures above room temperature. It is important that the mass of air inside the syringe does not change during your experiment. How can you ensure that the mass of air remains constant?

8. In your lab report, list each step you will perform in your experiment.

9. Before you carry out your experiment, your teacher must approve your plan.

Performing Your Experiment

10. After your teacher approves your plan, carry out your experiment.

11. Record your results in your data table.
    SAFETY CAUTION Use care when working with hot water; it can cause severe burns.

▶ Analysis

1. At each temperature you tested, calculate the average volume by adding the pull volume and push volume and dividing the sum by 2. Record the result in your data table.

2. Plot your data in your lab report in the form of a graph set up like the one at right. Draw a straight line on the graph that fits the data points best.

3. **Reaching Conclusions** How does the volume of a constant mass of air change as the temperature of the air increases? For the mass of air you used in your experiment, how much would the volume change if the temperature increased from 10°C to 60°C?

4. **Reaching Conclusions** Recall that the density of a substance equals the substance's mass divided by its volume. Do your results indicate that the density of air increases or decreases as the temperature of the air increases? Explain.

5. **Reaching Conclusions** Based on your results, would a body of air tend to rise or sink as it becomes colder than the surrounding air?

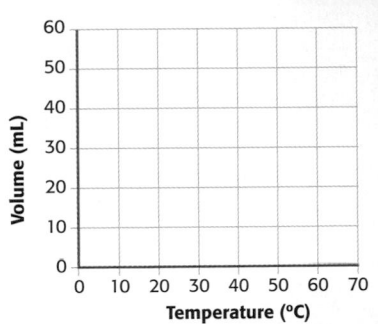

▶ Conclusions

6. Suppose someone tells you that your conclusions are invalid because some of your data points lie above or below the best-fit line you drew. How could you show that your conclusions are valid?

769

Linking Chapters

The widespread use of cars powered by internal combustion engines is often seen as one of the key factors in environmental deterioration. This feature points out that use of any technology will involve unexpected trade-offs that can cause harm to the marketplace, the environment, or both.

Answers

1. Critiquing Viewpoints

Students' analyses of a single argument's weak points may vary. Be sure that student responses clearly identify a weakness and provide support for their opinion. Possible answers include:

In response to Sheneah T.: The technology has not advanced far enough yet, so requiring it now is a bad idea. However, the requirement might spur companies to develop the technology.

In response to Megan B.: Requiring more expensive cars by law may not be fair to those who will now be unable to afford a car. However, the decreases in air pollution that may result will benefit all of society.

In response to Kathryn W.: We can't go back in time to fix the problem, so we should act now. However, if we act too soon, when the right technology is not available and not all of the information is in, we may create an even greater problem in the future.

In response to Margo K.: People don't always act on good ideas. However, it may not be right to force them to act by law.

viewpoints

Should Laws Require Zero-Emission Cars?

California law requires that by the year 2004, 10 percent of all cars sold in the state be zero-emission vehicles that produce no pollution as they are operated. Automobile companies are scrambling to develop electric cars and other technologies to meet this deadline.

Often these cars are substantially more expensive than gasoline-burning models. Is the pollution situation so desperate that this is necessary? Or is this a case of laws interfering with the car market?

> FROM: Sheneah T., Chicago, IL

As technology advances, we will be able to make better cars that won't depend on gas as much. If we were to cut down on the amount of pollution, this would make our environment better to live in. This goes back to an issue of public health. If cars emit less pollution, we would have fewer cases of respiratory disorders and a much cleaner environment.

Require These Cars Now

> FROM: Megan B., Houston, TX

A law requiring zero-emission vehicles after the year 2003 is probably the best way to prevent air pollution. The various ways companies are changing cars to be more environmentally safe just isn't cutting it. Why do we spend tens of thousands of dollars for fun but not to help save the world?

> FROM: Kathryn W., Rochester, MN

I think the government should definitely get involved in these issues. The laws can be changed later, if needed, but we can't go back in time and fix the problem.

770

In response to Marianne C.: Planting trees or carpooling may not be enough to undo the damage of many years. However, it is better for people to voluntarily choose what they want to do than to force them to do something by law.

In response to Amar T.: Damage being done to the environment should outweigh car performance issues. However, if the new cars do not compete well with the old ones, they are unlikely to be widely adopted.

2. Critiquing Viewpoints

Students' analyses of strong points in opposing arguments may vary. Be sure that student

responses clearly identify the strong points and respond in some way that provides support for the opposite point of view.

3. Interpreting and Communicating

Student answers may vary. Evaluate students' advertisements or brochures for organization, clarity and factuality, persuasiveness, creativity, and effort.

4. Understanding Systems

Student answers may vary. Evaluate students' paragraphs and ideas for practicality, thoroughness, probable ease of use, creativity, and effort.

> FROM: Margo K., Coral Springs, FL

Although zero-emission vehicles are better for the environment, there are many expenses that come along with them. I disagree with the law because of the cost of the new vehicles. Rather than making zero-emission vehicles mandatory, if the idea is spread, people will act upon it.

Don't Require These Cars Now

> FROM: Marianne C., Bowling Green, KY

Not everyone will be able to afford these expensive cars. I don't think people should be obligated to buy cars to save the planet. People should do other things instead, like planting trees or carpooling.

> FROM: Amar T., Palos Park, IL

From a car enthusiast's point of view, I feel that no state should make zero-emission cars mandatory for three main reasons: First, at this time zero-emission cars do not perform as well as cars with an internal-combustion engine. Second, zero-emission cars, like electrical cars, have small cruising ranges, and their fuel cells take up too much space. Finally, they are more expensive than gasoline-burning cars.

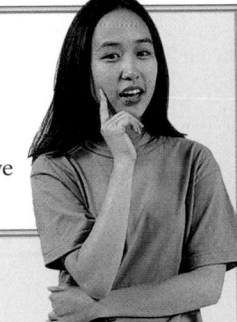

> Your Turn

1. **Critiquing Viewpoints** Select one of the statements on this page that you agree with. Identify and explain at least one weak point in the statement. What would you say to respond to someone who brought up this weak point as a reason you were wrong?

2. **Critiquing Viewpoints** Select one of the statements on this page that you disagree with. Identify and explain at least one strong point in the statement. What would you say to respond to someone who brought up this point as a reason they were right?

3. **Interpreting and Communicating** Imagine that you work for an advertising firm that has been hired to promote an expensive new zero-emission vehicle. Create an advertisement or brochure for the car that tries to persuade people to buy the new car.

4. **Understanding Systems** Other critics of such laws point out that zero-emission cars do not end the pollution entirely. Some toxic waste is made when these cars are manufactured. Write a paragraph in which you outline a method for deciding whether the pollution emitted by a regular car is worse for the environment than the waste made in making a zero-emission vehicle.

internet connect

TOPIC: Zero-Emission Vehicles
GO TO: go.hrw.com
KEYWORD: HK4 Zero-Emission

Should zero-emission vehicles be required? Why or why not? Share your views on this issue and learn about other viewpoints at the HRW Web site.

771

Using Natural Resources
Chapter Planning Guide

PACING	CLASSROOM RESOURCES	LABS, ACTIVITIES, AND DEMONSTRATIONS
BLOCK 1 · 45 min pp. 772–773 **Chapter Opener**		SE **Activity 1,** p. 773 SE **Activity 2,** p. 773
BLOCKS 2 & 3 · 90 min pp. 774–781 **Section 1** Organisms and Their Environment	CRF **Lesson Plan*** TT **Bellringer*** TT **Ecosystem Structure*** TM **Population Graph***	TE **Demonstration** Example of an Ecosystem, p. 775 GENERAL SE **Quick Lab** Why do seasons occur?, p. 779 GENERAL CRF **Datasheets for In-Text Labs** Why do seasons occur?* GENERAL
BLOCKS 4 & 5 · 90 min pp. 782–789 **Section 2** Energy and Resources	CRF **Lesson Plan*** TT **Bellringer*** TM **Energy Use*** TT **Carbon Cycle*** TT **Making Oil*** TM **Solar Cell Diagram***	TE **Opening Activity,** p. 782 GENERAL SE **Skills Practice Lab** Changing the Form of a Fuel, pp. 806–807 ◆ GENERAL CRF **Datasheets for SE Skills Practice Lab** Changing the Form of a Fuel* GENERAL
BLOCKS 6 & 7 · 90 min pp. 791–800 **Section 3** Pollution and Recycling	CRF **Lesson Plan*** TT **Bellringer*** TT **Concept Mapping***	SE **Quick Activity** Observing Air Pollution, p. 795 GENERAL CRF **Datasheets for In-Text Labs** Observing Air Pollution* GENERAL TE **Demonstration** Carbon Dioxide Analysis, p. 795 SE **Quick Lab** How can oil spills be cleaned up?, p. 797 GENERAL CRF **Datasheets for In-Text Labs** How can oil spills be cleaned up?* CRF **Observation Lab** Making Your Own Recycled Paper * ◆ BASIC CRF **CBL™ Probeware Lab** Investigating the Effects of Acid Rain* ◆ ADVANCED

BLOCKS 8 & 9 · 90 min

Chapter Review and Assessment Resources

SE **Chapter Review,** pp. 802–805
CRF **Chapter Tests*** GENERAL
OSP **Test Generator**
CRF **Standardized Test Practice with Guided Reading Development***
CRF **Test Item Listing for ExamView® Test Generator***
OSP **Scoring Rubrics and Classroom Management Checklists**

Online Resources

Visit the HRW Web site for a variety of free resources related to the text. Just type in the keyword **HK4 RES.**

Holt Science Spectrum: Physical Science: Online Edition

Students can access interactive problem solving help and active visual concept development with the *Holt Science Spectrum: Physical Science* Online Edition available at **www.hrw.com.**

student CNN **News**

cnnstudentnews.com

Find the latest news, lesson plans, and activities related to important scientific events.

KEY

TE Teacher Edition
SE Student Edition
OSP One-Stop Planner

CRF Chapter Resource File
TT Teaching Transparency
TM Transparency Master

* Also on One-Stop Planner
◆ Requires Advance Prep

PROBLEM SOLVING AND PRACTICE	SECTION REVIEW AND ASSESSMENT	STANDARDS CORRELATION
	CRF Pretest* GENERAL	
TE Inclusion Strategies, p. 775	**TE** Quiz, p. 781 BASIC **SE** Section 1 Review, p. 781 **CRF** Concept Review* GENERAL **CRF** Quiz* BASIC	LS 4c, 4d, 4e, 5b ES 1d UCP 1, 5 SAI 1 HNS 1 SPSP 3
CRF Cross-Disciplinary Worksheet Integrating Physics—Solar Cells* ADVANCED **CRF** Cross-Disciplinary Worksheet Connection to Social Studies—TVA: Finding Solutions* ADVANCED **CRF** Cross-Disciplinary Worksheet Connection to Language Arts—The Meaning of Efficiency* GENERAL **SE** Science and the Consumer Sun-Warmed Houses, p. 790 **CRF** Cross-Disciplinary Worksheet Science and the Consumer—Solar Energy* ADVANCED	**TE** Quiz, p. 789 BASIC **SE** Section 2 Review, p. 789 **CRF** Concept Review* GENERAL **CRF** Quiz* BASIC	PS 3b ES 1a UCP 2 SAI 1 SPSP 3, 4
TE Inclusion Strategies, p. 792 **CRF** Cross-Disciplinary Worksheet Integrating Chemistry—Pesticides* ADVANCED **SE** Science and the Consumer Recycling Codes: How Are Plastics Sorted?, p. 799 **CRF** Cross-Disciplinary Worksheet Integrating Biology—Composting* ADVANCED **CRF** Cross-Disciplinary Worksheet Science and the Consumer—Recycling Plastics* ADVANCED **SE** Study Skills Researching Information, p. 801 **SE** Science in Action Cars of the Future, pp. 808–809	**TE** Quiz, p. 800 BASIC **SE** Section 3 Review, p. 800 **CRF** Concept Review* GENERAL **CRF** Quiz* BASIC	LS 5e ST 2 SPSP 4, 5

SCiLINKS

www.scilinks.org

Topic: Fuel Cells
*Sci*Links code: HK4058

Topic: Ecosystem Factors
*Sci*Links code: HK4040

Topic: Ecosystems
*Sci*Links code: HK4041

Topic: Changes in Ecosystems
*Sci*Links code: HK4019

Topic: Alternative Energy Sources
*Sci*Links code: HK4004

Topic: Solar-heated Homes
*Sci*Links code: HK4159

Topic: Pollution
*Sci*Links code: HK4107

Topic: Solutions to Pollution Problems
*Sci*Links code: HK4130

Topic: Recycling Plastics
*Sci*Links code: HK4117

Topic: Global Warming
*Sci*Links code: HK4065

Topic: Electric Vehicles
*Sci*Links code: HK4115

Technology Resources

 One-Stop Planner
All of your printable resources and the Test Generator are on this convenient CD-ROM.

 CNN Science in the NEWS
each video segment is accompanied by a Critical Thinking Worksheet*

Segment 15
Coal-Oil Technology

Segment 11
Energy-Saving House

Segment 29
Electric Car

Overview

In this chapter, students learn what ecosystems are and how natural and human forces can affect ecosystems. Next, they learn about energy sources, including the advantages and disadvantages of different sources and the difference between renewable and nonrenewable resources. Finally, they learn about the causes of and ways to reduce different types of pollution.

Assessing Prior Knowledge

Be sure students understand the following concepts:

• solutions
• acids
• bases
• pH
• work
• energy
• heat
• temperature
• energy transfer
• the solar system
• planet Earth
• Earth's interior
• plate tectonics
• weather
• climate
• characteristics of the atmosphere

MISCONCEPTION ///ALERT\\\

Science education research has identified the following misconceptions about pollution and natural resources.

• Students believe that natural things cannot be pollution.
• Students think that oceans are a limitless resource.
• Students believe that humans are indestructible as a species.

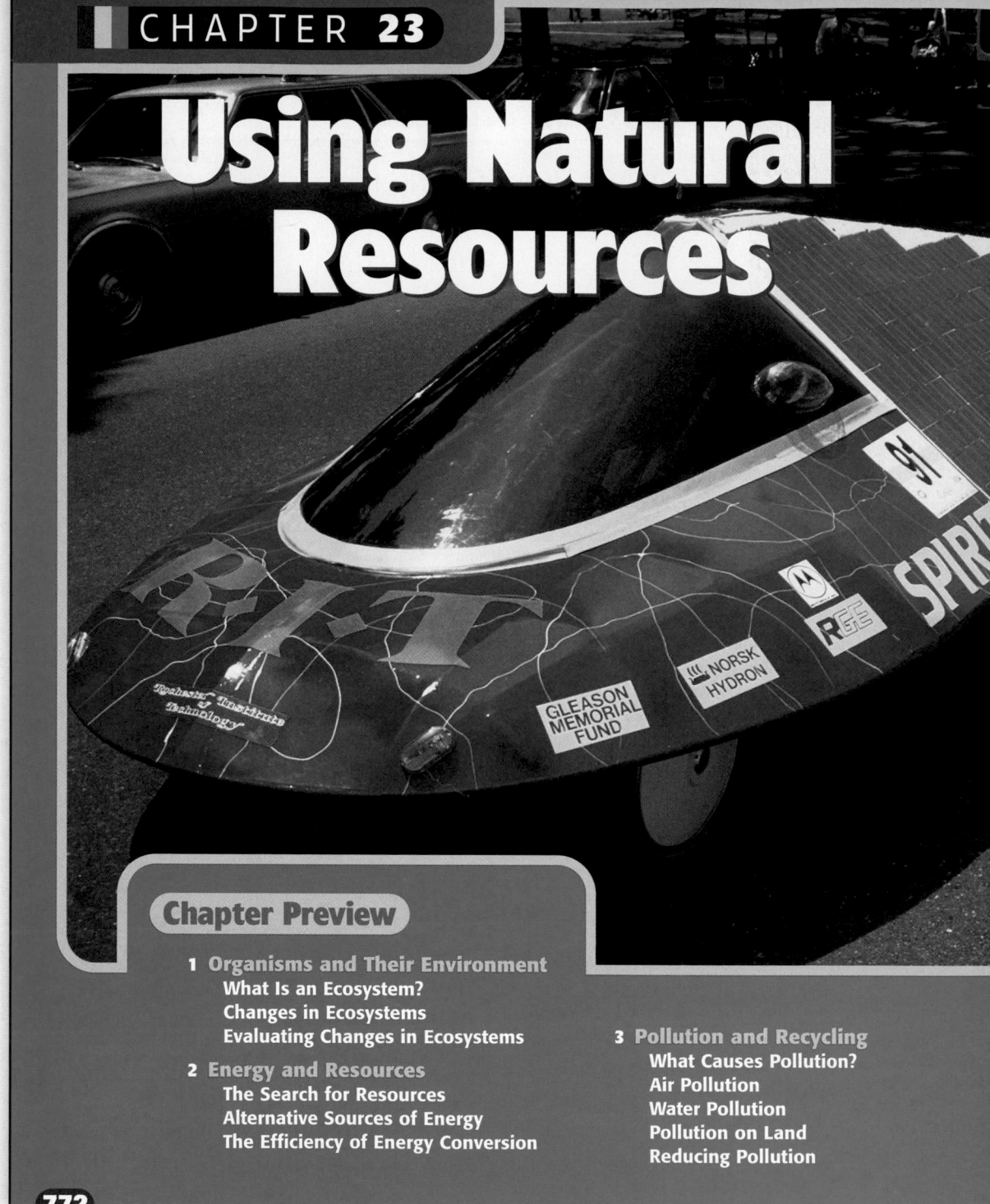

CHAPTER 23

Using Natural Resources

Chapter Preview

1 Organisms and Their Environment
What Is an Ecosystem?
Changes in Ecosystems
Evaluating Changes in Ecosystems

2 Energy and Resources
The Search for Resources
Alternative Sources of Energy
The Efficiency of Energy Conversion

3 Pollution and Recycling
What Causes Pollution?
Air Pollution
Water Pollution
Pollution on Land
Reducing Pollution

Standards Correlations

National Science Education Standards

The following descriptions summarize the National Science Standards that specifically relate to this chapter. For the full text of the standards, see the National Science Education Standards at the front of the book.

Section 1 Organisms and Their Environment
Life Science LS 4c–4e, 5b
Earth Science ES 1d
Unifying Concepts and Processes UCP 1, 5
Science as Inquiry SAI 1
History and Nature of Science HNS 1
Science in Personal and Social Perspectives SPSP 3

Section 2 Energy and Resources
Physical Science PS 3b
Earth and Space Science ES 1a
Unifying Concepts and Processes UCP 2
Science as Inquiry SAI 1
Science in Personal and Social Perspectives SPSP 3, 4

Section 3 Pollution and Recycling
Life Science LS 5e
Science and Technology ST 2
Science in Personal and Social Perspectives SPSP 4, 5

The World Solar Challenge is held each year to raise awareness of alternative energy sources. Drivers travel 3000 km across Australia using solar-powered cars.

Background The engines are started. The drivers check their gauges one last time. The flag rises, then falls. They're off!

Is this a typical race? Not quite. These race cars are powered by solar energy. Not a drop of gasoline is needed to make them run. Teams of college students designed and built the cars to compete in an annual event called Sunrayce.

Sunrayce began in 1990, and its goal is to raise awareness of alternative energy sources, such as the sun. Each summer, students travel more than 2093 km (1300 mi) across the United States in these solar-powered cars.

Solar-powered cars must be lightweight and efficient. Although the average race speed is about 35 mi/h, the cars can reach top speeds of 65 mi/h. Each car is powered by solar cells. These cells, which are made of thin layers of silicon, capture the sun's energy and store it in batteries.

Activity 1 Obtain a solar cell from an electronics store or your school's science lab. Using the solar cell, a low-wattage light bulb, and two pieces of insulated electrical wire, create a current in the cell and the bulb. Place the end of one wire on the metal bottom of the bulb and the end of the other wire on the side of the bulb. Place the other end of both wires on the solar cell. In your own words, describe the movement of charges between the cell and the bulb.

Activity 2 Make a list of the energy sources you use at home. Find out from your parents which source of energy costs the most per month. Make a pie chart showing your results. Call your local power company to find out what type of resource is used to generate electrical power.

☑ internet connect

www.scilinks.org
Topic: **Fuel Cells** SciLinks code: **HK4058**

SC*i*LINKS. Maintained by the National Science Teachers Association

Pre-Reading Questions
1. Where does energy come from?
2. What can be done to reduce pollution?

773

Background Solar-powered cars use light to create electrical energy, which in turn provides power. Although electric vehicles, or EVs, may seem futuristic, they are already in use in California and Arizona. One example is the EVI, an EV that can travel between 80 km and 145 km on a single charge. The EVI's rechargeable battery contains slightly more than two-dozen 12 V modules capable of holding more than 16 kW•h of energy.

Activity 1 Electrons will flow through the insulated wire from negative to positive until the materials in the solar cell are electrically neutral.
LS Kinesthetic

Activity 2 Answers may vary depending on the student. After students make their lists, have them classify their answers according to how the energy was produced. **LS Logical**

Answers to Pre-Reading Questions
1. Almost all energy on Earth comes from the sun.
2. Answers will vary. Students may state that pollution can be reduced by using alternative energy sources, by reducing consumption levels, and by reusing and recycling materials.

Chapter Resource File

• Pretest **GENERAL**
• Teaching Transparency Preview
• Answer Keys

SECTION
1

Focus

Overview

Before beginning this section, review with your students the Objectives listed in the Student Edition. In this section, students learn what an ecosystem is and how ecosystems are structured. They also study examples of changes in ecosystems due to both natural forces and human activities.

📀 Bellringer

Use the Bellringer transparency to prepare students for this section.

Motivate

Opening Discussion ——— GENERAL

Write the words *ecosystem* and *community* on the board. Ask students to come up with words and phrases that relate to the concepts and definitions of ecosystem and community. Have two students record the responses on the board under each word. Encourage students to use the section objectives and the figures on these two pages to develop definitions of the words and to understand how they relate to each other. Have the students use their new knowledge to illustrate how their school is an ecosystem with different communities of people in it. **LS Verbal**

Organisms and Their Environment

▶ **KEY TERMS**
ecosystem
community
succession

▶ **ecosystem** a community of organisms and their abiotic environment

internet connect

www.scilinks.org
Topic: Ecosystem Factors
SciLinks code: HK4040

Maintained by the
National Science
Teachers Association

OBJECTIVES

▶ **Explain** the structure of an ecosystem.
▶ **Describe** the effects one species can have on an ecosystem.
▶ **Discuss** two ways natural forces can change ecosystems.
▶ **Discuss** two ways humans can change ecosystems.

Should money be spent drilling for oil in the wilderness of Alaska, or would the money be better spent promoting solar power? Should a nearby swamp be drained to provide parking for a mall in your town, or is the swamp best left alone? No matter how you feel about these issues, you need to know how they will affect the world around you.

We humans, like all other living things, fill our needs by using natural resources. These resources are taken from the world around us. Every action taken, whether it is someone draining a swamp or a bird catching prey, affects all the other living things in the **ecosystem,** as shown in *Figure 1.*

To evaluate the effects of your decisions on the issues that cause change in your environment, you must first understand how the many parts of an ecosystem relate to one another.

Figure 1

Each living and nonliving element in this desert ecosystem directly affects the others.

774

LIFE SCIENCE
CONNECTION

Have a biology teacher visit your class to characterize the local ecosystem based on the native flora and fauna.

LANGUAGE ARTS
CONNECTION

In 1962, the scientist and naturalist Rachel Carson wrote *Silent Spring*, a book that drew wide attention to the effects of pollution on the living and nonliving elements of ecosystems. Choose relevant selections from the book and read them to your class.

What Is an Ecosystem?

Consider a squirrel in a city park. The squirrel gathers and stores food. It lives in a tree and drinks water from a sun-dappled pond. The soil, trees, sunshine, and water are natural resources found within the city park. The city park itself is an ecosystem.

Living elements in an ecosystem can include plants, animals, and people. Nonliving elements include sunlight, air, soil, water, and temperature. Cycles of energy and matter are at the base of an ecosystem. The interrelated elements within a desert ecosystem are shown in *Figure 1*. The cactuses, sagebrush, lizards, snakes, scorpions, and birds interact with one another and with their surroundings to make a stable, balanced ecosystem.

Not all ecosystems are the same size

Ecosystems can be large or small. The entire planet is one big ecosystem containing all the living and nonliving things on Earth—the land and water, the organisms, and the atmosphere. A shallow forest pool no bigger than a rain puddle is also an example of an ecosystem. This ecosystem is made up of water, mud, bacteria, mosquitoes, air, and larvae all living together.

Living things are adapted to their ecosystem

Living things are found almost everywhere: on land, in the air, and in water. Each organism has adapted to factors in its environment, such as temperature, humidity, and the other living things around it. For instance, polar bears are adapted to cold, wet places, and camels are adapted to hot, dry places. Neither animal could survive in the other's environment.

Ecosystems are divided into communities

All of the interacting animals and plants living in one area within an ecosystem form a **community.** There can be several communities in an ecosystem. The animals and plants of the tundra make up one community. The seals, fish, and algae of the nearby ocean make up another. Polar bears belong to *both* communities. *Figure 2* shows the different divisions within an ecosystem.

■ internet connect ≣

www.scilinks.org
Topic: Ecosystems
SciLinks code: HK4041

SC/LINKS. Maintained by the National Science Teachers Association

▶ **community** a group of species that live in the same habitat and interact with each other

Figure 2

Ecosystems are made up of communities that contain different populations of organisms.

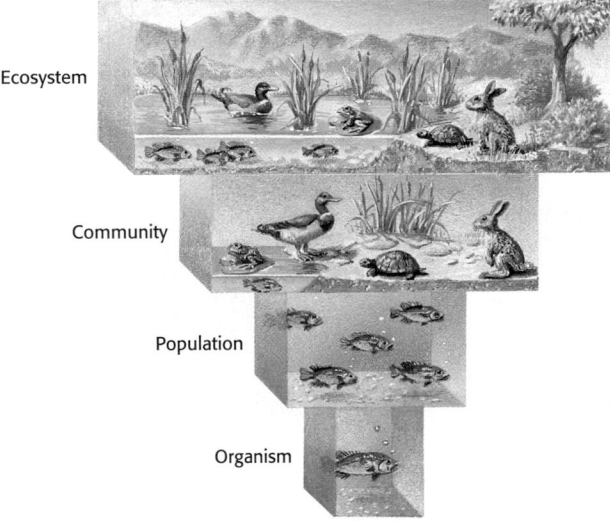

Ecosystem

Community

Population

Organism

Teach

Demonstration — GENERAL

Example of an Ecosystem

(Time: About 10 minutes)

Materials:

• aquarium or terrarium with assorted flora and fauna

(If you are not able to obtain an aquarium, simply collect some leaves and insects from outside and place them in a bowl covered with plastic wrap with air holes.)

Show the students an aquarium with plants and fish or a terrarium with plants and insects. Have students examine all aspects of the system closely.

Analysis

1. Is this an example of an ecosystem? Why or why not? (yes, because it is a distinct area where living and nonliving elements are interacting)

2. What are the living and nonliving elements of this ecosystem? (fish, plants, insects, sunlight, air, water, etc.)

3. How are the nonliving and living elements interacting? (Plants are using sunlight; animals are drinking water)

4. Could this ecosystem be considered part of a larger ecosystem? (yes, the classroom ecosystem or the school ecosystem) **LS** Logical

---

◉◉ INCLUSION Strategies BASIC

• *Developmentally Delayed*
• *Learning Disabled*
• *English Language Learners*

Have students use **Figure 2** as a model to design and draw an ecosystem from a different environment, such as a forest or a desert. Students can use pictures from magazines or original drawings. Tell them to label the organisms, populations, communities, and the entire ecosystem. The finished product and the oral presentation of the ecosystem can both be counted for a grade. **LS** Visual

Alternative Assessment — GENERAL

Finding Communities Take your class outside on the school grounds for a mini field trip. Tell them that the objective of the trip is to find as many communities of organisms as possible. As students identify these communities, have them list the living and nonliving elements of the environments where the communities are found. Are more communities found where there are more living or nonliving elements interacting? **LS** Kinesthetic

Chapter Resource File

• **Lesson Plan**

Transparencies

TT Bellringer
TT Ecosystem Structure

Teaching Tip

Characteristics of Ecosystems

All ecosystems have three fundamental characteristics: structure, processes, and changes over time. Ask students to draw a concept map with *ecosystem* as the central concept and the three subheadings *structure*, *processes*, and *changes over time*. Have students complete their concept maps as they read this section. You can discuss the following to help them get started.

Structure includes both the nonliving and the living elements of an ecosystem. Ask students for examples of nonliving elements. (Responses could include sunlight, water, rocks, air, and soil.) Also ask students to describe the living elements of an ecosystem (a set of interacting species that make up a community).

Processes involve things that cycle in and out of the ecosystem. Examples include energy flowing into and out of the system, and chemical elements cycling through the system.

Finally, all ecosystems change over time. One example is when an ecosystem develops through the process called succession. Ecosystem changes can be short-term (such as the seasons) or long-term (such as climate change).

Figure 3

The living and nonliving elements of this aquarium ecosystem help to keep it balanced.

Figure 4

The population of snowshoe hares directly affects the population of lynxes.

Changes in Ecosystems

Sunlight, water, air, soil, animals, plants—the elements that make up an ecosystem are numerous. Each element is balanced with the others so that the ecosystem can be maintained over a long period of time.

Look at the aquarium shown in **Figure 3**. The amount of salt in the water and the temperature of the water are at the right levels. The fish, snails, and plants all have sufficient space and food. In short, there are enough resources for every living thing in the aquarium ecosystem. The ecosystem is balanced.

Balanced ecosystems remain stable

When an ecosystem is balanced, the population sizes of the different species do not change much relative to one another. Overall, there is a natural balance between those that eat and those that are eaten. Other factors, such as disease or food shortage, prevent populations from growing too much. If the balance within an ecosystem is disturbed, change results.

The graph in **Figure 4** is based on a Canadian study of the snowshoe hare and its predator, the lynx. What happened to the lynx population when there were fewer snowshoe hares?

When the prey population of a particular place decreases, there is less food for the predators. As a result, some of the predators die. On the other hand, if the prey population increases, more predators may move into the area. A variety of factors, including population change, can affect the balance in an ecosystem. What happens when one factor changes?

Lynx and Hare Population Cycles

776

Alternative Assessment ——ADVANCED

Mount Saint Helens: Before and After

Writing Challenge students to write a report about the condition of Mount Saint Helens before the eruption and the process of succession that occurred after the eruption. Students should conduct research at the library or on the Internet for their reports. Research topics could include the following: What species moved in first? Is the ecosystem still recovering from the disturbance? **LS Verbal**

A change in one feature can affect the whole system

Throughout the 1990s, researchers closely watched a piece of land in Central America that was once a tropical rain forest. After the trees were logged, a species of wild pig vanished from the forest because the pigs no longer had enough food and shelter. Three species of frogs disappeared soon thereafter. Was this related to the loss of the wild pigs?

The pigs wallowed in mud, forming puddles that the frogs used for breeding. Suddenly there were no puddles, and the frogs had to find another place to breed. A change in one factor of an ecosystem can affect all the living and nonliving elements within the system.

The key to understanding ecosystems can be summed up in one word: *interrelatedness*. The elements that make up an ecosystem function together to keep the entire system stable. If something changes, time and natural forces often work to return the ecosystem to its previous state.

Ecosystems tend to gradually return to their original conditions

Yellowstone National Park is one of the largest tourist attractions in the United States. The park, located in northwestern Wyoming and southern Montana, covers about 2.2 million acres (3,472 mi^2) of land and is known for its active geysers and hot springs. Wildfire has had a role in the dynamics of this area for thousands of years, since long before it was a park.

But during the summer of 1988, large areas of the park were burned to the ground by fires, as shown in *Figure 5A*. The fires, which were started by lightning and careless human activity, spread quickly through the open forest during this particularly dry summer and left nearly a third of the park blackened with ash. Firefighters are often unsuccessful in putting out these fires. Frequently, it is rain or snow that eventually put the fires out.

The following spring, the appearance of the "dead" forest began to change. Large numbers of small, green plants began to flourish and replace areas that had been covered with fallen trees. Year after year, gradual developments took place in the recovering area, as shown in *Figure 5B*.

Figure 5

A Yellowstone National Park has been plagued with a series of uncontrollable fires. In fact, nearly all of Yellowstone's plant communities have burned at one time or another.

B The following spring, new plant life flourished in the recovering park.

777

REAL-LIFE CONNECTION — **GENERAL**

Writing **Ecosystems** Have students research a predator that lives in a local ecosystem, and write a short paper with their results. Ask them to focus on how the predator has adapted to the ecosystem. **LS Verbal**

Did You Know?

Yellowstone Fires In 2000, there were a total of 34 fires at Yellowstone. Four of the fires were caused by humans; the other 30 were caused by lightning. Students who want to learn more about the 2000 fires can visit the Yellowstone website: http://www.yellowstone.net/wildfires2000.htm

SKILL BUILDER — **BASIC**

Interpreting Visuals Have students examine **Figure 3** (on the facing page). Ask students to name some actions that may lead to the disruption of this balanced aquarium ecosystem. (insufficient food, insufficient sunlight, or too much salt) **LS Logical**

SKILL BUILDER — **BASIC**

Reading Skills Write the word *interrelatedness* on the board, and have students brainstorm all the words, phrases, and ideas that they associate with the word. List student contributions on the board and begin a discussion in which students seek clarification of the listed ideas. At the end of the discussion, have students make notes of everything they remember from the discussion. Encourage students to look over their notes to see what they know about the topic based on experience and the discussion. **LS Verbal**

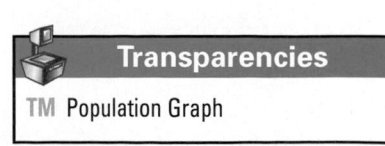

Transparencies

TM Population Graph

Teach, *continued*

Reading Strategy — BASIC

Reading Organizer Have students read the section titled *Evaluating Changes in Ecosystems,* and then organize the information presented in a two-column table. The heading for the left side of the table should read *Short-term changes,* and the heading for the right side of the table should read *Long-term changes.* Have students fill in the table with the appropriate information from the section.
LS Verbal

SKILL BUILDER — BASIC

Interpreting Visuals Have students examine **Figure 7.** Ask students how the living and nonliving elements of this ecosystem might differ with short-term ecosystem changes. (hibernation, migration, snow, loss of foliage) **LS** Visual

Teaching Tip — GENERAL

Succession Short-term ecosystem changes are easy to observe outside of the classroom during most times of the year. Take your class outside and ask them if they observe any evidence of short-term ecosystem changes. Ask them for evidence of long-term ecosystem changes as well. **LS** Logical

Figure 6
Different views of Glacier Bay, Alaska, show the types of change that took place over 200 years, as glaciers receded.

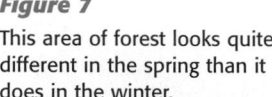 **succession** the replacement of one type of community by another at a single location over a period of time

Figure 7
This area of forest looks quite different in the spring than it does in the winter.

In time, a complex ecosystem will develop. This process, shown in *Figure 6,* is known as **succession.** The end product is a stable but complicated community where birth, death, growth, and decay take place continuously. This will keep the ecosystem stable if no major disruptions occur.

Evaluating Changes in Ecosystems

Ecosystems undergo both short-term and long-term changes. Short-term changes are usually easily reversed, but long-term changes can take many years to be reversed, and sometimes may never be reversed at all.

Short-term ecosystem changes include the seasons

During autumn, many trees and shrubs lose their leaves. In the winter, many birds migrate to warmer places. Other animals hibernate by lowering their metabolism. These animals can sleep through the winter in snug burrows or caves. In spring, the migrating birds return, animals come out of hibernation, buds open, and seeds begin to sprout. As *Figure 7* shows, an ecosystem can appear quite different during different times of the year.

Did You Know?

Amazing Migrations The arctic tern holds the record for the longest bird migration. Each year, the tern flies from the Arctic to the Antarctic and back, a trip that can be as long as 20,000 miles. Bar-headed geese might make the record for the highest flight. They have been observed flying across the Himalayas at 29,000 feet.

Trends in Life Science

The Great Backyard Bird Count Each year, scientists at the Cornell Lab of Ornithology collaborate with the National Audubon Society to sponsor a bird count. They ask for volunteers throughout North America to count the numbers and kinds of birds they observe in their area during one or all days of the count. To find out the dates of the next bird count and to learn more about how to involve your students, visit the following website: http://www.birdsource.com/gbbc/

Changes in climate cause long-term ecosystem changes

In your lifetime, the climate where you live probably hasn't changed much. Some years may be colder than others, but average monthly temperatures do not vary greatly from year to year. Throughout Earth's geologic history, there have been periods known as ice ages, when icy glaciers covered much of the continents. *Figure 8* shows the size of the glacier that covered much of North America during the last ice age. This period ended roughly 11 500 years ago.

During ice ages, temperatures are much colder than usual. These cold spells alternate with warmer periods. Scientists hypothesize that ice ages are caused by a variety of factors, including plate tectonics, changes in the tilt of Earth's axis, changes in the shape of Earth's orbit around the sun, and changes in the speed and pattern of the ocean's circulation.

The combined effect of these changes in Earth's position in space is difficult to predict. One thing we do know is that these changes cause temperature differences in ecosystems.

Long-term changes in ecosystems can also be caused by events such as volcanic eruptions. At other times, many small factors act together to cause change. In these cases, it may be hard to know how much each factor adds to the change. One example of this is the many and varied factors affecting global temperature change.

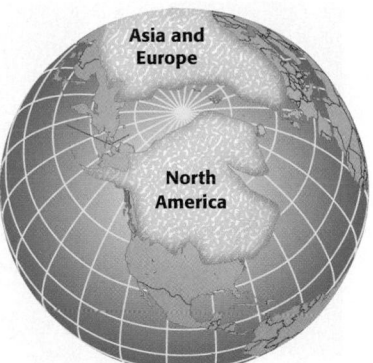

Figure 8
Icy glaciers covered much of North America and parts of Europe and Asia during the last ice age (approximately 20 000 years ago).

Quick Lab

Why do seasons occur?

Materials ✓ globe ✓ unshaded lamp

1. Place the lamp on a table, and turn the lamp on.
2. Stand about 2 m from the table, and hold the globe at arm's length, pointing it toward the lamp.
3. Tilt the globe slightly so that the bottom half—the Southern Hemisphere—is illuminated by the lamp.
4. Keeping the axis of Earth's rotation pointing in the same direction, walk halfway around the table.

Analysis

1. What part of the globe is lit by the lamp's light now? What season does this represent in this part of Earth?
2. Would there be any seasonal changes if the Earth's axis were not tilted? Explain your answer.
3. In addition to experiencing seasonal changes, ecosystems also experience short-term changes as day changes into night. What movement of Earth causes night and day to occur?

Trends in Ecology and Climatology

Climate Change and Disease Ecologists are working with climatologists to investigate the relationship between climatic changes and the outbreak of disease.

In 1993, a virus began killing young people in the southwestern United States. An unusually mild winter and a wet spring caused piñon trees to bloom vigorously, providing virus-carrying mice with a plentiful supply of pine nuts. The mice population increased tenfold. The mice found their way into people's homes and spread the virus to humans. Half the people infected with the virus died.

Researchers are carefully monitoring mice and other disease carriers in an effort to predict if, when, and where the next outbreak of disease might occur.

Teaching Tip

Long-Term Ecosystem Changes
Alert your students to the fact that long-term ecosystem changes are not limited to only destructive events such as volcanic eruptions. The construction of roads, shopping malls, and other human-related activities can be classified as long-term ecosystem changes as well.

Quick Lab

Why do seasons occur?

During spring and summer, the Northern Hemisphere tilts toward the sun and the Southern Hemisphere tilts away from the sun. Thus, sunlight strikes more nearly perpendicularly in the north than it does in the south. As a result, in the Southern Hemisphere the energy is spread out over a larger area. The opposite is true for both hemispheres during fall and winter.

Answers to Analysis

1. the Northern Hemisphere; summer
2. no, because there would be no variation in the amount of sunlight different parts of Earth receive
3. the rotation of the Earth on its own axis

Chapter Resource File

• **Quick Lab Datasheet** Why do seasons occur? GENERAL

Interpreting Visuals Have students look at **Figure 9.** The photos show various human activities and the changes that they cause to ecosystems. Ask students how these activities illustrate changed ecosystems. **LS** Logical

Reading Strategy — **BASIC**

Reading Organizer Have students read the section on changes to ecosystems caused by human activity, and have them organize the information presented in the form of a table. On the left side of the table, have students write *Positive aspects of human activities*, and on the right side of the table, have them write *Negative aspects of human activities*. Have students fill in the table with the appropriate information from the section. **LS** Verbal

Figure 9

Clearing trees, driving cars, constructing buildings, and farming are just a few human activities that cause changes in ecosystems.

internet connect

www.scilinks.org
Topic: Changes in
Ecosystems
SciLinks code: HK4019

SciLINKS Maintained by the
National Science
Teachers Association

Changes can be caused by human activity

Physical factors are not the only things that cause changes in ecosystems. People also alter the environment in a variety of ways, as shown in *Figure 9.* Activities such as logging, driving cars, growing crops, and constructing roads and buildings change the environment.

All of those activities have some benefits, but they also cause some problems. The benefits of some human activities, such as building dams, are numerous. For instance, the El Chocon Cerros Colorados project brought much-needed flood control to the foothills of the Andes Mountains in Argentina. Formerly, the rivers that flowed down from the mountains flooded twice each year. At other times the region was too dry to grow crops. The construction of a system of dams stopped this cycle. Excess water is now stored in large reservoirs behind the dams. This water is used for irrigation which allows farmers to grow crops year-round. In addition, the dam is used to generate hydroelectric power for much of the country.

Such large dams, however, can cause problems. Without the floods, rivers no longer deposit rich soil, so most farmers use chemical fertilizers on their crops instead. Runoff from these fertilizers can contaminate ground water and streams, making water supplies unsafe for humans and other living things.

Many of the adverse effects of dams constructed before the 1970s were not foreseen by their developers. Today scientists and engineers have a better understanding of how ecosystems work. Often, major projects such as constructing a dam must undergo an environmental analysis before construction begins. If the project is likely to destroy an entire ecosystem, it may have to be redesigned, relocated, or canceled altogether.

780

SOCIAL STUDIES
CONNECTION — **GENERAL**

Writing Have students conduct research to find out how their local ecosystem has changed over time. What did their ecosystem look like 50 years ago? 100 years ago? How have the living and nonliving elements of the ecosystem changed? What impact have humans had on the ecosystem and have they been positive or negative? How have natural forces changed the ecosystem? Ask students to write a short paper that summarizes the results of their research. **LS** Verbal

ECOLOGY
CONNECTION — **GENERAL**

Habitat loss as a result of human activities (destruction of forests, draining of wetlands, pollution) is one of the leading reasons why ecosystems are becoming unbalanced around the world today. Have students research how changes in their environment have resulted in habitat loss. **LS** Verbal

Changes can be caused by introduction of nonnative species

Some species move from one ecosystem to another on their own. Animals may migrate to new areas, and seeds can be carried by wind or water to different places. Humans often influence the spread of nonnative species to other ecosystems, sometimes accidentally and sometimes on purpose.

Starlings, for instance, were purposely brought to the United States. In the 1800s, a few dozen European starlings were released in New York City. The birds rapidly multiplied, and today there are millions of them across the United States.

Both starlings and native North American bluebirds nest in holes in tree trunks and fence posts, as shown in *Figure 10*. As a result, the two species compete for shelter. The bluebirds nearly lost the battle. Concerned citizens launched a multistate effort to build and distribute bird boxes. These boxes, specially designed with small entrances, provided nesting places for the bluebirds and kept the larger starlings out.

There are other ways in which unwanted organisms may be introduced to an ecosystem. For instance, small insects can be carried across borders accidentally. Often they are hidden in crates of fruit and vegetables.

In its new environment, the nonnative species may have no natural enemies to keep its numbers in check. As a result, its members can quickly multiply and modify an ecosystem. The new competing species may wipe out an existing native species and cause change in the ecosystem.

Figure 10

Starlings compete with bluebirds for nesting sites in the holes of tree trunks or fence posts.

SECTION 1 REVIEW

SUMMARY

▶ Living and nonliving elements form an ecosystem.

▶ The elements of an ecosystem maintain a balance. One change can affect an entire ecosystem.

▶ A disturbed ecosystem may return to its original state.

▶ Ecosystem changes can be short-term or long-term.

▶ Climatic changes, human activities, and nonnative species introduction can cause changes in ecosystems.

1. **List** two factors that keep populations stable.

2. **Analyze** the following statement: An ecosystem is like a fine-tuned car.

3. **Describe** how the loss of one species in a pond can affect other species.

4. **Define** *succession*.

5. **Predict** how a thunderstorm could lead to a long-term change in an ecosystem.

6. **List** two changes in the Earth's position in space that can affect climatic change in ecosystems.

7. **Critical Thinking** Describe a common human activity that can disrupt an ecosystem. Propose a solution to the problem.

8. **Critical Thinking** List several ways nonnative plants may be introduced to an environment.

781

Answers to Section 1 Review

1. disease and food shortage

2. There are many different aspects of an ecosystem that are interrelated, and all of those aspects must work together for success, just like a finely tuned car. If one part of the car breaks down, this will affect all of the other parts. If one part of the ecosystem is missing, the entire system is at risk of collapsing.

3. If the species that disappears is a food source for other species, then those species and any other species that depend on the food source will be impacted.

4. Succession is the gradual repopulating of a community by different species over a period of time.

5. A thunderstorm would entail lightning, which could result in forest fires that destroy trees and plants.

6. the tilt in Earth's axis; changes in Earth's orbit; Earth's distance from the sun

See "Continuation of Answers" at the end of the chapter.

Focus

Overview

Before beginning this section, review with your students the Objectives listed in the Student Edition. In this section, students study several energy sources, including fossil fuels, solar power, wind power, moving water, geothermal energy, and nuclear energy. They learn about the advantages and disadvantages of each source. The section concludes with a discussion of the efficiency of energy conversions.

🔔 Bellringer

Use the Bellringer transparency to prepare students for this section.

Motivate

Opening Activity — GENERAL

Write the following questions on the board:

• How do living things (plants, animals, and people) use energy?

• Where does this energy come from?

Before students read this section, group them in pairs and have them provide answers to these questions. After they have read the section, encourage them to discuss whether their answers changed or remained the same. Have students cite passages in the text that account for changes in or reinforcement of their original answers.
LS Interpersonal

Energy and Resources

▶ **KEY TERMS**
fossil fuel
nonrenewable resource
renewable resource
geothermal energy

OBJECTIVES

▶ **Identify** different sources of energy used by living things, and trace each source back to the sun.

▶ **Describe** the advantages and disadvantages of several energy sources.

▶ **Describe** the types of conversion processes necessary for different energy sources to produce electricity.

▶ **Identify** how efficient different conversion processes are.

How Energy Is Used in the United States

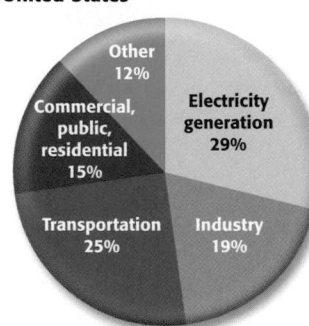

How Energy Is Used Worldwide

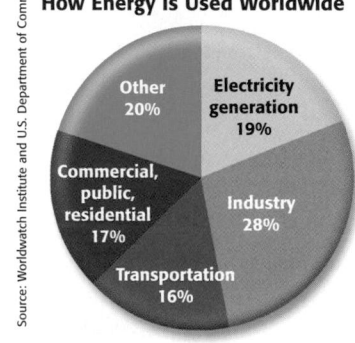

Source: Worldwatch Institute and U.S. Department of Commerce

Figure 11

These pie charts show how energy use in the United States compares with energy use worldwide. (Worldwide calculations include the United States.)

782

You probably use more energy than you think you do. Energy is needed to light streets and homes, heat water and buildings, cook food, power vehicles, run appliances, and make the products you use every day. Where does all this energy come from?

The Search for Resources

We rely on natural resources to meet our basic needs for food and shelter. We also depend on natural resources to provide the energy and raw materials needed at home, at work, and for growing food. *Figure 11* compares the patterns of energy use in the United States with the patterns of energy use worldwide.

The sun is the source of energy

Almost all energy comes from the sun. The sun sends out energy as radiation of various wavelengths.

Plants convert energy from the sun into chemical energy

Plants use some of the energy from sunlight to change the simple molecules of carbon dioxide and water into oxygen and more complex molecules of simple sugars. This process is called photosynthesis. It allows plants to change the sun's energy into stored chemical energy.

Some animals eat plants to obtain energy. Through a series of chemical reactions, the animals are able to convert the sugars in plants back to carbon dioxide and water. This process, which also produces energy, is known as cellular respiration. This cycle of energy transfer repeats itself continuously in nature. *Figure 12* shows the interrelatedness of organisms processing and moving carbon dioxide through an ecosystem.

Trends in Environmental Science

Gas Hydrates Most of the world's energy needs are met by fossil fuels. If current use continues and new reserves are not found, gas and oil supplies could run out by the year 2050. One potential solution is to extract molecules of gas trapped inside ice. Gas in this form is called gas hydrate, or methane hydrate. Estimates of the amount of gas hydrate beneath permafrost and the oceans suggest that there may be 730,000 times the world's current annual gas consumption. Although extraction may prove expensive, removing even a small percentage could vastly increase our energy supplies.

Did You Know ❓

Rising Demand In 1997, the world used a total of 73 million barrels of oil each day. The projected use for 2020 is 113 million barrels daily. Natural gas consumption is expected to be more than double its current value by 2020. In all, total global energy consumption is projected to rise 60% from 1997 to 2020.

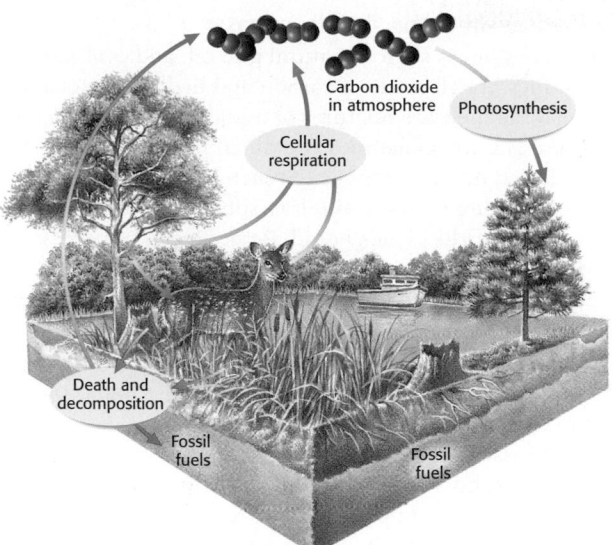

Figure 12
Carbon dioxide is a product of cellular respiration.

SKILL BUILDER —— BASIC
Reading Skills Write the phrase *fossil fuels* on the chalkboard, and have students come up with words that relate to the phrase. Encourage students to devise a hypothesis about the definition of the phrase based on their combined thoughts. **LS** Verbal

Fossil fuels form deep underground

Most living things cannot decompose and release their stored energy without air. For example, microscopic plants and animals living in the ocean die and are buried under layers of sediment where they are not in contact with air. Without air, the organic chemicals in the remains of the living things cannot combine with oxygen and change back to carbon dioxide and water. Instead, pressure and heat from the settled rock above the remains cause different chemical reactions. These reactions turn the organic chemicals into substances that contain mainly carbon and hydrogen, as shown in *Figure 13*. These substances are known as **fossil fuels.**

▶ **fossil fuel** a nonrenewable energy resource formed from the remains of organisms that lived long ago; examples include oil, coal, and natural gas

Teaching Tip —— GENERAL
Photosynthesis To demonstrate how plants change energy from the sun into chemical energy (by photosynthesis), lead your students to a patch of grass outside. Place a small waterproof container (plastic bowl or a shoe box covered with plastic wrap) over some of the grass, so that it is deprived of sunlight. Make sure there is a patch that will receive sunlight adjacent to the covered grass. Ask students to record their hypotheses about what the two patches will look like over the course of several days or a weekend. After a few days have passed, go back to the patches of grass and have students make observations. **LS** Logical

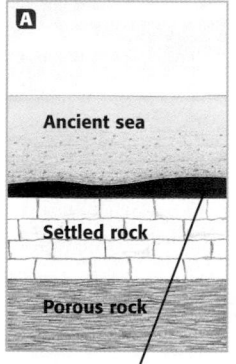

Layer of organic rich ooze

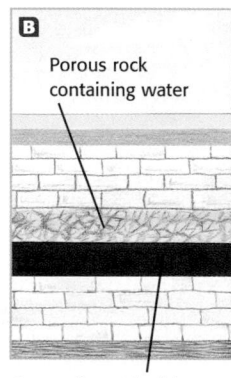

Layer of organic rich ooze buried and heated

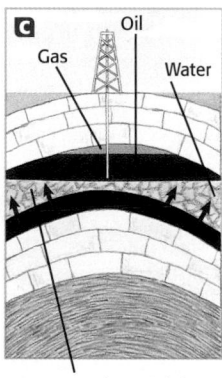

Porous rock containing water, oil, and gas

Figure 13
A Microscopic plants and animals collect in layers of mud forming an organic rich ooze.
B This organic rich layer is underlain and overlain by settled rock. Heat and pressure "cook" the ooze, causing oil to form.
C Geologic forces cause the rock layers to bend. The oil is forced out of its original layer, migrates into porous rocks, and is trapped.

783

Alternative Assessment —— ADVANCED
Fossil Fuels The combustion of fossil fuels for energy leads to several air pollution problems and contributes to global warming. Have students research the relationship between fossil fuel use, air pollution, and global warming in small groups, then have them present their results to the class. **LS** Interpersonal

Did You Know?

How Much Coal? It takes half a ton of coal to run a single electric stove for one year. A refrigerator uses about the same amount, while an electric water heater uses about 2 tons. It is easy to see how a family whose power comes from a fossil-fuel plant could use several tons of coal every year!

Chapter Resource File

• Lesson Plan

 Transparencies

TT Bellringer
TT Making Oil
TT Carbon Cycle
TM Energy Use

Air Pollutants Fumes from the burning of fossil fuels contain tiny particles of pollutants that circulate throughout the air that we breathe. Ask students if they can propose a hypothesis for a method of trapping some of these tiny air pollutants. (Answers may vary.)

For a demonstration, take pieces of plastic and smear petroleum jelly on them. Place the plastic pieces in various locations around the school and collect them in a week. Look for any particulates or debris that may have collected on the jelly, and discuss the results with your students.

|SKILL BUILDER — BASIC

Interpreting Visuals Have students study **Figure 15.** Ask: Which energy source is used most? (oil) Which is used least? (nuclear) What factors might change the estimated dates that each source will run out? (using more or less energy, discovering new reserves) You may wish to find the most current values from the Energy Information Administration and compare them with the 1999 values shown in the figure. **LS Visual**

Figure 14
Natural gas can provide the energy needed to prepare food.

Methane (CH₄)

▶ **nonrenewable resource**
a substance that is consumed faster than it forms and therefore can not be replaced within a human life span

Figure 15
Although 84% of the world's energy is supplied by fossil fuels, oil and natural-gas reserves will soon run out.

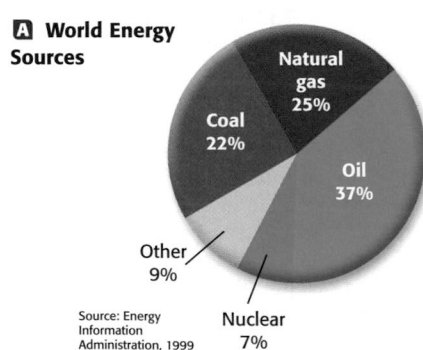

A World Energy Sources

Natural gas 25%

Coal 22%

Oil 37%

Other 9%

Nuclear 7%

Source: Energy Information Administration, 1999

B World Energy Reserves

Gas

Oil

Coal

1995 2045 2095 2145 2195 2245 2295
Years

Fossil fuels can be solids, liquids, or gases

Although substances such as natural gas, oil, and coal seem very different, they are all made of carbon and hydrogen molecules.

Some fossil fuels are mixtures of liquids and gases. Liquid oil and natural gas are sometimes contained in porous rocks. Wells must be drilled into these rocks to reach the fossil fuels.

Oil is a mixture of many different substances. It comes out of the ground as a tarlike black liquid. Before it can be used, it must be purified and separated. Refineries separate oil into gasoline, kerosene, diesel fuel, and other products. Oil formed in shallow oceans, as did natural gas.

Natural gas is made mostly of methane, a colorless, odorless, poisonous gas. It is often used for heating homes and cooking, as shown in *Figure 14.* In the United States, natural gas makes up 22% of the fossil fuels used.

Coal is a solid fossil fuel composed mainly of carbon. It formed in ancient swamps when the remains of large, fernlike plants were buried under layers of sediment.

The supply of fossil fuels is limited

When fossil fuels are burned, they form carbon dioxide and water and release energy. This energy is the energy of the sun that was trapped in plants hundreds of millions of years ago. Because fossil fuels take so long to form, they are considered **nonrenewable resources.** They are now being used by humans much faster than natural processes can replace them.

Figure 15A shows that the vast majority of the world's energy needs are met by the burning of fossil fuels. *Figure 15B* shows the estimated reserves of the world's supply of fossil fuels. If oil and natural gas use continue at current rates, the reserves may run out during your lifetime. What alternative sources of energy will provide the energy needed for everyday activities?

REAL-LIFE CONNECTION — GENERAL

Dependence on Cars For one week, have students keep track of how many miles they travel by car. At the end of the week, have them graph their travel data. Have each student assume that the vehicle(s) they traveled in used 1 gal of gasoline per 25 mi traveled. Have each student divide the distance traveled by 25 to determine how many gallons of fuel were used during the week. Students can then analyze the data and make comparisons with their classmates. Are there any alternatives that students can think of that will reduce their dependence on cars? **LS Logical**

Alternative Sources of Energy

Fossils fuels are not the only energy source around. We can harness energy from the sun, wind, water, and Earth. We can even obtain energy from atoms. The more these alternative sources of energy are used, the less we will rely on fossil fuels.

Another advantage of using alternative sources of energy is that many are **renewable resources.** This means they can be replaced by natural processes in a relatively short amount of time.

Solar power plants and solar cells can make electricity from sunlight

Every day, the sun makes more energy than is used to supply electricity to the United States for a year. But harnessing the sun's energy to supply electricity is not simple. Some parts of Earth do not get as much sunlight as other parts. Even when there is enough sunlight, tools are needed to convert the sun's energy and change it into a useful form.

In the 1990s, the first solar power plant capable of storing energy as heat was opened in the Mojave Desert, in California. The solar panels of the power plant shown in *Figure 16* store the sun's heat and use it for energy.

Another tool used to harness the sun's energy is a solar cell, shown in *Figure 17.* Solar cells are able to produce electricity from sunlight. Although their use has been limited, the level of efficiency of these devices has greatly improved over the last two decades.

The energy in wind can be used by windmills

Energy from the sun is a contributing factor in the production of another renewable energy resource —wind energy. Wind energy actually comes from the sun. Different places on Earth receive different amounts of sunlight, which causes variations in temperature. These temperature differences cause the movement of air, known as wind.

Wind is one of the oldest sources of renewable energy used by humans. It has been used to sail ships for thousands of years. As with other sources of energy, the use of wind energy has advantages and disadvantages. It can be unreliable. Even in exceptionally windy areas, wind doesn't blow steadily all the time. This can cause differences in the amount of power generated. However, the use of windmills is becoming increasingly popular in some areas because of their low cost.

▶ **renewable resource** a natural resource that can be replaced at the same rate as it is consumed, such as food production by photosynthesis

Figure 16
The Mojave Desert in California is home to the first solar power plant capable of storing heat.

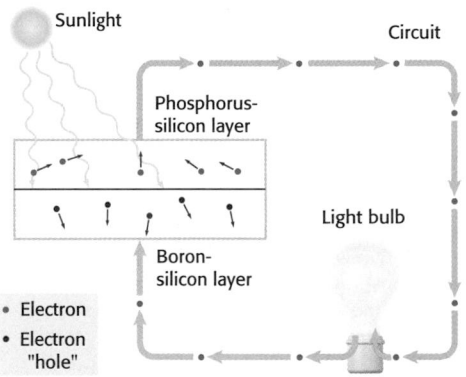

Sunlight
Circuit
Phosphorus-silicon layer
Light bulb
Boron-silicon layer
• Electron
• Electron "hole"

Figure 17
Solar cells convert sunlight directly into electricity.

Reading Strategy — BASIC

Reading Organizer Have students read the section titled *Alternative Sources of Energy,* and have them organize the information in a table like the one shown below. Have students fill in the table with the appropriate information from Section 2. **LS Verbal**

Teaching Tip — GENERAL

Solar Heating Have your students determine what materials are best for a solar heating system by gathering items made of different materials, such as aluminum, steel, copper, stone, wood, plastic, and glass. Tape a thermometer to the back of each item, and place the items outside in the sunlight. After 30 minutes, have your students record the temperature of each item.

Caution: *Be sure to warn students about touching very hot materials.*

Have your students make a list of the results, showing the hottest materials at the top of the list and the coolest at the bottom. Which material would work best in a solar heater? **LS Logical**

Alternative Energy Resources

Type/Source	Advantage	Disadvantage
Solar power/sun	Clean	High installation cost
Wind	Inexpensive	Unreliable
Hydroelectric/water	Renewable	May harm environment
Geothermal/Earth	Convenient in some areas	Limited by location

Chapter Resource File

• **Cross-Disciplinary Worksheet** Integrating Physics—Solar Cells ADVANCED

Transparencies

TT Solar Cell Diagram

Videos

CNN. Presents Chemistry

• **Segment 15** Coal-Oil Technology GENERAL

• **Critical Thinking** Worksheet 15

See the Science in the News video guide for more details.

Teaching Tip

Geothermal Energy Ask students if they have ever heard of Old Faithful. Tell them that Old Faithful is a geyser and that geysers are an example of geothermal energy. The ground water has been heated by molten rock within the Earth's crust, and the heat energy is released as steam.

|SKILL BUILDER| — GENERAL

Vocabulary Tell students that the prefix *hydr–* (as in hydroelectric) refers to water, while the prefix *geo–* (as in geothermal) refers to Earth. Ask them to use a dictionary to find other words that begin with the prefixes *hydr–* and *geo–*. Have them include definitions for each word. (Sample responses: hydrodynamics (the dynamics of fluids), hydrant (outlet from a water main with an upright pipe and one or more spouts), hydrated (chemically combined with water); geography (study of Earth and its features), geocentric (measured or observed from Earth's center), geology (study of the origin, history, and structure of Earth).)

SOCIAL STUDIES — CONNECTION

Making the Connection

Students' maps may vary but should show the entire area that the TVA covers.

Figure 18
The spinning blades of a windmill are connected to a generator. When the blades spin faster, the generator produces more energy.

▶ **geothermal energy** the energy produced by heat with the Earth

Connection to
SOCIAL STUDIES

To aid the nation's economic recovery following the Great Depression, President Franklin D. Roosevelt proposed a series of projects aimed at increasing employment and serving the public good. One such project was the Tennessee Valley Authority (TVA), a connected system of 40 dams spread over an area of 106 000 km². The TVA system still provides electricity to homes, farms, and factories in Tennessee, Kentucky, Virginia, North Carolina, Georgia, Alabama, and Mississippi.

Making the Connection

Go to your local library and find a map showing the area that the TVA covers. Draw your own map showing all 40 dams.

Windmills have been used for over 1000 years. Today wind farms generate electrical power, as seen in *Figure 18.*

Moving water produces energy

Falling water releases a lot of energy, but even the energy produced from water depends on sunlight. For example, the water in oceans evaporates as sunlight transfers energy by heat. This water rises into the atmosphere, falls back down later as rain or snow, and flows downhill into creeks and rivers. Once the water is in the creeks and rivers, it flows back to the ocean, and the cycle begins again.

In hydroelectric power plants, dams are built on fast-moving rivers to create large holding places for water. *Figure 19* illustrates how the stored water pours through turbines, making them spin. The turbines are connected to generators that produce hydroelectric power, another type of renewable resource.

Dams have already been built on most of the world's big rivers, so the potential for increasing the use of this energy source is limited.

Geothermal energy taps Earth's warmth

Under Earth's crust, reservoirs of steam or hot water produce **geothermal energy.** These holding pools usually lie near beds of molten magma, which heat the steam or water. Wells are drilled into the reservoirs, and the steam or hot water rises to the surface, where it is used to turn turbines to generate electrical power.

Cultural Awareness GENERAL

The Three Gorges Dam, in China, is currently being constructed in hopes of providing enough energy to meet the needs of the surrounding population. Have students research some of the pros and cons surrounding the construction of this dam. What will the benefits and losses be? What will the impacts of the dam be on the immediate environment and the cultural history of the area? **LS** Interpersonal

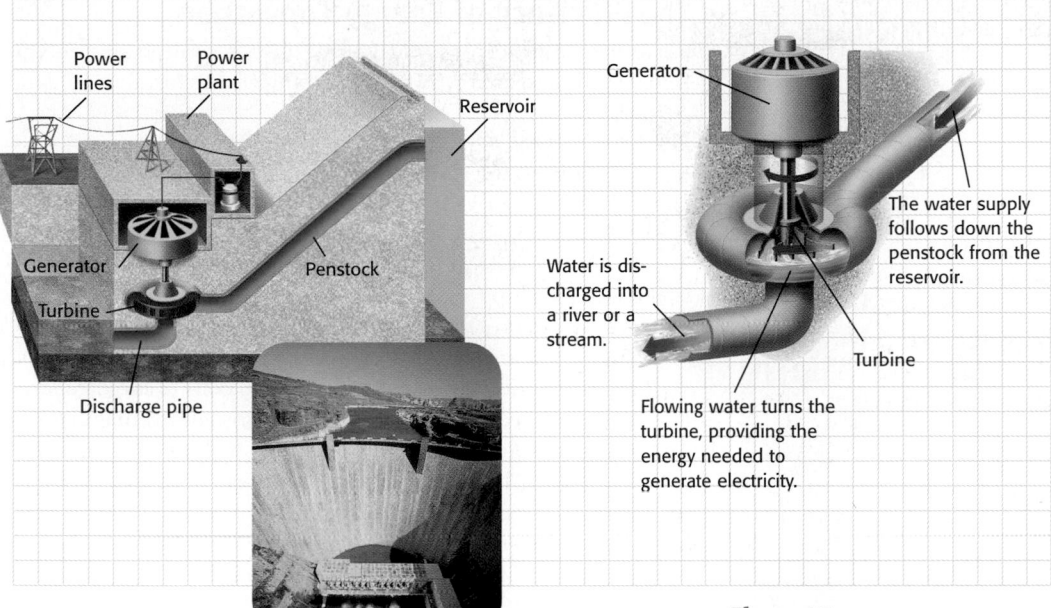

Generator

Power lines
Power plant

Reservoir

Generator

Penstock

Turbine

Discharge pipe

The water supply follows down the penstock from the reservoir.

Water is discharged into a river or a stream.

Turbine

Flowing water turns the turbine, providing the energy needed to generate electricity.

Figure 19

The incoming water causes the turbines in a dam to turn. The turbines run a generator that produces electricity.

Geothermal energy is a major source of electricity in volcanically active areas such as New Zealand, Iceland, and parts of California. Geothermal plants, such as the one shown in **Figure 20,** are common in these places. This type of energy works best when the beds of molten rock lie very close to Earth's surface. However, not many areas have magma beds close to the surface, so widespread use of this resource is unlikely.

Atoms produce nuclear energy

Atomic reactions produce a type of alternative energy resource called nuclear energy. Chain reactions involving nuclear fission produce a great deal of energy. This energy can be transferred as heat to water in a nuclear reactor. The heated water or steam can then be used to turn a turbine and generate electricity. Nuclear energy does not release pollutants into the air like the burning of fossil fuels does. However, harnessing nuclear energy produces highly radioactive waste.

Nuclear energy has seen limited use in the United States. It currently provides only 8% of our energy. Scientists working with nuclear fusion hope that it will someday be a useful and renewable resource. But so far, more energy is needed to sustain a fusion reaction than is produced by the reaction.

Figure 20

This geothermal plant in New Zealand produces electricity.

787

ENVIRONMENTAL SCIENCE — CONNECTION

Like all energy sources, nuclear power involves environmental trade-offs. One problem with nuclear power is determining how to safely store the spent fuel, which is radioactive. One proposal in the U.S. is to store the waste deep under Yucca Mountain, in Nevada. The safety of this option is a matter of debate. In the meantime, the spent fuel is currently being stored at the individual power plants where it is generated.

One environmental benefit of nuclear power is that it does not emit greenhouse gases into the atmosphere, like fossil-fuel plants do. Replacing all of the nuclear power plants in the U.S. with fossil-fuel plants would add about as many greenhouse gases to the air as putting 5 billion more cars on our roads.

Reading Skills Have students silently read the passage on the efficiency of energy conversion and have them mark on self-adhesive notes those sections that they do not understand. Be sure that students study **Figure 21** in order to clarify the relevant passages. Pair students together before or after the reading, and have them discuss the passages they found difficult.
LS Interpersonal

Interpreting Visuals Have students examine **Figure 21.** Ask them to answer the following: Where is the energy stored for this process, and where did it originate? (The energy is stored in the coal; the original source is our sun.) How is the energy released? (by burning the coal) What form is the energy in during steps A, B, and C? (A: heat energy; B: kinetic energy; C: electrical energy) **LS** Visual

Figure 21

This power plant converts coal into electrical energy.

The Efficiency of Energy Conversion

Regardless of which energy resource we use, some usable energy is lost each time energy is converted to another form.

Energy is wasted when input is greater than output

In a coal-fired power plant, chemical energy is released when coal burns with oxygen in the air in a combustion reaction. This heat energy is transferred to water, which forms steam. Some energy is lost in this conversion. In order to obtain a high pressure, the steam is heated, and more energy is lost. The steam must be at a high pressure to provide the force to turn the huge steam turbines. This changes the energy into kinetic energy of the moving turbines.

The spinning turbines are connected to coils of large generators. These coils carry current and act as large electromagnets. As they spin, they generate a high voltage in the fixed coils surrounding them. The energy is now in the form of electrical potential energy. This causes a current that carries the electricity to local consumers through a cable. *Figure 21* shows a coal-fired power plant and a diagram of how electricity is produced in a typical power station.

A large power station might be rated at 1000 MW. This means that it delivers 1 billion J of energy every second. But three times the fuel, equal to 3 billion J, is needed because two-thirds of the energy input is wasted. Most power stations that use fossil fuels for energy are only 30 to 40% efficient, though newer models recycle waste heat, which increases efficiency.

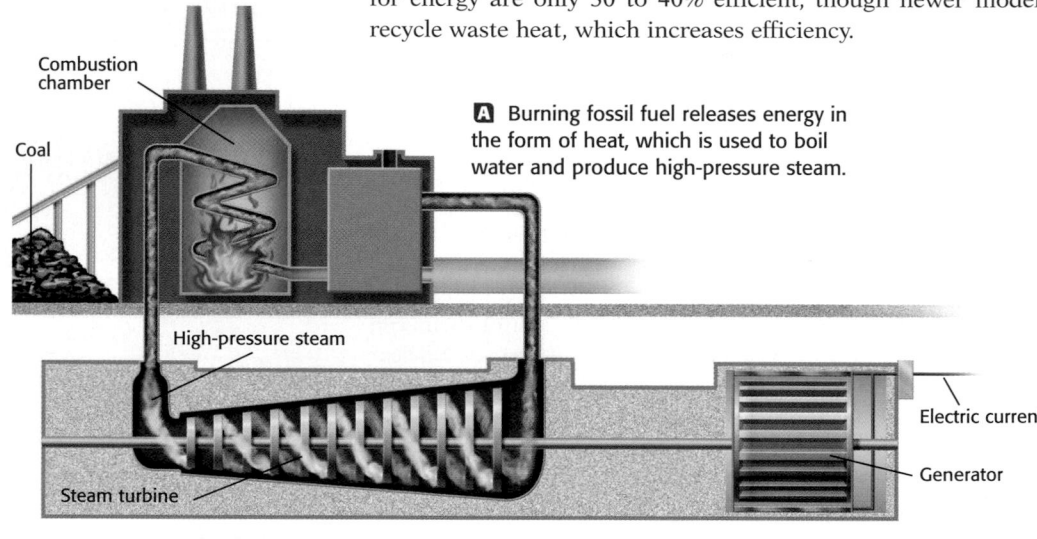

A Burning fossil fuel releases energy in the form of heat, which is used to boil water and produce high-pressure steam.

B The steam is directed against the blades of a turbine, which is set into motion.

C The turbine is connected to an electric generator. The turbine sets the generator in motion, generating electricity.

Trends in Technology

A New Way to Use Fossil Fuels Scientists at Los Alamos are developing a technique for producing electricity from fossil fuels with greater efficiency and less pollution than today's plants. The new technology involves converting a mixture of coal and water into hydrogen, which is then converted into electricity through a fuel cell.

The efficiency of this process is about 50%, compared to 30% or 40% for most fossil fuel plants today. The waste heat is captured and used to drive some of the reactions in the process. Compared to typical power plants, the process generates twice as much electricity per unit of fuel consumed. Another advantage is a significant reduction in greenhouse gases. The process produces less than half the carbon dioxide (per kilowatt-hour) of a conventional plant.

Wasted energy can be used

Some wasted energy occurs in the production of all types of energy sources. Nuclear power stations are roughly as efficient, or inefficient, as fossil-fuel power plants are. Both types of power plants produce waste. Even wind-powered plants have some inefficiency because of the energy conversions involved.

There are ways to make use of some of this wasted energy. For example, although the water from power stations is not hot enough to make electricity, it can be used to heat homes. In Germany, most towns have their own small power station. Rather than dump the warm water into a river, it is piped to people's homes to keep them heated. These are called combined heat and power schemes. They reduce wasted energy and make electricity less expensive.

internet connect

www.scilinks.org
Topic: Alternative Energy Sources
SciLinks code: HK4004

SCi LINKS. Maintained by the National Science Teachers Association

Close

Quiz ——————— BASIC

1. What are fossil fuels formed from? (plant and animal remains)

2. Are fossil fuels considered renewable or nonrenewable? (nonrenewable)

3. What are two examples of alternative energy sources? (any two of the following: solar power, wind power, moving water, geothermal energy, nuclear energy)

4. Some energy sources can generate electricity without wasting any energy in the conversion process. True or false? (false)

LS Logical

SECTION 2 REVIEW

SUMMARY

▶ Most of the energy used in the world comes from fossil fuels, which are nonrenewable resources.

▶ Energy from the sun produces solar energy and wind energy.

▶ Hydroelectric power can be generated from large reservoirs of water. Moving water produces energy.

▶ Geothermal energy is generated from underground reservoirs in volcanically active areas.

▶ The use of nuclear energy has been limited in the U.S.

▶ Energy is lost each time it is converted to another form. Some wasted energy has the potential for use.

1. **List** five possible ways that you might use electricity during lunch.

2. **Describe** the cycle of energy transfer among organisms. Name the process that plants use to convert energy from the sun into chemical energy.

3. **Compare** the amount of energy used in the United States with the amount used worldwide. Write a paragraph explaining the social and economic reasons for the difference.

4. **Describe** some advantages and disadvantages of solar energy.

5. **Explain** in a paragraph the energy conversions that occur when a drop of rain falls on a mountain, rolls downhill, passes through a hydroelectric plant, and turns a turbine to generate electricity.

6. **Predict** which of the following will be more efficient in terms of capturing the most energy. Justify your answer.
   a. burning paper to boil water to make steam to turn a turbine
   b. using wind to generate electricity

7. **Critical Thinking** A classmate says that geothermal power is perfectly efficient because it never runs out. What's wrong with this reasoning?

789

Answers to Section 2 Review

1. to boil water, to keep your drink cold, to turn the lights on in the kitchen, to open cans (electric can opener), to dispose of waste (garbage disposal)

2. The sun sends out energy as radiation, and this energy is used by animals and plants. Plants use this energy to make oxygen and simple sugars, and animals eat plants and other animals to get energy for themselves. The process plants use to convert the sun's energy into chemical energy is called photosynthesis.

3. The amount of energy used in the United States is much higher than in other places in the world. This is due to the fact that the United States is quite affluent compared to many other countries. This affluent lifestyle that many Americans enjoy contributes to high amounts of energy consumption.

4. advantages: solar energy is a nonpolluting renewable energy resource; disadvantages: solar energy is expensive, and some parts of Earth receive more sunlight than others

See "Continuation of Answers" at the end of the chapter.

Chapter Resource File

• **Cross-Disciplinary Worksheet**
Connection to Language Arts— The Meaning of Efficiency GENERAL

• **Concept Review** GENERAL

• **Section Quiz** BASIC

Science and the Consumer

Sun-Warmed Houses

Specially designed houses make use of the sun's heat in two different ways—passive solar heating and active solar heating.

How Does Passive Solar Heating Work?

In passive solar heating, no special devices are used. The house is simply built to take advantage of the sun's energy. For example, passive solar houses have large windows that face south, enabling them to receive a lot of sunlight throughout the day. Many have glass-enclosed fronts called sunspaces, which work to trap solar energy like glass in a greenhouse. During the winter, the energy that enters the sunspace keeps the room comfortably warm during the day. The floor is made of tiles that absorb heat and then radiate it out into the room throughout the evening.

How Is Energy Conserved in Passive Solar Heating?

The rest of the house still has to be heated in winter, but heating costs are generally much lower than they ordinarily would be. Some of the energy that would normally escape through an outside wall is kept inside because the sunspace acts as a good insulator.

Homes that take advantage of solar heating generally have lower electric bills.

On the north side of passive solar houses, there are only small windows in order to reduce energy loss. Also, the walls are built to be good insulators. Once the house is warm, its walls and furnishings act as a large "heat store," keeping the warmth inside.

How Does Active Solar Heating Work?

In active solar heating, houses use solar heaters—active solar-energy devices—to heat water or air. In solar heaters, a flat-plate collector is placed on the roof to gather sunlight. The collector is a flat box covered with glass or plastic. The bottom of the box is often painted black because dark colors absorb radiation better than light colors do. The collector gathers and traps heat, warming the air or water inside. The heated air or water flows into an insulated storage tank. Electric pumps or fans circulate the heated air or water throughout the house. Because they can store solar energy, active solar heating systems work 24 hours a day. However, during long periods of cloudy days, backup heating systems are necessary.

Your Choice

1. **Making Decisions** A house that uses fossil fuels for energy might cost its owners $1000 in heating expenses during the winter. An energy-efficient house could reduce this cost by 50 percent, but it might cost $10 000 more to build. Do you think the energy savings are worth the added construction cost? Explain your reasoning.

2. **Critical Thinking** What problem might people who live in houses with sunspaces have in the summer? How could they solve this problem?

internet connect

www.scilinks.org
Topic: Solar-heated Homes
SciLinks code: HK4159

SCiLINKS. Maintained by the National Science Teachers Association

790

Did You Know ?

Solar Power In 2000, solar cells produced around 280 megawatts (MW) of power worldwide. Of this, the U.S. used about 25 MW. Countries which rely more heavily on solar power include Japan and Germany. In Japan, about 5% of new homes have solar cells on their roofs.

Pollution and Recycling

INTEGRATING TECHNOLOGY and Society

OBJECTIVES

▶ **Identify** several pollutants caused by fossil fuel use.
▶ **Compare** the economic and environmental impacts of using various energy sources.
▶ **Describe** types of pollution in air, in water, and on land.
▶ **Identify** ways to reduce, reuse, and recycle.

▶ **KEY TERMS**
pollution
global warming
eutrophication
recycling

Think about the items in your classroom. You can probably identify ordinary items such as desks, lights, chalk, paper, pencils, backpacks, books, doors, and windows. Your classroom contains many products, all made from natural resources.

Making each product required energy. And the manufacture of nearly all of the products caused some kind of **pollution.** Whenever natural resources generate energy or become products, other things are usually made in the process. If these things are not used, they may be thrown out and cause pollution. Pollution is the contamination of the air, the water, or the soil, as shown in *Figure 22.*

▶ **pollution** an undesirable change in the natural environment that is caused by the introduction of substances that are harmful to living organisms, or by excessive wastes, heat, noise, or radiation

What Causes Pollution?

When you think of pollution, you may think of litter, such as that cluttering the water in *Figure 22A*. Or you may think of smog, the clouds of dust, smoke, and chemicals shown in *Figure 22B*.

Pollution can be as invisible as a colorless, odorless gas or as obvious as bad-smelling trash left by the side of the road. Most forms of pollution have several common features. Understanding these features will help you make better choices.

Figure 22

A Trash polluted the water off the coast of Oahu, Hawaii.
B Contaminants polluted the air in Mexico City.

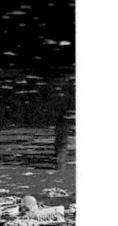

791

SOCIAL STUDIES
CONNECTION — ADVANCED

Before the Environmental Protection Agency (EPA) was proposed by President Nixon and approved by Congress in 1970, most environmental regulations were at the state and local levels. Have students research one example of federal legislation passed after 1970 to limit pollution, such as the Clean Air Act, the Clean Water Act, or the Pollution Prevention Act. Ask students to prepare a

report detailing the content of the legislation and the environmental conditions before and after the legislation was enacted. Other students may wish to report on legislation that is currently under consideration by Congress. Students with Internet access can begin their research at this EPA website: http://www.epa.gov/epahome/lawregs.htm
LS Verbal

Focus

Overview

Before beginning this section, review with your students the Objectives listed in the Student Edition. This section compares the economic and environmental impacts of different energy sources. Students learn about pollution in air, in water, and on land, including the pollution caused by the use of fossil fuels. The section concludes with a discussion of three ways to control pollution: reduce, reuse, and recycle.

⊙ Bellringer

Use the Bellringer transparency to prepare students for this section.

Motivate

Opening Discussion — GENERAL

Ask students to think of examples of pollution in your area. For each example, ask if any efforts are being made to control the source of the pollution. (You might want to research this topic before introducing the subject to the class.) For example, if you live in a large city, smog caused by heavy traffic might be a problem in the summer. To counter this, many cities promote the use of carpooling and city buses on high-ozone days. If students are unaware of pollution problems and solution attempts in their area, encourage them to research the subject. **LS** Verbal

Chapter Resource File

• **Lesson Plan**

Transparencies

TT Bellringer

Teach

Reading Strategy — BASIC

K-W-L Have students list what they know or what they think they know about the topic of pollution. After they have read the relevant portions of the text and studied the topic, have them look at their lists and write down what they have learned about the subject. Also have them write down any new questions they may have after reading the section. **LS Verbal**

SKILL BUILDER — BASIC

Interpreting Visuals Have students study **Table 1.** Students should compare and contrast the different pollutants, including their descriptions, their primary sources, and their effects. Lead students in a discussion about some of the similarities and differences among pollutants. **LS Logical**

Teaching Tip

Carbon Monoxide Carbon monoxide is a difficult pollutant to detect because it is odorless and colorless. Ask students if they have carbon monoxide detectors in their homes. Bring a detector in to show your students, and explain the benefits of having one.

Table 1 Air Pollutants

Pollutant	Description	Primary Sources	Effects
Carbon monoxide	CO is an odorless, colorless, poisonous gas.	It is produced by the incomplete burning of fossil fuels. Cars, trucks, buses, small engines, and some industrial processes are the major sources of CO.	◆ Interferes with the blood's ability to carry oxygen ◆ Slows reflexes and causes drowsiness ◆ Can cause death in high concentrations ◆ Can cause headaches and stress on the heart ◆ Can hamper the growth and development of the fetus
Nitrogen oxides (NO_x)	When combustion (burning) temperatures are greater than 538°C (1000°F), nitrogen and oxygen combine to form nitrogen oxides.	NO_x compounds come from the burning of fuels in vehicles, power plants, and industrial boilers.	◆ Can make the body vulnerable to respiratory infection, lung disease, and possibly cancer ◆ Contribute to the brownish haze often seen over congested areas and to acid rain ◆ Can cause metal corrosion and the fading and deterioration of fabrics
Sulfur dioxide (SO_2)	SO_2 is produced by chemical interactions between sulfur and oxygen.	SO_2 comes from the burning of fossil fuels. It is released from refineries, smelters, paper mills, and chemical plants.	◆ Contributes to acid rain ◆ Can harm plant life and irritate the respiratory systems of humans and animals
Volatile organic compounds (VOCs)	VOCs are organic chemicals that vaporize readily and produce toxic fumes. They include gasoline and paint thinner.	Cars are a major source of VOCs. They also come from solvents, paints, glues, and burning fuels.	◆ Contribute to the formation of smog ◆ Cause serious health problems, such as cancer ◆ May harm plants
Particulate matter (particulates or PM)	Particulates are tiny particles of liquid or solid matter. Some examples are smoke, dust, and acid droplets.	Construction, agriculture, forestry, and fires produce particulates. Industrial processes and motor vehicles that burn fossil fuels also produce particulates.	◆ Form clouds that reduce visibility and cause a variety of respiratory problems ◆ Are linked to cancer ◆ Corrode metals, erode buildings and sculptures, and soil fabrics

792

INCLUSION Strategies — BASIC

• *Developmentally Delayed* • *Learning Disabled*

Ask students to tell a story about a town or city with a pollution problem involving air pollution, litter, or water pollution. The students may wish to include a plan of how the people in the city began a clean-up effort. Be sure the story includes a beginning, middle, and end to show the sequence of events. The completed story may be written, drawn in a series of pictures, or dictated into a tape recorder. **LS Verbal**

Alternative Assessment — ADVANCED

Carbon Dioxide Levels For hundreds of thousands of years, atmospheric carbon dioxide levels have risen and fallen, with ice ages and interglacial periods being closely related to the changing levels. In the past 100 years, carbon dioxide levels have been rising quickly and are now exceeding the highest levels that have been measured previously. Does this mean that Earth is entering a warm phase? Have students conduct research to determine the answer. **LS Verbal**

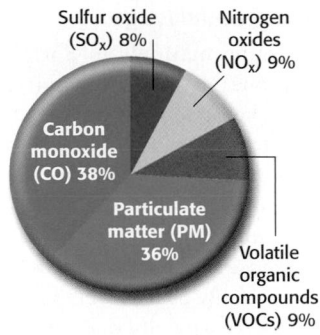

Sulfur oxide (SO$_x$) 8%
Nitrogen oxides (NO$_x$) 9%
Carbon monoxide (CO) 38%
Particulate matter (PM) 36%
Volatile organic compounds (VOCs) 9%

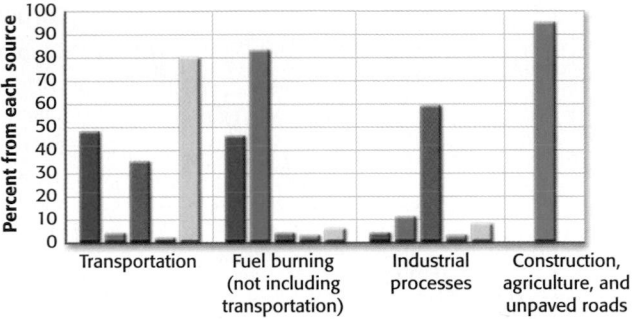

Figure 23

Carbon monoxide, produced mainly from the burning of fossil fuels, is the major air pollutant in the United States.

Some pollution has natural causes

Pollution can be caused by natural processes. For example, following an explosive volcanic eruption, dust and ash can be spread throughout the air. This can make it hard for some people to breathe. The dust and ash can also cover leaves of trees and plants, preventing them from absorbing sunlight.

Manmade pollution is more common

Most pollution is caused by human activities. Many chemical reactions can be used to make new materials or to release energy. But most chemical reactions produce two or more products. If the other products are not reused properly or recycled, they can add to the pollution problem.

Air Pollution

Air pollution comes in many forms, from individual molecules to clumps of dust and other matter, called particulates. *Figure 23* shows major air pollutants and sources of air pollutants in the United States. *Table 1* describes these different forms of pollution.

Combustion of fuels produces most air pollution

As you have learned, most of the energy we use to drive cars, heat and light buildings, and power machinery comes from the burning of fossil fuels. The burning process is known as combustion.

During combustion, the fuel, which contains carbon and hydrogen, reacts with oxygen to release energy. Along with this desirable product, two other products are formed: carbon dioxide gas and water vapor. These combustion products escape into the air as invisible gases. The reaction for burning methane from natural gas is shown below.

$$CH_4 + 2O_2 \longrightarrow CO_2 + 2H_2O$$

internet connect

www.scilinks.org
Topic: Pollution
SciLinks code: HK4107

SCI LINKS. Maintained by the National Science Teachers Association

Chapter 23 • Using Natural Resources   **793**

Interpreting Visuals Have students examine **Figure 24.** Levels of carbon dioxide in the atmosphere have been increasing for several decades. Ask students for hypotheses about future levels of carbon dioxide in the atmosphere; have them justify their answers. (Levels will continue to rise because of increasing numbers of people on the planet.) **LS Visual**

Teaching Tip — BASIC

Global Warming Write the phrase *global warming* on the board, and have students list all words, phrases, and ideas that they associate with the phrase. Begin a discussion in which students inquire about the listed ideas. At the end of the discussion, have students make notes of everything they remember from the discussion, then have them look over their notes to see what they know about the topic based on experience and the discussion. **LS Interpersonal**

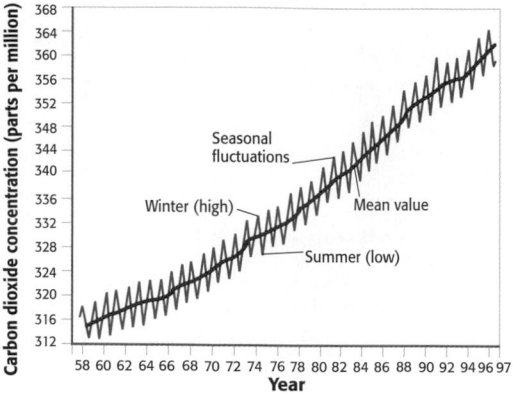

Figure 24

This graph shows an increase in atmospheric carbon dioxide from 1958 to 1997.

▶ **global warming** a gradual increase in the average global temperatures that is due to a higher concentration of gases such as carbon dioxide in the atmosphere

Figure 25

This coal-burning power plant provides power, but it also produces sulfur dioxide and nitrogen dioxide.

Carbon dioxide is a greenhouse gas

Carbon dioxide is found naturally in Earth's atmosphere. It is one of the greenhouse gases, which help keep temperatures on Earth balanced. Just as the glass of a greenhouse garden traps radiation that keeps the inside warm, greenhouse gases in the atmosphere trap radiation that keeps Earth warm. Without the greenhouse effect, temperatures on Earth would be roughly 33°C lower.

Humans release about twenty-one billion tons of carbon dioxide into the air every year. **Figure 24** shows how the amount of atmospheric carbon dioxide has changed since 1958. Some scientists hypothesize that by 2100 the amount of carbon dioxide in the atmosphere will be twice its level in 1880. Some records indicate that the average temperature of Earth is already showing a small increase. In the past 100 years, the average temperature in the United States has increased by about 0.7°C. It is estimated that if the level of atmospheric carbon dioxide doubles, global temperatures would increase by about 0.7°C. This **global warming** may not sound like a lot, but it could drastically affect Earth's climate. Weather patterns could change, bringing droughts to some areas and floods to others. Other scientists point out that ice ages and warming periods occurred in the past without large human sources of CO_2.

Combustion releases other pollutants

The burning of fossil fuels also releases sulfur dioxide and nitrogen dioxide. Once released, these gases react with other atmospheric gases and with water. Chemical reactions like these can make rain, sleet, or snow acidic. Normal precipitation is slightly acidic, with a pH of roughly 5.6. Acid precipitation typically has a pH of between 4 and 5—a very great difference.

Acid precipitation can harm or even kill aquatic life. It can also leach out nutrients from soil and damage large areas of forests. In addition, it can corrode metals and damage buildings by eroding stonework.

794

Did You Know?

"Soda-Pop" Rain In highly industrialized regions, acid rain can be extreme. The northeastern United States sometimes gets "soda-pop" rain with a pH near 4.0, similar to the pH of soft drinks.

Alternative Assessment — ADVANCED

Greenhouse Gases Have students conduct research to determine the distribution of the human sources of greenhouse gases in our environment, either in the United States or worldwide. For example, what percent comes from fossil-fuel power plants, and what percent comes from automobiles? Ask students to display their results in the form of a pie chart. **LS Logical**

Air pollution can cause breathing problems

When nitrogen oxide compounds in car exhaust react with sunlight, they can produce a cloud of chemicals called photochemical smog. The result is a brown haze that can make eyes sting and cause severe headaches and breathing difficulties.

Ozone is one of the harmful chemicals in photochemical smog. High up in the atmosphere, ozone blocks harmful ultraviolet radiation. Close to Earth's surface, however, ozone is a pollutant. It can cause problems for people who suffer from asthma or other conditions affecting the throat and lungs.

Photochemical smog is most common in sunny, densely populated cities, such as Los Angeles and Tokyo. In Tokyo, many people wear masks to protect themselves from the polluted air, and companies supply fresh-air dispensers to their employees. The smog also damages plants. Decreased yields of citrus fruits, such as oranges and lemons, may be linked to photochemical smog in areas not far from Los Angeles. To combat the problem, many cities have made concerted efforts to expand public transportation systems and to encourage people to carpool. These efforts help reduce the number of vehicles on the road and thus the amount of pollutants released into the air.

Water Pollution

All living things need water to survive. In fact, the bodies of most organisms, including humans, are made up mostly of water. Many people believe that water is our most valuable natural resource. Unpolluted water is even more important to aquatic organisms, which spend their entire life in a liquid environment.

On July 6, 1988, a load of aluminum sulfate was accidentally tipped into the water supply in Camelford, England. Before the problem was discovered, people became ill.

Accidents are responsible for some water pollution but not all. Most water pollution can be traced to industrial waste, agricultural fertilizers, and everyday human activities. A bucket of dirty, soapy water dumped down a kitchen drain can eventually make its way into the water supply. Flushing toilets, washing cars, and pouring chemicals down drains are actions that can contribute to water pollution.

In many countries, water is cleaned at water-purification plants before it is piped to consumers. But because many chemicals dissolve easily in water, it's difficult to remove all of the impurities from the water.

Quick ACTIVITY

Observing Air Pollution

1. Cut off a piece of masking tape about 8 cm long.
2. Place the sticky side of the tape against an outside wall, and press gently.
3. Remove the tape, and hold it against a sheet of white paper.
4. Did the tape pick up dust? If so, what might be the source of the dust?
5. Repeat the experiment on other walls in different places, and compare the amounts of dust observed.
6. Suggest reasons why some walls appear to have more dust than others.

Figure 26

The plastic six-pack ring around the neck of this herring gull is one example of how water pollution can affect the environment.

795

Demonstration
Carbon Dioxide Analysis
(Time: About 15 minutes)
Materials:
- bromothymol blue
- balloons (2)
- twist ties (4)
- straws (2)
- beakers (2)

Safety Caution: *Do not inhale gas from exhaust pipe of the vehicle.*

You can use bromothymol blue (BTB) to compare the amounts of carbon dioxide present in gasses. With the engine on, put a balloon around the small end of a funnel, and place the funnel over the tailpipe. Let the balloon fill, and close it with a twist tie. Exhale into another balloon. Close the balloon with a twist tie.

Insert a straw into the mouth of each balloon, and use a twist tie to tighten the balloon around the straw. (When the first twist tie is removed, the gas should come out of the straw.) Put 100 mL of BTB into two small beakers, and slowly introduce the samples of gas into each beaker. The sample that is more yellow contains a higher concentration of carbon dioxide.

Quick ACTIVITY

Observing Air Pollution
Materials:
- masking tape

Teacher's Notes: Before conducting this Quick Activity, brainstorm with your students:

- Where are the greatest amount of air pollutants?
- What causes air pollution?

Have students work where the differences in results will be greatest (near a bus stop, a chalkboard, an office with air conditioning, etc.)

LS Kinesthetic

HEALTH

CONNECTION — **GENERAL**

Smog and ozone are air pollutants that make it difficult to breathe. People with respiratory problems such as asthma and bronchitis are severely affected by air pollution. Have students research several respiratory-system conditions and explain how these conditions are affected by air pollution. **LS** Verbal

---

Chapter Resource File

- **Quick Activity Datasheet** Observing Air Pollution **GENERAL**

Reading Organizer Have students read the section on the different types of pollution (air, water, and land), and have them organize the information presented in the form of a table. The top heading should read *Types of Pollution* with three columns: *Air, Water,* and *Land.* There should be three rows across reading *Causes, Sources,* and *Effects.* Have students fill in the table with the appropriate information from the section.
LS Verbal

SKILL BUILDER — BASIC

Interpreting Visuals Have students examine **Figure 27,** and describe what the algal bloom looks like. Ask students what life might be like for a plant that lives on the bottom of a pond underneath an algal bloom. (The algal bloom will prevent the plant from receiving the proper amount of sunlight that it needs, and the plant will die.) **LS Visual**

Figure 27
This algal bloom is the result of an abundance of nitrates in the water.

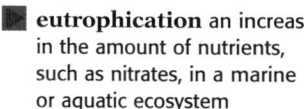
eutrophication an increase in the amount of nutrients, such as nitrates, in a marine or aquatic ecosystem

internet connect

www.scilinks.org
Topic: Solutions to Pollution Problems
SciLinks code: HK4130

SciLINKS Maintained by the National Science Teachers Association

796

Pesticides and fertilizers often end up polluting water

Most modern farms use chemical fertilizers. These fertilizers can get washed away by rain and end up in streams, rivers, lakes, ponds, or oceans. The fertilizers may contain nitrate ions, which encourage the growth of bacteria and algae.

Fed on the nitrates, the bacteria and algae grow so fast that they use up most of the oxygen in the water. The result is an algal bloom, such as the one shown in *Figure 27.* Fish and other aquatic wildlife suffer from the reduced oxygen and die. This process, known as **eutrophication,** is made worse if hot water from power stations or factories is discharged into the river or lake. The extra warmth makes the bacteria and algae multiply even faster.

Animals on land and in water can be affected by another group of chemicals that eventually make their way into bodies of water. These chemicals, called pesticides, are used to control crop-damaging pests.

Like fertilizers, pesticides can be washed by rain into streams, rivers, lakes, or ponds. There they are ingested by fish and other aquatic animals. As larger animals eat the fish, the chemicals are passed along the food chain. In the 1970s, a pesticide called DDT was widely used to control mosquitoes and other insects, leading to the eradication of malaria in the United States. DDT caused the eggs of fish-eating pelicans and other fish-eating birds to become thin and fragile. The pelicans nearly became extinct before the use of DDT was banned in the United States.

Alternative Assessment — ADVANCED

Aquatic Plants and Fertilizers

Materials:

• *Elodea* (or a similar aquatic plant)
• two small aquaria (or plastic containers)
• pond water
• fertilizer

Ask your students to predict what will happen if you take two equivalent samples of *Elodea* and expose them to similar amounts of sunlight, but give only one of the samples a daily amount of fertilizer. Ask students for hypotheses, and then conduct the experiment, comparing the results at the end of a week.

Challenge students to design an experiment to test the hypothesis: What effect does warm water have on *Elodea*? (Maintain one container at room temperature and the other at an elevated temperature. Compare the results at the end of the week or keep the experiment running longer to see more definitive results.)
LS Logical

Pollution on Land

Pollution that affects our land has many sources. In some cases, the source is obvious, as when trash is dumped illegally by a roadside. In other cases, the source is not so obvious. For example, dirt near many highways contains an unusually high amount of lead. This lead originally came from car exhaust. The lead was part of a compound added to gasoline to help car engines run more smoothly. Leaded gasoline was banned in the 1970s, but there are still greater amounts of lead in soil near busy highways.

Contaminants in soil are hard to remove

A common type of land-based pollution occurs when hazardous chemicals soak into the soil. For example, in 1983, the entire town of Times Beach, Missouri, was bought by the U.S. Environmental Protection Agency. The town's soil was contaminated by a highly toxic chemical compound called dioxin.

Exposure to dioxin, a chemical produced in the paper-making process, had been linked to an increased risk of cancer. The soil in the town had become contaminated because the waste oil used to keep the dust on the town's roads down contained small amounts of dioxin. After the roads were repeatedly sprayed over several years, the dioxin levels became very high. This resulted in the deaths of some livestock and other animals and it also adversely affected the health of some of the town's residents.

Dioxin, like many land-based pollutants, does not dissolve well in water. It does not break down easily. It is therefore very difficult to remove it from the soil. The EPA bought the town because it was less expensive than cleaning the entire town's soil.

Quick Lab

How can oil spills be cleaned up?

Materials ✓ cleaning materials ✓ cooking oil
✓ cold water ✓ rectangular baking pan

1. Fill the pan halfway with cold water.
2. Pour a small amount of cooking oil into the water.
3. Try to clean up the "oil spill" using at least four different cleaning materials.

Analysis

1. Evaluate the effectiveness of each material. Which worked best? Explain why.
2. Did any of the materials "pollute" the water with particles or residue? How might you clean up this pollution?

797

Did You Know ?

Phasing Out Leaded Gasoline Although leaded gasoline is banned in the United States, it is still used in some countries. About 22% of the gasoline sold worldwide contains lead. Currently, 36 countries have banned leaded gasoline, and 19 more countries are expected to do so by the year 2005. This will bring the total use of leaded gasoline to 16% or less by 2005.

Quick Lab

How can oil spills be cleaned up?

Students may benefit from wearing plastic gloves if they are available. This should prevent the classroom and materials from getting too oily. Ask students what other materials might be tested to help clean up the spill.

Answers to Analysis

1. Answers may vary. Students should compare the effectiveness of all materials used and suggest reasons for the differences they observed.
2. Answers may vary. Accept all reasonable responses.

Teaching Tip

Lead as a Health Hazard

Although it has long been known that exposure to high doses of lead is extremely dangerous, the effects of low-level exposure were not understood until recently. Throughout much of the twentieth century, lead was used in the U.S. as an additive in gasoline, in paint, and even in food cans. Today, scientists think that even small doses of lead can be dangerous, especially to young children.

Leaded gasoline and lead-based paint have been banned in the U.S. Since lead does not break down over time, most of this lead is still present in our soil, air, water, and in the bodies of living organisms. Lead-based paint can sometimes be found in homes built before 1978. This poses a health hazard for young children who may inhale or ingest lead from peeling or chipping paint.

K-W-L Have students list what they know or what they think they know about the topic of reducing pollution and the potential costs involved. After they have read the relevant portions of the text and studied the topic, have them look at their lists and write down what they have learned about the subject. Also have them write down any new questions they may have after reading the section. **LS Verbal**

INTEGRATING

BIOLOGY Explain to students the difference between the words *aerobes* and *anaerobes*. An aerobe is a microorganism that uses oxygen for decomposition, while an anaerobe is a microorganism that does not use oxygen for decomposition. Noting the presence of the prefix will help students understand the difference.

Figure 28
Many landfills are closing because of a lack of space.

INTEGRATING

BIOLOGY
When trash is buried at a landfill, microorganisms called aerobes use oxygen in the dirt to slowly decompose food, paper, and other biodegradable garbage. This process causes soil temperatures to rise, killing off the aerobes.

Anaerobes, microorganisms that thrive in oxygen-free environments, continue the long task of decomposing the refuse. But generally, anaerobes break down trash even more slowly than aerobes do. Nonbiodegradable trash, such as plastics, glass, and metals, stays essentially intact in landfills.

Landfill space is running out

Even when trash is taken to a landfill, like the one shown in *Figure 28,* and disposed of legally, it can still cause pollution. Each time an item is placed in a landfill, there is less space remaining for other materials. In some regions of the United States, landfills are closing because they are full. Few new landfills are opening.

Currently, each person in the United States throws away almost a half ton of garbage every year. If current trends continue, parts of the United States will soon run out of landfill space.

Reducing Pollution

The are many ways to reduce or limit pollution. Government regulation is one way. In the United States there are several laws that encourage clean water supplies and discourage the pollution of air, soil, and water. Countries may also work together to combat the problem. For instance, in December 1997, international representatives met in Kyoto, Japan, to negotiate an agreement to reduce greenhouse-gas emissions.

Choosing alternatives often involves trade-offs

Even greater improvements in the pollution problem come when individuals, communities, and companies make careful choices. For example, to reduce the air pollution caused by the burning of fossil fuels, people can make an effort to use alternative energy sources. Individuals can make a difference by conserving energy. However, even nonpolluting sources of energy, such as wind, solar, and hydroelectric power, require large amounts of land and are potentially disruptive to ecosystems.

Small-scale sources of energy, such as disposable batteries, have an environmental impact too. These batteries contain mercury and other potentially toxic chemicals. In the United States alone, more than 2 billion disposable batteries are discarded every year. The toxic chemicals can leak into the ground, polluting water supplies and soil.

Reducing the use of energy and products can cut down on pollution

Because of the trade-offs involved, many people believe that the best solution to the problem of pollution is to reduce our overall consumption. If less energy is used, less pollution is generated. Turning off lights and lowering thermostats are two simple ways to conserve energy. Carpooling or buying a car with higher-than-average gas mileage is another way to conserve energy.

798

REAL-LIFE
CONNECTION — GENERAL

Landfills Ask students to determine where their trash goes. Does it go to a landfill, and if so, how far away is the landfill from where students live? **LS Logical**

Alternative Assessment — GENERAL

Composting Landfills are so tightly packed with trash that materials that would ordinarily degrade—such as kitchen scraps like banana peels, eggshells, and coffee grounds—do not break down. One alternative is to compost biodegradable materials, rather than throwing them in the trash. Have students conduct research to learn more about composting. Ask them to display their results in a brochure. Their brochures should explain some of the advantages of composting and describe how individuals can make their own compost piles. **LS Visual**

Recycling Codes: How Are Plastics Sorted?

More than half of the states in the United States have enacted laws that require plastic products to be labeled with numerical codes that identify the type of plastic used in them. These codes are shown in the table below. Used plastic products can be sorted by these codes and properly recycled or processed. Only codes 1 and 2 are widely accepted for recycling. Codes 3 and 6 are rarely recycled. Knowing what the numerical codes mean will give you an idea of how successfully a given plastic product can be recycled. This may affect your decision to buy or not buy particular items.

Your Choice

1. **Making Decisions** Find out what types of plastic are recycled in your area. With this in mind, will you change your buying habits?

2. **Critical Thinking** How can consumers influence what types of plastic are recycled?

internet connect

www.scilinks.org
Topic: Recycling Plastics SciLinks code: HK4117

SCiLINKS. Maintained by the National Science Teachers Association

Recycling Codes for Plastic Products

Recycling code	Type of plastic	Physical properties	Example	Uses for recycled products
1	Polyethylene terephthalate (PET)	Tough, rigid; can be a fiber or a plastic; solvent resistant; sinks in water	Soda bottles, clothing, electrical insulation, automobile parts	Backpacks, sleeping bags, carpet, new bottles, clothing
2	High density polyethylene (HDPE)	Rough surface; stiff plastic; resistant to cracking	Milk containers, bleach bottles, toys, grocery bags	Furniture, toys, trash cans, picnic tables, park benches, fences
3	Polyvinyl chloride (PVC)	Elastomer or flexible plastic; tough; poor crystallization; unstable to light or heat; sinks in H_2O	Pipe, vinyl siding, automobile parts, clear bottles for cooking oil	Toys, playground equipment
4	Low density polyethylene (LDPE)	Moderately crystalline, flexible plastic; solvent resistant; floats on water	Trash bags, dry-cleaning bags, frozen-food and meat packaging	Trash cans, trash bags, compost containers
5	Polypropylene (PP)	Rigid, very strong; fiber or flexible plastic; lightweight; heat-and stress-resistant	Heat-proof containers, rope, appliance parts, outdoor carpet, luggage, diapers, automobile parts	Brooms, brushes, ice scrapers, battery cable, insulation, rope
6	Polystyrene (P/S, PS)	Somewhat brittle, rigid plastic; resistant to acids and bases but not to organic solvents; sinks in H_2O, unless it is a form	Fast-food containers, toys, videotape reels, electrical insulation, plastic utensils, disposable drinking cups, CD jewel cases	Insulated clothing, egg cartons, thermal insulation

799

Did You Know?

Recycling on the Rise In the United States, 1,511 million pounds of plastic bottles were recycled in 2000. This is a 368% increase over the amount recycled in 1990. Of this total, 758 million pounds were PET plastic and 745 million pounds were HDPE plastic.

REAL-LIFE
CONNECTION — BASIC

Water Conservation For one week, have students keep track of how they use water around their home. At the end of the week, have students present their results and brainstorm ways in which water can be conserved.
LS Intrapersonal

Science and the Consumer

Recycling Codes: How Are Plastics Sorted? Bring in different examples of plastics to display to your class, or ask your class to do the same. Ask your students to come up with hypotheses as to why Plastic Codes 3 and 6 are rarely recycled. (too rigid to break down, too expensive.)

To demonstrate how pervasive plastics are in our lives, ask students to make a list of all of the different types of plastics (and their codes) that are in their homes.

Answers to Your Choice

1. Answers may vary.
2. Answers may vary. Accept all well thought-out responses. Students may say that consumers can affect what plastics are recycled by choosing whether to buy products made of recycled plastics.

Chapter Resource File

• **Cross-Disciplinary Worksheet** Integrating Biology— Composting **ADVANCED**

• **Cross-Disciplinary Worksheet** Science and the Consumer— Recycling Plastics **ADVANCED**

Quiz ──────────── BASIC

1. What is the cause of most air pollution? (combustion of fuels)

2. All pollution is manmade. True or false? (false)

3. What is one way to limit pollution? (accept any of the following: choose alternatives, reduce, reuse, recycle)

LS Logical

Figure 29

This recycling plant in New York City helps reduce the amount of waste going into landfills.

▶ **recycling** the process of recovering valuable or useful materials from waste or scrap, or reusing some items

Some people conserve water by reusing rinse water from dishes and laundry to water gardens and lawns. This way, they use less water, so less energy is required to purify it and pump it to them. In addition, the water they use does not enter the sewer system. This reduces the amount of energy used by the water-treatment plant.

Recycling is the final way to prevent pollution

After you have tried to reduce how much you use and are successfully reusing it as much as possible, there's still one more thing you can do. When something is worn out and is no longer useful, instead of throwing it away, you can try **recycling.**

Recycling allows materials to be used again to make other products rather than being thrown away. **Figure 29** shows a recycling plant in New York City. Such plants commonly recycle paper products such as cardboard and newspapers; metal products such as copper, aluminum, and tin; and plastics such as detergent bottles and shopping bags.

Recycling these materials can make a huge difference in the amount of waste that ends up in a landfill. Paper products alone take up 41 percent of landfill space. Yet we currently recycle less than 30 percent of our paper.

SECTION 3 REVIEW

SUMMARY

▶ Pollution can be caused by natural events and by human activities.

▶ Acid rain is caused when sulfur dioxide and nitrogen oxides react with moisture in the air.

▶ Government regulation and recycling can lessen the pollution of air, water, and soil.

▶ Making an effort to reduce the energy products you use can cut down on pollution.

1. **List** an activity related to home, work, and growing food that can lead to water pollution.

2. **Distinguish** between the greenhouse effect and global warming.

3. **Explain** how riding a bike can reduce photochemical smog.

4. **Define** *eutrophication*. Explain how it occurs.

5. **Name** an alternative energy source, and describe one benefit and one drawback.

6. **Analyze** all of the possible polluting steps in the making of a pencil. Write a paragraph describing each step. *WRITING SKILL*

7. **Critical Thinking** Do you think recycling should be mandatory? Explain your reasoning.

Answers to Section 3 Review

1. Answers may vary. Sample response: using pesticides and fertilizers on the lawn or garden

2. The greenhouse effect is the natural warming of Earth by greenhouse gases, while global warming is an increase in Earth's temperature due to an increase in greenhouse gases.

3. Photochemical smog results from car exhaust. If you are riding a bike, then you are not producing any exhaust that can lead to the formation of photochemical smog.

4. Eutrophication is an increased amount of nutrients in an aquatic environment. It occurs when excess fertilizers are washed into streams, rivers, lakes, or ponds.

5. Sample answer: solar power; A benefit of solar power is that it does not directly produce pollutants. A drawback to solar power is that harnessing it requires large amounts of land and establishing that land may disrupt ecosystems.

6. The resources (wood, graphite, paint, rubber, and metal) must be obtained, shaped, and refined in the proper fashion.

7. Answers may vary.

Study Skills

Researching Information

Researching information is essential to learning more, including the latest information, about a topic. It is therefore important to know how to use resources, and to be able to decide for yourself if the information you receive is reliable.

1 **Begin research at your library using general reference materials.**

> We'll research the topic "global warming." Begin by using general references, such as a recent edition of *The New Encyclopaedia Britannica,* which may be found at most libraries. Start with a broader topic, such as "Climate and Weather," to find information specific to your research.

2 **Review articles written in periodicals.**

> Using *The Reader's Guide to Periodical Literature,* which may be found at your library, look up the topic for a given year. Note whether it is cross-referenced to another topic, and if there are subtopics listed underneath. For instance, if you want to know about the relationship between solar activity and climate, you would look under "Climate-Solar Relationships," which would refer you to "Sun and Meteorology." The citation "The Sun-Climate Connection, S. L. Baliurus and W. Soon, Sky and Telescope, v 92, p. 38–41, D' 96," indicates that the article may be found in volume 92 (December 1996) of *Sky and Telescope* magazine.

3 **Search the Internet.**

> Because of the wide range of quality and reliability of on-line sites, it is best to use sites maintained by scientific organizations. Among these are the large sites for the National Aeronautic and Space Administration (www.nasa.gov) and the National Oceanic and Atmospheric Administration (www.noaa.gov). Also available for research are websites for scientific periodicals, like *Scientific American* (www.sciam.com) and *Science* (www.scienceonline.org).

Practice

Use information from the chapter and the example above as a starting point to answer the questions below:

1. Research factors that affect the reflection and absorption of sunlight on Earth. Approximately how much solar energy is transferred to Earth's atmosphere through evaporation and condensation of water?

2. Research articles that suggest a connection between changes in solar activity and climate change on Earth. Use these to explain the Maunder Minimum and the "Little" Ice Age.

Teaching Tip

Locating Sources Choose a topic and have students locate information in three text sources (either books, encyclopedia articles, or articles in periodicals) and three online articles from valid sources. This will help them prepare for the Practice exercise.

Answers to Practice

1. The information may be found on p. 439 of the article "Climate and Weather," *The New Encyclopaedia Britannica, Vol. 16;* Encyclopaedia Britannica, Inc., Chicago, 1998, pp. 436–522. Evaporation and condensation account for about a fourth of the solar energy is transferred to the atmosphere.

2. The information may be found in the article "The Sun-Climate Connection," S. L. Baliunas and W. Soon, *Sky and Telescope,* vol. 92 (December 1996), pp. 38–41. The Maunder Minimum was a period during the seventeenth century where sunspot counts (solar activity) were especially low. The "Little" Ice Age describes the exceptionally cold climate of northern Europe during this time. The two are thought to be related.

Understanding Concepts

1. c

2. a

3. b

4. c

5. a

6. a

7. b

8. c

9. b

10. c

11. a

12. c

13. b

14. d

15. Interrelatedness describes how all the elements of an ecosystem function together to keep an ecosystem stable.

Using Vocabulary

16. Hydroelectric power means generating energy from water, which is a renewable resource. These generators must be built on large rivers, and most of the world's large rivers have been dammed already. Geothermal energy is energy derived from heated water within the Earth's crust. It is renewable as well but quite expensive, and is only practical in a few places around the world.

Chapter Resource File

• **Chapter Test** GENERAL

• **Test Item Listing for ExamView®
Test Generator**

Chapter Highlights

Before you begin, review the summaries of key ideas of each section, found at the end of each section. The vocabulary terms are listed on the first page of each section.

UNDERSTANDING CONCEPTS

1. Choose the list that is ordered from largest to smallest.
   a. population, organism, community, ecosystem
   b. organism, population, ecosystem, community
   c. ecosystem, community, population, organism
   d. ecosystem, population, community, organism

2. Starlings are an example of
   a. a non-native species introduced into the U.S.
   b. a climactic ecosystem change.
   c. a species that migrated to the U.S. on their own.
   d. a species threatened by ecosystem change.

3. Hydroelectric power is a(n) _____ energy source.
   a. nonrenewable   **c.** small-scale
   b. renewable   **d.** nuclear

4. An aquarium ecosystem is made up of
   a. cactuses, sagebrush, lizards, snakes, scorpions, and birds.
   b. grassland, termites, wild dogs, hyenas, and antelopes.
   c. water, fish, glass, and water plants.
   d. musk oxen, opposite-leaved saxifrage, wolves, and brown bears.

5. When fossil fuels are burned, they form carbon dioxide and
   a. water.   **c.** hydrocarbons.
   b. oxygen.   **d.** carbohydrates.

6. Fossil fuels are made mostly out of what elements?
   a. carbon and hydrogen
   b. carbon and calcium
   c. calcium and hydrogen
   d. potassium and chlorine

7. Earth's temperature is kept warmer by
   a. the surface of Earth.
   b. the greenhouse effect.
   c. the ozone layer.
   d. wind.

8. In power stations, _____ of the energy input is wasted.
   a. one-third   **c.** two-thirds
   b. one-half   **d.** all

9. Geothermal energy is produced by
   a. wind.
   b. underground reservoirs of steam or hot water.
   c. fast-moving water.
   d. solar radiation.

10. Usable energy is lost each time energy is
   a. gained.   **c.** converted.
   b. released.   **d.** trapped.

11. Which of the following is not true of recycling?
   a. Recycling does not affect the amount of waste that ends up in a landfill.
   b. Recycling reduces litter.
   c. Recycling reduces energy usage.
   d. Recycled materials can be used to make other products.

12. What causes eutrophication?
   a. CFCs   **c.** an increase in
   b. global warming   nitrates
                           d. recycling

13. The most abundant air pollutant is
   a. sulfur dioxides.
   b. carbon monoxide.
   c. nitrogen oxides.
   d. particulate matter.

---

17. The pollutants that come from driving cars are carbon dioxide, sulfur dioxide, and nitrogen dioxide. Sulfur dioxide and nitrogen dioxide can combine with other atmospheric gases to form acid precipitation.

18. The forest will recover as pioneer organisms move into the area. Hardy plants that can live in sub-optimum conditions will move in and establish themselves, which will lead to other communities eventually moving in as habitats return.

19. Fossil fuels are fuels that have formed from the remains of ancient plant and animal life. When those animals and plants were living, they needed energy to live, either by absorbing sunlight or by eating plants and other animals. The stored energy in these plants and animals did not have the chance to release while the organisms were alive. The burning of fossil fuels is actually the release of energy from the sun stored in the remains of ancient plant and animal life.

14. Atomic reactions can produce an alternative source of energy called
 a. fossil fuels.
 b. solar energy.
 c. geothermal energy.
 d. nuclear energy.

USING VOCABULARY

15. Describe how *interrelatedness* is crucial to the concept of ecosystems.

16. Compare hydroelectric power with geothermal energy. List advantages and disadvantages of each.

17. Driving cars releases pollutants into the atmosphere. State how these pollutants can lead to *acid precipitation.*

18. A forest *ecosystem* is wiped out by disease. Using the concept of *succession,* explain how the forest might recover.

19. Most energy on Earth comes from the sun. Describe how the burning of *fossil fuels* releases stored energy from the sun.

20. Is land a *renewable* or *nonrenewable* resource? Explain your reasoning.

21. Name one of the *greenhouse gases* and state its relationship to *global warming.*

22. Describe the process of *eutrophication.* List several sources of nitrates and name several consequences of *eutrophication.*

BUILDING MATH SKILLS

23. Estimating The world population in 1999 was approximately 5.8 billion. That number is expected to double in 40 years. Assuming a steady growth rate, estimate the world population in the years 2019, 2039, 2059, and 2079. Write a paragraph explaining how population changes might affect an ecosystem.

24. Calculating Each person in the United States produces roughly 2 kg (4.4 lb) of garbage per day. How long would it take you to fill a dump truck with a 16 000 lb capacity? If each person recycled one-half of his or her trash, how much less garbage would each person produce in one year?

THINKING CRITICALLY

25. Creative Thinking Green plants use sunlight to convert carbon dioxide and water into food in the process of photosynthesis. Explain how large-scale deforestation might affect global climate.

26. Applying Knowledge A factory is situated along the banks of a river. A large city is located farther upstream. On the outskirts of the city, farmers grow corn and wheat. Researchers are finding dead fish in the river. Write several paragraphs explaining the steps you would take to determine the cause.

WRITING SKILL

27. Creative Thinking In political and environmental circles, the phrase NIMBY stands for "Not In My Backyard." Judging from what you have learned about natural resources and pollution, what do you think this phrase means? Would you apply this sentiment to anything in your life?

28. Creative Thinking "Biological amplification" describes the process of the food chain becoming increasingly toxic as a substance travels upward. This can happen with a substance such as mercury. Why do you think it happens, and what can be done to prevent ill effects of eating contaminated seafood?

29. Applying Knowledge Have you ever had fish as pets? If so, you probably know that if you change the water without taking certain precautions the fish are likely to die. Knowing what you know about balanced ecosystems, why would this happen?

20. Land can be thought of as both a renewable and nonrenewable resource, depending on the practices that go on with the land. Land is a renewable resource because if it is cared for and pollutants do not destroy it, the land can be used for generations without diminishing returns. Land can also be thought of as a nonrenewable resource because it is not in perpetual supply. As people build on, excavate, and pollute land, it becomes unusable for future generations.

21. As a greenhouse gas, carbon dioxide helps keeps temperatures on Earth balanced. However, human activity has resulted in the release of billions of metric tons of non-naturally occurring carbon dioxide into the air each year. This increase in atmospheric carbon dioxide brings about an increase in Earth's temperature called global warming.

22. Eutrophication is the process in which an increase in nitrates and other nutrients in an environment results in a lethal algal bloom. The nitrates can come from chemical fertilizers used on farms. Eutrophication can result in reduced oxygen levels and the consequent death of fish and other aquatic life.

Building Math Skills

23. Answers may vary. One set of reasonable estimates follows.
2019: 9 billion
2039: 12 billion
2059: 18 billion
2079: 23 billion

24. 3600 days; you would produce 1600 lbs per year—but only 800 lbs per year if you recycled one-half of the trash.
$$\frac{16\ 000\ \text{lb}}{4.4\ \text{lb/day}} = 3600\ \text{days}$$
$$365\ \text{days} \times 4.4\ \text{lb/day} = 1600\ \text{lb}$$
$$\frac{1600\ \text{lb}}{2} - 800\ \text{lb}$$

803

Assignment Guide

Section	Questions
1	1, 2, 4, 15, 18, 23, 25, 29, 31, 36, 40
2	3, 5, 6, 8–10, 14, 16, 19, 20, 32, 34, 37, 43, 44
3	7, 11–13, 17, 21, 22, 24, 26–28, 30, 33, 35, 38, 39, 41, 42, 45, 46

Thinking Critically

25. Answers may vary. Cutting down trees could lead to an increase in the amount of carbon dioxide in the atmosphere. Carbon dioxide is a greenhouse gas that may contribute to global warming.

26. Have students make a sketch of the problem to help them visualize the scenario. Students would want to know the following: Are fertilizers and pesticides used on the farm? Is the factory releasing pollution into the river? Is eutrophication a problem in the river?

27. Answers may vary. Students may state that the phrase NIMBY means that people are reluctant to have things that they consider unhealthy (whether rational or not) built near where they live.

28. Answers may vary. Students may state that biological amplification occurs because toxic substances that survive one stage of the food chain will have unpredictable consequences in successive stages. The contamination of seafood can be prevented by reducing mercury levels and other pollution in our oceans and rivers.

29. As aquatic creatures, fish are highly susceptible to changes in their liquid environment, so any change in their ecosystem can create an imbalance, which may have adverse consequences.

30. Answers may vary.

31. A virgin forest has been untouched by human activities such as logging. There are still some virgin forests in the United States.

30. Acquiring and Evaluating Data The United States has a lot of space in comparison to certain other countries, like Japan. Using a search engine on the Internet, research what countries like Switzerland and Japan do instead of using landfills.

31. Applying Knowledge What does the phrase "virgin forest" mean? Using environmental texts from the library, determine whether there are there any virgin forests in the U.S.? What does one look like?

32. Applying Knowledge You have learned about solar powered cars. What other kinds of cars are under development in the United States? What are the government regulations regarding these things?

WRITING SKILL

DEVELOPING LIFE/WORK SKILLS

33. Teaching Others Using a desktop-publishing program, work in groups of three or four students to design a brochure to encourage recycling or energy conservation at your school. As a group, decide beforehand who your audience will be—do you want to reach younger students, your peers, or the school administration? List the benefits and possible drawbacks in your brochure. Explain how costs can be reduced without hindering the school's operation.

COMPUTER SKILL

WRITING SKILL

34. Applying Technology Research recent developments and advancements in solar-power technology. Then write a brief report analyzing the feasibility of the technology. Share your results with the class.

WRITING SKILL

804

35. Allocating Resources Three people live in the same neighborhood and work at the same office. One person spends $20 per week, one spends $25 per week, and one spends $30 per week for gasoline and parking. They work 51 weeks per year. If they formed a car pool, how much could each person save annually? Using a spreadsheet program, create a spreadsheet that will calculate how much each person would save in 5 years. (**Hint:** Assume that each person drives a total of 17 weeks.)

COMPUTER SKILL

36. Interpreting and Communicating Visit a local senior center and interview elderly people who have grown up in your area. Ask them to describe the natural habitats in and around your area as they were 50 years ago. Compare their descriptions with your observations of how your area's environment looks now. Then write a brief paragraph describing the changes that have occurred. Share your interview with your class.

WRITING SKILL

37. Improving Systems Use an atlas to find out the main sources of energy in your state. Based on the information in the atlas, identify alternative energy sources that might be used in different areas. For instance, are some places suitable for wind, hydroelectric, or solar power? Develop an alternative energy plan for your state.

38. Acquiring and Evaluating Data Call your local recycling company and find out where and how recycled materials are processed and reused. Also research how much money your city saves or spends by using recycled materials.

39. Acquiring and Evaluating Data Using an almanac, determine which five states have the most hazardous-waste sites and which five states have the fewest sites. What factors do you think might account for the number of hazardous-waste sites located in a state?

32. Answers may vary. Responses could include electric cars, gasoline-electric hybrids, and fuel-cell cars.

Developing Life/Work Skills

33. Student brochures will vary. Good brochures should give logical and persuasive arguments.

34. Answers may vary. Good reports should explain both pros and cons of technology using solar power.

35. The average annual cost of parking and gas is $1,275. Divided by three, the cost for each driver would be $425. Person A ($20 per week) would save $595 a year, Person B ($25 per week) would save $850 a year, and

Person C ($30 per week) would save $1,105 a year. Students' methods and spreadsheets will vary. Another possible method of calculation shows that Person A would save $680 a year, Person B would save $850 a year, and Person C would save $1,020 a year.

36. Answers may vary.

37. Answers may vary depending on location.

38. Answers may vary depending on location.

39. Population and types of businesses in the state will influence the number of hazardous waste sites in a state.

40. Answers may vary depending on location.

40. Acquiring and Evaluating Data Go to your local library and research any nonnative species that have been introduced into your area. What positive and negative effects have resulted from their arrival?

41. Acquiring and Evaluating Data Research the 1986 Safe Drinking Water Act and the 1987 Clean Water Act. What incentives are in place to encourage clean water supplies?

INTEGRATING CONCEPTS

42. Connection to Health Sunscreens, which protect people from ultraviolet rays, have different sun protection factors (SPFs). For someone who burns after 10 minutes in the sun, a sunscreen with an SPF of 8 would give that person 80 minutes of protection. If the same person wanted to stay in the sun for 2.5 hours without reapplying sunscreen, what SPF should he or she use?

43. Connection to Social Studies Research a particular region of the world other than the area where you live. Find out which energy resource is used to meet the energy needs of this particular region.

44. Connection to Social Studies One of the purported causes of global warming is the burning of fossil fuels. As a group, discuss how building energy-efficient houses might help counteract this effect. What incentives would you suggest to encourage people to make their homes more efficient? How would you judge if they're worth the cost?

45. Connection to History Human-caused environmental pollution has occurred throughout history. Research a historical example of environmental pollution, and write a short essay about it. Also research and include information about a historic conservation effort.

WRITING SKILL

46. Mapping Concepts Copy the unfinished concept map below onto a sheet of paper. Complete the map by writing the correct word or phrase in the lettered boxes.

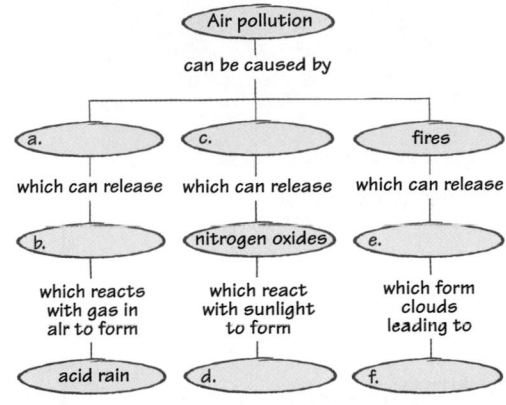

internet connect

www.scilinks.org
Topic: Global Warming SciLinks code: HK4065

SCILINKS Maintained by the National Science Teachers Association

Art Credits: Fig. 1, Robert Hynes/Mendola Artists; Fig. 2, Robert Hynes/Mendola Artists; Fig. 4, Robert Hynes/Mendola Artists; Fig. 8, Ortelius Design; Fig. 11, Leslie Kell; Fig. 12-13, Robert Hynes/Mendola Artists; Fig. 14, Kristy Sprott; Fig. 15, Leslie Kell; Fig. 17, Stephen Durke/Washington Artists; Figs. 19, 21, Uhl Studios, Inc.; Fig. 23-24, Leslie Kell; "Science and the Consumer" (recycling symbol), Uhl Studios, Inc.; "Section Review," Uhl Studios, Inc.

Photo Credits: Chapter Opener photo of blue solar car by William McCoy/Rainbow/Picture Quest; yellow car courtesy California State University, Los Angeles, College of Engineering and Technology; Fig. 3, M. Osf Gibbs/Animals Animals/Earth Scenes; Fig. 4, Tom J. Ulrich/Visuals Unlimited; Fig. 5A, Craig Fugii/©1988 The Seattle Times; Fig. 5B, Raymond Gehman/National Geographic Society Image Collection; Fig. 6(l), Ken M. Johns, The National Audubon Society Collection/Photo Researchers; Fig. 6(c), Glen M. Oliver/Visuals Unlimited; Fig. 6(r), Dr. E. R. Degginger/Color-Pic, Inc.; Fig.7, E. R. Degginger/Animals Animals/Earth Scenes; Fig. 9(trees), Richard Thom/Visuals Unlimited; Fig. 9 (cars), Mark Richards/Photo Edit; Fig. 9(house), Myrleen Ferguson/Photo Edit; Fig. 9 (farm), Peter Dean/Grant Heilman Photography; Fig. 10, Dr. E. R. Degginger/Color-Pic, Inc.; Fig. 14, Bruce Stoddard/Getty Images/FPG International; Fig. 16, Fred Bruemmer/Peter Arnold, Inc.; Fig. 18, Stefan Schott/Panoramic Images, Ltd.; Fig. 19, Telegraph Colour Library/Getty Images/FPG International; Fig. 20, G. R. Roberts Photo Library; Fig. 21, Grant Heilman Photography; "Science and the Consumer," Ulrike Welsch/Photo Edit; Fig. 22A, Warren Bolster/Getty Images/Stone; Fig. 22B, Travelpix/Getty Images/FPG International; Fig. 25, Corbis Images; Fig. 26, Patti Murray/Animals Animals/Earth Scenes; Fig. 27, John Sohlden/Visuals Unlimited; "Quick Lab," Victoria Smith/HRW; Fig. 28-29, Telegraph Colour Library/Getty Images/FPG International; "Science in Action"(fuel cell car), Laurent Gillieron/Keystone/AP/Wide World Photos; (traffic), Andrew Brown; Ecoscene/CORBIS; (plant), Todd Gipstein/CORBIS.

805

41. Incentives put in place to encourage clean water supplies include financial resources to clean up rivers and wells, legal support for citizens wanting to sue polluters, and governmental agencies responsible for regulation.

Integrating Concepts

42. SPF 15

43. Answers may vary.

44. An energy-efficient house conserves energy. As more energy is conserved, fewer fossil fuels are used and fewer amounts of greenhouse gases are emitted into the atmosphere. Incentives might include tax cuts and no- or low-interest loans from government or local utilities. Students should indicate which criteria would be used to judge whether such a program is feasible.

45. Students' research should be thorough and concise.

46. a. combustion of fossil fuels
   b. sulfur dioxide and/or nitrogen dioxide
   c. car exhaust
   d. photochemical smog
   e. particulates
   f. a reduction in visibility and/or respiratory problems

Transparencies

TT Concept Mapping

CHANGING THE FORM OF A FUEL

Teacher's Notes

Time Required 1 lab period

Ratings
```
        1      2      3      4
     EASY                  HARD
```

TEACHER PREPARATION	2
STUDENT SETUP	2
CONCEPT LEVEL	3
CLEANUP	3

Skills Acquired

- Classifying
- Collecting data
- Communicating
- Experimenting
- Identifying/recognizing patterns
- Inferring
- Interpreting
- Measuring
- Organizing and analyzing data

The Scientific Method

In this lab, students will:
- Make Observations
- Analyze the Results
- Draw Conclusions
- Communicate Results

Materials

Fresh wood splints made of pine will produce more condensate than dry wood splints will produce. Wood splints should be cut or broken to a length no more than half of the length of the test tubes used in the distillation to keep students from heating too close to the rubber stopper. Inserting the glass tubing through the rubber stoppers prior to the lab will minimize the risk that a student will force the tubing, break it, and be cut.

Skills Practice Lab

Introduction

Can you use your familiarity with products used in or near the home to help you identify some of the products of destructive distillation?

Objectives

▶ USING SCIENTIFIC METHODS **Observe** the process of destructive distillation.

▶ **Analyze** the amounts of products produced, and try to identify the products.

Materials

2 test tubes
one-hole stopper
two-hole stopper
bent glass tubing with fire-polished ends
20-cm long rubber tubing
gas burner
ringstand and 2 buret clamps
2 widemouth bottles
2 glass plates, 7 × 7 cm square
gas-collecting trough
pieces of wood splints
graduated cylinder
balance

806

Materials, *continued*

Because the destructive distillation leaves residues that are difficult to remove, the test tubes used may need to be replaced after the lab. The volumes of the gas bottles and test tubes can be measured by filling them with water, then pouring this water into a graduated cylinder and reading the level.

Changing the Form of a Fuel

▶ Procedure

Preparing for Your Experiment

Destructive distillation is the process of heating a material such as wood or coal in the absence of air. The material that is driven off as a gas is called volatile matter. When cooled, some of the matter remains as a gas. Much of the matter condenses to form a mixture of liquids. These liquids can be distilled to yield a number of different products. In this investigation, you (or your teacher) will heat wood to temperatures high enough to cause the wood to break down into different components, which you will try to identify.

1. On a clean sheet of paper, make a table like the one shown below.

2. Label your glassware as shown in the illustration on the next page.

3. Using the balance, determine the mass of test tube A. Record the value in your table.

4. Using the graduated cylinder, determine the volume of gas bottles 1 and 2. Record the values in your table.

5. Determine the mass of test tube B. Record the value in your table. Dry test tube B before setting up your equipment.

Data Table
Mass of test tube A (g)
Mass of test tube A with wood (g)
Mass of wood (g)
Mass of test tube A with solid residue (g)
Mass of solid residue (g)
Volume of gas bottle 1 (mL)
Volume of gas bottle 2 (mL)
Volume of gas produced (mL)
Mass of test tube B (g)
Mass of liquid produced (g)
Volume of liquid produced (mL)

Safety Cautions

- Review safety precautions about fire and burn procedures, and caution students about the hazards of using an open flame.
- Tongs must be used for hot test tubes.
- Buret clamps with protective rubber tips should be used if possible.
- Do not allow students to insert glass tubing through the rubber stopper by putting a hand behind the stopper and pushing the glass tube toward the hand.
- Inspect the distillation setup to make sure that the system is not plugged and that the tubing cannot become kinked and allow pressure to build in the system.

Destructive Distillation of Wood

SAFETY CAUTION Wear protective gloves when inserting the glass tubing through the stoppers. Rub glycerin on the tubing and the inside of the stopper holes before pushing the tubing through the stoppers. Rotate the tubing slowly, and push gently. If you have difficulty, ask your teacher for help.

6. Set up the equipment as illustrated below.

7. The gas bottle in the pan should be completely filled with water. Insert the delivery tube into the gas bottle.

SAFETY CAUTION Protect clothing, hair, and eyes when using a gas burner. The gases formed in the destructive distillation of wood are combustible.

8. Fill test tube A about two-thirds full with pieces of wood. Determine the mass of the test tube and the wood. Record the value in your table. Stopper the test tube, connect it, and heat the test tube. Move the gas burner frequently so that the entire mass of the wood is heated.

9. When all the water is driven from the gas bottle, place a glass plate over the mouth of the bottle, and remove the bottle from the pan. Set the gas bottle upright on the table, leaving it covered with the glass plate.

10. Place another water-filled bottle in the pan as before, and reinsert the gas delivery tube. Keep heating until the gas stops coming from test tube A.

▶ Analysis

1. How much gas was produced? How much gas was produced for 1 g of wood?

2. What happens when a burning splint is thrust into the gas?

3. Describe the contents of test tube B. What was the mass of the liquid produced? What was the volume? What about for 1 g of wood?

4. What does the solid material remaining in test tube A look like?

5. How much solid material was left? How much solid material remains for each 1 g of original wood?

6. Using insulated tongs, hold a piece of the solid material in the gas burner flame. How does it burn?

Test Tube A
Test Tube B
Gas Bottle 1

▶ Conclusions

7. Why would you expect charcoal to give off little or no flame when it is burned?

8. Why is this type of distillation called destructive?

9. What do you think the liquids can be used for?

807

Sample Data

Mass of test tube A with wood (g)	30.88	Volume of gas bottle 1 (mL)	515
Mass of test tube A (g)	23.60	Volume of gas bottle 2 (mL)	512
Mass of wood (g)	7.28	Volume of gas produced (mL)	903
Mass of test tube A with solid residue (g)	25.25	Volume of test tube B (mL)	36
Mass of solid residue (g)	1.65	Volume of liquid produced (mL)	4.08

Continuation of Answers

Continuation of Answers from p. 781
Section 1 Review

7. Answers may vary but may include damming waterways for hydro-electric power. Possible solutions may include analyzing the environmental impact of dams before their construction and redesigning or canceling them accordingly.

8. Answers could include the following: people transplant them for gardens; seeds are transported by wind; seeds are unknowingly transported by people; seeds are transported by ocean currents.

Continuation of Answers from p. 789
Section 2 Review

5. As rain rolls downhill, its kinetic energy increases, and as the rain moves through turbines in a hydroelectric plant, it makes the turbines spin. The turbines are connected to generators that produce hydroelectric power.

6. b. Using wind to generate electricity will be more efficient because there are fewer steps involved in the electricity production process; hence, there are fewer places for energy to be lost to the system.

7. No energy source can be perfectly efficient because some usable energy is lost every time energy is converted from one form to another.

Continuation of Answers from p. 807
Skills Practice Lab

3. The liquid in test tube B is brownish and watery. After a few minutes, the liquid may form two layers, one brown and one colorless. For each gram of wood, about 0.5 mL of liquid will be produced.

4. The material in test tube A is black and fragile. Students may recognize that the material is similar to common charcoal.

5. For each gram of wood, about 0.25 g of solid residue will be produced.

6. The solid material burns with a red glow.

Answers to Conclusions

7. Most of the original material in the wood has been changed into gases and liquids.

8. The wood cannot return to its original form.

9. The liquids in test tube B could be used as a stain. Some of the liquid in test tube B could be used as a liquid fuel.

Teaching Tip ——— GENERAL

Encourage interested students to conduct research to learn more about cars that use other energy sources in place of or in addition to gasoline. Research topics could include the following:

- What kind of alternative cars are available to consumers today?
- What are the advantages and disadvantages of each type?
- Should consumers be given extra incentives to purchase these cars when they are first produced?
- What are some of the hurdles scientists are trying to overcome for cars that are still in the development phase (such as fuel cell cars)?

You could also put students in groups, and have each group conduct research on one particular type of car that has recently been made available to consumers or is currently being developed.

Have students share the results of their research with the class in the form of a presentation, a poster, or a brochure. **LS** Logical

Science in ACTION

This fuel cell car looks futuristic, but alternatives to gas-burning vehicles are already available.

Cars of the Future

Nearly every engine in the world, from a lawnmower engine to the engines of a Boeing 747 airplane, depends on a petroleum-based product, such as gasoline, for fuel. But the world's supply of petroleum is limited. Emissions from gas-burning engines are a major cause of air pollution. Despite these problems, few better alternatives to gas engines have been developed until recently.

Traffic jams such as this are a major cause of air pollution.

Electric versus Hybrid

Because electric cars run on battery power, they do not release pollutants into the atmosphere. Although these zero-emission vehicles first hit the streets more than 100 years ago, they never became popular because they lacked power. Also, electric cars require frequent recharging. So how do we continue to use cars while conserving resources? The answer may be to use hybrid cars.

Hybrid cars use a combination of batteries and an efficient engine. Although hybrid cars need gasoline, they are more fuel-efficient than gas-powered cars. Hybrids are also convenient for drivers because the cars never need recharging. When the car brakes, the electric motor produces electricity and recharges the batteries.

808

The rapeseed plant is a source of cooking oil and is used in a process to make cleaner-burning diesel fuel.

Fuel Cells Burn Alternative Fuels

Alternative fuel sources may also be used to power our cars in the future. A fuel cell converts the energy of a chemical reaction into electricity that powers the car. Fuel cells are efficient because they contain no moving parts and do not lose heat. A typical car engine loses 80% of its energy as heat. Fuel cells are already being used in industry and on our space shuttles, and car makers are hurrying to make fuel cells for cars. Some people hope that in addition to reducing our use of petroleum and reducing emissions, fuel cells may also reduce groundwater pollution caused by discarded car batteries and spilled oil and gas.

Plants May Power Cars in the Future

Scientists are also looking to plant and animal materials, or *biomass*, as a potential source of fuel. The most common way to get energy from biomass is to burn it to make heat much like burning wood in a fireplace. But biomass can also be converted into liquid fuels for specially adapted car engines. Cars of the future may run on used vegetable oil from fast-food restaurants!

 Science and You

1. **Applying Knowledge** Describe the difference between an electric car and a hybrid car.

2. **Understanding Concepts** Describe two advantages that electric cars have over gas-powered cars.

3. **Critical Thinking** Many advances in alternative energy have come from NASA. Why do you think NASA and the space program is interested in alternative energy?

4. **Critical Thinking** Do you think wood and other organic materials could be recycled from landfills and used as biomass? Explain.

5. **Making Decisions** Electric cars rely on electricity produced by power plants that burn fossil fuels. Would you buy an electric car to help save natural resources? Explain.

internet connect

www.scilinks.org
Topic: Electric Vehicles
SciLinks code: HK4155

SCiLINKS. Maintained by the National Science Teachers Association

809

Reference Section

810

APPENDIX B

Figure B-3   The World: Physical

811

APPENDIX A

Practice

Answers

1. Answers may vary but should be concise. One possible answer is "Monet is considered the most influential impressionist."

2. Answers may vary. One possible answer is "Go pick up milk, butter, and syrup at store."

3. Answers may vary. One possible answer is "Venus is called the morning and the evening star."

Reading and Study Skills

As a science student, you are expected to understand the information you read in this textbook and hear from your teacher. To be successful, you also need to be able to organize the information you receive and to take good notes. This appendix is designed to help you learn these skills, which will help you become a more successful student.

The first section, which contains reading skills and study skills, is followed by sections on graphing skills, a math refresher, and lab skills. At the end of every chapter in this book, there is a skills practice page. The skills practice pages are designed to help you develop your skills. If you have difficulty in one particular area, such as graphing skills, this appendix is an excellent resource for extra explanation and practice.

Recognizing Key Words

To begin improving your understanding of what you read, you need to know how to recognize key words in a sentence and key ideas in a paragraph. Finding the key ideas or words may be difficult for several reasons. First, you may not understand what a key idea is. A key idea is the idea that the author is trying to help you understand. When you are reading, ask yourself what the author is trying to tell you in the paragraph or sentence. The answer to that question is often the key idea.

The second reason that finding the key word or idea may be difficult is that its location within a paragraph or sentence often changes. Sometimes the key idea appears at the beginning of a sentence, and sometimes it is in the middle or at the end. To recognize key words in a sentence, ask yourself the following question about every word in a sentence.

If this word was taken out of the sentence, would I still understand what the sentence is trying to say?

Consider the following example.

Find the key words in the following sentence.

You are to report to the counselor's office at 4:00 P.M., and don't forget to take your books with you.

Key words: you, report, counselor's office, 4:00 P.M., take, books.

If you communicated the key words to someone, he or she would understand what to do.

Practice

Identify the key words in the following sentences.

1. Claude Monet is considered by many to be one of the most influential impressionist painters that ever lived.

2. Please go to the store and pick up a container of milk, some butter, and maple syrup, so we can eat breakfast soon.

3. Venus, the second planet from the sun, appears close to the sun and is called both *the morning star* and *the evening star*.

Recognizing Key Ideas in a Paragraph

Recognizing key ideas in a sentence is much like recognizing key words in a paragraph. As you're reading, ask yourself whether you would still understand what the author is trying to say if the sentence was taken out of the paragraph.

Recognizing key ideas in a paragraph is very important to your success as a science student. It is usually the key ideas within a section or chapter that you are held accountable for and that will appear on tests and quizzes.

If you cannot identify the key ideas within a paragraph, it is important to reread the paragraph. After you have reread the section, try to identify particular sentences or parts of sentences that do not directly address the subject of the paragraph. Reread the paragraph again, and omit the less important passages. Then, ask yourself again if you understand what the author is trying to say. Repeat this process until you feel confident that you understand what the passage is trying to convey.

This process may take time in the beginning, but recognizing key ideas is an important skill to acquire if you want to become a good student. As you build your scientific vocabulary and become more familiar with this and other study skills, recognizing key ideas will become easier.

Consider the following example.

Some household products should never be combined because they react to produce harmful substances. Ammonia and bleach react to produce a poisonous substance called *chloramine*, NH_2Cl. Also, vinegar and bleach react to produce chlorine gas, Cl_2, another poisonous substance. To be safe, you should never combine household products.

Key idea: Household products react to produce poisonous substances, so you should never combine them.

The key idea is the most important information. In this paragraph, the specific examples of household chemicals were not a part of the key idea. They help explain the concept, but they can be left out of the summarzed key idea.

Practice

Identify the key idea in the following paragraph.

1. Soaps have traditionally been made from animal fats or vegetable oils. Soap can dissolve in both oil and water. Soaps are emulsifiers that let oil and water mix and keep the oil and water from separating. When you wash your face with soap, the oil on your face is suspended in the soapy water. The water you use to rinse your face with carries the soap and unwanted oil away to leave your face clean.

813

Practice

Answers

1. Answers may vary but should be concise. One possible answer is "Soap cleans by letting oil and water mix. When you wash off soap it carries the oil away."

Outlining

Outlining is one of the most widely used methods for taking notes. Taking good notes is a skill that is very important for understanding, comprehension, and achieving success on tests. The information in the chapters of *Holt Science Spectrum* is organized in a way to help you easily outline the key ideas in a chapter or section. Outlines you make from the key ideas can help you prepare for tests and can be used to check your comprehension of the chapter.

Most outlines follow the same structure. Main ideas or topics are listed first and usually follow a roman numeral. We will use the chapter titled "Chemical Reactions" as an example. The main topic is on the first page of the chapter and is the title of the section that is written in green. Because the section title is the main topic, we list it first along with the roman numeral I.

I. The Nature of Chemical Reactions

After we write the main topic, we will add major points that provide information about the main topic. These major points appear in red type under the section title. The two major points that we will add to our outline follow the letters A and B.

I. The Nature of Chemical Reactions

    A. Chemical Reactions Change Substances
    B. Energy and Reactions

The next step in outlining is to fill in the subpoints that describe or explain the major points. In this book, these subpoints will appear as sentences in blue type. We add the subpoints to our outline under the major points that they explain. The subpoints should follow numerals.

I. The Nature of Chemical Reactions

  A. Chemical Reactions Change Substances
    1. Production of gas and change of color are signs of a chemical reaction.
    2. Chemical reactions rearrange atoms.

  B. Energy and Reactions
    1. Energy must be added to break bonds.
    2. Forming bonds releases energy.
    3. Energy is conserved in chemical reactions.
    4. Reactions that release energy are exothermic.
    5. Reactions that absorb energy are endothermic.

The last step in outlining is to add the supporting details for the subpoints. In this book, these details will appear in the body of the text under the details subpoint. We add the details to our outline under the subpoints that they explain. Details should follow lowercase letters.

I. The Nature of Chemical Reactions

  A. Chemical Reactions Change Substances
    1. Production of gas and change of color are signs of a chemical reaction.
      a. An example of a change caused by a chemical reaction is seen in the process of baking bread. The dough rises because gas is produced, and the dough turns brown during baking.

2. Chemical reactions rearrange atoms.
   a. Reactants contain the same types of atoms as products but they are often rearranged.
   b. Atoms are neither created nor destroyed.
   c. Atoms are rearranged as bonds are broken and formed.

Remember that the major topics are always listed next to a roman numeral. The major points are listed under the main topics and follow a capital letter. Major points are followed by subpoints, which follow a numeral. Finally, supporting details are under subpoints and follow a lowercase letter.

Outlining Guidelines

I. The main topic is listed here.

  A. Major points, which provide information about the main topic are listed here.

    1. Subpoints, which provide information about or describe the major points are listed here.

      a. Supporting details of the subpoints are listed here.

Practice

1. Build an outline using the following topics, ideas, and supporting details taken from the chapter entitled "Planet Earth."
   A. Physical Weathering
   c. Acid precipitation causes damage to both living organisms and inorganic matter such as statues.
   1. Ice can break rocks.
   I. Weathering and Erosion
   a. Frost wegding occurs when water seeps into crack in rock and then freezes.
   2. Plants can also break rocks.
   B. Chemical Weathering
   1. Carbon dioxide can cause chemical weathering.
   2. Water plays a key role in chemical weathering.
   3. Acid precipitation can slowly dissolve minerals.
   a. Carbon dioxide from the air dissolves in rainwater to create carbonic acid.
   b. Carbonic acid reacts with minerals in rocks and then washes away carrying the rock with it.
   a. When fossil fuels, especially coal, are burned sulfur dioxide and nitrogen oxides are released into the air.
   2. Acid precipitation occurs when sulfur dioxide and nitrogen oxides in the air react with water in clouds to form weak acids that fall to Earth.
   a. Water reacts chemically with many minerals.
   b. Leaching occurs when water dissolves or reacts with minerals in rocks and then is transported to lower layers or rock.

2. Pick a section of one of the chapters in this book, and write an outline.

815

Practice

Answers

1. Answers may vary. Outlines should be well organized and ordered properly.

2. Outline should look like the following:

I. Weathering and Erosion
   A. Physical Weathering
      1. Ice can break rocks.
         a. Frost wedging occurs when water seeps into cracks in rocks and then freezes.
      2. Plants can also break rocks.
   B. Chemical Weathering
      1. Carbon dioxide dissolved in water can cause chemical weathering.
         a. Carbon dioxide from the air dissolves in rainwater to create carbonic acid.
         b. Carbonic acid reacts with minerals in rocks and then washes away carrying the rock with it.
      2. Water plays a key role in chemical weathering.
         a. Water reacts chemically with many minerals.
         b. Leaching occurs when water dissolves or reacts with minerals in rocks and then is transported to the lower layers of rock.
      3. Acid rain can slowly dissolve minerals.
         a. When fossil fuels (especially coal) are burned, sulfur dioxide and nitrogen oxides are released into the air.
         b. Acid precipitation occurs when sulfur dioxide and nitrogen oxides in the air react with water in clouds to form weak acids that fall to Earth.
         c. Acid precipitation causes damage to both living organisms and inorganic matter such as statues.

2. Answers may vary.

APPENDIX A

Teaching Tip

Many students have not developed note-taking skills in earlier grades. At first, they may try to simply write down everything you say. Introducing them to strategies like Power Notes will help them learn to organize the information provided in class.

Practice

Answers

1. Power 1: Ionization

Power 2: Electron lost
Power 3: Cation
Power 3: Positive charge
Power 2: Electron gained
Power 3: Anion
Power 3: Negative charge

Power Notes

Power notes help you organize the concepts you are studying by distinguishing main ideas from details and providing a framework of important concepts. Power notes are easier to use than outlines because their structure is simpler. You assign a power of *1* to each main idea and a *2, 3,* or *4* to each detail. You can use power notes to organize ideas while reading your text or to reorganize your class notes to study.

Start with a few boldfaced vocabulary terms. Later you can strengthen your notes by expanding these into more-detailed phrases. Use the following general format to help you structure your power notes.

Power 1 Main idea
  Power 2 Detail or support for power 1
    Power 3 Detail or support for power 2
      Power 4 Detail or support for power 3

1. Pick a Power 1 word.

We'll use the term *atom* found in the chapter on Atoms and The Periodic Table of your textbook.

Power 1 Atom

2. Using the text, select some Power 2 words to support your Power 1 word.

We'll use the terms *nucleus* and *electron cloud*, which are two parts of an atom.

Power 1 Atom
  Power 2 Nucleus
  Power 2 Electron Cloud

3. Select some Power 3 words to support your Power 2 words.

We'll use the terms *positive charge* and *negative charge,* two terms that describe the Power 2 words.

Power 1 Atom
  Power 2 Nucleus
    Power 3 Positive charge
  Power 2 Electron cloud
    Power 3 Negative charge

4. Continue to add powers to support and detail the main idea as necessary.

If you have a main idea that needs a lot of support, add as many powers as needed to describe the idea. You can use power notes to organize the material in an entire section or chapter of your textbook to study for classroom quizzes and tests.

Power 1 Atom
  Power 2 Nucleus
    Power 3 Positive charge
    Power 3 Protons
      Power 4 Positive charge
    Power 3 Neutrons
      Power 4 No charge
  Power 2 Electron cloud
    Power 3 Negative charge

Practice

1. Use the chapter entitled "Atoms and The Periodic Table" and the power notes structure below to organize the following terms: *electron lost, electron gained, ionization, anion, cation, negative charge,* and *positive charge.*

   Power 1
     Power 2
       Power 3
       Power 3
     Power 2
       Power 3
       Power 3

Two-column Notes

Two-column notes can be used to learn and review definitions of vocabulary terms or details of specific concepts. The two-column-note strategy is simple: write the term, main idea, or concept in the left-hand column. Then write the definition, example, or detail on the right.

One strategy for using two-column notes is to organize main ideas and their details. The main ideas from your reading are written in the left-hand column of your paper and can be written as questions, key words, or a combination of both. Key words can include boldface terms as well as any other terms you may have trouble remembering. Questions may include those the author has asked or any questions your teacher may have asked during class. Details describing these main ideas are then written in the right-hand column of your paper.

1. Identify the main ideas.

The main ideas for each chapter are listed in the section objectives. However, you decide which ideas to include in your notes. The table below shows some of the main ideas from the objectives in the first section of the chapter entitled Introduction to Science.

2. Divide a blank sheet of paper into two columns, and write the main ideas in the left-hand column.

Do not copy ideas from the book or waste time writing in complete sentences. Summarize your ideas using quick phrases that are easy to understand and remember. Decide how many details you need for each main idea, and include that number to help you to focus on the necessary information.

3. Write the detail notes in the right-hand column.

Be sure you list as many details as you designated in the main-idea column.

The two-column method of review is perfect for preparing for quizzes or tests. Just cover the information in the right-hand column with a sheet of paper, and after reciting what you know, uncover the notes to check your answers.

Practice

1. Make your own two-column notes using the periodic table. Include in the details the symbol and the atomic number of each of the following elements.

   a. neon   c. calcium   e. oxygen
   b. lead   d. copper   f. sodium

Main idea	Detail notes	
▶ Scientific theory (4 characteristic properties)	▶ tested experimentally ▶ possible explanation	▶ explains natural event ▶ used to predict
▶ Scientific law (3 characteristic properties)	▶ tested experimentally ▶ summary of an observation	▶ can be disproved
▶ Models (4 characteristic properties)	▶ represents an object or event ▶ physical	▶ computer ▶ mathematical

817

Teaching Tip

Two-column notes help students build analysis skills, because it requires them to distinguish between main ideas and details. Often, students are so overwhelmed by new information in science classes that they do not remember to make the distinction.

Practice

Main idea	Detail notes
a. neon	• Ne • 10
b. lead	• Pb • 82
c. calcium	• Ca • 20
d. copper	• Cu • 29
e. oxygen	• O • 8
f. sodium	• Na • 11

Teaching Tip

Pattern puzzles are an excellent strategy for reinforcing information that belongs in a specific order. Examples of the kinds of information that can be studied in this way include lab procedures, problem-solving steps, and historical developments of scientific models.

Pattern Puzzles

You can use pattern puzzles to help you remember information in the correct order. Pattern puzzles are not just a tool for memorization. They can also help you better understand a variety of scientific processes, from the steps in solving a mathematical conversion to the procedure for writing a lab report.

1. **Write down the steps of a process in your own words.**

   We'll use the Math Skills feature on converting amount to mass from the chapter entitled "Atoms and The Periodic Table." On a sheet of paper, write down one step per line, and do not number the steps. Also, do not copy straight from your text. Writing the steps in your own words helps you check your understanding of the process. You may want to divide the longer steps into two or three shorter steps.

   - List the given and unknown information.
   - Look at the periodic table to determine the molar mass of the substance.
   - Write the correct conversion factor to convert moles to grams.
   - Multiply the amount of substance by the conversion factor.
   - Solve the equation and check your answer.

2. **Cut the sheet of paper into strips with only one step per strip of paper.**

   Shuffle the strips of paper so that they are out of sequence.

   - Look at the periodic table to determine the molar mass of the substance.

   - Solve the equation and check your answer.

   - List the given and unknown information.

   - Multiply the amount of substance by the conversion factor.

   - Write the correct conversion factor to convert moles to grams.

Practice

Answers

3. **Place the strips in their proper sequence.**
Confirm the order of the process by checking your text or your class notes.

• List the given and unknown information.

• Look at the periodic table to determine the molar mass of the substance.

• Write the correct conversion factor to convert moles to grams.

• Multiply the amount of substance by the conversion factor.

• Solve the equation and check your answer.

Pattern puzzles can be used to help you prepare for a laboratory experiment. That way it will be easier for you to remember what you need to do when you get into the lab, especially if your teacher gives pre-lab quizzes.

You'll want to use pattern puzzles if your teacher is planning a lab practical exam to test whether you know how to operate laboratory equipment. That way you can study and prepare for such a test even though you don't have a complete set of lab equipment at home.

Pattern puzzles work very well with problem-solving. If you work a pattern puzzle for a given problem type several times first, you will find it much easier to work on the different practice problems assigned in your homework.

Pattern puzzles are especially helpful when you are studying for tests. It is a good idea to make the puzzles on a regular basis so that when test time comes you won't be rushing to make them. Bind each puzzle using paper clips, or store the puzzles in individual envelopes. Before tests, use your puzzles to practice and to review.

Pattern puzzles are also a good way to study with others. You and a classmate can take turns creating your own pattern puzzles and putting each other's puzzles in the correct sequence. Studying with a classmate in this way will help make studying fun and allow you and your classmate to help each other.

Practice

1. Write the following sentences describing the process of making pattern puzzles in the correct order.
   • Place the strips in their proper sequence.
   • Write down the steps of the process in your own words.
   • Shuffle the strips of paper.
   • Choose a multiple-step process from your text.
   • Using your text, confirm the order of the process.
   • Cut the paper into strips so that there is one step per strip.

Answers
1. • Choose a multiple-step process from your text.
   • Write down the steps of the process in your own words.
   • Cut the paper into strips so that there is one sentence per strip.
   • Shuffle the strips of paper.
   • Place the strips in their proper sequence.
   • Using your text, confirm the order of the process.

Teaching Tip

The KWL strategy provides a structure that helps students focus on what they should be learning as they read the textbook. As such, it works well with students at all ability levels.

KWL Notes

The KWL strategy is a helpful way to learn. It is different from the other learning strategies you have seen in this appendix. KWL stands for "what I **K**now—what I **W**ant to know—what I **L**earned." KWL prompts you to brainstorm about the subject matter before you read the assigned pages. This strategy helps you relate your new ideas and concepts with those you have already learned. This allows you to understand and apply new knowledge more easily. The objectives at the beginning of each section in your text are ideal for using the KWL strategy. Just read and follow the instructions in the example below.

1. **Read the section objectives.**

   You may also want to scan headings, bold-face terms, and illustrations in the section. We'll use a few of the objectives from the first section of the chapter entitled Matter.

   ▶ Explain the relationship between matter, atoms, and elements.

   ▶ Distinguish between elements and compounds.

   ▶ Categorize materials as pure substances or mixtures.

2. **Divide a sheet of paper into three columns, and label the columns "What I know," "What I want to know," and "What I learned."**

3. **Brainstorm about what you know about the information in the objectives, and write these ideas in the first column.**

   It is not necessary to write complete sentences. What's most important is to get as many ideas out as possible. In this way, you will already be thinking about the topic being covered. That will help you learn new information, because it will be easier for you to link it to recently-remembered knowledge.

4. **Think about what you want to know about the information in the objectives, and write these ideas in the second column.**

   You should want to know the information you will be tested over, so include information from both the section objectives and any other objectives your teacher has given you.

5. **While reading the section, or after you have read it, use the third column to write down what you learned.**

   While reading, pay close attention to any information about the topics you wrote in the "What I want to know" column. If you do not find all of the answers you are looking for, you may need to reread the section or find a second source for the information. Be sure to ask your teacher if you still cannot find the information after reading the section a second time.

What I know	What I want to know	What I learned

Practice

Answers

1. **a.** As written, this was a misconception according to row 2 of "What I learned" of the notes. Elements must be chemically combined to make a compound, not just physically combined.

   b. As written, this was a misconception. According to "What I learned" in rows 1 and 2 of the notes, a compound can be broken down into elements but diamond cannot. Diamond is a form of the element carbon because it contains only one type of atom.

   c. This is another misconception. According to row 1 of "What I learned," elements contain only one type of atom. Sodium chloride is a compound made up of two elements, sodium and chlorine.

   d. This is another misconception. Lemonade contains many substances. It is a mixture, just as grape juice is a mixture.

What I know	What I want to know	What I learned
▸ atoms are very small particles ▸ oxygen is an element ▸ elements are listed on the periodic table	▸ Explain the relationship between matter, atoms, and elements.	▸ matter is anything that occupies space ▸ atoms are the smallest particles with properties of an element ▸ elements cannot be broken down into simpler substances with the same properties ▸ atoms and elements are matter
▸ compounds are made of elements	▸ Distinguish between elements and compounds.	▸ elements combine chemically to make compounds ▸ compounds can be broken down into elements
▸ mixtures are combinations of more than one substance ▸ pure substances have only one component	▸ Categorize materials as pure substances or mixtures.	▸ pure substances have fixed compositions and definite properties ▸ mixtures are combinations of more than one pure substance ▸ elements and compounds are pure substances ▸ grape juice is a mixture

6. **It is also important to review your brainstormed ideas when you have completed reading the section.**

Compare your ideas in the first column with the information you wrote down in the third column. If you find that some of your brainstormed ideas are incorrect, cross them out. It is extremely important to identify and correct any misconceptions you had before you begin studying for your test.

Your completed KWL notes can make learning science much easier. First of all, this system of note-taking makes gaps in your knowledge easier to spot. That way you can focus on looking for the content you need easier, whether you look in the textbook or ask your teacher.

If you've identified the objectives clearly, the ideas you are studying the most are the ones that will matter most.

Practice

1. Use column 3 from the table above to identify and correct any misconceptions in the following brainstorm list.
   a. Physically mixing elements will form a compound.
   b. Diamond is a compound.
   c. Sodium chloride is an element.
   d. Lemonade is a pure substance.

821

Teaching Tip

At first, students may not like to make concept maps because they are not as structured as other forms of note taking. However, this freedom to create one's own method of organizing information usually wins students over.

At first, you may want to use concept maps as a review activity. Making the concept map in preparation for studying is actually a good way to reinforce the concepts of a lesson or chapter.

Concept Maps

Making concept maps can help you decide what material in a chapter is important and how best to learn that material. A concept map presents key ideas, meanings, and relationships for the concepts being studied. It can be thought of as a visual road map of the chapter.

Concept maps can begin with vocabulary terms. Vocabulary terms are generally labels for concepts, and concepts are generally nouns. Concepts are linked using linking words to form propositions. A proposition is a phrase that gives meaning to the concept. For example, "matter is changed by energy" is a proposition.

1. **Select a main concept for the map.**

   We will use *matter* as the main concept for this map.

2. **List all the important concepts.**

   We'll use some of the terms in the second section of the chapter entitled "Matter."

   energy   chemical change
   chemical property   physical change
   physical property   reactivity
   density

3. **Build the map by placing the concepts according to their importance under the main concept, and adding linking words to give meaning to the arrangement of concepts.**

   One way of arranging the concepts is shown in *Map A.* When adding the links, be sure that each proposition makes sense. To distinguish concepts from links, place your concepts in circles, ovals, or rectangles. Then add cross-links with lines connecting concepts across the map. *Map B* on the next page is a finished map covering the main ideas found in the vocabulary list in Step 1.

   Practice mapping by making concept maps about topics you know. For example, if you know a lot about a particular sport, such as basketball, you can use that topic to make a practice map. By perfecting your skills with information that you know very well, you will begin to feel more confident about making maps from the information in a chapter.

Map A

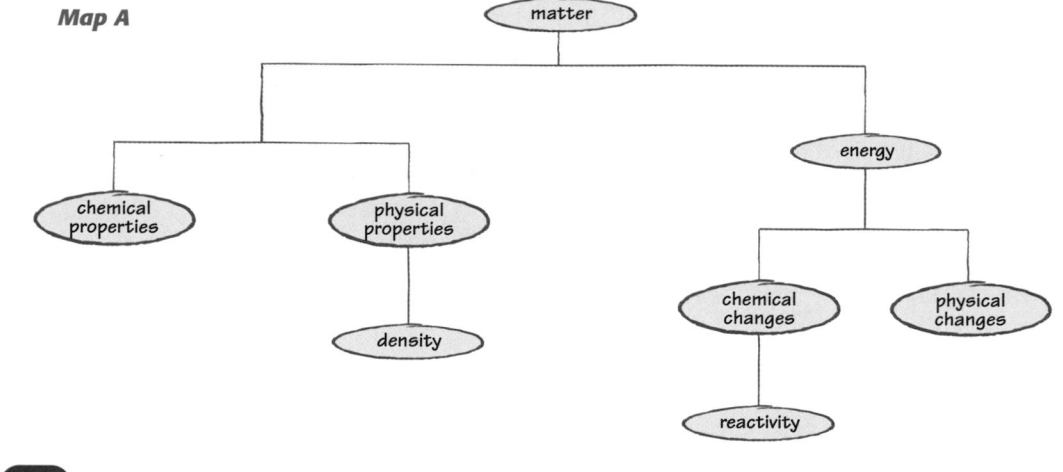

Making maps might seem difficult at first, but the process gets you to think about the meanings and relationships among concepts. If you do not understand those relationships, you can get help early on.

In addition, many people find it easier to study by looking at a concept map, rather than flipping through a chapter full of text because concept mapping is a visual way to organize the information in a chapter. Not only does it isolate the key concepts in a chapter, it also makes the relationships and linkages among those ideas easy to see and understand.

One useful strategy is to trade concept maps with a classmate. Everybody organizes information slightly differently, and something they may have done may help you understand the content better.

Remember, although concept mapping may take a little extra time, the time you spend mapping will pay off when it is time to review for a test or final exam.

Practice

Answers

1. **a.** concept
   b. linking words
   c. linking word
   d. linking words
   e. concept
   f. linking words
   g. concept
   h. linking word

2. Answers may vary. Check student responses against **Map B.** Some possible responses are the following:
   - Matter is described by chemical properties, which describe reactivity.
   - Matter is also described by physical properties such as density.
   - Matter is changed by energy, which causes chemical and physical changes.

> ### Practice
>
> 1. Classify each of the following as either a concept or linking word(s).
> | | |
> |---|---|
> | **a.** compound | **e.** element |
> | **b.** is classified as | **f.** reacts with |
> | **c.** forms | **g.** pure substance |
> | **d.** is described by | **h.** defines |
> 2. Write three propositions from the information in *Map B.*

Map B

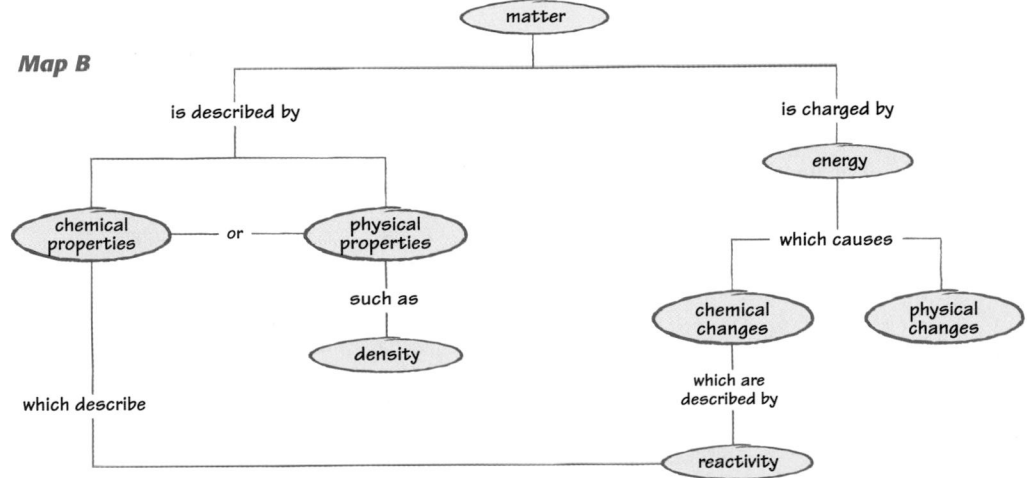

Answers

1. • Dams are built to harness energy
   • Water is forced through small channels at the top of the dam
   • As the water falls, it turns blades on a turbine fan
   • The fans are attached to a core that is wrapped with many loops of wire
   • The loops of wire rotate within a strong magnetic field
   • As the loops turn, electrical current (or energy) is produced

Process Flow Chart

A process flow chart is a special kind of concept map used for processes. The steps in a process almost always occur in the same order, so a process flow chart helps you to remember what order the steps occur in.

Unlike regular concept maps, process flow charts do not contain linking words. Instead, the arrows represent the next step in a process.

Examine the following process flow chart that shows the steps that occurred when the moon formed.

Another kind of process flow chart can be used to show the relationship between steps in a cycle, such as the one shown below that illustrates the steps of the water cycle.

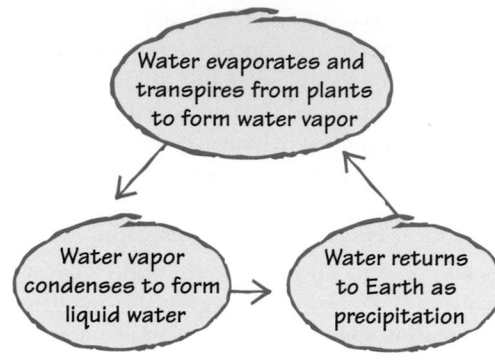

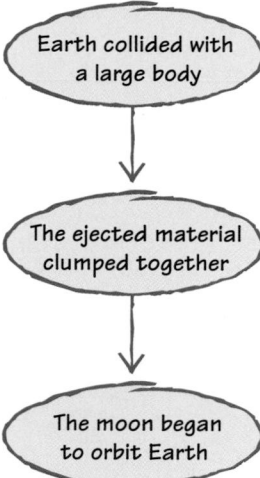

Notice that process flow charts that represent cycles have no beginnings or endings.

1. Create a process flow chart using the following scrambled steps.
   • As the water falls, it turns blades on a turbine fan
   • Dams are built to harness energy
   • As the loops turn, electrical current (or energy) is produced
   • Water is forced through small channels at the top of the dam
   • The fans are attached to a core that is wrapped with many loops of wire
   • The loops of wire rotate within a strong magnetic field
2. Use a cycle discussed in this book, and create your own process flow chart.

Interpreting Scientific Illustrations

Illustrations, figures, and photographs can be very useful when you are trying to understand a scientific idea. Many illustrations are included in this book to help you visualize relationships that are hard to visualize or understand. Some ideas or things are illustrated because they shown are too small or too large for you to see. Others are illustrated to help you remember relationships. The illustration on the right appears in the chapter entitled "Sound and Light."

When you are looking at a scientific illustration, refer to the text and remind yourself what the figure shows. The text that appears before the figure begins with the title "Objects have color because they reflect certain wavelengths of light." The title tells you the topic of the illustration.

Most figures have captions and labels that can also help you understand what the figure shows. First, examine the labels, and make sure that you understand what is being illustrated. Next, read the caption carefully, and restate it in your own words. If you can restate the caption you have a good idea of what the figure shows.

Practice

1. Look at Figure 24A. Write your own labels for Figure 24B by using labels similar to labels on Figure 24.
2. After you have examined the figure carefully, restate the captions in your own words.

Figure 24

A Rose in White and Red Light

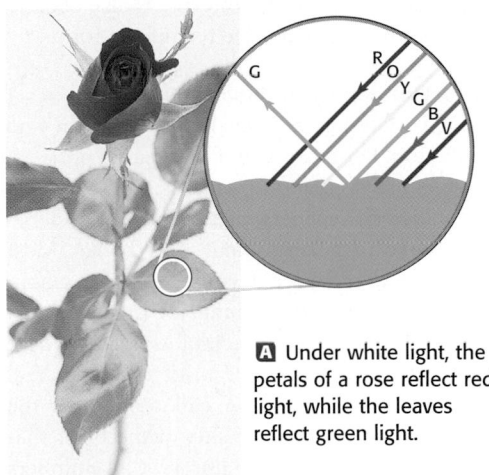

A Under white light, the petals of a rose reflect red light, while the leaves reflect green light.

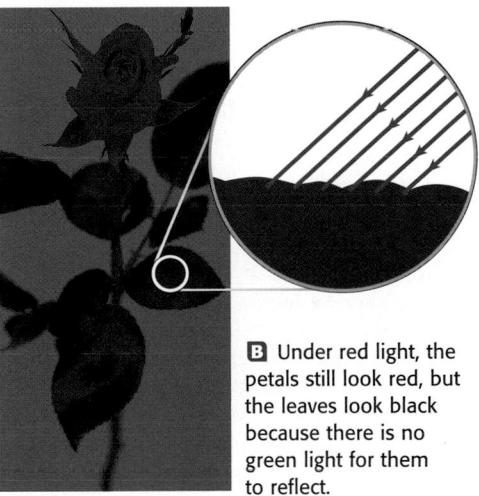

B Under red light, the petals still look red, but the leaves look black because there is no green light for them to reflect.

Practice

Answers

1. All arrows should be labeled with an R because they represent red light.
2. Answers may vary but they should express the figures accurately.

APPENDIX A

Practice

Answers

1. Answers may vary depending on your school library.
2. Answers may vary depending on your school library.
3. Answers may vary depending on your school library.
4. Answers may vary depending on your school library.
5. Answers may vary depending on your school library.

Researching Information

Many resources are available to help you research information. A wide variety of printed materials, such as newspapers, magazines, and books are available. The Internet is also becoming an important resource for information.

Using your library or media center

Printed materials are divided into fiction and nonfiction. For most scientific research projects, you will use nonfiction materials. These materials include newspapers, dictionaries, encyclopedias, some magazines, and other books. Most libraries and media centers use two main classification systems—the Library of Congress system and the Dewey decimal system. These classification systems are used to assign call numbers to the books. A call number is the series of numbers on the side of the book that help you to find it in the library. Call numbers are also listed in the card catalog or in the database your library uses to help you find books. *Table 1* shows a simplified Library of Congress system, which is used in most large libraries. *Table 2* shows a simplified Dewey decimal system. This system is often used in smaller libraries such as school libraries.

Practice

1. What system does your school library use to classify books?
2. Where are the encyclopedias in your school library?
3. Does your school library use a card catalog or a computer database to help you search for books?
4. List three nonfiction magazines that your school library has subscriptions to.
5. Name the title and call number for a book on science.

Table 1
Library of Congress classification system

Letter on book binding	Subject
A	General works
B	Philosophy, Psychology, and Religion
C–F	History
G–H	Geography and Social Sciences (e.g., Anthropology)
J	Political Science
K	Law
L	Education
M	Music
N	Fine Arts
P	Literature
Q	Science
R	Medicine
S	Agriculture
T	Technology
U–V	Military and Naval Science
Z	Bibliography and Library Science

Table 2
Dewey decimal system

Number on book binding	Subject
000–099	General works
100–199	Philosophy and Psychology
200–299	Religion
300–399	Social Sciences
400–499	Language
500–599	Pure Sciences
600–699	Technology
700–799	Arts
800–899	Literature
900–999	History

Using the Internet

Most research you do on a computer will involve the Internet and the World Wide Web. The Internet serves a variety of purposes but not all of these purposes involve providing accurate information, so it is important to be skeptical of information you get from the Internet.

To begin researching information on the Internet, you may want to begin with a search engine. A search engine is a Web site where you can search for other Web sites by subject or keyword. There are many different search engines that specialize in different kinds of information. When you are beginning a research project, find the address of a search engine and type that address into the address line like the one shown below.

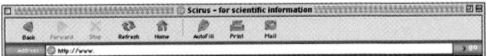

For example, use the search engine **www.scirus.com**. *Scirus* is a search engine that specializes in scientific information. After typing **www.scirus.com** into the address line, push the return or enter key. Spelling and punctuation become very important when you are trying to find a Web site. Because there are so many Web sites on the World Wide Web, misspellings may take you to a different site than you intended. The first page of the search engine will look something like this:

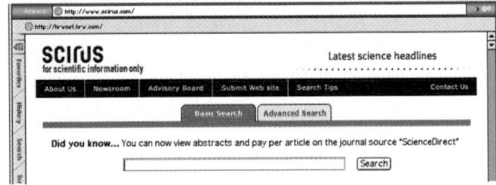

To search for information on black holes, enter the keyword "black hole" into the search line and hit the return or enter key. This process results in a search all the Web sites that are in the database for the keyword black holes and produces a list like the one shown below.

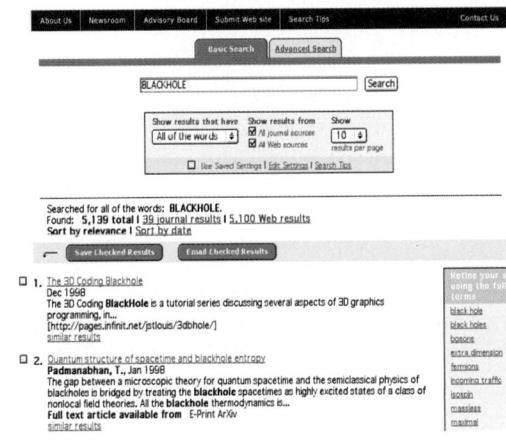

You should browse this list and pick Web sites that most closely fit your specific research topic. Another good search engine is **www.firstgov.gov**, which searches all government databases and Web sites. Government sites are a good source of reliable information.

Practice

1. Use an Internet search engine to search for Web sites about tectonic plates. List three Web sites, and discuss how reliable you think each one may be.
2. If you were looking for information about federal nutrition guidelines, where would you start? Explain your answer.
3. Why should you double check information you find through the Internet?

Practice

Answers

1. Answers may vary.
2. A good place to start would be the search engine at www.firstgov.gov because it would help you find good information about nutrition.
3. Answers may vary but should include information about ways in which the Internet can be unreliable including the fact that Web Sites often reflect individual points of view.

827

Fractions

Practice

Answers

1. a. $\dfrac{29}{24}$ or $1\dfrac{5}{24}$

   b. $\dfrac{7}{24}$

   c. $\dfrac{21}{8}$ or $2\dfrac{5}{8}$

   d. $\dfrac{13}{24}$

Percentages

Practice

1. 88.84%
2. saturated fat: 21%
   unsaturated fats: 79%

Math Skills Refresher

Fractions

Fractions represent numbers that are less than one. In other words, fractions are a way of numerically representing a part of a whole. For example, if you have a pizza with 8 slices, and you eat 2 of the slices, you have 6 out of the 8 slices, or $\dfrac{6}{8}$, of the pizza left. The top number in the fraction is called the numerator. The bottom number is the denominator.

There are special rules for adding, subtracting, multiplying, and dividing fractions. These rules are summarized in **Table 3**.

Table 3 Basic Operations for Fractions

	Rule and example
Multiplication	$\left(\dfrac{a}{b}\right)\left(\dfrac{c}{d}\right)=\dfrac{ac}{bd}$ $\left(\dfrac{2}{3}\right)\left(\dfrac{4}{5}\right)=\dfrac{8}{15}$
Division	$\dfrac{a}{b}\div\dfrac{c}{d}=\dfrac{\left(\dfrac{a}{b}\right)}{\left(\dfrac{c}{d}\right)}=\dfrac{ad}{bc}$ $\dfrac{2}{3}\div\dfrac{4}{5}=\dfrac{\left(\dfrac{2}{3}\right)}{\left(\dfrac{4}{5}\right)}=\dfrac{(2)(5)}{(3)(4)}=\dfrac{10}{12}$
Addition and subtraction	$\dfrac{a}{b}\pm\dfrac{c}{d}=\dfrac{ad\pm bc}{bd}$ $\dfrac{2}{3}-\dfrac{4}{5}=\dfrac{(2)(5)-(3)(4)}{(3)(5)}=-\dfrac{2}{15}$

Practice

1. Perform the following calculations:

   a. $\dfrac{7}{8}+\dfrac{1}{3}=$ c. $\dfrac{7}{8}\div\dfrac{1}{3}=$

   b. $\dfrac{7}{8}\times\dfrac{1}{3}=$ d. $\dfrac{7}{8}-\dfrac{1}{3}=$

Percentages

Percentages are no different from other fractions, except that in a percentage, the whole (or the number in the denominator) is considered to be 100. Any percentage, $x\%$, can be read as x out of 100. For example, if you have completed 50% of an assignment, you have completed $\dfrac{50}{100}$ or $\dfrac{1}{2}$ of the assignment.

Percentages can be calculated by dividing the part by the whole. When your calculator solves a division problem that is less than 1, it gives you a decimal value instead of a fraction. For example, 0.45 can be written as the fraction $\dfrac{45}{100}$. This is equal to 45%. An easy way to calculate percentages is to divide the part by the whole, then multiply by 100. This multiplication moves the decimal point two positions to the right, giving you the number that would be over 100 in a fraction. Try this example.

You scored 73 out of 92 problems on your last exam. What was your percentage score?

First divide the part by the whole to get a decimal value: $\dfrac{73}{92}$. Note that 0.7935 is equal to $\dfrac{79.35}{100}$.

Then multiply by 100 to yield the percentage: $0.7935 \times 100 = 79.35\%$.

Practice

1. Oxygen in water has a mass of 16.00 g. The water has a total mass of 18.01 g. What percentage of the mass of water is made up of oxygen?
2. A candy bar contains 14 g of fat. The total fat contains 3.0 g of saturated fat and 11 g of unsaturated fats. What are the percentages of saturated and unsaturated fat in the candy bar?

Exponents

An exponent is a number that is superscripted to the right of another number. The best way to explain how an exponent works is with an example. In the value 5^4, 4 is the exponent on 5. The number with its exponent means that 5 is multiplied by itself 4 times.

$$5^4 = 5 \times 5 \times 5 \times 5 = 625$$

You will frequently hear exponents referred to as powers. Using this terminology, the above equation could be read as *five to the fourth power equals 625*. Keep in mind that any number raised to the zero power is equal to one. Also, any number raised to the first power is equal to itself: $5^1 = 5$.

Just as there are special rules for dealing with fractions, there are special rules for dealing with exponents. These rules are summarized in **Table 4.**

You probably recognize the symbol for a square root, $\sqrt{}$. This means that a number times itself equals the value inside the square root. It is also possible to have roots other than the square root. For example, $\sqrt[3]{x}$ means that some number, n, times itself three times equals the number x, or $n \times n \times n = x$. We can turn our example of $5^4 = 625$ around to solve for the fourth root of 625.

$$\sqrt[4]{625} = 5$$

Taking the nth root of a number is the same as raising that number to the power of $1/n$. Therefore, $\sqrt[4]{625} = 625^{1/4}$.

A scientific calculator is a must for solving most problems involving exponents and roots. Many calculators have dedicated keys for squares and square roots. But what about different powers, such as cubes and cube roots? Most scientific calculators have a key shaped like a caret, $^\wedge$. If you type in "5^4," when you hit the equals sign or the enter key, the calculator will determine that $5^4 = 625$, and display that answer.

For roots, you enter the decimal equivalent of the fractional exponent. For example, to solve the problem of the fourth root of 625, instead of entering one-fourth as the exponent, enter "625^0.25," because 0.25 is equal to one-fourth.

Table 4   Rules for dealing with exponents

	Rule	Example
Zero power	$x^0 = 1$	$7^0 = 1$
First power	$x^1 = x$	$6^1 = 6$
Multiplication	$(x^n)(x^m) = x^{(n+m)}$	$(x^2)(x^4) = x^{(2+4)} = x^6$
Division	$\dfrac{x^n}{x^m} = x^{(n-m)}$	$\dfrac{x^8}{x^2} = x^{(8-2)} = x^6$
Exponents that are fractions	$x^{1/n} = \sqrt[n]{x}$	$4^{1/3} = \sqrt[3]{4} = 1.5874$
Exponents raised to a power	$(x^n)^m = x^{nm}$	$(5^2)^3 = 5^6 = 15\ 625$

Practice

1. Perform the following calculations:

   a. $9^1 =$ d. $(14^2)(14^3) =$

   b. $(3^3)^5 =$ e. $11^0 =$

   c. $\dfrac{2^8}{2^2} =$ f. $6^{1/6} =$

Practice

Answers

1. a. $9^1 = 9$

   b. $(3^3)5 \quad 3^{15} = 14\ 348\ 907$

   c. $\dfrac{2^8}{2^2} = 2^{(8-2)} = 2^6 = 64$

   d. $(14^2)(14^3) \quad 14^{(2+3)} = 14^5 \quad 537\ 824$

   e. $11^0 = 1$

   f. $6^{1/6} = \sqrt[6]{6} = 1.348$

829

Order of Operations

Answers

1. $2^3 \div 2 + 4 \times (9 - 2^2) =$
   $8 \div 2 + 4 \times (9 - 4) =$
   $8 \div 2 + 4 \times 5 =$
   $4 + 20 = 24$

2. $\dfrac{2 \times (6 - 3) + 8}{4 \times 2 - 6} =$

   $\dfrac{2 \times 3 + 8}{4 \times 2 - 6} =$

   $\dfrac{6 + 8}{8 - 6} =$

   $\dfrac{14}{2} = 7$

Geometry

Answers

1. $r = \dfrac{14cm}{2} = 7$ cm
   $volume = \pi r^2 h =$
   $\pi (7 \text{ cm})^2 (8 \text{ cm}) =$
   $volume = 1232 \text{ cm}^3$

2. surface area $= 2(lh + lw + hw) =$
   $2[(4 \text{ cm})2 + (4 \text{ cm})2 + (4 \text{ cm})2]$

3. volume of a sphere $=$
   $\dfrac{4}{3}\pi r3 = 76 \text{ cm}^3$
   $r^3 = 18 \text{ cm}^3$
   $r^3 = 2.6 \text{ cm}$
   diameter $= r \times 2 = 5.2$ cm
   The box must be at least 5.2 cm
   in each dimension in order for
   the sphere to fit. The box has
   one dimension that is too small.
   Therefore, the sphere will not fit.

APPENDIX A

Order of Operations

Use this phrase to remember the correct order for long mathematical problems: *Please Excuse My Dear Aunt Sally.* This phrase stands for *Parentheses, Exponents, Multiplication, Division, Addition, Subtraction.* These rules can be summarized in **Table 5.**

Table 5 **Order of Operations**

Step	Operation
1	Simplify groups inside parentheses. Start with innermost group and work out.
2	Simplify all exponents.
3	Perform multiplication and division in order from left to right.
4	Perform addition and subtraction in order from left to right.

Look at the following example:
$$4^3 + 2 \times [8 - (3 - 1)] = ?$$
First simplify the operations inside parentheses. Begin with the innermost parentheses:
$$(3 - 1) = 2$$
$$4^3 + 2 \times [8 - 2] = ?$$
Then move on to the next-outer parentheses:
$$[8 - 2] = 6$$
$$4^3 + 2 \times 6 = ?$$
Now, simplify all exponents:
$$4^3 = 64$$
$$64 + 2 \times 6 = ?$$
The next step is to perform multiplication:
$$2 \times 6 = 12$$
$$64 + 12 = ?$$
Finally, solve the addition problem:
$$64 + 12 = 76$$

1. $2^3 \div 2 + 4 \times (9 - 2^2) =$
2. $\dfrac{2 \times (6 - 3) + 8}{4 \times 2 - 6} =$

Geometry

Quite often, a useful way to model the objects and substances studied in science is to consider them in terms of their shapes. For example, many of the properties of a wheel can be understood by pretending that the wheel is a perfect circle.

For this reason, being able to calculate the area or the volume of certain shapes is a useful skill in science. **Table 6** provides equations for the area and volume of several geometric shapes.

Table 6 **Geometric Areas and Volumes**

Geometric Shape		Useful Equations
Rectangle		Area $= lw$
Circle		Area $= \pi r^2$ Circumference $= 2\pi r$
Triangle		Area $= \dfrac{1}{2}bh$
Sphere		Surface area $= 4\pi r^2$ volume $= \dfrac{4}{3}\pi r^3$
Cylinder		Volume $= \pi r^2 h$
Rectangular box		Surface area $= 2(lh + lw + hw)$ volume $= lwh$

1. What is the volume of a cylinder with a diameter of 14 cm and a height of 8 cm?
2. Calculate the surface area of a 4 cm cube.
3. Will a sphere with a volume of 76 cm^3 fit in a rectangular box that is 7 cm $\times$ 4 cm $\times$ 10 cm?

Algebraic Rearrangements

Algebraic equations contain *constants* and variables. Constants are simply numbers, such as *2, 5,* and *7.* Variables are represented by letters such as *x, y, z, a, b,* and *c.* Variables are unspecified quantities and are also called the unknowns.

Often, you will need to determine the value of a variable, but all you will be given will be an equation expressed in terms of algebraic expressions instead of a simple equation expressed in numbers only.

An algebraic expression contains one or more of the four basic mathematical operations: addition, subtraction, multiplication, and division. Constants, variables, or terms made up of both constants and variables can be involved in the basic operations.

The key to figuring out the value of a variable in an algebraic equation is that the quantity described on one side of the equals sign is equal to the quantity described on the other side of the equals sign.

If you are trying to determine the value of a variable in an algebraic expression, you would like to be able to rewrite the equation as a simple one that tells you exactly what x (or some other variable) equals.

But how do you get from a more complicated equation to a simple one?

Again, the key lies in the fact that both sides of the equation are equal. That means if you do the same operation on either side of the equation, the results will still be equal.

Look at the following simple problem:
$$8x = 32$$
If we wish to solve for x, we can multiply or divide each side of the equation by the same factor. You can add, subtract, multiply, or divide anything to or from one side of an equa-

tion as long as you do the same thing to the other side of the equation. In this case, if we divide both sides by 8, we have:

$$\frac{8x}{8} = \frac{32}{8}$$

The 8s on the left side of the equation cancel each other out, and the fraction $\frac{32}{8}$ can be reduced to give the whole number, 4. Therefore, $x = 4$.

Next consider the following equation:
$$x + 2 = 8$$
Remember, we can add or subtract the same quantity from each side. If we subtract 2 from each side, we get
$$x + 2 - 2 = 8 - 2$$
$$x + 0 = 6$$
$$x = 6$$

Now consider one more equation:
$$\frac{x}{5} = 9$$

If we multiply each side by 5, the 5 originally on the left side of the equation cancels out. We are left with x on the left by itself and 45 on the right:
$$x = 45$$
In all cases, *whatever operation is performed on the left side of the equals sign must also be performed on the right side.*

Practice

1. Rearrange each of the following equations to give the value of the variable indicated with a letter.

a. $8x - 32 = 128$ **d.** $-2(3m + 5) = 14$

b. $6 - 5(4a + 3) = 26$ **e.** $\left[8\dfrac{(8 + 2z)}{32}\right] + 2 = 5$

c. $-3(y - 2) + 4 = 29$ **f.** $\dfrac{(6b + 3)}{3} - 9 = 2$

831

---

f. $\dfrac{(6b + 3)}{3} - 9 = 2$

$\dfrac{(6b + 3)}{3} - 9 + 9 = 2 + 9$

$\dfrac{(6b + 3)}{3} = 11$

$\dfrac{(6b + 3)}{3} \times 3 = 11 \times 3$

$6b + 3 = 33$

$6b + 3 - 3 = 33 - 3$

$6b = 30$

$b = 5$

Practice

Answers

1. a. $8x - 32 = 128$
$9x - 32 + 32 = 128 + 32$
$8x = 160$

$x = \dfrac{160}{8} = 20$

b. $6 - 5 \times (4a + 3) = 26$
$6 - 6 - 5 \times (4a + 3) = 26 - 6$
$-5 \times (4a + 3) = 20$

$\dfrac{-5 \times (4a + 3)}{-5} = \dfrac{20}{-5}$

$4a + 3 = -4$
$4a + 3 - 3 = -4 - 3$
$4a - 7$
$a = -1.75$

c. $-3 \times (y - 2) + 4 = 29$
$-3 \times (y - 2) + 4 - 4 = 29 - 4$
$-3 \times (y - 2) = 25$

$\dfrac{-3 \times (y - 2)}{-3} = \dfrac{25}{-3}$

$y - 2 = -8.3$
$y - 2 + 2 = -8.3 + 2$
$y = -6.3$

d. $-2 \times (3m + 5) = 14$

$\dfrac{-2 \times (3m + 5)}{-2} = \dfrac{14}{-2}$

$(3m + 5) = -7$
$3m + 5) - 5 = -7 - 5$
$3m = -12$
$m = -4$

e. $\left[8\dfrac{(8 + 2z)}{32}\right] + 2 = 5$

$\left[8\dfrac{(8 + 2z)}{32}\right] + 2 - 2 = 5 - 2$

$8\dfrac{(8 + 2z)}{32} = 3$

$8\dfrac{(8 + 2z)}{32} \times 32 = 3 \times 32$

$8(8 + 2z) = 96$

$8\dfrac{(8 + 2z)}{8} = \dfrac{96}{8}$

$8 + 2z = 12$
$8 - 8 + 2z = 12 - 8$
$2z = 4$
$z = 2$

Math Skills   **831**

Practice

Answers

1. **a.** 1.23×10^7 m/s
   b. 4.5×10^{-12} kg
   c. 6.53×10^{-5} m
   d. 5.5432×10^{13} s
   e. 2.7315×10^2 K
   f. 6.2714×10^{-4} kg

Scientific Notation

Many quantities that scientists deal with have very large or very small values. For example, about 3 000 000 000 000 000 000 electrons' worth of charge pass through a standard light bulb in one second, and the ink required to make the dot over an *i* in this textbook has a mass of about 0.000 000 001 kg.

Obviously, it is very cumbersome to read, write, and keep track of numbers like these. We avoid this problem by using a method dealing with powers of the number 10.

Study the positive powers of ten shown in the chapter entitled Introduction to Science. You should be able to check those numbers using what you know about exponents. The number of zeros corresponds to the exponent on 10. The number for 10^4 is 10 000; it has 4 zeros.

But how can we use the powers of 10 to simplify large numbers such as the number of electron-sized charges passing through a light bulb? This large number is equal to $3 \times 1\ 000\ 000\ 000\ 000\ 000\ 000$. The factor of 10 has 18 zeros. Therefore, it can be rewritten as 10^{18}. This means that 3 000 000 000 000 000 000 can be expressed as 3×10^{18}.

That explains how to simplify really large numbers, but what about really small numbers, like 0.000 000 001 kg? Negative exponents can be used to simplify numbers that are less than 1.

Next, study the negative powers of 10. The exponent on 10 equals the number of decimal places you must move the decimal point to the right so that there is one digit just to the left of the decimal point. Using the mass of the ink in the dot on an i, the decimal point has to be moved 9 decimal places to the right for the numeral 1 to be just to the left of the decimal point. The mass of the ink, 0.000 000 001 kg, can be rewritten as 1×10^{-9} kg.

Numbers that are expressed as some power of 10 multiplied by another number with only one digit to the left of the decimal point are said to be written in scientific notation. For example, 5943 000 000 is 5.943×10^9 when expressed in scientific notation. The number 0.000 0832 is 8.32×10^{-5} when expressed in scientific notation.

When a number is expressed in scientific notation, it is easy to determine the order of magnitude of the number. The order of magnitude is the power of ten that the number would be rounded to. For example, in the number 5.943×10^9, the order of magnitude is 10^{10}, because 5.943 rounds to another 10, and 10 times 10^9 is 10^{10}. For numbers less than 5, the order of magnitude is just the power of ten when the number is written in scientific notation.

The order of magnitude can be used to help quickly estimate your answers. Simply perform the operations required, but instead of using numbers, use the orders of magnitude. Your final answer should be within two orders of magnitude of your estimate.

Practice

1. Rewrite the following values using scientific notation:

   a. 12 300 000 m/s
   b. 0.000 000 000 0045 kg
   c. 0.00 006 53 m
   d. 55 432 000 000 000 s
   e. 273.15 K
   f. 0.000 627 14 kg

Answers

1. **a.** 0.35 dm
   b. 5.24×10^6 cm^3
   c. 13.45 kg

SI

One of the most important parts of scientific research is being able to communicate your findings to other scientists. Today, scientists need to be able to communicate with other scientists all around the world. They need a common language in which to report data. If you do an experiment in which all of your measurements are in pounds, and you want to compare your results to a French scientist whose measurements are in grams, you will need to convert all of your measurements. For this reason, *Le Système International d'Unités*, or SI system was devised in 1960.

You are probably accustomed to measuring distance in inches, feet, and miles. Most of the world, however, measures distance in centimeters (abbreviated cm), meters (abbreviated m), and kilometers (abbreviated km). The meter is the official SI unit for measuring distance.

Notice that centi*meter* and kilo*meter* each contain the word *meter*. When dealing with SI units, you frequently use the base unit, in this case the meter, and add a prefix to indicate that the quantity you are measuring is a multiple of that unit. Most SI prefixes indicate multiples of 10. For example, the centimeter is 1/100 of a meter. Any SI unit with the prefix *centi-* will be 1/100 of the base unit. A centigram is 1/100 of a gram.

Table 7 **Some SI Units**

Quantity	Unit name	Abbreviation
Length	meter	m
Mass	kilogram	kg
Time	second	s
Temperature	kelvin	K
Amount of substance	mole	mol
Electric current	ampere	A
Pressure	pascal	Pa
Volume	meters^3	m^3

Table 8 **Some SI Prefixes**

Prefix	Abbreviation	Exponential factor
Giga-	G	10^9
Mega-	M	10^6
Kilo-	k	10^3
Hecto-	h	10^2
Deka-	da	10^1
Deci-	d	10^{-1}
Centi-	c	10^{-2}
Milli-	m	10^{-3}
Micro-	μ	10^{-6}
Nano-	n	10^{-9}
Pico-	p	10^{-12}
Femto-	f	10^{-15}

What about the *kilo*meter? The prefix *kilo-* indicates that the unit is 1000 times the base unit. A kilometer is equal to 1000 meters. Multiples of 10 make dealing with SI values much easier than values such as feet or gallons. If you wish to convert from feet to miles, you must remember a large conversion factor, 1.893939×10^{-4} miles = foot. If you wish to convert from kilometers to meters, you need only look at the prefix to know that you will multiply by 1000.

Table 7 lists the SI units. *Table 8* gives the possible prefixes and their meaning. When working with a prefix, simply take the unit abbreviation and add the prefix abbreviation to the front of the unit. For example, the abbreviation for kilometer is written km.

Practice

1. Convert each value to the requested units.
   a. 0.035 m to decimeters
   b. 5.24 m^3 to centimeters^3
   c. 13450 g to kilograms

833

Answers

1. a. 4
  b. 5
  c. 4
  d. 3
2. a. 0.004 dm + 0.12508 dm =
    0.12908 dm = 0.129 dm

  b. $\dfrac{340 \text{ m}}{0.1257 \text{ s}} =$

    2704.852824 m/s 2700 m/s

  c. 40.1 kg $\times$ 0.2453 m^2 =
    9.83653 kg•m^2 =
    9.84 kg•m^2

  d. 1.03 g $-$ 0.0456 g =
    0.9844 g = 0.98 g

Significant Figures

The following list can be used to review how to determine the number of significant figures in a reported value. After you have reviewed the rules, use *Table 9* to check your understanding of the rules. Cover up the second column of the table, and try to determine how many significant figures each number has. If you get confused, refer to the rule given.

Table 9 **Significant Figures**

Measurement	Number of significant figures	Rule
12 345	5	1
2400 cm	2	3
305 kg	3	2
2350. cm	4	4
234.005 K	6	2
12.340	5	6
0.001	1	5
0.002 450	4	5 and 6

Rules for Determining the Number of Significant Figures in a Measurement:

1. All nonzero digits are significant. **Example: 1246** has four significant figures (shown in red).
2. Any zeros between significant digits are also significant. **1206** has four significant figures.
3. If the value does not contain a decimal point, any zeros to the right of a nonzero digit are not significant. **1200** has only two significant figures.
4. Any zeros to the right of a significant digit and to the left of a decimal point are significant. **1200.** has four significant figures.
5. If a value has no significant digits to the left of a decimal point, any zeros to the right of the decimal point, and to the left of a significant digit, are not significant. **Example: 0.0012** has only two significant figures.
6. If a measurement is reported that ends with zeros to the right of a decimal point, those zeros are significant. **Example: 0.1200** has four significant figures.

If you are adding or subtracting two measurements, your answer can only have as many decimal positions as the value with the least number of decimal places. The final answer in the following problem has five significant figures. It has been rounded to two decimal places because 0.04 g only has two decimal places.

$$\begin{array}{r} 134.050 \text{ g} \\ - 0.04 \text{ g} \\ \hline 134.01 \text{ g} \end{array}$$

When multiplying or dividing measurements, your final answer can only have as many significant figures as the value with the least number of significant figures. Examine the following multiplication problem.

$$\begin{array}{r} 12.0 \text{ cm}^2 \\ \times 0.04 \text{ cm} \\ \hline 0.5 \text{ cm}^3 \end{array}$$

The final answer has been rounded to one significant figure because 0.04 cm has only one. When performing both types of operations (addition/subtraction vs. multiplication/division), complete one type, round, perform the other type, round, perform the other type, and round the result.

Practice

1. Determine the number of significant figures in each of the following measurements:

  a. 65.04 mL   **c.** 0.007504 kg
  b. 564.00 m   **d.** 1210 K

2. Perform each of the following calculations, and report your answer with the correct number of significant figures and units:

  a. 0.004 dm + 0.12508 dm
  b. 340 m ÷ 0.1257 s
  c. 40.1 kg × 0.2453 m^2
  d. 1.03 g − 0.0456 g

Graphing Skills

Line Graphs

In laboratory experiments, you will usually be controlling one variable and seeing how it affects another variable. Line graphs can show these relations clearly. For example, you might perform an experiment in which you measure the growth of a plant over time to determine the rate of the plant's growth. In this experiment, you are controlling the time intervals at which the plant height is measured. Therefore, time is called the *independent variable*. The height of the plant is the *dependent variable*. **Table 10** gives some sample data for an experiment to measure the rate of plant growth.

The independent variable is plotted on the *x*-axis. This axis will be labeled *Time (days)*, and will have a range from 0 days to 35 days. Be sure to properly label your axis including the units on the values.

The dependent variable is plotted on the *y*-axis. This axis will be labeled *Plant Height (cm)* and will have a range from 0 cm to 5 cm.

Think of your graph as a grid with lines running horizontally from the *y*-axis, and vertically from the *x*-axis. To plot a point, find the *x* (in this example time) value on the *x* axis. Follow the vertical line from the *x* axis until it

intersects the horizontal line from the *y*-axis at the corresponding *y* (in this case height) value. At the intersection of these two lines, place your point. **Figure 3** shows what a line graph of the data in **Table 10** might look like.

Figure 3

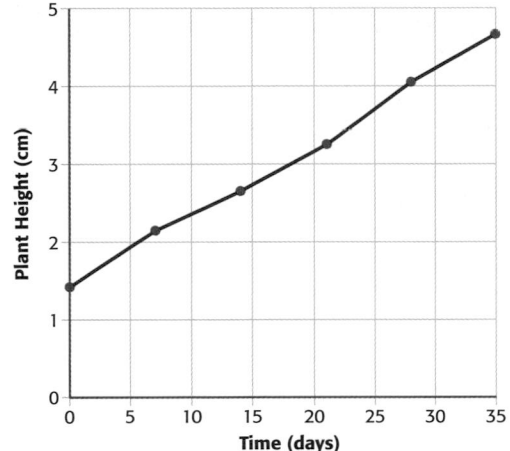

Table 10 Experimental Data for Plant Growth versus Time

Time (days)	Plant height (cm)
0	1.43
7	2.16
14	2.67
21	3.25
28	4.04
35	4.67

Practice

1. What does the line in **Figure 3** show, and what can you conclude about the plants used in the experiment?

2. Create a line graph of the following data.

Number of Days	Plant height (cm)
0	1.46
7	2.67
14	3.89
21	4.82

3. Compare the graph you made with **Figure 3**. What can you conclude about the two different groups of plants?

Teaching Tip

Be sure students understand that not all graphs are alike. Each type of graph showcases a certain aspect of a data set. A line graph reveals trends over constant ranges. Bar graphs work better for data that is organized by category. Pie charts easily show proportions making up a whole.

Practice

Answers

1. The line in **Figure 3** shows how the plant's height changed over time. It shows that the plants continued to grow at a steady rate over the course of the experiment.

2. Answers may vary. The axes of the graph should be clearly labeled and include units.

3. The plants in the graph from item 2 grew faster than those represented in **Figure 3**.

Scatter Plots

Practice

Answers

1. Answers may vary but should show a straight line.
2. The line represents an average number of magazine subscriptions, and how that average changed over time.
3. The city represented in item 1 had an average of more magazine subscriptions per 1000 households though the rate of change remained about the same as the city represented in **Figure 5.**

Some experiments or groups of data are best represented in a graph that is similar to a line graph and that is called a scatter plot. As in a line graph, the data points are plotted on the graph by using values on an x-axis and a y-axis. Scatter plots are often used to find trends in data. Instead of connecting the data points with a line, a trend can be represented by a best-fit line. A best-fit line is a line that represents all of the data points without necessarily going through all of them. To find a best-fit line, pick a line that is equidistant from as many data points as possible. Examine the graph below.

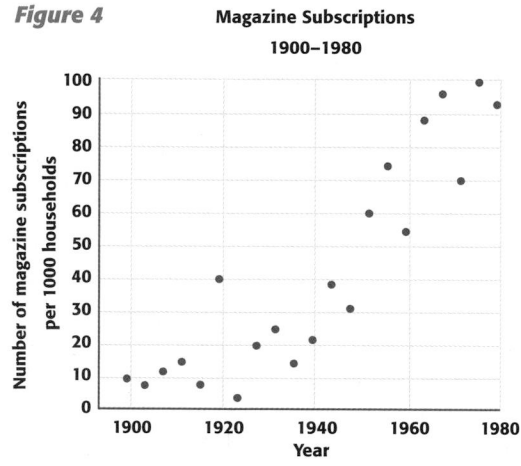

Figure 4 — Magazine Subscriptions 1900–1980

If we connected all of the data points with lines, the lines would create a zigzag pattern that would not tell us much about our data. But if we find a best-fit line, we can see a trend more clearly. Furthermore, if we pick two points on the best-fit line, we can estimate its slope. Examine the dotted lines on **Figure 5.**

The points can be estimated as 18 magazine subscriptions per 100 households in 1920, and 42 magazine subscriptions per 100 households in 1940. If we subtract 1920 from 1940, and 18

subscriptions from 42 subscriptions (using the point slope formula), we see that the line shows a trend of an increase of 24 subscriptions per 1000 households acres every 20 years. Scatter plots can also be used when there are two or more trends within one group of data or when there is no distinct trend at all.

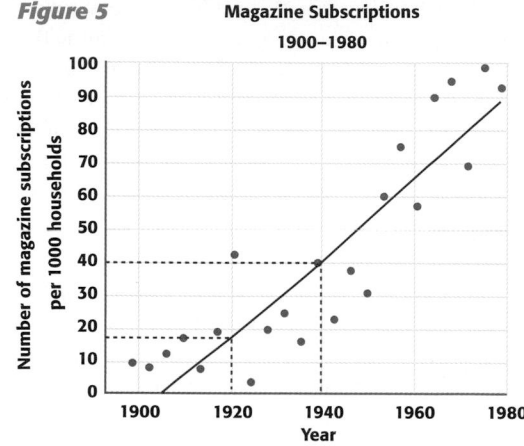

Figure 5 — Magazine Subscriptions 1900–1980

Practice

1. Copy the graph below, and draw a best-fit line.

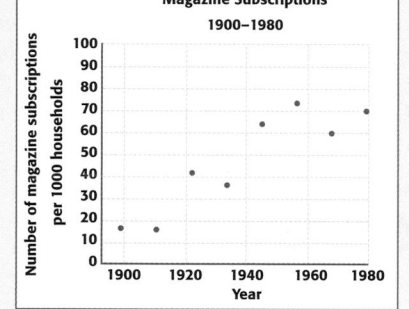

Magazine Subscriptions 1900–1980

2. What does that line represent?
3. If these were the data from a different city than the data in **Figure 5,** what conclusions could you draw about the two cities?

Bar Graphs

Figure 6

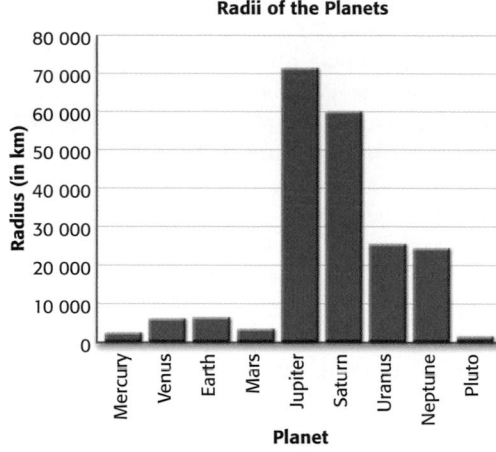

Radii of the Planets

Bar graphs make it easy to compare data quickly. We can see from *Figure 6* that Jupiter has the largest radius, and that Pluto has the smallest radius. We can also quickly arrange the planets in order of size.

Bar graphs can also be used to identify trends, especially trends among differing quantities. Examine *Figure 7* below.

Figure 7

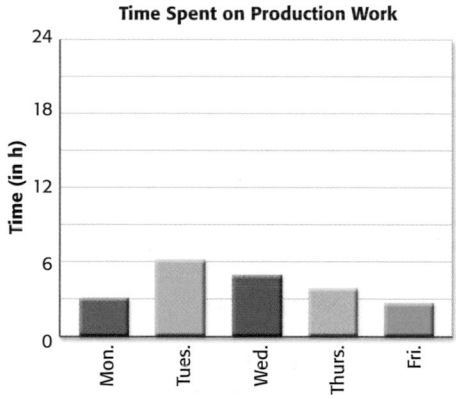

Time Spent on Production Work

The data are represented accurately, but it is not easy to draw conclusions quickly. Remember that when you are creating a graph, you want the graph to be as clear as possible. If we graph the exact same data on a graph with slightly different axes, as shown in *Figure 8*, it may be much easier to draw conclusions.

Figure 8

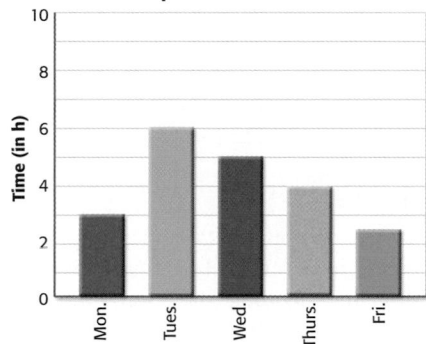

Time Spent on Production Work

Practice

1. What day of the week is most productive, according to *Figure 8*?
2. What day of the week is least productive?
3. Using the following data, create a clear and easily readable bar graph.

Fiscal period	Money spent (in millions)
First quarter	89
Second quarter	56
Third quarter	72
Fourth quarter	41

837

Practice

Answers

1. Tuesday
2. Friday
3. Answers may vary but student's graphs should have clearly labeled axes and have graphs that are easily interpreted.

Practice

Answers

1. Answers may vary but pie charts should be clearly labeled and easily interpreted.
2. Approximately half of the total area is used.

Pie Charts

Pie charts are an easy way to visualize how parts make up a whole. Frequently, pie charts are made from percentage data such as the data in *Table 11.*

Table 11 **Elemental Composition of Earth's Crust**

Element	Percentage of Earth's Crust
Oxygen	46%
Silicon	28%
Aluminum	8%
Iron	6%
Calcium	4%
Sodium	2%
Magnesium	2%
Potassium	2%
Titanium	1%
All remaining elements	1%

To create a pie chart, begin by drawing a circle. Imagine dividing the circle into 100 equal parts. Because 50 parts would be half of the circle, we know that 46% will be slightly less than half of the pie. We shade a piece that is less than half, and label it "Oxygen." Continue this process until the entire pie graph has been filled. Each element should be a different color to make the chart easy to read as in *Figure 9.*

Another way to construct a pie chart involves using a protractor. This method is especially helpful when your data can't be converted into simple fractions. First, convert the percentages to degrees by dividing each number by 100 and multiplying that result by 360. Next, draw a circle and make a vertical mark across the top of the circle. Using a protractor, measure the largest angle from your table and mark this angle along the circumference. For example, 32.9% would be 118° because 32.9/100 = .329 and .329 × 360 = 118.

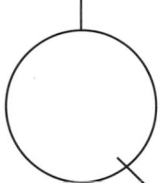

Next, measure a second angle from the second mark to make a third mark along the circumference. Continue this process until all of your slices are measured. Draw lines from the marks to the center of the circle, and label each slice.

Elemental Composition of Earth's Crust

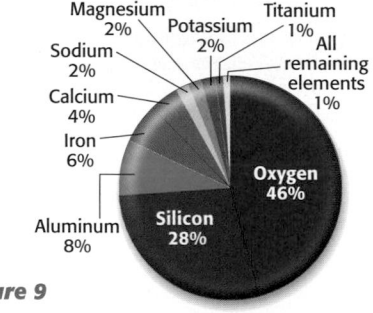

Magnesium 2% Potassium 2% Titanium 1%
Sodium 2%
All remaining elements 1%
Calcium 4%
Iron 6%
Oxygen 46%
Aluminum 8%
Silicon 28%

Figure 9

Practice

1. Use the data below to create a pie chart.

Kind of land use	Percentage of total land
Grassland and rangeland	29
Wilderness and parks	9
Urban	2
Wetlands and deserts	3
Forest	30
Cropland	17

2. If humans use half of forests and grasslands, plus all of croplands and urban areas, how much of the total land is used by humans?

Lab Skills

Making Measurements in the Laboratory

Reading a balance for mass

When a balance is required for determining mass, you will probably use a centigram balance like the one shown in *Figure 10.* The centigram balance is sensitive to 0.01 g. This means that your mass readings should all be recorded to the nearest 0.01 g.

Before using the balance, always check to see if the pointer is resting at zero. If the pointer is not at zero, check the slider weights. If all the slider weights are at zero, turn the zero adjust knob until the pointer rests at zero. The zero adjust knob is usually located at the far left end of the balance beam as shown in *Figure 10.* Note: The balance will not adjust to zero if the movable pan has been removed.

Figure 10

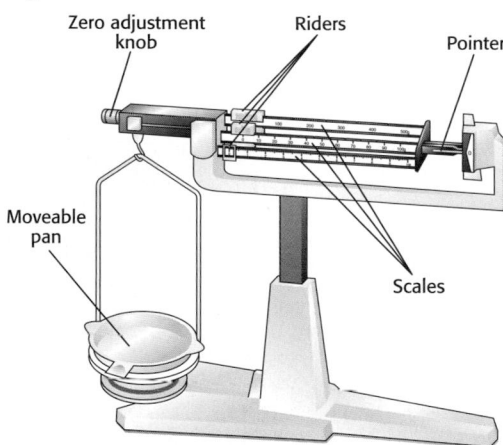

Zero adjustment knob · Riders · Pointer · Moveable pan · Scales

In many experiments you will be asked to obtain a specified mass of a solid. When measuring the mass of a chemical, place a piece of weighing paper on the balance pan. **Never place chemicals or hot objects directly on the balance pan.** They can permanently damage the surface of the balance pan and affect the accuracy of later measurements.

Determine the mass of the paper by adjusting the weights on the various scales. Record the mass of the weighing paper to the nearest 0.01 g. Then add the mass you wish to obtain by sliding over the appropriate weights on the balance. For example, if your weighing paper has a mass of 0.15 g, and you wish to obtain 13 g of table salt, the balance begins at 0.15 g. You then need to add 13 g to this mass. Do this by sliding the 10-gram scale to 10 and the 1-gram scale to 3. The balance is no longer balanced.

Slowly add the solid onto the weighing paper until the balance is once again balanced. Do not waste time trying to obtain *exactly* 13.00 g of a solid. Instead, read the mass when the pointer swings close to zero. Remember, you must read the final mass on the balance and subtract the mass of the weighing paper (0.15 g) from it to find the mass of the solid to two decimal places, as is appropriate for measurements that are made by using a centigram balance.

839

Teaching Tip

It is not practical to show every possible balance configuration in the textbook. If your balances are significantly different from the one shown in **Figure 10,** be sure to make clear to students what the differences are.

Teaching Tip

Because of the hazards due to spills, many school districts are eliminating use of mercury thermometers. While red alcohol thermometers are not quite as precise, the slight variation encountered is usually not significant for the high school science lab.

Measuring temperature with a thermometer

A thermometer is used to measure temperature. Examine your thermometer and the temperature range for the Celsius scale. You will probably be using an alcohol or a digital thermometer in your laboratory.

Mercury thermometers are hazardous and will probably not be available in your school laboratory, although you may still have a mercury fever thermometer at home. **If a mercury thermometer should ever break, immediately notify your teacher or parent. Your teacher or parent will clean up the spill. Do not touch the mercury.**

Alcohol thermometers, like mercury thermometers, have a column of liquid that rises in a glass cylinder depending on the temperature at the tip of the thermometer. One caution concerning alcohol thermometers is that they can burst at very high temperatures. Never let the thermometer be exposed to temperatures well above its range.

When working with any thermometer, it is especially important to pay close attention to the precision of the instrument. Most alcohol thermometers are marked in intervals of 1°C. The intervals are usually so close together that it is impossible to estimate temperature values measured with such a thermometer to any more precision than a half a degree, 0.5°C. Thus, if you are using this type of thermometer, it would be impossible to actually measure a temperature like 25.15°C.

It is also very important to keep your eye at about the same level as the colored fluid in the thermometer. If you are looking at the thermometer from below, the reading you see will appear a degree or two higher than it really is. Similarly, if you look at the thermometer from above, the reading will seem to be a degree or two lower than it really is.

Reading a graduated cylinder for volume

Many different types of laboratory glassware, from beakers to flasks contain markings indicating volume. However, these markings are merely approximate, and they are not consistently checked when the beaker or flask was made.

For truly accurate volume measurements, you should use a graduated cylinder, like the one shown in *Figure 11.* When a graduated cylinder is made, its accuracy is checked and rechecked. You will also notice that a graduated cylinder is marked in smaller increments than beakers are (usually individual milliliters, although some graduated cylinders are even more precise).

Most liquids have a concave surface that forms in a test tube or graduated cylinder. This concave surface is called a meniscus. When measuring the volume of a liquid, you must consider the meniscus, like the one labeled *Figure 11.* Always measure the volume from the bottom of the meniscus. The markings on a graduated cylinder are designed to take into account the little bits of water that extend up along the walls slightly above the marking lines.

It may be difficult to read a volume measurement, so if you need to, hold a piece of white paper behind the graduated cylinder. This should make the meniscus level easier to see.

Figure 11

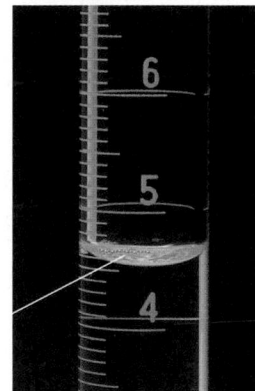

Meniscus, 4.5 mL

How to Write a Laboratory Report

In many of the laboratory investigations that you will be doing, you will be trying to support a hypothesis or answer a question by performing experiments following the scientific method. You will frequently be asked to summarize your experiments in a laboratory report. Laboratory reports should contain the following parts:

Title

This is the name of the experiment you are doing. If you are performing an experiment from a laboratory manual, the title will be the same as the title of the experiment.

Hypothesis

The hypothesis is what you think will happen during the investigation. It is often written as an "If . . . then" statement. When you conduct your experiment, you will be changing one condition, or variable, and observing and measuring the effect of this change. The condition that you are changing is called the *independent* variable and should follow the "If . . ." statement. The effect that you expect to observe is called the *dependent* variable and should follow the ". . . then" statement. For example, look at the following hypothesis:

If salamanders are reared in acidic water, then more salamanders will develop abnormally.

"If salamander are reared in acidic water" is the independent variable—salamanders normally live in nearly neutral water and you are changing this to acidic water. "Then more salamanders will develop abnormally" is the dependent variable—this is the change that you expect to observe and measure.

Materials

List of all the equipment and other supplies you will need to complete the experiment. If the investigation is taken from a laboratory manual, the materials are generally listed for you, but you will need to recopy them into your lab report. It is important that your lab report be complete enough for someone to use it to retest your results.

Procedure

The procedure is a step-by-step explanation of exactly what you did in the experiment. Investigations from laboratory manuals will have the procedure carefully written out for you, but you must write the procedure in your lab report EXACTLY as you performed it. This will not necessarily be an exact copy of the procedure in your laboratory manual.

Data

Your data are your observations. Data can include measurements, so it is important to record the correct units. They are often recorded in the form of tables, graphs, and drawings.

Analyses and Conclusions

This part of the report explains what you have learned. You should evaluate your hypothesis and explain any errors you made in the investigation. Keep in mind that not all hypotheses will be correct. Sometimes you will disprove your original hypothesis, rather than prove it. You simply need to explain why things did not work out the way you thought they would. In laboratory manual investigations, there will be questions to guide you in analyzing your data. You should use these questions as a basis for your conclusions.

841

Teaching Tip

While instructions for the traditional lab report are given here, this is not necessary for most labs. The decision whether to require full-fledged traditional lab reports is best left up to the teacher.

The *Holt Science Spectrum* package includes a system of *Datasheets*, which provide blank data tables, spaces for graphs, and room for answering the in-lab questions. These Datasheets are available as a booklet and also can be found on the *One-Stop Planner CD-ROM with Test Generator.*

Table 1 **SI Base Units**

Quantity	Unit	Abbreviation
Length	meter	m
Mass	kilogram	kg
Time	second	s
Temperature	kelvin	K
Electric current	ampere	A
Amount of substance	mole	mol
Luminous intensity	candela	cd

Table 2 **Other Commonly Used Units**

Quantity	Unit	Abbreviation	Conversion
Electric charge	coulomb	C	1 A·s
Temperature	degree Celsius	°C	1 K
Frequency	hertz	Hz	$1/\text{s}$
Work and energy	joule	J	$1 \dfrac{\text{kg·m}^2}{\text{s}^2} = 1 \text{ N·m}$
Force	newton	N	$1 \dfrac{\text{kg·m}}{\text{s}^2}$
Pressure	pascal	Pa	$1 \dfrac{\text{kg}}{\text{m·s}^2} = 1 \dfrac{\text{N}}{\text{m}^2}$
Angular displacement	radian	rad	(unitless)
Electric potential difference	volt	V	$1 \dfrac{\text{kg·m}^2}{\text{A·s}^3} = 1 \dfrac{\text{J}}{\text{C}}$
Power	watt	W	$1 \dfrac{\text{kg·m}^2}{\text{s}^3} = 1 \dfrac{\text{J}}{\text{s}}$
Resistance	ohm	Ω	$1 \dfrac{\text{kg·m}^2}{\text{A}^2\text{·s}^3} = 1 \dfrac{\text{V}}{\text{A}}$

Table 3 **Densities of Various Materials**

Material	Density (g/cm³)	Material	Density (g/cm³)
Air, dry	1.293×10^{-3}	Ice	0.917
Aluminum	2.70	Iron	7.86
Bone	1.7–2.0	Lead	11.3
Brick, common	1.9	Mercury	13.5336
Butter	0.86–0.87	Paper	0.7–1.15
Carbon (diamond)	3.5155	Rock salt	2.18
Carbon (graphite)	2.2670	Silver	10.5
Copper	8.96	Sodium	0.97
Cork	0.22–0.26	Stainless steel	8.02
Ethanol	0.783	Steel	7.8
Gasoline	0.7	Sugar	1.59
Gold	19.3	Water (at 25°C)	0.997 05
Helium	1.78×10^{-4}	Water (ice)	0.917

Table 4 **Specific Heats**

Material	c (J/kg·K)	Material	c (J/kg·K)
Acetic acid (CH_3COOH)	2070	Lead (Pb)	129
Air	1007	Magnetite (Fe_3O_4)	619
Aluminum (Al)	897	Mercury (Hg)	140
Calcium (Ca)	647	Methane (CH_4)	2200
Calcium carbonate ($CaCO_3$)	818	Neon (Ne)	1030
Carbon (C, diamond)	487	Nickel (Ni)	444
Carbon (C, graphite)	709	Nitrogen (N_2)	1040
Carbon dioxide (CO_2)	843	Oxygen (O_2)	918
Copper (Cu)	385	Platinum (Pt)	133
Ethanol (CH_3CH_2OH)	2440	Silver (Ag)	234
Gold (Au)	129	Sodium (Na)	1228
Helium (He)	5193	Sodium chloride (NaCl)	864
Hematite (Fe_2O_3)	650	Tin (Sn)	228
Hydrogen (H_2)	14 304	Tungsten (W)	132
Hydrogen peroxide (H_2O_2)	2620	Water (H_2O)	4186
Iron (Fe)	449	Zinc (Zn)	388

Values at 25°C and 1 atm pressure

843

Figure 1 **Periodic Table of the Elements**

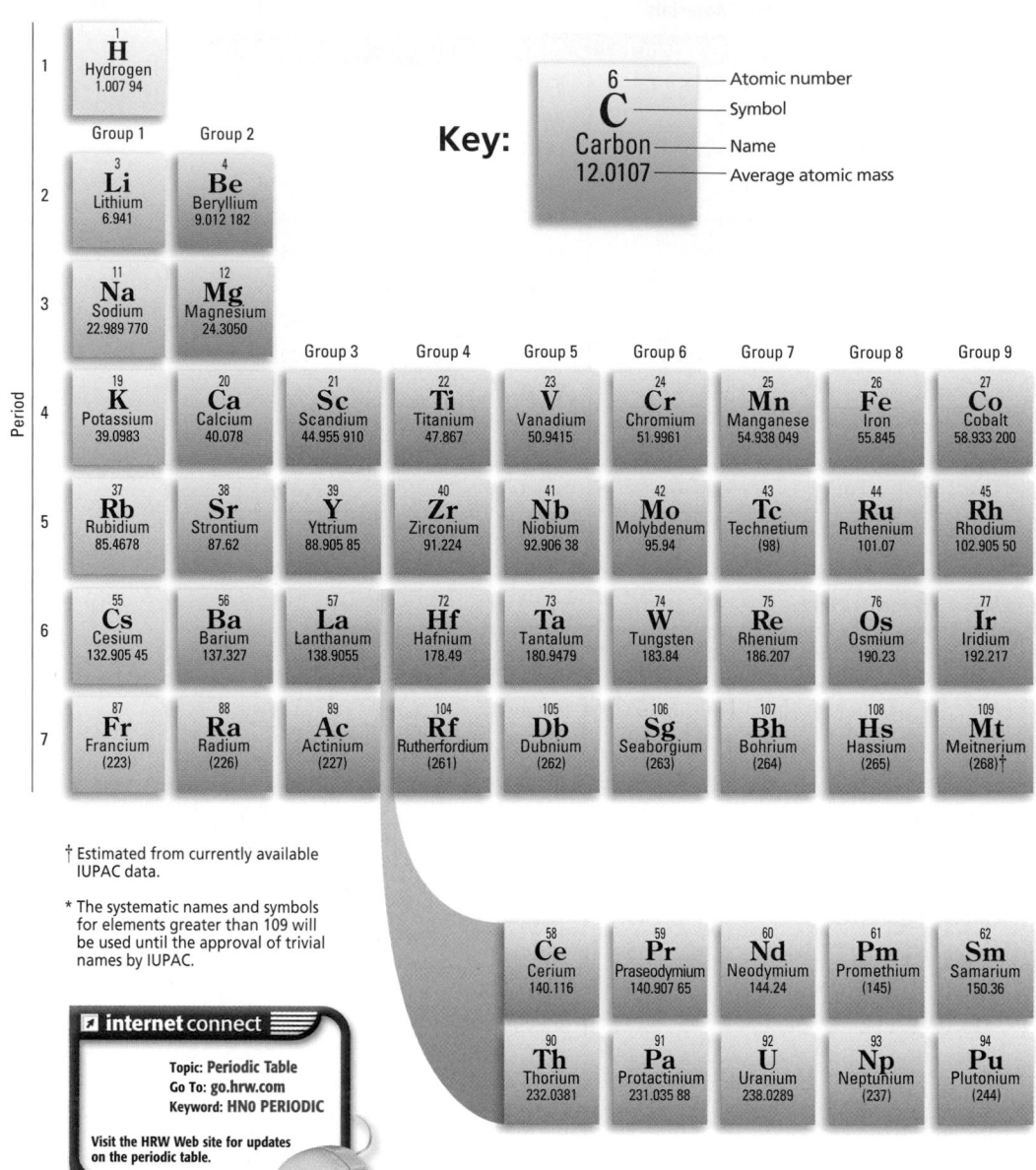

Key:

- 6 — Atomic number
- C — Symbol
- Carbon — Name
- 12.0107 — Average atomic mass

| | | | | Group 3 | Group 4 | Group 5 | Group 6 | Group 7 | Group 8 | Group 9 |

Period

1
1 H Hydrogen 1.007 94

Group 1 | Group 2

2
3 Li Lithium 6.941 | 4 Be Beryllium 9.012 182

3
11 Na Sodium 22.989 770 | 12 Mg Magnesium 24.3050

4
19 K Potassium 39.0983 | 20 Ca Calcium 40.078 | 21 Sc Scandium 44.955 910 | 22 Ti Titanium 47.867 | 23 V Vanadium 50.9415 | 24 Cr Chromium 51.9961 | 25 Mn Manganese 54.938 049 | 26 Fe Iron 55.845 | 27 Co Cobalt 58.933 200

5
37 Rb Rubidium 85.4678 | 38 Sr Strontium 87.62 | 39 Y Yttrium 88.905 85 | 40 Zr Zirconium 91.224 | 41 Nb Niobium 92.906 38 | 42 Mo Molybdenum 95.94 | 43 Tc Technetium (98) | 44 Ru Ruthenium 101.07 | 45 Rh Rhodium 102.905 50

6
55 Cs Cesium 132.905 45 | 56 Ba Barium 137.327 | 57 La Lanthanum 138.9055 | 72 Hf Hafnium 178.49 | 73 Ta Tantalum 180.9479 | 74 W Tungsten 183.84 | 75 Re Rhenium 186.207 | 76 Os Osmium 190.23 | 77 Ir Iridium 192.217

7
87 Fr Francium (223) | 88 Ra Radium (226) | 89 Ac Actinium (227) | 104 Rf Rutherfordium (261) | 105 Db Dubnium (262) | 106 Sg Seaborgium (263) | 107 Bh Bohrium (264) | 108 Hs Hassium (265) | 109 Mt Meitnerium (268)†

† Estimated from currently available IUPAC data.

* The systematic names and symbols for elements greater than 109 will be used until the approval of trivial names by IUPAC.

58 Ce Cerium 140.116 | 59 Pr Praseodymium 140.907 65 | 60 Nd Neodymium 144.24 | 61 Pm Promethium (145) | 62 Sm Samarium 150.36

90 Th Thorium 232.0381 | 91 Pa Protactinium 231.035 88 | 92 U Uranium 238.0289 | 93 Np Neptunium (237) | 94 Pu Plutonium (244)

internet connect

Topic: Periodic Table
Go To: go.hrw.com
Keyword: HN0 PERIODIC

Visit the HRW Web site for updates on the periodic table.

844

Metals
- Alkali metals
- Alkaline-earth metals
- Transition metals
- Other metals

Nonmetals
- Hydrogen
- Semiconductors (also known as *metalloids*)
- Other nonmetals
- Halogens
- Noble gases

Group 18

| 2 He Helium 4.002 602 |

Group 13	Group 14	Group 15	Group 16	Group 17	
5 B Boron 10.811	6 C Carbon 12.0107	7 N Nitrogen 14.006 74	8 O Oxygen 15.9994	9 F Fluorine 18.998 4032	10 Ne Neon 20.1797
13 Al Aluminum 26.981 538	14 Si Silicon 28.0855	15 P Phosphorus 30.973 761	16 S Sulfur 32.066	17 Cl Chlorine 35.4527	18 Ar Argon 39.948

Group 10	Group 11	Group 12						
28 Ni Nickel 58.6934	29 Cu Copper 63.546	30 Zn Zinc 65.39	31 Ga Gallium 69.723	32 Ge Germanium 72.61	33 As Arsenic 74.921 60	34 Se Selenium 78.96	35 Br Bromine 79.904	36 Kr Krypton 83.80
46 Pd Palladium 106.42	47 Ag Silver 107.8682	48 Cd Cadmium 112.411	49 In Indium 114.818	50 Sn Tin 118.710	51 Sb Antimony 121.760	52 Te Tellurium 127.60	53 I Iodine 126.904 47	54 Xe Xenon 131.29
78 Pt Platinum 195.078	79 Au Gold 196.966 55	80 Hg Mercury 200.59	81 Tl Thallium 204.3833	82 Pb Lead 207.2	83 Bi Bismuth 208.980 38	84 Po Polonium (209)	85 At Astatine (210)	86 Rn Radon (222)
110 Uun* Ununnilium (269)†	111 Uuu* Unununium (272)†	112 Uub* Ununbium (277)†		114 Uuq* Ununquadium (285)†				

A team at Lawrence Berkeley National Laboratories reported the discovery of elements 116 and 118 in June 1999. The same team retracted the discovery in July 2001. The discovery of element 114 has been reported but not confirmed.

63 Eu Europium 151.964	64 Gd Gadolinium 157.25	65 Tb Terbium 158.925 34	66 Dy Dysprosium 162.50	67 Ho Holmium 164.930 32	68 Er Erbium 167.26	69 Tm Thulium 168.934 21	70 Yb Ytterbium 173.04	71 Lu Lutetium 174.967
95 Am Americium (243)	96 Cm Curium (247)	97 Bk Berkelium (247)	98 Cf Californium (251)	99 Es Einsteinium (252)	100 Fm Fermium (257)	101 Md Mendelevium (258)	102 No Nobelium (259)	103 Lr Lawrencium (262)

The atomic masses listed in this table reflect the precision of current measurements. (Values listed in parentheses are those of the element's most stable or most common isotope.) In calculations throughout the text, however, atomic masses have been rounded to two places to the right of the decimal.

Figure 2   **The Electromagnetic Spectrum**

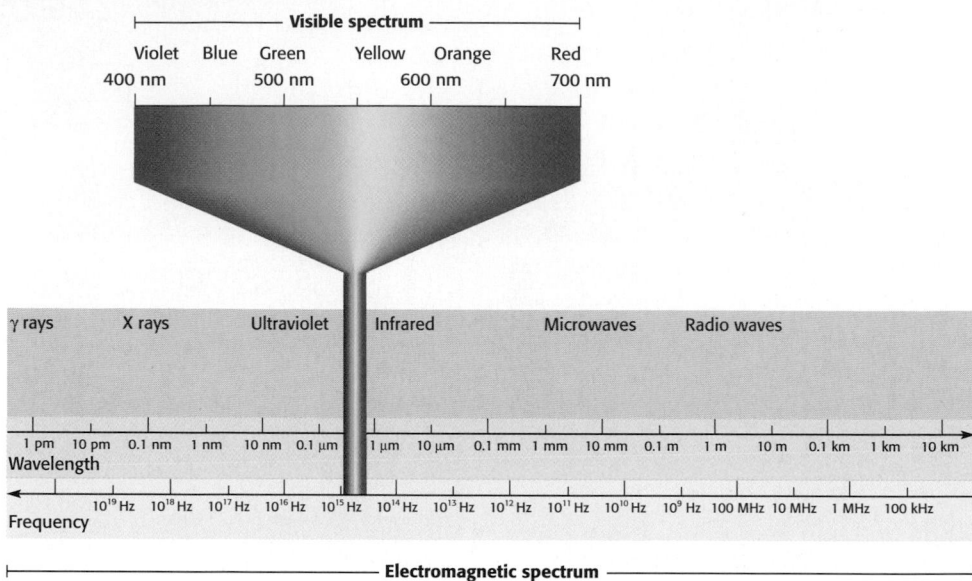

Table 5   **Properties of the Planets**

Planet	Diameter (km)	Average surface temperature (°C)	Number of moons	Atmosphere
Mercury	4879	350	0	Essentially none
Venus	12 104	460	0	Thick: carbon dioxide, sulfuric acid
Earth	12 756	20	1	nitrogen, oxygen
Mars	6790	−23	2	Thin: carbon dioxide
Jupiter	142 984	−120	16	Hydrogen, helium, ammonia, methane
Saturn	120 536	−180	18	Hydrogen, helium, ammonia, methane
Uranus	51 118	−210	20	Hydrogen, helium, ammonia, methane
Neptune	49 528	−220	8	Hydrogen, helium, methane
Pluto	2390	−230	1	Very thin: nitrogen, methane

Figure 3   Sky Maps for the Northern Hemisphere

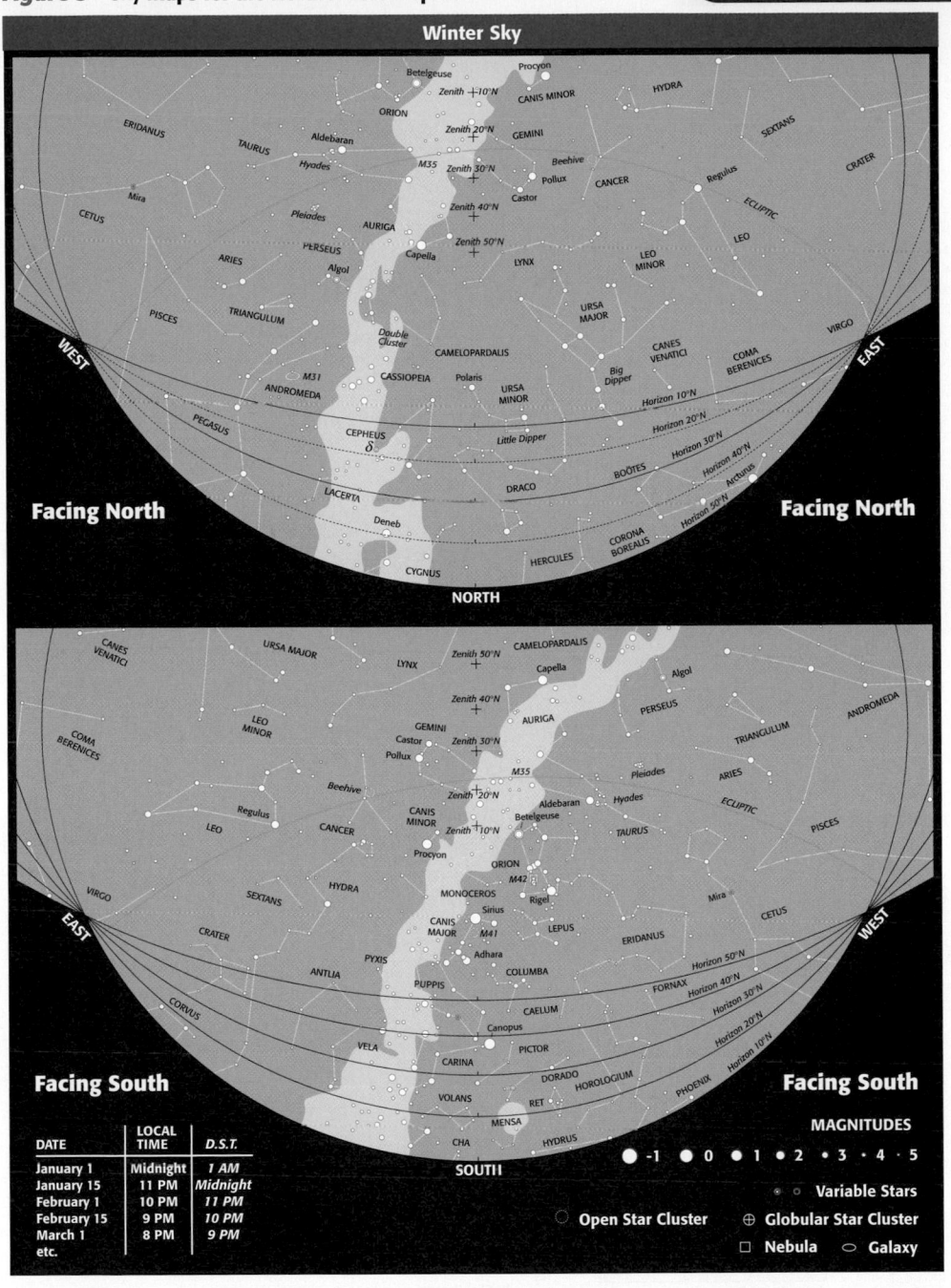

Winter Sky

DATE	LOCAL TIME	D.S.T.
January 1	Midnight	1 AM
January 15	11 PM	Midnight
February 1	10 PM	11 PM
February 15	9 PM	10 PM
March 1	8 PM	9 PM
etc.		

MAGNITUDES

● -1   ● 0   ● 1   ● 2   • 3   · 4   · 5

◦ ○ Variable Stars
○ Open Star Cluster   ⊕ Globular Star Cluster
□ Nebula   ◯ Galaxy

847

Figure 4 **The World: Physical**

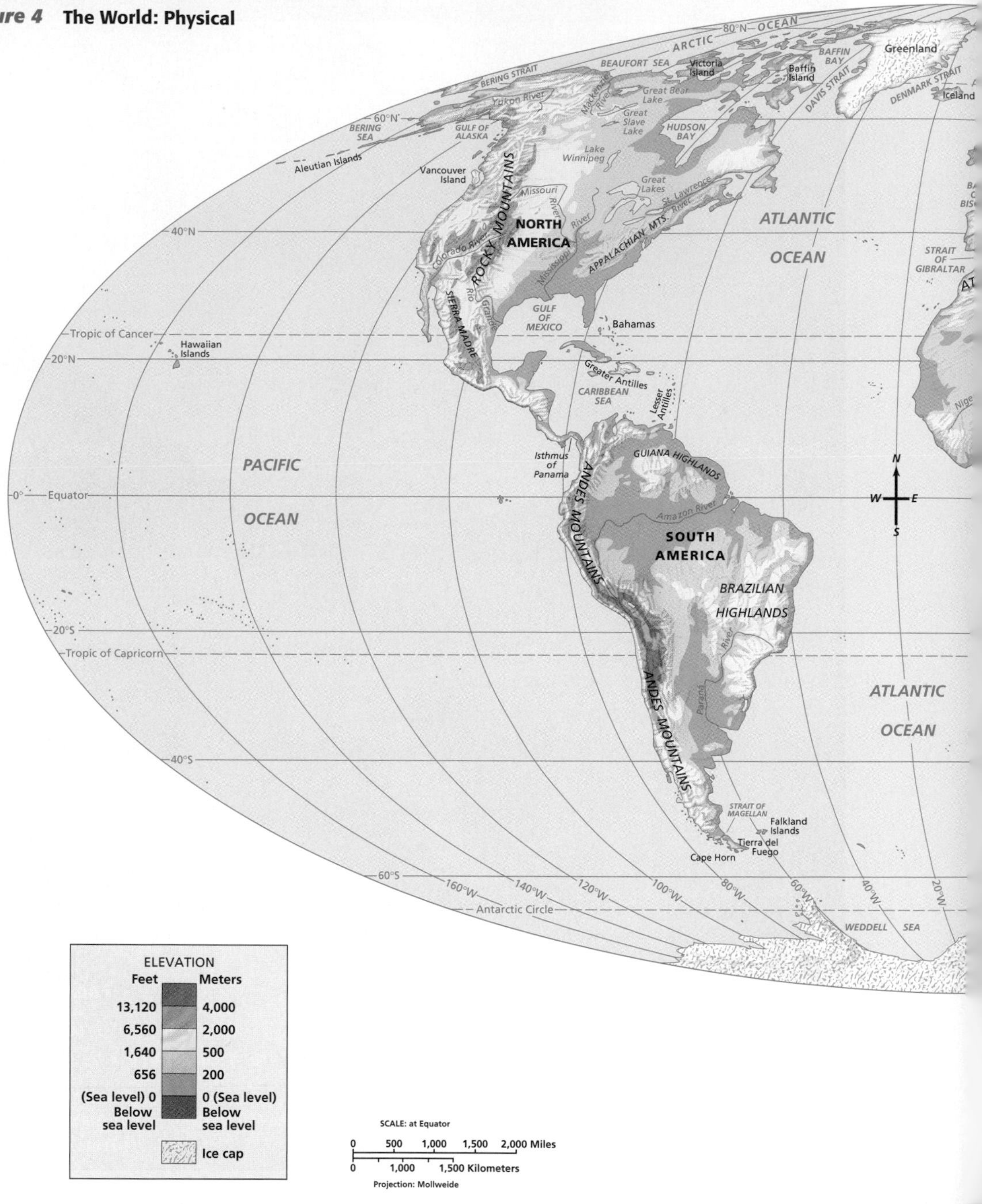

ELEVATION

Feet		Meters
13,120		4,000
6,560		2,000
1,640		500
656		200
(Sea level) 0		0 (Sea level)
Below sea level		Below sea level

Ice cap

SCALE: at Equator

0 500 1,000 1,500 2,000 Miles

0 1,000 1,500 Kilometers

Projection: Mollweide

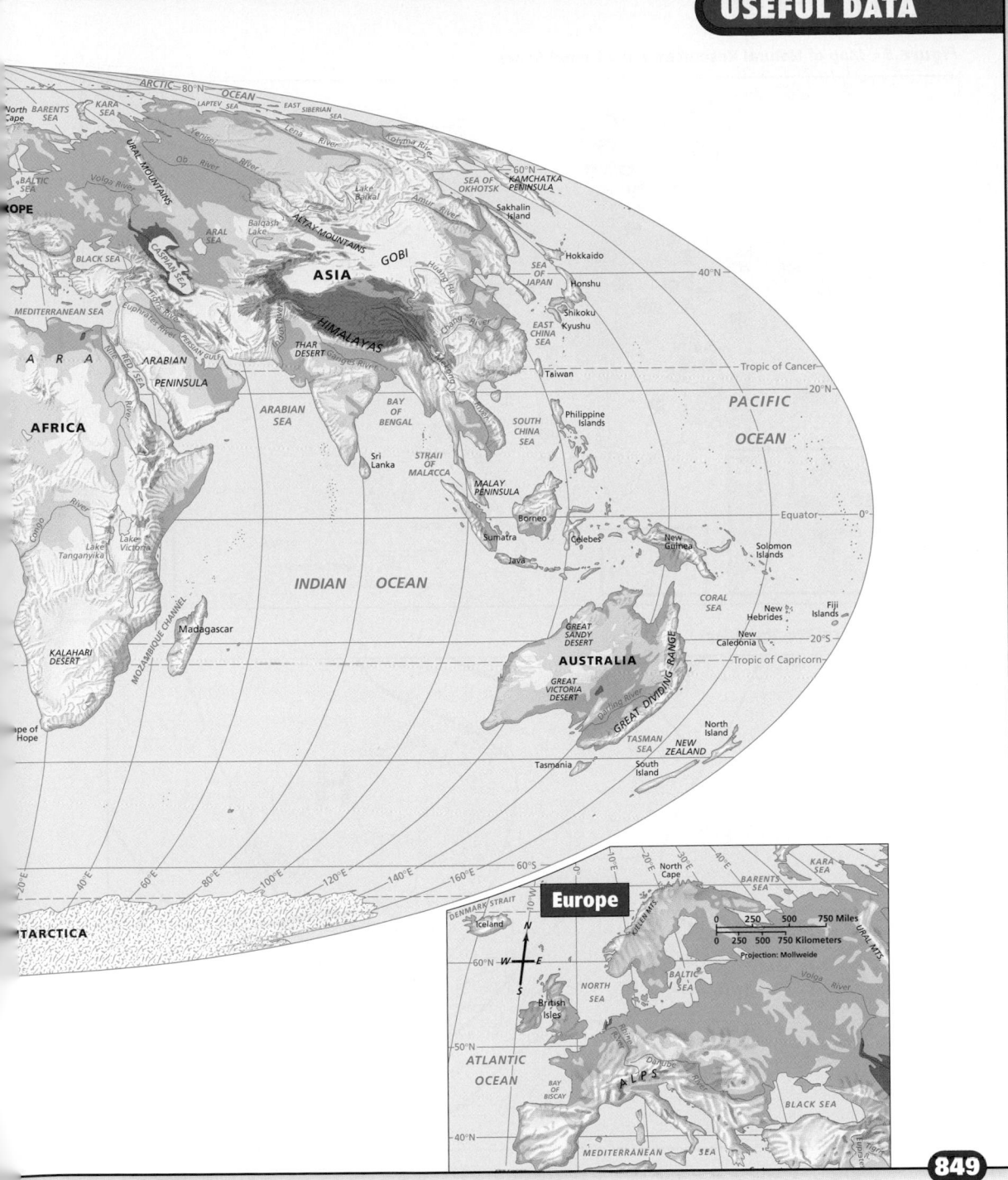

Europe

0   250   500   750 Miles
0   250   500   750 Kilometers
Projection: Mollweide

Figure 5 **Map of Natural Resources in the United States**

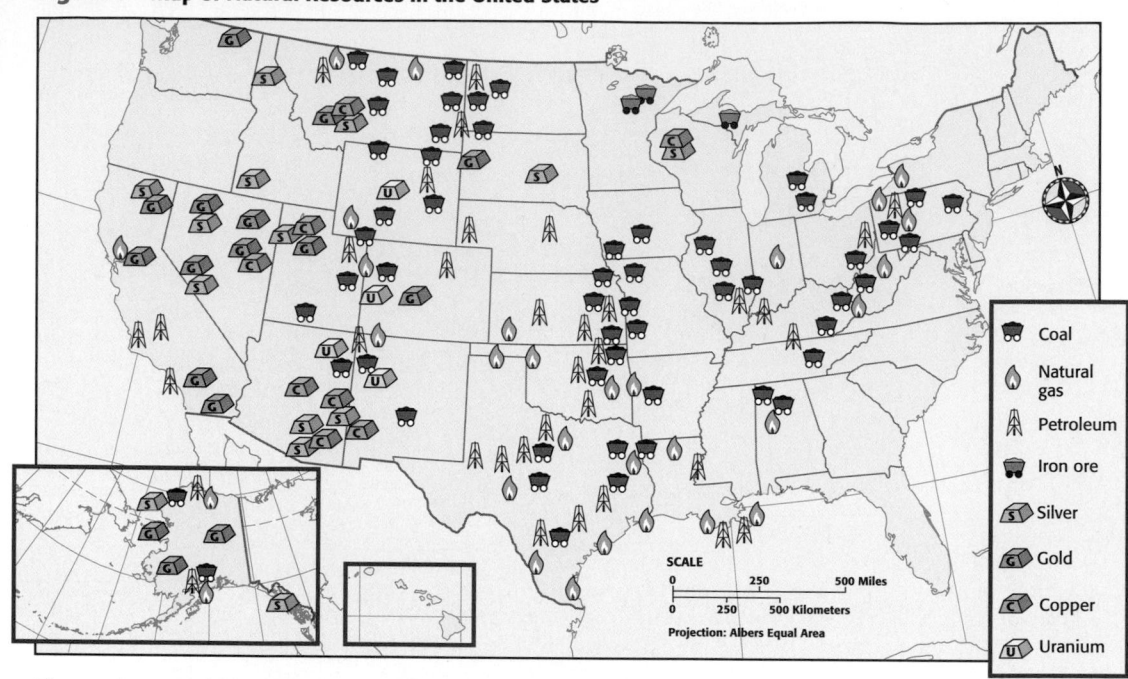

Figure 6 **Typical Weather Map**

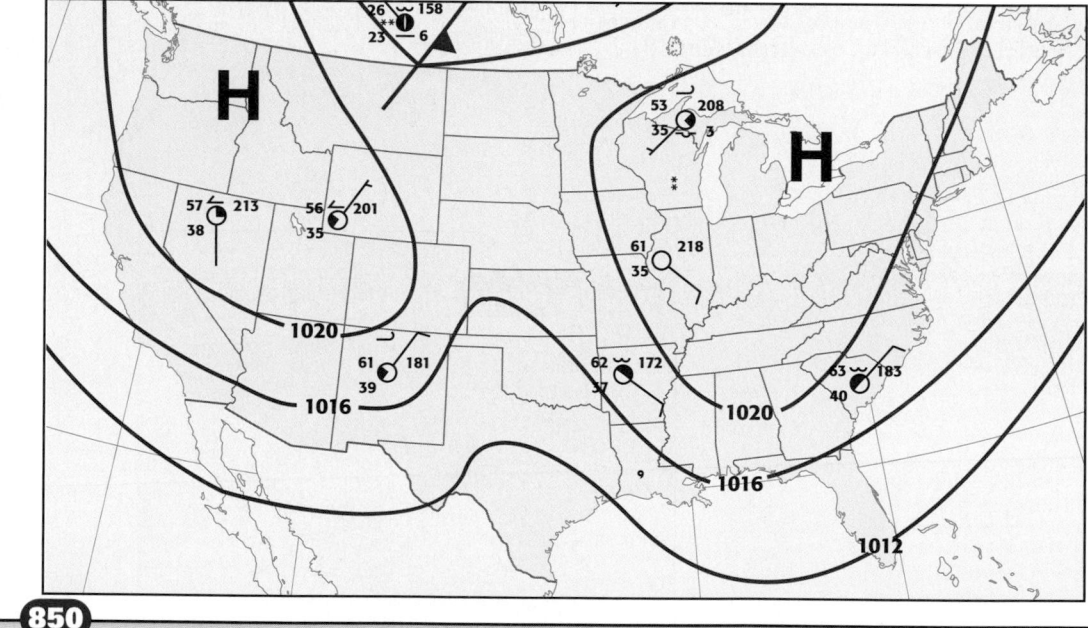

Table 6   **International Weather Symbols**

Current Weather

Hail	⊖	Light drizzle	ʼ	Light rain	•	Light snow	✳
Freezing rain	∿•	Steady, light drizzle	ʼ ʼ	Steady, light rain	• •	Steady, light snow	✳ ✳
Smoke	⌐∿	Intermittent, moderate drizzle	ʼ ʼ (stacked)	Intermittent, moderate rain	• • (stacked)	Intermittent, moderate snow	✳ ✳ (stacked)
Tornado	)(	Steady, moderate drizzle	ʼ ʼ ʼ	Steady, moderate rain	• • (over •)	Steady, moderate snow	✳ ✳ (over ✳)
Dust storms	⊸	Intermittent, heavy drizzle	ʼ ʼ ʼ (stacked)	Intermittent, heavy rain	• • • (stacked)	Intermittent, heavy snow	✳ ✳ ✳ (stacked)
Fog	≡	Steady, heavy drizzle	ʼ ʼ ʼ	Steady, heavy rain	• • •	Steady, heavy snow	✳ ✳ ✳
Thunder-storm	⊓						
Lightning	⟨						
Hurricane	၆						

Sky Coverage

No clouds	○	Two- to three-tenths covered	◔	Half covered	◑	Nine-tenths covered	◕•
One-tenth Coverage	⊕	Four-tenths covered	◑	Six-tenths covered	◑	Completely overcast	●

Clouds

Low:	Stratus	—	Cumulus	⌒	Cumulonimus calvus	⌂
	Stratocumulus	⌄	Cumulus congestus	⌒	Cumulonimbus with anvil	⊠
Middle:	Altostratus	∠	Altocumulus	⌣	Altocumulus castellanus	Ⳝ
High:	Cirrus	⌣	Cirrostratus	2	Cirrocumulus	⌇

Wind Speed (in km/h)

Calm	◎	4-13	⌐	24-33	⟍
1-3	—	14-23	∟	34-40	⟍⟍

Image Credits: Fig. 2, Kristy Sprott; Fig. 3-5, MapQuest.com. Inc., Fig. 6, Wil Tirion.

Problem Bank Solutions

Conversions

1. 1.738 Mm

2. **a.** $113 \text{ g} \times \dfrac{1000 \text{ mg}}{1 \text{ g}} =$
   113 000 mg

   b. $700 \text{ pm} \times \dfrac{10^{-12} \text{ m}}{1 \text{ pm}} \times$
   $\dfrac{1 \text{ nm}}{10^{-9} \text{ m}} = 0.7 \text{ nm}$

   c. $101.1 \text{ kPa} \times \dfrac{1000 \text{ Pa}}{1 \text{ kPa}} =$
   101 100 Pa

   d. $13 \text{ MA} \times \dfrac{10^6 \text{ A}}{1 \text{ MA}} =$
   13 000 000 A

3. 4989 000 cm

4. 341.11 m/s > 335.84 m/s
   Yes, he did break the sound barrier.

5. 15 000 000 000 000 000 Gg

Writing Scientific Notation

6. **a.** 1.1045×10^2 m
   b. 3.45×10^{-6} s
   c. $1.329 \, 48 \times 10^5$ kg
   d. $3.439 \, 00 \times 10^{-2}$ cm

7. 2.7×10^8 people

8. $7.029 \, 4601 \times 10^7$ airplanes

9. 5.685×10^{26} kg

10. 4.5×10^6 automobiles

Using Scientific Notation

11. **a.** 4.41×10^{-2}
    b. 5.79×10^{19}
    c. 3.01×10^5
    d. 6.572×10^{34}

12. $(8.850 \times 10^3 \text{ m}) + (1.0924 \times 10^4 \text{ m}) = (0.8850 \times 10^4 \text{ m}) + (1.0924 \times 10^4 \text{ m}) = 1.9774 \times 10^4$ m

13. by 8.8048 times

14. 5.2277×10^4 immigrants

15. 6.54×10^{17} m^3

Conversions

1. Earth's moon has a radius of 1738 km. Write this measurement in megameters.

2. Convert each of the following values as indicated:
   a. 113 g to milligrams
   b. 700 pm to nanometers
   c. 101.1 kPa to pascals
   d. 13 MA to amperes

3. Maryland has 49 890 m of coastline. What is this length in centimeters?

4. In 1997 Andy Green, a Royal Air Force pilot, broke the land speed record in Black Rock Desert, Nevada. His car averaged 341.11 m/s. The speed of sound in Black Rock at the time that he broke the record was determined to be 0.33584 km/s. Did Andy Green break the sound barrier?

5. The mass of the planet Pluto is 15 000 000 000 000 000 000 000 kg. What is the mass of Pluto in gigagrams?

Writing Scientific Notation

6. Express each of the following values in scientific notation:
   a. 110.45 m   **c.** 132 948 kg
   b. 0.000 003 45 s   **d.** 0.034 3900 cm

7. In 1998, the population of the United States was about 270 000 000 people. Write the estimated population in scientific notation.

8. In 1997, 70 294 601 airplanes either took off or landed at Chicago's O'Hare airport. What is the number of arrivals and departures at O'Hare given in scientific notation?

9. The planet Saturn has a mass of 568 500 000 000 000 000 000 000 000 kg. Express the mass of Saturn in scientific notation.

10. Approximately four-and-a-half million automobiles were imported into the United States last year. Write this number in scientific notation.

Using Scientific Notation

11. Perform the following calculations involving numbers that have been written in scientific notation:
    a. $3.02 \times 10^{-3} + 4.11 \times 10^{-2}$
    b. $(6.022 \times 10^{23}) \div (1.04 \times 10^4)$
    c. $(1.00 \times 10^2) \times (3.01 \times 10^3)$
    d. $6.626 \times 10^{34} - 5.442 \times 10^{32}$

12. Mount Everest, the tallest mountain on Earth, is 8.850×10^3 m high. The Mariana Trench is the deepest point of any ocean on Earth. It is 1.0924×10^4 m deep. What is the vertical distance from the highest mountain on Earth to the deepest ocean trench on Earth?

13. In 1950 Americans consumed nearly 1.4048×10^9 kg of poultry. In 1997, Americans consumed 1.2369×10^{10} kg of poultry. By what factor did America's poultry consumption increase between 1950 and 1997?

14. The following data were obtained for the number of immigrants admitted to the United States in major Texas cities in 1996. What is the total number of immigrants admitted in these Texas cities for that year?

City	Number of immigrants admitted
Houston	2.1387×10^4
Dallas	1.5915×10^4
El Paso	8.701×10^3
Ft. Worth/Arlington	6.274×10^3

15. The surface area of the Pacific Ocean is 1.66×10^{14} m^2. The average depth of the Pacific Ocean is 3.9395×10^3 m. If you could calculate the volume of the Pacific by simply multiplying the surface area by the average depth, what would the volume of the Pacific Ocean be?

Significant Figures

16. Determine the number of significant figures in each of the following values:
a. 0.026 48 kg   **c.** 1 625 000 J
b. 47.10 g   **d.** 29.02 cm

17. Solve the following addition problems, and round each answer to the correct number of significant figures:
a. 0.00241 g + 0.0123 g
b. 24.10 cm + 3.050 cm
c. 0.367 L + 2.51 L + 1.6004 L

18. Solve the following multiplication or division problems, and round each answer to the correct number of significant figures:
a. 129 g ÷ 29.20 cm^3
b. 120 mm × 355 mm × 12.1 mm
c. 45.4 g ÷ (0.012 cm × 0.444 cm × 0.221 cm)

19. Determine the volume of a cube whose width is 32.1 cm. Round your answer to the correct number of significant figures.

20. Solve the following subtraction problems, and round each answer to the correct number of significant figures:
a. 1.23 cm^3 − 0.044 cm^3
b. 89.00 kg − 0.1 kg
c. 780 mm − 64 mm

Density

21. Sugar has a density of 1.59 g/cm^3. What mass of sugar fits into a 140 cm^3 bowl?

22. The continent of North America has an area of 2.4346 × 10^{13} m^2. North America has a population of 3.01 × 10^8 people. What is the population density of North America?

23. The average density of Earth is 5.515 g/cm^3. The average density of Earth's moon is 3.34 g/cm^3. What is the difference in mass between 10.0 cm^3 of Earth and 10.0 cm^3 of Earth's moon?

24. A rubber balloon has a mass of 0.45 g, and can hold 1.78 × 10^{-3} m^3 of helium. If the density of helium is 0.178 kg/m^3, what is the balloon's total mass?

25. What is the density of a 0.996 g piece of graphite with a volume of 0.44 cm^3?

Pascal's Principle

26. A 6.50 × 10^{-3} m^2 piston compresses gas in a cylinder with a surface area of 9.75 × 10^{-2} m^2. What is the force on the cylinder walls if 50.0 N are applied to the piston?

27. A hydraulic lift raises a 1.0 × 10^4 N car on a 0.30 m^2 piston. If the compressor piston has an area of 0.015 m^2, what minimum force is needed to lift the car?

28. Air is blown into a trombone with a force of 3.5 N. A trombone's bore has a radius of 0.65 cm at the mouthpiece and 8.9 cm at the bell, where the air exits. What is the force on the exiting air?

29. A toy consists of a short, water-filled cylinder and a 28 cm^2 piston with a small hole in its center. A 4.5 N force applied to the piston causes water to flow through the hole with a force of 1.14 × 10^{-2} N. What is the area of the hole?

30. The air inside an automobile tire exerts a force that is 1.42 × 10^5 N greater than the force exerted by the outside air. The net force of air flowing through a hole in the tire is 2.72 N. If the tire's area is 0.656 m^2, what is the area of the hole?

Boyle's Law

31. A piston compresses gas in a 3.5 L cylinder to a volume of 2.1 L. If the gas pressure is initially 150 kPa, what is its pressure after it is compressed?

Boyle's Law

31. $\dfrac{(150\ \text{kPa})(3.5\text{L})}{2.1\ \text{L}} = 250\ \text{kPa}$

32. $\dfrac{(101.3\ \text{kPa})(15.0\ \text{L})}{15.5\ \text{L}} = 98.0\ \text{kPa}$

33. $\dfrac{(101\ \text{kPa})(0.250\ \text{L})}{0.108\ \text{L}} = 234\ \text{kPa}$

34. $\dfrac{(101\ \text{kPa})(0.75\ \text{L})}{85\ \text{kPa}} = 0.89\ \text{L}$

35. $\dfrac{(750\ \text{kPa})(1.3\ \text{cm}^3)}{125\ \text{kPa}} = 7.8\ \text{cm}^3$

Problem Bank Solutions
Significant Figures

16. a. 4
  b. 4
  c. 4
  d. 4

17. a. 0.0147 g
  b. 27.15 cm
  c. 4.48 L

18. a. $\dfrac{129\ \text{g}}{29.20\ \text{cm}^3}$
= 4.417 808 219 g/cm^3
= 4.42 g/cm^3
b. 120 mm × 355 mm × 12.1 mm =
5.1546 × 10^5 mm^3 =
5.2 × 10^5 mm^3
c. $\dfrac{45.4\ \text{g}}{0.012\ \text{cm} × 0.444\ \text{cm} × 0.221\ \text{cm}}$
= 3.855 665 62 × 10^4 g/cm^3
= 3.9 × 10^4 g/cm^3

19. (32.1 cm)^3 = 3.31 × 10^4 cm^3

20. a. 1.23 cm^3 − 0.044 cm^3 = 1.19 cm^3
b. 89.00 kg − 0.1 kg = 88.9 kg
c. 780 mm − 64 mm = 716 mm

Density

21. 140 cm^3 × 1.59 g/cm^3 = 220 g

22. $density = \dfrac{people}{area} =$
$\dfrac{4.64 × 10^8\ \text{people}}{2.4346 × 10^{13}\ \text{m}^2} =$
1.91 × 10^{-5} people/m^2

23. 55.2 g − 33.4 g = 21.8 g

24. m = 0.45 g + (1.78 × 10^{-3} m^3)(0.178 kg/m^3) =
7.7 × 10^{-4} kg or 0.77 g

25. $\dfrac{0.996\ \text{g}}{0.44\ \text{cm}^3} = 2.3\ \text{g/cm}^3$

Pascal's Principle

26. $F_2 = \left(\dfrac{50.0\ \text{N}}{6.5 × 10^{-3}\ \text{m}^2}\right)$
9.75 × 10^{-2} m^2 = 7.50 × 10^2 N

27. $F_2 = \left(\dfrac{1.0 × 10^4\text{N}}{0.30\ \text{m}^2}\right)$0.015 m^2 =
5.0 × 10^2 N

28. $F_2 = \left(\dfrac{3.5\ \text{N}}{\pi(0.65\ \text{cm})^2}\right)\pi(8.9\ \text{cm})^2 =$
660 N

29. $a_2 = \left(\dfrac{28\ \text{cm}^2}{4.5\ \text{N}}\right)1.14 × 10^{-2}\ \text{N} =$
0.071 cm^2

30. $a_2 = \left(\dfrac{0.656\ \text{m}^2}{1.42 × 10^5\ \text{N}}\right)2.72\ \text{N} =$
1.26 × 10^{-5} m^2 = 12.6 mm^2

Problem Bank Solutions
Conversion Factors

36. a. $\dfrac{2000\ \text{sheets}}{4\ \text{reams}} = \dfrac{500\ \text{sheets}}{1\ \text{ream}}$

b. $\dfrac{170\ \text{g Ga}}{2.5\ \text{mol Ga}} = \dfrac{68\ \text{g Ga}}{1\ \text{mol Ga}}$

c. $\dfrac{0.997\ \text{g H}_2\text{O}}{1.00\ \text{cm}^3\ \text{H}_2\text{O}}$

d. $\dfrac{2.24 \times 10^{10}\ \text{mol Ag}}{1.35 \times 10^{34}\ \text{atoms Ag}} =$

$\dfrac{1\ \text{mol Ag}}{6.03 \times 10^{23}\ \text{atoms Ag}}$

37. $\dfrac{4.184\ \text{kJ}}{1\ \text{Calorie}} \times 150\ \text{Calories} =$

$6.3 \times 10^2\ \text{kJ} = 6.3 \times 10^5\ \text{J}$

38. a. $m = dV =$

$19.3\ \text{g/cm}^3\ (10.0\ \text{cm} \times$

$26.0\ \text{cm} \times 8.0\ \text{cm}) =$

$4.0 \times 10^4\ \text{g}$

b. $(4.0 \times 10^4\ \text{g}) \times$

$\dfrac{0.0353\ \text{oz}}{1\ \text{g}} \times \dfrac{\$253.50}{1\ \text{oz}} =$

$\$3.6 \times 10^5$

39. $34.5\ \text{mol Cu} \times$

$\left(\dfrac{6.02 \times 10^{23}\ \text{atoms Cu}}{1\ \text{mol Cu}} \right) =$

$2.08 \times 10^{25}\ \text{atoms Cu}$

40. $295\ \text{min} \div 3\ \text{rev} =$

$98.3\ \text{min/rev for John Glenn}$

$\dfrac{\left(146\ \text{h} \times \dfrac{60\ \text{min}}{1\ \text{h}} \right) + 24\ \text{min}}{98.3\ \text{min/rev}} =$

$89.4\ \text{rev for Sally Ride}$

Converting Amount to Mass

41. a. $67.9\ \text{mol Si} \times \dfrac{28.09\ \text{g Si}}{1\ \text{mol Si}} =$

$1.91 \times 10^3\ \text{g Si}$

b. $1.45 \times 10^{-4}\ \text{mol Cd} \times$

$\dfrac{112.41\ \text{g Cd}}{1\ \text{mol Cd}} =$

$1.63 \times 10^{-2}\ \text{g Cd}$

c. $0.045\ \text{mol Au} \times$

$\dfrac{196.97\ \text{g Au}}{1\ \text{mol Au}} =$

$8.9\ \text{g Au}$

---

32. During a drive to the mountains, a tire's volume increases from 15.0 L to 15.5 L. If the pressure on the tire is initially 101.3 kPa, what is the final pressure?

33. An air-filled balloon has a volume of 0.250 L at 101 kPa. When immersed in a pool of mercury, the balloon's volume is 0.108 L. What pressure is exerted by the mercury on the balloon?

34. A toy balloon contains 0.75 L of helium at a pressure of 101 kPa. The balloon rises until the pressure on the balloon is 85 kPa. What is the balloon's volume?

35. An air bubble with a volume of 1.3 cm³ forms underwater, where the pressure is 750 kPa. What is the bubble's volume when the pressure is 125 kPa?

Conversion Factors

36. Give the correct factor to convert between each of the following values:
  a. 4 reams of paper → 2000 sheets of paper
  b. 2.5 mol of gallium, Ga → 170 g of Ga
  c. 1.00 cm³ of water → 0.997 g of water
  d. 1.35×10^{34} atoms of silver, Ag → 2.24×10^{10} mol of silver

37. A Calorie, as reported on nutritional labels, is equal to 4.184 kJ. A carbonated beverage contains about 150 Calories. What is the energy content in joules?

38. a. The density of gold is 19.3 g/cm³. What is the mass of a bar of gold with dimensions of 10.0 cm × 26.0 cm × 8.0 cm?
  b. Gold is priced by the ounce. One gram is equal to 0.0353 oz. If the price of gold is $253.50 per ounce, what is the value of the bar described in part (a)?

39. How many atoms of copper are there in a piece of copper tube that contains 34.5 mol of copper, Cu? (**Hint:** There are 6.02×10^{23} atoms in one mole.)

---

40. In February of 1962 John Glenn orbited Earth three times in 4 hours and 55 minutes. How long did it take him to make one revolution around Earth? In June of 1983 Sally Ride became the first U.S. woman in space. Her mission lasted 146 hours, 24 minutes. If each revolution of Sally Ride's mission took the same amount of time as each revolution of John Glenn's mission, how many times did Ride orbit Earth?

Converting Amount to Mass

41. Determine the mass in grams of each of the following:
  a. 67.9 mol of silicon, Si
  b. 1.45×10^{-4} mol of cadmium, Cd
  c. 0.045 mol of gold, Au
  d. 3.900 mol of tungsten, W

42. Fullerenes, also known as buckyballs, are a form of elemental carbon. One variety of fullerene has 60 carbon atoms in each molecule. What is the molar mass of 1 mol of this 60-carbon atom molecule? What is the mass of 5.23×10^{-2} mol of this fullerene?

43. An experiment requires 2.0 mol of cadmium, Cd, and 2.0 mol of sulfur, S. What mass of each element is required?

44. If there are 6.02×10^{27} mol of iron, Fe, in a portion of Earth's crust, what is the mass of iron present?

45. a. A certain molecule of polyester contains 1.00×10^5 carbon atoms. What is the mass of carbon in 1 mol of this polyester?
  b. The same polyester molecule contains 4.00×10^4 oxygen atoms. What is the mass of oxygen in 1 mol?

---

d. $3.900\ \text{mol W} \times \dfrac{183.84\ \text{g W}}{1\ \text{mol W}} =$

$7.170 \times 10^2\ \text{g W}$

42. $60 \times \dfrac{12.01\ \text{g C}}{1\ \text{mol C}} =$

$720.6\ \text{g/mol C}_{60}$

$5.23 \times 10^{-2}\ \text{mol C}_{60} \times$

$\dfrac{720.6\ \text{g C}_{60}}{1\ \text{mol C}_{60}} =$

$37.7\ \text{g C}_{60}$

43. $2.0\ \text{mol Cd} \times \dfrac{112.41\ \text{g Cd}}{1\ \text{mol Cd}} =$

$220\ \text{g Cd}$

$2.0\ \text{mol S} \times \dfrac{32.07\ \text{g S}}{1\ \text{mol S}} = 64\ \text{g S}$

44. $6.02 \times 10^{27}\ \text{mol Fe} \times$

$\dfrac{55.85\ \text{g Fe}}{1\ \text{mol Fe}} =$

$3.36 \times 10^{29}\ \text{g Fe}$

45. a. $1\ \text{mol polyester} \times \dfrac{6.02 \times 10^{23}\ \text{molecules}}{1\ \text{mol}} \times$

$\dfrac{10^5\ \text{atoms C}}{1\ \text{molecule polyester}} \times$

$\dfrac{12.01\ \text{g C}}{6.02 \times 10^{23}\ \text{atoms C}} = 1 \times 10^6\ \text{g C}$

Converting Mass to Amount

46. Imagine that you find a jar full of diamonds. You measure the mass of the diamonds and find that they have a mass of 45.4 g. Determine the amount of carbon in the diamonds.

47. A tungsten, W, filament in a light bulb has a mass of 2.0 mg. Calculate the amount of tungsten in this filament.

48. One liter of sea water contains 1.05×10^4 mg of sodium. How much sodium is in one liter of sea water?

49. Every kilogram of Earth's crust contains 282 g of silicon. How many moles of silicon are present in 2 kg of Earth's crust?

50. Chlorine rarely occurs in nature as Cl atoms. It usually occurs as gaseous Cl_2 molecules, molecules made up of two chlorine atoms joined together. What is the molar mass of gaseous Cl_2? Calculate the amount of chlorine molecules found in 4.30 g of chlorine gas.

Writing Ionic Formulas

51. Write the ionic formula for the salt made from potassium and bromine.

52. Calcium chloride is used by the canning industry to make the skin of fruit such as tomatoes more firm. What is the ionic formula for calcium chloride?

53. Write the ionic formulas formed by each of the following pairs:
a. lithium and oxygen
b. magnesium and oxygen
c. sodium and chlorine
d. magnesium and nitrogen

54. The active ingredient in most toothpaste is sodium fluoride. Write the ionic formula for this cavity-fighting compound.

55. What is the formula for the ionic compound formed from strontium and iodine?

Balancing Chemical Equations

56. Iron, Fe, combines with oxygen gas, O_2, to form iron(III) oxide. Balance the following equation for the synthesis of iron(III) oxide.
$$Fe + O_2 \rightarrow Fe_2O_3$$

57. Iron is often produced from iron ore by treating it with carbon monoxide in a blast furnace. Balance the equation for the production of iron.
$$Fe_2O_3 + CO \rightarrow Fe + CO_2$$

58. Zinc sulfide can be used as a white pigment. Balance the following equation for synthesizing zinc sulfide.
$$Na_2S + Zn(NO_3)_2 \rightarrow ZnS + NaNO_3$$

59. Kerosene, $C_{14}H_{30}$, is often used as a heating fuel or a jet fuel. When kerosene burns in oxygen, O_2, it produces carbon dioxide, CO_2, and water, H_2O. Write the balanced chemical equation for the reaction of kerosene and oxygen.

60. Plants use the process of photosynthesis to convert carbon dioxide and water into glucose and oxygen. This process helps remove carbon dioxide from the atmosphere. Balance the following equation for the production of glucose and oxygen from carbon dioxide and water.
$$CO_2 + H_2O \rightarrow C_6H_{12}O_6 + O_2$$

Molarity

61. Calculate the molarity of a hydrochloric acid (HCl) solution if 1.32 mol HCl are dissolved in water to form 5.28 L of solution.

62. Calculate the molarity of a potassium chloride (KCl) solution if 23.5 g of solute are dissolved in water to form 0.42 L of solution.

63. How many moles of sucrose, $C_{12}H_{22}O_{11}$, are needed to make 1.5 L of a 0.30 M sugar solution?

855

Problem Bank Solutions

b. 1 mol polyester $\times$

$$\frac{6.02 \times 10^{23} \text{ molecules}}{1 \text{ mol}} \times$$

$$\frac{4 \times 10^4 \text{ atoms O}}{1 \text{ molecule polyester}} \times$$

$$\frac{16.00 \text{ g O}}{6.02 \times 10^{23} \text{ atoms O}} =$$

6×10^3 g O

Converting Mass to Amount

46. $45.4 \text{ g C} \times \dfrac{1 \text{ mol C}}{12.01 \text{ g C}} =$
3.78 mol C

47. $2.0 \times 10^{-3} \text{ g W} \times \dfrac{1 \text{ mol W}}{183.84 \text{ g W}} =$
1.1×10^{-5} mol W

48. $1.05 \times 10^1 \text{ g Na} \times$
$\dfrac{1 \text{ mol Na}}{22.99 \text{ g Na}} =$
0.457 mol Na

49. $2 \text{ kg crust} \times \dfrac{282 \text{ g Si}}{1 \text{ kg crust}} \times$
$\dfrac{1 \text{ mol Si}}{28.09 \text{ g Si}} =$
20 mol Si

50. $\dfrac{2 \text{ mol Cl}}{1 \text{ mol Cl}_2} \times \dfrac{35.45 \text{ g Cl}}{1 \text{ mol Cl}} =$
70.90 g/mol Cl_2
$4.30 \text{ g Cl}_2 \times \dfrac{1 \text{ mol Cl}_2}{70.90 \text{ g Cl}_2} =$
6.06×10^{-2} mol Cl_2

Writing Ionic Formulas

51. KBr
52. $CaCl_2$
53. **a.** Li_2O
b. MgO
c. NaCl
d. Mg_3N_2
54. NaF
55. SrI_2

Balancing Chemical Equations

56. $2Fe_2O_3$
57. $Fe_2O_3 + 3CO \rightarrow 2Fe + 3CO_2$
58. $Na_2S + Zn(NO_3)_2 \rightarrow ZnS + 2NaNO_3$
59. $2C_{14}H_{30} + 43O_2 \rightarrow 30H_2O + 28CO_2$
60. $6CO_2 + 6H_2O \rightarrow C_6H_{12}O_6 + 6O_2$

Problem Bank Solutions

Molarity

61. $\dfrac{1.32 \text{ mol HCl}}{5.28 \text{ L solution}} = 0.250$ M HCl

62. $\dfrac{23.5 \text{ g KCl}}{(74.55 \text{ g/mol})(0.42 \text{ L solution})} =$

0.75 M KCl

63. $(0.30 \text{ M } C_{12}H_{22}O_{11})(1.5 \text{ L solution})$
$= 0.45 \text{ mol } C_{12}H_{22}O_{11}$

64. $(187.77 \text{ g/mol})(0.350 \text{ M AgBr})$
$(0.750 \text{ L solution}) = 49.3$ g AgBr

65. $\dfrac{450 \text{ g CuSO}_4}{(06.7 \text{ M CuSO}_4)(159.62 \text{ g/mol})} =$

4.2 L solution

Determining pH

66. 3

67. 3

68. $14 - 2 = 12$

69. 1×10^{-2} M HI $= 0.01$ M HI

70. pOH $= 14 - 11 = 3$;
1×10^{-3} M KOH $=$
0.001 M KOH

Nuclear Decay

71. a. $^{40}_{19}K \rightarrow {}^{40}_{20}Ca + {}^{0}_{-1}e$

b. Pa

72. nitrogen

73. a. beta emission

b. alpha emission

74. thorium

75. $^{222}_{86}Rn \rightarrow {}^{218}_{84}Po + {}^{4}_{2}He$

64. A 0.350 M solution of silver bromide, AgBr, in water, has a volume of 0.750 L. What is the mass of the solute?

65. A 0.67 M solution is made by dissolving 0.45 kg of copper(II) sulfate, $CuSO_4$, in water. What is the volume of the solution?

Determining pH

66. Calculate the pH of a 0.001 M solution of H_2SO_4, a strong acid.

67. A solution of HBr, a strong acid, has a concentration of 1.0×10^{-5} M. What is the pH of this solution?

68. Sodium hydroxide, NaOH, is a strong base. What is the pH of a 0.01 M solution of NaOH?

69. A solution of HI, a strong acid, has a pH of 2. What is the concentration of the solution?

70. A solution of KOH, a strong base, has a pH of 11. What is the concentration of the solution?

Nuclear Decay

71. Potassium undergoes nuclear decay by β-emission and other emissions. Complete the following equations for the nuclear decay of potassium.

a. $^{40}_{19}K \longrightarrow {}^{40}_{20}\underline{\quad} + {}^{0}_{-1}e$

b. $^{234}_{90}Th \longrightarrow {}^{234}_{91}\underline{\quad} + {}^{0}_{-1}e$

72. Carbon-14 decays by β-emission. What element is formed when carbon loses a β-particle?

73. Which type of emission would result in each of the following nuclear changes?

a. $^{15}_{6}C \longrightarrow {}^{15}_{7}N$

b. $^{147}_{62}Sm \longrightarrow {}^{143}_{60}Nd$

74. Uranium-238 decays by α-emission. What element is formed when a uranium atom loses an α-particle?

75. Complete the following equation for the decay of radon-222.

$$^{222}_{86}Rn \longrightarrow {}^{218}_{84}\underline{\quad} + {}^{4}_{2}He$$

Half-Life

76. The half-life of thorium-232, $^{232}_{90}Th$, is 1.4×10^{10} years. How much of a 50.0 g sample of thorium-232 will remain as thorium after 4.2×10^{10} years?

77. Radium, used in radiation treatment for cancer, has a half-life of 1.60×10^3 years. If you begin with a 0.25 g sample, what mass of radium will remain after 8.00×10^3 years?

78. The half-life of iodine-131 is 8.04 days. How long will it take for the mass of iodine present to drop to 1/16?

79. What is the half-life of an element if 1/8 of a sample remains after 12 years?

80. The half life of cobalt-60 is 10.47 minutes. What fraction of a sample will remain after 52.35 minutes?

Velocity

81. Amy Van Dyken broke the world record for a 50.0 m swim using the butterfly stroke in 1996. She swam 50.0 m in 26.55 seconds. What was her average velocity assuming that she swam the 50.0 m in a perfectly straight line?

82. If Amy Van Dyken swam her record-breaking 50 m by swimming to one end of the pool, then turning around and swimming back to her starting position, what would her average velocity be?

Half-life

76. $\dfrac{4.2 \times 10^{10} \text{ y}}{1.4 \times 10^{10} \text{ y}} = 3$ half-lives

$50.0 \text{ g} \times \dfrac{1}{2} \times \dfrac{1}{2} \times \dfrac{1}{2} = 6.25$ g Th

77. $\dfrac{8.00 \times 10^3 \text{ y}}{1.60 \times 10^3 \text{ y}} = 5$ half-lives

$0.25 \text{ g} \times \dfrac{1}{2} \times \dfrac{1}{2} \times \dfrac{1}{2} \times \dfrac{1}{2} \times \dfrac{1}{2} =$
7.8×10^{-3} g Ra

78. $\dfrac{1}{16} = \dfrac{1}{2} \times \dfrac{1}{2} \times \dfrac{1}{2} \times \dfrac{1}{2} = 4$ half-lives

$4 \times 8.04 \text{ days} = 32.2$ days

79. $\dfrac{1}{8} = \dfrac{1}{2} \times \dfrac{1}{2} \times \dfrac{1}{2} = 3$ half-lives

$\dfrac{12 \text{ years}}{3} = 4$ years

80. $\dfrac{52.35 \text{ min}}{10.47 \text{ min}} = 5$ half-lives

$\dfrac{1}{2} \times \dfrac{1}{2} \times \dfrac{1}{2} \times \dfrac{1}{2} \times \dfrac{1}{2} = \dfrac{1}{32}$

Velocity

81. $\dfrac{50.0 \text{ m}}{26.55 \text{ s}} = 1.88$ m/s straight ahead

82. $v = 0$ m/s (velocity is change in position over time, but at the end there was no change in position)

83. When Andy Green broke the land speed record, his vehicle was traveling across a flat portion of the desert with a forward velocity of 341.11 m/s. How long would it take him at that velocity to travel 4.500 km?

84. If a car moves along a perfectly straight road at a velocity of 24 m/s, how far will the car go in 35 minutes?

85. If you travel southeast from one city to another city that is 314 km away, and the trip takes you 4.00 hours, what is your average velocity?

Acceleration

86. While driving at an average velocity of 15.6 m/s down the road, a driver slams on brakes to avoid hitting a squirrel. The car stops completely in 4.2 seconds. What is the average acceleration of the car?

87. A sports car is advertised as being able to go from 0 to 60 in 6.00 seconds. If 60 mi/h is equal to 27 m/s, what is the sports car's average acceleration?

88. If a bicycle has an average acceleration of –0.44 m/s^2, and its initial forward velocity is 8.2 m/s, how long will it take the cyclist to bring the bicycle to a complete stop?

89. An airliner has an airborne velocity of 232 m/s. What is the plane's average acceleration if it takes the plane 15 minutes to reach its airborne velocity?

90. A school bus can accelerate from a complete stop at 1.3 m/s^2. How long will it take the bus to reach a velocity of 12.1 m/s?

Newton's Second Law

91. A peach falls from a tree with an acceleration of 9.8 m/s^2. The peach has a mass of 7.4 g. With what force does the peach strike the ground? (**Hint:** Convert g to kg.)

92. A group of people push a car from a resting position with a force of 1.99×10^3 N. The car and its driver have a mass of 831 kg. What is the acceleration of the car?

93. If the space shuttle accelerates upward at 35 m/s^2, what force will a 59 kg astronaut experience?

94. A soccer ball is kicked with a force of 15.2 N. The soccer ball has a mass of 2.45 kg. What is the ball's acceleration?

95. A person steps off a diving board and falls into a pool with an acceleration of 9.8 m/s^2, which causes the person to hit the water with a force of 637 N. What is the mass of the person?

Momentum

96. a. 703 kg car is traveling with a velocity of 20.1 m/s. What is the momentum of the car?
  b. If a 315 kg trailer is attached to the car, what is the new combined momentum?

97. You are traveling west on your bicycle at 4.2 m/s, and you and your bike have a combined mass of 75 kg. What is the momentum of you and your bicycle?

98. A runner, who has a mass of 52 kg, has a momentum of 218 kg · m/s along a trail. What is the runner's velocity?

99. A commercial airplane travels at a velocity of 234 m/s. The plane seats 253 people. If the average person on the plane has a mass of 68 kg, what is the momentum of the passengers on the plane?

100. A bowling ball has a mass of 5.44 kg. It is moving down the lane at 2.1 m/s when it strikes the pins. What is the momentum with which the ball hits the pins?

857

Problem Bank Solutions

83. $4500. \text{ m} \times \dfrac{1 \text{ s}}{341.11 \text{ m}} = 13.19 \text{ s}$

84. $35 \text{ min} \times \dfrac{60 \text{ s}}{1 \text{ min}} \times \dfrac{24 \text{ m}}{1 \text{ s}} = 5.0 \times 10^4 \text{ m}$

85. $\dfrac{314 \text{ km}}{4.00 \text{ h}} = 78.5$ km/h southeast

Acceleration

86. $a = \dfrac{\Delta v}{t} = \dfrac{0 \text{ m/s} - 15.6 \text{ m/s}}{4.2 \text{ s}} = -3.7 \text{ m/s}^2$

87. $a = \dfrac{\Delta v}{t} = \dfrac{27 \text{ m/s} - 0 \text{ m/s}}{6.00 \text{ s}} = 4.5 \text{ m/s}^2$

88. $\dfrac{-8.2 \text{ m/s}}{-0.44 \text{ m/s}^2} = 19 \text{ s}$

89. $\dfrac{232 \text{ m/s}}{15 \text{ min} \times \dfrac{60 \text{ s}}{1 \text{ min}}} = 0.26 \text{ m/s}^2$

90. $\dfrac{12.1 \text{ m/s}}{1.3 \text{ m/s}^2} = 9.3 \text{ s}$

Newton's Second Law

91. $F = ma = (7.4 \times 10^{-3} \text{ kg})(9.8 \text{ m/s}^2) = 7.3 \times 10^{-2} \text{ N}$

92. $\dfrac{1.99 \times 10^3 \text{ N}}{831 \text{ kg}} = 2.39 \text{ m/s}^2$

93. $59 \text{ kg} \times 35 \text{ m/s}^2 = 2.1 \times 10^3 \text{ N}$

94. $\dfrac{15.2 \text{ N}}{2.45 \text{ kg}} = 6.20 \text{ m/s}^2$

95. $\dfrac{637 \text{ N}}{9.8 \text{ m/s}^2} = 65 \text{ kg}$

Momentum

96. a. $p = mv = (703 \text{ kg})(20.1 \text{ m/s}) = 1.41 \times 10^4$ kg•m/s forward
  b. $p = mv = (703 \text{ kg} + 315 \text{ kg})(20.1 \text{ m/s}) = 2.05 \times 10^4$ kg•m/s forward

97. $p = mv = (75 \text{ kg})(4.2 \text{ m/s}) = 320$ kg•m/s west

98. $\dfrac{218 \text{ kg•m/s}}{52 \text{ kg}} = 4.2$ m/s along the trail

99. $253 \text{ people} \times \dfrac{68 \text{ kg}}{1 \text{ person}} \times 234 \text{ m/s} = 4.0 \times 10^6$ kg•m/s north

100. $5.44 \text{ kg} \times 2.1 \text{ m/s} = 11$ kg•m/s down the lane

Problem Bank Solutions

Work

101. $(1.99 \times 10^3 \text{ N})(2.1 \text{ m}) = 4.2 \times 10^3 \text{ J}$

102. $\dfrac{157 \text{ J}}{5.3 \text{ m}} = 3.0 \times 10^1 \text{ N}$

103. $(0.667 \text{ m})(3.2 \text{ kg})(3.2 \text{ m/s}^2) = 6.8 \text{ J}$

104. $\dfrac{1.56 \text{ J}}{(1.54 \text{ m/s}^2)(0.78 \text{ m})} = 1.3 \text{ kg}$

105. $\dfrac{686 \text{ J}}{(227 \text{ kg})(2.4 \text{ m})} = 1.3 \text{ m/s}^2$

Power

106. $P = \dfrac{W}{t} = \dfrac{686 \text{ J}}{3.1 \text{ s}} = 2.2 \times 10^2 \text{ W}$

107. $(60 \text{ W})(8 \text{ h})\left(\dfrac{60 \text{ min}}{1 \text{ h}}\right)\left(\dfrac{60 \text{ s}}{1 \text{ min}}\right) = 2 \times 10^6 \text{ J}$

108. $1.02 \times 10^{11} \text{ W}\left(\dfrac{24 \text{ h}}{1 \text{ day}}\right)$
$\left(\dfrac{60 \text{ min}}{1 \text{ h}}\right)\left(\dfrac{60 \text{ s}}{1 \text{ min}}\right) =$
$8.81 \times 10^{15} \text{ J} = 8.81 \times 10^9 \text{ MJ}$

109. $\dfrac{8.75 \times 10^3 \text{ J}}{350 \text{ W}} = 25 \text{ s}$

110. $\dfrac{(157 \text{ N})(2.3 \text{ km})\left(\dfrac{1000 \text{ m}}{1 \text{ km}}\right)}{20 \text{ min}\left(\dfrac{60 \text{ s}}{1 \text{ min}}\right)} =$
$3.0 \times 10^2 \text{ W per horse}$
$9.0 \times 10^2 \text{ W total}$

Mechanical Advantage

111. $\dfrac{1549 \text{ N}}{446 \text{ N}} = 3.47$

112. $\textit{input force} =$
$\dfrac{\textit{output force}}{\textit{mechanical advantage}} =$
$\dfrac{660 \text{ N}}{6} = 110 \text{ N}$

113. $(8650 \text{ N})(27) = 2.3 \times 10^5 \text{ N}$

114. $\textit{mechanical advantage}$
$\dfrac{87 \text{ cm}}{92 \text{ cm}} = 0.95$
If the knob were moved closer to the hinge, the mechanical advantage would be less.

Work

101. A car breaks down 2.1 m from the shoulder of the road. 1.99×10^3 N of force is used to push the car off the road. How much work has been done on the car?

102. Pulling a boat forward into a docking slip requires 157 J of work. The boat must be pulled a total distance of 5.3 m. What is the force with which the boat is pulled?

103. A box with a mass of 3.2 kg is pushed 0.667 m across a floor with an acceleration of 3.2 m/s². How much work is done on the box?

104. You need to pick up a book off the floor and place it on a table top that is 0.78 m above the ground. You expend 1.56 J of energy to lift the book. The book has an acceleration of 1.54 m/s². What is the book's mass?

105. A weight lifter raises a 227 kg weight above his head. The weight reaches a height of 2.4 m. The lifter expends 686 J of energy lifting the weight. What is the acceleration of the weight?

Power

106. A weight lifter does 686 J of work on a weight that he lifts in 3.1 seconds. What is the power with which he lifts the weight?

107. How much energy is wasted by a 60 W bulb if the bulb is left on over an 8 hour night?

108. A nuclear reactor is designed with a capacity of 1.02×10^8 kW. How much energy, in megajoules, should the reactor be able to produce in a day?

109. An electric mixer uses 350 W. If 8.75×10^3 J of work are done by the mixer, how long has the mixer run?

115. $MA = \dfrac{d_{in}}{d_{out}} =$
$\dfrac{2\pi(8.0 \times 10^{-2} \text{ m})750}{1.6 \times 10^3 \text{ m}} = 0.24$

110. A team of horses is hitched to a cart. The team pulls with a force of 471 N. The cart travels 2.3 km in 20 minutes. Calculate the power delivered by the horses.

Mechanical Advantage

111. A roofer needs to get a stack of shingles onto a roof. Pulling the shingles up manually uses 1549 N of force. The roofer decides that it would be easier to use a system of pulleys to raise the shingles. Using the pulleys, 446 N are required to lift the shingles. What is the mechanical advantage of the system of pulleys?

112. A dam used to make hydroelectric power opens and closes its gates with a lever. The gate weighs 660 N. The lever has a mechanical advantage of 6. Calculate the input force on the lever needed to move the gate.

113. A crane has a mechanical advantage of 27. An input force of 8650 N is used by the crane to lift a pile of steel girders. What is the weight of the girders?

114. A door that is 92 cm wide has a doorknob that is 87 cm from the door's hinge. What is the mechanical advantage of the door? Would the mechanical advantage be greater or less if the knob were moved 10 cm closer to the hinge?

115. A student pedals a bicycle to school. The gear on the bicycle has a radius of 8.0 cm. The student travels 1.6 km to school. During the journey, the pedals make 750 revolutions. What is the mechanical advantage of the bicycle?

Gravitational Potential Energy

116. A pear is hanging from a pear tree. The pear is 3.5 m above the ground and has a mass of 0.14 kg. What is the pear's gravitational potential energy?

117. A person in an airplane has a mass of 74 kg and has 6.6 MJ of gravitational potential energy. What is the altitude of the plane?

118. A high jumper jumps 2.04 m. If the jumper has a mass of 67 kg, what is his gravitational potential energy at the highest point in the jump?

119. A cat sits on the top of a fence that is 2.0 m high. The cat has a gravitational potential energy of 88.9 J. What is the mass of the cat?

120. A frog with a mass of 0.23 kg hops up in the air. At the highest point in the hop, the frog has a gravitational potential energy of 0.744 J. How high can it hop?

Kinetic Energy

121. A sprinter runs at a forward velocity of 10.9 m/s. If the sprinter has a mass of 72.5 kg, what is the sprinter's kinetic energy?

122. A car having a mass of 654 kg has a kinetic energy of 73.4 kJ. What is the car's speed?

123. A tennis ball with a mass of 51 g has a velocity of 9.7 m/s. What is the kinetic energy of the tennis ball?

124. A rock is rolling down a hill with a velocity of 4.67 m/s. It has a kinetic energy of 18.9 kJ. What is the mass of the rock?

125. Calculate the kinetic energy of an airliner with a mass of 7.6×10^4 kg that is flying at a speed of 524 km/h.

Efficiency

126. If a cyclist has 26.7% efficiency, how much energy is lost if 40.5 kJ of energy are put in by the cyclist?

127. What is the efficiency of a machine if 55.3 J of work are done on the machine, but only 14.3 J of work are done by the machine?

128. A microwave oven uses 89 kJ in one minute. The microwave has an output of 54 kJ per minute. What is the efficiency of the microwave?

129. A coal-burning power plant has an efficiency of 42%. If 4.99 MJ of energy are used by the power plant, how much useful energy is generated by the power plant?

130. A swimmer does 45 kJ of work while swimming. If the swimmer is wasting 42 kJ of energy while swimming, what is the efficiency for the activity?

Temperature Conversions

131. A normal body temperature is 98.6°F. What is this temperature in degrees Celsius?

132. Convert the following temperatures to the Kelvin scale.
   a. 214°F   **c.** 27°C
   b. 100°C   **d.** 32°F

133. What are the freezing point and boiling point of water in the Celsius, Fahrenheit, and Kelvin scales?

134. What is absolute zero in the Celsius scale?

135. If it is 315 K outside, is it hot or cold?

Specific Heat

See **Appendix B** for a table of specific heats.

136. How much energy is required to raise the temperature of 5.0 g of silver from 298 K to 334 K? (**Hint:** Convert g to kg.)

137. A burner transfers 45 J of energy to a small beaker with 5.3 g of water. If the water began at 27°C, what is the final temperature of the water?

138. A piece of aluminum foil is left on a burner until the temperature of the foil has risen from 27°C to 98°C. The foil absorbs 344 J of energy from the burner. What is the mass of the foil?

859

Problem Bank Solutions
Gravitational Potential Energy

116. $PE = mgh =$
$(0.14 \text{ kg})(9.8 \text{ m/s}^2)(3.5 \text{ m}) = 4.8 \text{ J}$

117. $\dfrac{6.6 \times 10^6 \text{ J}}{(74 \text{ kg})(9.8 \text{ m/s}^2)} = 9.1 \times 10^3 \text{ m}$

118. $PE =$
$(67 \text{ kg})(9.80 \text{ m/s}^2)(2.04 \text{ m})$
$= 1.3 \times 10^3 \text{ J}$

119. $\dfrac{88.9 \text{ J}}{(2.0 \text{ m})(9.80 \text{ m/s}^2)} = 4.5 \text{ kg}$

120. $\dfrac{0.744 \text{ J}}{(0.23 \text{ kg})(9.80 \text{ m/s}^2)} = 0.33 \text{ m}$

Kinetic Energy

121. $\dfrac{1}{2}(72.5 \text{ kg})(10.9 \text{ m/s})^2 =$
$4.31 \times 10^3 \text{ J}$

122. $v = \sqrt{\dfrac{2KE}{m}} =$
$\sqrt{\dfrac{2(7.34 \times 10^4 \text{ J})}{654 \text{ kg}}} = 15.0 \text{ m/s}$

123. $\dfrac{1}{2}(0.051 \text{ kg})(9.7 \text{ m/s})^2 = 2.4 \text{ J}$

124. $m = \dfrac{2(1.89 \times 10^4 \text{ J})}{(4.67 \text{ m/s})^2} =$
$1.73 \times 10^3 \text{ kg}$

125. $(524 \text{ km/h})(1000 \text{ m/km})$
$\left(\dfrac{1 \text{ h}}{60 \text{ min}}\right)\left(\dfrac{1 \text{ min}}{60 \text{ s}}\right) = 146 \text{ m/s}$
$\dfrac{1}{2}(7.6 \times 10^4 \text{ kg})(146 \text{ m/s})^2$
$= 8.1 \times 10^8 \text{ J}$

Efficiency

126. *energy lost = energy input* $\times$
$\left(\dfrac{100 - efficiency}{100}\right) =$
$40.5 \text{ kJ} \times \left(\dfrac{100 - 26.7}{100}\right) =$
29.7 kJ

127. $\dfrac{14.3 \text{ J}}{55.3 \text{ J}}(100) = 25.9\%$

128. *efficiency* $= \dfrac{54 \text{ kJ}}{89 \text{ kJ}}(100) = 61\%$

129. $4.99 \text{ MJ}\left(\dfrac{42}{100}\right) = 2.1 \text{ MJ}$

130. *efficiency* $=$
$\dfrac{45 \text{ kJ} - 42 \text{ kJ}}{45 \text{ kJ}}(100) = 7\%$

Temperature Conversions

131. $\dfrac{5}{9}(98.6°\text{F} - 32.0) = 37.0°\text{C}$

132. a. $T = \dfrac{5}{9}(214°\text{F} - 32.0) + 273 =$
   374 K
  b. $T = 100.°\text{C} + 273 = 373 \text{ K}$
  c. $T = 27°\text{C} + 273 =$
   $3.00 \times 10^2 \text{ K}$
  d. $\dfrac{5}{9}(32°\text{F} - 32.0) + 273 =$
   273 K

133. 0°C, 100°C, 32°F, 212°F, 273 K, 373 K

134. $t = 0 \text{ K} - 273 = -273°\text{C}$

135. 315 K − 273 = 42°C
As human body temperature is 37°C, this is hot.

Specific Heat

136. *energy* $= mc\Delta t = (5.0 \text{ g Ag})$
$(1 \text{ kg}/1000 \text{ g})(234 \text{ J/kg•K})$
$(334 \text{ K} - 298 \text{ K}) = 42 \text{ J}$

137. $\Delta t = \dfrac{energy}{mc} =$
$\dfrac{45 \text{ J}}{(5.3 \text{ g})\left(\dfrac{1 \text{ kg}}{1000 \text{ g}}\right)(4186 \text{ J/kg•K})} =$
$2.0° \text{ C}$
$t_f = 27°\text{C} + \Delta t = 29°\text{C}$

Problem Bank Solutions

138. $m = \dfrac{energy}{c \, \Delta t} =$

$\dfrac{344 \text{ J}}{(897 \text{ J/kg•K})(98°C - 27°C)} =$

5.4×10^{-3} kg

139. Because diamond has a lower specific heat (487 J/kg•K) than graphite (709 J/kg•K), it will heat up faster.

140. energy = $mc\Delta t$ =
(34 kg)(650 J/kg•K)(153°C) =
3.4×10^{6} J

Wave Speed

141. $\dfrac{698 \times 10^{-9} \text{ m}}{3.0 \times 10^{8} \text{ m/s}} = 2.3 \times 10^{-3}$ ps

142. $v = \dfrac{92 \text{ m}}{2.7 \times 10^{-1} \text{ s}} = 3.4 \times 10^{2}$ m/s

143. (1.3 m/s)(1.2 s) = 1.6 m

144. $t = \dfrac{7.78 \times 10^{11} \text{ m}}{3.0 \times 10^{8} \text{ m/s}} = 2.6 \times 10^{3}$ s

145. $\dfrac{3.0 \times 10^{8} \text{ m/s}}{5.08 \times 10^{-7} \text{ m}} = 5.9 \times 10^{14}$ Hz

Resistance

146. $R = \dfrac{V}{I} = \dfrac{6.0 \text{ V}}{1.4 \text{ A}} = 4.3 \ \Omega$

147. $\dfrac{120 \text{ V}}{12.0 \text{ A}} = 10. \ \Omega$

148. $V = (7.78 \times 10^{-3} \text{ A})(1150 \ \Omega) = 8.95$ V

149. $\dfrac{120 \text{ V}}{9.17 \text{ A}} = 13 \ \Omega$

150. $R = \dfrac{240 \text{ V}}{30 \text{ A}} = 8 \ \Omega$

139. If a piece of graphite and a diamond have the same mass and are placed on the same burner, which object will become hot faster? Why?

140. The iron ore hematite is heated until its temperature has risen by 153°C. If the piece of hematite has a mass of 34 kg, how much energy was required to raise the temperature this much?

Wave Speed

141. The speed of light in a vacuum is 3.0×10^{8} m/s. A red laser beam has a wavelength of 698 nm. How long, in picoseconds, will it take for one wavelength of the laser light to pass by a fixed point?

142. Two people are standing on opposite ends of a field. The field is 92 m long. One person speaks. It takes 270 ms for the person across the field to hear them. What is the speed of sound in the field?

143. A water wave has a speed of 1.3 m/s. A person sitting on a pier observes that it takes 1.2 s for a full wavelength to pass the edge of the pier. What is the wavelength of the water wave?

144. Jupiter is 7.78×10^{8} km from the sun. How long does it take the sun's light to reach Jupiter?

145. A green laser has a wavelength of 508 nm. What is the frequency of this laser light?

Resistance

146. What is the resistance of a wire that has a current of 1.4 A in it when it is connected to a 6.0 V battery?

147. An electric space heater is plugged into a 120 V outlet. A current of 12.0 A is in the coils in the space heater. What is the resistance of the coils?

148. A graphing calculator needs 7.78×10^{-3} A of current to function. The resistance in the calculator is 1150 Ω. What is the voltage required to operate the calculator?

149. A steam iron has a current of 9.17 A when plugged into a 120 V outlet. What is the resistance in the steam iron?

150. An electric clothes dryer requires a potential difference 240 V. The power cord that runs between the electrical outlet and the dryer supports a current of 30 A. What is the resistance in this power cord?

Electric Power

151. A flashlight uses a 3.0 V battery. The bulb has a current of 0.50 A. What is the electric power used by the flashlight?

152. What is the current in a 60 W light bulb when it is plugged into a 120 V outlet?

153. A student takes her hair dryer to Europe. In the United States, her hair dryer uses 1200 W of power when connected to a 120 V outlet. In Europe, the outlet has a potential difference of 240 V. When she uses her hair dryer in Europe, she notices that it gets very hot, and starts to smell as though it is burning. Determine the current in the hair dryer in the United States. Then calculate the resistance in the hair dryer. Calculate the current and the power in the hair dryer in Europe to explain why the hair dryer heats up when plugged into the European outlet.

154. A portable stereo requires a 12 V battery. It uses 43 W of power. Calculate the current in the stereo.

155. A microwave oven has a current of 12.3 A when operated using a 120 V power source. How much power does the microwave use?

Electric Power

151. (3.0 V)(0.50 A) = 1.5 W

152. $I = \dfrac{P}{V} = \dfrac{60 \text{ W}}{120 \text{ V}} = 0.5$ A

153. $\dfrac{1200 \text{ W}}{120 \text{ V}} = 1.0 \times 10^{1}$ A in the United States

$\dfrac{120 \text{ V}}{1.0 \times 10^{1} \text{ A}} = 12 \ \Omega$

$\dfrac{240 \text{ V}}{12 \ \Omega} = 2.0 \times 10^{1}$ A in Europe

(240 V)(2.0 × 10¹ A) = 4800 W in Europe

The hair dryer becomes hot in Europe because it is dissipating four times as much power as it does when used in the United States.

154. $\dfrac{43 \text{ W}}{12 \text{ V}} = 3.6$ A

155. (12.3 A)(120 V) = 1.5×10^{3} W

Selected Answers to Problems

Chapter 1
Introduction to Science
Math Skills
Practice, page 17
  2. 3500 ms
  4. 0.0025 kg
  6. 2.8 mol
  8. 3000 ng

Math Skills
Practice, page 23
  2. a. 4500 g
    b. 0.006 05 m
    c. 3 115 000 km
    d. 0.000 000 0199 cm

Math Skills
Practice, page 24
  2. a. 4.8×10^2 L/s
    b. 6.9 g/cm^3
    c. 5.5×10^5 cm^2
    d. 8.3×10^{-1} cm^3

Math Skills
Practice, page 25
  2. 1.6×10^2 cm^3
  4. 2.3 m/s

Section 3 Review
Math Skills, page 26
  6. a. 9.20×10^7 m^2
    b. 9.66×10^{-5} cm^2
    c. 6.70 g/cm^3

Chapter Review
Building Math Skills, page 29
  22. a. 69°C
    b. 3 minutes
    c. the first 3 minutes
    d. heating
  24. a. 2.2×10^4 mg
    b. 5×10^{-3} km
    c. 6.59×10^7 m
    d. 3.7×10^{-6} kg
    e. 7.22×10^{11} s
    f. 6.4×10^{-8} s
  26. a. 7.4 m
    b. 48 800 km or 4.88×10^4 km
    c. 0.09 g or 9×10^{-2} g
    d. 362.00 s
  28. a. 4 mg
    b. 75 km
    c. 325 Gg
    d. 3 ns
    e. 4.57 Ms

Chapter 2
Matter
Math Skills
Practice, page 48
  2. $D = \dfrac{163 \text{ g}}{50.0 \text{ cm}^3}$

    $D = \dfrac{3.26 \text{ g}}{\text{cm}^3}$

Section 2 Review
Math Skills, page 52
  6. $D = \dfrac{0.36 \text{ g}}{2500 \text{ cm}^3}$

    $D = \dfrac{1.4 \times 10^{-4} \text{g}}{\text{cm}^3}$

Chapter 3
States of Matter
Section 2 Review
Math Skills, page 86
  8. a. 15 N
    b. It will sink because its weight is greater than the buoyant force acting on it.

Math Skills
Practice, page 91
  2. $V_2 = \dfrac{P_1 V_1}{P_2}$

    $V_2 = \dfrac{(0.500 \text{ atm})(300 \text{ mL})}{0.750 \text{ atm}}$

    $V_2 = 200 \text{ mL}$

  4. $V_2 = \dfrac{P_1 V_1}{P_2}$

    $V_2 = \dfrac{(0.947 \text{ atm})(150 \text{ mL})}{1.000 \text{ atm}}$

    $V_2 = 140 \text{ mL}$

Chapter 4
Atoms and the Periodic Table
Math Skills
Practice, page 132
  2. 10 extra eggs

Math Skills
Practice, page 133
  2. 72.1 g Ca
  4. 203 g Cu

861

Section 4 Review
Math Skills, page 134
 6. 94 g Pt
 8. 0.39 mol Si

Chapter 5
The Structure of Matter
Math Skills
Practice, page 161
 2. $BeCl_2$
 4. $Co(OH)_3$

Section 3 Review
Math Skills, page 164
 6. The total charge of the compound must be zero. Each of the two cyanide ions has a charge of 1–. The charge of the cadmium ion must be 2+ to add to the 2×2 (1–) charge from the cyanide ions to equal zero.

Chapter 6
Chemical Reactions
Math Skills
Practice, page 213
 2. 322 g

Chapter Review
Building Math Skills, p. 215
 22. $2HgO \rightarrow 2Hg + O_2$
 24. a. The mole ratio of $Al_2(SO_4)_3$ and H_2SO_4 is 1:3. If 6 mol of H_2SO_4 are used, 2 mol of $Al_2(SO_4)_3$ are produced.
   b. The mole ratio of Al_2O_3 and H_2O is 1:3. Therefore, 3 mol of Al_2O_3 are needed for 9 mol of H_2O.
   c. 588 mol of $Al_2(SO_4)_3$ and 1764 mol of H_2O
 26. a. 130 g SO_2
   b. 128 g S
 28. 450 g

Chapter 7
Solutions
Section 3 Review
Math Skills, page 244
 6. 6. 0.5 M
 8. 0.373 M

Chapter Review
Building Math Skills, page 247
 26. 0.222 M
 28. 40.6 g

Chapter 8
Acids, Bases, Salts
Math Skills
Practice, page 262
 2. pH = 2
   pOH = 12

Chapter Review
Building Math Skills, page 277
 28. The concentration of OH^- ions is 0.001 M, or 1×10^{-3} M.
 30. The concentration of OH^- ions is $1 \times 10^{-(pOH)}$ M = 1×10^{-8} M

Chapter 9
Nuclear Changes
Math Skills
Practice, page 289
 2. $A = 4$
   $Z = 2$
   $X = He$
   Alpha decay occurs, and ${}^4_2He$ is produced.
 4. $212 = A + 4; A = 208$
   $83 = Z + 2; Z = 81$
   $X = Tl$
   Alpha decay occurs, and ${}^{208}_{81}Tl$ is produced.

Section 1 Review
Math Skills, page 292
 4. The product is ${}^{208}_{84}Po$.
 6. 64.4 minutes.
 8. 22 920 years old.

Chapter Review
Building Math Skills, page 309
 28. ${}^{130}_{52}Te \rightarrow {}^{130}_{53}I + {}^{0}_{-1}e$
 30. three half-lives
   1.72×10^4 years

Chapter 10
Motion
Math Skills
Practice, page 323
 2. 22 m/s toward first base

Section 1 Review
Math Skills, page 324
None

Math Skills
Practice, page 328
  2. 0.075 m/s^2 toward the shore
  4. 0.85 s

Section 2 Review
Math Skills, page 330
  4. 2.5 m/s^2 is the acceleration. The graph should be a straight line from 7.0 m/s at the zero line of time to 12 m/s at the 2.0 s line of time (with time on the x-axis and velocity on the y-axis).

Graphing Skills
Practice, page 337
  4. 8.0 m/s; from $t = 15.0$ s to $t = 17.5$ s; 0 m/s

Chapter Review
Building Math Skills, page 339
  22. 0.3 m/s^2
  24. 360 km/h northeast
  26. 6.6 h
  28. 27 s
  38. 11.1 s

Chapter 11
Forces
Math Skills
Practice, page 351
  2. 0.14 kg

Section 1 Review
Math Skills, page 351
  4. 0.26 m/s^2 forward
  6. 8.5 kg

Section 2 Review
Math Skills, page 359
  6. The force of gravity is inversely proportional to the square of distance. Since the distance is made twice as close, the force of gravity will be four times as great (the square of 2), or 4 million N.

Math Skills
Practice, page 363
  2. 6.3 m/s forward

Section 3 Review
Math Skills, page 366
  6. 12 kg • m/s eastward

Math Skills
Practice, page 367
  2. 3.5 m/s^2 in a backward direction (deceleration)

Chapter Review
Building Math Skills, page 369
  22. 10 m/s^2
  24. 2.1 kg
  26. 15 N
  28. 3.7 N
  30. 225 kg • m/s north
       (or 230 kg • m/s north using significant figures)
  32. $p = mv$
    a. 195 kg • m/s forward
    b. 660 kg • m/s west
    c. 0 kg • m/s
    d. 12 kg • m/s to the right

Chapter 12
Work and Energy
Math Skills
Practice, page 379
  2. 1 J
  4. 3960 J

Math Skills
Practice, page 381
  2. 900 MW
  4. a. 7.20×10^2 J

Math Skills
Practice, page 383
  2. MA = 66
  4. output force = 78 N

Section 1 Review
Math Skills, page 384
  6. MA = 2.40
  8. P = 2.0×10^4 W (27 hp)

Math Skills
Practice, page 393
  2. PE = 1.4×10^{15} J
  4. $m = 58$ kg

Math Skills
Practice, page 395
  2. $v = 3.3$ m/s

863

Section 3 Review
Math Skills, page 399

   8. KE = 900 J

Math Skills
Practice, page 407

   2. work input = 4800 J

Section 4 Review
Math Skills, page 408
   8. a. useful work output = 780 J
      b. P = 780 W

Chapter 13
Heat and Temperature
Math Skills
Practice, page 424
   2. Row 1: 294 K
      Row 2: 115°C, 239°F
      Row 3: –328°F, 73 K
      Row 4: 43°C, 316 K
   4. c

Section 1 Review
Math Skills, page 426
   6. 184K

Math Skills
Practice, page 434
   2. 28 000 J
   4. $T_f = 145°C$
   6. 480 J/kg • K

Section 2 Review
Math Skills, page 434
   6. 228 J/kg • K

Chapter Review
Understanding Concepts, page 446
   2. b
   4. b

Building Math Skills, page 447
   24. –454°F
   26. 4400 J

Chapter 14
Waves
Math Skills
Practice, page 468
   2. 3.00×10^{14} m/s
   4. 1.5 m

Section 2 Review
Math Skills, page 471
   8. 0.77 m

Chapter Review
Building Math Skills, page 481
   30. 3.0 m/s
   32. 1×10^9 Hz
   34. 440 Hz

Building Graphing Skills, page 482
   36. a. 9 cm = 0.09 m
      b. 20.0 cm = 0.20 m
      c. 5.00 m/s
      d. T = 0.0400 s

Chapter 15
Sound and Light
Math Skills
Practice, page 519
   2. 1.38×10^3 W/m^2

Chapter Review
Building Math Skills, page 521
   35. 41 m
   30. 5.5×10^{14} Hz

Chapter 16
Electricity
Math Skills
Practice, page 543
   2. 240 Ω
   4. 0.43 A

Section 2 Review
Math Skills, page 545
   8. 0.5 A

Math Skills
Practice, page 551
   2. 1.6×10^{-2} W
   4. 0.62 A

Section 3 Review
Math Skills, page 553
   8. The 75 W bulb has more current in it.

Chapter Review
Building Math Skills, page 555
24. 120 V
26. 0.99 W
28. 7.0×10^5 V

Chapter 17
Magnetism
none

Chapter 18
Communication Technology
none

Chapter 19
The Solar System
Chapter Review
Building Math Skills, page 657
32.

Planet	Wt in N
Mercury	267
Venus	615
Earth	684
Mars	258

Chapter 20
The Universe
none

Chapter 21
Planet Earth
none

Chapter 22
The Atmosphere
none

Chapter 23
Using Natural Resources
Chapter Review
Building Math Skills, page 803
24. 3600 days; you would produce 1600 lbs per year—but only 800 lbs per year if you recycled one-half of the trash.
42. SPF 15

865

Glossary

absolute zero the temperature at which molecular energy is at a minimum (0 K on the Kelvin scale or –273.16°C on the Celsius scale) (423)

acceleration the rate at which velocity changes over time; an object accelerates if its speed, direction, or both change (325)

accretion the accumulation of matter (649)

accuracy a description of how close a measurement is to the true value of the quantity measured (25)

acid any compound that increases the number of hydronium ions when dissolved in water; acids turn blue litmus paper red and react with bases and some metals to form salts (256)

acid precipitation precipitation, such as rain, sleet, or snow, that contains a high concentration of acids, often because of the pollution of the atmosphere (727)

aerobic describes a process that requires oxygen (747)

air mass a large body of air where temperature and moisture content are similar throughout (757)

air pressure the force with which air molecules push on a surface (83)

alkali metal one of the elements of Group 1 of the periodic table (lithium, sodium, potassium, rubidium, cesium, and francium) (121)

alkaline-earth metal one of the elements of Group 2 of the periodic table (beryllium, magnesium, calcium, strontium, barium, and radium) (122)

alloy a solid or liquid mixture of two or more metals (231)

alpha particle a positively charged atom that is released in the disintegration of radioactive elements and that consists of two protons and two neutrons (285)

alternating current an electric current that changes direction at regular intervals (abbreviation, AC) (578)

amino acid any one of 20 different organic molecules that contain a carboxyl and an amino group and that combine to form proteins (170)

amorphous solid a solid in which the particles are not arranged with periodicity or order (71)

amplitude the maximum distance that the particles of a wave's medium vibrate from their rest position (464)

analog signal a signal whose properties, such as amplitude and frequency, can change continuously in a given range (595)

angle of incidence the angle between a ray that strikes a surface and the perpendicular to that surface at the point of contact (507)

angle of reflection the angle formed by the line perpendicular to a surface and the direction in which a reflected ray moves (507)

anion an ion that has a negative charge (115)

antacid a weak base that neutralizes stomach acid (272)

antinode a point in a standing wave, halfway between two nodes; it indicates a position of maximum intensity (477)

Archimedes' principle the principle that states that the buoyant force on an object in a fluid is an upward force equal to the weight of the volume of fluid that the object displaces (81)

asteroid a small, rocky object that orbits the sun, usually in a band between the orbits of Mars and Jupiter (641)

asthenosphere the zone or layer of the mantle beneath the lithosphere where magma may be generated (704)

atmospheric pressure the pressure due to the weight of the atmosphere; also called air pressure or barometric pressure (753)

atmospheric transmission the passage of an electromagnetic wave signal through the atmosphere between a transmitter and a receiver (602)

atom the smallest unit of an element that maintains the properties of that element (39)

atomic mass unit a unit of mass that describes the mass of an atom or molecule; it is exactly one-twelfth of the mass of a carbon atom with mass number 12 (abbreviation, amu) (118)

atomic number the number of protons in the nucleus of an atom; the atomic number is the same for all atoms of an element (116)

autumnal equinox the moment when the sun passes directly above the equator from north to south; day and night are of equal length on the day that the autumnal equinox occurs (761)

average atomic mass the weighted average of the masses of all naturally occurring isotopes of an element (118)

Avogadro's constant equals 6.02×10^{23}; the number of particles in 1 mol (130)

background radiation the nuclear radiation that arises naturally from cosmic rays and from radioactive isotopes in the soil and air (299)

barometric pressure the pressure due to the weight of the atmosphere; also called air pressure or atmospheric pressure (753)

base any compound that increases the number of hydroxide ions when dissolved in water; bases turn red litmus paper blue and react with acids to form salts (258)

beat the interference of waves of slightly different frequencies traveling in the same direction (476)

Bernoulli's principle the principle that states that the pressure in a fluid decreases as the fluid's velocity increases (86)

beta particle a charged electron emitted during certain types of radioactive decay, such as beta decay (285)

big bang theory the theory that all matter and energy in the universe was compressed into an extremely small volume that 10 to 20 billion years ago exploded and began expanding in all directions (684)

biology the scientific study of living organisms and their interactions with the environment (6)

black hole an object so massive and dense that not even light can escape its gravity (672)

bleach a chemical compound used to whiten or make lighter, such as hydrogen peroxide or sodium hypochlorite (271)

blue shift an apparent shift toward shorter wavelengths of light caused when a luminous object moves toward the observer (682)

boiling point the temperature at which a liquid becomes a gas (46)

bond angle the angle formed by two bonds to the same atom (146)

bond length the distance between two bonded atoms at their minimum potential energy; the average distance between the nuclei of two bonded atoms (146)

botany the branch of biology that is the study of plants (6)

Boyle's law the law that states that for a fixed amount of gas at a constant temperature, the volume of the gas increases as the pressure of the gas decreases and the volume of the gas decreases as the pressure of the gas increases (89)

buoyant force the upward force exerted on an object immersed in or floating on a liquid (80)

C

carbohydrate any organic compound that is made of carbon, hydrogen, and oxygen and that provides nutrients to the cells of living things (170)

carrier in physics, a wave that can be modulated to send a signal (604)

catalyst a substance that changes the rate of a chemical reaction without being consumed or changed significantly (207)

cathode-ray tube a tube that uses an electron beam to create a display on a phosphorescent screen (606)

cation an ion that has a positive charge (115)

centripetal acceleration the acceleration directed toward the center of a circular path (326)

Charles's law the law that states that for a fixed amount of gas at a constant pressure, the volume of the gas increases as the temperature of the gas increases and the volume of the gas decreases as the temperature of the gas decreases (90)

chemical bond the attractive force that holds atoms or ions together (145)

chemical change a change that occurs when a substance changes composition by forming one or more new substances (55)

chemical energy the energy released when a chemical compound reacts to produce new compounds (397)

chemical equation a representation of a chemical reaction that uses symbols to show the relationship between the reactants and the products (198)

chemical equilibrium a state of balance in which the rate of a forward reaction equals the rate of the reverse reaction and the concentrations of products and reactants remain unchanged (210)

chemical formula a combination of chemical symbols and numbers to represent a substance (41)

chemical property a property of matter that describes a substance's ability to participate in chemical reactions (50)

chemical structure the arrangement of the atoms in a substance (146)

chemical weathering the process in which rocks break down as a result of chemical reactions (726)

chemistry the scientific study of the composition, structure, and properties of matter and the changes that matter undergoes (7)

cinder cone a steep-sloped deposit of solid fragments ejected from a volcano (715)

circuit breaker a switch that opens a circuit automatically when the current exceeds a certain value (552)

cirrus cloud a feathery cloud that is composed of ice crystals and that has the highest altitude of any cloud in the sky (752)

climate the average weather conditions in an area over a long period of time (760)

cluster a group of stars or galaxies bound by gravity (675)

code a set of rules used to interpret data that convey information (593)

cold front the front edge of a moving mass of cold air, usually accompanied by heavy rain (757)

colloid a mixture consisting of tiny particles that are intermediate in size between those in solutions and those in suspensions and that are suspended in a liquid, solid, or gas (226)

combustion reaction the oxidation reaction of an organic compound, in which heat is released (191)

comet a small body of ice, rock, and cosmic dust loosely packed together that follows an elliptical orbit around the sun and that gives off gas and dust in the form of a tail as it passes close to the sun (650)

community a group of species that live in the same habitat and interact with each other (775)

composite volcano a volcano made of alternating layers of lava and pyroclastic material; also called stratovolcano (714)

compound a substance made up of atoms of two or more different elements joined by chemical bonds (39)

compound machine a machine made of more than one simple machine (390)

compression a point of highest density in a longitudinal wave; corresponds to maximum amplitude (464)

computer an electronic device that can accept data and instructions, follow the instructions, and output the results (610)

concave mirror a mirror that is curved inward like the inside of a spoon (509)

concentration the amount of a particular substance in a given quantity of a mixture, solution, or ore (240)

condensation the change of a substance from a gas to a liquid (76)

constructive interference a superposition of two or more waves that produces a greater intensity than the sum of the intensities of the individual waves (475)

continental drift the hypothesis that states that the continents once formed a single landmass, broke up, and drifted to their present locations (702)

convection the movement of matter due to differences in density that are caused by temperature variations; can result in the transfer of energy as heat (428)

convection current the vertical movement of air currents due to temperature variations (429)

convergent boundary the border formed by the collision of two lithospheric plates (706)

conversion factor a ratio that is derived from the equality of two different units and that can be used to convert from one unit to the other (131)

core the central part of Earth below the mantle; also the center of the sun (701)

Coriolis effect the curving of the path of a moving object from an otherwise straight path due to Earth's rotation (755)

covalent bond a bond formed when atoms share one or more pairs of electrons (155)

crest the highest point of a wave (464)

critical mass the minimum mass of a fissionable isotope that provides the number of neutrons needed to sustain a chain reaction (297)

critical thinking the ability and willingness to assess claims critically and to make judgments on the basis of objective and supported reasons (12)

crude oil unrefined petroleum (230)

crust the thin and solid outermost layer of Earth above the mantle (700)

crystalline solid a solid that consists of crystals (71)

cumulus cloud the low-level, billowy cloud that often has a dark bottom and a top that resembles cotton balls (752)

cyclone an area in the atmosphere that has lower pressure than the surrounding areas and has winds that spiral toward the center (760)

D

decibel the most common unit used to measure loudness (abbreviation, dB) (492)

decomposition reaction a reaction in which a single compound breaks down to form two or more simpler substances (191)

density the ratio of the mass of a substance to the volume of the substance; often expressed as grams per cubic centimeter for solids and liquids and as grams per liter for gases (47)

dependent variable in an experiment, the variable that is changed or determined by manipulation of one or more factors (the independent variables) (21)

deposition the process in which material is laid down (728)

destructive interference a superposition of two or more waves whose intensity is less than the sum of the intensities of the individual waves (475)

detergent a water-soluble cleaner that can emulsify dirt and oil (270)

dew point the temperature at which air or a gas begins to condense to a liquid (752)

diffraction a change in the direction of a wave when the wave finds an obstacle or an edge, such as an opening (473)

diffusion the movement of particles from regions of higher density to regions of lower density (237)

digital signal a signal that can be represented as a sequence of discrete values (595)

diode an electronic device that allows electric charge to move more easily in one direction than in the other (602)

disinfectant a chemical substance that kills harmful bacteria or viruses (271)

dispersion in optics, the process of separating a wave (such as white light) of different frequencies into its individual component waves (the different colors) (517)

displacement the change in position of an object (319)

dissociation the separating of a molecule into simpler molecules, atoms, radicals, or ions (259)

distillation a process of separation in which a liquid is evaporated and then the vapor is condensed into a liquid (229)

divergent boundary the boundary between two tectonic plates that are moving away from each other (705)

Doppler effect an observed change in the frequency of a wave when the source or observer is moving (471)

double-displacement reaction a reaction in which a gas, a solid precipitate, or a molecular compound forms from the apparent exchange of atoms or ions between two compounds (195)

E

eclipse an event in which the shadow of one celestial body falls on another (635)

ecosystem a community of organisms and their abiotic environment (774)

efficiency a quantity, usually expressed as a percentage, that measures the ratio of useful work output to work input (406)

elastic potential energy the energy available for use when an elastic body returns to its original configuration (392)

electrical conductor a material in which charges can move freely and that can carry an electric current (532)

electrical energy the energy that is associated with charged particles because of their positions (550)

electrical insulator a material that does not transfer current easily (532)

electrical potential energy the ability to move an electric charge from one point to another (537)

electric field a region in space around a charged object that causes a stationary charged object to experience an electric force (535)

electric motor a device that converts electrical energy into mechanical energy (574)

electrolysis the process in which an electric current is used to produce a chemical reaction, such as the decomposition of water (191)

electrolyte a substance that dissolves in water to give a solution that conducts an electric current (257)

electromagnet a coil that has a soft iron core and that acts as a magnet when an electric current is in the coil (572)

electromagnetic induction the process of creating a current in a circuit by changing a magnetic field (576)

electromagnetic spectrum all of the frequencies or wavelengths of electromagnetic radiation (466)

electromagnetic wave a wave that consists of oscillating electric and magnetic fields, which radiate outward at the speed of light (455)

electron a subatomic particle that has a negative electric charge (106)

element a substance that cannot be separated or broken down into simpler substances by chemical means; all atoms of an element have the same atomic number (39)

empirical formula the composition of a compound in terms of the relative numbers and kinds of atoms in the simplest ratio (162)

emulsion any mixture of two or more immiscible liquids in which one liquid is dispersed in the other (227)

endothermic reaction a chemical reaction that requires heat (187)

energy the capacity to do work (73)

energy level the energy state of an atom (107)

enzyme a type of protein that speeds up metabolic reactions in plants and animals without being permanently changed or destroyed (207)

epicenter the point on Earth's surface directly above an earthquake's focus (710)

equivalence point the point at which the two solutions used in a titration are present in chemically equivalent amounts (267)

erosion a process in which the materials of Earth's surface are loosened, dissolved, or worn away and transported from one place to another by a natural agent, such as wind, water, ice, or gravity (728)

eutrophication an increase in the amount of nutrients, such as nitrates, in a marine or aquatic ecosystem (796)

evaporation the change of a substance from a liquid to a gas (75)

exosphere the outermost region of a planet's atmosphere in which the density is low enough that the lighter atmospheric atoms can escape into space (745)

exothermic reaction a chemical reaction in which heat is released to the surroundings (187)

F

fault a crack in Earth created when rocks on either side of a break move (707)

fission the process by which a nucleus splits into two or more fragments and releases neutrons and energy (295)

flammability the ability of a substance to react in the presence of oxygen and burn when exposed to a flame (50)

fluid a nonsolid state of matter in which the atoms or molecules are free to move past each other, as in a gas or liquid (80)

focal point the point on the axis of a mirror or lens at which all incident parallel light rays converge or diverge (516)

focus the area along a fault at which the first motion of an earthquake occurs (710)

force an action exerted on a body in order to change the body's state of rest or motion; force has magnitude and direction (331)

fossil fuel a nonrenewable energy resource formed from the remains of organisms that lived long ago; examples include oil, coal, and natural gas (783)

free fall the motion of a body when only the force of gravity is acting on the body (354)

freezing point the temperature at which a solid and liquid are in equilibrium at 1 atm pressure (76)

frequency the number of cycles or vibrations per unit of time; also the number of waves produced in a given amount of time (465)

friction a force that opposes motion between two surfaces that are in contact (332)

Glossary

front the boundary between air masses of different densities and usually different temperatures (757)

fuse an electrical device that contains a metal strip that melts when current in the circuit becomes too great (552)

fusion the process in which light nuclei combine at extremely high temperatures, forming heavier nuclei and releasing energy (298)

G

galaxy a collection of stars, dust, and gas bound together by gravity (674)

galvanometer an instrument that detects, measures, and determines the direction of a small electric current (573)

gamma ray the high-energy photon emitted by a nucleus during fission and radioactive decay (286)

gas giant a planet that has a deep, massive atmosphere, such as Jupiter, Saturn, Uranus, or Neptune (641)

Gay-Lussac's law the law that states that the pressure of a gas at a constant volume is directly proportional to the absolute temperature (92)

generator a machine that converts mechanical energy into electrical energy (578)

geocentric describes something that uses Earth as the reference point (647)

geology the study of the origin, history, and structure of Earth and the processes that shape Earth (7)

geostationary orbit a geosynchronous orbit in which a satellite moves in Earth's equatorial plane in the same direction as Earth's rotation such that the satellite remains at an altitude of 35,880 km above a fixed spot on the equator (600)

geothermal energy the energy produced by heat within Earth (786)

global warming a gradual increase in the average global temperatures that is due to a higher concentration of gases such as carbon dioxide in the atmosphere (749)

gravitational potential energy the potential energy stored in the gravitational fields of interacting bodies (392)

greenhouse effect the warming of the surface of Earth and the lower atmosphere as a result of carbon dioxide and water vapor, which absorb and reradiate infrared radiation (748)

group a vertical column of elements in the periodic table (also called family); elements in a group share chemical properties (114)

H

half-life the time required for half of a sample of a radioactive substance to disintegrate by radioactive decay or by natural processes (289)

halogen one of the elements of Group 17 of the periodic table (fluorine, chlorine, bromine, iodine, and astatine); halogens combine with most metals to form salts (126)

hardware the parts or pieces of equipment that make up a computer (614)

harmonic series a series of frequencies that includes the fundamental frequency and integral multiples of the fundamental frequency (494)

heat the energy transferred between objects that are at different temperatures; energy is always transferred from higher-temperature objects to lower-temperature objects (426)

heat engine a machine that transforms heat into mechanical energy, or work (442)

heliocentric sun-centered (647)

heterogeneous composed of dissimilar components (42)

homogeneous describes something that has a uniform structure or composition throughout (42)

humidity the amount of water vapor in the air (751)

hurricane a severe storm that develops over tropical oceans and whose strong winds of more than 120 km/h spiral in toward the intensely low-pressure storm center (760)

hydroelectric energy electrical energy produced by falling water (778)

hydrogen bond the intermolecular force occurring when a hydrogen atom that is bonded to a highly electronegative atom of one molecule is attracted to two unshared electrons of another molecule (234)

hydrosphere the portion of Earth that is water (639)

I

igneous rock rock that forms when magma cools and solidifies (719)

immiscible describes two or more liquids that do not mix with each other (226)

independent variable the factor that is deliberately manipulated in an experiment (21)

indicator a compound that can reversibly change color depending on the pH of the solution or other chemical change (256)

inertia the tendency of an object to resist being moved or, if the object is moving, to resist a change in speed or direction until an outside force acts on the object (346)

infrasound slow vibrations of frequencies lower than 20 Hz (493)

inhibitor a substance that slows down or stops a chemical reaction (207)

inner core the solid, dense center of Earth (701)

intensity the rate at which energy flows through a given area of space (502)

interference the combination of two or more waves of the same frequency that results in a single wave (474)

Internet a large computer network that connects many local and smaller networks all over the world (617)

interstellar matter the gas and dust located between the stars in a galaxy (676)

ion an atom, radical, or molecule that has gained or lost one or more electrons and has a negative or positive charge (115)

ionic bond a bond formed by the attraction between oppositely charged ions (152)

ionization the process of adding or removing electrons from an atom or molecule, which gives the atom or molecule a net charge (302)

ionosphere a region of the atmosphere that is above about 80 km and in which the air is ionized by solar radiation (745)

isotope an atom that has the same number of protons as other atoms of the same element do but that has a different number of neutrons (117)

K

kinetic energy the energy of a moving object due to its motion (73)

kinetic friction the force that opposes the movement of two surfaces that are in contact and are sliding over each other (333)

L

law of reflection the law that states that the angle of incidence is equal to the angle of reflection (507)

length a measure of the straight-line distance between two points (18)

lens a transparent object that refracts light waves such that they converge or diverge to create an image (515)

Le Système International d'Unités the International System of Units, which is the measurement system that is accepted worldwide (15)

light ray a line in space that matches the direction of the flow of radiant energy (506)

light-year the distance that light travels in one year; about 9.5 trillion kilometers (9.5×10^{12} km) (666)

lithosphere the solid, outer layer of Earth that consists of the crust and the rigid upper part of the mantle (703)

longitudinal wave a wave in which the particles of the medium vibrate parallel to the direction of wave motion (461)

loudness the extent to which a sound can be heard (491)

M

magma liquid rock produced under Earth's surface; igneous rocks are made of magma (705)

magnetic field a region where a magnetic force can be detected (567)

magnetic pole one of two points, such as the ends of a magnet, that have opposing magnetic qualities (566)

magnification a change in the size of an image compared with the size of an object (515)

mantle the layer of rock between Earth's crust and core (700)

maria large, dark areas of basalt on the moon (singular, mare) (634)

mass a measure of the amount of matter in an object (18)

mass defect the difference between the mass of an atom and the sum of the masses of the atom's protons, neutrons, and electrons (296)

mass number the sum of the numbers of protons and neutrons in the nucleus of an atom (116)

matter anything that has mass and takes up space (38)

mechanical advantage a quantity that measures how much a machine multiplies force or distance (383)

mechanical energy the sum of the kinetic and potential energy of large-scale objects in a system (396)

mechanical wave a wave that requires a medium through which to travel (455)

medium a physical environment in which phenomena occur (455)

melting point the temperature and pressure at which a solid becomes a liquid (46)

mesosphere the strong, lower part of the mantle between the asthenosphere and the outer core; also the coldest layer of the atmosphere between the stratosphere and the mesopause (745)

metallic bond a bond formed by the attraction between positively charged metal ions and the electrons around them (154)

metal an element that is shiny and that conducts heat and electricity well (121)

metamorphic rock a rock that forms from other rocks as a result of intense heat, pressure, or chemical processes (721)

meteorology the scientific study of Earth's atmosphere, especially in relation to weather and climate (757)

mineral a natural, usually inorganic solid that has a characteristic chemical composition, an orderly internal structure, and a characteristic set of physical properties (718)

miscible describes two or more liquids that can dissolve into each other in various proportions (42)

mixture a combination of two or more substances that are not chemically combined (41)

model a pattern, plan, representation, or description designed to show the structure or workings of an object, system, or concept (9)

modulate to change a wave's amplitude or frequency in order to send a signal (604)

molarity the concentration of a solution in moles of dissolved solute per liter of solution (243)

molar mass the mass in grams of 1 mol of a substance (130)

mole the SI base unit used to measure the amount of a substance whose number of particles is the same as the number of atoms of carbon in 12 g of carbon-12 (130)

molecular formula a chemical formula that shows the number and kinds of atoms in a molecule, but not the arrangement of the atoms (164)

molecule the smallest unit of a substance that keeps all of the physical and chemical properties of that substance; it can consist of one atom or two or more atoms bonded together (40)

mole ratio the relative number of moles of the substances required to produce a given amount of product in a chemical reaction (203)

momentum a quantity defined as the product of the mass and velocity of an object (362)

monomer a simple molecule that can combine with other like or unlike molecules to make a polymer (169)

motion an object's change in position relative to a reference point (318)

N

nebula a large cloud of dust and gas in interstellar space (648)

nebular model a model for the formation of the solar system in which the sun and planets condense from a cloud (or nebula) of gas and dust (648)

net force a single force whose external effects on a rigid body are the same as the effects of several actual forces acting on the body (334)

neutralization reaction the reaction of the ions that characterize acids (hydronium ions) and the ions that characterize bases (hydroxide ions) to form water molecules and a salt (264)

neutron a subatomic particle that has no charge and that is found in the nucleus of an atom (106)

neutron star a star that has collapsed under gravity to the point that the electrons and protons have smashed together to form neutrons (672)

noble gas an unreactive element of Group 18 of the periodic table (helium, neon, argon, krypton, xenon, or radon) that has eight electrons in its outer level (except for helium, which has two electrons) (127)

node in physics, a point in a standing wave that maintains zero amplitude (477)

nonmetal an element that conducts heat and electricity poorly and that does not form positive ions in an electrolytic solution (121)

nonpolar compound a compound whose electrons are equally distributed among its atoms (235)

nonrenewable resource a substance that is consumed faster than it forms and therefore cannot be replaced within a human life span (784)

nuclear chain reaction a continuous series of nuclear fission reactions (296)

nuclear radiation the particles that are released from the nucleus during radioactive decay, such as neutrons, electrons, and photons (284)

nucleus an atom's central region, which is made up of protons and neutrons (106)

O

operating system the software that controls a computer's activities (614)

optical fiber a transparent thread of plastic or glass that transmits light (597)

orbital a region in an atom where there is a high probability of finding electrons (109)

organic compound a covalently bonded compound that contains carbon, excluding carbonates and oxides (165)

oxidation reaction a chemical reaction in which a reactant loses one or more electrons such that the reactant becomes more positive in charge (195)

oxidation-reduction reaction any chemical change in which one species is oxidized (loses electrons) and another species is reduced (gains electrons); also called redox reaction (196)

ozone a gas molecule that is made up of three oxygen atoms (744)

ozone layer the thin layer of the atmosphere at an altitude of 15 to 40 km in which ozone absorbs ultraviolet solar radiation (744)

P

Pangea a single landmass that existed for about 40 million years before it began to break apart and form the continents that we know today (702)

parallel a circuit in which all of the components are connected to each other side by side (549)

pascal the SI unit of pressure; equal to the force of 1 N exerted over an area of 1 m^2 (abbreviation, Pa) (83)

Pascal's principle the principle that states that a fluid in equilibrium contained in a vessel exerts a pressure of equal intensity in all directions (84)

period in chemistry, a horizontal row of elements in the periodic table (114)

period the time that it takes a complete cycle or wave oscillation to occur (465)

periodic law the law that states that the repeating chemical and physical properties of elements change periodically with the atomic numbers of the elements (111)

pH a value used to express the acidity or alkalinity of a solution; it is defined as the logarithm of the reciprocal of the concentration of hydronium ions (261)

phase in astronomy, the change in the illuminated area of one celestial body as seen from another celestial body; phases of the moon are caused by the changing positions of Earth, the sun, and the moon (634)

photon a unit or quantum of light; a particle of electromagnetic radiation that has zero rest mass and carries a quantum of energy (500)

photosynthesis the process by which plants, algae, and some bacteria use sunlight, carbon dioxide, and water to produce carbohydrates and oxygen (746)

physical change a change of matter from one form to another without a change in chemical properties (53)

physical property a characteristic of a substance that does not involve a chemical change, such as density, color, or hardness (45)

physical science the scientific study of nonliving matter (7)

pitch a measure of how high or low a sound is perceived to be depending on the frequency of the sound wave (470)

pixel a picture element, the smallest element of a display image (607)

planet any of the nine primary bodies that orbit the sun; a similar body that orbits another star (630)

plasma a state of matter that starts as a gas and then becomes ionized; it consists of free-moving ions and electrons, it takes on an electric charge, and its properties differ from those of a solid, liquid, or gas (72)

plate tectonics the theory that explains how the outer parts of Earth change over time; explains the relationships between continental drift, sea-floor spreading, seismic activity, and volcanic activity (703)

polar molecule a molecule that has a negative charge on one side and a positive charge on the other (232)

pollution an undesirable change in the natural environment that is caused by the introduction of substances that are harmful to living organisms, or by excessive wastes, heat, noise, or radiation (791)

polyatomic ion an ion made of two or more atoms (156)

polymer a large molecule that is formed by more than five monomers, or small units (169)

potential difference between any two points, the work which must be done against electric forces to move a unit charge from one point to the other (538)

potential energy the stored energy resulting from the relative positions of objects in a system (392)

power a quantity that measures the rate at which work is done (380)

precipitation any form of water that falls to Earth's surface from the clouds; includes rain, snow, sleet, and hail (751)

precision the exactness of a measurement (24)

pressure the amount of force exerted per unit area of a surface (81)

prism a system that consists of two or more plane surfaces of a transparent solid at an angle with each other (517)

product a substance that forms in a chemical reaction (185)

projectile motion the curved path that an object follows when thrown, launched, or otherwise projected near the surface of Earth; the motion of objects that are moving in two dimensions under the influence of gravity (358)

protein an organic compound that is made of one or more chains of amino acids and that is a principal component of all cells (170)

proton a subatomic particle that has a positive charge and that is found in the nucleus of an atom (106)

pure substance a sample of matter, either a single element or a single compound, that has definite chemical and physical properties (41)

R

radar radio detection and ranging, a system that uses reflected radio waves to determine the velocity and location of objects (504)

radiation the energy that is transferred as electromagnetic waves, such as visible light and infrared waves (429)

radical an organic group that has one or more electrons available for bonding (196)

radioactive decay the disintegration of an unstable atomic nucleus into one or more different nuclides accompanied by either the emission of radiation, nuclear capture or ejection of electrons, or fission (124)

radioactive tracer a radioactive material that is added to a substance so that its distribution can be detected later (301)

radioactivity the process by which an unstable nucleus emits one or more particles or energy in the form of electromagnetic radiation (284)

Glossary

random-access memory a storage device that allows a computer user to write and read data (abbreviation, RAM) (613)

rarefaction the portion of a longitudinal wave in which the density and pressure of the medium are at a minimum (464)

reactant a substance or molecule that participates in a chemical reaction (185)

reactivity the capacity of a substance to combine chemically with another substance (50)

read-only memory a memory device that contains data that can be read but cannot be changed (abbreviation, ROM) (614)

real image an image of an object formed by light rays that actually come together at a specific location (509)

recycling the process of recovering valuable or useful materials from waste or scrap, or reusing some items (800)

red giant a large, reddish star late in its life cycle (671)

red shift an apparent shift toward longer wavelengths of light caused when a luminous object moves away from the observer (682)

reduction a chemical change in which electrons are gained, either by the removal of oxygen, the addition of hydrogen, or the addition of electrons (196)

reflection the bouncing back of a ray of light, sound, or heat when the ray hits a surface that it does not go through (472)

refraction the bending of a light ray as it passes from one substance to another one with a different density (474)

refrigerant a material used to cool an area or an object to a temperature that is lower than the temperature of the environment (440)

rem the quantity of ionizing radiation that does as much damage to human tissue as 1 roentgen of high-voltage X rays does (300)

renewable resource a natural resource that can be replaced at the same rate as it is consumed, such as food production by photosynthesis (785)

resistance the opposition posed by a material or a device to the flow of current (541)

resonance a phenomenon that occurs when two objects naturally vibrate at the same frequency (495)

respiration the interchange of oxygen and carbon dioxide between living cells and their environment; includes breathing and cellular respiration (747)

Richter scale a scale that expresses the magnitude of an earthquake (713)

rock cycle the series of processes in which a rock forms, changes from one type to another, is destroyed, and forms again by geological processes (722)

S

salt an ionic compound that forms when a metal atom or a positive radical replaces the hydrogen of an acid (265)

satellite a natural or artificial body that revolves around a planet (633)

saturated solution a solution that cannot dissolve any more solute under the given conditions (241)

schematic diagram a graphical representation of a circuit that uses lines to represent wires and different symbols to represent components (547)

science the knowledge gained by observing natural events and conditions in order to discover facts and formulate laws or principles that can be verified or tested (6)

scientific law a summary of many experimental results and observations; a law tells how things work (8)

scientific method a series of steps followed to solve problems including collecting data, formulating a hypothesis, testing the hypothesis, and stating conclusions (13)

scientific notation a method of expressing a quantity as a number multiplied by 10 to the appropriate power (22)

scientific theory an explanation for some phenomenon that is based on observation, experimentation, and reasoning (8)

sedimentary rock a rock formed from compressed or cemented layers of sediment (720)

seismic wave a vibration in rock that travels out from the focus of an earthquake in all directions; seismic waves can also be caused by explosions (710)

seismology the study of earthquakes, including their origin, propagation, energy, and prediction (711)

semiconductor an element or compound that conducts electric current better than an insulator but not as well as a conductor (121)

series the components of a circuit that form a single path for current (549)

shield volcano a large, gently sloped volcano that forms by eruptions of balsatic lava flows (714)

SI Le Système International d'Unités, or the International System of Units, which is the measurement system that is accepted worldwide (15)

signal anything that serves to direct, guide, or warn (592)

significant figure a prescribed decimal place that determines the amount of rounding off to be done based on the precision of the measurement (24)

simple harmonic motion a periodic motion whose path is formed by one or more vibrations that are symmetric about an equilibrium position (459)

simple machine one of the six basic types of machines of which all other machines are composed (385)

single-displacement reaction a reaction in which one element or radical takes the place of another element or radical in a compound (194)

soap a substance that is used as a cleaner and that dissolves in water (269)

software a set of instructions or commands that tells a computer what to do; a computer program (614)

solar system the sun and all of the planets and other bodies that travel around it (632)

solenoid a coil of wire with an electric current in it (571)

solubility the ability of one substance to dissolve in another at a given temperature and pressure; expressed in terms of the maximum amount of solute that will dissolve in a given amount of solvent (239)

soluble capable of dissolving in a particular solvent (239)

solute the substance that dissolves in the solvent (229)

solution a homogeneous mixture of two or more substances uniformly dispersed throughout a single phase (229)

solvent the substance in which the solute dissolves (229)

sonar sound navigation and ranging, a system that uses acoustic signals to determine the location of objects or to communicate (497)

sound wave a longitudinal wave that is caused by vibrations and that travels through a material medium (490)

specific heat the quantity of heat required to raise a unit mass of homogeneous material 1 K or 1°C in a specified way given constant pressure and volume (432)

spectator ion an ion that is present in a solution in which a reaction is taking place but that does not participate in the reaction (264)

speed the distance traveled by an object divided by the time interval during which the motion occurred (320)

standing wave a pattern of vibration that simulates a wave that is standing still (477)

star a large celestial body that is composed of gas and that emits light; the sun is a typical star (666)

static friction the force that resists the initiation of sliding motion between two surfaces that are in contact and at rest (333)

stationary front a front of air masses that moves either very slowly or not at all (758)

stratosphere the upper layer of the atmosphere, which lies immediately above the troposphere and extends from 10 km to about 50 km above Earth's surface (744)

stratus cloud a gray cloud that has a flat, uniform base and forms at very low altitudes (752)

strong acid an acid that ionizes completely in a solvent (257)

strong nuclear force the interaction that binds nucleons together in a nucleus (294)

structural formula a formula that indicates the location of the atoms, groups, or ions relative to one another in a molecule and that indicates the number and location of chemical bonds (146)

subduction the process by which one lithospheric plate moves beneath another as a result of tectonic forces (706)

sublimation the process in which a solid changes directly into a gas or a gas changes directly into a solid (75)

substrate the reactant in reactions catalyzed by enzymes (208)

succession the replacement of one type of community by another at a single location over a period of time (778)

summer solstice in the Northern Hemisphere, the moment in the year at which the sun appears to be at the greatest distance north of the equator; the first day of summer (761)

supernova a gigantic explosion in which a massive star collapses and throws its outer layers into space; plural supernovae (671)

surface wave a seismic wave that can move only through solids (711)

suspension a mixture in which particles of a material are more or less evenly dispersed throughout a liquid or gas (225)

synthesis reaction a reaction in which substances combine to form a new compound (190)

technology the application of science for practical purposes (7)

tectonic plate a block of lithosphere that consists of the crust and the rigid, outermost part of the mantle; also called lithospheric plate (703)

telecommunication the sending of visual or auditory information by electromagnetic means (594)

telescope an instrument that produces a magnified image of a distant object by using a system of lenses or mirrors (15)

temperature a measure of how hot (or cold) something is; specifically, a measure of the average kinetic energy of the particles in an object (420)

temperature inversion the atmospheric condition in which warm air traps cooler air near Earth's surface (743)

terminal velocity the constant velocity of a falling object when the force of air resistance is equal in magnitude and opposite in direction to the force of gravity (356)

terrestrial planet one of the highly dense planets nearest to the sun; Mercury, Venus, Earth, and Mars (637)

thermal conduction the transfer of energy as heat through a material (428)

thermal energy the kinetic energy of a substance's atoms (73)

thermometer an instrument that measures and indicates temperature (421)

Glossary

thermosphere the uppermost layer of the atmosphere, in which temperature increases as altitude increases; includes the ionosphere (745)

titration a method to determine the concentration of a substance in solution by adding a solution of known volume and concentration until the reaction is completed, which is usually indicated by a change in color (267)

topography the configuration of a land surface, including its relief (762)

total internal reflection the complete reflection that takes place within a substance when the angle of incidence of light striking the surface boundary is less than the critical angle (514)

transformer a device that increases or decreases the voltage of alternating current (581)

transform fault boundary the boundary between tectonic plates that are sliding past each other horizontally (707)

transition metal one of the metals that can use the inner shell before using the outer shell to bond (123)

transpiration the process by which plants release water vapor into the air through stomata; also the release of water vapor into the air by other organisms (751)

transverse wave a wave in which the particles of the medium move perpendicular to the direction the wave is traveling (461)

troposphere the lowest layer of the atmosphere, characterized by a constant drop of temperature with increasing altitude; the part of the atmosphere where weather conditions exist (743)

trough the lowest point of a wave (464)

typhoon a severe tropical cyclone that forms on the western Pacific Ocean and on the China Seas; a hurricane (760)

U

ultrasound any sound wave with frequencies higher than 20 000 Hz (493)

universe the sum of all space, matter, and energy that exists, that has existed in the past, and that will exist in the future (680)

unsaturated solution a solution that is able to dissolve additional solute (240)

V

vaccine a substance prepared from killed or weakened pathogens and introduced into an organism to produce immunity (223)

valence electron an electron that is found in the outermost shell of an atom and that determines the atom's chemical properties (110)

variable a factor that changes in an experiment in order to test a hypothesis (13)

velocity the speed of an object in a particular direction (322)

vent an opening at the surface of Earth through which volcanic material passes (714)

vernal equinox the moment when the sun passes directly above the equator from south to north; day and night are of equal length on the day that the vernal equinox occurs (761)

virtual image an image that forms at a location from which light rays appear to come but do not actually come (508)

viscosity the resistance of a gas or liquid to flow (85)

visible spectrum the portion of the electromagnetic spectrum that includes all of the wavelengths that are visible to the human eye (466)

volume a measure of the size of a body or region in three-dimensional space (18)

W

warm front a front that advances in such a way that warmer air replaces colder air (757)

water cycle the continuous movement of water from the ocean to the atmosphere to the land and back to the ocean (750)

watt the unit used to express power; equivalent to joules per second (abbreviation, W) (380)

wave a periodic disturbance in a solid, liquid, or gas as energy is transmitted through a medium (454)

wavelength the distance from any point on a wave to an identical point on the next wave (464)

weak acid an acid that releases few hydrogen ions in aqueous solution (257)

weathering the natural process by which atmospheric and environmental agents, such as wind, rain, and temperature changes, disintegrate, and decompose rocks (720)

weight a measure of the gravitational force exerted on an object (18)

white dwarf a small, hot, dim star that is the leftover center of an old star (671)

winter solstice in the Northern Hemisphere, the moment in the year at which the sun appears to be at the greatest distance south of the equator; the beginning of winter (761)

work the quantity of energy transferred by a force when it is applied to a body and causes that body to move in the direction of the force (378)

A

absolute zero/cero absoluto la temperatura a la que todo el movimiento molecular se detiene (0 K en la escala de Kelvin ó –273.16°C en la escala de Celsius (423)

acceleration/aceleración la tasa a la que la velocidad cambia con el tiempo; un objeto acelera si su rapidez cambia, si su dirección cambia, o si tanto su rapidez como su dirección cambian (325)

accuracy/exactitud término que describe qué tanto se aproxima una medida al valor verdadero de la cantidad medida (25)

acid/ácido cualquier compuesto que aumenta el número de iones de hidrógeno cuando se disuelve en agua; los ácidos cambian el color del papel tornasol a rojo y forman sales al reaccionar con bases y con algunos metales (256)

acid precipitation/precipitación ácida precipitación tal como lluvia, aguanieve o nieve, que contiene una alta concentración de ácidos debido a la contaminación de la atmósfera (727)

aerobic/aeróbico término que describe un proceso que requiere oxígeno (747)

air mass/masa de aire un gran volumen de aire que tiene una temperatura y contenido de humedad similar en toda su extensión (757)

air pressure/presión del aire la medida de la fuerza con la que las moléculas del aire empujan contra una superficie (83)

alkali metal/metal alcalino uno de los elementos del Grupo 1 de la tabla periódica (litio, sodio, potasio, rubidio, cesio y francio) (121)

alkaline-earth metal/metal alcalinotérreo uno de los elementos del Grupo 2 de la tabla periódica (berilio, magnesio, calcio, estroncio, bario y radio) (122)

alloy/aleación una mezcla sólida o líquida de dos o más metales (231)

alpha particle/partícula alfa un átomo cargado positivamente, liberado en la desintegración de elementos radiactivos, que está formado por dos protones y dos neutrones (285)

alpha particle/partícula alfa un átomo cargado positivamente, liberado en la desintegración de elementos radiactivos, que está formado por dos protones y dos neutrones (285)

alternating current/corriente alterna una corriente eléctrica que cambia de dirección en intervalos regulares (abreviatura: CA) (578)

amino acid/aminoácido cualquiera de las 20 distintas moléculas orgánicas que contienen un grupo carboxilo y un grupo amino y que se combinan para formar proteínas (170)

amorphous solid/sólido amorfo un sólido en el que las partículas no están ordenadas periódicamente o en orden (71)

amplitude/amplitud la distancia máxima a la que vibran las partículas del medio de una onda a partir de su posición de reposo (464)

analog signal/señal análoga una señal cuyas propiedades, tales como la amplitud y la frecuencia, cambian continuamente en un rango determinado (595)

angle of incidence/ángulo de incidencia el ángulo que se forma entre un rayo que choca contra una superficie y la línea perpendicular a esa superficie en el punto de contacto (507)

angle of reflection/ángulo de reflexión el ángulo formado por la línea perpendicular a la superficie y la dirección en la que se mueve un rayo reflejado (507)

anion/anión un ion que tiene carga negativa (115)

antacid/antiácido una base débil que neutraliza el ácido del estómago (0)

antinode/antinodo un punto en una onda estacionaria, ubicada en el punto medio entre dos nodos; indica una posición de intensidad máxima (477)

Archimedes' principle/principio de Arquímedes el principio que establece que la fuerza flotante de un objeto que está en un fluido es una fuerza ascendente cuya magnitud es igual al peso del volumen del fluido que el objeto desplaza (81)

asteroid/asteroide un objeto pequeño y rocoso que se encuentra en órbita alrededor del Sol, normalmente en una banda entre las órbitas de Marte y Júpiter (641)

asthenosphere/astenosfera la capa sólida y plástica del manto, que se encuentra debajo de la litosfera; está formada por roca del manto que fluye muy lentamente, lo cual permite que las placas tectónicas se muevan en su superficie (704)

atmospheric pressure/presión atmosférica la presión producida por el peso de la atmósfera (753)

atmospheric transmission/transmisión atmosférica el paso de la señal de una onda electromagnética a través de la atmósfera entre el transmisor y el receptor (602)

atom/átomo la unidad más pequeña de un elemento que conserva las propiedades de ese elemento (39)

atom/átomo la unidad más pequeña de un elemento que conserva las propiedades de ese elemento (104)

atomic mass unit/unidad de masa atómica una unidad de masa que describe la masa de un átomo o molécula; es exactamente 1/12 de la masa de un átomo de carbono con un número de masa de 12 (abreviatura: uma) (118)

atomic number/número atómico el número de protones en el núcleo de un átomo; el número atómico es el mismo para todos los átomos de un elemento (116)

Glosario

autumnal equinox/equinoccio otoñal el momento en el que el Sol pasa directamente encima del ecuador del Norte al Sur; el día y la noche tienen la misma duración en el día en que ocurre el equinoccio otoñal (761)

average atomic mass/masa atómica promedio el promedio ponderado de las masas de todos los isótopos de un elemento que se encuentran en la naturaleza (118)

Avogadro's number/número de Avogadro 6.02 ´ 1023, el número de átomos o moléculas que hay en 1 mol (130)

B

background radiation/radiación de fondo la radiación nuclear que surge naturalmente de los rayos cósmicos y de los isótopos radiactivos que están en el suelo y en el aire (299)

barometric pressure/presión barométrica la presión debida al peso de la atmósfera; también se llama presión del aire o presión atmosférica (753)

base/base cualquier compuesto que aumenta el número de iones de hidróxido cuando se disuelve en agua; las bases cambian el color del papel tornasol a azul y forman sales al reaccionar con ácidos (258)

beat/batido la interferencia de ondas que se desplazan en la misma dirección y que tienen frecuencias ligeramente distintas (476)

Bernoulli's principle/principio de Bernoulli el principio que establece que la presión de un fluido disminuye a medida que la velocidad del fluido aumenta (86)

beta particle/partícula beta un electrón con carga, emitido durante ciertos tipos de desintegración radiactiva, como por ejemplo, durante la desintegración beta (285)

big bang theory/teoría del Big Bang la teoría que establece que toda la materia y la energía del universo estaban comprimidas en un volumen extremadamente pequeño que explotó hace aproximadamente 10 a 20 mil millones de años y empezó a expandirse en todas direcciones (684)

biology/biología el estudio científico de los seres vivos y sus interacciones con el medio ambiente (6)

black hole/hoyo negro un objeto tan masivo y denso que ni siquiera la luz puede salir de su campo gravitacional (672)

bleach/blanqueador un compuesto químico que se usa para blanquear o aclarar, tal como el peróxido de hidrógeno o el hipoclorito de sodio (271)

boiling point/punto de ebullición la temperatura y presión a la que un líquido y un gas están en equilibrio (46)

boiling point/punto de ebullición la temperatura y presión a la que un líquido y un gas están en equilibrio (75)

bond angle/ángulo de enlace el ángulo formado por dos enlaces al mismo átomo (146)

bond length/longitud de enlace la distancia entre dos átomos que están enlazados en el punto en que su energía potencial es mínima; la distancia promedio entre los núcleos de dos átomos enlazados (146)

botany/botánica la rama de la biología que se ocupa del estudio de las plantas (6)

Boyle's law/ley de Boyle la ley que establece que para una cantidad fija de gas a una temperatura constante, el volumen del gas aumenta a medida que su presión disminuye y el volumen del gas disminuye a medida que su presión aumenta (89)

buoyant force/fuerza boyante la fuerza ascendente que hace que un objeto se mantenga sumergido en un líquido o flotando en él (80)

C

carbohydrate/carbohidrato cualquier compuesto orgánico que está hecho de carbono, hidrógeno y oxígeno y que proporciona nutrientes a las células de los seres vivos (170)

carrier/portador en física, una onda que puede modularse para enviar una señal (604)

catalyst/catalizador una substancia que cambia la tasa de una reacción química sin ser consumida ni cambiar significativamente (207)

cathode-ray tube/tubo de rayos catódicos un tubo que usa un haz de electrones para crear una representación en una pantalla fosforescente (606)

cation/catión un ion que tiene carga positiva (115)

centripetal acceleration/aceleración centrípeta la aceleración que se dirige hacia el centro de un camino circular (0)

Charles's law/ley de Charles la ley que establece que para una cantidad fija de gas a una presión constante, el volumen del gas aumenta a medida que su temperatura aumenta y el volumen del gas disminuye a medida que su temperatura disminuye (90)

chemical bond/enlace químico la fuerza de atracción que mantiene unidos a los átomos o iones (145)

chemical change/cambio químico un cambio que ocurre cuando una o más substancias se transforman en substancias totalmente nuevas con propiedades diferentes (55)

chemical energy/energía química la energía que se libera cuando un compuesto químico reacciona para producir nuevos compuestos (187)

chemical energy/energía química la energía que se libera cuando un compuesto químico reacciona para producir nuevos compuestos (397)

chemical equation/ecuación química una representación de una reacción química que usa símbolos para mostrar la relación entre los reactivos y los productos (198)

chemical equilibrium/equilibrio químico un estado de equilibrio en el que la tasa de la reacción directa es igual a la tasa de la reacción inversa y las concentraciones de los productos y reactivos no sufren cambios (210)

chemical formula/fórmula química una combinación de símbolos químicos y números que se usan para representar una substancia (41)

chemical property/propiedad química una propiedad de la materia que describe la capacidad de una substancia de participar en reacciones químicas (50)

chemical structure/estructura química la disposición de los átomos en una molécula (146)

chemical weathering/desgaste químico el proceso por medio del cual las rocas se fragmentan como resultado de reacciones químicas (726)

chemistry/química el estudio científico de la composición, estructura y propiedades de la materia y los cambios por los que pasa (7)

chemistry/química el estudio científico de la composición, estructura y propiedades de la materia y los cambios por los que pasa (38)

cinder cone/cono de escorias un depósito con pendiente empinada de fragmentos sólidos expulsados por un volcán (715)

circuit breaker/disyuntor un interruptor que abre un circuito automáticamente cuando la corriente excede un valor determinado (552)

cirrus cloud/nube cirro una nube liviana formada por cristales de hielo, la cual tiene la mayor altitud de todas las nubes en el cielo (752)

climate/clima las condiciones promedio del tiempo en un área durante un largo período de tiempo (760)

cluster/conglomerado un grupo de estrellas o galaxias unidas por la gravedad (675)

code/código un conjunto de reglas que se usan para interpretar datos y transmitir información (593)

cold front/frente frío el borde del frente de una masa de aire frío en movimiento, normalmente acompañado de fuertes lluvias (757)

colloid/coloide una mezcla formada por partículas diminutas que son de tamaño intermedio entre las partículas de las soluciones y las de las suspensiones y que se encuentran suspendidas en un líquido, sólido o gas (226)

combustion reaction/reacción de combustión la reacción de oxidación de un compuesto orgánico, durante la cual se libera calor (191)

comet/cometa un cuerpo pequeño formado por hielo, roca y polvo cósmico que sigue una órbita elíptica alrededor del Sol y que libera gas y polvo, los cuales forman una cola al pasar cerca del Sol (650)

community/comunidad un grupo de varias especies que viven en el mismo hábitat e interactúan unas con otras (775)

composite volcano/volcán compuesto un volcán formado por capas alternas de lava y material piroclástico; también se llama estratovolcán (714)

compound/compuesto una substancia formada por átomos de dos o más elementos diferentes unidos por enlaces químicos (39)

compound machine/máquina compuesta una máquina hecha de más de una máquina simple (390)

compression/compresión un punto de densidad máxima en una onda longitudinal; equivale a la amplitud máxima (464)

computer/computadora un aparato electrónico que acepta información e instrucciones, sigue instrucciones y produce una salida para los resultados (610)

concave mirror/espejo cóncavo un espejo que está curvado hacia adentro como la parte interior de una cuchara (509)

concentration/concentración la cantidad de una cierta substancia en una cantidad determinada de mezcla, solución o mena (240)

condensation/condensación el cambio de estado de gas a líquido (76)

constructive interference/interferencia constructiva una superposición de dos o más ondas que produce una intensidad mayor que la suma de las intensidades de las ondas individuales (475)

continental drift/deriva continental la hipótesis que establece que alguna vez los continentes formaron una sola masa de tierra, se dividieron y se fueron a la deriva hasta terminar en sus ubicaciones actuales (702)

convection/convección el movimiento de la materia debido a diferencias en la densidad que se producen por variaciones en la temperatura; puede resultar en la transferencia de energía en forma de calor (428)

convection current/corriente de convección el movimiento vertical de las corrientes de aire debido a variaciones en la temperatura (429)

convergent boundary/límite convergente el borde que se forma debido al choque de dos placas de la litosfera (706)

conversion factor/factor de conversión una razón que se deriva de la igualdad entre dos unidades diferentes y que se puede usar para convertir una unidad en otra (131)

core/núcleo la parte central de la Tierra, debajo del manto; también, el centro del Sol (701)

Coriolis effect/efecto de Coriolis la desviación de una línea recta que experimentan los objetos en movimiento debido a la rotación de la Tierra (755)

Glosario

covalent bond/enlace covalente un enlace formado cuando los átomos comparten uno más pares de electrones (155)

crest/cresta el punto más alto de una onda (464)

critical mass/masa crítica la cantidad mínima de masa de un isótopo fisionable que proporciona el número de neutrones que se requieren para sostener una reacción en cadena (297)

critical thinking/razonamiento crítico la capacidad y voluntad de evaluar declaraciones críticamente y de hacer juicios basados en razones objetivas y docum entadas (12)

crude oil/petróleo crudo petróleo no refinado (230)

crust/corteza la capa externa, delgada y sólida de la Tierra, que se encuentra sobre el manto (700)

crystalline solid/sólido cristalino un sólido formado por cristales (71)

cumulus cloud/nube cúmulo una nube esponjada ubicada en un nivel bajo, que normalmente es obscura en la parte inferior y cuya parte superior parece una bola de algodón (752)

cyclone/ciclón un área de la atmósfera que tiene una presión menor que la de las áreas circundantes y que tiene vientos que giran en espiral hacia el centro (760)

D

decibel/decibel la unidad más común que se usa para medir el volumen del sonido (abreviatura: dB) (492)

decomposition reaction/reacción de descomposición una reacción en la que un solo compuesto se descompone para formar dos o más substancias más simples (191)

density/densidad la relación entre la masa de una substancia y su volumen; comúnmente se expresa en gramos por centímetro cúbico para los sólidos y líquidos, y como gramos por litro para los gases (47)

dependent variable/variable dependiente en un experimento, la variable que se cambia o que se determina al manipular dos o más factores (las variables independientes) (21)

deposition/deposición el proceso por medio del cual un material se deposita (728)

destructive interference/interferencia destructiva una superposición de dos o más ondas cuya una intensidad menor que la suma de las intensidades de las ondas individuales (475)

detergent/detergente un limpiador no jabonoso, soluble en agua, que emulsiona la suciedad y el aceite (270)

dew point/punto de rocío la temperatura y presión a la que un gas se empieza a condensar para formar un líquido (752)

diffraction/difracción un cambio en la dirección de una onda cuando ésta se encuentra con un obstáculo o un borde, tal como una abertura (473)

diffusion/difusión el movimiento de partículas de regiones de mayor densidad a regiones de menor densidad (237)

digital signal/señal digital una señal que se puede representar como una secuencia de valores discretos (595)

diode/diodo un aparato electrónico que permite que la corriente eléctrica pase más fácilmente en una dirección que en otra (602)

disinfectant/desinfectante una substancia química que elimina bacterias dañinas o virus (271)

dispersion/dispersión en óptica, el proceso de separar una onda que tiene diferentes frecuencias (por ejemplo, la luz blanca) de las ondas individuales que la componen (los distintos colores) (517)

displacement/desplazamiento el cambio en la posición de un objeto (319)

dissociation/disociación la separación de una molécula en moléculas más simples, átomos, radicales o iones (259)

distillation/destilación un proceso de separación por medio del cual un líquido se evapora y, luego, el vapor se condensa en un líquido (229)

divergent boundary/límite divergente el límite entre dos placas tectónicas que se están separando una de la otra (705)

Doppler effect/efecto Doppler un cambio que se observa en la frecuencia de una onda cuando la fuente o el observador está en movimiento (471)

double-displacement reaction/reacción de doble desplazamiento una reacción en la que un gas, un precipitado sólido o un compuesto molecular se forma a partir del intercambio aparente de iones entre dos compuestos (195)

E

eclipse/eclipse un suceso en el que la sombra de un cuerpo celeste cubre otro cuerpo celeste (635)

ecosystem/ecosistema una comunidad de organismos y su ambiente abiótico (774)

efficiency/eficiencia una cantidad, generalmente expresada como un porcentaje, que mide la relación entre la entrada y la salida de trabajo (406)

elastic potential energy/energía potencial elástica la energía disponible para ser usada cuando un cuerpo elástico regresa a su configuración original (0)

electrical conductor/conductor eléctrico un material en el que las cargas se mueven libremente y que conduce una corriente eléctrica (532)

electrical energy/energía eléctrica la energía asociada con partículas que tienen carga debido a sus posiciones (550)

electrical insulator/aislante eléctrico un material que no transfiere corriente con facilidad (532)

electrical potential energy/ energía potencial eléctrica la capacidad de mover una carga eléctrica de un punto a otro (537)

electric charge/carga eléctrica una propiedad eléctrica de la materia que crea fuerzas e interacciones eléctricas y magnéticas (530)

electric field/campo eléctrico una región en el espacio alrededor de un objeto con carga experimente una fuerza eléctrica (535)

electric motor/motor eléctrico un aparato que transforma la energía eléctrica en energía mecánica (574)

electrolysis/electrólisis el proceso por medio del cual se utiliza una corriente eléctrica para producir una reacción química, como por ejemplo, la descomposición del agua (191)

electrolyte/electrolito una substancia que se disuelve en agua y crea una solución que conduce la corriente eléctrica (257)

electromagnet/electroimán una bobina que tiene un centro de hierro suave y que funciona como un imán cuando hay una corriente eléctrica en la bobina (572)

electromagnetic induction/inducción electromagnética el proceso de crear una corriente en un circuito por medio de un cambio en el campo magnético (576)

electromagnetic spectrum/espectro electromagnético todas las frecuencias o longitudes de onda de la radiación electromagnética (466)

electromagnetic wave/onda electromagnética una onda que está formada por campos eléctricos y magnéticos oscilantes, que irradia hacia fuera a la velocidad de la luz (455)

electron/electrón una partícula subatómica que tiene carga negativa (106)

element/elemento una substancia que no se puede separar o descomponer en substancias más simples por medio de métodos químicos; todos los átomos de un elemento tienen el mismo número atómico (39)

empirical formula/fórmula empírica la composición de un compuesto en función del número relativo y el tipo de átomos que hay en la proporción más simple (162)

emulsion/emulsión cualquier mezcla de dos o más líquidos inmiscibles en la que un líquido se encuentra disperso en el otro (227)

endothermic/endotérmico término que describe un proceso en que se absorbe calor del ambiente (?)

endothermic reaction/reacción endotérmica una reacción química que necesita calor (187)

energy/energía la capacidad de realizar un trabajo (73)

energy level/nivel de energía el estado de energía de un átomo (107)

enzyme/enzima un tipo de proteína que acelera las reacciones metabólicas en las plantas y animales, sin ser modificada permanentemente ni ser destruida (207)

epicenter/epicentro el punto de la superficie de la Tierra que queda justo arriba del punto de inicio, o foco, de un terremoto (710)

equilibrium/equilibrio en química, el estado en el que un proceso químico y el proceso químico inverso ocurren a la misma tasa, de modo que las concentraciones de los reactivos y los productos no cambian (?)

equivalence point/punto de equivalencia el punto en el que dos soluciones usadas en una titulación están presentes en cantidades químicas equivalentes (267)

erosion/erosión un proceso por medio del cual los materiales de la superficie de la Tierra se aflojan, disuelven o desgastan y son transportados de un lugar a otro por un agente natural, como el viento, el agua, el hielo o la gravedad (728)

eutrophication/eutrofización un aumento en la cantidad de nutrientes, tales como nitratos, en un ecosistema marino o acuático (796)

evaporation/evaporación el cambio de una substancia de líquido a gas (75)

exosphere/exosfera la porción más externa de la atmósfera de un planeta, en la cual la densidad es suficientemente baja como para permitir que los átomos atmosféricos más livianos escapen al espacio (745)

exothermic/exotérmico término que describe un proceso en el que un sistema libera calor al ambiente (?)

exothermic reaction/reacción exotérmica una reacción química en la que se libera calor a los alrededores (187)

F

fault/falla una grieta en un cuerpo rocoso a lo largo de la cual un bloque se desliza respecto a otro (707)

fin/aleta una estructura membranosa similar a un ala, que ayuda a los peces y a otros animales acuáticos a impulsarse hacia adelante, balancearse y guiar su cuerpo (730)

fission/fisión el proceso por medio del cual un núcleo se divide en dos o más fragmentos y libera neutrones y energía (295)

fluid/fluido un estado no sólido de la materia en el que los átomos o moléculas tienen libertad de movimiento, como en el caso de un gas o un líquido (80)

focal point/punto focal el punto en el eje de un espejo o lente en el que todos los rayos de luz paralelos e incidentes convergen o divergen (516)

focus/foco el punto a lo largo de una falla donde ocurre el primer movimiento de un terremoto (710)

force/fuerza una acción que se ejerce en un cuerpo con el fin de cambiar su estado de reposo o movimiento; la fuerza tiene magnitud y dirección (331)

fossil fuel/combustible fósil un recurso energético no renovable formado a partir de los restos de organismos que vivieron hace mucho tiempo; algunos ejemplos incluyen el petróleo, el carbón y el gas natural (783)

free fall/caída libre el movimiento de un cuerpo cuando la única fuerza que actúa sobre él es la fuerza de gravedad (354)

freezing point/punto de congelación la temperatura a la que un sólido y un líquido están en equilibrio a 1 atm de presión (76)

frequency/frecuencia el número de ciclos o vibraciones por unidad de tiempo; también, el número de ondas producidas en una cantidad de tiempo determinada (465)

friction/fricción una fuerza que se opone al movimiento entre dos superficies que están en contacto (0)

front/frente el límite entre masas de aire de diferentes densidades y, normalmente, diferentes temperaturas (757)

fuse/fusible un aparato eléctrico que contiene una tira de metal que se derrite cuando la corriente en el circuito es demasiado elevada (552)

fusion/fusión el proceso por medio del cual núcleos ligeros se combinan a temperaturas extremadamente altas formando núcleos más pesados y liberando energía (298)

G

galaxy/galaxia un conjunto de estrellas, polvo y gas unidos por la gravedad (674)

galvanometer/galvanómetro un instrumento que detecta, mide y determina la dirección de una corriente eléctrica pequeña (573)

gamma ray/rayo gamma el fotón de alta energía emitido por un núcleo durante la fisión y la desintegración radiactiva (286)

gas giant/gigante gaseoso un planeta con una atmósfera masiva y profunda, como por ejemplo, Júpiter, Saturno, Urano o Neptuno (641)

Gay-Lussac's law/ley de Gay-Lussac la ley que establece que la presión de un gas a volumen constante es directamente proporcional a la temperatura absoluta (92)

generator/generador una máquina que transforma la energía mecánica en energía eléctrica (578)

geocentric/geocéntrico término que describe algo que usa a la Tierra como punto de referencia (647)

geology/geología el estudio del origen, historia y estructura del planeta Tierra y los procesos que le dan forma (7)

geostationary orbit/órbita geoestacionaria una órbita geosincrónica en la que el satélite se mueve en el plano ecuatorial de la Tierra en la misma dirección que la rotación de la Tierra, de modo que el satélite permanece a una altitud de 35,880 km sobre un punto fijo en el ecuador (600)

geothermal energy/energía geotérmica la energía producida por el calor del interior de la Tierra (786)

global warming/calentamiento global un aumento gradual en las temperaturas globales promedio debido a una concentración más alta de gases (tales como dióxido de carbono) en la atmósfera (749)

gravitational potential energy/energía potencial gravitatoria la energía potencial almacenada en los campos gravitacionales entre cuerpos que interactúan (0)

greenhouse effect/efecto de invernadero el calentamiento de la superficie terrestre y de la parte más baja de la atmósfera, el cual se produce cuando el dióxido de carbono, el vapor de agua y otros gases del aire absorben radiación infrarroja y la vuelven a irradiar (748)

group/grupo una columna vertical de elementos de la tabla periódica; los elementos de un grupo comparten propiedades químicas (114)

H

half-life/vida media el tiempo que tarda la mitad de una muestra de una substancia radiactiva en desintegrarse por desintegración radiactiva o por procesos naturales (289)

halogen/halógeno uno de los elementos del Grupo 17 (flúor, cloro, bromo, yodo y ástato); se combinan con la mayoría de los metales para formar sales (126)

hardware/hardware las partes o piezas de equipo que forman una computadora (614)

harmonic series/serie armónica una serie de frecuencias que incluye la frecuencia fundamental y los múltiplos integrales de una frecuencia fundamental (494)

heat/calor la transferencia de energía entre objetos que están a temperaturas diferentes; la energía siempre se transfiere de los objetos que están a la temperatura más alta a los objetos que están a una temperatura más baja (426)

heat engine/motor térmico una máquina que transforma el calor en energía mecánica, o trabajo (442)

heliocentric/heliocéntrico centrado en el Sol (647)

heterogeneous/heterogéneo compuesto de componentes que no son iguales (42)

homogeneous/homogéneo término que describe a algo que tiene una estructura o composición global uniforme (42)

humidity/humedad la cantidad de vapor de agua que hay en el aire (751)

hurricane/huracán tormenta severa que se desarrolla sobre océanos tropicales, con vientos fuertes que soplan a más de 120 km/h y que se mueven en espiral hacia el centro de presión extremadamente baja de la tormenta (760)

hydroelectric energy/energía hidroeléctrica energía eléctrica producida por agua en caída (778)

hydrogen bond/enlace de hidrógeno la fuerza intermolecular producida por un átomo de hidrógeno que está unido a un átomo muy electronegativo de una molécula y que experimenta atracción a dos electrones no compartidos de otra molécula (234)

hydrosphere/hidrosfera la porción de la Tierra que es agua (639)

igneous rock/roca ígnea una roca que se forma cuando el magma se enfría y se solidifica (719)

immiscible/inmiscible término que describe dos o más líquidos que no se mezclan uno con otro (226)

independent variable/variable independiente el factor que se manipula deliberadamente en un experimento (21)

indicator/indicador un compuesto que puede cambiar de color de forma reversible dependiendo del pH de la solución o de otro cambio químico (256)

inertia/inercia la tendencia de un objeto a no moverse o, si el objeto se está moviendo, la tendencia a resistir un cambio en su rapidez o dirección hasta que una fuerza externa actúe en el objeto (346)

infrasound/infrasonido vibraciones lentas de frecuencias inferiores a 20 Hz (493)

inhibitor/inhibidor una substancia que desacelera o detiene una reacción química (207)

inner core/núcleo interno el centro sólido y denso de la Tierra (701)

intensity/intensidad la tasa a la que la energía fluye a través de un área determinada de espacio (502)

interference/interferencia la combinación de dos o más ondas de la misma frecuencias que resulta en una sola onda (474)

Internet/Internet una amplia red de computadoras que conecta muchas redes locales y redes más pequeñas por todo el mundo (617)

interstellar matter/materia interestelar el gas y polvo que están entre las estrellas de una galaxia (676)

ion/ion un átomo, radical o molécula que ha ganado o perdido uno o más electrones y que tiene una carga negativa o positiva (115)

ionic bond/enlace iónico una fuerza que atrae a los electrones de un átomo a otro y que transforma un átomo neutro a un ion (152)

ionization/ionización el proceso de añadir o quitar electrones de un átomo o molécula, lo cual da al átomo o molécula una carga neta (302)

ionosphere/ionosfera una región de la atmósfera que está a aproximadamente 80 km sobre la Tierra y en la que el aire está ionizado debido a la radiación solar (745)

isotope/isótopo un átomo que tiene el mismo número de protones (número atómico) que otros átomos del mismo elemento, pero que tiene un número diferente de neutrones (masa atómica) (117)

kinetic energy/energía cinética la energía de un objeto debido al movimiento del objeto (394)

kinetic friction/fricción cinética la fuerza que se opone al movimiento de dos superficies que están en contacto y que se deslizan una sobre la otra (333)

law of reflection/ley de la reflexión la ley que establece que el ángulo de incidencia es igual al ángulo de reflexión (507)

length/longitud una medida de la distancia en línea recta entre dos puntos (18)

lens/lente un objeto transparente que refracta las ondas de luz de modo que converjan o diverjan para crear una imagen (515)

light ray/rayo luz una línea en el espacio que corresponde con la dirección del flujo de energía radiante (506)

light-year/año luz la distancia que la luz viaja en un año; aproximadamente 9.5 trillones de kilómetros (9.5×10^{12} km) (666)

lithosphere/litosfera la capa externa y sólida de la Tierra que está formada por la corteza y la parte superior y rígida del manto (703)

longitudinal wave/onda longitudinal una onda en la que las partículas del medio vibran paralelamente a la dirección del movimiento de la onda (461)

longitudinal wave/onda longitudinal una onda en la que las partículas del medio vibran paralelamente a la dirección del movimiento de la onda (710)

loudness/volumen el grado al que se escucha un sonido (491)

Glosario

magma/magma roca líquida producida debajo de la superficie terrestre; las rocas ígneas están hechas de magma (705)

magnetic field/campo magnético una región donde puede detectarse una fuerza magnética (567)

magnetic pole/polo magnético uno de dos puntos, tales como los extremos de un imán, que tienen cualidades magnéticas opuestas (566)

magnification/magnificación el aumento del tamaño aparente de un objeto mediante el uso de lentes o espejos (515)

mantle/manto en las ciencias de la Tierra, la capa de roca que se encuentra entre la corteza terrestre y el núcleo (700)

maria/maria las áreas obscuras y grandes de basalto en la Luna (singular: mar) (634)

mass/masa una medida de la cantidad de materia que tiene un objeto; una propiedad fundamental de un objeto que no está afectada por las fuerzas que actúan sobre el objeto, como por ejemplo, la fuerza gravitacional (18)

mass defect/defecto de masa la diferencia entre la masa de un átomo y la suma de la masa de los protones, neutrones y electrones del átomo (296)

mass number/número de masa la suma de los números de protones y neutrones que hay en el núcleo de un átomo (116)

matter/materia cualquier cosa que tiene masa y ocupa un lugar en el espacio (38)

mechanical advantage/ventaja mecánica un número que dice cuántas veces una máquina multiplica una fuerza; se calcula dividiendo la fuerza de salida entre la fuerza de entrada (383)

mechanical energy/energía mecánica la cantidad de trabajo que un objeto realiza debido a las energías cinética y potencial del objeto (396)

mechanical wave/onda mecánica una onda que requiere un medio para desplazarse (455)

medium/medio un ambiente físico en el que ocurren fenómenos (455)

melting point/punto de fusión la temperatura y presión a la cual un sólido se convierte en líquido (46)

melting point/punto de fusión la temperatura y presión a la cual un sólido se convierte en líquido (75)

mesosphere/mesosfera la parte fuerte e inferior del manto que se encuentra entre la astenosfera y el núcleo externo; también, la capa más fría de la atmósfera que se encuentra entre la estratosfera y la termosfera, en la cual la temperatura disminuye al aumentar la altitud (745)

metallic bond/enlace metálico un enlace formado por la atracción entre iones metálicos cargados positivamente y los electrones que los rodean (154)

metalloid/metaloides elementos que tienen propiedades tanto de metales como de no metales; a veces de denominan semiconductores (121)

metamorphic rock/roca metamórfica una roca que se forma a partir de otras rocas como resultado de calor intenso, presión o procesos químicos (721)

meteorology/meteorología el estudio científico de la atmósfera de la Tierra, sobre todo en lo que se relaciona al tiempo y al clima (757)

mineral/mineral un sólido natural, normalmente inorgánico, que tiene una composición química característica, una estructura interna ordenada y propiedades físicas y químicas características (718)

miscible/miscible término que describe a dos o más líquidos que son capaces de disolverse uno en el otro en varias proporciones (42)

mixture/mezcla una combinación de dos o más substancias que no están combinadas químicamente (41)

model/modelo un diseño, plan, representación o descripción cuyo objetivo es mostrar la estructura o funcionamiento de un objeto, sistema o concepto (9)

modulate/modular cambiar la amplitud o la frecuencia de una onda con el fin de enviar una señal (604)

molarity/molaridad una unidad de concentración de una solución, expresada en moles de soluto disuelto por litro de solución (243)

molar mass/masa molar la masa en gramos de 1 mol de una substancia (130)

mole/mol la unidad fundamental del sistema internacional de unidades que se usa para medir la cantidad de una substancia cuyo número de partículas es el mismo que el número de átomos de carbono en exactamente 12 g de carbono-12 (130)

molecular formula/fórmula molecular una fórmula química que muestra el número y los tipos de átomos que hay en una molécula, pero que no muestra cómo están distribuidos (164)

molecule/molécula la unidad más pequeña de una substancia que conserva todas las propiedades físicas y químicas de esa substancia; puede estar formada por un átomo o por dos o más átomos enlazados uno con el otro (40)

mole ratio/razón molar el número relativo de moles de las substancias que se requieren para producir una cantidad determinada de producto en una reacción química (203)

momentum/momento una cantidad que se define como el producto de la masa de un objeto por su velocidad (362)

monomer/monómero una molécula simple que se puede combinar con otras moléculas parecidas o diferentes y formar un polímero (169)

motion/movimiento el cambio en la posición de un objeto respecto a un punto de referencia (318)

N

nebula/nebulosa una nube grande de polvo y gas en el espacio interestelar (648)

nebular model/teoría nebular un modelo de la formación del Sistema Solar en el que el Sol y los planetas se condensan a partir de una nube (o nebulosa) de gas y polvo (648)

net force/fuerza neta una fuerza única cuyos efectos externos en un cuerpo rígido son los mismos que los efectos de varias fuerzas reales ejercidas sobre el objeto (0)

neutralization reaction/reacción de neutralización la reacción de los iones que caracterizan a los ácidos (iones hidronio) y de los iones que caracterizan a las bases (iones hidróxido) para formar moléculas de agua y una sal (264)

neutron/neutrón una partícula subatómica que no tiene carga y que se encuentra en el núcleo de un átomo (106)

neutron star/estrella de neutrones una estrella que se ha colapsado debido a la gravedad hasta el punto en que los electrones y protones han chocado unos contra otros para formar neutrones (672)

noble gas/gas noble un elemento no reactivo del Grupo 18 de la tabla periódica; los gases nobles son: helio, neón, argón, criptón, xenón o radón (127)

node/nodo en física, un punto en una onda estacionaria que mantiene una amplitud de cero (477)

nonmetal/no metal un elemento que es mal conductor del calor y la electricidad y que no forma iones positivos en una solución de electrolitos (121)

nonpolar compound/compuesto no polar un compuesto cuyos electrones se encuentran distribuidos equitativamente entre los átomos (235)

nonrenewable resource/recurso no renovable un recurso que se forma a una tasa que es mucho más lenta que la tasa a la que se consume (784)

nuclear chain reaction/reacción nuclear en cadena una serie continua de reacciones nucleares de fisión (296)

nuclear radiation/radiación nuclear las partículas que el núcleo libera durante la desintegración radiactiva, tales como neutrones, electrones y fotones (284)

nucleus/núcleo la región central de un átomo, la cual está constituida por protones y neutrones (106)

O

operating system/sistema operativo el software (programas de computadora) que controla las actividades de una computadora (614)

optical fiber/fibra óptica una hebra transparente de plástico o vidrio que transmite luz (597)

orbital/orbital una región en un átomo donde hay una alta probabilidad de encontrar electrones (109)

organic compound/compuesto orgánico un compuesto enlazado de manera covalente que contiene carbono, excluyendo a los carbonatos y óxidos (165)

oxidation reaction/reacción de oxidación una reacción química en la que el reactivo pierde uno o más electrones, volviéndose más positivo en cuanto a su carga (195)

oxidation-reduction reaction/reacción de óxido-reducción cualquier cambio químico en el que una especie se oxida (pierde electrones) y otra especie se reduce (gana electrones); también se denomina reacción redox (196)

ozone/ozono una molécula de gas que está formada por tres átomos de oxígeno (744)

ozone layer/capa de ozono la capa de la atmósfera ubicada a una altitud de 15 a 40 km, en la cual el ozono absorbe la radiación solar (744)

P

Pangaea/Pangea una sola masa de tierra que existió durante aproximadamente 40 millones de años y luego comenzó a separarse para formar los continentes, tal como los conocemos en la actualidad (702)

parallel/paralelo cualquier círculo que va hacia el Este o hacia el Oeste alrededor de la Tierra y que es paralelo al ecuador; una línea de latitud (549)

pascal/pascal la unidad de presión del sistema internacional de unidades; es igual a la fuerza de 1 N ejercida sobre un área de 1 m2 (abreviatura: Pa) (83)

Pascal's principle/principio de Pascal el principio que establece que un fluido en equilibro que esté contenido en un recipiente ejerce una presión de igual intensidad en todas las direcciones (84)

period/período en química, una hilera horizontal de elementos en la tabla periódica (114)

period/período en física, el tiempo que se requiere para completar un ciclo o la oscilación de una onda (465)

periodic law/ley periódica la ley que establece que las propiedades químicas y físicas repetitivas de un elemento cambian periódicamente en función del número atómico de los elementos (111)

pH/pH un valor que expresa la acidez o la alcalinidad (basicidad) de un sistema; cada número entero de la escala indica un cambio de 10 veces en la acidez (261)

phase/fase en astronomía, el cambio en el área iluminada de la Luna según se ve desde la Tierra; las fases se producen como resultado de los cambios en la posición de la Tierra, el Sol y la Luna (634)

photon/fotón una unidad o quantum de luz; una partícula de radiación electromagnética que tiene una masa de reposo de cero y que lleva un quantum de energía (500)

photosynthesis/fotosíntesis el proceso por medio del cual las plantas, algas y algunas bacterias utilizan la luz solar, dióxido de carbono y agua para producir carbohidratos y oxígeno (746)

physical change/cambio físico un cambio de materia de una forma a otra sin que ocurra un cambio en sus propiedades químicas (53)

physical property/propiedad física una característica de una substancia que no implica un cambio químico, tal como la densidad, el color o la dureza (45)

physical science/ciencias físicas el estudio científico de la materia sin vida (7)

pitch/altura tona una medida de qué tan agudo o grave se percibe un sonido, dependiendo de la frecuencia de la onda sonora (470)

pitch/altura tona una medida de qué tan agudo o grave se percibe un sonido, dependiendo de la frecuencia de la onda sonora (492)

pixel/pixel el elemento más pequeño de una imagen de visualización (607)

planet/planeta cualquiera de los nueve cuerpos principales que giran en órbita alrededor del Sol; un cuerpo similar que gira en órbita alrededor de otra estrella (630)

plasma/plasma un estado de la materia que comienza como un gas y luego se vuelve ionizado; está formado por iones y electrones que se mueven libremente, tiene carga eléctrica y sus propiedades difieren de las de un sólido, líquido o gas (72)

plate tectonics/tectónica de placas la teoría que explica cómo cambian las partes externas de la Tierra con el tiempo; explica las relaciones entre la deriva continental, la expansión del suelo marino, la actividad sísmica y la actividad volcánica (703)

polar compound/compuesto polar un compuesto cuyas moléculas tienen una carga negativa en un lado y una carga positiva en el otro (232)

pollution/contaminación un cambio indeseable en el ambiente natural, producido por la introducción de substancias que son dañinas para los organismos vivos o por desechos, calor, ruido o radiación excesivos (791)

polyatomic ion/ion poliatómico un ion formado por dos o más átomos (156)

polymer/polímero una molécula grande que está formada por más de cinco monómeros, o unidades pequeñas (169)

potential difference/diferencia de potencial la diferencia de voltaje en el potencial entre dos puntos de un circuito (538)

potential energy/energía potencial la energía que tiene un objeto debido a su posición, forma o condición (392)

power/potencia una cantidad que mide la tasa a la que se realiza un trabajo o a la que se transforma la energía (380)

precipitation/precipitación cualquier forma de agua que cae de las nubes a la superficie de la Tierra; incluye a la lluvia, nieve, aguanieve y granizo (751)

precision/precisión la exactitud de una medición (24)

pressure/presión la cantidad de fuerza ejercida en una superficie por unidad de área (81)

prism/prisma un sistema formado por dos o más superficies planas de un sólido transparente ubicadas en un ángulo unas respecto a otras (517)

product/producto una substancia que se forma en una reacción química (185)

projectile motion/movimiento proyectil la trayectoria curva que sigue un objeto cuando es aventado, lanzado o proyectado de cualquier otra manera cerca de la superficie de la Tierra; el movimiento de objetos que se mueven en dos dimensiones bajo la influencia de la gravedad (358)

protein/proteína un compuesto orgánico que está hecho de una o más cadenas de aminoácidos y que es el principal componente de todas las células (170)

proton/protón una partícula subatómica que tiene una carga positiva y que se encuentra en el núcleo de un átomo (106)

pure substance/substancia pura una muestra de materia, ya sea un solo elemento o un solo compuesto, que tiene propiedades químicas y físicas definidas (41)

R

radar/radar detección y exploración a gran distancia por medio de ondas de radio; un sistema que usa ondas de radio reflejadas para determinar la velocidad y ubicación de los objetos (504)

radiation/radiación la energía que se transfiere en forma de ondas electromagnéticas, tales como las ondas de luz y las infrarrojas (429)

radicals/radicales un grupo orgánico que tiene uno o más electrones disponibles para formar enlaces (196)

radioactive decay/desintegración radiactiva la desintegración de un núcleo atómico inestable para formar uno o más nucleidos diferentes, lo cual va acompañado de la

emisión de radiación, la captura o expulsión nuclear de electrones, o fisión (124)

radioactive tracer/trazador radiactivo un material radiactivo que se añade a una substancia de modo que su distribución pueda ser detectada posteriormente (301)

radioactivity/radiactividad el proceso por medio del cual un núcleo inestable emite una o más partículas o energía en forma de radiación electromagnética (284)

random-access memory/memoria de acceso aleatorio un instrumento de almacenaje que permite que los usuarios de las computadoras escriban y lean datos (abreviatura: RAM, por sus siglas en inglés) (613)

rarefaction/rarefacción la porción de una onda sonora en la que la compresión del medio es mínima (464)

reactant/reactivo una substancia o molécula que participa en una reacción química (185)

reactivity/reactividad la capacidad de una substancia de combinarse químicamente con otra substancia (50)

read-only memory/memoria de sólo lectura un instrumento de memoria que contiene información que puede leerse pero que no puede cambiarse (abreviatura: ROM, por sus siglas en inglés) (614)

real image/imagen real la imagen de un objeto que se forma cuando pasan rayos de luz a través de un lente y se cruzan en un punto único (509)

recycling/reciclar el proceso de recuperar materiales valiosos o útiles de los desechos o de la basura; el proceso de reutilizar algunas cosas (800)

red giant/gigante roja una estrella grande de color rojizo que se encuentra en una etapa avanzada de su vida (671)

red shift/desplazamiento al rojo un aparente desplazamiento hacia una longitud de onda de luz mayor, que se origina cuando un objeto luminoso se aleja del observador (682)

reduction/reducción un cambio químico en el que se ganan electrones, ya sea por la remoción de oxígeno, la adición de hidrógeno o la adición de electrones (196)

reflection/reflexión el rebote de un rayo de luz, sonido o calor cuando el rayo golpea una superficie pero no la atraviesa (472)

refraction/refracción el curvamiento de un frente de ondas a medida que el frente pasa entre dos substancias en las que la velocidad de las ondas difiere (474)

refrigerant/refrigerante un material que se usa para enfriar un área o un objeto a una temperatura que es menor que la temperatura del ambiente (440)

rem/rem la cantidad de radiación ionizante que produce el mismo daño a los tejidos humanos que 1 roentgen de rayos X de alto voltaje (300)

renewable resource/recurso renovable un recurso natural que puede reemplazarse a la misma tasa a la que el se consume, como por ejemplo, el alimento que se produce por medio de la fotosíntesis (785)

resistance/resistencia en ciencias físicas, la oposición que un material o aparato presenta a la corriente (541)

resonance/resonancia un fenómeno que ocurre cuando dos objetos vibran naturalmente a la misma frecuencia (495)

respiration/respiración en química, el proceso por medio del cual las células producen energía a partir de los carbohidratos; el oxígeno atmosférico se combina con la glucosa para formar agua y dióxido de carbono (747)

Richter scale/escala de Richter una escala que expresa la magnitud de un terremoto (713)

rock cycle/ciclo de las rocas la serie de procesos por medio de los cuales una roca se forma, cambia de un tipo a otro, se destruye, y se forma nuevamente por procesos geológicos (722)

S

salt/sal un compuesto iónico que se forma cuando el átomo de un metal o un radical positivo reemplaza el hidrógeno de un ácido (265)

satellite/satélite un cuerpo natural o artificial que gira alrededor de un planeta (633)

saturated solution/solución saturada una solución que no puede disolver más soluto bajo las condiciones dadas (241)

schematic diagram/diagrama esquemático una representación gráfica de un circuito, la cual usa líneas para representar cables y diferentes símbolos para representar los componentes (547)

science/ciencia el conocimiento que se obtiene por medio de la observación natural de acontecimientos y condiciones con el fin de descubrir hechos y formular leyes o principios que puedan ser verificados o probados (6)

scientific method/método científico una serie de pasos que se siguen para solucionar problemas, los cuales incluyen recopilar información, formular una hipótesis, comprobar la hipótesis y sacar conclusiones (13)

scientific notation/notación científica un método para expresar una cantidad en forma de un número multiplicado por 10 a la potencia adecuada (22)

sedimentary rock/roca sedimentaria una roca que se forma a partir de capas comprimidas o cementadas de sedimento (720)

seismic wave/onda sísmica una vibración en las rocas que se aleja del epicentro de un terremoto en todas direcciones; las ondas sísmicas también pueden ser originadas por explosiones (710)

seismology/sismología el estudio de los terremotos, incluyendo su origen, propagación, energía y predicción (711)

semiconductor/semiconductor un elemento o compuesto que conduce la corriente eléctrica mejor que un aislante, pero no tan bien como un conductor (121)

series/serie los componentes de un circuito que forman un solo camino para la corriente (549)

shield volcano/volcán de escudo un volcán grande que tiene una pendiente suave y se forma por erupciones de flujos de lava basáltica (714)

SI/SI Le Système International d'Unités, o el Sistema Internacional de Unidades, que es el sistema de medición que se acepta en todo el mundo (15)

signal/señal cualquier cosa que sirve para dirigir, guiar o advertir (592)

significant figure/cifra significativa un lugar decimal prescrito que determina la cantidad de redondeo que se hará con base en la precisión de la medición (24)

simple harmonic motion/movimiento armónico simple un movimiento periódico cuya trayectoria se forma por una o más vibraciones que son simétricas respecto a una posición de equilibrio (459)

simple machine/máquina simple uno de los seis tipos fundamentales de máquinas, las cuales son la base de todas las demás formas de máquinas (385)

single-displacement reaction/reacción de sustitución simple una reacción en la que un elemento o radical toma el lugar de otro elemento o radical en el compuesto (194)

soap/jabón una sustancia que se usa como limpiador y que se disuelve en el agua (269)

software/software un conjunto de instrucciones o comandos que le dicen qué hacer a una computadora; un programa de computadora (614)

solar system/Sistema Solar el Sol y todos los planetas y otros cuerpos que se desplazan alrededor de él (632)

solenoid/solenoide una bobina de alambre que tiene una corriente eléctrica (571)

solubility/solubilidad la capacidad de una sustancia de disolverse en otra a una temperatura y presión dadas; se expresa en términos de la cantidad de soluto que se disolverá en una cantidad determinada de solvente (239)

soluble/soluble capaz de disolverse en un solvente determinado (239)

solute/soluto la sustancia que se disuelve en el solvente (229)

solution/solución una mezcla homogénea de dos o más sustancias dispersas de manera uniforme en una sola fase (229)

solvent/solvente la sustancia en la que se disuelve el soluto (229)

sonar/sonar navegación y exploración por medio del sonido; un sistema que usa señales acústicas y ondas de eco que regresan para determinar la ubicación de los objetos o para comunicarse (497)

sound wave/onda de sonido una onda longitudinal que se origina debido a vibraciones y que se desplaza a través de un medio material (490)

specific heat/calor específico la cantidad de calor que se requiere para aumentar una unidad de masa de un material homogéneo 1 K ó 1°C de una manera especificada, dados un volumen y una presión constantes (432)

spectator ions/iones espectadores iones que están presenten en una solución en la que está ocurriendo una reacción, pero que no participan en la reacción (264)

speed/rapidez la distancia que un objeto se desplaza dividida entre el intervalo de tiempo durante el cual ocurrió el movimiento (320)

standing wave/onda estacionaria un patrón de vibración que simula una onda que está parada (477)

star/estrella un cuerpo celeste grande que está compuesto de gas y emite luz; el Sol es una estrella típica (666)

static friction/fricción estática la fuerza que se opone a que se inicie el movimiento de deslizamiento entre dos superficies que están en contacto y en reposo (333)

stationary front/frente estacionario un frente de masas de aire que se mueve muy lentamente o que no se mueve (758)

stratosphere/estratosfera la capa de la atmósfera que se encuentra justo encima de la troposfera y se extiende de aproximadamente 10 km hasta 50 km sobre la superficie de la Tierra; ahí, la temperatura aumenta al aumentar la altitud; contiene la capa de ozono (744)

stratus cloud/nube stratus una nube gris que tiene una base plana y uniforme y que se forma a altitudes muy bajas (752)

strong acid/ácido fuerte un ácido que se ioniza completamente en un solvente (257)

strong nuclear force/fuerza fuerte la interacción que mantiene unidos a los nucleones en un núcleo (294)

structural formula/fórmula estructural una fórmula que indica la ubicación de los átomos, grupos o iones, unos respecto a otros en una molécula, y que indica el número y ubicación de los enlaces químicos (146)

subduction/subducción el proceso por medio del cual una placa de la litosfera se mueve debajo de otra como resultado de las fuerzas tectónicas (706)

sublimation/sublimación el proceso por medio del cual un sólido se transforma directamente en un gas o un gas se transforma directamente en un sólido (75)

substrate/sustrato el reactivo en reacciones que son catalizadas por enzimas (208)

succession/sucesión el reemplazo de un tipo de comunidad por otro en un mismo lugar a lo largo de un período de tiempo (778)

summer solstice/solsticio de verano el primer día del verano (761)

supernova/supernova una explosión gigantesca en la que una estrella masiva se colapsa y lanza sus capas externas hacia el espacio (671)

surface wave/onda superficial una onda sísmica que únicamente se puede mover a través de los sólidos (711)

suspension/suspensión una mezcla en la que las partículas de un material se encuentran dispersas de manera más o menos uniforme a través de un líquido o de un gas (225)

synthesis reaction/reacción de síntesis una reacción en la que dos o más sustancias se combinan para formar un compuesto nuevo (190)

T

technology/tecnología la aplicación de la ciencia con fines prácticos; el uso de herramientas, máquinas, materiales y procesos para satisfacer las necesidades de los seres humanos (7)

tectonic plate/placa tectónica un bloque de litosfera formado por la corteza y la parte rígida y más externa del manto; también se llama placa litosférica (703)

telecommunication/telecomunicación el envío de información visible o audible por medios electromagnéticos (594)

telescope/telescopio un instrumento que produce una imagen aumentada de un objeto distante por medio del uso de un sistema de lentes y espejos (15)

temperature/temperatura una medida de qué tan caliente (o frío) está algo; específicamente, una medida de la energía cinética promedio de las partículas de un objeto (420)

temperature inversion/inversión de la temperatura la condición atmosférica en la que el aire caliente retiene al aire frío cerca de la superficie terrestre (743)

terminal velocity/velocidad terminal la velocidad constante de un objeto en caída cuando la fuerza de resistencia del aire es igual en magnitud y opuesta en dirección a la fuerza de gravedad (356)

terrestrial planet/planeta terrestre uno de los planetas muy densos que se encuentran más cerca del Sol; Mercurio, Venus, Marte y la Tierra (637)

thermal conduction/conducción térmica la transferencia de energía en forma de calor a través de un material (428)

thermal energy/energía térmica la energía cinética de los átomos de una sustancia (73)

thermometer/termómetro un instrumento que mide e indica la temperatura (421)

thermosphere/termosfera la capa más alta de la atmósfera, en la cual la temperatura aumenta a medida que la altitud aumenta; incluye la ionosfera (745)

titration/titulación un método para determinar la concentración de una sustancia en una solución al añadir una solución de volumen y concentración conocidos hasta que se completa la reacción, lo cual normalmente es indicado por un cambio de color (267)

topography/topografía la configuración de una superficie de terreno, incluyendo su relieve (762)

total internal reflection/reflexión total interna

transformer/transformador un aparato que aumenta o disminuye el voltaje de la corriente alterna (581)

transform fault boundary/límite de transformación el límite entre placas tectónicas que se están deslizando horizontalmente una sobre otra (707)

transition metal/metal de transición uno de los metales que tienen la capacidad de usar su orbital interno antes de usar su orbital externo para formar un enlace (123)

transpiration/transpiración el proceso por medio del cual las plantas liberan vapor de agua al aire por medio de los estomas; también, la liberación de vapor de agua al aire por otros organismos (751)

transverse wave/onda transversal una onda en la que las partículas del medio se mueven perpendicularmente respecto a la dirección en la que se desplaza la onda (461)

troposphere/troposfera la capa inferior de la atmósfera, en la que la temperatura disminuye a una tasa constante a medida que la altitud aumenta; la parte de la atmósfera donde se dan las condiciones del tiempo (743)

trough/seno el punto más bajo de una onda (464)

typhoon/tifón un ciclón tropical severo que se forma en el océano Pacífico occidental y en los mares de China; un huracán (760)

U

ultrasound/ultrasonido cualquier onda de sonido que tenga frecuencias superiores a los 20,000 Hz (493)

universe/universo la suma de todo el espacio, materia y energía que existen, que han existido en el pasado y que existirán en el futuro (680)

unsaturated solution/solución no saturada una solución que contiene menos soluto que una solución saturada, y que tiene la capacidad de disolver más soluto (240)

V

vaccine/vacuna una sustancia que se prepara a partir de organismos patógenos muertos o debilitados y se introduce al cuerpo para producir inmunidad (223)

valence electron/electrón de valencia un electrón que se encuentra en el orbital más externo de un átomo y que determina las propiedades químicas del átomo (110)

Glosario

variable/variable un factor que se modifica en un experimento con el fin de probar una hipótesis (13)

velocity/velocidad la rapidez de un objeto en una dirección dada (322)

vent/chimenea una abertura en la superficie de la Tierra a través de la cual pasa material volcánico (714)

vernal equinox/equinoccio vernal el momento en el que el Sol pasa directamente encima del ecuador del Sur al Norte; el día en que ocurre un equinoccio vernal, el día y la noche tienen la misma duración (761)

virtual image/imagen virtual una imagen que se forma en un punto desde el cual parece que provienen los rayos de luz, pero en realidad no vienen de ahí (508)

viscosity/viscosidad la resistencia de un gas o un líquido a fluir (85)

visible spectrum/espectro visible la porción del espectro electromagnético que contiene todas las longitudes de onda que son visibles para el ojo humano (466)

volume/volumen una medida del tamaño de un cuerpo o región en un espacio de tres dimensiones (18)

W

warm front/frente cálido un frente que avanza de tal manera que el aire más cálido reemplaza al aire más frío (757)

water cycle/ciclo del agua el movimiento continuo del agua: del océano a la atmósfera, de la atmósfera a la tierra y de la tierra al océano (750)

watt/watt (o vatio) la unidad que se usa para expresar potencia; es equivalente a un joule por segundo (abreviatura: W) (380)

wave/onda una perturbación periódica en un sólido, líquido o gas que se transmite a través de un medio en forma de energía (454)

wavelength/longitud de onda la distancia entre cualquier punto de una onda y un punto idéntico en la onda siguiente (464)

weak acid/ácido débil un ácido que libera pocos iones de hidrógeno en una solución acuosa (257)

weathering/meteorización el proceso natural por el que los agentes atmosféricos y ambientales, tales como el viento, la lluvia y los cambios de temperatura, desintegran y descomponen las rocas (720)

weight/peso una medida de la fuerza gravitacional ejercida sobre un objeto; su valor puede cambiar en función de la ubicación del objeto en el universo (18)

white dwarf/enana blanca una estrella pequeña, caliente y tenue que es el centro sobrante de una estrella vieja (671)

winter solstice/solsticio de invierno el comienzo del invierno (761)

work/trabajo la transferencia de energía a un cuerpo por medio de la aplicación de una fuerza que hace que el cuerpo se mueva en la dirección de la fuerza; es igual al producto de la magnitud del componente de una fuerza aplicada en la dirección del desplazamiento por la magnitud del desplazamiento (378)

Index

Index

Index

Index

Acknowledgments continued from page iv.

Aaron Timperman, Ph.D.
Professor of Chemistry
Department of Chemistry
West Virginia University
Morgantown, West Virginia

Richard S. Treptow, Ph.D.
Professor of Chemistry
Department of Chemistry and
 Physics
Chicago State University
Chicago, Illinois

Martin VanDyke, Ph.D.
Professor of Chemistry,
 Emeritus
Front Range Community College
Westminister, Colorado

Text Reviewers

Dan Aude
Magnet Programs Coordinator
Montgomery Public Schools
Montgomery, Alabama

Robert Baronak
Science Teacher
Donegal High School
Mount Joy, Pennsylvania

David Blinn
Secondary Sciences Teacher
Wrenshall High School
Wrenshall, Minnesota

Robert Chandler
Science Teacher
Soddy-Daisey High School
Soddy-Daisey, Tennessee

Cindy Copolo, Ph.D.
Science Specialist
Summit Solutions
Bahama, North Carolina

Linda Culp
Science Teacher
Thorndale High School
Thorndale, Texas

Katherine Cummings
Science Teacher
Currituck County
Currituck, North Carolina

Donna Defrieze
Technical Communications
 Teacher
Soddy-Daisy High School
Soddy-Daisy, Tennessee

Chris Diehl
Science Teacher
Belleville High School
Belleville, Michigan

Benjamen Ebersole
Science Teacher
Donnegal High School
Mount Joy, Pennsylvania

Jeffrey L. Engel
Science Teacher
Madison County High School
Danielsville, Georgia

Randa Flinn
Science Teacher
Northeast High School
Fort Lauderdale, Florida

Sharon Harris
Science Teacher
Mother of Mercy High School
Cincinnati, Ohio

Gail Hermann
Science Teacher
Quincy High School
Quincy, Illinois

Donald R. Kanner
Physics Instructor
Lane Technical High School
Chicago, Illinois

Edward Keller
Science Teacher
Morgantown High School
Morgantown, West Virginia

Howard Knodle
Science Teacher
Maine South High School
Park Ridge, Illinois

Stewart Lipsky
Science Teacher
Seward Park High School
New York, New York

Mike Lubich
Science Teacher
Maple Town High School
Greensboro, Pennsylvania

Thomas Manerchia
Environmental Science Teacher,
 Retired
Archmere Academy
Claymont, Delaware

Tammie Niffenegger
Science Chair and Science
 Teacher
Port Washington High School
Waldo, Wisconsin

Donna Norwood
Science Teacher
Monroe High School
Charlotte, North Carolina

Jennifer Seelig-Fritz
Science Teacher
North Springs High School
Atlanta, Georgia

Aida Semerjibashian
Science Teacher
Pflugerville High School
Pflugerville, Texas

Bert Sherwood
Science/Health Specialist
Socorro ISD
El Paso, Texas

Linnaea Smith
Science Teacher
Bastrop High School
Bastrop, Texas

Dan Trockman
Science Teacher
Hopkins High School
Minnetonka, Massachusetts

Gabriela Waschesky, Ph.D.
Science and Math Teacher
Emery High School
Emeryville, California

Jim Watson
Science Teacher
Dalton High School
Dalton, Georgia

The Periodic Table of the Elements

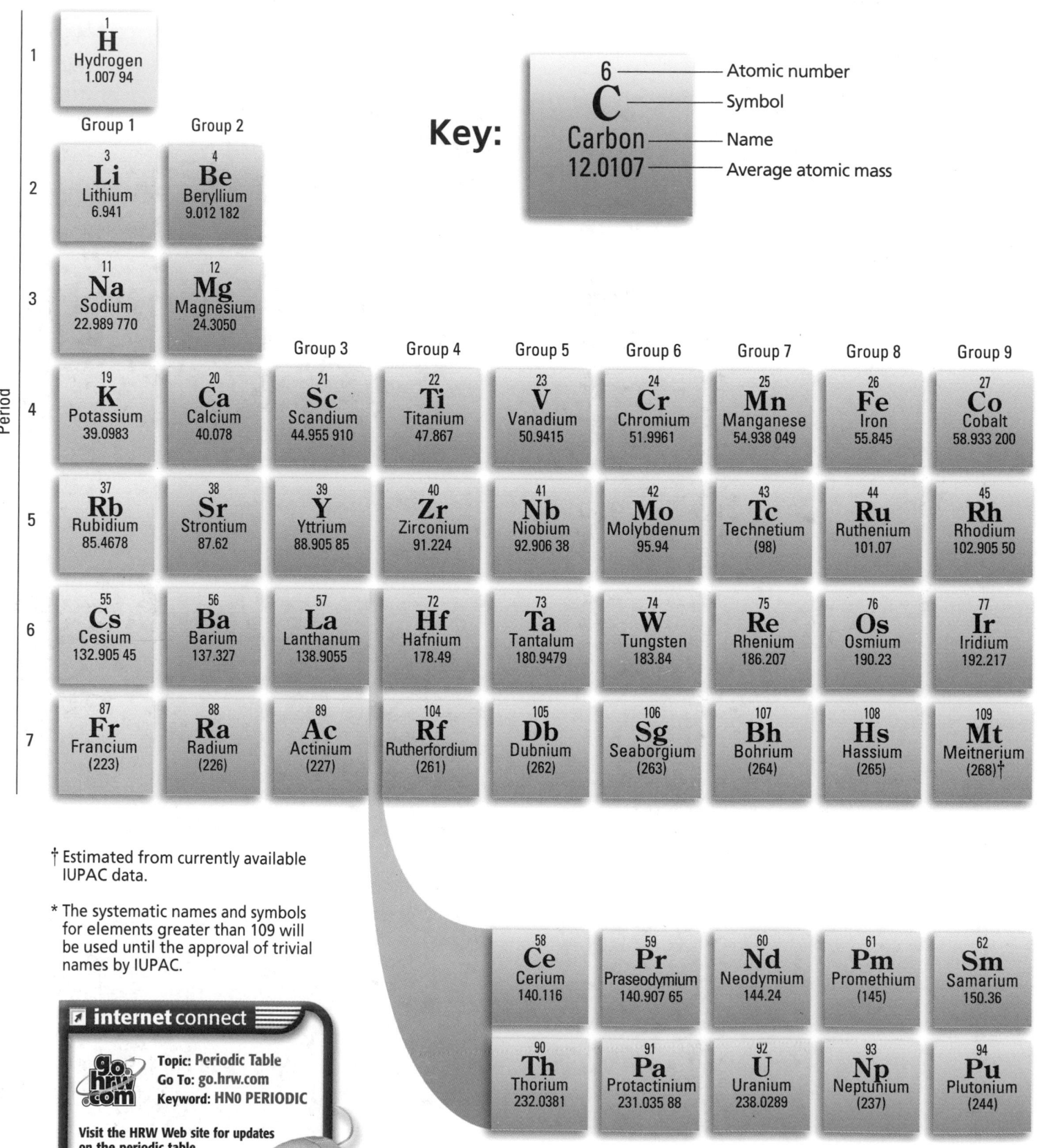

Key:

- 6 — Atomic number
- C — Symbol
- Carbon — Name
- 12.0107 — Average atomic mass

Period

Group 1

| 1 | H | Hydrogen | 1.007 94 |

| 3 | Li | Lithium | 6.941 |

| 11 | Na | Sodium | 22.989 770 |

| 19 | K | Potassium | 39.0983 |

| 37 | Rb | Rubidium | 85.4678 |

| 55 | Cs | Cesium | 132.905 45 |

| 87 | Fr | Francium | (223) |

Group 2

| 4 | Be | Beryllium | 9.012 182 |

| 12 | Mg | Magnesium | 24.3050 |

| 20 | Ca | Calcium | 40.078 |

| 38 | Sr | Strontium | 87.62 |

| 56 | Ba | Barium | 137.327 |

| 88 | Ra | Radium | (226) |

Group 3

| 21 | Sc | Scandium | 44.955 910 |

| 39 | Y | Yttrium | 88.905 85 |

| 57 | La | Lanthanum | 138.9055 |

| 89 | Ac | Actinium | (227) |

Group 4

| 22 | Ti | Titanium | 47.867 |

| 40 | Zr | Zirconium | 91.224 |

| 72 | Hf | Hafnium | 178.49 |

| 104 | Rf | Rutherfordium | (261) |

Group 5

| 23 | V | Vanadium | 50.9415 |

| 41 | Nb | Niobium | 92.906 38 |

| 73 | Ta | Tantalum | 180.9479 |

| 105 | Db | Dubnium | (262) |

Group 6

| 24 | Cr | Chromium | 51.9961 |

| 42 | Mo | Molybdenum | 95.94 |

| 74 | W | Tungsten | 183.84 |

| 106 | Sg | Seaborgium | (263) |

Group 7

| 25 | Mn | Manganese | 54.938 049 |

| 43 | Tc | Technetium | (98) |

| 75 | Re | Rhenium | 186.207 |

| 107 | Bh | Bohrium | (264) |

Group 8

| 26 | Fe | Iron | 55.845 |

| 44 | Ru | Ruthenium | 101.07 |

| 76 | Os | Osmium | 190.23 |

| 108 | Hs | Hassium | (265) |

Group 9

| 27 | Co | Cobalt | 58.933 200 |

| 45 | Rh | Rhodium | 102.905 50 |

| 77 | Ir | Iridium | 192.217 |

| 109 | Mt | Meitnerium | (268)† |

| 58 | Ce | Cerium | 140.116 |

| 59 | Pr | Praseodymium | 140.907 65 |

| 60 | Nd | Neodymium | 144.24 |

| 61 | Pm | Promethium | (145) |

| 62 | Sm | Samarium | 150.36 |

| 90 | Th | Thorium | 232.0381 |

| 91 | Pa | Protactinium | 231.035 88 |

| 92 | U | Uranium | 238.0289 |

| 93 | Np | Neptunium | (237) |

| 94 | Pu | Plutonium | (244) |

† Estimated from currently available IUPAC data.

* The systematic names and symbols for elements greater than 109 will be used until the approval of trivial names by IUPAC.

internet connect

go.hrw.com

Topic: Periodic Table
Go To: go.hrw.com
Keyword: HN0 PERIODIC

Visit the HRW Web site for updates on the periodic table.